Handbook of Cell Signaling

Volume 2

Handbook of Cell Signaling

Volume 2

Editors-in-Chief

Ralph A. Bradshaw
Department of Physiology and Biophysics
University of California Irvine
Irvine, California

Edward A. Dennis
Department of Chemistry and Biochemistry
University of California San Diego
La Jolla, California

ACADEMIC PRESS

An Imprint of Elsevier

Amsterdam   Boston   Heidelberg   London   New York   Oxford
Paris   San Diego   San Francisco   Singapore   Sydney   Tokyo

This book is printed on acid-free paper. ⊗

Copyright ©2004, Elsevier (USA).

All Rights Reserved.
No part of this publication may be reproduced or transmitted in any form or by any means, electronic or mechanical, including photocopy, recording, or any information storage and retrieval system, without permission in writing from the publisher.

Permissions may be sought directly from Elsevier's Science & Technology Rights Department in Oxford, UK: phone: (+44) 1865 843830, fax: (+44) 1865 853333, e-mail: permissions@elsevier.com.uk. You may also complete your request on-line via the Elsevier Science homepage (http://elsevier.com), by selecting "Customer Support" and then "Obtaining Permissions."

Academic Press
An imprint of Elsevier
525 B Street, Suite 1900, San Diego, California 92101-4495, USA
http://www.academicpress.com

Academic Press
84 Theobald's Road, London WC1X 8RR, UK
http://www.academicpress.com

Library of Congress Catalog Card Number: 2003103352

International Standard Book Number: 0-12-124546-2 (Set)
International Standard Book Number: 0-12-124547-0 (Volume 1)
International Standard Book Number: 0-12-124548-9 (Volume 2)
International Standard Book Number: 0-12-124549-7 (Volume 3)

PRINTED IN THE UNITED STATES OF AMERICA
04   05   06   07   7   6   5   4   3   2

Contents

VOLUME 2

Contributors   xlv

PART II
TRANSMISSION: EFFECTORS AND CYTOSOLIC EVENTS (CONTINUED FROM VOLUME 1)

Section C: Calcium Mobilization
Michael J. Berridge, Editor

CHAPTER 122
Phospholipase C   5
Hong-Jun Liao and Graham Carpenter

Introduction
PLC Anatomy
PLC Activation Mechanisms
PLC Physiology
References

CHAPTER 123
Inositol 1,4,5-trisphosphate 3-kinase and 5-phosphatase   11
Valérie Dewaste and Christophe Erneux

Introduction
Type I $InsP_3$ 5-phosphatase
$InsP_3$ 3-kinase
References

CHAPTER 124
Cyclic ADP-ribose and NAADP   15
Antony Galione and Grant C. Churchill

Introduction
References

CHAPTER 125
Sphingosine 1-phosphate   19
Kenneth W. Young and Stefan R. Nahorski

Introduction
Sphingolipid Metabolism
Activation of SPHK
Intracellular Target for SPP-mediated Ca^{2+} Release
Concluding Remarks
References

CHAPTER 126
Voltage-gated Ca^{2+} Channels   23
William A. Catterall

Introduction
Physiological Roles of Voltage-gated Ca^{2+} Channels
Ca^{2+} Current Types Defined by Physiological and Pharmacological Properties
Molecular Properties of Ca^{2+} Channels
Molecular Basis for Ca^{2+} Channel Function
Ca^{2+} Channel Regulation
Conclusion
References

CHAPTER 127
Store-operated Ca^{2+} Channels   31
James W. Putney, Jr.

Capacitative Calcium Entry
Store-operated Channels
Mechanism of Activation of Store-Operated Channels
Summary
References

CHAPTER 128
Arachidonic Acid-regulation Ca^{2+} Channel 35
Trevor J. Shuttleworth

Introduction
Identification and Characterization of ARC Channels
Specific Activation of ARC Channels by Low Agonist Concentrations
Roles of ARC Channels and SOC/CRAC Channels in [Ca^{2+}]$_i$ Signals: "Reciprocal Regulation"
Conclusions and Implications
References

CHAPTER 129
IP$_3$ Receptors 41
Colin W. Taylor

Introduction
References

CHAPTER 130
Ryanodine Receptors 45
David H. MacLennan and Guo Guang Du

Function and Structure
Activation of Ryanodine Receptor Ca^{2+} Release Channels
Molecular Biology of Ryanodine Receptors
References

CHAPTER 131
Intracellular Calcium Signaling 51
Martin D. Bootman, H. Llewelyn Roderick, Rodney O'Connor, and Michael J. Berridge

The "Calcium Signaling Toolkit" and Calcium Homeostasis
Multiple Channels and Messengers Underlie Ca^{2+} Increases
Temporal Regulation of Ca^{2+} Signals
Spatial Regulation of Ca^{2+} Signals
Modulation of Ca^{2+} Signal Amplitude
Ca^{2+} as a Signal within Organelles and in the Extracellular Space
References

CHAPTER 132
Calcium Pumps 57
Ernesto Carafoli

Introduction
Reaction Cycle of the SERCA and PMCA Pumps
The SERCA Pump
The PMCA Pump
Genetic Diseases Evolving Defects of Calcium Pumps
References

CHAPTER 133
Sodium/Calcium Exchange 63
Mordecai P. Blaustein

Introduction
Two Families of PM Na$^+$/Ca^{2+} Exchanges
Modes of Operation of the Na$^+$/Ca^{2+} Exchangers
Regulation of NCX
Inhibition of NCX
Localization of the NCX
Physiological Roles of the NCX
References

CHAPTER 134
Ca^{2+} Buffers 67
Beat Schwaller

Introduction
Relevant Parameters for Ca^{2+} Buffers
Ca^{2+} Buffers as One Component Contributing to Intracellular Ca^{2+} Homeostasis
Biological Effects of Ca^{2+} Buffers
References

CHAPTER 135
Mitochondria and Calcium Signaling, Point and Counterpoint 73
Michael R. Duchen

Introduction
Fundamentals
Machinery of Mitochondrial Ca^{2+} Movement
The Set Point
Quantitative Issues, Microdomains, and the Regulation of [Ca^{2+}]$_c$ Signals
Impact of Ca^{2+} Uptake on Mitochondrial Function
Mitochondrial Ca^{2+}, Disease, and Death
CODA
References

CHAPTER 136
EF-Hand Proteins and Calcium Sensing: The Neuronal Calcium Sensors 79
Jamie L. Weiss and Robert D. Burgoyne

Introduction
Class A. Neuronal Calcium Sensor 1 (Frequenin)
Class B. Neurocalcins (VILIPs) and Hippocalcin
Class C. Recoverins
Class D. Guanylate Cyclase Activating Proteins
Class E. K$^+$ Channel Interacting Proteins
Future Perspectives for the NCS Protein Family
References

CHAPTER 137
Calmodulin-Mediated Signaling 83
Anthony R. Means

Introduction
References

CHAPTER 138
The Family of S100 Cell Signaling Proteins 87
Claus W. Heizmann, Beat W. Schäfer, and Günter Fritz

Introduction
Protein Structures and Metal-Dependent Interactions with Target Proteins
Genomic Organization, Chromosomal Localization, and Nomenclature
Translocation, Secretion, and Biological Functions
Associations with Human Diseases
Conclusion and Perspectives
References

CHAPTER 139
C_2-Domains in Ca^{2+}-Signaling 95
Thomas C. Südhof and Josep Rizo

Structures of C_2-Domains
Ca^{2+}-Binding Mode of C_2-Domains
Phospholipid Binding Mechanism of C_2-Domains
Other Ligands of C_2-Domains
Functions of C_2-Domains
References

CHAPTER 140
Annexins and Calcium Signaling 101
Stephen E. Moss

Introduction
Annexins as Ca^{2+} Channel Regulators
Conclusions
References

CHAPTER 141
Calpain 105
Alan Wells and Anna Huttenlocher

Introduction
Calpain Family
Modes of Regulation
Calpain as a Signaling Intermediate: Potential Targets
Functional Roles
Future Considerations
References

CHAPTER 142
Regulation of Intracellular Calcium through Hydrogen Peroxide 113
Sue Goo Rhee

Introduction
Sources and Chemical Properties of ROS
Activation of Ryanodine and IP_3 Receptor Ca^{2+} Release Channels by H_2O_2
Enhancement of $[Ca^{2+}]_i$ through H_2O_2-mediated Inactivation of Protein Tyrosine Phosphatase and PTEN
References

Section D: Lipid-Derived Second Messengers
Lewis Cantley, Editor

CHAPTER 143
Historical Overview: Protein Kinase C, Phorbol Ester, and Lipid Mediators 119
Yasutomi Nishizuka and Ushio Kikkawa

Retrospectives of Phospholipid Research
Protein Kinase C and Diacylglycerol
Phorbol Ester and Cell Signaling
Structural Heterogeneity and Mode of Activation
Translocation and Multiple Lipid Mediators
Conclusion
References

CHAPTER 144
Type I Phosphatidylinositol 4-phosphate 5-kinases (PI4P 5-kinases) 123
K. A. Hinchliffe and R. F. Irvine

Introduction
Basic Properties
Regulation
Function
References

CHAPTER 145
Type 2 PIP4-Kinases 129
Lucia Rameh

Introduction
History
Structure
Type 2 PIP4-Kinase Isoforms
Regulation
Putative Models for the Function of the Type 2 PIP-Kinases
Conclusion
References

Chapter 146
Phosphoinositide 3-Kinases — 135
David A. Fruman

Introduction
The Enzymes
The Products
Lipid-Binding Domains
Effectors and Responses
Phosphatases
Genetics
Summary
References

Chapter 147
PTEN/MTM Phosphatidylinositol Phosphatases — 143
Knut Martin Torgersen, Soo-A Kim, and Jack E. Dixon

PTEN
Myotubularin: a Novel Family of Phosphatidylinositol Phosphatases
References

Chapter 148
SHIP Inositol Phosphate Phosphatases — 147
Larry R. Rohrschneider

Introduction
SHIP1 Structure, Expression, and Function
SHIP2 Structure, Expression, and Function
References

Chapter 149
Structural Principles of Lipid Second Messenger Recognition — 153
Roger L. Williams

Introduction
Phospholipid Second Messenger Recognition by Active Sites of Enzymes
Phosphoinositide-binding Domains
Non-phosphoinositide Lipid Messenger Recognition
Future Directions
References

Chapter 150
Pleckstrin Homology (PH) Domains — 161
Mark A. Lemmon

Identification and Definition of PH Domains
The Structure of PH Domains
PH Domains as Phosphoinositide-Binding Modules
Binding of PH Domains to Non-phosphoinositide Ligands
Possible Roles of Non-phosphoinositide PH Ligands
Conclusions
References

Chapter 151
PX Domains — 171
Hui Liu and Michael B. Yaffe

History and Overview of PX Domains
Lipid-Binding Specificity and the Structure of PX Domain
Function of PX Domain-containing Proteins
References

Chapter 152
FYVE Domains in Membrane Trafficking and Cell Signaling — 177
Christopher Stefan, Anjon Audhya, and Scott Emr

Introduction
Role for PtdIns(3)P in Membrane Trafficking and Identification of the FYVE Domain
Structural Basis for the FYVE Domain
Conservation of the FYVE Domain and Localization of PtdIns(3)P
FYVE Domains in Membrane Trafficking
FYVE Domains Involved in PtdIns(3)P Metabolism
FYVE Domains in Signaling
FYVE-like Domains
Conclusions
References

Chapter 153
Protein Kinase C: Relaying Signals from Lipid Hydrolysis to Protein Phosphorylation — 187
Alexandra C. Newton

Introduction
Protein Kinase C Family
Regulation of Protein Kinase C
Function of Protein Kinase C
Summary
References

Chapter 154
Role of PDK1 in Activating AGC Protein Kinase — 193
Dario R. Alessi

Introduction
Mechanism of Activation of PKB
PKB Is Activated by PDK1
Activation of Other Kinases by PDK1
Phenotype of PDK1 PKB- and S6K-Deficient Mice and Model Organisms
Hydrophobic Motif of AGC Kinases
Mechanism of Regulation of PDK1 Activity
Structure of the PDK1 Catalytic Domain
Concluding Remarks
References

Contents

CHAPTER 155
Modulation of Monomeric G Proteins by Phosphoinositides   203
Sonja Krugmann, Len Stephens, and Phillip T. Hawkins

Introduction
Rho Family Small GTPases
Arf Family GTPases
Modulation of Ras Family GTPases by PI3K
Conclusion
References

CHAPTER 156
Phosphoinositides and Actin Cytoskeletal Rearrangement   209
Paul A. Janmey, Robert Bucki, and Helen L. Yin

Historical Perspective
Stimulating Cellular Actin Polymerization
Actin-Membrane Linkers Localized or Activated by PIP2
Relation of Actin Assembly to Phsphoinositide-containing Lipid Rafts
Different Mechanisms of PPI-Actin Binding Protein Regulation
Effects on Lipid Membrane Structure
References

CHAPTER 157
The Role of PI3 Kinase in Directional Sensing during Chemotaxis in *Dictyostelium*, a Model for Chemotaxis of Neutrophils and Macrophages   217
Richard A. Firtel and Ruedi Meili

Introduction
Directional Movement
Localization of Cytoskeletal and Signaling Components
The Signaling Pathways Controlling Directional Movement
PI3K Effectors and their Roles in Controlling Chemotaxis
The Tumor Suppressor PTEN Regulates the Chemoattractant PI3K Pathways
Conclusions
References

CHAPTER 158
Phosphatidylinositol Transfer Proteins   225
Shamshad Cockcroft

Introduction
The Classical PITPs: α and β
RdgB Family of PITP Proteins
References

CHAPTER 159
Inositol Polyphosphate Regulation of Nuclear Function   229
John D. York

Introduction
Inositol Signaling and the Molecular Revolution
Links of Inositol Signaling to Nuclear Function
The Inositol Polyphosphate Kinase (IPK) Family
References

CHAPTER 160
Ins(1,3,4,5,6)P$_5$: A Signal Transduction Hub   233
Stephen B. Shears

Introduction
References

CHAPTER 161
Phospholipase D   237
Paul C. Sternweis

Introduction
Structural Domains and Requirements for Activity
Catalysis: Mechansim and Measurement
Modification of Mammalian PLDs
Regulatory Inputs for Mammalian PLD
Regulatory Pathways
Physiological Function of PA
Localization of PLD
Future Directions
References

CHAPTER 162
Diacylglycerol Kinases   243
M. K. Topham and S. M. Prescott

Introduction
The DGK Family
Regulation of DGKs
Paradigms of DGK Function
Conclusions
References

CHAPTER 163
Sphingosine-1-Phosphate Receptors   247
Michael Maceyka and Sarah Spiegel

Introduction
The S1PRs
S1P Signaling via S1PRs
Transactivation of S1PRs
Downstream Signaling from S1PRs
References

Chapter 164
SPC/LPC Receptors — 253
Linnea M. Baudhuin, Yijin Xiao, and Yan Xu

Introduction
Physiological and Pathological Functions of LPC and SPC
Identification of Receptors for SPC and LPC
Perspectives
References

Chapter 165
The Role of Ceramide in Cell Regulation — 257
Yusuf A. Hannun and L. Ashley Cowart

Ceramide-Mediated CellR
Biochemical Pathways of Ceramide Generation
Ceramide Targets
Conclusions
References

Chapter 166
Phospholipase A_2 Signaling and Arachidonic Acid Release — 261
Jesús Balsinde and Edward A. Dennis

Introduction
PLA_2 Groups
Cellular Function
Summary
References

Chapter 167
Prostaglandin Mediators — 265
Emer M. Smyth and Garret A. Fitzgerald

Introduction
The Cyclooxygenase Pathway
Prostanoid Receptors
Thromboxane A_2 (TxA_2)
Prostacyclin (PGI_2)
Prostaglandin D_2 (PGD_2)
Prostaglandin E_2 (PGE_2)
Prostaglandin $F_{2\alpha}$ ($PGF_{2\alpha}$)
Concluding Remarks
References

Chapter 168
Leukotriene Mediators — 275
Jesper Z. Haeggström and Anders Wetterholm

Introduction
Five-Lipoxygenase
Leukotriene A_4 Hydrolase
References

Chapter 169
Lipoxins and Aspirin-Triggered 15-epi-Lipoxins: Mediators in Anti-inflammation and Resolution — 281
Charles N. Serhan

Lipoxin Signals in the Resolution of Inflammation
Novel Anti-Inflammatory Signals and Pathways
Concluding Remarks
References

Chapter 170
Cholesterol Signaling — 287
Peter A. Edwards, Heidi R. Kast-Woelbern, and Matthew A. Kennedy

Introduction
Cholesterol Precursors
Cholesterol
Cholesterol Derivatives: Ligands for Nuclear Receptors
References

Section E: Protein Proximity Interactions
John D. Scott, Editor

Chapter 171
Protein Proximity Interactions — 293
John D. Scott

Introduction
Techniques for the Analysis of Protein-Protein Interactions
Subcellular Structures and Multiprotein Complexes
Kinase and Phosphatase Targeting Proteins

Chapter 172
Protein Interaction Mapping by Coprecipitation and Mass Spectrometric Identification — 295
Shao-En Ong and Matthias Mann

Introduction
General Considerations of the Coprecipitation Experiment
GST-Tagged Proteins
Antibodies
Epitope Tags
Mass Spectrometric Approaches
Outlook
References

Chapter 173
Proteomics, Fluorescence, and Binding Affinity — 301
Paul R. Graves and Timothy A. J. Haystead

Introduction
Isolation of Specific Proteomes

Affinity Chromatography for the Isolation of Protein Complexes
Specificity of Protein-Protein or Protein-Ligand Interactions
References

CHAPTER 174
FRET Analysis of Signaling Events in Cells — 305
Peter J. Verveer and Philippe I. H. Bastiaens

Introduction
Fluorescent Probes for FRET
FRET Detection Techniques
Conclusions and Prospects
References

CHAPTER 175
Peptide Recognition Module Networks: Combining Phage Display with Two-Hybrid Analysis to Define Protein-Protein Interactions — 311
Gary D. Bader, Amy Hin Yan Tong, Gianni Cesareni, Christopher W. Hogue, Stanley Fields, and Charles Boone

Introduction
References

CHAPTER 176
The Focal Adhesion: A Network of Molecular Interactions — 317
Benjamin Geiger, Eli Zamir, Yariv Kafri, and Kenneth M. Yamada

Introduction
Connectivity-Based Ordering of FA Components
Molecular Switches in FA
Future Challenges
References

CHAPTER 177
WASp/Scar/WAVE — 323
Charles L. Saxe

Introduction
WASp
Scar/WAVE
References

CHAPTER 178
Synaptic NMDA-Receptor Signaling Complex — 329
Mary B. Kennedy

Introduction
Structure of the NMDA Receptor Signaling Complex
Orchestration of Responses to Ca^{2+} Entering through the NMDA Receptor
References

CHAPTER 179
Toll Family Receptors — 333
Hana Bilak, Servane Tauszig-Delamasure, and Jean-Luc Imler

Introduction
Structure Function of Toll Receptors
Signaling by Toll Family Receptors
References

CHAPTER 180
Signaling and the Immunological Synapse — 339
Andrey S. Shaw

Introduction
Brief Introduction to T Cell Biology
Initiation of TCR Signaling
Definition of the Immunological Synapse
Immunological Synapses and T-Cell Development
Synapses and Different Kinds of T Cells
Natural Killer Cell Synapses
The Function of the Immunological Synapse
Immunological Synapses and TCR Downregulation
Conclusion
References

CHAPTER 181
The Ubiquitin-Proteasome System — 347
Mark Hochstrasser

Introduction
Overview of the Ubiquitin-Proteasome System
Components of the Ubiquitin Ligation and Deubiquitination Pathways
The 20S and 26S Proteasomes
Degradation Signals or Degrons
Examples of Regulation of Protein Ubiquitination
References

CHAPTER 182
Caspases: Cell Signaling by Proteolysis — 351
Guy S. Salvesen

Protease Signaling
Apoptosis and Limited Proteolysis
Caspase Activation
Regulation by Inhibitors
IAP Antagonists
References

CHAPTER 183
MAP Kinase in Yeast — 357
Elaine A. Elion

Introduction
Yeast Cells Use Multiple MAPKs to Respond to a Wide Variety of Stimuli
Functionally Defining *S. cerevisiae* MAPK Cascades

Major Regulatory Mechanisms that Control Specificity in
  S. cerevisai MAPK Cascades
References

CHAPTER 184
X^{c,v} Mammalian MAP Kinases — 365
Roger J. Davis

Introduction
The ERK Group of MAP Kinases
The p38 Group of MAP Kinases
The JNK Group of MAP Kinases
MAP Kinase-Related Protein Kinases
MAP Kinase Docking Interactions
Scaffold Proteins
References

CHAPTER 185
Subcellular Targeting of PKA through AKAPs: Conserved Anchoring and Unique Targeting Domains — 377
Mark L. Dell'Acqua

Introduction
Structurally Conserved PKA Anchoring Determinants
Unique Subcellular Targeting Domains
Probing Cellular Functions of AKAP-PKA Anchoring
Conclusions and Future Directions
References

CHAPTER 186
AKAP Signaling Complexes: The Combinatorial Assembly of Signal Transduction Units — 383
John D. Scott and Lorene K. Langeberg

Introduction
G-Protein Signaling Through AKAP Signaling
  Complexes
Kinase/Phosphatase Signaling Complexes
CAMP Signaling Units
Conclusions and Perspectives
References

CHAPTER 187
Protein Kinase C–Protein Interaction — 389
Peter J. Parker, Joanne Durgan, Xavier Iturrioz, and
Sipeki Szabolcs

Introduction
Priming
Activation
Substrates and Pathways
PKC Inactivation
Perspectives
References

CHAPTER 188
Dendrite Protein Phosphatase Complexes — 397
Roger J. Colbran

Introduction
The Importance of Dendritic Localization
Protein Phosphatase 1
Calcineurin (Protein Phosphatase 2B)
Dendritic Phosphatase Substrates
Role of Phosphatases in Synaptic Plasticity
Summary
References

CHAPTER 189
Protein Phosphatase 2A — 405
Adam M. Silverstein, Anthony J. Davis, Vincent A. Bielinski,
Edward D. Esplin, Nadir A. Mahmood, and Marc C. Mumby

Introduction
PP2A Regulatory Subunits Mediate Proximity
  Interactions
PP2A-Interacting Proteins
References

Section F: Cyclic Nucleotides
Jackie Corbin, Editor

CHAPTER 190
Adenylyl Cyclases — 419
Matt Whorton and Roger K. Sunahara

Introduction
Structure-Function
Regulation
Physiology
Summary
References

CHAPTER 191
Guanylyl Cyclases — 427
Ted D. Chrisman and David L. Garbers

Historic Perspectives
Guanylyl Cyclases
Guanylyl Cyclase Ligands
cGMP Effectors
Guanylyl Cyclases and Cell Growth Regulation
References

CHAPTER 192
Phosphodiesterase Families — 431
Jennifer L. Glick and Joseph A. Beavo

Introduction
The Gene Families
Implications of Multiple Gene Families/Splice Variants
PDE Inhibitors as Therapeutic Agents

Where Do We Go from Here?
References

CHAPTER 193
The cAMP-Specific Phosphodiesterases: A Class of Diverse Enzymes That Defines the Properties and Compartmentalization of the cAMP Signal — 437
Marco Conti

Structure of the cAMP-PDEs: Catalytic and Regulatory Domains
Subcellular Targeting of the cAMP-PDEs and cAMP Signal Compartmentalization
Regulation of cAMP-PDEs
References

CHAPTER 194
cAMP/cGMP Dual-Specificity Phosphodiesterases — 441
Marie C. Weston, Lena Stenson-Holst, Eva Degerman, and Vincent C. Manganiello

Introduction
PDE1 (Ca^{2+}/Calmodulin-dependenet PDE)
PDE2 (cGMP-stimulated PDE)
PDE3 (cGMP-inhibited cAMP PDE)
PDE10
PDE11
Conclusions
References

CHAPTER 195
Phosphodiesterase-5 — 447
Sharron H. Francis and Jackie D. Corbin

Introduction
Gene Organization and Regulation of Expression
General Structure
Concluding Remarks
References

CHAPTER 196
Structure, Function, and Regulation of Photoreceptor Phosphodiesterase (PDE6) — 453
Rick H. Cote

Introduction
Structure and Subcellular Localization of Rod PDE6
Regulation of Rod PDE6 Catalysis by γ
Catalytic Properties of Nonactivated and Activated PDE6P
Roles of the GAF Domains in PDE6 Regulation
Conclusion
References

CHAPTER 197
Spatial and Temporal Relationships of Cyclic Nucleotides in Intact Cells — 459
Manuela Zaccolo, Marco Mongillo, and Tullio Pozzan

The Complexity of Cyclic Nucleotides Signaling
Methodological Advances
Functional Compartments of cAMP in Heart Cells
Spatio-temporal Aspects of Cyclic Nucleotides Signaling in Neurons
Conclusions
References

CHAPTER 198
Regulation of Cyclic Nucleotide Levels by Sequestration — 465
Jackie D. Corbin, Jun Kotera, Venkatesh K. Gopal, Rick H. Cote, and Sharron H. Francis

Introduction
Sequestration of cGMP in Rod Photoreceptor Cells by PDE6
Sequestration of cGMP by PDE5
References

CHAPTER 199
cAMP-Dependent Protein Kinase — 471
Susan S. Taylor and Elzbieta Radzio-Andzelm

Introduction
Catalytic Subunit
Protein Kinase Inhibitor
Regulatory Subunits
References

CHAPTER 200
Cyclic GMP-Dependent Protein Kinase — 479
Thomas M. Lincoln

Introduction
Biochemical and Molecular Biology of PKG Isoforms
Physiologic Roles of PKG
Concluding Remarks
References

CHAPTER 201
Inhibitors of Cyclic Nucleotide-Dependent Protein Kinases — 487
Wolfgang R. G. Dostmann

Introduction
Cyclic Nucleotide Binding Site-Targeted Inhibitors
ATP Binding Site-Targeted Inhibitors
Peptide Binding in Site-Targeted Inhibitors
Conclusions
References

Chapter 202
Peptide Substrates of Cyclic Nucleotide-Dependent Protein Kinases — 495
Ross I. Brinkworth, Bostjan Kobe, and Bruce E. Kemp

Introduction
Peptide Substrate Recognition
Comparison of Kinase Substrate Acceptor Loci
Optimum Recognition Sequences
Comparison of PKA and PKG Specificity
Conclusions
References

Chapter 203
Physiological Substrates of PKA and PKG — 501
Kjetil Tasken, Anja Ruppelt, John Shabb, and Cathrine R. Carlson

Introduction
Abundance of PKA and PKG Phosphorylation Sites in the Human Proteome
Physiological Substrates
Concluding Remarks
References

Chapter 204
Effects of cGMP-Dependent Protein Kinase Knockouts — 511
Franz Hofmann, Robert Feil, Thomas Kleppisch, and Claudia Werner

Cyclic GMP-Dependent Protein Kinases: Genes and Knockouts
Outlook
References

Chapter 205
Cyclic Nucleotide-Regulated Cation Channels — 515
Martin Biel and Andrea Gerstner

Introduction
General Features of Cyclic Nucleotide-Regulated Cation Channels
CNG Channels
HCN Channels
References

Chapter 206
Epacs, cAMP-Binding Guanine Nucleotide Exchange Factors for Rap1 and Rap2 — 521
Holger Rehman, Johan de Rooij, and Johannes L. Bos

Introduction
The Epac Family
The cAMP-Binding Domain of Epac Closely Resembles Those of PKA and Channels
Epac Is Conserved Through Evolution
Properties of Epac
Expression and Subcellular Localization of Epacs
Cellular Function of Epacs
References

Chapter 207
Cyclic Nucleotide-Binding Phosphodiesterase and Cyclase GAF Domains — 525
Sergio E. Martinez, Xiao-Bo Tang, Stewart Turley, Wim G. J. Hol, and Joseph A. Beavo

Introduction
Atomic Structure
References

Chapter 208
cAMP Signaling in Bacteria — 531
J. M. Passner

Introduction and Significance
Background and History
Transcriptional Regulation by CAP
CAP Permits Differential Gene Regulation at Different cAMP Concentrations
A Second cAMP-Binding Site in a CAP Monomer
Perspectives and Conclusions
References

Chapter 209
Cyclic Nucleotide Signaling in Paramecium — 535
Jürgen U. Linder and Joachim E. Schultz

Introduction
cAMP Formation and Adenylyl Cyclase
Guanylyl Cyclase and cGMP Formation
Downstream of Cyclic Nucleotide Formation
References

Chapter 210
Cyclic Nucleotide Signaling in Trypanosomatids — 539
Roya Zoraghi and Thomas Seebeck

Introduction
Cyclic Nucleotide Signaling, Cell Proliferation, and Differentiation
Individual Components of the Cyclic Nucleotide Signaling Pathways
Cyclic Nucleotides and Host Parasite Intervention
Concluding Remarks
References

CHAPTER 211
Cyclic Nucleotide Specificity and Cross-Activation of Cyclic Nucleotide Receptors 545
Clay E. S. Comstock and John B. Shabb

cAMP Cross-Activation of PKG
cGMP Cross-Activation of PKA
Molecular Basis for cAMP/cGMP Selectivity of PKA and PKG
Other Cyclic Nucleotide Receptors
References

CHAPTER 212
Cyclic Nucleotide Analogs as Tools to Investigate Cyclic Nucleotide Signaling 549
Anne Elisabeth Christensen and Stein Ove Døskeland

Introduction
Use of cNMP Analogs: Guidelines and Examples
Chemistry and Properties of Cyclic Nucleotide Analogs
Future Developments
References

Section G: G Proteins
Heidi Hamm, Editor

CHAPTER 213
Signal Transduction by G Proteins — Basic Principles, Molecular Diversity, and Structural Basis of Their Actions 557
Lutz Birnbaumer

Introduction
Ras, the Prototypic Regulatory GTPases
Heterotrimeric G Proteins
Mechanism of G-Protein Activation by Receptors and Modulation of Activity
References

CHAPTER 214
Genetic Analysis of Heterotrimeric G-Protein Function 571
Juergen A. Knoblich

Introduction
Signaling by Heterotrimeric G Proteins in Yeast
Heterotrimeric G-Protein Function in *Drosophila*
Conclusions
References

CHAPTER 215
Heterotrimeric G Protein Signaling at Atomic Resolution 575
David G. Lambright

Introduction
Architecture and Switching Mechanism of the G_α Subunits
Insight into the GTP Hydrolytic Mechanism from an Unexpected Transition State Mimic
$G_{\beta\gamma}$ with and without G_α
Phosducin and $G_{\alpha\gamma}$
$G_{S\alpha}$ and Adenylyl Cyclase
Filling in the GAP
Visual Fidelity
What Structures May Follow
References

CHAPTER 216
In Vivo Functions of Heterotrimeric G Proteins 581
Stefan Offermanns

Introduction
Development
Central Nervous System
Immune System
Heart
Sensory Systems
Hemostasis
Conclusions
References

CHAPTER 217
Regulation of G Proteins by Covalent Modification 585
Jessica E. Smotrys and Maurine E. Linder

Introduction
N-Terminal Acylation of G_α
C-Terminal Modification of G_γ
Conclusions
References

CHAPTER 218
G-Protein-Coupled Receptors, Cell Transformation, and Signal Fidelity 589
Hans Rosenfeldt, Maria Julia Marinissen, and J. Silvio Gutkind

Introduction
Heptahelical Receptors and Tumorigenesis
G-Protein Signaling in Cancer
A Matrix of MAPK Cassettes Links GPCRs to Biological Outcomes
G-Protein-Independent Signaling
GPCR Effectors Are Organized by Scaffolding Molecules
Conclusion: GPCR Biology Requires Both Signal Integration and Separation
References

CHAPTER 219
Signaling through G_z 601
Jingwei Meng and Patrick J. Casey

General Properties
Receptors That Couple to G_z

Regulators of G_z Signaling: RGS Proteins
Effectors of G_z Signaling
$G\alpha_z$ Knockout Mice
Summary
References

CHAPTER 220
Effectors of $G_{\alpha 0}$ Signaling — 605
Prahlad T. Ram, J. Dedrick Jordan, and Ravi Iyengar

Introduction
Conclusions
References

CHAPTER 221
Phosphorylation of G Proteins — 609
Louis M. Luttrell and Deirdre K. Luttrell

Introduction
Serine Phosphorylation
Tyrosine Phosphorylation
Conclusions
References

CHAPTER 222
Mono-ADP-Ribosylation of Heterotrimeric G Proteins — 613
Maria Di Girolamo and Daniela Corda

Introduction
The Mono-ADP-Ribosylation Reaction
Bacterial Toxin-Induced ADP-Ribosylation
Endogenous Mono-ADP-Ribosylation
References

CHAPTER 223
Using Receptor-G-Protein Chimeras to Screen for Drugs — 619
Graeme Milligan, Richard J. Ward, Gui-Jie Feng, Juan J. Carrillo, and Alison J. McLean

Receptor-G-Protein Chimeras: An Introduction
Defining the Signal
Guanine Nucleotide Exchange Assays
Constitutive Activity and Inverse Agonism
Conclusions
References

CHAPTER 224
Specificity of G Protein βγ Dimer Signaling — 623
Janet D. Robishaw, William F. Schwindinger, and Carl A. Hansen

Introduction
Diversity of the β and γ Gene Families
Assembly of the βγ Dimer
Specificy of G Protein βγ Dimer Signaling
Conclusion
References

CHAPTER 225
The RGS Protein Superfamily — 631
David P. Siderovski and T. Kendall Harden

Introduction
The Signature RGS-Box as a Gα GAP
Gα GAP and Other Signaling Regulatory Activities of RGS Family Members
References

CHAPTER 226
Mechanism of βγ Effector Interaction — 639
Tohru Kozasa

Introduction
Effectors Interacting with βγ Subunits
Specificity of the Interaction between βγ Subunit and Effectors
References

CHAPTER 227
βγ Signaling in Chemotaxis — 645
Carol L. Manahan and Peter N. Devreotes

Introduction
Evidence that G Proteins Are Involved in Chemotaxis
PI3Ks — Role in Chemotaxis?
Lipid Phosphatases, PTEN and SHIP
References

CHAPTER 228
Reversible Palmitoylation in G-Protein Signaling — 651
Philip Wedegaertner

Introduction
Sites of Palmitoylation in G_α and RGS Proteins
Activation-Regulated Palmitoylation of Gα
Mechanisms of Reversible Palmitoylation
Functions of Reversible Palmitoylation
Conclusion
References

CHAPTER 229
G Proteins Mediating Taste Transduction — 657
Sami Damak and Robert F. Margolskee

Introduction
α-Gustducin
α-Transducin
Other G Protein α Subunits
βγ Subunits
G-Protein-Coupled Receptors
Second Messenger Pathways
Conclusion
References

CHAPTER 230
Regulation of Synaptic Fusion by Heterotrimeric G Proteins — 663
Simon Alford and Trillium Blackmer

Introduction
The Vesicle Fusion Machinery
G Protein-Coupled Receptor Mediated Modulation at the Presynaptic Terminal
Possible Mechanisms of Presynaptic Inhibition by G Proteins
Presynaptic Ca^{2+} Stores and Modulation of Neurotransmitter Release
G Proteins and Phosphorylation
References

CHAPTER 231
G Protein Regulation of Channels — 667
Wiser Ofer and Lily Yeh Jan

Interaction with K^+ Channels
The Gβγ Interacting Domain of GIRK
Coupling of GIRK Activation to Specific Receptors
Calcium Channel Interaction with G Proteins
G Protein Interacting Domains
The Gβγ Interacting Domain of HVA $Ca^{2\pm}$ Channels
Modulation of Gβγ Inhibition
Voltage-Independent G-Protein-Mediated Inhibition of Calcium Channels
References

CHAPTER 232
Ras and Cancer — 671
Frank McCormick

Introduction: Ras Activation in Cancer
Pathways Downstream of Ras
Mouse Models of Cancer
Prospects for Cancer Therapy Based on Ras
Referenccs

CHAPTER 233
The Influence of Cellular Location on Ras Function — 675
Janice E. Buss, Michelle A. Booden, and John T. Stickney

Cytosolic Ras Is not Functional
After Modifications by Endomembrane Enzymes, Ras Proteins Move Toward the Cell Surface
Destination-Cell Surface: Ras Proteins Distribute Among Several Plasma Membrane Domains
Ras Proteins Finally Become Active at the Plasma Membrane
Endocytosis — A New Stage for Ras Signaling
Drugs that Affect Ras Membrane Binding
References

CHAPTER 234
Role of R-Ras in Cell Growth — 681
Gretchen A. Murphy, Adrienne D. Cox, and Channing J. Der

Introduction
General Properties of R-Ras Proteins: Variations on Ras
R-Ras
TC21/R-Ras-2
M-Ras/R-Ras-3
Conclusions
References

CHAPTER 235
Molecular and Structural Organization of Rab GTPase Trafficking Networks — 689
Christelle Alory and William E. Balch

Introduction
Rab Proteins are Recycling GTPases
Rab Proteins: An Evolutionary Conserved Family
Structural Organization of the Rab Proteins
Posttranslational Modification and Localization
Effector Molecules: REP/CHM, GEF, Effectors (Motors/Tethers/Fusogens), GAP, and GDI
Rab Dysfunction and Disease
Perpective
References

CHAPTER 236
Cellular Roles of the Ran GTPase — 695
Jomon Joseph and Mary Dasso

Introduction
Introduction to the Ran Pathway
Structural Analysis of Ran Pathway Components
Ran's Role in Nuclear Transport
Ran's Function in Mitotic Progression
Ran's Function in Spindle Assembly
Ran's Role in Postmitotic Nuclear Assembly
Conclusions
References

CHAPTER 237
Rho Proteins and Their Effects on the Actin Cytoskeleton — 701
Anja Schmidt and Alan Hall

Introduction
Effects of Rho GTPases on the Actin Cytoskeleton
Signaling from Rho GTPases to the Actin Cytoskeleton
Conclusions
References

CHAPTER 238
Regulation of the NADPH Oxidase by Rac GTPase — 705
Becky A. Diebold and Gary M. Bokoch

Components and Regulation of the NADPH Oxidase
The Role of Rac in NADPH Oxidase Regulation
Current Models of Rac Function in NADPH Oxidase Regulation
Rac GTPase—A More General Role in Regulating Oxidant-Bases Signaling?
References

CHAPTER 239
The Role of Rac and Rho in Cell Cycle Progression — 711
Laura J. Taylor and Dafna Bar-Sagi

Introduction
Regulation of G1 Progression
The Function of Rac and Rho in Cell Cycle Progression and Transformation
Cell Cycle Targets of Rac and Rho
Future Perspectives
References

CHAPTER 240
Cdc42 and Its Cellular Functions — 715
Wannian Yang and Richard A. Cerione

Introduction
Biological Effects of Cdc42
Cell Adhesion and Migration
Cell Polarity
Molecular Mechanisms Underlying the Biological Activities of Cdc42
Conclusions
References

CHAPTER 241
Tissue Transglutaminase: A Unique GTP-Binding/GTPase — 721
Richard A. Cerione

Introduction
TGase as a GTP-Binding/GTPase
New Links to Biological Function
Future Directions
References

CHAPTER 242
The Role of ARF in Vesicular Membrane Traffic — 727
Melissa M. McKay and Richard A. Kahn

The ARF Family of Regulatory GTPases
ARF as a Regulator of Membrane Traffic
References

CHAPTER 243
Yeast Small G Protein Function: Molecular Basis of Cell Polarity in Yeast — 733
Hay-Oak Park and Keith G. Kozminski

Introduction
Conclusion
References

CHAPTER 244
Farnesyltransferase Inhibitors — 737
James J. Fiordalisi and Adrienne D. Cox

Introduction
Farnesylation and Protein Function
Ras—The Prototype of Farnesylated Proteins
Indentification and Development of FTIs
FTI Activity in Cell Culture and Animal Models
Alternative Prenylation in the Presence of FTIs
FTIs as Pharmacological Tools to Study Signaling and Biology
Targets of FTIs
Inhibition of Signaling by FTIs
Summary of Prospects
References

CHAPTER 245
Structure of Rho Family Targets — 745
Helen R. Mott and Darerca Owen

CRIB Proteins
Non-CRIB Rac Effectors
Rho Effectors
Concluding Remarks
References

CHAPTER 246
Structural Features of RhoGEFs — 751
Jason T. Snyder, Kent L. Rossman, David K. Worthylake, and John Sondek

Introduction
Structural Accomplishments
DH Domain Features
DH-Associated PH Domains
PH Domain Configurations
Mechanisms of Nucleotide Exchange
Molecular Recognition of Rho GTPase Substrates
External Regulation of the DH and PH Domains
References

CHAPTER 247
Structural Considerations of Small GTP-Binding Proteins — 757
Alfred Wittinghofer

Introduction
The G Domain Functional Unit

Guanine Nucleotide Exchange Factors
Effector B Via Switches and Others
Conclusions
References

CHAPTER 248
Conventional and Unconventional Aspects of Dynamin GTPases   763
Sandra L. Schmid

Introduction
Common and Unique Features of Dynamin as a GTPase
Dynamin's Function in Endocytic Vesicle Formation
Dynamin's Siblings: The Dynamin Subfamily of GTPases
Dynamin as a Signaling Molecule
Conclusion and Perspectives
References

CHAPTER 249
Mx Proteins: High Molecular Weight GTPases with Antiviral Activity   771
George Kochs, Othmar G. Engelhardt, and Otto Haller

Antiviral Activity of Mx GTPases
Mx Proteins Belong to the Superfamily of High Molecular Weight GTPases
Cellular Interaction Partners of Mx GTPases
References

Section H: Developmental Signaling
Geraldine Weinmaster, Editor

CHAPTER 250
Toll-Dorsal Signaling in Dorsal-Ventral Patterning and Innate Immunity   779
Ananya Bhattacharya and Ruth Steward

The Toll-Dorsal Pathway
Maturation of the Toll Ligand
Toll Signaling Establishes the Embryonic Dorsal Gradient
Dorsal Regulates the Function of Zygotic Genes
The Intracellular Pathway Is Conserved in the *Drosophila* Immune Response
Nuclear Import of Rel Proteins
References

CHAPTER 251
Developmental Signaling: JNK Pathway in *Drosophila* Morphogenesis   783
Beth E. Stronach and Norbert Perrimon

Introduction
The Paradigm of JNK Signaling: Dorsal Closure
Thorax Closure

Follicle Cell Morphogenesis
A New Paradigm: Planar Cell Polarity
Cellular Stress Response and Wound Healing
Perspectives
References

CHAPTER 252
Wnt Signaling in Development   789
Christian Wehrle, Heiko Lickert, and Rolf Kemler

Introduction
Wnt Signaling in Invertebrate Development
Wnt Signaling in Vertebrate Development
Wnt/β-Catenin Target Genes
References

CHAPTER 253
Hedgehog Signaling and Embryonic Development   793
Mark Merchant, Weilan Ye, and Frederic de Sauvage

The Hedgehog Proteins: Generation and Distribution
Transmitting the Hh Signal
Hh in Development and Disease
References

CHAPTER 254
Control of Left-Right (L/R) Determination in Vertebrates by the Hedgehog Signaling Pathway   799
Javier Capdevila and Juan Carlos Izpisúa Belmonte

Introduction
The Discovery of the First Molecular Asymmetries in Vertebrate Embryos and the Role of SHH
The Role of a Composite HH Signal during L/R Determination in the Mouse
References

CHAPTER 255
EGF-Receptor Signaling in *Caenorhabditis elegans* Vulval Development   805
Nadeem Moghal and Paul W. Sternberg

The Core LET-23 Signaling Pathway
Tissue Specificity
Positive and Negative Regulators
Prospects
References

CHAPTER 256
Induction and Lateral Specification Mediated by LIN-12/Notch Proteins   809
Sophie Jarriault and Iva Greenwald

The LIN-12/Notch Pathway
Cell-Cell Interactions Mediated by the LIN-12/Notch Pathway

The Role of LIN-12/Notch Proteins: Suppression of Differentiation versus Specification of Binary Cell Fate Decisions
References

CHAPTER 257
Notch Signaling in Vertebrate Development 813
Chris Kintner

Introduction
Components Mediating Vertebrate Notch Signaling
Notch Signaling in Vertebrate Development
Summary
References

CHAPTER 258
Reiterative and Concurrent Use of EGFR and Notch Signaling during *Drosophila* Eye Development 827
Raghavendra Nagaraj and Utpal Banerjee

Introduction
Establishment of the Eye Primordium
Proliferation and D/V Patterning
Morphogenetic Furrow and R8 Specification
R-Cell Specification
Sequential Linkage between Notch and EGFR Pathways
Parallel Linkage between EGFR and Notch
Pigment Cell Differentiation and Apoptosis
Conclusion
References

CHAPTER 259
BMPs in Development 833
Karen M. Lyons and Emmanuele Delot

Introduction
Gradients of BMP Activity
Establishing BMP Ligand Gradients
Extracellular Modifiers of BMP Activity
Interpreting the Gradient-Role of BMP Receptors
Differential Gene Activity in Response to BMP Signal Transduction
Intracellular Negative Regulation of BMP Signaling
Lessons from Loss-of-Function Studies in Mammals
Conclusions
References

CHAPTER 260
Neurotrophin Signaling in Development 839
Albert H. Kim and Moses V. Chao

Introduction
The Neurotrophin Ligands
Neurotrophin Receptors
Signaling Specificity during Development
The Importance of Tetrograde Transport
Interacting Proteins
References

CHAPTER 261
PDGF Receptor Signaling in Mouse Development 845
Richard A. Klinghoffer

Introduction
PDGTβR Signaling *In Vivo*
PDGFαR Signaling *In Vivo*
Specificity of PDGFR Signaling *In Vivo*
References

CHAPTER 262
VEGF and the Angiopoietins Activate Numerous Signaling Pathways that Govern Angiogenesis 849
Christopher Daly and Jocelyn Holash

Introduction
Endothelial Cell Proliferation
VEGF Promotes Vascular Permeability
Ang-1 Inhibits Vascular Permeabilitly
Vessel Destabilization and EC Migration
Regulation of EC Survival during Angiogenesis
Conclusion
References

CHAPTER 263
Vascular Endothelial Growth Factors and their Receptors in Vasculogenesis, Angiogenesis, and Lymphangiogenesis 855
Marja K. Lohela and Kari Alitalo

Vasculogenesis, Angiogenesis, and Lymphangiogenesis
The Vascular Endothelial Growth Factors and their Receptors
VEGF and VEGFR-1 and -2 are Essential for Vasculogenesis and Angiogenesis
Lymphangiogenesis is Regulated by VEGFR-3 and its Ligands VEGF-C and -D
Concluding Remarks
References

CHAPTER 264
Signaling from FGF Receptors in Development and Disease 861
Monica Kong and Daniel J. Donoghue

Introduction
Expression of FGFR during Development
Role of FGFR in Development
Syndromes Associated with FGFRs
Signaling Pathways Mediated by FGFRs
Summary
References

Chapter 265
The Role of Receptor Protein Tyrosine Phosphatases in Axonal Pathfinding 867
Andrew W. Stoker

Introduction
RPTPs and the Visual System
Neuromuscular System
Further Axon Growth and Guidance Roles
Axonal Signaling by RPTPs
References

Chapter 266
Attractive and Repulsive Signaling in Nerve Growth Cone Navigation 871
Guo-li Ming and Mu-ming Poo

Introduction
Netrin Signaling
Semaphorin Signaling
Slit Signaling
Ephrin Signaling
Nogo and Myelin-Associated Clycoprotein Signaling
Critical Roles of Modulatory Signals
Concluding Remarks
References

Chapter 267
Semaphorins and their Receptors in Vertebrates and Invertebrates 877
Eric F. Schmidt, Hideaki Togashi, and Stephen M. Strittmatter

The Semaphorin Family
Receptors for Semaphorins
Intracellular Signaling Pathways
Summary and Future Directions
References

Chapter 268
Signaling Pathways that Regulate Neuronal Specification in the Spinal Cord 883
Ann E. Leonard and Samuel L. Pfaff

Patterning along the Dorsoventral Axis
Dorsal Spinal Cord Development
Ventral Spinal Cord Development
Rostrocaudal Specification
References

Chapter 269
Cadherins: Interactions and Regulation of Adhesivity 889
Barbara Ranscht

Introduction
The Members of the Family
Multiple Modes for Regulating Cadherin Adhesive Activity
Conclusions and Perspectives
References

VOLUME 1

Contributors xlv
Preface lxvii

Chapter 1
Cell Signaling: Yesterday, Today, and Tomorrow 1
Ralph A. Bradshaw and Edward A. Dennis

Origins of Cell Signaling
Enter Polypeptide Growth Factors
Cell Signaling at the Molecular Level
Lipid Signaling
Cell Signaling Tomorrow
References

Part I
INITIATION: EXTRACELLULAR AND MEMBRANE EVENTS
James Wells, Editor

Section A: Molecular Recognition
Ian Wilson, Editor

Chapter 2
Structural and Energetic Basis of Molecular Recognition 11
Emil Alexov and Barry Honig

Introduction
Principles of Binding
Nonspecific Association with Membrane Surfaces
Protein–Protein Interactions
Prospects
References

Chapter 3
Computational Genomics: Prediction of Protein Functional Linkages and Networks 15
Todd O. Yeates and Michael J. Thompson

Introduction
Approaches to Analyzing Protein Functions on a Genome-Wide Scale

Current Issues and Future Prospects for Computing
   Functional Interactions
References

CHAPTER 4
Molecular Sociology   21
Irene M. A. Nooren and Janet M. Thornton

Transmembrane Signaling Paradigms
Structural Basis of Protein–Protein Recognition
Conclusion
References

CHAPTER 5
Free Energy Landscapes in Protein–Protein Interactions   27
Jacob Piehler and Gideon Schreiber

Introduction
Thermodynamics of Protein–Protein
   Interactions
Interaction Kinetics
The Transition State
Association of a Protein Complex
Dissociation of a Protein Complex
Summary
References

CHAPTER 6
Antibody–Antigen Recognition and Conformational Changes   33
Robyn L. Stanfield and Ian A. Wilson

Introduction
Antibody Architecture
Conformational Changes
Conclusion
References

CHAPTER 7
Binding Energetics in Antigen–Antibody Interfaces   39
Roy A. Mariuzza

Introduction
Thermodynamic Mapping of Antigen–
   Antibody Interfaces
Conclusions
References

CHAPTER 8
Immunoglobulin–Fc Receptor Interactions   45
Brian J. Sutton, Rebecca L. Beavil, and Andrew J. Beavil

Introduction
IgG–Receptor Interactions
IgE–Receptor Interactions
Summary
References

CHAPTER 9
Plasticity of Fc Recognition   51
Warren L. DeLano

Introduction
Structures of the Natural Fc Binding Domains
The Consensus Binding Site on Fc
Evolution of an Fc Binding Peptide
Factors Promoting Plasticity
Conserved and Functionally Important
   Molecular Interactions
Conclusion
References

CHAPTER 10
Ig-Superfold and Its Variable Uses in Molecular Recognition   57
Nathan R. Zaccai and E. Yvonne Jones

Introduction
The Immunoglobulin Superfamily
Ig-Superfold-Mediated Recognition
References

CHAPTER 11
T-Cell Receptor/pMHC Complexes   63
Markus G. Rudolph and Ian A. Wilson

TCR Generation and Architecture
Peptide Binding to MHC Class I and II
TCR/pMHC Interaction
Conclusions and Future Perspectives
References

CHAPTER 12
Mechanistic Features of Cell-Surface Adhesion Receptors   74
Steven C. Almo, Anne R. Bresnick, and Xuewu Zhang

Mechanosensory Mechanisms
Cell–Cell Adhesions/Adherens Junctions
T-Cell Costimulation
Axon Guidance and Neural Development
Conclusions
References

CHAPTER 13
The Immunological Synapse   79
Michael L. Dustin

Introduction
Migration and the Immunological Synapse
The Cytoskeleton and the Immunological
   Synapse
The Role of Self MHCp in T-Cell Sensitivity to
   Foreign MHCp

Integration of Adaptive and Innate Responses
Summary
References

Chapter 14
NK Receptors   83
Roland K. Strong

Introduction
Immunoreceptors
Natural Killer Cells
Ig-Type NK Receptors: KIR
C-Type Lectin-Like NK Receptors: Ly49A
C-Type Lectin-Like NK Receptors: NKG2D
References

Chapter 15
Carbohydrate Recognition and Signaling   87
James M. Rini and Hakon Leffler

Introduction
Biological Roles of Carbohydrate Recognition
Carbohydrate Structure and Diversity
Lectins and Carbohydrate Recognition
Carbohydrate-Mediated Signaling
Conclusions
References

Chapter 16
Rhinovirus–Receptor Interactions   95
Elizabeth Hewat

References

Chapter 17
HIV-1 Receptor Interactions   99
Peter D. Kwong

Molecular Interactions
Atomic Details
Recognition in the Context of a Humoral Immune Response
References

Chapter 18
Influenza Virus Neuraminidase Inhibitors   105
Garry L. Taylor

Introduction
Flu Virus: Role of NA
Structure of NA
Active Site
Inhibitor Development
Conclusion
References

Chapter 19
Signal Transduction and Integral Membrane Proteins   115
Geoffrey Chang and Christopher B. Roth

Introduction
Electrophysiology: Rapid Signal Transduction
Mechanosensation: How Do We Feel?
Active Transporters: Rapid Response and Energy Management
Receptors: Gate Keepers for Cell Signaling
References

Chapter 20
Structural Basis of Signaling Events Involving Fibrinogen and Fibrin   119
Russell F. Doolittle

References

Chapter 21
Structural Basis of Integrin Signaling   123
Robert C. Liddington

Introduction
Structure
Quaternary Changes
Tertiary Changes
Tail Interactions
Concluding Remarks
References

Chapter 22
Structures of Heterotrimeric G Proteins and Their Complexes   127
Stephen R. Sprang

Introduction
Gα Subunits
Gα-Effector Interactions
GTP Hydrolysis by Gα and Its Regulation by RGS Proteins
G$\beta\gamma$ Dimers
GPR/GoLoco Motifs
Gα-GPCR Interactions
References

Section B: Vertical Receptors
Henry Bourne, Editor

Chapter 23
Structure and Function of G-Protein-Coupled Receptors: Lessons from the Crystal Structure of Rhodopsin   139
Thomas P. Sakmar

Introduction
Introduction to Rhodopsin: a Prototypical G-Protein-Coupled Receptor

Molecular Structure of Rhodopsin
Molecular Mechanism of Receptor Activation
References

Chapter 24
Human Olfactory Receptors — 145
Orna Man, Tsviya Olender, and Doran Lancet

References

Chapter 25
Chemokines and Chemokine Receptors: Structure and Function — 149
Carol J. Raport and Patrick W. Gray

Introduction
Chemokine Structure and Function
Chemokine Receptors
References

Chapter 26
The Binding Pocket of G-Protein-Coupled Receptors for Biogenic Amines, Retinal, and Other Ligands — 155
Lei Shi and Jonathan A. Javitch

Introduction
The Binding Pocket of GPCRs
A Role of the Second Extracellular Loop in Ligand Binding
References

Chapter 27
Glycoprotein Hormone Receptors: A Unique Paradigm for Ligand Binding and GPCR Activation — 161
Gilbert Vassart, Marco Bonomi, Sylvie Claeysen, Cedric Govaerts, Su-Chin Ho, Leonardo Pardo, Guillaume Smits, Virginie Vlaeminck, and Sabine Costagliola

Introduction
Molecular Pathophysiology
Structure Function Relationships of the Glycoprotein Hormone Receptors
Conclusions and Perspectives
References

Chapter 28
Protease-Activated Receptors — 167
Shaun R. Coughlin

Introduction
Mechanisms of Activation
Protease-Activated Receptor Family
Roles of PARs *In Vivo*
References

Chapter 29
Constitutive and Regulated Signaling in Virus-Encoded 7TM Receptors — 173
Thue Schwartz

Virus-Encoded Proteins Are Developed through Targeted Evolution *In Vivo*
The Redundant Chemokine System Is an Optimal Target for Viral Exploitation
Multiple Virus-Encoded 7TM Receptors
Constitutive Signaling through Altered Pathways
Viral Receptors Recognize Multiple Ligands with Variable Function
Attempts to Identify the Function of Virus-Encoded Receptors *In Vivo*
References

Chapter 30
Frizzleds as G-Protein-Coupled Receptors for Wnt Ligands — 177
Sarah H. Louie, Craig C. Malbon, Randall T. Moon

Introduction
Wnt Signaling
Evidence for Frizzleds as G-Protein- Coupled Receptors
Perspective
References

Chapter 31
Agonist-Induced Desensitization and Endocytosis of G-Protein-Coupled Receptors — 181
Mark von Zastrow

Introduction
General Processes of GPCR Regulation
Mechanisms of GPCR Desensitization and Endocytosis
Functional Consequences of GPCR Endocytosis
References

Chapter 32
Functional Role(s) of Dimeric Complexes Formed from G-Protein-Coupled Receptors — 187
Marta Margeta-Mitrovic and Lily Yuh Jan

References

Chapter 33
The Role of Chemokine Receptors in HIV Infection of Host Cells — 191
Jacqueline D. Reeves and Robert W. Doms

Introduction
HIV Entry

Coreceptor Use *In Vivo*
Env Domains Involved in Coreceptor Interactions
Coreceptor Domains Involved in HIV Infection
Receptor Presentation and Processing
Role of Signaling in HIV Infection
Summary
References

CHAPTER 34
Chemotaxis Receptor in Bacteria: Transmembrane Signaling, Sensitivity, Adaptation, and Receptor Clustering — 197
Weiru Wang and Sung-Hou Kim

Signaling at Periplasmic Ligand Binding Domain
Signaling at the Cytoplasmic Domain
Adaptation
Clustering of the Chemoreceptor and Sensitivity
Future Studies
References

CHAPTER 35
Overview: Function and Three-Dimensional Structures of Ion Channels — 203
Daniel L. Minor, Jr.

Introduction
Studies of Full-Length Ion Channels
General Pore Features Revealed by Bacterial Channels
Pore Helices: Electrostatic Aids to Permeation
Open Channels
Eukaryotic Ion Channels at High Resolution: Divide and Conquer
Ion Channel Accessory Subunits: Soluble and Transmembrane
The Future: Ion Channels as Electrosomes
References

CHAPTER 36
How Do Voltage-Gated Channels Sense the Membrane Potential? — 209
Chris S. Gandhi and Ehud Y. Isacoff

Introduction
The Voltage-Sensing Gating Particle
S4 Is the Primary Voltage Sensor
Physical Models of Activation: Turning a Screw through a Bolt
Coupling Gating to S4 Voltage-Sensing Motions
References

CHAPTER 37
Ion Permeation: Mechanisms of Ion Selectivity and Block — 215
Bertil Hille

Aqueous Pore
Ion Selectivity
Block
References

CHAPTER 38
Agonist Binding Domains of Glutamate Receptors: Structure and Function — 219
Mark L. Mayer

References

CHAPTER 39
Nicotinic Acetylcholine Receptors — 223
Arthur Karlin

Function
Structure
References

CHAPTER 40
Small Conductance Ca^{2+}-Activated K^+ Channels: Mechanism of Ca^{2+} Gating — 227
John P. Adelman

Introduction
Clones Encoding SK Channels
Biophysical and Pharmacological Profiles
Mechanisms of Ca^{2+}-gating
Pantophobiac After All
References

CHAPTER 41
Regulation of Ion Channels by Direct Binding of Cyclic Nucleotides — 233
Edgar C. Young and Steven A. Siegelbaum

Introduction
The Cyclic Nucleotide-Gated Channels
Other Channels Directly Regulated by Cyclic Nucleotides
References

Section C: Horizontal Receptors
Robert Stroud, Editor

CHAPTER 42
Overview of Cytokine Receptors — 239
Robert M. Stroud

CHAPTER 43
Growth Hormone and IL-4 Families of Hormones and Receptors: The Structural Basis for Receptor Activation and Regulation — 241
Anthony A. Kossiakoff

Introduction
The Growth Hormone Family of Hormones and Receptors

Structural Basis for Receptor Homodimerization
Hormone Specificity and Cross-Reactivity Determine
    Physiological Roles
Hormone-Receptor Binding Sites
Receptor–Receptor Interactions
Hormone–Receptor Binding Energetics
Biological Implications of Transient Receptor Dimerization
A High-Affinity Variant of hGH (hGH$_v$) Reveals an
    Altered Mode for Receptor Homodimerization
Site1 and Site2 Are Structurally and Functionally Coupled
IL-4 Hormone-Induced Receptor Activation
IL-4–α-Chain Receptor Interface
Binding of the γ-Chain Receptor
Comparisons of IL-4 with GH(PRL)
Concluding Remarks
References

CHAPTER 44
Erythropoietin Receptor as a Paradigm for Cytokine Signaling — 251
Deborah J. Stauber, Minmin Yu, and Ian A. Wilson

Introduction
Biochemical Studies Supporting Preformed Dimers
Other Cytokine Receptor Superfamily Members
Conclusions
References

CHAPTER 45
A New Paradigm of Cytokine Action Revealed by Viral IL-6 Complexed to gp130: Implications for GCSF Interaction with GCSFR — 259
Dar-chone Chow, Lena Brevnova, Xiao-lin He, and K. Christopher Garcia

Introduction
Receptor/Ligand Interactions
The gp130 System
Viral Interleukin-6
GCSF and GCSFR
Structure of the Viral IL-6–gp130 Complex
Site 1
The Site 2 Interface
The Site 3 Interface
Implications of the vIL-6–gp130 Tetramer Structure
    for the Active GCSF–GCSFR Extracellular Signaling
    Complex
References

CHAPTER 46
The Fibroblast Growth Factor (FGF) Signaling Complex — 265
Fen Wang and Wallace L. McKeehan

Introduction
FGF Polypeptides
FGFR Tyrosine Kinases
Heparan Sulfate
Oligomeric FGF–FGFR–HS Signaling
    Complex
Intracellular Signal Transduction by
    the FGFR Complex
References

CHAPTER 47
Structure of IFN-γ and Its Receptors — 271
Mark R. Walter

References

CHAPTER 48
Structure and Function of Tumor Necrosis Factor at the Cell Surface — 275
Stephen R. Sprang

Introduction
Structure of Tumor Necrosis Factor
TNF Receptors
Extracellular (Ligand Binding) Domains of TNF
    Family Receptors
Ligand–Receptor Complexes
Consequences of Ligand–Receptor Complex
    Formation
Receptor Preassociation
Conclusion
References

CHAPTER 49
The Mechanism of NGF Suggested by the NGF–TrkA-D5 Complex — 281
Abraham M. de Vos and Christian Wiesmann

Introduction
Neurotrophins
Trks
NGF–TrkA-D5 Complex
p75^{NTR}
References

CHAPTER 50
The Mechanism of VEGFR Activation Suggested by the Complex of VEGF–flt1-D2 — 285
Christian Wiesmann and Abraham M. de Vos

Introduction
Heparin-Binding Domain of VEGF
Receptor-Binding Domain VEGF
VEGF Receptors
VEGF–flt1-D2 Complex
References

CHAPTER 51
Receptor–Ligand Recognition in the TGFβ Family as Suggested by the Crystal Structures of BMP-2–BR-IA$_{ec}$ and TGFβ3–TR-II$_{ec}$ 289
Matthias K. Dreyer

Introduction
Ligand and Receptor Structures
Receptor–Ligand Complexes
BMP-2–BR-IA$_{ec}$ Complex
Complex Formation with TGFβ Is Different than for BMP-2
References

CHAPTER 52
Insulin Receptor Complex and Signaling by Insulin   293
Lindsay G. Sparrow and S. Lance Macaulay

Introduction
Insulin Receptor Domain Structure
Binding Determinants of the IR
Insulin Signaling to Glucose Transport
References

CHAPTER 53
Structure and Mechanism of the Insulin Receptor Tyrosine Kinase   299
Steven R. Hubbard

Introduction
Structural/Mechanistic Studies
Prospects
References

CHAPTER 54
What Does the Structure of Apo2L/TRAIL Bound to DR5 Tell Us About Death Receptors?   305
Sarah G. Hymowitz and Abraham M. de Vos

Introduction
Novel Features in the Structure of Apo2L/TRAIL
Apo2L/TRAIL:DR5 Structures
Ligand-Independent Receptor Assembly
Intracellular Consequences of Ligand Binding
Conclusion
References

Section D: Membrane Proximal Events
Tom Alber, Editor

CHAPTER 55
TNF Receptor Associated Factors   311
Jee Y. Chung, Young Chul Park, Hong Ye, and Hao Wu

References

CHAPTER 56
Assembly of Signaling Complexes for TNF Receptor Family Molecules   315
Gail A. Bishop and Bruce S. Hostager

Introduction
Receptor Aggregation
Raft Recruitment
Ubiquitination
Receptor Interactions
Conclusions
References

CHAPTER 57
Mechanisms of CD40 Signaling in the Immune System   319
Aymen Al-Shamkhani, Martin J. Glennie, and Mark S. Cragg

Introduction
Signaling Pathways Triggered by CD40 Engagement
CD40 Signaling Is Mediated by TRAF-Dependent and TRAF-Independent Pathways
References

CHAPTER 58
Role of Lipid Domains in EGF Receptor Signaling   323
Linda J. Pike

Introduction
Localization of the EGF Receptor to Lipid Rafts
Rafts and EGF-Receptor-Mediated Signaling
The EGF Receptor and Caveolin
Summary
References

CHAPTER 59
Structure and Function of B-Cell Antigen Receptor Complexes   327
Michael Reth and Michael Huber

Introduction
The Structure of the B Cell Antigen Receptor
Initiation of BCR Signaling Is Controlled by Redox Regulation
References

CHAPTER 60
Lipid-Mediated Localization of Signaling Proteins   331
Maurine E. Linder

Introduction
Protein Lipidation
Summary
References

Chapter 61
G-Protein Organization and Signaling — 335
Maria R. Mazzoni and Heidi E. Hamm

Introduction
G-Protein Molecular Organization
Structural Features of G Protein Activation
Structural Determinants of Receptor–G-Protein Specificity
Gα Interactions with Effector Molecules
Gβγ Interactions with Effector Molecules
Conclusions
References

Chapter 62
JAK–STAT Signaling — 343
Rashna Bhandari and John Kuriyan

Introduction
Cytokine Signaling Proteins
JAK Structure and Localization
STAT Structure and Function
Inhibition of Cytokine Signaling
Summary
References

Chapter 63
Organization of Photoreceptor Signaling Complexes — 349
Susan Tsunoda

INAD Organizes Signaling Complexes
INAD-Signaling Complexes in Phototransduction
Assembly, Targeting, and Anchoring of Signaling Complexes
Signaling Complexes in Vertebrate Photoreceptors
References

Chapter 64
Protein Localization in Negative Signaling — 355
Jackson G. Egen and James P. Allison

Introduction
The Role of CD28 and CTLA-4 in T-Cell Activation
Expression and Localization of CTLA-4 and CD28: Consequences for Receptor Function
Mechanisms of CTLA-4-Mediated Negative Signaling
Conclusions
References

Chapter 65
Transmembrane Receptor Oligomerization — 361
Darren Tyson and Ralph A. Bradshaw

Introduction
Tyrosine Kinase-Containing Receptors
Cytokine Receptors
Guanylyl Cyclase-Containing Receptors
Serine/Threonine Kinase-Containing Receptors
Tumor Necrosis Factor Receptors
Heptahelical Receptors (G-Protein-Coupled Receptors)
Concluding Remarks
References

Part II
TRANSMISSION: EFFECTORS AND CYTOSOLIC EVENTS
Tony Hunter, Editor

Part II
Introduction — 369
Tony Hunter, Editor

Section A: Protein Phosphorylation
Tony Pawson

Chapter 66
Eukaryotic Kinomes: Genomic Cataloguing of Protein Kinases and Their Evolution — 373
Tony Hunter and Gerard Manning

Introduction
The Yeasts: *Saccharomyces cerevisiae* and *Schizosaccharomyces pombe*
Nematodes: *Caenorhabditis elegans*
Insects: *Drosophila melanogaster*
Vertebrates: *Homo sapiens*
Comparative Kinomics
Coda
References

Chapter 67
Modular Protein Interaction Domains in Cellular Communication — 379
Tony Pawson and Piers Nash

Introduction
Phosphotyrosine-Dependent Protein–Protein Interactions
Interaction Domains: A Common Theme in Signaling
Adaptors, Pathways, and Networks
Evolution of a Phospho-Dependent Docking Protein
Multisite Phosphorylation, Ubiquitination, and Switch-Like Responses
Summary
References

CHAPTER 68
Structures of Serine/Threonine and Tyrosine Kinases — 387
Matthew A. Young and John Kuriyan

Introduction
Structures of Protein Kinases
Structures of Inactive Protein Kinases
Summary
References

CHAPTER 69
Protein Tyrosine Kinase Receptor Signaling Overview — 391
Carl-Henrik Heldin

Introduction
PTK Subfamilies
Mechanism of Activation
Control of PTK Receptor Activity
Cross-Talk Between Signaling Pathways
PTK Receptors and Disease
References

CHAPTER 70
Signaling by the Platelet-Derived Growth Factor Receptor Family — 397
M. V. Kovalenko and Andrius Kazlauskas

Introduction
Platelet-Derived Growth Factors, Their Receptors, and Assembly of the PDGF Receptor Signaling Complex
Some Aspects of Regulation of the PDGF Receptor-Initiated Signaling
References

CHAPTER 71
EGF Receptor Family — 405
Mina D. Marmor and Yosef Yarden

Introduction
Domain Structure of ErbBs
Subcellular Localization of ErbB Proteins
ErbB-Induced Signaling Pathways
Negative Regulatory Pathways
Specificity of Signaling Through the ErbB Network
ErbB Proteins and Pathological Conditions
References

CHAPTER 72
IRS-Protein Scaffolds and Insulin/IGF Action — 409
Morris F. White

IRS-Proteins: The Beginnings
IRS-Proteins and Insulin Signaling
IRS-Protein Structure and Function
IRS-Protein Signaling in Growth, Nutrition, and Longevity
Interleukin-4 and IRS2 Signaling
Heterologous Regulation of IRS-Protein Signals
IRS2 and Pancreatic β-Cells
Summary
References

CHAPTER 73
Eph Receptors — 421
Rüdiger Klein

Introduction
Ephs and Ephrins
Eph Receptor Signaling Via Cytoplasmic Protein Tyrosine Kinases
Eph Receptor Signaling Via Rho Family GTPases
Effects on Cell Proliferation
Eph Receptor Signaling through PDZ-Domain-Containing Proteins
Eph Receptors and Cell Adhesion
Ephrin Reverse Signaling
EphrinB Reverse Signaling Via Phosphotyrosine
EphrinB Reverse Signaling Via PDZ Domain Interactions
Summary
References

CHAPTER 74
Cytokine Receptor Superfamily Signaling — 427
James N. Ihle

Cytokine Receptor Superfamily Signaling
References

CHAPTER 75
Negative Regulation of the JAK/STAT Signaling Pathway — 431
Joanne L. Eyles and Douglas J. Hilton

Introduction
The Phosphatases
STAT Phosphatases
PIAS (Protein Inhibitors of Activated STATS)
SOCS (Suppressors of Cytokine Signaling) Family
Concluding Comments
References

CHAPTER 76
Activation of Oncogenic Protein Kinases — 441
G. Steven Martin

Introduction
Physiological Regulation of Protein Kinases
Activation of Protein Kinases by Retroviruses
Activation of Protein Kinases in Human Cancer
Oncogenic Protein Kinases as Targets for Therapy
References

CHAPTER 77
Protein Kinase Inhibitors — 451
Alexander Levitzki

Signal Transduction Therapy
Protein Tyrosine Kinase Inhibitors
SER/THR Kinase Inhibitors
References

CHAPTER 78
Integrin Signaling: Cell Migration, Proliferation, and Survival — 463
J. Thomas Parsons, Jill K. Slack-Davis, and Karen H. Martin

Introduction
Integrins Nucleate the Formation of Multi-Protein Complexes
Cell Migration: A Paradigm for Studying Integrin Signaling
Integrin Regulation of Cell Proliferation and Survival: Links to Cancer
Concluding Remarks
References

CHAPTER 79
Downstream Signaling Pathways: Modular Interactions — 471
Bruce J. Mayer

Introduction
General Properties of Interaction Modules
Roles in Signaling
Prospects
References

CHAPTER 80
Non-Receptor Protein Tyrosine Kinases in T-Cell Antigen Receptor Function — 475
Kiminori Hasegawa, Shin W. Kang, Chris Chiu and Andrew C. Chan

Introduction
T-Cell Antigen Receptor Structure
Src PTKs
Csk (c-Src PTK)
ZAP-70/Syk PTKs
Tec PTKs
Summary
References

CHAPTER 81
Cbl: A Physiological PTK Regulator — 483
Wallace Y. Langdon

Introduction
Domains of Cbl Proteins
Sli-1: A Negative Regulator of RPTKs
PTK Downregulation by Polyubiquitylation
Cbl-Deficient Mice
Future Directions
References

CHAPTER 82
TGFβ Signal Transduction — 487
Jeffrey L. Wrana

Introduction
The Smad Pathway
Smads and the Ubiquitin–Proteasome System
Smad-Independent Signaling Pathways
Other Receptor Interaction Proteins
References

CHAPTER 83
MAP Kinases — 493
James R. Woodgett

Introduction
The ERK Module
Stress-Activated MAPKs, Part 1: SAPK/JNKs
Stress-Activated MAPKs, Part 2: p38 MAPKs
MAPKKs
MAPKKKs
MAPKKKKs
Summary
References

CHAPTER 84
Cytoskeletal Regulation: Small G-Protein–Kinase Interactions — 499
Ed Manser

Introduction
P21-Activated Kinases
Myotonic Dystrophy Kinase-Related Cdc42-Binding Kinase
Rho-Associated Kinase (ROK)
References

CHAPTER 85
Recognition of Phospho-Serine/Threonine Phosphorylated Proteins — 505
Stephen J. Smerdon and Michael B. Yaffe

Introduction
14-3-3 Proteins
FHA Domains
WW Domains
Leucine-Rich Repeats and WD40 Domains
Concluding Remarks
References

Chapter 86
Role of PDK1 in Activating AGC Protein Kinase — 513
Dario R. Alessi

Introduction
Mechanisms of Activation of PKB
PKB Is Activated by PDK1
Activation of Other Kinases by PDK1
Phenotype of PDK1 PKB- and S6K-Deficient Mice and Model Organisms
Hydrophobic Motif of AGC Kinases
Mechanisms of Regulation of PDK1 Activity
Structure of the PDK1 Catalytic Domain
Concluding Remarks
References

Chapter 87
Regulation of Cell Growth and Proliferation in Metazoans by mTOR and the p70 S6 Kinase — 523
Joseph Avruch

Introduction
Functions of TOR
Signaling from TOR
Regulation of mTOR Activity
References

Chapter 88
AMP-Activated Protein Kinase — 535
D. Grahame Hardie

Introduction
Structure of the AMPK Complex
Regulation of the AMPK Complex
Regulation in Intact Cells and Physiological Targets
Medical Implications of the AMPK System
References

Chapter 89
Principles of Kinase Regulation — 539
Bostjan Kobe and Bruce E. Kemp

Introduction
Protein Kinase Structure
General Principles of Control
Regulatory Sites in Protein Kinase Domains
Conclusions
References

Chapter 90
Calcium/Calmodulin-Dependent Protein Kinase II — 543
Mary B. Kennedy

Introduction
Structure of CaMKII
Regulation by Autophosphorylation
Regulatory Roles of CaMKII in Neurons
References

Chapter 91
Glycogen Synthase Kinase 3 — 547
Philip Cohen and Sheelagh Frame

Introduction
The Substrate Specificity of GSK3
The Regulation of GSK3 Activity by Insulin and Growth Factors
GSK3 as a Drug Target
The Role of GSK3 in Embryonic Development
GSK3 and Cancer
References

Chapter 92
Protein Kinase C: Relaying Signals from Lipid Hydrolysis to Protein Phosphorylation — 551
Alexandra C. Newton

Introduction
Protein Kinase C Family
Regulation of Protein Kinase C
Function of Protein Kinase C
Summary
References

Chapter 93
The PIKK Family of Protein Kinases — 557
Graeme C. M. Smith and Stephen P. Jackson

Introduction
Overview of PIKK Family Members
Overall Architecture of PIKK Family Proteins
MTOR: A Key Regulator of Cell Growth
DNA-Pkes: At the Heart of the DNA Nonhomologous End-Joining Machinery
ATM and ATR: Signalers of Genome Damage
SMG-1: A Regulator of Nonsense-Mediated mRNA Decay
TRRAP: A Crucial Transcriptional Co-Activator
PIKK Family Members as Guardians of Nucleic Acid Structure, Function, and Integrity?
References

Chapter 94
Histidine Kinases — 563
Fabiola Janiak-Spens and Ann H. West

References

Chapter 95
Atypical Protein Kinases: The EF2/MHCK/ChaK Kinase Family — 567
Angus C. Nairn

Introduction
Identification of an Atypical Family of Protein Kinases: EF2 Kinase, Myosin Heavy Chain Kinase and ChaK
The Structure of the Atypical Kinase Domain Reveals Similarity to Classical Protein Kinases and to Metabolic Enzymes with ATP-Grasp Domains
Substrate Specificity of Atypical Kinases
Regulation of Atypical Kinases
Functions of the Atypical Family of Protein Kinases
References

Chapter 96
Casein Kinase I and Regulation of the Circadian Clock — 575
Saul Kivimäe, Michael W. Young, and Lino Saez

Introduction
double-time: A Casein Kinase I Homolog in *Drosophila*
Casein Kinase I in the Mammalian Clock
Casein Kinase I in the *Neurospora* Clock
Similarities and Differences of CKI Function in Different Clock Systems
References

Chapter 97
The Leucine-Rich Repeat Receptor Protein Kinases of *Arabidopsis thaliana*: A Paradigm for Plant LRR Receptors — 579
John C. Walker and Kevin A. Lease

Introduction
LRR Receptor Protein Kinases: The Genomic Point of View
LRR Receptor Protein Kinases: The Functional View
Summary
References

Chapter 98
Engineering Protein Kinases with Specificity for Unnatural Nucleotides and Inhibitors — 583
Chao Zhang and Kevan M. Shokat

References

Section B: Protein Dephosphorylation
Jack E. Dixon, Editor

Chapter 99
Overview of Protein Dephosphorylation — 591
Jack E. Dixon

Chapter 100
Protein Serine/Threonine Phosphatases and the PPP Family — 593
Patricia T. W. Cohen

Current Classification of Protein Serine/Threonine Phosphatases
Background
Evolution and Conserved Features of the PPP Family
Catalytic Activities of the PPP Family Members
Eukaryotic PPP Subfamilies
Domain and Subunit Structure of PPP Family Members
Medical Importance of the PPP Family
References

Chapter 101
The Structure and Topology of Protein Serine/Threonine Phosphatases — 601
David Barford

Introduction
Protein Serine/Threonine Phosphatases of the PPP Family
Protein Serine/Threonine Phosphatases of the PPM Family
Conclusions
References

Chapter 102
Naturally Occurring Inhibitors of Protein Serine/Threonine Phosphatases — 607
Carol MacKintosh and Julie Diplexcito

Introduction
Effects of Inhibitors in Cell-Based Experiments
The Toxins Bind to the Active Sites of Protein Phosphatases
Chemical Synthesis of Protein Phosphatase Inhibitors
Microcystin Affinity Chromatography and Affinity Tagging
Avoiding the Menace of Toxins in the Real World Outside the Laboratory
References

Chapter 103
Protein Phosphatase 1 Binding Proteins — 613
Anna A. Depaoli-Roach

Introduction
Protein Phosphatase 1 (PP1)
PP1 Regulatory or Targeting Subunits
Conclusions
References

Chapter 104
Role of PP2A in Cancer and Signal Transduction — 621
Gernot Walter

Introduction
Structure of PP2A

Subunit Interaction
Association of PP2A with Cellular Proteins
Alteration or Inhibition of PP2A Is Essential in Human Cancer Development
Mutation of Aα and Aβ Isoforms in Human Cancer
Differences between Aα and Aβ Subunits
PP2A and Wnt Signaling
PP2A and MAP Kinase Pathway
Summary
References

CHAPTER 105
Serine/Threonine Phosphatase Inhibitor Proteins — 627
Shirish Shenolikar

Introduction
Protein Phosphatase 1 (PP1) Inhibitors
I-1, DARPP-32, and Other Phosphorylation-Dependent Phosphatase Inhibitors
Latent Phosphatase Complexes Activated by Inhibitor Phosphorylation
Inhibitors of Type-2 Serine/Threonine Phosphatases
Conclusions
References

CHAPTER 106
Calcineurin — 631
Claude B. Klee and Seun-Ah Yang

Introduction
Enzymatic Properties
Structure
Regulation
Distribution and Isoforms
Functions
Muscle Differentiation
Conclusion
References

CHAPTER 107
Protein Serine/Threonine-Phosphatase 2C (PP2C) — 637
Hisashi Tatabe and Kazuhiro Shiozaki

Introduction
Regulation of the Stress-Activated MAP Kinase Cascades
Control of the CFTR Chloride Channel by PP2C
Plant Hormone Abscisic Acid Signaling
Fem-2: A Sex-Determining PP2C in Nematode
Stress-Responsive PP2Cs in *Bacillus subtilis*
References

CHAPTER 108
Overview of Protein Tyrosine Phosphatases — 641
Nicholas K. Tonks

Background
Structural Diversity within the PTP Family
The Classical PTPs
The Dual Specificity Phosphates (DSPs)
Regulation of PTP Function
Oxidation of PTPs in Tyrosine Phosphorylation-Dependent Signaling
Substrate Specificity of PTPs
PTPs and Human Disease
Perspectives
References

CHAPTER 109
Protein Tyrosine Phosphatase Structure and Mechanisms — 653
Youngjoo Kim and John M. Denu

Introduction
Introduction to the Protein Tyrosine Phosphatase Family
Structure
Mechanism
Regulation
References

CHAPTER 110
Bioinformatics: Protein Tyrosine Phosphatases — 659
Niels Peter H. Møller, Peter Gildsig Jansen, Lars F. Iversen, and Jannik N. Andersen

Introduction to Bioinformatics
Amino Acid Homology Among PTP Domains and Structure-Function Studies
Identification of the Genomic Complement of PTPs
Functional Aspects of PTPs in Health and Disease: Bioinformatics
References

CHAPTER 111
PTP Substrate Trapping — 671
Andrew J. Flint

Introduction
Original C→S and D→A Substrate-Trapping Mutants
Second-Generation Trapping Mutants
Accessory or Noncatalytic Site Contributions to Substrate Recognition
New Twists on Trapping
Other Applications of Substrate Trapping Mutants
References

Chapter 112
Inhibitors of Protein Tyrosine Phosphatases — 677
Zhong-Yin Zhang

Introduction
Covalent PTP Modifiers
Oxyanions as PTP Inhibitors
PTyr Surrogates as PTP Inhibitors
Bidentate PTP Inhibitors
Other PTP Inhibitors
Concluding Remarks
References

Chapter 113
Regulating Receptor PTP Activity — 685
Erica Dutil Sonnenburg, Tony Hunter, and Joseph P. Noel

Introduction
Regulation by Dimerization
Regulation by Phosphorylation
Regulation by D2 Domain
References

Chapter 114
CD45 — 689
Zheng Xu, Michelle L. Hermiston, and Arthur Weiss

Introduction
Structure
Function
Regulation
References

Chapter 115
Properties of the Cdc25 Family of Cell-Cycle Regulatory Phosphatases — 693
William G. Dunphy

Introduction
Physiological Functions of Cdc25
Regulation of Cdc25
Concluding Remarks
References

Chapter 116
Cell-Cycle Functions and Regulation of Cdc14 Phosphatases — 697
Harry Charbonneau

Introduction
The Cdc14 Phosphatase Subgroup of PTPs
Budding Yeast Cdc14 is Essential for Exit from Mitosis
Fission Yeast Cdc14 Coordinates Cytokinesis with Mitosis
Potential Cell-Cycle Functions of Human Cdc14A and B
References

Chapter 117
MAP Kinase Phosphatases — 703
Marco Muda and Steve Arkinstall

Introduction
MAPK Phosphatases in Yeast
A MAPK Phosphatase in *C. elegans*
MAPK Phosphatases in *Drosophila melanogaster*
MAPK Phosphatases in Mammals
Summary
References

Chapter 118
SH2-Domain-Containing Protein–Tyrosine Phosphatases — 707
Benjamin G. Neel, Haihua Gu, and Lily Pao

History and Nomenclature
Structure, Expression, and Regulation
Biological Functions of Shps
Shp Signaling and Substrates
Determinants of Shp Specificity
Shps and Human Disease
Summary and Future Disease
References

Chapter 119
Insulin Receptor PTP: PTP1B — 729
Alan Cheng and Michel L. Tremblay

Introduction
PTP1B as a *Bona Fide* IR Phosphatase
PTP1B Gene Polymorphisms and Insulin Resistance
Insulin-Mediated Modulation of PTP1B
Genetic Evidence for Other PTP1B Substrates
Concluding Remarks
References

Chapter 120
Low-Molecular-Weight Protein Tyrosine Phosphatases — 733
Robert L. Van Etten

Introduction
Structures of LMW PTPases
Catalytic Mechanism
Inhibitors and Activators
Substrate Specificity, Regulation, and Biological Role
References

Chapter 121
STYX/Dead-Phosphatases 741
Matthew J. Wishart

Introduction
Gathering Styx: Structure Implies Function
The Gratefully Undead: STYX/Dead-Phosphatases Mediate Phosphorylation Signaling
Conclusions
References

VOLUME 3

Contributors xlv

Part III
NUCLEAR AND CYTOPLASMIC EVENTS: TRANSCRIPTIONAL AND POST-TRANSCRIPTIONAL REGULATION
Michael Karin, Editor

Part III
Introduction
Michael Karin

Section A: Nuclear Receptors
Michael G. Rosenfeld, Editor

Chapter 270
History of Nuclear Receptors 7
Elwood V. Jensen

Introduction
Discovery of Receptors and Shift in Research Direction
Receptor Forms and Physiological Action
Subsequent Discoveries Relevant to Receptor Structure and Function
References

Chapter 271
Regulation of Basal Transcription by RNA Polymerase II 11
Sohail Malik and Robert G. Roeder

Introduction
The Preinitiation Complex
Global Mechanisms of PIC Function
Gene-Specific Regulation of PIC Function by Transcriptional Activators
Conclusion
References

Chapter 272
Structural Mechanisms of Ligand-Mediated Signaling by Nuclear Receptors 21
H. Eric Xu and Millard H. Lambert

Introduction
Overall Structure of the LBD
Ligand-Binding Pockets
Ligand-Mediated Activation: Mouse Trap versus Charged Clamp
Ligand-Mediated Repression
Dimerization
Summary
References

Chapter 273
Nuclear Receptor Coactivators 25
Riki Kurokawa and Christopher K. Glass

Introduction
Mechanism of Coactivator Recruitment
General Classes of Coactivator Complexes
Coactivators as Targets of Signal Transduction Pathways
Conclusion
References

Chapter 274
Corepressors in Mediating Repression by Nuclear Receptors 29
Kristen Jepsen and Michael G. Rosenfeld

Introduction
N-CoR and SMRT in Repression by Nuclear Receptors
Purification of Corepressor Complexes
Other Nuclear Receptor and Transcription Factor Partners of N-CoR/SMRT
Multiple Mechanisms of N-CoR/SMRT Regulation
Roles in Development and Disease
Other Mediators of Nuclear Receptor Repression
Conclusion
References

Chapter 275
Steroid Hormone Receptor Signaling 35
Vincent Giguère

Introduction
Activation by the Hormone
Hormone-Independent Activation
Cross-Talk with Other Transcription Factors
Nongenomic Action of Steroid Hormones
Estrogen Related Receptors
Selective Steroid Hormone Receptor Modulators
References

Chapter 276
PPARγ Signaling in Adipose Tissue Development 39
Robert Walczak and Peter Tontonoz

Introduction
PPARγ: A Dominant Regulator of Adipose Tissue Development
Analysis of PPARγ Function in Animal Models
Transcriptional Networks in Adipose Tissue Development
Negative Regulation of Adipocyte Differentiation
PPARγ, TNF-α Signaling Antagonism and Insulin Resistance
PPARγ and Cell Cycle Regulation
References

Chapter 277
Orphan Nuclear Receptors 47
Barry Marc Forman

Classical Receptors versus Orphan Receptors
Orphan Receptors and Metabolite-Derived Signals
Orphan Receptors and Xenobiotic Signals
Future Directions
References

Chapter 278
Identification of Ligands for Orphan Nuclear Receptors 53
Steven A. Kliewer and Timothy M. Willson

Introduction
PPARs: Fatty Acid Sensors
LXRs: Cholesterol Sensors
FXR: Bile Acid Sensor
PXR and CAR: Xenobiotic Sensors
Ligands for Other Orphan Nuclear Receptors
Conclusion
References

Chapter 279
Orphan Receptor COUP-TFII and Vascular Development 57
Fabrice G. Petit, Sophia Y. Tsai, and Ming-Jer Tsai

Introduction
Vascular Development
PPARγ: Inhibitor of Angiogenesis
COUP-TFII: Positive Effector in Angiogenesis
Conclusion
References

Chapter 280
Crosss-Talk between Nuclear Receptors and Other Transcription Factors 61
Peter Herrlich

Introduction
Proliferation and Proinflammatory Pathways
Nuclear Receptors
Induced Expression of Inhibitory Molecules
Immediate Hormone Responses
Direct Modulation of Transcription Factors
Conclusion
References

Chapter 281
Drosophila Nuclear Receptors 69
Kirst King-Jones and Carl S. Thummel

Introduction
Nuclear Receptors and Embryonic Pattern Formation
Ecdysone Regulatory Hierarchies
The Neuronal Connection
References

Section B: Transcription Factors
Marc Montiminy, Editor

Chapter 282
JAK-STAT Signaling 77
Christian W. Schindler

Introduction
The JAK-STAT Paradigm
The JAK Family
The STAT Family
A Promising Future
References

Chapter 283
FOXO Transcription Factors: Key Targets of the PI3K-Akt Pathway That Regulate Cell Proliferation, Survival, and Organismal Aging 83
Anne Brunet, Hien Tran, and Michael E. Greenberg

Introduction
Identification of the FOXO Subfamily of Transcription Factors
Regulation of FOXO Transcription Factors by the PI3K-Akt Pathway
Other Regulatory Phosphorylation Sites in FOXOs
Mechanism of the Exclusion of FOXOs from the Nucleus in Response to Growth Factor Stimulation
Transcriptional Activator Properties of FOXOs
FOXOs and the Regulation of Apoptosis
FOXOs Are Key Regulators of Several Phases of the Cell Cycle
FOXOs in Cancer Development: Potential Tumor Suppressors
Role of FOXOs in the Response to Stress and Organismal Aging

FOXOs and the Regulation of Metabolism in
  Relation to Organismal Aging
Conclusion
References

Chapter 284
Multiple Signaling Routes to Histone Phosphorylation — 91
Claudia Crosio and Paolo Sassone-Corsi

Introduction
Histone Phosphorylation and Gene Activation
Histone Phosphorylation and DNA Repair
Histone Phosphorylation and Apoptosis
Histone Phosphorylation and Mitosis
Conclusions
References

Chapter 285
Multigene Family of Transcription Factor AP-1 — 99
Peter Angel

Introduction
General Structure of AP-1 Subunits
Transcriptional and Posttranslational Control of AP-1 Activity
Function of Mammalian AP-1 Subunits: Lessons from Loss-of-Function Approaches in Mice
References

Chapter 286
NFκB: A Key Integrator of Cell Signaling — 107
John K. Westwick, Klaus Schwamborn, and Frank Mercurio

References

Chapter 287
Transcriptional Regulation via the cAMP Responsive Activator CREB — 115
Marc Montminy and Keyong Du

The Transcriptional Response to cAMP
Mechanism of Transcriptional Activation via CREB
Signal Discrimination via CREB
Secondary Phosphorylation of CREB: Ser142
Methylation of the KIX Domain
Cooperative Binding with MLL
References

Chapter 288
The NFAT Family: Structure, Regulation, and Biological Functions — 119
Fernando Macian and Anjana Rao

Introduction
Structure and DNA-Binding
Regulation
Transcriptional Functions
Biological Programs Regulated by NFAT
Perspectives
References

Chapter 289
Transcriptional Control through Regulated Nuclear Transport — 125
Steffan N. Ho

Introduction
Regulated Nuclear Transport: Overview
Coordinate Regulation of Nuclear Import and Export: Calcium-Dependent Nuclear Localization of NFATc Transcription Factors
Regulated Nuclear Transport of Non-DNA-Binding Transcriptional Regulatory Proteins
Conclusion
References

Chapter 290
Proteasome/Ubiquitination — 129
Daniel Kornitzer and Aaron Ciechanover

Protein Degradation and the Ubiquitin/Proteasome System
Regulation of Ubiquitination by Substrate Modification
Regulation of Ubiquitin Ligase Activity
Protein Processing by the Ubiquitin System
Modulation of Kinase Activity by Ubiquitination
Conclusion
References

Chapter 291
Fluorescence Resonance Energy Transfer Microscopy and Nuclear Signaling — 135
Ty C. Voss and Richard N. Day

Introduction
References

Chapter 292
The Mammalian Circadian Timing System — 139
Ueli Schibler, Steven A. Brown, and Jürgen A. Ripperger

Introduction
The Molecular Oscillator
Photic Entrainment of the Central Pacemaker
Outputs of the SCN Pacemaker
Outputs via Subsidiary Clocks
Conclusions and Perspectives
References

CHAPTER 293
Protein Arginine Methylation 145
Michael David

Introduction
Arginine Methylation and Arginine-Methyltransferases
Function of Arginine Methylation
Role of Arginine Methylation in Signal Transduction
References

CHAPTER 294
Transcriptional Activity of Notch and CSL Proteins 149
Elise Lamar and Chris Kintner

Introduction
Components of the Notch Transcriptional Complex
Notch Transcriptional Activity *In Vivo*
Conclusion
References

CHAPTER 295
The β-Catenin: LEF/TCF Signaling Complex: Bigger and Busier than Before 161
Reiko Landry and Katherine A. Jones

Introduction
Regulated Proteolytic Turnover of β-Cat
Regulation of the Wnt-Assembled Enhancer Complex in the Nucleus
Enter Pygopus and Legless (hBcl9)
Perspectives
References

CHAPTER 296
Cubitus Interruptus 167
Sarah M. Smolik and Robert A. Holmgren

Introduction
Protein Structure and Expression Patterns of Ci
Regulation of Ci by Hedgehog
Regulation of Ci by PKA
Ci Transcriptional Regulation
References

CHAPTER 297
The Smads 171
Malcolm Whitman

Introduction
Families: R-Smads, Co-Smads, and I-Smads
Smad Oligomerization and Regulation by Receptors
Transcriptional Regulation by Smads
Down-Regulation and Cross-Regulation of Smads
Function *In Vivo*: Gain of Function Loss of Function
References

Section C: Damage/Stress Responses
Albert J. Fornace, Jr., Editor

CHAPTER 298
Complexity of Stress Signaling and Responses 179
Sally A. Amundson and Albert J. Fornace, Jr.

Introduction: A Variety of Stresses
Origin of Signals
Signal Transduction
Functional Genomics and Proteomics Approaches
References

CHAPTER 299
Signal Transduction in the *Escherichia coli* SOS Response 185
Penny J. Beuning and Graham C. Walker

SOS Response
LexA Cleavage and Other Self-Cleavage Reactions Regulating the SOS Response
Structures of Y-Family Polymerases
Conclusions
References

CHAPTER 300
Oxidative Stress and Free Radical Signal Transduction 191
Bruce Demple

Introduction: Redox Biology
Oxidative Stress Responses in Bacteria: Well-Defined Models of Redox Signal Transduction
Responses to Superoxide Stress and Nitric Oxide: SoxR Protein
Response to H_2O_2 and Nitrosothiols: OxyR Protein
Parallels in Redox and Free-Radical Sensing
Themes in Redox Sensing
References

CHAPTER 301
Budding Yeast DNA Damage Checkpoint: A Signal Transduction-Mediated Surveillance System 197
Marco Muzi-Falconi, Michel Giannattasio, Giordano Liberi, Achille Pelliccioli, Paolo Plevani, and Marco Foiani

Introduction
Sensing
Downstream Events
References

CHAPTER 302
Finding Genes That Affect Signaling and Toleration of DNA Damage, Especially DNA Double-Strand Breaks   203
Craig B. Bennet and Michael A. Resnick

Introduction
Nature of DSB and Repair and Genetic Consequences
Checkpoint Activation and Adaptation as Signaling Responses to DSBs
DNA Damage Signaling Networks
Identifying Checkpoint Defects by Screening Radiation-Sensitive Mutants
Checkpoint Mutants Revealed through Screening DNA Replication Mutants
Screening for Checkpoint Defects
Screen for Altered Checkpoint and Adaptation Responses to a Single DSB
Other Screens for DNA Damage Checkpoint Pathway Genes
Implications of DNA Damage Checkpoint Signaling
References

CHAPTER 303
Radiation Responses in *Drosophila*   213
Naoko Sogame and John M. Abrams

Introduction
Sensors and Transmitters
Effectors
Conclusions: What Can We Learn from the *Drosophila* Model?
References

CHAPTER 304
Double-Strand Break Recognition and Its Repair by Nonhomologous End Joining   219
Jane M. Bradbury and Stephen P. Jackson

Introduction
Repair of DSBs: Homologous Recombination and NHEJ
Recognition of DNA DSBs
Signal Transduction
DNA Repair
Other Sensors and Transducers of DNA Damage
New Factors in NHEJ
Future Prospects
References

CHAPTER 305
Role of ATM in Radiation Signal Transduction   225
Martin F. Lavin, Shaun Scott, Philip Chen, Sergei Kozlov, Nuri Gueven, and Geoff Birrell

Introduction
Sensing Radiation Damage in DNA
ATM Signaling: Recognition of Breaks in DNA
Checkpoint Activation
Role of ATM in More General Signaling
Perspective
References

CHAPTER 306
Signaling to the p53 Tumor Suppressor through Pathways Activated by Genotoxic and Nongenotoxic Stresses   237
Carl W. Anderson and Ettore Appella

Introduction
p53 Protein Structure
Posttransitional Modifications to p53
Regulation of p53 Activity
Activation of p53 by Genotoxic Stresses
Activation of p53 by Nongenotoxic Stresses
Conclusions
References

CHAPTER 307
Abl in Cell Signaling   249
Jean Y. J. Wang

Introduction
Functional Domains of Abl
Proteins that Interact with Abl
Abl in Signal Transduction
Future Prospects
References

CHAPTER 308
Radiation-Induced Cytoplasmic Signaling   257
Christine Blattner and Peter Herrlich

Introduction
Cytoplasmic Signaling Network
Redox Sensitivity and Metal Toxicity: Toxic Agents Activate Signaling Pathways
Activation of Signaling Components
Primary Radiation Targets: DNA Damage versus Cytoplasmic Signaling
Other Signaling-Initiating Principles
Conclusions
References

CHAPTER 309
Endoplasmic Reticulum Stress Responses   263
David Ron

Introduction
ER Stress Defined
The UPR in Yeast
The UPR Is Metazoans

Conclusions
References

CHAPTER 310
The Heat-Shock Response: Sensing the Stress of Misfolded Proteins 269
Richard I. Morimoto and Ellen A. A. Nollen

Introduction
Transcriptional Regulation of the Heat-Shock Response
Molecular Chaperones: Folding, Misfolding, and the Assembly of Regulatory Complexes
Neurodegenerative Diseases: When Aggregation-Prone Proteins Go Awry
References

CHAPTER 311
Hypoxia-Mediated Signaling Pathways 277
Albert C. Koong and Amato J. Giaccia

Introduction
HIF-1 Signaling
Unfolded Protein Response
Conclusions
References

CHAPTER 312
Regulation of mRNA Turnover by Cellular Stress 283
Myriam Gorospe

Introduction
mRNA Stability
Stress-Activated Signaling Molecules that Regulate mRNA Turnover
Conclusions
References

Section D: Post-Translational Control
Nahum Sonenberg, Editor

CHAPTER 313
RNA Localization and Signal Transduction 293
Vaughan Latham and Robert H. Singer

Introduction
Growth Factors Induce mRNA Localization
Signaling from the Extracellular Matrix Induces mRNA Localization
mRNAs Localized via the Cytoskeleton
mRNA Granule Movement in Neurons
Regulation of mRNA Localizing Proteins
GTPase Signals Regulating Actomyosin Interactions Are Involved in mRNA Localization

Conclusion
References

CHAPTER 314
Translational Control by Amino Acids and Energy 299
Tobia Schmelze, José L. Crespo, and Michael N. Hall

Introduction
GCN System
TOR Signaling Pathway
References

CHAPTER 315
Translational Control and Insulin Signaling 305
Thomas Radimerski and George Thomas

References

CHAPTER 316
Unfolded Protein Response: An Intracellular Signaling Pathway Activated by the Accumulation of Unfolded Proteins in the Lumen of the Endoplasmic Reticulum 311
Randal J. Kaufman

Introduction
UPR in *Saccharomyces cerevisiae*
UPR Transcriptional Activation in Metazoan Species
Physiological Role for the UPR in Mammals
Future Directions
References

CHAPTER 317
Regulation of mRNA Turnover 319
Perry J. Blackshear and Wi S. Lai

Introduction
Current Models of mRNA Stability in Vertebrate Cells
Presence of Instability Elements in Vertebrate mRNAs
Effects of ARE Binding Proteins on mRNA Turnover
Regulation of TTP Activity in Cells
Conclusion
References

CHAPTER 318
CPEB-Mediated Translation in Early Vertebrate Development 323
Joel D. Richter

Introduction
Mechanism of Translational Control
CPEB and Early Development
Conclusions
References

Chapter 319
Translational Control in Invertebrate Development 327
Paul Lasko

Introduction
Translational Control Targets Oskar to the Pole Plasm
Translational Control Targets Nanos to the Pole Plasm
Translational Control in the *Drosophila* Nervous System
Role for Translational Control in Regulation Growth
Translational Repression through MicroRNAs
References

Chapter 320
Role of Alternative Splicing During the Cell Cycle and Programmed Cell Death 331
Chanseok Shin and James L. Manley

Introduction
Apoptosis and Splicing
Cell Cycle and Splicing Regulation
References

Chapter 321
NF90 Family of Double-Stranded RNA-Binding Proteins: Regulators of Viral and Cellular Function 335
Trevor W. Reichman and Michael B. Mathews

Summary
Introduction
Members of the NF90 Protein Family
Domain Structure of NF90 Family Proteins
Proteins that Interact with NF90
Nucleic Acid Binding Properties of NF90
Functions of NF90 Homologs
Cellular Regulation of NF90 and NF45
Conclusions
References

Chapter 322
Signaling Pathways that Mediate Translational Control of Ribosome Recruitment of mRNA 343
Nahum Sonenberg and Emmanuel Petroulakis

Introduction
eIF4F Complex Formation
Repressors of Cap-Dependent Translation
Modulation of 4E-BP Phosphorylation FRAP/mTOR
Phosphorylation of eIF4G and eIF4B
Control of Cell Growth and Proliferation by eIF4E: Link to Cancer
Conclusions
References

PART IV
EVENTS IN INTRACELLULAR COMPARTMENTS
Marilyn Farquhar, Editor

Chapter 323
SREBPs: Gene Regulation through Controlled Protein Trafficking 353
Peter J. Espenshade, Joseph L. Goldstein, and Michael S. Brown

Introduction
SREBPs: Membrane-Bound Transcription Factors
SCAP: Sterol Sensor and Escorter of SREBP from ER to Golgi
Sterols Control Sorting of SCAP/SREBP into ER Vesicles
ER Retention of SCAP/SREBP
Conclusions
References

Chapter 324
Endoplasmic Reticulum Stress Responses 359
David Ron

Introduction
Conclusion
References

Chapter 325
Signaling Pathways from Mitochondria to the Nucleus 365
Zhengchang Liu and Ronald A. Butow

Introduction
Milestones in Mitochondrial Research
Mitochondrial Signaling
Aging and Retrograde Regulation
Conclusions
References

Chapter 326
Signaling During Exocytosis 375
Lee E. Eiden

Introduction
Functional, Morphological, and Historical Aspects of Exocytosis and Stimulus-Secretion Coupling
Secretion Begins with Secretagogues
Secretagogues Act at Target Cell Receptors
Calcium and Cyclic AMP: The Two Main Second Messengers for Secretion
Calcium and the Regulation of Exocytosis
Exocytosis and SNAREs
Calcium and cAMP Sensors for Exocytosis
Role of Signal Summation in Regulated Exocytosis
Role of PKC and Other PMA Targets in Regulated Secretion
Negative Regulation of Secretion

Upstream Regulation of Secretion
Far Upstream Regulation of Secretion
Conclusions and Future Outlook for Signaling in Exocytosis
References

CHAPTER 327
Nonclassical Pathways of Protein Export 393
Igor Prudovsky, Anna Mandinova, Cinzia Bagala, Raffaella Soldi, Stephen Bellum, Chiara Battelli, Irene Graziani, and Thomas Maciag

Introduction
Fibroblast Growth Factor Export Pathways
The Export of FGF-1 as a Multiprotein Complex
Interleukin-1 Export Pathways
Acidic Phospholipids and the Molten Globule Hypothesis
The Potential Pathophysiological Implication of Nonclassical Release
References

CHAPTER 328
Regulation of Cell Cycle Progression 401
Clare H. McGowan

Introduction
Being There: Cyclins Define Cell Cycle Phase
Signals to Slow Processes: Regulation of Cdks by Inhibitory Proteins
Cdks Are Positively and Negatively Regulated by Phosphorylation
Degradation: The Importance of Being Absent
Location, Location, Location
Checkpoint Signaling
References

CHAPTER 329
Endocytosis and Cytoskeleton 411
Pier Paolo Di Fiore and Giorgio Scita

Introduction
Actin Dynamics and Endocytosis
Role of Microtubule Cytoskeleton in Receptor Endocytosis
Physical and Functional Interactions of Dynamin and Dynamin-Interacting Proteins with the Actin Cytoskeleton
Integration of Signals in Endocytosis and Actin Dynamics by Small GTPases
Conclusions
References

CHAPTER 330
Molecular Basis for Nucleocytoplasmic Transport 419
Gino Cingolani and Larry Gerace

Introduction
Transport Signals
Transport Receptors
The Small GTPase Ran
Nuclear Pore Complex
Mechanism of Transport
Future Directions
References

CHAPTER 331
Apoptosis Signaling: A Means to an End 431
Lisa J. Pagliari, Michael J. Pinkoski, and Douglas R. Green

Introduction
The End of the Road
Caspase-8 Activation via Death Receptors
Mitochondria and the Activation of Caspase-9
Mitochondrial Outer Membrane Permeabilization
The Bcl-2 Family
Cell Cycle versus Apoptosis
Conclusions
References

CHAPTER 332
Signaling Down the Endocytic Pathway 441
Jeffrey L. Benovic and James H. Keen

Introduction
RTK Signaling from the Cell Surface
RTK Signaling from Endocytic Compartments
GPCR Signaling Paradigms and Desensitization
Control of RTK and GPCR Trafficking Leading to Degradation
GPCR Activation of MAP Kinases
Endocytic Signaling in Developmental Systems
Signaling between Neuronal Cell Body and Terminal
References

PART V
CELL-CELL AND CELL-MATRIX INTERACTIONS
E. Brad Thompson, Editor

PART V
Introduction
Brad Thompson

CHAPTER 333
Overview of Cell-Cell and Cell-Matrix Interactions 452
E. Brad Thompson and Ralph A. Bradshaw

References

CHAPTER 334
Angiogenesis: Cellular and Molecular Aspects of Postnatal Vessel Formation 455
Carla Mouta, Lucy Liaw, and Thomas Maciag

Introduction
Initiators of Angiogenesis: Cellular, Metabolic, and Mechanical
Vessel-Specific Requirements in Angiogenesis
Cellular and Soluble Regulators
Coordination of Angiogenesis by Cellular and Molecular Interactions
References

CHAPTER 335
Signaling Pathways Involved in Cardiogenesis 463
Deepak Srivastava

Introduction
Cardiomyocyte and Heart Tube Formation
Cardiac Looping and Left-Right Asymmetry
Patterning of the Developing Heart Tube
Myocardial Growth
Cardiac Valve Formation
Cardiac Outflow Tract and Aortic Arch Development
Conclusions
References

CHAPTER 336
Development and Regulatory Signaling in the Pancreas 471
Murray Korc

Introduction
Ontogeny of the Pancreas
Pancreatic Islet-Acinar Interactions
Cell-Cell and Matrix Interactions in the Endocrine Pancreas
Matrix and Cell-Cell Interactions in the Exocrine Pancreas
Conclusions
References

CHAPTER 337
Tropic Effects of Gut Hormones in the Gastrointestinal Tract 477
B. Mark Evers and Robert P. Thomas

Introduction
Tropic Effects of Gut Peptides in the Stomach, Small Bowel, and Colon
GI Hormone Receptors and Signal Transduction Pathways
Signaling Pathways Mediating the Effects of Intestinal Peptides
Conclusions
References

CHAPTER 338
Integrated Response to Neurotrophic Factors 485
J. Regino Perez-Polo

Introduction
Neural Cell Death
The Neurotrophic Hypothesis
Neurotrophins
Neurotrophin Receptors
Neurotrophin Signaling Pathways
Transcriptional Regulation
AP-1
NFκB Transcription Factor
Role of NFκB
Conclusions
References

CHAPTER 339
Cell-Cell and Cell-Matrix Interactions in Bone 497
L. F. Bonewald

Introduction
Diseases of Bone
Bone Cells and Their Functions
Mechanical Strain
Hormone Responsible for Bone Development, Growth, and Maintenance
Growth and Transcription Factors Responsible for Bone Development and Growth
Fibroblast Growth Factors
Bone Extracellular Matrix
Conclusions
References

CHAPTER 340
Cell-to-Cell Interactions in Lung 509
Joseph L. Alcorn

Introduction
Lung Organogenesis and Development
Soluble Factors of Cell-to-Cell Interactions Involved in Lung Injury
Conclusion
References

CHAPTER 341
Mechanisms of Stress Response Signaling and Recovery in the Liver of Young versus Aged Mice: The p38 MAPK and SOCS Families of Regulatory Proteins 515
John Papaconstantinou

Introduction
The p38 MAPK Pathway in Stress Response Signaling
SOCS Family of Negative Regulators of Inflammatory Response
Conclusions
References

Chapter 342
Cell-Cell Signaling in the Testis and Ovary — 531
Michael K. Skinner

Introduction
Cell-Cell Signaling in the Testis
Cell-Cell Signaling in the Ovary
Conclusions
References

Chapter 343
T Lymphocytes — 546
Rolf König and Wenhong Zhou

Introduction
Signaling Receptors in T Cells form Dynamic Macromolecular Signaling Complexes
Coreceptor and Costimulatory Proteins Modulate T-Cell Signaling Pathways
Intracellular Signaling Pathways Induced by Antigen Stimulation of T Cells
Conclusions
References

Chapter 344
Signal Transduction via the B-Cell Antigen Receptor: A Crucial Regulator of B-Cell Biology — 555
Louis B. Justement

Introduction
Initiation of Signal Transduction through the BCR
Propagation of Signal Transduction via the BCR
Conclusions
References

Chapter 345
Signaling Pathways in the Normal and Neoplastic Breast — 565
Danica Ramljak and Robert B. Dickson

Introduction
Signaling Molecules: A Class of Growth Factors
PI3K/Akt, MEK/Erk, and Stats: Major Proliferation/Survival Molecules Downstream of Growth Factor Receptors in Breast
Conclusions and Future Prospects
References

Chapter 346
Kidney — 573
Elsa Bello-Reuss and William J. Arendshorst

Overview of Kidney Functions and Cell-to-Cell Interactions
Vascular Endothelial Cells
Vascular Smooth Muscle Cells
Tubulovascular Interactions: The Juxtaglomerular Apparatus
Tubulovascular Interactions: The Juxtaglomerular Apparatus and Tubuloglomerular Feedback
Vasculotubular Communication
Tubule-Tubule Communication: Paracrine Agents Released from Epithelial Cells
Interstitial Cell-Tubule Communication
Conclusions
References

Chapter 347
Prostate — 591
Jean Closset and Eric Reiter

Introduction
Development of the Prostate during Fetal Life
The Adult Prostate
The Prostate during Aging
Conclusions
References

Chapter 348
Retrograde Signaling in the Nervous System: Dorsal Root Reflexes — 607
William D. Willis

Cell-to-Cell Signaling in the Nervous System
Retrograde Signaling
Neurogenic Inflammation
Dorsal Root Reflexes as Retrograde Signals
Conclusions
References

Chapter 349
Cytokines and Cytokine Receptors Regulating Cell Survival, Proliferation, and Differentiation in Hematopoiesis — 615
Fiona J. Pixley and E. Richard Stanley

General Aspects of Hematopoiesis
Signaling through Cytokine Receptors
Conclusions
References

Chapter 350
Regulation of Bartlett Endogenous Stem Cells in the Adult Mammalian Brain: Promoting Neuronal Repair — 625
Rodney L. Rietze and Perry F. Bartlett

Adult Neurogenesis Revealed
Isolation and Culture of Neural Stem Cells
Regulation of Stem Cell Differentiation into Neuron
References

Index

Contributors

Numbers in parentheses indicate the pages on which the authors' contributions begin.

John M. Abrams (3:213)
Department of Cell Biology, University of Texas Southwestern Medical Center, Dallas, Texas 75390

John P. Adelman (1:227)
Oregon Health Sciences University, Vollum Institute, Portland, Oregon 97201

Joseph L. Alcorn (3:509)
Department of Pediatrics, University of Texas Health Science Center, Houston, Texas 77030

Dario R. Alessi (1:513; 2:193)
Department of Biochemistry, MRC Unit, University of Dundee, Dundee DD1 5EH, Scotland, United Kingdom

Emil Alexov (1:11)
Department of Biochemistry and Molecular Biophysics, Columbia University, Howard Hughes Medical Institute, New York, New York 10032

Simon Alford (2:663)
Department of Biological Sciences, University of Illinois at Chicago, Chicago, Illinois 60637

Kari Alitalo (2:855)
Molecular/Cancer Biology Laboratory and Ludvig Institute for Cancer Research, Haartman Institute and Helsinki University Central Hospital, Biomedicum Helsinki, University of Helsinki, FIN-00014 Helsinki, Finland

James P. Allison (1:355)
Department of Molecular and Cell Biology, Howard Hughes Medical Institute, University of California, Berkeley, Berkeley, California 94720

Steven C. Almo (1:74)
Department of Biochemistry, Albert Einstein College of Medicine, Bronx, New York 10461-1975

Christelle Alory (2:689)
Departments of Cell and Molecular Biology and Institute for Childhood and Neglected Disease, The Scripps Research Institute, La Jolla, California 92037

Aymen Al-Shamkhani (1:319)
University of Southampton General Hospital, Tenovus Research Lab, Cancer Sciences Division, School of Medicine, Southampton SO16 6YD, United Kingdom

Sally A. Amundson (3:179)
Gene Response Section, Center for Cancer Research, National Cancer Institute, National Institutes of Health, Bethesda, Maryland 20892

Carl W. Anderson (3:237)
Biology Department, Brookhaven National Laboratory, Upton, New York 11973

Jannik N. Andersen (1:659)
Cold Spring Harbor Laboratory, Cold Spring Harbor, New York 11724

Peter Angel (3:99)
Department of Signal Transduction and Growth Control, Deutsches Krebsforschungszentrum, 69120 Heidelberg, Germany

Ettore Appella (3:237)
Laboratory of Cell Biology, National Cancer Institute, National Institutes of Health, Bethesda, Maryland 20892

William J. Arendshorst (3:573)
Department of Cell and Molecular Physiology, University of North Carolina at Chapel Hill, Chapel Hill, North Carolina 27599

Steve Arkinstall (1:703)
Departments of Microbiology and Medicine, Serono Reproductive Biology Institute, Rockland, Massachusetts 02370

Anjon Audhya (2:177)
Division of Cellular and Molecular Medicine, The Howard Hughes Medical Institute, University of California San Diego, School of Medicine, La Jolla, California 92093

Joseph Avruch (1:523)
Harvard Medical School, Massachusetts General Hospital, Boston, Massachusetts 02114

Gary D. Bader (2:311)
Samuel Lunenfeld Research Institute, Mount Sinai Hospital, Toronto, Ontario M5G 1X5, Canada

Cinzia Bagala (3:393)
Center for Molecular Medicine, Maine Medical Center Research Institute, Scarborough, Maine 04074

William E. Balch (2:689)
Departments of Cell and Molecular Biology and Institute for Childhood and Neglected Disease, The Scripps Research Institute, La Jolla, California 92037

Jesus Balsinde (2:261)
Institute of Molecular Biology and Genetics, University of Valladolid School of Medicine, E-47005 Valladolid, Spain

Utpal Banerjee (2:827)
Department of Molecular Cell and Developmental Biology, Biological Chemistry, Human Genetics, Molecular Biology Institute, University of California at Los Angeles, Los Angeles, California 90095

David Barford (1:601)
Chester Beatty Laboratories, Section of Structural Biology, Institute of Cancer Research, London, SW3 6JB, United Kingdom

Dafna Bar-Sagi (2:711)
Department of Molecular Genetics and Microbiology, State University of New York at Stony Brook, Stony Brook, New York 11794

Perry F. Bartlett (3:625)
Institute for Brain Research, The University of Queensland, Brisbane QLD 4072, Queensland, Australia

Philippe I. H. Bastiaens (2:305)
Cell Biology and Cell Biophysics Program, European Molecular Biology Laboratory, 69120 Heidelberg, Germany

Chiara Battelli (3:393)
Center for Molecular Medicine, Maine Medical Center Research Institute, Scarborough, Maine 04074

Linnea M. Baudhuin (2:253)
Department of Cancer Biology, The Lerner Research Institute, Cleveland Clinic Foundation, Cleveland, Ohio, and Department of Chemistry, Cleveland State University, Cleveland, Ohio 44195

Andrew J. Beavil (1:45)
The Randall Centre, King's College London, New Hunt's House, Guy's Campus, London SE1 1UL, United Kingdom

Rebecca L. Beavil (1:45)
The Randall Centre, King's College London, New Hunt's House, Guy's Campus, London SE1 1UL, United Kingdom

Joseph A. Beavo (2:431; 2:525)
Department of Pharmacology, University of Washington, Seattle, Washington 98195

Elsa Bello-Reuss (3:573)
Department of Internal Medicine, Division of Nephrology, and Department of Physiology and Biophysics, The University of Texas Medical Branch, Galveston, Texas 77555

Stephen Bellum (3:393)
Center for Molecular Medicine, Maine Medical Center Research Institute, Scarborough, Maine 04074

Juan Carlos Izpisúa Belmonte (2:799)
The Salk Institute for Biological Studies, Gene Expression Laboratory, La Jolla, California 92037

Craig B. Bennett (3:203)
Department of Surgery, Duke University Medical Center, Durham, North Carolina 27710

Jeffrey L. Benovic (3:441)
Cell Biology and Signaling Program, Kimmel Cancer Center, Thomas Jefferson University, Philadelphia, Pennsylvania 19107

Michael J. Berridge (2:51)
The Babraham Institute, Babraham, Cambridge CB2 4AT, United Kingdom

Penny J. Beuning (3:185)
Biology Department, Massachusetts Institute of Technology, Cambridge, Massachusetts 02139

Rashna Bhandari (1:343)
Departments of Molecular and Cell Biology and Chemistry, HHMI, University of California, Berkeley, Berkeley, California 94720-3202

Ananya Bhattacharya (2:779)
Waksman Institute, Department of Molecular Biology and Biochemistry, and Cancer Institute of New Jersey, Rutgers University, Piscataway, New Jersey 08901

Martin Biel (2:515)
Department Pharmzaie-Zentrum für Pharmaforschung, Ludwig-Maximilians-Universität München, 81377 München, Germany

Vincent A. Bielinski (2:405)
Department of Pharmacology, University of Texas Southwestern Medical Center, Dallas, Texas 75392

Hana Bilak (2:333)
National Center of Scientific Research, Institute of Molecular and Cellular Biology, 67084 Strasbourg, France

Lutz Birnbaumer (2:557)
Laboratory of Signal Transduction, National Institute of Environmental Health Sciences, Research Triangle Park, North Carolina 27709

Geoff Birrell (3:225)
Queensland Cancer Fund Research Laboratory, Queensland Institute of Medical Research, Brisbane 4029, Australia

Gail A. Bishop (1:315)
Departments of Microbiology and Internal Medicine, University of Iowa and the VA Medical Center, Iowa City, Iowa 52242

Trillium Blackmer (2:663)
Department of Biological Sciences, University of Illinois at Chicago, Chicago, Illinois 60637

Perry J. Blackshear (3:319)
Office of Clinical Research and Laboratory of Signal Transduction, National Institute of Environmental Health Sciences, Research Triangle Park, North Carolina 27709

and Departments of Medicine and Biochemistry, Duke University, Durham, North Carolina 27710

Christine Blattner (3:257)
Institute of Genetics and Toxicology, Forschungszentrum Karlsruhe, 76021 Karlsruhe, Germany

Mordecai P. Blaustein (2:63)
Department of Physiology, University of Maryland, School of Medicine, Baltimore, Maryland 21201

Gary M. Bokoch (2:705)
Departments of Immunology and Cell Biology, The Scripps Research Institute, La Jolla, California 92037

Lynda F. Bonewald (3:497)
University of Missouri at Kansas City, School of Dentistry, Kansas City, Missouri 64108

Marco Bonomi (1:161)
Faculty of Medicine, University of Brussels, Institut de Recherche Interdisciplinaire, Brussels B-1070, Belgium

Michelle A. Booden (2:675)
Lineberger Comprehensive Cancer Center, University of North Carolina at Chapel Hill, Chapel Hill, North Carolina 27599

Charles Boone (2:311)
Banting and Best Department of Medical Research and Department of Molecular and Medical Genetics, University of Toronto, Toronto, Ontario M5G 1X5, Canada

Martin D. Bootman (2:51)
The Babraham Institute, Babraham, Cambridge CB2 4AT, United Kingdom

Johannes L. Bos (2:521)
Department of Physiological Chemistry and Center for Biomedical Genetics, University Medical Center Utrecht, Utrecht, The Netherlands

Jane M. Bradbury (3:219)
Wellcome Trust and Cancer Research Campaign, Institute of Cancer and Developmental Biology, Cambridge, United Kingdom

Ralph A. Bradshaw (1:1; 1:361; 3:453)
Department of Physiology and Biophysics, Department of Anatomy and Neurobiology, University of California, Irvine, Irvine, California 92697-4560

Anne R. Bresnick (1:74)
Center for Synchrotron Biosciences, Albert Einstein College of Medicine, Bronx, New York 10461-1975

Lena Brevnova (1:259)
Department of Microbiology and Immunology, Department of Structural Biology, Stanford University School of Medicine, Stanford, California 94305-5124

Ross I. Brinkworth (2:495)
Department of Biochemistry and Molecular Biology, Institute for Molecular Bioscience, University of Queensland, St. Lucia, Brisbane QLD 4072, Queensland, Australia

Michael S. Brown (3:353)
Department of Molecular Genetics, University of Texas Southwestern Medical Center at Dallas, Dallas, Texas 75390

Steven A. Brown (3:139)
Department of Molecular Biology, Sciences II, University of Geneva, CH-1211 Geneva, Switzerland

Anne Brunet (3:83)
Division of Neuroscience, Children's Hospital and Department of Neurobiology, Harvard Medical School, Boston, Massachusetts 02115

Robert Bucki (2:209)
Department of Physiology, Institute of Medicine and Engineering, University of Pennsylvania, Philadelphia, Pennsylvania 19104

Robert D. Burgoyne (2:79)
The Physiological Laboratory, The University of Liverpool, Liverpool L69 3BX, United Kingdom

Janice E. Buss (2:675)
Department of Biochemistry, Biophysics, and Molecular Biology, Iowa State University, Ames, Iowa 50011

Ronald A. Butow (3:365)
Department of Molecular Biology, University of Texas Southwestern Medical Center, Dallas, Texas 75390

Javier Capdevila (2:799)
The Salk Institute for Biological Studies, Gene Expression Laboratory, La Jolla, California 92037

Ernesto Carafoli (2:52)
Department of Biological Chemistry, University of Maryland School of Medicine, Baltimore, Maryland 21201

Cathrine R. Carlson (2:501)
Department of Medical Biochemistry, Institute of Basic Medical Sciences, University of Oslo, N-0316 Oslo, Norway

Graham Carpenter (2:5)
Department of Biochemistry, Vanderbilt University School of Medicine, Nashville, Tennessee 37232

Juan J. Carrillo (2:619)
Molecular Pharmacology Group, Division of Biochemistry and Molecular Biology, University of Glasgow, Glasgow G12 8QQ, Scotland, United Kingdom

Patrick J. Casey (2:601)
Department of Pharmacology and Cancer Biology, Duke University Medical Center, Durham, North Carolina 27710

William A. Catterall (2:23)
Department of Pharmacology, University of Washington, Seattle, Washington 98195

Richard A. Cerione (2:715; 2:721)
Department of Molecular Medicine and Department of Chemistry and Chemical Biology, Cornell University, Ithaca, New York 14853

Gianni Cesareni (2:311)
Department of Biology, University of Rome Tor Vergata, 00153 Rome, Italy

Andrew C. Chan (1:475)
Department of Immunology, Genentech, Inc., South San Francisco, California 94080

Geoffrey Chang (1:115)
Department of Molecular Biology, The Scripps Research Institute, La Jolla, California 92037

Moses V. Chao (2:839)
Molecular Neurobiology Program, Skirball Institute of Biomolecular Medicine, New York University School of Medicine, New York, New York 10016

Harry Charbonneau (1:697)
Department of Biochemistry, Purdue University, West Lafayette, Indiana 47907-1153

Philip Chen (3:225)
Queensland Cancer Fund Research Laboratory, Queensland Institute of Medical Research, Brisbane 4029, Australia

Alan Cheng (1:729)
McGill Cancer Center, Department of Biochemistry, McGill University, Montreal, Quebec H3G 1Y6, Canada

Chris Chiu (1:475)
Department of Immunology, Genentech, Inc., South San Francisco, California 94080

Dar-chone Chow (1:259)
Department of Microbiology and Immunology, Department of Structural Biology, Stanford University School of Medicine, Stanford, California 94305-5124

Ted D. Chrisman (2:427)
Cecil H. and Ida Green Center for Reproductive Biology Sciences, Howard Hughes Medical Institute and Department of Pharmacology, University of Texas Southwestern Medical Center, Dallas, Texas 75390

Anne Elisabeth Christensen (2:549)
Department of Anatomy and Cell Biology, University of Bergen, N-5009 Bergen, Norway

Jee Y. Chung (1:311)
Department of Biochemistry, Cornell University, Weill Medical College, New York, New York 10021

Grant C. Churchill (2:15)
Department of Pharmacology, University of Oxford, Oxford OX1 2JD, United Kingdom

Aaron Ciechanover (3:129)
Department of Biochemistry, Bruce Rappaport Faculty of Medicine, Technion-Israel Institute of Technology, Haifa, Israel

Gino Cingolani (3:419)
Department of Cell Biology, The Scripps Research Institute, La Jolla, California 92037

Sylvie Claeysen (1:161)
Faculty of Medicine, University of Brussels, Institut de Recherche Interdisciplinaire, Brussels B-1070, Belgium

Jean Closset (3:591)
Department of Biochemistry, Faculty of Medicine, Institute of Pathology, University of Liège, B 4000 Liège, Belgium

Shamshad Cockcroft (2:225)
Department of Physiology, University College London, London WC1E 6JJ, United Kingdom

Patricia T. W. Cohen (1:593)
Medical Research Council Protein Phosphorylation Unit, School of Life Sciences, University of Dundee, Dundee DD1 5EH, Scotland

Philip Cohen (1:547)
MRC Protein Phosphorylation Unit, Department of Biochemistry, University of Dundee, Dundee DD1 5EH, Scotland

Roger J. Colbran (2:397)
Department of Molecular Physiology and Biophysics and the Center for Molecular Neuroscience, Vanderbilt University School of Medicine, Nashville, Tennessee 37232

Clay E. S. Comstock (2:545)
Department of Biochemistry and Molecular Biology, University of North Dakota School of Medicine and Health Sciences, Grand Forks, North Dakota 58203

Marco Conti (2:437)
Department of Gynecology and Obstetrics, Stanford University Medical Center, Stanford, California 94305

Jackie D. Corbin (2:447; 2:465)
Department of Molecular Physiology and Biophysics, Vanderbilt University School of Medicine, Nashville, Tennessee 37232

Daniela Corda (2:613)
Department of Cell Biology and Oncology, Instituto di Ricerche Farmacologiche "Mario Negri," Santa Maria Imbaro, Chieti, Italy

Sabine Costagliola (1:161)
Faculty of Medicine, University of Brussels, Institut de Recherche Interdisciplinaire, Brussels B-1070, Belgium

Rick H. Cote (2:453; 2:465)
Department of Biochemistry and Molecular Biology, University of New Hampshire, Durham, New Hampshire 03824

Shaun R. Coughlin (1:167)
Departments of Medicine and Cellular and Molecular Pharmacology, Cardiovascular Research Institute, University of California San Francisco, San Francisco, California 94920-0130

L. Ashley Cowart (2:257)
Department of Biochemistry and Molecular Biology, Medical University of South Carolina, Charleston, South Carolina 29425

Adrienne D. Cox (2:681; 2:737)
Department of Pharmacology, and Lineberger Comprehensive Cancer Center, University of North Carolina at Chapel Hill, Chapel Hill, North Carolina 27599

Mark S. Cragg (1:319)
University of Southampton General Hospital, Tenovus Research Laboratory, Cancer Sciences Division, School of Medicine, Southampton SO16 6YD, United Kingdom

José L. Crespo (3:299)
Division of Biochemistry, Biozentrum, University of Basel, 4056 Basel, Switzerland

Claudia Crosio (3:91)
Institute of Genetic and Molecular and Cellular Biology, 67084 Strasbourg, France

Christopher Daly (2:849)
Regeneron Pharmaceuticals, Inc., Tarrytown, New York 10591

Sami Damak (2:657)
Department of Physiology and Biophysics, The Mount Sinai School of Medicine, New York, New York 10029

Mary Dasso (2:695)
Laboratory of Gene Regulation and Development, National Institute of Child Health and Human Development, National Institutes of Health, Bethesda, Maryland 20892

Michael David (3:145)
Molecular Biology Section, Division of Biology, University of California San Diego, La Jolla, California 92093

Anthony J. Davis (2:405)
Department of Pharmacology, University of Texas Southwestern Medical Center, Dallas, Texas 75390

Roger J. Davis (2:365)
Howard Hughes Medical Institute and Program in Molecular Medicine, University of Massachusetts Medical School, Worcester, Massachusetts 01605

Richard N. Day (3:135)
Departments of Medicine and Cell Biology, NSF Center for Biological Timing, University of Virginia Health Sciences Center, Charlottesville, Virginia 22908

Eva Degerman (2:441)
Section for Molecular Signaling, Department of Cell and Molecular Biology, Lund University, SE-221 00 Lund, Sweden

Warren L. DeLano (1:51)
Sunesis Pharmaceuticals, Inc., San Francisco, California 94080

Mark L. Dell'Acqua (2:377)
Department of Pharmacology, University of Colorado Health Sciences Center, Denver, Colorado 80262

Emmanuèle Délot (2:833)
Department of Human Genetics, Department of Pediatrics, Department of Orthopedic Surgery, The David Geffen School of Medicine at the University of California Los Angeles, Los Angeles, California 90095

Bruce Demple (3:191)
Department of Cancer Cell Biology, Harvard School of Public Health, Boston, Massachusetts 02115

Edward A. Dennis (1:1; 2:261)
Department of Chemistry and Biochemistry, School of Medicine, University of California San Diego, La Jolla, California 92093

John M. Denu (1:653)
Department of Biochemistry and Molecular Biology, Oregon Health and Science University, Portland, Oregon 97201-3098

Anna A. DePaoli-Roach (1:613)
Department of Biochemistry and Molecular Biology, Indiana University School of Medicine, Indianapolis, Indiana 46202-5122

Channing J. Der (2:681)
Department of Pharmacology, and Lineberg Comprehensive Cancer Center, University of North Carolina at Chapel Hill, Chapel Hill, North Carolina 27599

Johan de Rooij (2:521)
Department of Physiological Chemistry and Center for Biomedical Genetics, University Medical Center Utrecht, Utrecht, The Netherlands

Frederic de Sauvage (2:793)
Department of Molecular Biology, Genentech, Inc., South San Francisco, California 94080

Peter N. Devreotes (2:645)
Department of Cell Biology, Johns Hopkins School of Medicine, Baltimore, Maryland 21201

Valérie Dewaste (2:11)
Interdisciplinary Research Institute (IRIBHN), Université Libre de Bruxelles, Brussels, Belgium

Robert B. Dickson (3:565)
Department of Oncology, Lombardi Cancer Center, Georgetown University, Washington, DC

Becky A. Diebold (2:705)
Department of Immunology, The Scripps Research Institute, La Jolla, California 92037

Pier Paolo Di Fiori (3:411)
Department of Experimental Oncology, Istituto Europeo di Oncologia, 20141 Milan, Italy, and Medical School, University of Milan, Milan Italy, and FIRC Institute for Molecular Oncology, 20134 Milan, Italy

Maria Di Girolamo (2:613)
Department of Cell Biology and Oncology, Instituto di Ricerche Farmacologiche "Mario Negri," Santa Maria Imbaro, Chieti, Italy

Julie Diplexcito (1:607)
School of Life Sciences, University of Dundee, MRC Protein Phosphorylation Unit, Dundee, Scotland DD1 5EH, United Kingdom

Jack E. Dixon (2:143)
The Life Science Institute and Department of Biological Chemistry, University of Michigan Medical School, Ann Arbor, Michigan 48109

Jack E. Dixon (1:591)
Department of Biological Chemistry, University of Michigan, Ann Arbor, Michigan 48109

Robert W. Doms (1:191)
Department of Microbiology, University of Pennsylvania, Philadelphia, Pennsylvania 19104

Daniel J. Donoghue (2:861)
Department of Chemistry and Biochemistry, Center for Molecular Genetics, University of California San Diego, La Jolla, California 92093

Russell F. Doolittle (1:119)
Center for Molecular Genetics, University of California San Diego, La Jolla, California 92093-0634

Stein Ove Døskeland (2:549)
Department of Anatomy and Cell Biology, University of Bergen, N-5009 Bergen, Norway

Wolfgang R. G. Dostmann (2:487)
Department of Pharmacology, University of Vermont, College of Medicine, Burlington, Vermont 05405

Matthias K. Dreyer (1:289)
Aventis Pharma Deutschland GmbH, Structural Biology, Frankfurt 65926, Germany

Guo Guang Du (2:45)
Banting and Best Department of Medical Research, University of Toronto, Charles H. Best Institute, Toronto, Ontario M5G 1X5, Canada

Keyong Du (3:115)
Peptide Biology Laboratories, The Salk Institute for Biological Studies, La Jolla, California 92037

Michael R. Duchen (2:73)
Department of Physiology and UCL Mitochondrial Biology Group, University College London, London WC1E 6JJ, United Kingdom

William G. Dunphy (1:693)
California Institute of Technology, Howard Hughes Medical Institute, Division of Biology, Pasadena, California 91125

Joanne Durgan (2:389)
Cancer Research UK, London Research Institute, Lincoln's Inn Fields Laboratories, WC2A 3PX London, United Kingdom

Michael L. Dustin (1:79)
NYU School of Medicine, Department of Pathology, Skirball Institute of Biomolecular Medicine, New York, New York 10016

Peter A. Edwards (2:287)
Departments of Biological Chemistry and Medicine, Molecular Biology Institute, University of California, Los Angeles, California 90095

Jackson G. Egen (1:355)
Department of Molecular and Cell Biology, Howard Hughes Medical Institute, University of California, Berkeley, Berkeley, California 94720

Lee E. Eiden (3:375)
Section on Molecular Neuroscience, Laboratory of Cellular and Molecular Regulation, National Institute of Mental Health Intramural Research Program, Bethesda, Maryland 20892

Elaine A. Elion (2:357)
Department of Biological Chemistry and Molecular Pharmacology, Harvard Medical School, Boston, Massachusetts 02115

Scott Emr (2:177)
Division of Cellular and Molecular Medicine, The Howard Hughes Medical Institute, University of California San Diego, School of Medicine, La Jolla, California 92093

Othmar G. Engelhardt (2:771)
Abteilung Virologie, Institut für Medizinische Mikrobiologie und Hygiene, Universität Freiburg, 79085 Freiburg, Germany

Christophe Erneux (2:11)
Interdisciplinary Research Institute (IRIBHN), Université Libre de Bruxelles, Brussels, Belgium

Peter J. Espenshade (3:353)
Department of Molecular Genetics, University of Texas Southwestern Medical Center at Dallas, Dallas, Texas 75390

Edward D. Esplin (2:405)
Department of Pharmacology, University of Texas Southwestern Medical Center, Dallas, Texas 75390

B. Mark Evers (3:477)
Department of Surgery, The University of Texas Medical Branch, Galveston, Texas 77555

Joanne L. Eyles (1:431)
The Walter and Eliza Hall Institute of Medical Research, The Cooperative Research Centre for Cellular Growth Factors, Victoria 3050, Australia

Sheelagh Fame (1:547)
MRC Protein Phosphorylation Unit, Department of Biochemistry, University of Dundee, Dundee DD1 5EH, Scotland

Marilyn Farquhar (3:351)
Department of Cellular and Molecular Medicine, University of California at San Diego, La Jolla, California 92093

Robert Feil (2:511)
Institut für Pharmakologie und Toxikologie, TU München, D-80802 München, Germany

Gui-Jie Feng (2:619)
Molecular Pharmacology Group, Division of Biochemistry and Molecular Biology, University of Glasgow, Glasgow G12 8QQ, Scotland, United Kingdom

Stanley Fields (2:311)
Howard Hughes Medical Institute, Departments of Genome Sciences and Medicine, University of Washington, Seattle, Washington 98195

James J. Fiordalisi (2:737)
Departments of Radiation Oncology and Pharmacology, and the Lineberger Comprehensive Cancer Center, University of North Carolina at Chapel Hill, Chapel Hill, North Carolina 27599

Richard A. Firtel (2:217)
Section of Cell Developmental Biology, Division of Biological Sciences, and Center for Molecular Genetics, University of California San Diego, La Jolla, California 92093

Garret A. Fitzgerald (2:265)
Center for Experimental Therapeutics, University of Pennsylvania, Philadelphia, Pennsylvania 19104

Andrew Flint (1:671)
CEPTYR, Inc., Bothell, Washington 98021

Marco Foiani (3:197)
Dipartimento di Genetica e Biologia dei Microrganismi, Universita' di Milano, 20133 Milano, Italy and Instituto FIRC di Oncologia Molecolare, 20139 Milano, Italy

Barry Marc Forman (3:47)
The Beckman Research Institute, City of Hope National Medical Center, Division of Molecular Medicine, The Gonda Diabetes and Genetic Research Center, Duarte California 91010

Albert J. Fornace, Jr. (3:179)
Gene Response Section, Center for Cancer Research, National Cancer Institute, Bethesda, Maryland 20892

Sharron H. Francis (2:447; 2:465)
Department of Molecular Physiology and Biophysics, Vanderbilt University School of Medicine, Nashville, Tennessee 37232

Günter Fritz (2:87)
Department of Pediatrics, Division of Clinical Chemistry and Biochemistry, University of Zürich, CH-8032 Zürich, Switzerland

David A. Fruman (2:135)
Department of Molecular Biology and Biochemistry, University of California Irvine, Irvine, California 92697

Antony Galione (2:15)
Department of Pharmacology, University of Oxford, Oxford OX1 2JD, United Kingdom

Chris S. Gandhi (1:209)
Department of Molecular and Cell Biology, University of California, Berkeley, Berkeley, California 94720

David L. Garbers (2:427)
Cecil H. and Ida Green Center for Reproductive Biology Sciences, Howard Hughes Medical Institute and Department of Pharmacology, University of Texas Southwestern Medical Center, Dallas, Texas 75390

K. Christopher Garcia (1:259)
Department of Microbiology and Immunology, Department of Structural Biology, Stanford University School of Medicine, Stanford, California 94305-5124

Benjamin Geiger (2:317)
Department of Molecular Cell Biology, Weizmann Institute of Science, Rehovot 76100, Isreal

Larry Gerace (3:419)
Department of Cell Biology, The Scripps Research Institute, La Jolla, California 92037

Andrea Gerstner (2:515)
Department Pharmzaie-Zentrum für Pharmaforschung, Ludwig-Maximilians-Universität München, 81377 München, Germany

Amato J. Giaccia (3:277)
Department of Radiation Oncology, Stanford University School of Medicine, Stanford, California 94305

Michele Giannattasio (3:197)
Dipartimento di Genetica e Biologia dei Microrganismi, Universita' di Milano, 20133 Milano, Italy

Vincent Giguère (3:35)
Molecular Oncology Group, McGill University Health Centre, Montréal, Québec H3G 1Y6, Canada

Christopher K. Glass (3:25)
Department of Cellular and Molecular Medicine, School of Medicine, University of California San Diego, La Jolla, California 92093

Martin J. Glennie (1:319)
University of Southampton General Hospital, Tenovus Research Laboratory, Cancer Sciences Division, School of Medicine, Southampton SO16 6YD, United Kingdom

Jennifer L. Glick (2:431)
Department of Pharmacology, University of Washington, Seattle, Washington 98195

Joseph L. Goldstein (3:353)
Department of Molecular Genetics, University of Texas Southwestern Medical Center at Dallas, Dallas, Texas 75390

Venkatesh Gopal (2:465)
Department of Molecular Physiology and Biophysics, Vanderbilt University School of Medicine, Nashville, Tennessee 37232

Myriam Gorospe (3:283)
Laboratory of Cellular and Molecular Biology, Gerontology Research Center, National Institute on Aging – IRP, National Institutes of Health, Baltimore, Maryland 21224

Cedric Govaerts (1:161)
Faculty of Medicine, University of Brussels, Institut de Recherche Interdisciplinaire, Brussels B-1070, Belgium

Paul R. Graves (2:301)
Department of Pharmacology and Cancer Biology, Center for Chemical Biology, Duke University, Durham, North Carolina 27710

Patrick W. Gray (1:149)
Macrogenics, Inc., Seattle, Washington 98103

Irene Graziani (3:393)
Center for Molecular Medicine, Maine Medical Center Research Institute, Scarborough, Maine 04074

Douglas R. Green (3:431)
Division of Cellular Immunology, La Jolla Institute for Allergy and Immunology, San Diego, California 92093

Michael E. Greenberg (3:83)
Division of Neuroscience, Children's Hospital and Department of Neurobiology, Harvard Medical School, Boston, Massachusetts 02115

Iva Greenwald (2:809)
Howard Hughes Medical Institute, Department of Biochemistry and Molecular Biophysics, Columbia University, New York, New York 10032

Haihua Gu (1:707)
Beth Israel-Deaconess Medical Center, Cancer Biology Program, Division of Hematology-Oncology, Harvard Medical School, Boston, Massachusetts 02215

Nuri Gueven (3:225)
Queensland Cancer Fund Research Laboratory, Queensland Institute of Medical Research, Brisbane 4029, Australia

J. Silvio Gutkind (2:589)
Oral and Pharyngeal Cancer Branch, National Institute of Dental and Craniofacial Research, National Institutes of Health, Bethesda, Maryland 20892

Jesper Z. Haeggström (2:275)
Department of Medical Biochemistry and Biophysics, Division of Chemistry II, The Scheele Laboratory, Karolinska Institutes, S-171 77 Stockholm, Sweden

Alan Hall (2:701)
Department of Biochemistry and Molecular Biology, University College London, London WC1E 6BT, United Kingdom

Michael N. Hall (3:299)
Division of Biochemistry, Biozentrum, University of Basel, 4056 Basel, Switzerland

Otto Haller (2:771)
Abteilung Virologie, Institut für Medizinische Mikrobiologie und Hygiene, Universität Freiburg, 79085 Freiburg, Germany

Heidi E. Hamm (1:335)
Vanderbilt University Medical Center, Department of Pharmacology, Nashville, Tennessee 37232-6600

Yusef A. Hannun (2:257)
Department of Biochemistry and Molecular Biology, Medical University of South Carolina, Charleston, South Carolina 29425

Carl A. Hansen (2:623)
Department of Biological and Allied Health Sciences, Bloomsburg University, Bloomsburg, Pennsylvania 17815

T. Kendall Harden (2:631)
Department of Pharmacology, Lineberger Comprehensive Cancer Center, and UNC Neuroscience Center, The University of North Carolina at Chapel Hill, Chapel Hill, North Carolina 27599

D. Grahame Hardie (1:535)
University of Dundee, Wellcome Trust Biocentre, Devision of Molecular Physiology, Dundee DD1 I, Scotland, United Kingdom

Kiminori Hasegawa (1:475)
Department of Immunology, Genentech, Inc., South San Francisco, California 94080

Phillip T. Hawkins (2:203)
Inositide Laboratory, Signaling Program, Babraham Institute, Cambridge CB2 4AT, United Kingdom

Timothy A. J. Haystead (2:301)
Serenex, Inc., Research Triangle Park, North Carolina 27709

Xiao-lin He (1:259)
Department of Microbiology and Immunology, Department of Structural Biology, Stanford University School of Medicine, Stanford, California 94305-5124

Claus W. Heizmann (2:87)
Department of Pediatrics, Division of Clinical Chemistry and Biochemistry, University of Zürich, CH-8032 Zürich, CH-8032 Switzerland

Carl-Henrik Heldin (1:391)
Biomedical Center, Ludwig Institute for Cancer Research, Uppsala S-751 24, Sweden

Michelle L. Hermiston (1:689)
Departments of Medicine, Microbiology and Immunology, Howard Hughes Medical Institute and Department of Pediatrics, University of California-San Francisco, San Francisco, California 94143-0795

Peter Herrlich (3:61; 3:257)
Forschungszentrum Karlsruhe, Institute of Genetics and Toxicology, 76021 Karlsruhe, Germany

Elizabeth A. Hewat (1:95)
Institut de Biologie Structurale J-P Ebel, Grenoble 38027, France

Bertil Hille (1:215)
Department of Physiology and Biophysics, University of Washington School of Medicine, Seattle, Washington 98195-7290

Douglas J. Hilton (1:431)
The Walter and Eliza Hall Institute of Medical Research, The Cooperative Research Centre for Cellular Growth Factors, Victoria 3050, Australia

K. A. Hinchliffe (2:123)
Department of Pharmacology, University of Cambridge, Cambridge CB2 1GA, United Kingdom

Steffan N. Ho (3:125)
Departments of Pathology and Cellular and Molecular Medicine, University of California San Diego, La Jolla, California 92093

Su-Chin Ho (1:161)
Faculty of Medicine, University of Brussels, Institut de Recherche Interdisciplinaire, Brussels B-1070, Belgium

Mark Hochstrasser (2:347)
Department of Molecular Biophysics and Biochemistry, Yale University, New Haven, Connecticut 06520

Franz Hofmann (2:511)
Institut für Pharmakologie und Toxikologie, TU München, D-80802 München, Germany

Christopher W. Hogue (2:311)
Department of Biochemistry, University of Toronto, Toronto, Ontario M5G 1X5, Canada

Wim G. J. Hol (2:525)
Department of Pharmacology, University of Washington, Seattle, Washington, and Department of Biochemistry and Biological Structure, Howard Hughes Medical Institute, University of Washington, Seattle, Washington 98195

Jocelyn Holash (2:849)
Regeneron Pharmaceuticals, Inc., Tarrytown, New York 10591

Robert A. Holmgren (3:167)
Department of Biochemistry Molecular Biology and Cell Biology, Robert H. Lurie Comprehensive Cancer Center, Northwestern University, Evanston, Illinois 60208

Barry Honig (1:11)
Department of Biochemistry and Molecular Biophysics, Howard Hughes Medical Institute, Columbia University, New York, New York 10032

Bruce S. Hostager (1:315)
Department of Microbiology, University of Iowa, Iowa City, Iowa 52242

Stevan R. Hubbard (1:299)
New York University, School of Medicine, Skirball Institute of Biomolecular Medicine, New York, New York 10016

Michael Huber (1:327)
Department of Molecular Immunology, Biology III, University of Freiburg and Max-Planck Institute for Immunobiology, Freiburg 79108, Germany

Tony Hunter (1:369; 1:373; 1:685)
Molecular and Cell Biology Laboratory, The Salk Institute for Biological Sciences, La Jolla, California 92037-1099

Anna Huttenlocher (2:105)
Departments of Pediatrics and Pharmacology, University of Wisconsin, Madison, Wisconsin 53716

Sarah G. Hymowitz (1:305)
Department of Protein Engineering, Genentech, Inc., South San Francisco, California 94080

James N. Ihle (1:427)
Department of Biochemistry, Howard Hughes Medical Institute, St. Jude Children's Research Hospital, Memphis, Tennessee 38105-2729

Jean-Luc Imler (2:333)
National Center of Scientific Research, Institute of Molecular and Cellular Biology, 67084 Strasbourg, France

R. F. Irvine (2:123)
Department of Pharmacology, University of Cambridge, Cambridge CB2 1GA, United Kingdom

Ehud Y. Isacoff (1:209)
Department of Molecular and Cell Biology, University of California, Berkeley, Berkeley, California 94720

Xavier Iturrioz (2:389)
Cancer Research UK, London Research Institute, Lincoln's Inn Fields Laboratories, WC2A 3PX London, United Kingdom

Lars F. Iversen (1:659)
Protein Chemistry, Novo Nordisk, Bagsvaerd DK 2880, Denmark

Ravi Iyengar (2:605)
Department of Pharmacology and Biological Chemistry, Mount Sinai School of Medicine, New York, New York 10029

Stephen P. Jackson (1:557; 3:219)
Wellcome Trust and Cancer Research Campaign, Institute of Cancer and Developmental Biology, Cambridge, United Kingdom, and Department of Zoology, University of Cambridge, Cambridge CB2 1GA, United Kingdom

Lily Yeh Jan (1:187; 2:667)
Howard Hughes Medical Institute, Departments of Physiology and Biochemistry, University of California San Francisco, San Francisco, California 94143

Fabiola Janiak-Spens (1:563)
Department of Chemistry and Biochemistry, University of Oklahoma, Norman, Oklahoma 73019

Paul A. Janmey (2:209)
Department of Physiology, Institute of Medicine and Engineering, University of Pennsylvania, Philadelphia, Pennsylvania 19104

Peter Gildsig Jansen (1:659)
Scientific Computing, Novo Nordisk, Måløv DK 2760, Denmark

Sophie Jarriault (2:809)
Howard Hughes Medical Institute, Department of Biochemistry and Molecular Biophysics, Columbia University, New York, New York 10032

Jonathan A. Javitch (1:155)
Columbia University Center for Molecular Recognition, and Departments of Psychiatry and Pharmacology, New York, New York 10032

Elwood V. Jensen (3:7)
Department of Cell Biology, University of Cincinnati College of Medicine, Cincinnati, Ohio 45267

Kristen Jepsen (3:29)
Howard Hughes Medical Institute, Department and School of Medicine, University of California San Diego, La Jolla, California 92093

E. Yvonne Jones (1:57)
Division of Structural Biology, Wellcome Trust Centre for Human Genetics, Cancer Research United Kingdom Receptor Stucture Research Group, Oxford, OX3 7BN, United Kingdom

Katherine A. Jones (3:161)
Regulatory Biology Laboratory Salk Institute for Biological Studies, La Jolla, California 92037

J. Dedrick Jordan (2:605)
Department of Pharmacology and Biological Chemistry, Mount Sinai School of Medicine, New York, New York 100329

Jomon Joseph (2:695)
Laboratory of Gene Regulation and Development, National Institute of Child Health and Human Development, National Institutes of Health, Bethesda, Maryland 20892

Louis B. Justement (3:555)
Division of Developmental and Clinical immunology, Department of Microbiology, University of Alabama at Birmingham, Birmingham, Alabama 35294

Yariv Kafri (2:317)
Department of Physics of Complex Systems, Weizmann Institute of Science, Rehovot 76100, Israel

Richard A. Kahn (2:727)
Department of Biochemistry, Emory University School of Medicine, Atlanta, Georgia 30322

Shin W. Kang (1:475)
Department of Immunology, Genentech, Inc., South San Francisco, California 94080

Arthur Karlin (1:223)
Departments of Biochemistry and Molecular Biophysics, College of Physicians and Surgeons, Center for Molecular Recognition, Columbia University, New York, New York 10032

Heidi R. Kast-Woelbern (2:287)
Department of Biological Chemistry, University of California, Los Angeles, California 90095

Randal J. Kaufman (3:311)
Department of Biological Chemistry, Howard Hughes Medical Institute, University of Michigan Medical Center, Ann Arbor, Michigan 48109

Andrius Kazlauskas (1:397)
Schepens Eye Research Institute, Harvard Medical School, Boston, Massachusetts 02114

James H. Keen (3:441)
Cell Biology and Signaling Program, Kimmel Cancer Center, Thomas Jefferson University, Philadelphia, Pennsylvania 19107

Rolf Kemler (2:789)
Department of Molecular Embryology, Max-Planck Institute of Immunobiology, 79085 Freiburg, Germany

Bruce E. Kemp (2:495)
St. Vincent's Institute of Medical Research, Fitzroy, Victoria, Australia

Bruce E. Kemp (1:539)
St. Vincent's Institute of Medical Research, Fitzroy, Victoria 3065, Australia

Mary B. Kennedy (1:543; 2:329)
Division of Biology, California Institute of Technology, Pasadena, California 91125

Matthew A. Kennedy (2:287)
Department of Biological Chemistry, University of California, Los Angeles, California 90095

Ushio Kikkawa (2:119)
Biosignal Research Center, Kobe University, Kobe, Japan

Albert H. Kim (2:839)
Molecular Neurobiology Program, Skirball Institute of Biomolecular Medicine, New York University School of Medicine, New York, New York 10016

Soo-A Kim (2:143)
The Life Science Institute and Department of Biological Chemistry, University of Michigan Medical School, Ann Arbor, Michigan 48109

Sung-Hou Kim (1:197)
Department of Chemistry and Lawrence Berkeley National Laboratory, University of California, Berkeley, Berkeley, California 94720

Youngjoo Kim (1:653)
Department of Biochemistry and Molecular Biology, Oregon Health and Science University, Portland, Oregon 97201-3098

Kirst King-Jones (3:69)
Howard Hughes Medical Institute, University of Utah, Salt Lake City, Utah 84112

Chris Kintner (2:813; 3:149)
Molecular Neurobiology Laboratory, Salk Institute for Biological Studies, La Jolla, California 92037

Saul Kivimäe (1:575)
Laboratory of Genetics, The Rockefeller University, New York, New York 10021

Claude B. Klee (1:631)
Laboratory of Biochemistry, National Cancer Institute, National Institutes of Health, Bethesda, Maryland 20892-4255

Rüdiger Klein (1:421)
Department of Molecular Neurobiology, Max-Planck Institute of Neurobiology, Martinsried D-82152, Germany

Thomas Kleppisch (2:511)
Institut für Pharmakologie und Toxikologie, TU München, D-80802 München, Germany

Steven A. Kliewer (3:53)
Nuclear Receptor Discovery Research, GlaxoSmithKline Research Triangle Park, North Carolina 27709

Richard A. Klinghoffer (2:845)
CEPTYR, Inc., Bothell, Washington 98021

Juergen A. Knoblich (2:571)
Research Institute of Molecular Pathology (I.M.P.), A-1030 Vienna, Austria

Bostjan Kobe (1:539; 2:495)
Department of Biochemistry and Molecular Biology, Institute for Molecular Bioscience, University of Queensland, St. Lucia, Brisbane QLD 4072, Queensland, Australia

George Kochs (2:771)
Abteilung Virologie, Institut für Medizinische Mikrobiologie und Hygiene, Universität Freiburg, 79085 Freiburg, Germany

Monica Kong-Beltran (2:861)
Genentech, Inc., San Francisco, California 94080

Rolf König (3:545)
Department of Microbiology and Immunology, Sealy Center for Molecular Science, The University of Texas Medical Branch, Galveston, Texas 77555

Albert C. Koong (3:277)
Division of Radiation Biology, Department of Radiation Oncology, Stanford University School of Medicine, Stanford, California 94305

Murray Korc (3:471)
Division of Endocrinology, Diabetes and Metabolism, University of California Irvine, Irvine, California 92697

Daniel Kornitzer (3:29)
Department of Molecular Microbiology, Bruce Rappaport Faculty of Medicine, Technion-Israel Institute of Technology, Haifa, Israel

Anthony A. Kossiakoff (1:241)
Cummings Life Sciences Center, University of Chicago, Department of Biochemistry and Molecular Biology, Chicago, Illinois 60637

Jun Kotera (2:465)
Department of Molecular Physiology and Biophysics, Vanderbilt University School of Medicine, Nashville, Tennessee 37232

M. V. Kovalenko (1:397)
Schepens Eye Research Institute, Harvard Medical School, Boston, Massachusetts 02114

Tohru Kozasa (2:639)
Department of Pharmacology, University of Illinois at Chicago, Chicago, Illinois 60637

Sergei Kozlov (3:25)
Queensland Cancer Fund Research Laboratory, Queensland Institute of Medical Research, Brisbane 4029, Australia

Keith G. Kozminski (2:733)
Departments of Biology and Cell Biology, University of Virginia, Charlottesville, Virginia 22904

Sonja Krugmann (2:203)
Inositide Laboratory, Signalling Programme, Babraham Institute, Cambridge CB2 4AT, United Kingdom

John Kuriyan (1:343; 1:387)
Physical Biosciences Division, Lawrence Berkeley National Laboratory, Berkeley, California 94720-3202

Riki Kurokawa (3:25)
Department of Cellular and Molecular Medicine, School of Medicine, University of California San Diego, La Jolla, California 92093

Peter D. Kwong (1:99)
National Institute of Allergy and Infectious Diseases, National Institute of Health Vaccine Research Center, Bethesda, Maryland 20892

Wi S. Lai (3:319)
Office of Clinical Research and Laboratory of Signal Transduction, National Institute of Environmental Health Sciences, Research Triangle Park, North Carolina 27709

Elise Lamar (3:149)
Molecular Neurobiology Laboratory, Salk Institute for Biological Studies, La Jolla, California 92037

Millard H. Lambert (3:21)
Nuclear Receptor Discovery Research, GlaxoSmithKline, Research Triangle Park, North Carolina 27709

David G. Lambright (2:575)
Program in Molecular Medicine and Department of Biochemistry and Molecular Pharmacology, University of Massachusetts Medical School, Worcester, Massachusetts 01605

Doron Lancet (1:145)
Department of Molecular Genetics, Weizmann Institute of Science, Rehovot 76100, Israel

Reiko Landry (3:161)
Regulatory Biology Laboratory Salk Institute for Biological Studies, La Jolla, California 92037

Wallace Y. Langdon (1:483)
Department of Pathology, University of Western Australia, Crawley, Western Australia 6009, Australia

Lorene K. Langeberg (2:383)
Howard Hughes Medical Institute, Vollum Institute, Oregon Health and Sciences University, Portland, Oregon 97239

Paul Lasko (3:327)
Department of Biology, McGill University, Montréal, Québec H3G 1Y6, Canada

Vaughn Latham (3:293)
Department of Anatomy and Structural Biology, Albert Einstein College of Medicine, Bronx, New York, and Dana-Farber Cancer Institute, Boston, Massachusetts 02115

Martin F. Lavin (3:225)
Queensland Cancer Fund Research Laboratory, Queensland Institute of Medical Research, Brisbane 4029, Australia, and Department of Surgery, University of Queensland, Brisbane QLD 4072, Australia

Kevin A. Lease (1:579)
Department of Biology, University of Virginia, Charlottesville, Virginia 22904

Hakon Leffler (1:87)
Section MIG, Institute of Laboratory Medicine, University of Lund, SE 223 62 Lund, Sweden

Mark A. Lemmon (2:161)
Department of Biochemistry and Biophysics, University of Pennsylvania Medical Center, Philadelphia, Pennsylvania 19104

Ann E. Leonard (2:883)
Gene Expression Laboratory, The Salk Institute for Biological Studies, La Jolla, California 92037

Alexander Levitzki (1:451)
Unit of Cell Signaling, Department of Biological Chemistry, The Alexander Silberman Institute of Life Sciences, The Hebrew University of Jerusalem, Jerusalem 91904, Israel

Hong-Jun Liao (2:5)
Department of Biochemistry, Vanderbilt University School of Medicine, Nashville, Tennessee 37232

Lucy Liaw (3:455)
Center for Molecular Medicine, Maine Medical Center Research Institute, Scarborough, Maine 04074

Giordano Liberi (3:197)
Dipartimento di Genetica e Biologia dei Microrganismi, Universita' di Milano, 20133 Milano, Italy and Instituto FIRC di Oncologia Molecolare, 20139 Milano, Italy

Heiko Lickert (2:789)
Department of Molecular Embryology, Max-Planck Institute of Immunobiology, 79085 Freiburg, Germany

Robert C. Liddington (1:123)
Program on Cell Adhesion, The Burnham Institute, La Jolla, California 92037

Thomas M. Lincoln (2:479)
Department of Pathology, University of Alabama at Birmingham, Birmingham, Alabama 35294

Jürgen U. Linder (2:535)
Department of Pharmaceutical Biochemistry, Pharmaceutical Institute, University of Tübingen, 72074 Tübingen, Germany

Maurine E. Linder (1:331; 2:585)
Department of Cell Biology and Physiology, Washington University School of Medicine, St. Louis, Missouri 63110

Hui Liu (2:171)
Center for Cancer Research, Massachusetts Institute of Technology, Cambridge, Massachusetts 02139

Zhengchang Liu (3:365)
Department of Molecular Biology, University of Texas Southwestern Medical Center, Dallas, Texas 75390

Marja K. Lohela (2:855)
Molecular/Cancer Biology Laboratory and Ludvig Institute for Cancer Research, Haartman Institute and Helsinki University Central Hospital, Biomedicum Helsinki, University of Helsinki, FIN-00014 Helsinki, Finland

Sarah H. Louie (1:177)
Department of Pharmacology and Center for Developmental Biology, Howard Hughes Medical Institute, University of Washington School of Medicine, Seattle, Washington 98195-7750

Deirdre K. Luttrell (2:609)
Department of High Throughput Biology, GlaxoSmithKline, Research Triangle Park, North Carolina 27709

Louis M. Luttrell (2:609)
The Geriatrics Research, Education and Clinical Center, Durham Veterans Affairs Medical Center, Durham, North Carolina, and Department of Medicine, Duke University Medical Center, Durham, North Carolina 27710

Karen M. Lyons (2:833)
Department of Molecular, Cell and Developmental Biology, Department of Biological Chemistry, Department of Orthopaedic Surgery, The David Geffen School of Medicine at the University of California Los Angeles, Los Angeles, California 90095

S. Lance Macaulay (1:293)
Health Sciences and Nutrition, CSIRO, Parkville, Victoria 3052, Australia

Michael Maceyka (2:247)
Department of Biochemistry, Medical College of Virginia Campus, Virginia Commonwealth University, Richmond, Virginia 23284

Thomas Maciag (3:393; 3:455)
Center for Molecular Medicine, Maine Medical Center Research Institute, Scarborough, Maine 04074

Fernando Macian (3:119)
Center for Blood Research and Department of Pathology, Harvard Medical School, Boston, Massachusetts 02115

Carol MacKintosh (1:607)
MRC Protein Phosphorylation Unit, School of Life Sciences, University of Dundee, Dundee, Scotland DD1 5EH, United Kingdom

David H. MacLennan (2:45)
Banting and Best Department of Medical Research, University of Toronto, Charles H. Best Institute, Toronto, Ontario M5G 1X5, Canada

Nadir A. Mahmood (2:405)
Department of Pharmacology, University of Texas Southwestern Medical Center, Dallas, Texas 75390

Craig C. Malbon (1:177)
Department of Molecular Pharmacology, Diabetes and Metabolic Research Center, University Medical Center, SUNY/Stony Brook, Stony Brook, New York 11794-8651

Sohail Malik (3:11)
Laboratory of Biochemistry and Molecular Biology, Rockefeller University, New York, New York 10021

Orna Man (1:145)
Department of Molecular Genetics, Weizmann Institute of Science, Rehovot 76100, Israel

Carol L. Manahan (2:645)
Department of Cell Biology, Johns Hopkins School of Medicine, Baltimore Maryland 21201

Anna Mandinova (3:393)
Center for Molecular Medicine, Maine Medical Center Research Institute, Scarborough, Maine 04074

Vincent C. Manganiello (2:441)
Department of Cell and Molecular Biology, Lund University, SE-22100 Lund, Sweden, and Pulmonary-Critical Care Medicine Branch, National Heart, Lung, and Blood Institute, National Institutes of Health, Bethesda, Maryland 20892

James L. Manley (3:331)
Department of Biological Sciences, Columbia University, New York, New York 10032

Matthias Mann (2:295)
Protein Interaction Laboratory, Center for Experimental Bioinformatics, University of Southern Denmark, DK-5230 Odense M, Denmark

Gerald Manning (1:373)
SUGEN, Inc., South San Francisco, California 94080

Ed Manser (1:499)
Institute of Molecular and Cell Biology, Glaxo–IMCB Group, Singapore 117609, Singapore

Marta Margeta-Mitrovic (1:187)
Department of Pathology, University of California, San Francisco, San Francisco, California 94143-0450

Robert F. Margolskee (2:657)
Department of Physiology and Biophysics, Howard Hughes Medical Institute, The Mount Sinai School of Medicine, New York, New York 10029

Julia Marinissen (2:589)
Oral and Pharyngeal Cancer Branch, National Institute of Dental and Craniofacial Research, National Institutes of Health, Bethesda, Maryland 20892

Roy A. Mariuzza (1:39)
Center for Advanced Research, W.M. Keck Laboratory for Structural Biology, University of Maryland, Biotechnology Institute, Rockville, Maryland 20850

Mina D. Marmor (1:405)
Department of Biological Regulation, Weizmann Institute of Science, Rehovot 76100, Israel

G. Steven Martin (1:441)
Department of Molecular and Cell Biology, University of California at Berkeley, Berkeley, California 94720-3204

Karen H. Martin (1:463)
Department of Microbiology, University of Virginia Health System, Charlottesville, Virginia 22908-0734

Sergio E. Martinez (2:525)
Department of Pharmacology, University of Washington, Seattle, Washington 98195

Michael B. Mathews (3:335)
Department of Biochemistry and Molecular Biology, New Jersey Medical School and the Graduate School of Biomedical Sciences, University of Medicine and Dentistry of New Jersey, Newark, New Jersey 07107

Bruce J. Mayer (1:471)
Department of Genetics and Developmental Biology, University of Connecticut Health Center, Farmington, Connecticut 06030-3301

Mark L. Mayer (1:219)
Laboratory of Cellular and Molecular Neurophysiology, National Institute of Child Health and Human Development, National Institutes of Health, Bethesda, Maryland 20892-4495

Maria R. Mazzoni (1:335)
Department of Pharmacology, Vanderbilt University Medical Center, Nashville, Tennessee 37232-6600

Frank McCormick (2:671)
Cancer Research Institute, University of California Comprehensive Cancer Center, San Francisco, California 94080

Clare H. McGowan (3:401)
Department of Molecular Biology and Department of Cell Biology, The Scripps Research Institute, La Jolla, California 92037

Melissa M. McKay (2:727)
Department of Biochemistry, Emory University School of Medicine, Atlanta, Georgia 30322

Wallace L. McKeehan (1:265)
Institute of Biosciences, Texas A&M University System Health Science Center, Center for Cancer Biology and Nutrition, Houston, Texas 77030-3303

Alison J. McLean (2:619)
Molecular Pharmacology Group, Division of Biochemistry and Molecular Biology, University of Glasgow, Glasgow G12 8QQ, Scotland, United Kingdom

Anthony R. Means (2:83)
Department of Pharmacology and Cancer Biology, Duke University Medical Center, Durham, North Carolina 27710

Ruedi Meili (2:217)
Section of Cell Developmental Biology, Division of Biological Sciences, and Center for Molecular Genetics, University of California San Diego, La Jolla, California 92093

Jingwei Meng (2:601)
Department of Pharmacology and Cancer Biology, Duke University Medical Center, Durham, North Carolina 27710

Mark Merchant (2:793)
Department of Molecular Biology, Genentech, Inc., South San Francisco, California 94080

Frank Mercurio (3:107)
Signal Research Division, Celgene, San Diego, California 92037

Graeme Milligan (2:619)
Molecular Pharmacology Group, Division of Biochemistry and Molecular Biology, University of Glasgow, Glasgow G12 8QQ, Scotland, United Kingdom

Guo-Li Ming (2:871)
The Salk Institute, La Jolla, California 92037

Daniel L. Minor (1:203)
Department of Biochemistry and Biophysics, Cardiovascular Research Institute, University of California, San Francisco, San Francisco, California 94143-0450

Nadeem Moghal (2:805)
HHMI and Division of Biology, California Institute of Technology, Pasadena, California 91125

Neils Peter H. Møller (1:659)
Signal Transduction, Novo Nordisk, Bagsvaerd, DK 2880, Denmark

Marco Mongillo (2:459)
Venetian Institute of Molecular Medicine, Department of Biomedical Studies, University of Padua, 35129 Padua, Italy

Marc Montminy (3:115)
Peptide Biology Laboratories, The Salk Institute for Biological Studies, La Jolla, California 92037

Randall T. Moon (1:177)
Department of Pharmacology and Center for Developmental Biology, Howard Hughes Medical Institute, University of Washington, Seattle, Washington 98195-7750

Richard I. Morimoto (3:269)
Department of Biochemistry, Molecular Biology and Cell Biology, Rice Institute for Biomedical Research, Northwestern University, Evanston, Illinois 60208

Stephen E. Moss (2:101)
Division of Cell Biology, Institute of Ophthalmology, University College London, London WC1E 6JJ, United Kingdom

Helen R. Mott (2:745)
Department of Biochemistry, University of Cambridge, Cambridge CB2 1GA, United Kingdom

Carla Mouta (3:455)
Center for Molecular Medicine, Maine Medical Center Research Institute, Scarborough, Maine 04074

Marco Muda (1:703)
Serono Reproductive Biology Institute, Rockland, Massachusetts 02370

Marc C. Mumby (2:405)
Department of Pharmacology, University of Texas Southwestern Medical Center, Dallas, Texas 75390

Gretchen A. Murphy (2:681)
Department of Pharmacology, and Lineberg Comprehensive Cancer Center, University of North Carolina at Chapel Hill, Chapel Hill, North Carolina 27599

Marco Muzi-Falconi (3:197)
Dipartimento di Genetica e Biologia dei Microrganismi, Universita' di Milano, 20133 Milano, Italy

Raghavendra Nagaraj (2:827)
Department of Molecular Cell and Developmental Biology, Biological Chemistry, Human Genetics, Molecular Biology Institute, University of California at Los Angeles, Los Angeles, California 90095

Stefan R. Nahorski (2:19)
Department of Cell Physiology and Pharmacology, University of Leicester, Leicester LE2 3QE, United Kingdom

Angus C. Nairn (1:567)
Department of Psychiatry, Yale University School of Medicine, New Haven, Connecticut 06508

Piers Nash (1:379)
Samuel Lunenfeld Research Institute, Mount Sinai Hospital, Toronto, ONT M5G 1X5, Canada

Benjamin G. Neel (1:707)
Cancer Biology Program, Division of Hematology-Oncology, Beth Israel Deaconess Medical Center, Harvard Medical School, Boston, Massachusetts 02215

Alexandra C. Newton (1:551; 2:187)
Department of Pharmacology, University of California at San Diego, La Jolla, California 92093

Yasutomi Nishizuka (2:119)
Biosignal Research Center, Kobe University, Kobe 657-8501, Japan

Joseph P. Noel (1:685)
Structural Biology Lab, The Salk Institute for Biological Sciences, La Jolla, California 92037

Ellen A. A. Nollen (3:269)
Department of Biochemistry, Molecular Biology and Cell Biology, Rice Institute for Biomedical Research, Northwestern University, Evanston, Illinois 60208

Irene M. A. Nooren (1:21)
European Bioinformatics Institute, Wellcome Trust Genome Campus, Hinxton, Cambridge, CB10 1SD, United Kingdom

Rodney O'Connor (2:51)
The Babraham Institute, Babraham, Cambridge CB2 4AT, United Kingdom

Stefan Offermanns (2:581)
Pharmakologisches Institut, Universität Heidelberg, 69120 Heidelberg, Germany

Tsviya Olender (1:145)
Department of Molecular Genetics, Weizmann Institute of Science, Rehovot 76100, Israel

Shao-En Ong (2:295)
Protein Interaction Laboratory, Center for Experimental Bioinformatics, University of Southern Denmark, DK-5230 Odense M, Denmark

Darerca Owen (2:745)
Department of Biochemistry, University of Cambridge, Cambridge CB2 1GA, United Kingdom

Lisa J. Pagliari (3:431)
Division of Cellular Immunology, La Jolla Institute for Allergy and Immunology, San Diego, California 92037

Lily Pao (1:707)
Cancer Biology Program, Division of Hematology-Oncology, Beth Israel Deaconess Medical Center, Harvard Medical School, Boston, Massachusetts 02215

John Papaconstantinou (3:515)
University of Texas Medical Branch, Department of Human Biological Chemistry and Genetics, Galveston, Texas 77555

Leonardo Pardo (1:161)
Unitat de Bioestadistica, Facultat de Medicina, Laboratori de Medicina Computacional, Universitat Autonoma de Barcelona, Bellaterra 08193, Spain

Hay-Oak Park (2:733)
Department of Molecular Genetics, The Ohio State University, Columbus, Ohio 43210

Young Chul Park (1:311)
Department of Biochemistry, Cornell University, Weill Medical College, New York, New York 10021

Peter J. Parker (2:389)
Cancer Research UK, London Research Institute, Lincoln's Inn Fields Laboratories, WC2A 3PX London, United Kingdom

J. Thomas Parsons (1:463)
Department of Microbiology, University of Virginia Health Systems Charlottesville, Virginia 22908-0734

J. M. Passner (2:531)
Department of Physiology and Biophysics, Mount Sinai School of Medicine, New York, New York 10029

Tony Pawson (1:379)
Samuel Lunenfeld Research Institute, Mount Sinai Hospital, Toronto, ONT M5G 1X5, Canada

Achille Pellicioli (3:197)
Dipartimento di Genetica e Biologia dei Microrganismi, Universita' di Milano, 20133 Milano, Italy and Instituto FIRC di Oncologia Molecolare, 20139 Milano, Italy

J. Regino Perez-Polo (3:485)
The University of Texas Medical Branch, Galveston Texas 77555

Norbert Perrimon (2:783)
Department of Genetics, Harvard Medical School, Howard Hughes Medical Institute, Boston, Massachusetts 02115

Fabrice G. Petite (3:57)
Department of Molecular and Cellular Biology, Baylor College of Medicine, Houston, Texas 77030

Emmanuel Petroulakis (3:343)
Department of Biochemistry and McGill Cancer Research Center, McGill University, Montréal, Québec H3G 1Y6, Canada

Samuel L. Pfaff (2:883)
Gene Expression Laboratory, The Salk Institute for Biological Studies, La Jolla, California 92037

Jacob Piehler (1:27)
Institute of Biochemistry, Goethe-University Frankfurt, Frankfurt am Main 60439, Germany

Linda J. Pike (1:323)
Department of Biochemistry and Molecular Biophysics, Washington University School of Medicine, St. Louis, Missouri 63110

Michael J. Pinkoski (3:431)
Division of Cellular Immunology, La Jolla Institute for Allergy and Immunology, San Diego, California 92037

Fiona J. Pixley (3:615)
Department of Developmental and Molecular Biology, Albert Einstein College of Medicine, Yeshiva University, Bronx, New York 10461

Paolo Plevani (3:197)
Dipartimento di Genetica e Biologia dei Microrganismi, Universita' di Milano, 20133 Milano, Italy

Mu-ming Poo (2:871)
Division of Neurobiology, Department of Molecular and Cell Biology, University of California Berkeley, Berkeley, California 94720

Tullioi Pozzan (2:459)
Venetian Institute of Molecular Medicine, Department of Biomedical Studies, University of Padua, 35129 Padua, Italy

Stephen M. Prescott (2:243)
Huntsman Cancer Institute, University of Utah, Salt Lake City, Utah 84112

Igor Prudovsky (3:393)
Center for Molecular Medicine, Maine Medical Center Research Institute, Scarborough, Maine 04074

James W. Putney, Jr. (2:31)
Calcium Regulation Section, Laboratory of Signal Transduction, National Institute of Environmental Health Sciences, National Institute of Health, Research Triangle Park, North Carolina 27709

Thomas Radimerski (3:305)
 Friedrich Miescher Institute for Biomedical Research, 4056 Basel, Switzerland
Elzbieta Radzio-Andzelm (2:471)
 Department of Chemistry and Biochemistry and the Howard Hughes Medical Institute, University of California San Diego, La Jolla, California 92093
Prahlad T. Ram (2:605)
 Department of Pharmacology and Biological Chemistry, Mount Sinai School of Medicine, New York, New York 10029
Lucia Rameh (2:129)
 Boston Biomedical Research Institute, Watertown, Massachusetts 02115
Danica Ramljak (3:565)
 Department of Oncology, Lombardi Cancer Center, Georgetown University, Washington, DC 20057
Barbara Ranscht (2:889)
 The Burnham Institute, Neurobiology Program, La Jolla, California 92037
Anjana Rao (3:119)
 Center for Blood Research and Department of Pathology, Harvard Medical School, Boston, Massachusetts 02115
Carol J. Raport (1:149)
 ICOS corporation, Bothell, Washington
Jacqueline D. Reeves (1:191)
 Department of Microbiology, University of Pennsylvania, Philadelphia, Pennsylvania 19104
Holger Rehman (2:521)
 Department of Physiological Chemistry and Center for Biomedical Genetics, University Medical Center Utrecht, Utrecht, The Netherlands, and Max-Planck Institute for Molecular Physiology, 44227 Dortmund, Germany
Trevor W. Reichman (3:335)
 Department of Biochemistry and Molecular Biology, New Jersey Medical School and the Graduate School of Biomedical Sciences, University of Medicine and Dentistry of New Jersey, Newark, New Jersey 07107
Eric Reiter (3:591)
 Station de Physiologie de la Reproduction et des Comportements, INRA/CNRS, University of Tours, 37380 Nouzilly, France
Michael A. Resnick (3:203)
 Laboratory of Molecular Genetics, National Institute of Environmental Health Sciences, National Institutes of Health, Research Triangle Park, North Carolina 27709
Michael Reth (1:327)
 Department of Molecular Immunology, Biology III, University of Freiburg and Max-Planck Institute for Immunobiology, Freiburg 79108, Germany
Sue Goo Rhee (2:113)
 Laboratory of Cell Signaling, National Heart, Lung, and Blood Institute, National Institutes of Health, Bethesda, Maryland 20892
Joel D. Richter (3:323)
 Program in Molecular Medicine, University of Massachusetts Medical School, Worcester, Massachusetts 01605

Rodney L. Rietze (3:625)
 Institute for Brain Research, The University of Queensland, Brisbane QLD 4072, Queensland, Australia
James M. Rini (1:87)
 Departments of Molecular and Medical Genetics and Biochemistry, University of Toronto, Toronto, ONT M5S 1A8, Canada
Jürgen A. Ripperger (3:139)
 Department of Molecular Biology, Sciences II, University of Geneva, CH-1211 Geneva, Switzerland
Josep Rizo (2:95)
 Departments of Biochemistry and Pharmacology, The University of Texas Southwestern Medical Center, Dallas, Texas 75390
Janet D. Robishaw (2:623)
 Weis Center for Research, Geisinger Clinic, Danville, Pennsylvania 17822
H. Llewelyn Roderick (2:51)
 The Babraham Institute, Babraham, Cambridge CB2 4AT, United Kingdom
Robert G. Roeder (3:11)
 Laboratory of Biochemistry and Molecular Biology, Rockefeller University, New York, New York 10021
Larry R. Rohrschneider (2:147)
 Fred Hutchinson Cancer Research Center, Division of Basic Sciences, Seattle, Washington 98109
David Ron (3:263; 3:359)
 Skirball Institute of Biomolecular Medicine, New York University School of Medicine, New York, New York 10016
Michael G. Rosenfeld (3:29)
 Howard Hughes Medical Institute, Department and School of Medicine, University of California San Diego, La Jolla, California 92093
Hans Rosenfeldt (2:589)
 Oral and Pharyngeal Cancer Branch, National Institute of Dental and Craniofacial Research, National Institutes of Health, Bethesda, Maryland 20892
Kent L. Rossman (2:751)
 Department of Pharmacology, Department of Biochemistry and Biophysics, and Lineberger Comprehensive Cancer Center, The University of North Carolina at Chapel Hill, Chapel Hill, North Carolina 27599
Christopher B. Roth (1:115)
 Department of Molecular Biology, The Scripps Research Institute, La Jolla, California 92037
Markus G. Rudolph (1:63)
 Department of Molecular Biology, Skaggs Institute for Chemical Biology, The Scripps Research Institute, La Jolla, California 92037
Anja Ruppelt (2:501)
 Department of Medical Biochemistry, Institute of Basic Medical Sciences, University of Oslo, N-0316 Oslo, Norway
Lino Saez (1:575)
 Laboratory of Genetics, The Rockefeller University, New York, New York 10021

Thomas P. Sakmar (1:139)
Laboratory of Molecular Biology and Biochemistry, Howard Hughes Medical Institute, Rockefeller University, New York, New York 10021

Guy S. Salvesen (2:351)
Program in Apoptosis and Cell Death Research, Burnham Institute, San Diego, California 92037

Paolo Sassone-Corsi (3:91)
Institute of Genetic and Molecular and Cellular Biology, 67084 Strasbourg, France

Charles L. Saxe (2:323)
Department of Cell Biology, Emory University School of Medicine, Atlanta, Georgia 30322

Beat W. Schäfer (2:87)
Department of Pediatrics, Division of Clinical Chemistry and Biochemistry, University of Zürich, CH-8032 Zürich, Switzerland

Ueli Schibler (3:139)
Department of Molecular Biology, Sciences II, University of Geneva, CH-1211 Geneva, Switzerland

Christian W. Schindler (3:77)
Department of Microbiology and Medicine, College of Physicians and Surgeons, Columbia University, New York, New York 10032

Tobias Schmelzle (3:299)
Division of Biochemistry, Biozentrum, University of Basel, 4056 Basel, Switzerland

Sandra L. Schmid (2:763)
Department of Cell Biology, The Scripps Research Institute, La Jolla, California 92037

Anja Schmidt (2:701)
MRC Laboratory for Molecular Cell Biology, CRC Oncogene and Signal Transduction Group, University College London, London WC1E 6BT, United Kingdom

Eric F. Schmidt (2:877)
Interdepartmental Science Program, Yale University School of Medicine, New Haven, Connecticut 06520

Gideon Schreiber (1:27)
Department of Biological Chemistry, Weizmann Institute of Science, Rehovot 76100, Israel

Joachim E. Schultz (2:535)
Department of Pharmaceutical Biochemistry, Pharmaceutical Institute, University of Tübingen, 72074 Tübingen, Germany

Beat Schwaller (2:67)
Division of Histology, Department of Medicine, University of Fribourg, Perolles, Fribourg, Switzerland

Klaus Schwamborn (3:107)
Signal Research Division, Celgene, San Diego, California 92037

Thue Schwartz (1:173)
Laboratory of Molecular Endocrinology, Ringshospitalet Copenhagen DK-2100, Denmark

William F. Schwindinger (2:623)
Weis Center for Research, Geisinger Clinic, Danville, Pennsylvania 17822

Giorgio Scita (3:411)
Department of Experimental Oncology, Istituto Europeo di Oncologia, 20141 Milan, Italy, and IFOM, The FIRC Institute of Molecular Oncology, 20134 Milan, Italy

John D. Scott (2:291; 2:383)
Howard Hughes Medical Institute, Vollum Institute, Oregon Health and Sciences University, Portland, Oregon 97239

Shaun Scott (3:225)
Queensland Cancer Fund Research Laboratory, Queensland Institute of Medical Research, Brisbane 4029, Australia

Thomas Seebeck (2:539)
Institute of Cell Biology, University of Bern, CH-3012 Bern, Switzerland

Charles N. Serhan (2:281)
Center for Experimental Therapeutics and Reperfusion Injury, Department of Anesthesiology, Perioperative and Pain Medicine, Brigham and Women's Hospital and Harvard Medical School, Boston, Massachusetts 02115

John B. Shabb (2:501; 2:545)
Department of Biochemistry and Molecular Biology, University of North Dakota School of Medicine and Health Sciences, Grand Forks, North Dakota 58203

Andrey S. Shaw (2:339)
Department of Pathology and Immunology, Washington University School of Medicine, St. Louis, Missouri 63110

Stephen B. Shears (2:233)
Laboratory of Signal Transduction, National Institute of Environmental Health Sciences, Research Triangle Park, North Carolina 27709

Shirish Shenolikar (1:627)
Department of Pharmacology and Cancer Biology, Duke University Medical Center, Durham, North Carolina 27710

Lei Shi (1:155)
Center for Molecular Recognition, Columbia University, New York, New York 10032

Chanseok Shin (3:331)
Department of Biological Sciences, Columbia University, New York, New York 10032

Kazuhiro Shiozaki (1:637)
Section of Microbiology, University of California, Davis, California 95616

Kevan M. Shokat (1:583)
Department of Cellular and Molecular Pharmacology, Department of Chemistry, University of California, Berkeley, California 94143-0450

Trevor J. Shuttleworth (2:35)
Department of Pharmacology and Physiology, University of Rochester Medical Center, Rochester, New York 14642

David P. Siderovski (2:631)
UNC Neuroscience Center, University of North Carolina at Chapel Hill, Chapel Hill, North Carolina 27599

Steven A. Siegelbaum (1:233)
Department of Pharmacology, Center for Neurobiology and Behavior, Howard Hughes Medical Institute, Columbia University, New York, New York 10032

Adam M. Silverstein (2:405)
Department of Pharmacology, University of Texas Southwestern Medical Center, Dallas, Texas 75390

Robert H. Singer (3:293)
Department of Anatomy and Structural Biology, Albert Einstein College of Medicine, Bronx, New York 10461

Michael K. Skinner (3:531)
Center for Reproductive Biology, School of Molecular Biosciences, Washington State University, Pullman, Washington 99164

Jill K. Slack-Davis (1:463)
Department of Microbiology, University of Virginia Health System, Charlottesville, Virginia 22908-0734

Stephen J. Smerdon (1:505; 2:171)
Division of Protein Structure, National Institute for Medical Research, London NW7 1AA, United Kingdom

Graeme C. M. Smith (1:557)
KuDOS Pharmaceuticals, Ltd, Cambridge CB4 0WG, United Kingdom

Guillaume Smits (1:161)
Faculty of Medicine, University of Brussels, Institut de Recherche Interdisciplinaire, Brussels B-1070, Belgium

Sarah M. Smolik (3:167)
Department of Medicine, Oregon Health Sciences University, Portland, Oregon 97239

Jessica E. Smotrys (2:585)
Department of Cell Biology and Physiology, Washington University School of Medicine, St. Louis, Missouri 63110

Emer M. Smyth (2:265)
Center for Experimental Therapeutics, University of Pennsylvania, Philadelphia, Pennsylvania 19104

Jason T. Snyder (2:751)
Department of Pharmacology, Department of Biochemistry and Biophysics, and Lineberger Comprehensive Cancer Center, The University of North Carolina at Chapel Hill, Chapel Hill, North Carolina 27599

Naoko Sogame (3:213)
Department of Cell Biology, University of Texas Southwestern Medical Center, Dallas, Texas 75390

Raffaella Soldi (3:393)
Center for Molecular Medicine, Maine Medical Center Research Institute, Scarborough, Maine 04074

John Sondek (2:751)
Department of Pharmacology, Department of Biochemistry and Biophysics, and Lineberger Comprehensive Cancer Center, The University of North Carolina at Chapel Hill, Chapel Hill, North Carolina 27599

Nahum Sonenberg (3:343)
Department of Biochemistry and McGill Cancer Research Center, McGill University, Montréal, Québec H3G 1Y6, Canada

Erica Dutil Sonneberg (1:685)
Structural Biology Lab, The Salk Institute for Biological Sciences, La Jolla, California 92037

Lindsay G. Sparrow (1:293)
Health Sciences and Nutrition, CSIRO, Parkville, Victoria 3052, Australia

Sarah Spiegel (2:247)
Department of Biochemistry, Medical College of Virginia Campus, Virginia Commonwealth University, Richmond, Virginia 23284

Stephen R. Sprang (1:127; 1:275)
Howard Hughes Medical Institute and Department of Biochemistry, The University of Texas Southwestern Medical Center, Dallas, Texas 75235-9050

Deepak Srivastava (3:463)
Department of Pediatrics and Molecular Biology, University of Texas Southwestern Medical Center, Dallas, Texas 75390

Robyn L. Stanfield (1:33)
Department of Molecular Biology, The Scripps Research Institute and The Skaggs Institute for Chemical Biology, La Jolla, California 92037

E. Richard Stanley (3:615)
Department of Developmental and Molecular Biology, Albert Einstein College of Medicine, Yeshiva University, Bronx, New York 10461

Deborah J. Stauber (1:251)
Department of Molecular Biology, The Scripps Research Institute and The Skaggs Institute for Chemical Biology, La Jolla, California 92037

Christopher Stefan (2:177)
Division of Cellular and Molecular Medicine, The Howard Hughes Medical Institute, University of California San Diego, School of Medicine, La Jolla, California 92093

Lena Stenson-Holst (2:441)
Section for Molecular Signaling, Department of Cell and Molecular Biology, Lund University, SE-22100 Lund, Sweden

Len Stephens (2:203)
Inositide Laboratory, Signalling Programme, Babraham Institute, Cambridge CB2 4AT, United Kingdom

Paul W. Sternberg (2:805)
HHMI and Division of Biology, California Institute of Technology, Pasadena, California 91125

Paul C. Sternweis (2:237)
Department of Pharmacology, University of Texas Southwestern Medical Center, Dallas, Texas 75390

Ruth Steward (2:779)
Waksman Institute, Department of Molecular Biology and Biochemistry, and Cancer Institute of New Jersey, Rutgers University, Piscataway, New Jersey 08901

John T. Stickney (2:675)
Department of Cell Biology, Neurobiology, and Anatomy, University of Cincinnati, Cincinnati, Ohio 45221

Andrew W. Stoker (2:867)
Neural Development Unit, Institute of Child Health, University College London, London WC1N 61EH, United Kingdom

Stephen M. Strittmatter (2:877)
Department of Neurology and Section of Neurobiology, Yale University School of Medicine, New Haven, Connecticut 06520

Beth E. Stronach (2:783)
Department of Biological Sciences, University of Pittsburgh, Pittsburgh, Pennsylvania 15261

Roland K. Strong (1:83)
Basic Sciences Division Fred Hutchinson Cancer Research Center, Seattle, Washington 98109-1024

Robert M. Stroud (1:239)
Department of Biochemistry and Biophysics, University of California, San Francisco, San Francisco, California 94143-0448

Thomas C. Südhof (2:95)
Center for Basic Neuroscience, Department of Molecular Genetics, and Howard Hughes Medical Institute, University of Texas Southwestern Medical Center, Dallas, Texas 75390

Roger K. Sunahara (2:419)
Department of Pharmacology, University of Michigan Medical School, Ann Arbor, Michigan 48109

Brian J. Sutton (1:45)
The Randall Centre, King's College London, New Hunt's House, Guy's Campus, London SE1 1UL, United Kingdom

Sipeki Szabolcs (2:389)
Cancer Research UK, London Research Institute, Lincoln's Inn Fields Laboratories, WC2A 3PX London, United Kingdom

Xiao-Bo Tang (2:525)
Department of Pharmacology, University of Washington, Seattle, Washington 98195

Kjetil Taskén (2:501)
Department of Medical Biochemistry, Institute of Basic Medical Sciences, University of Oslo, N-0316 Oslo, Norway

Hisashi Tatebe (1:637)
Section of Microbiology, University of California, Davis, California 95616

Servane Tauszig-Delamasure (2:333)
National Center of Scientific Research, Institute of Molecular and Cellular Biology, 67084 Strasbourg, France

Colin W. Taylor (2:41)
Department of Pharmacology, University of Cambridge, Cambridge CB2 1PD, United Kingdom

Garry L. Taylor (1:105)
Centre for Biomolecular Sciences, University of St. Andrews, St. Andrews, Fife KY16 9UA, Scotland

Laura J. Taylor (2:711)
Department of Molecular Genetics and Microbiology, State University of New York at Stony Brook, Stony Brook, New York 11794

Susan S. Taylor (2:471)
Department of Chemistry and Biochemistry and the Howard Hughes Medical Institute, University of California San Diego, La Jolla, California 92093

George Thomas (3:305)
Friedrich Miescher Institute for Biomedical Research, 4056 Basel, Switzerland

Robert P. Thomas (3:477)
Department of Surgery, The University of Texas Medical Branch, Galveston, Texas 77555

E. Brad Thompson (3:451)
Department of Human Biological Chemistry and Genetics, The University of Texas Medical Branch, Galveston, Texas 77555

Michael J. Thompson (1:15)
Protein Pathways, Inc., Woodland Hills, California 91367

Janet M. Thornton (1:21)
Wellcome Trust Genome Campus, European Bioinformatics Institute, Hinxton, Cambridge CB10 1SD, United Kingdom

Carl S. Thummel (3:69)
Howard Hughes Medical Institute, University of Utah, Salt Lake City, Utah 84112

Hideaki Togashi (2:877)
Department of Neurology and Section of Neurobiology, Yale University School of Medicine, New Haven, Connecticut 06520

Amy Hin Yan Tong (2:311)
Banting and Best Department of Medical Research and Department of Molecular and Medical Genetics, University of Toronto, Toronto, Ontario M5G 1X5, Canada

Nicholas K. Tonks (1:641)
Cold Spring Harbor Laboratory, Cold Spring Harbor, New York 11724-2208

Peter Tontonoz (3:39)
Howard Hughes Medical Institute, Department of Pathology and Laboratory Medicine, UCLA School of Medicine, Los Angeles, California 90095

M. K. Topham (2:243)
Huntsman Cancer Institute, University of Utah, Salt Lake City, Utah 84112

Knut Martin Torgersen (2:143)
The Life Science Institute and Department of Biological Chemistry, University of Michigan Medical School, Ann Arbor, Michigan 48109

Hien Tran (3:83)
Division of Neuroscience, Children's Hospital and Department of Neurobiology, Harvard Medical School, Boston, Massachusetts 02115

Michel L. Tremblay (1:729)
McGill Cancer Center, McGill University, Department of Biochemistry, Montreal, Quebec H3G 1Y6, Canada

Ming-Jer Tsai (3:57)
Department of Molecular and Cellular Biology, Baylor College of Medicine, Houston, Texas 77030

Sophia Y. Tsai (3:57)
Department of Molecular and Cellular Biology, Baylor College of Medicine, Houston, Texas 77030

Susan Tsunoda (1:349)
Department of Biology, Boston University, Boston, Massachusetts 02215

Stewart Turley (2:525)
Department of Biochemistry and Biological Structure, Howard Hughes Medical Institute, University of Washington, Seattle, Washington 98195

Darren Tyson (1:361)
Department of Physiology and Biophysics, University of California, Irvine, Irvine, California 92697-4560

Robert L. Van Etten (1:733)
Department of Chemistry, Purdue University, West Lafayette, Indiana 47907-1393

Gilbert Vassart (1:161)
Faculty of Medicine, University of Brussels, Institut de Recherche Interdisciplinaire and Department of Medical Genetics, Brussels B-1070, Belgium

Peter J. Verveer (2:305)
Cell Biology and Cell Biophysics Program, European Molecular Biology Laboratory, 69120 Heidelberg, Germany

Virginie Vlaeminck (1:161)
Faculty of Medicine, University of Brussels, Institut de Recherche Interdisciplinaire, Brussels B-1070, Belgium

Abraham M. de Vos (1:281; 1:285; 1:305)
Department of Protein Engineering, Genentech, Inc., South San Francisco, California 94080-4990

Ty C. Voss (3:135)
Departments of Medicine and Cell Biology, NSF Center for Biological Timing, University of Virginia Health Sciences Center, Charlottesville, Virginia 22908

Robert Walczak (3:39)
Howard Hughes Medical Institute, Department of Pathology and Laboratory Medicine, UCLA School of Medicine, Los Angeles, California 90095

Graham C. Walker (3:185)
Biology Department, Massachusetts Institute of Technology, Cambridge, Massachusetts 02139

John C. Walker (1:579)
Division of Biological Sciences, University of Missouri, Columbia, Missouri 65211

Gernot Walter (1:621)
Department of Pathology, University of California, San Diego, La Jolla, California 92093-0612

Mark R. Walter (1:271)
Department of Microbiology and Center for Macromolecular Crystallography, University of Alabama at Birmingham, Birmingham, Alabama 35294-0005

Fen Wang (1:265)
Center for Cancer Biology and Nutrition, Institute of Biosciences and Technology, Texas A&M University System Health Science Center, Houston, Texas 77030–3303

Jean Y. J. Wang (3:249)
Division of Biological Sciences and The Cancer Center, University of California at San Diego, La Jolla, California 92093

Weiru Wang (1:197)
Department of Chemistry and Lawrence Berkeley National Laboratory, University of California, Berkeley, Berkeley, California 94720

Richard J. Ward (2:619)
Molecular Pharmacology Group, Division of Biochemistry and Molecular Biology, University of Glasgow, Glasgow G12 8QQ, Scotland, United Kingdom

Philip Wedegaertner (2:651)
Department of Microbiology and Immunology, Kimmel Cancer Institute, Thomas Jefferson University, Philadelphia, Pennsylvania 19104

Christian Wehrle (2:789)
Department of Molecular Embryology, Max-Planck Institute of Immunobiology, 79085 Freiburg, Germany

Arthur Weiss (1:689)
Departments of Medicine, Microbiology and Immunology, Howard Hughes Medical Institute, University of California-San Francisco, San Francisco, California 94143-0795

Jamie L. Weiss (2:79)
The Physiological Laboratory, The University of Liverpool, Liverpool L69 3BX, United Kingdom

Alan Wells (2:105)
Department of Pathology, University of Pittsburgh, Pittsburgh, Pennsylvania 15261

Claudia Werner (2:511)
Institut für Pharmakologie und Toxikologie, TU München, D-80802 München, Germany

Ann H. West (1:563)
Department of Chemistry and Biochemistry, University of Oklahoma, Norman, Oklahoma 73019

Marie C. Weston (2:441)
Pulmonary-Critical Care Medicine Branch, National Heart, Lung, and Blood Institute, National Institutes of Health, Bethesda, Maryland 20892

John K. Westwick (3:107)
Signal Research Division, Celgene, San Diego, California 92037

Anders Wetterholm (2:275)
Department of Medical Biochemistry and Biophysics, Division of Chemistry II, The Scheele Laboratory, Karolinska Institutes, S-17177 Stockholm, Sweden

Morris F. White (1:409)
Joslin Diabetes Center, Howard Hughes Medical Institute, Harvard Medical School, Boston, Massachusetts 02215

Malcolm Whitman (3:171)
Department of Cell Biology, Harvard Medical School, Boston, Massachusetts 02115

Matt R. Whorton (2:419)
Department of Pharmacology, University of Michigan Medical School, Ann Arbor, Michigan 48109

Christian Wiesmann (1:281; 1:285)
Protein Engineering Department, Genetech, Inc., South San Francisco, California 94080

Christian Wiesmann (1:281; 1:285)
Sunesis Pharmaceuticals, South San Francisco, California 94080

Roger L. Williams (2:153)
Medical Research Council, Laboratory of Molecular Biology, Cambridge CB2 1GA, United Kingdom

William D. Willis (3:607)
Department of Anatomy and Neurosciences, Marine Biomedical Institute, The University of Texas Medical Branch, Galveston, Texas 77555

Timothy M. Willson (3:53)
Nuclear Receptor Discovery Research, GlaxoSmithKline Research Triangle Park, North Carolina 27709

Ian A. Wilson (1:33; 1:63; 1:251)
Department of Molecular Biology, The Scripps Research Institute and the Skaggs Institute for Chemical Biology, La Jolla, California 92037

Ofer Wiser (2:667)
Howard Hughes Medical Institute, Departments of Physiology and Biochemistry, University of California San Francisco, San Francisco, California 94080

Matthew J. Wishart (1:741)
Biological Chemistry Department, University of Michigan, Ann Arbor, Michigan 48109-0606

Alfred Wittinghofer (2:757)
Max-Planck Institute for Molecular Physiology, 44227 Dortmund, Germany

James R. Woodgett (1:493)
Ontario Cancer Institute, Toronto, Ontario M5G 2M9, Canada

David K. Worthylake (2:751)
Department of Pharmacology, Department of Biochemistry and Biophysics, and Lineberger Comprehensive Cancer Center, The University of North Carolina at Chapel Hill, Chapel Hill, North Carolina 27599

Jeffrey L. Wrana (1:487)
Samuel Lunenfeld Research Institute, Mount Sinai Hospital, Program in Molecular Biology and Cancer and Department of Medical Genetic and Microbiology, University of Toronto, Toronto, Ontario M5G 1X5, Canada

Hao Wu (1:311)
Department of Biochemistry, Cornell University, Weill Medical College, New York, New York 10021

Yijin Xiao (2:253)
Department of Cancer Biology, The Lerner Research Institute, Cleveland Clinic Foundation, Cleveland, Ohio 44195

H. Eric Xu (3:21)
Nuclear Receptor Discovery Research, GlaxoSmithKline, Research Triangle Park, North Carolina 27709

Yan Xu (2:253)
Department of Cancer Biology, The Lerner Research Institute, Cleveland Clinic Foundation, Cleveland, Ohio 44195

Zheng Xu (1:689)
Departments of Medicine, Microbiology and Immunology, Howard Hughes Medical Institute, University of California-San Francisco, San Francisco, California 94143-0795

Michael B. Yaffe (2:171)
Center for Cancer Research, Massachusetts Institute of Technology, Cambridge, Massachusetts 02139

Michael B. Yaffe (1:505)
Center for Cancer Research, Massachusetts Institute of Technology, Cambridge, Massachusetts 02139

Kenneth M. Yamada (2:317)
Craniofacial Developmental Biology and Regeneration Branch, National Institute of Dental and Craniofacial Research, National Intitutes of Health, Bethesda, Maryland 20892

Seun-Ah Yang (1:631)
Division of Hematology-Oncology, University of Pennsylvania School of Medicine, Philadelphia, Pennsylvania 19104

Wannian Yang (2:715)
Department of Molecular Medicine, Cornell University, Ithaca, New York 14853

Yosef Yarden (1:405)
Department of Biological Regulation, Weizmann Institute of Science, Rehovot 76100, Israel

Hong Ye (1:311)
Department of Biochemistry, Cornell University, Weill Medical College, New York, New York 10021

Weilan Ye (2:793)
Department of Molecular Biology, Genentech, Inc., San Francisco, California 94080

Todd O. Yeates (1:15)
UCLA-DOE Center for Genomics and Proteomics and UCLA Molecular Biology Institute, University of California, Los Angeles, 611 Charles Young Dr. East, Los Angeles, California 90095-1569

Helen L. Yin (2:209)
Department of Physiology, University of Texas Southwestern Medical Center, Dallas, Texas 75390

John D. York (2:229)
Departments of Pharmacology and Cancer Biology, and Biochemistry, Howard Hughes Medical Institute, Duke University Medical Center, Durham, North Carolina 27710

Edgar C. Young (1:233)
Department of Pharmacology, Center for Neurobiology and Behavior, Howard Hughes Medical Institute, Columbia University, New York, New York 10032

Kenneth W. Young (2:19)
Department of Cell Physiology and Pharmacology, University of Leicester, Leicester LE2 3QE, United Kingdom

Matthew A. Young (1:387)
Departments of Molecular and Cell Biology and Chemistry, Howard Hughes Medical Institute, University of California, Berkeley, Berkeley, California 94720-3202

Michael W. Young (1:575)
Laboratory of Genetics, The Rockefeller University, New York, New York 10021

Minmin Yu (1:251)
Department of Molecular Biology, The Scripps Research Institute and The Skaggs Institute for Chemical Biology, La Jolla, California 92037

Nathan R. Zaccai (1:57)
Division of Structural Biology, Wellcome Trust Centre for Human Genetics, Cancer Research United Kingdom Receptor Stucture Research Group, Oxford OX3 7BN United Kingdom

Manuela Zaccolo (2:459)
Venetian Institute of Molecular Medicine, and Dulbecco Telethon Institute, University of Padua, 35129 Padua, Italy

Eli Zamir (2:317)
Department of Molecular Cell Biology, Weizmann Institute of Science, Rehovot 76100, Israel

Mark von Zastrow (1:181)
Psychiatry, Box 0984 LPPI F A014, University of California, San Francisco, 513 Parnassus, Box 0450, San Francisco, California 94143-0450

Chao Zhang (1:583)
Department of Cellular and Molecular Pharmacology, University of California, San Francisco, California 94143-0450

Xuewu Zhang (1:74)
Department of Cell Biology, Albert Einstein College of Medicine, Bronx, New York 10461

Zhong-Yin Zhang (1:677)
Department of Molecular Pharmacology, Albert Einstein College of Medicine, Bronx, New York 10461

Wenhong Zhou (3:545)
Department of Microbiology and Immunology, Sealy Center for Molecular Science, The University of Texas Medical Branch, Galveston, Texas 77555

Roya Zoraghi (2:539)
Institute of Cell Biology, University of Bern, CH-3012 Bern, Switzerland

PART II

Transmission: Effectors and Cytosolic Events (Continued)

Tony Hunter, Editor

SECTION C

Calcium Mobilization

Michael J. Berridge, Editor

Phospholipase C

Hong-Jun Liao and Graham Carpenter
*Department of Biochemistry,
Vanderbilt University School of Medicine,
Nashville, Tennessee*

Introduction

The phospholipase C enzymes that hydrolyze phosphatidylinositol 4,5-bisphosphate in mammalian cells are subdivided into four families, denoted β, γ, δ, and ε, based on sequence similarities. Each family has a unique organization of regulatory sequence motifs or domains that facilitate protein:protein and/or protein:phospholipid interactions. Utilizing these motifs, each family responds to distinct hormonal signals or intracellular cues to produce the second messenger molecules inositol 1,4,5-trisphosphate and diacylglycerol. These metabolites in turn control intracellular levels of free Ca^{2+} and protein kinase C activity, respectively. This review, in addition to discussing molecular structure/function and activation mechanisms for phospholipase C enzymes, presents the physiologic consequences of PLC genetic knockouts.

This review is focused on the phosphoinositide-specific phospholipase C (PLC) isozymes expressed in mammalian cells. This family of isozymes is defined on the basis of sequence similarities and the capacity to mediate the hydrolysis of phosphatidylinositol 4,5-bisphosphate (PI 4,5-P_2) to the second messenger molecules inositol 1,4,5-trisphosphate and diacylglycerol. The former provokes mobilization of intracellular Ca^{2+} by regulating the release of stored Ca^{2+} from within intracellular organelles into the cytosol and nucleus. The latter functions as an endogenous and required activator of protein kinase C isozymes. Hence, this enzyme uniquely activates two second messengers, which in turn may control a variety of signaling pathways and thereby influence a panoply of cellular events. This review is constrained by space, and readers are referred to other recent reviews [1–3] for additional information and pertinent references.

PLC Anatomy

The eleven mammalian PI 4,5-P_2 specific PLCs are divided into four subgroups (designated β, γ, δ, ε) based on sequence similarities that produce an organization of structural motifs unique to each subgroup [1]. The organization of these motifs or domains is illustrated in Fig. 1. All PLC isozymes have motifs designated X and Y, which in the native protein are folded together to constitute the catalytic domain. An X-ray structure of PLC-δ1 provides the clearest picture of exactly how this enzymatic center is organized and suggests potential catalytic mechanisms [2]. In addition to conserved catalytic function, each PLC subgroup is characterized by additional motifs that are involved in regulating aspects of enzyme function, such as topological localization within the cell and sensitivity to protein:protein and protein:lipid interactions.

PLC Activation Mechanisms

PLC-β. The activity of four PLC-β isozymes is regulated by hormones that bind to G-protein coupled receptors (GPCRs) [1]. These receptors typically have multiple membrane spanning domains, have no intrinsic catalytic activity, and utilize heterotrimeric G proteins to communicate with downstream second messenger producing enzymes, such as PLC isozymes. When GPCRs are stimulated by hormone binding, G protein complexes containing α, β, γ subunits are activated with the following characteristics: GDP bound to the α subunit is replaced by GTP, dissociating the trimeric complex into two active species–a GTP-bound free α subunit and a βγ dimeric complex. Both of these act as signal transducers to activate PLC-β isoforms in a manner that may depend on both for maximal activation.

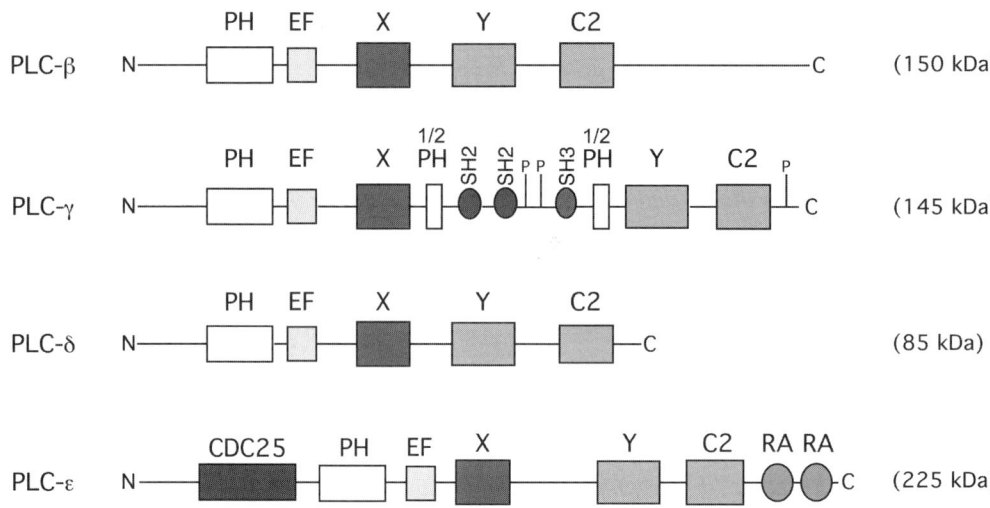

Figure 1 Schematized arrangement of domains within PLC isozymes. Functional domains or motifs (PH, EF, SH2, SH3, C2, RA, CDC25, X, Y) are explained in the text. Tyrosine phosphorylation sites are depicted by Y in the γ isozymes.

The α·GTP subunit interacts with a region in PLC-β that includes a portion of the C2 domain, while the βγ complex interacts with part of the PH domain. The region within PLC-β that binds the α·GTP subunit also appears to function as a dimerization interface, suggesting that PLC-β may function as a dimer [4,5]. Since α subunits and βγ complexes are constitutively anchored by lipid modifications to the cytoplasmic face of the plasma membrane, one consequence of these interactions is to promote a catalytically competent association of PLC-β with the plasma membrane. In unstimulated cells, PLC-βs are usually found loosely associated with the plasma membrane in what can be termed a catalytically incompetent association.

There is evidence for most PLCs that formation of a highly specific plasma membrane association is necessary for hydrolysis of the plasma membrane-localized substrate PI 4,5-P$_2$. Productive membrane association by PLC-β is further facilitated by interaction of its PH domain with phosphatidylinositol 3-phosphate. Separate regions of the PH domain accommodate this phosphoinositide and βγ complexes. Although the C2 domain in PLC-δ does mediate a Ca^{2+}-dependent phospholipid interaction, there is no evidence for this in PLC-β. While interactions of the C2 and PH domains of PLC-β with membrane-bound molecules might seem sufficient to explain formation of a productive membrane:enzyme complex, it is unclear whether this association *per se* is sufficient for increased catalytic activity or whether these interactions also provoke changes within the X/Y catalytic domain.

Signal transduction mechanisms are, by definition, reactions that are readily reversible. In the case of PLC-β, the most readily reversible component resides in the α·GTP subunit, which can be rapidly converted by intrinsic GTPase activity to α·GTP.

PLC-γ. This subgroup contains two members, γ1 and γ2 [1,3]. Initially, the cloning and sequencing of the isozymes indicated significant differences in the COOH terminal sequences; however, more recent data indicate that these apparent differences resulted from sequencing errors for γ2 [6]. As shown in Fig. 1, PLC-γ uniquely contains motifs known as SH2 and SH3 domains in addition to the PH and C2 domains present in other PLC subgroups. The SH2 motifs, in particular, are important to facilitate activation of γ isozymes by growth factor receptor tyrosine kinases (RTKs). RTKs possess a single transmembrane domain separating the ligand-binding ectodomain from a cytoplasmic domain that contains sequences encoding tyrosine kinase activity [7]. Ligand binding facilitates dimerization of RTKs and this in turn facilitates activation of the tyrosine kinase domain. Substrates for the tyrosine kinase include the receptor itself and other proteins, such as PLC-γ.

The initial step in growth factor-dependent activation of PLC-γ involves the recognition of autophosphorylation sites in a RTK by the SH2 domains of PLC-γ1 [1,3]. This recognition event is a prerequisite for tyrosine phosphorylation of PLC-γ by the RTK, which constitutes a major step in the activation mechanism. Receptor association may also relocalize PLC-γ1 from the cytosol to the cytoplasmic face of the plasma membrane, much like the association of PLC-β with membrane-anchored G protein subunits.

Compared to other PLCs, PLC-γ has an elongated linker segment between the X and Y domains. This linker contains not only the SH2 and SH3 domains, but also a split PH domain and at least two important tyrosine phosphorylation sites, which are close together between the C-SH2 and SH3 domains and are conserved in the γ1 and γ2 isoforms. It is possible that modulation of the structure of this linker region, by protein:protein interaction and/or by tyrosine phosphorylation, contributes to activation of the catalytic site [3]. One additional site of tyrosine phosphorylation is located C-terminal to C2 domain in both the γ1 and γ2 isoforms [1,3], while a fourth site in γ2 is located between the Y and C2 domains [8].

Evidence has been presented that in some cell systems PI-3 kinase activity and the formation of phosphatidylinositol 3,4,5-trisphosphate (PI 3,4,5-P_3) is necessary for maximal PLC-γ1 activation [1,3]. The site of action of PI 3,4,5-P_3 is most likely the N-terminal PH domain of PLC-γ1. However, not all reports are in agreement on this and data have been presented to support an interaction of PI 3,4,5-P_3 with the C-SH2 domain of PLC-γ1. The possible contributions of the SH3 and C2 domains of PLC-γ1 to the activation mechanism are unclear. Evidence has been presented to indicate that the SH3 domain of PLC-γ1 can associate with PIKE, a nuclear GTPase, and stimulate its activation [9].

In hematopoietic cells, adaptor proteins, such as LAT in T cells, are localized in specialized membrane microdomains termed lipid rafts and are also tyrosine phosphorylated following antigen activation [1,3]. In T cells, phosphorylated LAT becomes associated with PLC-γ1, and this interaction is necessary for PLC-γ1 activation. Whether similar membrane components are involved in PLC-γ1 activation in nonhematopoietic cells is unknown. However, there is evidence that RTK activation provokes the preferential association of tyrosine phosphorylated PLC-γ with caveolae [10,11], which resemble raft membrane microdomains but contain the protein caveolin.

PLC-δ. The activation mechanisms for this subgroup of PLC isozymes, which are not activated by GPCRs or RTKs, are perhaps least understood. Studies *in vitro* indicate that while all PLCs require free Ca^{2+}, the δ isozymes are the most sensitive to free Ca^{2+} levels [1]. This has led to the notion that this enzymes activity may be enhanced by increased levels of intracellular Ca^{2+}, and there is evidence consistent with this in a few cell-based systems. Available data also indicate that the PH and C2 domains facilitate membrane association of the δ isozymes. The PH domain of PLC-δ recognizes PI 4,5-P_2 and may not only tether the enzyme to the plasma membrane but may also facilitate a processive mechanism of hydrolysis. The C2 domain mediates membrane association by recognition of a Ca^{2+}-phospholipid complex in the membrane. It is interesting that the PH domain of PLC-δ also binds inositol 1,4,5-P_3, the product of PI 4,5-P_2 hydrolysis, and this may represent a mechanism to decrease PLC-δ activity when product levels become high.

PLC-ε. This is the most recent addition to the mammalian PLC family and was foreshadowed by the isolation of a *C. elegans* cDNA that has the same organization [1]. Within the PLC family, PLC-ε has novel protein:protein interaction motifs that indicate that its activation is directly controlled by the G protein Ras, which is a particularly important molecule in signal transduction pathways initiated by growth factor RTKs.

Near its N-terminus, PLC-ε contains a sequence identified as a CDC25 or a RasGEF (guanine nucleotide exchange factor) motif, which in other proteins facilitates the activation of Ras by mediating the exchange of GDP for GTP. Evidence indicates that expression of exogenous PLC-ε in cells does promote increased levels of Ras≅GTP. This would place PLC-ε upstream of Ras in a signal transduction pathway. PLC-ε also has RA or Ras association motifs near its C-terminus. This sequence motif allows recognition of PLC-ε by the GTP-bound, or activated, form of Ras. Evidence shows that PLC-ε is indeed recognized with high affinity by Ras·GTP and is not recognized by Ras·GTP. This predicts that Ras is an activator of PLC-ε and would place this PLC isoform downstream of activated Ras. Also, recognition of PLC-ε by activated Ras could also facilitate membrane translocation of PLC-ε, as Ras is constitutively membrane-localized by the presence of covalent lipid constituents. The complex relationship of PLC-ε to Ras is analogous to the reported observation that PLC-β1 acts as a GAP (GTPase activating protein) toward the α·GTP subunit that is its direct activator [12]. Ras is a prototype for a large family of single subunit GTPases, and there are data suggesting that PLC-ε also participates in signaling dependent on Rap 1 [13] and Rap2B [14], members of the Ras superfamily.

It also appears that PLC-ε can be activated by heterotrimeric G protein subunits, including βγ complexes [15] and at least one α·GTP subunit [16]. The former proceeds through recognition of a PH domain in PLC-ε. Hence, PLC-ε may be activated by growth factor RTKs through Ras or by GPCRs through heterotrimeric G proteins. This finding raises the prospect that the PLC activity downstream of these receptors may represent contributions by more than one PLC subgroup.

PLC Physiology

While structural and biochemical questions regarding PLC isozymes have yielded significant information regarding the molecular mechanisms by which these enzymes are activated in cells, there is much less information available regarding the role of these PLC isozymes in physiologic or pathophysiologic processes. Given the ubiquitous occurrence of PLC isozymes and the pleiotropic potential of the second messengers that they generate, this may seem either too obvious or too complex a question to resolve. In view of the multiplicity of isoforms in each PLC subgroup, one might expect substantial functional redundancy to exist, although this expectation is partially offset by the differing patterns of expression for each isoform. Also at play is the extent to which PLC-dependent signal transduction is necessary, sufficient, or dispensable for any given cell response. These issues can to some extent be addressed by selective abrogation of each PLC isozyme through targeted gene disruption technology. The contents of Table I describe results that have been obtained to date by the application of this technology to PLC genes. The results, in some cases, represent phenotypes obtained at the first crucial point in development when a particular PLC isoform becomes required for further development of the organism.

In the case of *Plcb3* knockouts, discordant results have been reported that may reflect the manner in which the gene was actually disrupted. When *Plcb3* genomic information corresponding to exons encoding the last one-third of the

Table I Phenotypes Resulting from Targeted Disruption of PLC Genes in Mice

Gene (Protein)	Phenotype	Reference
Plcb1 (PLC-β1)	Death 2–6 weeks after birth, increased level of recurrent seizures due to decrease in muscarinic acetylcholine signaling	17
Plcb2 (PLC-β2)	Normal life span, increased neutrophil chemotactic response	18
Plcb3 (PLC-β3)	Embryonic lethal E2.5	19
	Normal life span, decreased opoid-dependent behavioural responses	20
	Increased skin ulcers	21
Plcb4 (PLC-β4)	Normal life span, locomotor ataxia due to metabotropic glutamate receptor signaling	17
	Impaired visual response	22
	Decreased climbing fiber elimination	23
	Decreased long-term depression, decreased conditioned motor learning	24
Plcg1 (PLC-γ1)	Embryonic lethal E9.0, impaired erythrogenesis, vasculogenesis	25
	Chimeric mice (plcg1 −/− and +/+) mice, impaired hematopoiesis, polycystic kidney	26
Plcg2 (PLC-γ2)	Normal life span, decreased mast cell function, decreased B cell numbers	27
Plcd4 (PLC-δ4)	Normal life span, male infertility due to deficiency in acrosome reaction	28

X domain plus the first two-thirds of the Y domain was deleted, embryonic lethality was produced at approximately 2.5 days in gestation [19]. In the second knockout [20], a genomic deletion representing one exon encoding some residues in the X domain was produced and the mice were normal as far as development and growth are concerned. At this time the discrepancy between these two studies has not been resolved. It is possible that in one knockout a mutant protein was produced that acted as a dominant-negative molecule affecting other pathways. Alternatively, one of the knockouts may represent only a partial loss of PLC-β1 function.

References

1. Rhee, S. G. (2001). Regulation of phosphoinositide-specific phospholipase C. *Annu. Rev. Biochem.* **70**, 281–312.
2. Williams, R. L. (1999). Mammalian phosphoinositide-specific phospholipase C. *Biochim. Biophys. Acta* **1441**, 255–267.
3. Carpenter, G. and Ji, Q.-S. (1999). Phospholipase C-γ as a signal transducing element. *Exp. Cell Res.* **253**, 15–24.
4. Singer, A. U., Waldo, G. L., Harden, T. K., and Sondek, J. (2002). A unique fold of phospholipase C-β mediates dimerization and interaction with αq. *Nature Struc. Biol.* **9**, 32–36.
5. Ilkaeva, O., Kinch, L. N., Paulssen, R. H., and Ross, E. M. (2002). Mutations in the carboxyl-terminal domain of phospholipase C-β1 delineate the dimer interface and a potential $G\alpha_q$ interaction site. *J. Biol. Chem.* **277**, 4294–4300.
6. Ozdener, F., Kunapuli, S. P., and Daniel, J. L. (2001). Carboxyl terminal sequence of human phospholipase Cγ2. *Platelets* **12**, 121–123.
7. Schlessinger, J. and Ullrich, A. (1992). Growth factor signaling by receptor tyrosine kinases. *Neuron* **9**, 383–391.
8. Watanabe, D., Hashimoto, S., Ishiai, M., Matsushita, M., Baba, Y., Kishimoto, T., Kurosaki, T., and Tsukada, S. (2001). Four tyrosine residues in phospholipase C-γ2, identified as Btk-dependent phosphorylation sites, are required for B cell antigen receptor-coupled calcium signaling. *J. Biol. Chem.* **276**, 38595–38601.
9. Ye, K., Aghdasi, B., Luo, H. R., Moriarity, J. L., Wu, F. Y., Hong, J. J., Hurt, K. J., Bae, S. S., Suh, P.-G., and Snyder, S. H. (2002). Phospholipase Cγ1 is a physiological guanine nucleotide exchange factor for the nuclear GTPase PIKE. *Nature* **415**, 541–544.
10. Wang, X.-J., Liao, H.-J., Chattopadhyay, A., and Carpenter, G. (2001). EGF-dependent translocation of green fluorescent protein-tagged PLC-γ1 to the plasma membrane and endosomes. *Exp. Cell Res.* **267**, 28–36.
11. Jang, I.-H., Kim, J. H., Lee, B. D., Bae, S. S., Park, M. H., Suh, P.-G., and Ryu, S. H. (2001). Localization of phospholipase C-γ1 signaling in caveolae: importance of BGF-induced phosphoinositide hydrolysis but not in tyrosine phosphorylation. *FEBS Lett.* **491**, 4–9.
12. Berstein, G., Blank, J. L., Jhon, D.-Y., Exton, J. H., Rhee, S. G., and Ross, E. M. (1992). Phospholipase C-β is a GTPase-activating protein form Gq/11, its physiological regulator. *Cell* **70**, 411–418.
13. Jin, T.-G., Satoh, T., Liao, Y., Song, C., Gao, X., Kariya, K.-i., Hu, C.-D., and Kataoka, T. (2001). Role of the CDC25 homology domain of phospholipase Cε in amplification of Rap1-dependent signaling. *J. Biol. Chem.* **276**, 30301–30307.
14. Schmidt, M., Evellin, S., Weernink, P. A. O., vom Dorp, F., Rehmann, H., Lomasney, J. W., and Jakobs, K. H. (2001). A new phospholipase C-calcium signalling pathway mediated by cyclic AMP and a Rap GTPase. *Nature Cell Biol.* **3**, 1020–1024.
15. Wing, M. R., Houston, D., Kelley, G. G., Der, C. J., Siderovski, D. P., and Harden, T. H. (2001). Activation of phospholipase C-ε by heterotrimeric G protein βγ- subunits. *J. Biol. Chem.* **276**, 48257–48261.
16. Lopez, I., Mak, E. C., Ding, J., Hamm, H. E., and Lomasney, J. W. (2001). A novel bifunctional phospholipase C that is regulated by $G\alpha_{12}$ and stimulates the Ras/mitogen-activated protein kinase pathway. *J. Biol. Chem.* **276**, 2758–2765.
17. Kim, D., Jun, K. I. S., Lee, S. B., Kang, N.-G., Min, D. S., Kim, Y.-H., Ryu, S. H., Suh, P.-G., and Shin, H.-S. *Nature* **389**, 290–293.
18. Jiang, H., Kuang, Y., Wu, Y., Xie, W., Simon, M. I., and Wu, D. (1997). Roles of phospholipase Cβ2 in chemoattractant-elicited responses. *Proc. Natl. Acad. Sci. USA* **94**, 7971–7975.
19. Wang, S., Gebre-Medhin, S., Betsholtz, C., Stålberg, P., Zhou, Y., Larsson, C., Weber, G., Feinstein, R., Öberg, K., Gobl, A., and Skogseid, B. (1998). Targeted disruption of the mouse phospholipase Cβ3 gene results in early embryonic lethality. *FEBS Lett.* **441**, 261–265.
20. Xie, W., Samoriski, G. M., McLaughlin, J. P., Romoser, V. A., Smrcka, A., Hinkle, P. M., Bidlack, J. M., Gross, R. A., Jiang, H., and Wu, D. (1999). Genetic alteration of phospholipase Cβ3 expression modulates behavioral and cellular responses to µ opioids. *Proc. Natl. Acad. Sci. USA* **96**, 10385–10390.
21. Li, Z., Jiang, H., Xie, W., Zhang, Z., Smrcka, A. V., and Wu, D. (2000). Roles of PLC-β2 and -β3 and PI3Kγ in chemoattractant-mediated signal transduction. *Science* **287**, 1046–1049.

22. Jiang, H., Lyubarsky, A., Dodd, R., Vardi, N., Pugh, E., Baylor, D., Simon, M. I., and Wu, D. (1996). Phospholipase Cβ4 is involved in modulating the visual response in mice. *Proc. Natl. Acad. Sci. USA* **93**, 14598–14601.

23. Kano, M., Hashimoto, K., Watanabe, M., Kurihara, H., Offermanns, S., Jiang, H., Wu, Y., Jun, K., Shin, H.-S., Ionue, Y., Simon, M. I., and Wu, D. (1998). Phospholipase Cβ4 is specifically involved in climbing fiber synapse elimination in the developing cerebellum. *Proc. Natl. Acad. Sci. USA* **95**, 15724–15729.

24. Miyata, M., Kim, H.-T., Hashimoto, K., Lee, T.-W., Cho, S.-Y., Jiang, H., Wu, Y., Jun, K., Wu, D., Kano, M., and Shin, H.-S. (2001). Deficient long-term synaptic depression in the rostral cerebellum correlated with impaired motor learning in phospholipase Cβ4 mutant mice. *Eur. J. Neurosci.* **13**, 1945–1954.

25. Ji, Q.-S., Winnier, G. E., Niswender, K. D., Horstman, D., Wisdom, R., Magnuson, M. A., and Carpenter, G. (1997). Essential role of the tyrosine kinase substrate phospholipase C-γ1 in mammalian growth and development. *Proc. Natl. Acad. Sci. USA* **94**, 2999–3003.

26. Shirane, M., Sawa, H., Kobayashi, Y., Nakano, T., Ktajima, K., Shinkai, Y., Nagashima, K., and Negishi, I. (2001). Deficiency of phospholipase C-γ1 impairs renal development and hematopoiesis. *Development* **128**, 5173–5180.

27. Wang, D., Feng, J., Wen, R., Marine, J.-C., Sangster, M. Y., Parganas, E., Hoffmeyer, A., Jackson, C. W., Cleveland, J. L., Murray, P. J., and Ihle, J. N. (2000). Phospholipase Cγ2 is essential in the functions of B Cell and several Fc receptors. *Immunity* **13**, 25–35.

28. Fukami, K., Nakao, K., Inoue, T., Kataoka, Y., Kurokawa, M., Fissore, R. A., Nakamura, K., Katsuki, M., Mikoshiba, K., Yoshida, N., and Takenawa, T. (2001). Requirement of phospholipase Cδ4 for the zna pellucida-induced acrosome reaction. *Science* **292**, 920–923.

CHAPTER 123

Inositol 1,4,5-trisphosphate 3-kinase and 5-phosphatase

Valérie Dewaste and
Christophe Erneux

*Interdisciplinary Research Institute (IRIBHN)
Université Libre de Bruxelles,
Brussels, Belgium*

Introduction

Inositol 1,4,5-trisphosphate (InsP$_3$) 5-phosphatase and 3-kinase are the two major enzyme activities that metabolize InsP$_3$ in mammalian cells. Distinct forms of inositol and phosphatidylinositol 5-phosphatase selectively remove the phosphate from the 5-position of the inositol ring from both soluble and lipid substrates, i.e. InsP$_3$, inositol 1,3,4,5-tetrakisphosphate (InsP$_4$), phosphatidylinositol 4,5-bisphosphate (PtdInsP$_2$) or phosphatidylinositol 3,4,5-trisphosphate (PtdInsP$_3$). This reaction is often associated with the deactivation pathway of both soluble and lipid second messengers involved in Ca^{2+} signalling, proliferation, growth factor mediated events, and apoptosis [1–3]. Type I InsP$_3$ 5-phosphatase is unable to use the phosphoinositides as substrates and therefore is specific for InsP$_3$ and InsP$_4$. It is a quite different and unique enzyme as compared to the type II isoforms (sometimes also called type II, III, and IV and considered as inositol lipid phosphatases *in vivo*, for review, see [2]).

The InsP$_3$ 3-kinase catalyses the phosphorylation of InsP$_3$ to InsP$_4$. This enzyme is particularly interesting in view of the rapid increase in InsP$_4$ levels in cells upon stimulation by agonists [4] and of several second messenger functions that had been proposed for InsP$_4$ [5]. cDNAs encoding three human isoenzymes of InsP$_3$ 3-kinase (3-kinases A, B, and C) have been reported (Table I). A mammalian inositol polyphosphate multikinase has been shown to phosphorylate inositol 4,5-bisphosphate, thereby providing an alternative biosynthesis for InsP$_3$ [6].

Type I InsP$_3$ 5-phosphatase

Initially detected in human erythrocyte membranes [7], type I InsP$_3$ 5-phosphatase has a 10-fold higher affinity but 100-fold lower Vmax for InsP$_4$ than it does for InsP$_3$ [8]. Molecular cloning revealed that it is a 412 amino acid protein with a C-terminal isoprenylation site CCVVQ (Fig. 1) [9,10]. Type I InsP$_3$ 5-phosphatase is targeted to plasma membranes and perinuclear regions as shown in several transfection studies [12]. The activity of this enzyme could be modified by targeting mechanisms and by phosphorylation. Direct interaction with platelet proteins such as pleckstrin and 14-3-3ξ has been reported [13,14]. In several cell models (e.g. rat cortical astrocytes), this enzyme has been shown to be stimulated by an InsP$_3$ mobilizing agonist such as ATP or carbachol. This effect is triggered by Ca^{2+}/calmodulin kinase II protein phosphorylation, resulting in an inhibition of enzyme activity [15]. Underexpression of type I InsP$_3$ 5-phosphatase in rat kidney cells is associated with increasing levels of InsP$_3$ and InsP$_4$ and increasing levels of intracellular Ca^{2+} [16]. Antisense-transfected cells grow faster as compared to cells transfected with vector alone. Antisense-transfected cells formed colonies in soft agar and tumors in nude mice [17]. In over-expression studies in CHO cells, prenylation of type I InsP$_3$

Table I Characteristics of Type I InsP$_3$ 5-phosphatase and InsP$_3$ 3-kinase Isoforms

Protein (human)	Accession numbers	Pest regions	Stimulation of activity by Ca^{2+}/CaM	Tissue distribution
Type I InsP$_3$ 5-phosphatase	X77567	—	—	Heart, skeletal muscle, and brain [10]
InsP$_3$ 3-kinase A	X54938	2	2–3 fold	Brain and testis [11]
InsP$_3$ 3-kinase B	Y18024, AJ242780	5	8–10 fold	Ubiquitous [24]
InsP$_3$ 3-kinase C	AJ290975	4	Insensitive	Ubiquitous [25]

The PESTfind program (at *www.embnet.org*) developed by Rogers *et al* [26] was used. The tissue distribution was obtained by Northern blotting.

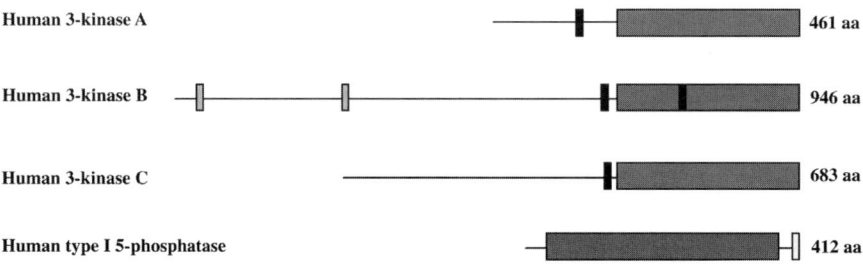

Figure 1 Schematic representation of the enzymes that metabolize InsP$_3$. Dark blue boxes represent the C-terminal catalytic domain of the InsP$_3$ 3-kinase enzymes and the red one of the type I InsP$_3$ 5-phosphatase. Potential phosphorylation sites are represented by black boxes. Two proline rich regions are found in the sequence of the InsP$_3$ 3-kinase B (light blue boxes). The yellow box shows the isoprenylation site of the type I InsP$_3$ 5-phosphatase enzyme.

5-phosphatase appears to be critical in the control of Ca^{2+} oscillations in response to agonists [18].

InsP$_3$ 3-kinase

This enzyme activity that specifically produced InsP$_4$ was initially reported in rat brain and T lymphocytes [19,20]. Purification of the protein and microsequence determination allowed the cloning of a 459 amino acid protein referred to as the A-isoform of 50 kDa. When it was expressed in bacteria, it showed an activity that was stimulated upon the addition of Ca^{2+} and calmodulin as shown for the native enzyme [21]. Transfection studies in HeLa cells show the importance of a 66-aminoacid N-terminal sequence in targeting the InsP$_3$ 3-kinase A to the actin cytoskeleton [22]. The first demonstration of two distinct sequences was reported in 1991 with the isolation of two different cDNAs from a cDNA library. These were referred to as human InsP$_3$ 3-kinases A and B, respectively [23]. The full-length sequence of human InsP$_3$ 3-kinase B has been recently reported and encodes a protein of 946 amino acids [24]. A third human isoenzyme, i.e. InsP$_3$ 3-kinase C, has been isolated following the screening of a human thyroid cDNA library. Its open reading frame encoding 683 amino acids has been expressed in *E. coli* and also in COS-7 cells [25]. Full-length sequences of human InsP$_3$ 3-kinases A, B, and C have been reported in databases (Table I).

The three human isoforms are members of a large family of inositol phosphate kinases present in mammalian but also in yeast that contain a C-terminal catalytic domain and specific residues involved in binding inositide substrates [3]. The mammalian InsP$_3$ 3-kinases A, B, and C show the presence of four conserved motifs in the catalytic C-terminal domain that are not present in inositol hexakisphosphate kinase or in the inositol multikinase [25]. Human InsP$_3$ 3-kinases A, B, and C contain potential PEST-sequences as identified by the PESTfind program (Table I). This suggests that the enzymes are particularly sensitive to proteolysis as noticed during purification of the native enzyme in several tissues.

Several data in the literature suggest that InsP$_4$ by itself shows second messenger function(s) in neurons or in endothelial cells [27,28]. One approach to address the function of InsP$_4$ is to look for the distribution and relative expression of the proteins responsible for its metabolism or action in cells. In this context, a particularly high level of the A isoform is found in the dendritic spines of neurons (Purkinje cells and hippocampal CA1 pyramidal cells) [22,29], thus supporting a role of InsP$_4$ in LTP. The A isoform was reported to associate with F-actin, whereas the B isoform was shown to exist in the cytosol and also to be associated to the cytosolic face of endoplasmic reticulum membranes [22,30]. The localization of the inositol type I 5-phosphatase in plasma membranes was reported to be critical in the control of Ca^{2+} oscillations [18]. The mammalian InsP$_3$ 3-kinases A, B, and C could also control the concentration of several inositol phosphates, InsP$_3$ isomers, and/or higher phosphorylated inositol phosphates such as InsP$_4$ isomers, InsP$_5$, or InsP$_6$. In particular, Ins(1,3,4)P$_3$, the product of InsP$_4$ dephosphorylation by type I inositol

5-phosphatase, is a potent inhibitor of Ins(3,4,5,6)P_4 1-kinase, resulting in an increase of Ins(3,4,5,6)P_4 and a decrease in chloride efflux [31]. The situation is also complicated by the fact that the three InsP_3 3-kinase isoforms in animals can be distinguished by their N-terminal sequence, presumably targeting sequences and sensitivity to Ca^{2+}/calmodulin [24,25].

References

1. Erneux, C., Govaerts, C., Communi, D., and Pesesse, X. (1998). The diversity and possible functions of the inositol polyphosphate 5-phosphatases. *Biochim. Biophys. Acta* **1436**, 185–199.
2. Majerus, P. W., Kisseleva, M. V., and Norris, F. A. (1999). The role of phosphatases in inositol signaling reactions. *J. Biol. Chem.* **274**, 10669–10672.
3. Irvine, R. F. and Schell, M. J. (2001). Back in the water: the return of the inositol phosphates. *Nature Rev. Mol. Cell Biol.* **2**, 327–338.
4. Batty, I. R., Nahorski, S. R., and Irvine, R. (1985). Rapid formation of inositol 1,3,4,5-tetrakisphosphate following muscarinic receptor stimulation of rat cerebral cortical slices. *Biochem. J.* **232**, 211–215.
5. Irvine, R. F., McNulty, T. J., and Schell, M. J. (1999). Inositol 1,3,4,5-tetrakisphosphate as a second messenger–a special role in neurones? *Chem. Phys. Lipids* **98**, 49–57.
6. Saiardi, A., Nagata, E., Luo, H. R., Sawa, A., Luo, X., Snowman, A. M., and Snyder, S. H. (2001). Mammalian inositol polyphosphate multikinase synthesizes inositol 1,4,5-trisphosphate and an inositol pyrophosphate. *Proc. Natl. Acad. Sci. USA* **98**, 2306–2311.
7. Downes, C. P., Mussat, M. C., and Michell, R. H. (1982). The inositol trisphosphate phosphomonoesterase of the human erythrocyte membrane. *Biochem. J.* **203**, 169–177.
8. Erneux, C., Lemos, M., Verjans, B., Vanderhaegen, P., Delvaux, A., and Dumont, J. E. (1989). Soluble and particulate Ins(1,4,5)P_3/Ins(1,3,4,5)P_4 5-phosphatase in bovine brain. *Eur. J. Biochem.* **181**, 317–322.
9. Verjans, B., De Smedt, F., Lecocq, R., Vanweyenberg, V., Moreau, C., and Erneux, C. (1994). Cloning and expression in *Escherichia coli* of a dog thyroid cDNA encoding a novel inositol 1,4,5-trisphosphate 5-phosphatase. *Biochem. J.* **300**, 85–90.
10. Laxminarayan, K. M., Chan, B. K., Tetaz, T., Bird, P. I., and Mitchell, C. A. (1994). Characterization of a cDNA encoding the 43-kDa membrane-associated inositol-polyphosphate 5-phosphatase. *J. Biol. Chem.* **269**, 17305–17310.
11. Vanweyenberg, V., Communi, D., D'Santos, C. S., and Erneux, C. (1995). Tissue- and cell-specific expression of Ins(1,4,5)P3 3-kinase isoenzymes. *Biochem. J.* **306**, 429–435.
12. De Smedt, F., Boom, A., Pesesse, X., Schiffmann, S. N., and Erneux, C. (1996). Post-translational modification of human brain type I inositol-1,4,5-trisphosphate 5-phosphatase by farnesylation. *J. Biol. Chem.* **1996**, 10419–10424.
13. Auethavekiat, V., Abrams, C. S., and Majerus, P. W. (1997). Phosphorylation of platelet pleckstrin activates inositol polyphosphate 5-phosphatase I. *J. Biol. Chem.* **272**, 1786–1790.
14. Campbell, J. K., Gurung, R., Romero, S., Speed, C. J., Andrews, R. K., Berndt, M. C., and Mitchell, C. A. (1997). Activation of the 43 kDa inositol polyphosphate 5-phosphatase by 14-3-3 zeta. *Biochemistry* **36**, 15363–15370.
15. Communi, D., Gevaert, K., Demol, H., Vandekerckhove, J., and Erneux, C. (2001). A novel receptor-mediated regulation mechanism of type I inositol polyphosphate 5-phosphatase by calcium/calmodulin-dependent protein kinase II phosphorylation. *J. Biol. Chem.* **276**, 38738–38747.
16. Speed, C. J., Neylon, C. B., Little, P. J., and Mitchell, C. A. (1999). Underexpression of the 43 kDa inositol polyphosphate 5-phosphatase is associated with spontaneous calcium oscillations and enhanced calcium responses following endothelin-1 stimulation. *J. Cell. Sci.* **112**, 669–679.
17. Speed, C., Little, P., Hayman, J. A., Mitchell, C. A. (1996). Underexpression of the 43 kDa inositol polyphosphate 5-phosphatase is associated with cellular transformation. *EMBO J.* **15**, 4852–4861.
18. De Smedt, F., Missiaen, L., Parys, J. B., Vanweyenberg, V., De Smedt, H., and Erneux, C. (1997). Isoprenylated human brain type I inositol 1,4,5-trisphosphate 5-phosphatase controls Ca^{2+} oscillations induced by ATP in Chinese hamster ovary cells. *J. Biol. Chem.* **272**, 17367–17375.
19. Irvine, R. F., Letcher, A. J., Heslop, J. P., and Berridge, M. J. (1986). The inositol tris/tetrkisphosphate pathway-demonstration of Ins(1,4,5)P_3 3-kinase activity in animal tissues. *Nature* **320**, 631–634.
20. Steward, S. J., Prpic, V., Powers, F. S., Bocckino, S. B., Isaacks, R. E., and Exton, J. H. (1986). Perturbation of the human T-cell antigen receptor-T3 complex leads to the production of inositol tetrakisphosphate: evidence for conversion from inositol trisphosphate. *Proc. Natl. Acad. Sci. USA* **83**, 6098–6102.
21. Takazawa, K., Vandekerckove, J., Dumont, J. E., and Erneux, C. (1990). Cloning and expression in *Escherichia coli* of a rat brain cDNA encoding a Ca^{2+}/calmodulin-sensitive inositol 1,4,5-trisphosphate 3-kinase. *Biochem. J.* **272**, 107–112.
22. Schell, M. J., Erneux, C., and Irvine, R. F. (2001). Inositol 1,4,5-trisphosphate 3-kinase A associates with F-actin and dendritic spines via its N-terminus. *J. Biol. Chem.* **276**, 37537–37546.
23. Takazawa, K., Perret, J., Dumont, J. E., and Erneux, C. (1991). Molecular cloning and expression of a new putative inositol 1,4,5-trisphosphate 3-kinase isoenzyme. *Biochem. J.* **278**, 883–886.
24. Dewaste, V., Roymans, D., Moreau, C., and Erneux, C. (2002). Cloning and expression of a full-length cDNA encoding human inositol 1,4,5-trisphosphate 3-kinase B. *Biochem. Biophys. Res. Commun.* **291**, 400–405.
25. Dewaste, V., Pouillon, V., Moreau, C., Shears, S., Takazawa, K., and Erneux, C. (2000). Cloning and expression of a cDNA encoding human inositol 1,4,5-trisphosphate 3-kinase C. *Biochem. J.* **352**, 343–351.
26. Rogers, S., Wells, R., and Rechsteiner, M. (1986). Amino acid sequences common to rapidly degraded proteins: the PEST hypothesis. *Science* **234**, 364–368.
27. Luckhoff, A., and Clapham, D. E. (1992). Inositol 1,3,4,5-tetrakisphosphate activates an endothelial Ca^{2+} permeable channel. *Nature* **355**, 356–358.
28. Tsubokawa, H., Oguro, K., Robinson, H. P. C., Masukawa, T., and Kawai, N. (1996). Intracellular Inositol 1,3,4,5-tetrakisphosphate enhances the Ca^{2+} current in hippocampal CA1 neurones of the gerbil after ischemia. *J. Physiol.* **497**, 67–78.
29. Mailleux, P., Takazawa, K., Erneux, C., and Vanderhaeghen, J. J. (1991). Inositol 1,4,5-trisphosphate 3-kinase distribution in rat brain. High levels in the hippocampal CA1pyramidal and cerebellar Purkinje cells suggest its involvement in some memory processes. *Brain Res.* **539**, 203–210.
30. Soriano, S., Thomas, S., High, S., Griffiths, G., D'Santos, C., Cullen, P., and Banting, G. (1997). Membrane association, localization and topology of rat inositol 1,4,5-trisphosphate 3-kinase B: implications for membrane traffic and Ca^{2+} homoeostasis. *Biochem. J.* **324**, 579–589.
31. Yang, X., Rudolf, M., Carew, M. A., Yoshida, M., Nerreter, V., Riley, A. M., Chung, S. K., Bruzik, K. S., Potter, B. V., Schultz, C., and Shears, S. B. (1999). Inositol 1,3,4-trisphosphate acts *in vivo* as a specific regulator of cellular signalling by inositol 3,4,5-tetrakisphosphate. *J. Biol. Chem.* **274**, 18973–18980.

CHAPTER 124

Cyclic ADP-ribose and NAADP

Antony Galione and Grant C. Churchill
*Department of Pharmacology, Oxford University, Mansfield Road,
Oxford, United Kingdom*

Introduction

Cyclic adenosine diphosphate ribose (cADPR) and nicotinic acid adenine dinucleotide phosphate (NAADP) are endogenous pyridine nucleotide metabolites with potent Ca^{2+} mobilizing activities. Although the Ca^{2+} mobilizing properties of these two molecules was first discovered in sea urchin eggs, their actions seem to extend to many mammalian, invertebrate, and plant systems where they may function as Ca^{2+} mobilizing intracellular messengers.

In a seminal study, Lee and colleagues reported in 1987 that not only could the established Ca^{2+} mobilizing messenger inositol 1,4,5 trisphosphate (IP_3) release Ca^{2+} from intracellular stores in sea urchin egg homogenates, but so too could the two pyridine nucleotides, NAD and NADP [1]. However, NAD and NADP were not Ca^{2+} mobilizing agents themselves. The active principles were subsequently identified as a cyclic metabolite of NAD, cADPR [2] (Fig. 1A) and a contaminant of commercially available NADP, NAADP [3] (Fig. 1B). Both cADPR and NAADP were shown directly to mobilize Ca^{2+} from intracellular stores by microinjection into intact sea urchin eggs [3,4].

A useful property of the sea urchin homogenate system is that Ca^{2+} release by different Ca^{2+} mobilizing agents displays homologous desensitization. After stimulation of maximal Ca^{2+} release by a given agent, Ca^{2+} stores become refractory to Ca^{2+} release by that same agents. Sequential additions of IP_3, cADPR and NAADP to the same aliquot of egg homogenate all evoked Ca^{2+} releases regardless of the order in which they were added, while a second addition of any of these agents failed to elicit any response. (see [5] for review.) From these data it was proposed that these three Ca^{2+} releasing agents mobilized Ca^{2+} stores by three distinct mechanisms. Further studies with selective pharmacological agents confirmed this view. Heparin, a competitive IP_3 receptor (IP_3R) antagonist, selectively inhibited IP_3-evoked Ca^{2+} release, while cADPR or NAADP-induced Ca^{2+} release were unaffected [4]. Ca^{2+} release by cADPR was found to be selectively blocked by ryanodine receptor (RyR) inhibitors [6] and chemically synthesized 8-substituted analogues of cADPR [7]. In contrast, NAADP-evoked Ca^{2+} release was neither affected by IP_3 or RyR antagonists nor cADPR analogues, but selectively blocked by inhibitors of voltage-gated Ca^{2+} and K^+ channels [8].

cADPR has now been implicated in the regulation of Ca^{2+} release via RyRs in many different cell types, while NAADP, which similarly has also been shown to mobilize Ca^{2+} in a number of cell types from different organisms [9], appears to act on a novel Ca^{2+} release channel. While IP_3Rs and RyRs are well-characterized Ca^{2+} release channels of intracellular organelles, the molecular nature of the NAADP-sensitive Ca^{2+} release channel is unknown. IP_3 and cADPR appear to predominantly mobilize Ca^{2+} from the endoplasmic reticulum [10,11] while NAADP releases Ca^{2+} from a distinct organelle [10,12], possibly an acidic compartment.

A key property of both IP_3Rs and RyRs is that both are modulated by Ca^{2+} itself. This property is responsible for Ca^{2+}-induced Ca^{2+} release (CICR), which serves to amplify normally locally restricted Ca^{2+} transients as global Ca^{2+} signals and is thought to be critical in determining the complex patterns of Ca^{2+} signals widely observed in cells such as repetitive Ca^{2+} spikes and regenerative Ca^{2+} waves [13]. Both IP_3 and cADPR appear to sensitize IP_3Rs and RyRs, respectively, to activation by Ca^{2+}, thereby promoting CICR. The molecular interactions of IP_3 with its receptors have been relatively well defined [14]; however, the molecular mechanisms by which cADPR activates RyRs are not, and the possibility exists that cADPR binds to an accessory protein that in turn modulates RyR openings [15,16]. Indeed, two known RyR associated proteins, calmodulin [17] and FKBP12.6 [18],

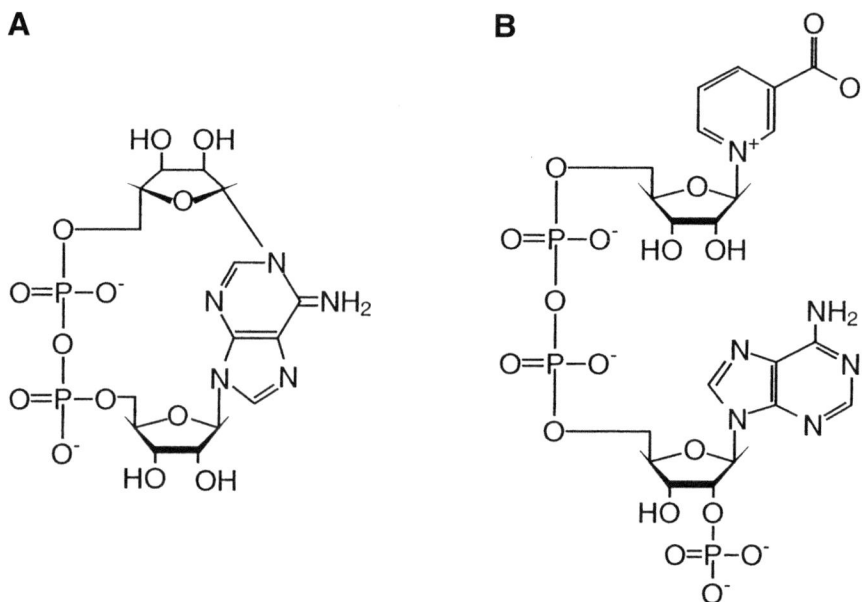

Figure 1 The chemical structures of cyclic adenosine dinucleotide phosphate ribose (cADPR) (A) and nicotinic acid adenine dinucleotide phosphate (NAADP) (B).

have been implicated in cADPR-induced Ca^{2+} release. However, NAADP elicits local Ca^{2+} signals unless amplified by triggering CICR by recruiting IP_3Rs and RyRs [19] or through its own diffusion through cells [20].

Two types of evidence exist to suggest that cADPR or NAADP function as intracellular messengers for Ca^{2+} signaling in cells and tissues. The first is that endogenous levels of these compounds are modulated by cellular stimuli and the second is that the selective block of cADPR or NAADP-sensitive Ca^{2+} release mechanisms can inhibit Ca^{2+} mobilization or functional responses to a range of hormones, transmitters, and other cell regulators. ADP-ribosyl cyclases are a class of enzyme that can synthesise cADPR and NAADP from alternate substrates NAD and NADP, respectively, with pH and phosphorylation state determining which product is produced [21,22]. The best-characterized example of such an enzyme is CD38, which has been implicated in cADPR-based Ca^{2+} signaling in a number of cell types [23]. It was originally thought to be an ectoenzyme, although several reports indicate that it is also present in several intracellular compartments. Studies from tissues and cells derived from CD38 knockout mice implicate CD38 in both cADPR synthesis and Ca^{2+} signaling [24,25], and perhaps also NAADP synthesis as well [21,26], although other synthetic pathways are possible. CD38 is a complex enzyme in that it also catalyzes the hydrolysis of cADPR as well, while NAADP may be metabolized by a calcium-dependent phosphatase [27]. A number of approaches have been employed to determine changes in cADPR levels in cells, including thin layer chromatography, hplc, radioimmunoassay, radioreceptor binding [28], and a new cycling assay [29]. A variety of cell stimuli have been shown to increase cADPR levels in cells [30], including G protein-linked receptors and those linked to tyrosine kinase activities [31], although the precise coupling mechanisms are unknown. There are no data at present on changes in NAADP levels in cells and tissues, although endogenous levels have been reported in plant tissues [32].

The role of cADPR in the generation of Ca^{2+} signals in response to cellular stimuli have been dissected by use of selective cADPR antagonists [30], and for NAADP by injecting high concentrations of NAADP into cells to desensitize NAADP-evoked Ca^{2+} release [19], since no selective antagonists for NAADP exist at present. Such studies have suggested a key role for the use of multiple messengers and multiple Ca^{2+} stores in dictating the complex patterns of Ca^{2+} signals observed in cells, which may be linked differentially to specific cellular responses. In T cells, IP_3 elicits a brief Ca^{2+} transient, while cADPR prolongs the Ca^{2+} signal [31]. In ascidian oocytes, different Ca^{2+} releasing messengers regulate different cell functions with IP_3 inducing Ca^{2+} spiking, cADPR regulating exocytosis, and NAADP modulating plasma membrane Ca^{2+} currents [33]. In pancreatic acinar cells, different Ca^{2+} mobilizing transmitters and hormones appear to be coupled to one or more types of Ca^{2+} mobilizing messengers [34]. The different Ca^{2+} release mechanisms appear also to be coupled in different ways depending on cell phenotype. For example, in pancreatic acinar cells and sea urchin eggs the NAADP mechanism couples to CICR channels to elicit global Ca^{2+} signals, while in T cells and ascidian oocytes, desensitization of NAADP receptors rather puzzlingly renders cells insensitive to IP_3 [9,33,35].

Although much work still remains to be done in elucidating many of the molecular details of the cADPR and NAADP signaling pathways, it is clear that through their involvement along with IP_3 in regulating Ca^{2+} signaling, they provide another layer of regulation in determining complex Ca^{2+} signaling patterns. They are thus likely to be important components of the Ca^{2+} code whereby a single ion can specifically regulate a diverse array of cellular functions.

References

1. Clapper, D. L., Walseth, T. F., Dargie, P. J., and Lee, H. C. (1987). Pyridine nucleotide metabolites stimulate calcium release from sea urchin egg microsomes desensitized to inositol trisphosphate. *J. Biol. Chem.* **262**, 9561–9568.
2. Lee, H. C., Walseth, T. F., Bratt, G. T., Hayes, R. N., and Clapper, D. L. (1989). Structural determination of a cyclic metabolite of NAD with intracellular calcium-mobilizing activity. *J. Biol. Chem.* **264**, 1608–1615.
3. Lee, H. C. and Aarhus, R. (1995). A derivative of NADP mobilizes calcium stores insensitive to inositol trisphosphate and cyclic ADP-ribose. *J. Biol. Chem.* **270**, 2152–2157.
4. Dargie, P. J., Agre, M. C., and Lee, H. C. (1990). Comparison of calcium mobilizing activities of cyclic ADP-ribose and inositol trisphosphate. *Cell Regul.* **1**, 279–290.
5. Genazzani, A. A. and Galione, A. (1997). A Ca^{2+} release mechanism gated by the novel pyridine nucleotide, NAADP. *Trends Pharmacol. Sci.* **18**, 108–110.
6. Galione, A., Lee, H. C., and Busa, W. B. (1991). Ca^{2+}-induced Ca^{2+} release in sea urchin egg homogenates: modulation by cyclic ADP-ribose. *Science* **253**, 1143–1146.
7. Walseth, T. F. and Lee, H. C. (1993). Synthesis and characterization of antagonists of cyclic ADP-ribose-induced calcium release. *Biochim. Biophys. Acta* **1178**, 235–242.
8. Genazzani, A. A., Mezna, M., Dickey, D. M., Michelangeli, F., Walseth, T. F., and Galione, A. (1997). Pharmacological properties of the Ca^{2+}-release mechanism sensitive to NAADP in the sea urchin egg. *Br. J. Pharmacol.* **121**, 1489–1495.
9. Patel, S., Churchill, G. C., and Galione, A. (2001). Coordination of Ca^{2+} signalling by NAADP. *Trends Biochem. Sci.* **26**, 482–489.
10. Lee, H. C. and Aarhus, R. (2000). Functional visualization of the separate but interacting calcium stores sensitive to NAADP and cyclic ADP-ribose. *J. Cell Sci.* **113**, 4413–4420.
11. Churchill, G. C. and Galione, A. (2001). NAADP induces Ca^{2+} oscillations via a two-pool mechanism by priming IP$_3$- and cADPR-sensitive Ca^{2+} stores. *Embo J.* **20**, 2666–2671.
12. Genazzani, A. A. and Galione, A. (1996). Nicotinic acid-adenine dinucleotide phosphate mobilizes Ca^{2+} from a thapsigargin-insensitive pool. *Biochem. J.* **315**, 721–725.
13. Berridge, M. J., Lipp, P., and Bootman, M. D. (2000). The versatility and universality of calcium signalling. *Nature Mol. Cell Biol. Rev.* **1**, 11–21.
14. Taylor, C. W. (1998). Inositol trisphosphate receptors: Ca^{2+}-modulated intracellular Ca^{2+} channels. *Biochim. Biophys. Acta* **1436**, 19–33.
15. Walseth, T. F., Aarhus, R., Kerr, J. A., and Lee, H. C. (1993). Identification of cyclic ADP-ribose-binding proteins by photoaffinity labeling. *J. Biol. Chem.* **268**, 26686–26691.
16. Thomas, J. M., Masgrau, R., Churchill, G. C., and Galione, A. (2001). Pharmacological characterization of the putative cADP-ribose receptor. *Biochem. J.* **359**, 451–457.
17. Lee, H. C., Aarhus, R., Graeff, R., Gurnack, M. E., and Walseth, T. F. (1994). Cyclic ADP ribose activation of the ryanodine receptor is mediated by calmodulin. *Nature* **370**, 307–309.
18. Noguchi, N., Takasawa, S., Nata, K., Tohgo, A., Kato, I., Ikehata, F., Yonekura, H., and Okamoto, H. (1997). Cyclic ADP-ribose binds to FK506-binding protein 12.6 to release Ca^{2+} from islet microsomes. *J. Biol. Chem.* **272**, 3133–3136.
19. Cancela, J. M., Churchill, G. C., and Galione, A. (1999). Coordination of agonist-induced Ca^{2+}-signalling patterns by NAADP in pancreatic acinar cells. *Nature* **398**, 74–76.
20. Churchill, G. C. and Galione, A. (2000). Spatial control of Ca^{2+} signaling by nicotinic acid adenine dinucleotide phosphate diffusion and gradients. *J. Biol. Chem.* **275**, 38687–38692.
21. Aarhus, R., Graeff, R. M., Dickey, D. M., Walseth, T. F., and Lee, H. C. (1995). ADP-ribosyl cyclase and CD38 catalyze the synthesis of a calcium-mobilizing metabolite from NADP. *J. Biol. Chem.* **270**, 30327–30333.
22. Wilson, H. L. and Galione, A. (1998). Differential regulation of nicotinic acid-adenine dinucleotide phosphate and cADP-ribose production by cAMP and cGMP. *Biochem. J.* **331**, 837–843.
23. Lee, H. C. (2000). Enzymatic functions and structures of CD38 and homologs. *Chem. Immunol.* **75**, 39–59.
24. Partida-Sanchez, S., Cockayne, D. A., Monard, S., Jacobson, E. L., Oppenheimer, N., Garvy, B., Kusser, K., Goodrich, S., Howard, M., Harmsen, A., Randall, T. D., and Lund, F. E. (2001). Cyclic ADP-ribose production by CD38 regulates intracellular calcium release, extracellular calcium influx and chemotaxis in neutrophils and is required for bacterial clearance in vivo. *Nat. Med.* **7**, 1209–1216.
25. Fukushi, Y., Kato, I., Takasawa, S., Sasaki, T., Ong, B. H., Sato, M., Ohsaga, A., Sato, K., Shirato, K., Okamoto, H., and Maruyama, Y. (2001). Identification of cyclic ADP-ribose-dependent mechanisms in pancreatic muscarinic Ca^{2+} signaling using CD38 knockout mice. *J. Biol. Chem.* **276**, 649–655.
26. Chini, E. N., Chini, C. C., Kato, I., Takasawa, S., and Okamoto, H. (2002). CD38 is the major enzyme responsible for synthesis of nicotinic acid-adenine dinucleotide phosphate in mammalian tissues. *Biochem. J.* **362**, 125–130.
27. Berridge, G., Galione, A., and Patel, S. P. (2002). Metabolism of the novel Ca^{2+} mobilising messenger, nicotinic acid adenine dinucleotide phosphate, via a 2'-specific Ca^{2+}-dependent phosphatase. *Biochem. J.* in press
28. Galione, A., Cancela, J.-M., Churchill, G., Genazzani, A., Lad, C., Thomas, J., Wilson, H. L., and Terrar, D. (2000). Methods in cADPR and NAADP research. In *Methods in Calcium Signaling* (J. Putney, ed.), pp. 249–296. CRC Press, Boca Raton.
29. Graeff, R. M. and Lee, H. C. (2002). A novel cycling assay for cellular cADP-ribose with nanomolar sensitivity. *Biochem. J.* **361**, 379–384.
30. Galione, A. and Churchill, G. (2000). Cyclic ADP-ribose as a calcium mobilizing messenger. *Science STKE*, www.stke.org/cgi/content/full/OC_sigtrans;2000/41/pe1,1–6.
31. Guse, A. H., Da Silva, C. P., Berg, I., Skapenko, A. L., Weber, K., Heyer, P., Hohenegger, M., Ashamu, G. A., Schulze-Koops, H., Potter, B. V., and Mayr, G. W. (1999). Regulation of calcium signalling in T lymphocytes by the second messenger cyclic ADP-ribose. *Nature* **398**, 70–73.
32. Navazio, L., Bewell, M. A., Siddiqua, A., Dickinson, G. D., Galione, A., and Sanders, D. (2000). Calcium release from the endoplasmic reticulum of higher plants elicited by the NADP metabolite nicotinic acid adenine dinucleotide phosphate. *Proc. Natl. Acad. Sci. USA* **97**, 8693–8698.
33. Albrieux, M., Lee, H. C., and Villaz, M. (1998). Calcium signaling by cyclic ADP-ribose, NAADP, and inositol trisphosphate are involved in distinct functions in ascidian oocytes. *J. Biol. Chem.* **273**, 14566–14574.
34. Cancela, J. M., Van Coppenolle, F., Galione, A., Tepikin, A. V., and Petersen, O. H. (2002). Transformation of local Ca^{2+} spikes to global Ca^{2+} transients: the combinatorial roles of multiple Ca^{2+} releasing messengers. *Embo J.* **21**, 909–919.
35. Berg, I., Potter, B. V., Mayr, G. W., and Guse, A. H. (2000). Nicotinic acid adenine dinucleotide phosphate (NAADP$^+$) is an essential regulator of T-lymphocyte Ca^{2+}-signaling. *J. Cell Biol.* **150**, 581–588.

Sphingosine 1-phosphate

Kenneth W. Young and Stefan R. Nahorski
*Department of Cell Physiology and Pharmacology,
University of Leicester,
Leicester, United Kingdom*

Introduction

Sphingosine 1-phosphate is a putative intracellular second messenger for Ca^{2+} release. The metabolic pathways controlling sphingosine 1-phosphate levels are beginning to be understood, and it is clear that intracellular levels of this molecule can be actively regulated. However, the signaling machinery through which sphingosine 1-phosphate releases Ca^{2+} from intracellular stores remains poorly characterised. This chapter addresses recent issues in this area of Ca^{2+} signaling.

It is well established that sphingolipids, which are ubiquitous structural membrane constituents, are also capable of giving rise to a number of lipid signaling metabolites. In particular, sphingosine 1-phosphate (SPP) (Fig. 1) has been implicated in a variety of processes at both the intracellular and extracellular levels [1–3]. The extracellular targets for SPP, which is present in serum and can be released into the extracellular environment by activated platelets [4], have been clearly identified as members of the endothelial differentiation gene (Edg) family of heptahelical G protein-coupled receptors (GPCRs) [2]. These receptors couple to a variety of heterotrimeric G proteins, and can stimulate intracellular Ca^{2+} release through activation of the inositol (1,4,5)-trisphosphate [$Ins(1,4,5)P_3$] signaling pathway. However, SPP has also a number of attributes associated with a role as an intracellular messenger molecule. Thus intracellular levels of SPP can be regulated, often within seconds of an appropriate extracellular stimulus, through the action of the enzyme sphingosine kinase (SPHK). In this case, the resultant increases in cytosolic SPP appear capable of directly activating cell growth and intracellular Ca^{2+} mobilization [1,3,5]; however, unlike the extracellular effects of SPP, the intracellular targets for SPP remain to be defined.

Sphingolipid Metabolism

SPP is formed from the membrane lipid sphingomyelin via a series of enzymatic reactions [see 2,6]. Hydrolysis of sphingomyelin produces ceramide (N-acyl sphingosine), and this appears to be a central molecule in the SPP metabolic pathway. Subsequent removal of the amide-linked fatty acid side chain of ceramide produces sphingosine, which can then be phosphorylated by SPHK to produce SPP. Out of the sphingolipid metabolites, only SPP appears to have any direct Ca^{2+} release activity [7,8], and there is now clear experimental evidence that a variety of extracellular stimuli can activate SPHK and increase intracellular SPP levels (see below). Consistent with a second messenger role, SPP can be rapidly degraded either by dephosphorylation back to sphingosine, or irreversibly removed from the sphingolipid cycle via the enzyme SPP lyase, which hydrolyses a carbon-carbon bond in the sphingosine backbone of the molecule [9]. It is interesting that platelets, a noted source of extracellular SPP, lack SPP lyase [9].

Activation of SPHK

Two distinct forms of human SPHK have been cloned. The type 1 form is a 49 kDa protein that contains putative phosphorylation sites for PKC, PKA, and casein kinase II, as well as Ca^{2+}/calmodulin and SH3 binding domains. The type 2 form is notably larger (65 kDa) and has a different tissue distribution [6, 10,11]. Both these proteins are predominantly cytosolic [5,12] (Fig. 2). SPHK activity has been quantified by measuring the ability of cell lysates (whole cell or fractionated, stimulated vs. unstimulated) to [$\gamma^{32}P$]-label added

Sphingosine 1-phosphate

Figure 1 Structural representation of sphingosine 1-phosphate.

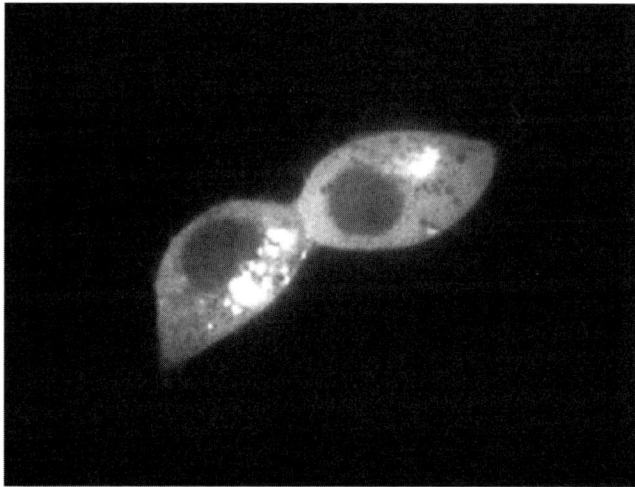

Figure 2 Cellular distribution of SPHK. HEK-293 cells, grown on coverslips, were transiently transfected with SPHK tagged with eGFP (gift from S. Spiegel) and imaged on a confocal microscope ($\times 100$ magnification). Similar to studies using HA-tagged SPHK, or involving SPHK1 antisera, SPHK-eGFP was predominantly cytosolic in location.

sphingosine from a pool of $[\gamma^{32}P]$-ATP. Such experiments have demonstrated SPHK activity both in cytosolic and membrane fractions [5]. Pretreatment of NIH 3T3 fibroblast cells with PDGF raises the Vmax of SPHK, with no apparent change in Km [13], and both cytosolic and membrane forms of SPHK are similarly activated [5]. SPHK activity can also be investigated indirectly by measuring changes in intracellular SPP. In such experiments, cells are pulse-labeled with $[^3H]$-sphingosine and the resultant $[^3H]$-SPP extracted. A complicating factor with this method is the lack of equilibrium labeling of $[^3H]$-sphingosine, hence making substrate availability a potential issue.

Using these methods, a number of extracellular stimuli, which result in Ca^{2+} mobilizing responses, have been shown to stimulate SPHK, hence supporting the possibility that intracellular SPP functions as a Ca^{2+} release mediator. This list of stimuli include PDGF [13], antigen stimulation [14], and a variety of recombinant and endogenous GPCRs (muscarinic M2 and M3 [15], lysophosphatidic acid (LPA), Edg-4 receptor [16]). Although investigating the role of SPP in Ca^{2+} mobilization is complicated by the fact that many of these extracellular stimuli also utilize the $Ins(1,4,5)P_3/Ca^{2+}$ release pathway, this is not always the case. Of particular note here is the Fcγ receptor-I in U937 monocytes. This receptor, which is an integral membrane glycoprotein, undergoes a molecular switching between $Ins(1,4,5)P_3$ mediated- and SPHK dependent-Ca^{2+}release according to the differentiation state of the cell [14], thus changing the temporal profile of the Ca^{2+} response. In addition, our own work in the SH-SY5Y human neuroblastoma cell line indicates an $Ins(1,4,5)P_3$-independent but SPHK-dependent Ca^{2+} mobilizing response to LPA [16,17].

The mechanism by which SPHK, a cytosolic protein, can be activated by cell surface stimuli remains unclear, though there is evidence for a number of possible signaling pathways. PDGF-mediated SPHK activation in TRMP canine kidney epithelial cells (recombinantly expressing PDGF-β receptors) is inhibited by addition of the Ca^{2+} chelator BAPTA [18]. A similar Ca^{2+}-sensitive SPP production has been shown for GPCRs [19]. In addition, receptor-independent increases in intracellular Ca^{2+} levels, either via Ca^{2+} ionophores [18] or voltage-gated Ca^{2+} channels [20], activate SPHK. However, although SPHK binds calmodulin with high affinity in the presence of Ca^{2+} [21], thus supporting a role for Ca^{2+}-mediated SPHK activation, it should be noted that purified SPHK is insensitive to Ca^{2+} in the range 1–100 μM [21]. A second activation pathway involves acidic phospholipids, which despite the lack of a recognized binding domain, increases SPHK activity *in vitro* [22]. In this way, a phospholipase D-mediated increase in phosphatidic acid may be the mechanism through which antigen activation of SPHK and subsequent Ca^{2+} release occurs [14]. Another possibility is that SPHK is activated via protein-protein interactions. Thus, although Ca^{2+} signals were not investigated, activated TNFα-receptors recruit the adaptor protein TRAF2, which in turn binds to and activates SPHK [23]. Such recruitment of SPHK to protein signaling scaffolds near the plasma membrane may have additional important consequences, as this would bring SPHK into contact with its substrate, sphingosine. It is of interest that antigen [24] and GPCR (unpublished data) stimulation of SPHK also involves recruitment of the kinase to the plasma membrane.

Intracellular Target for SPP-mediated Ca^{2+} Release

Although progress is clearly being made on the signaling pathways leading to SPP production, frustratingly little is known about the intracellular targets responsible for Ca^{2+} release. SPP-mediated Ca^{2+} mobilization was first noted in 1990 by Ghosh and co-workers [7]. Using a permeabilized cell preparation, this group demonstrated that sphingosine and the related compound sphingosylphosphorylcholine (SPC) could release $^{45}Ca^{2+}$ from the endoplasmic reticulum (ER). As the sphingosine response involved a short time lag, and required the presence of ATP, it was suggested that sphingosine was being converted to SPP, and it was SPP that possessed Ca^{2+} release activity [7,8]. Direct SPP-mediated Ca^{2+} release has now been shown by a number of groups, either in permeabilized cell preparations [8,16,25] or via microinjection [15]. SPP is active over the concentration range 1–100 μM, and although SPP utilizes ostensibly the same intracellular Ca^{2+} pool as $Ins(1,4,5)P_3$, there is no requirement for $InsP_3$- or ryanodine-receptor activation [17]. Functional studies indicate the presence of both SPHK and the putative release channel for SPP to be present in the ER [7,8]; however, the identity of the intracellular release channel remains elusive. A putative sphingolipid-mediated Ca^{2+} release channel, termed SCaMPER (sphingolipid Ca^{2+}-release mediating protein of the endoplasmic reticulum), has been cloned [26]. However, a more recent study suggests that SCaMPER is not an ion channel and is not associated with the ER [27]. Indeed it is not clear whether SCaMPER has any sort of interaction with SPP.

Due to the complicating effects of cell surface GPCRs for SPP, and because SPP production often occurs alongside increases in $Ins(1,4,5)P_3$, it has been difficult to look at the single cell and subcellular Ca^{2+} signals to SPP. Thus questions concerning the spatial and temporal aspects of SPP responses remain largely unanswered. It is of interest to note that the SPP dependent-, $Ins(1,4,5)P_3$ independent-, Ca^{2+} responses to antigen- [15] and LPA- [16] stimulation are similarly transient in nature. To date, there is no evidence that intracellular SPP is capable of producing Ca^{2+} oscillations; however, SPP-mediated Ca^{2+} release can stimulate subcellular Ca^{2+} puffs, presumably by a direct effect of the released Ca^{2+} on $InsP_3$ receptors [16].

Concluding Remarks

The role of SPP as an intracellular messenger will remain controversial until intracellular targets for SPP are clearly identified. Recent elegant work using a fluorescent bioassay for SPP demonstrates that PDGF-mediated increases in SPHK activity produce detectable increases in extracellular SPP that are capable of activating cell surface Edg-receptors [28]. Furthermore, SPP may itself be produced extracellularly via the extracellular export of SPHK [29], and so great care must be taken when investigating supposedly intracellular actions of SPP. As SCaMPER does not appear to be an appropriate intracellular target, attention needs to be focused on the identification and characterization of the intracellular Ca^{2+}-release channel for SPP. In addition, the use of molecular tools such as the recently described catalytically inactive form of SPHK [30] should allow for greater investigation into the role of SPP in mediating Ca^{2+} signals. Such experiments are essential for fully understanding the significance of SPP as an intracellular mediator for Ca^{2+} release.

Acknowledgments

K. W. Young is funded by The Wellcome Trust.

References

1. Spiegel, S. and Milstein, S. (1994). Sphingolipid metabolites: members of a new class of lipid messengers. *J. Membr. Biol.* **146**, 225–237.
2. Pyne, S. and Pyne, N. (2000). Sphingosine 1-phosphate signalling in mammalian cells. *Biochem. J.* **349**, 385–402.
3. Young, K. W. and Nahorski, S. R. (2001). Intracellular sphingosine 1-phosphate production: a novel pathway for Ca^{2+} release. *Semin. Cell Dev. Biol.* **12**, 19–25.
4. Yatomi, Y., Yamamura, S., Ruan, F., and Igarashi, Y. (1997). Sphingosine 1-phosphate induces platelet activation through an extracellular action and shares a platelet surface receptor with lysophosphatidic acid. *J. Biol. Chem.* **272**, 5291–5297.
5. Olivera, A., Kohama, T., Edsall, L., Nava, V., Cuvillier, O., Poulton, S., and Spiegel, S. (1999). Sphingosine kinase expression increases intracellular sphingosine-1-phosphate and promotes cell growth and survival. *J. Cell Biol.* **147**, 545–557.
6. Olivera, A. and Spiegel, S. (2001). Sphingosine kinase: a mediator of vital cell functions. *Prostaglan. Lipid. Meds.* **64**, 123–134.
7. Ghosh, T. K., Bian, J., and Gill, D. L. (1990). Intracellular calcium release mediated by sphingosine derivatives generated in cells. *Science* **248**, 1653–1656.
8. Ghosh, T. K., Bian, J., and Gill, D. L. (1994). Sphingosine 1-phosphate generated in the endoplasmic reticulum membrane activates release of stored calcium. *J. Biol. Chem.* **269**, 22628–22635.
9. Mandala, S. M. (2001). Sphingosine-1-phosphate phosphatases. *Prostaglan. Lipid. Meds.* **64**, 143–156.
10. Kohama, T., Olivera, A., Edsall, L., Nagiec, M. M., Dickson, R., and Spiegel, S. (1998). Molecular cloning and functional characterization of murine sphingosine kinase. *J. Biol. Chem.* **273**, 23722–23728.
11. Liu, H., Sugiura, M., Nava, V. E., Edsall, L. C., Kono, K., Poulton, S., Milstein, S., Kohama, T., and Spiegel, S. (2000). Molecular cloning and functional characterisation of a novel mammalian sphingosine kinase type 2 isoform. *J. Biol. Chem.* **275**, 19513–19520.
12. Murate, T., Banno, Y., T-Koizumi, K., Watanabe, K., Mori, N., Wada, A., Igarashi, Y., Takagi, A., Kojima, T., Asano, H., Akao, Y., Yoshida, S., Saito, H., and Nozawa, Y. (2001). Cell type-specific localization of sphingosine kinase 1a in human tissues. *J. Histochem. Cytochem.* **49**, 845–855.
13. Olivera, A. and Spiegel, S. (1993). Sphingosine-1-phosphate as second messenger in cell proliferation induced by PDGF and FCS mitogens. *Nature* **365**, 557–560.
14. Melendez, A., Floto, R. A., Cameron, A. J., Gillooly, D. J., Harnett, M. M., and Allen, J. M. (1998). A molecular switch changes the signalling pathway used by the FcγRI antibody receptor to mobilize Ca^{2+}. *Curr. Biol.* **8**, 210–211.
15. Meyer zu Heringdorf, D., Lass, H., Alemany, R., Laser, K. T., Neumann, E., Zhang, C., Schmidt, M., Rauen, U., Jakobs, K. H., and Van Koppen, C. J. (1998). Sphingosine kinase-mediated Ca^{2+} signalling by G-protein-coupled receptors. *EMBO J.* **17**, 2830–2837.
16. Young, K. W., Bootman, M. D., Channing, D. R., Lipp, P., Maycox, P. R., Meakin, J., Challiss, R. A. J., and Nahorski, S. R. (2000). Lysophosphatidic acid-induced Ca^{2+} mobilisation requires intracellular sphingosine 1-phosphate production. *J. Biol. Chem.* **275**, 38532–38539.
17. Young, K. W., Challiss, R. A. J., Nahorski, S. R., and Mackrill, J. J. (1999). Lysophosphatidic acid-mediated Ca^{2+} mobilization in human SH-SY5Y neuroblastoma cells is independent of phosphoinositide signalling, but dependent on sphingosine kinase activation. *Biochem. J.* **343**, 45–52.
18. Olivera, A., Edsall, L., Poulton, S., Kazlauskas, A., and Spiegel, S., (1999). Platelet-derived growth factor-induced activation of sphingosine kinase requires phosphorylation of the PDGF receptor tyrosine residue responsible for binding PLCγ. *FASEB J.* **13**, 1592–1600.
19. Alemany, R., Sichelschmidt, B., Meyer zu Heringdorf, D., Lass, H., Van Koppen, C. J., and Jakobs, K. H. (2000). Stimulation of sphingosine-1-phosphate formation by the $P2Y_2$ receptor in HL-60 cells: Ca^{2+} requirement and implication in receptor-mediated Ca^{2+} mobilisatization, but not MAP kinase activation. *Mol. Pharmacol.* **58**, 491–498.
20. Alemany, R., Kleuser, B., Ruwisch, L., Danneberg, K., Lass, H., Hashemi, R., Spiegel, S., Jakobs, K. H., and Meyer zu Heringdorf, D. (2001). Depolarisation induces rapid and transient formation of intracellular sphingosine 1-phosphate. *FEBS Letts.* **509**, 239–244.
21. Olivera, A., Kohama, T., Tu, Z., Milstien, S., and Spiegel, S. (1998). Purification and characterization of rat kidney sphingosine kinase. *J. Biol. Chem.* **273**, 12576–12583.
22. Olivera, A., Rosenthal, J., and Spiegel, S. (1996). Effect of acidic phospholipids on sphingosine kinase. *J. Cell. Biochem.* **60**, 529–537.
23. Xia, P., Wang, L., Moretti, P. A. B., Albanese, N., Chai, F., Pitson, S. M., D'Andrea, R. J., Gamble, J. R., and Vadas, M. A. (2002).

Sphingosine kinase interacts with TRAF2 and dissects tumor necrosis factor-α signaling. *J. Biol. Chem.* **277**, 7996–8003.

24. Mendelez, A. J. and Khaw, A. K. (2002). Dichotomy of Ca^{2+} signals triggered by different phospholipid pathways in antigen stimulation of human mast cells. *J. Biol. Chem.* **277**, 17255–17262.

25. Meyer zu Heringdorf, D., Niederdräing, N., Neumann, E., Fröde, R., Lass, H., Van Koppen, C. J., and Jakobs, K. H. (1998). Discrimination between plasma membrane and intracellular target sites of sphingosylphosphorylcholine. *Eur. J. Pharmacol.* **354**, 113–122.

26. Mao, C., Kim, S. H., Almenoff, J. S., Rudner, X. L., Kearney, D. M., and Kindman, L. A. (1996). Molecular cloning and characterization of SCaMPER, a sphingolipid Ca^{2+} release-mediating protein from endoplasmic reticulum. *Proc. Natl. Acad. Sci. USA* **93**, 1993–1996.

27. Schnurbus, R., De Pietri Tonelli, D., Grohavaz, F., and Zacchetti, D. (2002). Re-evaluation of primary structure, topology, and localization of Scamper, a putative intracellular Ca^{2+} channel activated by sphingosylphosphocholine. *Biochem. J.* **362**, 183–189.

28. Hobson, J. P., Rosenfeldt, H. M., Barak, L. S., Olivera, A., Poulton, S., Caron, M. G., Milstein, S., and Spiegel, S. (2001). Role of spingosine-1-phosphate receptor EDG-1 in PDGF-induced cell motility. *Science* **291**, 1800–1803.

29. Ancellin, N., Colmont, C., Su, J., Li, Q., Mittereder, N., Chae, S. S., Stefansson, S., Liau, G., and Hla, T. (2002). Extracellular export of sphingosine kinase-1 enzyme. *J. Biol. Chem.* **277**, 6667–6675.

30. Pitson, S. M., Moretti, P. A. B., Zebol, J. R., Xia, P., Gamble, J. R., Vadas, M. A., D'Andrea, R. J., and Wattenberg, B. W. (2000). Expression of a catalytically inactive sphingosine kinase mutant blocks agonist-induced sphingosine kinase activation. *J. Biol. Chem.* **275**, 33945–33950.

CHAPTER 126

Voltage-gated Ca^{2+} Channels

William A. Catterall
Department of Pharmacology,
University of Washington, Seattle, Washington

Introduction

Voltage-gated Ca^{2+} channels mediate calcium entry into cells in response to membrane depolarization. Electrophysiological studies reveal different Ca^{2+} currents designated L-, N-, P-, Q-, R-, and T-type. The high-voltage-activated Ca^{2+} channels that have been characterized biochemically are complexes of a pore-forming α1 subunit of about 190 to 250 kDa, a transmembrane, disulfide-linked complex of α2 and δ subunits, an intracellular β subunit, and in some cases a transmembrane γ subunit. Ten α1 subunits, four α2δ complexes, four β subunits, and several γ subunits are known. The Ca$_V$1 family of α1 subunits conduct L-type Ca^{2+} currents, which initiate muscle contraction, endocrine secretion, sensory transduction, cardiac pacemaking, and gene transcription and are regulated primarily by second messenger-activated protein phosphorylation pathways. The Ca$_V$2 family of α1 subunits conduct N-type, P/Q-type, and R-type Ca^{2+} currents, which initiate rapid synaptic transmission and are regulated primarily by direct interaction with G proteins and SNARE proteins and secondarily by protein phosphorylation. The Ca$_V$3 family of α1 subunits conduct T-type Ca^{2+} currents, which are activated and inactivated more rapidly and at more negative membrane potentials than other Ca^{2+} channel types. The distinct structures and patterns of regulation of these three families of Ca^{2+} channels provides a flexible array of Ca^{2+} entry pathways in response to changes in membrane potential and a range of possibilities for regulation of Ca^{2+} entry by second messenger pathways and interacting proteins.

Physiological Roles of Voltage-gated Ca^{2+} Channels

Ca^{2+} channels in many different cell types activate upon membrane depolarization and mediate Ca^{2+} influx in response to action potentials and subthreshold depolarizing signals. Ca^{2+} entering the cell through voltage-gated Ca^{2+} channels serves as the second messenger of electrical signaling, initiating many different cellular events (Fig. 1). In cardiac and smooth muscle cells, activation of Ca^{2+} channels initiates contraction directly by increasing cytosolic Ca^{2+} concentration and indirectly by activating ryanodine-sensitive Ca^{2+} release channels in the sarcoplasmic reticulum [1,2]. In skeletal muscle cells, voltage-gated Ca^{2+} channels in the transverse tubule membranes interact directly with ryanodine-sensitive Ca^{2+} release channels in the sarcoplasmic reticulum and activate them to initiate rapid contraction [3,4]. The same Ca^{2+} channels in the transverse tubules also mediate a slow Ca^{2+} conductance that increases cytosolic concentration and thereby regulates the force of contraction in response to high-frequency trains of nerve impulses. In endocrine cells, voltage-gated Ca^{2+} channels mediate Ca^{2+} entry that initiates secretion of hormones [5]. In neurons, voltage-gated Ca^{2+} channels initiate synaptic transmission [6,7,8]. In many different cell types, Ca^{2+} entering the cytosol via voltage-gated Ca^{2+} channels regulates enzyme activity, gene expression, and other biochemical processes [9,10]. Thus, voltage-gated Ca^{2+} channels are the key signal transducers of electrical excitability, converting the electrical signal of the action potential in the cell surface membrane to an intracellular Ca^{2+} transient. Signal transduction in different cell types involves different molecular subtypes of voltage-gated Ca^{2+} channels, which mediate voltage-gated Ca^{2+} currents with different physiological, pharmacological, and regulatory properties.

Ca^{2+} Current Types Defined by Physiological and Pharmacological Properties

Since the first recordings of Ca^{2+} currents in cardiac myocytes [11,12], it has become apparent that there are multiple types of Ca^{2+} currents as defined by physiological and

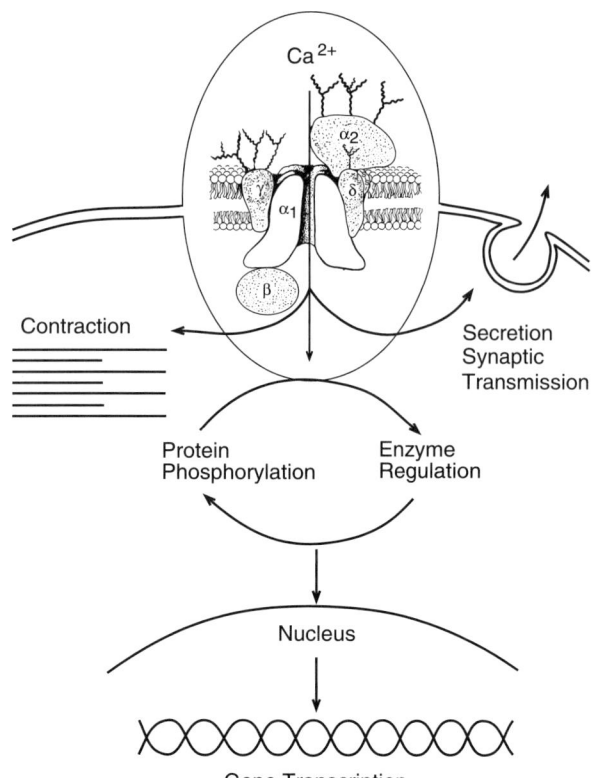

Figure 1 Ca^{2+} channels and signal transduction. Ca^{2+} entering cells initiates numerous intracellular events, including contraction, secretion, synaptic transmission, enzyme regulation, protein phosphorylation-dephosphorylation, and gene transcription. *Inset.* Subunit structure of voltage-gated Ca^{2+} channels. The five-subunit complex that forms high voltage-activated calcium channels is illustrated with a central, pore-forming α1 subunit, a disulfide-linked dimer of $α_2$ and δ glycoprotein subunits, an intracellular β subunit, and a transmembrane glycoprotein γ subunit.

pharmacological criteria [6,13–15]. In cardiac, smooth, and skeletal muscle, the major Ca^{2+} currents are distinguished by high voltage of activation, large single channel conductance, slow voltage-dependent inactivation, marked regulation by cAMP-dependent protein phosphorylation pathways, and specific inhibition by Ca^{2+} antagonist drugs, including dihydropyridines, phenylalkylamines, and benzothiazepines [16]. These Ca^{2+} currents have been designated L-type, as they are long-lasting when Ba^{2+} is the current carrier [17]. L-type Ca^{2+} currents are also recorded in endocrine cells, where they initiate release of hormones [18], and in neurons, where they are important in regulation of gene expression and in integration of synaptic inputs [9,10,14].

Voltage clamp studies of Ca^{2+} currents in starfish eggs [19] and recordings of Ca^{2+} action potentials in cerebellar Purkinje neurons [20] first revealed Ca^{2+} currents with different properties from L-type, and these were subsequently characterized in detail in voltage-clamped dorsal root ganglion neurons [17,21–23]. In comparison to L-type, these Ca^{2+} currents activate at much more negative membrane potentials, inactivate rapidly, deactivate slowly, have small single channel conductance, and are insensitive to Ca^{2+} antagonist drugs. They are designated low-voltage-activated Ca^{2+} currents for their negative voltage dependence [21] or T-type for their transient kinetics [17].

Whole-cell voltage clamp and single-channel recording from dissociated dorsal root ganglion neurons revealed an additional Ca^{2+} current, N-type [17]. In these initial experiments, N-type Ca^{2+} currents were distinguished by their intermediate voltage dependence and rate of inactivation—more negative and faster than L-type but more positive and slower than T-type [17]. They are insensitive to organic L-type Ca^{2+} channel blockers but blocked by the cone snail peptide ω-conotoxin GVIA [6,24]. This pharmacological profile has been the primary method to distinguish N-type Ca^{2+} currents, because the voltage dependence and kinetics of N-type Ca^{2+} currents in different neurons vary considerably.

Analysis of the effects of other peptide toxins revealed three additional Ca^{2+} current types. P-type Ca^{2+} currents, first recorded in Purkinje neurons [25], are distinguished by high sensitivity to the spider toxin ω-agatoxin IVA [26]. Q-type Ca^{2+} currents, first recorded in cerebellar granule neurons [27], are blocked by ω-agatoxin IVA with lower affinity. R-type Ca^{2+} currents in cerebellar granule neurons are resistant to the subtype-specific organic and peptide Ca^{2+} channel blockers [27] and may include multiple channel subtypes [28]. While L-type and T-type Ca^{2+} currents are recorded in a wide range of cell types, N-, P-, Q-, and R-type Ca^{2+} currents are most prominent in neurons.

Molecular Properties of Ca^{2+} Channels

Subunit Structure. Ca^{2+} channels purified from skeletal muscle transverse tubules are complexes of α1, β, and γ subunits, and the α1 and β subunits are substrates for cAMP-dependent protein phosphorylation [29,30]. More detailed biochemical analyses revealed an additional α2δ subunit co-migrating with the α1 subunit [31–34]. Analysis of the biochemical properties, glycosylation, and hydrophobicity of these five subunits led to a model comprising a principal transmembrane $α_1$ subunit of 190 kDa in association with a disulfide-linked $α_2δ$ dimer of 170 kDa, an intracellular phosphorylated β subunit of 55 kDa, and a transmembrane γ subunit of 33 kDa (Fig. 1, inset) [31].

The α1 subunit is a protein of about 2000 amino acid residues with an amino acid sequence and predicted transmembrane structure like the previously characterized, pore-forming α subunit of sodium channels [35] (Fig. 2). The amino acid sequence is organized in four repeated domains (I to IV), each of which contains six transmembrane segments (S1 to S6) and a membrane-associated loop between transmembrane segments S5 and S6. As expected from biochemical analysis [31], the intracellular β subunit has predicted alpha helices but no transmembrane segments [36] (Fig. 2), while the γ subunit is a glycoprotein with four transmembrane segments [37] (Fig. 2). The cloned α2 subunit has many glycosylation sites and several hydrophobic sequences [38], but biosynthesis studies indicate that it is an extracellular,

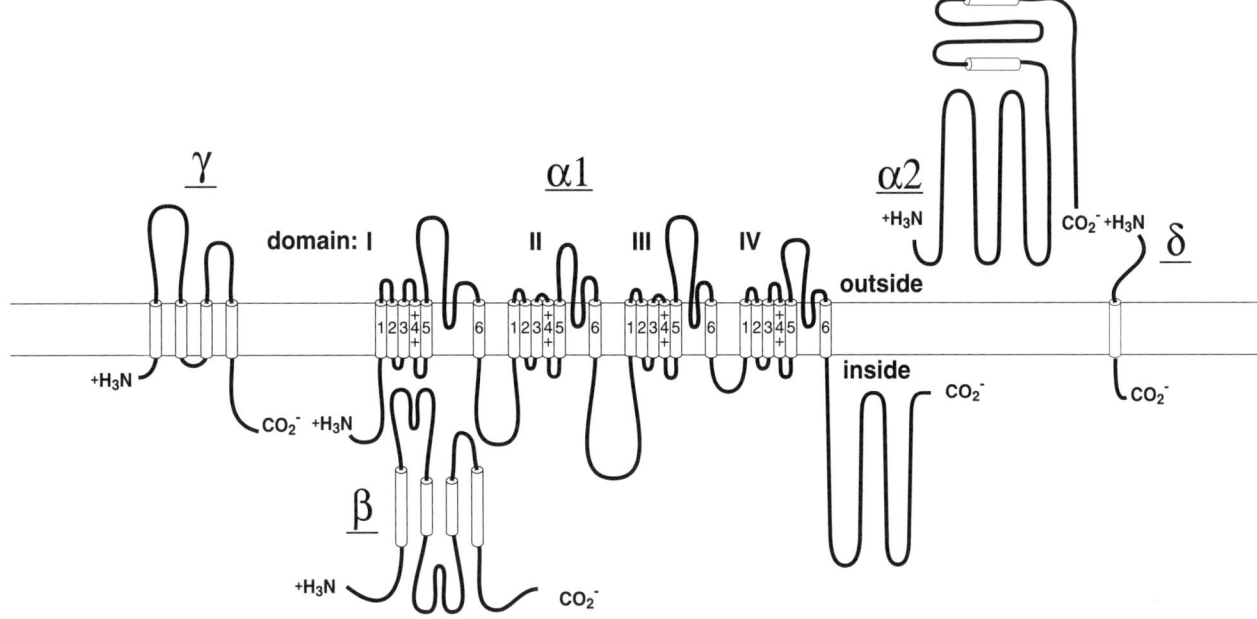

Figure 2 Transmembrane organization of voltage gated Ca^{2+} channels. The primary structures of the subunits of voltage-gated Ca^{2+} are illustrated. *Cylinders* represent probable alpha helical transmembrane segments. *Bold lines* represent the polypeptide chains of each subunit with length approximately proportional to the number of amino acid residues.

extrinsic membrane protein, attached to the membrane through disulfide linkage to the δ subunit [39] (Fig. 2). The δ subunit is encoded by the 3′ end of the coding sequence of the same gene as the α2 subunit, and the mature forms of these two subunits are produced by posttranslational proteolytic processing and disulfide linkage [40,41] (Fig. 2).

Purification of cardiac Ca^{2+} channels revealed subunits of the sizes of the α1, α2δ, β, and γ subunits of skeletal muscle Ca^{2+} channels [42–45], and immunoprecipitation of Ca^{2+} channels from neurons labeled by dihydropyridine Ca^{2+} antagonists revealed α1, α2δ, and β subunits but no γ subunit [46]. Together, these results suggest a similar subunit composition for L-type Ca^{2+} channels in cardiac and skeletal muscle and in neurons.

Purification and immunoprecipitation of N-type Ca^{2+} channels labeled by ω-conotoxin GVIA from brain membrane preparations revealed α1, α2δ, and β subunits [47,48]. Similarly, purified P/Q-type Ca^{2+} channels are composed of α1, α2δ, and β subunits [49–51]. In addition, more recent experiments have unexpectedly revealed a novel γ subunit, which is the target of the *stargazer* mutation in mice [52], and a related series of γ subunits expressed in brain and other tissues [53,54]. These γ-subunit-like proteins can modulate the voltage dependence of P/Q-type Ca^{2+} currents, so they may be associated with these Ca^{2+} channels *in vivo* [52]. If these new γ subunits are indeed associated with all neuronal Ca^{2+} channels, their subunit composition would be identical to that of skeletal muscle Ca^{2+} channels defined in biochemical experiments [31] (Fig. 2).

Functions of Ca^{2+} Channel Subunits. The initial analyses of functional expression of Ca^{2+} channel subunits were carried out with skeletal muscle Ca^{2+} channels. Expression of the α1 subunit is sufficient to produce functional skeletal muscle Ca^{2+} channels, but with low expression level and abnormal kinetics and voltage dependence of the Ca^{2+} current [55]. Co-expression of the α2δ subunit and especially the β subunit enhances the level of expression and confers more normal gating properties [56,57]. As for skeletal muscle Ca^{2+} channels, co-expression of β subunits has a large effect on the level of expression and the voltage dependence and kinetics of gating of cardiac and neuronal Ca^{2+} channels. In general, the level of expression is increased and the voltage dependence of activation and inactivation is shifted to more negative membrane potentials, and the rate of inactivation is increased. However, these effects are different for the individual β subunit isoforms (reviewed in [58,59]). For example, co-expression of the β2a subunit slows inactivation in most subunit combinations. In contrast, co-expression of α2δ subunits [58,59] and γ subunits [52] has much smaller functional effects.

Ca^{2+} Channel Diversity. The different types of Ca^{2+} currents are primarily defined by different α1 subunits. The primary structures of ten distinct Ca^{2+} channel α1 subunits have been defined by homology screening, and their function has been characterized by expression in mammalian cells or Xenopus oocytes. These subunits can be divided into three structurally and functionally related families (Ca_V1, Ca_V2, and Ca_V3) (Table I, [60]). L-type Ca^{2+} currents are mediated by the Ca_V1 type of α1 subunits, which have about 75% amino acid sequence identity among them [35,61,62]. The Ca_V2 type Ca^{2+} channels form a distinct subfamily with less than 40% amino acid sequence identity with Ca_V1 α1 subunits but greater than 70% amino acid sequence identity among themselves. Cloned $Ca_V2.1$ subunits [63,64] form

Table I Subunit Composition and Function of Ca^{2+} Channel Types

Ca^{2+} channel type	α_1 subunits	Specific blocker	Principal physiological functions	Inherited diseases
L	$Ca_V1.1$	DHPs	Excitation-contraction coupling in skeletal muscle Regulation of transcription	Hypokalemic periodic paralysis
	$Ca_V1.2$	DHPs	Excitation-contraction coupling in cardiac and smooth muscle Regulation of enzyme activity Regulation of transcription	
	$Ca_V1.3$	DHPs	Endocrine secretion Cardiac pacemaking Auditory transduction	
	$Ca_V1.4$	DHPs	Visual transduction	Stationary night blindness
N	$Ca_V2.1$	ω-CTx-GVIA	Neurotransmitter release Dendritic Ca^{2+} transients	Migraine Cerebellar ataxia Absence seizures (in mice)
P/Q	$Ca_V2.2$	ω-Agatoxin	Neurotransmitter release Dendritic Ca^{2+} transients	
R	$Ca_V2.3$	SNX-482	Neurotransmitter release Dendritic Ca^{2+} transients	
T	$Ca_V3.1$ $Ca_V3.2$ $Ca_V3.3$	None	Pacemaking and repetitive firing	

Abbreviations: DHP, dihydropyridine; ω-CTx-GVIA, ω-conotoxin GVIA from the cone snail *Conus geographus*; SNX-482, a synthetic version of a peptide toxin from venom of the tarantula *Hysterocrates gigas*.

P- or Q-type Ca^{2+} channels, which are inhibited by ω-agatoxin IVA [65–67]. $Ca_V2.2$ subunits form N-type Ca^{2+} channels with high affinity for ω-conotoxin GVIA [68,69]. Cloned $Ca_V2.3$ subunits form R-type Ca^{2+} channels, which are resistant to both organic Ca^{2+} antagonists specific for L-type Ca^{2+} currents and the peptide toxins specific for N-type or P/Q-type Ca^{2+} currents [27,70,71]. T-type Ca^{2+} currents are mediated by the Ca_V3 channels [72]. These $\alpha1$ subunits are only distantly related to the other known homologs, with less than 25% amino acid sequence identity. These results reveal a surprising structural dichotomy between the T-type, low-voltage-activated Ca^{2+} channels and the high-voltage-activated Ca^{2+} channels. Evidently, these two lineages of Ca^{2+} channels diverged very early in evolution of multicellular organisms.

The diversity of Ca^{2+} channel structure and function is substantially enhanced by multiple β subunits. Four β subunit genes have been identified, and each is subject to alternative splicing to yield additional isoforms (reviewed in [58,73]). In Ca^{2+} channel preparations isolated from brain, each Ca^{2+} channel $\alpha1$ subunit that has been investigated is associated with multiple β subunits, although there is a different rank order in each case [74,75]. The different β subunit isoforms cause different shifts in the kinetics and voltage dependence of gating, so association with different β subunits can substantially alter the physiological function of an $\alpha1$ subunit. Genes encoding four $\alpha2\delta$ subunits have been described [76], but the $\alpha2\delta$ isoforms produced by these different genes have relatively small functional effects on channel gating and expression. A new family of γ subunits has been recently described [52–54], which has small, but significant effects on the voltage dependence of Ca^{2+} channel gating.

Molecular Basis for Ca^{2+} Channel Function

Intensive studies of the structure and function of the related pore-forming subunits of Na^+, Ca^{2+}, and K^+ channels have led to identification of their principal functional components (reviewed in [77–80]). Each domain of the principal subunits consists of six transmembrane alpha helices (S1 through S6) and a membrane-associated loop between S5 and S6 (Fig. 2). The S4 segments of each homologous domain serve as the voltage sensors for activation, moving outward and rotating under the influence of the electric field and initiating a conformational change that opens the pore. The S5 and S6 segments and the membrane-associated pore loop between them form the pore lining of the voltage-gated ion channels. The narrow external pore is lined by the pore loop, which contains a pair of glutamate residues in each domain that are required for Ca^{2+} selectivity. Remarkably, substitution of only three amino acid residues in the pore loops between the S5 and S6 segments in domains II, III, and IV of sodium channels is sufficient to confer Ca^{2+} selectivity [81]. The inner pore is lined by the S6 segments, which form the receptor sites for the pore-blocking Ca^{2+} antagonist drugs specific for L-type Ca^{2+} channels [82–84]. All Ca^{2+} channels share these general structural features, but the amino acid residues that confer high affinity for the organic Ca^{2+} antagonists used in therapy of cardiovascular diseases are present only in the Ca_V1 family of Ca^{2+} channels, which conduct L-type Ca^{2+} currents.

Ca^{2+} Channel Regulation

The activity of voltage-gated Ca^{2+} channels is tightly regulated by second messenger signal transduction pathways through direct interactions with G proteins and other intracellular signaling proteins and through protein phosphorylation (reviewed in [85]). Regulation of L-type Ca^{2+} currents in cardiac, skeletal, and smooth muscle cells by the β-adrenergic receptor/cAMP pathway involves a signaling complex of the $Ca_V1.1$ or $Ca_V1.2$ channels, an A kinase anchoring protein designated AKAP-15 or AKAP-18, and PKA targeted to the channel by AKAP binding via a leucine zipper to a site in the C-terminal domain of the α1 subunit [86–89]. A functionally similar signaling complex in the brain includes the β-adrenergic receptor itself, adenylate cyclase, the AKAP MAP 250, and PKA [90]. This signaling pathway is also engaged by other G protein-coupled receptors that activate adenylate cyclase. This pathway regulates beating rate and contractility in the heart, vascular tone, skeletal muscle contractile force, and gene expression in neurons, myocytes, endocrine cells, and other cell types.

The activity of the Ca_V2 family of channels is regulated primarily by direct interaction with G proteins and other signaling proteins and secondarily by protein phosphorylation (reviewed in [8,91]). Many different neurotransmitters and hormones activate G protein-coupled receptors, which release Gβγ subunits that inhibit Ca^{2+} channel activity [92–94]. G protein inhibition can be reversed by strong depolarization, resulting in facilitation of Ca^{2+} channel activity, and by phosphorylation by protein kinase C [8,91,95–98]. In addition, the activity of the Ca_V2 family of channels is regulated by interaction with the SNARE proteins that are required in exocytosis [99–102]. This regulatory mechanism appears designed to focus Ca^{2+} entry on Ca^{2+} channels with exocytotic vesicles docked nearby.

Ca^{2+} itself also regulates the activity of both Ca_V1 and Ca_V2 channels. Low levels of Ca^{2+} entry cause facilitation of Ca^{2+} channel activity, and higher levels cause Ca^{2+}-dependent inactivation [103–107]. Both processes involve binding to calmodulin and interaction with specific calmodulin-binding sites in the C-terminal domain of the Ca^{2+} channel. In repetitively firing neurons and in cardiac myocytes, this mechanism allows integration of Ca^{2+} signals as a function of frequency of action potential generation. This mode of regulation also serves to tune the Ca^{2+} entry to the needs of intracellular regulatory processes and prevent inappropriately wide swings in local Ca^{2+} concentration.

Conclusion

Voltage-gated Ca^{2+} channels are essential signal transducers, converting cell surface electrical signals to intracellular Ca^{2+} transients that initiate many physiological and biochemical events. Recent research has defined their molecular properties, identified many genes that encode their subunits and provide diversity of function, and revealed their complex interaction with cellular regulatory pathways. Further work on this protein family will give essential insights into cellular signaling and its dysfunction in diseases as diverse as epilepsy, migraine, cardiac arrhythmia, hypertension, and diabetes.

References

1. Striessnig, J., Berger, W., and Glossman, H. (1993). Molecular properties of voltage-dependent Ca^{2+} channels in excitable tissues. *Cell. Physiol. Biochem.* **3**, 295–317.
2. Bers, D. M. (2002). Cardiac excitation-contraction coupling. *Nature* **415**, 198–205.
3. Catterall, W. A. (1991). Excitation-contraction coupling in vertebrate skeletal muscle: a tale of two calcium channels. *Cell* **64**, 871–874.
4. MacLennan, D. H. (2000). Ca^{2+} signalling and muscle disease. *Eur. J. Biochem.* **267**, 5291–5297.
5. Berggren, P. O. and Larsson, O. (1994). Ca^{2+} and pancreatic B-cell function. *Biochem. Soc. Trans.* **22**, 12–18.
6. Tsien, R. W., Lipscombe, D., Madison, D. V., Bley, K. R., and Fox, A. P. (1988). Multiple types of neuronal calcium channels and their selective modulation. *Trends Neurosci.* **11**, 431–438.
7. Dunlap, K., Luebke, J. I., and Turner, T. J. (1995). Exocytotic Ca^{2+} channels in mammalian central neurons. *TINS* **18**, 89–98.
8. Catterall, W. A. (1998). Structure and function of neuronal Ca^{2+} channels and their role in neurotransmitter release. *Cell Calcium* **24**, 307–323.
9. West, A. E., Chen, W. G., Dalva, M. B., Dolmetsch, R. E., Kornhauser, J. M., Shaywitz, A. J., Takasu, M. A., Tao, X., and Greenberg, M. E. (2001). Calcium regulation of neuronal gene expression. *Proc. Natl. Acad. Sci. USA* **98**, 11024–11031.
10. Hardingham, G. E., Cruzalegui, F. H., Chawla, S., and Bading, H. (1998). Mechanisms controlling gene expression by nuclear calcium signals. *Cell Calcium* **23**, 131–134.
11. Reuter, H. (1979). Properties of two inward membrane currents in the heart. *Annu. Rev. Physiol.* **41**, 413–424.
12. Reuter, H. (1967). The dependence of slow inward current in Purkinje fibres on the extracellular calcium-concentration. *J. Physiol. (London)* **192**, 479–492.
13. Hess, P. (1990). Calcium channels in vertebrate cells. *Annu. Rev. Neurosci.* **13**, 337–356.
14. Bean, B. P. (1989). Classes of calcium channels in vertebrate cells. *Annu. Rev. Physiol.* **51**, 367–384.
15. Llinas, R., Sugimori, M., Hillman, D. E., and Cherksey, B. (1992). Distribution and functional significance of the P-type, voltage-dependent Ca^{2+} channels in the mammalian central nervous system. *Trends Neurosci.* **15**, 351–355.
16. Reuter, H. (1983). Calcium channel modulation by neurotransmitters, enzymes and drugs. *Nature* **301**, 569–574.
17. Nowycky, M. C., Fox, A. P., and Tsien, R. W. (1985). Three types of neuronal calcium channel with different calcium agonist sensitivity. *Nature* **316**, 440–443.
18. Milani, D., Malgaroli, A., Guidolin, D., Fasolato, C., Skaper, S. D., Meldolesi, J., and Pozzan, T. (1990). Ca^{2+} channels and intracellular Ca^{2+} stores in neuronal and neuroendocrine cells. *Cell Calcium* **11**, 191–199.
19. Hagiwara, S., Ozawa, S., and Sand, O. (1975). Voltage clamp analysis of two inward current mechanisms in the egg cell membrane of a starfish. *J. Gen. Physiol.* **65**, 617–644.
20. Llinas, R. and Yarom, Y. (1981). Electrophysiology of mammalian inferior olivary neurones *in vitro*. Different types of voltage-dependent ionic conductances. *J. Physiol. (London)* **315**, 569–584.
21. Carbone, E. and Lux, H. D. (1984). A low voltage-activated, fully inactivating Ca channel in vertebrate sensory neurones. *Nature* **310**, 501–502.

22. Fedulova, S. A., Kostyuk, P. G., and Veselovsky, N. S. (1985). Two types of calcium channels in the somatic membrane of new-born rat dorsal root ganglion neurones. *J. Physiol.* **359**, 431–446.
23. Swandulla, D. and Armstrong, C. M. (1988). Fast deactivating calcium channels in chick sensory neurons. *J. Gen. Physiol.* **92**, 197–218.
24. McCleskey, E. W., Fox, A. P., Feldman, D. H., Cruz, L. J., Olivera, B. M., Tsien, R. W., and Yoshikami, D. (1987). ω-Conotoxin: direct and persistent blockade of specific types of calcium channels in neurons but not muscle. *Proc. Natl. Acad. Sci. USA* **84**, 4327–4331.
25. Llinás, R. R., Sugimori, M., and Cherksey, B. (1989). Voltage-dependent calcium conductances in mammalian neurons. The P channel. *Ann. N.Y. Acad. Sci.* **560**, 103–111.
26. Mintz, I. M., Adams, M. E., and Bean, B. P. (1992). P-type calcium channels in rat central and peripheral neurons. *Neuron* **9**, 85–95.
27. Randall, A. and Tsien, R. W. (1995). Pharmacological dissection of multiple types of Ca^{2+} channel currents in rat cerebellar granule neurons. *J. Neurosci.* **15**, 2995–3012.
28. Tottene, A., Moretti, A., and Pietrobon, D. (1996). Functional diversity of P-type and R-type calcium channels in rat cerebellar neurons. *J. Neurosci.* **16**, 6353–6363.
29. Curtis, B. M. and Catterall, W. A. (1984). Purification of the calcium antagonist receptor of the voltage-sensitive calcium channel from skeletal muscle transverse tubules. *Biochem.* **23**, 2113–2118.
30. Curtis, B. M. and Catterall, W. A. (1985). Phosphorylation of the calcium antagonist receptor of the voltage-sensitive calcium channel by cAMP-dependent protein kinase. *Proc. Natl. Acad. Sci. USA* **82**, 2528–2532.
31. Takahashi, M., Seagar, M. J., Jones, J. F., Reber, B. F., and Catterall, W. A. (1987). Subunit structure of dihydropyridine-sensitive calcium channels from skeletal muscle. *Proc. Natl. Acad. Sci. USA* **84**, 5478–5482.
32. Leung, A. T., Imagawa, T., and Campbell, K. P. (1987). Structural characterization of the 1,4-dihydropyridine receptor of the voltage-dependent Ca^{2+} channel from rabbit skeletal muscle. Evidence for two distinct high molecular weight subunits. *J. Biol. Chem.* **262**, 7943–7946.
33. Striessnig, J., Knaus, H. G., Grabner, M., Moosburger, K., Seitz, W., Lietz, H., and Glossmann, H. (1987). Photoaffinity labelling of the phenylalkylamine receptor of the skeletal muscle transverse-tubule calcium channel. *FEBS Lett.* **212**, 247–253.
34. Hosey, M. M., Barhanin, J., Schmid, A., Vandaele, S., Ptasienski, J., O'Callahan, C., Cooper, C., and Lazdunski, M. (1987). Photoaffinity labelling and phosphorylation of a 165 kilodalton peptide associated with dihydropyridine and phenylalkylamine-sensitive calcium channels. *Biochem. Biophys. Res. Commun.* **147**, 1137–1145.
35. Tanabe, T., Takeshima, H., Mikami, A., Flockerzi, V., Takahashi, H., Kangawa, K., Kojima, M., Matsuo, H., Hirose, T., and Numa, S. (1987). Primary structure of the receptor for calcium channel blockers from skeletal muscle. *Nature* **328**, 313–318.
36. Ruth, P., Röhrkasten, A., Biel, M., Bosse, E., Regulla, S., Meyer, H. E., Flockerzi, V., and Hofmann, F. (1989). Primary structure of the beta subunit of the DHP-sensitive calcium channel from skeletal muscle. *Science* **245**, 1115–1118.
37. Jay, S. D., Ellis, S. B., McCue, A. F., Williams, M. E., Vedvick, T. S., Harpold, M. M., and Campbell, K. P. (1990). Primary structure of the gamma subunit of the DHP-sensitive calcium channel from skeletal muscle. *Science* **248**, 490–492.
38. Ellis, S. B., Williams, M. E., Ways, N. R., Brenner, R., Sharp, A. H., Leung, A. T., Campbell, K. P., McKenna, E., Koch, W. J., Hui, A., Schwartz, A., and Harpold, M. M. (1988). Sequence and expression of mRNAs encoding the alpha 1 and alpha 2 subunits of a DHP-sensitive calcium channel. *Science* **241**, 1661–1664.
39. Gurnett, C. A., De Waard, M., and Campbell, K. P. (1996). Dual function of the voltage-dependent Ca^{2+} channel $\alpha_2\delta$ subunit in current stimulation and subunit interaction. *Neuron* **16**, 431–440.
40. De Jongh, K. S., Warner, C., and Catterall, W. A. (1990). Subunits of purified calcium channels. α2 and δ are encoded by the same gene. *J. Biol. Chem.* **265**, 14738–14741.
41. Jay, S. D., Sharp, A. H., Kahl, S. D., Vedvick, T. S., Harpold, M. M., and Campbell, K. P. (1991). Structural characterization of the dihydropyridine-sensitive calcium channel α_2-subunit and the associated δ peptides. *J. Biol. Chem.* **266**, 3287–3293.
42. Schneider, T. and Hofmann, F. (1988). The bovine cardiac receptor for calcium channel blockers is a 195-kDa protein. *Eur. J. Biochem.* **174**, 369–375.
43. De Jongh, K. S., Murphy, B. J., Colvin, A. A., Hell, J. W., Takahashi, M., and Catterall, W. A. (1996). Specific phosphorylation of a site in the full-length form of the α1 subunit of the cardiac L-type calcium channel by cAMP-dependent protein kinase. *Biochemistry.* **35**, 10392–10402.
44. Chang, F. C. and Hosey, M. M. (1988). Dihydropyridine and phenylalkylamine receptors associated with cardiac and skeletal muscle calcium channels are structurally different. *J. Biol. Chem.* **263**, 18929–18937.
45. Kuniyasu, A., Oka, K., Ide-Yamada, T., Hatanaka, Y., Abe, T., Nakayama, H., and Kanaoka, Y. (1992). Structural characterization of the dihydropyridine receptor-linked calcium channel from porcine heart. *J. Biochem. (Tokyo)* **112**, 235–242.
46. Ahlijanian, M. K., Westenbroek, R. E., and Catterall, W. A. (1990). Subunit structure and localization of dihydropyridine-sensitive calcium channels in mammalian brain, spinal cord, and retina. *Neuron* **4**, 819–832.
47. McEnery, M. W., Snowman, A. M., Sharp, A. H., Adams, M. E., and Snyder, S. H. (1991). Purified ω-conotoxin GVIA receptor of rat brain resembles a dihydropyridine-sensitive L-type calcium channel. *Proc. Natl. Acad. Sci. USA* **88**, 11095–11099.
48. Witcher, D. R., De Waard, M., Sakamoto, J., Franzini-Armstrong, C., Pragnell, M., Kahl, S. D., and Campbell, K. P. (1993). Subunit identification and reconstitution of the N-type Ca^{2+} channel complex purified from brain. *Science* **261**, 486–489.
49. Martin-Moutot, N., Leveque, C., Sato, K., Kato, R., Takahashi, M., and Seagar, M. (1995). Properties of omega conotoxin MVIIC receptors associated with α_{1A} calcium channel subunits in rat brain. *FEBS Lett.* **366**, 21–25.
50. Liu, H., De Waard, M., Scott, V. E. S., Gurnett, C. A., Lennon, V. A., and Campbell, K. P. (1996). Identification of three subunits of the high affinity ω-conotoxin MVIIC-sensitive Ca^{2+} channel. *J. Biol. Chem.* **271**, 13804–13810.
51. Martin-Moutot, N., Charvin, N., Leveque, C., Sato, K., Nishi, T., Kozaki, S., Takahashi, M., and Seagar, M. (1996). Interaction of SNARE complexes with P/Q-type calcium channels in rat cerebellar synaptosomes. *J. Biol. Chem.* **271**, 6567–6570.
52. Letts, V. A., Felix, R., Biddlecome, G. H., Arikkath, J., Mahaffey, C. L., Valenzuela, A., Bartlett, I. F. S., Mori, Y., Campbell, K. P., and Frankel, W. N. (1998). The mouse stargazer gene encodes a neuronal Ca^{2+}-channel γ subunit. *Nature Genet.* **19**, 340–347.
53. Burgess, D. L., Gefrides, L. A., Foreman, P. J., and Noebels, J. L. (2001). A cluster of three novel Ca^{2+} channel gamma subunit genes on chromosome 19q13.4: evolution and expression profile of the gamma subunit gene family. *Genomics* **71**, 339–350.
54. Klugbauer, N., Dai, S., Specht, V., Lacinova, L., Marais, E., Bohn, G., and Hofmann, F. (2000). A family of gamma-like calcium channel subunits. *FEBS Lett* **470**, 189–197.
55. Perez-Reyes, E., Kim, H. S., Lacerda, A. E., Horne, W., Wei, X. Y., Rampe, D., Campbell, K. P., Brown, A. M., and Birnbaumer, L. (1989). Induction of calcium currents by the expression of the alpha 1-subunit of the dihydropyridine receptor from skeletal muscle. *Nature* **340**, 233–236.
56. Singer, D., Biel, M., Lotan, I., Flockerzi, V., Hofmann, F., and Dascal, N. (1991). The roles of the subunits in the function of the calcium channel. *Science* **253**, 1553–1557.
57. Lacerda, A. E., Kim, H. S., Ruth, P., Perez-Reyes, E., Flockerzi, V., Hofmann, F., Birnbaumer, L., and Brown, A. M. (1991). Normalization of current kinetics by interaction between the α1 and β subunits of the skeletal muscle dihydropyridine-sensitive Ca^{2+} channel. *Nature* **352**, 527–530.

58. Hofmann, F., Biel, M., and Flockerzi, V. (1994). Molecular basis for Ca^{2+} channel diversity. *Annu. Rev. Neurosci.* **17**, 399–418.
59. Hosey, M. M., Chien, A. J., and Puri, T. S. (1996). Structure and regulation of L-type calcium channels–A current assessment of the properties and roles of channel subunits. *Trends Cardiovasc. Med.* **6**, 265–273.
60. Ertel, E. A., Campbell, K. P., Harpold, M. M., Hofmann, F., Mori, Y., Perez-Reyes, E., Schwartz, A., Snutch, T. P., Tanabe, T., Birnbaumer, L., Tsien, R. W., and Catterall, W. A. (2000). Nomenclature of voltage-gated calcium channels. *Neuron* **25**, 533–535.
61. Mikami, A., Imoto, K., Tanabe, T., Niidome, T., Mori, Y., Takeshima, H., Narumiya, S., and Numa, S. (1989). Primary structure and functional expression of the cardiac dihydropyridine-sensitive calcium channel. *Nature* **340**, 230–233.
62. Snutch, T. P., Tomlinson, W. J., Leonard, J. P., and Gilbert, M. M. (1991). Distinct calcium channels are generated by alternative splicing and are differentially expressed in the mammalian CNS. *Neuron* **7**, 45–47.
63. Starr, T. V. B., Prystay, W., and Snutch, T. P. (1991). Primary structure of a calcium channel that is highly expressed in the rat cerebellum. *Proc. Natl. Acad. Sci. USA* **88**, 5621–5625.
64. Mori, Y., Friedrich, T., Kim, M.-S., Mikami, A., Nakai, J., Ruth, P., Bosse, E., Hofmann, F., Flockerzi, V., Furuichi, T., Mikoshiba, K., Imoto, K., Tanabe, T., and Numa, S. (1991). Primary structure and functional expression from complementary DNA of a brain calcium channel. *Nature* **350**, 398–402.
65. Sather, W. A., Tanabe, T., Zhang, J.-F., Mori, Y., Adams, M. E., and Tsien, R. W. (1993). Distinctive biophysical and pharmacological properties of class A (BI) calcium channel α1 subunits. *Neuron* **11**, 291–303.
66. Bourinet, E., Soong, T. W., Sutton, K., Slaymaker, S., Matthews, E., Monteil, A., Samoni, G. W., Nargeot, J., and Snutch, T. P. (1999). Splicing of alpha 1A subunit gene generates phenotypic variants of P- and Q-type calcium channels. *Nat. Neurosci* **2**, 407–415.
67. Stea, A., Tomlinson, W. J., Soong, T. W., Bourinet, E., Dubel, S. J., Vincent, S. R., and Snutch, T. P. (1994). The localization and functional properties of a rat brain α$_{1A}$ calcium channel reflect similarities to neuronal Q- and P-type channels. *Proc. Natl. Acad. Sci. USA* **91**, 10576–10580.
68. Dubel, S. J., Starr, T. V. B., Hell, J., Ahlijanian, M. K., Enyeart, J. J., Catterall, W. A., and Snutch, T. P. (1992). Molecular cloning of the α–1 subunit of an ω-conotoxin-sensitive calcium channel. *Proc. Natl. Acad. Sci. USA* **89**, 5058–5062.
69. Williams, M. E., Brust, P. F., Feldman, D. H., Patthi, S., Simerson, S., Maroufi, A., McCue, A. F., Velicelebi, G., Ellis, S. B., and Harpold, M. M. (1992). Structure and functional expression of an omega-conotoxin-sensitive human N-type calcium channel. *Science* **257**, 389–395.
70. Soong, T. W., Stea, A., Hodson, C. D., Dubel, S. J., Vincent, S. R., and Snutch, T. P. (1994). Structure and functional expression of a member of the low voltage-activated calcium channel family. *Science* **260**, 1133–1136.
71. Zhang, J.-F., Randall, A. D., Ellinor, P. T., Horne, W. A., Sather, W. A., Tanabe, T., Schwarz, T. L., and Tsien, R. W. (1993). Distinctive pharmacology and kinetics of cloned neuronal Ca^{2+} channels and their possible counterparts in mammalian CNS neurons. *Neuropharmacology* **32**, 1075–1088.
72. Perez-Reyes, E., Cribbs, L. L., Daud, A., Lacerda, A. E., Barclay, J., Williamson, M. P., Fox, M., Rees, M., and Lee, J. H. (1998). Molecular characterization of a neuronal low-voltage-activated T-type calcium channel. *Nature* **391**, 896–900.
73. Perez-Reyes, E. and Schneider, T. (1995). Molecular biology of calcium channels. *Kidney International* **48**, 1111–1124.
74. Witcher, D. R., De Waard, M., Liu, H., Pragnell, M., and Campbell, K. P. (1995). Association of native Ca^{2+} channel β subunits with the α1 subunit interaction domain. *J. Biol. Chem.* **270**, 18088–18093.
75. Pichler, M., Cassidy, T. N., Reimer, D., Haase, H., Krause, R., Ostler, D., and Striessnig, J. (1997). β subunit heterogeneity in neuronal L-type Ca^{2+} channels. *J. Biol. Chem.* **272**, 13877–13882.
76. Klugbauer, N., Lacinová, L., Marais, E., Hobom, M., and Hofmann, F. (1999). Molecular diversity of the calcium channel α$_2$δ subunit. *J. Neurosci.* **19**, 684–691.
77. Catterall, W. A. (1995). Structure and function of voltage-gated ion channels. *Annu. Rev. Biochem.* **65**, 493–531.
78. Jan, L. Y. and Jan, Y. N. (1997). Cloned potassium channels from eukaryotes and prokaryotes. *Annu. Rev. Neurosci.* **20**, 91–123.
79. Stuhmer, W. and Parekh, A. B. (1992). The structure and function of Na$^+$ channels. *Curr. Opin. Neurobiol.* **2**, 243–246.
80. Hofmann, F., Lacinová, L., and Klugbauer, N. (1999). Voltage-dependent calcium channels: from structure to function. *Rev. Physiol. Biochem. Pharmacol.* **139**, 33–87.
81. Heinemann, S. H., Terlau, H., Stühmer, W., Imoto, K., and Numa, S. (1992). Calcium channel characteristics conferred on the sodium channel by single mutations. *Nature* **356**, 441–443.
82. Catterall, W. A. and Striessnig, J. (1992). Receptor sites for Ca^{2+} channel antagonists. *Trends Pharmacol. Sci.* **13**, 256–262.
83. Hockerman, G. H., Johnson, B. D., Scheuer, T., and Catterall, W. A. (1995). Molecular determinants of high affinity phenylalkyamine block of L-type calcium channels. *J. Biol. Chem.* **270**, 22119–22122.
84. Hockerman, G. H., Peterson, B. Z., Johnson, B. D., and Catterall, W. A. (1997). Molecular determinants of drug binding and action on L-type calcium channels. *Annu. Rev. Pharmacol. Toxicol.* **37**, 361–396.
85. Catterall, W. A. (2000). Structure and regulation of voltage-gated Ca^{2+} channels. *Annu. Rev. Cell Dev. Bio.* **16**, 521–555.
86. Gray, P. C., Tibbs, V. C., Catterall, W. A., and Murphy, B. J. (1997). Identification of a 15-kDa cAMP-dependent protein kinase-anchoring protein associated with skeletal muscle L-type calcium channels. *J. Biol. Chem.* **272**, 6297–6302.
87. Fraser, I. D. C., Tavalin, S. J., Lester, L. B., Langeberg, L. K., Westphal, A. M., Dean, R. A., Marrion, N. V., and Scott, J. D. (1998). A novel lipid-anchored A-kinase anchoring protein facilitates cAMP-responsive membrane events. *EMBO J.* **17**, 2261–2272.
88. Gray, P. C., Johnson, B. D., Westenbroek, R. E., Hays, L. G., Yates, I. J., Scheuer, T., Catterall, W. A., and Murphy, B. J. (1998). Primary structure and function of an A kinase anchoring protein associated with calcium channels. *Neuron* **20**, 1017–1026.
89. Hulme, J. T., Ahn, M., Hauschka, S. D., Scheuer, T., and Catterall, W. A. (2002). A novel leucine zipper targets AKAP15 and cyclic AMP-dependent protein kinase to the C terminus of the skeletal muscle Ca^{2+} channel and modulates its function. *J. Biol. Chem.* **277**, 4079–4087.
90. Davare, M. A., Avdonin, V., Hall, D. D., Peden, E. M., Burette, A., Weinberg, R. J., Horne, M. C., Hoshi, T., and Hell, J. W. (2001). A beta2 adrenergic receptor signaling complex assembled with the Ca^{2+} channel Ca$_v$1.2. *Science* **293**, 98–101.
91. Hille, B. (1994). Modulation of ion-channel function by G-protein-coupled receptors. *Trends Neurosci.* **17**, 531–536.
92. Ikeda, S. R. (1996). Voltage-dependent modulation of N-type calcium channels by G-protein βγ subunits. *Nature* **380**, 255–258.
93. Herlitze, S., Garcia, D. E., Mackie, K., Hille, B., Scheuer, T., and Catterall, W. A. (1996). Modulation of Ca^{2+} channels by G protein βγ subunits. *Nature* **380**, 258–262.
94. Ikeda, S. R. and Dunlap, K. (1999). Voltage-dependent modulation of N-type calcium channels: role of G protein subunits. *Adv. Second Messenger Phosphoprotein Res.* **33**, 131–151.
95. Bean, B. P. (1989). Neurotransmitter inhibition of neuronal calcium currents by changes in channel voltage dependence. *Nature* **340**, 153–156.
96. Swartz, K. J. (1993). Modulation of Ca^{2+} channels by protein kinase C in rat central and peripheral neurons: Disruption of G protein-mediated inhibition. *Neuron* **11**, 305–320.
97. Tsunoo, A., Yoshii, M., and Narahashi, T. (1986). Block of calcium channels by enkephalin and somatostatin in neuroblastoma-glioma hybrid NG108-115 cells. *Proc. Natl. Acad. Sci. USA* **83**, 9832–9836.
98. Zamponi, G. W., Bourinet, E., Nelson, D., Nargeot, J., and Snutch, T. P. (1997). Crosstalk between G proteins and protein kinase C mediated by the calcium channel α$_1$ subunit. *Nature* **385**, 442–446.
99. Sheng, Z.-H., Rettig, J., Takahashi, M., and Catterall, W. A. (1994). Identification of a syntaxin-binding site on N-type calcium channels. *Neuron* **13**, 1303–1313.

100. Bezprozvanny, I., Scheller, R. H., and Tsien, R. W. (1995). Functional impact of syntaxin on gating of N-type and Q-type calcium channels. *Nature* **378**, 623–626.
101. Wiser, O., Bennett, M. K., and Atlas, D. (1996). Functional interaction of syntaxin and SNAP-25 with voltage-sensitive L- and N-type Ca^{2+} channels. *EMBO J.* **15**, 4100–4110.
102. Zhong, H., Yokoyama, C., Scheuer, T., and Catterall, W. A. (1999). Reciprocal regulation of P/Q-type Ca^{2+} channels by SNAP-25, syntaxin and synaptotagmin. *Nat. Neurosci.* **2**, 939–941.
103. Zühlke, R. D. and Reuter, H. (1998). Ca^{2+}-sensitive inactivation of L-type Ca^{2+} channels depends on multiple cytoplasmic amino acid sequences of the α_{1C} subunit. *Proc. Natl. Acad. Sci. USA* **95**, 3287–3294.
104. Peterson, B. Z., DeMaria, C. D., and Yue, D. T. (1999). Calmodulin is the Ca^{2+} sensor for Ca^{2+}-dependent inactivation of L-type calcium channels. *Neuron* **22**, 549–558.
105. Zühlke, R. D., Pitt, G. S., Deisseroth, K., Tsien, R. W., and Reuter, H. (1999). Calmodulin supports both inactivation and facilitation of L-type calcium channels. *Nature* **399**, 159–162.
106. Lee, A., Wong, S. T., Gallagher, D., Li, B., Storm, D. R., Scheuer, T., and Catterall, W. A. (1999). Ca^{2+}/calmodulin binds to and modulates P/Q-type calcium channels. *Nature* **399**, 155–159.
107. Lee, A., Scheuer, T., and Catterall, W. A. (2000). Ca^{2+}-Calmodulin dependent inactivation and facilitation of P/Q-type Ca^{2+} channels. *Biophys. J.* **78**, 265A

CHAPTER 127

Store-operated Ca^{2+} Channels

James W. Putney, Jr.
*Calcium Regulation Section, Laboratory of Signal Transduction,
National Institute of Environmental Health Sciences,
National Institutes of Health,
Research Triangle Park, North Carolina*

Capacitative Calcium Entry

Many signaling pathways involve the generation of cytoplasmic Ca^{2+} signals. As reviewed elsewhere in this volume, in many instances these Ca^{2+} signals arise as a result of the Ca^{2+}-mobilizing actions of the intracellular messenger, inositol 1,4,5-trisphosphate (IP$_3$) (see also [1]). IP$_3$ binds to specific receptor/channels on the endoplasmic reticulum; the binding of IP$_3$ results in channel opening and release of stored Ca^{2+} to the cytoplasm. In most cell types, this release of Ca^{2+} from intracellular stores is accompanied by an accelerated entry of Ca^{2+} across the plasma membrane. A variety of mechanisms may be responsible for this entry of Ca^{2+} (reviewed in [2,3]). One mechanism that appears to be ubiquitous in nonexcitable cells, and is found in a number of excitable cell types, is *capacitative calcium entry*, also known at *store-operated calcium entry* [4–6]. The signal for capacitative calcium entry appears to be the fall in the concentration of Ca^{2+} in the endoplasmic reticulum or in a specialized subcompartment of it.

While the physiological mechanism for depleting stores and activating capacitative calcium entry generally involves IP$_3$-mediated discharge of Ca^{2+} stores, a number of experimental manipulations can bypass receptor activation to empty Ca^{2+} stores. Inhibitors of sarcoplasmic endoplasmic reticulum Ca^{2+} ATPases, such as thapsigargin, cause passive depletion of Ca^{2+} stores and are thus efficient activators of capacitative calcium entry. In electrophysiological studies, utilizing the patch clamp technique to examine whole-cell store-operated membrane currents, IP$_3$ can be included in the patch pipet, or Ca^{2+} stores can be depleted simply by high concentrations of a Ca^{2+} chelator. The first store-operated current to be described was the Ca^{2+} release-activated Ca^{2+} current (I_{crac}) characteristically found in hematopoetic cells [7]. Noise analysis indicates that the unitary conductance of single CRAC channels is likely to be too small to measure [8]. However, in other cell types, the electrophysiological profile of store-operated currents appears to differ significantly from I_{crac}, and in some instances single channels have been observed [9–12]. In these instances, the whole cell current resulting from store depletion is always less Ca^{2+} selective than I_{crac}. In some cases the whole cell currents appear to be nonselective cation currents [12–15]. This finding indicates that the molecular composition of store-operated channels differs among cell types, and it is also possible therefore that multiple mechanisms exist for gating these channels.

Store-operated Channels

The leading contenders for molecular components of store-operated channels are members of the *trp* gene superfamily [16]. In *Drosophila*, the *trp* gene encodes a subunit of a cation channel regulated by a light-sensitive phospholipase C [17]. Seven mammalian *trp* genes with 30–40% sequence similarity to *Drosophila trp* have been cloned, and the proteins they encode have been designated TRPC1 ... TRPC7, which fall into four groups based on structural similarities: TRPC1, TRPC2, TRPC3/6/7, TRPC4/5. While the results of transfection experiments sometimes vary from one laboratory to another, there are a gratifying number of instances in which these proteins appear to form or contribute to the formation of store-operated channels (TRPC1 [18,19], TRPC2 [20,21], TRPC3 [22], TRPC4 [23,24]). Generally, the channels formed in these expression studies are not highly calcium selective and thus may be candidates for the less selective channels found in nonhematopoetic cells. CaT1 (also known as TRPV6), a member of the *trp* superfamily, which is more distantly related to *trp* than the TRPC proteins, was shown in one study to express as a highly Ca^{2+}-selective store-operated channel [25]; however, these findings are at present controversial [26].

Mechanism of Activation of Store-operated Channels

There are two distinct proposals for the mechanism coupling Ca^{2+} store depletion to activation of capacitative calcium entry. The earliest idea was that a novel messenger molecule might be released from the endoplasmic reticulum, and this messenger would then diffuse to the plasma membrane and activate the store-operated channels [4]. A number of studies have published evidence for such a messenger, although its structure has not been elucidated [27–31]. The alternative proposal is based on analogy with the mechanism of coupling of L-type Ca^{2+} channels to ryanodine receptors in skeletal muscle. Thus, the conformational coupling model [6,32] proposes that IP_3 receptors in the endoplasmic reticulum interact directly with plasma membrane Ca^{2+} channels, perhaps members of the *trp* family. In support of this idea, TRPC3, when expressed in HEK293 cells, forms channels that are activated in a manner dependent on IP_3 and the IP_3 receptor [33–35]. However, in this expression system, TRPC3 is not a store-operated channel, rather its activation requires agonist activation of phospholipase C and production of IP_3. Also, in DT40 B lymphocytes, when IP_3 receptors were eliminated by gene disruption, store-operated Ca^{2+} entry was unaffected [36]. Expression of TRPC3 in these cells produced a store-operated channel whose activity was only partially reduced in the absence of IP_3 receptors [22]. Thus, conformational coupling may play a role in activation of store-operated channels in some situations, while a second messenger mode of signaling may be involved in others.

Summary

Capacitative calcium entry is a process whereby depletion of Ca^{2+} from intracellular stores leads to the activation of Ca^{2+} channels in the plasma membrane and accelerated entry of Ca^{2+} into the cytoplasm of cells. Current research focuses on the molecular nature of the channels and the mechanism of coupling the channels to Ca^{2+} store depletion. Leading contenders for the store-operated channels are members of the *trp* gene family, although no one gene has been definitively linked to a specific store-operated channel. The mechanism of activation of the channels may involve interactions between the plasma membrane channels and endoplasmic reticulum IP_3 receptors in some instances or a diffusible Ca^{2+} influx factor in others. Continued work is needed to clarify and resolve these important issues.

References

1. Berridge, M. J. (1993). Inositol trisphosphate and calcium signalling. *Nature* **361**, 315–325.
2. Meldolesi, J., Clementi, E., Fasolato, C., Zacchetti, D., and Pozzan, T. (1991). Ca^{2+} influx following receptor activation. *Trends Pharmacol. Sci.* **12**, 289–292.
3. Barritt, G. J. (1999). Receptor-activated Ca^{2+} inflow in animal cells: a variety of pathways tailored to meet different intracellular Ca^{2+} signalling requirements. *Biochem. J.* **337**, 153–169.
4. Putney, J. W., Jr. (1986). A model for receptor-regulated calcium entry. *Cell Calcium* **7**, 1–12.
5. Putney, J. W., Jr. (1997). *Capacitative Calcium Entry,* Landes Biomedical Publishing, Austin, TX.
6. Berridge, M. J. (1995). Capacitative calcium entry. *Biochem. J.* **312**, 1–11.
7. Hoth, M., and Penner, R. (1992). Depletion of intracellular calcium stores activates a calcium current in mast cells. *Nature* **355**, 353–355.
8. Zweifach, A. and Lewis, R. S. (1993). Mitogen-regulated Ca^{2+} current of T lymphocytes is activated by depletion of intracellular Ca^{2+} stores. *Proc. Nat. Acad. Sci. USA* **90**, 6295–6299.
9. Vaca, L. and Kunze, D. L. (1994). Depletion of intracellular Ca^{2+} stores activates a Ca^{2+}-selective channel in vascular endothelium. *Am. J. Physiol.* **267**, C920–C925.
10. Lückhoff, A. and Clapham, D. E. (1994). Ca^{2+} channels activated by depletion of internal calcium stores in A431 cells. *Biophys. J.* **67**, 177–182.
11. Zubov, A. I., Kaznacheeva, E. V., Alexeeno, V. A., Kiselyov, K., Muallem, S., and Mozhayeva, G. (1999). Regulation of the miniature plasma membrane Ca^{2+} channel I_{min} by IP_3 receptors. *J. Biol. Chem.* **274**, 25983–25985.
12. Trepakova, E. S., Gericke, M., Hirakawa, Y., Weisbrod, R. M., Cohen, R. A., and Bolotina, V. M. (2001). Properties of a native cation channel activated by Ca^{2+} store depletion in vascular smooth muscle cells. *J. Biol. Chem.* **276**, 7782–7790.
13. Zhang, H., Inazu, M., Weir, B., Buchanan, M., and Daniel, E. (1994). Cyclopiazonic acid stimulates Ca^{2+} influx through non-specific cation channels in endothelial cells. *Eur. J. Pharmacol.* **251**, 119–125.
14. Worley, J. F., III, McIntyre, M. S., Spencer, B., and Dukes, I. D. (1994). Depletion of intracellular Ca^{2+} stores activates a maitotoxin-sensitive nonselective cationic current in β cells. *J. Biol. Chem.* **269**, 32055–32058.
15. Krause, E., Pfeiffer, F., Schmid, A., and Schulz, I. (1996). Depletion of intracellular calcium stores activates a calcium conducting nonselective cation current in mouse pancreatic acinar cells. *J. Biol. Chem.* **271**, 32523–32528.
16. Birnbaumer, L., Zhu, X., Jiang, M., Boulay, G., Peyton, M., Vannier, B., Brown, D., Platano, D., Sadeghi, H., Stefani, E., and Birnbaumer, M. (1996). On the molecular basis and regulation of cellular capacitative calcium entry: roles for Trp proteins. *Proc. Nat. Acad. Sci. USA* **93**, 15195–15202.
17. Montell, C. (1999). Visual transduction in *Drosophila*. *Annu. Rev. Cell Dev. Biol.* **15**, 231–268.
18. Zitt, C., Zobel, A., Obukhov, A. G., Harteneck, C., Kalkbrenner, F., Lückhoff, A., and Schultz, G. (1996). Cloning and functional expression of a human Ca^{2+}-permeable cation channel activated by calcium store depletion. *Neuron* **16**, 1189–1196.
19. Liu, X., Wang, W., Singh, B. B., Lockwich, T., Jadlowiec, J., O'Connell, B., Wellner, R., Zhu, M. X., and Ambudkar, I. S. (2000). Trp1, a candidate protein for the store-operated Ca^{2+} influx mechanism in salivary gland cells. *J. Biol. Chem.* **275**, 3043–3411.
20. Vannier, B., Peyton, M., Boulay, G., Brown, D., Qin, N., Jiang, M., Zhu, X., and Birnbaumer, L. (1999). Mouse trp2, the homologue of the human trpc2 pseudogene, encodes mTrp2, a store depletion-activated capacitative Ca^{2+} channel. *Proc. Nat. Acad. Sci. USA* **96**, 2060–2064.
21. Jungnickel, M. K., Marreo, H., Birnbaumer, L., Lémos, J. R., and Florman, H. M. (2001). Trp2 regulates entry of Ca^{2+} into mouse sperm triggered by egg ZP3. *Nature Cell Biol.* **3**, 499–502.
22. Vazquez, G., Lièvremont, J.-P., Bird, G. St. J., and Putney, J. W., Jr. (2001). Trp3 forms both inositol trisphosphate receptor-dependent

and independent store-operated cation channels in DT40 avian B-lymphocytes. *Proc. Nat. Acad. Sci. USA* **98**, 11777–11782.
23. Philipp, S., Cavalié, A., Freichel, M., Wissenbach, U., Zimmer, S., Trost, C., Marguart, A., Murakami, M., and Flockerzi, V. (1996). A mammalian capacitative calcium entry channel homologous to *Drosophila* TRP and TRPL. *EMBO J.* **15**, 6166–6171.
24. Tomita, Y., Kaneko, S., Funayama, M., Kondo, H., Satoh, M., and Akaike, A. (1998). Intracellular Ca^{2+} store-operated influx of Ca^{2+} through TRP-R, a rat homolog of TRP, expressed in *Xenopus* oocytes. *Neurosci. Letters* **248**, 195–198.
25. Yue, L., Peng, J.-B., Hediger, M. A., and Clapham, D. E. (2001). CaT1 manifests the pore properties of the calcium release activated calcium channel. *Nature* **410**, 705–709.
26. Voets, T., Prenen, J., Fleig, A., Vennekens, R., Watanabe, H., Hoenderop, J. G. J., Bindels, R. J. M., Droogmans, G., Penner, R., and Nilius, B. (2001). CaT1 and the calcium release-activated calcium channel manifest distinct pore properties. *J. Biol. Chem.* **276**, 47767–47770.
27. Randriamampita, C. and Tsien, R. Y. (1993). Emptying of intracellular Ca^{2+} stores releases a novel small messenger that stimulates Ca^{2+} influx. *Nature* **364**, 809–814.
28. Thomas, D., and Hanley, M. R. (1995). Evaluation of calcium influx factors from stimulated Jurkat T-lymphocytes by microinjection into *Xenopus* oocytes. *J. Biol. Chem.* **270**, 6429–6432.
29. Rzigalinski, B. A., Willoughby, K. A., Hoffman, S. W., Falck, J. R., and Ellis, E. F. (1999). Calcium influx factor, further evidence it is 5,6-epoxyeicosatrienoic acid. *J. Biol. Chem.* **274**, 175–185.
30. Csutora, P., Su, Z., Kim, H. Y., Bugrim, A., Cunningham, K. W., Nuccitelli, R., Keizer, J. E., Hanley, M. R., Blalock, J. E., and Marchase, R. B. (1999). Calcium influx factor is synthesized by yeast and mammalian cells depleted of organellar calcium stores. *Proc. Nat. Acad. Sci. USA* **96**, 121–126.
31. Trepakova, E. S., Csutora, P., Hunton, D. L., Marchase, R. B., Cohen, R. A., and Bolotina, V. M. (2000). Calcium influx factor (CIF) directly activates store-operated cation channels in vascular smooth muscle cells. *J. Biol. Chem.* **275**, 26158–26163.
32. Irvine, R. F. (1990). "Quantal" Ca^{2+} release and the control of Ca^{2+} entry by inositol phosphates—a possible mechanism. *FEBS Lett.* **263**, 5–9.
33. Kiselyov, K., Xu, X., Mozhayeva, G., Kuo, T., Pessah, I., Mignery, G., Zhu, X., Birnbaumer, L., and Muallem, S. (1998). Functional interaction between $InsP_3$ receptors and store-operated Htrp3 channels. *Nature* **396**, 478–482.
34. Kiselyov, K., Mignery, G. A., Zhu, M. X., and Muallem, S. (1999). The N-terminal domain of the IP_3 receptor gates store-operated hTrp3 channels. *Mol. Cell* **4**, 423–429.
35. McKay, R. R., Szmeczek-Seay, C. L., Lièvremont, J.-P., Bird, G. St. J., Zitt, C., Jüngling, E., Lückhoff, A., and Putney, J. W., Jr. (2000). Cloning and expression of the human transient receptor potential 4 (TRP4) gene: localization and functional expression of human TRP4 and TRP3. *Biochem. J.* **351**, 735–746.
36. Sugawara, H., Kurosaki, M., Takata, M., and Kurosaki, T. (1997). Genetic evidence for involvement of type 1, type 2 and type 3 inositol 1,4,5-trisphosphate receptors in signal transduction through the B-cell antigen receptor. *EMBO J.* **16**, 3078–3088.

CHAPTER 128

Arachidonic Acid-regulated Ca^{2+} Channel

Trevor J. Shuttleworth

*Department of Pharmacology and Physiology,
University of Rochester Medical Center,
Rochester, New York*

Introduction

The arachidonic acid-regulated Ca^{2+} (ARC) channel is a recently identified Ca^{2+}-selective conductance that is distinct from the store-operated conductances (e.g. CRAC channels) discussed by Putney in the preceding chapter. These ARC channels play a key role in [Ca^{2+}]$_i$ signaling in nonexcitable cells in that they appear to provide the predominant pathway for the receptor-activated entry of Ca^{2+} at low, physiologically relevant, agonist concentrations. Under these conditions, [Ca^{2+}]$_i$ signals typically take the form of repetitive [Ca^{2+}]$_i$ oscillations, and the receptor-activated entry of Ca^{2+} acts to modulate the frequency of these oscillations [1–4]. The realization that during such signals intracellular Ca^{2+} stores are only transiently and/or partially depleted raised the question of whether a sufficient "capacitative signal" to activate the store-operated entry of Ca^{2+} would be generated under these conditions. Subsequent studies revealed that Ca^{2+} entry at these agonist concentrations displayed several features that were inconsistent with the capacitative model [5, 6] (see [6] for details), prompting a search for the basis of this apparent noncapacitative mechanism. The result was that in several different cell types such entry was found to be specifically dependent on the receptor-mediated generation of arachidonic acid [7–10]. Thus, arachidonic acid was shown to be generated by the same low agonist concentrations that induce the non-capacitative entry of Ca^{2+}, and inhibition of this arachidonic acid generation specifically and rapidly blocked the associated entry of Ca^{2+}. Moreover, the direct application of exogenous arachidonic acid activated an entry of Ca^{2+} (typically measured as Mn^{2+} quench rate) that was independent of any depletion of agonist-sensitive stores. Especially significant, the use of blockers of arachidonic acid metabolism and/or nonmetabolizable arachidonic acid analogues (e.g. ETYA) indicated that the observed effects reflected the actions of arachidonic acid itself, rather than any of its metabolites [8].

Identification and Characterization of ARC Channels

Although an agonist-activated, arachidonic acid-dependent entry of Ca^{2+} could be demonstrated in a variety of cells, it was unclear whether this involved some kind of "store-independent" activation of the already well-known capacitative Ca^{2+} entry channels (e.g. CRAC channels) or the activation of a novel channel type. This question was resolved by the identification of a novel Ca^{2+}-selective current that was specifically activated by low concentrations of arachidonic acid and that was entirely distinct from the endogenous "CRAC-like" store-operated Ca^{2+}-selective current recorded in the same cells [11]. This current was named I_{ARC} (for *a*rachidonate-*r*egulated *c*alcium current), and was first described in HEK293 cells [11], but similar currents have since been observed in mouse parotid cells, HeLa cells, and RBL cells (unpublished observations). When measured in either traditional whole-cell or perforated-patch modes, I_{ARC} is seen as a small inward current at negative holding potentials, with a current-voltage relationship displaying marked inward rectification and a reversal potential significantly greater than +30 mV (Fig. 1A) [11]. The current is potently blocked by 50 µM La^{3+} (Fig. 1A) and somewhat less effectively blocked by 50 µM Cd^{2+}. Substitution of

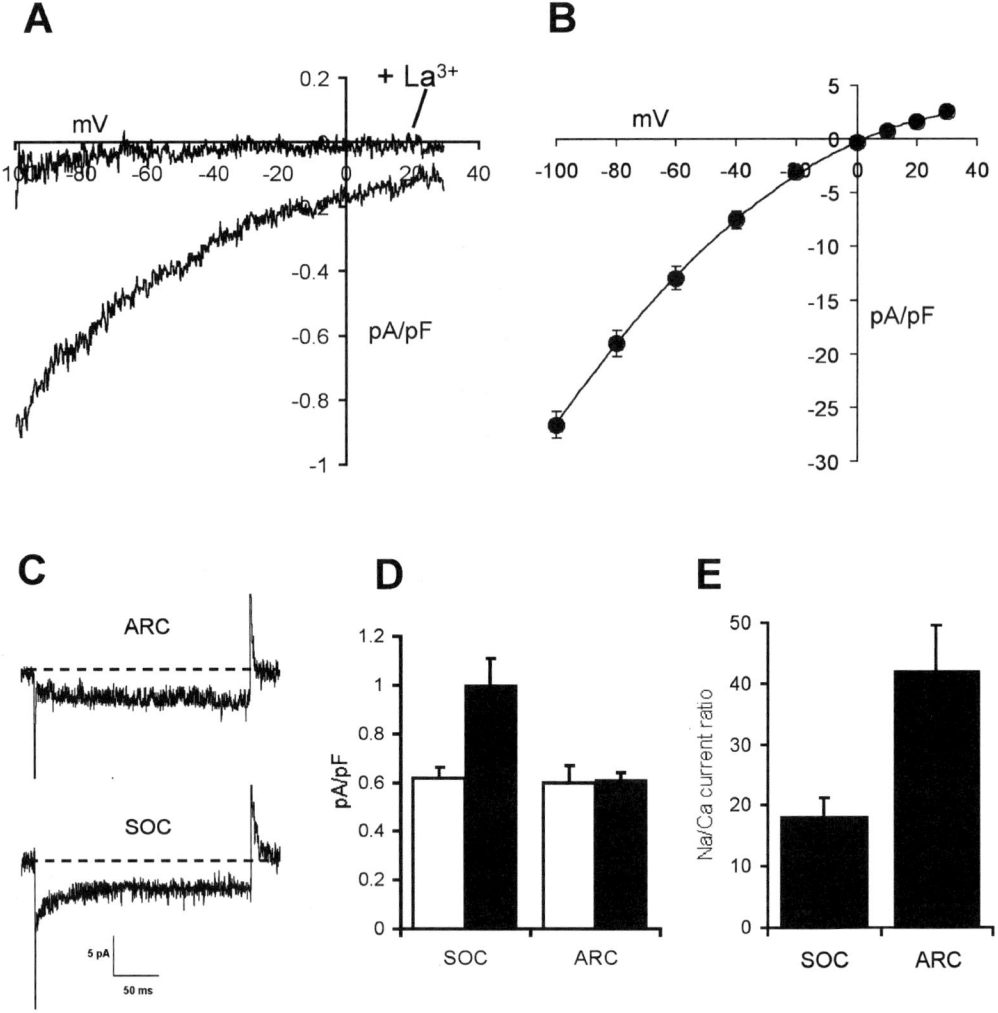

Figure 1 Characteristics of the arachidonic acid-activated current I_{ARC}. (A) Representative I/V curves for macroscopic I_{ARC} in the presence and absence of La^{3+} (50 µM). I_{ARC} was activated by exogenous addition of 8 µM arachidonic acid. Current-voltage relationships were recorded using 150 ms voltage ramps from −100 to +30 mV. External Ca^{2+} concentration was 20 mM. Adapted from [11]. (B) Mean I/V curve for the arachidonate-activated monovalent current recorded in the nominal absence of extracellular divalent cations. Internal $[Mg^{2+}]$ was 8 mm. ARC currents were activated and measured as described in A. (C) Comparison of fast inactivation in I_{ARC} (activated as described in A) and the endogenous CRAC-like store-operated current (I_{SOC}) in HEK293 cells. I_{SOC} was activated by inclusion of 2 µM adenophostin in the pipette solution. Representative recordings of the change in the current measured during a 250 ms pulse to −80 mV. Redrawn from [11]. (D) Effect of substituting internal cesium (open columns) with sodium (filled columns) on the magnitude of the endogenous store-operated currents (SOC) and the arachidonic acid-activated currents (ARC) determined at −80 mV. Store-operated and arachidonic acid-activated currents were determined as described. Data from [13]. (E) Sodium to calcium current ratios for store-operated currents (SOC) and arachidonic acid-activated currents (ARC). Maximal sodium current densities were measured at −80 mV by whole-cell patch clamp in the nominal absence of external divalent cations. SOC and ARC currents were activated as described above. Data from [13].

external Na^+ with $NMDG^+$ has negligible effects on the I/V, thus demonstrating that I_{ARC} is highly Ca^{2+}-selective.

Like other highly Ca^{2+}-selective conductances (including voltage-gated Ca^{2+} channels and CRAC channels), the ARC channels become permeable to monovalent cations on removal of external divalent ions (Fig. 1B). All these features of I_{ARC} are very similar to the archetypal store-operated conductance I_{CRAC} and the endogenous store-operated Ca^{2+}-selective current (I_{SOC}) in HEK293 cells. However, further examination revealed marked differences between I_{ARC} and these store-operated conductances. Unlike I_{CRAC} and I_{SOC}, I_{ARC} shows no Ca^{2+}-dependent fast-inactivation (Fig. 1C) and is largely insensitive to reductions in extracellular pH [11]. I_{ARC} is also insensitive to 2-APB (unpublished observations), which has been shown to potently inhibit I_{CRAC} in a variety of cell types in a manner independent of its originally reported actions on $InsP_3$ receptors [12]. In addition, substitution of the normal Cs^+-based internal (pipette) solution with a corresponding Na^+-based solution results in an approximate 70% increase in the magnitude of I_{SOC} [13], whereas similar substitution is without effect on the magnitude of I_{ARC} (Fig. 1D). Differences are also seen when the conductances are recorded in their monovalent-permeable modes in the nominal absence of external divalent cations [14]. For example, the ratio of the recorded monovalent (Na$^+$) currents relative to the normal Ca^{2+}-selective currents range from 5 to 20 for I_{CRAC} and I_{SOC}, whereas the corresponding ratio for I_{ARC} is greater than 40 (Fig. 1E). Additional differences are seen in the rates and apparent nature of the spontaneous decline in the monovalent currents [14]. Of course, the fundamental distinction between I_{ARC} and the store-operated conductances is that activation of I_{ARC} is specifically dependent on the generation or addition of arachidonic acid and is entirely independent of store-depletion. Moreover, I_{ARC} and I_{SOC} are additive in the same cell and I_{ARC} can be readily activated in cells whose Ca^{2+} stores have been maximally depleted, e.g. by treatment with thapsigargin or with the high-affinity InsP$_3$-receptor agonist adenophostin A [11]. Finally, the possibility that arachidonic acid was merely modifying the properties of endogenous store-operated conductances was eliminated by the demonstration that although Ca^{2+}-sensitive adenylyl cyclases are uniquely sensitive to Ca^{2+} entering via the capacitative pathway, they fail to respond to Ca^{2+} entering via the ARC channels [15]. Thus, the ARC channels are entirely distinct, both physically and spatially, from the store-operated channels.

Consistent with the previously observed arachidonic acid-dependent activation of noncapacitative Ca^{2+} entry, significant activation of I_{ARC} is detectable at concentrations of exogenous arachidonic acid as low as 2–3 µM. Such concentrations are likely to be physiologically meaningful as they lie within the typical range of the Km for the intracellular cyclo-oxygenases and lipoxygenases responsible for metabolizing arachidonic acid. It is important that use of higher concentrations be avoided, as it is clear that fatty acids such as arachidonic acid can have a variety of nonspecific effects (e.g. on membrane fluidity) at such concentrations. As demonstrated earlier for the arachidonic acid-dependent noncapacitative entry of Ca^{2+} (see above), experiments using the nonmetabolizable arachidonic acid analogue ETYA indicate that the activation of I_{ARC} is dependent on the fatty acid itself rather than any metabolite. Other poly-unsaturated fatty acids (e.g. linoleic acid) are also able to activate I_{ARC}, but none are as effective as arachidonic acid. Saturated (e.g. palmitic), and mono-unsaturated (e.g. oleic) fatty acids are completely without effect. Finally, the diacylgercerol analogue OAG fails to activate I_{ARC} even at concentrations as high as 100 µM (unpublished observations).

Specific Activation of ARC Channels by Low Agonist Concentrations

Together, the biophysical, biochemical, and pharmacological data demonstrate that the ARC channels represent a novel Ca^{2+} entry pathway entirely distinct from those activated by store depletion, and suggest that they are likely to be responsible for the arachidonic acid-dependent noncapacitative entry of Ca^{2+} seen in a variety of cells at low agonist concentrations. This suggestion was confirmed by the demonstration of the specific activation of I_{ARC} by the same low concentrations of agonists that had been shown to activate the noncapacitative entry of Ca^{2+}. Activation was achieved by using HEK 293 cells stably transfected with the m3 muscarinic receptor and a protocol based on the previous demonstration of the additive nature of the two Ca^{2+}-selective conductances (I_{ARC} and I_{SOC}) in the same cell [13]. Application of a low concentration (0.5 µM) of the muscarinic agonist carbachol to cells in which I_{SOC} had been maximally activated (using adenophostin A in the pipette solution) resulted in the development of an additional inward current at −80 mV (Fig. 2A). The I/V curve of this carbachol-activated current (after subtraction of the underlying I_{SOC}) showed marked inward rectification and a positive reversal potential (>+30 mV) (Fig. 2B). Development of the current was blocked by atropine and reversibly blocked by isotetrandrine, an inhibitor of the receptor-activation of arachidonic acid generation that does not affect the simultaneous stimulation of phospholipase C [8]. Thus, the current activated by carbachol under these conditions was dependent on the muscarinic receptor-mediated generation of arachidonic acid and was therefore likely to be I_{ARC}. This was confirmed by demonstrating that the carbachol-activated current showed no fast-inactivation, was unaffected by substituting Na$^+$ for Cs$^+$ in the pipette solution, and displayed a monovalent (Na$^+$) current in the nominal absence of external divalent ions that was more than 45 times larger than the corresponding normal Ca^{2+}-selective current [13]. As discussed above, these features are uniquely characteristic of I_{ARC}, confirming that the additional current activated by low concentrations of carbachol specifically reflects the activity of the ARC channels.

Activation of I_{ARC} by carbachol was measurable at concentrations that were just sufficient to initiate detectable [Ca^{2+}]$_i$ signals in the same cells (0.2 µM) and reached a maximum at 1 µM (Fig. 2C) [13]. [Ca^{2+}]$_i$ signals within this concentration range are typically oscillatory in nature, and previous evidence had indicated that the associated entry of Ca^{2+} is both noncapacitative and entirely dependent on the generation of arachidonic acid [8]. Thus, the data demonstrated that stimulation with low agonist concentrations results in the specific activation of I_{ARC}, which provides the predominant route for Ca^{2+} entry under these conditions.

Roles of ARC Channels and SOC/CRAC Channels in [Ca^{2+}]$_i$ Signals: "Reciprocal Regulation"

Although the ARC channels provide the major route for Ca^{2+} entry at low agonist concentrations, the additive nature of the two conductances I_{ARC} and I_{SOC} would lead to the prediction that entry at high agonist concentrations should be via a combination of both ARC and SOC channels. However, this is inconsistent with earlier evidence indicating that addition of an agonist to cells whose store-operated entry had been maximally activated by treatment with thapsigargin fails to induce any obvious increase in [Ca^{2+}]$_i$ [16]. Indeed, such data were widely used to support the proposition that the capacitative pathway was the sole mechanism responsible for the agonist-induced increase in the entry of Ca^{2+} in nonexcitable cells. Examination of the rate of Ca^{2+} entry at

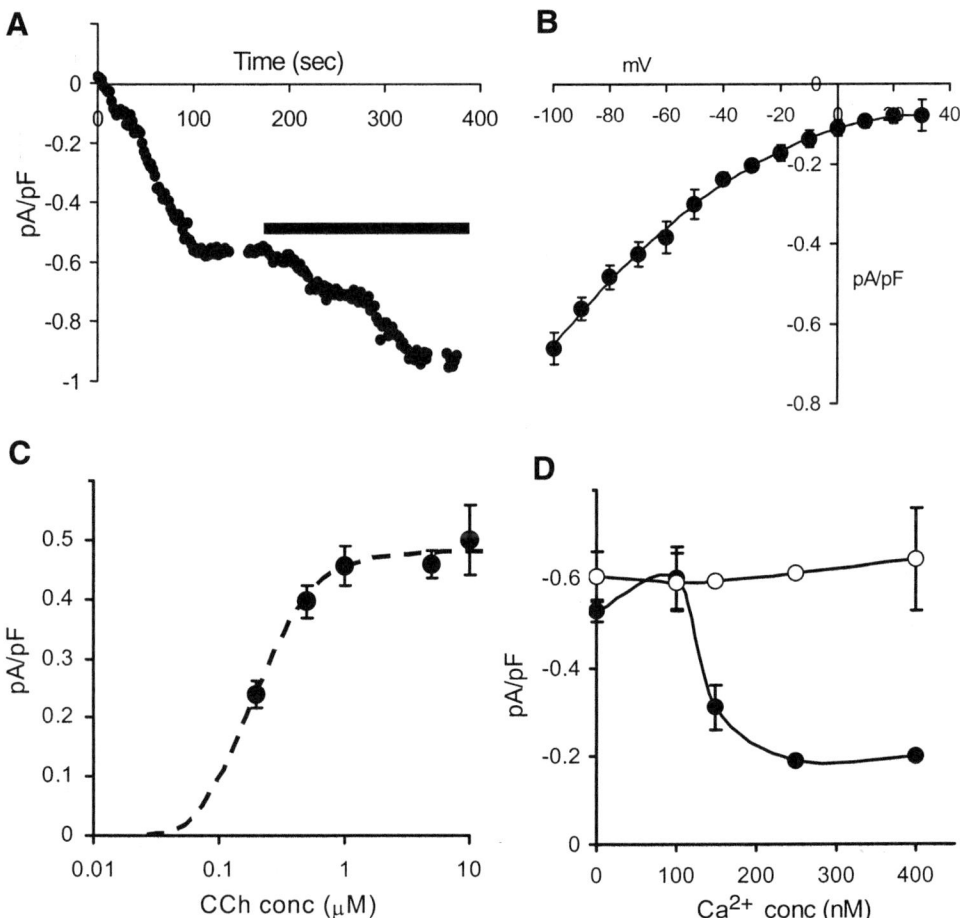

Figure 2 Activation of ARC currents by carbachol in m3-HEK cells. (A) Representative trace showing the activation of an additional inward current measured at −80 mV on addition of low concentrations of carbachol after maximal activation of store-operated currents. On going whole-cell (at time zero), inclusion of adenophostin A (2 μM) in the pipette solution rapidly depletes internal Ca^{2+} stores and maximally activates the endogenous store-operated current, I_{SOC}. Subsequent addition of carbachol (CCh, 0.5 μM, black bar) results in the development of additional inward current. Data taken from [13]. (B) Mean I/V curve for the current activated by carbachol (0.5 μM). Individual curves were obtained from voltage ramps after subtraction of the corresponding I/V curve for maximally activated I_{SOC} in the same cell. Taken from [13]. (C) The magnitude of the ARC current activated by different carbachol concentrations. Carbachol-activated ARC currents were measured after maximal activation of store-operated currents as described in B. Taken from [13]. (D) Effects of different buffered internal Ca^{2+} concentrations on the magnitude of SOC (maximally activated with 2 μM adenophostin A, open circles) and ARC currents (activated by 8 μM arachidonic acid, filled circles). Currents were measured at −80 mV. Taken from [13].

high agonist concentrations (using Mn^{2+} quench) confirmed that this was essentially entirely via the capacitative pathway and no significant arachidonic acid-dependent contribution could be detected, despite an increasing generation of arachidonic acid over the same agonist concentration range [13]. This apparent contradiction was resolved when it was revealed that I_{ARC} is potently inhibited by sustained increases in $[Ca^{2+}]_i$ above resting values [13] such as would be induced in cells stimulated with high agonist concentrations (Fig. 2D). As yet, the precise mechanism for this Ca^{2+}-dependent inhibition of I_{ARC} is unknown, but it is clear that it does not involve any action of Ca^{2+} entering through the channel itself as inhibition is seen even when I_{ARC} is carrying only monovalent cations (i.e. in the nominal absence of external divalent ions) [13]. Instead, it seems to reflect an effect of the general or global cytosolic Ca^{2+} concentration. This inhibition of the ARC channels by elevations in $[Ca^{2+}]_i$ develops only slowly, taking some two minutes to reach completion [13]. This means that the transient increases in $[Ca^{2+}]_i$ associated with oscillatory Ca^{2+} signals (each typically lasting only a few seconds) would have a negligible effect on the activity of I_{ARC} and would not impair the role of the ARC channels in the entry of Ca^{2+} under these conditions. Only with a prolonged elevation of $[Ca^{2+}]_i$, such as seen following the sustained depletion of the intracellular stores and activation of I_{CRAC} (or I_{SOC}), will I_{ARC} be significantly inhibited.

These findings lead to the interesting conclusion that the two coexisting, but independent, modes of receptor-stimulated Ca^{2+} entry, via the ARC channels and the SOC/CRAC channels, are regulated in a unique manner by increasing agonist concentrations—a process we have termed "reciprocal regulation" [13]. This is illustrated in Fig. 3. Low concentrations of

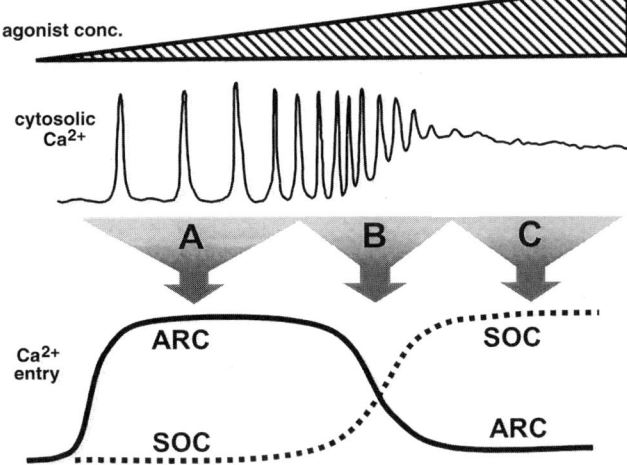

Figure 3 Reciprocal regulation of ARC and SOC channels. Diagram representing the regulation of I_{ARC} and I_{SOC} at different agonist concentrations and the corresponding changes in the nature of the resulting $[Ca^{2+}]_i$ signal (see text for details). Adapted from [13].

agonist result in the specific activation of the ARC channels which provide the main mode of Ca^{2+} entry under these conditions (see "A" in Fig. 3). This, together with the generation of low levels of $InsP_3$, initiates and modulates the cyclical transient discharge and refilling of intracellular Ca^{2+} stores resulting in the generation of the characteristic oscillatory $[Ca^{2+}]_i$ signals. SOC/CRAC channels fail to activate under these conditions because the transient and/or partial discharge of the intracellular Ca^{2+} stores is not able to generate an adequate "capacitative signal". As agonist concentrations increase, increasing levels of $InsP_3$ in the cytosol cause the discharge of the stores to become more complete and sustained, resulting in the activation of the SOC/CRAC channels and the development of a maintained elevated level of $[Ca^{2+}]_i$ ("C" in Fig. 3). This, in turn, inhibits the activity of the ARC channels. Thus, the transition from an oscillatory $[Ca^{2+}]_i$ signal to a sustained $[Ca^{2+}]_i$ signal ("B" in Fig. 3) is associated with a progressive switch in the predominant mode of Ca^{2+} entry from the ARC channels at low agonist concentrations to the SOC/CRAC channels at high ($\approx$ maximal) concentrations.

Conclusions and Implications

The demonstration that ARC channels represent a Ca^{2+} entry pathway that is spatially distinct from those activated by store depletion immediately raises the potential for the specific activation of different targets within the cell, as has already been demonstrated for the SOC channels and certain adenylyl cyclases [15]. Moreover, the fact that these pathways are independently activated at different agonist concentrations adds a new level of complexity to cellular Ca^{2+} signaling. Thus, the appropriate targeting of downstream Ca^{2+}-sensitive effectors to sites in close proximity to either ARC or SOC/CRAC channels may result in the specific selective regulation of these effectors at different agonist concentrations independent of any obvious overall changes in the spatial and/or temporal features of the induced $[Ca^{2+}]_i$ changes.

Obviously, the study of ARC channels is still only in its infancy and much remains to be discovered about their properties, regulation, and functions. Undoubtedly, as this novel Ca^{2+} entry pathway begins to receive more attention from researchers, a more complete picture of its distribution and specific roles will be revealed. Critical to this will be the identification of the molecular identity of the channels, which, to date, remains unknown. However, the demonstration that these channels provide the primary route for the receptor-activated entry of Ca^{2+} at physiologically relevant levels of stimulation [13] is of paramount significance, and the recent identification of currents identical to I_{ARC} in several different cell types suggests that this is a widespread phenomenon. Given this, it seems likely that the identification of this unique and specific function of the ARC channels will result in this channel becoming a prime target for possible pharmacological manipulation in any potential therapeutic strategies aimed at this key signaling system.

Acknowledgments

Studies from the author's laboratory described in this article were supported by grants from the National Institutes of Health (GM 40457).

References

1. Rooney, T. A., Sass, E. J., and Thomas, A. P. (1989). Characterization of cytosolic calcium oscillations induced by phenylephrine and vasopressin in single fura-2-loaded hepatocytes. *J. Biol. Chem.* **264**, 17131–17141.
2. Berridge, M. J. (1990). Calcium oscillations. *J. Biol. Chem.* **265**, 9583–9586.
3. Girard, S. and Clapham, D. (1993). Acceleration of intracellular calcium waves in *Xenopus* oocytes by calcium influx. *Science* **260**, 229–232.
4. Shuttleworth, T. J. and Thompson, J. L. (1996). Ca^{2+} entry modulates oscillation frequency by triggering Ca^{2+} release. *Biochem. J.* **313**, 815–819.
5. Shuttleworth, T. J. and Thompson, J. L. (1996). Evidence for a non-capacitative Ca^{2+} entry during $[Ca^{2+}]_i$ oscillations. *Biochem J.* **316**, 819–824.
6. Shuttleworth, T. J. (1999). What drives calcium entry during $[Ca^{2+}]_i$ oscillations? Challenging the capacitative model. *Cell Calcium* **25**, 237–246.
7. Shuttleworth, T. J. (1996). Arachidonic acid activates the noncapacitative entry of Ca^{2+} during $[Ca^{2+}]_i$ oscillations. *J. Biol. Chem.* **271**, 21720–21725.
8. Shuttleworth, T. J. and Thompson, J. L. (1998). Muscarinic receptor activation of arachidonate-mediated Ca^{2+} entry in HEK293 cells is independent of phospholipase C. *J. Biol. Chem.* **273**, 32636–32643.
9. Munaron, L., Antoiotti, S., Distasi, C., and Lovisolo, D. (1997). Arachidonic acid mediates calcium influx induced by basic fibroblast growth factor in Balb-c 3T3 fibrobalsts. *Cell Calcium* **22**, 179–188.
10. Broad, L. M., Cannon, T. R., and Taylor, C. W. (1999). A non-capacitative pathway activated by arachidonic acid is the major Ca^{2+} entry mechanism in rat A7r5 smooth muscle cells stimulated with low concentrations of vasopressin. *J. Physiol.* **517**, 121–134.
11. Mignen, O. and Shuttleworth, T. J. (2000). I_{ARC}, a novel arachidonate-regulated, noncapacitative Ca^{2+} entry channel. *J. Biol. Chem.* **275**, 9114–9119.
12. Prakriya, M. and Lewis, R. S. (2001). Potentiation and inhibition of Ca^{2+} release-activated Ca^{2+} channels by 2-aminoethyldiphenyl borate (2-APB) occurs independently of IP$_3$ receptors. *J. Physiol.* **536**, 3–19.

13. Mignen, O., Thompson, J. L., and Shuttleworth, T. J. (2001). Reciprocal regulation of capacitative and arachidonate-regulated non-capacitative Ca^{2+} entry pathways. *J. Biol. Chem.* **276**, 35676–35683.
14. Mignen, O. and Shuttleworth, T. J. (2001). Permeation of monovalent cations through the non-capacitative arachidonate-regulated Ca^{2+} channels in HEK293 cells. *J. Biol. Chem.* **276**, 21365–21374.
15. Shuttleworth, T. J. and Thompson, J. L. (1999). Discriminating between capacitative and arachidonate-activated Ca^{2+} entry pathways in HEK293 cells. *J. Biol. Chem.* **274**, 31174–31178.
16. Takemura, H., Hughes, A. R., Thastrup, O., and Putney, J. W. Jr. (1989). Activation of calcium entry by the tumor promoter thapsigargin in parotid acinar cells. *J. Biol. Chem.* **264**, 12266–12271.

IP$_3$ Receptors

Colin W. Taylor
*Department of Pharmacology,
University of Cambridge, Cambridge,
United Kingdom*

Introduction

Inositol 1,4,5-trisphosphate (IP$_3$) receptors are large proteins: the native receptor is some 20 nm across, extends more than 10 nm from the membrane of the endoplasmic reticulum (ER), and each of its four subunits comprises about 2,700 residues. Their close relatives, ryanodine receptors, are even bigger. Size is important for these intracellular Ca^{2+} channels because it allows opening of the channel to be controlled by many different intracellular stimuli and it allows IP$_3$ and ryanodine receptors to interact directly with proteins in other membranes, including Ca^{2+} channels in the plasma membrane. IP$_3$ receptors, for example, may interact directly with the trp channels that are thought to mediate store-regulated Ca^{2+} entry [1]. This role for IP$_3$ receptors remains controversial, but an essential role in linking receptors that stimulate IP$_3$ formation to release of Ca^{2+} from intracellular stores is accepted [2].

IP$_3$ receptors are expressed in most eucaryotic cells, with three genes encoding closely related subtypes in mammals and birds, but only a single subtype in each of *Xenopus*, *Drosophila*, and *C. elegans* [3]. At least two of the mammalian subtypes are also alternatively spliced [4]. Because the functional channel is a tetramer, which can assemble from the same or different subunits, and most mammalian cells express more than one receptor subtype, there is considerable scope for IP$_3$ receptor diversity [3]. The different subtypes and their splice variants are differentially expressed, respond differently to chronic stimulation, and their assembly into heterotetramers is itself regulated, but the physiological significance of IP$_3$ receptor heterogeneity is unclear. There are subtle differences in the affinities of the subtypes for IP$_3$ and in their modulation by various intracellular stimuli [5], but more striking than the differences are the properties shared by all IP$_3$ receptors. All are tetrameric intracellular Ca^{2+} channels with large conductances, they have similar primary structures, Ca^{2+} and IP$_3$ control their opening, and they are modulated by many additional intracellular signals.

Key structural features of the type 1 IP$_3$ receptor are shown in Fig. 1. Each subunit has an IP$_3$-binding site formed by two distinct domains lying close to the amino terminal and linked to each other by a short stretch of residues that includes the S1 splice site. Several conserved, positively charged residues are particularly important for recognition of IP$_3$; they probably interact with its phosphate groups. The core IP$_3$-binding region of just 350 residues, which can be expressed as a soluble protein with very high affinity for IP$_3$, has been used as an "IP$_3$-sponge" to define the role of IP$_3$ in intact cells [6]. This is a useful tool because the only other antagonists of IP$_3$ receptors, heparin, Xestospongin and 2-aminoethyldiphenyborane, are notorious for their side effects.

Toward the carboxy terminal of each subunit there are six membrane-spanning regions, the last two of which together with an intervening loop (the P-loop) line the pore of the channel [7]. In keeping with the similar ion permeation properties of the IP$_3$ receptor subtypes, the sequences within this pore region are conserved and are also similar in ryanodine receptors [8]. Although almost 1,700 residues separate the IP$_3$-binding site from the pore, the two regions are associated in the native receptor, with the IP$_3$-binding region of one subunit perhaps interacting directly with the pore region of a neighboring subunit to control its opening [9]. The long stretch of residues separating the IP$_3$-binding region from the pore has been described as the "modulatory domain": it certainly includes at least some of the sites through which channel opening is modulated by phosphorylation or binding of small molecules and proteins [4,10] (Fig. 1).

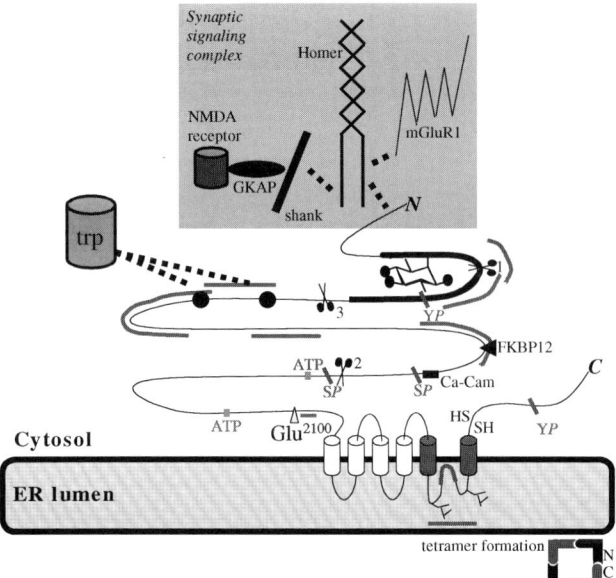

Figure 1 Structure of type 1 IP$_3$ receptor. The large cytosolic head of the receptor includes the IP$_3$-binding domain; seven regions (bold lines) to which Ca^{2+} has been shown to bind; the glutamate^{2100} residue shown to affect Ca^{2+}-sensitization; sites phosphoryated by PKA (SP) or tyrosine kinases (YP); the ATP-binding sites; the sites to which FKBP and trp are proposed to bind; and the three alternative splice sites (scissors). The pore is formed by the last two membrane-spanning regions together with part of the intervening loop. The loop also includes two glycosylation sites and a lumenal Ca^{2+}-binding site. The C-terminal tail includes conserved cysteine residues and another tyrosine kinase phosphorylation site. Assembly of the subunits into tetramers, which places the N-terminal of one subunit in close proximity to the channel region of another, requires residues downstream of the fourth membrane-spanning region. The box illustrates how Homer might function to link IP$_3$ receptors to various components of the synaptic signaling complex.

Despite controversy, it seems likely that all IP$_3$ receptors are biphasically regulated by cytosolic Ca^{2+}. Luminal Ca^{2+} has also been proposed to regulate channel opening, but it is difficult, for both ryanodine and IP$_3$ receptors [8], to resolve whether this really results from Ca^{2+} stimulating the receptor at its luminal surface or at its cytosolic surface after the Ca^{2+} has passed through the channel. There is a Ca^{2+}-binding site within a luminal loop of the type 1 IP$_3$ receptor (Fig. 1) and Ca^{2+}-binding proteins within the lumen of IP$_3$-sensitive organelles (calreticulin in ER; chromogranin in secretory vesicles) associate with IP$_3$ receptors [10], but none of these Ca^{2+}-binding sites has been shown to allow luminal Ca^{2+} to regulate channel opening.

Regulation of IP$_3$ receptors by luminal Ca^{2+} may be unresolved, but there is no such uncertainty about their regulation by cytosolic Ca^{2+}: all IP$_3$ receptors are stimulated by cytosolic Ca^{2+} and most (possibly all) can also be inhibited by cytosolic Ca^{2+} [5,11]. The details of how this biphasic regulation by cytosolic Ca^{2+} occurs have not been resolved; they are probably different for different receptor subtypes. It is accepted that IP$_3$ and Ca^{2+} must both bind to the IP$_3$ receptor before the channel can open and that binding of IP$_3$ regulates Ca^{2+} binding. IP$_3$, in other words, tunes the Ca^{2+} sensitivity of the IP$_3$ receptor. One scheme suggests that the major effect of IP$_3$ is to relieve Ca^{2+} inhibition by decreasing the affinity of an inhibitory Ca^{2+}-binding site [12], while another suggests that IP$_3$ binding reciprocally regulates two Ca^{2+}-binding sites, causing a stimulatory Ca^{2+}-binding site to be exposed and an inhibitory Ca^{2+}-binding site to be concealed [13]. It is not clear whether these Ca^{2+}-binding sites reside on the IP$_3$ receptor itself—it certainly has many cytosolic Ca^{2+}-binding sites (Fig. 1)—or on proteins associated with the receptor. A glutamate residue lying close to the C-terminal end of the modulatory domain (Fig. 1), and conserved within all IP$_3$ and ryanodine receptors, may be important in mediating the stimulatory effect of cytosolic Ca^{2+} [14]. The evidence that it is also involved in Ca^{2+} inhibition is less convincing [14] and difficult to reconcile with evidence suggesting that accessory proteins mediate Ca^{2+} inhibition [11]. Whether calmodulin is required for Ca^{2+} inhibition is hotly contested [15]. In summary, it seems likely that both IP$_3$ and Ca^{2+} must bind directly to the receptor for the channel to open, and that Ca^{2+} inhibition is via an accessory protein, whose binding to the receptor is probably regulated by IP$_3$. Both the regulation of IP$_3$ receptors by cytosolic Ca^{2+} and the ways in which different subtypes fine-tune that regulation are important in determining the complex spatio-temporal patterns of IP$_3$-evoked Ca^{2+} release in intact cells [16].

Besides Ca^{2+} and IP$_3$, there are many other modulators of IP$_3$ receptors. ATP binds to sites within the modulatory domain and increases IP$_3$ sensitivity [5]. Chemicals that modify sulphydryl groups, including reactive oxygen species, also increase IP$_3$ sensitivity, possibly by modifying conserved cysteine residues toward the C-terminal. These residues may thereby link the redox state of the cell to its IP$_3$ sensitivity. IP$_3$ receptors are phosphorylated by PKA, PKC, Ca^{2+}-calmodulin-dependent protein kinase II, and PKG [4,10]. The latter also phosphorylates a protein that is tightly associated with the IP$_3$ receptor (IRAG), causing inhibition of IP$_3$-evoked Ca^{2+} release [17]. Phosphorylation of tyrosine residues on IP$_3$ receptors can also set their sensitivity to IP$_3$ [18]. The immunophilin FKBP12, which certainly regulates ryanodine receptors, may regulate IP$_3$ receptors both directly and by anchoring the protein phosphatase, calcineurin, to them.

Most IP$_3$ receptors are found in the membranes of the ER, but they also occur in the Golgi, secretory vesicles, nuclear envelope, and plasma membrane. Even within the ER, the distribution of IP$_3$ receptors is far from uniform. On a molecular scale, IP$_3$ receptors occur in clusters [19], allowing the Ca^{2+} released by one receptor to rapidly influence its neighbors [16]. At a cellular level, they can be concentrated in discrete areas of ER: at the apical pole of pancreatic acinar cells, for example. There are also important functional associations between IP$_3$ receptors in the ER and other membranes: the possible link between IP$_3$ receptors and trp channels in the plasma membrane was mentioned earlier (Fig. 1), and there are many examples of IP$_3$ receptors in close association with mitochondria [20]. We are only just beginning to unravel the mechanisms responsible for putting IP$_3$ receptors into the right places, but scaffolding proteins are likely to be important.

An example illustrates the likely complexity of the interactions between scaffold proteins and IP_3 receptors. The Homer proteins are a family of dimeric scaffold proteins that assemble signaling proteins at excitatory synapses. The N-terminal of Homer binds to IP_3 receptors, type 1 metabotropic glutamate receptors, and to Shank, another scaffold protein that is targetted by its PDZ domain to the postsynaptic density and that itself binds further signaling proteins [21]. This chain of protein-protein interactions both targets IP_3 receptors to the dendritic spines of hippocampal neurones and brings them into intimate association with other signaling proteins, including receptors that stimulate IP_3 formation and channels that mediate Ca^{2+} entry (Fig. 1).

References

1. Zhang, Z., Tang, J., Tikunova, S., Johnson, J. D., Chen, Z., Qin, N., Dietrich, A., Stefani, E., Birnbaumer, L., and Zhu, M. X. (2001). Activation of Trp3 by inositol 1,4,5-trisphosphate receptors through displacement of inhibitory calmodulin from a common binding domain. *Proc. Natl. Acad. Sci. USA* **98**, 3168–3173.
2. Berridge, M. J. and Irvine, R. F. (1989). Inositol phosphates and cell signalling. *Nature* **341**, 197–205.
3. Taylor, C. W., Genazzani, A. A., and Morris, S. A. (1999). Expression of inositol trisphosphate receptors. *Cell Calcium* **26**, 237–251.
4. Patel, S., Joseph, S. K., and Thomas, A. P. (1999). Molecular properties of inositol 1,4,5-trisphosphate receptors. *Cell Calcium* **25**, 247–264.
5. Miyakawa, T., Maeda, A., Yamazawa, T., Hirose, K., Kurosaki, T., and Iino, M. (1999). Encoding of Ca^{2+} signals by differential expression of IP_3 receptor subtypes. *EMBO J.* **18**, 1303–1308.
6. Uchiyama, T., Yoshikawa, F., Hishida, A., Furuichi, T., and Mikoshiba, K. (2002). A novel recombinant hyper-affinity inositol 1,4,5-trisphosphate (IP_3) absorbent traps IP_3, resulting in specific inhibition of IP_3-mediated calcium signaling. *J. Biol. Chem.* **277**, 8106–8113.
7. Ramos-Franco, J., Galvan, D., Mignery, G. A., and Fill, M. (1999). Location of the permeation pathway in the recombinant type-1 inositol 1,4,5-trisphosphate receptor. *J. Gen. Physiol.* **114**, 243–250.
8. Balshaw, D., Gao, L., and Meissner, G. (1999). Luminal loop of the ryanodine receptor: a pore-forming segment? *Proc. Natl. Acad. Sci. USA* **96**, 3345–3347.
9. Boehning, D. and Joseph, S. K. (2000). Direct association of ligand-binding and pore domains in homo- and heterotetrameric inositol 1,4,5-trisphosphate receptors. *EMBO J.* **19**, 5450–5459.
10. Mackrill, J. J. (1999). Protein-protein interactions in intracellular Ca^{2+}-release channel function. *Biochem. J.* **337**, 345–361.
11. Taylor, C. W. (1998). Inositol trisphosphate receptors: Ca^{2+}-modulated intracellular Ca^{2+} channels. *Biochim. Biophys. Acta.* **1436**, 19–33.
12. Mak, D.-O., D., McBride, S., and Foskett, J. K. (1998). Inositol 1,4,5-trisphosphate activation of inositol trisphosphate receptor Ca^{2+} channel by ligand tuning of Ca^{2+} inhibition. *Proc. Natl. Acad. Sci. USA* **95**, 15821–15825.
13. Adkins, C. E. and Taylor, C. W. (1999). Lateral inhibition of inositol 1,4,5-trisphosphate receptors by cytosolic Ca^{2+}. *Curr. Biol.* **9**, 1115–1118.
14. Miyakawa, T., Mizushima, A., Hirose, K., Yamazawa, T., Bezprozvanny, I., Kurosaki, T., and Iino, M. (2001) Ca^{2+}-sensor region of IP_3 receptor controls intracellular Ca^{2+} signaling. *EMBO J.* **20**, 1674–1680.
15. Michikawa, T., Hirota, J., Kawano, S., Hiraoka, M., Yamada, M., Furuichi, T., and Mikoshiba, K. (1999). Calmodulin mediates calcium-dependent inactivation of the cerebellar type 1 inositol 1,4,5-trisphosphate receptor. *Neuron* **23**, 799–808.
16. Berridge, M. J., Lipp, P., and Bootman, M. D. (2000). The versatility and universality of calcium signalling. *Nature Rev. Mol. Cell Biol.* **1**, 11–21.
17. Ammendola, A., Geiselhöringer, A., Hofmann, F., and Schlossmann, J. (2001). Molecular determinants of the interaction between the inositol 1,4,5-trisphosphate receptor-associated cGMP kinase substrate (IRAG) and cGMP kinase Iβ. *J. Biol. Chem.* **276**, 24153–24159.
18. Yokoyama, K., Su, I., Tezuka, T., Yasuda, T., Mikoshiba, K., Tarakhovsky, A., and Yamamoto, T. (2002). BANK regulates BCR-induced calcium mobilization by promoting tyrosine phosphorylation of IP_3 receptor. *EMBO J.* **21**, 83–92.
19. Mak, D.-O., D., McBride, S., Raghiram, V., Yue, Y., Joseph, S. K., and Foskett, J. K. (2000). Single-channel properties in endoplasmic reticulum membrane of recombinant type 3 inositol trisphosphate receptor. *J. Gen. Physiol.* **115**, 241–255.
20. Rutter, G. A. and Rizzuto, R. (2000). Regulation of mitochondrial metabolism by ER Ca^{2+} release: an intimate connection. *Trends Biochem. Sci.* **25**, 215–221.
21. Sala, C., Piëch, V., Wilson, N. R., Passafaro, M., Liu, G., and Sheng, M. (2001). Regulation of dendritic spine morphology and synaptic function by shank and homer. *Neuron* **31**, 115–130.

CHAPTER 130

Ryanodine Receptors

David H. MacLennan and Guo Guang Du

*Banting and Best Department of Medical Research,
University of Toronto, Charles H. Best Institute,
Toronto, Ontario, Canada*

Function and Structure

The store from which signal Ca^{2+} is derived is either the extracellular space or the lumenal space of intracellular organelles, the source depending on the specialization of the cell. In muscle, Ca^{2+} is the major signaling molecule for excitation-contraction coupling, the process involving release of Ca^{2+} from the sarcoplasmic reticulum (SR) in response to depolarization of the sarcolemma and transverse tubules, and the subsequent activation of muscle contraction by the binding of Ca^{2+} to troponin, a component of the contractile apparatus. In highly specialized skeletal muscle, signal Ca^{2+} is released almost exclusively from a store located in the lumen of the SR by the activation of a class of Ca^{2+} release channels referred to as ryanodine receptors (RyR). In cardiac muscle, more than two-thirds of signal Ca^{2+} is derived from the SR, the remainder coming from extracellular spaces [1]. Ryanodine receptors are also expressed in other excitable and nonexcitable cells where their contributions to signal transduction may be less pronounced.

Three RyR isoforms have been characterized: RyR1, associated with skeletal muscle; RyR2, associated with cardiac muscle; and RyR3, which is expressed more ubiquitously. Isolated RyR type Ca^{2+} release channels have a high single channel conductance of 80 to 100 pS for Ca^{2+} and 400 to 800 pS for monovalent cations [2,3]. They are activated by micromolar Ca^{2+} and millimolar adenine nucleotides and inhibited by millimolar Ca^{2+} and Mg^{2+}. They are also modulated by calmodulin and cyclic ADP ribose. Pharmaceutical agents that open the channels include caffeine, 4-chloro-*m*-cresol, and halothane, which is a trigger for malignant hyperthermia (MH). Ryanodine binds to the open channel, converting the open state to a partially open subconductance state. Dantrolene, an antidote for MH, blocks the channel.

All three RyR isoforms are homotetramers formed from subunits of about 5000 amino acids, with subunit masses of about 565,000 Da [4,5]. Electron microscopic reconstruction of the tetrameric RyR1 molecule at about 30 Å resolution shows a cytoplasmic component with a square prism shape with dimensions of $28 \times 28 \times 12$ nm and a square transmembrane domain with an edge measuring 12 nm at the point of attachment to the cytoplasmic region and a depth of about 7 nm perpendicular to the membrane [6,7] (Fig. 1). Structures of these dimensions are observed in the junctional terminal cisternae of the SR and in corbular SR in cardiac muscle but are absent from *RYR1* null mice [8]. The N-terminal 85% of the molecule is predicted to form cytosolic domains, while 6 to 8 segments of the remaining C-terminal sequences contribute to the formation of the channel pore [4,5,5a].

The cytoplasmic component appears as a scaffold-like structure composed of at least 10 arbitrarily numbered, interconnected, globular domains, and provides a physical linkage between the SR and the transverse tubule while facilitating flow of Ca^{2+} from a central channel to the periphery (Fig. 1). This structure provides a framework for the identification of binding sites for specific regulatory proteins such as calmodulin (CaM) and FK506 binding protein (FKBP) [9]. CaM binds to an amino acid sequence containing residues 3614–3643 of the ryanodine receptor [10] and FKBP12 to Val^{2461} [11].

The structures of channels, closed in the absence of Ca^{2+} or opened in the presence of Ca^{2+} and other ligands, have been compared [12]. A small, central-axis opening with a diameter of 7 Å is revealed in transiently open channels and this opens to 18 Å in the ryanodine-modified channel. The channel runs through the whole transmembrane structure along a four-fold axis opening into the lumen. The process of opening is comparable to the opening of a camera diaphragm. In open channels, the clamp-shaped subdomains at the four

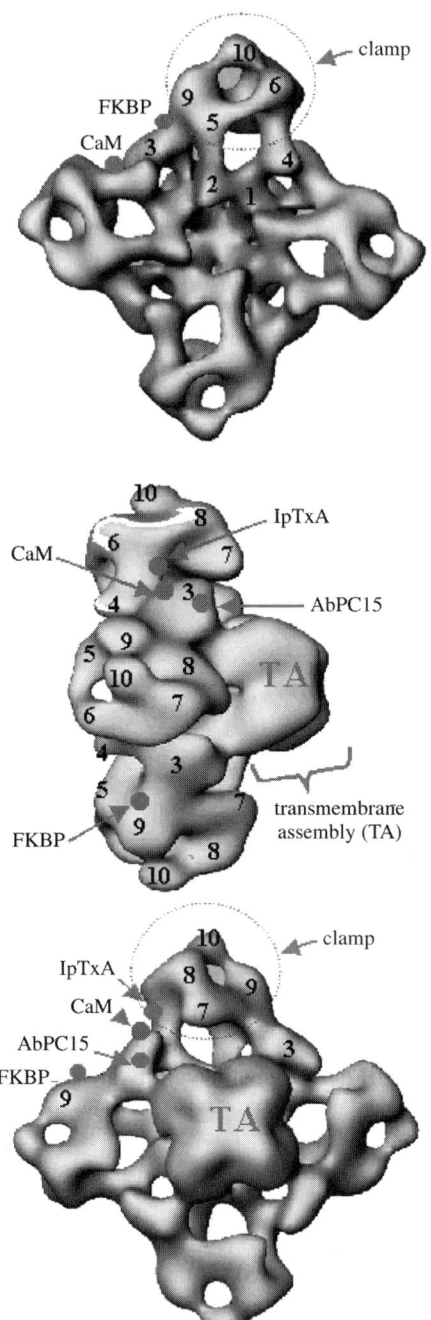

Figure 1 Solid body representation of a 3D reconstruction of RyR1 [7]. The numbers indicate distinct globular structures that correspond to structural domains, all of which are located in cytoplasmic regions of the protein. The filled circles indicate the locations of ligands as determined by reconstruction of RyR-ligand complexes. Abbreviations: CaM, calmodulin; FKBP, FK506-binding protein: IpTxA, imperatoxin A; AbPC15, monoclonal antibody against RyR residues 4425–4621; TA, transmembrane assembly. Reprinted by permission from *Eur. J. Biochem.* **267**, 5274–5279 (2000).

Activation of Ryanodine Receptor Ca^{2+} Release Channels

The location of RyR1 cytoplasmic domains in the junctional terminal cisternae of skeletal muscle SR suggests that they interact directly with the α1-subunit of the voltage-sensitive, dihydropyridine-modulated, slow or L-type Ca^{2+} channels (DHPR) located in closely apposed transverse tubules or plasma membranes [15]. A cluster of four DHPR molecules in the transverse tubule of skeletal muscle directly apposes every other ryanodine receptor molecule, with an individual DHPR molecule overlying an individual RyR1 subunit. Biochemical, physiological, and molecular genetic studies show that physical interactions occur between skeletal muscle RyR1 and DHPR isoforms, leading to both activation of the Ca^{2+} release channel (orthograde interaction) and modulation of the slow Ca^{2+} channel (retrograde interaction) [16]. The exact site of RyR/DHPR interaction is not clear [17]. Presumably, those Ca^{2+} release channels not opened by direct physical interaction are opened by Ca^{2+}-induced Ca^{2+} release. By contrast, there is no indication that direct physical interactions between cardiac RyR2 and DHPR α1-subunits lead to opening of the cardiac Ca^{2+} release channel [18]. In this case, entry of extracellular Ca^{2+} through the DHPR α1-subunit induces activation of RyR2 by Ca^{2+}-induced Ca^{2+} release.

Since no high-resolution structure is available and since it has proven difficult to carry out structure-function analysis of RyR molecules, evidence for a coherent mechanism of action is sketchy. If the ion pore corresponds to the model for a K^+ channel [19], then each of the four RyR subunits must contribute a hairpin-like structure with two transmembrane helices separated by an ion-selective pore-forming unit. In such a model, M8 and M10 [5] are the best candidates for the hairpin, while M9 is the best candidate for the selectivity filter [20]. M5 and M6 and possibly additional hairpin helices, such as M7a and M7b, may contribute to the periphery of the pore structure [5a]. Although interactions among the triggers that open and close this pore must be very complex [8], two triggers stand out as being of special significance [21]. In skeletal muscle, voltage-induced changes in the conformation of the "voltage sensor" DHPR α1-subunit undoubtedly drive conformational changes in the cytoplasmic segment of RyR1 that are transmitted over long ranges to activate Ca^{2+} release. In most tissues, and even in skeletal muscle, elevations in cytosolic Ca^{2+} trigger Ca^{2+} induced Ca^{2+} release. Indeed, Ca^{2+} may be the master trigger and the ability of all other agents, including protein-protein interactions, to activate the Ca^{2+} release channel may simply reflect an agonist-induced increase in the affinity of an RyR molecule for binding of Ca^{2+} to its trigger sites. A strong candidate for the site for binding of trigger Ca^{2+} is the "Ca^{2+} sensor" amino acid, Glu^{4032} (Glu^{3885} in RyR3), located within a hydrophobic sequence predicted earlier to form transmembrane helix M2 [22]. Other sites for Ca^{2+} binding might be located elsewhere [23].

ATP is a potent activator of the Ca^{2+} release channel in the presence of Ca^{2+} [3], but since cellular ATP concentrations

corners most distal from the membrane sector are in an open conformation and are slightly straightened toward the surface of the T-tubule membrane [13]. This feature supports the postulate that the four corners of the molecule interact with a specific protein in the transverse tubule. The N-terminus of RyR3 is located in the clamp region [14].

are rather constant, ATP is not likely to play a major regulatory role. The site of ATP binding is not defined. A transmembrane redox sensor exists within the RyR1 channel complex that confers tight regulation of channel activity in response to changes in transmembrane redox potential [24]. PO2 dynamically controls the redox state of several thiols in each RyR1 subunit and thereby tunes its response to NO [25]. At physiological pO2, nanomolar NO activates the channel by S-nitrosylating a single cysteine residue. S-nitrosylation is specific to RyR1 and its effect on the channel is CaM-dependent.

Caffeine activates Ca^{2+} release but appears to do so by increasing Ca^{2+} sensitivity [26]; ryanodine can drive the channel into an open subconductance state, but this state is Ca^{2+} dependent, with an exceptionally high Ca^{2+} affinity [27,28]. While ryanodine binds to C-terminal sequences [29], the binding site for caffeine is unknown. Most MH mutations alter the apparent affinity of the channel for caffeine and halothane [30], but these mutations are dispersed throughout two "hot spots" in the cytosolic domain [31] and one in the C-terminus [32]. The binding site for dantrolene, which closes the channel, is also not well defined [33].

FKBP [34], triadin [35], junctin[36], CaM [37], sorcin [38], and various protein kinases [39,40] may also regulate the function of ryanodine receptors. The interaction of FKBP with RyR increases channels to full conductance, decreases open probability after caffeine activation, increases mean open time, and coordinates opening of clusters of channels [11]. These observations, together with the 1:1 stoichiometry of FKBP12 with RyR1, suggest that FKBP is an RyR subunit. CaM is both an inhibitor and an activator of Ca^{2+} channel activity [37]. Triadin and junctin, which have single transmembrane sequences and positively charged lumenal sequences, form links to the lumenal, negatively charged Ca^{2+} buffering protein calsequestrin, so that RyR1, triadin, junctin, and calsequestrin form a quaternary complex that may be required for normal Ca^{2+} release. Sorcin acts as an inhibitor of RyR function. Phosphorylation of RyR1 at Ser^{2843} enhances open probability by increasing the sensitivity to Ca^{2+} and ATP [41]. Phosphorylation of Ser^{2809} in RyR2 by CaM kinase II reverses inhibition by CaM and restores prolonged channel opening [42]. Thus the RyR can be viewed as a massive protein with multiple protein and ligand binding sites that is designed to integrate complex signals for activation and inactivation from many different sites in the molecule [8].

Molecular Biology of Ryanodine Receptors

Three ryanodine receptors (*RYR*) genes have been identified: *RYR1* on human chromosome 19q13.1; *RYR2* on 1q42.1-43; and *RYR3* on 15q14-15 [31]. *RYR1* is expressed predominantly in fast and slow-twitch skeletal muscle and also in the esophagus and in cerebellar Purkinje cells in the brain. *RYR2* is the predominant isoform in cardiac muscle and brain. Its expression in the brain, brain stem, and spinal cord is widespread, but it is absent from the pituitary. *RYR3* is differentially expressed in the brain, T-lymphocytes, vas deferens, uterus, and testes [43]. It accounts for a small percentage of total *RYR* expression in mammalian skeletal muscle but is highly expressed in avian and amphibian skeletal muscles. RyR3 may flank RyR1 in the junctional terminal cisternae [44].

The disruption of *RYR1* is neonatally lethal [45]. The mutant mice resemble the mouse mutant *mdg*, which results from disruption in the DHPR α1-subunit gene *CACNA1S*, in that both disruptions lead to failure of excitation-contraction coupling in skeletal muscle [46]. The disruption of *RYR2* is lethal at embryonic day 10, with morphological abnormalities in the heart tube [47]. The disruption of *RYR3* does not cause gross abnormalities in mice, although *RYR3*-null mice have abnormal locomotor activity [48].

Mutations in *RYR1* cause MH, an autosomal dominant genetic abnormality in which susceptible individuals respond to potent inhalational anesthetics and depolarizing skeletal muscle relaxants with hypermetabolism, skeletal muscle rigidity, fever, and muscle cell damage [31]. Mutations in *RYR1* also cause central core disease (CCD), an autosomal dominant myopathy characterized by hypotonia and proximal muscle weakness. Central cores of skeletal muscle fibers lack oxidative or phosphorylase activity, and electron microscopy of the cores shows disintegration of the contractile apparatus and streaming of the Z lines, an increase in content of the sarcotubular system and depletion of mitochondria. CCD is usually closely associated with MH, but an exception has been found [49].

MH and CCD mutations are clustered in *RYR1* exons 2 to 17 (region 1), 34–46 (region 2), and 91–102 (region 3) [31,32]. The ratio of MH to CCD mutations in region 1 is 5:1, in region 2, 8:1, and in region 3, 1:8. MH mutations are more sensitive to caffeine and halothane activation than wild-type and are more "leaky": CCD mutant proteins are even more leaky than MH mutations [50,51]. CCD mutations may cause a more severe imbalance in Ca^{2+} regulation than those that cause MH. Elevation of resting Ca^{2+} by a very leaky CCD mutant channel may trigger the series of degenerative and compensatory events that lead to core formation in the center of the fiber without affecting the periphery of the muscle cell where Ca^{2+} homeostasis can be achieved through the intervention of plasma membrane Ca^{2+} pumps and exchangers. However, at least one CCD mutation is not leaky but, rather, uncouples excitation-contraction coupling by disrupting orthograde signaling between the DHPR and RyR1 proteins without disrupting retrograde signaling between these two proteins [52]. Mutations in the *CACNA1S* gene encoding the α1-subunit of the skeletal muscle DHPR have also been linked to MH, providing further support for strong functional interactions between these two proteins [53].

Mutations in *RYR2* have been linked to two autosomal dominant cardiac diseases [54,55]: catecholaminergic polymorphic ventricular tachycardia (CPVT), which occurs in response to stress and in the absence of either structural heart disease or prolonged QT interval; and arrhythmogenic right ventricular cardiomyopathy type 2 (ARVD2), which is characterized by partial degeneration of the myocardium of the right ventricle, electrical instability, and sudden death. The mutations are located in regions of the gene that correspond to MH regions 1 and 2 in *RYR1*. Since *RYR2* is not expressed in skeletal muscle, neither MH nor CCD manifest in these diseases. As a corollary, cardiac disease is not associated with MH or CCD mutations, since *RYR1* is not expressed in the heart. It is probable that CPVT and ARVD2, like MH and CCD, are differentiated on the basis of the severity of the alteration in RyR2 channel function.

References

1. Bers, D. M. (2001). *Excitation-Contraction Coupling and Cardiac Contractile Force*, Kluwer Academic Press, Amsterdam.
2. Coronado, R., Morrissette, J., Sukhareva, M., and Vaughan, D. M. (1994). Structure and function of ryanodine receptors. *Am. J. Physiol.* **266**, C1485–C1504.
3. Meissner, G. (1994). Ryanodine receptor/Ca^{2+} release channels and their regulation by endogenous effectors. *Annu. Rev. Physiol.* **56**, 485–508.
4. Takeshima, H., Nishimura, S., Matsumoto, T., Ishida, H., Kangawa, K., Minamino, N., Matsuo, H., Ueda, M., Hanaoka, M., Hirose, T., et al. (1989). Primary structure and expression from complementary DNA of skeletal muscle ryanodine receptor. *Nature* **339**, 439–445.
5. Zorzato, F., Fujii, J., Otsu, K., Phillips, M., Green, N. M., Lai, F. A., Meissner, G., and MacLennan, D. H. (1990). Molecular cloning of cDNA encoding human and rabbit forms of the Ca^{2+} release channel (ryanodine receptor) of skeletal muscle sarcoplasmic reticulum. *J. Biol. Chem.* **265**, 2244–2256.
5a. Du, G. G., Sandhu, B., Khanna, V. K., Guo, X., and MacLennan, D. H. (2002). Topology of the Ca^{2+} release channel of skeletal muscle sarcoplasmic reticulum (RyR1). *Proc. Natl. Acad. Sci. USA* **99**, 16725–16730.
6. Radermacher, M., Rao, V., Grassucci, R., Frank, J., Timerman, A. P., Fleischer, S., and Wagenknecht, T. (1994). Cryo-electron microscopy and three-dimensional reconstruction of the calcium release channel/ryanodine receptor from skeletal muscle. *J. Cell. Biol.* **127**, 411–423.
7. Stokes, D. L., and Wagenknecht, T. (2000). Calcium transport across the sarcoplasmic reticulum: structure and function of Ca^{2+}-ATPase and the ryanodine receptor. *Eur. J. Biochem.* **267**, 5274–5279.
8. Franzini-Armstrong, C., and Protasi, F. (1997). Ryanodine receptors of striated muscles: a complex channel capable of multiple interactions. *Physiol. Rev.* **77**, 699–729.
9. Wagenknecht, T., Radermacher, M., Grassucci, R., Berkowitz, J., Xin, H. B., and Fleischer, S. (1997). Locations of calmodulin and FK506-binding protein on the three-dimensional architecture of the skeletal muscle ryanodine receptor. *J. Biol. Chem.* **272**, 32463–32471.
10. Moore, C. P., Rodney, G., Zhang, J. Z., Santacruz-Toloza, L., Strasburg, G., and Hamilton, S. L. (1999). Apocalmodulin and Ca^{2+} calmodulin bind to the same region on the skeletal muscle Ca^{2+} release channel. *Biochemistry* **38**, 8532–8537.
11. Gaburjakova, M., Gaburjakova, J., Reiken, S., Huang, F., Marx, S. O., Rosemblit, N., and Marks, A. R. (2001). FKBP12 binding modulates ryanodine receptor channel gating. *J. Biol. Chem.* **276**, 16931–16935.
12. Serysheva, II, Schatz, M., van Heel, M., Chiu, W., and Hamilton, S. L. (1999). Structure of the skeletal muscle calcium release channel activated with Ca^{2+} and AMP-PCP. *Biophys. J.* **77**, 1936–1944.
13. Orlova, E. V., Serysheva, II, van Heel, M., Hamilton, S. L., and Chiu, W. (1996). Two structural configurations of the skeletal muscle calcium release channel. *Nat. Struct. Biol.* **3**, 547–552.
14. Liu, Z., Zhang, J., Sharma, M. R., Li, P., Chen, S. R., and Wagenknecht, T. (2001). Three-dimensional reconstruction of the recombinant type 3 ryanodine receptor and localization of its amino terminus. *Proc. Natl. Acad. Sci. USA* **98**, 6104–6109.
15. Block, B. A., Imagawa, T., Campbell, K. P., and Franzini-Armstrong, C. (1988). Structural evidence for direct interaction between the molecular components of the transverse tubule/sarcoplasmic reticulum junction in skeletal muscle. *J. Cell. Biol.* **107**, 2587–2600.
16. Nakai, J., Dirksen, R. T., Nguyen, H. T., Pessah, I. N., Beam, K. G., and Allen, P. D. (1996). Enhanced dihydropyridine receptor channel activity in the presence of ryanodine receptor. *Nature* **380**, 72–75.
17. Proenza, C., O'Brien, J., Nakai, J., Mukherjee, S., Allen, P. D., and Beam, K. G. (2002). Identification of a region of RyR1 that participates in allosteric coupling with the alpha(1S) (Ca(V)1.1) II-III loop. *J. Biol. Chem.* **277**, 6530–6535.
18. Nabauer, M., Callewaert, G., Cleemann, L., and Morad, M. (1989). Regulation of calcium release is gated by calcium current, not gating charge, in cardiac myocytes. *Science* **244**, 800–803.
19. Doyle, D. A., Morais Cabral, J., Pfuetzner, R. A., Kuo, A., Gulbis, J. M., Cohen, S. L., Chait, B. T., and MacKinnon, R. (1998). The structure of the potassium channel: molecular basis of K+ conduction and selectivity. *Science* **280**, 69–77.
20. Balshaw, D., Gao, L., and Meissner, G. (1999). Luminal loop of the ryanodine receptor: a pore-forming segment? *Proc. Natl. Acad. Sci. USA* **96**, 3345–3347.
21. Ebashi, S. (1991). Excitation-contraction coupling and the mechanism of muscle contraction. *Annu. Rev. Physiol.* **53**, 1–16.
22. Chen, S. R., Ebisawa, K., Li, X., and Zhang, L. (1998). Molecular identification of the ryanodine receptor Ca^{2+} sensor. *J. Biol. Chem.* **273**, 14675–14678.
23. Chen, S. R. and MacLennan, D. H. (1994). Identification of calmodulin-, Ca^{2+}-, and ruthenium red-binding domains in the Ca^{2+} release channel (ryanodine receptor) of rabbit skeletal muscle sarcoplasmic reticulum. *J. Biol. Chem.* **269**, 22698–22704.
24. Feng, W., Liu, G., Allen, P. D., and Pessah, I. N. (2000). Transmembrane redox sensor of ryanodine receptor complex. *J. Biol. Chem.* **275**, 35902–35907.
25. Eu, J. P., Sun, J., Xu, L., Stamler, J. S., and Meissner, G. (2000). The skeletal muscle calcium release channel: coupled O_2 sensor and NO signaling functions. *Cell* **102**, 499–509.
26. Herrmann-Frank, A., Luttgau, H. C., and Stephenson, D. G. (1999). Caffeine and excitation-contraction coupling in skeletal muscle: a stimulating story. *J. Muscle Res. Cell. Motil.* **20**, 223–237.
27. Du, G. G., Guo, X., Khanna, V. K., and MacLennan, D. H. (2001). Ryanodine sensitizes the cardiac Ca^{2+} release channel (ryanodine receptor isoform 2) to Ca^{2+} activation and dissociates as the channel is closed by Ca^{2+} depletion. *Proc. Natl. Acad. Sci. USA* **98**, 13625–13630.
28. Masumiya, H., Li, P., Zhang, L., and Chen, S. R. (2001). Ryanodine sensitizes the Ca^{2+} release channel (ryanodine receptor) to Ca^{2+} activation. *J. Biol. Chem.* **276**, 39727–39735.
29. Callaway, C., Seryshev, A., Wang, J. P., Slavik, K. J., Needleman, D. H., Cantu, C., 3rd, Wu, Y., Jayaraman, T., Marks, A. R., and Hamilton, S. L. (1994). Localization of the high and low affinity [3H]ryanodine binding sites on the skeletal muscle Ca^{2+} release channel. *J. Biol. Chem.* **269**, 15876–15884.
30. Tong, J., Oyamada, H., Demaurex, N., Grinstein, S., McCarthy, T. V., and MacLennan, D. H. (1997). Caffeine and halothane sensitivity of intracellular Ca^{2+} release is altered by 15 calcium release channel (ryanodine receptor) mutations associated with malignant hyperthermia and/or central core disease. *J. Biol. Chem.* **272**, 26332–26339.
31. Loke, J. and MacLennan, D. H. (1998). Malignant hyperthermia and central core disease: disorders of Ca^{2+} release channels. *Am. J. Med.* **104**, 470–486.
32. Monnier, N., Romero, N. B., Lerale, J., Landrieu, P., Nivoche, Y., Fardeau, M., and Lunardi, J. (2001). Familial and sporadic forms of central core disease are associated with mutations in the C-terminal domain of the skeletal muscle ryanodine receptor. *Hum. Mol. Genet.* **10**, 2581-9252.
33. Paul-Pletzer, K., Palnitkar, S. S., Jimenez, L. S., Morimoto, H., and Parness, J. (2001). The skeletal muscle ryanodine receptor identified as a molecular target of [3H]azidodantrolene by photoaffinity labeling. *Biochemistry* **40**, 531–542.

34. Jayaraman, T., Brillantes, A. M., Timerman, A. P., Fleischer, S., Erdjument-Bromage, H., Tempst, P., and Marks, A. R. (1992). FK506 binding protein associated with the calcium release channel (ryanodine receptor). *J. Biol. Chem.* **267**, 9474–9477.
35. Guo, W. and Campbell, K. P. (1995). Association of triadin with the ryanodine receptor and calsequestrin in the lumen of the sarcoplasmic reticulum. *J. Biol. Chem.* **270**, 9027–9030.
36. Jones, L. R., Zhang, L., Sanborn, K., Jorgensen, A. O., and Kelley, J. (1995). Purification, primary structure, and immunological characterization of the 26-kDa calsequestrin binding protein (junctin) from cardiac junctional sarcoplasmic reticulum. *J. Biol. Chem.* **270**, 30787–30796.
37. Tripathy, A., Xu, L., Mann, G., and Meissner, G. (1995). Calmodulin activation and inhibition of skeletal muscle Ca^{2+} release channel (ryanodine receptor). *Biophys. J.* **69**, 106–119.
38. Meyers, M. B., Pickel, V. M., Sheu, S. S., Sharma, V. K., Scotto, K. W., and Fishman, G. I. (1995). Association of sorcin with the cardiac ryanodine receptor. *J. Biol. Chem.* **270**, 26411–26418.
38a. Lokuta, A. J., Meyers, M. B., Sander, P. R., Fishman, G. I., and Valdivia, H. H. (1997). Modulation of cardiac ryanodine receptors by sorcin. *J. Biol. Chem.* **272**, 25333–25338.
39. Yang, J., Drazba, J. A., Ferguson, D. G., and Bond, M. (1998). A-kinase anchoring protein 100 (AKAP100) is localized in multiple subcellular compartments in the adult rat heart. *J. Cell. Biol.* **142**, 511–522.
40. Antos, C. L., Frey, N., Marx, S. O., Reiken, S., Gaburjakova, M., Richardson, J. A., Marks, A. R., and Olson, E. N. (2001). Dilated cardiomyopathy and sudden death resulting from constitutive activation of protein kinase A. *Circ. Res.* **89**, 997–1004.
41. Hain, J., Nath, S., Mayrleitner, M., Fleischer, S., and Schindler, H. (1994). Phosphorylation modulates the function of the calcium release channel of sarcoplasmic reticulum from skeletal muscle. *Biophys. J.* **67**, 1823–1833.
42. Hain, J., Onoue, H., Mayrleitner, M., Fleischer, S., and Schindler, H. (1995). Phosphorylation modulates the function of the calcium release channel of sarcoplasmic reticulum from cardiac muscle. *J. Biol. Chem.* **270**, 2074–2081.
43. Giannini, G., Conti, A., Mammarella, S., Scrobogna, M., and Sorrentino, V. (1995). The ryanodine receptor/calcium channel genes are widely and differentially expressed in murine brain and peripheral tissues. *J. Cell. Biol.* **128**, 893–904.
44. Protasi, F., Takekura, H., Wang, Y., Chen, S. R., Meissner, G., Allen, P. D., and Franzini-Armstrong, C. (2000). RYR1 and RYR3 have different roles in the assembly of calcium release units of skeletal muscle. *Biophys. J.* **79**, 2494–2508.
45. Takeshima, H., Iino, M., Takekura, H., Nishi, M., Kuno, J., Minowa, O., Takano, H., and Noda, T. (1994). Excitation-contraction uncoupling and muscular degeneration in mice lacking functional skeletal muscle ryanodine-receptor gene. *Nature* **369**, 556–559.
46. Beam, K. G., Knudson, C. M., and Powell, J. A. (1986). A lethal mutation in mice eliminates the slow calcium current in skeletal muscle cells. *Nature* **320**, 168–170.
47. Takeshima, H., Komazaki, S., Hirose, K., Nishi, M., Noda, T., and Iino, M. (1998). Embryonic lethality and abnormal cardiac myocytes in mice lacking ryanodine receptor type 2. *Embo. J.* **17**, 3309–3316.
48. Takeshima, H., Ikemoto, T., Nishi, M., Nishiyama, N., Shimuta, M., Sugitani, Y., Kuno, J., Saito, I., Saito, H., Endo, M., Iino, M., and Noda, T. (1996). Generation and characterization of mutant mice lacking ryanodine receptor type 3. *J. Biol. Chem.* **271**, 19649–19652.
49. Lynch, P. J., Tong, J., Lehane, M., Mallet, A., Giblin, L., Heffron, J. J., Vaughan, P., Zafra, G., MacLennan, D. H., and McCarthy, T. V. (1999). A mutation in the transmembrane/luminal domain of the ryanodine receptor is associated with abnormal Ca^{2+} release channel function and severe central core disease. *Proc. Natl. Acad. Sci. USA* **96**, 4164–4169.
50. Tong, J., McCarthy, T. V., and MacLennan, D. H. (1999). Measurement of resting cytosolic Ca^{2+} concentrations and Ca^{2+} store size in HEK-293 cells transfected with malignant hyperthermia or central core disease mutant Ca^{2+} release channels. *J. Biol. Chem.* **274**, 693–702.
51. Avila, G. and Dirksen, R. T. (2001). Functional effects of central core disease mutations in the cytoplasmic region of the skeletal muscle ryanodine receptor. *J. Gen. Physiol.* **118**, 277–290.
52. Avila, G., O'Brien, J. J., and Dirksen, R. T. (2001). Excitation-contraction uncoupling by a human central core disease mutation in the ryanodine receptor. *Proc. Natl. Acad. Sci. USA* **98**, 4215–4220.
53. Monnier, N., Procaccio, V., Stieglitz, P., and Lunardi, J. (1997). Malignant-hyperthermia susceptibility is associated with a mutation of the alpha 1-subunit of the human dihydropyridine-sensitive L-type voltage-dependent calcium-channel receptor in skeletal muscle. *Am. J. Hum. Genet.* **60**, 1316–1325.
54. Priori, S. G., Napolitano, C., Tiso, N., Memmi, M., Vignati, G., Bloise, R., Sorrentino, V. V., and Danieli, G. A. (2001). Mutations in the cardiac ryanodine receptor gene (hRyR2) underlie catecholaminergic polymorphic ventricular tachycardia. *Circulation* **103**, 196–200.
55. Tiso, N., Stephan, D. A., Nava, A., Bagattin, A., Devaney, J. M., Stanchi, F., Larderet, G., Brahmbhatt, B., Brown, K., Bauce, B., Muriago, M., Basso, C., Thiene, G., Danieli, G. A., and Rampazzo, A. (2001). Identification of mutations in the cardiac ryanodine receptor gene in families affected with arrhythmogenic right ventricular cardiomyopathy type 2 (ARVD2). *Hum. Mol. Genet.* **10**, 189–194.

CHAPTER 131

Intracellular Calcium Signaling

Martin D. Bootman, H. Llewelyn Roderick,
Rodney O'Connor, and Michael J. Berridge

*The Babraham Institute, Babraham,
Cambridge, United Kingdom*

The "Calcium Signaling Toolkit" and Calcium Homeostasis

Calcium (Ca^{2+}) is a ubiquitous intracellular messenger that controls a diverse range of cellular processes, such as gene transcription, muscle contraction, and cell proliferation. The ability of Ca^{2+} to play a pivotal role in cell biology results from the facility that cells have to shape Ca^{2+} signals in the dimensions of space, time, and amplitude. To generate and interpret the variety of observed Ca^{2+} signals, different cell types employ components selected from a "Ca^{2+} signaling toolkit," which comprises an array of homeostatic and sensory mechanisms (reviewed in [1]). Since many of the molecular components of this toolkit have multiple isoforms with subtly different properties, each specific cell type can exploit this large repertoire to construct highly versatile Ca^{2+} signaling networks. Thus by mixing and matching components from the toolkit, cells can obtain Ca^{2+} signals that suit their physiology.

In most cells, Ca^{2+} has its major signaling function when it is elevated in the cytosolic compartment. From there it can also diffuse into organelles such mitochondria and the nucleus. The Ca^{2+} concentration inside cells is regulated by the simultaneous interplay of multiple counteracting processes, which can be divided into Ca^{2+} "on" and "off" mechanisms depending on whether they serve to increase or decrease cytosolic Ca^{2+} (reviewed in [2–4]) (Fig. 1).

The Ca^{2+} "on" mechanisms include channels located at the plasma membrane that regulate the supply of Ca^{2+} from the extracellular space, and channels on the endoplasmic reticulum/sarcoplasmic reticulum (ER/SR, respectively), Golgi, secretory granules, and acidic stores (e.g. lysosomes), which release the finite intracellular Ca^{2+} stores. The "off" mechanisms include Ca^{2+}ATPases on the plasma membrane and ER/SR, and exchangers that utilize the electrochemical Na^{+} gradient to provide the energy to transport Ca^{2+} out of the cell. Occasionally, some of the "off" mechanisms contribute to cytosolic Ca^{2+} increases; examples are "slippage" of Ca^{2+} through Ca^{2+}ATPases and reverse-mode Na^{+}/Ca^{2+} exchange.

When cells are at rest, the balance lies in favour of the 'off' mechanisms, thus yielding an intracellular Ca^{2+} concentration of ~100 nM. However, when cells are stimulated by various means, e.g. depolarisation, mechanical deformation or hormones, the 'on' mechanisms are activated and the cytosolic Ca^{2+} concentration increases to levels of 1 μM or more.

As mentioned above, Ca^{2+} signals can be modulated in their temporal, amplitude, and spatial dimensions. Furthermore, Ca^{2+} signals can arise from different cellular sources, which appear to be regulated by a growing number of messengers (reviewed in [5]). The following sections describe the currently known messengers and channels and present examples of the versatility of Ca^{2+} signals.

Multiple Channels and Messengers Underlie Ca^{2+} Increases

Ca^{2+} Influx Channels

Cells utilize several different types of Ca^{2+} influx channels, which can be grouped on the basis of their activation mechanisms (reviewed in [2]). *Voltage-operated Ca^{2+} channels* (VOCs) are employed largely by excitable cell types such as muscle and neuronal cells, where they are activated by depolarization of the plasma membrane. Different types of VOCs,

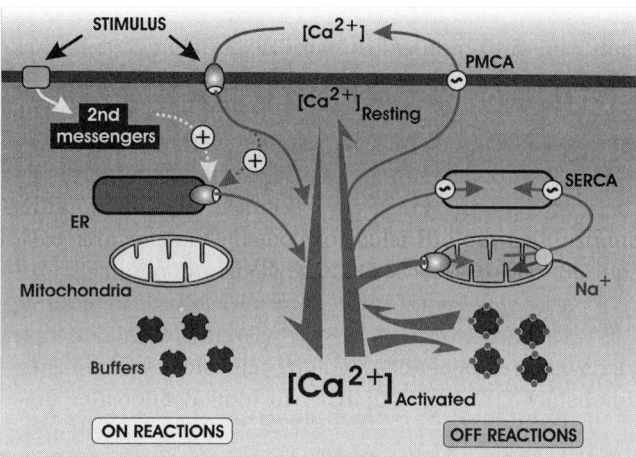

Figure 1 Calcium "on" and "off" mechanisms in cellular signaling and homeostasis. The figure illustrates various pathways and mechanisms by which cytosolic Ca^{2+} levels can increase or decline. At rest, cells generally have a free cytosolic Ca^{2+} concentration of around 100 nM. This can be increased by activation of channels at the plasma membrane and release from internal stores (denoted "ER"). Cytosolic Ca^{2+} signals are attenuated by passive Ca^{2+} buffering and actively reversed by mitochondria and Ca^{2+} ATPases on the ER and plasma membrane.

which are expressed in a tissue-specific manner, have been characterized on the basis of their gating characteristics and pharmacology. *Receptor-operated Ca^{2+} channels* (ROCs) comprise a range of structurally and functionally diverse channels that are particularly prevalent on secretory cells and at nerve terminals. Well-known ROCs include the nicotinic acetylcholine receptor and the N-methyl-D-aspartate (NMDA) receptor. ROCs are activated by the binding of an agonist to the extracellular domain of the channel. The different ROCs are activated by a wide variety of agonists, e.g. ATP, serotonin, glutamate, and acetylcholine. *Mechanically activated Ca^{2+} channels* are present on many cell types and respond to cell deformation. Such channels convey information into the cell concerning the stress/shape changes that a cell is experiencing. A particularly nice example of mechanically induced Ca^{2+} signaling was observed in epithelial cells from the trachea, where deformation of a single cell led to a radial Ca^{2+} wave that synchronized the Ca^{2+}-sensitive beating of cilia on many neighboring cells [6], which may serve to aid clearance of mucous or particles from the lungs. *Store-operated Ca^{2+} channels* (SOCs) are activated in response to depletion of the intracellular Ca^{2+} store. The mechanism by which the SOCs "sense" the filling status of the intracellular pool is unknown. At present, the best candidate for the molecular identity of SOCs are homologues of a protein named TRP (*t*ransient *r*eceptor *p*otential) that functions in *Drosophila* photoreception. Several mammalian TRP homologues have been identified and found to be expressed in almost all tissues (reviewed in [7]).

Ca^{2+} Release Channels

Inositol 1,4,5-trisphosphate Receptors (InsP$_3$Rs). The binding of many hormones and growth factors to specific receptors on the plasma membrane leads to the activation of an enzyme that catalyzes the hydrolysis of phospholipids to produce the intracellular messenger inositol 1,4,5-trisphosphate (InsP$_3$). InsP$_3$ is water-soluble and diffuses into the cell interior where it can engage InsP$_3$Rs on the ER/SR, allowing the Ca^{2+} stored at high concentrations to enter the cytoplasm. Three different isoforms of InsP$_3$Rs have been found, which appear to subtly differ in their characteristics, such as affinity for InsP$_3$. An important feature of InsP$_3$Rs is that they are actually co-regulated by InsP$_3$ and Ca^{2+}. Indeed, it seems that InsP$_3$ may simply serve to make InsP$_3$Rs responsive to an activating Ca^{2+} signal. InsP$_3$R opening is biphasically regulated by Ca^{2+}; 0.1–0.5 µM Ca^{2+} increases channel activity, whereas greater Ca^{2+} concentrations inhibit their gating (reviewed in [8,9]). This dependence of InsP$_3$R activity on cytosolic Ca^{2+} is crucial in the generation of the complex patterns of Ca^{2+} signals seen in many cells.

Ryanodine Receptors (RyRs) These receptors are structurally and functionally analogous to InsP$_3$Rs, although they have approximately twice the conductance and molecular mass of InsP$_3$Rs. Another property that RyRs share with InsP$_3$Rs is their sensitivity to cytosolic Ca^{2+} concentrations, although they are generally activated and inhibited by higher concentrations (activation at 1–10 µM; inhibition at >10 µM). In contrast to InsP$_3$Rs, which are almost ubiquitously expressed in mammalian tissues, RyRs are largely present in excitable cell types, such as muscle and neurons. As with InsP$_3$Rs, RyR subunits are encoded by three genes. However, these genes do not appear to have the same functional redundancy as observed with the InsP$_3$R isoforms. Instead, the different RyR proteins are often used for specific functions. For example, only type-1 RyRs are employed in triggering excitation of skeletal muscle, whereas only type-2 RyRs fulfil this role in cardiac muscle (reviewed in [1,10]).

Multiple Messengers

A wide range of messengers has been shown to mediate the activation of Ca^{2+} entry and Ca^{2+} release channels (reviewed in [5]). These messengers include InsP$_3$, cyclic adenosine 5′-diphosphoribose (cADPR), nitric oxide (NO), H$_2$O$_2$/O$_2^-$, nicotinic acid adenine dinucleotide phosphate (NAADP), diacylglycerol, arachidonic acid, sphingosine, sphingosine-1-phosphate (S-1-P), leukotrienes, and Ca^{2+} itself. From the specificities of the Ca^{2+}-releasing messengers we know that there must be several different types of intracellular Ca^{2+} release channel, although at present only InsP$_3$ receptors (InsP$_3$Rs) and ryanodine receptors (RyRs) have been characterized in detail. Exactly which messengers act in particular cells is far from clear. However, it is becoming apparent that Ca^{2+} signals can be activated by the simultaneous interplay of several factors. In pancreatic acinar cells, for example, the combined action of InsP$_3$, cADPR, and NAADP underlies the Ca^{2+} signals generated by physiological stimuli [11].

Temporal Regulation of Ca^{2+} Signals

Since prolonged elevation of cytoplasmic Ca^{2+} levels can be toxic, most cells do not usually respond with sustained Ca^{2+} signals. Rather, Ca^{2+} is commonly presented in a pulsatile manner [12,13]. A well-known example of such repetitive Ca^{2+} increases is the response of hepatocytes to stimulation with various hormones that, for example, regulate glycogen metabolism and mitochondrial respiration [14]. The Ca^{2+} increases that occur in hepatocytes during such stimulations are transient spikes, which arise from the cyclical activation of $InsP_3Rs$ (see below). The frequency of the Ca^{2+} spikes is directly proportional to the concentration of hormone applied to the cells, and they can persist for the duration of agonist application. These Ca^{2+} spikes are therefore essentially a frequency-modulated digital read-out of cell stimulation.

Another system in which pulsatile Ca^{2+} increases are critical is excitation-contraction coupling in striated muscle cells. The mechanism and channels underlying these signals are very different from those in hepatocytes. In cardiac muscle, for example, type 2 RyRs are activated following depolarization of the sarcolemma. The Ca^{2+} that enters the cell following voltage-operated Ca^{2+} channel activation triggers release of Ca^{2+} from ryanodine receptors (RyRs) by a process known as Ca^{2+}-induced Ca^{2+} release (CICR) [1]. This Ca^{2+} can then globally diffuse to the myofibrils and promote the interaction between actin and myosin that leads to contraction.

Although the Ca^{2+} signals observed in both hormonally stimulated hepatocytes and cardiomyocytes are repetitive Ca^{2+} transients, the periodicity and kinetics of these signals are very different. The Ca^{2+} spikes in hepatocytes (and many other nonelectrically excitable cells) typically have frequencies in the range of 0.1–0.01 Hz, a time-to-peak amplitude of several seconds, and a recovery phase lasting tens of seconds [15]. In contrast, cardiac Ca^{2+} signals are triggered at frequencies in the 1–10 Hz range (depending on the species of animal), reach peak within a few tens of milliseconds, and persist for only a few hundred milliseconds [16]. The distinct timescales of these Ca^{2+} responses reflect the very different mechanisms by which they are generated.

Pulsatile Ca^{2+} increases, such as those observed in hepatocytes, are generally considered to have a much higher fidelity of information transfer than simple tonic changes in Ca^{2+} concentrations, since they are much less prone to noisy fluctuations. The major sensors for these Ca^{2+} spikes are Ca^{2+} binding proteins such as calmodulin (reviewed in [17]). This ubiquitous protein is one of a family of proteins bearing structural Ca^{2+}-binding motifs known as EF-hands. The binding of Ca^{2+} to calmodulin has a K_d around 1 μM, making it an ideal receiver for the rapid transient Ca^{2+} increases seen with each spike. One of the best-known enzymes that uses calmodulin to help it "count" Ca^{2+} spikes is calmodulin-dependent protein kinase II, which can activate other proteins via phosphorylation. This enzyme is composed of many subunits that undergo variable degrees of activation depending on the frequency of Ca^{2+} spikes. Essentially, increasing the frequency or duration of Ca^{2+} spikes maintains this enzyme in an active state by trapping calmodulin and causing autophosphorylation [18,19].

There are many examples of cellular activities being modulated by the frequency of Ca^{2+} signals. The transcription of Ca^{2+}-regulated genes was sensitive to the frequency at which Ca^{2+} spikes occurred. Indeed, it appears that alternative transcription factors are tuned to distinct frequencies of Ca^{2+} spikes. Thus temporal modulation of Ca^{2+} signaling can underlie differential gene transcription [20].

Spatial Regulation of Ca^{2+} Signals

All Ca^{2+} signals derive initially from local sources such as the activation of Ca^{2+} channels. Both Ca^{2+} entry and Ca^{2+} release channels can give rise to brief pulses of Ca^{2+} that form a small plume around the mouth of the channel before diffusing into the cytoplasm (reviewed in [21]). In many situations, these local sources provoke global signals through regenerative CICR as described for cardiomyocytes above. However, there are numerous instances in which the spread of Ca^{2+} is constrained to a specific subcellular region. Such spatial regulation of Ca^{2+} provides perhaps the most elegant examples of how cells subtly modulate Ca^{2+} signals to control multiple, sometimes opposing, processes.

Nonelectrically Excitable Cells

A few different types of localized Ca^{2+} signals have been observed in various nonelectrically excitable cell types [22]. Probably the best know examples are "Ca^{2+} puffs" and the apical Ca^{2+} signals that occur in secretory cells (reviewed in [3]). Ca^{2+} puffs are local signals that derive from the activation of a cluster of $InsP_3Rs$. Typically Ca^{2+} puffs give a modest elevation of cytosolic Ca^{2+} (~50–600 nM) with a limited spatial spread (~2–6 μm) and are transient (duration of ~1 second) [23]. Such events were first observed in *Xenopus* oocytes [e.g. 24] but have subsequently been observed in many other cell types. The temporally and spatially coordinated recruitment of Ca^{2+} puffs is responsible for the generation of repetitive Ca^{2+} waves and oscillations observed during hormonal stimulation. Essentially Ca^{2+} waves reflect the progressive release of Ca^{2+} by Ca^{2+} puff sites distributed along the ER/SR. Ca^{2+} released by one puff site can diffuse to a neighboring site and activate it (providing $InsP_3$ is bound to the channels). Successive rounds of Ca^{2+} release and diffusion allow the initially local Ca^{2+} puffs to trigger global Ca^{2+} waves and oscillations (reviewed in [3,25]).

It is interesting that in HeLa cells [26] and *Xenopus* oocytes [27], it has been demonstrated that Ca^{2+} puff sites expressing a higher sensitivity to $InsP_3$ consistently trigger Ca^{2+} waves. What gives these pacemaking Ca^{2+} puff sites their enhanced sensitivity is unclear. In the case of somatic cells, the pacemaker sites tend to be distributed in a perinuclear region [28], thus raising the possibility that they can send signals specifically into the nucleus. In *Xenopus* oocytes, it has been shown that mitochondria can constrain the activity of Ca^{2+} puffs, and locations lacking these organelles may thus define pacemaking sites [29].

Another well-known local Ca^{2+} signal occurs in the apical region of secretory cells such as pancreatic acinar cells (reviewed in [30]). Similar to the pacemaker Ca^{2+} puffs (see above), the $InsP_3Rs$ that underlie the apical Ca^{2+} spikes are distinguished by a heightened sensitivity to $InsP_3$. Also like Ca^{2+} puffs, such apical Ca^{2+} spikes probably arise from the coordinated Ca^{2+} release from multiple Ca^{2+} release channels. Recent evidence has pointed to the apical spikes arising from a stimulus-dependent hierarchical activation of different types of Ca^{2+} release channel [11]. With low levels of cell stimulation, the Ca^{2+} spikes stay restricted to the apical pole of the acinar cells, where they can activate ion channels and trigger limited secretion. Greater stimulation causes the Ca^{2+} spikes to trigger Ca^{2+} waves that propagate toward the basal pole. It appears that the restriction of the Ca^{2+} signal in the apical pole is due in part to a "firewall" of mitochondria that buffer Ca^{2+} as it diffuses from the apical pole and prevent the activation of RyRs in the basal pole [31,32].

Electrically Excitable Cells

Spatial regulation of Ca^{2+} signaling is the forte of electrically excitable cells. For more detailed discussions, the reader is referred to recent reviews [3,33–35].

One of the best-known examples in which spatial regulation of Ca^{2+} signals can have diametrically opposing effects in the same cell is in the regulation of smooth muscle tone (reviewed in [35]). In these cells, global responses induce contraction by the activation of Ca^{2+}/calmodulin-dependent enzymes, whereas local subsarcolemmal Ca^{2+} signals promote relaxation by activating Ca^{2+}-dependent plasma membrane ion channels [36]. The subsarcolemmal Ca^{2+} signals are known as Ca^{2+} sparks; they are analogous to the Ca^{2+} puffs observed in nonelectrically excitable cells, but they arise from the activation of a cluster of RyRs. Ca^{2+} sparks are generally faster in onset and decline than Ca^{2+} puffs and have usually a more restricted spread (~1–3 µm).

In smooth muscle, the subsarcolemmal Ca^{2+} sparks activate K^+ and Cl^- conductances, giving rise to brief currents known as STOCs (spontaneous transient outward currents; K^+ current), STICs (spontaneous transient inward current; Cl^- current) and STOICs (mixed K^+ and Cl^- currents). STOCs have been measured in a wide variety of smooth muscle cell types and serve to hyperpolarize the cell membrane by ~20 mV, thus causing the muscle to relax. STOCs primarily arise due to the activation of large conductance Ca^{2+}-activated K^+ channels (BK channels). These Ca^{2+}-activated channels have a low sensitivity to cytosolic Ca^{2+}, requiring concentrations >1 µM for significant activity. It has been proposed that the BK channels sit in close apposition to Ca^{2+} spark sites and sense rapid step-like Ca^{2+} changes during RyR activation [37].

Spatio-temporal recruitment of Ca^{2+} sparks also underlies the global Ca^{2+} signals that activate skeletal and cardiac myocyte contraction. When the sarcolemma of these cells is depolarized by an action potential, VOCs open and allow a small influx of Ca^{2+} [38]. Through the process of CICR, this trigger Ca^{2+} signal is greatly amplified by clusters of closely apposed RyRs, thereby activating Ca^{2+} spark sites throughout the cell. The spatial overlap and temporal summation of the Ca^{2+} sparks gives rise to the global responses that ensure synchronized contraction in the muscle (reviewed in [3,34]).

The intricate morphologies of neurons means that these cells are well-suited to producing spatially regulated Ca^{2+} signals. Local Ca^{2+} changes in dendritic spines can underlie processes such as synaptic plasticity and neurite outgrowth, whereas more global Ca^{2+} signals can cause gene transcription and neuronal maturation within the brain (reviewed in [39–41]). An example of the necessity for precise spatial regulation of neuronal Ca^{2+} signals can be seen in the effects of activating synaptic versus nonsynaptic glutamate receptors on hippocampal neurons. At synaptic junctions, hippocampal neurons respond to the release of the neurotransmitter glutamate by activation of NMDA receptors. These ROCs are ligand-gated ion channels that allow the influx of Ca^{2+}. The Ca^{2+} that enters neurons through the synaptic NMDARs causes local activation of ERK1/2 [42] and promotes cell survival. When neurons and glia become anoxic, the glutamate released at synaptic terminals can diffuse to nonsynaptic NMDARs and cause Ca^{2+} signals that lead to cell death [43]. The drastically different effects of stimulating synaptic or nonsynaptic NMDARs explains the paradox that glutamate can be a physiological neurotransmitter, yet bath application of glutamate kills cultured neurons. Essentially, the spatial location of the Ca^{2+} signals determines which biochemical pathways will become activated and can switch cells from life to death.

Modulation of Ca^{2+} Signal Amplitude

Although many cell types can grade the amplitude of their Ca^{2+} signals, most control of Ca^{2+} signaling occurs through the types of spatial and temporal regulation described above. Consequently, there are only a few situations in which such modulation has been shown to have a physiological relevance. One well-known example is in muscle, where the amplitude of Ca^{2+} signals governs the force of contraction. In the case of cardiac muscle, inotropic agents (e.g. adrenaline) can alter the influx of Ca^{2+} through VOCs or Ca^{2+} release from RyRs, thus altering the capacity of the heart for pumping blood.

Since large, rapid increases in Ca^{2+} are easier to detect than small, graded changes, Ca^{2+} signals based on frequency modulation are believed to have greater fidelity than those occurring through amplitude modulation. However, it has been shown that cells may interpret modest changes in cytoplasmic concentration. For example, differential gene activation may occur by varying the amplitude of Ca^{2+} signals [44].

Ca^{2+} as a Signal within Organelles and in the Extracellular Space

The discussion above has largely considered the regulation of Ca^{2+} signals within the cytoplasm. However, it is important to point out that Ca^{2+} has crucial functions within organelles. Mitochondrial Ca^{2+} signals can enhance mitochondria respiration by activation enzymes of the citric acid cycle to stimulate production of NADH [45]. Elevation of nuclear Ca^{2+} appears to be important for transcription of specific genes [46]. Ca^{2+} has a diverse range of functions within the lumen of the ER. Depletion of ER Ca^{2+} leads to incorrect folding of nascent proteins and a stress response culminating in cell death [47].

Many cell types express receptors for Ca^{2+} on their surface, allowing them to sense changes in extracellular Ca^{2+} concentration (reviewed in [48]). These receptors can activate $InsP_3$ production to evoke intracellular Ca^{2+} changes. Through the action of such Ca^{2+}-sensing receptors, the Ca^{2+} that is extruded from a cell at the termination of a cytosolic signal can become an agonist for its neighbors [49], perhaps serving to coordinate the activity of adjacent cells.

References

1. Berridge, M. J., Lipp, P., and Bootman, M. D. (2000). The versatility and universality of calcium signaling. *Nature Reviews Mol. Cell. Biol.* **1**, 11–21.
2. Berridge, M. J. and Bootman, M. D. (1995). Calcium signalling. In *Modular Texts in Molecular and Cell Biology*. (R. A. Bradshaw and M. Purton, Eds.), Chapman and Hall, New York, pp. 205–221.
3. Bootman, M. D., Lipp, P., and Berridge, M. J. (2001). The organisation and functions of local Ca^{2+} signals. *J. Cell Sci.* **114**, 2213–2222.
4. Carafoli, E. (2002). Calcium signaling: a tale foe all seasons. *Proc. Natl. Acad. Sci. USA* **99**, 1115–1122.
5. Bootman, M. D., Berridge, M. J., and Roderick, H. L. (2002). Calcium signalling; more messengers, more channels, more complexity. *Curr. Biol.* In press.
6. Boitano, S., Dirksen, E. R., and Sanderson, M. J. (1992). Intercellular propagation of calcium waves mediated by inositol trisphosphate. *Science* **258**, 292–295.
7. Montell, C. (2001). Physiology, phylogeny, and functions of the TRP superfamily of cation channels. *Science's STKE*, http://stke.sciencemag.org/cgi/content/full/OC_sigtrans;2001/90/re1.
8. Taylor, C. W. (1998). Inositol trisphosphate receptors: Ca^{2+}-modulated intracellular Ca^{2+} channels. *Biochim. Biophys. Acta* **1436**, 19–33.
9. Patel, S., Joseph, S. K., and Thomas, A. P. (1999). Molecular properties of inositol 1,4,5-trisphosphate receptors. *Cell Calcium* **25**, 247–264.
10. Sorrentino, V. (1995). Molecular biology of ryanodine receptors. In *Ryanodine receptors: A CRC Pharmacology & Toxicology Series, Basic and Clinical Aspects* (V. Sorrentino, Ed.), pp. 85–100.
11. Cancela, J. M., Van Coppenolle, F., Galione, A., Tepikin, A. V., and Petersen, O. H. (2002). Transformation of local Ca^{2+} spikes to global Ca^{2+} transients: the combinatorial roles of multiple Ca^{2+} releasing messengers. *EMBO J.* **21**, 909–919.
12. Berridge, M. J. and Galione, A. (1988). Cytosolic calcium oscillators. *FASEB J.* **2**, 3074–3082.
13. Thomas, A. P., Bird, G. S., Hajnoczky, G., Robb-Gaspers, L. D., and Putney, J. W. Jr (1996). Spatial and temporal aspects of cellular calcium signalling. *FASEB J.* **10**, 1505–1517.
14. Hajnoczky, G., Robb-Gaspers, L. D., Seitz, M. B., and Thomas, A. P. (1995). Decoding of cytosolic calcium oscillations in the mitochondria. *Cell* **82**, 415–424.
15. Woods, N. M., Cuthbertson, K. S. R., and Cobbold, P. H. (1986). Repetitive transient rises in cytoplasmic free calcium in hormone-stimulated hepatocytes. *Nature* **319**, 600–602.
16. Mackenzie, L., Bootman, M. D., Berridge, M. J., and Lipp, P. (2001). Pre-determined recruitment of calcium release sites underlies excitation-contraction coupling in rat atrial myocytes. *J. Physiol.* **530**, 417–429.
17. Chin, D. and Means, A. R. (2000). Calmodulin: a prototypical calcium sensor. *Trends Cell Biol.* **10**, 322–328.
18. De Koninck, P. and Schulman, H. (1998). Sensitivity of CAM kinase II to the frequency of Ca^{2+} oscillations. *Science* **279**, 227–230.
19. Hudmon, A. and Schulman, H. (2002). Neuronal Ca^{2+}/calmodulin-dependent protein kinase II: the role of structure and autoregulation in cellular function. *Annu. Rev. Biochem.* **71**, 473–510.
20. Dolmetsch, R. E., Xu, K. L., and Lewis, R. S. (1998). Calcium oscillations increase the efficiency and specificity of gene expression. *Nature* **392**, 933–936.
21. Neher, E. (1998). Vesicle pools and Ca^{2+} microdomains: new tools for understand their roles in neurotransmitter release. *Neuron* **20**, 389–399.
22. Bootman, M. D. (1996) Hormone-evoked subcellular Ca^{2+} signals in HeLa cells. *Cell Calcium* **20**, 97–104.
23. Thomas, D., Lipp, P., Berridge, M. J., and Bootman, M. D. (1998). Hormone-stimulated calcium puffs in non-excitable cells are not stereotypic, but reflect activation of different size channel clusters and variable recruitment of channels within a cluster. *J. Biol. Chem.* **273**, 27130–27136.
24. Yao, Y., Choi, J., and Parker, I. (1995). Quantal puffs of intracellular Ca^{2+} evoked by inositol trisphosphate in *Xenopus* oocytes. *J. Physiol.* **482**, 533–553.
25. Berridge, M. J. (1997). Elementary and global aspects of calcium signaling. *J. Physiol.* **499**, 291–306.
26. Thomas, D., Lipp, P., Tovey, S. C., Berridge, M. J., Li, W. H., Tsien, R. Y., and Bootman, M. D. (2000). Microscopic properties of elementary Ca^{2+} release sites in non-excitable cells. *Curr. Biol.* **10**, 8–15.
27. Marchant, J. S. and Parker, I. (2001). Role of elementary Ca^{2+} puffs in generating repetitive Ca^{2+} oscillations. *EMBO J.* **20**, 65–76.
28. Lipp, P., Thomas, D., Berridge, M. J., and Bootman, M. D. (1997). Nuclear calcium signalling by individual cytoplasmic calcium puffs. *EMBO J.* **16**, 7166–7173.
29. Marchant, J. S., Ramos, V., and Parker, I. (2002). Structural and functional relationships between Ca^{2+} puffs and mitochondria in Xenopus oocytes. *Am. J. Physiol.* **282**, C1374–C1386.
30. Petersen, O. H., Burdakov, D., and Tepikin, A. Y. (1999). Polarity in intracellular calcium signaling. *Bioessays* **21**, 851–860.
31. Straub, S. V., Giovannucci, D. R., and Yule, D. I. (2000). Calcium wave propagation in pancreatic acinar cells. *J. Gen. Physiol.* **116**, 547–559.
32. Tinel, H., Cancela, J. M., Mogami, H., Gerasimenko, J. V., Gerasimenko, O. V., Tepikin, A. V., and Petersen, O. H. (2000). Active mitochondria surrounding the pancreatic acinar granule region prevent spreading of inositol trisphosphate-evoked local cytosolic Ca^{2+} signals. *EMBO J.* **18**, 4999–5008.
33. Cannell, M. B. and Soeller, C. (1998). Sparks of interest in cardiac excitation-contraction coupling. *TiPS* **19**, 16–20.
34. Niggli, E. (1999). Localized intracellular calcium signaling in muscle: Calcium sparks and calcium quarks. *Annu. Rev. Physiol.* **61**, 311–335.
35. Jaggar, J. H., Porter, V. A., Lederer, W. J., and Nelson, M. T. (2000). Ca^{2+} sparks in smooth muscle. *Am. J. Physiol.* **278**, C235–C256.
36. Nelson, M. T., Cheng, H., Rubart, M., Santana, L. F., Bonev, A. D. Knot, H. J., and Lederer, W. J. (1995). Relaxation of arterial smooth muscle by Ca^{2+} sparks. *Science* **270**, 633–637.
37. ZhuGe, R., Fogarty, K. E., Tuft, R. A., Lifshitz, L. M., Sayar, K., and Walsh, J. V. (2000). Dynamics of signaling between Ca^{2+} sparks and Ca^{2+}-activated K^+ channels studied with a novel image-based method for direct intracellular measurement of ryanodine receptor Ca^{2+} current. *J. Gen. Physiol.* **116**, 845–864.

38. Wang, S. Q., Song, L. S., Lakatta, E. G., and Cheng, H. P. (2001). Ca^{2+} signalling between single L-type Ca^{2+} channels and ryanodine receptors in heart cells. *Nature* **410**, 592–596.
39. Denk, W., Yuste, R., Svoboda, K., and Tank, D. W. (1996). Imaging Ca^{2+} dynamics in dendritic spines. *Curr. Opin. Neurobiol.* **6**, 372–378.
40. Berridge, M. J. (1998). Neuronal calcium signalling. *Neuron* **21**, 13–26.
41. Spitzer, N. C., Lautermilch, N. J., Smith, R. D., and Gomez, T. M. (2000). Coding of neuronal differentiation by calcium transients. *Bioesssays* **22**, 811–817.
42. Hardingham, G. E., Arnold, F. J. L., and Bading, H. (2001). A calcium microdomain near NMDA receptors: on switch for ERK-dependent synapse-to-nucleus communication. *Nat. Neurosci.* **4**, 565–566.
43. Hardingham, G. E., Fukunaga, Y., and Bading, H. (2002). Extrasynaptic NMDARs oppose synaptic NMDARs by triggering CREB shut-off and cell death pathways. *Nat Neurosci.* **5**, 405–414.
44. Dolmetsch, R. E., Lewis, R. S., Goodnow, C. C., and Healy, J. I. (1997). Differential activation of transcription factors induced by Ca^{2+} response amplitude and duration. *Nature* **386**, 855–858.
45. Robb-Gaspers, L. D., Burnett, P., Rutter, G. A., Denton, R. M., Rizzuto, R., and Thomas, A. P. (1998). Integrating cytosolic calcium signals into mitochondrial metabolic responses. *EMBO J.* **17**, 4987–5000.
46. Hardingham, G. E., Chawla, S., Johnson, C. M., and Bading, H. (1997). Distinct functions of nuclear and cytoplasmic calcium in the control of gene expression. *Nature* **385**, 260–265.
47. Roderick, H. L., Berridge, M. J., and Bootman, M. D. (2002). The Endoplasmic Reticulum: a central player in cell signalling and protein synthesis. In *Lecture Notes in Physics* (M. Falcke, Ed.), Springer Verlag, New York. In press.
48. Riccardi, D. (1999). Cell surface, Ca^{2+}(cation)-sensing receptor(s): one or many? *Cell Calcium* **26**, 77–83.
49. Hofer, A. M., Curci, S., Doble, M. A., Brown, E. M., and Soybel, D. I. (2000). Intercellular communication mediated by the extracellular calcium-sensing receptor. *Nat. Cell Biol.* **2**, 392–398.
50. Bootman, M. D., Collins, T. J., Peppiatt, C. M., Prothero, L. S., MacKenzie, L., De Smet, P., Travers, M., Tovey, S. C., Seo, J. T., Berridge, M. J., Ciccolini, F., and Lipp, P. (2001). Calcium signalling—an overview. *Seminars in Cell and Developmental Biology* **12**, 3–10.

CHAPTER 132

Calcium Pumps

Ernesto Carafoli
*Department of Biochemistry, University of Padova,
and Venetian Institute of Molecular Medicine (VIMM),
Padova, Italy*

Introduction

The plasma membrane controls the exchange of calcium between the intracellular and extracellular environments [1,2]. A limited and strictly controlled amount of Ca^{2+} is allowed to penetrate into the cells through a number of specific channels to trigger important cellular events, including the massive liberation of Ca^{2+} from membrane enclosed stores. An equivalent amount of Ca^{2+} must then be ejected to the extracellular spaces. Two systems preside over this function in animal cells: a large system that is particularly active in excitable cells exchanges electrogenically Na^+ for Ca^{2+} interacting with Ca^{2+} with low affinity. The other system is an ATPase (the PMCA pump [3]), which interacts instead with Ca^{2+} with high affinity but has low Ca^{2+} ejecting capacity: it is thus normally considered as the fine-tuner of cellular Ca^{2+}. Calcium is also exchanged between the cytoplasm and the internal space of the organelles, chiefly the mitochondria and the endo(sarco) plasmic reticulum (ER/SR). The latter contains an ATPase (the SERCA pump, [4]), which is similar in mechanism to the PMCA pump. The total Ca^{2+} transporting capacity of the reticulum depends on the amount of pump it contains, which is high in heart and skeletal muscle and low in nonmuscle tissues. The SERCA pump works in concert with channels in the ER/SR membrane that are activated by second messengers and return to the cytoplasm the calcium that the pump had transported to the ER/SR lumen. Since the ER/SR is located in close proximity to the mitochondria, the released calcium creates an ambient of high calcium concentration adequate to activate the low affinity electrophoretic uptake uniporter of the inner mitochondrial membrane [5]. Ca^{2+} accumulated in the mitochondrial matrix is released to the cytoplasm via two systems, a well-characterized a Na^+/Ca^{2+} exchanger [6] and a less well-characterized Ca^{2+}/H^+ antiporter.

Ca^{2+} pumps have also been described in lower eukaryotes. In yeasts, two pumps termed PMR1 and PMC1 [7–9] have been described in the Golgi complex [10] and the vacuoles [9], respectively. Their degree of sequence homology to the SERCA and PMCA pump does not exceed 40–50%, and in particular, the PMC1 pump does not contain the calmodulin binding domain that characterizes the PMCA pumps. Most bacteria extrude calcium via Ca^{2+}/H^+ or Ca^{2+}/Na^+ antiporters [11], but bona fide Ca^{2+} ATPases have also been described, e.g. in *Flavobacterium odoratum* [12] and in a cyanobacterium [13].

Reaction Cycle of the SERCA and PMCA Pumps

The basic enzyme cycle of the two calcium pumps is essentially the same [14] (Fig. 1). Ca^{2+} is bound on one side of the membrane in a reaction that does not require ATP, since Ca^{2+} binding can be measured in its absence. ATP is then bound and split to form an acyl-phosphate intermediate on an aspartic residue [15]. The formation of a phosphorylated intermediate has suggested the nomenclature of "P type" pumps [16,17]. After phosphorylation, the pump undergoes a conformational transition from a state termed E1 to one termed E2. In the E1 conformation the pump binds Ca^{2+} with high affinity to sites exposed to the cytosolic site, whereas in the E2 conformation the Ca^{2+} binding sites have lower affinity and are exposed to the ER/SR lumen or to the extracellular space. Ca^{2+} can thus be released. After releasing ATP and Ca^{2+} the enzyme becomes slowly dephosphorylated and returns to the E1 state. The SERCA and PMCA pumps differ in the Ca^{2+}/ATP transport stoichiometry, which is 2 in the former and 1 in the latter. Powerful inhibitors have been described. Lanthanum inhibits both pumps but with interesting differences. In the SERCA pump it decreases the steady state level of

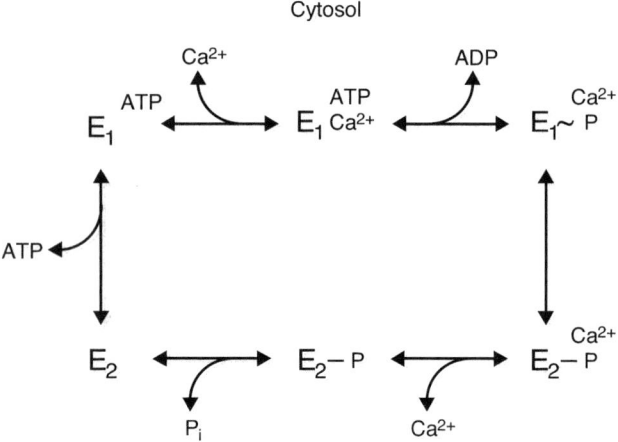

Figure 1 A simplified scheme of the reaction mechanism of calcium pumps. The pump, symbolized by E, is assumed to exist in two different conformations, E1 and E2. E1 binds calcium with high affinity at the cytoplasmic site of the membrane; E2 has lower affinity for calcium and releases it to the opposite site. Molecular details on the uptake and release path for calcium are discussed in the text. The energy of ATP is momentarily conserved in the enzyme as a phosphorylated intermediate (an aspartyl phosphate) that is formed prior to the translocation of calcium. The scheme shows a 1 to 1 stoichiometry between hydrolyzed ATP and transported calcium, which is that of the PMCA pump. The SERCA pump transports instead two Ca^{2+} per ATP hydrolyzed. See text for details.

the phosphorylated intermediate, whereas in the PMCA pump it greatly stimulates it. The phosphate analogue orthovanadate ($[VO_3(OH)]^{2-}$) inhibits both pumps with presumably identical mechanisms, whereas the inhibitors thapsigargin and thapsigarcin [18] and cyclopiazonic acid [19] only act on the SERCA pump by interacting with it with high affinity (K_d in the sub-nM range) [20].

The SERCA Pump

An enzyme that couples the hydrolysis of ATP to the transport of Ca^{2+} across the membrane of SR had been postulated about 40 years ago by Ebashi and Lipmann [21] and Hasselbach and Makinose [22]. Later work has identified the pump in the ER of nonmuscle cells as well. The ATPase, later termed the SERCA pump, was purified by MacLennan in 1970 [4] as a protein of about 100 kDa and cloned 15 years later [23]. The enzyme was predicted to be organized in the membrane of the reticulum with ten transmembrane domains and to protrude into the cytosol with three large units. The ATP binding domain and the catalytic aspartic acid are located in the cytosolic unit that protrudes between the fourth and the fifth transmembrane domains. The pump is the product of a multigene family: three basic gene products have so far been described with peculiar tissue distribution, additional isoform diversity being generated by alternative splicing of primary transcripts. SERCA1a is the major isoform of adult fast twitch muscle, whereas the transcripts of SERCA1b are detected in large amounts in neonatal fast twitch muscle.

The SERCA2 gene transcript is spliced to generate SERCA2a, which is found in slow twitch and heart muscles, whereas SERCA2b is found in smooth muscle and most nonmuscle cells. The SERCA2b protein is of particular interest because it replaces the last four residues of the SERCA2a isoform with a 49 amino acid stretch [24] that contains a hydrophobic sequence predicted to be the eleventh transmembrane domain [25]. Thus, the C-terminus of the SERCA2b isoform protrudes into the ER lumen. SERCA3 is only expressed in a limited range of nonmuscle cells [26].

Striking advances on the structure of the SERCA pump have recently extended our understanding of the molecular mechanism by which the enzyme couples the hydrolysis of ATP to the transport of Ca^{2+} across the protein. The pump has been crystallized in the Ca^{2+} bound E1 state by Toyoshima et al. [27] (Fig. 2). Its structure has been solved at 2.6 Å resolution, validating a number of previous suggestions on membrane topography and Ca^{2+} binding and transport. Specifically, the structure has confirmed that the number of transmembrane domains is 10 and has shown that the three large cytosolic domains (N for nucleotide binding; P, which contains the catalytic aspartic acid; and A, termed actuator or N anchoring domain) undergo large movements during ATP energized Ca^{2+} translocation. The movement of the three cytosolic units has been predicted by fitting the atomic structure to a low resolution structure (8 Å) derived from tubular crystals of the pump in the vanadate inhibited Ca^{2+} free (E2) conformation. The cytoplasmic portion of the E2 pump is more compact, suggesting that Ca^{2+} loosens the interactions between the cytosolic units. The N and P domains come close to each other whereas the A domain rotates by about 90° to bring a conserved, critically important sequence (TGES) next to the catalytic aspartic acid. The structure has also validated previous mutagenesis experiments [28] that had led to the conclusion that a number of residues in transmembrane domains 4, 5, 6, and 8 would form the two Ca^{2+} binding sites and the path of Ca^{2+} across the protein. The atomic structure has shown that transmembrane domain 5 is straight and extends to the center of the P domain, whereas transmembrane domains 4 and 6 are unwound in the middle to optimize Ca^{2+} coordination geometry. The two Ca^{2+} binding sites are separated by a distance of 5.7 Å, site I being formed essentially by transmembrane domains 5 and 6 (with a contribution of transmembrane domain 8) and site II by transmembrane domains 4 and 6. The two Ca^{2+} binding sites are stabilized by H bridges between coordinating residues and to residues on other transmembrane helices. The structure has also suggested the path for Ca^{2+} to the binding sites and from them to the lumenal space. The path to the sites may be a cavity opened to the cytoplasm formed by transmembrane domains 2, 4, and 6. Ca^{2+} would move along a row of hydrophilic carbonyl oxygens and would exit to the lumen of the ER through a zone ringed by hydrophilic oxygens surrounded by transmembrane domains 3, 4, and 5.

The SERCA pump is regulated by interaction with phospholamban (PLN, [29]), a small hydrophobic protein that has a strong tendency to form pentamers but that is active in

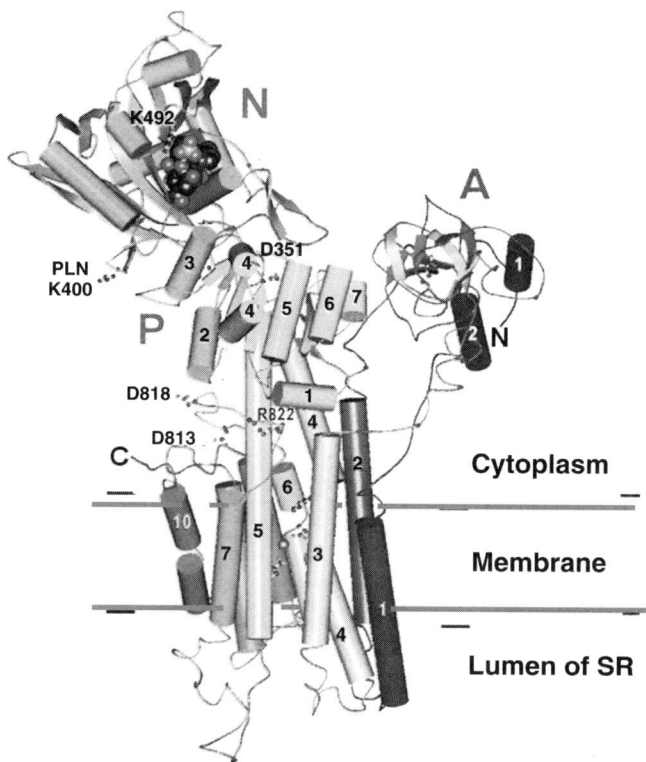

Figure 2 Crystal structure of the calcium bound (E1 form) of the SERCA pump [27]. The structure shows the predicted ten transmembrane domains and three units protruding into the cytoplasm, termed N (nucleotide binding), P (phosphorylation), and A (actuator, or N anchoring domain). Some residues important to the function of the pump are indicated, including K400, which is part of a loop that binds the cytosolic portion of phospholamban. Additional details of the structure and on the predicted motions of the cytosolic domains of the pump in the E1 to E2 conformational transition are discussed in the text.

the monomeric state [30]. Since PLN is only expressed in slow-twitch, heart, and smooth muscles, it only regulates the activity of the SERCA pump in these tissues. PLN is the substrate of two protein kinases, protein kinase A and a calmodulin-dependent kinase (protein kinase G may also phosphorylate it). In the unphosphorylated state it interacts with a cytosolic loop around Lys 400 [31], maintaining the pump inhibited. When phosphorylated on Ser16 and/or Thr17, PLN becomes detached from the binding loop freeing the pump from inhibition. Mutagenesis studies [32] have indicated that PLN also interacts with the intramembrane sector of the pump, specifically, with transmembrane domain 6. This indication has been recently supported by molecular modeling studies of the interaction of PLN [33] with the pump, based on the structure of the latter in the vanadate inhibited state and on the recently solved tertiary structure of PLN [34].

The PMCA Pump

The PMCA pump had been discovered by Schatzmann in 1966 as a system that ejected calcium from erythrocytes. The pump was purified in 1979 as a protein of about 135 kDa by Niggli et al. [35], and was cloned ten years later by Shull and Greb [36] and Verma et al. [37]. The membrane architecture of the protein resembles that of the SERCA pump, i.e., it is predicted to contain ten transmembrane domains and three large hydrophilic units protruding into the cytoplasm. One important difference with respect to the SERCA pump is the long C-terminal tail, which contains a calmodulin-binding domain [38]. Calmodulin is the most important regulator of the PMCA pump, although polyunsaturated fatty acids, acidic phospholipids, phosphorylation steps involving the C-terminal tail by protein kinase A, or protein kinase C may also activate the pump by lowering its K_m for calcium. Activation is also brought about by a dimerization process that occurs through the calmodulin-binding domain and by the proteolytic removal (e.g. by calpain) of most of the C-terminal tail of the pump [39]. At variance with the SERCA pump, the reaction cycle of the PMCA pump is not regulated by PLN but by a mechanism that has striking similarities to that of the SERCA pump. Specifically, the calmodulin-binding domain interacts in the resting state with two sites in the cytoplasmic portion of the pump, keeping it inhibited [40,41]. Calmodulin removes the binding domain from its "receptors" in the cytosolic portion of the pump, relieving the inhibition. Although in this case phosphorylation is not involved, the similarity to the reversible mechanism of inhibition of the SERCA pump by PLN is even more striking. The phosphorylation of the calmodulin-binding domain of the PMCA pump by protein kinase C impairs its ability to bind to the cytosolic portion of the pump [42,43].

The PMCA pump is the product of a multigene family, with four basic gene products. As in the case for the SERCA pump the number of isoforms is increased by the alternative splicing of primary transcripts. Two of the four basic isoforms (PMCA1 and 4) are expressed in all tissues, whereas PMCA2 and 3 are expressed in significant amounts only in neurons and in cells somehow related to them, e.g. the outer hair cells of the organ of Corti. Alternative splicing occurs at two sites. Site A is located upstream of the third transmembrane domain, next to a site that mediates the sensitivity of the pump to acidic phospholipids, site C within the calmodulin-binding domain itself. Information on the differential functional properties of the PMCA isoforms is very scarce, but it is known that PMCA2 has the highest sensitivity to calmodulin. The proximity of the splicing sites to domains that are important in regulation suggests different regulatory properties of the spliced isoforms. C-spliced variants of the pump may indeed interact with calmodulin with peculiar pH sensitivity [44], whereas a variant of the pump truncated C-terminally as a result of the insertion of a 154 bp hexon at site C [45] has decreased affinity for calmodulin.

An interesting development in the regulation of the PMCA pump has been the finding that its genes are transcriptionally regulated by Ca^{2+} itself [46,47]. The discovery has been made on maturing cultured cerebellar granular neurons, and reflect the regulation of PMCA gene expression within the cerebellum. The cultured granular neurons require a modest increase in cytosolic calcium (about three-fold) to switch off the apoptotic programs that would otherwise kill them in 3–5 days, and do so by rearranging the expression of PMCA isoforms to accommodate the changing requirements of calcium homeostasis necessary to set cell calcium at a higher level. Under these conditions, PMCA2 and 3 become strongly upregulated within days after the beginning of culture; PMCA1 experiences instead a splicing switch that favors a C-terminally truncated variant. PMCA4, by contrast, becomes rapidly and dramatically downregulated in a process that is mediated by the Ca^{2+}-dependent protein phosphatase calcineurin.

Genetic Diseases Evolving Defects of Calcium Pumps

Pathological phenotypes linked to genetic defects in the genes of both the SERCA and PMCA pumps have been described. In agreement with the distinct brain distribution of PMCA2, which appears to be specifically expressed in cerebellar Purkinje cells and in the outer hair cells of the inner ear, mice with defects in the gene of PMCA2 have been described that display vestibular/motor imbalance and are deaf [48,49]. A similar phenotype has also been described in PMCA2 knockout mice [50].

Pathological phenotypes have also been described as a result of inactivating mutations in the SERCA pump genes. Brody's disease, an autosomal recessive disorder of skeletal muscle characterized by muscle cramping and exercise-induced impairment of relaxation, has been traced back to three different mutations in the SERCA1 gene [51,52] that lead to a loss of SERCA1 activity (although not all cases of Brody's disease are linked to SERCA1 gene defects). Darier's disease, an autosomal dominant skin disorder, has been traced back to mutations in the SERCA2a gene. It has been suggested that the SERCA2 pump influences the adhesion between keratinocytes and thus cellular differentiation in the epidermis [53].

References

1. Carafoli, E., Santella, L., Branca, D., and Brini, M. (2001). Generation, control and processing of cellular calcium systems. *Crit. Rev. Biochem. Mol. Biol.* **36**, 107–260.
2. Carafoli, E. (2002). Calcium signaling: a tale for all seasons. *Proc. Natl. Acad. Sci. USA* **99**, 1115–1122.
3. Schatzmann, H. J. (1966). ATP-dependent Ca^{2+} extrusion from human red cells. *Experientia* **22**, 364–368.
4. MacLennan, D. H. (1970). Purification and properties of an adenosine triphosphatase from sarcoplasmic reticulum. *J. Biol. Chem.* **245**, 4508–4518.
5. Rizzuto, R., Simpson, A. W. M., Brini, M., and Pozzan, T. (1992). Rapid changes of mitochondrial Ca^{2+} revealed by specifically targeted recombinant aequorin. *Nature* **358**, 325–328.
6. Carafoli, E., Tiozzo, R., Lugli, G., Crovetti, F., and Kratzing, C. (1974). The release of calcium from heart mitochondria by sodium. *J. Mol. Cell. Cardiol.* **6**, 361–371.
7. Rudolph, H. K., Antebi, A., Fink, G. R., Buckley, C. M., Dorman, T. E., Le Vitre, J., Davidow, L. S., Mao, J. I., and Moir, D. T. (1989). The yeast secretory pathway is perturbed by mutations in PMR1, a member of a Ca^{2+} ATPase family. *Cell* **58**, 133–145.
8. Cunningham, K. W. and Fink, G. R. (1994a). Ca^{2+} transport in *Saccharomyces cerevisiae*. *J. Exp. Biol.* **196**, 157–166.
9. Cunningham, K. W. and Fink, G. R. (1994b). Calcineurin-dependent growth control in *Saccharomyces cerevisiae* mutants lacking PMC1, a homolog of plasma membrane Ca^{2+} ATPases. *J. Cell Biol.* **124**, 351–363.
10. Antebi, A. and Fink, G. R. (1992). The yeast Ca^{2+}-ATPase homologue, PMR1, is required for normal Golgi function and localizes in a novel Golgi-like distribution. *Mol. Biol. Cell* **3**, 633–654.
11. Rosen, B. P. (1987). Bacterial calcium transport. *Biochim. Biophys. Acta* **906**, 101–110.
12. Desrosiers, M. G., Gately, L. J., Gambel, A. M., and Menick, D. R. (1996). Purification and characterization of the Ca^{2+}-ATPase of Flavobacterium odoratum. *J. Biol. Chem.* **271**, 3945–3951.
13. Geisler, M., Richter, J., Schumann, J. (1993). Molecular cloning of a P-type ATPase gene from the cyanobacterium Synechocystis sp. PCC 6803. Homology to eukaryotic Ca^{2+}-ATPases. *J. Mol. Biol.* **234**, 1284–1289.
14. Makinose, M. (1973). Possible functional states of the enzyme of the sarcoplasmic calcium pump. *FEBS Lett.* **37**, 140–143.
15. Degani, C. and Boyer, P. D. (1973). Characterization of acyl phosphate in transport ATPase by a borohydride reduction method. *Ann. NY Acad. Sci.* **242**, 77–79.
16. Pedersen, P. L. and Carafoli, E. (1987). Ion motive ATPases. I. Ubiquity, properties, and significance for cell function. *Trends Biochem. Sci.* **12**, 146–150.
17. Pedersen, P. and Carafoli, E. (1987). Ion motive ATPases. II. Energy coupling and work output. *Trends Biochem. Sci.* **12**, 186–189.
18. Sagara, Y., Fernandez-Belda, F., de Meis, L., and Inesi, G. (1992). Characterization of the inhibition of intracellular calcium transport ATPases by thapsigargin. *J. Biol. Chem.* **267**, 12606–12613.
19. Inesi, G. and Sagara, Y. (1994). Specific inhibitors of intracellular Ca^{2+} transport ATPases. *J. Membr. Biol.* **141**, 1–6.
20. Sagara, D. and Inesi, G. (1991). Inhibition of the sarcoplasmic reticulum Ca^{2+} transport ATPases by thapsigargin at subnanomolar concentrations. *J. Biol. Chem.* **267**, 13503–13506.
21. Ebashi, S. and Lipmann, F. (1962). Adenosine triphosphate-linked concentration of calcium ions in a particulate fraction of rabbit muscle. *J. Cell. Biol.* **14**, 389–400.
22. Hasselbach, W. and Makinose, M. (1961). Die Calcium Pumpe der "Erschlaffungsgrana" des Muskles und ihre Abhangigkeit von der ATP-Spaltung. *Biochem. Z.* **333**, 518–528.
23. MacLennan, D. H., Brandl, C. J., Korczac, B., and Green, N. M. (1985). Amino-acid sequence of a $Ca^{2+}+Mg^{2+}$-dependent ATPase from rabbit muscle sarcoplasmic reticulum, deduced from its complementary DNA sequence. *Nature* **316**, 696–700.
24. Lytton, J. and MacLennan, D. H. (1988). Molecular cloning of cDNAs from human kidney coding for two alternatively spliced products of the cardiac Ca^{2+}-ATPase gene. *J. Biol. Chem.* **263**, 15024–15031.
25. Campbell, A. M., Kessler, P. D., and Fambrough, D. M. (1992). The alternative carboxyl termini of avian cardiac and brain sarcoplasmic reticulum/endoplasmic reticulum Ca^{2+}-ATPases are on opposite sides of the membrane. *J. Biol. Chem.* **267**, 9321–9325.
26. Bobe, R., Bredoux, R., Wuytack, F., Quarck, R., Kovacs, T., Papp, B., Corvazier, E., Magnier, C., and Enouf, J. (1994). The rat platelet 97-kDa Ca^{2+}ATPase isoform is the sarcoendoplasmic reticulum Ca^{2+}ATPase 3 protein. *J. Biol. Chem.* **269**, 1417–1424.
27. Toyoshima, C., Nakasako, M., Nomura, H., and Ogawa, H. (2000). Crystal structure of the calcium pump of sarcoplasmic reticulum at 2.6 Å resolution. *Nature* **405**, 647–655.
28. Clarke, D. M., Loo, T. W., Inesi, G., and MacLennan, D. H. (1989). Location of high affinity Ca^{2+}-binding sites within the predicted transmembrane domain of the sarcoplasmic reticulum Ca^{2+}-ATPase. *Nature* **339**, 476–478.
29. Tada, M., Kirchberger, M. A., and Katz, A. M. (1975). Phosphorylation of a 22,000-dalton component of the cardiac sarcoplasmic reticulum by adenosine 3′:5′-monophosphate-dependent protein kinase. *J. Biol. Chem.* **250**, 2640–2647.

30. Kimura, Y., Kurzydlowski, K., Tada, M., and MacLennan, D. H. (1997). Phospholamban inhibitory function is activated by depolimerization. *J. Biol. Chem.* **272**, 15061–15064.
31. James, P., Inui, M., Tada, .M., Chiesi, M., and Carafoli, E. (1989). Nature and site of phospholamban regulation of the Ca^{2+} pump of sarcoplasmic reticulum. *Nature*, **342**, 90–92.
32. Asahi, M., Kimura, Y., Kurzydlowski, K., Tada, M., and MacLennan, D. H. (1999). Transmembrane helix M6 in sarco(endo)plasmic reticulum Ca^{2+}-ATPase forms a functional interaction site with phospholamban. Evidence for physical interactions at other sites. *J. Biol. Chem.* **274**, 32855–32862.
33. Hutter, M. C., Krebs, J., Meiler, J., Griesinger, C., Carafoli, E., and Helms, V. (2002). A structural model of the complex between phospholamban and the calcium pump of sarcoplasmic reticulum obtained by molecular mechanics. *Submitted.*
34. Lamberth, S., Schmid, H., Muenchbach, M., Vorherr, T., Krebs, J., Carafoli, E., and Griesinger, C. (2000). NMR Solution structure of phospholamban. *Helvetica Chim. Acta* **83**, 2141–2152.
35. Niggli, V., Penniston, J. T., and Carafoli, E. (1979). Purification of the $(Ca^{2+}-Mg^{2+})$-ATPase from human erythrocyte membranes using a calmodulin affinity column. *J. Biol. Chem.* **254**, 9955–9958.
36. Shull, G. E. and Greeb, J. (1988). Molecular cloning of two isoforms of the plasma membrane Ca^{2+}-transporting ATPases from rat brain. *J. Biol. Chem.* **263**, 8646–8657.
37. Verma, A. K., Filoteo, A. G., Stanford, D. R., Wieben, E. D., Penniston, J. T., Strehler, E. E., Fischer, R., Heim, R., Vogel, G., and Mathews, S. (1988). Complete primary structure of a human plasma membrane Ca^{2+} pump. *J. Biol. Chem.* **263**, 14152–14159.
38. James, P., Maeda, M., Fischer, R., Verma, A. K., Penniston, J. T., and Carafoli, E. (1988). Identification and primary structure of a calmodulin binding domain of the Ca^{2+} pump of human erythrocytes. *J. Biol. Chem.* **263**, 2905–2910.
39. James, P., Vorherr, T., Krebs, J., Morelli, A., Castello, G., McCormick, D. J., Penniston, J. T., De Flora, A., and Carafoli, E. (1989). Modulation of erythrocyte Ca^{2+}-ATPase by selective calpain cleavage of the calmodulin-binding domain. *J. Biol. Chem.* **264**, 8289–8296.
40. Falchetto, R., Vorherr, T., Brunner, J., and Carafoli, E. (1991). The plasma membrane Ca^{2+} pump contains a site that interacts with its calmodulin-binding domain. *J. Biol. Chem.* **266**, 2930–2936.
41. Falchetto, R., Vorherr, T., and Carafoli, E. (1992). The calmodulin-binding site of the plasma membrane Ca^{2+} pump interacts with the transduction domain of the enzyme. *Protein Sci.* **1**, 1613–1621.
42. Hofmann, F., James, P., Vorherr, T., and Carafoli, E. (1993). The C-terminal domain of the plasma membrane Ca^{2+} pump contains three high affinity Ca^{2+} binding sites. *J. Biol. Chem.* **268**, 10252–10259.
43. Hofmann, F., Anagli, J., Carafoli, E., and Vorherr, T. (1994). Phosphorylation of the calmodulin binding domain of the plasma membrane Ca^{2+} pump by protein kinase C reduces its interaction with calmodulin and with its pump receptor site. *J. Biol. Chem.* **269**, 24298–24303.
44. Kessler, F., Falchetto, R., Heim, R., Meili, R., Vorherr, T., Strehler, E. E., and Carafoli, E. (1992). Study of calmodulin binding to the alternatively spliced C-terminal domain of the plasma membrane Ca^{2+} pump. *Biochemistry* **31**, 11785–11792.
45. Strehler, E. E., Strehler-Page, M. A., Vogel, G., and Carafoli, E. (1989). mRNAs for plasma membrane calcium pump isoforms differing in their regulatory domain are generated by alternative splicing that involves two internal donor sites in a single exon. *Proc. Natl. Acad. Sci. USA* **86**, 6908–6912.
46. Guerini, D., Garcia-Martin, E., Gerber, A., Volbracht, C., Leist, M., Gutierrez Merino, C., and Carafoli, E. (1999). The expression of plasma membrane Ca^{2+} pump isoforms in cerebellar granule neurons is modulated by Ca^{2+}. *J. Biol. Chem.* **274**, 1667–1676.
47. Guerini, D., Wang, X., Li, L., Genazzani, A., and Carafoli, E. (2000). Calcineurin controls the expression of isoform 4CII of the plasma membrane Ca^{2+} pump in neurons. *J. Biol. Chem.* **275**, 3706–3712.
48. Takahashi, K. and Kitamura, K. (1999). A point mutation in a plasma membrane Ca^{2+}-ATPase gene causes deafness in Wriggle Mouse Sagami. *Biochem. Biophys. Res. Commun.* **261**, 773–778.
49. Street, V. A., McKee-Johnson, J. W., Fonseca, R. C., Temperl, B. L. and Noben-Trauth, K. (1998). Mutations in a plasma membrane Ca^{2+}-ATPase gene cause deafness in deafwaddler mice. *Nat. Genet.* **19**, 390–394.
50. Kozel, P. J., Friedman, R. A., Eway, L. C., Yamoah, E. N., Liu, L. H., Riddle, T., Duffy J. J., Doetschman, T., Miller, M. L., Cardell, E. L., and Shull, G. E. (1998). Balance and hearing deficits in mice with a null mutation in the gene encoding plasma membrane Ca^{2+}-ATPase isoform 2. *J. Biol. Chem.* **273**, 18693–18696.
51. Brody, I. A. (1969). Muscle contracture induced by exercise. A syndrome attributable to decreased relaxing factor. *N. Engl. J. Med.* **281**, 187–192.
52. Karpati, G., Charuk, J., Carpenter, S., Jablecki, C., and Holland, P. (1986). Myopathy caused by a deficiency of Ca^{2+}-adenosine triphosphatase in sarcoplasmic reticulum (Brody's disease). *Ann. Neurol.* **20**, 38–49.
53. Clarke, D. M., Loo, T. W., and MacLennan, D. H. (1990). Functional consequences of alterations to polar amino acids located in the transmembrane domain of the Ca^{2+}-ATPase of sarcoplasmic reticulum. *J. Biol. Chem.* **265**, 6262–6267.

CHAPTER 133

Sodium/Calcium Exchange

Mordecai P. Blaustein

*Department of Physiology, University of Maryland School of Medicine
Baltimore, Maryland*

Introduction

The plasma membrane (PM) Na^+/Ca^{2+} exchanger (NCX) is one of the critical mechanisms involved in Ca^{2+} homeostasis and the regulation of Ca^{2+} signaling in most cells. The PM NCX was discovered about 35 years ago in mammalian cardiac muscle [1] and squid neurons [2]. It uses energy from the Na^+ electrochemical gradient, and not directly from ATP, to transport Ca^{2+}. Therefore, as we shall see, a critical aspect of the exchanger's function is that it may either export or import Ca^{2+}, depending upon the NCX coupling ratio and the prevailing membrane potential and Na^+ concentration gradient. The Na^+ gradient and membrane potential are maintained by the ATP-dependent, ouabain-sensitive Na^+ pump (Na^+, K^+-ATPase). A mitochondrial membrane Na^+/Ca^{2+} exchanger has also been identified [3] but has been less well characterized than the PM NCX; the mitochondrial exchanger will not be discussed here. Recent, more extensive reviews of NCX structure and function [3,4] should be consulted for details.

Two Families of PM Na^+/Ca^{2+} Exchangers

Early measurements suggested that the cardiac and neuronal exchangers both had coupling ratios of $3 Na^+ : 1 Ca^{2+}$, and that these two ion species were the only ones translocated by the exchanger [3,5]. Subsequently, a Na^+/Ca^{2+} exchanger was identified in the PM of photoreceptor cells. The photoreceptor exchanger was also dependent upon K^+ and appeared to have a coupling ratio of $4Na^+ : (1Ca^{2+} + 1K^+)$ [6,7]. This latter exchanger is therefore designated as the Na/(Ca, K) exchanger or NCKX.

Two families of Na^+/Ca^{2+} exchanger molecules have been cloned and sequenced [8,9]. One corresponds to the cardiac/neuronal NCX [8]; three members of this family, designated NCX1, NXC2, and NCX3, have been identified in mammals [10]. Each of these isoforms is the product of a different gene. NCX1 is the most prevalent, but they all have different tissue distributions. The functional significance of these different isoforms is unclear. In addition, there are several tissue-specific splice variants of NCX1; these, too, exhibit different tissue expression [11], but the functional significance has not been resolved.

The membrane topology of NCX1 is illustrated in Fig. 1. NCX has a molecular weight of 108 kDa (excluding glycosylation) and appears to have nine membrane-spanning segments [12]. A large cytoplasmic loop is located between the 5 N-terminal and 4 C-terminal transmembrane segments. This loop includes a calmodulin-like "exchanger inhibitory peptide" (XIP) binding site, a Ca^{2+} binding site that is involved in internal Ca^{2+}-dependent Ca^{2+} entry, and a peptide region that is alternatively spliced in different tissues (Fig. 1). A site that participates in intracellular Na^+-dependent inactivation may be included within the XIP region. The alpha helix repeat that occurs in helices 2–3 and 7 (gray regions in Fig. 1) has been postulated to participate in the binding and translocation of Na^+ and Ca^{2+}, but the evidence is inconclusive. Part of the second alpha repeat is a P loop-like region between transmembrane segments 7 and 8 that dips into the membrane from the cytoplasmic side but does not traverse the membrane.

Three mammalian members of the second exchanger family, the NCKX family, also have been cloned: NCKX1 is found in rod photoreceptors, NCKX2 is expressed in cones and neurons, and NCKX3 is expressed in the brain and smooth muscles [13,14]. The topology of the deduced NCKX proteins is similar to that of NCX. Nevertheless, the sequence homology of the two families of expressed proteins is limited to two of the putative membrane-spanning domains that

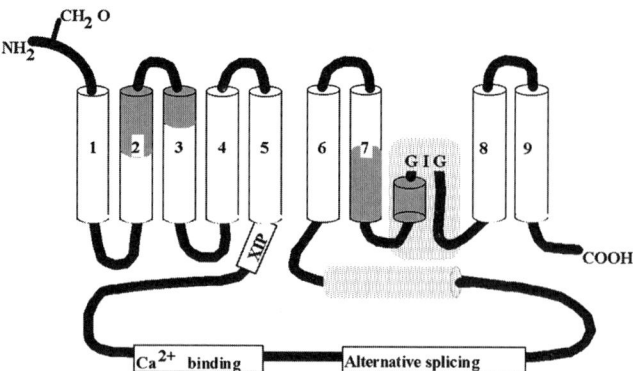

Figure 1 Diagram of Na$^+$/Ca^{2+} exchanger topology. The model shows the dog cardiac NCX1 (938 amino acids), which apparently has 9 transmembrane segments. The glycosylated N terminus is extracellular. The region between transmembrane segments 7 and 8 apparently forms a "P-type" loop that dips into the membrane (shaded area). The grey portions of segments 2–3 and 7 (and part of the P-type loop) are the alpha repeats. The large cytoplasmic loop (which actually contains nearly 550 amino acids) includes the XIP binding region (and internal Na$^+$-dependent inactivation site), the Ca^{2+} regulatory site, and the alternative splice site. This cytoplasmic loop also apparently includes a hydrophobic alpha helix region (shaded segment). Reproduced from Philipson and Nicoll [4] with permission.

may be involved in ion binding and translocation. Thus, the NCX and NCKX genes evolved independently.

Modes of Operation of the Na$^+$/Ca^{2+} Exchangers

As diagramed in Fig. 2, the NCX can mediate electroneutral Na$^+$/Na$^+$ exchange and electroneutral Ca^{2+}/Ca^{2+} exchange, as well as the Na$^+$ entry/Ca^{2+} exit and Na$^+$exit/Ca^{2+} entry exchange modes. The Ca^{2+}/Ca^{2+} exchange mode is activated by nontransported alkali metal ions. These partial reactions are consistent with a sequential transport mechanism (Fig. 2) [3] in which either one Ca^{2+} ion or three Na$^+$ ions are bound at one side of the membrane, translocated to the other side, and dissociated before the ion(s) from that side are bound. The reversal potential, $E_{Na/Ca}$, for an NCX with a coupling ratio of 3Na$^+$: 1Ca^{2+} is given by the equation [3]:

$$E_{Na/Ca} = 3E_{Na} - 2E_{Ca}$$

where $E_{Na} = (RT/F) \ln ([Na^+]_o/[Na^+]_i)$ and $E_{Ca} = (RT/2F) \ln ([Ca^{2+}]_o/[Ca^{2+}]_i)$, and the subscripts "o" and "i" refer to the extracellular and intracellular ion concentrations, respectively; R, T, and F have their usual meanings. If the membrane potential is more negative than $E_{Na/Ca}$, the NCX will extrude Ca^{2+}, and if more positive, the NCX will move Ca^{2+} into the cell.

The Na$^+$ entry/Ca^{2+} exit and Na$^+$exit/Ca^{2+} entry exchange modes are both rheogenic (i.e., they are associated with net current flow). The exchange of 3Na$^+$ for 1Ca^{2+} in both of these modes means that one positive charge enters the cells during each Ca^{2+} exit exchange and one positive charge exits the cells during Ca^{2+} entry exchange. Net Ca^{2+} transport mediated by the NCKX also is rheogenic, with one net charge transported per cycle. Consequently, NCX- and NCKX-mediated Ca^{2+} transport can both be measured electrically as ionic current flow across the plasma membrane in the direction opposite the net Ca^{2+} flux. Furthermore, the coupling ratio indicates that Na$^+$/Ca^{2+} exchange is voltage-sensitive: membrane hyperpolarization promotes Ca^{2+} exit via the exchanger, while depolarization promotes exchanger-mediated Ca^{2+} entry. This is counterintuitive, because hyperpolarization is normally expected to drive Ca^{2+} into cells, while depolarization should slow Ca^{2+} entry.

Regulation of NCX

Several regulatory sites have been identified in the large cytoplasmic loop of NCX. These sites play critical roles in exchanger function. When the cytoplasmic Na$^+$ concentration ([Na$^+$]$_i$) is increased, exchanger-mediated Ca^{2+} entry is increased almost instantly, but only transiently; the exchange then declines in a time- and [Na$^+$]$_i$-dependent manner [15]. This phenomenon, known as Na$_i$-dependent inactivation, might be expected to limit exchanger-mediated Ca^{2+} entry. However, binding of cytosolic Ca^{2+} to the activation site on the cytoplasmic loop not only is required to activate exchanger-mediated Ca^{2+} entry, but it also reduces Na$^+$-dependent inactivation [16].

Cardiac NCX is activated by phosphatidylinositol-4, 5-bisphosphate (PIP$_2$), which is generated from membrane-bound phosphatidylinositol by a mechanism that involves ATP hydrolysis [17]. The PIP$_2$ apparently binds to the XIP binding region of the large cytoplasmic loop [16]. This not only activates the NCX, but also eliminates Na$^+$-dependent inactivation.

Inhibition of NCX

NCX is highly selective for Na$^+$; other monovalent cations cannot substitute for Na$^+$. While Sr^{2+} and Ba^{2+} can be transported by the NCX, they are very poor substitutes for Ca^{2+} (i.e. maximum transport rates are much lower). Other divalent cations including Ni^{2+} and Cd^{2+}, and La^{3+} and some other lanthanides, inhibit NCX but are nonselective.

NCX inhibitory activity is displayed by various organic molecules. These include some hydrophobic amiloride analogs (e.g. 3,4-dichlorobenzamil), some antiarrhythmic agents (e.g. quinacrine and bepridil), and an isothiourea derivative ("compound 7943"). Unfortunately, none of these molecules is completely selective.

XIP is a synthetic calmodulin-like peptide that can be used as an experimental tool [16]. When introduced into the cytosol, it binds to the "XIP region" of the large cytoplasmic loop (Fig. 1) and inhibits NCX activity.

Localization of the NCX

The PM NCX functions in parallel with the ATP-driven PM Ca^{2+} pump (PMCA), and both transport systems are

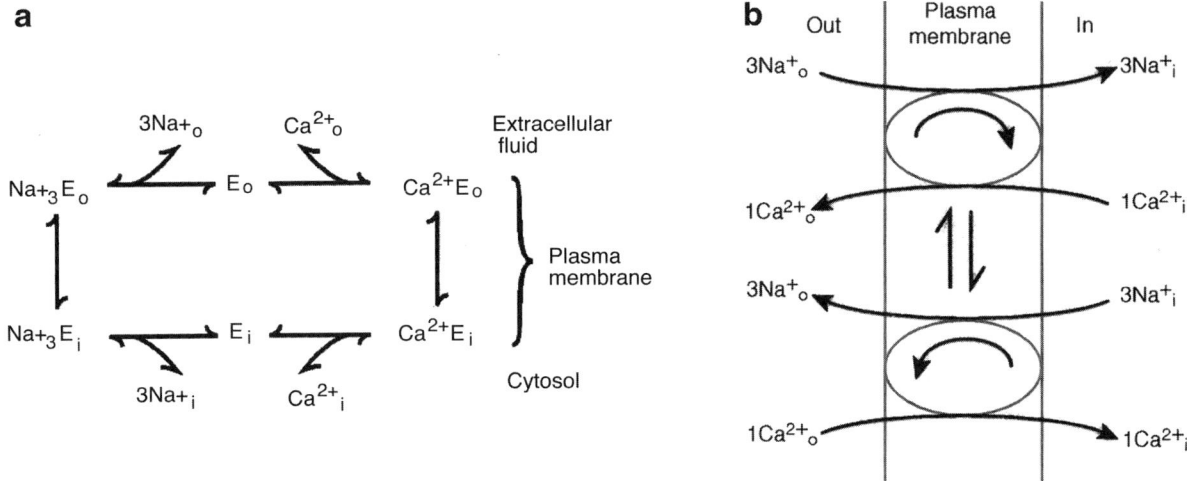

Figure 2 (a) State diagram illustrating the transport reactions mediated by the NCX ("E"). Subscripts "o" and "i" refer to the extracellular fluid or exofacial configuration of the carrier and cytosol or endofacial configuration of the carrer, respectively. Note that the carrier can switch between exofacial and endofacial conformations only when the carrier is loaded ($Na^+_3E_o$, $Na^+_3E_i$, $Ca^{2+}E_o$, or $Ca^{2+}E_i$); the unloaded carrier does not undergo conformational change (i.e. between E_o and E_i). (b). Diagram of net transport reactions mediated by the Na^+/Ca^{2+} exchanger. The exchanger can either move $3Na^+$ ions into the cell in exchange for one exiting Ca^{2+} ion (top) or move $3Na^+$ ions out of the cell in exchange for one entering Ca^{2+} ion (bottom). Reproduced from Blaustein et al. [33] with permission.

present in the PM of most cells. Moreover, the PMCA and NCX have very different kinetic properties—most notably, their affinities for cytosolic Ca^{2+} ($K_{Ca(cyt)} \approx 0.1\,\mu M$ for PMCA and $\approx 1.0\,\mu M$ for NCX1) and their turnover numbers ($\approx 30\,sec^{-1}$ for PMCA and $\approx 5{,}000\,sec^{-1}$ for NCX1). This implies that they have very different functions.

A further clue to their relative functions is their different distributions in the PM. The PMCA is very widely (uniformly?) distributed in the PM of several cell types, including astroglal cells, neurons, and smooth muscle cells [18,19]. In contrast, the NCX has a very much more limited distribution; indeed, in these same three cell types, NCX1 appears to be confined to microdomains of PM that overlie sub-PM ("junctional") elements of the endoplasmic or sarcoplasmic reticulum (jER or jSR) [18]. In skeletal muscle, NCX is localized primarily in T-tubule membranes. In cardiac muscle, too, the NCX is concentrated in T-tubule membranes [20,21]. However, there also is evidence (albeit controversial) of high levels of NCX expression in the peripheral PM [22] and some evidence that the NCX does not reside in the PM overlying jSR [21]. NCX is prevalent at presynaptic nerve terminals, but it appears to be excluded from transmitter release sites ("active zones") where the PMCA is concentrated [23].

Physiological Roles of the NCX

NCX (and NCKX) are expressed at high levels in cells with a large traffic of Ca^{2+} across the PM. Important examples are cardiac myocytes, neurons (especially nerve terminals), photoreceptor cells, and renal distal tubule epithelial cells [3,8,24]. The high level of activity in cardiac myocytes and neurons is consistent with the major role of the NCX in Ca^{2+} extrusion following periods of activity in these cells, and with the >100-fold difference in turnover number between NCX and PMCA. In the heart, the plateau of the action potential may help to maintain a high $[Ca^{2+}]_{CYT}$ during systole by temporarily reducing NCX-mediated Ca^{2+} extrusion. The possibility that NCX-mediated Ca^{2+} entry may contribute to cardiac excitation-contraction coupling has long intrigued investigators but is still controversial.

In the nervous system, the relative distribution of NCX has not yet been directly compared to that of NCKX. The specific roles of these two types of exchangers are not known, nor is it known whether members of both families are expressed in the same cells, but it is noteworthy that the two transporters have different coupling ratios and different reversal potentials.

The NCX is expressed in many epithelia, including gastrointestinal and renal epithelia, and in various endocrine and endocrine secretory cells. In renal distal tubules, the NCX is a key player in the reabsorption of Ca^{2+} and control of Ca^{2+} homeostasis.

NCX plays a role in the modulation of Ca^{2+} signaling in many types of cells. Indeed, this is the basis of the cardiotonic and vasotonic action of cardiotonic steroids [2,25–27]. In some cells, the NCX co-localizes with Na^+ pumps containing $\alpha 2$ or $\alpha 3$ subunits in PM microdomains [18] that are functionally coupled to the underlying jSR or jER [19,24,26]. These units ("PLasmERosomes"), which apparently help regulate Ca^{2+} signaling, contain a tiny diffusion-restricted volume of cytosol wedged between the PM and jSR or jER. Therefore, modulation of the Na^+ pump activity within the PM microdomains by hormones [27] or neurotransmitters [29,30] can alter the local (sub-PM) Na^+ and, via NCX, local Ca^{2+} concentrations. In this way, the Ca^{2+} content of the jSR or jER can be increased or decreased and can thus influence global Ca^{2+} signaling despite minimal change in the bulk $[Na^+]_i$. This resolves a long-standing dilemma about how low-dose cardiotonic steroids can exert their cardiotonic effect without altering bulk $[Na^+]_i$ [31]. Inhibition (by ouabain, for example) of just a small fraction of the total Na^+ pump molecules [26,32] should raise the local (sub-PM) $[Na^+]_i$. The PLasmERosome structure/function relationships then, in effect, enable the NCX to help translate and amplify the local $[Na^+]_I$ rise into an augmented global Ca^{2+} signal [26]. In other words, the NCX is not simply a "second" Ca^{2+} extrusion mechanism, even though Ca^{2+} extrusion may be a very important part of its function. In addition, the Na^+ pumps and NCX in the PLasmERosome work together to influence jSR/jER Ca^{2+} content; they thereby modulate Ca^{2+} signaling and all of the downstream consequences.

Acknowledgments

Supported by NIH grants NS-16106 and HL-45215.

References

1. Reuter, H. and Seitz, N. (1968). The dependence of calcium efflux from cardiac muscle on temperature and external ion composition. *J. Physiol. (London)* **195**, 451–470.
2. Baker, P. F., Blaustein, M. P., Hodgkin, A. L., and Steinhardt, R. A. (1969). The influence of calcium on sodium efflux in squid axons. *J. Physiol. (London)* **200**, 431–458.
3. Blaustein, M. P. and Lederer, W. J. (1999). Sodium/calcium exchange: its physiological implications. *Physiol. Rev.* **79**, 763–854.
4. Philipson, K. D. and Nicoll, D. A. (2000). Sodium-calcium exchange: a molecular perspective. *Annu. Rev. Physiol.* **62**, 111–133.
5. Reeves, J. P. and Hale, C. C. (1984). The stoichiometry of the cardiac sodium-calcium exchange system. *J. Biol. Chem.* **259**, 7733–7739.
6. Schnetkamp, P. P., Basu, D. K., and Szerencsei, R. T. (1989). Na^+–Ca^{2+} exchange in bovine rod outer segments requires and transports K^+. *Am. J. Physiol.* **275**, C153–C157.
7. Cervetto, L., Lagnado, L., Perry, R. J., Robinson, D. W., and McNaughton P. A. (1989). Extrusion of calcium from rod outer segments is driven by both sodium and potassium gradients. *Nature (London)* **337**, 740–743.
8. Nicoll, D. A., Longoni, S., and Philipson, K. D. (1990). Molecular cloning and functional expression of the cardiac sarcolemmal Na^+-Ca^{2+} exchanger. *Science* **250**, 62–65.
9. Reilander, H., Achilles, A., Friedel, U., Maul, G., Lottspeich, F., and Cook, N. J. (1992). Primary structure and functional expression of the Na/Ca,K-exchanger from bovine rod photoreceptors. *EMBO J.* **11**, 1689–1695.
10. Quednau, B. D., Nicoll, D. A., and Philipson, K. D. (1997). Tissue specificity and alternative splicing of the Na^+Ca^{2+} exchanger isoforms NCX1, NCX2, and NCX3 in rat. *Am. J. Physiol.* **272**, C1250–C1261.
11. Kofuji, P., Lederer, W. J., and Schulze, D. H. (1992). Mutually exclusive and cassette exons underlie alternatively spliced isoforms of the Na^+/Ca^{2+} exchanger. *J. Biol. Chem.* **269**, 5145–5149.
12. Nicoll, D. A., Ottolia, M., Lu, L., Lu, Y., and Philipson, K. D. (1999). A new topological model of the cardiac sarcolemmal Na^+-Ca^{2+} exchanger. *J. Biol. Chem.* **274**, 910–917.
13. Dong, H., Light, P. E., French, R. J., and Lytton, J. (2001). Electrophysiological characterization and ionic stoichiometry of the rat brain K^+-dependent Na^+/Ca^{2+} exchanger, NCKX2. *J. Biol. Chem.* **276**, 25919–25928.
14. Kraev, A., Quednau, B. D., Leach, S., Li, X. F., Dong, H., Winkfein, R., Perizzolo, M., Cai, X., Yang, R., Philipson, K. D., and Lytton, J. (2001). Molecular cloning of a third member of the potassium-dependent sodium-calcium exchanger gene family, NCKX3. *J. Biol. Chem.* **276**, 23161–23172.
15. Matsuoka, S. and Hilgemann D. W. (1994). Inactivation of outward Na^+–Ca^{2+} exchange current in guinea-pig ventricular myocytes. *J. Physiol. (London)* **476**, 443–458.
16. Matsuoka, S., Nicoll, D. A., He, Z., and Philipson K. D. (1997). Regulation of cardiac Na^+-Ca^{2+} exchanger by the endogenous XIP region. *J. Gen. Physiol.* **109**, 273–286.
17. Hilgemann, D. W. and Ball, R. (1996). Regulation of cardiac Na^+, Ca^{2+} exchange and KATP potassium channels by PIP2. *Science* **273**, 956–959.
18. Juhaszova, M. and Blaustein, M. P. (1997). Distinct distribution of different Na^+ pump alpha subunit isoforms in plasmalemma. Physiological implications *Ann. NY Acad. Sci.* **834**, 524–536.
19. Moore, E. D., Etter, E. F., Philipson, K. D., Carrington, W. A., Fogarty, K. E., Lifshitz, L. M., and Fay, F. S. (1993). Coupling of the Na^+ Ca^{2+} exchanger, Na^+ K^+ pump and sarcoplasmic reticulum in smooth muscle. *Nature (London)* **365**, 657–660.
20. Frank, J. S., Mottino, G., Reid, D., Molday, R. S., and Philipson, K. D. (1992). Distribution of the Na^+–Ca^{2+} exchange protein in mammalian cardiac myocytes: an immunofluorescence and immunocolloidal gold-labeling study. *J. Cell Biol.* **117**, 337–345.
21. Scriven, D. R., Dan, P., and Moore, E. D. (2000). Distribution of proteins implicated in excitation-contraction coupling in rat ventricular myocytes. *Biophys. J.* **79**, 2682–2691.
22. Kieval, R. S., Bloch, R. J., Lindenmayer, G. E., Ambesi, A., and Lederer, W. J. (1992). Immunofluorescence localization of the Na-Ca exchanger in heart cells. *Am. J. Physiol.* **263**, C545–C550.
23. Juhaszova, M., Church, P., Blaustein, M. P., and Stanley, E. F. (2000). Location of calcium transporters at presynaptic terminals. *Eur. J. Neurosci.* **12**, 39–846.
24. Blaustein, M. P. and Golovina, V. A. (2001). Structural complexity and functional diversity of endoplasmic reticulum Ca^{2+} stores. *Trends Neurosci.* **24**, 602–608.
25. Slodzinski, M. K., Juhaszova, M., and Blaustein, M. P. (1995). Antisense inhibition of Na^+/Ca^{2+} exchange in primary cultured arterial myocytes. *Am. J. Physiol.* **269**, C1340–C1345.
26. Arnon, A., Hamlyn, J. M., and Blaustein, M. P. (2000). Ouabain augments Ca^{2+} transients in arterial smooth muscle without raising cytosolic Na^+. *Am. J. Physiol.* **279**, H679–H691.
27. Reuter, H., Henderson, S. A., Han, T., Ross, R. S., Goldhaber, J. I., and Philipson, K. D. (2002). The Na^+–Ca^{2+} exchanger is essential for the action of cardiac glycosides. *Circ.Res.* **22**, 90:305–308.
28. Hamlyn, J. M., Lu, Z. R., Manunta, P., Ludens, J. H., Kimura, K., Shah, J. R., Laredo, J., Hamilton, J. P., Hamilton, M. J., and Hamilton, B. P. (1998). Observations on the nature, biosynthesis, secretion and significance of endogenous ouabain. *Clin. Exp. Hypertens.* **20**, 523–533.
29. Aperia, A. (2001). Regulation of sodium/potassium ATPase activity: impact on salt balance and vascular contractility. *Curr. Hypertens. Rep.* **3**, 165–171.
30. Mathias, R. T., Cohen, I. S., Gao, J., and Wang, Y. (2000). Isoform-specific regulation of the Na^+–K^+ pump in heart. *News Physiol. Sci.* **15**, 176–180.
31. Levi, A. J., Boyett, M. R., and Lee, C. O. (1994). The cellular actions of digitalis glycosides on the heart. *Prog. Biophys. Mol. Biol.* **62**, 1–54.
32. James, P. F., Grupp, I. L., Grupp, G., Woo, A. L., Askew, G. R., Croyle, M. L., Walsh, R. A., and Lingrel, J. B. (1999). Identification of a specific role for the Na,K-ATPase alpha 2 isoform as a regulator of calcium in the heart. *Mol. Cell.* **3**, 555–563.
33. Blaustein, M. P., Kao, J. P. Y., and Matteson, D. R. (2002). *Cellular Physiology*, Mosby, New York, in press.

CHAPTER 134

Ca^{2+} Buffers

Beat Schwaller
*Division of Histology, Department of Medicine,
University of Fribourg, Perolles, Fribourg, Switzerland*

Introduction

In principal, any molecule with several negatively charged groups can act as a chelator for Ca^{2+} ions if the negative charges are spatially distributed in such a manner as to satisfy the necessary geometrical considerations for coordination. In biological systems, these requirements are fulfilled by the carboxylic groups of small molecules such as citrate and more especially by the acidic side chain residues (e.g. glutamate, aspartate) or carbonyl groups of proteins. The possibilities for forming Ca^{2+}-binding sites are clearly numerous, and several protein families have been identified that contain different, evolutionarily well-conserved Ca^{2+}-binding domains. These include the EF-hand proteins [1], annexins, and C_2 domain proteins, each of which is described in a separate chapter (see Chapters 136, 140, and 141 of this volume). Almost all known proteins described as "Ca^{2+} buffers" belong to the family of EF-hand proteins [1,2]. An analysis of the human genome has revealed 242 proteins with EF-hand domains, which renders this one of the largest groups of proteins sharing a common motif [3]. EF-hand proteins have been somewhat arbitrarily designated as either "buffers" or "sensors" [4], the distinction being made on the basis that "sensors" undergo Ca^{2+}-dependant conformational changes, which permit them to interact with specific targets in a Ca^{2+}-regulated manner. Typical sensor proteins include calmodulin (see Chapter 137 by Means), some S100 proteins (see Chapter 138 by Heizmann *et al.*), and several others (see Chapter 136 by Bourgoyne and Weiss). Some so-called EF-hand "buffers" [e.g. calretinin (CR) and calbindin D-28k (CB28k)] also display Ca^{2+}-dependent conformational changes, but since no specific targets have as yet been identified, they are currently viewed as buffers. Whether an EF-hand protein can contribute to Ca^{2+}-buffering in a given cell depends largely upon its intracellular concentration.

Thus, all proteins classified as "sensors" could essentially act as "buffers" if present at sufficiently high levels.

Relevant Parameters for Ca^{2+} Buffers

In order to understand how a buffer will affect Ca^{2+} homeostasis within a cell, one first needs to consider the relevant parameters. These include (a) its cytosolic concentration, (b) its affinity for Ca^{2+} and possibly also for other metal ions, (c) the kinetics of Ca^{2+} binding and release, and (d) its mobility. But for no single protein have all of these parameters been determined with precision *in vivo*. During the past few years, most studies dealing with EF-hand Ca^{2+}-binding proteins have focused either on their metal-binding affinities (K_D values) or on elucidating their intracellular localization within specific cell types in a given tissue [1]. Proteins that will be discussed here include CB28k, CR, parvalbumin (PV), calbindin D-9k (CB9k; an S100-family protein), visinin-like protein III, and calmodulin (CaM).

Intracellular Concentration

With the exception of the ubiquitously expressed CaM, each of the aforementioned proteins is characterized by a very restricted pattern of expression within a given tissue, which renders an accurate determination of their intracellular concentrations extremely difficult. The proteins are frequently found in excitable cells (e.g. neurons), whose complex morphologies are prone to yield erroneous estimations of volume. Predictions of concentration in specific neurons usually fall within the range 1–50 µM, but the levels of PV in fast-twitch muscles and of CB9k or CB28k in specific cells of the kidney attain millimolar concentrations.

Metal-Binding Affinities

Two types of Ca^{2+}-binding sites in EF-hand proteins have been identified on the basis of differences in their selectivity and affinity for Ca^{2+} and Mg^{2+} ions [5]. The so-called Ca^{2+}-specific sites predominate, the affinity for this cation being much higher ($K_{Ca} = 10^{-3}$–10^{-7} M) than those for Mg^{2+} ($K_{Mg} = 10^{-1}$–10^{-2} M). Under basal conditions (free intracellular Ca^{2+} concentrations $[Ca^{2+}]_i = 40$–100 nM), the Ca^{2+}-specific sites of most of these proteins are assumed to be essentially vacant of metal ions and thus capable of binding Ca^{2+} rapidly, when $[Ca^{2+}]_i$ is raised. The second type, the mixed Ca^{2+}/Mg^{2+} site binds Ca^{2+} with high and Mg^{2+} with moderate affinity in a competitive manner (dissociation constants: $K_{Ca} = 10^{-7}$–10^{-9} M; $K_{Mg} = 10^{-3}$–10^{-5} M). Under basal conditions, these sites are occupied principally by Mg^{2+} ions, which must dissociate before Ca^{2+} binding can occur. EF-hand proteins have also been shown to contain allosteric effector, Mg^{2+}-specific binding sites, which can influence the affinities of the EF-hand Ca^{2+} binding sites [6]. Most EF-hand domains are paired to form a tandem domain consisting of two helix-loop-helix regions linked by a short stretch of 5–10 amino acid residues. Hence, the majority of these proteins have an even number of EF-hand domains (2, 4, or 6; for details, see Chapter 136). Not only are the tandem domains important for the structural stability of the individual EF-hand domains, but binding of Ca^{2+} ions to one site allosterically affects the affinity and probably also the binding kinetics of the second.

Metal-Binding Kinetics

Under physiological conditions, Ca^{2+}-binding kinetics (on-rates) can vary from $>10^8$ $M^{-1}s^{-1}$ for proteins with Ca^{2+}-specific sites implicated in very fast biological processes, such as muscle contraction (e.g. troponin C; TnC), down to an apparent on-rate of approximately 3×10^6 $M^{-1}s^{-1}$ for the slow-onset buffer PV. In the absence of Mg^{2+} ions, the on-rate of Ca^{2+}-binding to PV is very rapid (1.08×10^8 $M^{-1}s^{-1}$) [7]. But at the free intracellular concentration of magnesium ions $[Mg^{2+}]_i$, pertaining within neurons (0.3–0.6 mM) and rat myocytes (0.9 mM) [8,9], the rate for Ca^{2+}-binding, being determined by the rather slow Mg^{2+} off-rate [7,10] (Table I), will consequently be significantly slower.

During muscle contraction, PV does not compete with TnC for the binding of Ca^{2+} but helps increase the initial rate of $[Ca^{2+}]_i$ decay [11], thereby shortening the relaxation phase following very brief contractions. In this case, Mg^{2+} plays a role of almost equal importance to that of Ca^{2+}; it not only lowers Ca^{2+} affinity to within the physiological range but also exerts a considerable influence on the kinetics of Ca^{2+} binding to, and its release from, PV. The kinetics of Ca^{2+} binding for "fast" and "slow" buffer proteins are similar to those characterizing the synthetic chelators BAPTA and EGTA (Table I), respectively, which are thus often used experimentally to mimic endogenous buffer proteins. Owing to differences in their Mg^{2+}-buffering capacities, PV and EGTA are comparable only so far as their Ca^{2+}-binding kinetics are concerned— not with respect to the Ca^{2+}/Mg^{2+} antagonism (Table I).

Protein Mobility

In the cytosol of *Xenopus laevis* oocytes, only slowly mobile or immobile Ca^{2+} buffers exist. Accordingly, the rate of diffusion for Ca^{2+} ions under basal conditions ($D^* = 13$ $\mu m^2/s$) is much slower than that of another small molecule involved in cellular signaling, IP_3 (283 $\mu m^2/s$) [12]. Even when $[Ca^{2+}]_i$ is raised to 1 μM, with a view of saturating the immobile buffer sites, the diffusion coefficient remains relatively low (65 $\mu m^2/s$). The manner in which a Ca^{2+} transient is affected by the presence of a buffer is also linked to its intracellular localization, that is, whether the buffer is freely

Table I Properties of Ca^{2+}-binding Proteins and Artificial Ca^{2+} Buffers (Adapted from [24])

Buffer	K_D value(s)	k^+_{Ca2+} $M^{-1}s^{-1}$	Ca^{2+}/Mg^{2+} antagonism	No. of EF-hands (functional)	Refs.
PV	4–9 nM [1]	1.1×10^8	strong	3 (2)	[7]
	$K_{D,app}$ 50 nM [2]	1–2×10^7			[25]
	1 nM–100 nM [3]				
CB	$K_{D1} \approx 180$–240 nM [4]	$\approx 1.2 \times 10^7$	weak	6 (4)	[26]
	$K_{D2} \approx 410$–510 nM	$\approx 8.2 \times 10^7$			
CR	380–1500 nM	$\geq 10^8$ [5]	weak	6 (5)	[27,28]
BAPTA	130–800 nM	10^8–10^9	weak		[29,30]
EGTA	≈ 70 nM	3×10^6–1×10^7	weak		[26]

[1]This represents the K_D value in the absence of Mg^{2+}.

[2]The apparent dissociation constant ($K_{D,app}$) for Ca^{2+} depends heavily upon $[Mg^{2+}]_i$. The K_D value of 50 nM was obtained at a $[Mg^{2+}]_i$ of 0.16 mM. But at a $[Mg^{2+}]_i$ of 0.3–0.6 mM, which corresponds to the range encountered in neurons, $K_{D,app}$ lies around 80–150 nM. This will affect the on-rate of Ca^{2+} binding, lowering it to values of approximately 3–6×10^6 $M^{-1}s^{-1}$ (similar to that for EGTA).

[3]K_D values ranging from 1–100 nM were obtained under different experimental conditions (pH, ionic strength, etc.).

[4]CB contains two types of binding sites, which differ in their affinity for Ca^{2+} and their Ca^{2+}-binding on-rates.

[5]This is an approximation proposed by Edmonds *et al.* [31]. With its five Ca^{2+}-binding sites, CR would be expected to have different K_D and k^+_{Ca2+} values, as is the case with CB. The cited on-rate of 10^8 $M^{-1}s^{-1}$ most probably represents that of the fastest site(s).

diffusible or is bound to structures such as organelles, the plasma membrane, or cytoskeletal structures. The mobility effect may be further complicated if the buffer relocalizes as a result of changes in $[Ca^{2+}]_i$, as in the case for the Ca^{2+} "sensor" visinin-like protein III [13].

Ca^{2+} Buffers as One Component Contributing to Intracellular Ca^{2+} Homeostasis

Following an influx of Ca^{2+} ions into a cell, the role played by Ca^{2+} buffers is apparently a simple one, namely, to bind this cation and thereby lower $[Ca^{2+}]_i$. However, soluble buffers represent but one component of the intricate system implicated in Ca^{2+} homeostasis. A rise in $[Ca^{2+}]_i$ activates also the pumps involved in Ca^{2+} extrusion or Ca^{2+} uptake by organelles, such as the endoplasmic reticulum or mitochondria. These will remain operative until $[Ca^{2+}]_i$ has once again attained its steady-state level and the Ca^{2+} buffers have essentially reverted to their Ca^{2+}-free form, loading at this point being determined by their K_D and by basal $[Ca^{2+}]_i$. It is important to bear in mind that steady-state $[Ca^{2+}]_i$ is determined by the balance obtaining between Ca^{2+}-fluxes across the membranes surrounding the cytosol; it is not influenced by the presence of buffers *per se*. Neither the addition of a Ca^{2+} buffer such as PV or CB28k [7,14] to cells nor its elimination in knockout mice [11,15] affects basal $[Ca^{2+}]_i$, but rather prolongs the time ensuing until the steady-state level has been reattained (Fig.1). In the simplest case, the reduction in amplitude is inversely correlated to the lengthening of the transient, that is the time integral (the product of amplitude and time constant) remains unchanged by the presence of a Ca^{2+} buffer [16].

Intracellular Ca^{2+} transients are often characterized by highly complex patterns in time and space, since several relevant processes, such as Ca^{2+} entry via different pathways, Ca^{2+} binding to buffers (mobile and immobile), and sequestration by pumps, occur on the same temporal scale [7]. Furthermore, saturation of buffers may occur leading to nonlinear (e.g. supralinear) summation of Ca^{2+} signals as demonstrated in cerebellar Purkinje cells [17]. It is evident that the temporal and spatial aspects of Ca^{2+} signals are governed by an intricate interplay of the participating components. In biological systems, nonlinear summation of these signals is rather the rule than the exception, which makes it difficult to analyze the contribution of individual components [16].

Biological Effects of Ca^{2+} Buffers

CR, CB28k, and PV are three major representatives of EF-hand proteins that are classified as "buffers," and each is expressed within a specific subpopulation of neurons. The former two proteins possess, respectively, 5 and 4 Ca^{2+}-specific sites with presumably fast Ca^{2+}-binding kinetics (Table I), whereas PV has two Ca^{2+}/Mg^{2+}-mixed sites with a slow Ca^{2+}-onset rate. In knockout mice for any one of these proteins [11,15,18], the remaining two have been observed to be

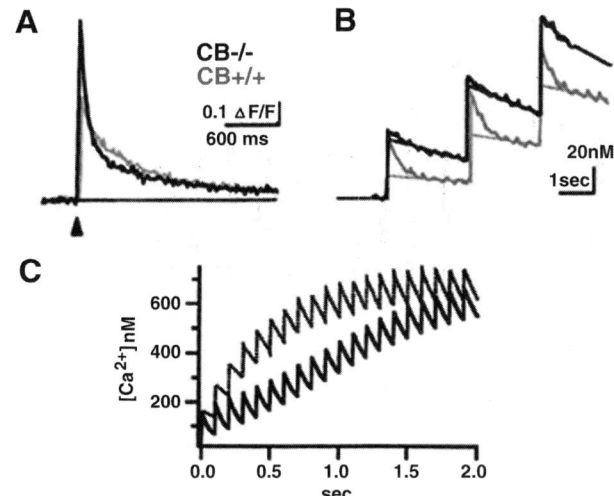

Figure 1 Effect of a fast and a slow buffer on Ca^{2+} transients. (A) Dendritic Ca^{2+} signals in Purkinje cells of CB+/+ (grey trace) and a CB-/- (black trace) mice, elicited by single-shock synaptic stimulation of the climbing fiber (arrowheads; for details, see [15]), CB markedly reduces the amplitude of a Ca^{2+} transient but prolongs the temporal decay of $[Ca^{2+}]_i$. (B) A series of Ca^{2+} transients evoked in a patched chromaffin cell by applying short (20 ms) depolarizing bursts just after break-in (black trace) and after loading with PV via a patch pipette (grey trace, for details, see [7]). PV does not affect the amplitude of the Ca^{2+} transients but increases the initial rate of decay of $[Ca^{2+}]_i$. (C) Simulated Ca^{2+} transients evoked in a neuron by 10 Hz stimulation in the absence (gray trace) or presence (black trace) of 200 μM PV (modified from [7]). Although the build-up of $[Ca^{2+}]_i$ is more rapid in the absence of PV, once the protein is Ca^{2+}-saturated, steady-state $[Ca^{2+}]_i$ is identical under both conditions. Hence, it is the time-course en route to the steady state that is significantly different. From this model, it is likewise evident that it is the metal-binding kinetic parameters that define the frequencies (stimulation intervals) at which a slow-onset buffer is effective in lowering the residual $[Ca^{2+}]_i$ between impulses.

neither upregulated nor expressed by any types of neurons other than those expressing them in wild-type animals. Hence, neurons are either incapable of inducing the expression of the other two buffers, or the relevant parameters (binding affinities, kinetics or diffusion) are unsuited to these acting as surrogates for the missing one.

Typical hallmarks of Ca^{2+} transients in excitable cells are their short duration (in the range of 10 to several 100 ms) and often restricted localization, within the axon of neurons, in the subplasmalemmal region of the soma, within parts of the dendrites or even within single spines only. Cytosolic Ca^{2+} buffers have a considerable influence on the spatiotemporal characteristics of such transients. Fast buffers such as CB or CR are able to buffer Ca^{2+} entering via channels from the extracellular space or being released from internal stores with virtually no delay. This reduces the peak amplitude of Ca^{2+} transients but prolongs the decay phase, since proteins such as CB28k or CR act as sources of Ca^{2+} at a later juncture (Fig. 1A). CB28k, on the one hand, is a fast enough buffer to slow down the Ca^{2+}-dependent inactivation of a Ca^{2+} channel and thereby even increases the total Ca^{2+} load [19]. PV, on the other hand, is too slow to affect the peak $[Ca^{2+}]_i$ in most cases, but it can significantly increase the rate of

$[Ca^{2+}]_i$ decay, as revealed in murine fast-twitch muscle fibers [11] or PV-injected chromaffin cells ([7], Fig. 1B).

In the presynaptic terminals or postsynaptic regions (soma, dendrites, and spines) of neurons, repetitive Ca^{2+} transients occurring at short time intervals are a typical physiological signaling event. Whether a particular Ca^{2+} buffer influences the spatiotemporal characteristics of these transients depends upon its concentration, Ca^{2+} affinity, binding kinetics, and diffusion rate. This is exemplified for PV during repetitive stimulations. At short pulse intervals (30 ms), paired-pulse modulation at the synapse between stellate or basket cells and Purkinje cells shifts from depression (Fig. 2A) to facilitation (Fig. 2B), if PV is absent (for example, in PV−/− mice). This phenomenon is attributable to the higher residual $[Ca^{2+}]_i$ obtaining in the absence of PV, which results in the second inhibitory postsynaptic current (IPSC) having a higher amplitude than that of the first (Fig. 2D). Clearly, if the pulses are delivered at longer intervals (300 ms), when residual $[Ca^{2+}]_i$ has decayed to basal levels irrespective of the presence of PV, paired-pulse depression is also observed in PV−/− mice (Fig. 2C). Steady state $[Ca^{2+}]_i$ level during burst-like action potentials (AP) depends upon $\Delta t/\tau$; Δt being the time interval between APs and τ the Ca^{2+} relaxation-time constant of individual Ca^{2+} transients. In the presence of PV, the time course until the equilibrium is reached is delayed, but eventually catches up if all PV molecules are saturated with Ca^{2+} (Fig. 1C; [7]).

In *Xenopus* oocytes, the injection or overexpression of PV induces elementary Ca^{2+} release events (Ca^{2+} puffs), which are elicited from discrete clusters of inositol 1,4,5 trisphosphate receptors (IP_3Rs) at low concentrations of IP_3 [20]. Ca^{2+} puff activity has also been detected after the injection of low concentrations of EGTA, but not after that of CB28k, which supports the idea that particular buffers are not simply interchangeable. This circumstance indicates that each buffer has distinct functions, which accord with its specific buffering properties. This is further illustrated by the finding that the changes in spine morphology of Purkinje cell dendrites (increased length and volume) observed in CB-deficient mice are not seen in PV-deficient ones [21]. In the fast-twitch muscles of PV-deficient mice, the volume of mitochondria, organelles also involved in Ca^{2+} sequestration helping to decrease $[Ca^{2+}]_i$ after Ca^{2+} transients (see Chapter 14 by Duchen), is almost twice as large as those in wild-type animals [22]. Ca^{2+} buffers thus constitute an integral part of the finely tuned system involved in Ca^{2+} homeostasis and have a profound effect on many aspects of Ca^{2+} signaling. The removal of such a buffer does not apparently trigger the obvious compensation mechanism (that is, the upregulation of another Ca^{2+} buffer), but leads rather to subtle changes in cell morphology or to discrete modulations in Ca^{2+} uptake or release systems. This may represent the cell's attempt to re-establish a "normal" state of Ca^{2+} homeostasis.

References

1. Celio, M., Pauls, T., and Schwaller, B. (1996). In Celio, M., Pauls T., and Schwaller, B., Eds., *Guidebook to the Calcium-Binding Proteins*, Oxford University Press, Oxford.
2. Kretsinger, R. H. (1980). Structure and evolution of calcium-modulated proteins. *CRC Crit. Rev. Biochem.* **8**, 119–174.
3. Lander, E. S. *et al.* (2001). Initial sequencing and analysis of the human genome. *Nature* **409**, 860–921.
4. Yap, K. L., Ames, J. B., Swindells, M. B., and Ikura, M. (1999). Diversity of conformational states and changes within the EF-hand protein superfamily. *Proteins* **37**, 499–507.
5. Celio, M., Pauls, T., and Schwaller, B. (1996). Introduction to EF-hand calcium-binding proteins, in Celio, M., Pauls, T., and Schwaller, B., Eds., Guidebook to the Calcium-Binding Proteins, pp. 15–20, Oxford University Press, Oxford.
6. Gilli, R., Lafitte. D., Lopez. C., Kilhoffer. M., Makarov, A., Briand. C., and Haiech J. (1998). Thermodynamic analysis of calcium and magnesium binding to calmodulin. *Biochemistry* **37**, 5450–5456.
7. Lee, S. H., Schwaller, B., and Neher. E. (2000). Kinetics of Ca^{2+} binding to parvalbumin in bovine chromaffin cells: implications for $[Ca^{2+}]$ transients of neuronal dendrites. *J. Physiol. (London)* **525**, 419–432.
8. Li-Smerin, Y., Levitan, E. S., and Johnson. J. W. (2001). Free intracellular $Mg^{(2+)}$ concentration and inhibition of NMDA responses in cultured rat neurons. *J Physiol. (London)* **533**, 729–743.
9. Watanabe, M. and Konishi, M. (2001). Intracellular calibration of the fluorescent Mg^{2+} indicator furaptra in rat ventricular myocytes. *Pflugers Arch.* **442**, 35–40.
10. Hou, T.-T., Johnson, J. D., and Rall, J. A. (1991). Parvalbumin content and Ca^{2+} and Mg^{2+} dissociation rates correlated with changes in relaxation rate of frog muscle fibres. *J. Physiol. (London)* **441**, 285–304.
11. Schwaller, B., Dick, J., Dhoot, G., Carroll, S., Vrbova, G., Nicotera, P., Pette, D., Wyss, A., Bluethmann, H., Hunziker, W., and Celio, M. R. (1999). Prolonged contraction-relaxation cycle of fast-twitch muscles in parvalbumin knockout mice. *Am. J. Physiol.* **276**, C395–403.

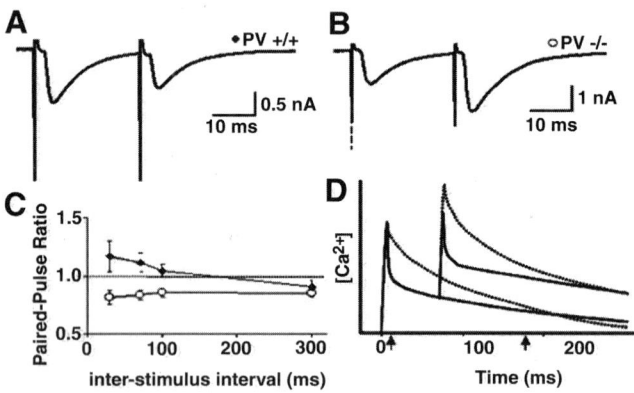

Figure 2 Parvalbumin affects short-term plasticity at the synapse between stellate or basket cells and Purkinje cells in the cerebellum. (A–B) Inhibitory postsynaptic currents (IPSCs) recorded from Purkinje cells during extracellular paired-pulse protocols (at an inter-stimulus interval (ISI) of 30 ms) of GABAergic interneurons. In the presence of PV (i. e. in PV+/+ mice), the second IPSC is depressed (A), whereas in its absence (i. e. in PV−/− mice), facilitation occurs (B). (C) The effect of PV on the paired-pulse ratio ($IPSC_2/IPSC_1$) is seen only when the ISI lies between 30 and 100 ms. When the interval is increased to 300 ms, the ratio does not differ between PV−/− and PV+/+ mice. (D) Within a presynaptic terminal, the peak $[Ca^{2+}]_i$ attained during the initial pulse is not affected by the absence or presence of PV, but the decay phase is slower in the former case (broken line). Hence, for a certain period of time (arrows), residual $[Ca^{2+}]_i$ will be elevated. If a second pulse is delivered during this period, then the peak $[Ca^{2+}]_i$ attained will be higher than during the first, a result that leads to enhanced facilitation. At the synapse between PV-containing stellate or basket cells and Purkinje cells, the effect of this Ca^{2+} buffer is maximal at a ISI of 30 ms (A, B, C: modified from [23]; D: modified from [24]).

12. Allbritton, N. L., Meyer, T., and Stryer, L. (1992). Range of messenger action of calcium ion and inositol 1,4,5,-trisphosphate. *Science* **258**, 1812–1815.
13. Spilker, C., Richter, K., Smalla, K. H., Manahan-Vaughan, D., Gundelfinger, E. D., and Braunewell, K. H. (2000). The neuronal EF-hand calcium-binding protein visinin-like protein-3 is expressed in cerebellar Purkinje cells and shows a calcium-dependent membrane association. *Neuroscience* **96**, 121–129.
14. Chard, P. S., Bleakman, D., Christakos, S., Fullmer, C. S., and Miller, R. J. (1993). Calcium buffering properties of calbindin D28k and parvalbumin in rat sensory neurones. *J. Physiol. (London)* **472**, 341–357.
15. Airaksinen, M. S., Eilers, J., Garaschuk, O., Thoenen, H., Konnerth, A., and Meyer, M. (1997). Ataxia and altered dendritic calcium signaling in mice carrying a targeted null mutation of the calbindin D28k gene. *Proc. Natl. Acad. Sci. USA* **94**, 1488–1493.
16. Neher, E. (1998). Usefulness and limitations of linear approximations to the understanding of Ca^{++} signals. *Cell Calcium* **24**, 345–357.
17. Maeda, H., Ellis-Davies, G. C., Ito, K., Miyashita, Y., and Kasai, H. (1999). Supralinear Ca^{2+} signaling by cooperative and mobile Ca^{2+} buffering in Purkinje neurons. *Neuron* **24**, 989–1002.
18. Schurmans, S., Schiffmann, S. N., Gurden, H., Lemaire, M., Lipp, H.-P., Schwam, V., Pochet, R., Imperato, A., Böhme, G. A., and Parmentier, M. (1997). Impaired LTP induction in the dentate gyrus of calretinin-deficient mice. *Proc. Natl. Acad. Sci. USA* **94**, 10415–10420.
19. Klapstein, G. J., Vietla, S., Lieberman, D. N., Gray, P. A., Airaksinen, M. S., Thoenen, H., Meyer, M., and Mody, I. (1998). Calbindin-D28k fails to protect hippocampal neurons against ischemia in spite of its cytoplasmic calcium buffering properties: evidence from calbindin-D28k knockout mice. *Neuroscience* **85**, 361–373.
20. John, L. M., Mosquera-Caro, M., Camacho, P., and Lechleiter, J. D. (2001). Control of $IP_{(3)}$-mediated Ca^{2+} puffs in *Xenopus laevis* oocytes by the Ca^{2+}-binding protein parvalbumin. *J Physiol.(London)* **535**, 3–16.
21. Vecellio, M., Schwaller, B., Meyer, M., Hunziker, W., and Celio, M. R. (2000). Alterations in Purkinje cell spines of calbindin D-28k and parvalbumin knock-out mice. *Eur. J. Neurosci.* **12**, 945–954.
22. Chen, G., Carroll, S., Racay, P., Dick, J., Pette, D., Traub, I., Vrbova, G., Eggli, P., Celio, M., and Schwaller, B. (2001). Deficiency in parvalbumin increases fatigue resistance in fast-twitch muscle and upregulates mitochondria. *Am. J. Physiol. (Cell Physiol.)* **281**, C114–C122.
23. Caillard, O., Moreno, H., Schwaller, B., Llano, I., Celio, M. R., and Marty, A. (2000). Role of the calcium-binding protein parvalbumin in short-term synaptic plasticity. *Proc. Natl. Acad. Sci. USA* **97**, 13372–13377.
24. Schwaller, B., Meyer, M., and Schiffmann, S. N. (2002). "New" functions for "old" proteins: The role of the calcium-binding proteins calbindin D-28k, calretinin and parvalbumin, in cerebellar physiology. Studies with knockout mice. *The Cerebellum* **1**, 248–251.
25. Eberhard, M. and Erne, P. (1994). Calcium and magnesium binding to rat parvalbumin. *Eur. J. Biochem.* **222**, 21–26.
26. Nagerl, U. V., Novo, D., Mody, I., Vergara, J. L. (2000). Binding kinetics of calbindin-D(28k) determined by flash photolysis of caged $Ca^{(2+)}$. *Biophys. J.* **79**, 3009–3018.
27. Schwaller, B., Durussel, I., Jermann, D., Herrmann, B., and Cox, J. A. (1997). Comparison of the Ca^{2+}-binding properties of human recombinant calretinin-22k and calretinin. *J. Biol. Chem.* **272**, 29663–29671.
28. Stevens, J. and Rogers, J. H. (1997). Chick calretinin: purification, composition, and metal binding activity of native and recombinant forms. *Protein Expr. Purif.* **9**, 171–181.
29. Tiffert, T. and Lew, V. L. (1997). Apparent Ca^{2+} dissociation constant of Ca^{2+} chelators incorporated non-disruptively into intact human red cells. *J. Physiol. (London)* **505**, 403–410.
30. Pethig, R., Kuhn, M., Payne, R., Adler, E., Chen, T. H., and Jaffe, L. F. (1989). On the dissociation constants of BAPTA-type calcium buffers. *Cell Calcium* **10**, 491–498.
31. Edmonds, B., Reyes, R., Schwaller, B., and Roberts, W. M. (2000). Calretinin modifies presynaptic calcium signaling in frog saccular hair cells. *Nat. Neurosci.* **3**, 786–790.

Mitochondria and Calcium Signaling, Point and Counterpoint

Michael R. Duchen

*Department of Physiology and UCL Mitochondrial Biology Group,
University College London, London, United Kingdom*

Introduction

Mitochondria can no longer be considered as static structures whose sole function is the unobtrusive manufacture of ATP. It is now clear that they also represent a storeroom for a number of potentially lethal proteins that are unleashed during programmed cell death and that they are significant participants in the detailed intracellular organization of cellular $[Ca^{2+}]_c$ signaling. While the expression in the mitochondrial membrane of Ca^{2+} transporting mechanisms was established years ago, the physiological significance of these pathways has only recently become apparent. There is now no question that mitochondria will take up and accumulate Ca^{2+} in all cells studied during the routine events of cellular $[Ca^{2+}]_c$ signaling, and that the pathway influences both mitochondrial function itself and the spatiotemporal and quantitative characteristics of the cellular $[Ca^{2+}]_c$ signal. As general issues relating to mitochondrial Ca^{2+} handling have recently been widely reviewed (e.g. [1–3]), I propose in this essay to highlight some of the more controversial and novel developments in the field over recent years and some of the mechanistic, quantitative, and comparative questions that remain.

Fundamentals

When energized mitochondria are exposed to raised $[Ca^{2+}]_c$, Ca^{2+} will move into the matrix. The accumulation of Ca^{2+} by mitochondria depends on the electrochemical gradient for Ca^{2+}, defined by the mitochondrial membrane potential, referred to as $\Delta\psi_m$, and by the intramitochondrial Ca^{2+} concentration ($[Ca^{2+}]_m$), which is kept low under resting conditions largely through the activity of a xNa^+/Ca^{2+} exchanger (see below). The mitochondrial potential is established and maintained by respiration and so requires a supply of oxygen and carbon substrate—collapse of $\Delta\psi_m$ due to anoxia, ischemia, damage to the respiratory chain, or the action of biochemical reagents, such as uncouplers, limit mitochondrial Ca^{2+} accumulation. One might ask whether any cell is ever really "at rest" *in situ* in the active organism in contrast to the artificial situation of the cell grown in culture—in which case, what would the normal $[Ca^{2+}]_m$ be in a cell in the living tissue and organism? The closest we get to that information comes from electron probe microanalysis (e.g. see [4]), but multiphoton imaging now holds the promise of being able to study mitochondria within intact tissues.

Machinery of Mitochondrial Ca^{2+} Movement

The Uniporter

Ca^{2+} is taken up through the mitochondrial inner membrane by a uniporter. Remarkably, we do not know the molecular identity or even the precise nature of this pathway. Is it a channel or a carrier? Flux rates are equivalent to those measured for fast gated pores, but rather slower than those seen for channels (see [3] for review). The activity of the uniporter shows little sensitivity to changes in temperature, and it also shows a wide spectrum of cation selectivity, together suggesting that it is a channel rather than a carrier. Ca^{2+} uptake via the uniporter is inhibited by ruthenium red (RuR), a compound that inhibits a variety of cation channels, including L-type plasmalemmal Ca^{2+} channels [5], ryanodine sensitive ER Ca^{2+} release channels [6], and vanilloid receptor operated channels [7], again suggesting that the uniporter may share channel properties.

One of the most interesting features of the uniporter is an apparent gating by $[Ca^{2+}]_c$, identified primarily through studies of the Ca^{2+} sensitivity of RuR-sensitive mitochondrial Ca^{2+} *efflux* in response to dissipation of $\Delta\psi_m$ [8,9]. Montero et al. [9], showed that, while collapse of $\Delta\psi_m$ prevents mitochondrial Ca^{2+} uptake, collapse of $\Delta\psi_m$ *after* the accumulation of mitochondrial Ca^{2+} inhibited mitochondrial efflux, i.e. all mitochondrial efflux pathways were inhibited by depolarization. Addition of Ca^{2+} to the depolarized Ca^{2+} loaded mitochondria then promoted mitochondrial Ca^{2+} release sensitive to RuR, suggesting release through the uniporter. This is consistent with suggestions that the uniporter is allosterically gated by $[Ca^{2+}]_o$ [8], an observation that may also explain why local $[Ca^{2+}]_c$ needs to be higher than one might expect from the behavior of a conducting Ca^{2+} channel in order to see significant increases in $[Ca^{2+}]_m$.

An uptake pathway with properties distinct from those of the uniporter has also been described [10,11] and dubbed the rapid uptake mode (RaM). This pathway has the capacity to transfer Ca^{2+} very rapidly into the mitochondria during the rising phase of a Ca^{2+} pulse. The properties of the pathway differ in different tissues [11], but in heart the pathway saturates quickly and is slow to reset after activation. Again, the functional significance of the pathway remains to be established.

VDAC

The mitochondrial outer membrane has been assumed to be permeant to small ions and so has been largely neglected in considerations of mitochondrial Ca^{2+} handling. However, the outer membrane may play a more significant role in modulating access of Ca^{2+} to the uniporter through the selectivity filter of the voltage-dependent anion channel (VDAC). It appears that VDAC is Ca^{2+} permeant and is regulated both by $[Ca^{2+}]$ and by RuR [12]. This finding raises questions about the extent to which the properties of the uptake pathway are defined by VDAC acting as a first filter. It is also tantalizing that VDAC appears to be part of the mitochondrial permeability transition pore (mPTP; see below), itself regulated by $[Ca^{2+}]_m$, as the mPTP provides a potential efflux pathway for Ca^{2+}, although the physiological relevance of this pathway is debated. Such studies point to the outer membrane as a significant permeability barrier that may itself be regulated.

Mitochondrial xNa$^+$/Ca^{2+} Exchange

The major route for Ca^{2+} efflux from mitochondria is a xNa$^+$/Ca^{2+} exchange. Identified about twenty years ago, it has a discrete pharmacology distinct from the plasmalemmal exchanger. The stoichiometry of the exchanger seems still to be controversial. Initially, it was thought to be an electroneutral 2Na$^+$/Ca^{2+} exchanger [13], but this has been questioned, as the exchanger can operate against a $[Ca^{2+}]$ gradient whose energy is over twice that of the Na$^+$ gradient [14]. Jung et al. [14] suggested a stoichiometry of 3Na$^+$/Ca^{2+}, in which case the operation of the exchanger will be dependent on $\Delta\psi_m$. The inhibition of mitochondrial Ca^{2+} efflux by mitochondrial depolarization (see above [9], and also [15]) supports this electrogenic stoichiometry. An electrogenic stoichiometry also predicts that Ca^{2+} efflux should be associated with mitochondrial depolarization. To my knowledge, this has not been documented.

The Set Point

Flux studies in isolated mitochondria revealed many years ago that mitochondria will take up Ca^{2+}. With small elevations of $[Ca^{2+}]_o$, the removal of Ca^{2+} from the matrix by the xNa$^+$/Ca^{2+} exchange may be sufficiently rapid so that net $[Ca^{2+}]_m$ changes little. As $[Ca^{2+}]_o$ rises above ~4–500 nM, the capacity of the exchanger is exceeded and mitochondria show net accumulation of Ca^{2+}. This was termed the "set point" for mitochondrial Ca^{2+} uptake by Nicholls and Crompton [16]. It is worth considering that Ca^{2+} flux into mitochondria is not necessarily synonymous with a net increase in $[Ca^{2+}]_m$, especially given our ignorance of the Ca^{2+} buffering capacity of the matrix. This is not purely semantic, as Ca^{2+} uptake by the uniporter is electrogenic and is therefore associated with small changes in $\Delta\psi_m$. Experimentally, changes in $\Delta\psi_m$ will reflect the rate of Ca^{2+} flux, and may therefore prove a more sensitive measurement of Ca^{2+} movement into mitochondria than measurement of $[Ca^{2+}]_m$. Further, net mitochondrial Ca^{2+} accumulation will be partly set by the activity of the xNa$^+$/Ca^{2+} exchanger—and we still know little about its regulation.

Many excitable cells respond to depolarization with a rise in $[Ca^{2+}]_c$, which rises rapidly and recovers with an initial rapid phase and a slower second phase that can even form a plateau [17–19]. It has been established in many preparations that the slow recovery phase reflects the redistribution of mitochondrial Ca^{2+} through the activity of the Na$^+$/Ca^{2+} exchanger, reflecting the set point, typically initiated at a $[Ca^{2+}]_c$ of ~500 nM. The operation of this system has functional consequences at presynaptic terminals, where the $[Ca^{2+}]_c$ plateau that follows repetitive stimulation, maintained by the reequilibration of mitochondrial Ca^{2+}, provides an elevated $[Ca^{2+}]_c$ baseline upon which subsequent stimulation initiates an enhanced synaptic response—the basis for post-tetanic potentiation of synaptic transmission [18,20]. It is also intriguing that the post stimulus plateau phase is not seen in nonexcitable cells following the transmission of $[Ca^{2+}]_c$ signals from ER to mitochondria. Certainly in astrocytes, $[Ca^{2+}]_m$ remains high for a very prolonged period after stimulation [21], suggesting that mitochondrial Ca^{2+} efflux must be very slow and perhaps the activity of the exchanger differs between tissues or cell types.

Quantitative Issues, Microdomains, and the Regulation of $[Ca^{2+}]_c$ Signals

There has been some debate about the quantitative relationships between ambient $[Ca^{2+}]$ and mitochondrial uptake.

In HeLa cells transfected with mitochondrially targetted aequorin and then permeabilized, net mitochondrial Ca^{2+} accumulation was only detectable if the added Ca^{2+} reached concentrations higher than 3 µM, while $[Ca^{2+}]_c$ signals evoked by IP_3 mobilizing agonists were far more effective at raising $[Ca^{2+}]_m$ even though the mean $[Ca^{2+}]_c$ signal might rise to <1 µM [22]. This led to the suggestion that mitochondria must be positioned at privileged sites close to the ER Ca^{2+} release sites where they would be exposed to microdomains of high local $[Ca^{2+}]_c$ sufficient to promote rapid Ca^{2+} uptake.

The proximity of mitochondria to SR or ER Ca^{2+} release sites has been further emphasized through evidence that focal, nonpropagating ER/SR Ca^{2+} release can cause a transient increase in $[Ca^{2+}]_m$ in mitochondria close to the release site. Thus, we found [23] that mitochondria in cardiomyocytes show spontaneous transient mitochondrial depolarizations that were dependent on local SR Ca^{2+} release and were blocked by inhibition of mitochondrial Ca^{2+} uptake. Hajnoczky et al. [24] have since shown that local $[Ca^{2+}]_c$ sparks may be associated with the direct transfer of Ca^{2+} to mitochondria visualized as transient increases in $[Ca^{2+}]_m$, which the group termed Ca^{2+} "marks." Further data from cardiomyocytes [25] strongly suggest that, in cardiomyocytes, mitochondria and SR must show very close coupling, as the transfer of Ca^{2+} to mitochondria in response to SR Ca^{2+} release with caffeine in permeabilized cells was sustained despite Ca^{2+} buffering by BAPTA sufficient to suppress the cytosolic signal. The transfer of Ca^{2+} was prevented by disrupting the cytoskeleton, suggesting that the maintained close apposition of mitochondria to SR was central to this signal.

The proximity of mitochondria to Ca^{2+} release sites has functional consequences for $[Ca^{2+}]_c$ signaling. Using $[Ca^{2+}]$ indicators in both mitochondria and ER in permeabilized cells, Csordas et al. [26] showed direct transfer of Ca^{2+} from ER to mitochondria and suggested that the proximity must be ~10–20 nm. This work was extended to show that mitochondrial Ca^{2+} uptake enhances the release of Ca^{2+} from the ER in response to IP_3 by acting as a local buffer [27]. Thus, by removing Ca^{2+} from the microdomain close to the IP_3 Ca^{2+} release channel, mitochondria prevent the Ca^{2+} dependent inactivation of the channel and facilitate ER Ca^{2+} release. This mechanism allows mitochondria to play a significant role in shaping the spatiotemporal patterning of $[Ca^{2+}]_c$ signals. In Xenopus oocytes, energization of mitochondria enhances the propagation and coordination of $[Ca^{2+}]_c$ waves [28], while in astrocytes, which express primarily IP_3 type 3 receptors, energized mitochondria serve as a spatial buffer that limit the rate and extent of propagation of $[Ca^{2+}]_c$ waves [21]. Microdomains of $[Ca^{2+}]_c$ regulated by mitochondria also play a significant role in the regulation of capacitative Ca^{2+} influx [29,30], suggesting that the mitochondria must be positioned close to the plasma membrane. The principle is very much as outlined above for the IP_3 receptor, as the Ca^{2+} influx channel is desensitized by Ca^{2+}. By keeping $[Ca^{2+}]_c$ low in microdomains close to the channels, mitochondria keep the channels open and facilitate Ca^{2+} influx through the channels.

In the blowfly salivary gland and in pituitary gonadotropes, mitochondrial Ca^{2+} uptake may even play a major role in defining the rates of oscillation of the IP_3 generated $[Ca^{2+}]_c$ signal [31,32], suggesting that the interplay between mitochondrial Ca^{2+} uptake and ER Ca^{2+} release contribute significantly to the temporal patterning of the $[Ca^{2+}]_c$ signal.

In pancreatic acinar cells, the mitochondria are concentrated into a band that isolates the secretory pole of these polarized cells, and they seem to act as a "firewall" that limits the spread of $[Ca^{2+}]_c$ signals from their initiation at the apical pole to the basal pole [33]. Furthermore, mitochondria localized close to the basal pole are more sensitive to local Ca^{2+} influx by capacitative entry, and so it seems that the positions of mitochondria within the cell may have a profound influence on their interaction with cellular $[Ca^{2+}]_c$ signals [34]. This issue alone is fascinating but unresolved—what dictates the positions mitochondria occupy within cells? Indeed, imaging mitochondria within cells shows that they move, and that the movement is erratic and unpredictable, and, to my knowledge, we have no idea what might be the functional significance of that movement.

Are $[Ca^{2+}]_c$ microdomains essential for mitochondria to sense changes in $[Ca^{2+}]_c$ associated with $[Ca^{2+}]_c$ signals? In permeabilized adrenal glomerulosa cells, Szabadkai et al. [35] found that graded additions of buffered external Ca^{2+} caused a graded but nonlinear increase in $[Ca^{2+}]_m$, showing a response even when the ambient $[Ca^{2+}]$ was only ~200–300 nM. Such data suggest that the close juxtaposition of mitochondria to Ca^{2+} sources is not an absolute requirement if they are to respond to $[Ca^{2+}]_c$ signals. A recent study in HeLa cells [36] also suggests that areas of maximal mitochondrial Ca^{2+} uptake may be divorced from areas of maximal proximity with ER in HELA cells. This study suggested that peripheral mitochondria had larger mitochondrial potentials and that this might provide a mechanism to enhance mitochondrial $[Ca^{2+}]_c$ accumulation into that mitochondrial population. The notion that mitochondria within a single cell may have different potentials remains contentious and the observation is critically dependent on the behavior of the fluorescent indicators used to measure $\Delta\psi_m$. This is probably not the place for further discussion of this issue, but my own view is that the question remains open and has not been satisfactorily resolved either way.

Impact of Ca^{2+} Uptake on Mitochondrial Function

In teleological terms, it seems that the major functional significance of mitochondrial Ca^{2+} uptake is in the regulation of mitochondrial metabolism. In the early 1990s it was shown that the three major rate-limiting enzymes of the citric acid cycle are all upregulated by Ca^{2+} (for review, see [37]). The question that remained was the functional issue—is mitochondrial Ca^{2+} uptake during physiological signaling sufficient for this mechanism to provide a functional regulation of metabolism? First suggestions that such a system operates in intact cells came from measurements of changes in mitochondrial redox state, reflected as changes in mitochondrial NADH and flavoprotein autofluorescence, in response to changes in $[Ca^{2+}]_c$ [5,38]. These observations showed clearly that (1) mitochondria must be taking up Ca^{2+} during $[Ca^{2+}]_c$ signals, and (2) that this was sufficient to activate the TCA cycle, causing increased net reduction of the coenzymes. More recently, transfection of cells with firefly luciferase allowed a clear and unequivocal demonstration that mitochondrial Ca^{2+} uptake increases mitochondrial ATP production [39]. The relative importance of this mechanism in the regulation of mitochondrial oxidative phosphorylation over the more traditional model, in which the rate of ATP generation is regulated largely by the ATP/ADP ratio, is not clear. It is very attractive to suggest that the transfer of Ca^{2+} from the cytosol to mitochondria during $[Ca^{2+}]_c$ signals represents a major mechanism to couple ATP supply with demand, as in almost all systems, increases in work are associated with increases in $[Ca^{2+}]_c$. Nevertheless, direct evidence for a significant role in intact systems is limited and there are conflicting data (see e.g. [40,41]).

The time course of the changes in $[Ca^{2+}]_m$ and in activation of the enzyme systems becomes crucial. $[Ca^{2+}]_c$ signals are typically brief, transient phenomena. Typically, it seems that the resultant mitochondrial activation is prolonged with respect to the change in $[Ca^{2+}]_c$ [5,42,43], and this in turn will be a function of the rate of mitochondrial Ca^{2+} efflux and the half-life of the activated states of the enzymes.

A further major question that is important in considering the impact of Ca^{2+} on mitochondrial function is, how high does $[Ca^{2+}]_m$ rise during these signals? There is not space here for a detailed discussion, but experiments using low-affinity variants of aequorin suggest that, at least in some mitochondria in some cells, $[Ca^{2+}]_m$ may rise into the millimolar region [44].

Mitochondrial Ca^{2+}, Disease, and Death

Most important, mitochondrial Ca^{2+} uptake may have profound consequences for mitochondrial function under pathological conditions. A combination of mitochondrial Ca^{2+} loading and oxidative stress and/or ATP depletion may promote opening of the mitochondrial permeability transition pore (mPTP). This appears to reflect a pathological conformation of a group of mitochondrial membrane proteins, notably the adenine nucleotide translocase (ANT) and VDAC, with the association of cyclophilin D, a regulatory protein that confers sensitivity of the complex to cyclosporin A, and a possible association of a number of other proteins, including the antiapoptotic Bcl-2, and the benzodiazepine receptor (see [45,46] for reviews). It is not clear whether the mPTP has any physiological function, but its complete opening under pathological conditions will lead inevitably to energetic collapse and cell death, and has been implicated in Ca^{2+} dependent cell death in reperfusion injury in the heart and in glutamate neurotoxicity in the CNS.

CODA

Our perception of the mitochondrion has changed radically over the last few years. Mitochondrial function is critical for the viability of the cell, mitochondria play an integral role in shaping cell signaling, and the dysfunction of the pathways necessary for these functions may trigger cell death. These are not trivial and peripheral functions but are central to cell life and cell death.

Acknowledgments

Work in my laboratory is supported by the Wellcome Trust, the Medical Research Council, and the Royal Society, whom I thank. I also thank Remi Dumollard, Jake Jacobson, and Laura Canevari for their invaluable discussion and comments on the manuscript. In so short an essay, it is not possible to describe all the fascinating activity in this field, and so I also apologize to those whose work is not cited here.

References

1. Rizzuto, R., Bernardi, P., and Pozzan, T. (2000). Mitochondria as all-round players of the calcium game. *J. Physiol.* **529**, 37–47.
2. Duchen, M. R. (2000). Mitochondria and calcium: from cell signaling to cell death. *J Physiol.* **529**, 57–68.
3. Gunter, T. E., Buntinas, L., Sparagna, G., Eliseev, R., and Gunter, K. (2000). Mitochondrial calcium transport: mechanisms and functions. *Cell Calcium* **28**, 285–296.
4. Isenberg, G., Han, S., Schiefer, A., and Wendt-Gallitelli, M. F. (1993). Changes in mitochondrial calcium concentration during the cardiac contraction cycle. *Cardiovasc Res.* **27**, 1800–1809.
5. Duchen, M. R. (1992). Ca^{2+}-dependent changes in the mitochondrial energetics in single dissociated mouse sensory neurons. *Biochem J.* **283**, 41–50.
6. Lukyanenko, V., Gyorke, I., Subramanian, S., Smirnov, A., Wiesner, T. F., and Gyorke, S. (2000). Inhibition of Ca^{2+} sparks by ruthenium red in permeabilized rat ventricular myocytes. *Biophys J.* **79**, 1273–1284.
7. Wood, J. N., Winter, J., James, I. F., Rang, H. P., Yeats, J., and Bevan, S. (1998). Capsaicin-induced ion fluxes in dorsal root ganglion cells in culture. *J. Neurosci,* **8**, 3208–3220.
8. Igbavboa, U. and Pfeiffer, D. R. (1988). EGTA inhibits reverse uniport-dependent Ca^{2+} release from uncoupled mitochondria. Possible regulation of the Ca^{2+} uniporter by a Ca^{2+} binding site on the cytoplasmic side of the inner membrane. *J. Biol. Chem.* **263**, 1405–1412.
9. Montero, M., Alonso, M. T., Albillos, A., Garcia-Sancho, J., and Alvarez, J. (2001). Mitochondrial Ca^{2+} induced Ca^{2+} release mediated by the Ca^{2+} uniporter. *Mol. Biol. Cell* **12**, 63–71.
10. Sparagna, G. C., Gunter, K. K., Sheu, S. S., and Gunter, T. E. (1995). Mitochondrial calcium uptake from physiological-type pulses of calcium. A description of the rapid uptake mode. *J. Biol. Chem.* **270**, 27510–27515.
11. Buntinas, L., Gunter, K. K., Sparagna, G. C., and Gunter, T. E. (2001). The rapid mode of calcium uptake into heart mitochondria (RaM): comparison to RaM in liver mitochondria. *Biochim Biophys Acta* **1504**, 248–261.
12. Gincel, D., Zaid, H., and Shoshan-Barmatz, V. (2001). Calcium binding and translocation by the voltage-dependent anion channel: a possible regulatory mechanism in mitochondrial function. *Biochem. J.* **358**, 147–155.
13. Brand, M. D. (1985). The stoichiometry of the exchange catalysed by the mitochondrial calcium/sodium antiporter. *Biochem. J.* **229**, 161–166.
14. Jung, D. W., Baysal, K., and Brierley, G. P. (1995). The sodium-calcium antiport of heart mitochondria is not electroneutral. *J. Biol. Chem.* **270**, 672–678.

15. Bernardi, P. and Azzone, G. F. (1982). A membrane potential-modulated pathway for Ca^{2+} efflux in rat liver mitochondria. *FEBS Lett.* **139**, 13–16.
16. Nicholls, D. G. and Crompton, M. (1980). Mitochondrial calcium transport. *FEBS Lett.* **111**, 261–268.
17. Thayer, S. A. and Miller, R. J. (1990). Regulation of the intracellular free calcium concentration in single rat dorsal root ganglion neurones in vitro. *J Physiol.* **425**, 85–115.
18. David, G., Barrett, J. N., and Barrett, E. F. (1998). Evidence that mitochondria buffer physiological Ca^{2+} loads in lizard motor nerve terminals. *J Physiol.* **509**, 59–65.
19. Babcock, D. F., Herrington, J., Goodwin, P. C., Park, Y. B., and Hille, B. (1997). Mitochondrial participation in the intracellular Ca^{2+} network. *J. Cell Biol.* **136**, 833–844.
20. Tang, Y. and Zucker, R. S. (1997). Mitochondrial involvement in post-tetanic potentiation of synaptic transmission. *Neuron* **18**, 483–491.
21. Boitier, E., Rea, R., and Duchen, M. R. (1999). Mitochondria exert a negative feedback on the propagation of intracellular Ca^{2+} waves in rat cortical astrocytes. *J. Cell Biol.* **145**, 795–808
22. Rizzuto, R., Brini, M., Murgia, M., and Pozzan, T. (1993). Microdomains with high Ca^{2+} close to IP_3-sensitive channels that are sensed by neighboring mitochondria. *Science* **262**, 744–747.
23. Duchen, M. R., Leyssens, A., and Crompton, M. (1998). Transient mitochondrial depolarizations reflect focal sarcoplasmic reticular calcium release in single rat cardiomyocytes. *J. Cell Biol.* **142**, 975–988.
24. Pacher, P., Thomas, A. P., and Hajnoczky, G. (2002). Ca^{2+} marks: miniature calcium signals in single mitochondria driven by ryanodine receptors. *Proc. Natl. Acad. Sci. USA* **99**, 2380–2385.
25. Sharma, V. K., Ramesh, V., Franzini-Armstrong, C., and Sheu, S. S. (2000). Transport of Ca^{2+} from sarcoplasmic reticulum to mitochondria in rat ventricular myocytes. *J. Bioenerg. Biomembr.* **32**, 97–104.
26. Csordas, G., Thomas, A. P., and Hajnoczky, G. (1999). Quasi-synaptic calcium signal transmission between endoplasmic reticulum and mitochondria. *EMBO J.* **18**, 96–108.
27. Hajnoczky, G., Hager, R., and Thomas, A. P. (1999). Mitochondria suppress local feedback activation of inositol 1,4,5-trisphosphate receptors by Ca^{2+}. *J. Biol. Chem.* **274**, 14157–14162.
28. Jouaville, L. S., Ichas, F., Holmuhamedov, E. L., Camacho, P., and Lechleiter, J. D. (1995). Synchronization of calcium waves by mitochondrial substrates in *Xenopus laevis* oocytes. *Nature* **377**, 438–441.
29. Hoth, M., Fanger, C. M., and Lewis, R. S. (1997). Mitochondrial regulation of store-operated calcium signaling in T lymphocytes. *J. Cell Biol.* **137**, 633–648.
30. Gilabert, J. A., Bakowski, D., and Parekh, A. B. (2001). Energized mitochondria increase the dynamic range over which inositol 1,4,5-trisphosphate activates store-operated calcium influx. *EMBO J.* **20**, 2672–2679.
31. Zimmermann, B. (2000). Control of $InsP_3$-induced Ca^{2+} oscillations in permeabilized blowfly salivary gland cells: contribution of mitochondria. *J. Physiol.* **525**, 707–719.
32. Kaftan, E. J., Xu, T., Abercrombie, R. F., and Hille, B. (2000). Mitochondria shape hormonally induced cytoplasmic calcium oscillations and modulate exocytosis. *J. Biol. Chem.* **275**, 25465–25470.
33. Tinel, H., Cancela, J. M., Mogami, H., Gerasimenko, J. V., Gerasimenko, O. V., Tepikin, A. V., and Petersen, O. H. (1999). Active mitochondria surrounding the pancreatic acinar granule region prevent spreading of inositol trisphosphate-evoked local cytosolic Ca^{2+} signals. *EMBO J.* **18**, 4999–5008.
34. Park, M. K., Ashby, M. C., Erdemli, G., Petersen, O. H., and Tepikin, A. V. (2001). Perinuclear, perigranular and sub-plasmalemmal mitochondria have distinct functions in the regulation of cellular calcium transport. *EMBO J.* **20**, 1863–1874.
35. Szabadkai, G., Pitter, J. G., and Spat, A. (2001). Cytoplasmic Ca^{2+} at low submicromolar concentration stimulates mitochondrial metabolism in rat luteal cells. *Pflugers Arch.* **441**, 678–685.
36. Collins, T. J., Berridge, M. J., Lipp, P., and Bootman, M. D. (2002). Mitochondria are morphologically and functionally heterogeneous within cells. *EMBO J.* **21**, 1616–1627.
37. McCormack, J. G., Halestrap, A. P., and Denton, R. M. (1990). Role of calcium ions in regulation of mammalian intramitochondrial metabolism. *Physiol Rev.* **70**, 391–425.
38. Pralong, W. F., Hunyady, L., Varnai, P., Wollheim, C. B., and Spat, A. (1992). Pyridine nucleotide redox state parallels production of aldosterone in potassium-stimulated adrenal glomerulosa cells. *Proc. Natl .Acad. Sci. USA* **89**, 132–136.
39. Jouaville, L. S., Pinton, P., Bastianutto, C., Rutter, G. A., and Rizzuto, R. (1999). Regulation of mitochondrial ATP synthesis by calcium: evidence for a long-term metabolic priming. *Proc. Natl. Acad. Sci. USA* **96**, 13807–13812.
40. Moravec, C. S., Desnoyer, R. W., Milovanovic, M., Schluchter, M. D., and Bond, M. (1997). Mitochondrial calcium content in isolated perfused heart: effects of inotropic stimulation. *Am. J. Physiol.* **273**, H1432–H1439.
41. Horikawa, Y., Goel, A., Somlyo, A. P., and Somlyo, A. V. (1998). Mitochondrial calcium in relaxed and tetanized myocardium. *Biophys. J.* **74**, 1579–1590.
42. Robb-Gaspers, L. D., Burnett, P., Rutter, G. A., Denton, R. M., Rizzuto, R., and Thomas, A. P. (1998). Integrating cytosolic calcium signals into mitochondrial metabolic responses. *EMBO J.* **17**, 4987–5000.
43. Hajnoczky, G., Robb-Gaspers, L. D., Seitz, M. B., and Thomas, A. P. (1995). Decoding of cytosolic calcium oscillations in the mitochondria. *Cell* **82**, 415–424.
44. Montero, M., Alonso, M. T., Carnicero, E., Cuchillo-Ibanez, I., Albillos, A., Garcia, A. G., Garcia-Sancho, J., and Alvarez, J. (2000). Chromaffin-cell stimulation triggers fast millimolar mitochondrial Ca^{2+} transients that modulate secretion. *Nature Cell Biol.* **2**, 57–61.
45. Crompton, M. (1999). The mitochondrial permeability transition pore and its role in cell death. *Biochem. J.* **341**, 233–249.
46. Jacobson, J. and Duchen, M. R. (2001). 'What nourishes me, destroys me': towards a new mitochondrial biology. *Cell Death Differ.* **8**, 963–966.

EF-Hand Proteins and Calcium Sensing: The Neuronal Calcium Sensors

Jamie L. Weiss and Robert D. Burgoyne
*The Physiological Laboratory,
The University of Liverpool,
Liverpool, United Kingdom*

Introduction

EF-hand calcium (Ca^{2+})-binding proteins play many important roles in Ca^{2+}-homeostasis and Ca^{2+}-signaling mechanisms. The EF-hand containing protein calmodulin has been most extensively studied but the EF-hand motif is the most prevalent and widely distributed protein domain in Ca^{2+} signaling. The neuronal Ca^{2+}-sensor (NCS) proteins are a closely-related family of Ca^{2+}-binding proteins whose physiological functions have begun to emerge in recent years.

EF-hand Ca^{2+}-binding proteins fall into two main categories: Ca^{2+} buffers, which act as Ca^{2+} chelators but do not undergo a conformational change, and Ca^{2+} sensors, which upon Ca^{2+} binding undergo a conformational change and transfer the signal to other proteins. This conformational change can alter interactions of EF-hand Ca^{2+} sensing proteins with target proteins or cause a conformation change in proteins already bound to the sensor concurrent with the Ca^{2+} concentration shift. Members of the neuronal Ca^{2+}-sensor (NCS) family (Table I) include proteins expressed only in the retina (e.g. recoverin) and others expressed mainly in neuronal and neuroendocrine cells such as the neurocalcins (*visinin-like proteins* [VILIPs]) and NCS-1 (frequenin). The NCS proteins are ~22 kDa, high-affinity Ca^{2+}-binding proteins, with ~30–70% protein sequence identity to each another, and most members of the family are N-terminally myristoylated. A Ca^{2+}-myristoyl switch mechanism has been proposed for some members of the family in which the proteins bind membranes in a Ca^{2+}-dependent manner via exposure of the myristoyl-group upon Ca^{2+} binding (Table I). The myristoyl group may also be important for protein-protein interactions. Two recent reviews [1,2] cover the NCS proteins in detail, and this chapter will emphasize recent developments in the field.

Class A. Neuronal Calcium Sensor 1 (Frequenin)

Drosophila mutants overexpressing frequenin were found to have frequency-dependent facilitation of evoked neurotransmission at the neuromuscular junction [3], and overexpression of NCS-1 results in enhancement of evoked exocytosis in neuroendocrine cells via an indirect effect of NCS-1 [4]. It is now known that NCS-1 regulates both voltage-gated (VG) Ca^{2+} channels and potassium (K^+) channels. NCS-1 functions in a G-protein-coupled receptor-mediated pathway of VG Ca^{2+} channel inhibition in chromaffin cells that relies on the activity of a Src-like tyrosine kinase [5,6]. NCS-1 has also been shown to increase the current amplitude for Kv4 (A-type) K^+ channels [7]. In a pathway involving glial-derived neurotrophic factor (GDNF), NCS-1 has been implicated in the facilitation of N-type channels in neuromuscular synapses [8]. NCS-1 knockouts and overexpression mutants in *C. elegans* demonstrate that NCS-1 has a role in learning and memory [9].

Table I General Properties of the NCS Family of Proteins

Protein name	Myristoyl switch mechanism	Structure determined	Ca^{2+} binding affinity	Tissue/cell localization	Subcellular localization	Pathways/ molecular targets	Possible physiological function(s)
NCS-1	No	Yes	0.3 µM	Brain, adrenal gland, COS cells, mouse inner ear, insulin secreting cells, heart, etc.	Golgi, synapses, dendrites, synaptic-like vesicles	PI4 kinase, calcineurin, ion channels	Control of neurotransmitter release, Ca^{2+} and K^+ channel regulation, learning and memory
Recoverin	Yes	Yes	2.1 µM	Photoreceptors	Photoreceptor membranes	Rhodopsin kinase	Retinal phototransduction
VILIP 1	Yes	No	1 µM	Brain-highest expression in cerebellar granule cells, and hippocampus, retina	Unknown	Actin, tubulin, ds RNA, cGMP	Cytoskeletal regulation, cGMP pathways
VILIP 2	Yes	No	Unknown	Brain-except cerebellum	Unknown	Unknown	Unknown
VILIP 3	Yes	No	Unknown	Highest expression in cerebellar purkinje cells	Unknown	Unknown	Unknown
Neurocalcin δ	Yes	Yes	0.6 µM	Highest expression in cerebellar purkinje cells	Plasma membrane, Golgi complex	Actin, tubulins and clathrin	Cytoskeletal regulation, protein trafficking
Hippocalcin	Yes	No	5 µM	Highest expression in hippocampus	Hippocampal pyramidal neurons-soma and dendrites	PLD, NAIP, Cdc42, MLK2	Apoptosis, MAP kinase pathway
GCAP 1	Yes	No	0.26 µM	Photoreceptors	Photoreceptor membranes	Guanylate cyclase	Retinal phototransduction
GCAP 2	Reversed	Yes	0.25 µM	Photoreceptors	Photoreceptor membranes	Guanylate cyclase	Retinal phototransduction
GCAP 3	Yes	No	0.25 µM	Photoreceptors	Photoreceptor membranes	Guanylate cyclase	Retinal phototransduction
KChIP 1	Unknown	No	Unknown	Brain	Plasma membrane	K^+ channels	K^+ channel regulation
KChIP 2	No myr. group	No	Unknown	Brain, heart	Plasma membrane	K^+ channels	K^+ channel regulation
KChIP 3	No myr. group	No	14 µM	Brain, testes	Plasma membrane	K^+ channels, DNA, Presenilins	K^+ channel regulation, transcriptional repressor for pain modulation
KChIP 4	No myr. group	No	Unknown	Brain	Membranes	K^+ channels	K^+ channel regulation, KchIP4b abolishes fast inactivation of Kv4 A-type currents.

Further evidence for a role of NCS-1 in learning and memory comes from the observation of an increase in brain NCS-1-mRNA levels following induction of long-term potentiation [10]. It has been demonstrated both biochemically and in cells that myristoylated NCS-1 binds membranes in a Ca^{2+}-independent manner [11,12]. It has also been reported that yeast frequenin does not need Ca^{2+} for interaction with its major target PI-4-kinase [13]. This suggests that NCS-1 does not rely on the Ca^{2+}-myristoyl switch mechanism for binding interactions. NCS-1 has multiple binding partners in adrenal chromaffin cell fractions; some of these interactions are Ca^{2+}-independent and others are Ca^{2+}-dependent [11].

Class B. Neurocalcins (VILIPs) and Hippocalcin

Neurocalcins (VILIPs) are expressed in certain classes of neurons (Table I). Both VILIPs and hippocalcin have been shown to have a classic Ca^{2+}-myristoyl switch mechanism in cells [12]. VILIP1 and neurocalcin δ have been shown to bind actin, tubulins, and, in the case of neurocalcin δ, clathrin as well [14,15]. These interactions appear to link neurocalcins to cytoskeletal regulation and possibly vesicle trafficking mechanisms. VILIP1 has also been shown to interact with double stranded RNA in a Ca^{2+}-dependent manner [16] and also increases cGMP levels in PC12 cells and cerebellar granule neurons [17]. VILIP1 expression promoted cell death, tau phosphorylation, and appeared to have a role in Ca^{2+}-mediated cytotoxicity in PC12 cells [18]. Hippocalcin is predominantly expressed in mammalian hippocampus and has been implicated in interactions with neuronal apotosis inhibitory protein (NAIP), phospholipase D (PLD)/Cdc42, and mixed lineage kinase 2 (MLK2) [19–21]. These studies link hippocalcin to endocytosis, MAP kinase, and apoptosis pathways.

Class C. Recoverins

Recoverins are the best-studied members of the NCS protein family. The structure of recoverin has been resolved both in the Ca^{2+}-free and Ca^{2+}-bound states. Analysis of its biochemical properties and structure has revealed that recoverin uses the Ca^{2+}-myristoyl switch mechanism. Recoverins are expressed in rod photoreceptors and have an important role in the Ca^{2+}-signaling of retinal phototransduction [22]. Recoverin regulates cGMP levels indirectly via a Ca^{2+}-dependent interaction with rhodopsin kinase inhibiting the kinase in the dark when Ca^{2+} levels are high [22].

Class D. Guanylate Cyclase Activating Proteins

Like recoverins, guanylate cyclase activating proteins (GCAPs) are Ca^{2+} sensors expressed in photoreceptors that play an important role in the signal transduction pathways of the retina by regulating guanylate cyclase (GC) activity. In the retina Ca^{2+} levels are low in the light. In such low Ca^{2+} conditions the GCAPs activate GC. This allows cGMP levels to be directly increased to allow activation of the cyclic-nucleotide gated channels. GCAP2 and GCAP3 directly activate GC in their Ca^{2+}-free state and at increased levels inhibit GC. GCAP2 is bound to membranes at low Ca^{2+} levels and as Ca^{2+} is elevated it dissociates from membranes [22]. EF-hand 1 in all the NCS family members is unable to bind Ca^{2+}. Recently it was determined that GCAP2 interacts with GC via this EF1 domain [22]. This may be an evolutionarily conserved mechanism in the NCS proteins, whereby the ability of EF1 to bind Ca^{2+} was lost in order to gain a protein interaction domain.

Class E. K^+ Channel Interacting Proteins

K^+ channel interacting proteins (KChIPs) [24,25] regulate A-type K^+ channel currents. KChIP1 is the only member of the KChIP subfamily that is myristoylated. Modulation of Kv4.2 K^+ channels in CHO cells and Kv4.2 and Kv4.3 K^+ channels in *Xenopus* oocytes via arachidonic acid has been shown to be dependent on KChIP1 [26]. KChIP2 also regulates KV4.2 and Kv4.3 K^+ channels, and KChIP2 knockout mice are highly susceptible to ventricular tachycardia due to the loss of a transient outward K^+ channel current in the heart [27]. KChIP3 is the same protein as calsenilin, which interacts with presenilin-1 and 2. Presenilin-1 mutations are the most common cause of familial Alzheimer's disease, and calsenilin interacts with the endogenous 20 kDa C-terminal fragment of presenilin 2 that is a product of regulated proteolytic cleavage [28]. KChIP3/calsenilin is also the same protein as downstream regulatory element antagonist modulator (DREAM), which is a Ca^{2+}-regulated transcriptional repressor involved in pain modulation [29,30]. An alternative splice variant of KChIP4 (KChIP4a) encodes a protein with a novel N-terminus called the KIS (K-channel inactivation suppressor) domain. The KIS domain appears to be important for the abolishment of fast inactivation of Kv4.3 channels [31].

Future Perspectives for the NCS Protein Family

The NCS proteins are already known to have multiple binding protein partners. More studies are needed to characterize these interactions so that we can place NCS proteins in known pathways. Further *in vivo* knockout and overexpression studies will reveal more about the roles of the NCS proteins and clarify their range of physiological functions in the regulation of neuronal activity.

References

1. Burgoyne, R. D. and Weiss, J. L. (2001). The neuronal calcium sensor family of Ca^{2+}-binding proteins. *Biochem. J.* **353**, 1–12.

2. Braunewell, K.-H. and Gundelfinger, E. D. (1999). Intracellular neuronal calcium sensor proteins: a family of EF-hand calcium-binding proteins in search of a function. *Cell Tissue Res.* **295**, 1–12.
3. Pongs, O., Lindemeier, J., Zhu, X. R., Theil, T., Endelkamp, D., Krah-Jentgens, I., Lambrecht, H.-G., Koch, K. W., Schwemer, J., Rivosecchi, R., Mallart, A., Galceran, J., Canal, I., Barbas, J. A., and Ferrus, A. (1993). Frequenin—a novel calcium-binding protein that modulates synaptic efficacy in the Drosophila nervous system. *Neuron* **11**, 15–28.
4. McFerran, B. W., Graham, M. E., and Burgoyne, R. D. (1998). NCS-1, the mammalian homologue of frequenin is expressed in chromaffin and PC12 cells and regulates neurosecretion from dense-core granules. *J. Biol. Chem.* **273**, 22768–22772.
5. Weiss, J. L., Archer, D. A., and Burgoyne, R. D. (2000). NCS-1/frequenin functions in an autocrine pathway regulating Ca^{2+} channels in bovine adrenal chromaffin cells. *J. Biol. Chem.* **275**, 40082–40087.
6. Weiss, J. L. and Burgoyne, R. D. (2001). Voltage-independent inhibition of P/Q-type Ca^{2+} channels in adrenal chromaffin cells via a neuronal Ca^{2+} sensor-1-dependent pathway involves Src-family tyrosine kinase. *J. Biol. Chem.* **276**, 44804–44811.
7. Nakamura, T. Y., Pountney, D. J., Ozaita, A., Nandi, S., Ueda, S., Rudy, B., and Coetzee, W. A. (2001). A role for frequenin, a Ca^{2+} binding protein, as a regulator of Kv4 K^+ currents. *Proc. Natl. Acad. Sci. USA* **98**, 12808–12813.
8. Wang, C.-Y., Yang, F., He, X., Chow, A., Du, J., Russell, J. T., and Lu, B. (2001). Ca^{2+} binding protein frequenin mediates GDNF-induced potentiation of Ca^{2+} channels and transmitter release. *Neuron* **32**, 99–112.
9. Gomez, M., De Castro, E., Guarin, E., Sasakura, H., Kuhara, A., Mori, I., Bartfai, T., Bargmann, C. I., and Nef, P. (2001). Ca^{2+} signalling via the neuronal calcium sensor-1 regulates associative learning and memory in C. elegans. *Neuron* **30**, 241–248.
10. Genin, A., Davis, S., Meziane, H., Doyere, V., Jeromin, A., Roder, J., Mallet, J., and Laroche, S. (2001). Regulated expression of the neuronal calcium sensor-1 gene during long-term potentiation in the dentate gyrus in vivo. *Neuroscience* **106**, 571–577.
11. McFerran, B. W., Weiss, J. L., and Burgoyne, R. D. (1999). Neuronal Ca^{2+}-sensor 1: characterisation of the myristoylated protein, its cellular effects in permeabilised adrenal chromaffin cells, Ca^{2+}-independent membrane-association and interaction with binding proteins suggesting a role in rapid Ca^{2+} signal transduction. *J. Biol. Chem.* **274**, 30258–30265.
12. O' Callaghan, D.W., Ivings, L., Weiss, J. L, Ashby, M. C., Tepikin, A. V., and Burgoyne, R. D. (2002). Differential use of myristoyl groups on neuronal calcium sensor proteins as a determinant of spatio-temporal aspects of Ca^{2+}-signal transduction. *J. Biol. Chem.* **277**, 14227–14237.
13. Hendricks, K. B., Wang, B. Q., Schnieders, E. A., and Thorner, J. (1999). Yeast homologue of neuronal frequenin is a regulator of phosphatidyl-inositol-4-OH kinase. *Nature Cell Biol.* **1**, 234–241.
14. Mornet, D. and Bonet-Kerrache, A. (2001). Neurocalcin-actin interaction. *Biochim. Biophys. Acta* **1549**, 197–203.
15. Ivings L., Pennington S. R., Jenkins, R., Weiss, J. L., and Burgoyne, R. D. (2002). Identification of calcium-dependent binding partners for the neuronal calcium sensor protein neurocalcin δ: interaction with actin, clathrin and tubulin. *Biochem. J.* **363**, 599–608.
16. Mathisen, P. M., Johnson, J. M., Kawczak, J. A., and Tuohy, V. K. (1999). Visinin-like protein (VILIP) is a neuron-specific calcium-dependent double-stranded RNA-binding protein. *J. Biol. Chem.* **274**, 31571–31576.
17. Braunwell, K.-H., Brackmann, M., Schaupp, M., Spilker, C., Anand, R., and Gundelfinger, E. D. (2001). Intracellular neuronal calcium sensor (NCS) protein VILIP-1 modulates cGMP signalling pathways in transfected neural cells and cerebellar granule neurons. *J. Neurochem.* **78**, 1277–1286.
18. Schnurra, I., Bernstein, H.-G., Riederer, P., and Braunewell, K.-H. (2001). The neuronal calcium sensor protein VILIP-1 is associated with amyloid plaques and promotes cell death and tau phosphorylation in vitro: a link between calcium sensors and Alzheimer's disease? *Neurobiol. Dis.* **8**, 900–909.
19. Mercer, W. A., Korhonen, L., Skoglosa, Y., Olssen, P.-A., Kukknen, J. P., and Lindholm, D. (2000). NAIP interacts with hippocalcin and protects neurons against calcium-induced cell death through caspase-3-dependent and -independent pathways. *EMBO J.* **19**, 3597–3607.
20. Hyun, J. K., Yon, C., Kim, Y. S., Lee, K. H., and Han, J. S. (2000). Role of hippocalcin in Ca^{2}-induced activation of phospholipase D. *Molecules Cells* **10**, 669–677.
21. Nagata, K., Puls, A., Futter, C., Aspenstrom, P., Schaefer, E., Nakata, T., Hirokawa, N., and Hall, A. (1998). The Map kinase kinase kinase MLK2 co-localizes with activated JNK along microtubultes and associates with kinesin superfamily motor KIF3. *EMBO J.* **17**, 149–158.
22. Palczewski, K., Polans, A., Baehr, W., and Ames, J. B. (2000). Ca^{2+}-binding proteins in the retina: structure, function and the etiology of human visual diseases. *BioEssays* **22**, 337–350.
23. Ermilov, A. N., E.V., O. and Dizhoor, A. M. (2001). Instead of binding calcium, one of the EF-hand structures in guanyl cyclase activating protein-2 is required for targeting photoreceptor guanylyl cylcase. *J. Biol. Chem.* **276**, 48143–48148.
24. An, W. F., Bowlby, M. R., Bett, M., Cao, J., Ling, H. P., Mendoza, G., Hinson, J. W., Mattsson, K. I., Strassle, B. W., Trimmer, J. S., and Rhodes, K. J. (2000). Modulation of A-type potassium channels by a family of calcium sensors. *Nature* **403**, 553–556.
25. Li, M., and Adelman, J. P. (2000). CHIPping away at potassium channel regulation. *Nature Neurosci.* **3**, 202–204.
26. Holmqvist, M. H., Cao, J., Knoppers, M. H., Jurman, M. E., Distefano, P. S., Rhodes, K. J., Xie, Y., and An, F. W. (2001). Kinetic modulation of Kv4-mediated A-current by arachidonic acid is dependent on potassium channel interacting proteins. *J. Neurosci.* **21**, 4154–4161.
27. Kuo, H.-C., Cheng, C.-F., Clark, R. B., Lin, J. J.-C., Lin, J. L.-C., Hoshijima, M., Nguyen-Tran, V. T. B., Gu, Y., Ikeda, Y., Chu, P.-H., Ross, J., Giles, W. R., and Chien, K. R. (2001). A defect in the Kv channel-interacting protein 2 (KChIP2) gene leads to a complete loss of I_{to} and confers susceptibility to ventricular tachycardia. *Cell* **107**, 801–813.
28. Choi, E.-K., Zaidi, N. F., Miller, J. S., Crowley, A. C., Merriam, D. E., Lilliehook, C., Buxbaum, J. D., and Wasco, W. (2001). Calsenilin is a substrate for caspase-3 that preferentially interacts with the familial Alzheimer's disease-associated C-terminal fragment of presenilin 2. *J. Biol. Chem.* **276**, 19197–19204.
29. Spreafico, F., Barski, J. J., Farina, C., and Meyer, M. (2001). Mouse DREAM/Calsenilin/KChIP3: gene structure, coding potential and expression. *Mol. Cell. Neurosci.* **17**, 1–16.
30. Cheng, H.-Y., Pitcher, G. M., Laviolette, S. R., Whishaw, I. Q., Tong, K. I., Kockeritz, L. K., Wada, T., Joza, N. A., Crackower, M., Goncalves, J., Sarosi, I., Woodgett, J. R., Oliveira-dos-Santos, A. J., Ikura, M., van der Kooy, D., Salter, M. W., and Penninger, J. M. (2002). DREAM is a critical transcriptional repressor for pain modulation. *Cell* **108**, 31–43.
31. Holmqvist, M. H., Cao, J., Hernandez-Pineda, R., Jacobson, M. D., Carroll, K. I., Sung, M. A., Beety, M., Ge, P., Gilbride, K. J., Brown, M. E., Jurman, M. E., Lawson, D., Silos-Santiago, I., Xie, Y., Covarrubias, M., Rhodes, K. J., Distefano, P. S., and An, W. F. (2002). Elimination of fast inactivation in Kv4 A-type potassium channels by an auxiliary subunit domain. *Proc. Natl. Acad. Sci. USA* **99**, 1035–1040.

CHAPTER 137

Calmodulin-Mediated Signaling

Anthony R. Means
*Department of Pharmacology and Cancer Biology,
Duke University Medical Center,
Durham, North Carolina*

Calmodulin (CaM) is a ubiquitous, essential protein that serves as the primary sensor of changes in intracellular Ca^{2+} in all eukaryotic cells [1]. In this way CaM is the Ca^{2+} receptor that orchestrates Ca^{2+}-initiated signal transduction cascades leading to changes in cell function. Some of these cascades are initiated by Ca^{2+} and proceed in a linear fashion that culminate in regulation of the vital cellular response. However, Ca^{2+}/CaM can also influence physiologically important processes indirectly via cross-talk with pathways initiated by other second messengers [2]. These pathways include those regulated by cyclic nucleotides, nitric oxide, and MAP kinases (Fig. 1). Thus, it is with good reason that Ca^{2+} is recognized as the most versatile of the second messengers and CaM as its pivotal receptor.

The exquisite design of CaM allows it to assume an almost limitless number of conformations, each one dictated by the amino acid sequence of the CaM-binding domain of the individual target proteins. Illustrative of the versatility of CaM, over 50 CaM-binding proteins have been described to date [1]. Whereas CaM is generally considered to be a Ca^{2+}-binding protein and binds to many of its targets only in the presence of Ca^{2+}, this is not the only mode of action of CaM in cells. Some physiologically relevant interacting proteins also bind CaM in a Ca^{2+}-independent way. For example, CaM is an integral subunit of phosphorylase kinase (PK). Whereas the subunit association of PK is Ca^{2+}-independent, CaM still serves to sense changes in the concentration of Ca^{2+} as Ca^{2+} can regulate the activity of the enzyme. Alternatively, CaM can bind to a partner in the absence of Ca^{2+} but be released from its binding partner upon Ca^{2+}-binding. CaM-binding proteins of this kind include the unconventional myosins, neurogranin and neuromodulin. However, this review will focus on the central role of CaM as a transducer of Ca^{2+} signals to Ca^{2+}/CaM-dependent effector proteins. In addition, because of space limitations, references are primarily restricted to other reviews that focus on different aspects of CaM structure and function.

Calmodulin is a 148 amino acid protein in which globular domains, each containing a pair of EF-hand Ca^{2+}-binding motifs, are separated by an 8-turn α-helix. Ca^{2+} binding causes a conformational change in CaM that exposes hydrophobic residues and generates a considerable amount of biochemical energy, which is then used to modify the activity of its targets. It is this remarkable ability to serve as a selective allosteric regulator of so many enzymes that allows CaM to react to both global and transient changes in Ca^{2+} concentration within the dynamic and physiologically relevant range experienced by a cell. A rise in Ca^{2+} markedly increases the affinity of CaM for its target proteins and promotes conformational changes in both CaM and the effector [1].

Although generally considered a soluble protein, almost 95% of the CaM in mammalian cells, such as those in smooth muscle, is immobile [3]. However, in neurons and pancreatic acinar cells CaM has been reported to specifically translocate from plasma membrane to nucleus in response to brief stimuli that result in a localized rise in Ca^{2+} [4,5]. Thus, although CaM may relocate in response to Ca^{2+} signals, it seems to primarily move around in cells by diffusion. Calmodulin is found in all subcellular compartments where CaM-binding proteins characteristic to a given organelle serve to tether the Ca^{2+} receptor in a manner appropriate to facilitate Ca^{2+}/CaM-dependent function [6].

In the absence of Ca^{2+}/CaM, the target enzymes for this protein are inactive. In most cases this is due to intrasteric

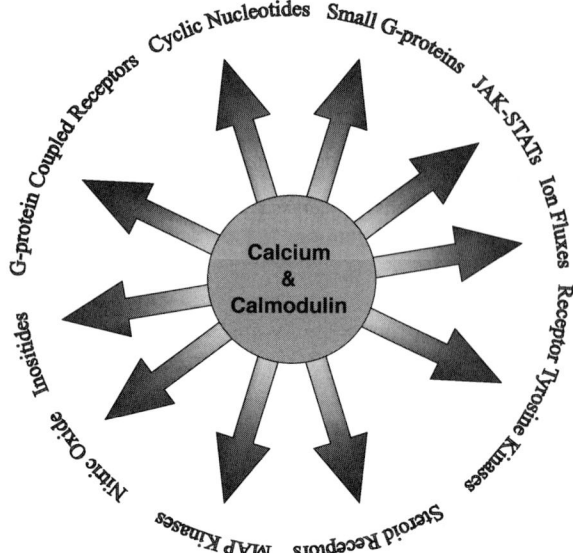

Figure 1 Signaling pathways influenced by Ca^{2+}/CAM.

autoinhibiton in which a portion of the protein is folded in a manner that prevents substrate access to its active site [7]. Ca^{2+}/CaM-binding to a region contiguous to the inhibitory segment relieves autoinhibition and thus promotes enzyme activity. Many of the autoinhibitory segments possess sequence similarity to the protein substrate, which led to the idea that autoinhibition was of the "pseudosubstrate" type. This concept has also been confirmed by both biochemical and structural approaches [7]. Ca^{2+}/CaM binding to a target enzyme can have one of three consequences. First, binding can activate the enzyme and inactivation follows the termination of the Ca^{2+} signal [1]. Enzymes that behave in this way include myosin light chain kinase, protein phosphatase 2B (calcineurin), CaMKI, and phosphodiesterase. Second, Ca^{2+}/CaM can activate the enzyme by promoting intersubunit phosphorylation that results in generation of constitutive activity. CaMKII is the prototypic example of this mode of regulation. As a CaMKII holoenzyme is composed of 10 to 12 subunits, this protein kinase can sense and record incremental increases in the intracellular concentration of Ca^{2+} [8]. Third, Ca^{2+}/CaM binding can enable phosphorylation of the target enzyme by an upstream enzyme in a signaling cascade of reactions. CaMKI and CaMKIV are examples of enzymes subject to this type of control. In response to Ca^{2+}/CaM binding these enzymes are phosphorylated by a CaMKK in a manner similar to the cascade of reactions that characterize MAPK pathways. This analogy has led to the concept of a CaMK cascade [9]. Like CaMKII, CaMKIV activity is maintained even after termination of the Ca^{2+} signal. The second and third control mechanisms provide means of continuing a Ca^{2+}-initiated reaction even after the Ca^{2+} signal has dissipated. Since in these cases the kinases are inactivated by dephosphorylation, kinase activity can also be terminated in the presence of a high level of intracellular Ca^{2+}, thus uncoupling the maintenance and inactivation of the response from the changes in Ca^{2+} required for its initiation.

One biological system that illustrates the multifaceted modes by which CaM can transduce a Ca^{2+} signal is the T lymphocyte. Stimulation of this cell through the T cell receptor results in a rapid and robust increase in intracellular Ca^{2+} that peaks about 2 min after receptor occupancy. The rise in Ca^{2+} is required for activation of two CaM-dependent pathways involved in expression of the interleukin 2 gene (IL-2) that encodes a mitogen necessary to expand this cohort of T cells. On the one hand, Ca^{2+}/CaM activates calcineurin in the cytoplasm. Calcineurin dephosphorylates a subunit of the NF-AT transcription factor, which allows the NF-AT/calcineurin complex to translocate into the nucleus and bind to the IL-2 gene promoter [10]. On the other hand, Ca^{2+}/CaM also activates CaMKIV, which resides in the nucleus and is involved in the phosphorylation of CREB [11], which is required for activation of immediate early genes such as those of the Fos and Jun families of transcription factors. Fos and Jun heterodimerize to form the AP1 transcription factor, which collaborates with NF-AT to activate the IL-2 gene promoter [10]. By 5 min after T cell receptor activation, the Ca^{2+} levels have largely decreased but remain elevated in the nucleus at a level about twice that present in resting T cells. This higher sustained Ca^{2+} concentration, which has to be maintained for approximately 2 hours, is required to maintain the activity of calcineurin and keep the NF-AT/calcineurin complexes in the nucleus [10]. However, because once fully activated CaMKIV becomes independent of Ca^{2+}/CaM, this enzyme is inactivated by dephosphorylation, which is catalyzed by PP2A that is in a complex with CaMKIV [11]. Hence, although a very similar Ca^{2+}/CaM-dependent mechanism is used to activate calcineurin and CaMKIV, the characteristics distinct to each enzyme permit maintenance of calcineurin activity and thus continued IL-2 gene transcription but inactivation of CaMKIV and the immediate early genes it regulates in the face of a sustained elevation of intranuclear Ca^{2+}.

Calcium is mandatory for cell proliferation and plays an important role in all cell cycle transitions [12]. In model organisms such as fungi, nematodes, and flies, the single CaM gene is essential for cell growth. It is interesting that the CaM effector proteins shown to be essential in *S. cerevisiae* are unique, as all bind CaM in the absence of Ca^{2+}. However, in the filamentous fungus *Aspergillus nidulans*, several essential CaM targets are Ca^{2+}-dependent and seem to be very similar to those required in mammalian cells. Thus, for the entry of quiescent cells into the cell cycle, both calcineurin and a CaMK are required and function in mid-G1 upstream of the activation of the cyclin/cdk complex that is a prelude to entry into DNA synthesis. In mammalian cells calcineurin is required to maintain stability of cyclin D, whereas the CaMK is required for activation of the cyclin D/cdk4 complex in the nucleus. Progression of *A. nidulans* and mammalian cells from G2 into mitosis requires a second CaMK. Again the CaMK is required prior to the activation of cyclin B/cdc2 and at least one target protein has been suggested to be the cdc25 phosphatase that activates the cdc2 complex. Finally, Ca^{2+}/CaM is important for mitosis. Presently available data suggest that Ca^{2+}/CaM is likely to be required for breakdown of the nuclear envelope, the transition from metaphase to anaphase and for cytokinesis. However, the Ca^{2+}/CaM-dependent target proteins appear to be different at all three points in mitotic progression, although the details are unclear and under intense investigation [12].

Based on studies in a variety of cell types except *S. cerevisiae*, it seems clear that all roles for CaM in the cell cycle are heralded by transient changes in the intracellular Ca^{2+} concentration, which emphasizes the ubiquitous role of CaM as a Ca^{2+} receptor [13,14]. In addition, at least six different Ca^{2+}/CaM-dependent effector proteins have been shown to be essential for cell proliferation. These observations illustrate the pleiotypic ability of CaM to relay Ca^{2+} signals to very different effectors. Thus, Ca^{2+}, CaM, and CaM-dependent target proteins are essential components of signaling cascades that monitor progression through the cell cycle.

References

1. Chin, D. and Means, A. R. (2000). Calmodulin: a prototypical calcium sensor. *Trends Cell Biol.* **10**, 322–328.
2. Soderling, T. R. (1999). The Ca^{2+}-calmodulin-dependent protein kinase cascade. *TIBS* **24**, 232–236.
3. Luby-Phelps, K., Hori, M., Phelps, J. M., and Won, D. (1995). Ca^{2+}-regulated dynamic compartmentalization of calmodulin in living smooth muscle cells. *J. Biol. Chem.* **270**, 21532–21538.
4. Deisseroth, K., Heist, E. K., and Tsien, R. W. (1998). Translocation of calmodulin to the nucleus supports CREB phosphorylation in hippocampal neurons. *Nature* **392**, 198–202.
5. Craske, M., Takeo, T., Gerasimenko, O., Vaillant, C., Torok, K., Petersen, O. H., and Tepikin, A. V. (1999). Hormone-induced secretory and nuclear translocation of calmodulin: oscillations of calmodulin concentration with the nucleus as an integrator. *Proc. Natl. Acad. Sci. USA* **96**, 4426–4431.
6. Liao, B., Pachal, B. M., and Luby-Phelps, K. (1999). Mechanism of Ca^{2+}-dependent nuclear accumulation of calmodulin. *Proc. Natl. Acad. Sci. USA* **96**, 6217–6222.
7. Kobe, B., Heierhorst, J., and Kemp, B. (1997). Intrasteric regulation of protein kinases. *Adv. Second Messenger Phosphoprotein Res.* **31**, 29–40.
8. DeKoninck, P. and Schulman, H. (1998). Sensitivity of CaM kinase II to the frequency of Ca^{2+} oscillations. *Science* **279**, 227–230.
9. Corcoran, E. E. and Means, A. R. (2001). Defining Ca^{2+}/calmodulin-dependent protein kinase cascades in transcriptional regulation. *J. Biol. Chem.* **276**, 2975–2978.
10. Crabtree, G. R. (2001). Calcium, calcineurin, and the control of transcription. *J. Biol. Chem.* **276**, 2313–2316.
11. Westphal, R. S., Anderson, K. A., Means, A. R., and Wadzinski, B. E. (1998). A signaling complex of Ca^{2+}-calmodulin-dependent protein kinase IV and protein phosphatase 2A. *Science* **280**, 1258–1261.
12. Means, A. R., Kahl, C. R., Crenshaw, D. G., and Dayton, J. S. (1999). Traversing the cell cycle: The calcium/calmodulin connection. In *Calcium as a Cellular Regulator* (E Carafoli and C Klee, eds.), Oxford University Press, New York, pp. 512–528.
13. Whitaker, M. and Larman, M. G. (2001). Calcium and mitosis. *Cell Dev.l Biol.* **12**, 53–58.
14. Carafoli, E., Santella, L., Branca, D., and Brini, M. (2001). Generation, control, and processing of cellular calcium signals. *Crit. Rev. Biochem. Mol. Biol.* **36**, 107–260.

CHAPTER 138

The Family of S100 Cell Signaling Proteins

Claus W. Heizmann, Beat W. Schäfer and Günter Fritz
Department of Pediatrics, Division of Clinical Chemistry and Biochemistry,
University of Zürich, Zürich, Switzerland

Introduction

S100 proteins attracted great interest in recent years due to their association with various human pathologies and their use in the diagnosis of these diseases. Twenty members have been discovered so far, and altogether S100 proteins represent the largest subgroup within the EF-hand Ca^{2+}-protein family. S100 proteins show a very divergent pattern of cell- and tissue-specific expression and of affinities for Ca^{2+}, Zn^{2+}, and Cu^{2+}, consistent with their pleiotropic intra- and extracellular functions. Several genetically engineered animal models have now been generated to study the roles of S100 proteins under normal and pathological conditions.

Many cellular events are regulated by oscillations of intracellular Ca^{2+} concentrations wherein the signal specificity is obtained through differences in location, duration, and frequency [1]. Participants in most Ca^{2+}-signaling pathways are members of the large family of Ca^{2+}-binding proteins characterized by the EF-hand structural motif [2]. Certain members, notably calbindin D28k and parvalbumin, serve as cytosolic Ca^{2+} buffers/shuttles whereas others such as calmodulin, troponin C, and the S100 proteins are Ca^{2+}-dependent regulatory proteins.

Unlike the ubiquitous calmodulin, most of the 20 members of the S100 protein family show cell- and tissue-specific expression, which is often deregulated in a number of human diseases including cancer, neurodegenerative disorders, inflammations and cardiomyopathy [3]. S100 proteins have a size of 10 to 12 kDa and form homo- and heterodimers. The monomer is composed of two helix-loop-helix (EF-hand) motifs connected by a central hinge region. The C-terminal EF-hand contains the canonical Ca^{2+}-binding loop, common to all EF-hand proteins. The N-terminal EF-hand consists of 14 amino acids and is characteristic for S100 proteins. Upon Ca^{2+}-binding S100 proteins undergo a conformational change required for target recognition and binding [32]. Generally, the dimeric S100 proteins bind four Ca^{2+} per dimer ($K_d = 20$–$500\,\mu M$). Besides Ca^{2+} a number of S100 proteins bind Zn^{2+} with a wide range of affinities ($K_d = 0.1$–$2000\,\mu M$). For S100B and S100A5 even Cu^{2+}-binding was reported ($K_d = 0.4$–$5\,\mu M$). This suggests that S100 protein target interactions and cellular functions may also be triggered by Zn^{2+} and Cu^{2+}.

Another unique feature is that individual members of S100 proteins are localized within specific cellular compartments from which some of them are able to relocate upon Ca^{2+} or Zn^{2+} activation [4], transducing the signal in a temporal and spatial manner by interacting with different targets specific for each S100 protein. Furthermore, some S100 proteins are even secreted from cells acting in a cytokine-like manner. The individual members seem to utilize distinct pathways (ER-Golgi route, tubulin- or actin-dependent) for their translocation/secretion into the extracellular space [5]. S100B and S100A12 specifically bind to the surface receptor RAGE (receptor for advanced glycation endproduct)—a multiligand member of the immunoglobulin superfamily [6]. The extracellular levels of S100B thereby play a crucial role in that nanomolar concentrations of S100B have trophic effects on cells whereas pathological levels (as found in Alzheimer's patients) induce apoptosis [7].

Unique to the S100 protein family is that most S100 genes are located in a gene cluster on human chromosome 1q21. Within this chromosomal region, several rearrangements and deletions have been reported during tumor development, probably linked to the deregulated S100 gene expression observed in various tumor types. Genetically manipulated mice (knockout and transgenics for S100 proteins and RAGE) and microarray technologies are now becoming available that will further advance our understanding of the diverse cell signaling activities of S100 proteins.

Protein Structures and Metal-Dependent Interactions with Target Proteins

The amount of detailed structural data for S100 proteins is growing rapidly. With the exception of calbindin D_{9K} all structures of S100 proteins revealed a tight homodimer whereby the dimerization plane is composed of strictly conserved hydrophobic residues, which are missing in the case of calbindin D_{9K} [2,3]. Each S100 monomer consists of two helix-loop-helix Ca^{2+}-binding domains termed EF-hands (Fig. 1). The N-terminal domain consisting of helices H_I and H_{II} connected by loop L_1 is different from the canonical EF-hand motif and is therefore called S100-specific or pseudo EF-hand, whereas the C-terminal domain (H_{III}-L_3-H_{IV}) contains the canonical EF-hand motif. Upon Ca^{2+}-binding almost all S100 proteins undergo a conformational change exposing a previously covered hydrophobic patch. The Ca^{2+}-dependent conformational change of S100 proteins was characterized by NMR and high-resolution X-ray studies. This conformational change of Ca^{2+}-bound S100 proteins is distinct from Ca^{2+}-dependent changes observed in other EF-hand proteins such as calmodulin or troponin C. In the C-terminal EF-hand (canonical EF-hand) there is a large change in the position of helix H_{III} upon Ca^{2+}-binding. The interhelical angle between helices H_{III} and H_{IV} changes by 90° in S100B compared to the Ca^{2+}-free structure, opening the structure and exposing the residues required for target recognition and binding. A similar change in conformation is observed for Ca^{2+}-bound S100A6 [30,31]. The crystal structures of Ca^{2+}-bound S100A7 [8], S100A8 [9], S100A11 [10], and S100A12 [11] confirmed the observations made for Ca^{2+}-bound S100B. All four structures revealed an open conformation suitable for target binding. A further interesting phenomenon was observed for S100A10, which is not able to bind Ca^{2+}. The crystal structure of S100A10 showed that the Ca^{2+}-free protein is already in a Ca^{2+}-bound like open conformation that enables S100A10 to interact Ca^{2+}-independently with its target molecule [12]. Recently a hexameric form of S100A12 was described [11]. Three S100A12 dimers assemble as a hexamer that is stabilized by six additional Ca^{2+} ions bound to the interface of two adjacent dimers.

S100A3 is a unique member of the S100 protein family; it has a low affinity for Ca^{2+} but a high affinity for Zn^{2+} ions. The crystal structure of S100A3 (Fig. 1A) [13] allowed the prediction of one putative Zn^{2+}-binding site (distinct from the EF-hand) in the C-terminus of each monomer, thus disturbing the hydrophobic interface of the homodimer.

Target Binding

So far three different S100-target complexes have been characterized: S100B complexed with a peptide of the regulatory domain of p53 [14], S100A10 with a peptide of annexin II, and S100A11 complexed with a peptide of annexin I [10]. All three peptides were located in a cavity formed by helices H_{III} and H_{IV} in the open conformation of the C-terminal canonical EF-hand. The binding of the target peptides with the protein matrix is accomplished by hydrophobic and ionic interactions. Furthermore the stoichiometry of the complex is two target peptides per S100 homodimer. However, the binding

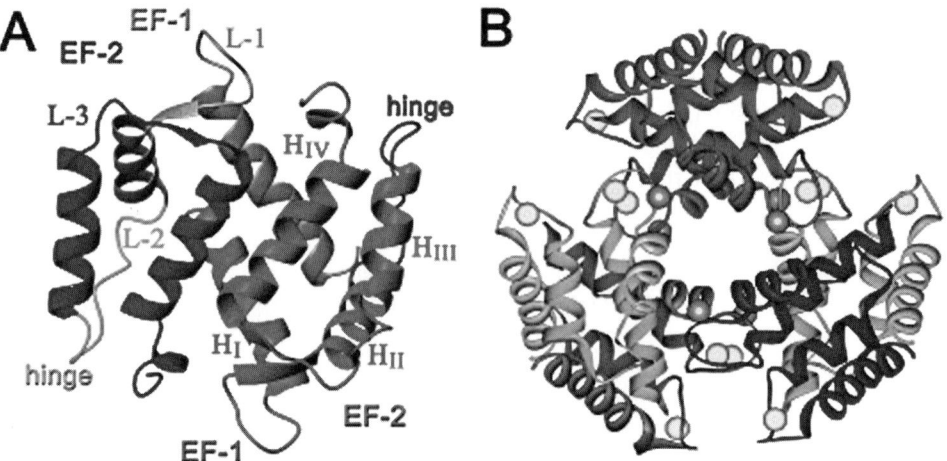

Figure 1 Structure of S100 proteins. (A) Dimeric structure of S100A3. The monomers are depicted in red and blue, respectively. Each monomer consists of two EF-hands connected by a hinge region. (B) Hexameric structure of S100A12 (pdb code 1GQM [11]). Three S100A12 dimers form a hexamer. The Ca^{2+} ions bound to the EF-hands are shown as bright yellow spheres. At the hexamer-forming interface six additional Ca^{2+} ions are located, which are shown as dark yellow spheres. A color representation of this figure is available on the CD version of the *Handbook of Cell Signaling*.

mode of the annexin peptides to S100A10 and S100A11 is strikingly different from that of the p53 peptide to S100B. In contrast to the p53 peptide the annexin peptides interact with both monomers whereby the required residues are located on the helices H_{III} and H_{IV} of one monomer and on helix H_I of the second monomer.

Based on these observations one can suppose that there are further modes of target binding to other S100 proteins. For example, it was proposed [11] that the hexameric form of S100 A12 (Fig. 1B) might interact with three extracellular domains of the RAGE receptor, bringing them together into large trimeric assemblies.

Zn^{2+} Binding

Although a number of S100 proteins bind Zn^{2+} with high affinity, only the structure of Zn^{2+} bound-S100A7 is known so far [8]. The Zn^{2+} binding sites in other S100 proteins, such as S100B, S100A2, S100A3, S100A4, or S100A6, are only preliminarily characterized [15]. The Zn^{2+}-binding residues were identified by spectroscopic and mutational analysis mainly as histidine and cysteine residues; however, a common Zn^{2+} binding motif has not yet been recognized in the sequence of these proteins.

Genomic Organization, Chromosomal Localization, and Nomenclature

The structural organization of S100 genes is highly conserved both within an organism and in different species [3]. A typical S100 gene consists of three exons whereby the first exon carries exclusively 5′ untranslated sequences. The second exon contains the ATG and codes for the N-terminal EF-hand, and the third exon encodes the carboxy-terminal canonical EF-hand. Only a few genes, such as S100A4, S100A5, and the newly identified S100A14 [16], are composed of four exons. In these genes, the first two exons can either be alternatively spliced (S100A4) or noncoding (S100A5), leaving the two exon-splitting of the coding region intact. It is interesting that for both S100A11 and S100A14 this region encoding the corresponding proteins is split into three exons. Whether this finding reflects a functional or evolutionarily close relationship between these two members of the S100 family remains to be seen.

Presently 14 bona fide S100 genes are found in a gene cluster on human chromosome 1q21 (Fig. 2), a finding that led to the introduction of the now widely accepted S100 nomenclature (Table I). Four additional S100 genes are found on other human chromosomes, including the newly discovered S100Z [17] likely to be localized on chromosome 5 (the 3-terminal sequence of the S100Z cDNA is part of a human BAC clone on region 5q12-q13). Hence, one can recognize at least four different subgroups of S100 genes located closely together (S100A1–S100A13–S100A14; S100A2 to S100A6; S100A8–S100A9–S100A12; S100A10–S100A11). This finding raises the question of whether each gene is regulated by its own promoter elements or by as yet uncharacterized locus control elements, as has been suggested for the epidermal differentiation genes. The evidence available today would rather suggest an individual regulatory mechanism for most S100 genes. It is striking, however, that for some of these subgroups of genes, similarities in the function of their encoded proteins have been recognized. Furthermore,

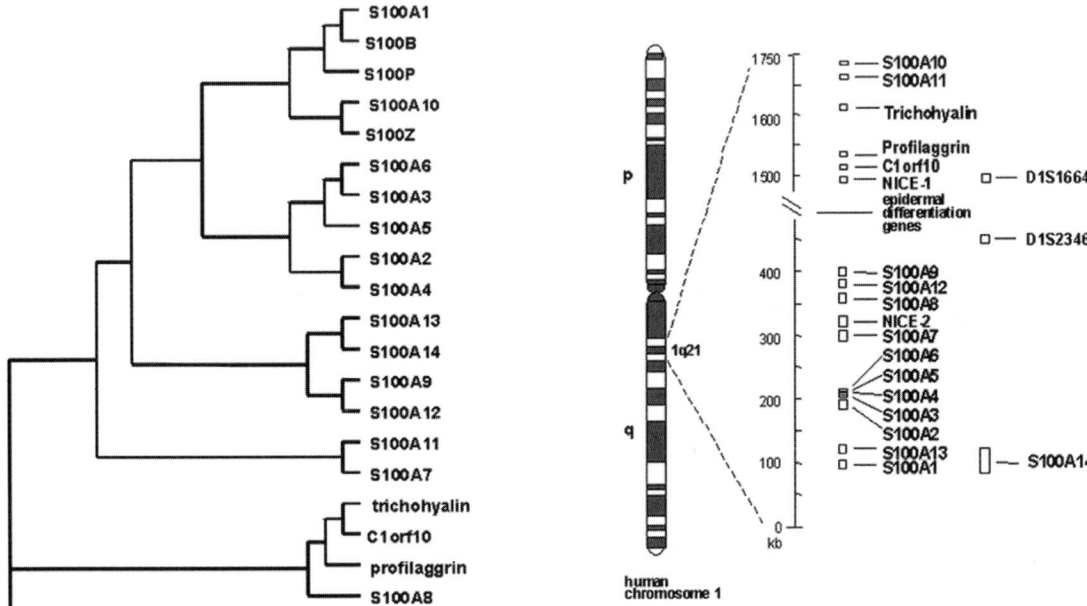

Figure 2 Phylogenetic tree of human S100 proteins in comparison with physical map of the S100 gene cluster on human chromosome 1q21. In the phylogenetic tree each node represents a putative gene duplication. The left-most nodes indicate the earliest gene duplications. Genes lying in the cluster region are indicated as well as two commonly used genomic markers (D1S1664 and D1S2346); p and q represent the long and the short arm of the chromosome, respectively.

Table I Nomenclature for S100 Genes Clustered on Human Chromosome 1q21

New name	Previous symbols/synonyms
S100A1	S100α
S100A2	S100L, CaN19
S100A3	S100E
S100A4	CAPL, p9ka, pEL98, mts 1, metastasin, calvasculin, murine placental calcium protein, Fsp1, 18A2
S100A5	S100D
S100A6	Calcyclin, CACY, 2A9, PRA, CaBP, 5B10
S100A7	Psoriasin, PSOR1, BDA11, CAAF2
S100A8	Calgranulin A, CAGA, CFAg, MRP8, p8, MAC387, 60B8Ag, L1Ag, CP-10, MIF, NIF, calprotectin
S100A9	Calgranulin B, CAGB, CFAg, MRP-14, p14, MAC 387, 60B8Ag, L1Ag, MIF, NIF, (S100A8/A9 dimer)
S100A10	Calpactin light chain, CAL12, CLP11, p11, p10, 42C
S100A11	Calgizzarin, S100C
S100A12	Calgranulin C, p6, CAAF1, CGRP, corned-associated antigen
S100A13	CAAF2
S100A14	—

Nomenclature for S100 Genes Located on Different Chromosomes

Name	Previous symbols/synonyms	Chromosomal location
Calbindin-D9K	9K CALB3, CaBP9k, I CaBP, Cholecalcin	Xp22
S100B	S100β, NEF	21q22
S100P	S100P	4p16
S100Z	—	5

a number of chromosomal abnormalities such as deletions, rearrangements or translocations in this region have been associated with neoplasias, suggesting that the expression of S100 genes might be altered in human cancer. It is also interesting that the clustered organization of the human genes seems to be evolutionarily conserved, at least in the mouse. In other species, S100 genes are less well characterized.

Comparison of a phylogenetic tree constructed on the basis of sequence alignments between the different human S100 proteins revealed a clustering of S100 proteins, which is mirrored in the physical map of the S100 genes on chromosome 1q21 (Fig. 2). For example, S100A2 through to S100A6 cluster tightly together both on the phylogenetic tree and on the chromosome. The small distances on the chromosome and the phylogenetic tree indicate that these S100 proteins most likely originate from late gene duplication events. Similarly, clusters are seen for S100A10 and S100A11, as well as for S100A13 and S100A14. Furthermore, there are three proteins encoded in 1q21 that carry in the N-terminus an S100-like domain, namely Trichohyalin, Profilaggrin, and C1or f10.

Translocation, Secretion, and Biological Functions

S100 proteins are generally involved in a large number of cellular activities such as signal transduction, cell differentiation, regulation of cell motility, transcription, and cell cycle progression (Table II) [3]. S100 proteins are thought to modulate the activity of target proteins in a Ca^{2+}- (and possibly also in a Zn^{2+}- and Cu^{2+}-) dependent manner. During the last decade, a large number of such possible interactions have been described involving enzymes, cytoskeletal elements, and transcription factors.

Apart from these intracellular functions, some S100 proteins, such as S100A8/A9, S100B, S100A4, and probably others, are secreted from cells and exhibit cytokine-like extracellular functions [3]. These include chemotactic activities related to inflammation (S100A8; S100A9 and S100A12), neurotrophic activities (S100B), and a recently described angiogenic affect (S100A4). In all cases, the mechanisms of secretion as well as the nature of high affinity surface receptors remain largely unknown. One candidate receptor to mediate at least some of the described extracellular functions is RAGE, which was shown to be activated upon binding of S100A12 and S100B [6]. It is currently not known whether RAGE is a universal S100 receptor.

Generation of some animal models has been initiated to study the physiological impact of S100 proteins [3]. Ectopic overexpression in the mouse has been described for S100B and S100A4. In the case of S100B, enhanced expression in the brain led to hyperactivity associated with an impairment of hippocampal function. Brains from S100B transgenic mice show a higher density of dendrites in the hippocampus postnatally compared to controls and a loss of dendrites by one year of age. In contrast to this mild phenotype, expression of S100A4 in oncogene-bearing transgenic mice is capable of inducing metastasis of mammary tumors, suggesting that S100A4 has an important role in the acquisition of the metastatic phenotype during tumor progression. While stimulation of angiogenesis might play a role, the exact mechanisms of this function are still under investigation.

Inactivation through homologous recombination in mouse embryonic stem cells has been achieved for S100B and S100A8. While inactivation of S100B has no obvious consequences for life, S100A8 null mice die via early resorption of the mouse embryo, a result that suggests a role for this protein in prevention of maternal rejection of the implanting embryo.

Since S100 proteins can form homo- and also heterodimers and usually more than one S100 protein is found to be expressed in a given cell type, functional redundancy or compensatory mechanisms might explain the lack of phenotype observed in some animal models. Clearly, more animal models inactivating single S100 proteins as well as combinations thereof are needed before their physiological roles can be clarified.

Table II S100 Proteins: Functions and Association with Human Diseases

Protein	Postulated functions	Disease association
S100A1	Regulation of cell motility, muscle contraction, phosphorylation, Ca^{2+} release channel, transcription	Cardiomyopathies
S100A2	Tumor suppression, nuclear functions, chemotaxis	Cancer, tumor suppression
S100A3	Hair shaft formation, tumor suppression, secretion, and extracellular functions	Hair damage, cancer
S100A4	Regulation of cell motility, secretion and extracellular functions, angiogenesis	Cancer (metastasis)
S100A5	Ca^{2+}-, Zn^{2+}-, and Cu^{2+}-binding protein in the CNS and other tissues; unknown function	Not known
S100A6	Regulation of insulin release, prolaction secretion, Ca^{2+} homeostasis, tumor progression	Amyotrophical lateral sclerosis
S100A7	S100A7-fatty acid binding protein complex regulates differentiation of keratinocytes	Psoriasis, cancer
S100A8/A9	Chemotactic activities, adhesion of neutrophils, myeloid cell differentiation, apoptosis, fatty acid metabolism	Inflammation, wound healing, cystic fibrosis
S100A10	Inhibition of phospholipase A2, neurotransmitter release, in connection with annexin II regulates membrane traffic, ion currents	Inflammation
S100A11	Organization of early endosomes inhibition of annexin I function, regulation of phosphorylation, physiological role in keratinocyte cornified envelope	Skin diseases, ocular melanoma
S100A12	Host-parasite interaction, differentiation of squamous epithelial cells and extracellular functions	Mooren's ulcer (autoimmune disease), inflammation
S100A13	Regulation of FGF-1 and synaptotagmin-1 release	—
S100A14	—	malignant transformation
S100B	Cell motility, proliferation, inhibition of phosphorylation, inhibition of microtubule assembly transcription, regulation of nuclear kinase, extracellular functions, e.g. neurite extension	Alzheimer's disease, Down's syndrome, melanoma, amyotrophic lateral sclerosis, epilepsy
S100P	Function in the placenta	Cancer
S100Z	Function in spleen and leukocytes	Aberrant in some tumors
Calbindin D_{9K}	Ca^{2+} buffer and Ca^{2+} transport	Vitamin D deficiency, abnormal mineralization

Associations with Human Diseases

As listed in Table II, various S100 proteins are closely associated with several human diseases. We will briefly discuss a few examples.

S100 A1 is mainly expressed in the human myocardium. Reduced levels measured in the left ventricles of patients with end stage heart failure may contribute to a reduced contractility [18]. This notion is in agreement with the reported interactions of S100A1 with SR proteins, regulating Ca^{2+}-induced Ca^{2+} release [19], and with SERCA2a, phospholamban (own results), and titin [20], modulating Ca^{2+} homeostasis and contractile performance [21]. Therefore, an S100A1 gene transfer to the heart *in vivo* might provide a new therapeutic approach to correct the altered Ca^{2+} signaling pathways that cause abnormal myocardial contractility.

S100A4 has been implicated in invasion and metastasis [22]. The prognostic significance of its selective expression in various cancers has been exploited. Identification of predictive markers of cancer is of major importance to the improvement of clinical management, therapeutic outcome, and survival of patients. In gastric cancer the inverse expression of S100A4 in relation to E-cadherin (a tumor supressor) was found to be a powerful aid in histological typing and in evaluating the metastatic potential/prognosis of patients with this type of cancer [23].

Extracellular functions of S100 proteins such as S100A4 have been described. Recently it was demonstrated [24] that S100A4 could act as an angiogenic factor and might induce tumor progression via an extracellular route stimulating angiogenesis. Inhibiting the process of tumor angiogenesis might be possible by either blocking S100A4 secretion or its extracellular function.

A prognostic significance of S100A2 in laryngeal squamous-cell carcinoma has also been found [25], thus allowing discrimination of high- and low-risk patients in the lymph-node negative subgroup to provide better therapy.

Human S100A8 and S100A9 are associated with chronic inflammatory diseases. Both proteins are also involved in wound repair by reorganizing the keratin cytoskeleton in the injured epidermis [26].

S100B and its interacting tau protein are individually affected in Alzheimer's disease (AD). S100B is overexpressed in AD, and hyperphosphorylated tau constitutes the primary component of neurofibrillary tangles. In addition, S100B interacts with other AD-associated proteins such as presenilin (PS1 and PS2) and with the amyloid precursor protein (APP). These interactions are altered by the phosphorylation state of tau and the overexpresssion of S100B, a finding that strongly indicates that S100B plays an important role in pathways associated with neurodegeneration [27–29].

Conclusion and Perspectives

S100 proteins have been implicated in pleiotropic Ca^{2+}-dependent cellular events, with specific functions for each of the family members. However, some S100 proteins have also physiologically relevant Zn^{2+} affinities, suggesting that Zn^{2+} rather than Ca^{2+} controls their biological activities. In order to understand how the biological functions of S100 proteins are regulated by Zn^{2+} and Ca^{2+} it will be necessary to determine the three-dimensional structures of the Zn^{2+} loaded S100 proteins and their interactions with target proteins.

S100 proteins are localized in specific cellular compartments from which some of them relocate upon cellular stimulation and even secrete exerting extracellular, cytokine-like activities. This finding suggests that translocation might be a temporal and spatial determinant of their interactions with different partner proteins. Our recent experiments suggest that different S100 proteins utilize distinct translocation pathways, which might lead them to certain subcellular compartments in order to perform their physiological tasks in the same cellular environment.

Some S100 proteins can play a crucial role in physiological responses through their paracrine effects on neighboring cells. The discovery of the surface receptor RAGE for two S100 proteins (S100B and S10012) has shed more light on the extracellular functions of these two proteins. Just how they are secreted and how they interact with RAGE still remains to be investigated. This could be done using animal models with inactivated single S100 proteins or deletion mutants of RAGE. Future research activities will also focus on the deregulated expression of S100 genes, which is a hallmark of a wide range of human diseases.

Acknowledgments

We thank Patricia McLoughlin for critical reading and D. Arévalo for help with the preparation of the manuscript. This work was supported in part by Wilhelm Sander-Stiftung (Germany), NCCR (National Competence Center for Research), Neuronal Plasticy and Repair, and the Swiss National Science Foundation.

References

1. Berridge, M. J., Lipp, P., and Bootman, M. D. (2000). The versatility and universality of calcium signaling. *Nature Rev. Mol. Cell Biol.* **1**, 11–21.
2. Kawasaki, H., Nakayama, S., and Kretsinger, R. H. (1998). Classification and evolution of EF-hand proteins. *BioMetals* **11**, 277–295.
3. Heizmann, C. W., Fritz, G., and Schäfer, B. W. (2002). S100 proteins: structure, functions and pathology. *Frontiers in Bioscience* **7**, 1356–1368.
4. Davey, G. E., Murmann, P., and Heizmann, C. W. (2001). Intracellular Ca^{2+} and Zn^{2+} levels regulate the alternative cell density-dependent secretion of S100B in human glioblastoma cells. *J. Biol. Chem.* **276**, 30819–30826.
5. Hsieh, H. L., Schäfer, B. W., Cox, J. A., and Heizmann, C. W. (2002). S100A13 and S100A6 exhibit distinct translocation pathways in endothelial cells. *J. Cell Sci.* **115**, 3149–3158.
6. Schmidt, A. M., Yan, S. D., Yan, S. F., and Stern, D. M. (2000). The biology of the receptor for advanced glycation end products and its ligands. *Biochim. Biophys Acta* **1498**, 99–111.
7. Huttunen, H. J., Kuja-Panulat, J., Sorci, G., Agneletti, A. L., Donato, R., and Rauvala, H. (2000). Coregulation of neurite outgrowth and cell survival by amphoterin and S100 proteins through receptor for advanced glycation end products (RAGE) activation. *J. Biol. Chem.* **275**, 40096–40105.
8. Brodersen, D. E., Etzerodt, M., Madsen, P., Celis, J. E., Thogersen, H. C., Nyborg, J., and Kjeldgaard, M. (1998). EF-hands at atomic resolution: the structure of human psoriasin (S100A7) solved by MAD phasing. *Structure* **6**, 477–489.
9. Ishikawa, K., Nakagawa, A., Tanaka, I., Suzuki, M., and Nishihira, J. (2000). The structure of human MRP8, a member of the S100 calcium-binding protein family, my MAD phasing at 1.9 A resolution. *Acta Crystallogr. D. Biol. Crystallogr.* **56**, 559–566.
10. Rety, S., Osterloh, D., Arie, J. P., Tabaries, S., Seemann, J., and Russo-Marie, F. (2000). Structural basis of the Ca^{2+}-dependent association between S100C (S100A11) and its target, the N-terminal part of annexin I. *Structure Fold. Des.* **8**, 175–184.
11. Moroz, O. V., Antson, A. A., Dodson, E. J., Burrell, H. J., Grist, S. J., Lloyd, R. M., Mait-land, N. J., Dodson, G. G., Wilson, K. S., Lukanidin, E., and Bronstein, I. B. (2002). The structure of S100A12 in a hexameric form and its proposed role in receptor signalling. *Acta Cryst.* D **58**, 407–413.
12. Rety, S., Sopkovo, J., Renouard, M., Osterloh, D., Gerke, V., Tabaries, S., Russo-Marie, F., and Lewit-Bentley, A. (1999). The crystal structure of a complex of p11 with the annexin II N-terminal peptide. *Nat. Struct. Biol.* **6**, 89–95.
13. Fritz, G., Mittl, P., Sargent, D. F., Vasak, M., Grütter, M. G., and Heizmann, C. W. (2002). The crystal structure of metal-free human EF-hand protein S100A3 at 1.7 A resolution. *J. Biol. Chem.* **277**, 33092–33098.
14. Rustandi, R. R., Baldisseri, D. M., and Weber, D. J. (2000). Structure of the negative regulatory domain of p53 bound to S100B ((ββ) *Nat. Struct. Biol.* **7**, 570–574.
15. Heizmann, C. W. and Cox, J. A. (1998). New perspective on S100 proteins: a multi-functional Ca^{2+}-, Zn^{2+}- and Cu^{2+}-binding protein family. *Biometals* **11**, 383–397.
16. Piétas, A., Schlüns, K., Marenholz, I., Schäfer, B. W., Heizmann, C. W., Petersen, I. (2002). Molecular cloning and characterization of a human gene encoding a novel member of the S100 family. *Genomics.* **79**, In press.
17. Gribenko, A. V., Hopper, J. E., and Makhatadze, G. I. (2001) Molecular characterization and tissue distribution of a novel member of the S100 family of EF-hand proteins. *Biochemistry* **40**, 15538–15548.
18. Remppis, A., Greten, T., Schäfer, B. W., Hunziker, P., Erne, P., Katus, H. A., and Heizmann, C. W. (1996). Altered expression of the Ca^{2+}-binding protein S100A1 in human cardiomyopathy. *Biochim. Biophys. Acta* **1313**, 253–257.
19. Treves, S., Scutari, E., Robert, M., Groh, S., Ottolia, M., Prestipino, G., Ronjat, M., and Zorzato, F. (1997). Interaction of S100A1 with the Ca^{2+} release channel (ryanodine receptor) of skeletal muscle. *Biochemistry* **36**, 11496–11503.
20. Yamasaki, R., Berri, M., Wu, Y. Trombitas, K., McNabb, M., Kellermayer, M. S. Z., Witt, C., Labeit, D., Labeit, S., Greaser, M., and Granzier, H. (2001). Titin-ctin interaction in mouse myocardium: passive tension modulation and its regulation by calcium/S100A1. *Biophys. J.* **81**, 2297–2313.
21. Most, P., Bernotat, J., Ehlermann, P., Pleger, S. T., Reppel, M., Boerries, M., Niroomand, F., Pieske, B., Janssen, P. M. L., Eschenhagen, T., Karczewski, P., Smith, G. L., Koch, W. J., Katus, H. A., and Remppis, A. (2001). S100A1: a regulator of myocardial contractility. *Proc. Natl. Acad. Sci.* **98**, 13889–13894.
22. Barraclough, R. (1998). Calcium-binding protein S100A4 in health and disease. *Biochim. Biophys. Acta* **1448**, 190–199.

23. Yonemura, Y., Endou, Y., Kimura, K., Fushida, S., Bandou, E., Taniguchi, K., Kinoshita, K., Ninomiya, I., Sugiyama, K., Heizmann, C. W., Schäfer, B. W., and Sasaki, T. (2000). Inverse expression of S100A4 and E-cadherin is associated with metastatic potential in gastric cancer. *Clin. Cancer Res.* **6**, 4234–4242.
24. Ambartsumian, N., Klingelhofer, J., Grigorian, M., Christensen, C., Kriajevska, M., Tulchinsky, E., Georgiev, G., Berezin, V., Bock, E., Rygaard, J., Cao, R., Cao, Y., and Lukanidin, E. (2001). The metastasis-associated Mts1 (S100A4) protein could act as an angiogenic factor. *Oncogene* **20**, 4685–4695.
25. Lauriola, L., Michetti, F., Maggiano, N., Galli, J., Cadoni, G., Schäfer, B. W., Heizmann, C. W., and Ranelletti, F. O. (2000). Prognostic significance of the Ca^{2+} binding protein S100A2 in laryngeal squamous-cell carcinoma. *Int. J. Cancer (Pred. Oncol.)* **89**, 345–349.
26. Kerkhof, C., Klempt, M., and Sorg, C. (1998). Novel insights into structure and function of MRP8 (S100A8) and MRP14 (S100A9). *Biochim. Biophys. Acta* **1448**, 200–211.
27. Baudier, J., and Cole, R. D. (1988). Interactions between the microtubule-associated tau proteins and S100b regulate tau phosphorylation by the Ca^{2+}/calmodulin-dependent protein kinase II. *J. Biol. Chem.* **263**, 5876–5883.
28. Sheng, J. G., Mrak, R. E., Rovnaghi, C. R., Kozlowska, E., Van Eldik, L. J., and Griffin, W. S. (1996). Human brain S100 beta and S100 beta mRNA expression increases with age: pathogenic implications for Alzheimer's disease. *Neurobiol. Aging* **17**, 359–363.
29. Sheng, J. G., Mrak, R. E., Bales, K. R., Cordell, B., Paul, S.M., Jones, R. A., Woodward, S., Zhou, X. Q., McGinness, J. M., and Griffin, W. S. (2000). Overexpression of the neuritotrophic cytokine S100 beta precedes the appearance of neuritic beta amyloid plaques in APPV717F mice. *J. Neurochem.* **74**, 295–301.
30. Otterbein, L. R., Kordowska, J., Witte-Hoffmann, C., Wang, C. L., and Dominguez, R. (2002). Crystal structures of S100A6 in the Ca^{2+}-free and Ca^{2+}-bound states. The calcium sensor mechanism of S100 proteins revealed at atomic resolution. *Structure* **10**, 557–567.
31. Maler, L., Sastry, M., and Chazin, W. J. (2002). A structural basis for S100 protein specificity derived from comparative analysis of apo and Ca^{2+}-calcyclin. *J. Mol. Biol.* **317**, 279–290.
32. Fritz, G. and Heizmann, C. W. (2003). 3D structures of the calcium- and zinc-binding S100 proteins *in* "Handbook of Metallo Proteins" (W., Bode, A., Messeischmidt and M., Cygler, eds.). Volume 3, Wiley and Sons: New York, in press.

CHAPTER 139

C_2-Domains in Ca^{2+}-Signaling

Thomas C. Südhof[1] and Josep Rizo[2]

[1]Center for Basic Neuroscience, Department of Molecular Genetics,
and Howard Hughes Medical Institute and
[2]Departments of Biochemistry and Pharmacology,
University of Texas Southwestern Medical Center,
Dallas, TX

Structures of C_2-Domains

C_2-domains are widespread protein modules of approximately 130 residues that are defined by a consensus sequence and a common three-dimensional fold [1]. Most C_2-domains bind Ca^{2+} and mediate interactions of their proteins with phospholipids. In the initial analysis of the human genome [2], C_2-domains were the second most abundant Ca^{2+}-regulatory domain (123 genes) after EF-hands (242 genes). C_2-domains primarily occur in membrane trafficking proteins such as synaptotagmin 1 and in signal transduction proteins such as protein kinase C (PKC). Although most C_2-domains bind Ca^{2+}, a significant number of C_2-domains do not bind Ca^{2+}.

As initially revealed in the structure of the synaptotagmin 1 C_2A-domain [3] and confirmed in more than ten C_2-domain structures (see Table I), C_2-domains are composed of a compact β-sandwich containing two β-sheets with four β-strands each. This is illustrated exemplarily for the two C_2-domains for synaptotagmin 1 in Fig. 1. Flexible loops at the top and bottom connect the eight β-strands. The β-strands are conserved between C_2-domains, whereas the loops vary and frequently contain additional secondary structure elements. The eight β-strands in the C_2-domain β-sandwich can be arranged in two distinct topologies. These topologies are circular permutations of each other, and are referred to as type 1 and type 2 (Fig. 2; reviewed in [1]). As a result, the N- and C-termini of the C_2-domains are either on top or at the bottom of the domains. However, C_2-domains with distinct topologies can nevertheless have similar three-dimensional structures. For example, the type 1 C_2A-domain from synaptotagmin 1 [3] and the type 2 C_2-domain from phospholipase Cδ1 [4] exhibit a root mean square deviation of 1.4 Å for 109 equivalent α-carbons. One reason for the different C_2-domain topologies may be that the topology influences the relative orientation of a C_2-domain and the neighboring domains.

The conserved core β-sandwich of C_2-domains serves as a scaffold for the emergence of variable loops at the top and bottom of the domain. In C_2-domains that bind Ca^{2+} and/or phospholipids, these exclusively bind to sites formed by the top loops. In contrast, the function of the bottom loops is unclear. The distinct Ca^{2+}- and phospholipid-binding properties of C_2-domains—as well as other activities—are probably encoded in the variable top and bottom loops which differ in size and sequence. C_2-domains thus are "janus-faced" modules in which a stable β-scaffold supports two variable surfaces: a generally Ca^{2+}-dependent top surface and a Ca^{2+}-independent bottom surface [5].

C_2-domains can be classified based on structural or functional properties. Structurally, in addition to the assignment of C_2-domains to the two principal types of β-strand topology, C_2-domains can be subdivided into classes that share common sequences. For example, the two C_2-domains of synaptotagmins (referred to as the C_2A- and C_2B-domains) exhibit conserved sequence differences. The fourth bottom loop of all C_2B-domains from synaptotagmins and related proteins forms a α-helix that is absent from all C_2A-domains and other C_2-domains [5]. In addition, the C_2B-domains from synaptotagmins 1, 2, and 8 but not other synaptotagmins include an extra C-terminal α-helix [6]. In contrast to this structural classification, C_2-domains can also be divided into functional classes. Again taking synaptotagmins as an example, synaptotagmins 1 and 2 bind Ca^{2+} but synaptotagmin 8

Table I Atomic Structures of C_2-Domains

Protein/C_2-domain	Method[a]	Ca^{2+}-binding state of structure[b]	Ca^{2+}-binding sites of C_2-domain[c]		References
			Number	Intrinsic affinity (in μM Ca^2)	
1. Trafficking proteins					
Synaptotagmin 1 C_2A-domain	X-ray & NMR	Ca^{2+}-free & Ca^{2+}-bound	3	Ca1 ≈ 50; Ca2 ≈ 500; Ca3 >10,000	3, 8–11, 18
Synaptotagmin 1 C_2B-domain	NMR	Ca^{2+}-bound	2	Ca1 ≈ 350; Ca2 ≈ 550	6
Rabphilin C_2B-domain	NMR	Ca^{2+}-bound	2	Ca1 ≈ 7; Ca2 ≈ 11	5
Synaptotagmin 3 C_2A/B-domain	X-ray	Ca^{2+}-free	n.d.	n.d.	41
2. Signal transduction proteins					
Protein kinase Cα C_2-domain	X-ray	Ca^{2+}-bound ± phospholipids	2	n.d.	17
Protein kinase Cβ C_2-domain	X-ray	Ca^{2+}-bound	3	n.d.	16
Protein kinase Cδ C_2-domain	X-ray	Ca^{2+}-independent		n.a.	23
Protein kinase Cε C_2-domain	X-ray	Ca^{2+}-independent		n.a.	24
Phospholipase Cδ1 C_2-domain	X-ray	Ca^{2+}-free & Ca^{2+}-bound	3	n.d.	4, 12
Phospholipase A_2 C_2-domain	X-ray & NMR	Ca^{2+}-bound	2	Ca1 ≈ 10; Ca2 ≈ 60	13, 14, 42
PTEN	X-ray	Ca^{2+}-independent		n.a.	21

[a]X-ray = X-ray crystallography; NMR = NMR spectroscopy.
[b]Number of Ca^{2+}-ions in structure.
[c]n.d. = not determined; n.a. = not applicable.

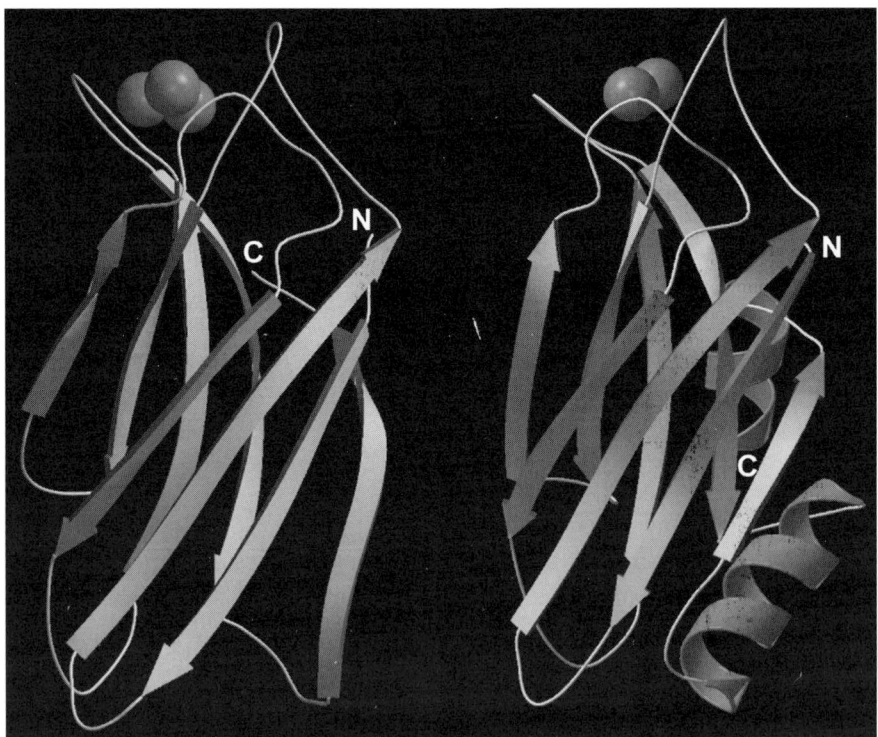

Figure 1 Structures of the synaptotagmin 1 C_2A- and C_2B-domains. Pictures show ribbon diagrams of the synaptotagmin C_2-domains (wide ribbons = β-strands; helices = α-helices; thin light strands = no secondary structure) in the Ca^{2+}-bound state. Three Ca^{2+}-ions are shown bound to the top loops for the C_2A-domain (left), and two Ca^{2+}-ions for the C_2B-domain (right). Note that only the C_2B-domain contains significant α-helices. Positions of N- and C-termini are indicated by N and C.

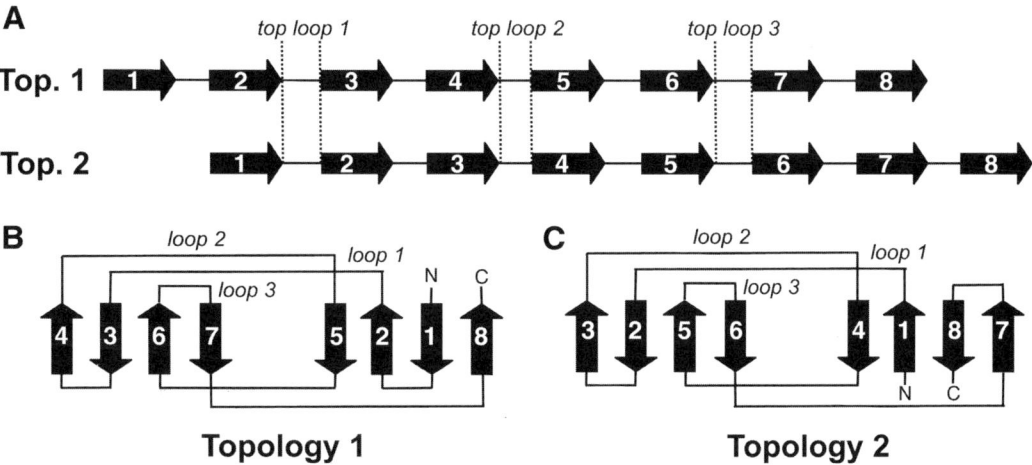

Figure 2 Diagram of the topography of β-strands in various types of C_2-domains. Two types of C_2-domain structures have been described that differ in the arrangement of β-strands; they are circular permutations of each other. In type 1 C_2-domains exemplified by the synaptotagmin C_2-domains (diagrams A and B; see also Fig. 1), N- and C-termini are on top of the C_2-domains, whereas in type 2 C_2-domains exemplified by the phospholipase Cδ1 structure (diagrams A and C; see ref. 4), the N- and C-termini are on the bottom. Note that the position of the top loops changes, but the actual localization of the loops within the three-dimensional structure does not.

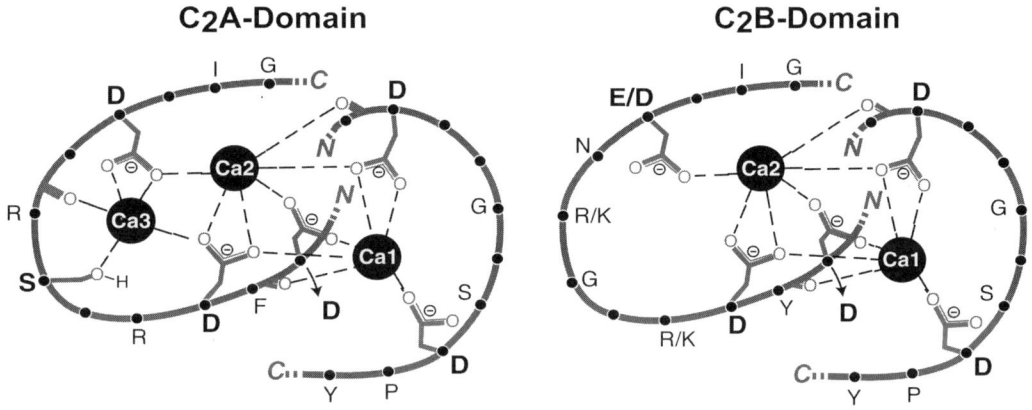

Figure 3 Model of the Ca^{2+}-binding sites of synaptotagmin C_2-domains. In both the C_2A- and C_2B-domains, the Ca^{2+}-binding sites are primarily formed by negatively charged residues located on top loops 1 and 3 (loop 2 is not shown). Most of the Ca^{2+}-coordinating negatively charged residues are multidentate ligands for the Ca^{2+}-ions. The sequences shown for the top loops are consensus sequences present in most but not all synaptotagmins (modified from [40]).

does not, in spite of the fact that the C_2B-domains from these three synaptotagmins share the same extra C-terminal α-helix, which is absent from other C_2B-domains [6,7]. Thus structural and functional classifications do not necessarily overlap.

Ca^{2+}-Binding Mode of C_2-Domains

In all Ca^{2+}-binding C_2-domains, Ca^{2+} binds exclusively to the top loops [3–6,8–17]. C_2-domains usually bind either two or three Ca^{2+}-ions at closely spaced sites that are primarily formed by aspartate residues, as illustrated in Fig. 3 for the two C_2-domains of synaptotagmin 1. The residues that form the Ca^{2+}-binding sites are widely separated in the primary sequences. The Ca^{2+}-binding residues often serve as bidentate ligands for multiple Ca^{2+}-ions [8,11,12]. The coordination spheres of the bound Ca^{2+} ions in the C_2-domain are incomplete in many C2-domains, resulting in low intrinsic Ca^{2+}-affinities (e.g. the synaptotagmin 1 C_2A-domain exhibits an affinity of >1.0 mM for complete Ca^{2+}-binding; [18]). As a result of this design, multiple Ca^{2+}-ions are concentrated in a small region on top of the C_2-domains and contain unsatisfied coordination sites that remain available for interaction with target molecules (see model in Fig. 3). Other C_2-domains, however, exhibit a much higher intrinsic Ca^{2+}-affinity (e.g. in the rabphilin C_2B-domain the two Ca^{2+}-ions bind with intrinsic affinities of 7 and 11 μM; [5]), and may have complete Ca^{2+}-coordination spheres.

The detailed characterization of the Ca^{2+}-binding sites for several C_2-domains, especially the C_2A-domain of synaptotagmin 1 [8–11,18] and the C_2-domain of phospholipase Cδ [4,12], revealed that the Ca^{2+}-binding sites are very similar, allowing a reasonably reliable prediction of whether a given C_2-domain is likely to bind Ca^{2+}. The aspartate residues involved in Ca^{2+} binding in the synaptotagmin I C_2A-domain are conserved in many C_2-domains, and the conserved Ca^{2+}-binding sequence that they form is referred to as the C_2-motif [8]. However, it has been difficult to predict the precise Ca^{2+}-binding properties of C_2-domains—for example their phospholipid specificities and apparent Ca^{2+}-affinities—presumably because they are determined by the variable sequences of their top loops that do not have defined conformations.

In all C_2-domains studied so far except for the piccolo C_2A-domain [19], Ca^{2+}-binding does not induce a substantial conformational change. For example, comparison of the Ca^{2+}-free and Ca^{2+}-bound forms of the synaptotagmin I C_2A-domain demonstrated that Ca^{2+} binding involves rotations of some side chains but causes no substantial backbone rearrangements [10]. The Ca^{2+}-binding region appears to be flexible in the absence of Ca^{2+}, and is stabilized after Ca^{2+} binding. Similar findings have been obtained for the C_2B-domain of synaptotagmin 1 [6] and the C_2-domain of phospholipase Cδ [4,12]. These results suggest that in most C_2-domains, Ca^{2+}-binding to a small patch on the top surface causes only a local effect that transduces the Ca^{2+}-binding signal. The nature of this effect probably depends on an electrostatic switch, since Ca^{2+}-binding causes a major change in the electrostatic potential of the top surface of the synaptotagmin 1 C_2A-domain [9].

The only C_2-domain that has been shown to undergo a major conformational change in response to Ca^{2+}-binding is the C_2A-domain of piccolo/aczonin [19]. Although this C_2-domain probably has "standard" C_2-domain Ca^{2+}-binding sites, Ca^{2+}-binding appears to induce a rearrangement of β-strands. The fact that a C_2-domain can undergo such a conformational change in response to Ca^{2+} indicates that C_2-domains are more versatile than suggested by the characterization of the initial C_2-domain structures.

Phospholipid Binding Mechanism of C_2-Domains

As first described for the Ca^{2+}-dependent binding of the synaptotagmin 1 C_2A-domain to phospholipids [20], the most common property of C_2-domains is phospholipid binding. This is true even for C_2-domains that do not bind Ca^{2+}; in fact, the function of most Ca^{2+}-independent C_2-domains appears to be to attach their resident proteins to phospholipid membranes [21–24]. Ca^{2+}-independent phospholipid binding is possibly best illustrated by the C_2-domain of PTEN, a tumor suppressor gene that is a phosphatase for the lipid phosphatidylinositol 3,4,5-trisphosphate [21,22]. The C_2-domain of PTEN positions its catalytic domain on top of the substrate. The C_2-domain not only recruits PTEN to the membrane, it also orients the catalytic domain with respect to the membrane substrate. Similar functions have been ascribed to the N-terminal C_2-domains of novel PKCs whose structures have been solved [23,24]. However, although significant evidence exists that phospholipid binding may be an even more general property of C_2-domains than Ca^{2+}-binding, it seems likely that not all C_2-domains bind phospholipids.

In all phospholipid-binding C_2-domains, phospholipids bind exclusively to the top loops similar to Ca^{2+}-binding. In spite of this similarity, however, the mechanism of phospholipid binding and the phospholipid specificity vary greatly among C_2-domains. Some C_2-domains (such as both C_2-domains of synaptotagmin 1) bind promiscuously negatively charged residues [6,20], whereas other (such as the C_2-domain of cytoplasmic phospholipase A2) bind neutral lipids [25]. Both hydrophobic and electrostatic interactions contribute to phospholipid binding but to different degrees in the various C_2-domains (see e.g. [25,26]). The contribution of different types of interactions has been described in detail for the double C_2-domain fragment of synaptotagmin 1 in which both C_2-domains contribute to the overall interaction [27,28]. Here, each C_2-domain separately participates in three types of interactions with the phospholipid bilayer, Ca^{2+}-mediated binding, hydrophobic attachment, and electrostatic interactions via positively charged residues.

Ca^{2+}-ions serve as a bridge that connect the C_2-domains to the phospholipid headgroups. The bound Ca^{2+}-ions are incompletely coordinated by the top loops of the C_2-domains. When phospholipids bind, they probably fill unsatisfied coordination sites on the bound Ca^{2+} ions. This results in a 100 to 1,000-fold increase in the apparent Ca^{2+}-affinity of the C_2-domains, and converts noncooperative intrinsic Ca^{2+}-binding into highly cooperative Ca^{2+}-binding by the C_2-domain/phospholipid complex [17,18,29]. In fact, at least for synaptotagmin 1, intrinsic Ca^{2+}-binding has an unphysiologically low affinity and probably never occurs in the absence of phospholipids, suggesting that the true signaling structure for synaptotagmins is the C_2-domain/phospholipid complex [18].

In addition to Ca^{2+}-ions, the synaptotagmin 1 C_2-domains are connected to the phospholipids by hydrophobic residues that insert into the bilayer [30,31] and by positively charged residues that form electrostatic interactions with negatively charged phospholipid headgroups [18,29]. All three forces contribute; in fact, the hydrophobic interactions, although constitutive, are essential for Ca^{2+}-triggered phospholipid binding [31]. Because the two C_2-domains are so closely spaced, the C_2-domains cooperate, resulting in a higher apparent Ca^{2+}-affinity of the double C_2-domain fragment than for the individual C_2-domains [28,32]. Furthermore, mutations that induce dramatic changes in the properties of isolated C_2-domains have unpredictable effects on the double C_2-domain fragment: Whereas some mutations induce the same change in the double C_2-domain fragment [18], others cause no change at all [32].

It seems likely that other C_2-domains, such as that of PKCα [17], bind phospholipids by a similar mechanism, although the contribution of the various types of interactions vary dramatically among C_2-domains [25]. In C_2-domains that bind phospholipids Ca^{2+} independently, the two Ca^{2+}-independent types of interactions are presumably sufficient to mediate constitutive binding. Such interactions could easily be modulated by other signaling pathways; for example, phosphorylation as shown for a C_2-domain from *Aplysia* PKC [33]. It would not be surprising if in Ca^{2+}-dependent and Ca^{2+}-independent C_2-domains membrane binding was further modulated by additional mechanisms.

Other Ligands of C_2-Domains

In addition to phospholipids, various C_2-domains have been reported to bind to many other molecules, primarily proteins (reviewed in [1]). Like all protein-protein interactions, the *in vivo* importance of these *in vitro* interactions is difficult to assess, and all of these interactions remain to be validated. Nevertheless, indirect evidence indicates that at least some of these interactions are important. First, some C_2-domains apparently do not bind to either phospholipids or Ca^{2+}. Although this finding does not exclude the possibility that the right lipids have not yet been tested, some of these C_2-domains strongly bind to other ligands that may mediate their functions. For example, the C_2B-domain of the active zone protein RIM does not bind Ca^{2+} or phospholipids but strongly interacts with proteins called α-liprins [34], suggesting that some C_2-domains might function as standard protein-protein interaction domains. Second, as janus-faced modules, C_2-domains have conserved sequence elements on their bottom surfaces that have no role in either Ca^{2+} or phospholipid binding [5]. It stands to reason that these sequences perform a function, although the nature of this activity remains obscure.

Function of C_2-domains

As is evident from their properties, the function of most C_2-domains is to attach their resident proteins to phospholipid membranes, although C_2-domains probably also connect their resident proteins to other ligands. The membrane-attachment function of C_2-domains may differ between trafficking and signal transduction proteins, the two classes that contain most of the C_2-domains in the genome. Signal transduction proteins usually contain a single C_2-domain that serves to position the catalytic domain of these proteins close to their substrates in the membrane. This is most obvious for enzymes that act on lipids such as phospholipase A2 and PTEN, where the C_2-domain is essential for placing the catalytic domain close to the phospholipid substrate either in a Ca^{2+}-dependent or constitutive manner [21,22,35,36]. However, this is also true for enzymes such as ras-GAP and PKC, where the C_2-domain brings the enzyme into proximity with membrane-bound ras (for ras-GAP) or diacylglycerol (for PKC) [37,38].

In contrast to signal transduction proteins, membrane trafficking proteins usually contain tandem C_2-domains (e.g. synaptotagmins), but some proteins include as many as six C_2-domains (e.g. ferlins; [39]). The properties of the tandem C_2-domain architecture has only been worked out for synaptotagmin 1 (reviewed in [40]). Here both C_2-domains bind Ca^{2+} and phospholipids, and both are essential for the function of the protein. Since synaptotagmins are intrinsic membrane proteins, their C_2-domains do not function to recruit these proteins to the membrane. The precise need for two C_2-domains is unknown, but a possible function of this configuration is to effect a physicochemical change in the target membranes to which they bind. For example, the role of synaptotagmin 1 in fast Ca^{2+}-triggered exocytosis may be mediated by a rapid Ca^{2+}-induced rearrangement of phospholipids during the fusion reaction, thereby opening the fusion pore that lets the transmitters escape. Although this is a plausible hypothesis for the need for two C_2-domains in synaptotagmin 1, it does not explain why so many other membrane trafficking proteins also contain two C_2-domains, even membrane proteins such as RIMs in which the C_2-domains do not appear to bind Ca^{2+} and/or phospholipids.

In the emerging universe of protein modules that are used to construct many of the eukaryotic signaling pathways, C_2-domains are remarkable for several reasons. A rigid core composed of a relatively invariant β-sandwich is used as a scaffold to form variable binding surfaces on the top and bottom of the module. The Ca^{2+}-binding sites formed in most C_2-domains are unusual because these sites are built from residues that are widely separated in the primary sequence and because Ca^{2+}-binding does not generally cause a conformational change but induces an electrostatic switch.

Although the progress in understanding C_2-domains has been significant over the last ten years, many questions remain to be addressed. For example, the mechanism and validity of the protein-protein interactions mediated by C_2-domains needs to be examined, the function of the bottom surface of C_2-domains needs to be elucidated, and the molecular basis for the Ca^{2+}-affinity and phospholipid-binding specificity of C_2-domains needs to be clarified. Before these important goals are realized, it will be difficult to postulate general conclusions about the functions of these domains.

References

1. Rizo, J. and Südhof, T. C. (1998). C_2-domains, structure of a universal Ca^{2+}-binding domain. *J. Biol. Chem.* **273**, 15879–15882.
2. International Human Genome Sequencing Consortium (2001). Initial sequencing and analysis of the human genome. *Nature* **409**, 860–921.
3. Sutton, A. B., Davletov, B. A., Berghuis, A. M., Südhof, T. C., and Sprang, S. R. (1995). Structure of the first C_2-domain of synaptotagmin I: A novel Ca^{2+}/phospholipids binding fold. *Cell* **80**, 929–938.
4. Essen, L.-O., Perisic, O., Cheung, R., Katan, M., and Williams, R. L. (1996). Crystal structure of a mammalian phosphoinositide-specific phospholipase C delta. *Nature* **380**, 595-602.
5. Ubach, J., Garcia, J., Nittler, M. P., Südhof, T. C., and Rizo, J. (1999). Structure of the janus-faced competence of glutamatergic synaptic vesicles. *Nature Cell Biol.* **1**, 106–112.
6. Fernandez, I., Arac, D., Ubach, J., Gerber, S. H., Shin, O.-K., Gao, Y., Anderson, R. G. W., Südhof, T. C., and Rizo, J. (2001). Three-dimensional

structure of the synaptotagmin 1 C$_2$B-domain: Synaptotagmin 1 as a phospholipid-binding machine. *Neuron* **23**, 1057–1069.
7. Li, C., Ullrich, B., Zhang, Z. Z., Anderson, R. G. W., Brose, N., and Südhof, T. C. (1995). Ca^{2+}-dependent and Ca^{2+}-independent activities of neural and nonneural synaptotagmins. *Nature* **375**, 594–599.
8. Shao, X., Davletov, B. A., Sutton, R. B., Südhof, T. C., and Rizo, J. (1996). Bipartite Ca^{2+}-binding motif in C$_2$-domains of synaptotagmin and protein kinase C. *Science* **273**, 248–251.
9. Shao, X., Li, C., Fernandez, I., Zhang, X., Südhof, T. C., and Rizo, J. (1997). Synaptotagmin-syntaxin interaction: the C$_2$-domain as a Ca^{2+}-dependent electrostatic switch. *Neuron* **18**, 133–142.
10. Shao, X., Fernandez, I., Südhof, T. C., and Rizo, J. (1998). Solution structures of the Ca^{2+}-free and Ca^{2+}-bound C$_2$A-domain of synaptotagmin I: does Ca^{2+} induce a conformational change? *Biochemistry* **37**, 16106–16115.
11. Ubach, J., Zhang, X., Shao, X., Südhof, T. C., and Rizo, J. (1998). Ca^{2+} binding to synaptotagmin: how many Ca^{2+} ions bind to the tip of a C$_2$-domain? *EMBO J.* **17**, 3921–3930.
12. Essen, L. O., Perisic, O., Lynch, D. E., Katan, M., Williams, R. L. (1997). A ternary metal binding site in the C$_2$ domain of phosphoinositide-specific phospholipase C-delta1. *Biochemistry* **36**, 2753–2762.
13. Perisic, O., Fong, S., Lynch, D. E., Bycroft, M., and Williams, R. L. (1998). Crystal structure of a calcium phospholipid binding domain from cytosolic phospholipase A2. *J. Biol. Chem.* **273**, 1596–1604.
14. Xu, G. Y., McDonagh, T., Yu, H. A., Nalefski, E. A., Clark, J. D., and Cumming, D. A. (1998). Solution structure and membrane interactions of the C2 domain of cytosolic phospholipase A2. *J. Mol. Biol.* **280**, 485–500.
15. Dessen, A., Tang, J., Schmidt, H., Stahl, M., Clark, J. D., Seehra, J., Somers, W. S. (1999). Crystal structure of human cytosolic phospholipase A2 reveals a novel topology and catalytic mechanism. *Cell* **97**, 349–360.
16. Sutton, R. B. and Sprang, S. R. (1998). Structure of the protein kinase C beta phospholipid-binding C$_2$ domain complexed with Ca^{2+}. *Structure* **6**, 1395–1405.
17. Verdaguer, N., Corbalan-Garcia, S., Ochoa, W. F., Fita, I., and Gomez-Fernandez, J. C. (1999). Ca^{2+} bridges the C$_2$ membrane-binding domain of protein kinase Cα directly to phosphatidylserine. *EMBO J.* **18**, 6329–6338.
18. Fernández-Chacón, R., Königstorfer, A., Gerber, S. H., García, J., Matos, M. F., Stevens, C. F., Brose, N., Rizo, J., Rosenmund, C., and Südhof, T. C. (2001). Synaptotagmin I functions as a Ca^{2+}-regulator of release probability. *Nature* **410**, 41–49.
19. Gerber, S. H., Garcia, J., Rizo, J., and Südhof, T. C. (2001). An unusual C$_2$-domain in the active zone protein piccolo: implications for Ca^{2+}-regulation of neurotransmitter release. *EMBO J.* **20**, 1605–1619.
20. Davletov, B. and Südhof, T. C. (1993). A single C$_2$-domain from synaptotagmin I is sufficient for high affinity Ca^{2+}/phospholipid-binding. *J. Biol. Chem.* **268**, 26386–26390.
21. Lee, J. O., Yang, H., Georgescu, M. M., Di Cristofano, A., Maehama, T., Shi, Y., Dixon, J. E., Pandolfi, P., and Pavletich, N. P. (1999). Crystal structure of the PTEN tumor suppressor: implications for its phosphoinositide phosphatase activity and membrane association. *Cell* **99**, 323–334.
22. Leslie, N. R. and Downes, C. P. (2002). PTEN: The down side of PI3-kinase signalling. *Cell Signal.* **14**, 285–295.
23. Pappa, H., Murray-Rust, J., Dekker, L. V., Parker, P. J., and McDonald, N. Q. (1998). Crystal structure of the C$_2$ domain from protein kinase C-delta. *Structure* **6**, 885–894.
24. Ochoa, W. F., Garcia-Garcia, J., Fita, I., Corbalan-Garcia, S., Verdaguer, N., Gomez-Fernandez, J. C. (2001). Structure of the C$_2$ domain from novel protein kinase C epsilon. A membrane binding model for Ca^{2+}-independent C$_2$ domains. *J. Mol. Biol.* **311**, 837–849.
25. Davletov, B., Perisic, O., and Williams, R. L. (1998). Calcium-dependent membrane penetration is a hallmark of the C$_2$ domain of cytosolic phospholipase A2 whereas the C$_2$A domain of synaptotagmin binds membranes electrostatically. *J. Biol. Chem.* **273**, 19093–19096.
26. Gerber, S. H., Rizo, J., and Südhof, T. C. (2001). The top loops of the C$_2$ domains from synaptotagmin and phospholipase A2 control functional specificity. *J. Biol. Chem.* **276**, 32288–32292.
27. Earles, C. A., Bai, J., Wang, P., and Chapman, E. R. (2001). The tandem C$_2$ domains of synaptotagmin contain redundant Ca^{2+} binding sites that cooperate to engage t-SNAREs and trigger exocytosis. *J. Cell Biol.* **17**, 1117–1123.
28. Shin, O.-K., Rizo, J., and Südhof, T. C. (2002). Synaptotagmin function in dense core vesicle exocytosis studied in cracked PC12 cells. *Nature Neurosci.* **5** 649–656.
29. Zhang, X., Rizo, R., and Südhof, T. C. (1998). Mechanism of phospholipid binding by the C$_2$A-domain of synaptotagmin. *Biochemistry* **37**, 12395–12403.
30. Chapman, E. R., and Davis, A. F. (1998). Direct interaction of a Ca^{2+}-binding loop of synaptotagmin with lipid bilayers. *J. Biol. Chem.* **273**, 13995–14001.
31. Gerber, S. H., Rizo, J., and Südhof, T. C. (2002). Role of electrostatic and hydrophobic interactions in Ca^{2+}-dependent phospholipid binding by the C$_2$A-domain of synaptotagmin 1. *Diabetes* **51**, S12–18.
32. Fernández-Chacón, R., Shin, O.-H., Königstorfer, A., Matos, M. F., Meyer, A. C., Garcia, J., Gerber, S. H., Rizo, J., Südhof, T. C., and Rosenmund, C. (2002). Structure/function analysis of Ca^{2+}-binding to the C$_2$A-domain of synaptotagmin 1. *J. Neurosci.* In press.
33. Pepio, A. M. and Sossin, W. S. (2001). Membrane translocation of novel protein kinase Cs is regulated by phosphorylation of the C$_2$ domain. *J. Biol. Chem.* **276**, 3846–3855.
34. Schoch, S., Castillo, P. E., Jo, T., Mukherjee, K., Geppert, M., Wang, Y., Schmitz, F., Malenka, R. C., and Südhof, T. C. (2002). RIM1α forms a protein scaffold for regulating neurotransmitter release at the active zone. *Nature* **415**, 321–326.
35. Perisic, O., Paterson, H. F., Mosedale, G., Lara-Gonzalez, S., and Williams, R. L. (1999). Mapping the phospholipid-binding surface and translocation determinants of the C2 domain from cytosolic phospholipase A2. *J. Biol. Chem.* **274**, 14979–14987.
36. Gijon, M. A., Spencer, D. M., Kaiser, A. L., and Leslie, C. C. (1999). Role of phosphorylation sites and the C$_2$ domain in regulation of cytosolic phospholipase A2. *J. Cell Biol.* **145**, 1219–1232.
37. Ponting, C. P. and Parker, P. J. (1996). Extending the C$_2$ domain family: C$_2$s in PKCs delta, epsilon, eta, theta, phospholipases, GAPs, and perforin. *Protein Sci.* **5**, 162–166.
38. Conesa-Zamora, P., Lopez-Andreo, M. J., Gomez-Fernandez, J. C., and Corbalan-Garcia, S. (2001). Identification of the phosphatidylserine binding site in the C$_2$ domain that is important for PKCα activation and in vivo cell localization. *Biochemistry* **40**, 13898–13905.
39. Britton, S., Freeman, T., Vafiadaki, E., Keers, S., Harrison, R., Bushby, K., and Bashir, R. (2000). The third human FER-1-like protein is highly similar to dysferlin. *Genomics* **68**, 313–321.
40. Südhof, T. C. (2002). Synaptotagmins: Why so many? *J. Biol. Chem.* **277**, 7629–7632.
41. Sutton, R. B., Ernst, J. A., and Brunger, A. T. (1999). Crystal structure of the cytosolic C$_2$A-C$_2$B domains of synaptotagmin III. Implications for Ca^{2+}-independent SNARE complex interaction. *J. Cell Biol.* **147**, 589–598.
42. Nalefski, E. A., Slazas, M. M., and Falke, J. J. (1997). Ca^{2+}-signaling cycle of a membrane-docking C$_2$ domain. *Biochemistry* **36**, 12011–12018.

CHAPTER 140

Annexins and Calcium Signaling

Stephen E. Moss

*Division of Cell Biology, Institute of Ophthalmology,
University College London,
London, United Kingdom*

Introduction

The vertebrate family of annexins comprises 12 calcium-binding proteins encoded by distinct genes. Although the functions of annexins are not yet fully elucidated, there is growing evidence that certain members of this family are involved in the homeostatic regulation of intracellular calcium ion concentration [1]. Annexins have a lower affinity for Ca^{2+} than E-F hand proteins, but this affinity is increased in the presence of negatively charged phospholipids. This defining biochemical property forms the basis of a generalized paradigm for annexin function in which elevation of intracellular Ca^{2+} concentration during cell stimulation is accompanied by translocation of annexins from the cytosol to the inner face of the plasma membrane. Ca^{2+}-dependent spatiotemporal regulation of subcellular localization is therefore likely to be a key aspect of annexin function, enabling annexins to influence the activities of other peripherally bound or integral membrane proteins in response to transient increases in cytosolic Ca^{2+} concentration. The precise question of what annexins do once membrane-bound has not been clearly answered, but it is probable that annexins are involved in membrane-associated events such as phospholipid clustering in lipid rafts, control of membrane fluidity, and structural changes to the membrane-cytoskeleton during endocytosis or phagocytosis (for review see [2]). Another possible role for membrane-bound annexins is in the generation and regulation of intracellular Ca^{2+} fluxes, and it is this topic that forms the focus of this chapter.

Annexins as Ca^{2+} Channels

The notion that annexins could function as Ca^{2+} channels first emerged in 1987, with the observation that purified annexin 7 (synexin) displays the properties of a voltage-gated Ca^{2+} channel when added to synthetic phospholipid bilayers [3]. Subsequent studies on the *in vitro* channel activities of various annexins, often supported by parallel structural analyses, have revealed this to be a general property of most members of the family. Annexin 5, which has been most extensively investigated with regard to the relationship between structure and Ca^{2+} channel activity, is approximately doughnut shaped with a slightly convex upper surface on which the Ca^{2+}-binding loops are located, and a slightly concave lower surface [4]. The proposed ion conductance pathway is lined with acidic residues, some of which have been demonstrated by mutagenesis studies to function as ion selectivity filter and voltage sensor [5,6]. Structural analysis of many other annexins has revealed almost superimposable tertiary architectures, and it is unsurprisingly that most family members exhibit Ca^{2+} channel activity *in vitro*.

The convincing structural basis for the Ca^{2+} channel activities of annexins is supported by electrophysiological and pharmacological correlates between the properties of putative annexin channels and as yet uncharacterized Ca^{2+} channels in mammalian cells [7–9]. For example, annexin 5 exhibits the properties of a classic voltage-gated Ca^{2+} channel, with unitary channel conductance values in the 10–20 pS range at both depolarizing and hyperpolarizing membrane potentials. The annexin 5 Ca^{2+} channel activity is inhibited by La^{3+}, which is known to block Ca^{2+} influx in most nonexcitable cells, whereas blockers of the L-, N-, P-, and T-type Ca^{2+} channels, such as nifedipine and Cd^{2+}, are without effect on the annexin 5 channel. Perhaps the most interesting pharmacological antagonist of annexin 5 is a cardioprotective benzothiazepine named K201, which exerts its effect by inhibiting Ca^{2+} influx into cardiomyocytes following ischemia-reperfusion injury [10]. Co-crystallization studies of K201 in complex with annexin 5 revealed the inhibitor to be tightly bound in a cleft at the proposed exit site of the Ca^{2+} conductance pathway [11].

Despite the weight of the structural, pharmacological, and electrophysiological evidence, the consensus view that annexins are either cytosolic or peripheral membrane-binding proteins presents a conceptual obstacle to the idea that such proteins could function as ion channels. There are few studies that have directly addressed this point, but investigations using bone-derived matrix vesicles [12] and chick DT40 cells containing a targeted disruption of the annexin 5 gene [13] add credibility to this theory. Mineralizing chondrocytes shed vesicles rich in phosphatidylserine and annexin 5 and take up Ca^{2+}, which forms crystals that embed in the collagen matrix during *de novo* bone deposition. The Ca^{2+} entry pathway in matrix vesicles was shown to be inhibited by Zn^{2+} and GTP, and increased by ATP. These characteristics mirror those observed in phospholipid vesicles containing annexin 5, suggesting that annexin 5 may be directly responsible for Ca^{2+} influx in mineralizing matrix vesicles. In annexin 5 null-mutant DT40 cells, Ca^{2+} signals elicited by both thapsigargin and activation of the B-cell receptor are normal, whereas the Ca^{2+} influx component of the biphasic response to hydrogen peroxide is absent. Cells lacking annexin 2 were normal with regard to peroxide-induced Ca^{2+} fluxes, showing that although both annexins exhibit Ca^{2+} channel activity *in vitro* [14], only annexin 5 seems to be involved in Ca^{2+} signaling *in vivo*. These data show that annexin 5 functions either as a Ca^{2+} channel, Ca^{2+} channel subunit, or signaling intermediate in the peroxide-activated Ca^{2+} influx pathway. Although the case for annexin 5 having a role as a Ca^{2+} channel is not yet proven, it would be a curious denouement if a protein that possesses so many of the structural and biophysical characteristics of a bona fide Ca^{2+} channel were to be shown to function as a Ca^{2+} channel regulator *in vivo*.

Annexins as Ca^{2+} Channel Regulators

A more readily acceptable modus operandi for annexins is in the regulation of intracellular Ca^{2+} fluxes. Annexins could influence Ca^{2+} signals by direct interaction with the cytoplasmic domains of proteins involved either in Ca^{2+} extrusion or in release of Ca^{2+} from the endoplasmic reticulum (ER) or sarcoplasmic reticulum (SR) Ca^{2+} stores. The first evidence for such roles came from studies in which annexin 6 was shown to increase the mean open time and opening probability of SR ryanodine-sensitive Ca^{2+} release channels in isolated membrane preparations [15]. However, annexin 6 was shown to exert this effect only when added to the lumenal side of the vesicles, and annexin 6 is generally considered to be a cytosolic protein. Also, these experiments were performed before it was known that annexin 6 itself has Ca^{2+} channel activity [16], raising the possibility that the changes observed in Ca^{2+} conductance may have been directly due to annexin 6. In A431 squamous epithelial carcinoma cells (which normally lack annexin 6), ectopic expression of annexin 6 was found to attenuate the sustained phase of the Ca^{2+} response to epidermal growth factor [17]. Other responses, such as to thapsigargin, were unaffected by annexin 6. An interesting finding is that only the larger of the two splice forms of annexin 6 exhibited this effect, which correlated with slower proliferative rate and tumor growth in nude mice [18].

Further studies in transgenic mice showed that targeted overexpression of an annexin 6 transgene in the heart led to cardiomyopathy, acute myocarditis, and fibrosis [19]. Experiments on isolated cardiomyocytes from these animals revealed lower resting Ca^{2+} levels, decreased amplitude of electrically evoked Ca^{2+} spikes, and impaired contractility. Similar studies on cardiomyocytes from annexin 6 knockout mice failed to identify any changes in resting cytosolic Ca^{2+} levels, but the contractile properties of these cells were significantly enhanced with regard to rate of contraction, extent of contraction, and rate of relaxation [20]. These mechanical changes correlated with accelerated diastolic clearance of Ca^{2+} from the cytosol, perhaps through enhanced activity of either the SR Ca^{2+} ATPase or the Na^+/Ca^{2+} exchanger. It is interesting to note that in humans with end-stage heart failure annexin 6 is downregulated [21]. Based on the studies of transgenic and null mutant mice, this would be predicted to enhance cardiomyocyte contractility, suggesting a negative inotropic role for annexin 6 in cardiomyocyte function.

Similar changes in cardiomyocyte function have been reported in mice lacking annexin 7 [22]. In normal mice, the degree of cell shortening increases with increasing frequency of stimulation, but this was not the case in the annexin 7 KO mice. In a separate study, targeted disruption of the annexin 7 gene in mice was reported to be lethal, and the heterozygous mice exhibited severe defects in insulin secretion that were apparently due to abnormally low expression of the inositol trisphosphate (IP_3) receptor [23]. Although these observations suggest a functional link between annexin 7 and IP_3 receptor expression, it is not clear why loss of a single annexin 7 allele should have such a striking effect on the expression of the IP_3 receptor.

Conclusions

The evidence that certain members of the annexin family have roles in Ca^{2+} signaling continues to grow. Most of the studies in this area have focused on annexins 5 and 6, and a combination of structural, electrophysiological, pharmacological, and genetic evidence tends to support the idea that annexin 5 functions as a Ca^{2+} channel. If this is indeed a physiological role of annexin 5, how does one account for the ability of a resident cytosolic protein to channel calcium ions? One study based on theoretical calculations predicted that membrane binding by annexin 5 would lead to foci of increased ion permeability caused by electrostatic destabilization of the lipid bilayer [24]. An alternative mechanism emerged from spin-labeling studies on annexin B12, which suggested that under certain conditions structural changes could occur that would lead to membrane insertion [25]. The availability of annexin null mutant cells and animals provides the opportunity to test these and other models of annexin function *in vivo*. Studies of this type should refine our understanding of the increasingly firm link between calcium signaling and annexin function.

References

1. Hawkins, T. E., Merrifield, C. J., and Moss, S. E. (2000). Calcium signaling and the annexins. *Cell Biochem. Biophys.* **33**, 275–296.
2. Gerke, V. and Moss, S. E. (2002). Annexins: from structure to function. *Physiol. Rev.* In press.
3. Rojas, E. and Pollard, H. B. (1987). Membrane capacity measurements suggest a calcium-dependent insertion of synexin into phosphatidylserine bilayers. *FEBS. Lett.* **217**, 25–31.
4. Huber, R., Romisch, J., and Paques, E. P. (1990). The crystal and molecular structure of human annexin V, an anticoagulant protein that binds to calcium and membranes. *EMBO J.* **9**, 3867–3874.
5. Burger, A., Voges, D., Demange, P., Perez, C. R., Huber, R., and Berendes, R. (1994). Structural and electrophysiological analysis of annexin V mutants. Mutagenesis of human annexin V, an in vitro voltage-gated calcium channel, provides information about the structural features of the ion pathway, the voltage sensor and the ion selectivity filter. *J. Mol. Biol.* **237**, 479–499.
6. Liemann, S., Benz, J., Burger, A., Voges, D., Hofmann, A., Huber, R., and Gottig, P. (1996). Structural and functional characterisation of the voltage sensor in the ion channel human annexin V. *J. Mol. Biol.* **258**, 555–561.
7. Demange, P., Voges, D., Benz, J., Liemann, S., Gottig, P., Berendes, R., Burger, A., and Huber, R. (1994). Annexin V: the key to understanding ion selectivity and voltage. *Trends. Biochem. Sci.* **19**, 272–276.
8. Rojas, E., Pollard, H. B., Haigler, H. T., Parra, C., and Burns, A. L. (1990). Calcium-activated endonexin II forms calcium channels across acidic phospholipid bilayer membranes. *J. Biol. Chem.* **265**, 21207–21215.
9. Berendes, R., Voges, D., Demange, P., Huber, R., and Burger, A. (1993). Structure-function analysis of the ion channel selectivity filter in human annexin V. *Science* **262**, 427–430.
10. Kaneko, N. (1994). New 1,4-benzothiazepine derivative, K201, demonstrates cardioprotective effects against sudden cardiac cell death and intracellular calcium blocking action. *Drug. Dev. Res.* **33**, 429–438.
11. Kaneko, N., Ago, H., Matsuda, R., Inagaki, E., and Miyano, M. (1997). Crystal structure of annexin V with its ligand K-201 as a calcium channel activity inhibitor. *J. Mol. Biol.* **274**, 16–20.
12. Arispe, N., Rojas, E., Genge, B. R., Wu, L. N., and Wuthier, R. E. (1996). Similarity in calcium channel activity of annexin V and matrix vesicles in planar lipid bilayers. *Biophys. J.* **71**, 1764–1775.
13. Kubista, H., Hawkins, T. E., Patel, D. R., Haigler, H. T., and Moss, S. E. (1999). Annexin 5 mediates a peroxide-induced Ca^{2+} influx in B cells. *Curr. Biol.* **9**, 1403–1406.
14. Burger, A., Berendes, R., Liemann, S., Benz, J., Hofmann, A., Gottig, P., Huber, R., Gerke, V., Thiel, C., Romisch, J., and Weber, K. (1996). The crystal structure and ion channel activity of human annexin II, a peripheral membrane protein. *J. Mol. Biol.* **257**, 839–847.
15. Diaz Munoz, M., Hamilton, S. L., Kaetzel, M. A., Hazarika, P., and Dedman, J. R. (1990). Modulation of Ca^{2+} release channel activity from sarcoplasmic reticulum by annexin VI (67-kDa calcimedin). *J. Biol. Chem.* **265**, 15894–15899.
16. Benz, J., Bergner, A., Hofmann, A., Demange, P., Gottig, P., Liemann, S., Huber, R., and Voges, D. (1996). The structure of recombinant human annexin VI in crystals and membrane-bound. *J. Mol. Biol.* **260**, 638–643.
17. Fleet, A., Ashworth, R., Kubista, H., Edwards, H. C., Bolsover, S., Mobbs, P., and Moss, S. E. (1999). Inhibition of EGF-dependent calcium influx by annexin VI is splice-form specific. *Biochem. Biophys. Res. Commun.* **260**, 540–546.
18. Theobald, J., Hanby, A., Patel, K., and Moss, S. E. (1995). Annexin VI has tumour-suppressor activity in human A431 squamous epithelial carcinoma cells. *Br. J. Cancer* **71**, 786–788.
19. Gunteski Hamblin, A. M., Song, G., Walsh, R. A., Frenzke, M., Boivin, G. P., Dorn, G. W. n., Kaetzel, M. A., Horseman, N. D., and Dedman, J. R. (1996). Annexin VI overexpression targeted to heart alters cardiomyocyte function in transgenic mice. *Am. J. Physiol.* **270**, H1091–1100.
20. Song, G., Harding, S. E., Duchen, M., Tunwell, R., O'Gara, P., Hawkins, T. E., and Moss, S. E. (2002). Altered mechanical properties and intracellular calcium transits in cardiomyocytes from mice with targeted disruption of the annexin 6 gene. *FASEB J.* In press.
21. Song, G., Campos, B., Wagoner, L. E., Dedman, J. R., and Walsh, R. A. (1998). Altered cardiac annexin mRNA and protein levels in the left ventricle of patients with end-stage heart failure. *J. Mol. Cell. Cardiol.* **30**, 443–451.
22. Herr, C., Smyth, N., Ullrich, S., Yun, F., Sasse, P., Hescheler, J., Fleischmann, B., Lasek, K., Brixius, K., Schwinger, R. H., Fassler, R., Schroder, R., and Noegel, A. A. (2001). Loss of annexin A7 leads to alterations in frequency-induced shortening of isolated murine cardiomyocytes. *Mol. Cell Biol.* **21**, 4119–4128.
23. Srivastava, M., Atwater, I., Glasman, M., Leighton, X., Goping, G., Caohuy, H., Miller, G., Pichel, J., Westphal, H., Mears, D., Rojas, E., and Pollard, H. B. (1999). Defects in inositol 1,4,5-trisphosphate receptor expression, $Ca^{(2+)}$ signaling, and insulin secretion in the anx7(+/−) knockout mouse. *Proc. Natl. Acad. Sci. USA* **96**, 13783–13788.
24. Karshikov, A., Berendes, R., Burger, A., Cavalie, A., Lux, H. D., and Huber, R. (1992). Annexin V membrane interaction: an electrostatic potential study. *Eur. Biophys. J.* **20**, 337–344.
25. Langen, R., Isas, J. M., Hubbell, W. L., and Haigler, H. T. (1998). A transmembrane form of annexin XII detected by site-directed spin labeling. *Proc. Natl. Acad. Sci. USA* **95**, 14060–14065.

Calpain

Alan Wells[1] and Anna Huttenlocher[2]
[1] Department of Pathology, University of Pittsburgh, Pittsburgh, Pennsylvania
[2] Department of Pediatrics and Pharmacology, University of Wisconsin, Madison, Wisconsin

Introduction

Four decades of study have provided much understanding of the calpain family of intracellular cysteine proteases [1]. Due to the limitations of investigative tools, earlier work focused on *in vitro* regulation and activities of these proteases, primarily the two ubiquitous isoforms μ- (calpain I) and m-calpain (calpain II) [2]. Structure-function studies have culminated in elucidating the molecular structure of the two ubiquitous calpains and deciphering how these enzymes might be activated and modulated. More recently, investigators have focused on connecting calpain function to physiology and pathology, particularly in regard to motility during wound repair [3,4], injury-mediated apoptosis in stroke and ischemia [5], protein degradation in muscular dystrophies [6], and susceptibility for non-insulin dependent diabetes mellitus [7,8].

Calpain Family

Thirteen distinct mammalian calpain gene products comprise the calpain gene family. The general structure is of a large subunit complexed to a single 30 kDa small subunit [6,9]. These isoforms differ in the length of N-terminal sequences, regulatory domain structures, and presence of calcium-binding domains. Ten calpain isoforms have been studied; most of these appear to be relatively selective for or enriched in cell and tissue types. The two ubiquitous calpains, μ- and m-calpain, are the best understood due to their high level of expression and primacy of discovery. These two isoforms were named according to their relative requirement for calcium *in vitro*, with μ-calpain requiring micromolar concentrations and m-calpain requiring near millimolar levels of calcium to elicit proteolytic activity *in vitro*.

Two other calpains, p94 calpain III and calpain X, also have a high level of interest due to their potential involvement in pathologies. The muscle-specific calpain III is characterized by two inserts, one within domain II and the other between domains III and IV [6]. It is interesting that p94 calpain appears to be independent of high calcium for activation but demonstrates low-level constitutive activity across a wide range of calcium concentrations. Calpain X has recently gained attention because of its identification in a linkage analysis study for type II or non-insulin-dependent diabetes [7,8]. This calpain is present in the β-cells of islets as well as in muscle and liver. Its structure is similar to that of the ubiquitous calpains.

Structure

The calpain molecule can be divided into five domains, initially described in protein structure-function studies and now by the crystal structure [10,11] (Fig. 1). Domain I contains a short 19 amino acid N-terminal sequence that is cleaved during autoproteolysis. The catalytic domain is divided into two parts, with the active site forming in the cleft between them. Domain III is a regulatory domain that has been shown to contain sites for attenuative phosphorylation [12] and a potential phospholipid-binding domain [13]. The fourth domain contains four calcium-binding EF-hand domains.

The crystal structure of calpain provides for the mechanism of activation. Unlike papains, the N-terminal domain is not a prodomain residing in the active site, a finding that confirms that autolysis of the N-terminal is not required for activation [14,15]. The most intriguing aspect is the active site itself.

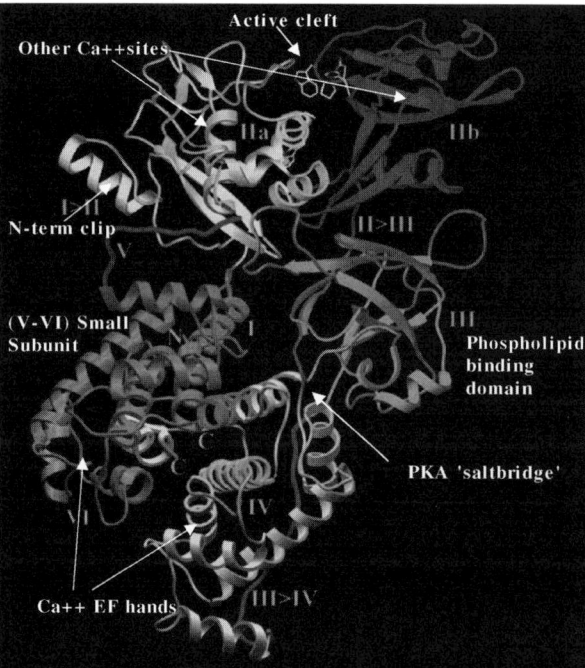

Figure 1 Structural domains of human m-calpain. Ribbon diagram of the crystal structure in the absence of calcium (domain V of the small subunit is absent) [10]. Denoted are sites of calcium binding in domains V and VI and in the active sites. Also noted are potential regulatory sites of phospholipid binding (the C2-like domain III), autoproteolysis of the N-terminal leader, and negative attenuation by PKA phosphorylation at serine/threonine 369/370. Adapted from [3,10].

In the inactive state, the catalytic residues (Cys105, His262, Asn286) are misaligned and too far apart to form a catalytic center [11,16]. Therefore, activating processes such as calcium binding, phospholipid binding, intramolecular cleavage, or phosphorylation must effect a realignment of these domains. Recent crystallography demonstrates that calcium loading at supraphysiological levels can accomplish such a shift. However, how these activating reorganization are effected *in vivo* remains a major challenge.

Modes of Regulation

An intricate strategy for the temporal and spatial regulation of calpain activity is necessary because calpain, which is abundant in the cytoplasm, cleaves many intracellular signaling and structural proteins. However, controversy still exists about how calpain activity is regulated *in vivo*. The lack of progress in this arena stems from many of the earlier studies having focused on calpain behavior *in vitro*. Whatever mechanisms are used to form an active site, the upstream signals triggering this activation and the downstream targets are critical in understanding the physiological roles of calpain. Furthermore, the ways that calpains are activated might vary not only by isoform, but also by subcellular localization as the ubiquitous isoforms are found throughout the cell, including in the nucleus. Therefore, multiple, potentially alternative or complementary mechanisms of activation have been proposed (Table I).

Table I Possible Mechanisms of Modulation

Activation	Inhibition
Calcium fluxes	Endogenous inhibitor Calpastatin
Phosphorylation	Phosphorylation
Proteolysis–limited	Proteolysis–extended
Phospholipids	
Protein cofactors	
DNA	

Calcium

Based on *in vitro* findings, calpains were proposed to be activated by intracellular calcium fluxes. That calcium can activate calpains is well supported *in vitro* [10,16]. *In vivo*, calcium chelation blocks activation of μ-calpain in response to chemokines in keratinocytes [17]. However, the need for seemingly supraphysiological levels of calcium has instigated searches for other modes of activation. Recent advances in calcium imaging suggest that levels high enough for μ-calpain activation could be achieved in highly localized calcium puffs (up to ~600 nM in nonexcitable cells) or sparks (excitable cells)[18]. During traumatic or ischemic compromise of the plasma membrane, calcium influx may reach levels that activate both μ- and m-calpain; but this level of calcium is not compatible with cell survival. Thus, the *in vitro* calcium levels required for m-calpain cannot be attained for other physiological responses. Therefore, a number of mechanisms have been suggested to lower the calcium requirement, even down to ambient cytosolic levels.

Phosphorylation

Most recently, an old standby of signal transduction cascades has been shown to regulate m-calpain. Early studies reported calpains not to be phosphorylated *in vivo* as determined by autoradiography due to the long half-life of calpains in unstimulated cells [19]. However, both m- and μ-calpain have been shown to be phosphorylated *in vivo* [20]. Under unstimulated conditions, there are three sites each of phospho-tyrosine, phospho-serine, and phospho-threonine phosphorylation, with the isolated calpains demonstrating varied sub-stoichiometric phosphorylation. Growth factors activate m-calpain downstream of ERK MAP kinase [21]; this is likely to occur by direct phosphorylation at amino acid S50 [22]. This is an intriguing finding, as p94 calpain III, which is considered constitutively active [23], presents a glutamic acid at this site.

Accessory Molecules

Mechanisms to reduce the requirement for calcium to the physiological range have been proposed. These include phospholipid binding, release of calpain from its inhibitor calpastatin, and binding of activator proteins. Phospholipids decrease the calcium requirement *in vitro* [24,25]. Calpain translocates to the plasma membrane in the presence of calcium, where it associates with phosphatidylinositol bis-phosphate [26]. Of particular interest, there is a putative phospholipid-binding activity in the regulatory domain III [13]. What is especially intriguing, DNA has been reported to lessen the requirement of m-calpain for select nucleoproteins [27], which may provide a mode of activation for the nuclear-localized pool of this protease.

A ubiquitous, endogenous inhibitor of calpains, calpastatin, provided hope that dissociation/re-association would be the mainstay of calpain regulation. Calpastatin binds and inactivates calpains through each of its four repetitive inhibitory domains. However, release of calpain from calpastatin, although it correlates with activity, is insufficient for activation. Furthermore, calpastatin is neither always present in excess molar levels, nor always co-localized with calpain. Calcium fluxes actually enhance calpastatin inhibition of calpains, suggesting that calpastatin might attenuate activated calpains rather than prevent activation [28,29]. Despite the conflicting evidence of physiological relevance, overexpression of this molecule can be employed to prevent calpain activation.

Other protein-protein interactions have been proposed to activate calpain. Select proteins co-purify with active calpains from many cell types. In rat skeletal muscle, bovine brain, and rat brain, activator proteins were found that increased autolysis and lowered the calcium requirement of μ-calpain [30–32]. Acyl-CoA-binding protein has been proposed as an activator for m-calpain [33]. Unfortunately, the association and activation of calpain *in vivo* by these proteins has not been demonstrated, and the mechanism by which these proteins would activate calpains remains unclear.

Inactivation

Key to all enzymes, especially those that cause irreversible signaling such as proteolytic cleavage, is an efficient system to prevent unintended activity. Calpain activity appears to be kept at minimal levels until signaled. How this occurs is still unknown, in part because the activation mechanisms are similarly unclear.

Calpain autoproteolysis and degradation rapidly remove active enzyme. The half-life of active calpain I or calpain II is shortened from almost a week [34] to just hours [21]. This autolysis was thought to activate calpain [35,36], since the N-terminal clipped intermediaries display increased activity. However, as intact calpain can be equally active [15,37,38], these are now considered just steps on the way to degradative removal.

Calpastatin can inhibit calpain activity by acting through each of four repeated domains. Expression of exogenously encoded calpastatin has been used successfully to block calpain activation [39,40]. Still, this does not address whether this endogenous protein acts as such *in vivo*. In fact, the reported discrepancies in subcellular localization [41] argue against this being the only inhibitory mechanism for preventing calpain activity. However, it is possible that calpastatin serves to attenuate activated calpain [42], whereas low calpain activity levels are maintained through lack of positive signals.

Phosphorylation of calpains may serve to either prevent activation or attenuate triggered enzyme. PKA phosphorylation of at least m-calpain limits the ability of growth factors to activate this isoform [12,43]. Whether this mechanism is operative in other isoforms is still an open question. The target serine at amino acid 369 is present in some of the other isoforms, but the recipient residues of the putative ensuing salt-bridge is lacking in μ-calpain.

Calpain as a Signaling Intermediate: Potential Targets

Evidence supports a critical role for calpain as a signaling intermediate downstream of both integrin and growth factor signaling pathways. The role for calcium, phospholipids, and phosphorylation in calpain regulation supports a central role for calpain in basic signal transduction mechanisms. However, the key question for understanding calpain function remains frustratingly unsolved—what are the operative targets of calpains? Cell behaviors dependent on calpain have provided hints as to what these targets might be, and many substrates have been identified both *in vitro* and *in vivo* (Table II). However, establishing whether proteolysis of these targets is either sufficient or required for the cellular responses has remained challenging due to the difficulties in generating calpain-resistant functional target molecules. The structure, primary or tertiary, of the proteolytic sites remains unknown, a fact that has confounded attempts to identify key target molecules or negative calpain cleavage of putative targets to allow assessment of functional role. Originally, calpain was proposed to cleave downstream of PEST sequences [9]; although further identification of targets demonstrated that the presence of a PEST sequence was not required [10].

The limited proteolysis of calpain suggests that it functions as an irreversible step in signaling cascades, generating constitutively active or dominant-negative versions of signaling proteins, rather than serving a degradative function. A number of structural and signaling molecules have been

Table II   Potential Targets of m- and μ-Calpain

Signaling	Adhesion	Proliferation/survival	Cytoskeletal components
EGF receptor	β-integrins	cyclin D1	spectrin
Protein kinase C	ezrin	caspases	MAP2
Src	talin	p53	filamin
Rho A	paxillin	p35	fodrin
Myosin light chain kinase	vinculin		tau
Focal adhesion kinase	α-actinin		

identified *in vitro* and in cells as targets of calpain. An early identified target of calpain was the EGF receptor, wherein calpain removes most of the carboxy-terminal domain that serves both as an autoinhibitory and a docking domain. Thus, it is not obvious whether calpain cleavage would increase or decrease EGF signaling or generate a signaling-restricted EGFR. Many of the other targets of the ubiquitous calpains are involved in cell adhesion and motility, being linked to the cytoskeletal machinery. These include FAK, ezrin, talin, paxillin, src, MLCK, RhoA, and the cytosolic tails of some of the β-integrins [3]. Recent studies with calpain-deficient embryonic fibroblasts adherent to fibronectin substrata demonstrate *in vivo* cleavage of talin but not FAK, paxillin, α-actinin, or vinculin, suggesting that talin may be a critical calpain substrate *in vivo* [44]. In accordance with previously published reports [45], Capn4-/-embryonic fibroblasts have reduced stress fibers, thus implicating a role for calpain in the formation of Rho-mediated stress fibers [44]. It must be mentioned that various investigators report different spectra of cellular targets in very similar systems, thereby suggesting that calpain targeting is likely to be plastic and redundant and possibly dependent on the mode of activation and analysis.

Functional Roles

Selective inhibitors for calpain have provided functional indications of calpain's roles in a wide range of physiological processes, including cell motility, cell proliferation, and apoptosis. However, confusion and controversy have existed regarding calpain's functional role, to a large extent, because of a lack of specificity of many of the cell-permeable inhibitors. More recent studies using calpain-deficient embryonic fibroblasts or ectopic expression of the endogenous calpain inhibitor calpastatin have helped clarify calpain's physiological role; however, the current efforts are also not isoform specific but target both m- and μ-calpain. These studies support a critical role for calpain in regulating the actin cytoskeleton and cell migration [44] but have called into question its role in other processes such as cell proliferation [24]. However, part of calpain's widespread functional profile is likely to be due to the various calpain isoforms and their cell-specific functions. In many cases these various functions of calpain can be seen in the same cell system, thus suggesting that these different functions may also be subserved by the different localized pools of calpain that exist throughout a cell.

The critical importance of ubiquitous m- and μ-calpain for normal development has been demonstrated by transgenic mice deficient in the regulatory subunit that eliminates detectable calpain activity [24]. These mice die during embryonic development with vascular defects, thus supporting a role for calpain in blood vessel formation.

Platelet Activation

Initial studies of the role of calpain were conducted in platelets, an interesting system in which calpain clearly plays a role in secretion, adhesion, and aggregation. Inhibition of calpain via overexpression of calpastatin prevents α-granule secretion, platelet aggregation, and spreading on glass surfaces [46]. Platelets uniquely express predominantly μ-calpain and have negligible levels of M-calpain. Therefore, molecular inhibition of μ-calpain is sufficient to down-regulate all detectable calpain activity, and antibodies to the autolyzed form of μ-calpain can yield meaningful results.

In this context, calpain was shown to be part of the signal transduction apparatus. Calpain is activated following signaling by the platelet integrin αIIbβ3 [47,48]. Calpain associates with focal adhesion proteins in platelets, regulates the attachment of αIIbβ3 to the cytoskeleton, and relaxes the retraction of fibrin clots. Activation of calpain by ionophore A23187 increased the proteolysis of pp60c-src and PTP-1B, which then dissociated from the cytoskeleton, thereby inactivating these proteins. This correlated with the inhibition of fibrin clot retraction observed in aggregated platelets in the presence of calcium. Calpain inhibition also blocked the cleavage of the actin-binding protein talin, whereas calpain activation caused the movement of both cleaved talin and integrin αIIbβ3 from the Triton X-100 insoluble fraction (cytoskeleton) to the Triton X-100 soluble fraction [49]. Calpain therefore functions as a signaling molecule in platelets by coordinating the cellular response of aggregation and clot formation.

Adhesion Modulation—Spreading and Motility

Calpains regulate cell adhesion to the substratum and thereby affect spreading and motility of many cell types. Cell spreading requires active remodeling and turnover of adhesion sites to enable cells to extend processes subsequent to attachment. It is also considered to be similar to forward protrusion during active cell locomotion. In bovine aortic endothelial cells, calpain enables spreading by allowing formation of Rac-induced adhesions under the extended lamellae [45,50]. Inhibition of calpain caused a marked reduction in cell spreading and adhesion formation, without affecting initial attachment. Calpain acting to enable new supramolecular assembly is also noted in T cells, in which integrin ligation activates calpain to promote integrin diffusion to form focal complexes and ultimately cell spreading [51,52]. Calpain inhibition may have very different effects on cell spreading in different cellular contexts. For example, enhanced membrane protrusion and filopodia formation is observed in calpain deficient embryonic fibroblasts ([44]; A. Huttenlocher, unpublished).

Calpain-mediated regulation of cell/substratum adhesion is critical not only during spreading and forward protrusion but also in rear release during productive motility [40,53]. Haptokinetic motility, signaled by adhesion receptors, primarily integrins, is calpain dependent. β1 and β3 integrin-mediated CHO cell migration is sensitive to calpain inhibition [40]. Calpain inhibition stabilized peripheral focal adhesions and decreased the detachment rate. If the effect of calpain was to alter adhesion to the substratum, one would predict a varied effect dependent on substrate density, with calpain being required most for migration over highly adhesive surfaces but only minimally involved for low adhesive regimens [40,54]. That calpain modulated cell motility dependent on adhesive strength identically to alterations in integrin affinity for fibronectin [54] indicates that calpain is acting effectively as a physiologic rheostat for adhesion.

Although the calpain isoform that functions downstream of integrin-mediated adhesion and migration has not been clearly identified, some evidence supports a role for μ-calpain in this regulation, most specifically during endothelial cell and platelet spreading. A growing body of evidence suggests that m-calpain may be involved in growth factor motility [53]. Growth factor–induced chemokinesis also requires de-adhesion [55], dependent on calpain [21]. It is interesting that this de-adhesion and motility occurs via ERK MAP kinase phosphorylating and enabling activation of m- but not μ-calpain in the absence of a calcium flux ([21]; A. Glading, unpublished). The site of phosphorylation appears to be S50, which is absent in μ-calpain. This finding provides an imposed rationale for the evolutionary duplication of the ubiquitous isoforms.

Calpain in Muscular Dystrophy

Calpain 3 is the skeletal muscle–specific calpain isoform. Defects in the human calpain 3 gene are responsible for a form of muscular dystrophy, limb girdle muscular dystrophy type 2A [56]. A calpain 3–deficient mouse model also shows a progressive muscular dystrophy with perturbations in membrane architecture and apoptosis-associated regulation of the IkappaB pathway [57]. It is interesting that in the mouse model of Duchenne's muscular dystrophy there is an increase in the expression and activity of the ubiquitous calpain isoforms, suggesting that perturbation of the muscle calpains, and not just enhanced proteolysis, may contribute to the pathogenesis of muscular dystrophy.

Apoptosis

Calpain has been implicated in necrotic and apoptotic cell death [58]. Previous reports have shown that calpain inhibitors have protective effects in *in vivo* models of CNS [58] and cardiac ischemia [59]. The combined treatment of neurons with both calpain and caspase inhibitors may have an additive protective effect against neuronal apoptosis [58]. These studies support the intriguing potential of calpain inhibitors as a therapeutic target to treat cerebral or cardiac ischemia. How calpain inhibitors exert anti-necrotic and anti-apoptotic effects remain unclear. During ischemic compromise, calcium influx may reach levels that activate both μ- and m-calpain. Under these conditions, calpain may cleave multiple substrates, including signaling, cytoskeletal proteins, and transcription factors. It is likely that calpain-mediated cleavage of focal adhesion and cytoskeletal proteins contributes to cell rounding and loss of focal adhesions during apoptosis and necrotic cell death. However, a direct modulation of apoptotic signaling pathways, i.e. by the cleavage and regulation of caspase activity, for example, may also contribute to calpain's role during apoptosis [60].

Proliferation

Substantial controversy exists about calpain's role during cell proliferation and cell cycle progression. Capn4-/- embryonic fibroblasts exhibit normal proliferation rates. However, ectopic expression of calpastatin reduces CHO cell proliferation [61] and Src-mediated transformation [62]. The calpastatin-induced inhibition of cell cycle progression in Src-transformed cells is associated with a decrease in pRb phosphorylation and reduced levels of cyclin A and D. However, although calpain may cleave cell cycle proteins such as cyclin D1 [40], a substrate for calpain's effects on cell cycle progression has not been identified. Defining calpain's role during cell cycle progression will be an important challenge for future investigation.

Future Considerations

Much is known about this ubiquitous family of limited intracellular proteases. Many investigators have defined the extended family and begun to establish structural bases of calpain activation and regulation. In addition, a number of functional roles have been established by calpain family–selective inhibitors, which in turn have provided potential target proteins. However, much remains to be learned about this complex family of enzymes.

A glaring gap in our knowledge is what precise roles the various members serve in cells, and how the different calpain localizations contribute to these cellular responses. For instance, if membrane-associated μ- and m-calpain contribute to rear detachment during motility [3,12,40], what do the majority of cytosolic and nuclear μ- and m-calpains do? Only by linking the function of isoform pools of calpain to specific cellular behaviors will we be able to understand the key targets of calpains and whether calpain clipping results in an active or dead molecule. To achieve this level of understanding will require significant advances in our tool sets. First, isoform-specific inhibitors have been attempted without widespread adoption. Second, calpain activity or activation needs to be imaged within subcellular compartments; the ubiquitous nature of calpain distribution throughout the cell renders simple localization and colocalization data of limited utility. That these advances will occur is ever more likely due to the increased realization that calpains function in motility during wound repair and tumor progression, in ischemia-induced apoptosis that aggravates stroke and myocardial infarction, and in myosin degradation of muscle-wasting syndromes. That calpain may prove a target for intervention in these major medical conditions ensures a burgeoning body of work on these fascinating molecules.

References

1. Guroff, G. (1964). A neutral, calcium-activated proteinase from the soluble fraction of rat brain. *J. Biol. Chem.* **239**, 149–155.
2. Murachi, T. (1989). Intracellular regulatory system involving calpain and calpastatin. *Biochem. Int.* **18**, 263–294.
3. Glading, A., Lauffenburger, D. A., and Wells, A. (2002). Cutting to the chase: calpain proteases in cell migration. *Trends Cell Biol.* **12**, 46–54.
4. Perrin, B. J. and Huttenlocher, A. (2002). Calpain. *Int. J. Biochem. Cell Biol.* **34**, 722–725.
5. Vanderklish, P. and Bahr, B. (2000). The pathogenic activation of calpain: a marker and mediator of cellular toxicity and disease states. *Int. J. Exp. Pathol.* **81**, 323–339.
6. Sorimachi, H. and Suzuki, K. (2001). The structure of calpain. *J. Biochem.* **129**, 653–664.
7. Horikawa, Y., Oda, N., Cox, N. J., Li, X., Orho-Melander, M., Hara, M., Hinokio, Y., Lindner, T. H., Mashima, H., Schwarz, P. E., delBosque-Plata, L., Horikawa, Y., Oda, Y., Yoshiuchi, I., Colilla, S., Polonsky, K. S., Wei, S., Concannon, P., Iwasaki, N., Schulze, J., Baier, L. J., Bogardus, C., Groop, L., Boerwinkle, E., Hanis, C. L., and Bell, G. I. (2000). Genetic variation in the gene encoding calpain-10 is associated with type 2 diabetes mellitus. *Nature Genet.* **26**, 163–175.
8. Sreenan, S. K., Zhou, Y. P., Otani, K., Hansen, P. A., Curie, K. P., Pan, C. Y., Lee, J. P., Ostrega, D. M., Pugh, W., Horikawa, Y., Cox, N. J., Hanis, C. L., Burant, C. F., Fox, A. P., Bell, G. I., and Polonsky, K. S. (2001). Calpains play a role in insulin secretion and action. *Diabetes* **50**, 2013–2020.
9. Sorimachi, H., Ishura, S., and Suzuki, K. (1997). Structure and physiological function of calpains. *Biochem. J.* **328**, 721–732.
10. Strobl, S., Fernandez-Catalan, C., Braun, M., Huber, R., Masumoto, H., Nakagawa, K., Irie, A., Sorimachi, H., Bourenkow, G., Bartunik, H., Suzuki, K., and Bode, W. (2000). The crystal structure of calcium-free human m-calpain suggests an electrostatic switch mechanism for activation by calcium. *Proc. Natl. Acad. Sci. USA* **97**, 588–592.
11. Hosfield, C. M., Elce, J. S., Davies, O. K., and Jia, Z. (1999). Crystal structure of calpain reveals the structural basis for Ca^{2+}-dependent protease activity and a novel model of enzyme activation. *EMBO J.* **18**, 6880–6889.
12. Shiraha, H., Glading, A., Chou, J., Jia, Z., and Wells, A. (2002). Activation of m-calpain (calpain II) by epidermal growth factor is limited by PKA phosphorylation of m-calpain. *Mol. Cell. Biol.* **22**, 2716–2727.
13. Tompa, P., Emori, Y., Sorimachi, H., Suzuki, K., and Friedrich, P. (2001). Domain III of calpain is a Ca^{+2}-regulated phospholipid-binding domain. *Biochem. Biophys. Res. Commun.* **280**, 1333–1339.
14. Guttmann, R. P., Elce, J. S., Bell, P. D., Isbell, J. C., and Johnson, G. V. (1997). Oxidation inhibits substrate proteolysis by calpain I, but not autolysis. *J. Biol. Chem.* **272**, 2005–2012.
15. Johnson, G. V. W. and Guttmann, R. P. (1997). Calpains: intact and active? *Bioessays* **19**, 1011–1018.
16. Moldoveanu, T., Hosfield, C. M., Lim, D., Elce, L. S., Jia, Z., and Davies, P. L. (2002). A $Ca^{(2+)}$ switch aligns the active site of calpain. *Cell* **108**, 649–660.
17. Satish, L., Yager, D., and Wells, A. (2003). ELR-negative CXC chemokine IP-9 as a mediator of epidermal-dermal communication during wound repair. *J. Invest. Derm.*, in press.
18. Bootman, M. D., Lipp, P., and Berridge, M. J. (2001). The organisation and functions of local Ca^{2+} signals. *J. Cell Sci.* **114**, 2213–2222.
19. Adachi, Y., Kobayashi, N., Murachi, T., and Hatanaka, M. (1986). Ca^{2+}-dependent cysteine proteinase, calpains I and II are not phosphorylated in vivo. *Biochem. Biophys. Res. Commun.* **136**, 1090–1096.
20. Cong, J. Y., Thompson, V. F., and Goll, D. E. (2000). Phosphorylation of the calpains. *Mol. Biol. Cell* **11**, S2003.
21. Glading, A., Chang, P., Lauffenburger, D. A., and Wells, A. (2000). Epidermal growth factor receptor activation of calpain is required for fibroblast motility and occurs via an ERK/MAP kinase signaling pathway. *J. Biol. Chem.* **275**, 2390–2398.
22. Glading, A., Reynolds, I. J., Shiraha, H., Blair, H. C., and Wells, A. (2003). M-calpain is activated by direct phosphorylation by ERK in response to EGF stimulation. Submitted.
23. Branca, D., Gugliucci, A., Bano, D., Brini, M., and Carafoli, E. (1999). Expression, partial purification and functional properties of the muscle-specific calpain isoform p94. *Eur. J. Biochem.* **265**, 839–846.
24. Arthur, J. S., Elce, J. S., Hegadorn, C., Williams, K., and Greer, P. A. (2000). Disruption of the murine calpain small subunit gene, Capn4: calpain is essential for embryonic development but not for cell growth and division. *Mol. Cell. Biol.* **20**, 4474–4481.
25. Melloni, E., Michetti, M., Salamino, F., Minafra, R., and Pontremoli, S. (1996). Modulation of the calpain autoproteolysis by calpastatin and phospholipids. *Biochem. Biophys.l Res. Communic.* **229**, 193–197.
26. Suzuki, K., Saido, T. C., and Hirai, S. (1992). Modulation of cellular signals by calpain. *Ann. NY Acad. Sci.* **674**, 218–227.
27. Mellgren, R. L., Song, K., and Mericle, M. T. (1993). m-Calpain requires DNA for activity on nuclear proteins at low calcium concentrations. *J. Biol. Chem.* **268**, 653–657.
28. Barnoy, S., Zipser, Y., Glaser, T., Grimberg, Y., and Kosower, N. S. (1999). Association of calpain (Ca^{2+}-dependent thiol protease) with its endogenous inhibitor calpastatin in myoblasts. *J. Cell Biochem.* **74**, 522–531.
29. Tullio, R. D., Passalacqua, M., Averna, M., Salamino, F., Melloni, E., and Pontremoli, S. (1999). Changes in intracellular localization of calpastatin during calpain activation. *Biochem. J.* **343**, 467–472.
30. Michetti, M., Viotti, P. L., Melloni, E., and Pontremoli, S. (1991). Mechanism of action of the calpain activator protein in rat skeletal muscle. *Eur. J. Biochem.* **202**, 1177–1180.
31. Melloni, E., Michetti, M., Salamino, F., and Pontremoli, S. (1998). Molecular and functional properties of a calpain activator protein specific for μ-isoforms. *J. Biol. Chem.* **273**, 12827–12831.
32. Salamino, F., DeTullio, R., Mengotti, P., Viotti, P. L., Melloni, E., and Pontremoli, S. (1993). Site-directed activation of calpain is promoted by a membrane-associated natural activator protein. *Biochem. J.* **290**, 191–197.
33. Melloni, E., Averna, M., Salamino, F., Sparatore, B., Minafra, R., and Pontremoli, S. (2000). Acyl-CoA-binding protein is a potent m-calpain activator. *J. Biol. Chem.* **275**, 82–86.
34. Zhang, W., Lane, R. D., and Mellgren, R. L. (1996). The major calpain isozymes are long-lived proteins. Design of an antisense strategy for calpain depletion in cultured cells. *J. Biol. Chem.* **271**, 18825–18830.
35. Fujitani, K., Kambayashi, J., Sakon, M., Ohmi, S. I., Kawashima, S., Yukawa, M., Yano, Y., Miyoshi, H., Ikeda, M., Shinoki, N., and Monden, M. (1997). Identification of μ–, m-calpains and calpastatin and capture of m-calpain activation in endothelial cells. *J. Cell. Chem.* **66**, 197–209.
36. Baki, A., Tompa, P., Alexa, A., Molnar, O., and Friedrich, P. (1996). Autolysis parallels activation of mu-calpain. *Biochem. J.* **318**, 897–901.
37. Cong, J., Goll, D. E., Peterson, A. M., and Kapprell, H. P. (1989). The role of autolysis in activity of the Ca^{2+}-dependent proteinases (μ-calpain and m-calpain). *J. Biol. Chem.* **264**, 10096–10103.
38. Molinari, M., Anagli, J., and Carafoli, E. (1994). Ca^{2+}-activated neutral protease is active in erythrocyte membrane in its nonautolyzed 80 kDa form. *J. Biol. Chem.* **269**, 27992–27995.
39. Potter, D. A., Tirnauer, J. S., Janssen, R., Croall, D. E., Hughes, C. N., Fiacco, K. A., Mier, J. W., Maki, M., and Herman, I. M. (1998). Calpain regulates actin remodeling during cell spreading. *J. Cell. Biol.* **141**, 647–662.
40. Huttenlocher, A., Palecek, S. P., Lu, Q., Zhang, W., Mellgren, R. L., Lauffenburger, D. A., Ginsburg, M. H., and Horwitz, A. F. (1997). Regulation of cell migration by the calcium-dependent protease calpain. *J. Biol. Chem.* **272**, 32719–32722.
41. Lane, R. D., Allan, D. M., and Mellgren, R. L. (1992). A comparison of the intracellular distribution of μ-calpain, m-calpain, and calpastatin in proliferating human A431 cells. *Exp. Cell Res.* **203**, 5–16.
42. Averna, M., deTullio, R., Passalacqua, M., Salamino, F., Pontremoli, S., and Melloni, E. (2001). Changes in intracellular calpastatin localization are mediated by reversible phosphorylation. *Biochem. J.* **354**, 25–30.

43. Shiraha, H., Gupta, K., Glading, A., and Wells, A. (1999). Chemokine transmodulation of EGF receptor signaling: IP-10 inhibits motility by decreasing EGF-induced calpain activity. *J. Cell Biol.* **146**, 243–253.
44. Dourdin, N., Bhatt, A. K., Greer, P. A., Arthur, J., Elce, J., and Huttenlocher, A. (2001). Reduced cell migration in calpain-deficient embryonic fibroblast. *J. Biol. Chem.* **276**, 48382–48388.
45. Kulkarni, S., Saido, T. C., Suzuki, K., and Fox, J. E. (1999). Calpain mediates integrin-induced signaling at a point upstream of rho family members. *J. Biol. Chem.* **274**, 21265–21275.
46. Croce, K., Flaumenhaft, R., Rivers, M., Furie, B., Furie, B. C., Herman, I. M., and Potter, D. A. (1999). Inhibition of calpain blocks platelet secretion, aggregation, and spreading. *J. Biol. Chem.* **274**, 36321–36327.
47. Fox, J. (1994). Transmembrane signaling across the platelet integrin glycoprotein IIb-IIIa. *Ann. NY Acad. Sci.* **714**, 75–87.
48. Inomata, M., Hayashi, M., Ohno-Iwashita, Y., Tsubuki, S., Saido, T. C., and Kawashima, S. (1996). Involvement of calpain in integrin-mediated signal transduction. *Arch. Biochem. Biophys.* **328**, 129–134.
49. Schoenwaelder, S. M., Yuan, Y., Cooray, P., Salem, H. H., and Jackson, S. P. (1997). Calpain cleavage of focal adhesion proteins regulates the cytoskeletal attachment of integrin αIIbβ3 (platelet glycoprotein IIb/IIIa) and the cellular retraction of fibrin clots. *J. Biol. Chem.* **272**, 1694–1702.
50. Bialkowska, K., Kulkarni, S., Du, X., Goll, D. E., Saido, T. C., and Fox, J. E. (2000). Evidence that β3 integrin-induced Rac activation involves the calpain-dependent formation of integrin clusters that are distinct from the focal complexes and focal adhesions that form as Rac and RhoA become active. *J. Cell Biol.* **151**, 685–695.
51. Stewart, M. P., McDowall, A., and Hogg, N. (1998). LFA-1-mediated adhesion is regulated by cytoskeletal restraint and by a Ca^{+2}-dependent protease, calpain. *J. Cell Biol.* **140**, 699–707.
52. Rock, M. T., Dix, A. R., Brooks, W. H., and Roszman, T. L. (2000). β1 integrin-mediated T cell adhesion and cell spreading are regulated by calpain. *Exp. Cell Res.* **261**, 260–270.
53. Wells, A., Gupta, K., Chang, P., Swindle, S., Glading, A., and Shiraha, H. (1998). Epidermal growth factor receptor-mediated motility in fibroblasts. *Microsc. Res. Techn.* **43**, 395–411.
54. Palecek, S., Huttenlocher, A., Horwitz, A. F., Lauffenburger, D. A. (1998). Physical and biochemical regulation of integrin release during rear detachment of migrating cells. *J. Cell Sci.* **111**, 929–940.
55. Xie, H., Pallero, M. A., Gupta, D., Chang, P., Ware, M. F., Witke, W., Kwiatkowski, D. J., Lauffenburger, D. A., Murphy-Ullrich, J. E., and Wells, A. (1998). EGF receptor regulation of cell motility: EGF induces disassembly of focal adhesions independently of the motility-associated PLCγ signaling pathway. *J. Cell Sci.* **111**, 615–624.
56. Tidball, J. G. and Spencer, M. J. (2000). Calpains and muscular dystrophies. *Int. J. Biochem. Cell Biol.* **32**, 1–5.
57. Richard, I., Roudaut, C., Marchand, S., Baghdiguian, S., Herasse, M., Stockholm, D., Ono, Y., Suel, L., Bourg, N., Sorimachi, H., Lefranc, G., Fardeau, M., Sebille, A., and Beckmann, J. S. (2000). Loss of calpain 3 proteolytic activity leads to muscular dystrophy and to apoptosis-associated IκBα/nuclear factor κB pathway perturbation in mice. *J. Cell Biol.* **151**, 1583–1590.
58. Wang, K. K. (2000). Calpain and caspase: can you tell the difference? *Trends Neurosci.* **23**, 20–26.
59. Reverter, D., Sorimachi, H., and Bode, W. (2001). The structure of calcium-free human m-calpain: implications for calcium activation and function. *Trends Cardiovasc. Med.* **11**, 222–229.
60. Carragher, N. O., Fincham, V. J., Riley, D., and Frame, M. C. (2001). Cleavage of focal adhesion kinase by different proteases during SRC-regulated transformation and apoptosis. Distinct roles for calpain and caspases. *J. Biol. Chem.* **276**, 4270–4275.
61. Xu, Y. and Mellgren, R. L. (2002). Calpain inhibition decreases the growth rate of mammalian cell colonies. *J. Biol. Chem.* In press.
62. Carragher, N. O., Westhoff, M. A., Riley, D., Potter, D. A., Dutt, P., Elce, J. S., Greer, P. A., and Frame, M. C. (2002). v-Src-induced modulation of the calpain-calpastatin proteolytic system regulates transformation. *Mol. Cell. Biol.* **22**, 257–269.

CHAPTER 142

Regulation of Intracellular Calcium through Hydrogen Peroxide

Sue Goo Rhee
Laboratory of Cell Signaling, National Heart, Lung, and Blood Institute, National Institutes of Health, Bethesda, Maryland

Introduction

H_2O_2 production, as a result of normal metabolism, environmental factors, and ligand-receptor interactions, is generally associated with an increase in cytoplasmic Ca^{2+} concentration. This Ca^{2+} increase can be attributed partly to the fact that H_2O_2 causes selective oxidation of certain reactive cysteine residues of the ryanodine receptor and the inositol(1,4,5)P_3 receptor, leading to enhanced Ca^{2+} channel activity of these two receptors. The Ca^{2+} elevation may also arise indirectly from inactivation of protein tyrosine phosphatase and PTEN, both of which contain an essential cysteine residue that is especially sensitive to H_2O_2-dependent oxidation.

Sources and Chemical Properties of ROS

Incomplete reduction of O_2 during respiration produces superoxide anion ($O_2^{•-}$), which is spontaneously or enzymatically dismutated to H_2O_2. H_2O_2 can be reduced further to hydroxyl radicals (HO•) in the presence of catalytic amounts of iron and electron donor molecules such as thiols and ascorbic acid (reviewed in [1,2]). These reactive oxygen species (ROS), $O_2^{•-}$, H_2O_2, and HO•, are also produced in response to environmental factors such as inflammation and UV radiation. Furthermore, substantial evidence suggests that $O_2^{•-}$ and H_2O_2 are generated transiently upon interaction of various ligand-cell surface receptor pairs and function as intracellular messengers (reviewed in [3]). Therefore, $O_2^{•-}$ and H_2O_2 are not merely damage-causing agents but are also mediators of physiological functions.

Calcium homeostasis is controlled by (1) Ca^{2+} channels such as the ryanodine receptor (RyR), inositol(1,4,5)P_3 receptor (IP$_3$R), dihydropyridine receptor (DHPR), and L-type voltage-sensitive channels, (2) Ca^{2+} pumps such as the sarcoplasmic reticulum Ca^{2+}ATPase (SERCA pump) and sarcolemmal Ca^{2+}ATPase, and (3) Na^+/Ca^{2+} exchangers [4]. A shift in the cellular redox status to a more oxidized state generally causes a rapid increase in the concentration of intracellular calcium ($[Ca^{2+}]_i$) (reviewed in [5–9]). The effect of ROS on $[Ca^{2+}]_i$, however, is variable, depending on the cell type, the type of ROS, the level of ROS production, and the duration of exposure to ROS. The effects of ROS on Ca^{2+} homeostasis have been studied extensively in vascular endothelial cells, smooth muscle cells, cardiomyocytes, and neuronal cells because of the pathophysiologic role of oxidative injury in myocardial ischemia-reperfusion, atherosclerotic lesion formation, and trauma. Despite abundant studies, the target molecules on which ROS act and the chemical nature of ROS-induced modification are largely unknown. Considerable differences in the chemical reactivity of $O_2^{•-}$, H_2O_2, and HO• also add complexity to such studies.

Hydroxyl radicals are extremely reactive, with a lifetime of several nanoseconds in the cellular milieu, and inflict indiscriminate damage on proteins, DNA, and lipids. Oxidation of membrane lipids by HO• alters the physical properties of membranes and membrane-associated proteins, leading to nonspecific ion leakage. It is therefore unlikely that HO• functions as a specific mediator of redox regulation.

$O_2^{\bullet-}$ and H_2O_2 are less reactive species that are known to display selective oxidation of particular target molecules. H_2O_2 is a mild oxidant that can oxidize the sulfur atom of methinone and cysteine residues in proteins. Cysteine is oxidized to cysteine sulfenic acid or disulfide, both of which are readily reduced back to cysteine by various cellular reductants. The pK_a (where K_a is the acid constant) of the sulfhydryl group (Cys–SH) of most cysteine residues is ~8.5. Because Cys–SH is less readily oxidized by H_2O_2 than is the cysteine thiolate anion (Cys–S$^-$), few proteins might be expected to possess a cysteine residue that is vulnerable to oxidation by H_2O_2 in cells [10]. However, certain protein cysteine residues have low pK_a values and exist as thiolate anions at neutral pH because of nearby positively charged amino acid residues that are available for interaction with the negatively charged thiolate. Proteins with low-pK_a cysteine residues can be the targets of specific oxidation by H_2O_2, and such oxidation can be reversed by thiol donors such as glutathione and thioredoxin. Methionine (Met) is more susceptible to oxidation by H_2O_2 and is converted to Met sulfoxide, which is reduced back to Met by specific enzymes called Met sulfoxide reductases (reviewed in [11]). Furthermore, reversible oxidation of Met residues can serve as a control switch for the regulation of protein function, as exemplified by calmodulin, which loses the ability to activate plasma membrane Ca^{2+}ATPases when its COOH-terminal Met residues are oxidized [12]. However, it is not known if H_2O_2 can effect selective oxidation of specific Met residues among many solvent-exposed Met residues.

Although $O_2^{\bullet-}$ is a poorer oxidant than H_2O_2, it specifically oxidizes certain metal ions bound to proteins and consequently modifies the function of these proteins (e.g. inactivation of calcineurin due to oxidation of its Fe-Zn center [13]). It seems that $O_2^{\bullet-}$ is also able to react selectively with certain proteins as the result of electrostatic attraction between the negatively charged $O_2^{\bullet-}$ molecules and positively charged amino acid residues of the targeted proteins. One such example is the vascular smooth muscle SR Ca^{2+}ATPase, which is inactivated by $O_2^{\bullet-}$ but not by H_2O_2 ([14]. However, the amino acid residues affected by $O_2^{\bullet-}$ have not been identified in this case. Furthermore, the cardiac muscle SR isoform, which shares 90% homology with the smooth muscle isoform, is insensitive to $O_2^{\bullet-}$.

Activation of Ryanodine and IP$_3$ Receptor Ca^{2+} Release Channels by H$_2$O$_2$

At the present time, H_2O_2-mediated oxidation of Cys residues residing within special microenvironments appears to provide the most well defined mechanism underlying the reversible and specific effects of ROS [3]. Good examples of this phenonmenon are RyR and IP$_3$R, both of which are activated when specific Cys residues are oxidized. RyRs, which are involved in Ca^{2+} release from the SR in skeletal and cardiac muscles, are composed of four subunits and form a complex with triadin. RyR contains about 21 cysteine residues per subunit. Some of the 21 cysteine residues have higher reactivity than others toward H_2O_2 and various sulfhydryl reagents, but these have not been mapped precisely (reviewed in [7,15]). Oxidation by H_2O_2 or modification by sulfhydryl reagents of the reactive Cys–SH residues decreases the K_d for ryanodine as well as the EC_{50} for Ca^{2+} activation [16]. Single-channel reconstitution experiments indicate that H_2O_2, at submicromolar concentrations, enhances the Ca^{2+} release that follows fusion of SR vesicles to planar lipid membranes [16,17]. Cysteine oxidation also contributes to the stabilization of a RyR/triadin complex during channel activation, probably through intermolecular disulfide bonding. The stimulatory effects of peroxide are reversed by thiol-reducing agents such as dithiothreitol and glutathione (GSH). Increased oxidative stress produced by ROS and nitric oxide is generally reflected by an increased ratio of GSSG to GSH in cells. GSSG is capable of forming a mixed disulfide with a reactive cysteine or converting two neighboring cysteines to a disulfide. In accordance with this capacity, GSSG, like H_2O_2, has been demonstrated to enhance the binding affinity of RyR to ryanodine and enhance its reconstituted single channel Ca^{2+} release [18].

Sensitivity to sulfhydryl oxidation also appears to be a property of the endoplasmic reticulum (ER) Ca^{2+} channel IP$_3$R. H_2O_2 and GSSG were shown to cause spontaneous release and oscillation of Ca^{2+} by sensitizing IP$_3$R to endogenous IP$_3$ [19–22]. Recent reports suggest that H_2O_2 generated intracellularly as the result of ligation of cell surface receptors also contributes to Ca^{2+} mobilization. For example, histamine produces H_2O_2 through activation of NADPH oxidase in endothelial cells, and the NADPH oxidase-derived H_2O_2 is critical for the generation of Ca^{2+} oscillations during histamine stimulation [23]. Many other agonists induce Ca^{2+} oscillations as well as H_2O_2 production. Therefore, receptor-mediated H_2O_2 production is likely to be a key process affecting Ca^{2+} signaling. However, care should be taken not to attribute the H_2O_2 effect entirely to the oxidation of IP$_3$R, as H_2O_2 is also known to cause the production of IP$_3$ (see below).

Many studies on other Ca^{2+} channels (dihydropyridine receptors, L-type voltage-sensitive channels), Ca^{2+}ATPases, and Na$^+$/Ca^{2+} exchangers also suggest that H_2O_2 affects their activity through cysteine oxidation (reviewed in [4]). However, the results remain inconclusive and, at times, controversial.

Enhancement of [Ca^{2+}]$_i$ through H$_2$O$_2$-mediated Inactivation of Protein Tyrosine Phosphatase and PTEN

The changes in Ca^{2+} homeostasis need not be entirely due to the modification of Ca^{2+} transporters (channels, pumps, and exchangers) but may also arise indirectly from the modification of other proteins. Candidates for such modification include protein tyrosine phosphatases (PTPs) and PTEN. All PTPs contain an essential cysteine residue (pK_a, 4.7 to 5.4) in the signature active site motif HCXXGXXRS/T (where X is any amino acid residue) that exists as a thiolate anion at neutral pH [24]. This active site cysteine is the target of specific oxidation by H_2O_2, and the ability of intracellularly produced H_2O_2 to inhibit PTP activity has been demonstrated in cells stimulated with EGF, PDGF, and insulin [25–27]. Furthermore, EGF- and PDGF-induced protein tyrosine phosphorylation of cellular proteins, including their respective receptor protein tyrosine kinases (RPTKs) and PLC-gamma, requires H_2O_2 production [28,29]. These results indicate that the activation of an RPTK *per se* by binding of the corresponding growth factor may not be sufficient to increase the steady state level of protein tyrosine phosphorylation in cells. Rather, the concurrent inhibition of PTPs by H_2O_2 may also be required. As such, H_2O_2 plays a major messenger role in the activation (tyrosine phosphorylation) of PLC-gamma and subsequent production of IP_3 in cells stimulated with PDGF and EGF. Exogenous H_2O_2 alone, in the absence of a growth factor, induces tyrosine phosphorylation of various cellular proteins including PLC-gamma and elicits IP_3 production [30]. This probably reflects the background activity of various protein tyrosine kinases, which is apparently sufficient to enhance the level of protein tyrosine phosphorylation when the activity of most PTPs is suppressed by H_2O_2.

PTEN is a member of the PTP family and reverses the action of phosphoinositide (PI) 3-kinase by catalyzing the removal of the 3′-phosphate of $PI(3,4,5)P_3$. H_2O_2 induces reversible inactivation of PTEN through specific oxidation of the catalytic site cysteine [31]. As with protein tyrosine phosphorylation, it is likely that the activation of PI 3-kinase in receptor-stimulated cells may not be sufficient to achieve the accumulation of $PI(3,4,5)P_3$ because of the opposing activity of PTEN; the concomitant inactivation of PTEN by H_2O_2 might thus be necessary to increase the abundance of $PI(3,4,5)P_3$ sufficiently to trigger downstream signaling events. However, production of $PI(3,4,5)P_3$ was shown to be necessary for PDGF-induced H_2O_2 production [32]. This is probably because $PI(3,4,5)P_3$ activates Rac, an essential component of the activated NADPH oxidase complex, by binding to the pleckstrin homology domains of the Rac guanine nucleotide exchange factors [33]. Thus, through its effect on the concentration of $PI(3,4,5)P_3$, the oxidation of PTEN by H_2O_2 constitutes a positive feedback loop that increases the production of H_2O_2. This positive feedback loop is expected to result in a rapid increase in Ca^{2+} concentration. Because many inositol polyphosphate phosphatases also contain a cysteine at their active site [34], degradation of IP_3 might be inhibited by H_2O_2. There are observations that support this possibility.

In all likelihood, PTPs and PTEN represent the first examples among many more proteins that connect H_2O_2 and Ca^{2+} signaling. Hence, we are merely taking our first steps in understanding how oxidants modulate Ca^{2+} signaling.

References

1. Stadtman, E. R. (1992). Protein oxidation and aging. *Science* **257**, 1220–1224.
2. Rhee, S. G. (1999). Redox signaling: hydrogen peroxide as intracellular messenger. *Exp. Mol. Med.* **31**, 53–59.
3. Rhee, S. G., Bae, Y. S., Lee, S.-R., and Kwon, J. (2000). Hydrogen peroxide: a key messenger that modulates protein phosphorylation through cysteine oxidation. *Science's STKE*. www.stke.org/cgi/contentfull/OC_sigtrans;2000/53/pe1.
4. Kourie, J. I. (1998). Interaction of reactive oxygen species with ion transport mechanisms. *Am. J. Physiol.* **275**, C1-24.
5. Wada, S. and Okabe, E. (1997). Susceptibility of caffeine- and Ins (1,4,5)P_3-induced contractions to oxidants in permeabilized vascular smooth muscle. *Eur. J. Pharmacol.* **320**, 51–59.
6. Suzuki, Y. J. and Ford, G. D. (1999). Redox regulation of signal transduction in cardiac and smooth muscle. *J. Mol. Cell Cardiol.* **31**, 345–353.
7. Pessah, I. N. and Feng, W. (2000). Functional role of hyperreactive sulfhydryl moieties within the ryanodine receptor complex. *Antioxid. Redox Signal.* **2**, 17–25.
8. Wang, H. and Joseph, J. A. (2000). Mechanisms of hydrogen peroxide-induced calcium dysregulation in PC12 cells. *Free Radic. Biol. Med.* **28**, 1222–1231.
9. Lounsbury, K. M., Hu, Q., and Ziegelstein, R. C. (2000). Calcium signaling and oxidant stress in the vasculature. *Free Radic. Biol. Med.* **28**, 1362–1369.
10. Kim, J. R., Yoon, H. W., Kwon, K. S., Lee, S. R., and Rhee, S. G. (2000). Identification of proteins containing cysteine residues that are sensitive to oxidation by hydrogen peroxide at neutral pH]. *Anal Biochem.* **283**, 214–221.
11. Hoshi, T. and Heinemann, S. (2001). Regulation of cell function by methionine oxidation and reduction. *J. Physiol.* **531**, 1–11.
12. Yao, Y., Yin, D., Jas, G. S., Kuczer, K., Williams, T. D., Schoneich, C., and Squier, T. C. (1996). Oxidative modification of a carboxyl-terminal vicinal methionine in calmodulin by hydrogen peroxide inhibits calmodulin-dependent activation of the plasma membrane Ca-ATPase. *Biochemistry* **35**, 2767–2787.
13. Wang, X., Culotta, V. C., and Klee, C. B. (1996). Superoxide dismutase protects calcineurin from inactivation. *Nature* **383**, 434–437.
14. Suzuki, Y. J. and Ford, G. D. (1991). Inhibition of $Ca^{(2+)}$-ATPase of vascular smooth muscle sarcoplasmic reticulum by reactive oxygen intermediates. *Am. J. Physiol.* **261**, H568–574.
15. Anzai, K., Ogawa, K., Ozawa, T., and Yamamoto, H. (2000). Oxidative modification of ion channel activity of ryanodine receptor. *Antioxid. Redox Signal.* **2**, 35–40.
16. Favero, T. G., Zable, A. C., and Abramson, J. J. (1995). Hydrogen peroxide stimulates the Ca^{2+} release channel from skeletal muscle sarcoplasmic reticulum. *J. Biol. Chem.* **270**, 25557–25563.
17. Boraso, A. and Williams, A. J. (1994). Modification of the gating of the cardiac sarcoplasmic reticulum $Ca^{(2+)}$-release channel by H_2O_2 and dithiothreitol. *Am. J. Physiol.* **267**, H1010–1016.
18. Zable, A. C., Favero, T. G., and Abramson, J. J. (1997). Glutathione modulates ryanodine receptor from skeletal muscle sarcoplasmic reticulum. Evidence for redox regulation of the Ca^{2+} release mechanism. *J. Biol. Chem.* **272**, 7069–7077.
19. Missiaen, L., Taylor, C. W., and Berridge, M. J. (1991). Spontaneous calcium release from inositol trisphosphate-sensitive calcium stores. *Natur.* **352**, 241–244.
20. Rooney, T. A., Renard, D. C., Sass, E. J., and Thomas, A. P. (1991). Oscillatory cytosolic calcium waves independent of stimulated inositol 1,4,5-trisphosphate formation in hepatocytes. *J. Biol. Chem.* **266**, 12272–12282.
21. Doan, T. N., Gentry, D. L., Taylor, A. A., and Elliott, S. J. (1994). Hydrogen peroxide activates agonist-sensitive $Ca^{(2+)}$-flux pathways in canine venous endothelial cells. *Biochem. J.* **297**, 209–215.
22. Hu, Q., Corda, S., Zweier, J. L., Capogrossi, M. C., and Ziegelstein, R. C. (1998). Hydrogen peroxide induces intracellular calcium oscillations in human aortic endothelial cells. *Circulation* **97**, 268–275.
23. Hu, Q., Zheng, G., Zweier, J. L., Deshpande, S., Irani, K., and Ziegelstein, R. C. (2000). NADPH oxidase activation increases the sensitivity of intracellular Ca^{2+} stores to inositol 1,4,5-trisphosphate in human endothelial cells. *J. Biol. Chem.* **275**, 15749–15757.
24. Denu, J. M. and Dixon, J. E. (1998). Protein tyrosine phosphatases: mechanisms of catalysis and regulation. *Curr. Opin. Chem. Biol.* **2**, 633–641.

25. Lee, S. R., Kwon, K. S., Kim, S. R., and Rhee, S. G. (1998). Reversible inactivation of protein-tyrosine phosphatase 1B in A431 cells stimulated with epidermal growth factor. *J. Biol. Chem.* **273**, 15366–15372.
26. Meng, T. C., Fukada, T., and Tonks, N. K. (2002). Reversible oxidation and inactivation of protein tyrosine phosphatases *in vivo*. *Mol. Cell.* **9**, 387–399.
27. Mahadev, K., Zilbering, A., Zhu, L., and Goldstein, B. J. (2001). Insulin-stimulated hydrogen peroxide reversibly inhibits protein-tyrosine phosphatase 1b in vivo and enhances the early insulin action cascade. *J. Biol. Chem.* **276**, 21938–21942.
28. Sundaresan, M., Yu, Z. X., Ferrans, V. J., Irani, K., and Finkel, T. (1995). Requirement for generation of H_2O_2 for platelet-derived growth factor signal transduction. *Science* **270**, 296–299.
29. Bae, Y. S., Kang, S. W., Seo, M. S., Baines, I. C., Tekle, E., Chock, P. B., and Rhee, S. G. (1997). Epidermal growth factor (EGF)-induced generation of hydrogen peroxide. Role in EGF receptor-mediated tyrosine phosphorylation. *J. Biol. Chem.* **272**, 217–221.
30. Wang, X. T., McCullough, K. D., Wang, X. J., Carpenter, G., and Holbrook, N. J. (2001). Oxidative stress-induced phospholipase C-gamma 1 activation enhances cell survival. *J. Biol. Chem.* **276**, 28364–28371.
31. Lee, S.-R., Yang, K.-S., Kwon, J., Lee, C., Jeong, W., and Rhee, S. G. (2002). Regulation of PTEN by superoxide and H_2O_2 through the reversible formation of a disulfide between Cys124 and Cys71. *J. Biol. Chem.* **277**, in press.
32. Bae, Y. S., Sung, J. Y., Kim, O. S., Kim, Y. J., Hur, K. C., Kazlauskas, A., and Rhee, S. G. (2000). Platelet-derived growth factor-induced H(2)O(2) production requires the activation of phosphatidylinositol 3-kinase. *J. Biol.Chem.* **275**, 10527–10531.
33. Welch, H. C., Coadwell, W. J., Ellson, C. D., Ferguson, G. J., Andrews, S. R., Erdjument-Bromage, H., Tempst, P., Hawkins, P. T., and Stephens, L. R. (2002). P-Rex1, a PtdIns(3,4,5)P_3- and Gbetagamma-regulated guanine-nucleotide exchange factor for Rac. *Cell* **108**, 809–821.
34. Majerus, P. W., Kisseleva, M. V., and Norris, F. A. (1999). The role of phosphatases in inositol signaling reactions. *J. Biol. Chem.* **274**, 10669–10672.

SECTION D

Lipid-Derived Second Messengers

Lewis Cantley, Editor

CHAPTER 143

Historical Overview: Protein Kinase C, Phorbol Ester, and Lipid Mediators

Yasutomi Nishizuka and Ushio Kikkawa
Biosignal Research Center, Kobe University, Kobe, Japan

Retrospectives of Phospholipid Research

Nearly 200 years ago, a French chemist, L. N. Vauquelin, found phosphorus in the brain material extracted with hot alcohol. This material was probably a mixture of crude phospholipids. Thirty years later, choline-containing phospholipid was obtained from the brain by F. Fremy (oleophosphoric acid) and from the egg yolk by M. Gobley (lecithin). Since then, during a period of more than 100 years, several phospholipids were isolated and structurally identified. The existence of inositol in plants was known in the nineteenth century but was unknown in animal tissues until 1941, when D. W. Woolley found it in the mammalian brain. In the next year, J. Folch and Woolley at Rockefeller Institute in New York fractionated several phospholipids and identified the chemical structure of inositol phospholipid. In the late 1940s Folch, then at Harvard University, noticed that additional phosphate was attached to the inositol moiety.

In the subsequent years many efforts were made to clarify the metabolic and synthetic pathways of various lipids, including inositol phospholipids. In parallel with these investigations, in the decade of 1960s, phospholipids were shown to be cofactors essential to the catalytic activity of enzymes such as β-hydroxybutyrate dehydrogenase (D. E. Green), Na^+/K^+ ATPase (T. Tanaka), NADH-cytochrome C reductase (S. J. Wakil), and many others. Nevertheless, with some exceptions such as the production of platelet-activating factor (D. J. Hanahan) and eicosanoid (S. K. Bergström and B. I. Samuelsson), membrane phospholipids were generally viewed as a biologically inert entity that provide a semipermeable barrier between exterior and interior compartments within and between cells.

In the early 1950s, with radioactive orthophosphate, Hokin and Hokin [1] observed that acetylcholine induced rapid labeling of acid-precipitable materials of some exocrine tissues such as pancreas. It became evident soon that the materials were inositol phospholipid and phosphatidic acid. Namely, the rapid labeling of these lipids resulted from the enhanced breakdown and resynthesis of inositol phospholipid, but its biological significance remained to be clarified for many years. In 1975, Michell postulated that this phospholipid hydrolysis may open the Ca^{2+} gate [2].

Protein Kinase C and Diacylglycerol

In 1977, when protein kinase C (PKC) was first found as an undefined protein kinase present in many mammalian tissues, the enzyme was activated by limited proteolysis with Ca^{2+}-dependent protease, and no obvious evidence was available for its role in signal transduction. Before long it became clear that without proteolysis the enzyme could be activated by a membrane factor in the presence of Ca^{2+}. The membrane factor was identified as anionic phospholipids, particularly phosphatidylserine. Curiously, crude phospholipids extracted from brain membranes could activate the enzyme in the absence of added Ca^{2+}, whereas pure phospholipids obtained from erythrocyte membranes could not produce any enzyme activation unless a higher concentration of Ca^{2+} was added to the reaction mixture. Analysis of the lipid impurities on a silicic acid column led us to conclude that diacylglycerol is an essential activator.

To explore the link of PKC activation to inositol phospholipid hydrolysis, we developed a procedure to activate this enzyme in intact cells by applying membrane-permeant diacylglycerols. Diacylglycerols having two long fatty acyl moieties could not be readily intercalated into the cell membrane. If, however, one of the fatty acids is replaced with a short chain, the resulting diacylglycerols, such as 1-oleoyl-2-acetyl-glycerol (OAG), obtain detergent-like properties and could be dispersed into the phospholipid bilayer and activate PKC directly. In the initial studies, human platelets were employed. Thrombin and collagen induce release of serotonin with the concomitant hydrolysis of inositol phospholipid and phosphorylation of two endogenous proteins with 20 and 47 kDa molecular size. It was already known that the 20 kDa protein is myosin light chain, and a specific calmodulin-dependent kinase is responsible for its phosphorylation. Before long we knew that the 47 kDa protein, pleckstrin we call it today, was a substrate specific to PKC. Thus, the phosphorylation of these two proteins served as excellent markers for the increase of Ca^{2+} and diacylglycerol-dependent activation of PKC, respectively. In 1980, we were able to show that both Ca^{2+} increase and PKC activation were essential and acted synergistically to elicit full activation of platelets and release of serotonin. Similarly, it was possible to show unequivocally that PKC activation is indispensable for neutrophil release reaction and T-cell activation, thus establishing the biological role of PKC in cellular responses. In 1983, Berridge and his colleges announced at Cambridge the important inositol 1,4,5-trisphosphate story [3].

Phorbol Ester and Cell Signaling

In the summer of 1981, M. Castagna visited our laboratory from Villejuif, France. We discussed a possible role of tumor-promoting phorbol ester in the PKC signaling pathway. It was already known that phorbol ester shows pleiotropic activities by mimicking hormone actions. When platelets were stimulated by 12-O-tetradecanoylphorbol-13-acetate (TPA), the 47 kDa protein was remarkably phosphorylated, but against our expectation diacylglycerol was not produced. It was extremely disappointing to us because this meant that our idea that diacylglycerol is the mediator for PKC activation was not correct. A few days later, however, an idea flashed: What would happen if TPA could activate PKC directly because the phorbol ester contains a diacylglycerol-like structure very similar to the membrane-permeant lipid molecule OAG that we had used? This insight occurred near the end of August. The following year several groups of investigators showed that PKC is the major target of phorbol ester. It was also shown that phorbol ester could cause translocation of PKC from the cytosol to the membrane. As a result, the traditional concept of tumor promotion originally proposed by I. Berenblum at Oxford in 1941 was replaced by an explicit biochemical explanation providing for an understanding the role of PKC. Along this line of study, phorbol esters and membrane-permeant diacylglycerols have since then been used as crucial tools for the manipulation of PKC in intact cells, and have allowed the determination of the wide range of cellular processes regulated by this enzyme [4]. It was realized much later, however, that phorbol ester can bind to other cellular proteins, such as chimaerin and RasGRP [5], and potentially affect cell functions through additional targets.

Structural Heterogeneity and Mode of Activation

Although PKC was once considered as a single entity, molecular cloning and enzymological studies in the mid 1980s revealed the existence of multiple isoforms of PKC. The mammalian PKC family consists of at least ten isoforms encoded by nine genes. These isoforms are divided into three subgroups based on their primary structures and biochemical properties: classical PKC isoforms (cPKC), novel PKC isoforms (nPKC), and atypical PKC isoforms (aPKC) [6]. The PKC isoforms are conserved in a variety of species, including yeast, nematoda, fly, fish, and frog. On the one hand, the serine-threonine protein kinase region that is located in the C-terminal half does not show much difference and exhibits similar enzymatic properties when tested in *in vitro* systems. On the other hand, the N-terminal half of the enzyme molecule contains multiple characteristic functional domains, such as the C1 domain, which binds diacylglycerol or phorbol ester; the C2 domain, which binds phospholipid in the presence of Ca^{2+}; and the OPR (octicosapeptide repeat) domain, which is involved in protein-protein interaction. The structural feature and multiple functional domains of the PKC isoforms are well investigated, as documented in excellent reviews [7,8]. In addition, several protein kinases that share kinase regions closely related to the PKC family are isolated and characterized [9]. These include protein kinase N (PKN or PRK), protein kinase D (PKD or PKCμ), and protein kinase B (PKB, Akt or rac-PK). The N-terminal regions of these enzymes contain multiple distinct functional domains such as PH and HR1 domains.

Structural analysis also made it clear that the mode of activation of the PKC family is far more complicated than we initially had thought. Newton [8] has shown that the newly synthesized kinase is catalytically inert and is regulated by phosphorylation by itself and also by other kinases, including PDK1 and related enzymes. A unique cross-talk thus emerged between the PKC signal pathway and one branch of the inositol phospholipid 3-kinase pathway that was described first by L. Cantley in the mid-1980s [10]. Another cross-talk with tyrosine kinase pathway for the activation of PKC is becoming clearer. PKC was initially recognized as an enzyme that can be activated by limited proteolysis, but later this proteolysis was recognized as a process of downregulation. More recently, however, the PKC δ-isoform is proposed to be a target of caspase 3 for its activation during apoptosis.

Translocation and Multiple Lipid Mediators

The specific functions of individual PKC isoform have been studied for many years, but whether the isoforms exert functional redundancy or functional specialization remains unclear, although some of them obviously play unique specific roles (see PKC minireview series, isoform-specific functions, *J. Biochem.* 2002–2003).

In addition to phospholipase C, phospholipase A2, phospholipase D, and sphingomyelinase appear to be indispensable players in signal transduction. In fact, it becomes increasingly clear that fatty acids, lysophospholipids, ceramide, and other lipid products may play roles in cell signaling, as described elsewhere [6] and in this chapter. In addition, the products of phosphatidylinositol 3-kinases play key roles in the transmembrane control of cellular processes as proposed by Toker and Cantley [10]. Multiple lipid mediators produced in membranes may recruit various protein kinases and other signal molecules to "lipid rafts" through lipid-lipid, lipid-protein, and protein-protein interactions. The lipid-mediated translocation of protein kinases to selective intracellular compartments such as plasma membrane, Golgi complex, and cell nucleus represents an essential step for stable access to their substrate proteins. It is attractive to surmise, then, that the N-terminal half of the enzyme molecule with multiple membrane-binding domains governs the enzymatic activity as well as the functional specificity of the C-terminal half. It is curious that such lipid-mediated translocation of PKC isoforms to membranes sometimes shows oscillation back and forth from the cytosol to the membrane. The mechanism of this oscillation is not clear, but lipid mediators appear to oscillate after receptor stimulation, presumably due to a repetitive feedback mechanism. The destination and reversibility of such translocation appear to differ with the isoform, lipid mediator, and cell type. The dynamic behavior of the PKC isoforms and related kinases can be visualized with the enzymes fused to green fluorescence protein.

Conclusion

In the last decade, our knowledge of the PKC family and related enzymes as well as the lipid mediators derived from membrane phospholipids has expanded enormously. It appears that the enzymes anchor to specific protein complexes through interaction with some adapter proteins, as directed by multiple lipid mediators. Such interactions may be essential for the function of each protein kinase at selective intracellular compartments. Further exploration of the dynamic aspects of such lipid mediators and identification of interacting proteins may unveil more of the transmembrane control of physiological and pathological cellular processes.

References

1. Hokin, M. R. and Hokin, L. E. (1953). Enzyme secretion and the incorporation of ^{32}P into phospholipids of pancreatic slices. *J. Biol. Chem.* **203**, 967–977.
2. Michell, R. H. (1975). Inositol phospholipids and cell surface receptor function. *Biochim. Biophys. Acta* **415**, 81–147.
3. Streb, H., Irvine, R. F., Berridge, M. J., and Schulz, I. (1983). Release of Ca^{2+} from a nonmitochondrial intracellular store in pancreatic acinar cells by inositol-1,4,5-trisphosphate. *Nature* **306**, 67–69.
4. Nishizuka, Y. (1984). The role of protein kinase C in cell surface signal transduction and tumour promotion. *Nature* **308**, 693–697.
5. Kazanietz, M. G. (2000). Eyes wide shut: protein kinase C isozymes are not the only receptors for the phorbol ester tumor promoters. *Mol. Carcinog.* **8**, 5–11.
6. Nishizuka, Y. (1995). Protein kinase C and lipid signaling for sustained cellular responses. *FASEB J.* **9**, 84–96.
7. Parekh, D. B., Ziegler, W., and Parker, P. J. (2000). Multiple pathways control protein kinase C phosphorylation. *EMBO J.* **19**, 496–503.
8. Newton, A. C. (2001). Protein kinase C: structural and spatial regulation by phosphorylation, cofactors, and macromolecular interactions. *Chem. Rev.* **101**, 2353–2364.
9. Mellor, H. and Parker, P. J. (1998). The extended protein kinase C superfamily. *Biochem. J.* **332**, 281–292.
10. Toker, A. and Cantley, L. C. (1997). Signalling through the lipid products of phosphoinositide-3-OH kinase. *Nature* **387**, 673–676.

CHAPTER 144

Type I Phosphatidylinositol 4-phosphate 5-kinases (PI4P 5-kinases)

K. A. Hinchliffe and R. F. Irvine
Department of Pharmacology
University of Cambridge,
Cambridge, United Kingdom

Introduction

Type I phosphatidylinositol 4-phosphate 5-kinases (PI4P5Ks) phosphorylate phosphatidylinositol 4-phosphate (PI4P) in the 5-position to form phosphatidylinositol 4,5-bisphosphate (PI45P$_2$). Because metabolic evidence suggests that *in vivo* the major route of synthesis of PI45P$_2$ in animal cells is by the 5-phosphorylation of PI4P, both in the plasma membrane [1,2] and the nucleus [3], type I PI4P5Ks are obviously the enzymes primarily responsible for regulating levels of this multifunctional lipid. In the test tube type I PI4P5Ks have been reported to catalyze other reactions. For example, both Iα and Iβ isoforms can convert PI into PI5P [4], and PI3P into PI34P$_2$ [4, 5] or PI35P$_2$ [4], or even eventually PI345P$_3$ [4,5]. A PI4P5K from *Arabadopsis* shows a similar flexibility when expressed in insect cells [6]. However, the 5-phosphorylation of PI4P is the major activity of the Type I enzymes, and the physiological significance (or even natural occurrence) of these other reactions remains unclear. The exception is the demonstration that an endogenous type I PI4P5K (isoform unknown) makes a physiologically significant contribution to the synthesis of PI345P$_3$ from PI34P$_2$ in response to cell stress [7].

Several fuller reviews have discussed these enzymes directly or indirectly (e.g. [8–10]).

Basic Properties

Cloning

Our current understanding of type I PI4P5Ks is that there are three distinct mammalian isoforms, and no other obvious candidate emerges from a scan of the current human genome database. Nomenclature is rather confusing, as the type Iβ cloned from mouse [11] and the human isoform called type Iα cloned shortly afterwards by Loijens and Anderson [12] are exact orthologues (and similarly, mouse type Iα and human type Iβ). As the type Iγ [13] has come from the same species (mouse) and lab as the original cloning of the Iα and Iβ, we have in Fig. 1 used the mouse nomenclature. The isoform that Carvajal *et al.* [14] identified as the STM7 gene, mapping close to the Friedrich's ataxia gene (X25), is the human type Iβ isoform.

Loijens and Anderson [12] reported two splice variants of the human type Iα and one of the type Iβ, and the mouse type Iγ also has at least two splice variants [13].

Structure

The lineup in Fig. 1 tells a superficially simple story of a highly conserved central core, which consists of the catalytic

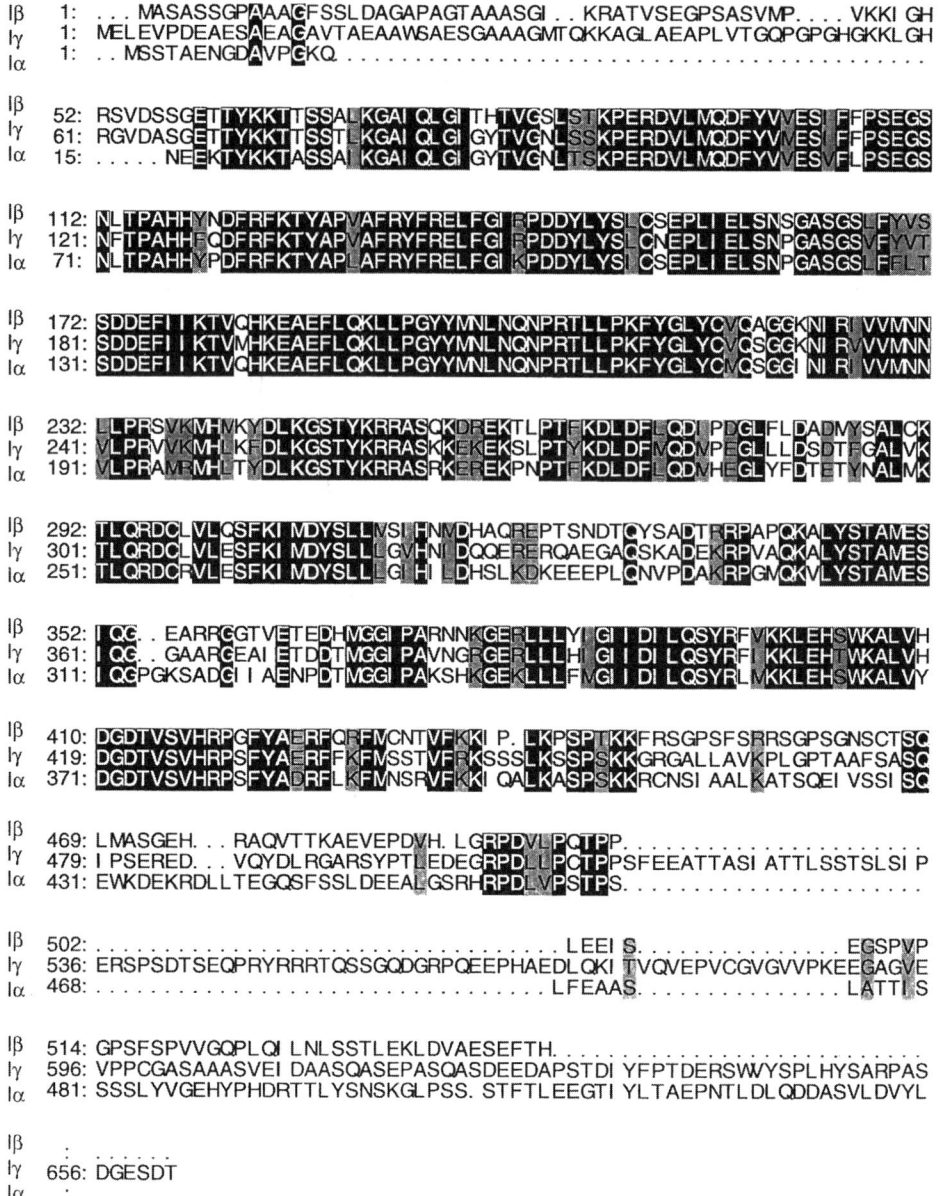

Figure 1 The sequences of the mouse type I PI4P5Ks are shown (Genebank accession numbers: Iβ, NM_008847; Iγ, NM_008844; Iα, NM_008846).

site interspersed with some loops of significant variation between isoforms, and then virtually no sequence similarity whatsoever between the isoforms at the C- and N-termini. The latter, in turn, implies diverse isoform-specific regulation, which has implications in the discussions about regulation, below.

There is also a close similarity in the catalytic "core" with the yeast gene Mss4 [11,15], but again no similarity in other parts of the sequence between the yeast and the mammalian enzymes. There is a limited amount of similarity with the members of the Fab1 family, both yeast and mammalian; these are PI3P 5-kinases, now given the name type III PIP kinases. Also, there is similarity in the catalytic core with the type II PI5P4Ks, and from this, and from the x-ray structure of the type IIβ PI5P4K [16], some deductions can been made about probable crucial residues for catalytic activity in the type I enzymes. Ishihara et al. showed that lysine 138 of the type Iα PI4P 5K, which they identified as being in the putative ATP-binding site, is essential for catalytic activity [13].

Substrate Specificity

Kunz et al. [17] have shown that the substrate specificity of the type I and II PIP kinases is dictated largely by their "activation loop," that is, transferring the activation loop from type IIβ PIPK into type Iβ (human) converted the type I enzyme into a PI5P4K activity (the activation loop of the orthologous mouse type Iα is residues 347–387 in Fig. 1). The converse (converting a type II PIPK into a PI4P5K activity by inserting a type I loop) was also observed. These observations have recently been taken a stage further by some elegant site-directed changes in this loop [18].

A remarkable finding is that changing a single residue, glutamate 362 in human type Iβ (equivalent to E362 in murine Iα, see Fig. 1), to an alanine transformed its substrate specificity to that resembling a type II activity in that it would use PI5P as a substrate (though its activity against PI4P was diminished rather than lost). Kunz et al. have suggested that the activation loop might fold into an α-helix in vivo [18]. The structure of type IIβ PI5P4K [16] suggests how the activation loop might lie adjacent to the presumed active site where the PI5P substrate head-group binds. Although the activation loop did not crystallize [16], it seems likely that it will influence the orientation of two loops that link contiguous β strands; both of these loops contribute residues that interact with the inositol 1,5 bisphosphate moiety of the substrate [16].

Localization

These latter studies on substrate specificity [18] have implications also for the localization of the type I enzymes. So far, localization studies have suggested that they are all primarily in the plasma membrane (though see below for some possible regulation of this). The combined data of the two papers on the influence of the activation loop [17,18] demonstrated that changing the substrate specificity from favoring PI4P to PI5P also changes the localization of the type Iβ enzyme from plasma membrane to cytosol. This in turn implies that the plasma membrane localization is governed primarily by interaction with the PI4P substrate. This conclusion is subject to the caveat that these studies use transfection, which might saturate endogenous (protein) binding sites in locations other than the plasma membrane, and it is then the "excess" that is being visualized, bound to its substrate.

Chatah and Abrams have reported a different localization of human type Iβ PI4P5K, that is, perinuclear, with a translocation to the plasma membrane after prolonged activation of the cells [19]. Our own experience is that there is some variation in subcellular localization of transfected type I PI4P5Ks between cell types, and also some dependence on culture conditions and length of transfection time. There is still a lot to learn about the localization in vivo of endogenous type I PI4P5Ks and how it is controlled.

Regulation

Given their self-evident role in cell regulation, it is not surprising that type I PI4P5Ks have been found to be subject to a variety of regulatory influences. Only a brief summary of the literature to the end of 2001 is possible here.

Phosphatidic Acid

This lipid has long been known to be a potent stimulator of type I (but not type II) PIPKs [20]. Under some circumstances it can be essential—for example, Honda et al. could only see the effects of Arf-6 (below) if PA was supplied [21]. Jones et al. [22] have produced evidence that endogenous PA may be a significant regulator of type I PI4P5Ks in vivo. PA is of course the product of PLD, itself an enzyme frequently tied in with $PI45P_2$ and with type I PI4P5Ks (e.g. [23]), and it may be that the two enzymes have a complex interregulatory relationship.

Monomeric G Proteins

There is abundant evidence that members of the Rho and Arf family can regulate type I PI4P5Ks, though to a significant degree we do not know the physiological veracity of these events, nor the isoform involved. The clear difference between the three isoforms (Fig. 1) raises the possibility that in vivo there may be significant specificity in the G-protein-PI4P5K interaction.

Arguably the strongest evidence supports regulation by members of the Arf family. For example, Honda et al. [21] purified from brain cytosol the major GTPγS-dependent activator of murine type Iα PI4P5K, and found it to be Arf-1. They went on to show that its localization in HeLa cells was not consistent with its being a natural regulator of type Iα PI4P5K (Arf-1 being predominantly in the Golgi in these cells), but rather that Arf-6 fitted the bill under all the criteria they addressed. Martin et al. [24] also thought that Arf and not Rho (see below) was the endogenous regulator of a type I PI4P5K (isoform unknown). Brown et al. [25] have implicated Arf-6 in endosome formation, and showed that human Type Iα PI4P5K can mimic the effects of a constitutively active Arf-6 (though again the endogenous Type I PI4P5K is unknown). Arf-1 may regulate $PI45P_2$ synthesis in the Golgi, though in these experiments it most likely recruited the type I PI4P5K from the cytosol [26,27].

There is also a reasonable case for type I PI4P5K activation by Rho family members, though it is sometimes confusing. Thus using the Rho-specific C3 Botulinum toxin, Chong et al. [28] implied that Rho regulates a type I PI4P5K activity in fibroblasts, whereas others have failed to see a Rho-type I PI4P5K interaction in experiments where it did interact directly with Rac [29]. Rac interaction with type I PI4P5Ks has been suggested in other experiments [30,31], and there are convincing data placing type Iα or Iβ PI4P5Ks in the signaling pathway from the thrombin receptor, via Rac, to actin polymerization [32]. For the most part the evidence for Rho involvement still remains indirect [33]. Some of these simplistic contradictions may be due to differences in isoforms, though Honda et al. [21] stated that this was unlikely to be the reason they could not see an effect of Rho in their experiments. An interaction with RhoGDI has also been reported [31], and we think that a fair summary of the state of play is that the involvement of monomeric G-proteins in regulation of type I PI4P5Ks is real, and important, but incompletely understood.

Phosphorylation

Several protein kinases have been reported to associate with or regulate type I PI4P5Ks; for example, casein kinase I in S. pombe [34], Rho-kinase [35] (which might explain some of the contradictions about Rho, though see ref [33]), and PKCμ (a.k.a. PKD) [36]. Also, Park *et al.* [37] showed that all three type I PI4P5Ks can be negatively regulated by PKA, and also suggested that receptor activation led to a dephosphorylation (and thus activation) by an uncharacterized mechanism that may involve PKC. Wenk *et al.* [38] have shown a stimulation-dependent dephosphorylation of type Iγ PI4P5K in synapses (where it is the major type I isoform). Another intriguing possible regulatory mechanism has been suggested by Itoh *et al.* [39]. All three isoforms of type I PI4P5K are capable of autophosphorylation, an activity that is stimulated by PI and that leads to an inhibition of the enzyme's activity against PI4P. The physiological relevance of this awaits further study, as does the even more intriguing (and as yet untested) possibility that, like some of the type I PI3Ks [40], type I PI4P5Ks might phosphorylate other proteins.

Other Regulation Mechanisms

Mejillano *et al.* [41] have suggested that human type Iα PI4P5K is cleaved by caspase during apoptosis, an event that, because they also suggest $PI45P_2$ to be anti-apoptotic, serves as part of the amplification of the apoptotic process once it has started. Recently, Barbieri *et al.* have shown an isomeric specificity for the involvement of type I PI4P5Ks in EGF receptor-mediated endocytosis [42], in that the mouse type Iβ PI4P5K was required but the type Iα was not. How the type Iβ is regulated in this process is an intriguing question for further exploration.

Function

The physiological role of type I PI4P 5Ks is self-evidently well established (in contrast with the more enigmatic type II PI5P 4-kinases; see the next chapter by Rameh), because their primary function is to synthesize $PI45P_2$. Thus the question, what is the function of type I PI4P 5Ks, is essentially the same as the question, what is the function of $PI45P_2$. This is now a huge topic, with upwards of 20 suggested physiological functions (e.g. see [8,10] for reviews) and therefore is outside the scope of this short review.

Acknowledgments

K.A.H. is supported by the MRC and R.F.I. by the Royal Society.

References

1. King, C. E., Stephens, L. R., Hawkins, P. T., Guy, G. R., and Michell, R. H. (1987). Multiple metabolic pools of phosphoinositides and phosphatidate in human erythrocytes incubated in a medium that permits rapid transmembrane exchange of phosphate. *Biochem. J.* **244**, 209–217.
2. Stephens, L. R., Hughes, K. T., and Irvine, R. F. (1991). Pathway of phosphatidylinositol(3,4,5)-trisphosphate synthesis in activated neutrophils. *Nature* **351**, 33–39.
3. Vann, L. R., Wooding, F. B., Irvine, R. F., and Divecha, N. (1997). Metabolism and possible compartmentalization of inositol lipids in isolated rat-liver nuclei. *Biochem. J.* **327**, 569–576.
4. Tolias, K. F., Rameh, L. E., Ishihara, H., Shibasaki, Y., Chen, J., Prestwich, G. D., Cantley, L. C., and Carpenter, C. L. (1998). Type I phosphatidylinositol-4-phosphate 5-kinases synthesize the novel lipids phosphatidylinositol 3,5-bisphosphate and phosphatidylinositol 5-phosphate. *J. Biol. Chem.* **273**, 18040–18046.
5. Zhang, X. *et al.* (1997). Phosphatidylinositol-4-phosphate 5-kinase isozymes catalyze the synthesis of 3-phosphate-containing phosphatidylinositol signaling molecules. *J. Biol. Chem.* **272**, 17756–17761.
6. Elge, S., Brearley, C., Xia, H. J., Kehr, J., Xue, H. W., and Mueller-Roeber, B. (2001). An Arabidopsis inositol phospholipid kinase strongly expressed in procambial cells: synthesis of PtdIns(4,5)P2 and PtdIns(3,4,5)P3 in insect cells by 5-phosphorylation of precursors. *Plant J.* **26**, 561–571.
7. Halstead, J. R., Roefs, M., Ellson, C. D., D'Andrea, S., Chen, C., D'Santos, C. S., and Divecha, N. (2001). A novel pathway of cellular phosphatidylinositol(3,4,5)-trisphosphate synthesis is regulated by oxidative stress. *Curr. Biol.* **11**, 386–395.
8. Hinchliffe, K. A., Ciruela, A., and Irvine, R. F. (1998). PIPkins, their substrates and their products: new functions for old enzymes. *Biochim. Biophys. Acta* **1436**, 87–104.
9. Anderson, R. A., Boronenkov, I. V., Doughman, S. D., Kunz, J., and Loijens, J. C. (1999). Phosphatidylinositol phosphate kinases, a multifaceted family of signaling enzymes. *J. Biol. Chem.* **274**, 9907–9910.
10. Martin, T. F. (1998). Phosphoinositide lipids as signaling molecules: common themes for signal transduction, cytoskeletal regulation, and membrane trafficking. *Annu. Rev. Cell Dev. Biol.* 14231–14264.
11. Ishihara, H., Shibasaki, Y., Kizuki, N., Katagiri, H., Yazaki, Y., Asano, T., and Oka, Y. (1996). Cloning of cDNAs encoding two isoforms of 68-kDa type I phosphatidylinositol-4-phosphate 5-kinase. *J. Biol. Chem.* **271**, 23611–23614.
12. Loijens, J. C. and Anderson, R. A. (1996). Type I phosphatidylinositol-4-phosphate 5-kinases are distinct members of this novel lipid kinase family. *J. Biol. Chem.* **271**, 32937–32943.
13. Ishihara, H., Shibasaki, Y., Kizuki, N., Wada, T., Yazaki, Y., Asano, T., and Oka, Y. (1998). Type I phosphatidylinositol-4-phosphate 5-kinases. Cloning of the third isoform and deletion/substitution analysis of members of this novel lipid kinase family. *J. Biol. Chem.* **273**, 8741–8748.
14. Carvajal, J. J. *et al.* (1996). The Friedreich's ataxia gene encodes a novel phosphatidylinositol-4-phosphate 5-kinase. *Nat. Genet.* **14**, 157–162.
15. Yoshida, S., Ohya, Y., Nakano, A., and Anraku, Y. (1994). Genetic interactions among genes involved in the STT4-PKC1 pathway of Saccharomyces cerevisiae. *Mol. Gen. Genet.* **242**, 631–640.
16. Rao, V. D., Misra, S., Boronenkov, I. V., Anderson, R. A., and Hurley, J. H. (1998). Structure of type II beta phosphatidylinositol phosphate kinase: a protein kinase fold flattened for interfacial phosphorylation. *Cell* **94**, 829–839.
17. Kunz, J., Wilson, M. P., Kisseleva, M., Hurley, J. H., Majerus, P. W., and Anderson, R. A. (2000). The activation loop of phosphatidylinositol phosphate kinases determines signaling specificity. *Mol. Cell* **5**, 1–11.
18. Kunz, J., Fuelling, A., Kolbe, L., and Andeson, R. A. (2002). Stereospecific substrate recognition by phosphatidylinositol phosphate kinases is swapped by changing a single amino acid residue. *J. Biol. Chem.* In press.
19. Chatah, N. E. and Abrams, C. S. (2001). G-protein-coupled receptor activation induces the membrane translocation and activation of phosphatidylinositol-4-phosphate 5-kinase I alpha by a Rac- and Rho-dependent pathway. *J. Biol. Chem.* **276**, 34059–34065.
20. Jenkins, G. H., Fisette, P. L., and Anderson, R. A. (1994). Type I phosphatidylinositol 4-phosphate 5-kinase isoforms are specifically stimulated by phosphatidic acid. *J. Biol. Chem.* **269**, 11547–11554.
21. Honda, A. *et al.* (1999). Phosphatidylinositol 4-phosphate 5-kinase alpha is a downstream effector of the small G protein ARF6 in membrane ruffle formation. *Cell* **99**, 521–532.
22. Jones, D. R., Sanjuan, M. A., and Merida, I. (2000). Type I alpha phosphatidylinositol 4-phosphate 5-kinase is a putative target for increased intracellular phosphatidic acid. *FEBS Lett* **476**, 160–165.
23. Divecha, N. *et al.* (2000). Interaction of the type I alpha PIP kinase with phospholipase D: a role for the local generation of

phosphatidylinositol 4,5-bisphosphate in the regulation of PLD2 activity. *EMBO J.* **19**, 5440–5449.
24. Martin, A., Brown, F. D., Hodgkin, M. N., Bradwell, A. J., Cook, S. J., Hart, M., and Wakelam, M. J. (1996). Activation of phospholipase D and phosphatidylinositol 4-phosphate 5-kinase in HL60 membranes is mediated by endogenous Arf but not Rho. *J. Biol. Chem.* **271**, 17397–17403.
25. Brown, F. D., Rozelle, A. L., Yin, H. L., Balla, T., and Donaldson, J. G. (2001). Phosphatidylinositol 4,5-bisphosphate and Arf6-regulated membrane traffic. *J. Cell Biol.* **154**, 1007–1017.
26. Godi, A. *et al.* (1999). ARF mediates recruitment of PtdIns-4-OH kinase-beta and stimulates synthesis of PtdIns(4,5)P2 on the Golgi complex. *Nat. Cell Biol.* **1**, 280–287.
27. Jones, D. H., Morris, J. B., Morgan, C. P., Kondo, H., Irvine, R. F., and Cockcroft, S. (2000). Type I phosphatidylinositol 4-phosphate 5-kinase directly interacts with ADP-ribosylation factor 1 and is responsible for phosphatidylinositol 4,5-bisphosphate synthesis in the golgi compartment. *J. Biol. Chem.* **275**, 13962–13966.
28. Chong, L. D., Traynor-Kaplan, A., Bokoch, G. M., and Schwartz, M. A. (1994). The small GTP-binding protein Rho regulates a phosphatidyl-inositol 4-phosphate 5-kinase in mammalian cells. *Cell* **79**, 507–513.
29. Tolias, K. F., Cantley, L. C., and Carpenter, C. L. (1995). Rho family GTPases bind to phosphoinositide kinases. *J. Biol. Chem.* **270**, 17656–17659.
30. Hartwig, J. H., Bokoch, G. M., Carpenter, C. L., Janmey, P. A., Taylor, L. A., Toker, A., and Stossel, T. P. (1995). Thrombin receptor ligation and activated Rac uncap actin filament barbed ends through phosphoinositide synthesis in permeabilized human platelets. *Cell* **82**, 643–653.
31. Tolias, K. F., Couvillon, A. D., Cantley, L. C., and Carpenter, C. L. (1998). Characterization of a Rac1- and RhoGDI-associated lipid kinase signaling complex. *Mol. Cell Biol.* **18**, 762–770.
32. Tolias, K. F., Hartwig, J. H., Ishihara, H., Shibasaki, Y., Cantley, L. C., and Carpenter, C. L. (2000). Type Ialpha phosphatidylinositol-4-phosphate 5-kinase mediates Rac-dependent actin assembly. *Curr. Biol.* **10**, 153–156.
33. Matsui, T., Yonemura, S., and Tsukita, S. (1999). Activation of ERM proteins in vivo by Rho involves phosphatidyl-inositol 4-phosphate 5-kinase and not ROCK kinases. *Curr. Biol.* **9**, 1259–1262.
34. Vancurova, I., Choi, J. H., Lin, H., Kuret, J., and Vancura, A. (1999). Regulation of phosphatidylinositol 4-phosphate 5-kinase from Schizosaccharomyces pombe by casein kinase I. *J. Biol. Chem.* **274**, 1147–1155.
35. Oude Weernink, P. A. *et al.* (2000). Stimulation of phosphatidyl-inositol-4-phosphate 5-kinase by Rho-kinase. *J. Biol. Chem.* **275**, 10168–10174.
36. Nishikawa, K., Toker, A., Wong, K., Marignani, P. A., Johannes, F. J., and Cantley, L. C. (1998). Association of protein kinase Cmu with type II phosphatidylinositol 4-kinase and type I phosphatidylinositol-4-phosphate 5-kinase. *J. Biol. Chem.* **273**, 23126–23133.
37. Park, S. J., Itoh, T., and Takenawa, T. (2001). Phosphatidylinositol 4-phosphate 5-kinase type I is regulated through phosphorylation response by extracellular stimuli. *J. Biol. Chem.* **276**, 4781–4787.
38. Wenk, M.R. *et al.* (2001). Pip kinase igamma is the major pi(4,5)p(2) synthesizing enzyme at the synapse. *Neuron* **32**, 79–88.
39. Itoh, T., Ishihara, H., Shibasaki, Y., Oka, Y., and Takenawa, T. (2000). Autophosphorylation of type I phosphatidylinositol phosphate kinase regulates its lipid kinase activity. *J. Biol. Chem.* **275**, 19389–19394.
40. Bondeva, T., Pirola, L., Bulgarelli Leva, G., Rubio, I., Wetzker, R., and Wymann, M. P. (1998). Bifurcation of lipid and protein kinase signals of PI3Kgamma to the protein kinases PKB and MAPK. *Science* **282**, 293–296.
41. Mejillano, M., Yamamoto, M., Rozelle, A. L., Sun, H. Q., Wang, X., and Yin, H. L. (2001). Regulation of apoptosis by phosphatidylinositol 4,5-bisphosphate inhibition of caspases, and caspase inactivation of phosphatidylinositol phosphate 5-kinases. *J. Biol. Chem.* **276**, 1865–1872.
42. Barbieri, M. A., Heath, C. M., Peters, E. M., Wells, A., Davis, J. N., and Stahl, P. D. (2001). Phosphatidylinositol-4-phosphate 5-kinase-1beta is essential for epidermal growth factor receptor-mediated endocytosis. *J. Biol. Chem.* **276**, 47212–47216.

CHAPTER 145

Type 2 PIP4-Kinases

Lucia Rameh
*Boston Biomedical Research Institute,
Watertown, Massachusetts*

Introduction

The type 2 PIP4-kinase family of enzymes appeared relatively late in the evolution of eukaryotes. Homologous to the type 1 PIP5-kinases, they also catalyze the synthesis of phosphatidylinositol-4,5-bisphosphate (PtdIns-4,5-P_2). However, the type 2 PIP4-kinases are 4-kinases that use phosphatidylinositol-5-phosphate (PtdIns-5-P) as substrate, while the type 1 are 5-kinases that use phosphatidylinositol-4-phosphate (PtdIns-4-P) as substrate. Unlike type 1 PIP5-kinases, type 2 PIP4-kinases are not found in yeast (*S. cerevisiae* and *S. pombe*), but they are present in lower multicellular eukaryotes (such as *C. elegans*). Thus, the type 2 PIP4-kinases probably diverged from the type 1 PIP5-kinase to fulfill a specialized but essential function in multicellular organisms. Despite the fact that the type 2 PIP4-kinases were the first phosphoinositide kinases to be isolated, cloned, and crystallized, their purpose in cells still remains elusive. Type 2 PIP4-kinases appear be regulated by extracellular factors, which suggests a role for these enzymes in cell-cell signaling. Here I review the history of the type 2 PIP4-kinases along with their structure and regulation. I present their potential roles in phosphoinositide metabolism and in the transduction of extracellular signals.

History

The kinases capable of synthesizing PtdIns-4,5-P_2 were first purified from erythrocytes in the late 1980s [1]. Two distinct activities, type 1 and type 2, were separated and initially distinguished from each other based on biochemical and immunogenic characteristics [2]. In the literature prior to 1997, type 1 and type 2 PIP-kinases were assumed to carry out the same reaction-conversion of PtdIns-4-P to PtdIns-4,5-P_2. In fact, they were first named PtdIns-4-P 5-kinases.

In 1997, a surprising observation led to the realization that the type 2 PIP-kinases actually produce PtdIns-4,5-P_2 by phosphorylating the 4 position of PtdIns-5-P (a contaminate in commercial PtdIns-4-P) [3]. This observation demonstrated that PtdIns-4,5-P_2 can be synthesized through two independent pathways. The pathway catalyzed by the type 1 PIP5-kinase uses PtdIns-4-P as intermediate and is referred to as the canonical pathway for PtdIns-4,5-P_2 synthesis. The pathway catalyzed by the type 2 PIP4-kinase uses PtdIns-5-P as intermediate and is referred to as the alternative pathway for PtdIns-4,5-P_2 synthesis, because it accounts for only a fraction of the total PtdIns-4,5-P_2 in cells. PtdIns-5-P levels in cells are very small when compared to PtdIns-4-P and cannot be easily detected via conventional HPLC separation protocols [3]. For this reason, PtdIns-5-P was not known to exist *in vivo* prior to this discovery. *In vitro*, the type 2 PIP4-kinases can also convert PtdIns-3-P to PtdIns-3,4-P_2, but PtdIns-5-P is the preferred substrate (50-fold better) [3]. In summary, it is now clear that the type 1 and 2 PIP-kinases have different biological and metabolic functions in cells, even though they both synthesize the same lipid product.

Structure

The domain structure of the type 2 PIP4-kinase protein is fairly simple (Fig. 1). Its predicted molecular weight is approximately 47 kDa, but the α and β isoforms migrate with apparent molecular weight of 55 kDa in SDS-polyacrylamide gels. The kinase domain is located in the carboxy-terminal portion of the protein and accounts for most of the protein. The amino-terminal portion of the protein is involved in dimerization. Crystals of the type 2β PIP4-kinase revealed that these enzymes have structures similar to protein kinases. The homodimer forms an elongated disc shape and a large flat surface containing the two catalytic pockets of the subunits [4].

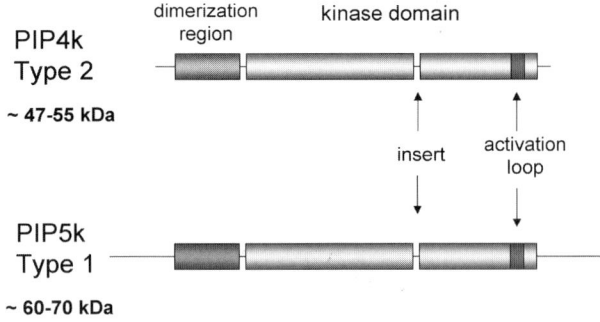

Figure 1 Structure of the type 2 and type 1 PIP-kinases.

A high concentration of positive residues on this flat surface suggests that this region is involved in membrane interaction through electrostatic forces. The substrate pocket is not as deep as protein kinase's substrate pocket, suggesting that the homodimer can float across the surface of membranes and phosphorylate PtdIns-5-P without a necessity for the lipid to significantly protrude from the membrane [4].

The kinase domain of the type 1 and the type 2 PIP-kinases are both interrupted by an insert that does not resolve in the type 2 crystal structure. The type 1 and type 2 are 35 percent identical at the kinase domain. However, the sequence of the type 1 and type 2 PIP-kinases are significantly divergent at a stretch of about 25 amino acids in the region of the kinase domain that corresponds to the activation loop of protein kinases. This region is highly conserved between the different isoforms of a given isotype. Anderson and collaborators showed that this activation loop region is sufficient to determine substrate specificity to the type 1 and type 2 PIP-kinases [5]. When the activation loop of the type 1 PIP5-kinase was swapped with the activation loop of the type 2, the chimeric enzymes lost their original substrate specificity and acquired the catalytic properties of the donor enzyme.

Type 2 PIP4-Kinase Isoforms

There are three isoforms of the type 2 PIP4-kinase in mammalian cells, namely the α, β, and γ isoforms [6–8]. At the protein level, the α and β isoforms are 83 percent identical and the γ isoform is about 60 percent identical to either one of them. All isoforms are ubiquitously expressed, but the α isoform is predominantly found in brain and platelets, the β isoform in brain and muscle, and the γ isoform in brain and kidney. Type 2 PIP4-kinase orthologs are present in the C. elegans (F535H12.4) and Drosophila (CG17471) genomes. Even though the biochemical properties of the products of the C. elegans and Drosophila genes have not yet been demonstrated, they are likely to be enzymatically active, based on conservation of critical residues in the active site. Because this gene family has been conserved from worms to humans, it is likely that the type 2 enzymes serve an important function in multicellular organisms. Similarities between the type 1 and type 2 PIP-kinases suggest that they have a common ancestor.

Regulation

The levels of PtdIns-4,5-P_2 in cells can be affected by extracellular signals, thereby suggesting that the activity of PIPkinases may be regulated [9–11]. Although the mechanisms by which the type 2 PIP4-kinases are regulated in cells are not completely clear, the existing data suggest that subcellular localization, interaction with membrane receptors, phosphorylation, and substrate availability are important factors.

Subcellular Localization

The first indication that the type 2 PIP4-kinases respond to extracellular factors came from studies in platelets. Thrombin-stimulated aggregation of platelets induced the redistribution of type 2 PIP4-kinase to the cytoskeleton [12]. This phenomenon correlated with increased cytoskeleton-associated PIP-kinase activity and increased levels of cytoskeleton-associated PtdIns-4,5-P_2. This thrombin-stimulated PIP4-kinase re-localization to the cytoskeleton was mediated by integrins, and the results suggested a role for this enzyme in controlling cell morphology and adhesion.

The subcellular localization of the type 2 PIP4-kinases (α and β) was also examined in fibroblasts by immunofluorescence and by expression of GFP-tagged fusion proteins. A surprising finding was that a fraction of these enzymes, together with the type 1 PIP5-kinases, was present in the nucleus, in structures that appear as nuclear speckles and contain pre-mRNA processing factors [13]. In a different study, the type 2β was found in the nucleus and cytosol, but the α was found exclusively in the cytosol [14,15]. Mutations in the β isoform revealed that α helix-7 of type 2 β is necessary for its nuclear localization. The function of the type 2 PIP4-kinase in the nucleus remains to be determined. Nonetheless, many studies have indicated that phosphoinositide metabolism in the nucleus is an active process.

Interaction with Membrane Receptors

The type 2β isoform was first cloned from a yeast two-hybrid screen by using the p55/tumor necrosis factor (TNF) receptor as bait [7]. Later, it was also shown to associate with the EGF receptor and with ErbB2 [16]. Association with the TNF receptor is specific for the p55 subunit and involves the juxta-membrane region of the receptor. Very little is known about how these interactions affect type 2 PIP4-kinase activity. It is possible that association with receptors may bring PIP4-kinase in close proximity with its substrate or with other regulatory proteins. Association with receptors is independent of ligand stimulation, thus it is not clear whether PIP4-kinase can be regulated by TNFα, EGF, or neuregulin stimulation or whether it participates in signaling by these growth factors.

More recently, the type 2α PIP4-kinase was shown to be present in bovine photoreceptor rod outer segments (ROS), a compartment of retinal photoreceptor cells in which phosphoinositide metabolism is active and responsive to light stimuli [17]. It is interesting that in tyrosine-phosphorylated ROS, the type 2 enzyme can be precipitated with anti-phosphotyrosine antibodies. The type 2 protein itself does not seem to be phosphorylated in these preparations, indicating rather that it associates with phosphotyrosine-containing proteins. These results suggest that receptor tyrosine kinases may regulate type 2 PIP4-kinase activity, although the phosphotyrosine containing partner of the type 2 PIP4-kinase has not been identified in these studies. Furthermore, these studies indicate that the type 2 enzymes may have a role in the transduction of signals initiated by light. Further analysis will be necessary to confirm these hypotheses.

Phosphorylation

The type 2α PIP4-kinase present in platelets was shown to be phosphorylated on serine and threonine residues [18]. Unlike other lipid kinases, such as the type 1 PIP5-kinase and PI3-kinases, the type 2 PIP4-kinase is not autophosphorylated [19]. The protein kinase CK2 was identified as a PIP4-kinase kinase and shown to phosphorylate Serine 304 (S304), a residue that is not conserved in the β and γ isoforms [20]. The phosphorylation state of S304 in resting versus activated platelets has not been determined. Since CK2 is a constitutively active kinase in cells, it is not clear whether S304 is involved in PIP4-kinase regulation in response to extracellular factors.

The type 2γ isoform was also found to be phosphorylated [8]. In polyacrylamide gels, this isoform migrates as a doublet, and the upper band was shown to be phosphatase-sensitive. *In vivo* labeling of cells with [^{32}-P]-phosphate demonstrated that this enzyme is phosphorylated on serine and threonine but not on tyrosine. The phosphorylation state of the type 2γ changes in response to various signals, including EGF and serum. This is strong evidence that the type 2 PIP4-kinases may be regulated by extracellular signals. However, the exact role of phosphorylation on the type 2γ activity in cells remains to be determined.

Substrate Availability

The levels of PtdIns-5-P in cells are comparable to the levels of 3′-phosphorylated phosphoinositides, such as PtdIns-3-P, but are much lower than the levels of PtdIns-4-P [3]. This suggests that the availability of PtdIns-5-P substrate may be the limiting step in the production of PtdIns-4,5-P_2 by the type 2 PIP4-kinases. Although the type 2 PIP4-kinases are subject to posttranslational modifications and protein-protein interactions, no direct effect on kinase activity was reported, as discussed above. In addition, expression of these PIP4-kinases in bacteria results in active enzymes (except for the type 2γ isoform) and indicates that the type 2 α and β may be constitutively active in cells. Therefore, it is possible that the local and temporal activation of the alternative pathway for PtdIns-4,5-P_2 synthesis is dependent upon PtdIns-5-P synthesis and the co-localization of type 2 enzyme with this substrate. PtdIns-5-P levels in cells were shown to be regulated by thrombin, by cell cycle progression, and by serum stimulation ([21,22] and personal unpublished data). However, the pathways for PtdIns-5-P synthesis *in vivo* have not been determined. *In vitro* PtdIns-5-P can be generated through phosphorylation of PtdIns by 5′-kinases, such as the type 1 PIP5kinase [23] and PIKfyve [24], or by dephosphorylation of PtdIns-4,5-P_2 by SHIP [3].

Putative Models for the Function of the Type 2 PIP-Kinases

Despite more than a decade of research on type 2 PIP4-kinases, the biological role of the alternative pathway for PtdIns-4,5-P_2 synthesis is not clear. Nevertheless, it is clear that the type 1 and the type 2 PIP-kinases have nonoverlapping biological functions. For example, overexpression of the type 1 PIP5-kinase, but not the type 2, leads to a dramatic reorganization of actin cytoskeleton [25].

Pulse labeling of phosphoinositides in cultured cells has indicated that the phosphate at the 5′ position of the inositol ring is incorporated last in the majority of PtdIns-4,5-P_2 synthesized *in vivo* [26]. Therefore, the type 2 PIP-kinase is not involved in maintaining the bulk of the PtdIns-4,5-P_2 in cells. At this point we can only speculate on the roles for this enzyme in phosphoinositide metabolism and cell signaling. Here are a few possibilities:

Model 1: to Regulate the Synthesis of Specific Pools of PtdIns-4,5-P_2 in Cells. One possibility is that the type 2 PIP4-kinases may contribute to PtdIns-4,5-P_2 synthesis in specific subcellular compartments where PtdIns-5-P is present (model 1, Fig. 2). This would permit the regulation of local synthesis of PtdIns-4,5-P_2 independent of the bulk of PtdIns-4,5-P_2 synthesis. Even though total PtdIns-4,5-P_2 levels in cells are high, there are reports that demonstrate that a large fraction of cellular PtdIns-4,5-P_2 is unavailable [27]. For instance, the PtdIns-4,5-P_2 synthesized through the alternative pathway could be the main source of substrate for the

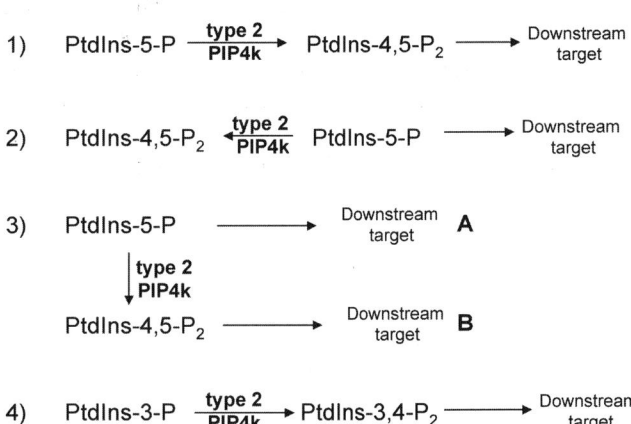

Figure 2 Putative models for the type 2 PIP4-kinase function in cells.

enzyme PLCγ to generate IP$_3$ and diacylglycerol, despite the availability of PtdIns-4,5-P$_2$ in the cell.

Model 2: to Regulate the Levels of PtdIns-5-P in Cells. As discussed above, PtdIns-5-P levels in cells are small and comparable to the levels of other signaling phosphoinositides. This observation raises the possibility that the important function of the type 2 PIP4-kinases is to get rid of PtdIns-5-P rather than generate PtdIns-4,5-P$_2$. This implies that PtdIns-5-P has a specific role in cells that may not be related to its role as an intermediate for PtdIns-4,5-P$_2$ synthesis. PtdIns-5-P could be a signaling molecule with specific downstream targets or a substrate for other enzymes such as PI 3-kinases. In this case, the function of the type 2 PIP4-kinase could be to assure that the levels of PtdIns-5-P are kept low and tightly regulated (model 2, Fig. 2). New data showing that PtdIns-5-P can be regulated by extracellular factors make this an attractive model.

Model 3: to Coordinate PtdIns-5-P Consumption with PtdIns-4,5-P$_2$ Synthesis. Models 1 and 2 are not mutually exclusive and it is possible that PtdIns-5-P and PtdIns-4,5-P$_2$ generated through PtdIns-5-P can trigger opposite cellular responses. In this model, the type 2 PIP4-kinase could serve as a switch between these two modes of signaling, necessary to assure that the termination of the signal generated by PtdIns-5-P is coupled to the initiation of the PtdIns-4,5-P$_2$ signal (model 3, Fig. 2).

Model 4: to Regulate the Synthesis of PtdIns-3,4-P$_2$ in Cells. The type 2 PIP-kinase could possibly be responsible for PtdIns-3,4-P$_2$ synthesis in cells, independent of PtdIns-3,4,5-P$_3$ (model 4, Fig. 2). In this case, PtdIns-3-P would be the major substrate for type 2 PIP4-kinases. This model is unlikely, based on the strong preference that these enzymes have for PtdIns-5-P.

Conclusion

Despite new biochemical, genetic, and structural information that has been acquired in recent years, the type 2 PIP4-kinase family remains a mystery to cell biologists who are trying to identify the physiologic role for the alternative pathway for PtdIns-4,5-P$_2$ synthesis, catalyzed by these lipid kinases. Future experiments involving inactivation or suppression of the type 2 activity in cell or animal models are likely to shed a light on this intriguing question.

References

1. Ling, L. E., Schulz, J. T., and Cantley, L. C. (1989). Characterization and purification of membrane-associated phosphatidylinositol-4-phosphate kinase from human red blood cells. *J. Biol. Chem.* **264**, 5080–5088.
2. Bazenet, C. E., Ruano, A. R., Brockman, J. L., and Anderson, R. A. (1990). The human erythrocyte contains two forms of phosphatidylinositol-4-phosphate 5-kinase which are differentially active toward membranes. *J. Biol. Chem.* **265**, 18012–18022
3. Rameh, L. E., Tolias, K. F., Duckworth, B. C., and Cantley, L. C. (1997). A new pathway for synthesis of phosphatidylinositol-4,5-bisphosphate [see comments]. *Nature* **390**, 192–196
4. Rao, V. D., Misra, S., Boronenkov, I. V., Anderson, R. A., and Hurley, J. H. (1998). Structure of type IIbeta phosphatidylinositol phosphate kinase: a protein kinase fold flattened for interfacial phosphorylation. *Cell* **94**, 829–839.
5. Kunz, J., Wilson, M. P., Kisseleva, M., Hurley, J. H., Majerus, P. W., and Anderson, R. A. (2000). The activation loop of phosphatidylinositol phosphate kinases determines signaling specificity. *Mol. Cell* **5**, 1–11.
6. Divecha, N., Truong, O., Hsuan, J. J., Hinchliffe, K. A., and Irvine, R. F. (1995). The cloning and sequence of the C isoform of PtdIns4P 5-kinase. *Biochem. J.* **309**, 715–719
7. Castellino, A. M., Parker, G. J., Boronenkov, I. V., Anderson, R. A., and Chao, M. V. (1997). A novel interaction between the juxtamembrane region of the p55 tumor necrosis factor receptor and phosphatidylinositol-4-phosphate 5-kinase. *J. Biol. Chem.* **272**, 5861–5870
8. Itoh, T., Ijuin, T., and Takenawa, T. (1998). A novel phosphatidylinositol-5-phosphate 4-kinase (phosphatidylinositol-phosphate kinase IIgamma) is phosphorylated in the endoplasmic reticulum in response to mitogenic signals. *J. Biol. Chem.* **273**, 20292–20299.
9. McNamee, H. P., Ingber, D. E., and Schwartz, M. A. (1993). Adhesion to fibronectin stimulates inositol lipid synthesis and enhances PDGF-induced inositol lipid breakdown. *J. Cell Biol.* **121**, 673–678.
10. Payrastre, B., Plantavid, M., Breton, M., Chambaz, E., and Chap, H. (1990). Relationship between phosphoinositide kinase activities and protein tyrosine phosphorylation in plasma membranes from A431 cells. *Biochem. J.* **272**, 665–670.
11. Halenda, S. P. and Feinstein, M. B. (1984). Phorbol myristate acetate stimulates formation of phosphatidyl inositol 4-phosphate and phosphatidyl inositol 4,5-bisphosphate in human platelets. *Biochem. Biophys. Res. Commun.* **124**, 507–513.
12. Hinchliffe, K. A., Irvine, R. F., and Divecha, N. (1996). Aggregation-dependent, integrin-mediated increases in cytoskeletally associated PtdInsP2 (4,5) levels in human platelets are controlled by translocation of PtdIns 4-P 5-kinase C to the cytoskeleton. *EMBO J.* **15**, 6516–6524.
13. Boronenkov, I. V., Loijens, J. C., Umeda, M., and Anderson, R. A. (1998). Phosphoinositide signaling pathways in nuclei are associated with nuclear speckles containing pre-mRNA processing factors. *Mol. Biol. Cell* **9**, 3547–3560.
14. Divecha, N., Rhee, S. G., Letcher, A. J., and Irvine, R. F. (1993). Phosphoinositide signalling enzymes in rat liver nuclei: phosphoinositidase C isoform beta 1 is specifically, but not predominantly, located in the nucleus. *Biochem. J.* **289**, 617–620.
15. Ciruela, A., Hinchliffe, K. A., Divecha, N., and Irvine, R. F. (2000). Nuclear targeting of the beta isoform of type II phosphatidylinositol phosphate kinase (phosphatidylinositol 5-phosphate 4-kinase) by its alpha-helix 7. *Biochem. J.* **346 Pt 3**, 587–591.
16. Castellino, A. M. and Chao, M. V. (1999). Differential association of phosphatidylinositol-5-phosphate 4-kinase with the EGF/ErbB family of receptors. *Cell Signal* **11**, 171–7.
17. Huang, Z., Guo, X. X., Chen, S. X., Alvarez, K. M., Bell, M. W., and Anderson, R. E. (2001). Regulation of type II phosphatidylinositol phosphate kinase by tyrosine phosphorylation in bovine rod outer segments. *Biochemistry* **40**, 4550–4559.
18. Hinchliffe, K. A., Irvine, R. F., and Divecha, N. (1998). Regulation of PtdIns4P 5-kinase C by thrombin-stimulated changes in its phosphorylation state in human platelets. *Biochem. J.* **329**, 115–119.
19. Itoh, T., Ishihara, H., Shibasaki, Y., Oka, Y., and Takenawa, T. (2000). Autophosphorylation of type I phosphatidylinositol phosphate kinase regulates its lipid kinase activity. *J. Biol. Chem.* **275**, 19389–19394.
20. Hinchliffe, K. A., Ciruela, A., Letcher, A. J., Divecha, N., and Irvine, R. F. (1999). Regulation of type IIalpha phosphatidylinositol phosphate kinase localisation by the protein kinase CK2. *Curr. Biol.* **9**, 983–986.
21. Morris, J. B., Hinchliffe, K. A., Ciruela, A., Letcher, A. J., and Irvine, R. F. (2000). Thrombin stimulation of platelets causes an increase in phosphatidylinositol 5-phosphate revealed by mass assay. *FEBS Lett.* **475**, 57–60.

22. Clarke, J. H., Letcher, A. J., D'Santos C. S., Halstead, J. R., Irvine, R. F., and Divecha, N. (2001). Inositol lipids are regulated during cell cycle progression in the nuclei of murine erythroleukaemia cells. *Biochem. J.* **357**, 905–910.
23. Tolias, K. F., Rameh, L. E., Ishihara, H., Shibasaki, Y., Chen, J., Prestwich, G. D., Cantley, L. C., and Carpenter, C. L. (1998). Type I phosphatidyl-inositol-4-phosphate 5-kinases synthesize the novel lipids phosphatidylinositol 3,5-bisphosphate and phosphatidylinositol 5-phosphate. *J. Biol. Chem.* **273**, 18040–18046.
24. Sbrissa, D., Ikonomov, O. C., and Shisheva, A. (1999). PIKfyve, a mammalian ortholog of yeast Fab1p lipid kinase, synthesizes 5-phosphoinositides. Effect of insulin. *J. Biol. Chem.* **274**, 21589–21597
25. Ishihara, H., Shibasaki, Y., Kizuki, N., Wada, T., Yazaki, Y., Asano, T., and Oka, Y. (1998). Type I phosphatidylinositol-4-phosphate 5-kinases. Cloning of the third isoform and deletion/substitution analysis of members of this novel lipid kinase family. *J. Biol. Chem.* **273**, 8741–8748.
26. Whiteford, C. C., Brearley, C. A., and Ulug, E. T. (1997). Phosphatidylinositol 3,5-bisphosphate defines a novel PI 3-kinase pathway in resting mouse fibroblasts. *Biochem. J.* **323**, 597–601
27. Cross, D. A., Watt, P. W., Shaw, M., van der Kaay, J., Downes, C. P., Holder, J. C., and Cohen, P. (1997). Insulin activates protein kinase B, inhibits glycogen synthase kinase-3 and activates glycogen synthase by rapamycin-insensitive pathways in skeletal muscle and adipose tissue. *FEBS Lett.* **406**, 211–215.

CHAPTER 146

Phosphoinositide 3-Kinases

David A. Fruman
Department of Molecular Biology and Biochemistry,
University of California,
Irvine, California

Introduction

Engagement of a great variety of cell surface receptors triggers the activation of phosphoinositide 3-kinase (PI3K). The lipid products of PI3K serve as second messengers by interacting with phosphoinositide-binding domains in certain cytoplasmic proteins, thereby recruiting these "PI3K effectors" to specific sites in cellular membranes. PI3K and its effectors have been implicated in diverse cellular functions, including vesicle trafficking, cell proliferation, survival, cytoskeletal remodeling, migration, and glucose uptake. The importance of PI3K signaling in cellular and organismal function has been illustrated by the striking phenotypes of animals with genetic perturbations of PI3K signaling.

This chapter is intended to provide a brief summary of PI3K signaling with an emphasis on recent advances. The reader is referred in the text to several excellent reviews that provide more detail on specific topics.

The Enzymes

Phosphoinositide 3-kinase (PI3K) isoforms have been divided into three classes that differ in subunit structure, substrate selectivity, and regulation [1,2]. Class I PI3Ks exist as heterodimers with a tightly bound regulatory subunit (Fig. 1). In addition to the kinase domain, each class I catalytic subunit possesses a Ras-binding domain, a C2 domain for interaction with phospholipid membranes, and a helical "PIK" domain that is also conserved in PtdIns-4-kinases. In a landmark paper describing the crystal structure of the p110γ isoform, Walker and colleagues showed that the helical domain acts as a scaffold or spine on which the other three functional domains are organized [3]. This report also confirmed that the kinase domain is similar in structure to protein kinases, as predicted from primary sequence comparison and limited protein kinase activity of PI3K enzymes. The shape of the substrate binding pocket helped explain the selectivity for phosphoinositide recognition and the likely basis for differential recognition of single and multiply phosphorylated substrates by different PI3K classes. Subsequent structural work from this group has clarified the mode of binding of various PI3K inhibitors to the active site of p110γ and has provided evidence for allosteric activation by Ras [4,5].

Class I PI3Ks are further subdivided by their modes of regulation. The class I_A subgroup (p110α, p110β, and p110δ) associates with regulatory subunits (p85α, p55α, p50α, p85β, or p55γ) that have multiple modular protein-protein interaction domains (Fig. 1). Class I_A PI3Ks function downstream of receptors with intrinsic or associated tyrosine kinase activity. Full activation of class I_A PI3K is thought to require occupancy of both Src-homology 2 (SH2) domains of the regulatory subunit by tyrosine phosphopeptides, along with the binding of catalytic subunit to Ras-GTP [6]. This normally occurs only in proximity with membrane-associated tyrosine kinases that also activate Ras. Other domains of the regulatory subunits may also contribute to activation or localization [1]. The class I_B enzyme (p110γ) interacts with a distinct regulatory subunit, p101, with no significant homology to other known proteins (Fig. 1). The class I_B enzyme is activated by βγ subunits of heterotrimeric G proteins [1,2] following engagement of G-protein-coupled receptors (GPCRs). The presence of a Ras binding domain within p110γ suggests that this isoform also integrates signals from tyrosine kinase pathways.

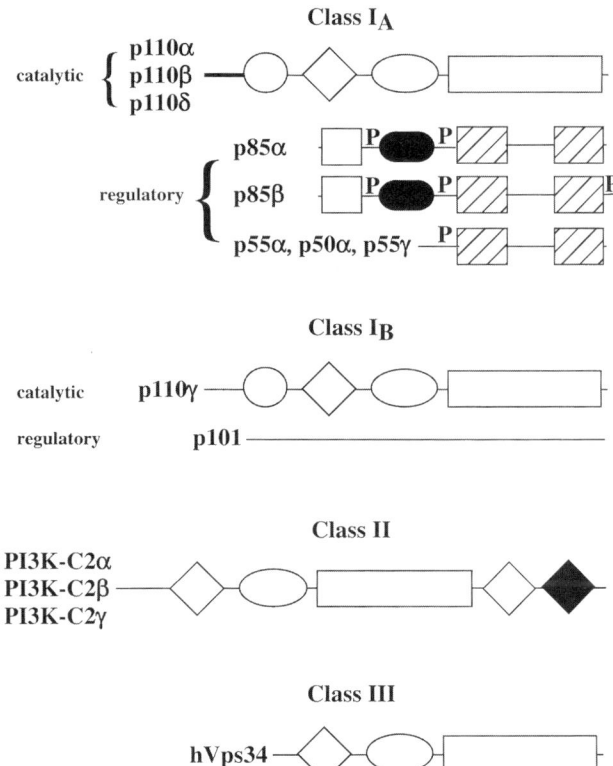

Figure 1 Schematic diagram of the domain structure of mammalian PI3Ks. The common names of the proteins are listed at the left of each structure (hVps34 = human homolog of class III PI3K [Vps34] first cloned from *S. cerevisiae*). The three class I_A catalytic subunits associate with each of the five regulatory subunits without any apparent preference. p85α, p55α, and p50α are alternative transcripts of a single gene. Open boxes, kinase domain; open ovals, PIK domain; open diamond, C2 domain; open circle, Ras-binding domain; open square, SH3 domain; hatched rectangle, SH2 domain; closed oval, RhoGAP-homology domain; closed diamond, PX domain; P, proline-rich motif.

Class II PI3Ks are distinguished by the presence of an additional C-terminal C2 domain and a PX domain (Fig. 1). Although comparatively little is known about class II PI3K regulation, there is growing evidence that these enzymes can be activated by extracellular signals [2]. Genes for class I and class II enzymes have been found in all multicellular animals. Class III PI3Ks are found in all eukaryotes from yeast to humans. These enzymes appear to have a housekeeping function related to vesicular transport and protein sorting [1,2]. They interact with an associated serine kinase in both yeast (vps15p) and humans (p150).

The Products

PI3Ks phosphorylate the 3′-hydroxyl of the D-*myo*-inositol ring of phosphatidylinositol (PtdIns) (Fig. 2A). Four D-3 phosphoinositides exist in mammalian cells: PtdIns(3)P, PtdIns(3,4)P_2, PtdIns(3,5)P_2, and PtdIns(3,4,5)P_3. The pathways of synthesis and degradation of these lipids have been reviewed recently in detail [2,7] and are summarized in Fig. 2B. PtdIns(3,4,5)P_3 is produced only by class I PI3Ks with PtdIns(4,5)P_2 as a substrate. PtdIns(3,4)P_2 can be generated from PtdIns(4)P by class I or class II PI3Ks, or by 5′-phosphatase action on PtdIns(3,4,5)P_3, or by a PtdIns(3) P-4-kinase. The signaling functions of PtdIns(3,4,5)P_3 and PtdIns(3,4)P_2 are well studied and will be discussed further below. Although PtdIns(3)P can be produced by all PI3Ks *in vitro*, the majority of PtdIns(3)P in cells appears to be made by class III PI3Ks and is detected primarily in endosomal vesicles [8]. PtdIns(3,5)P_2 is generated *in vivo* probably by a PtdIns(3)P-5-kinase and like PtdIns(3)P may be involved in vesicle trafficking.

Lipid-Binding Domains

Pleckstrin homology (PH) domains are small (~60aa) protein modules that mediate protein-lipid and protein-protein interactions. There is a growing list of PH domains shown to bind selectively to D-3 phosphoinositides [2,9]. It is important to note that PtdIns(4)P and PtdIns(4,5)P_2 are much more abundant in cellular membranes compared to D-3 phosphoinositides; thus, for a PH domain to be considered D-3-specific, the binding affinity for D-3 lipids must be considerably higher than the affinity for PtdIns(4)P or PtdIns(4,5)P_2. Subgroups have been identified that show greater affinity for either PtdIns(3,4,5)P_3 or PtdIns(3,4)P_2, with others that bind these lipids comparably [2,9]. A combination of biochemical and structural approaches has helped define features of PH domain primary sequence that determine selectivity for different D-3 phosphoinositides [2,10,11]. D-3-selective PH domains are found in a variety of proteins involved in signal transduction, some of which are discussed below.

PX domains are found in a diverse list of proteins involved in vesicle trafficking, protein sorting, and signal transduction [12,13]. Like PH domains, different PX domains exhibit selectivity for different phosphoinositides. Two components of the oxidative burst complex in phagocytes, p40phox and p47phox, possess PX domains that bind preferentially to PtdIns(3)P and PtdIns(3,4)P_2, respectively [14,15]. These interactions are thought to be important for targeting the cytosolic components of the NADPH oxidase complex to the phagolysosome, where they meet with the membrane-bound components p22phox and gp91phox to initiate the oxidative burst. The PX domain of class II PI3K (Fig. 1) binds selectively to PtdIns(4,5)P_2 [16]. The PX domain of cytokine-independent survival kinase (CISK) binds both PtdIns(3,5)P_2 and PtdIns(3,4,5)P_3 [17]. The crystal structure of the PX domain of p40phox bound to PtdIns(3)P shows that the lipid binds in a positively charged pocket and suggests how phosphoinositide binding specificity is determined [18].

The FYVE domain, originally identified in several yeast proteins, is a protein module that binds selectively to PtdIns(3)P. The structure of FYVE domains and their binding to PtdIns(3)P are distinct from the PH domain/D-3 lipid interaction [2,19]. Although some mammalian FYVE domain-containing proteins are involved in signal transduction,

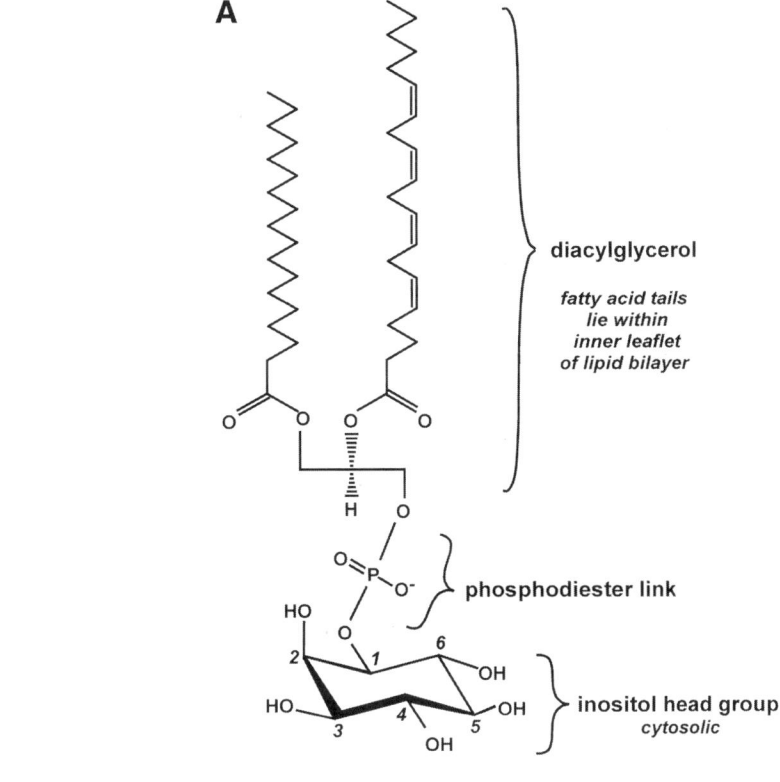

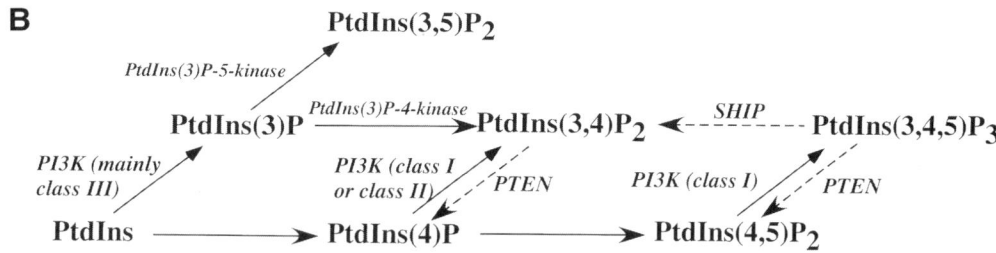

Figure 2 (A) Structure of D-*myo*-phosphatidylinositol (PtdIns). Note that the head group is positioned to interact with cytoplasmic molecules. Although free hydroxyls exist at positions 2–6, phosphorylation *in vivo* has only been detected at positions 3, 4, and 5. Reprinted with permission from the *Annual Review of Biochemistry*, Volume 70 ©2001 by Annual Reviews www.AnnualReviews.org. (B) Pathways of synthesis and degradation of D-3 phosphoinositides. The major enzymes responsible for particular reactions are indicated. For simplicity, many of the enzymes involved in metabolism of other phosphoinositides are omitted.

the primary role of most mammalian and all yeast proteins with this module is in membrane trafficking.

A recent study described the use of phosphoinositide affinity matrices to purify and clone a number of D-3 lipid-binding proteins [20]. Many of these were known proteins with PH or FYVE domains previously shown to bind to PI3K products, helping to validate the method. A novel protein with five PH domains, termed ARAP3, was found to possess distinct domains with GAP activity for Arf and Rho family G proteins. The ARAP family, along with other regulators of Arf and Rho function [2], may thus play an integral role in PI3K-regulated cytoskeletal changes (Fig. 3). This study also identified the Sec14 homology domain, originally identified in the yeast PtdIns transfer protein Sec14p, as a putative phosphoinositide-binding module.

Effectors and Responses

PI3K activation has been linked to distinct cellular responses downstream of different receptors. For example, PI3K is required for proliferation induced by numerous growth factors and cytokines, for glucose uptake triggered by insulin, and for cell migration in response to chemoattractants [1,2]. A major challenge in PI3K research has been to determine how specificity in signaling is achieved. With all the factors that can trigger increases in D-3 phosphoinositides, how is it that different stimuli evoke distinct responses through PI3K?

There are several answers to this puzzle. One level of specificity is conferred by differential expression of PI3K isoforms in distinct tissues and cell types. For example, the p110δ isoform is leukocyte specific, and antibody-blocking

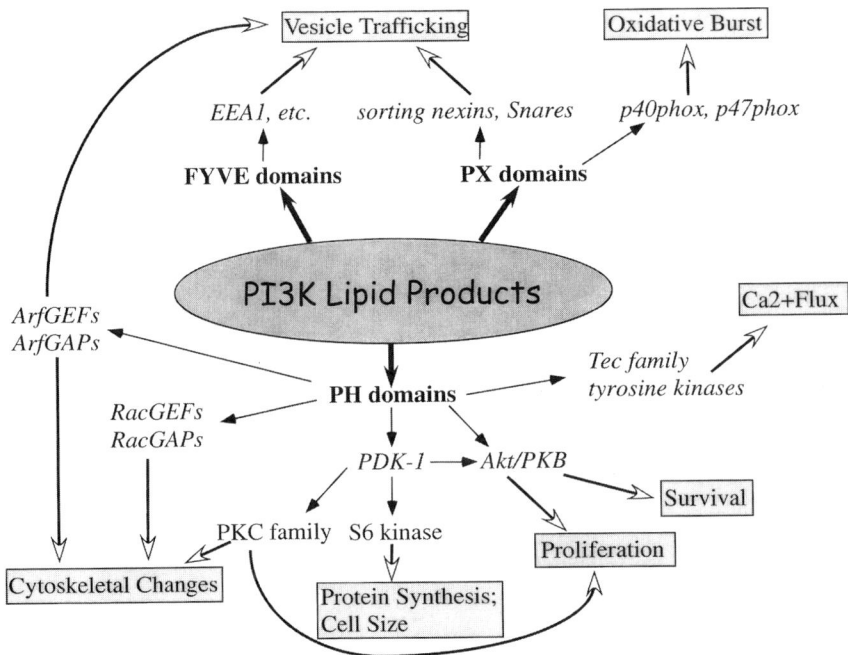

Figure 3 Overview of the effector proteins and signaling pathways regulated by PI3K lipid products. D-3 lipid-binding modules are in bold print. Functional responses are in boxes. The diagram shows selected effector proteins (in italics) whose lipid-binding domains have been well studied and whose activation has been linked to particular responses. GEF, guanine nucleotide exchange factor; GAP, GTPase-activating protein.

experiments suggest that in macrophages p110δ is required for migration whereas p110α is required for proliferation triggered by the CSF-1 receptor [21]. The regulatory isoform p50α is expressed at highest levels in the liver, and loss of p50α is associated with hepatocellular necrosis [22]. Similarly, some PI3K effectors are differentially expressed. An example is Btk, a PH domain-containing tyrosine kinase expressed primarily in B lymphocytes and mast cells. Mice lacking either Btk or the predominant class I_A regulatory isoform, p85α, exhibit similar defects in B cell development and function [23, 24]. Another factor in signaling specificity could be the compartmentalization of PI3K activation. In other words, distinct localization of receptors in membrane subdomains affects the pool of PI3K substrates and effectors utilized. D-3 lipids accumulate at the leading edge of cells migrating in response to chemoattractants, thus resulting in localized activation of PI3K effectors [25,26]. Finally, full activation of a given PI3K effector may require synergy with other signals, which may be differentially provided by distinct receptors. For example, full activation of Btk requires phosphorylation by Src family tyrosine kinases that are also activated by B cell antigen receptors [27].

Figure 3 summarizes current knowledge of the linkage of certain PI3K effectors to distinct responses. It is important to note that this diagram is simplified for clarity, and some effectors have been linked to additional functions. A central player in many responses to PI3K activation is phosphoinositide-dependent kinase-1 (PDK-1) [28]. This serine/threonine kinase has a PH domain that binds both PtdIns(3,4,5)P_3 and PtdIns(3,4)P_2. Current evidence suggests that PDK-1 is constitutively active but only gains access to substrates upon binding D-3 lipids. Phosphorylation by PDK-1 contributes to the activation of many downstream kinases, including Akt/PKB, S6kinase, and some protein kinase C isoforms.

Phosphatases

The membrane-targeting signal provided by D-3 phosphoinositides can be modulated by the action of phosphoinositide phosphatases (PPases). PTEN (*p*hosphatase and *ten*sin homology deleted on chromosome 10) hydrolyzes the 3′-phosphate of PtdIns(3,4,5)P_3 and PtdIns(3,4)P_2, effectively reversing the action of PI3K (Fig. 2A) [2,29]. Although the importance of PTEN is well established (see next section), it is not yet clear how PTEN is regulated or recruited to sites of PI3K activation. SHIP1 and SHIP2 are related 5′-PPases that contain N-terminal SH2 domains (*SH*2-containing *I*nositol polyphosphate 5-*p*hosphatase). SHIPs can remove the 5′-phosphate from PtdIns(3,4,5)P_3 to produce PtdIns(3,4)P_2 (Fig. 2A) [2,30]. Hence, these enzymes may alter the spectrum of PI3K effectors recruited to the membrane rather than simply turning the signal off. The SH2 domains of SHIP1 and SHIP2 are selective for phosphotyrosines within a particular sequence context known as the immunoreceptor tyrosine-based inhibitory motif (ITIM). ITIMs are found in a number of receptors (for example, FcγRIIB) whose ligation attenuates signaling through antigen receptors [30].

Genetics

Pharmacological inhibitors of PI3K enzyme activity impair proliferation in a variety of cell systems [1,2]. Natural and engineered mutations in PI3Ks and lipid phosphatases have further established the fundamental role of PI3K signaling in promoting growth of normal and transformed cells. The transforming oncogene of an avian sarcoma virus, ASV16, encodes a membrane-targeted variant of p110 whose expression in cells causes accumulation of D-3 phosphoinositides [31]. A truncated variant of p85α (termed p65), first isolated from a T-cell lymphoma, increases basal activity of class I_A catalytic subunits and promotes lymphoproliferation when expressed as a transgene in the T lineage [32,33]. Mice heterozygous for a disrupted PTEN gene exhibit a similar lymphoproliferative disorder [34,35]. Mice with homozygous loss of PTEN specifically in the T lineage develop autoimmune symptoms associated with spontaneously activated T cells that are resistant to apoptosis [36]. Inherited mutations in PTEN are the cause of three autosomal dominant cancer syndromes in humans: Cowden's disease, Lhermitte-Duclos disease, and Bannayan-Zonana syndrome [29]. Moreover, loss of PTEN function is seen in a large fraction of sporadic human cancers, especially glial, prostate, and endometrial tumors [29]. Mice lacking SHIP1 develop a myeloproliferative disorder and have lower activation thresholds for a variety of immune cell stimuli [30]. In addition to these examples of increased PI3K signaling promoting proliferation and tumorigenesis, there are also examples of decreased PI3K signaling causing impaired proliferation. Forced expression of PTEN in PTEN-deficient embryonic fibroblasts and tumor cells impairs growth by inducing cell cycle arrest and/or apoptosis [29]. Loss of the class I_A regulatory isoform p85α in mice abrogates B lymphocyte proliferation in response to antigen receptor engagement, diminishes interleukin-4-mediated B cell survival, and reduces stem cell factor-driven mast cell growth [23,24,37,38]. These mice show impaired immune responses to T-cell-independent antigens, bacteria, and parasitic worms [24,38].

Genetic studies have also implicated PI3K signaling in responses to insulin and insulin-like growth factors. In *C. elegans*, a class I PI3K functions downstream of the insulin receptor homolog in a pathway that regulates both dauer entry and lifespan [39]. This pathway involves the worm orthologs of PDK-1 and Akt and is antagonized by PTEN. In mice, SHIP2 phosphatase is a critical modulator of insulin signaling as SHIP2-deficient mice show increased insulin sensitivity [40]. Based on these findings and a wealth of cell culture experiments, it was expected that mice deficient in class I_A PI3K would show insulin resistance. However, in every case examined, the opposite result has been observed. Mice lacking p85α alone, or all p85α gene products (including p55α and p50α), exhibit hypoglycemia and decreased glucose tolerance [41,42]. Mice lacking p85α alone or p85β also show increased insulin sensitivity [41,43]. Fibroblast experiments suggest that class I_A regulatory isoforms are expressed in excess of catalytic subunits, producing a "buffer" effect that is overcome when regulatory subunit expression is reduced genetically [44]. However, it is not yet known whether this mechanism explains altered insulin sensitivity *in vivo*. Deletion of the mouse class I_A catalytic isoform p110α causes early embryonic lethality, preventing the analysis of insulin signaling in these animals [45].

In *Drosophila*, class I PI3K acts downstream of the insulin receptor ortholog in a pathway that controls cell size [2,46]. Overexpression of class I PI3K in wing imaginal discs increases cell size and yields enlarged wings in the adult fly. Mutation of *Drosophila* PTEN has a similar effect. Conversely, mutation of class I PI3K genes or expression of dominant-negative PI3K reduces cell and wing size. PI3K signaling was also shown to regulate the size of mouse cardiac myocytes [47]. The critical downstream effectors of PI3K in the *Drosophila* system are Akt and S6K, a serine/threonine kinase that regulates protein synthesis [2,46]. These kinases are also regulated by PI3K signaling in mammalian cells (Fig. 3), but their role in controlling the size of cardiac myocytes or other cells has not yet been reported.

p110γ, the class I_B isoform, is expressed primarily in leukocytes. Disruption of the mouse p110γ gene causes defects in inflammatory responses that correlate with defective chemotaxis to GPCR ligands such as f-Met-Leu-Phe and C5a [48–50].

Summary

Signaling through PI3K is an evolutionarily conserved process that enables reversible membrane localization of cytoplasmic proteins. Three modular domains (PH, PX, and FYVE) that interact with D-3 phosphoinositides are broadly distributed among proteins of different function. The recruitment and activation of specific subsets of PI3K effectors in a receptor-specific and cell type–specific manner allows PI3K activation to be linked to different functional responses. Given the pleiotropic effects of pharmacological PI3K inhibitors, therapeutic modulation of PI3K signaling is likely to require targeting of specific effectors that govern particular responses.

Note Added in Proof

Since submission of this chapter, new mouse genetic models have yielded a number of important advances in the PI3K field. Of particular interest are studies demonstrating lymphocyte defects in mice lacking functional p110δ [51-53], analysis of more tissue-specific PTEN knockouts (reviewed in ref. [54], also see [55]), studies showing a role for Akt and S6 kinase in regulation of mammalian cell and organ size [56,57], and a study establishing a role for GPCR signaling through the p110γ isoform in cardiac muscle contractility [58].

References

1. Fruman, D. A., Meyers, R. E., and Cantley, L. C. (1998). Phosphoinositide kinases. *Annu. Rev. Biochem.* **67**, 481–507.

2. Vanhaesebroeck, B., Leevers, S. J., Ahmadi, K., Timms, J., Katso, R., Driscoll, P. C., Woscholski, R., Parker, P. J., and Waterfield, M. D. (2001). Synthesis and function of 3-phosphorylated inositol lipids. *Annu. Rev. Biochem.* **70**, 535–602.

3. Walker, E. H., Perisic, O., Ried, C., Stephens, L., and Williams, R. L. (1999). Structural insights into phosphoinositide 3-kinase catalysis and signalling. *Nature* **402**, 313–320.

4. Walker, E. H., Pacold, M. E., Perisic, O., Stephens, L., Hawkins, P. T., Wymann, M. P., and Williams, R. L. (2000). Structural determinants of phosphoinositide 3-kinase inhibition by wortmannin, LY294002, quercetin, myricetin, and staurosporine. *Mol. Cell* **6**, 909–919.

5. Pacold, M. E., Suire, S., Perisic, O., Lara-Gonzalez, S., Davis, C. T., Walker, E. H., Hawkins, P. T., Stephens, L., Eccleston, J. F., and Williams, R. L. (2000). Crystal structure and functional analysis of Ras binding to its effector phosphoinositide 3-kinase gamma. *Cell* **103**, 931–943.

6. Rodriguez-Viciana, P., Warne, P. H., Vanhaesebroeck, B., Waterfield, M. D., and Downward, J. (1996). Activation of phosphoinositide 3-kinase by interaction with Ras and by point mutation. *EMBO J.* **15**, 2442–2451.

7. Tolias, K. F. and Cantley, L. C. (1999). Pathways for phosphoinositide synthesis. *Chem. Phys. Lipids* **98**, 69–77.

8. Gillooly, D. J., Morrow, I. C., Lindsay, M., Gould, R., Bryant, N. J., Gaullier, J. M., Parton, R. G., and Stenmark, H. (2000). Localization of phosphatidylinositol 3-phosphate in yeast and mammalian cells. *EMBO J.* **19**, 4577–4588.

9. Lemmon, M. A. and Ferguson, K. M. (1998). Pleckstrin homology domains. *Curr. Top. Microbiol. Immunol.* **228**, 39–74.

10. Lietzke, S. E., Bose, S., Cronin, T., Klarlund, J., Chawla, A., Czech, M. P., and Lambright, D. G. (2000). Structural basis of 3-phosphoinositide recognition by pleckstrin homology domains. *Mol. Cell* **6**, 385–394.

11. Ferguson, K. M., Kavran, J. M., Sankaran, V. G., Fournier, E., Isakoff, S. J., Skolnik, E. Y., and Lemmon, M. A. (2000). Structural basis for discrimination of 3-phosphoinositides by pleckstrin homology domains. *Mol. Cell* **6**, 373–384.

12. Wishart, M. J., Taylor, G. S., and Dixon, J. E. (2001). Phoxy lipids: revealing PX domains as phosphoinositide binding modules. *Cell* **105**, 817–820.

13. Sato, T. K., Overduin, M., and Emr, S. D. (2001). Location, location, location: membrane targeting directed by PX domains. *Science* **294**, 1881–1885.

14. Kanai, F., Liu, H., Field, S. J., Akbary, H., Matsuo, T., Brown, G. E., Cantley, L. C., and Yaffe, M. B. (2001). The PX domains of p47phox and p40phox bind to lipid products of PI(3)K. *Nat. Cell Biol.* **3**, 675–678.

15. Ellson, C. D., Gobert-Gosse, S., Anderson, K. E., Davidson, K., Erdjument-Bromage, H., Tempst, P., Thuring, J. W., Cooper, M. A., Lim, Z. Y., Holmes, A. B., Gaffney, P. R., Coadwell, J., Chilvers, E. R., Hawkins, P. T., and Stephens, L. R. (2001). PtdIns(3)P regulates the neutrophil oxidase complex by binding to the PX domain of p40phox. *Nat. Cell Biol.* **3**, 679–682.

16. Song, X., Xu, W., Zhang, A., Huang, G., Liang, X., Virbasius, J. V., Czech, M. P., and Zhou, G. W. (2001). Phox homology domains specifically bind phosphatidylinositol phosphates. *Biochemistry* **40**, 8940–8944.

17. Xu, J., Liu, D., Gill, G., and Songyang, Z. (2001). Regulation of cytokine-independent survival kinase (CISK) by the Phox homology domain and phosphoinositides. *J. Cell Biol.* **154**, 699–705.

18. Bravo, J., Karathanassis, D., Pacold, C. M., Pacold, M. E., Ellson, C. D., Anderson, K. E., Butler, P. J., Lavenir, I., Perisic, O., Hawkins, P. T., Stephens, L., and Williams, R. L. (2001). The crystal structure of the PX domain from p40(phox) bound to phosphatidylinositol 3-phosphate. *Mol. Cell* **8**, 829–839.

19. Fruman, D. A., Rameh, L. E., and Cantley, L. C. Phosphoinositide binding domains: embracing 3-phosphate. (1999). *Cell* **97**, 817–820.

20. Krugmann, S., Anderson, K. E., Ridley, S. H., Risso, N., McGregor, A., Coadwell, J., Davidson, K., Eguinoa, A., Ellson, C. D., Lipp, P., Manifava, M., Ktistakis, N., Painter, G., Thuring, J. W., Cooper, M. A., Lim, Z. Y., Holmes, A. B., Dove, S. K., Michell, R. H., Grewal, A., Nazarian, A., Erdjument-Bromage, H., Tempst, P., Stephens, L. R., and Hawkins, P. T. (2002). Identification of ARAP3, a novel PI3K effector regulating both Arf and Rho GTPases, by selective capture on phosphoinositide affinity matrices. *Mol. Cell* **9**, 95–108.

21. Vanhaesebroeck, B., Jones, G. E., Allen, W. E., Zicha, D., Hooshmand-Rad, R., Sawyer, C., Wells, C., Waterfield, M. D., and Ridley, A. J. Distinct PI(3)Ks mediate mitogenic signalling and cell migration in macrophages. (1999). *Nat. Cell Biol.* **1**, 69–71.

22. Fruman, D. A., Mauvais-Jarvis, F., Pollard, D. A., Yballe, C. M., Brazil, D., Bronson, R. T., Kahn, C. R., and Cantley, L. C. (2000). Hypoglycaemia, liver necrosis and perinatal death in mice lacking all isoforms of phosphoinositide 3-kinase p85alpha. *Nat. Genet.* **26**, 379–382.

23. Fruman, D. A., Snapper, S. B., Yballe, C. M., Davidson, L., Yu, J. Y., Alt, F. W., and Cantley, L. C. (1999). Impaired B cell development and proliferation in absence of phosphoinositide 3-kinase p85alpha. *Science* **283**, 393–397.

24. Suzuki, H., Terauchi, Y., Fujiwara, M., Aizawa, S., Yazaki, Y., Kadowaki, T., and Koyasu, S. (1999). Xid-like immunodeficiency in mice with disruption of the p85alpha subunit of phosphoinositide 3-kinase. *Science* **283**, 390–392.

25. Dekker, L. V. and Segal, A. W. (2000). Perspectives: signal transduction. Signals to move cells. *Science* **287**, 982–983, 985.

26. Stephens, L., Ellson, C., and Hawkins, P. (2002). Roles of PI3Ks in leukocyte chemotaxis and phagocytosis. *Curr. Opin. Cell Biol.* **14**, 203–213.

27. Li, Z., Wahl, M. I., Eguinoa, A., Stephens, L. R., Hawkins, P. T., and Witte, O. N. (1997). Phosphatidylinositol 3-kinase-gamma activates Bruton's tyrosine kinase in concert with Src family kinases. *Proc. Natl. Acad. Sci. USA* **94**, 13820–13825.

28. Toker, A. and Newton, A. C. (2000). Cellular signaling: pivoting around PDK-1. *Cell* **103**, 185–188.

29. Cantley, L. C. and Neel, B. G. (1999). New insights into tumor suppression: PTEN suppresses tumor formation by restraining the phosphoinositide 3-kinase/AKT pathway. *Proc. Natl. Acad. Sci. USA* **96**, 4240–4245.

30. Rohrschneider, L. R., Fuller, J. F., Wolf, I., Liu, Y., and Lucas, D. M. (2000). Structure, function, and biology of SHIP proteins. *Genes Dev.* **14**, 505–520.

31. Chang, H. W., Aoki, M., Fruman, D., Auger, K. R., Bellacosa, A., Tsichlis, P. N., Cantley, L. C., Roberts, T. M., and Vogt, P. K. (1997). Transformation of chicken cells by the gene encoding the catalytic subunit of PI 3-kinase. *Science* **276**, 1848–1850.

32. Jimenez, C., Jones, D. R., Rodríguez-Viciana, P., Gonzalez-García, A., Leonardo, E., Wennström, S., von Kobbe, C., Toran, J. L., R.-Borlado, L., Calvo, V., Copin, S. G., Albar, J. P., Gaspar, M. L., Diez, E., Marcos, M. A. R., Downward, J., Martinez, A. C., Mérida, I., and Carrera, A. C. (1998). Identification and characterization of a new oncogene derived from the regulatory subunit of phosphoinositide 3-kinase. *EMBO J.* **17**, 743–753.

33. Borlado, L. R., Redondo, C., Alvarez, B., Jimenez, C., Criado, L. M., Flores, J., Marcos, M. A., Martinez, A. C., Balomenos, D., and Carrera, A. C. (2000). Increased phosphoinositide 3-kinase activity induces a lymphoproliferative disorder and contributes to tumor generation in vivo. *FASEB J.* **14**, 895–903.

34. Podsypanina, K., Ellenson, L. H., Nemes, A., Gu, J., Tamura, M., Yamada, K. M., Cordon-Cardo, C., Catoretti, G., Fisher, P. E., and Parsons, R. (1999). Mutation of Pten/Mmac1 in mice causes neoplasia in multiple organ systems. *Proc. Natl. Acad. Sci. USA* **96**, 1563–1568.

35. Di Cristofano, A., Kotsi, P., Peng, Y. F., Cordon-Cardo, C., Elkon, K. B., and Pandolfi, P. P. (1999). Impaired Fas response and autoimmunity in Pten+/− mice. *Science* **285**, 2122–2125.

36. Suzuki, A., Yamaguchi, M. T., Ohteki, T., Sasaki, T., Kaisho, T., Kimura, Y., Yoshida, R., Wakeham, A., Higuchi, T., Fukumoto, M., Tsubata, T., Ohashi, P. S., Koyasu, S., Penninger, J. M., Nakano, T., and Mak, T. W. (2001). T cell-specific loss of Pten leads to defects in central and peripheral tolerance. *Immunity* **14**, 523–534.

37. Lu-Kuo, J. M., Fruman, D. A., Joyal, D. M., Cantley, L. C., and Katz, H. R. (2000). Impaired kit- but not FcepsilonRI-initiated mast

cell activation in the absence of phosphoinositide 3-kinase p85alpha gene products. *J. Biol. Chem.* **275**, 6022–6029.
38. Fukao, T., Yamada, T., Tanabe, M., Terauchi, Y., Ota, T., Takayama, T., Asano, T., Takeuchi, T., Kadowaki, T., Hata Ji, J., and Koyasu, S. (2002). Selective loss of gastrointestinal mast cells and impaired immunity in PI3K-deficient mice. *Nat. Immunol.* **3**, 295–304.
39. Guarente, L. and Kenyon, C. (2000). Genetic pathways that regulate ageing in model organisms. *Nature* **408**, 255–262.
40. Clement, S., Krause, U., Desmedt, F., Tanti, J. F., Behrends, J., Pesesse, X., Sasaki, T., Penninger, J., Doherty, M., Malaisse, W., Dumont, J. E., Le Marchand-Brustel, Y., Erneux, C., Hue, L., and Schurmans, S. (2001). The lipid phosphatase SHIP2 controls insulin sensitivity. *Nature* **409**, 92–97.
41. Terauchi, Y., Tsuji, Y., Satoh, S., Minoura, H., Murakami, K., Okuno, A., Inukai, K., Asano, T., Kaburagi, Y., Ueki, K., Nakajima, H., Hanafusa, T., Matsuzawa, Y., Sekihara, H., Yin, Y., Barrett, J. C., Oda, H., Ishikawa, T., Akanuma, Y., Komuro, I., Suzuki, M., Yamamura, K., Kodama, T., Suzuki, H., Kadowaki, T. *et al.* (1999). Increased insulin sensitivity and hypoglycaemia in mice lacking the p85 alpha subunit of phosphoinositide 3-kinase. *Nat. Genet.* **21**, 230–235.
42. Fruman, D. A., Mauvais-Jarvis, F., Pollard, D. A., Yballe, C. M., Brazil, D., Bronson, R. T., Kahn, C. R., and Cantley, L. C. (2000). Hypoglycaemia, liver necrosis and perinatal death in mice lacking all isoforms of phosphoinositide 3-kinase p85alpha. *Nat. Genet.* **26**, 379–382.
43. Ueki, K., Yballe, C. M., Brachmann, S. M., Vicent, D., Watt, J. M., Kahn, C. R., and Cantley, L. C. (2002). Increased insulin sensitivity in mice lacking p85beta subunit of phosphoinositide 3-kinase. *Proc. Natl. Acad. Sci. USA* **99**, 419–424.
44. Ueki, K., Fruman, D. A., Brachmann, S. M., Tseng, Y. H., Cantley, L. C., and Kahn, C. R. (2002). Molecular balance between the regulatory and catalytic subunits of phosphoinositide 3-kinase regulates cell signaling and survival. *Mol. Cell Biol.* **22**, 965–977.
45. Bi, L., Okabe, I., Bernard, D. J., Wynshaw-Boris, A., and Nussbaum, R. L. (1999). Proliferative defect and embryonic lethality in mice homozygous for a deletion in the p110alpha subunit of phosphoinositide 3-kinase. *J. Biol. Chem.* **274**, 10963–10968.
46. Weinkove, D. and Leevers, S. J. (2000). The genetic control of organ growth: insights from *Drosophila*. *Curr. Opin. Genet. Dev.* **10**, 75–80.
47. Shioi, T., Kang, P. M., Douglas, P. S., Hampe, J., Yballe, C. M., Lawitts, J., Cantley, L. C., and Izumo, S. (2000). The conserved phosphoinositide 3-kinase pathway determines heart size in mice. *EMBO J.* **19**, 2537–2548.
48. Sasaki, T., Irie-Sasaki, J., Jones, R. G., Oliveira-dos-Santos, A. J., Stanford, W. L., Bolon, B., Wakeham, A., Itie, A., Bouchard, D., Kozieradzki, I., Joza, N., Mak, T. W., Ohashi, P. S., Suzuki, A., and Penninger, J. M. (2000). Function of PI3Kgamma in thymocyte development, T cell activation, and neutrophil migration. *Science* **287**, 1040–1046.
49. Li, Z., Jiang, H., Xie, W., Zhang, Z., Smrcka, A. V., and Wu, D. (2000). Roles of PLC-beta2 and -beta3 and PI3Kgamma in chemoattractant-mediated signal transduction. *Science* **287**, 1046–1049.
50. Hirsch, E., Katanaev, V. L., Garlanda, C., Azzolino, O., Pirola, L., Silengo, L., Sozzani, S., Mantovani, A., Altruda, F., and Wymann, M. P. (2000). Central role for G protein-coupled phosphoinositide 3-kinase gamma in inflammation. *Science* **287**, 1049–1053.
51. Okkenhaug, K., Bilancio, A., Farjot, G., Priddle, H., Sancho, S., Peskett, E., Pearce, W., Meek, S. E., Salpekar, A., Waterfield, M. D., Smith, A. J., and Vanhaesebroeck, B. (2002). Impaired B and T cell antigen receptor signaling in p110delta PI 3-kinase mutant mice. *Science* **297**, 1031–1034.
52. Clayton, E., Bardi, G., Bell, S. E., Chantry, D., Downes, C. P., Gray, A., Humphries, L. A., Rawlings, D., Reynolds, H., Vigorito, E., and Turner, M. (2002). A crucial role for the p110delta subunit of phosphatidylinositol 3-kinase in B cell development and activation. *J. Exp. Med.* **196**, 753–763.
53. Jou, S. T., Carpino, N., Takahashi, Y., Piekorz, R., Chao, J. R., Wang, D., and Ihle, J. N. (2002). Essential, nonredundant role for the phosphoinositide 3-kinase p110delta in signaling by the B-cell receptor complex. *Mol. Cell. Biol.* **22**, 8580–8591.
54. Kishimoto, H., Hamada, K., Saunders, M. Backman, S., Sasaki, T., Nakano, T., Mak, T. W,. and Suzuki, A. (2003). Physiological functions of pten in mouse tissues. *Cell Struct. Funct.* **28**, 11–21.
55. Anzelon, A. N., Wu, H., and Rickert, R. C. (2003). Pten inactivation alters peripheral B lymphocyte fate and reconstitutes CD19 function. *Nat. Immunol.* **4**, 287–294.
56. Pende, M., Kozma, S. C., Jaquet, M., Oorschot, V., Burcelin, R., Le Marchand-Brustel, Y., Klumperman, J., Thorens, B., and Thomas, G. (2000). Hypoinsulinaemia, glucose intolerance and diminished beta-cell size in S6K1-deficient mice. *Nature* **408**, 994–997.
57. Shioi, T., McMullen, I. R., Kang, P. M., Douglas, P. S., Obata, T., Franke, T. F., Cantley, L. C., and Izumo, S. (2002). Akt/protein kinase B promotes organ growth in transgenic mice. *Mol. Cell Biol.* **22**, 2799–2809.
58. Crackower, M. A., Oudit, G. Y., Kozieradzki, I., Sarao, R., Sun, H., Sasaki, T., Hirsch, E., Suzuki, A., Shioi, T., Irie-Sasaki, J., Sah, R., Cheng, H. Y., Rybin, V. O., Lembo, G., Fratta, L., Oliveira-dos-Santos, A. J., Benovic, J. L., Kahn, C. R., Izumo, S., Steinberg, S. F., Wymann, M. P., Backx, P. H., and Penninger, J. M. (2002). Regulation of myocardial contractility and cell size by distinct PI3K-PTEN signaling pathways. *Cell* **110**, 737–749.

CHAPTER 147

PTEN/MTM Phosphatidylinositol Phosphatases

Knut Martin Torgersen, Soo-A Kim, and Jack E. Dixon

The Life Science Institute and Department of Biological Chemistry, University of Michigan Medical School, Ann Arbor, Michigan

PTEN

Introduction

PTEN (*p*hosphatase and *ten*sin homolog deleted on chromosome 10) was first identified as a tumor supressor gene localized on chromosome 10q23. PTEN mutations are found at high frequencies in certain tumors, including endometrial carcinomas, gliomas, and breast and prostate cancers. Furthermore, germline mutations in the *PTEN* gene are found in the related autosomal disorders Cowden disease and Lhermitte-Duclos and Bannayan-Zonana syndromes. Biochemical and genetic analyses of PTEN and its role in these diseases have placed it in a group of gatekeeper genes essential for controlling cell growth and development [1–3].

Activity and Function

PTEN is a member of the protein tyrosine phosphatase (PTP) superfamily of enzymes characterized by the invariant Cys-x_5-Arg (Cx_5R) active site motif (Table I). However, unlike other PTP superfamily enzymes, PTEN utilizes the lipid second messenger phosphatidylinositol 3,4,5-trisphosphate (PIP_3) as its substrate [4]. This places PTEN as a negative regulator of phosphatidyl inositol 3-kinase (PI3K) signaling [5]. PTEN has been reported to regulate signaling through Akt/PKB, PDK1, SGK1, and Rho GTPases and therefore as a modulator of a broad range of cellular processes [6,7]. Loss of PTEN function can lead to tumor development through defects in cell cycle regulation, apoptosis, or angiogenesis.

Homozygous $PTEN^{-/-}$; mice die before birth, and embryos display regions of increased proliferation and disturbed developmental patterning. Heterozygous $PTEN^{+/-}$ mice are viable but spontaneously develop various types of tumors [3]. Cells from both $PTEN^{-/-}$-mice and $PTEN^{+/-}$ have constitutively activated Akt and are resistant to apoptotic stimuli. A direct role of PTEN and its lipid phosphatase activity in the regulation of Akt has been demonstrated in several tumor cell lines [2,3]. Furthermore, $PTEN^{+/-}$ mice have a tendency of developing both T-cell lymphomas and autoimmune disorders. A role for PTEN in this postulated link between autoimmune disorders and cancers are further supported by studies of mice where PTEN is conditionally targeted in T cells [8].

Mice in which PTEN is conditionally deleted in neuronal brain cells develop macrocephaly as a result of increased cell numbers, decreased cell death, and enlarged soma size [9,10]. Targeted deletion early in brain development suggests a role for PTEN in controlling the proliferation and potency of stem cells, whereas restricted deletion of PTEN in postmitotic neurons does not result in increased cell proliferation, but rather causes a progressive enlargement of soma size resulting in enlarged cerebellum and seizures. It is of note that the abnormal phenotype of these mice resembles that of Lhermitte-Duclos disease, suggesting that loss of PTEN function is sufficient to cause this disease in humans.

Table I PTEN and Myotubularin-Related Genes in Human.

	Name	Active site	C-terminal domains	Length (aa)	Chromosome	Disease
PTEN	PTEN	IHCKAGKGRT	PDZ-binding	403	10q23.3	Multiple cancers, Cowden and Lhermitte-Duclos
	PTENP1	IHCKAGKGRT	PDZ-binding	>405	9p21	—
	PTENR1	IHCKGGKGRT	—	>408	—	—
	TPIP	IHCKGGKGRT	—	445	13	—
	TPTE	IHCKGGTDRT	—	551	21 and others	—
MTM	MTM1	VHCSDGWDRT	PDZ-binding	603	Xq28	Myotybular myopathy
	MTMR1	VHCSDGWDRT	PDZ-binding	669	Xq28	—
	MTMR2	VHCSDGWDRT	PDZ-binding	643	11q22	Charcot-Marie-Tooth 4B
	MTMR3	VHCSDGWDRT	FYVE	1199	22q12.2	—
	MTMR4	VHCSDGWDRT	FYVE	1195	17q22-23	—
	MTMR6	VHCSDGWDRT	—	>567	13q12	—
	MTMR7	VHCSDGWDRT	—	>574	8p22	—
	MTMR8	VHCSDGWDRT	—	704	Xq11.2-12	—
	MTMR5 (SBF1)	VGLEDGWDIT	PH	1930	22pter	—
	MTMR9 (LIP-STYX)	IHGTEGTDST	—	549	8	—
	MTMR10	LQEEEGRDLS	—	>451	15	—
	MTMR11	LQERGDRDLN	—	710	1	—
	MTMR12 (3-PAP)	LLEENASDLC	—	>637	5	—

Length of predicted protein products (amino acids) and chromosomal localization are listed for each gene. Conserved amino acids within predicted active site sequences are presented, including the catalytic cysteine (yellow) and arginine (light-blue), and non-catalytic basic (blue) and acidic (red) residues. Non-catalytic domains predicted for carboxy-terminal regions as well as related diseases are also listed.

A conserved role for PTEN as a PIP_3 phosphatase and negative regulator of PI3K signaling has also been demonstrated by genetic studies of *D. melanogaster* (dPTEN) and *C. elegans* (Daf-18). By balancing signals from the insulin receptor, dPTEN controls cell size and number in flies, whereas Daf-18 regulates metabolism and longevity in worms [11,12].

The crystal structure of PTEN has revealed several features that contribute to its unique substrate specificity [13]. A 4-residue insertion in one loop of the PTP domain results in the widening and extension of the catalytic pocket and enough space for the bulky PIP_3 headgroup. In addition, the two lysines (Lys125 and Lys128) within the Cx_5R active site sequence, as well as an upstream histidine (His93), coordinate the D1 and D5 phosphate groups of the inositol ring. Hence, the specificity of PTEN toward PIP_3 is generated by a larger active site pocket combined with the conserved residues within the Cx_5R active site. C-terminal to the PTP catalytic domain PTEN contains a Ca^{2+} independent C2 domain, two PEST sequences, and a PDZ-binding motif (Fig. 1). These domains are likely to play important roles in PTEN regulation (see below).

The human genome contains several PTEN-related genes, but so far little is known about their function. Most of these genes exhibit restricted expression pattern and/or subcellular localization different from PTEN and do not appear to regulate Akt phosporylation. In that respect it is interesting to note that these genes have a different active site sequence, which might suggest a different substrate specificity [14,15].

Regulation

The crystal structure of PTEN revealed an extensive interface between its PTP-domain and C2-domain, suggesting that membrane targeting and lipid phosphatase activity are interdependent [13]. This is further supported by the observation that mutations affecting this interface are frequently found in cancers [3]. *In vitro*, the C2-domain of PTEN binds phospholipids independent of Ca^{2+} and its structural characteristics predict a direct membrane association. Mutations in critical lipid binding residues inhibit the ability of PTEN to function as a tumor suppressor and cannot be rescued by artificial membrane targeting. Hence, both structural and functional analysis suggests that the C2-domain play a dual role of both membrane recruitment and positioning of the PTP-domain.

The extreme C-terminus of PTEN contains tandem PEST sequences and a consensus PDZ-binding domain. Whereas the regulatory role of the PEST sequences remain elusive, the PDZ-binding motif has been demonstrated to associate with several PDZ-domain containing proteins [3,16]. The identification of phosphorylation sites in the C-terminal tail of PTEN regulating PDZ-binding and

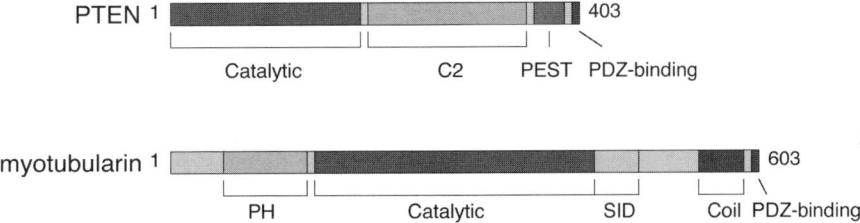

Figure 1 Structural features of PTEN and myotubularin phosphoinositide phosphatases. PTEN and myotubularin contain a catalytic domain that encompasses the CX_5R active site motif of PTP. In addition, both proteins possess several other domains/motifs that are likely to facilitate membrane association and protein-protein interactions. The C2 domain of PTEN is required for binding to lipid vesicles, whereas the carboxy-terminal PDZ-binding motif mediates interaction with PDZ-containing proteins. Phosphorylation in this region inhibits PDZ binding. Myotubularin contains a PH domain that may function to regulate membrane association. Furthermore, myotubularin contains a coiled coil motif as well as a putative PDZ-binding motif.

complex formation (Fig. 1) suggests additional levels of PTEN regulation [3,16].

Finally, several regulatory elements have recently been identified in the PTEN promoter, including binding sites for the tumor suppressors p53, early growth response-1 (Egr-1), and the perioxisome proliferator–activator receptor γ (PPARγ) [17–19]. The inducible transactivation of PTEN by these genes leads to reduced Akt activity and increased cell survival.

Myotubularin: a Novel Family of Phosphatidylinositol Phosphatases

Myotubularin-related proteins constitute one of the largest and most highly conserved protein tyrosine phosphatase (PTP) subfamilies in eukaryotes [14,19]. The MTM family includes at least eight catalytically active proteins as well as five catalytically inactive proteins in human [14,20]. Phylogenetic analysis of MTM family proteins allow a division of myotubularin family onto six subgroups, which include the catalytically active MTM1/MTMR1/MTMR2, MTMR3/MTMR4, and MTMR6/MTMR7/MTMR8 enzymes, as well as MTMR5 (Sbf1), MTMR9 (LIP-STYX), and MTMR10/MTMR11/MTMR12 (3-PAP) inactive forms [14,20]. One gene from *D. melanogaster* and *C. elegans* corresponding to each of these subfamilies has been identified [14].

Phosphatase Activity

Myotubularin (MTM1), the first characterized member of this novel family, utilizes the lipid second messenger phosphatidylinositol 3-phosphate (PI(3)P) as a physiological substrate [21,22]. In addition, recent findings demonstrate that other MTM-related phosphatases MTMR1, MTMR2, MTMR3, MTMR4, and MTMR6 also dephosphorylate PI(3)P, a finding that suggests that activity toward this substrate is common to all active myotubularin family enzymes [23,24].

The consensus CX_5R active site motif of PTP/DSP (dual specificity protein phosphatase) is found in the myotubularin family proteins, and the sequence "CSDGWDR" is invariant within all members of the active phosphatase subgroups. Unlike PTEN, in which two lysine residues within its active site (**CKAGKGR**) contribute to substrate specificity by interacting with the D1 and D5 phosphates of PIP_3, two aspartic acid residues are found in myotubularin family phosphatases. It is possible that interactions between the active site aspartic acid residues and phoshoryl groups at either the D4 or D5 position of the inositol ring may contribute to the high degree of specificity for PI(3)P found in MTM family emzymes.

One of the most notable characteristics of the human MTM family is the existence of at least five catalytically inactive forms, which contain germline substitution in catalytically essential residues within the PTP active site motif (Table I). Myotubularin-related inactive forms may function to regulate PI(3)P levels by opposing the actions of myotubularin phosphatases or directly affect the activity and/or subcellular localization of their active MTM counterparts [25].

Myotubularin Family and Human Diseases

To date, two myotubularin-related proteins have been associated with human disease. The myotubularin gene on chromosome Xq28, *MTM1*, is mutated in X-linked myotubular myopathy (XLMTM), a severe congenital muscular disorder characterized by hypotonia and generalized muscle weakness in newborn males [26]. Myogenesis in affected individuals is arrested at a late stage of differentiation/maturation following myotube formation, and the muscle cells have a characteristically large centrally located nuclei [26].

Mutations in a second MTM family member, *MTMR2* on chromosome 11q22, have recently been shown to cause the neurodegenerative disorder, type 4B Charcot-Marie-Tooth disease (CMT4B) [27]. CMT4B is an autosomal recessive demyelinating neuropathy characterized by abnormally folded myelin sheaths and Schwann cell proliferation in peripheral nerves.

Because these two highly similar genes, *MTM1* and *MTMR2* (64 percent identity, 76 percent similarity) utilize the same physiologic substrate, have a ubiquitous expression pattern, and are mutated in diseases with different target tissues and pathological characteristics, myotubularin and

MTMR2 may be subjected to differential regulatory mechanisms that preclude functional redundancy. Although their specific physiological roles are not known, a recent study has shown that developmental expression and subcellular localization of myotubularin and MTMR2 are differentially regulated, resulting in their utilization of specific cellular pools of PI(3)P [23].

Structural Features

In addition to the phosphatase domain, myotubularin-related proteins possess several motifs known to mediate protein-protein interactions and lipid binding. A PH domain, which was previously defined as a GRAM domain in myotubularin, is present in the N-terminal region of all myotubularin family members, including the catalytically inactive MTMs (Fig. 1). Although the physiologic relevance of this domain is not known, its presence in the myotubularin family lipid phosphatases suggests a role in membrane targeting of these proteins. A coiled coil motif is also present in all family members (Fig. 1) and may play a role in the regulation of MTM proteins through interactions with protein effectors and/or subcellular location. Some myotubularin family members have additional lipid-binding domains. For example, MTMR3 and MTMR4 contain a FYVE domain, and MTMR5 has an additional PH domain in its C-terminal region (Table I).

Although the role of the PH and FYVE domains in MTM function has yet to be determined, it is possible that they serve as targeting motifs to direct the lipid phosphatase domains to specific subcellular environments where PI(3)P is abundant. The physiologic function of myotubularin and related proteins in cell development and signaling processes remains unknown. Studies directed toward clarifying the regulation of myotubularin-related enzymes, as well as identifying downstream effectors, will be of significant value in understanding their roles in cell signaling and development.

References

1. Cantley, L. C. and Neel, B. G. (1999). New insights into tumor suppression: PTEN suppresses tumor formation by restraining the phosphoinositide 3-kinase/Akt pathway. *Proc. Natl. Acad. Sci. USA* **96**, 4240–4245.
2. Di Cristofano, A. and Pandolfi, P. P. (2000). The multiple roles of PTEN in tumor suppression. *Cell* **100**, 387–390.
3. Maehama, T., Taylor, G. S., Dixon, J. E. (2001). PTEN and myotubularin: novel phosphoinositide phosphatases. *Annu. Rev. Biochem.* **70**, 247–279.
4. Maehama, T. and Dixon, J. E. (1998). The tumor suppressor PTEN/MMAC1, dephosphorylates the lipid second messenger, phosphatidylinositol 3,4,5-trisphosphate. *J. Biol. Chem.* **273**, 13375–13378.
5. Vanhaesebroeck, B., Leevers, S. J., Ahmadi, K., Timms, J., Katso, R., Driscoll, P. C., Woscholski, R., Parker, P. J., Waterfield, M. D. (2001). Synthesis and function of 3-phosphorylated inositol lipids. *Annu. Rev. Biochem.* **70**, 535–602.
6. Datta, S. R., Brunet, A., Greenberg, M. E. (1999). Cellular survival: a play in three Akts. *Genes Dev.* **13**, 2905–2927.
7. Toker, A. and Newton, A. C. (2000). Cellular signaling: pivoting around PDK-1. *Cell* **103**, 185–188.
8. Suzuki, A., Yamaguchi, M. T., Ohteki, T., Sasaki, T., Kaisho, T., Kimura, Y., Yoshida, R., Wakeham, A., Higuchi, T., Fukumoto, M., Tsubata, T., Ohashi, P. S., Koyasu, S., Penninger, J. M., Nakano, T., Mak, T. W. (2001). T cell-specific loss of Pten leads to defects in central and peripheral tolerance. *Immunity* **14**, 523–534.
9. Penninger, J. M. and Woodgett, J. (2001). Stem cells. PTEN—coupling tumor suppression to stem cells? *Science* **294**, 2116–2118.
10. Morrison, S. J. (2002). Pten-uating neural growth. *Nat. Med.* **8**,1618.
11. Edgar B. A. (1999). From small flies come big discoveries about size control. *Nat. Cell Biol.* **1**, E191–193.
12. Guarente, L., Kenyon, C. (2000). Genetic pathways that regulate ageing in model organisms. *Nature* **408**, 255–262.
13. Lee, J.-O., Yang, H., Georgescu, M.-M., Di Cristofano, A., Maehama, T., Shi Y., Dixon, J. E., Pandolfi, P., and Pavletich, N. P. (1999). Crystal structure of the PTEN tumor suppressor: implications for its phosphoinositide phosphatase activity and membrane association. *Cell* **99**, 323–334.
14. Wishart, M. J., Taylor, G. S., Slama, J. T., Dixon, J. E. (2001). PTEN and myotubularin phosphoinositide phosphatases: bringing bioinformatics to the lab bench. *Curr. Opin. Cell Biol.* **13**, 172–181.
15. Leslie, N. R., Downes, C. P. (2002). PTEN: The down side of PI 3-kinase signalling. *Cell Signal.* **14**, 285–295.
16. Stambolic, V., MacPherson, D., Sas, D., Lin, Y., Snow, B., Jang, Y., Benchimol, S., Mak, T. W. (2001). Regulation of PTEN transcription by p53. *Mol. Cell.* **8**, 317–325.
17. Virolle, T., Adamson, E. D., Baron, V., Birle, D., Mercola, D., Mustelin, T., de Belle, I. (2001). The Egr-1 transcription factor directly activates PTEN during irradiation-induced signalling. *Nat. Cell Biol.* **3**, 1124–1128.
18. Patel, L., Pass, I., Coxon, P., Downes, C. P., Smith, S. A., Macphee, C. H. (2001). Tumor suppressor and anti-inflammatory actions of PPARγ agonists are mediated via upregulation of PTEN. *Curr. Biol.* **11**, 764–768.
19. Laporte, J., Blondeau, F., Buj-Bello, A., Tentler, D., Kretz, C., Dahl, N., and Mandel, J.-L. (1998). Characterization of the myotubularin dual specificity phosphatase gene family from yeast to human. *Hum. Mol. Genet.* **7**, 1703–1712.
20. Laporte, J., Blondeau, F., Buj-Bello, A., and Mandel, J.-L. (2001). The myotubularin family: from genetic disease to phosphoinositide metabolism. *Trends Genet.* **17**, 221–228.
21. Taylor, G. S., Maehama, T., and Dixon, J. E. (2000). Myotubularin, a protein tyrosine phosphatase mutated in myotubular myopathy, dephosphorylates the lipid second messenger, phosphatidylinositol 3-phosphate. *Proc. Natl. Acad. Sci. USA* **97**, 8910–8915.
22. Blondeau, F., Laporte, J., Bodin, S., Superti-Furga, G., Payrastre, B., and Mandel, J.-L. (2000). Myotubularin, a phosphatase deficient in myotubular myopathy, acts on phosphatidylinositol 3-kinase and phosphatidylinositol 3-phosphate pathway. *Hum. Mol. Genet.* **9**, 2223–2229.
23. Kim, S.-A., Taylor, G. S., Torgersen, K. M., and Dixon, J. E. (2002). Myotubularin and MTMR2, phosphatidylinositol 3-phosphatases mutated in myotubular myopathy and type 4B Charcot-Marie-Tooth disease. *J. Biol. Chem.* **277**, 4526–4531.
24. Laporte J., Liaubet, L., Blondeau, F., Tronchere, H., Mandel, J. L., Payrastre, B. (2002). Functional redundancy in the myotubularin family. *Biochem. Biophys. Res. Commun.* **291**, 305–312.
25. Wishart, M. J. (2002). Styx/Dead phosphatases, in *Handbook of CellSignaling*, vol II. Transmission: Effectors and Cytosolic Events, Section B: Protein Dephosphorylation, Academic Press, San Diego.
26. Laporte, J., Biancalana, V., Tanner, S. M., Kress, W., Schneider, V., Wallgren-Pettersson, C., Herger, F., Buj-Bello, A., Blondeau, F., Liechti-Gallati, S., and Mandel, J.-L. (2000). MTM1 mutations in X-linked myotubular myopathy. *Human Mutation* **15**, 393–409.
27. Bolino, A., Muglia, M., Conforti, F. L., LeGuern, E., Salih, M. A. M., Georgiou, D.-M., Christodoulou, K., Hausmanowa-Petrusewicz, I., Mandich, P., Schenone, A., Gambardella, A., Bono, F., Quattrone, A., Devoto, M., and Monaco, A. P. (2000). Charcot-Marie-Tooth type 4B is caused by mutations in the gene encoding myotubularin-related protein-2. *Nat. Genet.* **25**, 17–19.

CHAPTER 148

SHIP Inositol Phosphate Phosphatases

Larry R. Rohrschneider
Fred Hutchinson Cancer Research Center,
Division of Basic Sciences,
Seattle, Washington

Introduction

The SHIP (*SH2* domain-containing *i*nositol 5-*p*hosphatase) class of cytoplasmic signaling proteins in higher eucaryotes currently includes a pair of distinct gene products, each encoding an N-terminal SH2 domain, a central amino acid region with inositol 5-phosphatase enzymatic activity, and a C-terminal tail region. The two proteins, named SHIP1 and SHIP2, designating their domain structure and sequence of discovery (gene symbols INPP5D and INPPL1, respectively), are currently the only known members of this family. However, extended family members include many more proteins with inositol phosphatase activity (such as PTEN, type II IP5P, OCRL, and the synaptojanins). A search of the human genome yields only a single orthologue for SHIP2 (Ch. 11q13.3, contig. NT_030106.2) whereas the human genomic sequence for SHIP1 (Ch. 2q37, contig. NT_030597.1) is still incomplete. This review will focus on the principal structural, biochemical, and biological features and familial relationships of the two, so far identified, SHIP proteins. Additional details can be found in recent reviews [1–3].

SHIP1 Structure, Expression, and Function

The 27 exons encoding the SHIP1 protein stretch along an approximately 100 kb region of the murine genome (Ch1, 57.0 CM, C4 to band C5), and are spliced into an approximately 5 kb mRNA as shown in Fig. 1 [4]. The largest SHIP1 protein product, encoded by the co-linear expression of all genomic exons, results in a 1190-amino acid protein termed SHIP1α. This prototypical product contains an N-terminal SH2 domain, an ~450 amino acid inositol 5-phosphatase enzymatic domain, and a C-terminal tail containing multiple motifs for binding potential effector proteins with PTB, SH2, and/or SH3 domains. The SH2 domain has binding specificity for the Y-phosphorylated YxxL motif [3], and the 5′-phosphatase enzymatic activity of the central domain converts PtdIns 3,4,5-P3 to PtdIns 3,4-P2 [5,6]. Either inositol 1,3,4,5-P4 or phosphatidylinositol 3,4,5-P3 can serve as substrate but must contain phosphate at the 3′ position, suggesting that the substrate for SHIP1 is the end product of PI3K activity. Within the C-tail region, notable are the two NPXY motifs, which, when tyrosine phosphorylated, interact with the PTB domain of Shc [6,7]. The NPNY motif also interacts with p85/PI3K, and the Y within this motif plus the three adjacent amino acids comprise the canonical YIGM, which binds the C-terminal SH2 domain of the p85 most avidly [8,9]. The adapter protein, Grb2, contains two SH3 domains, and at least one interacts avidly with SHIP1, probably via one of the "PxxP" motifs in the C-terminal tail region [5,10].

The apparent molecular mass of SHIP1α on SDS acrylamide gel electrophoresis is 145 kDa; however, a large number of additional SHIP-related proteins are detectable (by immunoprecipitation, for example). These additional proteins may be ascribed to spliced isoforms [3], usage of an alternative internal SHIP promoter [11], and C-terminal proteolysis, which affects each of the above protein products [12].

Three SHIP1α isoforms result from three distinct splicing reactions (see Fig. 1), and two complete cDNAs and their

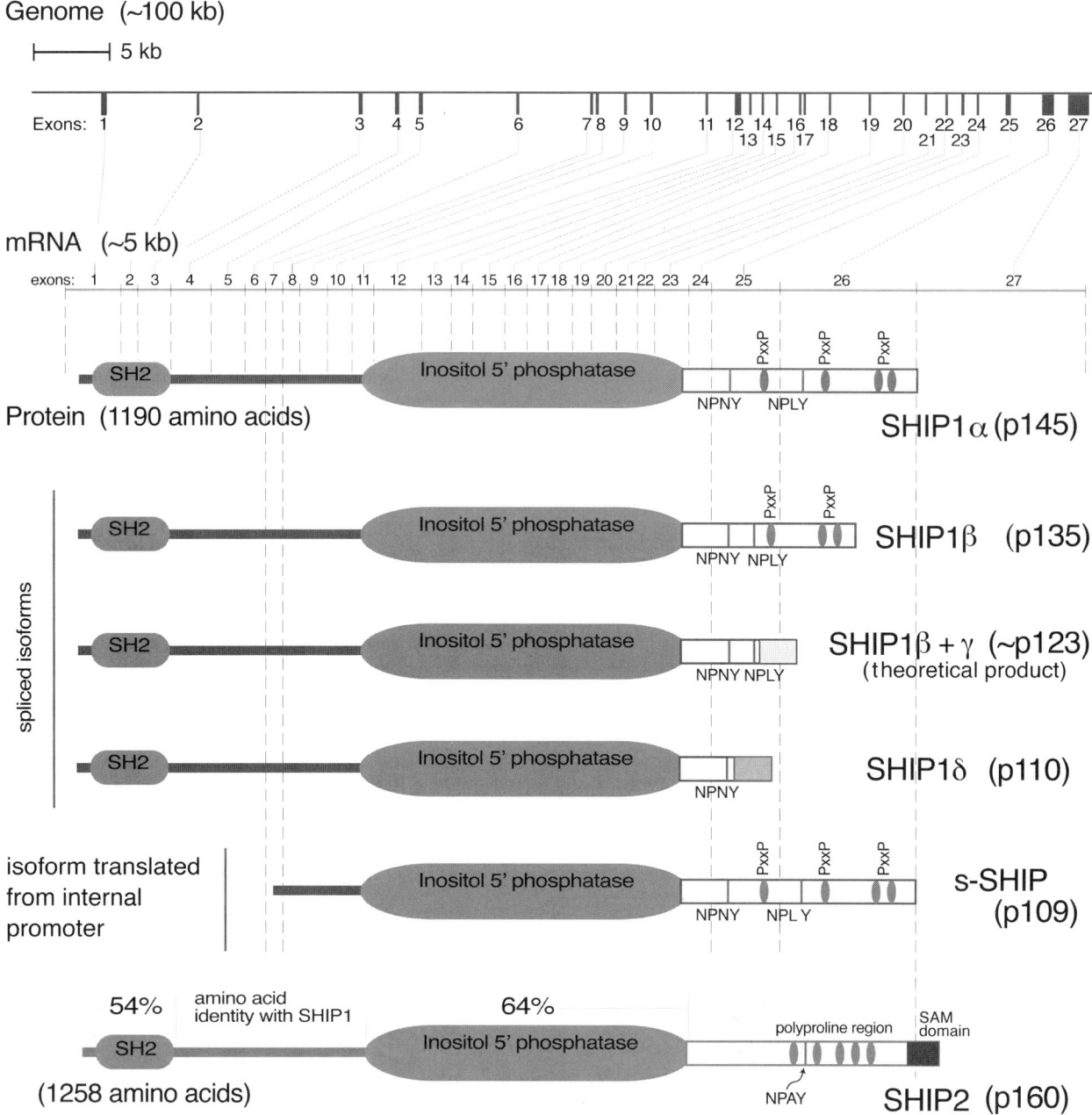

Figure 1 SHIP1 and SHIP2 proteins. The genomic organization of *ship1* is at the top with the mRNA and protein isoforms diagrammed below. At the bottom is the single known protein for SHIP2. See text for details.

protein products have been described for two of the spliced products (the β and δ isoforms) [4,9]. These splicing events, outlined in detail elsewhere [3], in general result in removal of one or several of the C-tail motifs required for binding PTB-, SH2-, or SH3-domain containing proteins. The splice within the β+γ and δ isoforms results in addition of new amino acid sequences at the C-terminal end. The biological function of each shorter isoform is not completely understood.

Recent experiments have established the existence of an SH2-less form of SHIP1 [11]; this isoform is termed s-SHIP (GenBank AF184912) for stem- or short-SHIP. This protein probably results from the usage of a potential promoter region within intron 5 [11]. Transcription of s-SHIP originates at least 44 nucleotides upstream of exon 6 and includes all downstream exons. Translation would probably not begin until exon 7, where the first ATG in the appropriate Kozak motifs is found. The β spliced product has been observed in s-SHIP (GenBank AF184913). In ES cells grown in LIF, s-SHIP is not tyrosine phosphorylated and is not associated with Shc. s-SHIP does, however, form a constitutive complex with Grb2 and is found in the membrane fraction of ES cells. The structure and expression (see below) of s-SHIP suggests it may have a function different from SHIP1 expressed in growth factor stimulated mature cells.

SHIP1 is expressed throughout hematopoietic cell development and tyrosine phosphorylated by a broad range of cytokines and growth factors of blood cells [3,13,14]. Adult uterus and kidney express SHIP1 detectable by RT-PCR [4], and immunohistochemistry shows strong testis expression [14]. SHIP1 is located within the seminiferous tubules of the testis and exhibits an interesting expressional relationship to both SHIP2, also in these structures of the testis, and the well-defined sequence for spermatozoa development in this organ (discussed in the SHIP2 section). Within the hematopoietic program for blood cell development, the largest SHIP1α product is found in more mature cells [9,13], especially in cell lines, and although numerous spliced isoforms are produced from the SHIP gene, their function and the cellular cues for their production are not understood. Some evidence suggests a developmental role [13]. The s-SHIP product is an exception, as good evidence exists for a function in primitive or stem cells of the blood [11]. s-SHIP is expressed only in very early progenitor or stem cells of the bone marrow and vanishes as cells mature. s-SHIP is also expressed in embryonic stem cell lines. The cDNA for s-SHIP predicts a "stem cell" promoter within intron 5 (see Fig. 1), and the correct size mRNA and protein are expressed in stem or progenitor cells. The exact function of s-SHIP in these cells is not known.

Gene knockout studies in mice and *in vitro* studies have convincingly reconfirmed the negative regulatory role of SHIP1 in myeloid cell development, mast cell activation, and antigen-induces B cell activation [15, 16]. SHIP1$^{-/-}$ mice exhibit a myeloproliferative disorder and inability to regulate mature blood cell functions [3]. Different molecular mechanisms can account for the negative regulatory role of SHIP in each cell type, but a few common themes are apparent (Fig. 2). One mechanism, initiated through receptor tyrosine kinases such as Kit or the M-CSF receptor, may regulate the activation of the survival factor Akt/PKB by eliminating the phosphatidylinositol lipid, PIP3, necessary for Akt/PKB activation (Fig. 2A). How SHIP is recruited into this pathway is not clear; however, one possibility is via the Gab-family of proteins. All three Gab proteins contain the consensus YxxL SHIP SH2 binding motif, and both Gab1 and Gab2 interact with SHIP after growth-factor receptor stimulation [17]. A second general mechanism is shown in Fig. 2B. Here the SHIP SH2 domain is recruited to an Ig-binding receptor (FcγRIIB in B cells and macrophages, FcεRI in mast cells). In B cells the FcγRIIB-SHIP complex terminates a positive signal from the B cell receptor (BCR) [18–20]. In mast cells aggregation of FcεRI is sufficient alone for degranulation, a step regulated by SHIP [21]. In contrast to the above mechanisms, the interaction of SHIP with DOK and RasGAP presents a negative regulatory mechanism altogether different [22,23]. RasGAP in this complex is sufficient to convert active RasGTP to the inactive GDP-bound form and attenuate the MAPK pathway (Fig. 2C). No doubt, additional negative regulatory mechanisms will be uncovered in the future, and it is unlikely that all will be mutually exclusive. Future tasks will be directed at understanding the cellular "when, where, and how" of these different mechanisms.

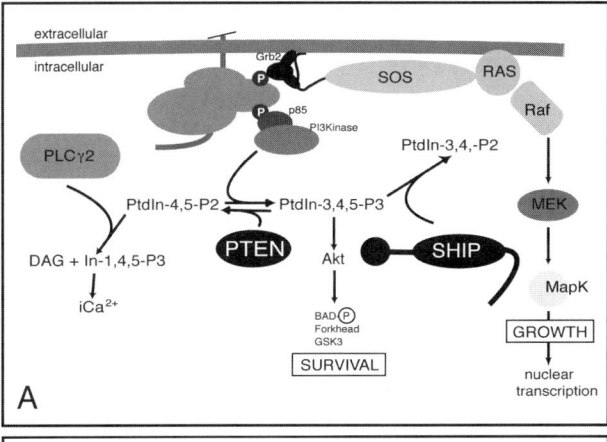

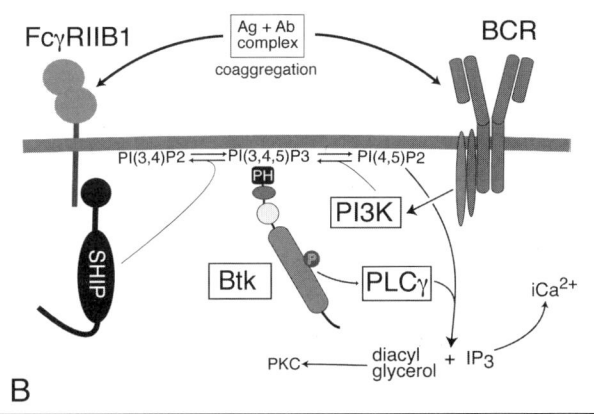

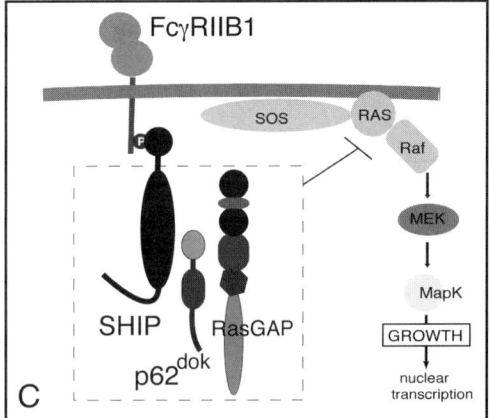

Figure 2 Mechanisms for the negative regulatory function of SHIP in cell signaling. See text for details.

SHIP2 Structure, Expression, and Function

The second member of the SHIP family, the SHIP2 protein, was first isolated by homology to the 51C protein [24]. 51C had been thought, incorrectly, to be a Fanconi anemia protein [25,26]. The 51C protein was also identified simultaneously with SHIP1 as containing an NPXY motif interacting with the PTB domain of Shc [6]. The murine SHIP2 gene is encoded in 29 exons [27]. The full sequence is complete for the mouse genome (AF162781). Translation is predicted to start in the second exon, which would encode part of the SH2 domain, and complete translation would produce a predicted protein of 142 kDa. Antibodies to the C-tail of the SHIP2 protein recognize a protein with the apparent mass of 160 kDa, a size corresponding roughly to the predicted full-length SH2 domain-containing protein [28].

Spliced isoforms of the SHIP2 protein have not been reported; however, if exons 3–29 alone were transcribed, they might encode an alternative protein product from this gene. The 51C protein might be such a product. The nucleotide sequence for the 51C cDNA (GenBank L36818) comprises exons 3–29 but contains 393 different nucleotides at the 5′ end. These nucleotides appear to be the intron immediately upstream of SHIP2 exon 3. If a protein were translated from this 51C mRNA, it would initiate translation at the methionine homologous with the same site in the s-SHIP protein. Therefore, the 51C cDNA could represent an s-SHIP version of the SHIP2 protein; however, it is also possible that this 51C cDNA is merely derived from incompletely-spliced mRNA.

SHIP2 contains, in general, a structure highly related to SHIP1 but the regions between the SH2 domain and the inositol 5-phosphatase domain and the C-tail region exhibit the least identity [24]. The specificity of the 5′-phosphatase enzymatic activity is similar to that in SHIP, but the inositol polyphosphates may be weaker substrates than the phosphatidylinositol analogues [29,30]. The C-tail region contains a single NPXY motif, several potential SH3-domain binding sites, and a C-terminal SAM (Sterile Alpha Motif) domain of yet unknown function in SHIP2 signaling. The amino acid sequence of the SH2 domain of SHIP2 suggests a binding specificity similar to that of SHIP1, and indeed, both are reported to interact with the phosphorylated immunoreceptor tyrosine-based inhibitory motif (ITIM) of FcγRIIB [31]. In addition, the SH2 domain of SHIP2 was found to bind tyrosine phosphorylated p130^{Cas} and therefore may have some role in the actin-based cytoskeletal reorganization accompanying cell spreading or migration [32].

Unlike SHIP1, SHIP2 is expressed in a broader range of cells and tissues of the mouse [24,26,27] and is present and constitutively tyrosine-phosphorylated in chronic myelogenous leukemia [30]. Brain and thymus exhibit the most prominent expression in both adult and 15.5 day embryos; liver expression was also highest in the embryo, but all other major embryonic or adult organs express some SHIP2. Testes express both SHIP1 and SHIP2 within the seminiferous tubules. Expression of SHIP2 is strongest at the periphery of the tubules where Sertoli cells and spermatogonia precursors reside. Also strongly positive are the mature spermatozoa occupying the inner portions of the seminiferous tubules. In contrast, SHIP1 expression is strongest in membranes of the developing spermatids located between the periphery and central core of the tubules. Therefore, the largely exclusive expression patterns of SHIP1 and SHIP2 in this tissue follow the developmental stages of spermatozoa production from the immature cells at the periphery of the seminiferous tubules to the mature cells in the central core. Expression of SHIP2 appears strongest in the most immature cells and decreases in spermatids while SHIP1 increases; expression levels again reverse in the mature spermatozoa.

Several growth factor receptors stimulate tyrosine phosphorylation of SHIP2 and activation of the Ras/Map kinase and Akt pathways [24,26,28]. The insulin receptor is extremely proficient at stimulating rapid and prolonged SHIP2 tyrosine phosphorylation and Akt/PKB activity. A negative regulatory role for SHIP2 in insulin-induced glucose uptake and glycogen synthesis has been shown by two independent methods. One study utilized wild-type and an inositol phosphatase-inactive mutant of SHIP2, demonstrating the requirement of the phosphatase activity in suppressing insulin-induced metabolic activities [33]. Another study generated SHIP2 knockout mice (lacking exons 18–29) and concluded that SHIP2 is necessary for the negative regulation of insulin signaling and sensitivity to insulin [34]. The homozygous mice lacking functional SHIP2 exhibited perinatal death, and heterozygous mice expressed symptoms of adult-onset diabetes mellitus. Additional abnormalities were not detected in the SHIP2 knockout mice, suggesting that negative regulation of insulin signaling may be a primary function of SHIP2.

References

1. Rohrschneider, L. R. et al. (2000). Structure, function, and biology of SHIP proteins. *Genes Dev.* **14**, 505–520.
2. Brauweiler, A. M., Tamir, I., and Cambier, J. C. (2000). Bilevel control of B-cell activation by the inositol 5-phosphatase SHIP. *Immunol. Rev.* **176**, 69–74.
3. Krystal, G. (2000). Lipid phosphatases in the immune system. *Sem. Immunol.* **12**, 397–403.
4. Wolf, I. et al. (2000). Cloning of the genomic locus of mouse SH2 containing inositol 5-phosphatase (SHIP) and a novel 110-kDa splice isoform, SHIPdelta. *Genomics* **69**, 104–112.
5. Damen, J. E. et al. (1996). The 145-kDa protein induced to associate with Shc by multiple cytokines is an inositol tetraphosphate and phosphatidylinositol 3,4,5-triphosphate 5-phosphatase. *Proc. Natl. Acad. Sci. USA* **93**, 1689–1693.
6. Lioubin, M. N. et al. (1996). p150Ship, a signal transduction molecule with inositol polyphosphate-5-phosphatase activity. *Genes Dev.* **10**, 1084–1095.
7. Lamkin, T. D. et al. (1997). Shc interaction with Src homology 2 domain containing inositol phosphatase (SHIP) *in vivo* requires the Shc-phosphotyrosine binding domain and two specific phosphotyrosines on SHIP. *J. Biol. Chem.* **272**, 10396–10401.
8. Gupta, N. et al. (1999). The SH2 domain-containing inositol 5′-phosphatase (SHIP) recruits the p85 subunit of phosphoinositide 3-kinase during FcgammaRIIb1-mediated inhibition of B cell receptor signaling. *J. Biol. Chem.* **274**, 7489–7494.
9. Lucas, D. M. and Rohrschneider, L. R. (1999). A novel spliced form of SH2-containing inositol phosphatase is expressed during myeloid development. *Blood* **93**, 1922–1933.
10. Kavanaugh, W. M. et al. (1996). Multiple forms of an inositol polyphosphate 5-phosphatase form signaling complexes with Shc and Grb2. *Curr. Biol.* **6**, 438–445.
11. Tu, Z. et al. (2001). Embryonic and hematopoietic stem cells express a novel SH2-containing inositol 5′-phosphatase isoform that partners with the Grb2 adapter protein. *Blood* **98**, 2028–2038.
12. Damen, J. E. et al. (1998). Multiple forms of the SH2-containing inositol phosphatase, SHIP, are generated by C-terminal truncation. *Blood* **92**, 1199–1205.
13. Geier, S. J. et al. (1997). The human SHIP gene is differentially expressed in cell lineages of the bone marrow and blood. *Blood* **89**, 1876–1885.

14. Liu, Q. *et al.* (1998). The SH2-containing inositol polyphosphate 5-phosphatase, ship, is expressed during hematopoiesis and spermatogenesis. *Blood* **91**, 2753–2759.
15. Helgason, C. D. *et al.* (1998). Targeted disruption of SHIP leads to hemopoietic perturbations, lung pathology, and a shortened life span. *Genes Dev.* **12**, 1610–1620.
16. Liu, Q. *et al.* (1999). SHIP is a negative regulator of growth factor receptor-mediated PKB/Akt activation and myeloid cell survival. *Genes Dev.* **13**, 786–791.
17. Liu, Y. *et al.* (2001). Scaffolding protein Gab2 mediates differentiation signaling downstream of Fms receptor tyrosine kinase. *Mol. Cell. Biol.* **21**, 3047–3056.
18. Chacko, G. W. *et al.* (1996). Negative signaling in B lymphocytes induces tyrosine phosphorylation of the 145-kDa inositol polyphosphate 5-phosphatase, SHIP. *J. Immunol.* **157**, 2234–2238.
19. Ono, M. *et al.* (1996). Role of the inositol phosphatase SHIP in negative regulation of the immune system by the receptor Fc(gamma)RIIB. *Nature* **383**, 263–266.
20. Tridandapani, S. *et al.* (1999). Protein interactions of Src homology 2 (SH2) domain-containing inositol phosphatase (SHIP): association with Shc displaces SHIP from FcgRIIb in B cells. *J. Immunol.* **162**, 1408–1414.
21. Huber, M. *et al.* (1999). The role of the SRC homology 2-containing inositol 5′-phosphatase in Fc epsilon R1-induced signaling. *Curr. Top. Micro. Immunol.* **244**, 29–41.
22. Yamanashi, Y. *et al.* (2000). Role of the rasGAP-associated docking protein p62(dok) in negative regulation of B cell receptor-mediated signaling. *Genes Dev.* **14**, 11–16.
23. Tamir, I. *et al.* (2000). The RasGAP-binding protein p62dok is a mediator of inhibitory FcgammaRIIB signals in B cells. *Immunity* **12**, 347–358.
24. Pesesse, X. *et al.* (1997). Identification of a second SH2-domain-containing protein closely related to the phosphatidylinositol polyphosphate 5-phosphatase SHIP. *Biochem. Biophys. Res. Comm.* **239**, 697–700.
25. Hejna, J. A. *et al.* (1995). Cloning and characterization of a human cDNA (INPPL1) sharing homology with inositol polyphosphate phosphatases. *Genomics* **29**, 285–287.
26. Habib, T. *et al.* (1998). Growth factors and insulin stimulate tyrosine phosphorylation of the 51C/SHIP2 protein. *J. Biol. Chem.* **273**, 18605–18609.
27. Schurmans, S. *et al.* (1999). The mouse SHIP2 (Inppl1) gene: complementary cDNA, genomic structure, promoter analysis, and gene expression in the embryo and adult mouse. *Genomics* **62**, 260–271.
28. Pesesse, X. *et al.* (2001). The Src homology 2 domain containing inositol 5-phosphatase SHIP2 is recruited to the epidermal growth factor (EGF) receptor and dephosphorylates phosphatidylinositol 3,4,5-trisphosphate in EGF-stimulated COS-7 cells. *J. Biol. Chem.* **276**, 28348–28355.
29. Pesesse, X. *et al.* (1998). The SH2 domain containing inositol 5-phosphatase SHIP2 displays phosphatidylinositol 3,4,5-triphosphate and inositol 1,3,4,5-tetrakisphosphate 5-phosphatase activity. *FEBS Lett.* **437**, 301–303.
30. Wisniewski, D. *et al.* (1999). The novel SH2-containing phosphatidylinositol 3,4,5-triphosphate 5-phosphatase (SHIP2) is constitutively tyrosine phosphorylated and associated with src homologous and collagen gene (SHC) in chronic myelogenous leukemia progenitor cells. *Blood* **93**, 2707–2720.
31. Bruhns, P. *et al.* (2000). Molecular basis of the recruitment of the SH2 domain-containing inositol 5-phosphatases SHIP1 and SHIP2 by fcgamma RIIB. *J. Biol. Chem.* **275**, 37357–37364.
32. Prasad, N., Topping, R. S., and Decker, S. J. (2001). SH2-containing inositol 5′-phosphatase SHIP2 associates with the p130(Cas) adapter protein and regulates cellular adhesion and spreading. *Mol. Cell. Biol.* **21**, 1416–1428.
33. Wada, T. *et al.* (2001). Overexpression of SH2-containing inositol phosphatase 2 results in negative regulation of insulin-induced metabolic actions in 3T3-L1 adipocytes via its 5′-phosphatase catalytic activity. *Mol. Cell. Biol.* **21**, 1633–1646.
34. Clement, S. *et al.* (2001). The lipid phosphatase SHIP2 controls insulin sensitivity. *Nature* **409**, 92–97.

CHAPTER 149

Structural Principles of Lipid Second Messenger Recognition

Roger L. Williams
*Medical Research Council,
Laboratory of Molecular Biology,
Cambridge, United Kingdom*

Introduction

Structural analyses have shown that domains with a variety of different folds can recognize a single type of lipid second messenger and that a single type of fold can evolve different binding sites and alternative modes of interaction for the same lipid second messenger. Specificity in lipid recognition is achieved by both electrostatic and shape complementarity. A common theme suggested by the structures of the lipid-modifying enzymes and the specific recognition modules is that secondary, nonspecific membrane interactions cooperate with specific lipid recognition to increase membrane avidity. Most binding domains have evolved mechanisms such as partial membrane penetration to bind lipids without removing them from the bilayer.

A wide range of lipid second messengers that are generated in response to external signals has been characterized in terms of their molecular biology. This review will focus on underlying structural principles involved in recognizing these messengers both by the enzymes that produce or consume them and by the downstream effector domains. In most cases, the lipid-modifying enzymes are structurally homologous to enzymes that catalyze an analogous reaction using soluble substrates, suggesting that the constraints imposed by the catalytic chemistry are a stronger determinant of fold than specific lipid binding. For example, the lipid kinases are homologous to protein kinases [1,2]. The phosphoinositide phosphatases are homologous to protein phosphatases and endonucleases [3,4]. The phosphoinositide-specific phospholipases C have a catalytic domain with a TIM-barrel fold similar to many other enzymes and an arrangement of catalytic residues similar to nucleases [5]. A variety of domains present in downstream effector proteins also specifically recognize the lipid second messengers. In contrast to the metabolizing enzymes, these effector domains typically bind lipids with higher affinity and have unique folds.

Phospholipid Second Messenger Recognition by Active Sites of Enzymes

The phosphoinositides are the most diverse family of lipid messengers. All of them share a phosphatidyl D-*myo*-inositol (PtdIns) scaffold that can be phosphorylated at all possible combinations of the 3-, 4-, and 5-hydroxyls to generate lipid messengers with specific roles in intracellular signaling. Several generalizations can be made regarding the recognition of phosphoinositides by proteins. The enzymes that recognize phosphoinositides as substrates tend to envelope the headgroup and make Van der Waals contacts with both faces of the inositol ring (Fig. 1). In contrast, the domains that have evolved simply to bind the phosphoinositides, such as PH domains, tend to form interactions with some or all of the phosphates but to leave one or both faces of the ring exposed (Fig. 2). Presumably, the tendency for the enzymes to more fully bury the headgroup arises from a necessity to exclude water from the active site or to more precisely position the reactive moieties in the active site. Within a family of domains, the affinity of phosphoinositide headgroup binding generally correlates with the number of hydrogen bonds between the

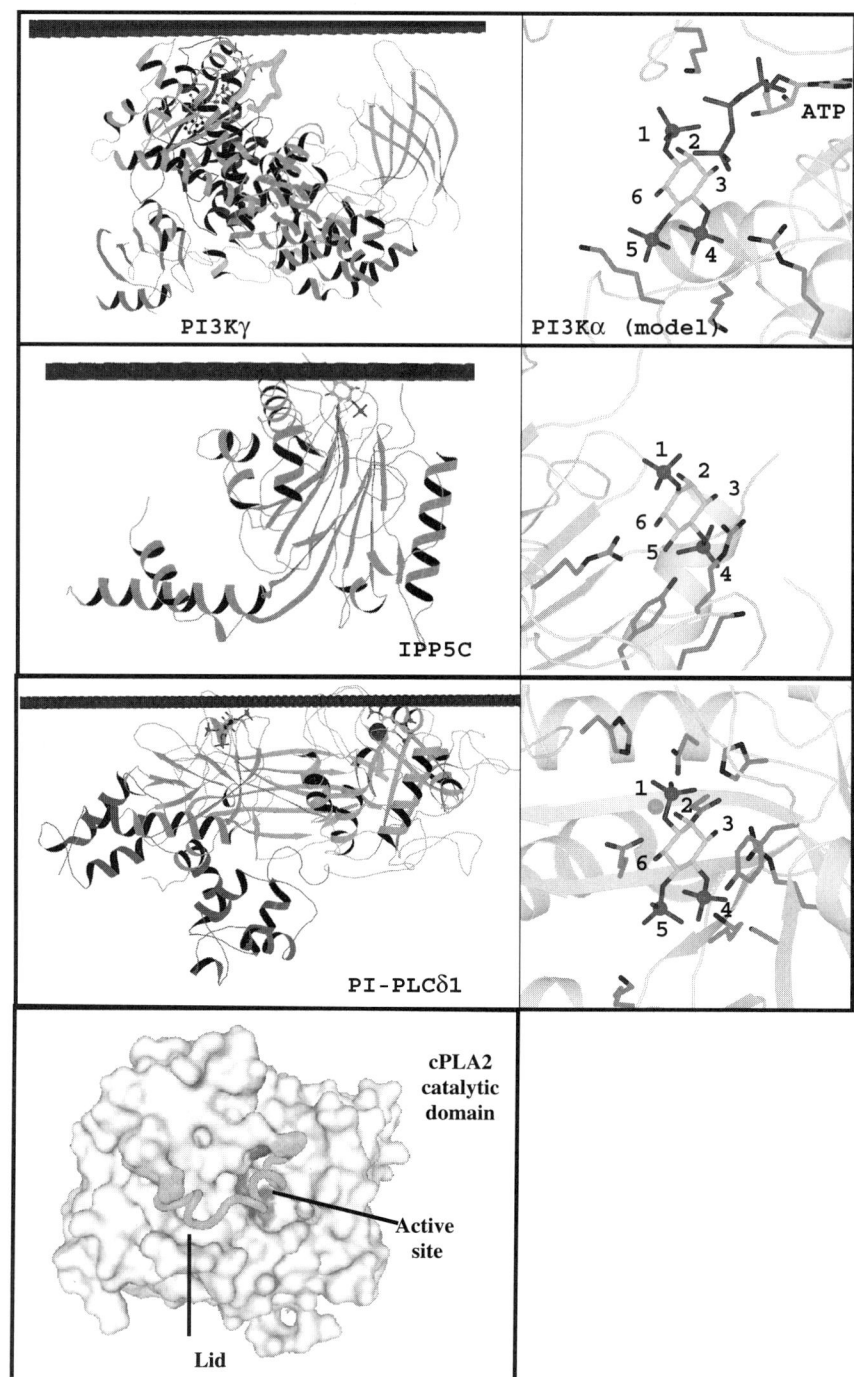

Figure 1 Lipid second messenger recognition by lipid-modifying enzymes. The left panels show the overall folds of the phosphoinositide-modifying enzymes with putative membrane-interacting regions placed in contact with a schematic membrane represented by a layer of spheres. The bound phosphoinositides are shown in stick representation. In the right panels, close-up views of the phosphoinositide/protein interactions are shown. The structures were optimally superimposed on the inositide moieties to present a common view. The molecular surface of cPLA$_2$'s catalytic domain is shown in the lower panel.

phosphoinositide and the protein. The enzymes generally have a lower affinity for the phosphoinositide headgroup than the highest affinity binding modules—as would be expected from the role of an enzyme to preferentially recognize the transition state rather than the substrate or the product.

Phosphoinositide 3-Kinase (PI3K). PI3Ks catalyze the phosphorylation of phosphoinositides at the 3-OH, giving rise to the second messengers PtdIns(3)P, PtdIns(3,4)P$_2$, and PtdIns(3,4,5)P$_3$. The structure of PI3Kγ, representative of both PI 3- and PI 4-kinases, has a catalytic domain with an

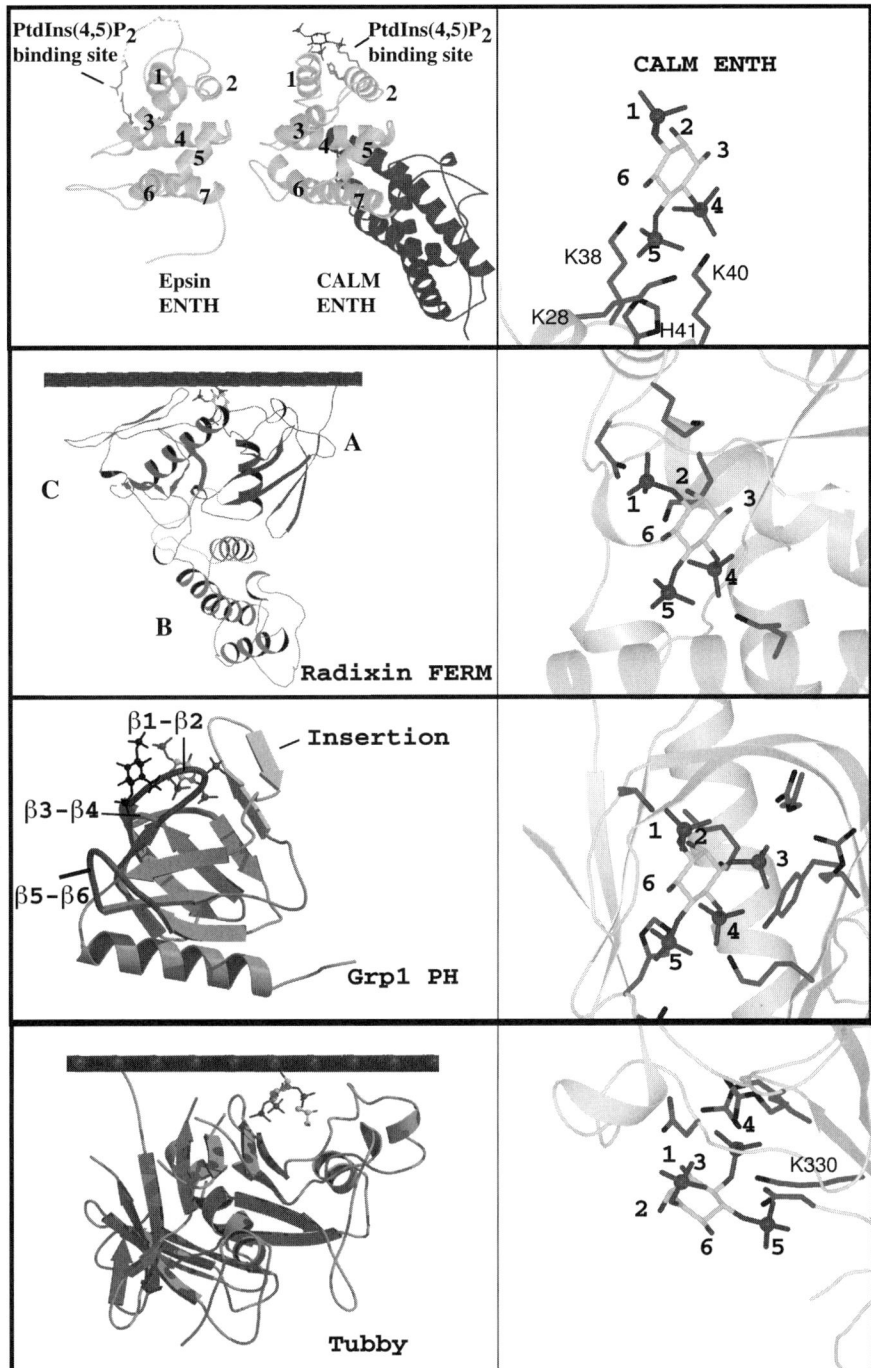

Figure 2 Recognition of polyphosphorylated phosphoinositides by specific binding modules. The representations are as in Fig. 1. In the first pair of panels, the PtdIns(4,5)P$_2$-binding sites of the epsin and CALM ENTH domains are illustrated. The structurally similar regions of the two domains are colored yellow. Part of the epsin PtdIns(4,5)P$_2$ site involves N-terminal residues that have been modeled (dotted lines). In the third pair of panels, Ins(1,3,4,5)P$_4$ bound to the Grp1 PH domain is shown (magenta phosphates). To illustrate the differences in the locations of the phosphoinositide binding pockets in β–spectrin and other PH domains such as Grp1, an Ins(1,4,5)P$_3$ (black) has been placed on the Grp1 domain in a location analogous to the β-spectrin binding pocket.

N-terminal lobe consisting of a five-stranded β-sheet closely related to protein kinases and a C-terminal lobe that is predominantly helical and more distantly related to protein kinases. The primary determinant of substrate preference for both PI3Ks is a region in the C-terminal lobe analogous to the activation loop of protein kinases [6,7]. Models of substrate binding proposed for PI3Ks place the phosphoinositide in a shallow pocket (Fig. 1) so that the 4- and 5-phosphates interact with basic residues in the activation loop, and the 1-phosphate contacts a Lys in a loop analogous to the glycine-rich loop of protein kinases (but without glycines in PI3Ks) [2,7]. As with PLCδ1, the location of the active site and accessory domains for membrane binding suggest that the enzyme interacts with the membrane in such a manner that substrate lipids do not have to be removed from the lipid bilayer (Fig. 1).

Phosphatidylinositol Phosphate 4- and 5-Kinases (PIPkins). PtdIns(4,5)P_2-mediated signal transduction is essential for cytoskeletal organization and dynamics, membrane trafficking, and apoptosis. Synthesis of PtdIns(4,5)P_2 is catalyzed by PIPkins [8]. The type IIβ PIPkin has an N-terminal lobe with a seven-stranded antiparallel β-sheet structurally related to protein kinases and a C-terminal lobe consisting of a smaller five-stranded β-sheet [1]. The PIPkins have a requirement for phosphorylated phosphoinositides due to a cluster of four conserved, basic residues in a putative phosphoinositide-binding pocket. The binding pocket is surprisingly shallow and open and suggests that there are few or no contacts with the 2- and 3-OH of the headgroup. The specificity of the enzyme for PtdIns(4)P *versus* PtdIns(5)P is completely dictated by a loop in the C-terminal lobe analogous to the activation loop of PI3K and protein kinases, and a single point mutation in this loop can swap the specificity [9].

PTEN, a 3-Phosphoinositde Phosphatase. Essential to any signal transduction system is a mechanism to produce second messengers and a mechanism to eliminate them. PTEN has a critical role in cells to antagonize the action of PI 3-kinases by catalyzing the dephosphorylation of the 3-phosphate from PtdIns(3,4,5)P_3. The structure of PTEN has a fold and active site configuration similar to the dual-specificity protein phosphatases [3]. A model for substrate binding places His 93 and Lys 128 as ligands of the 5-phosphate [3]. Although the 4-phosphate is deeply buried in the PTEN active site, there is no basic residue present with which it would associate. This is consistent with the ability of the enzyme to dephosphorylate PtdIns(3)P, PtdIns(3,4)P_2, and PtdIns(3,4,5)P_3. The 3-phosphate is also deeply buried, but, consistent with the presence of the scissile bond on this group, there is a basic residue, Arg 130, interacting with it.

Inositol Polyphosphate 5-Phosphatase (IPP5P). IPP5P plays an essential role in signaling by utilizing both inositol phosphates and phosphatidylinositol polyphosphates as substrates. The 5-phosphatases regulate the levels of both the soluble Ins(1,4,5)P_3 and the membrane-resident PtdIns(4,5)P_2. The structure of the catalytic domain of the synaptojanin IPP5P from *S. pombe* bound to the product of the reaction, Ins(4)P, shows an active site located at the bottom of a funnel-shaped depression containing the histidine essential for catalysis [4]. The catalytic mechanism is closely related to those of nucleases such as DNase I and DNase III. The 4-phosphate of the Ins(4)P interacts with three basic groups in the active site and makes water-mediated interactions with the divalent metal co-factor (Fig. 1). The product of the reaction binds in a catalytically nonproductive manner with the 4-phosphate remote from the catalytic histidine, thus showing why this family of enzymes is not able to use Ins(1,4)P_2 as a substrate.

Phosphoinositide-specific Phospholipase C (PI-PLC). PtdIns(4,5)P_2 is hydrolyzed by PI-PLC. The catalytic domain of the mammalian PLCδ1 consists of a $(β/α)_8$ barrel [5], a common architecture for enzymes in general. Principles of PtdIns(4,5)P_2 headgroup recognition by PI-PLC have been inferred from a complex of PLCδ1 with the product of the reaction, Ins(1,4,5)P_3. With the exception of the 6-OH of the headgroup, all of the hydroxyls of the bound inositide are stereospecifically recognized by the enzyme. The PtdIns(4,5)P_2 headgroup lodges edge-on in the binding pocket with the 3-OH at the bottom and the 1-OH at the top. This places the 1-OH at the level of the putative membrane-binding surface, suggesting that the enzyme does not remove substrate from the membrane during the catalytic cycle, similarly to most of the phosphoinositide-recognizing enzymes and binding domains (Fig. 1).

Cytosolic Phospholipase A_2. The phospholipase A_2 (PLA$_2$) family of enzymes hydrolyzes the *sn-2* bond of phospholipids to generate free fatty acids and lysophospholipids. The cytosolic PLA$_2$ (cPLA$_2$) selectively hydrolyzes phospholipids with an *sn-2* arachidonic acid and therefore has a key role in supplying the precursor for eicosanoid biosynthesis. cPLA$_2$ has an N-terminal C2 domain that is important for Ca^{2+}-dependent membrane translocation and a catalytic domain. The enzyme has a central β-sheet with an active-site nucleophile located in a portion of the structure analogous to the nucleophilic elbow of other phospholipases having an α/β hydrolase fold [10]. Apart from this feature, however, cPLA$_2$ has a quite divergent fold. Residues in the active site that are buried by a flexible lid accomplish recognition of the substrate. Upon binding to the membrane interface, this lid undergoes a conformational change to expose a wide hydrophobic platform surrounding a funnel-shaped pocket that cradles the substrate (Fig. 1). Even though the structure suggests that the catalytic domain partially penetrates into the hydrophobic portion of the lipid membrane, the cleft leading to the active-site nucleophile is deep enough to require that the substrate be removed from the lipid bilayer [10].

Phosphoinositide-binding Domains

Polyphosphorylated Phosphoinositide-binding Domains

ENTH Domain. Several proteins involved in endocytosis have an N-terminal domain of about 140 residues known as the ENTH domain, which is necessary for binding to PtdIns(4,5)P_2. The ENTH domains of CALM [11], AP180 [12], and epsin [13,14] consist of helices wound into a solenoid reminiscent of other helical domains such as armadillo and TPR. The PtdIns(4,5)P_2 binding sites in the CALM and epsin ENTH domains differ significantly (Fig. 2). The unique binding site of the CALM ENTH domain is on an exposed surface with the PtdIns(4,5)P$_2$ headgroup poised at the tips of three lysines (K28, K38, K40) and a histidine (H41) in helices α1 and α2 and the loop between them (Fig. 2). The residues involved in the interaction define a KX$_9$KX(K/R)(H/Y) motif that is present in other AP180 homologues but not in epsin [11]. The binding site in epsin involves basic residues in helices α3, α4 and a disordered N-terminal region that changes conformation upon lipid binding (Fig. 2) [14]. As was observed for PH domains, similarity in domain fold does not imply that the same region of the fold is used to interact with phosphoinositides.

The FERM Domain. The FERM domain is found in the ezrin/radixin/moesin (ERM) family of proteins as well as in talin, the erythrocyte band 4.1 protein, several tyrosine kinases and phosphatases, and the tumor suppressor merlin. Members of the ERM family of proteins have three structural domains, and the N-terminal FERM domain binds to PtdIns(4,5)P$_2$-containing membranes. Phospholipid binding is masked by an intra or intermolecular interaction between the C-terminal domain and the FERM domain. The FERM domain consists of three compact modules, A, B, and C [15–17]. Although the C module has an overall fold similar to PH domains that are known to bind PtdIns(4,5)P$_2$, the crystal structure of the radixin complex with the Ins(1,4,5)P$_3$ shows that the phosphoinositide binds between the A and C modules [18]. Two basic residues from the A module interact with the 4- and 5-phosphates and one from the C module interacts with the 1-phosphate (Fig. 2). The binding site is more open than most PtdIns(4,5)P$_2$ binding sites but less open than the ENTH-type PtdIns(4,5)P$_2$ binding site. PtdIns(4,5)P$_2$ binding causes conformational changes in the C module that prevent a self-association with the C-terminal tail of the protein and enable the N-terminal domain to interact with the cytosolic regions of integral membrane proteins. Mutagenesis suggests that the β5-β6 and β6-β7 loops in the C module may constitute a second PtdIns(4,5)P$_2$-binding site [19,20]. The phosphoinositide binding pocket defines part of a basic surface that is likely to be juxtaposed to the lipid bilayer, leaving an acidic groove between subdomains B and C free to interact with integral membrane adhesion proteins (Fig. 2) [18].

Tubby C-terminal DNA-binding Domain. A common feature among the tubby family proteins is the presence of a C-terminal DNA-binding domain with a unique fold consisting of a 12-stranded antiparallel β-barrel and a hydrophobic helix running through the barrel [21]. PtdIns(4,5)P$_2$ binding to the C-terminal domain causes Tubby to be localized to the plasma membrane until the levels of PtdIns(4,5)P$_2$ fall in response to receptor-mediated activation of PLC-β [21]. Loss of plasma-membrane localization is accompanied by nuclear translocation of the protein. The complex of the C-terminal domain of Tubby with glycerophosphoinositol 4,5-bisphosphate shows the PtdIns(4,5)P$_2$ headgroup in a shallow pocket that involves residues from three adjacent β-strands [21] and is located at one edge of the putative DNA-binding surface.

The side chain of a single Lys (330) intercalates between the 4- and 5-phosphates in a manner that is unique to Tubby and the CALM-N ENTH domains (Fig. 2). In these domains, Lys side chain approaches the 4- and 5-phosphates approximately parallel to the plane of the inositol ring. In Tubby, the Lys makes an unusually close (2.1 Å) contact with the 5-phosphate. An additional Arg that coordinates the 4-phosphate is also positioned so that 3-phosphorylated lipids could interact with it, which may account for the PtdIns(3,4,5)P$_3$ and PtdIns(3,4)P$_2$ binding observed *in vitro* [22].

PH Domains. PH domains are among the most common phosphoinositide-binding modules present in mammalian genomes and show a wide range of phosphoinositide affinities and specificities. They consist of two orthogonal β-sheets curving to form a barrel-like structure closed off by a C-terminal α-helix [8,23]. High-affinity binding to phosphoinositides is achieved using residues in the β1-β2 (VL1), β3-β4 (VL2), and β6-β7 (VL3) loops. The PtdIns(4,5)P$_2$-specific PH domain of PLC-δ1 [24] differs from other PH domains in that the orientation of the bound inositide is flipped by 180° so that the position occupied by the 5-phosphate in PLC-δ1 is occupied by the 3-phosphate in the 3-phosphoinositide-specific PH domains that have been characterized (Fig. 2).

Among PH domains recognizing 3-phosphoinositides, three types of specificities are apparent: PtdIns(3,4,5)P$_3$-specificity such as the PH domain of Grp1 and Btk, dual PtdIns(3,4,5)P$_3$/PtdIns(3,4)P$_2$-specificity such as the PH domain of DAPP1 and PKB and PtdIns(3,4)P$_2$-specificity such as the C-terminal PH domain of TAPP1. The PtdIns(3,4,5)P$_3$ specificity is achieved by enveloping the 5-phosphate by using insertions in either the β6-β7 loop (as in GRP1 [25,26]) or in the β1-β2 loop (as in Btk, [27]). DAPP1 makes more interactions with the 4-phosphate while the 5-phosphate is largely exposed. The PtdIns(3,4)P$_2$ specificity of the TAPP1 PH domain arises from steric clashes of the 5-phosphate with residues in the β1-β2 loop [28]. The analogous region of the closely related DAPP1 PH domain has a Gly that makes space to accommodate the 5-phosphate of PtdIns(3,4,5)P$_3$. Basic and hydrophobic residues in the β1-β2 loop of Grp1 and Btk suggest that these PH domains may have additional, nonspecific interactions with lipid bilayers that enhance membrane avidity (Fig. 2) [26].

Other modes of phosphoinositide binding have been shown for PH domains. The PH domain of β-spectrin uses the β5-β6 loop and the side of the β1-β2 loop opposite that used by PLC-δ1 to interact with Ins(1,4,5)P$_3$ [29], showing that the same fold can be adapted to several different binding modes (Fig. 2). The PH domain from β-spectrin is an example of a PH domain with low affinity and little specificity for lipid binding. More recent analyses of the genome suggest that this may be characteristic of the vast majority of PH domains [23].

PtdIns(3)P-binding Domains

PtdIns(3)P is present in mammalian cells at fairly high concentrations relative to such transient lipid second messengers as PtdIns(3,4,5)P$_3$. Its distribution in cells is restricted mainly to endosomal membranes. PtdIns(3)P levels can increase rapidly during certain processes such as receptor-mediated phagocytosis [30]. Two structurally unrelated domain types, FYVE and PX, are capable of specifically binding PtdIns(3)P [31].

FYVE Domains. The FYVE domains are found in many proteins involved in membrane transport [32]. The FYVE domains from Vps27 [33], Hrs [34], and EEA1 [35,36] consist of two small β-sheets stabilized by two Zn^{2+} ions and a C-terminal α-helix. The PtdIns(3)P forms hydrogen bonds with the protein by using the 1- and 3-phosphates and the 4-, 5-, and 6-OH groups [36]. The close approach of these hydrogen-bonding partners precludes polyphosphorylated phosphoinositides from binding (Fig. 3). The 3-phosphate forms a hydrogen bond with the last arginine in the (R/K) (R/K)HHCR signature motif characteristic of the FYVE domains. The 1-phosphate interacts with the protein *via* the first Arg of this motif. Like the PH domains, the FYVE domain buries only one face of the bound phosphoinositide. For EEA1, the face with the axial 2-OH is exposed to solution. The presence of the coiled-coil region preceding the EEA1 FYVE domain helps to unambiguously define the mode of membrane interaction and suggests that a loop flanking the PtdIns(3)P pocket, the "turret" loop, penetrates into the lipid bilayer (Fig. 3). Biophysical measurements indicate that this partial membrane penetration follows rather than precedes specific PtdIns(3)P binding [37].

PX Domains. PX domains are found in a wide range of proteins including many involved in lipid modification, intracellular signaling, and vesicle trafficking [38]. They consist of a three-stranded β-sheet subdomain and an α-helical subdomain that are joined by a conserved RR(Y/F) motif [39,40]. The structure of the PX domain from the p40 cytosolic subunit of the NADPH oxidase in a complex with PtdIns(3)P shows that the first Arg from the RR(Y/F) motif has a structural role in the core of the protein, while the second Arg and the Tyr residue interact with the 3-phosphate and the face of the inositide ring, respectively [40] (Fig. 3). The PX domain buries the face of the inositide adjacent to the axial 2-OH, leaving the opposite face largely exposed. The mode of membrane binding of the PX domain is suggested by the diacylglycerol moiety of the bound PtdIns(3)P and hydrophobic residues adjacent to the phosphoinositide binding pocket (Fig. 3).

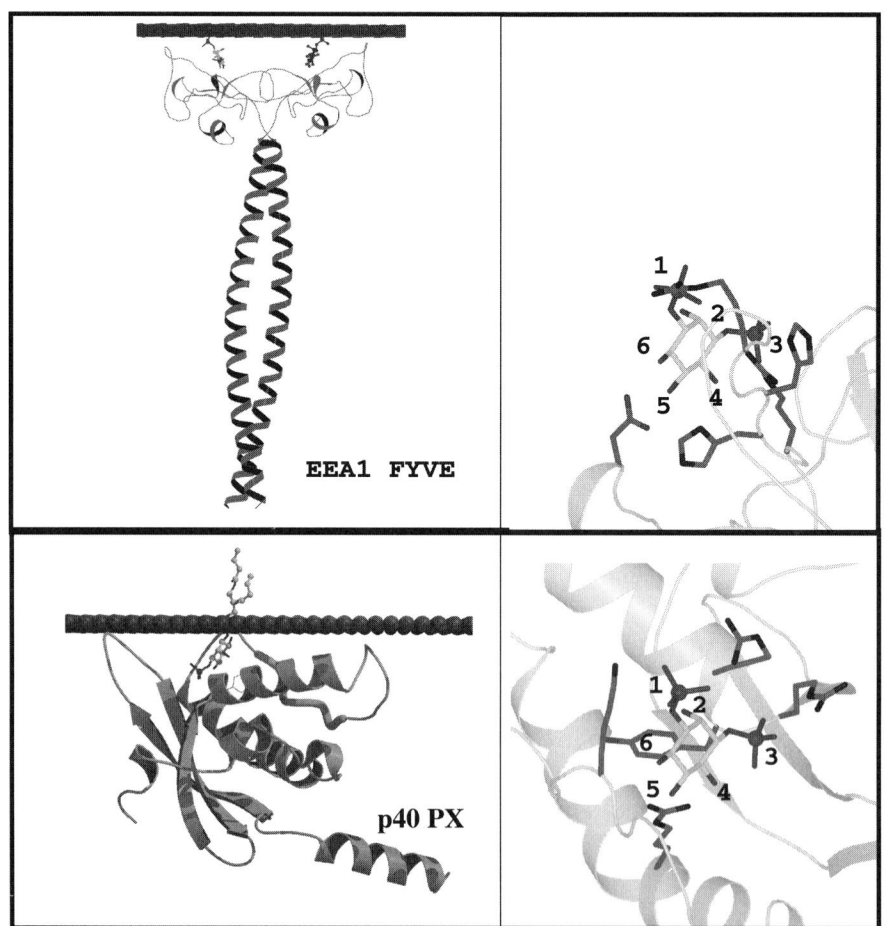

Figure 3 PtdIns(3)P recognition by specific binding modules. The representations are as in Fig. 1

Non-phosphoinositide Lipid Messenger Recognition

C1 Domains. The C1 domain is essential for membrane localization and activation of many proteins involved in signal transduction, including the protein kinase C isozymes [41]. The C1 domains are 50-residue modules containing two small β-sheets and a short C-terminal helix. The domains have been classified into two groups, the "typical" domains that fit a profile derived for phorbol ester or diacylglycerol (DAG) binding and the "atypical" domains that do not [42]. The phorbol ester sits in a groove that is formed by a splaying of adjacent β-strands in a sheet [43,44]. Hydrophilic groups on the phorbol ester intercalate between the strands and make backbone interactions with their exposed main-chain atoms. Once the phorbol ester is bound, the entire end of the domain presents a hydrophobic surface that penetrates into the lipid bilayer. Available binding data are consistent with a model in which the DAG fits into the same groove as the phorbol ester, forming hydrogen bonds with the main-chain atoms of the strands using its 3-OH.

Future Directions

Although much progress has been made in defining the nature of the interactions of lipid second messengers with proteins, many questions remain unanswered. Several lipid second messengers have been characterized for which there is no structural information about specific binding modules, e.g. PtdIns(3,5)P_2 and phosphatidic acid. A dimension of response to lipid-messenger recognition that remains largely unexplored is the effect of membrane binding on membrane structure during processes such as formation of multivesicular bodies. Many proteins use multiple weak interactions to bind to membranes in response to lipid second messengers, but an analysis of the energetics of the individual interactions is often lacking. Although membrane translocation in response to lipid second messengers is common, the nature and extent of allosteric responses mediated by membrane interactions are not clear. With methodologies that have emerged in the wake of genomic studies, we can look forward to answers to many of these questions in the near future.

Note Added in Proof

The details of the PtdIns(4,5) P_2-binding site of the epsin ENTH domain were described in the report of Ford, *et al* [45].

Acknowledgments

I apologize to colleagues whose work I was unable to cite given the wide scope of the review and the severe limitations on space. Marketa Zvelebil is thanked for coordinates of the PI3Kα model.

References

1. Rao, V. D., Misra, S., Boronenkov, I. V., Anderson, R. A., and Hurley, J. H. (1998). Structure of type IIβ phosphatidylinositol phosphate kinase: a protein kinase fold flattened for interfacial phosphorylation. *Cell* **94**, 829–839.
2. Walker, E. H., Perisic, O., Ried, C., Stephens, L., and Williams, R. L. (1999). Structural insights into phosphoinositide 3-kinase catalysis and signalling. *Nature* **402**, 313–320.
3. Lee, J. O., Yang, H., Georgescu, M. M., Di Cristofano, A., Maehama, T. *et al.* (1999). Crystal structure of the PTEN tumor suppressor: implications for its phosphoinositide phosphatase activity and membrane association. *Cell* **99**, 323–334.
4. Tsujishita, Y., Guo, S., Stolz, L. E., York, J. D., and Hurley, J. H. (2001). Specificity determinants in phosphoinositide dephosphorylation: crystal structure of an archetypal inositol polyphosphate 5-phosphatase. *Cell* **105**, 379–389.
5. Essen, L.-O., Perisic, O., Cheung, R., Katan, M., and Williams, R. L. (1996). Crystal structure of a mammalian phosphoinositide-specific phospholipase Cδ. *Nature,* **380**, 595–602.
6. Bondeva, T., Pirola, L., Bulgarelli-Leva, G., Rubio, I., Wetzker, R. *et al.* (1998). Bifurcation of lipid and protein kinase signals of PI3Kγ to the protein kinases PKB and MAPK. *Science* **282**, 293–296.
7. Pirola, L., Zvelebil, M. J., Bulgarelli-Leva, G., Van Obberghen, E., Waterfield, M. D. *et al.* (2001). Activation loop sequences confer substrate specificity to phosphoinositide 3-kinase α (PI3Kα). *J. Biol. Chem.* **276**, 21544–21554.
8. Hurley, J. H. and Misra, S. (2000). Signaling and subcellular targeting by membrane-binding domains. *Annu. Rev. Biophys. Biomol. Struct.* **29**, 49–79.
9. Kunz, J., Fuelling, A., Kolbe, L., and Anderson, R. A. (2002). Stereospecific substrate recognition by phosphatidylinositol phosphate kinases is swapped by changing a single amino acid residue. *J. Biol. Chem.* **277**, 5611–5619.
10. Dessen, A., Tang, J., Schmidt, H., Stahl, M., Clark, J. D. *et al.* (1999). Crystal structure of human cytosolic phospholipase A_2 reveals a novel topology and catalytic mechanism. *Cell* **97**, 349–360.
11. Ford, M. G. J., Pearse, B. M. F., Higgins, M. K., Vallis, Y., Owen, D. J. *et al.* (2001). Simultaneous binding of PtdIns(4,5)P_2 and clathrin by AP180 in the nucleation of clathrin lattices on membranes. *Science* **291**, 1051–1055.
12. Mao, Y., Chen, J., Maynard, J. A., Zhang, B., and Quiocho, F. A. (2001). A novel all helix fold of the AP180 amino-terminal domain for phosphoinositide binding and clathrin assembly in synaptic vesicle endocytosis. *Cell* **104**, 433–440.
13. Hyman, J., Chen, H., Di Fiore, P. P., De Camilli, P., and Brunger, A. T. (2000). Epsin 1 undergoes nucleocytosolic shuttling and its eps15 interactor NH(2)-terminal homology (ENTH) domain, structurally similar to Armadillo and HEAT repeats, interacts with the transcription factor promyelocytic leukemia $Zn^{(2)+}$ finger protein (PLZF). *J. Cell Biol.* **149**, 537–546.
14. Itoh, T., Koshiba, S., Kigawa, T., Kikuchi, A., Yokoyama, S. *et al.* (2001). Role of the ENTH domain in phosphatidylinositol-4,5-bisphosphate binding and endocytosis. *Science* **291**, 1047–1051.
15. Shimizu, T., Seto, A., Maita, N., Hamada, K., Tsukita, S. *et al.* (2001). Structural basis for neurofibromatosis type 2: Crystal structure of the merlin FERM domain. *J. Biol. Chem.* **277**, 10332–10336.
16. Edwards, S. D. and Keep, N. H. (2001). The 2.7 A crystal structure of the activated FERM domain of moesin: an analysis of structural changes on activation. *Biochemistry* **40**, 7061–7068.
17. Pearson, M. A., Reczek, D., Bretscher, A., and Karplus, P. A. (2000). Structure of the ERM protein moesin reveals the FERM domain fold masked by an extended actin binding tail domain. *Cell* **101**, 259–270.
18. Hamada, K., Shimizu, T., Matsui, T., Tsukita, S., Tsukita, S. *et al.* (2000). Structural basis of the membrane-targeting and unmasking mechanisms of the radixin FERM domain. *EMBO J.* **19**, 4449–4462.

19. Barret, C., Roy, C., Montcourrier, P., Mangeat, P., and Niggli, V. (2000). Mutagenesis of the phosphatidylinositol 4,5-bisphosphate (PIP(2)) binding site in the NH(2)-terminal domain of ezrin correlates with its altered cellular distribution. *J Cell Biol.* **151**, 1067–1080.
20. Niggli, V. (2001). Structural properties of lipid-binding sites in cytoskeletal proteins. *Trends Biochem. Sci.* **26**, 604–611.
21. Boggon, T. J., Shan, W. S., Santagata, S., Myers, S. C., and Shapiro, L. (1999). Implication of tubby proteins as transcription factors by structure-based functional analysis. *Science* **286**, 2119–2125.
22. Santagata, S., Boggon, T. J., Baird, C. L., Gomez, C. A., Zhao, J. *et al.* (2001). G-protein signaling through tubby proteins. *Science* **292**, 2041–2050.
23. Lemmon, M. A., Ferguson, K. M., and Abrams, C. S. (2002). Pleckstrin homology domains and the cytoskeleton. *FEBS Lett.* **513**, 71–76.
24. Ferguson, K. M., Lemmon, M. A., Schlessinger, J., and Sigler, P. B. (1995). Structure of the high affinity complex of inositol trisphosphate with a phospholipase C pleckstrin homology domain. *Cell* **83**, 1037–1046.
25. Ferguson, K. M., Kavran, J. M., Sankaran, V. G., Fournier, E., Isakoff, S. J. *et al.* (2000). Structural basis for discrimination of 3-phosphoinositides by pleckstrin homology domains. *Mol. Cell* **6**, 373–384.
26. Lietzke, S. E., Bose, S., Cronin, T., Klarlund, J., Chawla, A. *et al.* (2000). Structural basis of 3-phosphoinositide recognition by pleckstrin homology domains. *Mol. Cell* **6**, 385–394.
27. Baraldi, E., Carugo, K. D., Hyvonen, M., Surdo, P. L., Riley, A. M. *et al.* (1999). Structure of the PH domain from Bruton's tyrosine kinase in complex with inositol 1,3,4,5-tetrakisphosphate. *Structure Fold. Des.* **7**, 449–460.
28. Thomas, C. C., Dowler, S., Deak, M., Alessi, D. R., and van Aalten, D. M. (2001). Crystal structure of the phosphatidylinositol 3,4-bisphosphate-binding pleckstrin homology (PH) domain of tandem PH-domain-containing protein 1 (TAPP1): molecular basis of lipid specificity. *Biochem. J.* **358**, 287–294.
29. Hyvonen, M., Macias, M. J., Nilges, M., Oschkinat, H., Saraste, M. *et al.* (1995). Structure of the binding site for inositol phosphates in a PH domain. *EMBO J.* **14**, 4676–4685.
30. Vieira, O. V., Botelho, R. J., Rameh, L., Brachmann, S. M., Matsuo, T. *et al.* (2001). Distinct roles of class I and class III phosphatidylinositol 3-kinases in phagosome formation and maturation. *J. Cell Biol.* **155**, 19–25.
31. Misra, S., Miller, G. J., and Hurley, J. H. (2001). Recognizing phosphatidylinositol 3-phosphate. *Cell* **107**, 559–562.
32. Gillooly, D. J., Simonsen, A., and Stenmark, H. (2001). Cellular functions of phosphatidylinositol 3-phosphate and FYVE domain proteins. *Biochem. J.* **355**, 249–258.
33. Misra, S. and Hurley, J. H. (1999). Crystal structure of a phosphatidylinositol 3-phosphate-specific membrane-targeting motif, the FYVE domain of Vps27p. *Cell* **97**, 657–666.
34. Mao, Y., Nickitenko, A., Duan, X., Lloyd, T. E., Wu, M. N. *et al.* (2000). Crystal structure of the VHS and FYVE tandem domains of Hrs, a protein involved in membrane trafficking and signal transduction. *Cell* **100**, 447–456.
35. Kutateladze, T. and Overduin, M. (2001). Structural mechanism of endosome docking by the FYVE domain. *Science* **291**, 1793–1796.
36. Dumas, J. J., Merithew, E., Sudharshan, E., Rajamani, D., Hayes, S. *et al.* (2001). Multivalent endosome targeting by homodimeric EEA1. *Mol. Cell* **8**, 947–958.
37. Stahelin, R. V., Long, F., Diraviyam, K., Bruzik, K. S., Murray, D. *et al.* (2002). Phosphatidylinositol 3-phosphate induces the membrane penetration of the FYVE domains of Vps27p and Hrs. *J. Biol. Chem.* **277**, 26379–26388.
38. Wishart, M. J., Taylor, G. S., and Dixon, J. E. (2001). Phoxy lipids: revealing PX domains as phosphoinositide binding modules. *Cell* **105**, 817–820.
39. Hiroaki, H., Ago, T., Ito, T., Sumimoto, H., and Kohda, D. (2001). Solution structure of the PX domain, a target of the SH3 domain. *Nat. Struct. Biol.* **8**, 526–530.
40. Bravo, J., Karathanassis, D., Pacold, C. M., Pacold, M. E., Ellson, C. D. *et al.* (2001). The crystal structure of the PX domain from p40^{phox} bound to phosphatidylinositol 3-phosphate. *Mol. Cell* **8**, 829–839.
41. Ono, Y., Fujii, T., Igarashi, K., Kuno, T., Tanaka, C. *et al.* (1989). Phorbol ester binding to protein kinase C requires a cysteine-rich zinc-finger-like sequence. *Proc. Natl. Acad. Sci. USA* **86**, 4868–4871.
42. Hurley, J. H., Newton, A. C., Parker, P. J., Blumberg, P. M., and Nishizuka, Y. (1997). Taxonomy and function of C1 protein kinase C homology domains. *Protein Sci.* **6**, 477–480.
43. Zhang, G., Kazanietz, M. G., Blumberg, P. M., and Hurley, J. H. (1995). Crystal structure of the cys2 activator-binding domain of protein kinase C delta in complex with phorbol ester. *Cell* **81**, 917–924.
44. Xu, R. X., Pawelczyk, T., Xia, T. H., and Brown, S. C. (1997). NMR structure of a protein kinase C-gamma phorbol-binding domain and study of protein-lipid micelle interactions. *Biochemistry* **36**, 10709–10717.
45. Ford, M. G., Mills, I. G., Peter, B. J., Vallis, Y., Praefcke, G. J., Evans, P. R., McMahan, H. T. (2002). Curvature of clathrin-coated pits driven by epsin. *Nature* **419**, 361–366.

CHAPTER 150

Pleckstrin Homology (PH) Domains

Mark A. Lemmon
*Department of Biochemistry and Biophysics,
University of Pennsylvania Medical Center,
Philadelphia, Pennsylvania*

Identification and Definition of PH Domains

The 100 to 120-amino acid pleckstrin homology (PH) domain was first named in 1993 [1–3] as a region of sequence similarity that occurs twice in pleckstrin [4] and is shared by a large number of other proteins. Levels of sequence identity between PH domains are generally low, lying between around 10 (or less) to 30 percent, and there is no conserved motif that identifies PH domains. Rather, PH domains are defined by a pattern of sequence similarity that suggests a common fold, and may therefore share structural similarity in the absence of functional relatedness. The majority of PH domain-containing proteins require membrane association for some aspect of their function. These proteins participate in cellular signaling, cytoskeletal organization, membrane trafficking, and/or phospholipid modification. Sequences encoding PH domains occur in some 252 genes in the first draft of the human genome sequence [5], making this the eleventh most populous domain family in humans. PH domains occur in 77 genes in *D. melanogaster*, 71 genes in *C. elegans*, and 27 in *S. cerevisiae* [5]. Understanding the functions of these common domains has therefore been a subject of considerable interest.

The Structure of PH Domains

Structures of 15 different PH domains have been determined by NMR and/or X-ray crystallography [6–19]. At the core of each PH domain is the same seven-stranded β-sandwich of two near-orthogonal β-sheets containing four- and three-strands respectively (Fig. 1). A characteristic C-terminal α-helix (αC) closes off one "splayed" or open corner [20] of the β-sandwich (top in Fig. 1), while three interstrand loops (the most variable in PH domains) close off the opposite splayed corner (abutting the membrane surface in Fig. 1). This core fold has also been seen in several other classes of domain that share no significant sequence similarity with PH domains [21]. These include the phosphotyrosine binding (PTB) domain [22,23], the Enabled/VASP homology 1 (EVH1) domain [24,25], a Ran binding domain [26], and the FERM domain (for band *f*our-point-one, *e*zrin, *r*adixin, *m*oesin homology domain) [27]. The basic β-sandwich structure has been termed the PH domain "superfold" by Saraste and colleagues [28]. The frequent occurrence of this fold probably reflects its adaptability to multiple functions by creating a stable structural scaffold that can bear loops with quite different recognition properties.

Beyond the conserved β-sandwich fold, one characteristic shared by all PH domains of known structure (except the *C. elegans* Unc89 PH domain [7]) is a marked electrostatic sidedness. Each PH domain is electrostatically polarized, with a positively charged face that coincides with the three most variable loops in the PH domain [9,29]. This positively-charged face abuts the membrane in Fig. 1, and its existence provided part of the motivation for initial tests of PH domain binding to (negatively charged) membrane surfaces.

PH Domains as Phosphoinositide-Binding Modules

The Fesik laboratory was the first to point out that PH domains can bind membranes containing phosphoinositides [30]. Specifically, they showed that the N-terminal PH domain

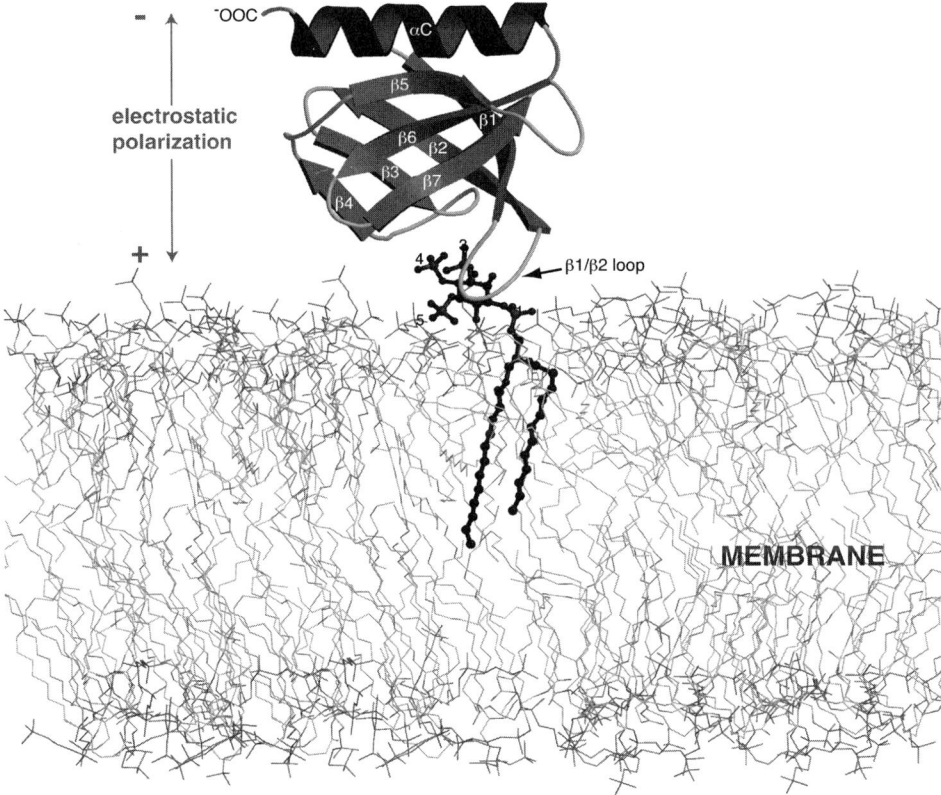

Figure 1 Hypothetical view of how the DAPP1 PH domain binds to PtdIns(3,4,5)P$_3$ in a membrane. The X-ray crystal structure of the DAPP1 PH domain [11] is shown in a ribbon representation with bound Ins(1,3,4,5)P$_4$. The β-sandwich structure of the PH domain can be seen, with strands β1 though β4 forming a sheet behind the plane of the paper, and strands 5 through 7 forming a β-sheet in front of the plane of the page. The characteristic C-terminal α-helix (αC) is also labeled (and caps the upper splayed corner of the β-sandwich). The direction of electrostatic polarization of PH domains is depicted schematically on the left. The positive face abuts the membrane in this orientation. A diacylglycerol molecule has been attached to the Ins(1,3,4,5)P$_4$ molecule to generate a hypothetical view of PtdIns(3,4,5)P$_3$ bound to the DAPP1 PH domain. The PtdIns(3,4,5)P$_3$ is embedded in a stick model of a phosphatidylcholine bilayer to guide thinking as to how the PH domain might bind the lipid headgroup in this context. MOLSCRIPT [98] was used to generate this figure.

from pleckstrin binds phosphatidylinositol-(4,5)-bisphosphate (PtdIns(4,5)P$_2$) with a K_D of approximately 30 μM. NMR analyses demonstrated that the positively charged face of the domain (shown to abut the membrane in Fig. 1) is the site at which the lipid binds [30]. A large number of subsequent studies have shown that phosphoinositide binding is a characteristic shared *in vitro* by nearly all PH domains, and a view has emerged that phosphoinositide binding is a conserved and likely physiologically relevant function for most PH domains [21,31,32]. For several PH domains, phosphoinositide binding has been convincingly demonstrated to be an important (and perhaps the only) function. In these cases the PH domain specifically recognizes the headgroup of a particular phosphoinositide, and this interaction plays an important role in targeting the PH domain-containing protein to cellular membranes [21,33]. However, PH domains in this category are rare. The majority—perhaps over 90 percent of PH domains—bind phosphoinositides with only low affinity and specificity [21,34–36]. How these PH domains participate in membrane targeting (if indeed they do) is not yet clear.

Highly Specific Recognition of Phosphoinositides (and Inositol Phosphates) by PH Domains

PH Domain Binding to Phosphatidylinositol-4,5-Bisphosphate

The phospholipase C-δ$_1$ (PLC-δ$_1$) PH domain was the first shown to recognize a specific phosphoinositide with high affinity [37–39]. The PLC-δ$_1$ PH domain recognizes both PtdIns(4,5)P$_2$ (which it binds with a K_D of approximately 2 μM) and its isolated soluble headgroup, inositol-(1,4,5)-trisphosphate (Ins(1,4,5)P$_3$), with which it forms a 1:1 complex (K_D=210 nM) [39]. An X-ray crystal structure of the Ins(1,4,5)P$_3$/PLC-δ$_1$ PH domain complex [10] showed that the three variable loops on the positively-charged face of the PH domain form the PtdIns(4,5)P$_2$/Ins(1,4,5)P$_3$ binding site. The detailed structure of this binding site also provided clear explanations for the strong Ins(1,4,5)P$_3$-specificity of the PLC-δ$_1$ PH domain (it binds Ins(1,4,5)P$_3$ at least 15-fold more strongly than any other inositol polyphosphate). When expressed as a green fluorescent protein (GFP) fusion, or analyzed by indirect immunofluorescence, the PLC-δ_1 PH domain shows clear plasma membrane localization [40–43]. GFP fusion proteins of this PH domain have been used to identify the location of PtdIns(4,5)P_2 in living cells, and to monitor PtdIns(4,5)P_2 dynamics and/or Ins(1,4,5)P_3 accumulation in response to different agonists [41–45].

RECOGNITION OF PHOSPHATIDYLINOSITOL 3-KINASE PRODUCTS

Following the realization that some PH domains recognize specific phosphoinositides, it was found that protein kinase B (PKB, also known as Akt), a serine/threonine kinase with an N-terminal PH domain, is a downstream effector of phosphatidylinositol 3-kinase (PI 3-kinase) [46,47]. Mutations in the PKB PH domain prevent its PI 3-kinase-dependent activation, indicating that the PH domain itself plays a critical role in this step [47]. The PKB PH domain specifically recognizes both PtdIns(3,4,5)P_3 and PtdIns(3,4)P_2, the major products of agonist-stimulated PI 3-kinase, but does not bind strongly to PtdIns(4,5)P_2 or other phosphoinositides [48–50]. As discussed elsewhere in this volume, PtdIns(3,4,5)P_3 and PtdIns(3,4)P_2 are all but undetectable in quiescent cells but accumulate transiently in the plasma membrane following stimulation of cells with a variety of agonists (to an estimated local concentration of 150 µM [51]). A PH domain that fails to bind PtdIns(4,5)P_2 (present constitutively in the plasma membrane), but which binds strongly to PtdIns(3,4,5)P_3 and/or PtdIns(3,4)P_2, will be recruited to the plasma membrane specifically when these PI 3-kinase-generated lipid second messengers are present. The PH domain from PKB has these binding characteristics, and can be shown (as a GFP fusion protein) to be recruited efficiently to the plasma membrane of mammalian cells following growth factor stimulation [52,53]. As discussed elsewhere in this volume, once recruited by its PH domain to PI 3-kinase products at the plasma membrane, PKB is activated at this location by phosphorylation at two sites [54]. One phosphorylation event is performed by a serine/threonine kinase named PDK1 (for phosphoinositide-dependent kinase-1), which also has a PH domain that can recruit it to the plasma membrane in a PI 3-kinase-dependent manner [55,56].

Other PH domains that specifically recognize PI 3-kinase products include that from Bruton's tyrosine kinase (Btk) [34,57–59] and the PH domain from the Arf-guanine nucleotide exchanger Grp1 (general receptor for phosphoinositides-1) [60]. Both of these PH domains bind exclusively (and strongly) to PtdIns(3,4,5)P_3 or its headgroup Ins(1,3,4,5)P_4 [35,59,60]. A point mutation (at arginine-28) in the Btk PH domain, which leads to agammaglobulinemia in humans and mice [61,62], abolishes PtdIns(3,4,5)P_3/Ins(1,3,4,5)P_4 binding [34,57,58]. The effects of this Btk mutation on B-cell signaling provided the first clue that PH domains may play a role in signal transduction. Like the PKB PH domain, the Btk and Grp1 PH domains are recruited directly to the plasma membrane upon PI 3-kinase activation [53,63–65].

Skolnik and colleagues identified more than 12 different PH domains capable of driving PI 3-kinase-dependent plasma membrane recruitment using a novel yeast-based assay [66]. Where studied, each of these PH domains binds *in vitro* to PtdIns(3,4,5)P_3 (or Ins(1,3,4,5)P_4) with a K_D in the 10–100 nM range, and selects for PtdIns(3,4,5)P_3 over PtdIns(4,5)P_2 by a factor of 20 or more [21]. PH domains in this group share a sequence motif centered around the $\beta 1/\beta 2$ loop that links the first two β-strands of the PH domain sandwich. Several crystal structures of PH domains bound to Ins(1,3,4,5)P_4 have shown how this motif defines a specific binding site for the PtdIns(3,4,5)P_3 [6,11,67] (Fig. 2). The structural details of the binding site are remarkably well conserved across different structures and bear a strong resemblance (in structure and sequence) to the Ins(1,4,5)P_3 binding site of the PLC-δ_1 PH domain. The sequence motif identified by Skolnik and colleagues [66] serves as a strong and reliable predictor of which PH domains specifically recognize PI 3-kinase products.

PTDINS(3,4,5)P_3 VERSUS PTDINS(3,4)P_2

Among the PH domains with the sequence motif shown in Fig. 2, some bind equally well to both PtdIns(3,4,5)P_3 and PtdIns(3,4)P_2 (e.g. the PKB and DAPP1 PH domains) while others bind only PtdIns(3,4,5)P_3 (e.g. the Grp1 and Btk PH domains). PH domains that recognize only PtdIns(3,4,5)P_3 tend either to have extended $\beta 1/\beta 2$ loops or (as in the Grp1 PH domain) insertions elsewhere in the structure that can make specific contacts with the 5-phosphate group. In the complex between the DAPP1 (dual-specific) PH domain and Ins(1,3,4,5)P_4 there are no hydrogen bonds between PH domain side chains and the 5-phosphate group [11], providing one explanation for why this PH domain binds equally well to Ins(1,3,4,5)P_4/PtdIns(3,4,5)P_3 and Ins(1,3,4)P_3/PtdIns(3,4,5)P_2.

There is no currently known PH domain that binds exclusively to PtdIns(3,4)P_2. Alessi and colleagues identified the C-terminal PH domain from TAPP1 (for *t*andem *P*H domain-containing *p*rotein-1) as a PH domain that prefers PtdIns(3,4)P_2 over other phosphoinositides according to protein-lipid overlay studies [68]. However, Ferguson *et al.* [11] showed clearly that this PH domain (called AA054961 in that study) binds with high affinity to the headgroups of both PtdIns(3,4)P_2 *and* PtdIns(3,4,5)P_3. In several other assays, using isolated headgroups and intact lipids, it has been shown that the C-terminal TAPP1 PH domain does prefer PtdIns(3,4)P_2, but binds to this phosphoinositide only four-fold more strongly than to PtdIns(3,4,5)P_3 [V. J. Sankaran and M. A. Lemmon, unpublished data]. In spite of this weak selectivity for PtdIns(3,4)P_2 over PtdIns(3,4,5)P_3, Alessi and colleagues have provided some evidence to suggest that the TAPP1 PH domain is recruited to the plasma membrane *in vivo* when PtdIns(3,4)P_2 production is stimulated but not when PtdIns(3,4,5)P_3 is thought to accumulate without PtdIns(3,4)P_2 production [69]. Whether other proteins exist that are regulated exclusively by PtdIns(3,4)P_2 remains to be seen.

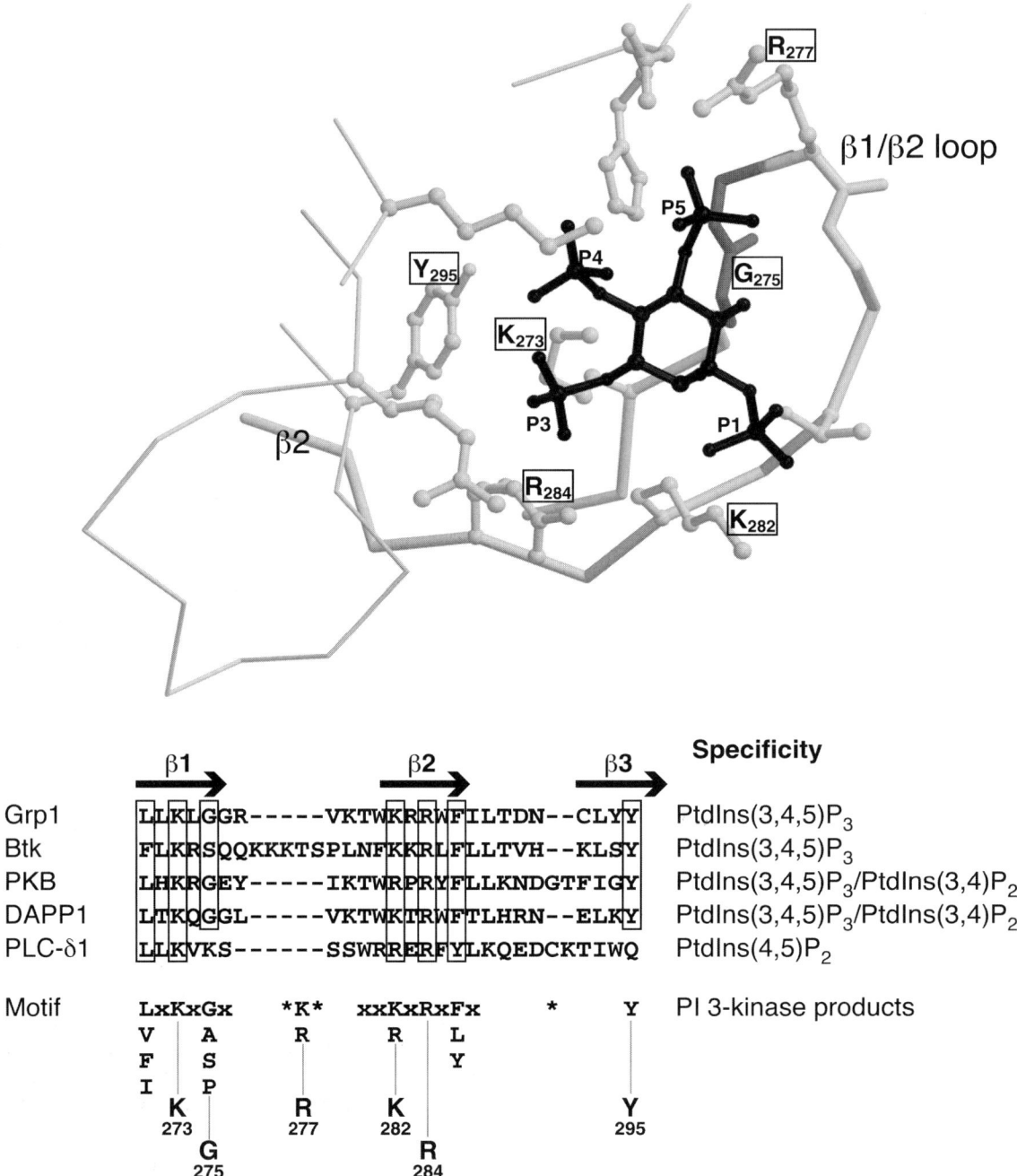

Figure 2 Close-up view of Ins(1,3,4,5)P$_4$ in the binding site of the Grp1 PH domain, depicting the side chains of residues in the sequence motif that predicts PI 3-kinase product specificity. The β1/β2 loop of the Grp1 PH domain is shown to "cradle" the Ins(1,3,4,5)P$_4$ molecule, with several residues marked forming side-chain hydrogen bonds with the bound lipid headgroup. The motif that predicts specificity for PI 3-kinase product binding is shown in the lower part of the figure, imposed upon the sequences of the N-terminal portions of the Grp1, Btk, PKB, and DAPP1 PH domains. The motif positions corresponding to the residues highlighted in the structural figure are also shown. K273, in strand β1 of Grp1, forms a hydrogen bond with both the 3- and 4-phosphates of Ins(1,3,4,5)P$_4$. G275 must be small in order to allow space for the inositol ring in this binding configuration. K282 forms a hydrogen bond with the Ins(1,3,4,5)P$_4$ 1-phosphate. R284 forms a critical hydrogen bond with the 3-phosphate. This is equivalent to the arginine at which mutations in Btk cause agammaglobulinemias. Y295, in strand β3, is a conserved feature of PH domains that bind PI 3-kinase products, and its side chain forms a hydrogen bond with the 4-phosphate group. These 5 motif characteristics are conserved in all PH domains that recognize PtdIns(3,4,5)P$_3$ and/or PtdIns(3,4)P$_2$, and several (but not all) are conserved in PtdIns(4,5)P$_2$ binding by the PLC-δ$_1$ PH domain [10]. Equivalents to the additional interaction of R277 with the 5-phosphate of Ins(1,3,4,5)P$_4$ are seen only in PtdIns(3,4,5)P$_3$-specific PH domains (Grp1 and Btk), and not those that also bind PtdIns(3,4)P$_2$ [11]. The extra long β1/β2 loop of the Btk PH domain contributes to 5-phosphate interactions, as does the β6/β7 loop insertion (not shown) of the Grp1 PH domain.

PH DOMAINS WITH OTHER PHOSPHOINOSITIDE-BINDING SPECIFICITIES

Dowler et al. [68] recently identified several PH domains that appear from protein-lipid overlay assays to have novel phosphoinositide specificities. Their strategy was to identify PH domains with sequences that match (or closely resemble) the PI 3-kinase product-binding motif presented in Fig. 2 and to assess phosphoinositide-binding specificity. The C-terminal TAPP1 PH domain was identified as a target for PI 3-kinase products with this approach, as were examples of PH domains that appear in overlay assays to recognize PtdIns-4-P, PtdIns-3-P, or PtdIns(3,5)P_2 specifically. It should be stressed that these phosphoinositide-binding specificities have not yet been confirmed via quantitative approaches *in vitro* or with localization studies *in vivo*, and that several of them appear from surface plasmon resonance (SPR) studies to have rather low affinities [68]. As well as PH domains with apparently novel specificities, Dowler et al. found several PH domains (with sequences related to the motif in Fig. 2) that interact with all phosphoinositides tested [68]. These results argue that the PtdIns(3,4,5)P_3-specific binding site depicted in Fig. 2 can be "remodeled" with only a handful of mutations to generate binding sites that instead recognize only the PtdIns(4,5)P_2 headgroup (the PLC-δ_1 PH domain), or perhaps only the PtdIns-4-P, PtdIns-3-P or PtdIns(3,5)P_2 headgroup. Remodeling of a different nature can alternatively generate a binding site that accommodates any phosphoinositide headgroup, so that binding is promiscuous.

Nonspecific Phosphoinositide Binding by PH Domains: The Majority Occupation?

Although this fact may not be immediately clear from a reading of the PH domain literature, by far the majority (>90 percent) of PH domains do *not* have a sequence that significantly resembles the motif shown in Fig. 2. Nonetheless, most PH domains lacking the motif do appear capable of phosphoinositide binding, although binding is weak and nonspecific in almost every case [21,35,36]. Where K_D values have been reported for phosphoinositide binding by this class of PH domains, they have ranged from around 30 μM to 4 mM or weaker [12,13,30,34,36,57,70–73]. In one case, that of the β-spectrin PH domain, a crystal structure of the PH domain with a weakly bound Ins(1,4,5)P_3 was reported [13]. Ins(1,4,5)P_3 binds to the surface of this electrostatically polarized PH domain, in the center of its positively charged face. NMR studies have similarly located the site of weak phosphoinositide binding in other PH domains to the variable loops on the positively-charged face [30,57,73,74], most likely driven by delocalized electrostatic attraction to the negatively charged ligand.

Although the physiological relevance of specific, high-affinity, phosphoinositide binding by PH domains has been well established in several cases, it remains unclear in most cases whether weak and promiscuous binding of phosphoinositides to the majority of PH domains plays any physiological role. It has been shown that the low-affinity, nonspecific binding of phosphoinositides to the PH domain of dynamin is essential for this protein's function in receptor-mediated endocytosis [75–77]. Similarly, the low affinity (and usually promiscuous) binding of PH domains from Dbl-family members to phosphoinositides [72] appears to be critical for their Rac/Rho exchange activity *in vivo* and their ability to transform cells [78–81]. Despite intensive study, and three crystal structures of DH/PH fragments from Dbl-family proteins [14,15,17], it remains unclear how low-affinity binding of phosphoinositides to the PH domains of these proteins influences the exchange activity of the adjacent DH (Dbl homology) domain. For many other proteins with PH domains in this class it has been demonstrated that the PH domain is critical for *in vivo* function, but it has not been established whether or not phosphoinositide binding is a physiologically relevant feature of the PH domain. Many more studies are required to address this question for other "promiscuous" PH domains.

Binding of PH Domains to Non-phosphoinositide Ligands

Since the first description of PH domains, many potential protein binding-partners have been reported. The first were βγ-subunits of heterotrimeric G-proteins, which were suggested to bind all PH domains [82] but now appear only to participate in membrane targeting of a small subset, which includes the PH domains from β-adrenergic receptor kinases (βARK's) [83,84]. Other reported protein targets for PH domains include protein kinase C (PKC) isoforms [85,86], the product of the *TCL1* (for T-cell leukemia) oncogene (which binds the PKB PH domain) [87–89], the receptor for activated PKC (RACK1) [90], $G_{12}\alpha$ [91], a protein called BAP-135 (reported to bind the Btk PH domain) [92], filamentous actin [93], acidic motifs found in proteins such as nucleolin (shown to bind the PH domains of IRS1 and IRS2) [94], and several others (reviewed in [21,95]). Although not all of these PH domain/protein interactions have been demonstrated to have physiological relevance, there is no doubt that some do, and that protein binding by PH domains cannot be ignored. Despite the relative wealth of reported protein targets, however, no common themes emerge from the described PH domain/protein interactions. This should not be surprising given the observed diversity in the modes of protein-target recognition by the structurally related EVH1, PTB, and Ran-binding domains [21].

Possible Roles of Non-phosphoinositide PH Ligands

PH domains for which protein targets have been reported include both examples that bind phosphoinositides weakly and promiscuously (e.g. the βARK, IRS-1, and dynamin PH domains), *as well as* PH domains that bind strongly and specifically to particular phosphoinositides (e.g. the Btk and PKB PH domains). It can therefore not be argued that the protein targets described for PH domains are simply alternatives to, or surrogates for, the well-studied (but rare) specific phosphoinositide ligands. Rather, it appears likely that some PH domains bind multiple ligands.

Cooperation of Multiple Ligands in Membrane Recruitment of PH Domains

A requirement for simultaneous PH domain binding to two different ligands was first demonstrated for membrane targeting by the βARK PH domain [83]. The βARK PH domain binds very weakly to PtdIns(4,5)P$_2$ ($K_D > 200 \mu M$) [12]. It also binds rather weakly to the βγ-subunits of heterotrimeric G-proteins [82]. Neither of these weak interactions alone is sufficient for high-affinity targeting of βARK to membranes, but the two interactions can cooperate to recruit βARK efficiently to relevant membrane surfaces [83].

Golgi Targeting of PH Domains by Multiple Ligands

The PH domain from oxysterol binding protein (OSBP), as well as several other related PH domains, is targeted specifically to the Golgi through interactions that appear to require both phosphoinositides and another unidentified (Golgi-specific) component [96]. These Golgi-targeted PH domains, which include those from FAPP1 and the Goodpasture antigen binding protein (GPBP), are highly promiscuous in their phosphoinositide binding (and are not PtdIns-4-P-specific) [34,96], arguing that phosphoinositide recognition alone cannot possibly determine their Golgi targeting. Phosphoinositide binding by these PH domains is several-fold weaker than PtdIns(4,5)P$_2$ binding by the PLC-δ$_1$ PH domain ([96] and D. Keleti, V. J. Sankaran, and M. A. Lemmon, unpublished), further suggesting that it may not be strong enough to drive membrane targeting of the OSBP PH domain independently. Studies in a series of yeast mutants have demonstrated that Golgi targeting of the OSBP and FAPP1 PH domains is dependent on PtdIns-4-P and not on PtdIns(4,5)P$_2$ production [96,97], but that the activity of Arf1p is also important [96]. It is therefore hypothesized that the presence of two binding partners in the Golgi is responsible for specific targeting of the OSBP, FAPP1, and GPBP PH domains to that organelle. On its own, phosphoinositide binding by these PH domains is not strong enough to drive membrane targeting *in vivo*, and would certainly not provide targeting specificity. The second (so far unidentified) target of these PH domains is thought to be Golgi-specific, but does not bind to the PH domains tightly enough to achieve Golgi targeting on its alone. Rather both phosphoinositide and this unknown component must be present in the same membrane (the Golgi) in order to recruit the OSBP and other PH domains to that compartment with high affinity and specificity. According to this model [96], PtdIns-4-P is implicated in Golgi targeting of the OSBP, FAPP1, and other PH domains not because of headgroup recognition, but because this happens to be the most abundant phosphoinositide in the membranes that contain the second PH domain ligand.

Conclusions

The eleventh most populous domain family in humans is now rather well understood structurally and lends its name to the PH domain superfold that includes proteins involved in binding to variety of phosphoinositide and protein ligands. Ligand binding by a small subgroup of PH domains—those that bind phosphoinositide headgroups with high affinity and specificity—is now understood rather well, although it remains possible that some PH domains from this class have additional, as yet unidentified, binding partners. PH domains that do *not* bind phosphoinositides with high affinity or specificity constitute the majority—perhaps 90 percent. The interactions driven by these PH domains are far less well understood. In many cases it even remains unclear whether phosphoinositide binding observed *in vitro* has any relevance *in vivo*. How weak and nonspecific phosphoinositide binding could contribute to membrane binding is a question that has yet to be fully addressed. It may do so though cooperation of multiple ligands that bind to a single PH domain (as discussed for the βARK and OSBP PH domains). Alternatively, the PH domain may be one of several domains within a multidomain protein or oligomer that cooperate with one another in driving membrane targeting. In these cases, specificity of membrane targeting may be defined not by the precise nature of the individual interactions (as with PH domains that bind PI 3-kinase products), but rather by the available combinations of interactions. Recruitment to a specific membrane may require that two or more PH domain targets coexist in that membrane.

References

1. Haslam, R. J., Koide, H. B., and Hemmings, B. A. (1993). Pleckstrin domain homology. *Nature* **363**, 309–310.
2. Mayer, B. J., Ren, R., Clark, K. L., and Baltimore, D. (1993). A putative modular domain present in diverse signaling molecules. *Cell* **73**, 629–630.
3. Musacchio, A., Gibson, T., Rice, P., Thompson, J., and Saraste, M. (1993). The PH domain: a common piece in a patchwork of signalling proteins. *Trends. Biochem. Sci.* **18**, 343–348.
4. Tyers, M., Rachubinski, R. A., Stewart, M. I., Varrichio, A. M., Shorr, R. G. L., Haslam, R. J., and Harley, C. B. (1988). Molecular cloning and expression of the major protein kinase C substrate of platelets. *Nature* **333**, 470–473.
5. Consortium, I. H. G. S. (2001). Initial sequencing and analysis of the human genome. *Nature* **409**, 860–921.
6. Baraldi, E., Djinovic Carugo, K., Hyvönen, M., Lo Surdo, P., Riley, A. M., Potter, B. V. L., O'Brien, R., Ladbury, J. E., and Saraste, M. (1999). Structure of the PH domain from Bruton's tyrosine kinase in complex with inositol 1,3,4,5-tetrakisphosphate. *Structure* **7**, 449–460.
7. Blomberg, N., Baraldi, E., Sattler, M., Saraste, M., and Nilges, M. (2000). Structure of a PH domain from the *C. elegans* muscle protein UNC-89 suggests a novel function. *Structure Fold Des.* **8**, 1079–1087.
8. Dhe-Paganon, S., Ottinger, E. A., Nolte, R. T., Eck, M. J., and Shoelson, S. E. (1999). Crystal structure of the pleckstrin homology-phosphotyrosine binding (PH-PTB) targeting region of insulin receptor substrate 1. *Proc. Natl. Acad. Sci. USA* **96**, 8378–8383.
9. Ferguson, K. M., Lemmon, M. A., Schlessinger, J., and Sigler, P. B. (1994). Crystal structure at 2.2 Å resolution of the pleckstrin homology domain from human dynamin. *Cell* **79**, 199–209.

10. Ferguson, K. M., Lemmon, M. A., Schlessinger, J., and Sigler, P. B. (1995). Structure of a high affinity complex between inositol-1,4,5-trisphosphate and a phospholipase C pleckstrin homology domain. *Cell* **83**, 1037–1046.
11. Ferguson, K. M., Kavran, J. M., Sankaran, V. G., Fournier, E., Isakoff, S. J., Skolnik, E. Y., and Lemmon, M. A. (2000). Structural basis for discrimination of 3-phosphoinositides by pleckstrin homology domains. *Mol. Cell* **6**, 373–384.
12. Fushman, D., Najmabadi-Kaske, T., Cahill, S., Zheng, J., LeVine, H., and Cowburn, D. (1998). The solution structure and dynamics of the pleckstrin homology domain of G protein-coupled receptor kinase 2 (β-adrenergic receptor kinase 1): A binding partner of $G_{\beta\gamma}$ subunits. *J. Biol. Chem.* **273**, 2835–2843.
13. Hyvönen, M., Macias, M. J., Nilges, M., Oschkinat, H., Saraste, M., and Wilmanns, M. (1995). Structure of the binding site for inositol phosphates in a PH domain. *EMBO J.* **14**, 4676–4685.
14. Rossman, K. L., Worthylake, D. K., Snyder, J. T., Siderovski, D. P., Campbell, S. L., and Sondek, J. (2002). A crystallographic view of interactions between Dbs and Cdc42: PH domain-assisted guanine nucleotide exchange. *EMBO J.* **21**, 1315–1326.
15. Soisson, S. M., Nimnual, A. S., Uy, M., Bar-Sagi, D., and Kuriyan, J. (1998). Crystal structure of the Dbl and pleckstrin homology domains from the human Son of Sevenless protein. *Cell* **95**, 259–268.
16. Thomas, C. C., Dowler, S., Deak, M., Alessi, D. R., and van Aalten, D. M. (2001). Crystal structure of the phosphatidylinositol 3,4-bisphosphate-binding pleckstrin homology (PH) domain of tandem PH-domain-containing protein 1 (TAPP1): molecular basis of lipid specificity. *Biochem. J.* **358**, 287–294.
17. Worthylake, D. K., Rossman, K. L., and Sondek, J. (2000). Crystal structure of Rac1 in complex with the guanine nucleotide exchange region of Tiam1. *Nature* **408**, 682–688.
18. Yoon, H. S., Hajduk, P. J., Petros, A. M., Olejniczak, E. T., Meadows, R. P., and Fesik, S. W. (1994). Solution structure of a pleckstrin-homology domain. *Nature* **369**, 672–675.
19. Zhang, P., Talluri, S., Deng, H., Branton, D., and Wagner, G. (1995). Solution structure of the pleckstrin homology domain of Drosophila beta-spectrin. *Structure* **3**, 1185–1195.
20. Chothia, C. (1984). Principles that determine the structure of proteins. *Annu. Rev. Biochem.* **53**, 537–572.
21. Lemmon, M. A., and Ferguson, K. M. (2000). Signal-dependent membrane targeting by pleckstrin homology (PH) domains. *Biochem. J.* **350**, 1–18.
22. Eck, M. J., Dhe-Paganon, S., Trüb, T., Nolte, R. T., and Shoelson, S. E. (1996). Structure of the IRS-1 PTB domain bound to the juxtamembrane region of the insulin receptor. *Cell* **85**, 695–705.
23. Zhou, M.-M., Ravichandran, K. S., Olejniczak, E. T., A.M., P., Meadows, R. P., Sattler, M., Harlan, J. E., Wade, W. S., Burakoff, S. J., and Fesik, S. W. (1995). Structure and ligand recognition of the phosphotyrosine binding domain of Shc. *Nature* **378**, 584–592.
24. Beneken, J., Tu, J. C., Xiao, B., Nuriya, M., Yuan, J. P., Worley, P. F., and Leahy, D. J. (2000). Structure of the Homer EVH1 domain-peptide complex reveals a new twist in polyproline recognition. *Neuron* **26**, 143–154.
25. Prehoda, K. E., Lee, D. J., and Lim, W. A. (1999). Structure of the enabled/VASP homology 1 domain-peptide complex: a key component in the spatial control of actin assembly. *Cell* **97**, 471–480.
26. Vetter, I. R., Nowak, C., Nishimotot, T., Kuhlmann, J., and Wittinghofer, A. (1999). Structure of a Ran-binding domain complexed with Ran bound to a GTP analogue: implications for nuclear transport. *Nature* **398**, 39–46.
27. Pearson, M. A., Reczek, D., Bretscher, A., and Karplus, P. A. (2000). Structure of the ERM protein moesin reveals the FERM domain fold masked by an extended actin binding tail domain. *Cell* **101**, 259–270.
28. Blomberg, N., Baraldi, E., Nilges, M., and Saraste, M. (1999). The PH superfold: A structural scaffold for multiple functions. *Trends Biochem. Sci.* **24**, 441–445.
29. Macias, M. J., Musacchio, A., Ponstingl, H., Nilges, M., Saraste, M., and Oschkinat, H. (1994). Structure of the pleckstrin homology domain from β-spectrin. *Nature* **369**, 675–677.
30. Harlan, J. E., Hajduk, P. J., Yoon, H. S., and Fesik, S. W. (1994). Pleckstrin homology domains bind to phosphatidylinositol 4,5-bisphosphate. *Nature* **371**, 168–170.
31. Hurley, J. H. and Misra, S. (2000). Signaling and subcellular targeting by membrane-binding domains. *Annu. Rev. Biophys. Biomol. Struct.* **29**, 49–79.
32. Bottomley, M. J., Salim, K., and Panayotou, G. (1998). Phospholipid-binding domains. *Biochim. Biophys. Acta* **1436**, 165–183.
33. Rameh, L. E. and Cantley, L. C. (1999). The role of phosphoinositide 3-kinase lipid products in cell function. *J. Biol. Chem.* **274**, 8347–8350.
34. Rameh, L. E., Arvidsson, A.-K., Carraway III, K. L., Couvillon, A. D., Rathbun, G., Cromptoni, A., VanRentergem, B., Czech, M. P., Ravichandran, K. S., Burakoff, S. J., Wang, D.-S., Chen, C.-S., and Cantley, L. C. (1997). A comparative analysis of the phosphoinositide binding specificity of pleckstrin homology domains. *J. Biol. Chem.* **272**, 22059–22066.
35. Kavran, J. M., Klein, D. E., Lee, A., Falasca, M., Isakoff, S. J., Skolnik, E. Y., and Lemmon, M. A. (1998). Specificity and promiscuity in phosphoinositide binding by pleckstrin homology domains. *J. Biol. Chem.* **273**, 30497–30508.
36. Takeuchi, H., Kanematsu, T., Misumi, Y., Sakane, F., Konishi, H., Kikkawa, U., Watanabe, Y., Katan, M., and Hirata, M. (1997). Distinct specificity in the binding of inositol phosphates by pleckstrin homology domains of pleckstrin, RAC-protein kinase, diacylglycerol kinase and a new 130 kDa protein. *Biochim. Biophys. Acta.* **1359**, 275–285.
37. Yagisawa, H., Hirata, M., Kanematsu, T., Watanabe, Y., Ozaki, S., Sakuma, K., Tanaka, H., Yabuta, N., Kamata, H., Hirata, H., and Nojima, H. (1994). Expression and characterization of an inositol 1,4,5-trisphosphate binding domain of phosphatidylinositol-specific phospholipase C-delta 1. *J. Biol. Chem.* **269**, 20179–20188.
38. Garcia, P., Gupta, R., Shah, S., Morris, A. J., Rudge, S. A., Scarlata, S., Petrova, V., McLaughlin, S., and Rebecchi, M. J. (1995). The pleckstrin homology domain of phospholipase C-delta 1 binds with high affinity to phosphatidylinositol 4,5-bisphosphate in bilayer membranes. *Biochemistry* **34**, 16228–16234.
39. Lemmon, M. A., Ferguson, K. M., O'Brien, R., Sigler, P. B., and Schlessinger, J. (1995). Specific and high-affinity binding of inositol phosphates to an isolated pleckstrin homology domain. *Proc. Natl. Acad. Sci. USA.* **92**, 10472–10476.
40. Paterson, H. F., Savopoulos, J. W., Perisic, O., Cheung, R., Ellis, M. V., Williams, R. L., and Katan, M. (1995). Phospholipase C delta 1 requires a pleckstrin homology domain for interaction with the plasma membrane. *Biochem. J.* **312**, 661–666.
41. Hirose, K., Kadowaki, S., Tanabe, M., Takeshima, H., and Iino, M. (1999). Spatiotemporal dynamics of inositol 1,4,5-trisphosphate that underlies complex Ca^{2+} mobilization patterns. *Science* **284**, 1527–1530.
42. Stauffer, T. P., Ahn, S., and Meyer, T. (1998). Receptor-induced transient reduction in plasma membrane $PtdIns(4,5)P_2$ concentration monitored in living cells. *Curr. Biol.* **8**, 343–346.
43. Varnai, P. and Balla, T. (1998). Visualization of phosphoinositides that bind pleckstrin homology domains: calcium- and agonist-induced dynamic changes and relationship to myo-[^{3}H]inositol-labeled phosphoinositide pools. *J. Cell Biol.* **143**, 501–510.
44. Tall, E. G., Spector, I., Pentyala, S. N., Bitter, I., and Rebecchi, M. J. (2000). Dynamics of phosphatidylinositol 4,5-bisphosphate in actin-rich structures. *Curr. Biol.* **10**, 743–746.
45. Botelho, R. J., Teruel, M., Dierckman, R., Anderson, R., Wells, A., York, J. D., Meyer, T., and Grinstein, S. (2000). Localized biphasic changes in phosphatidylinositol-4,5-bisphosphate at sites of phagocytosis. *J. Cell Biol.* **151**, 1353–1368.
46. Burgering, B. M. and Coffer, P. J. (1995). Protein kinase B (c-Akt) in phosphatidylinositol-3-OH kinase signal transduction. *Nature* **376**, 599–602.
47. Franke, T. F., Yang, S. I., Chan, T. O., Datta, K., Kazlauskas, A., Morrison, D. K., Kaplan, D. R., and Tsichlis, P. N. (1995). The protein kinase encoded by the Akt proto-oncogene is a target of the PDGF-activated phosphatidylinositol 3-kinase. *Cell* **81**, 727–736.

48. Franke, T. F., Kaplan, D. R., Cantley, L. C., and Toker, A. (1997). Direct regulation of the Akt proto-oncogene product by phosphatidylinositol-3,4-bisphosphate. *Science* **275**, 665–668.
49. Frech, M., Andjelkovic, M., Ingley, E., Reddy, K. K., Falck, J. R., and Hemmings, B. A. (1997). High affinity binding of inositol phosphates and phosphoinositides to the pleckstrin homology domain of RAC/protein kinase B and their influence on kinase activity. *J. Biol. Chem.* **272**, 8474–8481.
50. Klippel, A., Kavanaugh, W. M., Pot, D., and Williams, L. T. (1997). A specific product of phosphatidylinositol 3-kinase directly activates the protein kinase Akt through its pleckstrin homology domain. *Mol. Cell. Biol.* **17**, 338–344.
51. Stephens, L. R., Jackson, T. R., and Hawkins, P. T. (1993). Agonist-stimulated synthesis of phosphatidylinositol 3,4,5-trisphosphate: a new intracellular signaling system? *Biochim. Biophys. Acta.* **1179**, 27–75.
52. Watton, S. J. and Downward, J. (1999). Akt/PKB localisation and 3' phosphoinositide generation at sites of epithelial cell-matrix and cell-cell interaction. *Curr. Biol.* **9**, 433–436.
53. Gray, A., Van der Kaay, J., and Downes, C. P. (1999). The pleckstrin homology domains of protein kinase B and GRP1 (general receptor for phosphoinositides-1) are sensitive and selective probes for the cellular detection of phosphatidylinositol 3,4-bisphosphate and/or phosphatidylinositol 3,4,5-trisphosphate in vivo. *Biochem. J.* **344**, 929–936.
54. Vanhaesebroeck, B. and Alessi, D. R. (2000). The PI3K-PDK1 connection: more than just a road to PKB. *Biochem. J.* **346**, 561–576.
55. Anderson, K. E., Coadwell, J., Stephens, L. R., and Hawkins, P. T. (1998). Translocation of PDK-1 to the plasma membrane is important in allowing PDK-1 to activate protein kinase B. *Curr. Biol.* **8**, 684–691.
56. Currie, R. A., Walker, K. S., Gray, A., Deak, M., Casamayor, A., Downes, C. P., Cohen, P., Alessi, D. R., and Lucocq, J. (1999). Role of phosphatidylinositol 3,4,5-trisphosphate in regulating the activity and localization of 3-phosphoinositide-dependent protein kinase-1. *Biochem. J.* **337**, 575–583.
57. Salim, K., Bottomley, M. J., Querfurth, E., Zvelebil, M. J., Gout, I., Scaife, R., Margolis, R. L., Gigg, R., Smith, C. I. E., Driscoll, P. C., Waterfield, M. D., and Panayotou, G. (1996). Distinct specificity in the recognition of phosphoinositides by the pleckstrin homology domains of dynamin and Bruton's tyrosine kinase. *EMBO J.* **15**, 6241–6250.
58. Fukuda, M., Kojima, T., Kabayama, H., and Mikoshiba, K. (1996). Mutation of the pleckstrin homology domain of Bruton's tyrosine kinase in immunodeficiency impaired inositol 1,3,4,5-tetrakisphosphate binding capacity. *J. Biol. Chem.* **271**, 30303–30306.
59. Kojima, T., Fukuda, M., Watanabe, Y., Hamazato, F., and K., M. (1997). Characterization of the pleckstrin homology domain of Btk as an inositol polyphosphate and phosphoinositide binding domain. *Biochem. Biophys. Res. Commun.* **236**, 333–339.
60. Klarlund, J. K., Guilherme, A., Holik, J. J., Virbasius, A., and Czech, M. P. (1997). Signaling by 3,4,5-phosphoinositide through proteins containing pleckstrin and Sec7 homology domains. *Science* **275**, 1927–1930.
61. Rawlings, D. J., Saffran, D. C., Tsukada, S., Largaespada, D. A., Grimaldi, J. C., Cohen, L., Mohr, R. N., Bazan, J. F., Howard, M., and Copeland, N. G. (1993). Mutation of unique region of Bruton's tyrosine kinase in immunodeficient XID mice. *Science* **261**, 358–361.
62. Thomas, J. D., Sideras, P., Smith, C. I., Vorechovsky, I., Chapman, V., and Paul, W. E. (1993). Colocalization of X-linked agammaglobulinemia and X-linked immunodeficiency genes. *Science* **261**, 355–358.
63. Varnai, P., Rother, K. I., and Balla, T. (1999). Phosphatidylinositol 3-kinase-dependent membrane association of the Bruton's tyrosine kinase pleckstrin homology domain visualized in single living cells. *J. Biol. Chem.* **274**, 10983–10989.
64. Nagel, W., Zeitlmann, L., Schilcher, P., Geiger, C., Kolanus, J., and Kolanus, W. (1998). Phosphoinositide 3-OH kinase activates the beta2 integrin adhesion pathway and induces membrane recruitment of cytohesin-1. *J. Biol. Chem.* **273**, 14853–14861.
65. Venkateswarlu, K., Gunn-Moore, F., Oatey, P. B., Tavare, J. M., and Cullen, P. J. (1998). Nerve growth factor- and epidermal growth factor-stimulated translocation of the ADP-ribosylation factor-exchange factor GRP1 to the plasma membrane of PC12 cells requires activation of phosphatidylinositol 3-kinase and the GRP1 pleckstrin homology domain. *Biochem. J.* **335**, 139–146.
66. Isakoff, S. J., Cardozo, T., Andreev, J., Li, Z., Ferguson, K. M., Abagyan, R., Lemmon, M. A., Aronheim, A., and Skolnik, E. Y. (1998). Identification and analysis of PH domain-containing targets of phosphatidylinositol 3-kinase using a novel *in vivo* assay in yeast. *EMBO J.* **17**, 5374–5387.
67. Lietzke, S. E., Bose, S., Cronin, T., Klarlund, J., Chawla, A., Czech, M. P., and Lambright, D. G. (2000). Structural basis of 3-phosphoinositide recognition by pleckstrin homology domains. *Mol. Cell* **6**, 385–394.
68. Dowler, S., Currie, R. A., Campbell, D. G., Deak, M., Kular, G., Downes, C. P., and Alessi, D. R. (2000). Identification of pleckstrin-homology-domain-containing proteins with novel phosphoinositide-binding specificities. *Biochem. J.* **351**, 19–31.
69. Kimber, W. A., Trinkle-Mulcahy, L., Cheung, P. C., Deak, M., Marsden, L. J., Kieloch, A., Watt, S., Javier, R. T., Gray, A., Downes, C. P., Lucocq, J. M., and Alessi, D. R. (2002). Evidence that the tandem-pleckstrin-homology-domain-containing protein TAPP1 interacts with Ptd(3,4)P$_2$ and the multi-PDZ-domain-containing protein MUPP1 *in vivo*. *Biochem. J.* **361**, 525–536.
70. Klein, D. E., Lee, A., Frank, D. W., Marks, M. S., and Lemmon, M. A. (1998). The pleckstrin homology domains of dynamin isoforms require oligomerization for high affinity phosphoinositide binding. *J. Biol. Chem.* **273**, 27725–27733.
71. Koshiba, S., Kigawa, T., Kim, J. H., Shirouzu, M., Bowtell, D., and Yokoyama, S. (1997). The solution structure of the pleckstrin homology domain of mouse Son-of-sevenless 1 (mSos1). *J. Mol. Biol.* **20**, 579–591.
72. Snyder, J. T., Rossman, K. L., Baumeister, M. A., Pruitt, W. M., Siderovski, D. P., Der, C. J., Lemmon, M. A., and Sondek, J. (2001). Quantitative analysis of the effect of phosphoinositide interactions on the function of Dbl family proteins. *J. Biol. Chem.* **276**, 45868–45875.
73. Zheng, J., Cahill, S. M., Lemmon, M. A., Fushman, D., Schlessinger, J., and Cowburn, D. (1996). Identification of the binding site for acidic phospholipids on the PH domain of dynamin: Implications for stimulation of GTPase activity. *J. Mol. Biol.* **255**, 14–21.
74. Zheng, J., Chen, R.-H., Corbalan-Garcia, S., Cahill, S. M., Bar-Sagi, D., and Cowburn, D. (1997). The solution structure of the pleckstrin homology domain of human SOS1. A possible structural role for the sequential association of diffuse B cell lymphoma and pleckstrin homology domains. *J. Biol. Chem.* **272**, 30340–30344.
75. Achiriloaie, M., Barylko, B., and Albanesi, J. P. (1999). Essential role of the dynamin pleckstrin homology domain in receptor-mediated endocytosis. *Mol. Cell. Biol.* **19**, 1410–1415.
76. Lee, A., Frank, D. W., Marks, M. S., and Lemmon, M. A. (1999). Dominant-negative inhibition of receptor-mediated endocytosis by a dynamin-1 mutant with a defective pleckstrin homology domain. *Curr. Biol.* **9**, 261–264.
77. Vallis, Y., Wigge, P., Marks, B., Evans, P. R., and McMahon, H. T. (1999). Importance of the pleckstrin homology domain of dynamin in clathrin-mediated endocytosis. *Curr. Biol.* **9**, 257–260.
78. Booden, M. A., Campbell, S. L., and Der, C. J. (2002). Critical but distinct roles for the pleckstrin homology and cysteine-rich domains as positive modulators of Vav2 signaling and transformation. *Mol Cell Biol.* **22**, 2487–2497.
79. Russo, C., Gao, Y., Mancini, P., Vanni, C., Porotto, M., Falasca, M., Torrisi, M. R., Zheng, Y., and A., E. (2001). Modulation of oncogenic DBL activity by phosphoinositol binding to pleckstrin homology domains. *J. Biol. Chem.* **276**, 19524–19531.
80. Han, J., Luby-Phelps, K., Das, B., Shu, X., Xi, Y., Mosteller, R. D., Krishna, U. M., Falck, J. R., White, M. A., and Broek, D. (1998). Role of substrates and products of PI 3-kinase in regulating activation of Rac-related guanine triphosphatases by Vav. *Science* **279**, 558–560.
81. Nimnual, A. S., Yatsula, B. A., and Bar-Sagi, D. (1998). Coupling of the Ras and Rac guanine triphosphatases through the Ras exchanger Sos. *Science* **279**, 560–563.
82. Touhara, K., Inglese, J., Pitcher, J. A., Shaw, G., and Lefkowitz, R. J. (1994). Binding of G protein beta gamma-subunits to pleckstrin homology domains. *J. Biol. Chem.* **269**, 10217–10220.

83. Pitcher, J. A., Touhara, K., Payne, E. S., and Lefkowitz, R. J. (1995). Pleckstrin homology domain-mediated membrane association and activation of the β-adrenergic receptor kinase requires coordinate interaction with $G_{\beta\gamma}$ subunits and lipid. *J. Biol. Chem.* **270**, 11707–11710.
84. Jamora, C., Yamanouye, N., Van Lint, J., Laudenslager, J., Vandenheede, J. R., Faulkner, D. J., and Malhotra, V. (1999). $G_{\beta\gamma}$-mediated regulation of Golgi organization is through the direct activation of protein kinase D. *Cell* **98**, 59–68.
85. Yao, L., Suzuki, H., Ozawa, K., Deng, J., Lehel, C., Fukamachi, H., Anderson, W. B., Kawakami, Y., and Kawakami, T. (1997). Interactions between protein kinase C and pleckstrin homology domains. Inhibition by phosphatidylinositol 4,5-bisphosphate and phorbol 12-myristate 13-acetate. *J. Biol. Chem.* **272**, 13033–13039.
86. Konishi, H., Kuroda, S., Tanaka, M., Matsuzaki, H., Ono, Y., Kameyama, K., Haga, T., and Kikkawa, U. (1995). Molecular cloning and characterization of a new member of the RAC protein kinase family: association of the pleckstrin homology domain of three types of RAC protein kinase with protein kinase C subspecies and beta gamma subunits of G proteins. *Biochem. Biophys. Res. Comm.* **216**, 526–534.
87. Kunstle, G., Laine, J., Pierron, G., Kagami, S. S., Nakajima, H., Hoh, F., Roumestand, C., Stern, M. H., and Noguchi, M. (2002). Identification of Akt association and oligomerization domains of the Akt kinase coactivator TCL1. *Mol. Cell. Biol.* **22**, 1513–1525.
88. Laine, J., Kunstle, G., Obata, T., Sha, M., and Noguchi, M. (2000). The protooncogene TCL1 is an Akt kinase coactivator. *Mol. Cell* **6**, 395–407.
89. Pekarsky, Y., Koval, A., Hallas, C., Bichi, R., Tresini, M., Malstrom, S., Russo, G., Tsichlis, P., and Croce, C. M. (2000). Tcl1 enhances Akt kinase activity and mediates its nuclear translocation. *Proc. Natl. Acad. Sci. USA* **97**, 3028–3033.
90. Rodriguez, M. M., Ron, D., Touhara, K., Chen, C.-H., and Mochly-Rosen, D. (1999). RACK1, a protein kinase C anchoring protein, coordinates the binding of activated protein kinase C and select pleckstrin homology domains *in vitro*. *Biochemistry* **38**, 13787–13794.
91. Jiang, Y., Ma, W., Wan, Y., Kozasa, T., Hattori, S., and Huang, X. Y. (1998). The G protein $G_{\alpha 12}$ stimulates Bruton's tyrosine kinase and a rasGAP through a conserved PH/BM domain. *Nature* **395**, 808–813.
92. Yang, W. and Desiderio, S. (1997). BAP-135, a target for Bruton's tyrosine kinase in response to B cell receptor engagement. *Proc. Natl. Acad. Sci. USA* **94**, 604–609.
93. Yao, L., Janmey, P., Frigeri, L. G., Han, W., Fujita, J., Kawakami, Y., Apgar, J. R., and Kawakami, T. (1999). Pleckstrin homology domains interact with filamentous actin. *J. Biol. Chem.* **274**, 19752–19761.
94. Burks, D. J., Wang, J., Towery, H., Ishibashi, O., Lowe, D., Riedel, H., and White, M. F. (1998). IRS pleckstrin homology domains bind to acidic motifs in proteins. *J. Biol. Chem.* **273**, 31061–31067.
95. Maffucci, T. and Falasca, M. (2001). Specificity in pleckstrin homology (PH) domain membrane targeting: a role for a phosphoinositide-protein co-operative mechanism. *FEBS Letts.* **506**, 173–179.
96. Levine, T. P. and Munro, S. (2002). Targeting of Golgi-specific pleckstrin homology domains involves both PtdIns 4-kinase-dependent, and independent, components. *Curr. Biol.* **12**, 695–704.
97. Stefan, C. J., Audhya, A., and Emr, S. D. (2002). The yeast synaptojanin-like proteins control the cellular distribution of phosphatidylinositol (4,5)-bisphosphate. *Mol. Biol. Cell.* **13**, 542–557.
98. Kraulis, P. J. (1991). MOLSCRIPT: A program to produce both detailed and schematic plots of protein structures. *J. Appl. Crystallog* **24**, 946–950.

PX Domains

Hui Liu and Michael B. Yaffe

Center for Cancer Research,
Massachusetts Institute of Technology,
Cambridge, Massachusetts

History and Overview of PX Domains

The Phox (*ph*agocyte *o*xidase) homology (PX) domain, containing ~100 amino acids, was initially identified by sequence profiling as a conserved region present in the C2 domain-containing class of PI 3-kinases and in the N-terminal region of the p40[phox] and p47[phox] subunits of the NADPH oxidase [1]. For nearly five years following their discovery, the function of PX domains remained obscure. A conserved polyproline motif conforming to the consensus sequence PXXP, where X denotes any amino acid, was noted within most PX domain sequences. This observation, coupled with the presence of one or more SH3 domains in numerous PX domain-containing proteins (Fig. 1), led to speculation that one function of PX domains might involve binding to SH3 domains [1].

PX domain-containing proteins are found in all eukaryotes from yeast to human, and can be loosely divided into two groups (c.f. Fig. 1). The first group includes a large family of cytoplasmic and/or para-membrane proteins known as sorting nexins (SNX), including SNX1 through SNX11 and SNX17 in higher eukaryotes [2–5], and Vam7p, Vps5p, Mvp1p, and Grd19p in yeast [6–9]. Most sorting nexins contain no other recognizable domain other than the PX domain (Fig. 1). Collectively, all members of the SNX family are believed to be involved in vesicular trafficking (Table I, top). A second group of PX domain-containing proteins all contain one or more domains of known function in addition to the PX domain (Fig. 1). These co-associating domains include protein-protein interaction domains such as SH3 domains, PDZ domains, and RGS domains; protein-lipid binding domains such as C2 domains and PH domains; and catalytic domains such as the lipid kinase domain of PI 3-kinase or the phospholipase domain of Phospholipase D. These additional domains play an important role in defining the functions of their constituent proteins, whose intracellular localization is determined, in part, by the PX domain (Table I, bottom).

Lipid-Binding Specificity and the Structure of PX Domain

The observation that many PX domain-containing proteins were membrane associated suggested that their ligands might be specific phospholipids. This was experimentally verified via protein overlay assays on solid-phase immobilized phospholipids and by solution-phase binding assays using phosphoplipid-containing synthetic liposomes. These experiments demonstrated that the ligands for many PX domains were specific phosphoinositide products of PI 3-kinase [10–13]. Different PX domains show distinct specificity for different phosphoinositides. The PX domains of p47[phox] and p40[phox], for example, bound to phosphatidylinositol-3,4-bisphosphate [PtdIns $(3,4)P_2$] and PtdIns(3)P, respectively [12,13]. The PX domains of Vam7p and SNX3 bind specifically to PtdIns(3)P [10,11,14] whereas the PX domain of cytokine-independent survival kinase (CISK) interacts with PtdIns$(3,5)P_2$, PtdIns $(3,4,5)P_3$, and PtdIns$(4,5)P_2$ [15]. It appears that the vast majority of PX domains, however, including all of those in the budding yeast *Saccharomyces cerevesiae*, interact primarily with PtdIns(3)P [16].

A sequence alignment of PX domains that shows specificity for PtdIns(3)P binding, including that of p40[phox], SNX3, SNX4, MVP1p, Vam7p, and MDM1p, is shown in Fig. 2. Insight into the structural basis of lipid-binding specificity for PX domains is beginning to emerge from a recent NMR structure of the p47[phox] PX domain without a bound ligand [17], and the X-ray crystal structure of the p40[phox] PX domain

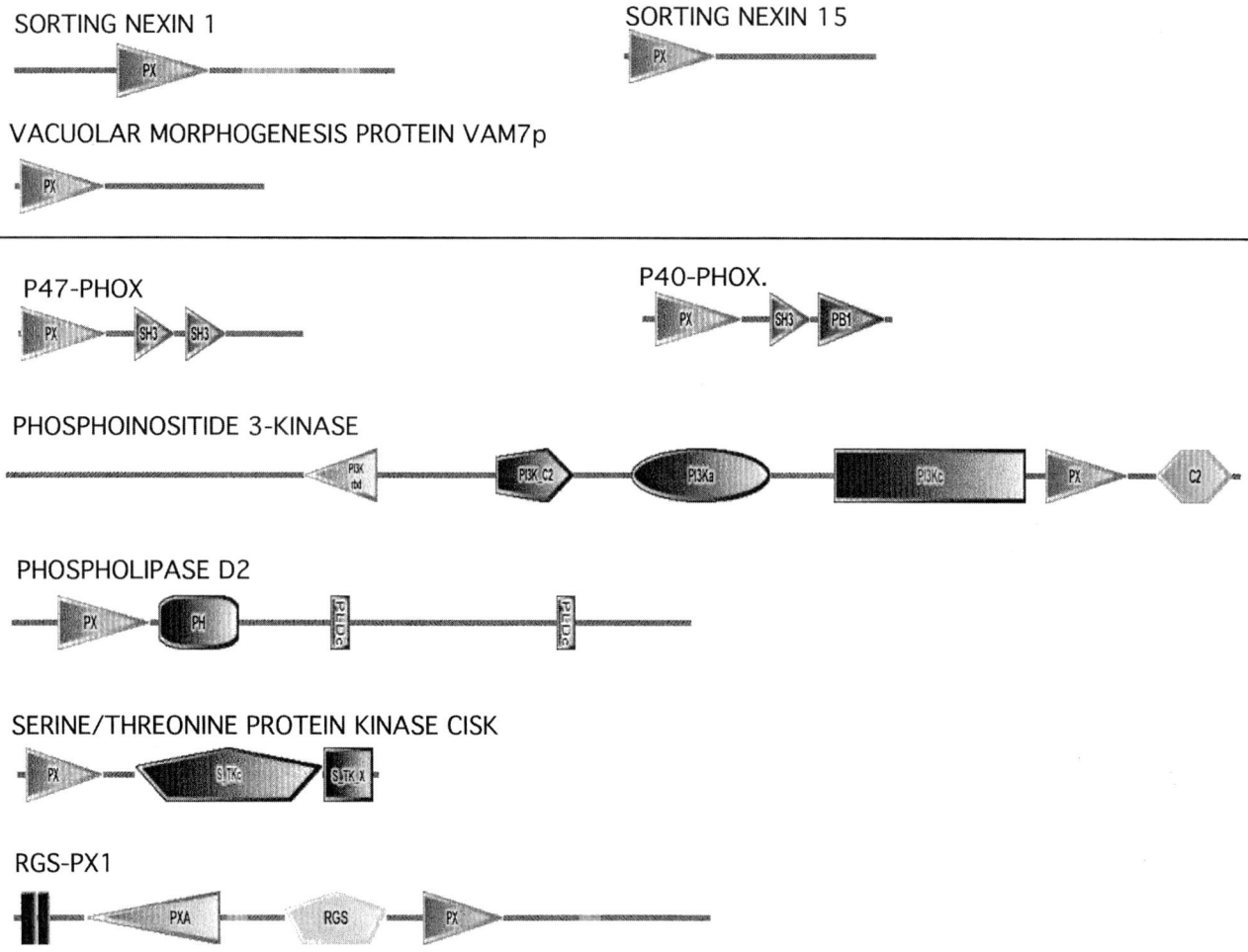

Figure 1 Domain architecture of PX domain-containing proteins. The upper panel shows the structures of sorting nexin proteins and their yeast homologues. The lower panel shows various PX domain-containing signaling molecules that contain additional co-associating domains.

Table I PX Domain-Containing Proteins

Protein	PX domain lipid target	Other domains	Protein function	Reference
SNX1	PtdIns (3)P	—	Binding to EGF receptor Targets EFGR to lysosome	4,7
SNX3	PtdIns (3)P	—	Regulates endosomal function	10
Vps5p/Mvp1p	PtdIns (3)P	—	Sorting carboxypeptidase Y to the vacuole	6
Vam7p	PtdIns (3)P	—	Golgi-to-vacuole transport	11
Grd19p	PtdIns (3)P	—	Retrieval of proteins from prevacuole to late Golgi	9
P40phox	PtdIns (3)P	SH3	Regulation of the NADPH oxidase	12,13
P47phox	PtdIns $(3,4)P_2$	SH3	Activation of NADPH oxidase	13
FISH	ND	SH3	Tyrosine kinase signaling	29
RGS-PX1	ND	RGS PXA	Gα-specific GAP Involved in vesicle trafficking	30
PLD1,2	ND	PH PLD Kinase	Cell division, signal transduction, and vesicle trafficking	31
Pi3K C2-γ	PtdIns$(4,5)P_2$	C2 Pi3 Kinase	EGF receptor signaling	14
CISK	PtdIns(3)P	Ser/Thr kinase	Cell survival	15
HS1BP3	ND	Leucine zipper	T and B cell development	32

The top section displays SNX proteins and their yeast homologs. The bottom section displays proteins containing other known modular signaling domains. (ND indicates Not Determined.)

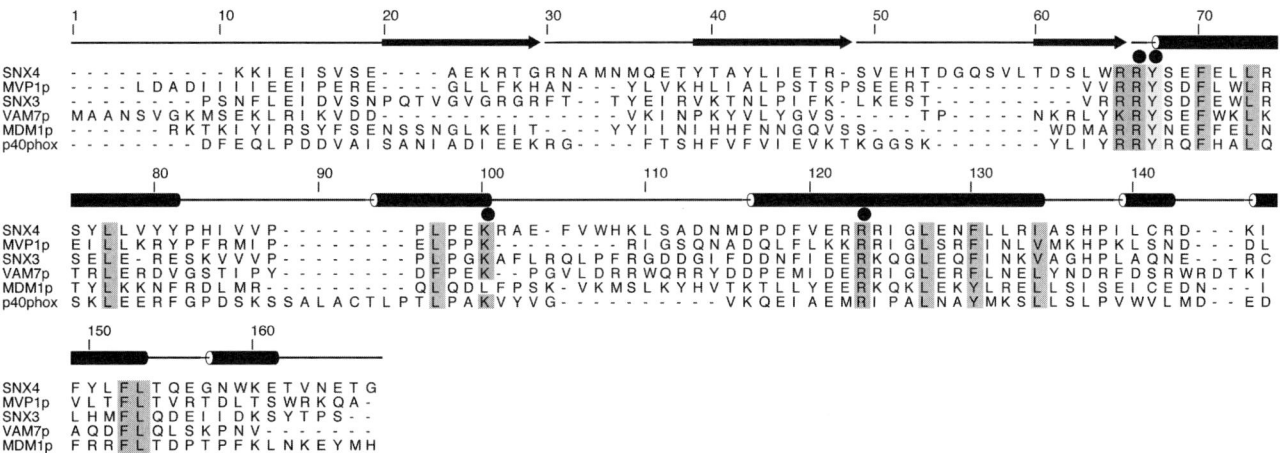

Figure 2 A structure-based sequence alignment of PX domains that bind to PtdIns(3)P. A general numbering scheme is indicated above the alignment, along with a cartoon indicating the positions of β-strands (arrows) and α-helixes (cylinders) based on the structure of the p40^phox PX domain. Conserved hydrophobic residues within the core of the domain are shaded in green, basic and aromatic residues that are involved in lipid binding are shaded in cyan and yellow, respectively. Sequence numbers for particular amino acid residues within individual PX domains are mentioned in the text.

bound to PtdIns(3)P [18]. The PX domain fold is a small three-stranded β-sheet packed against a helical subdomain containing four α helices and a short stretch of 3_{10} helix (Fig. 3). Both the p40^{phox} and p47^{phox} PX domain structures also contain a conserved PXXP motif that forms a type II polyproline (PP$_{II}$) helix. Since type II polyproline helices are well known to bind to SH3 domains, this structural observation suggests that some PX domains may, in fact, form intramolecular interactions with their co-associating SH3 domains.

Many PX domains contain a conserved Arg residue immediately preceding a conserved Tyr residue (Tyr-67 in the sequence alignment shown in Fig. 2). The preceeding Arg residue (Arg-66 in Fig. 2), which corresponds to R58 in the p40^{phox} PX domain, forms two salt bridges with the 3-phosphate of the lipid in the p40^{phox}PX:PtdIns(3)P crystal structure. The PX domain from the C2-containing PI 3-kinase, which binds to PtdIns(4,5)P$_2$, lacks an Arg residue at this position, suggesting that residues equivalent to R58 are specific to PX domains that bind to 3-phosphorylated phosphatidylinositols. The 4- and 5-hydroxyl groups of PtdIns(3)P form hydrogen bonds with another highly conserved residue corresponding to R105 in the p40phox structure (Fig. 3) and R123 in the alignment shown in Fig. 2. Phosphorylation on either the 4- or 5-hydroxyl would sterically impinge on the R105 side chain, rationalizing the PtdIns(3)P binding specificity of the p40^{phox} PX domain. However, R105 is also conserved in PX domains that bind phosphatidylinositols other than PtdIns(3)P, including those of CISK and p47phox. Presumably, alterations in the loops surrounding this residue relieve this steric clash and may allow direct interactions of R105 with the lipid phosphates in the 4- and 5- positions, in place of the interaction with the 4- and 5- hydroxyl groups seen in the p40^{phox} structure.

Tyrosine-59 in the p40^{phox}PX:PtdIns(3)P crystal structure (corresponding to Y67 in the alignment in Fig. 2) is another highly conserved amino acid, which forms the floor of the

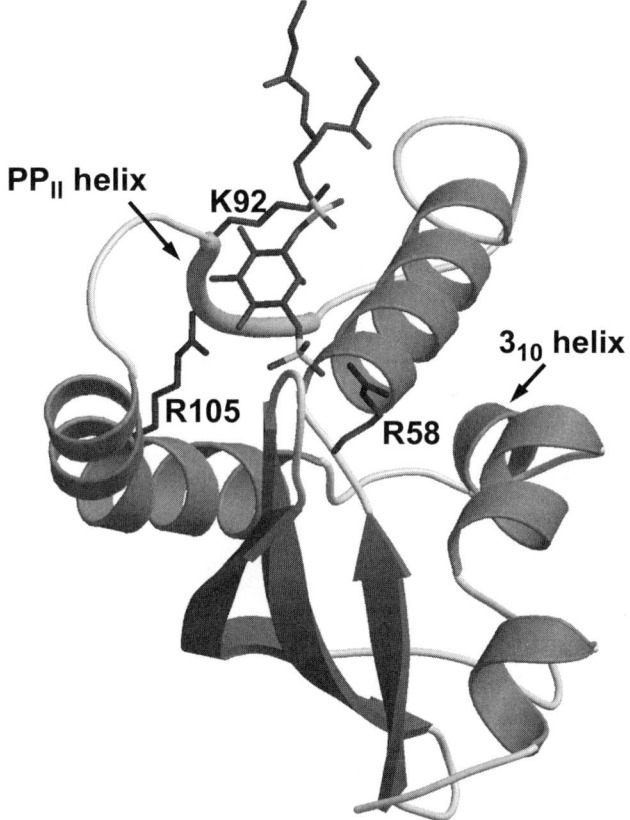

Figure 3 The structure of the p40^{phox}:PtdIns(3)P complex [18]. The lipid is depicted with the acyl chains at the top of the figure, as though it were protruding from a cell membrane. Amino acids that play key roles in PtdIns(3)P-binding are indicated, together with the polyproline-II helix and the 3_{10} helix.

lipid-binding pocket through interactions between its aromatic side chain and the inositol ring. In the p40^{phox} PX:PtdIns(3)P structure, the 1-phosphate forms salt bridges with the side chains of K92 (position 100 in Fig. 2) and R60 (position 68 in Fig. 2) to stabilize the interaction between the domain and the membrane proximal portion of the inositol lipid. Both of these residues are conserved in some, but not all, PX domains, suggesting that other residues also participate. Additional PX domain structures will clearly be necessary to fully understand the molecular determinants of phosphatidylinositol binding.

Function of PX Domain-containing Proteins

The function of several PX domain-containing proteins is reasonably well understood, although the exact role fulfilled by the PX domain is not yet well defined. The *phox* proteins, after which the PX domain was named, are subunits of the NADPH oxidase, the heme-containing enzyme responsible for superoxide production and killing of microorganisms by phagocytic cells. In resting phagocytes, the inactive NADPH oxidase is separated into a set of cytoplasmic subunits including p47^{phox}, p67^{phox}, and p40^{phox} and a membrane-bound heme-containing flavocytochrome b$_{558}$, which consists of gp91^{phox} and p22^{phox}. Upon phagocyte activation, the cytoplasmic components dock with the membrane-bound subunits to form a catalytically active enzyme that can transfer electrons from NADPH to oxygen to form reactive oxygen species (ROS), such as superoxide [19]. Production of superoxide in response to some stimuli requires the activity of PI 3-kinase, suggesting that specific PI 3-kinase lipid products may be directly involved in regulating oxidase assembly. The different lipid-binding specificities observed for the PX domains of p47^{phox} and p40^{phox} may target oxidase assembly to occur only within specific membrane compartments that contain the appropriate combination of PI 3-kinase-derived lipids [13,20,21].

Several PX domain-containing proteins in the budding yeast *Saccharomyces cerevesiae* participate in vesicular protein trafficking, including the Mvp1p and Vps1p proteins involved in vacuolar protein sorting [22] and the Vps17p and Vps5p proteins which translocate pre-vacuolar endosomes to the Golgi as part of the retromer protein complex [23]. In higher eukaryotes, SNX1, SNX2, SNX4, SNX6, and a splice variant of SNX1 (SNX1A) are known to associate with a variety of growth factor receptors, suggesting that they mediate receptor trafficking to vesicles [7,24].

CISK is a PX domain-containing Ser/Thr kinase, which functions in parallel with Ser/Thr kinase Akt/PKB to mediate IL-3 dependent cell survival in hematopoetic cells. CISK localizes to vesicular compartments, and this localization is dependent on the PX domain [15,25].

Phospholipase D (PLD) catalyzes hydrolysis of phospholipids to produce phosphatidic acid (PA) [26]. Both mammalian isoforms of PLD, PLD1 and PLD2, contain PX domains as well as PH domains that also bind to PI 3-kinase lipid products. The presence of two different lipid binding domains might target the lipase domain to specific phosphoinositide-containing regions of membranes, or might function as some type of lipid-regulated switch to control the activity of the lipase domain. The specificity of the PLD PX domain has not yet been reported.

Class II PI 3-kinases, defined by their *in vitro* usage of phosphatidylinositol and phosphatidylinositol 4-phosphate as substrates, also contain PX domains. The function of Class II PI 3-kinases is not well understood, though recent results suggest that they may participate in clathrin-mediated endocytosis [27,28].

In summary, PX domains join an expanding family of phosphoinositide-binding domains that includes C2 domains, PH domains, FYVE domains, ENTH domains, and tubby domains. The large differences in structure between these domains suggest that their lipid-binding function arose through the convergent evolution of different structures for the same biological function of lipid binding. It will be important for future work to further examine the structural foundation for lipid binding specificity by PX domains, and explore how their lipid-binding ability contributes to the overall function of PX domain-containing molecules.

References

1. Ponting, C. P. (1996). Novel domains in NADPH oxidase subunits, sorting nexins, and PtdIns 3-kinases: binding partners of SH3 domains? *Protein Sci.* **5**, 2353–2357.
2. Barr, V. A., Phillips, S. A., Taylor, S. I., and Haft, C. R. (2000). Overexpression of a novel sorting nexin, SNX15, affects endosome morphology and protein trafficking. *Traffic* **1**, 904–916.
3. Florian, V., Schluter, T., and Bohnensack, R. (2001). A new member of the sorting nexin family interacts with the C-terminus of P-selectin. *Biochem. Biophys. Res. Commun.* **281**, 1045–1050.
4. Kurten, R. C., Cadena, D. L., and Gill, G. N. (1996). Enhanced degradation of EGF receptors by a sorting nexin, SNX1. *Science* **272**, 1008–1010.
5. Teasdale, R. D., Loci, D., Houghton, F., Karlsson, L., and Gleeson, P. A. (2001). A large family of endosome-localized proteins related to sorting nexin 1. *Biochem. J.* **358**, 7–16.
6. Ekena, K. and Stevens, T. H. (1995). The Saccharomyces cerevisiae MVP1 gene interacts with VPS1 and is required for vacuolar protein sorting. *Mol. Cell Biol.* **15**, 1671–1678.
7. Haft, C. R., de la Luz Sierra, M., Barr, V. A., Haft, D. H., and Taylor, S. I. (1998). Identification of a family of sorting nexin molecules and characterization of their association with receptors. *Mol. Cell. Biol.* **18**, 7278–7287.
8. Sato, T. K., Darsow, T., and Emr, S. D. (1998). Vam7p, a SNAP-25-like molecule, and Vam3p, a syntaxin homolog, function together in yeast vacuolar protein trafficking. *Mol. Cell. Biol.* **18**, 5308–5319.
9. Voos, W. and Stevens, T. H. (1998). Retrieval of resident late-Golgi membrane proteins from the prevacuolar compartment of Saccharomyces cerevisiae is dependent on the function of Grd19p. *J. Cell Biol.* **140**, 577–590.
10. Xu, Y., Hortsman, H., Seet, L., Wong, S. H., and Hong, W. (2001). SNX3 regulates endosomal function through its PX-domain-mediated interaction with PtdIns(3)P. *Nat. Cell Biol.* **3**, 658–666.
11. Cheever, M. L., Sato, T. K., de Beer, T., Kutateladze, T. G., Emr, S. D., and Overduin, M. (2001). Phox domain interaction with PtdIns(3)P targets the Vam7 t-SNARE to vacuole membranes. *Nat. Cell Biol.* **3**, 613–618.
12. Ellson, C. D., Gobert-Gosse, S., Anderson, K. E., Davidson, K., Erdjument-Bromage, H., Tempst, P., Thuring, J. W., Cooper, M. A., Lim, Z. Y., Holmes, A. B. et al. (2001). PtdIns(3)P regulates the neutrophil oxidase complex by binding to the PX domain of p40(phox). *Nat. Cell Biol.* **3**, 679–682.
13. Kanai, F., Liu, H., Field, S. J., Akbary, H., Matsuo, T., Brown, G. E., Cantley, L. C., and Yaffe, M. B. (2001). The PX domains of p47phox and p40phox bind to lipid products of PI(3)K. *Nat. Cell Biol.* **3**, 675–678.

14. Song, X., Xu, W., Zhang, A., Huang, G., Liang, X., Virbasius, J. V., Czech, M. P., and Zhou, G. W. (2001). Phox homology domains specifically bind phosphatidylinositol phosphates. *Biochemistry* **40**, 8940–8944.
15. Xu, J., Liu, D., Gill, G., and Songyang, Z. (2001). Regulation of cytokine-independent survival kinase (CISK) by the Phox homology domain and phosphoinositides. *J. Cell Biol.* **154**, 699–705.
16. Yu, J. W. and Lemmon, M. A. (2001). All phox homology (PX) domains from *Saccharomyces cerevisiae* specifically recognize phosphatidylinositol 3-phosphate. *J. Biol. Chem.* **276**, 44179–44184.
17. Hiroaki, H., Ago, T., Ito, T., Sumimoto, H., and Kohda, D. (2001). Solution structure of the PX domain, a target of the SH3 domain. *Nat. Struct. Biol.* **8**, 526–530.
18. Bravo, J., Karathanassis, D., Pacold, C. M., Pacold, M. E., Ellson, C. D., Anderson, K. E., Butler, P. J., Lavenir, I., Perisic, O., Hawkins, P. T. *et al.* (2001). The crystal structure of the PX domain from p40(phox) bound to phosphatidylinositol 3-phosphate. *Mol. Cell* **8**, 829–839.
19. Babior, B. M. (1999). NADPH oxidase: an update. *Blood* **93**, 1464–1476.
20. Palicz, A., Foubert, T. R., Jesaitis, A. J., Marodi, L., and McPhail, L. C. (2001). Phosphatidic acid and diacylglycerol directly activate NADPH oxidase by interacting with enzyme components. *J. Biol. Chem.* **276**, 3090–3097.
21. Vieira, O. V., Botelho, R. J., Rameh, L., Brachmann, S. M., Matsuo, T., Davidson, H. W., Schreiber, A., Backer, J. M., Cantley, L. C., and Grinstein, S. (2001). Distinct roles of class I and class III phosphatidylinositol 3-kinases in phagosome formation and maturation. *J. Cell Biol.* **155**, 19–25.
22. Wilsbach, K. and Payne, G. S. (1993). Vps1p, a member of the dynamin GTPase family, is necessary for Golgi membrane protein retention in Saccharomyces cerevisiae. *EMBO J.* **12**, 3049–3059.
23. Seaman, M. N., McCaffery, J. M., and Emr, S. D. (1998). A membrane coat complex essential for endosome-to-Golgi retrograde transport in yeast. *J. Cell Biol.* **142**, 665–681.
24. Parks, W. T., Frank, D. B., Huff, C., Renfrew Haft, C., Martin, J., Meng, X., de Caestecker, M. P., McNally, J. G., Reddi, A., Taylor, S. I., *et al.* (2001). Sorting nexin 6, a novel SNX, interacts with the transforming growth factor-beta family of receptor serine-threonine kinases. *J. Biol. Chem.* **276**, 19332–19339.
25. Liu, D., Yang, X., and Songyang, Z. (2000). Identification of CISK, a new member of the SGK kinase family that promotes IL-3-dependent survival. *Curr. Biol.* **10**, 1233–1236.
26. Cockcroft, S. (2001). Signalling roles of mammalian phospholipase D1 and D2. *Cell. Mol. Life Sci.* **58**, 1674–1687.
27. Domin, J., Gaidarov, I., Smith, M. E., Keen, J. H., and Waterfield, M. D. (2000). The class II phosphoinositide 3-kinase PI3K-C2alpha is concentrated in the trans-Golgi network and present in clathrin-coated vesicles. *J. Biol. Chem.* **275**, 11943–11950.
28. Gaidarov, I., Smith, M. E., Domin, J., and Keen, J. H. (2001). The class II phosphoinositide 3-kinase C2alpha is activated by clathrin and regulates clathrin-mediated membrane trafficking. *Mol. Cell* **7**, 443–449.
29. Lock, P., Abram, C. L., Gibson, T., and Courtneidge, S. A. (1998). A new method for isolating tyrosine kinase substrates used to identify fish, an SH3 and PX domain-containing protein, and Src substrate. *Embo. J.* **17**, 4346–4357.
30. Zheng, B., Ma, Y. C., Ostrom, R. S., Lavoie, C., Gill, G. N., Insel, P. A., Huang, X. Y., and Farquhar, M. G. (2001). RGS-PXI, a GAP for GalphaS and sorting nexin in vesicular trafficking. *Science* **294**, 1939–1942.
31. Liscovitch, M., Czarny, M., Fiucci, G., and Tang, X. (2000). Phospholipase D: molecular and cell biology of a novel gene family. *Biochem. J.* **345**, 401–415.
32. Takemoto, Y., Furuta, M., Sato, M., Kubo, M., and Hashimoto, Y. (1999). Isolation and characterization of a novel HS1 SH3 domain binding protein, HS1BP3. *Int. Immunol.* **11**, 1957–1964.

CHAPTER 152

FYVE Domains in Membrane Trafficking and Cell Signaling

Christopher Stefan, Anjon Audhya, and Scott Emr

*Division of Cellular and Molecular Medicine,
The Howard Hughes Medical Institute, University of California,
San Diego, School of Medicine, La Jolla, California*

Introduction

The recruitment of cytoplasmic proteins to specific membrane compartments is important for a diverse spectrum of cellular processes, including intracellular protein trafficking, cytokine and growth factor receptor signaling, actin cytoskeleton organization, and apoptosis [1–4]. Many proteins are localized to membranes through tightly regulated interactions with membrane-associated factors. Derivatives of phosphatidylinositol (PtdIns) that can be reversibly phosphorylated at different positions of the inositol ring are ideally suited for this function. Through the action of a set of well-conserved specific lipid kinases [5], different phosphoinositide (PI) species phosphorylated at the 3′, 4′, or 5′ positions of the inositol headgroup are generated, each of which can recruit or activate a specific subset of cytoplasmic effector proteins. The activity of these target proteins can then be attenuated through the action of PI-specific phosphatases and lipases [6–9]. Several studies have now identified multiple, well-conserved PI-binding motifs, each of which can recognize particular PI isoforms with a high degree of specificity [10]. In this chapter, we discuss the structural basis of a novel zinc finger that binds PtdIns 3-phosphate [PtdIns(3)P], termed the FYVE domain, and the roles played by several proteins harboring this motif in membrane trafficking and cell signaling.

Role for PtdIns(3)P in Membrane Trafficking and Identification of the FYVE Domain

A role for PtdIns(3)P in vesicular transport was first discovered in the study of Golgi to vacuole transport in yeast [11]. *Saccharomyces cerevisiae* expresses one PtdIns 3-kinase isoform, Vps34 [5]. Deletion of the *VPS34* gene resulted in a lack of PtdIns(3)P synthesis and defects in endosomal membrane trafficking from the Golgi and plasma membrane to the lysosome-like vacuole [12]. Likewise, PtdIns(3)P has been shown to play important roles in several membrane trafficking pathways to mammalian lysosomes [13]. The fungal metabolite wortmannin, an inhibitor of PI 3-kinase activity, has been shown to impair homotypic endosome fusion *in vitro* and the transport of enzymes such as cathepsin D to lysosomes *in vivo* [14–16]. Accordingly, the human homolog of the yeast Vps34 PtdIns 3-kinase has been identified and found to be sensitive to wortmannin [5].

Several proteins have been implicated as downstream effectors of PtdIns(3)P in vesicle transport. One of these, mammalian EEA1 (early endosome antigen 1), has been shown to localize to endosomal membranes in a wortmannin-sensitive manner [17,18]. Consequently, deletion of the FYVE domain of EEA1 was shown to diminish its endosomal association, suggesting that this domain may directly bind PtdIns(3)P [19]. The FYVE domain, named after the first four proteins found to contain this motif (*F*ab1, *Y*OTB,

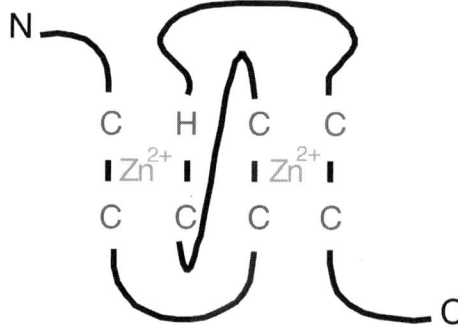

Fab1	WMK DESSKE CFS C GKT FNTFR RKHH C RI CX$_4$CX$_2$CX$_{15}$R$_V$ CYN CYE
YDR313	WQA DEEAHS CFQ C KTN FSFLV RRHH C RI CX$_4$CX$_2$CX$_{27}$ RT CNE CYD
Vac1	WRD DRSVLF C NI C SEP FGLLL RKHH C RL CX$_4$CX$_6$ CX$_{32}$ RL C SH C ID
EEA1	WAE DNEVQN CMA C GKG FSVTV RRHH C RQ CX$_4$CX$_2$CX$_{14}$ RV CDA CFN

Figure 1 The FYVE domain is a conserved RING finger domain. (Top) Schematic cartoon of the RING finger FYVE domain. The conserved cysteine/histidine residues that coordinate two Zn^{2+} atoms are shown. The highly conserved basic patch surrounding the third cysteine is indicated in blue. (Bottom) Sequence alignment of the FYVE domains of Fab1, Pib1 (YDR313c), Vac1, and EEA1. Identical residues are shown in boldface. The conserved cysteine residues are highlighted in gray. The highly conserved basic patch, R(R/K)HHCR, found in all FYVE domains is shown. The conserved hydrophobic region adjacent to the basic patch is indicated in red.

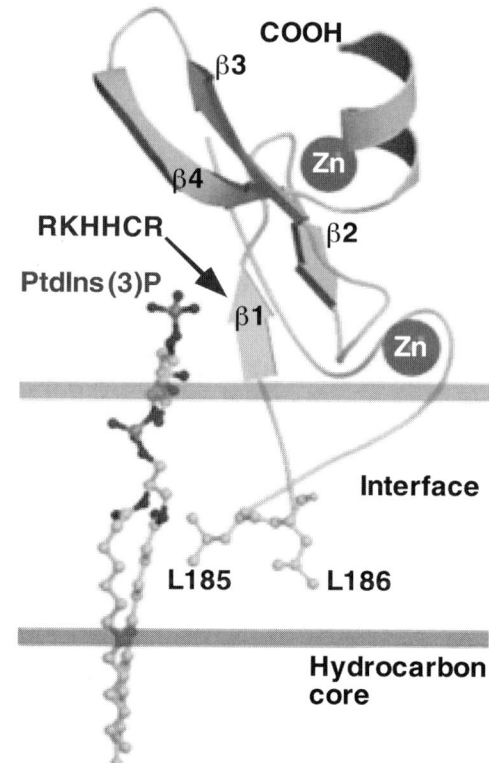

Figure 2 The FYVE domain is a modular PtdIns(3)P-binding motif. The crystal structure of the FYVE domain of yeast Vps27 [26]. The ribbon depicts four β strands followed by a carboxy-terminal α-helix. The two Zn^{2+} atoms are shown. The highly conserved positively charged residues (RKHHCR) in the β1 strand predicted to make contacts with the 3'-phosphate group of PtdIns(3)P are indicated. In addition, residues in a hydrophobic loop upstream of the first β-sheet predicted to penetrate into the membrane bilayer are indicated [26]. The membrane layer is divided into an interfacial region (lipid headgroups and the hydrophobic interface) and a hydrocarbon core (lipid acyl chains). (Reprinted from Hurley, Cell, 97, 657–666, 1999. With permission).

Vac1 and EEA1), was originally identified as a RING-finger family member that coordinates two Zn^{2+} ions through eight cysteine/histidine residues spaced in a conserved manner $[CX_2CX_{9-39}CX_{1-3}(C/H)X_{2-3}CX_2CX_{4-48}CX_2C]$ (Fig. 1) [19,20]. An important finding of subsequent studies was the demonstrated ability of FYVE domains to specifically bind PtdIns-3-P *in vitro*, as recombinant EEA1 FYVE sedimented with liposomes containing PtdIns(3)P but not other PI species [21–23]. The identification of modular protein domains that bind PtdIns(3)P with high affinity and specificity, such as the FYVE domain, has been a crucial step in further understanding the roles of this lipid in membrane trafficking events, as described in greater detail below.

Structural Basis for the FYVE Domain

Insight into the molecular mechanisms that mediate the interaction between FYVE domains and PtdIns(3)P is provided by several structural studies on this motif [24,25]. The FYVE domain, as mentioned, is an approximately 80-amino-acid sequence containing eight conserved cysteine/histidine residues that coordinate two Zn^{2+} ions. In addition, several other residues are conserved, most notably a highly basic R(R/K)HHCR patch adjacent to the third cysteine residue, an amino-terminal WxxD motif, and a conserved hydrophobic region upstream of the basic patch (Fig. 1). As first determined from the crystal structure of the yeast Vps27 FYVE domain, the basic patch is localized within β1 of two double-stranded antiparallel β-sheets (composed of β1/β2 and β3/β4), which are stabilized by the two zinc ions and a C-terminal α-helix [26]. Molecular modeling suggested that the inositol head group of PtdIns(3)P fits into a pocket created by the backbone of the first β-sheet, and the 3'-phosphate group contacts side groups of the final histidine and arginine residues found in the basic patch (Fig. 2) [26]. In addition, from this model, the 1'-phosphate of PtdIns(3)P is poised to form a salt bridge with the first arginine in the conserved basic patch [26]. In combination, these interactions are specific for PtdIns(3)P as additional or other phosphate groups on the inositol ring would prohibit interaction with the FYVE domain due to spatial constraints, consistent with previous *in vitro* binding studies indicating this motif does not bind to other phosphoinositides.

Although a similar structure was proposed for the FYVE domain of *Drosophila* Hrs, a homologue of Vps27, a different model for PtdIns(3)P binding was suggested [27]. The major difference involved an anti-parallel association of two Hrs FYVE monomers to generate a homodimer with two ligand-binding pockets. Residues from β1, including the conserved basic patch, together with a hydrophobic strand from β4 of the opposite FYVE monomer, line each pocket.

In addition, this model differed from that of the Vps27 FYVE structure in regard to the orientation of the FYVE domain with respect to the membrane. Thus, even though both models were consistent in regard to the overall structure of the FYVE domain, further studies were required to help resolve the true nature of the interaction between the FYVE domain and PtdIns(3)P containing membranes.

To gain further insight into the interaction of the FYVE domain with PtdIns(3)P, NMR studies of the EEA1 FYVE domain were performed [28,29]. As expected, these studies highlighted the importance of the basic residues found in the first β-sheet as they displayed large chemical shift changes in the presence of PtdIns(3)P. In addition, residues in a hydrophobic loop upstream of the first β-sheet displayed chemical shifts, but only in the presence of micelle-embedded PtdIns(3)P, suggesting that these residues contact the membrane nonspecifically. Similar to the membrane orientation predicted by the Vps27 FYVE structure (see Fig. 2), this hydrophobic loop may extend into the cytoplasmic side of the membrane bilayer, perhaps directly interacting with hydrophobic acyl chains.

However, when fused to GFP or GST, the EEA1 FYVE domain alone failed to efficiently localize to cellular membranes [21–23,30]. Residues from an additional coiled-coil region adjacent to the FYVE domain were required. Most recently, the crystal structure of the EEA1 FYVE domain including these additional residues has revealed the formation of stable homodimers that could bind two molecules of inositol 1,3-bisphosphate, a soluble mimic of PtdIns(3)P [31]. However, unlike the model proposed from studies of the Hrs FYVE domain, each EEA1 FYVE domain independently bound inositol 1,3-bisphosphate. Dimerization of the EEA1 FYVE domains was mediated primarily through interactions between the coiled-coil domains. Taken together, these data suggest that dimerization enhances binding of individual FYVE domains to membrane-restricted PtdIns(3)P. However, while the EEA1 FYVE structures provide an accurate model for PtdIns(3)P binding, additional factors may be involved in targeting and/or stabilization of FYVE domains at cellular membranes. It is interesting that residues adjacent to the EEA1 FYVE domain required for membrane localization are required for EEA1 to bind Rab5, a small GTPase that functions on membranes in the endocytic pathway [18,30]. Together, these data suggest that a combination of protein-PtdIns(3)P and protein-protein interactions is essential for specific and stable localization of FYVE domain-containing proteins to particular cellular membranes.

Conservation of the FYVE Domain and Localization of PtdIns(3)P

To date, analysis of the human genome has uncovered a total of approximately 30 FYVE domain-containing proteins, while the *Caenorhabditis elegans* genome contains 15 and the *S. cerevisiae* genome harbors 5. Thus, while the FYVE domain itself has been well conserved through the course of evolution, there appears to have been a significant expansion in the roles played by this lipid-binding motif. As described earlier, the major role for PtdIns(3)P, and by extension its FYVE domain-containing effectors, involves endocytic membrane transport. However, recent findings also show that PtdIns(3)P may have roles in growth factor signaling and actin cytoskeleton organization through the recruitment/activation of other FYVE domain-containing proteins. However, before further examining the role of the FYVE domain in cell signaling, localization of PtdIns(3)P itself must first be explored.

Initial studies of PtdIns(3)P localization were carried out via a GFP fusion to the FYVE domain of EEA1 in yeast [21]. Results indicated that the fusion co-localized with prevacuolar endosomes and weakly labeled the vacuolar membrane. An important finding is that this localization was dependent on Vps34 PtdIns 3-kinase activity, demonstrating a requirement for PtdIns(3)P in mediating membrane association *in vivo* [21]. Two FYVE domains in tandem fused to GFP similarly localized to endosomal structures in fibroblasts [32]. However, due to the limitations of light microscopy, a more detailed analysis of PtdIns(3)P localization required more extensive studies of the recombinant FYVE domain dimer. Using an electron microscopic labeling approach, PtdIns(3)P was found to be highly enriched on endosomes as expected from previous work, but the lipid was also observed in the nucleolus and in the internal vesicles of multivesicular bodies (MVBs) [32]. Consistent with the presence of PtdIns(3)P on vesicles inside the lumen of MVBs, the efficient turnover of PtdIns(3)P in yeast was shown to be dependent on hydrolase-mediated degradation in the vacuole [33]. It is likely that most, if not all, PtdIns(3)P effectors are recruited to and/or activated at endosomal/vacuolar membranes, the major sites of PtdIns(3)P accumulation in cells.

FYVE Domains in Membrane Trafficking

Studies of EEA1 have been instrumental in defining the localization of PtdIns(3)P. Closer examination of the protein reveals that it is a large coiled-coil protein (Fig. 3A) that can bind to the GTP-bound form of Rab5, a GTPase required for endosomal membrane fusion [18,34]. Together with PtdIns(3)P, Rab5-GTP recruits EEA1 to endosomal membranes where it functions in membrane fusion. Consistent with this hypothesis, depletion of EEA1 inhibits homotypic endosome fusion *in vitro*, while excess EEA1 stimulates fusion [18,34]. Furthermore, studies suggest that EEA1 may engage in oligomeric complexes during membrane fusion, which may tether Rab5-positive endosomes, thus facilitating pairing of SNARE proteins to drive membrane fusion [34,35].

Similar to EEA1, another FYVE domain-containing protein, Rabenosyn-5 (Table I), is an effector of Rab5-GTP. Its localization to the endosome is dependent on PtdIns(3)P binding and is required for endosome fusion [36]. While EEA1 appears to interact with specific SNARES including syntaxin-13 [35], Rabenosyn-5 directly interacts with the

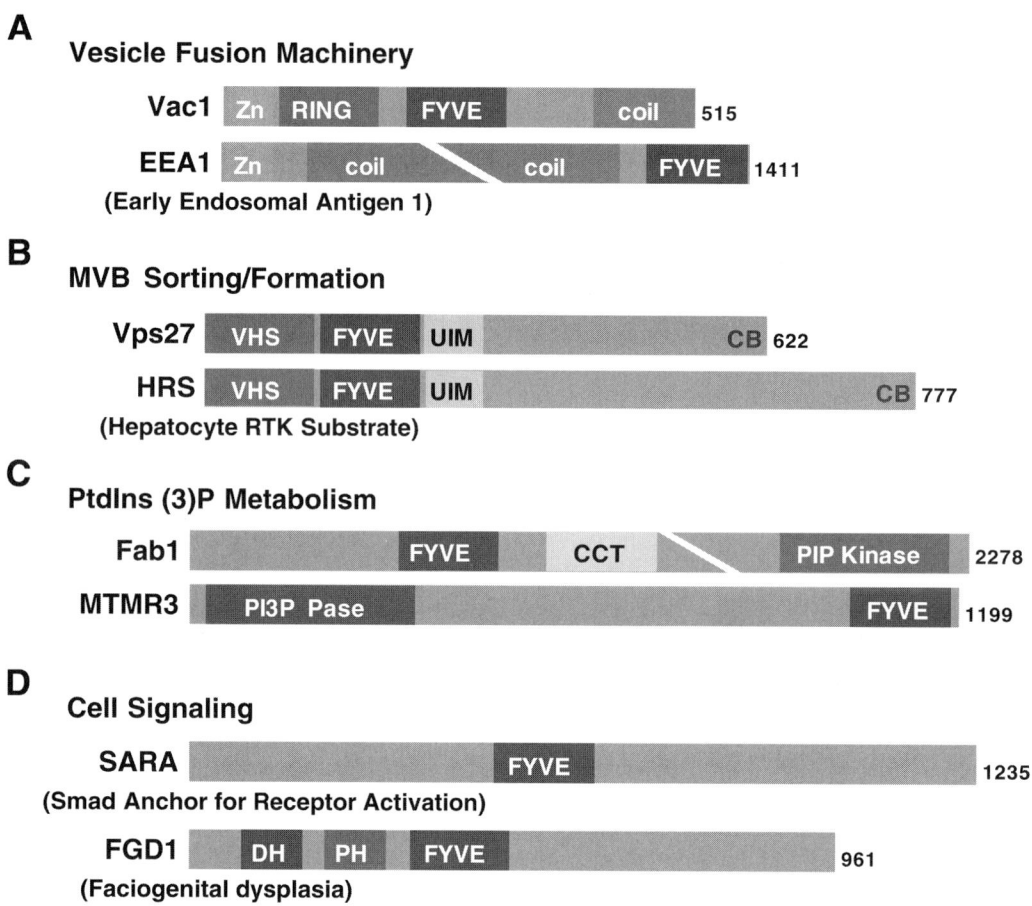

Figure 3 Schematic representation of protein motifs found within FYVE domain proteins. Several FYVE domain proteins have additional domains that bind protein targets. Together, these PtdIns(3)P-protein and protein-protein interactions define the specific function of each FYVE domain protein. Examples of yeast and mammalian FYVE domain proteins that act in various cellular processes are shown. (A) FYVE domain proteins implicated in vesicle targeting and fusion events. (B) FYVE domain proteins involved in endosomal/MVB sorting. (C) FYVE domain proteins containing enzymatic activities implicated in PI synthesis and turnover. (D) FYVE domain proteins involved in cell signaling. Other abbreviations: Zn, Zn^{2+} finger domain; RING, Zn^{2+} finger domain; coil, coiled-coil domain; VHS, conserved domain found in Vps27, Hrs, and STAM; SH3, Src homology 3 domain; UIM, ubiquitin-interacting motif; CB, clathrin box binding motif; CCT, chaperonin-like region; PIP kinase, PtdIns(3)P 5-kinase catalytic domain; PI3P Pase, myotubularin-related PtdIns(3)P phosphatase catalytic domain; DH, Dbl homology domain; PH, pleckstrin homology domain.

Sec1-like protein hVps45, suggesting distinct roles for these Rab5/PtdIns(3)P effectors in endosome fusion [36]. Similarly, the yeast homolog of Rabenosyn-5, Vac1, also contains a FYVE domain (Fig. 3A) and interacts with Vps21 and Vps45, homologs of Rab5 and Sec1 [37–39]. Deletion of *VAC1* results in an accumulation of vesicles destined for the endosome and a defect in protein sorting to the lysosome-like vacuole, suggesting that Vac1 is evolutionarily conserved and required for endocytic docking and/or fusion [38]. Substitutions in the FYVE domain of Vac1 also result in defects in vacuolar protein sorting, indicating a requirement for this domain in Vac1 function. However, these Vac1 mutants can still associate with membranes, suggesting that additional factors are involved in Vac1 localization [38]. Nevertheless, the FYVE domains found in EEA1 and Vac1/Rabenosyn-5 may still be required for concentration of these proteins on PtdIns(3)P-rich endosomes, while Rab binding may further drive specificity of these interactions. Once on endosomal membranes, EEA1 and Vac1/Rabenosyn-5 appear to intimately participate in the machinery that drives endosome fusion (Fig. 4).

In addition to Rab5, another small GTPase (Rab4) that has been implicated in the recycling of internalized receptors back to the plasma membrane regulates a FYVE domain-containing effector, Rabip4 (Table I). Like EEA1 and Vac1/Rabenosyn-5, Rabip4 localizes to endosomes and can affect endosomal morphology [40]. Moreover, overproduction of Rabip4 leads to the intracellular retention of normally recycled transporters such as Glut1 [40]. These data suggest that FYVE domains and thus PtdIns(3)P are not only involved in anterograde trafficking to lysosomes but also endosomal membrane recycling to the plasma membrane.

Another well conserved FYVE domain-containing protein that has been implicated in membrane trafficking is Hrs (Fig. 3B), a *h*epatocyte growth factor *r*eceptor tyrosine

Table I FYVE Domain-Containing Proteins Discussed in the Review

Yeast

Protein	Cellular function	Domains	Targets	Ref.
Vac1	Golgi to endosome transport	FYVE, RING, COIL	PtdIns(3)P, Vps21, Vps45, Pep12	[37–39]
Vps27	MVB sorting/formation	VHS, FYVE, UIM	PtdIns(3)P, ubiquitin	[42–44]
Fab1	PtdIns(3)P 5-synthesis, MVB sorting	FYVE, PtdIns(3)P 5-kinase	PtdIns(3)P	[44, 51]
Pib1	Ubiquitin ligase	FYVE, E3 ubiquitin ligase	PtdIns(3)P, unknown	[48]
Pib2	unknown	FYVE	PtdIns(3)P?	[21]

Mammalian

Protein	Cellular function	Domains	Targets	Ref.
EEA1	Endosome fusion	FYVE, COIL	PtdIns(3)P, Rab5, syntaxins	[17–23,35]
Rabenosyn-5	Endosome fusion	FYVE	PtdIns(3)P, Rab5, and hVps45	[36]
Rabip4	Endosomal membrane recycling	FYVE	PtdIns(3)P, Rab4	[40]
Hrs	Endosomal/MVB Sorting	VHS, FYVE, UIM, CB	PtdIns(3), ubiquitin, Clathrin	[45–47]
PIKfyve	Endosome morphology, PtdIns(3,5)P_2 synthesis	FYVE, PtdIns(3)P 5-kinase	PtdIns(3)	[52]
MTMR4	PtdIns(3)P turnover	FYVE, PtdIns(3)P phosphatase	PtdIns(3)P	[6,55]
endofin	Unknown	FYVE	PtdIns(3)P	[59]
Frabin	Actin cytoskeleton	FYVE, PH, GEF	PI isoforms, Cdc42, Rac	[61,62]
DFCP1	Unknown	FYVE-like	PtdIns(3)P?	[63]

As well as binding to PtdIns(3)P, FYVE domain-containing proteins have other domains that interact with protein targets. Together, these interactions likely play important roles in defining protein function.

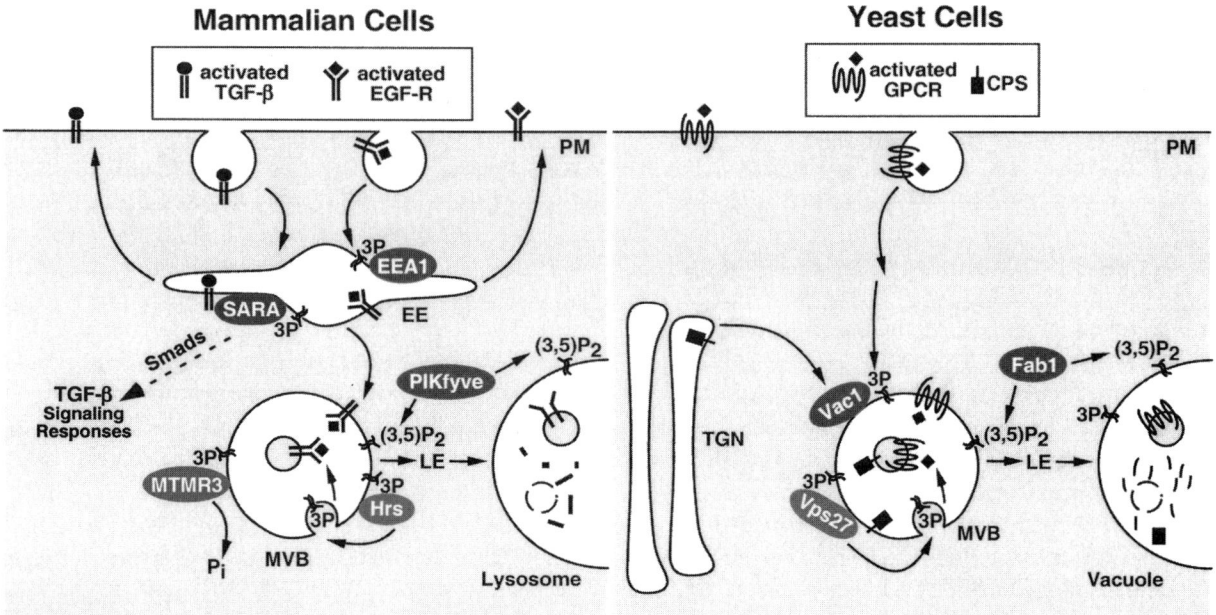

Figure 4 Cellular localization and functions of FYVE domain-containing proteins in mammalian and yeast cells. In mammalian and yeast cells, PtdIns(3)P (designated as 3P) recruits the Rab effectors EEA1 or Vac1 to endosomes where they participate in vesicle fusion in Golgi (TGN) to vacuole transport and endocytic trafficking. The FYVE domain-containing orthologs Hrs and Vps27 are required for MVB sorting pathways that transport PtdIns(3)P and cargo proteins, such as carboxypeptidase S (CPS) and internalized cell surface receptors, to the vacuole lumen where they are degraded. The PtdIns(3)P 5-kinases Fab1/PIKfyve and the PtdIns(3)P-specific phosphatase MTMR3 terminate PtdIns(3)P signaling by converting PtdIns(3)P to PtdIns(3,5)P_2 or PtdIns, respectively. PtdIns(3)P 5-kinase signaling via PtdIns(3,5)P_2 (designated as 3,5P_2) is also required for MVB sorting. In mammalian cells, the FYVE domain containing-protein SARA recruits Smad proteins, effectors of transforming growth factor beta (TGF-β) signaling, to the endosome.

kinase *s*ubstrate [41]. Studies of its yeast homolog Vps27 demonstrate a requirement for this PtdIns(3)P effector in protein trafficking, functioning after Vac1 in the endocytic pathway [42]. Specifically, inactivation of Vps27 results in a defect in the generation of intralumenal vesicles within MVBs and the vacuole [43,44]. Similarly, mouse embryos that lack Hrs exhibit defects in endosomal morphogenesis [45]. It is striking that Hrs and EEA1 are localized to different regions on endosomes. Specifically, Hrs colocalizes with clathrin and can bind to clathrin via a carboxy-terminal clathrin interacting motif [46]. Disruption of the PtdIns(3)P-FYVE interaction in Hrs by treatment with the PtdIns 3-kinase inhibitor wortmannin results in loss of both Hrs and clathrin localization to endosomes, again demonstrating the importance of the FYVE domain in endosomal function [47]. Further studies are required to precisely determine what other requirements may be necessary for Hrs/Vps27 to associate with endosomes in addition to PtdIns(3)P.

In addition to Vac1 and Vps27, yeast harbor another FYVE domain-containing protein that localizes to the endosome and vacuole, Pib1 (Table I) [21]. Localization of Pib1 to these structures is dependent on its FYVE domain through an interaction with PtdIns(3)P [48]. In addition to its FYVE domain, Pib1 contains a RING domain that possesses E2-dependent ubiquitin ligase activity *in vitro* [48]. In light of recent studies indicating a role for ubiquitin modification in the sorting of proteins into multivesicular bodies [43,49,50], the finding that Pib1 is an E3 RING-type ubiquitin ligase that localizes to the endosome via a PtdIns(3)P-FYVE interaction is especially interesting. However, deletion of *PIB1* fails to result in a defect in sorting of known ubiquitinated substrates through the MVB pathway [48], suggesting that other E3 ubiquitin ligases may act together with Pib1 in this process. Alternatively, Pib1 may be responsible for ubiquitination of a specific subset of cargo that have not yet been examined.

FYVE Domains Involved in PtdIns(3)P Metabolism

Additional FYVE domain-containing proteins found in both yeast and mammalian cells that appear to be involved in the formation of MVBs are the PtdIns(3)P 5-kinases, Fab1 and PIKfyve (Fig. 3C). Deletion of *FAB1* results in a loss of intralumenal vesicles and drastically enlarged vacuoles [44,51]. This abnormal vacuole morphology may also be in part due to defects in the recycling and/or turnover of membranes deposited at the vacuole. However, this remains to be demonstrated, since effectors of PtdIns(3,5)P$_2$ generated by Fab1 have yet to be identified. It is interesting that Fab1 has been shown to localize to both prevacuolar and vacuolar membranes, similar to the distribution of PtdIns(3)P, suggesting that its amino-terminal FYVE domain may have a role in localization and/or activity of Fab1 [51]. However, deletion of an amino-terminal fragment of Fab1 including its FYVE domain fails to significantly perturb its function, since this form of Fab1 can rescue a temperature-sensitive *fab1* mutant, suggesting that other determinants for Fab1 localization exist [51]. In contrast, the FYVE domain of mammalian PIKfyve is absolutely critical for its localization to membranes of the late endocytic pathway [52]. These studies highlight a surprising difference between certain yeast members of the FYVE domain family and their mammalian counterparts. Specifically, Vac1, Vps27, and Fab1 contain FYVE domains that are not entirely essential for membrane binding of the intact proteins. Nonetheless, this does not exclude a role for PtdIns(3)P-FYVE interactions for protein localization in yeast, but instead may emphasize the role of additional protein-protein interactions in precise subcellular targeting.

In addition to PtdIns kinases, recent studies have uncovered a set of PI phosphatases that contain a FYVE domain, such as MTMR3 (Fig. 3C) and MTMR4 (Table I). Both are members of the myotubularin family of phosphatases that were originally shown to dephosphorylate serine/threonine and tyrosine residues *in vitro* but have subsequently been shown to act upon PtdIns(3)P as their primary substrate [6,53]. It is interesting that various myotubularin members have been implicated in multiple disorders, including myotubular myopathy [53], which involves defects in muscle differentiation and Charcot-Marie-Tooth disease [54], a condition caused by defects in myelin development. MTMR3 and MTMR4 (also named FYVE-DSP1 and FYVE-DSP2) are localized in membrane fractions [54], but further studies are required to determine whether this localization is dependent on their FYVE domains. What is more important, however, the identification of both PtdIns(3)P 5-kinases and PtdIns(3)P phosphatases that contain FYVE domains raises the possibility that the FYVE domain may serve a regulatory role in the control of these enzyme activities when lipid is bound (Fig. 4). Recruitment of either a PtdIns(3)P 5-kinase or PtdIns(3)P-specific phosphatase could play a role in terminating PtdIns(3)P signaling by converting PtdIns(3)P to PtdIns(3,5)P$_2$ or PtdIns, respectively. Future work on these members of the FYVE domain family will be informative in shedding light on this question.

FYVE Domains in Signaling

In addition to membrane trafficking and phosphoinositide metabolism, FYVE domains are also found in proteins required for other cellular processes, such as growth factor signaling. The FYVE domain containing-protein SARA (Fig. 3D) recruits Smad proteins, effectors of transforming growth factor beta (TGF-beta) signaling, to the endosome (Fig. 4) [56]. There, bound TGF-beta receptors can phosphorylate Smad2 and Smad3 via their cytoplasmic serine/threonine kinase domain [56]. Phosphorylated Smads can then bind to Smad4, and this resulting complex is able to translocate to the nucleus and activate transcription of target genes [56]. SARA provides an excellent example in which the trafficking of cell-surface receptors is intimately coupled to intracellular signaling [56,57]. In this case, the FYVE domain of SARA spatially regulates TGF-beta signaling, restricting it to endosomes that contain both PtdIns(3)P and activated TGF-beta receptors. This spatial control permits the cell to prevent inappropriate activation of Smad signaling and allows for a large range of separation between the on and off states of this pathway. Consistent with this, treatment with the PtdIns 3-kinase inhibitor wortmannin results in mislocalization of SARA and leads to defects in Smad phosphorylation and downstream transcriptional activation [58]. Similar to SARA, a largely uncharacterized protein named endofin (Table I) also localizes to endosomes in a PtdIns(3)P-dependent manner [59]. Although 50% identical to SARA, endofin fails to interact with Smad2 and does not play a role in TGF-beta signaling [59]. This suggests that other yet to be defined signaling pathways may be regulated at the level of the endosome in a PtdIns(3)P-dependent manner.

Although less clear, FYVE domains are also found in proteins that regulate the actin cytoskeleton. These include Fgd1, a *faciogenital dysplasia* gene product implicated in the developmental disease Aarskog-Scott syndrome (Fig. 3D), and Frabin (Table I), which act as guanine nucleotide exchange factors (GEFs) for the small GTPases Cdc42 and Rac. Fgd1 specifically activates Cdc42, which in turn regulates actin cytoskeleton organization [60]. Frabin, through the action of Cdc42-dependent and independent pathways, has been implicated in filopodia and lamellipodia formation, respectively [61]. However, these events are not likely to involve endocytic trafficking, since early studies have indicated that this family of FYVE domain-containing proteins does not localize to the endosome [62]. Closer examination of these GEFs shows they also contain PH domains that can bind other PI species [61], and their FYVE domains lack a well-conserved tryptophan residue that is conserved in most other FYVE domains. Further studies are required to determine whether these regulators of actin cytoskeleton organization actually bind PtdIns(3)P through their FYVE domains or bind another ligand that may be structurally related to PtdIns(3)P.

FYVE-like Domains

In addition to the highly conserved FYVE domain, several other FYVE-like domains have recently been uncovered. For example, a protein identified from a human bone marrow cDNA library named DFCP1 contains two FYVE-like domains (Table I), but in both cases, the first conserved arginine in the conserved basic R(R/K)HHCR patch is replaced with a serine or threonine [63]. It remains to be determined how this might effect PtdIns(3)P binding or whether DFCP1 localization is dependent on PtdIns(3)P *in vivo*. Initial studies indicate that DFCP1 localizes to the endoplasmic reticulum, Golgi, and other intracellular vesicles, unlike what has been seen with bona fide PtdIns(3)P effectors, such as EEA1 and Hrs [63]. Future studies aimed at determining the ligand binding specificities of FYVE-like domains should help resolve these apparent discrepancies.

Conclusions

Within the span of a few years, the identification of the FYVE domain as a specific PtdIns(3)P binding motif has had a significant impact on the field of vesicular trafficking and has shed light on additional cellular functions of PtdIns(3)P. Through localization studies of the FYVE domain via both conventional light microscopy and high-resolution electron microscopy, PtdIns(3)P has been found to exist in endosomal membranes, on vesicles contained within endosomes, and in vacuolar/lysosomal membranes. More recently, GFP-FYVE fusions have been used to observe PtdIns(3)P on phagosomes [64]. However, additional PtdIns(3)P interacting proteins are involved in this process, as recent studies indicate that another lipid binding motif, the PX domain, specifically recognizes PtdIns(3)P [65]. Hence, it is likely that new effectors of this lipid will continue to emerge.

The question still remains whether the FYVE domain alone provides sufficient specificity for protein localization. Studies in yeast would favor a significant but not singular role for the FYVE domain in this regard. Instead, PtdIns(3)P-FYVE interactions coupled with protein-protein interactions are likely to ensure specific membrane recruitment of these proteins. This level of specificity would help ensure appropriate membrane-restricted responses and functions. Further studies are required to determine the validity of this concept and whether this is a general principle or may only apply to a certain subset of FYVE domain-containing proteins.

Acknowledgments

We thank members of the Emr lab for useful comments on the manuscript. C.J.S. is a fellow of the American Cancer Society supported by the Holland Peck Charitable Fund. S.D.E. is an investigator of the Howard Hughes Medical Institute.

References

1. Simonsen, A., Wurmser, A. E., Emr, S. D., and Stenmark, H. (2001). The role of phosphoinositides in membrane transport. *Curr. Opin. Cell Biol.* **13**, 485–492.
2. Odorizzi, G., Babst, M., and Emr, S. D. (2000). Phosphoinositide signaling and the regulation of membrane trafficking in yeast. *Trends Biochem. Sci.* **25**, 229–235.
3. Rameh, L. E. and Cantley, L. C. (1999). The role of phosphoinositide 3-kinase lipid products in cell function. *J. Biol. Chem.* **274**, 8347–8350.
4. Janmey, P. A. (1994). Phosphoinositides and calcium as regulators of cellular actin assembly and disassembly. *Annu. Rev. Physiol.*, **56**, 169–191.
5. Fruman, D. A., Meyers, R. E., and Cantley, L. C. (1998). Phosphoinositide kinases. *Annu. Rev. Biochem.* **67**, 481–507.
6. Wishart, M. J., Taylor, G. S., Slama, J. T., and Dixon, J. E. (2001). PTEN and myotubularin phosphoinositide phosphatases: bringing bioinformatics to the lab bench. *Curr. Opin. Cell Biol.* **13**, 172–181.
7. Hughes, W. E., Woscholski, R., Cooke, F. T., Patrick, R. S, Dove, S. K., McDonald, N. Q., and Parker, P. J. (2000). SAC1 encodes a regulated lipid phosphoinositide phosphatase, defects in which can be suppressed by the homologous Inp52p and Inp53p phosphatases. *J. Biol. Chem.* **275**, 801–808.
8. Majerus, P. W., Kisseleva, M. V., and Norris, F. A. (1999). The role of phosphatases in inositol signaling reactions. *J. Biol. Chem.* **274**, 10669–10672.

9. Berridge, M. J. (1981). Phosphatidylinositol hydrolysis: a multifunctional transducing mechanism. *Mol. Cell. Endocrinol*, **24**, 115–140.
10. Hurley, J. H. and Meyer, T. (2001). Subcellular targeting by membrane lipids. *Curr. Opin. Cell Biol.* **13**, 146–152.
11. Wurmser, A. E., Gary, J. D., and Emr, S. D. (1999). Phosphoinositide 3-kinases and their FYVE domain-containing effectors as regulators of vacuolar/lysosomal membrane trafficking pathways. *J. Biol. Chem.* **274**, 9129–9132.
12. Schu, P. V., Takegawa, K., Fry, M. J., Stack, J. H., Waterfield, M. D., and Emr, S. D. (1993). Phosphatidylinositol 3-kinase encoded by yeast *VPS34* gene essential for protein sorting. *Science* **12**, 88–91.
13. Corvera, S. (2001). Phosphatidylinositol 3-kinase and the control of endosome dynamics: new players defined by structural motifs. *Traffic* **2**, 859–866.
14. Brown, W. J., DeWald, D. B., Emr, S. D., Plutner, H., and Balch, W. E. (1995). Role for phosphatidylinositol 3-kinase in the sorting and transport of newly synthesized lysosomal enzymes in mammalian cells. *J. Cell Biol.* **130**, 781–796.
15. Li, G., D'Souza-Schorey, C., Barbieri, M. A., Roberts, R. L., Klippel, A., Williams, L. T., and Stahl, P. D. (1995). Evidence for phosphatidylinositol 3-kinase as a regulator of endocytosis via activation of Rab5. *Proc. Natl. Acad. Sci. USA* **92**, 10207–10211.
16. Davidson, H. W. (1995). Wortmannin causes mistargeting of procathepsin D. Evidence for the involvement of a phosphatidylinositol 3-kinase in vesicular transport to lysosomes. *J. Cell Biol.* **130**, 797–805
17. Patki, V., Virbasius, J., Lane, W. S., Toh, B. H., Shpetner, H. S., and Corvera, S. (1997). Identification of an early endosomal protein regulated by phosphatidylinositol 3-kinase. *Proc. Natl. Acad. Sci. USA* **94**, 7326–7330.
18. Simonsen, A., Lippe, R., Christoforidis, S., Gaullier, J. M., Brech, A., Callaghan, J., Toh, B. H., Murphy, C., Zerial, M., and Stenmark, H. (1998). EEA1 links PI(3)K function to Rab5 regulation of endosome fusion. *Nature* **394**, 494–498.
19. Stenmark, H., Aasland, R., Toh, B. H., and D'Arrigo, A. (1996). Endosomal localization of the autoantigen EEA1 is mediated by a zinc-binding FYVE finger. *J. Biol. Chem.* **271**, 24048–24054.
20. Mu, F. T., Callaghan, J. M., Steele-Mortimer, O., Stenmark, H., Parton, R. G., Campbell, P. L., McCluskey, J., Yeo, J. P., Tock, E. P, and Toh, B. H. (1995). EEA1, an early endosome-associated protein. EEA1 is a conserved alpha-helical peripheral membrane protein flanked by cysteine "fingers" and contains a calmodulin-binding IQ motif. *J. Biol. Chem.* **270**, 13503–13511.
21. Burd, C. G., and Emr, S. D. (1998). Phosphatidylinositol(3)-phosphate signaling mediated by specific binding to RING FYVE domains. *Mol. Cell* **2**, 157–162.
22. Patki, V., Lawe, D. C., Corvera, S., Virbasius, J. V., and Chawla, A. (1998). A functional PtdIns(3)P-binding motif. *Nature* **394**, 433–434.
23. Gaullier, J. M., Simonsen, A., D'Arrigo, A., Bremnes, B., Stenmark, H., and Aasland, R. (1998). FYVE fingers bind PtdIns(3)P. *Nature* **394**, 432–433.
24. Misra, S., Miller, G. J., and Hurley, J. H. (2001). Recognizing phosphatidylinositol 3-phosphate. *Cell* **107**, 559–562.
25. Fruman, D. A., Rameh, L. E., and Cantley, L. C. (1999). Phosphoinositide binding domains: embracing 3-phosphate. *Cell* **97**, 817–820.
26. Misra, S. and Hurley, J. H. (1999). Crystal structure of a phosphatidylinositol 3-phosphate-specific membrane-targeting motif, the FYVE domain of Vps27p. *Cell* **97**, 657–666.
27. Mao, Y., Nickitenko, A., Duan, X., Lloyd, T. E., Wu, M. N., Bellen, H., and Quiocho, F. A. (2000). Crystal structure of the VHS and FYVE tandem domains of Hrs, a protein involved in membrane trafficking and signal transduction. *Cell* **100**, 447–456.
28. Kutateladze T. G., Ogburn, K. D., Watson, W. T., de Beer, T., Emr, S. D., Burd, C. G., and Overduin, M. (1999). Phosphatidylinositol 3-phosphate recognition by the FYVE domain. *Mol. Cell* **3**, 805–811.
29. Kutateladze, T. and Overduin, M. (2001). Structural mechanism of endosome docking by the FYVE domain. *Science* **291**, 1793–1796.
30. Lawe, D. C., Patki, V., Heller-Harrison, R., Lambright, D., and Corvera, S. (2000). The FYVE domain of early endosome antigen 1 is required for both phosphatidylinositol 3-phosphate and Rab5 binding. Critical role of this dual interaction for endosomal localization. *J. Biol. Chem.* **275**, 3699–3705.
31. Dumas, J. J., Merithew, E., Sudharshan, E., Rajamani, D., Hayes, S., Lawe, D., Corvera, S., and Lambright, D. G. (2001). Multivalent endosome targeting by homodimeric EEA1. *Mol. Cell* **8**, 947–958.
32. Gillooly, D. J., Morrow, I. C., Lindsay, M., Gould, R., Bryant, N. J., Gaullier, J. M., Parton, R. G., and Stenmark, H. (2000). Localization of phosphatidylinositol 3-phosphate in yeast and mammalian cells. *EMBO J.* **19**, 4577–4588.
33. Wurmser, A. E. and Emr, S. D. (1998). Phosphoinositide signaling and turnover: PtdIns(3)P, a regulator of membrane traffic, is transported to the vacuole and degraded by a process that requires lumenal vacuolar hydrolase activities. *EMBO J.* **17**, 4930–4942.
34. Christoforidis, S., McBride, H. M., Burgoyne, R. D., and Zerial, M. (1999). The Rab5 effector EEA1 is a core component of endosome docking. *Nature* **397**, 621–625.
35. McBride, H. M., Rybin, V., Murphy, C., Giner, A., Teasdale, R., and Zerial, M. (1999). Oligomeric complexes link Rab5 effectors with NSF and drive membrane fusion via interactions between EEA1 and syntaxin 13. *Cell* **98**, 377–386.
36. Nielsen, E., Christoforidis, S., Uttenweiler-Joseph, S., Miaczynska, M., Dewitte, F., Wilm, M., Hoflack, B., and Zerial, M. (2000). Rabenosyn-5, a novel Rab5 effector, is complexed with hVPS45 and recruited to endosomes through a FYVE finger domain. *J. Cell Biol.* **151**, 601–612.
37. Peterson, M. R., Burd, C. G., and Emr, S. D. (1999). Vac1p coordinates Rab and phosphatidylinositol 3-kinase signaling in Vps45p-dependent vesicle docking/fusion at the endosome. *Curr. Biol.* **9**, 159–162.
38. Burd, C. G., Peterson, M., Cowles, C. R., and Emr, S. D. (1997). A novel Sec18p/NSF-dependent complex required for Golgi-to-endosome transport in yeast. *Mol. Biol. Cell* **8**, 1089–1104.
39. Tall, G. G., Hama, H., DeWald, D. B., and Horazdovsky, B. F. (1999). The phosphatidylinositol 3-phosphate binding protein Vac1p interacts with a Rab GTPase and a Sec1p homologue to facilitate vesicle-mediated vacuolar protein sorting. *Mol. Biol. Cell* **10**, 1873–1889.
40. Cormont, M., Mari M., Galmiche, A., Hofman, P., and Le Marchand-Brustel, Y. (2001). A FYVE-finger-containing protein, Rabip4, is a Rab4 effector involved in early endosomal traffic. *Proc. Natl. Acad. Sci. USA* **98**, 1637–1642.
41. Komada, M. and Kitamura N. (1995). Growth factor-induced tyrosine phosphorylation of Hrs, a novel 115-kilodalton protein with a structurally conserved putative zinc finger domain. *Mol. Cell. Biol.* **15**, 6213–6221.
42. Piper, R. C. and Cooper, A. A., Yang, H., Stevens, T. H. (1995). VPS27 controls vacuolar and endocytic traffic through a prevacuolar compartment in *Saccharomyces cerevisiae*. *J. Cell Biol.* **131**, 603–617.
43. Shih, S. C., Katzmann, D. J., Schnell, J. D., Sutanto, M., Emr, S. D., and Hicke, L. (2002). Epsins and Vps27p/Hrs contain ubiquitin-binding domains that function in receptor endocytosis. *Nat. Cell Biol.* **4**, 389–393.
44. Odorizzi. G., Babst, M., and Emr, S. D. (1998). Fab1p PtdIns(3)P 5-kinase function essential for protein sorting in the multivesicular body. *Cell* **95**, 847–858.
45. Komada, M. and Soriano, P. (1999). Hrs, a FYVE finger protein localized to early endosomes, is implicated in vesicular traffic and required for ventral folding morphogenesis. *Genes Dev.* **13**, 1475–1485.
46. Raiborg, C., Bache, K. G., Mehlum, A., Stang, E., and Stenmark, H. (2001). Hrs recruits clathrin to early endosomes. *EMBO J.* **20**, 5008–5021.
47. Raiborg, C., Bremnes, B., Mehlum, A., Gillooly, D. J., D'Arrigo, A., Stang, E., and Stenmark H. (2001). FYVE and coiled-coil domains determine the specific localisation of Hrs to early endosomes. *J. Cell Sci.* **114**, 2255–2263.
48. Shin, M. E., Ogburn, K. D., Varban, O. A., Gilbert, P. M., and Burd, C. G. (2001). FYVE domain targets Pib1p ubiquitin ligase to endosome and vacuolar membranes. *J. Biol. Chem.* **276**, 41388–41393.
49. Bishop, N., Horman, A., and Woodman, P. (2002). Mammalian class E vps proteins recognize ubiquitin and act in the removal of endosomal protein-ubiquitin conjugates. *J. Cell Biol.* **157**, 91–101.

50. Katzmann, D. J., Babst, M., and Emr, S. D. (2001). Ubiquitin-dependent sorting into the multivesicular body pathway requires the function of a conserved endosomal protein sorting complex, ESCRT-I. *Cell* **106**, 145–155.
51. Gary, J. D., Wurmser, A. E., Bonangelino, C. J., Weisman, L. S., and Emr, S. D. (1998). Fab1p is essential for PtdIns(3)P 5-kinase activity and the maintenance of vacuolar size and membrane homeostasis. *J. Cell Biol.* **143**, 65–79.
52. Sbrissa, D., Ikonomov, O. C., and Shisheva, A. (2002). Phosphatidylinositol 3-phosphate-interacting domains in PIKfyve. Binding specificity and role in PIKfyve. Endomembrane localization. *J. Biol. Chem.* **277**, 6073–6079.
53. Taylor, G. S., Maehama, T., and Dixon, J. E. (2000). Inaugural article: myotubularin, a protein tyrosine phosphatase mutated in myotubular myopathy, dephosphorylates the lipid second messenger, phosphatidylinositol 3-phosphate. *Proc. Natl. Acad. Sci. USA* **97**, 8910–8915.
54. Kim, S. A., Taylor, G. S., Torgersen, K. M., and Dixon, J. E. (2002). Myotubularin and MTMR2, phosphatidylinositol 3-phosphatases mutated in myotubular myopathy and type 4B Charcot-Marie-Tooth disease. *J. Biol. Chem.* **277**, 4526–4531.
55. Zhao, R., Qi, Y., Chen, J., and Zhao, Z. J. (2001). FYVE-DSP2, a FYVE domain-containing dual specificity protein phosphatase that dephosphorylates phosphotidylinositol 3-phosphate. *Exp. Cell Res.* **265**, 329–338.
56. Tsukazaki, T., Chiang, T. A., Davison, A. F., Attisano, L., and Wrana, J. L. (1998). SARA, a FYVE domain protein that recruits Smad2 to the TGFbeta receptor. *Cell* **95**, 779–791.
57. Miura, S., Takeshita, T., Asao, H., Kimura, Y., Murata, K., Sasaki, Y., Hanai. J. I., Beppu, H., Tsukazaki,T., Wrana, J. L., Miyazono, K., and Sugamura, K. (2000). Hgs (Hrs), a FYVE domain protein, is involved in Smad signaling through cooperation with SARA. *Mol. Cell. Biol.* **20**, 9346–9355.
58. Itoh, F., Divecha, N., Brocks, L., Oomen, L., Janssen, H., Calafat, J., Itoh, S., and Dijke, Pt. P. (2002). The FYVE domain in Smad anchor for receptor activation (SARA) is sufficient for localization of SARA in early endosomes and regulates TGF-beta/Smad signalling. *Genes Cells* **7**, 321–331.
59. Seet, L. F. and Hong, W. (2001). Endofin, an endosomal FYVE domain protein. *J. Biol. Chem.* **276**, 42445–42454.
60. Zheng, Y., Fischer, D. J., Santos, M. F., Tigyi, G., Pasteris, N. G., Gorski, J. L., and Xu ,Y. (1996). The faciogenital dysplasia gene product FGD1 functions as a Cdc42Hs-specific guanine-nucleotide exchange factor. *J. Biol. Chem.* **271**, 33169–33172.
61. Obaishi, H., Nakanishi, H., Mandai, K., Satoh, K., Satoh, A., Takahashi, K., Miyahara, M., Nishioka, H., Takaishi, K., and Takai, Y. (1998). Frabin, a novel FGD1-related actin filament-binding protein capable of changing cell shape and activating c-Jun N-terminal kinase. *J. Biol. Chem.* **273**, 18697–18700.
62. Kim, Y., Ikeda, W., Nakanishi, H., Tanaka, Y., Takekuni, K., Itoh, S., Monden, M., and Takai, Y. (2002). Association of frabin with specific actin and membrane structures. *Genes Cells* **7**, 413–420.
63. Ridley, S. H., Ktistakis, N., Davidson, K., Anderson, K. E., Manifava, M., Ellson, C. D., Lipp, P., Bootman, M., Coadwell, J., Nazarian, A., Erdjument-Bromage, H., Tempst, P., Cooper, M. A., Thuring, J. W., Lim, Z. Y., Holmes, A. B., Stephens, L. R., and Hawkins, P. T. (2001). FENS-1 and DFCP1 are FYVE domain-containing proteins with distinct functions in the endosomal and Golgi compartments. *J. Cell. Sci.* **114**, 3991–4000.
64. Ellson, C. D., Anderson, K. E., Morgan, G., Chilvers, E. R., Lipp, P., Stephens, L. R., and Hawkins, P. T. (2001). Phosphatidylinositol 3-phosphate is generated in phagosomal membranes. *Curr. Biol.* **11**, 1631–1635.
65. Sato, T. K., Overduin, M., and Emr, S. D. (2001). Location, location, location: membrane targeting directed by PX domains. *Science* **294**, 1881–1885.

CHAPTER 153

Protein Kinase C: Relaying Signals from Lipid Hydrolysis to Protein Phosphorylation

Alexandra C. Newton
*Department of Pharmacology, University of California at San Diego,
La Jolla, California*

Introduction

Protein kinase C (PKC) has been in the spotlight since the discovery a quarter of a century ago that, through its activation by diacylglycerol, it relays signals from lipid hydrolysis to protein phosphorylation [1]. The subsequent discovery that PKCs are the target of phorbol esters resulted in an avalanche of reports on the effects on cell function of phorbol esters, nonhydrolyzable analogs of the endogenous ligand, diacylglycerol [2–4]. Despite the enduring stage presence of PKC and tremendous advances in understanding the enzymology and regulation of this key protein, an understanding of the function of PKC in biology is still the subject of intense pursuit. Its uncontrolled signaling wreaks havoc in the cell, as epitomized by the potent tumor-promoting properties of phorbol esters. In fact, the pluripotent effects of phorbol esters, compounded with the existence of multiple isozymes of PKC, has made it difficult to uncover the precise cellular function of this key enzyme [5]. Studies with knockout mice have underscored the problem, with knockouts of most isozymes having only subtle phenotypic effects [6]. This chapter summarizes our current understanding of the molecular mechanisms of how protein kinase C transduces information from lipid mediators to protein phosphorylation.

Protein Kinase C Family

The 10 members of the mammalian PKC family are grouped into three classes based on their domain structure, which, in turn, dictates their cofactor dependence (Fig. 1). All members comprise a single polypeptide that has a conserved kinase core carboxyl-terminal to a regulatory moiety. This regulatory moiety contains two key functionalities: an autoinhibitory sequence (pseudosubstrate) and one or two membrane-targeting modules (C1 and C2 domains). The C1 domain binds diacylglycerol and phosphatidylserine specifically and is present as a tandem repeat in conventional and novel PKCs (C1A and C1B); the C2 domain nonspecifically binds Ca^{2+} and anionic phospholipids such as phosphatidylserine. Non-ligand-binding variants of each domain exist: atypical C1 domains do not bind diacylglycerol and novel C2 domains do not bind Ca^{2+}.

Conventional PKC isozymes (α, γ, and the alternatively spliced βI and βII) are stimulated by diacylglycerol and phosphatidylserine (C1 domain) and Ca^{2+} (C2 domain); novel PKC isozymes (δ, ε, η/L, θ) are stimulated by diacylglycerol and phosphatidylserine (C1 domain); and atypical PKC isozymes (ζ, ι/λ) are stimulated by phosphatidylserine (atypical C1 domain) [5–7]. (Note that PKC μ and ν were considered to constitute a fourth class of PKCs but are now generally regarded as members of a distinct family

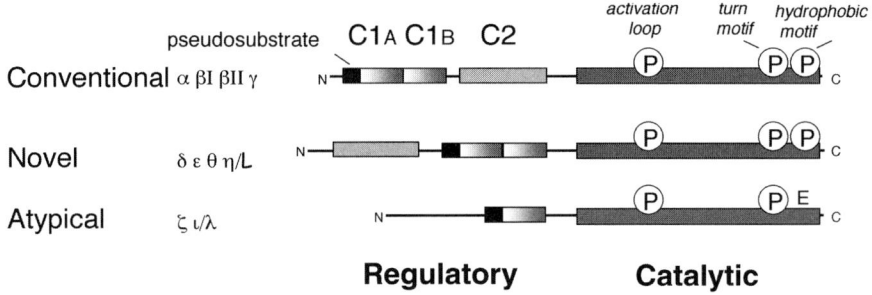

Figure 1 Domain composition of protein kinase C family members showing autoinhibitory pseudosubstrate, membrane-targeting modules (C1A and C1B and C2 domains), and kinase domain of the three subclasses: conventional, novel, and atypical isozymes. Also indicated are the positions of the three processing phosphorylation sites, the activation loop and two carboxyl-terminal sites, the turn motif, and hydrophobic motif. (Adapted from Newton, A. C. and Johnson, J. E., *Biochem. Biophys. Acta*, 1376, 155–172, 1998.)

called protein kinase D.) The role of the novel C2 domain in novel PKCs and that of the atypical C1 domain in atypical PKCs is not clear, but each may regulate the subcellular distribution of these isozymes through protein–protein interactions.

Regulation of Protein Kinase C

The normal function of PKC is under the coordinated regulation of three major mechanisms: phosphorylation/dephosphorylation, membrane targeting modules, and anchor proteins. First, the kinase must be processed by a series of ordered phosphorylations to become catalytically competent. Second, it must have its pseudosubstrate removed from the active site to be catalytically active, a conformational change driven by engaging the membrane-targeting modules with ligand. Third, it must be localized at the correct intracellular location for unimpaired signaling. Perturbation at any of these points of regulation disrupts the physiological function of PKC [7].

Phosphorylation/ Dephosphorylation

The function of PKC isozymes is controlled by phosphorylation mechanisms that are required for the maturation of the enzyme. In addition to the processing phosphorylations, the function of PKC isozymes is additionally fine-tuned by both Tyr and Ser/Thr phosphorylations [8,9]. The conserved maturation phosphorylations are described below.

PHOSPHORYLATION IS REQUIRED FOR THE MATURATION OF PROTEIN KINASE C

The majority of PKC in tissues and cultured cells is phosphorylated at two key phosphorylation switches: a loop near the active site, referred to as the *activation loop*, and a sequence at the carboxyl terminus of the kinase domain [10,11]. The carboxyl-terminal switch contains two sites: the turn motif, which by analogy with protein kinase A is at the apex of a turn on the upper lobe of the kinase domain, and the hydrophobic motif, which is flanked by hydrophobic residues (note that, in atypical PKCs, a Glu occupies the phospho-acceptor position of the hydrophobic motif). It is the phosphorylated species that transduces signals. While it had been appreciated since the late 1980s that PKC is processed by phosphorylation [12], the mechanism and role of these phosphorylations are only now being unveiled [7,9].

The first step in the maturation of PKC is phosphorylation by the phosphoinositide-dependent kinase, PDK-1, of the activation loop. This enzyme was originally discovered as the upstream kinase for Akt/protein kinase B [13] and was subsequently shown to be the activation loop kinase for a large number of AGC kinases, including all PKC isozymes [14–16]. The name PDK-1 was based on the phosphoinositide-dependence of Akt phosphorylation and is an unfortunate misnomer because the phosphorylation of other substrates (for example, the conventional PKCs) has no dependence on phosphatidylinositol 3-kinase (PI3K) lipid products [17]. Rather, PDK-1 appears to be constitutively active in the cell, with substrate phosphorylation regulated by the conformation of the substrate [18–20].

Completion of PKC maturation requires phosphorylation of the two carboxyl-terminal sites, the turn motif and hydrophobic motif. In the case of conventional PKCs, this reaction occurs by an intramolecular autophosphorylation mechanism [21]. Autophosphorylation also accounts for the hydrophobic motif processing of the novel PKCε [22]; however, it has been suggested that another member of this family, PKCδ, may be the target of a putative hydrophobic motif kinase [23].

Research in the past few years has culminated in the following model for PKC phosphorylation. Newly synthesized enzyme associates with the plasma membrane, where it adopts an open conformation with the pseudosubstrate exposed, thus unmasking the PDK-1 site on the activation loop [17,24]. It is likely held at the membrane by multiple weak interactions with the exposed pseudosubstrate, the C1 domain, and the C2 domain (because the C1 and C2 ligands are absent, these domains are weakly bound via their interactions with anionic phospholipids). PDK-1 docks onto the carboxyl terminus of PKC, where it is positioned to phosphorylate the activation loop [25]. This phosphorylation is the first and required step in the maturation of PKC; mutation of the phospho-acceptor position at the activation loop to Ala or Val prevents the maturation of PKC and results in accumulation of unphosphorylated, inactive species in the detergent-insoluble fraction of cells [26,27].

Completion of PKC maturation requires release of PDK-1 from its docking site on the carboxyl terminus. Physiological mechanisms for this release have not yet been elucidated, but it is interesting that over-expression of peptides that have a high affinity for PDK-1 promotes the maturation of PKC [25]. One such peptide is PIF, the carboxyl-terminus of PRK-2, which has a hydrophobic phosphorylation motif with a Asp at the phospho-acceptor position [28]. Release of PDK-1 unmasks the carboxyl terminus of PKC, allowing phosphorylation of the turn motif and the hydrophobic motif [7,10].

DEPHOSPHORYLATION: DEACTIVATION SIGNAL

While the phosphorylation of conventional PKCs is constitutive, the dephosphorylation appears to be agonist stimulated [29]. Both phorbol esters and ligands such as tumor necrosis factor α (TNFα) result in PKC inactivation and dephosphorylation [29–31]. In addition, serum selectively promotes the dephosphorylation of the activation loop site in conventional PKCs, thus uncoupling the phosphorylation of the activation loop from that of the carboxyl-terminal sites [17]. The hydrophobic site of PKCε has also been reported to be selectively dephosphorylated by a rapamycin-sensitive phosphatase [32]. It is likely that the uncoupling of the dephosphorylation of these sites has contributed to confusion as to whether the hydrophobic site is regulated by its own upstream kinase rather than autophosphorylation [23].

Membrane Translocation

The translocation from the cytosol to the membrane has served as the hallmark for PKC activation since the early 1980s [33,34]. The molecular details of this translocation have emerged from abundant biophysical, biochemical, and cellular studies showing that diacylglycerol acts like molecular glue to recruit PKC to membranes, an event that, for conventional PKCs, is facilitated by Ca^{2+} [35–37].

Both *in vitro* and *in vivo* data converge on the following model for the translocation of conventional PKC in response to elevated Ca^{2+} and diacylglycerol [35,38]. In the resting state, PKC bounces on and off membranes by a diffusion-limited reaction. However, its affinity for membranes is so low that its lifetime on the membrane is too short to be significant. Elevation of Ca^{2+} results in binding of Ca^{2+} to the C2 domain of this soluble species of PKC. This Ca^{2+}-bound species has a dramatically enhanced affinity for the membrane, with which it rapidly associates. The membrane-bound PKC then diffuses in the two-dimensional plane of the membrane, searching for the much less abundant ligand, diacylglycerol. This search for diacylglycerol is considerably more efficient from the membrane than one initiated from the cytosol. Following collision with, and binding to, diacylglycerol, PKC is bound to the membrane with sufficiently high affinity to allow release of the pseudosubstrate sequence and activation of PKC. Decreases in the level of either second messenger weaken the membrane interaction sufficiently to release PKC back into the cytosol. Note that if PMA is the C1 domain ligand, PKC can be retained on the membrane in the absence of elevated Ca^{2+} because this ligand binds PKC two orders of magnitude more tightly than diacylglycerol [39]. Similarly, if Ca^{2+} levels are elevated sufficiently, PKC can be retained at the membrane in the absence of a C1 ligand.

Novel PKC isozymes translocate to membranes much more slowly than conventional PKCs in response to receptor-mediated generation of diacylglycerol because they do not have the advantage of pre-targeting to the membrane by the soluble ligand, Ca^{2+} [40]. Atypical PKC isozymes do not respond directly to either diacylglycerol or Ca^{2+}.

Anchoring Proteins

The control of subcellular localization of kinases by scaffold proteins is emerging as a key requirement in maintaining fidelity and specificity in signaling by protein kinases [41]. PKC is no exception, and a battery of binding partners for members of this kinase family have been identified [42–45]. These proteins position PKC isozymes near their substrates, near regulators of activity such as phosphatases and kinases, or in specific intracellular compartments. Disruption of anchoring can impair signaling by PKC, and *Drosophila* photoreceptors provide a compelling example. Mislocalization of eye-specific PKC by abolishing its binding to the scaffold protein, ina D, disrupts phototransduction [46].

Unlike protein kinase A binding proteins (AKAPs) [47], there is no consensus binding mechanism for interaction of PKC with its anchor proteins. Rather, each binding partner identified to date interacts with PKC by unique determinants and unique mechanisms. Some binding proteins regulate multiple PKC isozymes, while others control the distribution of specific isozymes. There are binding proteins for newly synthesized unphosphorylated PKC, phosphorylated but inactive PKC, phosphorylated and activated PKC, and dephosphorylated, inactivated PKC [43,45]. Anchoring proteins for PKC have diverse functions—some positively regulate signaling while others negatively regulate it. An emerging theme is that many scaffolds bind multiple signaling molecules in a signaling complex; for example, AKAP 79 binds PKA, PKC, and the phosphatase calcineurin [48]. The physical coupling of kinases and phosphatases underscores the acute regulation that each must be under to maintain fidelity in signaling.

Model for Regulation of Protein Kinase C by Phosphorylation and Second Messengers

Figure 2 outlines a model for the regulation of PKC by phosphorylation, second messengers, and anchoring proteins. Newly synthesized PKC associates with the membrane in a conformation that exposes the pseudosubstrate

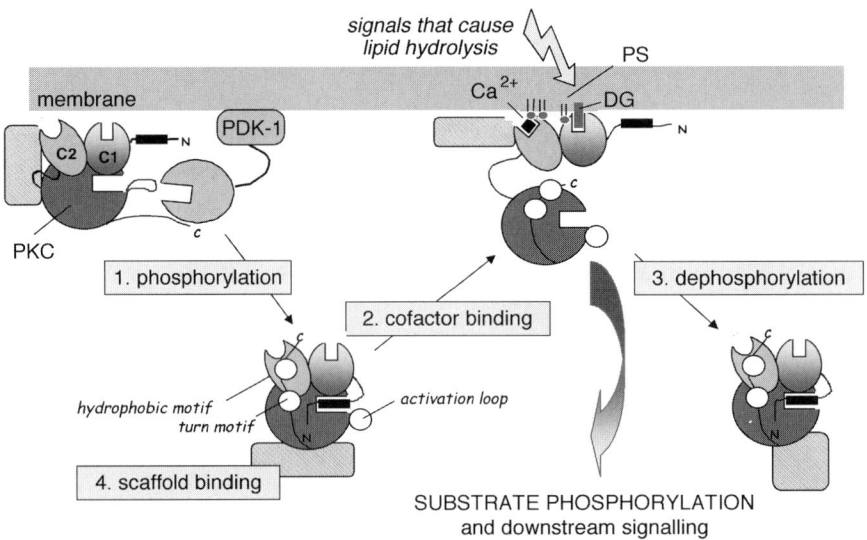

Figure 2 Model showing the major regulatory mechanisms for PKC function: (1) processing by phosphorylation, (2) activation by lipid mediators, (3) deactivation by dephosphorylation, and (4) spatial control by scaffold proteins. See text for details. (Adapted from Newton, A. C., *Chem. Rev.*, **101**, 2353–2364, 2001.)

(black rectangle), allowing access of the upstream kinase, PDK-1, to the activation loop. PDK-1 docks onto the carboxyl terminus of PKC. Following its phosphorylation of the activation loop and release from PKC, the turn motif and hydrophobic motif are autophosphorylated. The mature PKC is released into the cytosol, where it is maintained in an auto-inhibited conformation by the pseudosubstrate (middle panel), which has now gained access to the substrate-binding cavity (open rectangle in the large circle representing the kinase domain of PKC). It is this species that is competent to respond to second messengers. Generation of diacylglycerol and, for conventional PKCs, Ca^{2+} mobilization provide the allosteric switch to activate PKC. This is achieved by engaging the C1 and C2 domains on the membrane (Fig. 2, right panel), thus providing the energy to release the pseudosubstrate from the active site, allowing substrate binding and catalysis. In addition to the regulation by phosphorylation and cofactors, anchoring/scaffold proteins (stippled rectangle) play a key role in PKC function by positioning specific isozymes at particular intracellular locations [43,45]. Following activation, PKC is either released into the cytosol or, following prolonged activation, dephosphorylated and downregulated by proteolysis.

Function of Protein Kinase C

Despite over two decades of research on the effects of phorbol esters on cell function, a unifying mechanism for the role of PKC in the cell has remained elusive. An abundance of substrates have been identified, and the reader is referred to reviews summarizing these and potential signaling pathways involving PKC [5,6,49–52]. However, a unique role for PKC in defining cell function is lacking. This is epitomized by the finding that there is no severe phenotype associated with knocking-out specific PKC isozymes in mice.

Closer analysis of the phenotypes of knockout animals of various PKC isozymes does suggest a common theme: animals deficient in PKC are deficient in adaptive responses. For example, PKCε $^{-/-}$ mice have reduced anxiety and reduced tolerance to alcohol [53], PKCγ$^{-/-}$ mice have reduced pain perception [54], and PKCβII$^{-/-}$ mice have reduced learning abilities [55]. This theme carries over to the molecular level, where many of the substrates of PKC are receptors that become desensitized following PKC phosphorylation.

Summary

PKC plays a pivotal role in cell signalling by relaying information from lipid mediators to protein substrates. The relay of this information is under exquisite conformational, spatial, and temporal regulation, and extensive studies on the molecular mechanisms of this control have provided much insight into how PKC is regulated. With novel approaches in chemical genetics, analysis of crosses of PKC isozyme knockout mice, and proteomics, the PKC signaling field is poised to move to the next level of making headway into the *raison d'être* of this ubiquitous family of kinases.

Acknowledgments

This work was supported in part by National Institutes of Health Grants NIH GM 43154 and P01 DK54441.

References

1. Takai, Y. *et al.* (1979). Unsaturated diacylglycerol as a possible messenger for the activation of calcium-activated, phospholipid-dependent protein kinase system. *Biochem. Biophys. Res. Comm.* **91**, 1218–1224.

2. Castagna, M. et al. (1982). Direct activation of calcium-activated, phospholipid-dependent protein kinase by tumor-promoting phorbol esters. *J. Biol. Chem.* **257**, 7847–7851.
3. Blumberg, P. M. et al. (1984). Mechanism of action of the phorbol ester tumor promoters: specific receptors for lipophilic ligands. *Biochem. Pharmacol.* **33**, 933–940.
4. Nishizuka, Y. (1986). Studies and perspectives of protein kinase C. *Science* **233**, 305–312.
5. Nishizuka, Y. (1995). Protein kinase C and lipid signaling for sustained cellular responses. *FASEB J.* **9**, 484–496.
6. Mellor, H. and Parker, P. J. (1998). The extended protein kinase C superfamily. *Biochem. J.* **332**, 281–292.
7. Newton, A. C. (2001). Protein kinase C: structural and spatial regulation by phosphorylation, cofactors, and macromolecular interactions. *Chem. Rev.* **101**, 2353–2364.
8. Konishi, H. et al. (1997). Activation of protein kinase C by tyrosine phosphorylation in response to H_2O_2. *Proc. Natl. Acad. Sci.* **94**, 11233–11237.
9. Parekh, D. B., Ziegler, W., and Parker, P. J. (2000). Multiple pathways control protein kinase C phosphorylation. *EMBO J.* **19**, 496–503.
10. Keranen, L. M., Dutil, E. M., and Newton, A. C. (1995). Protein kinase C is regulated *in vivo* by three functionally distinct phosphorylations. *Curr. Biol.* **5**, 1394–1403.
11. Tsutakawa, S. E. et al. (1995). Determination of *in vivo* phosphorylation sites in protein kinase C. *J. Biol. Chem.* **270**, 26807–26812.
12. Borner, C. et al. (1989). Biosynthesis and posttranslational modifications of protein kinase C in human breast cancer cells. *J. Biol. Chem.* **264**, 13902–13909.
13. Alessi, D. R. et al. (1997). Characterization of a 3-phosphoinositide-dependent protein kinase which phosphorylates and activates protein kinase Bα. *Curr. Biol.* **7**, 261–269.
14. Chou, M. M. et al. (1998). Regulation of protein kinase C ζ by PI 3-kinase and PDK-1. *Curr. Biol.* **8**, 1069–1077.
15. Le Good, J. A. et al. (1998). Protein kinase C isotypes controlled by phosphoinositide 3-kinase through the protein kinase PDK1. *Science* **281**, 2042–2045.
16. Dutil, E. M., Toker, A., and Newton, A. C. (1998). Regulation of conventional protein kinase C isozymes by phosphoinositide-dependent kinase 1 (PDK-1). *Curr. Biol.* **8**, 1366–1375.
17. Sonnenburg, E. D., Gao, T., and Newton, A. C. (2001). The phosphoinositide dependent kinase, PDK-1, phosphorylates conventional protein kinase C isozymes by a mechanism that is independent of phosphoinositide-3-kinase. *J. Biol. Chem.* **28**, 28.
18. Toker, A. and Newton, A. (2000). Cellular signalling: pivoting around PDK-1. *Cell* **103**, 185–188.
19. Parker, P. J. and Parkinson, S. J. (2001). AGC protein kinase phosphorylation and protein kinase C. *Biochem. Soc. Trans.* **29**, 860–863.
20. Storz, P. and Toker, A. (2002). 3′-phosphoinositide-dependent kinase-1 (PDK-1) in PI 3-kinase signaling. *Front. Biosci.* **7**, D886–D902.
21. Behn-Krappa, A. and Newton, A. C. (1999). The hydrophobic phosphorylation motif of conventional protein kinase C is regulated by autophosphorylation. *Curr. Biol.* **9**, 728–737.
22. Cenni, V. et al. Regulation of novel protein kinase C epsilon by phosphorylation. *Biochem. J.* **363**, 537–545.
23. Ziegler, W. H. et al. (1999). Rapamycin-sensitive phosphorylation of PKC on a carboxyl-terminal site by an atypical PKC complex. *Curr. Biol.* **9**, 522–529.
24. Dutil, E. M. and Newton, A. C. (2000). Dual role of pseudosubstrate in the coordinated regulation of protein kinase C by phosphorylation and diacylglycerol. *J. Biol. Chem.* **275**, 10697–10701.
25. Gao, T., Toker, A., and Newton, A. C. (2001). The carboxyl terminus of protein kinase C provides a switch to regulate its interaction with the phosphoinositide-dependent kinase, PDK-1. *J. Biol. Chem.* **276**, 19588–19596.
26. Cazaubon, S., Bornancin, F., and Parker, P. J. (1994). Threonine-497 is a critical site for permissive activation of protein kinase C α. *Biochem. J.* **301**, 443–448.
27. Orr, J. W. and Newton, A. C. (1994). Requirement for negative charge on activation loop of protein kinase C. *J. Biol. Chem.* **269**, 27715–27718.
28. Balendran, A. et al. (1999). PDK1 acquires PDK2 activity in the presence of a synthetic peptide derived from the carboxyl terminus of PRK2. *Curr. Biol.* **9**, 393–404.
29. Hansra, G. et al. (1999). Multisite dephosphorylation and desensitization of conventional protein kinase C isotypes. *Biochem. J.* **342**, 337–344.
30. Lee, J.Y., Hannun, Y. A., and Obeid, L. M. (2000). Functional dichotomy of protein kinase C in TNFα signal transduction in L929 cells. *J Biol Chem*.
31. Sontag, E., Sontag, J. M., and Garcia, A. (1997). Protein phosphatase 2A is a critical regulator of protein kinase Cζ signaling targeted by SV40 small t to promote cell growth and NF-κB activation. *EMBO J.* **16**, 5662–5671.
32. England, K. et al. (2001). Signalling pathways regulating the dephosphorylation of Ser729 in the hydrophobic domain of PKC (ε) upon cell passage. *J. Biol. Chem.* **276**, 10437–10442.
33. Kraft, A. S. et al. (1982). Decrease in cytosolic calcium/phospholipid-dependent protein kinase activity following phorbol ester treatment of EL4 thymoma Cells. *J. Biol. Chem.* **257**, 13193–13196.
34. Kraft, A. S. and Anderson, W. B. (1983). Phorbol esters increase the amount of Ca^{2+}, phospholipid-dependent protein kinase associated with plasma membrane. *Nature* **301**, 621–623.
35. Newton, A. C. and Johnson, J. E. (1998). Protein kinase C: a paradigm for regulation of protein function by two membrane-targeting modules. *Biochem. Biophys. Acta* **1376**, 155–172.
36. Sakai, N. et al. (1997). Direct visualization of the translocation of the γ-subspecies of protein kinase C in living cells using fusion proteins with green fluorescent protein. *J. Cell Biol.* **139**, 1465–176.
37. Oancea, E. and Meyer, T. (1998). Protein kinase C as a molecular machine for decoding calcium and diacylglycerol signals. *Cell* **95**, 307–318.
38. Nalefski, E. A. and Newton, A. C. (2001). Membrane binding kinetics of protein kinase C βII mediated by the C2 domain. *Biochemistry* **40**, 13216–29.
39. Mosior, M. and Newton, A. C. (1996). Calcium-independent binding to interfacial phorbol esters causes protein kinase C to associate with membranes in the absence of acidic lipids. *Biochemistry*, **35**, 1612–1623.
40. Schaefer, M. et al. (2001). Diffusion-limited translocation mechanism of protein kinase C isotypes. *FASEB J.* **15**, 1634–1636.
41. Edwards, A. S. and Scott, J. D. (2000). A-kinase anchoring proteins: protein kinase A and beyond. *Curr. Opin. Cell Biol.* **12**, 217–21.
42. Kiley, S. C. et al. (1995). Intracellular targeting of protein kinase C isozymes: functional implications. *Biochem. Soc. Trans.* **23**, 601–605.
43. Mochly-Rosen, D. and Gordon, A. S. (1998). Anchoring proteins for protein kinase C: a means for isozyme selectivity. *FASEB J.* **12**, 35–42.
44. Colledge, M. and Scott, J. D. (1999). AKAPs: from structure to function. *Trends Cell Biol.* **9**, 216–221.
45. Jaken, S. and Parker, P. J. (2000). Protein kinase C binding partners. *Bioessays*, **22**, 245–254.
46. Tsunoda, S. et al. (1997). A multivalent PDZ-domain protein assembles signalling complexes in a G- protein-coupled cascade. *Nature*, **388**, 243–249.
47. Newlon, M. G. et al. (1999). The molecular basis for protein kinase A anchoring revealed by solution NMR. *Nat. Struct. Biol.* **6**, 222–227.
48. Klauck, T. M. et al. (1996). Coordination of three signalling enzymes by AKAP 79, a mammalian scaffold protein. *Science* **271**, 1589–1592.
49. Toker, A. (1998). Signaling through protein kinase C. *Front. Biosci.* **3**, D1134–D1147.
50. Black, J. D. (2000). Protein kinase C-mediated regulation of the cell cycle. *Front. Biosci.* **5**, D406–D423.
51. Newton, A. C. and Toker, A. (2001). Cellular regulation of protein kinase C, in Storey, K. B. and Storey, J. M., Eds., *Protein Adaptations and Signal Transduction*, pp. 163–173. Elsevier, Amsterdam.52.

Gokmen-Polar, Y. *et al.* (2001). Elevated protein kinase CβII is an early promotive event in colon carcinogenesis. *Cancer Res.* **61**, 1375–1381.
53. Hodge, C. W. *et al.* (1999). Supersensitivity to allosteric GABA(A) receptor modulators and alcohol in mice lacking PKCε. *Nat. Neurosci.* **2**, 997–1002.
54. Malmberg, A. B. *et al.* (1997). Preserved acute pain and reduced neuropathic pain in mice lacking PKCγ. *Science* **278**, 279–283.
55. Weeber, E. J. *et al.* (2000). A role for the beta isoform of protein kinase C in fear conditioning. *J. Neurosci.* **20**, 5906–5914.

CHAPTER 154

Role of PDK1 in Activating AGC Protein Kinase

Dario R. Alessi
MRC Protein Phosphorylation Unit,
School of Life Sciences, University of Dundee,
Dundee, United Kingdom

Introduction

Stimulation of cells with growth factors, survival factors, and hormones leads to recruitment to the plasma membrane of a family of lipid kinases known as class 1 phosphoinositide 3-kinases (PI 3-kinases, [1]). In this location PI 3-kinases phosphorylate the glycerophospholipid phosphatidylinositol 4,5-bisphosphate (PtdIns(4,5)P_2), at the D-3 position of the inositol ring, converting it to PtdIns(3,4,5)P_3, which is then converted to PtdIns(3,4)P_2 through the action of the SH2-containing inositol phosphatases (SHIP1 and SHIP2) or back to PtdIns(4,5)P_2 via the action of the lipid phosphatase PTEN (phosphatase and tensin homolog deleted on chromosome 10).

PtdIns(3,4,5)P_3 and perhaps PtdIns(3,4)P_2 play key roles in regulating many physiological processes, including controlling cell apoptosis and proliferation, most of the known physiological responses to insulin, and cell differentiation and cytoskeletal organization [2]. PtdIns(3,4,5)P_3 and PtdIns(3,4)P_2 exert their cellular effects by interacting with proteins that possess a certain type of pleckstrin homology domain (PH domain). A number of types of PH domain containing proteins that interact with PtdIns(3,4,5)P_3 and/or PtdIns(3,4)P_2 have now been identified. These include the serine/threonine protein kinases protein kinase B (PKB; also known as Akt) [3], tyrosine kinases of the Tec family [4,5], numerous adaptor molecules such as the Grb2-associated protein (GAB1 [6]), the dual adaptor of phosphotyrosine and 3-phosphoinositides (DAPP1 [7–10]), and the tandem PH-domain-containing proteins (TAPP1 and TAPP2 [11]), as well as guanosine triphosphate (GTP)/guanosine diphosphate (GDP) exchange [12–14] and GTPase-activating proteins [15,16] for the ARF/Rho/Rac family of GTP binding proteins (Fig 1). This chapter focuses on research aimed at understanding the mechanism by which PtdIns(3,4,5)P_3 regulates one branch of its downstream signaling pathways, namely enabling PDK1 to phosphorylate and activate a group of serine/threonine protein kinases that belong to the AGC subfamily of protein kinases. These include isoforms of PKB [3,17], p70 ribosomal S6 kinase (S6K) [18,19], serum- and glucocorticoid-induced protein kinase (SGK) [20], p90 ribosomal S6 kinase (RSK) [21], and protein kinase C (PKC) isoforms [22]. Once these diverse AGC kinase members are activated, they phosphorylate and change the activity and function of key regulatory proteins that control processes such as cell proliferation and survival as well as cellular responses to insulin [2,3,23].

Mechanism of Activation of PKB

The three isoforms of PKB (PKBα, PKBβ, and PKBγ) possess high sequence identity and are widely expressed in human tissues [17]. Stimulation of cells with agonists that activate PI 3-kinase induce a large activation of PKB isoforms within a few minutes. The activation of PKB is downstream of PI 3-kinase, as inhibitors of PI 3-kinase such as wortmannin or LY294002, or the over-expression of a dominant-negative regulatory subunit of PI 3-kinase inhibit the activation of PKB in cells by virtually all agonists tested [24–26]. Over-expression of a constitutively active mutant of PI 3-kinase induces PKB activation in unstimulated cells [27],

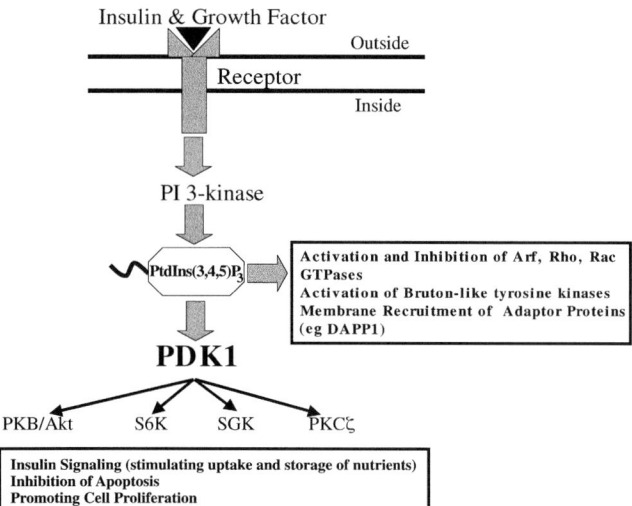

Figure 1 Overview of the PI 3-kinase signaling pathway. Insulin and growth factors induce the activation of PI 3-kinase and generation of PtdIns(3,4,5)P$_3$. In addition to leading to the activation of PKB/Akt, S6K, SGK, and atypical PKC isoforms such as PKCζ, PtdIns(3,4,5)P$_3$ also recruits a number of other proteins (outlined in the text) to the plasma membrane to trigger the activation of non-PDK1/AGC-kinase-dependent signaling pathways. Key challenges for future experiments are not only to define the specific cellular roles of the individual AGC kinase but also to understand the function and importance of other branches of signaling pathways activated by PI 3-kinase.

as does deletion of the PTEN phosphatase which also results in increased cellular levels of PtdIns(3,4,5)P$_3$ [28–32].

All PKB isoforms possess an N-terminal pleckstrin homology (PH domain) that interacts with PtdIns(3,4,5)P$_3$ and PtdIns(3,4)P$_2$ followed by a kinase catalytic domain and then a C-terminal tail. Stimulation of cells with agonists that activate PI 3-kinase induces the translocation of PKB to the plasma membrane, where PtdIns(3,4,5)P$_3$ as well as PtdIns(3,4)P$_2$ are located and, consistent with this, translocation of PKB is prevented by inhibitors of PI 3-kinase or by the deletion of the PH domain of PKB [33–35]. These findings strongly indicate that PKB interacts with PtdIns(3,4,5)P$_3$ and/or PtdIns(3,4)P$_2$ *in vivo*. The binding of PKB to PtdIns(3,4,5)P$_3$ or PtdIns(3,4)P$_2$ does not activate the enzyme but instead recruits PKB to the plasma membrane where it becomes phosphorylated at two residues at this location, namely Thr308 and Ser473. Inhibitors of PI 3-kinase and dominant-negative PI 3-kinase prevent phosphorylation of PKB at both residues following stimulation of cells with insulin and growth factors [17]. Thr308 is located in the T-loop (also known as *activation loop*) between subdomains VII and VIII of the kinase catalytic domain, situated at the same position as the activating phosphorylation sites found in many other protein kinases. As discussed later, Ser473 is located outside of the catalytic domain in a motif that is present in most AGC kinases and which has been termed the *hydrophobic motif*. The phosphorylation of PKBα at both Thr308 or Ser473 is likely to be required to activate PKBα maximally, as mutation of Thr308 to Ala abolishes PKBα activation, whereas mutation of Ser473 to Ala reduces the activation of PKBα by approximately 85%. The mutation of both Thr308 and Ser473 to Asp (to mimic the effect of phosphorylation by introducing a negative charge) increases PKBα activity substantially in unstimulated cells, and this mutant cannot be further activated by insulin [3]. Attachment of a membrane-targeting domain to PKBα results in it becoming highly active in unstimulated cells and induces a maximal phosphorylation of Thr308 and Ser473 [33,36]. These observations indicate that recruitment of PKB to the membrane of unstimulated cells is sufficient to induce the phosphorylation of PKBα at Thr308 and Ser473. Furthermore, there must be sufficient basal levels of PtdIns(3,4,5)P$_3$/PtdIns(3,4)P$_2$, T308 kinase, and Ser473 kinase located at the membrane to stimulate phosphorylation and activation of membrane-targeted PKB. PKBβ and PKBγ are activated by phosphorylation of the equivalent residues in their T-loops and hydrophobic motifs [37,38].

PKB Is Activated by PDK1

A protein kinase was purified [39,40] and subsequently cloned [41,42] that phosphorylated PKBα at Thr308 only in the presence of lipid vesicles containing PtdIns(3,4,5)P$_3$ or PtdIns(3,4)P$_2$. Because of these properties it was named 3-phosphoinositide-dependent protein kinase 1 (PDK1) and is composed of an N-terminal catalytic domain and a C-terminal PH domain which, like that of PKB, interacts with PtdIns(3,4,5)P$_3$ and PtdIns(3,4)P$_2$ [42,43]. The activation of PKB by PDK1 is stereospecific for the physiological D-enantiomers of these lipids, and neither PtdIns(4,5)P$_2$ nor any inositol phospholipid other than PtdIns(3,4)P$_2$ can replace PtdIns(3,4,5)P$_3$ in the PDK1-catalyzed activation of PKB [39,42].

Although co-localization of PKB and PDK1 at the plasma membrane through their mutual interaction with 3-phosphoinositides is likely to be important for PDK1 to phosphorylate PKB, the binding of PKB to PtdIns(3,4,5)P$_3$ or PtdIns(3,4)P$_2$ is also postulated to induce a conformational change in PKB, exposing Thr308 for phosphorylation by PDK1. This conclusion is supported by the observation that in the absence of 3-phosphoinositides, PDK1 is unable to phosphorylate wild-type PKB under conditions where it is able to efficiently phosphorylate a mutant form of PKB that lacks its PH domain, termed ΔPH-PKB [40,41]. Consistent with this, a PKB mutant in which a conserved Arg residue in the PH domain is mutated to abolish the ability of PKB to bind PtdIns(3,4,5)P$_3$ cannot be phosphorylated by PDK1 in the presence of lipid vesicles containing PtdIns(3,4,5)P$_3$ [40]. Moreover, artificially promoting the interaction of PDK1 with wild-type PKB and ΔPH-PKB by the attachment of a high-affinity PDK1 interaction motif to these enzymes is sufficient to induce maximal phosphorylation of Thr308 in ΔPH-PKB but not in wild-type PKB in unstimulated cells [44].

More recently, the three-dimensional structure of the isolated PH domain of PKB complexed with the head group of

PtdIns(3,4,5)P$_3$ has been solved [45]. Interestingly, the structure of the PH domain of PKB complexed to the inositol head group of PtdIns(3,4,5)P$_3$ revealed that the 3- and the 4-phosphate groups form numerous interactions with specific basic amino acids in the PKB PH domain, but in contrast the 5-phosphate group does not make any significant interaction with the protein backbone and is solvent exposed, thus providing the first structural explanation of why PKB interacts with both PtdIns(3,4,5)P$_3$ and PtdIns(3,4)P$_2$ with similar affinity [45].

The interaction of PDK1 with PtdIns(3,4,5)P$_3$ and PtdIns(3,4)P$_2$ is thought to be the primary determinant in enabling PDK1 and PKB to colocalize at membranes and permitting PDK1 to phosphorylate PKB efficiently. These conclusions are supported by the finding that the rate of activation of PKBα by PDK1 *in vitro*, in the presence of lipid vesicles containing PtdIns(3,4,5)P$_3$, is lowered considerably if the PH domain of PDK1 is deleted. Furthermore, the mutant of PKB that lacks its PH domain is also a very poor substrate for PDK1, compared to wild-type PKB, as it is unable to interact with lipid vesicles containing PtdIns(3,4,5)P$_3$.

Activation of Other Kinases by PDK1

The finding that the T-loop residues of PKB are very similar to those found on other AGC kinases suggested that PDK1 might phosphorylate and activate these members [46,47]. An alignment of the T-loop sequences of insulin and growth-factor-stimulated AGC kinases is shown in Fig. 2. It was found that the AGC kinases activated downstream of PI 3-kinase (namely, S6K1 [48,49], SGK isoforms [50–52], and atypical PKC isoforms [53,54]) were phosphorylated specifically at their T-loop residue by PDK1 *in vitro* or following the over-expression of PDK1 in cells. Moreover, AGC kinases that were not activated in a PI 3-kinase-dependent manner in cells—such as the p90 ribosomal S6K (p90RSK) isoforms [55,56], conventional and related PKC isoforms [57–60], PKA [61], and the non-AGC Ste20 family member PAK1 [62]—were also proposed to be physiological substrates for PDK1, as they could all be phosphorylated by PDK1 at their T-loop residue *in vitro* or following over-expression of PDK1 in cells.

Genetic evidence for the central role that PDK1 plays in mediating the activation of these AGC kinases was obtained from the finding that in PDK1$^{-/-}$ ES cells, isoforms of PKB, S6K, and RSK could not be activated by agonists that switch on these enzymes in wild-type cells [63]. In ES cells lacking PDK1, the intracellular levels of endogenously expressed PKCα, PKCβI, PKCγ, PKCδ, PKCε, and PRK1 are also vastly reduced compared to wild-type ES cells [64], consistent with the notion that PDK1 phosphorylation of these enzymes plays an essential role in post-translational stabilization of these kinases [65,66]. The levels of PKCζ were only moderately reduced in the PDK1$^{-/-}$ ES cells and PKCζ in these cells is not phosphorylated at its T-loop residue [64], providing genetic evidence that PKCζ is a physiological substrate for PDK1. In contrast, PKA was active and phosphorylated at its T-loop in PDK1$^{-/-}$ ES cells, to the same extent as in wild-type ES cells [63], thus arguing that PDK1 is not rate limiting for the phosphorylation of PKA in ES cells. It is possible that PKA phosphorylates itself at its T-loop residue *in vivo*, as it has been shown to possess the intrinsic ability to phosphorylate its own T-loop when expressed in bacteria. Thus far, we have no genetic data in PDK1-deficient cells as to whether or not PAK1 is active, but it should be noted that PAK1 can also phosphorylate itself at its T-loop in the presence of Cdc42-GTP or Rac-GTP, stimulating its own activation in the absence of PDK1 [67].

Phenotype of PDK1 PKB- and S6K-Deficient Mice and Model Organisms

PDK1$^{-/-}$ mouse embryos die at day E9.5, displaying multiple abnormalities that include a lack of somites, forebrain, and neural-crest-derived tissues, although the development of the hind- and midbrain proceeds relatively normally [68]. Other eukaryotic organisms also possess homologs of PDK1 that activate homologs of PKB and S6K in these species [69]. As in mice, knocking out PDK1 homologs in yeast [70–72], *Caenorhabditis elegans* [73], and *Drosophila* [74,75] results in nonviable organisms, confirming that PDK1 plays a key role in regulating normal development and survival of these organisms. Elegant genetic analysis of the PI 3-kinase/PDK1/AGC kinase pathway in *Drosophila* has demonstrated that this pathway plays a key role in regulating both cell size and number [76,77]. For example, the over-expression of dPI 3-kinase [78,79] or inactivation of the PtdIns(3,4,5)P$_3$ 3-phosphatase dPTEN [80–82] results in an increase in both the cell number as well as the cell size of *Drosophila*. Moreover, loss-of-function mutants of Chico, the fly homolog of insulin receptor substrate adaptor protein [83], dPI 3-kinase, or over-expression

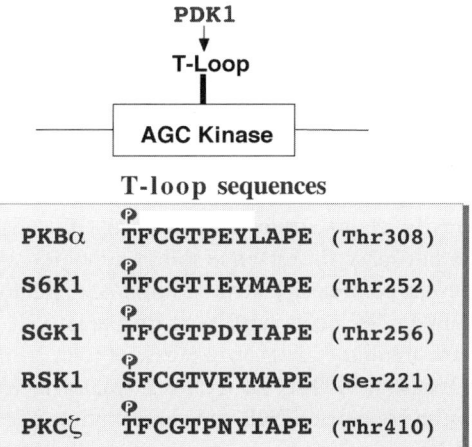

Figure 2 Alignment of the amino acid sequences surrounding the T-loop of insulin and growth-factor-stimulated AGC kinases.

of dPTEN results in a decrease in cell size and number. More recently, a partial loss-of-function mutation in dPDK1 was shown to cause a 15% reduction in fly body weight and a 7% reduction in cell number [74]. Loss of function mutants of dS6K1 [84] or dPKB reduce *Drosophila* cell size without affecting cell number [82,85]. PKB and S6K have also been knocked out in mice, but these studies are complicated by the presence of two isoforms of S6K (S6K1 and S6K2) and three isoforms of PKB (PKBα, PKBβ, and PKBγ) encoded for by distinct genes, in contrast to *Drosophila*, which have one isoform of these enzymes. Mice lacking S6K1 were viable, but adult mice were 15% smaller and possessed 10 to 20% reduced organ masses [86]. It was subsequently shown that S6K1 knockout mice possessed a reduced pancreatic islet β-cell size but the size of other cells types investigated was apparently unaffected [87]. Mice lacking PKBα were also reported to be 20% smaller than wild-type animals, but it was not determined whether the lack of PKBα resulted in a reduction of cell size or cell number [88,89]. In contrast, deletion of PKBβ caused insulin resistance without affecting mouse size [90].

PDK1 hypomorphic mutant mice that express only ≈10% of the normal level of PDK1 in all tissues have been generated [68]. These mice are viable and fertile, and despite the reduced levels of PDK1, injection of these mice with insulin induces the normal activation of PKB, S6K, and RSK in insulin-responsive tissues. Nevertheless, these mice have a marked phenotype, being 40 to 50% smaller than control animals. The volumes of the kidney, pancreas, spleen, and adrenal gland of the PDK1 hypomorphic mice are reduced proportionately. Furthermore, the volume of adrenal gland zona fasciculata cells is 45% lower than control cells, whereas the total cell number and the volume of the nucleus remains unchanged. Cultured embryonic fibroblasts from the PDK1 hypomorphic mice are also 35% smaller than control cells but proliferate at the same rate. Embryonic endoderm cells completely lacking PDK1 from E7.5 embryos were 60% smaller than wild-type cells [68]. These results establish that, as in *Drosophila*, PDK1 plays a key role regulating cell size in mammals. However, the finding that AGC kinases tested are still activated normally in the PDK1 hypomorphic mice may suggest that PDK1 regulates cell size by a pathway that is independent of PKB, S6K, and RSK, although this hypothesis requires further investigation. In this regard, Tian *et al.* [91] have recently reported that PDK1 can interact via its noncatalytic N terminus with the PI 3-kinase-regulated Ral GTP exchange factor, leading to its activation. The Ral GTPase has not been implicated in regulating cell size, but it will be important to investigate whether activation of Ral GTPases is defective in PDK1 hypomorphic or knockout cell lines or mice tissues.

Hydrophobic Motif of AGC Kinases

All insulin and growth-factor-activated AGC kinases, in order to become maximally activated, require phosphorylation of a residue located in a region of homology to the hydrophobic motif of PKBα that encompasses Ser473. This is located ≈160 amino acids C-terminal to the T-loop residue lying outside the catalytic regions of these enzymes. This hydrophobic motif is characterized by a conserved motif: Phe–Xaa–Xaa–Phe–Ser/Thr–Tyr/Phe (where Xaa is any amino acid and the Ser/Thr residue is equivalent to Ser473 of PKB). Atypical PKC isoforms (PKCζ, PKCλ, PKCτ) and the related PKC isoforms (PRK1 and PRK2), instead of possessing a Ser/Thr residue in their hydrophobic motifs, have an acidic residue. PKA, in contrast, possesses only the Phe–Xaa–Xaa–Phe moiety of the hydrophobic motif, as the PKA amino acid sequence terminates at this position [92]. PDK1 is the only AGC kinase member that does not appear to possess an obvious hydrophobic motif [92], and the implications of this are discussed below. A major outstanding challenge is to characterize the mechanism by which PKB and other AGC kinases are phosphorylated at their hydrophobic motifs. In spite of considerable effort to discover the kinases responsible for the phosphorylation of AGC kinase members, no convincing evidence has thus far been obtained. The extensive literature and considerable controversy in this area have been extensively reviewed [93]. The only exception is for RSK and conventional PKC isoforms. For RSK, the phosphorylation of the C-terminal non-AGC kinase domain of this enzyme by ERK1/ERK2 triggers this domain to phosphorylate the N-terminal AGC kinase domain at its hydrophobic motif [21]. In the case of conventional PKC isoforms, there is good evidence that these enzymes can autophosphorylate themselves at their hydrophobic motifs following phosphorylation of their T-loops by PDK1 [22].

Mechanism of Regulation of PDK1 Activity

An important question is to determine the mechanism by which the ability of PDK1 activity to phosphorylate its AGC kinase substrates is regulated by extracellular agonists. When isolated from unstimulated or cells stimulated with insulin or growth factors, PDK1 possesses the same activity toward PKB or S6K1 [41,49,94]. Furthermore, although PDK1 is phosphorylated at 5 serine residues in 293 cells, insulin or insulin-like growth factor 1 (IGF1) did not induce any change in the phosphorylation state of PDK1 [95]. Only one of these phosphorylation sites (namely, Ser241) was essential for PDK1 activity. Ser241 is located in the T-loop of PDK1, and, because PDK1 expressed in bacteria is stoichiometrically phosphorylated at Ser241, it is likely that PDK1 can phosphorylate itself at this residue [95]. Although PDK1 becomes phosphorylated on tyrosine residues following stimulation of cells with peroxovanadate (a tyrosine phosphatase inhibitor) or over-expression with a Src-family tyrosine kinase [96–98], no tyrosine phosphorylation of PDK1 has been detected following stimulation of cells with insulin [95,96].

Taken together, these observations suggest that PDK1 might not be activated directly by insulin/growth factors.

Instead, one possibility that might explain how PDK1 could phosphorylate a number of AGC kinases in a regulated manner is that PDK1, instead of being activated by an agonist, is constitutively active in cells and that it is the substrates that are converted into forms that can interact with PDK1 and thus become phosphorylated at their T-loops. In the case of PKB as discussed above, it is the interaction of PKB with PtdIns(3,4,5)P_3 that converts it into a substrate for PDK1. In the case of other AGC kinases that are activated downstream of PI 3-kinase, such as S6K, SGK, and PKC isoforms, which do not possess a PH domain and thus do not interact with PtdIns(3,4,5)P_3 and whose phosphorylation by PDK1 in vitro is not enhanced by PtdIns(3,4,5)P_3, it is not obvious how PtdIns(3,4,5)P_3 can regulate the phosphorylation of these enzymes in vivo. Recent studies indicate that a conserved motif located C-terminal to the catalytic domains of isoforms of most AGC kinases (the hydrophobic motif of S6K1, or SGK1 [44]) and atypical (PKCζ) and related PKC (PRK2) isoforms [57] can interact with a hydrophobic pocket in the kinase domain of PDK1 (the PIF pocket) [92]. Evidence indicates that this results in a docking interaction, which is required for the efficient T-loop phosphorylation of AGC kinases that do not interact with PtdIns(3,4,5)P_3/PtdIns(3,4)P_2. These experiments indicate that the interaction of S6K and SGK with PDK1 is significantly enhanced if these enzymes are phosphorylated at their hydrophobic motifs in a manner equivalent to that of the Ser473 phosphorylation site of PKB [44]. It is therefore possible that PtdIns(3,4,5)P_3 does not activate PDK1 but instead induces phosphorylation of S6K and SGK isoforms at their hydrophobic motifs, thereby converting these enzymes into forms that can interact with PDK1 and hence become activated. Consistent with this notion, the expression of mutant forms of S6K1 and SGK1 in which the hydrophobic motif phosphorylation site is altered to Glu to mimic phosphorylation is constitutively phosphorylated at their T-loop residues in unstimulated cells [50,99,100]. It is currently not clear how PtdIns(3,4,5)$_3$ could stimulate the phosphorylation of the hydrophobic motif, but it is possible that it could either activate the hydrophobic motif kinases or inhibit the hydrophobic motif phosphatases.

Frodin et al. [101] demonstrated that phosphorylation of the hydrophobic motif of p90RSK (which is induced following phosphorylation of p90RSK by ERK1/ERK2 [21]) strongly promotes its interaction with PDK1, therefore enhancing the ability of PDK1 to phosphorylate p90RSK at its T-loop motif. Thus, the phosphorylation of p90RSK by ERK1/ERK2 converts RSK into a form that can interact with and be activated by PDK1. Thus, the mechanism by which PDK1 recognizes isoforms of RSK is analogous to that by which it recognizes SGK/S6K, the only difference being the mechanism regulating phosphorylation of the hydrophobic motifs of these enzymes. The model of how isoforms of PKB, S6K, SGK, and RSK are activated by PDK1 is summarized in Fig. 3.

Related PKC isoforms (PRK1 and PRK2) and atypical PKC isoforms (PKCζ and PKCτ) possess a hydrophobic motif in which the residue equivalent to Ser473 is Asp or Glu, and these enzymes can in principle interact with PDK1 as soon as they are expressed in a cell [57]. However, it is possible that the interaction of related PKC isoforms and atypical PKC isoforms with PDK1 could be regulated through the interaction of these enzymes with other molecules. For example, the interaction of PRK2 with Rho-GTP [60] or PKCζ with hPar3 and hPar6 [102] might induce a conformational change in these enzymes that controls their interaction with PDK1.

PDK1 would be expected to activate PKB at the plasma membrane and its other non-3-phosphoinositide binding substrates in the cytosol. Consistent with this finding, PDK1 has been found to be localized in mainly the cytosol and plasma membrane of both stimulated and unstimulated cells [43,94]. It is controversial as to whether or not PDK1 translocates to the plasma membrane of cells in response to agonists that activate PI 3-kinase. Three reports [43,94,96] indicate that a small proportion of PDK1 is associated with the membrane of unstimulated cells, and they do not report any further translocation of PDK1 to membranes in response to agonists that activate PI 3-kinase and PKB. However, other groups have reported that PDK1 translocates to cellular membranes in response to agonists that activate PI 3-kinase [103,104]. Indeed, as mentioned earlier, there is evidence that at least some PDK1 is likely to be located at cell membranes of unstimulated cells as the expression of a membrane-targeted PKB construct in such cells is active and fully phosphorylated at Thr308 [33,36].

Structure of the PDK1 Catalytic Domain

Further insight into the mechanism by which PDK1 interacts with its AGC kinase substrates has been obtained recently from the high-resolution crystal structure of the human PDK1 catalytic domain. The structure defines the location of the PIF pocket on the small lobe of the catalytic domain—a marked hydrophobic pocket in the small lobe of the kinase domain [105] that corresponds to the region of the catalytic domain predicted from previous modeling and mutational analysis to form the PIF pocket [92]. Interestingly, mutation of several of the hydrophobic amino acids that make up the surface of this pocket abolish or significantly inhibit the ability of PDK1 to interact and activate S6K1 and SGK1 [44], indicating that this hydrophobic pocket does indeed represent the PIF pocket. As phosphorylation of the hydrophobic motif of S6K1 and SGK1 promotes the binding of S6K1 and SGK1 with PDK1, this suggests that a phosphate-interacting site is located near the PIF pocket. Interestingly, close to the PIF pocket in the PDK1 crystal structure, an ordered sulfate ion was interacting with four surrounding side chains (Lys76, Arg131, Thr148, and Gln150). Mutation of Lys76, Arg131, or Q150 to Ala reduces or abolishes the ability of PDK1 to interact with a phospho-peptide that encompasses the phosphorylated residues of the hydrophobic motif of S6K1, thereby

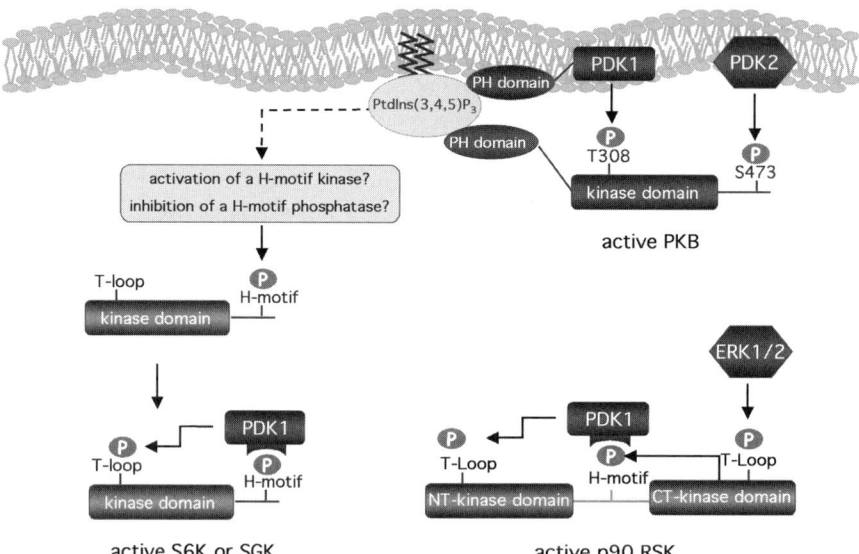

Figure 3 The mechanism by which phosphorylation of PKB, S6K SGK, and RSK by PDK1 is regulated. It should be noted that in this model of how PKB, S6K, SGK, and RSK are phosphorylated at their T-loop, PDK1 activity is not directly activated by insulin or growth factors, consistent with the experimental observation that PDK1 is constitutively active in cells. Instead, it is the substrates of PDK1 that are converted into forms that can be phosphorylated. In the case of PKB, it is the interaction of PKB with PtdIns(3,4,5)P_3 at the plasma membrane that colocalizes PDK1 and PKB and also induces a conformational change in PKB that converts it into a substrate for PDK1. In the case of S6K and SGK, which do not possess PH domains and cannot interact with PtdIns(3,4,5)P_3, this is achieved by the phosphorylation of these enzymes at their hydrophobic motif (H-motif) by an unknown mechanism, which thereby generates a docking site for PDK1. RSK isoforms possess two catalytic domains: an N-terminal AGC-kinase-like kinase domain and a C-terminal non-AGC kinase domain. The activation of RSK isoforms is initiated by the phosphorylation of these enzymes by the ERK1/ERK2 classical MAP kinases, which phosphorylate the T-loop of the C-terminal kinase domain. This activates the C-terminal kinase domain, which then phosphorylates the hydrophobic motif of the N-terminal AGC kinase. This creates a binding site for PDK1 to interact with RSK isoforms, leading to the phosphorylation of the T-loop of the N-terminal kinase domain and activating it. Phosphorylation of all RSK substrates characterized thus far is mediated by the N-terminal kinase domain; however, it is possible that the C-terminal domain of this enzyme will phosphorylate distinct substrates that have not as yet been identified.

suggesting that this region of PDK1 does indeed represent a phosphate docking site [105]. The only other AGC kinase for which the structure is known (namely, PKA) also possesses a hydrophobic pocket at a region of the kinase catalytic domain equivalent to that of PKA which is occupied by the four C-terminal residues of PKA(FXXF) and resembles the first part of the hydrophobic motif phosphorylation site of S6K and SGK (FXXFS/TY) in which the Ser/Thr is the phosphorylated residue [92]. Occupancy of this pocket of PKA by the FXXF residues is likely to be essential to maintaining PKA in an active and stable conformation, as mutation of either Phe residue drastically reduces PKA activity toward a peptide substrate, as well as reducing PKA stability [106,107]. In contrast to the PIF-pocket in the PDK1 structure, PKA does not possess a phosphate docking site located next to the hydrophobic FXXF binding pocket. Sequence alignments of the catalytic domains of AGC kinases, including PDK1, indicate that all AGC kinases possess a PIF pocket, and kinases such as isoforms of RSK, PKB, S6K, and SGK possess a phosphate docking site next to this pocket. The role of these pockets of the AGC kinases is probably to interact with their own hydrophobic motifs, and this interaction may account for the ability of these kinases to be activated following the phosphorylation of their hydrophobic motif. However, unlike other AGC kinases, PDK1 does not possess a hydrophobic motif C-terminal to its catalytic domain and therefore utilizes its empty PIF/phosphate binding pocket to latch onto its substrates that are phosphorylated at their hydrophobic motifs, thereby enabling PDK1 to phosphorylate these enzymes at their T-loop residue and activate them.

Concluding Remarks

Elucidation of the mechanism by which PKB was activated by PDK1 in cells provided the first example of how the second messenger PtdIns(3,4,5)P_3 could activate downstream signaling processes. However, there remain many major unsolved questions for future research to address. A major challenge will be to clarify the mechanism by which PtdIns(3,4,5)P_3 induces the phosphorylation of the hydrophobic motif of PKB and other AGC kinases members, which is a key trigger for the activation of these enzymes. The results discussed in this chapter also provide a framework within which drugs could be developed to inhibit the PDK1/AGC kinase pathway to treat forms of cancers in which this pathway may be constitutively activated. Indeed, it is now estimated that PTEN is mutated in up to 30% of all human tumors, resulting in elevated PtdIns(3,4,5)P$_3$ levels and hence PKB and S6K activity which are likely to contribute to the proliferation and survival of these tumors [108]. It could be envisaged that a PDK1 inhibitor would be effective at reducing the PKB and S6K activities that contribute to growth and survival of these tumors.

Acknowledgments

The work of the author is supported by the U.K. Medical Research Council, Diabetes UK, the Association for International Cancer Research, and the pharmaceutical companies supporting the Division of Signal Transduction Therapy unit in Dundee (AstraZeneca, Boehringer Ingelheim, GlaxoSmithKline, Novo-Nordisk, Pfizer).

References

1. Vanhaesebroeck, B., Leevers, S. J., Ahmadi, K., Timms, J., Katso, R., Driscoll, P. C., Woscholski, R., Parker, P. J., and Waterfield, M. D. (2001). Synthesis, and function of 3-phosphorylated inositol lipids. *Annu. Rev. Biochem.* **70**, 535–602.
2. Cantley, L. C. (2002). The phosphoinositide 3-kinase pathway. *Science* **296**, 1655–1657.
3. Brazil, D. P. and Hemmings, B. A. (2001). Ten years of protein kinase B signalling: a hard Akt to follow. *Trends Biochem. Sci.* **26**, 657–664.
4. Li, Z., Wahl, M. I., Eguinoa, A., Stephens, L. R., Hawkins, P. T., and Witte, O. N. (1997). Phosphatidylinositol 3-kinase-gamma activates Bruton's tyrosine kinase in concert with Src family kinases. *Proc. Natl. Acad. Sci. USA* **94**, 13820–13825.
5. Qiu, Y. and Kung, H. J. (2000). Signaling network of the BTK family kinases. *Oncogene* **19**, 5651–5661.
6. Rodrigues, G. A., Falasca, M., Zhang, Z., Ong, S. H., and Schlessinger, J. (2000). A novel positive feedback loop mediated by the docking protein Gab1 and phosphatidylinositol 3-kinase in epidermal growth factor receptor signaling. *Mol. Cell. Biol.* **20**, 1448–1459.
7. Dowler, S., Currie, R. A., Downes, C. P., and Alessi, D. R. (1999). DAPP1: a dual adaptor for phosphotyrosine and 3-phosphoinositides. *Biochem. J.* **342**, 7–12.
8. Dowler, S., Montalvo, L., Cantrell, D., Morrice, N., and Alessi, D. R. (2000). Phosphoinositide 3-kinase-dependent phosphorylation of the dual adaptor for phosphotyrosine and 3-phosphoinositides by the Src family of tyrosine kinase. *Biochem. J.* **349**, 605–610.
9. Marshall, A. J., Niiro, H., Lerner, C. G., Yun, T. J., Thomas, S., Disteche, C. M., and Clark, E. A. (2000). A novel B lymphocyte-associated adaptor protein, Bam32, regulates antigen receptor signaling downstream of phosphatidylinositol 3-kinase. *J. Exp. Med.* **191**, 1319–1332.
10. Rao, V. R., Corradetti, M. N., Chen, J., Peng, J., Yuan, J., Prestwich, G. D., and Brugge, J. S. (1999). Expression cloning of protein targets for 3-phosphorylated phosphoinositides. *J. Biol. Chem.* **274**, 37893–37900.
11. Dowler, S., Currie, R. A., Campbell, D. G., Deak, M., Kular, G., Downes, C. P., and Alessi, D. R. (2000). Identification of pleckstrin-homology-domain-containing proteins with novel phosphoinositide-binding specificities. *Biochem. J.* **351**, 19–31.
12. Gray, A., Van Der Kaay, J., and Downes, C. P. (1999). The pleckstrin homology domains of protein kinase B, and GRP1 (general receptor for phosphoinositides-1) are sensitive and selective probes for the cellular detection of phosphatidylinositol 3,4-bisphosphate and/or phosphatidylinositol 3,4,5-trisphosphate *in vivo*. *Biochem. J.* **344**, 929–936.
13. Klarlund, J. K., Rameh, L. E., Cantley, L. C., Buxton, J. M., Holik, J. J., Sakelis, C., Patki, V., Corvera, S., and Czech, M. P. (1998). Regulation of GRP1-catalyzed ADP ribosylation factor guanine nucleotide exchange by phosphatidylinositol 3,4,5-trisphosphate. *J. Biol. Chem.* **273**, 1859–1862.
14. Welch, H. C., Coadwell, W. J., Ellson, C. D., Ferguson, G. J., Andrews, S. R., Erdjument-Bromage, H., Tempst, P., Hawkins, P. T., and Stephens, L. R. (2002). P-Rex1, a PtdIns(3,4,5)P(3)- and Gβγ-regulated guanine-nucleotide exchange factor for Rac. *Cell* **108**, 809–821.
15. Venkateswarlu, K., Oatey, P. B., Tavare, J. M., Jackson, T. R., and Cullen, P. J. (1999). Identification of centaurin-alpha1 as a potential *in vivo* phosphatidylinositol 3,4,5-trisphosphate-binding protein that is functionally homologous to the yeast ADP-ribosylation factor (ARF) GTPase-activating protein, Gcs1. *Biochem. J.* **340**, 359–363.
16. Krugmann, S., Anderson, K. E., Ridley, S. H., Risso, N., McGregor, A., Coadwell, J., Davidson, K., Eguinoa, A., Ellson, C. D., Lipp, P., Manifava, M., Ktistakis, N., Painter, G., Thuring, J. W., Cooper, M. A., Lim, Z. Y., Holmes, A. B., Dove, S. K., Michell, R. H., Grewal, A., Nazarian, A., Erdjument-Bromage, H., Tempst, P., Stephens, L. R., and Hawkins, P. T. (2002). Identification of ARAP3, a novel PI3K effector regulating both Arf and Rho GTPases, by selective capture on phosphoinositide affinity matrices. *Mol. Cell* **9**, 95–108.
17. Vanhaesebroeck, B. and Alessi, D. R. (2000). The PI3K-PDK1 connection: more than just a road to PKB. *Biochem. J.* **346**, 561–576.
18. Avruch, J., Belham, C., Weng, Q., Hara, K., and Yonezawa, K. (2001). The p70 S6 kinase integrates nutrient and growth signals to control translational capacity. *Prog. Mol. Subcell. Biol.* **26**, 115–154.
19. Volarevic, S. and Thomas, G. (2001). Role of S6. phosphorylation, and S6 kinase in cell growth. *Prog. Nucleic Acid Res. Mol. Biol.* **65**, 101–127.
20. Lang, F. and Cohen, P. (2001). Regulation, and physiological roles of serum- and glucocorticoid-induced protein kinase isoforms. *Sci. STKE* **2001**, RE17.
21. Frodin, M. and Gammeltoft, S. (1999). Role and regulation of 90 kDa ribosomal S6 kinase (RSK). in signal transduction. *Mol. Cell. Endocrinol.* **151**, 65–77.
22. Newton, A. C. (2001). Protein kinase C: structural, and spatial regulation by phosphorylation, cofactors, and macromolecular interactions. *Chem. Rev.* **101**, 2353–2364.
23. Lawlor, M. A. and Alessi, D. R. (2001). PKB/Akt: a key mediator of cell proliferation, survival, and insulin responses? *J. Cell Sci.* **114**, 2903–2910.
24. Burgering, B. M. and Coffer, P. J. (1995). Protein kinase B (c-Akt) in phosphatidylinositol-3-OH kinase signal transduction. *Nature* **376**, 599–602.
25. Franke, T. F., Yang, S. I., Chan, T. O., Datta, K., Kazlauskas, A., Morrison, D. K., Kaplan, D. R., and Tsichlis, P. N. (1995). The protein kinase encoded by the Akt proto-oncogene is a target of the PDGF-activated phosphatidylinositol 3-kinase. *Cell* **81**, 727–736.
26. Kohn, A. D., Kovacina, K. S., and Roth, R. A. (1995). Insulin stimulates the kinase activity of RAC-PK, a pleckstrin homology domain containing ser/thr kinase. *EMBO J.* **14**, 4288–4295.
27. Klippel, A., Reinhard, C., Kavanaugh, W. M., Apell, G., Escobedo, M. A., and Williams, L. T. (1996). Membrane localization of phosphatidylinositol 3-kinase is sufficient to activate multiple signal-transducing kinase pathways. *Mol. Cell. Biol.* **16**, 4117–4127.
28. Haas-Kogan, D., Shalev, N., Wong, M., Mills, G., Yount, G., and Stokoe, D. (1998). Protein kinase B (PKB/Akt) activity is elevated in glioblastoma cells due to mutation of the tumor suppressor PTEN/MMAC. *Curr. Biol.* **8**, 1195–1198.
29. Li, D. M. and Sun, H. (1998). PTEN/MMAC1/TEP1 suppresses the tumorigenicity and induces G1 cell cycle arrest in human glioblastoma cells. *Proc. Natl. Acad. Sci. USA* **95**, 15406–15411.

30. Wu, X., Senechal, K., Neshat, M. S., Whang, Y. E., and Sawyers, C. L. (1998). The PTEN/MMAC1 tumor suppressor phosphatase functions as a negative regulator of the phosphoinositide 3-kinase/Akt pathway. *Proc. Natl. Acad. Sci. USA* **95**, 15587–15591.

31. Suzuki, A., de la Pompa, J. L., Stambolic, V., Elia, A. J., Sasaki, T., del Barco Barrantes, I., Ho, A., Wakeham, A., Itie, A., Khoo, W., Fukumoto, M., and Mak, T. W. (1998). High cancer susceptibility, and embryonic lethality associated with mutation of the PTEN tumor suppressor gene in mice. *Curr. Biol.* **8**, 1169–1178.

32. Myers, M. P., Pass, I., Batty, I. H., Van der Kaay, J., Stolarov, J. P., Hemmings, B. A., Wigler, M. H., Downes, C. P., and Tonks, N. K. (1998). The lipid phosphatase activity of PTEN is critical for its tumor supressor function. *Proc. Natl. Acad. Sci. USA* **95**, 13513–13518.

33. Andjelkovic, M., Alessi, D. R., Meier, R., Fernandez, A., Lamb, N. J., Frech, M., Cron, P., Cohen, P., Lucocq, J. M., and Hemmings, B. A. (1997). Role of translocation in the activation and function of protein kinase B. *J. Biol. Chem.* **272**, 31515–31524.

34. Goransson, O., Wijkander, J., Manganiello, V., and Degerman, E. (1998). Insulin-induced translocation of protein kinase B to the plasma membrane in rat adipocytes. *Biochem. Biophys. Res. Commun.* **246**, 249–254.

35. Watton, S. J. and Downward, J. (1999). Akt/PKB localisation and 3′ phosphoinositide generation at sites of epithelial cell–matrix and cell–cell interaction. *Curr. Biol.* **9**, 433–436.

36. Kohn, A. D., Takeuchi, F., and Roth, R. A. (1996). Akt, a pleckstrin homology domain containing kinase, is activated primarily by phosphorylation. *J. Biol. Chem.* **271**, 21920–21926.

37. Walker, K. S., Deak, M., Paterson, A., Hudson, K., Cohen, P., and Alessi, D. R. (1998). Activation of protein kinase B beta and gamma isoforms by insulin *in vivo* and by 3-phosphoinositide-dependent protein kinase-1 *in vitro*: comparison with protein kinase B alpha. *Biochem. J.* **331**, 299–308.

38. Brodbeck, D., Cron, P., and Hemmings, B. A. (1999). A human protein kinase Bγ with regulatory phosphorylation sites in the activation loop and in the C-terminal hydrophobic domain. *J. Biol. Chem.* **274**, 9133–9136.

39. Alessi, D. R., James, S. R., Downes, C. P., Holmes, A. B., Gaffney, P. R., Reese, C. B., and Cohen, P. (1997). Characterization of a 3-phosphoinositide-dependent protein kinase which phosphorylates and activates protein kinase Bα. *Curr. Biol.* **7**, 261–269.

40. Stokoe, D., Stephens, L. R., Copeland, T., Gaffney, P. R., Reese, C. B., Painter, G. F., Holmes, A. B., McCormick, F., and Hawkins, P. T. (1997). Dual role of phosphatidylinositol-3,4,5-trisphosphate in the activation of protein kinase B. *Science* **277**, 567–570.

41. Alessi, D. R., Deak, M., Casamayor, A., Caudwell, F. B., Morrice, N., Norman, D. G., Gaffney, P., Reese, C. B., MacDougall, C. N., Harbison, D., Ashworth, A., and Bownes, M. (1997). 3-Phosphoinositide-dependent protein kinase-1 (PDK1): structural and functional homology with the *Drosophila* DSTPK61 kinase. *Curr. Biol.* **7**, 776–789.

42. Stephens, L., Anderson, K., Stokoe, D., Erdjument-Bromage, H., Painter, G. F., Holmes, A. B., Gaffney, P. R., Reese, C. B., McCormick, F., Tempst, P., Coadwell, J., and Hawkins, P. T. (1998). Protein kinase B kinases that mediate phosphatidylinositol-3,4,5-trisphosphate-dependent activation of protein kinase B. *Science* **279**, 710–714.

43. Currie, R. A., Walker, K. S., Gray, A., Deak, M., Casamayor, A., Downes, C. P., Cohen, P., Alessi, D. R., and Lucocq, J. (1999). Role of phosphatidylinositol 3,4,5-trisphosphate in regulating the activity and localization of 3-phosphoinositide-dependent protein kinase-1. *Biochem. J.* **337**, 575–583.

44. Biondi, R. M., Kieloch, A., Currie, R. A., Deak, M., and Alessi, D. R. (2001). The PIF-binding pocket in PDK1 is essential for activation of S6K and SGK, but not PKB. *EMBO J.* **20**, 4380–4390.

45. Thomas, C. C., Deak, M., Kelly, S. M., Price, N. C., Alessi, D. R., and Van Aalten, D. M. (2002). High resolution structures of the pleckstrin homology domain of protein kinase B/Akt and a complex with phosphatidylinositol (3,4,5)-trisphosphate. *Curr. Biol.* **12**, 1256–1262.

46. Alessi, D. R. (2001). Discovery of PDK1, one of the missing links in insulin signal transduction. *Biochem. Soc. Trans.* **29**, 1–14.

47. Belham, C., Wu, S., and Avruch, J. (1999). Intracellular signalling: PDK1-a kinase at the hub of things. *Curr. Biol.* **9**, R93–96.

48. Alessi, D. R., Kozlowski, M. T., Weng, Q. P., Morrice, N., and Avruch, J. (1998). 3-Phosphoinositide-dependent protein kinase 1 (PDK1) phosphorylates and activates the p70 S6 kinase *in vivo* and *in vitro*. *Curr. Biol.* **8**, 69–81.

49. Pullen, N., Dennis, P. B., Andjelkovic, M., Dufner, A., Kozma, S. C., Hemmings, B. A., and Thomas, G. (1998). Phosphorylation and activation of p70s6k by PDK1. *Science* **279**, 707–710.

50. Kobayashi, T. and Cohen, P. (1999). Activation of serum- and glucocorticoid-regulated protein kinase by agonists that activate phosphatidylinositide 3-kinase is mediated by 3-phosphoinositide-dependent protein kinase-1 (PDK1) and PDK2. *Biochem. J.* **339**, 319–328.

51. Kobayashi, T., Deak, M., Morrice, N., and Cohen, P. (1999). Characterization of the structure and regulation of two novel isoforms of serum- and glucocorticoid-induced protein kinase. *Biochem. J.* **344**, 189–197.

52. Park, J., Leong, M. L., Buse, P., Maiyar, A. C., Firestone, G. L., and Hemmings, B. A. (1999). Serum and glucocorticoid-inducible kinase (SGK). is a target of the PI3-kinase-stimulated signaling pathway. *EMBO J.* **18**, 3024–3033.

53. Chou, M. M., Hou, W., Johnson, J., Graham, L. K., Lee, M. H., Chen, C. S., Newton, A. C., Schaffhausen, B. S., and Toker, A. (1998). Regulation of protein kinase C zeta by PI3-kinase and PDK-1. *Curr. Biol.* **8**, 1069–1077.

54. Le Good, J. A., Ziegler, W. H., Parekh, D. B., Alessi, D. R., Cohen, P., and Parker, P. J. (1998). Protein kinase C isotypes controlled by phosphoinositide 3-kinase through the protein kinase PDK1. *Science* **281**, 2042–2045.

55. Jensen, C. J., Buch, M. B., Krag, T. O., Hemmings, B. A., Gammeltoft, S., and Frodin, M. (1999). 90-kDa ribosomal S6 kinase is phosphorylated and activated by 3-phosphoinositide-dependent protein kinase-1. *J. Biol. Chem.* **274**, 27168–27176.

56. Richards, S. A., Fu, J., Romanelli, A., Shimamura, A., and Blenis, J. (1999). Ribosomal S6 kinase 1 (RSK1) activation requires signals dependent on and independent of the MAP kinase ERK. *Curr. Biol.* **12**, 810–820.

57. Balendran, A., Biondi, R. M., Cheung, P. C., Casamayor, A., Deak, M., and Alessi, D. R. (2000). A 3-phosphoinositide-dependent protein kinase-1 (PDK1) docking site is required for the phosphorylation of protein kinase Cζ (PKCζ) and PKC-related kinase 2 by PDK1. *J. Biol. Chem.* **275**, 20806–20813.

58. Dong, L. Q., Landa, L. R., Wick, M. J., Zhu, L., Mukai, H., Ono, Y., and Liu, F. (2000). Phosphorylation of protein kinase N by phosphoinositide-dependent protein kinase-1 mediates insulin signals to the actin cytoskeleton. *Proc. Natl. Acad. Sci. USA* **97**, 5089–5094.

59. Dutil, E. M., Toker, A., and Newton, A. C. (1998). Regulation of conventional protein kinase C isozymes by phosphoinositide-dependent kinase 1 (PDK-1). *Curr. Biol.* **8**, 1366–1375.

60. Flynn, P., Mellor, H., Casamassima, A., and Parker, P. J. (2000). Rho GTPase control of protein kinase C-related protein kinase activation by 3-phosphoinositide-dependent protein kinase. *J. Biol. Chem.* **275**, 11064–11070.

61. Cheng, X., Ma, Y., Moore, M., Hemmings, B. A., and Taylor, S. S. (1998). Phosphorylation and activation of cAMP-dependent protein kinase by phosphoinositide-dependent protein kinase. *Proc. Natl. Acad. Sci. USA* **95**, 9849–9854.

62. King, C. C., Gardiner, E. M., Zenke, F. T., Bohl, B. P., Newton, A. C., Hemmings, B. A., and Bokoch, G. M. (2000). p21-activated kinase (PAK1) is phosphorylated and activated by 3-phosphoinositide-dependent kinase-1 (PDK1). *J. Biol. Chem.* **275**, 41201–41209.

63. Williams, M. R., Arthur, J. S., Balendran, A., van der Kaay, J., Poli, V., Cohen, P., and Alessi, D. R. (2000). The role of 3-phosphoinositide-dependent protein kinase 1 in activating AGC kinases defined in embryonic stem cells. *Curr. Biol.* **10**, 439–448.

64. Balendran, A., Hare, G. R., Kieloch, A., Williams, M. R., and Alessi, D. R. (2000). Further evidence that 3-phosphoinositide-dependent

protein kinase-1 (PDK1) is required for the stability and phosphorylation of protein kinase C (PKC) isoforms. *FEBS Lett.* **484**, 217–223.

65. Bornancin, F. and Parker, P. J. (1997). Phosphorylation of protein kinase C-α on serine 657 controls the accumulation of active enzyme and contributes to its phosphatase-resistant state. *J. Biol. Chem.* **272**, 3544–3549 (erratum appears in *J. Biol. Chem.,* May 16, **272**(20), 13458, 1997).

66. Edwards, A. S. and Newton, A. C. (1997). Phosphorylation at conserved carboxyl-terminal hydrophobic motif regulates the catalytic and regulatory domains of protein kinase C. *J. Biol. Chem.* **272**, 18382–18390.

67. Manser, E., Huang, H. Y., Loo, T. H., Chen, X. Q., Dong, J. M., Leung, T., and Lim, L. (1997). Expression of constitutively active alpha-PAK reveals effects of the kinase on actin and focal complexes. *Mol. Cell. Biol.* **17**, 1129–1143.

68. Lawlor, M. A., Mora, A., Ashby, P. R., Williams, M. R., Murray-Tait, V., Malone, L., Prescott, A. R., Lucocq, J. M., and Alessi, D. R. (2002). Essential role of PDK1 in regulating cell size and development in mice. *Emboj.* **21**, 3728–3738.

69. Scheid, M. P. and Woodgett, J. R. (2001). Pkb/Akt: functional insights from genetic models. *Nat. Rev. Mol. Cell. Biol.* **2**, 760–768.

70. Casamayor, A., Torrance, P. D., Kobayashi, T., Thorner, J., and Alessi, D. R. (1999). Functional counterparts of mammalian protein kinases PDK1 and SGK in budding yeast. *Curr. Biol.* **9**, 186–197.

71. Inagaki, M., Schmelzle, T., Yamaguchi, K., Irie, K., Hall, M. N., and Matsumoto, K. (1999). PDK1 homologs activate the Pkc1-mitogen-activated protein kinase pathway in yeast. *Mol. Cell. Biol.* **19**, 8344–8352.

72. Niederberger, C. and Schweingruber, M. E. (1999). A *Schizosaccharomyces pombe* gene, *ksg1*, that shows structural homology to the human phosphoinositide-dependent protein kinase PDK1, is essential for growth, mating, and sporulation. *Mol. Gen. Genet.* **261**, 177–183.

73. Paradis, S., Ailion, M., Toker, A., Thomas, J. H., and Ruvkun, G. (1999). A PDK1 homolog is necessary and sufficient to transduce AGE-1 PI3 kinase signals that regulate diapause in *Caenorhabditis elegans*. *Genes Dev.* **13**, 1438–1452.

74. Rintelen, F., Stocker, H., Thomas, G., and Hafen, E. (2001). PDK1 regulates growth through Akt and S6K in *Drosophila*. *Proc. Natl. Acad. Sci. USA* **98**, 15020–15025.

75. Cho, K. S., Lee, J. H., Kim, S., Kim, D., Koh, H., Lee, J., Kim, C., Kim, J., and Chung, J. (2001). Drosophila phosphoinositide-dependent kinase-1 regulates apoptosis and growth via the phosphoinositide 3-kinase-dependent signaling pathway. *Proc. Natl. Acad. Sci. USA* **98**, 6144–6149.

76. Kozma, S. C. and Thomas, G. (2002). Regulation of cell size in growth, development and human disease: PI3K, PKB and S6K. *Bioessays* **24**, 65–71.

77. Coelho, C. M. and Leevers, S. J. (2000). Do growth and cell division rates determine cell size in multicellular organisms? *J. Cell Sci.* **113**, 2927–2934.

78. Leevers, S. J., Weinkove, D., MacDougall, L. K., Hafen, E., and Waterfield, M. D. (1996). The *Drosophila* phosphoinositide 3-kinase Dp110 promotes cell growth. *EMBO J.* **15**, 6584–6594.

79. Weinkove, D., Twardzik, T., Waterfield, M. D., and Leevers, S. J. (1999). The *Drosophila* class IA phosphoinositide 3-kinase and its adaptor are autonomously required for imaginal discs to achieve their normal cell size, cell number, and final organ size. *Curr. Biol.* **9**, 1019–1029.

80. Goberdhan, D. C., Paricio, N., Goodman, E. C., Mlodzik, M., and Wilson, C. (1999). *Drosophila* tumor suppressor PTEN controls cell size and number by antagonizing the Chico/PI3-kinase signaling pathway. *Genes Dev.* **13**, 3244–3258.

81. Huang, H., Potter, C. J., Tao, W., Li, D. M., Brogiolo, W., Hafen, E., Sun, H., and Xu, T. (1999). PTEN affects cell size, cell proliferation and apoptosis during *Drosophila* eye development. *Development* **126**, 5365–5372.

82. Scanga, S. E., Ruel, L., Binari, R. C., Snow, B., Stambolic, V., Bouchard, D., Peters, M., Calvieri, B., Mak, T. W., Woodgett, J. R., and Manoukian, A. S. (2000). The conserved PI3′K/PTEN/Akt signaling pathway regulates both cell size and survival in *Drosophila*. *Oncogene* **19**, 3971–3977.

83. Bohni, R., Riesgo-Escovar, J., Oldham, S., Brogiolo, W., Stocker, H., Andruss, B. F., Beckingham, K., and Hafen, E. (1999). Autonomous control of cell and organ size by CHICO, a *Drosophila* homolog of vertebrate IRS1-4. *Cell* **97**, 865–875.

84. Montagne, J., Stewart, M. J., Stocker, H., Hafen, E., Kozma, S. C., and Thomas, G. (1999). *Drosophila* S6. kinase: a regulator of cell size. *Science* **285**, 2126–2129.

85. Verdu, J., Buratovich, M. A., Wilder, E. L., and Birnbaum, M. J. (1999). Cell-autonomous regulation of cell and organ growth in *Drosophila* by Akt/PKB. *Nat. Cell Biol.* **1**, 500–506.

86. Shima, H., Pende, M., Chen, Y., Fumagalli, S., Thomas, G., and Kozma, S. C. (1998). Disruption of the p70(s6k)/p85(s6k) gene reveals a small mouse phenotype and a new functional S6 kinase. *EMBO J.* **17**, 6649–6659.

87. Pende, M., Kozma, S. C., Jaquet, M., Oorschot, V., Burcelin, R., Le Marchand-Brustel, Y., Klumperman, J., Thorens, B., and Thomas, G. (2000). Hypoinsulinaemia, glucose intolerance and diminished beta-cell size in S6K1-deficient mice. *Nature* **408**, 994–997.

88. Cho, H., Thorvaldsen, J. L., Chu, Q., Feng, F., and Birnbaum, M. J. (2001). Akt1/PKBα is required for normal growth but dispensable for maintenance of glucose homeostasis in mice. *J. Biol. Chem.* **276**, 38349–38352.

89. Chen, W. S., Xu, P. Z., Gottlob, K., Chen, M. L., Sokol, K., Shiyanova, T., Roninson, I., Weng, W., Suzuki, R., Tobe, K., Kadowaki, T., and Hay, N. (2001). Growth retardation and increased apoptosis in mice with homozygous disruption of the Akt1 gene. *Genes Dev.* **15**, 2203–2208.

90. Cho, H., Mu, J., Kim, J. K., Thorvaldsen, J. L., Chu, Q., Crenshaw, 3rd, E. B., Kaestner, K. H., Bartolomei, M. S., Shulman, G. I., and Birnbaum, M. J. (2001). Insulin resistance and a diabetes mellitus-like syndrome in mice lacking the protein kinase Akt2 (PKB beta). *Science* **292**, 1728–1731.

91. Tian, X., Rusanescu, G., Hou, W., Schaffhausen, B., and Feig, L. A. (2002). PDK1 mediates growth-factor-induced Ral-GEF activation by a kinase-independent mechanism. *EMBO J.* **21**, 1327–1338.

92. Biondi, R. M., Cheung, P. C., Casamayor, A., Deak, M., Currie, R. A., and Alessi, D. R. (2000). Identification of a pocket in the PDK1 kinase domain that interacts with PIF and the C-terminal residues of PKA. *EMBO J.* **19**, 979–988.

93. Leslie, N. R., Biondi, R. M., and Alessi, D. R. (2001). Phosphoinositide-regulated kinases and phosphoinositde phosphatases. *Chem. Rev.* **101**, 2365–2380.

94. Yamada, T., Katagiri, H., Asano, T., Tsuru, M., Inukai, K., Ono, H., Kodama, T., Kikuchi, M., and Oka, Y. (2002). Role of PDK1 in insulin-signaling pathway for glucose metabolism in 3T3-L1 adipocytes. *Am. J. Physiol. Endocrinol. Metab.* **282**, E1385–E1394.

95. Casamayor, A., Morrice, N., and Alessi, D. R. (1999). Phosphorylation of Ser 241 is essential for the activity of PDK1: identification of five sites of phosphorylation *in vivo*. *Biochem. J.* **342**, 287–292.

96. Grillo, S., Gremeaux, T., Casamayor, A., Alessi, D. R., Le Marchand-Brustel, Y., and Tanti, J. F. (2000). Peroxovanadate induces tyrosine phosphorylation of phosphoinositide-dependent protein kinase-1 potential involvement of Src kinase. *Eur. J. Biochem.* **267**, 6642–6649.

97. Prasad, N., Topping, R. S., Zhou, D., and Decker, S. J. (2000). Oxidative stress and vanadate induce tyrosine phosphorylation of phosphoinositide-dependent kinase 1 (PDK1). *Biochemistry* **39**, 6929–6935.

98. Park, J., Hill, M. M., Hess, D., Brazil, D. P., Hofsteenge, J., and Hemmings, B. A. (2001). Identification of tyrosine phosphorylation sites on 3-phosphoinositide-dependent protein kinase-1 and their role in regulating kinase activity. *J. Biol. Chem.* **276**, 37459–37471.

99. Balendran, A., Currie, R. A., Armstrong, C. G., Avruch, J., and Alessi, D. R. (1999). Evidence that PDK1 mediates the phosphorylation of p70 S6 kinase *in vivo* at Thr412 as well as Thr252. *J. Biol. Chem.* **274**, 37400–37406.

100. Weng, Q. P., Kozlowski, M., Belham, C., Zhang, A., Comb, M. J., and Avruch, J. (1998). Regulation of the p70 S6 kinase by phosphorylation *in vivo*. Analysis using site-specific anti-phosphopeptide antibodies. *J. Biol. Chem.* **273**, 16621–16629.

101. Frodin, M., Jensen, C. J., Merienne, K., and Gammeltoft, S. (2000). A phosphoserine-regulated docking site in the protein kinase RSK2 that recruits and activates PDK1. *EMBO J.* **19**, 2924–2934.
102. Brazil, D. P. and Hemmings, B. A. (2000). Cell polarity: scaffold proteins par excellence. *Curr. Biol.* **10**, R592–594.
103. Anderson, K. E., Coadwell, J., Stephens, L. R., and Hawkins, P. T. (1998). Translocation of PDK-1 to the plasma membrane is important in allowing PDK-1 to activate protein kinase B. *Curr. Biol.* **8**, 684–691.
104. Filippa, N., Sable, C. L., Hemmings, B. A., and Van Obberghen, E. (2000). Effect of phosphoinositide-dependent kinase 1 on protein kinase B translocation and its subsequent activation. *Mol. Cell. Biol.* **20**, 5712–5721.
105. Biondi, R. M., Komander, D., Thomas, C. C., Lizcano, J. M., Deak, M., Alessi, D. R., and Van Aalten, D. M. (2002). 2Å structure of human PDK1 catalytic domain defines the regulatory phosphate docking site. *Emboj.* **21**, 4214–4228.
106. Batkin, M., Schvartz, I., and Shaltiel, S. (2000). Snapping of the carboxyl terminal tail of the catalytic subunit of PKA onto its core: characterization of the sites by mutagenesis. *Biochemistry* **39**, 5366–5373.
107. Etchebehere, L. C., Van Bemmelen, M. X., Anjard, C., Traincard, F., Assemat, K., Reymond, C., and Veron, M. (1997). The catalytic subunit of dictyostelium cAMP-dependent protein kinase: role of the N-terminal domain and of the C-terminal residues in catalytic activity and stability. *Eur. J. Biochem.* **248**, 820–826.
108. Leslie, N. R. and Downes, C. P. (2002). PTEN: the down side of PI3-kinase signalling. *Cell Signal* **14**, 285–295.

CHAPTER 155

Modulation of Monomeric G Proteins by Phosphoinositides

Sonja Krugmann, Len Stephens, and Phillip T. Hawkins

*Inositide Laboratory, Signalling Programme, Babraham Institute,
Babraham, Cambridge, United Kingdom*

Introduction

Most, if not all, membranes in eukaryotic cells carry low mole percent phosphoinositides. These lipids act as regulatable scaffolds, dictating the localization and functions of many proteins on the membrane surface. A clear example of this principle is at the inner leaflet of the plasma membrane, where $PtdIns(3,4,5)P_3$ and $PtdIns(3,4)P_2$ are rapidly generated following cell-surface receptor activation of type I phosphoinositide 3OH-kinases (PI3K; Chapter 25 by D. Fruman) and act to recruit several PH domain containing proteins. These translocations are driven by the high specificity and affinity of specific PH domains for $PtdIns(3,4,5)P_3$ and $PtdIns(3,4)P_2$ (Chapter 29 by M. Lemmon). Many of the proteins that harbor PH domains are enzymes that regulate the activity of monomeric (small) GTPases: the guanine nucleotide exchange factors (GEFs), which promote GTP-loading of the GTPase, and GTPase activating proteins (GAPs), which enhance the endogenous GTPase activity of the GTPase. Small GTPases are considered to be active or "on" in their GTP-bound form and inactive or "off" in their GDP-bound form. Thus, historically, GEFs are generally considered to be activators and GAPs to be inactivators. More recent work suggests this is too simplistic. It appears that monomeric GTPases often need to cycle between on/off states to function effectively and that the GAPs can themselves be the target or effector of the monomeric GTPase or act as scaffolds to bring the monomeric GTPase together with targets and hence dictate the context of its activation. We summarize here recent evidence on the modulation of GEFs and GAPs by phosphoinositides with a focus on events regulated by PI3K.

Rho Family Small GTPases

Rho family GTPases are best known for their ability to modulate the actin cytoskeleton (where Rho regulates the formation of stress fibers, Rac regulates membrane ruffles, and Cdc42 regulates filopodia) but they are also involved in the control of such diverse processes as NADPH oxidase, transcriptional activation, G1 cell cycle progression, cell transformation, and secretion [1].

Direct Interactions of Rho family GTPases with Phosphoinositides

Both Rac and Cdc42 interact directly with $PtdIns(3,4,5)P_3$ and $PtdIns(3,4)P_2$ with an apparent affinity of 4.5 μM [2]. Lipid binding has been mapped to two sites containing abundant hydrophobic and positively charged sites on the GTPase, and leads to the enhanced dissociation of Rac-associated guanine nucleotides but not association of GTP and hence activation.

Modulation of Rac GEFs by Phosphoinositides

PI3K is thought to be involved in the activation of Rac by stimulating the following Dbl family GEFs: SOS, Vav1, Tiam-1, P-Rex, and possibly PIX. Like all Rho family GEFs [3] they contain the characteristic Dbl homology (DH)-PH domain module (see Fig. 1).

Apart from its well-characterized function as a Ras GEF, SOS functions also as a Rac GEF. The SOS PH domain translocates to the plasma membrane and preferentially to

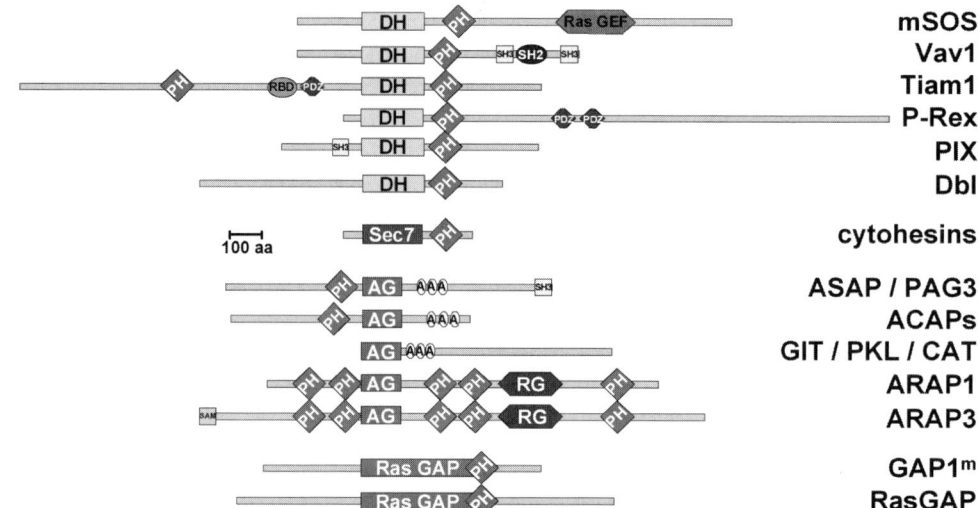

Figure 1 Alignment of all GEF and GAP proteins discussed along their relevant catalytic domains. Domains are drawn as determined via the SMART program. Abbreviations: A, ankyrin repeat; RBD, Ras binding domain; RG, Rho GAP; AG, Arf GAP domains.

leading edges after serum stimulation [4], and this is PI3K dependent. This phenomenon depends on the presence of key residues within the SOS PH domain mediating the interaction with PtdIns(3,4,5)P_3. The SOS DH domain alone functions as a Rac GEF both *in vitro* and *in vivo* [5], but a DH-PH construct is active only in the presence of an activating signal for PI3K. Current evidence suggests PtdIns(3,4,5)P_3-binding to the PH domain relieves an autoinhibitory constraint on the DH domain leading to an increase in catalyic activity toward Rac [6,7].

Vav proteins (Vavs 1–3) are essential for cytoskeletal, proliferative, and developmental pathways in lymphoid cells. Vav-1 is the only family member known to be regulated by PI3K; its regulation has been studied in-depth *in vitro*. Based on structural studies of Vav1 [8], the current model is that GEF activity of Vav1 is autoinhibited by the binding of its N-terminus to the DH domain. Phosphorylation of Y174 by Src-type tyrosine kinases [9] causes the release of the N-terminal inhibitory peptide and this transition is facilitated by PtdIns(3,4,5)P_3, but not PtdIns(4,5)P_2, binding to the PH domain [6,10]. The evidence is less clear-cut *in vivo*, but Gringhuis *et al.* [11] demonstrated convincingly, that in CD5 receptor signaling in T lymphocytes, Vav lies upstream of Rac and downstream of PI3K.

Tiam-1 is a broadly expressed, PI3K-regulated Rac GEF involved in cell adhesion and migration. It contains two PH domains. The N-terminal one binds with high affinity to PtdIns(3,4,5)P_3, is crucial for activation of Rac (ruffling), and plays a role in, but does not dictate, membrane localization [12,13]. Both *in vivo* and *in vitro* evidence suggests that Tiam-1 is regulated by threonine phosphorylation by CaMKII [14] and by binding of PtdIns(3,4,5)P_3 to the Tiam1 N-terminal PH domain [15], both of which moderately activate Tiam-1 catalytic activity, in a cooperative fashion [15]. Neither interactions with nor activation by PtdIns(3,4,5)P_3 of a Tiam1 DH–PH domain construct could be reproduced in a recent study [16].

P-Rex (PtdIns(3,4,5)P_3-dependent Rac exchanger) was purified from porcine neutrophils as a PtdIns(3,4,5)P_3-activated Rac GEF that would be responsible for regulating Rac downstream of activation of heterotrimeric G proteins [17]. Analysis of recombinant P-Rex1 shows that it interacts with lipid vesicles in a PtdIns(3,4,5)P_3-dependent fashion. P-Rex1 RacGEF activity is directly and substantially activated by G protein βγ subunits and PtdIns(3,4,5)P_3 in a synergistic fashion both *in vivo* and *in vitro*.

PIX was identified as a Pak-binding protein with GEF activity toward Rac and Cdc42 involved in the recruitment of PAK to focal adhesions. PIX binds directly to GIT, a potential PtdIns(3,4,5)P_3-stimulated Arf GAP ([18]; see s "PtdIns(3,4,5)P_3-regulated Arf GEFs: The Cytohesin Family," below). PIX interacts with the p85 regulatory subunit of type IB PI3K, and PIX GEF activity is very weakly stimulated by PtdIns(3,4,5)P_3 [19], but the mechanism is unclear.

Phosphoinositide Binding to Cdc42 GEFs

The PH domain of Dbl is reported to bind to PtdIns(4,5)P_2 and PtdIns(3,4,5)P_3. Both phosphoinositides inhibit Dbl GEF activity. Dbl is partially localized to the plasma membrane in a PH domain dependent fashion [20]. It is interesting that Dbl PH domain point mutants that do not bind phosphoinositides and confer increased Cdc42 GEF activity do not promote focus formation. Another study reports binding of PtdIns(4,5)P_2 to DH–PH constructs of Dbs (Dibl's big sister) and intersectin, but neither GEF activity was found to be affected by its inclusion into assays *in vitro* [16].

Arf Family GTPases

Arf family G proteins [21] differ from other small G proteins in that they do not have any detectable intrinsic GTPase activity; therefore they have an absolute requirement for both GEFs and GAPs to cycle rapidly between GTP- and GDP-bound states. The closely related Arf family members are grouped into class I (Arfs 1–3), class II (Arfs 4 and 5) and Class III (Arf 6). Class I Arfs are primarily involved in trafficking in the ER-Golgi and endosomal systems. Little is known about class II Arfs. Arf 6 functions in endosomal and plasma membranes to regulate secretion and coordinate cytoskeletal changes (ruffling) in collaboration with Rac, which it transports to the plasma membrane [22]. Phosphoinositides have regulatory inputs into Arfs 6 and 1 by acting on both their GEFs and GAPs.

PtdIns(3,4,5)P_3-regulated Arf GEFs: The Cytohesin Family

Out of the growing number of Arf GEFs [23] only the cytohesin family (comprising the highly homologous Cytohesin 1 and 4, ARNO and Grp1) are regulated by phosphoinositides. Cytohesins share with other Arf GEFs the catalytic Sec7 domain and contain further a characteristic C-terminal PH domain (Fig. 1), which selectively binds specific phosphoinositides. The cytohesins exhibit characteristically dramatic translocations from the cytosol to the plasma membrane that are both dependent on PI3K activity and a functional PH domain. This translocation probably represents their main mode of activation. The field is littered with controversy regarding cytohesin family lipid-binding specificities, an issue that may be explained by the presence of allelic variants of all cytohesins bar cytohesin 4 [24]. The variants are distinguished by the insertion of a single glycine residue into the lipid-binding specificity-determining region of the PH domain. Binding and selectivity of diglycine PH domains to PtdIns(3,4,5)P_3 over PtdIns(4,5)P_2 or PtdIns-(3,4)P_2 is exquisite; tri-glycine PH domains are less selective and bind less tightly. Hence, isolated diglycine PH domain constructs can translocate to membranes containing PtdIns-(3,4,5)P_3, whereas triglycine PH domains translocate only in the context of the full-length GEFs [25], when a polybasic stretch C-terminally adjacent of the PH domain enhances membrane association [26]. ARNO is phosphorylated by PKC on serine 392, which lies in the polybasic stretch adjacent to the PH domain. The negative charge conferred by the phosphate reduces catalytic activity and binding to membranes, suggesting that PKC may act to switch off ARNO GEF activity [27]. It is interesting that 80% of GRP-1 is expressed in the diglyine version whereas Arno and cytohesin-1 are predominently in the triglycine form [24], possibly conferring differential sensitivity to PI3K activation.

A second issue of controversy concerns Grp1 and ARNO substrate specificities. Their abilities to use Arf6 as a substrate *in vitro* appear to depend on the precise assay conditions used, whereas Arfs 1 and 5 act as more robust substrates [28,29].

In vivo, all cytohesins have been shown very convincingly to translocate from a cytoplasmic location to the plasma membrane in a PI3K-dependent fashion ([30] and references therein). This coincides with Arf6 distribution, which cycles between endosomal compartments and plasma membrane, but not Arf1, which is confined to intracellular membranes. Indeed, Arf6 and ARNO have been shown to co-localize in ruffles in a PtdIns(3,4,5)P_3-, GEF activity-, and PH domain-dependent fashion [30].

PtdIns(4,5)P_2 and PtdIns(3,4,5)P_3-regulated Arf GAPs

Arf GAPs are characterized by a Zn-finger containing Arf GAP domain and adjacent ankyrin repeats. Our knowledge of Arf GAPs has expanded dramatically over the last few years [23]. Many Arf GAPs are regulated by phosphoinositides. The Pap/ASAP/ACAP family is activated by binding to PtdIns(4,5)P_2 whereas the GIT/CAT/PKL and ARAP families bind to and are activated by PtdIns(3,4,5)P_3 (Fig. 1). Both Pap (PAG3)/ASAP/ACAPs and GIT/CAT/PKLs function in cytoskeletal remodeling and associate with paxillin (reviewed in [31]).

The Pap, ASAP, and ACAPs ([31] and references therein; [32,33]) are all robustly activated in their Arf GAP activities by PtdIns(4,5)P_2 (in the context of phosphatidic acid), which they bind to using their PH domains. ASAP and Pap/PAG3 function as Arf 1 and 5 GAPs *in vitro*, but there is some *in vivo* evidence for Pap/PAG3 acting on Arf 6 also. ACAPs function as PtdIns(4,5)P_2-stimulated Arf 6 GAPs *in vitro* and *in vivo*. There is an intriguing link between PDGF and ASAP/ACAP Arf GAPs, which are reported to localize to PDGF-induced dorsal ruffles ([34]; although this has not been demonstrated by video imaging) and inhibit their formation when overexpresssed [35,36]. The role of the PH domain has been investigated further in the context of ASAP [34]. It doesn't bind preferentially to PtdIns(3,4,5)P_3 and supports but is not vital for the observed translocation; it is crucial for catalytic Arf GAP activity, leading to a model of ASAP regulation, where Arf GAP and PH domains interact until lipid binding by the PH domain allows freeing of the Arf GAP catalytic site.

GIT family proteins have been identified in various screens as binding partners for paxillin [see 31], the Rho GEF PIX [18,35], and G-protein coupled receptor kinases (GRK; [36]). Several family members exist in addition to multiple spliced variants, which appear to be expressed tissue-specifically [36]. GIT1 and 2 were analyzed in terms of their GAP activities and both were found to use Arfs 1, 5, and 6 as substrates. Their catalytic activity was enhanced moderately by high concentrations (200 μM) of PtdIns(3,4,5)P_3 but not PtdIns5P, PtdIns(4,5)P_2, or diacylglycerol [37]. We do not know how PtdIns(3,4,5)P_3 interacts with GIT proteins, since they lack domains known to mediate interactions with lipids.

The second class of PtdIns(3,4,5)P_3 regulated Arf GAPs are ARAPs. Their unusual domain structure comprising 5 PH domains, as well as Arf- and Rho GAP domains, predicts that these proteins are ideally suited to mediate cross-talk between small G-protein families. ARAP1 and 2 were identified by homology-based cloning [38] and ARAP3 on the basis of its binding to PtdIns(3,4,5)P_3 [39]. ARAP1 is a PtdIns-(3,4,5)P_3-dependent Arf GAP specific for Arf 1 and Arf 5 *in vitro*. *In vivo*, ARAP1 localizes to Golgi structures and regulates cell spreading and the formation of filopodia via the regulation of Arf 1/5 and Cdc42. In contrast, ARAP3 is a promiscuous Rho GAP *in vitro* and is a very specific, PtdIns-(3,4,5)P_3-dependent Arf 6-GAP *in vitro* and *in vivo*. *In vivo*, ARAP3 causes dynamic remodeling of the actin cytoskeleton and a striking loss of adhesion in a PI3K dependent manner.

Modulation of Ras Family GTPases by PI3K

The best understood pathway regulated by Ras family G proteins is the Raf-Erk-MAPK cascade. In addition, Ras has a well-established role in the activation of class I PI3K [40], but there are now numerous, cell-type specific examples in which PI3K has a regulatory role upstream of Ras. One hypothesis is that low amounts of PtdIns(3,4,5)P_3 (generated in most attached cells via integrin-engagement) may fulfill a permissive role in the activation of Ras, possibly allowing the recruitment of Shc-Grb2-SOS RasGEF complexes to the plasma membrane in the absence of significant recruitment via the phosphorylated tails of activated growth-factor receptors [41]. Alternatively, PI3K might modulate some RasGAPs (p120RasGAP, GAP1^m) as they clearly possess PH domains capable of binding PtdIns(3,4,5)P_3. Insulin-stimulation of Swiss 3T3 adipocytes causes a PI3K-dependent inhibition of RasGAP, leading to activation of Ras [42]. Similarly, in U937 cells, PI3K inhibitors act to inhibit Ras, which is mediated via an increase in GTP hydrolysis on Ras, indicating negative control of a RasGAP by PI3K [43]. GAP1^m has been shown to be recruited to the plasma membrane in a PH-domain and PI3K-dependent fashion [44]. Curiously, neither PtdIns-(3,4,5)P_3 nor its soluble headgroup Ins(1,3,4,5)P_4 could significantly influence GAP1^m RasGAP activity *in vivo* or *in vitro* [44,45], thus raising the question as to whether it may influence an effector function of the RasGAP rather than the GAP activity.

Conclusion

Monomeric GTPases are molecular switches that drive many complex processes involving cellular membranes. They are characteristically lipid modified in their active states and hence retained at the lipid bilayer. Their activities are controlled by the actions of "GTP-loading" GEFs and "GTP-hydrolyzing" GAPs whose activities appear to be regulated by the presence of phosphoinositides in the lipid surface in which they reside. In some cases, it is clear that the phosphoinositides act predominantly to localize and hence concentrate the GEF/GAP with the GTPase (usually via direct binding to an appropriate PH domain); perhaps the clearest example is in PtdIns(3,4,5)P_3-dependent recruitment of cytohesin family Arf GEFs from the cytosol to the plasma membrane. In other examples of GEF/GAP regulations by phosphoinositides, bulk translocation is probably irrelevant or only part of the regulatory mechanism; in these cases phosphoinositide binding allows some form of allosteric change leading to an increase in catalytic activity. In the best-studied examples, phosphoinositide binding to a PH domain relieves an autoinhibitory constraint on a neighboring catalytic domain, and given the almost universal positioning of PH domains adjacent to GEF and GAP catalytic modules (see Fig. 1), this may prove to be a general principle. Both of these types of mechanisms can be seen as "acutely regulatory" (where the levels of the appropriate phosphoinositides are changed rapidly, e.g. PtdIns(3,4,5)P_3 synthesis via receptor regulated PI3Ks) or more "permissive" (where the levels of the phosphoinositides are more constant, e.g. arguably in examples of PtdIns(4,5)P_2 regulation). Clearly more focused work needs to be done before we have a satisfactory explanation of how phosphoinositides regulate the activity of individual GEFs and GAPs, but it is already clear just how universal this form of regulation appears to be and hence how important phosphoinositides are for coordinating the regulation of this class of molecules.

Acknowledgments

S. K. is supported by a Deutsche Forschungsgemeinschaft research fellowship; P.T.H. is a BBSRC senior research fellow.

References

1. Bishop, A. L. and Hall, A. (2000). Rho GTPases and their effector proteins. *Biochem. J.* **348**, 241–245
2. Missy, K., Van Poucke, V., Raynal, P., Viala, C., Mauco, G., Plantavid, M., Chap, H., and Payrastre, B. (1998). Lipid products of phosphoinositide 3-kinase interact with Rac1 GTPase and stimulate GDP dissociation. *J. Biol. Chem.* **273**, 30279–30286.
3. Zheng, Y. (2001). Dbl family guanine nucleotide exchange factors. *Trends Biochem. Sci.* **26**, 724–732.
4. Chen, R. H., Corbalan-Garcia, S., and Bar-Sagi, D. (1997). The role of the PH domain in the signal dependent membrane targetting of Sos. *EMBO J.* **16**, 1351–1359.
5. Nimnual, A. S., Yatsula, B. A., and Bar-Sagi, D. (1998). Coupling of Ras and Rac guanosine triphosphatases through the Ras exchanger Sos. *Science* **279**, 560–563.
6. Das, B., Shu, X., Day, G. J., Han, J., Krishna, U. M., Falck, J. R., and Broek, D. (2000). Control of intermolecular interactions between the pleckstrin homology and Dbl homology domains of Vav and Sos1 regulates Rac binding. *J. Biol. Chem.* **275**, 15074–15081.
7. Soisson, S. M., Nimnual, A. S., Uy, M., Bar-Sagi, D., and Kuriyan, J. (1998). Crystal structure of the Dbl and pleckstrin homology domains from the human Son of sevenless protein. *Cell* **95**, 259–268.
8. Aghazadeh, B., Lowry, W. E., Huang, X. Y., and Rosen, M. K. (2000). Structural basis for relief of autoinhibition of the Dbl homology domain of proto-oncogene Vav by tyrosine phosphorylation. *Cell* **102**, 625–633.
9. Crespo, P., Schuebel, K. E., Pstrom, A. A., Gutkind, J. S., and Bustelo, X.R. (1997). Phosphotyrosine-dependent activation of Rac-1 GDP/GTP exchange by the vav proto-oncogene product. *Nature* **385**, 169–172.
10. Han, J., Luby-Phelps, K., Das, B., Shu, X., Xia, Y. L., Mosteller, R. D., Krishna, U. M., Falck, J. R., White, M. A., and Broek, D. (1998). Role of substrates and products of PI3-kinase in regulating activation of Rac-related guanosine trisphosphates by Vav. *Science* **279**, 558–560.

11. Gringhuis, S. I., de Leij, L. F., Coffer, P. J., and Vellenga, E. (1998). Signaling through CD5 activates a pathway involving phosphatidylinositol 3-kinase, Vav, and Rac1 in human mature T lymphocytes. *Mol. Cell Biol.* **18**, 1725–1735.
12. Michiels, F., Stam, J. C., Horgijk, P. L., van der Kammen, R. A., Ruuls-Van Stalle, L., Feltkamp, C. A., and Collard, J. G. (1997). Regulated membrane localization of Tiam1, mediated by the NH_2-terminal pleckstrin homology domain, is required for Rac-dependent membrane ruffling and C-Jun NH_2-terminal kinase activation. *J. Cell Biol.* **137**, 387–398.
13. Stam, J. C., Sander, E. E., Michiels, F., van Leuwen, F. N., Kain, H. E. T., van der Kammen, R. A., and Collard, J. G. (1997). Targeting of Tiam1 to the plasma membrane requires the cooperative function of the N-terminal pleckstrin homology domain and an adjacent protein interaction domain. *J. Biol. Chem.* **272**, 28447–28454.
14. Buchanan, F. G., Elliot, C. M., Gibbs, M., and Exton, J. H. (2000). Translocation of the Rac1 guanine nucleotide exchange factor Tiam1 induced by platelet-derived growth factor and lysophosphatidic acid. *J. Biol. Chem.* **275**, 9742–9748.
15. Fleming, I. N., Gray, A., and Downes, C. P. (2000). Regulation of the Rac1-specific exchange factor Tiam1 involves both phosphoinositide 3-kinase-dependent and -independent components. *Biochem J.* **351**, 173–182.
16. Snyder, J. T., Rossman, K. L., Baumeister, M. A., Pruitt, W. M., Siderovski, D. P., Der, C. J., Lemmon, M. A., and Sondek, J. (2001). Quantitative analysis of the effect of phosphoinositide interactions on the function of Dbl family proteins. *J. Biol. Chem.* **276**, 45868–45875.
17. Welch, H. C. E., Ellson, C. D., Coadwell, J., Erdjument-Bromage, H., Tempst, P., Hawkins, P. T., and Stephens, L. R. (2002). P-Rex1, a novel $PtdIns(3,4,5)P_3$- and $G\beta\gamma$-regulated guanine-nucleotide exchange factor for Rac. *Cell.* **108**, 1–20.
18. Bragodia, S., Bailey, D., Lenard, Z., Hart, M., Guan, J. L., Premont, R. T., Taylor, S. J., and Cerione, R. A. (1999). A tyrosine-phosphorylated protein that binds to an important regulatory region on the cool family of p21-activated kinase-binding proteins. *J. Biol. Chem.* **274**, 22393–22400.
19. Yoshii, S., Tanaka, M., Otsuki, Y., Wang, D. Y., Guo, R. J., Zhu, Y., Takeda, R., Hanai, H., Kaneko, E., and Sugimura, H. (1999). αPIX nucleotide exchange factor is activated by interaction with phosphatidylinositol 3-kinase. *Oncogene* **18**, 5680–5690.
20. Russo, C., Gao, Y., Mancini, P., Vanni, C., Porotto, M., Falasca, M., Torrisi, M. R., Zheng, Y., and Eva, A. (2001). Modulation of oncogenic DBL activity by phosphoinositol phosphate binding to pleckstrin homology domain. *J. Biol. Chem.* **276**, 19524–19531.
21. Moss, J. and Vaughan, M. (1998). Molecules in the ARF orbit. *J. Biol. Chem.* **273**, 21431–2144.
22. Radhakrishna, H., Al-Awar, O., Khachikian, Z., and Donaldson, J. G. (1999). ARF6 requirement for Rac ruffling suggests a role for membrane trafficking in cortical actin rearrangements. *J. Cell Sci.* **112**, 855–866.
23. Donaldson, J. G. and Jackson, C. L. (2000). Regulators and effectors of the ARF GTPases. *Curr. Opin. Cell Biol.* **12**, 475–482.
24. Ogasawara, M., Kim, S. C., Adamik, R., Togawa, A., Ferrans, V. J., Takeda, K., Kirby, M., Moss, J., and Vaughan, M. (2000). Similarities in function and gene structure of cytohesin-4 and cytohesin-1, guanine nucleotide-exchange proteins for ADP-ribosylation factors. *J. Biol. Chem.* **275**, 3221–3230.
25. Klarlund, J. K., Tsiaras, W., Holik, J., Chawla, A., and Czech, M. P. (2000). Distinct polyphosphoinositide binding selectivities for pleckstrin homology domains of GRP1-like proteins based on diglycine versus triglycine motifs. *J. Biol. Chem.* **275**, 32816–32821.
26. Macia, E., Paris, S., and Franco, M. (2000). Binding of the PH and polybasic C-terminal domains of ARNO to phosphoinositides and to acidic lipids. *Biochemistry* **39**, 5893–5901.
27. Santy, L. C., Frank, S. R., Hatfield, J. C., and Casanova, J. E. (1999). Regulation of ARNO nucleotide exchange by a PH domain electrostatic switch. *Curr. Biol.* **9**, 1173–1176.
28. Frank, S., Upender, S., Hansen, S. H., and Casanova, J. E. (1998). ARNO is a guanine nucleotide exchange factor for ADP-ribosylation factor 6. *J. Biol. Chem.* **273**, 23–27.
29. Langille, S. E., Patki, V., Klarlund, J. K., Buxton, J. M., Holik, J. J., Chawla, A., Corvera, S., and Czech, M. P. (1999). ADP-ribosylation factor 6 as a target of guanine nucleotide exchange factor GRP1. *J. Biol. Chem.* **274**, 27099–27104.
30. Venkateswarlu, K. and Cullen, P. J. (2000). Signalling via ADP-ribosylation factor 6 lies downstream of phosphatidylinositide 3-kinase. *Biochem. J.* **345**, 719–724.
31. Turner, C. E., West, K. A., and Brown, M. C. (2001). Paxillin-ARF GAP signaling and the cytoskeleton. *Curr. Opin. Cell Biol.* **13**, 593–599.
32. Brown, M. T., Andrade, J., Radhakrishna, H., Donaldson, J. G., Cooper, J. A., and Randazzo, P. A. (1998). ASAP1, a phospholipid-dependent Arf GTPase-activating protein that associates with and is phosphorylated by Src. *Mol. Cell Biol.* **18**, 7038–7051.
33. Andreev, J., Simon, J. P., Sabatini, D. D., Kam, J., Plowman, G., Randazzo, P. A., and Schlessinger, J. (1999). Identification of a new Pyk2 target protein with Arf-GAP activity. *Mol. Cell Biol.* **19**, 2338–2350.
34. Kam, J. L., Miura, K., Jackson T. R., Gruschus, J., Roller, P., Stauffer, S., Clark, J., Aneja, R., and Randazzo, P. A. (2000). Phosphoinositide-dependent activation of the ADP-ribosylation factor GTPase-activating protein ASAP1. Evidence for the pleckstrin homology domain functioning as an allosteric site. *J. Biol. Chem.* **275**, 53–63.
35. Zhao, Z. S., Manser, E., Loo, T. H., and Lim, L. (2000). Coupling of PAK-interacting exchange factor PIX to GIT1 promotes focal complex disassembly. *Mol. Cell Biol.* **20**, 6354–6363.
36. Premont, R. T., Claing, A., Vitale, N., Freeman, J. L., Pitcher, J. A., Patton, W. A., Moss, J., Vaughan, M., and Lefkowitz, R. J. (1998). 2-Adrenergic receptor regulation by GIT1, a G protein-coupled receptor kinase-associated ADP ribosylation factor GTPase-activating protein. *Proc. Natl. Acad. Sci. USA* **95**, 14082–14087.
37. Vitale, N., Patton, W. A., Moss, J., Vaughan, M., Lefkowitz, R. J., and Premont, R. T. (2000). GIT proteins, A novel family of phosphatidylinositol 3,4,5-trisphosphate-stimulated GTPase-activating proteins for ARF6. *J. Biol. Chem.* **275**, 13901–13906.
38. Miura, K., Jacques, K. M., Stauffer, S., Kubosaki, A., Zhu, K., Hirsch, D. S., Resau, J., Zheng, Y., and Randazzo, P. A. (2002). ARAP1: A point of convergence for Arf and Rho signalling. *Mol. Cell.* **9**, 109–119.
39. Krugmann, S., Anderson, K. E., Ridley, S. H., Risso, N., McGregor, A., Coadwell, J., Davidson, K., Eguinoa, A., Ellson, C. D., Lipp, P., Manifava, M., Ktistakis, N., Painter, G., Thuring, J. W., Cooper, M. A, Lim, Z.-Y., Holmes, A. B., Dove, S. K., Michell, R. H., Grewal, A., Nazarian, A., Erdjument-Bromage, H., Tempst, P., Stephens, L.R., and Hawkins, P. T (2002). Identification of ARAP-3, a novel PI3K effector regulating both Arf and Rho GTPases by selective capture on phosphoinositide affinity matrices. *Molecular Cell.* **9**, 95–108.
40. Rodriguez-Viciana, P., Warne, P. H., Khwaja, A., Marte, B. M., Pappin, D., Das, P., Waterfield, M. D., Ridley, A., Downward, J. (1997). Role of phosphoinositide 3-OH kinase in cell transformation and control of the actin cytoskeleton by Ras. *Cell* **89**, 457–467.
41. Wennström, S. and Downward, J. (1999). Role of phosphoinositide 3-kinase in activation of ras and mitogen-activated protein kinase by epidermal growth factor. *Mol. Cell Biol.* **19**, 4279–4288.
42. DePaolo, D., Reusch, J. E., Carel, K., Bhuripanyo, P., Leitner, J. W., and Draznin, B. (1996). Functional interactions of phosphatidylinositol 3-kinase with GTPase-activating protein in 3T3-L1 adipocytes. *Mol. Cell Biol.* **16**, 1450–1457.
43. Rubio, I. and Wetzker, R. (2000). A permissive function of phosphoinositide 3-kinase in Ras activation mediated by inhibition of GTPase-activating proteins. *Curr. Biol.* **10**, 1225–1228.
44. Lockyer, P. J., Wennström, S., Kupzig, S., Venkateswarlu, K., Downward, J. and Cullen, P. J. (1999). Identification of the ras GTPase-activating protein $GAP1^m$ as a phosphatidylinositol-3,4,5-trisphosphate-binding protein in vivo. *Curr. Biol.* **9**, 265–268.
45. Fukuda, M. and Mikoshiba, K. (1996). Structure-function relationships of the mouse $Gap1^m$. Determination of the inositol 1,3,4,5-tetrakisphosphate-binding domain. *J. Biol. Chem.* **271**, 18838–18842.

CHAPTER 156

Phosphoinositides and Actin Cytoskeletal Rearrangement

Paul A. Janmey,[1] Robert Bucki,[1] and Helen L. Yin[2]

[1]Department of Physiology, Institute for Medicine and Engineering,
University of Pennsylvania, Philadelphia, Pennsylvania and
[2]Department of Physiology, University of Texas,
Southwest Medical Center, Dallas, Texas

Historical Perspective

Cytoskeletal proteins were the first proteins shown to be regulated by phosphoinositides (PPIs), beginning with the report by Lassing and Lindberg that PIP2 dissociated profilin-actin complexes *in vitro* and promoted actin polymerization [1]. This finding suggested that increases in cellular PIP2 would drive the polymerization of cytoskeletal actin. At that time products of the PI3-kinase pathway had not yet been implicated in cell signaling, and it was thought that PI(4,5)P2 was the primary lipid responsible for direct effects on profilin, a hypothesis that is largely supported by many subsequent studies of actin-binding proteins. Since that time dozens of actin-binding proteins have been found to be either activated or inhibited by PPIs *in vitro*, usually by PIP2, and studies in the last few years confirm that at least some of these reactions occur in a similar way in cytoplasm. The fundamental predictions that increased PPI synthesis leads to actin assembly and depletion of PPIs triggers actin depolymerization have been borne out in studies in which PPI levels are altered in cells either by manipulation of expression of lipid kinases [2–5] or phosphatases [3,6–10] or by introduction of constructs such as PH domains [8,11] or PIP2-binding peptides [12–15] that sequester the lipids and may mimic the effects of endogenous proteins whose cellular role appears to involve sequestration of membrane phosphoinositides [16–18].

In the last several years the number of PPI-binding proteins has increased greatly, and actin-binding proteins are now a minority of the total ligands proposed for these lipids. Many of the newly reported proteins were identified by their possession of PH, FYVE, PX, or other PPI-binding motifs and the lipid-binding potential measured after this identification. Often these protein modules bind specifically to PPIs generated by PI3-kinases rather than to PI(4)P or PI(4,5)P2. In contrast, most PPI-binding cytoskeletal proteins were first identified biochemically to interact with PIP2 and the specific binding sites sought only afterward. It is noteworthy that the structures of PPI binding sites in cytoskeletal proteins are less well characterized than the motifs listed above, and it is perhaps not a coincidence that the PPI-binding domains common to proteins involved in vesicle traffic or spatial localization of signaling are conspicuously missing from most actin-binding proteins.

This review will focus on recent advances that demonstrate how PPIs are involved in stimulation of actin polymerization *in vivo*, or activation of proteins involved in the formation of cytoskeleton and membrane links, and how binding of cytoskeletal proteins to membrane PPIs may relate to the lateral or transverse movement of lipids to affect raft formation or lipid asymmetry.

Stimulating Cellular Actin Polymerization

There is increasing evidence for a localized increase in PIP2 at sites of actin polymerization and remodeling using the fluorescent chimera GFP-PH-PLCδ1 as a PIP2 reporter [19] (see section entitled Relation of Actin Assembly to

Phosphoinosite-containing Lipid Rafts). Localized PIP2 increase may depend on small GTPases in the Rho family (Rac and Rho) and the ADP ribosylation factor family (Arf6, Arf1). These GTPases have profound effects on the actin cytoskeleton, and can either alter the activity of type I phosphatidylinositol 4 phosphate 5 kinases (PIP5K), the enzymes that convert PI4P to PIP2, or recruit them to sites of actin polymerization [5]. For example, Rac and Rho bind PIP5Ks and recruit them to the plasma membrane [20] while Arf6 acts downstream of Rac to activate PIP5Ks [21].

The Mechanisms of Actin Polymerization

Since actin polymerization *in vivo* occurs primarily through the rapid growing end (+) of actin nuclei, the mechanisms by which (+) ends are generated are of intense interest. Three mechanisms have been proposed (reviewed in [22,23]): first, *de novo* actin nucleation by the Arp2/3 complex as a result of activation by the Wiskott-Aldrich syndrome family proteins (WASP, N-WASP etc.) or other proteins; second, severing of preexisting filaments by cofilin/actin depolymerizing protein (cofilin/ADF) or gelsolin family proteins; third, uncapping of the (+) end by capping proteins such as CapZ. Once the (+) ends are liberated, actin monomer delivery is accelerated by profilin and funneling of actin monomers to the favored sites by (+) end capping at other sites. The supply of actin monomer is sustained by severing and facilitated depolymerization from the (−) end by cofilin/ADF.

PIP2 alters *in vitro* the activity of critical proteins in each of these steps. It activates N-WASP, synergistically with Cdc42 [24] or with SH3 adapters such as Nck independently of Cdc42 [25], to promote *de novo* nucleation from the Arp2/3 complexes by an unmasking mechanism depicted in Fig. 1C. In contrast, PIP2 inactivates cofilin/ADF, CapZ, profilin and gelsolin-related severing and capping proteins (reviewed in [18]). The mechanism for PIP2 inhibition of these proteins is not completely understood but may involve the changes depicted in Figs. 1A and B (see section entitled Different Mechanisms of PPI-Actin Binding Protein Regulation).

Several recent studies implicate PIP2 in control of the cytoskeleton *in vivo*. Genetic disruption of *Mss4m*, which encodes the single PIP5K in yeast, or *skittles* (one of two PIP5K) in *Drosophila* produce cytoskeletal defects. Since mammalian cells have multiple PIP5Ks, and knockout animals are not yet available, other approaches have been used to examine the role of PIP5K *in vivo*. These include microinjection of an anti-PIP2 antibody [26], introduction of a cell-permeant gelsolin PIP2-binding peptide [13,15], addition of PIP2 or a PIP2-binding peptide to semi-intact cells [5], overexpression of actin regulatory proteins with defective PIP2 binding [27], and manipulation of the expression levels of PIP5Ks [9] and the phosphoinositide phosphatases that dephosphorylate PIP2 [10].

PIP5K overexpression induces dramatic actin phenotypes, establishing a causal relation between PIP2 and actin cytoskeletal dynamics. The responses vary depending on the cell

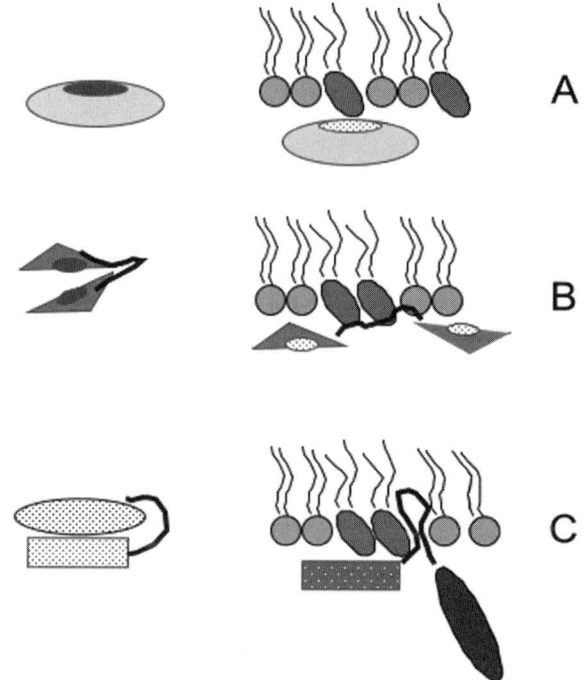

Figure 1 Three models for regulation of protein function by membrane-bound phosphoinositides. Actin-binding sites are shown in blue. Solid patches denote active sites and dotted patches inhibited sites. PIP2 is shown as large-headed lipids within one leaflet of a bilayer composed of neutral lipid. (A) The actin site is occluded by the lipid without other structural change. (B) PIP2 binding reorients two protein domains such that structures required for actin binding can no longer function cooperatively. (C) Protein binds and inserts in membrane to simultaneously stabilize membrane association and expose sites for actin and membrane proteins.

types used and most likely the extent of overexpression. They include NWASP-dependent actin comet tail formation [9], Rho and Rho-kinase dependent actin stress fiber formation [28], and the arrest of Arf6-regulated plasma membrane-endosome recycling [29]. The multiple phenotypes are not surprising, given that PIP5Ks may be regulated by several small GTPases that induce distinct actin structures in a sometimes sequential and at other times mutually exclusive manner. Furthermore, the site of PIP2 generation as well as the subset of actin regulatory proteins at those sites will dictate the dominant response. The actin-modulating proteins that contribute to these phenotypes have been identified in the first two cases, thus providing mechanistic insight into how PIP2 regulates the actin cytoskeleton *in vivo*.

PIP5K Overexpression Induces Actin Comet Formation

Actin comets formed around pathogens, such as *Listeria*, *Shigella*, and vaccinia, have contributed significantly to our understanding of the mechanism of cellular actin polymerization because they either introduce their own membrane protein or hijack the host's N-WASP to initiate actin assembly by the same mechanism used at the cell membrane. Since N-WASP also stimulates actin comet formation around intracellular vesicles and lipid vesicles in cell extracts [30] and in *Xenopus* eggs [31], the possibility that PIP2 promotes vesicle trafficking by generating actin comets is particularly attractive. Indeed, tiny comets that form spontaneously have been sighted, and much more robust comets are found in cells that overexpress PIP5K [9]. Overexpressed PIP5K is enriched at the head of the comets, establishing that the *in situ* generation of PIP2 at the vesicle may recruit and activate N-WASP to promote *de novo* actin nucleation by Arp2/3. Dynamin, another PIP2-activated protein that is involved in vesicle trafficking, is also recruited to the head of the PIP5K-induced comets, and overexpression of dominant negative dynamin mutants inhibits comet formation [4,32].

Actin comets are formed from endocytic and Golgi-derived exocytic vesicles, particularly those with cholesterol and sphinogolipid-enriched membrane microdomains (rafts), and raft disruption reduces the number of comets dramatically [9]. Rafts have previously been shown to contain an agonist-sensitive pool of phosphoinositides that responds to PLC signaling, and they are the primary sites of PIP2 synthesis [17]. The preferential formation of actin comets in raft domains establishes that rafts are platforms for the integration of PIP2 signaling and actin polymerization, reinforcing the concept that specialized regions of the membrane are closely related to specific cytoskeletal structures discussed in the section on mechanisms of PPI-actin binding protein regulation.

PIP5K Overexpression Induces Actin Stress Fiber Formation

PIP5K overexpression in CV1 cells induces robust actin stress fiber formation and inhibition of membrane ruffling in response to growth factors due to an inability to generate (+) end actin nuclei [28]. These two effects are consistent with activation of Rho and inhibition of Rac-dependent cytoskeletal pathways, respectively, and are remarkably similar to that observed in fibroblasts isolated from gelsolin knockout animals [34]. Gelsolin binding to actin is inhibited in PIP5K-overexpressing cells, suggesting that inhibition of severing by gelsolin and perhaps by cofilin/ADF may account for the formation of long actin filaments and the inability to generate (+) end nuclei to mount an actin polymerization response. Furthermore, profilin and CapZ are also inhibited, while ezrin/radixin/moesin, the membrane-linker proteins (see below), are activated. Together, these changes can amplify the consequences of severing inhibition. In conclusion, these studies show that several PIP2-sensitive actin modulating proteins behave *in vivo* as predicted from their well-characterized behavior *in vitro*.

Relation Among PIP5K, Arf6, and Actin Remodeling

Arf6 overexpression induces actin comets [35] and stimulates membrane ruffling [29]. However, although one study finds comet formation is not inhibited by an antibody to PIP2 [35], other studies show that Arf6 activates PIP5K *in vitro* and *in vivo* [21,29]. Overexpression of a constitutively active Arf6 or PIP5K induces PIP2 generation and actin polymerization around recycling endosomes, eventually trapping them into an aggregate that cannot recycle back to the plasma membrane [29]. The phenotype suggests that cycling of Arf6 between the GTP- and GDP-bound forms is important for actin regulation, and this is likely to be achieved through activation and inactivation of PIP5K. Sustained high-level PIP2 production due to constitutive Arf6 overexpression or PIP5K overexpression promotes polymerization and inhibits depolymerization to generate abnormal actin structures around the vesicles. The abnormally large actin comets found in other cellular contexts [9] are another manifestation of sustained actin polymerization. The requirement for transient changes in PIP2 levels highlights the importance of dissipating PIP2 in a spatially and temporally defined manner. Although PIP2 is hydrolyzed by PLC during agonist signaling, PIP2 is likely to be cleared primarily by phosphoinositide phosphatases.

Effects of Manipulating the Level of Phosphoinositide Phosphatases

Phosphatases that dephosphorylate phosphoinositides or inositol polyphosphates at the 5′ position are classified into four groups, according to their substrate specificity [10]. The type II phosphatases that hydrolyze PIP2 have been used to study the effect on the actin cytoskeleton. Overexpression of synaptojanin or other type II phosphatases decreases actin stress fibers [6,36] or induces actin arborization [37]. However, it is not known whether the effects are due specifically to a decrease in PIP2 or water-soluble inositol phosphates.

The most definitive evidence is obtained by disruption of the synaptojanin 1 gene [38], which results in an accumulation of clathrin-coated vesicles and polymerized actin in the endocytic zones of nerve terminals. These changes are correlated with an increase in PIP2 concentration. Synaptojanin and PIP5K are both concentrated at synapses, and they antagonize each other in the recruitment of clathrin coats to lipid membranes *in vitro* [3]. These results strongly suggest that the PIP2 level at the synapse is critically dependent on the balance of phosphoinositide kinase and phosphatases, and that PIP2 has a pivotal role in the regulation of actin and endocytic vesicle formation at the synapse [39].

Actin-Membrane Linkers Localized or Activated by PIP2

In contrast to actin monomer-binding or severing proteins that are generally inactivated by PPIs, proteins that crosslink actin filaments to each other or link them to the cell membrane are usually activated to bind actin or directed to link actin to transmembrane receptors by the lipids [40]. Evidence from mutational analyses suggests how this activating switch occurs, and interfering with PPI binding disrupts this linking process in cells.

Alpha Actinin

Recent studies of alpha actinin provide a good example of how activation of actin or other ligand binding may occur. In this case, an actin- and titin-binding motif of the antiparallel alpha actinin dimer is occluded in the inactive state because it binds a complementary domain within the same homodimer [41]. When PIP2 binds to alpha actinin, self-association is disrupted, exposing the actin- and titin-binding motifs so that they can bind their targets (see Fig. 1C). Similar activation switches have been proposed for band 4.1/ezrin/radixin/moesin (FERM) protein family members and for the focal adhesion proteins talin and vinculin.

Ezrin/Radixin/Moesin

ERM proteins are among the best currently characterized PPI-activated proteins. Both actin and membrane protein binding sites are inactive in the dormant state of the protein because of self-association between the two domains responsible for these separate activities, and the structural basis for this self-inactivation is now clear from protein structures determined by crystallography for radixin [42] and moesin [43,44]. This self-inactivation is a common feature of several actin-membrane linkers, including talin and vinculin, and in retrospect explains why the *in vitro* actin binding of these proteins was so difficult to characterize compared to proteins such as cofilin or filamin, where the actin-binding sites appear to be constitutively exposed. Biochemical and cell localization studies show that ERM proteins colocalize with transmembrane proteins in activated cells and that *in vitro* this association is stimulated by PPIs. The PIP2-dependent linkage of ezrin to ICAMs [45,46] involves reordering of the FERM domain, which contains an acidic loop distinct from the IP3-binding site that may also participate in binding to the basic juxta-membrane regions present in adhesion receptors such as CD44 [42]. The importance of the PIP2-binding regions for cellular localization and function of ERM proteins has been increasingly well demonstrated in recent studies. Mutation of four basic residues found in the PIP2-binding site prevented localization of ezrin to actin-rich membrane structures [47].

Talin

Like ezrin, talin is also activated by PIP2 to increase membrane association. In this case one consequence of PIP2 binding is an increased affinity of the intact protein for the cytoplasmic domain of beta 1 integrins [48]. The relevance of this interaction in a cellular context is reinforced by evidence that PIP2 cosediments after immunoprecipitation with anti-talin antibodies, and the amount of PIP2 shows a strong transient increase after suspended cells are plated on fibronectin, reaching a maximum 15 minutes after engagement of integrins that is five times higher than the initial state or the levels 1 hour after plating. The finding that only PIP2, but not PIP or PI, shows this transient change rules out the possibility that the lipid in the immunoprecipitates results from nonspecific contaminating membranes but also raises the question of the state of the lipid in these lipid-protein complexes.

Relation of Actin Assembly to Phosphoinositide-containing Lipid Rafts

The finding that the specialized regions of the plasma membranes such as caveoli or lipid rafts are potentially enriched in PPI [49] and that several cytoskeletal proteins do not bind PIP2 unless it is present in bilayers at approximately 10 mol percent [50] suggests that clustering of inositol lipids in specialized regions of lipid monolayers is an important aspect of their ability to modify specific cellular processes. Experiments with liposomes show that in the presence of multivalent cations PIP2 organizes into domains [51] and domains in PIP2-containing monolayers bind and are reorganized by peptides based on the PIP2-binding site of gelsolin [52]. Evidence that areas of local PIP2 concentration form in cells and are associated with actin assembly has emerged from several recent studies. In fibroblasts, the PH domains of PLCδ1 fused with GFP localized to actin-rich membrane ruffles and the selective concentration of the PLCδ1-PH-GFP in highly dynamic regions of the plasma membrane, which are rich in F-actin, supports the hypothesis that local synthesis and lateral segregation of PI(4,5)P2 spatially restrict actin polymerization [53]. In macrophages, the PLCδ1-PH-GFP protein localizes transiently to the phagosomal cups along with PLCγ2, PI(4)P 5-kinase and actin [54,55]. The dissociation of a PLCδ1-PH-CFP fusion protein was accompanied by recruitment of a C1-PKCδ-C1-YFP fusion protein, which suggests that PLCγ2 mediates the conversion of PI(4,5)P2 to diacylglycerol upon sealing of the phagosomal cup.

An immunocytochemical study with PI(4,5)P2 antibodies in PC12 cells, COS-7 cells, and hippocampal neurons has visualized clusters of PIP2 that co-localize with plasmalemma-associated PKC substrates that affect actin cytoskeleton: GAP43, myristoylated alanine-rich C kinase substrate (MARCKS), and CAP23 (GMC proteins) [17]. These clusters are interpreted to be raft domains, which were dispersed by membrane cholesterol extraction by using cyclodextrin. Cells that overexpress MARCKS exhibited larger macroscopic PIP2 clusters, whereas expression of MARCKS lacking basic effector domain exhibited reduced PIP2 clusters. These results suggest that GMC proteins regulate the availability of PIP2 for interaction with actin-binding proteins. A focal pattern of labeling suggesting colocalization of actin with PIP2 at the membrane has also been recently observed in NIH3T3 fibroblast plasma membranes treated with a fluorescent gelsolin-derived, cell permeant phosphoinositide–binding peptide [13].

Different Mechanisms of PPI-Actin Binding Protein Regulation

There are at least three distinct mechanisms and several variations by which membranes containing PPIs can alter actin-binding protein function. The simplest mechanism shown in Fig. 1A is that an actin-binding site coincides with a PIP2-binding site and therefore targeting of the protein to PPI-rich membranes dissociates actin competitively without necessarily changing the protein structure. However, even for small monomer-sequestering proteins such as cofilin/ADF/actophorin, careful mapping of the actin- and PIP2-binding sites shows that they are not precisely coincident, and that specific residues can be altered to perturb one but not the other activity [27,56]. The recent finding that PIP2 promotes oligomerization of cofilin and subsequent actin filament bundling [57] further complicates the model of simple competition. Profilin likewise appears to have an extensive surface that interacts with PIP2, and binding to the lipid promotes increased alpha-helix in the protein [58].

A different model for inhibition of actin-binding function shown in Fig. 1B is that the binding to PIP2 causes a rearrangement of actin-binding domains or a local unfolding of polypeptide within these domains to derange the surface required to bind actin. This model appears to account for effects on gelsolin and related proteins [50]. This type of allosteric regulation may occur either with or without the protein inserting into the hydrophobic domain of the membrane.

The third mode of binding (Fig. 1C) involves docking of the protein to the membrane in a manner that disrupts interactions between domains within monomers or homo-oligomers that mask binding sites for actin or membrane anchors. This model, which may apply to ERM proteins, talin, alpha actinin, N-WASP, and vinculin, would result in activation rather than inhibition of the protein function. In the model drawn, both sites are activated after PIP2 binding, but it is also plausible that one of the sites remains occupied by the lipid and is then available only after PIP2 hydrolysis or reorganization of the membrane. Such a mechanism would explain how PIP2 binding can work to activate vinculin *in vivo* whereas purified PIP2 may inhibit vinculin-actin binding *in vitro* [59] and would allow for sequential activation of two sites as PIP2 is turned over at the cell membrane.

Effects on Lipid Membrane Structure

Binding of protein to membranes containing PPIs can have a number of effects on the membrane depending on the charge of the protein docking site, the degree of penetration into the hydrophobic domain, and the number of lipids to which the protein binds. In contrast to the extensive documentation of changes in protein function or structure, changes in lipid structure are less often studied but are likely to be equally important in a cellular context. Although it is often assumed that protein docking to PPIs in a membrane mainly localizes the protein to the surface, there are many specific changes that can occur, such as the recently documented extended conformation [60], especially for a lipid such as PIP2, which is relatively unstable in a bilayer. One interesting possibility is that some forms of binding to PIP2 can destabilize bilayer packing to the extent that loss of membrane asymmetry occurs. Resent observations indicate that PIP2 is a positive regulator of Ca^{2+}-induced lipid scrambling due to PIP2-enriched domain formation [61]. The possibility to reorganize membranes by specific interaction with PIP2 has been documented by electron spin resonance measurements that show disordering of lipid bilayer vesicles by myristoylated ARF6 only when the vesicles contain PIP2 [62]. Phospholipid scrambling was also observed in platelets and lipid vesicles containing PIP2 treated with a gelsolin-derived peptide [12]. These observations suggest that the binding of cytoskeletal proteins to membrane phosphoinositides has a vast potential for regulation and reorganization of cells that is now beginning to be appreciated.

References

1. Lassing, I. and Lindberg, U. (1985). Specific interaction between phosphatidylinositol 4,5-bisphosphate and profilactin. *Nature* **314**, 472–474.
2. Oude Weernink, P. A., Schulte, P., Guo, Y., Wetzel, J., Amano, M., Kaibuchi, K., Haverland, S., Voss, M., Schmidt, M., Mayr, G. W., and Jakobs, K. H. (2000). Stimulation of phosphatidylinositol-4-phosphate 5-kinase by Rho-kinase. *J. Biol. Chem.* **275**, 10168–10174.
3. Wenk, M. R., Pellegrini, L., Klenchin, V. A., Di Paolo, G., Chang, S., Daniell, L., Arioka, M., Martin, T. F., and De Camilli, P. (2001). PIP kinase Igamma is the major PI(4,5)P(2) synthesizing enzyme at the synapse. *Neuron* **32**, 79–88.
4. Lee, E. and De Camilli, P. (2002). Dynamin at actin tails. *Proc. Natl. Acad. Sci. USA* **99**, 161–166.
5. Tolias, K. F., Hartwig, J. H., Ishihara, H., Shibasaki, Y., Cantley, L. C., and Carpenter, C. L. (2000). Type Ialpha phosphatidylinositol-4-phosphate 5-kinase mediates Rac-dependent actin assembly. *Curr. Biol.* **10**, 153–156.
6. Ijuin, T., Mochizuki, Y., Fukami, K., Funaki, M., Asano, T., and Takenawa, T. (2000). Identification and characterization of a novel inositol polyphosphate 5- phosphatase. *J. Biol. Chem.* **275**, 10870–10875.
7. Payrastre, B., Missy, K., Giuriato, S., Bodin, S., Plantavid, M., and Gratacap, M. (2001). Phosphoinositides: key players in cell signalling, in time and space. *Cell Signal* **13**, 377–387.
8. Raucher, D., Stauffer, T., Chen, W., Shen, K., Guo, S., York, J. D., Sheetz, M. P., and Meyer, T. (2000). Phosphatidylinositol 4,5-bisphosphate functions as a second messenger that regulates cytoskeleton-plasma membrane adhesion. *Cell* **100**, 221–228.
9. Rozelle, A. L., Machesky, L. M., Yamamoto, M., Driessens, M. H., Insall, R. H., Roth, M. G., Luby-Phelps, K., Marriott, G., Hall, A., and Yin, H. L. (2000). Phosphatidylinositol 4,5-bisphosphate induces actin-based movement of raft-enriched vesicles through WASP-Arp2/3. *Curr. Biol.* **10**, 311–320.
10. Takenawa, T. and Itoh, T. (2001). Phosphoinositides, key molecules for regulation of actin cytoskeletal organization and membrane traffic from the plasma membrane. *Biochim. Biophys. Acta* **1533**, 190–206.
11. Raucher, D. and Sheetz, M. P. (2001). Phospholipase C activation by anesthetics decreases membrane-cytoskeleton adhesion. *J. Cell Sci.* **114**, 3759–3766.
12. Bucki, R., Janmey, P. A., Vegners, R., Giraud, F., and Sulpice, J. C. (2001). Involvement of phosphatidylinositol 4,5-bisphosphate in phosphatidylserine exposure in platelets: use of a permeant phosphoinositide-binding peptide. *Biochemistry* **40**, 15752–15761.
13. Cunningham, C. C., Vegners, R., Bucki, R., Funaki, M., Korde, N., Hartwig, J. H., Stossel, T. P., and Janmey, P. A. (2001). Cell permeant phosphoinositide-binding peptides that block cell motility and actin assembly. *J. Biol. Chem.* **276**, 43390–43399.
14. Glogauer, M., Hartwig, J., and Stossel, T. (2000). Two pathways through Cdc42 couple the N-formyl receptor to actin nucleation in permeabilized human neutrophils. *J. Cell Biol.* **150**, 785–796.

15. Guttman, J., Janmey, P., and Vogl, A. (2002). Gelsolin—evidence for a role in turnover of junction-related actin filaments in sertoli cells. *J. Cell Sci.* **115**, in press.
16. Caroni, P. (2001). New EMBO members' review: actin cytoskeleton regulation through modulation of PI(4,5)P(2) rafts. *EMBO J.* **20**, 4332–4336.
17. Laux, T., Fukami, K., Thelen, M., Golub, T., Frey, D., and Caroni, P. (2000). GAP43, MARCKS, and CAP23 modulate PI(4,5)P(2) at plasmalemmal rafts, and regulate cell cortex actin dynamics through a common mechanism. *J. Cell Biol.* **149**, 1455–1472.
18. Lanier, L. M. and Gertler, F. B. (2000). Actin cytoskeleton: thinking globally, actin locally. *Curr. Biol.* **10**, R655–R657.
19. Stauffer, T. P., Ahn, S., and Meyer, T. (1998). Receptor-induced transient reduction in plasma membrane PtdIns(4,5)P2 concentration monitored in living cells. *Curr. Biol.* **8**, 343–346.
20. Chatah, N. E. and Abrams, C. S. (2001). G-protein-coupled receptor activation induces the membrane translocation and activation of phosphatidylinositol-4-phosphate 5- kinase I alpha by a Rac- and Rho-dependent pathway. *J. Biol. Chem.* **276**, 34059–34065.
21. Honda, A., Nogami, M., Yokozeki, T., Yamazaki, M., Nakamura, H., Watanabe, H., Kawamoto, K., Nakayama, K., Morris, A. J., Frohman, M. A., and Kanaho, Y. (1999). Phosphatidylinositol 4-phosphate 5-kinase alpha is a downstream effector of the small G protein ARF6 in membrane ruffle formation. *Cell* **99**, 521–532.
22. Yin, H. L. and Stull, J. T. (1999). Proteins that regulate dynamic actin remodeling in response to membrane signaling minireview series. *J. Biol. Chem.* **274**, 32529–32530.
23. Condeelis, J. (2001). How is actin polymerization nucleated in vivo? *Trends Cell Biol.* **11**, 288–293.
24. Rohatgi, R., Ho, H. Y., and Kirschner, M. W. (2000). Mechanism of N-WASP activation by CDC42 and phosphatidylinositol 4,5-bisphosphate. *J. Cell Biol.* **150**, 1299–1310.
25. Rohatgi, R., Nollau, P., Ho, H. Y., Kirschner, M. W., and Mayer, B. J. (2001). Nck and phosphatidylinositol 4,5-bisphosphate synergistically activate actin polymerization through the N-WASP-Arp2/3 pathway. *J. Biol. Chem.* **276**, 26448–26452.
26. Fukami, K., Matsuoka, K., Nakanishi, O., Yamakawa, A., Kawai, S., and Takenawa, T. (1988). Antibody to phosphatidylinositol 4,5-bisphosphate inhibits oncogene-induced mitogenesis. *Proc. Natl. Acad. Sci. USA* **85**, 9057–9061.
27. Ojala, P. J., Paavilainen, V., and Lappalainen, P. (2001). Identification of yeast cofilin residues specific for actin monomer and PIP2 binding. *Biochemistry* **40**, 15562–15569.
28. Yamamoto, M., Hilgemann, D. H., Feng, S., Bito, H., Ishihara, H., Shibasaki, Y., and Yin, H. L. (2001). Phosphatidylinositol 4,5-bisphosphate induces actin stress-fiber formation and inhibits membrane ruffling in CV1 cells. *J. Cell Biol.* **152**, 867–876.
29. Brown, F. D., Rozelle, A. L., Yin, H. L., Balla, T., and Donaldson, J. G. (2001). Phosphatidylinositol 4,5-bisphosphate and Arf6-regulated membrane traffic. *J. Cell Biol.* **154**, 1007–1017.
30. Ma, L., Rohatgi, R., and Kirschner, M. W. (1998). The Arp2/3 complex mediates actin polymerization induced by the small GTP-binding protein Cdc42. *Proc. Natl. Acad. Sci. USA* **95**, 15362–15367.
31. Taunton, J., Rowning, B. A., Coughlin, M. L., Wu, M., Moon, R. T., Mitchison, T. J., and Larabell, C. A. (2000). Actin-dependent propulsion of endosomes and lysosomes by recruitment of N-WASP. *J. Cell Biol.* **148**, 519–530.
32. Orth, J. D., Krueger, E. W., Cao, H., and McNiven, M. A. (2002). The large GTPase dynamin regulates actin comet formation and movement in living cells. *Proc. Natl. Acad. Sci. USA* **99**, 167–172.
33. Pasolli, H. A., Klemke, M., Kehlenbach, R. H., Wang, Y., and Huttner, W. B. (2000). Characterization of the extra-large G protein alpha-subunit XLalphas. I. Tissue distribution and subcellular localization. *J. Biol. Chem.* **275**, 33622–33632.
34. Azuma, T., Witke, W., Stossel, T. P., Hartwig, J. H., and Kwiatkowski, D. J. (1998). Gelsolin is a downstream effector of rac for fibroblast motility. *EMBO J.* **17**, 1362–1370.
35. Schafer, D. A., D'Souza-Schorey, C., and Cooper, J. A. (2000). Actin assembly at membranes controlled by ARF6. *Traffic* **1**, 892–903.
36. Sakisaka, T., Itoh, T., Miura, K., and Takenawa, T. (1997). Phosphatidylinositol 4,5-bisphosphate phosphatase regulates the rearrangement of actin filaments. *Mol. Cell Biol.* **17**, 3841–3849.
37. Asano, T., Mochizuki, Y., Matsumoto, K., Takenawa, T., and Endo, T. (1999). Pharbin, a novel inositol polyphosphate 5-phosphatase, induces dendritic appearances in fibroblasts. *Biochem. Biophys. Res. Commun.* **261**, 188–195.
38. Cremona, O., Di Paolo, G., Wenk, M. R., Luthi, A., Kim, W. T., Takei, K., Daniell, L., Nemoto, Y., Shears, S. B., Flavell, R. A., McCormick, D. A., and De Camilli, P. (1999). Essential role of phosphoinositide metabolism in synaptic vesicle recycling. *Cell* **99**, 179–188.
39. Cremona, O. and De Camilli, P. (2001). Phosphoinositides in membrane traffic at the synapse. *J. Cell Sci.* **114**, 1041–1052.
40. Sechi, A. S. and Wehland, J. (2000). The actin cytoskeleton and plasma membrane connection: PtdIns(4,5)P(2) influences cytoskeletal protein activity at the plasma membrane. *J. Cell Sci.* **113 Pt 21**, 3685–3695.
41. Young, P. and Gautel, M. (2000). The interaction of titin and alpha-actinin is controlled by a phospholipid-regulated intramolecular pseudoligand mechanism. *EMBO J.* **19**, 6331–6340.
42. Hamada, K., Shimizu, T., Matsui, T., Tsukita, S., and Hakoshima, T. (2000). Structural basis of the membrane-targeting and unmasking mechanisms of the radixin FERM domain. *EMBO J.* **19**, 4449–4462.
43. Pearson, M. A., Reczek, D., Bretscher, A., and Karplus, P. A. (2000). Structure of the ERM protein moesin reveals the FERM domain fold masked by an extended actin binding tail domain. *Cell* **101**, 259–270.
44. Edwards, S. D. and Keep, N. H. (2001). The 2.7 A crystal structure of the activated FERM domain of moesin: an analysis of structural changes on activation. *Biochemistry* **40**, 7061–7068.
45. Serrador, J. M., Vicente-Manzanares, M., Calvo, J., Barreiro, O., Montoya, M. C., Schwartz-Albiez, R., Furthmayr, H., Lozano, F., and Sanchez-Madrid, F. (2002). A novel serine-rich motif in the intercellular adhesion molecule 3 is critical for its ERM-directed subcellular targeting. *J. Biol. Chem.* **9**, 9.
46. Heiska, L., Alfthan, K., Gronholm, M., Vilja, P., Vaheri, A., and Carpen, O. (1998). Association of ezrin with intercellular adhesion molecule-1 and -2 (ICAM-1 and ICAM-2). Regulation by phosphatidylinositol 4, 5-bisphosphate. *J. Biol. Chem.* **273**, 21893–21900.
47. Barret, C., Roy, C., Montcourrier, P., Mangeat, P., and Niggli, V. (2000). Mutagenesis of the phosphatidylinositol 4,5-bisphosphate (PIP(2)) binding site in the NH(2)-terminal domain of ezrin correlates with its altered cellular distribution. *J. Cell Biol.* **151**, 1067–1080.
48. Martel, V., Racaud-Sultan, C., Dupe, S., Marie, C., Paulhe, F., Galmiche, A., Block, M. R., and Albiges-Rizo, C. (2001). Conformation, localization, and integrin binding of talin depend on its interaction with phosphoinositides. *J. Biol. Chem.* **276**, 21217–21227.
49. Hope, H. R. and Pike, L. J. (1996). Phosphoinositides and phosphoinositide-utilizing enzymes in detergent-insoluble lipid domains. *Mol. Biol. Cell* **7**, 843–851.
50. Tuominen, E. K., Holopainen, J. M., Chen, J., Prestwich, G. D., Bachiller, P. R., Kinnunen, P. K., and Janmey, P. A. (1999). Fluorescent phosphoinositide derivatives reveal specific binding of gelsolin and other actin regulatory proteins to mixed lipid bilayers. *Eur. J. Biochem.* **263**, 85–92.
51. Denisov, G., Wanaski, S., Luan, P., Glaser, M., and Mclaughlin, S. (1998). Binding of basic peptides to membranes produces lateral domains enriched in the acidic lipids phosphatidylserine and phosphatidylinositol 4,5-bisphosphate—an electrostatic model and experimental results. *Biophys. J.* **74**, 731–744.
52. Foster, W. J. and Janmey, P. A. (2001). The distribution of phosphoinositides in lipid films. *Biophys. Chem.* **91**, 211–218.
53. Tall, E. G., Spector, I., Pentyala, S. N., Bitter, I., and Rebecchi, M. J. (2000). Dynamics of phosphatidylinositol 4,5-bisphosphate in actin-rich structures. *Curr. Biol.* **10**, 743–746.
54. Botelho, R. J., Teruel, M., Dierckman, R., Anderson, R., Wells, A., York, J. D., Meyer, T., and Grinstein, S. (2000). Localized biphasic changes in phosphatidylinositol-4,5-bisphosphate at sites of phagocytosis. *J. Cell Biol.* **151**, 1353–1368.

55. Bajno, L., Peng, X. R., Schreiber, A. D., Moore, H. P., Trimble, W. S., and Grinstein, S. (2000). Focal exocytosis of VAMP3-containing vesicles at sites of phagosome formation. *J. Cell Biol.* **149**, 697–706.
56. Van Troys, M., Dewitte, D., Verschelde, J. L., Goethals, M., Vandekerckhove, J., and Ampe, C. (2000). The competitive interaction of actin and PIP2 with actophorin is based on overlapping target sites: design of a gain-of-function mutant. *Biochemistry* **39**, 12181–12189.
57. Pfannstiel, J., Cyrklaff, M., Habermann, A., Stoeva, S., Griffiths, G., Shoeman, R., and Faulstich, H. (2001). Human cofilin forms oligomers exhibiting actin bundling activity. *J. Biol. Chem.* **276**, 49476–49484.
58. Raghunathan, V., Mowery, P., Rozycki, M., Lindberg, U., and Schutt, C. (1992). Structural changes in profilin accompany its binding to phosphatidylinositol, 4,5-bisphosphate. *FEBS Lett.* **297**, 46–50.
59. Steimle, P. A., Hoffert, J. D., Adey, N. B., and Craig, S. W. (1999). Phosphoinositides inhibit the interaction of vinculin with actin filaments. *J. Biol. Chem.* **274**, 18414–18420.
60. Tuominen, E. K., Wallace, C. J., and Kinnunen, P. K. (2002). Phospholipid-cytochrome c interaction: Evidence for the extended lipid anchorage. *J. Biol. Chem.* **277**, 8822–8826.
61. Bucki, R., Giraud, F., and Sulpice, J. C. (2000). Phosphatidylinositol 4,5-bisphosphate domain inducers promote phospholipid transverse redistribution in biological membranes. *Biochemistry* **39**, 5838–5844.
62. Ge, M., Cohen, J. S., Brown, H. A., and Freed, J. H. (2001). ADP ribosylation factor 6 binding to phosphatidylinositol 4,5-bisphosphate-containing vesicles creates defects in the bilayer structure: an electron spin resonance study. *Biophys. J.* **81**, 994–1005.

CHAPTER 157

The Role of PI3 Kinase in Directional Sensing during Chemotaxis in *Dictyostelium*, a Model for Chemotaxis of Neutrophils and Macrophages

Richard A. Firtel and Ruedi Meili

Section of Cell and Developmental Biology, Division of Biological Sciences, and Center for Molecular Genetics, University of California, San Diego, La Jolla, California

Introduction

Chemotaxis, or directional movement of eukaryotic cells toward a small molecular attractant, is highly conserved evolutionarily and is regulated by a variety of ligands (including chemoattractants, chemokines, and growth factors) that activate G-protein-coupled and receptor tyrosine kinase effector pathways. Concentration differences of only a few percent over the length of a cell are sufficient to be recognized and converted into directional motility. Chemotaxis of *Dictyostelium* cells toward the extracellular chemoattractant cyclic AMP (cAMP) provides a model for the mechanisms underlying chemotaxis in a number of mammalian cell types, including neutrophils and macrophages. The ability to employ genetic, biochemical, and single-cell assays makes *Dictyostelium* an exceptional system for finding new genes involved in chemotaxis and understanding the interplay of known and novel gene products in the signal transduction processes underlying chemotactic responses. Furthermore, the simplicity of *Dictyostelium* chemotaxis combined with a vast body of experimental knowledge makes this organism an ideal system for testing predictions made by theoretical models of the chemotactic response.

Here, we discuss the signaling events that control the ability of a cell to sense the direction of a chemoattractant gradient. The phosphatidylinositol-3 kinase (PI3K) pathway plays a pivotal role in the conversion of such a shallow external gradient into the extreme cytoskeletal polarization necessary for directed cell movement of neutrophils, macrophages, and *Dictyostelium* cells.

Directional Movement

When cells are placed in a chemoattractant gradient, there is a rapid polymerization of actin on the side of the cell closest to the chemoattractant source resulting in the formation of a new leading edge [1–6]. Polymerization of F-actin leads to protrusion of the plasma membrane [7]. Assembly of conventional, nonmuscle myosin II and F-actin at the posterior of the cell enables actin-myosin contractility, which lifts off the posterior (also called a uropod) and promotes a rapid contraction of the posterior of the cell [3,6].

Localization of Cytoskeletal and Signaling Components

Even in the absence of an external gradient, many chemotaxis-competent cells exhibit a polarized distribution of cytoskeletal components, with the majority of F-actin found at the leading edge and myosin II and the remaining F-actin found at the cell's posterior. Only when cells are placed in a gradient is there a dynamic activation of localized F-actin polymerization and myosin assembly resulting in directional cell movement. When the chemoattractant gradient changes direction, the cells respond by dismantling the old actin/myosin cytoskeleton and forming a new leading edge and uropod, thereby realigning their axis with the external gradient [1,3,4,6]. In *Dictyostelium*, neutrophils, and macrophages, members of the Rac/Cdc42 and RhoA family of small GTPases regulate WASP proteins (e.g. WASP, Scar/WAVE), members of the PAK family of serine/threonine protein kinases, and myosin kinases are required for chemoattractant-mediated actin polymerization, myosin assembly, and directional movement (Fig. 2). For a more detailed discussion of this topic see Nobes et al., 1999 [8]. In contrast to cytoskeletal elements, upstream signal transduction components exhibit a uniform distribution in the absence of an external signal. In response to stimulation, some of these components undergo a dramatic redistribution and polarization, reflecting an underlying signal processing that can recognize and amplify a shallow external gradient enabling the chemotactic response. Recent experimental progress has identified some of these proteins and has shed light on some of the mechanistic aspects regulating their change in subcellular localization.

Analysis of the subcellular localization of known cell surface sensors such as G-protein-coupled chemoattractant receptors has demonstrated that these receptors in *Dictyostelium* and neutrophils are uniformly localized along the plasma membrane [1,4–6], even in the presence of an external signal. Although the concentration of G$\beta\gamma$ subunits is highest at the anterior of the cell, the gradient is extremely shallow, comparable to the gradient of the external signal, and cannot account for the steep intracellular gradient of other signaling components [9].

The Signaling Pathways Controlling Directional Movement

Activation of PI3K at the leading edge appears to play a predominant role in controlling directional movement in many amoeboid cell types such as neutrophils, macrophages, and *Dictyostelium* cells [1,4–6]. Class I members of the PI3K family phosphorylate $PI(4,5)P_2$ and $PI(4)P$ at the 3′ position in the inositol ring to produce the membrane lipids $PI(3,4,5)P_3$ and $PI(3,4)P_2$ (Fig. 1). $PI(3,4)P_2$ is also derived through the dephosphorylation of $PI(3,4,5)P_3$ by the 5′ inositol lipid phosphatase SHIP [10].

Studies in which PI3K function is abrogated through gene knockouts, use of PI3K-specific inhibitors, or PI3K isoform-specific antibodies demonstrate that PI3K is required for proper chemotaxis in neutrophils, macrophages, and *Dictyostelium* cells (see [1] for references). *Dictyostelium pi3k1/2* null cells, in which two of the three genes encoding the Class I PI3Ks PI3K1 and PI3K2 have been disrupted by homologous recombination, exhibit strong chemotaxis defects [11–15]. *Dictyostelium* wild-type cells chemotaxing toward cAMP are highly polarized, preferentially form a single pseudopod at the leading edge, and produce few if any lateral pseudopodia. In contrast, *pi3k1/2* null cells or wild-type cells treated with the PI3K inhibitor LY294002 exhibit a significant loss of polarity, move more slowly than wild-type cells, and produce multiple pseudopodia simultaneously along the periphery of the cell, although the cells still move predominantly toward the chemoattractant source [14]. These results indicate that PI3K is an important component of chemotaxis regulating directionality and polarity; however, these findings also imply that there are pathways parallel to those regulated by PI3K1 and PI3K2 contributing to directional movement. The nature of this pathway is currently unknown.

These observations make the PI3K pathway a likely candidate to be involved in polarization amplification. This notion is further supported by the subcellular localization of PI3K pathway components. Initial evidence for the localized activation of PI3K has derived from the demonstration, first in *Dictyostelium* and then in neutrophils and fibroblasts, of the preferential localization of a subfamily of PH-domain-containing proteins that preferentially bind the PI3K products $PI(3,4,5)P_3$ and $PI(3,4)P_2$ [14,16–19]. GFP fusions of these proteins function as reporters for local accumulation of $PI(3,4,5)P_3/PI(3,4)P_2$ corresponding to localized activation of Class I PI3Ks. These proteins are cytosolic in unstimulated cells and rapidly and transiently move to the plasma membrane in response to cells being globally stimulated by a chemoattractant (cells being rapidly bathed in chemoattractant) so that all of the receptors around the cell are uniformly activated. When cells are placed in a chemoattractant gradient, these proteins preferentially localize to the leading edge, suggesting that PI3K is preferentially localized in this region of the cell (Fig. 3). The localization of these PH-domain-containing proteins, which in *Dictyostelium* peaks at ~6–8 seconds after global stimulation with a chemoattractant, is one of the most rapid responses that has been described [14,16,17]. In *Dictyostelium*, three PH-domain-containing proteins exhibit identical patterns of spatial localization in chemotaxing cells and in response to global stimulation by a chemoattractant. These include CRAC, a PH-domain-containing protein required for chemoattractant-mediated activation of adenylyl cyclase; the serine/threonine protein kinase Akt/PKB, which is a PI3K effector important in

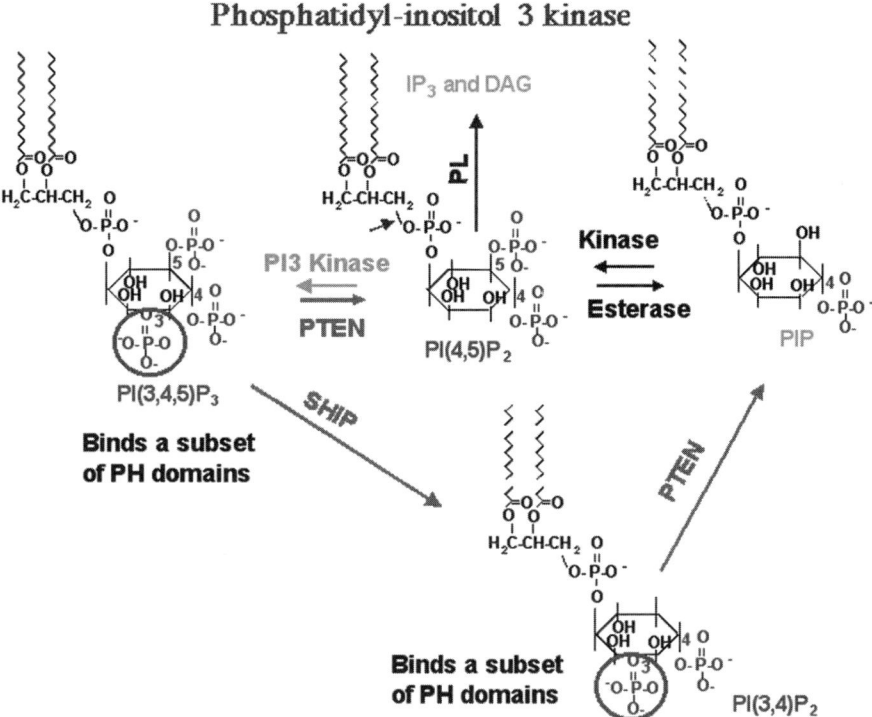

Figure 1 The phosphatidylinositol kinase/PTEN biochemical pathway. The figure illustrates the biochemical pathway for the production and degradation of $PI(3,4,5)P_3$ and $PI(3,4)P_2$, PI3K products that preferentially bind a subclass of PH-domain-containing proteins involved in chemotaxis. PI3K phosphorylates on the 3′ position of inositol, $PI(4,5)P_2$, also a substrate of phospholipase C (PLC), or $PI(4)P$. $PI(3,4,5)P_3$ can be degraded by the 5′ inositol phosphatase SHIP to produce $PI(3,4)P_2$. The tumor suppressor PTEN dephosphorylates $PI(3,4,5)P_3$ and $PI(3,4)P_2$ on the 3′ position of the inositol ring, thereby removing the binding sites for the PH domains on the plasma membrane.

regulating cell growth and cell survival; and PhdA, which is involved in the spatial-temporal control of F-actin polymerization at the leading edge. In neutrophils, the PH domain of Akt/PKB exhibits changes in its spatial distribution in response to chemoattractant stimulation similar to those of its counterpart in *Dictyostelium* cells [18].

Genetic and biochemical studies have demonstrated that these localizations occur in response to activation of PI3K. First, the localization is inhibited by the PI3K-specific inhibitor LY294002. Second, in *Dictyostelium* cells carrying disruptions of the genes encoding two of the Class I PI3Ks (PI3K1 and PI3K2), chemoattractant-mediated membrane localization is not observed. Finally, point mutations in the PH domain that abrogate the ability of the PH domain to bind PI3K lipid products prevent the PH domains from localizing to the leading edge in response to global chemoattractant stimulation (see [1] for references). These findings indicate that PH domain localization and thus PI3K activation respond to one of the primary downstream responses to chemoattractant stimulation. These findings have been reproduced in mammalian leukocytes [18].

The activation of this response at the leading edge is reflected by a very steep intracellular gradient of the PH domain localization at the plasma membrane from the front to the back of the cell. Since this occurs in a very shallow chemoattractant gradient (as low as a 2–5% difference between the front and back of the cell), there must be a mechanism by which an external, shallow gradient gives rise to a very steep intracellular gradient of the second messengers $PI(3,4,5)P_3$ and $PI(3,4)P_2$. Recent studies have demonstrated that PI3K in *Dictyostelium* preferentially localizes to the leading edge (Fig. 3). GFP fusions of PI3K1 or PI3K2 localize to the leading edge when cells are placed in a chemoattractant gradient and are transiently localized to the plasma membrane in response to global stimulation [20]. This provides for the localized activation of PI3K at the leading edge and thus the localized production of $PI(3,4,5)P_3/PI(3,4)P_2$. What localizes PI3K to the leading edge is presently unknown. However, as already mentioned, PI3K's localization is not due to a corresponding localization of either the chemoattractant receptor or the coupled heterotrimeric G protein.

PI3K Effectors and their Roles in Controlling Chemotaxis

pi3k1/2 null cells exhibit strong chemoattractant defects. *Dictyostelium pi3k1/2* null cells chemotaxing toward cAMP exhibit a significant loss of polarity and move more slowly than wild-type cells. Whereas wild-type cells preferentially form a single pseudopod at the leading edge and produce few if any lateral pseudopodia, *pi3k1/2* null cells produce multiple pseudopodia simultaneously along the periphery of the cell, although the cells still move predominantly toward the chemoattractant source. Results from neutrophils and macrophages in which the function of PI3Kγ and PI3Kδ, respectively, are abrogated are consistent with the observations in *Dictyostelium* (see [1] for references).

Biochemical analysis of *Dictyostelium pi3k1/2* null cells has provided insight into the downstream effector pathways. *pi3k1/2* null cells exhibit a reduced chemoattractant-mediated F-actin assembly and a more severe defect in the spatial-temporal regulation of F-actin assembly. When cells are chemotaxing toward a micropipette emitting a chemoattractant and the location of the micropipette is changed, resulting in a change in direction, the kinetics of the directional change in *pi3k1/2* null cells are delayed compared to those of wild-type cells. *Dictyostelium* cells carrying a disruption of the gene encoding PhdA, a PH domain-containing protein with PI3K-dependent localization to the leading edge, exhibit actin phenotypes similar to those observed for *pi3k1/2* null cells, suggesting that this function of PI3K is mediated, at least in part, through PhdA [14].

Akt/PKB is activated in response to chemoattractants in both *Dictyostelium* and neutrophils and probably in other cell types as well, and this activation is lost in *Dictyostelium* and mammalian *pi3k* null cells (see [1] for references), indicating these pathways are probably conserved between *Dictyostelium* and mammalian cells. Ligand-regulated Akt/PKB activity is required for proper chemotaxis in *Dictyostelium* and has been linked to the migration of mammalian endothelial cells [17,21].

In *Dictyostelium*, a gene knockout of Akt/PKB exhibits a subset of the *pi3k* null defects, including a reduction in cell movement, directionality, and chemoattractant-mediated myosin II assembly, providing a link between the activation of PI3K at the front of the cell and the regulation of myosin assembly at the cell's posterior [17,22] (Fig. 2). Myosin assembly and disassembly in *Dictyostelium* is regulated by phosphorylation of myosin heavy chain kinase (MHCK). Phosphorylation of myosin II by MHCK leads to myosin II disassembly, whereas phosphatase treatment leads to assembly [23]. Genetic and biochemical evidence demonstrates that PAKa, a p21-activated serine/threonine kinase related to mammalian PAK1 and yeast Cla4 and Ste20, is essential for myosin II assembly during cytokinesis and chemotaxis; *paka* null cells exhibit phenotypes similar to those of cells lacking myosin II. Moreover, PAKa colocalizes with myosin at the posterior of chemotaxing cells and at the contractile ring during cytokinesis. Studies have demonstrated that PAKa is a direct substrate, both *in vivo* and *in vitro*, of Akt/PKB. Phosphorylation of PAKa by Akt/PKB leads to its activation [22]. PAKa then is thought to function by inhibiting MHCK, leading to myosin II assembly. It is interesting that one of the MHCKs, MHCK-A, preferentially localizes to the leading edge in chemotaxing cells, where it is presumably activated, causing disassembly of myosin and more efficient pseudopod extension [24].

The Tumor Suppressor PTEN Regulates the Chemoattractant PI3K Pathways

PTEN, the tumor suppressor that acts as a negative regulator of the PI3K cell growth and cell survival pathway by dephosphorylating $PI(3,4,5)P_3$ and $PI(3,4)P_2$ on the 3' position [25], also functions as a negative regulator of chemotaxis. *Dictyostelium* cells overexpressing PTEN exhibit a reduced activation of Akt/PKB and chemotaxis defects consistent with a reduced level of PI3K pathway activity [20]. Hypomorphs, cells expressing a lower level of PTEN, exhibit higher levels of Akt/PKB activation. The strongest link between PTEN and chemotaxis derives from the analysis of *Dictyostelium* PTEN null cells [26]. These cells exhibit prolonged localization of PH-domain-containing proteins in response to chemoattractant stimulation. Moreover, when chemotaxing cells are examined, PH domain protein localization extends around the side of the cell, including some localization to the cell's posterior. These phenotypes are very similar to those obtained by expressing myr-tagged PI3K. These results demonstrate that PTEN restricts the domain of $PI(3,4,5)P_3/PI(3,4)P_2$ localization and regulates the function of the PI3K pathway. Further linkage of the PI3K pathway to chemotaxis was demonstrated by the finding that PTEN null cells exhibit an elevated and prolonged level of chemoattractant-mediated F-actin assembly, indicating a linkage between PI3K and F-actin assembly. This observation suggests that PI3K activation at the leading edge may be one of the mechanisms that drives the expected activation of Rho exchange factors and F-actin polymerization at these sites. The results are consistent with the phenotypes of *pi3k* null cells and those of PhdA, a PI3K effector.

The subcellular localization of PTEN is also consistent with a role as a negative regulator of the PI3K pathway during chemotaxis: PTEN is uniformly localized around the plasma membrane of unstimulated cells and rapidly delocalizes from the plasma membrane in response to chemoattractant stimulation [20] (Fig. 3). In polarized, chemotaxing cells, PTEN is preferentially excluded from the leading edge, while it remains along the sides of the cell. This finding suggests that PTEN localization, like that of PI3K, is dynamic and is complementary to that of PI3K. Presumably, PTEN exclusion from the leading edge allows an amplification of the PI3K activity at this site of the cell, whereas its localization on the plasma membrane on the sides of the cell helps restrict PI3K activity and sharpen the boundary of PIP_3/PIP_2 localization.

Conclusions

Results in *Dictyostelium* provide a model of the mechanism controlling directional movement (Fig. 2). The localization of PI3K to the leading edge results in the production of the second messengers $PI(3,4,5)P_3$ and $PI(3,4)P_2$, causing a localization of PH-domain-containing proteins and regulation of downstream effector pathways. The delocalization of

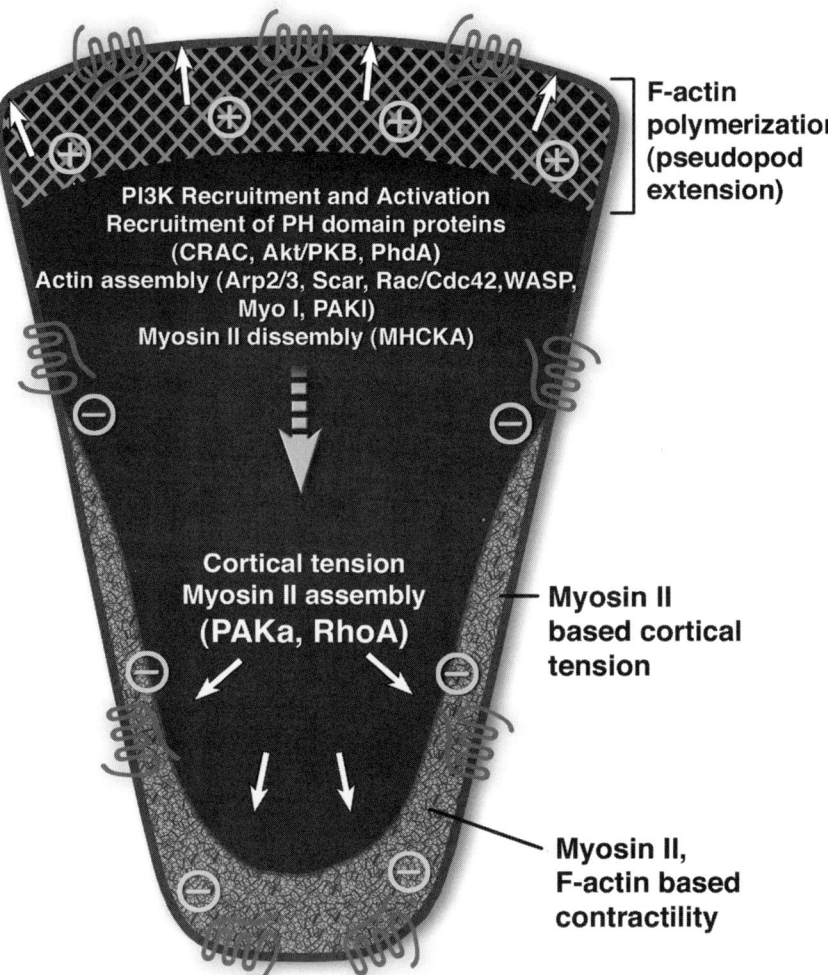

Figure 2 Spatial regulation of signaling components and the actin/myosin cytoskeleton. The image shows a polarized, chemotaxing amoeboid cell. Preferential activation in a series of signaling pathways at the leading edge results in F-actin polymerization and extension of the pseudopod. The pathways activated include phosphatidylinositol-3 kinase (PI3K), leading to the recruitment and activation of PH-domain-containing proteins. Other components that are localized to the leading edge include those regulating actin assembly (the Arp2/3 complex, the Wiskott-Aldrich Syndrome protein WASP and its relative Scar/WAVE, and the small GTPases Rac and Cdc42), and myosin I, which may regulate the translocation of the WASP/Arp2/3 complex along the F-actin filaments. Mammalian PAK1 is also vital in regulating pseudopod extension. Myosin assembly and contractility at the posterior of the cell is required for uropod contraction. In *Dictyostelium*, these processes are mediated by myosin heavy chain kinase A (MHCKA), which localized to the front of the cell, and PAKa found at the posterior of the cell. In mammalian cells, it is regulated by the small GTPase RhoA and downstream pathways. See text for additional details.

PTEN from the leading edge while PTEN remains along the sides of the cell helps localize and restrict the PI3K pathway. Downstream effector pathways include F-actin polymerization at the leading edge and myosin assembly at the cell's posterior. Other studies have implicated this pathway in controlling cell polarization, which is necessary for effective chemotaxis. Parts of this pathway have also been described in neutrophils and macrophages, including the essential role of PI3K in directional sensing and the localization of PH-domain-containing proteins at the leading edge. Because of the increased recognition of the importance of chemotaxis in a variety of cellular processes, including cell polarization, metastasis, and embryonic cell movement, understanding these mechanisms is paramount to providing mechanistic insights into basic biological processes and many aspects of human disease.

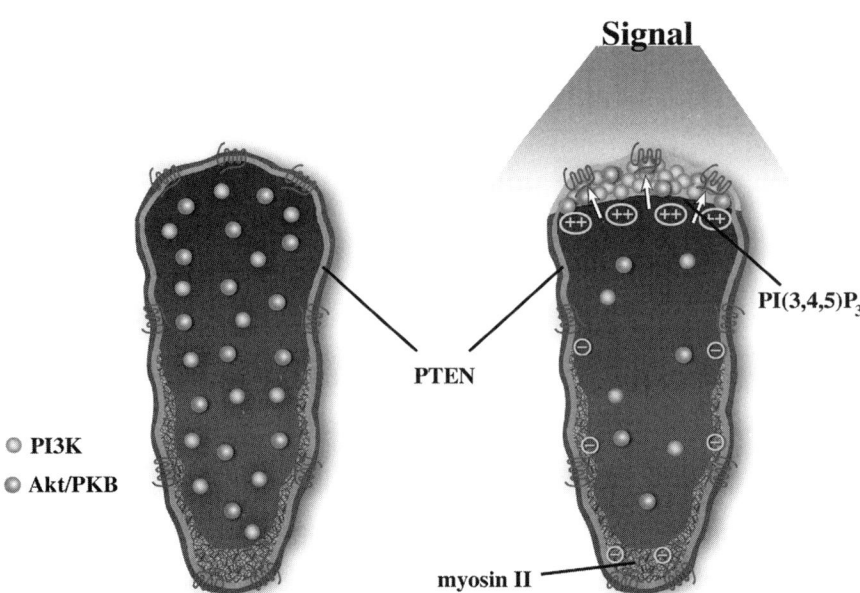

Figure 3 Differential localization of PI3K and PTEN in a chemotaxing cell. The figure on the right shows that in a resting cell, PI3K and PH-domain-containing proteins Akt/PKB, CRAC, and PhdA are uniformly distributed in the cytosol, whereas the tumor suppressor PTEN is localized uniformly at the edge of the cell. When cells are placed in a chemoattractant gradient, as illustrated in the panel on the right, PI3K preferentially localizes to the leading edge, causing the production of $PI(3,4,5)P_3$ and the localization of the PH-domain-containing proteins. PTEN is preferentially lost from the leading edge. This loss is thought to help sharpen the gradient by preferentially limiting the site of $PI(3,4,5)P_3$ membrane localization.

References

1. Chung, C., Funamoto, S., and Firtel, R. (2001). Signaling pathways controlling cell polarity and chemotaxis. *Trends Biochem. Sci.* **26**, 557–566.
2. Katanaev, V. L. (2001). Signal transduction in neutrophil chemotaxis. *Biochemistry* **66**, 351–368.
3. Sanchez-Madrid, F. and del Pozo, M. A. (1999). Leukocyte polarization in cell migration and immuneinteractions. *EMBO J.* **18**, 501–511.
4. Parent C. A. and Devreotes, P. N. (1999). A cell's sense of direction. *Science* **284**, 765–770.
5. Rickert, P., Weiner, O., Wang, F., Bourne, H., and Servant, G. (2000). Leukocytes navigate by compass: roles of PI3K-gamma and its lipid products. *Trends Cell Biol.* **10**, 466–473.
6. Firtel, R. A. and Chung, C. Y. (2000). The molecular genetics of chemotaxis: Sensing and responding to chemoattractant gradients. *BioEssays* **22**, 603–615.
7. Borisy, G. G. and Svitkina, T. M. (2000). Actin machinery: pushing the envelope. *Curr. Opinion Cell Biol.* **12**, 104–112.
8. Nobes, C. and Hall, A. (1999). Rho GTPases control polarity, protrusion, and adhesion during cell movement. *J. Cell Biol.* **144**, 1235–1244.
9. Jin, T., Zhang, N., Long, Y., Parent, C. A., and Devreotes, P. N. (2000). Localization of the G protein beta gamma complex in living cells during chemotaxis [see comments]. *Science* **287**, 1034–1036.
10. Rameh, L. E. and Cantley, L. C. (1999). The role of phosphoinositide 3-kinase lipid products in cell function. *J. Biol. Chem.* **274**, 8347–8350.
11. Li, Z., Jiang, H., Xie, W., Zhang, Z., Smrcka, A. V., and Wu, D. (2000). Roles of PLC-beta2 and -beta3 and PI3Kgamma in chemoattractant-mediated signal transduction [see comments]. *Science* **287**, 1046–1049.
12. Hirsch, E., Katanaev, V. L., Garlanda, C., Azzolino, O., Pirola, L., Silengo, L., Sozzani, S., Mantovani, A., Altruda, F., and Wymann, M. P. (2000). Central role for G protein-coupled phosphoinositide 3-kinase gamma in inflammation [see comments]. *Science* **287**, 1049–1053.
13. Sasaki, T., Irie-Sasaki, J., Jones, R. G., Oliveira-dos-Santos, A. J., Stanford, W. L., Bolon, B., Wakeham, A., Itie, A., Bouchard, D., Kozieradzki, I., Joza, N., Mak, T. W., Ohashi, P. S., Suzuki, A., and Penninger, J. M. (2000). Function of PI3Kgamma in thymocyte development, T cell activation, and neutrophil migration [see comments]. *Science* **287**, 1040–1046.
14. Funamoto, S., Milan, K., Meili, R., and Firtel, R. (2001). Role of phosphatidylinositol 3' kinase and a downstream pleckstrin homology domain-containing protein in controlling chemotaxis in *Dictyostelium*. *J. Cell Biol.* **153**, 795–810.
15. Vanhaesebroeck, B., Jones, G. E., Allen, W. E., Zicha, D., Hooshmand-Rad, R., Sawyer, C., Wells, C., Waterfield, M. D., and Ridley, A. J. (1999). Distinct PI(3)Ks mediate mitogenic signalling and cell migration in macrophages. *Nature Cell Biol.* **1**, 69–71.
16. Parent, C. A., Blacklock, B. J., Froehlich, W. M., Murphy, D. B., and Devreotes, P. N. (1998). G protein signaling events are activated at the leading edge of chemotactic cells. *Cell* **95**, 81–91.
17. Meili, R., Ellsworth, C., Lee, S., Reddy, T. B., Ma, H., and Firtel, R. A. (1999). Chemoattractant-mediated transient activation and membrane localization of Akt/PKB is required for efficient chemotaxis to cAMP in *Dictyostelium*. *EMBO J.* **18**, 2092–2105.
18. Servant, G., Weiner, O. D., Herzmark, P., Balla, T., Sedat, J.W., and Bourne, H. R. (2000). Polarization of chemoattractant receptor signaling during neutrophil chemotaxis. *Science* **287**, 1037–1040.
19. Haugh, J. M., Codazzi, F., Teruel, M., and Meyer, T. (2000). Spatial sensing in fibroblasts mediated by 3' phosphoinositides. *J. Cell Biol.* **151**, 1269–1280.
20. Funamoto, S., Meili, R., Lee, S., Parry, L., and Firtel, R. A. (2002). Spatial and temporal regulation of 3-phosphoinositides by PI3 kinase and PTEN mediates chemotaxis. *Cell.* **In press**.
21. Morales-Ruiz, M., Fulton, D., Sowa, G., Languino, L. R., Fujio, Y., Walsh, K., and Sessa, W. C. (2000). Vascular endothelial growth factor–stimulated actin reorganization and migration of endothelial cells is regulated via the serine/threonine kinase Akt. *Circ. Res.* **86**, 892–896.

22. Chung, C. Y., Potikyan, G., and Firtel, R. A. (2001). Control of cell polarity and chemotaxis by Akt/PKB and PI3 kinase through the regulation of PAKa. *Mol. Cell* **7**, 937–947.
23. Chung, C. Y. and Firtel, R. A. (2000). *Dictyostelium*: a model experimental system for elucidating the pathways and mechanisms controlling chemotaxis. In P. M. Conn and A. Means, Ed., *Principles of Molecular Regulation*. (The Humana Press, Totowa, N.J., 99–114.)
24. Steimel, P. A., Yumura, S. Y., Cote, G. P., Medley, Q. G., Polyakov, M. V., Leppert, B., and Egelhoff, T. T. (2001). Recruitment of a myosin heavy chain kinase to actin-rich protrusions in *Dictyostelium*. *Curr. Biol.* **11**, 708–713.
25. Maehama, T. and Dixon, J. (1998). The tumor suppressor, PTEN/MMAC1, dephosphorylates the lipid second messenger, phosphatidylinositol 3,4,5-trisphosphate. *J. Biol. Chem.* **273**, 13375–13378.
26. Iijima, M. and Devreotes, P. (2002). Tumor suppressor PTEN mediates sensing of chemoattractant gradients. *Cell.* In press.

CHAPTER 158

Phosphatidylinositol Transfer Proteins

Shamshad Cockcroft
*Department of Physiology,
University College London,
London, United Kingdom*

Introduction

The phosphatidylinositol transfer protein (PITP) family is defined by its ability to bind one molecule of either phosphatidylinositol (PtdIns) or phosphatidylcholine (PtdCho) and facilitate lipid transfer between separate membrane compartments [1]. PITPs have now emerged as critical regulators of phosphoinositide metabolism in specific cellular compartments where they participate in signal transduction and membrane traffic [2]. PITP was originally purified as a soluble 35 kDa protein, which is now known to contain a single structural domain [3]. The PITP domain has now been found in the larger RdgB proteins, originally identified in *Drosophila* as *r*etinal *d*egeneration (Class *B*) mutants. Today, the mammalian PITP family includes five proteins divided into three subgroups, all containing a PITP domain: the classical PITPs, α and β (35 kDa), two larger related proteins M-rdgBα1 and M-rdgBα2 (160 kDa), and the soluble M-rdgBβ protein (38 kDa) [4]. In addition to mammals, proteins with a PITP domain are found in *Caenorhabditis elegans* (worms), *Drosophila melanogaster* (flies) and *Dictyostelium discoideum* (soil amoebae), but not yeast or plants. The larger rdgB proteins are not found in *Dictyostelium*, however. The yeast Sec14p and its related family members form a separate group of PtdIns transfer proteins that, although they share lipid binding properties and transfer function with mammalian PITPs, have no sequence or structural similarity [3,5].

The Classical PITPs: α and β

PITPα and PITPβ are expressed ubiquitously in all tissues, are abundant proteins, and share 77% identity and 94% similarity in amino acid sequence. In addition to PtdIns and PtdCho transfer, PITPβ can also transfer sphingomyelin [6]. Transfer occurs down a concentration gradient without input of energy *in vitro*. Thus PITPs solubilize specific lipids from membranes and can facilitate their movement through the aqueous phase (Fig. 1). Although PITPs are defined by their ability to bind either one molecule of PtdIns or PtdCho, the affinity of PITPα for PtdIns is 16-fold greater compared to PtdCho. This reflects the lower levels of PtdIns compared to PtdCho in cells, and typically 30 to 40% of the PITPα and β proteins are PC-bound compared to 60 to 70% that are loaded with PtdIns.

PITPα and PITPβ are localized in different compartments. PITPα is present in the cytosol and the nucleus whereas PITPβ is localized at the Golgi and in the cytosol. In mammalian cells, the function of PITP was first identified in biochemical studies involving reconstitution of phospholipase C-signaling and exocytosis in cytosol-depleted cells [7,8]. Phospholipase C hydrolyses phosphatidylinositol(4,5)bisphosphate (PIP$_2$) to generate the second messengers, diacylglycerol and inositol(1,4,5)trisphosphate. Activation of G-protein-coupled receptors or receptor tyrosine kinases is responsible for increasing phospholipase C activity, and PITPα was identified as an essential component in ensuring PIP$_2$ supply for the enzyme [9,10]. Exocytosis could be similarly recovered in

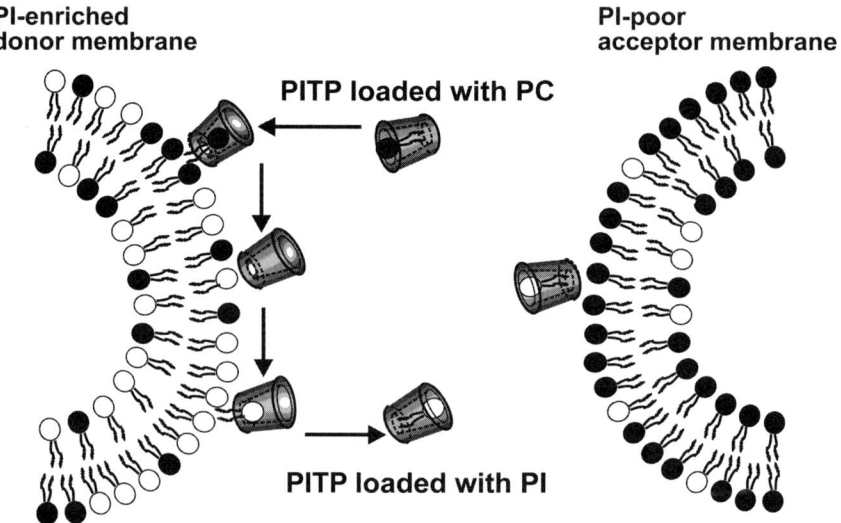

Figure 1 PITPs bind and transfer PtdIns and PtdCho between membrane compartments. Phosphatidylinositol transfer proteins (PITPs) were first purified based on their ability to transfer PtdIns between two membrane compartments *in vitro*.

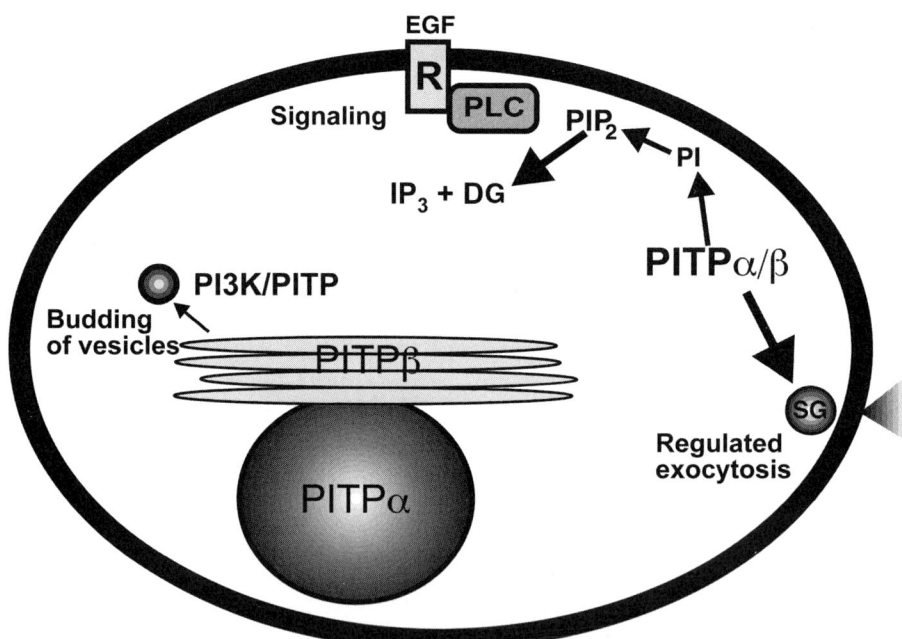

Figure 2 Functions and location of PITPα and PITPβ. PITPα is primarily localized in the cytosol and the nucleus. PITPα is required for supplying the substrate, PtdIns for PIP_2 synthesis utilized by phospholipase C and for maintaining a pool of PIP_2 for exocytosis. PITPβ is primarily localized at the Golgi and cytosol and is involved in the budding of vesicles by making available a pool of phosphoinositides at the Golgi. The function of PITPα in the nucleus is probably in making substrate available for phosphorylation.

permeabilized cells where PITPα and a PIP 5-kinase worked in synergy to make PIP_2 [11,12]. Finally, biogenesis of vesicles from Golgi was also dependent on cytosolic proteins, and PITP was thus purified [13,16]. In all these studies, both PITPα and PITPβ were equally capable of restoring function. Several of these functions are summarized in Fig. 2.

Studies aimed at elucidating the mechanism of action of PITP in each of these seemingly disparate functions have yielded a singular theme. The activity of PITP stems from its ability to transfer PtdIns from its site of synthesis (ER) to sites of cellular activity and to stimulate the local synthesis of phosphorylated forms of PtdIns, including PtdIns(4)P, $PtdIns(4,5)P_2$, PtdIns(3)P, and $PtdIns(3,4,5)P_3$ [14,15]. It is speculated that PITP could present PtdIns to the lipid kinases within a signaling complex. This concept is supported by observations that PITPα does associate with the EGF receptor phopsholipase Cγ and PtdIns 4-kinase following stimulation with EGF [9].

A reduction in the expression levels of PITPα, as seen in the *vibrator* mutation in mice leads, to neurodegeneration [17] in the presence of normal concentrations of PITPβ, suggesting that although these proteins share transfer activity and can substitute for each other in reconstitution assays [8,13,18], they do have distinct functions *in vivo*. PITPβ is an essential protein, since mice carrying mutations in PITPβ die early in embryogenesis, and PITPβ may be essential for stem cell viability [19]. PITPβ was originally cloned from a rat brain cDNA library by its ability to rescue *Sec14* mutants in yeast, *S. cerevisiae*. Mutations in *Sec14* lead to a defect in the formation of secretory vesicles destined for the plasma membrane. Despite the absence of sequence or structural homology, PITPβ was able to rescue the temperature-sensitive *Sec14* mutants. In mammalian cells, PITPβ also localizes to the Golgi, and considerable data have accumulated that suggest that PITPβ may be involved in vesicle budding in this compartment by maintaining a pool of phosphorylated PtdIns [16].

The amino acid sequence of the PITP domain is highly conserved in all isoforms, and no characteristic short sequence motifs have been identified. From the crystal structure of PITPα bound to PtdCho, the PITP domain comprises an amino-terminal lipid-binding region that contains an eight-stranded, concave, mostly anti-parallel ß-sheet and two helices, the carboxy-terminal helical region, and the intervening regulatory loop region [3]. Upon stimulation with receptor-directed agonists or PMA, PITPα is phosphorylated at Ser164, which resides in the regulatory loop region. This is an important regulatory control as mutation of the serine residue to glutamate (which mimics phosphorylation) inhibits transfer function as well as the ability to provide substrate for phospholipase C signaling.

The mechanism of how PITP can abstract a lipid from a bilayer and facilitate exchange can be conjectured from the extensive biochemical and structural analysis of PITPα. For PITPα to perform its task, a change in affinity for membranes has to occur for it to associate with membranes to exchange its bound lipid, but this change in affinity has to be reversed so that the protein can move rapidly away from the membrane. Deletions of the C-terminus induce a more relaxed conformation and enhances its affinity for membranes without affecting its lipid binding properties. Thus movement of the carboxy-terminal helical region very likely governs change in membrane affinity and also exposes the lipid tails toward the membrane. The lipid is now able to move out of its cavity, and lipid exchange can then occur. Following exchange the protein has to return to its compact structure to be released from the membrane.

One of the major roles of PITPα is to provide PtdIns for PLC signaling. PLC signaling is thought to occur in inositol-lipid enriched membranes rafts, since destruction of rafts inhibits PLC activation [20,21]. It may be speculated that the localized depletion of the inositol lipids in membrane rafts could result in changes of membrane bilayer curvature. Since the activity of PITPα is sensitive to membrane curvature, this would mean that the transfer activity is regulated by changes in the local membrane environment. This conclusion is supported by the observation that upon stimulation with EGF, PITPα is part of a signaling complex, which includes the EGF receptor, Type II PI-4-kinase, and phospholipase Cγ [9]. Thus in cells, PITPα does not randomly transfer lipids but does so at specific sites of active consumption of phosphoinositides.

RdgB Family of PITP Proteins

As already mentioned, the RdgB acronym is derived from a *r*etinal *deg*eneration mutant phenotype (type **B**) in *Drosophila* where this family of PITP proteins were first identified. The *D-rdgB* mutation causes abnormal photoreceptor responses and light-enhanced retinal degeneration, and genetic evidence indicates that the D-rdgB product acts within the light-triggered phosphoinositide cascade responsible for phototransduction. The *D-rdgB* gene encodes a 160 kDa protein that has an N-terminal PITP domain, an acidic Ca^{2+}-binding domain, and an extended hydrophobic region; in mammals, there are of two homologues, M-RdgBα1 and M-RdgBα2. In addition, a smaller protein of 38 kDa (RdgBβ) was identified by homology to rdgBα and this isoform is also found in both flies and mammals [4]. Transgenic expression of murine rdgBα1 or 2 rescues *rdgB* null *Drosophila*. However, deletion of RdgBα2 in mice has no obvious phenotype and phototransduction and photoreceptor survival is unaffected, whereas deletion of RdgBα1 is embryonically lethal [22].

M-RdgB1 specifically associates with phosphatidylinositol 4-kinase (α-isoform) and can increase the kinase activity [23]. These data are consistent with our proposal that PITP proteins mediate spatially restricted synthesis of phosphorylated inositol lipids. In conclusion, proteins with a PITP domain all appear to function in many aspects of biology by virtue of its ability to regulate phosphoinositides synthesis. Phosphoinositides play important roles not only for providing substrate for signaling pathways but also as ligands for proteins containing specific domains, including PH domains, PX domains, ENTH domains.

References

1. Wirtz, K. W. A. (1997). Phospholipid transfer proteins revisited. *Biochem. J.* **324**, 353–360.
2. Cockcroft, S. (2001). Phosphatidylinositol transfer proteins couple lipid transport to phosphoinositide synthesis. *Semin.Cell Dev. Biol.* **12**, 183–191.
3. Yoder, M. D., Thomas, L. M., Tremblay, J. M., Oliver, R. L., Yarbrough, L. R., and Helmkamp, G. M., Jr. (2001). Structure of a multifunctional protein. Mammalian phosphatidylinositol transfer protein complexed with phosphatidylcholine. *J Biol Chem.* **276**, 9246–9252.
4. Hsuan, J. and Cockcroft, S. (2001). The PITP family of phosphatidylinositol transfer proteins. *Genome Biol.* **2**, 3011.1–3011.8.
5. Sha, B., Phillips, S. E., Bankaitis, V., and Luo, M. (1998). Crystal structure of the *Saccharomyces cerevisiae* phosphatidylinositol transfer protein. *Nature* **391**, 506–510.
6. De Vries, K. J., Heinrichs, A. A. J., Cunningham, E., Brunink, F., Westerman, J., Somerharju, P. J., Cockcroft, S., Wirtz, K. W. A., and Snoek, G. T. (1995). An isoform of the phosphatidylinositol transfer protein transfers sphingomyelin and is associated with the golgi system. *Biochem. J.* **310**, 643–649.

7. Thomas, G. M. H., Cunningham, E., Fensome, A., Ball, A., Totty, N. F., Troung, O., Hsuan, J. J., and Cockcroft, S. (1993). An essential role for phosphatidylinositol transfer protein in phospholipase C-mediated inositol lipid signalling. *Cell* **74**, 919–928.
8. Hay, J. C. and Martin, T. F. J. (1993). Phosphatidylinositol transfer protein required for ATP-dependent priming of Ca^{2+}-activated secretion. *Nature* **366**, 572–575.
9. Kauffmann-Zeh, A., Thomas, G. M. H., Ball, A., Prosser, S., Cunningham, E., Cockcroft, S., and Hsuan, J. J. (1995). Requirement for phosphatidylinositol transfer protein in epidermal growth factor signalling. *Science* **268**, 118–1190.
10. Cunningham, E., Thomas, G. M. H., Ball, A., Hiles, I., and Cockcroft, S. (1995). Phosphatidylinositol transfer protein dictates the rate of inositol trisphosphate production by promoting the synthesis of PIP_2. *Curr. Biol.* **5**, 775–783.
11. Hay, J. C., Fisette, P. L., Jenkins, G. H., Fukami, K., Takenawa, T., Anderson, R. E., and Martin, T. F. J. (1995). ATP-dependent inositide phosphorylation required for Ca^{2+}-activated secretion. *Nature* **374**, 173–177.
12. Fensome, A., Cunningham, E., Prosser, S., Tan, S. K., Swigart, P., Thomas, G., Hsuan, J., and Cockcroft, S. (1996). ARF and PITP restore GTPγS-stimulated protein secretion from cytosol-depleted HL60 cells by promoting PIP_2 synthesis. *Curr. Biol.* **6**, 730–738.
13. Ohashi, M., Jan de Vries, K., Frank, R., Snoek, G., Bankaitis, V., Wirtz, K., and Huttner, W. B. (1995). A role for phosphatidylinositol transfer protein in secretory vesicle formation. *Nature* **377**, 544–547.
14. Kular, G., Loubtchenkov, M., Swigart, P., Whatmore, J., Ball, A., Cockcroft, S., and Wetzker, R. (1997). Co-operation of phosphatidylinositol transfer protein with phosphoinositide 3-kinase(gamma) in the formylmethionyl-leucylphenylalanine-dependent production of phosphatidylinositol 3,4,5 trisphosphate in human neutrophils. *Biochem. J.* **325**, 299–301.
15. Panaretou, C., Domin, J., Cockcroft, S., and Waterfield, M. D. (1997). Characterization of p150, an adaptor protein for the human phosphatidylinositol (PtdIns) 3-kinase. Substrate presentation by phosphatidylinositol transfer protein to the p150-PtdIns 3-kinase complex. *J. Biol. Chem.* **272**, 2477–2485.
16. Jones, S. M., Alb, J. G., Jr., Phillips, S. E., Bankaitis, V. A., and Howell, K. E. (1998). A phosphatidylinositol 3-kinase and phosphatidylinositol transfer protein act synergistically in formation of constitutive transport vesicles from the trans-golgi network. *J. Biol. Chem.* **273**, 10349–10354.
17. Hamilton, B. A., Smith, D. J., Mueller, K. L., Kerrebrock, A. W., Bronson, R. T., Berkel, V. v., Daly, M. J., Kroglyak, L., Reeve, M. P., Nernhauser, J. L., Hawkins, T. L., Rubin, E. M., and Lander, E. S. (1997). The *vibrator* mutation causes neurogeneration via reduced expression of PITPα: positional complementation cloning and extragenic suppression. *Neuron* **18**, 711–722.
18. Cunningham, E., Tan, S. W., Swigart, P., Hsuan, J., Bankaitis, V., and Cockcroft, S. (1996). The yeast and mammalian isoforms of phosphatidylinositol transfer protein can all restore phospholipase C-mediated inositol lipid signalling in cytosol-depleted RBL-2H3 and HL60 cells. *Proc. Natl. Acad. Sci. USA* **93**, 6589–6593.
19. Bankaitis, V. A. (2002). Cell biology. Slick recruitment to the Golgi. *Science* **295**, 290–291.
20. Pike, L. J. and Casey, L. (1996). Localization and turnover of phosphatidylinositol 4,5-bisphosphate in caveolin-enriched membrane domains. *J. Biol. Chem.* **271**, 26453–26456.
21. Waugh, M. G., Lawson, D., Tan, S. K., and Hsuan, J. J. (1998). Phosphatidylinositol 4-phosphate synthesis in immunoisolated caveolae-like vesicles and low bouyant density non-caveolar membranes. *J. Biol. Chem.* **273**, 17115–17121.
22. Lu, C., Peng, Y. W., Shang, J., Pawlyk, B. S., Yu, F., and Li, T. (2001). The mammalian retinal degeneration B2 gene is not required for photoreceptor function and survival. *Neuroscience* **107**, 35–41.
23. Aikawa, Y., Kuraoka, A., Kondo, H., Kawabuchi, M., and Watanabe, T. (1999). Involvement of PITPnm, a mammalian homologue of *Drosophila rdgB*, in phosphoinositide synthesis on Golgi membranes. *J. Biol. Chem.* **274**, 20569–20577.

CHAPTER 159

Inositol Polyphosphate Regulation of Nuclear Function

John D. York
*Departments of Pharmacology and Cancer Biology and of
Biochemistry, Howard Hughes Medical Institute,
Duke University Medical Center,
Durham, North Carolina*

Introduction

As several chapters in this Handbook attest, inositol signaling pathways have emerged as a multifaceted ensemble of cellular switches that regulate a number of processes well beyond calcium release, including membrane trafficking, channel activity, and nuclear function. Over 30 inositol messengers are found in eukaryotic cells that may be generally grouped into two classes: (1) inositol lipids or phosphoinositides (PIPs) and (2) water-soluble inositol polyphosphates (IPs). Insights into the roles of these messengers have come through the characterization of numerous gene products that control the metabolism of PIPs and IPs, over eighty in humans and twenty-six in budding yeast. This review will discuss in brief a small subset of the overall inositol signaling pathway, namely higher IPs, generally defined as having four or more phosphates. Two important concepts have emerged: (1) the higher IPs discussed here are derived from phospholipase C-dependent activation, thus IP_3 is both a messenger and a precursor to others, and (2) the higher IPs have been linked to the regulation of several nuclear processes. Emphasis will be placed on the gene products that synthesize IP_4, IP_5, IP_6 and diphosphoryl IPs and the processes they have been found to regulate. Several of these kinases appear to localize within the nucleus, and their activities are necessary for proper gene expression, mRNA export, and DNA metabolism. The breadth of nuclear processes regulated and the evolutionary conservation of the genes involved in their synthesis have sparked renewed interest in higher IPs as important intracellular messengers.

Inositol Signaling and the Molecular Revolution

The molecular revolution has left an indelible mark on inositol signaling pathways, fueling an expansion of our thinking [1–8]. As the roles of inositol 1,4,5-trisphosphate (IP_3) and 1,2-diacylglycerol were forged as intracellular messengers, many researchers questioned whether other inositol lipids and inositol polyphosphates, some 30 in all, had important roles in cell signaling. The cloning and characterization of kinases, phosphatases, lipases, and effectors has made it clear that the functions of inositol phosphate derivatives are numerous. Over the past decades, dozens of gene products have been characterized as players that act in concert to generate a combinatorial ensemble of distinct chemical messengers with instructions for the cell. Nature may have utilized *myo*-inositol as a signaling scaffold because of its elegant chemistry—a six-carbon asymmetric cyclitol that is readily modified by combinatorial phosphorylation—and because it may be formed through two metabolic steps from glucose 6-phosphate. Ancestral relationships have been identified at the sequence and structural level among proteins involved in inositol signaling and those involved in nucleotide, protein, and carbohydrate metabolism, thereby providing clues as to how signaling machinery evolved. Examples of genetic economy are found among several promiscuous IP kinases and phosphatases harboring multiple specificities. Remarkably, a dual-functional gene product has been identified, conserved from yeast to man, which has two distinct autonomously folded inositol lipid phosphatase domains that together are capable of dephosphorylating all known PIPs. Three inositol lipid phosphatases—OCRL-1, MTM, and PTEN/MMAC—have been identified in which a loss of function results in human disease. Together these findings have generated much new excitement within the signaling community, and as we look to the future there is every expectation that we are in for many more surprises.

Links of Inositol Signaling to Nuclear Function

A recurring theme in intracellular signaling is the spatial restriction of pathways to selective compartments. In the past 15 years, discrete nuclear specific pathways of inositol metabolism have been identified that may provide a provocative mechanism by which extracellular stimuli may ultimately elicit nuclear responses. Initially it was demonstrated that phosphoinositides are present in nuclear membranes and that activities required for their synthesis and breakdown are within nuclear fractions [9–11]. The functional importance of such pathways were then suggested through studies of insulin-like growth factor I (IGF-I), which stimulates nuclear but not cytoplasmic phosphoinositide metabolism [12,13]. Studies by Crabtree and coworkers [14] have found that PIP_2 regulates chromatin-remodeling complexes. Many other studies have been recently reviewed by Divecha and coworkers [13], and hint that inositols influence nuclear processes such as DNA synthesis, cell cycle, nuclear calcium, chromatin structure, gene expression, and messenger RNA export.

Genetic and biochemical studies of a phospholipase C-dependent pathway in the budding yeast have provided compelling functional evidence for regulation of three distinct nuclear processes by higher IPs. The budding yeast genome contains a single phosphoinositide-specific phospholipase C gene (PLC1) whose activation induces a kinase pathway that sequentially converts $I(1,4,5)P_3$ to higher IPs, including $I(1,4,5,6)P_4$, $I(1,3,4,5,6)P_5$, IP_6, and PP-IPs (see Fig. 1) [15]. Examination of IP metabolism in a variety of yeast strains reveal that activation of Plc1 results in the production of IP_3, which is then sequentially phosphorylated by two kinases, Ipk2 and Ipk1, to IP_6 [15–17]. A third kinase, Kcs1, has been identified as a diphosphoryl inositol synthase, which generates PP-IP branches from IP_5 and IP_6 substrates [18,19].

Individual mutations in plc1, ipk2, or ipk1 result in defects in the production of IP_6 as well as defects in efficient mRNA export [15]. In contrast, mutation in kcs1 and hence PP-IP production does not appear to alter mRNA export [19]. Furthermore, induction of the pathway through overexpression of Plc1 results in suppression of defects in a gle1-1 mRNA export mutant [15]. These data suggest that phospholipase C and kinase-dependent higher IP production, possibly IP_6 or some yet identified product of the Ipk1 2-kinase, regulates mRNA export. The cloning of Ipk1 orthologs from plants and

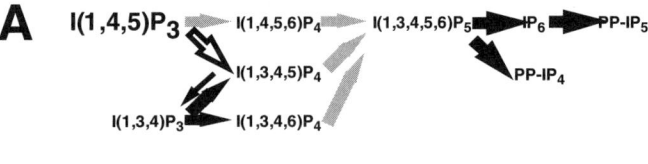

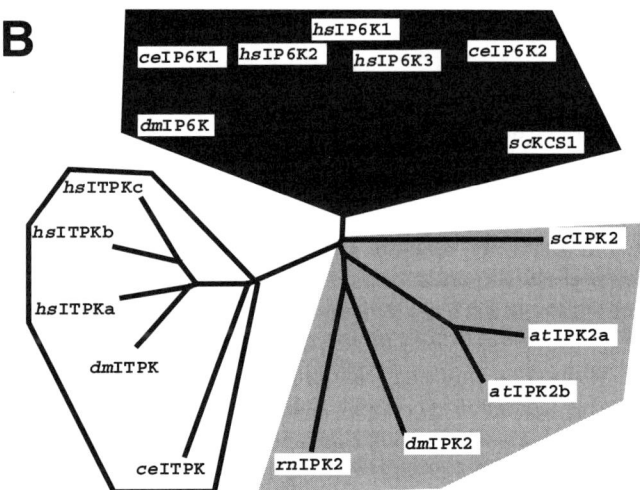

Figure 1 Higher inositol polyphosphate synthesis pathways. (A) An abridged description of IP_3 metabolism—most phosphatase activities and several higher IP intermediates have been omitted for clarity. (B) Dendogram showing three branches of the IPK family, including IPK2, ITPK, and IP6K kinases. Species are abbreviated: hs, Homo sapiens; rn, Rattus norvegicus; dm, Drosophila melanogaster; at, Arabidopsis thaliana; sc Saccharomyces cerevisiae; ce, Caenorhabditis elegans. Color legend: red—Ipk2 family members having 6-/3-/5-kinase activities; green—$I(1,4,5)P_3$ 3-kinase (ITPK); blue—diphosphoryl inositol synthetase, which generate PP-IPs (IP6K); black—IP_5 2-kinase; magenta—$I(1,3,4)P_3$ 5-/6-kinase; and cyan—IP_3 5-phosphatase.

mammals, and a recent report that reduction of higher IPs in mammalian cells also affects mRNA export suggests this pathway is conserved throughout eukaryotes (J. Stevenson-Paulik, R. A. Frye, and J. D. York, unpublished; [20,21]).

Second, a role for IP_4 and/or IP_5 in the regulation of gene expression has come from studies of a yeast IP_3/IP_4 kinase, Ipk2, which found it is identical to Arg82 [16,17]. Messenguy and co-workers [22,23] have studied Arg82 as a regulator of gene expression through the ArgR-Mcm1 transcription complex. Ipk2 is a dual-specificity 6-/3-kinase that sequentially converts IP_3 to IP_5, is localized within the nucleus, and is required to assemble protein complexes on DNA-promoter elements. Both Plc1 activity and Ipk2-mediated IP_4/IP_5 production are required for ArgR-Mcm1 transcriptional activation. Our results indicate that Ipk2 influences transcriptional responses through a two-step mechanism. First, Ipk2 protein but not IP synthesis is needed to enable formation of ArgR-Mcm1 complexes on DNA promoter elements. Second, production of IP_4 and possibly IP_5 through both phospholipase C and Ipk2 kinase activity is required to properly execute transcriptional control. While Messenguy and colleagues [24] have recently suggested that higher IPs are not required for ArgR-Mcm1 transcription, we find in using both genome-array analysis and transcriptional reporter assays that both Plc1 and Ipk2 kinase activities are required for appropriate gene expression [A. R. Odom and J. D. York, unpublished]. Because Ipk1 is not required for complex formation or transcription control [16], we conclude that these two IP kinases generate distinct nuclear messengers.

It is also important to mention earlier studies of Henry and Coworkers that found that changes in cellular levels of *myo*-inositol regulate the transcription of *INO1*, whose gene product converts glucose 6-phosphate to D-inositol 3-phosphate (reviewed in [25]). This enables cells to initiate *de novo* synthesis of inositol under conditions in which it is unavailable in the growth medium. This transcriptional regulation is accomplished through defined *cis*-acting DNA elements and *trans*-acting factors. An important future area of study will be to determine how the cell detects changes in inositol.

A third role for phospholipase C pathway in nuclear function has come from studies of the Kcs1, a disphosphoryl synthase that generates PP-IPs from IP_5 and IP_6 substrates (referred to as an IP_6 kinase by Snyder and coworkers). *KCS1* was originally identified on a genetic screen as one of two genes that when mutated overcome a hyper-recombination phenotype found in certain mutant alleles of protein kinase C [26]. Snyder and coworkers [27] have demonstrated that Kcs1 has IP_6 kinase activity and find that a point mutation in the kinase domain results in rescue of hyper-recombination, indicating that PP-IPs play a role in DNA metabolism. Of note, Shears and co-workers [19,28] have suggested a role for PP-IPs in binding components involved in membrane trafficking and have recently reported that *kcs1* mutant yeast strains have aberrant vacuole morphology. Thus, it is possible that PP-IPs have distinct compartment specific functions.

An additional role for higher IPs in DNA metabolism is suggested by the work of West and coworkers [29]. This group finds that IP_6 is a regulator of non-homologous end-joining (NHEJ), a DNA repair pathway mediated through the DNA-dependent protein kinase (DNA-PK). Through an elegant biochemical purification of a cellular regulator of NHEJ, IP_6 was identified [29]. Subsequently it has been shown that inding of IP_6 occurs via the *KU* heterodimer, the noncatalytic subunit of DNA-PK [30,31]. While both *KU* heterodimer and NHEJ pathways are found in yeast, the yeast *KU* heterodimer does not bind IP_6, and unpublished studies of Llorente and Symington (referred to in [31]) have indicated that loss of IP_6 production does not impair NHEJ in yeast. It will be important to show that changes in IP_6 levels, or one of its metabolites such as PP-IP_5, within the cell regulate NHEJ *in vivo*.

The Inositol Polyphosphate Kinase (IPK) Family

The sequence motif "PxxxDxKxG" is conserved among a growing family of inositol polyphosphate kinases, which in general we have called IPKs, depicted by the dendogram in Fig. 1B. This motif was originally described as common to IP_3 3-kinases (reviewed in [32]), and subsequently it was shown by several research groups to be a hallmark of IPK family members discussed in the preceding section, which include IP_3 3-kinases, IP_3/IP_4 dual-specificity 6-/3-kinases, and diphosphoryl IP synthase. This motif is not found in IP kinases that phosphorylate the axial second position of the inositol ring [15], nor in I(1,3,4)P_3 5/6 kinases [33], suggesting these kinases evolved from different ancestors. Thus there appear to be three branches on the tree, each of which encodes kinases that regulate distinct processes within the cell.

The Ipk2 and diphosphoryl inositol synthase (IP6K) branches are conserved from yeast to man, but the IP_3 3-kinase (ITPK) branch is not. Two groups have found that certain Ipk2 proteins exhibit diphosphoryl synthase activity [34,35]. These data indicate that the Ipk2 branch may be the oldest and raises a question related to the origins of soluble inositol polyphosphate signaling and higher IP function. What is remarkable, all eukaryotes have pathways in which the activation of phospholipase C results in cleavage of PI(4,5)P_2 to produce inositol 1,4,5-trisphosphate (IP_3) and diacylglycerol. However, the commonly viewed cellular function of these two messengers in releasing intracellular calcium and stimulation of protein kinase C has not yet been described in lower eukaryotes. Budding yeast do not appear to have IP_3-mediated calcium release pathways (and no evidence of the IP_3 receptor in their genome), nor does diacylglycerol appear to activate yeast Pkc1. I will leave you with a final question: does this indicate that the primordial role of phospholipase C induced production of IP_3 is to serve as fuel for production of higher IPs and regulation of nuclear processes?

Acknowledgments

I wish to thank members of the lab, past and present, and numerous colleagues for helpful discussions. I would also like to apologize to the authors of numerous studies whose work had to be omitted from discussion due to extreme space limitations.

References

1. Hokin, L. E. (1985). Receptors and phosphoinositide-generated second messengers. *Annu. Rev. Biochem.* **54**, 205–235.
2. Berridge, M. J. (1993). Inositol trisphosphate and calcium signaling. *Nature* **361**, 315–325.
3. Kapeller, R. and Cantley, L. C. (1994). Phosphatidylinositol 3-kinase. *BioEssays*, **16**, 565–576.
4. Majerus, P. W. (1992). Inositol phosphate biochemistry. *Annu. Rev. Biochem.* **61**, 225–250.
5. Shears, S. B. (1998). The versatility of inositol phosphates as cellular signals. *Biochimica et Biophysica Acta* **1436**, 49–67.
6. Irvine, R. F. and Schell, M. J. (2001). Back in the water: the return of the inositol phosphates. *Nat. Rev. Mo. Cell Biol.* **2**, 327–338.
7. York, J. D., Xiong, J. P., and Spiegelberg, B. (1997). Nuclear inositol signaling: a structural and functional approach. *Advances in Enz. Reg.* **38**, 365–374.
8. Odorizzi, G., Markus, B., and Emr, S. D. (2000). Phosphoinositide signaling and the regulation of membrane trafficking in yeast. *Trends in Biol. Sci.* **25**, 229–235.
9. Smith, C. and Wells, W. (1983). Phosphorylation of rat liver nuclear envelopes. *J. Biol. Chem*, **258**, 9368–9373.
10. Cocco, L., Gilmour, R. S., Ognibene, A., Manzoli, F. A., and Irvine, R. F. (1987). Synthesis of polyphosphoinositides in nuclei of Friend cells.

Evidence for polyphosphoinositide metabolism inside the nucleus which changes with cell differentiation. *Biochem. J.* **248**, 765–770.
11. Payrastre, B., Nievers, M., Boonstra, J., Breton, M., Verkleij, A. J., and Van Bergenen Henegouwen, P. M. (1992). A differential location of phosphoinositide kinases, diacylglycerol kinase, and phospholipase C in the nuclear matrix. *J. Biol. Chem.* **267**, 5078–5084.
12. Cocco, L., Martelli, A., Gilmour, R. S., Ognibene, A., Manzoli, F., and Irvine, R. (1988). Rapid changes in phospholipid metabolism in the nuclei of Swiss 3t3 cells induced by treatment of the cells with insulin-like growth factor I. *Biochem. Biophys. Res. Commun.* **154**, 1266–1272.
13. Divecha, N., Banfic, H., Treagus, J., Vann, L., Irvine, R., and D'Santos, C. (1997). Nuclear diacylglycerol, the cell cycle, the enzymes and a red herring (or how we can to love phosphatidylcholine). *Biochem.l Soc. Trans.* **25**, 571–575.
14. Zhao, K., Wang, W., Rando, O., Xue, Y., Swiderek, K., Kuo, A., and Crabtree, G. (1998). Rapid and phosphoinositol-dependent binding of the SWI/SNK-like BAF complex to chromatin after T lymphocyte receptor signaling. *Cell* **95**, 625–636.
15. York, J. D., Odom, A. R., Murphy, R., Ives, E. A., and Wente, S. R. (1999). A phospholipase C-dependent inositol polyphosphate kinase pathway required for efficient mRNA export. *Science* **285**, 96–100.
16. Odom, A. R., Stahlberg, A., Wente, S. R., and York, J. D. (2000). A role for nuclear inositol 1,4,5-trisphosphate kinase in transcriptional control. *Science* **287**, 2026–2029.
17. Saiardi, A., Caffrey, J. J., Snyder, S. H., and Shears, S. B. (2000). Inositol polyphosphate multikinase (ArgRIII) determines nuclear mRNA export in *Saccharomyces cerevisiae*. *FEBS Lett.* **468**, 28–32.
18. Saiardi, A., Erdjument-Bromage, H., Snowman, A. M., Tempst, P., and Snyder, S. H. (1999). Synthesis of diphosphoinositol pentakisphosphate by a newly identified family of higher inositol polyphosphate kinases. *Curr. Biol.* **9**, 1323–1326.
19. Saiardi, A., Caffrey, J. J., Snyder, S. H., and Shears, S. B. (2000). The inositol hexakisphosphate kinase family. Catalytic flexibility and function in yeast vacuole biogenesis. *J. Biol. Chem.* **275**, 24686–24692.
20. Verbsky, J. W., Wilson, M. P., Kisseleva, M. V., Majerus, P. W., and Wente, S. R. (2002). The synthesis of inositol hexakisphosphate: characterization of human inositol 1,3,4,5,6-pentakisphosphate 2-kinase. *J. Biol. Chem.* [epub ahead of print].
21. Feng, Y., Wente, S. R., and Majerus, P. W. (2001). Overexpression of the inositol phosphatase SopB in human 293 cells stimulates cellular chloride influx and inhibits nuclear mRNA export. *Proc. Natl. Acad. Sci. USA* **98**, 875–879.
22. Dubois, E., Bercy, J., and Messenguy, F. (1987). Characterization of two genes, ARGRI and ARGRIII required for specific regulation of arginine metabolism in yeast. *Mol. Gen. Genet.* **207**, 142–148.
23. Messenguy, F. and Dubois, E. (1993). Genetic evidence for a role for MCM1 in the regulation of arginine metabolism in *Saccharomyces cerevisiae*. *Mol. Cell. Biol.* **13**, 2586–2592.
24. Dubois, E., Dewaste, V., Erneux, C., and Messenguy, F. (2000). Inositol polyphosphate kinase activity of Arg82/ArgRIII is not required for the regulation of the arginine metabolism in yeast. *FEBS Lett.* **486**, 300–304.
25. Carman, G. M. and Henry, S. A. (1999). Phospholipid biosynthesis in the yeast *Saccharomyces cerevisiae* and interrelationship with other metabolic processes. *Prog Lipid Res.* **38**, 361–399.
26. Huang, K. N. and Symington, L. S. (1995). Suppressors of a *Saccharomyces cerevisiae* pkc1 mutation identify alleles of the phosphatase gene PTC1 and of a novel gene encoding a putative basic leucine zipper protein. *Genetics* **141**, 1275–1285.
27. Luo, H. R., Saiardi, A., Yu, H., Nagata, E., Ye, K., and Snyder, S. H. (2002). Inositol pyrophosphates are required for DNA hyperrecombination in protein kinase c1 mutant yeast. *Biochemistry* **41**, 2509–2515.
28. Dubois, E., Scherens, B., Vierendeels, F., Ho, M. M., Messenguy, F., and Shears, S. B. (2002). In *Saccharomyces cerevisiae*, the inositol polyphosphate kinase activity of Kcs1p is required for resistance to salt stress, cell wall integrity, and vacuolar morphogenesis. *J Biol Chem.* **277**, 23755–23763.
29. Hanakahi, L. A., Bartlet-Jones, M., Chappell, C., Pappin, D., and West, S. C. (2000). Binding of inositol phosphate to DNA-PK and stimulation of double-strand break repair. *Cell* **102**, 721–729.
30. Ma, Y. and Lieber, M. R. (2002). Binding of inositol hexakisphosphate (IP6) to Ku but not to DNA-PKcs. *J. Biol. Chem.* **277**, 10756–10759.
31. Hanakahi, L. A. and West, S. C. (2002). Specific interaction of IP6 with human Ku70/80, the DNA-binding subunit of DNA-PK. *EMBO J.* **21**, 2038–2044.
32. Communi, D., Vanweyenberg, V., and Erneux, C. (1995). Molecular study and regulation of D-myo-inositol 1,4,5-trisphosphate 3-kinase. *Cell Signal.* **7**, 643–650.
33. Wilson, M. P. and Majerus, P. W. (1996). Isolation of inositol 1,3, 4-trisphosphate 5/6-kinase, cDNA cloning, and expression of recombinant enzyme. *J. Biol. Chem.* **271**, 11904–11910.
34. Saiardi, A., Nagata, E., Luo, H. R., Sawa, A., Luo, X., Snowman, A. M., and Snyder, S. H. (2001). Mammalian inositol polyphosphate multikinase synthesizes inositol 1,4,5-trisphosphate and an inositol pyrophosphate. *Proc. Natl. Acad. Sci. USA* **98**, 2306–2311.
35. Zhang, T., Caffrey, J. J., and Shears, S. B. (2001). The transcriptional regulator, Arg82, is a hybrid kinase with both monophosphoinositol and diphosphoinositol polyphosphate synthase activity. *FEBS Lett.* **494**, 208–212.

CHAPTER 160

Ins(1,3,4,5,6)P$_5$: A Signal Transduction Hub

Stephen B. Shears
*Laboratory of Signal Transduction,
National Institute of Environmental Health Sciences,
Research Triangle Park, North Carolina*

Introduction

All nucleated cells contain approximately 15 to 50 μM Ins(1,3,4,5,6)P$_5$ [1]. In this review, I will illustrate how Ins-(1,3,4,5,6)P$_5$ serves a number of signaling roles, both by itself, and also as a precursor pool for other physiologically active inositol polyphosphates (Fig. 1).

For example, Ins(1,3,4,5,6)P$_5$ dephosphorylation by a receptor-regulated 1-phosphatase [2] generates Ins(3,4,5,6)P$_4$, which inhibits CaMKII-dependent activation of a family of Cl$^-$ channels in the plasma membrane [3–5]. This carefully controlled regulation of ion channel conductance, through a dynamic balance between competing stimulatory and inhibitory signals, permits a high degree of signal amplification. Enhancement of the signaling process is aided by the precipitous dose-response curve that describes the highly cooperative manner with which Ins(3,4,5,6)P$_4$ inhibits Cl$^-$ channels [3,4,6]. Specificity is another of the hallmarks of an efficient cellular signal; this is certainly the case here. Cl$^-$ channels are unaffected by Ins(1,3,4)P$_3$, Ins(3,4,5)P$_3$, Ins(3,4,6)P$_3$, Ins(4,5,6)P$_3$, Ins(3,5,6)P$_3$, Ins(1,3,4,6)P$_4$, Ins(1,3,4,5)P$_4$, Ins(1,4,5,6)P$_4$, and Ins(1,3,4,5,6)P$_5$ [2–4,7]. The efficacy of Ins(3,4,5,6)P$_4$ (IC$_{50}$ = 3–7 μM) reflects a physiologically relevant concentration range (1–10 μM; [8]).

Ca^{2+} also directly activates some Cl$^-$ channels, independently of CaMKII; this effect of Ca^{2+} is also blocked by Ins(3,4,5,6)P$_4$ in certain situations [7] although not in others [5]. Both CaMKII- and Ca^{2+}-regulated Cl$^-$ channels regulate salt and fluid secretion [8,9], cell volume homeostasis [10], and electrical excitability in neurones and smooth muscle [11]. Recently [12] we showed that at least one member of a molecularly distinct Cl$^-$ channel family, ClC, is also inhibited by Ins(3,4,5,6)P$_4$. These channels are located in secretory vesicles, endosomes, and lysosomes, where they act as a charge shunt that facilitates functionally indispensable vesicle acidification by ATP-driven H$^+$ pumps [13]. Inhibition of insulin granule acidification by Ins(3,4,5,6)P$_4$ attenuates Ca^{2+}-dependent insulin secretion [12]. We anticipate that further cellular functions for Ins(3,4,5,6)P$_4$ will emerge from its effect upon vesicle acidification. Thus, receptor-activated Ins(1,3,4,5,6)P$_5$ dephosphorylation regulates a versatile range of physiological activities that depend upon Cl$^-$ channel activity.

The Ins(1,3,4,5,6)P$_5$ 1-phosphatase is of particular interest because it is reversible *in vivo* [2]; in fact, this enzyme was originally identified as an Ins(3,4,5,6)P$_4$ 1-kinase [14]. Furthermore, the 1-phosphate group that is removed from Ins(1,3,4,5,6)P$_5$ can be directly transferred to the 6-OH of Ins(1,3,4)P$_3$ [2]. By accepting this phosphate group, Ins(1,3,4)P$_3$ enhances the Ins(1,3,4,5,6)P$_5$ 1-phosphatase activity [2]. This is how PLC-initiated increases in levels of Ins(1,3,4)P$_3$ elevate Ins(3,4,5,6)P$_4$ levels [15]. Thus, the 1-kinase and 1-phosphatase activities of a single enzyme switches Ins(3,4,5,6)P$_4$ signaling on and off. These opposing reactions offer an alternative to general doctrine that intracellular signals are regulated by integrating multiple, distinct phosphatases and kinases [16].

A different role for Ins(1,3,4,5,6)P$_5$ concerns its interaction with PTEN, a tumor suppressor [17]. PTEN has classically been recognized as a PtdIns(3,4,5)P$_3$ 3-phosphatase that downregulates the lipid's enhancement of cell proliferation and Akt-dependent cell survival [17]. My laboratory

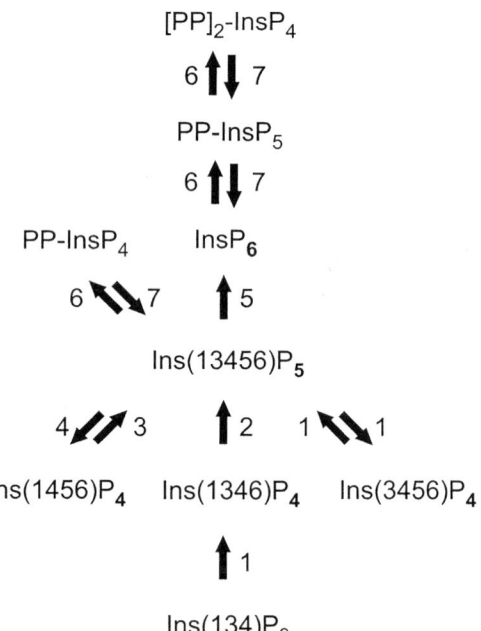

Figure 1 Enzymes that synthesize and metabolize Ins(1,3,4,5,6)P_5. (1) The multifunctional Ins(1,3,4)P_3 6-kinase/Ins(3,4,5,6)P_4 1-kinase/Ins(1,3,4,5,6)P_5 1-phosphatase [2]. (2) Ins(1,3,4,6)P_4 5-kinase [1]. (3) Ins(1,4,5,6)P_4 3-kinase [1,21,23]. (4) PTEN [18] and MIPP [26]. (5) Ins(1,3,4,5,6)P_5 2-kinase [27]. (6) Diphosphoinositol polyphosphate synthase (a.k.a. "InsP$_6$ kinase") [30,31]. (7) Diphosphoinositol polyphosphate phosphatase [33].

recently discovered that PTEN is also a high-affinity Ins(1,3,4,5,6)P_5 3-phosphatase [18]. Competition from soluble Ins(1,3,4,5,6)P_5 will temper the ability of PTEN to bind and dephosphorylate PtdIns(3,4,5)P_3 and further restrict PTEN's already weak protein phosphatase activity [19]. Ins(1,3,4,5,6)P_5 may therefore be viewed as "clamping" PTEN activity, the significance being that overall regulation of a signaling system is much tighter, and permits greater amplification, if it has to be de-inhibited (i.e. when PTEN escapes Ins(1,3,4,5,6)P_5) as well as activated (i.e. when PTEN locates PtdIns(3,4,5)P_3). Ins(1,3,4,5,6)P_5 that is dephosphorylated by PTEN will be replenished by Ins(1,4,5,6)P_4 3-kinase activity; we discovered this kinase over 10 years ago, and we also demonstrated the dynamic nature of Ins(1,3,4,5,6)P_5 3-phosphatase/Ins(1,4,5,6)P_4 3-kinase metabolic cycling *in vivo* [1,20]. Perhaps this cycling is simply the metabolic "price" for regulation of PTEN. Alternately, this cycle may itself have specific functions, for example in the nucleus, since that is where most mammalian Ins(1,4,5,6)P_4 3-kinase is located [21], together with some PTEN [22]. Furthermore, in yeast there is evidence that Ins(1,4,5,6)P_4 synthesis regulates transcription [23]. Salmonella's SopB and SopE virulence factors activate rapid Ins(1,3,4,5,6)P_5 hydrolysis to Ins(1,4,5,6)P_4 as an obligatory part of the process by which the bacteria activate host cell Rac/Rho GTPases, which promotes cellular invasion [24,25]. There is a mammalian enzyme (MIPP; multiple inositol polyphosphate phosphatase) that further dephosphorylates Ins(1,4,5,6)P_4 to Ins(1,4,5)P_3 [26]. Thus, Ins(1,3,4,5,6)P_5 potentially provides a PLC-independent source of Ins(1,4,5)P_3, although to date, only *Dictyostelium* MIPP is proven to generate Ins(1,4,5)P_3 in this manner [26].

Ins(1,3,4,5,6)P_5 is also phosphorylated by a 2-kinase [27], yielding InsP$_6$, to which a number of diverse functions have been attributed, but unfortunately, *in vitro* experiments with InsP$_6$ provide many opportunities for nonphysiological artifacts, so the significance of many of these studies has been criticized [28]. This is why genetic manipulations of InsP$_6$ levels inside cells are more likely to uncover useful information. Finally, there are enzymes that convert Ins(1,3,4,5,6)P_5 to a diphosphorylated derivative (PP-InsP$_4$) [29]. This kinase family generally receives more attention for phosphorylating InsP$_6$ [30], but Ins(1,3,4,5,6)P_5 is also a substrate [31]. Indeed, metabolic cycling between Ins(1,3,4,5,6)P_5 and PP-InsP$_4$ is at least as extensive in intact cells as is the cycling between InsP$_6$ and its diphosphorylated derivatives (PP-InsP$_5$ and [PP]$_2$-InsP$_4$ [29]). All of the diphosphorylated inositol phosphates are considered "high-energy" phosphate donors that apparently regulate vesicle trafficking and possibly other energy-demanding processes [31,32].

References

1. Oliver, K. G., Putney, J. W., Jr., Obie, J. F., and Shears, S. B. (1992). The interconversion of inositol 1,3,4,5,6-pentakisphosphate and inositol tetrakisphosphates in AR4-2J cells. *J. Biol. Chem.* **267**, 21528–21534.
2. Ho, M. W., Yang, X., Carew, M. A., Zhang, T., Hua, L., Kwon, Y.-U., Chung, S.-K., Adelt, S., Vogel, G., Riley, A. M., Potter, B. V. L., and Shears, S. B. (2002). Regulation of Ins(3456)P4 signaling by a reversible kinase/phosphatase. *Curr. Biol.* **12**, 477–482.
3. Xie, W., Kaetzel, M. A., Bruzik, K. S., Dedman, J. R., Shears, S. B., and Nelson, D. J. (1996). Inositol 3,4,5,6-tetrakisphosphate inhibits the calmodulin-dependent protein kinase II-activated chloride conductance inT84 colonic epithelial cells. *J. Biol. Chem.* **271**, 14092–14097.
4. Ho, M. W. Y., Shears, S. B., Bruzik, K. S., Duszyk, M., and French, A. S. (1997). Inositol 3,4,5,6-tetrakisphosphate specifically inhibits a receptor-mediated Ca^{2+}-dependent Cl$^-$ current in CFPAC-1 cells. *Am. J. Physiol.*, **272**, C1160–C1168.
5. Ho, M. W. Y., Kaetzel, M. A., Armstrong, D. L., and Shears, S. B. (2001). Regulation of a human chloride channel: a paradigm for integrating input from calcium, CaMKII and Ins(3,4,5,6)P_4. *J. Biol. Chem.* **276**, 18673–18680.
6. Xie, W., Solomons, K. R. H., Freeman, S., Kaetzel, M. A., Bruzik, K. S., Nelson, D. J., and Shears, S. B. (1998). Regulation of Ca^{2+}-dependent Cl$^-$ conductance in T84 cells: cross-talk between Ins(3,4,5,6)P_4 and protein phosphatases. *J. Physiol. (London)*, **510**, 661–673.
7. Ismailov, I. I., Fuller, C. M., Berdiev, B. K., Shlyonsky, V. G., Benos, D. J., and Barrett, K. E. (1996). A biologic function for an "orphan" messenger: D-*myo*-Inositol 3,4,5,6-tetrakisphosphate selectively blocks epithelial calcium-activated chloride current. *Proc. Nat. Acad. Sci. USA* **93**, 10505–10509.
8. Vajanaphanich, M., Schultz, C., Rudolf, M. T., Wasserman, M., Enyedi, P., Craxton, A., Shears, S. B., Tsien, R. Y., Barrett, K. E., and Traynor-Kaplan, A. E. (1994). Long-term uncoupling of chloride secretion from intracellular calcium levels by Ins(3,4,5,6)P_4. *Nature*, **371**, 711–714.
9. Carew, M. A., Yang, X., Schultz, C., and Shears, S. B. (2000). Ins(3,4,5,6)P_4 inhibits an apical calcium-activated chloride conductance in polarized monolayers of a cystic fibrosis cell-line. *J. Biol. Chem.*, **275**, 26906–26913.
10. Nilius, B., Prenen, J., Voets, T., Eggermont, J., Bruzik, K. S., Shears, S. B., and Droogmans, G. (1998). Inhibition by inositoltetrakisphosphates of calcium- and volume-activated Cl$^-$ currents in macrovascular endothelial cells. *Pflügers Arch. Eur. J. Physiol.* **435**, 637–644.

11. Frings, S., Reuter, D., and Kleene, S. J. (2000). Neuronal Ca^{2+}-activated Cl^- channels—homing in on an elusive channel species. *Prog. Neurobiol.* **60**, 247–289.
12. Renström, E., Ivarsson, R., and Shears, S. B. (2002). Ins(3,4,5,6)P4 inhibits insulin granule acidification and fusogenic potential. *J. Biol. Chem.*, **277**. In press.
13. Nishi,T. and Forgac, M. (2002). The vacuolar (H^+)-ATPases—nature's most versatile proton pumps. *Nat. Rev. Mol. Cell Biol.* **3**, 94–103.
14. Stephens, L. R., Hawkins, P. T., Morris, A. J., and Downes, P. C. (1988). L-*myo*-Inositol 1,4,5,6-tetrakisphosphate (3-hydroxy)kinase. *Biochem. J.*, **249**, 283–292.
15. Yang, X., Rudolf, M., Yoshida, M., Carew, M. A., Riley, A. M., Chung, S.-K., Bruzik, K. S., Potter, B. V. L., Schultz, C., and Shears,S. B. (1999). Ins(1,3,4)P_3 acts *in vivo* as a specific regulator of cellular signaling by Ins(3,4,5,6)P_4. *J. Biol. Chem.*, **274**, 18973–18980.
16. Woscholski, R. and Parker, P. J. (2000). Inositol phosphatases: constructive destruction of phosphoinositides and inositol phosphates. In S. Cockcroft (Ed.), *Biology of Phosphoinositides*, pp. 320–338. Oxford University Press, Oxford.
17. Di Cristofano, A. and Pandolfi, P. P. (2000). The multiple roles of PTEN in tumor suppression. *Cell* **100**, 387–390.
18. Caffrey, J. J., Darden, T., Wenk, M. R., and Shears, S. B. (2001). Expanding coincident signaling by PTEN through its inositol 1,3,4,5,6-pentakisphosphate 3-phosphatase activity. *FEBS Lett.*, **499**, 6–10.
19. Myers, M. P., Stolarov, J. P., Eng, C., Li, J., Wang, S. I., Wigler, M. H., Parsons, R., and Tonks, N. K. (1997). PTEN, the tumor suppressor from human chromosome 10q23, is a dual specificity phosphatase. *Proc. Nat. Acad. Sci. USA* **94**, 9052–9057.
20. Menniti, F. S., Oliver, K. G., Nogimori, K., Obie, J. F., Shears, S. B., and Putney, J. W., Jr. (1990). Origins of *myo*-inositol tetrakisphosphates in agonist-stimulated rat pancreatoma cells. Stimulation by bombesin of *myo*-inositol (1,3,4,5,6) pentakisphosphate breakdown to *myo*-inositol (3,4,5,6) tetrakisphosphate. *J. Biol. Chem.* **265**, 11167–11176.
21. Nalaskowski, M. M., Deschermeier, C., Fanick, W., and Mayr, G. W. (2002). The human homologue of yeast ArgRIII protein is an inositol phosphate multikinase with predominantly nuclear localization. *Biochem. J.* In press.
22. Perren, A., Komminoth, P., Saremaslani, P., Matter, C., Feurer, S., Lees, J. A., Heitz, P. U., and Eng, C. (2000). Mutation and expression analysis reveal differential subcellular compartmentalization of PTEN in endocrine pancreatic tumors compared to normal islet cells. *Am. J. Pathol.* **157**, 1097–1103.
23. Odom, A. R., Stahlberg, A., Wente, S. R., and York, J. D. (2000). A role for nuclear inositol 1,4,5-trisphosphate kinase in transcriptional control. *Science* **287**, 2026–2029.
24. Eckmann, L., Rudolf, M. T., Ptasznik, A., Schultz, C., Jiang, T., Wolfson, N., Tsien, R., Fierer, J., Shears, S. B., Kagnoff, M. F., and Traynor-Kaplan, A. (1997). D-*myo*-inositol 1,4,5,6-tetrakisphosphate produced in human intestinal epithelial cells in response to *Salmonella* invasion inhibits phosphoinositide 3-kinase signaling pathways. *Proc. Nat. Acad. Sci. USA* **94**, 14456–14460.
25. Zhou, D., Chen, L.-M., Hernandez, L., Shears, S. B., and Galán, J. E. (2001). A *Salmonella* inositol polyphosphatase acts in conjunction with other bacterial effectors to promote host-cell actin cytoskeleton rearrangements and bacterial internalization. *Mol. Microbiol.*, **39**, 248–259.
26. Van Dijken, P., de Haas, J.-R., Craxton, A., Erneux, C., Shears, S. B., and van Haastert, P. J. M. (1995). A novel, phospholipase C-independent pathway of inositol 1,4,5-trisphosphate formation in Dictyostelium and rat liver. *J. Biol. Chem.* **270**, 29724–29731.
27. Verbsky, J. W., Wilson, M. P., Kisseleva, M. V., Majerus, P. W., and Wente, S. R. (2002). The synthesis of inositol hexakisphosphate: characterization of human inositol 1,3,4,5,6-pentakisphosphate 2-kinase. *J. Biol. Chem.* **277**, in press.
28. Shears, S. B. (2001). Assessing the omnipotence of inositol hexakisphosphate. *Cell. Signal.* **13**, 151–158.
29. Menniti, F. S., Miller, R. N., Putney, J. W., Jr., and Shears, S. B. (1993). Turnover of inositol polyphosphate pyrophosphates in pancreatoma cells. *J. Biol. Chem.*, **268**, 3850–3856.
30. Saiardi, A., Erdjument-Bromage, H., Snowman, A., Tempst, P., and Snyder, S. H. (1999). Synthesis of diphosphoinositol pentakisphosphate by a newly identified family of higher inositol polyphosphate kinases. *Curr. Biol.* **9**, 1323–1326.
31. Saiardi, A., Caffrey, J. J., Snyder, S. H., and Shears, S. B. (2000). The inositol hexakisphosphate kinase family: catalytic flexibility, and function in yeast vacuole biogenesis. *J. Biol. Chem.* **275**, 24686–24692.
32. Dubois, E., Scherens, B., Vierendeels, F., Ho, M. W. Y., Messenguy, F., and Shears, S. B. (2002). In *Saccharomyces cerevisiae*, the inositol polyphosphate kinase activity of Kcs1p is required for resistance to salt stress, cell wall integrity and vacuolar morphogenesis. *J. Biol. Chem.*, **277**, 23755–23763.
33. Safrany, S. T., Caffrey, J. J., Yang, X., Bembenek, M. E., Moyer, M. B., Burkhart, W. A., and Shears, S. B. (1998). A novel context for the "MutT" module, a guardian of cell integrity, in a diphosphoinositol polyphosphate phosphohydrolase. *EMBO J.* **17**, 6599–6607.

CHAPTER 161

Phospholipase D

Paul C. Sternweis
*Department of Pharmacology, University of Texas,
Southwestern Medical Center at Dallas,
Dallas, Texas*

Introduction

Phospholipase D (PLD) activity is found throughout the biological world, and enzymes have been characterized from a broad spectrum of organisms. Phosphatidic acid is a key molecule in the metabolic pathways for phospholipid synthesis and degradation. One pathway for the production of this lipid is hydrolysis of glycerophospholipids by PLD enzymes, which produce phosphatidic acid (PA) and the associated base or headgroup.

Numerous hormones, growth factors, neurotransmitters, and other cellular stimuli regulate the generation of PA by PLD in mammalian cells. This has led to an emerging role for the mammalian enzymes and PA in signal transduction. Phosphatidic acid is hypothesized to act as a second messenger through direct interaction with a variety of targets, which are discussed elsewhere [1,2]. In addition, PA is a precursor for formation of diacylglycerol (DAG) and lysophosphatidic acid (LPA). DAG can act as a second messenger to regulate the activity of protein kinase C (PKC) isozymes; LPA can function as an autocoid or paracrine through interaction with receptors in the edg family.

This brief overview summarizes the properties and functions of PLD enzymes with a focus on the mammalian proteins. More detailed information can be found in several reviews that provide excellent depth [1] and earlier perspectives [2–4], as well as description of the enzymes in plants [5,6] and yeast [7].

Structural Domains and Requirements for Activity

Two mammalian PLD isozymes have been identified and characterized at the molecular and biochemical levels. A schematic representation of the domain structure of these and several PLD enzymes from other organisms is presented in Fig. 1A. The phospholipase D enzymes belong to a larger family of proteins that includes several endonucleases, cardiolipin synthases, and phosphatidylserine synthases. These enzymes are characterized by the presence of HKD motifs (consensus sequence, HxKxxxxD). Structures have been determined for two members of this superfamily, the PLD from *Streptomyces* sp. Strain PMF [8] (MMDB, 15995; PDB, 1FO1) and the dimer of the endonuclease, Nuc, from *Salmonella typhimurium* [9] (MMDM, 11347; PDB, 1BYR). The structures clearly show that two HKD motifs come together to form a single functional catalytic site with the two histidines as likely nucleophiles for catalysis. Whereas most of the enzymes in this superfamily contain both HKD motifs in a single polypeptide, Nuc contains only a single motif and dimerizes to form its active site (Fig. 1B). Mutations of conserved residues in the HKD motifs have indicated their requirement for catalytic activity in the mammalian enzymes. The C-terminus is a second region required for activity by the mammalian enzymes. Evidence from Liu and colleagues [10] indicates that C-terminal residues and especially the C-terminal α-carboxyl group were required for activity. The actual role of these residues in the catalytic reaction is unknown.

Figure 1A identifies other potential regulatory features that distinguish PLD enzymes. The classical plant PLDs, represented by the enzyme from maize, contain an N-terminal C2 domain, which provides for binding of Ca^{2+} and phospholipids [6]. In contrast, the mammalian enzymes contain putative PX and PH domains that may be important for interaction with regulatory molecules. The presence of the latter domains extends to PLDs from many organisms, including *D. melanogaster*, *C. elegans*, and yeast. Recently, two PLD

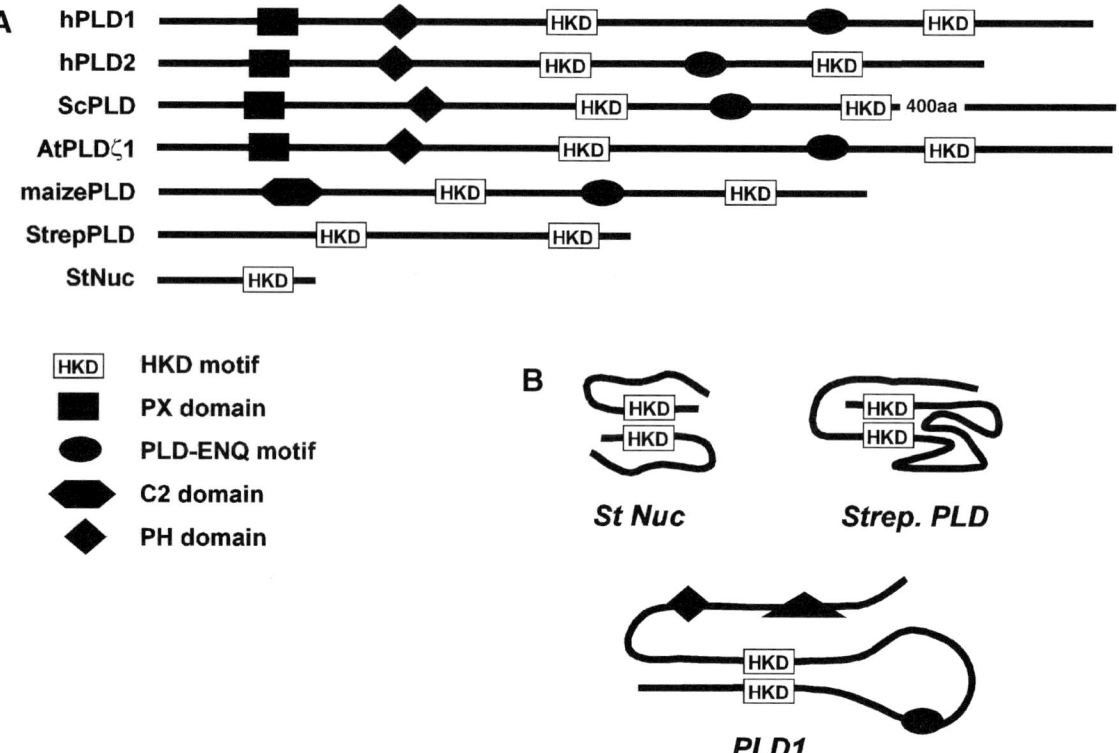

Figure 1 A. Schematic representation of several members of the PLD superfamily: hPLD1, human PLD1 (GI:4505873); hPLD2, human PLD2 (GI:20070141); ScPLD, SPO14 gene from *Saccharomyces cerevisiae* (GI:1174406); AtPLDζ1, isozyme from *Arabidopsis thaliana* (GI:20139230); maizePLD, PLDα1 from *Zea mays* (GI:2499708); StrepPLD, enzyme from *Streptomyces septatus* (GI:15823702); StNuc, nuclease from *Salmonella typhimurium* (GI:6435643). B. Dual HKD motifs cooperate to form a single active site. In the case of Nuc, which only contains one HKD motif, the protein dimerizes to form an active catalytic site.

enzymes that closely resemble the mammalian structural paradigm were identified in *Arabidopsis thaliana* (see AtPLDζ1, Fig. 1). This departure from the classical plant PLDs suggests a much broader scope for regulation of PLD activity in plants [11].

Catalysis: Mechanism and Measurement

It has been appreciated for some time that the enzymatic cleavage carried out by PLD is a two-step process (Fig. 2, see [12] for details). The preferred substrate for the mammalian enzymes is phosphatidylcholine (PC) [13]. The initial reaction involves the formation of a phosphatidylated enzyme with the concomitant release of choline. This attachment is presumably to one of the histidines of the HKD motifs. Under physiological conditions, water is used to release phosphatidic acid and regenerate the active enzyme. A common substrate for assessment of PLD activity *in vitro* is [^{3}H-choline]-PC; the reaction is easily followed by measurement of released choline. Optimal assay procedures, which use phospholipid vesicles containing the labeled PC, have been described in detail [14].

Primary alcohols are much better acceptors for the phosphatidic acid than water. In solutions containing 1% ethanol, the primary product of PLD is phosphatidylethanol rather than PA. This property, referred to as transphosphatidylation, is unique to PLD enzymes and is exploited to measure PLD activity *in vivo*. The pool of PC in cultured cells can be preferentially labeled with [^{3}H]-myristate or labeled lysophosphatidylcholine. The exposure of cells to ethanol or lower concentrations of n-butanol results in the formation of the labeled phosphatidyl alcohol, which can be uniquely distinguished after separation by thin later chromatography or other resolving techniques. An extensive discussion of this and other methods to measure PLD activity is available [13].

Modification of Mammalian PLDs

Two types of modification have been observed in the mammalian enzymes. Two cysteines (Cys240 and Cys241) in the PH domain of PLD1 can be acylated [15]. Elimination of this acylation in PLD1 [15] and PLD2 [16] results in altered cellular location and modest reduction in membrane association. However, a definitive role of this modification in physiological regulation remains to be elucidated.

The PLD isozymes can be phosphorylated by PKCα, *in vitro*, and on Ser/Thr and Tyr, *in vivo*. While these modifications are putative mechanisms for modulation of PLD

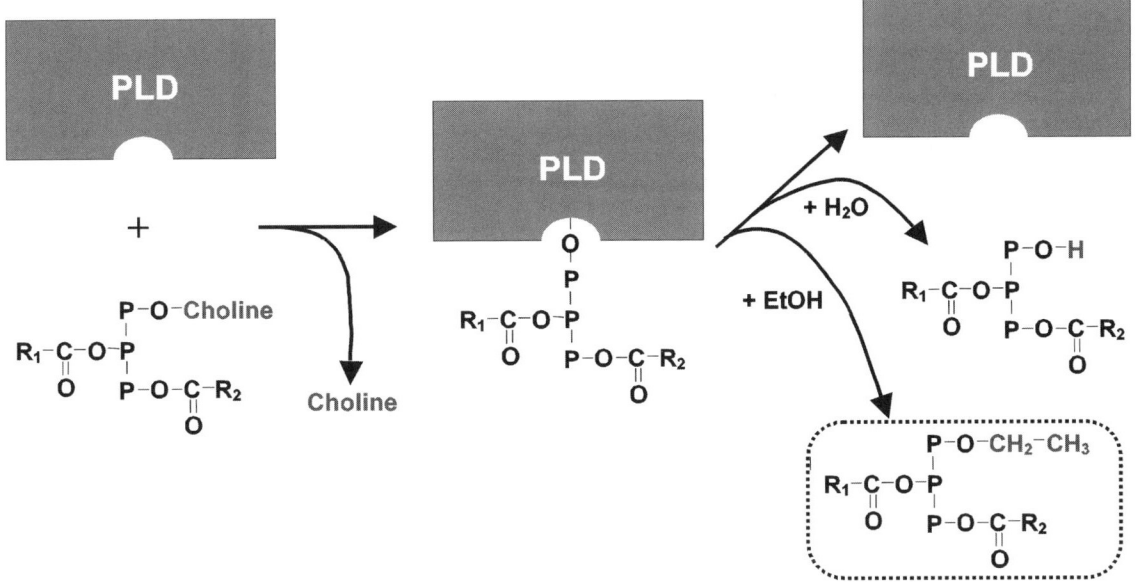

Figure 2 Catalytic mechanism for PLD. Hydrolysis of PC involves two steps. The first cleavage releases choline while forming an intermediate phosphatidylated enzyme. Subsequent hydrolysis with water yields the normal product, phosphatidic acid. In the presence of low concentrations of primary alcohols, the reaction can be shunted to form the phosphatidyl alcohol.

activity by G-protein-coupled receptors and growth factors, the evidence for stimulation of PLD activity by direct phosphorylation is not clear [1].

Regulatory Inputs for Mammalian PLD

Phospholipase D activity in mammalian cells can be stimulated by numerous hormones that act through G-protein-coupled receptors, growth factor receptors, and other stimuli (see [1,2] for surveys). Potential mechanisms for regulation include at least four direct activators that have been identified and characterized, *in vitro*; these are two families of monomeric GTPases (Arf and Rho), protein kinase C (PKC), and phosphatidylinositol 4,5-bisphosphate (PIP_2) (Fig. 3). The lipid PIP_2 was originally identified as a key component of substrate vesicles used to assay the mammalian enzyme [17,18]. Subsequently, this lipid was shown to be an efficacious activator of both PLD1 and PLD2, SPO14 from yeast, and most recently two isoforms of PLD from *Arabidopsis* [11]. While many reviews and papers have stated that this lipid is a cofactor or essential for phospholipase activity, the mammalian PLD1 is clearly active and regulated by Arf in the absence of PIP_2 [14]. Thus, this lipid should be considered a bonafide regulator of the enzyme. Phosphatidylinositol 3,4,5-trisphosphate has also been shown to activate PLD, but the low abundance of this lipid in cells suggest this is unlikely to be a physiological mechanism.

Members of the Arf family of monomeric GTPases were the first identified protein regulators of PLD activity [17,19]. Although they are potent activators of PLD1, Arfs provide only slight modulation of wild type PLD2 [20]; this becomes more efficacious if the N-terminus of PLD2 is truncated [21]. The physiological relevance of this is not known. A second family of monomeric GTPases, the Rho proteins, can also directly activate PLD1 but not PLD2. All subgroups of the Rho family (Rho, Rac, and Cdc42) have proven effective in this function. For both Arf and Rho proteins, regulation of PLD activity occurs through the activated form of the GTPase and thus downstream of the regulation of these proteins.

A more traditional pathway for regulation of PLD is via the classical forms of PKC. It had been noted for some time that direct stimulators of PKC (e.g. phorbol myristate acetate, PMA) increased the activity of PLD in cells. The unusual feature of this regulation, when assessed *in vitro*, is that direct stimulation of PLD by activated PKCα does not require phosphorylation, but rather may occur through the regulatory domain [22]. There is evidence, however, that phosphorylation mechanisms may be required *in vivo* [1].

An intriguing property of these activators is their synergistic action when combined *in vitro* [22,23]. Such action is especially noted among the protein activators, as well as in conjunction with PIP_2. This finding suggests that the most efficacious stimulation of PLD activity *in vivo* may require the action of multiple regulatory pathways.

Regulatory Pathways

Evidence has been presented to implicate these stimulatory molecules in the regulation of PLD activity in cells. Extensive discussions are found elsewhere [1,2]. One clear mechanism for G-protein-coupled receptors is through the stimulation of phospholipase Cβ and activation of protein kinase C (Fig. 3). Similarly, growth factors can stimulate PKC via PLCγ.

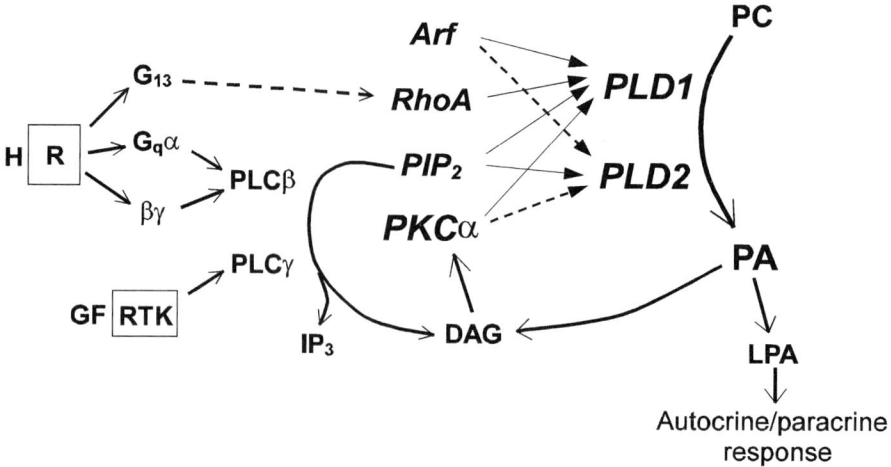

Figure 3 Potential pathways for regulation of mammalian PLD isozymes. See text for discussion of interactions and pathways.

Use of inhibitors of PKC and other strategies to reduce PKC activity in cells is generally effective for attenuation of PLD activity due to direct stimulation of PKC with PMA. In contrast, hormone responses are often retained. The use of PKC to stimulate PLD is intriguing because it potentially leads to an autocatalytic cycle in which the product of PLD activity could lead to further activation of PKC via conversion of PA to DAG. The ineffectiveness of PKC inhibitors in many systems (including some hormones that stimulate PLC activity) indicates that this regulation is more complex and that other mechanisms for hormonal stimulation are operative.

Evidence for the role of Rho proteins in regulation of PLD activity derives largely from the use of C3 toxin from *C. botulinum* and expression of dominant negative forms of the GTPases (summarized in [1]). The interpretation of these data is complicated by the potential regulation of PIP_2 synthesis by Rho proteins and the potential for nonspecific effects of disruption of cytoskeletal architecture through attenuation of these GTPases. The Rho proteins offer an attractive mechanism for regulation of PLD by receptors coupled to the G_{12} and G_{13} proteins. At this time, evidence for this regulation by endogenous receptors and G proteins is lacking.

Physiological regulation of PLD by PIP_2 is still an open question. Various studies, especially in permeabilized cells, that correlate PLD activity with manipulations to vary PIP_2 are suggestive [1,2]. However, the multiple actions of PIP_2 in cells suggests that global change in the concentration of PIP_2 is an unlikely regulatory mechanism. Recent studies that show increased activity of PLD2 when coexpressed with Type1α PIPkinase and the potential association between this kinase and the PLDs [24] support hypotheses that regulation could occur through localized and coordinated changes in the lipid.

The emerging picture predicts that several pathways regulate PLD activity *in vivo*. The primary pathways used in any one system will probably depend on the cell type, the stimulus used, and the local and temporal environment, such as the state of other pathways being utilized in the cells.

Physiological Function of PA

The number of stimuli that regulate PLD activity strongly indicate the importance of PA in signal tranduction. The known activators of PLD have given rise to hypotheses that include roles in basic hormone signaling (PKC), vesicle trafficking (Arf), cytoskeletal rearrangements (Rho), and exocytosis (Arf and Rho) [1,2]. Yet the various roles proposed are largely unproven.

Evidence for prolonged production of DAG via PLD activity is clear, and downstream regulation is probable. This pathway offers a mechanism for regulation by DAG that is independent of Ca^{2+} and selective for a spectrum of PKC isozymes different from those activated by the action of phospholipase C. The potential role of PA in facilitation of membrane rearrangements required for vesicle budding and fusion is attractive. Thus, production of PA, and subsequent products, DAG or LPA, could profoundly affect curvature and stability of the membranes involved. However, the evidence for these events is controversial. Finally, numerous proteins have been shown to interact with or be affected directly by PA *in vitro* [1,2]. These represent potential targets for regulation by the lipid, but definitive demonstration of these putative pathways has been elusive.

Localization of PLD

Phospholipase D activity is found primarily in particulate fractions of mammalian cells and tissues. Subcellular distribution of PLD isozymes within the cell is still in question and has been inferred largely from studies with exogenous expression of tagged proteins [1,2]. In several studies, overexpression of PLD1 resulted in localization to perinuclear regions, frequently with a punctate appearance. In contrast, overexpressed PLD2 was found in the plasma membrane and potential endosomal vesicles near the surface of cells. One report, which examined endogenous PLD1, found the enzyme enriched in the Golgi apparatus, but diffuse staining also indicated a more diverse distribution [25].

Future Directions

Which PLD isozyme is responsible for activity observed in response to hormones? PLD1 was initially favored because of its reponsiveness *in vitro* to regulatory molecules in the known hormone pathways (PKC and Rho in particular). Yet the presumed main site for regulation of PLD activity by hormones is the plasma membrane, and by this criterion, localization studies suggest PLD2 as the potential target. Studies with overexpressed enzymes demonstrate that PLD2 is responsive to activation by PKC in the cellular environment and can enhance stimulations observed with some hormonal stimuli. In contrast, overexpressed PLD1 can be unresponsive to agonists that stimulate putative activators of the enzyme. The discordance between regulation observed *in vitro* and *in vivo* indicates there is still much to be learned about the molecular mechanisms of PLD regulation.

In addition to understanding the responsiveness of individual isozymes, basic questions about the putative roles of PA and its metabolites abound. Fundamental to this are clear determinations of the specific regulation of putative effector proteins and the potential role for these lipids as mediators of membrane restructuring. As for the enzyme itself, little is known about its three-dimensional structure or the molecular mechanisms by which multiple activators can synergistically stimulate its activity. A clear understanding of these mechanisms *in vitro* should facilitate investigation of hormonal pathways *in vivo*.

Acknowledgments

This effort was supported in part by grants from the NIH (GM31954) and the Robert A Welch foundation.

References

1. Exton, J. H. (2002). *Rev. Physiol. Biochem. Pharmacol.* **144**, 1–94.
2. Cockcroft, S. (2001). *Cell Mol. Life Sci.* **58**, 1674–1687.
3. Singer, W. D., Brown, H. A., and Sternweis, P. C. (1997). *Annu. Rev. Biochem.* **66**, 475–509.
4. Frohman, M. A., Sung, T. C., and Morris, A. J. (1999). *Biochim. Biophys. Acta* **1439**, 175–186.
5. Munnik, T. and Musgrave, A. (2001). *Science STKE.* **2001**, E42.
6. Pappan, K. and Wang, X. (1999). *Biochim. Biophys. Acta* **1439**, 151–166.
7. Rudge, S. A. and Engebrecht, J. (1999). *Biochim. Biophys. Acta* **1439**, 167–174.
8. Leiros, I., Secundo, F., Zambonelli, C., Servi, S., and Hough, E. (2000). *Structure. Fold. Des.* **8**, 655–667.
9. Stuckey, J. A. and Dixon, J. E. (1999). *Nat. Struct. Biol.* **6**, 278–284.
10. Liu, M. Y., Gutowski, S., and Sternweis, P. C. (2001). *J. Biol. Chem.* **276**, 5556–5562.
11. Qin, C. and Wang, X. (2002). *Plant Physiol.* **128**, 1057–1068.
12. Waite, M. (1999). *Biochim. Biophys. Acta* **1439**, 187–197.
13. Morris, A. J., Frohman, M. A., and Engebrecht, J. (1997). *Anal. Biochem.* **252**, 1–9.
14. Jiang, X., Gutowski, S., Singer, W. D., and Sternweis, P. C. (2002). *Methods Enzymol.* **345**, 328–334.
15. Sugars, J. M., Cellek, S., Manifava, M., Coadwell, J., and Ktistakis, N. T. (1999). *J. Biol. Chem.* **274**, 30023–30027.
16. Xie, Z., Ho, W. T., and Exton, J. H. (2002). *Biochim. Biophys. Acta* **1580**, 9–21.
17. Brown, H. A., Gutowski, S., Moomaw, C. R., Slaughter, C., and Sternweis, P. C. (1993). *Cell* **75**, 1137–1144.
18. Brown, H. A. and Sternweis, P. C. (1995). *Methods Enzymol.* **257**, 313–324.
19. Cockcroft, S., Thomas, G. M., Fensome, A., Geny, B., Cunningham, E., Gout, I., Hiles, I., Totty, N. F., Truong, O., and Hsuan, J. J. (1994). *Science* **263**, 523–526.
20. Lopez, I., Arnold, R. S., and Lambeth, J. D. (1998). *J. Biol. Chem.* **273**, 12846–12852.
21. Sung, T. C., Altshuller, Y. M., Morris, A. J., and Frohman, M. A. (1999). *J. Biol.Chem.* **274**, 494–502.
22. Singer, W. D., Brown, H. A., Jiang, X., and Sternweis, P. C. (1996). *J. Biol. Chem.* **271**, 4504–4510.
23. Hammond, S. M., Jenco, J. M., Nakashima, S., Cadwallader, K., Gu, Q., Cook, S., Nozawa, Y., Prestwich, G. D., Frohman, M. A., and Morris, A. J. (1997). *J. Biol. Chem.* **272**, 3860–3868.
24. Divecha, N., Roefs, M., Halstead, J. R., D'Andrea, S., Fernandez-Borga, M., Oomen, L., Saqib, K. M., Wakelam, M. J., and D'Santos, C. (2000). *EMBO J.* **19**, 5440–5449.
25. Freyberg, Z., Sweeney, D., Siddhanta, A., Bourgoin, S., Frohman, M., and Shields, D. (2001). *Mol. Biol. Cell* **12**, 943–955.

CHAPTER 162

Diacylglycerol Kinases

M. K. Topham and S. M. Prescott
*The Huntsman Cancer Institute and
Department of Internal Medicine,
University of Utah, Salt Lake City, Utah*

Introduction

Many signaling cascades are initiated by phospholipase C (PLC) isozymes. One product of this reaction is diacylglycerol (DAG), a prolific second messenger that activates proteins involved in a variety signaling cascades. The protein kinase Cs (PKCs) are the best-characterized DAG-activated proteins, but diacylglycerol also activates other proteins [1], including RasGRP and two guanine nucleotide exchange factors (GEFs), CalDAG GEFs I and III. The chimaerins, which are GTPase-activating proteins (GAPs) for Rac, and the *Unc-13* gene product from *Caenorhabditis elegans* also bind to DAG. Because it can associate with a diverse set of proteins, DAG potentially activates numerous signaling cascades. Thus, its accumulation needs to be strictly regulated. Diacylglycerol kinases (DGKs), which phosphorylate DAG, are widely considered to be responsible for terminating diacylglycerol signaling [2,3]. But the product of the DGK reaction, phosphatidic acid (PA), also can be a signal: it can activate phosphatidylinositol 4-phosphate 5-kinases and PKCζ, participates in recruiting Raf1 to the plasma membrane, and is involved in vesicle trafficking. Because they manipulate both DAG and PA signaling, the DGKs can regulate numerous signaling events.

The DGK Family

DGKs have been identified in most organisms that have been studied, and it appears that they have gained specialization in more complex species. For example, bacteria express only one DGK, which is an integral membrane protein capable of phosphorylating DAG and other lipids such as ceramide. This DGK does not appear to have structural elements that allow regulation of its activity, indicating that the limiting factor is access to its substrates. With the exception of yeast, in which no DGKs have been identified, higher organisms appear to have several DGKs that can be grouped by common structural elements into five subfamilies. The mammalian DGKs are the best characterized, and nine of them have been identified [2,3]. All of these DGKs have two common structural features: a catalytic domain and at least two C1 domains, which are thought to bind diacylglycerol (Fig. 1). Other structural domains, which form the basis of the five subtypes, appear to have regulatory roles. For example, type I DGKs, α, β, and γ, have calcium-binding EF hand motifs that make these enzymes more active in the presence of calcium. Type II DGKs, δ and η, have pleckstrin homology (PH) domains near their amino termini. DGKδ also has a sterile alpha motif (SAM) at its carboxy terminus that may allow protein-protein interactions. The only type III DGK, ε, does not have identifiable structural motifs outside its C1 and catalytic domains. It is interesting that this is the only DGK that displays specificity toward acyl chains of DAG—it dramatically prefers DAGs with an arachidonoyl group at the *sn*-2 position. Type IV DGKs, ζ and ι, have a motif enriched in basic amino acids that acts as a nuclear localization signal and is a substrate for conventional PKCs. This motif is homologous to the phosphorylation site domain of the myristoylated alanine rich C kinase substrate (MARCKS) protein. Type IV DGKs also have four ankyrin repeats at their carboxy termini that may be sites of protein-protein interactions. The only type V DGK, θ, is distinguished by three C1 domains, a PH domain, and a Ras-association (RA) domain. To date, no binding partners for the PH and RA domains have been identified. Based on their structural diversity, the mammalian DGKs likely have specific roles dictated by their unique structural motifs.

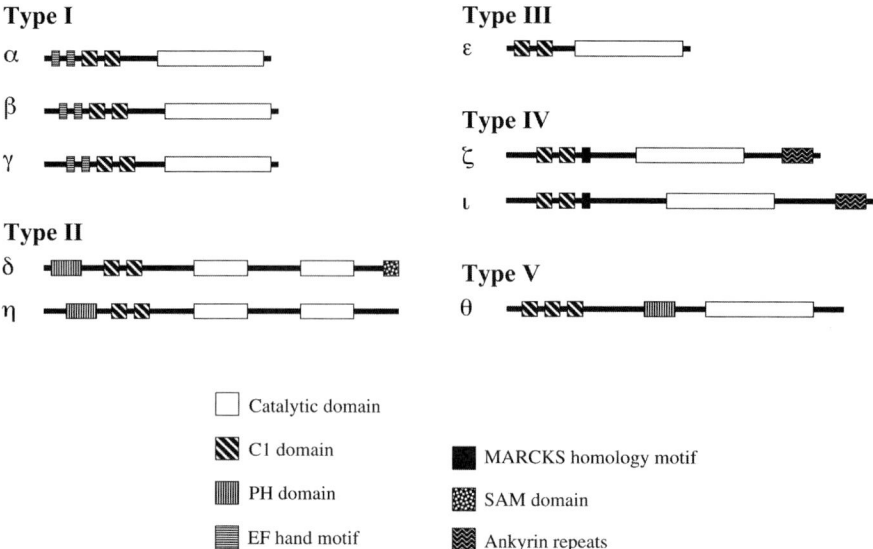

Figure 1 The nine members of the mammalian diacylglycerol kinase family are grouped by sequence homology into five subtypes. Shown are protein motifs common to several DGKs.

Regulation of DGKs

Activation of DGKs is complex, requiring translocation to a membrane compartment as well as binding to appropriate cofactors. Additional regulation of their activity occurs by posttranslational modifications. This complexity allows tissue- or cell-specific regulation depending on the availability of cofactors and the type of stimulus that the cell receives. Tissue-specific alternative splicing of DGKs β and ζ—and probably other isotypes—adds additional opportunities for regulation.

DGKα demonstrates the complex regulation of DGKs. In T lymphocytes, it translocates to at least two membrane compartments depending upon the agonist used to activate the cells. For example, stimulation of T cells with IL-2 causes DGKα to translocate from the cytosol to a perinuclear region [4]. But activation of the antigen receptor causes DGKα to translocate to the plasma membrane [5]. At the membrane, the activity of DGKα can be modified by the availability of several co-factors. Calcium is known to bind to the EF hand structures and stimulates DAG kinase activity *in vitro*, and lipids modify its activity: phosphatidylserine and sphingosine activate DGKα *in vitro* and probably *in vivo*. Finally, DGKα can be phosphorylated by several protein kinases, including PKC isoforms and Src. Although the consequences of these phosphorylations are not clear, evidence suggests that phosphorylation by Src enhances DAG kinase activity [6]. Thus, several events can modify the activity of DGKα, and combinations of them likely allow titration of its activity depending upon the cellular context.

Like DGKα, other DGK isotypes appear to be regulated by access to DAG through membrane translocation and by the availability of lipid or protein co-factors. Members of each DGK subfamily are likely to be regulated similarly, although there probably are subtle differences between subfamily members due to tissue-specific expression patterns, unique binding partners, alternative splicing, and subcellular localization. Type II DGKs, for example, have a PH domain, and this motif in DGKδ binds to phosphatidylinositols. Sakane *et al.* using an *in vitro* system, could not detect activation of DGKδ by phosphatidylinositols [2], suggesting that binding to these lipids instead provides a localization cue. The activity of types III and IV DGKs can be modified by lipids: DGKε is inhibited by phosphatidylinositols and by phosphatidylserine, while type IV DGKs are activated by phosphatidylserine. Type IV DGKs are also strongly regulated by subcellular translocation. They are imported into the nucleus, which requires their MARCKS homology domain, a nuclear localization signal that is regulated by PKC phosphorylation [7]. There is also evidence that the syntrophin family of scaffolding proteins regulates nuclear import of DGKζ by associating with its carboxy-terminal PDZ binding domain to sequester DGKζ in the cytoplasm [8]. And we have observed that DGKζ has a strong nuclear export signal (M. K. Topham, unpublished). Thus, nuclear accumulation of type IV DGKs is exquisitely regulated, suggesting an important nuclear function for these isozymes. Finally, DGKθ, a type V DGK, can be regulated through its association with active RhoA [3]. Binding to RhoA completely inhibits its DAG kinase activity and is the only example of regulation of a DGK through a protein-protein interaction.

Paradigms of DGK Function

Although there is substantial information regarding their regulation, little is known of the biologic functions of the individual DGKs. But recently, a few paradigms have emerged.

Spatial Regulation of DAG Signaling

Evidence suggests that DGKs selectively associate with and regulate DAG-activated proteins. Van der Bend et al. [2] initially tested this concept by either initiating spatially restricted DAG synthesis through receptor activation or by causing nonspecific, global DAG generation with exogenous PLC. They observed DAG kinase activity—measured by generation of PA—following receptor activation, but not after treating the cells with exogenous PLC. Their data demonstrate that DGKs are active only in spatially restricted compartments following physiologic generation of DAG. Recently, more specific examples of spatially restricted DAG kinase function have emerged. We found that DGKζ associated with RasGRP, a guanine nucleotide-exchange factor for Ras [9]. Their association was enhanced in the presence of phorbol esters, which are slowly metabolized DAG analogues. Since RasGRP requires DAG to function, we hypothesized that DGKζ associated with it to spatially metabolize DAG and consequently to regulate the function of RasGRP. Indeed, we found that kinase-dead DGKζ did not affect the function of RasGRP. Demonstrating the specificity of this regulation, we found that five other DAG kinases did not significantly inhibit RasGRP activity. Thus, our data demonstrate that in some cases DAG kinase activity is spatially restricted and serves to specifically regulate DAG-activated proteins.

Nurrish et al. presented another example of compartmentalized DGK function [3]. They isolated a *Caenorhabditis elegans* strain resistant to serotonin-induced inhibition of locomotion. The mutated gene responsible for the effect, *dgk-1*, is homologous to DGKθ. Their data suggested a model in which serotonin signaling activated DGK-1 to reduce local accumulation of DAG. This resulted in inhibition of UNC-13, a protein activated by DAG that may mediate acetylcholine release. Thus, their results represent another example of compartmentalized DGK function that modulates the activity of a DAG-activated protein.

Regulation of Signaling Through Fatty Acid Specificity

Inositol phospholipids, including PIP_2, a precursor of DAG, are enriched in arachidonate at the *sn-2* position. Some DAG targets, including PKCs, are specifically activated by diacylglycerol-containing unsaturated fatty acids, such as arachidonate. So to maintain the integrity of some DAG-activated signaling cascades, it is important that phosphatidylinositols maintain a proper fatty acid composition. Because DGKε selectively phosphorylates arachidonoyl-DAG, the first step in resynthesis of phosphatidylinositols, DGKε may be responsible for their enrichment with arachidonate. Inositol lipid signaling is an important component of neuronal transmission. In a collaborative effort we examined seizure susceptibility in mice with targeted deletion of DGKε [10] and found that the null mice were resistant to seizures induced by electroconvulsive shock. Examination of brain lipids revealed that compared to wild-type mice, DGKε-deficient mice had reduced levels of arachidonate in both PIP_2 and DAG. This lipid profile demonstrated a critical role for DGKε in maintaining a proper balance of arachidonate-enriched inositol phospholipids. Thus, through its selectivity for arachidonoyl-DAG, DGKε regulates lipid signaling events and, consequently, seizure susceptibility.

Nuclear DGKs

There is a nuclear phosphatidylinositol cycle regulated separately from its plasma membrane/cytosolic counterpart [7]. DAG is present in nuclear preparations and appears to fluctuate with the cell cycle, but the specific pattern of its accumulation is not clear because of the many different methods used to isolate nuclei. DAG kinases have also been observed in nuclei and appear to have a prominent role there. Some DGKs, like DGKs α, ζ, and ι, translocate to the nucleus, while others, like DGKθ, are constitutively located there [3]. These DGKs are confined to specific compartments within the nucleus. For example, both DGKθ and DGKζ appear in a speckled pattern within the nucleus, while DGKα associates with the nuclear envelope [3]. This compartmentalization suggests that DGK isotypes have specific roles in the nucleus. Movement of proteins in the nucleus largely occurs by random diffusion, so overexpression of one DGK isotype may interfere with the function of another DGK. This fact, combined with the lack of specific DGK inhibitors, has made it difficult to study the nuclear function of the different DGK isotypes. But they are likely to affect nuclear signaling either by terminating DAG signals or by generating PA. For example, in T lymphocytes, the PA produced by nuclear DGKα appears to be necessary for IL-2-mediated progression to S phase of the cell cycle [4]. Conversely, nuclear DGKζ inhibits exit from G1 phase of the cell cycle by metabolizing DAG [7]. These data indicate both the complexity and importance of lipid signaling and DGK function in the nucleus.

Visual Signal Transduction

A clear role for DGK function has been demonstrated in *Drosophila*. A mutant strain, *rdgA*, undergoes rapid retinal degeneration after birth. The defect is due to a deficiency in retinal DGK activity because of a point mutation that inactivates dDGK2, a DAG kinase very similar to mammalian type IV DGKs. Although there are differences in photoreceptor signaling between *Drosophila* and vertebrates, these observations indicate that DGKs may be functionally important in the vertebrate retina. Three mammalian DGK isoforms, γ, ε, and ι, have been definitively localized to the retina, but their functions there have not yet been identified [2]. However, there is evidence of light-dependent activation of PIP_2 hydrolysis and generation of PA in vertebrate retina, indicating a role there for DGK activity.

Conclusions

Diacylglycerol kinases are expressed in all multicellular organisms that have been studied. Their structural diversity and complexity indicate that they are functionally important in a variety of cellular signaling events. Since they can affect both DAG and PA signals, DGK activity plays a central role in many lipid signaling pathways.

References

1. Ron, D. and Kazanietz, M. G. (1999). New insights into the regulation of protein kinase C and novel phorbol ester receptors. *FASEB J.* **13**, 1658–1676.
2. Topham, M. K. and Prescott, S. M. (1999). Mammalian diacyglycerol kinases, a family of lipid kinases with signaling functions. *J. Biol. Chem.* **274**, 11447–11450.
3. van Blitterswijk, W. J. and Houssa, B. (2000). Properties and functions of diacylglycerol kinases. *Cell. Signal.* **12**, 595–605.
4. Flores, I., Casaseca, T., Martinez-A. C., Kanoh, H., and Merida, I. (1996). Phosphatidic acid generation through interleukin 2 (IL-2)-induced α-diacylglycerol kinase activation is an essential step in IL-2-mediated lymphocyte proliferation. *J. Biol. Chem.* **271**, 10334–10340.
5. Sanjuan, M. A., Jones, D. R., Izquierdo, M., and Merida, I. (2001). Role of diacylglycerol kinase a in attenuation of receptor signaling. *J. Cell Biol.* **153**, 207–219.
6. Cutrupi, S., Baldanzi, G., Gramaglia, D., Maffe, A., Schaap, D., Giraudo, E., van Blitterswijk, W. J., Bussolino, F., Comoglio, P. M., and Graziani, A. (2000). Src-mediated activation of α-diacylglycerol kinase is required for hepatocyte growth factor-induced cell motility. *EMBO J.* **19**, 4614–4622.
7. Topham, M. K., Bunting, M., Zimmerman, G. A., McIntyre, T. M., Blackshear, P. J., and Prescott, S. M. (1998). Protein kinase C regulates the nuclear localization of diacylglcyerol kinase-ζ. *Nature* **394**, 697–700.
8. Hogan, A., Shepherd, L., Chabot, J., Quenneville, S., Prescott, S. M., Topham, M. K. and Gee, S. H. (2001). Interaction of γ1-syntrophin with diacylglycerol kinase-ζ. Regulation of nuclear localization by PDZ interactions. *J Biol Chem.* **276**, 26526–26533.
9. Topham, M. K. and Prescott, S. M. (2001). Diacylglycerol kinase ζ regulates ras activation by a novel mechanism. *J. Cell Biol.* **152**, 1135–1143.
10. Rodriguez de Turco, E. B., Tang, W., Topham, M. K., Sakane, F., Marcheselli, V. L., Chen, C., Taketomi, A., Prescott, S. M., and Bazan, N. G. (2001). Diacylglycerol kinase ε regulates seizure susceptibility and long-term potentiation through arachidonoyl-inositol lipid signaling. *Proc. Natl. Acad. Sci. USA* **98**, 4740–4745.

Sphingosine-1-Phosphate Receptors

Michael Maceyka and Sarah Spiegel
Department of Biochemistry, Medical College of Virginia Campus, Virginia Commonwealth University, Richmond, Virginia

Introduction

The endothelial differentiation gene (EDG) family of G-protein coupled receptors (GPCRs) comprises high-affinity receptors for the lysophospholipids, lysophosphatidic acid (LPA), and sphingosine-1-phosphate (S1P) [1,2]. The homologous EDG receptors are clearly divided into two classes: three that bind LPA (EDG-2/LPA$_1$, EDG-4/LPA$_2$, EDG-7/LPA$_3$) and five that bind S1P (EDG-1/S1P$_1$, EDG-3/S1P$_3$, EDG-5/S1P$_2$, EDG-6/S1P$_4$, EDG-8/S1P$_5$). Molecular modeling and targeted mutagenesis have shown that S1PRs and LPARs use very similar motifs for binding of ligands, with one amino acid primarily determining the difference in specificity [3]. As several excellent reviews have recently appeared on LPARs and our studies have concentrated on dissecting molecular signaling pathways regulated by S1P [4,5], we have focused in this chapter on lipid signaling to and through S1PRs.

S1P is formed by sphingosine kinase (SphK), of which there are two known mammalian isoforms (for review, see [6]). SphKs are evolutionarily conserved and catalyze the ATP-dependent phosphorylation of the primary hydroxyl of sphingosine, the common backbone of mammalian sphingolipids. S1P is an interesting molecule that is an intercellular messenger and an intracellular second messenger [7]. This greatly complicates interpretation of results when adding exogenous S1P to cells: Is the response observed due to cell surface receptors, effects on intracellular targets, or both? A preponderance of studies have indicated that many of the biological effects of S1P are mediated by specific S1PRs and the lack of confirmed intracellular targets appears to bolster these claims. However, others have suggested that certain results are better explained by receptor-independent intracellular effects of S1P. First, the well-known S1PRs typically have K_ds in the 2–30 nM range [2,8], whereas effects of S1P on growth and suppression of apoptosis usually require micromolar concentrations [9]. In addition, dihydrosphingosine-1-phosphate (dhS1P), which has the same structure as S1P but only lacks the 4,5-*trans* double bond, binds to and activates all of the S1PRs. However, dhS1P does not mimic the effects of S1P on growth and survival [10], thus suggesting that these effects are likely to be mediated by intracellular actions of S1P.

The S1PRs

S1P was identified as the natural high affinity ligand of S1P$_1$ [2], which was shown to be highly specific, only binding S1P and dhS1P [11,12]. S1P$_1$ is coupled to G$_{\alpha i}$ and G$_{\alpha o}$ [13] but not G$_{\alpha s}$, G$_{\alpha q}$, or G$_{\alpha 12/13}$ [14]. Thus, pertussis toxin, which inhibits G$_{\alpha i/o}$ proteins, is a useful tool for dissecting signaling through S1P$_1$. *s1p$_1$* deleted mice died *in utero* between E12.5 and E14.5 due to massive hemorrhaging [15]. Although vasculogenesis and angiogenesis are normal in the *s1p$_1$*-/- mice, vascular smooth muscle cells failed to completely surround and seal the vasculature, thereby leading to hemorrhage. On a cellular level, the defect was linked to an inability of S1P$_1$ null fibroblasts to migrate toward S1P, likely due to dysfunctional Rac activation, and indicated the important role of S1P/S1P$_1$ signaling in motility.

S1P$_2$ is unique in being the only one of the S1PRs with a significantly poorer affinity for dhS1P than S1P [16]. S1P$_2$ has a wide tissue distribution [17] and a K_d for S1P of 20–30 nM [11]. In addition to G$_{\alpha i/o}$, S1P$_2$ couples to G$_{\alpha q}$ and G$_{\alpha 12/13}$ [14]. S1P$_2$ has been linked to increases in cAMP levels and thus may couple weakly to G$_{\alpha s}$ in some cell types depending on the pattern of expression of both GPCRs and

G proteins [18]. S1P$_2$ regulates diverse signaling pathways, including calcium mobilization, stimulation of NF-κB, and inhibition of Rac-dependent cell migration in certain cell types [19,20]

S1P$_2$ has also been knocked out in mice [21] and in contrast to *s1p$_1$-/-* mice, these mice have no obvious anatomical or physiological phenotypes. It is interesting that mammalian S1P$_2$ is highly homologous to the zebrafish gene *miles apart* [22]. Two inactivating mutations in *miles apart* prevent the normal migration of heart primordia, thus resulting in abnormal cardiac development. The heart precursor cells from the *miles apart* mutants migrated normally when transplanted into wild-type embryos, but wild-type cells failed to migrate in mutant embryos, a result that suggests that the zebrafish S1P$_2$ homologue is required for generating a migration-permissive environment.

S1P$_3$ was shown to be activated by S1P [23] with a K_d of 20–30 nM [11,12] and to couple to $G_{\alpha i/o}$, $G_{\alpha q}$, and $G_{\alpha 12/13}$, but not $G_{\alpha s}$ [14]. S1P$_3$-null mice have been generated and also have no obvious phenotype [24]. S1P$_3$ has been linked to many signaling pathways, including calcium mobilization, stimulation of NF-κB, and NO production [25,26].

The two remaining S1PRs have a more narrow tissue distribution. S1P$_4$ is expressed almost exclusively in lymphoid and hematopoietic cells, as well as in the lung [27]. S1P$_4$ has a K_d for S1P of 12–63 nM, as determined by different groups [28,29], and couples to $G_{\alpha i/o}$. The final S1PR, S1P$_5$, previously named EDG-8 and *nrg-1*, is expressed predominantly in the central nervous system and to a lesser extent in lymphoid tissue [8,30]. S1P$_5$ couples to $G_{\alpha i/o}$ and $G_{\alpha 12/13}$, but not to $G_{\alpha s}$ or $G_{\alpha q}$ [31] and has a K_d for S1P of 2–6 nM [8,31].

S1P Signaling via S1PRs

Intriguing questions concerning lysolipid messengers are what regulates the levels of these amphipathic molecules and how do they get to their target cells? Platelets are known to store S1P and release it upon stimulation (reviewed in [32]). HUVECs and C6 glioma cells release S1P to the extracellular milieu [33,34]. Moreover, even when S1P release from cells is below detectable limits, co-culturing cells expressing S1P$_1$ with cells producing S1P due to overexpression of SphK induced activation of S1P$_1$ on adjacent as well as distant cells, thus indicating either that vanishingly small amounts of S1P are released or that it can be transferred from one cell to another by cell–cell interactions, or both [35]. Thus, S1P can act in an autocrine and/or paracrine manner. In support of this concept, the chemoattractant PDGF recruits SphK to the plasma membrane, where S1PRs are located, and especially to structures known as lamellipodia [36]. Given the importance of lamellipodia and S1PRs in chemotaxis, this finding suggests that S1P is produced and released from the cell in a spatially restricted manner, providing cells with a sense of direction. In addition, a recent report claims that type 1 SphK is secreted from cells in a catalytically active form and may catalyze the formation of SIP at or near the plasma membrane [33]. Further studies are necessary to confirm a role for extracellular SphK.

Transactivation of S1PRs

An intriguing aspect of the *s1p$_1$-/-* phenotype is that it appears to be nearly identical to that of the PDGF-BB and PDGFR-β knockouts [15], as these embryos also die because of a vascular smooth muscle cell migration defect. Because PDGF stimulates SphK and increases S1P [10], it therefore appeared possible that S1P$_1$ and PDGF signaling pathways are linked. Indeed, embryonic fibroblasts from *s1p$_1$-/-* mice, in contrast to wild-type cells, failed to migrate toward both S1P and PDGF [35]. Moreover, enforced expression of S1P$_1$ in HEK 293 cells, which express low basal levels of S1P$_1$, increased their ability to migrate toward PDGF, and antisense ablation of S1P$_1$ significantly inhibited migration toward PDGF [36]. A specific inhibitor of SphK also blocked PDGF-induced motility. Taken together, these results suggest a transactivation pathway linking PDGF though SphK to the autocrine and/or paracrine release of S1P that then stimulates S1P$_1$ to regulate motility. Furthermore, it was independently shown that S1P$_1$ potentiated the response to PDGF in HEK 293 cells overexpressing PDGFR [37]. However, in this case, these effects appeared to be independent of SphK, and it was suggested that PDGFR and S1P$_1$ were tethered in a complex that was activated independently of S1P.

Downstream Signaling from S1PRs

Because the S1PRs are coupled to heterotrimeric G proteins, the types of signals transduced are many and varied, depending on the specific isoforms of G_α and $G_{\beta\gamma}$ that are present. Thus, signals linked to a S1PR in one cell type may not be linked in the same manner in a second cell type. For example, transfection with S1P$_1$ increases S1P-induced calcium mobilization in CHO cells [38] but not in COS-7 cells [39]. Determining which specific S1PR is involved in a particular response is difficult because most cells express multiple S1PRs. To date, S1PR specific agonists or antagonists have not been developed. Thus, to elucidate the role of a particular S1PR, either transfection of receptor negative or knockout cells or antisense approaches have been used. Given the diversity of GPCR signaling, it is not surprising that results from these experiments demonstrate that S1PRs control the major lipid-mediated signaling pathways, as discussed below.

Phospholipase C. Many of the responses linked to S1PR signaling involve increases in intracellular calcium. Generation of the second messenger inositol trisphosphate (IP$_3$) by activation of phospholipase C (PLC) is the major pathway leading to intracellular calcium increases. CHO cells transfected with S1P$_1$, S1P$_2$, S1P$_3$, or S1P$_4$, but not vector controls, had increased IP$_3$ production and calcium release in an

S1P-dependent manner [28]. In contrast, in Jurkat T cells, S1P$_2$ and S1P$_3$, but not S1P$_1$, elicited IP$_3$-mediated calcium responses [25]. On a more physiological level, in HUVECs, which express S1P$_1$, and to a lesser extent S1P$_3$, S1P stimulated nitric oxide (NO) production by calcium-dependent epithelial nitric oxide synthase (eNOS) [40]. NO production was blocked by the PLC inhibitor U73122, the calcium chelator BAPTA-AM, and antisense oligonucleotides to S1P$_1$ or S1P$_3$, thus demonstrating a role for both S1PRs in activation of PLC. Furthermore, fibroblasts from S1P$_3$-null mice, but not littermate controls, failed to activate PLC upon S1P addition [24].

Phospholipase D. Another important lipid second messenger is phosphatidic acid (PA), which is generated by activation of phospholipase D (PLD). Overexpression of S1P$_1$ in HEK 293 or NIH 3T3 cells did not result in activation of PLD [9]. However, in C2C12 skeletal muscle cells, S1P stimulated PLD via either S1P$_1$, S1P$_2$, or S1P$_3$ in a pertussis toxin–sensitive manner [41]. Transfection of either S1P$_1$ or S1P$_2$ in C6 glioma cells conferred S1P-dependent PLD stimulation and PA formation [42]. S1P$_3$ also induced production of PA, specifically through activation of PLD2 in CHO cells [43].

Phosphatidylinositol-3-kinase. Activation of phosphatidylinositol-3-kinase (PI3K) promotes cell survival, cytoskeletal remodeling, and vesicular trafficking [44]. PI3K also promotes activation of the protein kinase Akt in two ways: translocation of Akt to the membrane by binding phosphatidylinositol-3,4-bisphosphate and activation of phosphoinositide-dependent kinases, which phosphorylate and activate Akt (reviewed in [45]). Though the S1PR(s) involved were not identified, S1P induced chemotaxis and angiogenesis of endothelial cells both *in vivo* and *in vitro* in a PI3K- and Akt-dependent manner [46,47]. S1P$_1$ transiently transfected in COS-7 cells led to activation of Akt, which was inhibited by the PI3K inhibitor wortmannin [48]. Further work from this group implicated G$_{\beta\gamma}$ stimulation of the PI3Kβ isoforms in S1P-dependent signaling to PI3K [49]. On a more physiological level, ventricular cardiomyocyte hypertrophy induced by S1P was inhibited by both wortmannin and by S1P$_1$ antibody [50]. S1P$_1$, S1P$_2$, and S1P$_3$ transfected into CHO cells each activated PI3K in response to S1P [43,51]. It is interesting that in this system, S1P$_1$ and S1P$_3$ promoted S1P-induced chemotaxis, while S1P$_2$ inhibited it.

Sphingosine Kinase. S1P has been demonstrated to release calcium from non-IP$_3$ releasable microsomal stores, though the intracellular receptor(s) are unknown [52–54]. Meyer zu Heringdorf and colleagues demonstrated that HEK 293 cells endogenously expressing S1P$_1$, S1P$_2$, and S1P$_3$ mobilized calcium in response to S1P [55]. However, in these cells, PLC was not activated and there was no measurable production of IP$_3$. What is especially interesting, they found that S1P stimulated SphK and S1P production, and the increase in S1P levels, as well as calcium release, was reduced by inhibitors of SphK. S1P production was also completely blocked by pertussis toxin, indicating the involvement of G$_i$-linked GPCRs in the process. Thus, the remarkable observation was made that extracellular S1P regulates intracellular S1P formation [55].

Acknowledgments

This work was supported by research grants from the National Institutes of Health (GM43880 and CA61774) and the Department of the Army (DAMD17-02-1-0060) (SS) and a postdoctoral fellowship (DAMD 7-02-1-0240) to M. M.).

References

1. Hecht, J. H., Weiner, J. A., Post, S. R., and Chun, J. (1996). Ventricular zone gene-1 (vzg-1) encodes a lysophosphatidic acid receptor expressed in neurogenic regions of the developing cerebral cortex. *J. Cell Biol.* **135**, 1071–1083.
2. Lee, M. J., Van Brocklyn, J. R., Thangada, S., Liu, C. H., Hand, A. R., Menzeleev, R., Spiegel, S., and Hla, T. (1998). Sphingosine-1-phosphate as a ligand for the G protein-coupled receptor EDG-1. *Science* **279**, 1552–1555.
3. Parrill, A. L., Wang, D., Bautista, D. L., Van Brocklyn, J. R., Lorincz, Z., Fischer, D. J., Baker, D. L., Liliom, K., Spiegel, S., and Tigyi, G. (2000). Identification of Edg1 receptor residues that recognize sphingosine 1-phosphate. *J. Biol. Chem.* **275**, 39379–39384.
4. Hla, T., Lee, M. J., Ancellin, N., Paik, J. H., and Kluk, M. J. (2001). Lysophospholipids-receptor revelations. *Science* **294**, 1875–1878.
5. Spiegel, S. and Milstien, S. (2002). Sphingosine-1-phosphate a key cell signaling molecule. *J. Biol. Chem.* **277**, 25851–25854.
6. Liu, H., Chakravarty, D., Maceyka, M., Milstien, S., and Spiegel, S. (2002). Sphingosine kinases: a novel family of lipid kinases. *Prog. Nucl. Acid Res.* **71**, 493–511.
7. Maceyka, M., Payne, S. G., Milstien, S., and Spiegel, S. (2002). Sphingosine kinase, sphingosine-1-phosphate, and apoptosis. *Biochim. Biophys. Acta* **1585**, 193–201.
8. Im, D. S., Heise, C. E., Ancellin, N., O'Dowd, B. F., Shei, G. J., Heavens, R. P., Rigby, M. R., Hla, T., Mandala, S., McAllister, G., George, S. R., and Lynch, K. R. (2000). Characterization of a novel sphingosine 1-phosphate receptor, Edg-8. *J. Biol. Chem.* **275**, 14281–14286.
9. Van Brocklyn, J. R., Lee, M. J., Menzeleev, R., Olivera, A., Edsall, L., Cuvillier, O., Thomas, D. M., Coopman, P. J. P., Thangada, S., Hla , T., and Spiegel, S. (1998). Dual actions of sphingosine-1-phosphate: extracellular through the G$_i$-coupled orphan receptor edg-1 and intracellular to regulate proliferation and survival. *J. Cell Biol.* **142**, 229–240.
10. Olivera, A. and Spiegel, S. (1993). Sphingosine-1-phosphate as a second messenger in cell proliferation induced by PDGF and FCS mitogens. *Nature* **365**, 557–560.
11. Van Brocklyn, J. R., Tu, Z., Edsall, L. C., Schmidt, R. R., and Spiegel, S. (1999). Sphingosine 1-phosphate-induced cell rounding and neurite retraction are mediated by the G protein-coupled receptor H218. *J. Biol. Chem.* **274**, 4626–4632.
12. Kon, J., Sato, K., Watanabe, T., Tomura, H., Kuwabara, A., Kimura, T., Tamama, K., Ishizuka, T., Murata, N., Kanda, T., Kobayashi, I., Ohta, H., Ui, M., and Okajima, F. (1999). Comparison of intrinsic activities of the putative sphingosine 1-phosphate receptor subtypes to regulate several signaling pathways in their cDNA-transfected Chinese hamster ovary cells. *J. Biol. Chem.* **274**, 23940–23947.
13. Lee, M.-J., Evans, M., and Hla, T. (1996). The inducible G protein-coupled receptor *edg-1* signals via the G$_i$/mitogen-activated protein kinase pathway. *J. Biol. Chem.* **271**, 11272–11282.
14. Windh, R. T., Lee, M. J., Hla, T., An, S., Barr, A. J., and Manning, D. R. (1999). Differential coupling of the sphingosine 1-phosphate receptors edg-1, edg-3, and H218/Edg-5 to the g(i), g(q), and G(12) families of heterotrimeric G proteins. *J. Biol. Chem.* **274**, 27351–27358.

15. Liu, Y., Wada, R., Yamashita, T., Mi, Y., Deng, C. X., Hobson, J. P., Rosenfeldt, H. M., Nava, V. E., Chae, S. S., Lee, M. J., Liu, C. H., Hla, T., Spiegel, S., and Proia, R. L. (2000). Edg-1, the G protein-coupled receptor for sphingosine-1-phosphate, is essential for vascular maturation. *J. Clin. Invest.* **106**, 951–961.
16. Hla, T. (2001). Sphingosine 1-phosphate receptors. *Prostaglandins* **64**, 135–142.
17. Okazaki, H., Ishizaka, N., Sakurai, T., Kurokawa, K., Goto, K., Kumada, M., and Takuwa, Y. (1993). Molecular cloning of a novel putative G protein-coupled receptor expressed in the cardiovascular system. *Biochem. Biophys. Res. Commun.* **190**, 1104–1109.
18. Gonda, K., Okamoto, H., Takuwa, N., Yatomi, Y., Okazaki, H., Sakurai, T., Kimura, S., Sillard, R., Harii, K., and Takuwa, Y. (1999). The novel sphingosine 1-phosphate receptor AGR16 is coupled via pertussis toxin-sensitive and -insensitive G-proteins to multiple signalling pathways. *Biochem. J.* **337**, 67–75.
19. Meacci, E., Cencetti, F., Formigli, L., Squecco, R., Donati, C., Tiribilli, B., Quercioli, F., Zecchi Orlandini, S., Francini, F., and Bruni, P. (2002). Sphingosine 1-phosphate evokes calcium signals in C2C12 myoblasts via Edg3 and Edg5 receptors. *Biochem. J.* **362**, 349–357.
20. Ryu, Y., Takuwa, N., Sugimoto, N., Sakurada, S., Usui, S., Okamoto, H., Matsui, O., and Takuwa, Y. (2002). Sphingosine-1-phosphate, a platelet-derived lysophospholipid mediator, negatively regulates cellular Rac activity and cell migration in vascular smooth muscle cells. *Circ. Res.* **90**, 325–332.
21. MacLennan, A. J., Carney, P. R., Zhu, W. J., Chaves, A. H., Garcia, J., Grimes, J. R., Anderson, K. J., Roper, S. N., and Lee, N. (2001). An essential role for the H218/AGR16/Edg-5/LP(B2) sphingosine 1-phosphate receptor in neuronal excitability. *Eur. J. Neurosci.* **14**, 203–209.
22. Kupperman, E., An, S., Osborne, N., Waldron, S., and Stainier, D. Y. (2000). A sphingosine-1-phosphate receptor regulates cell migration during vertebrate heart development. *Nature* **406**, 192–195.
23. An, S., Bleu, T., Huang, W., Hallmark, O. G., Coughling, S. R., and Goetzl, E. J. (1997). Identification of cDNAs encoding two G protein-coupled receptors for lysosphingolipids. *FEBS Lett.* **417**, 279–282.
24. Ishii, I., Friedman, B., Ye, X., Kawamura, S., McGiffert, C., Contos, J. J., Kingsbury, M. A., Zhang, G., Heller Brown, J., and Chun, J. (2001). Selective loss of sphingosine 1-phosphate signaling with no obvious phenotypic abnormality in mice lacking its G protein-coupled receptor, LP(B3)/EDG-3. *J. Biol. Chem.* **276**, 33697–33704.
25. An, S., Bleu, T., and Zheng, Y. (1999). Transduction of intracellular calcium signals through G protein-mediated activation of phospholipase C by recombinant sphingosine 1-phosphate receptors. *Mol. Pharmacol.* **55**, 787–794.
26. Siehler, S., Wang, Y., Fan, X., Windh, R. T., and Manning, D. R. (2001). Sphingosine 1-phosphate activates nuclear factor-kappa B through Edg receptors. Activation through Edg-3 and Edg-5, but not Edg-1, in human embryonic kidney 293 cells. *J. Biol. Chem.* **276**, 48733–48739.
27. Gräler, M. H., Bernhardt, G., and Lipp, M. (1998). EDG6, a novel G-protein-coupled receptor related to receptors for bioactive lysophospholipids, is specifically expressed in lymphoid tissue. *Genomics* **53**, 164–169.
28. Yamazaki, Y., Kon, J., Sato, K., Tomura, H., Sato, M., Yoneya, T., Okazaki, H., Okajima, F., and Ohta, H. (2000). Edg-6 as a putative sphingosine 1-phosphate receptor coupling to Ca(2+) signaling pathway. *Biochem. Biophys. Res. Commun.* **268**, 583–589.
29. Van Brocklyn, J. R., Graler, M. H., Bernhardt, G., Hobson, J. P., Lipp, M., and Spiegel, S. (2000). Sphingosine-1-phosphate is a ligand for the G protein-coupled receptor EDG-6. *Blood* **95**, 2624–2629.
30. Glickman, M., Malek, R. L., Kwitek-Black, A. E., Jacob, H. J., and Lee, N. H. (1999). Molecular cloning, tissue-specific expression, and chromosomal localization of a novel nerve growth factor-regulated G-protein-coupled receptor, nrg-1. *Mol. Cell. Neurosci.* **14**, 141–152.
31. Malek, R. L., Toman, R. E., Edsall, L. C., Wong, S., Chiu, J., Letterle, C. A., Van Brocklyn, J. R., Milstien, S., Spiegel, S., and Lee, N. H. (2001). Nrg-1 belongs to the endothelial differentiation gene family of G protein-coupled sphingosine-1-phosphate receptors. *J. Biol. Chem.* **276**, 5692–5699.
32. Yatomi, Y., Ohmori, T., Rile, G., Kazama, F., Okamoto, H., Sano, T., Satoh, K., Kume, S., Tigyi, G., Igarashi, Y., and Ozaki, Y. (2000). Sphingosine 1-phosphate as a major bioactive lysophospholipid that is released from platelets and interacts with endothelial cells. *Blood* **96**, 3431–3438.
33. Ancellin, N., Colmont, C., Su, J., Li, Q., Mittereder, N., Chae, S. S., Steffansson, S., Liau, G., and Hla, T. (2002). Extracellular export of sphingosine kinase-1 enzyme: sphingosine 1-phosphate generation and the induction of angiogenic vascular maturation. *J. Biol. Chem.* **277**, 6667–6675.
34. Vann, L. R., Payne, S. G., Edsall, L. C., Twitty, S., Spiegel, S., and Milstien, S. (2002). Involvement of sphingosine kinase in TNF-alpha-stimulated tetrahydrobiopterin biosynthesis in C6 glioma cells. *J. Biol. Chem.* **277**, 12649–12656.
35. Hobson, J. P., Rosenfeldt, H. M., Barak, L. S., Olivera, A., Poulton, S., Caron, M. G., Milstien, S., and Spiegel, S. (2001). Role of the sphingosine-1-phosphate receptor EDG-1 in PDGF-induced cell motility. *Science* **291**, 1800–1803.
36. Rosenfeldt, H. M., Hobson, J. P., Maceyka, M., Olivera, A., Nava, V. E., Milstien, S., and Spiegel, S. (2001). EDG-1 links the PDGF receptor to Src and focal adhesion kinase activation leading to lamellipodia formation and cell migration. *FASEB J.* **15**, 2649–2659.
37. Alderton, F., Rakhit, S., Choi, K. K., Palmer, T., Sambi, B., Pyne, S., and Pyne, N. J. (2001). Tethering of the platelet-derived growth factor beta receptor to G-protein coupled receptors: a novel platform for integrative signaling by these receptor classes in mammalian cells. *J. Biol. Chem.* **276**, 28578–28585.
38. Okamoto, H., Takuwa, N., Gonda, K., Okazaki, H., Chang, K., Yatomi, Y., Shigematsu, H., and Takuwa, Y. (1998). EDG1 is a functional sphingosine-1-phosphate receptor that is linked via a Gi/o to multiple signaling pathways, including phospholipase C activation, Ca^{2+} mobilization, ras-mitogen-activated protein kinase activation, and adenylate cyclase inhibition. *J. Biol. Chem.* **273**, 27104–27110.
39. Zondag, G. C. M., Postma, F. R., Etten, I. V., Verlaan, I., and Moolenaar, W. H. (1998). Sphingosine 1-phosphate signalling through the G-protein-coupled receptor Edg-1. *Biochem. J.* **330**, 605–609.
40. Kwon, Y. G., Min, J. K., Kim, K. M., Lee, D. J., Billiar, T. R., and Kim, Y. M. (2001). Sphingosine 1-phosphate protects human umbilical vein endothelial cells from serum-deprived apoptosis by nitric oxide production. *J. Biol. Chem.* **276**, 10627–10633.
41. Meacci, E., Vasta, V., Donati, C., Farnararo, M., and Bruni, P. (1999). Receptor-mediated activation of phospholipase D by sphingosine 1-phosphate in skeletal muscle C2C12 cells. A role for protein kinase C. *FEBS Lett.* **457**, 184–188.
42. Sato, K., Ui, M., and Okajima, F. (2000). Differential roles of Edg-1 and Edg-5, sphingosine 1-phosphate receptors, in the signaling pathways in C6 glioma cells. *Brain Res. Mol. Brain. Res.* **85**, 151–160.
43. Banno, Y., Takuwa, Y., Akao, Y., Okamoto, H., Osawa, Y., Naganawa, T., Nakashima, S., Suh, P. G., and Nozawa, Y. (2001). Involvement of phospholipase D in sphingosine 1-phosphate-induced activation of phosphatidylinositol 3-kinase and Akt in Chinese hamster ovary cells overexpressing EDG3. *J. Biol. Chem.* **276**, 35622–35628.
44. Sotsios, Y. and Ward, S. G. (2000). Phosphoinositide 3-kinase: a key biochemical signal for cell migration in response to chemokines. *Immunol. Rev.* **177**, 217–235.
45. Wymann, M. P., Bulgarelli-Leva, G., Zvelebil, M. J., Pirola, L., Vanhaesebroeck, B., Waterfield, M. D., and Panayotou, G. (1996). Wortmannin inactivates phosphoinositide 3-kinase by covalent modification of Lys-802, a residue involved in the phosphate transfer reaction. *Mol. Cell. Biol.* **16**, 1722–1733.
46. Morales-Ruiz, M., Lee, M. J., Zollner, S., Gratton, J. P., Scotland, R., Shiojima, I., Walsh, K., Hla, T., and Sessa, W. C. (2001). Sphingosine 1-phosphate activates Akt, nitric oxide production, and chemotaxis through a Gi protein/phosphoinositide 3-kinase pathway in endothelial cells. *J. Biol. Chem.* **276**, 19672–19677.
47. Rikitake, Y., Hirata, K., Kawashima, S., Ozaki, M., Takahashi, T., Ogawa, W., Inoue, N., and Yokoyama, M. (2002). Involvement of

endothelial nitric oxide in sphingosine-1-phosphate-induced angiogenesis. *Arterioscler. Thromb. Vasc. Biol.* **22**, 108–114.

48. Igarashi, J. and Michel, T. (2000). Agonist-modulated targeting of the EDG-1 receptor to plasmalemmal caveolae. eNOS activation by sphingosine 1-phosphate and the role of caveolin-1 in sphingolipid signal transduction. *J. Biol. Chem.* **275**, 32363–32370.

49. Kou, R., Igarashi, J., and Michel, T. (2002). Lysophosphatidic acid and receptor-mediated activation of endothelial nitric-oxide synthase. *Biochemistry* **41**, 4982–4988.

50. Mazurais, D., Robert, P., Gout, B., Berrebi-Bertrand, I., Laville, M. P., and Calmels, T. (2002). Cell type-specific localization of human cardiac S1P receptors. *J. Histochem. Cytochem.* **50**, 661–670.

51. Okamoto, H., Takuwa, N., Yokomizo, T., Sugimoto, N., Sakurada, S., Shigematsu, H., and Takuwa, Y. (2000). Inhibitory regulation of Rac activation, membrane ruffling, and cell migration by the G protein-coupled sphingosine-1-phosphate receptor EDG5 but not EDG1 or EDG3. *Mol. Cell Biol.* **20**, 9247–9261.

52. Ghosh, T. K., Bian, J., and Gill, D. L. (1990). Intracellular calcium release mediated by sphingosine derivatives generated in cells. *Science* **248**, 1653–1656.

53. Mattie, M., Brooker, G., and Spiegel, S. (1994). Sphingosine-1-phosphate, a putative second messenger, mobilizes calcium from internal stores via an inositol trisphosphate-independent pathway. *J. Biol. Chem.* **269**, 3181–3188.

54. Ghosh, T. K., Bian, J., and Gill, D. L. (1994). Sphingosine 1-phosphate generated in the endoplasmic reticulum membrane activates release of stored calcium. *J. Biol. Chem.* **269**, 22628–22635.

55. Meyer zu Heringdorf, D., Lass, H., Kuchar, I., Lipinski, M., Alemany, R., Rumenapp, U., and Jakobs, K. H. (2001). Stimulation of intracellular sphingosine-1-phosphate production by G-protein-coupled sphingosine-1-phosphate receptors. *Eur. J. Pharmacol.* **414**, 145–154.

CHAPTER 164

SPC/LPC Receptors

Linnea M. Baudhuin[1,2], Yijin Xiao[1], and Yan Xu[1,2,3]

[1]Department of Cancer Biology, Cleveland Clinic Foundation, Cleveland, Ohio,
[2]Department of Chemistry, Cleveland State University, Cleveland, Ohio,
[3]Department of Gynecology and Obstetrics, Cleveland Clinic Foundation,
9500 Euclid Avenue, Cleveland, Ohio

Introduction

Among LPLs, the biological effects and signaling mechanisms of lysophosphatidic acid (LPA), sphingosine-1-phosphate (S1P), and their receptors (LPA_{1-3} and $S1P_{1-5}$) have been studied most extensively [1–4]. The signaling mechanisms of their corresponding choline derivatives, lysophosphatidylcholine (LPC) and sphingosylphosphorylcholine (SPC), however, have been examined to a much lower extent, although their extracellular existence and evidence of their signaling properties have long been recognized.

Addition of a positively charged choline group to the negatively charged phosphate group provides LPC and SPC with zwitterionic and detergent-like properties. In fact, LPC is cell lytic at concentrations >30 μM when bovine serum albumin (BSA) is absent [5]. Moreover, the specific receptors for LPC and SPC were not previously identified. Thus, controversy has arisen as to whether LPC, and possibly SPC, act as specific signaling molecules or molecules modulating cellular functions nonspecifically, and whether their actions are receptor-mediated. This situation has been changed recently with the identification of three G-protein-coupled receptors (GPCRs)—OGR1, GPR4, and G2A—as receptors for LPC and SPC [6–8]. These discoveries provide an intriguing and novel opportunity to study the pathophysiological and functional roles of SPC, LPC, and their receptors.

Physiological and Pathological Functions of LPC and SPC

The potential physiological and pathological functions of LPC and SPC have been recently reviewed [9,10]. While LPA, S1P, LPC, and SPC may share similar, overlapping, or opposing effects in some cellular systems, each of these lipids may also have its own unique functions. For example, all four of these LPLs have been shown to play some role in wound healing and some inflammatory processes [11–15]. LPA and its receptors are involved in nervous system development, and S1P has been implicated in cardiovascular development [12,16,17]. LPC and SPC are implicated more specifically in diseases involving immunological and inflammatory processes, such as atherosclerosis and systemic lupus erythematosus [18–21].

The metabolic pathways involved in synthesis and release of LPC and SPC are closely related to those of LPA and potentially S1P. A lysophospholipase-D (lysoPLD) activity, which directly converts LPC to LPA, has been reported previously [22,23]. We have recently observed that when sterile, cell-free ovarian cancer ascites samples, but not non-malignant ascites, were incubated at 37°C, LPA levels were increased over time (Fig. 1A). The LPA production was completely abolished when EDTA or EGTA was added to the ascites, indicating that the LPA production in ovarian cancer ascites was probably due to a soluble enzymatic activity that requires bivalent metal ions and calcium. It is interesting that during the same time course, LPC levels were decreased (Fig. 1B), suggesting that a lysoPLD-like activity may be responsible for LPA production in ovarian cancer ascites. Furthermore, SPC is a substrate for bacterial PLD (unpublished observations), and thus SPC may be converted to S1P in mammalian cells *in vivo*, although such an endogenous activity has not been identified. These data support the notion that the physiological roles of LPLs may be closely related and intertwined and therefore may play a more complex role *in vivo* than what is observed *in vivo* when a single LPL is tested.

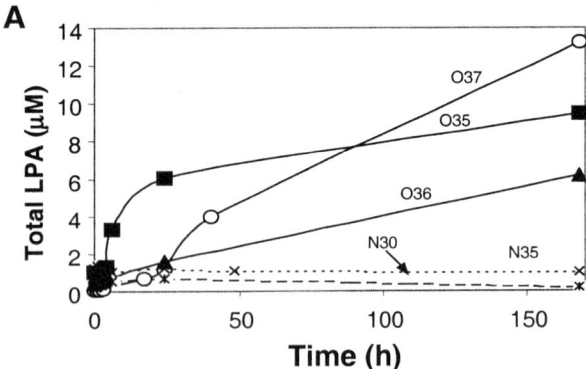

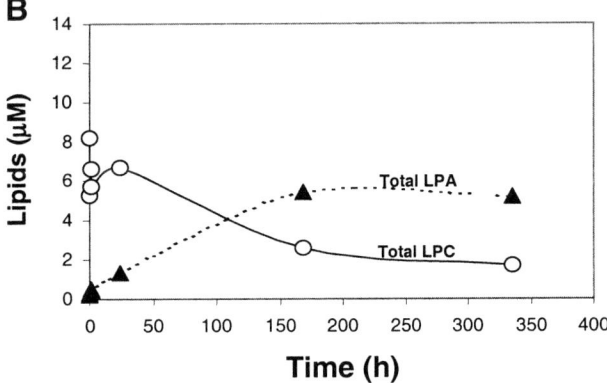

Figure 1 LPA production and LPC degradation in ascites samples. The ascites samples were incubated at 37°C for different durations as indicated, and then quantitatively analyzed for LPA and LPC content by ESI-MS [34,35]. (A) LPA production in ascites from patients with ovarian cancer (O35, O36, and O37) and nonmalignant diseases (N30 and N35). (B) LPC reduction and LPA production in ascites from a patient with ovarian cancer.

Identification of Receptors for SPC and LPC

The identification of receptors for SPC and LPC first began with the cloning of OGR1 from the HEY ovarian cancer cell line by using a PCR-based cloning strategy with primers based on the sequences of receptors for platelet-activating factor (PAF) and thrombin [24]. OGR1 shares approximately 30% sequence homology with the PAF receptor, a finding that indicated that the ligand for OGR1 may be also a lipid molecule and may also contain a choline group. Functional analyses were performed, which provided evidence for OGR1 as the first high-affinity receptor identified for SPC [6]. Similar studies were performed to determine whether SPC and/or LPC are ligands for GPR4 and G2A [7,8]. OGR1, GPR4, and G2A have no or very low affinity to LPA, S1P, PAF, lyso-PAF, lysophosphatidylinositol (LPI), PAF, sphingomyelin, ceramide, psychosine, glucosyl-β1,1′-sphingosine (Glu-Sph), galactosyl-β1,1′-ceramide (Gal-Cer), and lactosyl-β1, 1′-ceramide (Lac-Cer) [6–8]. TDAG8, a fourth related receptor that is 36% homologous to OGR1, has been recently identified as a receptor for a glycosphingolipid, psychosine (galactosyl-β1,1′-sphingosine) [25]. Due to the relative high homologies of these receptors, the potentials exist for TDAG8 to also be a LPC/SPC receptor and/or for OGR1, GPR4, and G2A to be receptors for psychosine.

Although LPA/S1P and SPC/LPC subfamily receptors are GPCRs for structurally related lysolipids, the two receptor subfamilies share little sequence homology and may prefer different G-protein coupling in certain signaling pathways. Evidence supports that three major G protein families (G_i, G_q, and $G_{12/13}$) are coupled to these receptors. Compared to the majority of GPCRs, which employ G_q as a mediator for calcium mobilization, SPC- and LPC-induced calcium release from intercellular stores are mediated through G_i in MCF10A cells, although other cell lines remained to be tested. Furthermore, while ERK activation via GPCRs is mainly mediated through G_i, SPC-induced PI3K and ERK activation via OGR1 appear to be mediated by a PTX-insensitive G protein ([6] and unpublished observations). Nonetheless, these differences are not restricted to LPC/SPC receptors. G_i-mediated calcium release and G_q-mediated PI3K and ERK activation have been reported previously [26–29].

The K_d values for LPC/SPC ligand binding are about one to two orders of magnitude higher than those for LPA/S1P receptors. Likewise, the serum and plasma concentrations of LPC are usually one to two orders of magnitude higher than those of LPA. Thus, the K_d values of their receptors may reflect a physiological adaptation to their concentrations. At the normal physiological concentrations of LPC (5–180 μM), if all of the LPC were in an active form, then its receptors would be saturated, downregulated, and/or desensitized. However, *in vivo*, the functionally available concentration of LPC may be affected by such conditions as percentage of LPC bound to albumin or lipoprotein and compartmentalization (i.e., tissue, cellular, and subcellular distribution) of LPC [30–32]. The concentrations of both S1P and SPC in physiological fluids are in the nM to sub-μM range. The lower affinity of SPC for its receptors may suggest (1) a physiological adaptation for a lower response to SPC; (2) the presence of a different, higher-affinity receptor(s) for SPC, which cannot be ruled out; and (3) these receptors (LPC/SPC subfamily receptors) may have different endogenous ligand(s). These issues remain to be further investigated.

Perspectives

Until now, very limited information has been accumulated regarding the pathophysiological functions of SPC, LPC, and their receptors. G2A-null mice develop a late-onset autoimmune disease [33]. Some of the effects of G2A may be compensated by other LPC receptors. Generation of OGR1- and GPR4-null mice is in progress and will provide important information about the physiological functions of these receptors. Comparative studies between LPA/S1P and LPC/SPC may generate interesting data to advance our understanding of these lipids, since: (1) LPA and S1P are prototypes of bioactive extracellular lipid signaling molecules; (2) LPC and SPC share similar, yet distinct signaling pathways as those induced by LPA and S1P; (3) the metabolic pathways LPA/S1P and LPC/SPC; are linked between and (4) all of these LPLs are present in serum and plasma. LPA, SPC, and

LPC levels are elevated under pathological conditions. It can be foreseen that the identification of their receptors will facilitate our understanding of the roles these LPLs play in physiological and pathological processes.

References

1. Hla, T., Lee, M. J., Ancellin, N., Paik, J. H., and Kluk, M. J. (2001). Lysophospholipids—receptor revelations. *Science* **294**, 1875–1888.
2. Goetzl, E. J. and An, S. (1998). Diversity of cellular receptors and functions for the lysophospholipid growth factors lysophosphatidic acid and sphingosine 1-phosphate. *FASEB J.* **12**, 1589–1598.
3. Spiegel, S. (1999). Sphingosine 1-phosphate: a prototype of a new class of second messengers. *J. Leukoc. Biol.* **65**, 341–344.
4. Moolenaar, W. H. (1999). Bioactive lysophospholipids and their G protein-coupled receptors. *Exp. Cell. Res.* **253**, 230–238.
5. Jalink, K., van Corven, E. J., and Moolenaar, W. H. (1990). Lysophosphatidic acid, but not phosphatidic acid, is a potent $Ca^{2(+)}$-mobilizing stimulus for fibroblasts. Evidence for an extracellular site of action. *J. Biol. Chem.* **265**, 12232–12239.
6. Xu, Y., Zhu, K., Hong, G., Wu, W., Baudhuin, L. M., Xiao, Y., and Damron, D. S. (2000). Sphingosylphosphorylcholine is a ligand for ovarian cancer G-protein-coupled receptor 1. *Nat. Cell Biol.* **2**, 261–267.
7. Zhu, K., Baudhuin, L. M., Hong, G., Williams, F. S., Cristina, K. L., Kabarowski, J. H., Witte, O. N., and Xu, Y. (2001). Sphingosylphosphorylcholine and lysophosphatidylcholine are ligands for the G protein-coupled receptor GPR4. *J. Biol. Chem.* **276**, 41325–41335.
8. Kabarowski, J. H., Zhu, K., Le, L. Q., Witte, O. N., and Xu, Y. (2001). Lysophosphatidylcholine as a ligand for the immunoregulatory receptor G2A. *Science* **293**, 702–705.
9. Chisolm, G. M. III and Chai, Y. (2000). Regulation of cell growth by oxidized LDL. *Free Radic. Biol. Med.* **28**, 1697–1707.
10. Prieschl, E. E. and Baumruker, T. (2000). Sphingolipids: second messengers, mediators and raft constituents in signaling. *Immunol. Today* **21**, 555–560.
11. Lee, H., Goetzl, E. J., and An, S. (2000). Lysophosphatidic acid and sphingosine 1-phosphate stimulate endothelial cell wound healing. *Am. J. Physiol. Cell Physiol.* **278**, C612–C618.
12. Lynch, K. R. and Macdonald, T. L. (2001). Structure activity relationships of lysophospholipid mediators. *Prostaglandins Other Lipid Mediat.* **64**, 33–45.
13. Igarashi, Y. and Yatomi, Y. (1998). Sphingosine 1-phosphate is a blood constituent released from activated platelets, possibly playing a variety of physiological and pathophysiological roles. *Acta Biochim. Pol.* **45**, 299–309.
14. Sun, L., Xu, L., Henry, F. A., Spiegel, S., and Nielsen, T. B. (1996). A new wound healing agent—sphingosylphosphorylcholine. *J. Invest. Dermatol.* **106**, 232–237.
15. Murugesan, G. and Fox, P. L. (1996). Role of lysophosphatidylcholine in the inhibition of endothelial cell motility by oxidized low density lipoprotein. *J. Clin. Invest.* **97**, 2736–2744.
16. Fukushima, N., Ishii, I., Contos, J. J., Weiner, J. A., and Chun, J. (2001). Lysophospholipid receptors. *Annu. Rev. Pharmacol. Toxicol.* **41**, 507–534.
17. Tigyi, G. (2001). Physiological responses to lysophosphatidic acid and related glycero-phospholipids. *Prostaglandins Other Lipid Mediat.* **64**, 47–62.
18. Lusis, A. J. (2000). Atherosclerosis. *Nature* **407**, 233–241.
19. Koh, J. S., Wang, Z., and Levine, J. S. (2000). Cytokine dysregulation induced by apoptotic cells is a shared characteristic of murine lupus. *J. Immunol.* **165**, 4190–4201.
20. Murata, Y., Ogata, J., Higaki, Y., Kawashima, M., Yada, Y., Higuchi, K., Tsuchiya, T., Kawainami, S., and Imokawa, G. (1996). Abnormal expression of sphingomyelin acylase in atopic dermatitis: an etiologic factor for ceramide deficiency? *J. Invest. Dermatol.* **106**, 1242–1249.
21. Sugiyama, E., Uemura, K., Hara, A., and Taketomi, T. (1993). Metabolism and neurite promoting effect of exogenous sphingosylphosphocholine in cultured murine neuroblastoma cells. *J. Biochem (Tokyo)* **113**, 467–472.
22. Tokumura, A., Miyake, M., Nishioka, Y., Yamano, S., Aono, T., and Fukuzawa, K. (1999). Production of lysophosphatidic acids by lysophospholipase D in human follicular fluids. *Biol. Repro.* **61**, 195–199.
23. Tokumura, A., Yamano, S., Aono, T., and Fukuzawa, K. (2000). Lysophosphatidic acids produced by lyosphospholipase D in mammalian serum and body fluid. *Ann. NY Acad. Sci.* **905**, 347–350.
24. Xu, Y. and Casey, G. (1996). Identification of human OGR1, a novel G protein-coupled receptor that maps to chromosome 14. *Genomics* **35**, 397–402.
25. Im, D. S., Heise, C. E., Nguyen, T., O'Dowd, B. F., and Lynch, K. R. (2001). Identification of a molecular target of psychosine and its role in globoid cell formation. *J. Cell Biol.* **16**, 429–434.
26. Ulloa-Aguirre, A. and Conn, P. M. (2000). G protein-coupled receptors and G proteins. *In* "Principles of Molecular Regulation" (P. M. Conn and A. R. Means, Eds.) Humana Press, Totowa, NJ.
27. Bogoyevitch, M. A., Clerk, A., and Sugden, P. H. (1995). Activation of the mitogen-activated protein kinase cascade by pertussis toxin-sensitive and—insensitive pathways in cultured ventricular cardiomyocytes. *Biochem. J.* **309**, 437–443.
28. Hawes, B. E., van Biesen, T., Koch, E. J., Luttrell, L. M., and Lefkowitz, R. H. (1995). Distinct pathways of G_i- and G_q-mediated mitogen-activated protein kinase activation. *J. Biol. Chem.* **270**, 17148–17153.
29. Cobb, M. H. and Goldsmith, E. J. (1995). How MAP kinases are regulated. *J. Biol. Chem.* **270**, 14843–14846
30. Croset, M., Brossard, N., Polette, A., and Lagarde, M. (2000). Characterization of plasma unsaturated lysophosphatidylcholines in human and rat. *Biochem. J.* **345**, 61–67.
31. Carson, M. J. and Lo, D. (2001). Immunology. The push-me pull-you of T cell activation. *Science* **293**, 618–619.
32. Mochizuki, M., Zigler, J. S. Jr, Russell, P., and Gery, I. (1982–1983). Serum proteins neutralize the toxic effect of lysophosphatidyl choline. *Curr. Eye Res.* **2**, 621–624.
33. Le, L. Q., Kabarowski, J. H., Weng, Z., Satterthwaite, A. B., Harvill, E. T., Jensen, E. R., Miller, J. F., and Witte, O. N. (2001). Mice lacking the orphan G protein-coupled receptor G2A develop a late-onset autoimmune syndrome. *Immunity.* **14**, 561–571.
34. Xiao, Y., Chen, Y., Kennedy, A. W., Belinson, J., and Xu, Y. (2000). Evaluation of plasma lysophospholipids for diagnostic significance using electrospray ionization mass spectrometry (ESI-MS) analyses. *Ann. NY Acad. Sci.* **905**, 242–259.
35. Xiao, Y-J., Schwartz, B., Washington, M., Kennedy, A., Webster, K., Belinson, J., and Xu, Y. (2001). Electrospray ionization mass spectrometry analysis of lysophospholipids in human ascitic fluids: comparison of the lysophospholipid contents in malignant vs. nonmalignant ascitic fluids. *Anal. Biochem.* **290**, 312–313.

CHAPTER 165

The Role of Ceramide in Cell Regulation

Yusuf A. Hannun and L. Ashley Cowart
*Department of Biochemistry and Molecular Biology,
Medical University of South Carolina,
Charleston, South Carolina*

During the past several years there has been a dramatic increase in information on the role of ceramide in many cellular events. Many excellent and thorough reviews have been written on ceramide; therefore, the purpose of this chapter is to offer the reader a starting point in the body of literature focused on the role of ceramide in cell regulation.

Ceramide-Mediated Cell Regulation

Ceramide is involved in several cellular processes, including apoptosis, cell senescence, differentiation, and cell stress. Many cytokines and stress agents, such as tumor necrosis factor (TNF) heat, and chemotherapy agents induce the production of ceramide, and several studies suggest critical roles for ceramide in regulating specific cell responses to these agents (Fig. 1). For example, augmentation of ceramide levels can lead to apoptosis in many cancer cells, whereas inhibition of ceramide formation can attenuate apoptosis in many, but not all, cell types and in response to several agonists. In other cell types, such as endothelial cells and fibroblasts, ceramide is more closely related to cell cycle arrest, differentiation, and senescence rather than apoptosis [1].

Biochemical Pathways of Ceramide Generation

De Novo Biosynthesis

The sphingolipid class of cell membrane lipids includes the sphingoid bases, ceramides (sphingoid bases with N-linked acyl groups), and complex sphingolipids based on ceramide and containing polar head groups. The committed step in sphingolipid biosynthesis is the condensation of palmitate with serine, which is catalyzed by serine palmitoyltransferase (SPT), a pyridoxal 5′-phosphate-dependent enzyme composed of two subunits and requiring (at least in yeast) another small subunit for maximal activity [2]. SPT generates 3-ketosphinganine, which is then reduced to dihydrosphingosine (DHS). The N-linked addition of a fatty acid to DHS, catalyzed by dihydroceramide synthase, yields dihydroceramide, which is subsequently desaturated to form ceramide in mammalian cells, or hydroxylated to form phytoceramide in yeast. From ceramide a variety of complex sphingolipids are derived. For example, complex glycosphingolipids including cerebrosides are formed by the addition of sugar groups to the ceramide backbone, and gangliosides are formed by the further addition of sialic acid. Sphingomyelin (SM) is generated by the transfer of a phosphocholine headgroup from phosphatidylcholine to ceramide. Each of these sphingolipid classes has distinct structural and/or functional roles [3].

Importantly, Fas, TNF, B-cell receptor stimulation, angiotensin II, palmitate loading, several chemotherapeutic agents (such as etoposide, CPT-11, and daunorubicin), phorbol esters, and UV radiation have been shown to activate the *de novo* pathway, leading to ceramide accumulation. Inhibition of this pathway with either myriocin, an SPT inhibitor, or with Fumonisin B_1, an inhibitor of dihydroceramide synthase, blocks formation of ceramide and attenuates the apoptotic response to these agents, thus implicating this pathway in the regulation of apoptosis [2]. Also, overexpression of glucosyl ceramide synthase (GCS), which clears

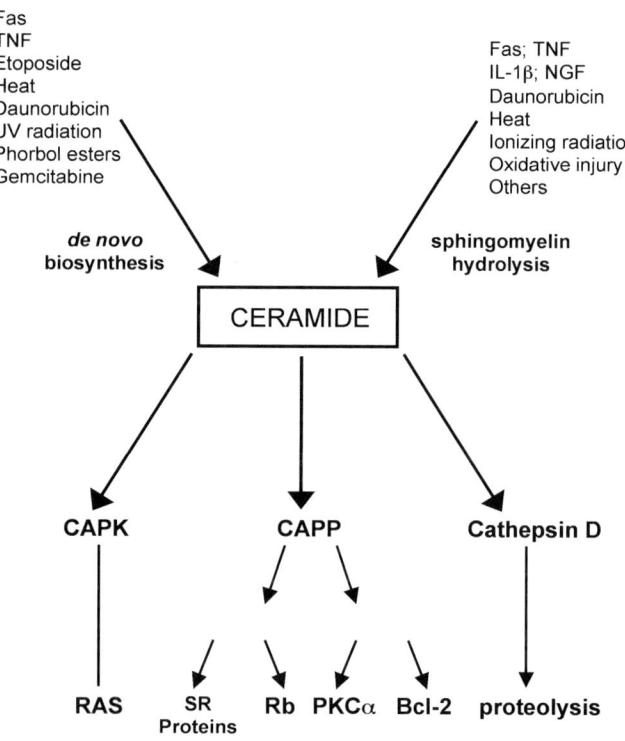

Figure 1 Several mechanisms of ceramide mediation of cell responses to extracellular signals.

ceramide (possibly with preference to *de novo* generated ceramide), attenuates apoptotic responses.

The Sphingomyelin Cycle

Ceramide can also be formed from hydrolysis of complex sphingolipids by stepwise removal of headgroups. Thus, sphingomyelinases (SMases) catalyze hydrolysis of SM, generating ceramide and phosphorylcholine. SM-derived ceramide can then act as a lipid mediator, undergo cleavage into sphingoid bases, or be reincorporated into SM, the latter completing the SM cycle. There are several different sphingomyelinase isoforms, displaying different pH optima, subcellular localization, and cofactor requirements [4].

SMase-derived ceramide has been shown to be an important component of the cell response to factors such as TNF, FAS, IL-1β, Ara-C, heat, and ionizing radiation. Cellular activities of ceramide derived from SM hydrolysis depend on the subcellular colocalization of the SMase as well as the SM pool, and studies suggest that the topology of ceramide production by SMase is a key factor in determining ceramide's ultimate cellular effects [4]. For example, recent results suggest that SM hydrolysis in the mitochondrion is sufficient to promote apoptosis [5], whereas ceramide at the membrane inhibits protein kinase C (PKC) translocation [6] and augments Fas action [7].

Moreover, as subsequent resynthesis of sphingomyelin from SMase-generated ceramide consumes phosphatidylcholine, yielding diacylglycerol (DAG), the dual action of SMase and SM synthase plays a role in the regulation of both ceramide and DAG, another important lipid mediator, primarily through activation of PKC, whose effects are often antagonistic to those of ceramide [4].

Ceramide Targets

Several important biochemical signaling pathways are regulated by ceramide, including pathways of stress and apoptosis. There are several key enzymes regulated by ceramide *in vitro*; thus serving as direct targets of ceramide, mediating at least some of its effects on these pathways.

CAPK

A ceramide-activated kinase activity which phosphorylates Raf has been shown to be the same as the kinase suppressor of Ras (KSR). KSR is activated by cytokine-induced ceramide production, coupling agents such as TNFα to apoptosis through regulation of MAP kinase pathways [8]. Other protein kinases for which there is evidence for direct ceramide-activation include PKCζ [9] and Raf [10].

CAPP

The discovery that ceramide specifically increased phosphatase activity in crude cell extracts resulted in the purification of the activity and subsequent identification of protein phosphatase 2A (PP2A) and protein phosphatase 1 (PP1) as ceramide-activated protein phosphatases (CAPP) [11]. Indeed, ceramide causes dephosphorylation of several key phosphoproteins with important roles in cell regulation. For example, inhibitors of ceramide biosynthesis blocked the TNF-induced, PP2A-dependent, dephosphorylation of PKCα, thus providing strong evidence for the involvement of *de novo* synthesized ceramide in PKCα regulation. Ceramide generation has also been shown to couple TNFα to the dephosphorylation of c-jun via PP2A. Furthermore, ceramide activation of PP2A caused dephosphorylation of bcl-2 and blocked its anti-apoptotic effects. Akt, a kinase involved in insulin signaling, mitogenesis, and apoptosis, has been shown to be dephosphorylated and inhibited in response to ceramide. Also, ceramide induces the PP1-mediated dephosphorylation of the retinoblastoma gene product (Rb), a key regulator of the cell cycle. Additionally, data indicate that the FAS-mediated dephosphorylation of SR proteins, which play important roles in mRNA splicing, is mediated by ceramide activation of PP1.

Cathepsin D

Cathepsin D is a lysosomal protease that has recently been shown to be activated by ceramide *in vitro*. Association of ceramide with the pre-pro cathepsin D causes autocatalytic cleavage to produce the 32-kDa active protease [12], which is thought to mediate several apoptotic events. This is particularly interesting as it presents a novel mechanism of ceramide-mediated responses independent of protein phosphorylation, and it highlights the compartment-specific functions of ceramide.

Mechanisms of Ceramide-Protein Interaction

Mechanisms of ceramide activation of these proteins are far from fully understood, however, some proteins that interact with ceramide have cysteine-rich domains which have been hypothesized to accommodate protein-ceramide interaction [13], but this has not been demonstrated. On the other hand, many ceramide-interacting proteins, including PP1 and PP2A, lack such domains. Therefore more research is needed to define ceramide-binding motifs.

Physical Properties of Ceramide

Because sphingolipid-enriched lipid microdomains such as rafts and caveolae exhibit distinct biophysical properties, it is possible that some of the described effects of ceramide are mediated via organization of such structures and/or assembly of signaling complexes [14]. Furthermore, ceramide is located in the membrane fractions of cells, and, based on its biochemical and biophysical properties, ceramide should not diffuse freely throughout the cell [15].

Conclusions

The field of ceramide signaling has rapidly expanded as studies reveal diverse roles for ceramide in cell biology and mechanisms of its regulation and action. Priorities for further study include determining the downstream actions of ceramide on protein targets that mediate specific cellular responses as well as determining the upstream events leading to ceramide production. Other key areas in need of further investigation include the subcellular localization of the substrates, enzymes, and protein targets of these sphingolipid signaling cascades.

Acknowledgments

The authors are partially supported by NIH grants GM 43825 and CA 87584.

References

1. Perry, D. K. and Hannun, Y. A. (1998). The role of ceramide in cell signaling. *Biochim. Biophys. Acta* **1436**, 233–243.
2. Linn, S., Kim, H., Keane, E., Andras, L., Wang, E., and Merrill, A. H., Jr. (2001). Regulation of *de novo* sphingolipid biosynthesis and the toxic consequences of its disruption. *Biochem. Soc. Trans.* **29**, 831–835.
3. Hannun, Y., Luberto, C., and Argraves, K. (2001). Enzymes of Sphingolipid Metabolism: From Modular to Integrative Signaling. *Biochemistry* **40**, 4893–4903.
4. Levade, T., Andrieu-Abadie, N., Segui, B., Auge, N., Chatelut, M., Jaffrezou, J.-P., and Salvayre, R. (1999). Sphingomyelin-degrading pathways in human cells. Role in cell singalling. *Chem. Phys. Lipids* **102**, 167–178.
5. Birbes, H., El Bawab, S., Hannun, Y. A., and Obeid, L. M. (2001). Selective hydrolysis of a mitochondrial pool of sphingomyelin induced apoptosis. *FASEB J.* **15**, 2669–2679.
6. Signorelli, P., Luberto, C., and Hannun, Y. A. (2001). Ceramide inhibition of NF-kappaB activation involves reverse translocation of classical protein kinase C (PKC) isoenzymes: requirement for kinase activity and carboxyl-terminal phosphorylation of PKC for the ceramide response. *FASEB J.* **15**, 2401–2414.
7. Cremesti, A., Paris, F., Grassmé, H., Holler, N., Tschopp, J., Fuks, Z., Gulbin, E., and Kolesnick, R. (2001). Ceramide enables Fas to cap and kill. *J. Biol. Chem.* **276**, 23954–23961.
8. Zhang, Y., Yao, B., Delikat, S., Bayoumy, S., Lin, X.-H., Basu, S., McGinley, M., Cahan-Hui, P.-Y., Lichenstein, H., and Kolesnick, R. (1997). Kinase suppressor of Ras is ceramide-activated protein kinase. *Cell* **89**, 63–72.
9. Lozano, J., Berra, E., Municio, M. M., Diaz-Meco, M. T., Dominguez, D, Sanz, L., and Moscat, J. (1994). Protein kinase C zeta isoform is critical for kappa B-dependent promoter activation by sphingomyelinase. *J. Biol. Chem.* **269**, 19200–19202.
10. Huwiler, A., Brunner, J., Hummel, R., Vervoordeldonk, M., Stabel, S., Van Den Bosch, H., and Pfeilschifter, J. (1996). Ceramide-binding and activation defines protein kinase c-Raf as a ceramide-activated protein kinase. *Proc. Natl. Acad. Sci. USA* **93**, 6959–6963.
11. Chalfant, C. and Hannun, Y. (2001). The role of serine/threonine protein phosphatases in ceramide signaling. In Futerman, T., Eds, *Ceramide Signaling*, Chap. 2, Eurekah. com.
12. Heinrich, M., Wickel, M., Schneider-Brachert, W., Sandberg, C., Gahr, J., Schwandner, R., Weber, T., Brunner, J., Krönke, M., and Schütze, S. (1999). Cathepsin D targeted by acid sphingomyelinase-derived ceramide. *EMBO J.* **18**, 5252–5263.
13. Van Blitterswijk, W. J. (1998). Hypothesis: Ceramide conditionally activates atypical protein kinases C, Raf-1, and KSR through binding to their cysteine-rich domains. *Biochem. J.* **331**, 679–680.
14. Dobrowsky, R. (2000). Sphingolipid signalling domains. Floating on rafts or buried in caves? *Cell. Signal.* **12**, 81–90.
15. Venkataraman, K. and Futerman, A. (2000). Ceramide as a second messenger: Sticky solutions to sticky problems. *Trends Cell Biol.* **10**, 408–412.

CHAPTER 166

Phospholipase A$_2$ Signaling and Arachidonic Acid Release

Jesús Balsinde[1] and Edward A. Dennis[2]

[1]Institute of Molecular Biology and Genetics,
University of Valladolid School of Medicine, Valladolid, Spain;
[2]Department of Chemistry and Biochemistry, School of Medicine and Revelle College,
University of California at San Diego, La Jolla, California

Introduction

Phospholipase A$_2$ (PLA$_2$) has attracted considerable interest in view of its role in lipid signaling and its involvement in a variety of inflammatory conditions. PLA$_2$ cleaves the sn-2 ester bond of cellular phospholipids, producing a free fatty acid and a lysophospholipid, both of which are implicated in lipid signaling. The free fatty acid produced is frequently arachidonic acid (AA, 5,8,11,14-eicosatetraenoic acid), the biosynthetic precursor of the eicosanoid family of potent inflammatory mediators that includes the prostaglandins, thromboxane, leukotrienes, and lipoxins. The other product of PLA$_2$ action on phospholipids is a lysophospholipid, which, depending on its molecular composition, may be converted into platelet-activating factor, another potent inflammatory mediator.

Phospholipase A$_2$ (PLA$_2$) consists of a superfamily of enzymes that catalyze the hydrolysis of the sn-2 ester bond in phospholipids, generating a free fatty acid and a lysophospholipid. This reaction is of the utmost importance in the context of cellular signaling, since it constitutes the main pathway by which arachidonic acid (AA) is liberated from phospholipids. Free AA is the precursor of a large family of compounds known as the eicosanoids, which includes the cyclooxygenase-derived prostaglandins and the lipoxygenase-derived leukotrienes [1]. If the other product of PLA$_2$ action on phospholipids is a choline-containing lysophospholipid possessing an alkyl linkage in the sn-1 position, then an acetyltransferase can act upon it to produce platelet-activating factor (PAF, 1-O-alkyl-2-acetyl-sn-3-phosphocholine).

The importance of the eicosanoids and platelet-activating factor as key mediators of inflammation as well as other pathophysiological conditions is now universally accepted [1]. Aspirin and other nonsteroidal anti-inflammatory drugs (NSAIDs) are well established as cyclooxygenase inhibitors, and are widely used in clinical practice. Similarly, the pharmaceutical industry has been actively pursuing lipoxygenase inhibitors and receptor antagonists for both leukotrienes and PAF. Note that since prostaglandins, leukotrienes, and PAF all derive from the action of a PLA$_2$, direct inhibition of such an enzyme would have the potential of blocking all three of the pathways at once, which could be of therapeutic advantage in certain settings. This is why the pharmaceutical industry has been actively pursuing the design of drugs with potential anti-PLA$_2$ effects, and some compounds are now in advanced clinical trials. Furthermore, cPLA$_2$ knockouts show distinct advantages in certain diseases [2,3].

The above approach is hampered, however, by the fact that cells in general contain multiple PLA$_2$ enzymes. For example, in humans, no less than 15 different proteins have been identified to possess PLA$_2$ activity [4]. Thus a first step in a rational PLA$_2$ drug design strategy is to define the different PLA$_2$ classes present in cells, as well as their putative roles in eicosanoid and PAF synthesis.

PLA$_2$ Groups

PLA$_2$s have been systematically classified according to their nucleotide sequence [4]. The latest update to this classification, published in October 2000, included eleven groups, most of them with several subgroups [4], but new PLA$_2$ enzymes have been described since, leading to a twelfth group [5–7]. Only PLA$_2$s whose nucleotide sequence has been determined should be included in the classification. However, this obvious criterion is the cause of some confusion, since many reports have appeared that erroneously link certain enzyme activities and functions to particular PLA$_2$ groups without it having been verified that such an association actually exists.

A parallel classification of the PLA$_2$s on the basis of biochemical properties is also frequently used, and it has value in describing PLA$_2$ activities for which sequence data are unavailable. This classification contemplates three main PLA$_2$ classes based on whether the enzyme is secreted (sPLA$_2$), cytosolic Ca^{2+}-dependent (cPLA$_2$), or cytosolic Ca^{2+}-independent (iPLA$_2$). One must be aware of the fact that this classification is not devoid of problems either, e.g. the Group IVC PLA$_2$ is generally referred to as cPLA$_2$-γ, despite its being a Ca^{2+}-independent enzyme. In addition, the PAF acetylhydrolase PLA$_2$s (Groups VII and VIII) also distribute among these categories.

Generally, the sPLA$_2$s (Groups I, II, III, V, IX, X, XI, XII) require millimolar levels of Ca^{2+} for activity, have low molecular masses, and lack specificity for arachidonate-containing phospholipids. The cPLA$_2$s (Group IV, comprising three subgroups) have higher molecular masses, require Ca^{2+} for translocation to membranes but not for activity, and are selective for arachidonate-containing phospholipids. Finally, the iPLA$_2$s (Group VI, and also Group IVC; see above) have high molecular masses but are not selective for arachidonate-containing phospholipids [4].

Cellular Function

A key determinant of the role of PLA$_2$s in a given cellular function is the mechanism of PLA$_2$ regulation/activation during such a process. A myriad of agents that exert effects on cells via receptor-dependent or independent pathways elicit a series of signals that ultimately lead to increased PLA$_2$ activity. Elucidation of these signals has been the subject of much effort for the last ten years [8]. The situation is further complicated by the evidence that most cells contain several PLA$_2$ forms and that all of them may eventually participate in the signaling process. Figure 1 shows the PLA$_2$ signal transduction mechanism developed over the years in our laboratory for the P388D$_1$ macrophage-like cells.

The scheme shown in Fig. 1 has been generally confirmed by many other laboratories and thus can be regarded as the currently accepted paradigm of PLA$_2$ signaling in immunoinflammatory cells. The cells may respond to two different kinds of signals that generate either a delayed response (bacterial lipopolysaccharide, LPS) or an immediate one (PAF). LPS acts primarily by inducing the cells to synthesize new proteins involved in the process. However, PAF acts on preexisting proteins. In either case, the foremost event is the translocation and activation of the cPLA$_2$ in an intracellular compartment. The mechanism of activation of this enzyme has been the subject of many studies, and generally involves the concerted action of the mitogen-activated protein kinase (MAPK) cascade and transient elevations of the intracellular Ca^{2+} concentration [9]. There are a few exceptions however, in which cPLA$_2$ activation has been described as being activated in a Ca^{2+}-independent manner [10] and/or

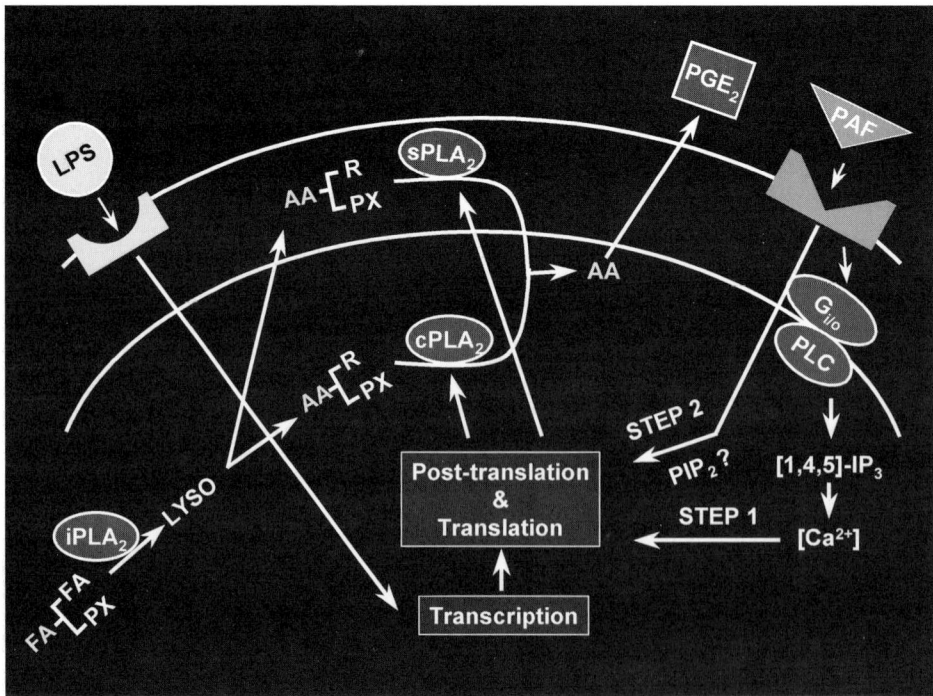

Figure 1 Signal transduction mechanism in P388D$_1$ macrophages. Adapted from [13].

phosphorylated by kinases not of the mitogen-activated protein kinase family [11].

Activation of the cPLA$_2$ is followed by activation of a sPLA$_2$, which, depending on cellular type, may belong to Groups IIA, V, or perhaps other groups. Depending on the stimulation conditions, the cPLA$_2$ modulation of sPLA$_2$ cellular activity may occur at a gene regulatory level (delayed responses) [12] or at the level of regulation of enzyme activity itself (immediate responses) [13]. In the latter case, a variety of cellular mechanisms may account for this activation, from cPLA$_2$-induced rearrangement of membrane phospholipids that enables further sPLA$_2$ attack to more sophisticated biochemical mechanisms such as inactivation of endogenous sPLA$_2$ inhibitors or Ca^{2+} fluxes. While it is clear that the cPLA$_2$ acts on perinuclear membranes, the precise site of action of the sPLA$_2$ has been the subject of numerous recent studies. The enzyme appears to be released to the extracellular medium, from which it re-associates with the outer cellular surface, where it hydrolyzes phospholipids. However, recent studies have suggested that the enzyme is re-internalized deep into the cell, probably via the caveolin system to the vicinity of nuclear membranes [14]. Whether the enzyme is still active in the cellular interior or this represents a signal termination mechanism is unclear at present [15]. This is currently an area of active study.

Free arachidonate generated by both cPLA$_2$ and sPLA$_2$ should be readily converted to prostaglandins and other eicosanoids by the cyclooxygenases and/or lipoxygenases. These eicosanoids are subsequently secreted to the extracellular medium, where they can act in both autocrine and paracrine manners.

Summary

The model depicted in Fig. 1 contemplates a scenario where the concerted action of two distinct PLA$_2$s leads to a full AA release response. The cPLA$_2$ appears to initiate the response and plays primarily a regulatory role, whereas the sPLA$_2$ acts in a second "wave" to amplify the response by providing the bulk of the AA liberated. Needless to say, in those cells that do not express a sPLA$_2$, the cPLA$_2$ would be the only one responsible for the release.

Cells usually also contain measurable levels of iPLA$_2$. This enzyme is frequently suggested to play a role in AA mobilization and eicosanoid production [16]. However, such an involvement remains controversial because practically all the evidence favoring this view has been inferred from studies utilizing bromoenol lactone, a compound that manifests high selectivity for the iPLA$_2$ *in vitro* but fails to do so *in vivo*. For example, a recent report has shown that the cPLA$_2$ counts among the cellular targets of bromoenol lactone in cells [17]. The iPLA$_2$ may also be involved indirectly in AA mobilization by modulating fatty acid reacylation reactions. The lysophospholipids produced by the iPLA$_2$ may be used to re-incorporate part of the fatty acids (including AA) that have previously been released by its Ca^{2+}-dependent counterparts. Thus, by regulating AA reacylation reactions, the iPLA$_2$ may participate in the formation of the cellular AA pools. Thus, all three types of PLA$_2$ (sPLA$_2$, cPLA$_2$, iPLA$_2$) appear to serve important but distinct functions in cells.

References

1. Smith, W. L., De Witt, D. L., and Garavito, R. M. (2000). Cyclooxygenases: structural, cellular, molecular biology. *Annu. Rev. Biochem.* **69**, 145–182.
2. Bonventre, J. V., Huang, Z., Taheri, M. R., O'Leary, E., Li, E., Moskowitz, M. A., and Sapirstein, A. (1997). Reduced fertility and postischaemic brain injury in mice deficient in cytosolic phospholipase A$_2$. *Nature* **390**, 622–625.
3. Uozumi, N., Kume, K., Nagase, T., Nakatani, N., Ishii, S., Tashiro, F., Komagata, Y., Maki, K., Ikuta, K., Ouichi, Y., Miyazaki, J., and Shimizu, T. (1997). Role of cytosolic phospholipase A$_2$ in allergic response and parturition. *Nature* **390**, 618–622.
4. Six, D. A. and Dennis, E. A. (2000). The expanding superfamily of phospholipase A$_2$ enzymes: classification and characterization. *Biochim. Biophys. Acta* **1488**, 1–19.
5. Gelb, M. H., Valentin, E., Ghomashchi, F., Lazdunski, M., and Lambeau, G. (2000). Cloning and recombinant expression of a structurally novel human secreted phospholipase A$_2$. *J. Biol. Chem.* **275**, 39823–39826.
6. Ho, I. C., Arm. J. P., Bingham, C. O., Choi, A., Austen, K. F., and Glimcher, L. F. (2001). A novel group of phospholipase A2s preferentially expressed in type 2 helper T cells. *J. Biol. Chem.* **276**, 18321–18326.
7. Mizenina, O., Musatkina, E., Yanushevich, Y., Rodina, A., Krasilnikov, M., de Gunzburg, J., Camonis, J. H., Travitian, A., and Tatosyan, A. (2001). A novel Group IIA phospholipase A$_2$ interacts with v-src oncoprotein from RSV-transformed hamster cells. *J. Biol. Chem.* **276**, 34006–34012.
8. Balsinde, J., Balboa, M. A., Insel, P. A., and Dennis, E. A. (1999). Regulation and inhibition of phospholipase A$_2$. *Annu. Rev. Pharmacol. Toxicol.* **39**, 175–189.
9. Dessen, A. (2000). B Structure and mechanism of human cytosolic phospholipase A$_2$. *Biochim. Biophys. Acta* **1488**, 40–47.
10. Balsinde, J., Balboa, M. A., Li, W., Llopis, J., and Dennis, E. A. (2000). Cellular regulation of cytosolic group IV phospholipase A$_2$ by phosphatidylinositol bisphosphate levels. *J. Immunol.* **164**, 5398–5402.
11. Leslie, C. C. (1997). Properties and regulation of cytosolic phospholipase A$_2$. *J. Biol. Chem.* **272**, 16709–16712.
12. Balsinde, J., Shinohara, H., Lefkowitz, L. J., Johnson, C. A., Balboa, M. A., and Dennis, E. A. (1999). Group V phospholipase A$_2$-dependent induction of cyclooxygenase-2 in macrophages. *J. Biol. Chem.* **274**, 25967–25970.
13. Balsinde, J. and Dennis, E. A. (1996). Distinct roles in signal transduction for each of the phospholipase A$_2$ enzymes present in P388D$_1$ macrophages. *J. Biol. Chem.* **271**, 6758–6765.
14. Murakami, M., Nakatani, Y., Kuwata, H., and Kudo, I. (2000). Cellular components that functionally interact with signaling phospholipase A$_2$s. *Biochim. Biophys. Acta* **1488**, 159–166.
15. Cho, W. (2000). Structure, function, and regulation of group V phospholipase A$_2$. *Biochim. Biophys. Acta* **1488**, 48–58.
16. Winstead, M., Balsinde, J., and Dennis, E. A. (2000). Ca^{2+}-independent phospholipase A$_2$: structure and function. *Biochim. Biophys. Acta* **1488**, 28–39.
17. Farooqui, A. A., Horrocks, L. A., and Farooqui, T. (2000). Deacylation and reacylation of neural membrane glycerophospholipids. *J. Mol. Neurosci.* **14**, 123–135.

CHAPTER 167

Prostaglandin Mediators

Emer M. Smyth and Garret A. FitzGerald
Center for Experimental Therapeutics, University of Pennsylvania
Philadelphia, Pennsylvania

Introduction

Arachidonic acid (AA), a 20-carbon unsaturated fatty acid containing four double bonds ($\Delta 5,8,11,14$:C20:4), circulates in plasma in both free and esterified forms and is a natural constituent of the phospholipid domain of cell membranes. AA is mobilized for release by phospholipases (PLs) A_2, particularly type IV cytosolic (c) PLA_2, [1] following its calcium-dependent translocation to the nuclear membrane and the endoplasmic reticulum (Fig. 1). Three major groups of enzymes, prostaglandin G/H synthase (PGHS), lipoxygenase, or cytochrome p450, then catalyze the formation of the prostaglandins (PGs) and thromboxane A_2 (TxA_2), the leukotrienes, or the epoxyeicosatrienoic acids, respectively. Collectively, these products are known as eicosanoids. A parallel family of free radical catalyzed isomers, the isoeicosanoids, are formed by direct peroxidation of AA *in situ* in cell membranes [2]. This chapter will focus on the PGs and TxA_2, collectively termed the prostanoids.

The Cyclooxygenase Pathway

Prostanoids are formed by the action of PGHS, or cyclooxygenase (COX), on AA to form bisenoic products containing two double bonds, denoted by a subscript 2, (e.g. PGE_2 [3]). COX-1 or COX-2 dimers [4], homotypically inserted into the ER membrane, contain both cyclooxygenase and hydroperoxidase activities [3]. AA is sequentially transformed into the unstable cyclic endoperoxides, PGG_2 and PGH2, for delivery to downstream isomerases and synthases to generate TxA_2 and D, E, F, and I series PGs (Fig. 1). It is presently not understood either how AA is delivered specifically to COX or how PGH_2 is presented to downstream enzymes. Two COX genes have been identified: COX-1 is expressed constitutively in most cells while COX-2 is upregulated by cytokines, shear stress, and tumor promoters [3]. These observations suggest housekeeping functions, such as gastric epithelial cytoprotection and hemostasis, for COX-1-derived prostanoids, although it appears that both enzymes contribute to the generation of autoregulatory prostanoids. Conversely, the inducible COX-2 is considered the dominant source of prostanoid formation in inflammation and cancer, although both isozymes can contribute to prostanoid formation in syndromes of human inflammation, including atherosclerosis [5] and rheumatoid arthritis [6].

COX-1 and COX-2 are closely related in their amino acid sequence [7] and crystal structure [8]. Although both isozymes demonstrate similar subcellular distribution [9], preference for downstream enzymes is sometimes evident in heterologous expression systems and apparently *in vivo*. COX-1 preferentially couples with TxS, PGFS [10], and the cytosolic (c) PGES isozymes [11]. COX-2 prefers PGIS [10] and the microsomal (m) PGES isozymes, which are induced by cytokines and tumor promoters [12]. Two forms of PGDS [13,14] and PGFS [15,16] have been identified, underscoring the diversity of the isomerases and synthases.

COX Deletion Deletion of COX-2 results in multiple defects of implantation and reproduction, leading to breeding difficulties [17]. Offspring have variably revealed cardiac fibrosis, renal defects and impaired inflammatory responses; however, the extent to which these phenotypes are modulated by genetic background is presently unclear. Impaired inflammatory responses [18] and delayed parturition [19] secondary to COX-1 deletion have been reported. Deletion of the COX-2 but not the COX-1 gene increases the frequency of patent ductus arteriosus (PDA) in newborn pups [20]. Coincidental deletion of COX-1 increases the frequency of the COX-2 knockout PDA phenotype [20]. It is the absence of PGE_2 that underlies this phenotype; deletion of 15 PGE_2 deydrogenase, the major inactivating enzyme of PGE_2, produces sustained high levels of PGE_2 throughout the perinatal period and results in ductal closure [21]. Both COX-1 and

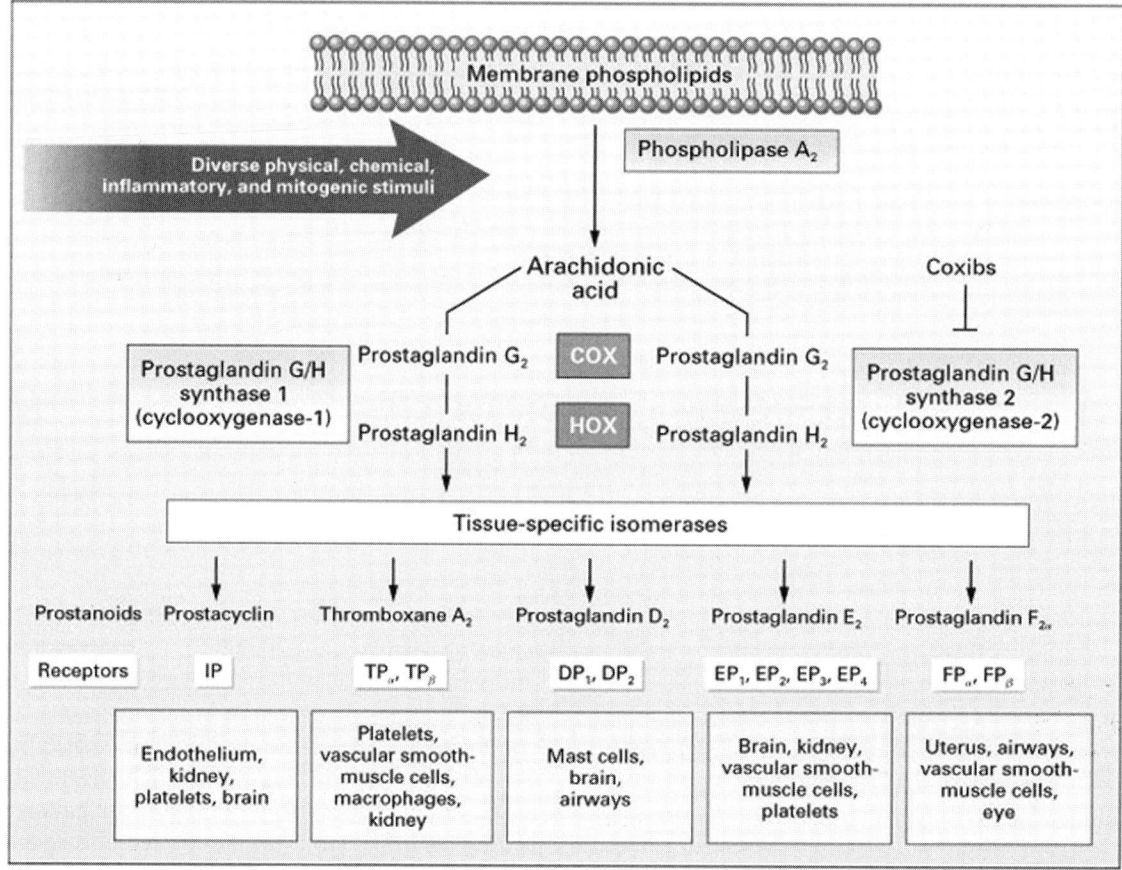

Figure 1 Production and actions of prostanoids: Arachidonic acid, a 20-carbon fatty acid containing four double bonds, is liberated from the *sn*2 position in membrane phospholipids by PLA_2. COX converts arachidonic acid to the unstable intermediate prostaglandin H2, which is converted by tissue-specific isomerases to multiple prostanoids. These bioactive lipids activate specific cell-membrane receptors of the superfamily of GPCRs.

COX-2 are subject to developmental regulation, and interference with their developmental expression may condition adult phenotypes [22]. COX-1 and COX-2 are expressed in a spatially and temporaly segregated manner during thymic development, where they influence T-cell maturation [23].

Prostanoid Receptors

Due to their short half lives (seconds to minutes), prostanoids act as autacoids, rather than circulating hormones, by activating membrane receptors at, or close to, the site of their formation. Specific G-protein-coupled receptors (GPCRs) have been cloned for all the prostanoids [24]. A single gene product has been identified for prostacyclin (the IP), $PGF_{2\alpha}$ (the FP), and TxA_2 (the TP), while four distinct PGE_2 receptors (the EP_{1-4}) and two PGD_2 (DP_1 and DP_2) have been cloned. The prostanoid receptors appear to derive from an ancestral EP receptor and share high homology [24]. Phylogenetic comparison of this family reveals three subclusters: first the EP_2, EP_4, IP, and DP_1, the relaxant receptors, which increase cAMP generation; second EP_1, FP, and TP, the contractile receptors, which increase intracellular calcium levels; and third the EP_3, which elevates intracellular

calcium and decreases cAMP [24,25]. The DP_2, a member fMLP receptor superfamily [26,27], is the exception to this characterization. Differential mRNA splicing gives rise to additional isoforms of the TP (α and β) [28], FP (A and B) [29], and EP_3 (A–D) [30]. The prostanoid receptors have been reviewed thoroughly elsewhere [24,31].

Thromboxane A_2 (TxA_2)

TxA_2, the major product of platelet COX-1, is a potent vasoconstrictor [32], mitogen [33] and platelet activator [34]. Despite the diversity of platelet agonists, inhibition of platelet TxA_2 formation apparently accounts for cardioprotection from aspirin [35], reflecting the importance of TxA_2 as an amplification signal for more potent agonists, such as thrombin and ADP [34]. TxA_2, also a major product of macrophage COX-2, contributes to atherogenesis in mouse models [36]. Analogous to its role in vascular proliferation (see below), TxA_2 may also mediate cellular hypertrophy [37].

Two forms of the platelet TP have been segregated pharmacologically, one mediating shape change, the other aggregation [38]. However, the cloned human TP splice variants (splice variants of the mouse TP are not apparent), do not account for this distinction, and TPα is apparently the sole isoform expressed in platelets [39,40]. Recognized differences between the splice variants are limited to G-protein activation in heterologous expression systems [41,42] and agonist-induced desensitization and sequestration [43,44]. Given the identification of distinct low homology GPCRs mediating ADP-induced platelet shape change and aggregation [45], it seems likely that at least one more distinct TP remains to be identified. In this regard, evidence suggests that iPF$_{2\alpha}$-III, an isoprostane, acts *in vivo* at the TP [46] but does not bind to either isoform *in vitro* [47], further suggesting the existence of another TP. Distinct receptor sites can be generated through GPCR heterodimerization [48]. It is interesting that TPα and TPβ appear to dimerize, and their coexpression augments iPF$_{2\alpha}$-III signaling compared to either receptor alone [49]. The extent to which associations between TPα with TPβ, and/or other prostanoid receptors, might contribute to the family of prostanoid receptors remains to be examined. The cloned TPs couple via G_q, G_{11}, $G_{12/13}$, and G_h (which is also tissue transglutaminase II) to activate PLC-dependent inositol phosphate generation and elevate intracellular calcium [24,31]. Activation of the TP isoforms may also activate or inhibit adenylyl cyclase, via G_s (TPα) or G_i (TPβ), respectively, and signal via G_q and related proteins to MAP kinase signaling pathways.

TP mRNAs are expressed widely in lung, liver, kidney, heart, uterus, and vascular cells with TPα usually the predominant isoform [50]. Despite reports of abundant TP expression in thymus, the role of TxA$_2$ in lymphocyte development and function is presently unclear. A naturally occurring mutation in the first intracellular loop of the TP is associated with a mild bleeding disorder and platelet resistance to TP agonists [40], while a polymorphism in the TP has been linked to bronchodilator resistance in asthma [51].

TP Deletion, Overexpression Deletion of the TP reveals a mild haemostatic defect and resistance to AA-induced platelet activation [52], reflecting the role of TxA$_2$ in vascular biology. TP null mice have reduced proliferative responses to vascular injury, while the opposite is true of mice engineered to overexpress TPβ in the vasculature [53]. TPβ overexpressors also develop a syndrome reminiscent of intrauterine growth retardation, probably secondary to placental ischemia [54].

Prostacyclin (PGI$_2$)

A major product of COX-2 in healthy individuals [55], PGI$_2$ is a potent vasodilator, inhibitor of platelet aggregation by all recognized agonists [56], and an inhibitor of cell proliferation *in vitro* [57]. PGI$_2$ biosynthesis is increased in syndromes of platelet activation [58,59], perhaps as a homeostatic response to accelerated platelet-vascular interactions. Chronic PGI$_2$ treatment can reduce pulmonary vascular resistance in patients with primary pulmonary hypertension [60]. Delivery of the gene for PGIS *in vivo* diminishes vascular smooth muscle cell proliferation and migration in response to injury [61]. PGIS polymorphs have been associated with essential hypertension [62] and myocardial infarction [63], while PGI$_2$ attenuates angiotensin II-induced renal vasoconstriction and systemic hypertension [64]. PGI$_2$ also attenuates the response to thrombotic stimuli in dogs [65] and specifically limits the effects on TxA$_2$ on platelets and the vessel wall in mice [53].

The sole identified IP couples to activation of adenylyl cyclase via G_s, although it can also activate phospholipase C via G_q. IP mRNA is abundantly expressed in kidney, where PGI$_2$ may regulate renal blood flow, renin release, and glomerular filtration rate and in lung, where PGI$_2$ can modulate vascular tone [24,31]. The IP is also expressed in the spinal column, where PGI$_2$ plays a role in pain perception, and in the liver, where its role is unknown. Expression within the cardiovascular system is most abundant in the aorta, consistent with the major biological role of PGI$_2$ in platelet and macrovascular homeostasis. IP expression in the heart, together with reports that COX-2-dependent PGI$_2$ formation limits oxidant-induced injury in cardiomyocytes [66], suggests a possible protective role for PGI$_2$ in cardiac tissue. A major difference between humans and rodents is the marked expression of IP in the thymus [67], although the functional relevance of this observation is not clear.

It is likely that at least one other IP remains to be identified. Pharmacologically distinct IP sites in the brain and kidney are not attributable to the cloned IP [68,69]. In addition, PGI$_2$ may activate the peroxisome proliferator activated receptors (PPARs). However, while both PGI$_2$ and iloprost activated PPARα and PPARδ *in vitro*, another PGI$_2$ analog, cicaprost, did not [70], and it is as yet unclear whether PGI$_2$ activation of PPARs occurs *in vivo*. The loss of PGI$_2$-mediated PPARδ activation was thought to underlie the implantation defect in COX-2-deficient mice, although no implantation defect was evident in PPARδ-deficient mice [71]. Two IP polymorphs, with some alterations in ligand binding and signaling in overexpression systems, have been identified [72].

IP Deletion Although results in IP-deficient mice have implicated PGI$_2$ in the mediation of pain and inflammation [73], these consequences seem conditioned by genetic background. Platelets of IP-null mice are resistant to disaggregation by IP agonists [73] and the thrombotic and proliferative response to vascular injury is enhanced [53], as is hypoxia-induced pulmonary hypertension and remodeling [74]. Despite its expression in murine thymus, disordered T-cell function secondary to IP deletion has not been reported, and the IP appears to play a minor role, if any, in murine T cell maturation [23]. Deletion of the IP undermines the atheroprotective effect of female gender in LDL receptor deficient mice [75], a possible consequence of estrogen-mediated upregulation of PGIS [76] and/or its protection from free radical attack. Interestingly, unlike their normotensive IP knock out counterparts, mice deficient in PGIS are hypertensive [77]. However, while this supports the possibility of a second IP, formation of both PGE$_2$ and TxA$_2$ are increased in the PGIS knockouts, perhaps due to rediversion of PGH$_2$.

Prostaglandin D$_2$ (PGD$_2$)

PGD$_2$, the major COX product formed by mast cells, is released during allergic responses, including asthma and systemic mastocytosis [78,79]. Infusion of PGD$_2$ in humans results in flushing, nasal stuffiness, and hypotension [80]. In mice, overexpression of lipocain-like PGDS increases response to bronchial challenge with ovalbumin [81]. The hematopoietic PGDS is expressed abnormally in patients with coronary disease [82] and a polymorphic variant has been linked to human asthma [83]. PGD$_2$, an abundant COX product in brain, is considered an important regulator of sleep-wake cycles [84]. Deletion of PGDS abolishes allodynia (touch-evoked pain) in mice [85], demonstrating a role for PGD$_2$ in pain perception.

The DP$_1$ is coupled positively to adenylyl cyclase through G$_s$ [24,31], which directs PDG$_2$-induced inhibition of platelet aggregation, bronchodilation, and vasodilation. Among the prostanoid receptors the DP$_1$ is the least abundant, with minor expression reported in mouse ileum, lung, stomach, and uterus and expression in the CNS limited specifically to the leptomeninges. Recently, a chemoattractant receptor–homologous molecule (CRTH2), expressed on T-helper (H) type-2 cells [27], was classified as the DP$_2$. This receptor is distinct from other prostanoid receptors, couples to increased intracellular Ca^{2+}, and directs PGD$_2$-induced chemotaxis and migration of TH2 cells. Both DPs integrate coordinately the effects of PGD$_2$ on eosinophils, modulating chemokinesis, degranulation, and apoptosis [86]. DP$_2$ and PGDS are coordinately expressed at the fetal/maternal interface in human deciduas, where they may participate in lymphocyte recruitment [87].

PGD$_2$ may be metabolized to PGJ$_2$ and its metabolite, 15-deoxy Δ (12,14) PGJ$_2$ [88], a possible natural ligand for PPARγ, regulating adipogenesis, inflammation, tumorigenesis, and immunity [89]. However, while PGJ$_2$ and its metabolite can activate the nuclear receptor *in vitro* [90], it is presently unclear whether sufficient concentrations are formed *in vivo*.

DP Deletion Deletion of the DP$_1$ sharply reduces ovalbumin-induced lymphocytes and eosinophils infiltration and airway hyperreactivity, reflecting PGD$_2$'s apparent role in asthma [91]. Work with these mice demonstrates the action of PDG$_2$ on arachnoid trabecular cells in the basal forebrain to increase extracellular adenosine, which in turn facilitates induction of sleep [92]. DP$_2$ null mice have not yet been generated.

Prostaglandin E$_2$ (PGE$_2$)

PGE$_2$ regulates diverse biological processes, including cell growth, inflammation, reproduction, sodium homeostasis and blood pressure [93]. Its biological effects are complex and often opposing; vasodilation in the arterial and venous systems [94] but constriction of smooth muscle in the trachea, gastric fundus, and ileum [95]. Like COX-2, the inducible mPGES isoforms [96] may contribute to the increase in PGE$_2$ associated with inflammatory and pyretic responses. The COX-1-cPGES axis is considered the predominant source of homeostatic PGE$_2$ [11], although mPGES and COX-1 seen coupled in the mouse kidney [97]. However, COX-2-derived prostanoids may differentially regulate salt excretion and glomerular circulation in volume overload or depletion [98]. PGE$_2$, along with PGI$_2$, apparently derived from COX-2, maintains renal blood flow and salt excretion [99], effects that may be counterbalanced by COX-1-derived TxA$_2$ [64].

Both the EP$_2$ and the EP$_4$ activate adenylyl cyclase via G$_s$ [24,31]. Differences in agonist-induced desensitization may be one reason for the presence of such similar receptors for PGE$_2$. The EP$_1$, via an unclassified G-protein, and the EP$_{3D}$, via G$_q$, are coupled to PLC activation [24,31]. A splice variant of the EP$_1$ in the rat may antagonize coupling of other EPs [100]. The EP$_{3B}$/EP$_{3C}$ couple to G$_s$-mediated activation, and the EP$_{3D}$/EP$_{3A}$ to G$_i$-mediated inhibition, of adenylyl cyclase.

The mRNAs for all four EPs are widely expressed; however, the limited distribution of EP$_1$ and EP$_2$, compared with EP$_3$ and EP$_4$, together with induction of EP$_2$ in response to inflammatory stimuli [101], suggests specialized functions for the different EPs [24,31]. The biological actions of PGE$_2$ may be conditioned by this differential receptor expression and/or PGE$_2$ levels. EP$_4$ directs platelet inhibition at low PGE$_2$ concentrations, while increased PGE$_2$ levels in, for example, inflammation, lead to EP$_3$-mediated platelet aggregation [102]. Indeed, high concentrations of PGE$_2$ condition platelet responses through EP$_3$- and IP-mediated regulation of intracellular cAMP [102]. Despite higher renal EP$_4$ expression [31], evidence supports EP$_2$-mediated renal vasodilation and salt handling [103], while an EP$_1$-directed increase salt excretion may contribute to PGE$_2$-dependent natriuresis [31]. PGE$_2$ may also directly stimulate renin and angiotensin II generation in the kidney [104], or directly constrict the renal vasculature [105] leading to hypertension. In the gastrointestinal tract, cytoprotective effects are mediated by EP$_1$ in stomach [106] but by EP$_3$ and EP$_4$ in the intestine [107]. EP$_1$ and EP$_3$ receptors appear responsible for myometrial contractility caused by PGE analogs, such as misoprost, used to induce labor [108], while selective EP$_2$-mediated inhibition of myometrial contractility [109] may be useful against preterm labor. EP$_2$- [110] and EP$_3$-mediated [111] interactions with growth factors may underlie the proliferative and angiogenic actions of PGE$_2$ in cancer. In the immune system, activation of the EP$_2$ inhibits T-cell proliferation, while both EP$_2$ and EP$_4$ receptors regulate antigen-presenting function *in vivo* [112]. Circulating levels of interleukin-1β induce coordinate COX-2-mPGES expression at the blood brain barrier, permitting activation of the central EPs [113]. Localized infusions of PGE$_2$ into the third ventricle induce wakefulness via the EP$_1$ and EP$_2$ receptors, while EP$_4$ activation in subarachnoid space induces sleep [114]. Pyrexial responses may mediated through the EP$_3$ [115], while the EP$_1$ and the EP$_3$ increase neuronal excitability and pain perception [115,116].

EP Deletion Knockout mouse models have been generated for all the EPs. EP_1-deficient mice have reduced nociceptive perception, while male, but not female, knockouts have reduced systolic blood pressure accompanied by elevated renin-angiotensin activity [116]. EP_2-deficient mice are normotensive at baseline but demonstrate increased salt- and pressor hormone-induced hypertension, although this is modified by genetic background [103]. The EP_2 knockouts also demonstrate a preimplantation defect, which may underlie some of the breeding difficulties seen in the COX-2 knockouts (see above). EP_3-deficient mice are resistant to pyrogen-induced fever [115]. However, despite its abundant expression in the kidney, there is no renal phenotype in EP_3 knockouts [117]. Deletion of the EP_4 results in PDA and neonatal death [118].

Prostaglandin $F_{2\alpha}$ ($PGF_{2\alpha}$)

$PGF_{2\alpha}$ actions include luteolysis [119] and smooth muscle contraction across a variety of tissues [120,121]. PGFS catalyzes the reduction of PGH_2 to $PGF_{2\alpha}$, PGD_2 to $9\alpha\ 11\beta$ $PGF_{2\alpha}$ [122], and retinal to retinol [123]. It exists in at least two isoforms, identified initially in liver [16] and lung [122], and is also expressed in lymphocytes [122] and spinal chord [124]. $PGF_{2\alpha}$ induces cardiac myocyte hypertrophy and induction of myofibrillar genes, independent of muscle contraction [125], suggesting a role for this eicosanoid during development, in compensatory hypertrophy, and/or in recovery of the heart from injury.

Thus far, one GPCR for $PGF_{2\alpha}$, the FP, which couples via G_q activation of PLC, has been cloned [24,31]. Stimulation of FP also activates Rho kinase, leading to the formation of actin stress fibers, phosphorylation of p125 focal adhesion kinase, and cell rounding [126]. Similar to the EP_3 and TP, carboxy terminal splice variants, FP_A and FP_B [29], have been identified. These are indistinguishable in their ligand-binding properties and signaling but may differ in their constitutive activity [29] and rates of desensitization [127]. FP_A and FP_B also differ in coupling to the Tcf/β catenin-signaling pathway, which may underlie the prolonged cytoskeletal effects mediated thought FP_B [128]. The FP is expressed in kidney, heart, lung, and stomach; however, it is most abundant in the corpus luteum, where its expression varies during the estrus cycle, consistent with the role for $PGF_{2\alpha}$ in luteolysis. The FP is also expressed in the ciliary body of the eye, where FP agonists have clinical utility in the treatment of raised intraocular pressure in patients with glaucoma [129]. Although activation of the FP results in vaso- and broncho-constriction [121], cell proliferation [130], and cardiomyocyte hypertrophy [131], the role of this prostanoid in cardiopulmonary disease is poorly characterized. Similarly, activation of the FP blocks preadipocyte differentiation *in vitro* [132], but the role of the FP, if any, in obesity is poorly understood.

FP Deletion Mice deficient in the FP do not deliver at term, resulting from a failure to induce the oxytocin receptor and lack of the normal decline in elevated progesterone levels [133]. Ovariectomy restores responsiveness to oxytocin and permits successful parturition. COX-1-derived $PGF_{2\alpha}$ in these mice appears important for luteolysis, consistent with delayed parturition in COX-1-deficient mice [19]. Subsequent upregulation of COX-2 generates prostanoids, including $PGF_{2\alpha}$ and TxA_2, important in the final stages of parturition [134]. Mice lacking both COX-1 and oxytocin underwent normal parturition, demonstrating the critical interplay between $PGF_{2\alpha}$ and oxytocin in onset of labor [19].

Concluding Remarks

The cyclooxygenase pathway of arachidonic acid metabolism generates a family of evanescent mediators with wide and varied physiological and pathophysiological actions. Understanding the biological role of the prostanoids requires examination of the biosynthetic pathways that lead to their temporal and tissue-specific generation together with the array of signaling pathways activated by their multiple receptors.

References

1. Leslie, C. C. (1997). Properties and regulation of cytosolic phospholipase A2. *J. Biol. Chem.* **272**, 16709–16712.
2. Patrono, C. and FitzGerald, G. A. (1997). Isoprostanes: potential markers of oxidant stress in atherothrombotic disease. *Arterioscler. Thromb. Vasc. Biol.* **17**, 2309–2315.
3. Herschman, H. R. (1996). Prostaglandin synthase 2. *Biochim. Biophys. Acta* **1299**, 125–140.
4. Garavito, R. M., Picot, D., and Loll, P. J. (1995). The 3.1 A X-ray crystal structure of the integral membrane enzyme prostaglandin H2 synthase-1. *Adv Prostaglandin Thromboxane Leukot. Res.* **23**, 99–103.
5. Wijeyaratne, S. M., Abbott, C. R., Homer-Vannisinkam, S., Mavor, A. I., and Gough, M. J. (2001). Differences in the detection of cyclooxygenase 1 and 2 proteins in symptomatic and asymptomatic carotid plaques. *Br. J. Surg.* **88**, 951–957.
6. Iniguez, M. A., Pablos, J. L., Carreira, P. E., Cabre, F., and Gomez-Reino, J. J. (1998). Detection of COX-1 and COX-2 isoforms in synovial fluid cells from inflammatory joint diseases. *Br. J. Rheumatol.* **37**, 773–778.
7. Smith, W. L., Garavito, R. M., and DeWitt, D. L. (1996). Prostaglandin endoperoxide H synthases (cyclooxygenases)-1 and -2. *J. Biol. Chem.* **271**, 33157–33160.
8. FitzGerald, G. A. and Loll, P. (2001). COX in a crystal ball: current status and future promise of prostaglandin research. *J. Clin. Invest.* **107**, 1335–1337.
9. Spencer, A. G., Woods, J. W., Arakawa, T., Singer, II, and Smith, W. L. (1998). Subcellular localization of prostaglandin endoperoxide H synthases-1 and -2 by immunoelectron microscopy. *J. Biol. Chem.* **273**, 9886–9893.
10. Ueno, N., Murakami, M., Tanioka, T., Fujimori, K., Tanabe, T., Urade, Y., and Kudo, I. (2001). Coupling between cyclooxygenase, terminal prostanoid synthase, and phospholipase A2. *J. Biol. Chem.* **276**, 34918–34927.
11. Tanioka, T., Nakatani, Y., Semmyo, N., Murakami, M., and Kudo, I. (2000). Molecular identification of cytosolic prostaglandin E2 synthase that is functionally coupled with cyclooxygenase-1 in immediate prostaglandin E2 biosynthesis. *J. Biol. Chem.* **275**, 32775–32782.
12. Murakami, M., Naraba, H., Tanioka, T., Semmyo, N., Nakatani, Y., Kojima, F., Ikeda, T., Fueki, M., Ueno, A., Oh, S., and Kudo, I. (2000).

Regulation of prostaglandin E2 biosynthesis by inducible membrane-associated prostaglandin E2 synthase that acts in concert with cyclooxygenase-2. *J. Biol. Chem.* **275**, 32783–32792.
13. Nagata, A., Suzuki, Y., Igarashi, M., Eguchi, N., Toh, H., Urade, Y., and Hayaishi, O. (1991). Human brain prostaglandin D synthase has been evolutionarily differentiated from lipophilic-ligand carrier proteins. *Proc. Natl. Acad. Sci. USA* **88**, 4020–4024.
14. Kanaoka, Y., Ago, H., Inagaki, E., Nanayama, T., Miyano, M., Kikuno, R., Fujii, Y., Eguchi, N., Toh, H., Urade, Y., and Hayaishi, O. (1997). Cloning and crystal structure of hematopoietic prostaglandin D synthase. *Cell* **90**, 1085–1095.
15. Watanabe, K., Fujii, Y., Nakayama, K., Ohkubo, H., Kuramitsu, S., Kagamiyama, H., Nakanishi, S., and Hayaishi, O. (1988). Structural similarity of bovine lung prostaglandin F synthase to lens epsilon-crystallin of the European common frog. *Proc. Natl. Acad. Sci. USA* **85**, 11–15.
16. Suzuki, T., Fujii, Y., Miyano, M., Chen, L. Y., Takahashi, T., and Watanabe, K. (1999). cDNA cloning, expression, and mutagenesis study of liver-type prostaglandin F synthase. *J. Biol. Chem.* **274**, 241–248.
17. Lim, H., Paria, B. C., Das, S. K., Dinchuk, J. E., Langenbach, R., Trzaskos, J. M., and Dey, S. K. (1997). Multiple female reproductive failures in cyclooxygenase 2-deficient mice. *Cell* **91**, 197–208.
18. Langenbach, R., Morham, S. G., Tiano, H. F., Loftin, C. D., Ghanayem, B. I., Chulada, P. C., Mahler, J. F., Lee, C. A., Goulding, E. H., Kluckman, K. D. *et al.* (1995). Prostaglandin synthase 1 gene disruption in mice reduces arachidonic acid-induced inflammation and indomethacin-induced gastric ulceration. *Cell*, **83**, 483–492.
19. Gross, G. A., Imamura, T., Luedke, C., Vogt, S. K., Olson, L. M., Nelson, D. M., Sadovsky, Y., and Muglia, L. J. (1998). Opposing actions of prostaglandins and oxytocin determine the onset of murine labor. *Proc. Natl. Acad. Sci. USA* **95**, 11875–11879.
20. Loftin, C. D., Trivedi, D. B., Tiano, H. F., Clark, J. A., Lee, C. A., Epstein, J. A., Morham, S. G., Breyer, M. D., Nguyen, M., Hawkins, B. M., Goulet, J. L., Smithies, O., Koller, B. H., and Langenbach, R. (2001). Failure of ductus arteriosus closure and remodeling in neonatal mice deficient in cyclooxygenase-1 and cyclooxygenase-2. *Proc. Natl. Acad. Sci.USA* **98**, 1059–1064.
21. Coggins, K. G., Latour, A., Nguyen, M. S., Audoly, L., Coffman, T. M., and Koller, B. H. (2002). Metabolism of PGE2 by prostaglandin dehydrogenase is essential for remodeling the ductus arteriosus. *Nat. Med.* **8**, 91–92.
22. Grosser, T., Yusuff, S., Cheskis, E., Pack, M. A., and FitzGerald, G. A. (2002). Developmental expression of functional cyclooxygenases in zebrafish. *Proc. Natl. Acad. Sci. USA* **14**, 14.
23. Rocca, B., Spain, L. M., Pure, E., Langenbach, R., Patrono, C., and FitzGerald, G. A. (1999). Distinct roles of prostaglandin H synthases 1 and 2 in T-cell development. *J. Clin. Invest.* **103**, 1469–1477.
24. Narumiya, S., Sugimoto, Y., and Ushikubi, F. (1999). Prostanoid receptors: structures, properties, and functions. *Physiol. Rev.* **79**, 1193–1226.
25. Boie, Y., Sawyer, N., Slipetz, D. M., Metters, K. M., and Abramovitz, M. (1995). Molecular cloning and characterization of the human prostanoid DP receptor. *J. Biol. Chem.* **270**, 18910–18916.
26. Nagata, K., Hirai, H., Tanaka, K., Ogawa, K., Aso, T., Sugamura, K., Nakamura, M., and Takano, S. (1999). CRTH2, an orphan receptor of T-helper-2-cells, is expressed on basophils and eosinophils and responds to mast cell-derived factor(s). *FEBS. Lett.* **459**, 195–199.
27. Hirai, H., Tanaka, K., Yoshie, O., Ogawa, K., Kenmotsu, K., Takamori, Y., Ichimasa, M., Sugamura, K., Nakamura, M., Takano, S., and Nagata, K. (2001). Prostaglandin D2 selectively induces chemotaxis in T helper type 2 cells, eosinophils, and basophils via seven-transmembrane receptor CRTH2. *J. Exp. Med.* **193**, 255–261.
28. Raychowdhury, M. K., Yukawa, M., Collins, L. J., McGrail, S. H., Kent, K. C., and Ware, J. A. (1994). Alternative splicing produces a divergent cytoplasmic tail in the human endothelial thromboxane A2 receptor. *J. Biol. Chem.* **269**, 19256–19261.
29. Pierce, K. L., Bailey, T. J., Hoyer, P. B., Gil, D. W., Woodward, D. F., and Regan, J. W. (1997). Cloning of a carboxyl-terminal isoform of the prostanoid FP receptor. *J. Biol. Chem.* **272**, 883–887.
30. Schmid, A., Thierauch, K. H., Schleuning, W. D., and Dinter, H. (1995). Splice variants of the human EP3 receptor for prostaglandin E2. *Eur. J. Biochem.* **228**, 23–30.
31. Breyer, R. M., Bagdassarian, C. K., Myers, S. A., and Breyer, M. D. (2001). Prostanoid receptors: subtypes and signaling. *Annu. Rev. Pharmacol. Toxicol.* **41**, 661–690.
32. Dorn, G. W. 2nd, Sens, D., Chaikhouni, A., Mais, D., and Halushka, P. V. (1987). Cultured human vascular smooth muscle cells with functional thromboxane A2 receptors: measurement of U46619-induced 45calcium efflux. *Circ. Res.* **60**, 952–956.
33. Pakala, R., Willerson, J. T., and Benedict, C. R. (1997). Effect of serotonin, thromboxane A2, and specific receptor antagonists on vascular smooth muscle cell proliferation. *Circulation* **96**, 2280–2286.
34. FitzGerald, G. A. (1991). Mechanisms of platelet activation: thromboxane A2 as an amplifying signal for other agonists. *Am. J. Cardiol.* **68**, 11B–15B.
35. Patrono, C. (1994). Aspirin as an antiplatelet drug. *N. Engl. J. Med.* **330**, 1287–1294.
36. Cayatte, A. J., Du, Y., Oliver-Krasinski, J., Lavielle, G., Verbeuren, T. J., and Cohen, R. A. (2000). The thromboxane receptor antagonist S18886 but not aspirin inhibits atherogenesis in apo E-deficient mice: evidence that eicosanoids other than thromboxane contribute to atherosclerosis. *Arterioscler. Thromb. Vasc. Biol.* **20**, 1724–1728.
37. Ali, S., Davis, M. G., Becker, M. W., and Dorn, G. W. 2nd. (1993). Thromboxane A2 stimulates vascular smooth muscle hypertrophy by up-regulating the synthesis and release of endogenous basic fibroblast growth factor. *J. Biol. Chem.* **268**, 17397–17403.
38. Dorn, G. W. 2nd and DeJesus, A. (1991). Human platelet aggregation and shape change are coupled to separate thromboxane A2-prostaglandin H2 receptors. *Am. J. Physiol.* **260**, H327–334.
39. Habib, A., FitzGerald, G. A., and Maclouf, J. (1999). Phosphorylation of the thromboxane receptor alpha, the predominant isoform expressed in human platelets. *J. Biol. Chem.* **274**, 2645–2651.
40. Hirata, T., Kakizuka, A., Ushikubi, F., Fuse, I., Okuma, M., and Narumiya, S. (1994). Arg60 to Leu mutation of the human thromboxane A2 receptor in a dominantly inherited bleeding disorder. *J. Clin. Invest.* **94**, 1662–1667.
41. Vezza, R., Habib, A., and FitzGerald, G. A. (1999). Differential signaling by the thromboxane receptor isoforms via the novel GTP-binding protein, Gh. *J. Biol. Chem.* **274**, 12774–12779.
42. Hirata, T., Ushikubi, F., Kakizuka, A., Okuma, M., and Narumiya, S. (1996). Two thromboxane A2 receptor isoforms in human platelets. Opposite coupling to adenylyl cyclase with different sensitivity to Arg60 to Leu mutation. *J. Clin. Invest.* **97**, 949–956.
43. Yukawa, M., Yokota, R., Eberhardt, R. T., von Andrian, L., and Ware, J. A. (1997). Differential desensitization of thromboxane A2 receptor subtypes. *Circ. Res*, **80**, 551–556.
44. Parent, J. L., Labrecque, P., Orsini, M. J., and Benovic, J. L. (1999). Internalization of the TXA2 receptor alpha and beta isoforms. Role of the differentially spliced COOH terminus in agonist-promoted receptor internalization. *J. Biol. Chem.* **274**, 8941–8948.
45. Takasaki, J., Kamohara, M., Saito, T., Matsumoto, M., Matsumoto, S., Ohishi, T., Soga, T., Matsushime, H., and Furuichi, K. (2001). Molecular cloning of the platelet P2T(AC) ADP receptor: pharmacological comparison with another ADP receptor, the P2Y(1) receptor. *Mol. Pharmacol.* **60**, 432–439.
46. Audoly, L. P., Rocca, B., Fabre, J. E., Koller, B. H., Thomas, D., Loeb, A. L., Coffman, T. M., and FitzGerald, G. A. (2000). Cardiovascular responses to the isoprostanes iPF(2alpha)-III and iPE(2)-III are mediated via the thromboxane A(2) receptor in vivo. *Circulation* **101**, 2833–2840.
47. Pratico, D., Smyth, E. M., Violi, F., and FitzGerald, G. A. (1996). Local amplification of platelet function by 8-Epi prostaglandin F2alpha is not mediated by thromboxane receptor isoforms. *J. Biol. Chem.* **271**, 14916–14924.
48. Devi, L. A. (2001). Heterodimerization of G-protein-coupled receptors: pharmacology, signaling and trafficking. *Trends Pharmacol. Sc.* **22**, 532–537.

49. Sullivan, P. and Smyth, E. M. (2002). Heterodimerization of the a and b isoforms of the human thromboxane receptor. *Arterioscler. Thromb. Vasc. Biol.* **22**, 878.
50. Miggin, S. M. and Kinsella, B. T. (1998). Expression and tissue distribution of the mRNAs encoding the human thromboxane A2 receptor (TP) alpha and beta isoforms. *Biochim. Biophys. Acta* **1425**, 543–559.
51. Leung, T. F., Tang, N. L., Lam, C. W., Li, A. M., Chan, I. H., and Ha, G. (2002). Thromboxane A2 receptor gene polymorphism is associated with the serum concentration of cat-specific immunoglobulin E as well as the development and severity of asthma in Chinese children. *Pediatr. Allergy Immunol.* **13**, 10–17.
52. Thomas, D. W., Mannon, R. B., Mannon, P. J., Latour, A., Oliver, J. A., Hoffman, M., Smithies, O., Koller, B. H., and Coffman, T. M. (1998). Coagulation defects and altered hemodynamic responses in mice lacking receptors for thromboxane A2. *J. Clin. Invest.* **102**, 1994–2001.
53. Cheng, Y., Austin, S. C., Rocca, B., Koller, B. H., Coffman, T. M., Grosser, T., Lawson, J. A., and FitzGerald, G. A. (2002). Role of prostacyclin in the cardiovascular response to thromboxane A2. *Science* **296**, 539–541.
54. Rocca, B., Loeb, A. L., Strauss, J. F. 3rd, Vezza, R., Habib, A., Li, H., and FitzGerald, G. A. (2000). Directed vascular expression of the thromboxane A2 receptor results in intrauterine growth retardation. *Nat. Med.* **6**, 219–221.
55. Catella-Lawson, F., McAdam, B., Morrison, B. W., Kapoor, S., Kujubu, D., Antes, L., Lasseter, K. C., Quan, H., Gertz, B. J., and FitzGerald, G. A. (1999). Effects of specific inhibition of cyclooxygenase-2 on sodium balance, hemodynamics, and vasoactive eicosanoids. *J. Pharmacol. Exp. Ther.* **289**, 735–741.
56. Moncada, S. and Vane, J. R. (1981). Prostacyclin: homeostatic regulator or biological curiosity? *Clin. Sci. (Colch)* **61**, 369–372.
57. Zucker, T. P., Bonisch, D., Hasse, A., Grosser, T., Weber, A. A., and Schror, K. (1998). Tolerance development to antimitogenic actions of prostacyclin but not of prostaglandin E1 in coronary artery smooth muscle cells. *Eur. J. Pharmacol.* **345**, 213–220.
58. Fitzgerald, D. J., Roy, L., Catella, F., and FitzGerald, G. A. (1986). Platelet activation in unstable coronary disease. *N. Engl. J. Med.* **315**, 983–989.
59. Fitzgerald, D. J., Doran, J., Jackson, E., and FitzGerald, G. A. (1986). Coronary vascular occlusion mediated via thromboxane A2 prostaglandin endoperoxide receptor activation in vivo. *J. Clin. Invest.* **77**, 496–502.
60. McLaughlin, V. V., Genthner, D. E., Panella, M. M., and Rich, S. (1998). Reduction in pulmonary vascular resistance with long-term epoprostenol (prostacyclin) therapy in primary pulmonary hypertension. *N. Engl. J. Med.* **338**, 273–277.
61. Numaguchi, Y., Naruse, K., Harada, M., Osanai, H., Mokuno, S., Murase, K., Matsui, H., Toki, Y., Ito, T., Okumura, K., and Hayakawa, T. (1999). Prostacyclin synthase gene transfer accelerates reendothelialization and inhibits neointimal formation in rat carotid arteries after balloon injury. *Arterioscler. Thromb. Vasc. Biol.* **19**, 727–733.
62. Nakayama, T., Soma, M., Rahmutula, D., Tobe, H., Sato, M., Uwabo, J., Aoi, N., Kosuge, K., Kunimoto, M., Kanmatsuse, K., and Kokubun, S. (2002). Association study between a novel single nucleotide polymorphism of the promoter region of the prostacyclin synthase gene and essential hypertension. *Hypertens. Res.* **25**, 65–68.
63. Nakayama, T., Soma, M., Rehemudula, D., Takahashi, Y., Tobe, H., Satoh, M., Uwabo, J., Kunimoto, M., and Kanmatsuse, K. (2000). Association of 5' upstream promoter region of prostacyclin synthase gene variant with cerebral infarction. *Am. J. Hypertens.* **13**, 1263–1267.
64. Qi, Z., Chuan-Ming, H., Langenbach, R. I., Breyer, R. M., Redha, R., Morrow, J. D., and Breyer, M. D. (2002). Opposite effects of cyclooxygenases 1 and 2 activity on the pressor response to angiotensin II. *J. Clin. Invest.* **110**, 61–69.
65. Hennan, J. K., Huang, J., Barrett, T. D., Driscoll, E. M., Willens, D. E., Park, A. M., Crofford, L. J., and Lucchesi, B. R. (2001). Effects of selective cyclooxygenase-2 inhibition on vascular responses and thrombosis in canine coronary arteries. *Circulation* **104**, 820–825.
66. Adderley, S. R. and Fitzgerald, D. J. (1999). Oxidative damage of cardiomyocytes is limited by extracellular regulated kinases 1/2-mediated induction of cyclooxygenase-2. *J. Biol. Chem.* **274**, 5038–5046.
67. Namba, T., Oida, H., Sugimoto, Y., Kakizuka, A., Negishi, M., Ichikawa, A., and Narumiya, S. (1994). cDNA cloning of a mouse prostacyclin receptor. Multiple signaling pathways and expression in thymic medulla. *J. Biol. Chem.* **269**, 9986–9992.
68. Hebert, R. L., Regnier, L., and Peterson, L. N. (1995). Rabbit cortical collecting ducts express a novel prostacyclin receptor. *Am. J. Physiol.* **268**, F145–154.
69. Takechi, H., Matsumura, K., Watanabe, Y., Kato, K., Noyori, R., and Suzuki, M. (1996). A novel subtype of the prostacyclin receptor expressed in the central nervous system. *J. Biol. Chem.* **271**, 5901–5906.
70. Forman, B. M., Chen, J., and Evans, R. M. (1997). Hypolipidemic drugs, polyunsaturated fatty acids, and eicosanoids are ligands for peroxisome proliferator-activated receptors alpha and delta. *Proc. Natl. Acad. Sci. USA* **94**, 4312–4317.
71. Peters, J. M., Lee, S. S., Li, W., Ward, J. M., Gavrilova, O., Everett, C., Reitman, M. L., Hudson, L. D., and Gonzalez, F. J. (2000). Growth, adipose, brain, and skin alterations resulting from targeted disruption of the mouse peroxisome proliferator-activated receptor beta(delta). *Mol. Cell Biol.* **20**, 5119–5128.
72. Stitham, J., Stojanovic, A., and Hwa, J. (2002). Impaired receptor binding and activation associated with a human prostacyclin receptor polymorphism. *J. Biol. Chem.*, **277**, 15439–15444.
73. Murata, T., Ushikubi, F., Matsuoka, T., Hirata, M., Yamasaki, A., Sugimoto, Y., Ichikawa, A., Aze, Y., Tanaka, T., Yoshida, N., Ueno, A., Oh-ishi, S., and Narumiya, S. (1997). Altered pain perception and inflammatory response in mice lacking prostacyclin receptor. *Nature* **388**, 678–682.
74. Hoshikawa, Y., Voelkel, N. F., Gesell, T. L., Moore, M. D., Morris, K. G., Alger, L. A., Narumiya, S., and Geraci, M. W. (2001). Prostacyclin receptor-dependent modulation of pulmonary vascular remodeling. *Am. J. Respir. Crit. Care Med.* **164**, 314–318.
75. Egan, K., Austin, S., Smyth, E. M., and FitzGerald, G. A. (2000). Accelerated atherogenesis in prostacyclin receptor deficient mice. *Circulation* **102**, 234.
76. Seeger, H., Mueck, A. O., and Lippert, T. H. (1999). Effect of estradiol metabolites on prostacyclin synthesis in human endothelial cell cultures. *Life Sci.* **65**, L167–L170.
77. Yokoyama, C., Yabuki, T., Shimonishi, M., Wada, M., Hatae, T., Takeda, J., Okabe, M., and Tanabe, T. (2001). Prostacyclin deficiency in mice induces vascular disorders in kindey. *Ist Takeda Science Foundation Symposium on Pharma Sciences—Lipids in Signaling and Related Diseases.*, Tokyo, Japan.
78. Sladek, K., Sheller, J. R., FitzGerald, G. A., Morrow, J. D., and Roberts, L. J. 2nd. (1991). Formation of PGD2 after allergen inhalation in atopic asthmatics. *Adv. Prostaglandin Thromboxane Leukot. Res.* 433–436.
79. Roberts, L. J. 2nd, Sweetman, B. J., Lewis, R. A., Austen, K. F., and Oates, J. A. (1980). Increased production of prostaglandin D2 in patients with systemic mastocytosis. *N. Engl. J. Med.* **303**, 1400–1404.
80. Heavey, D. J., Lumley, P., Barrow, S. E., Murphy, M. B., Humphrey, P. P., and Dollery, C. T. (1984). Effects of intravenous infusions of prostaglandin D2 in man. *Prostaglandins* **28**, 755–767.
81. Fujitani, Y., Kanaoka, Y., Aritake, K., Uodome, N., Okazaki-Hatake, K., and Urade, Y. (2002). Pronounced eosinophilic lung inflammation and Th2 cytokine release in human lipocalin-type prostaglandin D synthase transgenic mice. *J Immunol*, **168**, 443–449.
82. Inoue, T., Takayanagi, K., Morooka, S., Uehara, Y., Oda, H., Seiki, K., Nakajima, H., and Urade, Y. (2001). Serum prostaglandin D synthase level after coronary angioplasty may predict occurrence of restenosis. *Thromb. Haemost.* **85**, 165–170.
83. Noguchi, E., Shibasaki, M., Kamioka, M., Yokouchi, Y., Yamakawa-Kobayashi, K., Hamaguchi, H., Matsui, A., and Arinami, T. (2002). New polymorphisms of haematopoietic prostaglandin D synthase and human prostanoid DP receptor genes. *Clin. Exp. Allergy* **32**, 93–96.
84. Hayaishi, O. (2000). Molecular mechanisms of sleep-wake regulation: a role of prostaglandin D2. *Philos. Trans. R. Soc. London B Biol. Sci.* **355**, 275–280.

85. Eguchi, N., Minami, T., Shirafuji, N., Kanaoka, Y., Tanaka, T., Nagata, A., Yoshida, N., Urade, Y., Ito, S., and Hayaishi, O. (1999). Lack of tactile pain (allodynia) in lipocalin-type prostaglandin D synthase-deficient mice. *Proc. Natl. Acad. Sci. USA* **96**, 726–730.
86. Monneret, G., Gravel, S., Diamond, M., Rokach, J., and Powell, W. S. (2001). Prostaglandin D2 is a potent chemoattractant for human eosinophils that acts via a novel DP receptor. *Blood* **98**, 1942–1948.
87. Michimata, T., Tsuda, H., Sakai, M., Fujimura, M., Nagata, K., Nakamura, M., and Saito, S. (2002). Accumulation of CRTH2-positive T-helper 2 and T-cytotoxic 2 cells at implantation sites of human decidua in a prostaglandin D(2)-mediated manner. *Mol. Hum. Reprod.* **8**, 181–187.
88. Fitzpatrick, F. A. and Wynalda, M. A. (1983). Albumin-catalyzed metabolism of prostaglandin D2. Identification of products formed in vitro. *J. Biol. Chem.* **258**, 11713–11718.
89. Harris, S. G., Padilla, J., Koumas, L., Ray, D., and Phipps, R. P. (2002). Prostaglandins as modulators of immunity. *Trends Immunol.* **23**, 144–150.
90. Forman, B. M., Tontonoz, P., Chen, J., Brun, R. P., Spiegelman, B. M., and Evans, R. M. (1995). 15-Deoxy-delta 12, 14-prostaglandin J2 is a ligand for the adipocyte determination factor PPAR gamma. *Cell* **83**, 803–812.
91. Matsuoka, T., Hirata, M., Tanaka, H., Takahashi, Y., Murata, T., Kabashima, K., Sugimoto, Y., Kobayashi, T., Ushikubi, F., Aze, Y., Eguchi, N., Urade, Y., Yoshida, N., Kimura, K., Mizoguchi, A., Honda, Y., Nagai, H., and Narumiya, S. (2000). Prostaglandin D2 as a mediator of allergic asthma. *Science* **287**, 2013–2017.
92. Mizoguchi, A., Eguchi, N., Kimura, K., Kiyohara, Y., Qu, W. M., Huang, Z. L., Mochizuki, T., Lazarus, M., Kobayashi, T., Kaneko, T., Narumiya, S., Urade, Y., and Hayaishi, O. (2001). Dominant localization of prostaglandin D receptors on arachnoid trabecular cells in mouse basal forebrain and their involvement in the regulation of non-rapid eye movement sleep. *Proc. Natl. Acad. Sci. USA* **98**, 11674–11679.
93. Dubois, R. N., Abramson, S. B., Crofford, L., Gupta, R. A., Simon, L. S., Van De Putte, L. B., and Lipsky, P. E. (1998). Cyclooxygenase in biology and disease. *FASEB J.* **12**, 1063–1073.
94. Lydford, S. J., McKechnie, K. C., and Dougall, I. G. (1996). Pharmacological studies on prostanoid receptors in the rabbit isolated saphenous vein: a comparison with the rabbit isolated ear artery. *Br. J. Pharmacol.* **117**, 13–20.
95. Coleman, R. A., Kennedy, I., Humphrey, P. P. A., Bunce, K., and Lumley, P. (1990). Prostanoids and their Receptors. In Hansch, C., Ed., *Comprehensive Medicinal Chemistry*. Pergamon Press, New York, pp. 643–714.
96. Jakobsson, P. J., Thoren, S., Morgenstern, R., and Samuelsson, B. (1999). Identification of human prostaglandin E synthase: a microsomal, glutathione-dependent, inducible enzyme, constituting a potential novel drug target. *Proc. Natl. Acad. Sci. USA* **96**, 7220–7225.
97. Guan, Y., Zhang, Y., Schneider, A., Riendeau, D., Mancini, J. A., Davis, L., Komhoff, M., Breyer, R. M., and Breyer, M. D. (2001). Urogenital distribution of a mouse membrane-associated prostaglandin E(2) synthase. *Am. J. Physiol. Renal Physiol.* **281**, F1173–F1177.
98. Yang, T., Singh, I., Pham, H., Sun, D., Smart, A., Schnermann, J. B., and Briggs, J. P. (1998). Regulation of cyclooxygenase expression in the kidney by dietary salt intake. *Am. J. Physiol.* **274**, F481–F489.
99. Breyer, M. D. and Breyer, R. M. (2000). Prostaglandin E receptors and the kidney. *Am. J. Physiol. Renal Physiol.* **279**, F12–F23.
100. Okuda-Ashitaka, E., Sakamoto, K., Ezashi, T., Miwa, K., Ito, S., and Hayaishi, O. (1996). Suppression of prostaglandin E receptor signaling by the variant form of EP1 subtype. *J. Biol. Chem.* **271**, 31255–31261.
101. Katsuyama, M., Nishigaki, N., Sugimoto, Y., Morimoto, K., Negishi, M., Narumiya, S., and Ichikawa, A. (1995). The mouse prostaglandin E receptor EP2 subtype: cloning, expression, and northern blot analysis. *FEBS Lett.* **372**, 151–156.
102. Fabre, J. E., Nguyen, M., Athirakul, K., Coggins, K., McNeish, J. D., Austin, S., Parise, L. K., FitzGerald, G. A., Coffman, T. M., and Koller, B. H. (2001). Activation of the murine EP3 receptor for PGE2 inhibits cAMP production and promotes platelet aggregation. *J. Clin. Invest.* **107**, 603–610.
103. Kennedy, C. R., Zhang, Y., Brandon, S., Guan, Y., Coffee, K., Funk, C. D., Magnuson, M. A., Oates, J. A., Breyer, M. D., and Breyer, R. M. (1999). Salt-sensitive hypertension and reduced fertility in mice lacking the prostaglandin EP2 receptor. *Nat. Med.* **5**, 217–220.
104. Jensen, B. L., Schmid, C., and Kurtz, A. (1996). Prostaglandins stimulate renin secretion and renin mRNA in mouse renal juxtaglomerular cells. *Am. J. Physiol.* **271**, F659–F669.
105. Inscho, E. W., Carmines, P. K., and Navar, L. G. (1990). Prostaglandin influences on afferent arteriolar responses to vasoconstrictor agonists. *Am. J. Physiol.* **259**, F157–F163.
106. Araki, H., Ukawa, H., Sugawa, Y., Yagi, K., Suzuki, K., and Takeuchi, K. (2000). The roles of prostaglandin E receptor subtypes in the cytoprotective action of prostaglandin E2 in rat stomach. *Aliment Pharmacol. Ther.* **14** (Suppl 1), 116–124.
107. Kunikata, T., Tanaka, A., Miyazawa, T., Kato, S., and Takeuchi, K. (2002). 16,16-Dimethyl prostaglandin E2 inhibits indomethacin-induced small intestinal lesions through EP3 and EP4 receptors. *Dig. Dis. Sci.* **47**, 894–904.
108. Asboth, G., Phaneuf, S., Europe-Finner, G. N., Toth, M., and Bernal, A. L. (1996). Prostaglandin E2 activates phospholipase C and elevates intracellular calcium in cultured myometrial cells: involvement of EP1 and EP3 receptor subtypes. *Endocrinology* **137**, 2572–2579.
109. Tani, K., Naganawa, A., Ishida, A., Egashira, H., Sagawa, K., Harada, H., Ogawa, M., Maruyama, T., Ohuchida, S., Nakai, H., Kondo, K., and Toda, M. (2001). Design and synthesis of a highly selective EP2-receptor agonist. *Bioorg. Med. Chem. Lett.* **11**, 2025–2028.
110. Sonoshita, M., Takaku, K., Sasaki, N., Sugimoto, Y., Ushikubi, F., Narumiya, S., Oshima, M., and Taketo, M. M. (2001). Acceleration of intestinal polyposis through prostaglandin receptor EP2 in Apc(Delta 716) knockout mice. *Nat. Med.* **7**, 1048–1051.
111. Pai, R., Soreghan, B., Szabo, I. L., Pavelka, M., Baatar, D., and Tarnawski, A. S. (2002). Prostaglandin E2 transactivates EGF receptor: a novel mechanism for promoting colon cancer growth and gastrointestinal hypertrophy. *Nat. Med.* **8**, 289–293.
112. Nataraj, C., Thomas, D. W., Tilley, S. L., Nguyen, M. T., Mannon, R., Koller, B. H., and Coffman, T. M. (2001). Receptors for prostaglandin E(2) that regulate cellular immune responses in the mouse. *J. Clin. Invest.* **108**, 1229–1235.
113. Ek, M., Engblom, D., Saha, S., Blomqvist, A., Jakobsson, P. J., and Ericsson-Dahlstrand, A. (2001). Inflammatory response: pathway across the blood-brain barrier. *Nature* **410**, 430–431.
114. Yoshida, Y., Matsumura, H., Nakajima, T., Mandai, M., Urakami, T., Kuroda, K., and Yoneda, M. (2000). Prostaglandin E (EP) receptor subtypes and sleep: promotion by EP4 and inhibition by EP1/EP2. *Neuroreport* **11**, 2127–2131.
115. Ushikubi, F., Segi, E., Sugimoto, Y., Murata, T., Matsuoka, T., Kobayashi, T., Hizaki, H., Tuboi, K., Katsuyama, M., Ichikawa, A., Tanaka, T., Yoshida, N., and Narumiya, S. (1998). Impaired febrile response in mice lacking the prostaglandin E receptor subtype EP3. *Nature* **395**, 281–284.
116. Stock, J. L., Shinjo, K., Burkhardt, J., Roach, M., Taniguchi, K., Ishikawa, T., Kim, H. S., Flannery, P. J., Coffman, T. M., McNeish, J. D., and Audoly, L. P. (2001). The prostaglandin E2 EP1 receptor mediates pain perception and regulates blood pressure. *J. Clin. Invest.* **107**, 325–331.
117. Fleming, E. F., Athirakul, K., Oliverio, M. I., Key, M., Goulet, J., Koller, B. H., and Coffman, T. M. (1998). Urinary concentrating function in mice lacking EP3 receptors for prostaglandin E2. *Am. J. Physiol.* **275**, F955–F961.
118. Nguyen, M., Camenisch, T., Snouwaert, J. N., Hicks, E., Coffman, T. M., Anderson, P. A., Malouf, N. N., and Koller, B. H. (1997). The prostaglandin receptor EP4 triggers remodelling of the cardiovascular system at birth. *Nature* **390**, 78–81.

119. Horton, E. W. and Poyser, N. L. (1976). Uterine luteolytic hormone: a physiological role for prostaglandin F2alpha. *Physiol. Rev.* **56**, 595–651.
120. Dong, Y. J., Jones, R. L., and Wilson, N. H. (1986). Prostaglandin E receptor subtypes in smooth muscle: agonist activities of stable prostacyclin analogues. *Br. J. Pharmacol.* **87**, 97–107.
121. Barnard, J. W., Ward, R. A., and Taylor, A. E. (1992). Evaluation of prostaglandin F2 alpha and prostacyclin interactions in the isolated perfused rat lung. *J. Appl. Physiol.* **72**, 2469–2474.
122. Suzuki-Yamamoto, T., Nishizawa, M., Fukui, M., Okuda-Ashitaka, E., Nakajima, T., Ito, S., and Watanabe, K. (1999). cDNA cloning, expression and characterization of human prostaglandin F synthase. *FEBS Lett.* **462**, 335–340.
123. Endo, K., Fukui, M., Mishima, M., and Watanabe, K. (2001). Metabolism of vitamin A affected by prostaglandin F synthase in contractile interstitial cells of bovine lung. *Biochem. Biophys. Res. Commun.* **287**, 956–961.
124. Vanegas, H. and Schaible, H. G. (2001). Prostaglandins and cyclooxygenases [correction of cycloxygenases] in the spinal cord. *Prog. Neurobiol.* **64**, 327–363.
125. Adams, J. W., Migita, D. S., Yu, M. K., Young, R., Hellickson, M. S., Castro-Vargas, F. E., Domingo, J. D., Lee, P. H., Bui, J. S., and Henderson, S. A. (1996). Prostaglandin F2 alpha stimulates hypertrophic growth of cultured neonatal rat ventricular myocytes. *J. Biol. Chem.* **271**, 1179–1186.
126. Pierce, K. L., Fujino, H., Srinivasan, D., and Regan, J. W. (1999). Activation of FP prostanoid receptor isoforms leads to Rho-mediated changes in cell morphology and in the cell cytoskeleton. *J. Biol. Chem.* **274**, 35944–35949.
127. Fujino, H., Srinivasan, D., Pierce, K. L., and Regan, J. W. (2000). Differential regulation of prostaglandin F(2alpha) receptor isoforms by protein kinase C. *Mol. Pharmacol.* **57**, 353–358.
128. Fujino, H. and Regan, J. W. (2001). FP prostanoid receptor activation of a T-cell factor/beta-catenin signaling pathway. *J. Biol. Chem.* **276**, 12489–12492.
129. Kunapuli, P., Lawson, J. A., Rokach, J., and FitzGerald, G. A. (1997). Functional characterization of the ocular prostaglandin f2alpha (PGF2alpha) receptor. Activation by the isoprostane, 12-iso-PGF2alpha. *J. Biol. Chem.* **272**, 27147–27154.
130. Hesketh, T. R., Moore, J. P., Morris, J. D., Taylor, M. V., Rogers, J., Smith, G. A., and Metcalfe, J. C. (1985). A common sequence of calcium and pH signals in the mitogenic stimulation of eukaryotic cells. *Nature* **313**, 481–484.
131. Kunapuli, P., Lawson, J. A., Rokach, J. A., Meinkoth, J. L., and FitzGerald, G. A. (1998). Prostaglandin F2alpha (PGF2alpha) and the isoprostane, 8,12-isoprostane F2alpha-III, induce cardiomyocyte hypertrophy. Differential activation of downstream signaling pathways. *J. Biol. Chem.* **273**, 22442–22452.
132. Casimir, D. A., Miller, C. W., and Ntambi, J. M. (1996). Preadipocyte differentiation blocked by prostaglandin stimulation of prostanoid FP2 receptor in murine 3T3-L1 cells. *Differentiation* **60**, 203–210.
133. Sugimoto, Y., Yamasaki, A., Segi, E., Tsuboi, K., Aze, Y., Nishimura, T., Oida, H., Yoshida, N., Tanaka, T., Katsuyama, M., Hasumoto, K., Murata, T., Hirata, M., Ushikubi, F., Negishi, M., Ichikawa, A., and Narumiya, S. (1997). Failure of parturition in mice lacking the prostaglandin F receptor. *Science* **277**, 681–683.
134. Tsuboi, K., Sugimoto, Y., Iwane, A., Yamamoto, K., Yamamoto, S., and Ichikawa, A. (2000). Uterine expression of prostaglandin H2 synthase in late pregnancy and during parturition in prostaglandin F receptor-deficient mice. *Endocrinology* **141**, 315–324.

CHAPTER 168

Leukotriene Mediators

Jesper Z. Haeggström and Anders Wetterholm
*Department of Medical Biochemistry and Biophysics,
Division of Chemistry II, Karolinska Institutet,
S-171 77 Stockholm, Sweden*

Introduction

The leukotrienes (LT) constitute a group of bioactive lipids derived from the metabolism of polyunsaturated fatty acids, e.g., arachidonic acid [1]. In two consecutive reactions, arachidonic acid is transformed into an unstable epoxide compound, LTA$_4$. This intermediate is either hydrolyzed into the dihydroxy acid LTB$_4$ or conjugated with glutathione to form LTC$_4$. The latter compound together with its metabolites LTD$_4$ and LTE$_4$ are referred to as the cysteinyl-containing leukotrienes (cys-LTs).

Leukotrienes possess a wide range of biological activities elicited via specific, G-protein-coupled, cell surface receptors [2]. LTB$_4$ is a very potent chemoattractant for neutrophils and recruits inflammatory cells to the site of injury. This compound also induces chemokinesis and increases leukocyte adhesion to the endothelial cells of the vessel wall. The cys-LTs are potent constrictors of smooth muscles, particularly in the airways, leading to bronchoconstriction. In the microcirculation, the cys-LTs constrict arterioles and increase the permeability of the postcapillary venules, which results in extravasation of plasma. Due to their potent biological activities, leukotrienes are considered to be chemical mediators in a number of inflammatory and allergic disorders, e.g. rheumatoid arthritis, inflammatory bowel disease, and bronchial asthma [3].

Five-Lipoxygenase

Five-lipoxygenase (5-LO) catalyzes the first two steps in leukotriene biosynthesis [4] (Fig. 1). Free arachidonic acid is oxygenated into the hydroperoxide 5-HPETE, which is subsequently dehydrated to yield the unstable epoxide intermediate LTA$_4$. The enzyme, which predominantly is found in bone marrow derived cells, is stimulated by Ca^{2+} and ATP. Furthermore, it contains one atom of non-heme iron that is involved in catalysis [5]. Mutagenetic analysis has demonstrated that His-372, His-550, and the C-terminal isoleucin Ile-673 are iron ligands [6]. The gene encoding human 5-LO, as well as the promoter, has been characterized, and some important features are listed in Table I, together with data for the other key enzymes in the leukotriene cascade.

The only crystal structure of a mammalian lipoxygenase that has been determined is rabbit 15-LO [7]. This enzyme contains an N-terminal β-barrel domain, a structure also found in the C-terminal domain of lipases. The role of this domain for lipoxygenases is presently unclear but for 5-LO it has been shown to bind Ca^{2+}, which stimulates enzyme activity and presumably facilitates its association of 5-LO with membranes during catalysis (see following section) [8]. 5-LO is also a substrate for p38 kinase-dependent MAPKAP kinases *in vitro*, suggesting that phosphorylation may be one additional factor, which determines 5-LO translocation and enzyme activity [9].

Five-Lipoxygenase Activating Protein (FLAP) and Cellular Leukotriene Biosynthesis

Cellular 5-LO activity is dependent on a small membrane protein, five lipoxygenase activating protein (FLAP), which presumably presents or transfers arachidonic acid to 5-LO [10].

Early studies showed that upon cell stimulation leading to an increase in Ca^{2+}, 5-LO is activated and translocates to a membrane compartment [11]. Of particular interest was the discovery that FLAP is localized to the nuclear envelope of

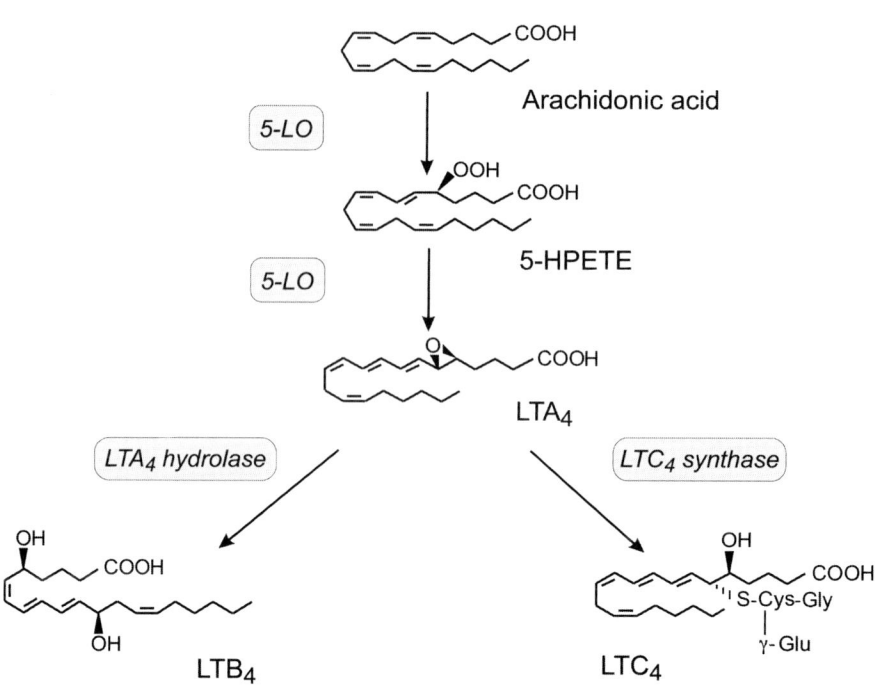

Figure 1 Enzymes and intermediates in the leukotriene cascade.

Table I Properties of Enzymes and Receptors in Leukotriene Biosynthesis and Action[a]

Protein	Protein size (no. of amino acids)[b]	Prosthetic group[c]	Gene size (kb)	Exon no.	Putative cis-elements of promoter regions	Chromosomal location	Gene deficient mice
5-Lipoxygenase	673	Fe	>82	14	Sp1, AP-2, NF-κB	10	+
FLAP	160	—	>31	5	TATA, AP-2, GRE	13	+
LTA$_4$ hydrolase	610	Zn	>35	19	XRE, AP-2	12	+
LTC$_4$ synthase	149	—	2.5	5	Sp1, AP-1, AP-2	5	+
BLT$_1$	351	—	5.5	3	Sp1, CpG site, NFκB AP-1	14	+
BLT$_2$[d]	357	—	ND[e]	1[e]		14	−
CysLT$_1$	336	—	ND	ND		X	+
CysLT$_2$	345	—	ND	ND		13	−

[a]Data refer to human proteins. ND, not determined.
[b]Initial methionine excluded.
[c]1 mol metal per mol protein.
[d]The ORF of BLT$_2$ is included in the promoter of the BLT$_1$ gene.

neutrophils and that 5-LO, upon cell activation, translocates to the same compartment [12] (Fig. 2). Further analysis revealed that 5-LO can also be present in the nucleus of *resting* cells associated with the nuclear euchromatin, a site from which it translocates to the nuclear envelope. In addition, 5-LO has been shown to associate with growth factor receptor-binding protein 2 (Grb2), an "adaptor" protein for tyrosine kinase-mediated cell signaling, through Src homology 3 (SH3) domain interactions [13]. It is interesting that inhibitors of tyrosine kinase activity, a determinant of SH3 interactions, also inhibited the catalytic activity of 5-LO and its translocation during cellular activation. In addition, an internal bipartite nuclear localization sequence, spanning Arg-638–Lys-655, has been shown to be necessary for the redistribution of 5-LO to the nuclear compartment [14,15]. Moreover, recent data indicate that also the N-terminal β-barrel domain in 5-LO plays a role in this process [16].

Not only 5-LO and FLAP are associated with the cell nucleus and nuclear membrane. Thus, LTC$_4$ synthase (see section entitled Leukotriene C$_4$ Synthase) resides in this compartment, and recent data suggest that the enzyme is located on the *outer* nuclear membrane and peripheral endoplasmic reticulum [17]. It is interesting that the soluble LTA$_4$ hydrolase (see section entitled Leukotriene A$_4$ Hydrolase) was also reported to reside in the nucleus of rat basophilic leukemia cells and rat alveolar macrophages [18].

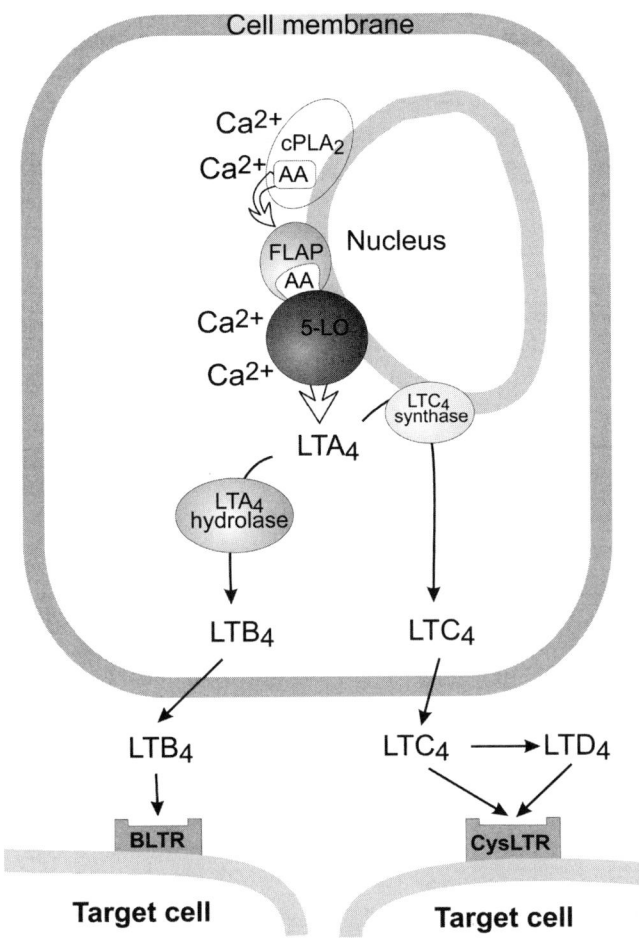

Figure 2 Leukotriene biosynthesis at the nuclear membrane of an activated leukocyte.

Together, these findings imply that leukotriene biosynthesis is carried out by a complex of enzymes assembled at the nuclear membrane (cf. Fig. 2). This conclusion in turn suggests that these enzymes and/or their products may have additional intracellular and intranuclear functions, perhaps related to signal transduction or gene regulation. In line with this notion, it has been reported that LTB_4 is a natural ligand to the nuclear orphan receptor PPARα, suggesting that LTB_4 may have intranuclear functions [19]. It was also reported that 5-LO can interact with several cellular proteins, including coactosine-like protein (CLP) and transforming growth factor type β-receptor-I-associated protein (TRAP-1) [20]. In addition, 5-LO interacts with a human homologue of the protein "Dicer," a member of the RNase III family of nucleases, which is implicated in the RNA interference mechanism of gene regulation [20,21].

Leukotriene A_4 Hydrolase

Leukotriene A_4 hydrolase catalyzes the final step in the biosynthesis of the proinflammatory compound LTB_4 (Fig. 1). In contrast to 5-LO, LTA_4 hydrolase is widely distributed and has been detected in almost all mammalian cells, organs, and tissues examined. The enzyme has been purified from several mammalian sources, and cDNAs encoding the human, mouse, rat, and guinea-pig enzymes have been cloned and sequenced [22].

Sequence comparison with certain zinc metalloenzymes revealed the presence of a zinc-binding motif (**HEXXH-X_{18}-E**) in LTA_4 hydrolase [23]. Accordingly, LTA_4 hydrolase was found to contain a catalytic zinc. The three proposed zinc-binding ligands, His-295, His-299, and Glu-318, were verified by mutagenetic analysis. Furthermore, the enzyme was found to exhibit a chloride-activated peptidase activity. Based on its zinc signature, sequence homology, and aminopeptidase activity, LTA_4 hydrolase has been classified as a member of the M1 family of the MA clan of metallopeptidases [24].

Identification of Catalytically Important Amino Acid Residues and Crystal Structure of LTA_4 Hydrolase

In addition to the zinc-binding ligands, several amino acid residues of catalytic importance have been identified by site-directed mutagenesis. Thus, mutagenetic replacements of Glu-296 in LTA_4 hydrolase abrogated only the peptidase activity, a finding that suggests a direct catalytic role for Glu-296 in the peptidase reaction, possibly as a general base [25]. Furthermore, sequence comparisons and mutational analysis have demonstrated that Tyr-383 plays an important role in the peptidase reaction of LTA_4 hydrolase, presumably as a proton donor [26].

Typically, LTA_4 hydrolase undergoes "suicide" inactivation with a concomitant covalent modification of the enzyme by its substrate LTA_4 [27]. Mutational analysis has demonstrated that Tyr-378 is a major structural determinant for suicide inactivation [28]. Mutated proteins, carrying a Gln or Phe residue in position 378, were neither inactivated nor covalently modified by LTA_4.

Recently, the X-ray crystal structure of LTA_4 hydrolase in complex with the competitive inhibitor bestatin was determined [29]. The protein molecule is folded into an N-terminal, a catalytic, and a C-terminal domain, packed in a flat triangular arrangement. Although the three domains pack closely and make contact with each other, a deep cleft is created between them. At the bottom of the interdomain cleft, the zinc site is located. As predicted from previous work, the metal is bound to the three amino acid ligands, His-295, His-299, and Glu-318. In the vicinity of the prosthetic zinc, the catalytic residues Glu-296 and Tyr-383 are located at positions that are commensurate with their proposed roles as general base and proton donor in the aminopeptidase reaction.

Close to the catalytic zinc, a glutamic acid residue (Glu-271), belonging to a conserved G*X*MEN motif in the M1 family of zinc peptidases, was identified [29]. By mutational analysis and crystallography it was shown that Glu-271 is necessary for *both* catalytic activities of LTA_4 hydrolase [30]. Presumably, the carboxylate of the glutamic acid residue participates in the opening of the epoxide moiety of LTA_4

and formation of a carbocation intermediate. In the peptidase reaction, the role of Glu-271 may be to serve as an N-terminal recognition site and to stabilize the transition state during turnover of peptide substrates.

The crystal structure, in combination with site-directed mutagenesis studies, also suggested that Asp-375 is a critical determinant for the introduction of the 12R-hydroxyl group of LTB4 [31].

Leukotriene C_4 Synthase

Leukotriene C_4 synthase catalyzes the committed step in the biosynthesis of cys-LTs through conjugation of LTA_4 with glutathione. The enzyme is a membrane-bound homodimer with a subunit molecular mass of 18 kDa [32]. LTC_4 synthase has been cloned and sequenced [33,34]. Two consensus sequences for protein kinase C phosphorylation were found, and subsequent studies have shown that phosphorylation reduces the LTC_4 synthase activity [35]. Sequence comparisons of LTC_4 synthase and FLAP demonstrated a surprising 31% identity between the two proteins. In addition, recent work has identified two microsomal GSH transferases (MGST2 and MGST3) that possess LTC_4 synthase activity and exhibit a high degree of similarity to both LTC_4 synthase and FLAP [36,37].

Leukotriene Receptors

For LTB4, two types of surface receptors are known (BLT_1 and BLT_2). The BLT_1-receptor has been cloned and characterized as a 43 kDa, G-protein-coupled receptor with seven transmembrane-spanning domains (7TM) [38]. The BLT_1 receptor is only expressed in inflammatory cells [39] and shows a high degree of specificity for LTB_4 with a K_d of 0.15–1 nM [38,40].

A second G-protein-coupled 7TM receptor for LTB_4, BLT_2, has recently been identified [40–42]. This receptor is homologous to the BLT_1 receptor but has a higher K_d value for LTB_4 (23 nM) [43]. In contrast to the BLT_1 receptor, BLT_2 is ubiquitously expressed in various tissues.

The cys-LTs are recognized by at least two receptor types ($CysLT_1$ and $CysLT_2$), both of which have been cloned and characterized as G-protein-coupled 7TM receptors [44–48]. The $CysLT_1$ receptor mRNA is found in, for example, spleen, peripheral blood leukocytes, lung tissue, smooth muscle cells, and tissue macrophages [45,47]. The preferred ligands for the $CysLT_1$ receptor are LTD_4 followed by LTC_4 and LTE_4 in decreasing order of potency.

The $CysLT_2$ receptor contains 345 amino acids with approximately 40% sequence identity to the $CysLT_1$ receptor [44,46,48]. This receptor binds LTC_4 and LTD_4 equally well, whereas LTE_4 shows low affinity to the receptor. Studies on the tissue distribution of the $CysLT_2$ receptor show high levels of mRNA in heart, brain, peripheral blood leukocytes, spleen, placenta, and lymph nodes, whereas only small amounts are found in the lung. The functional role(s) of the $CysLT_2$ receptor is presently unclear, but its wide tissue distribution suggests many possibilities, including regulation of brain and/or cardiac functions.

Gene Targeting of Enzymes and Receptors in the Leukotriene Cascade

The roles of the key enzymes and two of the receptors (BLT_1 and $CysLT_1$) in the leukotriene cascade have been studied by gene targeting. 5-LO-deficient mice are more resistant to lethal effects of PAF-induced shock and also show a marked reduction in the ear inflammatory response to exogenous arachidonic acid [49]. Furthermore, 5-LO null mice are more susceptible to infections with *Klebsiella pnemoniae* [50], exhibit a reduced airway reactivity in response to methacholine, and have lower levels of serum immunoglobulins [51].

FLAP deficient mice, like the 5-LO (−/−) mice, showed a blunted response to topical arachidonic acid, had increased resistance to PAF induced shock, and responded with less edema in zymosan-induced peritonitis [52]. Furthermore, the severity of collagen-induced arthritis was substantially reduced in FLAP (−/−) mice, thereby indicating a role for leukotrienes in this model of inflammation [53].

LTA_4 hydrolase (−/−) mice are resistant to the lethal effects of systemic shock induced by PAF, thus identifying LTB_4 as a key mediator of this reaction [54]. In zymosan A-induced peritonitis, LTB_4 modulates only the cellular component of the response, whereas the LTC_4 synthase (−/−) mice displayed a reduced plasma protein extravasation in this type of inflammation [55]. Furthermore, the LTC_4 synthase (−/−) mice were less prone to develop passive cutaneous anaphylaxis. Recently, the role of LTC_4 in plasma protein extravasation following zymosan A–induced peritonitis and IgE-mediated passive cutaneous anaphylaxis was confirmed in mice lacking the $CysLT_1$ receptor gene [56].

Finally, the role of the BLT_1 receptor has also been studied by targeted gene disruption [57,58]. The receptor was necessary to elicit the physiological effects of LTB_4 (e.g. chemotaxis, calcium mobilization, and adhesion to endothelium) and important for the recruitment of leukocytes in an *in vivo* model of peritonitis. As also observed in mice lacking 5-LO, FLAP, or LTA_4 hydrolase, BLT_1 (−/−) mice were protected from the lethal effects of PAF-induced anaphylaxis.

Acknowledgments

This work was supported by the Swedish Medical Research Council (O3X-10350), the European Union (QLG1-CT-2001-01521), the Vårdal Foundation, the Swedish Foundation for Strategic Research, and Konung Gustav V:s 80-Årsfond.

References

1. Funk, C. D. (2001). Prostaglandins and leukotrienes: Advances in eicosanoid biology. *Science* **294**, 1871–1875.
2. Izumi, T., Yokomizu, T., Obinata, H., Ogasawara, H., and Shimizu, T. (2002). Leukotriene receptors: Classification, gene expression, and signal transduction. *J. Biochem.* **132**, 1–6.

3. Lewis, R. A., Austen, K. F., and Soberman, R. J. (1990). Leukotrienes and other products of the 5-lipoxygenase pathway. *New Engl. J. Med.* **323**, 645–655.
4. Rouzer, C. A., Matsumoto, T., and Samuelsson, B. (1986). Single protein from human leukocytes possesses 5-lipoxygenase and leukotriene A_4 synthase activities. *Proc. Natl. Acad. Sci. USA* **83**, 857–861.
5. Percival, M. D. (1991). Human 5-lipoxygenase contains an essential iron. *J. Biol. Chem.* **266**, 10058–10061.
6. Rådmark, O. (2000). Mutagenesis studies of mammalian lipoxygenases. In *Molecular and Cellular Basis of Inflammation*, C. N. Serhan and P. A. Ward, Eds., pp. 93–108. Humana Press Inc., Totowa, NJ.
7. Gillmor, S. A., Villaseñor, A., Fletterick, R., Sigal, E., and Browner, M. (1997). The structure of mammalian 15-lipoxygenase reveals similarity to the lipases and the determinants of substrate specificity. *Nature Struct. Biol.* **4**, 1003–1009.
8. Hammarberg, T., Provost, P., Persson, B., and Radmark, O. (2000). The N-terminal domain of 5-lipoxygenase binds calcium and mediates calcium stimulation of enzyme activity. *J. Biol. Chem.* **275**, 38787–38793.
9. Werz, O., Klemm, J., Samuelsson, B., and Radmark, O. (2000). 5-lipoxygenase is phosphorylated by p38 kinase-dependent MAPKAP kinases. *Proc. Natl. Acad. Sci. USA* **97**, 5261–5266.
10. Ford-Hutchinson, A. W., Gresser, M., and Young, R. N. (1994). 5-Lipoxygenase. *Annu. Rev. Biochem.* **63**, 383–417.
11. Rouzer, C. A., and Kargman, S. (1988). Translocation of 5-lipoxygenase to the membrane in human leukocytes challenged with ionophore A23187. *J. Biol. Chem.* **263**, 10980–10988.
12. Peters-Golden, M. and Brock, T. G. (2001). Intracellular compartmentalization of leukotriene synthesis: unexpected nuclear secrets. *FEBS Lett.* **487**, 323–326.
13. Lepley, R. A. and Fitzpatrick, F. A. (1994). 5-Lipoxygenase contains a functional Src homology 3-binding motif that interacts with the Src homology 3 domain of Grb2 and cytoskeletal proteins. *J. Biol. Chem.* **269**, 24163–24168.
14. Lepley, R. A. and Fitzpatrick, F. A. (1998). 5-Lipoxygenase compartmentalization in granulocytic cells is modulated by an internal bipartite nuclear localizing sequence and nuclear factor kappa B complex formation. *Arch. Biochem. Biophys.* **356**, 71–76.
15. Healy, A. M., Peters-Golden, M., Yao, J. P., and Brock, T. G. (1999). Identification of a bipartite nuclear localization sequence necessary for nuclear import of 5-lipoxygenase. *J. Biol. Chem.* **274**, 29812–29818.
16. Chen, X. S. and Funk, C. D. (2001). The N-terminal "beta-barrel" domain of 5-lipoxygenase is essential for nuclear membrane translocation. *J. Biol. Chem.* **276**, 811–818.
17. Christmas, P., Weber, B. M., McKee, M., Brown, D., and Soberman, R. J. (2002). Membrane localization and topology of leukotriene C_4 synthase. *J. Biol. Chem.* **277**, 28902–28908.
18. Brock, T. G., Maydanski, E., McNish, R. W., and Peters-Golden, M. (2001). Co-localization of leukotriene A_4 hydrolase with 5-lipoxygenase in nuclei of alveolar macrophages and rat basophilic leukemia cells but not neutrophils. *J. Biol. Chem.* **276**, 35071–35077.
19. Devchand, P. R., Keller, H., Peters, J. M., Vazquez, M., Gonzalez, F. J., and Wahli, W. (1996). The PPARα-leukotriene B_4 pathway to inflammation control. *Nature* **384**, 39–43.
20. Provost, P., Samuelsson, B., and Rådmark, O. (1999). Interaction of 5-lipoxygenase with cellular proteins. *Proc. Natl. Acad. Sci. USA* **96**, 1881–1885.
21. Bernstein, E., Caudy, A. A., Hammond, S. M., and Hannon, G. J. (2001). Role for a bidentate ribonuclease in the initiation step of RNA interference. *Nature* **409**, 363–366.
22. Wetterholm, A., Blomster, M., and Haeggström, J. Z. (1996). Leukotriene A_4 hydrolase: a key enzyme in the biosynthesis of leukotriene B_4. In *Eicosanoids: From Biotechnology to Therapeutic Applications*, G. Folco, B. Samuelsson, J. Maclouf, and G. P. Velo, Eds., pp. 1–12. Plenum Press, New York.
23. Haeggstrom, J. Z. (2000). Structure, function, and regulation of leukotriene A_4 hydrolase. *Am. J. Resp. Crit. Care Med.* **161**, S25–31.
24. Barret, A. J., Rawlings, N. D., and Woessner, J. F. (1998). Family M1 of membrane alanyl aminopeptidase. In *Handbook of Proteolytic Enzymes*, A. J. Barret, N. D. Rawlings, and J. F. Woessner, Eds., pp. 994–996. Academic Press, London, San Diego.
25. Wetterholm, A., Medina, J. F., Rådmark, O., Shapiro, R., Haeggström, J. Z., Vallee, B. L., and Samuelsson, B. (1992). Leukotriene A_4 hydrolase: Abrogation of the peptidase activity by mutation of glutamic acid-296. *Proc. Natl. Acad. Sci. USA* **89**, 9141–9145.
26. Blomster, M., Wetterholm, A., Mueller, M. J., and Haeggström, J. Z. (1995). Evidence for a catalytic role of tyrosine 383 in the peptidase reaction of leukotriene A_4 hydrolase. *Eur. J. Biochem.* **231**, 528–534.
27. Orning, L., Gierse, J., Duffin, K., Bild, G., Krivi, G., and Fitzpatrick, F. A. (1992). Mechanism-based inactivation of leukotriene A_4 hydrolase/aminopeptidase by leukotriene A_4. Mass spectrometric and kinetic characterization. *J. Biol. Chem.* **267**, 22733–22739.
28. Mueller, M. J., Blomster, M., Oppermann, U. C. T., Jörnvall, H., Samuelsson, B., and Haeggstrom, J. Z. (1996). Leukotriene A_4 hydrolase—protection from mechanism-based inactivation by mutation of tyrosine-378. *Proc. Natl. Acad. Sci. USA* **93**, 5931–5935.
29. Thunnissen, M. G. M., Nordlund, P., and Haeggström, J. Z. (2001). Crystal structure of human leukotriene A_4 hydrolase, a bifunctional enzyme in inflammation. *Nature Str. Biol.* **8**, 131–135.
30. Rudberg, P. C., Tholander, F., Thunnissen, M. M. G. M., and Haeggström, J. Z. (2002). Leukotriene A_4 hydrolase/aminopeptidase: Glutamate 271 is a catalytic residue with specific roles in two distinct enzyme mechanisms. *J. Biol. Chem.* In press.
31. Rudberg, P. C., Tholander, F., Thunnissen, M. M. G. M., Samuelsson, B., and Haeggström, J. Z. (2002). Leukotriene A_4 hydrolase: selective abrogation of leukotriene formation by mutation of aspartic acid 375. *Proc. Natl. Acad. Sci. USA* **99**, 4215–4220.
32. Nicholson, D. W., Ali, A., Vaillancourt, J. P., Calaycay, J. R., Mumford, R. A., Zamboni, R. J., and Ford-Hutchinson, A. W. (1993). Purification to homogeneity and the N-terminal sequence of human leukotriene C_4 synthase: a homodimeric glutathione S-transferase composed of 18-kDa subunits. *Proc. Natl. Acad. Sci. USA* **90**, 2015–2019.
33. Lam, B. K., Penrose, J. F., Freeman, G. J., and Austen, K. F. (1994). Expression cloning of a cDNA for human leukotriene C_4 synthase, an integral membrane protein conjugating reduced glutathione to leukotriene A_4. *Proc. Natl. Acad. Sci. USA* **91**, 7663–7667.
34. Welsch, D. J., Creely, D. P., Hauser, S. D., Mathis, K. J., Krivi, G. G., and Isakson, P. C. (1994). Molecular cloning end expression of human leukotriene C_4 synthase. *Proc. Natl. Acad. Sci. USA* **91**, 9745–9749.
35. Ali, A., Ford-Hutchinson, A. W., and Nicholson, D. W. (1994). Activation of protein kinase C down-regulates leukotriene C_4 synthase activity and attenuates cysteinyl leukotriene production in an eosinophilic substrain of HL-60 cells. *J. Immunol.* **153**, 776–788.
36. Jakobsson, P. J., Mancini, J. A., and Ford-Hutchinson, A. W. (1996). Identification and characterization of a novel human microsomal glutathione S-transferase with leukotriene C_4 synthase activity and significant sequence identity to 5-lipoxygenase-activating protein and leukotriene C_4 synthase. *J. Biol. Chem.* **271**, 22203–22210.
37. Jakobsson, P. J., Mancini, J. A., Riendeau, D., and Ford-Hutchinson, A. W. (1997). Identification and characterization of a novel microsomal enzyme with glutathione-dependent transferase and peroxidase activities. *J. Biol. Chem.* **272**, 22934–22939.
38. Yokomizo, T., Izumi, T., Chang, K., Takuwa, Y., and Shimizu, T. (1997). A G-protein-coupled receptor for leukotriene B_4 that mediates chemotaxis. *Nature* **387**, 620–624.
39. Kato, K., Yokomizo, T., Izumi, T., and Shimizu, T. (2000). Cell-specific transcriptional regulation of human leukotriene B(4) receptor gene. *J. Exp. Med.* **192**, 413–420.
40. Yokomizo, T., Kato, K., Terawaki, K., Izumi, T., and Shimizu, T. (2000). A second leukotriene B_4 receptor, BLT_2: a new therapeutic target in inflammation and immunological disorders. *J. Exp. Med.* **192**, 421–431.
41. Kamohara, M., Takasaki, J., Matsumoto, M., Saito, T., Ohishi, T., Ishii, H., and Furuichi, K. (2000). Molecular cloning and characterization of another leukotriene B_4 receptor. *J. Biol. Chem.* **275**, 27000–27004.

42. Tryselius, Y., Nilsson, N. E., Kotarsky, K., Olde, B., and Owman, C. (2000). Cloning and characterization of cDNA encoding a novel human leukotriene B_4 receptor. *Biochem. Biophys. Res. Commun.* **274**, 377–382.
43. Yokomizo, T., Kato, K., Hagiya, H., Izumi, T., and Shimizu, T. (2001). Hydroxyeicosanoids bind to and activate the low affinity leukotriene B_4 receptor, BLT_2. *J. Biol. Chem.* **276**, 12454–12459.
44. Heise, C. E., O'Dowd, B. F., Figueroa, D. J., Sawyer, N., Nguyen, T., Im, D.-S., Stocco, R., Bellefeuille, J. N., Abramovitz, M., Cheng Jr., Williams, R., Zeng, Z., Liu, Q., Ma, L., Clements, M. K., Coulombe, N., Liu, Y., Austin, C. P., George, S. R., O'Neill, G. P., Metters, K. M., Lynch, K. P., and Evans, J. F. (2000). Characterization of the human cysteinyl leukotriene 2 ($CysLT_2$) receptor. *J. Biol. Chem.* **275**, 30531–30536.
45. Lynch, K. R., O'Neill, G. P., Liu, Q., Im, D. S., Sawyer, N., Metters, K. M., Coulombe, N., Abramovitz, M., Figueroa, D. J., Zeng, Z., Connolly, B. M., Bai, C., Austin, C. P., Chateauneuf, A., Stocco, R., Greig, G. M., Kargman, S., Hooks, S. B., Hosfield, E., Williams, D. L. Jr., Ford-Hutchinson, A. W., Caskey, C. T., and Evans, J. F. (1999). Characterization of the human cysteinyl leukotriene $CysLT_1$ receptor. *Nature* **399**, 789–793.
46. Nothacker, H. P., Wang, Z. W., Zhu, Y. H., Reinscheid, R. K., Lin, S. H. S., and Civelli, O. (2000). Molecular cloning and characterization of a second human cysteinyl leukotriene receptor: discovery of a subtype selective agonist. *Mol. Pharmacol.* **58**, 1601–1608.
47. Sarau, H. M., Ames, R. S., Chambers, J., Ellis, C., Elshourbagy, N., Foley, J. J., Schmidt, D. B., Muccitelli, R. M., Jenkins, O., Murdock, P. R., Herrity, N. C., Halsey, W., Sathe, G., Muir, A. I., Nuthulaganti, P., Dytko, G. M., Buckley, P. T., Wilson, S., Bergsma, D. J., and Hay, D. W. (1999). Identification, molecular cloning, expression, and characterization of a cysteinyl leukotriene receptor. *Mol. Pharmacol.* **56**, 657–663.
48. Takasaki, J., Kamohara, M., Matsumoto, M., Saito, T., Sugimoto, T., Ohishi, T., Ishii, H., Ota, T., Nishikawa, T., Kawai, Y., Masuho, Y., Isogai, T., Suzuki, Y., Sugano, S., and Furuichi, K. (2000). The molecular characterization and tissue distribution of the human cysteinyl leukotriene $CysLT_2$ receptor. *Biochem. Biophys. Res. Commun.* **274**, 316–322.
49. Chen, X. S., Sheller, J. R., Johnson, E. N., and Funk, C. D. (1994). Role of leukotrienes revealed by targeted disruption of the 5-lipoxygenase gene. *Nature* **372**, 179–182.
50. Bailie, M. B., Standiford, T. J., Laichalk, L. L., Coffey, M. J., Strieter, R., and Peters-Golden, M. (1996). Leukotriene-deficient mice manifest enhanced lethality from *Klebsiella pneumonia* in association with decreased alveolar macrophage phagocytic and bactericidal activities. *J. Immunol.* **157**, 5221–5224.
51. Irvin, C. G., Tu, Y. P., Sheller, J. R., and Funk, C. D. (1997). 5-lipoxygenase products are necessary for ovalbumin-induced airway responsiveness in mice. *Am. J. Physiol.* **16**, L1053–L1058.
52. Byrum, R. S., Goulet, J. L., Griffiths, R. J., and Koller, B. H. (1997). Role of the 5-lipoxygenase-activating protein (FLAP) in murine acute inflammatory responses. *J. Exp. Med.* **185**, 1065–1075.
53. Griffiths, R. J., Smith, M. A., Roach, M. L., Stock, J. L., Stam, E. J., Milici, A. J., Scampoli, D. N., Eskra, J. D., Byrum, R. S., Koller, B. H., and McNeish, J. D. (1997). Collagen-induced arthritis is reduced in 5-lipoxygenase-activating protein-deficient mice. *J. Exp. Med.* **185**, 1123–1129.
54. Byrum, R. S., Goulet, J. L., Snouwaert, J. N., Griffiths, R. J., and Koller, B. H. (1999). Determination of the contribution of cysteinyl leukotrienes and leukotriene B_4 in acute inflammatory responses using 5-lipoxygenase- and leukotriene A_4 hydrolase-deficient mice. *J. Immunol.* **163**, 6810–6819.
55. Kanaoka, Y., Maekawa, A., Penrose, J. F., Austen, K. F., and Lam, B. K. (2001). Attenuated zymosan-induced peritoneal vascular permeability and IgE dependent passive cutaneous anaphylaxis in mice lacking leukotriene C_4 synthase. *J. Biol. Chem.* **276**, 22608–22613.
56. Maekawa, A., Austen, K. F., and Kanaoka, Y. (2002). Targeted gene disruption reveals the role of cysteinyl leukotriene 1 receptor in the enhanced vascular permeability of mice undergoing acute inflammatory responses. *J. Biol. Chem.* **277**, 20820–20824.
57. Haribabu, B., Verghese, M. W., Steeber, D. A., Sellars, D. D., Bock, C. B., and Snyderman, R. (2000). Targeted disruption of the leukotriene B_4 receptor in mice reveals its role in inflammation and platelet-activating factor-induced anaphylaxis. *J. Exp. Med.* **192**, 433–438.
58. Tager, A. M., Dufour, J. H., Goodarzi, K., Bercury, S. D., von Andrian, U. H., and Luster, A. D. (2000). BLTR mediates leukotriene B_4-induced chemotaxis and adhesion and plays a dominant role in eosinophil accumulation in a murine model of peritonitis. *J. Exp. Med.* **192**, 439–446.

Lipoxins and Aspirin-Triggered 15-epi-Lipoxins: Mediators in Anti-inflammation and Resolution

Charles N. Serhan

*Center for Experimental Therapeutics and Reperfusion Injury,
Department of Anesthesiology, Perioperative and Pain Medicine,
Brigham and Women's Hospital and Harvard Medical School,
Boston, Massachusetts*

ASA	acetylsalicylic acid
ATL	aspirin-triggered lipoxin, 15R-LXA$_4$
COX-2	cyclooxygenase 2
EPA	eicosapentaenoic acid
HEPE	hydroxyeicosapentaenoic acid
HETE	hydroxyeicosatetraenoic acid
LT	leukotriene
LX	lipoxin
LXA$_4$	5S, 6R,15S-trihydroxy-7,9,13-*trans*-11-*cis*-eicosatetraenoic acid
15-epi-LXA$_4$	5S,6R,15R-trihydroxy-7,9,13-*trans*-11-*cis*-eicosatetraenoic acid
NSAID	non-steroidal anti-inflammatory drug
PUFA	polyunsaturated fatty acid
PMN	polymorphonuclear leukocytes

Lipoxin Signals in the Resolution of Inflammation

Biosynthesis

Cell-cell interactions and transcellular biosynthesis of mediators are now well recognized as important means of generating new signals [1]. In humans, lipoxin (LX) biosynthesis is an example of LO-LO interactions via transcellular circuits. Lipoxins are a separate class of mediators produced from arachidonic acid in that they contain a conjugated tetraene and trihydroxy structure, a feature that departs from the other structural classes of eicosanoids (see [2] and chapters therein) and that gives them distinct biological roles. The lipoxins are generated by two main routes (Fig. 1): the first involves initial lipoxygenation by 15-LO that inserts molecular oxygen in predominantly the S configuration at carbon 15 followed by 5-LO based transformation. This route is particularly relevant when polymorphonuclear leukocytes (PMN) interact with mucosal surfaces. A second route, which occurs predominantly as a major intravascular origin within blood vessels when, for example, platelet intracellular glutathione is depleted, involves the conversion of 5-LO-derived LTA$_4$ that is released from leukocytes and subsequently converted to lipoxins. On their own, human platelets do not generate lipoxins but become an important source of LX as they interact with leukocytes (reviewed in [3]).

Relation of Lipoxins to Diseases and Bioactions

In humans during disease processes, lipoxins are generated by airway, kidney, joints [see reviews [3,4] and references therein) and liver [5]. Their production within exudates is both temporally and spatially separated from the formation

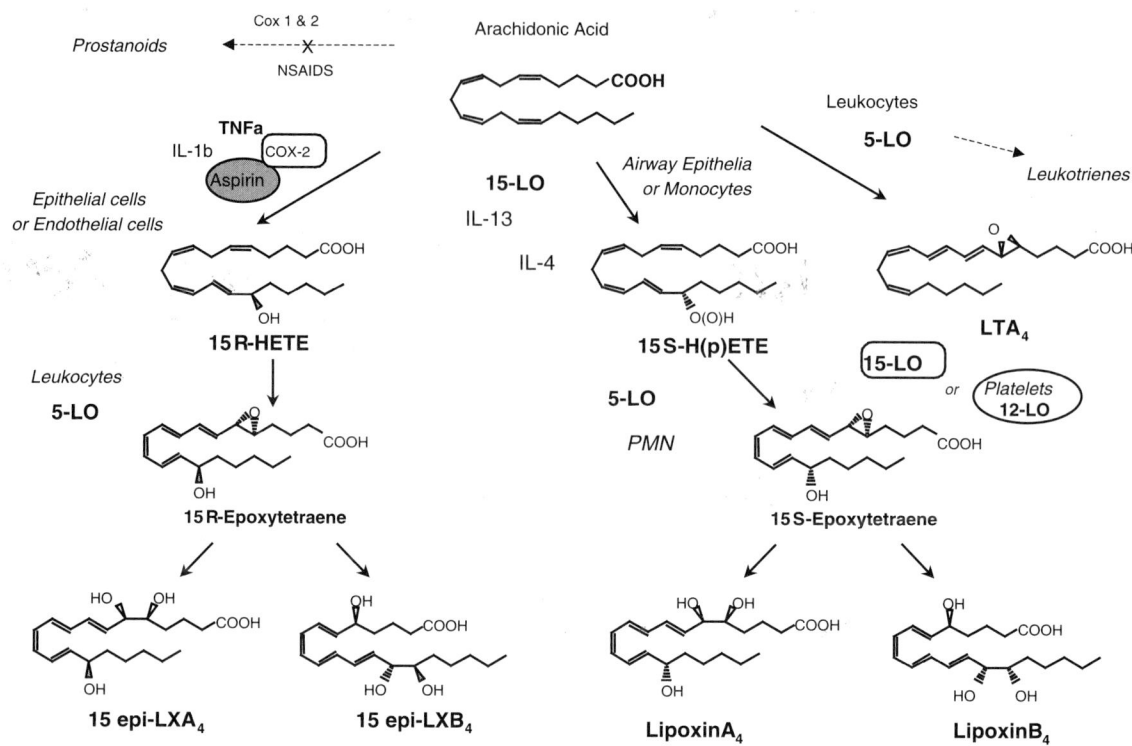

Figure 1 Biosynthesis of lipoxins and aspirin-triggered 15-epi-lipoxins. *Right*: 15-lipoxygenase initiated pathway and 5-lipoxygenase-12-lipoxygenase pathway. *Left*: ASA triggered pathway; irreversible acetylation of COX-2 by aspirin changes the enzyme's product from prostaglandin intermediate to precursors of ATL. The acetylated COX-2 remains catalytically active (see text).

of leukotrienes or prostaglandins [6], and the main actions of lipoxins stand apart from other known eicosanoids and lipid mediators (Table I). Acting in the nanomolar range, native lipoxins selectively regulate the motility of PMN, eosinophils, and monocytes in a stereospecific fashion, a result that raised our awareness to the possibility that lipoxins and related compounds could serve as endogenous "stop signals" of select leukocytes to help resolve local inflammation [3,4]. Much as lipoxins serve as endogenous suppressors of leukocyte-mediated tissue injury, their epimeric form generated with aspirin treatment (Fig. 1), namely aspirin-triggered lipoxins (ATL) 15-epi-LX (stereoisomers at carbon 15 position of native LX) and related compounds, may be the effectors of well-established anti-inflammatory therapies.

Lipoxins are rapidly generated within seconds to minutes, act locally, and are swiftly inactivated via enzymatic routes. Based on knowledge of LX routes of inactivation and the identity of a receptor for LXA_4 [7], metabolically stable LX and ATL analogs that resist rapid metabolic inactivation and are potent regulators of leukocyte traffic (*in vitro* and *in vivo*) were designed and synthesized by total organic synthesis [8]. Some of these more potent LX-mimetics are also topically active inhibitors of acute inflammation (Table I) and are potent inhibitors of TNF_α signals as well as IL-8 formation [9]. ATL and its active analogs compete with LXA_4 at its own receptor on leukocytes and act as agonists that stimulate and induce intracellular "stop signaling" that can have both rapid and gene transcriptional associated events [10]. These recent findings support the notion that ATL and native lipoxin prevent tissue damage by serving as endogenous anti-inflammatory molecules that also stimulate macrophage clearance of spent PMN and resolution of edema [11,12].

Table I Main Anti-Inflammatory and Resolving Actions of Lipoxins and Novel Aspirin-Triggered Lipoxins[a]

Compound/mediator	Response/action
LX and/or ATL[b]	• Regulate leukocyte traffic in acute inflammation & injury ("stop" PMN and eosinophils, "go" monocytes non-phlogistic activation) [3, 4]
	• Redirect chemokine-cytokine axis (gene expression, i.e. IL-8, IL-1) [10]
	• Reduce edema [11]
	• Stimulate clearance and phagocytosis of apoptotic PMN [12]
	• Turn down pain signals: downregulate PMN in neuropathic pain [25]
18R-EPA series (i.e. 18R,5,12-tri-HEPE) and 15-epi-LXA_5 series	• Inhibit PMN transmigration and block cytokine-stimulated inflammation *in vivo* [19]

[a]For further details, see text. For further details and original citations of bioactions in isolated cell systems and *in vivo* with disease models, please see [3].

[b]See abbreviations list in text.

Aspirin-Triggered Lipoxins and Other Polyunsaturated Fatty Acid–Derived Mediators

Despite nearly 100 years of wide use, the therapeutic impact of acetylsalicylic acid (ASA) is still evolving, and new beneficial effects are still being uncovered [13,14]. The irreversible acetylation of both cyclooxygenase 1 and 2 (COX-1 and COX-2) with subsequent inhibition of prostaglandin biosynthesis is well appreciated and explains some, but not all, of ASA's pharmacological actions [15], and until recently the mechanism for ASA's impact *in vivo* on PMN recruitment in inflammation remained largely unknown. In 1995, Clària and Serhan found that ASA treatment triggers formation of novel series of lipid mediators termed the aspirin-triggered lipoxins (ATL). Their formation relies on cell-cell interactions (15-epi-LX; Fig. 1). Co-activation of neutrophils with either endothelial cells treated with ASA or certain epithelial cells generates a novel class of 15R-containing lipoxins (ATL) that in turn downregulate PMN-endothelial cell interactions as well as epithelial function [8,9,16].

In most clinical arenas ASA is held to act strictly as an inhibitor of prostaglandins. However, the ASA-acetylated form of COX-2 *is still active* and converts arachidonate to 15-HETE, which carries its C15 alcohol in the R configuration [16]. The COX-2 substrate channel [17] is larger in this isoenzyme (cf. crystal structures for COX-2 [18]) and gives rise to an unusual L-shaped binding of arachidonic acid that gets oxygenated in the 15R position. 15-epi-LXA$_4$ is more potent and longer acting than its 15S-containing form because it is not as rapidly inactivated [8]. It appears that ASA triggers formation of endogenous eicosanoids and related substances that could mediate some of the many beneficial actions of ASA by pirating the native pathway of lipoxin production and signaling. It is important to note that the biosynthesis of 15-epi-lipoxins does not arise from a simple pathway shunt, but rather represents the effect of ASA on the oxygenating function of COX-2 at foci of inflammation.

The biological importance of this difference in the enzyme structure (COX-1 versus COX-2) is not clear, but the presence of an additional binding pocket in COX-2 for NSAID was exploited to make COX-2-inhibitors [18]. Acetylation of COX-1 by ASA does not permit substantial amounts of arachidonate conversion to 15R-HETE. Once formed, 15R-HETE is rapidly esterified in inflammatory cells, altering signal transduction as well as priming the supply of LX precursors [3]. The endothelial cell production of 15-HETE is highly effective *in situ* [19] and *in vivo* at sites of inflammation. Given the vast size of the vasculature and its role in host defense and inflammation, the vascular endothelium is likely to contain focal regions or "hot spots" under stress that express COX-2 and can generate substantial amounts of COX-2-derived products with ASA treatment.

Novel Anti-Inflammatory Signals and Pathways

Over the past 25 years, numerous studies reported that dietary supplementation with omega-3 polyunsaturated fatty acids (ω-3 PUFA) has beneficial effects in disease. Recent reviews discuss potential antithrombotic, immunoregulatory, and anti-inflammatory responses relevant in arteriosclerosis, arthritis, and asthma as well as anti-tumor and anti-metastatic effects [20]. The possible preventative or therapeutic actions of ω-3 PUFA supplementation in infant nutrition, for cardiovascular diseases, and for mental health led an international workshop to call for recommended dietary intakes [20], and data from one large trial (GISSI—Prevenzione, which included over 11,300 subjects) that evaluated the benefits of aspirin with or without ω-3 PUFA supplementation for patients surviving myocardial infarction found a significant decrease in death in the group taking the supplement [21].

Fish oils or n-3 PUFA per se are proposed to act by one or several possible mechanisms [20]. None of the proposed explanations are widely accepted, largely because of the supra-pharmacologic amounts, usually milligram to microgram range of ω-3 PUFA, that are required *in vitro* to achieve the supposed beneficial effects. Because compelling molecular evidence has been lacking and in view of beneficial profiles attributed to dietary ω-3 PUFA and those of aspirin in a variety of diseases, we sought evidence for possible new lipid-derived signals that could explain the epidemiological findings from humans.

A Protective Role for Vascular COX-2 in Micro-inflammation

Inflammatory exudates formed in murine dorsal pouches treated with ω-3 and ASA generate several novel compounds [19], including 18R-hydroxy-eicosapentaenoic acid (18R-HEPE) and several trihydroxy-containing compounds derived from the ω-3 fish oil eicosapentaenoic acid (EPA) (C20:5) used as an n-3 PUFA prototype. Human cells also generate these new 18R and 15R series of compounds from EPA, which carry intriguing bioactivities. When human endothelial cells expressing COX-2 are pulsed with EPA and treated with ASA, they generate 18R-HEPE or a mixture of 18R-HEPE and 15R-HEPE. A role for COX-2 in this biosynthetic pathway was confirmed with recombinant human COX-2, in which acetylation by ASA dramatically increased the production of both 18R-HEPE and 15R-HEPE, findings that could be of clinical significance [19].

When engaged in phagocytosis, activated human polymorphonuclear leukocytes (PMN) process the intermediates derived from acetylated recombinant COX-2 to produce two series of trihydroxy-containing compounds; one series carries an 18R-position hydroxyl group, and the other series in the 15R position that are related to 15-epi-LX$_5$. Trout macrophages and human leukocytes can indeed convert endogenous EPA to 15S-containing LX also denoted as 5-series LX$_5$ [22]. Briefly, we found that human PMN take up and convert 18R-HEPE via 5-lipoxygenation to insert molecular oxygen and, in subsequent steps, form 5-hydro(peroxy)-18R-DiH(p)EPE and a more labile intermediate 5(6)epoxide that gives rise to 5,12,18-triHEPE. In a similar biosynthetic pathway, 15R-HEPE released by endothelial cells is converted by activated PMN via 5-lipoxygenation to a 5-series LXA$_5$ analog that also retains their C15 R or *epi* configuration, namely 15-epi-LXA$_5$ [19]. The stereochemistry of compounds in this pathway is different from those of the LO-LO driven pathways that give predominantly C15 S containing LX$_5$ structures (so-called 5-series of five double bonds) as with endogenous sources of EPA in trout macrophages (cf. [22] and references therein). The chirality of the precursor with ASA-COX-2 (predominantly R) is retained when converted by human PMN to give 15-epi-LXA$_5$ [19].

The new 18R-series members might serve as dampers for inflammatory responses, since 18R-HEPE gave some inhibitory activity and its product 5,12,18R-triHEPE potently inhibits PMN transmigration and infiltration [19]. These results raise the question of whether arachidonate is the sole substrate for COX-2 in physiologic settings in human tissues or whether EPA or other PUFA are important as well [19]. Despite the many reports of possible beneficial impacts of ω-3 PUFAs and EPA in humans [20,21], oxygenation by COX-2 to generate bioactive compounds, as referenced herein, has not been addressed. In fish leukocytes and platelets, EPA (C20:5) and arachidonic acid are both mobilized and converted to both 5-series and 4-series eicosanoids (including PG, LT, and LX) with roughly equal abundance [22]. Given the gram amounts of ω-3 PUFA taken as dietary supplements by humans, as in [20,21], and the large area of the vasculature that can express COX-2 (vascular "hot spots" during local inflammation), the conversion of EPA by vascular endothelial cells and neighboring cells could represent a significant *in vivo* source.

Concluding Remarks

Inappropriate control of inflammation and its resolution is now recognized to contribute to many diseases. Aspirin as well as other NSAIDs that affect these signaling systems (Fig. 1) are in wide use, yet these agents are not without *unwanted* side effects, particularly in kidney and stomach. The discovery of the second isoform of COX (reviewed in [23]) sparked a large-scale search for safer aspirin-like drugs, namely COX-2 inhibitors, that would bypass the unwanted side effects. Results reviewed here indicate that lipoxins, their aspirin-triggered epimers (ATL), and broader arrays of aspirin-triggered lipid mediators derived from omega-3 PUFA reveal previously unappreciated *endogenous* anti-inflammation and pro-resolution signaling mechanisms (Table I) that could offer new treatment approaches. The finding that lipoxin counters inflammatory events led to more general concepts, namely that aspirin-triggered lipid mediators could serve as local mediators of anti-inflammation or endogenous agonists that favor resolution of inflammation. Additional support for this notion that lipoxins are protective and that ATLs share this property [3,19] comes from finding that LXA$_4$ stimulates macrophages to clear apoptotic PMN [12], and that LXA$_4$ receptors regulate gene expression, cytokines, and metalloproteases (see [24] and references within). These signaling pathways add a new dimension to the well-established use of low-dose ASA as a specific COX-1 inhibitor in platelets, which also triggers COX-2 generated protective products, thus underscoring the importance of transcellular biosynthetic signaling pathways.

Acknowledgments

These studies were supported in part by National Institutes of Health grants no. GM38765 and P01-DE13499 (C.N.S.). A full reference list appears at http://etherweb.bwh.harvard.edu/research/overview/serhan.php.

References

1. Marcus, A. J. (1999). Platelets: their role in hemostasis, thrombosis, and inflammation. In "Inflammation: Basic Principles and Clinical Correlates" (Gallin, J. I., and Snyderman, R., Eds.), pp. 77–95. Lippincott Williams & Wilkins, Philadelphia.
2. Funk, C. D. (2001). Prostaglandins and leukotrienes: Advances in eicosanoid biology. *Science* **294**, 1871–1875.
3. Serhan, C. N. and Chiang, N. (2001). Lipid-derived mediators in endogenous anti-inflammation and resolution: lipoxins and aspirin-triggered 15-epi-lipoxins. *The Scientific World*, on-line (www.thescientificworld.com).
4. McMahon, B., Mitchell, S., Brady, H. R., and Godson, C. (2001). Lipoxins: revelations on resolution. *Trends Pharmacol. Sci.* **22**, 391–395.
5. Clària, J., Titos, E., Jiménez, W., Ros, J., Ginès, P., Arroyo, V., Rivera, F., and Rodés, J. (1998). Altered biosynthesis of leukotrienes and lipoxins and host defense disorders in patients with cirrhosis and ascites. *Gastroenterology* **115**, 147–156.
6. Levy, B. D., Clish, C. B., Schmidt, B., Gronert, K., and Serhan, C. N. (2001). Lipid mediator class switching during acute inflammation: signals in resolution. *Nature Immunol.* **2**, 612–619.
7. Fiore, S., Maddox, J. F., Perez, H. D., and Serhan, C. N. (1994). Identification of a human cDNA encoding a functional high affinity lipoxin A$_4$ receptor. *J. Exp. Med.* **180**, 253–260.
8. Serhan, C. N., Maddox, J. F., Petasis, N. A., Akritopoulou-Zanze, I., Papayianni, A., Brady, H. R., Colgan, S. P., and Madara, J. L. (1995). Design of lipoxin A$_4$ stable analogs that block transmigration and adhesion of human neutrophils. *Biochemistry* **34**, 14609–14615.
9. Gewirtz, A. T., McCormick, B., Neish, A. S., Petasis, N. A., Gronert, K., Serhan, C. N., and Madara, J. L. (1998). Pathogen-induced chemokine secretion from model intestinal epithelium is inhibited by lipoxin A$_4$ analogs. *J. Clin. Invest.* **101**, 1860–1869.
10. Qiu, F.-H., Devchand, P. R., Wada, K., and Serhan, C. N. (2001). Aspirin-triggered lipoxin A$_4$ and lipoxin A$_4$ up-regulate transcriptional corepressor NAB1 in human neutrophils. *FASEB J.*, **15**, 2736–2738.
11. Bandeira-Melo, C., Serra, M. F., Diaz, B. L., Cordeiro, R. S. B., Silva, P. M. R., Lenzi, H. L., Bakhle, Y. S., Serhan, C. N., and Martins, M. A. (2000). Cyclooxygenase-2-derived prostaglandin E$_2$ and lipoxin A$_4$ accelerate resolution of allergic edema in *Angiostrongylus costaricensis*-infected rats: relationship with concurrent eosinophilia. *J. Immunol.* **164**, 1029–1036.
12. Godson, C., Mitchell, S., Harvey, K., Petasis, N. A., Hogg, N., and Brady, H. R. (2000). Cutting edge: Lipoxins rapidly stimulate nonphlogistic phagocytosis of apoptotic neutrophils by monocyte-derived macrophages. *J. Immunol.* **164**, 1663–1667.
13. Gum, P. A., Thamilarasan, M., Watanabe, J., Blackstone, E. H., and Lauer, M. S. (2001). Aspirin use and all-cause mortality among patients being evaluated for known or suspected coronary artery disease: a propensity analysis. *J.A.M.A.* **286**, 1187–1194.
14. Ridker, P. M., Cushman, M., Stampfer, M. J., Tracy, R. P., and Hennekens, C. H. (1997). Inflammation, aspirin, and the risk of cardiovascular disease in apparently healthy men. *N. Engl. J. Med.* **336**, 973–979.

15. Vane, J. R. (1982). Adventures and excursions in bioassay: the stepping stones to prostacyclin. *In* "Les Prix Nobel: Nobel Prizes, Presentations, Biographies and Lectures", pp. 181–206. Almqvist & Wiksell, Stockholm.
16. Clària, J. and Serhan, C. N. (1995). Aspirin triggers previously undescribed bioactive eicosanoids by human endothelial cell-leukocyte interactions. *Proc. Natl. Acad. Sci. USA* **92**, 9475–9479.
17. Rowlinson, S. W., Crews, B. C., Goodwin, D. C., Schneider, C., Gierse, J. K., and Marnett, L. J. (2000). Spatial requirements for 15-(R)-hydroxy-5Z,8Z,11Z,13E-eicosatetraenoic acid synthesis within the cyclooxygenase active site of murine COX-2. *J. Biol. Chem.* **275**, 6586–6591.
18. Kurumbail, R. G., Stevens, A. M., Gierse, J. K., McDonald, J. J., Stegeman, R. A., Pak, J. Y., Gildehaus, D., Miyashiro, J. M., Penning, T. D., Seibert, K., Isakson, P. C., and Stallings, W. C. (1996). Structural basis for selective inhibition of cyclooxygenase-2 by anti-inflammatory agents. *Nature* **384**, 644–648.
19. Serhan, C. N., Clish, C. B., Brannon, J., Colgan, S. P., Chiang, N., and Gronert, K. (2000). Novel functional sets of lipid-derived mediators with antiinflammatory actions generated from omega-3 fatty acids via cyclooxygenase 2-nonsteroidal antiinflammatory drugs and transcellular processing. *J. Exp. Med.* **192**, 1197–1204.
20. Simopoulos, A. P., Leaf, A., and Salem, N., Jr. (1999). Workshop on the essentiality of and recommended dietary intakes for omega-6 and omega-3 fatty acids. *J. Am. Coll. Nutr.* **18**, 487–489.
21. GISSI-Prevenzione Investigators (1999). Dietary supplementation with n-3 polyunsaturated fatty acids and vitamin E after myocardial infarction: results of the GISSI-Prevenzione trial. Gruppo Italiano per lo Studio della Sopravvivenza nell'Infarto miocardico. *Lancet* **354**(9177), 447–455.
22. Hill, D. J., Griffiths, D. H., and Rowley, A. F. (1999). Trout thrombocytes contain 12- but not 5-lipoxygenase activity. *Biochim. Biophys. Acta* **1437**, 63–70.
23. Herschman, H. R. (1998). Recent progress in the cellular and molecular biology of prostaglandin synthesis. *Trends Cardiovasc. Med.* **8**, 145–150.
24. Sodin-Semrl, S., Taddeo, B., Tseng, D., Varga, J., and Fiore, S. (2000). Lipoxin A_4 inhibits IL-1 beta-induced IL-6, IL-8, and matrix metalloproteinase-3 production in human synovial fibroblasts and enhances synthesis of tissue inhibitors of metalloproteinases. *J. Immunol.* **164**, 2660–2666.
25. Serhan, C. N., Fierro, I. M., Chiang, N., and Pouliot, M. (2001). Nociceptin stimulates neutrophil chemotaxis and recruitment: Inhibition by aspirin-triggered-15-epi-lipoxin A_4. *J. Immunol.* **166**, 3650–3654.

CHAPTER 170

Cholesterol Signaling

Peter A. Edwards,[1,2] Heidi R. Kast-Woelbern,[1] and
Matthew A. Kennedy[1]

[1]Departments of Biological Chemistry and Medicine,
[2]Molecular Biology Institute,
University of California,
Los Angeles, California

SREBP, sterol regulatory element binding protein
LXR, liver X-activated receptor
FXR, farnesoid X-activated receptor
PXR, pregnane X receptor
VDR, vitamin D receptor
GR, glucocorticoid receptor
ER, estrogen receptor
MR, mineralocorticoid receptor
PR, progesterone receptor
AR, androgen receptor
FPP, farnesyl diphosphate
GGPP, geranylgeranyl diphosphate
PE, phosphatidylethanolamine

Introduction

Numerous intermediates are formed either during the biosynthesis or catabolism of cholesterol that function as important components in many cell signaling events (Fig. 1). In this review we will briefly discuss some of the recent studies that have revealed the importance of these newly identified signaling molecules.

Cholesterol Precursors

HMG-CoA reductase is the rate-limiting enzyme of cholesterol biosynthesis. The expression level of this membrane-bound enzyme is controlled by many factors that in turn regulate cholesterol synthesis and cellular cholesterol homeostasis (reviewed in [1]). However, the most important mechanisms involve those that control the stability of the protein and transcription of the gene. Studies utilizing both mammalian cells and yeast have shown that increased degradation of HMG-CoA reductase occurs when cellular levels of either the 15-carbon isoprenoid farnesyl diphosphate (FPP), farnesol (dephosphoryled FPP), or an unidentified derivative of FPP are increased [2–4]. The isoprenoid-dependent increase in degradation of both mammalian [2] and yeast [5] HMG-CoA reductase also requires an oxysterol.

Some of the many biologically active oxysterols that are synthesized from cholesterol are illustrated in Fig. 1 (reviewed in [6] and discussed below). However, 24(S),25-epoxycholesterol is synthesized in many tissues from squalene, and not from cholesterol (Fig. 1) [7]. Recent studies have shown that this epoxysterol is one of the most potent activators of the nuclear receptor LXR (discussed below) [8]. Nonetheless, the physiological importance of the endogenous pathway that generates this epoxysterol is currently unknown.

FPP lies at a critical branch point in the cholesterol biosynthetic pathway, since it is a precursor of several important compounds (Fig. 1) (reviewed in [1]). One such compound is the 20-carbon isoprenoid geranylgeranyl diphosphate (GGPP) (Fig. 1). Both FPP and GGPP are important isoprenoid donors that are subsequently covalently linked, via a thioether bond, to a cysteine, located at or near the carboxy terminus of many proteins. This prenylation reaction is necessary for both the intracellular location and function of many proteins, including many members of the ras, rab, or rho family of small G proteins, kinases, and G-protein-coupled receptors [1].

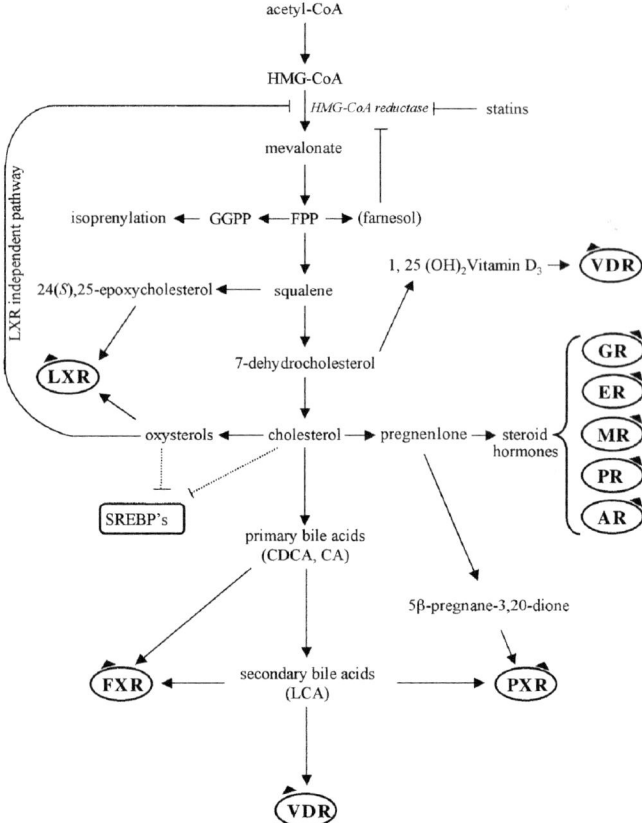

Figure 1 *Signaling molecules generated during the synthesis and catabolism of cholesterol. Only some of the intermediates in the synthesis and catabolism of cholesterol are shown. The regulatory enzyme HMG-CoA reductase is indicated in italics. The catabolism of cholesterol to primary bile acids, chenodeoxycholic acid (CDCA), and cholic acid (CA) occurs only in the liver. The subsequent synthesis of the secondary bile acid lithocholic acid (LCA) from CDCA occurs as a result of a 7α-dehydroxylation pathway present in certain intestinal bacteria. These bile acids activate FXR, PXR, or VDR, as indicated. The synthesis of steroid hormones that activate the steroid receptors (GR, ER, MR, PR, and AR) primarily occurs in steroidogenic tissues. The conversion of 7-dehydrocholesterol to 1,25 $(OH)_2$ Vitamin D_3, the ligand for the vitamin D receptor (VDR), is discussed in the text. Oxysterols derived from cholesterol or squalene can either activate the nuclear receptor LXR or inhibit the cleavage and maturation of SREBPs by a process that is independent of LXR. Ligands/hormones that function to activate members of the nuclear receptor family are indicated (▲).*

7-Dehydrocholesterol, one of the late intermediates in the cholesterol biosynthetic pathway, is found in high concentrations in the skin. Exposure of the skin to UV radiation results in cleavage of the B ring of 7-dehydrocholesterol to produce cholecalciferol. The latter is subsequently converted in the liver and kidney to $1,25(OH)2$ vitamin D_3, the biologically active form of vitamin D. Until very recently, $1,25(OH)2$ vitamin D_3 was considered to be the major endogenous ligand that activates the vitamin D receptor (VDR). This activated nuclear receptor is essential for the normal absorption of dietary calcium and for the control of calcium homeostasis. However, the secondary bile acid lithocholic acid has recently been shown to function as a potent agonist of VDR [9]. Secondary bile acids are synthesized by bacteria in the intestinal lumen from the primary bile acids that are secreted from the liver. These results imply that secondary bile acids, via activation of VDR in enterocytes, may have a heretofore unrecognized role in maintaining calcium absorption and metabolism.

Cholesterol

In May 1953, Gould *et al.* reported that hepatic cholesterol synthesis decreased when dogs were fed a cholesterol-rich diet [10]. Forty-nine years later we know much about the mechanisms involved in this feedback inhibition. One mechanism involves the accelerated degradation of pre-formed HMG-CoA reductase protein by a process that requires oxysterols and a derivitive of FPP (see preceding section, Cholesterol Precursors).

A second mechanism involves transcriptional repression. Studies initiated in the 1970s demonstrated that the enzymatic activity of many enzymes involved in cholesterol synthesis, including HMG-CoA reductase [11], HMG-CoA synthase, and FPP synthase, was repressed when cells were exposed to oxysterols but not pure cholesterol. It is now clear that cellular accumulation of cholesterol and/or oxysterols results in decreased transcription of these and many other genes that encode enzymes involved in cholesterol biosynthesis. In addition, transcription of the low-density lipoprotein receptor is also repressed by these sterols (reviewed in [12–14]). Goldstein and Brown and colleagues have identified a novel mechanism that controls the transcription of these genes. Transcription is dependent on the nuclear localization of a transcription factor termed sterol regulatory element binding protein (SREBP) (reviewed in [12–15]). In brief, there are two mammalian SREBP genes that, as a result of the use of alternative promoters and splicing, encode three proteins: SREBP1a, SREBP1c, and SREBP2. Each of these proteins is synthesized as a larger precursor that is embedded in the endoplasmic reticulum via a central hairpin loop containing two transmembrane domains [12]. When levels of cellular sterols are reduced, SREBP is escorted from the endoplasmic reticulum to the Golgi by the membrane-bound chaperone SCAP (SREBP-cleavage activating protein) [16–19]. Once in the Golgi, the SREBPs are sequentially cleaved by the site 1 and then the site 2 proteases (S1P and S2P, respectively) to release the mature amino terminal fragment of SREBP [19–21]. This protein fragment translocates to the nucleus, binds to SREBP response elements (SREs) in the promoters of target genes, and activates transcription [12]. These target genes encode enzymes that control the synthesis of cholesterol, fatty acids, triacylglycerides, phospholipids, and NADPH [13,14]. Space limitations prevent further discussion of this area, but the reader is referred to several reviews [1,12,15,32].

The cleavage and maturation of SREBPs is prevented by specific oxysterols but is relatively unaffected by pure cholesterol [22]. However, since exogenously added oxysterols have been shown to enhance the translocation of cholesterol from the plasma membrane to the endoplasmic reticulum [23], it is still unclear whether cholesterol or oxysterols are the active lipid that interferes with the movement of SREBPs and SCAP out of the endoplasmic reticulum.

A recent publication has shed light onto the possible mechanism by which cholesterol/oxysterols inhibit this translocation in mammalian cells; Dobrosotskaya *et al.* reported that the *Drosophila* SREBP protein undergoes a similar translocation and cleavage prior to the entry of the mature protein into the nucleus [24]. However, in contrast to the sterol-regulated translocation and cleavage of SREBPs in mammalian cells, the translocation of the *Drosophila* SREBP was prevented by phosphatidylethanolamine (PE) [24]. This effect specifically required PE containing the saturated fatty acid palmitate [25]. A surprising finding is that cholesterol had no effect on the processing of the *Drosophila* SREBP [24]. Based on these studies, the authors suggest a mechanism by which cholesterol and PE regulate the translocation of SREBP in mammals and flies, respectively; they propose that excess cellular cholesterol or PE may alter or distort the lipid phase of the endoplasmic reticulum membrane [24]. Such a change in the lipid phase may be sensed by SCAP (possibly via its "sterol sensing domains") and result in altered conformation of the SCAP protein, such that it can no longer bind and chaperone SREBP to the Golgi [24,26]. Since three other membrane-bound proteins (HMG-CoA reductase, Neimann-Pick C and Patched) are also reported to contain homologous sterol sensing domains [27], these data suggest that the function of all four proteins may be dependent upon their ability to "sense" the fluidity of the membranes in which they reside.

In addition to cholesterol/oxysterols, long-chain unsaturated fatty acids also repress the maturation of mammalian SREBPs [28–31]. In contrast, unsaturated fatty acids do not regulate the maturation of the *Drosophila* SREBP [25]. Taken together, these data suggest that a phospholipid containing one or more unsaturated fatty acids may function to regulate the maturation of mammalian SREBP, whereas in flies, this process is regulated by PE containing the saturated fatty acid palmitate.

Cholesterol Derivatives: Ligands for Nuclear Receptors

As shown in Fig. 1, cholesterol can be metabolized to many steroid hormones, including estrogen, testosterone, dihydrotestosterone, progesterone, aldosterone, and glucocorticoids. Each of these steroids activates specific members of the steroid receptor family (Fig. 1) (reviewed in [33,34]). Cholesterol can also be metabolized to a number of other biologically active compounds that function as potent agonists for other members of the nuclear receptor superfamily (Fig. 1). These agonists, which include oxysterols, primary bile acids (chenodeoxycholic acid and cholic acid), secondary bile acids (lithocholic acid), and 5β-pregnane,3,20-dione, activate LXR, FXR, PXR, and/or VDR, as illustrated in Fig. 1 [35,36]. Each of these latter nuclear receptors form functional heterodimers with RXR and bind to specific DNA sequences termed hormone response elements (reviewed in [34,37]). In general, agonists must bind to these DNA-bound nuclear receptor factors in order to activate transcription (reviewed in [38]).

The recent identification of novel nuclear receptors, which include FXR, LXR, CAR, and PXR, their natural ligands (see Fig. 1), their activated target genes [35,39–41], and the generation of nuclear receptor null mice, has led to a wealth of information about the physiological importance of each receptor. The results indicate that these nuclear receptors may represent useful targets for pharmacological intervention. For example, activation of LXR results in decreased cholesterol absorption [42]; activation of FXR results in decreased plasma triglyceride levels [43–45]; and activation of PXR results in the catabolism of a myriad of drugs, xenobiotics, and natural compounds (reviewed in [41]). The recent observations that St. John's Wort (taken as an antidepressant) contains a potent agonist of PXR [46] and that gugulipid (taken as a hypocholesterolemic agent) contains guggulsterone, an antagonist of FXR [47], are particularly intriguing. Based on these reports, it seems likely that many other natural compounds will be discovered that function as agonists or antagonists for these nuclear receptors.

When mevinolin, a natural fungal metabolite, was discovered in 1976 and shown to inhibit HMG-CoA reductase [48] (Fig. 1), few investigators would have guessed that this would lead to a class of drugs called statins that are used extensively worldwide to lower plasma cholesterol levels. Perhaps new agonists-antagonists of the recently discovered nuclear receptors that are activated by derivatives of cholesterol will prove to be equally useful in the clinical arena in the next few years.

Acknowledgments

We apologize to all those investigators whose work we should have referenced but were unable to do so because of severe space limitations. This work was funded by grants from the National Institutes of Health (HL30568 and HL68445 to P.A.E.), the Laubisch fund (to P.A.E.), and a National Institutes of Health Postdoctoral Fellowship (M.A.K.).

References

1. Edwards, P. A. and Ericsson, J. (1999). *Annu. Rev. Biochem.* **68**(1), 157–185.
2. Correll, C. C., Ng, L., and Edwards, P. A. (1994). *J. Biol. Chem.* **269**(26), 17390–17393.
3. Gardner, R. G. and Hampton, R. Y. (1999). *J. Biol. Chem.* **274**(44), 31671–31678.
4. Meigs, T. E., Roseman, D. S., and Simoni, R. D. (1996). *J. Biol. Chem.* **271**(14), 7916–7922.
5. Gardner, R. G., Shan, H., Matsuda, S. P., and Hampton, R. Y. (2001). *J. Biol. Chem.* **276**(12), 8681–8694.
6. Russell, D. W. (2000). *Biochim. Biophys. Acta* **1529**(1–3), 126–135.
7. Spencer, T. A. (1994). *Acc. Chem. Res.* **27**, 83–90.
8. Lehmann, J. M., Kliewer, S. A., Moore, L. B., Smith-Oliver, T. A., Oliver, B. B., Su, J. L., Sundseth, S. S., Winegar, D. A., Blanchard, D. E.,

Spencer, T. A., and Willson, T. M. (1997). *J. Biol. Chem.* **272**(6), 3137–3140.

9. Makishima, M., Lu, T. T., Xie, W., Whitfield, G. K., Domoto, H., Evans, R. M., Haussler, M. R., and Mangelsdorf, D. J. (2002). *Science* **296**(5571), 1313–1316.
10. Gould, R. G., Taylor, C. B., Hagerman, J. S., Warner, I., and Campbell, D. J. (1953). *J. Biol. Chem.* **201**, 498–501.
11. Kandutsch, A. A., Chen, H. W., and Heiniger, H. J. (1978). *Science* **201**(4355), 498–501.
12. Brown, M. S. and Goldstein, J. L. (1997). *Cell* **89**(3), 331–340.
13. Horton, J. D., Goldstein, J. L., and Brown, M. S. (2002). *J. Clin. Invest.* **109**(9), 1125–1131.
14. Edwards, P. A., Tabor, D., Kast, H. R., and Venkateswaran, A. (2000). *Biochim. Biophys. Acta* **1529**, 103–113.
15. Osborne, T. F. (2000). *J. Biol. Chem.* **275**(42), 32379–32382.
16. DeBose-Boyd, R. A., Brown, M. S., Li, W. P., Nohturfft, A., Goldstein, J. L., and Espenshade, P. J. (1999). *Cell* **99**(7), 703–712.
17. Sakai, J., Rawson, R. B., Espenshade, P. J., Cheng, D., Seegmiller, A. C., Goldstein, J. L., and Brown, M. S. (1998). *Mol. Cell* **2**(4), 505–514.
18. Rawson, R. B., Zelenski, N. G., Nijhawan, D., Ye, J., Sakai, J., Hasan, M. T., Chang, T. Y., Brown, M. S., and Goldstein, J. L. (1997). *Mol. Cell* **1**(1), 47–57.
19. Brown, M. S., Ye, J., Rawson, R. B., and Goldstein, J. L. (2000). *Cell* **100**(4), 391–398.
20. Espenshade, P. J., Cheng, D., Goldstein, J. L., and Brown, M. S. (1999). *J. Biol. Chem.* **274**(32), 22795–22804.
21. Ye, J., Dave, U. P., Grishin, N. V., Goldstein, J. L., and Brown, M. S. (2000). *Proc. Natl. Acad. Sci. USA* **97**(10), 5123–5128.
22. Wang, X., Sato, R., Brown, M. S., Hua, X., and Goldstein, J. L. (1994). *Cell* **77**(1), 53–62.
23. Lange, Y., Ye, J., Rigney, M., and Steck, T. L. (1999). *J. Lipid Res.* **40**(12), 2264–2270.
24. Dobrosotskaya, I. Y., Seegmiller, A. C., Brown, M. S., Goldstein, J. L., and Rawson, R. B. (2002). *Science* **296**(5569), 879–883.
25. Seegmiller, A. C., Dobrosotskaya, I., Goldstein, J. L., Ho, Y. K., Brown, M. S., and Rawson, R. B. (2002). *Dev. Cell* **2**(2), 229–238.
26. Nohturfft, A. and Losick, R. (2002). *Science* **296**(5569), 857–858.
27. Carstea, E. D., Morris, J. A., Coleman, K. G., Loftus, S. K., Zhang, D., Cummings, C., Gu, J., Rosenfeld, M. A., Pavan, W. J., Krizman, D. B., Nagle, J., Polymeropoulos, M. H., Sturley, S. L., Ioannou, Y. A., Higgins, M. E., Comly, M., Cooney, A., Brown, A., Kaneski, C. R., Blanchette-Mackie, E. J., Dwyer, N. K., Neufeld, E. B., Chang, T. Y., Liscum, L., Tagle, D. A. et al. (1997). *Science* **277**(5323), 228–231.
28. Yahagi, N., Shimano, H., Hasty, A. H., Amemiya-Kudo, M., Okazaki, H., Tamura, Y., Iizuka, Y., Shionoiri, F., Ohashi, K., Osuga, J., Harada, K., Gotoda, T., Nagai, R., Ishibashi, S., and Yamada, N. (1999). *J. Biol. Chem.* **274**(50), 35840–35844.
29. Worgall, T. S., Sturley, S. L., Seo, T., Osborne, T. F., and Deckelbaum, R. J. (1998). *J. Biol. Chem.* **273**(40), 25537–25540.
30. Thewke, D. P., Panini, S. R., and Sinensky, M. (1998). *J. Biol. Chem.* **273**(33), 21402–21407.
31. Tabor, D. E., Kim, J. B., Spiegelman, B. M., and Edwards, P. A. (1999). *J. Biol. Chem.* **274**(29), 20603–20610.
32. Brown, M. S. and Goldstein, J. L. (1999). *Proc. Natl. Acad. Sci. USA* **96**(20), 11041–11048.
33. Beato, M., Herrlich, P., and Schütz, G. (1995). *Cell* **83**(6), 851–857.
34. Mangelsdorf, D. J., Thummel, C., Beato, M., Herrlich, P., Schütz, G., Umesono, K., Blumberg, B., Kastner, P., Mark, M., Chambon, P., and Evans, R. M. (1995). *Cell* **83**(6), 835–839.
35. Kliewer, S. and Willson, T. (2002). *J. Lipid Res.* **43**(3), 359–364.
36. Repa, J. J. and Mangelsdorf, D. J. (2000). *Annu. Rev. Cell Dev. Biol.* **16**(20), 459–481.
37. Mangelsdorf, D. J. and Evans, R. M. (1995). *Cell* **83**(6), 841–850.
38. Glass, C. K. and Rosenfeld, M. G. (2000). *Genes Dev.* **14**(2), 121–141.
39. Edwards, P. A., Kast, H. R., and Anisfeld, A. M. (2002). *J. Lipid Res.* **43**(1), 2–12.
40. Goodwin, B., Moore, L. B., Stoltz, C. M., McKee, D. D., and Kliewer, S. A. (2001). *Mol. Pharm.* **60**(3), 427–431.
41. Goodwin, B. and Kliewer, S. A. (2002). *Am. J. Physiol. Gastrointest. Liver Physiol.* **282**(6), G926–931.
42. Schultz, J. R., Tu, H., Luk, A., Repa, J. J., Medina, J. C., Li, L., Schwendner, S., Wang, S., Thoolen, M., Mangelsdorf, D. J., Lustig, K. D., and Shan, B. (2000). *Genes Dev.* **14**(22), 2831–2838.
43. Maloney, P. R., Parks, D. J., Haffner, C. D., Fivush, A. M., Chandra, G., Plunket, K. D., Creech, K. L., Moore, L. B., Wilson, J. G., Lewis, M. C., Jones, S. A., and Willson, T. M. (2000). *J. Med. Chem.* **43**(16), 2971–2974.
44. Kast, H. R., Nguyen, C. M., Sinal, C. J., Jones, S. A., Laffitte, B. A., Reue, K., Gonzalez, F. J., Willson, T. M., and Edwards, P. A. (2001). *Mol. Endo.* **15**(10), 1720–1728.
45. Sinal, C., J., Tohkin, M., Miyata, M., Ward, J. M., Lambert, G., and Gonzalez, F. J. (2000). *Cell* **102**, 731–744.
46. Moore, L. B., Goodwin, B., Jones, S. A., Wisely, G. B., Serabjit-Singh, C. J., Willson, T. M., Collins, J. L., and Kliewer, S. A. (2000). *Proc. Natl. Acad. Sci. USA* **97**(13), 7500–7502.
47. Urizar, N. L., Liverman, A. B., Dodds, D. T., Silva, F. V., Ordentlich, P., Yan, Y., Gonzalez, F. J., Heyman, R. A., Mangelsdorf, D. J., and Moore, D. D. (2002). *Science* **2**, 2.
48. Endo, A., Kuroda, M., and Tsujita, Y. (1976). *J. Antibiot. (Tokyo)* **29**(12), 1346–1348.

SECTION E

Protein Proximity Interactions

John D. Scott, Editor

CHAPTER 171

Protein Proximity Interactions

John D. Scott

*Howard Hughes Medical Institute, Vollum Institute,
Oregon Health and Sciences University,
Portland, Oregon*

Introduction

Location, location, and location are the three most important aspects of real estate. The same may be true for intracellular signaling as the subcellular localization of protein kinases and phosphatases are key determinants that control the response time and specificity of many signal transduction pathways. This is equally true for growth factor, phosphotyrosine, and second messenger regulated signaling events where the localization of enzymes in close proximity to substrates ensures an efficient relay of information and directs the signal toward a subset of target proteins. This "protein proximity" section of the handbook highlights a range of cellular mechanisms that contribute to the compartmentalization of signaling enzymes and the assembly of multiprotein networks. Three general areas are covered:

1. Techniques for the analysis of protein-protein interactions,
2. Subcellular structures and multiprotein complexes that contribute to cell signaling and
3. Kinase and phosphatase targeting proteins.

Techniques for the Analysis of Protein-Protein Interactions

The first four chapters describe some of the methods that are used to define protein signaling neworks. Advances in mass spectrometry techniques have revolutionized the analysis of multiprotein complexes. The opening chapter by Shao-En Ong and Mathais Mann (chapter 172) introduces this approach and provides a practical step by step explanation of how signaling complexes are dissected. Paul Graves and Tim Haystead (chapter 173) expand on this theme by discussing some innovative approaches that use affinity chromatography and fluorescent tagging to isolate and characterize native signaling complexes. Peter Verver and Philippe Bastiaens (chapter 174) introduce Fluorescence Resonance Energy Transfer (FRET) and optical techniques that provide sensitive means to quantify and detect protein interactions in living cells. Finally, Gary Bader and colleagues (chapter 175) describe a combined phage display and yeast two-hybrid approach that has been successfully used to map protein-protein interactions within the yeast proteome.

Subcellular Structures and Multiprotein Complexes

The next six chapters define some of the specialized subcellular structures where signaling networks are organized and describe certain cellular events that are controlled by multi protein transduction units. Benjamin Geiger and colleagues (chapter 176) describe the molecular architecture of focal adhesions, where cells attach to the extracellular matrix and transfer signals to the actin cytoskeleton. Karl Saxe (chapter 177) presents evidence for SCAR/Wave and WASP proteins in the coordination of signals from Rho family GTPases to the actin remodeling machinery. Mary Kennedy (chapter 178) defines NMDA receptor signaling complexes at the postsynaptic densities of neurons. Hana Bilak and colleagues (chapter 179) describe the Toll family of receptors, which detect pathogen-associated materials and activate the innate immune response. Andrey Shaw (chapter 180) introduces a protein/membrane compartment called the immune synapse that provides a molecular basis for T cell signaling. Mark Hochstrasser (chapter 181) outlines the ubiquitin-proteosome system, an important group of cellular proteins and enzymes that target degradation, and removal of selected proteins. Finally Guy Salvesen (chapter 182) describes the role of caspase cascades in the interleukin mediated apoptosis.

Kinase and Phosphatase Targeting Proteins

The final seven chapters of this section introduce some of the anchoring, adapter and scaffolding proteins that organize broad specifity protein kinases and phosphoprotein phosphatases. Elaine Elion (chapter 183) provides a historical perspective by describing the identification and analyses of scaffolding proteins that maintain Mitogen Activated Protein (MAP) kinase cascades in yeast. A complementary chapter by Roger Davis (chapter 184) defines MAP kinase and Jun kinase scaffolds in mammalian cells. Mark Dell'Acqua introduces the A-kinase Anchoring Proteins (AKAPs) that localize the cAMP-dependent protein kinase at specific sites inside cells in chapter 185. This theme is expanded in chapter 186 by Lorene Langeberg and myself as we explore the cellular roles of AKAP signaling complexes containing PKA, PKC and signal termination enzymes such as phosphatases and phosphodiesterases. Peter Parker and colleagues (chapter 187) cover the topic of PKC binding partners and document the various classes of interacting proteins that compartmentalize this enzyme family. Roger Colbran (chapter 188) reviews the extensive literature on protein phosphatase localization with special emphasis of the role of targeting subunits that direct their enzymes to synaptic sites. Finally, in chapter 189, Marc Mumby and colleagues catalog the numerous families of targeting subunits that compartmentalize and modulate the activity of protein phosphatase 2A.

CHAPTER 172

Protein Interaction Mapping by Coprecipitation and Mass Spectrometric Identification

Shao-En Ong and Matthias Mann[1]
[1]Protein Interaction Laboratory, Center for Experimental Bioinformatics,
University of Southern Denmark,
Odense M, Denmark

MS   mass spectrometry
Y2H yeast-two-hybrid
GST glutathione S-transferase
TAP tandem affinity purification
TEV tobacco etch virus

Introduction

The complex web of signaling pathways and their mechanisms of action rely heavily on protein-protein interactions. These interactions often involve large signaling complexes containing many different protein kinases, protein phosphatases, their substrates, and scaffold proteins. For instance, the scaffold proteins, kinase suppressor of Ras (KSR) and Mek partner 1 (MP-1), bind members of signaling cascades in close proximity and help maintain the stoichiometric binding of these molecular scaffolds to kinases, which can determine the amplitude of the transduced signal[1].

The classical approach to the study of protein-protein interactions has been to employ antibodies and glutathione S-transferase (GST) fusions for the coprecipitation of proteins. These studies often make use of Western blotting as readout and as such are limited to use in verification rather than discovery, since antibodies or epitope tags specific to each potential interaction partner are required.

As studies of protein-protein interactions developed, new genetic approaches like the yeast two hybrid (Y2H) method [2] appeared that allowed the screening of large libraries of proteins for interacting partners. This approach is now commonly used to test for protein-protein interactions in functional characterization of proteins and has also been used in several large-scale Y2H studies (see Chapter 53, this volume).

With the advent of sensitive mass spectrometric technologies that allow the rapid identification of proteins in complex mixtures, protein interaction mapping on the scale of whole organisms has now become technologically feasible. In this chapter, we discuss the general principles that govern the study of multiprotein complexes by coprecipitations and subsequent analysis by mass spectrometry (MS), highlighting the role of MS as an especially useful and powerful tool in the field of cellular signaling.

General Considerations of the Coprecipitation Experiment

The coprecipitation experiment consists of three fundamental steps—presentation of the bait, providing the "hook" or affinity tag, and employing a particular mass spectrometric technology for binding partner identification (Table I). The examples described below (and in references cited therein)

Table I The Three Tiers of the Coprecipitation and Mass Spectrometric Approach to the Study of Protein-Protein Interactions

	Introduction of bait	
In vitro	*In vivo* with transient/ inducible overexpression	*In vivo* with endogenous promoter
Separate expression of bait (e.g. with GST vector in bacterial expression, incubation of lysate with GST-protein for pulldown)	Bait is epitope tagged expressed directly in system studied under the control of a inducible promoter (e.g. *GAL1*)	Bait is epitope tagged, homologous recombination of tagged gene to achieve genomic replacement.

	Choice of affinity separation/elution	
Single tag	Single tag with enzymatic cleavage	Multiple tags
Many options available: Can be large tags (GST), small moieties (biotin) or specific reactivities (6xHis). Monoclonal antibody specific epitopes (Flag, Myc, Hemagglutinin)	Affinity selection is similar to single tag, release from column is performed with an enzyme specific to the tag to increase specificity. (for example, Protein A or 3xMyc and TEV cleavage [29])	Several variants now exist incorporating different affinity tags and often an internal enzymatic cleavage site. Increased specificity, though more steps perhaps resulting in sample loss.

	Mass spectrometric methods	
Gel with MALDI	Gel with tandem MS	Gel free with LC-MS/MS
Matrix assisted desorption ionization time-of-flight MS (MALDI-TOF MS) and peptide mass fingerprinting. Analysis of bands containing only one protein. Less specific than tandem MS.	Prefractionation of sample with SDS-PAGE Nanoelectrospray and tandem MS and MS/MS. Analysis of simple mixtures. Sequence information allows unambiguous protein identifications.	Liquid chromatography and MS/MS allows analysis of complex mixtures. Separation of peptides in time as well as automated data acquisition will make this tool highly suited for large-scale approaches.

Depending on the experimenter's needs and the system studied, some combinations may prove more useful. Refer to text for further discussion.

---

illustrate the use of coprecipitation and mass spectrometry in a variety of ways and are not intended to be an exhaustive review. The pros and cons of each choice (bait format, type of tag, type of MS used) should be evaluated for each study. Important variables that govern these choices include the amount and type of starting material available (cell lysates versus tissue); the type of complexes to be purified, the strength of their interactions; the inherent tradeoff between specificity of purification and the loss of weaker interacting proteins; and the degree of prefractionation and the scale of the experiment (single bait versus proteome-wide).

GST-Tagged Proteins

Glutathione S-transferase (GST) fusions with bait proteins are well established as a means for coprecipitating interacting proteins. The major advantage of GST-fusions is that large amounts of the bait can often be easily made in bacterial expression systems. These fusion proteins can be used for coprecipitation studies in an *in vitro* format; that is, large volumes of cell lysate are incubated with the GST-fusion and subsequently purified over glutathione-agarose. Fig. 1 shows an example in which GST-fusions with an interaction domain like SH2 yielded promising results in coprecipitation experiments. Other examples of coprecipitation experiments with GST-fusions and mass spectrometry include the identification of novel binding partners for Shc [3] and the Cdk5 activator protein p35[nck5a] [4].

Although the use of GST fusions is widespread and it is a tremendously useful technique, there are some disadvantages in its application to large-scale proteomics studies. First, the large size of GST (220 aa) may sterically inhibit complex formation. Second, the production of GST fusion proteins in bacterial expression systems may also mean that the protein may not be folded in its native state. Third, bacterially expressed recombinant proteins are sometimes unstable and/or less soluble and do not carry mammalian posttranslational modifications. Fourth, bait presentation is not done within the cell and so the physiological relevance of the protein interactions is less certain than with *in vivo* methods.

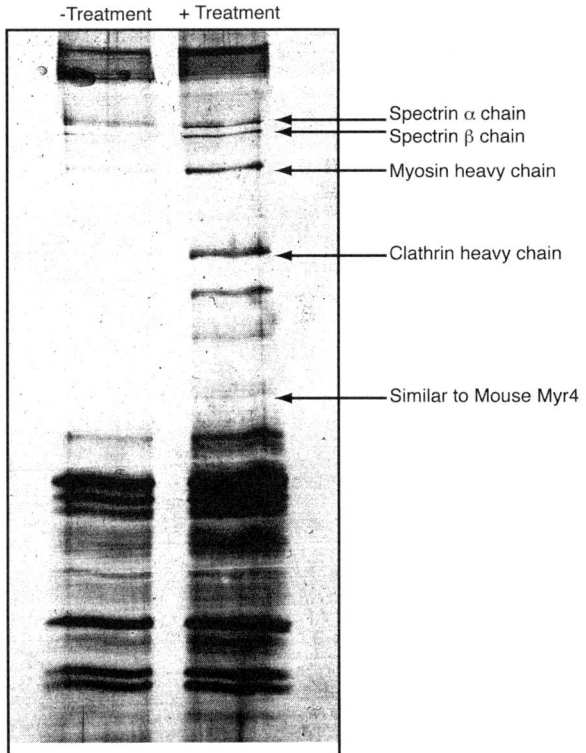

Figure 1 GST-SH2 coprecipitation experiment in mammalian cell lysates. Proteins in the right lane (marked with an arrow) are pulled down by the GST-fusion with SH2 following treatment, suggestive that these bands are specific. Gel bands were excised and then analyzed by Nano ES tandem MS.

For the above reasons, this approach is most useful as a validation method for protein interactions suggested by other evidence. For large-scale fishing strategies where coprecipitation is used as a discovery tool, we suggest considering other tags (refer to Table I for a summary of features in some of the more commonly used tags).

Antibodies

In discovery-oriented approaches—that is, where the interacting partners were not already suspected—the coupling of coimmunoprecipitation and mass spectrometry has proven to be an elegant and effective approach for finding novel substrates and novel proteins in growth factor and cytokine receptor signaling [5]. In addition to directing an antibody against a single "bait" protein, the antibody can also be directed against a common feature of a group of proteins, such as a phospho-residue. Pandey et al. were able to identify novel substrates of both the epidermal growth factor receptor (EGFR) and the platelet derived growth factor receptor (PDGFR) signaling pathway with the use of antiphosphotyrosine antibodies to immunoprecipitate phosphoproteins from cell lysates [6–8]. In these pull-down experiments, gel bands resolved by one-dimensional SDS-PAGE from immunoprecipitates of experimental samples treated with cytokine or growth factor were compared with untreated controls. Gel bands specific to treated cells were excised, digested, and analyzed by tandem mass spectrometry. In the example of the analysis of immunoprecipitated proteins from EGFR signaling, nine proteins were identified. Of these two were previously undescribed substrates in EGFR signaling and seven others were previously known.

The use of antibodies against endogenous bait protein is attractive because the bait is in its *in vivo* state, which makes this experiment the closest to the *in vivo* situation. However, it should be noted that it can be quite time consuming to develop antibodies against a large set of proteins and that the subset of antibodies capable of immunoprecipitating proteins is small. The specificity and ability to pull down proteins is also exceedingly variable (for example, the antiphosphoserine and antiphosphothreonine antibodies are not as effective as currently available antiphosphotyrosine antibodies). Large-scale collections of validated antibodies against all human proteins would solve these problems.

Epitope Tags

Epitope tags were historically derived by mapping the interaction site of monoclonal antibody (linear epitope of less than ten amino acids), and then fusing this epitope to bait proteins of interest. A commonly used epitope tag is the FLAG peptide [9,10]. Its salient features are its small size (8 aa), its hydrophilicity, and the availability of good monoclonal antibodies for highly specific purification. The size and hydrophilic nature of the tag were specifically designed to minimize the tag's effect on the formation of native protein complexes. Plasmid vectors for adding N- and C-terminal versions of the FLAG tag to the bait of interest exist, and each have specific conjugated monoclonal antibodies raised against them. Although these antibodies are highly specific, they do not have a large binding capacity. In MS-coupled approaches, this high specificity is a desirable tradeoff versus the lower cost of the GST-based approach, and the high sensitivity of MS may mean that large preparations are unnecessary. As shown in Table II, other epitope tags, such as Myc and HA, have also been used, but they may not be appropriate for use for their own reasons: Myc tagged proteins are difficult to elute from the monoclonal antibody (9E10) and should be used in conjunction with enzymatic cleavage (see below); anti-HA (12CA5) antibodies are not as specific as anti-Myc or anti-Flag antibodies.

In an effort to increase specificity of the affinity purification without the use of monoclonal antibodies, the combination of multiple epitopes in a single tag has recently been popularized (for example, TAP for tandem affinity purification [11]) and has been used successfully in several recent studies [12,13]. In the TAP example, the tag consists of a calmodulin-binding peptide and an IgG-binding region (ProtA) with a tobacco etch virus (TEV) cleavage site separating the two epitopes. Briefly, the tagged protein and its associated proteins are first purified on IgG beads, washed, and then eluted by

Table II A Summary Table of Common Tags Used in coprecipitation Studies

Types of Tags[a]	GST	Biotinylated	6xHis	Epitope	Affinity tag with enzymatic cleavage
Size of tag	Large, 26 kDa protein (220 aa residues)	Biotin plus linker region is small	Small	Less than 10 amino acids; (FLAG, Myc; HA)	Large, bulky (e.g. TAP is 184 aa residues)
Features	Many commercial vectors available in different expression systems	Biotin as a reactive hook for fishing with avidin conjugated beads Requires a chemical modification that attaches biotinylated tag to reactive moieties on peptide chain	Polyanionic	FLAG has an enterokinase cleavage site, hydrophilic tag; Myc binds tightly and should be combined with enzymatic release; HA is less specific	ProtA and TEV (29); TAP tag as a modification-Multiple affinity tags Protein A, Calmodulin binding peptide, Internal cleavage site (11, 30)
Purification	Fairly specific, glutathione conjugated agarose for affinity purification	Avidin binding is extremely strong, fairly specific depending on the type of avidin	Commonly Immobilized Metal Affinity Chromatography (IMAC) like Ni-NTA resin, not very specific.	Monoclonal antibodies-very specific	Enzymatic cleavage is one of the first steps to elute bound material. In TAP tagging, first bound to one affinity matrix (IgG beads), cleave purification tag internally to release, bind to second affinity matrix (calmodulin beads) elute with EGTA
Ease of elution	Competitive elution with glutathione; enzyme cleavage (e.g. thrombin)	Requires harsh elution conditions, typically organic solvent/acid.	Mild elution conditions with 100 mM imidazole or with L-histidine	Gentle–competitive elution with the peptide constituting the epitope	Enzymatic cleavage is generally specific. Dual elution steps– both fairly mild but might be biased against specific classes of proteins (EGTA)
Potential Pitfall	Large size of tag may interfere with native protein or protein complexes	Chemical modification is not specific–requires purification of complex to homogeneity which is generally impractical; harsh treatment of sample.	Affinity matrix is not very specific to His-tagged proteins. Might also pull down other proteins.		More complicated purification procedure; Complex needs to be stable under enzymatic release

[a]See also [31,32], for a good discussion of available tags.

incubating with the TEV protease. The enzyme will cleave the tag, resulting in the elution of the protein complex. The eluate is affinity purified in the next step over calmodulin beads and subsequently eluted with EGTA. The TAP tag and other purification methods based on two orthogonal principles significantly reduce background. Often, a single tag, for example Protein A or n x Myc, achieves similar specificity when combined with enzymatic release of the complex and is simpler to perform. Limitations of the TAP method include the large size of the tag (184 aa residues), the requirement for stability of the complex during enzymatic cleavage, and the fact that some mammalian proteins contain the recognition site for the TEV protease.

Mass Spectrometric Approaches

Since the mid-1990s, mass spectrometry (MS) has been the definitive method used for protein identification. Technological developments have brought significant advances in the sensitivity and throughput of MS approaches [14,15]. Quantitative approaches using stable isotopes together with MS greatly enchance the utility of such experiments [15].

A well-established and still commonly used approach is the identification of proteins following separation by gel electrophoresis. Proteins are visualized by staining with either silver or Coomassie-based stains. Gel bands from one-dimensional SDS-PAGE gels or spots from two-dimensional

IEF-SDS-PAGE gels are excised, reduced, and alkylated, then digested with a proteolytic enzyme (typically trypsin) in order to yield peptides that are more easily characterized by mass spectrometric methods [16].

The MS method used for the identification of proteins may vary depending on the degree of prefractionation of the sample. For instance, it may be acceptable to use matrix-assisted laser desorption ionization time-of-flight (MALDI-TOF) instruments and peptide mass fingerprinting (PMF) for protein identification of well-resolved proteins on two-dimensional gels. Analyses of large proteins (and correspondingly large numbers of peptides) or mixtures of proteins may require the use of tandem mass spectrometry in order to yield unambiguous identifications from femtomole (nanogram) amounts of protein.

Two recent studies in yeast involving the use of epitope tags for the study of multiprotein complexes by mass spectrometry have demonstrated the feasibility of large-scale approaches [17–19]. Notably, up to 25% of the yeast genome was identified in immunoprecipitations of tagged proteins in both studies. An important finding was that datasets from both of these interaction mapping studies did not show considerable overlap with known protein interaction data obtained from Y2H experiments [20]. Datasets from Y2H experiments have also shown that these derived interaction maps show variability even when the same approach is used [21,22]. This may be an indication that such studies, though large in scale, still only sample a small area of the interaction space in the overall scope of protein interactions.

It is important to bear in mind the principal differences between the Y2H approach and the pull-down experiment. Y2H experiments are more suited to the discovery of binary or tertiary (with yeast-three hybrid) of moderate to strong binding strengths. The pull-down experiment is more likely to yield protein complexes that interact strongly within a complex but may consist of a large number of interacting proteins that individually might have a much weaker binding affinity for other proteins within the complex. Another attractive feature of the coprecipitation approach is that the complex is pre-assembled *in vivo* and subsequently affinity purified. It is likely that the maintenance of the physiological conditions where these interactions form would have a positive impact on the biological relevance of the interaction data obtained. In contrast, the Y2H experiment is based on the inherent assumption that the cDNAs will express in yeast and that the interactions persist (or form) in possibly nonphysiological contexts.

Outlook

The combination of affinity purification of proteins and subsequent detection by mass spectrometry has become a very powerful tool over the last few years. Although different tags are continually being developed, the goal remains the same: the specific enrichment of certain proteins either in a modification-dependent fashion or as a member of a larger protein complex. The choice of the tag depends on the experimental context.

Recently, the move toward high-throughput analyses of complex protein mixtures has led to the increased use of liquid chromatographic separation as an orthogonal method of separation before mass spectrometric analyses [23,24]. In this approach, complex mixtures of proteins (e.g. a cell lysate) are digested in-solution and loaded onto a reverse phase C_{18} column, desalted, and eluted with a gradient of organic solvent directly into the mass spectrometer for analysis. Present approaches require that one determine the validity of a coprecipitating protein by visualizing the differences with a "control" sample (Fig. 1). It would therefore be necessary to incorporate quantitative information directly in protein complexes studied. This is also necessary because mass spectrometry is becoming ever more sensitive and can detect trace amounts of proteins in a pull-down. Labeling of protein complexes before mass spectrometric analyses with chemical reagents such as ICAT [25] or other derivatizing agents [26,27] may serve this function. In addition, recent methods for the labeling of cells *in vivo* [28,33] may be more suitable as proteins are quantitatively coded at the amino acid level even before any purification steps are performed. By using quantitatively labeled proteins from two different states in coprecipitation experiments, specific protein interactions can be easily distinguished from non-specific 'background' with a high degree of confidence and sensitivity [34,35]. Furthermore, this quantitative data can facilitate the determination of stoichiometries of protein binding within complexes.

Acknowledgment

We thank Leonard Foster for critical reading of the manuscript, Blagoy Blagoev for the figure, and other members of the Protein Interaction Laboratory for useful discussions and support. Work done in PIL is supported by a generous grant from the Danish Research Foundation.

References

1. Ferrell, J. E. Jr. (2000). *Sci STKE* **2000**, PE1.
2. Fields, S. and Song, O. (1989). *Nature* **340**, 245–246.
3. Thomas, D., Patterson, S. D., and Bradshaw, R. A. (1995). *J. Biol. Chem.* **270**, 28924–28931.
4. Qu, D., Li, Q., Lim, H. Y., Cheung, N. S., Li, R., Wang, J. H., and Qi, R. Z. (2002). *J. Biol. Chem.* **277**, 7324–7332.
5. Pandey, A., Andersen, J. S., and Mann, M. (2000). *Sci STKE* **2000**, PL1.
6. Pandey, A., Fernandez, M. M., Steen, H., Blagoev, B., Nielsen, M. M., Roche, S., Mann, M., and Lodish, H. F. (2000). *J. Biol. Chem.* **275**, 38633–38639.
7. Pandey, A., Podtelejnikov, A. V., Blagoev, B., Bustelo, X. R., Mann, M., and Lodish, H. F. (2000). *Proc. Natl. Acad. Sci. USA* **97**, 179–184.
8. Steen, H., Kuster, B., Fernandez, M., Pandey, A., and Mann, M. (2002). *J. Biol. Chem.* **277**, 1031–1039.
9. Hopp, T. P. P. K., Price, V. L., Libby, R. T., March, C. J., Cerretti, D. P., Urdal, D. L., and Conlon, P. J. (1988). *BIO-TECHNOLOGY* **6**, 1204–1210.
10. Einhauer, A. and Jungbauer, A. (2001). *J. Biochem. Biophys. Methods* **49**, 455–465.
11. Rigaut, G., Shevchenko, A., Rutz, B., Wilm, M., Mann, M., and Seraphin, B. (1999). *Nat. Biotechnol.* **17**, 1030–1032.
12. Chen, C. Y., Gherzi, R., Ong, S. E., Chan, E. L., Raijmakers, R., Pruijn, G. J., Stoecklin, G., Moroni, C., Mann, M., and Karin, M. (2001). *Cell* **107**, 451–464.

13. Deshaies, R. J., Seol, J. H., McDonald, W. H., Cope, G., Lyapina, S., Shevchenko, A., Shevchenko, A., Verma, R., and Yates, J. R., III (2002). *Mol Cell Proteomics* **1**, 3–10.
14. Aebersold, R. and Goodlett, D. R. (2001). *Chem. Rev.* **101**, 269–295.
15. Mann, M., Hendrickson, R. C., and Pandey, A. (2001). *Annu. Rev. Biochem.* **70**, 437–473.
15a. Aebersold, R. and Mann, M. (2003). *Nature* **422**, 198–207.
16. Shevchenko, A., Wilm, M., Vorm, O., and Mann, M. (1996). *Anal. Chem.* **68**, 850–858.
17. Ho, Y., Gruhler, A., Heilbut, A., Bader, G. D., Moore, L., Adams, S. L., Millar, A., Taylor, P., Bennett, K., Boutilier, K., Yang, L., Wolting, C., Donaldson, I., Schandorff, S., Shewnarane, J., Vo, M., Taggart, J., Goudreault, M., Muskat, B., Alfarano, C., Dewar, D., Lin, Z., Michalickova, K., Willems, A. R., Sassi, H., Nielsen, P. A., Rasmussen, K. J., Andersen, J. R., Johansen, L. E., Hansen, L. H., Jespersen, H., Podtelejnikov, A., Nielsen, E., Crawford, J., Poulsen, V., Sorensen, B. D., Matthiesen, J., Hendrickson, R. C., Gleeson, F., Pawson, T., Moran, M. F., Durocher, D., Mann, M., Hogue, C. W., Figeys, D., and Tyers, M. (2002). *Nature* **415**, 180–183.
18. Gavin, A. C., Bosche, M., Krause, R., Grandi, P., Marzioch, M., Bauer, A., Schultz, J., Rick, J. M., Michon, A. M., Cruciat, C. M., Remor, M., Hofert, C., Schelder, M., Brajenovic, M., Ruffner, H., Merino, A., Klein, K., Hudak, M., Dickson, D., Rudi, T., Gnau, V., Bauch, A., Bastuck, S., Huhse, B., Leutwein, C., Heurtier, M. A., Copley, R. R., Edelmann, A., Querfurth, E., Rybin, V., Drewes, G., Raida, M., Bouwmeester, T., Bork, P., Seraphin, B., Kuster, B., Neubauer, G., and Superti-Furga, G. (2002). *Nature* **415**, 141–147.
19. Shevchenko, A., Schaft, D., Roguev, A., Pijnappel, W. W. M. P., Stewart, A. F., and Shevchenko, A. (2002). *Mol Cell Proteomics* **1**, 204–212.
20. von Mering, C., Krause, R., Snel, B., Cornell, M., Oliver, S. G., Fields, S., and Bork, P. (2002). *Nature* **417**, 399–403.
21. Ito, T., Chiba, T., Ozawa, R., Yoshida, M., Hattori, M., and Sakaki, Y. (2001). *Proc. Natl. Acad. Sci. USA* **98**, 4569–4574.
22. Uetz, P., Giot, L., Cagney, G., Mansfield, T. A., Judson, R. S., Knight, J. R., Lockshon, D., Narayan, V., Srinivasan, M., Pochart, P., Qureshi-Emili, A., Li, Y., Godwin, B., Conover, D., Kalbfleisch, T., Vijayadamodar, G., Yang, M., Johnston, M., Fields, S., and Rothberg, J. M. (2000). *Nature* **403**, 623–627.
23. McCormack, A. L., Schieltz, D. M., Goode, B., Yang, S., Barnes, G., Drubin, D., and Yates, J. R., 3rd (1997). *Anal. Chem.* **69**, 767–776.
24. Mermall, V., Post, P. L., and Mooseker, M. S. (1998). *Science* **279**, 527–533.
25. Gygi, S. P., Rist, B., Gerber, S. A., Turecek, F., Gelb, M. H., and Aebersold, R. (1999). *Nat. Biotechnol.* **17**, 994–999.
26. Regnier, F. E., Riggs, L., Zhang, R., Xiong, L., Liu, P., Chakraborty, A., Seeley, E., Sioma, C., and Thompson, R. A. (2002). *J. Mass Spectrom.* **37**, 133–145.
27. Cagney, G. and Emili, A. (2002). *Nat. Biotechnol.* **20**, 163–170.
28. Ong, S. E., Blagoev, B., Kratchmarova, I., Kristensen, D. B., Steen, H., Pandey, A., and Mann, M. (2002). *Mol. Cell Proteomics* **1**, 376–386.
29. Senger, B., Simos, G., Bischoff, F. R., Podtelejnikov, A., Mann, M., and Hurt, E. (1998). *EMBO J.* **17**, 2196–2207.
30. Honey, S., Schneider, B. L., Schieltz, D. M., Yates, J. R., and Futcher, B. (2001). *Nucl. Acids Res.* **29**, E24.
31. Hearn, M. T. and Acosta, D. (2001). *J. Mol. Recognit.* **14**, 323–369.
32. Harlow, E. and Lane, D. (1999). *Using Antibodies: A Laboratory Manual*. Cold Spring Harbor Laboratory Press, Cold Spring Harbor.
33. Ong, S. E., Kratchmarova, I., and Mann, M. (2003). *J. Proteome Res.* **2**, 173–181.
34. Blagoev, B., Kratchmarova, I., Ong, S. E., Nielsen, M., Foster, L. J., and Mann, M. (2003). *Nat. Biotechnol.* **21**, 315–318.
35. Ranish, J. A., Yi, E. C., Leslie, D. M., Purvine, S. O., Goodlett, D. R., Eng, J., and Aebersold, R. (2003). *Nat. Genet.* **33**, 349–355.

CHAPTER 173

Proteomics, Fluorescence, and Binding Affinity

Paul R. Graves[1] and Timothy A. J. Haystead[2]

[1]Department of Pharmacology and Cancer Biology, Center for Chemical Biology,
Duke University, Durham, North Carolina
[2]Serenex Inc., Research Triangle Park, North Carolina

Introduction

In this chapter, we discuss how affinity chromatography can be used to purify protein complexes or isolate specific proteomes for further analysis. This approach can facilitate the investigation of protein-protein interactions, enable the identification of novel binding proteins, or allow changes in protein expression or modification to be monitored under different conditions.

Isolation of Specific Proteomes

Proteomics, the study of all proteins expressed in a cell, can provide insight into complex cellular processes. However, studying the entire proteome of a typical eukaryotic cell can be difficult because (1) the number of proteins may exceed the capacity of the systems used to analyze them, and (2) abundantly expressed proteins can dominate the analysis, obscuring less abundant yet potentially interesting proteins. Therefore, we have used affinity chromatography to select for specific types of proteins to simplify and direct the analysis.

Gamma-Phosphate Linked ATP-Sepharose

Several years ago we developed gamma-phosphate linked ATP-sepharose for the affinity purification of protein kinases from complex mixtures [1]. The design of this resin was based upon the orientation of ATP when bound in the active site of a protein kinase. The crystal structure of cyclic-AMP-dependent protein kinase revealed that the adenine portion of ATP is buried within the binding pocket of the enzyme while the gamma-phosphate group of ATP is exposed to the solvent [2]. Therefore, we linked ATP to sepharose via its gamma-phosphate group to enable binding of protein kinases. An important property is that if ATP is linked through adenosine at N6, the resin becomes nonfunctional [P. R. Graves et al., in preparation].

In addition to protein kinases, ATP-sepharose is capable of binding a large number of other purine-utilizing enzymes including the NAD+/NADP+ utilizing dehydrogenases, DNA ligases, nonprotein kinases, mononucleotide ATPases, and nonconventional purine utilizing enzymes. Indeed, the number of proteins that utilize purines (the purine binding proteome) is quite large and has been estimated to represent about 4% of the human genome [3]. Characterization of proteins that specifically bound ATP-sepharose from a whole mouse lysate revealed that a significant portion of the purine-binding proteome was captured (P. R. Graves et al., in preparation).

The following experiment illustrates how ATP-sepharose (or another type of affinity matrix specific for certain types of proteins) can be used (Fig. 1). Two sets of cells are prepared, one as a control, and one that undergoes some form of treatment (stimulation with growth factors, drug treatments, and so forth). The cells are lysed, and the extracts passed over a column of ATP-sepharose to capture the purine-binding proteome from each sample. After washing to remove nonspecific proteins, proteins are eluted with ATP and resolved by one- or two-dimensional gel electrophoresis. Proteins that are altered in their expression levels or have undergone

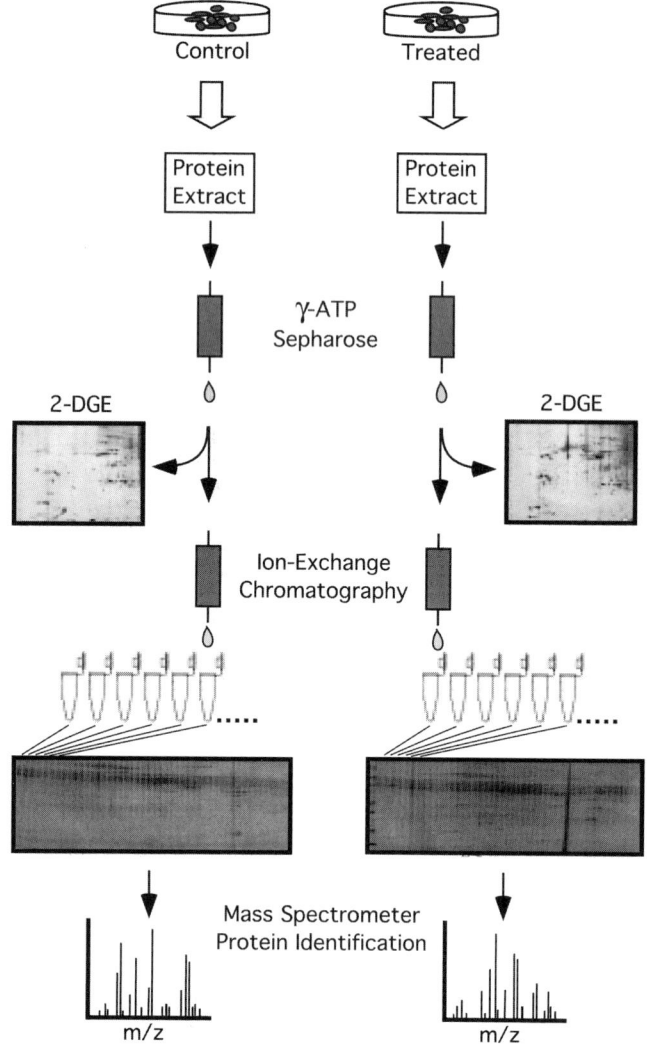

Figure 1

unclear whether low-abundance phosphoproteins can be recovered [4,5]. We developed a method to label phosphoserine in proteins or peptides with a fluorophore [7] based upon the work of Meyer and colleagues [8]. The labeling of phosphorylated amino acids with fluorescent moieties could serve as an alternative to labeling with the radioisotope ^{32}P and be used to study phosphorylation in animals, intact tissues, and humans [7].

Affinity Chromatography for the Isolation of Protein Complexes

Microcystin-Sepharose Chromatography for the Study of Protein-Protein Interactions

Microcystin is a cyclic heptapeptide known to inhibit type 1 and 2 protein phosphatases by binding to their catalytic subunits [9]. Microcystin-sepharose was developed by Sugimura and colleagues [10] and has been used for the purification of PP-1 and PP-2A [11,12]. We have used this resin for the identification of novel protein phosphatase interacting proteins [11,12]. Skeletal muscle lysates were applied to microcystin-sepharose and bound proteins analyzed by one-dimensional gel electrophoresis and protein sequencing. Using this approach, 36 protein phosphatase 1-binding proteins were detected and sequenced [12]. In addition to the recovery of many known protein phosphatase binding proteins, a novel PP-1 regulatory subunit was defined [12].

Gamma-Phosphate Linked ATP-Sepharose

Because ATP-sepharose is capable of binding protein kinases, it can also be used to characterize protein kinase complexes. For example, the subunit structure of AMP-activated protein kinase was established by purifying native AMP-kinase with ATP-sepharose [13]. ATP-sepharose was also used for the purification of Cdc28p kinase complexes from budding yeast [14]. Since ATP-sepharose binds protein kinases in the catalytic domain, it offers another method for the isolation of protein kinase interacting proteins.

changes in modification (such as phosphorylation) are excised from the gel and identified by mass spectrometry (Fig. 1). Alternatively, proteins isolated by ATP-sepharose can be further purified by more specific techniques before analysis.

Phosphoprotein Enrichment

Protein phosphorylation is a reversible and widespread mechanism for regulating protein function. One goal of proteomics is to allow study of the phosphoproteome, or all the phosphorylated proteins expressed in the cell, under different conditions. However, many phosphorylated proteins are in low abundance and cannot be detected in a whole cell extract. Therefore, a number of techniques have emerged to enrich for phosphorylated proteins. Two methods were recently reported that capture phosphorylated proteins by converting the phosphoamino acids in proteins to groups that allow attachment of affinity ligands for their subsequent purification [4–6]. However, because of the chemistry and the purification strategy involved in these methods, it is still

Specificity of Protein-Protein or Protein-Ligand Interactions

When starting with a cell extract, the recovery of proteins with an affinity matrix will depend on the solubility of the protein target in the lysis buffer, the binding constant between the protein and the ligand, the abundance of the protein, and the stability of the protein in the cellular extract. Regardless of which affinity approach is used, proteins will be isolated that are not specific for the protein or ligand used because of interaction with the affinity ligand support. Therefore, in each case the appropriate controls need to be performed. If two samples are being compared, ideally, an identical affinity column should be used for both samples. If a "bait" protein is being used to isolate interacting proteins, then it is a good idea to use a scrambled or irrelevant protein of the same mass as a control. In some cases, specific mutations can be introduced into the bait protein to prevent protein interactions. Double epitope tagging (also known as tandem affinity purification, or TAP) has also been used to increase the stringency of protein complex isolation [15].

One of the most important steps in affinity chromatography is the elution step. Elutions should be performed with an excess of the ligand used for the affinity matrix. In this way, the risk of nonspecific protein elutions is likely to be eliminated. As an example, to increase specificity, we linked microcystin to biotin for the capture of phosphatase interacting proteins [11]. This provides two advantages. First, since mild conditions can be used to elute the bound proteins, holoenzyme complexes remain intact. Second, since the elution is performed with biotin, only proteins that bind to microcystin are eluted, eliminating proteins that bind to the column matrix. Some additional factors to be considered are (1) Can the binding of the recovered proteins to an affinity column be rationalized? For example, if a protein is recovered with ATP-sepharose, is it a purine-utilizing enzyme or does it have a binding site for a molecule resembling ATP? (2) Does pretreatment of the extract with free ligand prevent binding of proteins to the column? (3) Does the protein binding survive stringent washing conditions? (4) Can the result be confirmed by other methods such as far western or co-immunoprecipitation experiments? Because of the complexity of the proteome of a eukaryotic cell and the fact that many proteins of low abundance cannot be detected in a cell lysate, affinity chromatography will continue to be an important method for the isolation of specific types of proteins.

References

1. Haystead, C. M., Gregory, P., Sturgill, T. W., and Haystead, T. A. (1993). Gamma-phosphate-linked ATP-sepharose for the affinity purification of protein kinases. Rapid purification to homogeneity of skeletal muscle mitogen-activated protein kinase kinase. *Eur. J. Biochem.* **214**, 459–467.
2. Knighton, D. R., Zheng, J. H., Ten Eyck, L. F., Xuong, N. H., Taylor, S. S., and Sowadski, J. M. (1991). Structure of a peptide inhibitor bound to the catalytic subunit of cyclic adenosine monophosphate-dependent protein kinase. *Science* **253**, 414–420.
3. Lander, E. S. *et al.* (2001). Initial sequencing and analysis of the human genome. *Nature* **409**, 860–921.
4. Oda, Y., Nagasu, T., and Chait, B. T. (2001). Enrichment analysis of phosphorylated proteins as a tool for probing the phosphoproteome. *Nat. Biotechnol.* **19**, 379–382.
5. Zhou, H., Watts, J. D., and Aebersold, R. (2001). A systematic approach to the analysis of protein phosphorylation. *Nat. Biotechnol.* **19**, 375–378.
6. Adamczyk, M., Gebler, J. C., and Wu, J. (2001). Selective analysis of phosphopeptides within a protein mixture by chemical modification, reversible biotinylation and mass spectrometry. *Rapid Commun. Mass Spectrom.* **15**, 1481–1488.
7. Fadden, P. and Haystead, T. A. (1995). Quantitative and selective fluorophore labeling of phosphoserine on peptides and proteins: characterization at the attomole level by capillary electrophoresis and laser-induced fluorescence. *Anal. Biochem.* **225**, 81–88.
8. Meyer, H. E., Hoffmann-Posorske, E., and Heilmeyer, L. M., Jr. (1991). Determination and location of phosphoserine in proteins and peptides by conversion to S-ethylcysteine. *Methods Enzymol.* **201**, 169–185.
9. MacKintosh, C. and MacKintosh, R. W. (1994). Inhibitors of protein kinases and phosphatases. *Trends Biochem. Sci.* **19**, 444–448.
10. Nishiwaki, S., Fujiki, H., Suganuma, M., Nishiwaki-Matsushima, R., and Sugimura, T. (1991). Rapid purification of protein phosphatase 2A from mouse brain by microcystin-affinity chromatography. *FEBS Lett.* **279**, 115–118.
11. Campos, M., Fadden, P., Alms, G., Qian, Z., and Haystead, T. A. (1996). Identification of protein phosphatase-1-binding proteins by microcystin-biotin affinity chromatography. *J. Biol. Chem.* **271**, 28478–28484.
12. Damer, C. K., Partridge, J., Pearson, W. R., and Haystead, T. A. (1998). Rapid identification of protein phosphatase 1-binding proteins by mixed peptide sequencing and data base searching. Characterization of a novel holoenzymic form of protein phosphatase 1. *J. Biol. Chem.* **273**, 24396–24405.
13. Davies, S. P., Hawley, S. A., Woods, A., Carling, D., Haystead, T. A., and Hardie, D. G. (1994). Purification of the AMP-activated protein kinase on ATP-gamma-sepharose and analysis of its subunit structure. *Eur. J. Biochem.* **223**, 351–357.
14. Shellman, Y. G., Svee, E., Sclafani, R. A., and Langan, T. A. (1999). Identification and characterization of individual cyclin-dependent kinase complexes from *Saccharomyces cerevisiae*. *Yeast* **15**, 295–309.
15. Puig, O., Caspary, F., Rigaut, G., Rutz, B., Bouveret, E., Bragado-Nilsson, E., Wilm, M., and Seraphin, B. (2001). The tandem affinity purification (TAP) method: a general procedure of protein complex purification. *Methods* **24**, 218–229.

CHAPTER 174

FRET Analysis of Signaling Events in Cells

Peter J. Verveer and Philippe I. H. Bastiaens
Cell Biology and Cell Biophysics Program,
European Molecular Biology Laboratory,
Heidelberg, Germany

Introduction

Detection of fluorescence resonance energy transfer (FRET) by optical techniques provides a sensitive means of detecting and quantifying molecular interactions and protein modifications in cells. Several strategies are available to develop sensors for use in FRET assays based on fluorescent labeling or green fluorescent protein (GFP) fusions. By using these sensors, techniques such as ratio imaging, sensitized emission measurements, photobleaching methods, or fluorescence lifetime imaging can be employed to spatially and temporally resolve the occurrence of FRET in cells. In this contribution, the strengths and weaknesses of the different sensors and measurement methods are discussed and compared and their use illustrated by reviewing the recent literature. We conclude that the spatially and temporally resolved measurement of FRET in cells has opened new opportunities to image biochemistry in intact cells and expect that these techniques will play an increasingly important role in cell biology.

Optical microscopy provides a sensitive, specific, and noninvasive approach to localize fluorescently labeled macromolecules in cells with high spatial and temporal resolution. Moreover, the spectroscopic properties of fluorescence probes can be used to obtain information on their molecular environment. With the advent of genetically encoded variants of green fluorescent proteins, the observation of biochemistry in cells has become feasible *in vivo* [1–3]. Fluorescence resonance energy transfer is one photophysical phenomenon that has been put to good use to detect and quantify molecular interactions and protein modifications [4]. FRET cannot be measured directly, but the resulting changes in the fluorescence properties of the fluorophores can be detected by several optical techniques with spatial and temporal resolution inside cells. FRET is a photophysical effect whereby energy is transferred from an excited donor fluorophore to an acceptor fluorophore. This does not involve the emission and subsequent absorption of a photon but occurs by a direct electromagnetic interaction. The efficiency of transfer depends on the spectral properties of donor and acceptor, and on their relative orientation and distance. An important factor is that the energy transfer efficiency has an inverse sixth order dependence on the distance between the two fluorophores, and typically FRET only occurs when the distance is less than 10 nm, which is generally only achieved when donor and acceptor are attached to the same macromolecule or to interacting molecules. For this reason FRET can be applied to specifically image such events as molecular interactions or conformational changes. FRET can be observed by its consequences, which are reflected by a change in the fluorescence kinetics of both the donor and the acceptor. Due to the transfer of energy the rate at which the donor returns to its ground state increases, and hence its fluorescence lifetime decreases. As a consequence the quantum yield of the donor, and therefore its steady-state fluorescence intensity, also decreases. The steady-state fluorescence of the acceptor, however, is increased by the sensitized emission that is emitted when the acceptor returns to its ground state.

Fluorescent Probes for FRET

The design of the fluorescent probes for FRET measurements must match the problem at hand. Single component molecule sensors, consisting of two fluorophores flanking a protein domain or subunit, change FRET efficiency upon a change of conformation or cleavage. These types of sensors are commonly constructed by fusing cyan and yellow variants of GFP to the reporter domains. They can be used to detect physiologically relevant ions [5–7] and small organic compounds [8–10], or report on protein activity or conformational or covalent state [11–14]. They can be targeted to specific cellular compartments by incorporating suitable localization signals. Another class of sensors is based upon the interaction between different compounds tagged with a donor or acceptor fluorophore [15–20]. One application is the detection of covalent modification of a protein, in which case the protein is tagged by a donor, and an acceptor-tagged reagent interacting with the conjugated group is present in large excess. The state of donor-tagged proteins can then be probed by FRET. An example is the use of generic phosphoamino-acid-specific antibodies to probe the phosphorylation state of a protein, whereby the protein is tagged with a donor fluorophore (e.g. by fusing a GFP molecule) and the antibody with the acceptor fluorophore (e.g. Cy3). Another application, not necessarily employing an excess of one species, is using cyan and yellow GFP variants to measure interactions between different proteins, or homo- or hetero-dimerization *in vivo*. The use of GFP variants is attractive for live cell applications, but FRET measurements can also be done via labeled antibodies. This may seem potentially problematic due to the antibody size, but depending on the donor/acceptor configuration and the type of FRET measurement this may in fact be an advantage. Consider an assay to probe the state of a protein based on the observation of the donor fluorophore (e.g. a GFP) only and an antibody labeled with multiple acceptor fluorophores (e.g. Cy3; see [20]). In this case, the increased density of acceptor fluorophores implies a higher probability that upon antibody binding, the donor is in close proximity of an acceptor. However, using an antibody labeled with multiple donor fluorophores would lead to an increased probability of detecting a donor that is too far away from an acceptor for efficient FRET and therefore a lower average signal. In general, the choice of sensor, fluorophore (GFP or fluorescent dye), and where donor and acceptor are located is also determined by the measurement method that is used. The different techniques are described in the next section, where the types of sensors that are appropriate for the respective techniques will also be discussed.

FRET Detection Techniques

Ratio Imaging

Upon the occurrence of FRET the steady-state fluorescence of the donor is quenched, since part of the excited state energy is transferred to the acceptor rather than emitted as photons. Simultaneously, emitting this energy as photons increases the steady-state fluorescence of the acceptor. Therefore, a change in FRET is reflected in an increase in the ratio of sensitized emission over donor emission. This type of measurement is straightforward, as it requires only measurements at two filter settings, and is therefore well suited to live cell imaging.

In an ideal situation, the sensitized emission and donor emission would be directly measured by choosing appropriate combinations of excitation and emission filters, exciting the donor specifically, and detecting the intensities through filters specific for donor and acceptor emissions. Indeed, the donor emission can be imaged specifically; however, in practice the measured acceptor channel contains significant contributions of leak-through of the donor emission and of direct excitation of the acceptor. This implies that the signal in the acceptor channel is strongly dependent on the relative concentrations of donor and acceptor, and that interpretation of the ratio as a diagnostic for FRET is problematic. However, if the relative concentrations of donor and acceptor are fixed, then a change in the donor/acceptor fluorescence ratio can be attributed to a change in FRET. Therefore this approach should mainly be used with sensors that consist of donor and acceptor fluorophores attached to a single molecule. A change in FRET due to a conformational change can then be reliably detected. Another situation is presented by a sensor where donor and acceptor disassociate, for instance by proteolytic activity. In such a case the relative concentrations are known before cleavage and, if no significant differential relocation of the cleaved products is expected, the ratio may still be a good measure of FRET. Also, if one is only interested in the total signal integrated over a cell, the relative total concentrations can be considered to remain the same, although this assumption may be incorrect in a confocal microscope, where molecules may relocate to a position outside the focal plane.

In general, quantification of ratio measurements in terms of the concentrations of the species that exhibit FRET is not straightforward, due to the unknown contributions of leak-through and direct excitation. However, it is possible to externally calibrate the ratio values to physiologically relevant quantities by using reference samples where a known concentration of the species of interest can be related to the measured ratio.

Ratio measurements of FRET have been utilized in a wide spectrum of applications. They have been used as an indicator for adenosine 3′,5′-cyclic monophosphate (cAMP) [8,10], guanosine 3′,5′-cyclic monophosphate (cGMP) [9], and Ca^{2+} [5,6]. More recently sensors for protein kinase [11,12,14], and GTPase [13] activity have been described.

Sensitized Emission Measurements

Whereas ratio measurements are difficult to quantify, it is possible to make use of reference measurements to further quantify results. Generally three measurements are made: (1) using a filter set where the donor is excited ands the donor emission is measured (donor filter-set); (2) using a filter set where the acceptor is excited and the acceptor emission is measured (acceptor filter-set); (3) using a filter set where the donor is excited and the acceptor emission is measured (FRET filter-set). The images from the donor and acceptor filter sets are multiplied with correction factors and subtracted from the image taken with the FRET filter-set to obtain the sensitized emission, corrected for contributions of donor leak-through and direct excitation. The correction factors are determined via reference samples that contain only donor or only acceptor molecules. The resulting estimation of the intensity of the sensitized emission is then normalized for donor and/or acceptor concentration by using an expression for apparent energy transfer, for which several approaches have been taken [21–23]. Ideally, quantification should provide the relative fraction of molecules that are exhibiting FRET. Up to a scalar factor, it is indeed possible to determine the fraction of acceptor molecules that are exhibiting FRET, since the acceptor concentration can be directly related to the acceptor fluorescence. Determining the fraction of donor molecules that exhibit FRET is much more difficult, since the donor fluorescence is not proportional to the donor concentration, owing to the quenching of the donor fluorescence by FRET. The donor concentration can, however, be measured using acceptor photobleaching, as described below.

In contrast to ratio measurements, this approach is suited to applications where donor and acceptor are not on the same molecule, due to the correction for leak-through and direct excitation. They are also suitable to live cell imaging since only three measurements need to be made. Mostly these approaches are used when donor- and acceptor-tagged molecules have concentrations that are in the same order of magnitude [17]. This FRET method may become less effective in cases where the acceptor is in large excess, since the corrections for direct excitation become large and the result is more susceptible to noise. Similarly, a large excess of donor leads to large corrections for donor leak-through, with associated problems to estimate the sensitized emission. Thus this approach is likely to be less suitable for sensors where a protein state is determined by a large excess of a probe, although sensors to probe the state of a protein have been reported [18]. Even if donor and acceptor concentrations are similar, the corrections can be large, depending on the spectral properties of the donor/acceptor pair that is employed. The different variants of GFP are examples in which the corrections may be substantial, and in such cases this method works best with a high energy transfer efficiency between donor and acceptor, implying a large relative contribution of sensitized emission.[18]. We note that the correction factors are generally determined from averages of images of reference samples, and are therefore scalar factors. It is not known how much these factors change as a function of the environment or whether scalar correction factors are sufficient.

Sensitized emission measurements have been used to study the interactions between different molecules. Examples are the studies by Sorkin and colleagues, who looked at the interaction of the EGF receptor with Grb2 in living cells [17], and by Mahajan et al., who investigated interaction of Bcl-2 with Bax [24]. The sensitized emission method was also used to localize the activity of the GTPase Rac [18].

Methods Based on Photobleaching

Photobleaching of either the donor or acceptor molecules can be utilized to detect the effects of FRET on the kinetics of the fluorescence of either. Photobleaching is the process whereby a fluorophore is converted to a nonfluorescent species, for instance in the presence of oxygen. This essentially happens only when the donor is in the excited state and therefore the rate at which photobleaching occurs is proportional to the average time it spends in the excited state, which in turn is inversely proportional to the rate at which the molecule returns to its ground state. Hence, an increase in the latter due to FRET can be detected by a decrease in the photobleaching rate. Thus, one approach to measure and quantify FRET is to measure the kinetics of photobleaching of the donor [25,26]. Generally, the kinetics are described by a sum of exponentials, and it is possible to quantify the fraction of donor molecules exhibiting FRET. Although this is a potentially precise approach, it is not in much use nowadays for several reasons. First, the requirement to photobleach the donor over extended time periods leads to long acquisition times. Therefore the method is mostly useful on fixed samples, although presumably live cells could be used if one is only interested in the integrated response of a complete cell. Second, the mechanisms of photobleaching are not understood very well, although the assumption of a multi-exponential model is reasonable in first approximation.

Rather than examining the photobleaching kinetics of the donor, one can utilize photobleaching of the acceptor [16,27]. Acceptor molecules can be excited specifically, since their absorption spectrum generally does not overlap with that of the donor. One makes use of the simple property that destroying all acceptor fluorophores abolishes FRET. Thus, after photobleaching the acceptor, the donor intensity should increase, since it is not quenched anymore by FRET. Therefore, comparing the intensity of the donor before and after acceptor photobleaching should indicate whether FRET was occurring. Dividing the difference of the donor intensity before and after photobleaching by the intensity after photobleaching yields an apparent energy transfer measure that is directly proportional to the fraction of donor molecules exhibiting FRET. Obviously this approach is attractive since it is experimentally easy, and also interpretation does not require extensive analysis. Acceptor photobleaching does suffer from the same drawback as donor photobleaching in that it is a time-consuming approach less suitable for imaging of FRET in live cells. A point of possible concern that has not been addressed much so far is that photobleaching of the acceptor may create a different species of molecule that fluoresces in the donor channel, leading to an overestimation of the unquenched donor signal. Alternatively, a dark species could be created that absorbs light and still acts as an acceptor, leading to an underestimation of the unquenched donor fluorescence.

Acceptor photobleaching has been used as an independent technique to measure FRET. For instance, it was used to study the localization of the A- and B-subunits of cholera toxin [16], and recently to image the three-dimensional distribution of receptor tyrosine kinases interacting with protein tyrosine phosphatase 1B [28]. It is also increasingly being used in combination with other techniques as a standard control for the occurrence of FRET [10,14,29,30].

Fluorescence Lifetime Imaging Microscopy

The kinetics of fluorescence can be measured by fluorescence lifetime imaging microscopy (FLIM). This makes it possible to detect whether the rate at which an excited donor molecule returns to the ground state increases by FRET, since the fluorescence lifetime of a fluorophore is inversely proportional to the sum of rates of all possible pathways by which an excited molecule returns to the ground state [31].

FLIM has been applied in a qualitative fashion where an average lifetime is measured that decreases upon FRET [31]. In such a case, photobleaching the acceptor may serve as an internal control since after destruction of the acceptor fluorophore the lifetime of the donor should attain its normal value [30]. In this case, the acceptor photobleaching can also be applied in live cell imaging by photobleaching after FLIM measurements are made. Since the fluorescence lifetime of the donor is independent of concentration and light path-length, the average donor lifetime can then serve as the control value. FLIM has also been applied quantitatively by resolving the multi-exponential decay kinetics of the donor to determine the fractions of donor molecules exhibiting FRET [20].

So far, FLIM has been applied mostly to donor-only imaging. It is therefore well suited to applications in which the acceptor is present in excess, e.g. to probe the state of a donor-tagged molecule [19,20]. However, it is also suited to cases in which the donor and acceptor are available in comparable concentrations [28]. FLIM has also been applied by imaging both donor and acceptor simultaneously. In this case, the kinetics of the acceptor come into play, and it becomes possible to use fluorophores that are difficult to separate spectrally, such as GFP and YFP [32]. In principle, it is then possible to quantitatively determine both the fractions of donor and acceptor that are participating in FRET, but this has not been demonstrated yet. FLIM requires the acquisition of multiple images but is rapid enough to enable live cell imaging. One drawback of the method in comparison with other FRET methods is that it is more technically involved and equipment cost is higher.

FLIM has been used in a wide variety of applications, among others to probe the phosphorylation state of proteins such as PKCα [19] and the EGF-receptor [20,30] and to study the proteolysis of PCKβ1 [15], the oligomerization of EGF-receptors [33], and the interactions between PKCα and ezrin [29]. In a rather different type of application, Murata et al. [34] studied the organization of DNA in cell nuclei by visualizing FRET between the AT-specific donor Hoechst 33258 and the GC-specific acceptor 7-aminoactinomycin D.

Conclusions and Prospects

Exploitation of the physical phenomenon of FRET in biomolecular systems has opened new ways to image biochemistry in live cells. Several optical techniques to measure FRET have been developed in recent years and the number of applications of these techniques to relevant biological systems has been increasing steadily. All of these techniques have their strengths and weaknesses, and it is to be expected that the technological developments will continue for some time. In addition, the development of novel sensors that are based on FRET measurements promises to open more fields of applications, not in the least due to the large variety of GFPs that is becoming available [35,36]. FRET is complementary to biochemical approaches for investigating the complex signaling systems that are encountered at the cellular level in the fundamental biological sciences. We therefore expect that FRET measurements will play an increasingly important role in cell biology in the future.

References

1. Heim, R. and Tsien, R. Y. (1996). Engineering green fluorescent protein for improved brightness, longer wavelengths and fluorescence resonance energy transfer. *Curr. Biol.* **6**, 178–182.
2. Matz, M. V., Fradkov, A. F., Labas, Y. A., Savitsky, A. P., Zaraisky, A. G., Markelov, M. L., and Lukyanov, S. A. (1999). Fluorescent proteins from nonbioluminescent Anthozoa species. *Nat. Biotechnol.* **17**, 969–973.
3. Tsien, R.Y. (1998). The green fluorescent protein. *Annu. Rev. Biochem.* **67**, 509–544.
4. Clegg, R. M. (1996). Fluorescence resonance energy transfer. In *Fluorescence Imaging Spectroscopy and Microscopy*, X. F. Wang and B. Herman, Eds., pp. 179–252. Wiley, New York.
5. Miyawaki, A., Llopis, J., Heim, R., McCaffery, J. M., Adams, J. A., Ikura, M., and Tsien, R. Y. (1997). Fluorescent indicators for Ca^{2+} based on green fluorescent proteins and calmodulin. *Nature* **388**, 882–887.
6. Miyawaki, A., Griesbeck, O., Heim, R., and Tsien, R. Y. (1999). Dynamic and quantitative Ca^{2+} measurements using improved cameleons. *Proc. Natl. Acad. Sci. USA* **96**, 2135–2140.
7. Truong, K., Sawano, A., Mizuno, H., Hama, H., Tong, K. I., Mal, T. K., Miyawaki, A., and Ikura, M. (2001). FRET-based in vivo Ca^{2+} imaging by a new calmodulin-GFP fusion molecule. *Nat. Struct. Biol.* **8**, 1069–1073.
8. Adams, S. R., Harootunian, A. T., Buechler, Y. J., Taylor, S. S., and Tsien, R. Y. (1991). Fluorescence ratio imaging of cyclic AMP in single cells. *Nature* **349**, 694–697.
9. Honda, A., Adams, S. R., Sawyer, C. L., Lev Ram, V. V., Tsien, R. Y., and Dostmann, W. R. (2001). Spatiotemporal dynamics of guanosine 3′,5′-cyclic monophosphate revealed by a genetically encoded, fluorescent indicator. *Proc. Natl. Acad. Sci. USA* **98**, 2437–2442.
10. Zaccolo, M., De Giorgi, F., Cho, C. Y., Feng, L., Knapp, T., Negulescu, P. A., Taylor, S. S., Tsien, R. Y., and Pozzan, T. (1999). A genetically encoded, fluorescent indicator for cyclic AMP in living cells. *Nat. Cell Biol.* **2**, 25–29.
11. Nagai, Y., Miyazaki, M., Aoki, R., Zama, T., Inouye, S., Hirose, K., Iino, M., and Hagiwara, M. (2000). A fluorescent indicator for visualizing cAMP-induced phosphorylation in vivo. *Nat. Biotechnol.* **18**, 313–316.

12. Ting, A. Y., Kain, K. H., Klemke, R. L., and Tsien, R. Y. (2001). Genetically encode fluorescent reporters of protein tyrosine kinase activities in living cells. *Proc. Natl. Acad. Sci. USA* **18**, 15003–15008.
13. Mochizuki, N., Yamashita, S., Kurokawa, K., Ohba, Y., Nagai, T., Miyawaki, A., and Matsuda, M. (2001). Spatio-temporal images of growth-factor-induced activation of Ras and Rap1. *Nature* **411**, 1065–1068.
14. Zhang, J., Ma, Y., Taylor, S. S., and Tsien, R. Y. (2001). Generically encode reporters of protein kinase A activity reveal impact of substrate thetering. *Proc. Natl. Acad. Sci. USA* **18**, 14997–15002.
15. Bastiaens, P. I. H. and Jovin, T. M. (1996). Microspectroscopic imaging tracks the intracellular processing of a signal transduction protein: fluorescent-labeled protein kinase C βI. *Proc. Natl. Acad. Sci. USA* **93**, 8407–8412.
16. Bastiaens, P. I. H., Majoul, I. V., Verveer, P. J., Söling, H.-D., and Jovin, T. M. (1996). Imaging the intracelllar trafficking and state of the AB_5 quaternary structure of cholera toxin. *EMBO J.* **15**, 4246–4253.
17. Sorkin, A., McClure, M., Huang, F., and Carter, R. (2000). Interaction of EGF receptor and Grb2 in living cells visualized by fluorescence resonance energy transfer (FRET) microscopy. *Curr. Biol.* **10**, 1395–1398.
18. Kraynov, V. S., Chamberlain, C., Bokoch, G. M., Schwartz, M. A., Slabaugh, S., and Hahn, K. M. (2000). Localized Rac activation dynamics visualized in living cells. *Science* **290**, 333–337.
19. Ng, T., Squire, A., Hansra, G., Bornancin, F., Prevostel, C., Hanby, A., Harris, W., Barnes, D., Schmidt, S., Mellor, H., Bastiaens, P. I. H., and Parker, P. J. (1999). Imaging protein kinase Cα activation in cells. *Science* **283**, 2085–2089.
20. Verveer, P. J., Wouters, F. S., Reynolds, A. R., and Bastiaens, P. I. H. (2000). Quantitative imaging of lateral ErbB1 receptor signal propagation in the plasma membrane. *Science* **290**, 1567–1570.
21. Gordon, G. W., Berry, G., Liang, X. H., Levine, B., and Herman, B. (1998). Quantitative fluorescence energy transfer measurements using fluorescence microscopy. *Biophys. J.* **74**, 2702–2713.
22. Nagy, P., Vámosi, G., Bodnár, A., Locket, S. J., and Szöllösi, J. (1998). Intensity-based energy transfer measurements in digital imaging microscopy. *Eur. Biophys. J.* **27**, 377–389.
23. Xia, Z. and Liu, Y. (2001). Reliable and global measurement of fluorescence resonance energy transfer using fluorescence microscopes. *Biophys. J.* **81**, 2395–2402.
24. Mahajan, N., Linder, K., Berry, G., Gordon, G. W., Heim, R., and Herman, B. (1998). Bcl-2 and Bax interactions in mitochondria probed with green fluorescnct protein and fluorescence resonance energy transfer. *Nat. Biotechnol.* **16**, 547–552.
25. Jovin, T. M. and Arndt-Jovin, D. J. (1989). FRET microscopy: digital imaging of fluorescence resonance energy transfer. In *Cell Structure and Function by Microspectrofluorometry*, E. Kohen and J. G. Hirschberg, Eds., pp. 99–115. Academic Press, San Diego.
26. Jovin, T. M. and Arndt-Jovin, D. J. (1989). Luminescence digital imaging microscopy. *Annu. Rev. Biophys. Biophys. Chem.* **18**, 271–308.
27. Bastiaens, P. I. H. and Jovin, T. M. (1998). FRET microscopy. In *Cell Biology: A Laboratory Handbook*, J. E. Celis, Ed., pp. 136–146. Academic Press, New York.
28. Haj, F. G., Verveer, P. J., Squire, A., Neel, B. G., and Bastiaens, P. I. H. (2002). Imaging sites of receptor dephosphorylation by PTP1B on the surface of the endoplasmic reticulum. *Science* **295**, 1708–1711.
29. Ng, T., Parsons, M., Hughes, W. E., Monypenny, J., Zicha, D., Gautreau, A., Arpin, M., Gschmeissner, S., Verveer, P. J., Bastiaens, P. I. H., and Parker, P. J. (2001). Ezrin is a downstream effector of trafficking PKC-integrin complexes involved in the control of cell motility. *EMBO J.* **20**, 2723–2741.
30. Wouters, F. S. and Bastiaens, P. I. H. (1999). Fluorescence lifetime imaging of receptor tyrosine kinase activity in cells. *Curr. Biol.* **9**, 1127–1130.
31. Bastiaens, P. I. H. and Squire, A. (1999). Fluorescence lifetime imaging microscopy: spatial resolution of biochemical processes in the cell. *Trends Cell Biol.* **9**, 48–52.
32. Harpur, A. G., Wouters, F. S., and Bastiaens, P. I. H. (2001). Imaging FRET between spectrally similar GFP molecules in single cells. *Nat. Biotechnol.* **19**, 167–169.
33. Gadella, T. W. J. Jr. and Jovin, T. M. (1995). Oligomerization of epidermal growth-factor receptors on A431 cells studied by time-resolved fluorescence imaging microscopy—a stereochemical model for tyrosine kinase receptor activation. *J. Cell Biol.* **129**, 1543–1558.
34. Murata, S., Herman, P., Lin, H. J., and Lakowicz, J. R. (2000). Fluorescence lifetime imaging of nuclear DNA: effect of fluorescence resonance energy transfer. *Cytometry* **41**, 178–185.
35. Griesbeck, O., Baird, G. S., Campbell, R. E., Zacharias, D. A., and Tsien, R. Y. (2001). Reducing the environmental sensitivity of yellow fluorescent protein. Mechanism and applications. *J. Biol. Chem.* **276**, 29188–29194.
36. Nagai, T., Ibata, K., Park, E. S., Kubota, M., Mikoshiba, K., and Miyawaki, A. (2002). A variant of yellow fluorescent protein with fast and efficient maturation for cell-biological applications. *Nat. Biotechnol.* **20**, 87–90.

CHAPTER 175

Peptide Recognition Module Networks: Combining Phage Display with Two-Hybrid Analysis to Define Protein-Protein Interactions

Gary D. Bader,[4] Amy Hin Yan Tong,[1] Gianni Cesareni,[2] Christopher W. Hogue,[5] Stanley Fields,[3] and Charles Boone[1]

[1]*Banting and Best Department of Medical Research and Department of Molecular and Medical Genetics, University of Toronto, Toronto, Ontario, Canada*
[2]*Department of Biology, University of Rome Tor Vergata, 00133 Rome, Italy*
[3]*Howard Hughes Medical Institute, Departments of Genome Sciences and Medicine, University of Washington, Seattle, Washington*
[4]*Samuel Lunenfeld Research Institute, Mount Sinai Hospital, Toronto, Ontario, Canada, and*
[5]*Department of Biochemistry, University of Toronto, Toronto, Canada*

Introduction

Many of the protein-protein interactions of macromolecular signaling complexes are mediated by domains that function as recognition modules to bind specific peptide sequences found in their partner proteins [1]. For example, SH3, WW, and EVH1 domains bind to proline-rich peptides [2–4], EH domains bind to peptides containing the NPF motif [5,6], and SH2 and FHA domains bind to peptides phosphorylated on Tyr and Thr, respectively [7,8]. For particular modules within the same family, specificity is determined by critical residues in the binding partner flanking the core peptide motif [9,10]. A major challenge is to construct protein-protein interaction networks in which every module within the predicted proteome of a sequenced organism is linked to its cognate partner.

To address this problem, we developed a four-step strategy for the derivation of protein-protein interaction networks mediated by peptide recognition modules [11–13]. First, the consensus sequences for preferred ligands for each peptide recognition module are defined by isolating 10 to 20 different peptide ligands from screens of phage display libraries. Second, the consensus sequences resulting from the phage display experiments are used to computationally derive a protein-protein interaction network that links each peptide recognition module to proteins containing a preferred peptide ligand. Third, a protein-protein interaction network is experimentally derived via large-scale two-hybrid analysis.

Fourth, the intersection of the predicted and experimental networks is determined.

As a test of this strategy, we constructed a protein interaction network for the SH3 domains of the yeast *Saccharomyces cerevisiae*. The SH3 domain is one of the more commonly used protein recognition modules. In fact, over 1500 different SH3 domains have been identified in the protein databases of eukaryotic organisms [14]. The yeast proteome contains a total 28 SH3 domains, found in 24 different proteins [15], the majority of which had been implicated in signal transduction (Bem1, Boi1, Boi2, Cdc25, Sdc25, and Sho1) or reorganization of the actin cytoskeleton (Abp1, Bud14, Cyk3, Hof1, Myo3, Myo5, Rvs167, Sla1) [16,17]. A set of eight SH3 proteins remained to be characterized (Bbc1, Bzz1, Nbp2, Yfr024c, Ygr136w, Yhl002w, Ypr154w, and Ysc84).

We were able to express 24 different SH3 domains in a soluble form as glutathione S-transferase (GST)-SH3 fusion proteins in *Escherichia coli*. Because some of the SH3 domains did not select a ligand from the nonapeptide library, we were able to obtain a consensus sequence for only a subset of 20 different SH3 domains. Most SH3 domains bind to a core PxxP ligand motif (P = proline, x = any amino acid), with particular residues that occur on either side of the core determining binding specificity. Two general classes of SH3 ligands have been defined; class I peptides conform to the general consensus RxxPxxP (R = arginine) and class II peptides conform to PxxPxR [2], Most of the yeast SH3 domains selected peptides that aligned to yield a class I or class II consensus ligand, with one to six domain-specific residues constrained outside the PxxP motif (Table I).

The consensus sequences were used to search the yeast proteome for proteins that contained potential SH3 ligands. Because hundreds of the predicted yeast proteins contain an SH3 class I and class II consensus ligand, we used a position specific scoring matrix (PSSM) to rank the peptides present in yeast proteins based upon their similarity to the peptides selected from the phage display libraries. The peptides within the top 20% of the PSSM scores captured most of the literature-validated SH3 domain interactions, and therefore this set was considered as potential ligands. The predicted protein-protein interactions were imported into the Biomolecular Interaction Network Database (BIND) [18] and formatted for visualization in the Pajek package [19], a program originally designed for visualization of social networks. The resulting phage display protein-protein interaction network contained 394 interactions among 206 proteins (Fig. 1A). Proteins are represented as nodes on the graph and the interactions represented as edges connecting the nodes.

Proteins found within highly connected subgraphs can be extracted from more complex networks by using graph

Table I   Consensus Sequence of Yeast SH3 Peptide Ligands

	Class I								Class II							Unusual
Bem1-1																P P x V x P Y
Fus1																R x x R st st S l
Abp1										rk x x p x		x P	x rk P x w #			
Myo3	P	x	@	p	P	P	x	x	P							
Myo5	P	x	@	p	P	P	x	x	P							
Pex13		R	x	l	P	x	#		P							
Sla1-3	h	R	x	p	P	x	p		P							
Sho1	s	kr	x	L	P	x	x		P							
Ygr136w		R	x	rk	#@	x	l		P	P	x	#P	x R p			
Ypr154w	@	kr R	P	p	#	x	l		P	P P	#P	x R P				
Yhl002w	y	R	p	#	P	x	x		P							f R x x x h Y t
Ysc84										P x	LP	x R				
Yfr024c										P p	LP	x R P				
Rvs167		R	x	#	P	x	p		P	P P	#P	P R				
Bzz1-1	K	kr	x	P	P	p	x		P							
Bzz1-2	kr	kr	p	P	P	P	p	#	P							
Bbc1		R	kr	x	P	x	p		P	P kr	#P	x R P				
Boi1		R	x	x	P	x	x		P	p P R	x P	r R #				
Boi2		R								p p R	n P	x R #				
Nbp2	P	x	R	P	a	P	x	x	P							

The consensus peptides were derived from an alignment of the selected phage-display peptides (x, any amino acid; lowercase letters, residues conserved in 50 to 80% of the selected peptides; uppercase letters, residues conserved in more than 80% of the selected peptides). Abbreviations for the amino acid residues are as follows: A, Ala; H, His; K, Lys; L, Leu; N, Asn; P, Pro; R, Arg; S, Ser; T, Thr; V, Val; W, Trp; Y, Tyr; #, hydrophobic residues; @, aromatic residues. The consensus sequences corresponding to Class I peptides, first column, Class II peptides, second column; unaligned, third column.

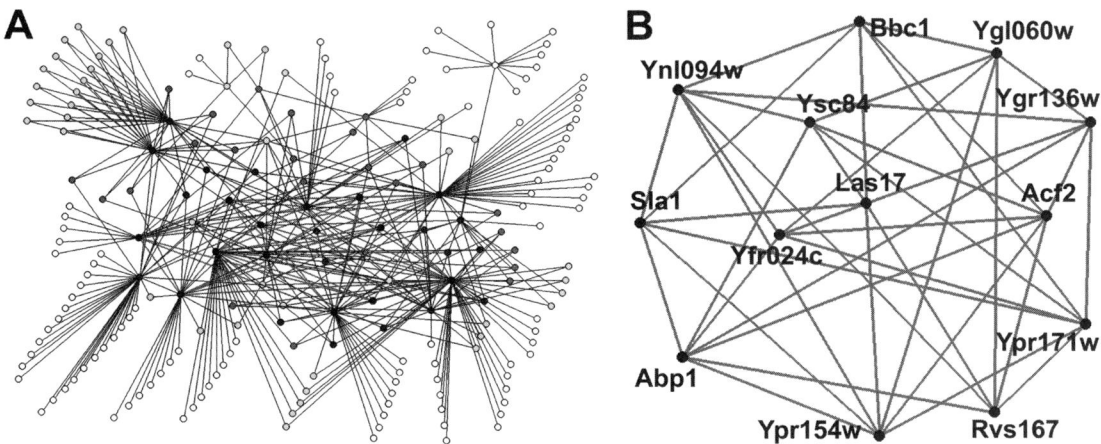

Figure 1 (A) Yeast SH3 domain protein-protein interaction network predicted via phage display selected peptides; 394 interactions and 206 proteins are shown; a network with each gene name labeled is included in the supplementary material [7]. The proteins are colored according to their k-core value (six-core=black, five-core=cyan, four-core=blue, three-core=red, two-core=green, one-core=yellow), identifying subsets of interconnected proteins in which each protein has at least k interactions. By definition, lower core numbers encompass all higher core numbers (e.g. a four-core includes all the nodes in the four-core, five-core, and six-core). The interactions of the six-core subgraph are highlighted in red. (B) The six-core subgraph derived from the phage display protein-protein interaction network, expanded to allow identification of individual proteins. The six-core subset contains eight SH3 domain proteins (Abp1, Bbc1, Rvs167, Sla1, Yfr024c, Ysc84, Ypr154w, and Ygr136w) and five proteins predicted to bind at least six different SH3 domains (Las17, Acf2, Ypr171w, Ygl060w, and Ynl094w).

theoretical algorithms. The phage display network contained a highly connected six-core subgraph, in which each protein has at least six interactions with the other proteins in the subgraph (Fig. 1B). Because the phage display network represents an integration of all potential interactions and does not take into account temporal expression or protein localization information, the six-core is subject to various biological interpretations. It may represent a single complex, provided all the proteins are co-expressed *in vivo* and all of the interactions occur simultaneously; however, it may represent multiple dimers or other oligomers, each of which forms independently under some cellular state. In any case, the presence of a highly connected core suggests a functional association between the interacting proteins. We examined 1,000 random model networks, in which a similar number of random proteins were linked to each SH3 domain. The model networks were not as highly connected as the phage display network and at most contained a four-core subgraph, indicating that the six-core within the phage display network was unlikely to occur by chance. Indeed, the six-core contains a number of functionally related proteins. At the center of the six-core is Las17, the yeast homolog of human Wiscott-Aldrich syndrome protein, which binds and activates the Arp2/3 actin nucleation complex and assembles the filamentous actin of yeast cortical actin patches [20–23]. The six-core also contains Acf2, a protein required for Las17-dependent reconstitution of actin assembly *in vitro* [24] and a set of proteins that were either implicated previously in the endocytotic role of cortical actin patches (Abp1, Sla1, Rvs167) [25–27] or found to localize to cortical patches (Bbc1, Ysc84, Ynl094w, and Ypr171w) [11,28]. Thus, the construction of a protein-protein interaction network from *in vitro* peptide binding information and the graphical analysis of its connectivity revealed known components of the yeast cortical actin patch complex.

To construct a two-hybrid protein-protein interaction network for comparison to the phage display network, we screened 18 SH3 domain baits against conventional two-hybrid libraries and an ordered genome-wide array of yeast Gal4 activation domain–open reading frame fusions [29]. The results from these screens were assembled into a network containing 233 interactions and 145 proteins (Fig. 2A). Only a subset of the interactions within the phage-display network and the two-hybrid network are expected to overlap. In particular, the phage display and two-hybrid methodologies will lead to different sets of false positives, which should exclude them from the overlap network. A total of 59 interactions in the phage display network also occurred in the two-hybrid network (Fig. 2B). All of the overlap interactions are mediated directly by SH3 domains; the precise ligand of the binding partner was predicted by the phage display analysis. Three lines of evidence suggest that the interactions within the overlap network are meaningful. First, the phage display network was highly enriched for overlap interactions when compared to the random model networks, which contained an average of 0.84 overlap interactions (SD=1.01). Second, the overlap network was enriched for interactions validated previously in the literature, over three-fold compared to the two-hybrid network and over five-fold compared to the phage display network. Third, a focused analysis of the proline-rich peptides within Las17 revealed that the phage display ligand analysis consistently predicted the ligand fragment that showed the strongest binding.

Future experiments of this type may be able to achieve better results by optimizing specific steps. For example, some false positives in the phage display approach undoubtedly arise because the predicted ligand peptide is, in fact, buried in the core of the protein. This aspect of the analysis could be improved by assessing surface accessibility with a program

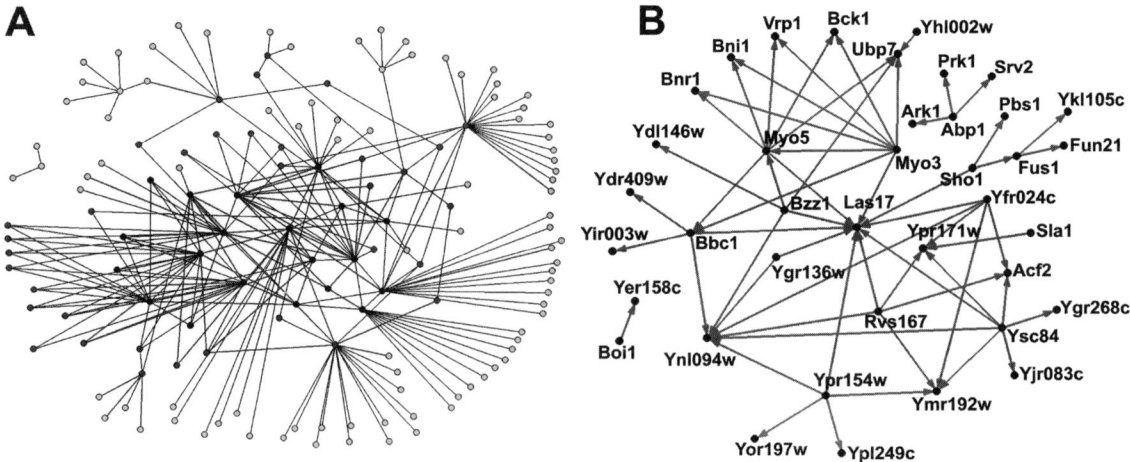

Figure 2 (A) Two-hybrid SH3 domain protein-protein interaction network. Two-hybrid results, based largely on screens with SH3 domains as baits, generated a network containing 233 interactions and 145 proteins. Proteins are colored according to their k-core (see Fig. 1A). The largest core of the two-hybrid network is a single four-core (blue nodes). Interactions common to the phage display network are highlighted in red. (B) Overlap of the protein-protein interaction networks derived from phage display and two-hybrid analysis. Expanded view of the common elements of the phage display and two-hybrid protein-protein interaction networks, 59 interactions, and 39 proteins. All of these interactions are predicted to be mediated directly by SH3 domains. The arrows point from an SH3 domain protein to the target protein. Additional evidence to support the relevance of several of these interactions is provided as supplementary material.

such as PHDacc [30], or homology models [31] of the protein could be scanned. Another means to improve proteome scanning would use a specificity and sensitivity analysis to assess what PSSM score threshold would retain the largest number of physiologically relevant interactions (true positives) and discard as many potential false positive interactions as possible. In this case, false positives can be defined operationally as those not identified within the literature or the yeast two-hybrid network. Thus, the optimization could be based on maximizing overlap with the yeast two-hybrid network or a set of confirmed interactions from a literature-based benchmark.

The overlap step could be improved in a number of ways. While the reasons for the false-positives and false-negatives of yeast two-hybrid screens seem satisfyingly orthogonal to those of the phage display predicted network, other protein interaction experimental methods, such as co-immunoprecipitation coupled with mass spectrometry [32,33], should also be evaluated. The current network representation, with a single node corresponding to a protein and a single edge corresponding to an interaction, could be much improved by making it probabilistic. The attachment of a probability value as a weight on the edges could enter into the overlap calculation to result in a more realistic model. For instance, a weight value on an edge could be high if the interaction has been characterized by several different methods, or found by multiple laboratories. These highly probable edges could be made to appear in the weighted combination of networks; in this fashion, "textbook" interactions would be included even if they were not found by both the phage display and two-hybrid derived networks. A review by Gerstein et al. (2002) addresses some of these points in more detail [12]. A better visualization tool that could draw networks with probabilistic information and allow one to examine parameter changes (for example, in the PSSM score threshold) in real-time would complement these method improvements and facilitate evaluation of the results.

Many of these future improvements depend on the availability of a literature-based benchmark, a manually curated collection of high-quality, expert-validated interactions. Sources of more stringently validated interactions are MIPS [17], YPD [34] and PreBIND [33]. Collecting these together in a nonredundant set creates a benchmark of over 3,300 protein-protein interactions for yeast. Because some experimental methods are more likely to yield physiologically relevant information (for example, interactions detected with full length proteins expressed at native levels), the literature benchmark could also include a reliability score for each record.

A set of over 15,000 unique protein interactions collected for yeast from the literature and from all available large-scale studies contained 519 interactions involving 364 proteins in which one interaction partner has an SH3 domain [18]. Because many of these proteins are highly conserved, it will be of interest to determine the extent to which the connectivity of the network is conserved. The prospects for applying this interaction network mapping approach to other organisms are reasonable; for example, *Caenorhabditis elegans* has only 99 SH3 domains in 77 proteins, according to the SMART database, whereas the mouse has on the order of 327 SH3 domains in 172 proteins. A map of peptide-binding module-mediated interaction networks across organisms will provide a powerful dataset to study the specificity of domain-mediated interactions, the evolution of complexity, and the biology that these interactions dictate. Finally, the systematic analysis of binding properties and protein-protein interaction networks for peptide recognition modules will enable the development of sets of dominant interfering small molecules for systematic functional interrogation of the network [35].

References

1. Pawson, T. and Scott, J. D. (1997). Signaling through scaffold, anchoring, and adaptor proteins. *Science* **278**(5346), 2075–2080.
2. Cesareni, G. *et al.* (2002). Can we infer peptide recognition specificity mediated by SH3 domains? *FEBS Lett.* **513**(1), 38–44.
3. Fedorov, A. A. *et al.* (1999). Structure of EVH1, a novel proline-rich ligand-binding module involved in cytoskeletal dynamics and neural function. *Nat. Struct. Biol.* **6**(7), 661–665.
4. Macias, M. J., Wiesner, S., and Sudol, M. (2002). WW and SH3 domains, two different scaffolds to recognize proline-rich ligands. *FEBS. Lett.* **513**(1), 30–37.
5. de Beer, T. *et al.* (1997). Molecular mechanism of NPF recognition by EH domains. *Nat. Struct. Biol.* **7**(11), 1018–1022.
6. Salcini, A. E. *et al.* (1997). Binding specificity and in vivo targets of the EH domain, a novel protein-protein interaction module. *Genes Dev.* **11**(17), 2239–2249.
7. Moran, M. F. *et al.* (1990). Src homology region 2 domains direct protein-protein interactions in signal transduction. *Proc. Natl. Acad. Sci. USA* **87**(21), 8622–8626.
8. Durocher, D. and Jackson, S. P. (2002). The FHA domain. *FEBS Lett.* **513**(1), 58–66.
9. Paoluzi, S. *et al.* (1998). Recognition specificity of individual EH domains of mammals and yeast. *EMBO J.* **17**(22), 6541–6550.
10. Panni, S., Dente, L., and Cesareni, G. (2002). In vitro evolution of recognition specificity mediated by SH3 domains reveals target recognition rules. *J. Biol. Chem.* **277**(24), 21666–21674.
11. Tong, A. H. *et al.* (2002). A combined experimental and computational strategy to define protein interaction networks for peptide recognition modules. *Science* **295**(5553), 321–324.
12. Gerstein, M., Lan, N., and Jansen, R. (2002). Proteomics integrating interactomes. *Science* **295**(5553), 284–287.
13. Legrain, P. (2002). Protein domain networking. *Nat. Biotechnol.* **20**(2), 128–129.
14. Mayer, B. J. SH3 domains: complexity in moderation. *J. Cell Sci.* **114**(Pt. 7), 1253–1263.
15. Letunic, I. *et al.* (2002). Recent improvements to the SMART domain-based sequence annotation resource. *Nucl. Acids Res.* **30**(1), 242–244.
16. *http://genome-www.stanford.edu/Saccharomyces/*.
17. Mewes, H. W. *et al.* (2002). MIPS: a database for genomes and protein sequences. *Nucl. Acids Res.* **30**(1), 31–34.
18. Bader, G. D. *et al.* (2001). BIND—the Biomolecular Interaction Network Database. *Nucl. Acids Res.* **29**(1), 242–245.
19. *http://vlado.fmf.uni-lj.si/pub/networks/pajek/*.
20. Winter, D., Lechler, T., and Li, R. (1999). Activation of the yeast Arp2/3 complex by Bee1p, a WASP-family protein. *Curr. Biol.* **9**(9), 501–504.
21. Madania, A. *et al.* (1999). The *Saccharomyces cerevisiae* homologue of human Wiskott-Aldrich syndrome protein Las17p interacts with the Arp2/3 complex. *Mol. Biol. Cell* **10**(10), 3521–3538.
22. Lechler, T., Shevchenko, A., and Li, R. (2000). Direct involvement of yeast type I myosins in Cdc42-dependent actin polymerization. *J. Cell Biol.* **148**(2), 363–373.
23. Evangelista, M. *et al.* (2000). A role for myosin-I in actin assembly through interactions with Vrp1p, Bee1p, and the Arp2/3 complex. *J. Cell Biol.* **148**(2), 353–362.
24. Lechler, T. and Li, R. (1997). In vitro reconstitution of cortical actin assembly sites in budding yeast. *J.* **138**(1), 95–103.
25. Lila, T. and Drubin, D. G. (1997). Evidence for physical and functional interactions among two *Saccharomyces cerevisiae* SH3 domain proteins, an adenylyl cyclase-associated protein and the actin cytoskeleton. *Mol. Biol. Cell* **8**(2), 367–385.
26. Colwill, K. *et al.* (1999). In vivo analysis of the domains of yeast Rvs167p suggests Rvs167p function is mediated through multiple protein interactions. *Genetics* **152**(3), 881–893.
27. Ayscough, K. R. *et al.* (1999). Sla1p is a functionally modular component of the yeast cortical actin cytoskeleton required for correct localization of both Rho1p-GTPase and Sla2p, a protein with talin homology. *Mol. Biol. Cell* **10**(4), 1061–1075.
28. Drees, B. L. *et al.* (2001). A protein interaction map for cell polarity development. *J. Cell Biol.* **154**(3), 549–571.
29. Uetz, P. *et al.* (2000). A comprehensive analysis of protein-protein interactions in *Saccharomyces cerevisiae*. *Nature* **403**(6770), 623–627.
30. Rost, B., Sander, C., and Schneider, R. (1994). PHD—an automatic mail server for protein secondary structure prediction. *Comput. Appl. Biosci.* **10**(1), 53–60.
31. Pieper, U. *et al.* (2002). MODBASE, a database of annotated comparative protein structure models. *Nucl. Acids Res.* **30**(1), 255–259.
32. Gavin, A. C. *et al.* (2002). Functional organization of the yeast proteome by systematic analysis of protein complexes. *Nature* **415**(6868), 141–147.
33. Ho, Y. *et al.* (2002). Systematic identification of protein complexes in *Saccharomyces cerevisiae* by mass spectrometry. *Nature* **415**(6868), 180–183.
34. Costanzo, M. C. *et al.* (2001). YPD, PombePD and WormPD: model organism volumes of the BioKnowledge library, an integrated resource for protein information. *Nucl. Acids Res.* **29**(1), 75–79.
35. Oneyama, C., Nakano, H., and Sharma, S. V. (2002). UCS15A, a novel small molecule, SH3 domain-mediated protein-protein interaction blocking drug. *Oncogene* **21**(13), 2037–2050.

CHAPTER 176

The Focal Adhesion: A Network of Molecular Interactions

Benjamin Geiger,[1] Eli Zamir,[1] Yariv Kafri,[2] and Kenneth M. Yamada[3]

[1]Department of Molecular Cell Biology, Weizmann Institute of Science, Rehovot, Israel;
[2]Department of Physics of Complex Systems, Weizmann Institute of Science, Rehovot, Israel;
[3]Craniofacial Developmental Biology and Regeneration Branch, National Institute of Dental and Craniofacial Research, National Institutes of Health, Bethesda, Maryland

Introduction

Focal adhesions (FA) and related structures are specialized subcellular sites through which cells in culture or in tissues attach to the extracellular matrix (ECM). The molecular structure and dynamics of these plaque-like adhesion sites [1–4], as well as their roles in adhesion, motility, and signaling, have been established in numerous studies over the past decades [5–8]. Although various types of cell-ECM adhesions are formed, affecting anchorage, motility, or ECM assembly, FA are still the best-studied type of adhesion.

The overall structural organization of FA consists of three major domains: transmembrane receptors, the attached cytoskeleton, and interconnecting plaque molecules. These three domains are structurally and functionally interlinked, and mutations that perturb their interactions disrupt the adhesive process. For example, mutations of integrin receptors that truncate the cytoplasmic domains through which they interact with cytoplasmic partners produce functional inactivation. Similarly, disruption of the actin cytoskeleton by use of drugs that inhibit actin polymerization or block actomyosin contractility leads to the loss of FA.

At the molecular level, both the transmembrane and cytoskeletal domains (but not the plaque domain) are relatively simple, with limited numbers of molecular components. The primary molecular constituents of the membrane domain are members of the integrin receptor family that bind specifically to the ECM molecule to which the cell is adhering. For example, when a fibroblast adheres to fibronectin, clusters of both the fibronectin receptor $\alpha_5\beta_1$ and the multipurpose receptor $\alpha_V\beta_3$ accumulate initially in FA, whereas only $\alpha_V\beta_3$ accumulates on a vitronectin substrate [4,9–11].

The cytoskeletal domain of FA consists mainly of actin filaments, with a few associated proteins, such as α-actinin and filamin. In contrast, the submembrane plaque between integrins and cytoskeleton is considerably more complex, consisting of a large number of linker and signaling molecules (more than 50 reported components). Some major actin-associated proteins are also potent linkers of actin to the membrane at FA. For example, α-actinin and filamin may bind to an integrin and also serve as an actin-integrin linker. Because the molecular components of FA have been extensively described and discussed in a number of recent reviews [8,12], they will not be addressed here in detail. Instead, we describe some key properties of the various constituents of the plaque and their interactions.

FA plaque constituents include both structural molecules, which can physically cross-link different FA molecules to each other and stabilize the adhesion complex, and signaling molecules, such as a variety of kinases, adapter proteins, and other signal transduction proteins. Some FA proteins contain specific motifs involved in protein-protein interactions. The best-characterized involve the SH2 domains, which interact with tyrosine-phosphorylated motifs, and the SH3 domains, which bind proline-rich motifs. Other types of interactions can be mediated by LIM, PDZ, and PH domains. Many of these proteins can bind to more than one partner, and thus they can potentially serve as "nodes" in the formation of multi-molecular adhesion complexes.

Although the binding affinity between many of these components might be weak, they form aggregates or molecular complexes that cumulatively generate high-avidity interactions. Their low intrinsic affinity may facilitate dynamic rearrangements of molecular complexes, for example during cell migration. Information on the modes of interaction of the different FA molecules with each other *in vivo* is quite limited. Consequently, "wiring diagrams" [12,13] or "network hierarchy schemes" (Fig. 1) can provide a general idea about potential interactions. Yet future research is needed to determine which interactions actually occur *in vivo* and in various physiological states (see below). It is also noteworthy that the actual complexity of FA in specific cell types is not clear, and that it may be less than expected by simply combining all information on the many plaque components in various cell types described in the scientific literature.

Connectivity-Based Ordering of FA Components

Although network schemes currently involve all potential interactions, rather than only interactions that actually occur in adhesion sites, they can shed novel light on molecular connectivity within FA. For example, Fig. 1A shows that actin and integrins can interact with many partners. High connectivity is also seen for several plaque proteins, such as FAK, vinculin, paxillin, α-actinin, and talin, which have at least nine potential interactions each. Less connected are a variety of signaling molecules, which have relatively few links. It is interesting that there are a number of molecules that have only one known link to another FA molecule, suggesting that even though they may be associated with the protein network, their interaction is not as an integral component.

This network type of analysis is presently preliminary, since the published data used for the construction of the network are incomplete and of variable quality. Nevertheless, integration of the currently available data on the players and their interactions into a network presentation allows one to gain some new information. For example, some insight can be gained about the effects of gene knockout (or mRNA inactivation), overexpression, or inhibition of specific certain signaling processes on the network. In Fig. 1B, a vinculin-null version of the network is shown. Such analysis predicts that in vinculin-null cells, vinexin and ponsin will be excluded from FA whereas other vinculin-binding molecules (e.g. talin, α-actinin, and paxillin) may remain via interactions with other components. Similarly, based on a FAK-null network, PLC-γ is predicted to be lost from FA (Fig. 1C). Obviously, such predictions can be challenged experimentally.

In order to illustrate a situation in which one molecular interaction dominates others, we have tested restricting the interactions mediated by the cytoplasmic domain of the integrin to only one molecule: talin (Fig. 1D). As a consequence, some molecules might lose their specific anchorage in FA (DRAL and uPAR) and many others are downshifted in their connectivity. Other scenarios presented here are the changes in molecular connectivity in networks in which phosphotyrosine-SH2-based interactions are excluded, either with or without the presence of FAK (Fig. 1E and 1F, respectively). Again, in these networks some proteins lose their connection to any FA component whereas others are lowered in their connectivity.

Despite its rather speculative and preliminary nature, this approach provides a novel tool for molecular modeling of FA interactions. It makes a number of specific predictions that can be directly tested experimentally. Note that even the deletion of the most highly connected proteins, namely FAK or vinculin, resulted in a surprisingly minor loss of other components (at most two). Furthermore, none of these network modifications caused a split into two unconnected sub-networks. These findings emphasize the robust nature of the connections between most components of FA.

Molecular Switches in FA

An obvious limitation of a network scheme, such as that shown in Fig. 1, is that all of the intermolecular links are presented as though they can all take place simultaneously. In fact, the structure and function of FA, and hence the interactions between its components, need to be regulated physiologically in many processes, including cell spreading, migration, and response to different growth factors and cytokines. By using different modes of regulation (defined here as "switches"), cells can employ a number of potentially important regulatory mechanisms, some of which are illustrated in Fig. 2.

Transcriptional Switches The level of a specific component can be regulated at the level of gene transcription, which in turn affects FA composition. For example, mRNA levels of vinculin can be modulated by matrix adhesiveness.

Protein Stability Switches The dynamic turnover of FA can be initiated by proteolytic cleavage of specific plaque components such as talin and FAK by proteases like calpain.

Tyrosine Phosphorylation Switches Dynamic changes in phosphorylation of plaque components such as FAK are thought to be involved in FA assembly and turnover.

Other Posttranslational Switches Changes in serine/threonine phosphorylation can also regulate integrin and plaque protein (e.g. paxillin) functions [14,15].

Conformational Switches The molecular conformation of proteins or their clustering can be regulated by phosphorylation or binding of other molecules. For example, vinculin can be transformed from a folded state to an open state with exposed intermolecular interaction sites after binding the signaling lipid PIP_2. Tensin conformation may also be regulated by tyrosine phosphorylation, and integrin clustering might trigger its binding to cytoplasmic partners and activate FAK autophosphorylation.

CHAPTER 176 The Focal Adhesion: A Network of Molecular Interactions 319

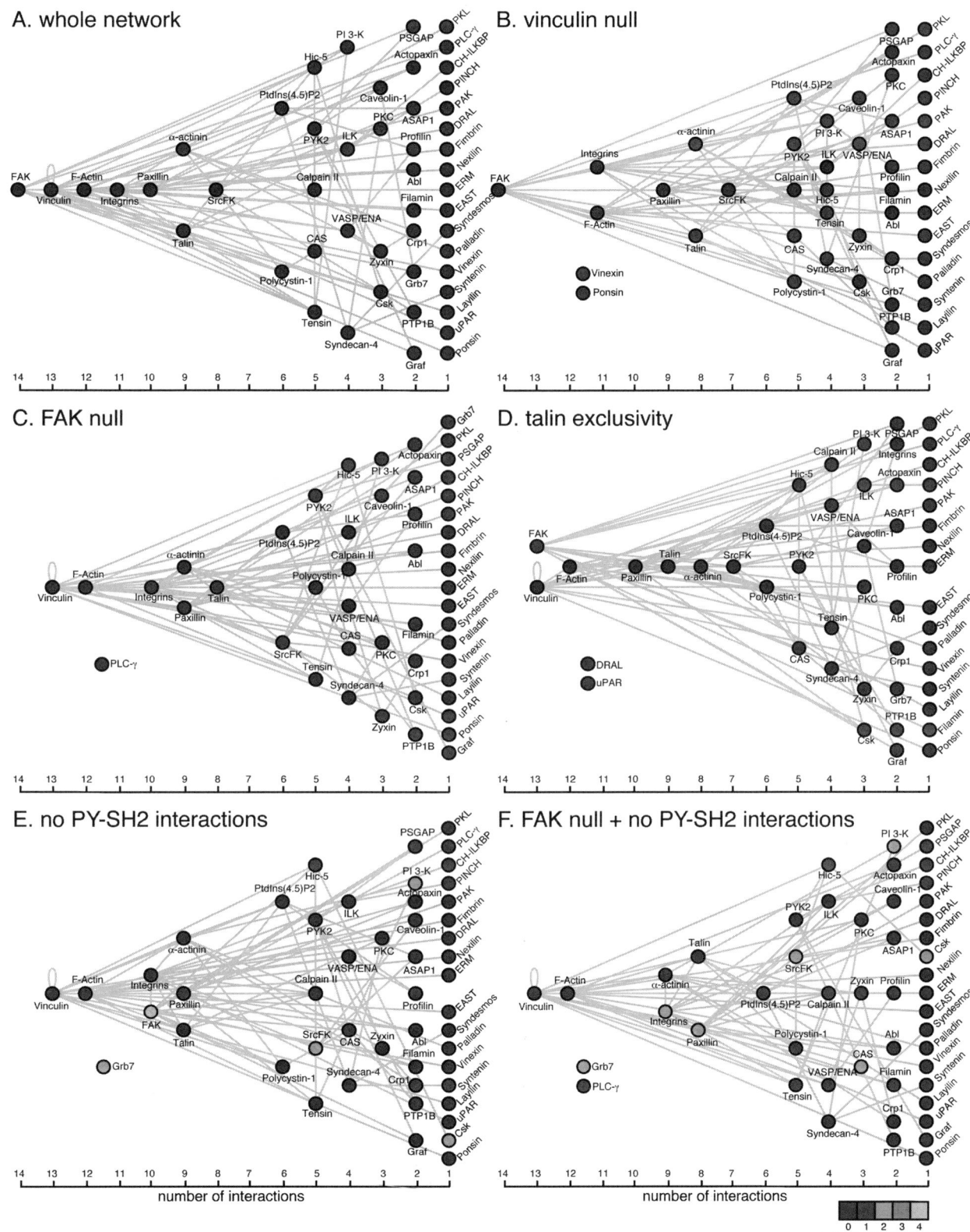

Figure 1 Connectivity-based ordering of FA components. In each of the network diagrams, the proteins were ordered from the most connected to the least connected, as indicated by the scale on the horizontal axis. Proteins listed in the lower left corner of each diagram have no known connections with the network. The connectivity data were taken from previous reviews [12,13] and analyzed via the Pajek program (http://vlado.fmf.uni-lj.si/pub/networks/pajek). (A) The whole network. (B) The network from which vinculin was removed. (C) The network from which FAK was removed. (D) The network in which talin dominates over all of the other cytoplasmic interactions with integrin. Thus, integrin is connected exclusively to talin and to membrane caveolin. (E) The whole network after removing all connections that are mediated by phosphotyrosine-SH2 motifs. (F) The same as (E) with the exclusion of FAK. The color of each vertex (circle) indicates the number of connections lost in comparison to the whole network; For each component, the number of lost connections can be realized by comparison with the whole network, or by the color of its node (a color representation of this figure is available on the CD-ROM version of *Handbook of Cell Signaling*).

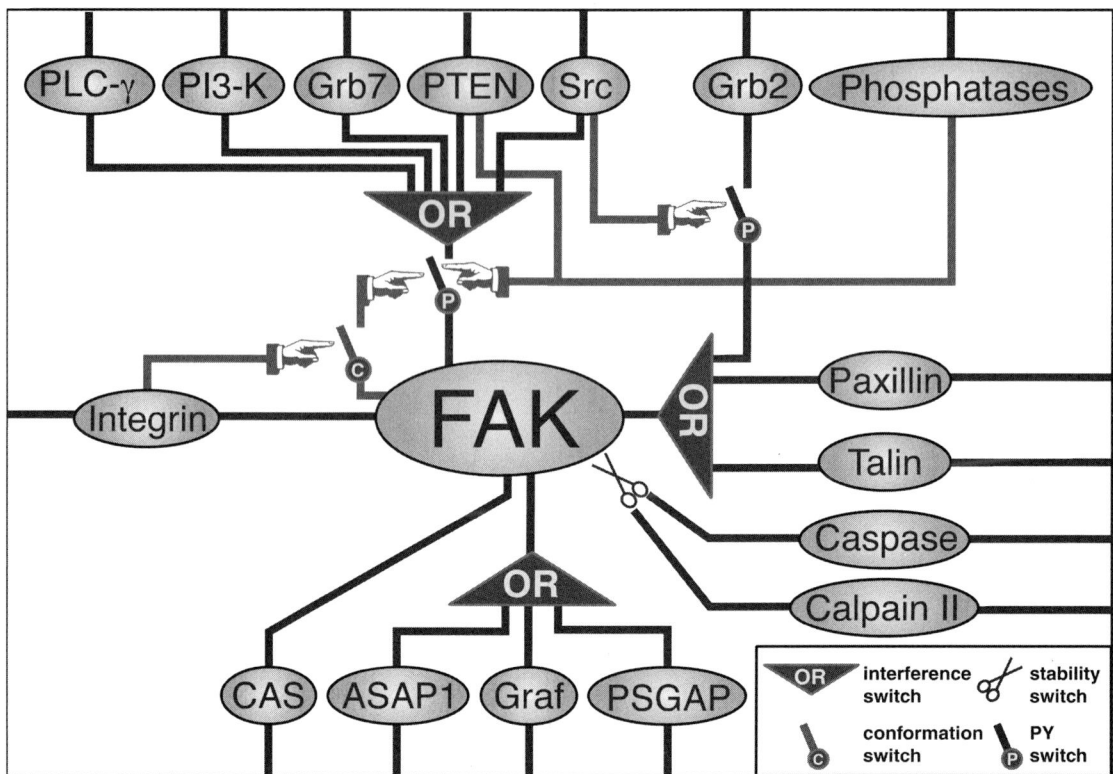

Figure 2 Molecular switches that might regulate protein-protein interactions in FA. The example shown here involves FAK (for general reviews about FAK see [20,21]), which can interact with multiple FA molecules (Fig. 1A). It is proposed that several types of "switches" can significantly constrain the possible combination of interactions with individual FAK. Molecular interference prevents more than one protein from interacting simultaneously with the same site or a nearby site. Thus, an exclusive "OR gate" can restrict the interactions of FAK with Src, Grb7, PTEN, PLC-γ, and PI3-K (all bind FAK Tyr397). Similarly, ASAP1, Graf, and PSGAP interact via their SH3 domain with the proline-rich motif located between amino acids 778–881 of FAK, and thus may interfere with each other. However, CAS interacts with a distinct proline-rich motif, located between amino acids 715–718 of FAK, and thus may be unaffected by other FAK interactions. An additional interference is predicted among paxillin, talin (both interact with the FAT domain of FAK), and Grb2 (which interact with Tyr925 at the FAT domain of FAK). Tyrosine phosphorylation of FAK at residues Tyr397 and Tyr925 is an additional "switch" activating binding sites for several SH2-containing proteins such as Src or Grb7 (Tyr397) and Grb2 (Tyr925) ("P-switch"). The phosphorylation at Tyr397 is performed by FAK itself (autophosphorylation) and can be induced by the clustering of integrin receptors, which aggregates FAK molecules together ("conformational (C) switch"). Phosphatases such as PTEN and Shp2 have been reported to dephosphorylate FAK, presumably targeting Tyr397, which could turn off all FAK interactions mediated by that phosphorylated residue. The phosphorylation of Tyr925 is mediated by Src and may play a role in the recruitment of Grb2, which, in turn, can interfere with FA localization. Finally, caspases and calpain II, in response to various signals, can mediate the proteolytic cleavage of FAK ("stability switch").

Tension Switches Although the mechanisms are still poorly understood at the molecular level, external tension on cells and intracellular actomyosin contractility can dramatically regulate the initiation and sizes of FA [16–18]. Shear stress can also trigger FAK tyrosine phosphorylation [19].

Molecular-interference Switches The binding of one component to a site may block the binding of other molecules to the same or neighboring sites in a dominant-negative fashion. A related switch involves the dominant-negative inhibition of FAK binding of its target molecules by a truncated form of the molecule, termed FRNK, the levels of which are regulated by alternative splicing.

In conclusion, it appears that in constructing meaningful interaction networks, such as those shown in Fig. 1, the various switches described above should be taken into account. We provide here one example how a set of switches can affect the molecular interactions of one of the busiest components of FA, namely FAK (Fig. 2). We show, for example, that many of FAK's partners compete with each other for binding to the same or adjacent sites, and that conformational, phosphorylation, and stability switches can all affect the actual connectivity of this molecule and its signaling activity. Obviously, it would be highly desirable to incorporate such switches into molecular interaction maps.

Future Challenges

Focal adhesions have been studied intensively for the past decade, and there is now a large body of literature on their composition and functions. Even so, our current understanding of the mechanisms involved in FA organization and function is still highly preliminary and speculative. Moreover, even though FA are thought to serve as signaling centers initiating signaling cascades (ranging from activation of MAP kinases to phosphoinositol lipid signaling), the evidence that signaling comes directly from interacting components located within FA is still minimal. Rigorous studies will be needed to establish which integrin-triggered signals come directly from FA or other cell-matrix adhesions.

A further challenge comes from the fact that many FA components and adapters have multiple potential binding sites. Therefore, it seems likely that FA contain complex mixtures of interacting components with variable binding to different components. That is, a given protein may be bound to multiple combinations of five to ten different proteins, and each of these complexes may behave differently. In addition, local heterogeneity of protein complexes within FA will also need to be evaluated. The approaches for probing these different sources of complexity will have to include experimental molecular perturbation of each component alone and then in various combinations. Obvious methods will include regulated overexpression, anti-sense and RNAi methods, enzymatic activity modulation, and dominant-negative inhibition approaches. However, because of the complexity of FA, computer-based network analysis seems essential. A combination of "wet" biochemical and molecular biology approaches with computerized analyses and generation of specific predictions for experimental testing should provide a powerful approach to understanding the role of these adhesive structures in regulating cell signaling and function.

References

1. Cukierman, E., Pankov, R., Stevens, D. R., and Yamada, K. M. (2001). Taking cell-matrix adhesions to the third dimension. *Science* **294**, 1708–1712.
2. Smilenov, L. B., Mikhailov, A., Pelham, R. J., Marcantonio, E. E., and Gundersen, G. G. (1999). Focal adhesion motility revealed in stationary fibroblasts. *Science* **286**, 1172–1174.
3. Zamir, E., Katz, B. Z., Aota, S., Yamada, K. M., Geiger, B., and Kam, Z. (1999). Molecular diversity of cell-matrix adhesions. *J. Cell Sci.* **112**, 1655–1669.
4. Zamir, E., Katz, M., Posen, Y., Erez, N., Yamada, K. M., Katz, B. Z., Lin, S., Lin, D. C., Bershadsky, A., Kam, Z., and Geiger, B. (2000). Dynamics and segregation of cell-matrix adhesions in cultured fibroblasts. *Nat. Cell Biol.* **2**, 191–196.
5. Izzard, C. S. and Lochner, L.R. (1976). Cell-to-substrate contacts in living fibroblasts: an interference reflexion study with an evaluation of the technique. *J. Cell Sci.* **21**, 129–159.
6. Izzard, C. S. and Lochner, L. R. (1980). Formation of cell-to-substrate contacts during fibroblast motility: an interference-reflexion study. *J. Cell Sci.* **42**, 81–116.
7. Burridge, K. and Chrzanowska-Wodnicka, M. (1996). Focal adhesions, contractility, and signaling. *Annu. Rev. Cell Dev. Biol.* **12**, 463–518.
8. Geiger, B., Bershadsky, A., Pankov, R., and Yamada, K. M. (2001). Transmembrane extracellular matrix—cytoskeleton crosstalk. *Nat. Rev. Mol. Cell Biol.* **2**, 793–805.
9. Katz, B. Z., Zamir, E., Bershadsky, A., Kam, Z., Yamada, K. M., and Geiger, B. (2000). Physical state of the extracellular matrix regulates the structure and molecular composition of cell-matrix adhesions. *Mol. Biol. Cell* **11**, 1047–1060.
10. Pankov, R., Cukierman, E., Katz, B. Z., Matsumoto, K., Lin, D. C., Lin, S., Hahn, C., and Yamada, K. M. (2000). Integrin dynamics and matrix assembly: tensin-dependent translocation of $\alpha_5\beta_1$ integrins promotes early fibronectin fibrillogenesis. *J. Cell Biol.* **148**, 1075–1090.
11. Singer, II, Scott, S., Kawka, D. W., Kazazis, D. M., Giant, J., and Ruoslahti, E. (1988). Cell surface distribution of fibronectin and vitronectin receptors depends on substrate composition and extracellular matrix accumulation. *J. Cell Biol.* **106**, 2171–2182.
12. Zamir, E. and Geiger, B. (2001). Molecular complexity and dynamics of cell-matrix adhesions. *J. Cell Sci.* **114**, 3583–3590.
13. Zamir, E. and Geiger, B. (2001). Components of cell-matrix adhesions. *J. Cell Sci.* **114**, 3577–3579.
14. Brown, M. C., Perrotta, J. A., and Turner, C. E. (1998). Serine and threonine phosphorylation of the paxillin LIM domains regulates paxillin focal adhesion localization and cell adhesion to fibronectin. *Mol. Biol. Cell* **9**, 1803–1816.
15. Han, J., Liu, S., Rose, D. M., Schlaepfer, D. D., McDonald, H., and Ginsberg, M. H. (2001). Phosphorylation of the integrin α_4 cytoplasmic domain regulates paxillin binding. *J. Biol. Chem.* **276**, 40903–40909.
16. Balaban, N. Q., Schwarz, U. S., Riveline, D., Goichberg, P., Tzur, G., Sabanay, I., Mahalu, D., Safran, S., Bershadsky, A., Addadi, L., and Geiger, B. (2001). Force and focal adhesion assembly: a close relationship studied using elastic micropatterned substrates. *Nat. Cell Biol.* **3**, 466–472.
17. Geiger, B. and Bershadsky, A. (2001). Assembly and mechanosensory function of focal contacts. *Curr. Opin. Cell Biol.* **13**, 584–592.
18. Riveline, D., Zamir, E., Balaban, N. Q., Schwarz, U. S., Ishizaki, T., Narumiya, S., Kam, Z., Geiger, B., and Bershadsky, A. D. (2001). Focal contacts as mechanosensors: externally applied local mechanical force induces growth of focal contacts by an mDia1-dependent and ROCK-independent mechanism. *J. Cell Biol.* **153**, 1175–1186.
19. Li, S., Kim, M., Hu, Y. L., Jalali, S., Schlaepfer, D. D., Hunter, T., Chien, S., and Shyy, J. Y. (1997). Fluid shear stress activation of focal adhesion kinase. Linking to mitogen-activated protein kinases. *J. Biol. Chem.* **272**, 30455–30462.
20. Parsons, J. T., Martin, K. H., Slack, J. K., Taylor, J. M., and Weed, S. A. (2000). Focal adhesion kinase: a regulator of focal adhesion dynamics and cell movement. *Oncogene* **19**, 5606–5613.
21. Schaller, M. D. (2001). Biochemical signals and biological responses elicited by the focal adhesion kinase. *Biochim. Biophys. Acta* **1540**, 1–21.

CHAPTER 177

WASp/Scar/WAVE

Charles L. Saxe
*Department of Cell Biology, Emory University,
School of Medicine, Atlanta, Georgia*

Introduction

The regulation of actin dynamics is central to a variety of biological events, including cell motility, chemotaxis, nerve growth cone extension, and establishment of cell polarity. A number of signaling mechanisms control these events and play a role in determining the position and extent of new actin assembly. Our understanding of the details is still incomplete, but many of these signals involve receptor stimulation and activation of members of the rho family of small GTPases [1]. One mechanism by which signaling pathways regulate actin polymerization is by controlling the formation of new actin filaments. Activation of the Arp2/3 complex appears to be critical in the nucleation step of this process. In turn, activation of members of the Wiskott-Aldrich protein (WASp) family leads to activation of the Arp2/3 complex (reviewed in [2]). WASp family members appear to play the role of signal integrator, responding to inputs from a variety of extracellular and intracellular signals and converting those signals into effects on actin dynamics.

The WASp protein family has two major branches, WASp and Scar/WAVE. WASp and Scar/WAVE proteins share a distinct modular architecture (Fig. 1). Conserved among all members of the family is a central proline-rich region, followed by a monomeric actin-binding domain (alternatively referred to as a Wiskott-Aldrich homology 2 domain, WH2, or verprolin homology domain, V; Fig. 1). Also common to all members of the family is a C-terminal acidic region shown to be essential for binding the Arp2/3 complex. The precise number and position of the acidic amino acids at the C-terminus are believed to influence the activity of the protein [3]. Between the WH2 and Acidic domains is a region that connects the two (C) that is important for WASp family proteins to activate Arp2/3. What distinguishes WASp (and the closely related N-WASp) from the Scar/WAVE subfamily is the N-terminal region. WASp and N-WASp have, at their most N-terminus, a region termed a WH1 domain, with structural similarities to the EVH1 domain of Ena/VASP proteins [4]. WASp and N-WASp also have in their N-terminus a basic region followed immediately by a small GTPase-binding domain (GBD or CRIB motif). The former, basic, region interacts with acidic phospholipids, particularly phosphatidylinositol 4,5-bisphosphate (PIP_2). The latter CRIB domain binds Cdc42 (and weakly rac1), and both regions are thought to be important in regulating the activation of WASp (see below). Scar proteins have neither a WH1 domain nor a conserved GBD, and though a distinct basic region is present, the role of PIP_2 in Scar function remains unclear. The difference in N-terminal domain structure has led to speculation that WASp and Scar proteins are regulated by different mechanisms. Recent evidence supports this idea (see below). For a more complete discussion of the role(s) and regulation of WASp/Scar/WAVE proteins see the extensive reviews of Higgs and Pollard [2] and Takenawa and Miki [5].

WASp

WASp was originally described as the protein defective in patients with Wiskott-Aldrich syndrome (WAS), an X-linked hematopoietic disease characterized by abnormalities in platelets and lymphocytes [6]. Hematopoietic cells from WAS patients have defects in actin-associated structures such as podosomes and microvilli and motility defects in cells such as macrophages. Independently, a related, more ubiquitously expressed protein, N-WASp, was identified as a binding partner for Ash/Grb2 [7]. It was recognized that both proteins contained a Cdc42-binding site, suggesting that they might provide a connection between the rho family of small GTPases and actin polymerization. More distantly related proteins, Las17p/bee1p and Wsp1p, were subsequently identified in *Saccharomyces* and *Schizosaccharomyces*, respectively,

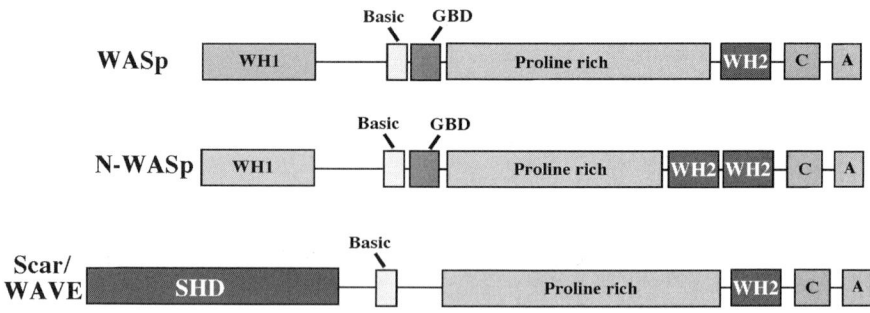

Figure 1 Diagram of domain structure of WASp family proteins. Abbreviations: WH1, Wiskott-Aldrich homology region 1; GBD, GTPase binding domain (also called CRIB motif); WH2, Wiskott-Aldrich homology region 2 (also called verprolin (V) domain); C, connecting domain (also called cofilin-like domain); A, acidic region.

and shown to be essential for actin patch formation and endocytosis [8,9].

The connection between WASp and the Arp2/3 complex came when it was determined that the C-terminal acidic domain of WASp (and hScar1) bound, in a yeast two hybrid screen, to the p21-Arc subunit [10]. Subsequently it was shown that C-terminal fragments of WASp, N-WASp, and Scar could all facilitate the ability of the Arp2/3 complex to nucleate new actin filament formation in vitro (reviewed in [2]). Further studies showed that full length WASp (and N-WASp) were significantly less effective at activating Arp2/3 than a C-terminal fragment. Addition of activated Cdc42 enhanced the ability of full-length N-WASp to activate Arp2/3. These data and other studies in cell culture led to a basic model whereby WASp/N-WASp is normally autoinhibited in the cell due to interaction of the C region of the protein with the GBD domain. Activation of WASp/N-WASp involves binding of Cdc42 and/or PIP_2 and unmasking of the carboxyl-terminal region, allowing it to bind actin and Arp2/3 (with an ancillary role for profilin binding to the proline rich region). There are subtle, but important, differences in the way WASp and N-WASp react in the presence of GTP-Cdc42 and PIP_2 suggesting that the model is incomplete. This notion is discussed in more detail in Higgs and Pollard [2]. In addition to Cdc42, a number of other proteins bind to various regions of WASp and/or N-WASp and are believed to be critical in regulating WASp/N-WASp function (Table I). The WH1 region provides a binding site for WASp-interacting protein (WIP) and two related proteins, CR16 and WICH (also called WIRE) [11–14]. These proteins bind not only WASp and N-WASp but also actin and isoforms of the adaptor Nck. The cellular role of these proteins is still unclear, but ectopic expression studies suggest they may couple WASp and N-WASp to SH3-containing proteins such as Nck and regulate subcellular localization, as well as affecting actin polymerization. The protein WISH (WASp-interacting SH3 protein) also binds to the proline-rich region of N-WASp and stimulates actin polymerization in an N-WASp, Arp2/3-dependent, but Cdc42 independent manner [15]. Cdc42 and PIP_2-independent signaling also has been reported in Drosophila where rescue of a WASp null mutation can be accomplished by expressing a form of WASp missing both the PIP_2- and Cdc42-binding domains [16]. SH3-containing tyrosine kinases also bind WASp and/or N-WASp (Table I). The importance of these interactions is incompletely understood. It appears, however, that there are many layers to WASp/N-WASp regulation. How this relates to specific function(s) is still unknown.

Some aspects of WASp function are relatively clear. WASp and N-WASp affect Arp2/3 activation and thus actin polymerization, and they can be found localized to filopodia, consistent with their being regulated by Cdc42. In humans the lack of WASp results in defects in filamentous actin formation that leads, among other things, to abnormal T-cell activation, abnormal formation of microvilli and podosomes, and cell motility defects in macrophage. Not so clear is why constitutively activating WASp should result in a severe neutropenia and monocytopenia [17]. Also, GTP-Cdc42 is still able to induce filopodia in fibroblasts derived from N-WASp null mice [18]. Likewise, the loss of WASp in Drosophila results in errors in cell fate determination, largely as a result of defects in asymmetric cell division. Defects in other fundamental aspects of actin function are not disturbed [19]. Clearly there is much more about WASP/N-WASp function and regulation that needs to be understood.

Scar/WAVE

Scar family proteins were originally identified in a genetic suppressor screen of a G-protein-coupled receptor mutation in Dictyostelium (Suppressor of cAMP receptor defect; [20]). Though the original report described the existence of Scar-like proteins in a variety of other organisms, the first study of a vertebrate ortholog was by Machesky and Insall [10] followed shortly by Miki et al. [21]. The latter group identified the protein, based on a database search, as WAVE (WASp family, Verprolin homology protein). All three early reports provided evidence that, like WASp, Scar proteins affect actin polymerization. In Dictyostelium, C. elegan, and Drosophila single isoforms of Scar exist. In the vertebrates reported to date there are (at least) three isoforms, Scar1, Scar2, and Scar3 (WAVE1, WAVE2, and WAVE3, respectively). The distribution

Table I Know Binding Interactions of WASp, N-WASp, and Scar/WAVE Proteins

WASp/Scar member	Binding partner	Binding region	Effect on WASp/Scar	References
WASp	Cdc42	GBD domain	"Activate"	2, 5
	PIP_2	Basic domain	"Activate"	2, 5
	WIP	WH1 domain	"Retards"	11
	WIRE(WICH)	WH1	Localization	14
	Ash/Grb2	Proline-rich	?	5
	Src family kinases	Proline-rich	?	2
	Btk family kinases	Proline-rich	?	2
	Profilin	Proline-rich	"Accelerates"	2, 5
	p21-Arc	Acidic domain	"Facilitates"	10
N-WASp	Cdc42	GBD domain	"Activate"	2, 5
	PIP_2	Basic domain	"Activate"	2, 5
	Calmodulin	IQ domain	?	5
	Ash/Grb2	Proline-rich	?	5
	WISH	Proline-rich	Activate	15
	Nck	Proline-rich	Stimulates	2, 5
	Profilin	Proline-rich	"Accelerates"	2, 5
	PSTPIP	Proline-rich	?	2
	Syndapin	Proline-rich	?	2
	p21-Arc	Acidic domain	"Facilitates"	10
	WIP/CR16/WICH	WH1	"Retards/?"	11–14
Scar/WAVE (interactions may not occur with all isoforms)	PKA	Partial WH2	?	26
	Abl kinase	Proline-rich	?	26
	NCKAP1	?	"Inhibits"	24
	PIR121	?	"Inhibits"	24
	HSPC300	?	?	24
	IRSp53	Proline-rich	"Activates"	23
	WRP	Proline-rich	"Terminates"?	25
	Profilin	Proline-rich	"Accelerates"	21
	p21-Arc	Acidic domain	"Facilitates"	10

Binding protein and domain on WASp family member are indicated. Domains are as shown in Fig. 1. Effects are meant as a broad description of the consequence based on either *in vitro* or *in vivo* interaction.

of each is somewhat different, though none seems as restricted as WASp. Although all three isoforms share the same basic domain structure there is evidence that their cellular roles and regulation may be somewhat different (see below). Mutations in Scar genes in *Dictyostelium* and *Drosophila* have established the importance of Scar in regulating actin assembly. In *Dictyostelium*, basic cell motility is abnormal in Scar mutants, and actin assembly in response to chemoattractant stimulation is severely impaired (Steiner *et al.* submitted). In *Drosophila*, Scar mutants are impaired in several aspects of development and oogenesis owing to defects in filamentous actin assembly. Scar, rather than WASp, appears to be the mediator of most Arp2/3-dependent events in this organism [22].

The presence of Scar proteins at the leading edge of cells suggests that they might be regulated by rac-like small GTPases, and Miki *et al.* showed that transvection of a nonfunctional Scar into cells blocked the activity of an activated rac1 but not an activated Cdc42 [21]. There is no obvious rac-binding region in Scar, and there is no evidence that Scar proteins directly bind rac proteins. Evidence does exist for at least two indirect mechanisms by which rac proteins regulate Scar activity. Using a yeast two-hybrid screen Miki *et al.* found that the rac-binding protein IRSp53 binds directly to the proline-rich region of hScar/WAVE proteins [23]. They further found that the interaction of IRSp53 and Scar2 was necessary for the formation of membrane ruffles induced by ectopic expression of activated rac1 in COS7 cells. It is interesting that interaction with IRSp53 seems to be largely (though perhaps not exclusively) restricted to the Scar2/WAVE2 isoform, and IRSp53 can associate with Scar2 in the absence of rac. In another recent study, Eden *et al.* reported the existence of a Scar1-containing complex that includes the rac1-binding proteins NCKAP1 and PIR121, as well as

Scar and the 8 KDa protein HSPC300 [24]. When a purified form of this complex is added to Arp2/3 complex and actin monomers *in vitro*, little actin polymerization occurs unless GTPγS-rac1 is added to the reaction. In the same assay purified Scar alone activates Arp2/3 complex to the maximal extent. The data suggest that in the cell Scar1 is retained in an inactive complex until GTP-bound rac1 binds either, or both, NCKAP1 and PIR121. The complex then separates into two components, one containing NCKAP1 and PIR121 and the other containing Scar and HSPC300. In this form Scar is able to activate Arp2/3 complex and facilitate formation of new actin polymers [24]. It should be noted that Nck can substitute for rac1 in this experiment. Whether this same inhibitory mechanism is common to all the isoforms of Scar is unknown, but it appears to be a possibility [24]. The relationship between the complexes containing IRSp53 and the one containing NCKAP1/PIR121/HSPC300 is not yet clear. It may be that different isoforms of Scar are retained in different complexes or that Scar proteins can each be sequestered in multiple ways. Further work is necessary to clarify this issue. What is true is that activation of WASp/N-WASp and Scar by small GTPases is performed by different mechanisms. Scar proteins are indirectly regulated by rac-like GTPases and can involve disassembly of a multiprotein complex that inhibits Scar function. WASp is autoinhibited and it is the direct binding of Cdc42 (and or other proteins) that can relieve the inhibition. The difference in regulation may provide a basis for the difference in localization seen for Scar and WASp proteins.

Another mechanism by which Scar function may be regulated involves a novel rac-GAP, WRP [25]. This GTPase-activating protein was identified as a binding partner of Scar through immunoprecipitation of rat brain extracts. The interaction is direct and involves the SH3 domain of WRP and the proline-rich region of WAVE-1. The direct interaction between a racGAP and Scar/WAVE proteins may provide at least part of the mechanism for terminating Scar-mediated signaling. It is interesting that in the same report, a number of other signaling/cytoskeleton proteins were reported to immunoprecipitate with WAVE-1. These included the Abl kinase interacting proteins, abi-1 and abi-2, α-tubulin, and SNAP-25 interacting protein. The *in vivo* significance of these interactions awaits further study.

Another level at which Scar/WAVE proteins may be regulated involves cAMP-dependent protein kinase (PKA). Scar1 binds directly to the regulatory subunit (RII) of PKA, and the binding site overlaps the region of monomeric actin binding [26] (Table I). Actin and PKA compete for this binding site, and the interaction is unique to Scar1 and does not occur with Scar2 or Scar3. Scar1 may, therefore, act as an A-kinase anchoring protein (AKAP) and be involved in regulating PKA activity at the sites of new actin assembly. The same study found that the tyrosine kinase abl also bound Scar. To date, there is no report of tyrosine phosphorylation of Scar (though phosphorylation downstream of MEK has been found; [27]), so the significance of abl binding remains to be determined.

Whereas WASp and N-WASp are enriched throughout extending lamellipods, in the general cell periphery and in the tips of filopodia, Scar proteins are localized in a very narrow region at the edge of a cell and at the tips of pseudopodia and lamellipodia ([28,29], Steiner *et al.*, submitted). The use of Scar-GFP in *Dictyostelium* has shown that Scar does not localize to the rim of the whole pseudopod but only to subregions that presumably reflect points of active actin filament assembly. The localization is thus transient and dynamic. Much has been learned about WASp and Scar/WAVE proteins during the last several years, but much remains to be determined. Even though they share a basic structure at the C-terminal end, how similar is the mechanism of activation of Arp2/3 complex? How do the mechanisms affect the branched structure of actin filaments associated with Arp2/3 activation? Are there proteins besides PKA that can compete for binding by the core components to that region of these molecules? What is the mechanistic significance of being able to bind multiple SH3-containing proteins? Does binding of each have a subtly different effect on WASp/Scar activation? Is binding of individual partners competitive or is it used to integrate responses via WASp/Scar proteins? How does all of this relate to the different localizations of Scar/WASp proteins and their potentially different roles in the formation of lamellipodia and filopodia. Although the field has progressed very rapidly, there is still a long way to go.

References

1. Ridley, A. J. (2001). Rho GTPases and cell migration. *J. Cell Sci.* **114**, 2713–2722.
2. Higgs, H. N. and Pollard, T. D. (2001). Regulation of actin filament network formation through Arp2/3 complex: activation by a diverse array of proteins. *Annu. Rev. Biochem.* **70**, 649–676.
3. Zalevsky, J., Lempert, L., Kranitz, H., and Mullins, R. D. (2001). Different WASP family proteins stimulate different Arp2/3 complex-dependent actin-nucleating activities. *Curr. Biol.* **11**, 1903–9113.
4. Rong, S.-B. and Vihinen, M. (2000). Structural basis of Wiskott-Aldrich syndrome causing mutations in the WH1 domain. *J. Mol. Med.* **78**, 530–537.
5. Takenawa, T. and Miki, H. (2001). WASP and WAVE family proteins: key molecules for rapid rearrangement of cortical actin filaments and cell movement. *J. Cell Sci.* **114**, 1801–1809.
6. J. M. J., Ochs, H. D., and Franke, U. (1994). Isolation of a novel gene mutated in Wiskott-Aldrich syndrome. *Cell.* **78**, 635–644.
7. Miki, H., Miura, K., and Takenawa, T. (1996). N-WASP, a novel actin-depolymerizing protein, regulates the cortical cytoskeleton rearrangement in a PIP_2-dependent manner downstream of tyrosine kinases. *EMBO J.* **15**, 5326–5335.
8. Li, R. (1997). Bee1, a yeast protein with homology to Wiskott-Aldrich syndrome protein, is critical for the assembly of the cortical actin cytoskeleton. *J. Cell Biol.* **136**, 649–658.
9. Lee, W.-L., Bezanilla, M., and Pollard, T. D. (2000). Fission yeast myosin-I, Myo1p, stimulates actin assembly by Arp2/3 complex and shares functions with WASp. *J. Cell Biol.* **151**, 78–799.
10. Machesky, L. M. and Insall, R. I. (1998). Scar1 and the related Wiskott-Aldrich syndrome protein, WASP, regulate the actin cytoskeleton through the Arp2/3 complex. *Curr. Biol.* **8**, 1347–1356.
11. Ramesh, N., Anton, I. M., Hartwig, J. H., and Geha, R. S. (1997). WIP, a protein associated with Wiskott-Aldrich syndrome protein, induces actin polymerization and redistribution in lymphoid cells. *Proc. Natl. Acad. Sci. USA* **94**, 14671–14676.

12. Ho, H. Y., Rohatgi, R., Ma, L., and Kirschner, M. W. (2001). CR16 forms a complex with N-WASp in brain and is a novel member of a conserved proline-rich actin-binding protein family. *Proc. Natl. Acad. Sci. USA* **98**, 11306–11311.
13. Kato, M., Miki, H., Kurita, S., Endo, T., Nakagawa, H., Miyamamoto, S., and Takenawa, T. (2002). WICH, a novel verprolin homology domain-containing protein that functions cooperatively with N-WASP in actin-microspike formation. *Biochem. Biophys. Res. Comm.* **291**, 41–47.
14. Aspenstrom, P. (2002). The WASP-binding protein WIRE has a role in the regulation of the actin filament system downstream of the platelet-derived growth factor receptor. *Exp. Cell Res.* **279**, 21–33.
15. Fukuoka, M., Suetsugu, S., Miki, H., Fukami, K., Endo, T., and Takenawa, T. (2001). A novel N-WASP binding protein, WISH, induced Arp2/3 complex activation independent of Cdc42. *J. Cell Biol.* **152**, 471–482.
16. Tal, T., Vaizel-Ohayon, D., and Schejter, E. D. (2002). Conserved interactions with cytoskeletal but not signaling elements are an essential aspect of *Drosophila* WASp function. *Dev Biol.* **243**, 260–271.
17. Devriendt, K., Kim, A. S., Mathijs, G., Frints, S. G. M., Schwartz, M., Van den Oord, J. J., Verhoef, E. G., Boogaerts, M. A., Fryns, J.-P., You, D., Rosen, M. K., and Vandenberghe, P. (2001). Constitutive activating mutation in WASP causes X-linked severe congenital neutropenia. *Nature Genetics* **27**, 313–317.
18. Lommel, S., Benesch, S., Rottner, K., Franz, T., Wehland, J., and Kuhn, R. (2001). Actin pedestal formation by enteropathogenic *Escherichia coli* and intracellular motility of *Shigella flexneri* are abolished in N-WASP defective cells. *EMBO Rep.* **2**, 850–857.
19. Ben-Yaacov, S., Le Borgne, R., Abramson, I., Schweisguth, F., and Schejter, E. D. (2001). *Wasp*, the *Drosophila* Wiskott-Aldrich syndrome protein homologue, is required for cell fate decisions mediated by *Notch* signaling. *J. Cell Biol.* **152**, 1–13.
20. Bear, J. E., Rawls, J. F., and Saxe, C. L. (1998). SCAR, a WASP-related protein isolated as a suppressor of receptor defects in late *Dictyostelium* development. *J. Cell Biol.* **142**, 1325–1335.
21. Miki, H., Suetsugu, S., and Takenawa, T. WAVE, a novel WASP-family protein involved in actin reorganization induced by Rac. *EMBO J.* **17**, 6932–6941.
22. Zallen, J. A., Cohen, Y., Hudson, A. M., Cooley, L., Wieschaus, E., and Schejter, E. D. (2002). SCAR is a primary regulator of Arp2/3-dependent morphological events in *Drosophila*. *J. Cell Biol.* **156**, 689–701.
23. Miki, H., Yamaguchi, H., Suetsugu, S., and Takenawa, T. (2000). IRSp53 is an essential intermediate between Rac and WAVE in the regulation of membrane ruffling. *Nature* **408**, 732–735.
24. Eden, S., Rohatgi, R., Podtelejnikov, A. V., Mann, M., and Kirschner, M. W. (2002). Mechanism of regulation of WAVE1-induced actin nucleation by rac1 and nck. *Nature* **418**, 790–793.
25. Soderling, S. H., Binns, K. L., Wayman, G. A., Davee, S. M., Ong, S. H., Pawson, T., and Scott, J. D. (2002). The WRP component of the WAVE-1 complex attenuates Rac-mediated signaling. *Nat. Cell Biol.* **4**, 970–975.
26. Westphal, R. S., Soderling, S. H., Alto, N. M., Langeberg, L. K., and Scott, J. D. (2000). Scar/WAVE-1, a Wiskott-Aldrich syndrome protein, assembles an actin-associated multi-kinase scaffold. *EMBO J.* **19**, 4589–4600.
27. Miki, H., Fukuda, M., Nishida, E., and Takenawa, T. (1999). Phosphorylation of WAVE downstream of mitogen-activated protein kinase signaling. *J. Biol. Chem.* **274**, 27605–27609.
28. Hahne, P., Sechi, A., Benesch, S., and Small, J. V. (2001). Scar/WAVE is localized at the tips of protruding lamellipodia in living cells. *FEBS Lett.* **49**, 215–220.
29. Nakagawa, H., Miki, H., Ito, M., Ohashhi, K., and Takenawa, T. (2001). N-WASP, WAVE, and Mena play different roles in the organization of actin cytoskeleton in lamellipodia. *J. Cell Sci.* **114**, 1555–1565.

CHAPTER 178

Synaptic NMDA-Receptor Signaling Complex

Mary B. Kennedy
Division of Biology, California Institute of Technology, Pasadena, California

Introduction

Glutamate is the neurotransmitter released from presynaptic terminals at most excitatory synapses in the central nervous system. It binds to two major classes of postsynaptic ligand-gated ion channels called AMPA (α-amino-3-hydroxy-5-methylisoxazole 4-propionic acid)-type glutamate receptors and NMDA (*N*-methyl-D-aspartate)-type glutamate receptors (further abbreviated NMDA receptor). Each of these receptors can trigger influx of ions across the membrane to produce an excitatory postsynaptic potential, but the NMDA receptor is also linked to a complex signaling pathway that produces biochemical changes in the postsynaptic neuron. The linkage is both physical [1], through direct interactions with cytosolic signaling molecules and scaffolds, and metabolic, through influx of Ca^{2+} ion through its ion channel [2,3].

The NMDA receptor is specialized to initiate and control changes in synaptic strength, based on the firing patterns of the synapse. When its channel opens, a large portion of the current across the membrane is carried by influx of Ca^{2+} ions [4,5]. The Ca^{2+} interacts with a variety of signaling molecules present just below the postsynaptic membrane. Opening of its channel is delicately controlled by the combination of glutamate-binding and concurrent depolarization of the postsynaptic membrane. At resting membrane potentials, the pore of the channel is blocked by Mg^{2+} ions. The block is relieved in a graded way by depolarization of the membrane. Thus, the NMDA receptor can "titrate" the amount of Ca^{2+} influx through its channel depending on how much the membrane is depolarized when glutamate is bound to it [6–8]. The required membrane depolarization can come from "back-propagating" action potentials that spread into the dendrites when the neuron fires an action potential, or from activation of several nearby synapses at the same time [9]. Ca^{2+} influx is highest when the synapse is activated a few milliseconds after the neuron has already fired an action potential that is spreading back into the dendrites [10]. Because of this "coincidence detection" property [11], the NMDA receptor can trigger synaptic changes when the presynaptic neuron and postsynaptic neuron are activated concurrently, fulfilling the prediction of Donald Hebb for a synaptic learning mechanism [12].

Here, I will discuss the structure and function of the signaling complex assembled in the cytosol around the NMDA receptor.

Structure of the NMDA Receptor Signaling Complex

Like other ligand-gated channels, the NMDA receptor is assembled from four (or five) distinct subunits that fall into structural and functional classes. However, it differs significantly from other ligand-gated channels because one major class of subunits, the NR2A-D subunits, have long (~300 residues) carboxyl terminal tails that extend into the cytosol and associate with a variety of proteins. The signaling complex that assembles around this tail does not appear to be a rigid structure with a fixed stoichiometry. Instead, it is believed to be assembled stochastically by associations with a set of scaffold proteins and, in some instances, by direct association of the tails of the NMDA receptor with signaling molecules. The meshwork of proteins that results is often referred to as the postsynaptic density, a structure that can be observed at synapses in the electron microscope [13].

Scaffold Proteins

At least three classes of scaffold molecules participate in organizing the interactions of signaling molecules associated with NMDA receptors in the postsynaptic density. A fifth class, the GRIPS/ABPs, associate primarily with AMPA receptors.

PSD-95

PSD-95 is the only known scaffold protein that associates directly with the NMDA receptor in the postsynaptic density. It was discovered as a prominent component of the "postsynaptic density fraction" recovered after detergent extraction of synaptosomes [14]. PSD-95 comprises three amino terminal "PDZ-domains" named after three proteins in which they were first recognized (PSD-95, Discs-large, and ZO-1), an SH3 domain, and a carboxyl terminal guanylate kinase domain. The first two PDZ domains interact directly with a terminal T/SXV motif at the carboxyl terminus of the NR2 class of NMDA receptor subunits [15,16]. The SH3 domain and guanylate kinase domains are fused into a structure that also appears to mediate a variety of protein interactions [17,18].

Immunocytochemical studies show that PSD-95 is highly concentrated in postsynaptic densities at glutamatergic synapses where it strongly co-localizes with the NMDA receptor [16,19]. Nearly all synapses that contain NMDA receptors appear to also contain PSD-95, indicating that it is an ubiquitous scaffold and associates with a variety of NMDA-receptor complexes. Several cytosolic signaling molecules have been shown to bind to PSD-95, including neuronal nitric oxide (NO) synthase; the protein kinase scaffold AKAP (next section); synGAP, a synaptic Ras GTPase activating protein; and GKAP, a protein that appears to link PSD-95 to an additional scaffold protein, shank (see below). At least one family of transmembrane proteins, the neuroligins, also binds to PSD-95.

Nitric Oxide Synthase. The neuronal form of NO synthase (nNOS) contains a PDZ domain near its amino terminus. Bredt and co-workers have shown that this PDZ domain associates directly with the second PDZ domain of PSD-95 by an atypical PDZ/PDZ interaction [20,21]. Thus, PSD-95 can link nNOS to the NMDA receptor. Neuronal NOS is activated by Ca^{2+}/calmodulin, and in cerebellar synapses nNOS is preferentially activated by Ca^{2+} flowing through NMDA receptors [22]. It seems likely that association of nNOS with PSD-95 plays an important role in controlling the specificity of its activation in cerebellar and other synapses.

SynGAP. SynGAP was discovered both as a prominent component of the PSD fraction [23] and in a two-hybrid screen for proteins that interact with SAP-102, a homologue of PSD-95 [24]. Its GTPase-activating (GAP) domain is highly homologous to that of the canonical p120 RasGAP, and the protein has RasGAP activity. Thus, it is assumed that its principal function is to inactivate Ras that has been activated by GTP exchange factors (GEFs) or by the action of protein tyrosine kinases such as Trks or Ephrin receptors. Although it is not a transmembrane protein, SynGAP nevertheless interacts strongly with the membrane via a combined PH/C2 domain near its amino terminus. It interacts with PSD-95 through a T/SXV motif at its carboxyl terminus. Immunocytochemical studies have shown that SynGAP is almost as highly concentrated at synaptic sites as PSD-95 itself [23,25], but it is also located on small vesicles throughout the cytosol and dendrite. Its precise functions are still mysterious. However, there are some clues. SynGAP is a prominent target for phosphorylation by Ca^{2+}/calmodulin-dependent protein kinase II (CaMKII) in the postsynaptic density (see below) [23]. Furthermore, deletion of synGAP is lethal to mice a few days after birth, producing abnormalities in development of certain brain areas and altering the number and size of synaptic sites in cultured hippocampal neurons (H.-J. Chen, L. Vazquez, Knuesel, and M. B. Kennedy, unpublished).

C. Neuroligins. Neuroligins are a large family of alternatively spliced transmembrane molecules that are located primarily in postsynaptic membranes, where they are believed to interact with the presynaptic neurexin proteins to mediate heterophilic adhesion [26–28]. Neuroligins were found to bind in a two-hybrid screen to PSD-95 through a terminal T/SXV motif [29]. Recent evidence suggests an important role for neuroligins in synaptic differentiation [30,31]. However, it is not yet known how their interaction with PSD-95 contributes to their function.

GKAP. GKAP (Guanylate kinase associated protein), a 70 kD protein with no identified functional domains, was isolated from a two-hybrid screen for proteins that bind to PSD-95 [32]. GKAP interacts specifically with the guanylate kinase homology domain of the PSD-95 family of proteins. GKAP was soon found to interact directly with an additional scaffold molecule termed Shank (or ProSAP; see below) [33,34]. Thus, one function of GKAP appears to be the formation of a link between the two scaffold proteins PSD-95 (and its family members) and Shank.

AKAPs (A-Kinase Interacting Proteins)

The AKAPs are a family of scaffold proteins that bind to the regulatory subunit of the cAMP-dependent protein kinase (PKA) and direct the kinase holoenzyme to particular subcellular compartments [35]. Many of the AKAPs also bind other enzymes in the cAMP pathway, forming discretely localized signaling complexes. AKAP79/150 provides a scaffold for PKA, protein kinase C, and calcineurin, a calcium-dependent protein phosphatase, positioning them adjacent to one another. AKAP79/150 binds to the SH3 and GK domains of both PSD-95 and SAP-97 (a relative of PSD-95 that associates specifically with the AMPA-receptor rather than with the NMDA receptor) [36]. It is localized together with glutamate receptors in most, but not all, excitatory synapses in hippocampal neurons in culture, and evidence suggests that it helps orchestrate phosphorylation of the AMPA receptor by PKA [36]. Calcineurin, a Ca^{2+}-dependent phosphatase that is localized to the PSD by its association with AKAP79/150, is a likely target of Ca^{2+} flowing through the NMDA receptor and may play a crucial role in control of LTP and LTD (see below).

SHANKS

The shank family was discovered in a two-hybrid screen for proteins that interact with GKAP (mentioned above) [33,34,37]. They are a particularly interesting family because they interact with other scaffold proteins and appear to be "scaffolds of scaffolds" linking the NMDA receptor complex, metabotropic glutamate receptor complexes [38], and perhaps also the AMPA receptor complex [37]. The shank proteins vary in size from ~240 kD to ~120 kD. The largest shank contains multiple ankyrin repeats, an SH3 domain, a PDZ domain, a long proline-rich region that occupies more than half the protein, and a sterile alpha motif (SAM) domain. Many of these domains are known to bind to specific proteins; the SH3 domain binds GRIP (AMPA receptor scaffold), the PDZ domain binds GKAP (linking to PSD-95 scaffold), and the proline rich domain binds homer (metabotropic glutamate receptor scaffold). A different region of the proline rich domain binds cortactin, an actin-associated protein. Thus, shank may provide a link between the postsynaptic density complexes and the actin cytoskeleton of the spine [33]. Overexpression of shank in hippocampal neurons in culture results in early formation of enlarged spines and enhanced recruitment of homer, the IP3 receptor, PSD-95, GKAP, and the NMDA receptor into these spines [39]. Thus, shank plays an important role in shaping the size and structure of spines.

Ca^{2+}/Calmodulin-Dependent Protein Kinase II (CaMKII) Binds Directly to the NMDA Receptor

CaMKII is highly enriched in the PSD [40,41] and is an important target for Ca^{2+} flowing through activated NMDA receptors [42–44]. Mice with a deletion mutation in the α-subunit of CaMKII are epileptic, exhibit deranged long-term potentiation at their hippocampal synapses, and perform poorly in learning tests [45,46].

CaMKII is anchored in the PSD through direct binding to at least two PSD components, the NMDA receptor itself and a transmembrane PSD protein termed densin. Binding of the CaMKII holoenzyme to the tail of the NR2A or NR2B subunits is strengthened by autophosphorylation of CaMKII that occurs upon activation of the kinase [47–49]. Binding to the tail of NR2B *in vitro* stabilizes the kinase in its activated state, suggesting that it may also do so *in vivo* [50].

Association of CaMKII with the PSD is highly dynamic in hippocampal neurons. Application of glutamate to neuronal cultures causes massive movement of CaMKII holoenzymes from dendritic shafts into the PSD over a period of a few minutes [51]. The process can be reversed by removal of Ca^{2+} from the medium. Thus, both activation of CaMKII and its localization at the postsynaptic site are delicately regulated by synaptic activity. Activation of CaMKII in the PSD can lead to phosphorylation and upregulation of AMPA-type glutamate receptors [43], or addition of new AMPA receptors to the synapse through a second process that does not require direct phosphorylation of the AMPA receptor [52].

Orchestration of Responses to Ca^{2+} Entering Through the NMDA Receptor

Two features of activity-dependent plasticity at excitatory synapses in the central nervous system are central to learning mechanisms in the brain and have been the focus of many recent studies. One is the tight dependence of synaptic plasticity on "spike-timing" of pre- and postsynaptic neurons [53]. The second is "metaplasticity" or the adjustment of the sensitivity and sign (LTP or LTD) of synaptic plasticity controlled by patterns of prior activity [54]. These mechanisms depend on the amplitude, time course, and location of origin of Ca^{2+} influx. Thus, the spatial arrangement of calcium-dependent signaling enzymes such as CaMKII, calcineurin, and nitric oxide synthase are likely to be critical determinants of both. Our understanding of the arrangement of signaling molecules at the postsynaptic site will set the stage for a precise quantitative understanding of these mechanisms of learning.

References

1. Kennedy, M. B. (2000). Signal-processing machines at the postsynaptic density. *Science* **290**, 750–754.
2. Lynch, G., Larson, J., Kelso, S., Barrionuevo, G., and Schottler, F. (1983). Intracellular injections of EGTA block induction of hippocampal long-term potentiation. *Nature* **305**, 719–721.
3. Malinow, R., Schulman, H., and Tsien, R. W. (1989). Inhibition of post-synaptic PKC or CaMKII blocks induction but not expression of LTP. *Science* **245**, 862–866.
4. Ascher, P. and Nowak, L. (1988). The role of divalent cations in the N-methyl-D-aspartate responses of mouse central neurones in culture. *J. Physiol.* **399**, 247–266.
5. MacDermott, A. B., Mayer, M. L., Westbrook, G. L., Smith, S. J., and Barker, J. L. (1986). NMDA-receptor activation increases cytoplasmic calcium concentration in cultured spinal cord neurones. *Nature* **321**, 519–522.
6. Nowak, L., Bregestovski, P., Ascher, P., Herbet, A., and Prochiantz, A. (1984). Magnesium gates glutamate-activated channels in mouse central neurones. *Nature* **307**, 462–465.
7. Mayer, M. L., Westbrook, G. L., and Guthrie, P. B. (1984). Voltage-dependent block by Mg^{2+} of NMDA responses in spinal cord neurones. *Nature* **309**, 261–263.
8. Mayer, M. L. and Westbrook, G. L. (1987). Permeation and block of N-methyl-D-aspartic acid receptor channels by divalent cations in mouse central neurones. *J. Physiol.* **394**, 501–528.
9. Magee, J. C. and Johnston, D. (1997). A synaptically controlled, associative signal for Hebbian plasticity in hippocampal neurons. *Science* **275**, 209–213.
10. Bi, G. and Poo, M.-M. (1998). Synaptic modifications in cultured hippocampal neurons: dependence on spike timing, synaptic strength, and postsynaptic cell type. *J. Neurosci.* **18**, 10464–10472.
11. Bourne, H. R. and Nicoll, R. (1993). Molecular machines integrate coincident synaptic signals. *Cell* **72**, 65–75.
12. Hebb, D. O. (1949). *The Organization of Behavior.* John Wiley & Sons, New York.
13. Kennedy, M. B. (1997). The postsynaptic density at glutamatergic synapses. *Trends Neurosci.* **20**, 264–268.

14. Cho, K.-O., Hunt, C. A., and Kennedy, M. B. (1992). The rat brain postsynaptic density fraction contains a homolog of the Drosophila discs-large tumor suppressor protein. *Neuron* **9**, 929–942.
15. Doyle, D. A., Lee, A., Lewis, J., Kim, E., Sheng, M., and MacKinnon, R. (1996). Crystal structures of a complexed and peptide-free membrane protein-binding domain: molecular basis of peptide recognition by PDZ. *Cell* **85**, 1067–1076.
16. Kornau, H.-C., Schenker, L. T., Kennedy, M. B., and Seeburg, P. H. (1995). Domain interaction between NMDA receptor subunits and the postsynaptic density protein PSD-95. *Science* **269**, 1737–1740.
17. McGee, A. W., Dakoji, S. R., Olsen, O., Bredt, D. S., Lim, W. A., and Prehoda, K. E. (2001). Structure of the SH3-guanylate kinase module from PSD-95 suggests a mechanism for regulated assembly of MAGUK scaffolding proteins. *Mol. Cell* **8**, 1291–1301.
18. Tavares, G. A., Panepucci, E. H., and Brunger, A. T. (2001). Structural characterization of the intramolecular interaction between the SH3 and guanylate kinase domains of PSD-95. *Mol. Cell* **8**, 1313–1325.
19. Hunt, C. A., Schenker, L. J., and Kennedy, M. B. (1996). PSD-95 is associated with the postsynaptic density and not with the presynaptic membrane at forebrain synapses. *J. Neurosci.* **16**, 1380–1388.
20. Christopherson, K. S., Hillier, B. J., Lim, W. A., and Bredt, D. S. (1999). PSD-95 assembles a ternary complex with the N-methyl-D-aspartic acid receptor and a bivalent neuronal NO synthase PDZ domain. *J. Biol. Chem.* **274**, 27467–27473.
21. Hillier, B. J., Christopherson, K. S., Prehoda, K. E., Bredt, D. S., and Lim, W. A. (1999). Unexpected modes of PDZ domain scaffolding revealed by structure of nNOS-syntrophin complex. *Science* **284**, 812–815.
22. Garthwaite, J., Charles, S. L., and Chess-Williams, R. (1988). Endothelium-derived relaxing factor release on activation of NMDA receptors suggests role as intracellular messenger in the brain. *Nature* **336**, 385–388.
23. Chen, H.-J., Rojas-Soto, M., Oguni, A., and Kennedy, M. B. (1998). A synaptic Ras-GTPase activating protein (p135 SynGAP) inhibited by CaM Kinase II. *Neuron* **20**, 895–904.
24. Kim, J. H., Liao, D., Lau, L.-F., and Huganir, R. L. (1998). SynGAP: a synaptic RasGAP that associates with the PSD-95/SAP90 protein family. *Neuron* **20**, 683–691.
25. Zhang, W., Vazquez, L., Apperson, M., and Kennedy, M. B. (1999). Citron binds to PSD-95 at glutamatergic synapses on inhibitory neurons in the hippocampus [In Process Citation]. *J. Neurosci.* **19**, 96–108.
26. Ichtchenko, K., Nguyen, T., and Südhof, T. C. (1996). Structures, alternative splicing, and neurexin binding of multiple neuroligins. *J. Biol. Chem.* **271**, 2676–2682.
27. Nguyen, T. and Sudhof, T. C. (1997). Binding properties of neuroligin 1 and neurexin 1-beta reveal function as heterophilic cell adhesion molecules. *J. Biol. Chem.* **272**, 26032–26039.
28. Song, J. Y., Ichtchenko, K., Sudhof, T. C., and Brose, N. (1999). Neuroligin 1 is a postsynaptic cell-adhesion molecule of excitatory synapses. *Proc. Natl. Acad. Sci. USA* **96**, 1100–1105.
29. Irie, M., Hata, Y., Takeuchi, M., Ichtchenko, A., Toyoda, A., Hirao, K., Takai, Y., Rosahl, T. W., and Sudhof, T. C. (1997). Binding of neuroligins to PSD-95. *Science* **277**, 1511–1515.
30. Scheiffele, P., Fan, J., Choih, J., Fetter, R., and Serafini, T. (2000). Neuroligin expressed in nonneuronal cells triggers presynaptic development in contacting axons. *Cell* **101**, 657–669.
31. Rao, A., Harms, K. J., and Craig, A. M. (2000). Neuroligation: building synapses around the neurexin-neuroligin link. *Nature Neurosci.* **3**, 747–749.
32. Kim, E., Naisbitt, S., Hsueh, Y. P., Rao, A., Rothschild, A., Craig, A. M., and Sheng, M. (1997). GKAP, a novel synaptic protein that interacts with the guanylate kinase-like domain of the PSD-95/SAP90 family of channel clustering molecules. *J. Cell Biol.* **136**, 669–678.
33. Naisbitt, S., Kim, E., Tu, J. C., Xiao, B., Sala, C., Valtschanoff, J., Weinberg, R. J., Worley, P. F., and Sheng, M. (1999). Shank, a novel family of postsynaptic density proteins that binds to the NMDA receptor/PSD-95/GKAP complex and cortactin. *Neuron* **23**, 569–582.
34. Boeckers, T. M., Kreutz, M. R., Winter, C., Zuschratter, W., Smalla, K. H., Sanmarti-Vila, L., Wex, H., Langnaese, K., Bockmann, J., Garner, C. C., and Gundelfinger, E. D. (1999). Proline-rich synapse-associated protein-1/cortactin binding protein 1 (ProSAP1/CortBP1) is a PDZ-domain protein highly enriched in the postsynaptic density. *J. Neurosci.* **19**, 6506–6518.
35. Colledge, M. and Scott, J. D. (1999). AKAPs: from structure to function. *Trends Cell Biol.* **9**, 216–221.
36. Colledge, M., Dean, R. A., Scott, G. K., Langeberg, L. K., Huganir, R. L., and Scott, J. D. (2000). Targeting of PKA to glutamate receptors through a MAGUK-AKAP complex. *Neuron* **27**, 107–119.
37. Sheng, M. and Kim, E. (2000). The Shank family of scaffold proteins. *J. Cell Sci.* **113**, 1851–1856.
38. Tu, J. C., Xiao, B., Naisbitt, S., Yuan, J. P., Petralia, R. S., Brakeman, P., Doan, A., Aakalu, V. K., Lanahan, A. A., Sheng, M., and Worley, P. F. (1999). Coupling of mGluR/Homer and PSD-95 complexes by the Shank family of postsynaptic density proteins. *Neuron* **23**, 583–592.
39. Sala, C., Piech, V., Wilson, N. R., Passafaro, M., Liu, G., and Sheng, M. (2001). Regulation of dendritic spine morphology and synaptic function by Shank and Homer. *Neuron* **31**, 115–130.
40. Kennedy, M. B., Bennett, M. K., and Erondu, N. E. (1983). Biochemical and immunochemical evidence that the "major postsynaptic density protein" is a subunit of a calmodulin-dependent protein kinase. *Proc. Natl. Acad. Sci. USA* **80**, 7357–7361.
41. Kelly, P. T., McGuinness, T. L., and Greengard, P. (1984). Evidence that the major postsynaptic density protein is a component of a Ca^{2+}/calmodulin-dependent protein kinase. *Proc. Natl. Acad. Sci. USA* **81**, 945–949.
42. Fukunaga, K., Stoppini, L., Miyamoto, E., and Muller, D. (1993). Long-term potentiation is associated with an increased activity of Ca^{2+}/calmodulin-dependent protein kinase II. *J. Biol. Chem.* **268**, 7863–7867.
43. Barria, A., Muller, D., Derkach, V., Griffith, L. C., and Soderling, T. R. (1997). Regulatory phosphorylation of AMPA-type glutamate receptors by CaMKII during long term potentiation. *Science* **276**, 2042–2045.
44. Ouyang, Y., Rosenstein, A., Kreiman, G., Schuman, E. M., and Kennedy, M. B. (1999). Tetanic stimulation leads to increased accumulation of Ca(2+)/calmodulin-dependent protein kinase II via dendritic protein synthesis in hippocampal neurons. *J. Neurosci.* **19**, 7823–7833.
45. Silva, A. J., Stevens, C. F., Tonegawa, S., and Wang, Y. (1992). Deficient hippocampal long-term potentiation in α-calcium-calmodulin kinase II mutant mice. *Science* **257**, 201–206.
46. Silva, A. J., Paylor, R., Wehner, J. M., and Tonegawa, S. (1992). Impaired spatial learning in α-calcium-calmodulin kinase II mutant mice. *Science* **257**, 206–211.
47. Strack, S. and Colbran, R. J. (1998). Autophosphorylation-dependent targeting of calcium/calmodulin-dependent protein kinase II by the NR2B subunit of the N-methyl-D-aspartate receptor. *J. Biol. Chem.* **273**, 20689–20692.
48. Leonard, A. S., Lim, I. A., Hemsworth, D. E., Horne, M. C., and Hell, J. W. (1999). Calcium/calmodulin-dependent protein kinase II is associated with the N-methyl-D-aspartate receptor. *Proc. Natl. Acad. Sci. USA* **96**, 3239–3244.
49. Gardoni, F., Schrama, L. H., van Dalen, J. J., Gispen, W. H., Cattabeni, F., and Di Luca, M. (1999). AlphaCaMKII binding to the C-terminal tail of NMDA receptor subunit NR2A and its modulation by autophosphorylation. *FEBS Lett.* **456**, 394–398.
50. Bayer, K. U., De Koninck, P., Leonard, A. S., Hell, J. W., and Schulman, H. (2001). Interaction with the NMDA receptor locks CaMKII in an active conformation. *Nature* **411**, 801–805.
51. Shen, K. and Meyer, T. (1999). Dynamic control of CaMKII translocation and localization in hippocampal neurons by NMDA receptor stimulation. *Science* **284**, 162–166.
52. Shi, S. H., Hayashi, Y., Petralia, R. S., Zaman, S. H., Wenthold, R. J., Svoboda, K., and Malinow, R. (1999). Rapid spine delivery and redistribution of AMPA receptors after synaptic NMDA receptor activation [see comments]. *Science* **284**, 1811–1816.
53. Sjostrom, P. J. and Nelson, S. B. (2002). Spike timing, calcium signals and synaptic plasticity. *Curr. Opin. Neurobiol.* **12**, 305–314.
54. Abraham, W. C., Mason-Parker, S. E., Bear, M. F., Webb, S., and Tate, W. P. (2001). Heterosynaptic metaplasticity in the hippocampus in vivo: a BCM-like modifiable threshold for LTP. *Proc. Natl. Acad. Sci. USA* **98**, 10924–10929.

CHAPTER 179

Toll Family Receptors

Hana Bilak, Servane Tauszig-Delamasure, and Jean-Luc Imler

*Centre National de le Recherche Scientifique,
Institut de Biologie Moléculaire et Cellulaire,
Strasbourg, France*

Introduction

In recent years, Toll receptors have emerged as an important family of molecules activating the innate immune response. Mammalian Toll receptors detect pathogen-associated molecular patterns (PAMPs) such as lipopolysaccharides (LPS) from Gram-negative bacteria or peptidoglycan (PGN) from Gram-positive bacteria, and activate signaling cascades that induce the transcription factors NF-κB and AP1.

Structure Function of Toll Receptors

Discovery of Toll and Toll-like Receptors

The *Toll* gene was first identified in the early 1980s by Nusslein-Volhard and Anderson during a mutagenesis screen to identify the genes controlling the establishment of the dorsoventral (DV) axis of the *Drosophila* embryo, and was found to encode a new kind of type I transmembrane receptor [1]. The eleven remaining genes identified in this screen encode factors acting upstream and downstream of Toll in the signaling pathway. Activation of Toll on the ventral side of the embryo results in a ventral to dorsal gradient of nuclear translocation of the transcription factor Dorsal, thus establishing embryonic polarity. It was also through studies in *Drosophila* that Toll was assigned an immune function in the control of the inducible expression of antimicrobial peptides. The transcriptional activation of the genes encoding these peptides requires transcriptional activators of the Rel family, to which NF-κB belongs. At the time, Dorsal was the only identified member of this family in *Drosophila*. This prompted analysis of the known mutant strains of the Toll pathway for immunodeficiency phenotypes, which led to the demonstration by Hoffmann and collaborators that the Toll pathway controls the response to fungal and Gram-positive bacterial infections [2,3]. These results were rapidly followed by the first description by Medzhitov and coworkers of a mammalian Toll homologue (now known as TLR4) capable of activating NF-κB and the synthesis of cytokines and co-stimulatory molecules [4]. Shortly after, Beutler and colleagues showed that mice from the LPS-hyporesponsive strains C57BL/10ScCr and C3H/HeJ carry mutations in their *tlr4* gene, indicating that Toll-like receptors (TLRs) play an important role in the control of infection in mammals [5]. Since these initial discoveries, a family of 10 TLRs has been described in mammals, the properties of which are described below.

Structure of Toll Family Receptors

The cytoplasmic domain of Toll bears striking similarities to that of the Interleukin-1 type I receptor (IL-1R), and is referred to as the TIR (Toll/IL-1R) homology domain. TIR domains are also present in intracellular signaling molecules such as MyD88 or TIRAP/MAL (see below). Plants also express TIR domain–containing factors [6]. The structures of the TIR domains of human TLR1 and TLR2 have been solved and shown to be composed of a central five-stranded parallel β-sheet surrounded by five α-helices [7], (Fig. 1).

The extracellular domain of Toll family receptors does not contain Ig domains, like the IL-1R, but comprises several leucine-rich repeats flanked by characteristic cysteine-rich motifs. This feature is shared by a number of membrane receptors such as the pattern recognition receptor CD14, the adhesion molecule GpIbα, or the members of the Trk family of neurotrophins receptors. Leucine-rich repeats are generally recognized as a protein-protein interaction domain [8].

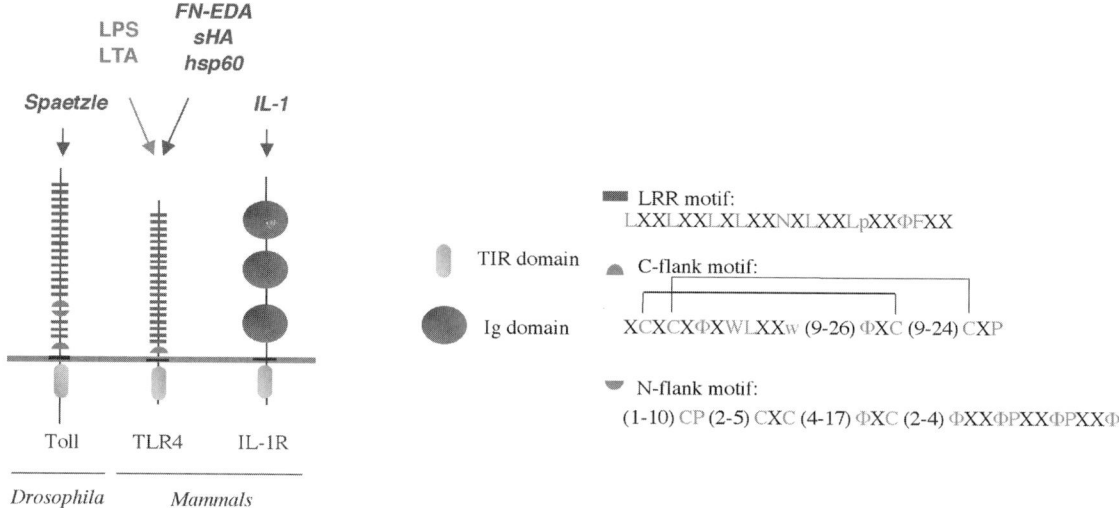

Figure 1 TIR domain receptors in *Drosophila* and mammals. (A) Domain architecture of members of the three groups of TIR domain membrane receptors. TIR domains are represented in yellow, LRR motifs as green rectangles, and flanking cystein-rich regions as orange half-circles. Microbe-derived ligands are in blue, and endogenous ligands in mauve. FN-EDA, fibronectine extra domain A; sHA, soluble Hyaluronic Acid. (B) Amino-acid sequence of the motifs found in the ectodomains of Toll family receptors. Φ, hydrophobic residue; X, any amino-acid [8].

Despite their similarity of structure and function, members of the family of Toll receptors can be subdivided in two different classes, based on phylogenetic analysis of their TIR domain [9]. All mammalian TLRs cluster to one of these subfamilies, together with a single *Drosophila* receptor, Toll-9. The eight other *Drosophila* Toll family members cluster to the second subfamily, together with the *Caenorhabditis elegans Tol-1* gene product. Close examination of the ectodomain of these molecules also reveals differences, the most evident being the presence of a single C-flank cysteine rich motif in TLRs and *Drosophila* Toll-9, as opposed to other Tolls, which most often contain additional C-flank motifs and always share at least one N-flank cysteine rich motif. Thus it appears that most *Drosophila* Tolls have evolved independently from mammalian TLRs, possibly to fulfill different functions. In agreement with this hypothesis, TLRs do not appear to be required for murine development, in contrast to Toll in *Drosophila* [10].

Activation of Toll and Toll-like Receptors

In *Drosophila* embryos, a cascade of proteases is activated on the ventral side of the embryo and promotes proteolytic maturation of the cysteine-knot growth factor Spaetzle, which is structurally related to neurotrophins. The cleaved Spaetzle is thought to bind to the Toll receptor, prompting its dimerization and activation [1]. In adult flies, activation of Toll in response to infections also requires a processed form of Spaetzle, although genetic experiments indicate that the proteases of the embryonic cascade upstream of Spaetzle are largely dispensable for the immune response [2]. Recognition of Gram-positive bacteria is mediated by a member of the PGN recognition protein (PGRP) family, PGRP-SA, which is thought to activate a protease(s) leading to Spaetzle activation [11].

No Spaetzle homologues have been described so far in mammals. Rather, TLRs appear to be directly activated by PAMPs. Analysis of macrophages derived from TLR knockout (KO) mice has revealed that (1) TLR4 is required for the response to LPS and lipoteichoic acid (LTA) derived from Gram-positive bacteria; (2) TLR2 is necessary for activation by several ligands, including PGN, bacterial lipopeptide (BLP), and the mycobacterial lipopeptide MALP2; (3) TLR3-deficient cells do not respond to dsRNA from viruses; (4) TLR5 recognizes flagellin, the principal constituent of bacterial flagella; (5) TLR6 deficient cells do not respond to MALP2; (6) TLR9 is required for cell activation by bacterial DNA containing unmethylated CpG motifs (reviewed in [10,12–14]).

Despite the genetic evidence, biochemical data for a direct interaction between most TLRs and their ligands are still lacking. There is good pharmacological evidence for a direct interaction between LPS and TLR4, or CpG and TLR9. In addition, biochemical experiments using radiolabeled LPS demonstrated that it can be specifically cross-linked to TLR4, thereby indicating that the two molecules directly interact [15].

Some TLRs associate with coreceptors to interact with PAMPs. For example, in the case of LPS, CD14 plays an essential role in the receptor complex. This protein is also required for maximal cell activation by PGN, a TLR2 agonist. In addition, plasma membrane expression of TLR4 requires the coexpression of MD2, a secreted accessory molecule that also seems to interact with LPS (Fig. 2a). In some cases, recognition of PAMPs involves heterodimerization of TLRs. For instance, activation of cells by MALP2 requires both TLR2 and TLR6. TLR6 coimmunoprecipitates with TLR2 in cell lines, and heterodimerization of the TIR domains of TLR2 and TLR6 elicits a cellular response. TLR2 is believed

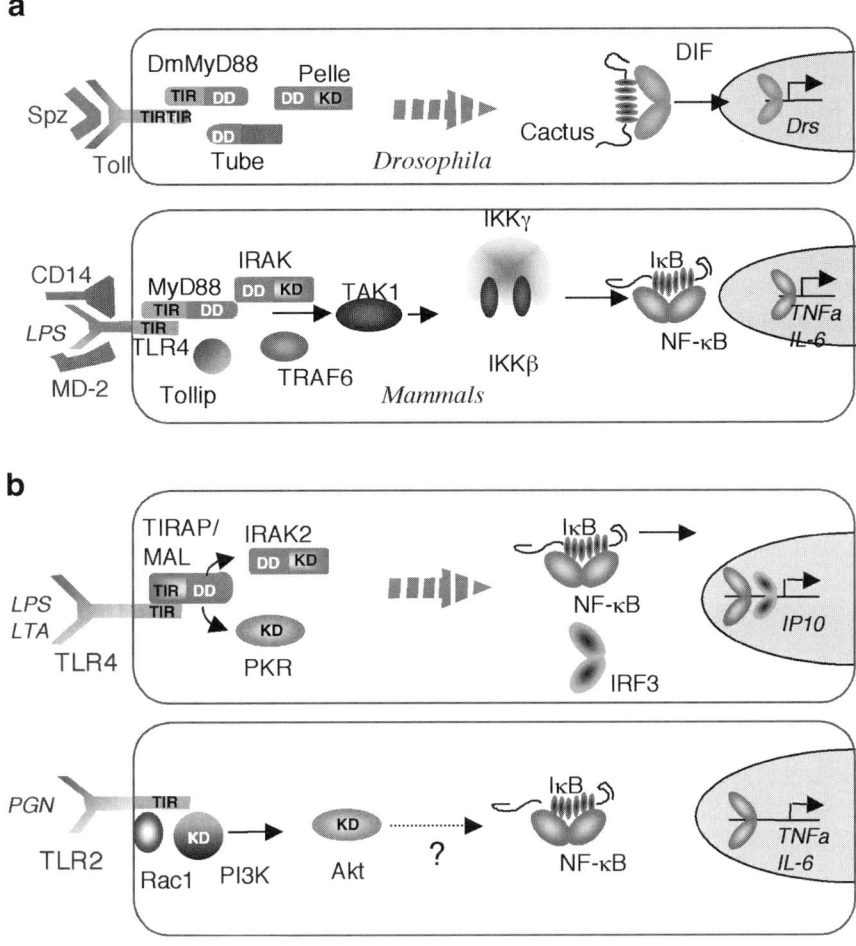

Figure 2 Signaling pathways activated by Toll family receptors. (a) Classical pathway activated by Toll family receptors in *Drosophila* and mammals. IkB and its homologue Cactus are inhibitory factors that retain the Rel proteins DIF and NF-kB in the cytoplasm. Drs, Drosomycin. (b) Examples of subunit-specific alternative signaling pathways downstream of TLR4 and TLR2. IRF3, interferon response factor 3; IP10, chemokine gene induced by TLR4 in a MyD88-independant fashion; DD, death-domain; KD, kinase domain.

to associate with another TLR to recognize BLP. It so appears that cooperation between TLRs is a means of broadening the recognition repertoire of these receptors [10,12,13].

One critical question about regulation of the innate immune response is the distinction by immune cells between harmless commensal microorganisms and infectious microbes, which often share common PAMPs. Cellular distribution of TLRs may contribute to this discrimination. For example, TLR5 is asymmetrically distributed in gut epithelial cells: it is absent from the apical side of the cells, which are exposed to the commensal bacterial flora of the gut lumen, and present on the basolateral side of the cells, to which noninvading bacteria do not have access [16]. In addition, it is becoming clear that some TLRs can be activated by host-derived molecules. These include hsp60, oligosaccharides from hyaluronic acid, and a fibronectin fragment [10,12,13]. A common feature of these molecules is that they are produced during stress responses in general and inflammation in particular. One intriguing possibility is that these endogenous products generated in response to infection synergize with PAMPs to activate TLRs and stimulate an efficient innate immune response.

Signaling by Toll Family Receptors

Activation of Toll family receptors results in the transcriptional induction of a number of genes involved in host defense such as antimicrobial molecules or cytokines. A common feature of these genes is the presence of binding sites for the transcription factor NF-κB and AP-1 in their promoters. As described below, a pathway leading from TIR domain receptors to the nucleus has been deciphered. The various ligands that activate TLRs induce overlapping but distinct sets of genes during infection, thus revealing the existence of receptor-specific alternative signaling pathways (Fig. 2b).

The Classical Toll/IL-1R Signaling Pathway

TIR domain receptors are associated with an intracytoplasmic plurimolecular platform in which the MyD88 factor plays a central role. MyD88 is composed of an amino-terminal death domain (DD) and a carboxy-terminal TIR domain. It interacts in a ligand-dependent manner with the receptors through homophilic TIR-TIR domain interactions. The DD of MyD88 interacts with the DD of the serine-threonine kinase IRAK (IL-1R associated kinase) [10,12,13]. Another adaptor protein, Tollip, is also recruited transiently to the receptor complex upon activation [17]. Tollip does not contain a TIR domain, and its function may be to bring IRAK to the receptor complex. Interaction of IRAK with MyD88 triggers autophosphorylation of the kinase. This progressive phosphorylation affects interaction with the receptor complex and allows IRAK to interact with TRAF6.

TRAF6 belongs to a family of signal tranducers originally identified for their role downstream of receptors of the TNFα family. TRAF6 associates with a dimeric ubiquitin-conjugating enzyme complex composed of Ubc13 and Uev1A. This complex mediates activation of the TAK-1 kinase through a nonclassical ubiquitination mechanism, which does not involve the proteasome. TAK1 then phosphorylates the IκB kinase (IKK) and MKK6 kinases, which in turn activate NF-κB and AP-1 [18].

Significant differences exist in the Toll pathway in flies. The receptor platform contains an additional DD component named Tube, which interacts with the IRAK homologue Pelle. Three TRAF factors are encoded by the *Drosophila* genome, but their role in the Toll pathway is not clear. Curiously, the *Drosophila* homologues of TAK1, Uev1a, Ubc13, IKKβ and γ are not required downstream of Toll during the antifungal response in adults or during embryonic development [19].

Subunit-Specific Alternative Pathways

Analysis of MyD88 KO mice confirmed the importance of this central factor in the signaling pathway downstream of the cytokines IL-1 and IL-18 and several PAMPs including LPS. However, the response to LPS differs between TLR4 and MyD88-deficient cells. Notably, activation of NF-κB and AP1 is only delayed in MyD88$^{-/-}$ cells whereas it is abolished in TLR4$^{-/-}$ cells. Furthermore, maturation of dendritic cells (DC) in response to LPS stimulation, though abolished in TLR4$^{-/-}$ cells, is still functional in MyD88$^{-/-}$ cells [20]. These data clearly point to the existence of a MyD88-independent pathway of cell activation, which also seems to operate downstream of TLR3. Nevertheless, this alternative signaling pathway is not activated by all TLRs. For example, in response to the TLR9 and TLR2 agonists CpG and MALP2, respectively, induction of NF-κB and DC maturation is completely abolished in both TLR(9 or 2) and MyD88 KO cells [10,12].

A new cytosolic factor has recently been identified and is likely to play a role in the MyD88-independent pathway of LPS signal transduction. This factor (MAL for MyD88 adaptor like protein, or TIRAP for TIR associated protein) contains a TIR domain but no DD [21,22]. TIRAP/MAL associates with TLR4 through its own TIR domain but not with TLR9. In addition, a dominant-negative version of the molecule has been shown to block activation of NF-κB by TLR4 but not by TLR9, IL-1R, or IL-18R. Treatment of DCs with a synthetic TIRAP-blocking peptide inhibits their LPS-induced maturation in both wild-type and MyD88$^{-/-}$ backgrounds [22]. These data strongly suggest that TIRAP/MAL controls the MyD88-independent pathway activated by TLR4. By contrast to MyD88, this factor does not interact with IRAK but with the related kinase IRAK2 through which it activates TRAF6. TIRAP/MAL can also associate with the dsRNA binding kinase PKR.

Another subunit-specific pathway seems to operate downstream of TLR2. This pathway is initiated by a ligand-dependent tyrosine phosphorylation of TLR2, which leads to recruitment of the p85 regulatory subunit of phosphatidyl-inositol 3 kinase (PI3K) and the small GTPase Rac-1, and to activation of the protein kinase Akt (protein kinase B) [23]. Although the effectors downstream of Akt remain to be identified, it is clear that this pathway regulates phosphorylation of NF-κB, rather than targeting its inhibitor IκB. Because this phosphorylation is required for the transactivation properties of NF-κB, it is possible that other TLRs will activate this pathway. However, the PI3-kinase binding motif, which contains the phosphorylated tyrosine residue is only conserved in TLR1, 2, and 6, thus suggesting specificity of this Rac-1 dependent pathway.

The RIP2 serine-threonine kinase was also recently shown to function downstream of some TLRs. Induction of cytokines in RIP2$^{-/-}$ macrophages is reduced upon stimulation with LPS, LTA, PGN, and dsRNA but not CpG. This result indicates that RIP2 is involved in TLR2, TLR3, TLR4 but not TLR9 signaling [24,25].

It is therefore becoming obvious that different combinations of transducing factors can dock to the receptor platform, depending on its molecular composition. This finding probably explains why some TLRs, such as TLR4, TLR3, and TLR9, can signal as homodimers, whereas others, such as TLR2 or TLR6, cannot. These distinct transduction factors initiate discrete signaling events, the combination of which may explain the panel of different responses elicited by the various TLR agonists. The understanding of these regulatory events holds promise for improved vaccination strategies and better control of infectious diseases.

Note Added in Proof

Recent work, including characterization of TIRAP/MAL deficient mice indicate that the MyD88-independent pathway does not involve this molecule, but may rely on a third TIR domain adaptor, TRIF/TICAM-1 (reviewed in Imler and Hoffmann (2003). *Nature Immunol.* **4**,105–106).

Acknowledgments

We thank Jules Hoffmann for continuous interest and support, and Petros Ligoxygakis and Sophie Rutschmann for critical reading of the manuscript.

References

1. Belvin, M. P. and Anderson, K. V. (1996). A conserved signaling pathway: the Drosophila toll-dorsal pathway. *Annu. Rev. Cell Dev. Biol.* **12**, 393–416.
2. Lemaitre, B., Nicolas, E., Michaut, L., Reichhart, J., and Hoffmann, J. (1996). The dorsoventral regulatory gene cassette spätzle/Toll/cactus controls the potent antifungal response in Drosophila adults. *Cell* **86**, 973–983.
3. Rutschmann, S., Kilinc, A., and Ferrandon, D. (2002). The Toll pathway is required for resistance to gram-positive bacterial infections in *Drosophila*. *J. Immunol.* **168**, 1542–1546.
4. Medzhitov, R., Preston-Hurlburt, P., and Janeway, C. (1997). A human homologue of the Drosophila Toll protein signals activation of adaptive immunity. *Nature*, **388**, 394–397.
5. Poltorak, A., He, X., Smirnova, I., Liu, M., Huffel, C., Du, X., Birdwell, D., Alejos, E., Silva, M., and Galanos, C. *et al.* (1998). Defective LPS signaling in C3H/HeJ and C57BL/10ScCr mice: mutations in Tlr4 gene. *Science* **282**, 2085–2088.
6. Imler, J. and Hoffmann, J. A. (2001). Toll receptors in innate immunity. *Trends Cell Biol.* **11**, 304–311.
7. Xu, Y., Tao, X., Shen, B., Horng, T., Medzhitov, R., Manley, J. L., and Tong, L. (2000). Structural basis for signal transduction by the Toll/interleukin-1 receptor domains. *Nature* **408**, 111–115.
8. Kajava, A. V. (1998). Structural diversity of leucine-rich repeat proteins. *J. Mol. Biol.* **277**, 519–527.
9. Du, X., Poltorak, A., Wei, Y., and Beutler, B. (2000). Three novel mammalian toll-like receptors: gene structure, expression, and evolution. *Eur. Cytokine Netw.* **11**, 362–371.
10. Akira, S., Takeda, K., and Kaisho, T. (2001). Toll-like receptors: critical proteins linking innate and acquired immunity. *Nat. Immunol.* **2**, 675–680.
11. Michel, T., Reichhart, J. M., Hoffmann, J. A., and Royet, J. (2001). Drosophila Toll is activated by Gram-positive bacteria through a circulating peptidoglycan recognition protein. *Nature* **414**, 756–759.
12. Medzhitov, R. (2001). Toll-like receptors and innate immunity. *Nature Rev. Immunol.* **1**, 135–145.
13. Underhill, D. M. and Ozinsky, A. (2002). Toll-like receptors: key mediators of microbe detection. *Curr. Opin. Immunol.* **14**, 103–110.
14. Alexopoulou, L., Holt, A. C., Medzhitov, and R., Flavell, R. A. (2001). Recognition of double-stranded RNA and activation of NF-kappaB by Toll-like receptor 3. *Nature* **413**, 732–738.
15. da Silva Correia, J., Soldau, K., Christen, U., Tobias, P. S., and Ulevitch, R. J. (2001). Lipopolysaccharide is in close proximity to each of the proteins in its membrane receptor complex transfer from CD14 to TLR4 and MD-2. *J. Biol. Chem.* **276**, 21129–21135.
16. Gewirtz, A. T., Navas, T. A., Lyons, S., Godowski, P. J., and Madara, J. L. (2001). Cutting edge: bacterial flagellin activates basolaterally expressed TLR5 to induce epithelial proinflammatory gene expression. *J. Immunol.* **167**, 1882–1885.
17. Burns, K., Clatworthy, J., Martin, L., Martinon, F., Plumpton, C., Maschera, B., Lewis, A., Ray, K., Tschopp, J., and Volpe, F. (2000). Tollip, a new component of the IL-1RI pathway, links IRAK to the IL-1 receptor. *Nat. Cell Biol.* **2**, 346–351.
18. Wang, C., Deng, L., Hong, M., Akkaraju, G. R., Inoue, J., and Chen, Z. J. (2001). TAK1 is a ubiquitin-dependent kinase of MKK and IKK. *Nature* **412**, 346–351.
19. Silverman, N. and Maniatis, T. (2001). NF-kappaB signaling pathways in mammalian and insect innate immunity. *Genes Dev.* **15**, 2321–2342.
20. Kawai, T., Takeuchi, O., Fujita, T., Inoue, J., Muhlradt, P. F., Sato, S., Hoshino, K., and Akira, S. (2001). Lipopolysaccharide stimulates the MyD88-independent pathway and results in activation of IFN-regulatory factor 3 and the expression of a subset of lipopolysaccharide-inducible genes. *J. Immunol.* **167**, 5887–5894.
21. Fitzgerald, K. A., Palsson-McDermott, E. M., Bowie, A. G., Jefferies, C. A., Mansell, A. S., Brady, G., Brint, E., Dunne, A., Gray, P., and Harte, M. T. *et al.* (2001). Mal (MyD88-adapter-like) is required for Toll-like receptor-4 signal transduction. *Nature* **413**, 78–83.
22. Horng, T., Barton, G. M., and Medzhitov, R. (2001). TIRAP: an adapter molecule in the Toll signaling pathway. *Nat. Immunol.* **2**, 835–841.
23. Arbibe, L., Mira, J., Teusch, N., Kline, L., Guha, M., Mackman, N., Godowski, P., Ulevitch, R., and Knaus, U. (2000). Toll-like receptor 2-mediated NF-κB activation requires a Rac1-dependent pathway. *Nat. Immunol.* **1**, 533–540.
24. Chin, A. I., Dempsey, P. W., Bruhn, K., Miller, J. F., Xu, Y., and Cheng, G. (2002). Involvement of receptor-interacting protein 2 in innate and adaptive immune responses. *Nature* **416**, 190–194.
25. Kobayashi, K., Inohara, N., Hernandez, L. D., Galan, J. E., Nunez, G., Janeway, C. A., Medzhitov, R., and Flavell, R. A. (2002). RICK/Rip2/CARDIAK mediates signalling for receptors of the innate and adaptive immune systems. *Nature* **416**, 194–199.

CHAPTER 180

Signaling and the Immunological Synapse

Andrey S. Shaw
*Department of Pathology and Immunology,
Washington University School of Medicine,
Saint Louis, Missouri*

Introduction

Although considerable progress has been achieved over the last decade in the field of T-cell activation, many interesting questions remain. We now understand the basic biochemical pathways that underlie the basis of T-cell receptor (TCR) signaling, but we don't know how these signals are used to regulate T-cell function. For example, how does the T cell recognize antigens with such high sensitivity and specificity? How are these signals modulated to mediate such diverse processes as T-cell development, T-cell survival, and T-cell effector functions? Understanding these processes will require new approaches that incorporate principles of biochemistry, cell biology, immunology, and systems biology. Recent work using state of the art methods of digital imaging have revealed a process of membrane protein rearrangement known as immunological synapse formation. Immunological synapse formation is intimately associated with the T-cell activation process. Here, I review recent progress in this field (also recently reviewed in [1–4]). Because the study of immunological synapse addresses issues pertaining mainly to the field of cellular immunology, and to put a discussion of immune synapses into a broader context, a brief summary of T cell biology is given first.

Brief Introduction to T Cell Biology

One of the central challenges today is to understand how the TCR can specifically and sensitively sense foreign antigens. The TCR binds to short peptide fragments that are themselves bound to membrane proteins known as major histocompatibility antigens (MHC). These short peptide fragments are generated by antigen-presenting cells via the proteolytic degradation of all proteins (intracellular and extracellular) in the cellular milieu. Thus, at any given time, a large variety of peptides compete for binding to a limited number of MHC molecules expressed on the surface of cells. The ability of the T cell to discern a rare antigenic peptide from an ocean of other, nonrelevant peptides attests to the incredible accuracy of TCR sensitivity. Recent studies verify this sensitivity by demonstrating that for CD8+ T cells, even a single peptide-MHC molecule is sufficient to activate a T cell [5]. The specificity of TCR recognition is also impressive. Conservative changes in the peptide sequence can change a peptide from being a strong agonist to one that is ignored by the T cell [6].

The trafficking of naïve T cells and professional antigen-presenting cells (APC) is designed so that both cells can interact with each other in secondary lymphoid organs like the lymph node or spleen. Although professional APCs include macrophages and B cells, the most important professional APC is the dendritic cell. Professional APCs function to not only carry foreign proteins from the periphery to the secondary lymphoid organs, but each also have specialized roles in the activation of T cells. The expression of membrane proteins like B7 on professional APCs functions to control the activation of naïve T cells via interaction with co-stimulator molecules such as CD28 expressed on the surface of naïve T cells.

Activated naïve T cells proliferate and differentiate into effector T cells. In the CD4+ T cell lineage, the most important effector cells are known as Th1 and Th2 cells and can be distinguished by the cytokines that they secrete. Th1 cells secrete IL-2 and γ-interferon and mediate delayed-type hypersensitivity (DTH) responses by activating macrophages. Th2 cells secrete IL-4 and function to stimulate humoral immune responses by controlling B-cell activation and immunoglobulin switching. Similar pathways operate in the differentiation of cells in the CD8 lineage with the development of Tc1 and Tc2 cells. The primary function of CD8 cells is to kill infected target cells. Finally, T cells differentiate into memory T cells, long-lived cells that enhance secondary responses to pathogens.

Initiation of TCR Signaling

How T-cell receptors are initiated by binding to antigenic ligands is still a controversial area [7]. Although receptor clustering is the trigger that initiates signaling in most signaling systems, the small numbers of antigenic ligands present on the surface of the APC *in vivo* and the ability of very small numbers of peptide/MHC molecules to activate a T cell makes receptor clustering an unlikely mechanism. A recent study suggests that ligand binding results in a conformational change that promotes the assembly of signaling molecules with the TCR [8], but such changes have not been seen by structural studies [9]. Another model proposes that the association of accessory molecules with the TCR may be the trigger for T cell activation [10]. Higher-affinity binding would be discerned from low-affinity binding because longer-lived complexes would allow accessory molecules to be recruited to the TCR. The ability of T cells, however, to become activated in the absence of accessory and co-stimulatory molecules suggests that this mechanism is not required for signal initiation [11]. Finally, it has been proposed that T-cell receptors may form microclusters [12,13]. This is supported by structural studies demonstrating that MHC molecules, as well as accessory molecules such as CD4 and CD28, can form dimers [14], but these studies are controversial [15]. One problem with this model is that given the low abundance of antigenic peptides on the surface of the APC, the chances that both halves of an MHC dimer would contain the same peptide are extremely unlikely. But it has also been shown that non-agonist peptides may also play a role in triggering T-cell activation [16]. Thus, it is possible that TCR dimerization could be achieved by combinations of antigenic and non-antigenic ligands.

Definition of the Immunological Synapse

Kupfer and coworkers were the first to demonstrate a specific reorganization of proteins in the contact area between the T cell and the APC [17]. Using digital reconstruction of confocal microscopic images, they noted that T-cell membrane proteins segregated into a bull's-eye pattern during T-cell activation. The center of the bull's-eye, dubbed the C-SMAC, is notable for the clustering of T-cell receptor, PKC-theta, and CD28. The outer ring consists of a zone containing the adhesion molecule, lymphocyte function-associated antigen (LFA-1). In a third zone, proteins excluded from the C-SMAC and P-SMAC, such as CD43 and CD45, are found. Recent work suggests that at least for CD43, there is an active process that targets it outside of the immunological synapse [18–20]. This pattern of a peripheral ring of LFA-1 surrounding a central zone of T-cell receptors is referred to as a classical or mature synapse.

Excitement about this concept and its rapid adoption by those studying T-cell activation is based on the possibility that the morphology of synapses may give important clues about how interactions with APCs lead to distinct T-cell responses (Fig. 1). Since interactions with other cells are fundamental to the T-cell activation process, one possibility is that different synapse morphologies may lead to different outcomes. Thus, recent publications have analyzed synapses during T-cell development [21,22], the activation of Th1 or Th2 cells [23], the synapses of CD8 [24,25] and NK cells [26–30], and those formed using different types of APCs [12,31].

Initial experiments were performed using CD4+ T cell clones and B-cell tumor lines as APCs [17]. These experiments showed that formation of a classical synapse occurs quickly, usually within minutes. Many other groups obtained similar results with T-cell hybridomas and T-cell tumor lines. Using fluorochrome-labeled MHC and adhesion molecules embedded in a planar lipid bilayer, Grakoui *et al.* were able to study the kinetics of synapse formation. Their studies suggested a distinct pattern that preceded mature synapse formation [32]. At early time points, TCRs were first recruited to the periphery of the contact, surrounding a central zone of LFA-1/ICAM complexes. This pattern, the opposite of that seen in the classical synapse has been referred to as an immature synapse. Over the next twenty minutes, this pattern inverts with movement of TCRs into the center of the synapse now surrounded by LFA-1 at the periphery of the contact.

Because mature synapses form very rapidly using T cell clones and cell lines, and the immature pattern is almost never seen, there was some question about the validity of the immature synapse pattern. However, the two-step formation of the immunological synapse was recently confirmed using naïve T cell/APC conjugates [31]. In this system, naïve T cells were incubated with primary APCs (T and B cell depleted splenocytes). Synapses formed very slowly, with the immature pattern achieved between 5–15 minutes after cell contact followed by the pattern of the mature synapse between 15–30 minutes. Because naïve T cells form synapses on lipid bilayers with kinetics similar to T cell blasts [33], the APC is likely to control the rate and pattern of synapse formation. In fact Richie *et al.* showed this directly when they tested the morphologies of synapses formed using blasted T cells or thymocytes with a variety of different APCs [21]. This suggests that synapse formation is strongly influenced by the APC.

Live-cell imaging studies further suggest that interactions with dendritic cells may be distinct from the interactions with other antigen presenting cells [34]. When T cells were

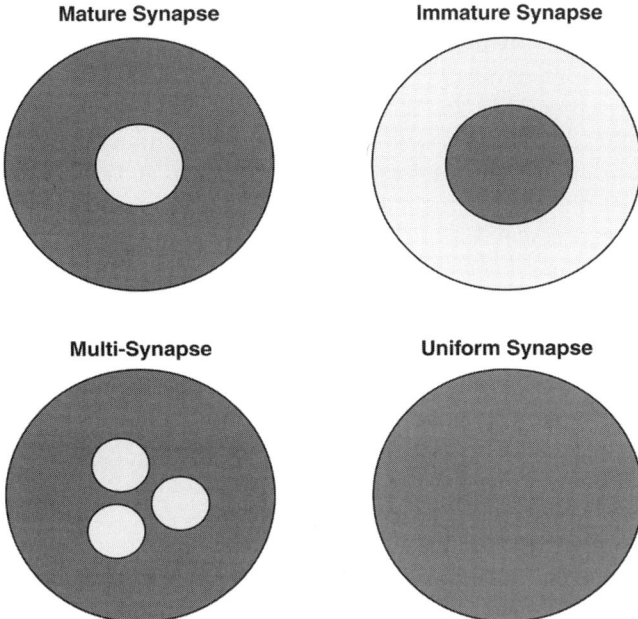

Figure 1 Basic synapse morphologies. Depicted are four basic synapse morphologies. The classical synapse has a central zone known as a C-SMAC surrounded by a peripheral ring of LFA-1 called a P-SMAC. The immature synapse is seen in some systems before the classical synapse. It is also seen in thymocytes undergoing negative selection. The multisynapse consists of multiple clusters of TCRs that are dynamic and stable for long periods. The uniform synapse represents contact areas where there is no visible segregation of membrane protein components. The significance of such synapses is not clear.

incubated with dendritic cells in a collagen matrix, T cell interactions with dendritic cells were transient (minutes) suggesting that a sustained interaction with an antigen-presenting cell might not be required for an activation response. T cells can also from synapses with dendritic cells in the absence of antigenic peptide [35]. But two recent studies imaging T cell/dendritic cell interactions in lymph nodes gave conflicting results. Stoll et al. injected fluorescently labeled dendritic cells and T cells into a mouse and imaged movement of T cells in a lymph node using standard one-photon confocal microscopy [36]. They found that T cells formed long-lived stable contacts for about 20 hours after which the T cells become highly motile and started to proliferate. In contrast, Miller et al. using two-photon fluorescent microscopy, found instead that the vast majority of antigen-specific T cells were highly motile in the lymph node [37]. T cells were described to be "swarming". Whether differences in the technique of microscopy or immunization procedure account for this difference is unclear. Clearly, this is an area that will require further investigation.

Another active area of interest has been the morphology of synapses formed with APCs displaying altered peptide ligands (APLs). APLs are modified antigenic peptides, in which single amino acids have been changed. These changes can result in a variety of different effects ranging from changing the peptide to a null, or completely non-antigenic peptide, to converting the peptide into an antagonist to having no effect at all. Early interest in the synapse focused on the morphology of synapses formed with APLs. Grakoui et al. found that weak agonist peptides still formed mature immunological synapses but accumulated lower levels of MHC-peptide complexes [32]. This suggests that MHC-peptide accumulation is related to the strength of the antigen. Antagonist peptides were able to accumulate MHC-peptide complexes in the junction between the T cell and the APC, however, a C-SMAC did not form and the T cells were unable to stop moving. Consequently, synapses were short lived. In a more recent study, Zal et al. used energy transfer techniques to analyze the interaction of the co-receptor, CD4, with the TCR after stimulation with agonist or antagonist peptides [38]. They found that CD4 and the TCR interact in the immunological synapse when stimulated with agonist. Antagonist peptides, interestingly, still recruited CD4 and TCR to the complex, but an interaction between the two never occurred. This is likely related to the shorter half-lives of antagonist peptide/MHC complexes with the TCR preventing association of CD4 with the TCR.

Immunological Synapses and T-Cell Development

During development, T cells must be "educated" to distinguish between self versus foreign antigens. It is thought that T cells learn first to recognize self-MHC in a process known as positive selection. Only T cells that are able to bind to self-MHC molecules are allowed to survive as well as expand. To rid this immature T-cell population of potentially self-reactive, autoimmune cells, cells that react strongly to self-MHC are deleted in a process known as negative selection. This results, at the end of development, in T cells that are weakly reactive against self and that are strongly reactive against foreign antigens. This property is important, for the low self-reactivity is required for the long-term survival of the T-cell population. The foreign reactivity results in the activation, proliferation, and differentiation of T cells into effector T cells. Are the processes of positive and negative selection distinct at the level of the immunological synapse?

Davis and coworkers developed an *in vitro* system to study thymocyte-positive and negative selection and were able to image contacts between thymocytes and thymic epithelial cells during negative selection [21]. They found that thymocytes undergoing negative selection form a variety of synaptic patterns that is most notable by the absence of a C-SMAC. For the most part, the synapses of negative selection are long-lived and are reminiscent of the immature synapse with peripheral localization of the TCR. This pattern was unique to synapses with thymic epithelial cell, as synapses using thymocytes with other types of antigen presenting cells had different morphologies. This suggests that the antigen-presenting cell plays a key role in synapse formation. Hailman et al., using a planar bilayer system, also noted a distinct synapse pattern for negative selection [39]. They found that the synapses of thymocytes undergoing negative selection were dynamic, long-lived, and notable for sustained TCR signaling.

The morphology was different from the Richie *et al.* study [21] with synapses showing multiple clusters of TCRs. But common to both studies is the lack of a single, central C-SMAC and the long-lived nature of the synapses. Neither study was able to study positive selection because of inefficient conjugate formation. However, Buosso *et al.* recently imaged cells undergoing positive selection by using two-photon microscopy and a three dimensional organ culture system [22]. They found cell-cell interactions were both stable and highly dynamic. Clearly much more work is necessary to understand whether there is a relationship among cell contacts, synapse morphology and T-cell development.

Synapses and Different Kinds of T Cells

Understanding T cell activation is further complicated by the fact that different types of T cells have different activation requirements. Naïve T cells have relatively high thresholds for T-cell activation, require secondary signals mediated by receptors classified as co-stimulators, and can only be activated by professional antigen-presenting cells in secondary lymphoid organs such as lymph node and spleen. Memory T cells, in contrast, have much lower thresholds for T-cell activation, do not require co-stimulation, and can be activated outside of secondary lymphoid organs. The activation requirements for CD4 T cells versus CD8 T cells also appear to be distinct. Although CD4 T cells appear to require sustained periods of TCR engagement, CD8 T cells can apparently make the decision to fire or not within minutes. Whereas CD4 activation requires 100–1000 peptide-MHC complexes [40], CD8 activation requires only a single peptide-MHC complex [5,41]. Important subgroups of CD4 T cells, the so-called TH1 and TH2 T cells, also appear to have distinct signaling requirements.

Although several groups have imaged synapses of both naïve and previously activated T-cell blasts, there is relatively little analysis about any possible differences. This issue is complicated by the fact that a careful analysis will also require use of a variety of types of antigen-presenting cells. This requirement is critical because *in vivo*, different types of T cells are stimulated by different types of APCs. Both naïve and blasted T cells can form immature and classical synapses, but the kinetics may be slower in naïve T cells [31]. The specific differences between the two will need to be examined closely in the future.

One area that has seen progress in recent years is the nature of CD8+ T-cell synapses. This is an interesting topic because CD8 T-cell activation occurs very quickly and can be stimulated by a single peptide-MHC complex. In addition, multiple targets can be killed within a 10 to 20 minute period. Kupfer and coworkers demonstrated that naïve CD8 T cells form classical synapses even at extremely low levels of peptide-MHC complex [24]. These synapses have a normal P-SMAC but surprisingly have no detectable enrichment of TCRs in the C-SMAC at very low levels of antigen. The C-SMAC, however, can be identified by the recruitment and concentration of PKC-theta in a zone that is central to the P-SMAC. At higher concentrations of antigen, however, visible recruitment of the TCR is seen. This suggests that recruitment and enrichment of the TCR in the C-SMAC is not required for T-cell activation, at least in CD8 cells. A beautiful high-resolution study by Griffiths and coworkers confirmed these findings and also demonstrated that the contact contains a small cleft into which secretory granules empty their contents [25]. This suggests that one important function of the synapse is to help form a target zone for the secretion of cytolytic granules. Electron microscopic studies demonstrated that there is fusion between the plasma membranes of the T cell and the target cell in the synapse. Combined with the tight adhesion of the P-SMAC, the synapse might be important in preventing the spillage of cytolytic proteins into the extracellular space.

Natural Killer Cell Synapses

Three groups have also imaged synapses formed by natural killer (NK) cells and their targets [26,27,39,30,42]. This is an interesting area because NK cells like CD8+ cells must quickly make a decision to kill or not kill a target cell. Unlike CD8+ cells, this decision is not solely dependent on antigen but is negatively regulated by recognition of normal membrane proteins. Thus, the NK cells must integrate information from both stimulatory and inhibitory receptors before they can make a decision to fire or not. First, Leibson and coworkers imaged the recruitment of lipid rafts to synapses in the presence of absence of inhibitory signaling [42]. They found that rafts were recruited when a target cell was killed but that inhibitory signaling blocked raft recruitment. The Strominger and Dupont groups have analyzed these cytolytic and noncytolytic synapses in greater detail. They found that in cytolytic synapses, NK cells formed a P-SMAC as defined as an outer ring of LFA-1 and talin and a C-SMAC as defined by the central accumulation of PKC-theta and other signaling molecules [26,27]. In noncytolytic synapses, while a contact surface is generated, no clearly defined P-SMAC and C-SMAC can be seen, nor is there recruitment of lipid rafts [29]. This is an active process because addition of an inhibitory ligand to the system results in the loss of receptor segregation in the contact area as well as lipid raft recruitment. Thus, in this system, classical synapse formation appears to be linked to not only to antigen recognition but also to the absence of inhibitory receptor signaling.

The Function of the Immunological Synapse

Early studies proposed that the synapse functioned to initiate T-cell receptor signal transduction [32]. This was based on the fact that T-cell receptors were clustered and concentrated in the C-SMAC. In addition, it was noted that lipid rafts, a rich source of signaling molecules, also clustered in the synapse [43]. Thus it seemed logical that the C-SMAC might function as a way to crosslink and aggregate T-cell receptors. Although early studies suggested that calcium signaling might precede formation of immunological synapse, in many systems, especially those using T-cell clones and T-cell hybridomas, the formation of the immunological synapse occurred quickly, within minutes of contact formation. Thus it was difficult to say definitively whether synapse formation preceded or followed the initiation of signal transduction.

Support for the idea that the C-SMAC serves is an active area of signal transduction was the finding that lipid rafts are enriched in immunological synapses [43]. Since lipid rafts are a rich source of signaling components [44], it was logical to assume that formation of the C-SMAC accompanied by lipid raft recruitment would form a strong trigger and an amplifier of TCR signal transduction. Burack et al. were the first to show that lipid rafts are enriched in the C-SMAC and depleted in the P-SMAC [45]. However, the level of raft enrichment is low, about two- to three-fold over basal levels and likely to be due to displacement from the P-SMAC rather than active recruitment from other areas of the plasma membrane.

As discussed above, Lee et al., using naïve T cells and primary antigen-presenting cells, were able to demonstrate that synapse formation occurs in two distinct phases [31]. Up to 15 minutes after conjugate formation, the synapses have an immature morphology; TCRs are found mainly at the periphery of the contact, surrounding a central zone of LFA-1/ICAM-1. Between 15 and 30 minutes, the zones invert, with T-cell receptors moving to the center of the contact and LFA-1/ICAM-1 complexes moving to the periphery. The slow development of the mature synapse in this system allowed a careful analysis of the kinetics of TCR signal transduction via phospho-specific antibodies. The activation of LCK and ZAP-70 occurred in the immature synapse and had largely abated by the time the mature synapse had formed. Not only does this suggest that immunological synapse formation is not involved in initiating T-cell receptor signaling, it also suggests that large-scale aggregation of T-cell receptors is not required to initiate T cell receptor signaling. This in turn suggests that TCR signaling is initiated before mature synapse formation.

Work from Krummel et al. examining the recruitment of Lck and CD4 to the synapse is consistent with these findings [46]. They found that CD4 and CD3 zeta, components of the TCR, were both recruited rapidly to the contact area. Their recruitment was coincident with the initiation of calcium signaling. Although CD3 zeta was recruited to the C-SMAC, CD4 surprisingly moved out of the contact area. This result is consistent with the idea that the initiation of TCR signaling does not require the C-SMAC, nor does a significant portion of signaling occur there, and suggests that the C-SMAC is involved in some other biological activity occurring between T cell/ APC conjugates.

Perhaps the immunological synapse functions to sustain T-cell receptor signaling. Several groups have now showed that sustained contact for over two hours is required before a naïve T cell is committed to proliferate [3,47]. The synapse might function, therefore, to help sustain signaling. What signals are required for two hours remains unclear, however. Studies using phospho-specific antibodies suggest that TCR signaling has largely abated by 30 minutes after contact formation [31]. This scenario does not rule out that low-level TCR signaling continues at levels below the level of detection. It is also possible that the immunological synapse may function to facilitate engagement of receptors in addition to the TCR. Important candidates include co-stimulatory molecules as well as cytokine receptors [48].

The engagement of co-stimulatory molecules such as CD28 is required for the activation of naïve T cells, whose sustained signaling may be required in the synapse. The ligand for CD28, B7, is expressed on professional antigen-presenting cells, and CD28/B7 complexes are recruited to the C-SMAC [33]. Although the exact signals transduced by CD28 are still controversial, activation almost certainly will involve more than one signaling pathway. CD28 functions by directly potentiating TCR signaling, enhancing cell survival, and increasing the metabolism of the T cells. One important signaling pathway mediated by CD28 is the PI-3 kinase pathway, and prolonged signaling by the PI-3 kinase/AKT pathways may be required for T-cell commitment [49]. It is also interesting to note that other co-stimulatory molecules, such as CTLA-4 and ICOS, are upregulated hours after the initiation of TCR signaling [50,51]. This finding emphasizes the important role for prolonged co-stimulatory signaling after initiation of TCR signaling. In addition, long-term stability of synapses may play roles in regulating the differentiation of naïve T cells into effector T cells. Cytokine released by the APCs may be more effective if the cells are attached to each other [48]. Thus, it seems logical to consider that the persistence of the immunological synapse would enhance engagement and signal transduction by these accessory molecules.

Immunological Synapses and TCR Downregulation

Because a few peptide MHC molecules are able to downregulate thousands of T-cell receptors, it has been proposed that T-cell activation occurs via the serial triggering of T-cell receptor molecules [52,53]. The immunological synapse might therefore serve to allow multiple receptors to be engaged by the same few MHC peptide molecules clustered into the synapse. Although a provocative model, the serial triggering model is still controversial. First, it has been shown that T-cell receptors can internalize their MHC peptide ligands in the APC plasma membrane [54]. If this occurs in vivo, it is difficult to explain how antigen can be maintained in the immunological synapse for many hours. Long-term stability of the peptide-MHC complex in the synapse is also complicated by the half-life of peptide/MHC complexes; sustained signaling could not be maintained for 20 hours if the stability of the peptide-MHC complex is much shorter than this. Finally, it has been shown that

T cells expressing engineered receptors that have very high affinities and that should therefore limit serial engagement are still able to efficiently activate T cells [55]. Thus, whether serial engagement plays an important role in T-cell activation is not currently known.

Still, the formation of a classical immunological synapse might function to direct the internalization and degradation of engaged TCR complexes. In resting T cells, the TCR is recycled continuously from the surface to an early endosomal compartment and back [56]. During activation, TCRs are internalized and instead of being recycled to the plasma membrane are directed for degradation in lysosomes [57,58]. Since the C-SMAC has been shown to be the spot where secretory granules are inserted into the membrane [25], it is likely that it is also the spot where membranes are retrieved. To maintain membrane homeostasis, much of the inserted membrane will need to be retrieved. Thus, it seems possible that the C-SMAC denotes an area of the T-cell plasma membrane where membrane dynamics are particularly active.

Conclusion

More than a contact surface, the immunological synapse is a specific arrangement of proteins in the contact area. Many different patterns of the immunological synapse have been discovered, and it is suggested that these patterns are related to the biological outcome of T-cell activation. Although exactly what determines the morphology of the immunological synapse remains unknown, it almost certainly is due to differences in the antigen-presenting cell as well as differences in the particular T cell studied. It will be important as this field grows to analyze this phenomenon in greater detail. Although the immunological synapse was first proposed to function to initiate T-cell signaling, it is clear now that TCR signaling can occur in the absence of organized synapse formation. Rather, the formation of the immunological synapse appears to be a consequence of activation rather than a requirement for activation. The jury is still out on whether the synapse functions to sustain TCR signaling, but it seems very likely that the synapse is required for several hours to commit the naïve T cell to enter cell cycle. Although much remains to be learned, all can agree that the immunological synapse does represent the polarization of the T cell, a process that is required to allow T-cell effector functions to be accomplished—be that cytolysis or cytokine release. It seems likely, given the rapid progress in this field, that much more will be known about immunological synapses in the very near future.

References

1. Delon, J. and Germain, R. N. (2000). Information transfer at the immunological synapse. *Curr. Biol.* **10**, R923–933.
2. Krummel, M. F. and Davis, M. M. (2002). Dynamics of the immunological synapse: finding, establishing and solidifying a connection. *Curr. Opin. Immunol.* **14**, 66–74.
3. Lanzavecchia, A. and Sallusto, F. (2000). From synapses to immunological memory: the role of sustained T cell stimulation. *Curr. Opin. Immunol.* **12**, 92–98.
4. Lanzavecchia, A. and Sallusto, F. (2001). Antigen decoding by T lymphocytes: from synapses to fate determination. *Nat. Immunol.* **2**, 487–492.
5. Sykulev, Y., Joc, M., Vturina, I., Tsomides, T. J., and Eisen, H. N. (1996). Evidence that a single peptide-MHC complex on a target cell can elicit a cytolytic T cell response. *Immunity* **4**, 565–571.
6. Evavold, B. D., Sloan-Lancaster, J., and Allen, P. M. (1993). Tickling the TCR: selective T-cell functions stimulated by altered peptide ligands. *Immunol. Today* **14**, 602–609.
7. van der Merwe, P. A. (2001). The TCR triggering puzzle. *Immunity* **14**, 665–668.
8. Gil, D., Schamel, W. W., Montoya, M., Sanchez-Madrid, F., and Alarcon, B. (2002). Recruitment of Nck by CD3epsilon reveals a ligand-induced conformational change essential for T cell receptor signaling and synapse formation. *Cell* **109**, 901–912.
9. Garcia, K. C., Teyton, L., and Wilson, I. A. (1999). Structural basis of T cell recognition. *Annu. Rev. Immunol.* **17**, 369–397.
10. Holdorf, A. D., Lee, K. H., Burack, W. R., Allen, P. M., and Shaw, A. S. (2002). Regulation of Lck activity by CD4 and CD28 in the immunological synapse. *Nat. Immunol.* **3**, 259–264.
11. Shahinian, A., Pfeffer, K., Lee, K. P., Kundig, T. M., Kishihara, K., Wakeham, A., Kawai, K., Ohashi, P. S., Thompson, C. B., and Mak, T. W. (1993). Differential T cell costimulatory requirements in CD28-deficient mice. *Science* **261**, 609–612.
12. Reich, Z., Boniface, J. J., Lyons, D. S., Borochov, N., Wachtel, E, J., and Davis, M. M. (1997). Ligand-specific oligomerization of T-cell receptor molecules. *Nature* **387**, 617–620.
13. Boniface, J. J., Rabinowitz, J. D., Wulfing, C., Hampl, J., Reich, Z., Altman, J. D., Kantor, R. M., Beeson, C., McConnell, H. M., and Davis, M. M. (1998). Initiation of signal transduction through the T cell receptor requires the multivalent engagement of peptide/MHC ligands [corrected]. *Immunity* **9**, 459–466.
14. Schafer, P. H., Pierce, S. K., and Jardetzky, T. S. (1995). The structure of MHC class II: a role for dimer of dimers. *Semin. Immunol.* **7**, 389–398.
15. Baker, B. M. and Wiley, D. C. (2001). Alpha beta T cell receptor ligand-specific oligomerization revisited. *Immunity* **14**, 681–692.
16. Wulfing, C., Sumen, C., Sjaastad, M. D., Wu, L. C., Dustin, M. L., and Davis, M. M. (2002). Costimulation and endogenous MHC ligands contribute to T cell recognition. *Nat. Immunol.* **3**, 42–47.
17. Monks, C. R., Freiberg, B. A., Kupfer, H., Sciaky, N., and Kupfer, A. (1998). Three-dimensional segregation of supramolecular activation clusters in T cells. *Nature* **395**, 82–86.
18. Allenspach, E. J., Cullinan, P., Tong, J., Tang, Q., Tesciuba, A. G., Cannon, J. L., Takahashi, S. M., Morgan, R., Burkhardt, J. K., and Sperling, A. I. (2001). ERM-dependent movement of CD43 defines a novel protein complex distal to the immunological synapse. *Immunity* **15**, 739–750.
19. Delon, J., Kaibuchi, K., and Germain, R. N. (2001). Exclusion of CD43 from the immunological synapse is mediated by phosphorylation-regulated relocation of the cytoskeletal adaptor moesin. *Immunity* **15**, 691–701.
20. Roumier, A., Olivo-Marin, J. C, Arpin, M., Michel, F., Martin, M., Mangeat, P., Acuto, O., Dautry-Varsat, A., and Alcover, A. (2001). The membrane-microfilament linker ezrin is involved in the formation of the immunological synapse and in T cell activation. *Immunity* **15**, 715–728.
21. Richie, L. I., Ebert, P. J., Wu, L. C., Krummel, M. F., Owen, J. J., and Davis, M. M. (2002). Imaging synapse formation during thymocyte selection: inability of CD3zeta to form a stable central accumulation during negative selection. *Immunity* **16**, 595–606.
22. Bousso, P., Bhakta, N. R., Lewis, R. S., and Robey, E. (2002). Dynamics of thymocyte-stromal cell interactions visualized by two-photon microscopy. *Science* **296**, 1876–1880.
23. Balamuth, F., Leitenberg, D., Unternaehrer, J., Mellman, I., and Bottomly, K. (2001). Distinct patterns of membrane microdomain partitioning in Th1 and Th2 cells. *Immunity* **15**, 729–736.

24. Potter, T. A., Grebe, K., Freiberg, B. A., and Kupfer, A. (2001). Formation of supramolecular activation clusters on fresh ex vivo CD8+ T cells after engagement of the T cell antigen receptor and CD8 by antigen-presenting cells. *Proc. Natl. Acad. Sci. USA* **98**, 12624–12629.
25. Stinchcombe, J. C., Bossi, G., Booth, S., and Griffiths, G. M. (2001). The immunological synapse of CTL contains a secretory domain and membrane bridges. *Immunity* **15**, 751–761.
26. Vyas, Y. M., Mehta, K. M., Morgan, M., Maniar, H., Butros, L., Jung, S., Burkhardt, J. K., and Dupont, B. (2001). Spatial organization of signal transduction molecules in the NK cell immune synapses during MHC class I-regulated noncytolytic and cytolytic interactions. *J. Immunol.* **167**, 4358–4367.
27. Vyas, Y. M., Maniar, H., and Dupont, B. (2002). Cutting edge: differential segregation of the SRC homology 2-containing protein tyrosine phosphatase-1 within the early NK cell immune synapse distinguishes noncytolytic from cytolytic interactions. *J. Immunol.* **168**, 3150–3154.
28. Davis, D. M. (2002). Assembly of the immunological synapse for T cells and NK cells. *Trends Immunol.* **23**, 356–363.
29. Fassett, M. S., Davis, D. M., Valter, M. M., Cohen, G. B., and Strominger, J. L. (2001). Signaling at the inhibitory natural killer cell immune synapse regulates lipid raft polarization but not class I MHC clustering. *Proc. Natl. Acad. Sci. USA* **98**, 14547–14552.
30. Davis, D. M., Chiu, I., Fassett, M., Cohen, G. B., Mandelboim, O., and Strominger, J. L. (1999). The human natural killer cell immune synapse. *Proc. Natl. Acad. Sci. USA* **96**, 15062–15067.
31. Lee, K. H., Holdorf, A. D., Dustin, M. L., Chan, A. C., Allen, P. M., and Shaw, A. S. (2002). T cell receptor signaling precedes immunological synapse formation. *Science.* **295**, 1539–1542.
32. Grakoui, A., Bromley, S. K., Sumen, C., Davis, M. M., Shaw, A. S., Allen, P. M., and Dustin, M. L. (1999). The immunological synapse: a molecular machine controlling T cell activation. *Science* **285**, 221–227.
33. Bromley, S. K., Iaboni, A., Davis, S. J., Whitty, A., Green, J. M., Shaw, A. S., Weiss, A., and Dustin, M. L. (2001). The immunological synapse and CD28-CD80 interactions. *Nat. Immunol.* **2**, 1159–1166.
34. Gunzer, M., Schafer, A., Borgmann, S., Grabbe, S., Zanker, K. S., Brocker, E. B., Kampgen, E., and Friedl, P. (2000). Antigen presentation in extracellular matrix: interactions of T cells with dendritic cells are dynamic, short lived, and sequential. *Immunity* **13**, 323–332.
35. Revy, P., Sospedra, M., Barbour, B., and Trautmann, A. (2001). Functional antigen-independent synapses formed between T cells and dendritic cells. *Nat. Immunol.* **2**, 925–931.
36. Stoll, S., Delon, J., Brotz, T. M., and Germain, R. N. (2002). Dynamic imaging of T cell-dendritic cell interactions in lymph nodes. *Science* **296**, 1873–1876.
37. Miller, M. J., Wei, S. H., Parker, I., and Cahalan, M. D. (2002). Two-photon imaging of lymphocyte motility and antigen response in intact lymph node. *Science* **296**, 1869–1873.
38. Zal, T., Zal, M. A., and Gascoigne, N. R. (2002). Inhibition of T cell receptor-coreceptor interactions by antagonist ligands visualized by live FRET imaging of the T-hybridoma immunological synapse. *Immunity* **16**, 521–534.
39. Hailman, E., Burack, W. R., Shaw, A. S., Dustin, M. L., and Allen, P. M. (2002). Immature CD4(+)CD8(+) thymocytes form a multifocal immunological synapse with sustained tyrosine phosphorylation. *Immunity* **16**, 839–848.
40. Harding, C. V. and Unanue, E. R. (1990). Quantitation of antigen-presenting cell MHC class II/peptide complexes necessary for T-cell stimulation. *Nature* **346**, 574–576.
41. Delon, J., Gregoire, C., Malissen, B., Darche, S., Lemaltre, F., Kourilsky, P., Abastado, J.-P., and Trautmann, A. (1998). CD8 expression allows T cell signaling by monomeric peptide-MHC complexes. *Immunity* **9**, 467–473.
42. Lou, Z., Jevremovic, D., Billadeau, D. D., and Leibson, P. J. (2000). A balance between positive and negative signals in cytotoxic lymphocytes regulates the polarization of lipid rafts during the development of cell-mediated killing. *J. Exp. Med.* **191**, 347–354.
43. Viola, A., Schroeder, S., Sakakibara, Y., and Lanzavecchia, A. (1999). T lymphocyte costimulation mediated by reorganization of membrane microdomains. *Science* **283**, 680–682.
44. Simons, K. and Toomre, D. (2000). Lipid rafts and signal transduction. *Nat. Rev. Mol. Cell Biol.* **1**, 31–39.
45. Burack, W. R., Lee, K. H., Holdorf, A. D., Dustin, M. L., and Shaw, A. S. (2002). Cutting edge: quantitative imaging of raft accumulation in the immunological synapse. *J. Immunol.* In press.
46. Krummel, M. F., Sjaastad, M. D., Wulfing, C., and Davis, M. M. (2000). Differential clustering of CD4 and CD3zeta during T cell recognition. *Science* **289**, 1349–1352.
47. van Stipdonk, M. J., Lemmens, E. E., and Schoenberger, S. P. (2001). Naive CTLs require a single brief period of antigenic stimulation for clonal expansion and differentiation. *Nat. Immunol.* **2**, 423–429.
48. Miyamoto, S., Teramoto, H., Gutkind, J. S., and Yamada, K. M. (1996). Integrins can collaborate with growth factors for phosphorylation of receptor tyrosine kinases and MAP kinase activation: roles of integrin aggregation and occupancy of receptors. *J. Cell Biol.* **135**, 1633–1642.
49. Rudd, C. E. (1996). Upstream-downstream: CD28 cosignaling pathways and T cell function. *Immunity* **4**, 527–534.
50. Hutloff, A., Dittrich, A. M., Beier, K. C., Eljaschewitsch, B., Kraft, R., Anagnostopoulos, I., and Kroczek, R. A. (1999). ICOS is an inducible T-cell co-stimulator structurally and functionally related to CD28. *Nature* **397**, 263–266.
51. Linsley, P. S., Greene, J. L., Tan, P., Bradshaw, J., Ledbetter, J. A., Anasetti, C., and Damle, N. K. (1992). Coexpression and functional cooperation of CTLA-4 and CD28 on activated T lymphocytes. *J. Exp. Med.* **176**, 1595–1604.
52. Valitutti, S., Muller, S., Cella, M., Padovan, E., and Lanzavecchia, A. (1995). Serial triggering of many T-cell receptors by a few peptide-MHC complexes. *Nature* **375**, 148–151.
53. Viola, A. and Lanzavecchia, A. (1996). T cell activation determined by T cell receptor number and tunable thresholds. *Science* **273**, 104–106.
54. Huang, J. F., Yang, Y., Sepulveda, H., Shi, W., Hwang, I., Peterson, P. A., Jackson, M. R., Sprent, J., and Cai, Z. (1999). TCR-mediated internalization of peptide-MHC complexes acquired by T cells. *Science* **286**, 952–954.
55. Holler, P. D., Lim, A. R., Cho, B. K., Rund, L. A., and Kranz, D. M. (2001). CD8(−) T cell transfectants that express a high affinity T cell receptor exhibit enhanced peptide-dependent activation. *J. Exp. Med.* **194**, 1043–1052.
56. Liu, H., Rhodes, M., Wiest, D. L., and Vignali, D. A. (2000). On the dynamics of TCR:CD3 complex cell surface expression and down-modulation. *Immunity* **13**, 665–675.
57. Alcover, A. and Alarcon, B. (2000). Internalization and intracellular fate of TCR-CD3 complexes. *Crit. Rev. Immunol.* **20**, 325–346.
58. D'Oro, U., Vacchio, M. S., Weissman, A. M., and Ashwell, J. D. (1997). Activation of the Lck tyrosine kinase targets cell surface T cell antigen receptors for lysosomal degradation. *Immunity* **7**, 619–628.

The Ubiquitin-Proteasome System

Mark Hochstrasser
Department of Molecular Biophysics and Biochemistry,
Yale University,
New Haven, Connecticut

Introduction

The ubiquitin-proteasome system provides the major route of degradation for most short-lived intracellular proteins in eukaryotes. Ubiquitin is a extraordinarily well-conserved 76-residue polypeptide that is found either free or covalently joined through its C-terminus to a variety of cytoplasmic, nuclear, and membrane proteins. Although the best defined function of ubiquitin is to direct substrate proteins to their destruction by the 26S proteasome, this is not its only role. In recent years, protein ubiquitination has also been shown to regulate endocytosis and intracellular trafficking of membrane proteins, modify proteins in signal transduction pathways, and alter activity of the ribosome. Moreover, we have become aware of a broader set of structurally related proteins—the *ub*iquitin-*l*ike proteins or Ubls—that are also covalently ligated to and removed from other macromolecules by specific enzymatic pathways. These enzymes are generally similar in both mechanism and sequence to those that act on ubiquitin. The purpose of this short review is to give a general description of the ubiquitin-proteasome system, while highlighting some of the key regulatory pathways that depend on it.

Overview of the Ubiquitin-Proteasome System

Ubiquitin is joined reversibly to proteins by an amide (isopeptide) linkage between the C-terminus of ubiquitin and lysine ε-amino groups of the acceptor proteins. A simplified view of the ubiquitin pathway, which is highly conserved among diverse eukaryotes, is depicted in Fig. 1 [1,2]. The C-terminus of ubiquitin must be activated before it can form isopeptide bonds with other proteins. Initially, ubiquitin is adenylated by the ubiquitin activating enzyme, E1. The activated carbonyl in the ubiquitin~AMP intermediate is then attacked by a sulfhydryl group of the E1 enzyme, yielding an E1-ubiquitin thioester. Ubiquitin is subsequently passed to one of a large number of distinct ubiquitin-conjugating enzymes or E2s to form an E2-ubiquitin thioester. The E2s almost always catalyze substrate ubiquitination in conjunction with a specificity factor known as ubiquitin-protein ligase or E3. For proteolytic substrates, assembly of a multiubiquitin chain(s) on the protein is generally necessary for degradation. Ubiquitinated proteins are in a dynamic state, subject to either further rounds of ubiquitin addition, ubiquitin removal by deubiquitinating enzymes (DUBs), or degradation by the 26S proteasome (Fig. 1). The proteasome specifically recognizes multiubiquitin-protein conjugates; it then unfolds the substrate moiety and degrades it into small peptides, while releasing intact ubiquitin for further rounds of protein tagging.

A series of ubiquitin-related proteins (Ubls) that can be ligated to target molecules has also been uncovered, primarily as a result of genomic-scale DNA sequencing efforts. At present, over a dozen proven or putative Ubls are known [3,4]. Some, such as Rub1/NEDD8 and Smt3/SUMO, are known to make important contributions to cell signaling.

Components of the Ubiquitin Ligation and Deubiquitination Pathways

Ubiquitin-Activating Enzyme (E1) The E1 proteins are well-conserved proteins that are about 100 kD in size and

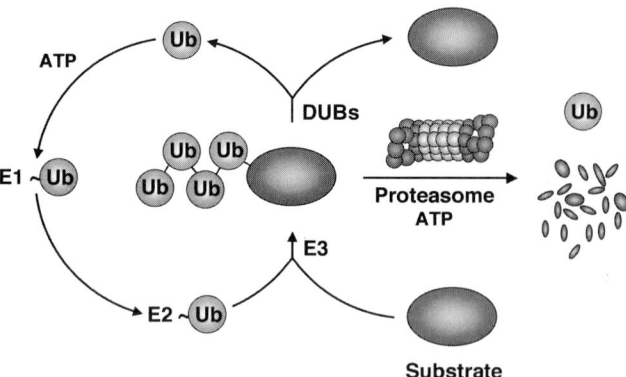

Figure 1 Outline of the ubiquitin-proteasome system. See text for details.

have a recognizable nucleotide-binding motif for ATP binding and a conserved cysteine residue that serves as the site of ubiquitin thioester formation. Most organisms have a single E1-encoding gene, which is essential for survival. No three-dimensional structure for a ubiquitin-activating enzyme has been solved, but a set of structures for an evolutionarily related bacterial protein required for molybdenum cofactor synthesis has been reported recently [5]; the structures suggest a mechanism for ubiquitin adenylation as well.

Ubiquitin-Conjugating Enzyme (E2) There is considerable diversity among the E2 isozymes but they share a core domain of ~120–150 residues with roughly 35 percent sequence identity. A cysteine in this domain accepts ubiquitin from the E1-ubiquitin thioester in a transthiolation reaction. Multiple E2 crystal structures have been reported. Different E2s are specifically required with the ubiquitination of particular substrates. This requirement appears to reflect both specific E2 association with a subset of E3 ligases and direct E2 contacts with the substrate [1].

Ubiquitin-Protein Ligase (E3) The E3s generally make the greatest contribution to substrate recognition and are usually essential for efficient ubiquitin transfer. They can be either single polypeptides or multisubunit complexes. Two mechanistically and structurally distinct classes of E3s have been reported to date. One class appears to function primarily as an adaptor between E2 and substrate, and the most recent structural data are consistent with this view [6]. All E3s of this mechanistic class share a common structural feature, namely, a RING domain. The RING motif utilizes a characteristic arrangement of Cys and His residues to coordinate two zinc ions, and this structural domain can bind directly to E2s. (In a structurally related motif, called the U-box, the same RING-like fold is predicted, but without the coordination of zinc ions; several E3s with U-boxes have been reported [7].) A distinct domain or subunit of the E3 associates with substrate, and the E2-E3-substrate ternary complex is thought to optimize the position or orientation of the E2-linked ubiquitin for attack by a substrate lysine.

In the second class of E3s, an additional ubiquitin transthiolation occurs, in this case between the E2 and E3, and the ubiquitin is then finally transferred from the E3 thiol to the substrate lysine. All E3s of this type bear a conserved ~350-residue stretch called the HECT domain, the key feature of which is the conserved cysteine that forms the thioester with ubiquitin. A crystal structure of a HECT E3 ligase-E2 complex is available, but the relevant cysteines in the E2 and E3 are over 40 Å apart, so very little can be inferred about how ubiquitin is transferred between the two thiols [8]. Although there are many fewer HECT E3s than RING E3s, it was a mutation in a HECT E3, the E6-AP gene mutated in Angelman's Syndrome, that provided the first direct link between the ubiquitin system and a heritable human disease [9].

Deubiquitinating Enzymes (DUBs) In terms of individual physiological functions, most DUBs remain enigmatic. Nevertheless, both the broad requirements for such enzymes and some of their basic molecular features are fairly well understood [10,11]. DUBs are required first of all to process the precursor forms of ubiquitin insofar as all ubiquitin genes encode either head-to-tail fusions of multiple ubiquitin moieties or ubiquitin sequences fused to ribosomal peptides. Because ubiquitin ligation involves intermediates of ubiquitin susceptible to nucleophilic attack, adventitious conjugation to abundant intracellular nucleophiles such as lysine or glutathione is believed to be common, and this must be reversed by DUBs to maintain adequate free ubiquitin pools. DUBs also can negatively regulate ubiquitin-dependent processes by removing the protein modification; for example, they can prevent or limit ubiquitin-dependent proteolysis by the proteasome. Finally, an important positive regulatory function of DUBs is the recycling of ubiquitin from proteins following their commitment to proteolysis by either the proteasome or lysosome/vacuole [10,11].

There are two known classes of DUBs: the ubiquitin C-terminal hydrolases (UCHs) and the ubiquitin-specific processing proteases (UBPs). Both are specialized cysteine proteases. The UCHs, which are usually <40 kD in size, are the less common class, and generally can only cleave ubiquitin from small adducts or disordered protein segments. This substrate constraint reflects the presence of a loop in the UCH enzymes that lies over the active site; segments of the ubiquitin-substrate conjugate may need to be threaded through this loop for efficient cleavage [12]. Less is known about the structure and enzymatic mechanism of the UBPs, which are generally larger than the UCHs. The UBPs are extremely heterogeneous in sequence and are primarily classified together by virtue of two short conserved sequence elements, the Cys box and the His box, which appear to constitute part of the active site of these enzymes. *Saccharomyces cerevisiae*, for example, has 16 such enzymes, and the only regions that can be unequivocally aligned in all of them are the Cys and His boxes. Surprisingly, none of the individual yeast enzymes is required for viability [13].

The 20S and 26S Proteasomes

The 26S proteasome is both the largest and most complex of ubiquitin system components (reviewed in [14]). The most rigorous measurements put its mass at ~2.5 MDa, which is on the same order as ribosome subunits. The complex comprises some 32–35 distinct polypeptides, with additional proteins that either interact with only a subset of proteasomes or associate only transiently with proteasomes. In the absence of ATP, the 26S proteasome readily dissociates into a 19S regulatory particle and a 20S core particle. The 20S proteasome core is a cylindrical structure made of a set of 28 subunits arranged in four coaxially stacked rings. A protected interior chamber houses the protease active sites; access to this hydrolytic chamber is restricted by a set of narrow channels that extend from the two ends of the particle, which at their narrowest are ~13 Å across. Therefore, for a protein to be translocated into the catalytic core, it must be completely unfolded or nearly so. Proteolysis proceeds processively until short peptides are generated and released from the inner chambers.

Unfolding and translocation into the 20S core are controlled by the 19S regulatory particles that are positioned over each axial pore. The 19S complex includes six distinct AAA-type ATPase subunits; these are believed to form a ring that interacts directly with the outer heptameric ring of the 20S proteasome. Functions of the 19S complex include recognition of polyubiquitinated proteins (and less frequently, of nonubiquitinated substrates), protein unfolding, translocation of protein into the 20S proteasome, and recycling of ubiquitin from protein substrates bound to the proteasome. Under some conditions, the 19S regulatory complex can be further dissociated into lid and base subcomplexes. The base, which interacts directly with the 20S proteasome, includes all the ATPase subunits and the two largest non-ATPase subunits.

Degradation Signals or Degrons

A fundamental question about protein degradation is exactly what structural features render a particular protein into a target for ubiquitin-dependent proteolysis [15]. Despite its importance, this remains one of the least well understood aspects of ubiquitin-dependent degradation. Short peptide stretches can sometimes target a protein to an E2/E3 ubiquitin ligase complex, and in some cases their ability to function as targeting signals or degrons is controlled by their phosphorylation. In other examples, it is clear that a higher order structure, which may be made up of discontinuous sequence elements, is required for E2/E3 recognition [15,16]. Virtually any protein that is misfolded or misassembled can become a substrate for the ubiquitin-proteasome system. This implies that there are common features that help a cell distinguish a correctly folded protein from one that is not. The most straightforward idea, for which there is experimental evidence, is that surface-exposed hydrophobic structures, which are normally buried in subunit interfaces or in the hydrophobic core of a protein, can function as folding-dependent degradation signals.

Examples of Regulation by Protein Ubiquitination

Protein ubiquitination is now recognized to be almost as pervasive as protein phosphorylation, and it would be impossible to enumerate here all the known instances in which ubiquitination is an important component of a cellular regulatory mechanism. Instead, a few illustrative examples will be briefly described. Perhaps the most widely appreciated function of ubiquitin-dependent proteolysis is in cell cycle control. Multiple, irreversible transitions in the cell cycle, including G1-to-S, metaphase-to-anaphase, and mitotic exit, require the timed degradation of specific positive or negative cell-cycle regulators. For instance, separation of sister chromatids to initiate anaphase depends on the ubiquitin-dependent destruction of an inhibitor called securin [17]. Securin binds to and keeps inactive a caspase-related protease, called separase or separin. Degradation of securin frees separase to cleave a subunit of the cohesin complexes, which prior to subunit cleavage maintain connections between sister chromatids even while they are under tension from the mitotic spindle.

Another well-studied example of ubiquitin-dependent regulation is in the NFκB signaling pathway. The NFκB transcription factor initiates transcription of multiple genes in response to an array of different environmental signals [18]. Most of these signals work by inactivating an inhibitor of NFκB called IκB, which sequesters NFκB in the cytoplasm by enhancing its export from the nucleus. Signaling leads to the site-specific phosphorylation of IκB, which triggers its multiubiquitination by an ubiquitin-protein ligase and rapid degradation by the proteasome. An interesting finding is that protein multiubiquitination appears to serve an additional, nonproteolytic signaling function upstream of the kinase that phosphorylates IκB, but the mechanism of this regulation remains to be fully elaborated [18].

A final example of how ubiquitin is deployed as a regulator of signaling is the remarkable finding that the ubiquitination of certain transcription factors is both a requirement for their activator function and a prelude to their silencing by proteasomal degradation [19]. A single ubiquitin moiety is sufficient for activation, but a multiubiquitin chain is necessary for subsequent degradation. This latter finding is consistent with the fact that proteasomes only bind efficiently to ubiquitin chains with at least four ubiquitin moieties [20]. Monoubiquitination, however, is also a known signal in other processes, particularly in the endocytosis and trafficking of cell surface receptors [21].

Acknowledgments

I wish to thank David Schwartz for comments on the manuscript. Work from my laboratory has been supported by grants from the National Institutes of Health.

References

1. Hochstrasser, M. (1996). Ubiquitin-dependent protein degradation. *Annu. Rev. Genet.* **30**, 405–439.
2. Pickart, C. M. (2001). Mechanisms underlying ubiquitination. *Annu. Rev. Biochem.* **70**, 503–533.
3. Hochstrasser, M. (2000). Evolution and function of ubiquitin-like protein-conjugation systems. *Nat. Cell Biol.* **2**, E153–E157.
4. Jentsch, S. and Pyrowolakis, G. (2000). Ubiquitin and its kin: how close are the family ties? *Trends Cell Biol.* **10**, 335–342.
5. Lake, M. W., Wuebbens, M. M., Rajagopalan, K. V., and Schindelin, H. (2001). Mechanism of ubiquitin activation revealed by the structure of a bacterial MoeB-MoaD complex. *Nature* **414**, 325–329.
6. Zheng, N., Schulman, B. A., Song, L., Miller, J. J., Jeffrey, P. D., Wang, P., Chu, C., Koepp, D. M., Elledge, S. J., Pagano, M., Conaway, R. C., Conaway, J. W., Harper, J. W., and Pavletich, N. P. (2002). Structure of the Cul1-Rbx1-Skp1-F boxSkp2 SCF ubiquitin ligase complex. *Nature* **416**, 703–709.
7. Hatakeyama, S., Yada, M., Matsumoto, M., Ishida, N., and Nakayama, K. I. U box proteins as a new family of ubiquitin-protein ligases. *J. Biol. Chem.* **276**, 33111–33120.
8. Huang, L., Kinnucan, E., Wang, G., Beaudenon, S., Howley, P. M., Huibregtse, J. M., and Pavletich, N. P. (1999). Structure of an E6AP-UbcH7 complex: insights into ubiquitination by the E2-E3 enzyme cascade. *Science* **286**, 1321–1326.
9. Kishino, T., Lalande, M., and Wagstaff, J. (1997). UBE3A/E6-AP mutations cause Angelman syndrome. *Nat. Genet.* **15**, 70–73.
10. Wilkinson, K. D. and Hochstrasser, M. (1998). The deubiquitinating enzymes, in Peters, J. M., Finley, D. Eds., *Ubiquitin and the Biology of the Cell*, pp. 99–125, Plenum Press, New York.
11. Amerik, A. Y., Nowak, J., Swaminathan, S., and Hochstrasser, M. (1999). The Doa4 deubiquitinating enzyme is functionally linked to the vacuolar protein-sorting and endocytic pathways. *Mol. Biol. Cell* **11**, 3365–3380.
12. Johnston, S. C., Riddle, S. M., Cohen, R. E., and Hill, C. P. (1999). Structural basis for the specificity of ubiquitin C-terminal hydrolases. *EMBO J.* **18**, 3877–3887.
13. Amerik, A. Y., Li, S. J., and Hochstrasser, M. (2000). Analysis of the deubiquitinating enzymes of the yeast *Saccharomyces cerevisiae*. *Biol. Chem.* **381**, 981–992.
14. Voges, D., Zwickl, P., and Baumeister, W. (1999). The 26S proteasome: a molecular machine designed for controlled proteolysis. *Annu. Rev. Biochem.* **68**, 1015–1068.
15. Laney, J. D. and Hochstrasser, M. (1999). Substrate targeting in the ubiquitin system. *Cell* **97**, 427–430.
16. Gardner, R. G. and Hampton, R. Y. (1999). A 'distributed degron' allows regulated entry into the ER degradation pathway. *EMBO J.* **18**, 5994–6004.
17. Nasmyth, K., Peters, J. M., and Uhlmann, F. (2000). Splitting the chromosome: cutting the ties that bind sister chromatids. *Science* **288**, 1379–1385.
18. Ghosh, S. and Karin, M. (2002). Missing pieces in the NF-kB puzzle. *Cell* **109**, S81–96.
19. Salghetti, S. E., Caudy, A. A., Chenoweth, J. G., and Tansey, W. P. (2001). Regulation of transcriptional activation domain function by ubiquitin. *Science* **293**, 1651–1653.
20. Pickart, C. M. Ubiquitin in chains. *Trends Biochem. Sci.* **25**, 544–548.
21. Hicke, L. (2001). Protein regulation by monoubiquitin. *Nat. Rev. Mol. Cell Biol.* **2**, 195–201.

CHAPTER 182

Caspases: Cell Signaling by Proteolysis

Guy S. Salvesen
*Program in Apoptosis and Cell Death Research,
Burnham Institute,
San Diego, California*

Protease Signaling

Proteolytic enzymes, including the cell's degrading machine the proteasome, calpains, and integral membrane proteases such as γ-secretase and rhomboid, participate in several intracellular signaling processes (Table I). But given that the human genome contains in excess of 500 genes that encode proteases, it seems odd that only the caspases constitute a formal multistep pathway able to transmit intracellular signals by proteolysis. In contrast, extracellular multistep signaling pathways frequently use the principle of proteolysis extensively for coagulation, fibrinolysis, complement activation in mammals, and gastrulation in flies. What is it about the caspases that makes them so suitable to transmit intracellular signals?

The consensus view of caspases places them in two main camps. First are the cytokine activators related to caspase 1, probably including mouse caspase 11 and its close homologs caspases 4 and 5 in humans. Their main role is to respond to bacterial infection by rapidly converting active cytokines (IL-1β, IL-18) from intracellular stores. Confirmation of the important roles of the caspases in the inflammatory cytokine response comes from gene ablation experiments in mice. Animals ablated in caspase 1 or 11 are deficient in cytokine processing [1,2] but without any overt apoptotic phenotype. The second camp constitutes the apoptotic caspases that transduce and execute death signals. The phenotypes of these knockouts are very gross, evidently anti-apoptotic, and vary from early embryonic lethality (caspase 8) to perinatal lethality (caspases 3 and 9) [3–5] to relatively mild with defects in the process of normal oocyte ablation [6]. Techniques in biochemistry and cell biology have allowed us to place the apoptotic caspases in two converging pathways, such that some are activated by others (Fig. 1). This core pathway probably represents a minimal apoptotic program, and certainly their simplicity is complicated by cell-specific additions that help fine-tune individual cell fates. Nevertheless, the basic order and at least some of the essential functions and, especially important, endogenous regulators of the caspases are now known.

Apoptosis and Limited Proteolysis

Apoptosis is a mechanism to regulate cell number and is vital throughout the life of all metazoan animals. Though several different types of biochemical events have been recognized as important in apoptosis, perhaps the most fundamental is the participation of the caspases [7–9].

The name caspase is a contraction of *c*ysteine-dependent *asp*artate specific prote*ase* [10], thus their enzymatic properties are governed by a dominant specificity for protein substrates containing Asp, and by the use of a Cys side chain for catalyzing peptide bond cleavage. The use of a Cys side chain as a nucleophile during peptide bond hydrolysis is common to several protease families. However, the primary specificity for Asp turns out to be very rare among proteases throughout biotic kingdoms. Of all known mammalian proteases only the caspase activator granzyme B, a serine protease, has the same primary specificity [11,12]. Caspases cleave a number of cellular proteins [13], and the process is one of limited proteolysis whereby a small number of cuts, usually only one, are made. Sometimes cleavage results in activation of the protein, sometimes in inactivation [14], but never in degradation, since their substrate specificity distinguishes the caspases as among the most restricted of endopeptidases.

Table I Proteases Involved in Intracellular Signaling

Protease	Signaling function
Caspases	Apoptosis
	Pro-inflammatory cytokine activation
Proteasome	Cell cycle progression
	NFκB activation
Rhomboid	EGF signaling
γ-Secretase	Toll receptor signaling
SREB Site2 Protease	Upregulation of sterol synthesis genes
Separase	Anaphase
Calpains	Various signaling events

While not an exhaustive survey, the table highlights the principle of proteolysis as a mechanism of signal transmission.

This is an important distinction from the proteasome, which permits signaling by wholesale destruction of regulatory proteins such as IκB in NFκB signaling and PDS1 in anaphase promotion [see Chapter 291 in Volume 3].

The most primitive organism with a bona fide caspase appears to be *Caenhorabditis elegans*. Indeed, the first apoptotic caspase, Ced3, was identified in this organism, a finding that galvanized the apoptotic research field [15]. As the complexity of primitive cell death pathways developed, so apparently did the number of caspases. *Drosophila* have at least seven caspases [16] and humans have at least 11. Mapping the inherent substrate specificity of caspases has allowed some broad consensuses to be recognized [17]. These consensuses also allow apoptotic caspases to be distinguished from pro-inflammatory caspases, since the latter have a rather distinct specificity that presumably allows them to carry out their job without threatening cell viability. What is interesting, there seems to have been a parallel evolution of apoptotic caspases along with their substrates. Consensus caspase targets in humans such as nuclear lamins and poly (ADP)ribose polymerase have easily recognizable caspase cleavage sites in *Drosophila* but apparently not in organisms such as yeast and plants, which lack an apoptotic pathway.

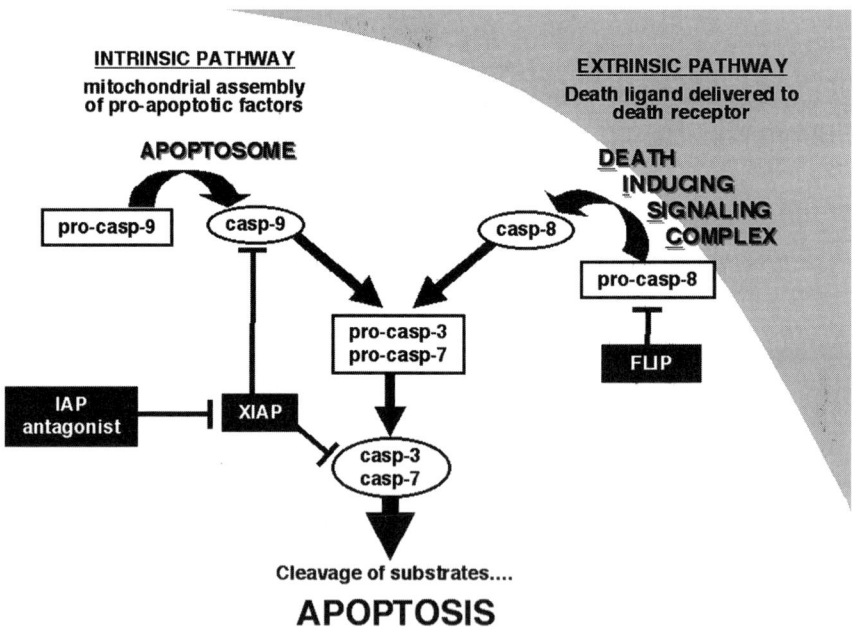

Figure 1 The framework of apoptosis. Death may be signaled by direct ligand-enforced clustering of receptors at the cell surface, which leads to the activation of initiator caspases 8 [46]. This caspase then directly activates the executioner caspases 3 and 7 (and possibly 6), which are predominantly responsible for the limited proteolysis that characterizes apoptotic dismantling of the cell. Alternatively, irreparable damage to the genome caused by mutagens, pharmaceuticals that inhibit DNA repair, or ionizing radiation—transmitted by a mechanism thought to involve the release of cytochrome c from mitochondria—engages the same executioner caspases [47]. The latter events progress through the initiator caspase 9 and its cofactor Apaf-1 [21]. Activation of the extrinsic pathway is regulated by FLIP, which serves to modulate the recruitment of caspase 8 to its adapters [48]. The common execution phase is regulated through direct caspase inhibition by IAPs, some of which can also regulate the active form of caspase 9. In turn, the IAPs are under the influence of antagonist proteins that compete with caspases for IAPs [38]. Though other modulators may regulate the apoptotic pathway in a cell-specific manner, this framework is considered common to most mammalian cells.

Caspase Activation

Induced Proximity

The seminal discovery that death receptor signaling requires, in its most basic form, simply a transmembrane receptor, an adapter molecule, and a caspase [18,19] revealed a solution to the perplexing problem of how the first proteolytic signal was generated during apoptosis, since it implicated a caspase directly in the triggering event. Prior to this work receptors were thought to signal by either altering the phosphorylation status of key signaling molecules or by functioning as ion channels. Death receptors such as Fas signal by direct recruitment and activation of a protease (caspase 8). Concomitantly with this work, groundbreaking studies showed that the intrinsic pathway was activated by a cofactor known as apoptosis protease activating factor (Apaf-1) [20]. Subsequently, Apaf-1 was found to recruit and activate caspase 9, forming the "apoptosome" [21].

In common with most proteolytic enzymes, caspases reside as latent forms that are usually activated by limited proteolysis. It is relatively easy to imagine that the caspases operating at the bottom of the pathway are activated by ones above. Until recently the question of how the first caspase in a pathway became activated, how the first death signal was generated, was a perplexing issue. Now several groups have focused on this issue and a consensus has been arrived at to describe the intriguing operation of the initiation of the proteolytic pathways that execute apoptosis (reviewed in [22,23]). This mechanism is known as the "induced proximity hypothesis" and although it originally referred to death induced by caspase 8 (the extrinsic pathway), it has now been extended to death induced by caspase 9 (the intrinsic pathway); see Fig. 1. Notably, this mechanism does not apply to the executioner caspases.

How exactly does a recruited zymogen become active? To understand this, as a basis for formulating an adequate hypothesis, one must understand the unusual properties of caspase zymogens that set them apart from most other proteases. Unlike most other proteases, simple expression of caspase zymogens in *E. coli* usually results in their activation by limited proteolysis within a "linker segment" that separates the large (~20 kD) and small (~10 kD) subunits of the catalytic domain [24,25]. This activation results from processing that is a consequence of intrinsic proteolytic activity residing in the caspase zymogens. It is not due to *E. coli* proteases, since catalytically disabled C285A (caspase 1 numbering convention) mutants fail to undergo processing.

Initiator Caspases

At the cytosolic concentration in human cells (<50 nM), pro-caspase 9 is a monomer [26] and requires oligomerization within the apoptosome to become active [27,28]. Significantly, unlike the executioner caspases 3 and 7, pro-caspase 9 does not need to be cleaved in the linker region to become active [29,30]. Not only is cleavage unnecessary, but also it is insufficient to produce an active enzyme. Instead, caspase 9 is activated by small-scale rearrangements of surface loops that define the substrate cleft and catalytic residues [26]. In the simplest model, this is achieved by dimerization of caspase 9 monomers within the apoptosome [31], with the dimer interface providing surfaces compatible with catalytic organization of the active site. More recently, it has been demonstrated that a similar dimerization mechanism activates the caspose 8 Zymogen to trigger the extrinsic pathway [51,52].

Executioner Caspases

Once an initiator caspase has become active, the ensuing activation of the executioners is more straightforwardly explained. At cytosolic concentration in human cells, the cxaspase-3 and 7 zymogens are already dimers, but cleavage within their respective linker segments is required for activation [32,33]. The same re-ordering of catalytic and substrate binding residues occurs in caspase 7 as seen in caspase 9, so the fundamental mechanism of zymogen activation is equivalent. Only the driving forces are distinct, since the linker segment of pro-caspase 7 blocks ordering of the active site, and upon cleavage the new N- and C-terminal sequences so generated aid in active site stabilization. The property that allows the distinct driving forces to converge on the same activation mechanism seems to be the unusual plasticity of the residues constituting the caspase active site, which, rather unusually for proteases, are predominantly placed on flexible loops and not on an ordered secondary structure.

Regulation by Inhibitors

The first level of regulating proteolytic pathways is by zymogen activation, but an equally important level is achieved by the use of specific inhibitors that can govern the activity of the active components. The endogenous inhibitors of caspases, those present in mammalian cells, are members of the inhibitor of apoptosis (IAP) family. In addition to these endogenous regulators are the virally encoded cowpox virus CrmA and baculovirus p35 proteins that are produced early in infection to suppress caspase-mediated host responses. Each of the inhibitors has a characteristic specificity profile against human caspases, as determined *in vitro*, and these profiles, with few caveats [34], agree with the biologic function of the inhibitors (reviewed in [35]). Though IAPs and CrmA would be expected to regulate mammalian caspases *in vivo*, p35 would never be present normally in mammals because it is expressed naturally by baculoviruses.

The best-characterized endogenous caspase inhibitor is the X-linked IAP (XIAP), a member of the IAP family. The IAPs are broadly distributed and, as their name indicates, the founding members are capable of selectively blocking apoptosis, having initially been identified in baculoviruses (reviewed in [36]). Eight distinct IAPs have been identified in humans. XIAP (which is the human family paradigm) has been found by multiple research groups to be a potent but restricted inhibitor targeting caspases-3, 7, and 9 (reviewed in [37]). Similarly, evidence implicates human cIAPs 1 and 2, ML-IAP, *Drosophila* DIAP-1, as well as ILP-2, as caspase inhibitors (reviewed in [38]). IAPs certainly have functions in addition to caspase inhibition because they have been found in organisms such as yeast, which neither contain caspases nor undergo apoptosis [39].

IAPs contain one, two, or three baculovirus IAP repeat (BIR) domains, which represent the defining characteristic of the family. Currently there is no known function for BIR1; however, domains closely related to the second BIR domain (BIR2) of XIAP specifically target caspases 3 and 7 ($K_i \approx 0.1$–1 nM), and regions closely related to the third BIR domain (BIR3) specifically target caspase 9 ($K_i \approx 10$ nM). This led to the general assumption that the BIR domain itself was important for caspase inhibition. The recent structures of BIR2 in complex with caspases 3 and 7 have surprisingly revealed the BIR domain to have almost no direct role in the inhibitory mechanism. All the important inhibitory contacts are made by the flexible region preceding the BIR domain [40–42]. It is interesting that the mechanism of inhibition of caspase 9 by the BIR3 domain requires cleavage in the inter-subunit linker to generate the new sequence NH_2-ATPF [30]. In part this explains the cleavage of caspase 9 during apoptosis, which as described above is not required for its activation. Paradoxically, it seems required for its inactivation by XIAP.

Neither CrmA-like nor p35-like inhibitors, which operate by mechanism-based inactivation [35], have been chosen for endogenous caspase regulation; rather, IAPs have been adapted to regulate the executioner caspases. Although the reason for this is not certain, it seems likely that the IAP solution provides a degree of specificity that mechanism-based inhibitors cannot achieve. Thus, XIAP inhibition of caspases 3 and 7 requires a nonstandard interaction with the extended 381 loop that is specific to these two caspases (reviewed in [35]). Possibly the 381 loop has evolved to achieve substrate specificity in the executioner caspases [43]. But an equally likely possibility is that the 381 loop has been generated to enable the IAP scaffold to provide a unique control level over the execution phase of apoptosis. Adding to this level of sophistication, IAPs, but not CrmA or p35-like proteins, are subject to negative regulation by IAP antagonists that go by the names of Hid, Grim, Reaper, and Sickle in *Drosophila* and Smac/Diablo, and HtrA2/Omi in mammals (reviewed in [38]).

IAP Antagonists

Biochemical studies led to the identification of a potentially important IAP-interacting protein known as second mitochondrial activator of caspases (Smac) in humans [44] and Diablo in mice [45]. Smac is a nuclear-encoded mitochondrial protein whose 55 amino terminal residues are removed by a proteolytic process during translocation.

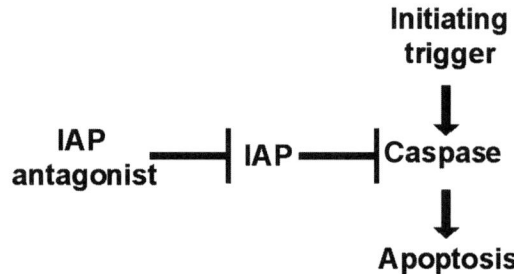

Figure 2 The basis of IAP antagonists as pro-apoptotic proteins. Genetic screens in *Drosophila* initially revealed a genomic region encoding three proteins—head involution defective (Hid), Grim, and Reaper—that together seemed responsible for almost all developmental apoptosis in the fly [49]. These proteins, along with the more recently discovered Sickle, constitute the currently known IAP antagonists in flies. In flies there is evidence for a continuous low-level production of caspases, which are neutralized by *Drosophila* IAP-1 [50]. In this scenario simple upregulation of one or more fly IAP antagonists could send the system into apoptosis, and to this extent the system is transcriptionally regulated. Input from the left of the diagram may be the most important event. In contrast to flies, the currently known mammalian IAP antagonists are mitochondrial proteins and require translocation before they can influence the inhibitory activity of IAPs. This implies a more complex role for IAP antagonists in mammals, since both positive initiator caspase signaling and mitochondrial fluxes would presumably be required, and the system could only be transcriptionally regulated in an indirect manner. In mammals, input from the top of the diagram could be more important.

Smac can be released, apparently along with cytochrome c, in response to apoptotic stimuli. Upon its release from mitochondria, mature Smac binds XIAP and probably several other IAPs, in a manner that displaces caspases from XIAP. Thus, Smac is a negative regulator of IAPs and therefore an apoptosis-enhancing molecule. Residues 56–59 of Smac and DIABLO are homologous to the exposed amino terminal motif utilized by caspase 9 to bind BIR3 of XIAP (see preceding section), and Smac has been found to bind to the same pocket in XIAP, thereby attenuating the affinity and displacing caspase 9 from the complex. The remarkable similarity in binding mode of caspase 9 and Smac to XIAP is not limited to these two proteins. In *Drosophila* the death-inducing proteins Hid, Grim, Reaper, and Sickle all contain homologous IAP binding motifs at their N-terminus (sometimes called the RHG or IBM motif). In distinction to the known mammalian IAP antagonists, the *Drosophila* ones do not have transit peptides and are therefore presumably fully functional following synthesis and removal of their initiator methionine. The mitochondrial versus nonmitochondrial disposition of IAP antagonists between mammals and flies suggests a rather different IAP regulation process (Fig. 2), which may have profound consequences for the interpretation of IAP function.

References

1. Kuida, K., Lippke, J. A., Ku, G., Harding, M. W., Livingston, D. J., Su, M. S. S., and Flavell, R. A. (1995). Altered cytokine export and apoptosis in mice deficient in interleukin-1-beta converting enzyme. *Science* **267**, 2000–2003.

2. Wang, S., Miura, M., Jung, Y.-K., Zhu, H., and Yuan, J. (1998). Murine caspase-11, an ice-interacting protease, is essential for the activation of ice. *Cell* **92**, 501–509.
3. Kuida, K., Haydar, T. F., Kuan, C. Y., Gu, Y., Taya, C., Karasuyama, H., Su, M. S., Rakic, P., and Flavell, R. A. (1998). Reduced apoptosis and cytochrome c-mediated caspase activation in mice lacking caspase 9. *Cell* **94**, 325–337.
4. Kuida, K., Zheng, T. S., Na, S., Kuan, C.-y., Yang, D., Karasuyama, H., Rakic, P., and Flavell, R. A. (1996). Decreased apoptosis in the brain and premature lethality in CPP32-deficient mice. *Nature* **384**, 368–372.
5. Varfolomeev, E. E., Schuchmann, M., Luria, V., Chiannilkulchai, N., Beckmann, J. S., Mett, I. L., Rebrikov, D., Brodianski, V. M., Kemper, O. C., Kollet, O., Lapidot, T., Soffer, D., Sobe, T., Avraham, K. B., Goncharov, T., Holtmann, H., Lonai, P., and Wallach, D. (1998). Targeted disruption of the mouse caspase 8 gene ablates cell death induction by the tnf receptors, fas/apo1, and dr3 and is lethal prenatally. *Immunity* **9**, 267–276.
6. Morita, Y., Maravei, D. V., Bergeron, L., Wang, S., Perez, G. I., Tsutsumi, O., Taketani, Y., Asano, M., Horai, R., Korsmeyer, S. J., Iwakura, Y., Yuan, J., and Tilly, J. L. (2001). Caspase-2 deficiency prevents programmed germ cell death resulting from cytokine insufficiency but not meiotic defects caused by loss of ataxia telangiectasia-mutated (ATM) gene function. *Cell Death Differen.* **8**, 614–620.
7. Salvesen, G. S. and Dixit, V. M. (1997). Caspases: intracellular signaling by proteolysis. *Cell* **91**, 443–446.
8. Cohen, G. M. (1997). Caspases: the executioners of apoptosis. *Biochem. J.* **326**, 1–16.
9. Thornberry, N. A. and Lazebnik, Y. (1998). Caspases: enemies within. *Science* **281**, 1312–1316.
10. Alnemri, E. S., Livingston, D. J., Nicholson, D. W., Salvesen, G., Thornberry, N. A., Wong, W. W., and Yuan, J. (1996). Human ICE/Ced-3 protease nomenclature. *Cell* **87**, 171.
11. Odake, S., Kam, C. M., Narasimhan, L., Poe, M., Blake, J. T., Krahenbuhl, O., Tschopp, J., and Powers, J. C. (1991). Human and murine cytotoxic t lymphocyte serine proteases: subsite mapping with peptide thioester substrates and inhibition of enzyme activity and cytolysis by isocoumarins. *Biochemistry USA* **30**, 2217–2227
12. Harris, J. L., Backes, B. J., Leonetti, F., Mahrus, S., Ellman, J. A., and Craik, C. S. (2000). Rapid and general profiling of protease specificity by using combinatorial fluorogenic substrate libraries. *Proc. Natl. Acad. Sci. USA* **97**, 7754–7759.
13. Nicholson, D. W. (1999). Caspase structure, proteolytic substrates, and function during apoptotic cell death. *Cell Death Differen.* **6**, 1028–1042.
14. Karran, L. and Dyer, M. J. (2001). Proteolytic cleavage of molecules involved in cell death or survival pathways: a role in the control of apoptosis? *Crit. Rev. Eukaryot. Gene Expr.* **11**, 269–277.
15. Yuan, J., Shaham, S., Ledoux, S., Ellis, H. M., and Horvitz, H. M. (1993). The *C. elegans* cell death gene Ced-3 encodes a protein similar to mammalian interleukin-1β-converting enzyme. *Cell* **75**, 641–652.
16. Kumar, S. and Doumanis, J. (2000). The fly caspases. *Cell Death Differen.* **7**, 1039–1044.
17. Thornberry, N. A., Rano, T. A., Peterson, E. P., Rasper, D. M., Timkey, T., Garcia-Calvo, M., Houtzager, V. M., Nordstrom, P. A., Roy, S., Vaillancourt, J. P., Chapman, K. T., and Nicholson, D. W. (1997). A combinatorial approach defines specificities of members of the caspase family and granzyme B. Functional relationships established for key mediators of apoptosis. *J. Biol. Chem.* **272**, 17907–17911.
18. Boldin, M. P., Goncharov, T. M., Goltsev, Y. V., and Wallach, D. (1996). Involvement of mach, a novel MORT1/FADD-interacting protease, in Fas/Apo-1-and TNF receptor-induced cell death. *Cell* **85**, 803–815.
19. Muzio, M., Stockwell, B. R., Stennicke, H. R., Salvesen, G. S., and Dixit, V. M. (1998). An induced proximity model for caspase-8 activation. *J. Biol. Chem.* **273**, 2926–2930.
20. Zou, H., Henzel, W. J., Liu, X., Lutschg, A., and Wang, X. (1997). Apaf-1, a human protein homologous to *C. elegans* Ced-4, participates in cytochrome c-dependent activation of caspase-3. *Cell* **90**, 405–413.
21. Li, P., Nijhawan, D., Budihardjo, I., Srinivasula, S. M., Ahmad, M., Alnemri, E. S., and Wang, X. (1997). Cytochrome c and dATP-dependent formation of Apaf-1/caspase-9 complex initiates an apoptotic protease cascade. *Cell* **91**, 479–489.
22. Salvesen, G. S. and Dixit, V. M. (1999). Caspase activation: the induced-proximity model. *Proc. Natl. Acad. Sci. USA* **96**, 10964–10967.
23. Steller, H. (1998). Artificial death switches: induction of apoptosis by chemically induced caspase multimerization. *Proc. Natl. Acad. Sci. USA* **95**, 5421–5422.
24. Orth, K., O'Rourke, K., Salvesen, G. S., and Dixit, V. M. (1996). Molecular ordering of apoptotic mammalian Ced-3/ICE-like proteases. *J. Biol. Chem.* **271**, 20977–20980.
25. Stennicke, H. R. and Salvesen, G. S. (1997). Biochemical characteristics of caspases-3, -6, -7, and -8. *J. Biol. Chem.* **272**, 25719–25723.
26. Renatus, M., Stennicke, H. R., Scott, F. L., Liddington, R. C., and Salvesen, G. S. (2001). Dimer formation drives the activation of the cell death protease caspase 9. *Proc. Natl. Acad. Sci. USA* **98**, 14250–14255.
27. Cain, K., Brown, D. G., Langlais, C., and Cohen, G. M. (1999). Caspase activation involves the formation of the aposome, a large (approximately 700 KDa) caspase-activating complex. *J. Biol. Chem.* **274**, 22686–22692.
28. Zou, H., Li, Y., Liu, X., and Wang, X. (1999). An Apaf-1, cytochrome c multimeric complex is a functional apoptosome that activates procaspase-9. *J. Biol. Chem.* **274**, 11549–11556.
29. Stennicke, H. R., Deveraux, Q. L., Humke, E. W., Reed, J. C., Dixit, V. M., and Salvesen, G. S. (1999). Caspase-9 can be activated without proteolytic processing. *J. Biol. Chem.* **274**, 8359–8362.
30. Srinivasula, S. M., Hegde, R., Saleh, A., Datta, P., Shiozaki, E., Chai, J., Lee, R. A., Robbins, P. D., Fernandes-Alnemri, T., Shi, Y., and Alnemri, E. S. (2001). A conserved XIAP-interaction motif in caspase-9 and Smac/Diablo regulates caspase activity and apoptosis. *Nature* **410**, 112–116.
31. Acehan, D., Jiang, X., Morgan, D. G., Heuser, J. E., Wang, X., and Akey, C. W. (2002). Three-dimensional structure of the apoptosome: implications for assembly, procaspase-9 binding and activation. *Mol. Cell* **9**, 423–432.
32. Riedl, S. J., Fuentes-Prior, P., Renatus, M., Kairies, N., Krapp, R., Huber, R., Salvesen, G. S., and Bode, W. (2001). Structural basis for the activation of human procaspase-7. *Proc. Natl. Acad. Sci. USA* **98**, 14790–14795.
33. Chai, J., Wu, Q., Shiozaki, E., Srinivasula, S. M., Alnemri, E. S., and Shi, Y. (2001). Crystal structure of a procaspase-7 zymogen. Mechanisms of activation and substrate binding. *Cell* **107**, 399–407.
34. Ryan, C. A., Stennicke, H. R., Nava, V. E., Lewis, J., Hardwick, J. M., and G.S., S. (2002). Inhibitor specificity of recombinant and endogenous caspase 9. *Biochem. J.* **366**, 595–601.
35. Stennicke, H. R., Ryan, C. A., and Salvesen, G. S. (2002). Reprieval from execution: the molecular basis of caspase inhibition. *Trends Biochem. Sci.* **27**, 94–101.
36. Verhagen, A. M., Coulson, E. J., and Vaux, D. L. (2001). Inhibitor of apoptosis proteins and their relatives: IAPs and other BIRPs. *Genome Biol.* **2**
37. Deveraux, Q. L. and Reed, J. C. (1999). IAP family proteins—suppressors of apoptosis. *Genes Dev.* **13**, 239–252.
38. Salvesen, G. S. and Duckett, C. S. (2002). IAP proteins: Blocking the road to death's door. *Nat. Rev. Mol. Cell Biol.* **3**, 401–410.
39. Uren, A. G., Coulson, E. J., and Vaux, D. L. (1998). Conservation of baculovirus inhibitor of apoptosis repeat proteins (BIRPs) in viruses, nematodes, vertebrates and yeasts. *Trends Biochem. Sci.* **23**, 159–162.
40. Chai, J., Shiozaki, E., Srinivasula, S. M., Wu, Q., Dataa, P., Alnemri, E. S., and Yigong Shi, Y. (2001). Structural basis of caspase-7 inhibition by XIAP. *Cell* **104**, 769–780.
41. Huang, Y., Park, Y. C., Rich, R. L., Segal, D., Myszka, D. G., and Wu, H. (2001). Structural basis of caspase inhibition by XIAP: differential roles of the linker versus the bir domain. *Cell* **104**, 781–790.
42. Riedl, S. J., Renatus, M., Schwarzenbacher, R., Zhou, Q., Sun, S., Fesik, S. W., Liddington, R. C., and Salvesen, G. S. (2001). Structural basis for the inhibition of caspase-3 by XIAP. *Cell* **104**, 791–800.

43. Rotonda, J., Nicholson, D. W., Fazil, K. M., Gallant, M., Gareau, Y., Labelle, M., Peterson, E. P., Rasper, D. M., Tuel, R., Vaillancourt, J. P., Thornberry, N. A., and Becher, J. W. (1996). The three-dimensional structure of apopain/cpp32, a key mediator of apoptosis. *Nature Struct. Biol.* **3**, 619–625.
44. Du, C., Fang, M., Li, Y., Li, L., and Wang, X. (2000). Smac, a mitochondrial protein that promotes cytochrome c-dependent caspase activation by eliminating IAP inhibition. *Cell* **102**, 33–42.
45. Verhagen, A. M., Ekert, P. G., Pakusch, M., Silke, J., Connolly, L. M., Reid, G. E., Moritz, R. L., Simpson, R. J., and Vaux, D. L. (2000). Identification of Diablo, a mammalian protein that promotes apoptosis by binding to and antagonizing iap proteins. *Cell* **102**, 43–53.
46. Ashkenazi, A. and Dixit, V. M. (1998). Death receptors: signaling and modulation. *Science* **281**, 1305–1308.
47. Green, D. R. and Reed, J. C. (1998). Mitochondria and apoptosis. *Science* **281**, 1309–1312.
48. Chang, D. W., Xing, Z., Pan, Y., Algeciras-Schimnich, A., Barnhart, B. C., Yaish-Ohad, S., Peter, M. E., and Yang, X. (2002). cFLIP(l) is a dual function regulator for caspase-8 activation and cd95-mediated apoptosis. *EMBO J.* **21**, 3704–3714.
49. White, K., Grether, M. E., Abrams, J. M., Young, L., Farrell, K., and Steller, H. (1994). Genetic control of programmed cell death in *Drosophila*. *Science* **264**, 677–683.
50. Rodriguez, A., Chen, P., Oliver, H., and Abrams, J. M. (2002). Unrestrained caspase-dependent cell death caused by loss of DIAP1 function requires the *Drosophila* Apaf-1 homolog, dark. *EMBO J.* **21**, 2189–2197.
51. Boatright, K. M., Renatus, M., Scott, F. L., Sperandio, S., Shin, H., Pedersen, I., Ricci, J.-E., Edris, W. A., Sutherlin, D. P., Green, D. R., and Salvesen, G. S. (2003). A Unified Model for Apical Caspase Activation. *Mol. Cell* **11**, 529–541.
52. Donepudi, M., Mac Sweeney, A., Briand, C., and Gruetter, M. G. (2003). Insights into the regulatory mechanism for caspase-8 activation. *Mol. Cell* **11**, 543–549.

CHAPTER 183

MAP Kinase in Yeast

Elaine A. Elion
*Department of Biological Chemistry and
Molecular Pharmacology,
Harvard Medical School,
Boston, Massachusetts*

Introduction

Yeast cells use multiple mitogen-activated protein (MAP) kinases to respond to a wide variety of external stimuli that regulate proliferation, differentiation, survival, and response to stress. As in mammalian cells, yeast MAPKs are activated within MAPK cascades that form the cores of larger signal transduction cascades. Activation of MAPKs leads to the phosphorylation of a variety of effector proteins, including many nuclear transcription factors. This chapter presents a brief overview of the different MAPK kinases in yeast, discusses how physical interactions with regulatory proteins such as scaffolds, activating kinases, and substrates regulate pathway specificity, and elucidates how their function is dynamically controlled at the level of localization and the strength and duration of activation.

Yeast Cells Use Multiple MAPKs to Respond to a Wide Variety of Stimuli

All eukaryotic cells have multiple MAPK cascades that are activated by one or more environmental stimulus and that mediate a variety of cellular responses [1,2]. Much of our knowledge about how MAPK cascades function has come from analysis in yeast. The proliferation, differentiation, morphogenesis, and response to stress of budding yeast *S. cerevisiae* is governed by five different MAPKs (Fig. 1), which function in six MAPK cascades that respond to different nutritional and hormonal inputs (Fig. 2). Three of the MAPKs are expressed in dividing haploid and diploid cells (Kss1, Mpk1/Slt2, Hog1), the fourth MAPK (Fus3) is expressed in dividing haploid cells (Fus3), and the fifth MAPK (Smk1) is expressed in cells undergoing meiosis and sporulation. MAPKs Fus3 and Kss1 function in the mating pathway, which is activated by peptide pheromones and causes cell cycle arrest in G1 phase along with polarized growth and additional differentiative events that prepare cells for cell attachment and fusion (reviewed in [3]). The Kss1 MAPK also functions in the filamentous growth pathways that respond to altered nutritional conditions and induce pseudohyphal diploid cells with higher surface-to-volume ratio and altered budding pattern and invasive haploid cells that can digest the substratum (reviewed in [4]). The Mpk1 MAPK functions in the PKC-regulated MAPK cascade, which is essential for mitotic growth, particularly the G1 to S phase transition, and plays a key role in maintaining cell integrity through cell wall synthesis and the actin cytoskeleton. The PKC-regulated pathway is activated under conditions that perturb the cell wall or plasma membrane (e.g. heat shock) (reviewed in [5,6]). The Kss1 MAPK cascade plays a lesser role in regulating cell integrity in parallel with the PKC-regulated MAPK cascade and is also thought to be activated under conditions of cell wall stress [7]. The Hog1 MAPK functions in the high osmolarity glycerol (HOG) pathway, which responds to hypertonic stress by increasing intracellular pools of glycerol (reviewed in [8]). This pathway is the closest functional equivalent of mammalian p38/JNK pathways. Finally, the Smk1 MAPK functions in a less-defined MAPK cascade that regulates sporulation and meiosis [9]. As in other eukaryotes, the yeast MAPKs phosphorylate many different proteins, including transcription factors in the nucleus (Fig. 2).

The yeast MAPKs are activated by a wide variety of means, including (1) a serpentine receptor and heterotrimeric G protein in the mating pathway, (2) a Ras-linked mechanism for the filamentous growth pathways, (3) a plasma membrane sensor and two component system in the HOG pathway, and (4) integrin-like sensors that physically link to Rho guanine exchange factors in the PKC-regulated pathway (Fig. 2). Additional signal transduction enzymes that regulate pathway activity include G-protein regulators, adapter proteins, and phosphatases. G-protein regulators include the RGS protein Sst2 that positively regulates Gpa1, the Gα subunit of the mating pathway G protein [10], Cdc25 (GEF) and Ira1, Ira2 (GAPs) that regulate Ras2 [4], and Rom1, 2 (GEFs) that regulate Rho1 [6]. Adapter proteins include the Ste5 scaffold in the mating pathway [11], Pbs2 MAPKK in the HOG pathway (reviewed in [12]), and the Ste11 MAPKKK regulator Ste50 [3] (Fig. 3).

Functionally Defining *S. cerevisiae* MAPK Cascades

The *S. cerevisiae* MAPKs and their activating kinases were identified through genetic screens for mutations or high copy plasmids that affect growth and differentiation. The mating MAPK cascade was the first to be identified and is the best characterized. Full assessment of MAPK function should involve analysis of null alleles as well as catalytically inactive mutants. Useful mutations include a lysine→arginine change in the binding site for ATP, which does not block phosphorylation by MKK or protein substrate binding, and threonine→alanine or tyrosine→phenylalanine changes in the T-X-Y sequence, which block phosphorylation by MKK and can affect protein substrate binding. A combined use of these different types of alleles has shown, for example, that certain MAPKs are functionally redundant (e.g. Fus3 and Kss1 for mating, [3]; Fig. 2) and that inactive kinases serve regulatory functions (e.g. Kss1, [13], Fig. 3B).

The core elements of a *S. cerevisiae* MAPK cascade module are three sequentially activated protein kinases [i.e. MAPKK kinase (MKKK) →MAPK kinase (MKK) →MAPK, Fig. 2]. *S. cerevisiae* MAPK modules show remarkably high conservation with MAPK modules in humans [14]. Fus3 and

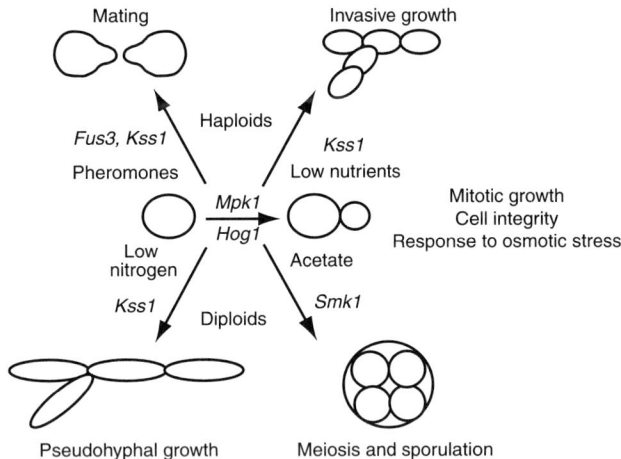

Figure 1 *S. cerevisiae* life cycle is regulated by 5 MAPKs.

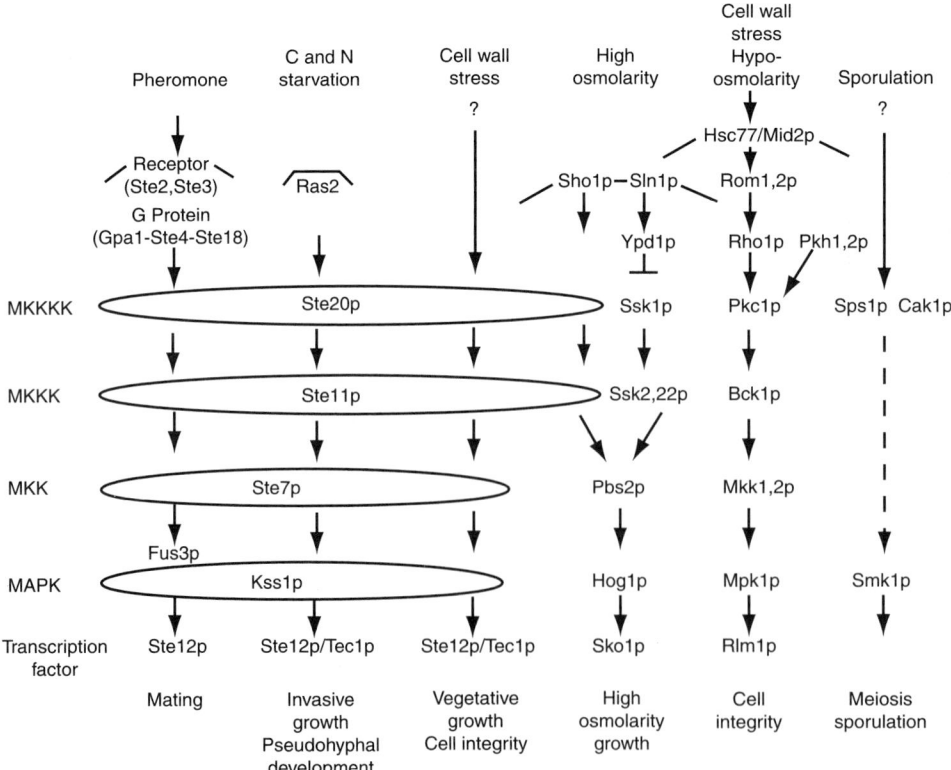

Figure 2 *S. cerevisiae* MAPK cascades.

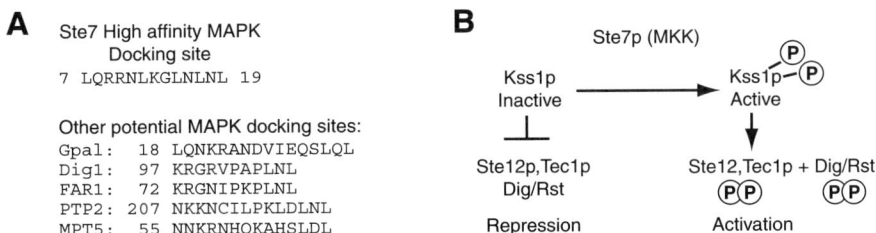

Figure 3 MAPK interactions with activating kinases and targets.

Kss1 are most similar to ERK1, 2, and Hog1 is most similar to p38 MAPK, whereas Mpk1 and Smk1 are distinct [14]. The activation mechanism for the *S. cerevisiae* MAPKs is highly conserved and involves phosphorylation of conserved threonine and tyrosine residues in the signature T-X-Y sequence within the activation (or T) loop at the catalytic cleft by the appropriate dual specificity MKK. Similar to ERK1, Fus3 autophosphorylates on tyrosine 182; however, dual phosphorylation is required for full activation. The yeast MAPKs are tightly regulated and are highly active only in the presence of a stimulus. However, the basal activities of Mpk1, Kss1, and Fus3 perform specific functions in the absence of stimulus [7–8,13]. Biochemical and genetic data support a series of direct phosphorylation events within each MAPK module, leading to serial activation of the MKK and MAPK tiers. A third upstream activating kinase (a putative MKKKK) is thought to activate the uppermost kinase (the MKKK) in a number of MAPK cascades. This third upstream activating kinase includes two subtypes, Ste20 p21-activated kinases (PAK) and protein kinase C (Fig. 2). Ste20 has been shown to directly phosphorylate MKKKK Ste11; its activation involves derepression of a negative regulatory domain and association with a conserved protein called Ste50 through mutual SAM domains [3]. MKK Ste7 may be activated by dual phosphorylation of serine-threonine residues within a conserved domain near the activation loop [15].

Major Regulatory Mechanisms that Control Specificity in *S. cerevisae* MAPK Cascades

A major biological question is how specificity is achieved to assure that the appropriate response takes place for a given stimulus. The individual *S. cerevisiae* MAPK cascades exhibit exquisite specificity based on the phenotypes of cells with mutations in the individual MAPKs. This specificity is remarkable because four of the pathways use subsets of the same kinases (Fig. 2). Pathway specificity occurs through multiple layers of control, which include preferred kinase-kinase and kinase-substrate interactions, the use of scaffold or adapter-like proteins to route signals through specific MAPK modules, cross-regulation between MAPK modules, controlled localization of MAPKs, and control of the strength of the activation. Common themes of signal transduction through yeast MAPK cascades are that they form molecular assemblies within cells and spatial organization at specific locations plays a critical role in mediating efficient activation and specificity [16,17]. The pathways are also highly dynamic at the level of movement of signal transduction proteins [18–20].

Kinase Interactions within a MAPK Cascade

MAPKs prefer to bind to specific activating kinases within a module, providing one level of pathway specificity. For example, the yeast MKK Ste7 preferentially binds to and activates MAPKs Fus3 and Kss1 of the mating and invasive growth and pseudohyphal development pathways, but does not bind to or activate MAPKs Mpk1 or Hog1. Consistent with this, *ste7Δ* null mutants only have defects in mating and invasive growth and pseudohyphal development. A docking site of charged residues has been defined for the interaction of Ste7 with Fus3 and Kss1 [21]; this docking site defines a consensus sequence that is generally applicable for other MAPK binding partners (Fig. 3A; [22]).

Functional analysis in yeast demonstrates that MAPK cascade specificity is enhanced through the use of different MKKs and MAPKs that recognize specific substrates. The mating, invasive growth and pseudohyphal development, and high osmolarity growth pathways use some of the same MAPK module components (i.e. MKKKK Ste20, MKKK Ste11, MKK Ste7, MAPK Kss1), yet elicit very different responses by using a specific MKK (Pbs2) and MAPK (Hog1) for high osmolarity growth and a specific MAPK (Fus3) for mating (Fig. 2). For example, during mating, MAPK Fus3 specifically regulates G1 arrest and mating through phosphorylation of the Far1 protein [23], which is a poor substrate for MAPK Kss1 [24].

Yeast MAPKs form stable complexes with their targets, either as active or inactive kinases, suggesting that such complexes may play a regulatory role [23,25]. The functional importance of catalytically inactive MAPK/substrate complexes was first demonstrated for MAPK Kss1, which exerts both negative and positive control over pseudohyphal development and invasive growth (Fig. 3B; [26,27]). Kss1 forms a stable complex with the Ste12 transcription factor in its unphosphorylated inactive form and acts as a transcriptional repressor. When phosphorylated, Kss1 becomes a positive regulator, presumably through dissociation from Ste12 together with phosphorylation of Ste12, Tec1, and the

Dig1,2 repressors (Fig. 3B). The fact that the same MAPK provides both negative and positive functions for signaling means that a *kss1* null mutation does not block signaling whereas catalytically inactive Kss1 derivatives do. These findings argue that it is important to analyze catalytically inactive kinase mutations in addition to null mutations in assigning function.

Scaffold Proteins

A variety of biochemical and genetic evidence suggests that simple activation of the kinases within a MAPK module is not sufficient for function *in vivo*. Scaffold proteins have been found to play key roles in maintaining MAPK pathway function and fidelity by linking individual components to each other, to upstream activators, downstream targets, and specific cellular locales [16,17]. Many of our ideas about scaffolds are based upon work on the prototype MAPK scaffold Ste5 of the mating pathway (Fig. 4, reviewed in [11]). The criteria for assigning a scaffold function to Ste5 included (1) definition of separable binding sites for each of the three tiers of protein kinases (i.e. Ste11, Ste7, and Fus3/Kss1) and (2) biochemical evidence for a Ste5-multikinase complex that enhances Ste11/Ste7 interactions and increases the specific activity of MAPK Fus3. In addition to tethering kinases, Ste5 plays a direct role in the activation of MKKKK Ste11 by MKKKK Ste20, which is normally associated with the Rho-type G protein Cdc42 at the cell cortex (Fig. 4). Ste11 must be properly targeted to Ste20 for activation to take place. This function is tied to proper recruitment of Ste5 through its conserved RING-H2 domain to Gβγ dimers at the plasma membrane, which also bind to Ste20 upon release from Gα after pheromone induction of the receptor. Formation of an active signaling complex requires oligomerization of Ste5; glycerol sedimentation is consistent with Ste5 functioning as a dimer.

The MKK Pbs2, of the yeast high osmolarity growth (HOG) pathway, also appears to serve a scaffold function based on pair-wise interactions between Pbs2 and the other two kinases in the MAPK module and separate binding sites for these kinase (reviewed in [12]). Pbs2 has a large regulatory domain in addition to its catalytic domain. A variety of evidence suggests that Pbs2 tethers the Hog1 MAPK and Ste11 MKKK to a transmembrane sensor called Sho1 that senses changes in osmolarity. This interaction is thought to occur through a proline-rich domain in Pbs2. Much like Ste5, Pbs2 is thought to recruit Ste11 to Ste20 while it is linked through its CRIB domain to Cdc42, which is asymmetrically enriched at the plasma membrane , with the end result of phosphorylation and activation of Ste11 by Ste20. Thus, the common theme of Ste5 and Pbs2 is that they link the cytosolic kinases to a plasma membrane-linked G protein or sensor and an activating GTPase-linked PAK. This arrangement allows a proximal sense of the status of the environment and provides a spatial arrangement that promotes phosphorylation of the uppermost kinase of each MAPK module.

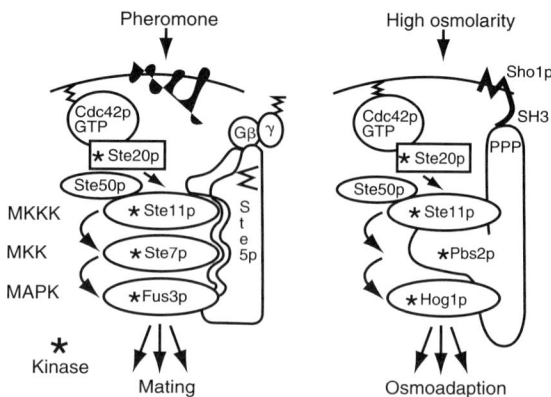

Figure 4 Ste5 and Pbs2 scaffold proteins regulate MAPK cascades of overlapping components.

It is interesting that a common regulatory scheme is used for the two pathways that share signaling components and respond to stimulus in a polarized manner—perhaps the scaffolds allow for a highly localized activation mechanism that facilitates a polarized response, such as by targeting the MAPK to substrates, and also prevent cross-activation.

MAPK cascade scaffolds are now being described in mammalian cells [17]. None bear homology to each other, suggesting that they have evolved independently for the individual requirements of individual subsets of kinases and upstream activating events. It is interesting, however, that the JIP and KSR scaffolds dimerize and are localized to the cell periphery through interactions with membrane-linked receptors or GTPases, thus suggesting that aspects of their function are similar to Ste5's. It has been proposed that a three-tier MAPK cascade has ultrasensitive switch-like responses [28]. The existence of MAPK cascade scaffolds raises the important question of how they affect signal propogation [11,17]. For example, co-localization of three tiers of kinases by a scaffold may prevent amplification by physical sequestration or provide a means to ensure feedback control and prevent cross-activation of other pathways that utilize the same kinases, localize the MAPK to targets, or protect it from phosphatases. Alternatively, a MAPK cascade scaffold could promote kinase-kinase phosphorylation and dissociation after activation and could enhance amplification.

Dynamic Localization of MAPK Cascades

Specificity is also regulated by proper localization of the MAPKs. MAPKs are stably associated with a variety of subcellular structures, including tubulin and centromeres (reviewed in [29,30]). MAPKs can shuttle between the nucleus and cytoplasm and be retained in the cytoplasm by a cognate MAPKK or undergo nuclear translocation as a result of activation and dimerization [30]. In *S. cerevisiae*, the MAPK Hog1 translocates to the nucleus in response to osmotic stress [18], then forms a stable complex with the transcription apparatus at specific promoters and activates transcription through derepression of Sln1 ([32], Fig. 5). MAPKs Fus3 and Kss1 are predominantly nuclear; however, Fus3 shuttles continuously between the nucleus and cytoplasm and is translocated to the projection tip of pheromone treated cells along with MKK Ste7, possibly through Ste5 [19,20]. This localization event may serve to promote polarized growth and attenuation by co-localizing Fus3 with targets such as Far1 or other components of the morphogenesis apparatus in addition to upstream signaling components such as Ste5, Sst2, and Gα (Gpa1) (Fig. 6). Once activated, Fus3 dissociates rapidly from Ste5 and translocates to the nucleus, where it phosphorylates transcription factors. The Ste5 scaffold also shuttles through the nucleus, and this localization event positively regulates its recruitment and activation of MAPK Fus3 [19]. The Far1 and Cdc24 signaling proteins involved in morphogenesis are anchored together in the nucleus and exported to the cell cortex as a result of pheromone activation (Fig. 6, [3]). Thus, a majority of components in the mating MAPK cascade are dynamic. These findings may be generally applicable to other pathways.

Signal Strength

DOWNREGULATION

The duration and strength of activation of a given MAPK cascade may also affect the ultimate outcome of a response. For example, transient activation of the Ras/Raf/ERK pathway by epidermal growth factor in PC12 cells causes cell proliferation, whereas sustained activation of the same pathway by nerve growth factor (NGF) causes terminal differentiation as shown by neurite outgrowth [33]. Multiple factors can regulate the duration of activation of a MAPK cascade. The *S. cerevisiae* MAPKs are inactivated by several phosphatases that may form stable associations with their target MAPKs [30,34]. These include a dual-specificity phosphatase Msg5 that inactivates Fus3, protein tyrosine phosphatases Ptp2 and Ptp3 that inactivate Hog1, Fus3, and Mpk1 with different specificities, and a serine-threonine phosphatase Ptc1 that inactivates Hog1 MAPK (reviewed in [35]). Yeast MAPK phosphatases can be induced by their cognate MAPK either by direct phosphorylation or increased expression [34].

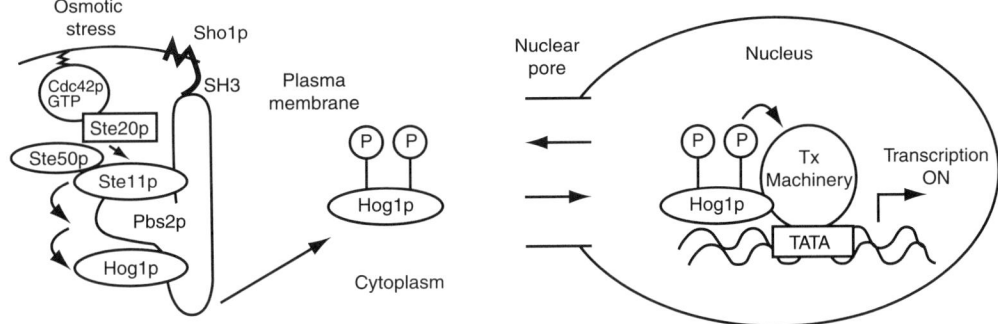

Figure 5 Hog1 forms a stable part of the transcriptional activation machinery.

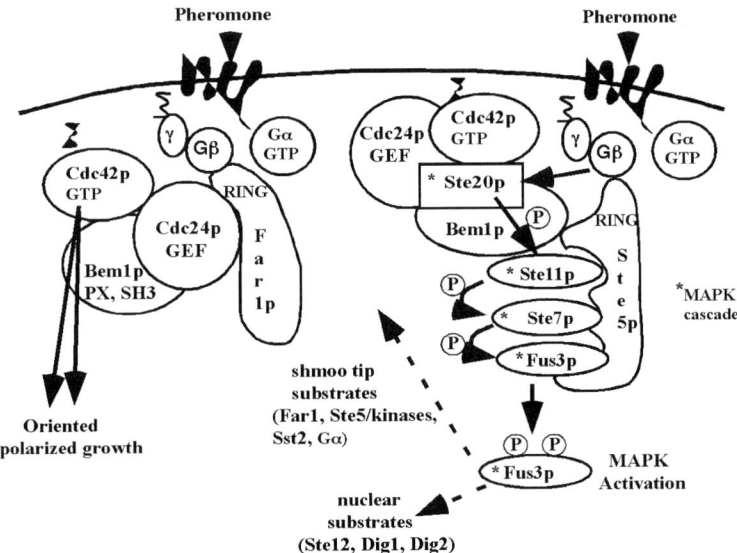

Figure 6 MAPK Fus3 at cell cortex may promote polarized growth fusion.

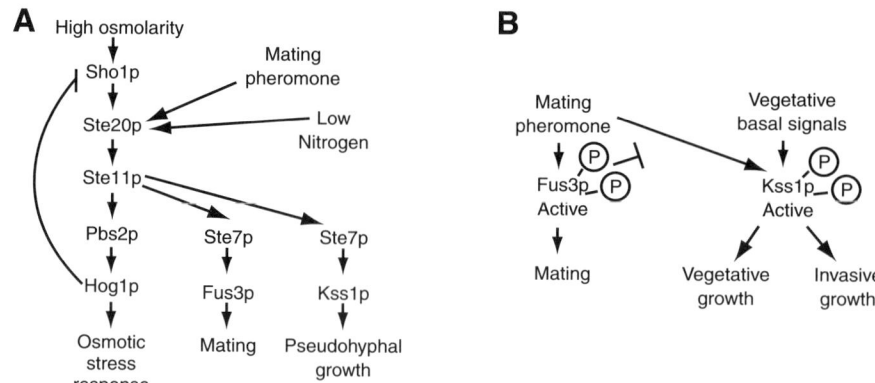

Figure 7 Yeast MAPKs crossregulate one another to maintain specificity.

Because there is considerable sharing of signal transduction components between the MAPK cascades, inputs into one MAPK cascade have the potential to affect multiple pathways. An interesting feature of the yeast MAPKs is their propensity to act as attenuation factors and prevent cross-talk with MAPKs within the same or other pathways. In haploid cells, the mating MAPK Fus3 inhibits the invasive growth and vegetative growth pathways and prevents their activation by mating pheromone (Fig. 4B, [7,13]). Activated Fus3 attenuates the activity of MAPK Kss1 [36], thereby reducing the strength and duration of its activation by pheromone, presumably preventing hyperactivation of the invasive growth pathway. The osmoregulatory pathway MAPK Hog1 inhibits the Fus3 and Kss1 MAPKs both in the absence and presence of hyperosmotic stress (Fig. 7, [37,38]). This cross-regulation requires Hog1 kinase activity and may involve attenuation of the transmembrane sensor Sho1 to prevent persistent activation of the MKKK Ste11. A variety of evidence suggests that MAPKs attenuate the activity of their own pathway through direct phosphorylation of upstream signaling components. For example, MAPK Fus3 phosphorylates Ste5, Ste7, and Ste11 *in vitro* and could directly regulate signal amplification. Fus3 also associates Gpa1-GTP (Gα) *in vivo,* and this interaction appears to attenuate pathway activity [22].

CELL ARCHITECTURE

The strength and nature of the MAPK cascade activation process is dependent upon proper cellular architecture. Regulators of the actin cytoskeleton, such as the Rho family member Cdc42 and an SH3-domain morphogenesis protein Bem1, have been demonstrated to modulate the activity of the *S. cerevisiae* mating and invasive growth and pseudohyphal development MAPK cascades [39]. A MAPK cascade can also be secondarily activated as a consequence of changes in cellular architecture that arise from activation of a primary MAPK cascade. For example, activation of the mating MAPK by pheromone induces polarized growth, which in turn activates the PKC1-regulated MAPK cascade [40].

Acknowledgments

This work was supported by grants from the National Institutes of Health (GM46962) and the American Heart Association.

References

1. Ip, Y. T. and Davis, R. J. (1998). Signal transduction by the c-Jun N-terminal kinase (JNK)-from inflammation to development. *Curr. Opin. Cell Biol.* **10**, 205–219.
2. Garrington, T. P. and Johnson, G. L. (1999). Organization and regulation of mitogen-activated protein kinase signaling pathways. *Curr. Opin. Cell Biol.* **11**, 211–218.
3. Elion, E. A. (2000). Pheromone response, mating and cell biology. *Curr. Opin. Microbiol.* **3**, 573–581.
4. Pan, X., Harashima, T., and Heitman J. (2000). Signal transduction cascades regulating pseudohyphal differentiation of *Saccharomyces cerevisiae. Curr. Opin Microbiol.* **3**, 567–572.
5. Kamada, Y., Jung, U., Piotrowaki, J., and Levin, D. E. (1995). The protein kinase C-activated MAP kinase pathway of *Saccharomyces cerevisiae* mediates a novel aspect of the heat shock response. *Genes Dev.* **9**, 1–13.
6. Schmelzle, T., Helliwell, S. B., and Hall, M. N. (2002). Yeast protein kinases and the RHO1 exchange factor TUS1 are novel of the cell integrity pathway in yeast. *Mol. Cell. Biol.* **22**, 1329–1339.
7. Lee, B. N. and Elion, E. A. (1999). The MAPKKK Ste11 regulates vegetative growth through a kinase cascade of shared signaling components. *Proc. Natl. Acad. Sci. USA* **96**, 12679–12684.
8. Gustin, M. C., Albertyn, J., Alexander, M., and Davenport, K. (1998). MAP kinase pathways in the yeast *Saccharomyces cerevisiae. Microbiol. Mol. Biol. Rev.* **62**, 1264–1300.
9. Schaber, M., Lindgren, A., Schindler, K., Bungard, D., Kaldis, P., and Winter, E. (2002). CAK1 promotes meiosis and spore formation in *Saccharomyces cerevisiae* in a CDC28-independent fashion. *Mol. Cell. Biol.* **22**, 57–68.
10. Dohlman, H. G. and Thorner, J. W. (2001). Regulation of G protein-initiated signaltransduction in yeast: paradigms and principles. *Annu. Rev. Biochem.* **70**, 703–754.
11. Elion, E. A. (2001). The Ste5 scaffold. *J. Cell Sci.* **114**, 3967–3978.
12. Raitt, D., Posas, F., and Saito, H. (2000). Yeast Cdc42 GTPase and Ste20 PAK-like kinase regulate Sho1-dependent activation of the Hog1 MAPK pathway. *EMBO J.* **19**, 4623–4631.
13. Madhani, H. D. and Fink, G. R. (1998). The riddle of MAP kinase signaling specificity. *Trends Genet.* **14**, 151–155.
14. Caffrey, D. R., O'Neill, L. A. J., and Shields, D. C. (1999). The evolution of the MAP kinase pathways: Coduplication of interacting proteins leads to new signaling cascades. *J. Mol. Evol.* **49**, 567–582.

15. Guan, K. L. (1994). The mitogen activated protein kinase signal transduction pathway from the cell surface to the nucleus. *Cell. Signal.* **6**, 581–589.
16. Pawson, T. and Scott, J. D. (1997). Signaling through scaffold, anchoring, and adaptor proteins. *Science* **278**, 2075–2080.
17. Burack, W. R. and Shaw, A. S. (2000). Signal transduction: hanging on a scaffold. *Curr. Opin. Cell Biol.* **12**, 211–216.
18. Hood, J. K. and Silver, P. A. (1999). In or out? Regulating nuclear transport. *Curr. Opin. Cell Biol.* **11**, 241–247.
19. Mahanty, S. K., Wang, Y., Farley, F. W., and Elion, E. A. (1999). Nuclear shuttling of yeast scaffold Ste5 is required for its recruitment to the plasma membrane and activation of the mating MAPK cascade. *Cell* **98**, 501–512.
20. van Drogen, F., Stucke, V. M., Jorritsma, G., and Peter, M. (2001). MAP kinase dynamics in response to pheromones in budding yeast. *Nature Cell Biol.* **3**, 1051–1059.
21. Bardwell, A. J., Flatauer, L. J., Matsukuma, K., Thorner, J., and Bardwell, L. (2001). A conserved docking site in MEKs mediates high-affinity binding to MAP kinases and cooperates with a scaffold protein to enhance signal transmission. *J. Biol. Chem.* **276**, 10374–10386.
22. Metodiev, M. V., Matheos, D., Rose, M. D., and Stone, D. E. (2002). Regulation of MAPK function by direct interaction with the mating-specific Gα in yeast. *Science* **296**, 1483–1486.
23. Elion, E. A., Satterberg, B., and Kranz, J. (1993). FUS3 phosphorylates components of the mating signal transduction cascade: Evidence for STE12 and FAR1. *Mol. Biol. Cell* **4**, 495–510.
24. Breitkreutz, A. and Tyers, M. (2002). MAPK signaling: it takes two to tango. *Trends Cell Biol.* **12**, 254–257.
25. Kranz, J., Satterberg, B., and Elion, E. A. (1994). The MAP kinase Fus3 associates with and phosphorylates the upstream signaling component Ste5. *Genes Devel.* **8**, 313–327.
26. Madhani, H. D., Styles, C. A., and Fink, G. R. (1997). MAP kinases with distinct inhibitory functions impart signaling specificity during yeast differentiation. *Cell* **91**, 673–684.
27. Bardwell, L., Cook, J. G., Voora, D., Baggott, D. M., Martinez, A. R., and Thorner, J. (1998). Repression of yeast Ste12 transcription factor by direct binding of unphosphorylated Kss1 MAPK and its regulation by the Ste7 MEK. *Genes Devel.* **12**, 2887–2898.
28. Ferrell, J. E. and Machleder, E. M. (1998). The biochemical basis of an all-or-none cell fate switch in *Xenopus* oocytes. *Science* **280**, 895–898.
29. Cobb, M. H. (1999). MAP kinase pathways. *Progr. Biophys. Mol. Biol.* **71**, 479–500.
30. Schaeffer, H. J. and Weber, M. J. (1999). Mitogen-activated protein kinases: Specific messages from ubiquitous messengers. *Mol. Cell. Biol.* **19**, 2435–2444.
31. Cobb, M. H. and Goldsmith, E. J. (2000). Dimerization in MAP-kinase signaling. *Trends Biochem. Sci.* **25**, 7–9.
32. Alepuz, P. M., Jovanovic, A., Reiser, V., and Ammerer, G. (2001). Stress-induced MAP kinase Hog1 is part of transcription activation complexes. *Mol. Cell* **7**, 767–777.
33. Marshall, C. J. (1995). Specificity of receptor tyrosine kinase signaling: transient versus sustained extracellular signal-regulated kinase activation. *Cell* **80**, 179–185.
34. Keyse, S. M. (2000). Protein phosphatases and the regulation of mitogen-activated protein kinase signalling. *Curr. Opin. Cell Biol.* **12**, 186–192.
35. Warmka, J., Hanneman, J., Lee, J., Amin, D., and Ota, I. (2001). Ptc1, a type 2C ser/thr phosphatase, inactivates the HOG pathway by dephosphorylating themitogen-activated protein kinase Hog1. *Mol. Cell. Biol.* **21**, 51–60.
36. Sabbagh, W. Jr., Flatauer, L. J., Bardwell, A. J., and Bardwell L. (2001). Specificity of MAP kinase signaling in yeast differentiation involves transient versus sustained MAPK activation. *Mol. Cell* **8**, 683–691.
37. Hall, J. P., Cherkasova, V., Elion, E., Gustin, M. C., and Winter, E. (1996). The osmoregulatory pathway represses mating pathway activity in *Saccharomyces cerevisiae*: Isolation of a Fus3 mutant that is insensitive to the repression mechanism. *Mol. Cell. Biol.* **16**, 6715–6723.
38. O'Rourke, S. and Herskowitz, I. (1998). The Hog1 MAPK prevents cross talk between the HOG and pheromone response MAPK pathways in *Saccharomyces cerevisiae*. *Genes Devel.* **12**, 2874–2886.
39. Lyons, D. L., Mahanty, S. M., Choi, K.-Y., Manandhar, M., and Elion, E. A. (1996). The SH3-domain protein Bem1 coordinates MAP kinase cascade activation with cell cycle control during mating in *S. cerevisiae*. *Mol. Cell. Biol.* **16**, 4095–4106.
40. Buehrer, B. M. and Errede, B. (1997). Coordination of the mating and cell integrity mitogen-activated protein kinase cascades in *Saccharomyces cerevisiae*. *Mol. Cell. Biol.* **17**, 6517–6525.

Mammalian MAP Kinases

Roger J. Davis

*Howard Hughes Medical Institute and
Program in Molecular Medicine,
University of Massachusetts Medical School,
Worcester, Massachusetts*

Introduction

MAP kinases are critical mediators of signal transduction in mammalian cells [17,43,54,58,95]. The MAP kinases are encoded by a group of genes that are related in sequence. These enzymes are also functionally related. For example, MAP kinases phosphorylate substrate proteins on conserved Ser-Pro and Thr-Pro motifs [22]. However, the substrate specificity of MAP kinases is different for individual MAP kinase isoforms. This substrate specificity is mediated, in part, by the selective docking of MAP kinases to substrate proteins [29]. A second functional similarity between MAP kinases is the mechanism of activation. MAP kinases are activated within conserved protein kinase signaling modules composed of a MAP kinase, a MAP kinase kinase, and a MAP kinase kinase kinase. Phosphorylation and activation of MAP kinases by MAP kinase kinases occurs on two residues within a tripeptide motif (Thr-Xaa-Tyr) located within the T loop that controls the conformation of the MAP kinase active site [18].

There are three major groups of mammalian MAP kinases, the extracellular signal regulated protein kinases (ERK), the p38 MAP kinases, and the c-Jun NH2-terminal kinases (JNK) (Table I). These groups of MAP kinases can be distinguished by the sequence of the dual phosphorylation motif that mediates MAP kinase activation: Thr-Glu-Tyr (ERK), Thr-Gly-Tyr (p38), and Thr-Pro-Tyr (JNK). In addition, there are several protein kinases that are related to the MAP kinases with similar dual phosphorylation motifs: Thr-Glu-Tyr (MAK and MOK), Thr-Asp-Tyr (ICK, KKIAMRE, KKIALRE), and Thr-His-Glu (NLK). Here I review the properties of these MAP kinases and MAP kinase-related protein kinases.

The ERK Group of MAP Kinases

ERK1 and ERK2

The ERK1 and ERK2 protein kinases are activated by phosphorylation of the Thr-Glu-Tyr motif located in the T loop by the MAP kinase kinases MEK1 and MEK2 (Table II). Structural analysis of nonphosphorylated and inactive ERK2 [114] and phosphorylated and activated ERK2 [13] by X-ray crystallography reveals that the mechanism of activation involves conformational changes that remodel the T loop and the active site. The activated form of ERK2 has been identified as a homodimer [50] and it is likely that this dimeric form of ERK2 is critical for nuclear accumulation [19].

The ERK1 and ERK2 protein kinases are major targets of the Ras signaling pathway. Substrates of these MAP kinases include a wide array of proteins, including the epidermal growth factor receptor [22], cytoplasmic phospholipase A2 [59], and Ets family transcription factors [64]. The ERK1 and ERK2 protein kinases are implicated in proliferation [119], tumorigenesis [63], and differentiation [21]. It appears that the time course of ERK1 and ERK2 activation may be critical for specifying the biological outcome of signal transduction: transient activation correlates with growth and sustained activation correlates with growth arrest and differentiation [21].

Recent studies indicate that the targeted disruption of the *Erk1* gene in mice causes defects in thymocyte development [83] and causes increased synaptic plasticity associated with improved striatal-mediated learning and memory [67]. It is likely that these limited phenotypes reflect the partial complementation of ERK1-deficiency by ERK2. Further studies of ERK2-deficiency and compound mutations in ERK1 and ERK2 will be important to establish the function of these protein kinases *in vivo* (Table III).

Table I Nomenclature of Human MAP Kinases

Name	Alternative name	Gene name	Chromosomal location	LocusID/accession number
ERK1	p44 MAPK	MAPK3	16pll.2	5595
ERK2	p42 MAPK	MAPK1	22qll.21	5594
ERK3-related	p63 MAPK	MAPK4	18ql2-q21	5596
ERK3	p97 MAPK	MAPK6	15q21	5597
ERK5	Big MAP kinase, BMK	MAPK7	17pll.2	5598
ERK7		ERK7		XM 139448
ERK8		ERK8	Chromosome 8	225689
p38α	CSBP, RK, SAPK2A	MAPK14	6p21.3-p21.2	1432
p38β	p38-2, p38β2, SAPK2B	MAPK11	22ql3.33	5600
p38γ	SAPK3, ERK6	MAPK12	22ql3.33	6300
p38δ	SAPK4	MAPK13	6p21.1	5603
JNK1	SAPKγ, SAPK1C	MAPK8	10q21.1	5599
JNK2	SAPKα, SAPK1A	MAPK9	5q35	5601
JNK3	SAPKβ, SAPK1B	MAPK10	4q22.1-q23	5602

Human MAP kinases corresponding to the ERK group (ERK1, ERK2, ERK3, ERK3-related, ERK5, ERK7, and ERK8), the p38 MAP kinase group (p38α, p38β, p38γ, p38δ), and the JNK group (JNK1, JNK2, and JNK3) are presented with their alternative names, the gene name, the chromosomal localization, and the LocusID or Accession Number.

Table II Nomenclature of Human MAP Kinase Kinases

Name	Alternative name	Gene name	Chromosomal location	LocusID
MEK1	MAPKK1, MKK1	MAP2K1	15q22.1-q22.33	5604
MEK2	MAPKK2, MKK2	MAP2K2	7q32	5605
MKK3	MEK3, SKK2	MAP2K3	17qll.2	5606
MKK4	MEK4, SEK1, JNKK1, SKK1	MAP2K4	17pll.2	6416
MEK5	MKK5	MAP2K5	15q22.1	5607
MKK6	MEK6, SKK3	MAP2K6	17q25.1	5608
MKK7	MEK7, SEK2, JNKK2, SKK4	MAP2K7	19pl3.3-pl3.2	5609

Human MAP kinase kinases are presented with their alternative names, the gene name, the chromosomal localization, and the LocusID.

ERK3

The ERK3 subgroup of MAP kinases is encoded by two genes [10,36]. These protein kinases are not true members of the ERK group because they contain the sequence Ser-Glu-Gly instead of the dual phosphorylation motif Thr-Glu-Tyr. A MAP kinase kinase activity for ERK3 has been described, but the regulation of the ERK3 protein kinases is not understood [17]. The physiological function of the ERK3 protein kinases is also unclear [17].

ERK5

The ERK5 protein kinase is activated by the MAP kinase kinase MEK5 by dual phosphorylation on the conserved T-loop motif Thr-Glu-Tyr [28,137]. Substrates of ERK5 include members of the MEF2 family of transcription factors [47]. The signaling pathways that lead to the activation of the ERK5/MEK5 module are poorly understood. However, ERK5 is critical for survival signal transduction by retrograde NGF receptors localized in endosomes [115] and for proliferation caused by epidermal growth factor [48]. Recent studies indicate that ERK5 is essential for viability during murine embryonic development because ERK5-deficiency causes defects in cardiovascular development [87].

ERK7 and ERK8

The ERK7 and ERK8 protein kinases form a subgroup of ERK-related protein kinases that contain the dual phosphorylation motif Thr-Glu-Tyr [2,3]. ERK7 and ERK8 are not activated by MEK1 and MEK2. Recent studies indicate that ERK7 is constitutively activated by autophosphorylation [1]. However, ERK8 appears to have a low basal activity that is increased in cells with activated Src or following

Table III Targeted Disruption of Genes in Mice

	Phenotype	Reference
ERK1	Viable. Defects in thymocyte maturation. Increased synaptic plasticity and improved striatal-mediated learning and memory.	[83,67]
ERK5	Lethal. Embryos die during mid-gestation. Defects in cardiovascular development.	[87]
p38α	Lethal. Defects in formation of the placenta and expression of erythropoietin.	[4,5,104]
JNK1	Viable. Defects in CD4 effector T cell differentiation and function. Defects in CD8 T cell activation.	[25,24,92,20]
JNK2	Viable. Defects in CD4 effector T cell differentiation and function. Increased CD8 T cell activation.	[90,24,92,20]
JNK1 + JNK2	Lethal. Embryos die during mid-gestation with neural tube closure defects and increased fore-brain apoptosis. Primary fibroblasts are resistant to stress-induced apoptosis.	[53,91,112]
JNK3	Viable. Defects in neuronal apoptosis in response to excitotoxic stress.	[130]
MEK1	Lethal. Defects in placental development *in vivo* and cell migration *in vitro*.	[35]
MKK3	Viable. Defects in cytokine secretion by CD4 T cells and macrophages. Defects in activation-induced cell death of peripheral CD4 T cells.	[60,125,105]
MKK4	Lethal. Embryos die during mid-gestation with liver apoptosis.	[79,80,128,33,81, 102,82,111]
MKK6	Viable. Defects in thymocyte apoptosis.	[105]
MKK7	Lethal. Embryos die during mid-gestation. Defects in TNF-stimulated JNK activation.	[24,93,111]
MKK4 + MKK7	Lethal. Embryos die during mid-gestation. JNK is not activated in cultured fibroblasts.	[111]

The effect of disruption of the murine genes that encode MAP kinases and MAP kinase kinases is summarized.

serum stimulation [3]. Interestingly, ERK7 appears to be associated with an intracellular chloride channel [85]. The physiological function of ERK7 and ERK8 is unclear and further studies of these MAP kinases are warranted.

The p38 Group of MAP Kinases

The regulation of the p38 group of MAP kinases differs from the ERK group of MAP kinases [95,54]. The ERK1 and ERK2 protein kinases are activated by many growth factors by a Ras-dependent mechanism. In contrast, the p38 MAP kinases are activated by inflammatory cytokines and by the exposure of cells to environmental stress. It is thought that members of the Rho family of small GTPases, rather than the Ras family of GTPases, are critical mediators of p38 MAP kinase activation. The p38 group of MAP kinases is activated by three different MAP kinase kinases [43]. Two of these MAP kinase kinases specifically activate the p38 MAP kinases (MKK3 and MKK6). In contrast, the third MAP kinase kinase (MKK4) activates both the JNK and p38 MAP kinases.

p38α and p38β MAP Kinases

The p38α and p38β MAP kinases are structurally related protein kinases that contain the dual phosphorylation motif Thr-Gly-Tyr [54,95]. The structure of p38α MAP kinase has been determined by x-ray crystallography [14,122]. This structure is similar to ERK2. However, there are significant differences that distinguish p38α MAP kinase from ERK2. Included among these differences is the structure of the active site of p38α MAP kinase. This difference has allowed the discovery of active site-directed inhibitors that are extremely selective for p38α and p38β MAP kinases [27]. These drugs are useful for functional dissection of the p38α and p38β MAP kinase signaling pathway. In addition, it is likely that such drugs may be useful for therapeutic intervention. In particular, a major role for p38α and p38β MAP kinases in inflammatory responses has been identified. The p38α and p38β MAP kinases appear to be critical for the expression of several inflammatory cytokines, including interleukin-1, interleukin-6, and tumor necrosis factor [56]. These p38 MAP kinases regulate several steps in cytokine expression including transcription, mRNA stability, and translation. The transcriptional effects of the p38α and p38β MAP kinases are mediated by phosphorylation of several transcription factors, including ATF2 and members of the MEF2 and Ets families [39,86,121]. The mechanisms that account for the effects of p38α and p38β MAP kinases on mRNA stability and translation have not been defined. However, recent studies have implicated proteins that bind to the 3' ARE element of regulated mRNAs and MAPKAP2, a protein kinase that is phosphorylated and activated by p38α and p38β MAP kinases [52,62,77,123].

Mice with targeted disruption of the p38α MAP kinase gene have been described [4,5,104]. These mice die during mid-gestation because of defects in the formation of the placenta. p38α MAP kinase-deficient cells derived from these embryos exhibit severe defects in the expression of inflammatory cytokines [5].

p38γ and p38δ MAP Kinases

The p38γ and p38δ MAP kinases are related enzymes that represent a separate subgroup of p38 MAP kinases [54,95].

Like other members of the p38 family, the p38γ and p38δ MAP kinases contain a Thr-Gly-Tyr dual phosphorylation motif. However, the p38γ and p38δ MAP kinases are not inhibited by drugs that selectively inhibit the p38α and p38β MAP kinases [27]. The regulation of the p38γ and p38δ MAP kinases is similar to the p38α and p38β MAP kinases, and all four p38 MAP kinases are activated by inflammatory cytokines and exposure to environmental stress. Interestingly, p38γ MAP kinase interacts with the PDZ domain of α-syntrophin, although the functional significance of this interaction is unclear [40]. The physiological role of p38α MAP kinase is unclear. However, recent studies have demonstrated an important role for p38δ MAP kinase in the regulation of translation by eEF2 kinase [51].

The JNK Group of MAP Kinases

The JNK group of MAP kinases is activated by many of the same stimuli that cause activation of the p38 MAP kinases, including the exposure of cells to inflammatory cytokines and environmental stress [23]. This similarity in regulation indicates that the JNK and p38 MAP kinases may be functionally related. Indeed, both the JNK and p38 MAP kinases are collectively named stress-activated protein kinases (SAPK). However, the mechanism of JNK activation differs from the p38 MAP kinases. First, the dual phosphorylation motif located in the T loop of JNK is Thr-Pro-Tyr. Second, the signaling module that activates JNK includes the MAP kinase kinases MKK4 and MKK7 [23]. The protein kinase MKK7 is a specific activator of the JNK pathway, while MKK4 can activate both p38 MAP kinase and JNK [23]. Biochemical studies demonstrate that MKK4 and MKK7 can cooperate to activate JNK by selectively phosphorylating JNK on Tyr and Thr, respectively, and gene disruption experiments in mice demonstrate functional cooperation of MKK4 and MKK7 *in vivo* [30,55,111].

The JNK protein kinases are encoded by three genes. Two of the genes (*JNK1* and *JNK2*) are expressed ubiquitously and one gene (*JNK3*) is expressed in a limited number of tissues, including the brain, heart, and testis [23]. Alternative splicing creates ten different JNK protein kinases: four JNK1 isoforms, four JNK2 isoforms, and two JNK3 isoforms [38]. Analysis of JNK3 by x-ray crystallography indicates that this MAP kinase is structurally related to both ERK2 and p38α MAP kinase [127].

The physiological role of JNK appears to be complex. It is established that JNK is required for the normal regulation of AP-1 transcription activity [23]. This is mediated, in part, by the phosphorylation of the transcription factors ATF2, c-Jun, JunB, and JunD on two sites within the NH2-terminal activation domain [23]. JNK can also regulate other transcription factors by phosphorylation. Nevertheless, although it is known that JNK causes increased AP-1-dependent gene expression, the physiological consequence of JNK activation appears to be both cell-type and context dependent. JNK can cause apoptosis by a mechanism that involves the mitochondrial pathway [57,112]. JNK can also signal cell survival [41,84]. The mechanism that accounts for these markedly divergent cellular responses to JNK activation is unclear. However, it is likely that the cellular response to JNK activation reflects the activation state of other signal transduction pathways within the cell. A reasonable hypothesis is that the program of gene expression that is induced by JNK-stimulated AP-1 activity depends on the cooperation of AP-1 with other transcription factors bound to the promoters of relevant genes. Further studies will be required to test this hypothesis.

Gene disruption studies in mice demonstrate that JNK1, JNK2, and JNK3 are not essential for viability [20,25,90,92,129,130]. However, these mice exhibit defects in apoptosis and immune responses. Compound mutations in JNK1 and JNK2 cause early embryonic lethality associated with neural tube defects and markedly increased apoptosis in the developing forebrain [53,91]. Studies of *Jnk1-/- Jnk2-/-* CD4 T cells isolated from *Rag1 -/-* chimeric mice indicate that JNK is required for effector CD4 T-cell function, but is not required for CD4 T-cell activation [24]. Primary embryo fibroblasts isolated from *Jnk1 -/- Jnk2 -/-* embryos exhibit severe growth defects with premature senescence associated with increased expression of ARF and p53. These JNK-deficient fibroblasts also exhibit marked resistance to stress-induced apoptosis [112]. This resistance to apoptosis was also observed in *Mkk4 -/- Mkk7 -/-* fibroblasts that contain JNK, but lack a mechanism to allow JNK activation [111]. Together, these data indicate that JNK contributes to multiple physiological processes, including differentiation, survival, and apoptosis.

MAP Kinase-Related Protein Kinases

There are three major groups of mammalian MAP kinases (ERK, p38, and JNK). Several additional human protein kinases have been identified that exhibit similarities with these MAP kinases (Table IV). Thus, two protein kinases (MAK and MOK) contain the same dual phosphorylation motif that is found in the ERK group of MAP kinases (Thr-Glu-Tyr). It is established that MOK is regulated by phosphorylation on this motif, but the mechanism that causes this phosphorylation is unclear [73]. Similar studies of MAK have not been reported [66]. The physiological role of MAK is unclear because MAK-deficient mice lack an obvious phenotype [100]. Three protein kinases have been identified (ICK, p42 KKIALRE, p56 KKIAMRE) that contain a related dual phosphorylation motif (Thr-Asp-Tyr). The role of the dual phosphorylation motif in these kinases is unclear. Mutational analysis indicates that the dual phosphorylation motif is required for the regulation of ICK [110], but it is not essential for the regulation of p42 KKIALRE and p56 KKIAMRE [103]. The protein kinase (NLK) contains the divergent motif Thr-His-Glu. Phosphorylation of this motif in response to the MAP kinase kinase kinase TAK1 is required for NLK activation [99]. Further studies are warranted to discover whether these protein kinases (and other related protein kinases) represent additional members of the MAP kinase family.

Table IV Human MAP Kinase-related Protein Kinases

Name	TXY motif	Gene name	Chromosomal location	LocusID
ICK	TDY	*ICK*	6pl2.3-pll.2	22858
p42 KKIALRE	TOY	*CDKL1*	14q21.3	8814
p56 KKIAMRE	TDY	*CDKL2*	4q21.1	8999
MAK	TEY	*MAK*	6q22	4117
MOK	TEY	*RAGE*	14q32	5891
NLK	THE	*NLK*	17qll.2	51701

Human protein kinases that are related to MAP kinases are presented with the sequence of the TXY dual phosphorylation motif, the gene name, the chromosomal location, and the LocusID.

MAP Kinase Docking Interactions

Although MAP kinases can phosphorylate Ser-Pro and Thr-Pro sites on substrate proteins, the substrate specificities of individual MAP kinases are distinct. One mechanism that can dictate the substrate specificity of a MAP kinase is the requirement for a docking site on the substrate [29]. The docking site is physically separate from the site of phosphorylation and is required for efficient substrate phosphorylation by MAP kinases. Mutational removal of the docking site prevents substrate phosphorylation by MAP kinases. Examples of docking sites include the δ domain on c-Jun that interacts with JNK [23], the D domain on the Elk-1 transcription factor that binds ERK and JNK [132,133], the D domain on MEF transcription factors that binds p38α MAP kinase [131], and the FXF motif on the SAP-1 transcription factor that binds ERK and p38δ MAP kinase [31,45]. The δ and D domains are similar and consist of a Leu-Xaa-Leu motif separated from several basic residues [29]. In contrast, the FXF domain contains the motif Phe-Xaa-Phe-(Pro) [29].

Interestingly, these docking domains are conserved in many proteins that interact with MAP kinases. Thus, the NH2-terminal region of MAP kinase kinases contains a D domain [7]. Disruption of this MAP kinase docking site on MAP kinase kinases is caused by the anthrax lethal factor protease and prevents MAP kinase activation [26]. Docking sites are also observed in MAP kinase phosphatases and scaffold proteins [29]. It therefore appears that many proteins that interact with MAP kinases contain conserved motifs that mediate this interaction [29]. Mutational analysis indicates that these motifs bind MAP kinases at a common site [106–108]. A recent study has provided structural insight into the mechanism of protein docking to MAP kinases. This study used x-ray crystallographic analysis to determine the structures of the complexes of p38α MAP kinase with the D domains of MKK3 and MEF2 [14]. This analysis demonstrated that the Leu-Xaa-Leu motif is directly involved in the protein-protein contact and that the basic residues in the D domain may interact with acidic residues in the common docking site [116]. The site of interaction on the surface of p38α MAP kinase is not located close to the T loop or the active site. Thus, the active site and the T loop of MAP kinases are available for interaction with docked proteins, including MAP kinase kinases, MAP kinase phosphatases, and substrates.

Scaffold Proteins

The protein kinases that form MAP kinase signaling modules can interact via a series of sequential binary interactions to create a protein kinase cascade. One example is represented by MKK4, which can dock to both an upstream kinase (MEKK1) and to a downstream MAP kinase (JNK) [126]. Alternatively, the protein kinases may simultaneously interact with a common component of the cascade. Thus, MEKK2 can synergistically interact with both MKK4 and JNK [16] and MEKK1 can bind c-Raf-1, MEK1, and ERK2 [46]. MEKK1 can also bind JNK in a phosphorylation-dependent manner [32]. These interactions may lead to the assembly of a functional signalling module [118].

Functional MAP kinase signaling modules can also be created by the interaction of the protein kinases with scaffold molecules that serve to assemble the protein kinases [118]. These putative scaffold proteins include KSR, MP1, the JIP proteins, β-arrestin-2, and SKRP1/MKPX (Table V). In addition, several other molecules have been proposed to function as MAP kinase scaffolds, including Filamin [65], Crk II [34], and IKAP [42].

KSR

Kinase suppressor of Ras (KSR) shares structural similarity with the MAP kinase kinase kinase c-Raf-1 [74,89]. One major difference between KSR and c-Raf-1 is that while KSR does have a protein kinase-like domain, KSR does not function as a protein kinase [70]. KSR binds to c-Raf-1, MEK1/2, and ERK1/2 and appears to function as a scaffold for the activation of the ERK1/2 signaling module that is activated by growth factors [74,89]. Following the activation of growth factor receptors, KSR is recruited to the cell surface by a phosphorylation-dependent mechanism that involves the

Table V Nomenclature of Human MAP Kinase Scaffold Proteins

Name	Alternative name	Gene name	Chromosomal location	LocusID
KSR-1	KSR	*KSR*	17qll.1	8844
KSR-2			Chromosome 12	125806
MP-1		*MAP2K1IP1*	4q22.3	8649
JIP-1	IB1	*MAPK8IP1*	Ilpl2-pll.2	9479
JIP-2	IB2	*MAPK8IP2*	22ql3.33	23542
JIP-3	JSAP1, SYD2	*MAPK8IP3*	16pl3.3	23162
β-Arrestin-2		*ARRB2*	17pl3	409
SKRP1	JKAP	*SKRP1*	2q32.1	142679
MKPX	VHX, JSP-1	*MKPX*	6p24.3	56940

Human MAP kinase scaffold proteins are presented with their alternative names, the gene name, the chromosomal localization, and the LocusID.

C-TAK1 protein kinase [75]. Gene disruption studies in mice have examined the requirement of KSR for growth-factor-stimulated ERK1/2 activation. KSR-deficient cells derived from these mice were found to exhibit partial defects in ERK activation [78]. This partial defect may reflect a specialized role for KSR under certain conditions. Alternatively, it is possible that the functions of KSR are redundant with the product of another gene that encodes a KSR-like protein (Table V). Further studies are required to define the role of these KSR proteins. Nevertheless, it is very likely that the mammalian KSR proteins do function as an essential scaffold for ERK1/2 activation because RNAi experiments have demonstrated an important role for KSR in the activation of ERK in *Drosophila* [88].

MP1

The MP1 scaffold protein binds to MEK1 and ERK1 [94]. Transfection assays demonstrate that MP1 potentiates the activation of ERK1 caused by MEK1 [94]. Recent studies demonstrate that MP1 also binds to the late endosomal protein p!4 [124]. MP1 therefore localizes a MEK1/ERK1 signaling module on the cytoplasmic surface of late endosomes. It is possible that MP1 contributes to ERK activation following ligand-induced endocytosis of growth factor receptors. This role of MP1 on late endosomes serves to distinguish MP1 from the KSR scaffold proteins that appear to function at the cell surface. The possible functional interaction between the MP1 and KSR scaffold proteins warrants further study. In addition, gene disruption studies are required to establish the physiological function of MP1 in ERK activation *in vivo*.

JIP

Three genes encode the JIP group of scaffold proteins [23]. The JIP1 and JIP2 proteins are structurally similar and contain an SH3 domain and a PTB domain in the COOH-terminal region [9,76,117,134]. The PTB domain can interact with p190 RhoGEF and with members of the low density lipoprotein receptor family, including ApoER2 [37,69,101]. In addition, JIP2 has been reported to bind the Rac exchange factor Tiaml, Ras-GRF, and members of the fibroblast growth factor homologous protein family [12,96,97]. The JIP3 protein is structurally distinct and consists of an extended coiled-coil domain [44,49]. All of these JIP proteins share several common properties. Each JIP isoform binds to JNK, MKK7, and members of the mixed-lineage group of MAP kinase kinase kinases. An interaction of JIP3 with MEKK1 and MKK4 has also been described [44]. Transfection studies demonstrate that the JIP proteins potentiate the activation of JNK [44,49,117,134]. Some studies have demonstrated that JIP2 can also activate p38 MAP kinases under some circumstances [96,97].

The JIP1, JIP2, and JIP3 proteins bind to kinesin light chain and are transported by the microtubule motor protein kinesin [11,113,120]. This interaction with motor proteins accounts for the accumulation of the JIP proteins in the growth cones of developing neurons. In mature neurons, the JIP1 and JIP2 proteins accumulate at synapses and JIP3 is mostly localized to perinuclear vesicular structures. The JIP proteins may act as adapter molecules for the transport of cargo by the kinesin motor protein. In addition, the JIP proteins may act to locally regulate JNK activation in response to specific stimuli.

Two groups have reported the phenotype of mice with targeted disruption of the *Jip1* gene. One group reported that the JIP1-deficiency causes very early embryonic lethality prior to implantation [109]. In contrast, a second group reported that JIP1-deficient mice are viable [120]. It is possible that the difference in viability reflects an effect of the mouse strain background. The viable JIP1-deficient mice were found to exhibit defects in stress-induced JNK activation in hippocampal neurons following exposure to stress. Studies of JIP2- and JIP3-deficient mice will be required to identify the redundant and nonredundant functions of these JIP scaffold proteins.

β-Arrestin

Ligand-induced activation of seven transmembrane-spanning receptors causes receptor phosphorylation, recruitment of arrestin molecules, and subsequent downregulation of heterotrimeric G protein signaling [71]. The arrestin molecules can also serve as a platform for the recruitment of additional molecules to the receptor [71]. Evidence has been presented that indicates that the ubiquitously expressed isoform β-arrestin-1 may serve as a scaffold for components of the ERK pathway [61]. More detailed studies have been performed on β-arrestin-2. This scaffold protein contains a D domain that selectively binds to JNK3 [68,72]. Interestingly, both β-arrestin-2 and JNK3 are selectively expressed in the brain and the heart. The β-arrestin-2 scaffold also binds the MAP kinase kinase kinase ASK1 [68]. The signaling module assembled by β-arrestin-2 also contains MKK4, which interacts with both JNK3 and ASK1, but does not directly contact β-arrestin-2 [68]. Biochemical assays demonstrate that β-arrestin-2 is essential for the activation of JNK3 caused by the angiotensin II receptor [68]. Interestingly, the activated receptor bound to the β-arrestin-2 scaffold complex is localized to endosomal structures. The β-arrestin-2 scaffold provides a mechanism for the activation of JNK by seven transmembrane-spanning receptors. In addition, the signaling module assembled by β-arrestin-2 provides a mechanism for the selective activation of the JNK3 isoform of JNK. Studies of β-arrestin-2-deficient mice have been reported [8]. Further studies of these mice to investigate defects in the activation of JNK3 are warranted.

SKRP1/MKPX

SKRP1 and MKPX are two related small phosphatases that belong to the MAP kinase phosphatase (MKP) family. Like other MKPs, these phosphatases can inactivate MAP kinases [6,135]. However, it appears that the normal function of these phosphatases is to activate JNK [15,98,136]. Interestingly, the phosphatase catalytic activity is required for these MKPs to activate JNK, but the physiologically relevant substrates have not been identified. It appears that these phosphatases interact with MKK7 [15,136] and may also interact with ASK1 [136]. Gene disruption studies in mice demonstrated that SKRP1 is not an essential gene, but SKRP1 is required for JNK activation caused by tumor necrosis factor-α and transforming growth factor-β, but not for JNK activation caused by ultraviolet radiation [15]. Together, these data suggest that these MKPs may act as scaffold proteins for the JNK signaling pathway.

References

1. Abe, M. K., Kahle, K. T., Saelzler, M. P., Orth, K., Dixon, J. K, and Rosner, M. R. (2001). ERK7 is an autoactivated member of the MAPK family. *J. Biol. Chem.* **276**, 21272–21279.
2. Abe, M. K., Kuo, W. L., Hershenson, M. B., and Rosner, M. R. (1999). Extracellular signal-regulated kinase 7 (ERK7), a novel ERK with a C-terminal domain that regulates its activity, its cellular localization, and cell growth. *Mol. Cell. Biol.* **19**, 1301–1312.
3. Abe, M. K., Saelzler, M. P., Espinosa, R., 3rd, Kahle, K. T., Hershenson, M. B., Le Beau, M. M., and Rosner, M. R. (2002). ERK8, a new member of the mitogen-activated protein kinase family. *J. Biol. Chem.* **277**, 16733–16743.
4. Adams, R. H., Porras, A., Alonso, G., Jones, M., Vintersten, K., Panelli, S., Valladares, A., Perez, L., Klein, R., and Nebreda, A. R. (2000). Essential role of p38alpha MAP kinase in placental but not embryonic cardiovascular development. *Mol. Cell* **6**, 109–116.
5. Alien, M., Svensson, L., Roach, M., Hambor, J., McNeish, J., and Gabel, C. A. (2000). Deficiency of the stress kinase p38alpha results in embryonic lethality: Characterization of the kinase dependence of stress responses of enzyme-deficient embryonic stem cells. *J. Exp. Med.* **191**, 859–870.
6. Alonso, A., Merlo, J. J., Na, S., Kholod, N., Jaroszewski, L., Kharitonenkov, A., Williams, S., Godzik, A., Posada, J. D., and Mustelin, T. (2002). Inhibition of T cell antigen receptor signaling by VHR-related MKPX (VHX), a new dual specificity phosphatase related to VH1 related (VHR). *J. Biol. Chem.* **277**, 5524–5528.
7. Bardwell, L. and Thorner, J. (1996). A conserved motif at the amino termini of MEKs might mediate high-affinity interaction with the cognate MAPKs. *Trends Biochem. Sci.* **21**, 373–374.
8. Bohn, L. M., Lefkowitz, R. J., Gainetdinov, R. R., Peppel, K., Caron, M. G., and Lin, F. T. (1999). Enhanced morphine analgesia in mice lacking beta-arrestin 2. *Science* **286**, 2495–2498.
9. Bonny, C., Nicod, P., and Waeber, G. (1998). IB1, a JIP-1-related nuclear protein present in insulin-secreting cells. *J. Biol. Chem.* **273**, 1843–1846.
10. Boulton, T. G., Nye, S. H., Robbins, D. J., Ip, N. Y., Radziejewska, E., Morgenbesser, S. D., DePinho, R. A., Panayotatos, N., Cobb, M. H., and Yancopoulos, G. D. (1991). ERKs: A family of protein-serine/threonine kinases that are activated and tyrosine phosphorylated in response to insulin and NGF. *Cell* **65**, 663–675.
11. Bowman, A. B., Kamal, A., Ritchings, B. W., Philp, A. V., McGrail, M., Gindhart, J. G., and Goldstein, L. S. (2000). Kinesin-dependent axonal transport is mediated by the Sunday driver (SYD) protein. *Cell* **103**, 583–594.
12. Buchsbaum, R. J., Connolly, B. A., and Feig, L. A. (2002). Interaction of Rac exchange factors Tiaml and Ras-GRFl with a scaffold for the p38 mitogen-activated protein kinase cascade. *Mol. Cell. Biol.* **22**, 4073–4085.
13. Canagarajah, B. J., Khokhlatchev, A., Cobb, M. H., and Goldsmith, E. J. (1997). Activation mechanism of the MAP kinase ERK2 by dual phosphorylation. *Cell* **90**, 859–869.
14. Chang, C. I., Xu, B. E., Akella, R., Cobb, M. H., and Goldsmith, E. J. (2002). Crystal structures of MAP kinase p38 complexed to the docking sites on its nuclear substrate MEF2A and activator MKK3b. *Mol. Cell* **9**, 1241–1249.
15. Chen, A. J., Zhou, G., Juan, T., Colicos, S. M., Cannon, J. P., Cabriera-Hansen, M., Meyer, C. F., Jurecic, R., Copeland, N. G., Gilbert, D. J., Jenkins, N. A., Fletcher, F., Tan, T. H., and Belmont, J. W. (2002). The Dual Specificity JKAP Specifically Activates the c-Jun N-terminal Kinase Pathway. *J. Biol. Chem.* **277**, 36592–36601.
16. Cheng, J., Yang, J., Xia, Y., Karin, M., and Su, B. (2000). Synergistic interaction of MEK kinase 2, c-Jun N-terminal kinase (JNK) kinase 2, and JNK1 results in efficient and specific JNK1 activation. *Mol. Cell. Biol.* **20**, 2334–2342.
17. Cobb, M. H. (1999). MAP kinase pathways. *Prog. Biophys. Mol. Biol.* **71**, 479–500.
18. Cobb, M. H. and Goldsmith, E. J. (1995). How MAP kinases are regulated. *J. Biol. Chem.* **270**, 14843–14846.
19. Cobb, M. H. and Goldsmith, E. J. (2000). Dimerization in MAP-kinase signaling. *Trends Biochem. Sci.* **25**, 7–9.
20. Conze, D., Krahl, T., Kennedy, N., Weiss, L., Lumsden, J., Hess, P., Flavell, R. A., Le Gros, G., Davis, R. J., and Rincon, M. (2002). c-Jun NH(2)-terminal kinase (JNK)l and JNK2 have distinct roles in CD8(+) T cell activation. *J. Exp. Med.* **195**, 811–823.
21. Cowley, S., Paterson, H., Kemp, P., and Marshall, C. J. (1994). Activation of MAP kinase kinase is necessary and sufficient for PC 12

differentiation and for transformation of NIH 3T3 cells. *Cell* **77**, 841–852.
22. Davis, R. J. (1993). The mitogen-activated protein kinase signal transduction pathway. *J. Biol. Chem.* **268**, 14553–14556.
23. Davis, R. J. (2000). Signal transduction by the JNK group of MAP kinases. *Cell* **103**, 239–252.
24. Dong, C., Yang, D. D., Tournier, C., Whitmarsh, A. J., Xu, J., Davis, R. J., and Flavell, R. A. (2000). JNK is required for effector T-cell function but not for T-cell activation. *Nature* **405**, 91–94.
25. Dong, C., Yang, D. D., Wysk, M., Whitmarsh, A. J., Davis, R. J., and Flavell, R. A. (1998). Defective T cell differentiation in the absence of Jnk1. *Science* **282**, 2092–2095.
26. Duesbery, N. S., Webb, C. P., Leppla, S. H., Gordon, V. M., Klimpel, K. R., Copeland, T. D., Ahn, N. G., Oskarsson, M. K., Fukasawa, K., Paull, K. D., and Vande Woude, G. F. (1998). Proteolytic inactivation of MAP-kinase-kinase by anthrax lethal factor. *Science* **280**, 734–737.
27. English, J. M. and Cobb, M. H. (2002). Pharmacological inhibitors of MAPK pathways. *Trends Pharmacol. Sci.* **23**, 40–45.
28. English, J. M., Vanderbilt, C. A., Xu, S., Marcus, S., and Cobb, M. H. (1995). Isolation of MEK5 and differential expression of alternatively spliced forms. *J. Biol. Chem.* **270**, 28897–28902.
29. Enslen, H. and Davis, R. J. (2001). Regulation of MAP kinases by docking domains. *Biol. Cell* **93**, 5–14.
30. Fleming, Y., Armstrong, C. G., Morrice, N., Paterson, A., Goedert, M., and Cohen, P. (2000). Synergistic activation of stress-activated protein kinase 1/c-Jun N-terminal kinase (SAPK1/JNK) isoforms by mitogen-activated protein kinase kinase 4 (MKK4) and MKK7. *Biochem. J.* **352**, Pt 1, 145.
31. Galanis, A., Yang, S. H., and Sharrocks, A. D. (2001). Selective targeting of MAPKs to the ETS domain transcription factor SAP-1. *J. Biol. Chem.* **276**, 965–973.
32. Gallagher, E. D., Xu, S., Moomaw, C., Slaughter, C. A., and Cobb, M. H. (2002). Binding of JNK/SAPK to MEKK1 is regulated by phosphorylation. *J. Biol. Chem.* **12**, 12.
33. Ganiatsas, S., Kwee, L., Fujiwara, Y., Perkins, A., Ikeda, T., Labow, M. A., and Zon, L. I. (1998). SEK1 deficiency reveals mitogen-activated protein kinase cascade crossregulation and leads to abnormal hepatogenesis. *Proc. Natl. Acad. Sci. USA* **95**, 6881–6886.
34. Girardin, S. E. and Yaniv, M. (2001). A direct interaction between JNK1 and CrkII is critical for Rac1- induced JNK activation. *EMBO J.* **20**, 3437–3446.
35. Giroux, S., Tremblay, M., Bernard, D., Cardin-Girard, J. F., Aubry, S., Larouche, L., Rousseau, S., Huot, J., Landry, J., Jeannotte, L., and Charron, J. (1999). Embryonic death of Mek1-deficient mice reveals a role for this kinase in angiogenesis in the labyrinthine region of the placenta. *Curr. Biol.* **9**, 369–372.
36. Gonzalez, F. A., Raden, D. L., Rigby, M. R., and Davis, R. J. (1992). Heterogeneous expression of four MAP kinase isoforms in human tissues. *FEBS Lett.* **304**, 170–178.
37. Gotthardt, M., Trommsdorff, M., Nevitt, M. F., Shelton, J., Richardson, J. A., Stockinger, W., Nimpf, J., and Herz, J. (2000). Interactions of the low density lipoprotein receptor gene family with cytosolic adaptor and scaffold proteins suggest diverse biological functions in cellular communication and signal transduction. *J. Biol. Chem.* **275**, 25616–25624.
38. Gupta, S., Barrett, T., Whitmarsh, A. J., Cavanagh, J., Sluss, H. K., Derijard, B., and Davis, R. J. (1996). Selective interaction of JNK protein kinase isoforms with transcription factors. *EMBO J.* **115**, 2760–2770.
39. Han, J., Jiang, Y., Li, Z., Kravchenko, V. V., and Ulevitch, R. J. (1997). Activation of the transcription factor MEF2C by the MAP kinase p38 in inflammation. *Nature* **386**, 296–299.
40. Hasegawa, M., Cuenda, A., Spillantini, M. G., Thomas, G. M., Buee-Scherrer, V., Cohen, P., and Goedert, M. (1999). Stress-activated protein kinase-3 interacts with the PDZ domain of alpha 1-syntrophin. A mechanism for specific substrate recognition. *J. Biol. Chem.* **274**, 12626–12631.
41. Hess, P., Pihan, G., Sawyers, C. L., Flavell, R. A., and Davis, R. J. (2002). Survival signaling mediated by c-Jun NH(2)-terminal kinase in transformed B lymphoblasts. *Nat. Genet.* **32**, 201–205.
42. Holmberg, C., Katz, S., Lerdrup, M., Herdegen, T., Jaattela, M., Aronheim, A., and Kallunki, T. (2002). A novel specific role for I kappa B kinase complex-associated protein in cytosolic stress signaling. *J. Biol. Chem.* **277**, 31918–31928.
43. Ip, Y. T. and Davis, R. J. (1998). Signal transduction by the c-Jun N-terminal kinase (JNK)—From inflammation to development. *Curr. Opin. Cell Biol.* **10**, 205–219.
44. Ito, M., Yoshioka, K., Akechi, M., Yamashita, S., Takamatsu, N., Sugiyama, K., Hibi, M., Nakabeppu, Y., Shiba, T., and Yamamoto, K. I. (1999). JSAP1, a novel jun N-terminal protein kinase (JNK)-binding protein that functions as a Scaffold factor in the JNK signaling pathway. *Mol. Cell. Biol.* **19**, 7539–7548.
45. Jacobs, D., Glossip, D., Xing, H., Muslin, A. J., and Kornfeld, K. (1999). Multiple docking sites on substrate proteins form a modular system that mediates recognition by ERK MAP kinase. *Genes Dev.* **13**, 163–175.
46. Karandikar, M., Xu, S., and Cobb, M. H. (2000). MEKK1 binds Raf-1 and the ERK2 cascade components. *J. Biol. Chem.* **275**, 40120–40127.
47. Kato, Y., Kravche, Tapping, R. I., Han, J., Ulevitch, R. J., and Lee, J. D. (1997). BMK1/ERK5 regulates serum-induced early gene expression through transcription factor MEF2C. *EMBO J.* **16**, 7054–7066.
48. Kato, Y., Tapping, R. I., Huang, S., Watson, M. H., Ulevitch, R. J., and Lee, J. D. (1998). Bmk1/Erk5 is required for cell proliferation induced by epidermal growth factor. *Nature* **395**, 713–716.
49. Kelkar, N., Gupta, S., Dickens, M., and Davis, R. J. (2000). Interaction of a mitogen-activated protein kinase signaling module with the neuronal protein JIP3. *Mol. Cell. Biol.* **20**, 1030–1043.
50. Khokhlatchev, A. V., Canagarajah, B., Wilsbacher, J., Robinson, M., Atkinson, M., Goldsmith, E., and Cobb, M. H. (1998). Phosphorylation of the MAP kinase ERK2 promotes its homodimerization and nuclear translocation. *Cell* **93**, 605–615.
51. Knebel, A., Morrice, N., and Cohen, P. (2001). A novel method to identify protein kinase substrates: eEF2 kinase is phosphorylated and inhibited by SAPK4/p38delta. *EMBO J.* **20**, 4360–4369.
52. Kotlyarov, A., Neininger, A., Schubert, C., Eckert, R., Birchmeier, C., Volk, H. D., and Gaestel, M. (1999). MAPKAP kinase 2 is essential for LPS-induced TNF-alpha biosynthesis. *Nat. Cell Biol.* **1**, 94–97.
53. Kuan, C. Y., Yang, D. D., Samanta Roy, D. R., Davis, R. J., Rakic, P., and Flavell, R. A. (1999). The Jnk1 and Jnk2 protein kinases are required for regional specific apoptosis during early brain development. *Neuron* **22**, 667–676.
54. Kyriakis, J. M. and Avruch, J. (2001). Mammalian mitogen-activated protein kinase signal transduction pathways activated by stress and inflammation. *Physiol. Rev.* **81**, 807–869.
55. Lawler, S., Fleming, Y., Goedert, M., and Cohen, P. (1998). Synergistic activation of SAPK1/JNK1 by two MAP kinase kinases in vitro. *Curr. Biol.* **8**, 1387–1390.
56. Lee, J. C., Laydon, J. T., McDonnell, P. C., Gallagher, T. F., Kumar, S., Green, D., McNulty, D., Blumenthal, M. J., Keys, J. R., Landvatter, S. W. et al. (1994). A protein kinase involved in the regulation of inflammatory cytokine biosynthesis. *Nature* **372**, 739–746.
57. Lei, K., Nimnual, A., Zong, W. X., Kennedy, N. J., Flavell, R. A., Thompson, C. B., Bar-Sagi, D., and Davis, R. J. (2002). The Bax subfamily of Bcl2-related proteins is essential for apoptotic signal transduction by c-Jun NH(2)-terminal kinase. *Mol. Cell. Biol.* **22**, 4929–4942.
58. Lewis, T. S., Shapiro, P. S., and Ahn, N. G. (1998). Signal transduction through MAP kinase cascades. *Adv. Cancer Res.* **74**, 49–139.
59. Lin, L. L., Wartmann, M., Lin, A. Y., Knopf, J. L., Seth, A., and Davis, R. J. (1993). cPLA2 is phosphorylated and activated by MAP kinase. *Cell* **72**, 269–278.
60. Lu, H. T., Yang, D. D., Wysk, M., Gatti, E., Mellman, I., Davis, R. J., and Flavell, R. A. (1999). Defective IL-12 production in mitogen-activated protein (MAP) kinase kinase 3 (Mkk3)-deficient mice. *EMBO J.* **18**, 1845–1857.
61. Luttrell, L. M., Roudabush, F. L., Choy, E. W., Miller, W. E., Field, M. E., Pierce, K. L., and Lefkowitz, R. J. (2001). Activation and targeting of extracellular signal-regulated kinases by beta-arrestin scaffolds. *Proc. Natl. Acad. Sci. USA* **98**, 2449–2454.

62. Mahtani, K. R., Brook, M., Dean, J. L., Sully, G., Saklatvala, J., and Clark, A. R. (2001). Mitogen-activated protein kinase p38 controls the expression and posttranslational modification of tristetraprolin, a regulator of tumor necrosis factor alpha mRNA stability. *Mol. Cell. Biol.* **21**, 6461–6469.

63. Mansour, S. J., Matten, W. T., Hermann, A. S., Candia, J. M., Rong, S., Fukasawa, K., Vande Woude, G. F., and Ahn, N. G. (1994). Transformation of mammalian cells by constitutively active MAP kinase kinase. *Science* **265**, 966–970.

64. Marais, R., Wynne, J., and Treisman, R. (1993). The SRF accessory protein Elk-1 contains a growth factor-regulated transcriptional activation domain. *Cell* **73**, 381–393.

65. Marti, A., Luo, Z., Cunningham, C., Ohta, Y., Harrwig, J., Stossel, T. P., Kyriakis, J. M., and Avruch, J. (1997). Actin-binding protein-280 binds the stress-activated protein kinase (SAPK) activator SEK-1 and is required for tumor necrosis factor-alpha activation of SAPK in melanoma cells. *J. Biol. Chem.* **272**, 2620–2628.

66. Matsushime, H., Jinno, A., Takagi, N., and Shibuya, M. (1990). A novel mammalian protein kinase gene (mak) is highly expressed in testicular germ cells at and after meiosis. *Mol. Cell. Biol.* **10**, 2261–2268.

67. Mazzucchelli, C., Vantaggiato, C., Ciamei, A., Fasano, S., Pakhotin, P., Krezel, W., Welzl, H., Wolfer, D. P., Pages, G., Valverde, O., Marowsky, A., Porrazzo, A., Orban, P. C., Maldonado, R., Ehrengruber, M. U., Cestari, V., Lipp, H. P., Chapman, P. F., Pouyssegur, J., and Brambilla, R. (2002). Knockout of ERK1 MAP kinase enhances synaptic plasticity in the striatum and facilitates striatal-mediated learning and memory. *Neuron* **34**, 807–820.

68. McDonald, P. H., Chow, C. W., Miller, W. E., Laporte, S. A., Field, M. E., Lin, F. T., Davis, R. J., and Lefkowitz, R. J. (2000). Beta-arrestin 2: A receptor-regulated MAPK scaffold for the activation of JNK3. *Science* **290**, 1574–1577.

69. Meyer, D., Liu, A., and Margolis, B. (1999). Interaction of c-Jun aminoterminal kinase interacting protein-1 with p190 rhoGEF and its localization in differentiated neurons. *J. Biol. Chem.* **274**, 35113–35118.

70. Michaud, N. R., Therrien, M., Cacace, A., Edsall, L. C., Spiegel, S., Rubin, G. M., and Morrison, D. K. (1997). KSR stimulates Raf-1 activity in a kinase-independent manner. *Proc. Natl. Acad. Sci. USA* **94**, 12792–12796.

71. Miller, W. E. and Lefkowitz, R. J. (2001). Expanding roles for beta-arrestins as scaffolds and adapters in GPCR signaling and trafficking. *Curr. Opin. Cell Biol.* **13**, 139–145.

72. Miller, W. E., McDonald, P. H., Cai, S. F., Field, M. E., Davis, R. J., and Lefkowitz, R. J. (2001). Identification of a motif in the carboxyl terminus of beta-arrestin2 responsible for activation of JNK3. *J. Biol. Chem.* **276**, 27770–27777.

73. Miyata, Y., Akashi, M., and Nishida, E. (1999). Molecular cloning and characterization of a novel member of the MAP kinase superfamily. *Genes Cells* **4**, 299–309.

74. Morrison, D. K. (2001). KSR: A MAPK scaffold of the Ras pathway? *J. Cell Sci.* **114**, 1609–1612.

75. Muller, J., Ory, S., Copeland, T., Piwnica-Worms, H., and Morrison, D. K. (2001). C-TAK1 regulates Ras signaling by phosphorylating the MAPK scaffold, KSR1. *Mol. Cell* **8**, 983–993.

76. Negri, S., Oberson, A., Steinmann, M., Sauser, C., Nicod, P., Waeber, G., Schorderet, D. F., and Bonny, C. (2000). cDNA cloning and mapping of a novel islet-brain/JNK-interacting protein. *Genomics* **64**, 3.

77. Neininger, A., Kontoyiannis, D., Kotlyarov, A., Winzen, R., Eckert, R., Volk, H. D., Holtmann, H., Kollias, G., and Gaestel, M. (2002). MK2 targets AU-rich elements and regulates biosynthesis of tumor necrosis factor and interleukin-6 independently at different post-transcriptional levels. *J. Biol. Chem.* **277**, 3065–3068.

78. Nguyen, A., Burack, W. R., Stock, J. L., Kortum, R., Chaika, O. V., Afkarian, M., Muller, W. J., Murphy, K. M., Morrison, D. K., Lewis, R. E., McNeish, J., and Shaw, A. S. (2002). Kinase suppressor of Ras (KSR) is a scaffold which facilitates mitogen-activated protein kinase activation in vivo. *Mol. Cell. Biol.* **22**, 3035–3045.

79. Nishina, H., Bachmann, M., Oliveira-dos-Santos, A. J., Kozieradzki, L., Odermatt, B., Wakeham, A., Shahinian, A., Takimoto, H., Bernstein, A., Mak, T. W., Woodgett, J. R., Ohashi, P. S., and Penninger, J. M. (1997). Impaired CD28-mediated interleukin 2 production and proliferation in stress kinase SAPK/ERK1 kinase (SEKl)/mitogen-activated protein kinase kinase 4 (MKK4)-deficient T lymphocytes. *J. Exp. Med.* **186**, 941–953.

80. Nishina, H., Fischer, K. D., Radvanyi, L., Shahinian, A., Hakem, R., Rubie, E. A., Bernstein, A., Mak, T. W., Woodgett, J. R., and Penninger, J. M. (1997). Stress-signalling kinase Sek1 protects thymocytes from apoptosis mediated by CD95 and CDS. *Nature* **385**, 350–353.

81. Nishina, H., Radvanyi, L., Raju, K., Sasaki, T., Kozieradzki, L., and Penninger, J. M. (1998). Impaired TCR-mediated apoptosis and Bcl-XL expression in T cells lacking the stress kinase activator SEK1/MKK4. *J. Immunol.* **161**, 3416–3420.

82. Nishina, H., Vaz, C., Billia, P., Nghiem, M., Sasaki, T., De la Pompa, J. L., Furlonger, K., Paige, C., Hui, C., Fischer, K. D., Kishimoto, H., Iwatsubo, T., Katada, T., Woodgett, J. R., and Penninger, J. M. (1999). Defective liver formation and liver cell apoptosis in mice lacking the stress signaling kinase SEK1/MKK4. *Development* **126**, 505–516.

83. Pages, G., Guerin, S., Grail, D., Bonino, F., Smith, A., Anjuere, F., Auberger, P., and Pouyssegur, J. (1999). Defective thymocyte maturation in p44 MAP kinase (Erk 1) knockout mice. *Science* **286**, 1374–1377.

84. Potapova, O., Gorospe, M., Dougherty, R. H., Dean, N. M., Gaarde, W. A., and Holbrook, N. J. (2000). Inhibition of c-Jun N-terminal kinase 2 expression suppresses growth and induces apoptosis of human tumor cells in a p53-dependent manner. *Mol. Cell. Biol.* **20**, 1713–1722.

85. Qian, Z., Okuhara, D., Abe, M. K., and Rosner, M. R. (1999). Molecular cloning and characterization of a mitogen-activated protein kinase-associated intracellular chloride channel. *J. Biol. Chem.* **274**, 1621–1627.

86. Raingeaud, J., Whitmarsh, A. J., Barrett, T., Derijard, B., and Davis, R. J. (1996). MKK3-and MKK6-regulated gene expression is mediated by the p38 mitogen-activated protein kinase signal transduction pathway. *Mol. Cell. Biol.* **16**, 1247–1255.

87. Regan, C. P., Li, W., Boucher, D. M., Spatz, S., Su, M. S., and Kuida, K. (2002). Erk5 null mice display multiple extraembryonic vascular and embryonic cardiovascular defects. *Proc. Natl. Acad. Sci. USA* **99**, 9248–9253.

88. Roy, F., Laberge, G., Douziech, M., Ferland-McCollough, D., and Therrien, M. (2002). KSR is a scaffold required for activation of the ERK/MAPK module. *Genes Dev.* **16**, 427–438.

89. Roy, F. and Therrien, M. (2002). MAP Kinase Module: The Ksr Connection. *Curr. Biol.* **12**, R325–327.

90. Sabapathy, K., Hu, Y., Kallunki, T., Schreiber, M., David, J. P., Jochum, W., Wagner, E. F., and Karin, M. (1999). JNK2 is required for efficient T-cell activation and apoptosis but not for normal lymphocyte development. *Curr. Biol.* **9**, 116–125.

91. Sabapathy, K., Jochum, W., Hochedlinger, K., Chang, L., Karin, M., and Wagner, E. F. (1999). Defective neural tube morphogenesis and altered apoptosis in the absence of both JNK1 and JNK2. *Mech. Dev.* **89**, 115–124.

92. Sabapathy, K., Kallunki, T., David, J. P., Graef, I., Karin, M., and Wagner, E. F. (2001). c-Jun NH2-terminal kinase (JNK)1 and JNK2 have similar and stage-dependent roles in regulating T cell apoptosis and proliferation. *J. Exp. Med.* **193**, 317–328.

93. Sasaki, T., Wada, T., Kishimoto, H., Irie-Sasaki, J., Matsumoto, G., Goto, T., Yao, Z., Wakeham, A., Mak, T. W., Suzuki, A., Cho, S. K., Zuniga-Pflucker, J. C., Oliveira-dos-Santos, A. J., Katada, T., Nishina, H., and Penninger, J. M. (2001). The stress kinase mitogen-activated protein kinase kinase (MKK)7 is a negative regulator of antigen receptor and growth factor receptor-induced proliferation in hematopoietic cells. *J. Exp. Med.* **194**, 757–768.

94. Schaeffer, H. J., Catling, A. D., Eblen, S. T., Collier, L. S., Krauss, A., and Weber, M. J. (1998). MP1: A MEK binding partner that enhances enzymatic activation of the MAP kinase cascade. *Science*, 1668–1671.

95. Schaeffer, H. J. and Weber, M. J. (1999). Mitogen-activated protein kinases: specific messages from ubiquitous messengers. *Mol. Cell. Biol.* **19**, 2435–2444.

96. Schoorlemmer, J. and Goldfarb, M. (2001). Fibroblast growth factor homologous factors are intracellular signaling proteins. *Curr. Biol.* **11**, 793–797.
97. Schoorlemmer, J. and Goldfarb, M. (2002). FGF homologous factors and the islet brain-2 scaffold protein regulate activation of a stress-activated protein kinase. *J. Biol. Chem.* **18**, 18.
98. Shen, Y., Luche, R., Wei, B., Gordon, M. L., Diltz, C. D., and Tonks, N. K. (2001). Activation of the Jnk signaling pathway by a dual-specificity phosphatase, JSP-1. *Proc. Natl. Acad. Sci. USA* **98**, 13613–13618.
99. Shin, T. H., Yasuda, J., Rocheleau, C. E., Lin, R., Soto, M., Bei, Y., Davis, R. J., and Mello, C. C. (1999). MOM-4, a MAP kinase kinase kinase-related protein, activates WRM-I/LIT-1 kinase to transduce anterior/posterior polarity signals in C. elegans. *Mol. Cell* **4**, 275–280.
100. Shinkai, Y., Satoh, H., Takeda, N., Fukuda, M., Chiba, E., Kato, T., Kuramochi, T., and Araki, Y. (2002). A testicular germ cell-associated serine-threonine kinase, MAK, is dispensable for sperm formation. *Mol. Cell. Biol.* **22**, 3276–3280.
101. Stockinger, W., Brandes, C., Fasching, D., Hermann, M., Gotthardt, M., Herz, J., Schneider, W. J., and Nimpf, J. (2000). The reelin receptor ApoER2 recruits JNK-interacting proteins-1 and -2. *J. Biol. Chem.* **275**, 25625–25632.
102. Swat, W., Fujikawa, K., Ganiatsas, S., Yang, D., Xavier, R. J., Harris, N. L., Davidson, L., Ferrini, R., Davis, R. J., Labow, M. A., Flavell, R. A., Zon, L. I., and Alt, F. W. (1998). SEK1/MKK4 is required for maintenance of a normal peripheral lymphoid compartment but not for lymphocyte development. *Immunity* **8**, 625–634.
103. Taglienti, C. A., Wysk, M., and Davis, R. J. (1996). Molecular cloning of the epidermal growth factor-stimulated protein kinase p56 KKIAMRE. *Oncogene* **13**, 2563–2574.
104. Tamura, K., Sudo, T., Senftleben, U., Dadak, A. M., Johnson, R., and Karin, M. (2000). Requirement for p38alpha in erythropoietin expression: A role for stress kinases in erythropoiesis. *Cell* **102**, 221–231.
105. Tanaka, N., Kamanaka, M., Enslen, H., Dong, C., Wysk, M., Davis, R. J., and Flavell, R. A. (2002). Differential involvement of p38 mitogen-activated protein kinase kinases MKK3 and MKK6 in T-cell apoptosis. *EMBO Rep.* **3**, 785–791.
106. Tanoue, T., Adachi, M., Moriguchi, T., and Nishida, E. (2000). A conserved docking motif in MAP kinases common to substrates, activators and regulators. *Nat. Cell Biol.* **2**, 110–116.
107. Tanoue, T., Maeda, R., Adachi, M., and Nishida, E. (2001). Identification of a docking groove on ERK and p38 MAP kinases that regulates the specificity of docking interactions. *EMBO J.* **20**, 466–479.
108. Tanoue, T., Yamamoto, T., and Nishida, E. (2002). Modular structure of a docking surface on MAPK phosphatases. *J. Biol. Chem.* **277**, 22942–22949.
109. Thompson, N. A., Haefliger, J. A., Senn, A., Tawadros, T., Magara, F., Ledermann, B., Nicod, P., and Waeber, G. (2001). Islet-brain 1/JNK-interacting protein-1 is required for early embryogenesis in mice. *J. Biol. Chem.* **276**, 27745–27748.
110. Togawa, K., Yan, Y. X., Inomoto, T., Slaugenhaupt, S., and Rustgi, A. K. (2000). Intestinal cell kinase (ICK) localizes to the crypt region and requires a dual phosphorylation site found in map kinases. *J. Cell. Physiol.* **183**, 129–139.
111. Tournier, C., Dong, C., Turner, T. K., Jones, S. N., Flavell, R. A., and Davis, R. J. (2001). MKK7 is an essential component of the JNK signal transduction pathway activated by proinflammatory cytokines. *Genes Dev.* **15**, 1419–1426.
112. Toumier, C., Hess, P., Yang, D. D., Xu, J., Turner, T. K., Nimnual, A., Bar-Sagi, D., Jones, S. N., Flavell, R. A., and Davis, R. J. (2000). Requirement of JNK for stress-induced activation of the cytochrome c- mediated death pathway. *Science* **288**, 870–874.
113. Verhey, K. J., Meyer, D., Deehan, R., Blenis, J., Schnapp, B. J., Rapoport, T. A., and Margolis, B. (2001). Cargo of kinesin identified as JIP scaffolding proteins and associated signaling molecules. *J. Cell Biol.* **152**, 959–970.
114. Wang, Z., Harkins, P. C., Ulevitch, R. J., Han, J., Cobb, M. H., and Goldsmith, E. J. (1997). The structure of mitogen-activated protein kinase p38 at 2.1-A resolution. *Proc. Natl. Acad. Sci. USA* **94**, 2327–2332.
115. Watson, F. L., Heerssen, H. M., Bhattacharyya, A., Klesse, L., Lin, M. Z., and Segal, R. A. (2001). Neurotrophins use the Erk5 pathway to mediate a retrograde survival response. *Nat. Neurosci.* **4**, 981–988.
116. Weston, C. R., Lambright, D. G., and Davis, R. J. (2002). Signal transduction. MAP kinase signaling specificity. *Science* **296**, 2345–2347.
117. Whitmarsh, A. J., Cavanagh, J., Tournier, C., Yasuda, J., and Davis (1998). A mammalian scaffold complex that selectively mediates MAP kinase activation. *Science* **281**, 1671–1674.
118. Whitmarsh, A. J. and Davis, R. J. (1998). Structural organization of MAP-kinase signaling modules by scaffold proteins in yeast and mammals. *Trends Biochem. Sci.* **23**, 481–485.
119. Whitmarsh, A. J. and Davis, R. J. (2000). A central control for cell growth. *Nature* **403**, 255–256.
120. Whitmarsh, A. J., Kuan, C. Y., Kennedy, N. J., Kelkar, N., Haydar, T. F., Mordes, J. P., Appel, M., Rossini, A. A., Jones, S. N., Flavell, R. A., Rakic, P., and Davis, R. J. (2001). Requirement of the JIPl scaffold protein for stress-induced JNK activation. *Genes Dev.* **15**, 2421–2432.
121. Whitmarsh, A. J., Yang, S. H., Su, M. S., Sharrocks, A. D., and Davis, R. J. (1997). Role of p38 and JNK mitogen-activated protein kinases in the activation of ternary complex factors. *Mol. Cell. Biol.* **17**, 2360–2371.
122. Wilson, K. P., Fitzgibbon, M. J., Caron, P. R., Griffith, J. P., Chen, W., McCaffrey, P. G., Chambers, S. P., and Su, M. S. (1996). Crystal structure of p38 mitogen-activated protein kinase. *J. Biol. Chem.* **271**, 27696–27700.
123. Winzen, R., Kracht, M., Ritter, B., Wilhelm, A., Chen, C. Y., Shyu, A. B., Muller, M., Gaestel, M., Resch, K., and Holtmann, H. (1999). The p38 MAP kinase pathway signals for cytokine-induced mRNA stabilization via MAP kinase-activated protein kinase 2 and an AU-rich region-targeted mechanism. *EMBO J.* **18**, 4969–4980.
124. Wunderlich, W., Fialka, I., Teis, D., Alpi, A., Pfeifer, A., Parton, R. G., Lottspeich, F., and Huber, L. A. (2001). A novel 14-kilodalton protein interacts with the mitogen-activated protein kinase scaffold mpl on a late endosomal/lysosomal compartment. *J. Cell Biol.* **152**, 765–776.
125. Wysk, M., Yang, D. D., Lu, H. T., Flavell, R. A., and Davis, R. J. (1999). Requirement of mitogen-activated protein kinase kinase 3 (MKK3) for tumor necrosis factor-induced cytokine expression. *Proc. Natl. Acad. Sci. USA* **96**, 3763–3768.
126. Xia, Y., Wu, Z., Su, B., Murray, B., and Karin, M. (1998). JNKK1 organizes a MAP kinase module through specific and sequential interactions with upstream and downstream components mediated by its amino-terminal extension. *Genes Dev.* **12**, 3369–3381.
127. Xie, X., Gu, Y., Fox, T., Coll, J. T., Fleming, M. A., Markland, W., Caron, P. R., Wilson, K. P., and Su, M. S. (1998). Crystal structure of JNK3: a kinase implicated in neuronal apoptosis. *Structure* **6**, 983–991.
128. Yang, D., Tournier, C., Wysk, M., Lu, H. T., Xu, J., Davis, R. J., and Flavell, R. A. (1997). Targeted disruption of the MKK4 gene causes embryonic death, inhibition of c-Jun NH2-terminal kinase activation, and defects in AP-1 transcriptional activity. *Proc. Natl. Acad. Sci. USA* **94**, 3004–3009.
129. Yang, D. D., Conze, D., Whitmarsh, A. J., Barrett, T., Davis, R. J., Rincon, M., and Flavell, R. A. (1998). Differentiation of CD4+ T cells to Thl cells requires MAP kinase JNK2. *Immunity* **9**, 575–585.
130. Yang, D. D., Kuan, C. Y., Whitmarsh, A. J., Rincon, M., Zheng, T. S., Davis, R. J., Rakic, P., and Flavell, R. A. (1997). Absence of excitotoxicity-induced apoptosis in the hippocampus of mice lacking the Jnk3 gene. *Nature* **389**, 865–870.
131. Yang, S. H., Galanis, A., and Sharrocks, A. D. (1999). Targeting of p38 mitogen-activated protein kinases to MEF2 transcription factors. *Mol. Cell. Biol.* **19**, 4028–4038.

132. Yang, S. H., Whitmarsh, A. J., Davis, R. J., and Sharrocks, A. D. (1998). Differential targeting of MAP kinases to the ETS-domain transcription factor Elk-1. *EMBO J.* **17**, 1740–1749.
133. Yang, S. H., Yates, P. R., Whitmarsh, A. J., Davis, R. J., and Sharrocks, A. D. (1998). The Elk-1 ETS-domain transcription factor contains a mitogen-activated protein kinase targeting motif. *Mol. Cell. Biol.* **18**, 710–720.
134. Yasuda, J., Whitmarsh, A. J., Cavanagh, J., Sharma, M., and Davis, R. J. (1999). The JIP group of mitogen-activated protein kinase scaffold proteins. *Mol. Cell. Biol.* **19**, 7245–7254.
135. Zama, T., Aoki, R., Kamimoto, T., Inoue, K., Ikeda, Y., and Hagiwara, M. (2002). A novel dual specificity phosphatase SKRPl interacts with the MAPK kinase MKK7 and inactivates the JNK MAPK pathway. Implication for the precise regulation of the particular MAPK pathway. *J. Biol. Chem.* **277**, 23909–23918.
136. Zama, T., Aoki, R., Kamimoto, T., Inoue, K., Ikeda, Y., and Hagiwara, M. (2002). Scaffold role of a mitogen-activated protein kinase phosphatase, SKRPl, for the JNK signaling pathway. *J. Biol. Chem.* **277**, 23919–23926.
137. Zhou, G., Bao, Z. Q., and Dixon, J. E. (1995). Components of a new human protein kinase signal transduction pathway. *J. Biol. Chem.* **270**, 12665–12669.

CHAPTER 185

Subcellular Targeting of PKA Through AKAPs: Conserved Anchoring and Unique Targeting Domains

Mark L. Dell'Acqua

*Department of Pharmacology,
University of Colorado Health Sciences Center,
Denver, Colorado*

Introduction

Regulation of the opposing actions of adenylyl cyclases (AC) and phosphodiesterases (PDE) that control levels of the diffusible intracellular second messenger cAMP is central to many signaling responses. The major effector of cAMP in eukaryotic cells is the cAMP-dependent protein kinase (PKA). PKA is a heterotetrameric enzyme consisting of two regulatory subunits (R) that dimerize and each bind and inhibit a single catalytic subunit (C) to form an R_2C_2 holoenzyme (Fig. 1A) [1]. Activation of the inactive PKA holoenzyme by cAMP occurs when two molecules of cAMP bind to each R subunit, resulting in a conformational change that releases the active C-subunits from the inhibitory R_2 dimer (Fig. 1A). The released active C-subunits phosphorylate serine and threonine residues commonly found in sequence contexts of RRXS/T or KRXXS/T to regulate target protein function. The catalytic subunit (C) of PKA can phosphorylate target proteins rapidly near the site of release and in time diffuse to more distant location such as the nucleus, where additional targets are phosphorylated [2]. It is remarkable that this ubiquitous signaling pathway is used in different cell types to tranduce signals to myriad different yet very specific target proteins in a variety of cellular compartments. Thus, elucidating how this PKA signaling versatility is achieved without compromising specificity is fundamental to understanding cAMP signal transduction pathways.

Over the last decade we have learned that both diversity and specificity in cAMP signaling is in large part achieved by targeting the PKA holoenzyme to discrete subcellular locations such that, upon release of the C subunit, phosphorylation of co-localized target substrates is greatly enhanced (Fig. 1B) [2]. This subcellular targeting of PKA is mediated by a class of anchoring-scaffolding proteins called AKAPs (A-kinase anchoring proteins) [3–5]. AKAPs anchor PKA by binding the R-subunit dimer through a structurally conserved anchoring domain and then target the holoenzyme to specific locations within the cell through unique targeting domains (Fig. 1B). AKAPs frequently contain additional protein-protein interaction motifs that serve as tethering sites for the target substrates, as well as additional signaling proteins such as other kinases, phosphatases, scaffold, and adapter proteins. Thus, many AKAPs function as signal-integrating scaffolds that coordinate both subcellular localization and complex assembly of multiprotein signal tranduction machines. This chapter will focus on the structurally conserved R-subunit

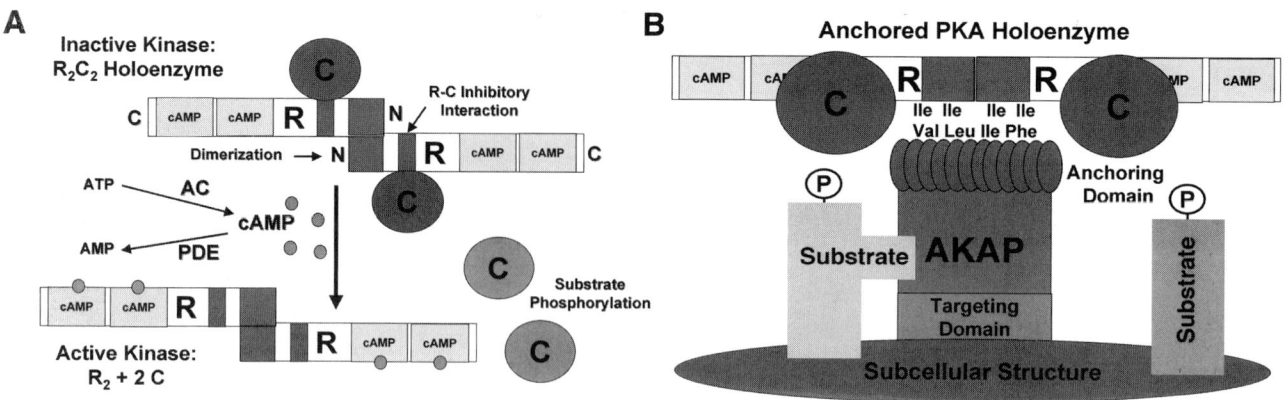

Figure 1 Anchoring and subcellular targeting of PKA by AKAPs. (A) Structure of the PKA holoenzyme and regulation of catalytic activity by cAMP. (B) AKAP anchoring of the PKA holoenzyme near specific cellular substrates. A co-localized substrate and a tethered substrate (see Chapter 186 by Scott) are shown.

anchoring domain and examples of unique subcellular targeting domains that localize anchored-PKA for regulation of localized target substrates. The following chapter will deal more with the issue of multi-enzyme signaling complexes that are maintained by AKAPs.

Structurally Conserved PKA Anchoring Determinants

There are three PKA C subunit isoforms (Cα, Cβ, Cγ) and four R subunit isoforms (RIα, RIβ, RIIα, RIIβ) [1]. Although the C subunit isoforms share very similar properties, the RIα,β and RIIα,β subunit groups show significant differences in both cAMP-binding affinity [1] and AKAP anchoring [6]. Thus, due to these differences, PKA is divided into type I and type II holoenzymes based on the identity of the R subunit dimer present. It is clear that both type I and type II PKA holoenzymes can bind to AKAPs and thus be targeted to specific subcellular domains, but the majority of AKAP proteins bind RII dimers with 100 to 1000-fold higher affinity than RI dimers [6–10]. This difference in AKAP binding affinity is reflected in observations that in many cell types the type II PKA is more discretely localized to specific cellular structures whereas type I PKA is diffusely distributed in the cytoplasm. The differences in cAMP-binding affinity between RI and RII fit well with these differing distributions. Type I holoenzymes have higher cAMP-binding affinities and thus may be adapted to generate more prolonged responses to lower cAMP signals encountered in the bulk cytoplasm. In contrast, type II holoenzymes bind cAMP with lower affinity, a possible adaptation to being targeted by AKAPs to local environments where cAMP can be much higher due to compartmentalized regulation of either AC or PDE [11,12]. One important function of reduced cAMP sensitivity might be to maintain low activity of type II holoenzyme anchored in local environments, where even basal cAMP levels would fully activate the type I enzyme, thus leading to complete loss of C subunits.

A number of mutagenesis studies have shown that AKAPs bind hydrophobic residues in the N-terminus of the R subunits in a manner that depends on formation of the R-R dimer [13–15]. Indeed, recent NMR structural studies show that both the RI and RII N-terminal dimerization domains form anti-parallel four-helix bundles in which dimerization creates an extended hydrophobic surface that binds the AKAP (Fig. 2C) [16–18]. This hydrophobic surface is somewhat less extensive and more sterically hindered in RI compared to RII dimers, a result that possibly explains why in general most AKAPs bind RII with higher affinity than RI. Accordingly, the AKAPs all bind to the R-subunit N-terminal dimerization domain through hydrophobic interactions [18–21]. However, PKA-anchoring domains from different AKAPs have very little primary amino acid sequence similarity yet share a conserved hydrophobic character and secondary structure (Fig. 2A,B) [18,20,21]. Thus, AKAPs are a family of functionally related proteins arising from convergent evolution as opposed to diverging from a common ancestral AKAP protein.

The common secondary structure in AKAP PKA-anchoring domains is seen as a conserved spacing of hydrophobic residues that map to one side of an amphipathic alpha-helix of about 18 residues in length (Fig. 2A,B) [18,20]. One exception to this conservation of an amphipathic-helix anchoring domain is seen in Pericentrin, which anchors PKA-R subunits through a unique, more extended hydrophobic motif of unknown structure [22]. For the conserved anchoring motif found in all other AKAPs, mutagenesis studies have lent support to the amphipathic helix structural model by showing that substitution of hydrophobic residues with either hydrophilic residues to change polarity or proline residues to break helical structure causes inhibition of R-subunit binding [19,20]. Recent NMR structures have been obtained showing clearly that two AKAP-anchoring domain peptides, Ht31 (493–515) and AKAP79(392–413), with divergent primary sequences both bind to RII in similar helical conformations with extensive contacts made between the hydrophobic face of the helices and the large hydrophobic surface on the RII dimer (Fig. 2C) [18]. These extensive hydrophobic interactions explain why these two AKAPs bind PKA-RII with such high affinities. High-affinity R-subunit

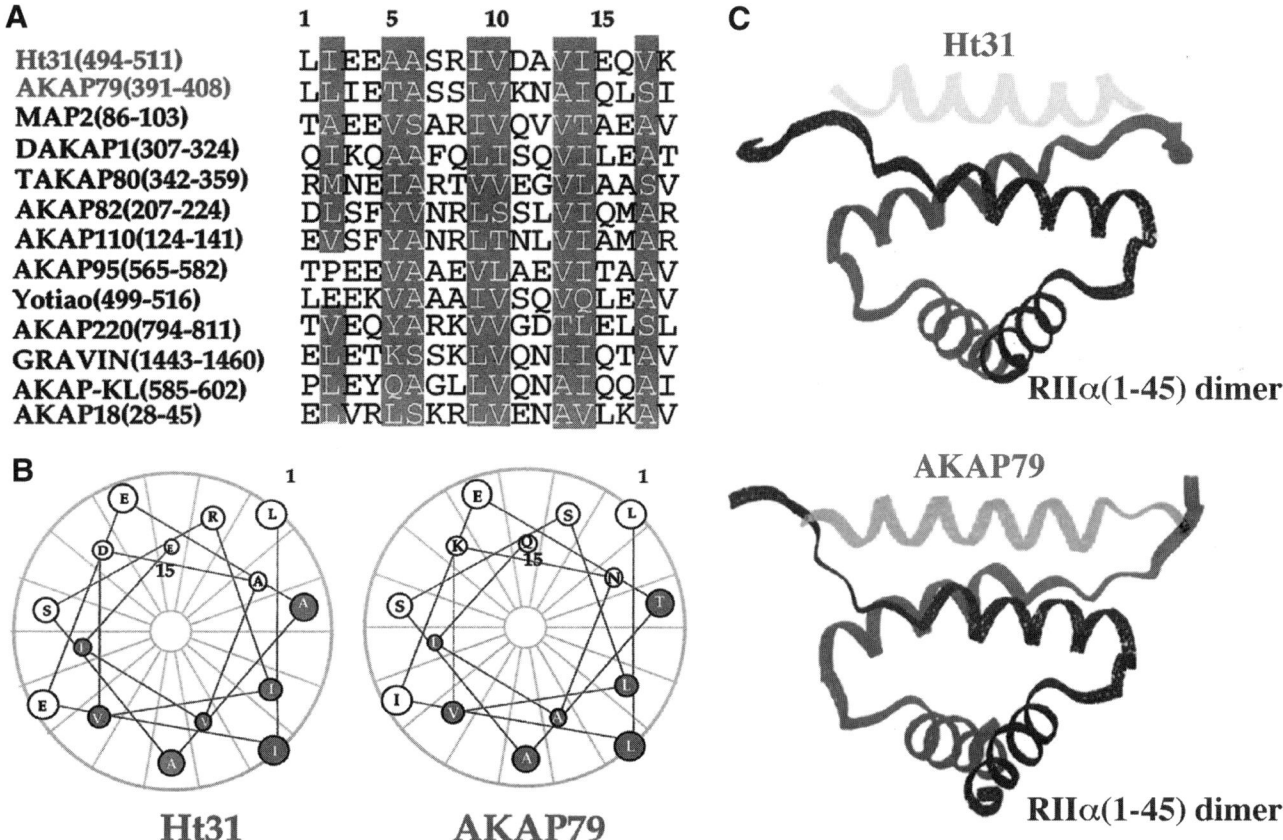

Figure 2 Structurally conserved AKAP-PKA anchoring domains. (A) Divergent primary structures for PKA-anchoring in AKAP family members. Hydrophobic residues are blocked. (B) Conserved amphipathic alpha-helical secondary structures for AKAP PKA-anchoring domains. Hydrophobic residues are shaded on helical wheel diagrams. (C) Conserved mechanisms of PKA anchoring revealed by NMR solution structures of AKAP-RII complexes. The structures show AKAP anchoring domain peptides from Ht31(493-515) and AKAP79 (392-413) in green bound to an RIIα(1-45) N-terminal domain dimer. This figure was adapted from Newlon et al. (2002) EMBO J. 20, 1651–1662 (18) with permission from P. A. Jennings and Oxford University Press.

binding of isolated AKAP-anchoring domain peptides has allowed them to be used as competitive inhibitors of PKA-anchoring in cells to probe numerous cellular functions of AKAP-PKA complexes in cAMP signaling (discussed more below).

Unique Subcellular Targeting Domains

The variety and specificity that is made possible by AKAP-PKA anchoring is in large part a function of unique targeting domains in different AKAP molecules (Fig. 1B). AKAP molecules have been identified at many distinct subcellular locations (Fig. 3A,B), including the plasma membrane, intracellular vesicles [23,24], actin and microtubule cytoskeletons, mitochondria, endoplasmic reticulum (ER), Golgi, and centrosomes [25,26]. For example, MAP2 is targeted to dendritic microtubules in neurons by direct binding to tubulin [27–29]. Scar/Wave1, an AKAP that also anchors the abl-Tyrosine kinase, binds to actin both in focal adhesions (Fig. 3B) and membrane ruffles in fibroblasts, where it regulates the actin polymerization activity of the Arp2/3 complex (see next chapter) [30]. The skeletal and cardiac muscle-enriched mAKAP protein, which also anchors PDE4D3 (see next chapter) [11], is targeted by a series of spectrin repeats to perinuclear ER/SR membranes, including most prominently the nuclear membrane (Fig. 3B) [31,32]. D-AKAP1/sAKAP84 can be targeted either to the ER (in the liver) or mitochondria (in most other cells) by two different N-terminal targeting sequences produced by alternate mRNA splicing [33–35]. One AKAP, AKAP95, has even been shown to associated with the nuclear matrix and chromatin, even though PKA holoenzyme/R-subunits are excluded from the nucleus during interphase (Fig. 3A) [36,37]. However, interactions of AKAP95 with PKA and chromatin could have important functions in regulating chromosome condensation during mitosis when the nuclear envelope is absent [38,39].

In certain cell types precise localization to discrete plasma membrane domains is seen, such as localization to apical or basolateral membrane domains in polarized epithelial cells and synaptic membrane specializations in neurons. AKAP75/79/150, an AKAP scaffold protein that also binds protein kinase C and calcineurin-protein phosphatase 2B(CaN-PP2B) (see next chapter), is targeted to the plasma membrane/cortical cytoskeleton (Fig. 3A) by three N-terminal

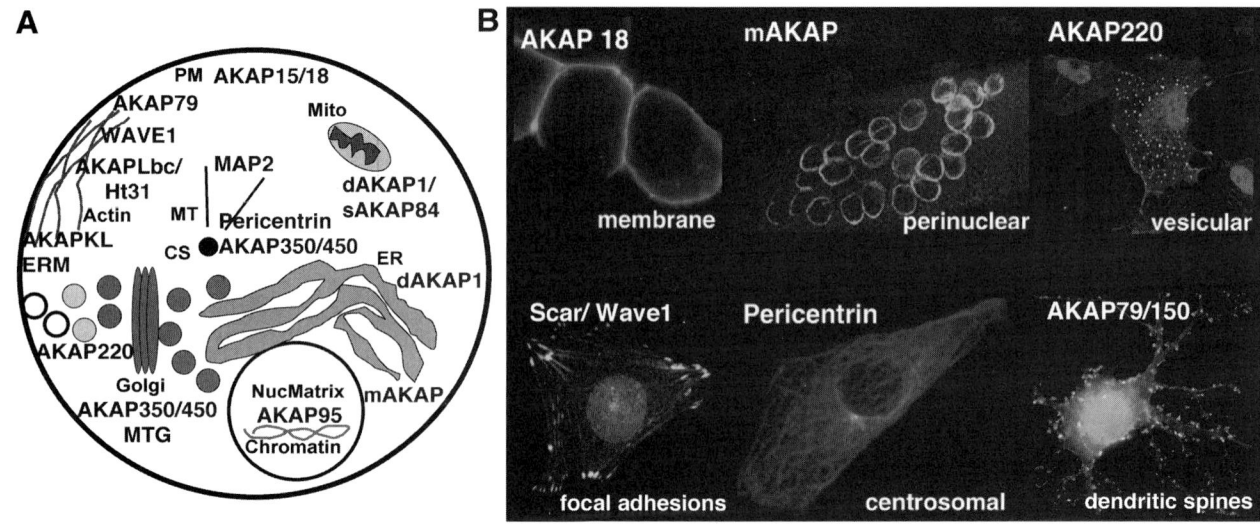

Figure 3 Unique AKAP targeting domains and subcellular localizations. (A) Subcellular compartmentalization of different AKAP family members. References for AKAPLbc/Ht31 [63] and MTG Localization [64]. For other references see the text. (B) Specific subcellular localizations observed for selected AKAPs.

polybasic domains that bind to acidic phospholipids, including phosphatidylinositol-4,5-bisphosphate, as well as F-actin [40–42]. This same N-terminal basic domain mediates targeting to excitatory postsynaptic membrane specializations located on actin-rich dendritic spines in neurons (Fig. 3B) [42]. The low-molecular-weight AKAP15/18α is targeted to the plasma membrane in HEK-293 cells by N-terminal lipid modifications of myristoylation of Gly-1 and dual palmitoylation of Cys-4, Cys-5 (Fig. 3B) [43]. In MDCK epithelial cells this AKAP15/18α isoform selectively targets to the basolateral membrane. However, an alternate splice variant, AKAP15/18β, which contains an additional exon coding for a 24 amino acid insert, localizes to the apical membrane [44]. It is interesting that two additional AKAP families, the AKAP-KL and ERM proteins (Ezrin/Radixin/Moesin), are also specifically targeted to apical membranes, where they bind to cortical actin (Fig. 3A) [45,46].

Probing Cellular Functions of AKAP-PKA Anchoring

The importance of compartmentalized pools of PKA has been implicated in numerous cellular responses, including transcription [47,48], secretion [43,49,50], and cell cycle-regulation [39,51,52]. However, functional roles for AKAP-PKA targeting have been studied in the greatest detail for cAMP regulation of membrane ion channels [53,54]. These studies of ion channel regulation have either used anchoring inhibitor peptides such as Ht31(493-515) to displace PKA from endogenous AKAPs or heterologous co-expression of an AKAP with the ion channel of interest. Some of the best examples of both of these approaches come from studies of PKA-regulation of neuronal ionotropic glutamate receptors and skeletal and cardiac muscle L-type calcium channels. In each case, use of anchoring inhibitor peptides to disrupt PKA-R-subunit anchoring was first shown to block cAMP regulation of endogenous plasma membrane channel activity similar to inhibition of PKA-C subunit catalytic activity [55–58]. Thus, these studies suggested that endogenous AKAPs were important for targeting PKA to the plasma membrane in close proximity to the regulated channels. Subsequent studies confirmed these results by showing that co-expression of these channels with an appropriate membrane-targeted AKAP partner, such as AKAP15/18α or AKAP79, could reconstitute cAMP-PKA regulation of channel activity in heterologous HEK-293 cell expression systems [57,59,60]. In contrast, heterologous expression of the channels by themselves was characterized by a complete lack of PKA regulation or abnormal PKA regulation relative to that seen for endogenous channels. Very recently it has been appreciated that AKAPs and their anchored pools of PKA can be even more directly targeted to membrane ion channel substrates through protein-protein interactions (see next chapter; Fig. 1B). In heterologous systems, AKAP79 and AKAP15/18α were shown to be able to substitute for each other in some aspects of PKA channel regulation [43,57,60]; however, more recent biochemical and electrophysiological studies clearly indicate that these AKAPs serve more specific functions *in vivo* as well as in heterologous systems. In particular, AKAP15/18α is very likely to serve specifically in PKA regulation of skeletal and cardiac L-type channels through forming a complex with channel proteins [61]. In contrast, AKAP79 seems to be adapted for PKA and CaN-PP2B regulation of postsynaptic AMPA-subtype ionotropic glutamate receptors linked through additional protein-protein interactions with PSD-95 family MAGUK scaffold proteins (see next chapter) [60,62].

Conclusions and Future Directions

In summary, subcellular targeting of PKA by AKAPs appears to be a very efficient mechanism for adapting the versatile cAMP signal transduction pathway for highly selective local regulatory phosphorylation events. The key factors in maintaining both the diversity and specificity of this system are the unique subcellular targeting domains present in different AKAP family members. Thus, it is these targeting interactions that might serve in the future as targets for the development of novel therapeutics. The attractiveness of this approach is already supported by studies described in the next chapter that show that selective disruption of interactions between AKAPs and ion channel substrates has the same functional effect as disrupting PKA anchoring to the AKAPs with the nonselective anchoring inhibitor peptides [61]. Furthermore, recent studies of AKAP79/150 targeting to excitatory synapses suggest that regulation of AKAP-targeting domains by cell signaling pathways may serve as important endogenous mechanisms that control PKA signaling [42]. Thus, further dissection of the mechanisms of AKAP targeting, as well as the substrate binding and scaffolding interactions discussed in the following chapter, will continue to be active and important areas of AKAP research in the next few years.

References

1. Taylor, S. S., Buechler, J. A., and Yonemoto, W. (1990). cAMP-dependent protein kinase: framework for a diverse family of regulatory enzymes. *Annu. Rev. Biochem.* **59**, 971–1005.
2. Zhang, J., Ma, Y., Taylor, S. S., and Tsien, R. Y. (2001). Genetically encoded reporters of protein kinase A activity reveal impact of substrate tethering. *Proc. Natl. Acad. Sci. USA* **98**, 14997–15002.
3. Colledge, M. and Scott, J. D. (1999). AKAPs: from structure to function. *Trends Cell Biol.* **9**, 216–221.
4. Feliciello, A., Gottesman, M. E., and Avvedimento, E. V. (2001). The biological functions of A-kinase anchor proteins. *J. Mol. Biol.* **308**, 99–114.
5. Carlisle Michel, J. J. and Scott, J. D. (2002). AKAP mediated signal transduction. *Annu. Rev. Pharmacol. Toxicol.* **42**, 235–257.
6. Burton, K. A., Johnson, B. D., Hausken, Z. E., Westenbroek, R. E., Idzerda, R. L. et al. (1997). Type II regulatory subunits are not required for the anchoring-dependent modulation of Ca^{2+} channel activity by cAMP-dependent protein kinase. *Proc. Natl. Acad. Sci. USA* **94**, 11067–11072.
7. Huang, L. J., Durick, K., Weiner, J. A., Chun, J., and Taylor, S. S. (1997). Identification of a novel dual specificity protein kinase A anchoring protein, D-AKAP1. *J. Biol. Chem.* **272**, 8057–8064.
8. Miki, K. and Eddy, E. M. (1999). Single amino acids determine specificity of binding of protein kinase A regulatory subunits by protein kinase A anchoring proteins. *J. Biol. Chem.* **274**, 29057–29062.
9. Herberg, F. W., Maleszka, A., Eide, T., Vossebein, L., and Tasken, K. (2000). Analysis of A-kinase anchoring protein (AKAP) interaction with protein kinase A (PKA) regulatory subunits: PKA isoform specificity in AKAP binding. *J. Mol. Biol.* **298**, 329–339.
10. Angelo, R. G. and Rubin, C. S. (2000). Characterization of structural features that mediate the tethering of *Caenorhabditis elegans* protein kinase A to a novel A kinase anchor protein. Insights into the anchoring of PKAI isoforms. *J. Biol. Chem.* **275**, 4351–4362.
11. Dodge, K. L., Khouangsathiene, S., Kapiloff, M. S., Mouton, R., Hill, E. V. et al. (2001). mAKAP assembles a protein kinase A/PDE4 phosphodiesterase cAMP signaling module. *EMBO J.* **20**, 1921–1930.
12. Tasken, K. A., Collas, P., Kemmner, W. A., Witczak, O., Conti, M. et al. (2001). Phosphodiesterase 4D and protein kinase a type II constitute a signaling unit in the centrosomal area. *J. Biol. Chem.* **276**, 21999–22002.
13. Hausken, Z. E., Coghlan, V. M., Hasting, C. A. S., Reimann, E. M., and Scott, J. D. (1994). Type II regulatory subunit (RII) of the cAMP dependent protein kinase interaction with A-kinase anchor proteins requires isoleucines 3 and 5. *J. Biol. Chem.* **269**, 24245–24251.
14. Li, Y. and Rubin, C. S. (1995). Mutagenesis of the regulatory subunit (RIIβ) of cAMP-dependent protein kinase IIβ reveals hydrophobic amino acids that are essential for RIIβ dimerization and/or anchoring RIIβ to the cytoskeleton. *J. Biol. Chem.* **270**, 1935–1944.
15. Hausken, Z. E., Dell'Acqua, M. L., Coghlan, V. M., and Scott, J. D. (1996). Mutational analysis of the A-kinase anchoring protein (AKAP)-binding site on RII. *J. Biol. Chem.* **271**, 29016–29022.
16. Newlon, M. G., Roy, M., Morikis, D., Hausken, Z. E., Coghlan, V. et al. (1999). The molecular basis for protein kinase A anchoring revealed by solution NMR. *Nat. Struct. Biol.* **6**, 222–227.
17. Banky, P., Newlon, M. G., Roy, M., Garrod, S., Taylor, S. S. et al. (2000). Isoform-specific differences between the type Ialpha and IIalpha cyclic AMP-dependent protein kinase anchoring domains revealed by solution NMR. *J. Biol. Chem.* **275**, 35146–35152.
18. Newlon, M. G., Roy, M., Morikis, D., Carr, D. W., Westphal, R. et al. (2001). A novel mechanism of PKA anchoring revealed by solution structures of anchoring complexes. *EMBO J.* **20**, 1651–1662.
19. Glantz, S. B., Li, Y., and Rubin, C. S. (1993). Characterization of distinct tethering and intracellular targeting domains in AKAP75, a protein that links cAMP-dependent protein kinase IIβ to the cystoskeleton. *J. Biol. Chem.* **268**, 12796–12804.
20. Carr, D. W., Stofko-Hahn, R. E., Fraser, I. D. C., Bishop, S. M., Acott, T. S. et al. (1991). Interaction of the regulatory subunit (RII) of cAMP-dependent protein kinase with RII-anchoring proteins occurs through an amphipathic helix binding motif. *J. Biol. Chem.* **266**, 14188–14192.
21. Carr, D. W., Hausken, Z. E., Fraser, I. D., Stofko-Hahn, R. E., and Scott, J. D. (1992). Association of the type II cAMP-dependent protein kinase with a human thyroid RII-anchoring protein. Cloning and characterization of the RII-binding domain. *J. Biol. Chem.* **267**, 13376–13382.
22. Diviani, D., Langeberg, L. K., Doxsey, S. J., and Scott, J. D. (2000). Pericentrin anchors protein kinase A at the centrosome through a newly identified RII-binding domain. *Curr. Biol.* **10**, 417–420.
23. Lester, L. B., Coghlan, V. M., Nauert, B., and Scott J. D. (1996). Cloning and characterization of a novel A-kinase anchoring protein: AKAP220, association with testicular peroxisomes. *J. Biol. Chem.* **272**, 9460–9465.
24. Schillace, R. V. and Scott, J. D. (1999). Association of the type 1 protein phosphatase PP1 with the A-kinase anchoring protein AKAP220. *Curr. Biol.* **9**, 321–324.
25. Schmidt, P. H., Dransfield, D. T., Claudio, J. O., Hawley, R. G., Trotter, K. W. et al. (1999). AKAP350: a multiply spliced A-kinase anchoring protein associated with centrosomes. *J. Biol. Chem.* **274**, 3055–3066.
26. Witczak, O., Skalhegg, B. S., Keryer, G., Bornens, M., Tasken, K. et al. (1999). Cloning and characterization of a cDNA encoding an A-kinase anchoring protein located in the centrosome, AKAP450. *EMBO J.* **18**, 1858–1868.
27. Lewis, S. A., Wang, D., and Cowan, N. J. (1988). Microtubule-associated protein MAP2 shares a microtubule binding motif with tau protein. *Science* **242**, 936–939.
28. Rubino, H. M., Dammerman, M., Shafit-Sagardo, B., and Erlichman, J. (1989). Localization and characterization of the binding site for the regulatory subunit of type II cAMP-dependent protein kinase on MAP2. *Neuron* **3**, 631–638.
29. Luo, Z., Shafit-Zagardo, B., and Erlichman, J. (1990). Identification of the MAP2- and P75-binding domain in the regulatory subunit (RIIβ) of Type II cAMP-dependent protein kinase. *J. Biol. Chem.* **265**, 21804–21810.

30. Westphal, R. S., Soderling, S. H., Alto, N. M., Langeberg, L. K., and Scott, J. D. (2000). Scar/WAVE-1, a Wiskott-Aldrich syndrome protein, assembles an actin-associated multi-kinase scaffold. *EMBO J.* **19**, 4589–4600.
31. Kapiloff, M. S., Schillace, R. V., Westphal, A. M., and Scott, J. D. (1999). mAKAP: an A-kinase anchoring protein targeted to the nuclear membrane of differentiated myocytes. *J. Cell Sci.* **112**, 2725–2736.
32. Kapiloff, M. S., Jackson, N., and Airhart, N. (2001). mAKAP and the ryanodine receptor are part of a multi-component signaling complex on the cardiomyocyte nuclear envelope. *J. Cell Sci.* **114**, 3167–3176.
33. Lin, R.-Y., Moss, S. B., and Rubin, C. S. (1995). Characterization of S-AKAP84, a novel developmentally regulated A kinase anchor protein of male germ cells. *J. Biol. Chem.* **270**, 27804–27811.
34. Chen, Q., Reigh-Yi, L., and Rubin, C. (1997). Organelle-specific targeting of protein kinase AII (PKA). *J. Biol. Chem.* **272**, 15247–15257.
35. Huang, L. J., Wang, L., Ma, Y., Durick, K., Perkins, G. et al. (1999). NH2-Terminal targeting motifs direct dual specificity A-kinase-anchoring protein 1 (D-AKAP1) to either mitochondria or endoplasmic reticulum. *J. Cell Biol.* **145**, 951–959.
36. Coghlan, V. M., Langeberg, L. K., Fernandez, A., Lamb, N. J. C., and Scott, J. D. (1994). Cloning and characterization of AKAP95, a nuclear protein that associates with the regulatory subunit of type II cAMP-dependent protein kinase. *J. Biol. Chem.* **269**, 7658–7665.
37. Eide, T., Coghlan, V., Orstavik, S., Holsve, C., Solberg, R. et al. (1997). Molecular cloning, chromosomal localization and cell cycle-dependent subcellular distribution of the A-kinase anchoring protein, AKAP95. *Exp. Cell Res.* **238**, 305–316.
38. Collas, P., Le Guellec, K., and Tasken, K. (1999). The A-kinase-anchoring protein AKAP95 is a multivalent protein with a key role in chromatin condensation at mitosis. *J. Cell Biol.* **147**, 1167–1180.
39. Landsverk, H. B., Carlson, C. R., Steen, R. L., Vossebein, L., Herberg, F. W. et al. (2001). Regulation of anchoring of the RIIalpha regulatory subunit of PKA to AKAP95 by threonine phosphorylation of RIIalpha: implications for chromosome dynamics at mitosis. *J. Cell Sci.* **114**, 3255–3264.
40. Li, Y., Ndubuka, C., and Rubin, C. S. (1996). A kinase anchor protein 75 targets regulatory (RII) subunits of cAMP-dependent protein kinase II to the cortical actin cytoskeleton in non-neuronal cells. *J. Biol. Chem.* **271**, 16862–16869.
41. Dell'Acqua, M. L., Faux, M. C., Thorburn, J., Thorburn, A., and Scott, J. D. (1998). Membrane-targeting sequences on AKAP79 bind phosphatidylinositol-4,5-bisphosphate. *EMBO J.* **17**, 2246–2260.
42. Gomez, L. L., Alam, S., Horne, E., Smith, K. E., and Dell'Acqua, M. L. (2002). Regulation of AKAP79/150-PKA postsynaptic targeting by NMDA receptor activation of calcineurin and remodeling of dendritic actin. *J. Neurosci.*, in press.
43. Fraser, I. D., Tavalin, S. J., Lester, L. B., Langeberg, L. K., Westphal, A. M. et al. (1998). A novel lipid-anchored A-kinase anchoring protein facilitates cAMP-responsive membrane events. *EMBO J.* **17**, 2261–2272.
44. Trotter, K. W., Fraser, I. D., Scott, G. K., Stutts, M. J., Scott, J. D. et al. (1999). Alternative splicing regulates the subcellular localization of A-kinase anchoring protein 18 isoforms. *J. Cell Biol.* **147**, 1481–1492.
45. Dong, F., Felsmesser, M., Casadevall, A., and Rubin, C. S. (1998). Molecular characterization of a cDNA that encodes six isoforms of a novel murine A kinase anchor protein. *J. Biol. Chem.* **273**, 6533–6541.
46. Dransfield, D. T., Bradford, A. J., Smith, J., Martin, M., Roy, C. et al. (1997). Ezrin is a cyclic AMP-dependent protein kinase anchoring protein. *EMBO J.* **16**, 101–109.
47. Feliciello, A., Li, Y., Avvedimento, E. V., Gottesman, M. E., and Rubin, C. S. (1997). A-kinase anchor protein 75 increases the rate and magnitude of cAMP signaling to the nucleus. *Curr. Biol.* **7**, 1011–1014.
48. Paolillo, M., Feliciello, A., Porcellini, A., Garbi, C., Bifulco, M. et al. (1999). The type and the localization of cAMP-dependent protein kinase regulate transmission of cAMP signals to the nucleus in cortical and cerebellar granule cells. *J. Biol. Chem.* **274**, 6546–6552.
49. Lester, L. B., Langeberg, L. K., and Scott, J. D. (1997). Anchoring of protein kinase A facilitates hormone-mediated insulin secretion. *Proc. Natl. Acad. Sci. USA* **94**, 14942–14947.
50. Lester, L. B., Faux, M. C., Nauert, J. B., and Scott, J. D. (2001). Targeted protein kinase A and PP-2B regulate insulin secretion through reversible phosphorylation. *Endocrinology* **142**, 1218–1227.
51. Keryer, G., Yassenko, M., Labbe, J. C., Castro, A., Lohmann, S. M. et al. (1998). Mitosis-specific phosphorylation and subcellular redistribution of the RIIalpha regulatory subunit of cAMP-dependent protein kinase. *J. Biol. Chem.* **273**, 34594–34602.
52. Carlson, C. R., Witczak, O., Vossebein, L., Labbe, J. C., Skalhegg, B. S. et al. (2001). CDK1-mediated phosphorylation of the RIIalpha regulatory subunit of PKA works as a molecular switch that promotes dissociation of RIIalpha from centrosomes at mitosis. *J. Cell Sci.* **114**, 3243–3254.
53. Gray, P. C., Scott, J. D., and Catterall, W. A. (1998). Regulation of ion channels by cAMP-dependent protein kinase and A-kinase anchoring proteins. *Curr. Opin. Neurobiol.* **8**, 330–334.
54. Fraser, I. D. and Scott, J. D. (1999). Modulation of ion channels: a "current" view of AKAPs. *Neuron* **23**, 423–436.
55. Johnson, B. D., Scheuer, T., and Catterall, W. A. (1994). Voltage-dependent potentiation of L-type Ca^{2+} channels in skeletal muscle cells requires anchored cAMP-dependent protein kinase. *Proc. Natl. Acad. Sci. USA* **91**, 11492–11496.
56. Gray, P. C., Tibbs, V. C., Catterall, W. A., and Murphy, B. J. (1997). Identification of a 15-kDa cAMP-dependent protein kinase-anchoring protein associated with skeletal muscle L-type calcium channels. *J. Biol. Chem.* **272**, 6297–6302.
57. Gao, T., Yatani, A., Dell'Acqua, M. L., Sako, H., Green, S. A. et al. (1997). cAMP-dependent regulation of cardiac L-type Ca^{2+} channels requires membrane targeting of PKA and phosphorylation of channel subunits. *Neuron* **19**, 185–196.
58. Rosenmund, C., Carr, D. W., Bergeson, S. E., Nilaver, G., Scott, J. D. et al. (1994). Anchoring of protein kinase A is required for modulation of AMPA/kainate receptors on hippocampal neurons. *Nature* **368**, 853–856.
59. Gray, P. C., Johnson, B. D., Westenbroek, R. E., Hays, L. G., Yates, J. R. et al. (1998). Primary structure and function of an A kinase anchoring protein associated with calcium channels. *Neuron* **20**, 1017–1026.
60. Tavalin, S. J., Colledge, M., Hell, J. W., Langeberg, L. K., Huganir, R. L. et al. (2002). Regulation of GluR1 by the A-kinase anchoring protein 79 (AKAP79) signaling complex shares properties with long-term depression. *J Neurosci.* **22**, 3044–3051.
61. Hulme, J. T., Ahn, M., Hauschka, S. D., Scheuer, T., and Catterall, W. A. (2002). A novel leucine zipper targets AKAP15 and cyclic AMP-dependent protein kinase to the C-terminus of the skeletal muscle Ca^{2+} channel and modulates its function. *J. Biol. Chem.* **277**, 4079–4087.
62. Colledge, M., Dean, R. A., Scott, G. K., Langeberg, L. K., Huganir, R. L. et al. (2000). Targeting of PKA to glutamate receptors through a MAGUK-AKAP complex. *Neuron* **27**, 107–119.
63. Diviani, D., Soderling, J., and Scott, J. D. (2001). AKAP-Lbc anchors protein kinase A and nucleates Galpha 12-selective Rho-mediated stress fiber formation. *J. Biol. Chem.* **276**, 44247–44257.
64. Schillace, R. V., Andrews, S. F., Liberty, G. A., Davey, M. P., and Carr, D. W. (2002). Identification and characterization of myeloid translocation gene 16b as a novel A kinase anchoring protein in T lymphocytes. *J. Immunol.* **168**, 1590–1599.

AKAP Signaling Complexes: The Combinatorial Assembly of Signal Transduction Units

John D. Scott and Lorene K. Langeberg
Howard Hughes Medical Institute,
Vollum Institute, Oregon Health and Sciences University,
3181 S.W. Sam Jackson Park Road,
Portland, Oregon

Understanding the molecular organization of intracellular signaling pathways is a topic of considerable research interest. Multiprotein signaling complexes create focal points of enzyme activity to disseminate the intracellular action of many hormones and neurotransmitters. The spatio-temporal activation of protein kinases and/or phosphatases is important in controlling where and when phosphorylaton events occur. Anchoring proteins and targeting subunits provide a molecular framework that orients protein kinases and phosphatases toward selected substrates. Prototypic examples of these "signal-directing molecules" are A-kinase anchoring proteins (AKAPs) that sustain multicomponent signaling complexes of the cAMP dependent protein kinase (PKA) and G proteins and other enzymes. These protein-protein interactions not only focus PKA toward certain substrates but also spatially segregate parallel signaling pathways.

Introduction

The efficient transmission of cellular signals often involves the positioning of signaling proteins in proximity to their upstream activators and downstream targets. In fact, the clustering of receptors, G proteins, and enzymes with their substrates is believed to contribute significantly to the specifcity of signaling. This sophisticated degree of organization may also help prevent the indiscriminate activation of related signaling complexes that are close by. This is of particular importance for second messenger dependant signaling pathways that lead to the activation of broad specificity enzymes such as the cAMP dependent protein kinase (PKA), protein kinase C (PKC), and a variety of calmodulin dependant kinases (CaM kinases). Compartmentalization of these enzymes is often achieved through their association with scaffolding proteins that simultaneously coordinate the location of several enzymes [1–4].

A-kinase anchoring proteins (AKAPs) are a growing family of scaffolding proteins that package PKA and other signaling enzymes into multiprotein complexes [5,6]. As discussed in the previous chapter, each AKAP contains both a conserved amphipathic helix that binds to the R subunit dimer with high affinity and a targeting domain that directs the PKA-AKAP complex to specific subcellular compartments [7,8]. Another important role for AKAPs is to place PKA in the proximity of enzymes such as phosphatases and phosphodiesterases that terminate cAMP signaling events [9–11]. The focus of this chapter is to highlight advances in our understanding of AKAP signaling complexes and their role in facilitating this bi-directional control of various signaling events.

G-Protein Signaling Through AKAP Signaling Complexes

A shared property of several AKAPs is to position enzymes in microenvironments where they can respond to upstream signals. Clearly, there are potential advantages of anchoring PKA in close proximity to primary transduction elements such as G-protein coupled receptors and the cAMP synthesis machinery. In fact, two anchoring proteins, gravin/AKAP250 and AKAP79/150, maintain kinase complexes that bind to the β2-adrenergic (β2-AR) receptor [12,13]. The AKAP79 complex binds to regions within the third cytoplasmic loop and C-terminal tail of the β2-AR in an agonist-independent manner, whereas gravin/AKAP250 recruits PKA and PKC to the receptor in an agonist-dependent manner [13–15] (Fig. 1). These receptor based AKAP complexes also contribute to β2-AR phosphorylation, desensitization, and indirect activation of MAP kinase pathways that emanate from the receptor [16]. Furthermore, dephosphorylation of β2-AR and the receptor kinase GRK2 are likely to be important signal termination events in this process and could be mediated by an anchored pool of PP-2B that associates with AKAP79. Another anchoring protein MAP2 seems to nucleate a membrane associated signaling complex that includes β2-AR, adenylyl cyclase, PKA, PP-2B, and a substrate for the kinase of the class C L-type Ca^{2+} channel [17,18]. The identification of such a signaling complex emphasizes the notion that receptors, effectors, kinases, and their substrates are spatially coordinated. However, it also confirms the view put forward by a number of investigators that in some cases cAMP may not have to diffuse very far from its site of synthesis to activate the PKA holoenzyme.

Other classes of G proteins have been implicated in the channeling of signals through AKAP complexes, although not necessarily via cAMP dependant mechanisms [4]. Scar/WAVE-1 is a member of the Wiskott-Aldrich syndrome family of scaffolding proteins that binds PKA, the Abl tyrosine kinase, and the Arp2/3 complex, a group of seven proteins that control actin remodeling [19–21] (Fig. 1). The dynamic assembly of this complex at sites of lamellapodial extension occurs in response to growth factor signals that activate the low-molecular-weight GTPase Rac [19]. Consequently, Scar/WAVE may direct PKA and Abl toward cytoskeletal substrates and synchronize cell movement by ensuring efficient transmission of Rac-mediated signals to the actin remodeling machinery. Analogous AKAP signaling networks participate in the formation of actin stress fibers. AKAP-Lbc, a splice variant of the Lbc oncogene, encodes a chimeric molecule that anchors PKA and functions as a Rho-selective guanine nucleotide exchange factor [22]. Application of lysophosphatydic acid or selective expression of Gα12 enhances cellular AKAP-Lbc activation and leads to the formation of actin stress fibers in fibroblasts [22]. This provides an example where the spatial organization of heterotrimeric and small molecular weight G proteins may involve interactions with the same AKAP. Finally, certain unconventional modes of signaling to PKA may also be governed by G-protein recruitment to AKAP-signaling complexes. For example, the testis specific anchoring protein, AKAP110, has been reported to interact with the heterotrimeric

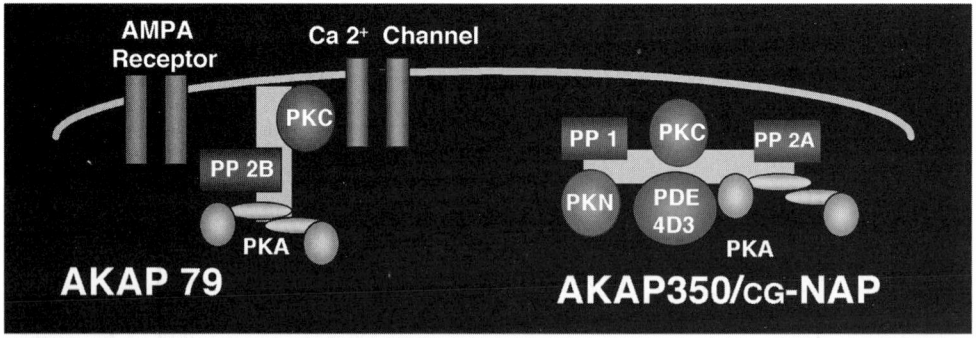

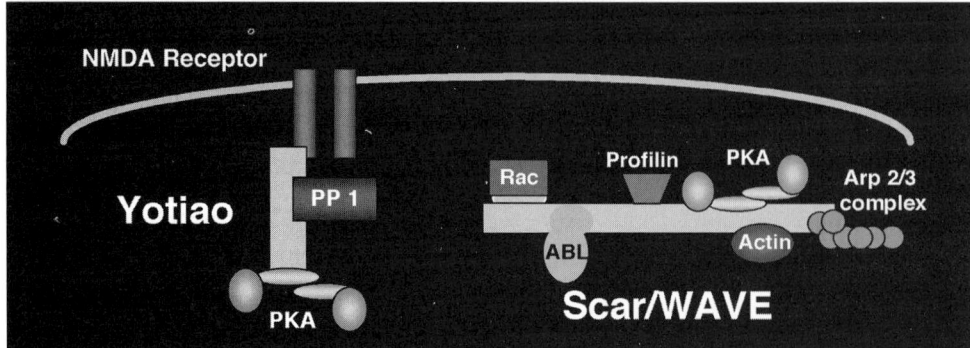

Figure 1 AKAP signaling complexes. A schematic representation of certain AKAP signaling complexes discussed in this chapter. Each interacting protein is labeled.

G-protein subunit Gα13 that activates AKAP110-associated PKA via a cAMP-independent mechanism [23]. Each of these examples underscores the notion that AKAP-signaling complexes can respond to G-protein signaling events in a variety of pathways.

Kinase/Phosphatase Signaling Complexes

Several AKAP signaling complexes include both signal transduction and signal termination enzymes. This generates a locus to regulate the forward and backward steps of a given signaling process. One example of this, mentioned in other chapters, is the clustering of second messenger regulated kinases and phosphatases at the excitatory synapses of neurons by the AKAP79/150 family of anchoring proteins [24].

The human form AKAP79 and its bovine and murine counterparts AKAP75 and AKAP150, respectively, are enriched in the synaptosomal and postsynaptic densitiy fractions of neuronal lysates and are present in dendritic spines [25,26]. In 1995 the A subunit of the calcium/calmodulin dependent phosphatase PP2B was identified in a two-hybrid screen with AKAP79 as bait [27]. Subsequent biochemical and cellular analyses defined the phosphatase-binding site and demonstrated that both enzymes simultaneously associate with AKAP79/150 in neurons [28]. A year later it was demonstrated that PKC is also a component of the AKAP79 signaling complex [29]. At that time it was postulated that the simultaneous anchoring of these three signaling enzymes generated a locus for the integration of distinct second messenger signals at the postsynaptic membranes [30]. Functional studies have largely confirmed this notion by showing that the AKAP79/150-signaling complex controls the phosphorylation status and facilitates the regulation of a variety of ion channels, including L-type calcium channels, KCNQ potassium channels, aquiporin water channels, and AMPA type glutamate receptor ion channels [30–33] (Fig. 1).

The most detailed studies have dissected the phosphorylation events that occur on the cytoplasmic tail of AMPA type glutamate receptors. This channel is present at the terminals of excitatory synapses and is gated by the release of glutamate across the synaptic cleft [34,35]. A series of reports have shown that the AKAP79 signaling complex is recruited into a larger transduction unit with the GluR1 subunit of the AMPA type glutamate receptor (reviewed by Dodge and Scott [24]). Simultaneous association with the membrane-associated guanylate kinase bridging protein SAP97 brings the channel and the signaling complex together [36]. Functional studies indicate that AKAP79-bound PKA enhances GluR1 phosphorylation on serine 845 in the cytoplasmic tail of the channel subunit, an important site for the regulation of channel function during the induction of long-term synaptic depression (LTD) [37–39]. These findings extend an earlier report showing that perfusion of anchoring inhibitor peptides into cultured hippocampal neurons antagonizes PKA anchoring and causes rundown of synaptic AMPA-type glutamate receptor activity [40]. Since this phenomenon occurs with a time-course that is similar to the inhibition of the kinase, it was initially assumed that disruption of PKA anchoring displaced the kinase from the proximity of the AMPA receptor. However, more recent studies indicate that the phosphatase PP2B may play a prominent role in the downregulation of channel activity. Electrophysiological recordings suggest the proximity of the AKAP79-bound phosphatase to sites of calcium entry ensures that the enzyme is rapidly activated upon synaptic elevation of intracellular calcium and is responsible for the dephosphorylation of serine 845. It is interesting that serine 845 is phosphorylated by PKA upon elevation of synaptic cAMP levels [41,42]. Thus AKAP79/150 maintains a kinases and a phosphatase in close proximity to the channel in a manner that allows second messenger dependant changes in the phosphorylation status and activity of GluR1 (Fig. 1).

Other synaptic AKAPs also maintain kinase/phosphatase complexes. For example, yotiao interacts with the NR1a subunit of synaptic NMDA glutamate receptor ion channels and anchors PKA and protein phosphatase 1 (PP-1) [10,43–46]. The modulation of NMDA receptors containing NR1a requires interactions with the scaffolding protein as peptide-mediated displacement of either PKA or PP-1 causes changes in the modulation of channel activity [10]. Thus yotiao maintains a signaling complex that is directly attached to the substrate, the NMDA receptor (Fig. 1). Although this example certainly highlights the role of AKAPs to ensure the rapid preferential phosphorylation of substrates, the compartmentalization of enzymes in the yotiao complex may also contribute to the segregation of signals at excitatory synapses, where the GluR1/AKAP79 complex is also in the immediate vicinity.

Two AKAP additional signaling complexes are found at the centrosome. AKAP350/CG-NAP, a large centrosomal AKAP of unknown function, has been reported to bind three kinases (PKA, PKC, and PKN) and two phosphatases (PP-1 and PP2A; Fig. 1) [20,47–49]. Likewise, pericentrin, an integral component of the centreolar machinery, anchors PKA and other enzymes, presumably for a role in the coordination of centrosomal phosphorylation events. An interesting finding is that both AKAP350. CG-NAP and pericentrin contain a c-terminal PACT domain that is responsible for targeting each anchoring protein to the centrosome. Expression of this 100 amino acid region alone is sufficient to promote centrosomal targeting of GFP [50]. This raises the intriguing possibility that the PACT domains of these anchoring proteins might interact with the same structure in the centrosome in a mutually exclusive manner. This could represent one mechanism to generate greater diversity, as distinct signaling complexes could be tethered to the same cellular locus. Thus, the possibilities for coordinated phosphorylation and dephosphorylation events mediated by association with AKAPs are increased.

cAMP Signaling Units

Another way to exert tight control of PKA phosphorylation events is to compartmentalize the kinase with enzymes that control the intracellular concentrations of its activator, cAMP. In fact, two recent reports have demonstrated that phosphodiesterases, the enzymes that catalyze cAMP degradation, are components of AKAP/PKA signaling complexes [51,52]. These findings add to the complexity of cAMP signaling as they point toward a role for anchored pools of phosphodiesterase in the tight control of local second messenger concentrations. This in turn controls where and when PKA becomes active. For example, a muscle-selective anchoring protein mAKAP directly binds PKA and a splice variant of the cAMP-specific, type 4 phosphodiesterase PDE4D3 and targets them to the perinuclear membranes of cardiomyocytes [51]. Yet in Sertoli cells the PDE4D3 interacts with AKAP350/CG-NAP, one of the large centrosomal AKAPs discussed above (Fig. 1) [52].

Two important regulatory factors that are built into these cAMP-signaling modules favor the signal termination process. First, the tethered PDE is constitutively active and will rapidly restore basal cAMP levels when the flow of second messenger is turned off from its site of synthesis at the plasma membrane. Second, elegant experiments have demonstrated that PKA phosphorylation of PDE4D3 on serine 54 increases the Vmax of the enzyme two- to threefold over basal conditions [53–57]. Phosphorylation of PDE4D3 increases cAMP degradation to favor reformation of the PKA holoenzyme. PKA anchoring is a unique and critical element in this PKA-PDE4D3 feedback loop, as displacement of the kinase with the anchoring inhibitor peptide Ht31 prevents cAMP-dependent stimulation of the mAKAP associated PDE4D activity [51]. This finding emphasizes the importance of PDE localization to maintain the balance of intracellular cAMP levels. This notion is also supported by recent imaging studies using intermolecular FRET that have shown that micro-gradients of cAMP emanate from sites of synthesis at the plasma membrane. Hormonal stimulation of cardiomyocytes induced changes in the rate and magnitude of local cAMP gradients with the concomitant effect on the activation of anchored PKA pools [58]. Thus multiple regulatory processes are involved in controlling where and when cellular PKA activation occurs.

Although PDE4D3 is a substrate for the kinase, it is clear that there are other PKA substrates associated with the mAKAP scaffold. For example, the regulation of ryanodine receptor (RyR) phosphorylation is important for maintaining contractility in response to β-adrenergic signaling and increases in intracellular Ca^{2+} concentration in the heart. Hyperphosphorylation of sarcoplasmic reticulum RyR leads to increased Ca^{2+} sensitivity of the channel and decreased sensitivity to β-adrenergic stimulation [59–61]. These changes are manifest in human heart tissue undergoing heart failure where changes in RyR phosphorylation are also detected [59]. Atypical regulation of RyR function may be due to several factors that regulate cAMP/PKA signaling in heart, including loss of phosphatase activity from the RyR complex [59] and defects in regulation of cAMP levels by PDE activity associated with the complex [51]. It is interesting that two groups have detected PP1 and PP2A phosaphatase subunits in the mAKAP signaling complex. Given the myriad binding partners for mAKAP, it is plausible to suggest that the composition of this signaling network may be altered in response to different intracellular stimuli and in disease states.

Conclusions and Perspectives

AKAPs provide the platforms for the assembly of multiprotein signaling complexes in a variety of cellular compartments. Two factors contribute to the diversity of these signaling units. First, individual AKAP complexes may control distinct signaling events within the same subcellular compartment. This may be best exemplified by WAVE-1 and AKAP-Lbc that nucleate the formation intracellular signaling cascades to catalyze distinct forms of cytoskeletal reorganization. In both cases receptor occupancy at the plasma membrane triggers the assembly of signaling complexes that transmit distinct signals to the actin cytoskeleton [19,22]. Likewise, AKAP350 and pericentrin may synchronize different phosphorylation events at the centrosome, and three anchoring proteins, D-AKAP-1/sAKAP82, D-AKAP-2, and Rab32, are involved in mitochondrial signaling processes [62–64]. Second, the recruitment and release of individual enzymes from the signaling complex provides a dynamic component to the composition of a given protein network. For example, Ca^{2+}/calmodulin antagonizes PKC anchoring by competing for binding to AKAP79. The calcium influx from ion channels within the synaptic membrane releases PKC from its anchor, changing the activity status of this kinase. Presumably the soluble enzyme is more available to propagate calcium/phospholipid signaling events, as it has a less restricted access to its substrates. An additional level of complexity may be present, as biochemical and proteomic experiments have detected each PKC isoform in AKAP79/150 complexes isolated from rat brain [65,66]. This implies that at these synapses a variety of AKAP79/150 signaling complexes exist that contain conventional, novel, or atypical PKC isozymes. Again this adds to the diversity of AKAP signaling, as each PKC class responds to different combinations of phospholipid activator. These examples not only highlight the sophisticated degree of spatial organization achieved by anchoring proteins but also emphasize the degree of specificity that can be generated through the combinatorial assembly of unique AKAP signaling complexes.

Acknowledgments

J.D.S. and L.K.L. are supported in part by NIH grant GM48231.

References

1. Pawson, T. and Scott, J. D. (1997). Signaling through scaffold, anchoring, and adaptor proteins. *Science* **278**, 2075–2080.
2. Jordan, J. D., Landau, E. M., and Iyengar, R. (2000). Signaling networks: the origins of cellular multitasking. *Cell* **103**, 193–200.

3. Hunter, T. (2000). Signaling—2000 and beyond. *Cell* **100**, 113–127.
4. Bauman, A. L. and Scott, J. D. (2002). Kinase- and phosphatase-anchoring proteins: harnessing the dynamic duo. *Nat. Cell Biol.* **4**, E203–206.
5. Colledge, M. and Scott, J. D. (1999). AKAPs: from structure to function. *Trends Cell Biol.* **9**, 216–221.
6. Feliciello, A., Gottesman, M. E., and Avvedimento, E. V. (2001). The biological functions of A-kinase anchor proteins. *J. Mol. Biol.* **308**, 99–114.
7. Carr, D. W., Stofko-Hahn, R. E., Fraser, I. D. C., Bishop, S. M., Acott, T. S., Brennan, R. G., and Scott, J. D. (1991). Interaction of the regulatory subunit (RII) of cAMP-dependent protein kinase with RII-anchoring proteins occurs through an amphipathic helix binding motif. *J. Biol. Chem.* **266**, 14188–14192.
8. Newlon, M. G., Roy, M., Morikis, D., Carr, D. W., Westphal, R., Scott, J. D., and Jennings, P. A. (2001). A novel mechanism of PKA anchoring revealed by solution structures of anchoring complexes. *EMBO J.* **20**, 1651–1662.
9. Smith, F. D. and Scott, J. D. (2002). Signaling complexes: junctions on the intracellular information super highway. *Curr. Biol.* **12**, R32–40.
10. Westphal, R. S., Tavalin, S. J., Lin, J. W., Alto, N. M., Fraser, I. D., Langeberg, L. K., Sheng, M., and Scott, J. D. (1999). Regulation of NMDA receptors by an associated phosphatase-kinase signaling complex. *Science* **285**, 93–96.
11. Tavalin, S. J., Colledge, M., Hell, J. W., Langeberg, L. K., Huganir, R. L., and Scott, J. D. (2002). Regulation of GluR1 by the A-kinase anchoring protein 79 (AKAP79) signaling complex shares properties with long-term depression. *J. Neurosci.* **22**, 3044–3051.
12. Shih, M., Lin, F., Scott, J. D., Wang, H. Y., and Malbon, C. C. (1999). Dynamic complexes of beta2-adrenergic receptors with protein kinases and phosphatases and the role of gravin. *J. Biol. Chem.* **274**, 1588–1595.
13. Fraser, I., Cong, M., Kim, J., Rollins, E., Daaka, Y., Lefkowitz, R., and Scott, J. (2000). Assembly of an AKAP/beta2-adrenergic receptor signaling complex facilitates receptor phosphorylation and signaling. *Curr. Biol.* **10**, 409–412.
14. Lin, F., Wang, H., and Malbon, C. C. (2000). Gravin-mediated formation of signaling complexes in beta 2-adrenergic receptor desensitization and resensitization. *J. Biol. Chem.* **275**, 19025–19034.
15. Ferguson, S. S. (2001). Evolving concepts in G protein-coupled receptor endocytosis: the role in receptor desensitization and signaling. *Pharmacol. Rev.* **53**, 1–24.
16. Daaka, Y., Luttrell, L. M., and Lefkowitz, R. J. (1997). Switching of the coupling of the beta2-adrenergic receptor to different G proteins by protein kinase A. *Nature* **390**, 88–91.
17. Davare, M. A., Avdonin, V., Hall, D. D., Peden, E. M., Burette, A., Weinberg, R. J., Horne, M. C., Hoshi, T., and Hell, J. W. (2001). A beta2 adrenergic receptor signaling complex assembled with the Ca^{2+} channel Cav1.2. *Science* **293**, 98–101.
18. Davare, M. A., Dong, F., Rubin, C. S., and Hell, J. W. (1999). The A-kinase anchor protein MAP2B and cAMP-dependent protein kinase are associated with class C L-type calcium channels in neurons. *J. Biol. Chem.* **274**, 30280–30287.
19. Westphal, R. S., Soderling, S. H., Alto, N. M., Langeberg, L. K., and Scott, J. D. (2000). Scar/WAVE-1, a Wiskott-Aldrich syndrome protein, assembles an actin-associated multi-kinase scaffold. *EMBO J.* **19**, 4589–4600.
20. Diviani, D. and Scott, J. D. (2001). AKAP signaling complexes at the cytoskeleton. *J. Cell Sci.* **114**, 1431–1437.
21. Machesky, L. M. and Insall, R. H. (1998). Scar1 and the related Wiskott-Aldrich syndrome protein, WASP, regulate the actin cytoskeleton through the Arp2/3 complex. *Curr. Biol.* **8**, 1347–1356.
22. Diviani, D., Soderling, J., and Scott, J. D. (2001). AKAP-Lbc anchors protein kinase A and nucleates Galpha 12-selective Rho-mediated stress fiber formation. *J. Biol. Chem.* **276**, 44247–44257.
23. Niu, J., Vaiskunaite, R., Suzuki, N., Kozasa, T., Carr, D. W., Dulin, N., and Voyno-Yasenetskaya, T. A. (2001). Interaction of heterotrimeric G13 protein with an A-kinase-anchoring protein 110 (AKAP110) mediates cAMP-independent PKA activation. *Curr. Biol.* **11**, 1686–1690.
24. Dodge, K. and Scott, J. D. (2000). AKAP79 and the evolution of the AKAP model. *FEBS Lett.* **476**, 58–61.
25. Carr, D. W., Stofko-Hahn, R. E., Fraser, I. D. C., Cone, R. D., and Scott, J. D. (1992). Localization of the cAMP-dependent protein kinase to the postsynaptic densities by A-kinase anchoring proteins: characterization of AKAP79. *J. Biol. Chem.* **24**, 16816–16823.
26. Sik, A., Gulacsi, A., Lai, Y., Doyle, W. K., Pacia, S., Mody, I., and Freund, T. F. (2000). Localization of the A kinase anchoring protein AKAP79 in the human hippocampus, *Eur J Neurosci.* **12**, 1155–1164.
27. Coghlan, V. M., Perrino, B. A., Howard, M., Langeberg, L. K., Hicks, J. B., Gallatin, W. M., and Scott, J. D. (1995). Association of protein kinase A and protein phosphatase 2B with a common anchoring protein. *Science* **267**, 108–112.
28. Klauck, T. M., Faux, M. C., Labudda, K., Langeberg, L. K., Jaken, S., and Scott, J. D. (1996). Coordination of three signaling enzymes by AKAP79, a mammalian scaffold protein. *Science* **271**, 1589–1592.
29. Faux, M. C. and Scott, J. D. (1996). Molecular glue: kinase anchoring and scaffold proteins. *Cell* **70**, 8–12.
30. Fraser, I. D. and Scott, J. D. (1999). Modulation of ion channels: a "current" view of AKAPs. *Neuron* **23**, 423–426.
31. Gao, T., Yatani, A., Dell'Acqua, M. L., Sako, H., Green, S. A., Dascal, N., Scott, J. D., and Hosey, M. M. (1997). cAMP-dependent regulation of cardiac L-type Ca^{2+} channels requires membrane targeting of PKA and phosphorylation of channel subunits. *Neuron* **19**, 185–196.
32. Jo, I., Ward, D. T., Baum, M. A., Scott, J. D., Coghlan, V. M., Hammond, T. G., and Harris, H. W. (2001). AQP2 is a substrate for endogenous PP2B activity within an inner medullary AKAP-signaling complex. *Am. J. Physiol. Renal Physiol.* **281**, F958–965.
33. Potet, F., Scott, J. D., Mohammad-Panah, R., Escande, D., and Baro, I. I. (2001). AKAP proteins anchor cAMP-dependent protein kinase to KvLQT1/IsK channel complex. *Am. J. Physiol. Heart Circ. Physiol.* **280**, H2038–H2045.
34. Mayer, M. L. and Westbrook, G. L. (1987). The physiology of excitatory amino acids in the vertebrate central nervous system. *Prog. Neurobiol.* **28**, 197–276.
35. Jahr, C. E. and Lester, R. A. J. (1992). Synaptic excitation mediated by glutamate-gated ion channels. *Curr. Opin. Neurobiol.* **2**, 395–400.
36. Colledge, M., Dean, R. A., Scott, G. K., Langeberg, L. K., Huganir, R. L., and Scott, J. D. (2000). Targeting of PKA to glutamate receptors through a MAGUK-AKAP complex. *Neuron* **27**, 107–119.
37. Kameyama, K., Lee, H. K., Bear, M. F., and Huganir, R. L. (1998). Involvement of a postsynaptic protein kinase A substrate in the expression of homosynaptic long-term depression. *Neuron* **21**, 1163–1175.
38. Lee, H. K., Barbarosie, M., Kameyama, K., Bear, M. F., and Huganir, R. L. (2000). Regulation of distinct AMPA receptor phosphorylation sites during bidirectional synaptic plasticity. *Nature* **405**, 955–959.
39. Banke, T. G., Bowie, D., Lee, H., Huganir, R. L., Schousboe, A., and Traynelis, S. F. (2000). Control of GluR1 AMPA receptor function by cAMP-dependent protein kinase. *J Neurosci,* **20**, 89–102.
40. Rosenmund, C., Carr, D. W., Bergeson, S. E., Nilaver, G., Scott, J. D., and Westbrook, G. L. (1994). Anchoring of protein kinase A is required for modulation of AMPA/kainate receptors on hippocampal neurons. *Nature* **368**, 853–856.
41. Raymond, L. A., Blackstone, C. D., and Huganir, R. L. (1993). Phosphorylation and modulation of recombinant GluR6 glutamate receptors by cAMP-dependent protein kinase. *Nature* **361**, 637–641.
42. Swope, S. L., Moss, S. I., Raymond, L. A., and Huganir, R. L. (1999). Regulation of ligand-gated ion channels by protein phosphorylation. *Adv. Second Messenger Phosphoprotein Res.* **33**, 49–78.
43. Lin, J. W., Wyszynski, M., Madhavan, R., Sealock, R., Kim, J. U., and Sheng, M. (1998). Yotiao, a novel protein of neuromuscular junction and brain that interacts with specific splice variants of NMDA receptor subunit NR1. *J. Neurosci.* **18**, 2017–2027.

44. Raman, I. M., Tong, G., and Jahr, C. E. (1996). β-adrenergic regulation of synaptic NMDA receptors by cAMP-dependent protein kinase. *Neuron* **16**, 415–421.
45. Wang, L.-Y., Orser, B. A., Brautigan, D. L., and Macdonald, J. F. (1994). Regulation of NMDA receptors in cultured hippocampal neurons by protein phosphatases 1 and 2A. *Nature* **369**, 230–232.
46. Snyder, G. L., Fienberg, A. A., Huganir, R. L., and Greengard, P. (1998). A dopamine/D1 receptor/protein kinase A/dopamine- and cAMP-regulated phosphoprotein (Mr 32 kDa)/protein phosphatase-1 pathway regulates dephosphorylation of the NMDA receptor. *J. Neurosci.* **18**, 10297–10303.
47. Takahashi, M., Mukai, H., Oishi, K., Isagawa, T., and Ono, Y. (2000). Association of immature hypo-phosphorylated protein kinase C epsilon with an anchoring protein CG-NAP. *J. Biol. Chem.*
48. Schmidt, P. H., Dransfield, D. T., Claudio, J. O., Hawley, R. G., Trotter, K. W., Milgram, S. L., and Goldenring, J. R. (1999). AKAP350: a multiply spliced A-kinase anchoring protein associated with centrosomes. *J. Biol. Chem.* **274**, 3055–3066.
49. Witczak, O., Skalhegg, B. S., Keryer, G., Bornens, M., Tasken, K., Jahnsen, T., and Orstavik, S. (1999). Cloning and characterization of a cDNA encoding an A-kinase anchoring protein located in the centrosome, AKAP450. *EMBO J.* **18**, 1858–1868.
50. Gillingham, A. K. and Munro, S. (2000). The PACT domain, a conserved centrosomal targeting motif in the coiled-coil proteins AKAP450 and pericentrin. *EMBO Rep.* **1**, 524–529.
51. Dodge, K. L., Khouangsathiene, S., Kapiloff, M. S., Mouton, R., Hill, E. V., Houslay, M. D., Langeberg, L. K., and Scott, J. D. (2001). mAKAP assembles a protein kinase A/PDE4 phosphodiesterase cAMP signaling module. *EMBO J.* **20**, 1921–1930.
52. Tasken, K. A., Collas, P., Kemmner, W. A., Witczak, O., Conti, M., and Tasken, K. (2001). Phosphodiesterase 4D and protein kinase a type II constitute a signaling unit in the centrosomal area. *J. Biol. Chem.* **276**, 21999–22002.
53. Sette, C. and Conti, M. (1996). Phosphorylation and activation of a cAMP-specific phosphodiesterase by the cAMP-dependent protein kinase. *J. Biol. Chem.* **271**, 16526–16534.
54. Lim, J., Pahlke, G., and Conti, M. (1999). Activation of the cAMP-specific phosphodiesterase PDE4D3 by phosphorylation. Identification and function of an inhibitory domain. *J. Biol. Chem.* **274**, 19677–19685.
55. Oki, N., Takahashi, S. I., Hidaka, H., and Conti, M. (2000). Short term feedback regulation of cAMP in FRTL-5 thyroid cells. Role of PDE4D3 phosphodiesterase activation. *J. Biol. Chem.* **275**, 10831–10837.
56. Conti, M. (2000). Phosphodiesterases and cyclic nucleotide signaling in endocrine cells. *Mol. Endocrinol.* **14**, 1317–1327.
57. Hoffmann, R., Wilkinson, I. R., McCallum, J. F., Engels, P., and Houslay, M. D. (1998). cAMP-specific phosphodiesterase HSPDE4D3 mutants which mimic activation and changes in rolipram inhibition triggered by protein kinase A phosphorylation of Ser-54: generation of a molecular model. *Biochem. J.* **333**, 139–149.
58. Zaccolo, M. and Pozzan, T. (2002). Discrete microdomains with high concentration of cAMP in stimulated rat neonatal cardiac myocytes. *Science* **295**, 1711–1715.
59. Marx, S. O., Reiken, S., Hisamatsu, Y., Jayaraman, T., Burkhoff, D., Rosemblit, N., and Marks, A. R. (2000). PKA phosphorylation dissociates FKBP12.6 from the calcium release channel (ryanodine receptor): defective regulation in failing hearts. *Cell* **101**, 365–376.
60. Fink, M. A., Zakhary, D. R., Mackey, J. A., Desnoyer, R. W., Apperson-Hansen, C., Damron, D. S., and Bond, M. (2001). AKAP-mediated targeting of protein kinase a regulates contractility in cardiac myocytes. *Circ. Res.* **88**, 291–297.
61. Zakhary, D. R., Moravec, C. S., and Bond, M. (2000). Regulation of PKA binding to AKAPs in the heart: alterations in human heart failure. *Circulation* **101**, 1459–1464.
62. Huang, L. J., Durick, K., Weiner, J. A., Chun, J., and Taylor, S. S. (1997). Identification of a novel dual specificity protein kinase A anchoring protein, D-AKAP1. *J. Biol. Chem.* **272**, 8057–8064.
63. Huang, L. J., Durick, K., Weiner, J. A., Chun, J., and Taylor, S. S., (1997). D-AKAP2, a novel protein kinase A anchoring protein with a putative RGS domain. *Proc. Natl. Acad. Sci. USA* **94**, 11184–11189.
64. Alto, N. M., Soderling, J., and Scott, J. D. (2002). Rab32 is an A-kinase anchoring protein and participates in mitochondrial dynamics. *J. Cell Biol.* **158**, 659–668.
65. Faux, M. C., Rollins, E. N., Edwards, A. S., Langeberg, L. K., Newton, A. C., and Scott, J. D. (1999). Mechanism of A-kinase-anchoring protein 79 (AKAP79) and protein kinase C interaction. *Biochem. J.* **343**, 443–452.
66. Husi, H., Ward, M. A., Choudhary, J. S., Blackstock, W. P., and Grant, S. G. (2000). Proteomic analysis of NMDA receptor-adhesion protein signaling complexes. *Nat. Neurosci.* **3**, 661–669.

CHAPTER 187

Protein Kinase C Protein Interactions

Peter J. Parker, Joanne Durgan,
Xavier Iturrioz, and Sipeki Szabolcs

*Cancer Research UK, London Research Institute,
Lincoln's Inn Fields Laboratories,
London, United Kingdom*

Introduction

Protein kinase C (PKC) was initially identified in screens for broad specificity kinases that could, like protein kinase A (PKA), respond to second messengers and so integrate agonist-induced responses (see [1]). There is no doubt that the PKC family, comprising the classical (α,β,γ), novel ($\delta,\varepsilon,\eta,\theta$), and atypical ($\zeta,\iota/\lambda$) isoforms, fulfil both elements of this definition, that is, they are responsive to second messengers and they are of broad specificity (at least *in vitro*). This latter property has triggered the question of how specificity is imposed on this rather promiscuous group of proteins. In part such considerations have led to the identification of binding partners and substrates that provide specificity to the system. However, this is not all that is achieved by such PKC-binding partners; interactions are seen to operate at a number of levels that can be related to the "life cycle" of the kinase. These can be summarized as follows (see Fig. 1):

PKC priming
PKC activation
PKC substrate engagement
PKC inactivation

Although a number of reviews on this subject have described classes of binding proteins or specific examples (see [2] and references therein), here we discuss examples of PKC-binding proteins with reference to the above properties conferred on PKC proteins.

Priming

Although considered to be under acute allosteric control by diacylglycerol (DAG) and possibly other bioactive lipids [3], PKC proteins all appear to require multisite phosphorylation for their optimum function (see [4]). In some respects this can be viewed as a priming device for subsequent DAG-dependent (or other) activation.

The dephosphorylated form of PKCα is intrinsically insoluble [5], and it is likely that the unphosphorylated primary translation product is associated with an as yet unidentified chaperone. This is suggested by the observation that the newly synthesized protein is soluble, unlike its phosphorylated and then dephosphorylated counterpart [5]. There is to date no evidence that PKCα is a client protein for known chaperones such as Hsp90, and the identity of its "chaperone" remains to be determined. However, for PKCε, which undergoes a similar array of phosphorylation events, the anchoring protein CG-NAP (centrosome and Golgi localized PKN-associated protein) binds the hypophosphorylated form of the protein. It has been proposed that CG-NAP acts as a scaffold to promote phosphorylation of PKCε [6], taking the chaperone role for this member of the family.

The subsequent steps in this priming process are the phosphorylations themselves. There is general agreement that a key step in this process involves the action of the upstream kinase PDK1 ([7,8] and see Newton, this volume). This PtdIns3,4,5P$_3$-activated kinase can bind directly to all

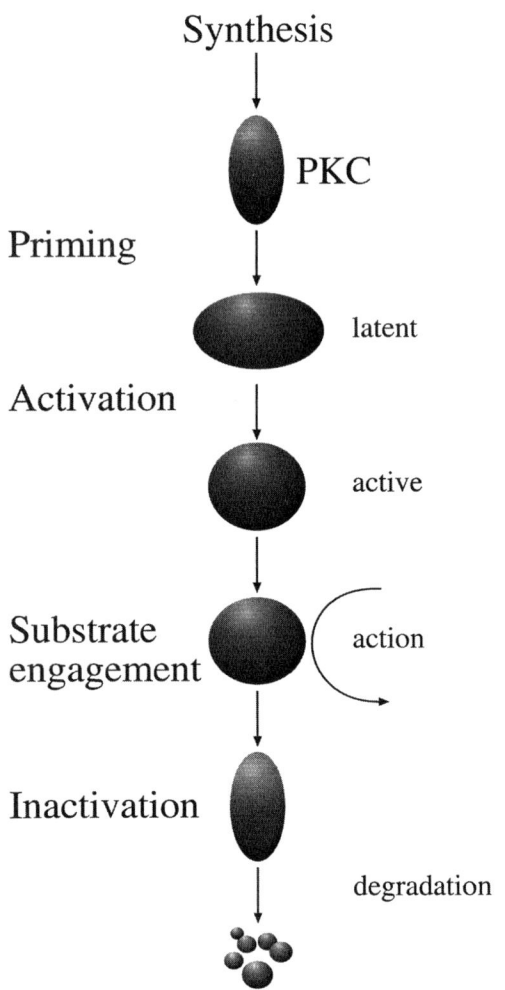

Figure 1 The life and death of PKC. Various states along the pathway of PKC synthesis, modification, activation, and proteolysis are indicated. The generic events associated with the different phases of existence are denoted on the left and referred to in the text in respect of PKC-binding proteins. These events take place in distinct compartments in part defined by binding partners (see text).

members of the PKC superfamily (α, β, γ, δ, ϵ, ζ, η, θ, ι/λ) [7]. PDK1-PKC interaction appears to be mediated through the PKC kinase domain [7], and based upon the PRK2-PDK1 interaction studies it would appear to involve the conserved phosphorylated C-terminal hydrophobic motif FXXF{S(P),T(P),E}F/Y [9]. Thus PDK1 can engage these PKC substrates at a site distal to its target site of phosphorylation, which resides in the activation loop of the kinases at the conserved TFCGTP motif. It is unclear whether these PKC-PDK1 interactions confer specificity, efficiency, or some other property to the system. In principle this may reflect an ordering of events at least for those PKCs requiring hydrophobic motif phosphorylation.

14-3-3 proteins act as dimers and engage phosphorylated proteins with a preference for RXXSXP or RXXXSXP motifs [10]. It is known that PKCs can bind 14-3-3 proteins *in vitro* and also *in vivo*, although the role of the interaction is not clear. However, the precedents set by the c-Raf1 protein kinase and its functional interaction with 14-3-3 (as established genetically [11]) indicate that this is likely to be important also in PKC function. Perhaps as for c-Raf1 [12], 14-3-3 facilitates stabilization of phosphorylated conformers of PKCs, thereby contributing to the priming process.

Though not a priming device, tyrosine phosphorylation of PKC, in particular PKCδ, has critical effects on cellular responses (see for example [13]). The distinct functions and reported properties for different phosphorylated forms of PKCδ (reviewed [14]) indicate that some of these tyrosine phosphorylations may be protein autonomous; however, there remains the probability that some involve direct recognition, with the likely recruitment of partners via SH2 domain (http://www.ncbi.nlm.nih.gov/Structure/cdd/cddsrv.cgi?uid=smart00252) interactions.

Activation

An intrinsic property of the classical and novel PKC allosteric activator DAG is that it is membrane-limited. It is the ability of PKCs to "sample" the membrane environment that provides the opportunity for activation. This feature of PKC activation lends itself to the compartmental targeting of PKC to provide for selectivity of activation.

Targeting of PKC prior to activation is clearly exemplified in the *Drosophila* compound eye, where the scaffolding protein InaD is responsible for assembling the eye-specific phospholipase C NorpA, the eye-specific PKC InaC, and the ion channel protein TRPL (recently reviewed [15]). This assembly contributes to the efficiency with which photoreception is relayed and by which it is terminated. In this instance the interaction of the scaffolding protein InaD with the downstream effectors occurs via PDZ domains (http://www.ncbi.nlm.nih.gov/Structure/cdd/cddsrv.cgi?uid=smart00228), with the different proteins binding to distinct PDZ domains. A similar PDZ-dependent interaction is seen for PKCα and the binding partner PICK1 [16]; PKCα has a canonical PDZ binding motif (..SAV) at its C-terminus. The physiological role of this partnership is not understood.

AKAPs (A-kinase associated proteins) represent a distinct class of scaffolds that were first identified as protein kinase A (PKA) binding proteins [17]. Elegant studies by Scott and colleagues have provided evidence for the ability of AKAPs to direct PKA to sites of activation and action (see Chapters 62, 63, this volume). Two of these proteins, AKAP79 and Gravin (AKAP250), independently bind PKC as well as PKA and other proteins [18,19]. In particular, Gravin, which has been isolated in a number of different guises, scaffolds a broad spectrum of signal transducers and cell-cycle regulators (reviewed [20]). One other AKAP (AKAP78) has been identified as ezrin [21], which has been shown recently also to bind PKCα [22] (see further below). Although the specific roles for this group of PKC-binding proteins remain largely unknown, the precedents set by the studies on PKA provide a clear paradigm, and it is anticipated that these scaffolds serve to place PKCs at sites where their action is required. The ability of this class of scaffolds to assemble multiple transducers also provides for integration (as for InaD) and duplication of signal outputs (for example, a common target for distinct upstream pathways, such as PKA and PKC).

For the atypical PKCs (ι/λ, ζ), activation is not DAG-dependent but can be effected by the assembly of a cdc42/PAR6/PAR3/aPKC complex (recently reviewed [23]). The importance of such an assembly was demonstrated in *C. elegans*, where it was shown that PKC3 (the only aPKC in *C. elegans*) ablation produced a polarity phenotype related to that documented for PAR3 and PAR6 mutants [24]. Subsequent studies have shown that the mammalian homologues of these proteins are involved in establishing the polarity of epithelial cells, where they assemble at and control the formation of tight junctions (see [23]). The sites of interaction for these proteins have been mapped (see Fig. 2). For aPKC the binding site for PAR6 resides in the aminoterminal domain whereas that for PAR3 is in the carboxyterminal, catalytic domain. It also remains possible that subcomplexes form, for example between the brain-specific PKMζ protein (catalytic domain protein from the PKCζ gene involved in memory; see [25]) and PAR3; however, this possibility requires further investigation.

Transmembrane proteins (including receptors) also interact with PKCs, targeting them to the resident compartment. These include β_1-integrin [26], syndecan 4 [27], and tetraspanins [28]. In the case of β_1-integrin, the interaction with PKCα occurs between the cytoplasmic C-terminal integrin tail and the central variable region (V3) of PKCα [29]. This interaction has been reported to be involved in β_1-integrin migratory responses by controlling the traffic of β_1-integrin [26]. There is a complex two-way relationship between β_1-integrin and PKCα, a part of which represents an integrin-scaffolding role in the local assembly with, and PKC phosphorylation of the protein ezrin (see below).

A broad class of transmembrane proteins long-associated with PKC function are ion channels (see [30]). Although it has become apparent that in fact DAG can directly control the TrpL proteins [31], DAG also acts via PKC in other contexts. For example, PKC forms a complex with the GABA(A) receptor β_1 and β_3 subunits [32]. This association, like that conferred by InaD, predisposes the receptor to localized and efficient control by PKC—in this instance through phosphorylation of sites within a conserved motif shared by $\beta1$, $\beta2$, and $\beta3$ subunits [33,34].

The interaction of PKCα with syndecan is distinctive in being promoted by PtdIns4,5,P$_2$ [35], and the association is negatively regulated by phosphorylation of syndecan on S183, possibly by PKCδ [36]. In the syndecan 4/PtdIns4,5P$_2$ complex, PKCα is in an active conformation resembling the interaction with RACKs (see below). Syndecan 4 is implicated in responses to FGF2, and the association and activation of PKCα via syndecan 4 is thought to contribute to FGF2-induced endothelial cell migration and proliferation [37].

There is a well-documented class of PKC-binding proteins that interact selectively with the activated conformer. These are typified by the RACKs (receptors for activated c-kinases) first documented by Mochly-Rosen and colleagues [38]. The two characterized RACKs are RACK1 [39] and β-COP (RACK2) [40]. Although the specific roles for these RACKs in the context of PKC action remain incompletely understood, their interactions with PKCs have proved informative. In particular, two features are notable here. First, interaction of PKC with RACK1 has been shown to lock the protein in an active conformation, providing a mechanism for DAG-triggered but DAG-independent sustained activation. Second, the site of interaction of RACKs with PKCs appears to be through the PKC C2/C2-like domain [41]. The latter property has been exploited to demonstrate that short

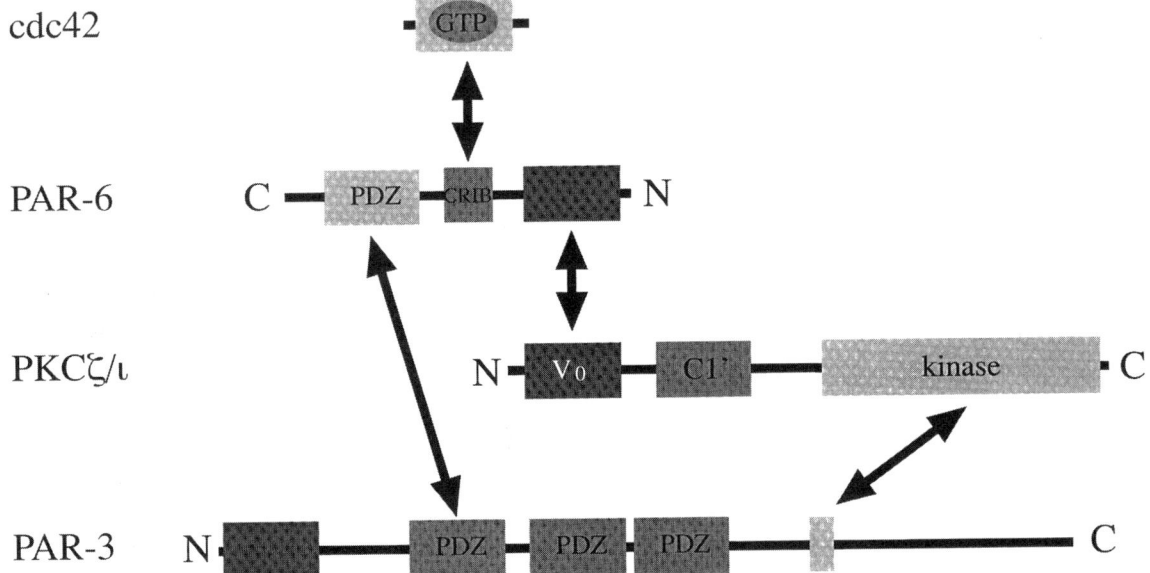

Figure 2 PAR-aPKC associations. The various interactions that compose the aPKC-PAR complex are illustrated (adapted from [23]). The defined domains are as indicated. The PKCζ C1 domain is a C1'-related domain that does not bind DAG/phorbol esters. The association of the small GTP-binding protein cdc42 with PAR6 requires its GTP-bound state.

oligopeptides corresponding to the sites of interaction can block PKC translocation and PKC effects (see [41]). Thus, interactions through these regions of PKC (whether with RACKs or other proteins) appear to be important in PKC action. Corroboration of this has come from studies showing the inhibition of preconditioning in cardiac myocytes with a PKCε inhibitory peptide [42] and the subsequent demonstration in knockout mice that PKCε is essential for one component of preconditioning *in vivo* [43].

Annexin VI is another example of a PKC-binding protein that interacts with PKC (PKCα and β) but is not itself a substrate for the kinase [44]. This interaction is of interest because both proteins bind phospholipids in a Ca^{2+}-dependent fashion and their interaction is dependent upon lipid binding. Nevertheless, there is evidence for direct protein-protein contact in this association [44].

Substrates and Pathways

The distinctions between scaffolds and substrates are somewhat blurred where the scaffolds themselves are substrates. However, conceptually there is a clear distinction between those proteins that bind PKC in its inactive conformation predisposing it to localized responses and those that are recruited postactivation. This latter category is exemplified by the STICK proteins(substrates that interact with c-kinase) [45]. These proteins have been identified by far-western analysis and include the PKC substrates clone72/SSecKS and γ-adducin (see [2]).

The STICK adducin is associated with actin and forms specific heterotetramers of the types α–β or α–γ (reviewed [46]). The PKC-dependent phosphorylation of γ-adducin (at serine 660) is associated with its release from this cytoskeletal location [47]. SSecKS binds a number of other signaling proteins (see above). It is interesting that like the PKC substrates MARCKS and GAP-43, the phosphorylation of SSeCKS by PKC occurs at a calmodulin-binding site, interfering with calmodulin interaction [48]. SSeCKS also undergoes an interaction with cyclinD, which is attenuated by PKC phosphorylation of SSeCKS [48]. Phosphorylated SSeCKS is associated with membrane protrusions and ruffles, thus implying an involvement in the reorganization of cortical actin. How this relates to the nonbinding/release of calmodulin and cyclin D is as yet unclear. The cortical actin association and its control by phosphorylation is also observed for another PKCα-interacting protein, ezrin [22]. In this latter case, there is evidence for the phosphorylation of ezrin by PKCα playing an essential role in β1-integrin dependent directional movement of cells [Ng, 2001 #6991]. As noted above, this also involves integrins (see Fig. 3).

In the context of substrates and pathways, phospholipase D (PLD) is a very interesting example of a PKC-binding protein. PKCα will activate PLD1 *in vitro*, and, as first evidenced in membrane reconstitutions [49], the complex formed between PKCα and PLD1 is sufficient for activation in the presence of other factors; phosphorylation is not required.

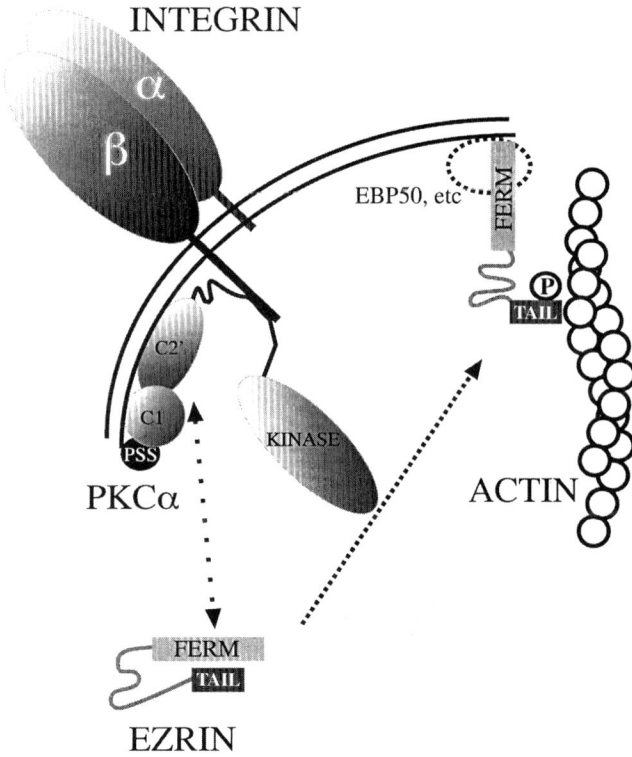

Figure 3 Integrin-PKCα-ezrin associations. PKCα and integrinβ1 form a complex through the V3 domain of PKCα and the C-terminal tail of integrinβ1 [29]. Ezrin can be recruited into this complex, and this leads to ezrin phosphorylation on T567. This phosphorylation alters the conformation by reducing the FERM-tail domain interaction and so permitting distinct contacts to be made through the FERM domain (for example with EBP50) and through the tail domain with actin (see [74]). The domains of PKCα and ezrin are indicated. PSS refers to the PKC pseudosubstrate site, which, along with the C1 and C2 domains, can interact with membranes.

This interaction occurs through the regulatory domain of PKCα and requires the open/active conformer [50]; how the activity of PKC relates to the downstream pathways in this context remains to be determined.

Mammalian genetic analysis of PKC functions has provided a number of insights. Of note here, the mouse knockout of the PKCβ gene leads to a B-cell phenotype reminiscent of Btk loss-of-function [51]. This implies that there is a functional relationship between these two kinases, and indeed they have been shown to interact [52]. This interaction occurs through the PH domain of Btk [52]. Other PH domain-PKC interactions have been documented, including that for PKCη -$PH^{PKC\mu/PKD}$ [53]; PKCμ/PKD, like Btk, is also a downstream target for PKC [54].

PKC Inactivation

A characteristic property of classical and novel forms of PKC is that their chronic activation frequently leads to their inactivation and/or degradation. This feature is often observed on phorbol ester (or functionally related pharmacological agonist) stimulation of cells. For PKCα this process of activation-induced degradation has been documented to proceed via caveolae/raft-dependent traffic, dephosphorylation, and then degradation [55]. PKCα can interact with caveolin 1 through its scaffold domain [56]. This interaction may well contribute to its traffic through the caveolae compartment, although the serum deprivation protein, sdr, has also been proposed as a caveolae targeting device for PKCα [57]. It is possible that a related role is played by p62/ZIP, a PKCζ-interacting protein [58,59] that has been reported to recruit PKCζ and traffic through endosomal fractions in response to EGF [59] and NGF [60].

Dephosphorylation of PKC appears to be effected through protein phosphatase 2A (PP2A) (for example [61]) and there is evidence that PP2A can associate with certain PKC isoforms [62], as has been described for other AGC kinase family proteins [63]. As noted above, dephosphorylated PKC is often found to be associated with the neutral detergent-insoluble fraction. Although this may reflect an intrinsic property, it is of note that the p32 protein recovered from the detergent-insoluble fraction of hepatocytes has been shown to interact with certain conformers of PKC isoforms and hence may contribute to this behavior [64].

The process of degradation of PKC is a ubiquitin-dependent one [65,66], implying that E3 component(s) recognize and interact with PKCs to facilitate ubiquitination. To date the only such component identified that binds PKC isoforms is the VHL tumor suppressor gene product [67]. Whether this is the only interacting E3 protein remains to be determined.

Perspectives

There is an increasing need to be able to monitor at a subcellular level the interactions of PKCs and their binding partners. It will be important to be able to follow such events in real time and in the context of catalytic activity, for example by using antisera that monitor phosphorylation events [68,69]. The elucidation of the spatio-temporal behavior of complexes promises to be very informative in defining how individual ones contribute to responses. Beyond this, application of accumulating molecular knowledge to the development of improved health care is increasingly tractable. Indeed, our understanding of PKC targets has already started to have an impact on clinical trials [70].

In a broader context, it is of interest that comprehensive interaction maps have been compiled for yeast (for example [71]) and other organisms (see web sites in [71]). To date these generic activities have had limited novel input into our appreciation of PKC and its partners in higher eukaryotes; in yeast the defined partner for PKC1, a Rho family member, in fact reflects a property of the PKC-related kinases (PRKs/PKNs) [72]. However such mapping approaches promise much for the future.

It will be evident to those in the field that this review has really only touched on what is a substantial collection of PKC-interacting proteins. Indeed, the diversity of these PKC-associated proteins means that we must acknowledge that the majority have not been discussed in this commentary, and we hope colleagues will accept our apologies if their favorite PKC partner is missing.

References

1. Nishizuka, Y. (1986). Studies and perspectives of protein kinase C. *Science* **233**, 305–312.
2. Jaken, S. and Parker, P. J. (2000). Protein kinase C binding partners. *Bioessays* **22**, 245–254.
3. Nishizuka, Y. (1995). Protein kinase C and lipid signaling for sustained cellular responses. *FASEB J.* **9**, 484–496.
4. Parekh, D. B., Ziegler, W., and Parker, P. J. (2000). Multiple pathways control protein kinase C phosphorylation. *EMBO J.* **19**, 496–503.
5. Bornancin, F. and Parker, P. J. (1997). Phosphorylation of protein kinase C-α on serine 657 controls the accumulation of active enzyme and contributes to its phosphatase-resistant state. *J. Biol. Chem.* **272**, 3544–3549.
6. Takahashi, M., Mukai, H., Oishi, K., Isagawa, T., and Ono, Y. (2000). Association of immature hypophosphorylated protein kinase C-epsilon with an anchoring protein CG-NAP. *J. Biol. Chem.* **275**, 34592–34596.
7. Le Good, J. A., Ziegler, W. H., Parekh, D. B., Alessi, D. R., Cohen, P., and Parker, P. J. (1998). Protein kinase C isotypes controlled by phosphoinositide 3-kinase through the protein kinase PDK1. *Science* **281**, 2042–2045.
8. Dutil, E. M., Toker, A., and Newton, A. C. (1998). Regulation of conventional protein kinase C isozymes by phosphoinositide-dependent kinase 1 (PDK-1). *Curr. Biol.* **8**, 1366–1375.
9. Balendran, A., Casamayor, A., Deak, M., Paterson, A., Gaffney, P., Currie, R., Downes, C. P., and Alessi, D. R. (1999). PDK1 acquires PDK2 activity in the presence of a synthetic peptide derived from the carboxyl terminus of PRK2. *Curr. Biol.* **9**, 393–404.
10. Tzivion, G. and Avruch, J. (2002). 14-3-3 proteins: active cofactors in cellular regulation by serine/threonine phosphorylation. *J. Biol. Chem.* **277**, 3061–3064.
11. Hsu, V., Zobel, C. L., Lambie, E. J., Schedl, T., and Kornfeld, K. (2002). Caenorhabditis elegans lin-45 raf is essential for larval viability, fertility and the induction of vulval cell fates. *Genetics* **160**, 481–492.
12. Yip-Schneider, M. T., Miao, W., Lin, A., Barnard, D. S., Tzivion, G., and Marshall, M. S. (2000). Regulation of the Raf-1 kinase domain by phosphorylation and 14-3-3 association. *Biochem. J.* **351**, 151–159.
13. Kronfeld, I., Kazimirsky, G., Lorenzo, P. S., Garfield, S. H., Blumberg, P. M., and Brodie, C. (2000). Phosphorylation of protein kinase Cdelta on distinct tyrosine residues regulates specific cellular functions. *J. Biol. Chem.* **275**, 35491–35498.
14. Gschwendt, M. (1999). Protein kinase C delta. *Eur. J. Biochem.* **259**, 555–564.
15. Tsunoda, S. and Zuker, C. S. (1999). The organization of INAD-signaling complexes by a multivalent PDZ domain protein in *Drosophila* photoreceptor cells ensures sensitivity and speed of signaling. *Cell Calcium* **26**, 165–171.
16. Staudinger, J., Lu, J., and Olson, E. N. (1997). Specific interaction of the PDZ domain protein PICK1 with the COOH terminus of protein kinase C-alpha. *J. Biol. Chem.* **272**, 32019–32024.
17. Dell'Acqua, M. L. and Scott, J. D. (1997). Protein kinase A anchoring. *J. Biol. Chem.* **272**, 12881–12884.
18. Klauck, T. M., Faux, M. C., Labudda, K., Langeberg, L. K., Jaken, S., and Scott, J. D. (1996). Coordination of three signaling enzymes by AKAP79, a mammalian scaffold protein. *Science* **271**, 1589–1592.
19. Nauert, J. B., Klauck, T. M., Langeberg, L. K., and Scott, J. D. (1997). Gravin, an autoantigen recognized by serum from myasthenia gravis patients, is a kinase scaffold protein. *Curr. Biol.* **7**, 52–62.
20. Gelman, I. H. (2002). The role of SSeCKS/Gravin/AKAP12 scaffolding proteins in the spaciotemporal control of signaling pathways in oncogenesis and development. *Front. Biosci.* **1**, D1782–1797.

21. Dransfield, D. T., Bradford, A. J., Smith, J., Martin, M., Roy, C., Mangeat, P. H., and Goldenring, J. R. (1997). Ezrin is a cyclic AMP-dependent protein kinase anchoring protein. *EMBO J.* **16**, 35–43.
22. Ng, T., Parsons, M., Hughes, W. E., Monypenny, J., Zicha, D., Gautreau, A., Arpin, M., Gschmeissner, S., Verveer, P. J., Bastiaens, P. I., and Parker, P. J. (2001). Ezrin is a downstream effector of trafficking PKC-integrin complexes involved in the control of cell motility. *EMBO J.* **20**, 2723–2741.
23. Ohno, S. (2001). Intercellular junctions and cellular polarity: the PAR-aPKC complex, a conserved core cassette playing fundamental roles in cell polarity. *Curr. Opin. Cell Biol.* **13**, 641–648.
24. Tabuse, Y., Izumi, Y., Piano, F., Kemphues, K. J., Miwa, J., and Ohno, S. (1998). Atypical protein kinase C cooperates with PAR-3 to establish embryonic polarity in *Caenorhabditis elegans*. *Development* **125**, 3607–3614.
25. Drier, E. A., Tello, M. K., Cowan, M., Wu, P., Blace, N., Sacktor, T. C., and Yin, J. C. (2002). Memory enhancement and formation by atypical PKM activity in *Drosophila melanogaster*. *Nat. Neurosci.* **5**, 316–324.
26. Ng, T., Shima, D., Squire, A., Bastiaens, P. I. H., Gschmeissner, S., Humphries, M. J., and Parker, P. J. (1999). PKCα regulates β1 integrin-dependent cell motility through association and control of integrin traffic. *EMBO J.* **18**, 3309–3923.
27. Oh, E. S., Woods, A., Lim, S. T., Theibert, A. W., and Couchman, J. R. (1998). Syndecan-4 proteoglycan cytoplasmic domain and phosphatidylinositol 4,5-bisphosphate coordinately regulate protein kinase C activity. *J. Biol. Chem.* **273**, 10624–10629.
28. Zhang, X. A., Bontrager, A. L., and Hemler, M. E. (2001). Transmembrane-4 superfamily proteins associate with activated protein kinase C (PKC) and link PKC to specific beta(1) integrins. *J. Biol. Chem.* **276**, 25005–25013.
29. Parsons, M., Keppler, M. D., Kline, A., Messent, A., Humphries, M. J., Gilchrist, R., Hart, I. R., Quittau-Prevostel, C., Hughes, W. E., Parker, P. J., and Ng, T. (2002). Site-directed perturbation of protein kinase C-integrin interaction blocks carcinoma cell chemotaxis. *Mol. Cell Biol.* **22**, 5897–5911.
30. Catterall, W. A. (1997). Modulation of sodium and calcium channels by protein phosphorylation and G proteins. *Adv. Second Messenger Phosphoprotein Res.* **31**, 159–181.
31. Estacion, M., Sinkins, W. G., and Schilling, W. P. (2001). Regulation of *Drosophila* transient receptor potential-like (TrpL) channels by phospholipase C-dependent mechanisms. *J. Physiol.* **530**, 1–19.
32. Brandon, N. J., Uren, J. M., Kittler, J. T., Wang, H., Olsen, R., Parker, P. J, and Moss, S. J. (1999). Subunit-specific association of protein kinase C and the receptor for activated C kinase with GABA type A receptors. *J. Neurosci.* **19**, 9228–9234.
33. Moss, S. J., Doherty, C. A., and Huganir, R. L. (1992). Identification of the cAMP-dependent protein kinase and protein kinase C phosphorylation sites within the major intracellular domains of the beta 1, gamma 2S, and gamma 2L subunits of the gamma-aminobutyric acid type A receptor. *J. Biol. Chem.* **267**, 14470–14476.
34. McDonald, B. J. and Moss, S. J. (1997). Conserved phosphorylation of the intracellular domains of GABA(A) receptor beta2 and beta3 subunits by cAMP-dependent protein kinase, cGMP-dependent protein kinase protein kinase C and Ca^{2+}/calmodulin type II-dependent protein kinase. *Neuropharmacology* **36**, 1377–1385.
35. Oh, E. S., Woods, A., Lim, S. T., Theibert, A. W., and Couchman, J. R. (1998). Syndecan-4 proteoglycan cytoplasmic domain and phosphatidylinositol 4,5-bisphosphate coordinately regulate protein kinase C activity. *J. Biol. Chem.* **273**, 10624–10629.
36. Murakami, M., Horowitz, A., Tang, S., Ware, J. A., and Simons, M. (2002). Protein kinase C (PKC) delta regulates PKCalpha activity in a Syndecan-4-dependent manner. *J. Biol. Chem.* **277**, 20367–20371.
37. Horowitz, A., Tkachenko, E., and Simons, M. (2002). Fibroblast growth factor-specific modulation of cellular response by syndecan-4. *J. Cell Biol.* **157**, 715–725.
38. Mochly-Rosen, D., Khaner, H., and Lopez, J. (1991). Identification of intracellular receptor proteins for activated protein-kinase-C. *Proc. Natl. Acad. Sci. USA* **88**, 3997–4000.
39. Ron, D., Chen, C. H., Caldwell, J., Jamieson, L., Orr, E., and Mochly-Rosen, D. (1994). Cloning of an intracellular receptor for protein kinase C: a homolog of the beta subunit of G proteins. *Proc. Natl. Acad. Sci. USA* **91**, 839–843.
40. Csukai, M., Chen, C. H., De Matteis, M. A., and Mochly-Rosen, D. (1997). The coatomer protein beta'-COP, a selective binding protein (RACK) for protein kinase Cepsilon. *J. Biol. Chem.* **272**, 29200–29206.
41. Ron, D., Luo, J., and Mochly-Rosen, D. (1995). C2 region-derived peptides inhibit translocation and function of beta protein kinase C in vivo. *J. Biol. Chem.* **270**, 24180–24187.
42. Liu, G. S., Cohen, M. V., Mochly-Rosen, D., and Downey, J. M. (1999). Protein kinase C-epsilon is responsible for the protection of preconditioning in rabbit cardiomyocytes. *J. Mol. Cell Cardiol.* **31**, 1937–1948.
43. Saurin, A. T., Pennington, D. J., Raat, N. J. H., Owen, M. J., and Marber, M. S. (2002). Targeted disruption of the protein kinase C epsilon gene abolishes the infarct size reduction that follows ischaemic preconditioning of isolated buffer-perfused mouse hearts. *Cardiovasc. Res.* **55**, 672–680.
44. Schmitz-Peiffer, C., Browne, C. L., Walker, J. H., and Biden, T. J. (1998). Activated protein kinase C alpha associates with annexin VI from skeletal muscle. *Biochem. J.* **330**, 675–681.
45. Chapline, C., Ramsay, K., Klauck, T., and Jaken, S. (1993). Interaction cloning of protein-kinase-c substrates. *J. Biol. Chem.* **268**, 6858–6861.
46. Matsuoka, Y., Li, X., and Bennett, V. (2000). Adducin: structure, function and regulation. *Cell Mol. Life Sci.* **57**, 884–895.
47. Dong, L., Chapline, C., Mousseau, B., Fowler, L., Ramsay, K., Stevens, J. L., and Jaken, S. (1995). 35H, a sequence isolated as a protein kinase C binding protein, is a novel member of the adducin family. *J. Biol. Chem.* **270**, 25534–25540.
48. Lin, X. and Gelman, I. H. (2002). Calmodulin and cyclin D anchoring sites on the Src-suppressed C kinase substrate, SSeCKS. *Biochem. Biophys. Res. Commun.* **290**, 1368–1375.
49. Conricode, K. M., Brewer, K. A., and Exton, J. H. (1992). Activation of phospholipase D by protein kinase C. Evidence for a phosphorylation-independent mechanism. *J. Biol. Chem.* **267**, 7199–7202.
50. Singer, W. D., Brown, H. A., Jiang, X., and Sternweis, P. C. (1996). Regulation of phospholipase D by protein kinase C is synergistic with ADP-ribosylation factor and independent of protein kinase activity. *J. Biol. Chem.* **271**, 4504–4510.
51. Leitges, M., Schmedt, C., Guinamard, R., Davoust, J., Schaal, S., Stabel, S., and Tarakhovsky, A. (1996). Immunodeficiency in protein kinase Cbeta-deficient mice. *Science* **273**, 788–791.
52. Yao, L., Kawakami, Y., and Kawakami, T. (1994). The pleckstrin homology domain of Bruton tyrosine kinase interacts with protein kinase C. *Proc. Natl. Acad. Sci. USA* **91**, 9175–9179.
53. Waldron, R. T., Iglesias, T., and Rozengurt, E. (1999). The pleckstrin homology domain of protein kinase D interacts preferentially with the eta isoform of protein kinase C. *J. Biol. Chem.* **274**, 9224–9230.
54. Zugaza, J. L., Sinnett-Smith, J., Van Lint, J., and Rozengurt, E. (1996). Protein kinase D (PKD) activation in intact cells through a protein kinase C-dependent signal transduction pathway. *EMBO J.* **15**, 6220–6230.
55. Prevostel, C., Joubert, D., Alice, V., and Parker, P. J. (2000). Protein kinase Cα actively downregulates through caveolae-dependent traffic to an endosomal compartment. *J. Cell Sci.* **113**, 2575–2584.
56. Oka, N., Yamamoto, M., Schwencke, C., Kawabe, J., Ebina, T., Ohno, S., Couet, J., Lisanti, M. P., and Ishikawa, Y. (1997). Caveolin interaction with protein kinase C. Isoenzyme-dependent regulation of kinase activity by the caveolin scaffolding domain peptide. *J. Biol. Chem.* **272**, 33416–33421.
57. Mineo, C., Ying, Y. S., Chapline, C., Jaken, S., and Anderson, R. G. (1998). Targeting of protein kinase Calpha to caveolae. *J. Cell Biol.* **141**, 601–610.
58. Puls, A., Schmidt, S., Grawe, F., and Stabel, S. (1997). Interaction of protein kinase C zeta with ZIP, a novel protein kinase C-binding protein. *Proc. Natl. Acad. Sci.* **94**, 6191–6196.

59. Sanchez, P., De Carcer, G., Sandoval, I. V., Moscat, J., and Diaz-Meco, M. T. (1998). Localization of atypical protein kinase C isoforms into lysosome-targeted endosomes through interaction with p62. *Mol. Cell Biol.* **18**, 3069–3080.
60. Samuels, I. S., Seibenhener, M. L., Neidigh, K. B., and Wooten, M. W. (2001). Nerve growth factor stimulates the interaction of ZIP/p62 with atypical protein kinase C and targets endosomal localization: evidence for regulation of nerve growth factor-induced differentiation. *J. Cell Biochem.* **82**, 452–466.
61. Hansra, G., Bornancin, F., Whelan, R., Hemmings, B. A., and Parker, P. J. (1996). 12-O-Tetradecanoylphorbol-13-acetate-induced dephosphorylation of protein kinase Calpha correlates with the presence of a membrane-associated protein phosphatase 2A heterotrimer. *J. Biol. Chem.* **271**, 32785–32788.
62. Srivastava, J., Goris, J., Dilworth, S. M., and Parker, P. J. (2002). Dephosphorylation of PKCdelta by protein phosphatase 2Ac and its inhibition by nucleotides. *FEBS Lett.* **516**, 265–269.
63. Westphal, R. S., Coffee, R. L., Jr., Marotta, A., Pelech, S. L., and Wadzinski, B. E. (1999). Identification of kinase-phosphatase signaling modules composed of p70 S6 kinase-protein phosphatase 2A (PP2A) and p21-activated kinase-PP2A. *J. Biol. Chem.* **274**, 687–692.
64. Robles-Flores, M., Rendon-Huerta, E., Gonzalez-Aguilar, H., Mendoza-Hernandez, G., Islas, S., Mendoza, V., Ponce-Castaneda, M. V., Gonzalez-Mariscal, L., and Lopez-Casillas, F. (2002). p32 (gC1qBP) is a general protein kinase C (PKC)-binding protein; interaction and cellular localization of P32-PKC complexes in ray hepatocytes. *J. Biol. Chem.* **277**, 5247–5255.
65. Lee, H. W., Smith, L., Pettit, G. R., Vinitsky, A., and Smith, J. B. (1996). Ubiquitination of protein kinase C-alpha and degradation by the proteasome. J. Biol. Chem. **271**, 2097–20976.
66. Lu, Z., Liu, D., Hornia, A., Devonish, W., Pagano, M., and Foster, D. A. (1998). Activation of protein kinase C triggers its ubiquitination and degradation. *Mol. Cell Biol.* **18**, 839–845.
67. Okuda, H., Saitoh, K., Hirai, S., Iwai, K., Takaki, Y., Baba, M., Minato, N., Ohno, S., and Shuin, T. (2001). The von Hippel-Lindau tumor suppressor protein mediates ubiquitination of activated atypical protein kinase C. *J. Biol. Chem.* **276**, 43611–43617.
68. Ng, T., Squire, A., Hansra, G., Bornancin, F., Prevostel, C., Hanby, A., Harris, W., Barnes, D., Schmidt, S., Mellor, H., Bastiaens, P. I., and Parker, P. J. (1999). Imaging protein kinase Calpha activation in cells. *Science* **283**, 2085–2089.
69. Kiley, S. C., Clark, K. J., Duddy, S. K., Welch, D. R., and Jaken, S. (1999). Increased protein kinase C delta in mammary tumor cells: relationship to transformtion and metastatic progression. *Oncogene* **18**, 6748–6757.
70. Sausville, E. A., Arbuck, S. G., Messmann, R., Headlee, D., Bauer, K. S., Lush, R. M., Murgo, A., Figg, W. D., Lahusen, T., Jaken, S., Jing, X., Roberge, M., Fuse, E., Kuwabara, T., and Senderowicz, A. M. (2001). Phase I trial of 72-hour continuous infusion UCN-01 in patients with refractory neoplasms. *J. Clin. Oncol.* **19**, 2319–2333.
71. Ho, Y., Gruhler, A., Heilbut, A., Bader, G. D., Moore, L., Adams, S. L., Millar, A., Taylor, P., Bennett, K., Boutilier, K., Yang, L., Wolting, C., Donaldson, I., Schandorff, S., Shewnarane, J., Vo, M., Taggart, J., Goudreault, M., Muskat, B., Alfarano, C., Dewar, D., Lin, Z., Michalickova, K., Willems, A. R., Sassi, H., Nielsen, P. A., Rasmussen, K. J., Andersen, J. R., Johansen, L. E., Hansen, L. H., Jespersen, H., Podtelejnikov, A., Nielsen, E., Crawford, J., Poulsen, V., Sorensen, B. D., Matthiesen, J., Hendrickson, R. C., Gleeson, F., Pawson, T., Moran, M. F., Durocher, D., Mann, M., Hogue, C. W., Figeys, D., and Tyers, M. (2002). Systematic identification of protein complexes in *Saccharomyces cerevisiae* by mass spectrometry. *Nature* **415**, 180–183.
72. Sayers, L. G., Katayama, S., Nakano, K., Mellor, H., Mabuchi, I., Toda, T., and Parker, P. J. (2000). Rho-dependence of schizosaccharomyces pombe pck2. *Genes Cells* **5**, 17–27.

CHAPTER 188

Dendritic Protein Phosphatase Complexes

Roger J. Colbran
*Department of Molecular Physiology and Biophysics and the Center for Molecular Neuroscience,
Vanderbilt University School of Medicine, Nashville, Tennessee*

AKAP	A-kinase (PKA) anchoring protein
AMPA	a-amino-3-hydroxy-5-methyl-4-isoxazole-propionic acid
AMPAR	AMPA-type glutamate receptor
CaM	calmodulin
CaMKII	Ca^{2+}/CaM-dependent protein kinase II
CREB	cyclic AMP-response element binding protein
DARPP-32	dopamine and cyclic AMP-regulated phosphoprotein of 32 kDa
I1	inhibitor 1
LTD	long-term depression
LTP	long-term potentiation
NMDA	N-methyl-D-aspartate
NMDAR	NMDA-type glutamate receptor
PKA	protein kinase A, A-kinase, or cyclic AMP-dependent protein kinase
PP1	protein phosphatase 1
PP2B	protein phosphatase 2B or calcineurin
PSD	postsynaptic density

Introduction

Fast excitatory glutamatergic synaptic transmission is a primary contributor to many normal behaviors such as learning and memory and is often disrupted in complex neural disorders. The primary mediators of excitatory transmission are dendritic ionotropic AMPA-type and NMDA-type glutamate receptors, which are phosphorylated by many protein kinases. The modulatory actions of these kinases are antagonized by dendritic protein phosphatases, although there are incomplete, often contradictory, data implicating specific enzymes in modulating synaptic function, because only a relatively limited number of these enzymes often display promiscuous *in vitro* activity (see chapter 91, Volume 1 by Cohen). This review focuses on the nature and roles of dendritic protein phosphatase (PP1) and calcineurin (PP2B), the two phosphatases that have been best implicated in postsynaptic signaling.

The Importance of Dendritic Localization

The glutamate receptors involved in excitatory transmission are localized to the synapse by association with other proteins in a cytoskeletal structure called the postsynaptic density (PSD), which is often located at the tip of dendritic spines (see Chapter 55 by Kennedy). "Extra-synaptic" glutamate receptors are found in the membrane of dendritic shafts and the cell body. Synaptic and extra-synaptic receptors may be differentially regulated depending on the localization of enzymes involved in their regulation. Although specific receptors may be *in vitro* substrates for kinases and phosphatases or be modulated in whole-cell electrophysiological studies, this may not reflect synaptic regulation. Thus, understanding mechanisms that localize the relevant kinases and phosphatases to PSDs is a critical part of this puzzle.

Protein Phosphatase 1

Three mammalian genes encode four distinct but highly homologous PP1 catalytic subunits (PP1α, PP1β, PP1γ$_1$, and PP1γ$_2$), which interact with about 50 divergent proteins. Most PP1-binding proteins contain a binding motif defined by a consensus sequence, R/K-I/V-X-F. The interacting proteins modulate the activity or subcellular location of $PP1_C$ (see chapters 103 and 105 by DePaoli-Roach and Shenolikar in Volume 1, and [1]).

All four PP1 isoforms are expressed in mammalian brain, but with variable cellular distribution [2–5]: PP1β and PP1γ$_1$ also exhibit differential subcellular distribution. For example, PP1β is enriched in microtubule fractions whereas PP1γ$_1$ is selectively abundant in F-actin enriched extracts [3]. In addition, PP1γ$_1$, but not PP1β, is highly enriched in PSDs [6,7], which also contain F-actin. Consistent with these observations, immunohistochemistry shows that PP1γ$_1$ is enriched in synaptic layers and dendritic spines of neurons in brain slices [2,3], whereas PP1β is enriched in cell body layers [3]. Moreover, PP1γ$_1$ is localized to synapses in cultured cortical neurons, whereas PP1β colocalizes with microtubules in the cell body [3]. PP1α exhibits similar distribution to PP1γ$_1$ [2,6,7], but the subcellular localization of PP1γ$_2$ is unknown. Thus, PP1 isoforms are differentially targeted, suggesting that they have at least partially distinct neuronal functions. Furthermore, targeting appears to be dynamically regulated by synaptic NMDAR activation, which recruits additional unidentified PP1 isoforms to synapses [8]. Understanding the isoform selectivity and modulation of PP1 localization, and thus PP1 functions, demands a thorough understanding of the mechanisms involved. Several dendritic PP1-binding proteins (see Fig. 1) may play critical roles, as discussed below.

Inhibitor Proteins

PP1 interacts with I1 and DARPP-32, primarily soluble proteins that both inhibit PP1 activity when phosphorylated by PKA at homologous Thr residues (Thr35 in I1 and Thr34 in DARPP-32). Inhibition is mediated by interactions involving an R/K-I/V-X-F consensus motif (approximately residues 6–12) and residues surrounding Thr34/Thr35 [9,10]. Phospho-Thr34/35 in I1/DARPP-32 are effective substrates for calcineurin, the Ca^{2+}/CaM-dependent phosphatase [11]. Thus, cAMP and Ca^{2+} signaling modulate PP1 activity by antagonistic regulation of Thr34/35 phosphorylation [12]. In addition, regulation of PP1 by both DARPP-32 and I1 is modulated by phosphorylation at distinct sites by other kinases (reviewed in [13] and chapter 105 in Volume 1 by Shenolikar). Although there is no evidence indicating that I1 or DARPP32 distinguish between PP1 isoforms, their differential but overlapping expression suggests that they have different roles. For example, each protein appears to regulate dendritic PP1, but DARPP-32 is more abundant in the striatum, where as I1 is prevalent in hippocampus [14].

Targeting Proteins

Four major PP1-binding proteins (216,175,134, and 75 kDa) were initially detected in isolated PSDs [15]. The 175 and 134 kDa proteins selectively bound PP1γ$_1$ over PP1β, and were identified as *neurabin and spinophilin* [16], proteins previously identified as F-actin- and PP1α-binding proteins [17,18], respectively. Spinophilin also was isolated as an F-actin binding protein and termed neurabin II [19]. Both spinophilin and neurabin are selectively associated with PP1α and PP1γ$_1$ in brain extracts, but not PP1β [16,20], and may contribute to the differential localization of PP1β and PP1γ$_1$ in neurons (see above). An R/K-I/V-X-F motif is critical for binding and inhibition of PP1 by both proteins *in vitro*, but additional residues are also important

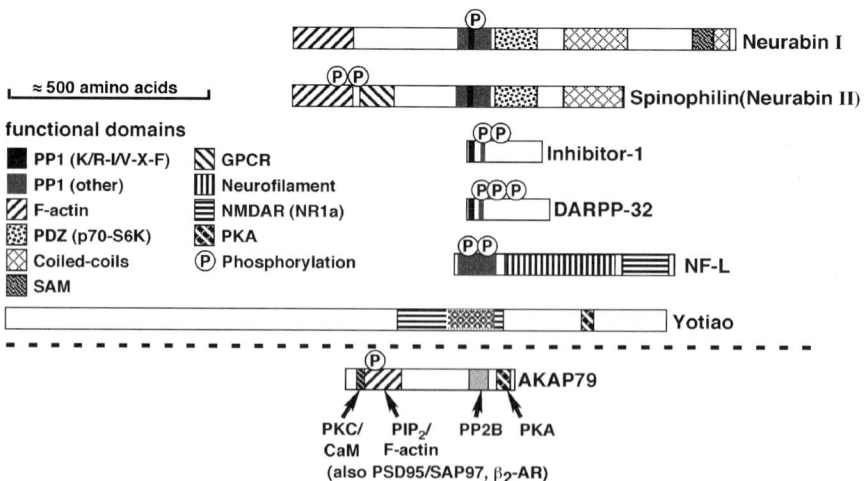

Figure 1 Dendritic PP1- and calcineurin-associated proteins. Amino acid sequences of the selected proteins are represented by open bars drawn to approximate scale. Many interact with PP1 via K/R-I/V-X-F motifs (black boxes) or other domains (gray boxes) (above dashed line), whereas AKAP79 interacts with calcineurin (PP2B). Most of these proteins also interact with protein kinases, cytoskeletal proteins, or other signaling proteins, probably serving as signaling scaffolds, and are phosphorylated, as indicated by additional shaded boxes (see legend). GPCR, G-protein coupled receptor; PDZ, PSD95/Dlg/ZO-1 domain; SAM, sterile alpha motif.

for inhibition [21,22]. However, it should be noted that PP1 activity is detected in neurabin complexes immunoprecipitated from brain extracts [21]. Although the activity of these complexes was not compared to that of free PP1, it seems that endogenous PP1 complexes with spinophilin or neurabin may possess phosphatase activity.

Subsequently, neurabin and spinophilin were shown to interact with many additional proteins that play critical roles in signal transduction, subcellular trafficking, and cytoskeletal dynamics (Fig. 1). Despite functional homology (e.g., interaction with PP1, F-actin, p70 S6 kinase), spinophilin and neurabin are likely to serve different cellular functions. For example, only spinophilin binds D2 dopamine and α2 adrenergic receptors [23,24], although it is not clear whether these interactions occur in neurons. In addition, reduction of neurabin expression in cultured neurons blocks neurite outgrowth [17], whereas neurons cultured from mice lacking spinophilin prematurely develop processes and the neurons from mature animals have an overabundance of spines [25].

The 216 and 75 kDa binding proteins in PSDs [15] may be yotiao and the neurofilament-L protein (NF-L), respectively. *Yotiao*, a protein first identified by binding NR1a subunits of NMDARs, was shown to be a PP1-targeting protein in addition to an AKAP [26]. It is interesting that small PP1-binding fragments of yotiao lack a recognizable K/R-I/V-X-F motif, although peptides containing the consensus motif compete for binding. Unusually for a PP1-binding protein, yotiao did not significantly inhibit PP1. High constitutive phosphatase activity in the NR1a-yotiao-PP1 complex may maintain the NMDAR in a dephosphorylated state under basal conditions [26]. PP1 also binds to the head domain of NF-L, resulting in inhibition of PP1, although this domain also lacks a K/R-I/V-X-F consensus motif. NF-L was identified in isolated PSDs [7] and also binds to the NMDAR NR1a subunit [27]. Thus, both yotiao and NF-L may contribute to PP1 targeting to NMDARs, but their isoform selectivity and specific contributions to subcellular localization remain unknown.

Calcineurin (Protein Phosphatase 2B)

Calcineurin was originally identified as an abundant Ca^{2+}/CaM-binding protein in brain extracts and was later shown to possess Ca^{2+}/CaM-dependent phosphatase activity identical to PP2B. The holoenzyme is a heterodimer of a Ca^{2+}/CaM-binding catalytic A subunit with a Ca^{2+}-binding regulatory B subunit (see chapter 106 in Volume 1 by Klee & Yang). Calcineurin is largely soluble in brain extracts, but it is also present in isolated PSDs [6] and colocalizes with F-actin in dendritic spines [28]. Calcineurin activity is inhibited by AKAP79, a PSD- and actin-associated dendritic spine protein that also interacts with many other synaptic proteins ([29–31] and see Fig. 1). Ca^{2+} influx via NMDA receptors stimulates calcineurin-dependent remodeling of actin in dendritic spines [28] and also the re-distribution of AKAP79, calcineurin, and PKA [29].

Dendritic Phosphatase Substrates

NMDARs and AMPARs

Phosphorylation may regulate the conductance, open probability, or trafficking of AMPARs and NMDARs. Initially, relatively nonselective inhibitors implicated PP1, PP2A, and PP2B in the regulation of AMPARs [32,33] and NMDARs [34]. Later studies in mice with a DARPP-32 knockout or overexpressing a constitutively active I1 mutant specifically implicated PP1. PKA-mediated regulation of extrasynaptic AMPAR and NMDAR is disrupted in striatal neurons from DARPP-32 knockout mice [35–37]. Although PP1 activities in extracts from wild-type and DARPP-32 knockout mice were not directly compared, these data suggest that PP1 inhibition by PKA-phosphorylated DARPP-32 is essential for phosphorylation of other PKA substrates. In contrast, expression of constitutively active I1 reduced hippocampal PP1 activity by 68 percent and enhanced phosphorylation of hippocampal AMPARs, as well as that of CaMKII and CREB [38]. In combination these data implicate I1/DARPP-32 modulation of PP1 activity in normal AMPAR and NMDAR regulation.

PP1 targeting has also been implicated in modulation of glutamate receptors. Intracellular perfusion of peptides containing the R/K-I/V-X-F PP1-binding motif disrupts dopamine D1 receptor regulation of extrasynaptic AMPARs in striatal neurons [39] but does not affect the basal activities of synaptic AMPARs and NMDARs in hippocampal neurons [8]. However, interpretation of these data is complicated because similar peptides disrupt PP1 interactions with spinophilin, neurabin, yotiao, NF-L, DARPP-32, and I1 *in vitro*. Subsequently, it was shown that striatal AMPARs and hippocampal NMDARs were abnormally regulated in spinophilin knockout mice [25]. These studies suggest that in wild-type animals spinophilin targets PP1 to appropriate subcellular locations to permit efficient regulation (dephosphorylation) of extrasynaptic AMPARs and NMDARs. However, the role of spinophilin in regulation of synaptic receptors is not as clearly defined.

Yotiao and NF-L also may play a role in PP1 modulation of NMDARs. PP1-binding to yotiao and NF-L is disrupted by R/K-I/V-X-F peptides; thus, disruption of these complexes may account for some of the effects of similar peptides in cells (see above). However, disruption of PP1 complexes with yotiao and NF-L would not be expected in spinophilin knockout animals. Thus, although ternary complexes containing NMDAR subunits, PP1, and either spinophilin, yotiao, or NF-L have not been reported, it seems that the mechanism of PP1 targeting to NMDARs may depend on the cell type, developmental stage, or other factors.

Calcineurin is also strongly implicated in NMDAR regulation [40–42] and mediates Ca^{2+}-dependent rundown of AMPARs in hippocampal neurons [43]. Association of calcineurin with AMPARs via AKAP79 and SAP97 promotes Ca^{2+}-stimulated rundown of AMPAR currents in HEK293 cells, an effect requiring an intact PKA site (Ser845) in the AMPAR [43]. However, despite evidence for PSD targeting of calcineurin by AKAP79, direct actions of calcineurin on synaptic AMPARs or NMDARs have not been clearly established. Calcineurin-dependent regulation of PP1 via I1 or DARPP-32 may contribute to some of the observed effects.

Ca^{2+}/CaM-Dependent Protein Kinase II (CaMKII)

Autophosphorylation at Thr286 generates a Ca^{2+}/calmodulin-independent form of CaMKII [44–46] and is critical for stable association of CaMKII with PSDs *in vitro* and in brain slices [47], as well as in cultured neurons [48]. Some forms of synaptic plasticity, learning, and memory require an intact Thr286 autophosphorylation site in CaMKII [49]. Thus, protein phosphatases have a potentially important role in regulating CaMKII.

Initially, PP1 was identified as the major CaMKII phosphatase in PSDs, with a minor role for PP2A and no significant direct role for calcineurin or PP2C [50,51]. Subsequently, PSD-associated CaMKII was shown to be primarily dephosphorylated by PP1, but PP2A was the major activity toward soluble CaMKII [6,47]. Thus, CaMKII translocation to PSDs may modulate its availability to cellular phosphatases. Although mechanism(s) accounting for this effect are unclear, it could play a role in determining the half-life of autophosphorylated CaMKII in cells. Addition of okadaic acid to cells enhances CaMKII autophosphorylation [52], consistent with a dominant role for PP1 and PP2A in intact cells. Moreover, inhibition of hippocampal PP1 by induced overexpression of constitutively active I1 results in enhanced autophosphorylation of CaMKII at Thr286 [38]. Thus, calcineurin may play a role in CaMKII regulation by modulating PP1 (see above) and PP2C-related phosphatases may be involved in some cells [53].

Role of Phosphatases in Synaptic Plasticity

Synaptic plasticity describes the long-lasting adaptations of synaptic function that are thought to underlie certain forms of memory. The best-studied forms of long-term potentiation (LTP) and long-term depression (LTD) in hippocampal CA1 neurons require dendritic Ca^{2+} influx via the NMDARs, and modulate the phosphorylation, activity, and subcellular trafficking of AMPARs. LTP requires Thr286 autophosphorylation of CaMKII and the activation of several other kinases (reviewed in [54–56]). The actions of protein phosphatases oppose the kinases, thus depressing or depotentiating synaptic transmission (see Fig. 2).

Initially, calcineurin and PP1 were implicated in LTD induction [57,58], in part because LTD required Ca^{2+} influx and the Ca^{2+}/CaM-dependent calcineurin regulates PP1 via I1 or DARPP-32 or both. One relevant substrate appears to be GluR1 subunits of AMPARs. LTD of naïve cells correlates with dephosphorylation of Ser845 (a PKA site), whereas depotentiation (LTP reversal) induces dephosphorylation of Ser831 (a CaMKII site) [59]. In contrast, LTP of naïve cells enhances Ser831 phosphorylation, and potentiation of "previously depressed" synapses enhances Ser845 phosphorylation [59]. However, these changes occur in the total AMPAR pool, not just in synaptic receptors, and the identity of the relevant phosphatases remains unclear.

Genetic manipulation of calcineurin activities in mice provided important insight into mechanisms of synaptic plasticity. A surprising finding was that mice expressing either constitutively active calcineurin or a calcineurin inhibitory peptide exhibited no detectable defect in LTD induction [60,61]. Rather, both transgenic models implicate calcineurin as exerting a negative effect on LTP [60,61], consistent with some prior pharmacological data [57,62]. It is interesting that mice lacking the calcineurin α gene, but retaining the minor calcineurin β gene, exhibit normal LTD and LTP but are defective in depotentiation [63]. Given the known changes in AMPAR phosphorylation during LTP, LTD, and depotentiation (see above), these data implicate calcineurin in the dephosphorylation of Ser831 but not Ser845 in synaptic AMPARs. However, this could be due to direct dephosphorylation of Ser831 by calcineurin or, indirectly, to PP1-mediated dephosphorylation/inactivation of CaMKII. Although this model is inconsistent with data implicating calcineurin in Ser845 dephosphorylation ([43]; see above), the synaptic AMPARs relevant to synaptic plasticity may be regulated differently.

Synaptic plasticity may involve modulation of PP1 activity by DARPP-32 or I1 or targeting of PP1 to AMPARs by spinophilin or other PP1-binding proteins. R/K-I/V-X-F motif peptides disrupt PP1 interactions with I1, DARPP-32, spinophilin, neurabin, and yotiao *in vitro* and block induction of some forms of LTD [8]. More specifically, spinophilin knockout mice are deficient in the induction of LTD but not LTP [25]. Moreover, PP1 plays a role in selective dephosphorylation of AMPARs at Ser845 but not Ser831 in striatal neurons [35]. Thus, AMPAR dephosphorylation at Ser845 associated with LTD may be due to activation of PP1 associated with spinophilin. Dephosphorylation of I1 by calcineurin may liberate PP1 to interact with spinophilin [8] and dephosphorylate Ser845, whereas dephosphorylation of Ser831 associated with depotentiation may be due to a direct action of calcineurin (Fig. 2).

In addition to a role for PP1 activation in LTD induction, inhibition of hippocampal PP1 appears important for LTP induction. LTP induces PKA-mediated phosphorylation of I1 and inhibition of PP1, thereby "gating" Thr286 autophosphorylation of CaMKII and promoting phosphorylation of AMPARs (Ser831) and other substrates [64,65]. Consistent with this model, Thr286 phosphorylation of CaMKII and Ser831 phosphorylation of AMPARs were enhanced in transgenic mice induced to overexpress a constitutively active mutant I1, correlating with enhanced learning and memory [38]. Thus, at least some forms of LTP require PKA-dependent inhibition of PP1 activity via the phosphorylation of Thr35 in I1. This contrasts with data obtained from I1 and spinophilin knockout mice in which LTP at the hippocampal CA3-CA1 synapses is normal [25,66]. However, these observations may reflect adaptations to chronic protein deficiency

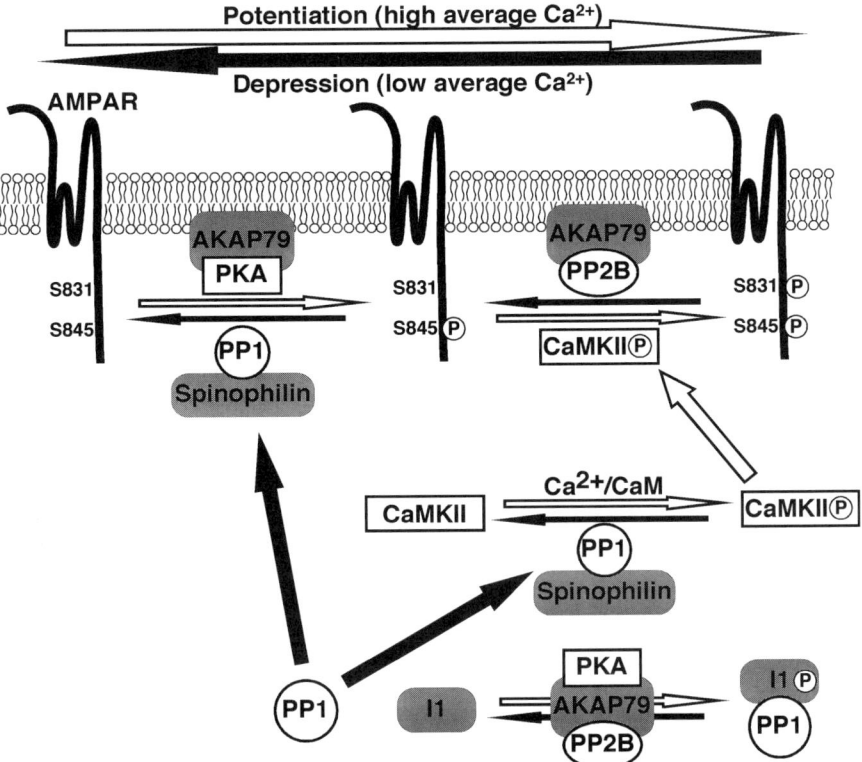

Figure 2 Working model implicating dendritic protein phosphatase complexes in hippocampal CA1 synaptic plasticity. Phosphorylation-dephosphorylation of AMPARs (black line) in synaptic membranes and other locations is associated with synaptic plasticity. *Unfilled arrows* indicate reactions and promoted by intense synaptic stimulation (e.g., high-frequency stimulation) that potentiate synaptic transmission. Activation of NMDARs results in high-average Ca^{2+} concentrations that favor activation of protein kinases (open rectangles), such as CaMKII and PKA (via Ca^{2+}/CaM-dependent adenylyl cyclases). PKA substrates include AMPARs (Ser845) and inhibitor 1 (I1). Phosphorylated I1 directly inhibits PP1 activity and may compete PP1 away from spinophilin. PP1 inhibition promotes the accumulation of Ser845 phosphorylated AMPARs and autophosphorylated (activated) CaMKII, which phosphorylates AMPARs at Ser831. In contrast, *solid arrows* indicate reactions promoted by low-intensity synaptic stimulation (e.g., prolonged low-frequency stimulation). NMDAR activation again provides the primary signal, but lower average Ca^{2+} concentrations favor phosphatase (open circles) activation, since calcineurin (PP2B) has a higher affinity for Ca^{2+}/CaM than CaMKII or adenylyl cyclases. Dephosphorylation of I1 by calcineurin liberates free PP1 that may associate with spinophilin and dephosphorylate (inactivate) CaMKII. Calcineurin and PP1 may dephosphorylate Ser831 and Ser845 in AMPARs, respectively. Assembly of phosphatases and kinase in complexes with anchoring targeting or regulatory proteins (gray) plays a critical role in dictating signaling specificity and fidelity.

in these knockout animals. Thus, the precise role of specific PP1 complexes in CaMKII regulation and LTP induction remains unclear.

Summary

There is compelling evidence that multiprotein complexes containing PP1 and calcineurin are critical for synaptic regulation, but specific roles of their molecular components still need to be determined. Roles for other serine/threonine and tyrosine phosphatases are also likely to be better described. Apparently contradictory data in the cited literature may be due to cell-specific issues related to differences in cellular functions, prior synaptic activity or developmental stage analyzed, and variations in experimental conditions. The challenge is to carefully control all the variables and use emerging animal models and specific molecular tools to more precisely understand the roles of these phosphatase complexes in synaptic regulation.

Acknowledgments

I appreciate the critical review of an initial draft by Eric D. Norman, Brian E. Wadzinski, Danny G. Winder, and members of my laboratory. I apologize to colleagues whose significant related contributions may not have been cited due to space constraints. Work in my laboratory was funded by the NIMH, NINDS, and AHA.

References

1. Bollen, M. (2001). Combinatorial control of protein phosphatase-1. *Trends Biochem. Sci.* **26**, 426–431.

2. Ouimet, C. C., da Cruz e Silva, E. F., and Greengard, P. (1995). The alpha and gamma 1 isoforms of protein phosphatase 1 are highly and specifically concentrated in dendritic spines. *Proc. Natl. Acad. Sci. USA* **92**, 3396–3400.
3. Strack, S., Kini, S., Ebner, F. F., Wadzinski, B. E., and Colbran, R. J. (1999). Differential cellular and subcellular localization of protein phosphatase 1 isoforms in brain. *J. Comp. Neurol.* **413**, 373–384.
4. da Cruz e Silva, E. F., Fox, C. A., Ouimet, C. C., Gustafson, E., Watson, S. J., and Greengard, P. (1995). Differential expression of protein phosphatase 1 isoforms in mammalian brain. *J. Neurosci.* **15**, 3375–3389.
5. Sakagami, H., Ebina, K., and Kondo, H. (1994). Localization of phosphatase inhibitor-1 mRNA in the developing and adult rat brain in comparison with that of protein phosphatase-1 mRNAs. *Brain Res. Mol. Brain. Res.* **25**, 7–18.
6. Strack, S., Barban, M. A., Wadzinski, B. E., and Colbran, R. J. (1997). Differential inactivation of postsynaptic density-associated and soluble Ca2+/calmodulin-dependent protein kinase II by protein phosphatases 1 and 2A. *J. Neurochem.* **68**, 2119–2128.
7. Terry-Lorenzo, R. T., Inoue, M., Connor, J. H., Haystead, T. A., Armbruster, B. N., Gupta, R. P., Oliver, C. J., and Shenolikar, S. (2000). Neurofilament-L is a protein phosphatase-1-binding protein associated with neuronal plasma membrane and post-synaptic density. *J. Biol. Chem.* **275**, 2439–2446.
8. Morishita, W., Connor, J. H., Xia, H., Quinlan, E. M., Shenolikar, S., and Malenka, R. C. (2001). Regulation of synaptic strength by protein phosphatase 1. *Neuron* **32**, 1133–1148.
9. Kwon, Y. G., Huang, H. B., Desdouits, F., Girault, J. A., Greengard, P., and Nairn, A. C. (1997). Characterization of the interaction between DARPP-32 and protein phosphatase 1 (PP-1): DARPP-32 peptides antagonize the interaction of PP-1 with binding proteins. *Proc. Natl. Acad. Sci. USA* **94**.
10. Endo, S., Zhou, X., Connor, J., Wang, B., and Shenolikar, S. (1996). Multiple structural elements define the specificity of recombinant human inhibitor-1 as a protein phosphatase-1 inhibitor. *Biochemistry* **35**, 5220–5228.
11. King, M. M., Huang, C. Y., Chock, P. B., Nairn, A. C., Hemmings, H. C., Jr., Chan, K. F., and Greengard, P. (1984). Mammalian brain phosphoproteins as substrates for calcineurin. *J. Biol. Chem.* **259**, 8080–8083.
12. Halpain, S., Girault, J. A., and Greengard, P. (1990). Activation of NMDA receptors induces dephosphorylation of DARPP-32 in rat striatal slices. *Nature* **343**, 369–372.
13. Greengard, P. (2001). The neurobiology of slow synaptic transmission. *Science* **294**, 1024–1030.
14. Hemmings, H. C., Jr., Girault, J. A., Nairn, A. C., Bertuzzi, G., and Greengard, P. (1992). Distribution of protein phosphatase inhibitor-1 in brain and peripheral tissues of various species: comparison with DARPP-32. *J. Neurochem.* **59**, 1053–1061.
15. Colbran, R. J., Bass, M. A., McNeill, R. B., Bollen, M., Zhao, S., Wadzinski, B. E., and Strack, S. (1997). Association of brain protein phosphatase 1 with cytoskeletal targeting/regulatory subunits. *J. Neurochem.* **69**, 920–929.
16. MacMillan, L. B., Bass, M. A., Cheng, N., Howard, E. F., Tamura, M., Strack, S., Wadzinski, B. E., and Colbran, R. J. (1999). Brain actin-associated protein phosphatase 1 holoenzymes containing spinophilin, neurabin, and selected catalytic subunit isoforms. *J. Biol. Chem.* **274**, 35845–35854.
17. Nakanishi, H., Obaishi, H., Satoh, A., Wada, M., Mandai, K., Satoh, K., Nishioka, H., Matsuura, Y., Mizoguchi, A., and Takai, Y. (1997). Neurabin—a novel neural tissue-specific actin filament-binding protein involved in neurite formation. *J. Cell Biol.* **139**, 951–961.
18. Allen, P. B., Ouimet, C. C., and Greengard, P. (1997). Spinophilin, a novel protein phosphatase 1 binding protein localized to dendritic spines. *Proc. Natl. Acad. Sci. USA* **94**, 9956–9961.
19. Satoh, A., Nakanishi, H., Obaishi, H., Wada, M., Takahashi, K., Satoh, K., Hirao, K., Nishioka, H., Hata, Y., Mizoguchi, A., and Takai, Y. (1998). Neurabin-II/spinophilin: an actin-filament-binding protein with one PDZ domain localized at cadherin-based cell-cell adhesion sites. *J. Biol. Chem.* **273**, 3470–3475.
20. Terry-Lorenzo, R. T., Carmody, L. C., Voltz, J. W., Connor, J. H., Li, S., Smith, F. D., Milgram, S. L., Colbran, R. J., and Shenolikar, S. (2002). The neuronal actin-binding proteins, neurabin I and neurabin II, recruit specific isoforms of protein phosphatase-1 catalytic subunits. *J. Biol. Chem.* **277**, 27716–27724.
21. Oliver, C. J., Terry-Lorenzo, R. T., Elliott, E., Bloomer, W. A., Li, S., Brautigan, D. L., Colbran, R. J., and Shenolikar, S. (2002). Targeting protein phosphatase 1 (PP1) to the actin cytoskeleton: the neurabin I/PP1 complex regulates cell morphology. *Mol. Cell. Biol.* **22**, 4690–4701.
22. Hsieh-Wilson, L. C., Allen, P. B., Watanabe, T., Nairn, A. C., and Greengard, P. (1999). Characterization of the neuronal targeting protein spinophilin and its interactions with protein phosphatase-1. *Biochemistry* **38**, 4365–4373.
23. Smith, F. D., Oxford, G. S., and Milgram, S. L. (1999). Association of the D2 dopamine receptor third cytoplasmic loop with spinophilin, a protein phosphatase-1-interacting protein. *J. Biol. Chem.* **274**, 19894–19900.
24. Richman, J. G., Brady, A. E., Wang, Q., Hensel, J. L., Colbran, R. J., and Limbird, L. E. (2001). Agonist-regulated interaction between alpha2-adrenergic receptors and spinophilin. *J. Biol. Chem.* **276**, 15003–15008.
25. Feng, J., Yan, Z., Ferreira, A., Tomizawa, K., Liauw, J. A., Zhuo, M., Allen, P. B., Ouimet, C. C., and Greengard, P. (2000). Spinophilin regulates the formation and function of dendritic spines. *Proc. Natl. Acad. Sci. USA* **97**, 9287–9292.
26. Westphal, R. S., Tavalin, S. J., Lin, J. W., Alto, N. M., Fraser, I. D., Langeberg, L. K., Sheng, M., and Scott, J. D. (1999). Regulation of NMDA receptors by an associated phosphatase-kinase signaling complex. *Science* **285**, 93–96.
27. Ehlers, M. D., Fung, E. T., Obrien, R .J., and Huganir, R. L. (1998). Splice variant-specific interaction of the NMDA receptor subunit NR1 with neuronal intermediate filaments. *J. Neurosci.* **18**, 720–730.
28. Halpain, S., Hipolito, A., and Saffer, L. (1998). Regulation of F-actin stability in dendritic spines by glutamate receptors and calcineurin. *J. Neurosci.* **18**, 9835–9844.
29. Gomez, L. L., Alam, S., Smith, K. E., Horne, E., and Dell'Acqua, M. L. (2002). Regulation of A-kinase anchoring protein 79/150-cAMP-dependent protein kinase postsynaptic targeting by NMDA receptor activation of calcineurin and remodeling of dendritic actin. *J. Neurosci.* **22**, 7027–7044.
30. Coghlan, V. M., Perrino, B. A., Howard, M., Langeberg, L. K., Hicks, J. B., Gallatin, W. M., and Scott, J. D. (1995). Association of protein kinase A and protein phosphatase 2B with a common anchoring protein. *Science* **267**, 108–111.
31. Dell'Acqua, M. L., Dodge, K. L., Tavalin, S. J., and Scott, J. D. (2002). Mapping the protein phosphatase-2B anchoring site on AKAP79: binding and inhibition of phosphatase activity are mediated by residues 315–360. *J. Biol. Chem.*
32. Figurov, A., Boddeke, H., and Muller, D. (1993). Enhancement of AMPA-mediated synaptic transmission by the protein phosphatase inhibitor calyculin A in rat hippocampal slices. *Eur. J. Neurosci.* **5**, 1035–1041.
33. Wyllie, D. J. and Nicoll, R. A. (1994). A role for protein kinases and phosphatases in the Ca(2+)-induced enhancement of hippocampal AMPA receptor-mediated synaptic responses. *Neuron* **13**, 635–643.
34. Wang, L. Y., Orser, B. A., Brautigan, D. L., and MacDonald, J. F. (1994). Regulation of NMDA receptors in cultured hippocampal neurons by protein phosphatases 1 and 2A. *Nature* **369**, 230–232.
35. Snyder, G. L., Allen, P. B., Fienberg, A. A., Valle, C. G., Huganir, R. L., Nairn, A. C., and Greengard, P. (2000). Regulation of phosphorylation of the GluR1 AMPA receptor in the neostriatum by dopamine and psychostimulants in vivo. *J. Neurosci.* **20**, 4480–4488.
36. Snyder, G. L., Fienberg, A. A., Huganir, R. L., and Greengard, P. (1998). A dopamine/D1 receptor/protein kinase A/dopamine- and cAMP-regulated phosphoprotein (Mr 32 kDa)/protein phosphatase-1 pathway regulates dephosphorylation of the NMDA receptor. *J. Neurosci.* **18**, 10297–10303.

37. Fienberg, A. A., Hiroi, N., Mermelstein, P. G., Song, W., Snyder, G. L., Nishi, A., Cheramy, A., O'Callaghan, J. P., Miller, D. B., Cole, D. G., Corbett, R., Haile, C. N., Cooper, D. C., Onn, S. P., Grace, A. A., Ouimet, C. C., White, F. J., Hyman, S. E., Surmeier, D. J., Girault, J., Nestler, E. J., and Greengard, P. (1998). DARPP-32: regulator of the efficacy of dopaminergic neurotransmission. *Science* **281**, 838–842.
38. Genoux, D., Haditsch, U., Knobloch, M., Michalon, A., Storm, D., and Mansuy, I. M. (2002). Protein phosphatase 1 is a molecular constraint on learning and memory. *Nature* **418**, 970–975.
39. Yan, Z., Hsieh-Wilson, L., Feng, J., Tomizawa, K., Allen, P. B., Fienberg, A. A., Nairn, A. C., and Greengard, P. (1999). Protein phosphatase 1 modulation of neostriatal AMPA channels: regulation by DARPP-32 and spinophilin. *Nat. Neurosci.* **2**, 13–17.
40. Umemiya, M., Chen, N., Raymond, L. A., and Murphy, T. H. (2001). A calcium-dependent feedback mechanism participates in shaping single NMDA miniature EPSCs. *J. Neurosci.* **21**, 1–9.
41. Lieberman, D. N. and Mody, I. (1994). Regulation of NMDA channel function by endogenous $Ca^{(2+)}$-dependent phosphatase. *Nature* **369**, 235–239.
42. Krupp, J. J., Vissel, B., Thomas, C. G., Heinemann, S. F., and Westbrook, G. L. (2002). Calcineurin acts via the C-terminus of NR2A to modulate desensitization of NMDA receptors. *Neuropharmacology* **42**, 593–602.
43. Tavalin, S. J., Colledge, M., Hell, J. W., Langeberg, L. K., Huganir, R. L., and Scott, J. D. (2002). Regulation of GluR1 by the A-kinase anchoring protein 79 (AKAP79) signaling complex shares properties with long-term depression. *J. Neurosci.* **22**, 3044–3051.
44. Thiel, G., Czernik, A. J., Gorelick, F., Nairn, A. C., and Greengard, P. (1988). Ca^{2+}/calmodulin-dependent protein kinase II: identification of threonine-286 as the autophosphorylation site in the alpha subunit associated with the generation of Ca2+-independent activity. *Proc. Natl. Acad. Sci. USA* **85**, 6337–6341.
45. Schworer, C. M., Colbran, R. J., Keefer, J. R., and Soderling, T. R. (1988). Ca^{2+}/calmodulin-dependent protein kinase II. Identification of a regulatory autophosphorylation site adjacent to the inhibitory and calmodulin-binding domains. *J. Biol. Chem.* **263**, 13486–13489.
46. Miller, S. G., Patton, B. L., and Kennedy, M. B. (1988). Sequences of autophosphorylation sites in neuronal type II CaM kinase that control Ca2(+)-independent activity. *Neuron* **1**, 593–604.
47. Strack, S., Choi, S., Lovinger, D. M., and Colbran, R. J. (1997). Translocation of autophosphorylated calcium/calmodulin-dependent protein kinase II to the postsynaptic density. *J. Biol. Chem.* **272**, 13467–13470.
48. Shen, K., Teruel, M. N., Connor, J. H., Shenolikar, S., and Meyer, T. (2000). Molecular memory by reversible translocation of calcium/calmodulin-dependent protein kinase II. *Nat. Neurosci.* **3**, 881–886.
49. Giese, K. P., Fedorov, N. B., Filipkowski, R. K., and Silva, A. J. (1998). Autophosphorylation at Thr(286) of the alpha calcium-calmodulin kinase II In LTP and learning. *Science* **279**, 870–873.
50. Dosemeci, A. and Reese, T.S. (1993). Inhibition of endogenous phosphatase in a postsynaptic density fraction allows extensive phosphorylation of the major postsynaptic density protein. *J. Neurochem.* **61**, 550–555.
51. Shields, S. M., Ingebritsen, T. S., and Kelly, P. T. (1985). Identification of protein phosphatase 1 in synaptic junctions: dephosphorylation of endogenous calmodulin-dependent kinase II and synapse-enriched phosphoprotein*s. J. Neurosci.* **5**, 3414–3422.
52. Molloy, S. S. and Kennedy, M. B. (1991). Autophosphorylation of type II Ca2+/calmodulin-dependent protein kinase in cultures of postnatal rat hippocampal slices. *Proc. Natl. Acad. Sci. USA* **88**, 4756–4760.
53. Fukunaga, K., Kobayashi, T., Tamura, S., and Miyamoto, E. (1993). Dephosphorylation of autophosphorylated Ca2+/calmodulin-dependent protein kinase II by protein phosphatase 2C. *J. Biol. Chem.* **268**, 133–137.
54. Malenka, R. C. and Nicoll, R. A. (1999). Long-term potentiation—a decade of progress? *Science* **285**, 1870–1874.
55. Malinow, R. and Malenka, R. C. (2002). AMPA receptor trafficking and synaptic plasticity. *Annu. Rev. Neurosci.* **25**, 103–126.
56. Lisman, J., Schulman, H., and Cline, H. (2002). The molecular basis of CaMKII function in synaptic and behavioural memory. *Nat. Rev. Neurosci.* **3**, 175–190.
57. Mulkey, R. M., Endo, S., Shenolikar, S., and Malenka, R. C. (1994). Involvement of a calcineurin/inhibitor-1 phosphatase cascade in hippocampal long-term depression. *Nature* **369**, 486–488.
58. Mulkey, R. M., Herron, C. E., and Malenka, R. C. (1993). An essential role for protein phosphatases in hippocampal long-term depression. *Science* **261**, 1051–1055.
59. Lee, H. K., Barbarosie, M., Kameyama, K., Bear, M. F., and Huganir, R. L. (2000). Regulation of distinct AMPA receptor phosphorylation sites during bidirectional synaptic plasticity. *Nature* **405**, 955–959.
60. Malleret, G., Haditsch, U., Genoux, D., Jones, M. W., Bliss, T. V., Vanhoose, A. M., Weitlauf, C., Kandel, E. R., Winder, D. G., and Mansuy, I. M. (2001). Inducible and reversible enhancement of learning, memory, and long-term potentiation by genetic inhibition of calcineurin. *Cell* **104**, 675–686.
61. Winder, D. G., Mansuy, I. M., Osman, M., Moallem, T. M., and Kandel, E. R. (1998). Genetic and pharmacological evidence for a novel, intermediate phase of long-term potentiation suppressed by calcineurin. *Cell* **92**, 25–37.
62. Wang, J. H. and Kelly, P. T. (1997). Postsynaptic calcineurin activity downregulates synaptic transmission by weakening intracellular Ca2+ signaling mechanisms in hippocampal CA1 neurons. *J. Neurosci.* **17**, 4600–4611.
63. Zhuo, M., Zhang, W., Son, H., Mansuy, I., Sobel, R. A., Seidman, J., and Kandel, E. R. (1999). A selective role of calcineurin alpha in synaptic depotentiation in hippocampus. *Proc. Natl. Acad. Sci. USA* **96**, 4650–4655.
64. Blitzer, R. D., Connor, J. H., Brown, G. P., Wong, T., Shenolikar, S., Iyengar, R., and Landau, E. M. (1998). Gating of CaMKII by cAMP-regulated protein phosphatase activity during LTP. *Science* **280**, 1940–1942.
65. Brown, G. P., Blitzer, R. D., Connor, J. H., Wong, T., Shenolikar, S., Iyengar, R., and Landau, E. M. (2000). Long-term potentiation induced by theta frequency stimulation is regulated by a protein phosphatase-1-operated gate. *J. Neurosci.* **20**, 7880–7887.
66. Allen, P. B., Hvalby, O., Jensen, V., Errington, M. L., Ramsay, M., Chaudhry, F. A., Bliss, T. V., Storm-Mathisen, J., Morris, R. G., Andersen, P., and Greengard, P. (2000). Protein phosphatase-1 regulation in the induction of long-term potentiation: heterogeneous molecular mechanisms. *J. Neurosci.* **20**, 3537–3543.

Protein Phosphatase 2A

Adam M. Silverstein, Anthony J. Davis,
Vincent A. Bielinski, Edward D. Esplin,
Nadir A. Mahmood, and Marc C. Mumby

*Department of Pharmacology, University of Texas Southwestern Medical Center,
Dallas, Texas*

Introduction

Serine/threonine phosphatases are integral components of many signal transduction pathways. There are eight classes of serine/threonine phosphatases in vertebrates. Protein serine/threonine phosphatases 1, 2A, 2B/calcineurin, 4, 5, 6, and 7 are members of the PPP gene family that contain a conserved serine/threonine phosphatase domain. Protein phosphatase 2A (PP2A) is a ubiquitously expressed member of the PPP gene family that accounts for a substantial portion of the total serine/threonine phosphatase activity in many cell types. PP2A is an essential enzyme that functions in fundamental cellular processes, including metabolism and the cell cycle. Like the other signaling molecules discussed in this chapter, proximity interactions play a primary role in regulating PP2A.

Once thought of as a single, broad-specificity phosphatase, PP2A is actually many different enzymes composed of complexes between catalytic subunits, scaffold subunits, regulatory subunits, and interacting proteins [1–3]. The catalytic and scaffold subunits bind tightly to form a core dimer that is the common component of most, but not all, forms of PP2A. The core dimer interacts with an array of regulatory subunits to generate multiple heterotrimeric holoenzymes. Additional interactions between PP2A and a variety of interacting proteins generate additional diversity. The regulatory subunits and interacting proteins target PP2A to specific substrates and intracellular locations. The existence of many different forms of PP2A accounts for the ability of the enzyme to regulate a wide variety of biological processes.

Interaction of the core dimer with regulatory subunits is critical for PP2A function. The regulatory subunits bind to the core dimer through interactions with both the scaffold and the catalytic subunits. The scaffold subunit is composed entirely of 15 copies of a conserved motif termed the HEAT repeat [4]. HEAT repeats 1–10 mediate interactions with regulatory subunits whereas repeats 11–15 mediate interaction with the C subunit [5]. The regulatory subunits must form contacts with both the scaffold and the catalytic subunits to generate stable heterotrimers [5,6]. The regulatory subunits bind to the core dimer in a mutually exclusive manner. Although some sites of interaction are conserved, there are unique amino acids within the scaffold subunit that are involved in the interaction with individual regulatory subunits [7] (Fig. 1).

PP2A Regulatory Subunits Mediate Proximity Interactions

Regulatory subunits play a primary role in specifying the proximity interactions of PP2A. Three families of PP2A regulatory subunits have been identified in vertebrates by biochemical and genetic methods. A list of PP2A subunits is presented in Table I. In order to avoid confusion, we have used a nomenclature for the PP2A subunits derived from their official human gene symbols. In contrast to the scaffold and catalytic subunits, which are ubiquitously expressed, the PP2A regulatory subunits are expressed in a cell- and tissue-specific manner. PP2A regulatory subunits are also differentially expressed during development and have distinct subcellular localizations. Neither the structural basis for interaction of regulatory subunits with the PP2A core dimer nor the biochemical effects of these interactions have been clearly elucidated. The PP2A regulatory subunit families have little overall amino acid sequence similarity. Several regulatory

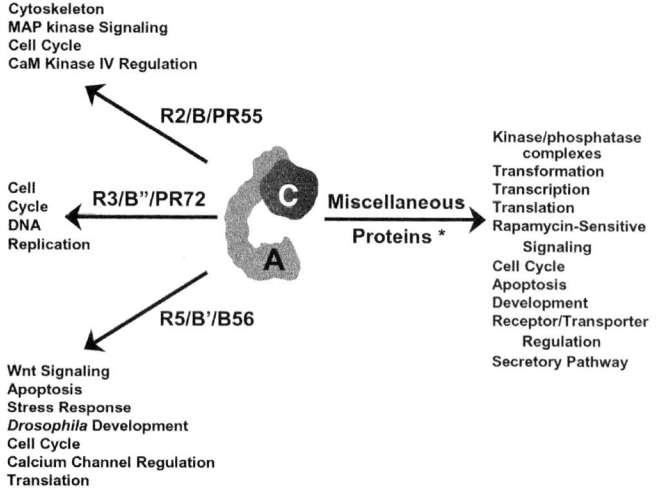

Figure 1 PP2A is a family of enzymes with multiple cellular functions. The PP2A holoenzyme consists of a common core dimer (AC) that complexes with a wide variety of regulatory molecules to generate a diversity of enzyme forms. These regulatory molecules include three regulatory subunit families (R2, R3, and R5) and a variety of miscellaneous proteins that interact with the core dimer or the free catalytic subunit. The regulatory molecules target PP2A to distinct substrates and intracellular locations, allowing the enzyme to participate in numerous cellular functions. The types of functions targeted by the individual regulatory subunits and miscellaneous proteins are listed.

subunits contain WD domains, which have been proposed as a conserved motif responsible for the interaction with the core dimer [3,8–10]. Recently a loosely conserved A-subunit binding domain has been identified in each of the regulatory subunit families [11]. The regulatory subunits have effects on the kinetics of dephosphorylation that are consistent with a role in controlling the binding of substrates to PP2A [12–15]. This model is consistent with the notion that regulatory subunit-mediated proximity interactions play a role in targeting PP2A to phosphoprotein substrates. In contrast to simple enzyme-substrate interactions, the interaction of PP2A with many substrates involves a stable interaction involving regions of the enzyme removed from the active site. These stable interactions serve to maintain a high effective concentration of PP2A in the vicinity of the substrate.

Consistent with roles in defining PP2A specificity, different families of PP2A regulatory subunits have non-overlapping functions. The stress-induced growth arrest caused by mutations in the R5 subunit gene (RTS1) in yeast can be rescued by introduction of wild-type versions of either the yeast R5 gene or the human R5γ gene [16]. In contrast, wild-type R5 cannot rescue the cold-sensitive phenotype resulting from mutations in the yeast R2 subunit gene (CDC55). Knockdown of individual PP2A regulatory subunits in *Drosophila* S2 cells by RNA interference causes distinct defects. Loss of the R2 subunit

Table I Nomenclature of Mammalian PP2A Subunits

Name	Gene symbol	Aliases	Chromosomal location	LocusID/accession number[a]
Catalytic subunits				
α isoform	PPP2CA	Cα, PP2A$_{Cα}$	5q23–31	*5515*
β isoform	PPP2CB	Cβ, PP2A$_{Cβ}$	8p21–12	*5516*
Scaffold subunits				
α isoform	PPP2R1A	Aα, PR65α	19	*5518*
β isoform	PPP2R1B	Aβ, PR65β	11q23	*5519*
R2 subunits				
α isoform	PPP2R2A	Bα, PR55α	8	*5520*
β isoform	PPP2R2B	Bβ, PR55β	5q31–5q33	*5521*
γ isoform	PPP2R2C	Bγ	4p16	*5522*
δ isoform		Bδ		*AF180350*
R3 subunits				
α isoform	PPP2R3	PR72, B″	3	*5523*
β isoform		PR59		*AF050165*
γ isoform		PR48		*28227*
R5 subunits				
α isoform	PPP2R5A	B′α, B56α	1q41	*5525*
β isoform	PPP2R5B	B′β, B56β	11q12	*5526*
γ isoform	PPP2R5C	B′γ, B56γ	3p21	*5527*
δ isoform	PPP2R5D	B′δ, B56δ	6p21.1	*5528*
ε isoform	PPP2R5E	B′ε, B56ε	7p11.1–12	*5529*

[a]Entries in this column include the LocusID, when available, for the *NCBI LocusLink* entry for the corresponding the protein, or the GenBank/EMBL Accession number.

causes an increase in insulin-dependent MAP kinase signaling, whereas loss of both R5 isoforms induces apoptosis [17]. These data suggest that PP2A holoenzymes containing the R2 subunit play a negative regulatory role in MAP kinase signaling whereas holoenzymes containing the R5 subunit function in cell survival.

The R2 Family

The R2 family comprises a set of proteins present in a form of PP2A originally designated PP2A$_1$ [18]. This family currently contains four known isoforms (Table I) that are 79–87 percent identical. R2α mRNA is ubiquitously expressed and is the most abundant PP2A regulatory subunit in many cells and tissues. The R2β and R2γ isoforms are only expressed at high levels in brain and testis. Although R2α and R2β are both expressed in the brain, they are present at different levels in different types of neurons [19]. R2α is distributed mainly in neuronal cell bodies and is localized in both the cytosol and nucleus. In contrast, the β isoform is excluded from the nucleus and is localized in axons and dendrites in addition to the cell body. Expression of R2 subunit mRNA is also differentially regulated during development. The differential expression and localization of R2 subunits support the idea that different members of this family play distinct roles in regulating PP2A functions.

Information about the functions of PP2A regulatory subunits has been derived from genetic analysis in yeast, *Drosophila*, and *C. elegans*. The pleiotropic phenotypes of mutant alleles of the R2 subunit in yeast and its numerous genetic interactions indicate that the R2 (cdc55p) protein plays multiple roles during mitosis, including the bud morphogenetic checkpoint and the mitotic spindle-assembly checkpoint [20–22]. The genetic results suggest that R2/cdc55p is involved in promoting activation of the yeast cell cycle regulatory kinase CDC2 (cyclin B/Cdc28 in *S. cerevisiae*) via dephosphorylation of the inhibitory tyrosine 19 phosphorylation site. Since PP2A does not directly dephosphorylate tyrosine, a likely target of R2 action is the cdc25 dual-specificity phosphatase, which is responsible for dephosphorylating tyrosine 19 in *S. cerevisiae* cdc28p.

Reduced levels of the R2 subunit in *Drosophila* result in varied phenotypes depending on the severity of the alleles. The *aar1* allele (for abnormal anaphase resolution) contains a P-element insertion in the R2 gene [23]. Mutant *aar1* flies die as larvae or early adults with overcondensed chromosomes and abnormal anaphase figures in larval brain cells. These defects can be rescued by reintroduction of the wild-type R2 gene. The *aar1* phenotype is reminiscent of the mitotic spindle-assembly checkpoint defects seen in the yeast R2 mutants. Another P-element mutant allele of *Drosophila* R2 (*twins^P*) causes death at an early pupal stage and shows pattern duplication of wing imaginal discs [24]. Flies harboring a weaker allele, *twins^{55}*, survive but have duplicated bristles in sensory neurons [25]. The effects of the *twins* mutation are consistent with a role for the R2 subunit in *Drosophila* embryonic cell fate determination.

Both the *aar1* and *twins^P* mutant larvae have a specific reduction in phosphatase activity toward substrates of cyclin-dependent kinases, suggesting the R2 subunit directs PP2A toward these substrates.

The R2 subunit targets PP2A to pathways that regulate MAP kinase activity. Overexpression of the small-t antigen of SV40 virus disrupts endogenous PP2A complexes containing the R2 subunit. This leads to enhanced activation of MAP kinase in response to growth factors in some but not all cell types [26,27]. The small-t antigen effects may involve protein kinase C and the PI-3 kinase pathway [28]. Depletion of the R2 subunit in *Drosophila* S2 cells via RNA interference also leads to a prolonged activation of MAP kinase in response to insulin [17]. These studies indicate that the R2 subunit plays a negative role in regulating MAP kinase activity, presumably by targeting PP2A to a component that is activated by phosphorylation. The *C. elegans* R2 subunit (*sur-6*) was isolated as a suppressor of the multivulval phenotype caused by an activated ras mutation [29]. Sur-6 mutations do not cause defects in vulval development by themselves but enhance the effects of weak mutant alleles of the *C. elegans* Raf protein kinase. These genetic interactions indicate that *sur-6* mutations reduce signaling through the Ras pathway and may act with the kinase suppressor of raf (KSR) protein in a common pathway to positively regulate signaling through the Ras-Raf-MAP kinase pathway. The PP2A core dimer can associate with the Raf-1 protein kinase (Table II). This interaction appears to mediate the dephosphorylation of inhibitory phosphorylation sites and enhance activation of Raf-1 during mitogenic stimulation [30]. This interaction does not appear to be mediated by the R2 subunit, since neither R2α nor R2β were detected in Raf-1 complexes. These studies suggest that PP2A is targeted to components of signaling pathways that regulate MAP kinase in both positive and negative ways. At least some of this targeting is mediated by R2 subunits. Multiple roles in MAP kinase signaling are consistent with genetic studies showing that mutations in the PP2A catalytic subunit have both positive and negative effects on MAP kinase activation in *Drosophila* [31]. The multiple actions in MAP kinase signaling are likely to be due to different forms of PP2A acting at distinct sites in this regulatory network.

Another R2-mediated proximity interaction regulates the microtubule cytoskeleton. A population of PP2A is associated with microtubules in neuronal and non-neuronal cells [32]. The association of PP2A with microtubules in brain is specific for R2α- and R2β-containing isoforms, and can be enhanced by a heat-labile anchoring factor [15]. PP2A holoenzymes containing R2α or R2β also interact with the neuronal microtubule-associated protein tau (Table II) and act as potent tau phosphatases [33]. The microtubule-binding and organizing activity of tau is regulated by phosphorylation. Hypophosphorylated forms of tau bind to microtubules, leading to increased microtubule stability. In contrast, hyperphosphorylated tau dissociates from microtubules, leading to a decrease in microtubule stability. Tau-dependent stabilization of microtubules is important for formation and maintenance

Table II PP2A Interacting Proteins

Protein	Comments	Refs
Signaling proteins/transcription factors		
Adenomatous polyposis coli (APC)	APC binds to R5 subunits in yeast two-hybrid assays. This interaction may target PP2A to the Wnt signaling pathway, but physical complexes between PP2A and APC have not been demonstrated. Overexpression of R5 subunits decrease β-catenin levels and suppress Wnt signaling.	[49]
Axin	Axin forms complexes with the C and R5 subunits. The interaction targets PP2A to a complex of axin, APC, GSK3, and β-catenin and plays a role in regulating Wnt signaling.	[50, 51]
Cas (p130 Crk-associated substrate)	Cas is a Src substrate that has increased association with PP2A when Src is activated. PP2A dephosphorylates serine residues on Cas *in vitro*.	[60]
E-cadherin/β-catenin	The C_α but not the C_β subunit is required for stabilization of E-cadherin/β-catenin complexes at the Plasma membrane.	[61]
Heat shock transcription factor 2 (HSF2)	HSF2 interacts with the A-subunit in two-hybrid and co-immunoprecipitation assays. HSF2 may displace the catalytic subunit from PP2A holoenzymes.	[62, 63]
HOX11	HOX 11 is homeobox transcription factor that controls development of the spleen. HOX11 binds to the PP2A catalytic subunit and inhibits phosphatase activity. HOX 11 also interacts with protein phosphatase 1.	[64]
HRX	HRX binds to PP2A through the SET/I_2^{PP2A} inhibitor protein. HRX is commonly mutated in acute leukemias.	[65]
Sex combs reduced (SCR)	SCR is a *Drosophila* homeobox transcription factor that interacts with the *Drosophila* R5 subunit in two-hybrid assays. SCR is homologous to human HOX5 and HOX6. PP2A may control phosphorylation and DNA binding activity of SCR.	[57]
RelA	RelA interacts with the scaffold subunit *in vitro*. The association may be transient since cross-linking is required to isolate a PP2A/RelA complex. RelA is dephosphorylated by PP2A *in vitro*.	[66]
Shc	PP2A associates with the PTB domain of Shc in the basal state and dissociates in response to insulin- and EGF-induced tyrosine phosphorylation. Expression of SV40 small-t antigen also causes dissociation of this complex.	[67]
Sp1	The Sp1 transcription factor interacts with the catalytic subunit in dividing T lymphocytes.	[68, 69]
Stat5	Stat5 associates with PP2A in an IL-3-dependent manner in the cytoplasm but not the nucleus.	[70]
Cell cycle related proteins		
Anaphase-promoting complex/cyclosome (APC/C)	APC/C binds to the adenovirus E4orf4-PP2A complex. E4orf4 may target PP2A to APC/C, leading to its inactivation. This interaction may play a role in E4orf4-mediated cell cycle arrest and apoptosis.	[71]
Cdc6	Cdc6 binds to the R3γ/PR48 subunit and interacts with the AC-R3γ heterotrimer. The interaction may regulate Cdc6 phosphorylation and DNA replication. Overexpression of R3γ causes G1 arrest.	[43]
Cdc25c dual-specificity phosphatase	Cdc25c co-immunoprecipitates with PP2A following cross-linking of cell lysates. The interaction requires the R2 subunit and results in dephosphorylation of cdc25c. The interaction is enhanced by the HIV-1 Vpr protein, suggesting that dephosphorylation and inactivation of cdc25c is involved in Vpr-mediated G2 arrest.	[72]
Cyclin G2	The association of cyclin G2 with PP2A catalytic and R5 subunits correlates with its ability to inhibit cell cycle progression.	[54]
DNA polymerase α-primase	PP2A is recovered with the hypophosphorylated form of DNA polymerase α-primase in G1. PP2A dephosphorylates DNA polymerase α-primase and restores its origin-dependent initiation activity *in vito*.	[73]
p107	p107 (a retinoblastoma-related protein) binds the R3β/PR59 subunit-containing holoenzyme. Overexpression of R3β/PR59 causes p107 dephosphorylation and G1 arrest.	[41]
Membrane receptors/transporters		
Beta$_2$-adrenergic receptor	The association of PP2A with this G-protein coupled receptor is dependent upon agonist stimulation, receptor internalization, and acidification of endosomes. PP2A dephosphorylation is important for receptor resensitization and recycling to plasma membrane.	[74]
Biogenic amine transporters	Dopamine, norepinephrine, and serotonin transporters associate with PP2A. Transporter phosphorylation results in disruption of the PP2A association. The interaction may be involved in the regulation of the surface expression of transporters.	[75]
Class C L-type calcium channel ($Ca_v1.2$)	PP2A binds to the pore-forming α_{1C} subunit of this channel and reverses PKA-catalyzed serine phosphorylation. The interaction is selective for R5γ-containing PP2A complexes.	[58]
CXCR2 chemokine receptor	The chemokine receptor CXCR2 is a G-protein coupled receptor involved in chemotaxis. CXCR2 interacts with the AC core dimer. The interaction is dependent on internalization of the receptor following agonist stimulation.	[76]
NMDA receptor	PP2A forms a stable complex with NR3A subunit of the NMDA receptor. The association increases phosphatase activity and dephosphorylation of the NR1 subunit. Stimulation of the receptor leads to dissociation of PP2A and a reduction in phosphatase activity.	[77]

Table II *continued*

Protein	Comments	Refs
Protein kinases		
CaM kinase IV (CaMKIV)	CaMKIV binds to the AC-R2 form and is dephosphorylated by PP2A.	[78]
Casein kinase II (CK2)	CK2 binds to the AC core dimer. CK2 can phosphorylate and stimulate PP2A activity *in vitro*.	[79]
JAK2	There is a transient association of JAK2 and PP2A upon interleukin-11 stimulation of adipocytes.	[80]
p21-Activated kinase (PAK1)	PAK1 interacts with and is a substrate of PP2A.	[81]
p70 S6 kinase	p70 S6 kinase is a PP2A substrate.	[81]
PKCα	The PP2A catalytic subunit co-immunoprecipitates with PKCα. PKCα is dephosphorylated by PP2A. This association may be involved in the regulation of mast cell IL-6 production.	[82]
PKCδ	PKCδ is a substrate for PP2A.	[83]
PKR (Double-stranded RNA-dependent protein kinase)	PKR binds to and phosphorylates the R5α regulatory subunit. Phosphorylation of R5α enhances PP2A activity and may alter the activity of the translation initiation factor eIF4.	[84]
RAF-1	RAF-1 interacts with the AC core dimer. PP2A dephosphorylates inhibitory sites on RAF-1.	[30]
Src	PP2A binds to the SH2, SH3, and catalytic domains of Src. This interaction decreases Src tyrosine kinase activity.	[85]
Apoptotic proteins		
Cyclin G1	Cyclin G1 binds to R5 subunits and the association is dependent on the induction of p53. Cyclin G1 plays a role in enhancing apoptosis.	[53, 55]
Bcl-2	Bcl-2 interacts with the PP2A isoform containing the R5α subunit. PP2A dephosphorylates Bcl-2 and regulates the function of Bcl-2 in apoptosis.	[86–88]
Cytoskeletal proteins		
CG-NAP (AKAP 350/450/CG-NAP)	This 450-kDa centrosome and Golgi localized PKN-associated protein coimmunoprecipitates with PP2A in R3α-130 expressing cells. CG-NAP is involved in regulation of centrosome dynamics during the cell cycle.	[40]
Mid-1	Mid-1 binds to the PP2A interacting protein alpha 4 at a site independent from the C-subunit binding site. This interaction may regulate mid-1 binding to microtubules and formation of the midline during embryonic development.	[89]
Myosin	PP2A associates with myosin following mast cell activation. This interaction may play a role in regulating cytoskeletal remodeling and mast cell secretion.	[90]
Neurofilament proteins (NFs)	The AC-R2 complex associates with NF proteins. PP2A dephosphorylates sites in all three NF proteins (NF-L, NF-M, and NF-H). Dephosphorylation by PP2A promotes assembly of NF-L into filaments.	[91, 92]
Paxillin	Paxillin interacts with C-subunit and R5γ regulatory subunit. R5γ1 co-localizes with paxillin at focal adhesions and may target PP2A to paxillin.	[59]
Tau	Tau specifically interacts with R2-containing trimers. AC-R2 trimers dephosphorylate tau, promote microtubule binding, and stabilize microtubules.	[33, 93]
Vimentin	The AC-R2 complex associates with and dephosphorylates vimentin in an interaction mediated by the R2 subunit. Depletion of R2 by antisense RNA causes hyperphosphorylation of vimentin and reorganization of intermediate filaments.	[94]
Secretory pathway proteins		
Carboxypeptidase D (CPD)	PP2A binds to and dephosphorylates the cytoplasmic tail of this secretory pathway protein. PP2A may play a role in the intracellular trafficking of CPD between the cell surface and the trans-Golgi network.	[95]
Mannose-6-phosphate receptor (cation-independent)	PP2A binds to the cytoplasmic tail of this secretory pathway protein.	[95]
Peptidylglycine-a-amidating mono-oxygenase (PAM)	PP2A binds to the cytoplasmic tail of this secretory pathway protein.	[95]
TGN38	PP2A binds to the cytoplasmic tail of this secretory pathway protein.	[95]
Translation		
Eukaryotic termination factor-1 (eRF1)	eRF1 binds to the AC core dimer through C subunit. This interaction may target PP2A to ribosomes.	[96]
α4/Tap42 (IGBP1)	Alpha 4 interacts directly with the C subunit and decreases phosphatase acitivity toward eIF4E-BP1 that has been phosphorylated by the mTOR kinase.	[97–100]

continues

Table II *continued*

Protein	Comments	Refs
Viral proteins		
Adenovirus E4orf4 protein	E4orf4 binds to the AC-R2 and AC-R5 complexes. Formation of a complex with AC-R2 is required for E4orf4-mediated apoptosis.	[56, 101, 102]
HIV Vpr protein	Vpr binds to AC-R2 complex and mediates Vpr-induced G2 arrest. This interaction regulates the Cdc25 dual-specificity phosphatase and Wee1 kinase.	[68, 72, 103]
Polyomavirus middle tumor antigen	Middle-T antigen binds to the AC core dimer and targets PP2A to the signaling complex assembled by middle-T antigen. The role of this interaction in middle-T mediated transformation is not clear.	[104–106]
Polyomavirus small tumor antigen	Similar to SV40 small-t antigen. Binds to the AC core dimer.	[104–106]
SV40 small tumor antigen	Binds to AC core dimer, displacing the R2 subunits and inhibiting PP2A acitivity toward some substrates. This interaction enhances MAP kinase signaling and viral transformation.	[104, 105]
Other cellular proteins		
I_1^{PP2A} (PHAP1, mapmodulin)	I_1^{PP2A} can inhibit PP2A activity *in vitro*, but its physiological function is unknown.	[107]
I_2^{PP2A} (SET)	I_2^{PP2A} can inhibit PP2A *in vitro*, but its function is unknown.	[108]
Phosphotyrosyl phosphatase activator (PTPA)	PTPA displays a weak interaction with PP2A and can enhance the low activity of the AC core dimer toward phosphotyrosine.	[3]
Protein phosphatase 5 (PP5)	PP5 interacts with the scaffold subunit of PP2A and may replace the catalytic subunit. The interaction appears to involve the R3α subunit, which co-immunoprecipitates with PP5.	[39]
Protein phosphatase methylesterase (PME-1)	Associates with catalytically inactive C-subunit point mutants. Demethylates the catalytic subunit *in vitro*.	[109]
SG2NA	SG2NA binds to the AC core dimer. The protein is localized in nucleus. SG2NA contains WD repeats, such as R2 subunits and striatin, and binds calmodulin. The function of SG2NA is currently unknown.	[10]
Striatin	Striatin binds to the AC core dimer. The protein contains WD repeats, such as R2 subunits and SG2NA, and binds to calmodulin. The function of Striatin is currently unknown.	[10]

of axons in the central nervous system [34]. Disruption of the PP2A-tau interaction by expression of SV40 small-t antigen (which disrupts interaction of R2 subunits with the core dimer) causes hyperphosphorylation of tau and its dissociation from microtubules [33]. These observations suggest that proximity interactions among R2-containing forms of PP2A, microtubules, and tau play important roles in maintaining tau in a hypophosphorylated state. The targeted dephosphorylation of tau is important for axonal integrity, since inhibition of PP2A leads to tau hyperphosphorylation, loss of organized microtubules, and axonal degeneration in cultured neuronal cells [35]. The R2-mediated interactions of PP2A with microtubules and tau may have implications in neurodegenerative diseases, including Alzheimer's disease, where tau becomes hyperphosphorylated.

Expansion of a novel CAG trinucleotide repeat within the human R2β gene (PPPR2B) is associated with a form of autosomal dominant spinocerebellar ataxia termed SCA12 [36]. SCA12 is caused by neurodegeneration with atrophy of the cortex and cerebellum. The CAG expansion lies near the transcription start site of the R2β gene and could alter expression of this brain-specific isoform. The presence of the CAG expansion in affected individuals and its absence in non-affected family members suggest that altered expression of R2β may cause this disease. Although the mechanism of R2β loss in SCA12 is unknown, these data suggest that R2β may play a role in maintenance of neuronal viability.

The R3 Family

The second family of regulatory subunits identified by molecular cloning was the R3 family (Table I). The R3 subunit was first identified as a 74 kDa protein present in a PP2A holoenzyme termed PCS_M [37]. Current evidence indicates that this family plays a role in targeting PP2A to proteins involved in cell cycle regulation, including Cdc6, p107, and CG-NAP (Table II). The gene encoding the R3 subunit (designated R3α in Table I) produces two alternatively spliced transcripts encoding proteins of 72 and 130 kDa [38]. R3α-72 and R3α-130 contain the same C-terminal protein sequence, but PR130 contains a 665-amino-acid N-terminal extension. Both the 72 and 130 kDa variants are selectively but not exclusively expressed in skeletal muscle and heart. *In vitro*, the R3α subunit suppresses the activity of the AC dimer toward exogenous substrates and increases sensitivity of the enzyme to polycations [37]. The functions of R3α-72 or R3α-130 subunits have not been identified. Protein phosphatase 5 (another member of the PPP gene family) can interact with PP2A. Immunoprecipitated PP5 is associated with R3α-72 but not other regulatory subunits [39]. Although the significance of this interaction is not known, the data suggest that PP5 can be present in a PP2A oligomer containing the scaffold and R3α-72 subunits and that PP5 might act as the catalytic subunit in this heterocomplex. R3α-130 interacts with the giant scaffolding protein CG-NAP (centrosome and

Golgi localized PKN-associated protein). CG-NAP anchors a signaling complex containing protein kinase-A, protein kinase-N, protein kinase-Cε, PP2A (R3α-130), and protein phosphatase 1 to the centrosome and Golgi apparatus in a cell-cycle-dependent manner [40]. The CG-NAP signaling complex may mediate some of the complex phosphorylation-based regulation of the centrosome that occurs during the cell cycle. One potential substrate for PP2A in this complex is protein kinase-N.

The R3 family contains additional isoforms that function in cell cycle regulation through unique proximity interactions. The R3β (PR59) protein was discovered in a yeast two-hybrid screen via the retinoblastoma-related protein p107 as bait [41]. R3β forms complexes with the PP2A core dimer when expressed in cells. Although R3β shares 56 percent identity with R3α-72, the interaction with p107 is specific. Furthermore, although R3β binds to p107, it fails to interact with the retinoblastoma protein. Forced overexpression of R3β results in dephosphorylation of p107 and cell cycle arrest in the G1 phase. R3β-mediated cell cycle arrest may be the result of hypophosphorylation of p107 (due to increased PP2A targeting) and its association with the E2F transcription factor. Binding of p107 to E2F would repress expression of genes required for entry into S phase. R3β may be targeted to dephosphorylate p107 in response to UV irradiation [42].

The R3γ regulatory subunit (PR48) was discovered in a yeast two-hybrid screen with the Cdc6 protein as bait [43]. Cdc6 is required for formation of pre-replication complexes during DNA replication. Phosphorylation of Cdc6 by S-phase cyclin-dependent kinases is the rate-limiting step for initiation of DNA replication. In mammalian cells, phosphorylation of Cdc6 at the beginning of S phase causes its dissociation from chromatin and triggers replication. In addition, Cdc6 phosphorylation induces its nuclear export and ubiquitin-dependent degradation. R3γ shares 50 and 68 percent sequence identity with R3α and R3β, respectively. R3γ localizes to the nucleus in mammalian cells and, like PR59, forced overexpression of PR48 results in cell cycle arrest at G1.

The R5 Family

The R5 regulatory subunits are a complex family of proteins that are components of a PP2A holoenzyme originally termed $PP2A_0$ [18,44]. There are at least five isoforms (Table I) that have distinct patterns of expression [45–47]. The α and γ isoforms are expressed predominantly in muscle, the β and δ isoforms in brain, and the ε isoform in brain and testis. In cardiac muscle, nearly all of the PP2A holoenzyme is composed of the R5α subunit [44]. *In vitro*, the R5 subunits suppress phosphatase activity toward multiple substrates [14]. This implies that the R5 subunits target PP2A by disfavoring interactions with some substrates while favoring interactions with others. The R5 family has been subdivided into cytosolic and nuclear types based on localization of transiently expressed proteins [46,48]. The R5α, R5β, and R5ε isoforms are cytoplasmic whereas R5γ and R5δ are present in both the cytoplasm and nucleus. Ectopically expressed R5 subunits are also phosphorylated in intact cells. Thus, the regulation of PP2A or interaction with other proteins may be modulated by covalent modification of R5 family members.

The R5 subunits mediate interactions between PP2A and components of the Wnt signaling pathway involved in cell growth and transformation. Members of the R5 family were identified in a yeast two-hybrid screen by using the adenomatous polyposis coli (APC) protein as bait [49]. APC forms a signaling complex with axin and glycogen synthase kinase 3β that mediates the phosphorylation and proteasome-dependent degradation of β-catenin. A basal level of β-catenin degradation normally prevents transcription of β-catenin target genes involved in cell growth and transformation. Stimulation of the Wnt pathway causes inhibition of β-catenin phosphorylation and degradation, leading to increased transcription of β-catenin target genes. Ectopic expression of R5 subunits in mammalian cells causes a reduction in β-catenin levels and a decrease in expression of β-catenin target genes. Further supporting a role for PP2A in the Wnt/β-catenin pathway, the catalytic subunit of PP2A interacts with axin in two-hybrid assays and can be co-immunoprecipitated with axin [50]. Subsequent studies have shown that the scaffold subunit, the catalytic subunit, and R5 subunits can be immunoprecipitated with axin from *Xenopus* embryos [51,52]. Ectopic expression of the PP2A scaffold subunit, the catalytic subunit, or R5 subunits all have ventralizing activity in *Xenopus* embryos, consistent with a negative role in Wnt/β-catenin signaling. The R5 subunits appear to interact directly with axin at a site that is distinct from the sites that interact with APC, GSK-3β, and β-catenin [51]. The data are all consistent with an important role for R5 subunits in targeting PP2A to the axin/GSK-3/APC complex and regulating the Wnt signaling pathway.

The R5 subunits are also linked to cell survival and apoptosis. Cyclin G1, cyclin G2, and cyclin I are members of a unique family of cyclin-related proteins that are expressed in brain and muscle. R5 subunits interact with both cyclin G1 [53] and cyclin G2 [54]. Cyclin G1 and R5 subunits can be co-immunoprecipitated from neurons whereas cyclin G2-R5-catalytic subunit complexes can be isolated from cultured cells [54]. Although the function of the cyclin G1 is not known, the p53 tumor suppressor protein regulates its transcription. Ectopic expression of cyclin G1 enhances apoptosis in response to multiple stimuli in cultured cells [55]. Similarly, forced overexpression of cyclin G2 causes formation of aberrant nuclei and cell cycle arrest [54]. These observations raise the possibility that the cyclin G1-PP2A interaction could be involved in cell cycle arrest and apoptosis. The interaction of R5 subunits with the adenovirus E4orf4 protein is essential for E4orf4-mediated apoptosis [56]. Finally, the use of RNA interference in *Drosophila* cells has shown that loss of both of the *Drosophila* R5 subunits results in apoptosis [17].

R5 subunits interact with a variety of other proteins, thus indicating roles for this family in other signaling pathways (Table II). A *Drosophila* homolog of R5 interacts with a homeodomain-containing transcription factor called Sex Combs Reduced. This interaction positively modulates transcriptional activity [57]. The R5γ subunit is associated with L-type calcium channels, where it appears to target PP2A to regulatory sites phosphorylated by protein kinase A [58]. R5α interacts with the double-stranded RNA-dependent protein kinase PKR. PKR phosphorylates R5α, leading to an increase in PP2A phosphatase activity. PKR-enhanced PP2A activity may lead to decreased phosphorylation of eIF4E and altered protein synthesis. R5-containing PP2A may also be targeted to focal adhesions through interaction with paxillin [59].

PP2A-Interacting Proteins

Proximity interactions are the most important mechanism for regulating the activity of PP2A. Association with interacting proteins mediates many proximity interactions of PP2A, and allows targeting of this phosphatase to a wide variety of signaling pathways. PP2A interacting proteins include phosphoproteins that are PP2A substrates, scaffold proteins, and components of the cytoskeleton. As discussed above, many of these interactions occur with PP2A holoenzymes and are mediated by specific regulatory subunits. However, interacting proteins have been identified that interact directly with the PP2A core dimer and the catalytic subunit. PP2A-interacting proteins include virally encoded proteins and a host of cellular proteins that participate in interesting aspects of signal transduction. A compilation of the currently identified PP2A-interacting proteins is presented in Table II. Although many of the proteins listed in the table are substrates for PP2A, others act to target PP2A to specific signaling complexes, and some alter signaling by disrupting endogenous PP2A complexes. These proteins have been grouped into categories based on functional similarities. Brief descriptions of individual proteins and their interaction with PP2A are presented in the table.

References

1. Millward, T. A., Zolnierowicz, S., and Hemmings, B. A. (1999). Regulation of protein kinase cascades by protein phosphatase 2A. *Trends Biochem. Sci.* **24**, 186–191.
2. Virshup, D. M. (2000). Protein phosphatase 2A: a panoply of enzymes. *Curr. Opin. Cell Biol.* **12**, 180–185.
3. Janssens, V. and Goris, J. (2001). Protein phosphatase 2A: a highly regulated family of serine/threonine phosphatases implicated in cell growth and signalling. *Biochem. J.* **353**, 417–439.
4. Groves, M. R., Hanlon, N., Turowski, P., Hemmings, B. A., and Barford, D. (1999). The structure of the protein phosphatase 2A PR65/A subunit reveals the conformation of its 15 tandemly repeated HEAT motifs. *Cell* **96**, 99–110.
5. Ruediger, R., Hentz, M., Fait, J., Mumby, M., and Walter, G. (1994). Molecular model of the A subunit of protein phosphatase 2A: interaction with other subunits and tumor antigens. *J. Virol.* **68**, 123–129.
6. Kamibayashi, C., Lickteig, R. L., Estes, R., Walter, G., and Mumby, M. C. (1992). Expression of the A subunit of protein phosphatase 2A and characterization of its interactions with the catalytic and regulatory subunits. *J. Biol. Chem.* **267**, 21864–21872.
7. Ruediger, R., Fields, K., and Walter, G. (1999). Binding specificity of protein phosphatase 2A core enzyme for regulatory B subunits and T antigens. *J. Virol.* **73**, 839–842.
8. Neer, E. J., Schmidt, C. J., Nambudripad, R., and Smith, T. F. (1994). The ancient regulatory-protein family of WD-repeat proteins. **371**, 297–300.
9. Griswold-Prenner, I., Kamibayashi, C., Maruoka, E. M., Mumby, M. C., and Derynck, R. (1998). Physical and functional interactions between type I transforming growth factor beta receptors and B-alpha, a WD-40 repeat subunit of phosphatase 2A. *Mol. Cell. Biol.* **18**, 6595–6604.
10. Moreno, C. S., Park, S., Nelson, K., Ashby, D., Hubalek, F., Lane, W. S., and Pallas, D. C. (2000). WD40 repeat proteins striatin and S/G(2) nuclear autoantigen are members of a novel family of calmodulin-binding proteins that associate with protein phosphatase 2A. *J. Biol. Chem.* **275**, 5257–5263.
11. Li, X. and Virshup, D. M. (2002). Two conserved domains in regulatory B subunits mediate binding to the A subunit of protein phosphatase 2A. *Eur. J. Biochem.* **269**, 546–552.
12. Chen, S.-C., Kramer, G., and Hardesty, B. (1989). Isolation and partial characterization of an M_r 60,000 subunit of a type 2A phosphatase from rabbit reticulocytes. *J. Biol. Chem.* **264**, 7267–7275.
13. Imaoka, T., Imazu, M., Usui, H., Kinohara, N., and Takeda, M. (1983). Resolution and reassociation of three distinct components from pig heart phosphoprotein phosphatase. *J. Biol. Chem.* **258**, 1526–1535.
14. Kamibayashi, C., Estes, R., Lickteig, R. L., Yang, S.-I., Craft, C., and Mumby, M. C. (1994). Comparison of heterotrimeric protein phosphatase 2A containing different B subunits. *J. Biol. Chem.* **269**, 20139–20148.
15. Price, N. E. and Mumby, M. C. (2000). Effects of regulatory subunits on the kinetics of protein phosphatase 2A. *Biochemistry* **39**, 11312–11318.
16. Zhao, Y., Boguslawski, G., Zitomer, R. S., and DePaoli-Roach, A. A. (1997). *Saccharomyces cerevisiae* homologs of mammalian B and B' subunits of protein phosphatase 2A direct the enzyme to distinct cellular functions. *J. Biol. Chem.* **272**, 8256–8262.
17. Silverstein, A. M., Barrow, C. A., Davis, A. J., and Mumby, M. C. (2002). Actions of PP2A on the MAP kinase pathway and apoptosis are mediated by distinct regulatory subunits. *Proc. Natl. Acad. Sci. USA* **99**, 4221–4226.
18. Tung, H. Y. L., Alemany, S., and Cohen, P. (1985). The protein phosphatases involved in cellular regulation 2. Purification, subunit structure and properties of protein phosphatases-2A0, 2A1, and 2A2 from rabbit skeletal muscle. *Eur. J. Biochem.* **148**, 253–263.
19. Strack, S., Zaucha, J. A., Ebner, F. F., Colbran, R. J., and Wadzinski, B. E. (1998). Brain protein phosphatase 2A: developmental regulation and distinct cellular and subcellular localization by B subunits. *J. Comp. Neurol.* **392**, 515–527.
20. Healy, A. M., Zolnierowicz, S., Stapelton, A. E., Goebl, M., DePaoli-Roach, A. A., and Pringle, J. R. (1991). CDC55, a *Saccharomyces cerevisiae* gene involved in cellular morphogenesis: identification, characterization, and homology to the B subunit of mammalian type 2A protein phosphatase. *Mol. Cell. Biol.* **11**, 5767–5780.
21. Minshull, J., Straight, A., Rudner, A. D., Dernburg, A. F., Belmont, A., and Murray, A. W. (1996). Protein phosphatase 2A regulates MPF activity and sister chromatid cohesion in budding yeast. *Curr. Biol.* **6**, 1609–1620.
22. Wang, Y. and Burke, D. J. (1997). Cdc55p, the B-type regulatory subunit of protein phosphatase 2A, has multiple functions in mitosis and is required for the kinetochore/spindle checkpoint in *Saccharomyces cerevisiae*. *Mol. Cell. Biol.* **17**, 620–626.
23. Mayer-Jaekel, R. E., Ohkura, H., Gomes, R., Sunkel, C. E., Baumgartner, S., Hemmings, B. A., and Glover, D. M. (1993). The 55 kd regulatory subunit of *Drosophila* protein phosphatase 2A is required for anaphase. *Cell* **72**, 621–633.

24. Uemura, T., Shiomi, K., Togashi, S., and Takeichi, M. (1993). Mutation of *twins* encoding a regulator of protein phosphatase 2A leads to pattern duplication in *Drosophila* imaginal disks. *Genes Dev.* **7**, 429–440.
25. Shiomi, K., Takeichi, M., Nishida, Y., Nishi, Y., and Uemura, T. (1994). Alternative cell fate choice induced by low-level expression of a regulator of protein phosphatase 2A in the *Drosophila* peripheral nervous system. *Development* **120**, 1591–1599.
26. Sontag, E., Fedorov, S., Kamibayashi, C., Robbins, D., Cobb, M., and Mumby, M. (1993). The interaction of SV40 small tumor antigen with protein phosphatase 2A stimulates the MAP kinase pathway and induces cell proliferation. *Cell* **75**, 887–897.
27. Frost, J. A., Alberts, A. S., Sontag, E., Guan, K., Mumby, M. C., and Feramisco, J. R. (1994). SV40 small t antigen cooperates with mitogen activated kinases to stimulate AP-1 activity. *Mol. Cell. Biol.* **14**, 6244–6252.
28. Sontag, E., Sontag, J. M., and Garcia, A. (1997). Protein phosphatase 2A is a critical regulator of protein kinase C zeta signaling targeted by SV40 small t to promote cell growth and NF-kappaB activation. *EMBO J.* **16**, 5662–5671.
29. Sieburth, D. S., Sundaram, M., Howard, R. M., and Han, M. (1999). A PP2A regulatory subunit positively regulates Ras-mediated signaling during *Caenorhabditis elegans* vulval induction. *Genes Devel.* **13**, 2562–2569.
30. Abraham, D., Podar, K., Pacher, M., Kubicek, M., Welzel, N., Hemmings, B. A., Dilworth, S. M., Mischak, H., Kolch, W., and Baccarini, M. (2000). Raf-1-associated protein phosphatase 2A as a positive regulator of kinase activation. *J. Biol. Chem.* **275**, 22300–22304.
31. Wassarman, D. A., Solomon, N. M., Chang, H. C., Karim, F. D., Therrien, M., and Rubin, G. M. (1996). Protein phosphatase 2A positively and negatively regulates Ras1-mediated photoreceptor development in *Drosophila. Genes Dev.* **10**, 272–278.
32. Sontag, E., V., Nunbhakdi-Craig, G. S., Bloom, and Mumby, M. C. (1995). A novel pool of protein phosphatase 2A is associated with microtubules and is regulated during the cell cycle. *J. Cell Biol.* **128**, 1131–1144.
33. Sontag, E., Nunbhakdi-Craig, V., Lee, G., Bloom, G. S., and Mumby, M. C. (1996). Regulation of the phosphorylation state and microtubule-binding activity of tau by protein phosphatase 2A. *Neuron* **17**, 1201–1207.
34. Billingsley, M. L. and Kincaid, R. L. (1997). Regulated phosphorylation and dephosphorylation of tau protein: effects on microtubule interaction, intracellular trafficking and neurodegeneration. *Biochem. J.* **323**, 577–591.
35. Merrick, S. E., Trojanowski, J. Q., and Lee, V. M. Y. (1997). Selective destruction of stable microtubules and axons by inhibitors of protein serine/threonine phosphatases in cultured human neurons (NT2N cells). *J. Neurosci.* **17**, 5726–5737.
36. Holmes, S. E., O'Hearn, E. E., McInnis, M. G., Gorelick-Feldman, D. A., Kleiderlein, J. J., Callahan, C., Kwak, N. G., Ingersoll-Ashworth, R. G., Sherr, M., Sumner, A. J., Sharp, A. H., Ananth, U., Seltzer, W. K., Boss, M. A., Vieria-Saecker, A. M., Epplen, J. T., Riess, O., Ross, C. A., and Margolis, R. L. (1999). Expansion of a novel CAG trinucleotide repeat in the 5' region of PPP2R2B is associated with SCA12. *Nature Genetics* **23**, 391–392.
37. Waelkens, E., Goris, J., and Merlevede, W. (1987). Purification and properties of polycation-stimulated phosphorylase phosphatases from rabbit skeletal muscle. *J. Biol. Chem.* **262**, 1049–1059.
38. Hendrix, P., Mayer-Jaekel, R. E., Cron, P., Goris, J., Hofsteenge, J., Merlevede, W., and Hemmings, B. A. (1993). Structure and expression of a 72-kDa regulatory subunit of protein phosphatase 2A. Evidence for different size forms produced by alternative splicing. *J. Biol. Chem.* **268**, 15267–15276.
39. Lubert, E. J., Hong, Y., and Sarge, K. D. (2001). Interaction between protein phosphatase 5 and the A subunit of protein phosphatase 2A: evidence for a heterotrimeric form of protein phosphatase 5. *J. Biol. Chem.* **276**, 38582–38587.
40. Takahashi, M., Shibata, H., Shimakawa, M., Miyamoto, M., Mukai, H., and Ono, Y. (1999). Characterization of a novel giant scaffolding protein, CG-NAP, that anchors multiple signaling enzymes to centrosome and the Golgi apparatus. *J. Biol. Chem.* **274**, 17267–17274.
41. Voorhoeve, P. M., Hijmans, E. M., and Bernards, R. (1999). Functional interaction between a novel protein phosphatase 2A regulatory subunit, PR59, and the retinoblastoma-related p107 protein. *Oncogene* **18**, 515–524.
42. Voorhoeve, P. M., Watson, R. J., Farlie, P. G., Bernards, R., and Lam, E. W. (1999). Rapid dephosphorylation of p107 following UV irradiation. *Oncogene* **18**, 679–688.
43. Yan, Z., Fedorov, S. A., Mumby, M. C., and Williams, R. S. (2000). PR48, a novel regulatory subunit of protein phosphatase 2A, interacts with Cdc6 and modulates DNA replication in human cells. *Mol. Cell Biol.* **20**, 1021–1029.
44. Zolnierowicz, S., Csortos, C., Bondor, J., Verin, A., Mumby, M. C., and DePaoli-Roach, A. A. (1994). Diversity in the regulatory B-subunits of protein phosphatase 2A: identification of a novel isoform highly expressed in brain. *Biochemistry* **33**, 11858–11867.
45. McCright, B. and Virshup, D. M. (1995). Identification of a new family of protein phosphatase 2A regulatory subunits. *J. Biol. Chem.* **270**, 26123–26128.
46. Ahmadian-Tehrani, M., Mumby, M. C., and Kamibayashi, C. (1996). Identification of a novel protein phosphatase 2A regulatory subunit highly expressed in muscle. *J. Biol. Chem.* **271**, 5164–5170.
47. Csortos, C., Zolnierowicz, S., Bako, E., Durbin, S. D., and DePaoli-Roach, A. A. (1996). High complexity in the expression of the B' subunit of protein phosphatase 2A0. Evidence for the existence of at least seven novel isoforms. *J. Biol. Chem.* **271**, 2578–2588.
48. McCright, B., Rivers, A. M., Audlin, S., and Virshup, D. M. (1996). The B56 family of protein phosphatase 2A (PP2A) regulatory subunits encodes differentiation-induced phosphoproteins that target PP2A to both nucleus and cytoplasm. *J. Biol. Chem.* **271**, 22081–22089.
49. Seeling, J. M., Miller, J. R., Gil, R., Moon, R. T., White, R., and Virshup, D. M. (1999). Regulation of beta-catenin signaling by the B56 subunit of protein phosphatase 2A. *Science* **283**, 2089–2091.
50. Hsu, W., Zeng, L. and Costantini, F. (1999). Identification of a domain of axin that binds to the serine/threonine protein phosphatase 2A and a self-binding domain. *J. Biol. Chem.* **274**, 3439–3445.
51. Yamamoto, H., Hinoi, T., Michiue, T., Fukui, A., Usui, H., Janssens, V., Van Hoof, C., Goris, J., Asashima, M., and Kikuchi, A. (2001). Inhibition of the Wnt signaling pathway by the PR61 subunit of protein phosphatase 2A. *J. Biol. Chem.* **276**, 26875–26882.
52. Li, X., Yost, H. J., Virshup, D. M., and Seeling, J. M. (2001). Protein phosphatase 2A and its B56 regulatory subunit inhibit Wnt signaling in Xenopus. *EMBO J.* **20**, 4122–4131.
53. Okamoto, K., Kamibayashi, C., Serrano, M., Prives, C., Mumby, M. C., and Beach, D. (1996). p53-dependent association between cyclin G and the B' subunit of protein phosphatase 2A. *Mol. Cell. Biol.* **16**, 6593–6602.
54. Bennin, D. A., Arachchige Don, A. S., Brake, T., McKenzie, J. L., Rosenbaum, H., Ortiz, L., DePaoli-Roach, A. A., and Horne, M. C. (2002). Cyclin G2 associates with protein phosphatase 2A catalytic and regulatory B' subunits in active complexes and induces nuclear aberrations and a G1/S phase cell cycle arrest. *J. Biol. Chem.*
55. Okamoto, K. and Prives, C. (1999). A role of cyclin G in the process of apoptosis. *Oncogene* **18**, 4606–4615.
56. Shtrichman, R., Sharf, R., and Kleinberger, T. (2000). Adenovirus E4orf4 protein interacts with both B alpha and B' subunits of protein phosphatase 2A, but E4orf4-induced apoptosis is mediated only by the interaction with B alpha. *Oncogene* **19**, 3757–3765.
57. Berry, M. and Gehring, W. (2000). Phosphorylation status of the SCR homeodomain determines its functional activity: essential role for protein phosphatase 2A,B'. *EMBO J.* **19**, 2946–2957.
58. Davare, M. A., Horne, M. C., and Hell, J. W. (2000). Protein phosphatase 2A is associated with class C L-type calcium channels (Cav1.2) and antagonizes channel phosphorylation by cAMP-dependent protein kinase. *J. Biol. Chem.* **275**, 39710–39717.
59. Ito, A., Kataoka, T. R., Watanabe, M., Nishiyama, K., Mazaki, Y., Sabe, H., Kitamura, Y., and Nojima, H. (2000). A truncated isoform

of the PP2A B56 subunit promotes cell motility through paxillin phosphorylation. *EMBO J.* **19**, 562–571.
60. Yokoyama, N. and Miller, W. T. (2001). Protein phosphatase 2A interacts with the Src kinase substrate p130(CAS). *Oncogene* **20**, 6057–6065.
61. Gotz, J., Probst, A., Mistl, C., Nitsch, R. M., and Ehler, E. (2000). Distinct role of protein phosphatase 2A subunit C alpha in the regulation of E-cadherin and beta-catenin during development. *Mechan. Devel.* **93**, 83–93.
62. Hong, Y. L. and Sarge, K. D. (1999). Regulation of protein phosphatase 2A activity by heat shock transcription factor 2. *J. Biol. Chem.* **274**, 12967–12970.
63. Hong, Y. L., Lubert, E. J., Rodgers, D. W., and Sarge, K. D. (2000). Molecular basis of competition between HSF2 and catalytic subunit for binding to the PR65/A subunit of PP2A. *Biochem. Biophys. Res. Commun.* **272**, 84–89.
64. Kawabe, T., Muslin, A. J., and Korsmeyer, S. J. (1997). Hox11 interacts with protein phosphatases PP2A and PP1 and disrupts a G2/M cell-cycle checkpoint. *Nature* **385**, 454–458.
65. Adler, H. T., Nallaseth, F. S., Walter, G., and Tkachuk, D. C. (1997). HRX leukemic fusion proteins form a heterocomplex with the leukemia-associated protein SET and protein phosphatase 2A. *J. Biol. Chem.* **272**, 28407–28414.
66. Yang, J., Fan, G. H., Wadzinski, B. E., Sakurai, H., and Richmond, A. (2001). Protein phosphatase 2A interacts with and directly dephosphorylates RelA. *J. Biol. Chem.* **276**, 47828–47833.
67. Ugi, S., Imamura, T., Ricketts, W., and Olefsky, J. M. (2002). Protein phosphatase 2A forms a molecular complex with Shc and regulates Shc tyrosine phosphorylation and downstream mitogenic signaling. *Mol. Cell Biol.* **22**, 2375–2387.
68. Elder, R. T., Yu, M., Chen, M., Zhu, X., Yanagida, M., and Zhao, Y. (2001). HIV-1 Vpr induces cell cycle G2 arrest in fission yeast (*Schizosaccharomyces pombe*) through a pathway involving regulatory and catalytic subunits of PP2A and acting on both Wee1 and Cdc25. *Virology* **287**, 359–370.
69. Lacroix, I., Lipcey, C., Imbert, J., and Kahn-Perles, B. (2002). Sp1 transcriptional activity is upregulated by phosphatase 2A in dividing T lymphocytes. *J. Biol. Chem.* **277**, 9598–9605.
70. Yokoyama, N., Reich, N. C., and Miller, W. T. (2001). Involvement of protein phosphatase 2a in the interleukin-3-stimulated jak2-stat5 signaling pathway. *J. Interferon Cytokine Res.* **21**, 369–378.
71. Kornitzer, D., Sharf, R., and Kleinberger, T. (2001). Adenovirus E4orf4 protein induces PP2A-dependent growth arrest in *Saccharomyces cerevisiae* and interacts with the anaphase-promoting complex/cyclosome. *J. Cell Biol.* **154**, 331–344.
72. Hrimech, M., Yao, X. J., Branton, P. E., and Cohen, E. A. (2000). Human immunodeficiency virus type 1 Vpr-mediated G(2) cell cycle arrest: Vpr interferes with cell cycle signaling cascades by interacting with the B subunit of serine/threonine protein phosphatase 2A. *EMBO J.* **19**, 3956–3967.
73. Dehde, S., Rohaly, G., Schub, O., Nasheuer, H. P., Bohn, W., Chemnitz, J., Deppert, W., and Dornreiter, I. (2001). Two immunologically distinct human DNA polymerase alpha-primase subpopulations are involved in cellular DNA replication. *Mol. Cell Biol.* **21**, 2581–2593.
74. Krueger, K. M., Daaka, Y., Pitcher, J. A., and Lefkowitz, R. J. (1997). The role of sequestration in G protein-coupled receptor resensitization. Regulation of beta2-adrenergic receptor dephosphorylation by vesicular acidification. *J. Biol. Chem.* **272**, 5–8.
75. Bauman, A. L., Apparsundaram, S., Ramamoorthy, S., Wadzinski, B. E., Vaughan, R. A., and Blakely, R. D. (2000). Cocaine and antidepressant-sensitive biogenic amine transporters exist in regulated complexes with protein phosphatase 2A. *J. Neurosci.* **20**, 7571–7578.
76. Fan, G. H., Yang, W., Sai, J., and Richmond, A. (2001). Phosphorylation-independent association of CXCR2 with the protein phosphatase 2A core enzyme. *J. Biol. Chem.* **276**, 16960–16968.
77. Chan, S. F. and Sucher, N. J. (2001). An NMDA receptor signaling complex with protein phosphatase 2A. *J. Neurosci.* **21**, 7985–7992.
78. Westphal, R. S., Anderson, K. A., Means, A. R., and Wadzinski, B. E. (1998). A signaling complex of Ca^{2+}-calmodulin-dependent protein kinase IV and protein phosphatase 2A. *Science* **280**, 1258–1261.
79. Heriche, J. K., Lebrin, F., Rabilloud, T., Leroy, D., Chambaz, E. M., and Goldberg, Y. (1997). Regulation of protein phosphatase 2A by direct interaction with casein kinase 2α. *Science* **276**, 952 955.
80. Fuhrer, D. K., and Yang, Y. C. (1996). Complex formation of JAK2 with PP2A, P13K, and Yes in response to the hematopoietic cytokine interleukin-11. *Biochem. Biophys. Res. Commun.* **224**, 289–296.
81. Westphal, R. S., Coffee, R. L., Marotta, A., Pelech, S. L., and Wadzinski, B. E. (1999). Identification of kinase-phosphatase signaling modules composed of p70 S6 kinase-protein phosphatase 2A (PP2A) and p21-activated kinase-PP2A. *J. Biol. Chem.* **274**, 687–692.
82. Boudreau, R. T., Garduno, R., and Lin, T. J. (2002). Protein phosphatase 2A and protein kinase Calpha are physically associated and are involved in *Pseudomonas aeruginosa*-induced interleukin 6 production by mast cells. *J. Biol. Chem.* **277**, 5322–5329.
83. Srivastava, J., Goris, J., Dilworth, S. M., and Parker, P. J. (2002). Dephosphorylation of PKCdelta by protein phosphatase 2Ac and its inhibition by nucleotides. *FEBS Lett.* **516**, 265–269.
84. Xu, Z. and Williams, B. R. (2000). The B56alpha regulatory subunit of protein phosphatase 2A is a target for regulation by double-stranded RNA-dependent protein kinase PKR. *Mol. Cell. Biol.* **20**, 5285–5299.
85. Yokoyama, N. and Miller, W. T. (2001). Inhibition of Src by direct interaction with protein phosphatase 2A. *FEBS Lett.* **505**, 460–464.
86. Deng, X. M., Ito, T., Carr, B., Mumby, M., and May, W. S. (1998). Reversible phosphorylation of Bcl2 following interleukin 3 or bryostatin 1 is mediated by direct interaction with protein phosphatase 2A. *J. Biol. Chem.* **273**, 34157–34163.
87. Ruvolo, P. P., Deng, X., Ito, T., Carr, B. K., and May, W. S. (1999). Ceramide induces bcl2 dephosphorylation via a mechanism involving mitochondrial PP2A. *J. Biol. Chem.* **274**, 20296–20300.
88. Ruvolo, P. P., Clark, W., Mumby, M., Gao, F., and May, W. S. (2002). A functional role for the B56 alpha-subunit of protein phosphatase 2A in ceramide-mediated regulation of Bcl2 phosphorylation status and function. *J. Biol. Chem.* **277**, 22847–22852.
89. Liu, J., Prickett, T. D., Elliott, E., Meroni, G., and Brautigan, D. L. (2001). Phosphorylation and microtubule association of the Opitz syndrome protein mid-1 is regulated by protein phosphatase 2A via binding to the regulatory subunit alpha 4. *Proc. Natl. Acad. Sci. USA* **98**, 6650–6655.
90. Holst, J., Sim, A. T., and Ludowyke, R. I. (2002). Protein phosphatases 1 and 2A transiently associate with myosin during the peak rate of secretion from mast cells. *Mol. Biol. Cell* **13**, 1083–1098.
91. Saito, T., Shima, H., Osawa, Y., Nagao, M., Hemmings, B. A., Kishimoto, T., and Hisanaga, S. (1995). Neurofilament-associated protein phosphatase 2A: its possible role in preserving neurofilaments in filamentous states. *Biochemistry* **34**, 7376–7384.
92. Strack, S., Westphal, R. S., Colbran, R. J., Ebner, F. F., and Wadzinski, B. E. (1997). Protein serine/threonine phosphatase 1 and 2A associate with and dephosphorylate neurofilaments. *Brain Res. Mol. Brain Res.* **49**, 15–28.
93. Sontag, E., Nunbhakdi-Craig, V., Lee, G., Brandt, R., Kamibayashi, C., Kuret, J., White, C. L. III, Mumby, M. C., and Bloom, G. S. (1999). Molecular interactions among protein phosphatase 2A, tau, and microtubules: implications for the regulation of tau phosphorylation and the development of tauopathies. *J. Biol. Chem.* **274**, 25490–25498.
94. Turowski, P., Myles, T., Hemmings, B. A., Fernandez, A., and Lamb, N. C. (1999). Vimentin dephosphorylation by protein phosphatase 2A is modulated by the targeting subunit B55. *Mol. Biol. Cell* **10**, 1997–2015.
95. Varlamov, O., Kalinina, E., Che, F. Y., and Fricker, L. D. (2001). Protein phosphatase 2A binds to the cytoplasmic tail of carboxypeptidase D and regulates post-trans-Golgi network trafficking. *J. Cell Sci.* **114**, 311–322.

96. Andjelkovic, N., Zolnierowicz, S., Van Hoof, C., Goris, J., and Hemmings, B. A. (1996). The catalytic subunit of protein phosphatase 2A associates with the translation termination factor eRF1. *EMBO J.* **15**, 156–167.
97. Chen, J., Peterson, R. T., and Schreiber, S. L. (1998). Alpha 4 associates with protein phosphatases 2A, 4, and 6. *Biochem. Biophys. Res. Commun.* **247**, 827–832.
98. Maeda, K., Inui, S., Tanaka, H., and Sakaguchi, N. (1999). A new member of the alpha 4-related molecule (alpha 4-b) that binds to the protein phosphatase 2A is expressed selectively in the brain and testis. *Eur. J. Biochem.* **264**, 702–706.
99. Nanahoshi, M., Nishiuma, T., Tsujishita, Y., Hara, K., Inui, S., Sakaguchi, N., and Yonezawa, K. (1998). Regulation of protein phosphatase 2A catalytic activity by alpha4 protein and its yeast homolog tap42. *Biochem. Biophys. Res. Commun.* **251**, 520–526.
100. Jiang, Y. and Broach, J. R. (1999). Tor proteins and protein phosphatase 2A reciprocally regulate Tap42 in controlling cell growth in yeast. *EMBO J.* **18**, 2782–2792.
101. Kleinberger, T. and Shenk, T. (1993). Adenovirus E4orf4 protein binds to protein phosphatase 2A, and the complex down regulates E1A-enhanced *junB* transcription. *J. Virol.* **67**, 7556–7560.
102. Shtrichman, R., Sharf, R., Barr, H., Dobner, T., and Kleinberger, T. (1999). Induction of apoptosis by adenovirus E4orf4 protein is specific to transformed cells and requires an interaction with protein phosphatase 2A. *Proc. Natl. Acad. Sci.USA* **96**, 10080–10085.
103. Tung, H. Y., De Rocquigny, H., Zhao, L. J., Cayla, X., Roques, B. P., and Ozon, R. (1997). Direct activation of protein phosphatase-2A0 by HIV-1 encoded protein complex NCp7:vpr. *FEBS Lett.* **401**, 197–201.
104. Mumby, M. (1995). Regulation by tumour antigens defines a role for PP2A in signal transduction. *Sem. Cancer Biol.* **6**, 229–237.
105. Pallas, D. C., Shahrik, L. K., Martin, B. L., Jaspers, S., Miller, T. B., Brautigan, D. L., and Roberts, T. M. (1990). Polyoma small and middle T antigens and SV40 small t antigen form stable complexes with protein phosphatase 2A. *Cell* **60**, 167–176.
106. Cayla, X., Ballmer-Hofer, K., Merlevede, W., and Goris, J. (1993). Phosphatase 2A associated with polyomavirus small-T or middle-T antigen is an okadaic acid-sensitive tyrosyl phosphatase. *Eur. J. Biochem.* **214**, 281–286.
107. Li, M., Makkinje, A., and Damuni, Z. (1996). Molecular identification of I1PP2A, a novel potent heat-stable inhibitor protein of protein phosphatase 2A. *Biochemistry* **35**, 6998–7002.
108. Li, M., Makkinje, A., and Damuni, Z. (1996). The myeloid leukemia-associated protein SET is a potent inhibitor of protein phosphatase 2A. *J. Biol. Chem.* **271**, 11059–11062.
109. Ogris, E., Du, X. X., Nelson, K. C., Mak, E. K., Yu, X. X., Lane, W. S., and Pallas, D. C. (1999). A protein phosphatase methylesterase (PME-1) is one of several novel proteins stably associating with two inactive mutants of protein phosphatase 2A. *J. Biol. Chem.* **274**, 14382–14391.

SECTION F

Cyclic Nucleotides

Jackie Corbin, Editor

CHAPTER 190

Adenylyl Cyclases

Matt R. Whorton and Roger K. Sunahara
Department of Pharmacology,
University of Michigan Medical School, Ann Arbor, Michigan

Introduction

The second messenger cyclic AMP is a key component in intracellular signaling pathways in both prokaryotes and eukaryotes[2]. The enzyme responsible for its synthesis, adenylyl cylcase, is found in these organisms as either membrane-bound or soluble forms. Ten genes have been identified in mammals that encode either membrane-bound (AC1 to AC9) or soluble forms (sAC) of AC, and may be regulated by various factors [3–5]. Soluble and membrane-bound forms of AC encoded by genes from various genera have also been identified, although their modes of regulation are not fully appreciated [6,7].

In eukaryotes the primary role cAMP plays is to activate protein kinase A (PKA), however, cAMP also directly activates exchange factors of small molecular weight G proteins (RaplA)[8,9], activates cyclic nucleotide-gated channels, and regulates the activity of some cGMP-specific phosphodiesterases [10]. Through protein phosphorylation activated PKA can regulate a plethora of enzymes, secondary kinases, transcription factors, receptors, and channels [11]. The actions of PKA may support mechanisms of feed-forward (activation) or feedback (desensitization and downregulation). In bacteria, cAMP binds directly to transcription factors and is responsible for repression of expression of genes involved in metabolism, also serving as a feedback mechanism [12].

In higher eukaryotes such as mammals, receptor-activated G proteins, Ca^{2+}-activated calmodulin (CaM), protein kinases, and bicarbonate ions appear to be the native modulators of AC activity [3,4,13,14]. AC function may also be affected by cellular stress [15,16], as well as by exogenous small molecules such as adenosine analogs and the diterpene, forskolin. The responses to these regulators are exquisitely AC-subtype specific. Although several AC isoforms may be expressed together in the same cell, each isoform may be selectively regulated by specific factors. Even though this complicates studies evaluating the physiological role of ACs, overexpression studies, gene knockouts, and gene mutations have been developed to elucidate these roles. Furthermore, aberrant AC is implicated in several human diseases, making this a very important molecule to study.

In recent years the biochemical characterization of ACs has been the subject of intense research. Through structural and functional approaches, researchers have gained a firm understanding of how ACs are regulated and even elucidated the mechanism of catalysis [17–19]. Moreover, improvements in biochemical approaches have allowed scientists to study the function of AC at cellular and even atomic detail. The following few pages will place emphasis on the mammalian forms of AC but highlight some differences in the ACs from other genera. We will summarize our current understanding of the mechanism of regulation of ACs, summarize the catalytic mechanism, and discuss the physiological roles this regulation plays in the function of AC.

Structure-Function

Ten AC isoforms, nine membrane-bound and one soluble, have been identified and cloned in mammals [3,4,6,13]. The membrane-bound forms share the same topology in that they are composed of 12 transmembrane (TM) segments and 2 large cytoplasmic domains (Cl and C2). These proteins exist in the membrane as tandem repeats of 6-TM regions followed by a large cytoplasmic loop (Fig. 1A and Fig. 2A). The sequence similarity between the different ACs is about 60% with the most conserved residues residing in the cytoplasmic domains. These two loops also share considerable sequence similarity with guanylyl cyclases (GC), to the degree that as few as two point mutations may be introduced into AC to convert it to a functional GC.

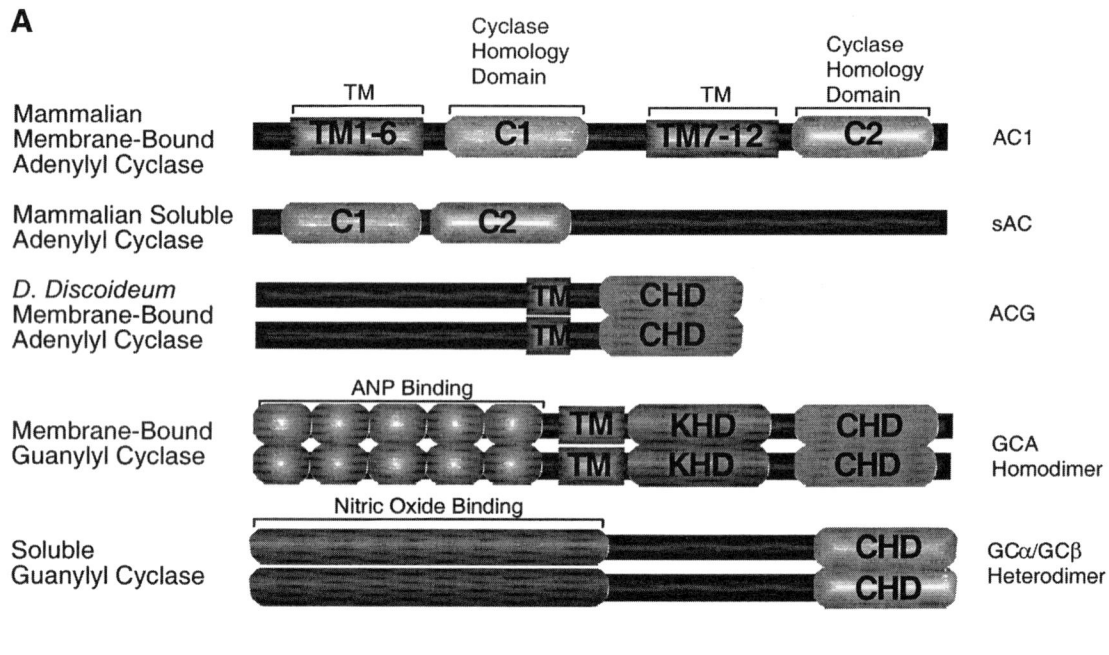

Figure 1 (**A**) This is the alignment of the domain structure of adenylyl and guanylyl cyclases. The putative domain structures of the cyclase homology domain (CHD), green and light blue; transmembrane domain (TM), yellow; ligand binding domain for atrial natriuretic peptide (ANP), pink; heme binding domain where nitric oxide binds, silver; and the kinase homology domain (KHD), purple. Note that the functional enzymes are organized as homo- or heterodimers of the catalytic domains. Membrane-bound and soluble ACs have both domains contained within one polypeptide, whereas the other cyclases require two proteins to have activity. (**B**) This is the alignment of the amino acid sequences of the adenylyl and guanylyl cyclases. Amino acid sequences bovine AC1 (GI: 162612), rat sAC (GI: 11067412), *Dictyostelium discoideum* ACG (GI: 167661) and ACA (GI: 457431), rat GC-A (GI: 204265), rat soluble GCα1 (GI: 1655846), and rat soluble GCβ1 (GI: 6980995). The mammalian ACs (AC1 and sAC) are divided into the C1 and C2 domains. Only the C2 domain from the *D. discoideum* ACA was included. The selected cyclase sequences were chosen as representative of eukaryotic nucleotide cyclase and were not singled out based on regulatory or mechanistic attributes. Residues are color coded to outline residue conservation. Residues that are important to enzyme function are also color coded as follows: lysine (K), glutamate (E), aspartate, (D) and cysteine (C) residues which contribute toward substrate specificity are boxed in red; conserved aspartate (D) residues which coordinate the two magnesium ions are boxed in royal blue; residues (arginine, R; asparagine, N; and lysine, K) which contribute toward stabilizing the transition state and that coordinate polyphosphate binding are boxed in black.

Several other unique forms of AC have been found in invertebrates as well as pathogenic bacteria. In the slime mold *Dictyostelium discoideum,* two diverse ACs containing a single TM region (ACA and ACR) have been identified, in addition to the canonical 12-TM form [20,21]. The bacteria *Bordella pertussis, Bacillus anthracis,* and *Pseudomonas aeruginosa* each excrete exotoxins, which possess AC activity [22–24]. These soluble ACs are taken up by host cells where they are then activated and begin producing very high levels of cAMP, thereby disrupting intracellular signaling pathways.

The X-ray crystal structure of the cytoplasmic C1 and C2 domains has recently been solved and has provided much information about the relevant active sites as well as the

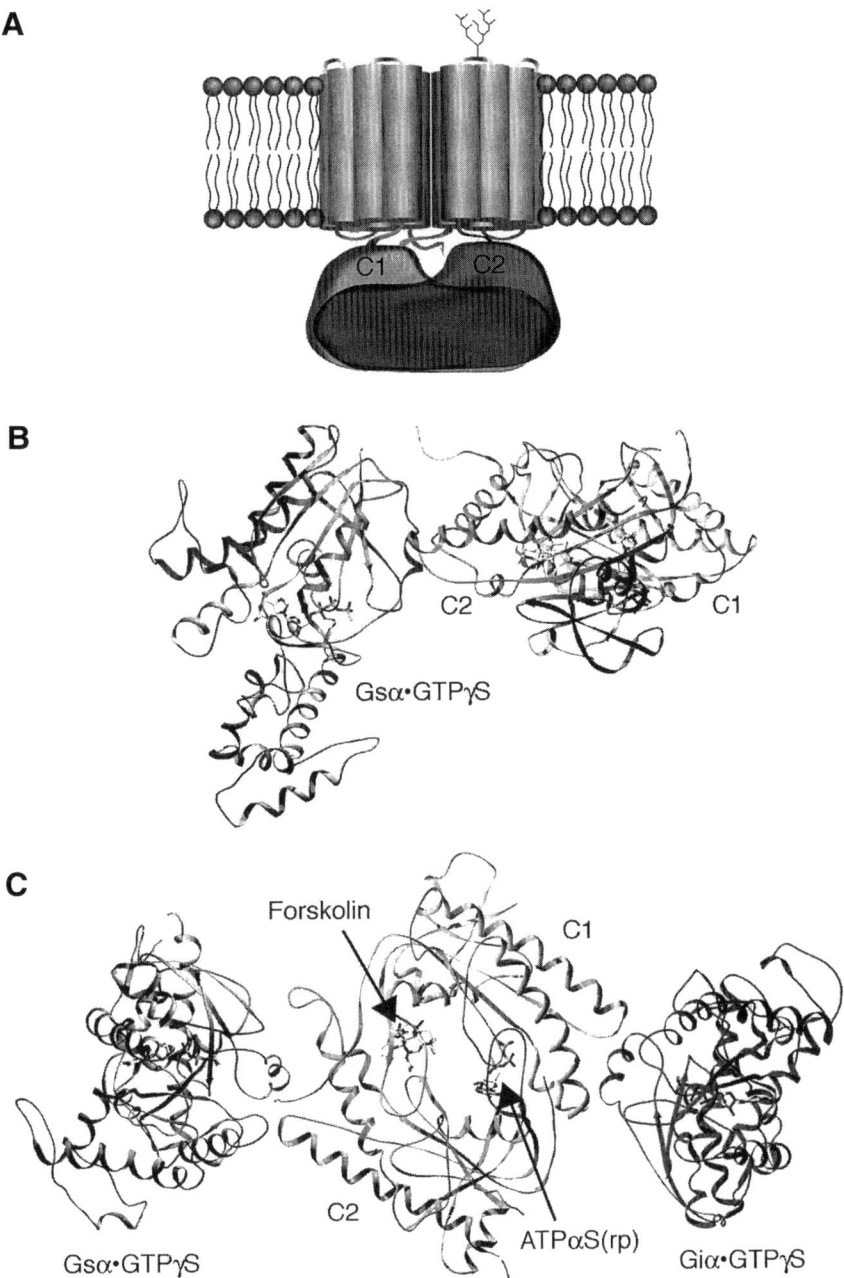

Figure 2 (A) This is an illustration of the membrane topology of a typical mammalian membrane-bound adenylyl cyclase. The 12-TM domain polypeptide also contains two large and homologous cytoplasmic domains each approximately 40 kDa in size. The Cl domain (lime green) and the C2 domain (sky blue) are represented by pseudosymmetrically related globular proteins. An asparagine-linked glycosylation site is also depicted in green. (B) This is an X-ray crystal structure of the stimulatory G protein, Gsα, bound to the catalytic core of adenylyl cyclase. GTPγS-activated Gsα (gray steel) is shown bound to the Cl(lime green) and C2 domain (sky blue) of adenylyl cyclase in the presence of forskolin and substrate inhibitor, ATPαS(rp). Illustrations were generated using SwissPDBViewer [73] and rendered with POV-RAY™ using the coordinates for the Gsα-GTPγ·Cl·C2·Fsk complex with ATPαS(rp) in the presence of Mn^{2+} and Mg^{2+} (PDB id: 1CJK)[26]. (C) Rotated view of the Gsα-GTPγS·Cl·C2·Fsk structure in (B) and with Giα-GTPγS (PDB id:lGIA) modeled into the pseudosymmetrically related Gsα-binding site. Visible also is the diterpene activator forskolin and the substrate inhibitor, ATPαS(rp). Note the twofold pseudosymmetry in the C1·C2 complex.

mechanism of catalysis (Fig. 2C) [17,18]. The catalytic core of AC is composed of a heterodimer of the Cl and C2 domains, which are related to each other by a twofold pseudosymmetry (Fig. 2). Forskolin binds to a hydrophobic pocket at the interface of the two domains, while G proteins bind on the surface, contacting both domains. The active site of catalysis, where nucleotides bind, is also located at the interface of the Cl and C2 domains and is pseudosymmetrically related to the forskolin binding site. The residues in this active site that are responsible for coordinating the binding of the nucleotide as well as two magnesium ions are highly conserved across all isoforms of both AC and GC [25,26]. In a manner similar to DNA and RNA polymerases, RNA spliceosomes, and reverse transcriptases, ACs utilize the metals to both stabilize the transition state of the reaction and also to deprotonate the 3' hydroxyl moiety of the ribose ring of ATP [19]. This is a key step that is necessary for the nucleophilic attack on the alpha phosphate by the newly formed oxyanion. The products are cAMP and the leaving group in the reaction, pyrophosphate (PPi).

The structure of AC from *Trypanosoma brucie* was recently solved and shares a similar protein-fold to the mammalian forms [27]. In sharp contrast, the structure of the catalytic AC domain of the exotoxin from *B. anthracis* recently delineated by the Tang laboratory, portrays a highly divergent protein-fold and a completely different catalytic mechanism [28]. AC from *B. anthraxis* utilizes the traditional catalytic triad consisting of histidine, serine, and aspartic acid residues to stabilize the transition state and deprotonate the 3'-OH.

Regulation

In invertebrates or vertebrates, neurotransmitter and hormonal regulation of ACs occurs primarily through heterotrimeric G proteins [29] (Table I and Fig. 3 for summary). G-protein-coupled receptor (GPCR) activation by these extracellular stimuli in turn leads to activation of bound G proteins by initiating the exchange of GDP for GTP. The α subunit of the stimulatory G protein (Gsα) activates all nine membrane-bound isoforms of AC in a nucleotide-dependent fashion, preferring the GTP-bound form to the GDP-bound form by a factor of 10 [4,13,30,31]. Gsα activation of ACs is terminated by GTP hydrolysis to GDP, a reaction that is accelerated by RGS proteins [32,33]. The α subunits of the inhibitory family of G proteins, $Gi_{1,2,3}$, Go, and Gz [34–36], inhibit AC activity, as the name would indicate, in an isoform-dependent manner [3,4,13,30]. Giαcl-3, Goα, and Gzα inhibit AC5 and AC6, and Goα also inhibits AC1 and possibly AC8. For AC5 and AC6 Giα-inhibition does not occur by competition with Gsα; in fact, mutagenesis experiments suggest that Giα binds to a site pseudosymmetrically related to the Gsα site, on the opposite side of AC (see Fig. 2C) [37]. Gβγ subunits are also important modulators of AC activity. They can potently stimulate the activity of AC2, AC4, and AC7, but in a manner that is dependent on co-activation by Gsα [34,38]. Gβγ subunits are also potent inhibitors of AC1 and AC8 [34].

CaM is a ubiquitous Ca^{2+} sensor protein and is a potent activator of several mammalian membrane-bound AC isoforms: AC1 [39], AC8 [40], and perhaps AC3 [41]. The primary source of calcium ions is thought to be derived from capacitative entry through Ca^{2+} channels, rather than the G-protein-regulated and inositol triphosphate (IP3) sensitive release of Ca^{2+} from intracellular stores [42,43]. CaM is also implicated in the pathology of the bacterial exotoxins mentioned above, as it is the principal AC activator [44,45]. CaM activation of edema factor (EF; the exotoxin from *B. anthraxis*) yields a catalytic rate 1000-fold higher than that of CaM-activated mammalian ACs.

CaM also inhibits AC1 and AC3 indirectly through the activity of CaM-dependent protein kinase II and IV (CaMKII, IV)[46,47]. Phosphorylation of AC1 and AC3 by CaM kinases inhibits cyclase activity by blocking the binding of activators. In this sense, posttranslational modification of ACs by phosphorylation is generally inhibitory and can also be caused by the PKA as well as protein kinase C (PKC). PKA supports a negative-feedback mechanism whereby the more cAMP that is produced by ACs, the more PKA is activated, and thus the more ACs that are phosphorylated and inhibited.

The effects of Ca^{2+} have also been shown to be quite inhibitory on AC5 and anthrax [48,49]. Low micromolar concentrations of Ca^{2+}, well below the toxic levels and certainly within the physiological dynamic range found in a cell, effectively and specifically inhibit these two isoforms. While all isoforms of AC are inhibited by higher concentrations (mM), the effect on these isoforms are consistent with levels derived from capacitative entry, similar to the CaM-dependent stimulatory effect.

The small molecule forskolin (isolated from the plant *Choleus forskohlii*) is a potent activator of all mammalian membrane-bound isoforms of AC except for AC9, which is weakly activated [50]. AC isoform-specific forskolin analogs have been discovered using structural-based drug design, and it has been hypothesized that endogenous forskolin-like molecules may exist [51]. While the binding site of forskolin is within the conserved catalytic domain, the stimulatory actions appear to be selective for the membrane-bound vertebrate and invertebrate forms [50].

In contrast, ACs are inhibited by a class of adenosine analogs known as P site inhibitors [52]. These small molecules act by binding to a conformation of the enzyme that closely resembles the product-bound state or the posttransition state [53,54]. Inhibition is enhanced with the presence of PPi. The potency of P-site inhibitors is therefore increased by higher levels of AC activity [53]. Several adenosine analogs have been developed that appear to display some isoform preference [51,55]. In addition, polyphosphorylated acyclic nucleosides (such as 9-(2-triphosphonylmethoxyethyl) adenine, PMEApp) and foscarnet (phosphonoformic acid), also inhibit ACs [55,56]. Both drugs are used clinically as antiviral and antifungal agents and share a similar proposed mechanism of action as the P-site inhibitors on AC. It should be noted

Table I Summary of the Regulatory Properties of the Mammalian Adenylyl Cyclases*

AC isoform	Tissue distribution	Gαs	Gβγ	Gαi	Protein kinases	Calcium	Forskolin	Notes
AC1	Brain, adrenal (medulla)	↑	↓	↓(Gαo)	↑ PKC ↓ CaMKIV	↑ CaM	↑	
AC2	Brain, skeletal muscle, Lung (heart)	↑	↑*		↑ PKC		↑	
AC3	Brain, olfactory epithelium	↑			↑ PKC ↓ CaMKII	↑ CaM	↑	
AC4	Brain (heart, kidney, liver, lung, BAT, uterus)	↑	↑*		↑ PKC		↑	
AC5	Heart, brain, kidney, liver, lung, Uterus, adrenal, BAT	↑	↓	↓	↑ PKCα,ζ ↓ PKA	↓	↑	
AC6	Ubiquitous	↑	↓	↓	↑ PKC ↓ PKA	↓	↑	
AC7	Ubiquitous, High in brain	↑	↑*		↑ PKC		↑	
AC8	Brain, lung (testis, adrenal, uterus, heart)	↑				↑ CaM	↑	
AC9	Brain, skeletal muscle	↑						
sAC	Testis							↑ Bicarbonate

*Gβγ stimulation of AC isoforms is conditional upon Gsα co-activation.

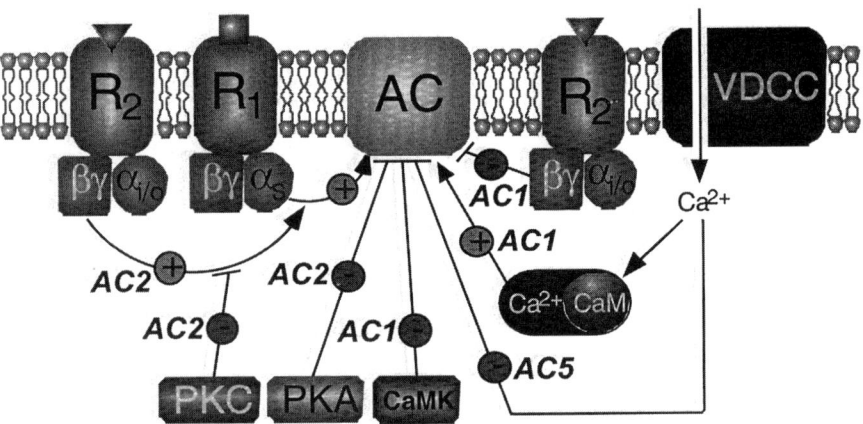

Figure 3 This is an illustration of the complex modes of regulation of membrane-bound mammalian adenylyl cyclase. Summarized are the stimulatory (green circles with pluses) or inhibitory (red circles with minuses) influences of hormone-receptor-mediated G-protein regulation, protein kinase regulation, and Ca^{2+} and/or CaM effects on AC activity. AC isoform-specific effects are demarcated in italics beside the stimulatory or inhibitory signs. For example, the Gβγ effects are inhibitory on the AC1 family of cyclases (AC1, AC8, and presumably AC3) and stimulatory on the AC2 family of cyclases (AC2, AC4, and AC7). The stimulatory effects of Gβγ are dependent on prior activation by Gsα. Ca^{2+}-CaM directly and potently activates the AC1 family of cyclases, whereas Ca^{2+} alone effectively inhibits the AC5 family of cyclase (AC5 and AC6). Not illustrated is forskolin, which activates all membrane-bound isoforms except AC9.

that the primary therapeutic target of these compounds is the mechanistically similar viral polymerases and transcriptases.

The lone soluble mammalian AC isoform (sAC) is as unique in overall structure as it is unique in regulation [6,7,57]. sAC activity is not affected at all by the classic AC modulators: G proteins, CaM, or forskolin [57]. Instead, it is activated *in vivo* as well as *in vitro* by bicarbonate ions [14]. Soluble forms of AC in prokaryotes are also sensitive to bicarbonate, suggesting that sAC is an evolutionarily conserved bicarbonate sensor [15]. The single TM domain AC in *Dictyostelium* is unique in the same way. It is regulated by osmotic stress; however, it is not yet clear if it actually has intrinsic osmosensing activity [16]. Although the mechanisms of regulation of these isoforms are quite different, the sequence of the catalytic domains are similar and the putative catalytic residues are conserved.

Physiology

Adenylyl cyclases are studied in many systems and have been implicated in numerous physiological roles. At the least, it is known that all mammalian AC isoforms are expressed in the central nervous system and in excitable tissues; but, for the most part, AC is expressed in nearly every tissue (see Table I). More precise patterns of expression have been difficult to obtain due to relatively low levels of expression as well as a general lack of highly specific antibodies. Exceptions in expression patterns do exist, most notably in sAC, which is most highly expressed in the testis [6].

The precise roles of specific AC isoforms have been difficult to assess because most cells express multiple isoforms. The specific contributions of these ACs have only recently been segregated from the remaining isoforms. Researchers have taken advantage of genetic mutations in AC or gene disruption using homologous recombination in mice. Several studies have investigated the function of AC in the *Drosophila* mutant *rutabaga* [58]. These mutants are deficient in a calcium-activated AC, which is quite similar to the mammalian AC1. Deficiency of this AC causes these flies to avoid a trained odor, indicating that AC1 is important in memory and learning [59]. Likewise, specific disruption of the AC1 gene in mutant mice or the spontaneous mutation of AC1 in the *barrelless* mouse, have a negative effect on long-term potentiation (LTP) [60,61]. AC1 and AC8 are both necessary for both late-LTP (L-LTP) as well as long-term memory (LTM) [62]. Knockouts of either AC gene by itself yields normal L-LTP and LTM; however, double knockout mice exhibit no L-LTP or LTM. This effect can be reversed by infusion of forskolin into the hippocampus, which may compensate the null AC1 and AC8 by producing cAMP through other AC isoforms.

AC3 has been demonstrated to be involved in transmitting olfactory responses in mice [63]. AC isoforms 2, 3, and 4 are all present in olfactory cilia; however, it is interesting that a knockout of just AC3 is sufficient to completely ablate responses to odorants. It has also been shown that ACs are important in developing drug dependencies. Following chronic opiate treatment, several ACs are upregulated and become supersensitized to additional stimulation by either Gsα or forskolin. Specifically, AC1, AC5, AC6, and AC8 are sensitized, while AC2, AC3, AC4, and AC7 are not [64–66]. Depending on the system, upregulation of AC may or may not involve a transcriptional step.

Upregulation of AC isoforms is also important for cell differentiation. AC2, AC5, and AC6 are upregulated in differentiation of pluripotent PI9 cells. Additionally, upregulation of AC2, AC5, and AC8 accompany neuronal differentiation [67]. It is also interesting that ectopic expression of AC2 in NIH3T3 cells inhibits cell cycle progression [68]. One resulting hypothesis is that for cell differentiation to occur, upregulation of AC2 is necessary to induce a temporary arrest of cell proliferation.

Summary

Adenylyl cyclase is clearly an incredibly important molecule as it is intimately involved in the very complex signaling pathways that regulate the numerous facets of life itself. The identification and characterization of ACs has come a long way since the initial discovery of cAMP nearly forty years ago, especially with recent advances applying molecular genetic and structural biology approaches. Nevertheless, many important questions still remain unanswered. For instance, what is the function of the 12-TM domain structure, other than localizing ACs to the membrane? Can the 12-TM structure support transport of molecules across the plasma membrane, as originally proposed when the first AC cDNA [39] was reported? Membrane-bound isoforms of ACs have recently been shown to homodimerize [69]. Although the relevance of AC dimerization is unknown, it is particularly intriguing with regard to the specter of heterodimerization. Heterodimers between different AC isoforms would add a new dimension to the already complex network of AC regulation. In any case, much more research is needed to more fully understand the precise regulation and physiological roles of ACs.

References

1. Hepler, J. R., Biddlecome, G. H., Kleuss, C., Camp, L. A., Hofmann, S. L., Ross, E. M., and Gilman, A. G. (1996). Functional importance of the amino terminus of Gq alpha. *J. Biol. Chem.* **271**, 496–504.
2. Robison, G. A., Butcher, R. W., and Sutherland, E. W. (1968). Cyclic AMP. *Annu. Rev. Biochem.* **37**, 149–174.
3. Sunahara, R., Dessauer, C., and Gilman, A. (1996). Complexity and diversity of mammalian adenylyl cyclases. *Annu. Rev. Pharmacol. Toxicol.* **36**, 461–480.
4. Smit, M. J. and Iyengar, R. (1998). Mammalian adenylyl cyclases. *Adv. Second Messenger Phosphoprot. Res.* **32**, 1–21.
5. Hanoune, J., Pouille, Y., Tzavara, E., Shen, T., Lipskaya, L., Miyamoto, N., Suzuki, Y., and Defer, N. (1997). Adenylyl cyclases: Structure, regulation and function in an enzyme superfamily. *Mol. Cell. Endocrinol.* **128**, 179–194.
6. Buck, J., Sinclair, M. L., Schapal, L., Cann, M. J., and Levin, L. R. (1999). Cytosolic adenylyl cyclase defines a unique signaling molecule in mammals. *Proc. Natl. Acad. Sci. USA* **96**, 79–84.
7. Jaiswal, B. S. and Conti, M. (2001). Identification and functional analysis of splice variants of the germ cell soluble adenylyl cyclase. *J. Biol. Chem.* **276**, 31698–31708.
8. de Rooij, J., Zwartkruis, F. J., Verheijen, M. H., Cool, R. H., Nijman, S. M., Wittinghofer, A., and Bos, J. L. (1998). Epac is a Rap1 guanine-nucleotide-exchange factor directly activated by cyclic AMP. *Nature* **396**, 474–477.
9. Kawasaki, H., Springett, G. M., Mochizuki, N., Toki, S., Nakaya, M., Matsuda, M., Housman, D. E., and Graybiel, A. M. (1998). A family of cAMP-binding proteins that directly activate Rap1. *Science* **282**, 2275–2279.
10. Broillet, M. C. and Firestein, S., (1999). Cyclic nucleotide-gated channels. Molecular mechanisms of activation. *Ann. N. Y. Acad. Sci.* **868**, 730–740.
11. Taylor, S. S., Buechler, J. A., and Yonemoto, W. (1990). cAMP-dependent protein kinase: framework for a diverse family of regulatory enzymes. *Annu. Rev. Biochem.* **59**, 971–1005.
12. Daniel, P. B., Walker, W. H., and Habener, J. F. (1998). Cyclic AMP signaling and gene regulation. *Annu. Rev. Nutr.* **18**, 353–383.
13. Patel, T. B., Du, Z., Pierre, S., Cartin, L., and Scholich, K. (2001). Molecular biological approaches to unravel adenylyl cyclase signaling and function. *Gene* **269**, 13–25.
14. Chen, Y., Cann, M. J., Litvin, T. N., Lourgenko, V., Sinclair, M. L., Levin, L. R., and Buck, J. (2000). Soluble adenylyl cyclase as an evolutionarily conserved bicarbonate sensor. *Science* **289**, 625–628.

15. Zippin, J. H., Levin, L. R., and Buck, J. (2001). CO(2)/HCO(3)(-)-responsive soluble adenylyl cyclase as a putative metabolic sensor. *Trends Endocrinol. Metab.* **12**, 366–370.
16. van Es, S., Virdy, K., Pitt, G., Meima, M., Sands, T., Devreotes, P., Cotter, D., and Schaap, P. (1996). Adenylyl cyclase G, an osmosensor controlling germination of Dictyostelium spores. *J. Biol. Chem.* **271**, 23623–23625.
17. Zhang, G., Liu, Y., Ruoho, A. E., and Hurley, J. H. (1997). Structure of the adenylyl cyclase catalytic core. *Nature* **386**, 247–253.
18. Tesmer, J. J. and Sunahara, R. K., Gilman, A.G., and Sprang, S. R. (1997). Crystal structure of the catalytic domains of adenylyl cyclase in a complex with Gsalpha.GTPgammaS. *Science* **278**, 1907–1916.
19. Tesmer, J. J. and Sprang, S. R. (1998). The structure, catalytic mechanism and regulation of adenylyl cyclase. *Curr. Opin. Struct. Biol.* **8**, 713–719.
20. Soderbom, F., Anjard, C., Iranfar, N., Fuller, D., and Loomis, W. F. (1999). An adenylyl cyclase that functions during late development of Dictyostelium. *Development* **126**, 5463–5471.
21. Pitt, G., Milona, N., Borleis, J., Lin, K., Reed, R., and Devreotes, P. (1992). Structurally distinct and stage-specific adenylyl cyclase genes play different roles in Dictyostelium development. *Cell* **69**, 305–315.
22. Ladant, D. and Ullmann, A. (1999). Bordatella pertussis adenylate cyclase: a toxin with multiple talents. *Trends Microbiol.* **7**, 172–176.
23. Baillie, L. and Read, T. D. (2001). Bacillus anthracis, a bug with attitude! *Curr. Opin. Microbiol.* **4**, 78–81.
24. Yahr, T. L., Vallis, A. J., Hancock, M. K., Barbieri, J. T., and Frank, D. W. (1998). ExoY, an adenylate cyclase secreted by the Pseudomonas aeruginosa type III system. *Proc. Natl. Acad. Sci. USA* **95**, 13899–13904.
25. Liu Y., Rao, R. A. VD, and Hurley, J. H. (1997). Catalytic mechanism of the adenylyl and guanylyl cyclases: modeling and mutational analysis. *Proc. Natl. Acad. Sci. USA* **94**, 13414–13419.
26. Tesmer, J. J., Sunahara, R. K., Johnson, R. A., Gosselin, G., Gilman, A. G., and Sprang, S. R. (1999). Two-metal-Ion catalysis in adenylyl cyclase. *Science* **285**, 756–760.
27. Bieger, B. and Essen, L. O. (2001). Structural analysis of adenylate cyclases from Trypanosoma brucei in their monomeric state. *EMBO J.* **20**, 433–445.
28. Drum, C. L., Yan, S. Z., Bard, J., Shen, Y. Q., Lu, D., Soelaiman, S., Grabarek, Z., Bohm, A., and Tang, W. J. (2002). Structural basis for the activation of anthrax adenylyl cyclase exotoxin by calmodulin. *Nature* **415**, 396–402.
29. Gilman, A. G. (1990). Regulation of adenylyl cyclase by G proteins. *Adv. Second Messenger Phosphoprot. Res.* **24**, 51–57.
30. Hanoune, J. and Defer, N. (2001). Regulation and role of adenylyl cyclase isoforms. *Annu. Rev. Pharmacol. Toxicol.* **41**, 145–174.
31. Sunahara, R. K., Dessauer, C. W., Whisnant, R. E., Kleuss, C., and Gilman, A. G. (1997). Interaction of $G_{s\alpha}$ with the cytosolic domains of mammalian adenylyl cyclase. *J. Biol. Chem.* **272**, 22265–22271.
32. De Vries, L., Zheng, B., Fischer, T., Elenko, E., and Farquhar, M. G. (2000). The regulator of G protein signaling family. *Annu. Rev. Pharmacol. Toxicol.* **40**, 235–271.
33. Zheng, B., Ma, Y. C., Ostrom, R. S., Lavoie, C., Gill, G. N., Insel, P. A., Huang, X. Y., and Farquhar, M. G. (2001). RGS-PX1, a GAP for GalphaS and sorting nexin in vesicular trafficking. *Science* **294**, 1939–1942.
34. Tang, W.-J. and Gilman, A. G. (1991). Type-specific regulation of adenylyl cyclase by G protein beta gamma subunits. *Science* **254**, 1500–1503.
35. Taussig, R., Tang, W. J., Hepler, J. R., and Gilman, A. G. (1994). Distinct patterns of bidirectional regulation of mammalian adenylyl cyclases. *J. Biol. Chem.* **269**, 6093–6100.
36. Kozasa, T., and Gilman, A. (1995). Purification of recombinant G proteins from Sf9 cells by hexahistidine tagging of associated subunits. Characterization of alpha 12 and inhibition of adenylyl cyclase by alpha z. *J. Biol. Chem.* 1734–1741.
37. Dessauer, C. W., Tesmer, J. J., Sprang, S. R., and Gilman, A. G. (1998). Identification of a Gialpha binding site on type V adenylyl cyclase. *J. Biol. Chem.* **273**, 25831–25839.
38. Gao, B. N. and Gilman, A. G. (1991). Cloning and expression of a widely distributed (type IV) adenylyl cyclase. *Proc. Natl. Acad. Sci. USA* **88**, 10178–10182.
39. Krupinski, J., Coussen, F., Bakalyar, H. A., Tang, W. J., Feinstein, P. O. K. Orth, K., Slaughter, C., Reed, R. R., and Gilman, A. G. (1989). Adenylyl cyclase amino acid sequence: Possible channel- or transporter-like structure. *Science* **244**, 1558–1564.
40. Cali, J. J., Zwaagstra, J. C., Mons, N., Cooper, D. M., and Krupinski, J. (1994). Type VIII adenylyl cyclase. A Ca^{2+}/calmodulin-stimulated enzyme expressed in discrete regions of rat brain. *J. Biol. Chem.* **269**, 12190–12195.
41. Choi, E. J., Xia, Z., and Storm, D. R. (1992). Stimulation of the type III olfactory adenylyl cyclase by calcium and calmodulin. *Biochemistry* **31**, 6492–6498.
42. Pagan, K. A., Mahey, R., and Cooper, D. M. (1996). Functional co-localization of transfected Ca(2+)-stimulable adenylyl cyclases with capacitative Ca^{2+} entry sites. *J. Biol. Chem.* **271**, 12438–12444.
43. Pagan, K. A., Graf, R. A., Tolman, S., Schaack, J., and Cooper, D. M. (2000). Regulation of a Ca2+-sensitive adenylyl cyclase in an excitable cell. Role of voltage-gated versus capacitative Ca2+ entry. *J. Biol. Chem.* **275**, 40187–40194.
44. Leppla, S. H. (1984). Bacillus anthracis calmodulin-dependent adenylate cyclase: Chemical and enzymatic properties and interactions with eucaryotic cells. *Adv. Cyclic Nucleotide Protein Phosphoryl. Res.* **17**, 189–198.
45. Oldenburg, D., Gross, M., Wong, C., and Storm, D. (1992). High-affinity calmodulinbinding is required for the rapid entry of Bordetella pertussis adenylyl cyclase into neuroblastoma cells. *Biochemistry* **31**, 8884–8891.
46. Wayman, G. A., Wei, J., Wong, S., and Storm, D. R. (1996). Regulation of type I adenylyl cyclase by calmodulin kinase IV in vivo. *Mol. Cell. Biol.* **16**, 6075–6082.
47. Wei, J., Wayman, G., and Storm, D. R. (1996). Phosphorylation and inhibition of type III adenylyl cyclase by calmodulin-dependent protein kinase II in vivo. *J. Biol. Chem.* **271**, 24231–24235.
48. Gu, C. and Cooper, D. M. (2000). Ca(2+), Sr(2+), and Ba(2+) identify distinct regulatory sites on adenylyl cyclase (AC) types VI and VIII and consolidate the apposition of capacitative cation entry channels and Ca(2+)-sensitive ACs. *J. Biol. Chem.* **275**, 6980–6986.
49. Cooper, D. M. (1991). Inhibition of adenylate cyclase by Ca(2+)-a counterpart to stimulation by Cas+/calmodulin. *Biochem. J.* **278**, 903–904.
50. Hacker, B. M., Tomlinson, I. E., Wayman, G. A., Sultana, R., Chan, G. Villacres, E., Disteche, C., and Storm, D. R. (1998). Cloning, chromosomal mapping, and regulatory properties of the human type 9 adenylyl cyclase (ADCY9). *Genomics* **50**, 97–104.
51. Onda, T., Hashimoto, Y., Nagai, M. *et al.* (2001). Type-specific regulation of adenylyl cyclase. Selective pharmacological stimulation and inhibition of adenylyl cyclase isoforms. *J. Biol. Chem.* **276**, 47785–47793.
52. Londos, C. and Wolff, J. (1977). Two distinct adenosine-sensitive sites on adenylate cyclase. *Proc. Natl. Acad. Sci. USA* **74**, 5482–5486.
53. Dessauer, C. W. and Gilman, A. G. (1997). The catalytic mechanism of mammalian adenylyl cyclase. Equilibrium binding and kinetic analysis of P-site inhibition. *J. Biol. Chem.* **272**, 27787–27795.
54. Dessauer, C. W., Tesmer, J. J., Sprang, S. R., and Gilman, A. G. (1999). The interactions of adenylate cyclases with P-site inhibitors. *Trends Pharmacol. Sci.* **20**, 205–210.
55. Johnson, R., Desaubry, L., Bianchi, G. *et al.* (1997). Isozyme-dependent sensitivity of adenylyl cyclases to P-site-mediated inhibition by adenine nucleosides and nucleoside 3'-polyphosphates. *J. Biol. Chem.* **272**, 8962–8966.
56. Kudlacek, O., Mitterauer, T., Nanoff, C., Hohenegger, M., Tang, W. J., Freissmuth, M., and Kleuss, C. (2001). Inhibition of adenylyl and guanylyl cyclase isoforms by the antiviral drug foscarnet. *J. Biol. Chem.* **276**, 3010–3016.
57. Neer, E. J. (1978). Physical and functional properties of adenylate cyclase from mature rat testis. *J. Biol. Chem.* **253**, 5808–5812.

58. Levin, L. R., Han, P. L., Hwang, P. M., Feinstein, P. O., Davis, R. L., and Reed, R. R. (1992). The Drosophila learning and memory gene rutabaga encodes a Ca2+/Calmodulin-responsive adenylyl cyclase. *Cell* **68**, 479–489.

59. Zars, T., Fischer, M., Schulz, R., and Heisenberg, M. (2000). Localization of a short-term memory in Drosophila. *Science* **288**, 672–675.

60. Wu, Z. L., Thomas, S. A., Villacres, E. G., Xia, Z., Simmons, M. L., Chavkin, C., Palmiter, R. D., and Storm, D. R. (1995). Altered behavior and long-term potentiation in type I adenylyl cyclase mutant mice. *Proc. Natl. Acad. Sci. USA* **92**, 220–224.

61. Abdel-Majid, R. M., Leong, W. L., Schalkwyk, L. C. *et al.* (1998). Loss of adenylyl cyclase I activity disrupts patterning of mouse somatosensory cortex. *Nat. Genet.* **19**, 289–291.

62. Schaefer, M. L., Wong, S. T., Wozniak, D. F. *et al.* (2000). Altered stress-induced anxiety in adenylyl cyclase type VHI-deficient mice. *J. Neurosci.* **20**, 4809–4820.

63. Wong, S. T., Trinh, K., Hacker, B., Chan, G. C., Lowe, G., Gaggar, A., Xia, Z., Gold, G. H., and Storm, D. R. (2000). Disruption of the type III adenylyl cyclase gene leads to peripheral and behavioral anosmia in transgenic mice. *Neuron* **27**, 487–497.

64. Avidor-Reiss, T., Nevo, I., Saya, D., Bayewitch, M., and Vogel, Z. (1997). Opiate-induced adenylyl cyclase superactivation is isozyme-specific. *J. Biol. Chem.* **272**, 5040–5047.

65. Watts, V. J. and Neve, K. A. (1996). Sensitization of endogenous and recombinant adenylate cyclase by activation of D2 dopamine receptors. *Mol. Pharmacol.* **50**, 966–976.

66. Thomas, J. M. and Hoffman, B. B. (1996). Isoform-specific sensitization of adenylylcyclase activity by prior activation of inhibitory receptors: role of beta gamma subunits in transducing enhanced activity of the type VI isoform. *Mol. Pharmacol.* **49**, 907–914.

67. Lipskaia, L., Djiane, A., Defer, N., and Hanoune, J. (1997). Different expression of adenylyl cyclase isoforms after retinoic acid induction of PI 9 teratocarcinoma cells. *FEBS Lett.* **415**, 275–280.

68. Smit, M. J., Verzijl, D., and Iyengar, R. (1998). Identity of adenylyl cyclase isoform determines the rate of cell cycle progression in NIH 3T3 cells. *Proc. Natl. Acad. Sci. USA* **95**, 15084–15089.

69. Gu, C., Cali, J. J., and Cooper, D. M. (2002). Dimerization of mammalian adenylate cyclases. *Eur. J. Biochem.* **269**, 413–421.

70. Guex, N. and Peitsch, M. C. (1997). SWISS-MODEL and the Swiss-PdbViewer: An environment for comparative protein modeling. *Electrophoresis* **18**, 2714–2723.

CHAPTER 191

Guanylyl Cyclases

Ted D. Chrisman and David L. Garbers

*Cecil H. & Ida Green Center for Reproductive Biology Sciences,
Howard Hughes Medical Institute and Department of Pharmacology,
University of Texas Southwestern Medical Center, Dallas, Texas*

Receptor guanylyl cyclases and their ligands together with guanosine-3′,5′-monophosphate (cGMP), and its effectors compose signal transduction pathways regulating essential tissue and cell functions. For example, cGMP is the documented second messenger for NO- and atrial natriuretic peptide (ANP)-induced vascular smooth muscle relaxation, and in the kidney cGMP mediates ANP-induced natriuresis and diuresis. Likewise, pathologic elevations of cGMP in intestinal mucosal cells result in severe diarrhea in response to secretion of a heat-stable peptide (Sta) from pathogenic strains of *Escherichia coli*; Sta is a potent agonist of a guanylyl cyclase in these cells. And disruption of the murine C-type natriuretic peptide (CNP) gene, a ligand for a chondrocyte guanylyl cyclase, results in dwarfism and early death, while disruption of a photoreceptor guanylyl cyclase results in cone-specific dystrophy in the mouse. Therefore, the various guanylyl cyclase/cGMP signaling pathways are physiologically important, but the relevance of the interactions with other signaling pathways has been less clear. Recently, cGMP signaling pathways have been shown to impact on signaling systems that regulate cell proliferation and differentiation.

Historic Perspectives

Guanylyl cyclases catalyzing the formation of cGMP and pyrophosphate (P~P) from MgGTP or MnGTP were identified in crude extracts of mammalian tissues shortly after the discovery of cGMP in rat urine [1]; importantly those studies established that the guanylyl cyclases were distinct from adenylyl cyclases implying different functions for each of the families of cyclases. Differences in subcellular distribution and kinetics of guanylyl cyclases in mammalian tissue homogenates prompted the suggestion that different forms of the cyclase exist rather than there being a differential distribution of a single form of the enzyme. Several decades later through the use of recombinant DNA technology, the early speculations were confirmed, and two categories of guanylyl cyclases were identified through cloning: single-pass plasma membrane or particulate guanylyl cyclases (pGC) and cytosolic or soluble guanylyl cyclases (sGC). P~P has not been shown to signal, and thus cGMP is considered the second messenger following ligand activation and no initial signaling pathway other than the generation of cGMP has yet been firmly documented for any of the cyclases (Fig. 1).

Guanylyl Cyclases

Seven single-pass plasma membrane (pGC) and four cytosolic or soluble (sGC) guanylyl cyclase subunits have been identified in mammals. Many more (23 putative guanylyl cyclase genes) have been identified in *Caenorhabditis elegans*, and multiple-pass plasma membrane cyclases, similar to the mammalian membrane forms of adenylyl cyclase, have been reported in *Dictyostelium discoideum, Plasmodium falciparum, Paramecium tetraurelia,* and *Tetrahymena pyriformis* [1].

The seven mammalian pGCs (GCA through GCG) are expressed in many different tissues and cultured cells (Table I). The structurally similar 120- to 140-kDa proteins contain an amino terminal extracellular domain (BCD; the apparent ligand-binding domain is the least similar within the family), a single-pass transmembrane domain (TMD), a protein kinase homology domain (KHD; 30% homologous to protein kinase catalytic domains [2], and a carboxyl-terminal cyclase catalytic domain (CCD; the most conserved and the most highly similar to adenylyl cyclases [1]). Homodimeric plasma membrane cyclases appear to be preferentially expressed *in vivo* even though more than one cyclase is expressed

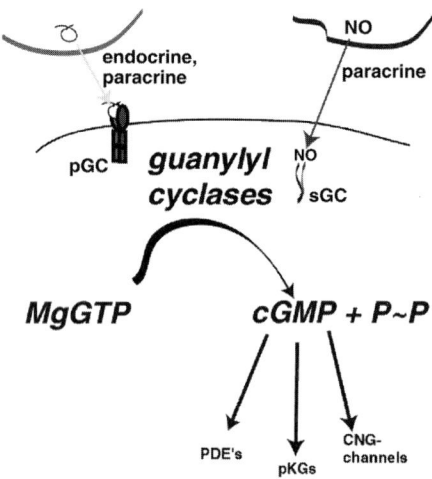

Figure 1 The two general forms of guanylyl cyclase found in mammals. The plasma membrane (pGC) forms exist as minimal homodimers and the soluble (sGC) forms exist as minimal heterodimers to display catalytic activity. The increased concentrations of cGMP found as a result of stimulation of these forms then acts on phosphodiesterases (PDEs) that are stimulated or inhibited by cGMP, ion channels (CNG) directly gated by cGMP, or on cGMP-dependent protein kinases (PKGs) to elicit a cell behavioral response.

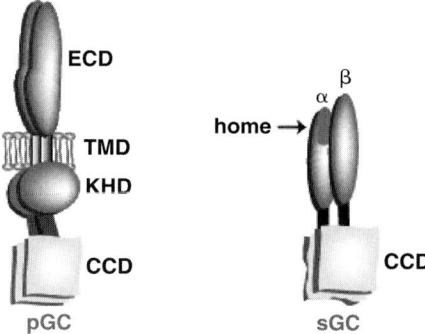

Figure 2 The general domain structure of the membrane forms (pGC) or the soluble forms (sGC) of guanylyl cyclase. BCD represents the extracellular ligand binding domain, TMD the transmembrane segment, KHD the protein kinase homology domain, and CCD the cyclase catalytic domain. The KHD appears to bind ATP as a regulatory molecule. For sGC, the subunits of the heterodimer have been arbitrarily named a and p. Heme binds to the amino terminal region, probably to the p subunit.

simultaneously by a single cell. The minimum catalytic unit of the pGCs appears to be that of a homodimer [3] (Fig. 2).

The sGCs are heterodimeric proteins composed of subunit isoforms a (cti, 0.2, both 82 kDa) and P (pi,70 kDa; 02,76 kDa) and are expressed in many of the same cells and tissues as the pGCs (Table II). A heme moiety noncovalently bound to the p subunit amino terminus confers ligand (NO) sensitivity to the cyclase. Two human subunits, named 0:3 and ps are orthologs of rat oti and pi. The aipi heterodimer is more commonly found *in vivo* but aapi has been detected as well. The $2 subunit also contains a potential geranylgeranylation site [4] raising the possibility of plasma membrane localization of this form and the possibility therefore of NO-stimulated cGMP elevations at the level of the membrane. The minimum active mammalian soluble cyclase appears to be a heterodimer as expression of a or p subunits alone results in no detectable guanylyl cyclase activity while co-expression of various subunits produces high-soluble guanylyl cyclase activity. The studies of Sunhara *et al.* [5] and of Liu *et al.* [3] also support the existence of a dimer as the minimal unit for catalytic activity.

Guanylyl Cyclase Ligands

Ligands have been identified for some but not all mammalian GCs (Table II). Human ANP (28 amino acids), brain natriuretic peptide (BNP, 32 amino acids), and CNP(22 or 53 amino acids) compose a family of distinct and structurally similar oligopeptides having a highly conserved 17-member ring required for biological activity. ANP and BNP are endocrine ligands released from the heart that promote natriuresis, diuresis, and vasorelaxation by direct activation of GCA in the kidney, in vascular smooth muscle, or the adrenal gland. CNP, classified as a natriuretic peptide by virtue of its structure and the only known ligand for GCB, has no known tissue depots and only marginally mimics the vascular and renal actions of ANP and BNP. CNP is generally considered a paracrine ligand involved in regulation of cell proliferation and differentiation [6,7].

ANP, BNP, and CNP bind with near equal avidity to a third cell surface receptor, the natriuretic peptide clearance receptor (NPR-C). NPR-C has a BCD similar to GCA and GCB, a short cytosolic domain devoid of the KHD and CCD, and is expressed by most mammalian cells. Internalization of the peptide-receptor complex serves to remove or "clear" natriuretic peptides from the extracellular space thus buffering their effects on cell function. There is some evidence suggesting that NPR-C may also signal via a pertussis-toxin-sensitive pathway to inhibit adenylyl cyclase [8].

GCC was identified initially as the receptor for the enterotoxin, Sta. Subsequently three mammalian ligands for GCC, guanylin, uroguanylin, and lymphoguanylin were isolated. All are small peptides with structural homology and sequence identity and expressed in the intestine and other tissues.

NO, the ligand for sGCs, has diverse actions on cardiovascular, renal, and immune cell function and is expressed by many cell types [1]. NO has effects similar to the natriuretic peptide signaling pathways in those cells where GCA, GCB, and sGC are expressed.

Extracellular ligands for GCD, GCE, GCF, and GCD have not been identified and remain orphan receptors. GCE and GCF (human RetGC-1 and RetGC-2), expressed in photoreceptors, are activated intracellularly by Ca^{2+}-free forms of guanylyl cyclase activating proteins (GCAPs) which bind to the coiled-coil region linking the KHD and CCD [9,10]. Although GCE and GCF may not require extracellular ligands based on these observations, the conservation of Cys within the BCD compared to the guanylyl cyclases with known ligands, and the conservation of the BCD across all vertebrates that have been studied suggests heavy evolutionary pressure is being exerted to retain the BCD structure. One source for this pressure would be the need to recognize a ligand.

cGMP Effectors

At least three classes of cGMP-binding proteins amplify and mediate changes in intracellular cGMP levels in mammalian tissues: cGMP-dependent protein kinases (PKG), cyclic nucleotide-gated (CNG) ion channels, and cyclic nucleotide phosphodiesterases (PDE). The serine/threonine protein kinases PKG1 and PKG2 mediate most of the known effects of cGMP. The cytosolic PKG1 is the more widely distributed form, highly expressed in vascular smooth muscle, cerebellum, and platelets. Gene disruption of this kinase in the mouse results in vascular, intestinal, and erectile dysfunctions [1]. PKG2, abundant in intestine, bone, lung, and brain, contains a myristolated site and is localized to the plasma membrane, and gene disruption in the mouse results in resistance to Sta-induced diarrhea, intestinal secretory defects, and dwarfism [11].

The 11-gene PDE family functions to decrease signaling levels of both cGMP and cAMP and also provides a point at which both cGMP and cAMP signaling pathways can intersect. PDE5, -6, and -9 are cGMP-specific [12], cGMP-stimulated PDE-2 hydrolyzes both cGMP and cAMP, and PDE3A is a cGMP-stimulated, cAMP-specific family member.

Cyclic nucleotide-gated channels are ubiquitously expressed, the prototypical CNG channel being the photoreceptor, relatively nonselective cation channel.

Guanylyl Cyclases and Cell Growth Regulation

Although NO, ANP, and BNP appear to be important counterbalances to the renin/angiotensin/aldosterone axis in the cardiovascular system, there is considerable and convincing evidence that various guanylyl cyclases also regulate cell proliferation and differentiation. The molecular basis of such regulation remains principally at the descriptive level in that a number of growth factors, including serum, act rapidly to desensitize either GCA or GCB [1,13] and likewise, cGMP, possibly in the same manner as cAMP, inhibits growth factor activation of the MAP kinase pathway. Interestingly, this apparent adversarial relationship between various mitogens and the guanylyl cyclases appears to primarily involve the membrane forms of the enzyme [13]. hi fibroblasts, whereas serum, basic fibroblast growth factor, or platelet-derived growth factor decrease CNP-stimulated GCB activity, they fail to alter the activity of NO-stimulated guanylyl cyclase [13]. Identification of the pathway by which the various mitogens, including serum, communicate with the membrane forms of guanylyl cyclase, in particular GCA and GCB, remains unknown.

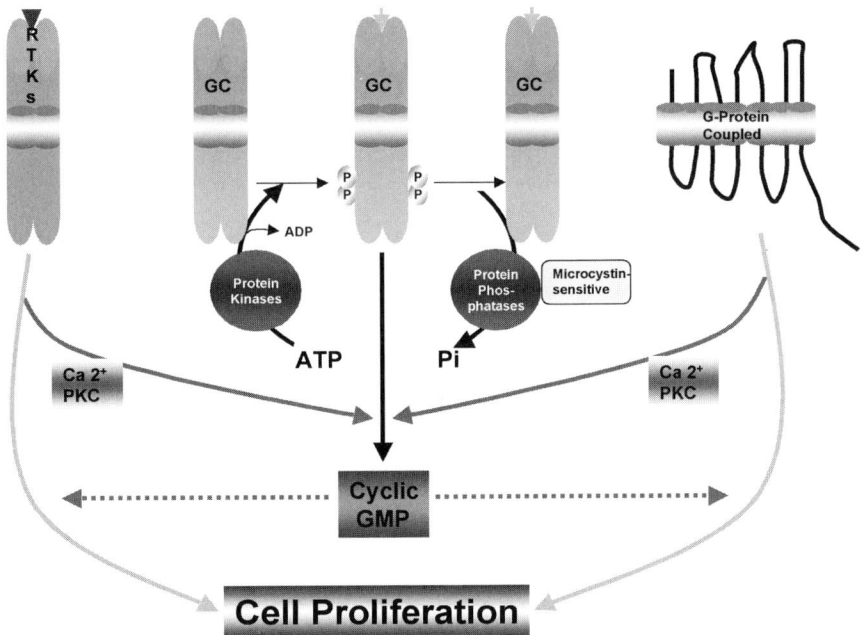

Figure 3 Schematic showing the adversarial relationship between various growth factor receptor signaling pathways and guanylyl cyclases A and B. By mechanisms not yet understood, various growth factors including serum act on receptor tyrosine kinase receptors or G-protein-coupled receptors to rapidly desensitize the ligand-stimulated forms of GCA or GCB. Both Ca^{2+} and protein kinase C(PKC) have been suggested as important components of the desensitization pathway. The dephosphorylation of the guanylyl cyclase receptor, leading in part or totally to desensitization, is mediated by a microcystin-sensitive protein phosphatase. Likewise, cGMP inhibits activation of the MAP kinase pathway in response to growth factors or serum. The subsequent rate of cell proliferation is dictated by whether the guanylyl cyclase or mitogen-stimulated pathways are dominant.

Both GCA and GCB appear to exist in a phosphorylated state in the absence of ligand, and it is the phosphorylated form that is most sensitive to the addition of ligand. Based on the work of Potter and Hunter [14], 6 principal sites of phosphorylation exist, all located within or just to the amino terminal side of the consensus protein kinase homology domain. A number of reports have suggested that dephosphorylation at these sites leads to desensitization of either GCA or GCB [14]. However, neither the protein kinase(s) nor the protein phosphatase(s) responsible for the apparent regulation of these receptors have been identified, although the protein phosphatase(s) responsible for dephosphorylation have been shown to be particularly sensitive to inhibition by microcystin [15,16] (Fig. 3). A significant number of reports suggest that activation of protein kinase C is one of the important upstream events that lead to desensitization of the natriuretic peptide receptors [14,17], where even the dephosphorylation of a single serine through activation of the protein kinase C pathway appears to desensitize GCB [18]. However, the recent work of Abbey and Potter [17] suggest that not only protein kinase C but also the other arm of the phospholipase C pathway (IPS/Ca^{2+}) is capable of causing desensitization of GCB. Their work in A10 smooth muscle cells further suggests that it is the elevations of Ca^{2+} that are required for the desensitization. Interestingly, however, the degree of desensitization obtained through elevations of Ca^{2+} alone are not equivalent to those seen through activation of the phospholipase C pathway [17]. Identification of the guanylyl cyclase protein kinase(s) and phosphatase(s) now seems essential for the understanding of the mechanisms by which the various mitogens, acting through either G-protein-coupled or receptor tyrosine kinase receptors mediate a rapid and marked decrease in guanylyl cyclase activity, where Ca^{2+} or protein kinase C may often serve as upstream regulators.

Acknowledgments

This work was supported in part by the Cecil H. & Ida Green Center for Reproductive Biology Sciences, the Cecil H. & Ida Green Distinguished Chair in Reproductive Biology Sciences, the Howard Hughes Medical Institute, and the Sandier Program for Asthma Research.

References

1. Wedel, B. and Garbers, D. (2001). The guanylyl cyclase family at Y2K. *Annu. Rev. Physiol.* **63**, 215–233.
2. Chinkers, M. and Garbers, D. L.(1989). The protein kinase domain of the ANP receptor is required for signaling. *Science* **245**(4924), 1392–1394.
3. Liu, Y., Ruoho, A. E., Rao, V. D., and Hurley, J. H. (1997). Catalytic mechanism of the adenylyl and guanylyl cyclases: Modeling and mutational analysis. *Proc. Natl. Acad. Sci. USA* **94**(25), 13414–13419.
4. Yuen, P. S., Potter, L. R., and Garbers, D. L. (1990). A new form of guanylyl cyclase is preferentially expressed in rat kidney. *Biochemistry* **29**(49), 10872–10878.
5. Sunahara, R. K., Beuve, A., Tesmer, J. J., Sprang, S. R., Garbers, D. L., and Oilman, A. G. (1998). Exchange of substrate and inhibitor specificities between adenylyl and guanylyl cyclases. *J. Biol. Chem.* **273**(26), 16332–16338.
6. Garbers, D. L. (1992). Guanylyl cyclase receptors and their endocrine, paracrine, and autocrine ligands. *Cell.* **71**(1), 1–4.
7. Suga, S., Nakao, K., Itoh, H., Komatsu, Y., Ogawa, Y., Kama, N., and Imura, H. (1992). Endothelial production of C-type natriuretic peptide and its marked augmentation by transforming growth factor-beta. Possible existence of "vascular natriuretic peptide system." *J. Clin. Invest.* **90**(3), 1145–1149.
8. Drewett, J. G. and Garbers, D. L. (1994). The family of guanylyl cyclase receptors and their ligands. *Endocr. Rev.* **15**(2), 135–162.
9. Ramamurthy, V., Tucker, C., Wilkie, S. E., Daggett, V., Hunt, D. M., and Hurley, J. B. (2001). Interactions within the coiled-coil domain of RetGC-1 guanylyl cyclase are optimized for regulation rather than for high affinity. *J. Biol. Chem.* **276**(28), 26218–26229.
10. Ames, J. B., Dizhoor, A. M., Doira, M., Palezewski, K., and Stryer L. (1999). Three-dimensional structure of guanylyl cyclase activating protein-2, a calcium-sensitive modulator of photoreceptor guanylyl cyclases. *J. Biol. Chem.* **274**(27), 19329–19337.
11. Pfeifer, A., Aszodi, A., Seidler, U., Ruth, P., Hofmann, F., and Fassler, R. (1996). Intestinal secretory defects and dwarfism in mice lacking cGMP-dependent protein kinase II. *Science* **274**(5295), 2082–2086.
12. Koyama, H., Bornfeldt, K. E., Fukumoto, S., and Nishizawa, Y. (2001). Molecular pathways of cyclic nucleotide-induced inhibition of arterial smooth muscle cell proliferation. *J. Cell. Physiol.* **186**(1), 1–10.
13. Chrisman, T. D. and Garbers, D. L. (1999). Reciprocal antagonism coordinates C-type natriuretic peptide and mitogen-signaling pathways in fibroblasts. *J. Biol. Chem.* **274**(7), 4293–4299.
14. Potter, L. R. and Hunter, T. (2001). Guanylyl cyclase-linked natriuretic peptide receptors: Structure and regulation. *J. Biol. Chem.* **276**(9), 6057–6060.
15. Foster, D. C. and Garbers, D. L. (1998). Dual role for adenine nucleotides in the regulation of the atrial natriuretic peptide receptor, guanylyl cyclase-A. *J. Biol. Chem.* **273**(26), 16311–16318.
16. Bryan, P. M. and Potter, L. R. (2002). The atrial natriuretic peptide receptor (NPR-A/GC-A) is dephosphorylated by distinct microcystin-sensitive and magnesium-protein phosphatases. *J. Biol. Chem.* **277** (18), 16041–16047.
17. Abbey, S. E. and Potter, L. R. (in press). Vasopressin-dependent inhibition of the C-type natriuretic peptide receptor, NPR-B/GC-B, requires elevated intracellular calcium concentrations. *J. Biol. Chem.*
18. Potter, L. R. and Hunter, T. (2000). Activation of protein kinase C stimulates the dephosphorylation of natriuretic peptide receptor-B at a single serine residue: A possible mechanism of heterologous desensitization. *J. Biol. Chem.* **275**(40), 31099–31106.

CHAPTER 192

Phosphodiesterase Families

Jennifer L. Glick and Joseph A. Beavo
Department of Pharmacology, University of Washington, Seattle, Washington

Introduction

The cyclic nucleotides are ubiquitous second messengers that regulate a large number of processes, including proliferation, chemotaxis, differentiation, contraction, gene transcription, and inflammation. These second messengers are produced and utilized by nearly all eukaryotes from amoebae to man. Regulation of the intracellular levels of cGMP or cAMP, both in the resting state and in response to stimuli, is therefore critical for the proper functioning and survival of many organisms. The levels of a cyclic nucleotide in the cell are determined by the relative rates of synthesis by the cyclases, adenylate cyclase, and guanylate cyclase and degradation by the phosphodiesterases (PDEs). Although the generation of cyclic nucleotide signals in many systems has been the subject of intense study, the understanding of the mechanism by which these signals are terminated by the PDEs is, in most cases, still in its infancy. This is curious as the cyclase and phosphodiesterase activities were discovered within a few years of one another [1,25].

"Cyclic nucleotide phosphodiesterase" was first described as a widely distributed enzyme that could catalyze the hydrolysis of cAMP and cGMP to their respective 5′ monophosphates [2]. The initial studies of PDE activity used either tissue homogenates or partially purified preparations of these enzymes from various tissues. The characteristics of PDE activity from these studies varied greatly depending on the PDE source. It was unclear whether these differences were a consequence of the purification scheme of these enzymes (that is, the presence of different contaminating proteins) or of the existence of multiple forms of PDE. PDEs were therefore referred to in terms of the tissue from which they were purified (for example, rat liver PDE or bovine brain PDE). Later anion exchange chromatography experiments demonstrated the presence of several PDE activities in an individual tissue or cell type. These observations were later confirmed in experiments by use of immunocytochemistry, immunoblotting, and *in situ* hybridization. With the purification of multiple enzymes to apparent homogeneity and more stringent characterization of their properties, PDEs were subsequently named according to their regulatory properties and substrate specificities (for example, calcium/calmodulin-stimulated PDE or cGMP-stimulated PDE). With the advent of molecular biology, there has been a virtual explosion of new information, including the cloning of the previously known and many new PDE genes, as well the identification of a number of new splice variants. Nucleotide sequence data for the PDEs have also allowed for the organization of them into gene families according to homology. As more data emerge regarding the distribution, characteristics, and roles of the many PDE isozymes, it is clear that the regulation of cyclic nucleotide signaling by PDEs is far more complex than could have been imagined when they were first studied in the 1960s.

The Gene Families

There currently exist 11 PDE gene families. The current nomenclature for a PDE contains, in order, two letters to indicate species, followed with a number indicating gene family, a letter to represent an individual gene, and finally a letter to identify the splice variant. For example, HSPDE7A1 represents *Homo sapiens* PDE gene family 7, gene A, splice variant 1. For a correlation between older and current PDE nomenclature, see Beavo *et al.* [3]. The kinetic properties, substrate specificities and drug sensitivities of these families (Table I) have been described extensively elsewhere [3,23,26,27] and will not be discussed at length here. In general, the phosphodiesterases share the same organizational structure. Each protein has an N-terminal domain that confers regulatory properties to the protein, followed by a more C-terminal ~270 amino acid catalytic domain and a

Table I Characteristics of the PDE Families

PDE family	Genes	# Splice variants	Regulatory domains, role	Phosphorylation	Substrate(s)	Commonly used inhibitors
PDE1	1A, 1B, 1C	9	CaM, activation	PKA	cGMP, cAMP	KS-505
PDE2	2A	3	GAF, activation	Unknown	cAMP, cGMP	EHNA
PDE3	3A, 3B	1 each	Transmembrane domains, membrane targeting	PKB	cAMP	Milrinone
PDE4	4A, 4B, 4C, 4D	>20	UCR1, UCR2, unclear	ERK, PKA	cAMP	Rolipram
PDE5	5A	3	GAF, unclear	PKA, PKG	cGMP	Sildenafil, Dipyrimadole, Zaprinast
PDE6	6A, 6B, 6C	1 each	GAF, activation	PKC, PKA	cGMP	Dipyrimadole, Zaprinast
PDE7	7A, 7B	6	Unknown	Unknown	cAMP	None Identified
PDE8	8A, 8B	6	PAS, unknown	Unknown	cAMP	None Identified
PDE9	9A	4	Unknown	Unknown	cGMP	None Identified
PDE10	10A	2	GAF, unknown	Unknown	cAMP, cGMP	None Identified
PDE11	11A	4	GAF, unknown	Unknown	cAMP, cGMP	None Identified

short C-terminal tail. The sequence identity in the catalytic domain between genes is only about 35 percent, yet all PDEs possess the signature sequence H–D–X$_2$–H–X$_4$–N [4]. The substrate specificities of the different PDE families run the gamut from dual-specificity PDEs to those that are highly specific for either cAMP or cGMP. Further, the relative substrate specificity can vary even between members of a gene family. For example, within the PDE1 family, PDE1A2 has a K_m for cGMP that is approximately 20-fold higher than that for cAMP, yet PDE1C2 has a K_m that is equal for both. In addition to variation in specificity, the activity of a PDE toward one nucleotide may depend upon the concentration of the other. For instance, PDE2s hydrolyze cAMP and cGMP with relatively similar K_m values. However, the presence of a small amount of cGMP (which binds allosterically) stimulates the enzymes' catalytic activity toward cAMP several-fold [5]. To make things more complex, there are also PDEs for which the cyclic nucleotides are competitive inhibitors for one another. Cyclic AMP is a competitive inhibitor of cGMP hydrolysis by PDE10 [6], and cGMP is a potent competitive inhibitor of cAMP hydrolysis for PDE3 [7]. This variety and flexibility in substrate specificities of the PDEs makes them a family of enzymes with tremendous diversity, suitable for the fine tuning of many cyclic nucleotide-mediated signaling systems.

As mentioned above, most of the PDEs also possess domains within their N-termini that regulate the activity of the catalytic site (Fig. 1). The PDE1 proteins have two Ca^{2+}/calmodulin binding domains, and binding of calmodulin to these PDEs stimulates their activity [8]. The PDE2, PDE5, PDE6, PDE10, and PDE11 proteins all have allosteric, cGMP-binding domains known to be part of the larger GAF domain family [9]. The consequence of binding of cGMP to these domains varies with the PDE. As discussed above, cGMP binding to PDE2 stimulates activity. For PDE5, binding of cGMP to this domain alters the protein's susceptibility to phosphorylation, but a direct effect on activity

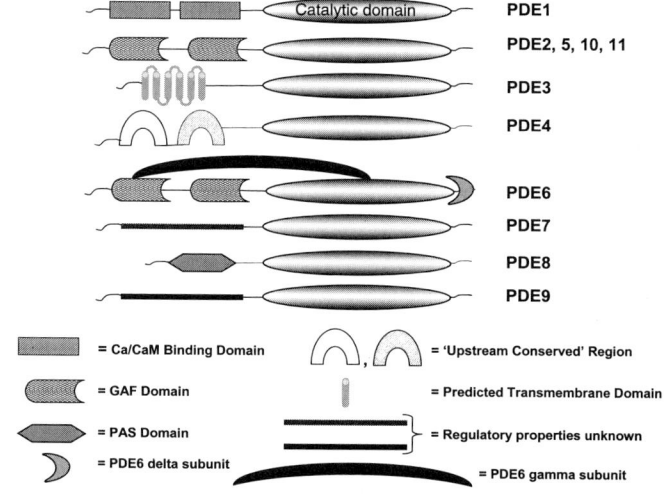

Figure 1 Domain organization of the PDE families.

has not been demonstrated. For further discussion of the GAF domains in PDEs, see Martinez, Tang et al. (Chapter 207) in this volume. The PDE3 family proteins have six predicted transmembrane segments in their amino terminal domains, consistent with the observation that PDE3 activity is at least partially membrane-associated. The PDE4 family, a large family of enzymes with four genes and many splice variants, is responsible for the majority of basal cAMP-hydrolyzing activity in most cells. For further discussion of the PDE4 family, see Conti (Chapter 71) in this volume. The N-terminus of PDE8 contains a PAS domain, a domain that generally is found in proteins that are involved in sensing and responding to the cellular environment (for example, redox state, light or energy levels) [10]. The PAS domains in many proteins bind small molecules such as heme, NAD, or chromaphores and can also serve as a site for homodimerization. It will be interesting to see whether some small molecule also binds the PDE8 PAS domain, what effect that may have on activity, and also whether the PAS domain of PDE8 serves to dimerize the protein. PDE7 and PDE9 have substantial N-terminal segments that bear no resemblance to known proteins. What role these portions of PDE7 and PDE9 have in the protein remains to be seen.

Thus, with great variation in nucleotide specificity and regulatory properties, the PDE superfamily comprises a complex set of enzymes that can provide cross talk between the cGMP and cAMP pathways, with Ca^{2+}/CaM dependent pathways, and allow the cell exquisite control of cyclic nucleotide dynamics.

Implications of Multiple Gene Families/Splice Variants

Another remarkable feature of the PDE superfamily is the expression and localization patterns of its members. Individual PDEs, within gene families, and even between splice variants, have unique expression patterns. For example, the PDE1 genes PDE1A, PDE1B, and PDE1C are all expressed in brain. However, *in situ* localization studies have shown that PDE1A is expressed primarily in the cortex, PDE1B in the striatum, and PDE1C in the cerebellum [11]. The expression pattern of PDE1C splice variants has been further broken down. PDE1C5 is highly expressed in testis; PDE1C1 is more generally distributed, found in heart, testis, cerebellum, and olfactory epithelium; and PDE1C2 is primarily expressed in the olfactory neuroepithelium.

The PDE3 genes provide another example of the complex localization of family members. PDE3A and PDE3B are generally found in distinct cell types, with PDE3A expressed in platelets and PDE3B in adipose cells and hepatocytes. Both are apparently expressed in vascular smooth muscle cells, although probably in different compartments. For further discussion of the PDE3 family, see Weston *et al.* (Chapter 72) of this volume.

The ability to produce multiple N-terminal variants allows for specific differential targeting of the PDE2 family members. The three PDE2A splice variants differ only at their extreme N-terminus. However, the PDE2A2 N-terminus contains a putative transmembrane domain, and the PDE2A3 variant contains an N-terminal myrisoylation site, both of which probably allow for targeting to membrane compartments of the cell and are responsible for the membrane associated forms of PDE2 activity observed in tissue homogenates [12].

All of the above examples of differential localization of PDE genes/splice variants imply that each of these PDEs probably plays specialized roles in the regulation of cyclic nucleotide signaling in cells. Perhaps the definitive example of precise localization of a PDE to achieve specialized function is in the case of the PDE6 gene family. In the photoreceptor cells of the retina, a visual signal is generated through the activation of a cascade of proteins that ultimately result in the activation of PDE6. PDE6 rapidly hydrolyzes the cGMP in the cell, the resident cGMP-gated cation channels close, and the cells become hyperpolarized. All of the proteins involved in this cascade, including PDE6, are expressed primarily in the retina and specifically targeted to the membrane disks of the photoreceptors. Although this PDE family is the first known and best characterized example of a highly specialized PDE, there are certainly others to follow. A prime candidate example is PDE1C2. As mentioned previously, PDE1C2 is highly expressed in the olfactory neuroepithelium. The other major PDE expressed there is PDE4A. However, expression of PDE1C2 is restricted to the cilia of the epithelium, where it co-localizes with adenylate cyclase III. PDE4A is not expressed in the cilia of the neurons, but rather throughout the remainder of the neuronal layer [13]. Clearly, in the olfactory neuroepithelium, PDE1C2 and PDE4A are playing different roles in regulating cAMP during olfaction.

Another demonstration of compartmentalization of PDE activities was accomplished by Dousa and colleagues while investigating the effects of cAMP-elevating agents on kidney mesanglial cells. In these cells, mitogenesis and superoxide production are both stimulated through the production of cAMP. Rolipram (a PDE4-specific inhibitor) inhibited the production of superoxide in these cells while having no effect on mitogenesis. Likewise, cilostamide (a PDE3-specific inhibitor) inhibited mitogenesis while having no effect on superoxide production. These results indicate that these two PDEs have access to different cAMP pools within the cell [14].

The phenomenon of differential localization of PDEs speaks to the idea of compartmentalization of cyclic nucleotide signals. Compartmentalization was originally proposed to explain how a second messenger as ubiquitous as cAMP can mediate different effects within a cell [15,20]. The idea that there are different cyclic nucleotide "pools" available only to certain kinases, channels, exchange factors, or PDEs is a compelling one. However, the details of how these compartments are established are only beginning to emerge. For the cAMP-dependent protein kinases, differential localization is accomplished through their association with anchoring proteins, AKAPs. It is becoming increasingly clear that signaling "modules" are built through the association of receptors, mediators, and targets with each other, AKAPs, and other scaffolding proteins. To that end, in recent years the AKAPs have been shown to associate with receptors, channels, G proteins, and PDEs [16,18,28,29,30]. PDEs also may have their own scaffolding partners. Recently, it was shown that PDE4D5 interacts with RACK1, a WD-repeat protein that also serves as a scaffold for PKC, Src, and integrin [17]. In addition, PDE4D interacts with myomegalin, a protein that appears to target PDE4 to the Golgi and centrosomes of cells [18]. As better tools become available to investigate PDE targeting, we are likely to see more examples of this type of compartmentalization of PDE function.

PDE Inhibitors as Therapeutic Agents

The wide variety of PDE isozymes implies that PDEs have tremendous therapeutic potential. Indeed, some of the oldest drugs man has known (for example, caffeine, ginseng) are PDE inhibitors! Some PDE inhibitors such as theophylline, papaverine, and dipyridamole were in fact used before their mechanism of action was known. However, with a new appreciation of the complexity of the PDE signaling systems and the availability of more sophisticated endpoint assays, a new generation of PDE-inhibiting drugs is being developed.

PDE inhibitors have long been known to have anti-inflammatory properties and have been used for the treatment of asthma, stroke, and chronic obstructive pulmonary disease (COPD). Early treatments for these diseases with PDE inhibitors have unfortunately been hampered by the side effect of emesis. Novel PDE4 inhibitors (Ariflo, GlaxoSmithKline; and Roflumilast, Byk Gulden) are now in clinical trial for the treatment of asthma and COPD with milder emetic side effects [19]. Milrinone, a PDE3 inhibitor that has been used to treat patients in congestive heart failure, has recently been shown *in vitro* to increase conductance of the CFTR transporter, indicating promise for the treatment of cystic fibrosis [20,31]. Cilostazol, also a PDE3 inhibitor, is currently being applied to the treatment of patients with intermittent claudication, and clinical trials suggest that cilostazol may also be useful in the prevention of restenosis after angioplasty [21,32]. Dipyridamole inhibits PDEs in platelets and is commonly used in combination with aspirin (ASA) to reduce clotting, despite the initial lack of clinical data demonstrating an added benefit of dipyridamole over ASA [22]. However, the recent European Stroke Prevention Study (ESPS2) clearly demonstrated an additive effect of dipyridamole and ASA in the prevention of second stroke [23]. Thus, dipyridamole will likely continue to be prescribed in combination with ASA. And finally, Viagra, a PDE5-specific inhibitor, has been used successfully for the treatment of male erectile dysfunction with minimal side effects.

The successful treatment of patients with specific PDE inhibitors is encouraging but not surprising given the specific, diverse distribution of many of the PDEs. In addition, the recent identification of the PDE8, PDE9, PDE10, and PDE11 should lead to the development of novel inhibitors with as yet unknown application. As we begin to fully appreciate the full complement of PDEs in a system of interest and how they are compartmentalized and regulated, new therapeutics (as well as novel application of older drugs) will no doubt be applied to a wide variety of disease states.

Where Do We Go from Here?

Although it is certainly possible that more PDEs exist that bear little resemblance in linear sequence to the currently known proteins, it is also certain that most if not all of the proteins responsible for the major PDE activities in tissues have been identified and their genes cloned. With the upcoming completion of the sequencing of multiple genomes, the identification of new PDE gene families by the "traditional" method of BLAST searching with current known sequences is probably drawing to a close. One of the new challenges in the field lies in determining exactly how many PDEs exist. Many of the newly identified splice variants are based on EST searching or RACE from cDNA libraries. It is often difficult to distinguish alternative splicing from library construction artifacts or mis-splicing. It is therefore important to confirm the existence of these variants by using a variety of techniques, including RNAse protection and immunoblotting. There are currently few good antibodies available for PDEs, especially the newly discovered ones. It will take the development of a full complement of good, sensitive antibodies specific for all PDEs and spice variants to accomplish this goal.

Another challenge in the field lies in the potential differences in PDE activities and their regulation *in vitro* versus '*in cellulo.*' Only a few of the PDEs have potent, highly selective, cell-permeable inhibitors. Therefore, current methods for studying PDE activity/regulation require either purification of the protein or overexpression in heterologous systems. Localization of PDEs is determined through subcellular fractionation or immunocytochemistry by using tissues that have been fixed and mounted to a slide. Little is known about how well these studies correlate to the situation in the cell. Removing a PDE from its native environment also removes it from other regulatory factors. It is possible that factors that mildly stimulate a PDE *in vitro* have much greater effects *in situ*. In addition, PDEs are also vulnerable to proteolysis in the *in vitro* assays, which can affect their activities. Two things are in great need. One is selective inhibitors for all the PDE families, preferably for the different splice variants as well. This need is exemplified by the discovery that the newest PDEs are insensitive to IBMX, which has been used for decades as a nonspecific PDE inhibitor. It is not clear whether the development of splice variant-specific PDE inhibitors is possible, but efforts in this area are under way. A second need is the accurate, sensitive, real-time measurement of PDE activity in live, intact cells. New techniques have already been developed for the real-time measurement of cyclic nucleotide levels in cells [24,50,51]. Although further refinement of these technologies is needed, the real-time measurement of cyclic nucleotide levels, combined with specific PDE inhibitors, will prove to be a powerful step in furthering the understanding of PDE functions in cells.

References

1. Sutherland, E. W. and Rall, T. W. (1958). *J. Biol. Chem.* **232**, 1077.
2. Cheung, W. Y. (1970). *Adv. Biochem. Psychopharmacol.* **3**, 51–65.
3. Beavo, J. A. (1995). *Physiol. Rev.* **75**(4), 725–748.
4. Charbonneau, H., Beier, N., Walsh, K. A., and Beavo, J. A. (1986). *Proc. Natl. Acad. Sci. USA* **83**(24), 9308–9312.
5. Martins, T. J., Mumby, M. C., and Beavo, J. A. (1982). *J. Biol. Chem.* **257**(4), 1973–1979.
6. Soderling, S. H., Bayuga, S. J., and Beavo, J. A. (1999). *Proc. Natl. Acad. Sci. USA* **96**(12), 7071–7076.
7. Degerman, E., Belfrage, P., and Manganiello, V. C. (1997). *J. Biol. Chem.* **272**(11), 6823–6826.
8. Kakkar, R., Raju, R. V., and Sharma, R. K. (1999). *Cell. Mol. Life Sci.* **55**(8–9), 1164–1186.

9. Aravind, L. and Ponting, C. P. (1997). *Trends Biochem. Sci.* **22**(12), 458–459.
10. Taylor, B. L. and Zhulin, I. B. (1999). *Microbiol. Mol. Biol. Rev.* **63**(2), 479–506.
11. Yan, C., Bentley, J. K., Sonnenburg, W. K., and Beavo, J. A. (1994). *J. Neurosci.* **14**(3 Pt 1), 973–984.
12. Beavo, J. A., Hardman, J. G., and Sutherland, E. W. (1971). *J. Biol. Chem.* **246**(12), 3841–3846.
13. Juilfs, D. M., Fulle, H. J., Zhao, A. Z., Housley, M. D., Garbers, D. L., and Beavo, J. A. (1997). *Proc. Natl. Acad. Sci. USA* **94**(7), 3388–3395.
14. Chini, C. C., Grande, J. P., Chini, E. N., and Dousa, T. P. (1997). *J. Biol. Chem.* **272**(15), 9854–9859.
15. Hayes, J. S., Brunton, L. L., and Mayer, S. E. (1980). *J. Biol. Chem.* **255**(11), 5113–5119.
16. Dodge, K. and Scott, J. D. (2000). *FEBS Lett.* **476**(1–2), 58–61.
17. Yarwood, S. J., Steele, M. R., Scotland, G., Houslay, M. D., and Bolger, G. B. (1999). *J. Biol. Chem.* **274**(21), 14909–14917.
18. Verde, I., Pahlke, G., Salanova, M., Zhang, G., Wang, S., Coletti, D., Onuffer, J., Jin, S. L., and Conti, M. (2001). *J. Biol. Chem.* **276**(14), 11189–11198.
19. Hele, D. J. (2001). *Respir. Res.* **2**(5).
20. Kelley, T. J., Thomas, K., Milgram, L. J., and Drumm, M. L. (1997). *Proc. Natl. Acad. Sci. USA* **94**(6), 2604–2608.
21. Eberhardt, R. T. and Coffman, J. D. (2000). *Heart Dis.* **2**(1), 62–74.
22. Gibbs, C. R. and Lip, G. Y. (1998). *Br. J. Clin. Pharmacol.* **45**(4), 323–328.
23. Redman, A. R. and Ryan, G. J. (2001). *Clin. Ther.* **23**(9), 1391–1408.
24. Zhang, J., Ma, Y., Taylor, S. S., and Tsien, R. Y. (2001). *Proc. Natl. Acad. Sci. USA* **98**(26), 14997–5002.
25. Sutherland, E. W., Rall, T. W., and Menon, T. (1962). *J. Biol. Chem.* **237**(4), 1220–1227.
26. Houslay, M. D. (2001). *Prog. Nucleic Acid Res. Mol. Biol.* **69**, 249–315.
27. Soderling, S. H. and Beavo, J. A. (2000). *Curr. Opin. Cell Biol.* **12**, 174–179.
28. Cong, M., Perry, S. J., Lin, F., Fraser, I. D., Hu, L. A., Chen, W., Pitcher, J. A., Scott, J. D., and Lefkowitz, R. J. (2001). *J. Biol. Chem.* **276**(18), 15192–15199.
29. Dodge, K. L., Khouangsathiene, S., Kapiloff, M. S., Mouton, R., Hill, E. V., Houslay, M. D., Langeberg, L. K., and Scott, J. D. (2001). *EMBO J.* **20**(8), 1921–1930.
30. Niu, J., Vaiskunaite, R., Suzuki, N., Kozasa, T., Carr, D. W., Dulin, N., and Voyno-Yasenetskaya, T. A. (2001). *Curr. Biol.* **11**, 1686–1690.
31. Smith, S. N., Middleton, P. G., Chadwick, S., Jafe, A., Bush, K. A., Rolleston, S., Farley, R., Delaney, S. J., Wainwright, B., Geddes, D. M., and Alton, E. W. F. W. (1999). *Am. J. Respir. Cell Mol. Biol.* **20**, 129–134.
32. El-Beyrouty, C. and Spinler, S. A. (2001). *Ann. Pharmacother.* **35**, 1108–1113.

The cAMP-Specific Phosphodiesterases: A Class of Diverse Enzymes That Defines the Properties and Compartmentalization of the cAMP Signal

Marco Conti
Division of Reproductive Biology, Department of Gynecology and Obstetrics, Stanford University School of Medicine, Stanford, California

The second messenger cAMP generated by adenylyl cyclases either interacts with and activates intracellular effectors or is degraded and inactivated by phosphodiesterases (PDEs). Of the eleven PDE families thus far identified, three families of isoenzymes are specific for cAMP, including family 4, 7, and 8 (see chapter *xx*. for the nomenclature). Given their high affinity for cAMP in the submicromolar to micromolar range, they are the primary enzymes involved in the inactivation of this cyclic nucleotide under resting or stimulated conditions, thus playing a critical role in signaling.

Isoenzymes that belong to the PDE4 family were the first to be identified on the basis of their pharmacological and biochemical properties. Indeed, rolipram, a PDE inhibitor developed in the fifties as an antidepressant that specifically targets PDE4, has been of tremendous value in dissecting the properties and functions of PDE4. Because of their potential therapeutic use in inflammatory disorders, drugs specific for PDE7 and PDE8 are also under development.

The cloning of the PDE4 genes by virtue of their homology with the *Drosophila* PDE [1] was soon followed by the identification of the other two families using either a strategy of complementation of yeasts deficient in phosphodiesterases for PDE7 [2], or database homologous searches for PDE8 [3]. It is now established that four genes code for the PDE4s, whereas two genes each encode the PDE7s and PDE8s. The presence of multiple genes represents only a first layer of complexity in the cAMP-PDE enzymes, as multiple transcripts originate from these genes either by alternate splicing or the use of different promoters, greatly expanding the number of cAMP-PDE proteins expressed in mammalian cells. Although the number likely will be corrected upward, 15–18 different PDE4, 4 different PDE7, and potentially 6 PDE8 proteins thus far have been identified [4]. A total of 28 different proteins specifically hydrolyze cAMP in the cell (Fig. 1). Given this extreme heterogeneity, it has been proposed that cAMP-PDEs have a modular structure with

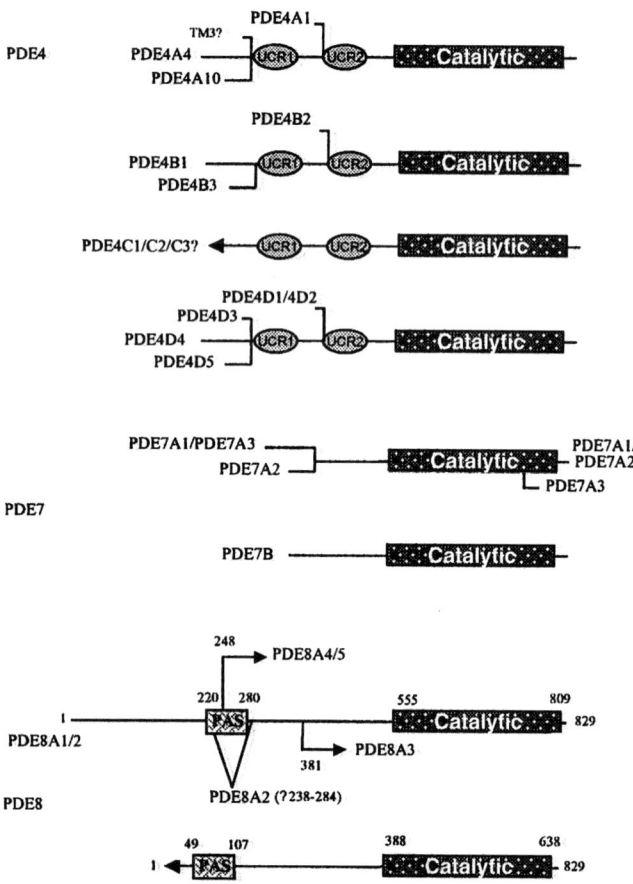

Figure 1 Schematic representation of the different cAMP-PDE expressed in human cells. The gene family is reported on the left. The open reading frame is indicated as a continuous line with the catalytic region indicated as a box. The point of divergence of a splicing variant is indicated by a bracket. The conserved UCR1 and UCR2 regions present in all PDE4 are designated as ovals. The PAS domain in PDE8 is indicated as a cross-hatched box.

different regulatory cassettes specific for each variant, and that cells use distinct cAMP-PDE proteins with subtle differences in their properties to adapt cAMP signaling to their specialized functions.

Structure of the cAMP-PDEs: Catalytic and Regulatory Domains

Cyclic AMP-PDEs are proteins with molecular weights ranging between 50 and 130 kDa and are composed of a catalytic domain surrounded by domains with regulatory functions. Deletion and site-directed mutagenesis, as well as analysis of sequence conservation, have defined the boundaries of the catalytic domain and mapped it to the carboxyl-terminus end of the cAMP-PDE. The structure of this domain in a PDE4 has been recently resolved at the atomic level, demonstrating a compact globular bundle of 17 helices subdivided in 3 subdomains that define a catalytic pocket [5]. Two metal ion binding sites are present in this pocket: one of these metal binding sites is thought to be permanently occupied by $Zn^{\wedge}$, whereas binding of the second metal ion at the second site, likely Mg^{2+}, can be rapidly exchanged. Both metals are essential for substrate binding and for the catalytic activity of PDE4 and presumably for the other cAMP PDEs. Changes in affinity for $Mg^{\wedge}$ follow posttranslational modification of PDE4 and may impact the hydrolytic capacity of the enzyme [6].

Domains flanking the catalytic domain of the cAMP-PDEs have been identified by different strategies and can roughly be divided into regulatory and targeting domains. Two regions highly conserved from worm and fly to mammals are present at the amino terminus of the PDE4 long forms and are termed upstream conserved regions 1 and 2 (UCR1/UCR2). These domains, signatures for PDE4, serve regulatory functions because a phosphorylation site for PKA has mapped to the amino-terminus end of UCR1. Yeast two-hybrid or pulldown assays have indicated that UCR1 and UCR2 interact with each other and that phosphorylation modulates this interaction [7].

Targeting domains have been identified in several PDE4s as well as at the aminoterminus of PDE7 (see the following). In addition, motifs corresponding to a PKA pseudosubstrate are present at the amino terminus of PDE7A1 and PDE7A2 [2]. Although their function has not been confirmed, they may play a role in the regulation of PKA. Finally, PDE8 contains a PAS domain likely involved in protein/protein interaction [3].

Subcellular Targeting of the cAMP-PDEs and cAMP Signal Compartmentalization

The cAMP PDEs are not uniformly distributed throughout the cell. Conversely, they are targeted to discrete compartments via mechanisms that are only partially known. This nonrandom distribution of PDEs is described, for instance, in the olfactory sensory neurons. The cAMP PDE 4A is present in the body of the neuron in a region surrounding the nucleus, whereas a PDE1 is targeted to the cilia, suggesting that the two enzymes control different cAMP pools [8].

Targeting domains have been identified in PDE4s, and in most instances they coincide with domains that mediate protein/protein interaction. The interaction of a PDE4D5 with RACK-1 [9], a scaffold protein initially identified as a PKC binding protein, is the best characterized, even though the significance of this interaction is unclear. The localization of PDE4D to the Golgi/centrosome structures is probably mediated by the PDE4 interaction with myomegalin, a large coiled-coil protein discovered by yeast two-hybrid screening [10]. Interestingly, in skeletal muscle where it is expressed at high levels, myomegalin colocalizes with PDE in a region where β adrenergic receptors, adenylyl cyclases, PKAs, and AKAPs are also localized [11,12]. A PDE4D3 is targeted to the centrosome or the perinuclear region through interaction with AKAP450 or mAKAP, two scaffold proteins that also bind PKA (see the following). PDE4A5 binds to SH3 domain-containing proteins and is localized at the plasma membrane [13]. An additional mechanism of targeting may be dependent on the direct interaction with the lipid bilayer, as the amino terminus of PDE4A1 is highly hydrophobic and is sufficient for membrane localization of this PDE isoform [14]. One of the splicing variants of PDE7, PDE7A2, contains a unique hydrophobic amino terminus which likely targets it to insoluble structures of the skeletal muscle [15].

The physiological consequence of this distribution of PDEs in different cell districts is still a matter of debate. However, in one case, it is clear that interaction of PDE with scaffold protein serves to create a signaling unit with PKA. The PDE4D3 variant co-immunoprecipitates with the RII regulatory subunit of PKA and with two AKAPs, indicating the presence of a PKA/PDE signaling complex organized on the AKAP scaffold [16,17]. AKAP350/450, also termed Yotiao, is a true signaling scaffold because it complexes PKA, a PDE, PP1, and in some cases, is anchored to receptors [18]. Given the rapid phosphorylation and activation of this PDE isoform by PKA, this colocalization allows the rapid termination of the cAMP signal in a discrete domain of the cell.

Regulation of cAMP-PDEs

The activity of cAMP-PDEs is tightly regulated and integrated in several signaling pathways. Two major mechanisms of regulation have been described. These include regulation by phosphorylation and regulation at the level of transcription/translation of the genes [19]. In addition, interaction with other proteins or with lipids may affect the conformation and activity of these enzymes, even though this area of research is in its infancy.

Regulation of cAMP PDEs by Phosphorylation: Feedback Regulation of cAMP

Using a thyroid cell line as a model of PDE regulation, it was demonstrated that TSH rapidly activates a PDE4D and that this activation is dependent on cAMP accumulation and mediated byPKA [4]. The phosphorylation sites in PDE4D3 have been mapped by site-directed mutagenesis and by phosphopeptide analysis. These data, together with experiments in intact cells, have established that a short-term feedback regulation of PDE4 is operating in the cell to dampen or maintain cAMP levels within a narrow range of concentrations. PKA phosphorylation sites are also present in PDE7, though their significance is unclear.

In addition to a PKA-mediated phosphorylation at the amino terminus, all PDE4s, except PDE4A, are phosphorylated by MAPK at a carboxyl-terminus site [20]. This phosphorylation decreases the activity of the long forms or increases that of the short forms. These effects are overridden by the PKA-mediated phosphorylation. It has been proposed that this regulation is a means to terminate the MAPK kinase activation.

Other kinases may use PDE4s as substrates, as an increase in PDE4A activity follows S6 kinase activation by GH or monocytic cell line activation by lipopolysaccharides (LPS) [13]. PDE4B may be the target of kinases functioning in the T-cell-receptor-activated pathway [13].

Regulation of cAMP PDE Expression During Cell Adaptation and Differentiation

In the seventies, it was observed that changes in cAMP produced an increase in PDE activity and that this increase required protein synthesis. The cloning of PDE4 provided the tools to demonstrate that the regulation of transcription of the PDE4D gene and mRNA accumulation are regulated by cAMP. hi the Sertoli cell of the testis, FSH, which increases cAMP, produces more than a 100-fold increase in PDE4D mRNA and accumulation of PDE4D2 protein. This upregulation and consequent increase in cAMP hydrolysis contributes to the state of desensitization that follows hormonal stimulation [19]. Similar findings have been reported for T cells and several cell lines. PDE4B is subject to a similar cAMP-mediated regulation, eventhough both transcription and mRNA stabilization may contribute to the regulation of the corresponding mRNA [4]. This long-term feedback regulation of cAMP-PDE may play a particularly important role in neuronal cell adaptation and in gonadal cell function. This is inferred by recent observations on the phenotype of mice deficient in a PDE4 where several functions including fertility and behavioral effects follow inactivation of PDE4D [13,21].

An increase in PDE4 expression is also associated with the activation of other signaling pathways, including activation of T-cell receptors by mechanisms that are mostly unknown. In the same vein, PDE7 and PDE8 expression are induced by activation of T lymphocytes by CD3/CD28 antibodies [22]. This induction seems to be critical for the activation of these cells because PDE7 mRNA antisense treatment blocks replication and IL-2 production. PDE4B2 expression is induced in monocytes by LPS activation of the Toll-like receptor pathway [23], a regulation that may be critical for cytokine production, since PDE4 inhibitors block TNF-α production. These regulations in immune cells provide the rationale for the development of cAMP-PDE inhibitors for the treatment of inflammatory disorders.

The above summarized properties and regulations indicate that cAMP-PDEs play an important role in cAMP signaling as well as acting as integrators of multiple signaling pathways. Although much needs to be done to fully understand their functions, these PDEs should be regarded as homeostatic regulators of signaling.

References

1. Conti, M., Jin, S. L., Monaco, L., Repaske, D. R., and Swinnen, J. V. (1991). Hormonal regulation of cyclic nucleotide phosphodiesterases. *Endocr. Rev.* **12**, 218–234.
2. Michaeli, T., Bloom, T. J., Martins, T., Loughney, K., Ferguson, K., Riggs, M., Rodgers, L., Beavo, J. A., and Wigler, M. (1993). Isolation and characterization of a previously undetected human cAMP phosphodiesterase by complementation of cAMP phosphodiesterase-deficient Saccharomyces cerevisiae. *J. Biol. Chem.* **268**, 12925–12932.

3. Soderling, S. H., Bayuga, S. J., and Beavo, J. A. (1998). Cloning and characterization of a cAMP-specific cyclic nucleotide phosphodiesterase. *Proc. Natl. Acad. Sci. USA* **95**, 8991–8996.
4. Conti, M. and Jin, S. L. (1999). The molecular biology of cyclic nucleotide phosphodiesterases. *Prog. Nucleic Acid Res. Mol. Biol.* **63**, 1–38.
5. Xu, R. X., Hassell, A. M., Vanderwall, D., Lambert, M. H., Holmes, W. D., Luther, M. A., Rocque, W. J., Milbum, M. V., Zhao, Y., Ke, H., and Nolte, R.T. (2000). Atomic structure of PDE4: Insights into phosphodiesterase mechanism and specificity. [In Process Citation]. *Science*, 1822–1825.
6. Alvarez, R., Sette, C., Yang, D., Eglen, R., Wilhelm, R., Shelton, E. R., and Conti, M. (1995). Activation and selective inhibition of a cyclic AMP-specific phosphodiesterase, PDE4D3. *Mol. Pharmacol.* **48**, 616–622.
7. Lim, J., Pahlke, G., and Conti, M. (1999). Activation of the cAMP-specific Phosphodiesterase PDE4D3 by Phosphorylation. Identification and function of an inhibitory domain. *J. Biol. Chem.* **274**, 19677–19685.
8. Juilfs, D. M., Fulle, H. J., Zhao, A. Z., Housley, M. D., Garbers, D. L., and Beavo, J. A. (1997). A subset of olfactory neurons that selectively express cGMP- stimulated phosphodiesterase (PDE2) and guanylyl cyclase-D define a unique olfactory signal transduction pathway. *Proc. Natl. Acad. Sci. USA* **94**, 3388–3395.
9. Yarwood, S. J., Steele, M. R., Scotland, G., Houslay, M. D., and Bolger, G. B. (1999). The RACK1 signaling scaffold protein selectively interacts with the cAMP-specific phosphodiesterase PDE4D5 isoform. *J. Biol. Chem.* **274**, 14909–14917.
10. Verde, I., Pahlke, G., Salanova, M., Zhang, G., Wang, S., Coletti, D., Onuffer, J., Jin, S. L., and Conti. M. (2001). Myomegalin is a novel protein of the golgi/centrosome that interacts with a cyclic nucleotide phosphodiesterase. *J. Biol. Chem.* **276**, 11189–11198.
11. Gao, T., Puri, T. S., Gerhardstein, B. L., Chien, A. J., Green, R. D., and Hosey, M. M. (1997). Identification and subcellular localization of the subunits of L-type calcium channels and adenylyl cyclase in cardiac myocytes. *J. Biol. Chem.* **272**, 19401–19407.
12. Yang, J., Drazba, J. A., Ferguson, D. G., and Bond, M. (1998). A-kinase anchoring protein 100 (AKAP100) is localized in multiple subcellular compartments in the adult rat heart. *J. Cell. Biol.* **142**, 511–522.
13. Houslay, M. D. (2001). PDE4 cAMP-specific phosphodiesterases. *Prog. Nucleic Acid Res. Mol. Biol.* **69**, 249–315.
14. Smith, K. J., Scotland, G., Beattie, J., Trayer, I. P., and Houslay, M. D. (1996). Determination of the structure of the N-terminal splice region of the cyclic AMP-specific phosphodiesterase RD1 (RNPDE4A1) by 1H NMR and identification of the membrane association domain using chimeric constructs. *J. Biol. Chem.* **271**, 16703–16711.
15. Han, P., Zhu, X., and Michaeli, T. (1997) Alternative splicing of the high affinity cAMP-specific phosphodiesterase (PDE7A) mRNA in human skeletal muscle and heart. *J. Biol. Chem.* **272**, 16152–16157.
16. Tasken, K. A., Collas, P., Kemmner, W. A., Witczak, O., Conti, M., and Tasken, K. (2001) Phosphodiesterase 4D and protein kinase a type II constitute a signaling unit in the centrosomal area. *J. Biol. Chem.* **276**, 21999–22002.
17. Dodge, K. L., Khouangsathiene, S., Kapiloff, M. S., Mouton, R., Hill, E. V., Houslay, M. D., Langeberg, L. K., and Scott, J. D. (2001). mAKAP assembles a protein kinase A/PDE4 phosphodiesterase cAMP signaling module. *EMBO J.* **20**, 1921–1930.
18. Scott, J. D., Dell'Acqua, M. L., Fraser, I. D., Tavalin, S. J., and Lester, L. B. (2000). Coordination of c AMP signaling events through PKA anchoring. *Adv. Pharmacol* **47**, 175–207.
19. Conti, M., Nemoz, G., Sette, C., and Vicini, E. (1995). Recent progress in understanding the hormonal regulation of phosphodiesterases. *Endocr. Rev.* **16**, 370–389.
20. Hoffmann, R., Baillie, G. S., MacKenzie, S. J., Yarwood, S. J., and Houslay, M. D. (1999). The MAP kinase ERK2 inhibits the cyclic AMP-specific phosphodiesterase HSPDE4D3 by phosphorylating it at Ser579. *EMBO J.* **18**, 893–903.
21. Conti, M. (2000). Phosphodiesterases and cyclic nucleotide signaling in endocrine cells. *Mol. Endocrinol.* **14**, 1317–1327.
22. Glavas, N. A., Ostenson, C., Schaefer, J. B., Vasta, V., and Beavo, J. A. (2001). T cell activation up-regulates cyclic nucleotide phosphodiesterases 8A1 and 7A3. *Proc. Natl. Acad. Sci. USA* **98**, 6319–6324.
23. Wang, P., Wu, P., Ohleth, K. M., Egan, R. W., and Billah, M. M. (1999). Phosphodiesterase 4B2 is the predominant phosphodiesterase species and undergoes differential regulation of gene expression in human monocytes and neutrophils. *Mol. Pharmacol.* **56**, 170–174.

CHAPTER 194

cAMP/cGMP Dual-Specificity Phosphodiesterases

Marie C. Weston,[2] Lena Stenson Holst, Eva Degerman,[1] and Vincent C. Manganiello[1,2]

[1]*Department of Cell and Molecular Biology, Lund University, Lund, Sweden and*
[2]*Pulmonary-Critical Care Medicine Branch, National Heart, Lung, and Blood Institute, National Institutes of Health, Bethesda, Maryland*

Introduction

The mammalian PDE superfamily contains at least 11 functionally distinct, highly regulated, and structurally related gene families. PDE families differ in their primary sequences, substrate affinities and catalytic properties, sensitivity to effectors and inhibitors, responses to regulatory molecules, and cellular functions [1–5]. Intracellular cAMP and cGMP pools are tightly regulated and seem to be temporally, spatially, and functionally compartmentalized [6]. Most cells contain representatives of more than one PDE gene family (and different variants of the same family) but in different amounts, proportions, and subcellular locations. By virtue of their distinct intrinsic characteristics, their intracellular targeting to different subcellular locations, and their interactions with molecular scaffolds, cellular structural elements, and regulatory partners, different PDEs integrate multiple cellular inputs and modulate the intracellular diffusion and functional compartmentalization, as well as the amplitude, duration, termination, and specificity of cyclic nucleotide signals and actions [6–8]. PDEs are critical determinants of the unique cellular phenotypes (or "fingerprints") that characterize cyclic nucleotide signaling pathways.

Some PDE families are rather specific for cAMP hydrolysis (PDEs 4,7,8); others are cGMP-specific (PDEs 5,6,9); some hydrolyze both cGMP and cAMP (PDEs 1,2,3,10,11).

This chapter will briefly discuss various aspects of the molecular diversity, structure/function, regulation, and functions of these dual-specificity PDEs.

The catalytic core of mammalian PDEs (~270 amino acids near the carboxy terminus) is more highly conserved among members of the same gene family than between different gene families [1,9]. This core contains common structural elements responsible for hydrolysis of cAMP and cGMP, as well as family-specific sequences responsible for differences in substrate affinities, catalytic properties, and sensitivities to family-selective inhibitors, such as SCH51866 and methoxymethyl-isobutylxanthine (PDE1), EHNA (erythro-9-(2-hydroxy-3-nonyl)-adenine) (PDE2), and milrinone, cilostazol, and cilostamide (PDE3) [1,9]. Selective inhibitors—that is, drugs that target individual PDE families with 10 to 100-fold greater potency than for other PDE families—are not yet available for the more recently identified PDEs 10 and 11, although dipyridamole (IC_{50}, ~0.37 µM) is a potent inhibitor of PDE11.

Widely divergent N-terminal portions of PDEs contain information that determine responses of different PDEs to specific regulatory signals, such as binding sites for Ca^{2+}/calmodulin and autoinhibitory domains (PDE1), membrane-targeting domains (PDEs 2,3), and sites for phosphorylation by cAMP-, cGMP- and Ca^{2+}/calmodulin-dependent protein kinases, protein kinase B (PKB/Akt), and protein kinase C (PDEs 1,2,3,10,11) [1–5].

PDE2s contain two homologous, noncatalytic, cGMP-binding regions upstream of the catalytic core. These domains, conserved in PDE5, PDE6, PDE10, and PDE11 and a wide variety of proteins, are referred to as GAF domains (because of proteins containing these domains: cGMP-binding PDEs, Anabena adenylyl cyclase, and the *E. coli* transcriptional regulator, fhl A) [5,10]. In PDE2, PDE5, and PDE6, GAF domains bind cGMP with high affinity but with different functional consequences. Binding of cGMP to GAF domains results in allosteric activation of PDE2, enhances PKG-induced phosphorylation/activation of PDE5, and enhances interactions between PDE6 catalytic and inhibitory subunits and transducin. Functional consequences of interactions with GAF domains in PDEs 10 and 11 are unknown. Since the K_d for cGMP binding to the PDE10A GAF domains (>9 μM) is much higher than physiological cGMP concentrations, cGMP binding is probably not the primary function of these GAF domains [5,10]. Other small molecules can bind GAF domains; for example, fhl A binds formate in an N-terminal region containing two GAF domains [11].

PDE1 (Ca^{2+}/Calmodulin-dependent PDE)

Of the three PDE1 subfamilies (1A–C), PDE1A and PDE1B have higher affinity for cGMP (K_m ~5 μM and 2.7 μM, respectively) than for cAMP (K_m ~113 μM and 24 μM, respectively) [15], whereas PDE1C has high affinity for both cAMP and cGMP (K_m ~1 μM). PDEs1B and 1C hydrolyze cAMP and cGMP at similar rates, with V_{max} for cAMP by PDE1A twice that for cGMP [15]. There are five PDE1A [16,17], two PDE1B [17,18], and five PDE1C [15,19] splice variants; N-terminal, not C-terminal, diversity apparently accounts for functional and regulatory differences. Alternative splicing generates structural changes in calmodulin-binding domains, with PDE1A1 and PDE1C2 having greater affinity for calmodulin than PDE1A2 [20] or other PDE1C isoforms [15], respectively. *In vitro* phosphorylation of PDE1A1 and PDE1A2 by PKA [21,22], or PDE1B by calmodulin-dependent protein kinase II [23], renders the enzymes less sensitive to Ca^{2+}/calmodulin. PDE1A possesses an N-terminal PEST recognition motif for m-calpain which generates an activated cleavage fragment that is independent of calmodulin [24] and could provide an alternative intracellular mechanism for cyclic nucleotide regulation.

PDE1s may mediate cross-talk between Ca^{2+}, lipid, and cyclic nucleotide signals. Stimulation of CHO cells with PMA or agonists that activate lipid-signaling pathways, or transfection with specific PKC isoforms, rapidly induces PDE1 activity and expression [25,26]. In mammalian brain, PDE1 is relatively highly expressed, with distinct isozyme-specific distributions. PDE1B distribution in the striatum correlates with that of dopamine receptors, inferring a role in modulating dopamine-mediated cAMP signaling [27]. PDE1C2 and adenylyl cyclase AC3 colocalize to olfactory sensory neuronal cilia that extend to nasal epithelium, where they may regulate transient cAMP responses to odorants [27].

PDE1A upregulation has been implicated in developing tolerance to vasodilator effects of chronic nitroglycerin treatment [28]. Quiescent smooth muscle cells (SMCs) from intact normal human aorta express PDEs 1A and 1B, whereas proliferating SMCs in human arterial primary cultures and cultures from atherosclerotic lesions contain high levels of PDE1C(12).

In pancreatic islets [29,30] and cultured β-cell lines [31], PDE1, especially PDE1C, may regulate cAMP and glucose-induced insulin secretion, since PDE1 inhibition augments glucose-stimulated insulin release from βTC3 cells and exposure to glucose activates PDE1 [31].

PDE1 may be elevated in certain tumors [32]. PDE1B1 is induced in activated T cells and expressed in human lymphoblastoid cell lines but not in normal, quiescent, peripheral blood lymphocytes [33]. PDE1 inhibitors, 8-methoxymethyl-IBMX and vinpocetine, can attenuate IL-13 production and induce apoptosis of lymphoma cells [34], suggesting the potential for PDE1-targeted therapy of leukemia and inflammatory disorders.

PDE2 (cGMP-stimulated PDE)

Unique N-terminal regions in three variants of the single PDE2A gene, most likely generated via alternative exon splicing [4,35–41], may localize PDE2 to soluble (PDE2A1) and particulate (PDE2A2, PDE2A3) subcellular fractions. In general, PDE2A mRNA expression is similar in human [38], rat [42], and bovine [39] tissues, with relatively high levels in brain and intermediate levels in heart, liver, skeletal muscle, and pancreas.

PDE2 isoforms exhibit similar catalytic properties, hydrolyzing cAMP (K_m, ~30 μM) and cGMP (K_m, ~10 μM) at similar maximal velocities (100–160 μmol/min/mg) with positively cooperative kinetics. At physiological concentrations, cGMP is the preferred substrate and effector for PDE2, and can stimulate PDE2-mediated cAMP hydrolysis up to 50-fold with a K_d ~0.35–0.5 μM. By interaction with non-catalytic cGMP-binding regions in the two GAF domains upstream of the catalytic domain [43], cGMP induces a conformational change and converts PDE2 from a ligand-free state, displaying low affinity for cAMP and positively cooperative or sigmoidal kinetics, to an activated high-affinity conformation, displaying classical Michaelis-Menton kinetics without any change in V_{max}, [36]. Because of this unique allosteric regulation, PDE2 is referred to as the cGMP-stimulated cAMP PDE. At higher concentrations, cGMP inhibits cAMP hydrolysis, due to competition at the catalytic site.

Physiologically, PDE2 is a locus for cross-talk between cAMP- and cGMP-signaling pathways and may play an important role in cells where cAMP and cGMP regulate opposing functions. Regulation of PDE2 by cGMP may be more important when intracellular cAMP is elevated, rather than in the basal state [4,36]. The relatively recent development of a selective PDE2 inhibitor, EHNA [44], has been critical in elucidation of putative pathways regulated by PDE2s. However, although EHNA potently inhibits PDE2 (IC50, ~0.8–2 μM, at least 50-fold less than for other PDEs), it also inhibits adenosine deaminase.

In heart and platelets of different animal species, effects of NO and cAMP and cGMP seem to be regulated by the interplay of different PDEs, especially PDE2, PDE3, and PDE5. In frog ventricular myocytes and human atrial myocytes, NO-induced activation of guanylyl cyclase increases cGMP, which can activate PDE2, resulting in increased hydrolysis of cAMP and decreased L-type channel Ca^{2+} current I_{ca} [45–48]. PDE2 may be located in the same subcellular compartment as PKA and the L-type channel [49], and might specifically regulate L-type channel phosphorylation. In rabbit platelets, nitrovasodilators and prostacyclins act synergistically to inhibit platelet aggregation via increased cAMP, generated by NO-induced activation of guanylyl cyclase, cGMP accumulation, and subsequent inhibition of cAMP hydrolysis by PDE3 [50]. In human platelets, however, although PDE3 may be important in the absence of nitrovasodilators or at low cAMP concentrations, increases in cAMP seem to be restricted by cGMP-induced activation of PDE2 [51]. [Stimulation by nitrovasodilators, via increases in cGMP, results in inhibition of PDE3 as well as activation of PDE2, resulting in net cAMP hydrolysis. Thus, clinical effectiveness of nitrovasodilators as platelet inhibitors, which could be compromised by activation of PDE2, may be enhanced by PDE2 inhibitors [50–52].

In cultured bovine adrenal glomerulosa cells, ANP inhibits ACTH-stimulated aldosterone secretion, presumably via cGMP-induced activation of PDE2 and hydrolysis of cAMP [53]. PDE2 may also be important in regulation of phytohemaglutinin-induced activation (but not by anti CD3) of the T-cell receptor [54], the hypoxic pressor response in rat lung [55], NO-induced inhibition of prolactin release from the pituitary gland [56], NaCl absorption by the rat thick ascending tubule [57], olfactory neuroepithelial signaling [58,59], and effects of NGF on cAMP-induced PC12 cell differentiation [60,61].

PDE3 (cGMP-inhibited cAMP PDE)

Both PDE3 subfamilies, PDE3A and PDE3B, generated from separate genes located on chromosome 11 and 12, respectively [62,63], hydrolyze cAMP and cGMP with high affinity (K_m values of ~0.1–0.8 μM) in a mutually competitive manner, with V_{max} for cAMP higher (~4–10-fold) than for cGMP [14,64]. Both are specifically inhibited by compounds such as milrinone, enoximone, and cilostazol [64].

PDE3A and B exhibit cell-specific differences in expression. PDE3B is relatively highly expressed in cells important in energy metabolism, such as white and brown adipocytes, pancreatic β cells, and hepatocytes [64–67]; PDE3A is highly expressed primarily in the cardiovascular system, for example platelets, smooth muscle, and cardiac myocytes [64,68]. Full-length PDE3s (Mw~135 kDA) are found in association with membranes; smaller PDE3A forms are found in cytosolic fractions [68].

The structural organization of PDE3A and PDE3B proteins is identical [14] with the catalytic domain found in all PDEs located in the C-terminal portions of the molecules. The catalytic domains of PDE3A and B are highly conserved, except for an insertion of 44 unique amino acids that is not found in the catalytic domains of other PDE families and that also differs in, and thus distinguishes, PDE3A and B isoforms. PDE3A and B N-terminal portions are quite divergent, consisting of large hydrophobic regions containing 5 to 6 transmembrane helical segments and separated from the catalytic domain by a regulatory domain with consensus sites for phosphorylation by PKA and PKB [14,64].

Studies with specific, cell-permeable, PDE3 inhibitors (some of which are used in clinical situations [64,65]) suggest that PDE3s regulate several important physiological functions, including adipose tissue lipolysis, insulin secretion, platelet aggregation, vessel relaxation, cardiac function, oocyte maturation, and cell proliferation [14,64]. In intact cells, insulin/IGF-1 and cAMP-increasing agents activate (and presumably phosphorylate) PDE3s in adipocytes [14], hepatocytes [67,69], platelets [64], oocytes [70], and pancreatic β cells [71].

In adipocytes, insulin-induced phosphorylation and activation of membrane-associated PDE3B, via insulin receptor substrates (IRS)/phosphatidylinositol-3 kinase (PI3K)/PKB signaling pathways [72–77], is a major mechanism whereby insulin acutely antagonizes catecholamine-induced lipolysis and release of fatty acids. Activation of PDE3B leads to increased degradation of cAMP and consequent lowering of PKA activity, net dephosphorylation of hormone-sensitive lipase, and reduced lipolysis [78]. Protein phosphatase 2A catalyzes PDE3B dephosphorylation [79]. cAMP-elevating hormones also phosphorylate and activate PDE3 in adipocytes and several other cells, including platelets, as a feedback mechanism to limit excess production of cAMP.

Several studies suggest that reduced PDE3B gene expression in adipocytes is associated with insulin resistance. PDE3B activity and gene expression are decreased in adipose tissue from JCR: LA-cp rats, a strain that develops obesity, insulin resistance, and vasculopathy [80], and from obese, insulin-resistant, diabetic KKAy mice [81,82]. Administration of pioglitazone to KKAy mice increased adipocyte PDE3B expression and restored its responsiveness to insulin. Long-term incubation (~24 hr) of 3T3-L1 adipocytes with TNFα and ceramides results in increased lipolysis that is associated with downregulation of PDE3B; troglitazone reverses the effect on PDE3B [83]. Whether downregulation of PDE3B is important in development of TNFα-induced insulin resistance is not certain.

In pancreatic islets, insulinoma cells, and clonal β cells [30,31,71,84,85], PDE3, most likely PDE3B, seems to be important in regulation of insulin secretion. Activation of PDE3B by insulin-like growth factor I (IGF-1) [71] or leptin [88] attenuates insulin release stimulated by the cAMP-elevating hormone glucagon-like peptide 1 (GLP-1).

Selective PDE3 inhibitors increased plasma insulin in normal rats [86,87]. In pancreatic islets and insulinoma cells, they enhanced insulin secretion stimulated by glucose [30,84–86], indicating that PDE3-mediated control of cAMP may also be important for nutrient-stimulated release of insulin (although inhibition of the cAMP/PKA pathway has been reported not to adversely affect glucose-induced release [89,90]).

In hepatocytes, activation of PDE3B is thought to be important in the antiglycogenolytic actions of insulin and leptin [69,91]. In frog *(Xenopus laevis)* and murine oocytes and in intact rodents, specific PDE3, not PDE4, inhibitors blocked oocyte maturation [92,93]. In the frog oocyte, IGF-1-induced meiotic maturation is associated with PKB-induced activation of oocyte PDE3A and reduction in cAMP [70]. Taken together, these studies suggest that activation of PDE3B (perhaps via PI3-K/PKB-signaling) is important in counterregulatory actions of insulin, IGF-1, leptin, and other cytokines on certain cAMP-mediated processes.

PDE10

Two variants of the single PDE10 gene, PDE10A1 and PDE10A2, with unique N-termini but otherwise identical sequences, and several rat variants, PDE10A3, 10A4, 10A5, and 10A6, have been cloned and characterized [94–97]. PDE10 activities were detected in extracts of rat striatum and testis [98,99]; human PDE10A1 transcripts, in brain, testis, heart, kidney, lung, liver, and pancreas; PDE10A2 in brain, kidney, and placenta. Recombinant PDE10 hydrolyzes both cAMP (K_m~0.05–0.26 µM) and cGMP (K_m~3–8 µM), with V_{max} for cGMP ~ five-fold higher than cAMP. Since cGMP hydrolysis is potently inhibited by cAMP, PDE10 may serve as a cAMP-inhibited cGMP PDE *in vivo*, but characterization of PDE10 function(s) awaits development of specific inhibitors.

Two PDE10 N-terminal GAF domains, although similar to GAF domains in PDE2, PDE5, and PDE6, are probably not important in functional cGMP-binding [96]. The unique N-terminal region of PDE10A2 contains a consensus site for PKA phosphorylation, but effects of phosphorylation on activity in intact cells and *in vitro* are unknown.

PDE11

Several rat [100] and human [100–102] splice variants of the single PDE11 gene, with unique N-termini, have been cloned and characterized. Recombinant PDE11A hydrolyzes both cGMP and cAMP, with K_m values of 0.52 µM and 1.04 µM, respectively [103]. With similar V_{max} values for both cAMP and cGMP, despite the two-fold higher affinity for cGMP, PDE11A may function as a genuine dual substrate PDE at physiologically relevant concentrations. PDE11A splice variants exhibit differences in V_{max} (PDE11A4 >> PDE11A2 > PDE11A3 >> PDE11A1), and in their sensitivities to inhibitors, some of which were two- to three-fold more potent against PDE11A3 than PDE11A4 [102], indicating that the N-terminus of PDE11A may affect the conformation of the protein and thus regulate catalytic activity. The splice variants possess different GAF domains in their N-terminal regions: PDE11A4, with two GAF domains; PDE11A3, one complete and one incomplete GAF domain; PDE11A2, one complete GAF domain; and PDE11A1, an incomplete GAF domain. Although PKA and PKG phosphorylate PDE11A4, but not other variants, it is uncertain whether phosphorylation occurs *in vivo* or serves a regulatory function [102].

PDE11A transcripts are highly expressed in prostate and testis, with moderate expression in salivary gland, pituitary gland, thyroid gland, and liver. PDE11A3 transcripts are specifically expressed in testis, whereas PDE11A4 is particularly abundant in prostate [102], suggesting a potential role for PDE11 in these tissues.

Conclusions

In general, dual-specificity PDEs seem to be regulators of many cyclic nucleotide signaling pathways, including proliferation of vascular smooth muscle (PDE1) [12,13], myocardial contractility and platelet aggregation (PDE2 and PDE3) [4,14], adrenal steroidogenesis (PDE2), and insulin/IGF-1 action (PDE3) [3,4,12–14]. In addition, because of their intrinsic characteristics and regulatory properties, dual-specificity PDEs can serve as a locus for cross-talk among Ca^{2+}, cAMP, and cGMP signaling pathways, since Ca^{2+}, calmodulin, and calmodulin kinase regulate PDE1, and since, depending on physiological cyclic nucleotide concentrations, cGMP can modulate intracellular cAMP concentrations by either stimulating or inhibiting cAMP hydrolysis by activating PDE2 or inhibiting PDE3. Little is known of intracellular functions of recently identified PDE10 and 11, but PDE10 might be expected to function as a cAMP-inhibitable cGMP PDE.

Acknowledgments

We thank Carol Kosh for her secretarial assistance; E.D. is supported in part by Swedish MRC grant 3362.

References

1. Conti, M. and Jin, S. L. (1999). *Prog. Nucleic Acids Res. Mol. Biol.* **66**, 1–38.
2. Dousa, T. P. (1999). *Kidney Int.* **55**, 29–62.
3. Mehats, C., Andersen, C. B., Filopanti, M., Jin, S.-L. C., and Conti, M. (2002). *Trends Endocrinol. Med.* **13**, 29–33.
4. Juilfs, D. M., Soderling, S., Burns, F., and Beavo, J. A. (1999). *Rev. Physiol. Biochem. Pharmacol.* **135**, 67–104.
5. Soderling, S. H. and Beavo, J. A. (2000). *Curr. Opin. Cell Biol.* **12**, 174–179.
6. Houslay, M. D. and Milligan, G. (1997). *Trends Biochem. Sci.* **22**, 217–224.
7. Jurevicuis, J. and Fischmeister, R. (1996). *Proc. Natl. Acad. Sci. USA* **93**, 295–299.

8. Rich, T. C., Fagan, K. A., Tse, T. E., Schaack, J., Cooper, D. M., and Karpen, J. W. (2001). *Proc. Natl. Acad. Sci. USA* **98**, 13049–13054.
9. Francis, S. H., Turko, I. V., and Corbin, J. D. (2001). *Prog. Nucleic Acids Res. Mol. Biol.* **65**, 1–52.
10. Aravind, L. and Ponting, C. P. (1997). *Trends Biochem. Sci.* **22**, 458–459.
11. Korsa, I. and Bock, A. (1997). *J. Bacteriol.* **179**, 41–45.
12. Rybalkin, S. D., Bornfeldt, K. E., Sonnenburg, W. K., Rybalkina, I. G., Kwak, K. S., Hanson, K., Krebs, E. G., and Beavo, J. A. (1997). *J. Clin. Invest.* **100**, 2611–2621.
13. Kayama, H., Bornfeldt, K. E., Fukimoto, S., and Nishizawa, Y. (2001). *J. Cell. Physiol.* **186**, 1–10.
14. Degerman, E., Belfrage, P., and Manganiello, V. C. (1997). Structure, localization, and regulation of cGMP-inhibited phosphodiesterase (PDE3). *J. Biol. Chem.* **272**, 6823–6826.
15. Yan, C., Zhao, A. Z., Bentley, J. K., and Beavo, J. A. (1996). *J. Biol. Chem.* **271**, 25699–25706.
16. Snyder, P. B., Florio, V. A., Ferguson, K., and Loughney, K. (1999). *Cell Signal.* **11**, 535–544.
17. Fidock M., Miller, M., and Lanfear, J. (2002). *Cell Signal.* **14**, 53–60.
18. Yu, J., Wolda, S. L., Frazier, A. L. B., Florio, V. A., Martins, T. J., Snyder, P. B., Harris, E. A. S., McCaw, K. N., Farrell, C. A., Steiner, B., Bentley, J. K., Beavo, J. A., Ferguson, K., and Gelinas, R. (1997). *Cell Signal.* **9**, 519–529.
19. Loughney, K., Martins, T. J., Harris, E. A. S., Sadhu, K., Hicks, J. B., Sonnenburg, W. K., Beavo, J. A., and Ferguson, K. (1996). *J. Biol. Chem.* **271**, 796–806.
20. Sonnenburg, W. K., Seger, D., Kwak, K. S., Huang, J., Charbonneau, H., and Beavo, J. A. (1995). *J. Biol. Chem.* **270**, 30989–31000.
21. Sharma, R. K. (1991). *Biochemistry* **30**, 5964–5968.
22. Florio, V. A., Sonnenburg, W. K., Johnson, R., Kwak, K. S., Jensen, G. S., Walsh, K. A., and Beavo, J. A. (1994). *Biochemistry* **33**, 8948–8954.
23. Hashimoto, Y., Sharma, R. K., and Soderling, T. R. (1989). *J. Biol. Chem.* **264**, 10884–10887.
24. Kakkar, R., Raju, R. V. S., and Sharma, R. K. (1998). *Arch. Biochem. Biophys.* **358**, 320–328.
25. Spence, S., Rena, G., Sweeney, G., and Houslay, M. D. (1995). *Biochem. J.* **310**, 975–982.
26. Spence, S., Rena, G., Sullivan, M., Erdogan, S., and Houslay, M. D. (1997). *Biochem. J.* **321**, 157–163.
27. Zhao, A. Z., Yan, C., Sonnenburg, W. K., and Beavo, J. A. (1997). *Adv. Second Messenger Phosphop. Res.* **31**, 237–250.
28. Kim, D., Rybalkin, S. D., Pi, X., Wang, Y., Zhang, C., Munzel, T., Beavo, J. A., Berk, B. C., and Yan, C. (2001). *Circulation* **104**, 2338–2343.
29. Capito, K., Hedeskov, C., and Thams, P. (1986). *Acta Endocrinol.* **111**, 533–538.
30. Shafiee-Nick, R., Pyne, N. J., and Furman, B. L. (1995). *Br. J. Pharmacol.* **115**, 1–7.
31. Han, P., Werber, J., Surana, M., Fleischer, N., and Michaeli, T. (1999). *J. Biol. Chem.* **274**, 22337–22344.
32. Sharma, R. K. and Hickie, R. A. (1996). In Dent, G., Rabe, K., and Schudt, C., Eds., *Phosphodiesterase Inhibitors*, pp. 65–79.
33. Jiang, X., Jianping, L., Paskind, M., and Epstein, P. M. (1996). *Proc. Natl. Acad. Sci. USA* **93**, 11236–11241.
34. Kanda, N. and Watanabe, S. (2001). *Biochem. Pharmacol.* **62**, 495–507.
35. Beavo, J. A., Hardman, J. G., and Sutherland, E. W. (1970). *J. Biol. Chem.* **245**, 5649–5655.
36. Manganiello, V. C., Tanaka, T., and Murashima, S. (1990). In Beavo, J. A., and Houslay, M. D., Eds., *Cyclic Nucleotide Phosphodiesterase: Structure, Regulation, and Drug Action*, vol. 2, pp. 61–86.Wiley, Chichester, U.K.
37. Le Trong, H., Beier, N., Sonnenburg, W. K., Stroop, S. D., Walsh, K. A., Beavo, J. A., and Charbonneau, H. (1990). *Biochemistry* **29**, 10280–10288.
38. Rosman, G. J., Martins, T. J., Sonenburg, W. K., Beavo, J. A., Fergusin, K., and Loughney, K. (1997). *Gene* **191**, 89–95.
39. Sonnenburg, W. K., Mullaney, P. J., and Beavo, J. A. (1991). *J. Biol. Chem.* **266**, 17655–17661.
40. Yang, Q., Paskin, M., Bolger, G., Thompson, W. J., Repaske, D. R., Cutler, L. S., and Epstein, P. M. (1994). *Biochem. Biophys. Res. Commun.* **205**, 1850–1858.
41. Tanaka, T., Hockman, S., Moos, M. Jr., Taira, M., Meacci, E., Murashima, S., and Manganiello, V. C. (1991). *Second Messengers Phosphoproteins* **13**, 87–98.
42. Repaske, D. R., Corbin, J. G., Conti, M., and Goy, M. F. (1993). *Neuroscience* **56**, 673–686.
43. Charbonneau, H., Prusti, R. K., LeTrong, H., Sonnenburg, W. K., Mullaney, P. J., Walsh, K. A., and Beavo, J. A. (1990). *Proc. Natl. Sci. USA* **87**, 288–292.
44. Podzuweit, T., Nennsteil, P., and Muller, A. (1995). *Cell Signal.* **7**, 733–738.
45. Hartzell, H. C. and Fischmeister, R. (1986). *Nature* **323**, 273–275.
46. Mery, P. F., Pavolne, C., Belhassen, L., Pecker, F., and Fischmeister, R. (1993). *J. Biol. Chem.* **268**, 26286–26295.
47. Mery, P. F., Pavolne, C., Pecker, F., and Fischmeister, R. (1995). *Mol. Pharmacol.* **48**, 121–130.
48. Vandecasteele, G., Verde, I., Rucker-Martin, C., Donzeau-Gouge, P., and Fischmeister, R. (2001). *J. Physiol.* **533**, 329–340.
49. Dittrich M., Jurevicius, J., Georget, M., Rochais, F., Fleischmann, B. K., Hescheler, J., and Fischmeister, R. (2001). *J. Physiol.* **534**, 109–121.
50. Maurice, D. H. and Haslam, R. J. (1990). *Mol. Pharmacol.* **37**, 671–681.
51. Dickinson, N. T., Jang, E. K., and Haslam, R. J. (1997). *Biochem. J.* **323**, 371–377.
52. Haslam, R. J., Dickinson, N. T., and Jang, E. K. (1999). *Thromb. Haemost.* **82**, 412–423.
53. MacFarland R. T., Zelus B. D., and Beavo J. A. (1991). *J. Biol. Chem.* **266**, 136–142.
54. Michie, A. M., Lobban, M., Muller, T., Harnett, M. M., and Houslay, M. D. (1996). *Cell Signal.* **8**, 97–110.
55. Haynes, J., Killilea, D., Peterson, P. D., and Thompson, W. J. (1996). *J. Pharmacol. Exp. Therap.* **276**, 752–757.
56. Velardez, M. O., De Laurentiis, A., del Carmen Diaz, M., Lasaga, M., Pisera, D., Seilicovich, A., and Duvilanski, B. H. (2000). *Eur. J. Endocrinol.* **143**, 279–284.
57. Oritz, P. A. and Garvin, J. L. (2001). *Hypertension* **37**, 467–471.
58. Juilfs, D. M., Fulle, H. J., Zhao, A. Z., Houslay, M. D., Garbers, D. L., and Beavo, J. A. (1997). *Proc. Natl. Acad. Sci. USA* **94**, 3388–3395.
59. Meyer, M. R., Angele, A., Kremmer, E., Kaupp, U. B., and Muller, F. (2000). *Proc. Natl. Acad. Sci. USA* **97**, 10595–10600.
60. Whalin, M. W., Scammell, J. G., Strada, S. J., and Thompson, W. J. (1991). *Mol. Pharmacol.* **39**, 711–717.
61. Bentley, J. K., Juilfs, D. M., and Uhler, M. D. (2001). *J. Neurochem.* **76**, 1252–1263.
62. Miki, T., Taira, M., Hockman, S., Shimada, F., Lieman, J., Napolitano, M., Ward, T., Makino, H., and Manganiello, V. C. (1996). *Genomics* **36**, 476–485.
63. Lobbert, R. W., Winterpacht, A., Seipel, B., and Zabel, B. U. (1996). *Genomics* **37**, 211–218.
64. Degerman, E. and Manganiello, V. C. (2001). *Prog. Nucleic Acids Res. Mol. Biol.* **66**, 241–277.
65. Snyder, P. (1999). *Emerging Therapeutic Targets* **3**, 587–599.
66. Conti, M. (2000). *Mol. Endocrinol.* **14**, 1317–1327.
67. Houslay, M. D. and Kilgour, E. (1990). In Beavo, J. A., and Houslay, M. D., Eds., *Cyclic Nucleotide Phosphodiesterases: Structure, Regulation, and Drug Action*, pp. 185–224. Wiley, Chichester, U.K.
68. Choi, Y. H., Ekholm, D., Krall, J., Ahmad, F., Degerman, E., Manganiello, V. C., and Movsesian, M. A. (2001). *Biochem. J.* **353**, 41–50.
69. Zhao, A., Shinohara, M. M., Huang, D., Shimizu, N., Elder-Finkelman, H., Krebs, E. G., Beavo, J. A., and Bornfeldt, K. E. (2000). *J. Biol. Chem.* **275**, 11348–11354.
70. Andersen, C. E., Roth, R. A., and Conti, M. (1998). *J. Biol. Chem.* **273**, 18705–18708.
71. Zhao, A. Z., Zhao, H., Teague, J., Fujimoto, W., and Beavo, J. A. (1997). *Proc. Natl. Acad. Sci. USA* **94**, 3223–3228.
72. White, M. F. (1998). *Recent Prog. Horm. Res.* **53**, 119–138.

73. Shepherd, P. R., Withers, D. J., and Siddle, K. (1998). *Biochem. J.* **333**, 471–490.
74. Rahn, T., Ronnstrand, L., Leroy, M. J., Wernstedt, C., Tornqvist, H., Manganiello, V. C., Belfrage, P., and Degerman, E. (1996). *J. Biol. Chem.* **271**, 11575–11580.
75. Kitamura, T., Kitamura, Y., Kuroda, S., Hino, Y., Ando, M., Kotani, K., Konishi, H., Matsuzaki, H., Kikkawa, U., Ogawa, W., and Kasuga, M. (1999). *Mol. Cell. Biol.* **19**, 6286–6296.
76. Ahmad, F., Cong, L. N., Stenson-Holst, L., Wang, L. M., Rahn-Landstrom, T., Pierce, J. H., Quon, M. J., Degerman, E., and Manganiello, V. C. (2000). *J. Immunol.* **164**, 4678–4688.
77. Wijkander, J., Landstrom, T. R., Manganiello, V. C., Belfrage, P., and Degerman, E. (1998). *Endocrinology* **139**, 219–227.
78. Holm, C., Langin, D., Manganiello, V. C., Belfrage, P., and Degerman, E. (1997). *Methods Enzymol.* **286**, 45–67.
79. Resjo, S., Okinianska, A., Zolnierowicz, S., Manganiello, V. C., and Degerman, E. (1999). *Biochem. J.* **341**, 839–845.
80. Russell, J. C., Shillabeer, G., Bar-Tana, J., Lau, D. C., Richardson, M., Wenzel, L. M., Graham, S. E., and Dolphin, P. J. (1998). *Diabetes* **47**, 770–778.
81. Tang, Y., Osawa, H., Onuma, H., Nishimiya, T., Ochi, M., and Makino, H. (1999). *Diabetes* **48**, 1830–1835.
82. Tang, Y., Osawa, H., Onuma, H., Hasegawa, M., Nishimiya, T., Ochi, M., and Makino, H. (2001). *Eur. J. Endocrinol.* **145**, 93–99.
83. Rahn-Landstrom, T., Mei, J., Karlsson, M., Manganiello, V. C., and Degerman, E. (2000). *Biochem. J.* **346**, 337–343.
84. Ahmad, M., Abdel-Wahab, Y. H., Tate, R., Flatt, P. R., Pyne, N. J., and Furman, B. L. (2000). *Br. J. Pharmacol.* **129**, 1228–1234.
85. Parker, J. C., Van Volkenburg, M. A., Ketchum, R. J., Brayman, K. L., and Andrews, K. M. (1995). *Biochem. Biophys. Res. Commun.* **217**, 916–923.
86. El-Metwally, M., Shafiee-Nick, R., Pyne, N. J., and Furman, B. L. (1997). *Eur. J. Pharmacol.* **324**, 227–232.
87. Parker, J. C., Van Volkenburg, M. A., Nardone, N. A., Hargrove, D. M., and Andrews, K. M. (1997). *Biochem. Biophys. Res. Commun.* **236**, 665–669.
88. Zhao, A. Z., Bornfeldt, K. E., and Beavo, J. A. (1998). *J. Clin. Invest.* **102**, 869–873.
89. Persaud, S. J., Jones, P. M., and Howell, S. L. (1990). *Biochem. Biophys. Res. Commun.* **173**, 833–839.
90. Harris, T. E., Persaud, S. J., and Jones, P. M. (1997). *Biochem. Biophys. Res. Commun.* **232**, 648–651.
91. Beebe, S. J., Redman, J. B., Blackmon, P. F., and Corbin, J. D. (1985). *J. Biol. Chem.* **260**, 15781–15788.
92. Sadler, S. E. (1991). *Mol. Endocrinol.* **5**, 1939–1946.
93. Wiersma, A., Hirsch, B., Tsafiri, A., Hanssen, R. G., Van de Kant, M., Kloosterboer, H. J., Conti, M., and Hsueh, I. (1998). *J. Clin. Invest.* **102**, 532–537.
94. Fujishige, K., Kotera, J., Michibata, H., Yuasa, K., Takebayashi, S., Okumura, K., and Omori, K. (1999). *J. Biol. Chem.* **274**, 18438–18445.
95. Loughney, K., Snyder, P. B., Uher, L., Rosman, G. J., Ferguson, K., and Florio, V. A. (1999). *Gene* **234**, 109–117.
96. Soderling, S. H., Bayuga, S., and Beavo, J. A. (1999). *Proc. Natl. Acad. Sci. USA* **96**, 7071–7076.
97. Kotera, J., Fujishige, K., Yuasa, K., and Omori, K. (1999). *Biochem. Biophys. Res. Commun.* **261**, 551–557.
98. Fujishige, K., Kotera, J., and Omori, K. (1999). *Eur. J. Biochem.* **266**, 1118–1127.
99. Fujishige, K., Kotera, J., Yuasa, K., and Omori, K. (2000). *Eur. J. Biochem.* **267**, 5943–5951.
100. Yuasa, K., Kanoh, Y., Okumura, K., and Omori, K. (2001). *Eur. J. Biochem.* **268**, 168–178.
101. Hetman, J. M., Robas, N., Baxendale, R., Fidock, M., Phillips, S. C., Soderling, S. H., and Beavo, J. A. (2000). *Proc. Natl. Acad. Sci. USA* **97**, 12891–12895.
102. Yuasa, K., Kotera, J., Fujishige, K., Michibata, H., Sasaki, T., and Omori, K. (2000). *J. Biol. Chem.* **275**, 31469–31479.
103. Fawcett, L., Baxendale, R., Stacey, P., McGrouther, C., Harrow, I., Soderling, S., Hetman, J., Beavo, J. A., and Phillips, S. C. (2000). *Proc. Natl. Acad. Sci. USA* **97**, 3702–3707.

CHAPTER 195

Phosphodiesterase-5

Sharron H. Francis and Jackie D. Corbin
*Department of Molecular Physiology and Biophysics,
Vanderbilt University Medical School, Nashville, Tennessee*

Introduction

Cyclic GMP-specific cGMP-binding phosphodiesterase (PDE5) is a class I PDE, and it is abundant in smooth muscle, platelets, Purkinje cells, and renal tissues. It is the pharmacological target for sildenafil citrate (Viagra™), vardenafil (Levitra™), and Fadalafil (Cialis™), all of which are treatments for male erectile dysfunction. It is also an emerging target for treatment of other disorders involving smooth muscle, such as pulmonary hypertension. PDE5 is dimeric; each 100-kDa monomer contains a catalytic domain and a regulatory domain that includes two GAF domains associated with the allosteric cGMP-binding function and a single phosphorylation site for either cGMP-dependent protein kinase (PKG) or cAMP-dependent protein kinase (PKA). The role of the allosteric cGMP-binding in enzyme regulation is poorly understood. There is no apparent homology between the allosteric cGMP-binding site(s) and the catalytic site; they have distinctly different analog specificities, and neither is homologous with the cyclic nucleotide-binding sites of PKA and PKG. There are three splice variants that differ at their amino-termini, and they are derived from a single gene. Expression of PDE5 is subject to long-term and developmental regulation. Phosphorylation of the single phosphorylation site requires binding of cGMP to the regulatory domain and results in activation of catalysis as well as increased cGMP-binding affinity at the allosteric sites. Increased phosphorylation occurs in intact cells when cGMP is elevated in response to atrial natriuretic peptide. PDE5 plays a central role in selectively controlling cGMP levels in numerous tissues through breakdown of cGMP at the catalytic site and binding of cGMP at the allosteric sites.

PDE5 has recently gained prominence with the realization that the activity of this enzyme in combination with that of guanylyl cyclases plays a key role in modulating cGMP levels in smooth muscle cells and therefore in determining contractile tone in this tissue (Fig. 1). PDE5 is highly abundant in corpus cavernosum of the penis and is inhibited by several drugs that are a highly effective are for treatment of male erectile dysfunction. The demonstrated role of PDE5 activity as a major determinant in modulating cGMP signaling in numerous tissues has led to renewed interest in this enzyme and its potential as an important pharmacological target.

Gene Organization and Regulation of Expression

The single human PDE5 gene is located on chromosome 4q26, contains 23 exons, and encodes mRNA for three variants: PDE5A1, A2, and A3. These proteins, 100 kDa, 95 kDa and 95 kDa, respectively, share similar properties and have different amino-termini, a result of alternative mRNA splicing [1–4]. Alternative first exons arranged in the order A1-A3-A2 account for the three forms [5]. PDE5 mRNA is present in many tissues [3,4,6], but PDE5 protein level is relatively low in most tissues. PDE5 protein occurs in relative abundance in smooth muscle, platelets, and Purkinje cells and has been detected in proximal renal tubules, collecting renal ducts, and epithelial cells of pancreatic ducts [2,7]. It is by far the major cGMP-hydrolyzing PDE in penile corpora cavernosae of human, dog, and rabbit. PDE5A3 appears to predominate in smooth muscle, and its expression may only occur in smooth muscle [4,8]. PDE5A1 and PDE5A2 are co-expressed in a variety of tissues, but PDE5A2 is the major form found in most human tissues [4,9].

PDE5 is subject to long-term and developmental regulation [1,9–11]. Both cAMP and cGMP stimulate PDE5A promoter activity, and cAMP increases PDE5A2 expression [9,12]. PDE5A1 promoter contains multiple Sp1- and AP2-binding sequences and involves a portion of the first exon. PDE5A2 utilizes an intronic promoter located downstream from the

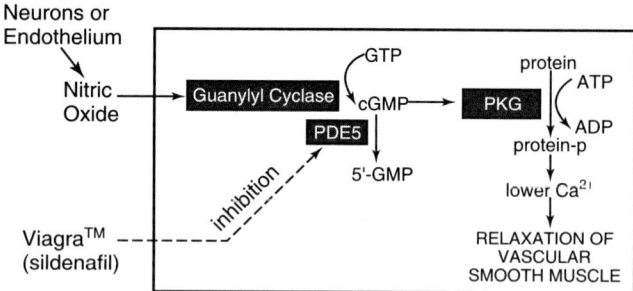

Figure 1 Nitric oxide and cGMP signaling in smooth muscle cells. Cyclic GMP level in smooth muscle cells is determined by the balance between the rate of cGMP synthesis by guanylyl cyclases and the rate of cGMP breakdown by phosphodiesterases (PDE). When cGMP is elevated, it binds to cGMP-dependent protein kinase (PKG) and activates that enzyme to bring about relaxation of smooth muscle. PDE5 is abundant in vascular smooth muscle and contributes to the downregulation of cGMP levels. Inhibitory agents such as sildenafil (Viagra™), Vardenafil (Levitra™), or tadalafil (Cialis™) that block catalytic activity of PDE5 enhance elevation of cGMP in response to stimuli that activate guanylyl cyclases.

first exon of PDE5A1. This promoter also contains sequences that bind Sp1 and AP2 transcription factors [5]. Several of the Sp1 sites are important for promoting PDE5A2 gene transcription and in providing for cGMP stimulation of transcription [13].

General Structure

PDE5 is a dimer of identical subunits each with $M_r \sim 100,000$ (Fig. 2). A regulatory domain is located in the amino-terminal portion of each monomer, and a catalytic domain occupies a more carboxyl-terminal region. The regulatory domain of PDE5 contains a phosphorylation site (Ser-92 in bovine PDE5A1 and Ser-102 in human PDE5A1) for PKG or PKA [14]. The regulatory domain binds cGMP at highly specific, allosteric binding sites that show kinetic heterogeneity [15]. The catalytic site is also highly specific for cGMP and, like all known PDEs, contains binding sites for divalent cation(s) that are required to support catalysis. The amino acid sequence of the PDE5 allosteric cGMP-binding region is not homologous with the catalytic site, and neither of these is homologous with cyclic nucleotide-binding sites in PKA, PKG, cyclic nucleotide-gated cation channels, or cyclic nucleotide-regulated guanine nucleotide exchange factors [16].

Phosphorylation of a single serine near the aminoterminus of PDE5 by PKG or PKA requires cGMP binding to the regulatory region and would be predicted to occur only when cellular cGMP is significantly increased and PKG is concomitantly activated (Fig. 3). PDE5 is phosphorylated in intact cells in response to stimuli that elevate cGMP. Elevation of cAMP does not mimic this. PDE5 is at least a ten-fold better substrate for PKG than for PKA. This preference results from a phenylalanine located at P^{+4}, which is a negative determinant for PKA. PKG has ~20-fold higher affinity for PDE5 regulatory domain as substrate ($K_m \sim 2.7\,\mu M$) than

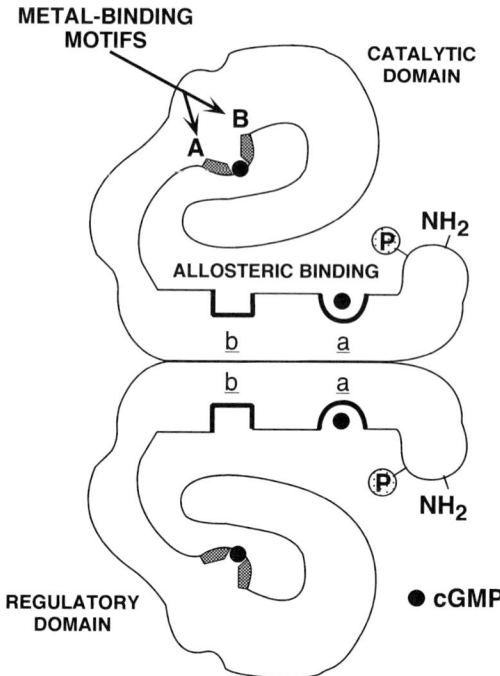

Figure 2 Working model of PDE5. The regulatory domain in the amino-terminal portion of PDE5 contains the phosphorylation site (Ser-102), two GAF domains (*a* and *b*) associated with allosteric cGMP-binding and the dimerization domain. Cyclic GMP binding to the isolated GAF domain *a* has been demonstrated, but whether the GAF domain *b* binds cGMP remains to be determined. The catalytic domain in the carboxyl-terminal portion of PDE5 contains a cGMP-binding substrate site and two Zn^{2+}-binding motifs (A and B), portions of which form a novel metal-binding site.

for the peptide containing the phosphorylation site ($K_m \sim 68\,\mu M$), indicating that additional features in the PDE5 regulatory domain contribute to high-affinity interaction with PKG [17].

PDE5 Catalytic Domain

Catalytic domains of all mammalian PDEs include ~270 residues that are 25 to 50 percent conserved and contain 18 invariant amino acids [16,18]. PDE5 is highly specific for cGMP and exhibits a k_{cat} of 4.3 sec^{-1} and K_m of 1–6 μM for that nucleotide. Catalytic site affinity for cAMP is at least 100-fold lower than that for cGMP, but V_{max} for cAMP is actually higher than that for cGMP. Cyclic nucleotide analog specificity of the catalytic site differs from that for cGMP-binding to the regulatory domain, and the catalytic site tolerates more modifications to the cGMP molecule. Cyclic GMP analogs with modifications at the 2'-OH or at N1 and C2 are accommodated well at the catalytic site. High catalytic site affinity vardenafil and PDE5 inhibitors such as sildenafil, tadalafil which exhibit slight resemblance to cGMP, further indicate the promiscuous nature of this site. Recombinant PDE5 containing only the catalytic domain is monomeric, and the catalytic properties (K_m, V_{max} for cGMP, and IC_{50} for inhibitors) compare well to those of full-length PDE5, indicating that requirements for catalysis exist within a single catalytic domain [16,19].

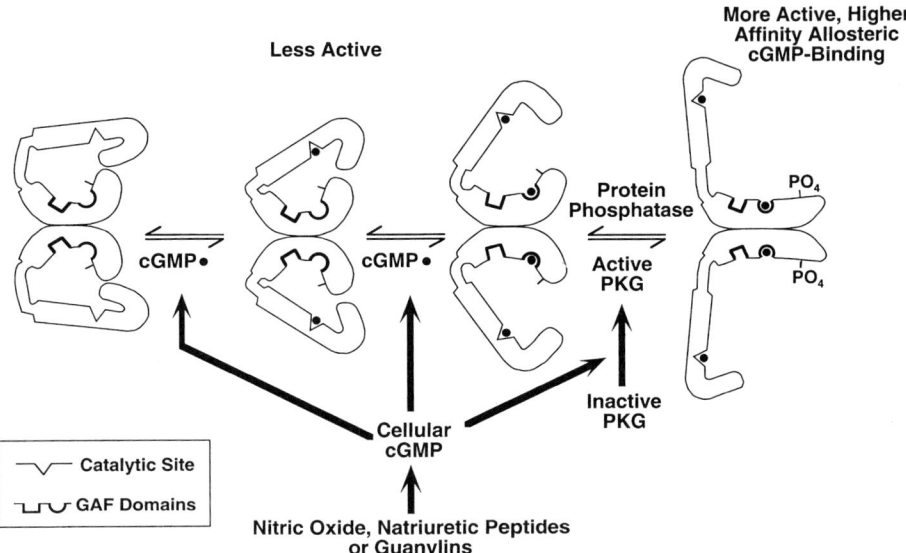

Figure 3 Working model of conformational changes that occur in PDE5. Upon elevation of cellular cGMP, cGMP interacts with its major intracellular receptors, PDE5 and PKG. For PDE5, evidence suggests that cGMP interacts first with the catalytic site where it is hydrolyzed. That initial interaction induces a conformational change in PDE5 to increase cGMP binding to the regulatory domain. The possibility that cGMP binding to the regulatory domain increases the affinity of the catalytic site for cGMP, and maximum catalytic activity is also increased. Cyclic GMP-binding to the regulatory domain produces yet another change in the amino-terminal portion of PDE5 to expose the serine that can be phosphorylated by PKG. Under these circumstances, PKG would be activated due to specific binding of cGMP to its allosteric sites. Evidence suggests that the phosphorylated PDE5 is the most active form and exhibits highest affinity for cGMP binding in the regulatory domain. Existence of each conformation is based on extensive biochemical studies.

Cloning of PDE5 cDNA led to identification of two conserved Zn^{2+}-binding motifs (A and B) in catalytic domains of all mammalian PDEs. The importance of Zn^{2+} in catalytic function of PDEs was first demonstrated experimentally with PDE5. Zinc binds to PDE5 with high stoichiometry (~3 mol/mol PDE5 monomer), is not competed by 10,000-fold excess of either Mn^{2+} or Mg^{2+}, and is effective at far lower concentration (submicromolar) than either Mn^{2+} or Mg^{2+} in supporting catalytic activity. Site-directed mutagenesis of PDE5 confirms the catalytic importance of residues in Zn^{2+}-binding motifs A and B and reveals involvement of residues from each motif to form a novel Zn^{2+} binding site [20,21]. Two conserved downstream aspartates are also implicated as critical for efficient catalysis. Mutation of two other residues, Tyr-602 and Glu-775, causes marked changes in K_m for cGMP.

A variety of PDE inhibitors block catalysis by competing with cGMP at the catalytic site. The order of potency, in descending order, of some common PDE inhibitors for PDE5 is vardenafil, tadalafil ≥ sildenafil, zaprinast, dipyridamole, 3-isobutyl-1-methylxanthine (IBMX), cilostamide, theophylline, caffeine, rolipram [19]. PDE5 is the specific target of sildenafil citrate, a potent competitive inhibitor (IC_{50} of 1–4 nM) marketed as Viagra™. Affinity of sildenafil for PDE5 is 10- and 100-fold greater than that for PDE6 and PDE1, respectively, and much greater than that for other PDE families. Other commercialized compounds [Levitra™ (vardenafil), Cialis™ (tadalafil)] have now been produced that inhibit PDE5 with equal or better potencies than sildenafil and that show even stronger selectivity than sildenafil for inhibition of PDE5 compared to other PDEs [22,23].

Occupation of PDE5 catalytic site by cGMP, cGMP analogs, or inhibitors such as sildenafil increases cGMP binding to allosteric cGMP-binding sites in the regulatory domain (Fig. 3). This increaes catalytic activity and also leads to exposure of the phosphorylation site, thereby allowing for phosphorylation, further activation of PDE5, and enhanced cGMP-binding affinity of the allosteric sites. Inhibitors such as sildenafil bind exclusively to the PDE5 catalytic site and do not compete with cGMP for binding in the regulatory domain sites [16,19].

PDE5 Regulatory Domain

Allosteric sites in unphosphorylated PDE5 bind cGMP with a K_d ~0.2–1.5 µM; affinity varies depending on conditions. Cyclic GMP binding is associated with one or two GAF domains comprised of ~110 residues each [29]. The GAF acronym derives from a group of apparently unrelated proteins that contain these ancient domains, bind diverse ligands, and share low amino acid sequence similarity: cGMP-binding PDEs, cyanobacterial *Anabaena* adenylyl cyclase, and *Escherichia coli* transcriptional factor fhlA. Whether both GAF domains in a single PDE5 monomer bind cGMP is not known. Two kinetically distinct cGMP-binding sites with ~four-fold difference in affinity exist in the enzyme. Over the years, much evidence has been presented and interpreted as supporting the likelihood that cGMP binds to both GAF domains in PDE5. However, stoichiometry of cGMP binding to the regulatory region is best estimated at 1 mol cGMP per monomer. A recombinant protein containing only the more amino-terminal GAF domain of PDE5a binds cGMP with an affinity similar to that of the high-affinity component in native and wild-type PDE5 [17]. It is clear that at least the cGMP-binding associated with the more amino-terminal GAF domain is a very stable and self-contained entity (Figs. 2 and 3).

Functional roles of allosteric cGMP-binding in PDE5 has been partly clarified (see below), but the biological advantage of having two GAF domains in most cGMP-binding PDEs (PDE2, PDE5, PDE6, PDE10, PDE 11) is a mystery [16]. Site-directed mutagenesis of residues in the more amino-terminal GAF or in both GAFs diminishes or compromises cGMP binding and causes a three- to four-fold increase in K_m of PDE5 for cGMP. Integrity of both GAFs and retention of cGMP binding by the regulatory domain influences exposure of the phosphorylatable serine for PKG or PKA action (Fig. 3). The phosphorylation in turn increases the affinity with which the regulatory domain binds cGMP, and the K_m for cGMP as substrate is lowered. Whether this latter effect results from phosphorylation alone or from a concomitant increase in the cGMP-binding affinity of the regulatory domain is not known.

Dimerization occurs through stable interactions located near or within the allosteric cGMP-binding sites. The nature of the contacts and the advantages of a dimeric enzyme are not known. Dimerization could contribute to stability or function of the cGMP-binding domains. It is clearly not required for catalytic function.

Concluding Remarks

Studies of PDE5 have provided major insights into molecular characteristics of that enzyme and have advanced our understanding of general properties shared by all members of the PDE superfamily. PDE5 has been successfully targeted for a widely marketed treatment for male erectile dysfunction. The success of sildenafil (Viagra™) in this regard has renewed interest in PDEs as potential pharmacological targets. The relatively restricted expression of PDE5 along with its well-defined biochemical properties enhances its attraction as a target for future pharmacological interventions.

References

1. Kotera, J., Yanaka, N., Fujishige, K., Imai, Y., Akatsuka, H., Ishizuka, T., Kawashima, K., and Omori, K. (1997). Expression of rat cGMP-binding cGMP-specific phosphodiesterase mRNA in Purkinje cell layers during postnatal neuronal development. *Eur. J. Biochem.* **249**, 434–442.
2. Loughney, K., Hill, T. R., Florio, V. A., Uher, L., Rosman, G. J., Wolda, S. L., Jones, B. A., Howard, M. L., McAllister-Lucas, L. M., Sonnenburg, W. K., Francis, S. H., Corbin, J. D., Beavo, J. A., and Ferguson, K. (1998). Isolation and characterization of cDNAs encoding PDE5A, a human cGMP-binding cGMP-specific 3',5'-cyclic nucleotide phosphodiesterase. *Gene* **216**, 137–147.
3. Yanaka, N., Kotera, J., Ohtsuka, A., Akatsuka, H., Imai, Y., Michibata, H., Fujishige, K., Kawai, E., Takebayashi, S., Okumura, K., and Omori, K. (1998). Expression, structure and chromosomal localization of the human cGMP-binding cGMP-specific phosphodiesterase PDE5A gene. *Eur. J. Biochem.* **255**, 391–399.
4. Lin, C. S., Lau, A., Tu, R., and Lue, T. F. (2000). Expression of three isoforms of cGMP-binding cGMP-specific phosphodiesterase (PDE5) in human penile cavernosum. *Biochem. Biophys. Res. Commun.* **268**, 628–635.
5. Lin, C. S., Lau, A., Tu, R., and Lue, T. F. (2000). Identification of three alternative first exons and an intronic promoter of human PDE5A gene. *Biochem. Biophys. Res. Commun.* **268**, 596–602.
6. Kotera, J., Fujishige, K., Akatsuka, H., Imai, Y., Yanaka, N., and Omori, K. (1998). Novel alternative splice variants of cGMP-binding cGMP-specific phosphodiesterase. *J. Biol. Chem.* **273**, 26982–26990.
7. Kotera, J., Fujishige, K., and Omori, K. (2000). Immunohistochemical localization of cGMP-binding cGMP-specific phosphodiesterase (PDE5) in rat tissues. *J. Histochem. Cytochem.* **48**, 685–693.
8. Gopal, V. K., Francis, S. H., and Corbin, J. D. (2001). Allosteric sites of phosphodiesterase-5 (PDE5). A potential role in negative feedback regulation of cGMP signaling in corpus cavernosum. *Eur. J. Biochem.* **268**, 3304–3312.
9. Kotera, J., Fujishige, K., Imai, Y., Kawai, E., Michibata, H., Akatsuka, H., Yanaka, N., and Omori, K. (1999). Genomic origin and transcriptional regulation of two variants of cGMP-binding cGMP-specific phosphodiesterases. *Eur. J. Biochem.* **262**, 866–872.
10. Bakre, M. M. and Visweswariah, S. S. (1997). Dual regulation of heat-stable enterotoxin-mediated cGMP accumulation in T84 cells by receptor desensitization and increased phosphodiesterase activity. *FEBS Lett.* **408**, 345–349.
11. Sanchez, L. S., de la Monte, S. M., Filippov, G., Jones, R. C., Zapol, W. M., and Bloch, K. D. (1998). Cyclic-GMP-binding, cyclic-GMP-specific phosphodiesterase (PDE5) gene expression is regulated during rat pulmonary development. *Pediatr .Res.* **43**, 163–168.
12. Lin, C. S., Chow, S., Lau, A., Tu, R., and Lue, T. F. (2001). Regulation of human PDE5A2 intronic promoter by cAMP and cGMP: identification of a critical Sp1-binding site. *Biochem. Biophys. Res. Commun.* **280**, 693–699.
13. Lin, C. S., Chow, S., Lau, A., Tu, R., and Lue, T. F. (2001). Identification and regulation of human PDE5A gene promoter. *Biochem. Biophys. Res. Commun.* **280**, 684–692.
14. Thomas, M. K., Francis, S. H., and Corbin, J. D. (1990). Substrate- and kinase-directed regulation of phosphorylation of a cGMP-binding phosphodiesterase by cGMP. *J. Biol. Chem.* **265**, 14971–14978.
15. McAllister-Lucas, L. M., Haik, T. L., Colbran, J. L., Sonnenburg, W. K., Seger, D., Turko, I. V., Beavo, J. A., Francis, S. H., and Corbin, J. D. (1995). An essential aspartic acid at each of two allosteric cGMP-binding sites of a cGMP-specific phosphodiesterase. *J. Biol. Chem.* **270**, 30671–30679.
16. Francis, S. H., Turko, I. V., and Corbin, J. D. (2000). Cyclic nucleotide phosphodiesterases: relating structure and function. *Nucl. Acid Res. Mol. Biol.* **65**, 1–52.
17. Liu, L., Underwood, T., Li, H., Pamukcu, R., and Thompson, W. J. (2001). Specific cGMP binding by the cGMP binding domains of cGMP-binding cGMP-specific phosphodiesterase. *Cell. Signal.* **13**, 1–7.
18. Soderling, S. H. and Beavo, J. A. (2000). Regulation of cAMP and cGMP signaling: new phosphodiesterases and new functions. *Curr. Opin. Cell Biol.* **12**, 174–179.
19. Corbin, J. D. and Francis, S. H. (1999). Cyclic GMP phosphodiesterase 5: target for sildenafil. *J. Biol. Chem.* **274**, 13729–13732.
20. Turko, I. V., Francis, S. H., and Corbin, J. D. (1998). Potential roles of conserved amino acids in the catalytic domain of the cGMP-binding cGMP-specific phosphodiesterase. *J.Biol.Chem.* **273**, 6460–6466.
21. Francis, S. H., Turko, I. V., Grimes, K. A., and Corbin, J. D. (2000). Histidine-607 and histidine-643 provide important interactions for metal support of catalysis in phosphodiesterase-5. *Biochemistry* **39**, 9591–9596.
22. Ukita, T., Nakamura, Y., Kubo, A., Yamamoto, Y., Moritani, Y., Saruta, K., Higashijima, T., Kotera, J., Takagi, M., Kikkawa, K., and Omori, K. (2001).

Novel, potent, and selective phosphodiesterase 5 inhibitors: synthesis and biological activities of a series of 4-aryl-1-isoquinolinone derivatives. *J. Med. Chem.* **44**, 2204–2218.

23. Padma-Nathan, H., McMurray, J. G., Pullman, W. E., Whitaker, J. S., Saoud, J. B., Ferguson, K. M., and Rosen, R. C. (2001). On-demand IC351 (Cialis) enhances erectile function in patients with erectile dysfunction. *Int. J. Impotence Res.* **13**, 2–9.

24. Turko, I. V., Francis, S. H., and Corbin, J. D. (1998). Binding of cGMP to both allosteric sites of cGMP-binding cGMP-specific phosphodiesterase (PDE5) is required for its phosphorylation. *Biochem. J.* **329**, 505–510.

25. Corbin, J. D., Turko, I. V., Beasley, A., and Francis, S. H. (2000). Phosphorylation of phosphodiesterase-5 by cyclic nucleotide-dependent protein kinase alters its catalytic and allosteric cGMP-binding activities. *Eur. J. Biochem.* **267**, 2760–2767.

26. Rybalkin, S. D., Rybalkina, I. G., Shimizu-Albergine, M., Tang, X. B., and Beavo, J. A. (2003). PDE5 is converted to an activated state upon cGMP binding to the GAF A domain. *EMBO J.* **22**, 469–478.

27. Mullershausen, F., Friebe, A., Feil, R., Thompson, W. J., Hofmann, F., Koesling, D. (2003). Direct activation of PDE5 by cGMP: long-term effects within NO/cGMP signaling. *J Cell Biol* **160**, 719–727.

28. Corbin, J. D., Blount, M. A., Weeks, J. L., Beasley, A., Kuhn, K. P., Yew, S. J. H., Saidi, L. F., Hurley, J. H., Kotera, J., and Francis, S. H. (2003). [^{3}H]Sildenafil binding to phosphodiesterase-5 is specific, kinetically heterogeneous, and stimulated by cGMP. *Mol Pharm* **63**, In Press.

29. Aravind, L. and Ponting, C. P. (1997). The GAF domain: an evolutionary link between diverse phototransducing proteins. *Trends Biochem. Sci.* **22**, 458–459.

Structure, Function, and Regulation of Photoreceptor Phosphodiesterase (PDE6)

Rick H. Cote

*Department of Biochemistry and Molecular Biology,
University of New Hampshire,
Durham, New Hampshire*

Introduction

The remarkable single-photon sensitivity of the visual pathway in rod photoreceptors requires that a photon of light be converted into a cascade of biochemical reactions. This process starts with activation of the visual pigment molecule (rhodopsin, R^*). During the lifetime of one R^*, ~100 heterotrimeric G-proteins (transducin) can be catalytically activated per second. Each activated transducin (specifically, the GTP-loaded α subunit, α_t^*-GTP) then binds to and activates the effector enzyme, cGMP phosphodiesterase (PDE). The hydrolytic action of PDE rapidly reduces the cytoplasmic cGMP concentration, causing dissociation of cGMP from cyclic GMP-gated ion channels. A single R^* can suppress over 1 pA of current that normally flows through these plasma membrane channels in the dark. Termination of the rod photoresponse, along with modulation of photoresponse sensitivity during dark and light adaptation, requires biochemical reactions distinct from the excitation pathway described above. Cone photoreceptors have phototransduction pathways similar to rods, but the biochemical machinery of cone phototransduction must differ significantly to account for physiological differences in the light sensitivity and adaptational properties of cones. For reviews of rod and cone phototransduction, see [1–4].

Photoreceptor cells express a single type of PDE (classified as PDE6). In rods, the PDE6 holoenzyme consists of a catalytic heterodimer ($\alpha\beta$) to which two inhibitory γ subunits bind ($\alpha\beta\gamma_2$). Different classes of cone photoreceptors exist in mammalian retina, and each expresses cone-specific PDE6 catalytic (α') and inhibitory (γ') subunits. There is some evidence that PDE6 subunits—particularly the γ subunit—are expressed and function in nonretinal tissues [5,6], but the level of expression is low compared to photoreceptor cells (20 μM). The PDE6 family is one of five families of PDEs that contain noncatalytic, cGMP-binding GAF domains (see Chapter 192 by Glick and Beavo, this volume). The catalytic and regulatory features of the PDE6 family make it uniquely suited as the central effector of the visual transduction pathway.

Structure and Subcellular Localization of Rod PDE6

The primary sequence of the catalytic subunits of PDE6 largely consists of highly conserved regulatory and catalytic domains along with a C-terminal sequence that is the site of posttranslational modifications conferring membrane attachment (Fig. 1). The N-terminal half of the sequence contains homologous GAFa and GAFb domains. The GAF domain

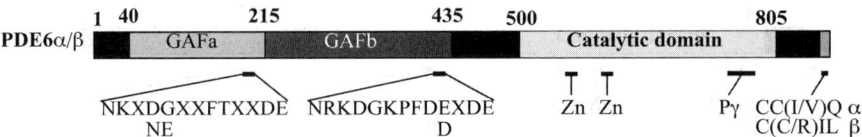

Figure 1 Domain organization of the catalytic subunits of rod PDE6. The bovine rod α (859 a.a.) or β (853 a.a.) subunit each contain two tandem GAF domains (blue and red), a catalytic domain (yellow), and unique C-terminal isoprenylation motifs (green). The consensus sequences for GAFa and GAFb represent amino acid identity [except for positions occupied by two (alternate shown below) or more (denoted by X) different residues]; sequence alignment was performed with human, bovine, mouse, dog, and frog α and β subunits. For the isoprenylation motifs, the frog sequences were omitted.

is a structural module responsible for binding allosteric regulatory molecules [ref. [7,8]; see review by Hurley, this volume]. Based on the crystal structure of the related PDE2 GAF domains [9] and amino acid sequence analysis of the known cGMP-binding phosphodiesterases (PDE2, PDE5, and PDE6), it is likely that the GAFa domains of PDE6 contain high-affinity cGMP binding sites (Fig. 1). An interaction site of the γ subunit with GAFa has been identified by cross-linking experiments [10]. The fact that the PDE6 holoenzyme has 4 GAF domains but only 2 high-affinity cGMP binding sites [11,12,13] raises intriguing questions about the function of the GAFb domains. Whereas 26% of the residues in GAFa are identical in known rod and cone PDE6 sequences, 60% of the GAFb amino acids are identical, including a unique 24 amino acid insert present in PDE6 GAFb but not in other GAF-containing phosphodiesterases.

Progress in elucidating the structure-function relationship of the catalytic domain of PDE6 has relied on its structural similarity to the PDE5 catalytic domain [14], the ability to express recombinant PDE5/cone PDE6 chimeras [15–17], and homology modeling of the catalytic domain based on the PDE4 crystal structure [18]. Two metal ion binding sites critical for catalysis are present in all PDE families [19,20], and substitution of PDE6-specific residues from this region into the PDE5 sequence partially restores PDE6 catalytic efficiency [17]. Using a similar approach, PDE6-specific residues near the catalytic pocket of the enzyme that interact with the Pγ subunit have been delineated [16,17,21].

The arrangement of the GAF and catalytic domains within the quaternary structure of PDE6 has been obtained at 2.8 nm resolution by using electron microscopy and image analysis of individual PDE6 molecules [22]. Three distinct domains, probably representing two GAF domains and the larger catalytic domain, were observed. The molecular organization of PDE5 was highly homologous. The isolated GAFa domain [23] or the catalytic domain [24] of PDE5 have been expressed and retain their ability to bind or hydrolyze cGMP, respectively. These results strongly suggest that individual domains fold into functional units that can interact to generate the final quaternary structure of the catalytic dimer.

Regulation of Rod PDE6 Catalysis by γ

The γ subunit has multiple sites of interaction with the PDE6 αβ dimer and with the α_t^*-GTP subunit of transducin;

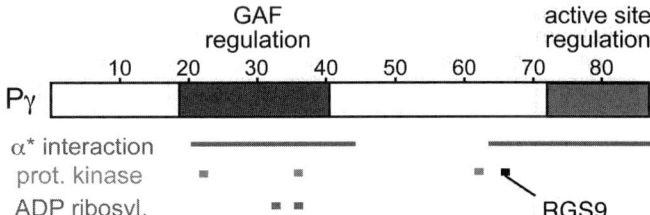

Figure 2 Sites of interaction and regulation of the inhibitory γ subunit of PDE6. The rod γ subunit (87 a.a.) binds to PDE6 at its central GAF interaction domain (red) and its C-terminal inhibitory domain (green). Two major sites of interaction with activated transducin α_t^* subunits are shown in green, and interaction with RGS9 at Val[66] is in black [50]. Phosphorylation sites are shown in orange. ADP ribosylation at either Arg[33] or Arg[36] is in purple [51].

in addition, it serves as a potential substrate for posttranslational modifications (Fig. 2). The C-terminus of γ binds at the entrance to the active site and physically blocks entry of substrate [16,17,25,26]. The central, polycationic region of γ also interacts with the catalytic dimer, and with 50-fold higher affinity than the C-terminal γ region [27]. Binding of this region of γ to the catalytic dimer enhances cGMP binding to PDE6. The combined interactions of the central and C-terminal domains of γ with PDE6 result in very high binding affinity of full-length γ [28]. Recent work has revealed two nonidentical γ binding sites on αβ, one of which is regulated by cGMP binding to the GAF domain [27].

Post-translational modifications of γ have been postulated as feedback mechanisms for PDE6 regulation (Fig. 2). Protein kinases present in rod outer segments can phosphorylate γ at Thr[22] [29–31] or Thr[35] [31–33]. The primary effect of phosphorylation at either site is to markedly reduce γ interaction with activated transducin [29,31,33]. However, feedback regulation of PDE6 or transducin by phosphorylated γ is unlikely, based on the low stoichiometry of γ phosphorylation observed in the above-cited studies. Phosphorylation of γ at Thr[62] in nonretinal tissue culture cells has been linked to regulation of the MAP kinase signaling cascade [6], suggesting that γ might be involved in regulating other signaling pathways in photoreceptor cells.

Catalytic Properties of Nonactivated and Activated PDE6

In rod outer segments, there is equimolar γ subunit relative to the PDE6 catalytic subunit concentration [34].

The very high (K_D~1pM) binding affinity of γ for the catalytic dimer ensures that only 1 out of 2200 αβ dimers lacks bound γ and are activated in the dark-adapted condition [35], in good agreement with electrophysiological estimates of spontaneous PDE6 activation [36].

Removing both γ subunits from the PDE6 catalytic dimer activates the enzyme more than 300-fold, with a catalytic constant (k_{cat}) of 5500 s^{-1} for bovine rod PDE6 [27,37] and close to 8000 s^{-1} for frog PDE6 [34,35]. The catalytic efficiency ($k_{cat}/K_M = 4 \times 10^8$ M^{-1}s^{-1}) for cGMP approaches the diffusion-controlled limit. cAMP is a relatively poor substrate ($k_{cat}/K_M = \sim 3 \times 10^6$ M^{-1}s^{-1}), primarily because of its reduced affinity for the active site [27,35]. Under conditions tested to date, the GAF domains of PDE6 do not allosterically activate or inhibit the active site [27,38].

During visual excitation, PDE6 activation results from binding of $α_t$*-GTP to the PDE6 holoenzyme, thereby releasing the inhibitory constraint of γ. However, the maximum extent of activation of cGMP hydrolysis ($k_{cat} = 4400$ s^{-1} for frog PDE6. [34,35,39,40]) is about half of the fully activated value. In combination with observations that full activation of the PDE6 holoenzyme requires a single $α_t$*-GTP [41], it is likely that only one of the two catalytic sites on the PDE6 dimer is activated by transducin during phototransduction.

Roles of the GAF Domains in PDE6 Regulation

The GAF-containing PDE2, PDE5, and PDE6 families each utilize the GAF domain in a different manner. cGMP binding to PDE2 allosterically stimulates catalysis directly [42,43], whereas cGMP binding to PDE5 induces a conformational change enhancing phosphorylation of the enzyme that subsequently increases catalysis [44].

For PDE6, cGMP occupancy at the GAF domains enhances the affinity of γ for nonactivated PDE6 [31,41]; conversely, binding of γ to the PDE6 catalytic dimer enhances cGMP binding to the two nonidentical GAF domains [35,45]; conversely, binding of γ to the PDE6 catalytic dimer enhances cGMP binding to the two non-identical GAF domains [34,46]. This positive cooperativity of cGMP and γ binding serves to lower the basal catalytic activity of PDE6, and reduces cGMP metabolic flux in the dark-adapted state. Binding of cGMP to the GAF domains also serves to sequester a large majority of the total cellular cGMP, because the concentration of PDE6 is so large (see Chapter 195 by Francis and Corbin, this volume, for detailed discussion).

Upon transducin activation of frog PDE6, displacement of one γ subunit on PDE6 not only fully stimulates catalysis but also lowers cGMP-binding affinity at one GAF domain. Accelerated dissociation of cGMP at this site is accompanied by release of γ from the PDE6 holoenzyme. Rebinding of cGMP to PDE6 causes re-association of γ to the enzyme (see Fig. 8 of [34]).

It has been proposed that the primary purpose of the GAF domains is to release sequestered cGMP to assist guanylyl cyclase in restoring cytoplasmic cGMP levels during the recovery phase of the photoresponse [13,47]. However, the kinetics of cGMP dissociation from the GAF domain are too slow, and the rates of cGMP hydrolysis by activated PDE6 too fast to allow bound cGMP to significantly affect cytoplasmic cGMP levels during recovery to the dark-adapted state.

A more likely role for the GAF domains is to regulate the state of association of γ to catalytic dimer [46]. Free γ has been shown to act in concert with RGS9 to accelerate inactivation of $α_t$*-GTP [48,49]. Only when the GAF domains are unoccupied can γ dissociate from PDE6 and serve in this capacity [38,50]. Dissociation of cGMP (and hence γ) from PDE6 is likely only during prolonged light exposures when PDE6 is persistently activated and cytoplasmic cGMP levels remain low. This negative feedback mechanism involving cGMP, the GAF domains of PDE6, and the γ subunit acting as a GTPase accelerating factor for transducin could underlie the accelerated kinetics and reduced amplitude of the photoresponse characteristic of light-adapted rod photoreceptors.

Conclusion

The lifetime of activated rod PDE6 during rod phototransduction is a highly regulated process. Much work remains to fully describe the molecular events within the PDE holoenzyme and its interactions with transducin that precisely regulate the extent and duration of PDE6 activation. The interaction of PDE6 with other proteins such as the "δ subunit" [51] or GARP2 [52] is poorly understood, but may represent novel mechanisms for regulating the subcellular localization or catalytic activity of PDE6. Finally, it is possible that cross-talk between the visual transduction pathway and as yet uncharacterized signaling pathways may be fundamental to a complete understanding of rod PDE6 regulation during daytime illumination conditions when rod photoreceptors are near or at response saturation.

Acknowledgments

Work from the author's laboratory cited in this article was supported by the National Institutes of Health (National Eye Institute, EY-05798). The author gratefully acknowledges the contributions of past and present members of the lab.

References

1. Pugh, E. N. Jr. and Lamb, T. D. (2000). Phototransduction in vertebrate rods and cones: molecular mechanisms of amplification, recovery and light adaptation, in Stavenga, D. G., DeGrip, W. J., and Pugh, E. N. Jr., Eds., pp. 183–255. Elsevier Science B.V., New York.
2. Burns, M. E. and Baylor, D. A. (2001). Activation, deactivation, and adaptation in vertebrate photoreceptor cells. *Annu. Rev. Neurosci.* **24**, 779–805.
3. Fain, G. L., Matthews, H. R., Cornwall, M. C., and Koutalos, Y. (2001). Adaptation in vertebrate photoreceptors. *Physiol Rev.* **81**, 117–151.
4. Arshavsky, V. Y., Lamb, T. D., and Pugh, E. N., Jr. (2002). G Proteins and phototransduction. *Annu. Rev. Physiol.* **64**, 153–187.
5. Lochhead, A., Nekrasova, E., Arshavsky, V. Y., and Pyne, N. J. (1997). The regulation of the cGMP-binding cGMP phosphodiesterase by proteins that are immunologically related to the γ subunit of the photoreceptor cGMP phosphodiesterase. *J. Biol. Chem.* **272**, 18397–18403.

6. Wan, K. F., Sambi, B. S., Frame, M., Tate, R., and Pyne, N. J. (2001). The inhibitory γ subunit of the Type 6 retinal cyclic guanosine monophosphate phosphodiesterase is a novel intermediate regulating p42/p44 mitogen-activated protein kinase signaling in human embryonic kidney 293 cells. *J. Biol. Chem.* **276**, 37802–37808.

7. Aravind, L. and Ponting, C. P. (1997). The GAF domain: an evolutionary link between diverse phototransducing proteins. *Trends Biochem. Sci.* **22**, 458–459.

8. Anantharaman, V., Koonin, E. V., and Aravind, L. (2001). Regulatory potential, phyletic distribution and evolution of ancient, intracellular small-molecule-binding domains. *J. Mol. Biol.* **307**, 1271–1292.

9. Martinez, S. E., Wu, A. Y., Glavas, N. A., Tang, X. B., Turley, S., Hol, W. G. J., and Beavo, J. A. (2002). The two GAF domains in phosphodiesterase 2A have distinct roles in dimerization and in cGMP binding, *Proc. Natl. Acad. Sci. U. S. A.* **99**, 13260–13265.

10. Muradov, K. G., Granovsky, A. E., Schey, K. L., and Artemyev, N. O. (2002). Direct interaction of the inhibitory γ-subunit of rod cGMP phosphodiesterase (PDE6) with the PDE6 GAFa domains. *Biochemistry* **41**, 3884–3890.

11. Yamazaki, A., Sen, I., Bitensky, M. W., Casnellie, J. E., and Greengard, P. (1980). Cyclic GMP-specific, high affinity, noncatalytic binding sites on light-activated phosphodiesterase. *J. Biol. Chem.* **255**, 11619–11624.

12. Gillespie, P. G. and Beavo, J. A. (1989). cGMP is tightly bound to bovine retinal rod phosphodiesterase. *Proc. Natl. Acad. Sci. USA* **86**, 4311–4315.

13. Cote, R. H. and Brunnock, M. A. (1993). Intracellular cGMP concentration in rod photoreceptors is regulated by binding to high and moderate affinity cGMP binding sites. *J. Biol. Chem.* **268**, 17190–17198.

14. McAllister-Lucas, L., Sonnenburg, W. K., Kadlecek, A., Seger, D., Le Trong, H., Colbran, J. L., Thomas, M. K., Walsh, K. A., Francis, S. H., Corbin, J. D., and Beavo, J. A. (1993). The structure of a bovine lung cGMP-binding, cGMP-specific phosphodiesterase deduced from a cDNA clone. *J. Biol. Chem.* **268**, 22863–22873.

15. Granovsky, A. E., Natochin, M., McEntaffer, R. L., Haik, T. L., Francis, S. H., Corbin, J. D., and Artemyev, N. O. (1998). Probing domain functions of chimeric PDE6α′/PDE5 cGMP-phosphodiesterase. *J. Biol. Chem.* **273**, 24485–24490.

16. Granovsky, A. E. and Artemyev, N. O. (2000). Identification of the γ-subunit interacting residues on photoreceptor cGMP phosphodiesterase, PDE6α′. *J. Biol. Chem.* **275**, 41258–41262.

17. Granovsky, A. E. and Artemyev, N. O. (2001). Partial reconstitution of photoreceptor cGMP phosphodiesterase characteristics in cGMP phosphodiesterase-5. *J Biol. Chem.* **276**, 21698–21703.

18. Xu, R. X., Hassell, A. M., Vanderwall, D., Lambert, M. H., Holmes, W. D., Luther, M. A., Rocque, W. J., Milburn, M. V., Zhao, Y., Ke, H., and Nolte, R. T. (2000). Atomic structure of PDE4: insights into phosphodiesterase mechanism and specificity. *Science* **288**, 1822–1825.

19. Francis, S. H., Colbran, J. L., McAllister-Lucas, L. M., and Corbin, J. D. (1994). Zinc interactions and conserved motifs of the cGMP-binding cGMP-specific phosphodiesterase suggest that it is a zinc hydrolase. *J. Biol. Chem.* **269**, 22477–22480.

20. He, F., Seryshev, A. B., Cowan, C. W., and Wensel, T. G. (2000). Multiple zinc binding sites in retinal rod cGMP phosphodiesterase, PDE6αβ. *J. Biol. Chem.* **275**, 20572–20577.

21. Granovsky, A. E. and Artemyev, N. O. (2001). A conformational switch in the inhibitory γ-subunit of PDE6 upon enzyme activation by transducin. *Biochemistry* **40**, 13209–13215.

22. Tcheudji, J. F., Lebeau, L., Virmaux, N., Maftei, C. G., Cote, R. H., Lugnier, C., and Schultz, P. (2001). Molecular organization of bovine rod cGMP-phosphodiesterase 6. *J. Mol. Biol.* **310**, 781–791.

23. Ho, Y.-S. J., Burden, L. M., and Hurley, J. H. (2000). Structure of the GAF domain, a ubiquitous signaling motif and a new class of cyclic GMP receptor. *EMBO J.* **19**, 5288–5299.

24. Fink, T. L., Francis, S. H., Beasley, A., Grimes, K. A., and Corbin, J. D. (1999). Expression of an active, monomeric catalytic domain of the cGMP-binding cGMP-specific phosphodiesterase (PDE5). *J. Biol. Chem.* **274**, 34613–34620.

25. Artemyev, N. O., Natochin, M., Busman, M., Schey, K. L., and Hamm, H. E. (1996). Mechanism of photoreceptor cGMP phosphodiesterase inhibition by its gamma-subunits. *Proc. Natl. Acad. Sci. USA* **93**, 5407–5412.

26. Granovsky, A. E., Natochin, M., and Artemyev, N. O. (1997). The γ subunit of rod cGMP-phosphodiesterase blocks the enzyme catalytic site. *J. Biol. Chem.* **272**, 11686–11689.

27. Mou, H. and Cote, R. H. (2001). The catalytic and GAF domains of the rod cGMP phosphodiesterase (PDE6) heterodimer are regulated by distinct regions of its inhibitory γ subunit. *J. Biol. Chem.* **276**, 27527–27534.

28. Wensel, T. G. and Stryer, L. (1986). Reciprocal control of retinal rod cyclic GMP phosphodiesterase by its gamma subunit and transducin. *Prot. Struct. Funct. Genet.* **1**, 90–99.

29. Tsuboi, S., Matsumoto, H., Jackson, K. W., Tsujimoto, K., Williams, T., and Yamazaki, A. (1994). Phosphorylation of an inhibitory subunit of cGMP phosphodiesterase in *Rana catesbiana* rod photoreceptors. I. Characterization of the phosphorylation. *J. Biol. Chem.* **269**, 15016–15023.

30. Matsuura, I., Bondarenko, V. A., Maeda, T., Kachi, S., Yamazaki, M., Usukura, J., Hayashi, F., and Yamazaki, A. (2000). Phosphorylation by cyclin-dependent protein kinase 5 of the regulatory subunit of retinal cGMP phosphodiesterase: I. Identification of the kinase and its role in the turnoff of phosphodiesterase in vitro. *J. Biol. Chem.* **275**, 32950–32957.

31. Paglia, M. J., Mou, H., and Cote, R. H. (2002). Regulation of photoreceptor phosphodiesterase (PDE6) by phosphorylation of its inhibitory γ subunit re-evaluated. *J. Biol. Chem.*, in press.

32. Udovichenko, I. P., Cunnick, J., Gonzales, K., and Takemoto, D. J. (1993). Phosphorylation of bovine rod photoreceptor cyclic GMP phosphodiesterase. *Biochem. J.* **295**, 49–55.

33. Xu, L. X., Tanaka, Y., Bonderenko, V. A., Matsuura, I., Matsumoto, H., Yamazaki, A., and Hayashi, F. (1998). Phosphorylation of the gamma subunit of the retinal photoreceptor cGMP phosphodiesterase by the cAMP-dependent protein kinase and its effect on the gamma subunit interaction with other proteins. *Biochemistry* **37**, 6205–6213.

34. Norton, A. W., D'Amours, M. R., Grazio, H. J., Hebert, T. L., and Cote, R. H. (2000). Mechanism of transducin activation of frog rod photoreceptor phosphodiesterase: allosteric interactions between the inhibitory γ subunit and the noncatalytic cGMP binding sites. *J. Biol. Chem.* **275**, 38611–38619.

35. D'Amours, M. R. and Cote, R. H. (1999). Regulation of photoreceptor phosphodiesterase catalysis by its noncatalytic cGMP binding sites. *Biochem. J.* **340**, 863–869.

36. Rieke, F. and Baylor, D. A. (1996). Molecular origin of continuous dark noise in rod photoreceptors. *Biophys. J.* **71**, 2553–2572.

37. Mou, H., Grazio, H. J., Cook, T. A., Beavo, J. A., and Cote, R. H. (1999). cGMP binding to noncatalytic sites on mammalian rod photoreceptor phosphodiesterase is regulated by binding of its γ and δ subunits. *J. Biol. Chem.* **274**, 18813–18820.

38. Arshavsky, V. Y., Dumke, C. L., and Bownds, M. D. (1992). Noncatalytic cGMP binding sites of amphibian rod cGMP phosphodiesterase control interaction with its inhibitory γ-subunits. A putative regulatory mechanism of the rod photoresponse. *J. Biol. Chem.* **267**, 24501–24507.

39. Dumke, C. L., Arshavsky, V. Y., Calvert, P. D., Bownds, M. D., and Pugh, E. N. Jr. (1994) Rod outer segment structure influences the apparent kinetic parameters of cyclic GMP phosphodiesterase. *J. Gen. Physiol.* **103**, 1071–1098.

40. Leskov, I. B., Klenchin, V. A., Handy, J. W., Whitlock, G. G., Govardovskii, V. I., Bownds, M. D., Lamb, T. D., Pugh, E. N. Jr., and Arshavsky, V. Y. (2000). The gain of rod phototransduction: Reconciliation of biochemical and electrophysiological measurements. *Neuron* **27**, 525–537.

41. Melia, T. J., Malinski, J. A., He, F., and Wensel, T. G. (2000). Enhancement of phototransduction protein interactions by lipid surfaces. *J. Biol. Chem.* **275**, 3535–3542.

42. Martins, T. J., Mumby, M. C., and Beavo, J. A. (1982). Purification and characterization of a cyclic GMP-stimulated cyclic nucleotide phosphodiesterase from bovine tissues. *J. Biol. Chem.* **257**, 1973–1979.
43. Yamamoto, T., Manganiello, V. C., and Vaughan, M. (1983). Purification and characterization of cyclic GMP-stimulated cyclic nucleotide phosphodiesterase from calf liver. *J. Biol. Chem.* **258**, 12526–12533.
44. Corbin, J. D., Turko, I. V., Beasley, A., and Francis, S. H. (2000). Phosphorylation of phosphodiesterase-5 by cyclic nucleotide-dependent protein kinase alters its catalytic and allosteric cGMP-binding activities. *Eur. J. Biochem.* **267**, 2760–2767.
45. Yamazaki, A., Hayashi, F., Tatsumi, M., Bitensky, M. W., and George, J. S. (1990). Interactions between the subunits of transducin and cyclic GMP phosphodiesterase in *Rana catesbiana* rod photoreceptors. *J. Biol. Chem.* **265**, 11539–11548.
46. Cote, R. H., Bownds, M. D., and Arshavsky, V. Y. (1994). cGMP binding sites on photoreceptor phosphodiesterase: role in feedback regulation of visual transduction. *Proc. Natl. Acad. Sci. USA* **91**, 4845–4849.
47. Yamazaki, A., Bondarenko, V. A., Dua, S., Yamazaki, M., Usukura, J., and Hayashi, F. (1996). Possible stimulation of retinal rod recovery to dark state by cGMP release from a cGMP phosphodiesterase noncatalytic site. *J. Biol. Chem.* **271**, 32495–32498.
48. Arshavsky, V. Y. and Bownds, M. D. (1992). Regulation of deactivation of photoreceptor G protein by its target enzyme and cGMP. *Nature* **357**, 416–417.
49. He, W., Cowan, C. W., and Wensel, T. G. (1998). RGS9, a GTPase accelerator for phototransduction. *Neuron* **20**, 95–102.
50. Calvert, P. D., Ho, T. W., LeFebvre, Y. M., and Arshavsky, V. Y. (1998). Onset of feedback reactions underlying vertebrate rod photoreceptor light adaptation. *J. Gen. Physiol.* **111**, 39–51.
51. Florio, S. K., Prusti, R. K., and Beavo, J. A. (1996). Solubilization of membrane-bound rod phosphodiesterase by the rod phosphodiesterase recombinant δ subunit. *J. Biol. Chem.* **271**, 1–12.
52. Körschen, H. G., Beyermann, M., Müller, F., Heck, M., Vantler, M., Koch, K. W., Kellner, R., Wolfrum, U., Bode, C., Hofmann, K. P., and Kaupp, U. B. (1999). Interaction of glutamic-acid-rich proteins with the cGMP signalling pathway in rod photoreceptors. *Nature* **400**, 761–766.
53. Slep, K. C., Kercher, M. A., He, W., Cowan, C. W., Wensel, T. G., and Sigler, P. B. (2001). Structural determinants for regulation of phosphodiesterase by a G protein at 2.0 A. *Nature* **409**, 1071–1077.
54. Bondarenko, V. A., Desai, M., Dua, S., Yamazaki, M., Amin, R. H., Yousif, K. K., Kinumi, T., Ohashi, M., Komori, N., Matsumoto, H., Jackson, K. W., Hayashi, F., Usukura, J., Lipkin, V. M., and Yamazaki, A. (1997). Residues within the polycationic region of cGMP phosphodiesterase gamma subunit crucial for the interaction with transducin α subunit: identification by endogenous ADP-ribosylation and site-directed mutagenesis. *J. Biol. Chem.* **272**, 15856–15864.

Spatial and Temporal Relationships of Cyclic Nucleotides in Intact Cells

Manuela Zaccolo,[1] Marco Mongillo,[2] and Tullio Pozzan[2]

Venetian Institute of Molecular Medicine, Padua, Italy,
[1]*Dulbecco Telethon Institute and* [2]*Department of Biomedical Sciences,*
University of Padua, Italy

The Complexity of Cyclic Nucleotides Signaling

Protein phosphorylation is the most common posttranslational mechanism for regulating cellular functions. Regulation by this covalent modification of proteins is not a simple on-off mechanism occurring homogeneously within cells. On the contrary, activation and inhibition of protein kinases and phosphatases are tightly controlled both in time and in space. The exact timing of signal kinetics, the quantitative determination of the speed of signaling molecule diffusion, and their precise location within living cells are necessary prerequisites to a better understanding of cell physiology and pathology [1]. Protein kinases and phosphatases are often discretely localized within the cell in close proximity to receptors and targets, and this location appears critical for speed and specificity of response. The finding that signaling components are highly organized at the plasma membrane, in the cytoplasm, and in the nucleus has led to the proposal of "signaling domains" [2], that is, specific compartments within the three-dimensional matrix of the cell where the signal is generated and/or specifically targeted. cAMP-mediated signaling, for example, complies with tight local control, and the structural basis for such compartmentation is being uncovered. In hippocampal neurons the presence of a macromolecular signaling complex including the β2-adrenergic receptor, a G protein, adenylyl cyclase (AC), protein kinase A (PKA), the phosphatase PP2A, and the ultimate effector of the receptor, the L-type Ca^{2+}, channel has been documented [3], and a highly localized signal transduction from the receptor to the channel has been demonstrated. A crucial role in the nucleation of such signaling domains is played by A kinase anchoring proteins (AKAPs) [4], a family of functionally related proteins that anchor the regulatory subunit of PKA and possess unique targeting sequences that direct the PKA-AKAP complex to specific subcellular compartments. Within the "signaling domain" not only receptors, effectors and their ultimate targets are found in close proximity, but evidence is also accumulating that compartmentation involves the small diffusible molecules that act as intracellular second messengers and that second messenger limited diffusion can contribute to the efficiency and specificity of signal transmission [5].

The cyclic nucleotides cAMP and cGMP are freely diffusible intracellular second messengers that mediate specific responses to a staggering number of extracellular stimuli by activating downstream effectors, mainly the cAMP- and cGMP-dependent protein kinases (PKA and PKG, respectively). PKA and PKG in turn can potentially phosphorylate hundreds of different cellular targets. The enormous load of specific information that enters the cell and is funnelled through these two second messengers is relayed to the final effector systems without any loss of informational detail. The issue raised is how a single molecule can generate the appropriate response to so many different stimuli. One possible way that cyclic nucleotides could expand their signaling capability enough to accomplish this task is by adopting multiple signaling codes. According to this view, transmission of information is not exclusively dependent on the amplitude of the response but can be frequency-dependent [6], similar to what has been shown for Ca^{2+} oscillations, or space-dependent [7]; that is, signal transmission can take place in restricted subcellular compartments, while other signals are excluded. Several features of the cyclic nucleotides signaling cascades provide a potential molecular basis for such diversity in signaling modes [8]. First, intracellular cyclic nucleotide concentrations appear to be highly flexible and transient in nature because of a tightly controlled equilibrium of synthesis and breakdown. Second, both cyclases (the enzymes that make cyclic nucleotides) and phosphodiesterases (the enzymes that degrade cyclic nucleotides) exist in several isoforms with different tissue distribution, different intracellular targeting, specific regulation, and cross-talk with other signaling pathways. Third, Ca^{2+}, for which a frequency-encoded and a space-encoded mode of signaling has been clearly established, modulates both the synthesis and the breakdown of cyclic nucleotides by regulating, in an isoform-specific manner, both cyclases and phosphodiesterases [8].

A thorough and accurate analysis of the spatio-temporal aspects of cyclic nucleotides signaling appears to be crucial for our understanding of cell physiology [9,10]. So far, such analysis has been very difficult due to technical limitations. Lately, however, there have been methodological advances that allow monitoring of cyclic nucleotide in real time in live cells, thus opening the way to a detailed description in space and time of cAMP and cGMP biochemistry *in vivo*.

Methodological Advances

Traditionally, cyclic nucleotide concentration has been measured by immunoassay on cell lysates. Such a destructive method has several limitations: it is rather insensitive and often phosphodiesterase inhibition is necessary to achieve adequate sensitivity [11,12]; only the average concentration of the cyclic nucleotide in the cell population is recorded; the total, rather then the free cyclic nucleotide content, is detected; the temporal resolution is rather limited; any topographical information on the location of the response is cancelled out. A few attempts have been made to localize cyclic nucleotides in fixed tissue by immunocytochemistry [13] but accurate quantification is a problem, and time-courses are difficult to record.

Two nondestructive approaches have been pursued to monitor dynamically and in real time the levels of cyclic nucleotides in single live cells: (1) recombinant cAMP-cGMP sensitive channels, and (2) fluorescent probes sensitive to cAMP-cGMP and/or cAMP-cGMP dependent phosphorylation. The first method utilizes cyclic nucleotide-gated (CNG) ion channels genetically engineered to be especially sensitive to either cGMP [14] or cAMP [15]. One version of such methodology artificially introduces CNG channels into various cell types via the "patch-cramming" technique: an excised, inside-out membrane patch containing the channels is "crammed" into a recipient cell to measure cGMP concentration in the cytosol near the tip of the pipette [16]. The advantages of this approach are the possibility of easily calibrating the system and the minimal alteration or buffering of the intraclualar level of cGMP. One limitation is that detection of the cyclic nucleotide level is confined to the restricted area where the patch pipette is inserted. Also, given the relatively large size of the patch pipette, this approach is applicable only to relatively large cells (above about $40\,\mu m$ in diameter). Alternatively, CNG channels have been used to study the membrane-localized, cAMP-signaling pathway by expressing the channel at the plasma membrane, where the adenylyl cyclase is known to reside, and by measuring either cAMP-induced currents [17] or cAMP-induced Ca^{2+} influx via Ca^{2+}-sensitive fluorescent dyes [18].

The second approach for real-time monitoring of cyclic nucleotide levels in live cells relies on imaging of cyclic nucleotide-activated protein kinases (or of their target peptides) that have been fluorescently labeled and are competent for fluorescence resonance energy transfer (FRET) [19]. FRET is a physico-chemical phenomenon whereby the excited state energy of a donor fluorophore is non-radiatively transferred to a nearby acceptor fluorophore. For FRET to occur, the following requirements must be satisfied: (1) substantial overlap between the emission spectrum of the donor and the excitation spectrum of the acceptor, (2) alignment relative to each other in space of the fluorophore transition dipoles, and (3) distance between donor and acceptor fluorophores within 1 to 10 nm. The efficiency of FRET decreases with the sixth power of the distance between the fluorophores [20], making FRET a phenomenon that is exquisitely suitable for detecting conformational changes.

Most FRET-based probes for cyclic nucleotides exploit either PKA or PKG as the natural sensors for cAMP and cGMP, respectively. They take advantage of the conformational change that results from binding of the cyclic nucleotide to the kinase and that leads to a change in the distance between the FRET-donor and acceptor fluorescent labels.

Inactive PKA is a heterotetramer made of a dimer of regulatory subunits (R_2) and two catalytic subunits (C). When cAMP levels rise, two molecules of the second messenger bind to each of the R subunits, thus inducing a conformational change that leads to dissociation of two active C subunits. The prototype FRET-based sensor for cAMP consists of a fluorescein-labeled C subunit and a rhodamine-labeled R subunit [21]. This probe requires *in vitro* labeling of purified C and R subunits and subsequent microinjection in the cell of interest, which limits its use only to a restricted number of applications. Moreover, this very laborious approach suffers from several limitations [22,23], including probe instability, nonspecific compartmentation, and some toxic effect on the injected cell. A variant of this sensor has been generated that is entirely genetically encoded and in which the donor and acceptor fluorophores are the blue (BFP) and green (GFP) variants of the green fluorescent protein [24]. When cAMP is low PKA is mainly in the holotetrameric conformation R-BFP_2C-GFP_2 and the two fluorophores are close enough to generate FRET. Upon excitation of BFP at its proper excitation wavelength (380 nm), part of its excited state energy is emitted as blue light (460 nm) and part is transferred to the nearby acceptor GFP, which in turn is excited and emits green light (510 nm). When cAMP levels increase, C-GFP subunits are released, the two fluorophores diffuse apart, and FRET is abolished. In this condition excitation of BFP at 380 nm only generates emission of blue light and no green light is emitted. FRET can be conveniently measured as the ratio of donor to acceptor emission with the advantage of cancelling out fluorescence intensity variations due to probe concentration, optical path length, and excitation intensity [25]. More recently this probe was modified by changing the BFP-GFP couple with the more convenient mutants CFP-YFP [3].

Based on the same general principle, sensors for cGMP have been generated [26,27]. In the probe denominated "cygnet," the cyan (CFP) and yellow (YFP) variants of the green fluorescent protein have been genetically fused to the amino- and carboxy-terminus of PKG, respectively. CFP and YFP act as donor and acceptor for FRET. The relatively large conformational change generated by cGMP binding to the PKG regulatory domain moves the donor CFP away from the acceptor YFP, thus reducing FRET.

Two other genetically encoded probes have been recently generated to monitor the cAMP dynamics in living cells. These two probes are also based on FRET between two mutants of GFP. Rather than directly probing cAMP levels, these sensors report the PKA-mediated phosphorylation of a substrate peptide. In the first case the two variants of GFP are joined by the kinase-inducible domain (KID) of the transcription factor CREB (cAMP-responsive element binding protein) [28]. The phosphorylation of KID by PKA decreases FRET among the flanking GFPs. The cAMP probe described by Zhang and coworkers is based on a similar principle. In this latter case the fusions of CFP, a phosphoamino-acid-binding domain (14-3-3τ), a consensus substrate for PKA, and YFP results in a change of FRET dependent on PKA-mediated phosphorylation [29].

Functional Compartments of cAMP in Heart Cells

Mayer and coworkers [30] were the first to propose, more than 20 years ago, the existence of spatially restricted domains of cAMP. The hypothesis was formulated to explain a series of data obtained in cardiac myocytes. In particular, the authors found that the positive inotropic and lusitropic effects induced by β-adrenergic stimulation correlated with activation of a particulate, membrane-bound fraction of PKA; similar elevations of cAMP due to prostaglandin 1 resulted in the activation of a cytosolic form of PKA and in minimal functional consequences [31]. Similarly, stimulation of cardiac myocytes with glucagon-like peptide-1 (GLP-1) generates a rise in $[cAMP]_i$ comparable to that elicited by isoproterenol, a β-AR agonist, but causes modest negative inotropy and no lusitropic effect, in sharp contrast with isoproterenol [32]. Moreover, in contrast to $β_1$-AR, $β_2$-AR stimulation fails to induce a cAMP-dependent phosphorylation of non-sarcolemmal proteins (such as phospholamban, or the myofilament proteins troponin I and C protein), but it does activate sarcolemmal L-type Ca^{2+} channels [33], suggesting a differential spatial organization of the PKA activated by the two β-ARs.

More convincing, yet indirect evidence, for localized domains of cAMP was provided by Jurevicius and Fishmeister [34]. In a series of elegant experiments carried out by whole-cell patch-clamp recordings of Ca^{2+} currents in frog heart cells they showed that a local stimulus of the β-adrenergic receptor causes a local activation of I_{Ca}, whereas local application of the adenylyl cyclase activator forskolin induces activation of I_{Ca} throughout the cell. The authors suggested that β-adrenergic stimulation generates a localized accumulation of cAMP sufficient to efficiently activate only nearby located Ca2+ channels, whereas nonspecific activation of AC induces a generalized rise of cAMP concentration [34]. Due to the experimental setup, the spatial resolution of these experiments was in the order of 20 μm, making it difficult to anticipate the physiological relevance of such cAMP compartments in a 10 to 15 μm wide mammalian cell. New insight into this issue came from experiments in which cAMP dynamics were monitored in real time in rat neonatal cardiac myocytes by using a FRET-based, genetically encoded sensor for cAMP [24]. In these cells, β-adrenergic stimulation generates multiple microdomains with increased concentration of cAMP in correspondence of the T tubular system (Fig. 1). T tubules in heart cells run along sarcomeric Z lines whose distance from one another is about 2 μm. The local gradients of the second messenger spread for less than half such a distance, implying a range of action for cAMP as small as about 1 μm [35].

Local control and limited diffusion of cAMP is not a prerogative of heart cells. In fact, in human embryonic kidney cells it has been demonstrated, by using recombinant cyclic nucleotide-gated channels as a cAMP sensor, that a uniform extracellular stimulus with PGE_1 initiates a transient increase in cAMP concentration near the plasma membrane and a sustained rise of the second messenger in the bulk cytosol [18], suggesting that spatially and temporally distinct cAMP signals can coexist within simple cells.

An important role in cAMP diffusional restriction seems to be played by phosphodiesterases, the cAMP degrading enzymes. In fact, PDE inhibition allows uniform spreading of the second messenger from the restricted microdomains along the T tubular network into the bulk cytosol [35] (Fig. 1). It is interesting that certain PDE inhibitors seem to cause "spillover" of PGE_1-induced cAMP from the soluble into the particulate compartment of ventricular myocytes [7]. A reduction in the localized cAMP effects upon inhibition of PDE was observed also in the frog heart cells locally stimulated with isoproterenol [34], and a role for PDE in shaping the local cAMP response underneath the plasma membrane has been uncovered in embryonic kidney cells [18]. In support of the notion that PDE can control cAMP spreading is the finding that PDEs have also been found to be targeted to specific subcellular compartments [36] and in cardiac myocytes PDEs have been found to be anchored to AKAPs [31,37]. Alternative mechanisms, however, cannot be excluded. In the case of the $β_2$-AR, in fact, local signaling has been suggested to result from the coupling of $β_2$-AR to Gi proteins that may act locally, possibly through a protein phosphatase-dependent mechanism [38].

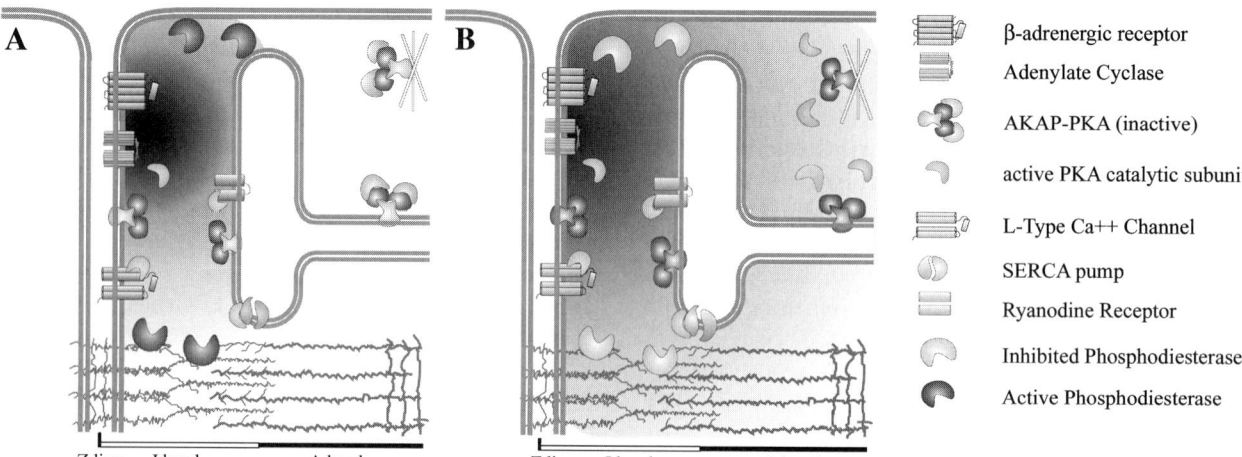

Figure 1 PKA, the main effector of cAMP, is a key regulator of excitation-contraction coupling in muscle cells. In the heart, sympathetic control of the frequency and strength of contraction is exerted by β-AR stimulation, activation of G proteins, and in turn activation of adenylyl cyclase and synthesis of cAMP. The second messenger activates PKA, which by phosphorylating L-type Ca^{2+} channels and the ryanodine receptor increases the amount of Ca^{2+} ions available for contraction of the sarcomere, and by phosphorylating phospholamban (PLB) it favors Ca^{2+} reuptake by the calcium pump SERCA-2. In heart cells, G proteins [44], AC [44], PKA [45], A kinase anchoring proteins (AKAPs) [37,45], the L-type Ca^{2+} channels [44], and the ryanodyne receptor [46] have all been found within the T tubular network/junctional SR membrane, thus forming a "signaling domain" that provides a potential anatomical basis for the confined activation of the cAMP signaling pathway.

When the β-adrenergic agonist interacts with the β-AR, a series of G-protein-mediated changes leads to activation of adenylyl cyclase and synthesis of cAMP. The latter acts by means of PKA to phosphorylate the L-type Ca^{2+} channel, which increases the inward movement of Ca^{2+} ions through the membrane of the T tubule. These Ca^{2+} ions release more Ca^{2+} from the sarcoplasmin reticulum (SR) through the ryanodine receptor, and more Ca^{2+} is thus available for binding to sarcomeric proteins for contraction. PKA also activates phospholamban (PLB), thus increasing Ca^{2+} reuptake through the Ca^{2+} pump SERCA-2, located on the SR membrane. All the molecular elements involved in the β-adrenergic signaling pathway are clustered in the region of the T tubular network/junctional SR membrane. PKA itself is anchored in this area through binding to AKAPs [45]. Activation of the β-AR generates in these cells a localized increase of cAMP, which in turn activates a subpopulation of PKA that is confined within the same signaling domain [35] (panel A). Compartmentation of cAMP is due to the activity of phosphodiesterases. Indeed, inhibition of PDE allows homogeneous diffusion of cAMP in the cytosol [35] (panel B). The bars at the bottom of panels A and B indicate sarcomeric elements.

Spatio-temporal Aspects of Cyclic Nucleotides Signaling in Neurons

Although both cAMP and cGMP have been shown to be involved in the molecular processes that govern certain types of learning and memory and, in particular, in activity-induced synaptic plasticity [39,40], most of the spatiotemporal intricacies of cyclic nucleotide signaling in neural circuits still remain to be elucidated [41,42]. Some of the available data, however, point to an important role of the spatial organization of the second messenger in controlling the output. cAMP is involved in both short- and long-term synaptic changes, and functional compartmentation of the second messenger has been proposed as one of the possible mechanisms that control which effect, either short- or long-term, is put in place [42]. In the monosynaptic connection between sensory and motor neurons of the gill-withdrawal reflex in the marine snail *Aplysia* the generalized application of serotonin very rapidly produces (20 sec) a spatial gradient of free cAMP, with the higher concentration being recorded at the distal neuronal processes and the lower concentration at the cell body. On the contrary, only a very prolonged stimulation (2 hrs) is accompanied by translocation of the PKA catalytic subunit to the nucleus [43]. The interpretation of these results is that the observed gradients confine cAMP where it is most needed for short-term plasticity, that is at the tip of the processes where most of the presynaptic terminals reside. The attenuation of the cAMP signal at the cell body could represent a safety mechanism to ensure that only repeated stimulation leads to active PKA C subunits entering the nucleus, activating transcriptional programs, and therefore generating long-term effects [43].

Conclusions

In the last decades it has become clear that signal transduction does not rely on stochastic events but rather on the sophisticated organization of molecular interactions that are precisely regulated both in time and space. Cyclic nucleotides are key players in the relay of intracellular signals, and are bound to be subject to the same strict control. The fine description of cAMP and cGMP dynamics has been hampered in the past by the lack of adequate technology for real-time monitoring of the levels of these second messengers in live cells. The recent development of imaging-based methodologies that allow single cell analysis and very high spatial and temporal resolution will certainly be crucial in our understanding of cAMP and cGMP biochemistry *in vivo*. At present, the large majority of the data available on the spatio-temporal aspects of cyclic nucleotide signaling refers to cAMP, and although some evidence for subcellular heterogeneity in cGMP has been presented in the past, no direct single cell data on this topic have yet been published (at the moment this manuscript was submitted). Certainly, in the coming years we can expect to see the uncovering of a finely detailed map of the molecular mechanisms that orchestrate both cAMP and cGMP signal transduction.

References

1. Hunter, T. (2000). Signaling-2000 and beyond. *Cell* **100**, 113–127.
2. Pawson, T. and Scott, J. D. (1997). Signaling through scaffold, anchoring and adaptor proteins. *Science* **278**, 2075–2080.
3. Davare, M. A., Avdonin, V., Hall, D. D., Peden, E. M., Burette, A., Weimberg, R. J., Horne, M. C., Hoshi, T., and Hell, J. W. (2001). A β$_2$ adrenergic receptor signaling complex assembled with the Ca^{2+} channel Ca$_v$1.2. *Science* **293**, 98–101.
4. Colledge, M. and Scott, J. D. (1999). AKAPs: from structure to function. *Trends Cell Biol.* **9**, 216–221.
5. Zaccolo, M., Magalhaes, P., and Pozzan, T. (2002). Compartmentation of cAMP and Ca^{2+} signals. *Curr. Opin. Cell. Biol.* **14**, 160–166.
6. Cooper, D. M., Mons, N., and Karpen, J. K. (1995). Adenylyl cyclases and the interaction between calcium and cAMP signaling. *Nature* **374**, 421–424.
7. Steinberg, S. F. and Brunton, L. L. (2001). Compartmentation of G protein-coupled signaling pathways in cardiac myocytes. *Annu. Rev. Pharmacol. Toxicol.* **41**, 751–773.
8. Houslay, M. D. and Milligan, G. (1997). Tailoring cAMP-signaling responses through isoform multiplicity. *Trends Biochem Sci.* **22**, 217–224.
9. Kasai, H. and Petersen, O. H. (1994). Spatial dynamics of second messengers:IP3 and cAMP as long-range and associative messengers. *Trends Neurosci.* **17**, 95–101.
10. Jordan, J. D., Landau, E. M., and Iyengar, R. (2000). Signaling networks: the origins of cellular multitasking. *Cell* **103**, 193–200.
11. Leitman, D. C. and Murad, F. (1986). Comparison of binding and cyclic GMP accumulation by atrial natriuretic peptides in endothelial cells. *Biochim. Biophys. Acta* **885**, 74–79.
12. Ledbetter, J. A., Parsons, M., Martin, P. J., Hansen, J. A., Rabinovitch, P. S., and June, C. H. (1986). Antibody binding to CD5 (Tp67) and Tp44 T cell surface molecules: effects on cyclic nucleotides, cytoplasmic free calcium, and cAMP-mediated suppression. *J. Immunol.* **137**, 3299–3305.
13. Borsani, J. and Marx, S. J. (1990). Immunocytology on microwave-fixed cells reveals rapid and agonist-specific changes in subcellular accumulation patterns for cAMP or cGMP. *Proc. Natl. Acad. Sci USA* **87**, 1188–1192.
14. Goulding, E. H., Tibbs, G. R., and Siegelbaum, S. A. (1994). Molecular mechanism of cyclic nucleotide-gated channel activation. *Nature* **372**, 369–374.
15. Rich T. C., Tse, T. E., Rohan, J. G., Schaack, J., and Karpen, J. W. (2001). In vivo assessment of local phosphodiesterase activity using tailored cyclic nucleotide-gated channels as cAMP sensors. *J. Gen. Physiol.* **118**, 63–78.
16. Triverdi, B. and Kramer, R. H. (1998). Real-time patch-cram detection of intracellular cGMP reveals long-term suppression of responses to NO and muscarinic agonists. *Neuron* **21**, 895–906.
17. Rich, T. C., Fagan, K. A., Nakata, H., Schaack, J., Cooper, D. M., and Karpen, J. W. (2000). Cyclic nucleotide-gated channels colocalize with adenylyl cyclase in regions of restricted cAMP diffusion. *J. Gen. Physiol.* **116**, 147–161.
18. Rich, T. C., Fagan, K. A., Tse, T. E., Schaack, J., Cooper, D. M., and Karpen, J. W. (2001). A uniform extracellular stimulus triggers distinct cAMP signals in different compartments of a simple cell. *Proc. Natl. Acad. Sci. USA* **98**, 13049–13054.
19. Zaccolo, M., Filippin, L., Magalhães, P., and Pozzan, T. (2001). Heterogeneity of second messenger levels in living cells. *Novartis Found. Symp.* **239**, 85–93.
20. Clegg, R. M. (1996). Fluorescence resonance energy transfer (FRET), in Wang, X. F., Herman, B., Eds., *Fluorescence Imaging, Spetroscopy and Microscopy*. Wiley, New York.
21. Adams, S. R., Harootunian, A. T., Buechler, Y. J., Taylor, S. S., and Tsien, R. Y. (1991). Fluorescence ratio imaging of cyclic AMP in single cells. *Nature* **249**, 694–697.
22. Goaillard, J. M., Vincent, P., and Fischmeister, R. (2001). Simultaneous measurements of intracellular cAMP and L-type Ca^{2+} current in single frog ventricular myocytes. *J. Physiol.* **530**, 79–91.
23. Webb, R. J., Bains, H., Cruttwell, C., and Carroll, J. (2002). Gap-junctional communication in mouse cumulus-oocyte complexes: implications for the mechanism of meiotic maturation. *Reproduction* **123**, 41–52.
24. Zaccolo, M., De Giorgi, F., Cho, C. Y., Feng, L., Knapp, T., Negulescu, P. A., Taylor, S. S., Tsien, R. Y., and Pozzan, T. (2000). A genetically encoded, fluorescent indicator for cyclic AMP in living cells. *Nat. Cell Biol.* **2**, 25–29.
25. Tsien, R. Y. (1989). Fluorescent probes of cell signaling. *Annu. Rev. Neurosci.* **12**, 227–253.
26. Sato, M., Hida, N., Ozawa, T., and Umezawa, Y. (2000). Fluorescent indicators for cyclic GMP based on cyclic GMP-dependent protein kinase Ialpha and green fluorescent proteins. *Anal. Chem.* **72**, 5918–5924.
27. Honda, A., Adams, S. R., Sawyer, C. L., Lev-Ram, V., Tsien, R. Y., and Dostmann, W. R. G. (2001). Spatiotemporal dynamics of guanosine 3′,5′-cyclic monophosphate revealed by a genetically encoded, fluorescent indicator. *Proc Natl. Acad. Sci. USA* **98**, 2437–2442.
28. Nagai, Y., Miyazaki, M., Aoki, R., Zama T., Inouye, S., Hirose, K., Iino, M., and Hagiwara, M. (2000). A fluorescent indicator for visualizing cAMP-induced phosphorylation in vivo. *Nat Biotechnol.* **18**, 262–263.
29. Zhang, J., Ma, Y., Taylor, S. S., and Tsien, R. Y. (2001). Genetically encoded reporters of protein kinase A activity reveal impact of substrate tethering. *Proc. Natl. Acad. Sci. USA* **98**, 14997–5002.
30. Brunton, L. L., Hayes, J. S., and Mayer, S. E. (1981). Functional compartmentation of camp and protein kinase in heart. *Adv. Cyclic Nucleotide Res.* **14**, 391–397.
31. Hayes, J. S., Brunton L. L., and Mayer, S. E. (1980). Selective activation of particulate camp-dependent protein kinase by isoproterenol and PGE$_1$. *J Biol. Chem.* **255**, 5113–5119.
32. Vila Petroff, M. G., Egan, J. M., Wang, X., and Sollot, S. J. (2001). Glucagon-like peptide-1 increases camp but fails to augment contraction in adult rat cardiac myocytes. *Cir. Res.* **89**, 445–452.
33. Zhou, Y. Y., Cheng, H., Bogdanov, K. Y., Hohl, C., Altshuld, R., Lakatta, E. G., and Xiao, R. P. (1997). Localized camp-dependent signaling mediates beta 2-adrenergic modulation of cardiac excitation-contraction coupling. *Am. J. Physiol.* **273**, 611–618.
34. Jurevicius, J. and Fischmeister, R. (1996). cAMP compartmentation is responsible for a local activation of Ca^{2+} channels by b-AR agonists. *Proc. Natl. Acad. Sci. USA* **93**, 295–299.
35. Zaccolo, M. and Pozzan, T. (2002). Discrete microdomains with high concentration of cAMP in stimulated rat neonatal cardiac myocytes. *Science* **295**, 1711–1715.
36. Houslay, M. D., Sullivan, M., and Bolger G. B. (1998). The multienzyme PDE4 cyclic adenosine monophosphate-specific phosphodiesterase family: intracellular targeting, regulation, and selective inhibition by compounds exerting anti-inflammatory and antidepressant actions. *Adv. Pharmacol.* **44**, 225–342.
37. Verde, I., Pahlke, G., Salanova, M., Zhang, G., Wang, S., Coletti, D., Onuffer, J., Jin, S. L. C., and Conti, M. (2001). Myomegalin is a novel protein of the Golgi/centrosome that interacts with a cyclic nucleotide phosphodiesterase. *J. Biol. Chem.* **276**, 11189–11198.
38. Kuschel, M., Zhou, Y. Y., Cheng, H., Zhang, S. J., Chen, Y., Lakatta, E. G., and Xiao, R. P. (1999). G$_i$ protein-mediated fubctional compartmentalization of cardiac b$_2$-adrenergic signaling. *J. Biol. Chem.* **274**, 22048–22052.

39. Bailey, C. H., Bartsch, D., and Kandel, E. R. (1996). Toward a molecular definition of long-term memory storage. *Proc. Natl. Acad. Sci. USA* **93**, 13445–13452.
40. Zhuo, M. and Hawkins, R. D. (1995). Long-term depression: a learning-related type of synaptic plasticity in the mammalian central nervous system. *Rev. Neurosci.* **6**, 259–277.
41. Lev-Ram, V., Jiang, T., Wood, J., Lawrance, D. S., and Tsien, R. Y. (1997). Synergies and coincidence requirements between NO, cGMP, and Ca^{2+} in the induction of cerebellar long-term depression. *Neuron* **18**, 1025–1038.
42. Hempel, C. M., Vincent, P., Adams, S. R., Tsien, R. Y., and Selverston, A. I. (1996). Spatio-temporal dynamics of cyclic AMP signals in an intact neural circuit. *Nature* **384**, 166–169.
43. Bacskai, B. J., Hochner, B., Mahaut-Smith, M., Adams, S. R., Kaang, B. K., Kandel, E. R., and Tsien, R. Y. (1993). Spatially resolved dynamics of cAMP and protein kinase A subunits in *Aplysia* sensory neurons. *Science* **260**, 222–226.
44. Laflamme, M. A. and Becker, P. L. (1999). Gs and adenylyl cyclase in transverse tubules of heart: implications for cAMP-dependent signaling. *Am. J. Physiol.* **46**, H1841–H1848.
45. Yang, J., Drazba, J. A., Ferguson, D. G., and Bond, M. (1998). A-kinase anchoring protein 1000 (AKAP100) is localized in multiple subcellular compartments in adult rat heart. *J. Cell Biol.* **142**, 511–522.
46. Carl, S. L. K., Felix, K., Caswell, A. H., Brandt, N. R., Ball, W. J., Vaghy, P. L., Meissner, G., and Ferguson, D. J. (1995). Immunolocalization of sarcolemmal dihydropyridine receptor and sarcoplasmic reticular triadin and ryanodyne receptor in rabbit ventricle and atrium. *J. Cell. Biol.* **126**, 673–682.

CHAPTER 198

Regulation of Cyclic Nucleotide Levels by Sequestration

Jackie D. Corbin,[1] Jun Kotera,[1] Venkatesh K. Gopal,[1]
Rick H. Cote,[2] and Sharron H. Francis[1]

[1]*Department of Molecular Physiology and Biophysics,
Vanderbilt University School of Medicine, Nashville, Tennessee and
[2]Department of Biochemistry and Molecular Biology,
University of New Hampshire, Durham, New Hampshire*

Introduction

Cyclic nucleotides bind to specific proteins in cells and are thereby sequestered into different functional compartments. Some of the bound cyclic nucleotide serves to activate target proteins such as protein kinases or ion channels while another portion is bound to sites in other proteins and therefore inactive for stimulating protein kinases or ion channels. Sequestration into or release from such inactive sites could be modulated in response to certain stimuli. In order for sequestration to be physiologically relevant, cyclic nucleotide-binding proteins should have ample affinity and be present in sufficient amounts to bind a significant portion of the total cellular cyclic nucleotide. Moreover, the sequestration should be a regulated process. These conditions appear to be met for cGMP binding to the allosteric (noncatalytic) sites of phosphodiesterase-5 in penile corpus cavernosum or of phosphodiesterase-6 in retina.

It is usually emphasized that extracellular signals modulate levels of cyclic nucleotides in cells by regulation of the activities of adenylyl and guanylyl cyclases. Changes in concentrations of signals presented to cells are thus critical determinants for rate of formation of cAMP or cGMP. Rate of loss, or inactivation, of cAMP or cGMP also contributes importantly to determining the cellular concentration of these nucleotides, but this aspect has been studied much less. There are at least three possible routes of loss of active cellular cyclic nucleotides. These are illustrated for cGMP in Fig. 1.

First, loss of cellular cAMP into the extracellular space can occur by efflux such as through leakage or specific transport processes [1,2]. This energy-dependent transport process for cAMP and cGMP may involve an organic anion transport process that utilizes multidrug resistance proteins in some tissues [3–5]. The rate of cyclic nucleotide efflux appears to be dependent on intracellular cyclic nucleotide levels and independent of direct signal regulation. Furthermore, loss of cyclic nucleotides by hydrolysis greatly exceeds that achieved by efflux [1,6–9].

Second, loss of cyclic nucleotides through degradation by cAMP and cGMP phosphodiesterases (PDEs) has been widely investigated [10–16]. PDEs have been shown to be regulated by a variety of signaling pathways [10–16]. See several chapters in this volume for additional information on specific PDE families.

Third, sequestration of cyclic nucleotide, which is the focus of this chapter, is also a possible route of loss of the cytoplasmic, active pool of cyclic nucleotide. In contrast to loss of cyclic nucleotide by either efflux or degradation, this process would be largely reversible.

Physiological sequestration of second messengers and other signaling agents is not a new concept. For example, Ca^{2+} sequestered in cardiomyocyte sarcoplasmic reticulum is unavailable to stimulate contraction of heart muscle until its release due to depolarization of the cardiomyocyte [17]. Compartmentalization and regulated trafficking of protein kinases and myriad other proteins among plasma membrane, cytoplasm, nucleus, and other organelles is commonplace.

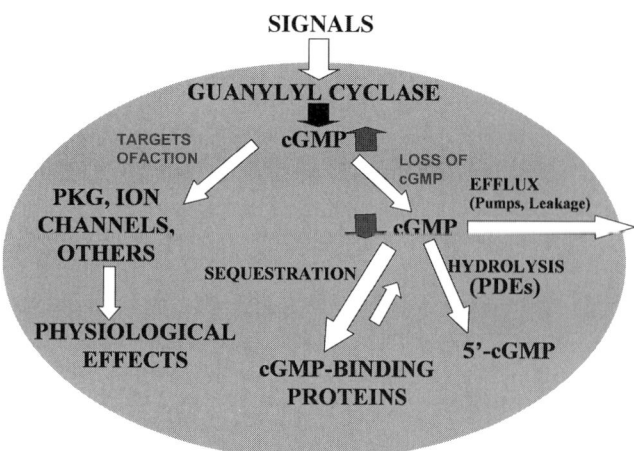

Figure 1 Routes of formation and loss of cGMP in cells.

Likewise, there is increasing evidence for compartmentalization of cAMP into subcellular microdomains that impose diffusional restriction on the nucleotide [18,19]. A portion of the cellular cyclic nucleotide pool can be bound as inert forms to certain cellular proteins. As such, this pool is diffusionally restricted and less available for degradation or efflux, but upon dissociation, this cyclic nucleotide can join the pool of free cyclic nucleotides or traffic among cyclic nucleotide-binding proteins. Thus, some proteins could act as sinks for cyclic nucleotides, not only providing for inhibition of action by sequestration but also as a source of active cyclic nucleotide after release by certain stimuli.

Localization of cyclic nucleotide-binding proteins to particular compartments within cells would concentrate both the proteins and bound cyclic nucleotide within these locales [18]. Compartmentalization of cyclic nucleotides with their target proteins such as PKA and PKG, or with PDEs, almost certainly occurs and is likely to be physiologically advantageous. Co-localization of cyclic nucleotides with PDE activity could also be brought about evolutionarily by fusing noncatalytic cyclic nucleotide-binding domains with catalytic domains responsible for cyclic nucleotide hydrolysis. This would effectively concentrate the cyclic nucleotide in the microenvironment of the catalytic site. Upon dissociation from the binding site, the nucleotide would most likely be hydrolyzed rather than reentering the active pool.

Conclusive evidence for sequestration of cAMP into inactive pools by cellular proteins has not been demonstrated. Type I regulatory subunit (R subunit) of PKA occurs in stoichiometric excess of catalytic subunit (C subunit) in some tissues [20]. This R subunit contains bound cAMP. However, the apparent excess of R subunit in some cases could be an artifact, since the instability of free C subunit makes quantification difficult. If excess R subunit is conclusively demonstrated, it might act as a buffer to dampen cAMP responses or act as a reservoir for cAMP. R subunit is in slight excess of cAMP in unstimulated rat heart [18].

Cellular sequestration of cGMP in certain instances seems entirely plausible. Of particular emphasis is the likelihood that cGMP could be sequestered by a group of cGMP-binding phosphodiesterases (PDEs) (PDE2, PDE5, PDE6, PDE10, PDE11) that contain noncatalytic GAF domains. Several of these families of PDEs (PDE2, PDE5, PDE6) degrade cGMP or cAMP at catalytic sites and have been shown to bind cGMP at noncatalytic (allosteric) sites associated with the GAF domains. These binding sites provide the potential for sequestration of cGMP away from its targets such as cGMP-dependent protein kinase (PKG) and cyclic nucleotide-gated channels. Some signals could cause release of this latent form of cGMP for stimulation of certain pathways, or alternatively, they could release the nucleotide near the catalytic site for its efficient breakdown and contribute to negative feedback regulation of the cGMP pathway. If sequestration of cGMP by a cGMP-binding PDE is to be physiologically meaningful, the affinity of cGMP for this PDE should be high. Furthermore, the stoichiometric amount of this PDE should be significant relative to the level of cGMP itself. There should also be mechanisms for modulating the affinity state of the cGMP-binding sites of the PDE. Whether the levels of PDE2, PDE10, or PDE11 are sufficient or capable to bind a significant amount of total cGMP in any cell has not been investigated [21–24]. It should be emphasized that cellular sequestration by any of the cGMP-binding PDEs would be enhanced by appropriate co-compartmentalization with cGMP production. This also applies to cGMP-gated channels, although it is unlikely that number and affinity of binding sites of these proteins are sufficient to bind a significant portion of total cGMP in most cells.

Sequestration of cGMP in Rod Photoreceptor Cells by PDE6

Cyclic GMP is the primary intracellular second messenger for visual transduction and is present in high concentrations in photoreceptor cells. Light stimulation activates PDE6 more than 100-fold, causing free cytoplasmic cGMP to drop to sub-micromolar levels. Recovery of the dark-adapted level of cGMP is aided by PDE inactivation in concert with guanylyl cyclase activation. cGMP-gated cation channels in the plasma membrane close in response to the light-induced decrease in cGMP, resulting in hyperpolarization of the cell (for reviews, see [25, 26]).

Sequestration of cGMP is central to this pathway, since the dark-adapted level of cytoplasmic cGMP ($2-4\,\mu M$) inferred from electrophysiological studies [27,28] is less than 10 percent of total cGMP concentration ($60\,\mu M$) in the signal-transducing outer segment portion of the photoreceptor [29]. Both PDE6 holoenzyme ($\alpha\beta\gamma_2$) and the cGMP-gated channel bind cGMP and could sequester it [30,31]. However, the relative number of cGMP-binding sites and the affinity with which each protein binds cGMP (Table I) indicate that the cGMP-gated ion channel plays a minor role in sequestration of cGMP in the outer segment.

PDE6 is present in high concentrations ($20\,\mu M$ holoenzyme concentration) in rod outer segments [30,32]. Two high-affinity

Table I Calculated Intracellular Concentrations in Frog Rod Outer Segments

Component	concentration[1] (µM)	[cGMP site] (µM)	Binding affinity (µM)	[cGMP] bound[3] (µM)
PDE6 (high affinity)	20	40	0.06	39.0
Low affinity		75	7.0	18.4
cGMP-gated channel	1	4	30[2]	0.3
cGMP	Total: 60 free: 2.3			

1. Concentrations in the rod outer segment are referenced to the cytoplasmic volume (see text).
2. The cited value is the half-maximal activation constant for this binding reaction.
3. Bound and free values were estimated using EQCAL (Elsevier Biosoft) and assuming non-interacting, non-cooperative binding.

cGMP-binding sites per holoenzyme are occupied in the inactive PDE6 [34], binding two-thirds of total cellular cGMP in amphibian rod outer segments [35]. A second class of cGMP-binding sites with 100-fold lower affinity has been detected [35]. Assignment of this binding to the second GAF domain within each catalytic subunit of PDE6 is largely based on the fact that no other high-abundance cGMP-binding protein has been reported in photoreceptors. Together, the high- and low-affinity sites sequester 95 percent of total cellular cGMP, bringing the free cGMP into the low micromolar range (Table I).

Light activation of PDE6 causes a sub-second decline in free cytoplasmic cGMP to sub-micromolar levels, causing closure of cGMP-gated ion channels in the plasma membrane. Notably, total extractable cGMP is decreased only approximately 50 percent (to approximately 30 µM). Activation of PDE6 converts one of its high-affinity cGMP-binding sites to a lower affinity that allows for rapid dissociation of this cGMP. The other high-affinity site on PDE6 undergoes a more modest loss in affinity, and cGMP probably remains bound to this site—and therefore protected from hydrolysis—during excitation and recovery [33,36–38]. Termination of the photoresponse requires inactivation of PDE6 catalysis by its γ subunit, restoration of the second high-affinity binding site on PDE6, acceleration of cGMP synthesis by guanylyl cyclase, and rebinding of cGMP to the cGMP-gated channel and to both sites on PDE6.

The functional significance of light-induced changes in the amount of sequestered cGMP is unknown. One hypothesis is that cGMP released from light-activated PDE6 could elevate cytoplasmic cGMP and assist guanylyl cyclase during the recovery phase of the response [35,36]. However, sequestered cGMP that is released upon illumination will be quickly hydrolyzed and is unlikely to contribute to restoring free cGMP levels during recovery [37]. A more plausible hypothesis is that cGMP release from activated PDE6 serves as a negative feedback mechanism to inactivate PDE6 during bright light adaptation of photoreceptors [33]. Cyclic GMP dissociation from PDE6 exerts an allosteric effect on inhibitory γ subunit [33,38,39], causing γ to accelerate transducin GTPase (in concert with RGS9; [37,40,41]). In this view, cGMP-binding sites on activated PDE6 would respond to a persistent reduction in free cGMP levels (that is, bright, continuous illumination) by releasing bound cGMP, causing transducin to more rapidly inactivate itself and allow the γ subunit to inhibit PDE6 once more. This is consistent with the speeded recovery of the photoreceptor light response under conditions of bright light adaptation.

In summary, cGMP sequestration by PDE6 meets the criteria for physiological relevance: (1) concentrations of PDE6 cGMP-binding sites and cGMP in photoreceptor cells are comparable, (2) GMP-binding to PDE6 is regulated, and (3) affinity of the cGMP-binding sites is sufficient for quantitatively significant cGMP binding in these cells. Cyclic GMP binding by PDE6 is likely to serve multiple functions in regulating the signal transduction in photoreceptors. In the dark-adapted state, PDE6 binding of cGMP maintains free cytoplasmic cGMP at the low levels needed for opening a small fraction of cGMP-gated ion channels in the plasma membrane. Following light activation, release of cGMP from PDE6 may shorten the lifetime of excitation and provide a mechanism for adapting photoreceptor cells to sustained illumination.

Sequestration of cGMP by PDE5

PDE5 was the first recognized mammalian cyclic nucleotide receptor other than PKA and PKG [42,43]. It is highly specific for hydrolysis of cGMP at a single catalytic site on each subunit, and it contains two potential cGMP-binding sites (GAF domains) in the regulatory domain of each subunit [44,45] with an average cGMP-binding affinity of about 200 nM. Phosphorylation of PDE5 by PKG at Ser-92 (bovine enzyme) increases cGMP-binding affinity ten-fold [46,47] and also stimulates the catalytic activity of the enzyme [47]. Furthermore, occupation of the catalytic site of bovine PDE5 by cGMP or cGMP analogs stimulates binding of cGMP to the allosteric sites, which in turn stimulates phosphorylation at Ser-92 [43,48]. It follows from these reciprocal, allosteric effects that elevation of cGMP in cells would initiate PDE5-mediated negative feedback by stimulation of both cGMP breakdown at its catalytic site and sequestration of cGMP at its allosteric sites. These two mechanisms for lowering free

Table II Calculated Intracellular Concentrations in Rabbit Corpus Cavernosum

Cell conditions		Free [cGMP] nM[2]	Bound [cGMP] nM[2]	
[cGMP] nM	PDE5		To PDE5	To PKG
Basal	Unphospho-	7	6.7	4.0
Elevated	Unphospho-	23	20.0	11.0
Elevated	Phospho-	8	41.1	4.5
Basal	Phospho-	2	14.2	1.4

1. Basal total cGMP concentration 18 ± 4 nM. Nitric oxide-stimulated cGMP concentration, ~54 nM. Total PDE5 concentration (referenced to cGMP binding sites), 188 ± 6 nM. Total PKG concentration (referenced to cGMP binding sites), 58 ± 6 nM. Calculations based on specific enzyme activity of pure proteins, and intracellular water of 0.5 g per g tissue.
2. Calculated assuming the following K_D values: PKG = 100 nM (form Km of PKGIα at 20°C), unphospho-PDE5 = 200 nM at 0°, phospho-PDE5 = 30 nM, using EQCAL and assuming noninteracting, noncooperative cGMP binding.

cGMP levels should diminish cGMP binding to PKG or other cGMP receptor proteins.

In order for sequestration of cGMP to play a significant role in modulating the free cGMP levels in PDE5-regulated signaling pathways, the number and affinity of cGMP-binding sites must be sufficient to buffer cGMP, and the rate of binding/sequestration must be faster than the rate of loss via hydrolysis. At least in some tissues, these conditions appear to be met. In rabbit corpus cavernosum smooth muscle cells (Table II), PDE5 allosteric binding site concentration is more than five times higher than basal cGMP [49]. The binding affinity of these sites is great enough (even for the unphosphorylated PDE5) to sequester a significant portion of cGMP, even after elevation of this nucleotide by various agents. Loss of active cellular cGMP by this sequestration could be quantitatively meaningful since loss of total cellular cGMP by PDE5 catalytic breakdown of this nucleotide is calculated to require about 16 sec [49].

Under basal intracellular conditions, most cGMP molecules would be bound to PDE5 or PKG given the high affinities of these two proteins for cGMP. Using the K_D value of 200 nM that we determined for unphospho-PDE5 [48,50], PKG K_D for cGMP of 100 nM [51], and the intracellular values of 188 nM PDE5, 58 nM PKG and 18 nM for cGMP (Table II), we used EQCAL (Elsevier Biosoft) to calculate free cGMP and bound cGMP under various conditions.

Figure 2 summarizes our model for regulation of cGMP signaling pathways by sequestration of this nucleotide by allosteric sites of PDE5. In the basal condition (low cGMP, unphosphorylated PDE5), both PDE5 and PKG contribute to binding about half of the total cellular cGMP (free concentration = 7 nM). A three-fold increase in total cGMP levels by guanylyl cyclase activation initially would cause corresponding increases in free cGMP and the amount bound to both PKG and PDE5. Increased PKG activation resulting from cGMP binding might phosphorylate PDE5, enhancing ten-fold its binding affinity at the noncatalytic binding sites. This would immediately reduce the free cGMP and reverse

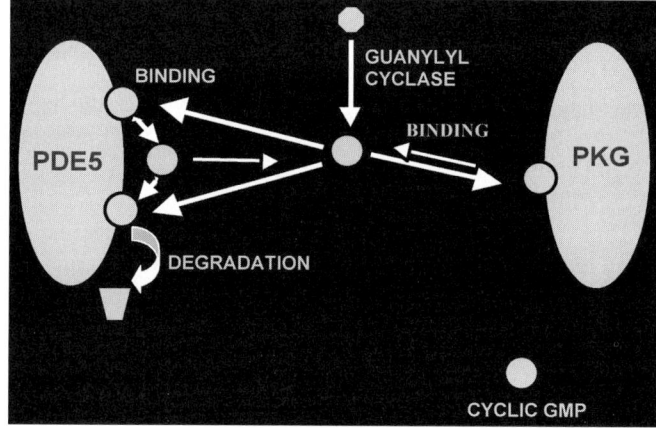

Figure 2 Proposal for regulation of some signaling pathways by sequestration of cGMP by allosteric sites of PDE5.

PKG activity to its basal state. If PDE5 dephosphorylation were relatively slow, free cGMP levels as well as PKG bound to cGMP might both drop several-fold lower than their basal values (Table II). In summary, cGMP elevation would cause increased sequestration, resulting in dampening of the cGMP signal and facilitating termination of this signal.

The process described above could be part of a concert of negative feedback processes for cGMP that have evolved for fine regulation of cGMP signaling in a number of tissues: (1) increased PDE5 catalytic activity due to mass action of elevated cGMP, (2) increased cGMP binding to PDE5 allosteric sites due to mass action of elevated cGMP, (3) increased PDE5 catalytic activity due to phosphorylation and activation of PDE5 by activated PKG [50], and (4) increased cGMP binding to PDE5 allosteric sites due to this phosphorylation. A fifth possible process is direct stimulation of the PDE5 catalytic site by allosteric cGMP binding to the enzyme, which would be predicted by the principle of reciprocity [52]. A final possibility is that concentrating cGMP near the catalytic site by its release from proximal cGMP-binding sites could increase efficiency of cGMP hydrolysis. The presence of such an array of mechanisms for negative feedback of the cGMP pathway suggests that cells cannot readily tolerate excessive activation of PKG or other target proteins. Thus, for both PDE5- and PDE6-containing cells, sequestration may act to (1) buffer cGMP levels in the basal state and (2) act as negative feedback regulator to prevent overstimulation and accelerate response termination of the signaling pathway.

References

1. Barber, R. and Butcher, R. W. (1981). The quantitative relationship between intracellular concentration and egress of cyclic AMP from cultured cells. *Mol. Pharmacol.* **19**, 38–43.
2. Heasley, L. E. and Brunton, L. L. (1985). Prostaglandin A1 metabolism and inhibition of cyclic AMP extrusion by avian erythrocytes. *J. Biol. Chem.* **260**, 11514–11519.
3. Schultz, C., Vaskinn, S., Kildalsen, H., and Sager, G. (1998). Cyclic AMP stimulates the cyclic GMP egression pump in human erythrocytes: effects of probenecid, verapamil, progesterone, theophylline, IBMX, forskolin, and cyclic AMP on cyclic GMP uptake and association to inside-out vesicles. *Biochemistry* **37**, 1161–1166.
4. Jedlitschky, G., Burchell, B., and Keppler, D. (2000). The multidrug resistance protein 5 functions as an ATP-dependent export pump for cyclic nucleotides. *J. Biol. Chem.* **275**, 30069–30074.
5. Chen, Z. S., Lee, K., and Kruh, G. D. (2001). Transport of cyclic nucleotides and estradiol 17-beta-D-glucuronide by multidrug resistance protein 4. Resistance to 6-mercaptopurine and 6-thioguanine. *J. Biol. Chem.* **276**, 33747–33754.
6. Barber, R., Ray, K. P., and Butcher, R. W. (1980). Turnover of adenosine 3',5'-monophosphate in WI-38 cultured fibroblasts. *Biochemistry* **19**, 2560–2567.
7. Nemecek, G. M., Wells, J. N., and Butcher, R. W. (1980). Inhibition of fibroblast cyclic AMP escape and cyclic nucleotide phosphodiesterase activities by xanthines. *Mol. Pharmacol.* **18**, 57–64.
8. Fehr, T. F., Dickinson, E. S., Goldman, S. J., and Slakey, L. L. (1990). Cyclic AMP efflux is regulated by occupancy of the adenosine receptor in pig aortic smooth muscle cells. *J. Biol. Chem.* **265**, 10974–10980.
9. Mercapide, J., Santiago, E., Alberdi, E., and Martinez-Irujo, J. J. (1999). Contribution of phosphodiesterase isoenzymes and cyclic nucleotide efflux to the regulation of cyclic GMP levels in aortic smooth muscle cells. *Biochem. Pharmacol.* **58**, 1675–1683.
10. Francis, S. H., Turko, I. V., and Corbin, J. D. (2000). Cyclic nucleotide phosphodiesterases: relating structure and function. *Nucl. Acid Res. Mol. Biol.* **65**, 1–52.
11. Conti, M., Nemoz, G., Sette, C., and Vicini, E. (1995). Recent progress in understanding the hormonal regulation of phosphodiesterases. [Review] [194 refs]. *Endocrine Rev.* **16**, 370–389.
12. Beavo, J. A. (1995). Cyclic nucleotide phosphodiesterases: functional implications of multiple isoforms. *Physiol. Rev.* **75**, 725–748.
13. Polson, J. B. and Strada, S. J. (1996). Cyclic nucleotide phosphodiesterases and vascular smooth muscle. [Review] [87 refs]. *Annu. Rev. Pharmacol. Toxicol.* **36**, 403–427.
14. Stryer, L. (1996). Vision: from photon to perception. *Proc. Natl. Acad. Sci. USA* **93**, 557–559.
15. Degerman, E., Belfrage, P., and Manganiello, V. C. (1997). Structure, localization, and regulation of cGMP-inhibited phosphodiesterase (PDE3). [Review] [61 refs]. *J. Biol. Chem.* **272**, 6823–6826.
16. Houslay, M. D., Sullivan, M., and Bolger, G. B. (1998). The multienzyme PDE4 cyclic adenosine monophosphate-specific phosphodiesterase family: intracellular targeting, regulation, and selective inhibition by compounds exerting anti-inflammatory and antidepressant actions. [Review] [250 refs]. *Adv. Pharmacol. NY* **44**, 225–342.
17. Langer, G. A. (1992). Calcium and the heart: exchange at the tissue, cell, and organelle levels. [Review] [68 refs]. *FASEB J.* **6**, 893–902.
18. Corbin, J. D., Sugden, P. H., Lincoln, T. M., and Keely, S. L. (1977). Compartmentalization of adenosine 3':5'-monophosphate and adenosine 3':5'-monophosphate-dependent protein kinase in heart tissue. *J. Biol. Chem.* **252**, 3854–3861.
19. Rich, T. C., Fagan, K. A., Nakata, H., Schaack, J., Cooper, D. M., and Karpen, J. W. (2000). Cyclic nucleotide-gated channels colocalize with adenylyl cyclase in regions of restricted cAMP diffusion. [see comments]. *J. Gen. Physiol.* **116**, 147–161.
20. Beebe, S. J. and Corbin, J. D. (1986). Cyclic nucleotide-dependent protein kinases, in P. D. Boyer, and E. G. Krebs, Eds., *The Enzymes*, Vol. 17A, pp. 43–111. Academic Press, Orlando, Florida.
21. Soderling, S. H., Bayuga, S. J., and Beavo, J. A. (1998). Identification and characterization of a novel family of cyclic nucleotide phosphodiesterases. *J. Biol. Chem.* **273**, 15553–15558.
22. Loughney, K., Hill, T. R., Florio, V. A., Uher, L., Rosman, G. J., Wolda, S. L., Jones, B. A., Howard, M. L., McAllister-Lucas, L. M., Sonnenburg, W. K., Francis, S. H., Corbin, J. D., Beavo, J. A., and Ferguson, K. (1998). Isolation and characterization of cDNAs encoding PDE5A, a human cGMP-binding cGMP-specific 3',5'-cyclic nucleotide phosphodiesterase. *Gene* **216**, 137–147.
23. Fujishige, K., Kotera, J., Michibata, H., Yuasa, K., Takebayashi, S., Okumura, K., and Omori, K. (1999). Cloning and characterization of a novel human phosphodiesterase that hydrolyzes both cAMP and cGMP (PDE10A). *J. Biol. Chem.* **274**, 18438–18445.
24. Fawcett, L., Baxendale, R., Stacey, P., McGrouther, C., Harrow, I., Soderling, S., Hetman, J., Beavo, J. A., and Phillips, S. C. (2000). Molecular cloning and characterization of a distinct human phosphodiesterase gene family: PDE11A. *Proc. Natl. Acad. Sci. USA* **97**, 3702–3707.
25. Pugh, E. N. J., Nikonov, S., and Lamb, T. D. (1999). Molecular mechanisms of vertebrate photoreceptor light adaptation. [Review] [61 refs]. *Curr. Opinion Neurobiol.* **9**, 410–418.
26. Ebrey, T. and Koutalos, Y. (2001). Vertebrate photoreceptors. [Review] [395 refs]. *Prog. Retinal Eye Res.* **20**, 49–94.
27. Nakatani, K. and Yau, K. W. (1988). Guanosine 3',5'-cyclic monophosphate-activated conductance studied in a truncated rod outer segment of the toad. *J. Physiol.* **395**, 731–753.
28. Pugh, E. N. and Lamb, T. D. (1993). Amplification and kinetics of the activation steps in phototransduction. [Review] [203 refs]. *Biochim. Biophys. Acta* **1141**, 111–149.
29. Cote, R. H., Biernbaum, M. S., Nicol, G. D., and Bownds, M. D. (1984). Light-induced decreases in cGMP concentration precede changes in membrane permeability in frog rod photoreceptors. *J. Biol. Chem.* **259**, 9635–9641.
30. Yamazaki, A., Sen, I., Bitensky, M. W., Casnellie, J. E., and Greengard, P. (1980). Cyclic GMP-specific, high affinity, noncatalytic binding sites on light-activated phosphodiesterase. *J. Biol. Chem.* **255**, 11619–11624.
31. Cook, N. J., Zeilinger, C., Koch, K. W., and Kaupp, U. B. (1986). Solubilization and functional reconstitution of the cGMP-dependent cation channel from bovine rod outer segments. *J. Biol. Chem.* **261**, 17033–17039.
32. Dumke, C. L., Arshavsky, V. Y., Calvert, P. D., Bownds, M. D., and Pugh, E. N. Jr. (1994). Rod outer segment structure influences the apparent kinetic parameters of cyclic GMP phosphodiesterase. *J. Gen. Physiol.* **103**, 1071–1098.
33. Cote, R. H., Bownds, M. D., and Arshavsky, V. Y. (1994). cGMP binding sites on photoreceptor phosphodiesterase: role in feedback regulation of visual transduction. *Proc. Natl. Acad. Sci. USA* **91**, 4845–4849.
34. Gillespie, P. G. and Beavo, J. A. (1989). cGMP is tightly bound to bovine retinal rod phosphodiesterase. *Proc. Natl. Acad. Sci. USA* **86**, 4311–4315.
35. Cote, R. H. and Brunnock, M. A. (1993). Intracellular cGMP concentration in rod photoreceptors is regulated by binding to high and moderate affinity cGMP binding sites. *J. Biol. Chem.* **268**, 17190–17198.

36. Yamazaki, A., Bondarenko, V. A., Dua, S., Yamazaki, M., Usukura, J., and Hayashi, F. (1996). Possible stimulation of retinal rod recovery to dark state by cGMP release from a cGMP phosphodiesterase noncatalytic site. *J. Biol. Chem.* **271**, 32495–32498.
37. Calvert, P. D., Ho, T. W., LeFebvre, Y. M., and Arshavsky, V. Y. (1998). Onset of feedback reactions underlying vertebrate rod photoreceptor light adaptation. *J. Gen. Physiol.* **111**, 39–51.
38. Norton, A. W., D'Amours, M. R., Grazio, H. J., Hebert, T. L., and Cote, R. H. (2000). Mechanism of transducin activation of frog rod photoreceptor phosphodiesterase. Allosteric interaction between the inhibitory gamma subunit and the noncatalytic cGMP-binding sites. *J. Biol. Chem.* **275**, 38611–38619.
39. Yamazaki, A., Bartucca, F., Ting, A., and Bitensky, M. W. (1982). Reciprocal effects of an inhibitory factor on catalytic activity and noncatalytic cGMP binding sites of rod phosphodiesterase. *Proc. Natl. Acad. Sci. USA* **79**, 3702–3706.
40. Arshavsky, V. Y. and Bownds, M. D. (1992). Regulation of deactivation of photoreceptor G protein by its target enzyme and cGMP. *Nature* **357**, 416–417.
41. He, W., Cowan, C. W., and Wensel, T. G. (1998). RGS9, a GTPase accelerator for phototransduction. *Neuron* **20**, 95–102.
42. Lincoln, T. M., Hall, C. L., Park, C. R., and Corbin, J. D. (1976). Guanosine 3':5'-cyclic monophosphate binding proteins in rat tissues. *Proc. Natl. Acad. Sci. USA* **73**, 2559–2563.
43. Francis, S. H., Lincoln, T. M., and Corbin, J. D. (1980). Characterization of a novel cGMP binding protein from rat lung. *J. Biol. Chem.* **255**, 620–626.
44. McAllister-Lucas, L. M., Sonnenburg, W. K., Kadlecek, A., Seger, D., LeTrong, H., Colbran, J. L., Thomas, M. K., Walsh, K. A., Francis, S. H., Corbin, J. D., and Beavo, J. A. (1993). The structure of a bovine lung cGMP-binding, cGMP-specific phosphodiesterase deduced from a cDNA clone. *J. Biol. Chem.* **268**, 22863–22873.
45. Corbin, J. D. and Francis, S. H. (1999). Cyclic GMP phosphodiesterase 5: target for sildenafil. *J. Biol. Chem.* **274**, 13729–13732.
46. Thomas, M. K., Francis, S. H., and Corbin, J. D. (1990). Characterization of a purified bovine lung cGMP-binding cGMP phosphodiesterase. *J. Biol. Chem.* **265**, 14964–14970.
47. Corbin, J. D., Turko, I. V., Beasley, A., and Francis, S. H. (2000). Phosphorylation of phosphodiesterase-5 by cyclic nucleotide-dependent protein kinase alters its catalytic and allosteric cGMP-binding activities. *Eur. J. Biochem.* **267**, 2760–2767.
48. Thomas, M. K., Francis, S. H., and Corbin, J. D. (1990). Substrate- and kinase-directed regulation of phosphorylation of a cGMP-binding phosphodiesterase by cGMP. *J. Biol. Chem.* **265**, 14971–14978.
49. Gopal, V. K., Francis, S. H., and Corbin, J. D. (2001). Allosteric sites of phosphodiesterase-5 (PDE5). A potential role in negative feedback regulation of cGMP signaling in corpus cavernosum. *Eur. J. Biochem.* **268**, 3304–3312.
50. Corbin, J. D., Turko, I. V., Beasley, A., and Francis, S. H. (2000). Phosphorylation of phosphodiesterase-5 by cyclic nucleotide-dependent protein kinase alters its catalytic and allosteric cGMP-binding activities. *Eur. J. Biochem.* **267**, 2760–2767.
51. Corbin, J. D. and Doskeland, S. O. (1983). Studies of two different intrachain cGMP-binding sites of cGMP-dependent protein kinase. *J. Biol. Chem.* **258**, 11391–11397.
52. Weber, G. (1975). Energetics of ligand binding to protein. *Adv. Protein Chem.* **29**, 1–83.

CHAPTER 199

cAMP-dependent Protein Kinase

Susan S. Taylor and Elzbieta Radzio-Andzelm
*Department of Chemistry and Biochemistry and the Howard Hughes Medical Institute,
University of California, San Diego, La Jolla, California*

Introduction

cAMP-dependent protein kinase (PKA) is one of the best characterized members of the large protein kinase superfamily. The catalytic subunit serves as a structural prototype for the entire family. The inactive holoenzyme comprises a regulatory (R) subunit dimer and two catalytic subunits. Binding of cAMP to the R subunits unleashes the active C subunits. The structure of the C subunit is described and correlated with its function. The structure of the dimerization/docking domain of RIIα and the cAMP binding domains of RIα and RIIβ are also described and correlated with the dynamic properties of the R subunits.

cAMP-dependent protein kinase (PKA) was one of the first protein kinases to be discovered [1], the first to be sequenced [2], the first to be cloned [3], and the first protein kinase for which a crystal structure was solved [4]. It thus serves in many ways as a prototype for the entire protein kinase superfamily, which represents approximately 2 percent of the human genome. cAMP is an ancient stress response signal; for example, it is a universal indicator of glucose deprivation. Whereas in bacteria, the cAMP second messenger is linked to the catabolite gene activator protein, in mammals it is linked primarily to the activation of PKA. PKA is ubiquitous in mammalian cells and regulates many diverse pathways.

The inactive holoenzyme complex consists of a regulatory (R) subunit dimer and two catalytic (C) subunits. Binding of cAMP to the R subunits unleashes the active C subunits, thus allowing them to phosphorylate a variety of protein substrates, both cytosolic and nuclear [5,6]. In addition to serving as inhibitors of PKA activity and receptors for cAMP, the R subunits also serve as adapters that tether the C subunit to specific cellular locations by binding to A kinase anchoring proteins (AKAPs) [7]. PKI, another inhibitor of the C subunit that is independent of cAMP [8], also contributes to trafficking of the free C subunits between the cytoplasm and the nucleus [9]. The inhibitors of PKA activity are both modular and multifunctional proteins. A review of PKA structure thus must include the diverse set of proteins that contribute overall to PKA regulation. The structures of the C subunit and its inhibitors, both the R subunits and PKI, are described here.

Catalytic Subunit

In mammals three isoforms of the C subunit have been identified: α, β, and γ [3,10,11]. The Cα subunit is expressed constitutively in all cells, whereas expression of Cβ is tissue specific and especially prominent in brain. Cγ is found primarily in testes. Several splice variants of both Cα [12] and Cβ [13] also exist; all differ in the first exon. In the primary form of Cα, β, and γ, exon I codes for 14 amino acids that include an N-terminal myristylation site [14]. The other C subunit splice variants are typically not myristylated. In addition to co-translational myristylation, the C subunit is phosphorylated posttranslationally at two essential sites [15]. Phosphorylation at Thr197 in the activation loop is essential for efficient catalysis [16]. Phosphorylation at Ser 338 is essential for stability and is very likely to be an important part of the maturation of the initial transcript into an active enzyme [17]. Phosphorylation at Ser10 [18] and deamidation of Asn2 [19] are other posttranslational modifications that have been identified.

Catalytic Properties

The C subunit is a highly concerted enzyme; all its energy is focused on transferring the γ–phosphate of ATP to an appropriate substrate protein [20]. There are two general recognition motifs for PKA substrates [21]: Arg-Arg-X-Ser/Thr-Hyd and Arg-X-X-Arg-X-X-Ser/Thr-Hyd, where X is any residue and Hyd is a hydrophobic residue. The mechanism for catalysis has been carefully defined by Adams [22]. Pre-steady state kinetics established that the actual rate of phosphoryl transfer is very fast (>500/s) whereas the kcat is only 20/sec. For PKA, the kcat correlates, in general, with the release of ADP and the conformational changes that allow for its release [23]. The K_m for the heptapeptide, kemptide (Leu-Arg-Arg-Ala-Ser-Leu-Gly), is 10–20 µM; however, this K_m does not reflect a true binding affinity. The K_d (200–300 µM) more accurately reflects affinity [24].

Structure

Multiple forms of the C subunit have been crystallized, and these structures provide a molecular understanding of nucleotide binding, peptide binding, and conformational flexibility [4,25–27]. The C subunit comprises a highly conserved core containing a smaller ATP binding domain (residues 40–126) that is dominated by β structure and a larger, mostly helical lobe (residues 127–300) that provides a docking site for peptides/proteins as well as several essential residues that contribute to catalysis (Fig. 1). The adenine ring of ATP is buried at the base of the cleft between the two lobes, and the peptide docks to the surface of the large lobe at the edge of the cleft. This core is conserved in all protein kinases that phosphorylate Ser, Thr, or Tyr [28].

In PKA, as seen in Fig. 2, the core is flanked by 40 additional residues at the N-terminus that begin with a myristyl moiety attached to the N-terminal Gly. This is followed by an amphipathic helix that is anchored by hydrophobic interactions to both the small and large lobe of the core [29]. The core is followed by a 50 residue "tail" that is anchored to the large lobe (residues 301–318), has a flexible anionic "gate" that draws basic peptides to the active site cleft, and terminates with a hydrophobic motif at the C terminus (Phe-Ser-Glu-Phe) [30,31]. This hydrophobic motif is anchored to a hydrophobic pocket on the small lobe and probably helps orient the C-helix into its active conformation.

CONSERVED CORE

As recognized initially by Hanks and Hunter [32], the conserved kinase core consists of a set of sequence motifs that span the entire core (Fig. 1). Although affinity labeling provided clues about the roles of some of these motifs [5,33,34], the first crystal structure revealed the unique architecture that brings most of these conserved motifs to the active site cleft where they contribute primarily to the binding of ATP and phosphoryl transfer [35]. The detailed characterization of the C subunit is reviewed in Johnson *et al.* [36].

SMALL LOBE

In general, the small lobe is more "loosely" structured than the large lobe. One of the most essential features of this enzyme is the glycine-rich loop that links β strands 1 and 2. In most of the crystal structures, this loop is disordered or ordered poorly [31]. Only in the ternary complex where ATP, or an ATP analog, and an inhibitor peptide, PKI (5–24), are bound [26,37] and in a recently solved aluminum fluoride complex that mimics a transition state intermediate, is the tip of the loop firmly anchored [38].

The hydrogen bond between the backbone amide of Ser53 and the γ-phosphate of ATP is probably the driving force for catalysis [39]. The two other essential residues in the small lobe are Lys72 in β strand 3, which anchors the α- and β-phosphates of ATP, and Glu91 in the C helix, which interacts with Lys72. All crystal structures of the C subunit so far have been of the active, fully phosphorylated protein.

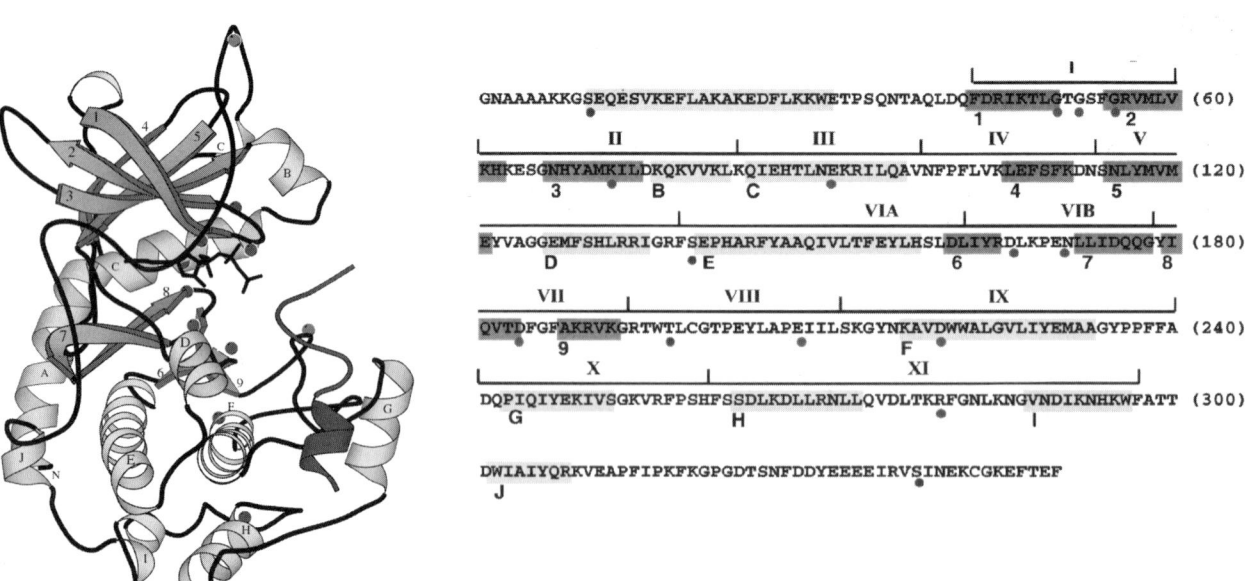

Figure 1 Structure and sequence of the catalytic subunit of PKA. A ribbon diagram of the mouse C subunit bound to ATP and an inhibitor peptide PKI(5-24) is on the left [26]. β strands are in green; α helices are yellow. PKI(5-24) is red. Conserved residues are indicated as red balls, phosphorylation sites as purple balls. On the right is the sequence with the same color coding.

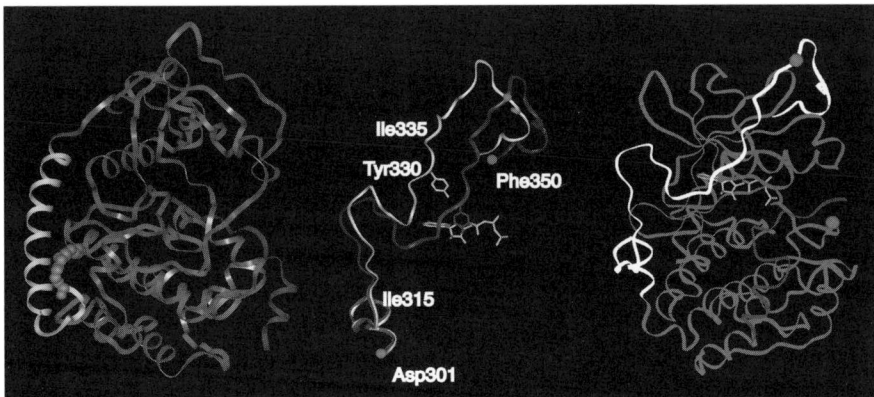

Figure 2 The N and C terminal tails of the catalytic subunit. The structure of the myristylated N terminus (residues 1–40) was observed in the mammalian C subunit (left); this structure represents an open conformation. On the right is a structure of a ternary complex of the recombinant C subunit with the C terminal tail highlighted in white. In the center is shown the conformation of the C terminal tail in an "open" and "closed" conformation. Tyr330 in the closed conformation forms a nucleation site by interacting with the ribose of ATP, the linker through Glu127, and the P-3 Arg through a water molecule. Replacement of Tyr330 with Ala leads to significant loss of activity. In the absence of ATP, the tail tends to be disordered.

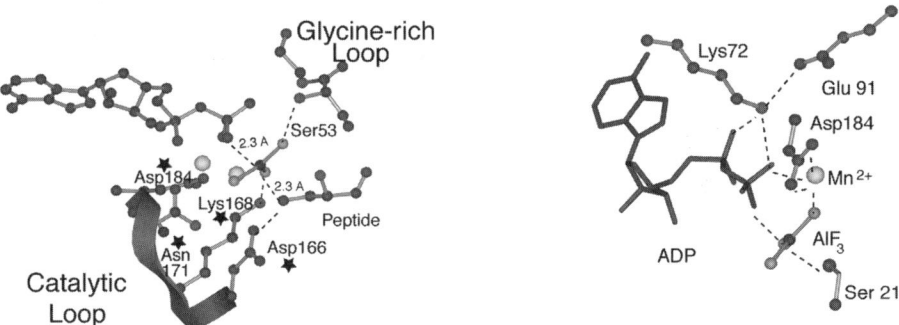

Figure 3 Aluminum fluoride complex mimics a transition state intermediate. On the left is the active site cleft in the presence of MgATP and PKA (5-24) [26]. On the right is the convergence of active site residues in a complex of C subunit with ADP, AlF$_3$, and a substrate peptide. This structure reveals how many of the conserved residues cluster around the active site cleft and contribute to the transfer of the γ phosphate of ATP to the peptide substrate. Conserved residues are highlighted with a yellow dot.

Phosphorylation decreases the K_M (ATP) 50-fold and increases the rate of phosphoryl transfer from 500s^{-1} to 20s^{-1} [16]. For many other protein kinases the proper orientation of the C helix is dependent upon phosphorylation of the activation loop in the large lobe [40]. Most likely, this is also true for the C subunit when the protein is unphosphorylated.

LARGE LOBE

Most of the conserved residues in the large lobe are localized on a β sheet that is anchored firmly through hydrophobic interactions to the large lobe (Fig. 3). The catalytic loop linking β strands 6 and 7 contains 3 conserved residues; Asp166 and Asn171 are universally conserved whereas Lys168 is conserved in all Ser/Thr specific kinases. Although Asp166 is positioned to be a catalytic base, it contributes only minimally to phosphoryl transfer and is thought to be used primarily for orienting the peptide hydroxyl moiety rather than contributing significantly to the nucleophillic properties of the attacking group [41]. Asn171 binds to the second metal ion that interacts with the α and γ phosphates of ATP. It also hydrogen bonds to the backbone carbonyl of Asp166, thereby stabilizing the backbone of the catalytic loop. The magnesium-positioning loop, residues 184–187, links β strands 8 and 9. Asp184 binds the activating Mg ion that bridges the β and γ phosphates of ATP. β strand 9 is followed by the activation loop, which is positioned for optimal phosphorylation by the phosphorylation of Thr197. When expressed in *E. coli*, Thr 197 can be autophosphorylated. However, Thr197 is also an excellent substrate for PDK1 [42], and in mammalian cells it is more likely that the C subunit is phosphorylated by a heterologous protein kinase, not by autophosphorylation [43]. Thr197 is followed by the P+1 loop, named because three residues (Leu198, Pro202, and Leu205) fold inward and form a docking site for hydrophobic P+1 residue. In fact, however, this loop can be more appropriately referred to as the peptide-organizing loop, since almost every residue contributes to some aspect of peptide recognition. Gly200 and Thr201 are essential and conserved in all Ser/Thr protein kinases. Gly200 abuts the backbone of the P-site residue and forms a hydrogen bond to the P+1 backbone amide. In contrast, the side chain of Thr201 interacts directly with catalytic loop residues, where it is wedged between and hydrogen bonds to the side chains of Lys168 that positions the γ-phosphate of ATP and Asp166 that positions the hydroxyl acceptor in the peptide substrate. The bridging role of Thr201 is seen most clearly in the structure of ADP, AlF$_3$, and a substrate peptide (Fig. 3) [38]. Glu203 in PKA provides a docking site for the P-6 Arg and Tyr204 hydrogen bonds to Glu230, a primary recognition site for the P-2 Arg. The aromatic ring of Tyr204 also contributes to peptide binding.

Protein Kinase Inhibitor

PKI contains a 20 residue inhibitor domain that binds to the free subunit ($K_d = 2$ nM). In solution, PKI, which contains 75 amino acids, is mostly disordered with the exception of two helical regions [44]. The first helical region provides high-affinity binding for PKI to the C subunit [8]. This amphipathic helix precedes the consensus site, which for PKI contains an Ala at the P site. The high-affinity binding of PKI requires ATP. While the consensus site segment of PKI binds to the active site cleft region, the high-affinity binding of PKI requires the amphipathic helix that docks to a hydrophobic pocket composed of Tyr^{235}-Pro-Pro-Phe-Phe [25]. The second helix in the C subunit lies in the region that harbors the nuclear export signal [9].

Regulatory Subunits

As seen in Fig. 4, the R subunits are modular proteins that are multifunctional and highly flexible [45]. There are two major isoforms, types I and II, with α and β subtypes in each class. RIα and RIIα are expressed in most mammalian tissues whereas the expression of the β isoforms is more tissue specific. There are also unique isoform distribution patterns: RIα is expressed predominantly in growing and transformed cells and RIIα predominates in differentiated cells [46]. The isoforms are clearly not functionally redundant. The only R subunit that is essential is RIα. Deletion of RIα is embryonically lethal and leads to cardiac defects [47]. Knockouts of other isoforms give unique phenotypes but are not lethal, and RIα tends to compensate when other R subunits are deleted [48]. For example, deletion of RIIβ gives a lean phenotype with a resistance to alcohol toxicity [47,49]. Myxomas and Carney disease are caused by premature stop codons in RIα [50].

Clearly, there is still much to be learned about the physiological importance of the PKA isoforms. RIα requires ATP and 2 Mg^{2+} ions to form a tight complex with the C subunit [51]; the high-affinity binding of ATP (60 nM) and C (0.2 nM) are synergistic. Type I holoenzyme is activated at lower levels of cAMP than type II holoenzyme [52,53]. RII binding to the C subunit is independent of MgATP; instead RII subunits are autophosphorylated at the consensus inhibitor site by the C subunit.

Molecular Architecture

All mammalian R subunits share the same organization. At the N-terminus is a dimerization/docking domain that locks the enzyme into a stable dimer. In the RI subunits the two protomers are actually linked by a disulfide bond [54].

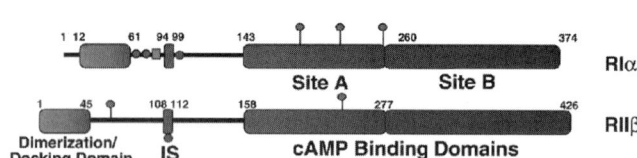

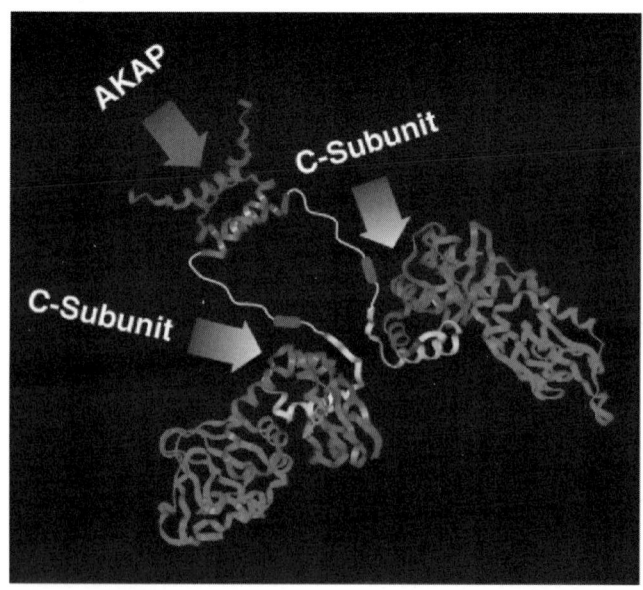

Figure 4 Domain organization of the regulatory subunits. The modular organization of RI and RII is shown on the left and a model of the subunit showing a flexible linker is on the right. Figure on the left is done by Ashton D. Taylor.

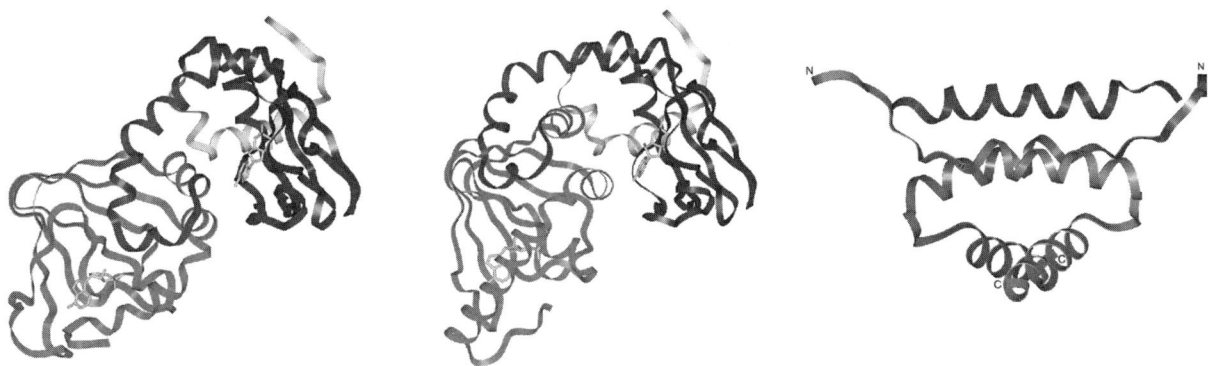

Figure 5 Structures of the regulatory subunits. On the left is the structure of the cAMP-binding domains of RIα [61] and RIIβ [62]. On the right is the dimerization/docking domain of RIα.

This is followed by a flexible and variable linker region that also contains a psuedo-substrate inhibitor site. In the absence of cAMP, this inhibitor site binds to the active site cleft of the C subunit, thus blocking access of other substrates. At the C-terminus lie two stable, tandem cAMP-binding domains. In RIα the first cAMP-binding domain also contributes to the docking of the C subunit [55]. Domain A thus shuttles between two conformations: a C bound form associated with the holoenzyme and a cAMP-bound conformation. The second cAMP-binding domain serves as a gatekeeper and regulates access of cAMP to site A [56].

cAMP is an ancient signaling molecule that has been conserved from bacteria to man. The cAMP-binding domain that serves to shield the cyclic phosphate from solvent and from phosphodiesterases is also ancient [57]. In bacteria the cAMP-binding domain is linked to a DNA-binding domain in the catabolite gene activator protein, whereas in mammals it is linked to protein kinase activation and is also found in cyclic nucleotide gated channels [58] and in a cGMP exchange factor, EPAC [59]. The highly conserved phosphate-binding cassette that surrounds the cyclic phosphate is the recognition motif for this [57]. One side of this motif interacts with cAMP while the other side interacts with the rest of the domain and is the heart of an extended network of interactions that reach to both the C-subunit docking site and the B domain [60]. Although the motif and the overall domain are highly conserved in RI and RII, the network of interactions (Fig. 5) that lead to the cooperative binding of cAMP and the release of the catalytic subunit are remarkably different in RIα and RIIβ [61,62].

The D/D domains are composed of a very stable four-helix bundle (Fig. 4), but once again there are striking differences between RI and RII [63]. The AKAP-binding surface is formed at the dimer interface, and dimerization is essential for AKAP binding. An amphipathic helix from the AKAP docks to this dimer interface [64].

Dynamics

In the absence of C subunit, the region that links the D/D domain to cAMP-binding domain A is quite mobile, as demonstrated by time-resolved fluorescence anisotropy [65]. Even in the holoenzyme the linker remains quite mobile. Small angle X-ray scattering reveals a highly asymmetric structure [66]. Hydrogen/deuterium exchange in the presence and absence of C-subunit and in the presence and absence of cAMP has provided a glimpse of the dynamic network that links cAMP binding to the release of the C-subunit [60].

Acknowledgments

This work was supported by grants from the National Institutes of Health.

References

1. Walsh, D. A., Perkins, J. P., and Krebs, E. G. (1968). An adenosine 3',5'-mono-phosphate-dependent protein kinase from rabbit skeletal muscle. *J. Biol. Chem.* **243**, 3763–3765.
2. Shoji, S., Ericsson, L. H., Walsh, D. A., Fischer, E. H., and Titani, K. (1983). Amino acid sequence of the catalytic subunit of bovine type II adenosine cyclic 3',5'-phosphate dependent protein kinase. *Biochemistry* **22**, 3702–3709.
3. Uhler, M. D., Carmichael, D. F., Lee, D. C., Chivia, J. C., Krebs, E. G., and McKnight, G. S. (1986). Isolation of cDNA clones for the catalytic subunit of mouse cAMP-dependent protein kinase. *Proc. Natl. Acad. Sci. USA* **83**, 1300–1304.
4. Knighton, D. R., Zheng, J., Ten Eyck, L. F., Ashford, V. A., Xuong, N.-h., Taylor, S. S., and Sowadski, J. M. (1991). Crystal structure of the catalytic subunit of cAMP-dependent protein kinase. *Science* **253**, 407–414.
5. Taylor, S. S., Buechler, J. A., and Yonemoto, W. (1990). cAMP-dependent protein kinase: framework for a diverse family of regulatory enzymes. *Annu. Rev. Biochem.* **59**, 971–1005.
6. Francis, S. H. and Corbin, J. D. (1994). Structure and function of cyclic nucleotide-dependent protein kinases. *Annu. Rev. Physiol.* **56**, 237–272.
7. Michel, J. J. and Scott, J. D. (2002). AKAP mediated signal transduction. *Annu. Rev. Pharmacol. Toxicol.* **42**, 235–257.
8. Walsh, D. A., Angelos, K. L., Van Patten, S. M., Glass, D. B., and Garetto, L. P. (1990). In B. E. Kemp, Ed., *Peptides and Protein Phosphorylation*, pp. 43–84. CRC Press, Boca Raton.
9. Wen, W., Meinkoth, J. L., Tsien, R. Y., and Taylor, S. S. (1995). Identification of a signal for rapid export of proteins from the nucleus. *Cell* **82**(3), 463–473.
10. Showers, M. O. and Maurer, R. A. (1986). A cloned bovine cDNA encodes an alternate form of the catalytic subunit of cAMP-dependent protein kinase. *J. Biol. Chem.* **261**(35), 16288–16291.

11. Beebe, S., Oyen, O., Sandberg, M., Froysa, A., Hansson, V., and Jahnsen, T. (1990). Molecular cloning of a tissue-specific protein kinase (C Gamma) from human testis—representing a third isoform for the catalytic subunit of cAMP-dependent protein kinase. *Mol. Endocrinol.* **4**, 465–475.
12. San Agustin, J. T., Wilkerson, C. G., and Witman, G. B. (2000). The unique catalytic subunit of sperm cAMP-dependent protein kinase is the product of an alternative Cα mRNA expressed specifically in spermatogenic cells. *Mol. Biol. Cell* **11**, 3031–3044.
13. Guthrie, C., Skalhegg, B. S., and McKnight, G. S. (1997). Two novel brain-specific splice variants of the murine Cbeta gene of cAMP-dependent protein kinase. *J. Biol. Chem.* **272**(47), 29560–295605.
14. Carr, S. A., Biemann, K., Shoji, S., Parmalee, D. C., and Titani, K. (1982). n-Tetradecanoyl in the NH2 terminal blocking group of the catalytic subunit of the cyclic AMP-dependent protein kinase from bovine cardiac muscle. *Proc. Natl. Acad. Sci. USA* **79**(20), 6128–6131.
15. Shoji, S., Titani, K., Demaille, J. G., and Fischer, E. H. (1979). Sequence of two phosphorylated sites in the catalytic subunit of bovine cardiac muscle adenosine 3':5'-monophosphate-dependent protein kinase. *J. Biol. Chem.* **254**, 6211–6214.
16. Adams, J. A., McGlone, M. L., Gibson, R. M., and Taylor, S. S. (1995). Phosphorylation modulates catalytic function and regulation in the cAMP-dependent protein kinase. *Biochemistry* **34**, 2447–2454.
17. Yonemoto, W., McGlone, M. L., Grant, B., and Taylor, S. S. (1997). Autophosphorylation of the catalytic subunit of cAMP-dependent protein kinase in *Escherichia coli*. *Protein Eng.* **10**(8), 915–925.
18. Toner-Webb, J., van Patten, S. M., Walsh, D. A., and Taylor, S. S. (1992). Autophosphorylation of the catalytic subunit of cAMP-dependent protein kinase. *J. Biol. Chem.* **267**(35), 25174–25180.
19. Kinzel, V., Konig, N., Pipkorn, R., Bossemeyer, D., and Lehmann, W. (2000). The amino terminus of PKA catalytic subunit-a site for introduction of postranslational heterogeneities by deamidation: D-Asp2 and D-isoAsp2 containing isozymes. *Protein Sci.* **11**, 2269–2277.
20. Li, F., Gangal, M., Juliano, C., Gorfain, E., Taylor, S. S., and Johnson, D. A. (2002). Evidence for an internal entropy contribution to phosphoryl transfer: a study of domain closure, backbone flexibility, and the catalytic cycle of cAMP-dependent protein kinase. *J. Mol. Biol.* **315**(3), 459–469.
21. Zetterqvist, Ö. Z., Ragnarsson, U., and Engstrom, L. (1990). In B. E. Kemp, Ed., *Peptides and Protein Phosphorylation*, pp. 171–187. CRC Press, Boca Raton.
22. Adams, J. A. (2001). Kinetic and catalytic mechanisms of protein kinases. *Chem. Rev.* **101**(8), 2271–2290.
23. Adams, J. A. and Taylor, S. S. (1992). The energetic limits of phosphotransfer in the catalytic subunit of cAMP-dependent protein kinase as measured by viscosity experiments. *Biochemistry* **31**(36), 8516–8522.
24. Adams, J. A. and Taylor, S. S. (1993). Effects of pH on the phosphorylation of peptide substrates for the catalytic subunit of cAMP-dependent protein kinase. *J. Biol. Chem.* **268**(11), 7747–7752.
25. Knighton, D. R., Zheng, J., Ten Eyck, L. F., Xuong, N.-h., Taylor, S. S., and Sowadski, J. M. (1991). Structure of a peptide inhibitor bound to the catalytic subunit of cyclic adenosine monophosphate-dependent protein kinase. *Science* **253**, 414–420.
26. Zheng, J., Knighton, D. R., Ten Eyck, L. F., Karlsson, R., Xuong, N.-h., Taylor, S. S., and Sowadski, J. M. (1993). Crystal structure of the catalytic subunit of cAMP-dependent protein kinase complexed with MgATP and peptide inhibitor. *Biochemistry* **32**(9), 2154–2161.
27. Zheng, J., Knighton, D. R., Xuong, N.-h., Taylor, S. S., Sowadski, J. M., and Ten Eyck, L. F. (1993). Crystal structures of the myristylated catalytic subunit of cAMP-dependent protein kinase reveal open and closed conformations. *Protein Sci.* **2**, 1559–1573.
28. Taylor, S. S., Knighton, D. R., Zheng, J., Ten Eyck, L. F., and Sowadski, J. M. (1992). Structural framework for the protein kinase family. *Annu. Rev. Cell Biol.* **8**, 429–462.
29. Veron, M., Radzio-Andzelm, E., Tsigelny, I., Ten Eyck, L. F., and Taylor, S. S. (1993). A conserved helix motif complements the protein kinase core. *Proc. Natl. Acad. Sci. USA* **90**, 10618–10622.
30. Batkin, M., Schvartz, I., and Shaltiel, S. (2000). Snapping of the carboxyl terminal tail of the catalytic subunit of PKA onto its core: characterization of the sites by mutagenesis. *Biochemistry* **39**(18), 5366–5373.
31. Narayana, N., Cox, S., Xuong, N.-h., Ten Eyck, L. F., and Taylor, S. S. (1997). A binary complex of the catalytic subunit of cAMP-dependent protein kinase and adenosine further defines conformational flexibility. *Structure* **5**, 921–935.
32. Hanks, S. K. and Hunter, T. (1995). Protein kinases 6. The eukaryotic kinase superfamily: kinase (catalytic) domain structure and classification. *FASEB J.* **8**, 576–596.
33. Zoller, M. J. and Taylor, S. S. (1979). Affinity labeling of the nucleotide binding site of the catalytic subunit of cAMP-dependent protein kinase using p-fluorosulfonyl-[^{14}C]benzoyl 5'-adenosine: Identification of a modified lysine residue. *J. Biol. Chem.* **254**, 8363–8368.
34. Buechler, J. A. and Taylor, S. S. (1988). Identification of Asp 184 as an essential residue in the catalytic subunit of cAMP-dependent protein kinase. *Biochemistry* **27**, 7356–7361.
35. Taylor, S. S., Knighton, D. R., Zheng, J., Sowadski, J. M., Gibbs, C. S., and Zoller, M. J. (1993). A template for the protein kinase family. *Trends Biochem. Sci.* **18**(3), 84–89.
36. Johnson, D. A., Akamine, P., Radzio-Andzelm, E., Madhusudan, and Taylor, S. S. (2001). Dynamics of cAMP-dependent protein kinase. *Chem. Rev.* **101**(8), 2243–2270.
37. Bossemeyer, D., Engh, R. A., Kinzel, V., Ponstingl, H., and Huber, R. (1993). Phosphotransferase and substrate binding mechanism of the cAMP-dependent protein kinase catalytic subunit from porcine heart as deduced from the 2.0 Å structure of the complex with Mn^{2+} adenyl imidodiphosphate and inhibitor peptide PKI(5-24) *EMBO J.* **12**(3), 849–859.
38. Madhusudan, Akamine, P., Xuong, N.-h., and Taylor, S. S. (2002). Crystal structure of a transition state mimic of the catalytic subunit of cAMP-dependent protein kinase. *Nat. Struct. Biol.* **9**(4), 273–277.
39. Aimes, R. T., Hemmer, W., and Taylor, S. S. (2000). Serine-53 at the tip of the glycine-rich loop of cAMP-dependent protein kinase: role in catalysis, P-site specificity, and interaction with inhibitors. *Biochemistry* **39**(28), 8325–8332.
40. Johnson, L., Noble, M., and Owen, D. (1996). Active and inactive protein kinases: structural basis for regulation. *Cell* **85**(2), 149–158.
41. Zhou, J. and Adams, J. A. (1997). Is there a catalytic base in the active site of cAMP-dependent protein kinase? *Biochemistry* **10**, 2977–2984.
42. Cheng, X., Ma, Y., Moore, M., Hemmings, B. A., and Taylor, S. S. (1998). Phosphorylation and activation of cAMP-dependent protein kinase by phosphoinositide-dependent protein kinase. *Proc. Natl. Acad. Sci.* **95**, 9849–9854.
43. Cauthron, R. D., Carter, K. B., Liauw, S., and Steinberg, R. A. (1998). Physiological phosphorylation of protein kinase A at Thr-197 is by a protein kinase A kinase. *Mol. Cell Biol.* **18**, 1416–1423.
44. Hauer, J. A., Barthe, P., Taylor, S. S., Parello, J., and Padille, A. (1999). Two well defined motifs in the cAMP-dependent protein kinase inhibitior (PKIa) correlate with inhibitory and nuclear export function. *Protein Sci.* **8**, 545–553.
45. Li, F., Gangal, M., Jones, J. M., Deich, J., Lovett, K. E., Taylor, S. S., and Johnson, D. A. (2000). Consequences of cAMP and catalytic-subunit binding on the flexibility of the A-kinase regulatory subunit. *Biochemistry* **39**(50), 15626–15632.
46. Stratakis, C. A., Miller, W. R., Severin, E., Chin, K. V., Bertherat, J., Amieux, P. S., Eng, C., Kammer, G. M., Dumont, J. E., Tortora, G., Beaven, M. A., Puck, T. T., Jan De Beur, S. M., Weinstein, L. S., and Cho-Chung, Y. S. (2002). Protein-kinase A and human disease: the core of cAMP-dependent signaling in health and disease. *Horm. Metab. Res.* **34**(4), 169–175.
47. Cummings, D. E., Brandon, E. P., Planas, J. V., Motamed, K., Idzerda, R. L., and McKnight, G. S. (1996). Genetically lean mice result from targeted disruption of the RII beta subunit of protein kinase A [see comments]. *Nature* **382**(6592), 622–626.
48. Amieux, P. S., Cummings, D. E., Motamed, K., Brandon, E. P., Wailes, L. A., Le, K., Idzerda, R. L., and McKnight, G. S. (1997).

Compensatory regulation of RIa protein levels in protein kinase A mutant mice. *J. Biol. Chem.* **272**, 3993–3998.

49. Thiele, T. E., Willis, B., Stadler, J., Reynolds, J. G., Bernstein, I. L., and McKnight, G. S. (2000). High ethanol consumption and low sensitivity to ethanol-induced sedation in protein kinase A-mutant mice. *J. Neurosci.* **20**, 1–6.

50. Kirschner, L. S., Carney, J. A., Svetlana, D. P., Taymans, S. E., Giatzkis, C., Cho, Y. S., Cho-Chung, Y. S., and Stratakis, C. A. (2000). Mutations of the gene encoding the protein kinase A type I-a regulatory subunit in patients with Carney complex. *Nature Genet.* **26**, 89.

51. Herberg, F. W., Doyle, M. L., Cox, S., and Taylor, S. S. (1999). Dissection of the nucleotide and metal-phosphate binding sites in cAMP-dependent protein kinase. *Biochemistry* **38**, 6352–6360.

52. Cadd, G. G., Uhler, M. D., and McKnight, G. S. (1990). Holoenzymes of cAMP-dependent protein kinase containing the neural form of type I regulatory subunit have an increased sensitivity to cyclic nucleotides. *J. Biol. Chem.* **265**(32), 19502–19506.

53. Steinberg, S. F. and Brunton, L. L. (2001). Compartmentation of G protein-coupled signaling pathways in cardiac myocytes. *Annu. Rev. Pharmacol. Toxicol.* **41**, 751–773.

54. Bubis, J., Neitzel, J. J., Saraswat, L. D., and Taylor, S. S. (1988). A point mutation abolishes binding of cAMP to site A in the regulatory subunit of cAMP-dependent protein kinase. *J. Biol. Chem.* **263**, 9668–9673.

55. Huang, L. J. and Taylor, S. S. (1998). Dissecting cAMP binding domain A in the RIa subunit of cAMP-dependent protein kinase: distinct subsites for recognition of cAMP and the catalytic subunit. *J. Biol. Chem.* **273**, 26739–26746.

56. Øgreid, D. and Døskeland, S. O. (1983). Cyclic nucleotides modulate the release of [^3H]adenosine Cyclic 3',5'-phosphate bound to the regulatory moiety of protein kinase I by the catalytic subunit of the kinase. *Biochemistry* **22**, 1686–1696.

57. Canaves, J. M. and Taylor, S. S. (2002). Classification and phylogenetic analysis of cAMP-dependent protein kinase regulatory subunit family. *J. Mol. Evol.* **54**(1), 17–29.

58. Bonigk, W. *et al.* (1999). The native rat olfactory cyclic nucleotide-gated channel is composed of three distinct subunits. *J. Neurosci.* **19**, 5332–5347.

59. de Rooij, J., Zwawrtkruis, F. J. T., Verheijen, M. H. G., Cool, R. H., Nijman, S. M. B., Wittinghofer, A., and Bos, J. L. (1998). Epac is a Rap1 guanine-nucleotide-exchange factor directly activated by cyclic AMP. *Nature* **396**(6710), 474–477.

60. Anand, G., Hughes, C., Jones, J., Taylor, S., and Komives, E. (2002). Amide H/^{2}H exchange reveals communication between the cAMP- and catalytic subunit-binding sites in the R$^I\alpha$ subunit of protein kinase A. *J. Mol. Biol.* In press.

61. Su, Y., Dostmann, W. R. G., Herberg, F. W., Durick, K., Xuong, N.-h., Ten Eyck, L. F., Taylor, S. S., and Varughese, K. I. (1995). Regulatory (RIa) subunit of protein kinase A: structure of deletion mutant with cAMP binding domains. *Science* **269**, 807–819.

62. Diller, T. C., Xuong, N.-h., and Taylor, S. S. (2000). Type IIB regulatory subunit of cAMP-dependent protein kinase: purification strategies to optimize crystallization. *Protein Express. Purif.* **20**, 357–364.

63. Banky, P., Newlon, M. G., Roy, M., Garrod, S., Taylor, S. S., and Jennings, P. A. (2000). Isoform-specific differences between the type Ia and IIa cyclic AMP-dependent protein kinase anchoring domains revealed by solution NMR. *J. Biol. Chem* **275**, 35146–35152.

64. Newlon, M. G., Roy, M., Morikis, D., Carr, D. W., Westphal, R., Scott, J. D., and Jennings, P. A. (2001). A novel mechanism of PKA anchoring revealed by solution structures of anchoring complexes. *EMBO J.* **7**, 1651–1662.

65. Gangal, M., Li, F., Jones, J. M., Deich, J., Lovett, K., Taylor, S. S., and Johnson, D. A. (2000). Consequences of cAMP and catalytic subunit binding on the flexibility of the A-kinase of the regulatory subunit. *Biochemistry* **39**(50), 15626–15632.

66. Tung, C. S., Walsh, D. A., and Trewhella, J. (2002). A structural model of the catalytic subunit-regulatory subunit dimeric complex of the cAMP-dependent protein kinase. *J. Biol. Chem.* **277**(14), 12423–12431.

Cyclic GMP-Dependent Protein Kinase

Thomas M. Lincoln
*Department of Pathology,
University of Alabama at Birmingham,
Birmingham, Alabama*

Introduction

The emergence of nitric oxide (NO) signaling between cells in biological systems over the past two decades has opened up new fields of exploration in such areas as vascular biology and neuroscience. Because NO signaling involves one of the more established intracellular signaling molecules, namely cyclic GMP, there has been a renewed interest among investigators to learn more about this signaling pathway. In this review, I will highlight one major player in the cyclic GMP signaling pathway, the cyclic GMP-dependent protein kinase (PKG). Although this particular protein kinase was among the first half-dozen protein kinases discovered more than 30 years ago, its role in biological systems is still not as well-defined as some of the more recently discovered protein kinases. I will review a few of the earlier seminal findings regarding the function of this enzyme and examine more thoroughly information that has come to light recently.

Biochemical and Molecular Biology of PKG Isoforms

Up to two genes in the genomes that have been studied to date encode cyclic GMP-dependent protein kinase [1,2]. In vertebrates, the products of the genes are termed the type I and the type II cyclic GMP-dependent protein kinase (referred to hereafter as PKG-I and PKG-II, respectively). In the mammalian tissues studied, PKG-I seems to be the more widely expressed form and is present at easily measured levels in smooth muscle cells, platelets, cerebellar Purkinje cells, and cardiomyocytes [3–6]. Other cells contain measurable but lower levels of the enzyme, and these include endothelial cells, leukocytes, and many endocrine-secretory cells. Still other cells do not appear to express measurable levels of PKG-I, and these include erythrocytes and skeletal muscle myocytes. PKG-II was first discovered in intestinal epithelial cells, where it is highly expressed and localized to the particulate fractions of these cells [7,8]. The presence of PKG-II has been confirmed in brain tissue [9], juxtaglomerular cells [10], and chondrocytes through molecular and genetic studies [11].

PKG is a member of the serine/threonine kinase branch of protein kinases and is a member of the group of cyclic nucleotide-dependent protein kinases. The protein kinases in this group have as their distinguishing feature an N-terminal regulatory component (either a domain of the enzyme or a separate subunit) that binds with high affinity to either cyclic AMP or cyclic GMP. The PKG members bind cyclic GMP on their regulatory domain with an affinity approximately 20–50 times that of cyclic AMP, whereas the cyclic AMP-dependent protein kinases (PKA) bind cyclic AMP to their regulatory subunits with the greater affinity [12,13]. The catalytic domain of the PKG members reside in the C-terminal half of the protein. The PKG isoforms can be divided into four regions: (1) an extreme N-terminal domain consisting of a leucine/isoleucine zipper dimerization motif and an autoinhibitory sequence (the autoinhibitory sequence contains serine and threonine residues that undergo phosphorylation that modulates activity of the enzyme once the enzyme has been activated by the binding of cyclic GMP [12–15]);

(2) tandem, high-affinity cyclic GMP binding sites probably created through the duplication of the DNA sequences encoding specific amino acid residues in this region [16]; (3) a catalytic domain having homology with all members of the entire protein kinase family; and (4) an extreme C-terminal region with unknown function.

PKG-I is expressed in cells as two isoforms that are derived by the alternate mRNA splicing for the first two exons encoding the N-terminal region of the enzyme [3–6]. The smaller of the two forms (PKG-Iα) has the first exon expressed as the N-terminal region and is the form most abundant in platelets, most vascular smooth muscle cells, and cerebellum. The larger of the two forms (PKG-Iβ) expresses the second exon in the protein and is abundant in vascular and nonvascular smooth muscle cells. However, there appears to be much variability in the expression of these two isoforms even in the same cell type (e.g., aortic smooth muscle cells). Furthermore, mRNA levels usually appear more abundant for the Iβ isoform in cells where the level of Iα protein may be more than twice that of the Iβ protein [17–19]. These results imply that the Iα mRNA may be more unstable than the Iβ, but there are few studies to date published on regulation of the expression of the mRNAs encoding these enzymes.

The concentration of cGMP necessary for half-maximal activation of the purified PKG-Iα isoform has been measured at approximately $0.1\,\mu M$, while the concentration of cGMP necessary for half-maximal activation of the purified Iβ isoform is 1 to $2\,\mu M$. This is despite both isoforms' having identical sequences that encode the two tandem cGMP binding sites. The reason given for this functional difference is that the N-terminal domain encoding the Iβ isoform contains an autoinhibitory domain that is either "more efficient" at inhibiting the holoenzyme or has a higher affinity for inhibiting the catalytic domain of the holoenzyme [20]. Predictably, therefore, PKG-Iα is activated at lower cGMP levels in the cell than PKG-Iβ. Corbin and colleagues have in fact shown that PKG-I exists as a partially-activated PKG especially after autophosphorylation [13,15]. The two cGMP-binding sites have different affinities for cGMP, and it has been shown that occupation of the high-affinity cGMP-binding site confers partial activation to PKG-I [13–16,20]. Upon elevation of cGMP in the cell, occupation of the second, lower-affinity binding site leads to further activation of the enzyme and autophosphorylation. The role of autophosphorylation of PKG-I has been investigated. Early studies suggested that autophosphorylation may prevent reassociation of regulatory and catalytic domains of PKG, similar to the role of autophosphorylation of the PKA regulatory subunit II. However, it is now clear that autophosphorylation may serve a more complex role for PKG and may somehow stabilize the activated enzyme in the active state even after the dissociation of cGMP. This mechanism has been shown to be operational for the calmodulin-activated protein kinase, CAM Kinase II, where the kinase remains active even after dissociation of the active calcium-calmodulin complex [21]. The molecular regulation of CAM kinase II in this fashion has been suggested to be the basis for synaptic facilitation and of memory and learning in the nervous system. For PKG, it is possible that autophosphorylation is a mechanism to maintain PKG activity and relaxation of the smooth muscle cell even after elevated NO and cGMP have dissipated and returned to baseline levels.

There may be different functional roles for PKG-Iα and Iβ, at least in smooth muscle cells. There has been speculation that PKG-Iβ, with its lower "affinity" for cGMP, may serve to buffer cGMP concentrations in the cell. This seems unlikely given the importance of PKG-Iβ in intracellular calcium regulation (see below). An attractive hypothesis for the functionally different roles of PKG-I isoforms is based on the fact that the sequence difference between the two enzymes is only at the N-terminus. In PKA, the N-terminus of the regulatory subunit determines what targeting proteins PKA binds to in the cell, the AKAPs (A kinase anchoring proteins) [22]. Hence, it has been suggested that the different N-terminal domains of PKG-I allow the kinases to bind to different targeting proteins in the cell. Experimental evidence for this hypothesis has recently been provided and demonstrates that PKG-Iα appears to localize to the perinuclear regions of smooth muscle cells [23,24] whereas PKG-Iβ is more widely distributed in smooth muscle cells [25]. Functionally, the different subcellular distribution could affect which isoform of PKG regulates intracellular calcium mobilization, for example. The PKG-Iβ isoform, but not PKG-Iα, has been shown to bind to a protein termed IRAG [for inositol (1,4,5) P_3-receptor-associated cGKI substrate] [26]. The function of the IRAG protein is to compartmentalize PKG-Iβ with the inositol 1,4,5-trisphosphate (IP_3) receptor in the endoplasmic reticulum (ER) compartment of the smooth muscle cell. PKG has been shown to catalyze the phosphorylation of the IP_3 receptor, and its localization with this substrate is necessary for phosphorylation [27,28]. The role of IP_3 receptor phosphorylation has not been unequivocally defined although there is speculation that PKG-dependent phosphorylation regulates calcium release from the ER [26,28].

Binding proteins specific for PKG-Iα have also been described, and these include a male germ cell 42 kDa protein [29] and the myosin light chain phosphatase binding subunit (MBS) [30]. Of these two potential targeting proteins, the binding to MBS has been most extensively characterized. Surks et al. [30] have shown that PKG-Iα, but not PKG-Iβ or Type II PKG, binds via the leucine zipper domain to MBS. As a consequence of PKG binding, the MBS is phosphorylated, resulting in the activation of the phosphatase. In contrast to PKG, both Rho-kinase A and protein kinase C (PKC)-catalyzed phosphorylation of MBS lead to the inhibition of the phosphatase activity [31,32]. Therefore, multisite phosphorylation mechanisms regulate the phosphorylation of myosin light chain and hence contractility in smooth muscle cells. The concepts of PKG-dependent regulation of smooth muscle contractility will be addressed again below. The purpose of this discussion here is to illustrate the importance of the N-terminal domains of PKG-I in targeting the kinase to subcellular compartments.

Other studies have suggested that various PKG isoforms bind to cytoskeletal compartments. PKG-binding proteins such as myosin [33], vimentin [34], and actin-binding proteins [24,25] have been reported, but except for VASP the physiologic significance is unknown. Recently, fluorescent indicators have been used to determine the localization of PKG in cells. In HEK-293 and A549 epithelial cells, PKG was found to be localized with the best-characterized PKG substrate to date, VASP [24]. VASP is an actin-binding protein that is widely expressed and preferentially localized to focal adhesions of cells [35,36]. The precise anchoring protein for PKG in focal adhesions is not known, however.

PKG-II, as stated earlier, has a different distribution compared with PKG-I. PKG-II was first identified in intestinal epithelial cells as a phosphorylated protein. Since then, PKG-II has been cloned from brain and epithelial tissues. PKG-II clearly is the key kinase that regulates intestinal chloride transport due to phosphorylation and activation of the CFTR [37–39]. PKG-II has also been shown to inhibit renin release from the juxtaglomerular cells in the kidney [10,40]. PKG-II exists as a homodimer, is N-terminally myristolated and anchored in the cell membrane, and has tandem cGMP-binding sites like the PKG-I isoforms, although differences in activation mechanisms exist [41,42]. The sequence homology between PKG-I and PKG-II is approximately 50 percent overall, but the PKG-II isoform possesses unique cGMP-binding properties and activation by cGMP compared with PKG-I. PKG-II also has a very low affinity for cAMP and therefore does not appear to be cross-activated by cAMP in the cell. These findings suggest that drugs may be devised that could selectively target PKG-II and PKG-I in the cell, thus providing better insight into the roles of these two isoforms of PKG in cell function.

Physiologic Roles of PKG

Both PKG-I and PKG-II have been genetically deleted in mice and the phenotypes of the knockout mice studied. PKG-I deficient mice demonstrate a loss of NO-dependent relaxation of smooth muscle and acquire hypertension at approximately four weeks of age [43]. The animals also demonstrate abnormalities in platelet function [44]. These phenotypic properties are predictable based on the known roles for cGMP in relaxing vascular and nonvascular smooth muscle, inhibiting platelet adhesion and activation, and relaxing cavernosal smooth muscle in the penis. The animals succumb at an early age apparently due to digestive and colorectal dysfunction. Recently, Wegener *et al.* [45], using PKG-I conditional knockout animals, demonstrated a clear role for PKG in mediating the negative inotropic actions of NO and cGMP in mouse heart . The role of NO and cGMP in regulating cardiac contractility has been the subject of much controversy over the past two decades, in part because the antiadrenergic effects of muscarinic agonists have been difficult to distinguish from those related directly to cardiomyocyte cGMP, such as the inhibition of voltage-gated Ca^{2+} channels. This study provides a rather conclusive role for PKG in directly mediating the actions of cardiomyocyte cGMP.

PKG-II deficient mice show predictable changes in salt and water absorption from intestinal epithelium but also produced an unexpected phenotype of dwarfism. Lack of PKG-II causes the epiphyses of the long bones to close and harden prematurely, suggesting that PKG-II is required for chondrocyte proliferation and matrix production.

The classic roles of PKG in cell function were established in smooth muscle and platelets. Cyclic GMP was identified as the intracellular mediator of nitrovasodilator-drug dependent smooth muscle relaxation, and then shown to mediate epidermal growth factor receptor (EDRF) (that is, NO)-dependent relaxation. Many studies in the 1980s and 1990s established that PKG mediated relaxation by diverse mechanisms. Many of the details of the mechanisms responsible for smooth muscle relaxation have been reviewed elsewhere and will not be repeated here [5,6,46]. However, what should be emphasized is that smooth muscle cell relaxation can be accomplished by a number of very different mechanisms. For example, active contraction of the VSMC is achieved and regulated by two pathways: increases in cytosolic free calcium with the activation of myosin light chain kinase (MLCK), and activation of the rho-rho kinase pathway preventing myosin light chain dephosphorylation by inhibition of myosin light chain phosphatase [47,48]. The latter pathway constitutes the mechanism of calcium-sensitization of smooth muscle [49]. PKG has been shown to catalyze the phosphorylation of proteins that regulate both pathways. For example, PKG-dependent phosphorylation of protein components of calcium-activated potassium channels (K_{Ca} channels) leads to the activation of this channel and hyperpolarization of the cell [50–54]. Hyperpolarization inhibits calcium entry into the cell and hence allows less MLCK activation. PKG also catalyzes the phosphorylation of components of the calcium-sequestering mechanism in smooth muscle cells such as phospholamban [23]. Phospholamban phosphorylation leads to its dissociation from and the activation of the endoplasmic reticulum (ER) Ca-ATPase pump. Not only does this mechanism decrease cytosolic free calcium levels, it loads up the ER (or sarcoplasmic reticulum, the SR) with calcium. More calcium is then available for release near the plasma membrane compartment, where it activates K_{Ca} channels [55]. The increases in the number of "calcium sparks" by PKG is a novel mechanism for regulation of cytosolic calcium levels. Furthermore, the PKG-dependent phosphorylation of the SR calcium release channel, the IP_3 receptor, could contribute to calcium sparks and relaxation by either increasing calcium release toward the plasma membrane or by inhibiting calcium release, thus allowing greater filling of the SR with calcium for release through the ryanodine receptor [27,56,58]. What seems to be the case is that in different types of smooth muscle cells, the regulation of intracellular calcium levels relies on any one of these diverse mechanisms more than the others. Hence, cGMP-dependent regulation of K_{Ca} channels, for instance, might be more important in the relaxation of one type of smooth muscle compared to another. Nonetheless, each calcium-lowering pathway has apparently evolved the capacity to be regulated by cGMP/PKG-dependent protein phosphorylation. The occurrence of redundant mechanisms controlling cytosolic calcium in smooth muscle apparently provides evolutionary or selective advantage to the cell and organism to prevent excessive calcium accumulation and contractile activity in the cell.

The role of PKG in regulating smooth muscle myosin light chain phosphatase activity followed earlier findings demonstrating that cGMP produced calcium desensitization in smooth muscle cells [49,59,63]. The detailed mechanism by which PKG activates the phosphatase is still uncertain, but it is clear that phosphorylation of the MBS due to PKG localization and phosphorylation of the protein is necessary [30]. The reports of PKG-dependent activation of myosin light chain phosphatase, which is a Type I serine/threonine protein phosphatase, follow other reports in which cGMP and PKG have been implicated in protein phosphatase regulation in the nervous system. In Purkinje cells, a protein substrate for PKG, originally discovered in Paul Greengard's laboratory and termed the G-substrate, is an inhibitor of protein phosphatase 1 [64] and protein phosphatase 2A [65] when phosphorylated. DARPP-32, an inhibitor for protein phosphatase 1 when phosphorylated by PKA, is also phosphorylated by the NO/cGMP/PKG pathway [66], thereby suggesting that both cAMP and cGMP signaling pathways regulate DARPP-32 activity.

In addition to serine/threonine protein phosphatase regulation, protein tyrosine phosphatases have also been reported to be regulated by PKG. In smooth muscle, PTP-1, a soluble form of protein tyrosine phosphatase, is both phosphorylated and activated by NO/cGMP/PKG [67]. The role of the NO/cGMP signaling pathway in smooth muscle cell proliferation has been extensively studied, but specific mechanisms by which PKG regulates proliferation are unknown. Perhaps the regulation of protein tyrosine phosphorylation by phosphatases contributes to the antiproliferative actions of NO and cGMP.

A final mechanism for cyclic nucleotide-dependent smooth muscle relaxation is the regulation of thin filaments and particularly the phosphorylation of small heat shock protein HSP-20 and HSP-27 [68–70]. HSP-20 is a actin-binding protein in smooth muscle whose phosphorylation is catalyzed in the intact tissue by PKA and PKG. Tissues deficient in HSP-20 are less sensitive to cyclic nucleotide-dependent relaxation, thus suggesting an important role for HSP-20 in active relaxation [71,72]. It is interesting that PKG not only catalyzes the phosphorylation of HSP-20 but also induces its expression in VSMC [73].

PKG has been implicated in the regulation of MAP kinase pathways, gene expression and transcriptional activity, and in the regulation of VSMC phenotypic modulation. Although these findings have been reviewed recently and will not be discussed here [5,46], one newer role for cGMP/PKG signaling is in tumor cell apoptosis. It has been known that the nonsteroidal, antiinflammatory sulindac derivatives have potent anticancer activity [74,75]. One derivative, exisulind, induces apoptosis in cells derived from a number of different tumors, and in clinical studies exisulind prevents colorectal polyp formation in patients with familial adenomatous polyposis. Exisulind exerts its actions on tumor cells by inhibiting the type V cGMP phosphodiesterase (PDE-V) [76,77]. PDE-V, which is also the target for sildenafil (trade name, Viagra), is a specific cGMP phosphodiesterase that when phosphorylated by PKG hydrolyzes cGMP at a high rate. Exisulind produces sustained inhibition of PDE-V in tumor cells, leading to increases in cGMP and activation of PKG. PKG promotes the phosphorylation of β-catenin either through direct catalysis of phosphorylation of the protein or through activation of other kinase pathways that lead to β-catenin phosphorylation. When phosphorylated in response to PKG activation, β-catenin is targeted for ubiquination in the cytoplasm of the tumor cell. β-catenin combines with T-cell factor (TCF) to form a transcriptional complex that activates gene expression in tumor cells, resulting in resistance to apoptosis. By virtue of its phosphorylation by PKG activation, anti-apoptotic pathways are apparently turned off while pro-apoptotic pathways are turned on (e.g., caspase genes), leading to tumor cell death [78,79]. The pro-apoptotic effects of PKG may be dependent on activation of the c-Jun NH_2-terminal kinase (JNK) pathway [78–80]. Clearly, the roles for PKG in regulating gene expression, cell growth and differentiation, and apoptosis are only beginning to be uncovered.

Concluding Remarks

The family of protein kinases is huge, with some estimates of the number of gene products expressed in cells that belong to this family at around 5 percent. And protein phosphorylation, being the important and widespread regulator of cell function that it is, is a major area of interest in every biological process. As already mentioned, PKG was perhaps the fourth protein kinase discovered, following phosphorylase kinase, casein kinase, and PKA. Yet until about 15 years ago, understanding and solving the roles of PKG in biological function lagged the understanding of most other kinases' roles because upstream pathways leading to PKG activation were not well-defined. This changed rather dramatically with the uncovering of the biological role of NO. Now it seems that PKG has many important roles in various biological processes—even more than could be imagined by those investigators that began studying PKG from the time of its discovery. As discussed here, PKG plays a central role in smooth muscle cell contraction (e.g., blood pressure regulation, erectile function, gastrointestinal motility), smooth muscle cell gene expression in diseases such as atherosclerosis, salt and water absorbtion, skeletal growth, cardiac contractility, memory and learning, and tumor apoptosis. In the world of fruit flies and honey bees, whether the insects are a "stay-at-home couch potatoes" or "active food foraging providers" seems to depend on whether or not PKG is expressed (the food foragers express PKG and the couch potatoes don't) [81,82]. While I suspect that there is not any message for humans in these findings, one predication is that there will continue to be active pursuit of therapeutic agents that interact with the PKG pathway. With PKG involvement in so many different biological processes, and with losses in its expression or activity correlated with human diseases, new agents directed to enhance PKG expression or activity may find their way into future therapies.

References

1. Orstavik, S., Natarajan, V., Tasken, K., Jahnsen, T., and Sandberg, M. (1997). Characterization of the human gene coding for the type I alpha and type I beta cGMP-dependent protein kinase. *Genomics* **42**, 311–318.
2. Kalderon, D. and Ruben, G. M. (1989). cGMP-dependent protein kinase genes in *Drosophila*. *J. Biol. Chem.* **264**, 10738–10748.
3. Butt, E., Geiger, J., Jarchau, T., Lohmann, S. M., and Walter, U. (1993). The cGMP-dependent protein kinase—gene, protein and function. *Neurochem. Res.* **18**, 27–42.
4. Lincoln, T. M. and Cornwell, T. L. (1993). Intracellular cyclic GMP receptor proteins. *FASEB J.* **7**, 328–338.
5. Eigenthaler, M., Lohmann, S. M., Walter, U., and Pilz, R. B. (1999). Signal transduction by cGMP-dependent protein kinases and their emerging roles in the regulation of cell adhesion and gene expression. *Rev. Physiol. Biochem. Pharmacol.* **135**, 173–209.
6. Hofmann, F., Ammendola, A., and Schlossmann, J. (2000). Rising behind NO: cGMP-dependent protein kinases. *J. Cell Sci.* **113**, 1671–1676.
7. DeJonge, H. R. (1981). Cyclic GMP-dependent protein kinase in intestinal brush borders. *Adv. Cyclic Nucleotide Res.* **14**, 315–333.
8. Jarchau, T., Hausler, C., Markert, T., Pohler, D., Vanderkerckhove, J., De Jonge, H. R., Lohmann, S. M., and Walter, U. (1994). Cloning, expression, and in situ localization of rat intestinal cGMP-dependent protein kinase II. *Proc. Natl. Acad. Sci. USA* **91**, 9426–9430.
9. Uhler, M. D. (1993). Cloning and expression of a novel cyclic GMP-dependent protein kinase from mouse brain. *J. Biol. Chem.* **268**, 13586–13591.
10. Gambaryan, S., Wagner, C., Smolenski, A., Walter, U., Poller, W., Hasse, W., Kurtz, A., and Lohmann, S. M. (1998). Endogenous or overexpressed cGMP-dependent protein kinase inhibits cAMP-dependent renin release from rat isolated perfused kidney, microdissected glomeruli, and isolated juxtaglomerular cells. *Proc. Natl. Acad. Sci. USA* **95**.
11. Pfeifer, A., Aszodi, A., Seidler, U., Ruth, P., Hofmann, F., and Fassler, R. (1996). Intestinal secretory defects and dwarfism in mice lacking cGMP-dependent protein kinase II. *Science* **274**, 2082–2086.
12. Hofmann, F., Gensheimer, H. P., and Gobel, C. (1985). *Eur. J. Biochem.* **147**, 361–365.
13. Smith, J. A., Francis, S. H., Walsh, K. A., Kumar, S., and Corbin, J. D. (1996). Autophosphorylation of type Iβ cGMP-dependent protein kinase increases basal catalytic activity and enhances allosteric activation by cGMP or cAMP. *J. Biol. Chem.* **271**, 20756–20762.
14. Francis, S. H., Smith, J. A., Colbran, J. L., Grimes, K., Walsh, K. A., Kumar, S., and Corbin, J. D. (1996). Arginine 75 in the pseudosubstrate sequence of type Iβ cGMP-dependent protein kinase is critical for autoinhibition, although autophosphorylated serine 63 is outside this sequence. *J. Biol. Chem.* **271**, 20748–20755.
15. Chu, D. M., Francis, S. H., Thomas, J. W., Maksymovitch, E. A., Fosler, M., and Corbin, J. D. (1998). Activation by autophosphorylation or cGMP binding produces a similar apparent conformational change in cGMP-dependent protein kinase. *J. Biol. Chem.* **273**, 14649–14656.
16. Francis, S. H. and Corbin, J. D. (1999). Cyclic nucleotide-dependent protein kinases: intracellular receptors for cAMP and cGMP action. *Crit. Rev. Clin. Lab. Sci.* **36**, 275–328.
17. Lincoln, T. M., Thompson, M., and Cornwell, T. L. (1988). Purification and characterization of two forms of cyclic GMP-dependent protein kinase from bovine aorta. *J. Biol. Chem.* **263**, 17632–17637.
18. Keilbach, A., Ruth, P., and Hofmann, F. (1992). Detection of cGMP-dependent protein kinase isozymes by specific antibodies. *Eur. J. Biochem.* **208**, 467–473.
19. Sandberg, M., Natarajan, V., Ronander, I., Kalderon, D., Walter, U., Lohmann, S. M., and Jahnsen, T. (1989). Molecular cloning and predicted full-length amino acid sequence of the type Iβ isozyme of cGMP-dependent protein kinase from human placenta. *FEBS Lett.* **255**, 321–329.
20. Francis, S. H., Poteet-Smith, C., Busch, J. L., Richie-Jannetta, R., and Corbin, J. D. (2002). Mechanisms of autoinhibition in cyclic nucleotide-dependent protein kinases. *Front. Biosci.* **7**, 580–592.
21. Soderling, T. R., Chang, B., and Brickey, D. (2001). Cellular signaling through multifunctional Ca^{2+}/calmodulin-dependent protein kinase II. *J. Biol. Chem.* **276**, 3719–3722.
22. Michel, J. J. and Scott, J. D. (2002). AKAP mediated signal transduction. *Annu. Rev. Pharmacol. Toxicol.* **42**, 235–257.
23. Cornwell, T. L., Pryzwansky, K. B., Wyatt, T. A., and Lincoln, T. M. (1991). Regulation of sarcoplasmic reticulum phosphorylation by localized cyclic GMP-dependent protein kinase in vascular smooth muscle cells. *Mol. Pharmacol.* **40**, 23–931.
24. Browning, D. D., McShane, M., Marty, C., and Ye, R. D. (2001). Functional analysis of type Iα cGMP-dependent protein kinase using green fluorescent fusion proteins. *J. Biol. Chem.* **276**, 13039–13048.
25. Feil, R., Gappa, N., Rutz, M., Schlossmann, J., Rose, C. R., Konnerth, A., Brummer, S., Kubandner, S., and Hofmann, F. (2002). Functional reconstitution of vascular smooth muscle cells with cGMP-dependent protein kinase I isoforms. *Circ. Res.* **90**, 1080–1085.
26. Schlossmann, J., Ammendola, A., Ashman, K., Zong, X., Huber, A., Neubauer, G., Wang, G. X., Allescher, H. D., Korth, M., Wilm, M., Hofmann, F., and Ruth, P. (2000). Regulation of intracellular calcium by a signaling complex of IRAG, IP_3 receptor, and cGMP kinase Iβ. *Nature* **404**, 197–201.
27. Komalavilas, P. and Lincoln, T. M. (1996). Phosphorylation of the inositol 1,4,5-trisphosphate receptor: cyclic GMP-dependent protein kinase mediates cAMP and cGMP dependent phosphorylation in the intact rat aorta. *J. Biol. Chem.* **271**, 21933–21938.
28. Koga, T., Yoshida, Y., Cai, J. Q., and Imai, S. (1994). Purification and characterization of a 240-kDa cGMP-dependent protein kinase substrate of vascular smooth muscle: close resemblance to inositol 1,4,5-trisphosphate receptor. *J. Biol. Chem.* **269**, 11640–11647.
29. Yuasa, K., Omori, K., and Yanaka, N. (2000). Binding and phosphorylation of a novel male germ cell-specific cGMP-dependent protein kinase anchoring protein by cGMP-dependent protein kinase Iα. *J. Biol. Chem.* **275**, 4897–4905.
30. Surks, H. K., Mochizuki, N., Kasai, Y., Georgescu, S. P., Tang, K. M., Ito, M., Lincoln, T. M., and Mendelsohn, M. E. (1999). Regulation of myosin phosphatase by a specific interaction with cGMP-dependent protein kinase Iα. *Science* **286**, 1583–1587.
31. Shinokawa, H. (2002). Rho-kinase as a novel therapeutic target in treatment of cardiovascular disease. *J. Cardiovasc. Pharmacol.* **39**, 319–327.
32. Toth, A., Kiss, E., Gergely, P., Walsh, M. P., Hartshorne, D. J., and Erdodi, F. (2001). Phosphorylation of MYPT1 by protein kinase C attenuates interaction with PP1 catalytic subunit and the 20 kDa light chain of myosin. *FEBS Lett.* **484**, 113–117.
33. Vo, N. K., Gettemy, J. M., and Coghlan, V. M. (1998). Identification of cGMP-dependent protein kinase anchoring proteins (GKAPs). *Biochem. Biophys. Res. Commun.* **246**, 831–835.
34. MacMillan-Crow, L. A. and Lincoln, T. M. (1994). High affinity binding and localization of cyclic GMP-dependent protein kinase with the intermediate filament protein vimentin. *Biochemistry* **33**, 8035–8043.
35. Reinhard, M., Halbrugge, M., Scheer, U., Wiegand, C., Jockusch, B. M., and Walther, U. (1992). The 46/50 kDa phosphoprotein VASP purified from human platelets is a novel protein associated with actin filaments and focal contacts. *EMBO J.* **11**, 2063–2070.
36. Reinhard, M., Giehl, K., Abel, K., Haffner, C., Jarchau, T., Hoppe, V., Jockusch, B. M., and Walter, U. (1995). The proline rich focal adhesion and microfilament protein VASP is a ligand for profilins. *EMBO J.* **14**, 1583–1589.

37. French, P. J., Bijman, J., Edixhoven, M., Vaandrager, A. B., Scholte, B. J., Lohmann, S. M., Nairn, A. C., and DeJonge, H. R. (1995). Isotype-specific activation of cystic fibrosis transmembrane conductance regulator-chloride channel by cGMP-dependent protein kinase II. *J. Biol. Chem.* **270**, 26626–26631.
38. Vaandrager, A. B., Tilly, B. C., Smolenski, A., Schneider-Rasp, S., Bot, A. G. M., Edixhoven, M., Scholte, B. J., Jarchau, T., Walter, U., Lohmann, S. M., Poller, W. C., and DeJonge, H. R. (1997). cGMP-stimulation of cystic fibrosis transmembrane conductance regulator Cl channels co-expressed with cGMP-depedndnet protein kinase type II but not type Iβ. *J. Biol. Chem.* **272**, 4195–4200.
39. Vaandrager, A. B., Smolenski, A., Tilly, B. C., Housmuller, A. B., Ehlert, E. M. E., Bot, A. G. M., Edixhoven, M., Boomaars, W. E. M., Lohmann, S. M., and DeJonge, H. R. (1998). Membrane targeting of cGMP-dependent protein kinase is required for cystic fibrosis transmembrane conductance regulator Cl channel activation. *Proc. Natl. Acad. Sci. USA* **95**, 1466–1471.
40. Wagner, C., Pfeifer, A., Ruth, P., Hofmann, F., and Kurtz, A. (1998). Role of cGMP-kinase II in the control of renin secretion and renin expression. *J. Clin. Invest.* **102**, 1576–1582.
41. Gamm, D. M., Francis, S. H., Angelotti, T. P., Corbin, J. D., and Uhler, M. D. (1995). The type II isoform of cGMP-dependent protein kinase is dimeric and possesses regulatory and catalytic properties distinct from type I isoforms. *J. Biol. Chem.* **270**, 27380–27388.
42. Taylor, M. K. and Uhler, M. D. (2000). The amino-terminal cyclic nucleotide binding site of the type II cGMP-dependent protein kinase is essential for full cyclic nucleotide-dependent activation. *J. Biol. Chem.* **275**, 28053–28062.
43. Pfeifer, A., Klatt, P., Massberg, S., Ny, L., Sausbier, M., Hirneiss, C., Wang, G. X., Korth, M., Aszodi, A., Andersson, K., Krombach, F., Mayerhofer, A., Ruth, P., Fassler, R., and Hofmann, F. (1998). Defective smooth muscle regulation in cGMP kinase I-deficient mice. *EMBO J.* **11**, 3045–3051.
44. Massberg, S., Sausbier, M., Klatt, P., Baurer, M., Pfeifer, A., Siess, W., Fassler, R., Ruth, P., Krombach, F., and Hofmann, F. (1999). Increased adhesion and aggregation of platelets lacking cGMP kinase I. *J. Exp. Med.* **189**, 1255–1263.
45. Wegener, J. W., Nawrath, H., Wolfsgruber, W., Kuhbandner, S., Werner, C., Hofmann, F., and Feil, R. (2002). cGMP-dependent protein kinase I mediates the negative inotropic effect of cGMP in the murine myocardium. *Circ. Res.* **90**, 18–20.
46. Lincoln, T. M., Dey, N., and Sellak, H. (2001). Signal transduction in smooth muscle (invited review): cGMP-dependent protein kinase signaling mechanisms in smooth muscle: from the regulation of tone to gene expression. *J. Appl. Physiol.* **91**, 1421–1430.
47. Gong, M. C., Iizuka, K., Nixon, G., Browne, J. P., Hall, A., Eccleston, J. F., Sugai, M., Kobayashi, S., Somlyo, A. V., and Somlyo, A. P. (1996). Role of guanosine nucleotide-binding proteins—ras family or trimeric proteins or both—in Ca^{2+} sensitization in smooth muscle. *Proc. Natl. Acad. Sci. USA* **93**, 1340–1345.
48. Kimura, K., Ito, M., Amano, M., Chihara, K., Fukata, Y., Nakafuku, M., Yamamori, B., Feng, J., Nakano, T., Okawa, K., Iwamatsu, A., and Kaibuchi, K. (1996). Regulation of myosin phosphatase by rho and rho-associated kinase (Rho-kinase). *Science* **273**, 245–248.
49. Somlyo, A. P. and Somlyo, A. V. (1998). From pharmacomechanical coupling to G-proteins and myosin phosphatase. *Acta Physiol. Scand.* **164**, 437–448.
50. Alioua, A., Huggins, P., and Rousseau, E. (1995). PKG-I phosphorylates the α-subunit and upregulates reconstituted GK_{Ca} channels from tracheal smooth muscle cells. *Am. J. Physiol.* **268**, L1057–L1063.
51. Archer, S. L., Huang, J. M. C., Hampl, V., Nelson, D. P., Shultz, P. J., and Weir, E. K. (1994). Nitric oxide and cGMP cause vasorelaxation by activation of a charybodotoxin-sensitive K channel by cGMP-dependent protein kinase. *Proc. Natl. Acad. Sci. USA* **91**, 7583–7587.
52. Fukao, M., Mason, H. S., Britton, F. C., Kenyon, J. L., Horowitz, B., and Keef, K. D. (1999). Cyclic GMP-dependent protein kinase activates cloned BK_{Ca} channels expressed in mammalian cells by direct phosphorylation at serine 1072. *J. Biol. Chem.* **274**, 10927–10935.
53. White, R., Lee, A., Scherbatko, A. D., Lincoln, T. M., Schonbrunn, A., and Armstrong, D. L. (1993). Potassium channel stimulation by natriuretic peptides through cGMP-dependent dephosphorylation. *Nature* **361**, 263–266.
54. Zhou, X. B., Ruth, P., Schlossmann, J., Hofmann, F., and Korth, M. (1996). Protein phosphatase 2A is essential for the activation of Ca^{2+}-activated K^+-currents by cGMP-dependent protein kinase in tracheal smooth muscle and Chinese hamster ovary cells. *J. Biol. Chem.* **271**, 19760–19767.
55. Porter, V. A., Bonev, A. D., Knot, H. J., Heppner, T. J., Stevenson, A. S., Kleppisch, T., Lederer, W. J., and Nelson, M. T. (1998). Frequency modulation of Ca sparks is involved in regulation of arterial diameter by cyclic nucleotides. *Am. J. Physiol.* **274**, C1346–C1355.
56. Murthy, K. S. and Makhlouf, G. M. (1995). Interaction of cAM-kinase and cG-kinase in mediating relaxation of dispersed smooth muscle cells. *Am. J. Physiol.* **268**, C171–C180.
57. Murthy, K. S., Severi, C., Grider, J. R., and Makhlouf, G. M. (1993). Inhibition of IP_3 and IP_3-dependent Ca^{2+} mobilization by cyclic nucleotides in isolated gastric muscle cells. *Am. J. Physiol.* **264**, C967–C974.
58. Nelson, M. T., Cheng, H., Rubart, M., Santana, L. F., Bonev, A. D., Knot, H. J., and Lederer, W. J. (1995). Relaxation of arterial smooth muscle by calcium sparks. *Science* **270**, 633–637.
59. Nishimura, N. and van Breemen, C. (1989). Direct regulation of smooth muscle contractile elements by second messengers. *Biochem. Biophys. Res. Commun.* **163**, 929–935.
60. Lee, M. R., Li, L., and Kitazawa, T. (1997). Cyclic GMP causes Ca desensitization in vascular smooth muscle by activating the myosin light chain phosphatase. *J. Biol. Chem.* **272**, 5063–5068.
61. Pfitzer, G., Merkel, L., Ruegg, J. C., and Hofmann, F. (1986). Cyclic GMP-dependent protein kinase relaxes skinned fibers from guinea pig taenia coli. *Pflug. Arch.* **407**, 87–91.
62. Wu, X., Somlyo, A. V., and Somlyo, A. P. (1996). Cyclic GMP-dependent stimulation reverses G-protein coupled inhibition of smooth muscle myosin light chain phosphatase. *Biochem. Biophys. Res. Commun.* **220**, 658–663.
63. Wu, X., Haystead, T. A. J., Nakamoto, R. K., Somlyo, A. V., and Somlyo, A. P. (1998). Acceleration of myosin light chain dephosphorylation and relaxation of smooth muscle by telokin: synergism with cyclic nucleotide-activated kinase. *J. Biol. Chem.* **273**, 11362–11369.
64. Hall, K. U., Collins, S. P., Gamm, D. M., Massa, E., DePaoli-Roach, A. A., and Uhler, M. D. (1999). Phosphorylation-dependent inhibition of protein phosphatase-1 by G-substrate: a Purkinje cell substrate of the cyclic GMP-dependent protein kinase. *J. Biol. Chem.* **274**, 3485–3495.
65. Endo, S., Suzuki, M., Sumi, M., Nairn, A. C., Morita, R., Yamakawa, K., Greengard, P., and Ito, M. (1999). Molecular identification of human G-substrate, a possible downstream component of the cGMP-dependent protein kinase cascade in cerebellar Purkinje cells. *Proc. Natl. Acad. Sci. USA* **96**, 2467–2472.
66. Tsou, K., Snyder, G. L., and Greengard, P. (1993). Nitric oxide/cGMP pathway stimulates phosphorylation of DARPP-32, a dopamine and cAMP-regulated phosphoprotein, in the substantia nigra. *Proc. Natl. Acad. Sci. USA* **90**, 3462–3465.
67. Hassid, A., Yao, J., and Huang, S. (1999). NO alters cell shape and motility in aortic smooth muscle cells via protein phosphatase 1B activation. *Am. J. Physiol.* **277**, H1014–H1026.
68. Beall, A., Bagwell, D., Woodrum, D., Stoming, T. A., Kato, K., Suzuki, A., Rasmussen, H., and Brophy, C. M. (1999). The small heat shock-related protein, HSP20, is phosphorylated on serine 16 during cyclic nucleotide-dependent relaxation. *J. Biol. Chem.* **274**, 11344–11351.
69. Rembold, C. M., Foster, D. B., Strauss, J. D., Wingard, C. J., and Eyk, J. E. (2000). cGMP-mediated phosphorylation of heat shock protein 20 may cause smooth muscle relaxation without myosin light chain dephosphorylation in swine carotid artery. *J. Physiol London* **524**, 865–878.
70. Yamboliev, I. A., Hedges, J. C., Mutnick, J. L., Adam, L. P., and Gerthoffer, W. T. (2000). Evidence for modulation of smooth muscle

force by the p38 MAP kinase/HSP27 pathway. *Am. J. Physiol.* **278**, H1899–H1907.
71. Beall, A. C., Kato, K., Goldenring, J. R., Rasmussen, H., and Brophy, C. M. (1997). Cyclic nucleotide-dependent vasorelaxation is associated with the phosphorylation of a small heat shock-related protein. *J. Biol. Chem.* **272**, 11282–11287.
72. Brophy, C. M., Dickinson, M., and Woodrum, D. (1999). Phosphorylation of the small heat shock-related protein, HSP20, in vascular smooth muscles is associated with changes in the macromolecular associations of HSP20. *J. Biol. Chem.* **274**, 6324–6329.
73. Brophy, C. M., Woodrum, D. A., Pollock, J., Dickinson, M., Komalavilas, P., Cornwell, T. L., and Lincoln, T. M. (2002). cGMP-dependent protein kinase expression restores contractile function in cultured vascular smooth muscle cells. *J. Vasc. Res.* **39**, 95–103.
74. Haanen, C. (2001). Sulindac and its derivatives: a novel class of anticancer agents. *Curr. Opin. Invest. Drugs* **5**, 677–683.
75. Goluboff, E. T. (2001). Exisulind, a selective apoptotic antineoplastic drug. *Expert Opin. Invest. Drugs* **10**, 1875–1882.
76. Thompson, W. J., Piazza, G. A., Li, H., Liu, L., Fetter, J., Zhu, B., Sperl, G., Ahnen, D., and Pamukcu, R. (2000). Exisulind induction of apoptosis involves guanosine 3′,5′-cyclic monophosphate phosphodiesterase inhibition, protein kinase G activation, and attenuated β-catenin. *Cancer Res.* **60**, 3338–3342.
77. Liu, L., Han, L., Underwood, T., Lloyd, M., David, M., Sperl, G., Pamukcu, R., and Thompson, W. J. (2001). Cyclic GMP-dependent protein kinase activation and induction by exisulind and CP461 in colon tumor cells. *J. Pharmacol. Exp. Therap.* **299**, 583–592.
78. Soh, J. W., Mao, Y., Kim, M. G., Pamukcu, R., Han, L., Piazza, G. A., Thompson, W. J., and Weinstein, I. B. (2000). Cyclic GMP mediates apoptosis induced by sulindac derivatives via activation of c-Jun NH_2-terminal kinase 1. *Clin. Cancer Res.* **6**, 4136–4141.
79. Soh, J. W., Mao, Y., Liu, L., Thomposn, W. J., Pamukcu, R., and Weinstein, I. B. (2001). Protein kinase G activates the JNK 1 pathway via phosphorylation of MEKK1. *J. Biol. Chem.* **276**, 16406–16410.
80. Komalavilas, P., Shah, P. K., Jo, H., and Lincoln, T. M. (1999). Activation of mitogen-activated protein kinase pathways by cyclic GMP and cyclic GMP-dependent protein kinase in contractile vascular smooth muscle cells. *J. Biol. Chem.* **274**, 34301–34309.
81. Osborne K. A., Robichon, A., Burgess, E., Butland, S., Shaw, R. A., Coulthard, A., Pereira, H. S., Greenspan, R. J., and Sokolowski, M. B. (1997). Natural behavior polymorphism due to a cGMP-dependent protein kinase of *Drosophila*. *Science* **277**, 834.
82. Ben-Sharar, Y., Robichon, A., Sokolowski, M. B., and Robinson, G. E. (2002). Influence of gene action across different time scales on behavior. *Science* **296**, 741–744.

CHAPTER 201

Inhibitors of Cyclic Nucleotide-Dependent Protein Kinases

Wolfgang R. G. Dostmann

*Department of Pharmacology,
University of Vermont, College of Medicine,
Burlington, Vermont*

Introduction

Inhibitors of cyclic nucleotide-dependent protein kinases have served as valuable tools in identifying the pivotal roles of cAMP-dependent protein kinase (PKA) and cGMP-dependent protein kinase (PKG) in cellular biology by allowing the elucidation of basic molecular mechanisms of kinase structure and function and the dissection of the specific roles for each kinase in intracellular signaling. The remarkably similar multidomain structures of PKA and PKG simultaneously provide an opportunity and pose a challenge for the design of potent and selective inhibitors. A set of inhibitors now available target one of three distinct regions found in all subforms of cyclic nucleotide-dependent protein kinases: the cyclic nucleotide binding sites, the ATP-binding domains, and the peptide/substrate binding regions. This review will present the apparent experimental advantages and pitfalls of each inhibitor class and will provide a guide for identifying the inhibitor best suited for a given experiment, whether a reconstituted enzyme assay or intact cell or tissue preparation is involved.

The cyclic nucleotide-dependent protein kinases PKA and PKG, primary targets for the second messengers cAMP and cGMP, respectively, have served as rosetta stones in our understanding of a vast number of intracellular signaling mechanisms ranging from smooth muscle cell relaxation to neuronal synaptic plasticity (for reviews see [1–7]). Therefore, the search for potent inhibitors of these kinases has been extensively investigated ever since their discoveries. However, the structural similarities of PKA and PKG have posed a formidable obstacle in the design of selective inhibitors that specifically target cyclic nucleotide-dependent protein kinases and show little inhibitory potency to other more distant Ser/Thr-kinase relatives.

The domain structures of PKA and PKG dictate key target sites for putative inhibitors. Figure 1 compares the domain structures of PKA and PKG and defines three distinct classes of inhibitors and their various sites of actions. The regulatory components of cyclic nucleotide-dependent protein kinases each harbors two tandem cyclic nucleotide binding sites that allow allosteric and cooperative control of kinase activity (for reviews see [1,6,7]). A particular class of cAMP/cGMP derivatives, (R_P)-phosphorothioates, are the only known inhibitors that bind to the cyclic nucleotide binding sites [8–10]. Although their mode of action is still not completely understood, studies have indicated that the binding of these derivatives fails to induce the conformational changes essential for releasing catalytic activity [11,12]. The catalytic components of cyclic nucleotide-dependent protein kinases contain two target sites for inhibitors: the ATP-binding site and the substrate-binding site. Compounds mimicking ATP represent a diverse class of inhibitors, as has been known for

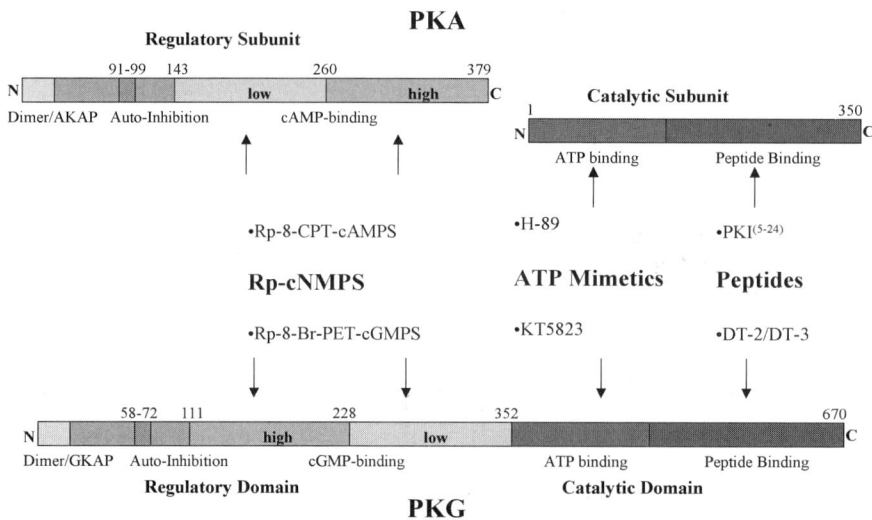

Figure 1 Domain structures of PKA and PKG with target sites for inhibitors and prominent inhibitor examples. Rp-cNMPS denotes (R_p)-diastereomers of cAMP/cGMP phosphorothioate derivatives.

all other major families of protein kinases [13,14]. In contrast, peptide inhibitors designed to block the substrate-binding site of PKA have been relatively straightforward [15–17]. PKA contains a well-defined substrate consensus sequence [18] that serves as an ideal template for peptide inhibitor design [19]. However, selective PKG peptide inhibitors have long remained elusive, partly because the sequence requirements for PKG inhibition do not follow a classic consensus sequence and the kinase appears to nonspecifically favor positively charged amino acids [18,20–26]. Only recently has a new class of potent and cell membrane permeable PKG peptide inhibitors based on combinatorial peptide libraries emerged [27–29].

Cyclic Nucleotide Binding Site-Targeted Inhibitors

The cyclic-nucleotide binding sites of PKA and PKG show remarkable sequence similarity with respect to the recognition motif of the nucleotide phosphodiester (FGE…RAA and FGE…RTA for all PKA and PKG isoforms, respectively). The crystal structures of PKA RIIβ [30] and RIα [31] revealed the architecture of the cyclic nucleotide binding pocket with an invariant arginine situated at its base. It is thought that chelation of the phosphate moiety by the invariant arginine is the first step in cAMP binding. The discovery that sulfur substitution of the axial exocyclic phosphate oxygen (Rp-) of cyclic nucleotides resulted in inhibitors of PKA and PKG, while cyclic nucleotide analogs carrying the equivalent apical sulfur substitution (Sp-) act as kinase agonists [8,32–35], prompted the synthesis and subsequent analysis of an entirely new class of cyclic nucleotide analogs [9,10,36–40] in an effort to identify selective and isozyme-specific inhibitors (for a comprehensive overview and a complete list of references visit *www.biolog.de*). Tables I and II give examples of PKA- and PKG-specific Rp-cyclic nucleotide phosphorothioates (Rp-cNMPS). These studies revealed that the competitive binding of Rp-cAMPS to both cAMP-binding sites in the PKA holoenzyme is thought to prevent dissociation of the regulatory subunits [11,12]. Likewise, it is believed that binding of Rp-cGMPS and analogs to PKG are unable to induce the conformational change needed to expose the enzyme's catalytic cleft.

Recent studies have demonstrated that Rp-cAMPS analogs can function as partial agonists dependent on the presence of MgATP and enzyme concentration [11,41]. These findings raise questions concerning the usefulness of inhibition constants when dealing with Rp-cAMPS analogs. However, this phenomenon is not observed for PKG partly because the holoenzyme complex does not dissociate into regulatory and catalytic subunits. Another important finding is that the sulfur substitution in Rp-cNMPS increases the relative lipophilicity and hence cell-membrane permeability compared to their cNMP counterparts [42], (see also: *www.biolog.de/logkw.html*). In addition, Rp-cNMPS exhibits complete resistance to phosphodiesterases [43]. These properties have markedly contributed to the diverse applications that cyclic nucleotide binding site-targeted inhibitors have enjoyed in intracellular signaling research [10]. It should be noted that Rp-cGMPS derivatives [37,38], in particular Rp-8-Br-PET-cGMPS, as a PKG inhibitor with high selectivity, low toxicity, and superior membrane permeability, has gained supremacy in its class [39].

ATP Binding Site-Targeted Inhibitors

Synthetic protein kinase inhibitors that are competitive with ATP and specific for PKA and/or PKG represent a structurally diverse group of small ligand compounds [44,45]. Polycyclic aromatics, such as Isoquinolinesulfonyl and Naphtalenesulfonyl compounds ("H-series"), and naturally occurring molecules, such as staurosporine analoga ("K-series"), have served primarily as valuable inhibitors of

Table I Potencies of Selected Cyclic Nucleotide-Dependent Protein Kinase Inhibitors

PKA inhibitors	K_i/IC_{50} (μM)	Isoform selectivity	Comments	Refs.	PKG inhibitors	K_i/IC_{50} (μM)	Isoform selectivity	Comments	Refs.
Kemptide		PKA Iα	pep, PKA	60	H2B[(29-35)]	86, PKG Iα	PKG Iα, Iβ	pep, non	24
Rp-cAMPS	12.5	PKA I	cyc, PKA, PDE	33	Rp-cGMPS	20, PKG Iα	PKG Iα, Iβ	cyc, PKG, PDE	38
H-8	1.2	PKA I	ATP, lip	47	WW21	7.5, PKG Iα	PKG Iα	pep, PKG	28
Rp-8-Br-cAMPS		PKA I	cyc, lip, PDE	9	HA1004	1.4, PKG Iα	—	ATP, lip, AGC, CaMK CMGC	48
Rp-8-CPT-cAMPS		PKA II	cyc, lip, PDE	9	Rp-8-Br-cGMPS	4.0, PKG Iα	PKG Iα, Iβ	cyc, lip, PKG, PDE	81
Rp-8-PIP-cAMPS		PKA II, site B	cyc, lip, PDE	40	Rp-8-CPT-cGMPS	0.5, PKG Iα	PKG II	cyc, lip, PKG, PDE	37
4-Cyano-3-methyl-isoquinoline	0.03, PKA	—	ATP, AGC	50	H-8	0.5, PKG Iα	—	ATP, lip, PKG	47
Balanol	0.004, PKA	PKA I	ATP, PKA	54, 55	KT5823	0.234, PKG Iα	—	ATP, lip, PKG	44, 46
Staurosporine	0.008, PKA	—	ATP, non	83–85	K-252b	0.1	—	ATP, lip, AGC, CaMK	46
H-89	0.048, PKA; 0.48, PKG	—	ATP, lip	44, 46	Rp-8-Br-PET-cGMPS	0.035, PKG Iα; 0.03, PKG Iβ	PKG Iα	cyc, lip, PKG, PDE	39
KT5720	0.06, PKA	—	ATP, lip	29	DT-3	0.025, PKG Iβ	PKG Iα	pep, MTS, PKG	29
K-252a	0.018	—	ATP, lip	46	K-252a	0.02	—	ATP, lip, AGC, CaMK	46
PKI[(5-24)]	0.002, PKA; 150, PKG	PKA Iα	pep, PKA	62	DT-2	0.012, PKG Iα	PKG Iα	pep, MTS, PKG	29

The following abbreviations are used: AGC specificity for AGC, subfamily protein kinases; ATP, ATP-binding site inhibitor; CaMK, specificity for CaMK subfamily protein kinases; CMGC, specificity for CMGC subfamily protein kinases; cyc, cyclic nucleotide binding site inhibitor; lip, lipophilic, cell permeable; MTS, membrane translocation signal; non, nonspecific; PDE, resistant against PDE hydrolysis; pep, peptide binding site inhibitor; PKA; PKC, high selectivity for PKC; PKG, high selectivity for PKG.

Table II Representative Commercially Available Inhibitors with Increasing PKA/PKG Selectivity

	$K_i(\mu M)$	PKA/PKG selectivity	References
Cyclic Nucleotide Inhibitors			
Rp-cAMPS	7.9 PKA I/II	1.5×10^{-1}	33, 35, 38, 82
	52, PKG Iα		
Rp-cGMPS	20, PKA II	1×10^0	37, 38
	20, PKG Iα		
Rp-8-CPT-cGMPS	8.3, PKA II	1.7×10^1	37
	0.5, PKG Iα		
Rp-8-Cl-cGMPS	100, PKA II	6.7×10^1	38
	1.5, PKG		
Rp-8-Br-PET-cGMPS	11, PKA II	3.1×10^2	39
	0.035, PKG Iα		
	0.030, PKG Iβ	3.7×10^2	
ATP Analogs			
KT5720	0.06, PKA I	3.0×10^{-2}	46
	>2, PKG Iα		
H89	0.048, PKA I	1.0×10^{-1}	45, 49
	0.48, PKG Iα		
H8	1.2, PKA I	2.5×10^1	45, 47
	0.48, PKG Iα		
KT5823	>10, PKA I	$>4.2 \times 10^1$	46, 44
	0.234, PKG Iα		
Peptide Inhibitors			
PKI(5–24)	0.002, PKA	1.8×10^{-5}	16, 20
TTYADFIASGRTGRRNAIHD	111, PKG		
PKI(14–22)	0.073, PKA	1.5×10^{-3}	17, 28
GRTGRRNAI	47, PKG		
Ala-Kemptide	376, PKA	4.7×10^{-1}	24
LRRAALG	800, PKG		
[A^{32}]-H2B(29–35)	550, PKA	6.4×10^0	20
RKRARKE	86, PKG		
WW21	750, PKA	1.0×10^2	28
TQAKRKKALAMA	7.5, PKG		
DT-2	16.5, PKA	1.32×10^3	29, 80
YGRKKRRQRRRPPLRKKKKKH	0.012, PKG		
DT-3	493, PKA	1.97×10^4	29, 80
RQIKIWFQNRRMKWKKLRKKKKKH	0.025, PKG		

AGC-type protein kinases, notably PKC [44–50]. In fact, the inhibitory potency against PKC is a defining property of most ATP-site inhibitors of the above series. However, a subset of compounds, including the H-series H89 and KT5720 and the K-series H8 and KT5823, are moderately specific inhibitors for PKA and PKG, respectively (see Tables I and II). The relatively simple chemical modifications of isoquinoline-derivatives in particular produced a wealth of selective protein kinase inhibitors with potential for clinical applications [45,50]. In addition, the cell-membrane permeability of these compounds further amplified their versatility in dissecting signaling pathways involving protein kinase signaling. However, concerns regarding toxicity and reports of problems using *in vivo* conditions, specifically with KT5823 as specific PKG inhibitors, have questioned their usefulness in intact cell preparations [51–53] (for a detailed discussion see: www.biolog.de/ti1003.html).

Recently it was observed that the natural product balanol inhibits protein kinases of the AGC-subfamily with high potency [54,55]. The crystal structure of balanol in complex with the catalytic subunit of PKA [56] confirmed a structural peculiarity of most ATP-site targeted inhibitors: the molecules satisfy essential interaction within the ATP-binding site but utilize unique interactions with the enzyme, thus gaining selectivity and specificity. It has been shown that analogs of balanol display variability in protein kinase inhibition, and the structural determinants of their protein kinase selectivity can now be elucidated with computational methods [57–59].

Peptide Binding Site-targeted Inhibitors

The observation that relatively short peptides corresponding to the regulatory subunit's autophosphorylation site were effective substrates for PKA [60] and the discovery of protein kinase inhibitor (PKI) [15,61] prompted a comprehensive search of PKA inhibitory peptides [15–17,19,62–65] and presented a prime example for the concept of "consensus sequences" [18,66,67]. However, full appreciation of the intricate structural web existing between kinase and inhibitor occurred only after the crystal structure of the catalytic subunit of the PKA:PKI adduct was solved [68]. Peptides derived from PKI isoforms α and β [69–71], namely PKI$^{5-24}$ and PKI$^{10-22}$, are still the most potent and, more important, the most selective PKA inhibitors known today (Tables I and II). However, their use in intact cell studies is limited, for example in patch-clamp techniques, due to their inability to cross the plasma membrane. Unfortunately, fusion peptides of PKI$^{15-22}$ with membrane translocation signal (MTS) peptides derived from Antennapedia-homeo domain or *HIV-1* tat [28,72] showed a profound loss in PKA selectivity (W. Dostmann, unpublished results).

Attempts to identify PKG-selective inhibitor peptides based on the auto-inhibitory domain of the enzyme or *in vivo* substrates have been tedious at best, due to the lack of a well-defined consensus sequence [20–24,73,74]. Only a relative preference for basic residues surrounding the phosphate acceptor site has been established [18,21,22,24–26]. Various synthetic peptides have been used with limited success to analyze and optimize the sequence requirements for PKG substrates and inhibitors [75–79]. Recently, the identification of selective inhibitors of PKG by a novel peptide library screen specifically designed to select for tight binding peptides was reported [27–29]. Cellular internalization of the peptides was accomplished by N-terminal fusion to the membrane translocation sequences from either the HIV-1 Tat protein [47–59], DT-2 or the *Drosophila* Antennapedia homeo-domain [43–58], DT-3 [29,80]. A surprising finding is that these fusion peptides result in an extraordinary synergism with respect to PKG inhibition (Tables I and II). It was shown that DT-2 and DT-3 effectively inhibit NO-induced vasodilation, further emphasizing the central role for PKG in the modulation of vascular contractility [80]. These results suggest that the cell membrane permeability of DT-2 and DT-3, combined with enormous PKG selectivity, will significantly advance our experimental ability to dissect PKG-mediated intracellular pathways from PKA and other kinases.

Conclusions

R_P-phosphorotioate derivatives of cAMP and cGMP competitively inhibit cyclic nucleotide-dependent protein kinases by 'freezing' the enzymes in their inactive holoenzyme states. A large pool of derivatives, moderate selectivity and cell membrane permeability are regarded as their advantages as tools in intact cell studies. However, partial antagonism and limited potencies restrict their versatility. ATP-analogs are a highly resourceful group of protein kinase inhibitors. Cell membrane permeability and limited selectivity highlight their advantages and disadvantages, respectively. Peptide-derived inhibitors present the most potent and selective group of PKA and PKG blockers. Low cell membrane permeability remains their main obstacle in cellular research. Recently, a subset of PKG-selective peptide inhibitors employing MTS sequences as a means of membrane translocation has overcome this problem.

References

1. Johnson, D. A., Akamine, P., Radzio-Andzelm, E., Madhusudan, M., and Taylor, S. S. (2001). Dynamics of cAMP-dependent protein kinase. *Chem. Rev.* **101**, 2243–2270.
2. Michel, J. J. and Scott, J. D. (2002). AKAP mediated signal transduction. *Annu. Rev. Pharmacol. Toxicol.* **42**, 235–257.
3. Hofmann, F., Ammendola, A., and Schlossmann, J. (2000). Rising behind NO: cGMP-dependent protein kinases. *J. Cell Sci.* **113**, 1671–1676.
4. Lincoln, T. M., Dey, N., and Sellak, H. (2001). cGMP-dependent protein kinase signaling mechanisms in smooth muscle: from the regulation of tone to gene expression. *J. Appl. Physiol.* **91**, 1421–1430.
5. Lohmann, S. M., Vaandrager, A. B., Smolenski, A., Walter, U., and De Jonge, H. R. (1997). Distinct and specific functions of cGMP-dependent protein kinases. *Trends Biochem. Sci.* **22**, 307–312.
6. Pfeifer, A., Ruth, P., Dostmann, W., Sausbier, M., Klatt, P., and Hofmann, F. (1999). Structure and function of cGMP-dependent protein kinases. *Rev. Physiol. Biochem. Pharmacol.* **135**, 105–149
7. Francis, S. H. and Corbin, J. D. (1994). Structure and function of cyclic nucleotide-dependent protein kinases. *Annu. Rev. Physiol.* **56**, 237–272.
8. Botelho, L. H., Rothermel, J. D., Coombs, R. V., and Jastorff, B. (1988). cAMP analog antagonists of cAMP action. *Methods Enzymol.* **159**, 159–172.
9. Dostmann, W. R., Taylor, S. S., Genieser, H. G., Jastorff, B., Doskeland, S. O., and Ogreid, D. (1990). Probing the cyclic nucleotide binding sites of cAMP-dependent protein kinases I and II with analogs of adenosine 3',5'-cyclic phosphorothioates. *J. Biol. Chem.* **265**, 10484–10491.
10. Schwede, F., Maronde, E., Genieser, H., and Jastorff, B. (2000). Cyclic nucleotide analogs as biochemical tools and prospective drugs. *Pharmacol. Ther.* **87**, 199–226.
11. Dostmann, W. R. and Taylor, S. S. (1991). Identifying the molecular switches that determine whether (Rp)-cAMPS functions as an antagonist or an agonist in the activation of cAMP-dependent protein kinase I. *Biochemistry* **30**, 8710–8716.
12. Dostmann, W. R. (1995). (Rp)-cAMPS inhibits the cAMP-dependent protein kinase by blocking the cAMP-induced conformational transition. *FEBS Lett.* **375**(3), 231–234.
13. Toledo, L. M., Lydon, N. B., and Elbaum, D. (1999). The structure-based design of ATP-site directed protein kinase inhibitors. *Curr. Med. Chem.* **6**, 775–805.
14. Garcia-Echeverria, C., Traxler, P., and Evans, D. B. (2000). ATP site-directed competitive and irreversible inhibitors of protein kinases. *Med. Res. Rev.* **20**, 28–57.

15. Scott, J. D., Fischer, E. H., Demaille, J. G., and Krebs, E. G. (1985). Identification of an inhibitory region of the heat-stable protein inhibitor of the cAMP-dependent protein kinase. *Proc. Natl. Acad. Sci. USA* **82**, 4379–4383.
16. Cheng, H. C., Kemp, B. E., Pearson, R. B., Smith, A. J., Misconi, L., Van Patten, S. M., and Walsh, D. A. (1986). A potent synthetic peptide inhibitor of the cAMP-dependent protein kinase. *J. Biol. Chem.* **261**, 989–992.
17. Glass, D. B., Cheng, H. C., Mende-Mueller, L., Reed, J., and Walsh, D. A. (1989). Primary structural determinants essential for potent inhibition of cAMP-dependent protein kinase by inhibitory peptides corresponding to the active portion of the heat-stable inhibitor protein. *J. Biol. Chem.* **264**, 8802–8810.
18. Kennelly, P. J. and Krebs, E. G. (1991). Consensus sequences as substrate specificity determinants for protein kinases and protein phosphatases. *J. Biol. Chem.* **266**, 15555–15558.
19. Kemp, B. E., Pearson, R. B., and House, C. M. (1991). Pseudosubstrate-based peptide inhibitors. *Methods Enzymol.* **201**, 287–304.
20. Glass, D. B., Cheng, H. C., Kemp, B. E., and Walsh, D. A. (1986). Differential and common recognition of the catalytic sites of the cGMP-dependent and cAMP-dependent protein kinases by inhibitory peptides derived from the heat-stable inhibitor protein. *J. Biol. Chem.* **261**, 12166–12171.
21. Glass, D. B. (1983). Differential responses of cyclic GMP-dependent and cyclic AMP-dependent protein kinases to synthetic peptide inhibitors. *Biochem. J.* **213**, 159–164.
22. Glass, D. B. and Krebs, E. G. (1979). Comparison of the substrate specificity of adenosine 3′:5′-monophosphate- and guanosine 3′:5′-monophosphate-dependent protein kinases. *J. Biol. Chem.* **254**, 9728–9738.
23. Glass, D. B. (1990). Substrate specificity of the cyclic GMP-dependent protein kinase, in B. E. Kemp, Ed., *Peptides and Protein Phosphorylation*, pp. 209–238. CRC Press, Boca Raton, Florida.
24. Glass, D. B. and Smith, S. B. (1983). Phosphorylation by cyclic GMP-dependent protein kinase of a synthetic peptide corresponding to the autophosphorylation site in the enzyme. *J. Biol. Chem.* **258**, 14797–14803.
25. Zeilig, C. E., Langan, T. A., and Glass, D. B. (1981). Sites in histone H1 selectively phosphorylated by guanosine 3′:5′-monophosphate-dependent protein kinase. *J. Biol. Chem.* **256**, 994–1001.
26. Glass, D. B. and Krebs, E. G. (1982). Phosphorylation by guanosine 3′:5′-monophosphate-dependent protein kinase of synthetic peptide analogs of a site phosphorylated in histone H2B. *J. Biol. Chem.* **257**, 1196–1200.
27. Tegge, W., Frank, R., Hofmann, F., and Dostmann, W. R. (1995). Determination of cyclic nucleotide-dependent protein kinase substrate specificity by the use of peptide libraries on cellulose paper. *Biochemistry* **34**, 10569–10577.
28. Dostmann, W. R., Nickl, C., Thiel, S., Tsigelny, I., Frank, R., and Tegge, W. J. (1999). Delineation of selective cyclic GMP-dependent protein kinase Ialpha substrate and inhibitor peptides based on combinatorial peptide libraries on paper. *Pharmacol. Ther.* **82**, 373–387.
29. Dostmann, W. R., Taylor, M. S., Nickl, C. K., Brayden, J. E., Frank, R., and Tegge, W. J. (2000). Highly specific, membrane-permeant peptide blockers of cGMP-dependent protein kinase Ialpha inhibit NO-induced cerebral dilation. *Proc. Natl. Acad. Sci. USA* **97**, 14772–14777.
30. Diller, T. C., Madhusudan, Xuong, N. H., and Taylor, S. S. (2001). Molecular basis for regulatory subunit diversity in cAMP-dependent protein kinase: crystal structure of the type II beta regulatory subunit. *Structure (Camb).* **10**, 73–82.
31. Su, Y., Dostmann, W. R., Herberg, F. W., Durick, K., Xuong, N. H., Ten Eyck, L., Taylor, S. S., and Varughese, K. I. (1995). Regulatory subunit of protein kinase A: structure of deletion mutant with cAMP binding domains. *Science* **269**, 807–813.
32. de Wit, R. J., Hoppe, J., Stec, W. J., Baraniak, J., and Jastorff, B. (1982). Interaction of cAMP derivatives with the 'stable' cAMP-binding site in the cAMP-dependent protein kinase type I. *Eur. J. Biochem.* **122**, 95–99.
33. de Wit, R. J., Hekstra, D., Jastorff, B., Stec, W. J., Baraniak, J., Van Driel, R., and Van Haastert, P. J. (1984). Inhibitory action of certain cyclophosphate derivatives of cAMP on cAMP-dependent protein kinases. *Eur. J. Biochem.* **142**, 255–260.
34. Rothermel, J. D. and Parker Botelho, L. H. (1988). A mechanistic and kinetic analysis of the interactions of the diastereoisomers of adenosine 3′,5′-(cyclic)phosphorothioate with purified cyclic AMP-dependent protein kinase. *Biochem. J.* **251**, 757–762.
35. Hofmann, F., Gensheimer, H. P., Landgraf, W., Hullin, R., and Jastorff, B. (1985). Diastereomers of adenosine 3′,5′-monothionophosphate (cAMP[S]) antagonize the activation of cGMP-dependent protein kinase. *Eur. J. Biochem.* **150**, 85–88.
36. Genieser, H.-G., Dostmann, W., Bottin, U., Butt, E., and Jastorff, B. (1988). Synthesis of nucleoside-3′,5′-cyclic phosphorothioates by cyclothiophosphorylation of unprotected nucleosides. *Tetrahedron Lett.* **29**, 2803–2804.
37. Butt, E., Eigenthaler, M., and Genieser, H. G. (1994). (Rp)-8-pCPT-cGMPS, a novel cGMP-dependent protein kinase inhibitor. *Eur. J. Pharmacol.* **269**, 265–268.
38. Butt, E., van Bemmelen, M., Fischer, L., Walter, U., and Jastorff, B. (1990). Inhibition of cGMP-dependent protein kinase by (Rp)-guanosine 3′,5′-monophosphorothioates. *FEBS Lett.* **263**, 47–50.
39. Butt, E., Pohler, D., Genieser, H. G., Huggins, J. P., and Bucher, B. (1995). Inhibition of cyclic GMP-dependent protein kinase-mediated effects by (Rp)-8-bromo-PET-cyclic GMPS. *Br. J. Pharmacol.* **116**, 3110–3116.
40. Ogreid, D., Dostmann, W., Genieser, H. G., Niemann, P., Doskeland, S. O., and Jastorff, B. (1994). (Rp)- and (Sp)-8-piperidino-adenosine 3′,5′-(cyclic)thiophosphates discriminate completely between site A and B of the regulatory subunits of cAMP-dependent protein kinase type I and II. *Eur. J. Biochem.* **221**, 1089–1094.
41. Gjertsen, B. T., Mellgren, G., Otten, A., Maronde, E., Genieser, H. G., Jastorff, B., Vintermyr, O. K., McKnight, G. S., and Doskeland, S. O. (1995). Novel (Rp)-cAMPS analogs as tools for inhibition of cAMP-kinase in cell culture. Basal cAMP-kinase activity modulates interleukin-1 beta action. *J. Biol. Chem.* **270**, 20599–20607.
42. Kraß, J., Jastorff, B., and Genieser, H.-G. (1997). Determination of lipophilicity by gradient elution high-performance liquid chromatography. *Anal. Chem.* **69**, 2575–2581.
43. Erneux, C. and Miot, F. (1988). Cyclic nucleotide analogs used to study phosphodiesterase catalytic and allosteric sites. *Methods Enzymol.* **159**, 520–530.
44. Hidaka, H. and Kobayashi, R. (1992). Pharmacology of protein kinase inhibitors. *Annu. Rev. Pharmacol. Toxicol.* **32**, 377–397.
45. Ono-Saito, N., Niki, I., and Hidaka, H. (1999). H-series protein kinase inhibitors and potential clinical applications. *Pharmacol. Ther.* **82**, 123–131.
46. Kase, H., Iwahashi, K., Nakanishi, S., Matsuda, Y., Yamada, K., Takahashi, M., Murakata, C., Sato, A., and Kaneko, M. (1987). K-252 compounds, novel and potent inhibitors of protein kinase C and cyclic nucleotide-dependent protein kinases. *Biochem. Biophys. Res. Commun.* **142**, 436–440.
47. Hidaka, H., Inagaki, M., Kawamoto, S., and Sasaki, Y. (1984). Isoquinolinesulfonamides, novel and potent inhibitors of cyclic nucleotide dependent protein kinase and protein kinase C. *Biochemistry* **23**, 5036–5041.
48. Ishikawa, T., Inagaki, M., Watanabe, M., and Hidaka, H. (1985). Relaxation of vascular smooth muscle by HA-1004, an inhibitor of cyclic nucleotide-dependent protein kinase. *J. Pharmacol. Exp. Ther.* **235**, 495–499.
49. Chijiwa, T., Mishima, A., Hagiwara, M., Sano, M., Hayashi, K., Inoue, T., Naito, K., Toshioka, T., and Hidaka, H. (1990). Inhibition of forskolin-induced neurite outgrowth and protein phosphorylation by a newly synthesized selective inhibitor of cyclic AMP-dependent protein kinase, N-[2-(p-bromocinnamylamino)ethyl]-5-isoquinolinesulfonamide (H-89), of PC12D pheochromocytoma cells. *J. Biol. Chem.* **265**, 5267–5272.

50. Lu, Z. X., Quazi, N. H., Deady, L. W., and Polya, G. M. (1996). Selective inhibition of cyclic AMP-dependent protein kinase by isoquinoline derivatives. *Biol. Chem. Hoppe Seyler* **377**, 373–384.
51. Burkhardt, M., Glazova, M., Gambaryan, S., Vollkommer, T., Butt, E., Bader, B., Heermeier, K., Lincoln, T. M., Walter, U., and Palmetshofer, A. (2000). KT5823 inhibits cGMP-dependent protein kinase activity in vitro but not in intact human platelets and rat mesangial cells. *J. Biol. Chem.* **275**, 33536–33541.
52. Komalavilas, P., Shah, P. K., Jo, H., and Lincoln, T. M. (1999). Activation of mitogen-activated protein kinase pathways by cyclic GMP and cyclic GMP-dependent protein kinase in contractile vascular smooth muscle cells. *J. Biol. Chem.* **274**, 34301–34309.
53. Wyatt, T. A., Pryzwansky, K. B., and Lincoln, T. M. (1991). KT5823 activates human neutrophils and fails to inhibit cGMP-dependent protein kinase phosphorylation of vimentin. *Res. Commun. Chem. Pathol. Pharmacol.* **74**, 3–14.
54. Koide, K., Bunnage, M. E., Gomez Paloma, L., Kanter, J. R., Taylor, S. S., Brunton, L. L., and Nicolaou, K. C. (1995). Molecular design and biological activity of potent and selective protein kinase inhibitors related to balanol. *Chem. Biol.* **2**, 601–608.
55. Setyawan, J., Koide, K., Diller, T. C., Bunnage, M. E., Taylor, S. S., Nicolaou, K. C., and Brunton, L. L. (1999). Inhibition of protein kinases by balanol: specificity within the serine/threonine protein kinase subfamily. *Mol. Pharmacol.* **56**, 370–376.
56. Narayana, N., Diller, T. C., Koide, K., Bunnage, M. E., Nicolaou, K. C., Brunton, L. L., Xuong, N. H., Ten Eyck, L. F., and Taylor, S. S. (1999). Crystal structure of the potent natural product inhibitor balanol in complex with the catalytic subunit of cAMP-dependent protein kinase. *Biochemistry* **38**, 2367–2376.
57. Hunenberger, P. H., Helms, V., Narayana, N., Taylor, S. S., and McCammon, J. A. (1999). Determinants of ligand binding to cAMP-dependent protein kinase. *Biochemistry* **38**, 2358–2366.
58. Gustafsson, A. B. and Brunton, L. L. (1999). Differential and selective inhibition of protein kinase A and protein kinase C in intact cells by balanol congeners. *Mol. Pharmacol.* **56**, 377–382.
59. Wong, C. F., Hunenberger, P. H., Akamine, P., Narayana, N., Diller, T., McCammon, J. A., Taylor, S. S., and Xuong, N.-H. (2001). Computational analysis of PKA-balaonol interactions. *J. Med. Chem.* **44**, 1530–1539.
60. Kemp, B. E., Graves, D. J., Benjamini, E., and Krebs, E. G. (1977). Role of multiple basic residues in determining the substrate specificity of cyclic AMP-dependent protein kinase. *J. Biol. Chem.* **252**, 4888–4894.
61. Scott, J. D., Fischer, E. H., Takio, K., Demaille, J. G., and Krebs, E. G. (1985). Amino acid sequence of the heat-stable inhibitor of the cAMP-dependent protein kinase from rabbit skeletal muscle. *Proc. Natl. Acad. Sci. USA* **82**, 5732–5736.
62. Scott, J. D., Glaccum, M. B., Fischer, E. H., and Krebs, E. G. (1986). Primary-structure requirements for inhibition by the heat-stable inhibitor of the cAMP-dependent protein kinase. *Proc. Natl. Acad. Sci. USA* **83**, 1613–1616.
63. Bhatnagar, D., Glass, D. B., Roskoski, R. Jr., Lessor, R. A., and Leonard, N. J. (1988). Synthetic peptide analogues differentially alter the binding affinities of cyclic nucleotide dependent protein kinases for nucleotide substrates. *Biochemistry* **27**, 1988–1994.
64. Glass, D. B., Lundquist, L. J., Katz, B. M., and Walsh, D. A. (1989). Protein kinase inhibitor-(6-22)-amide peptide analogs with standard and nonstandard amino acid substitutions for phenylalanine 10. Inhibition of cAMP-dependent protein kinase. *J. Biol. Chem.* **264**, 14579–14584.
65. Kemp, B. E. and Pearson, R. B. (1991). Intrasteric regulation of protein kinases and phosphatases. *Biochim. Biophys. Acta* **1094**, 67–76.
66. Pearson, R. B. and Kemp, B. E. (1991). Protein kinase phosphorylation site sequences and consensus specificity motifs: tabulations. *Methods Enzymol.* **200**, 62–81.
67. Kemp, B. E., Parker, M. W., Hu, S., Tiganis, T., and House, C. (1994). Substrate and pseudosubstrate interactions with protein kinases: determinants of specificity. *Trends Biochem. Sci.* **19**, 440–444.
68. Knighton, D. R., Zheng, J. H., Ten Eyck, L. F., Xuong, N. H., Taylor, S. S., and Sowadski, J. M. (1991). Structure of a peptide inhibitor bound to the catalytic subunit of cyclic adenosine monophosphate-dependent protein kinase. *Science* **253**, 414–420.
69. Van Patten, S. M., Ng, D. C., Th'ng, J. P., Angelos, K. L., Smith, A. J., and Walsh, D. A. (1991). Molecular cloning of a rat testis form of the inhibitor protein of cAMP-dependent protein kinase. *Proc. Natl. Acad. Sci. USA* **88**, 5383–5387.
70. Kumar, P., Van Patten, S. M., and Walsh, D. A. (1997). Multiplicity of the beta form of the cAMP-dependent protein kinase inhibitor protein generated by post-translational modification and alternate translational initiation. *J. Biol. Chem.* **272**, 20011–20020.
71. Kumar, P. and Walsh, D. A. (2002). A dual-specificity isoform of the protein kinase inhibitor PKI produced by alternate gene splicing. *Biochem. J.* **362**(Pt 3), 533–537.
72. Wadia, J. S. and Dowdy, S. F. (2002). Protein transduction technology. *Curr. Opin. Biotechnol.* **13**, 52–56.
73. Mitchell, R. D., Glass, D. B., Wong, C. W., Angelos, K. L., and Walsh, D. A. (1995). Heat-stable inhibitor protein derived peptide substrate analogs: phosphorylation by cAMP-dependent and cGMP-dependent protein kinases. *Biochemistry* **34**, 528–534.
74. Poteet-Smith, C. E., Corbin, J. D., and Francis, S. H. (1997). The pseudo-substrate sequences alone are not sufficient for potent autoinhibition of cAMP- and cGMP-dependent protein kinases as determined by synthetic peptide analysis. *Adv. Second Messenger Phosphoprotein Res.* **31**, 219–235.
75. Butt, E., Abel, K., Krieger, M., Palm, D., Hoppe, V., Hoppe, J., and Walter, U. (1994). cAMP- and cGMP-dependent protein kinase phosphorylation sites of the focal adhesion vasodilator-stimulated phosphoprotein (VASP) in vitro and in intact human platelets. *J. Biol. Chem.* **269**, 14509–14517.
76. Wood, J. S., Yan, X., Mendelow, M., Corbin, J. D., Francis, S. H., and Lawrence, D. S. (1996). Precision substrate targeting of protein kinases. *J. Biol. Chem.* **271**, 174–179.
77. Werner, D. S., Lee T. R., and Lawrence, D. S. (1996). Is protein kinase substrate efficacy a reliable barometer for successful inhibitor design? *J. Biol. Chem.* **271**, 180–185.
78. Yan, X., Corbin, J. D., Francis, S. H., and Lawrence, D. S. (1996). Precision targeting of protein kinases. *J. Biol. Chem.* **271**, 1845–1848.
79. Lev-Ram, V., Jiang, T., Wood, J., Lawrence, D. S., and Tsien, R. Y. (1997). Synergies and coincidence requirements between NO, cGMP, and Ca^{2+} in the induction of cerebellar long-term depression. *Neuron*, **18**, 1025–1038.
80. Dostmann, W. R. G., Tegge, W., Frank, R., Nickl, C. K., Taylor, M. S., and Brayden, J. E. (2002). Exploring the mechanisms of vascular smooth muscle tone with highly specific, membrane-permeable inhibitors of cyclic GMP-dependent protein kinase Iα. *Pharm. Ther.* **93**, 203–215.
81. Wei, J. Y., Cohen, E. D., Genieser, H. G., and Barnstable, C. J. (1998). Substituted cGMP analogs can act as selective agonists of the rod photoreceptor cGMP-gated cation channel. *J. Mol. Neurosci.* **10**, 53–64.
82. Van Haastert, P. J., Van Driel, R., Jastorff, B., Baraniak, J., Stec, W. J., and De Wit, R. J. (1984). Competitive cAMP antagonists for cAMP-receptor proteins. *J. Biol. Chem.* **259**, 10020–10024.
83. Nakano, H., Kobayashi, E., Takahashi, I., Tamaoki, T., Kuzuu, Y., and Iba, H. (1987). Staurosporine inhibits tyrosine-specific protein kinase activity of Rous sarcoma virus transforming protein p60. *J. Antibiot. (Tokyo)* **40**, 706–708.
84. Tamaoki, T., Nomoto, H., Takahashi, I., Kato, Y., Morimoto, M., and Tomita, F. (1986). Staurosporine, a potent inhibitor of phospholipid/Ca^{++}dependent protein kinase. *Biochem. Biophys. Res. Commun.* **135**, 397–402.
85. Ruegg, U. T. and Burgess, G. M. (1989). Staurosporine, K-252 and UCN-01: potent but nonspecific inhibitors of protein kinases. *Trends Pharmacol. Sci.* **10**, 218–220.

//CHAPTER 202

Peptide Substrates of Cyclic Nucleotide-Dependent Protein Kinases

Ross I. Brinkworth,[1] Bostjan Kobe,[1] and Bruce E. Kemp[2]

[1]Department of Biochemistry and Molecular Biology, Institute for Molecular Bioscience,
University of Queensland, St. Lucia, Brisbane, Queensland, Australia;
[2]St. Vincent's Institute of Medical Research,
Fitzroy, Victoria, Australia

Introduction

The cyclic nucleotide-dependent protein kinases cAMP and cGMP-dependent protein kinases (PKA and PKG) are closely related enzymes, with approximately 50 percent sequence identity. These two kinases have similar substrate specificities; in particular, both show a strong preference for Arg residues at positions (−3) and (−2) of peptide substrates. The crystal structure of PKA revealed multiple contacts between these substrate Arg side chains and the enzyme. There are differences in specificity between the two enzymes particularly at the (+1) site, with PKA having a higher preference for hydrophobic residues. However, hydrophobic residues at the (+1) site are only found in approximately 40 percent of PKA substrates, possibly because the phosphorylation sites must be hydrophilic enough to be located on the surface of the protein. Although PKG is more stringent in its requirement for residues at the (−1), (+2) and (+3) positions than PKA, there is substantial overlap in specificity, and the two kinases are expected to share some substrates, as is the case for the cystic fibrosis transmembrane conductance regulator.

The cAMP and cGMP-dependent protein kinases (PKA and PKG, respectively) belong to subgroup 1 of the AGC group of Ser/Thr protein kinases, the group that also contains protein kinase C (PKC) [1]. Early studies on PKA specificity over 25 years ago established that residues in the local phosphorylation site sequence were the primary specificity determinants (reviewed in [2]). Synthetic peptides corresponding to these sites were effective substrates. They have been used extensively to study the substrate specificity of protein kinases and have proved effective to measure PKA and PKG [3] activities in many systems. More recently synthetic peptides have been exploited as substrates in high-throughput screening for inhibitors and activators of protein kinases, including PKG [4]. This chapter focuses primarily on new developments in our understanding of the similarities and differences in the recognition of peptide substrates by PKA and PKG and the identification of important residues in enzyme substrate recognition.

Peptide Substrate Recognition

The catalytic domains [1] of PKA and PKG1 share 48 percent sequence identity; the strongest similarity is found in the central part of the catalytic domain, containing the majority of the substrate-binding residues as well as those involved in catalysis. We can therefore expect that the PKG catalytic domain will have a structure similar to the crystal structure of PKA [5–7]. The PKA crystal structures show the binding of a peptide inhibitor or substrate to PKA [5–9]. By examining these and other structures, we identified a constellation of 20 PKA residues (called acceptor loci)

Table I Acceptor Loci (Residues) in the Substrate-Binding Cleft of PKA, PKG1α, and PKG1β

Acceptor loci	PKA (Swiss-Prot # P17612)	PKG1α (Swiss-Prot # Q13976)	PKG1β (Swiss-Prot # P14619)	Substrate side chain subsite(s)
1	Gly 52	Gly 368	Gly 384	(−1)
2	**Ser 53**[a]	**Gly 369**	**Gly 385**	(−1)/(+2)
3	Phe 54	Phe 370	Phe 386	(+2)
4	**Leu 82**	**Thr 399**	**Thr 415**	(+2)
5	Gln 84	Gln 401	Gln 417	(+2)
6	Glu 127	Glu 444	Glu 460	(−3)
7	**Phe 129**	**Trp 446**	**Trp 462**	(−3)
8	**Ser 130**	**Thr 447**	**Thr 463**	(−3)
9	Asp 166, catalytic Asp	Asp 483, catalytic Asp	Asp 499, catalytic Asp	(0)
10	Glu 170	Glu 487	Glu 503	(−3)/(−2)
11	Phe 187	Phe 504	Phe 520	(+1)/(+2)
12	pThr 197	pThr 516	pThr 532	(+3)
13	**Leu 198**	**Phe 517**	**Phe 533**	(+1)/(+3)
14	Thr 201	Thr 520	Thr 536	(−2)
15	Pro 202	Pro 521	Pro 537	(+1)
16	Glu 203	Glu 522	Glu 537	(−2)
17	Tyr 204	Tyr 523	Tyr 538	(−2)
18	**Leu 205**	**Val 524**	**Val 539**	(+1)
19	Glu 230	Glu 549	Glu 565	(−2)
20	**Tyr 330**	**Phe 651**	**Phe 667**	(−3)

[a]Where the acceptor loci differ between PKA and PKG, the residues are shown in **bold**. Phosphothreonine, pThr.

predicted to interact with the side chains of peptide substrates between subsites (−3) and (+3) (the specificity determining residues in the phosphorylation site sequence are notated −n to +n, depending on their position on the N or C-terminal side of the phosphorylation site, 0; Table I). Other important peptide substrate enzyme interactions can occur outside the subsites (−3) and (+3) [6] but are not considered here for reasons of brevity. Although acceptor loci determine specificity, historically, residues in substrate phosphorylation site sequences that facilitate peptide substrate binding have been called specificity determinants, and for this reason acceptor loci are used here to notate contact residues on the kinase surface. A substrate Arg at (−3) (see Table I) binds to acceptor loci 6,7,8,10, and 20, plus the 3′-OH and 4′-OH of the ATP ribose moiety. An Arg at (−2), however, binds to acceptor loci 10,16,17, and 19 (acceptor locus 16 corresponding to Glu203 is diagnostic of the AGC group of protein kinase subfamily). The substrate (−3) and (−2) Arg residues make more contacts with the PKA catalytic core than any other substrate side chains, and this accounts for the strong specificity-determining role of Arg at these two positions. The only other subsite approaching this number of interactions is the (+1) hydrophobic pocket comprising acceptor loci 11,13,15, and 18. Acceptor locus 18 (Leu205) is situated at the end of this pocket, facilitating optimal binding to Leu or Ile residues (Val is suboptimal at this subsite). Acceptor loci 1,4, and 13 are located in the (−1), (+2), and (+3) side chain pockets, respectively, and suggest a modest preference for hydrophobic residues at these subsite pockets.

Comparison of Kinase Substrate Acceptor Loci

Due to the similarities in the acceptor loci, PKA and PKG share strong substrate specificity for Arg residues in peptide substrates at (−3) and (−2) positions [2]. By contrast, members of the PKC subgroup have an additional strong specificity for Arg or Lys at the (+2) position [10]. Of the 20 PKA acceptor loci, 13 are shared with PKG1 (Table I), whereas seven are different (2,4,7,8,13,18, and 20). The key difference appears to be acceptor locus 18 at the bottom of the (+1) pocket; the acceptor locus 18 of PKG is a Val, whose side chain is too short to contact a Leu residue bound at (+1). Thus, unlike PKA, a (+1) Val or Leu would make the same number of contacts with PKG. These differences in acceptor loci residues dictate the overall differences in the substrate specificities of PKA and PKG.

Optimum Recognition Sequences

Extensive studies of PKA over the past 30 years have identified a large number of substrates, in fact the largest

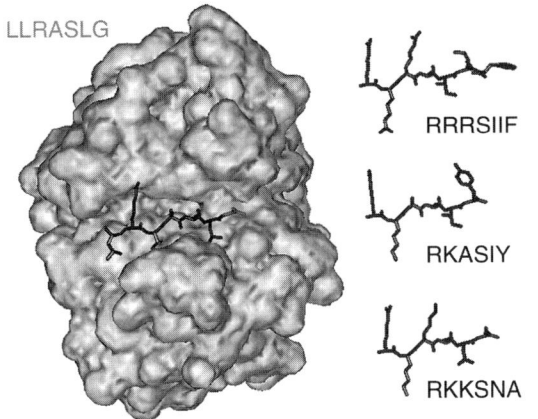

Figure 1 Surface structure of the catalytic domain of PKA with the "Kemptide" substrate (LRRASLG) bound. The structures of peptides representing the optimized recognition sequences for PKA (RRRSIIF [12], RKASIY [13]) and PKG (RKKSNA [13]) are illustrated.

Table II Abundance of Particular Residues in PKA and PKG Substrate Phosphorylation Sites (PhosphoBase), LIMVF, Leu/Ile/Met/Val/Phe; GASP, Gly/Ala/Ser/Pro

Subsite	Residue	PKA	PKG
(−3)	Arg	65.4	69.7
	Lys	16.7	6.1
(−2)	Arg	56.4	45.4
	Lys	16.4	42.1
(−1)	Arg	12.7	24.2
	Leu	9.7	16.2
	LIMVF	23.0	42.4
	GASP	35.2	12.1
(+1)	Leu	19.4	6.1
	LIMVF	36.9	24.2
	Ser	13.9	6.1
	GASP	29.4	27.3
	Arg	7.9	15.1
(+2)	LIMVF	12.4	18.2
	Ser	14.5	12.1
	Ala	19.1	21.2
	GASP	38.6	37.4
(+3)	LIMVF	15.8	9.1
	Pro	6.1	18.2
	GASP	32.9	24.2
	Glu	7.0	27.3

number for any protein kinase thus far; the number of phosphorylation sites currently listed in PhosphoBase [11] is 165, including redundancies such as entries for the same protein from different species. The results of oriented degenerate peptide library experiments have enabled estimates to be made of the optimal substrate sequence for protein kinases. The optimum recognition sequence for PKA [12], RRRSIIF (phosphorylation site underlined), is consistent with conclusions derived from both inspection of the acceptor loci and the early synthetic peptide studies using the Kemptide (LRRASLG). Optimal substrate sequences for PKA, RKASIY- and PKG, RKKSNA-have also been studied by using peptide libraries on cellulose paper [13]. The binding of the "Kemptide" substrate to PKA together with the structures of the optimized peptide substrates for PKA and PKG are illustrated in Fig. 1. Examination of substrate sequences for PKA (Table II) shows that specificity (for Arg) is strong at (−3) and (−2), but the specificity is broad at other positions, particularly at (+2) and (+3). Only a minority of substrate sequences conform reasonably well to the peptide library motif (e.g., RRNSILT, RKVSLAP, RRGSVPI, RRDSLFV, and RRQSVLV). The consensus sequence inferred from known PKA substrates would be RR(R/S)SLSS. Our structural analysis and the peptide library results indicate that the (+1) site should strongly favor hydrophobic residues, however, less than 40 percent of sites in known substrates contain hydrophobic residues at this position. There are a number of smaller residues [Gly, Ala, Ser, and Pro (GASP)] found at (−1), (+1), (+2), and (+3) in substrate sequences, which would be expected to make few, if any, contacts with the enzyme. The presence of these small residues accounts for the wide range of PKA substrate phosphorylation site sequences. This raises the question of whether within the cell the AKAPs (A kinase anchoring proteins) [14] may increase the probability of PKA binding by these suboptimal sequence-containing substrates, and therefore compensating for this suboptimal recognition, as well as localizing the substrate to a signaling complex.

One reason the optimal recognition sequence defined by peptide library experiments is not more highly represented in substrates is that it is rather hydrophobic and therefore difficult to accommodate in an exposed loop of a protein where the phosphorylation site would usually be located. The protein substrate's function also needs to be taken into account. Phosphorylation is required to induce structural and functional changes in the substrate, and these might not necessarily be compatible with a single highly conserved phosphorylation site sequence motif. Substitution of one or more hydrophobic residues in the motif by smaller residues such as Ser, Ala, Gly, or Pro would render the peptide less hydrophobic and increase the probability that the phosphorylation site is in a loop region. In addition, the smaller residues may facilitate structural changes in the substrate triggered by the phosphorylation event that are essential for altering protein function. The major implication of these considerations is that the binding of Arg at (−3) and (−2) provides the bulk of the interactions needed to bind the substrate to PKA, and this permits considerable flexibility in accommodating surrounding residues. A further reason for sequence diversity in local phosphorylation site sequences for PKA is that they may have a kinetic function in signaling, determining the order in which protein substrates are phosphorylated, that otherwise could not be achieved if there was a common conserved optimum motif [15]. By contrast, the peptide library motif for PKCα (RRR**S**LRK) [10,16] is more hydrophilic and therefore a better guide for PKC substrates such as MARCKS [17].

Comparison of PKA and PKG Specificity

Despite its close similarity to PKA, PKG is more restricted in its known substrates, with 33 phosphorylation sites listed in PhosphoBase. Some of these phosphorylation sites are common to both PKA and PKG. The consensus motif based on substrate sequences of PKG is R(R/K)(R/L)**S**(R/A)AE (Table II). Although largely similar to the PKA substrates, the PKG substrates show distinct differences. The (−2) site has a similar preference for Arg and Lys, whereas in PKA, Arg is heavily favored over Lys. The (−3) specificity is basically equal in PKA and PKG, despite the substitution of the acceptor locus 20 of PKA (Tyr to Phe) resulting in a loss of the acceptor group for the Arg side chain. PKG is more specific at the (−1), (+2), and (+3) sites than PKA, and Gly, Ala, Ser, and Pro occur more rarely at (−1) and (+3) positions. The moderate preference for hydrophobic residues at (+1) of PKA is not observed with PKG. Indeed the most dramatic differences in specificity between PKA and PKG were observed in peptide analogs (LRRRRF-aminoalcohol) containing α-substituted alcohols. PKG is able to phosphorylate the hydroxyl group in these peptide analogs in either the equivalent of a D or L-amino acid configuration, whereas PKA has a strict requirement for the L-configuration [18]. The PKG Val acceptor site 18 for the (+1) subsite is likely to explain its capacity to phosphorylate different isomers. The highest preference at (+1) is for Arg or Ala. PKG does display some preference for Arg/Leu, Ala/Ser, and Glu at (−1), (+2), and (+3), respectively. In summary, the specificity of PKG is narrower when compared to PKA, and a correspondingly smaller number of substrates have been identified. The additional specificity constraints for PKG may mean that anchoring proteins play less of a role in PKG signaling.

Overlapping substrate motifs for PKA and PKG suggest that there will be phosphorylation sites that the two protein kinases have in common, as is the case with RRL^{737}**S**LVP from the cystic fibrosis transmembrane conductance regulator [19,20]. By contrast, RTL^{7}**S**VSS from glycogen synthase is a substrate for at least eight different protein kinases, including PKA, but not PKG [11,21] possibly because the Ser residues at both (+2) and (+3) do not provide favorable contacts for PKG. The results show how two protein kinases such as PKA and PKG, with considerable sequence similarity, nevertheless exhibit significant diversity in substrate recognition properties.

Conclusion

The use of synthetic peptide analogs and peptide libraries has played a critical role in revealing the major specificity determinants in PKA and PKG substrate phosphorylation site sequences. Crystal structures of PKA with peptide substrates and inhibitors bound to the enzyme have provided a road map to understanding the important acceptor loci residues on the kinase surface responsible for peptide substrate binding. Molecular modeling of these structures has allowed us to identify important similarities and differences in PKA and PKG substrate specificity.

References

1. Hanks, S. K. and Quinn, A. M. (1991). Protein kinase catalytic domain sequence database: identification of conserved features of primary structure and classification of family members. *Methods Enzymol.* **200**, 38–62.
2. Kemp, B. E. and Pearson, R. B. (1990). Protein kinase recognition sequence motifs. *Trends Biochem. Sci.* **15**, 342–346.
3. Kemp, B. E. and Pearson, R. B. (1991). Design and use of peptide substrates for protein kinases. *Methods Enzymol.* **200**, 121–134.
4. Bader, B., Butt, E., Palmetshofer, A., Walter, U., Jarchau, T., and Drueckes, P. (2001). A cGMP-dependent protein kinase assay for high throughput screening based on time-resolved fluorescence resonance energy transfer. *J. Biomol. Screen* **6**, 255–264.
5. Knighton, D. R., Zheng, J., Ten Eyck, L. F., Ashford, V. A., Xuong, N.-H., Taylor, S. S., and Sowadski, J. M. (1991). Crystal structure of the catalytic subunit of cyclic adenosine monophosphate-dependent protein kinase. *Science* **253**, 407–414.
6. Knighton, D. R., Zheng, J., Ten Eyck, L. F., Xuong, N.-H., Taylor, S. S., and Sowadski, J. M. (1991). Structure of a peptide inhibitor bound to the catalytic subunit of cyclic adenosine monophosphate-dependent protein kinase. *Science* **253**, 414–420.
7. Bossemeyer, D., Engh, R. A., Kinzel, V., Postingl, H., and Huber, R. (1993). Phosphotransferase and substrate binding mechanism of the cAMP-dependent protein kinase catalytic subunit from porcine heart as deduced from the 2.0 Å structure of the complex with Mn^{2+} adenylyl imidodiphosphate and inhibitor peptide PKI(5-24). *EMBO J.* **12**, 849–859.
8. Zheng, J., Knighton, D. R., ten Eyck, L. F., Karlsson, R., Xuong, N., Taylor, S. S., and Sowadski, J. M. (1993). Crystal structure of the catalytic subunit of cAMP-dependent protein kinase complexed with MgATP and peptide inhibitor. *Biochemistry* **32**, 2154–2161.
9. Madhusudan, Trafny, E. A., Xuong, N. H., Adams, J. A., Ten Eyck, L. F., Taylor, S. S., and Sowadski, J. M. (1994). cAMP-dependent protein kinase: crystallographic insights into substrate recognition and phosphotransfer. *Protein Sci.* **3**, 176–187.
10. Nishikawa, K., Toker, A., Johannes, F. J., Songyang, Z., and Cantley, L. C. (1997). Determination of the specific substrate sequence motifs of protein kinase C isozymes. *J. Biol. Chem.* **272**, 952–960.
11. Kreegipuu, A., Blom, N., and Brunak, S. (1999). PhosphoBase, a database of phosphorylation sites: release 2.0. *Nucleic Acids Res.* **27**, 237–239.
12. Songyang, Z., Blechner, S., Hoagland, N., Hoekstra, M. F., Piwnica-Worms, H., and Cantley, L. C. (1994). Use of an oriented peptide library to determine the optimal substrates of protein kinases. *Curr. Biol.* **4**, 973–982.
13. Tegge, W., Frank, R., Hofmann, F., and Dostmann, W. R. (1995). Determination of cyclic nucleotide-dependent protein kinase substrate specificity by the use of peptide libraries on cellulose paper. *Biochemistry* **34**, 10569–10577.
14. Pawson, T. and Scott, J. D. (1997). Signaling through scaffold, anchoring, and adaptor proteins. *Science* **278**, 2075–2080.
15. Walsh, D. Newsholme, A. P., Cawley, K. C., van Patten, S. M., and Angelos, K. L. (1991). Motifs of protein phosphorylation and mechanisms of reversible covalent regulation. *Physiol. Rev.* **71**, 285–304.
16. Songyang, Z., Lu, K. P., Kwon, Y. T., Tsai, L. H., Filhol, O., Cochet, C., Brickey, D. A., Soderling, T. R., Bartleson, C., Graves, D. J., DeMaggio, A. J., Hoekstra, M. F., Blenis, J., Hunter, T., and Cantley, L. C. (1996). A structural basis for substrate specificities of protein Ser/Thr

kinases: primary sequence preference of casein kinases I and II, NIMA, phosphorylase kinase, calmodulin-dependent kinase II, CDK5, and Erk1. *Mol. Cell. Biol.* **16**, 6486–6493.

17. Aderem, A. (1992). The MARCKS brothers: a family of protein kinase C substrates. *Cell* **71**, 713–716.

18. Wood, J. S., Yan, X., Mendelow, M. J., Corbin, D., Francis, S. H., and Lawrence, D. S. (1996). Precision substrate targeting of protein kinases. The cGMP- and cAMP-dependent protein kinases. *J. Biol. Chem.* **271**, 174–179.

19. Picciotto, M. R., Cohn, J. A., Bertuzzi, G., Greengard, P., and Nairn, A. C. (1992). Phosphorylation of the cystic fibrosis transmembrane conductance regulator. *J. Biol. Chem.* **267**, 12742–12752.

20. Tien, X. Y., Brasitus, T. A., Kaetzel, M. A., Dedman, J. R., and Nelson, D. J. (1994). Activation of the cystic fibrosis transmembrane conductance regulator by cGMP in the human colonic cancer cell line, Caco-2. *J. Biol. Chem.* **269**, 51–54.

21. Roach, P. J. (1981). Glycogen synthase and glycogen synthase kinases. *Curr. Top. Cell Regul.* **20**, 45–105.

# Physiological Substrates of PKA and PKG

Kjetil Taskén, Anja Ruppelt, John Shabb,[1] and Cathrine R. Carlson

Department of Medical Biochemistry, Institute of Basic Medical Sciences, University of Oslo, Oslo, Norway and
[1]*Department of Biochemistry and Molecular Biology, University of North Dakota School of Medicine and Health Sciences, Grand Forks, North Dakota*

Introduction

The cAMP- and cGMP-kinases (PKA and PKG, respectively) belong to the ACG-subclass of Ser/Thr-specific protein kinases and generally prefer the phosphate acceptor residue preceded by a row of basic residues. S is the favored phosphate acceptor also when taking into account the 12-fold higher frequency of S over T in eukaryotic proteins. Based on an extensive body of work with peptide substrates *in vitro* (see Chapter 80, this volume) and mapping of phosphorylation sites in physiological substrates *in vivo*, PKA is well known to phosphorylate substrates with the general motif R(R/K)X(S/T) [1–6], whereas the consensus for PKG substrates is $(R/K_{2-3})(X/K)(S/T)$ and includes more basic residues than the PKA consensus [7]. However, considerable overlap of sites phosphorylated by both kinases is observed *in vivo*. By analysis of physiological substrates (Table II), the preference for PKA in the P^{-3} and P^{-2} positions is RR≥RK>>KR≥KK, and there are weaker preferences for small residues (S, G, P) at P^{-1}, for a basic residue (R) at P^{-4} to P^{-7}, and for a hydrophobic residue (F, I, L, V) at P^{+1} [8]. Substrate specificity of PKG is similar to that of PKA, but in physiological substrates (Table III) a stronger preference for R>>K at P^{-3} (at position P^{-2} K=R), a slight preference for basic or neutral residues (K, R, S) in P^{-4}, and an increased frequency of neutral and hydrophobic residues (S, V, A) at P^{-1} and (S, L, A) at P^{+1} is observed (this chapter).

Here we present data on the total availability of PKA and PKG consensus sites in the human proteome, estimate frequencies of phosphorylation of different motifs, and attempt to give an overview of physiological substrates of both kinases that meet a set of eligibility criteria. However, mechanisms whereby the phosphorylation event alters function of each individual substrate and thereby regulates its physiological role are not included in this short overview.

Abundance of PKA and PKG Phosphorylation Sites in the Human Proteome

We have searched available protein databases (January 2002), including all translated and indexed sequences from the full draft of the human genome, with all the permutations of the canonical PKA and PKG motif, (R/K)(R/K)X(S/T). A little more than 35,000 motifs were found in human proteins, which is approximately 15 percent of all motifs identified (Table Ia). Based on the total abundance of canonical motifs (phosphorylated and nonphosphorylated) in a limited set of approximately 100 substrates for PKA in which the phosphorylation sites have been mapped, Shabb estimated the probability of phosphorylation of permutations of the canonical sequence, which was RRXS (0.8) > RRXT, RKX(S/T) (0.5–0.3) > KKX(S/T), KRX(S/T) (<0.2) [8]. A similar analysis of frequency of PKG phosphorylation (this chapter, from substrates listed in Table III) shows that probability of phosphorylation *in vivo* by PKG is estimated as R(R/K)XS (0.7) > RKXT (0.5) > KRXS, RRXT (0.3) > KKXS (0.2).

Table I Abundance of PKA and PKG motifs[a]

a. Frequency and probability of phosphorylation of canonical subtrate sequences

Consensus	Homo sapiens	Estimated phosphorylated motifs of PKA[b]	Estimated phosphorylated motifs of PKG[c]
RRXS	7078	5662	5096
RKXS	5233	2460	3506
KRXS	4670	841	1541
KKXS	5194	831	883
RRXT	3682	1399	921
RKXT	3106	932	1553
KRXT	2783	390	0
KKXT	3801	266	0

b. Frequency of high affinity motifs for PKA [d,e] and PKG [e] identified in physiological substrates[d] or by peptide library screens[e]

PKA Consensus	Homo sapiens	PKG Consensus	Homo sapiens
RRSS(L, V, I, F) [a]	163	RKKS	407
RRGS(L, V, I, F) [a]	106	KRKKS	68
RRPS(L, V, I, F) [a]	93	KARKXS	19
RR(S, G, P)SF [a]	42	KARKKS	3
RRAS [c]	450	AKRKKS	4
AERRAS [c]	3	KRKKSL	5
RAERRASI [c]	0	KXRKKSL	3
LRRASLG [c] (kemptide)	0	KARKKSL	1
		TQAKRKKSLA	0

[a]All available human protein sequences including predicted proteins translated from the full draft of the human genome were analyzed (January 2002: 54687 sequences in the International Protein Index (IPI) version 2.4 at EMBL-European Bioinformatics Institute, Hinxton, Cambridge, UK, which provides a minimally redundant yet maximally complete set of human proteins assembled from SWISS-PROT, TrEMBL, RefSeq and Ensembl, see (http://www.ebi.ac.uk/IPI/IPIhelp.html).

[b,c]Probability for the motif being phosphorylated by PKA[b] ([8]) or PKG[c] (this study) in vivo.

[d] Preferred in vivo motifs according to [8].

[e]High-affinity peptides [9,10,17]. The authors acknowledge the kind help of Dr. Paul Kersey at EBI with searching IPI.

Using these probabilities, we estimate that PKA, PKG, or both can phosphorylate approximately 15,000 sites in the human proteome *in vivo*. When both frequency of different sites and probability of phosphorylation are taken into account, RRXS followed by RKXS stands out as the most abundant *in vivo* sites for both kinases (Table Ia). Notably, however, this analysis also predicts phosphorylation of significant numbers of substrates with less prevalent motifs such as K(R/K)X(S/T) with lower affinity but which may be physiologically relevant especially in contexts where kinase and substrate are colocalized [8]. In contrast, preferred substrates defined by more detailed analysis of physiological substrates (Tables II and III) or by *in vitro* phosphorylation of generations of peptide libraries with PKA and PKG are clearly less abundant due to their more restricted motifs (Table Ib). High-affinity substrates such as LRRASLG (Kemptide, [9]) and RAERRASI [7] for PKA and TQAKRKKSLA for PKG [10] used *in vitro* enzyme assays were not found in the human proteome.

Physiological Substrates

General criteria for identification of physiological substrates of protein kinases were originally outlined by Krebs and Beavo [11], recently reviewed by Shabb [8], and can be summarized thus:

1. The target protein should be phosphorylated stoichometrically and dephosphorylated by phosphatase *in vitro* at significant kinetic rates.
2. Functional properties of the substrate should change in correlation with the degree of phosphorylation.
3. Phosphorylation of the substrate should be demonstrated *in vivo* or in intact cells with accompanying functional changes.
4. The cellular levels of protein kinase should correspond to the extent of phosphorylation of the substrate.
5. The *in situ* phosphorylation sequence should be identified (new, adds stringency).

Table II Physiological Substrates of PKA

Substrate	In vivo/in situ site	Sequence	Accession no./species	Ref
Autophosphorylation				
cAMP-dependent protein kinase regulatory subunit type IIα	Ser-95[a]	pgrfdrrvsvcaet	P00515 (bovine)	[8]
Receptor mediated signaling				
β$_2$-adrenergic receptor	Ser-262[c]	tghglrrsskfclk	P07550 (human)	[8]
	Ser-345[c]	qellclrrsslkay		
	Ser-346[c]	ellclrrsslkayg		
Prostacyclin receptor (IP)	Ser-357[c]	qaplsrpasgrrdp	A54416 (mouse)	[13]
Regulator of protein signaling RGS9-1	Ser-427[c]	epqgttrkssslpf	O46469 (bovine)	[14]
	Ser-428[c]	pqgttrkssslpfm		
Regulator of G protein signaling (RGS10)	Ser-168[c]	aqtaakrasriynt	Q9CQE5 (mouse)	[15]
cAMP signaling				
cAMP-specific phosphodiesterase PDE4D3	Ser-54[a,c]	fvhsqrresflyrs	P14270 (rat)	[8]
cGMP inhibited phosphodiesterase PDE3B	Ser-302[a,c]	sgkmfrrpslpcis	Q63085 (rat)	[8]
D1 dopamine receptor	Thr-268[c]	fkmsfkretkvlkt	P18901 (rat)	[8]
Metabotropic glutamate receptor subunit mGluR2	Ser-843[a,c]	fgsaaprasanlgq	P31421 (rat)	[8]
Phosphodiesterase 4D5 (PDE4D5)	Ser-126[ac]	fvhsqrresflyrs	AAC00069 (human)	[16]
cGMP signaling				
Phosducin	Ser-73[a]	kermsrkmsiqeye	P20942 (rat)	[8]
Phosphoinositide and calcium signaling				
Elongation factor-2 kinase	Ser-499[c]	srlhlprpsavale	P70531 (rat)	[8]
Inositol 1,4,5-trisphosphate Type I receptor	Ser-1589[a]	arnaarrdsvlaas	P29994 (rat)	[8]
	Ser-1755[a]	irpsgrresltsfg		
Phospholipase C-γ1	Ser-1248[a]	fhvraregsfeary	P10686 (rat)	[8]
Phospholipase C-β3	Ser1105[a,c]	ildrkrhnsiseak	Q01970 (human)	[8]
Thromboxane A$_2$ receptor TPα	Ser-329[c]	prlstrprslslqp	NP 001051 (human)	[8]
CaM-kinase kinase α	Thr-108[c]	sprawrrptieshh	BAA75246 (rat)	[17]
	Ser-458[c]	vksmlrkrsfgnpf		
Rho signaling				
RhoA small GTP-binding protein	Ser-188[a,c]	qarrgkkksgclvl	P06749 (human)	[8]
T cell receptor signaling				
COOH-terminal Src kinase (Csk)	Ser-364[a,c]	ealrekkfstksdv	P41240 (human)	[8]
Mitogen-activated protein kinase signaling				
Hematopoeietic protein tyrosine phosphatase (hePTP)	Ser-23[a,c]	vrlqerrgsnvalm	P35236 (human)	[8]
Mammalian STE20-like kinase 3 b isoform (MST3b)	Thr-18[c]	lalnkrratlphpg	AAD42039 (human)	[8]
v-Mos	Ser-56[a,c]	psvdsrscsiplva	P00538 (maloney murine sarcoma virus)	[8]
	Ser-102[a,c]	vclmhrlqsggfgs		
	Ser-263[a,c]	qdlrgrqaspphig		
GTPase activating protein specific for Rap1 (rap1GAP)	Ser-490[a,c]	gksptrkksgpfgs	P47736 (human)	[8]
	Ser-499[a,c]	gpfgsrrssaigie		
Guanine nucleotide exchange factor Ras-GRF1	Ser-916[a,c]	nkevfrrmslantg	P27671 (mouse)	[8]
Protein tyrosine phosphatase-SL (PTP-SL)	Ser-231[c]	iglqerrgsnvslt	NP 035347 (human)	[8]
Raf-1 serine/threonine protein kinase	Ser-43[a,c]	qfgyqrrasddgkl	P11345 (rat)	[8]
Rap1b low molecular weight GTP/GDP-binding protein	Ser-179[c]	vpgkarkksscqll	P09526 (human)	[8]
Striatal-enriched protein tyrosine phosphatase, 61 kDa (STEP)	Ser-160[a,c]	lppedrrgsvsrqp	P54830 (mouse)	[8]
STEP$_{61}$ and 46 kDa STEP$_{46}$ splice variant	Ser221[a,c,e]	mglqerrgsnvslt		

continues

Table II *continued*

Substrate	In vivo/in situ site	Sequence	Accession no./species	Ref
Modulators of protein phosphatase 1				
Glycogen binding (G) subunit of protein phosphatase 1	Ser-46[a]	spqpsrrgsdssed	NP 002702 (human)	[8]
	Ser-65[a]	pssgtrrvsfadsf		
Inhibitor-1of (I-1) protein phosphatase I	Thr-35[a]	eqirrrrptpatlv	P01099 (rabbit)	[8]
Dopamine and cAMP-regulated phosphoprotein, 32 kDa (DARPP-32)	Thr-34[a,b]	emirrrrptpamlf	P07516 (bovine)	[8]
Transcriptional regulation				
Cyclic AMP response element binding protein (CREB)	Ser133[a,b,c]	reilsrrpsyrkil	P15337 (rat)	[8]
Cyclic AMP responsive element modulator (CREMτ)	Ser-117[c]	reilsrrpsyrkil	P27699 (mouse)	[8]
Nuclear factor of activated T cells 3 (NFAT3)	Ser-272[a,c]	spcgkrryssgtp	Q14934 (human)	[8]
	Ser-289[a,c]	spalsrrgslgeeg		
Nuclear factor κB (NF-κB)	Ser-276[a,c]	vsmqlrrpsdrels	Q04207 (mouse)	[8]
Retinoic acid receptor-α (RARα)	Ser-369[a,c]	vyvrkrrpsrphmf	P11416 (mouse)	[8]
Retinoid X receptor-α (RXRα)	Ser-27[c]	ltsptgrgsmaaps	P19793 (human)	[8]
Sex determining region of Y gene product (SRY protein)	Ser-32[c]	nipalrrsssflct	Q05066 (human)	[8]
SRY-box related transcription factor SOX9	Ser-64[c]	gepdlkkeseedkf	P48436 (human)	[8]
	Ser-181[b,c]	kyqprrrksvkngq		
Steroidogenic factor-1 (SF-1)	Ser-430[c]	clvevralsmqake	P33242 (mouse)	[8]
Thyroid hormone receptor α1	Ser-28[a,c]	ldgkrkrkssqclv	P04625 (chick)	[8]
	Ser-29[a,c]	dgkrkrkssqclvk		
Vasoactive intestinal polypeptide receptor transcriptional repressor protein (VIPR-RP)	Ser-245[c]	ktkkarkdseeges	AAC40192 (rat)	[8]
	Ser-361[c]	kgsptkresvsped		
Class II transactivator (CIITA)	Ser-834[c]	vqelpgrlsflgtr	AAA88861 (human)	[18]
	Ser1050[c]	laasllrlslynnc		
Thyroid transcription factor (TTF1)	Ser-337[c]	pdlahhaaspaalq	P23441 (rat)	[19]
Histones				
Histone H1c	Ser-37[a]	pagvrrkasgppvs	P15864 (mouse)	[8]
Histone H3	Ser-10[a,b,c]	rtkqtarkstggka	P16106 (human)	[8]
Apoptosis and cell survival				
Bcl-2/Bcl-X$_L$-antagonist, causing cell death (BAD)	Ser155[a,b,c]	ygrelrrmsdefeg	Q61337 (mouse)	[8]
Glycogen synthase kinase 3α (GSK-3α)	Ser-21[b]	gsgrartssfaepg	P18265 (rat)	[8]
Glycogen synthase kinase-3β (GSK-3β)	Ser-9[b,c]	msgrprttsfaesc	P18266 (rat)	[8]
Interleukin receptor-3 β$_c$ chain	Ser-585[b,c]	ylgpphsrslpdil	NP_000386 (human)	[8]
Ligand-gated ion channels				
GABA$_A$ receptor β1 subunit	Ser-409[c]	kgrirrrasqlkvk	P50571 (mouse)	[8]
GABA$_A$ receptor β3 subunit	Ser-408[c]	Kthlrrrssqlkik	P15433 (mouse)	[8]
	Ser-409[c]	kkthlrrrssqlki		
Glutamate receptor GluR1 subunit (AMPA receptor)	Ser845[a,b,c]	rtstlprnsgagas	P19490 (rat)	[8]
Glutamate receptor GluR4 subunit (AMPA receptor)	Ser-842[a,c]	airnkarlsitgsv	P48058 (human)	[8]
Glutamate receptor GluR6 subunit (kainate receptor)	Ser-684[c]	afmssrrqsvlvks	P42260 (rat)	[8]
Glutamate receptor NR1A subunit (NMDA receptor)	Ser897[a,b,c]	ssfkrrrsskdtst	P35439 (rat)	[8]
Nicotinic acetylcholine receptor δ subunit	Ser-361[a,c]	ndlklrrsssvgyi	P02718	[8]
	Ser-362[a,c]	dlklrrsssvgyis	(*T. californica*)	
P$_{2X2}$ purinoreceptor	Ser-431[c]	avqsprpcsisalt	P49653 (rat)	[8]
Sodium ion movement				
Na$^+$ H$^+$ exchanger 3 (NHE3)	Ser-552[a,c]	vaegerrgslafir	P26433 (rat)	[8]
	Ser-605[a,c]	qsleqrrrsirdte		

Table II *continued*

Substrate	In vivo/in situ site	Sequence	Accession no./species	Ref
Na+,K+ ATPase α1 subunit	Ser943[a,b,c]	vicktrrnsvfqqg	P06685 (rat)	[8]
Serum and glucocorticoid regulated kinase (Sgk)	Thr-369[c]	ddlinkkitppfnp	O00141 (human)	[8]
Voltage sensitive Na+ channel (Rat brain type IIA) α subunit	Ser-573[a,c]	slfsprrnsraslf	P04775 (rat)	[8]
	Ser-610[a,c]	edndsrrdslfvph		
	Ser-623[a,c]	hrhgerrpsnvsqa		
	Ser-687[a,c]	teirkrrsssyhvs		
Voltage-sensitive Na+ channel (cardiac type H1) α subunit	Ser526[a,b,c]	rtsmrprssrgsif	P15389 (rat)	[8]
	Ser529[a,b,c]	mrprssrgsiftfr		
Chloride conductance				
Cystic fibrosis transmembrane conductance regulator (CFTR)	Ser-660[a,c]	qfsaerrnsiltet	P13569 (human)	[8]
	Ser-700[a]	efgekrknsilnpi		
	Ser-737[c]	deplerrlslvpds		
	Ser-795[c]	ttastrkvslapqa		
	Ser-813[c]	idiysrrlsqetgl		
Phospholemman	Ser-68[a,c]	frssirrlstrrr	O08589 (rat)	[8]
Potassium channels				
Shaker K+ channel	Ser-507[c]	tlgqhmkksslses	P08511 (*Drosophila*)	[8]
	Ser-508[c]	lgqhmkksslsess		
Slo K$_{Ca}$ channel splice variant A1C2E1G3I0	Ser-942[b,c]	pivlqrrgsvygan	JH0697[d] (*Drosophila*)	[8]
hSlo BK$_{Ca}$ α subunit of large conductance Ca^{2+}-dep. K+ channel (maxi-K)	Ser-869[c]	vhgmlrqpsittgv	NP 002238 (human)	[8]
Kv1.1 α subunit of the *Shaker* RCK1 Voltage-gated K+ channel	Ser-446[c]	dsdlsrrsssttisk	P10499 (rat)	[8]
Kvβ1.3 subunit of the Kv1.5 K+ channel Ik$_{ur}$	Ser-24[c]	entklrrqsgfsva	AAC41926.1 (human)	[8]
Kv4.2 α subunit of the *Shal*-type K+ channel	Thr-38[a,b]	pprqerkrtqdali	NP 062671 (mouse)	[8]
	Ser-552[a,b]	nvsgshrgsvqels		
Kir 1.1 Renal outer medullary K+ channel 1, 2 (ROMK1, ROMK2)	Ser-25[a,c]	srqrarlvskegrc	P35560 (rat)	[8]
	Ser-200[a,c]	irvanlrkslligs		
	Ser-294[a,c]	satcqvrtsyvpee		
Kir2.1	Ser-425[c]	eprplrresei	Q64273 (rat)	[8]
Kir2.3 Inward rectifier K+ channel (IRK)	Ser-440[a,c]	dnisyrresai	P48050 (human)	[8]
Kir6.2 subunit of the ATP-sensitive K+ channel (K$_{ATP}$)	Ser-372[c]	argplrkrsvpmak	Q14654 (human)	[8]
	Thr-224[c]	hmqvvrkttspege		
SUR1 subunit of the ATP-sensitive K+ channel (K$_{ATP}$)	Ser-1571[c]	ekllsrkdsvfasf	Q09428 (human)	[8]
Water homeostasis				
Aquaporin-2	Ser-256[b,c]	erevrrrgsvelhs	P34080 (rat)	[8]
Other transporters				
P-glycoprotein mdr1b	Ser-665[a,c]	sksplirrsiyrsv	P06795 (mouse)	[8]
	Ser-681[a,c]	kqdqerrlsmkeav		
Steroidogenic acute regulatory protein (StAR)	Ser-57[c]	inqvrrrssllgsr	P49675 (human)	[8]
	Ser-195[c]	vrcakrrgstcvla		
Extracellular proteins				
Atrial natriuretic peptide	Ser-104[a]	gprslrrsscfggr	P01161 (rat)	[8]
Vitronectin	Ser-378[a]	rnqnsrrpsratwl	P04004 (human)	[8]
Trafficking and motility				
Actin bundling protein L-plastin	Ser-5[a,c]	margsvsdeemmel	P13796 (human)	[8]
Low-density lipoprotein receptor-related protein (LRP)	Ser-4520[c]	mgghgsrhslastd	Q07954 human	[8]
Myosin light-chain kinase (MLCK) and telokin splice variant	Ser-1005[a,c,f]	sglsgrksstgspt	P29294 (rabbit)	[8]

continues

Table II *continued*

Substrate	*In vivo/in situ* site	Sequence	Accession no./species	Ref
Protein tyrosine phosphatase-PEST (PTP-PEST)	Ser-39[a]	dfmrlrrlstkyrt	NP 002826 (human)	[8]
	Ser-435[a]	dkklernlsfeikk		
Small heat shock-related protein HSP20	Ser-16[a,b]	qpswlrrasaplpg	O14558 (human)	[8]
Synapsin I	Ser-9[a]	mnylrrrlsdsnfm	P09951 (rat)	[8]
Vasodilator-stimulated phosphoprotein (VASP)	Ser-157[a,c]	sehierrvsnaggp	P50552 (human)	[8]
	Ser-239[a,b,c]	agaklrkvskqeea		
	Thr-278[a,c]	mlarrrkatqvgek		
Snapin	Ser-50[c]	shvhavresqvelr	XP_057189 (human)	[20]
Sso1 t-SNARE	Ser-79[c]	eqashlrhsldnfv	NP_015092 (yeast)	[21]
Cysteine string protein (csp)	Ser-10[c]	dcqrqrslstsges	Q29455 (bovine)	[22]
Striated muscle contraction				
Myosin-binding protein-C cardiac isoform	Ser-275[a,c]	llsafrrtslaggg	Q14896 (human)	[8]
	Ser-284[a,c]	lagggrrisdshed		
	Ser-304[a,c]	ssllkkrdsfrtpr		
Phospholamban	Ser-16[a,b]	trsairrastiemp	P26678 (human)	[8]
Ryanodine receptor type 2 (sarcoplasmic reticulum Ca^{2+} release channel)	Ser-2809[c]	lynrtrrisqtsqv	P30957 (rabbit)	[8]
Troponin I	Ser-23[a]	apapirrrssnyra	P19429 (human)	[8]
	Ser-24[a]	papirrrssnyray		
Voltage-sensitive L-type Ca^{2+} channel (skeletal muscle) α1 Subunit	Ser-1757[a,c]	perggrrtsltgsl	P07293 (rabbit)	[8]
	Ser-1854[a,c]	pgslsrrsslgsld		
Voltage-sensitive L-type Ca^{2+} channel (cardiac) α1 subunit	Ser-1928[a,b,c]	saslgrrasfhlec	P15381 (rabbit)	[8]
Voltage-sensitive L-type Ca^{2+} channel (cardiac) $β_{2a}$ subunit	Ser-459[a,c]	drsaprsasqaeee	A42044 (rat)	[8]
	Ser-478[a,c]	vkksqhrsssathq		
	Ser-479[a,c]	kksqhrsssathqn		
Metabolism and respiration				
ATP citrate lyase	Ser-454[a]	tpapsrtasfsesr	P16638 (rat)	[8]
Cytochrome P450 CYP2E1	Ser-129[a,c]	twkdvrrfslsilr	P05182 (rat)	[8]
Glycogen synthase (muscle type)	Ser-7[a,c]	mplsrtlsvsslpg	AAB69872 (rabbit)	[8]
	Ser-697[a,c]	apewprrasctsss		
	Ser-710[a,c]	sssggskrsnsvdt		
Hormone-sensitive lipase	Ser-563[a,c]	rltesmrrsvseaa	P15304 (rat)	[8]
	Ser-659[a,c]	pdgfhprrssqgvl		
	Ser-660[a,c]	dgfhprrssqgvlh		
Phenylalanine hydroxylase	Ser-16[a,b]	npglgrklsdfgqe	P00439 (human)	[8]
Phosphorylase kinase α subunit (muscle type)	Ser-1018[a]	kqvefrrlsistes	P18688 (rabbit)	[8]
Phosphorylase kinase β subunit (muscle type)	Ser-26[a]	rartkrsgsvyepl	P12798 (rabbit)	[8]
6-phosphofructo-2-kinase-fructose-2,6-bisphosphatase liver isozyme 1	Ser-32[a]	svlqrrrgssipqf	P07953 (rat)	[8]
6-phosphofructo-1-kinase, isozyme A (muscle type)	Ser-376[a]	eamklrgrsfmnnw	P00511 (rabbit)	[8]
Pyruvate kinase (liver type)	Ser-43[a]	pagylrrasvaqlt	P12928 (rat)	[8]
Tyrosine hydroxylase	Ser-40[a,b,c]	prfgirrqslieda	P04177 (rat)	[8]
Nuclear-encoded subunit of complex I (NDUFS4)	Ser-131[a]	anfswnkrtrvstk	Q02375 (bovine)	[23]
Micellaneous				
cAMP-regulated phosphoprotein 16/19 kDa (ARPP16/19)	Ser-104[a,b,g]	qdlpqrkpslvask	P56211 (human)	[8]
cAMP-regulated phosphoprotein, 21 kDa (ARPP-21)	Ser-55[a,b]	aqnqerrksksgag	A34957 (bovine)	[8]
Serine/threonine protein kinase LKB1	Ser-431[b,c]	ssnkirrlsackqq	NP_035622 (mouse)	[8]

Table II continued

Substrate	In vivo/in situ site	Sequence	Accession no./species	Ref
Phogrin	Ser-680[c]	gphtsrinsvssql	CAA90600 (rat)	[24]
	Thr-699[c]	pspsarsstsswse		
rabphilin	Ser-234[a]	hgpptrrasearm	P47709 (rat)	[25]

[a] Direct sequencing and/or phosphopeptide mapping.
[b] Phospho/dephospho-specific antibodies.
[c] Site-directed mutagenesis.
[d] Ser-952 in this splice variant.
[e] $STEP_{61}$ Ser-221 is equivalent to $STEP_{46}$ Ser-49.
[f] MLCK Ser-1005 is equivalent to telokin Ser-13.
[g] ARPP-19 Ser-104 is equivalent to ARPP-16 Ser-88.

Table III Physiological Substrates of PKG

Substrate	In vivo/in situ site	Sequence	Accession no./species	Ref
Autophosphorylation				
Autophosphorylation	Ser-63	atqqagkqsastlq	P14619 (human)	[26]
	Ser-79	prtkrgaisaepta		
Regulation of smooth muscle tone				
Cardiac Troponin I	Ser-23	apapirrrssnyray	P19429 (human)	[27]
(cTN1)	Ser-24	apapirrrssnyray		[28]
(CRP2) Cystein rich Protein 2	Ser-104	vrteerktsgppkgp	P36201 (rat)	[29,30]
Hsp20 (Heat-shock 20 kD like protein P20)	Ser-16	qpswlrrasaplpg	P97541 (rat)	[31]
Regulation of smooth muscle tone (by regulation of intracell. Ca level)				
Cardiac phospholamban	Ser-16	trsairrastiemp	P26678 (human)	[32]
L-type Ca2+ channel α1c subunit	Ser-533	hriskskfsrywrr	P15381 (rabbit)	[33]
Calcium-acivated maxi K+ channel (BK_{Ca}) (α-subunit)	Ser-1072[a]	sqssskkssshvs	AAA84000 (canine)	[34,35]
Ins (1,4,5) P3 receptor type I	Ser-1756	irpsgrresltsfg	P29994 (rat)	[36]
Platelet Aggregation				
Heat shock protein 27 (Hsp27)	Ser-15	pfsllrgpswdpfr	XP_004991 (homo sapiens)	[37]
	Ser-78	apaysralsrqlssg		
	Ser-82	sralsrqlssgvsei		
	Thr-143	srcftrkytlppgv		
VASP (vasodilator stimulated phosphoprotein)	Ser-157	sehierrvsnaggp	CAA86523 (human)	[38,39]
	Ser-239	agaklrkvskqeea		
	Thr-278	mlarrrkatqvgek		
Neuronal function				
G-septin	Ser-91	ksqvsrkasswnre	AAD21035 (rat)	[40,41]
$GABA_A$ Receptor β2 subunit	Ser-410[a]	ksrlrrrasqlkit	P15432 (mouse)	[42]
$GABA_A$ Receptor β3-subunit	Ser-409	kthlrrrssqlkik	1095220 (mouse)	[42]
G-substrate	Thr-68	qkkprrkdtpalhi	AAD13030 (human)	[43]
	Thr-119	qkkprrkdtpalhm		
Dopamin/DARPP-32	Thr-34	emirrrrptpamlf	P07516 (bovine)	[44]
Thromboxane receptor alpha (TPα)	Ser-331[a]	lstrprslslqpql	NP_001051 (human)	[45,46]
Nucleus				
Splicing factor SF1	Ser-20	fpskkrkrsrwnqd	CAA03883 (human)	[47]

continues

Table III *continued*

Substrate	*In vivo/in situ* site	Sequence	Accession no./species	Ref
Histone H2B	Ser-32	kdgkkrkrsrkesy	XP_059791(human)	[48]
germ cell development				
GKAP42	Ser-106[a]	spnpagkesreenw	BAA92254(mouse)	[49]
Metabolic enzymes				
Tyrosine Hydroxylase	Ser-40	prfigrrgslieda	P04177(rat)	[50]
Nitric-oxide synthase NOS-III	Ser-633	swrrkrkessntds	P29474 (human)	[51]
	Ser-1177	vtsrirtgsfslqe		
6-phospho-fructo-2-kinase	Ser-32	svlqrrrgssipqf	P07953(rat)	[52]
Regulation of other signalling pathways				
CGMP-binding cGMP-specific hosphodiesterase (CGB-PDE)	Ser-92	pgtptrkisasefd	Q28156 (bovine)	[53]
c-Raf1	Ser-43[a]	qfgyqrrasddgkl	P04049 (human)	[54]
PLC-beta3	Ser-26[a]	vvktlrrgskfikw	P51432 (mouse)	[55]
	Ser-1105[a]	ildrkrnnsiseak		
HIV-1 replication and infectivity				
Vif (one of the HIV-1 Proteins)	Ser-144	qaghnkvgslqyla	AAA44202 (HIV type I)	[56]
	Thr-188	tkghrgshtmngh		
Filamentprotein				
Vimetin	Ser-26[b]	pgtasrpsstrsyv	P48616 (bovine)	[57,58]

[a] Site-directed mutagenesis.
[b] Potential phosphorylation site because the serine residue meets the consensus sequence

With the present level of available technologies such as deletion and mutation mapping of the site and mass spectrometry, current consensus is that the primary evidence to consider is that *in vivo* phosphorylation occurs in response to elevated cAMP or cGMP (criteria 3–5), and this takes priority over *in vitro* mapping of PKA and PKG sites (typically the method of use in older literature) where a pitfall is that less stringent phosphorylations may occur. However, application of these criteria excludes phosphorylations occuring *in vivo* that are either silent (that is, no functional change) or whose function has not yet been mapped. Conversely, substrates implicated in physiological pathways but where the precise mechanism or site has not yet been identified are also excluded.

Here we present a comprehensive list of PKA substrates that meet the above eligibility criteria identified through review of the literature (Table II). This list is based on a recently published review by Shabb [8], and has been revised by adding new published data, bringing the number of identified and listed physiological substrates of PKA that meet the eligibility criteria to 116 and the number of analyzed motifs to 162. Furthermore, we present a corresponding list of 29 PKG substrates, representing 38 analyzed motifs, that have been identified by systematic literature search and application of the same set of criteria (Table III). It is interesting that 13 of those motifs are also phosphorylated by PKA (Tables III and II), indicating at least 30 percent overlap in substrates. During these searches we also revealed a number of potential substrates that do not fulfil a convincing combination of the general criteria yet. Furthermore, the discrepancy between the identified substrates that meet the criteria (Tables II and III) and the estimated number of substrates (Table I) indicates that more than 99 percent of human PKA and PKG substrates are still unidentified, and this constitutes a major task to address in the future.

Concluding Remarks

The specificity of a substrate is determined not only by the primary sequence, but also by several other factors that affect the degree of phosphorylation of a given target. The tertiary structure of the substrate affects function and kinetics of the kinase, such as for example the catalytic subunit of cAMP-kinase that in part acquires its substrate specificity from the conserved F at position $P11^{-1}$[12]. The organization of the microenvironment around a phosphorylation event has a clear impact. In that respect, anchoring proteins (AKAPs, GKAPs) play an important role by locating PKA and PKG in close vicinity to their substrates and demonstrate how low-affinity substrates may become physiologically relevant.

References

1. Glass, D. B. and Krebs, E. G. (1979). Comparison of the substrate specificity of adenosine 3':5'-monophosphate- and guanosine

3':5'-monophosphate-dependent protein kinases. Kinetic studies using synthetic peptides corresponding to phosphorylation sites in histone H2B. *J. Biol. Chem.* **254**, 9728–9738.
2. Glass, D. B. (1983). Differential responses of cyclic GMP-dependent and cyclic AMP-dependent protein kinases to synthetic peptide inhibitors. *Biochem. J.* **213**, 159–164.
3. Glass, D. B., Cheng, H. C., Mende-Mueller, L., Reed, J., and Walsh, D. A. (1989). Primary structural determinants essential for potent inhibition of cAMP-dependent protein kinase by inhibitory peptides corresponding to the active portion of the heat-stable inhibitor protein. *J. Biol. Chem.* **264**, 8802–8810.
4. Kennelly, P. J. and Krebs, E. G. (1991). Consensus sequences as substrate specificity determinants for protein kinases and protein phosphatases. *J. Biol. Chem.* **266**, 15555–15558.
5. Kemp, B. E., Pearson, R. B., and House, C. M. (1991). Pseudosubstrate-based peptide inhibitors. *Methods Enzymol.* **201**, 287–304.
6. Kemp, B. E., Faux, M. C., Means, A. R., House, C., Tiganis, T., Hu, S. H., and Mitchelhill, K. I. (1994). Structural aspects: pseudosubstrate and substrate interactions, in J. R. Woodgett, Ed., *Protein Iinases*, pp. 30–67. IRL Press and Oxford University Press, Oxford, U.K.
7. Tegge, W., Frank, R., Hofmann, F., and Dostmann, W. R. (1995). Determination of cyclic nucleotide-dependent protein kinase substrate specificity by the use of peptide libraries on cellulose paper. *Biochemistry* **34**, 10569–10577.
8. Shabb, J. B. (2001). Physiological substrates of cAMP-dependent protein kinase. *Chem. Rev.* **101**, 2381–2411.
9. Kemp, B. E., Graves, D. J., Benjamini, E., and Krebs, E. G. (1977). Role of multiple basic residues in determining the substrate specificity of cyclic AMP-dependent protein kinase. *J. Biol. Chem.* **252**, 4888–4894.
10. Dostmann, W. R., Nickl, C., Thiel, S., Tsigelny, I., Frank, R., and Tegge, W. J. (1999). Delineation of selective cyclic GMP-dependent protein kinase Ialpha substrate and inhibitor peptides based on combinatorial peptide libraries on paper. *Pharmacol. Ther.* **82**, 373–387.
11. Krebs, E. G. and Beavo, J. A. (1979). Phosphorylation-dephosphorylation of enzymes. *Annu. Rev. Biochem.* **48**, 923–959.
12. Knighton, D. R., Zheng, J. H., Ten Eyck, L. F., Xuong, N. H., Taylor, S. S., and Sowadski, J. M. (1991). Structure of a peptide inhibitor bound to the catalytic subunit of cyclic adenosine monophosphate-dependent protein kinase. *Science* **253**, 414–420.
13. Lawler, O. A., Miggin, S. M., and Kinsella, B. T. (2001). Protein kinase A-mediated phosphorylation of serine 357 of the mouse prostacyclin receptor regulates its coupling to G(s)-, to G(i)-, and to G(q)-coupled effector signaling. *J. Biol. Chem.* **276**, 33596–33607.
14. Balasubramanian, N., Levay, K., Keren-Raifman, T., Faurobert, E., and Slepak, V. Z. (2001). Phosphorylation of the regulator of G protein signaling rgs9-1 by protein kinase A is a potential mechanism of light- and Ca $^{2+}$-mediated regulation of g protein function in photoreceptors. *Biochemistry* **40**, 12619–12627.
15. Burgon, P. G., Lee, W. L., Nixon, A. B., Peralta, E. G., and Casey, P. J. (2001). Phosphorylation and nuclear translocation of a regulator of G protein signaling (RGS10). *J. Biol. Chem.* **276**, 32828–32834.
16. Baillie, G., MacKenzie, S. J., and Houslay, M. D. (2001). Phorbol 12-myristate 13-acetate triggers the protein kinase A-mediated phosphorylation and activation of the PDE4D5 cAMP phosphodiesterase in human aortic smooth muscle cells through a route involving extracellular signal regulated kinase (ERK). *Mol. Pharmacol.* **60**, 1100–1111.
17. Kitani, T., Okuno, S., and Fujisawa, H. (2001). Regulation of Ca(2+)/calmodulin-dependent protein kinase kinase alpha by cAMP-dependent protein kinase: II. mutational analysis. *J. Biochem. (Tokyo)* **130**, 515–525.
18. Li, G., Harton, J. A., Zhu, X., and Ting, J. P. (2001). Downregulation of CIITA function by protein kinase a (PKA)-mediated phosphorylation: mechanism of prostaglandin E, cyclic AMP, and PKA inhibition of class II major histocompatibility complex expression in monocytic lines. *Mol. Cell Biol.* **21**, 4626–4635.
19. Feliciello, A., Allevato, G., Musti, A. M., De Brasi, D., Gallo, A., Avvedimento, V. E., and Gottesman, M. E. (2000). Thyroid transcription factor 1 phosphorylation is not required for protein kinase A-dependent transcription of the thyroglobulin promoter. *Cell Growth Differ.* **11**, 649–654.
20. Chheda, M. G., Ashery, U., Thakur, P., Rettig, J., and Sheng, Z. H. (2001). Phosphorylation of Snapin by PKA modulates its interaction with the SNARE complex. *Nat. Cell Biol.* **3**, 331–338.
21. Marash, M. and Gerst, J. E. (2001). t-SNARE dephosphorylation promotes SNARE assembly and exocytosis in yeast. *EMBO J.* **20**, 411–421.
22. Evans, G. J., Wilkinson, M. C., Graham, M. E., Turner, K. M., Chamberlain, L. H., Burgoyne, R. D., and Morgan, A. (2001). Phosphorylation of cysteine string protein by PKA: implications for the modulation of exocytosis. *J. Biol. Chem.*
23. Technikova-Dobrova, Z., Sardanelli, A. M., Speranza, F., Scacco, S., Signorile, A., Lorusso, V., and Papa, S. (2001). Cyclic adenosine monophosphate-dependent phosphorylation of mammalian mitochondrial proteins: enzyme and substrate characterization and functional role. *Biochemistry* **40**, 13941–13947.
24. Wasmeier, C. and Hutton, J. C. (2001). Secretagogue-dependent phosphorylation of the insulin granule membrane protein phogrin is mediated by cAMP-dependent protein kinase. *J. Biol. Chem.* **276**, 31919–31928.
25. Lonart, G. and Sudhof, T. C. (2001). Characterization of rabphilin phosphorylation using phospho-specific antibodies. *Neuropharmacology* **41**, 643–649.
26. de Jonge, H. R. and Rosen, O. M. (1977). Self-phosphorylation of cyclic guanosine 3':5'-monophosphate-dependent protein kinase from bovine lung. Effect of cyclic adenosine 3':5'-monophosphate, cyclic guanosine 3':5'-monophosphate and histone. *J. Biol. Chem.* **252**, 2780–2783.
27. Vallins, W. J., Brand, N. J., Dabhade, N., Butler-Browne, G., Yacoub, M. H., and Barton, P. J. (1990). Molecular cloning of human cardiac troponin I using polymerase chain reaction. *FEBS Lett.* **270**, 57–61.
28. Yuasa, K., Michibata, H., Omori, K., and Yanaka, N. (1999). A novel interaction of cGMP-dependent protein kinase I with troponin T. *J. Biol. Chem.* **274**, 37429–37434.
29. Okano, I., Yamamoto, T., Kaji, A., Kimura, T., Mizuno, K., and Nakamura, T. (1993). Cloning of CRP2, a novel member of the cysteine-rich protein family with two repeats of an unusual LIM/double zinc-finger motif. *FEBS Lett.* **333**, 51–55.
30. Huber, A., Neuhuber, W. L., Klugbauer, N., Ruth, P., and Allescher, H. D. (2000). Cysteine-rich protein 2, a novel substrate for cGMP kinase I in enteric neurons and intestinal smooth muscle. *J. Biol. Chem.* **275**, 5504–5511.
31. Beall, A. C., Kato, K., Goldenring, J. R., Rasmussen, H., and Brophy, C. M. (1997). Cyclic nucleotide-dependent vasorelaxation is associated with the phosphorylation of a small heat shock-related protein. *J. Biol. Chem.* **272**, 11283–11287.
32. Raeymaekers, L., Hofmann, F., and Casteels, R. (1988). Cyclic GMP-dependent protein kinase phosphorylates phospholamban in isolated sarcoplasmic reticulum from cardiac and smooth muscle. *Biochem. J.* **252**, 269–273.
33. Jiang, L. H., Gawler, D. J., Hodson, N., Milligan, C. J., Pearson, H. A., Porter, V., and Wray, D. (2000). Regulation of cloned cardiac L-type calcium channels by cGMP-dependent protein kinase. *J. Biol. Chem.* **275**, 6135–6143.
34. Alioua, A., Huggins, J. P., and Rousseau, E. (1995). PKG-I alpha phosphorylates the alpha-subunit and upregulates reconstituted GKCa channels from tracheal smooth muscle. *Am. J. Physiol* **268**, L1057–L1063.
35. Fukao, M., Mason, H. S., Britton, F. C., Kenyon, J. L., Horowitz, B., and Keef, K. D. (1999). Cyclic GMP-dependent protein kinase activates cloned BKCa channels expressed in mammalian cells by direct phosphorylation at serine 1072. *J. Biol. Chem.* **274**, 10927–10935.
36. Komalavilas, P. and Lincoln, T. M. (1996). Phosphorylation of the inositol 1,4,5-trisphosphate receptor. Cyclic GMP-dependent protein kinase mediates cAMP and cGMP dependent phosphorylation in the intact rat aorta. *J. Biol. Chem.* **271**, 21933–21938.

37. Butt, E., Immler, D., Meyer, H. E., Kotlyarov, A., Laass, K., and Gaestel, M. (2001). Heat shock protein 27 is a substrate of cGMP-dependent protein kinase in intact human platelets: phosphorylation-induced actin polymerization caused by HSP27 mutants. *J. Biol. Chem.* **276**, 7108–7113.
38. Butt, E., Abel, K., Krieger, M., Palm, D., Hoppe, V., Hoppe, J., and Walter, U. (1994). cAMP- and cGMP-dependent protein kinase phosphorylation sites of the focal adhesion vasodilator-stimulated phosphoprotein (VASP) *in vitro* and in intact human platelets. *J. Biol. Chem.* **269**, 14509–14517.
39. Haffner, C., Jarchau, T., Reinhard, M., Hoppe, J., Lohmann, S. M., and Walter, U. (1995). Molecular cloning, structural analysis and functional expression of the proline-rich focal adhesion and microfilament-associated protein VASP. *EMBO J.* **14**, 19–27.
40. Xue, J., Wang, X., Malladi, C. S., Kinoshita, M., Milburn, P. J., Lengyel, I., Rostas, J. A., and Robinson, P. J. (2000). Phosphorylation of a new brain-specific septin, G-septin, by cGMP-dependent protein kinase. *J. Biol. Chem.* **275**, 10047–10056.
41. Ammendola, A., Geiselhoringer, A., Hofmann, F., and Schlossmann, J. (2001). Molecular determinants of the interaction between the inositol 1,4,5-trisphosphate receptor-associated cGMP kinase substrate (IRAG) and cGMP kinase Ibeta. *J. Biol. Chem.* **276**, 24153–24159.
42. McDonald, B. J. and Moss, S. J. (1997). Conserved phosphorylation of the intracellular domains of GABA(A) receptor beta2 and beta3 subunits by cAMP-dependent protein kinase, cGMP-dependent protein kinase protein kinase C and Ca^{2+}/calmodulin type II-dependent protein kinase. *Neuropharmacology* **36**, 1377–1385.
43. Endo, S., Suzuki, M., Sumi, M., Nairn, A. C., Morita, R., Yamakawa, K., Greengard, P., and Ito, M. (1999). Molecular identification of human G-substrate, a possible downstream component of the cGMP-dependent protein kinase cascade in cerebellar Purkinje cells. *Proc. Natl. Acad. Sci. USA* **96**, 2467–2472.
44. Hemmings, H. C., Jr., Williams, K. R., Konigsberg, W. H., and Greengard, P. (1984). DARPP-32, a dopamine- and adenosine 3':5'-monophosphate-regulated neuronal phosphoprotein. I. Amino acid sequence around the phosphorylated threonine. *J. Biol. Chem.* **259**, 14486–14490.
45. Wang, G. R., Zhu, Y., Halushka, P. V., Lincoln, T. M., and Mendelsohn, M. E. (1998). Mechanism of platelet inhibition by nitric oxide: in vivo phosphorylation of thromboxane receptor by cyclic GMP-dependent protein kinase. *Proc. Natl. Acad. Sci. USA* **95**, 4888–4893.
46. Yamamoto, S., Yan, F., Zhou, H., and Tai, H. H. (2001). Serine 331 is the major site of receptor phosphorylation induced by agents that activate protein kinase G in HEK 293 cells overexpressing thromboxane receptor alpha. *Arch. Biochem. Biophys.* **393**, 97–105.
47. Wang, X. and Robinson, P. J. (1997). Cyclic GMP-dependent protein kinase and cellular signaling in the nervous system. *J. Neurochem.* **68**, 443–456.
48. Glass, D. B. and Krebs, E. G. (1982). Phosphorylation by guanosine 3':5'-monophosphate-dependent protein kinase of synthetic peptide analogs of a site phosphorylated in histone H2B. *J. Biol. Chem.* **257**, 1196–1200.
49. Yuasa, K., Omori, K., and Yanaka, N. (2000). Binding and phosphorylation of a novel male germ cell-specific cGMP-dependent protein kinase-anchoring protein by cGMP-dependent protein kinase Ialpha. *J. Biol. Chem.* **275**, 4897–4905.
50. Rodriguez-Pascual, F., Ferrero, R., Miras-Portugal, M. T., and Torres, M. (1999). Phosphorylation of tyrosine hydroxylase by cGMP-dependent protein kinase in intact bovine chromaffin cells. *Arch. Biochem. Biophys.* **366**, 207–214.
51. Butt, E., Bernhardt, M., Smolenski, A., Kotsonis, P., Frohlich, L. G., Sickmann, A., Meyer, H. E., Lohmann, S. M., and Schmidt, H. H. (2000). Endothelial nitric-oxide synthase (type III) is activated and becomes calcium independent upon phosphorylation by cyclic nucleotide-dependent protein kinases. *J. Biol. Chem.* **275**, 5179–5187.
52. Murray, K. J., el Maghrabi, M. R., Kountz, P. D., Lukas, T. J., Soderling, T. R., and Pilkis, S. J. (1984). Amino acid sequence of the phosphorylation site of rat liver 6-phosphofructo-2-kinase/fructose-2,6-bisphosphatase. *J. Biol. Chem.* **259**, 7673–7681.
53. Thomas, M. K., Francis, S. H., and Corbin, J. D. (1990). Substrate- and kinase-directed regulation of phosphorylation of a cGMP-binding phosphodiesterase by cGMP. *J. Biol. Chem.* **265**, 14971–14978.
54. Suhasini, M., Li, H., Lohmann, S. M., Boss, G. R., and Pilz, R. B. (1998). Cyclic-GMP-dependent protein kinase inhibits the Ras/Mitogen-activated protein kinase pathway. *Mol. Cell Biol.* **18**, 6983–6994.
55. Xia, C., Bao, Z., Yue, C., Sanborn, B. M., and Liu, M. (2001). Phosphorylation and regulation of G-protein-activated phospholipase C-beta 3 by cGMP-dependent protein kinases. *J. Biol. Chem.* **276**, 19770–19777.
56. Yang, X., Goncalves, J., and Gabuzda, D. (1996). Phosphorylation of Vif and its role in HIV-1 replication. *J. Biol. Chem.* **271**, 10121–10129.
57. Wyatt, T. A., Lincoln, T. M., and Pryzwansky, K. B. (1991). Vimentin is transiently co-localized with and phosphorylated by cyclic GMP-dependent protein kinase in formyl-peptide-stimulated neutrophils. *J. Biol. Chem.* **266**, 21274–21280.
58. MacMillan-Crow, L. A. and Lincoln, T. M. (1994). High-affinity binding and localization of the cyclic GMP-dependent protein kinase with the intermediate filament protein vimentin. *Biochemistry* **33**, 8035–8043.

CHAPTER 204

Effects of cGMP-Dependent Protein Kinase Knockouts

**Franz Hofmann, Robert Feil,
Thomas Kleppisch, and Claudia Werner**
Institut für Pharmakologie und Toxikologie, TU München, München, Germany

Both mammalian cGMP-dependent protein kinase genes, cGKI and cGKII, have been inactivated in mice. The major phenotypes observed in cGKI knockout mice are decreased life span, impaired relaxation of vascular, visceral, and penile smooth muscle, disturbed platelet function, and altered neuronal functions. cGKII knockout mice have a normal life span, decreased longitudinal bone growth, decreased intestinal chloride secretion, and altered renin secretion.

Cyclic GMP-Dependent Protein Kinases: Genes and Knockouts

Cyclic GMP-dependent protein kinases (cGKs) are serine/threonine kinases which are activated by the second messenger cGMP. Two cGK genes, cGKI and cGKII, have been identified in mammals. The amino terminus (approximately amino acids 1–100) of cGKI is encoded by two alternatively used exons, resulting in the production of two cGKI isoforms, cGKIα and cGKIβ. To generate cGKI and cGKII gene targeted mouse lines, the exon encoding part of the ATP-binding site in the catalytic domain (cGKI) or the first part of the cGMP-binding pocket (cGKII) was destroyed by homologous recombination in embryonic stem cells. These strategies resulted in loss-of-function mutants of the cGKII gene [1] and of the cGKI gene with complete inactivation of both the Iα and Iβ isozyme [2]. Since cGKI-null mutants have a decreased viability (most animals die until six weeks of age), a conditional cGKI mouse line was generated in which the ATP-binding site encoding exon was flanked by loxP sequences allowing for Cre-mediated tissue-specific inactivation of the cGKI gene [3].

Vasorelaxation and Hypertension

Nitric oxide (NO) and atrial natriuretic peptide (ANP) stimulate cGMP synthesis and relax small arteries and arterioles resulting in a decreased blood pressure. Targeted inactivation of the NO-synthase III (NOS III), ANP, or ANP receptor gene causes hypertension (reviewed in reference [4]). Juvenile (4 to 5 weeks old) cGKI knockout mice have an elevated blood pressure [2] indicating that the anti-hypertensive effects of NO and ANP are at least in part mediated by activation of cGKI. Potential *in vivo* targets for cGKI are the Ca^{2+}-activated K^+ (BK_{Ca}) channel [5] and IRAG [6], a protein involved in the intracellular calcium release mechanism. Phosphorylation of these two proteins could reduce the cytosolic calcium concentration, thereby leading to vasorelaxation. Vascular smooth muscle cells (VSMCs) isolated from wild-type mice endogenously express cGKIα, cGKIβ, and IRAG. NO/cGMP inhibited noradrenaline-induced Ca^{2+}-transients in wild-type but not in cGKI-deficient VSMCs. Interestingly, the defective Ca^{2+} regulation in cGKI-deficient VSMCs can be rescued by transfection of the cGKIα isozyme, but not the β isozyme [7]. These results suggest that cGKIα relaxed smooth muscle by decreasing the cytosolic Ca^{2+} level. The role of cGKIβ in VSMCs is unclear at present, but may be more related to the cGKI effects on smooth muscle proliferation, differentiation, and gene expression [8]. The results described above do not preclude the possibility that cGKI decreases vessel tone by additional Ca^{2+}-independent mechanisms including the activation of myosin phosphatase [9], of phosphorylation of RhoA [10], and of telokin [11] resulting in dephosphorylation of the myosin light chains without affecting the cytosolic calcium level. Blood pressure may be regulated also by cGKII via inhibition of renin secretion [12].

The interpretation of the pathophysiology of the conventional cGKI knockout animals was complicated by the finding that 7-week-old, cGKI-null mutants had a normal or only slightly elevated blood pressure [2] indicating that the lack of cGKI can be bypassed in older animals. However, cGKI-null mutants develop multiple phenotypes with increasing age including infections and inflammation [13], which are known to induce massive NO synthesis. High concentrations of NO can increase cGMP levels to extreme values in vascular smooth muscle [5]. Thus, it is tempting to speculate that the apparent "normalization" of blood pressure in older cGKI-null mutants may be due to crossactivation of cAMP-dependent protein kinase by the high cGMP levels that are potentially generated in these diseased mice [5]. Furthermore, cGMP levels should be increased in cGKI knockout animals since the cGMP hydrolysing phosphodiesterase 5 (PDE 5) is not phosphorylated in cGKI-deficient VSMCs [14]. It has been suggested that *in vivo* activation of PDE 5 requires phosphorylation by cGK [15].

Further analysis of the cGKI-null mutants was limited by their low viability and by the fact that the cGKI gene was inactivated in the germline and thus in every cell of these mice. Therefore, it was difficult to study the role of cGKI in adults and whether the age-related hypertension of these mice reflected a function of cGKI in VSMCs, endothelial cells, or other cell types such as cardiomyocytes. To overcome these limitations, a mouse line has been generated which allows the conditional inactivation of the cGKI gene in somatic cells [3]. Cardiomyocyte-specific cGKI mutants are fully viable and can be studied throughout adulthood. The combined analysis of conventional and cardiomyocyte-specific cGKI knockout mice demonstrated that cGKI mediates the negative inotropic effect of cGMP in the juvenile as well as in the adult murine heart [3]. In line with results obtained in older NOS III knockout mice [16,17], the NO/cGMP/cGKI pathway does not appear to be involved in the negative inotropic action of acetylcholine [3].

Platelet Function

NO is of major importance for the homeostasis of platelet-endothelium and platelet-platelet interactions by inhibiting the adhesion of platelets to injured endothelium, platelet activation, and aggregation [18]. *In vitro* and *in vivo* studies with cGKI-deficient platelets proved that these effects are mediated by activation of cGKI [19]. Platelet adhesion and aggregation during ischemia/reperfusion of the microcirculation was analyzed by intravital video microscopy in wild-type mice infused with cGKI-deficient platelets and cGKI-deficient mice infused with wild-type platelets. These experiments clearly showed that platelet cGKI but not endothelial or smooth muscle cGKI is essential to prevent intravascular adhesion and aggregation of platelets after ischemia, probably by inhibiting the activation of the platelet fibrinogen receptor, glycoprotein IIb–IIIa [19].

Gastrointestinal and Urogenital Function

Both cGKI and cGKII knockout mice show severe gastrointestinal malfunctions. cGKI-deficient mice have disturbed bowel movement leading to delayed passage of food [2]. Apparently, NO-dependent relaxation of stomach smooth muscle is disturbed leading to an increased tonus of the pylorus [2,20]. At present it is not clear whether or not the lack of cGKI in a subpopulation of the interstitial cells of Cajal [21] contributes to the motility defect. Deletion of the cGKII gene abolished CFTR-mediated intestinal water secretion after stimulation of particulate guanylyl cyclase by guanylin or the heat-stable toxin from *Escherichia coli* [1]. This disturbance is confined to the small intestine. Ion transport was normal in the colon of cGKII-deficient mice [22].

Deletion of the cGKI gene impaired the NO/cGMP-induced relaxation of penile smooth muscle leading to erectile dysfunction, but did not affect the motility and fertility of sperm [23]. Furthermore, the NO/cGMP-dependent relaxation of urinary tract smooth muscle is abolished in cGKI-null mutants [24].

Nervous System

Over the past few years a number of neuronal functions of cGKs have been identified. All three cGKs are expressed in specific brain regions of mammals. Analysis of naturally occurring *Drosophila* variants in food-searching behavior, combined with the introduction of transgenes and transposable P elements into the *Drosophila* genome, led to the identification of the *foraging* gene as the *dg2* gene, which encodes cGKII [25]. Behavioral analysis of the cGKII knockout mice revealed increased anxiety and increased consumption of alcohol at first contact [26]. cGKII is expressed in the amygdala and basal forebrain, structures thought to be involved in anxiety and addiction.

cGKIα is expressed in dorsal root ganglia and cerebellar Purkinje cells, whereas cGKIβ predominates in the hippocampus and olfactory bulb. Deletion of the cGKI gene led to impaired pathfinding of sensory neurons in the spinal cord and to a substantial reduction of nociceptive flexion reflexes [27], suggesting that cGKIα is required for the correct guidance of sensory neurons during development. The production of the cGKI and cGKII knockout mouse lines has also allowed to test whether or not these protein kinases are involved in hippocampal long-term potentiation (LTP) [28,29] and cerebellar long-term depression (LTD) [30]. These analyses showed that hippocampal LTP was not affected in juvenile (4 to 5 weeks old) conventional knockout mice lacking cGKI, cGKII, or both enzymes [31]. However, LTP was impaired partially in adult (at least 3 months old) mutants, in which the cGKI gene was inactivated specifically in the hippocampus. In agreement with the conventional cGKI knockout mice [31], LTP was unchanged in juvenile hippocampus-specific cGKI mutants (TK, RF, and FH, unpublished data). These phenotypes suggest that cGKs may regulate discrete neural functions in an age-dependent manner.

Preliminary experiments with Purkinje cell-specific cGKI mutants support the hypothesis that cGKI is an essential part in the regulation of LTD in the cerebellar Purkinje cells.

Bone

The particulate guanylyl cyclases, GC-A and GC-B, which generate cGMP upon binding of peptide ligands (ANP, BNP, and CNP) are expressed abundantly in mouse tibial epiphysis and vertebrae [32]. Cultivation of mouse tibia in the presence of BNP induced a significant increase in total bone length. Transgenic mice overexpressing BNP exhibited skeletal overgrowth that was restricted to bones with endochondral ossification [32]. cGKI and cGKII are expressed in the growth zones of bones [1]. The deletion of cGKI had no apparent effect on the growth of the skeleton. In contrast, cGKII-deficient mice are dwarfs, having limbs 16–30% shorter than normal [1]. Targeted expression of CNP in the growth plate chondrocytes failed to rescue the skeletal defect of the cGKII knockout mice, suggesting that cGKII plays a critical role in the CNP-mediated endochondral ossification [33].

Outlook

Taken together, the analysis of cGK-deficient mice has demonstrated the physiological relevance of these protein kinases as major mediators of cGMP signaling in diverse organs and tissues. cGKI plays an important role for cardiovascular and gastrointestinal homeostasis and has discrete functions in the central and peripheral nervous system. cGKII regulates longitudinal bone growth, intestinal ion transport, and renin release in the kidney, and it may be involved in the generation of complex behaviors like anxiety and addiction. In the future, the combined analysis of conventional and conditional cGK knockout mice and of cGK-deficient primary cells will further advance our understanding of the specific functions of cGKIα, cGKIβ, and cGKII in the mammalian body in health and disease. These studies may also help to develop new strategies for the treatment of cardiovascular, gastrointestinal, skeletal, and neuronal disorders.

Acknowledgment

Work in the authors' laboratory was supported by grants from Deutsche Forschungsgemeinschaft and Volkswagen Stiftung.

References

1. Pfeifer, A., Aszodi, A., Seidler, U., Ruth, P., Hofmann, F., and Fässler, R. (1996). Intestinal secretory defects and dwarfism in mice lacking cGMP-dependent protein kinase II. *Science* **274**, 2082–2086.
2. Pfeifer, A., Klatt, P., Massberg, S., Ny, L., Sausbier, M., Hirneiss, C., Wang, G. X., Korth, M., Aszodi, A., Andersson, K. E., Krombach, F., Mayerhofer, A., Ruth, P., Fassler, R., and Hofmann, F. (1998). Defective smooth muscle regulation in cGMP kinase I-deficient mice. *EMBO J.* **17**, 3045–3051.
3. Wegener, J. W., Nawrath, H., Wolfsgruber, W., Kühbandner, S., Werner, C., Hofmann, F., and Feil, R. (2002). cGMP-dependent protein kinase I mediates the negative inotropic effect of cGMP in the murine myocardium. *Circ. Res.* **90**, 18–20.
4. Garbers, D. L. and Dubois, S. K. (1999). The molecular basis of hypertension. *Annu. Rev. Biochem.* **68**, 127–155.
5. Sausbier, M., Schubert, R., Voigt, V., Hirneiss, C., Pfeifer, A., Korth, M., Kleppisch, T., Ruth, P., and Hofmann, F. (2000). Mechanisms of NO/cGMP-dependent vasorelaxation. *Circ. Res.* **87**, 825–830.
6. Schlossmann, J., Ammendola, A., Ashman, K., Zong, X., Huber, A., Neubauer, G., Wang, G. X., Allescher, H. D., Korth, M., Wilm, M., Hofmann, F., and Ruth, P. (2000). Regulation of intracellular calcium by a signalling complex of IRAG, IP3 receptor and cGMP kinase Ibeta. *Nature* **404**, 197–201.
7. Feil, R., Gappa, N., Rutz, M., Schlossmann, J., Rose, C. R., Konnerth, A., Brummer, S., Kühbandner, S., and Hofmann, F. (2002). Functional reconstitution of vascular smooth muscle cells with cGMP-dependent protein kinase I isoforms. *Circ. Res.* **90**, 1080–1086.
8. Lincoln, T. M., Dey, N., and Sellak, H. (2001). Invited Review: cGMP-dependent protein kinase signaling mechanisms in smooth muscle: from the regulation of tone to gene expression. *J. Appl. Physiol.* **91**, 1421–1430.
9. Surks, H. K., Mochizuki, N., Kasai, Y., Georgescu, S. P., Tang, K. M., Ito, M., Lincoln, T. M., and Mendelsohn, M. E. (1999). Regulation of myosin phosphatase by a specific interaction with cGMP-dependent protein kinase Ialpha. *Science* **286**, 1583–1587.
10. Sauzeau, V., Le Jeune, H., Cario-Toumaniantz, C., Smolenski, A., Lohmann, S. M., Bertoglio, J., Chardin, P., Pacaud, P., and Loirand, G. (2000). Cyclic GMP-dependent protein kinase signaling pathway inhibits RhoA-induced Ca^{2+} sensitization of contraction in vascular smooth muscle. *J. Biol. Chem.* **275**, 21722–21729.
11. Walker, L. A., MacDonald, J. A., Liu, X., Nakamoto, R. K., Haystead, T. A., Somlyo, A. V., and Somlyo, A. P. (2001). Site-specific phosphorylation and point mutations of telokin modulate its Ca^{2+}-desensitizing effect in smooth muscle. *J. Biol. Chem.* **276**, 24519–24524.
12. Wagner, C., Pfeifer, A., Ruth, P., Hofmann, F., and Kurtz, A. (1998). Role of cGMP-kinase II in the control of renin secretion and renin expression. *J. Clin. Invest.* **102**, 1576–1582.
13. Werner, C., Pryzwansky, K. B., and Hofmann, F. (2001). Cyclic GMP kinase I affects murine neutrophil migration and superoxide production. *N-S Arch. Pharmacol.* **363**, R81.
14. Rybalkin, S. D., Rybalkina, I. G., Feil, R., Hofmann, F., and Beavo, J. A. (2002). Regulation of cGMP-specific phosphodiesterase (PDE5) phosphorylation in smooth muscle. *J. Biol. Chem.* **277**, 3310–3317.
15. Wyatt, T. A., Naftilan, A. J., Francis, S. H., and Corbin, J. D. (1998). ANF elicits phosphorylation of the cGMP phosphodiesterase in vascular smooth muscle cells. *Am. J. Physiol.* **274**, H448–H455.
16. Vandecasteele, G., Eschenhagen, T., Scholz, H., Stein, B., Verde, I., and Fischmeister, R. (1999). Muscarinic and beta-adrenergic regulation of heart rate, force of contraction and calcium current is preserved in mice lacking endothelial nitric oxide synthase. *Nat. Med.* **5**, 331–334.
17. Godecke, A., Heinicke, T., Kamkin, A., Kiseleva, I., Strasser, R. H., Decking, U. K., Stumpe, T., Isenberg, G., and Schrader, J. (2001). Inotropic response to beta-adrenergic receptor stimulation and antiadrenergic effect of ACh in endothelial NO synthase-deficient mouse hearts. *J. Physiol.* **532**, 195–204.
18. Lloyd-Jones, D. M. and Bloch, K. D. (1996). The vascular biology of nitric oxide and its role in atherogenesis. *Annu. Rev. Med.* **47**, 365–375.
19. Massberg, S., Sausbier, M., Klatt, P., Bauer, M., Pfeifer, A., Siess, W., Fassler, R., Ruth, P., Krombach, F., and Hofmann, F. (1999). Increased adhesion and aggregation of platelets lacking cyclic guanosine 3′,5′-monophosphate kinase I. *J. Exp. Med.* **189**, 1255–1264.
20. Ny, L., Pfeifer, A., Aszodi, A., Ahmad, M., Alm, P., Hedlund, P., Fassler, R., and Andersson, K. E. (2000). Impaired relaxation of stomach smooth muscle in mice lacking cyclic GMP-dependent protein kinase I. *Br. J. Pharmacol.* **129**, 395–401.
21. Salmhofer, H., Neuhuber, W. L., Ruth, P., Huber, A., Russwurm, M., and Allescher, H. D. (2001). Pivotal role of the interstitial cells of

Cajal in the nitric oxide signaling pathway of rat small intestine. Morphological evidence. *Cell Tissue Res.* **305**, 331–340.
22. Vaandrager, A. B., Bot, A. G., Ruth, P., Pfeifer, A., Hofmann, F., and De Jonge, H. R. (2000). Differential role of cyclic GMP-dependent protein kinase II in ion transport in murine small intestine and colon. *Gastroenterology* **118**, 108–114.
23. Hedlund, P., Aszodi, A., Pfeifer, A., Alm, P., Hofmann, F., Ahmad, M., Fassler, R., and Andersson, K. E. (2000). Erectile dysfunction in cyclic GMP-dependent kinase I-deficient mice. *Proc. Natl. Acad. Sci. USA* **97**, 2349–2354.
24. Persson, K., Pandita, R. K., Aszodi, A., Ahmad, M., Pfeifer, A., Fassler, R., and Andersson, K. E. (2000). Functional characteristics of urinary tract smooth muscles in mice lacking cGMP protein kinase type I. *Am. J. Physiol. Regul. Integr. Comp. Physiol.* **279**, R1112–R1120.
25. Osborne, K. A., Robichon, A., Burgess, E., Butland, S., Shaw, R. A., Coulthard, A., Pereira, H. S., Greenspan, R. J., and Sokolowski, M. B. (1997). Natural behavior polymorphism due to a cGMP-dependent protein kinase of *Drosophila. Science* **277**, 834–836.
26. Werner, C., Sillaber, I., Spanagel, R., and Hofmann, F. (2000). Reduced ethanol sensitivity and enhanced ethanol consumption in cGMP-Kinase 2-deficient mice. *N-S Arch. Pharmacol.* **361**, R105.
27. Schmidt, H., Werner, M., Heppenstall, P. A., Henning, M., Moré, M. I., Kühbandner, S., Lewin, G. R., Hofmann, F., Feil, R., and Rathjen, F. G. (2002). cGMP-mediated signalling via cGKIα is required for the guidance and connectivity of sensory axons. *J. Cell Biol.* **159**, 489–498.
28. Zhuo, M., Hu, Y., Schultz, C., Kandel, E. R., and Hawkins, R. D. (1994). Role of guanylyl cyclase and cGMP-dependent protein kinase in long-term potentiation. *Nature* **368**, 635–639.
29. Arancio, O., Antonova, I., Gambaryan, S., Lohmann, S. M., Wood, J. S., Lawrence, D. S., and Hawkins, R. D. (2001). Presynaptic role of cGMP-dependent protein kinase during long-lasting potentiation. *J. Neurosci.* **21**, 143–149.
30. Lev-Ram, V., Jiang, T., Wood, J., Lawrence, D. S., and Tsien, R. Y. (1997). Synergies and coincidence requirements between NO, cGMP, and Ca2+ in the induction of cerebellar long-term depression. *Neuron* **18**, 1025–1038.
31. Kleppisch, T., Pfeifer, A., Klatt, P., Ruth, P., Montkowski, A., Fässler, R., and Hofmann, F. (1999). Long-term potentiation in the hippocampal CA1 region of mice lacking the cGMP-dependent kinase is normal and susceptible to inhibition of NO synthase. *J. Neurosci.* **19**, 48–55.
32. Suda, M., Ogawa, Y., Tanaka, K., Tamura, N., Yasoda, A., Takigawa, T., Uehira, M., Nishimoto, H., Itoh, H., Saito, Y., Shiota, K., and Nakao, K. (1998). Skeletal overgrowth in transgenic mice that overexpress brain natriuretic peptide. *Proc. Natl. Acad. Sci. USA* **95**, 2337–2342.
33. Miyazawa, T., Ogawa, Y., Chusho, H., Yasoda, A., Tamura, N., Komatsu, Y., Pfeifer, A., Hofmann, F., and Nakao, K. (2002). Cyclic GMP-dependent protein kinase II plays a critical role in C-type natriuretic peptide-mediated endochondral ossification. *Endocrinology* **143**, 3604–3610.

Cyclic Nucleotide-Regulated Cation Channels

Martin Biel and Andrea Gerstner

*Department Pharmazie–Zentrum für Pharmaforschung,
Ludwig-Maximilians-Universität München,
München, Germany*

Introduction

Cyclic nucleotides exert their physiological effects by binding to four major classes of cellular receptors: cAMP- and cGMP-dependent protein kinases [1], cyclic GMP-regulated phosphodiesterases [2], cAMP-binding guanine nucleotide exchange factors [3], and cyclic nucleotide-regulated cation channels. Cyclic nucleotide-regulated cation channels are unique among these receptors because their activation is directly coupled to the influx of extracellular cations into the cytoplasm and to the depolarization of the plasma membrane. Two families of channels regulated by cyclic nucleotides have been identified, the cyclic nucleotide-gated (CNG) channels and the hyperpolarization-activated cyclic nucleotide-gated (HCN) channels [4–7]. The two channel classes differ from each other with regard to their mode of activation. CNG channels are opened by direct binding of cAMP or cGMP. In contrast, HCN channels are principally operated by voltage. These channels open at hyperpolarized membrane potentials and close on depolarization. Apart from their voltage sensitivity, HCN channels are also activated directly by cyclic nucleotides, which act by increasing the channel open probability.

General Features of Cyclic Nucleotide-Regulated Cation Channels

Structurally, both CNG and HCN channels are members of the superfamily of voltage-gated cation channels. Like other subunits encoded by this large gene family, CNG and HCN channel subunits are believed to assemble to tetrameric complexes. The proposed structure and the phylogenetic relationship of mammalian CNG and HCN channel subunits is shown in Fig. 1 (see also Table I for recent nomenclature). The transmembrane channel core consists of six α-helical segments (S1–S6) and an ion-conducting pore loop between the S5 and S6. The amino- and carboxy-termini are localized in the cytosol. CNG and HCN channels contain a positively charged S4 helix carrying three to nine regularly spaced arginine or lysine residues at every third position. In HCN channels, as in most other members of the channel superfamily, the S4 helix functions as "voltage-sensor" conferring voltage-dependent gating [8,9]. In CNG channels, which are not gated by voltage, the specific role of S4 is not known.

CNG and HCN channels reveal different ion selectivities. CNG channels conduct both Ca^{2+} and monovalent cations with permeability ratios P_{Ca}/P_{Na} ranging from about 2 to 25 depending on the respective channel type and the cyclic nucleotide concentration [10,11]. By providing an entry pathway for Ca^{2+}, CNG channels control a variety of cellular processes that are triggered by this cation. In contrast, HCN channels are not permeable to Ca^{2+}. These channels pass Na^+ and K^+ ion with a relative permeability ratio P_{Na}/P_K of about 0.15 to 0.25 [12–14].

In the carboxy-terminus, CNG and HCN channels contain a cyclic nucleotide-binding domain (CNBD) that has significant sequence similarity to the CNBDs of most other

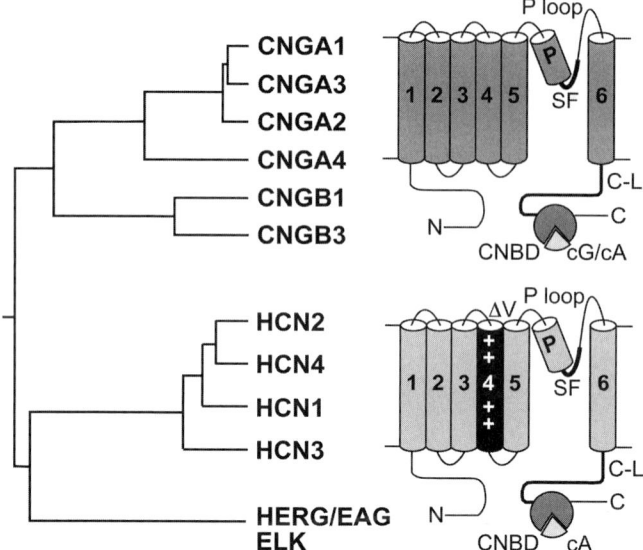

Figure 1 Phylogenetic tree and structural model of cyclic nucleotide-regulated cation channels. The CNG channel family comprises six members, which are classified into α subunits (CNGA1-4) and β subunits (CNGB1 and CNGB3). A "CNGB2" subunit does not exist. The HCN channel family comprises four members (HCN1-4). The channels are distantly related to EAG/HERG and ELK K^+ channels [49]. CNG and HCN channels share a common transmembrane topology, consisting of six transmembrane segments (1–6), a pore loop, and a cyclic nucleotide-binding domain (CNBD). The pore loop comprises the pore helix (P) and an uncoiled strand of 4 to 5 amino acid residues forming the selectivity filter (SF). CNG channels are activated *in vivo* by binding of either cAMP (cA) or cGMP (cG), depending on the channel type. HCN channels activate on membrane hyperpolarization (ΔV) and are enhanced by binding of cAMP. The positively charged amino acid residues in the S4 segment of HCN channels are indicated by "+" symbols. CL, C-linker involved in activation gating of CNG and HCN channels.

Table I Adopted Nomenclature for Cyclic Nucleotide-Regulated Ion Channel Subunits [5,48]

Adopted nomenclature	Previous designations
CNGA1	CNG1/CNGα1/RCNC1
CNGA2	CNG2/CNGα3/OCNC1
CNGA3	CNG3/CNGα2/CCNC1
CNGA4	CNG5/CNGα4/OCNC2/CNGB2
CNGB1	CNG4/CNGβ1/RCNC2
CNGB3	CNG6/CNGβ2/CCNC2
HCN1	HAC2/BCNG1
HCN2	HAC1/BCNG2
HCN3	HAC3/BCNG4
HCN4	HAC4/BCNG3

channel activation to more positive membrane potentials and thereby facilitate voltage-dependent channel activation [12–14]. Despite the fact that the CNBDs of HCN and CNG channels show significant sequence homology, the two channel classes reveal different selectivities for cyclic nucleotides. HCN channels display an approximately ten-fold higher affinitiy for cAMP than for cGMP whereas CNG channels select cGMP over cAMP [4,5].

CNG Channels

CNG channels are expressed in retinal photoreceptors and olfactory neurons and play a key role in visual and olfactory signal transduction [16]. CNG channels are also found at low density in some other cell types and tissues such as brain, testis, and kidney (for recent compilation of tissue expression see [17]). Although the function of CNG channels in sensory neurons has been unequivocally demonstrated, the role of these channels in other cell types remains to be established. Based on phylogenetic relationship, the six CNG channels identified in mammals are divided in two subfamilies, the α subunits (CNGA1-4) and the β subunits (CNGB1 and CNGB3) (Fig. 1). CNG channel α subunits (with the only exception of CNGA4) form functional homomeric channels in various heterologous expression systems. In contrast, β subunits do not give rise to functional channels when expressed alone. However, together with CNGA1-3 they confer novel properties (e.g. single channel flickering, increased sensitivity for cAMP and L-cis diltiazem) that are characteristic of native CNG channels [4]. The physiological role and subunit composition are known for three native CNG channels: the rod and cone photoreceptor channel and the olfactory channel. The CNG channel of rod photoreceptors consists of the CNGA1 subunit [18] and a long isoform of the CNGB1 subunit (CNGB1a [19]). The cone photoreceptor channel consists of the CNGA3 [20] and the CNGB3 [21] subunit. CNG channels control the membrane potential and the calcium concentration of photoreceptors. In the dark both channels are maintained in the open state by a high concentration of cGMP. The resulting influx of Na^+ and Ca^{2+} ("dark current") depolarizes the photoreceptor and promotes synaptic transmission. Light-induced hydrolysis of cGMP leads to the closure of the CNG channel. As a result the photoreceptor hyperpolarizes and shuts off synaptic glutamate release. Mutations in CNG channel genes have been linked to retinal diseases. Mutations in the CNGA1 [22] and CNGB1 [23] subunits have been identified in the genome of patients suffering from retinitis pigmentosa. The functional loss of either the CNGA3 [24,25] or the CNGB3 [26] subunit causes total color blindness (achromatopsia) and degeneration of cone photoreceptors.

The CNG channel expressed in cilia of olfactory neurons consists of three different subunits: CNGA2 [27,28], CNGA4 [29,30], and a short isoform of the CNGB1 subunit (CNGB1b) [31,32]. The channel is activated *in vivo* by cAMP, which is synthesized in response to the binding of

types of cyclic nucleotide receptors (Fig. 2). In CNG channels, the binding of cyclic nucleotides to the CNBD initiates a sequence of allosteric transitions that lead to the opening of the ion-conducting pore [15]. In HCN channels, the binding of cyclic nucleotides is not required for activation. However, cyclic nucleotides shift the voltage-dependence of

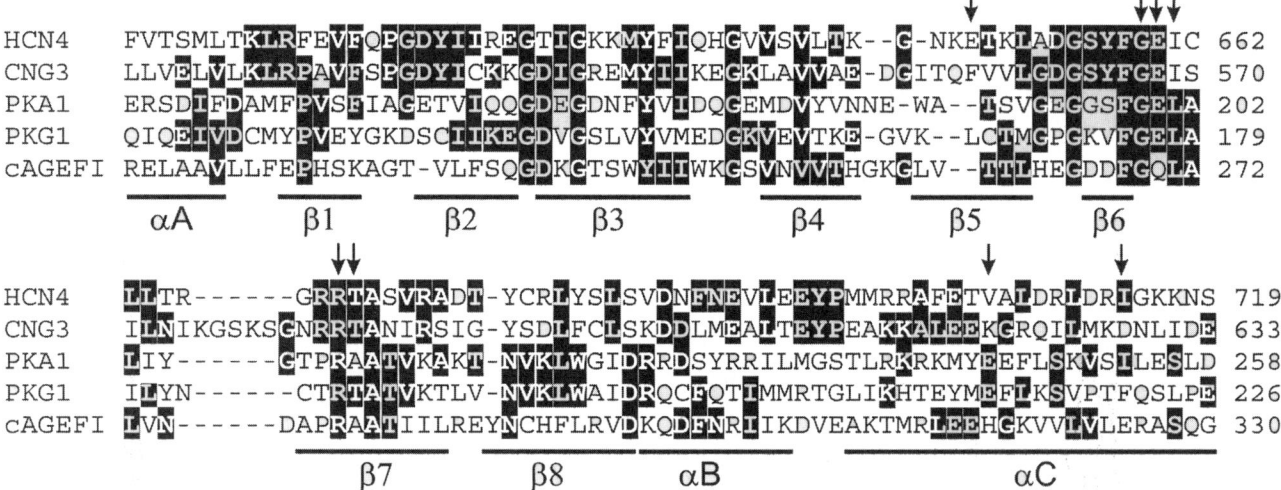

Figure 2 Amino acid sequences of CNBDs of cyclic nucleotide-binding proteins. Residues identical in at least two sequences are highlighted by a black background. Amino acids that are supposed to interact with cyclic nucleotides in CNG channels are indicated by arrows [50]. HCN4 (human HCN4, accession NP_005468), CNG3 (bovine CNG3, accession Q16281), PKA1 (cAMP binding site 1 of bovine cAMP-dependent protein kinase, α-catalytic subunit, accession P00517), PKG1 (cGMP-binding site 1 of bovine cGMP-dependent protein kinase type I α, accession P00516), cAGEF1 (human cAMP-GEF1, accession NP_006096).

odorants to their cognate receptors. The olfactory CNG channel is thought to conduct almost exclusively Ca^{2+} under physiological ionic conditions [33]. The resulting increase in cellular Ca^{2+} activates a Ca^{2+}-activated Cl^- channel, which further depolarizes the cell membrane. Ca^{2+} is not only a permeating ion of the olfactory CNG channel, it also represents an important modulator of this channel. By forming a complex with calmodulin that binds to the CNGA2 subunit, Ca^{2+} decreases sensitivity of the CNG channel to cAMP [34]. The resulting inhibition of channel activity is the principal mechanism underlying odorant adaptation [35].

HCN Channels

A cation current that is slowly activated by membrane hyperpolarization (termed I_h, I_f, or I_q) is found in a variety of excitable cells, including neurons, cardiac pacemaker cells, and photoreceptors [36]. The best understood function of I_h is to control heart rate and rhythm by acting as "pacemaker current" in the sinoatrial (SA) node [37]. I_h is activated during the membrane hyperpolarization following the termination of an action potential and provides an inward Na^+ current that slowly depolarizes the plasma membrane. Sympathetic stimulation of SA node cells raises cAMP levels and increases I_h by a positive shift of the current activation curve, thus accelerating diastolic depolarization and heart rate. Stimulation of muscarinic receptors slows down heart rate by the opposite action. In the brain I_h fulfills diverse functions: it controls the activity of spontaneously spiking neurons ("neuronal pacemaking" [36]), it is involved in the determination of resting potential [36], in photoreceptors it provides rebound depolarizations in response to pronounced hyperpolarizations [36], it is involved in the transduction of sour taste [38], and it is involved in the control of synaptic plasticity [39].

HCN channels represent the molecular correlate of the I_h current [5–7]. In mammals, the HCN channel family comprises four members (HCN1-4) that share about 60 percent sequence identity to each other and about 25 percent sequence identity to CNG channels. The highest degree of sequence homology between HCN and CNG channels is found in the CNBD. When expressed in heterologous systems all four HCN channels generate currents displaying the typical features of native I_h: (1) activation by membrane hyperpolarization, (2) permeation of Na^+ and K^+, (3) positive shift of the voltage-dependence of channel activation by direct binding of cAMP, and (4) channel blockade by extracellular Cs^+. HCN1–4 mainly differ from each other with regard to their speed of activation and the extent by which they are modulated by cAMP. HCN1 is the fastest channel, followed by HCN2, HCN3, and HCN4. Unlike HCN2 and HCN4, whose activation curves are shifted by about +15 mV by cAMP, HCN1 is only weakly affected by cAMP (shift of less than +5 mV). Site-directed mutagenesis experiments have provided initial insight into the complex mechanism underlying dual HCN channel activation by voltage and cAMP. Like in other voltage-gated cation channels, activation of HCN channels is initiated by the movement of the positively charged S4 helix in the electric field [8,9]. The resulting conformational change in the channel protein is allosterically coupled by other channel domains to the opening of the ion-conducting pore. Major determinants affecting channel activation are the intracellular S4–S5 loop [40], the S1 segment [41], and the extracellular S1–S2 loop [41]. The CNBD fulfills the role of an auto-inhibitory channel domain. In the absence of cAMP the cytoplasmic carboxy-terminus inhibits HCN channel gating by interacting with the channel core and thereby shifting the activation curve to more hyperpolarizing voltages [42]. Binding of cAMP to the CNBD relieves this inhibition. Differences in the magnitude of the response to cAMP among the four HCN channel isoforms are largely due to differences in the exent to which the CNBD inhibits basal gating. It remains to be determined whether the inhibitory effect of the CNBD is conferred by a direct physical interaction with the channel core domain or by some indirect pathway. There is initial evidence that the so-called C-linker, a peptide of about 80 amino acids that connects the last transmembrane helix (S6) to the CNBD, plays an important role in this process [43] (Fig. 1). The C-linker was also shown to play a key role in the gating of CNG channels, suggesting that the functional role of this domain has been conserved during channel evolution [15].

HCN channels are found in neurons and heart cells. In SA node cells, HCN4 represents the predominantly expressed HCN channel isoform [44,45]. In mouse brain all four HCN isoforms have been detected [46,47]. The expression levels and the regional distribution of the HCN channel mRNAs vary profoundly between the respective channel types. HCN2 is the most abundant neuronal channel and is found almost ubiquitously in the brain. In contrast, HCN1 and HCN4 are enriched in specific regions of the brain such as thalamus (HCN4) or hippocampus (HCN1). HCN3 is uniformly expressed throughout the brain, however at very low levels. HCN channels have also been detected in the retina and some peripheral neurons such as dorsal root ganglion neurons [45].

Acknowledgment

This work was supported by the Deutsche Forschungsgemeinschaft.

References

1. Pfeifer, A., Dostmann, W. R. G., Sausbier, M., Klatt, P., Ruth, P., and Hofmann, F. (1999). cGMP-dependent protein kinases: structure and function. *Rev. Physiol. Biochem. Pharmacol.* **135**, 105–149.
2. Soderling, S. H. and Beavo, J. A. (2000). Regulation of cAMP and cGMP signaling: new phosphodiesterases and new functions. *Curr. Opin. Cell. Biol.* **12**, 174–179.
3. de Rooij, J., Zwartkruis, F. J., Verheijen, M. H., Cool, R. H., Nijman, S. M., Wittinghofer, A., and Bos, J. L. (1998). Epac is a Rap1 guanine-nucleotide exchange factor directly activated by cyclic AMP. *Nature* **396**, 474–477.
4. Biel, M., Zong, X., Ludwig, A., Sautter, A., and Hofmann, F. (1999). Structure and function of cyclic nucleotide-gated channels. *Rev. Physiol. Biochem. Pharmacol.* **135**, 151–171.
5. Biel, M., Ludwig, A., Zong, X., and Hofmann, F. (1999). Hyperpolarization-activated cation channels: a multi-gene family. *Rev. Physiol. Biochem. Pharmacol.* **136**, 165–181.
6. Kaupp, U. B. and Seifert, R. (2001). Molecular diversity of pacemaker ion channels. *Annu. Rev. Physiol.* **63**, 235–257.
7. Santoro, B. and Tibbs, G. R. (1999). The HCN gene family: molecular basis of the hyperpolarization-activated pacemaker channels. *Ann. N.Y. Acad. Sci.* **868**, 741–764.
8. Chen, J., Mitcheson, J. S., Lin, M., and Sanguinetti, M. C. (2000). Functional roles of charged residues in the putative voltage sensor of the HCN2 pacemaker channel. *J. Biol. Chem.* **275**, 36465–36471.
9. Vaca, L., Stieber, J., Zong, X., Ludwig, A., Hofmann, F., and Biel, M. (2000). Mutations in the S4 domain of a pacemaker channel alter its voltage dependence. *FEBS Lett.* **479**, 35–40.
10. Frings, S., Seifert, R., Godde, M., and Kaupp, U. B. (1995). Profoundly different calcium permeation and blockage determine the specific function of distinct cyclic nucleotide-gated channels. *Neuron* **15**, 169–179.
11. Hackos, D. H. and Korenbrot, J. I. (1999). Divalent cation selectivity is a function of gating in native and recombinant cyclic nucleotide-gated ion channels from retinal photoreceptors. *J. Gen. Physiol.* **113**, 799–818.
12. Gauss, R., Seifert, R., and Kaupp, U. B. (1998). Molecular identification of a hyperpolarization-activated channel in sea urchin sperm. *Nature* **393**, 583–587.
13. Ludwig, A., Zong, X., Jeglitsch, M., Hofmann, F., and Biel, M. (1998). A family of hyperpolarization-activated mammalian cation channels. *Nature* **393**, 587–591.
14. Santoro, B., Liu, D. T., Yao, H., Bartsch, D., Kandel, E. R., Siegelbaum, S. A., and Tibbs, G. R. (1998). Identification of a gene encoding a hyperpolarization-activated pacemaker channel of brain. *Cell* **93**, 717–729.
15. Flynn, G. E., Johnson, J. P., and Zagotta, W. N. (2001). Cyclic nucleotide-gated channels: shedding light on the opening of a channel pore. *Nature Rev.* **2**, 643–651.
16. Finn, J. T., Grunwald, M. E., and Yau, K. W. (1996). Cyclic nucleotide-gated ion channels: an extended family with diverse functions. *Annu. Rev. Physiol.* **58**, 395–426.
17. Richards, M. J. and Gordon, S. E. (2000). Cooperativity and cooperation in cyclic nucleotide-gated ion channels. *Biochemistry* **39**, 14003–14011.
18. Kaupp, U. B., Niidome, T., Tanabe, T., Terada, S., Bönigk, W., Stühmer, W., Cook, N. J., Kangawa, K., Matsuo, H., Hirose, T., Miyata, T., and Numa, S. (1989). Primary structure and functional expression from complementary DNA of the rod photoreceptor cyclic GMP-gated channel. *Nature* **342**, 762–766.
19. Körschen, H. G., Illing, M., Seifert, R., Sesti, F., Williams, A., Gotzes, S., Colville, C., Müller, F., Dosé, A., Godde, M., Molday, L., Kaupp, U. B., and Molday, R. S. (1995). A 240 kDa protein represents the complete beta subunit of the cyclic nucleotide-gated channel from rod photoreceptor. *Neuron* **15**, 627–636.
20. Bönigk, W., Altenhofen, W., Müller, F., Dosé, A., Illing, M., Molday, R. S., and Kaupp, U. B. (1993). Rod and cone photoreceptor cells express distinct genes for cGMP-gated channels. *Neuron* **10**, 865–877.
21. Gerstner, A., Zong, X., Hofmann, F., and Biel, M. (2000). Molecular cloning and functional characterization of a new modulatory cyclic nucleotide-gated channel subunit from mouse retina. *J Neurosci.* **20**, 1324–1332.
22. Dryja, T. P., Finn, J. T., Peng, Y. W., McGee, T. L., Berson, E. L., and Yau, K. W. (1995). Mutations in the gene encoding the alpha subunit of the rod cGMP-gated channel in autosomal recessive retinitis pigmentosa. *Proc. Natl. Acad. Sci. USA* **92**, 10177–10181.
23. Bareil, C., Hamel, C. P., Delague, V., Arnaud, B., Demaille, J., and Claustres, M. (2001). Segregation of a mutation in CNGB1 encoding the β-subunit of the rod cGMP-gated channel in a family with autosomal recessive retinitis pigmentosa. *Hum. Genet.* **108**, 328–334.
24. Kohl, S., Marx, T., Giddings, I., Jägle, H., Jacobson, S. G., Apfelstedt-Sylla, E., Zrenner, E., Sharpe, L. T., and Wissinger, B. (1998). Total colourblindness is caused by mutations in the gene encoding the alpha-subunit of the cone photoreceptor cGMP-gated cation channel. *Nat. Genet.* **19**, 257–259.
25. Biel, M., Seeliger, M., Pfeifer, A., Kohler, K., Gerstner, A., Ludwig, A., Jaissle, G., Fauser, S., Zrenner, E., and Hofmann, F. (1999). Selective loss of cone function in mice lacking the cyclic nucleotide-gated channel CNG3. *Proc. Natl. Acad. Sci. USA* **96**, 7553–7557.
26. Sundin, O. H., Yang, J. M., Li, Y., Zhu, D., Hurd, J. N., Mitchell, T. N., Silva, E. D., and Maumenee, I. H. (2000). Genetic basis of total colourblindness among the Pingelapese islanders. *Nat. Genet.* **25**, 289–293.
27. Dhallan, R. S., Yau, K. W., Schrader, K. A., and Reed, R. R. (1990). Primary structure and functional expression of a cyclic nucleotide-activated channel from olfactory neurons. *Nature* **347**, 184–187.
28. Ludwig, J., Margalit, T., Eismann, E., Lancet, D., and Kaupp, U. B. (1990). Primary structure of cAMP-gated channel from bovine olfactory epithelium. *FEBS Lett.* **270**, 24–29.

29. Bradley, J., Li, J., Davidson, N., Lester, H. A., and Zinn, K. (1994). Heteromeric olfactory cyclic nucleotide-gated channels: a subunit that confers increased sensitivity to cAMP. *Proc. Natl. Acad. Sci. USA* **91**, 8890–8894.
30. Liman, E. R. and Buck, L. B. (1994). A second subunit of the olfactory cyclic nucleotide-gated channel confers high sensitivity to cAMP. *Neuron* **13**, 611–621.
31. Sautter, A., Zong, X., Hofmann, F., and Biel, M. (1998). An isoform of the rod photoreceptor cyclic nucleotide-gated channel beta subunit expressed in olfactory neurons. *Proc. Natl. Acad. Sci. USA* **95**, 4696–4701.
32. Bönigk, W., Bradley, J., Müller, F., Sesti, F., Boekhoff, I., Ronnett, G. V., Kaupp, U. B., and Frings, S. (1999). The native rat olfactory cyclic nucleotide-gated channel is composed of three distinct subunits. *J. Neurosci.* **19**, 5332–5347.
33. Dzeja, C., Hagen, V., Kaupp, U. B., and Frings, S. (1999). Ca^{2+} permeation in cyclic nucleotide-gated channels. *EMBO J.* **18**, 131–144.
34. Liu, M., Chen, T. Y., Ahamed, B., Li, J., and Yau, K. W. (1994). Calcium-calmodulin modulation of the olfactory cyclic nucleotide-gated cation channel. *Science* **266**, 1348–1354.
35. Kurahashi, T. and Menini, A. (1997). Mechanism of odorant adaptation in the olfactory receptor cell. *Nature* **385**, 725–729.
36. Pape, H. C. (1996). Queer current and pacemaker: the hyperpolarization-activated cation current in neurons. *Annu. Rev. Physiol.* **58**, 299–327.
37. DiFrancesco, D. (1993). Pacemaker mechanisms in cardiac tissue. *Annu. Rev. Physiol.* **55**, 455–472.
38. Stevens, D. R., Seifert, R., Bufe, B., Muller, F., Kremmer, E., Gauss, R., Meyerhof, W., Kaupp, U. B., and Lindemann, B. (2001). Hyperpolarization-activated channels HCN1 and HCN4 mediate responses to sour stimuli. *Nature* **413**, 631–635.
39. Mellor, J., Nicoll, R. A., and Schmitz, D. (2002). Mediation of hippocampal mossy fiber long-term potentiation by presynaptic I_h channels. *Science* **295**, 143–147.
40. Chen, J., Mitcheson, J. S., Tristani-Firouzi, M., Lin, M., and Sanguinetti, M. C. (2001). The S4-S5 linker couples voltage sensing and activation of pacemaker channels. *Proc. Natl. Acad. Sci. USA* **98**, 11277–11282.
41. Ishii, T. M., Takano, M., and Ohmori, H. (2001). Determinants of activation kinetics in mammalian hyperpolarization-activated cation channels. *J. Physiol.* **537**, 93–100.
42. Wainger, B. J., DeGennaro, M., Santoro, B., Siegelbaum, S. A., and Tibbs, G. R. (2001). Molecular mechanism of cAMP modulation of HCN pacemaker channels. *Nature* **411**, 805–810.
43. Wang, J., Chen, S., and Siegelbaum, S. A. (2001). Regulation of hyperpolarization-activated HCN channel gating and cAMP modulation due to interactions of COOH terminus and core transmembrane. *J. Gen. Physiol.* **118**, 237–250.
44. Ishii, T. M., Takano, M., Xie, L. H., Noma, A., and Ohmori, H. (1999). Molecular characterization of the hyperpolarization-activated cation channel in rabbit heart sinoatrial node. *J. Biol. Chem.* **274**, 12835–12839.
45. Moosmang, S., Stieber, J., Zong, X., Biel, M., Hofmann, F., and Ludwig, A. (2001). Cellular expression and functional characterization of four hyperpolarization-activated pacemaker channels in cardiac and neuronal tissues. *Eur. J. Biochem.* **268**, 1646–1652.
46. Santoro, B., Chen, S., Lüthi, A., Pavlidis, P., Shumyatsky, G. P., Tibbs, G. R., and Siegelbaum, S. A. (2000). Molecular and functional heterogeneity of hyperpolarization-activated pacemaker channels in the mouse CNS. *J. Neurosci.* **20**, 5264–5275.
47. Moosmang, S., Biel, M., Hofmann, F., and Ludwig, A. (1999). Differential distribution of four hyperpolarization-activated cation channels in mouse brain. *Biol. Chem.* **380**, 975–980.
48. Bradley, J., Frings, S., Yau, K. W., and Reed, R. (2001). Nomenclature for ion channel subunits. *Science* **294**, 2095–2096.
49. Warmke, J. W. and Ganetzky, B. (1994). A family of potassium channel genes related to *eag* in *Drosophila* and mammals. *Proc. Natl. Acad. Sci. USA* **91**, 3438–3442.
50. Scott, S. P., Harrison, R. W., Weber, I. T., and Tanaka, J. C. (1996). Predicted ligand interactions of 3′5′-cyclic nucleotide-gated channel binding sites: comparison of retina and olfactory binding site models. *Protein Eng.* **9**, 333–344.

CHAPTER 206

Epacs, cAMP-Binding Guanine Nucleotide Exchange Factors for Rap1 and Rap2

Holger Rehman,[1,2] Johan de Rooij,[1] and Johannes L. Bos[1]

[1]Department of Physiological Chemistry and Center for Biomedical Genetics,
University Medical Center Utrecht, STR. 3.233, Universiteitsweg 100,
P.O. Box 85060, 3584 CG Utrecht, The Netherlands, and
[2]Max-Planck Institut für Molekulare Physiologie,
Otto-Hahn-Straße 11, Dortmund, Germany

Introduction

In 1998 two independent groups identified a novel family of cAMP-binding proteins, Epacs or cAMP-GEFs [1,2]. These highly conserved and rather ubiquitously expressed proteins serve as guanine nucleotide exchange factors (GEFs) for the small GTPases Rap1 and Rap2 and as such induce signal transduction pathways independent from protein kinase A (PKA). One process among others that is controlled by Rap is integrin-mediated cell adhesion [3].

The Epac Family

In humans the Epac (exchange protein directly activated by cAMP) family consist of two members, Epac1 and Epac2 (Fig. 1). Epac1 has in its N-terminal part a DEP (dishevelled, egl-10, pleckstrin) domain, responsible for membrane localization, followed by a cAMP-binding site that closely resembles the cAMP-binding domains in PKA and cyclic nucleotide gated ion channels (Fig. 2). The C-terminal part consists of the REM domain and the catalytic domain. These domains closely resemble similar domains in GEFs for other members of the Ras family, such as Sos and RasGRP. The CDC25-homology domain mediates the exchange of guanine nucleotides bound to the target GTPase in the case of

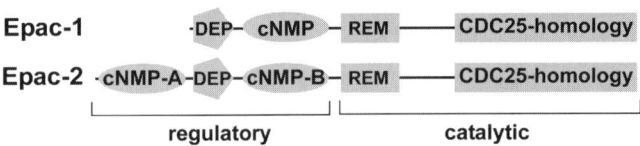

Figure 1 Domain organization of Epac. The regulatory and the catalytic part of the protein are indicated. Dishevelled, egl-10, pleckstrin (DEP) cyclic nucleotide binding domain (cNMP), Ras exchange motif (REM), CDC25 homology domain (CDC25 homology).

Epac Rap1 and Rap2. The REM domain is responsible for the stabilization of the CDC-25 homology domain [4] but is not directly involved in the catalysis of nucleotide exchange. Epac2 has in addition a second N-terminal cAMP domain.

The cAMP-Binding Domain of Epac Closely Resembles Those of PKA and Channels

Sequence comparison between various cAMP-binding sites as well as the crystal structure of cAMP-binding sites of PKA in the presence of cAMP identifies various consensus sequences specific for cAMP–binding, including the phosphate binding cassette (PBC) [5]. These motifs are all present in the cNMP-binding domain of Epac1, as well as in the second cNMP-binding domain of Epac2 (Fig. 2)

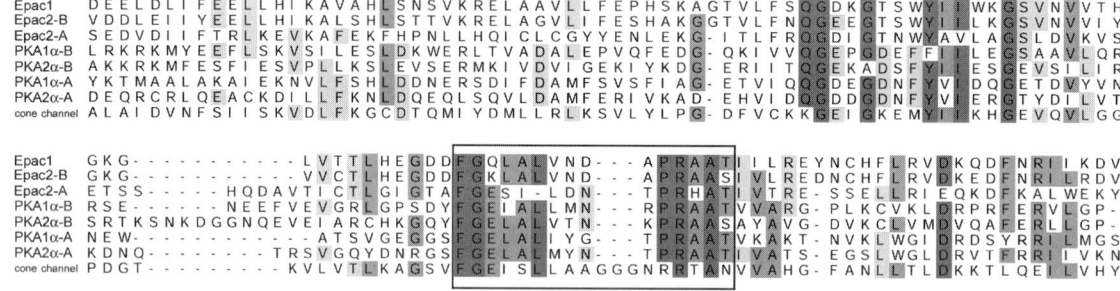

Figure 2 The phosphate binding cassette (PBC) is clearly conserved in Epac. The PBC is marked by a box. All sequences shown are human.

Epac Is Conserved Through Evolution

In higher organisms two Epac proteins, Epac1 and Epac2 (Fig. 1), are found, which show strong sequence homology not only among species (human Epac2 and mouse Epac2) but also between each other (human Epac1 and human Epac2) (Fig. 3). The major difference between Epac1 and Epac2 is an additional N-terminal cNMP-binding domain. In lower organisms only one Epac is present that is more related to Epac2, due to its additional putative cNMP-binding domain. However, this domain is less conserved and lacks PBC, which makes cAMP binding to these domains highly unlikely. Note also that sequence conservation is strongest in the CDC25 homology domain, which mediates the catalytic activity, and in the more C-terminal cNMP binding domain, which is the (main) regulatory element. A minor role in regulation for the first cNMP-binding domain fits the biochemical data (see below).

Properties of Epac

In vitro experiments have shown that in the absence of cAMP, Epac is unable to increase intrinsic exchange of nucleotide on the small GTPase Rap1. However, cAMP accelerates the exchange activity by at least two orders of magnitude, proving the responsiveness of Epac to cAMP. By using isothermal titration calorimetry, the affinity of cAMP has been determined for each of the single domains of Epac1 and Epac2, as well as for the whole regulatory domain of Epac2 (Table I). These data indicate that there is no significant cooperativity in cAMP binding to the two cAMP-binding sites in the regulatory part of Epac2 [6]. In addition, deletion of the first cAMP-binding domain from Epac2 does not abolish the requirement of cAMP for activity [6]. The second cAMP-binding domain, which is conserved in the domain structure of Epac1 and Epac2, is therefore sufficient to maintain the inactive state of Epac2 in the absence of cAMP. This is in agreement with the relative low conservation of the first cAMP domain between species (see above).

The value of half maximal activity (IC_{50}-value) of Epac was determined by measuring the concentration dependency of cAMP-induced activation of Epac. For Epac1 an IC_{50} value of 40 µM is found (H. Rehman, unpublished data). This value reflects the affinity of cAMP for full-length Epac, and is clearly lower than the affinity of cAMP for the separate regulatory domain. This implies that the presence of the catalytic part of the protein reduces the cAMP affinity. Apparently, a conformational change induced by cAMP binding "consumes" parts of the binding energy.

Expression and Subcellular Localization of Epacs

Epac1 is expressed in a large number of tissues tested but most notably in kidney, ovary, thyroid, heart, and brain, where it is confined to the septum and the thalamus of the neonatal brain. Epac2 is predominantly expressed in the cerebral cortex, the hippocampus, the habenula, and the cerebellum of the brain and in the adrenal gland [2]. Subcellularly, Epac 1 is attached to membranes through its DEP domain [6], and an Epac-GFP fusion protein is located predominantly in the preinuclear region, where Rap1 and Rap2 are also located (F. J. Zwartkruis, unpublished observation).

Cellular Function of Epacs

Epacs mediate the activation of Rap1 and Rap2 by receptors that elevate the levels of cAMP. However, the biological effects are still unclear. A number of reports indicate cAMP-mediated but PKA-independent events that may be mediated by Epac. For instance, the following events have been noted: (1) cAMP-induced translocation of CFTR-like chloride channels in renal cells [7], (2) cAMP regulation of Glut-1 translocation to plasma membrane through PI3-kinase-dependent and PKA-independent signaling pathways [8], (3) cAMP-induced calcium influx in PBL cells [9], (4) cAMP-induced activation of PKB in thryroid cells [10], and (5) cAMP-induced transcription in glioma cells [11].

cAMP-induced Epac-mediated regulated secretion in pancreatic B-cells [12,13] is the only effect directly assigned to Epacs, but a role for Rap in this process was not established.

Figure 3 Sequence alignment of Epac from various species. The variable N-terminus is excluded form the alignment and given as sequence only. Domains are indicated according to Fig. 1 by bars over the sequence. The conserved phosphate binding cassette (PBC) is highlighted by boxes in both cNMP binding domains (h, *homo sapiens*; m, *mus musculus*; d, *Drosophila melanogaster*; ce, *Caenorhabditis elegans*).

Tabel I Affinities of cAMP Binding Domains

Epac1	Epac2		
	cAMP-A	cAMP-B	RD
4 μM	90 μM	1 μM	80 μM ; 0.5 μM

Rap1 has been implicated in a variety of cellular functions, most notably in the regulation of ERK activity. However, these effects are mediated by PKA and Epacs are not involved. Whether this implies that a different pool of Rap1 is activated by PKA or that Rap1 is not involved in this process at all is still a matter of debate, although we favor the second alternative.

References

1. de Rooij, J. *et al.* (1998). Epac is a Rap1 guanine-nucleotide-exchange factor directly activated by cyclic AMP. *Nature* **396**, 474–477.
2. Kawasaki, H. *et al.* (1998). A family of cAMP-binding proteins that directly activate Rap1. *Science* **282**, 2275–2279.
3. Reedquist, K. A. *et al.* (2000). The small GTPase, Rap1, mediates CD31-induced integrin adhesion. *J. Cell Biol.* **148**, 1151–1158.
4. Boriack-Sjodin, P. A., Margarit, S. M., Bar-Sagi, D., and Kuriyan, J. (1998). The structural basis of the activation of Ras by Sos. *Nature* **394**, 337–343.
5. Canaves, J. M. and Taylor, S. S. (2002). Classification and phylogenetic analysis of the cAMP-dependent protein kinase regulatory subunit family. *J. Mol. Evol.* **54**, 17–19.
6. de Rooij, J. *et al.* (2000). Mechanism of regulation of the Epac family of cAMP-dependent RapGEFs. *J. Biol. Chem.* **275**, 20829–20836.
7. Shintani, Y. and Marunaka, Y. (1996). Regulation of chloride channel trafficking by cyclic AMP via protein kinase A-independent pathway in A6 renal epithelial cells. *Biochem. Biophys. Res. Commun.* **223**, 234–239.
8. Samih, N., Hovsepian, S., Aouani, A., Lombardo, D., and Fayet, G. (2000). Glut-1 translocation in FRTL-5 thyroid cells: role of phosphatidylinositol 3-kinase and N-glycosylation. *Endocrinology* **141**, 4146–4155.
9. de la Rosa, L. A., Vilarino, N., Vieytes, M. R., and Botana, L. M. (2001). Modulation of thapsigargin-induced calcium mobilisation by cyclic AMP-elevating agents in human lymphocytes is insensitive to the action of the protein kinase A inhibitor H-89. *Cell Signal* **13**, 441–449.
10. Cass, L.A. *et al.* (1999). Protein kinase A-dependent and -independent signaling pathways contribute to cyclic AMP-stimulated proliferation. *Mol. Cell Biol.* **19**, 5882–5891.
11. Skoglund, G., Hussain, M. A., and Holz, G. G. (2000). Glucagon-like peptide 1 stimulates insulin gene promoter activity by protein kinase A-independent activation of the rat insulin I gene cAMP response element. *Diabetes* **49**, 1156–1164.
12. Renstrom, E., Eliasson, L., and Rorsman, P. (1997). Protein kinase A-dependent and -independent stimulation of exocytosis by cAMP in mouse pancreatic B-cells. *J. Physiol.* **502**, 105–118.
13. Ozaki, N. *et al.* (2000). cAMP-GEFII is a direct target of cAMP in regulated exocytosis. *Nat. Cell Biol.* **2**, 805–811.

Cyclic Nucleotide-Binding Phosphodiesterase and Cyclase GAF Domains

Sergio E. Martinez,[1] Xiao-Bo Tang,[1] Stewart Turley,[2] Wim G. J. Hol[1,2], and Joseph A. Beavo[1]

[1]Department of Pharmacology, University of Washington, Seattle, Washington and
[2]Department of Biochemistry and Biological Structure, Howard Hughes Medical Institute, University of Washington, Seattle, Washington

Introduction

The GAF domains are one of the largest and most widespread of the small-molecule binding domains [1]. They are distantly related to PAS domains, another superfamily with a similar fold [1]. The acronym GAF was coined from the first three protein families in which they were identified: cyclic cGMP-specific and -regulated cyclic nucleotide phosphodiesterases, Anabaena adenylyl cyclase, and E. coli transcription factor FhlA [2]. Over 890 GAF domain containing proteins have now been identified in a non-redundant database [3]. They are involved in many signal transduction pathways, protein regulation, and sensory systems [1,2]. This review will focus on the structure and function of cyclic nucleotide binding GAF domains found in PDEs and also include a short discussion of several others to which they are related.

cAMP and cGMP are used as second messengers in many cellular signal transduction pathways. The intracellular levels of cyclic nucleotides are determined by the relative activities of adenylyl and guanylyl cyclases, the enzymes that catalyze their synthesis, and cyclic nucleotide phosphodiesterases (PDEs), which hydrolyze 3′,5′ cyclic nucleotides to their respective inactive 5′-nucleotide monophosphates. Eleven PDE families have been identified in mammalian tissues [4]. PDEs are nearly all homodimers with monomer molecular weights ranging from 50 to 135 kDa. The PDEs constitute the major pathway for the elimination of the cyclic nucleotide signal from the cells to precisely maintain their levels within a narrow range.

PDEs are multidomain proteins with a C-terminal conserved catalytic domain and one or more N-terminal regulatory segments. The regulatory segment from five of the PDE families (PDE2, 5, 6, 10, 11) contain one or two GAF domain modules. For the first three families, cGMP-binding to at least one GAF domain in the regulatory segment modifies the PDE's catalytic activity. In PDE2, cGMP bound to the GAF domain stimulates the Vmax by more than ten-fold [5–8]. For PDE5A, cGMP binding to GAF A stimulates the phosphorylation by PKG of a Ser just N-terminal to GAF A [9] and increases the catalytic activity.

In rod and cone photoreceptor cells, the catalytic activity of PDE6 is inhibited by binding of a small 9 kDa inhibitory gamma subunit whose affinity is regulated by binding of cGMP to the regulatory segment. The apparent binding affinity of frog rod Pγ for PDE6 was 28 pM in the presence of cGMP and 16-fold weaker with the GAF sites unoccupied [10]. Unlike PDE2A and 5A, there is no evidence for this enzyme showing direct cooperativity between the GAF domains and catalytic sites. cGMP binding by GAF A does not act allosterically on the catalytic domain. Instead, it binds the central polycationic region of Pγ and increases the affinity of the C-terminal region for the catalytic site [11]. In turn, binding or dissociation of cGMP to GAF A may also influence the interaction with transducin or its cognate GAP protein, RGS9. A photoactive peptide containing the polycationic region of Pγ cross-linked to two residues in GAF A [12]. There is circumstantial evidence for gamma-like subunits associated with PDE5A in lung [13].

Two more recently discovered and less well-characterized PDE families, PDE10A and PDE11A, are homologous to PDE 2, 5, and 6 in their regulatory segment and contain GAF domains; however, careful binding studies have not yet been reported. A preliminary study suggests that PDE10A lacks a high-affinity cGMP binding site but may contain a low affinity site [14]. One unique structural feature of this family is the presence of splice variants that contain truncated GAF domains in the N-terminal region. These tissue-specific splice variants have been postulated mostly from Northern blot or RT-PCR data, and their existence as active proteins in the cell is not yet fully confirmed [15–18]. However, a genomics study of the PDE11A gene shows separate promoters for both splice variants with an incomplete GAF A (PDE11A3) or GAF B domain (PDE11A1) [19]. It is interesting that a similar series of structures have been seen in a PDE family from Trypanosomes [20,21]. The initial studies of PDE11A1 suggest that neither cAMP nor cGMP are allosteric effectors [18]. However, the splice variant tested lacked the GAF A of the full-length enzyme (PDE11A4) and has an incomplete GAF B. PDE11A4, with both GAF A and B, had no significant cGMP or cAMP binding activity in another preliminary study [17]. A "half-a-PAS" splice variant has been reported for PDE8A, the only PDE to contain a PAS domain [22].

Atomic Structure

The first atomic structure of a GAF domain came recently from a structural genomics project for signal transduction domains. A hypothetical GAF protein, YKG9, from the yeast genome was characterized at high resolution. However, yeast do not make cGMP and this protein did not bind the nucleotide [23]. The first atomic structure of a PDE GAF domain, in this case with bound cGMP, was recently determined (Martinez et al., PNAS). The dimeric regulatory segment of mouse PDE2A, containing both the GAF A and B domains, was determined at 2.9 Å resolution. GAF A and B have very similar folds to YKG9. Although both the catalytic and GAF domains of PDEs bind cyclic nucleotides, and some homology exists, their folding structures are quite different and they appear to be examples of convergent evolution. For example, the catalytic domain of PDE4 is a bundle of 17 alpha helices in three subdomains [24], whereas the GAF domain is a beta sheet packed on the back side with two to four alpha helices and on the other side with a mixture of short alpha helices and loops that form the sides of the ligand-binding pocket [23]. In addition, the zinc-binding motif of the catalytic domain [25] is not present in the GAF domain.

These two new GAF domain structures from PDE2A can now be compared with that of YKG9. The overall folds of the three GAF domains with known structure are very similar, with the yeast variant having an additional N-terminal helix compared to the PDE2A GAF A. Yeast YKG9 (PDB 1F5M, monomer A) is distantly related to PDE2A GAF A with 136 Cα atoms superimposing within 4.0 Å and a sequence identity of 10 percent, and to GAF B with 130 equivalent Cα atoms that superimpose within an r.m.s deviation of 3.3 Å and a sequence identity of 16 percent.

In PDE2A, GAF A and B are connected through short linkers of several residues to a 32 residue long connecting helix of nine turns (Fig. 1). Unexpectedly, GAF A is engaged in numerous inter-subunit interactions, whereas the GAF B domains are well separated. The first five turns of the connecting helix also provide inter-subunit contacts but afterwards diverge. This occurs after C386, a residue that appears (at this somewhat limited resolution) to form a disulfide bond. A C386S mutant has an essentially identical structure. Therefore, the presence or absence of this disulfide bond has little effect on conformation. Whether this disulfide is integral to the activation or positive cooperativity of PDE2A remains to be determined. GAF A has a surprising, completely different dimer interface than the YKG9 dimer, with contacts from a completely different set of secondary structure elements. Recently, the overall shapes of the bovine PDE6A/B and PDE5A have been visualized at 28 Å resolution from uranyl-stained electron microscopy images [26]. These data show that both PDEs are dimers with the N-terminal regions interacting, which is clearly consistent with the dimerization interface between GAF A domains elucidated at higher resolution in PDE2A. An intriguing finding is that the dimerizing connecting helices that separate GAF A and B in PDE2A appear in the micrographs to be holes surrounded by connections of electron density. Although this could be an artifact of the low resolution and/or the uranyl acetate negative stain, the sequence of the PDE2A connecting helix is poorly conserved among the GAF PDEs.

In the 2.9 Å resolution crystal structure of PDE2A, electron density was found for a molecule of cGMP bound to GAF B. The binding site corresponds to the PAS pocket, showing the conservation of ligand binding function. It is interesting that the cGMP is completely buried, along with three bound waters. This strongly suggests that the protein must also exist in another more open state in which the initial binding occurs. The most distinctive feature of cGMP, its 3′,5′ cyclic phosphate group, has no side chain contacts. Two backbone amides (I458, Y481, A459) and a water make hydrogen bonds to the exocylic oxygens of the phosphate. In addition, the positive end of the helix dipole of helix α3 points directly at the negatively charged phosphate group. The guanine ring and ribose group make a total of six polar contacts and two hydrophobic contacts. One residue, D439, makes both a side chain contact (to the N2 nitrogen) and a backbone amide contact (to the O6 carbonyl). We believe this residue is a main specificity determinant for cGMP over cAMP.

CHAPTER 207 Cyclic Nucleotide-Binding Phosphodiesterase and Cyclase GAF Domains

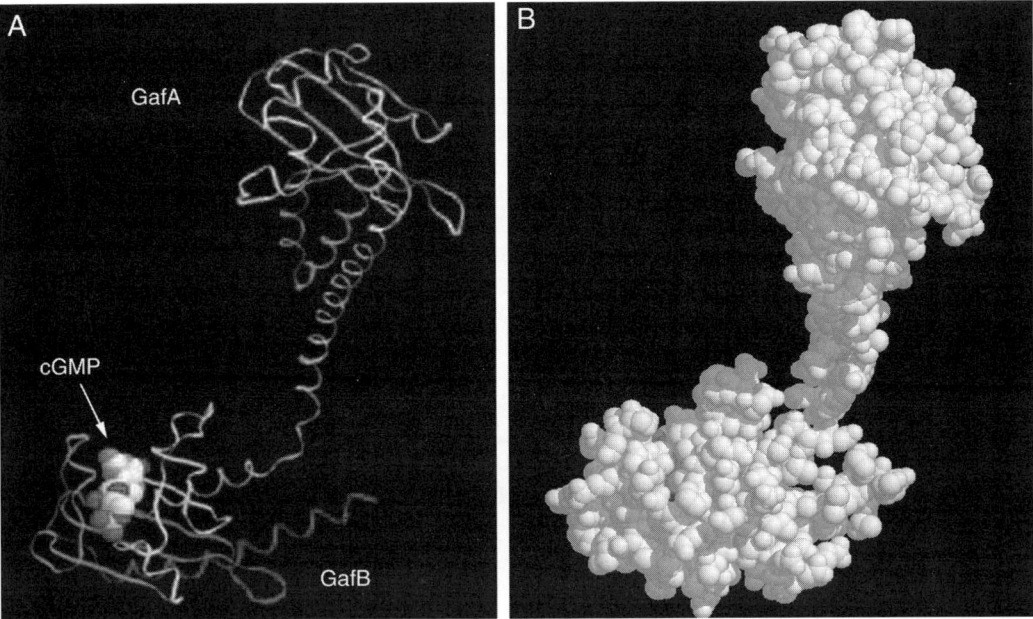

Figure 1 Structure of the PDE2A GAF A/GAF B monomer plus cGMP. Shown on the left is a ribbon structure of the monomer with a space-filling model of cGMP bound to GAF B. Shown on the right is a full space-filling model of both. Notice that the cGMP is almost completely buried. Less than 1 percent of the molecule is exposed to solvent. Presumably therefore the protein must go through some other conformation in order to bind the cyclic nucleotide.

Aside from this work, published cGMP binding to individual PDE GAF domains have only been done for PDE5A, which shows binding to the GAF A domain [23,27]. A comparison of the 2A GAF B structure to the 5A GAF A sequence suggests that the binding motif is SX(13-18)FDX(18-22)IAX(21)[Y/N]X(2)VDX(2)TX(3)TX(19)[E/Q]. The [Y/N] and IA are backbone contacts not expected to be conserved. It is a theoretical possibility that PDE2A GAF A and PDE5A GAF B may be able to form additional cGMP binding sites, but the poor conservation of the motif and biochemical data suggest otherwise. Given the variety of ligands bound by PAS and GAF domains, it raises the intriguing possibility of unknown small molecule ligands regulating GAF PDEs. PDE6 is the PDE whose stoichiometry of cGMP is best documented, one 1 cG/monomer. The binding motif is best conserved in GAF A.

Although the YKG9 structure has no bound ligand and does not bind cGMP [23], it does contain a buried pocket with a volume of 590 Å^3 corresponding to the location of the cGMP-binding site of our GAF domain. A structural superposition of YKG9 onto GAF B shows approximately 11 ordered waters, which are scattered about the volume occupied by the cGMP molecule in PDE2A GAF B. It is of possible significance that the YKG9 GAF pocket contains a buried acidic cluster (E132, D149, and D151). This may imply that YKG9 could bind an as yet unknown, positively charged ligand. One of these residues (D149) superimposes with E512 in PDE2A GAF B, the only charged residue buried in that GAF pocket.

The PDE2A GAF A+B structure does not suggest an obvious explanation for the activation mechanism of PDE2A.

In a recent paper, a cAMP-activated cyanobacterial adenylyl cyclase was enzymologically characterized and shown to contain GAF A and GAF B domains, like PDE2A. Binding of cAMP but not cGMP to this domain activated the cyclase. However, the organization of the cyclase differed in that it also contains a PAS domain interspersed in between GAF B and the cyclase domain [28]. Although the overall sequence homology of the cyanobacterial and mammalian GAF domains is not particularly high, it appears that the key residues necessary for binding in PDE2A GAF B are present (Fig. 2). Substitution of the mammalian PDE2 GAF A/GAF B domain cassette into the cyanobacterial cyclase amazingly allowed a fully functional cyclase to be formed, but one that now could be activated by cGMP. These data strongly suggest that the GAF domains function as a general, highly conserved, cyclic nucleotide switch for activating adjacent catalytic domains. In this case two organisms that are over two billion years removed from each other in evolution can utilize the other's GAF domain structural motif. The intervening PAS domain in the adenylyl cyclase apparently does not block this mechanism. This result argues against an intramolecular association between the GAF A and B assembly and the catalytic domain of either enzyme. More likely, binding of cGMP creates a conformational change in GAF B that is transmitted mechanically to the catalytic domain. This may involve a loop that blocks the active site. Alternatively, if the catalytic domains contact each other, the active site may be partially occluded in the inactive form by the opposite domain. A third possibility is conformational change within the active site itself.

In nearly all PDE GAF domains, as well as in the adenylyl cyclase cyaB1 of Anabaena, there is a conserved

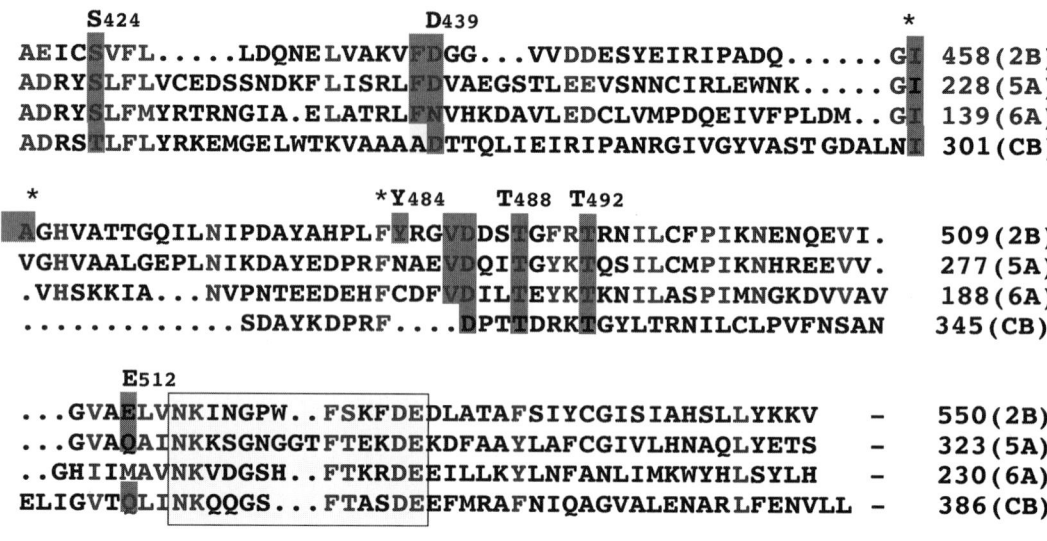

Figure 2 Comparison of cNMP binding GAF domain sequences. Clustal W sequence alignment of the four GAF domains known to bind cyclic nucleotide. Gaps were further aligned by eye. 2B, mouse PDE2A GAF *B*; 5A, human PDE5A GAF *A*; 6A, human PDE6A GAF *A*; and CB, Anabaena Adenylyl Cyclase GAF **B**. Notice that in this alignment nearly all of the major cGMP contact points (shading) present in the PDE2A GAF B crystal structure are present in all of the other GAF domains that bind cyclic nucleotide. Backbone contacts are shown by asterisks. The original GAF domain signature sequence NKxxxFxxDE sequence found in the PDE GAF domains is also present in the Anabaena cyclase.

N[KR]X$_n$FX$_3$DE motif (Fig. 2) [29]. This motif straddles a loop on the other side of the beta sheet and approximately 17 Å from the cGMP binding pocket in PDE2A GAF B. Nevertheless, point mutations in the N[KR]X$_n$FX$_3$DE motif of GAF A, but not GAF B, of PDE5 greatly weaken cGMP binding [30]. In the structure of both GAF A and GAF B of PDE2, a salt bridge appears to be present between the K and D residues. This suggests that this motif may be important for conformation, or for a conformational change upon cGMP binding. Similar mutations in the Asp of this motif in GAF B but not GAF A of Anabaena AC disabled cAMP-dependent activation of cyclase activity [28].

The first crystal structures of GAF domains have solved many questions regarding regulation of the proteins but have also raised many questions. There may well be other ligands for each of the non-cGMP binding GAF domains in PDE2, PDE5, PDE6, and possibly the poorly characterized PDE10 and PDE11. What is the role of the GAF domains that don't bind cGMP? Do they serve only as dimerization domains for the holoenzyme? Is the common regulatory mechanism hinted at by the Anabaena cyclase/ratPDE2 chimera also conserved among the five GAF PDE families, yet affects the catalytic domain of three different PDE in such different ways? Ultimately, crystal or NMR structures will be needed for the various PDE domains with and without bound ligands to visualize the conformational changes that occur upon binding. Structures of the full-length holoenzymes and possibly also chimeras may also be required. Since these small molecule-binding GAF domains are intimately involved in regulation of so many different enzymes, they are also likely to become good targets for drug development.

Acknowledgments

The work presented in this manuscript was supported by grants DK 21723 and HL 44948 to J.A.B. and CA 65656 to W.G.J.H. W.G.J.H. acknowledges a major equipment grant to the Biomolecular Structure Center by the Murdock Charitable Trust. S.E.M. was supported for part of the work by NIH training grant T32 HL07312-23. We would like to thank Joachim Schultz for providing us with data on Anabaena adenylyl cyclase GAF domains prior to publication.

References

1. Anantharaman, V., Koonin, E. V., and Aravind, L. (2001). *J. Mol. Biol.* **307**, 1271–1292.
2. Aravind, L. and Ponting, C. P. (1997). *Trends Biochem. Sci.* **22**, 458–459.
3. Schultz, J., Milpetz, F., Bork, P., and Ponting, C. P. (1998). *Proc. Natl. Acad. Sci. USA* **95**, 5857–5864.
4. Soderling, S. H. and Beavo, J. A. (2000). *Curr. Opin. Cell Biol.* **12**, 174–179.
5. Yamamoto, T., Manganiello, V. C., and Vaughan, M. (1983). *J. Biol. Chem.* **258**, 12526–12533.
6. Moss, J., Manganiello, V. C., and Vaughan, M. (1977). *J. Biol. Chem.* **252**, 5211–5215.
7. Erneux, C., Couchie, D., Dumont, J. E., Baraniak, J., Stec, W. J., Abbad, E. G., Petridis, G., and Jastorff, B. (1981). *Eur. J. Biochem.* **115**, 503–510.
8. Wada, H., Osborne, J. J., and Manganiello, V. C. (1987). *Biochemistry* **26**, 6565–6570.
9. Corbin, J. D., Turko, I. V., Beasley, A., and Francis, S. H. (2000). *Eur. J. Biochem.* **267**, 2760–2767.
10. D'Amours, M. R. and Cote, R. H. (1999). *Biochem. J.* **340**, 863–869.
11. Mou, H. M., Grazio, H. J., Cook, T. A., Beavo, J. A., and Cote, R. H. (1999). *J. Biol. Chem.* **274**, 18813–18820.
12. Muradov, K., Granovsky, A., Schey, K., and Artemyev, N. (2002). *Biochemistry* **41**, 3884–3890.

13. Lochhead, A., Nekrasova, E., Arshavsky, V. Y., and Pyne, N. J. (1997). *J. Biol. Chem.* **272**, 18397–18403.
14. Soderling, S. H., Bayuga, S. J., and Beavo, J. A. (1999). *Proc. Natl. Acad. Sci. USA* **96**, 7071–7076.
15. Hetman, J. M., Robas, N., Baxendale, R., Fidock, M., Phillips, S. C., Soderling, S. H., and Beavo, J. A. (2000). *Proc. Natl. Acad. Sci. USA* **97**, 12891–12895.
16. Yuasa, K., Ohgaru, T., Asahina, M., and Omori, K. (2001). *Eur. J. Biochem.* **268**, 4440–4448.
17. Yuasa, K., Kotera, J., Fujushige, K., Michibata, H., Sasaki, T., and Omori, K. (2000). *J. Biol. Chem.* **275**, 31469–31479.
18. Fawcett, L., Baxendale, R., Stacey, P., McGrouther, C., Harrow, I., Soderling, S., Hetman, J., Beavo, J. A., and Phillips, S. C. (2000). *Proc. Natl. Acad. Sci. USA* **97**, 3702–3707.
19. Yuasa, K., Kanoh, Y., Okumura, K., and Omori, K. (2001). *Eur. J. Biochem.* **268**, 168–178.
20. Zoraghi, R. and Seebeck, T. (2002). *Proc. Natl. Acad. Sci. USA* **99**, 4343–4348.
21. Rascon, A., Soderling, S. H., Schaefer, J. B., and Beavo, J. A. (2002). *Proc. Natl. Acad. Sci. USA* **99**, 4714–4719.
22. Wang, P., Wu, P., Egan, R. W., and Billah, M. M. (2001). *Gene* **280**, 183–194.
23. Ho, Y.-S. J., Burden, L. M., and Hurley, J. H. (2000). *EMBO J.* **19**, 1–12.
24. Xu, R. X., Hassell, A. M., Vanderwall, D., Lambert, M. H., Holmes, W. D., Luther, M. A., Rocque, W. J., Milburn, M. V., Zhao, Y., Ke, H., and Nolte, R. T. (2000). *Science* **288**, 1822–1825.
25. Omburo, G. A., Jacobitz, S., Torphy, T. J., and Colman, R. W. (1998). *Cell. Signal.* **10**, 491–497.
26. Kameni Tcheudji, J. F., Lebeau, L., Virmaux, N., Maftei, C. G., Cote, R. H., Lugnier, C., and Schultz, P. (2001) *J. Mol. Biol.* **310**, 781–791.
27. Liu, L., Underwood, T., Li, H., Pamukcu, R., and Thompson, W. J. (2002). *Cell. Signal.* **14**, 45–51.
28. Kanacher, T., Schultz, A., Linder, J. U., and Schultz, J. E. (2002). *EMBO J.* **21**, 1–9.
29. Charbonneau, H., Prusti, R. K., LeTrong, H., Sonnenburg, W. K., Mullaney, P. J., Walsh, K. A., and Beavo, J. A. (1990). *Proc. Natl. Acad. Sci. USA* **87**, 288–292.
30. Turko, I. V., Haik, T. L., McAllister-Lucas, L. M., Burns, F., Francis, S. H., and Corbin, J. D. (1996). *J. Biol. Chem.* **271**, 22240–22244.

CHAPTER 208

cAMP Signaling in Bacteria

J. M. Passner
*Department of Physiology and Biophysics,
Mount Sinai School of Medicine,
New York, New York*

Introduction and Significance

Our understanding of cyclic AMP (cAMP) signaling in bacteria has resulted primarily from studying the *Escherichia coli* catabolite gene activator protein (CAP), also known as the cAMP receptor protein (CRP). CAP is a 45 kDa homodimer that positively regulates the expression of over 150 genes [1]. Transcriptional activation of RNA polymerase by CAP requires that CAP bind cAMP, undergo an allosteric conformational change, and bind to a specific DNA sequence near the polymerase binding site.

Kolb *et al.* [2] write that "[t]he title of paradigm has been awarded to CRP on many occasions and for many of its functions." This is well-deserved. CAP is one of the most studied transcriptional regulators [3]. The conformational change of CAP upon cAMP binding has become an important paradigm for allostery [4]. The CAP DNA–binding domain contains a helix-turn-helix motif conserved in a large number of DNA-binding proteins; CAP is a founding member of this "First Family" of DNA-binding proteins, which include the homeodomain and the "winged helix" subfamilies [5,6]. The cAMP-binding domain is homologous to the regulatory subunit of cAMP-dependent protein kinases and the cyclic nucleotide binding domains of cyclic nucleotide-gated channels [7,8]. This homology to CAP has furthered our understanding of cyclic nucleotide-gated channels [8,9].

Background and History

In an environment containing glucose and other carbon sources, *E. coli* metabolizes glucose while metabolism of other sources is inhibited. Upon glucose depletion, the enzymes that metabolize the remaining carbon sources are induced. This phenomenon has been called the "glucose effect" or "catabolite or glucose repression" and was studied extensively by Jacques Monod over 60 years ago [10,11]. This effect is mediated by cAMP. *E. coli* growing in glucose-containing media have low cAMP levels; but once glucose is removed, cAMP concentrations increase rapidly [12]. Furthermore, exogenously added cAMP also relieves glucose repression [13,14]. Glucose indirectly inhibits adenylate cyclase, which synthesizes cAMP from ATP [15]. CAP was discovered over thirty years ago and was shown to bind to cAMP and mediate its effects [16,17].

The CAP-cAMP complex structure was originally determined in the absence of the amino acid sequence from a 2.9 Å resolution isomorphous map [18]. The amino acid sequence was subsequently incorporated into the structure [19], which was later refined to 2.5 Å resolution [20] and most recently to 2.1 Å resolution [21]. Each subunit folds into two domains, a large (residues 1–130) N-terminal cAMP-binding domain and a small (residues 140–209) DNA-binding domain. The residues between the two domains form a hinge that accommodates movement of one domain relative to the other.

The cAMP conformation when bound to CAP was a matter of disagreement. In the crystal structure, the cAMP molecules, which are among the best-ordered parts of the structure, are in an *anti* conformation. Just as unambiguously, and approximately when the CAP-cAMP structure was initially solved, a *syn* conformation for the cAMP was inferred from NMR experiments [22]. This apparent paradox was not resolved for 15 years, as discussed below.

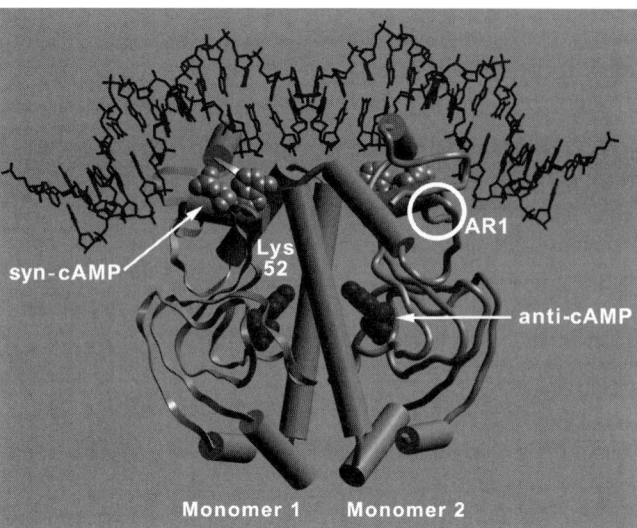

Figure 1 The cAMP-CAP-DNA dimer crystal structure [25]. The activating region 1 and Lys52 are labeled in the respective monomers where each is closer to the viewer. The *anti* and the *syn* cAMP molecules are shown as dark and light gray, respectively.

The crystal structure of CAP-cAMP complexed with a 30 base-pair DNA was solved initially at 3 Å resolution [23] and subsequently refined to 2.5 Å resolution [24]. The DNA in the complex is bent by 90°, which results almost entirely from two 40° kinks. A second CAP-DNA crystal form using a 46 base-pair DNA was solved at 2.2 Å resolution (Fig. 1) [25]. In this crystal form the DNA ends are disordered and unencumbered by crystal packing forces. The second crystal form structure confirms the major results of the original structure.

Transcriptional Regulation by CAP

The architecture of CAP-dependent promoters is amazingly varied. CAP can be either an activator or a repressor, and it can act alone or with other factors. CAP-binding site positions vary considerably among promoters. CAP can stimulate transcription alone when the DNA-binding sites are centered at −41.5, −61.5, and −70.5, as in the *gal*, *lac*, and *malT* promoters, respectively [26]. When the CAP site(s) is (are) further upstream, CAP cooperates with other transcriptional activators, such as MalT and AraC [27,28]. CAP represses transcription of itself and adenylate cyclase [29,30]. An example more reminiscent of eukaryotic systems is a set of CytR-regulated promoters: CytR repression requires two CAP dimers bound exactly 53 base-pairs apart. CytR binds to this nucleoprotein complex, disrupting transcription activation by CAP [31]. It is beyond the scope of this review to discuss the different mechanisms of CAP action at various promoters. However, these few examples offer a sampling of many ways *E. coli* uses CAP in transcription regulation.

Several CAP regions that are involved in transcriptional activation, and that presumably contact RNA polymerase, have been identified (Fig. 1). One such contact site, activating region 1 (AR1), is a surface-exposed loop (residues 158–162). Positive control mutations—that is, mutations that disrupt transcription activation but do not affect DNA-binding—have been found in AR1 [32–34]. An elegant experiment using heterodimers—one monomer has mutations in AR1, while the other has a mutation that alters the DNA sequence it recognizes—explored which activation loop is required at different promoters [35,36]. Zhou *et al.* demonstrated that when CAP is centered at −61.5, a functional AR1 is essential in the downstream subunit but dispensable in the upstream subunit; the reverse is the case when CAP is centered at −41.5. Suppressors of AR1 mutations map to a second surface-exposed loop containing Lys52 [37]. However, suppression only occurs when the CAP site is centered at −41.5. By studying RNA polymerase alpha subunit deletion and point mutants, it has been shown that the C terminus contains a CAP contact site, when CAP binds at −61.5 but not at −41.5 [38]. Promoter architecture thus plays an important role in determining the exact nature of the CAP-polymerase interaction.

CAP Permits Differential Gene Regulation at Different cAMP Concentrations

Bacteria have two major pathways to control transcription of a global network of genes in response to environmental stimuli and signals. One is to use a transcription factor such as CAP that responds to a signal and undergoes a conformational change, allowing it to bind DNA specifically and thereby regulate gene expression. The other pathway is to use alternative sigma factors, which recognize different promoter sequences, thereby providing the cell with an easy way to turn on a gene family in response to a signal [39]. For example, alternative sigma factors are involved in response to heat shock, sporulation, and flagellar synthesis. Why does *E. coli* use binding to CAP, and not the induction of an alternative sigma factor, in response to cAMP? The answer appears to lie in the flexibility of CAP in regulating gene expression, a flexibility that an alternative sigma factor would be incapable of achieving.

The use of an alternative sigma factor is a digital response. By using a different sigma factor, the cell can turn on and off a whole set of genes in unison. This type of response is critical for discrete changes in developmental or environmental states, such as sporulation or heat shock, or when the cell requires a complex cellular machine, such as flagella, where it is most efficient to produce all the components together.

As noted above, *E. coli* employs CAP in multiple ways; accordingly, the response of CAP to cAMP is analog. The regulation of the expression of different operons can be turned on or off at different cAMP concentrations, with different thresholds for a cAMP response. The threshold is set by adjusting the DNA sequence to change the affinity of the DNA for CAP-cAMP, so that different cAMP concentrations lead to differing affinities for various promoters. For example, the cAMP-CAP affinity for the *lac* promoter is about an order of magnitude higher than for the *gal* promoter [40]. CAP binds to a consensus sequence 450 times better than to the *lac* site, which is one of the strongest *in vivo*; no natural site is even close in binding affinity to that of the consensus site [41]. This is presumably because a consensus site will always be bound at all physiological cAMP concentrations, and hence no regulation would be possible at promoters containing a consensus-binding site [42].

What advantage does setting different cAMP thresholds at different promoters confer on the cell? Some carbon sources cost more energy to metabolize than others. Therefore, the cell will resort to certain metabolites at different levels of glucose depletion and hence different cAMP concentrations. Similarly, in the presence of multiple nonglucose carbon sources, the cell can utilize its metabolites most efficiently.

A Second cAMP-Binding Site in a CAP Monomer

Various experiments have shown that cAMP binding to CAP induces a conformational change in CAP [4]. It has been observed, with several conformational probes, that CAP has a biphasic dependence on cAMP concentration. These probes—which include proteolytic digestion rates, Cys178 modification, tryptophan fluorescence, fluorescence of an extrinsic probe, and DNA affinity—display two cAMP concentration-dependent behaviors. One set of behaviors is seen up to approximately 200 micromolar cAMP concentrations and the other at millimolar concentrations [43–45]. These observations have been explained by the presence of three conformational states: free cAMP, and CAP dimers with one and two molecules bound, respectively. However, the structure of CAP bound to a 46 base-pair DNA revealed a second cAMP molecule bound to each protein monomer (Fig. 1) [25]. This second cAMP molecule was in the *syn* conformation, and resolved the long-standing discrepancy between NMR and crystallographic observations on the cAMP conformation bound to CAP. Crothers and Steitz [46] postulated the existence of a second cAMP-binding site, when they suggested that at the millimolar concentrations in which the NMR experiments were done there was "a second weakly bound and rapidly exchanging cAMP which is in the *syn* conformation." The presence of a second cAMP-binding site in each monomer suggests that the experiments showing the biphasic dependence on cAMP concentration be reinterpreted in terms of a new model involving three conformational states: free CAP, CAP with two cAMP molecules bound to the *anti* binding site, and CAP with two cAMP molecules bound to the *anti* and two to the *syn* binding sites.

Whether the *syn*-cAMP binding site is physiologically relevant remains unresolved. *In vivo*, does the cAMP concentration near CAP reach the millimolar concentrations necessary for physiological relevance? Alternatively, is *syn*-cAMP binding an artifact only observed *in vitro*?

Perspectives and Conclusions

Great strides have been made in understanding the molecular mechanism of action by CAP and its response to cAMP; however, many questions remain. Despite extensive efforts to crystallize apo-CAP, its structure has not been determined. The nature of the conformational changes that occur upon cAMP binding has been the subject of speculation and investigation [4,21,47]. Recent progress has included both the crystal structure solution of a transcription factor in the CAP family, CooA, without its effector CO [48], and a low-resolution NMR apo-CAP structure [49]. However, a high-resolution apo-CAP structure would be invaluable.

Over the past two decades, structural biology has contributed a great deal to understanding transcription and its regulation. RNA polymerase structures have provided a wealth of information about the basic mechanisms of transcription. Meanwhile, progress in solving the structures of a large number of transcription factor DNA-binding domains bound to their specific sites has elucidated how transcriptional regulators bind DNA elements within promoters. Structural biology has only begun to address how transcriptional regulators, once bound to DNA, interact with the basic transcription apparatus. Understanding the molecular mechanism of CAP activation is an attainable goal and will provide a useful model for transcription activation in general. The structures of both RNA polymerase and the entire CAP protein are known, including the CAP regions thought to contact RNA polymerase. In addition, since there is no chromatin in prokaryotes, issues of DNA accessibility and chromatin remodeling are not relevant for transcription activation by CAP. Solving the prokaryotic DNA-CAP-polymerase structure would make an enormous contribution to understanding transcription activation even in eukaryotes.

Acknowledgments

The author is supported by a special fellowship from the Leukemia and Lymphoma Society. I wish to acknowledge A. K. Aggarwal for support and encouragement. I am grateful to K. Borden, S. Garges, and D. Possen for helpful comments on the manuscript.

References

1. Botsford, J. L. and Harman, J. G. (1992). Cyclic AMP in prokaryotes. *Microbiol Rev.* **56**, 100–122.
2. Kolb, A., *et al.* (1993). Transcriptional regulation by cAMP and its receptor protein. *Annu. Rev. Biochem.* **62**, 749–795.
3. Busby, S. and Ebright, R. H. (1999). Transcription activation by catabolite activator protein (CAP). *J. Mol. Biol.* **293**, 199–213.
4. Harman, J. G. (2001). Allosteric regulation of the cAMP receptor protein. *Biochim. Biophys. Acta* **1547**, 1–17.
5. Harrison, S. C. and Aggarwal, A. K. (1990). DNA recognition by proteins with the helix-turn-helix motif. *Annu. Rev. Biochem.* **59**, 933–969.
6. Steitz, T. A. (1990). Structural studies of protein-nucleic acid interaction: the sources of sequence-specific binding. *Q. Rev. Biophys.* **23**, 205–280.
7. Su, Y., *et al.* (1995). Regulatory subunit of protein kinase A: structure of deletion mutant with cAMP binding domains. *Science* **269**, 807–813.

8. Zagotta, W. N. and Siegelbaum, S. A. (1996). Structure and function of cyclic nucleotide-gated channels. *Annu. Rev. Neurosci.* **19**, 235–263.
9. Flynn, G. E., Johnson, J. P., Jr., and Zagotta, W. N. (2001). Cyclic nucleotide-gated channels: shedding light on the opening of a channel pore. *Nat. Rev. Neurosci.* **2**, 643–651.
10. Monod, J. (1947). The phenomenon of enzymatic adaptation. *Growth.* **11**, 223–289.
11. Magasanik, B. (1961). Catabolite repression. *Cold Spring Harbor Symp. Quant. Biol.* **26**, 249–262.
12. Makman, R. S. and Sutherland, E. W. (1965). Adenosine 3′,5′-phosphate in *Escherichia coli*. *J. Biol. Chem.* **240**, 1309–1314.
13. Ullmann, A. and Monod, J. (1969). Cyclic AMP as an antagonist of catabolite repression in *Escherichia coli*. *FEBS Lett.* **2**, 57–60.
14. Perlman, R. L., De Crombrugghe, B., and Pastan, I. (1969). Cyclic AMP regulates catabolite and transient repression in *E. coli*. *Nature* **223**, 810–812.
15. Peterkofsky, A. and Gazdar, C. (1974). Glucose inhibition of adenylate cyclase in intact cells of *Escherichia coli* B. *Proc. Natl. Acad. Sci. USA* **71**, 2324–2328.
16. Zubay, G., Schwartz, D., and Beckwith, J. (1970). Mechanism of activation of catabolite-sensitive genes: a positive control system" *Proc. Natl. Acad. Sci. USA* **66**, 104–110.
17. Emmer, M. *et al.* (1970). Cyclic AMP receptor protein of *E. coli*: its role in the synthesis of inducible enzymes. *Proc. Natl. Acad. Sci. USA* **66**, 480–487.
18. McKay, D. B. and Steitz, T. A. (1981). Structure of catabolite gene activator protein at 2.9-Å resolution suggests binding to left-handed B-DNA. *Nature* **290**, 744–749.
19. McKay, D. B., Weber, I. T., and Steitz, T. A. (1982). Structure of catabolite gene activator protein at 2.9-Å resolution. *J. Biol. Chem.* **257**, 9518–9524.
20. Weber, I. T. and Steitz, T. A. (1987). Structure of a complex of catabolite gene activator protein and cyclic AMP refined at 2.5 Å resolution. *J. Mol. Biol.* **198**, 311–326.
21. Passner, J. M., Schultz, S. C., and Steitz, T. A. (2000). Modeling the cAMP-induced allosteric transition using the crystal structure of CAP-cAMP at 2.1 Å resolution. *J. Mol. Biol.* **304**, 847–859.
22. Gronenborn, A. M. *et al.* (1981). Conformational selection of *Syn*-cAMP upon binding to the cAMP receptor protein. *FEBS Lett.* **136**, 160–164.
23. Schultz, S. C., Shields, G. C., and Steitz, T. A. (1991). Crystal structure of a CAP-DNA complex: the DNA is bent by 90 degrees. *Science* **253**, 1001–1007.
24. Parkinson, G. *et al.* (1996). Structure of the CAP-DNA complex at 2.5 Angstroms resolution: a complete picture of the protein-DNA interface. *J. Mol. Biol.* **260**, 395–408.
25. Passner, J. M. and Steitz, T. A. (1997). The structure of a CAP-DNA complex having two cAMP molecules bound to each monomer. *Proc. Natl. Acad. Sci. USA* **94**, 2843–2847.
26. Gaston, K. *et al.* (1990). Stringent spacing requirements for transcription activation by CRP. *Cell* **62**, 733–743.
27. Schwartz, M. (1987). In Neidhardt, F. C., Ed., Escherichia coli *and* Salmonella typhimurium. *Cellular and Molecular Biology,* Vol. 2, pp. 1482–1502. American Society for Microbiology, Washington, D.C.
28. Schleif, R. (1992). In McKnight, S. L., and Yamamoto, K. R., Eds., *Transcriptional Regulation*, Vol. 2, pp. 643–665. CSH Press, Cold Spring Harbor, Massachusetts.
29. Aiba, H. (1983). Autoregulation of the *Escherichia coli* CRP gene: CRP is a transcriptional repressor for its own gene. *Cell* **32**, 141–149.
30. Aiba, H. (1985). Transcription of the *Escherichia coli* adenylate cyclase gene is negatively regulated by cAMP-cAMP receptor protein. *J. Biol. Chem.* **260**, 3063–3070.
31. Kallipolitis, B. H., Norregaard-Madsen, M., and Valentin-Hansen, P. (1997). Protein-protein communication: structural model of the repression complex formed by CytR and the global regulator CRP. *Cell* **89**, 1101–1109.
32. Bell, A. *et al.* (1990). Mutations that alter the ability of the *Escherichia coli* cyclic AMP receptor protein to activate transcription. *Nucl. Acids Res.* **18**, 7243–7250.
33. Eschenlauer, A. C. and Reznikoff, W. S. (1991). *Escherichia coli* catabolite gene activator protein mutants defective in positive control of lac operon transcription. *J. Bacteriol.* **173**, 5024–5029.
34. Zhou, Y., Zhang, X., and Ebright, R. H. (1993). Identification of the activating region of catabolite gene activator protein (CAP): isolation and characterization of mutants of CAP specifically defective in transcription activation. *Proc. Natl. Acad. Sci. USA* **90**, 6081–6085.
35. Zhou, Y., Busby, S., and Ebright, R. H. (1993). Identification of the functional subunit of a dimeric transcription activator protein by use of oriented heterodimers. *Cell* **73**, 375–379.
36. Zhou, Y. *et al.* (1994). The functional subunit of a dimeric transcription activator protein depends on promoter architecture. *EMBO J.* **13**, 4549–4557.
37. Williams, R. *et al.* (1991). The role of two surface exposed loops in transcription activation by the *Escherichia coli* CRP and FNR proteins. *Nucl. Acids Res.* **19**, 6705–6712.
38. Ishihama, A. (1992). Role of the RNA polymerase alpha subunit in transcription activation. *Mol. Microbiol.* **6**, 3283–3288.
39. Gross, C. A., Lonetto, M., and Losick, R. (1992). In McKnight, S. L., and Yamamoto, K. R., Eds., *Transcription Regulation*, Vol. 1, pp. 129–176. CSH Press, Cold Spring Harbor, Massachusetts.
40. Kolb, A. *et al.* (1983). On the different binding affinities of CRP at the Lac, Gal, and MalT promoter Regions. *Nucl. Acids Res.* **11**, 7833–7852.
41. Ebright, R. H., Ebright, Y. W., and Gunasekera, A. (1989). Consensus DNA site for the *Escherichia coli* catabolite gene activator protein (CAP): CAP exhibits a 450-fold higher affinity for the consensus DNA than for the *E. coli* lac DNA site. *Nucl. Acids Res.* **17**, 10295–10305.
42. Gaston, K., Kolb, A., and Busby, S. (1989). Binding of the *Escherichia coli* cyclic AMP receptor protein to DNA fragments containing consensus nucleotide sequences. *Biochem. J.* **261**, 649–653.
43. Heyduk, T. and Lee, J. C. (1990). Application of fluorescence energy transfer and polarization to monitor *Escherichia coli* cAMP receptor protein and lac promoter interaction. *Proc. Natl. Acad. Sci. USA* **87**, 1744–1748.
44. Heyduk, E., Heyduk, T., and Lee, J. C. (1992). Intersubunit communications in *Escherichia coli* cyclic AMP receptor protein: studies of the ligand binding domain. *Biochemistry* **31**, 3682–3688.
45. Pyles, E. A. and Lee, J. C. (1996). Mode of selectivity in cyclic AMP receptor protein-dependent promoters in *Escherichia coli*. *Biochemistry* **35**, 1162–1172.
46. Crothers, D. M. and Steitz, T. A. (1992). In McKnight, S. L. and Yamamoto, K. R., Eds., *Transcriptional Regulation*, Vol. 1, pp. 501–534. CSH Press, Cold Spring Harbor, Massachusetts.
47. Adhya, S., Ryu, S., and Garges, S. (1995). Role of allosteric changes in cyclic AMP receptor protein function. *Subcell. Biochem.* **24**, 303–321.
48. Lanzilotta, W. N. *et al.* (2000). Structure of the CO sensing transcription activator CooA. *Nat. Struct. Biol.* **7**, 876–880.
49. Won, H. S. *et al.* (2000). Structural understanding of the allosteric conformational change of cyclic AMP receptor protein by cyclic AMP binding. *Biochemistry* **39**, 13953–13962.

Cyclic Nucleotide Signaling in Paramecium

Jürgen U. Linder and Joachim E. Schultz
*Department of Pharmaceutical Biochemistry,
Pharmaceutical Institute, University of Tübingen,
Tübingen, Germany*

Introduction

The unicellular protozoan *Paramecium* uses both cyclic nucleotide second messengers, cAMP and cGMP, for signal transduction. Contrary to mammalian cells, in which cyclic nucleotide formation is under hormonal control, the primary input in *Paramecium* appears to be membrane potential and ion currents. Studies showed that regulation of cyclic nucleotide levels takes place at the site of formation, that is, by direct modulation of adenylyl cyclase (AC) and guanylyl cyclase (GC) activities. Molecular analyses revealed that the architecture of the respective cyclase enzymes differs from their mammalian congeners, leading to new concepts for the coupling of signals to second messenger generation. This chapter concentrates on the features of *Paramecium* AC and GC. Events downstream of cyclic nucleotide formation will be discussed only briefly, as they have been investigated to a much lesser extent. See Fig. 1 for a summary.

cAMP Formation and Adenylyl Cyclase

In the 1980s regulation of cAMP levels in *Paramecium* has been investigated by exposing cells to sudden changes of the extracellular ion milieu [1]. Increases in cAMP content are strictly dependent on hyperpolarization, e.g. elicited by a dilution of external K^+. Upon an eight-fold dilution of K^+ cAMP levels increase up to four-fold within 5 to 10 seconds and then decline to a new steady-state level within few minutes. This behavior has the properties of an adaptive response. cAMP formation is strictly coupled to a K^+-conductance by which K^+-ions exit the cell upon dilution [2] because (1) specific K^+-channel blockers such as tetraethylammonium or Cs^+ abolish the cAMP response; (2) the cAMP response correlates with the K^+ concentration that the cells have been adapted to prior to the stimulus; upon dilution from 16 mM K^+ in the equilibration buffer where the K^+ resting conductance is high, the increase in cAMP is maximal whereas upon dilution from 1 mM K^+ in the equilibration medium where the resting conductance is low, the cAMP response is negligible; (3) The mutant *restless* which cannot control its K^+ resting conductance [3], shows an exaggerated cAMP response even when cells are adapted to low external K^+.

Subsequently the stimulation of cAMP formation has been investigated biochemically via purification of *Paramecium* AC [2,4]: *Paramecium* AC activity is exclusively membrane-bound and requires detergents for solubilization. AC activity is high in ciliary membranes (0.5 nmol cAMP/(mg·min)) and increases to 25 μmol/(mg·min) upon purification to homogeneity. Paramecium AC is a 97 kDa protein with ion pore-forming properties: upon reconstitution into black lipid bilayers an intrinsic cation-specific conductance of 320 ± 60 pS has been observed. The pore displays a slight preference for K^+ and is impermeable to tetraethylammonium. The unique bifunctional property of the *Paramecium* enzyme as an ion channel-AC suggests that the AC activity is regulated by an intrinsic K^+-conductance. This AC may be involved in setting the resting membrane potential and thus behaves like a biological ammeter. Such a view is compatible with the observation that cAMP formation upon K^+-dilution is

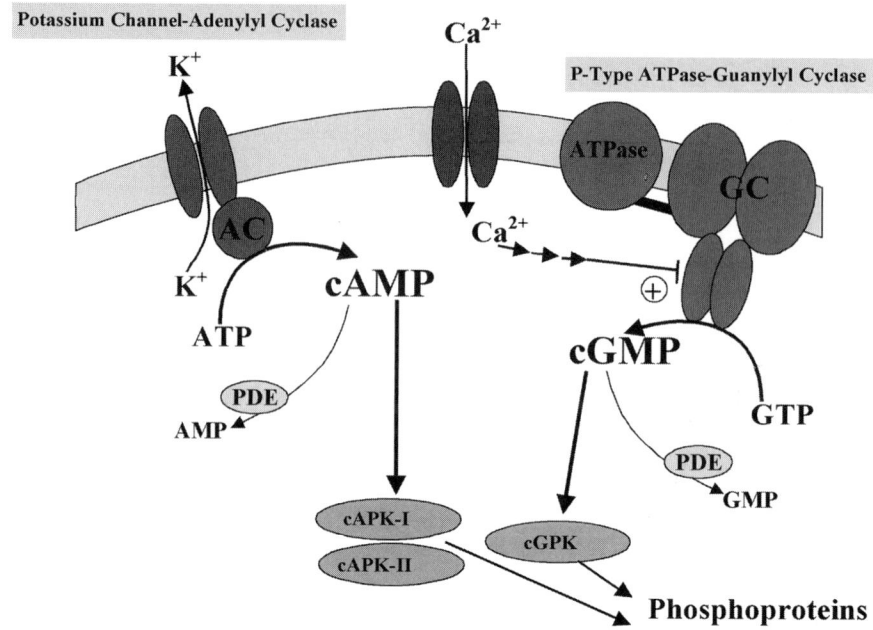

Figure 1 Overview of cyclic nucleotide signaling in *Paramecium*. PDE: cyclic nucleotide phosphodiesterase; cAPK/cGPK: cyclic AMP/GMP-dependent protein kinase; AC/GC: adenylyl/guanylyl cyclase. The light grey ribbon represents the cell membrane with incorporated channels and integral membrane domains. The P-type ATPase-like domain of *Paramecium* GC is connected by a cytosolic peptide linker to the GC-domain. The *Paramecium* AC most likely consists of a K^+-channel and a catalytic domain with a yet unknown multimeric composition.

dependent on the K^+ resting conductance set by adaptation to the respective equlibration buffer (see above).

Recently we cloned a cDNA from *Paramecium* coding for a 98 kDa protein consisting of three domains (unpublished): an N-terminal ion channel domain linked to an AC catalytic domain and a single C-terminal tetratricopeptide unit. Thus the predicted protein correlates with the previous physiological and biochemical data. The functional properties of the protein have yet to be established.

Some chemoattractants cause hyperpolarization and thus elicit a transient cAMP response [5], e.g. L-glutamate. However, hyperpolarization brought about by addition of NH_4^+ or acetate ions has no effect on cAMP levels. Currently, it is unclear whether the response to L-glutamate is via stimulation of AC or via inhibition of cAMP-phosphodiesterase.

Guanylyl Cyclase and cGMP Formation

Paramecium contains a membrane-bound GC activity about 10 percent of which localize to ciliary membranes [6,7]. Soluble, that is cytosolic, GC activity has never been detected. Ca^{2+}-ions play a prominent role in the regulation of the protozoan GC as removal of Ca^{2+} greatly diminishes activity [4,8]. The inhibition is specifically relieved by readdition of Ca^{2+} ions. Sr^{2+} is less effective and Ba^{2+} ions fail to reactivate [8]. The Ca^{2+}-dependence of the GC *in vitro* fits reasonably well the boost-like increase of cGMP levels *in vivo* upon a stimulated Ca^{2+} influx. One route by which Ca^{2+} can enter *Paramecium* is a depolarization-activated, voltage-gated Ca^{2+} channel activated e.g. by an increase of extracellular K^+. Channel inactivation is rapid within a few milliseconds, and the global cGMP content rises only by about 45 percent under these conditions [9]. In contrast, a depolarization caused by Ba^{2+}-addition results in a more sustained Ca^{2+}-entry and a substantial, yet short-lived increase in cGMP [9]. Mutants with defects in the depolarization-activated Ca^{2+} conductance show altered cGMP responses. *Pawn* mutants, which lack the depolarization-gated Ca^{2+}-influx, do not display a significant cGMP increase upon treatment with Ba^{2+} [9,10]. On the other hand, *dancer* cells in which the Ca^{2+} channel is very slowly inactivating after opening, show a sustained cGMP response toward Ba^{2+}-depolarization and, in contrast to the wild-type, cGMP levels in *dancer* cells rise three-fold upon a K^+-depolarization [11]. An alternate route of Ca^{2+} entry is activated by chemical hyperpolarization, e.g. by the drug amiloride, which itself does not affect GC activity *in vitro* [12]. Upon amiloride addition cGMP-levels increase 6 to 30-fold within 10 to 20 sec, dependent on extracellular Ca^{2+}. *Pawn* mutants show an identical cGMP-response to amiloride, suggesting that the hyperpolarization-activated conductance is physically distinct from the depolarization-gated Ca^{2+}-channel opened by Ba^{2+}.

The molecular architecture of *Paramecium* GC has been elucidated via a homology-cloning approach [13]. A 7.2 kb cDNA was obtained coding for a 280 kDa integral membrane protein of two domains. The N-terminal domain of 155 kDa is similar to P-type ion transport ATPases and is joined by a cytoplasmic linker to the C-terminal 115 kDa guanylyl cyclase domain. The calculated topology of the GC domain is identical with that of mammalian membrane-bound ACs (see previous chapters). It contains two sets of six transmembrane helices, which are linked to the two catalytic subdomains, called C1a and C2. The C1a subdomain contains the crucial amino acids required for substrate recognition and transition state stabilization and the C2 subdomain contains the two metal–cofactor binding aspartates. Thus the functions of C1a and C2 in *Paramecium* GC are inverse compared to mammalian ACs, where C1a binds Mg^{2+} and C2 takes over the other functionalities. The P-type ATPase-like domain contains ten transmembrane helices and harbors the invariant DKTGT(L/I)T signature motif of this class of transporters. However, several crucial deviations from functionally characterized consensus motifs indicate that this domain is not an active transporter but probably has adopted a new, as yet unrecognized function [13].

Due to the unorthodox genetic code in use in *Paramecium* a heterologous expression of the GC requires that the expression cassettes are adjusted to universal codon usage [13]. This way, a specific GC activity of up to 1 nmol cGMP/(mg·min) is detected in Sf9 cells expressing the C-terminal GC-domain. Cells expressing the holoenzyme yield about 0.5 nmol cGMP/(mg·min). A Western blot revealed, however, that in Sf9 cells the holoenzyme is cleaved into two large polypeptides.

The C1a and C2 catalytic subdomains of the *Paramecium* GC can be individually expressed in *E. coli* as soluble proteins, as previously demonstrated for the C1a and C2 domains of the mammalian ACs [14–16]. Alone the affinity-purified subdomains are inactive, yet a mixture of C1a and C2 reconstitutes a robust GC activity [17]. This system was then used for a further characterization of the catalyst. Based on x-ray and mutagenesis data from the mammalian AC/GC enzymes [18–20], Glu1681 and Ser1748 of *Paramecium* C1a were predicted to be key residues for GTP substrate specificity. Indeed mutation of both residues that specify the ATP substrate in the mammalian ACs, i.e. E1681K and S1748D, converts the *Paramecium* GC-C1a/C2 heterodimer into a specific AC [17].

Taken together the *Paramecium* GC is a unique protein assembly wherein an ion pump-like domain is fused to a classical mammalian AC domain in which the substrate specificity was modified during its evolution.

Downstream of Cyclic Nucleotide Formation

Intracellular levels of cyclic nucleotides reflect the balance between synthesis by the cyclases and hydrolysis by phosphodiesterases (PDE). In *Paramecium* PDE activity is high [21]. So far, the regulation of PDE activity has not been investigated in sufficient detail. However, the sharp rise and rapid fall in cGMP levels upon Ba^{2+}-depolarization vividly demonstrates the highly dynamic equilibrium of cGMP-formation and hydrolysis. Indeed it is estimated that the cellular cGMP turns over every two seconds in *Paramecium* [22].

The only known primary effectors for cyclic nucleotides in *Paramecium* are cAMP- and cGMP-dependent protein kinases. Two forms of cAMP-dependent protein kinases, cAPK-I (70 kDa) and cAPK-II (220 kDa), have been partially purified [23–25]. Both forms contain a single catalytic (C) and regulatory subunit (R). They differ in their R-subunits, which are subject to autophosphorylation *in vitro* [26]. The cloned subunit of cAPK-I has up to 38 percent identity to known R-subunits from other unicells and mammalian tissues. In agreement with the molecular weight of the protozoan cAPK-I an N-terminal dimerization region is missing. This explains RC-dimer composition in contrast to the R_2C_2 tetrameric structure in mammals [27]. Further, a cGMP-dependent protein kinase (cGPK) has been partially purified and correlated to a single 77 kDa protein [28]. It appears to autophosphorylate *in vitro* and uses ATP as well as GTP as phosphoryl donors. Numerous proteins appear to be phosphorylated in *Paramecium* in a cGMP and/or cAMP-dependent manner *in vitro* whose identity and function largely remains to be identified [29,30]. Some work has concentrated on a 29 kDa dynein light chain that is a substrate for cAMP-dependent phosphorylation *in vitro* and appears to be a component of the 22S axonemal dynein [31–34]. It has been suggested that this protein may regulate dynein function in *Paramecium* as hyperpolarization increases swimming speed and cAMP formation and because phosphorylation of the 29 kDa protein regulates microtubule translocation *in vitro*. It is interesting that the cAMP response to hyperpolarization is rather transient whereas the increase in swimming speed clearly extends to longer periods of time, demonstrating a lasting signal transduction.

With the advance of several genome projects targeting protozoans of the parvkingdom of Alveolata, including *Tetrahymena* and the apicomplexan parasites *Plasmodium* and *Cryptosporidium*, it is noteworthy that similar cGMP and cAMP signal transduction machineries seem to exist in all members of this parvkingdom, among those several of extremely high pathogenic potential. The studies with *Paramecium* may, therefore, turn out to be a valuable starting point for investigating signal transdcution in this group of protozoans [35].

Acknowledgments

Our work is funded by the Deutsche Forschungsgemeinschaft and the Fonds der Deutschen Industrie.

References

1. Schultz, J. E., Grünemund, R., von Hirschhausen, R., and Schönefeld, U. (1984). Ionic regulation of cyclic AMP levels in *Paramecium tetraurelia* in vivo. *FEBS Lett.* **167**, 113–116.
2. Schultz, J. E., Klumpp, S., Benz, R., Schürhoff-Goeters, W. J., and Schmid, A. (1992). Regulation of adenylyl cyclase from *Paramecium* by an intrinsic potassium conductance. *Science* **255**, 600–603.
3. Richard, E. A., Hinrichsen, R. D., and Kung, C. (1985). A single gene mutation that affects a potassium conductance and resting membrane potential in *Paramecium*. *J. Neurogenet.* **2**, 239–252.

4. Klumpp, S., Gierlich, D., and Schultz, J. E. (1984). Adenylate cyclase and guanylate cyclase in the excitable ciliary membrane from *Paramecium*: separation and regulation. *FEBS Lett.* **171**, 95–99.
5. Yang, W. Q., Braun, C., Plattner, H., Purvee, J., and Van Houten, J. L. (1997). Cyclic nucleotides in glutamate chemosensory signal transduction of *Paramecium*. *J. Cell Sci.* **110**, 2567–2572.
6. Schultz, J. E. and Klumpp, S. (1980). Guanylate cyclase in the excitable ciliary membrane of *Paramecium*. *FEBS Lett.* **122**, 64–66.
7. Schultz, J. E. and Klumpp, S. (1991). Calcium-regulated guanylyl cyclases from *Paramecium* and *Tetrahymena*. *Methods Enzymol.* **195**, 466–474.
8. Klumpp, S. and Schultz, J. E. (1982). Characterization of a Ca^{2+}-dependent guanylate cyclase in the excitable ciliary membrane from *Paramecium*. *Eur. J. Biochem.* **124**, 317–324.
9. Schultz, J. E., Pohl, T., and Klumpp, S. (1986). Voltage-gated Ca^{2+} entry into *Paramecium* linked to intraciliary increase in cyclic GMP. *Nature* **322**, 271–273.
10. Oertel, D., Schein, S. J., and Kung, C. (1977). Separation of membrane currents using a *Paramecium* mutant. *Nature* **268**, 120–124.
11. Hinrichsen, R. D. and Saimi, Y. (1984). A mutation that alters properties of the calcium channel in *Paramecium tetraurelia*. *J. Physiol.* **351**, 397–410.
12. Schultz, J. E., Guo, Y., Kleefeld, G., and Völkel, H. (1997). Hyperpolarization- and depolarization-activated Ca^{2+} currents in *Paramecium* trigger behavioral changes and cGMP formation independently. *J. Membr. Biol.* **156**, 251–259.
13. Linder, J. U., Engel, P., Reimer, A., Krüger, T., Plattner, H., Schultz, A., and Schultz, J. E. (1999). Guanylyl cyclases with the topology of mammalian adenylyl cyclases and an N-terminal P-type ATPase-like domain in *Paramecium*, *Tetrahymena* and *Plasmodium*. *EMBO J.* **18**, 4222–4232.
14. Tang, W. J. and Gilman, A. G. (1995). Construction of a soluble adenylyl cyclase activated by Gs alpha and forskolin. *Science* **268**, 1769–1772.
15. Dessauer, C. W. and Gilman, A. G. (1996). Purification and characterization of a soluble form of mammalian adenylyl cyclase. *J. Biol. Chem.* **271**, 16967–16974.
16. Whisnant, R. E., Gilman, A. G., and Dessauer, C. W. (1996). Interaction of the two cytosolic domains of mammalian adenylyl cyclase. *Proc. Natl. Acad. Sci. USA* **93**, 6621–6625.
17. Linder, J. U., Hoffmann, T., Kurz, U., and Schultz, J. E. (2000). A guanylyl cyclase from *Paramecium* with 22 transmembrane spans. Expression of the catalytic domains and formation of chimeras with the catalytic domains of mammalian adenylyl cyclases. *J. Biol. Chem.* **275**, 11235–11240.
18. Tesmer, J. J., Sunahara, R. K., Gilman, A. G., and Sprang, S. R. (1997). Crystal structure of the catalytic domains of adenylyl cyclase in a complex with $G_{s\alpha}$·GTPγS. *Science* **278**, 1907–1916.
19. Zhang, G., Liu, Y., Ruoho, A. E., and Hurley, J. H. (1997). Structure of the adenylyl cyclase catalytic core. *Nature* **386**, 247–253.
20. Sunahara, R. K., Beuve, A., Tesmer, J. J., Sprang, S. R., Garbers, D. L., and Gilman, A. G. (1998). Exchange of substrate and inhibitor specificities between adenylyl and guanylyl cyclases. *J. Biol. Chem.* **273**, 16332–16338.
21. von Hirschhausen, R. (1986). "Die Phosphodiesterasen in *Paramecium tetraurelia*." MD Thesis, University of Tübingen.
22. Schultz, J. E. and Schade, U. (1989). Calcium channel activation and inactivation in *Paramecium* biochemically measured by cyclic GMP production. *J. Membr. Biol.* **109**, 251–258.
23. Mason, P. A. and Nelson, D. L. (1989). Cyclic AMP-dependent protein kinases of *Paramecium*. I. Chromatographic and physical properties of the enzymes from cilia. *Biochim. Biophys. Acta* **1010**, 108–115.
24. Mason, P. A. and Nelson, D. L. (1989). Cyclic AMP-dependent protein kinases of *Paramecium*. II. Catalytic and regulatory properties of type II kinase from cilia. *Biochim. Biophys. Acta* **1010**, 116–121.
25. Hochstrasser, M. and Nelson, D. L. (1989). Cyclic AMP-dependent protein kinase in *Paramecium tetraurelia*. Its purification and the production of monoclonal antibodies against both subunits. *J. Biol. Chem.* **264**, 14510–14518.
26. Hochstrasser, M., Carlson, G. L., Walczak, C. E., and Nelson, D. L. (1996). *Paramecium* has two regulatory subunits of cyclic AMP-dependent protein kinase, one unique to cilia. *J. Eukaryot. Microbiol.* **43**, 356–362.
27. Carlson, G. L. and Nelson, D. L. (1996). The 44-kDa regulatory subunit of the *Paramecium* cAMP-dependent protein kinase lacks a dimerization domain and may have a unique autophosphorylation site sequence. *J. Eukaryot. Microbiol.* **43**, 347–356.
28. Miglietta, L. A. and Nelson, D. L. (1988). A novel cGMP-dependent protein kinase from *Paramecium*. *J. Biol. Chem.* **263**, 16096–16105.
29. Bonini, N. M. and Nelson, D. L. (1990). Phosphoproteins associated with cyclic nucleotide stimulation of ciliary motility in *Paramecium*. *J. Cell Sci.* **95**, 219–230.
30. Lewis, R. M. and Nelson, D. L. (1981). Biochemical studies of the excitable membrane of *Paramecium tetraurelia* VI. Endogenous protein substrates for in vitro and in vivo phosphorylation in cilia and ciliary membranes. *J. Cell Biol.* **91**, 167–174.
31. Walczak, C. E. and Nelson, D. L. (1993). In vitro phosphorylation of ciliary dyneins by protein kinases from *Paramecium*. *J. Cell. Sci.* **106**, 1369–1376.
32. Barkalow, K., Hamasaki, T., and Satir, P. (1994). Regulation of 22S dynein by a 29-kD light chain. *J. Cell Biol.* **126**, 727–735.
33. Hamasaki, T., Barkalow, K., Richmond, J., and Satir, P. (1991). cAMP-stimulated phosphorylation of an axonemal polypeptide that copurifies with the 22S dynein arm regulates microtubule translocation velocity and swimming speed in *Paramecium*. *Proc. Natl. Acad. Sci. USA* **88**, 7918–7922.
34. Wang, H. and Satir, P. (1998). The 29 kDa light chain that regulates axonemal dynein activity binds to cytoplasmic dyneins. *Cell Motil. Cytoskeleton* **39**, 1–8.
35. Cavalier-Smith, T. (1993). Kingdom protozoa and its 18 phyla. *Microbiol. Rev.* **57**, 953–994.

CHAPTER 210

Cyclic Nucleotide Signaling in Trypanosomatids

Roya Zoraghi* and Thomas Seebeck
*Institute of Cell Biology, University of Bern,
Bern, Switzerland*

Introduction

The trypanosomatids (order: kinetoplastida) include the causative agents of human sleeping sickness and of the cattle disease nagana in Sub-Saharan Africa (*Trypanosoma brucei*), a host of diseases of camels, water buffaloes, and horses (*T. evansi, T. equinum* and *T. equiperdum*); Chagas disease in South and Middle America (*T. cruzi*); and Kala Azar (*Leishmania donovani*) and a host of other human Leishmanial infections worldwide.

Cyclic nucleotide signaling in parasitic protozoa has lately become an attractive field of research, not at the least because many experimental approaches have been greatly facilitated by the advanced stage of several of the genome projects (e.g. *T. brucei*: http://parsun1.path.cam.ac.uk; *T. cruzi*: http://www.dbbm.fiocruz.br/TcruziDB; *Leishmania major*: http://www.ebi.ac.uk/parasites/leish.html).

Cyclic Nucleotide Signaling, Cell Proliferation, and Differentiation

During *in vitro* differentiation of *T. brucei* from bloodstream forms to the insect stage (procyclics),[1] two peaks of adenylyl cyclase (AC) activity were observed. The first peak occurred 6–10 hr after triggering differentiation, before the first cell division. A second peak was observed when the cells emerged from the first division and before they began to proliferate [1,2]. However, an independent study using a different trypanosome strain found no involvement of cAMP in differentiation *in vitro* [3]. During animal infections, the intracellular cAMP levels of the trypanosomes increased from the early stages of the infection to the peak of parasitaemia and then decreased as differentiation from long slender to short stumpy forms began [4,5].[2,3] These early observations were recently corroborated by *in vitro* experiments [6]. When cultured bloodstream form trypanosomes reached a threshold density, they secreted a low-molecular-mass factor (SIF: stumpy-inducing factor), which induced an increase in intracellular cAMP. This increase was followed by cell cycle arrest, and resulted in a high efficiency of differentiation to stumpy forms. Membrane-permeable cAMP analogs and the PDE inhibitor etazolate mimicked SIF activity, indicating that SIF acts via cAMP and that the elevated intracellular cAMP is a signal for differentiation. Similarly, membrane-permeable cAMP analogs, PDE inhibitors, or the inactivation of specific PDEs by RNA interference were shown to elevate intracellular cAMP and to concomitantly inhibit proliferation of cultured bloodstream forms of *T. brucei* [7,8].

---

*Current address: Dept. of Molecular Physiology and Biophysics, Vanderbilt University, Nashville, TN 37232-0615.
[1]Procyclics: proliferative form of *T. brucei* adapted to the environment of the tsetse fly midgut.
[2]Long slender forms of *T. brucei*: proliferative form of the trypanosomes in the blood of mammalian hosts.
[3]Short stumpy forms of *T. brucei*: long slender forms eventually differentiate into short stumpy forms, which are pre-adapted to survival in the tsetse fly vector. Short stumpy forms do no longer proliferate and have a limited lifetime in the mammalian bloodstream.

In *T. cruzi*, increased cAMP was also found to downregulate cell proliferation. High intracellular cAMP levels inhibited DNA, RNA, and protein synthesis [9]. Fetal calf serum, which is mitogenic for *T. cruzi*, was shown to decrease intracellular cAMP [10].

However, increased cAMP appears to play a role in differentiation of *T. cruzi*. Prior to differentiation to metacyclic forms, the intracellular cAMP levels increased four-fold.[4] In agreement with these findings, exogenous cAMP, cAMP analogs, and the PDE inhibitor papaverine induced differentiation *in vitro* [11]. In *T. cruzi*, the steady-state level of the mRNA of an unidentified gene, TC26, was elevated during differentiation to metacyclics or after induction by exogenous cAMP analogs. In a strain of *T. cruzi* that was unable to undergo differentiation, TC26 was not expressed [12]. Recent advances in the *T. cruzi* genome project have demonstrated that TC26 represents a family of repetitive elements.

In *L. donovani*, cAMP and PDE inhibitors blocked the transformation from the intracellular amastigote stage to the extracellular promastigote form.[5,6] In *L. tropica*, caffeine was shown to increase intracellular cAMP levels and to reduce proliferation [13]. Yet binding of a specific component of human serum, C-reactive protein, to the surface of *L. mexicana* promastigotes induced differentiation to the amastigote form, possibly through a cAMP mediated signaling cascade [14].

Individual Components of the Cyclic Nucleotide Signaling Pathways

Cyclases

In striking contrast to the situation in higher organisms, the very small genomes of *T. brucei* and *T. cruzi* (about 40 Mbp) contain a large number of genes for different adenylyl cyclases [15,16]. In *T. brucei*, two types of AC genes can be discerned in terms of their genomic organization. Each telomeric VSG expression site contains a number of expression-site associated genes (ESAGs), one of which is always an AC [17].[7] All other AC genes (GRESAGs, genes related to ESAG4) are scattered throughout the genome and are present either as single copy genes or as small gene families.

---

[4]Metacyclic forms: the mammalian-infective forms of *T. brucei* and *T. cruzi*, which are deposited by the respective insect vectors in the blood of the mammalian host.

[5]Amastigote: intracellular, nonflagellated forms of *T. cruzi* and *Leishmania ssp.*, which replicate within mammalian host cells.

[6]Promastigote: extracellular, flagellated forms of *T. cruzi* and *Leishmania ssp.*

[7]VSG (variable surface glycoprotein) of *T. brucei*: a tightly packed layer of single type of such a protein covers the entire surface of *T. brucei*. The genome contains several hundred genes for different VSGs. Only one of these is active at any given time, and activation switches stochastically between individual genes with a low frequency.

All ACs from *T. brucei* [18–21], *T. cruzi* [16], *L. donovani* [22], and *T. equiperdum* [23] for which sequence information is available exhibit a similar predicted structure. A short N-terminal leader sequence is followed by a large extracellular domain that bears no significant similarity to other proteins or among the different trypanosomal ACs. This putative extracellular receptor domain is connected via a single transmembrane helix to the intracellular catalytic domain [24]. The catalytic domains are strongly conserved, not only between kinetoplastid ACs, but also between the kinetoplastid and mammalian ACs [25,26] and are presumably activated by dimerization [22,24,26]. The AC activity of *T. brucei* is insensitive to agents known to activate mammalian cyclases, such as GTP or GTP analogs, forskolin, or cholera and pertussis toxins [1]. The AC activity in bloodstream form *T. brucei* can be transiently activated by Ca^{2+}, whereas no such effect could be detected in procyclic form [27].

The overall structure of the trypanosomal ACs is reminiscent of that of the mammalian membrane-bound receptor guanylyl cyclases [28]. The large number of different trypanosomal ACs, all with different N-terminal (extracellular) domains, and their structural similarity with the receptor guanylyl cyclases suggest that the trypanosomal ACs may serve as extracellular receptors [15]. As trypanosomatids contain neither G-protein-coupled receptors nor heterotrimeric G proteins (see below), the ACs may serve as enzyme-linked receptors in an alternative paradigm for chemical sensing. The functional correlate to the structural similarity between trypanosomal ACs and the metazoan receptor GCs might be reflected by the vertebrate GC-D olfactory neurons. In this subpopulation of the olfactory receptor neurons, membrane-bound receptor GCs serve as olfactory receptors, instead of the G-protein coupled receptors expressed in the majority of olfactory receptor neurons [29].

Cyclic Nucleotide-Specific Phosphodiesterases

In *T. brucei*, PDE activity in lysates of bloodstream forms was first demonstrated over 20 years ago [30]. More recent work has now identified at least two different PDE families in *T. brucei*. TbPDE1 is coded for by a single-copy gene and represents a class I PDE with an unusually high K_m for its specific substrate, cAMP (Kunz *et al.*, submitted). This PDE is not essential for proliferation of *T. brucei* in culture, nor for the infection of the tsetse fly vector [31]. Three members of a second class I family of cAMP-specific PDEs of *T. brucei* have recently been identified and characterized (TbPDE2A [7]; TbPDE2B: [7a] TbPDE2C [8]). They share highly conserved catalytic domains but differ in their N-terminal regulatory regions [15], each of which contain one or two GAF domains [32]. All TbPDE2 family members characterized so far are highly specific for cAMP and exhibit a low K_m, and their activity is not affected by cGMP. Several broad-spectrum PDE inhibitors such as the methyl-xanthines are completely inactive toward the TbPDE2 enzymes. Dipyridamole, trequinsin, sildenafil, and ethaverine inhibit the recombinant enzymes and block proliferation of bloodstream trypanosomes in culture [7]. RNA interference experiments confirmed that inactivation of TbPDE2 family members is lethal to bloodstream trypanosomes in culture [8], suggesting that this PDE family might constitute an interesting drug target.

Two kinds of cAMP-specific PDE activities were detected in *T. cruzi*. One enzyme was a soluble, cAMP-specific PDE with a K_m of 40 μM, and it exhibited full activity at pH 8.0 and in the presence of 5 mM Mn^{2+}. The enzymatic activity was not Ca^{2+} dependent, and was not inhibited by theophylline and caffeine [33]. A second PDE activity was purified by affinity chromatography on a brain calmodulin-Sepharose column. Activation required micromolar concentrations of Ca^{2+}, and activity was blocked by EGTA and by calmodulin inhibitors [34].

A cAMP PDE activity was also detected in *L. tropica* and *L. donovani*, which required Mg^{2+} for full activation [35]. A cAMP-specific PDE activity from *L. mexicana* was recently characterized [36]. This PDE activity was found both in the cytoplasm and on the outer cell surface. It was identified as a 60 kDa protein on SDS-PAGE, exhibited an unusually high K_m of 277 μM for cAMP, and required Mg^{2+} for maximal activity. Based on its high K_m, the enzyme might represent a class II PDE, or it may represent the leishmanial homolog of TbPDE1 of *T. brucei*.

Protein Kinase A

In *T. brucei*, three genes were predicted to code for different catalytic subunits of PKA homologs [37]. The single-copy gene of a regulatory subunit of PKA has recently been identified and characterized. Although the overall sequence organization of the protein is well conserved with respect to its mammalian homologs, the PKA holoenzyme from *T. brucei* could only be activated by cGMP but not by cAMP. When expressed as recombinant proteins, both cyclic-nucleotide-binding domains of the regulatory subunit did bind cGMP, and this binding was not competed by cAMP [38]. The K_ds for cGMP binding to both of the domains were in the 10 μM range, and its physiological significance remains to be established.

In *T. cruzi*, both the regulatory and catalytic subunits of PKA have been purified from cultured cells. The putative catalytic subunit appeared to be similar to the bovine heart PKA catalytic subunit and cross-reacted with an antibody against the bovine enzyme. The putative regulatory subunit inhibited the catalytic subunit of bovine heart PKA, and this inhibition could be reversed by cAMP. Reconstitution experiments with the two purified proteins resulted in a holoenzyme with an activity similar to that of bovine heart PKA holoenzyme. The molecular weight of the reconstituted holoenzyme suggested a tetrameric structure [39,40].

A presumptive catalytic subunit of PKA was purified from *L. donovani*. Its activity could be inhibited by the regulatory subunit of bovine heart PKA, and the inhibition could be reversed by cAMP. A heat-stable porcine heart PKA inhibitor also inhibited its activity [41].

Cyclic Nucleotides and Host Parasite Intervention

In the bloodstream of the host, *T. brucei* cells are faced with a massive antibody response. As a potential countermeasure, they have acquired a mechanism for active disaggregation when agglutinated by VSG-specific antibodies [42]. The disaggregation mechanism is energy-dependent, does not result in the proteolysis of the bound antibody, and does not involve the shedding of VSG from the cell surface. The mechanism is modulated by PKA and cAMP. This suggests that a cAMP-signaling cascade might be triggered by antibody binding to the VSG coat. This cascade eventually results in the evasion of *T. brucei* from the onslaught of the host's anti-VSG antibody response.

Infection of mammalian cells by *T. cruzi* involves the mobilization of intracellular Ca^{2+} and the elevation of cAMP in the host cells. Parasites that lack the oligopeptidase B required to trigger the Ca^{2+} release in the host cell are still able to induce the elevation of cAMP in the host cell, and to infect them, albeit more slowly [43]. In addition, cGMP was also shown to play a role in modulating macrophage susceptibility to *T. cruzi* infection. Increasing the intracellular cGMP level in macrophages resulted in a marked increase in the number of parasites associated with the cells and in the percentage of infected cells. Similar pretreatments of the parasites had no effect on the host-parasite interaction [44]. The first step of the invasion of *T. cruzi* into its host cells consists in the adhesion to the host membrane. Attachment of infective trypanosomes to host muscle cell sarcolemma resulted in the rapid inhibition of parasite AC activity and significantly reduced intracellular levels of cAMP in the parasite [45].

L. tropica and *L. donovani* secrete a soluble factor that inhibited mammalian AC activity whereas it had no effect on the AC activity of the parasite [46]. cAMP was reported to modulate the intracellular superoxide dismutase activity in *L. donovani*, resulting in a reduction of parasite survival in the intraphagolysosomal environment [47].

Concluding Remarks

The current state of the study of cyclic nucleotide signaling in trypanosomatids still is very patchy. Nevertheless, the available results already provide a glimpse on the contours of the larger picture. One important conclusion that already can be drawn from the available data is that the trypanosomatids differ widely from their mammalian hosts in some, though not all, aspects of their cyclic nucleotide signaling pathways. Although some enzymes involved in these pathways, such as the PDEs that were investigated in great detail in mammals, can similarly be detected in trypanosomatids, amazing differences are seen with others. No G-protein-coupled receptors, heterotrimeric G proteins, or G-protein-activated adenylyl cylases were found in trypanosomatids. The structure of their adenylyl cyclases, of which they express a large number of different isoenzymes, is entirely different from that of their mammalian counterparts, and it closely resembles mammalian receptor guanylyl cyclases. In trypanosomatids, the receptor adenylyl cyclases may replace the G-protein-coupled receptors of higher eukaryotes as the major sensory system.

Such novel structures in old pathways might result from the evolutionary adaptation of the trypanosomatids to their parasitic lifestyles, or from the evolutionary distance between them and their mammalian and insect hosts. The three groups of organisms are evolutionarily far apart; the trypanosomatids are among the oldest of the extant eukaryotes, whereas insects and mammals represent much more recent developments. They also might simply reflect a degree of evolutionary diversity that has remained largely undiscovered and unexplored due to an overly narrow focusing of contemporary cell biological studies to a very small number of model organisms such as man, mice, or *Drosophila*.

Studying the parasite-specific wrinkles of cyclic nucleotide signaling pathways not only is a fascinating challenge for the adventurous scientific mind, but it also holds great promise from a practical standpoint. For many of the major parasitic diseases, the current state of the art of chemotherapy is deplorable. The identification of new signaling molecules that differ significantly between the cyclic nucleotide signaling pathways of hosts and parasites might open up a treasure trove of new potential drug targets, not only for trypanosomatid-caused diseases, but just as well for many other protozoal diseases such as malaria, toxoplasmosis, or amebiasis.

However, the pharmacology of the PDEs, a class of enzymes that are similar between mammals and trypanosomatids, has been highly developed for the human enzymes. PDE inhibitors are currently on the market or in the pipelines as medication for an increasing number of clinical applications, and potency as well as isoenzyme and subtype specificity are ever increasing. This extensive know-how on developing inhibitors for mammalian PDEs could and should be applied to their homologs from trypanosomatids and other protozoal pathogens, possibly resulting in an assortment of effective and highly selective antiparasitic compounds.

References

1. Rolin, S., Hanocq-Quertier, J., Paturiax-Hanocq, F., Nolan, D., Salmon, D., Webb, H., Carrington, M., Voorheis, P., and Pays, E. (1996). Simultaneous but independent activation of adenylate cyclase and glycosylphosphatidylinositol-phospholipase C under stress conditions in *Trypanosoma brucei*. *J. Biol. Chem.* **271**, 10844–10852.
2. Rolin, S., Paindavoine, P., Hanocq-Quertier, J., Hanocq, F., Claes, Y., Le Ray, D., Overath, P., and Pays, E. (1993). Transient adenylate cyclase activation accompanies differentiation of *Trypanosoma brucei* from bloodstream to procyclic forms. *Mol. Biochem. Parasitol.* **61**, 115–125.
3. Bass, K. E. and Wang, C. C. (1991). The in vitro differentiation of pleomorphic *Trypanosoma brucei* from bloodstream into procyclic form requires neither intermediary nor short-stumpy stage. *Mol. Biochem. Parasitol.* **44**, 261–270.
4. Mancini, P. F. and Patton, C. L. (1981). Cyclic 3',5'-adenosine monophosphate levels during the developmental cycle of *Trypanosoma brucei brucei* in the rat. *Mol. Biochem. Parasitol.* **3**, 19–31.
5. Strickler, J. E. and Patton, C. L. (1975). Adenosine 3',5'-monophosphate in reproducing and differentiated *trypanosomes*. *Science* **190**, 1110–1112.
6. Vassella, E., Reuner, B., Yuzi, B., and Boshart, M. (1997). Differentiation of African trypanosomes is controlled by a density sensing mechanism which signals cell cycle arrest via the cAMP pathway. *J. Cell Sci.* **110**, 2661–2671.
7. Zoraghi, R., Kunz, S., Gong, K., and Seebeck, T. (2001). Characterization of TbPDE2A, a novel cyclic nucleotide-specific phosphodiesterase from the protozoan parasite *Trypanosoma brucei*. *J. Biol. Chem.* **276**, 11559–11566.
7a. Rascon, A., Soderling, S. H., Schaefer, J. B., and Beavo, J. A. (2002). Cloning and characterization of a cAMP-specific phosphodiesterase (TbPDE2B) from Trypanosoma brucei. *Proc. Natl. Acad. Sci. U.S.A.* **99**, 4714–4719.
8. Zoraghi, R. and Seebeck, T. (2002). The cAMP-specific phosphodiesterase TbPDE2C is an essential enzyme in bloodstream form *Trypanosoma brucei*. *Proc. Natl. Acad. Sci. USA*, 99, 4343–4348.
9. Santos, D. O. and Oliveira, M. O. (1988). Effect of cAMP on macromolecule synthesis in the pathogenic protozoon *Trypanosoma cruzi*. *Mem. Inst. Oswaldo. Cruz.* **83**, 287–292.
10. Oliveira, M. M., Rocha, E. D., Rondinelli. E., Arnholdt, A. V., and Scharfstein, J. (1993). Signal transduction in *Trypanosoma cruzi*: opposite effects of adenylcyclase and phospholipase C systems in growth control. *Mol. Cell. Biochem.* **124**, 91–99.
11. Rangel-Aldao, R., Triana, F., Fernandez, V., Comach, G., Abate, T., and Montoreano, R. (1988). Cyclic AMP as an inducer of the cell differentiation of *Trypanosoma cruzi*. *Biochem Int.* **17**, 337–344.
12. Heath, S., Hieny, S., and Sher, A. (1990). A cyclic AMP inducible gene expressed during the development of infective stages of *Trypanosoma cruzi*. *Mol. Biochem. Parasitol.* **43**, 133–141.
13. Walter, R.D., Buse, E., and Ebert, F. (1978). Effect of cyclic AMP on transformation and proliferation of *Leishmania* cells. *Tropenmed. Parasitol.* **29**, 439–442.
14. Bee, A., Culley, F. J., Alkhalife, I. S., Bodman-Smith, K. B., Raynens, J. G., and Bates, P. A. (2001). Transformation of *Leishmania mexicana* metacyclic promastigotes to amastigote-like forms mediated by binding of human C-reactive protein. *Parasitology* **122**, 521–529.
15. Seebeck, T., Gong, K. W., Kunz, S., Schaub, R., Shalaby, T., and Zoraghi, R. (2001). cAMP signalling in *Trypanosoma brucei*. *Int. J. Parasitol.* **31**, 491–498.
16. Taylor, M. C., Muhia, D. K., Baker, D. A., Mondragon, A., Schaap, P. B., and Kelly, J. M. (1999). *Trypanosoma cruzi* adenylyl cyclase is encoded by a complex multigene family. *Mol. Biochem. Parasitol.* **104**, 205–217.
17. Pays, E., Tebabi, P., Pays, A., Coquelet, H., Revelard, P., Salmon, D., and Steinert, M. (1989). The genes and transcripts of an antigen gene expression site from *T. brucei*. *Cell* **57**, 835–845.
18. Paindavoine, P., Rolin, S., Van Assel, S., Geuskens, M., Jauniaux, J. C., Dinsart, C., Huet, G., and Pays, E. (1992). A gene from the variant surface glycoprotein expression site encodes one of several transmembrane adenylate cyclases located on the flagellum of *Trypanosoma brucei*. *Mol. Cell Biol.* **12**, 1218–1225.
19. Alexandre, S., Paindavoine, P., Hanocq-Quertier, J., Paturiaux-Hanocq, F., Tebabi, P., and Pays, E. (1996). Families of adenylate cyclase genes in *Trypanosoma brucei*. *Mol. Biochem. Parasitol.* **77**, 173–182.
20. Alexandre, S., Paindavoine, P., Tebabi, P., Pays, A., Halleux, S., Steinert, M., and Pays, E. (1990). Differential expression of a family of putative adenylate/guanylate cyclase genes in *Trypanosoma brucei*. *Mol. Biochem. Parasitol.* **43**, 279–288.
21. Naula, C., Schaub, R., Leech, V., Melville, S., and Seebeck, T. (2001). Spontaneous dimerization and leucine-zipper induced activation of the recombinant catalytic domain of a new adenylyl cyclase of *Trypanosoma brucei*, GRESAG4.4B. *Mol. Biochem. Parasitol.* **112**, 19–28.
22. Sanchez, M. A., Zeoli, D., Klamo, E. M., Kavanaugh, M. P., and Landfear, S. M. (1995). A family of putative receptor-adenylate cyclases from *Leishmania donovani*. *J. Biol. Chem.* **270**, 17551–17558.
23. Ross, D. T., Raibaud, A., Florent, I. C., Sather, S., Gross, M. K., Storm, D. R., and Eisen, H. (1991). The *trypanosome* VSG expression site

encodes adenylate cyclase and a leucine-rich putative regulatory gene. *EMBO J.* **10**, 2047–2053.
24. Naula, C. and Seebeck, T. (2000). Cyclic AMP signaling in trypanosomatids. *Parasitol. Today* **16**, 35–38.
25. Bieger, B. and Essen, L. O. (2001). Structural analysis of adenylate cyclases from *Trypanosoma brucei* in their monomeric state. *EMBO J.* **20**, 433–445.
26. Liu, Y., Ruoho, S. A. E., Rao, V. D., and Hurley, J. H. (1997). Catalytic mechanism of the adenylyl and guanylyl cyclases: modeling and mutational analysis. *Proc. Natl. Acad. Sci. USA* **94**, 13414–13419.
27. Rolin, S,. Halleux, S., Van Sande, J., Dumont, J., Pays, E., and Steinert, M. (1990). Stage-specific adenylate cyclase activity in *Trypanosoma brucei. Exp. Parasitol.* **71**, 350–352.
28. Garbers, D. L. (1999). The guanylyl cyclase receptors. *Methods* **19**, 477–484.
29. Zufall, F. and Munger, S. D. (2001). From odor and pheromone transduction to the organization of the sense of smell. *Trends Neurosci.* **24**, 191–193.
30. Walter, R. D. (1974). 3':5'-cyclic-AMP phosphodiesterase from *Trypanosoma gambiense. Hoppe Seylers. Z. Physiol. Chem.* **355**, 1443–1450.
31. Gong, K. W., Kunz, S., Zoraghi, R., Kunz Renggli, C., Brun, R., and Seebeck, T. (2001). cAMP-specific phosphodiesterase TbPDE1 is not essential in *Trypanosoma brucei* in culture or during midgut infection of tsetse flies. *Mol. Biochem. Parasitol.* **116**, 229–232.
32. Ho, Y., Burden, L., and Hurley, J. H. (2000). Structure of the GAF domain, a ubiquitous signaling motif and a new class of cyclic GMP receptor. *EMBO J.* **19**, 5288–5299.
33. Goncalves, M. F., Zingales, B., and Colli, W. (1980). cAMP phosphodiesterase and activator protein of mammalian cAMP phosphodiesterase from *Trypanosoma cruzi. Mol. Biochem. Parasitol.* **1**, 107–118.
34. Tellez-Inon, M. T., Ulloa, R. M., Torruella, M., and Torres, H. N. (1985). Calmodulin and Ca^{2+}-dependent cyclic AMP phosphodiesterase activity in *Trypanosoma cruzi. Mol. Biochem. Parasitol.* **17**, 143–153.
35. Al-Chalabi, K. A., Ziz, L. A., and Al-Khayat, B. (1989). Presence and properties of cAMP phosphodiesterase from promastigote forms of *Leishmania tropica* and *Leishmania donovani. Comp. Biochem. Physiol.* **93**, 789–792.
36. Rascon, A., Viloria, M. F., De-Chiara, L., and Dubra, M. F. (2000). Characterization of cyclic AMP phosphodiesterases in *Leishmania mexicana* and purification of a soluble form. *Mol. Biochem. Parasitol.* **106**, 283–292.
37. Klöckner, T. (1996). "cAMP-Signaltransduktion in *Trypanosoma brucei*: Klonierung und Charakterisierung von Proteinkinase A und Phosphodiesterase Homologen." Ph.D Thesis, Ludwig Maximilian Universität, Munich.
38. Shalaby, T., Liniger, M., and Seebeck, T. (2001). The regulatory subunit of a cGMP regulated protein kinase A of *Trypanosoma brucei. Eur. J. Biochem.* **268**, 6197–6206.
39. Ulloa, R. M., Mesri, E., Esteva, M., Torres, H. N., and Tellez-Inon, M. T. (1988). Cyclic AMP-dependent protein kinase activity in *Trypanosoma cruzi. Biochem. J.* **255**, 319–326.
40. Ochatt, C. M., Ulloa, R. M., Torres, H. N., and Tellez-Inon, M. T. (1993). Characterization of the catalytic subunit of *Trypanosoma cruzi* cyclic AMP-dependent protein kinase. *Mol. Biochem. Parasitol.* **57**, 73–81.
41. Banerjee, C. and Sarkar, D. (1992). Isolation and characterization of a cyclic nucleotide-independent protein kinase from *Leishmania donovani. Mol. Biochem. Parasitol.* **52**, 195–205.
42. O'Beirne, C., Lowry, C. M., and Voorheis, H. P. (1998). Both IgM and IgG anti-VSG antibodies initiate a cycle of aggregation-disaggregation of bloodstream forms of *Trypanosoma brucei* without damage to the parasite. *Mol. Biochem. Parasitol.* **91**, 165–193.
43. Caler, E. V., Morty, R. E., Buriegh, B. A., and Andrews, N.W. (2000). Dual role of signaling pathways leading to Ca^{2+} and cyclic AMP elevation in host cell invasion *by Trypanosoma cruzi. Infect. Immunol.* **68**, 6602–6610.
44. Wirth, J. J. and Kierszenbaum, F. (1983). Modulatory effect of guanosine-3':5' cyclic monophosphate on macrophage susceptibility to *Trypanosoma cruzi* infection. *J. Immunol.* **31**, 3028–3031.
45. Von Kreuter, B. F., Walton, B. L., and Santos-Buch, C. (1995). Attenuation of parasite cAMP levels in *T. cruzi*-host cell membrane interactions in vitro. *J. Eukaryot. Microbiol.* **42**, 20–26.
46. Walter, R.D., Slutzky, G. M., and Greenblat, C. L. (1982). Effect of leishmanial excreted factor on the activities of adenylate cyclase from hamster liver and *Leishmania tropica. Tropenmed. Parasitol.* **33**, 137–139.
47. Dey, R., Mitra, S., and Data, S. C. (1995). Cyclic AMP mediates change in superoxide dismutase activity to monitor host-parasite interaction in *Leishmania donovani. J. Parasitol.* **81**, 683–686.

CHAPTER 211

Cyclic Nucleotide Specificity and Cross-Activation of Cyclic Nucleotide Receptors

Clay E. S. Comstock and John B. Shabb

*Department of Biochemistry and Molecular Biology,
University of North Dakota School of Medicine and Health Sciences,
Grand Forks, North Dakota*

The homologous cAMP- and cGMP-dependent protein kinases (PKA and PKG) are generally believed to be activated relatively specifically by cAMP and cGMP. Certain physiological and pathophysiological conditions may exist in which elevation of one cyclic nucleotide may lead to cross-activation of the opposing kinase [1–3]. This chapter summarizes the current evidence supporting cAMP and cGMP cross-activation and the structural basis for cyclic nucleotide specificity of the cyclic nucleotide dependent protein kinases. The interaction of cyclic nucleotides with other eukaryotic cyclic nucleotide receptors is also discussed.

cAMP Cross-Activation of PKG

Though there is some evidence for cAMP cross-activation of PKG in gastric smooth muscle [4], the physiological paradigm for this type of cross-activation is vascular smooth muscle relaxation. In one line of experiments, doses of cyclic nucleotides sufficient to relax pig coronary arteries correlate with their potencies for PKG activation but not PKA activation [5]. Consistent with this, a mere twofold elevation of intracellular cAMP in this same tissue is sufficient to cause vasorelaxation and activation of both PKA and PKG [6]. A second and more frequently tried line of evidence correlates the pharmacological inhibition of PKG with the disruption of cAMP-induced functions such as vasorelaxation [7], phosphorylation of the IPs receptor in rat aorta [8], gating of the L-type Ca^{2+} channel in rabbit portal vein myocytes [9], and gating of the Ca^{2+}-activated potassium (BKca) channel in porcine coronary artery myocytes [10]. A third experimental approach demonstrates that cAMP reduces vasopressin-induced Ca^{2+} mobilization in primary cultures of rat aortic myocytes, but not in myocytes depleted of PKG [11]. This last experimental model suggests cAMP-induced vascular smooth muscle relaxation is mediated solely by PKG. Other models suggest, however, that cAMP can also relax vascular smooth muscle through PKA activation [7,9]. The latter is further supported by the observation that PKG I-deficient mice still undergo cAMP-induced smooth muscle relaxation even though these mice already suffer severe vascular and intestinal disregulation [12].

Although the affinity of PKG for cAMP is 50- to 100-fold lower than its affinity for cGMP [13], two factors may make PKG susceptible to cAMP activation in the cell. First, the concentration of cAMP is estimated to be fivefold higher than cGMP in some tissues [5]. Second, autophosphorylation of PKG improves its affinity for cAMP *in vitro* [13–15]. For example, autophosphorylation at Ser-79 of the bovine type Iβ isoform of PKG results in a three- to fourfold increase in basal kinase activity and a twofold decrease in its K_a for cyclic nucleotides [16]. Though *in vivo* autophosphorylation at this site has not been determined directly, replacement of Ser-79 with aspartic acid to mimic the effect of autophosphorylation results in a mutant kinase with constitutive activity when expressed in CV-1 cells [17].

cGMP Cross-Activation of PKA

The best evidence for the cross-activation of PKA by cGMP is from studies with intestinal epithelial cell lines. The *Escherichia coli* heat-stable enterotoxin causes supraphysiological accumulation of cGMP and increased Cl⁻ permeability in the human intestinal T84 cell line, which expresses little or no PKG, but plenty of PKA [18]. Inhibitors of PKA activation also suppress increased Cl⁻ secretion in T84 cells upon stimulation with cGMP analogs or guanylin, an endogenous guanylyl cyclase activator peptide [19]. Similarly, induction of Cl⁻ conductance in the human colonic carcinoma cell line, Caco-2, by cAMP or cGMP analogs is suppressed by the PKA inhibitor PKI_{5-24} but not the PKG inhibitor KT5823 [20]. Other studies suggest that cGMP-dependent activation of PKA may also occur during porcine carotid artery relaxation [21] and during atrial natriuretic peptide-induced testosterone production in mouse Leydig cells [22].

The cross-activation of PKA may occur indirectly through cGMP inhibition of the cAMP-specific phosphodiesterase PDE3. Treatment of rat small intestine synaptosomal preparations with inhibitors of PKA, but not PKG, block the NO-induced release of bombesin-like immunoreactivity (BLI), considered to be important in enteric smooth muscle contractility [23]. Treatment with trequinsin, a blocker of PDE3, results in increased cAMP and mimics the NO-induced release of BLI. The effects of trequinsin and NO are not additive suggesting that the two agents act through a common mechanism.

Molecular Basis for cAMP/cGMP Selectivity of PKA and PKG

Central to the theme of cyclic nucleotide-dependent protein kinase cross-activation is the degree to which the cyclic nucleotide-binding domain (CNBD) can select for the appropriate cyclic nucleotide. The typical CNBD contains about 124 residues and is characterized by three α-helices, an eight-stranded anti-parallel (β-barrel), and a half-dozen invariant residues. Three invariant glycines are critical for proper folding of the β-barrel, an invariant glutamic acid forms a hydrogen bond with the ribose 2′-OH, and an invariant arginine interacts electrostatically with the cyclic phosphate of the cyclic nucleotide. The importance of these residues in cyclic nucleotide-binding has been demonstrated by site-directed mutagenesis of the type Iα regulatory subunit (RIα) of PKA, which contains two CNBDs termed the A and B domains (summarized in [24]). The contribution of the sixth invariant residue, an alanine, to high affinity cyclic nucleotide binding is not fully understood, though its mutagenesis destroys cAMP-binding activity [25]. Since all of these signature residues are present in cAMP- or cGMP-selective CNBDs, they are not involved in cyclic nucleotide selectivity.

The characteristic RAA and R(S/T)A motifs found in cAMP- and cGMP-selective CNBDs, respectively, have been identified as cAMP/cGMP selectivity determinants. The middle residues in each motif are bracketed by the invariant Arg and Ala residues described above. Mutagenesis of the first Ala in the RAA motif in the A and B domains of RIα increases its affinity 200-fold for cGMP with minor effects on cAMP affinity [26–28]. Likewise, mutation of Ser/Thr residues to Ala in the R(S/T)A motif in the A and B domains of PKG Iα [29], PKG Iβ [30], and PKG II [31] reduces their affinities for cGMP relative to cAMP. These site-directed mutagenesis studies along with cyclic nucleotide analog studies [32–34], molecular modeling [35,36], and determination of the structure of the RIα cAMP-binding domains [37] all support the interpretation that the selectivity of cGMP-binding domains is due to hydrogen bonding between the 2-amino position of cGMP and the Ser/Thr side chain.

Whereas cGMP selectivity can be explained in a large part by a single ligand-receptor interaction, identification of residues responsible for selective, high-affinity cAMP-binding has met with limited success. The RIα structure suggests that aromatic stacking between the A domain Trp-260 or the B domain Tyr-371 and the purine moiety of cAMP may contribute to cAMP/cGMP selectivity. Mutagenesis studies, however, demonstrate that Tyr-371, though important for general high-affinity cyclic nucleotide binding, does not affect cAMP/cGMP selectivity [38]. Other residues in the binding pocket close to the 6-position of the cyclic nucleotide have been implicated as potential selectivity determinants. A likely candidate is the peptide backbone carbonyl group of Asn-372, in the B domain of RIα. Mutagenesis of Asn-372 suggests that it contributes modestly to the cAMP selectivity in RIα (Comstock and Shabb, unpublished).

Another RIα B domain residue that may contribute to recognition of the 6-amino group is Ile-368. Conversion of Ile-368 to Tyr increases the affinity three- to fourfold for N^6-modified cAMP analogs [39]. The only fully converted cGMP-sensitive, cAMP-dependent protein kinase is a chimera where both of the cAMP-binding domains of RIα have been replaced by the cGMP-binding domains of PKG I [40].

Other Cyclic Nucleotide Receptors

Although cross-activation events are usually mediated by protein kinases, the potential involvement of other cyclic nucleotide receptors must be considered. For example, the homomeric expression of the olfactory cyclic nucleotide-gated (CNG) ion channel α subunit is more sensitive to cGMP than to cAMP [41,42]. However, heteromeric association with its β subunit tends to equalize the channel's affinity for cAMP and cGMP [43,44] suggesting that either cyclic nucleotide can act as its physiological ligand.

The CNBDs of the CNG ion channels are homologous to those found in the protein kinases. Like the kinases, mutagenesis of the invariant Arg in the ion channels results in reduced affinity for cGMP [45]. Substitution of the Thr in the R(S/T)A motif to an Ala in photoreceptor and olfactory CNG ion channels results in a reduction of cGMP-responsiveness without affecting cAMP-induced channel activation [41]. Other mutagenesis [46,47], molecular modeling [48], and analog studies [49] suggest that the 2-amino group of cGMP may not always interact with the CNG ion channel R(S/T)A motif, but may, under certain conditions, bind cyclic nucleotides in a conformation distinct from that found with the protein kinases.

The recently identified cAMP-binding protein EPAC (exchange protein directly activated by cAMP) or cAMP-GEF (guanine-nucleotide exchange factor) [50,51] represents a new family of CNBD-containing proteins. Preliminary characterization of EPAC-1 cyclic nucleotide selectivity has been done [52]. Further studies of this and other cyclic nucleotide receptors should continue to shed light on the complex interrelationships of intracellular cyclic nucleotide signaling pathways.

References

1. Jiang, H., Shabb, J. B., and Corbin, J. D. (1992). Cross-activation: overriding cAMP/cGMP selectivities of protein kinases in tissues. *Biochem. Cell Biol.* **70**, 1283–1289.
2. Torphy, T. J. (1994). Beta-adrenoceptors, cAMP and airway smooth muscle relaxation: challenges to the dogma. *Trends Pharmacol. Sci.* **15**, 370–374.
3. Carvajal, J. A., Germain, A. M., Huidobro-Toro, J. P., and Weiner, C. P. (2000). Molecular mechanism of cGMP-mediated smooth muscle relaxation. *J. Cell. Physiol* **184**, 409–420.
4. Murthy, K. S. and Makhlouf, G. M. (1995). Interaction of cA-kinase and cG-kinase in mediating relaxation of dispersed smooth muscle cells. *Am. J. Physiol.* **268**, C171–C180.
5. Francis, S. H., Noblett, B. D., Todd, B. W., Wells, J. N., and Corbin, J. D. (1988). Relaxation of vascular and tracheal smooth muscle by cyclic nucleotide analogs that preferentially activate purified cGMP-dependent protein kinase. *Mol. Pharmacol.* **34**, 506–517.
6. Jiang, H., Colbran, J. L., Francis, S. H., and Corbin, J. D. (1992). Direct evidence for cross-activation of cGMP-dependent protein kinase by cAMP in pig coronary arteries. *J. Biol. Chem.* **267**, 1015–1019.
7. Eckly-Michel, A., Martin, V., and Lugnier, C. (1997). Involvement of cyclic nucleotide-dependent protein kinases in cyclic AMP-mediated vasorelaxation. *Br. J. Pharmacol.* **122**, 158–164.
8. Komalavilas, P. and Lincoln, T. M. (1996). Phosphorylation of the inositol 1,4,5-trisphosphate receptor. Cyclic GMP-dependent protein kinase mediates cAMP and cGMP dependent phosphorylation in the intact rat aorta. *J. Biol. Chem.* **271**, 21933–21938.
9. Ruiz-Velasco, V., Zhong, J., Hume, J. R., and Keef, K. D. (1998). Modulation of Ca^{2+} channels by cyclic nucleotide cross activation of opposing protein kinases in rabbit portal vein. *Circ. Res.* **82**, 557–565.
10. Han, G., Kryman, J. P., McMillin, P. J., White, R. E., and Carrier, G. O. (1999). A novel transduction mechanism mediating dopamine-induced vascular relaxation: opening of BKCa channels by cyclic AMP-induced stimulation of the cyclic GMP-dependent protein kinase. *J. Cardiovasc. Pharmacol.* **34**, 619–627.
11. Lincoln, T. M., Cornwell, T. L., and Taylor, A. E. (1990). cGMP-dependent protein kinase mediates the reduction of $Ca2+$ by cAMP in vascular smooth muscle cells. *Am. J. Physiol* **258**, C399–C407.
12. Pfeifer, A., Klatt, P., Massberg, S., Ny, L., Sausbier, M., Hirneiss, C., Wang, G. X., Korth, M., Aszodi, A., Andersson, K. E., Krombach, F., Mayerhofer, A., Ruth, P., Fassler, R., and Hofmann, F. (1998). Defective smooth muscle regulation in cGMP kinase I-deficient mice. *EMBO J.* **17**, 3045–3051.
13. Wolfe, L., Corbin, J. D., and Francis, S. H. (1989). Characterization of a novel isozyme of cGMP-dependent protein kinase from bovine aorta. *J. Biol. Chem.* **264**, 7734–7741.
14. Foster, J. L., Guttmann, J., and Rosen, O. M. (1981). Autophosphorylation of cGMP-dependent protein kinase. *J. Biol. Chem.* **256**, 5029–5036.
15. Landgraf, W., Hullin, R., Gobel, C., and Hofmann, F. (1986). Phosphorylation of cGMP-dependent protein kinase increases the affinity for cyclic AMP. *Eur. J. Biochem.* **154**, 113–117.
16. Smith, J. A., Francis, S. H., Walsh, K. A., Kumar, S., and Corbin, J. D. (1996). Autophosphorylation of type Ibeta cGMP-dependent protein kinase increases basal catalytic activity and enhances allosteric activation by cGMP or cAMP. *J. Biol. Chem.* **271**, 20756–20762.
17. Collins, S. P. and Uhler, M. D. (1999). Cyclic AMP- and cyclic GMP-dependent protein kinases differ in their regulation of cyclic AMP response element-dependent gene transcription. *J. Biol. Chem.* **274**, 8391–8404.
18. Forte, L. R., Thorne, P. K., Eber, S. L., Krause, W. J., Freeman, R. H., Francis, S. H., and Corbin, J. D. (1992). Stimulation of intestinal Cl-transport by heat-stable enterotoxin: activation of cAMP-dependent protein kinase by cGMP. *Am. J. Physiol.* **263**, C607–C615.
19. Chao, A. C., de Sauvage, F. J., Dong, Y. J., Wagner, J. A., Goeddel, D. V., and Gardner, P. (1994). Activation of intestinal CFTR Cl-channel by heat-stable enterotoxin and guanylin via cAMP-dependent protein kinase. *EMBO J.* **13**, 1065–1072.
20. Tien, X. Y., Brasitus, T. A., Kaetzel, M. A., Dedman, J. R., and Nelson, D. J. (1994). Activation of the cystic fibrosis transmembrane conductance regulator by cGMP in the human colonic cancer cell line, Caco-2. *J. Biol. Chem.* **269**, 51–54.
21. van Riper, D. A., McDaniel, N. L., and Rembold, C. M. (1997). Myosin light chain kinase phosphorylation in nitrovasodilator induced swine carotid artery relaxation. *Biochim. Biophys. Acta* **1355**, 323–330.
22. Schumacher, H., Muller, D., and Mukhopadhyay, A. K. (1992). Stimulation of testosterone production by atrial natriuretic peptide in isolated mouse Leydig cells results from a promiscuous activation of cyclic AMP-dependent protein kinase by cyclic GMP. *Mol. Cell. Endocrinol.* **90**, 47–52.
23. Kurjak, M., Fritsch, R., Saur, D., Schusdziarra, V., and Allescher, H. D. (1999). NO releases bombesin-like immunoreactivity from enteric synaptosomes by cross-activation of protein kinase A. *Am. J. Physiol.* **276**, G1521–G1530.
24. Shabb, J. B. and Corbin, J. D. (1992). Cyclic nucleotide-binding domains in proteins having diverse functions. *J. Biol. Chem.* **267**, 5723–5726.
25. Zorn, M., Fladmark, K. E., Ogreid, D., Jastorff, B., Doskeland, S. O., and Dostmann, W. R. (1995). Ala335 is essential for high-affinity cAMP-binding of both sites A and B of cAMP-dependent protein kinase type I. *FEBS Lett.* **362**, 291–294.
26. Shabb, J. B., Ng, L., and Corbin, J. D. (1990). One amino acid change produces a high affinity cGMP-binding site in cAMP-dependent protein kinase. *J. Biol. Chem.* **265**, 16031–16034.
27. Shabb, J. B., Buzzeo, B. D., Ng, L., and Corbin, J. D. (1991). Mutating protein kinase cAMP-binding sites into cGMP-binding sites. Mechanism of cGMP selectivity. *J. Biol. Chem.* **266**, 24320–24326.
28. Muhonen, W. W. and Shabb, J. B. (2000). Resonant mirror biosensor analysis of type l alpha cAMP-dependent protein kinase B domain–cyclic nucleotide interactions. *Protein Sci.* **9**, 2446–2456.
29. Reed, R. B., Sandberg, M., Jahnsen, T., Lohmann, S. M., Francis, S. H., and Corbin, J. D. (1997). Structural order of the slow and fast intra-subunit cGMP-binding sites of type I alpha cGMP-dependent protein kinase. *Adv. Second Messenger Phosphoprot. Res.* **31**, 205–217.
30. Reed, R. B., Sandberg, M., Jahnsen, T., Lohmann, S. M., Francis, S. H., and Corbin, J. D. (1996). Fast and slow cyclic nucleotide-dissociation sites in cAMP-dependent protein kinase are transposed in type Ibeta cGMP-dependent protein kinase. *J. Biol. Chem.* **271**, 17570–17575.
31. Taylor, M. K. and Uhler, M. D. (2000). The amino-terminal cyclic nucleotide binding site of the type II cGMP-dependent protein kinase is essential for full cyclic nucleotide-dependent activation. *J. Biol. Chem.* **275**, 28053–28062.
32. Doskeland, S. O., Ogreid, D., Ekanger, R., Sturm, P. A., Miller, J. P., and Suva, R. H. (1983). Mapping of the two intrachain cyclic

nucleotide binding sites of adenosine cyclic 3',5'-phosphate dependent protein kinase I. *Biochemistry* **22**, 1094–1101.

33. Ogreid, D., Ekanger, R., Suva, R. H., Miller, J. P., and Doskeland, S. O. (1989). Comparison of the two classes of binding sites (A and B) of type I and type II cyclic-AMP-dependent protein kinases by using cyclic nucleotide analogs. *Eur. J. Biochem.* **181**, 19–31.

34. Corbin, J. D., Ogreid, D., Miller, J. P., Suva, R. H., Jastorff, B., and Doskeland, S. O. (1986). Studies of cGMP analog specificity and function of the two intrasubunit binding sites of cGMP-dependent protein kinase. *J. Biol. Chem.* **261**, 1208–1214.

35. Weber, I. T., Steitz, T. A., Bubis, J., and Taylor, S. S. (1987). Predicted structures of cAMP binding domains of type I and II regulatory subunits of cAMP-dependent protein kinase. *Biochemistry* **26**, 343–351.

36. Weber, I. T., Shabb, J. B., and Corbin, J. D. (1989). Predicted structures of the cGMP binding domains of the cGMP-dependent protein kinase: a key alanine/threonine difference in evolutionary divergence of cAMP and cGMP binding sites. *Biochemistry* **28**, 6122–6127.

37. Su, Y., Dostmann, W. R., Herberg, F. W., Durick, K., Xuong, N. H., Ten Eyck, L., Taylor, S. S., and Varughese, K. I. (1995). Regulatory subunit of protein kinase A: Structure of deletion mutant with cAMP binding domains. *Science* **269**, 807–813.

38. Kapphahn, M. A. and Shabb, J. B. (1997). Contribution of the carboxyl-terminal regional of the cAMP-dependent protein kinase type I alpha regulatory subunit to cyclic nucleotide interactions. *Arch. Biochem. Biophys.* **348**, 347–356.

39. Huq, I., Dostmann, W. R., and Ogreid, D. (1996). Isoleucine 368 is involved in low-affinity binding of N6-modified cAMP analogues to site B of the regulatory subunit of cAMP-dependent protein kinase I. *Biochem. J.* **316** (Ptl), 337–343.

40. Wild, N., Herberg, F. W., Hofmann, F., and Dostmann, W. R. (1995). Expression of a chimeric, cGMP-sensitive regulatory subunit of the cAMP-dependent protein kinase type I alpha. *FEBS Lett.* **374**, 356–362.

41. Altenhofen, W., Ludwig, J., Eismann, E., Kraus, W., Bonigk, W., and Kaupp, U. B. (1991). Control of ligand specificity in cyclic nucleotide-gated channels from rod photoreceptors and olfactory epithelium. *Proc. Natl. Acad. Sci. USA* **88**, 9868–9872.

42. Dhallan, R. S., Yau, K. W., Schrader, K. A., and Reed, R. R. (1990). Primary structure and functional expression of a cyclic nucleotide-activated channel from olfactory neurons. *Nature* **347**, 184–187.

43. Bradley, J., Li, J., Davidson, N., Lester, H. A., and Zinn, K. (1994). Heteromeric olfactory cyclic nucleotide-gated channels: a subunit that confers increased sensitivity to cAMP. *Proc. Natl. Acad. Sci. USA* **91**, 8890–8894.

44. Liman, E. R. and Buck, L. B. (1994). A second subunit of the olfactory cyclic nucleotide-gated channel confers high sensitivity to cAMP. *Neuron* **13**, 611–621.

45. Tibbs, G. R., Goulding, E. H., and Siegelbaum, S. A. (1997). Allosteric activation and tuning of ligand efficacy in cyclic-nucleotide-gated channels. *Nature* **386**, 612–615.

46. Varnum, M. D., Black, K. D., and Zagotta, W. N. (1995). Molecular mechanism for ligand discrimination of cyclic nucleotide-gated channels. *Neuron* **15**, 619–625.

47. Shapiro, M. S. and Zagotta, W. N. (2000). Structural basis for ligand selectivity of heteromeric olfactory cyclic nucleotide-gated channels. *Biophys. J.* **78**, 2307–2320.

48. Scott, S. P., Harrison, R. W., Weber, I. T., and Tanaka, J. C. (1996). Predicted ligand interactions of 3',5'-cyclic nucleotide-gated channel binding sites: comparison of retina and olfactory binding site models. *Protein Eng.* **9**, 333–344.

49. Scott, S. P., Cummings, J., Joe, J. C., and Tanaka, J. C. (2000). Mutating three residues in the bovine rod cyclic nucleotide-activated channel can switch a nucleotide from inactive to active. *Biophys. J.* **78**, 2321–2333.

50. de Rooij, J., Zwartkruis, F. J., Verheijen, M. H., Cool, R. H., Nijman, S. M., Wittinghofer, A., and Bos, J. L. (1998). Epac is a Rapl guanine-nucleotide-exchange factor directly activated by cyclic AMP. *Nature* **396**, 474–477.

51. Kawasaki, H., Springett, G. M., Mochizuki, N., Toki, S., Nakaya, M., Matsuda, M., Housman, D. E., and Graybiel, A. M. (1998). A family of cAMP-binding proteins that directly activate Rapl. *Science* **282**, 2275–2279.

52. Enserink, J. M., Christensen, A. E., de Rooij, J., van Triest, M., Schwede, F., Genieser, H. G., Døskeland, S. O., Blank, J. L., Bos, J. L. (2002). A Novel Epac-specific cAMP analogue Demonstrates independent Regulation of Rapl and ERK. *Nat. Cell. Biol.* **4**, 901–906.

CHAPTER 212

Cyclic Nucleotide Analogs as Tools to Investigate Cyclic Nucleotide Signaling

Anne Elisabeth Christensen and
Stein Ove Døskeland

*Department of Anatomy and Cell Biology,
University of Bergen,
Bergen, Norway*

Introduction

Early after the discovery of the two naturally occurring cyclic nucleotide monophosphates (cNMPs) cAMP and cGMP, the lipophilic analog N^6, 2′-O-dibutyryl cAMP was synthesized [1] and used to elicit cAMP responses in intact cells. Several hundred cNMP analogs have since been synthesized. See [2] for a review.

The effects of cAMP in vertebrates appear to be mediated mainly through activation of cAMP-dependent protein kinase isozyme I (cA-PKI) and cA-PKII, but also through activation of small GTPase exchange factors [3,4] and direct binding to ion channels [5,6]. Specialized extracellular cAMP receptors exist on the surface of *Dictyostelium discoideum* [7], and bacteria have a cAMP receptor [8] acting as a catabolite gene activator protein (CAP). The known receptors of cGMP are the cG-kinases (types Iα, Iβ, and II) and ion channels [5], but cGMP can also bind directly to allosteric sites on cyclic nucleotide phosphodiesterases and thereby modulate their action [9]. In the first section of this chapter we will give guidelines and examples of the use of cNMP analogs. In the second section we will describe the chemistry and some properties of commonly used cNMP analogs.

Use of cNMP Analogs: Guidelines and Examples

Activating cNMP Analogs

In order to pinpoint the receptor responsible for a given cyclic nucleotide effect, a dominant positive form of the receptor (not readily available for most cNMP receptors) can be introduced by transfection or by microinjection into cells. The alternative is to use cNMP analogs activating the cNMP receptor. The analog approach does not depend on artificially overexpressed gene products and can be used when transfection or microinjection is not applicable, e.g. in blood platelets. An additional advantage of cNMP analogs is that they generally can be removed by washing the cells and act within minutes, rather than hours or days.

A major problem has been to discriminate between cAMP effects mediated through the EPAC (exchange protein directly activated by cAMP) family of cAMP-regulated GTP exchange factors, cA-PKI and cA-PKII, since all commercially available analogs that activate EPAC also activate cA-PKI and cA-PKII [10]. Recently, novel 2′-O-modified cAMP analogs have been synthesized that have about three orders of magnitude higher affinity (relative to cAMP) for EPAC than for the R subunits of cA-PKI or cA-PKII (Table I), and show EPAC

Table I Relative Affinities of Some EPAC-1 Selective cAMP Analogs

cA analog	Rel. aff. EPAC	Rel. aff. RI: $\sqrt{(AI)(BI)}$	Rel. aff. RI: $\sqrt{(AII)(BII)}$	EPAC/RI	EPAC/RII
cA	1.0	1.0	1.0	1.0	1.0
8-CPT-cA	65	2.57	1.01	25.3	64.3
2′-O-Me-cA	0.12	0.0048	0.0052	24.8	23.3
8-CPT-2′-O-Me-cA	4.6	0.0089	0.0028	517	1626

The table lists the analog affinity for the single cNMP binding site of EPAC-1, and the average affinity for site AI and BI of the RI subunit of cA-PKI and for site AII and BII of cA-PKII. The two right-hand columns give the preference for binding to EPAC relative to RI and RII. The data are from [35].

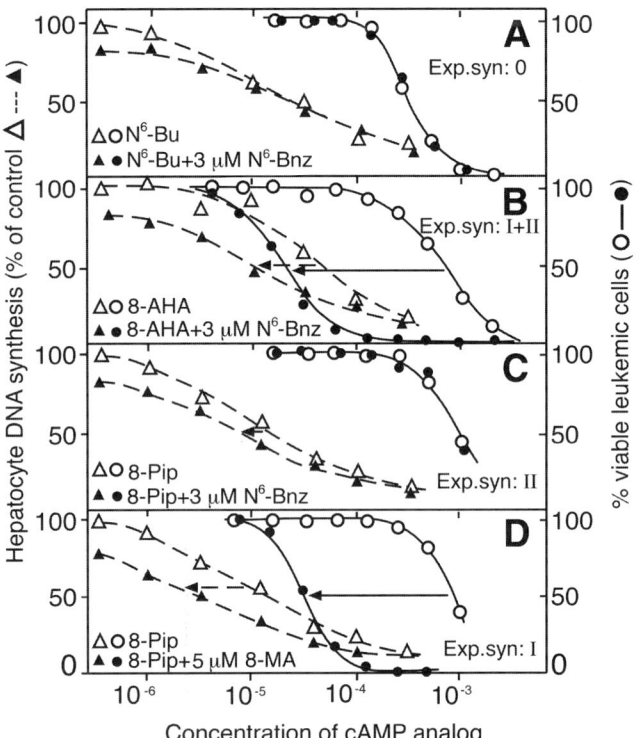

Figure 1 Effects of cAMP analog pairs selected for synergistic activation of cA-PKI and cA-PKII. Panel A shows that the two analogs N^6-benzoyl-cAMP and N^6-butyryl-cAMP (see Table III and Figure 2 for details of analog structure) fail to synergize in affecting hepatocyte DNA replication (Δ,▲) or leukemic cell apoptosis (○,●), as expected from lack of synergistic activation of either cA-PKI or cA-PKII (Table II). Panel B shows that N^6-benzoyl-cAMP synergized strongly with 8-aminohexyl-amino-cAMP, indicating that either cA-PKI or cA-PKII or both isozymes were able to induce the observed effects. Panel C shows that N^6-benzoyl-cAMP combined with 8-piperidino-cAMP, which only synergize for activation of cA-PKII (Table I), failed to synergize in induction of leukemic cell death but had a weak synergy in inhibiting hepatocyte DNA replication. Panel D shows that 8-piperidino-cAMP combined with 8-methylamino-cAMP, which selectively synergize for activation of cA-PKI (Table IIB), were strongly synergistic for leukemic cell death induction and moderately so for inhibition of hepatocyte DNA replication. The conclusion is that leukemic cell death can be induced by cA-PKI only, whereas hepatocyte DNA replication can be induced to a moderate extent by activation of cA-PKII, to a higher extent by cA-PKI, and in full by combined activation of cA-PKI and cA-PKII.

selectivity in intact cells [11]. N^6-modified cAMP analogs such as N^6-monobuturyl-cAMP and N^6-benzoyl-cAMP appear to activate cA-PKI and cA-PKII without activating EPAC [11a,35]. Such analogs do not bind to the retinal rod ion channel [12] and discriminate better than 8-Br-cAMP against cGMP-kinases [13].

No single cAMP analog is capable of selectively activating either cA-PKI or cA-PKII, but this can be achieved by analog pairs [9,14]. Examples are given in Fig. 1, where the end-point was either inhibition of hepatocyte DNA replication or induction of leukemic cell death by apoptosis [14]. The theoretical background for the approach is that the two cAMP-binding sites of cA-PKI (AI, BI) and cA-PKII (AII, BII) can be distinguished by cNMP analogs, and that both site A and B must be occupied in a given isozyme to achieve biologically relevant activation [15]. cNMP analogs preferring site AI and BI will therefore synergize in activation of cA-PKI, whereas AII- and BII-preferring analogs will activate cA-PKII synergistically (Table II). Synergistic activation is weaker for cG-PK [9,13], but N^2-substituted cGMP analogs can discriminate between the α and the β forms of cG-PKI [9]. The cyclic nucleotide-gated ion channels are less well characterized with respect to cNMP analog specificity; the photoreceptor channel was 30 times more sensitive to 8-Fl-cGMP (a fluorescein derived analog) than to cGMP itself [16]. Presumably, use of a battery of cGMP analogs, including 8-Fl-cGMP and analogs with cG-PK specificity, should allow a conclusion of whether an ion channel or cG-PK is responsible for a certain biological phenomenon.

Inhibitory cNMP Analogs

Considering the high latent energy of the cyclic phosphate bond it is no surprise that this part of the cNMP molecule is essential for biological action and that even minor modifications of the cyclic phosphate, as when either the axial (Sp) or equatorial (Rp) oxygen is replaced by sulphur, may interfere with activation. Sp-cNMPS analogs are generally agonistic, with the notable exception of the rod photoreceptor cGMP-gated cation channel, for which Sp-8-Br-PET-cGMPS

Table IIA Relative Affinities of Selected cN-analogs to cA-PK Binding Sites

cA analog	Rel. aff. AI	Rel. aff. BI	Rel. aff. AII	Rel. aff. BII
cA	1.0	1.0	1.0	1.0
N^6-Bu-cA	3.6	0.093	0.71	0.041
N^6-Bnz-cA	4.0	0.26	3.8	0.037
8-AHA-cA	0.056	4.1	0.010	0.39
8-Pip-cA	2.1	0.06	0.047	2.7
8-MA-cA	0.07	3.3	0.026	1.6
N^6-BC-cA	0.50	0.086	13	0.066
Sp-5,6diCl-cBIMPS	0.022	0.13	0.034	14

Table IIB Predicted Synergy between Pairs of cN-analogs

cA analog x + y	Predicted synergy cA-PKI $\dfrac{\sqrt{(AI^x+AI^y)(BI^x+BI^y)}}{\sqrt{(AI^x)(BI^x)}+\sqrt{(AI^y)(BI^y)}}$	Predicted synergy cA-PKII $\dfrac{\sqrt{(AII^x+AII^y)(BII^x+BII^y)}}{\sqrt{(AII^x)(BII^x)}+\sqrt{(AII^y)(BII^y)}}$
N^6-Bu-cA + N^6-Bnz-cA	1.0	1.1
8-AHA-cA + N^6-Bnz-cA	2.8	2.9
8-Pip-cA + N^6-Bnz-cA	1.0	4.4
8-Pip-cA + 8-MA-cA	3.2	1.0
N^6-BC-cA + Sp-5,6diCl-cBIMPS	1.3	8.3

Panel A shows the affinity of selected cAMP analogs for the binding sites of cA-PKI (AI, BI) and cA-PKII (AII, BII). Data are from Øgreid et al. [36] or the authors' unpublished data. Panel B shows the formulas to predict synergy between analog pairs for activation of cA-PKI or cA-PKII. Numbers above 1.0 signify synergism. For details of cAMP analog structures, see Table III and Figure 2.

is an antagonist [17]. The Rp-cNMPS analogs are generally antagonistic or partially agonistic, with the exception of the cGMP-gated cation channel, where Rp-cGMPS analogs such as Rp-8-Br-cGMPS are agonists [18].

For practical purposes the comparative use of Rp-8-Br-cAMPS, Rp-cAMPS, and Rp-8-Br-cGMPS may be useful. Since these analogs must be used at high concentrations (0.1–1 mM in the extracellular medium) to be able to compete efficiently with cAMP or cGMP in the cell, it is important to test them against submaximal concentrations of the cNMP receptor activator. If Rp-8-Br-cAMPS is the more efficient inhibitor one can suspect that cA-PKI is the main mediator; whereas when Rp-cAMPS is the better inhibitor cA-PKII is more likely; and if Rp-8-Br-cGMPS is the most efficient inhibitor one can suspect cG-PK involvement. Further proof of cyclic nucleotide protein kinase involvement can be obtained by use of cell-permeable inhibitors directed against the kinase moiety of these enzymes, either broadly acting ATP antagonists [19] or more specific peptide inhibitors (Chapter 79 by Dostmann, this volume). If applicable, dominant negative forms of RI or RII (with deficient cAMP-binding sites) can be introduced by transfection or microinjection to further prove cA-PK involvement.

Chemistry and Properties of Cyclic Nucleotide Analogs

Cyclic AMP and cGMP are composed of four rings: the cyclic phosphate, the ribose, and a substituted aromatic purine double ring (pyrimidine and imidazole). The structure of most cNMP analogs can be described (Table III) as variously substituted derivatives of purine riboside-3′,5′-cyclic monophosphate (cPuMP), shown in Fig. 2A. The parent cAMP and cGMP molecules are virtually unable to penetrate cell membranes by diffusion and are readily broken down by cyclic nucleotide phosphodiesterases. Most cNMP analogs that are synthesized for use in cell biology have enhanced lipophilicity (Table III) and improved resistance toward cyclic nucleotide phosphodiesterase attack.

The imidazole ring of cPuMP has been subjected to a number of substitutions at the 8 position. Substituent charge, hydrogen bonding potential, hydrophobicity, and bulk determine the specificity toward cAMP-binding sites type A and B of cA-PKI and cA-PKII [20]. Bulky 8-substituents will force the cNMP molecule into a *syn* conformation about the torsion angle between C4-N9 of the purine ring and the O1′-C1′ of the ribose ring (Fig. 2A). This will deter cN binding to the

Table III Compounds Resulting from Substitution of cPUMP (fig. 2A)

Compound	R2'	R2	R6	R8	Relative lipophilicity	Sp/Rp isomers commercially available
cPUMP	–	–	–	–	–	–
cAMP	OH	–	NH_2	–	1.0	+
8-MA-cAMP	OH	–	NH_2	$NHCH_3$	1.0	–
N^6-Bu-cA	OH	–	NHC_4OH_7	–	3.5	–
N^6-Bnz-cA	OH	–	NH-benzoyl	–	6.5	–
N^6-BC-cA	OH	–	NH-t.-butyl-carbamoyl	–	32	–
8-AHA-cA	OH	–	NH_2	$NH-(CH_2)_6-NH_3^+$	–	–
8-Pip-cA	OH	–	NH_2	piperidino	12	+
8-CPT-cA	OH	–	NH_2	Chloro-phenyl-thio	36	+
8-Br-cA	OH	–	NH_2	Br	1.8	+
8-Azido-cA	OH	–	NH_2	N_3	–	+
2'-O-butyrate-cA	$-O-C_4OH_7$	–	NH_2	–	–	+
8-CPT-2'-O-Me-cA	$-O-CH_3$	–	NH_2	Chloro-phenyl-thio	–	–
cGMP	OH	NH_2	=O	–	0.48	+
8-Br-cG	OH	NH_2	=O	Br	1.2	+
8-CPT-cG	OH	NH_2	=O	Chloro-phenyl-thio	27	+
8-Fluo-cG	OH	NH_2	=O	Fluo	–	+

Compounds resulting from substitution of cPUMP (Fig. 2A). For numbering of substituents, see Fig. 2A. 8-Fluo-cGMP is 8-[[2-[(Fluoresceinylthioureido)amino]ethyl]thio]-cGMP. The lipophilicity data are adapted from [2] and www.biolog.de.

E. coli CAP protein that binds cAMP in the *anti*-conformation [21] and to the surface cAMP receptor of *Dictyostelium* [22], but not to cA-PK [23] and cG-PK, which bind cAMP in the *syn* conformation. 8-azido-cAMP can be used to photo-affinity label cA-PK in broken cell preparations [24]. This compound may not work in intact cells, since it is labile in a reducing environment, where it quickly decomposes to 8-NH_2-cAMP [25]. 8-Fluo-cAMP and 8-NBD-cAMP are both useful fluorescent cAMP analogs (for more information, see www.biolog.de). 8-Cl-cAMP is slowly metabolized intracellularly to toxic 8-Cl-AMP and 8-Cl-adenosine, and its antitumor properties [26,27] may be related to this fact [28,29]. The 7-deaza-cAMP molecule has intact ability to bind to cA-PK, but is metabolized to the highly toxic [30] compound 7-deaza-adenosine (tubercidine).

Most pyrimidine ring substituents are in the 6 or 2 position, in which cAMP and cGMP differ (Table III). The introduction of an electron-withdrawing group (-Cl, -CF_3) in position 2 will enhance binding to site B of cA-PK, but decrease binding to cG-PK [13,31]. The introduction into cGMP of certain bulky NH-substituents in 2-position can greatly enhance the specificity of binding to cG-PKIα compared to cG-PKIβ [9]. The introduction of bulky hydrophobic NH-substituents in 6-position will produce molecules with preference for site A of cA-PK, low affinity for cG-PK, and resistance to degradation by cyclic nucleotide phosphodiesterases.

Modification of the purine ring structure itself has resulted in compounds with enhanced affinity for site B of cA-PKII (such as Sp-5,6-diCl-cBIMPS; Fig. 2B) or cG-PKI (such as 8-Br-PET-cGMP; Fig. 2C). Fluorescent analogs can also be produced by ring extension, as in 1,N^6-etheno-cAMP [32].

The ribose moiety of cAMP and cGMP is tolerant to modification regarding binding to the *Dictyostelium* surface receptor [22] and EPAC (Table I), but not regarding binding to cA-PK and cG-PK. The 2'-O-butyryl-substituents give at least 1000-fold decreased binding activity but are useful to enhance membrane penetration. Since the butyrate is removed quickly by ubiquitous esterases in the cell, the parent compound will be formed intracellularly (trapping effect).

In the cyclic phosphate moiety the equatorial (Rp) or axial (Sp) oxygen (O) can be substituted with a sulphur (S) (see also Fig. 2B and 2E). This modification greatly increases the stability toward phosphodiesterases. Sp-cAMPS and Sp-cGMPS and their derivatives are used as hydrolysis-resistant agonists of cA-PK and cG-PK. It is interesting that Sp-8-Br-PET-cGMPS is an antagonist for the retinal type ion channels and an activator of cG-PK [17]. Rp-cAMPS and Rp-cGMPS and their derivatives act as partial agonists or pure antagonists for cA-PK and cG-PK. Rp-8-Br-cAMPS is an excellent inhibitor of cA-PKI, whereas Rp-cAMPS preferentially inhibits cA-PKII. Rp-8-Br-cGMPS is an antagonist of cG-PK and an activator of the retinal cGMP-gated ion channel [18]. The ability of cNMP analogs to penetrate cell membranes is highly improved [33] by masking the cyclic phosphate charge by esterification (cNMP-acetoxymethyl ester, e.g. Sp-cAMPS-AM, Fig. 2E) or coupling to a coumarin derivative. The cNMP-AM analogs will be trapped intracellularly due to esterase cleavage of the AM ester. The coumarin

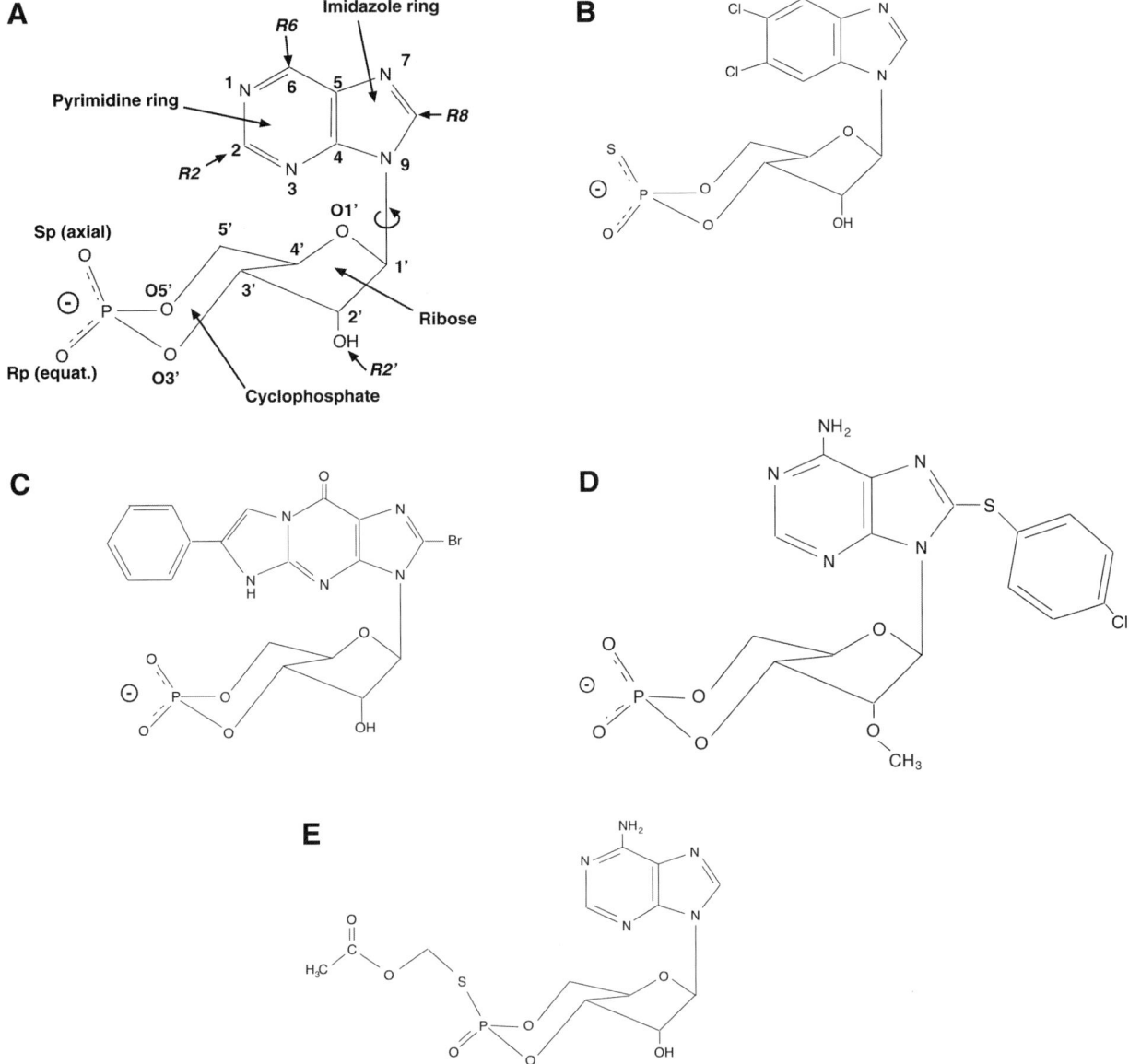

Figure 2 The structure of some cyclic nucleotide analogs. Panel A shows the structure of cyclic purine monophosphate (cPuMP), which can be considered the backbone of cAMP, cGMP, and the other analogs listed in Table III. The most common sites for substitution are indicated (R2′, R2, R6, and R8). Panel B shows Sp-cAMPS with modified pyrimidine ring (Sp-5, 6-di-chloro-cBIMPS), panel C shows 8-Br-cGMP with a phenyl-etheno extension of the pyrimidine ring, panel D shows 8-CPT-cAMP with a modified 2′-position (8-CPT-2′-O-Me-cAMP), and panel E shows the acetoxymethyl-modified Sp-cAMPS.

derivatives are caged, and in principle they are immediately converted to the active parent compound upon irradiation. Other modifications of the cyclic phosphate ring are replacements of the O3′ and O5′ by NH [22].

Future Developments

Currently available cNMP analogs have proven to be very useful tools in cell biology. There is nevertheless room for considerable progress regarding both the synthesis of more specific and potent analogs with improved pharmacokinetic properties and mapping of the cNMP receptors. An interesting novel approach is to synthesize polymer-linked cNMPs to simultaneously occupy two binding sites in the same receptor complex [34].

Acknowledgments

We are grateful to Dr. Erik Maronde for critically reviewing the manuscript.

References

1. Robison, G. A., Butcher, R. W., and Sutherland, E. W. (1968). Cyclic AMP. *Annu. Rev. Biochem.* **37**, 149–174.
2. Schwede, F., Maronde, E., Genieser, H., and Jastorff, B. (2000). Cyclic nucleotide analogs as biochemical tools and prospective drugs. *Pharmacol. Ther.* **87**, 199–226.

3. de Rooij, J., Zwartkruis, F. J., Verheijen, M. H., Cool, R. H., Nijman, S. M., Wittinghofer, A., and Bos, J. L. (1998). Epac is a Rap1 guanine-nucleotide-exchange factor directly activated by cyclic AMP. *Nature* **396**, 474–477.

4. Kawasaki, H., Springett, G. M., Mochizuki, N., Toki, S., Nakaya, M., Matsuda, M., Housman, D. E., and Graybiel, A. M. (1998). A family of cAMP-binding proteins that directly activate Rap1. *Science* **282**, 2275–2279.

5. Nakamura, T. and Gold, G. H. (1987). A cyclic nucleotide-gated conductance in olfactory receptor cilia. *Nature* **325**, 442–444.

6. Zufall, F., Firestein, S., and Shepherd, G. M. (1994). Cyclic nucleotide-gated ion channels and sensory transduction in olfactory receptor neurons. *Annu. Rev. Biophys. Biomol. Struct.* **23**, 577–607.

7. Hereld, D. and Devreotes, P. N. (1992). The cAMP receptor family of Dictyostelium. *Int. Rev. Cytol.* **137B**, 35–47.

8. Ullmann, A. and Danchin, A. (1983). Role of cAMP in bacteria. *Adv. Cyclic Nucl. Res.* **15**, 1–53.

9. Francis, S. H. and Corbin, J. D. (1994). Structure and function of cyclic nucleotide-dependent protein kinases. *Annu. Rev. Physiol.* **56**, 237–272.

10. Christensen, A. E., Dao, K. K., Nilsen, O. K., deRooij, J., Bos, J. L., and Døskeland, S. O. (2001). Comparison of the cAMP binding sites of EPAC-1 and cAMP-kinase. *FASEB J.* **15**, A9.

11. Enserink, J. M., Christensen, A. E., de Rooij, J., van Triest, M., Schwede, F., Genieser, H. G., Døskeland, S. O., Blank, J. L., and Bos, J. L. (2002). A novel Epac-specific cAMP analogue reveals independent regulation of Rap1 and ERK by cAMP. *Nat. Cell. Biol.*, **4**, 901–906.

11a. Kopperud, R., Krakstad, C., Selheim, F., and Døskeland, S. O. (2003). *Febs. Lett.* (in press).

12. Scott, S. P. and Tanaka, J. C. (1995). Molecular interactions of 3′,5′-cyclic purine analogues with the binding site of retinal rod ion channels. *Biochemistry* **34**, 2338–2347.

13. Corbin, J. D., Øgreid, D., Miller, J. P., Suva, R. H., Jastorff, B., and Døskeland, S. O. (1986). Studies of cGMP analog specificity and function of the two intrasubunit binding sites of cGMP-dependent protein kinase. *J. Biol. Chem.* **261**, 1208–1214.

14. Døskeland, S. O., Bøe, R., Bruland, T., Vintermyr, O. K., Jastorff, B., and Lanotte, M. (1991). *Cell Signal. Exp. Strat.* **21**, 103–114.

15. Døskeland, S. O., Maronde, E., and Gjertsen, B. T. (1993). The genetic subtypes of cAMP-dependent protein kinase-functionally different or redundant? *Biochim. Biophys. Acta* **1178**, 249–258.

16. Tanaka, J. C., Eccleston, J. F., and Furman, R. E. (1989). Photoreceptor channel activation by nucleotide derivatives. *Biochemistry* **28**, 2776–2784.

17. Wei, J. Y., Cohen, E. D., Yan, Y. Y., Genieser, H. G., and Barnstable, C. J. (1996). Identification of competitive antagonists of the rod photoreceptor cGMP-gated cation channel: beta-phenyl-1,N2-etheno-substituted cGMP analogues as probes of the cGMP-binding site. *Biochemistry* **35**, 16815–16823.

18. Wei, J. Y., Cohen, E. D., Genieser, H. G., and Barnstable, C. J. (1998). Substituted cGMP analogs can act as selective agonists of the rod photoreceptor cGMP-gated cation channel. *J. Mol. Neurosci.* **10**, 53–64.

19. Davies, S. P., Reddy, H., Caivano, M., and Cohen, P. (2000). Specificity and mechanism of action of some commonly used protein kinase inhibitors. *Biochem. J.* **351**, 95–105.

20. Schwede, F., Christensen, A., Liauw, S., Hippe, T., Kopperud, R., Jastorff, B., and Doskeland, S. O. (2000). 8-Substituted cAMP analogues reveal marked differences in adaptability, hydrogen bonding, and charge accommodation between homologous binding sites (AI/AII and BI/BII) in cAMP kinase I and II. *Biochemistry* **39**, 8803–8812.

21. Passner, J. M. and Steitz, T. A. (1997). The structure of a CAP-DNA complex having two cAMP molecules bound to each monomer. *Proc. Natl. Acad. Sci. USA* **94**, 2843–2847.

22. Theibert, A., Palmisano, M., Jastorff, B., and Devreotes, P. (1986). The specificity of the cAMP receptor mediating activation of adenylate cyclase in Dictyostelium discoideum. *Dev. Biol.* **114**, 529–533.

23. Su, Y., Dostmann, W. R., Herberg, F. W., Durick, K., Xuong, N. H., Ten Eyck, L., Taylor, S. S., and Varughese, K. I. (1995). Regulatory subunit of protein kinase A: structure of deletion mutant with cAMP binding domains. *Science* **269**, 807–813.

24. Taylor, S. S., Buechler, J. A., and Yonemoto, W. (1990). cAMP-dependent protein kinase: framework for a diverse family of regulatory enzymes. *Annu. Rev. Biochem.* **59**, 971–1005.

25. Øgreid, D. and Døskeland, S. O. (1982). Activation of protein kinase isoenzymes under near physiological conditions. Evidence that both types (A and B) of cAMP binding sites are involved in the activation of protein kinase by cAMP and 8-N3-cAMP. *FEBS Lett.* **150**, 161–166.

26. Ally, S., Tortora, G., Clair, T., Grieco, D., Merlo, G., Katsaros, D., Øgreid, D., Døskeland, S. O., Jahnsen, T., and Cho-Chung, Y. S. (1988). Selective modulation of protein kinase isozymes by the site-selective analog 8-chloroadenosine 3′,5′-cyclic monophosphate provides a biological means for control of human colon cancer cell growth. *Proc. Natl. Acad. Sci. USA* **85**, 6319–6322.

27. Srivastava, R. K., Srivastave, A. R., and Cho-Chung, Y. S. (2000). Synergistic effects of 8-Cl-cAMP and retinoic acids in the inhibition of growth and induction of apoptosis in ovarian cancer cells: induction of retinoic acid receptor beta. *Mol. Cell. Biochem.* **204**, 1–9.

28. Vintermyr, O. K., Bøe, R., Brustugun, O. T., Maronde, E., Aakvaag, A., and Døskeland, S. O. (1995). Cyclic adenosine monophosphate (cAMP) analogs 8-Cl- and 8-NH$_2$-cAMP induce cell death independently of cAMP kinase-mediated inhibition of the G1/S transition in mammary carcinoma cells (MCF-7). *Endocrinology* **136**, 2513–2520.

29. Taylor, C. W. and Yeoman, L. C. (1992). Inhibition of colon tumor cell growth by 8-chloro-cAMP is dependent upon its conversion to 8-chloro-adenosine. *Anticancer Drugs* **3**, 485–491.

30. Kozlowska, M., Smolenski, R. T., Makarewicz, W., Hoffmann, C., Jastorff, B., and Swierczynski, J. (1999). ATP depletion, purine riboside triphosphate accumulation and rat thymocyte death induced by purine riboside. *Toxicol. Lett.* **104**, 171–181.

31. Døskeland, S. O., Øgreid, D., Ekanger, R., Sturm, P. A., Miller, J. P., and Suva, R. H. (1983). Mapping of the two intrachain cyclic nucleotide binding sites of adenosine cyclic 3′,5′-phosphate dependent protein kinase I. *Biochemistry* **22**, 1094–1101.

32. White, H. D., Smith, S. B., and Krebs, E. G. (1983). Use of 1,N^6-etheno-cAMP as a fluorescent probe to study cAMP-dependent protein kinase. *Methods Enzymol.* **99**, 162–167.

33. Maronde, E., Korf, H. W., Niemann, P., and Genieser, H. G. (2001). Direct comparison of the potency of three novel cAMP analogs to induce CREB-phophorylation in rat pinealocytes. *J. Pineal. Res.* **31**, 183–185.

34. Kramer, R. H. and Karpen, J. W. (1998). Spanning binding sites on allosteric proteins with polymer-linked ligand dimers. *Nature* **395**, 710–713.

35. Christensen, A. E., Selheim, F., deRooij, J., Dremier, S., Schwede, F., Dao, K., Martinez, A., Maenhaut, C., Bos, J. L., Genieser, H.-G., and Døskeland, S. O. (2003). cAMP analog mapping of Epac1 and cAMP-kinase. Discriminating analogs demonstrate that Epac and cAMP-kinase act synergistically to promote PC-12 cell neurite extension. *J. Biol. Chem.* (in press).

36. Øgreid, D., Ekanger, R., Suva, R. H., Miller, J. P., and Døskeland, S. O. (1989). Comparison of the two classes of binding sites (A and B) of type I and type II cyclic-AMP-dependent protein kinases by using cyclic nucleotide analogs. *Eur. J. Biochem.* **181**, 19–31.

SECTION G

G Proteins

Heidi Hamm, Editor

CHAPTER 213

Signal Transduction by G Proteins—Basic Principles, Molecular Diversity, and Structural Basis of Their Actions

Lutz Birnbaumer

Laboratory of Signal Transduction,
National Institute of Environmental Health Sciences,
Research Triangle Park, North Carolina

Introduction

Cells do not live in isolation and respond to extracellular stimuli either with specific tasks such as neurotransmission, activation of substrate uptake, fatty acid release, or with adaptive changes that ensure homeostatic cohabitation with other cell types as is needed in multicellular organisms to ensure better survival (stress responses). Extracellular cues include nutrients, intoxicants, and of course, signaling molecules such as autacoids, growth factors, and hormones. Signaling molecules are either membrane-permeant (e.g., steroid hormones) or -impermeant (peptide hormones, biogenic amines) for which cells have evolved separate response mechanisms. For membrane-permeant signals, receptors are for the most part intracellular, cytosolic, or nuclear, and the changes they elicit are frequently modulations of gene expression with direct participation of the receptor in the regulatory complex that transcribes response genes. For membrane-impermeant signals, nature has evolved a repertoire of mechanisms by which binding of the ligand to its receptor on the cell surface leads to intracellular changes in one or more enzymatic activities and to activation or inhibition of regulatory signaling pathways. In many instances the action(s) of a receptor involve promotion or disruption of multimeric protein complexes. Some of the signaling pathways activated by receptors in response to membrane-impermeant ligands are wholly cytosolic or submembranous; others affect nuclear gene expression. The process by which the extracellular ligand-receptor interaction leads to changes inside the cell is commonly referred to as "signal transduction"—the ligand is the extracellular signal whose message is transduced into an intracellular signal of a different chemical nature.

Among the signals generated in this way are second messengers, such as ions entering through ligand-gated ion channels; signaling molecules derived from the receptor itself, such as the cytosolic domain of Notch; products of the activation of the receptors' intrinsic enzymatic activities, such as cGMP formed by the guanylyl cyclase activity of ANF and guanylin receptors; tyrosine phosphorylation, first of self and then of proteins recruited to the phosphorylated receptor, as happens when epidermal growth factor (EGF), platelet-derived growth factor (PDGF), insulin, or nerve growth factor (NGF) bind to their respective homodimeric receptors; and sequential serine/threonine phosphorylation, also first of self (receptor II phosphorylating receptor I) and then of cytosolic transducing proteins such as R-SMADs by phosphorylated receptor I, as it happens in response to interaction of the transforming growth factor β (TGFβ) and

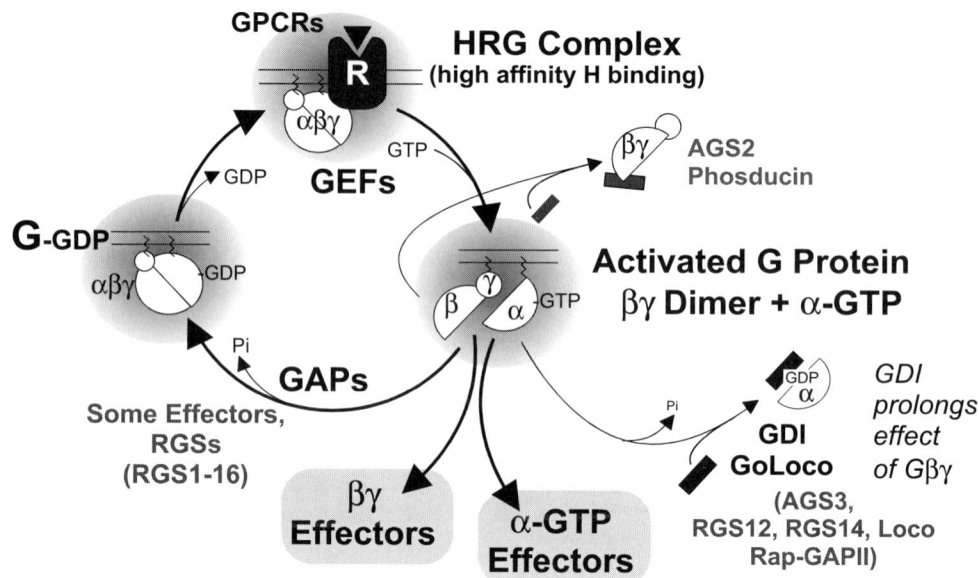

Figure 1 Hormone receptor complexes act as guanine nucleotide exchange factors (GEFs) to catalyze activation of the trimeric G protein with formation of α-GTP plus βγ, both of which modulate effector functions. Spontaneous decay of α-GTP to α-GDP can be modulated by GTPase activating proteins (GAPs) such as RGSs and some effectors, and by activators of G protein signaling (AGSs) that tilt the balance between effects through α-GTP versus those through βγ.

bone morphogenic protein (BMP) superfamily of signaling molecules with their heteromeric receptors. In the case of R-SMADs, their phosphorylation exposes a nuclear localization signal as well as a heterodimerization domain, leading ultimately to translocation into the nucleus and changes in gene expression. For general references see [1].

The use of heterotrimeric G proteins as signal transducers constitutes a completely different type of signal transduction process. In this case evolution has lead a structurally related superfamily of receptors that recognize peptide and protein hormones, neurotransmitters, autacoids, and sensory signals such as light, odor, and taste, to acquire the ability to recruit and regulate the intrinsic signaling capacity of a large family of structurally related heterotrimeric regulatory GTPases (Fig. 1). This chapter deals with the mechanism of activation/deactivation and the action of the heterotrimeric G proteins responsible for transducing receptor activation signals into cellular responses. Activation of a G protein by a receptor, first proposed by Rodbell and collaborators to be a two-step process in which a GTP-dependent transducer regulates an enzyme or amplifier without clear knowledge of the number of molecular components involved [2], is now known to be a multistep process that includes GDP/GTP exchange and subunit dissociation of the αβγ G protein, followed by spontaneous deactivation—hydrolysis of the activating GTP to GDP—and reassociation of the separated members in preparation for reactivation by a new round of GDP/GTP exchange promoted by the activating hormone receptor (HR) complex (bottom panel of Fig. 1). As shown in this figure, the basic cycle is affected by modulators of G-protein signaling, the RGSs, which increase the rate at which α-GTP deactivates and the AGS2- and AGS3-type modulators which sequester either α-GDP or Gβγ and thereby bias signaling through α-GTP versus signaling through Gβγ (see the following sections).

Ras, the Prototypic Regulatory GTPases

As the name indicates, regulatory GTPases are proteins that bind and hydrolyze GTP. Their regulatory power lies in the fact that their conformation differs when occupied by GTP or GDP. They are also referred to as molecular switches. Due to their intrinsic GTPase activity they carry a built-in inactivating timer that prevents the GTP state from being long-lived. The crystal structure of the 180 amino acid regulatory GTPase *ras*, in its GTP- and GDP-liganded forms, shows two principal regions that differ in the GTP state as compared to the GDP state, referred to as switch I and switch II [3]. Switch I, amino acids 32 through 40, is a large loop connecting α-helix 1 (α1) to β-strand 2 (β2). Mutations in this region interfere with activation of several of the downstream effectors of *ras*. Switch II, amino acids 60 through 75, changes conformation even more drastically than switch I, to the extent that upon binding of GTP, amino acids 66 through 74 rearrange into a well-ordered α-helix (α2). Switches I and II are required for activation of downstream effector(s) by *ras*-GTP, the activated form of *ras*. *Ras*-GDP appears to be neutral. Mutations of Gln-61, (at the start of switch II after β3), and of Gly-12 (at the base of α1), reduce the intrinsic GTPase activity of *ras* prolonging the life span of the activated GTP state. As is the case for most, if not all regulatory GTPases, the *ras* GTPase is under regulation of two types of proteins, a GEF or guanine nucleotide

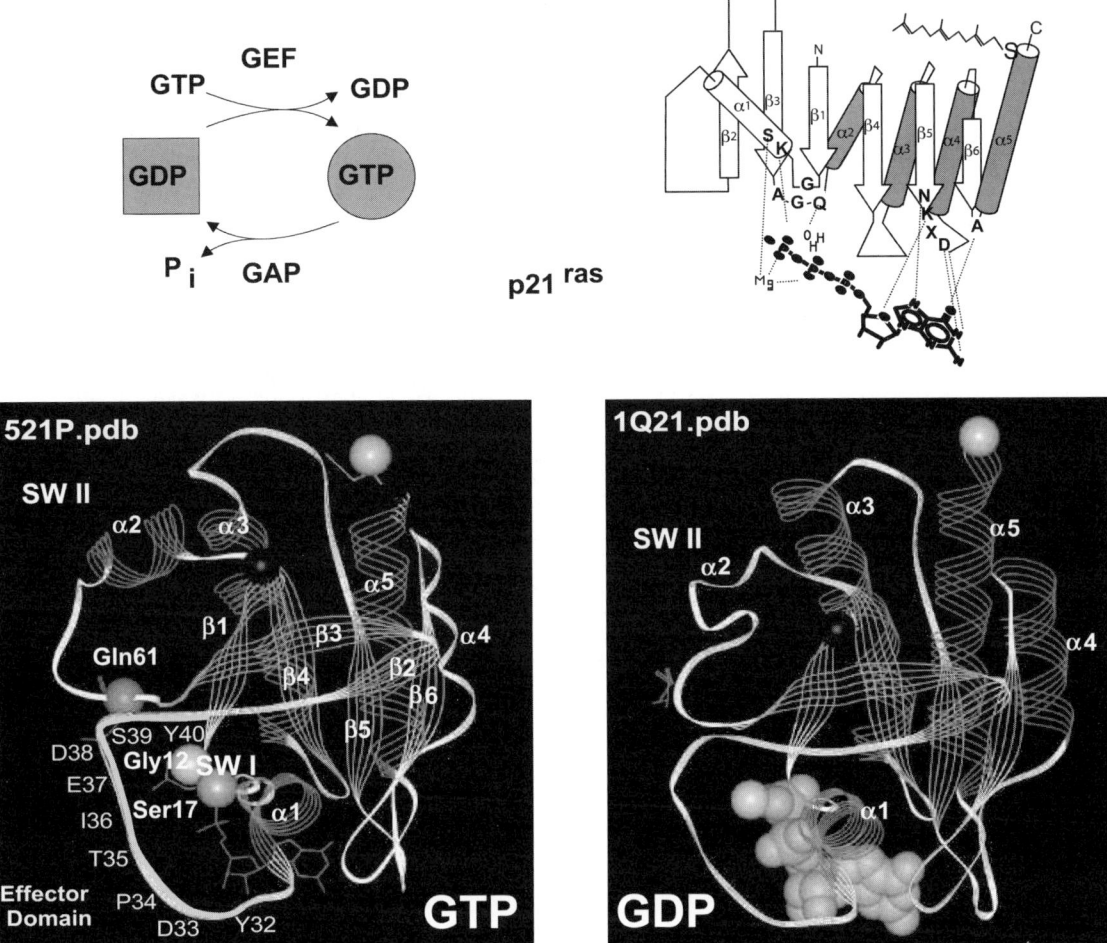

Figure 2 Model of the crystal structure of *ras* in its GTP and GDP states (PDB accession #521P and 1Q21) highlighting secondary structures (α-helices and β-strands as well as switch 1 (effector domain) and switch 2 regions. Bottom left illustrates the basic GTPase regulatory cycle and its two principal modulators: GEFs, responsible for activation, and GAPs, responsible for deactivation of the regulatory GTPases. The bottom right diagram is a two-dimensional representation of the three-dimensional features of the basic GTPase fold found in all regulatory GTPases.

exchange factor, responsible for promoting the transition of *ras*-GDP to *Ras*-GTP, and several GAPs or GTPase activating proteins, which, as their name indicates, shorten the life span of *ras*-GTP by increasing the catalytic efficacy of its intrinsic GTPase (see [4,5], and references therein). SOS (son of sevenless) and *ras*-GAP1, are the prototypes of *ras* GEF and *ras* GAP, respectively. Mutation of Ser-17 to Asn-17, roughly in the middle of α2, interferes with the nucleotide exchange reaction and locks *ras*-GDP in an inactive conformation in which it may still bind to proteins regulated by *ras*-GTP, but without affecting their activity. By reason of occupying the site to which *ras*-GTP should bind to activate effectors, Asn-17 *ras* interferes with signaling by wild-type *ras*-GTP and is referred to as a dominant negative form of *ras* [6]. Figure 2 depicts *ras* in its GTP and GDP conformations and highlights the locations of Gly-12 (G12), Ser-17 (S17), and Gln-61 (Q61), as well as switch I (in blue) and switch II. The upper left panel of Fig. 2 presents this basic GTPase cycle in schematic form with GEF and GAP regulating the life-times of the GTP and GDP sates. The upper right panel is a schematic two-dimensional representation of the

main three-dimensional features of this regulatory GTPase (adapted from reference [3])

Heterotrimeric G Proteins

Trimeric G proteins are responsible for transducing the effects of the seven-transmembrane superfamily of receptors, and are regulatory GTPases which, while more complex than *ras*, nevertheless preserve the basic features of the *ras*-type regulatory GTPases. G proteins activated by seven-transmembrane receptors are α, β, γ trimers, of which the α subunit is the GTPase-bearing subunit. The β and γ subunits form a dimer that exists either free or in association with α-GDP. Gβ and Gγ have never been found isolated as individual proteins.

Subunit Structure

α-Subunits were crystallized in the laboratories of PaulSiegler at Yale University and of Steven Sprang at

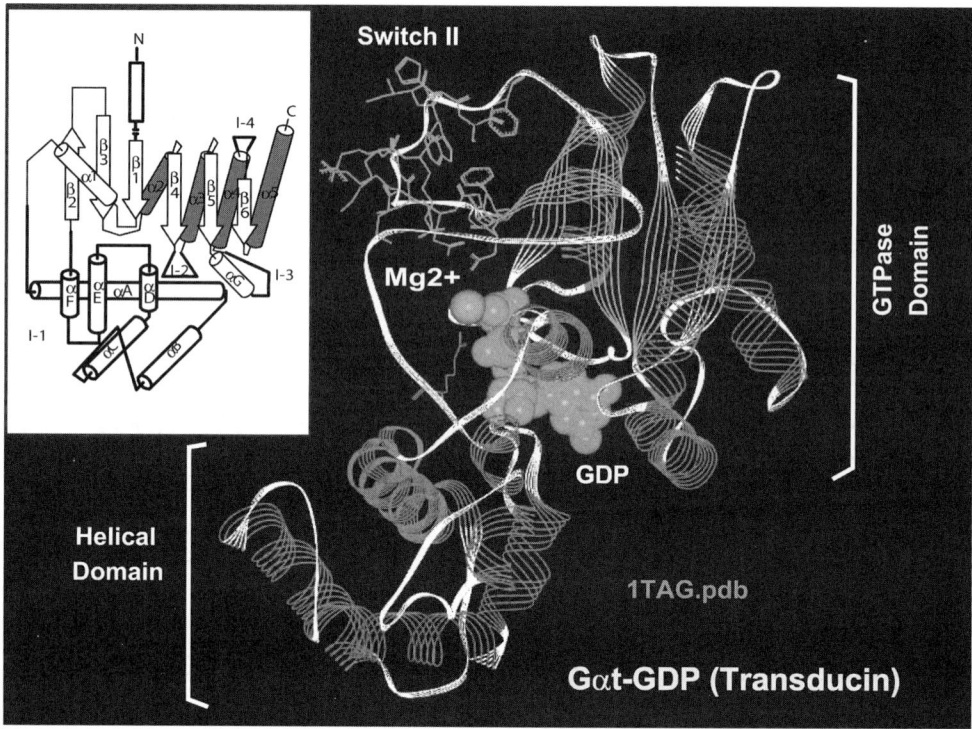

Figure 3 Crystal structure of an α subunit of a heterotrimeric G protein in its GDP state (PDB #1TAG) Its GTPase domain is oriented in the same way as *ras* in Fig. 2. Atoms of amino acid side chains of switch II two are displayed as sticks; space-filling CPK atoms were used to represent GDP. Mg is a cyan ball. Upper right inset: Two-dimensional diagram of three-dimensional features of heterotrimeric G-protein α subunit. Additions to the *ras* structure are in heavy lines. GTPase and helical domains are highlighted.

the University of Texas at Dallas, in collaboration with Heidi Hamm and Alfred Gilman, respectively [7,8]. From a structural viewpoint, α subunits of heterotrimeric G proteins are made up of a *ras*-like GTPase domain and a helical domain with 6 α-helices (αA through αF) inserted in the center of what would be switch I of the *ras* (Figs. 3 and 5). In its GDP-liganded form, Gαs have a disorganized switch II region and exhibit high affinity for Gβγ. Gα-GDPs are therefore found associated with Gβγs as heterotrimers. Gβγ locks GDP into its binding site on Gα, causing its dissociation rate to drop by a factor of 10 [9]. At the same time, Gβγ also shields the switch II region from interacting with possible effectors (Fig. 5). Consequently, activation of a heterotrimeric G protein requires two events: (1) exchange of GTP for GDP, and (2) dissociation of Gβγ from Gα. Subunit dissociation has not been measured in intact cells in an unequivocal way, but it is readily seen *in vitro*. Dissociation exposes the regulating surface of Gα-GTP and allows for effector regulation with direct involvement of switch II ([10]; PDB accession #1CS4). The converse also applies. That is, Gβγ is a signaling molecule able to regulate effector functions and its association with Gα-GDP resulting in occlusion of the signaling surface of its Gβγ.

The structure of Gβγ deserves separate comment ([11,12]; Figs. 4 and 5). Gβ is a seven-bladed propeller of which each blade is made up of four anti-parallel β-strands running from the center to the periphery. The innermost β-strand runs parallel to the axis of rotation of the propeller. The next two change pitch to approach the orientation of the outermost, which runs along the periphery of the propeller and is coplanar with the circle described by the rotating propeller. Blade seven is made up of three β-strands contributed by the very C terminus of Gβ and a fourth (outermost) "zipping" β-strand recruited from the sequence immediately preceding those that create blade I. Preceding the zipping β-strand is an extended N terminus with a long α-helix that interacts with the Gγ by forming a coiled-coil. Gβγ dimers have so far been crystallized only in association with Gα subunits or with a retinal regulatory protein, phosducin, but not in isolation. Conformational changes in Gβ that occur upon dissociating from Gα, if they occur, have not been observed as yet.

LIPID MODIFICATIONS

Gα and Gβγ subunits engaged in signal transduction are membrane bound by virtue of lipid modifications. Gβγ dimers are anchored to membranes through a C15 or C20 polyisoprene attached to the Cys of a C-terminal CAAX motif. Posttranslational processing not only attaches the polyisoprene, but also removes the last three amino acids and methylates the new C terminus. Prenylation is not necessary for association of Gγ with Gβ, but is required for association of Gβγ to Gα-GDP and for regulation of effector, e.g., adenylyl cyclase. Further, prenylation contributes to the association of the Gγ dimer to membranes.

CHAPTER 213 Signal Transduction by G Proteins—Basic Principles, Molecular Diversity, and Structural Basis of Their Actions   561

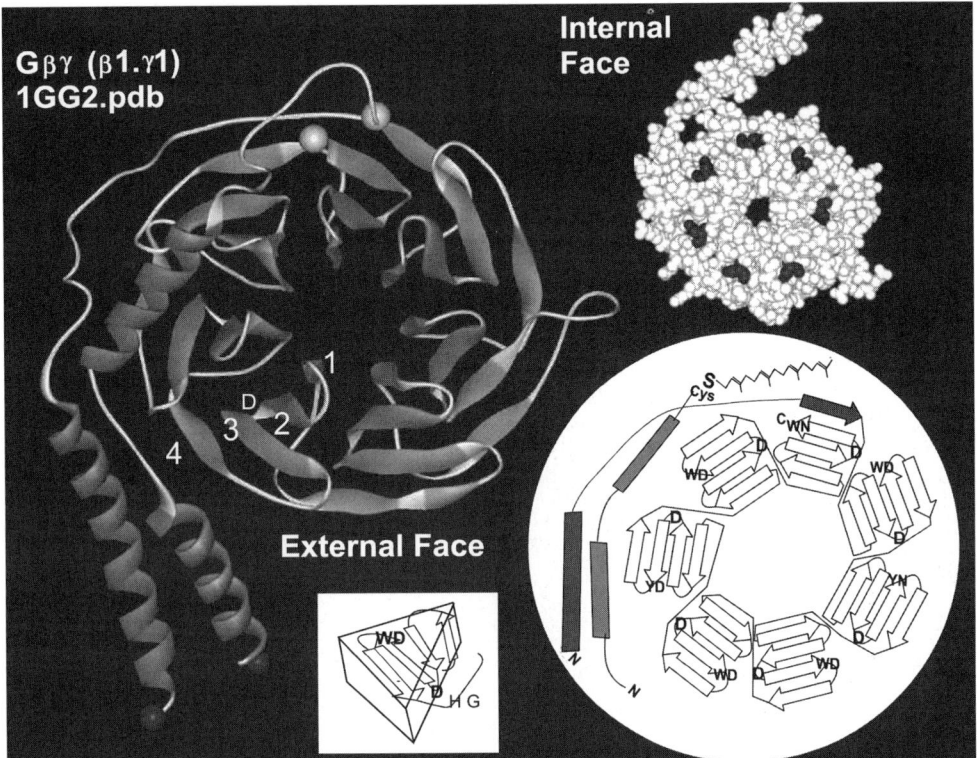

Figure 4 View of a seven-bladed Gβ propeller from the "external" side determined by its orientation when complexed with α-GDP. Upper right inset: CPK representation of the same propeller viewed from the side facing α-GDP in its trimeric form. Blue: Conserved aspartic acids (D) present in each propeller blade. Bottom right: Two-dimensional diagram of the three-dimensional features of a Gβ propeller highlighting the "zipping," outermost β-strand of blade seven, and location of conserved D residues shown in blue in the space-filling representation of the model. Note the γ coiled-coil interaction of the γ subunit with the N terminus of the Gβ subunit.

Most Gα subunits engaged in signal transduction are palmitoylated near their N terminus. Palmitoylation facilitates their anchoring to the plasma membrane. Gα subunits of the Gi/Go family are also myristolated at N-terminal glycines (Gly-2 of the primary transcript). Gα myristylation increases affinity for Gβγ dimers. Removal of myristic acid by Gly-2 to Ala-2 mutation renders Giα subunits inactive as inhibitors of adenylyl cyclase. Some but not all nonpalmitoylated Gα subunits fail to localize to membranes and are found in the cytosol. Lipid modification of Gα and Gγ subunits are therefore essential for their normal biological activity (Reviewed in reference[13]).

Molecular Diversity of G Proteins

Each of the subunits that make up a heterotrimeric G protein is encoded by a family of structurally homologous genes. There are 16 Gα (one with two splice variants), 5 Gβ, and 11 Gγ genes, raising the theoretical possibility of close to 1000 distinct heterotrimers. Gα subunits are the longest of the three subunits, ranging from 350 to 390. Their sequence similarities vary from almost identical when Gi1α is compared to Gi3α (86% identical) to up to 60% different when Gsα is compared to G16α. Amino acid sequence alignments show Gβs to be structurally a very closely related family with β1-4 being 350 amino acids long and differing by no more than 17% in their amino acid sequence. Gβ5 with 395 amino acids, exhibits the same degree of similarity, differing primarily by a 45 amino acid N-terminal extension. Gγs are the shortest (68–75 amino acids) and are the most diverse, differing in amino acid sequence between 40 and 65%. GGL (Gγ-like) domains of RGS 6, 7, 9, and 11 (for RGS see below) constitute an additional group of Gγ subunits that interact with the atypical N-terminal extension of Gβ5′.

The actual number of G-protein isoforms in any given cell is much lower than 1000 for two reasons: (1) there is no cell known that expresses all G-protein subunit genes, and (2) there are structural limitations that do not allow all βγ dimers to form, e.g., while β1 interacts with γ1, γ2, and γ3, β3 does not interact with γ1 or γ2, but partners with γ4. Vice versa, γ1 partners with β1 but not with either β2 or β3; and γ2 partners with β1 and β2, failing to do so with β3. The complete spectrum of permissible interactions among the 5β subunits and 11 γ subunits still needs to be worked out. One Gβ, Gβ5, interacts preferentially with the GGL domain of 6, 7, 9, and 11. It is quite possible that even for biochemically permissible interactions there may be βγ dimers that never form because they are not co-expressed in the same cell.

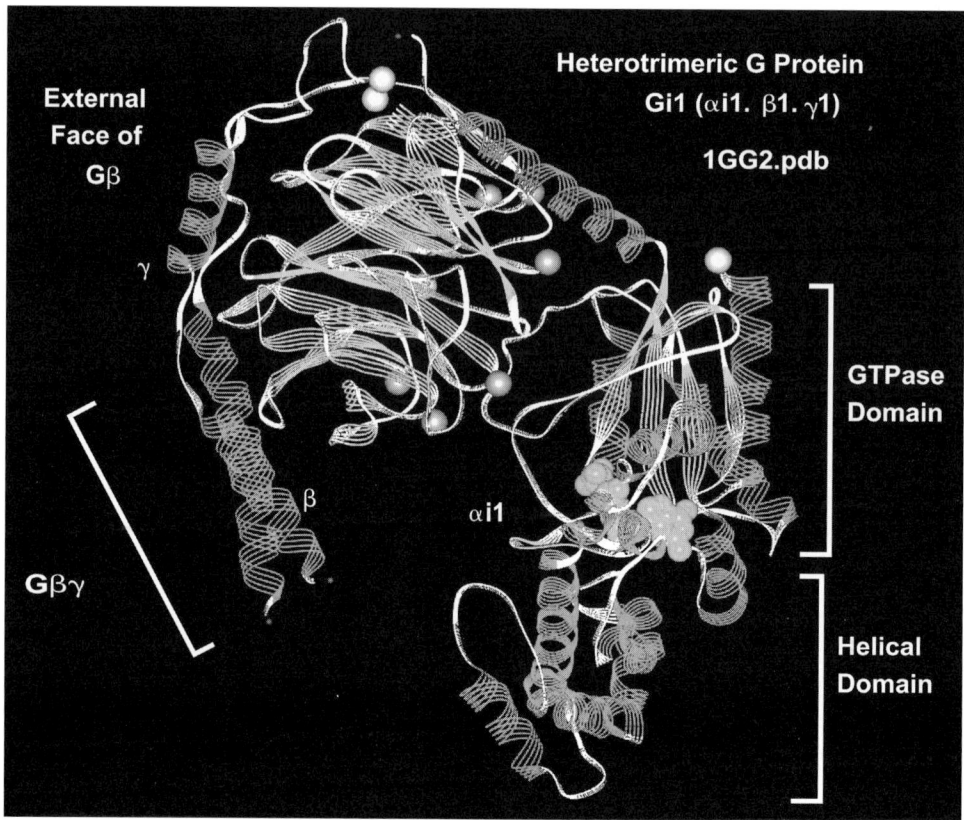

Figure 5 Model of Gi1 heterotrimer with bound GDP deduced from crystal x-ray diffraction studies (PDB accession #1GG2). Note that the internal face of Gβ shields the switch II of Gα subunit from being accessible to effectors. Blue balls, N-terminal α carbons; brown balls, C-terminal α carbons; gray balls on Gβ, α carbons of conserved aspartic acids.

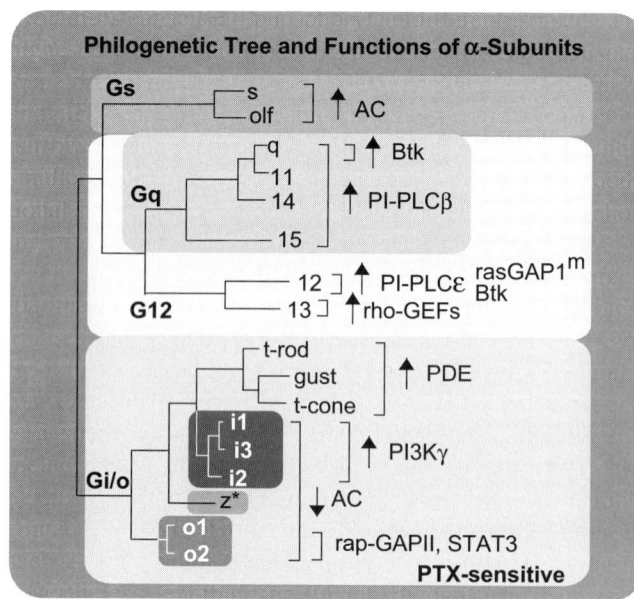

Figure 6 Phylogenetic analysis of G-protein α subunits reveals structural subdivisions that have functional correlates.

G proteins are named after their α subunit. This has its origin in the fact that for the first two G proteins discovered, Gs and transducin (Gt), it was established that the major (then sole) signaling function resided in their α subunits.

Gαs, activates adenylyl cyclases (ACs), and Gαt subunits activate visual phosphodiesterase (PDE, a tetramer of one α, one β, and two inhibitory γs). Activation of visual PDE results from the association of αt-GTPs with the PDEγs, thereby suppressing their inhibitory effects on PDEαβ. A phylogenetic tree of G-protein α subunits (Fig. 6), clusters their sequences into four subfamilies: Gs, Gq, G12/13, and the pertussis toxin (PTX)-sensitive Gα subunit subfamily. The latter includes three α subunits that play roles in light and taste perception, plus the α subunits of the Gi/Go family. PTX-sensitive α subunits not only show higher sequence similarity to each other than to the remaining α subunits, but as their name indicates, they are substrates for the ADP-ribosyl-transferase activity of the S1 subunit of pertussis toxin. The ADP-ribosylated amino acid is a Cys at position −4 from their C termini. PTX uncouples this group of G proteins from activation by receptors by virtue of creating of a steric hindrance to the G-protein::receptor interaction. Included in this structural group is the α subunit of Gz (αz), which lacks a Cys at -4 from the C terminus and is PTX-insensitive. It is functionally a Gi, its closest structural homologue. As illustrated in Fig 6, and in more detail in Fig. 7, α subunits modulate the activity of a large and diverse group of enzymes, including ACs, phospholipase Cβs (PLCβs), phosphatidylinositol 3-OH kinases (PI3Ks), as well as type 6 visual and type 3 gustatory PDEs and a Rho-GEF for regulation of cytoskeletal remodeling. An analysis of the evolution of the

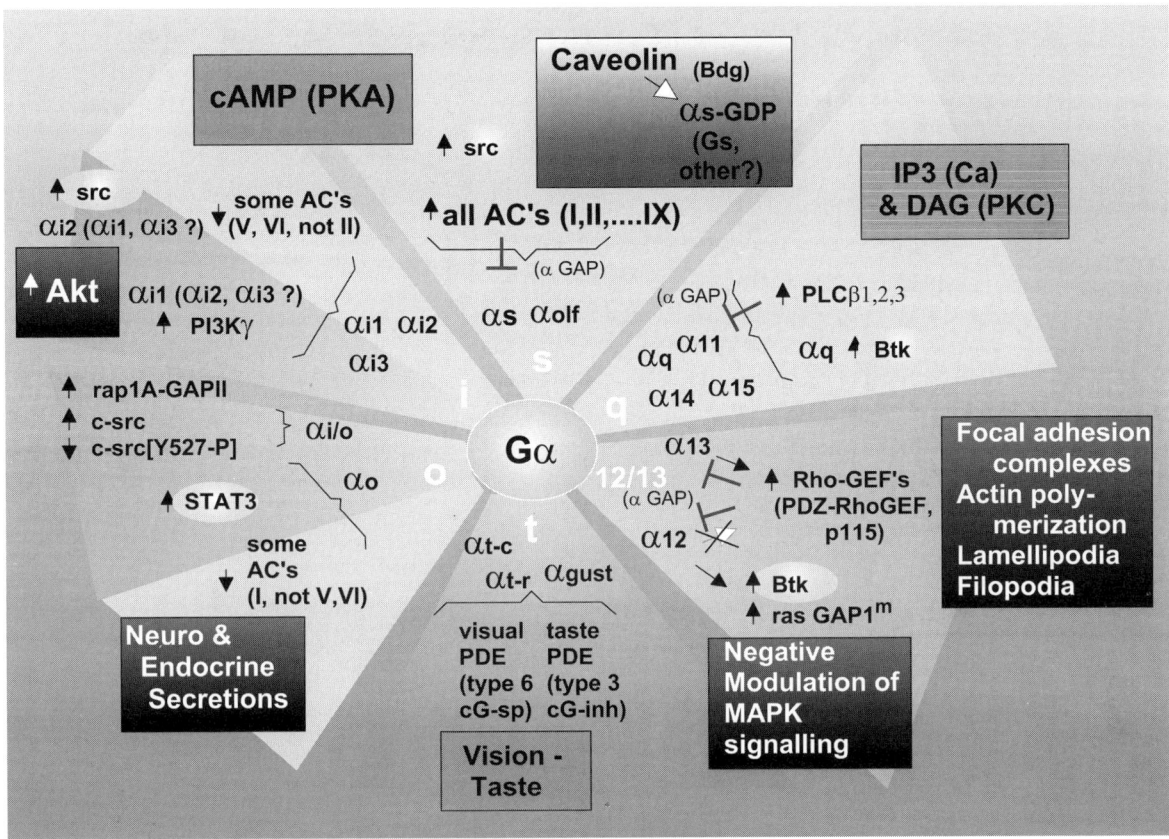

Figure 7 Effectors of activated α subunits. This figure shows the different signaling systems affected by α subunits at the periphery, with ever increasing detail toward the center. —|, inhibition.

intron-exon structure of G protein α subunits (Fig. 8) clusters the genes into the same groups: Gs, Gq/11, G12/13, and Gi/o and G-taste/vision. This type of analysis places Gz in the Gi/Go group solely on the basis of open reading frame sequence similarity, but shows no similarity in its intron-exon organization to those of any G protein as it has only three exons (two coding) instead of the eight to nine exons of the Gi/o family of α subunits. The loss of introns is highly suggestive of reactivation of a pseudogene with attendant chance mutation of the Cys at −4 from the C terminus.

Both Gα subunits and Gβγ dimers modulate, positively or negatively, a diverse set of cellular functions. In some instances, the effects of Gβγ dimers are in concert with those of α subunits, in others regulation of effector by Gβγ is unrelated to regulation by a Gα. It is not clear at this time whether effectors distinguish Gβγ isoforms. As illustrated on Fig. 9 the gamut of Gβγ effectors is as complex and as diverse as that of the α subunits. Among the regulated functions worth mentioning are inhibition of type 1 AC, but co-stimulation with Gsα of type 2 and type 4 ACs; stimulation of type 3 PLCβ independently of co-existing stimulation by the Gq/11 group of α subunits; co-stimulation with Gαi2 (possibly also αi1 and αi3) of PI3Kγ; and co-stimulation of PI3Kβ with tyrosine-phosphorylated p85, the regulatory subunit of PI3K. *In vitro* reconstitution experiments in which PI3Kβ was incubated with Gβγ and a tyrosine-phosphorylated peptide corresponding to the tyrosine-phosphorylated sequence of p85, showed that stimulation by each was three to fivefold, but became 100-fold when the tyrosine-phosphorylated peptide and the Gβγ were added together. This type of cross-dependence on dual inputs is highlighted in Figs. 7 and 9. Gβγ also modulates, sometimes positively and sometimes negatively, a variety of ion channels and thereby provides a nexus between regulation of second messenger formation by enzymes and regulation of cell excitability by voltage, which it can augment (activation of potassium channels) or dampen (inhibition of presynaptic Ca^{2+} channels) (Fig. 9).

In conclusion, signal transduction by G proteins is the result of structurally similar receptors activating structurally very similar G proteins, which then regulate positively or negatively the activity of a diverse gamut of structurally unrelated cellular functions that affect intracellular levels of cAMP, inositol tris-phosphate (IP3), Ca^{2+}, diacylglycerol (DAG), and phosphatidyl inositol 3 phosphates (PIP3s) as well as ion channel activity and formation of lamellipodia and filopodia and the attendant cytoskeletal changes. Moreover, because second messengers such as cAMP, DAG, and Ca^{2+} affect protein kinases, it is unlikely that there is a cellular function that in one way or another is not under the controlling or modulatory influence that emanates from activation of heterotrimeric G proteins. Indeed, as indicated in Figs. 7 and 9, cellular responses included are the PI3K-PDK-Akt-NF.kB anti-apoptotic response of cells to extracellular

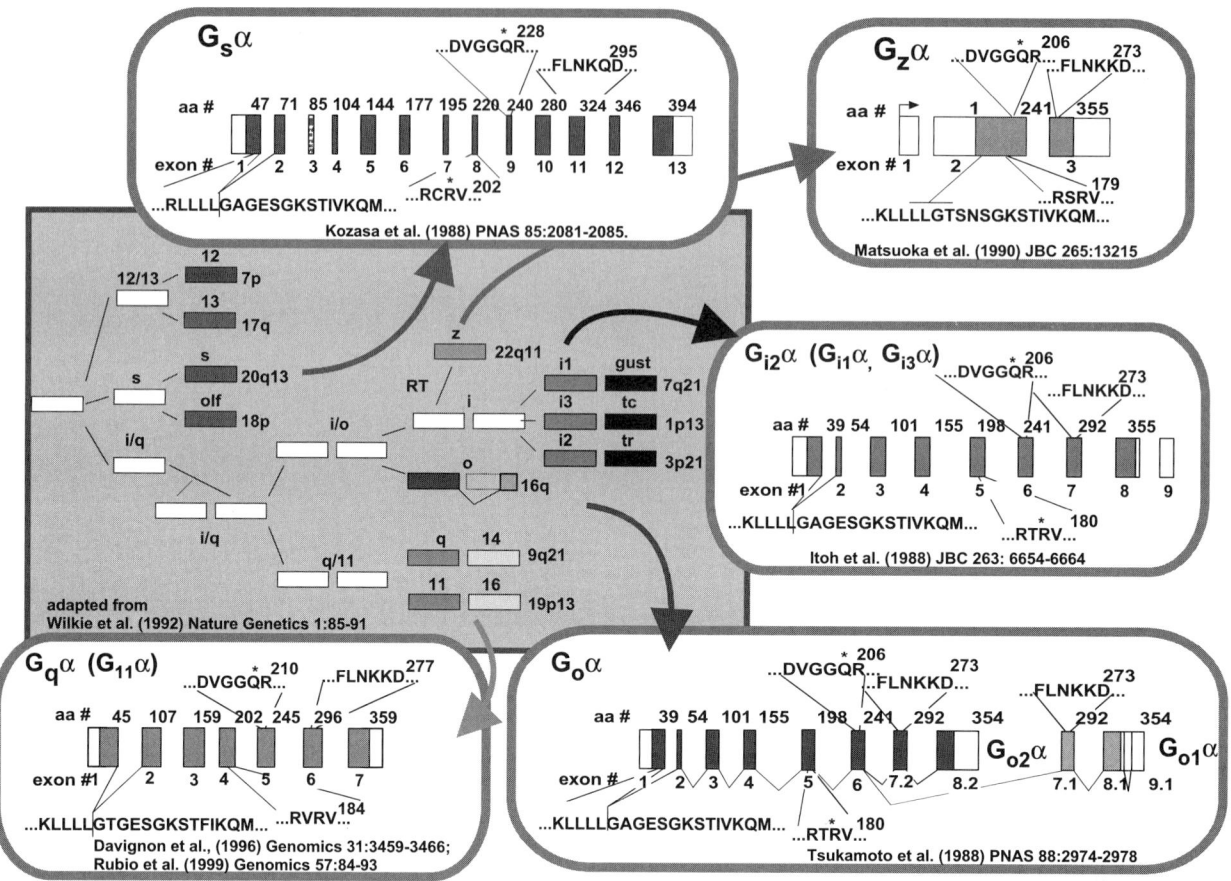

Figure 8 Analysis of the intron-exon structure and sequence similarities among proteins encoded in the exons, allow for development of model for the evolution of α subunit genes in which a series of gene duplications, loss of one intergenic region including most of the exons of the downstream gene followed by rescue of a pseudogene lead to the present day G-protein α subunits. The general layout of this figure was adapted from a similar layout in reference [26]. (Adapted from Wilkie et al., Nat. Genet., 1, 85–91, 1992. With permission). Other references on the figure refer to the publications in which the intron-exon structures shown were described.

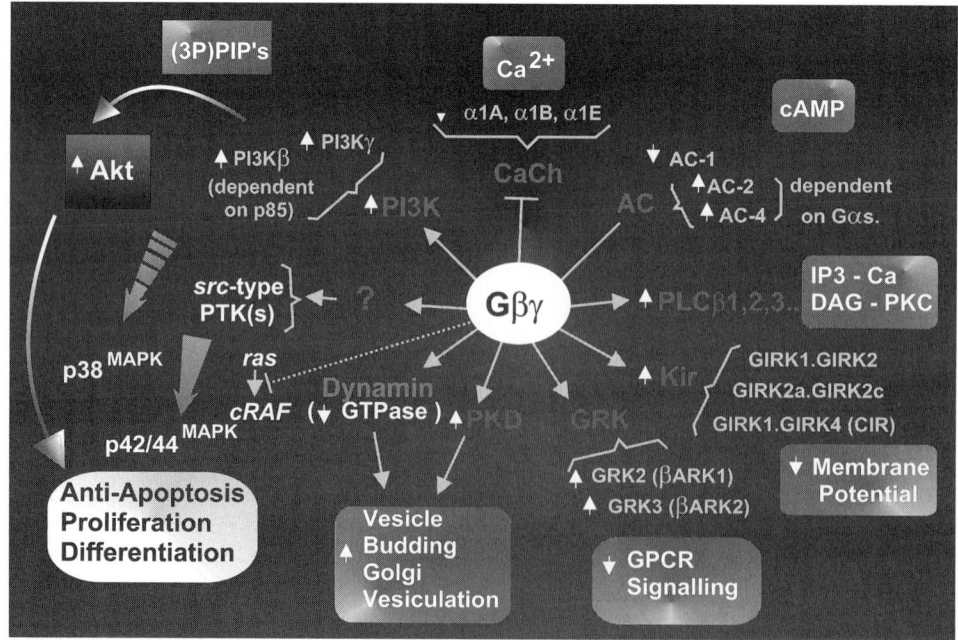

Figure 9 Effectors of βγ dimers. This figure shows the different intracellular trignaling systems affected by Gβg dimers at the periphery, with ever increasing detail toward the center. →, stimulation; —|, inhibition; ——, stimulation or inhibition.

signals and the effects of Gβγ on dynamin, Golgi vesiculation, and vesicle budding.

Mechanism of G-Protein Activation by Receptors and Modulation of Activity

Receptors acting through heterotrimeric G proteins are referred to as GPCRs or G-protein-coupled receptors. They play the role of GEFs in the regulatory GTPase cycle. Ligand binding to GPCRs, all of which belong to the rhodopsin superfamily of seven-transmembrane receptors, has as its final effect the GDP-GTP exchange with attendant subunit dissociation into Gα-GTP plus Gβγ (Fig. 1). The final effect of a hormone acting through a GPCR on any given cell depends on the type of G protein activated by the receptor, and the repertoire of effectors, i.e., regulatable enzymes, ion channels, and other affected molecules in the target cell.

Mechanism of Activation of a G Protein by a Receptor

At the molecular level, activation of a G protein by a rhodopsin-like receptor is still poorly understood. This is because of lack of knowledge of which amino acids of the receptor make contact with which amino acids of the G protein. In contrast, the regions of each molecule important for productive interaction are well known, as are some of the kinetic and molecular state changes that occur when a receptor under the influence of an activating ligand, i.e., agonist, interacts with and activates a G protein. Thus, binding of an agonist to a GPCR in the absence of guanine nucleotide (GTP or GDP), as can be done in vitro with purified membranes, has two consequences: (1) a shift of the equilibrium between two states of the receptor, from being mostly in state I (inactive) characterized by having low affinity for agonist as well as for the G protein(s), to mostly in state II (active) having higher affinity for the activating ligand, and (2) the stable association of the agonist-receptor complex to the G protein. The latter causes the G protein to reduce its affinity for GDP. Bound GDP, or prebound [3H]GDP, will thus dissociate under these conditions. Mg ion has to be present if the receptor is to cause GDP dissociation. Addition at this point of GTP or a GTP analog such as GTPγS or GMP-P(NH)P, leads to its binding in place of GDP and to the activation of the G protein as seen by stimulation of the activity of an effector such as adenylyl cyclase or visual phosphodiesterase. For most of the cases where this has been studied, a high concentration of Mg ion, about 50 mM, mimics the action of the agonist-activated receptor. With a purified G protein, incubation with Mg ion and GTPγS or GMP-P(NH)P leads not only to accumulation of Gα-bound guanine nucleotide, but also to subunit dissociation, i.e., formation of Gα-GTPγS plus free Gβγ, Dissociation is evident in several ways, the easiest being by a shift in sedimentation velocity from that corresponding to an approximate M_r of 100,000 protein (Gαβγ) to that of two co-sedimenting proteins of an approximate M_r of 50,000 (Gα-GTPγS + Gβγ). M_rs of α subunits are in the 40,000–50,000 range and those of Gβγ complexes are also approximately 50,000.

In intact membranes, where activation of a G protein of the αβγ type by agonist occupancy of a receptor can be measured in terms of stimulation of the activity of an effector, e.g., adenylyl cyclase, phospholipase C, visual phosphodiesterase, or an inwardly rectifying potassium channel, the net effect of receptor activity is thus facilitation of the activation of the G proteins by Mg ion. This comes about as a consequence of a receptor-induced shift in the apparent K_m for Mg ion from high millimolar to low micromolar. In other words, a receptor appears to act by reducing the concentration of Mg required for activation of the G protein by GTP from being above physiologic to being below physiologic. Free cytosolic Mg is in the order of 0.5 mM. The effect of glucagon (receptor) shifting the concentration of Mg ion required for activation of liver Gs, (the stimulatory regulatory component of adenylyl cyclase) by GTPγS, is illustrated in Fig. 10.

Structural Determinants

At the molecular level, mutational analysis has shown that amino acids in the third intracellular loop of GPCRs are involved in the ability of a GPCR to activate a G protein. Swapping intracellular loops between receptors of different G-protein preference, such as between M1 and M2 muscarinic receptors, or β and α1 adrenergic receptors, also points to the third intracellular loop as responsible for defining G-protein specificity. Further, most of the receptor mutations that are of the gain-of-function type are in the distal (C-terminal) end of the third intracellular loop. It is not known, however,

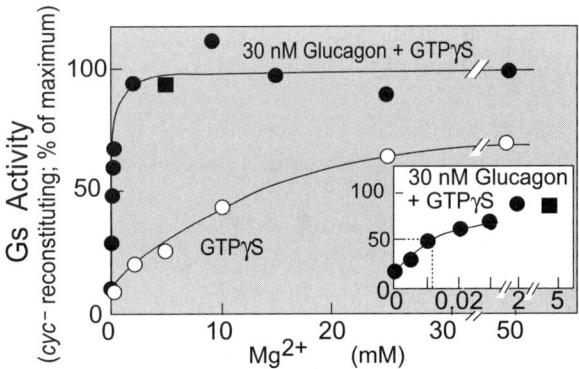

Figure 10 GPCRs reduce the concentration of Mg ion required for G-protein activation by GTP. The results of a three-step reaction are shown. In the first step (not shown), the adenylyl cyclase enzyme of liver membranes was inactivated by treatment with N-ethyl maleimide, leaving an intact receptor-G-protein system in its natural membrane environment. The second step tested for the effect of varying Mg ion on GDP/GTPγS exchange in the absence and presence of the hormone glucagon. In the third step, the Gsα-GTPγS complexes formed were extracted and quantified in a standard reconstitution assay. This figure shows the effect of the glucagon-activated receptor on the Mg required for G-protein activation. Note that (1) hormone was not necessary for Gs activation as long as high enough (supraphysiologic) Mg was present during incubation with GTPγS, and (2) that in the presence of the hormone, the Mg required to activate Gs was approximately 1000-fold lower than in its absence. Inset: Same as main panel but with an expanded Mg concentration scale. (Adapted from Iyengar, R. and Birnbaumer, L., *Proc. Natl. Acad. Sci. USA*, **79**, 5179–5183, 1982. With permission).

why some mutations are activating and others inactivating. Moreover, it is also not known whether these amino acids actually contact the G protein, and if so, which G-protein subunit (for further discussion see references [14–17,141]).

Mutational analysis and sequence swapping experiments with G protein α subunits indicate that receptors interact with the very C terminus of the G protein α subunit. Indeed, swapping as few as three of the last ten amino acids between two α subunits can lead to a switch in the type of receptor that activates the G protein. The C terminus is not the only region of interaction of an α subunit with a receptor. Multiple sites have been identified by mutational analysis, including the α3β5 and α4β6 loops of the GTPase domain (Figs. 2 and 12; references [16] and [18]). The α subunit C terminus and the α3β5 and α4β6 loops are part of the same face of the molecule presumed to be immediately juxtamembranous. Receptors not only interact with the α subunit of a trimeric G protein. Free Gα subunits are not recognized by receptors; they are only recognized in the context of the heterotrimer. In agreement with this conclusion, injection of subunit specific antisense oligonucleotides, or subunit specific antibodies, leads to loss of receptor-mediated effector regulation, not only upon suppressing Gα but also either Gβ or Gγ subunits, all in a gene-specific manner. It has been shown that in pituitary cells the M4 muscarinic receptor activates a Go G protein of subunit composition αo1β3γ4, while the somatostatin receptor activates a Go of subunit composition αo2β1γ3. It follows that receptors "proofread" the subunit subtypes that make up the particular trimer with which they enter in contact. Figure 5 shows that only one of the two faces of the Gβ propeller is exposed to the milieu, while the other one faces the switch II region of the α subunits. The exposed face and the sides of the propeller are therefore available for interaction with a receptor. In turn, because the receptor interacts with Gβ and Gγ, it can reasonably be expected that receptors may affect the Gβγ interaction with Gα. One model, based on these considerations, as well as the results shown in Fig. 10, is that while the interaction of the receptor with the C terminus of Gα is essential for selection of the type of Gαβγ that will be activated through the GDP-GTP exchange reaction, the "activating" effect of the receptor may in fact be mediated by the Gβγ. In this sense the Gβγ dimers would have a receptor dependent GEF activity. Thus, while Gβγ alone prevents GDP dissociation by acting as guanine nucleotide dissociation inhibitor (GDI), the Gβγ receptor complex has GEF activity facilitating binding of GTP to its site on α. Artificial "bending" of the N terminus, mimicking what Gβγ might do if it were acting as GEF, does indeed lead to an apparently constitutively active G protein [19]. It has been speculated that Gβγ may be the site of action of Mg ion [20]. If so, the role of receptor would be simply to promote binding of Mg ion to Gβγ. Mg ion would then be responsible for changing the activity of Gβγ from GDI to GEF.

Regardless of the final outcome regarding the events responsible for G-protein activation by a receptor at the submolecular level, the overall reaction for a receptor activating a G protein is facilitation of the action of Mg ion to promote GDP-GTP exchange, followed by dissociation of the trimer into Gα-GTP plus Gβγ.

RGSs or Regulators of G-Protein Signaling

As there are receptors that by virtue of their GEF activity promote activation of the heterotrimeric G proteins, there are also GAPs that accelerate the GTPase activity of activated, GTP-liganded Gα subunits. Two types of Gα GAPs have been identified. One type are the RGSs. RGSs accelerate GTPase activity by 100-fold or more and exhibit Gα subunit selectivity. Sixteen RGSs are known and many of them are multidomain, multifunction proteins, and thus not only affect the Gα subunits but also aid in the organization of multicomponent "signaling complexes" [21]. The second type of GAPs are some of the effectors regulated by the α subunits. GAP activities of effectors increase k_{cat} of the Gα-GTP complexes by 10- to 20-fold only. In both instances, increased GTP hydrolysis ensures not only prompt turn-off of the signaling protein, but also a faster approach to equilibrium, therefore increasing the rate at which responses to an extracellular stimulus can be obtained. Indeed for RGS proteins, faster on/off rates of the regulated function may be a primary raison d'être. In contrast, when effectors act as GAPs, the primary purpose may be to ensure that they are indeed affected by the activated α-GTP complex. The intrinsic GTPase activity of Gα subunits is very low, in the order of 4–8 per s, giving them a rather long half-life, ensuring that they "find" their effector(s) while still in their GTP state. Once the effector has been found and the receptor message is delivered, the recipient of the message "kills" the messenger through activation of its deactivating mechanism. Thus, continued stimulation of effector, if this is desirable, requires continued presence of receptor agonist and constant reactivation of the G protein. Figure 11, illustrates the effect of GAPs on accelerating the rate at which equilibrium is established in a simple on/off reaction, such as binding of a hormone H to its receptor R. The three panels show: (1) the basic rate at which a bimolecular reaction reaches equilibrium; (2) the effect of increasing the k_{off} rate on the rate at which equilibrium is reached, expressed as HR formed in absolute concentrations; and (3) the fact that while the number of complexes at equilibrium decreases with increasing values of k_{off}, the rate at which equilibrium is reached increases with increases in k_{off}. It follows that, given the low intrinsic GTPase activity of Gα subunits, the need for a rapid response can only be satisfied by both existence of a GAP and a very high concentration of reactants so that the amplitude of the read-out signal (regulated effector) is large enough upon activation of only a small fraction of the GTPase. This is, in fact, the case with the activation of transducin (heterotrimeric G protein) by rhodopsin. The GTPase of transducin is stimulated by RGS9, a GGL RGS. This ensures that the physiological rapid turn-off occurs within a tenth of a second as opposed to having a half-life 10 s ($t_{1/2} = \ln 2/k_{off} = 10$ s, where k_{off} corresponds to published intrinsic GTPase activity of Gα subunits of 4 per min) [22]. However, to ensure that sufficient

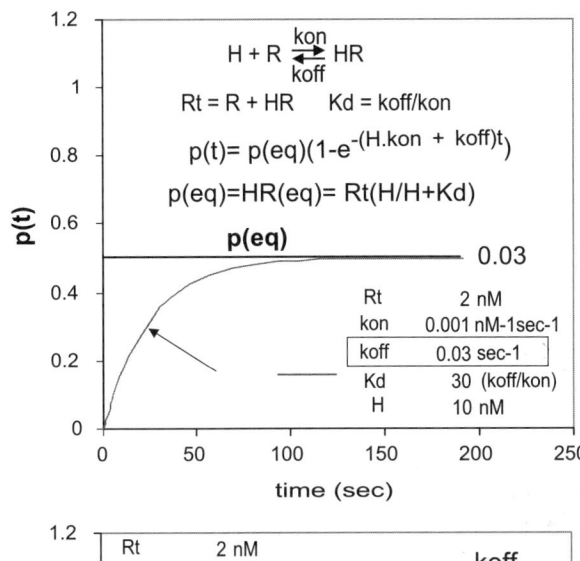
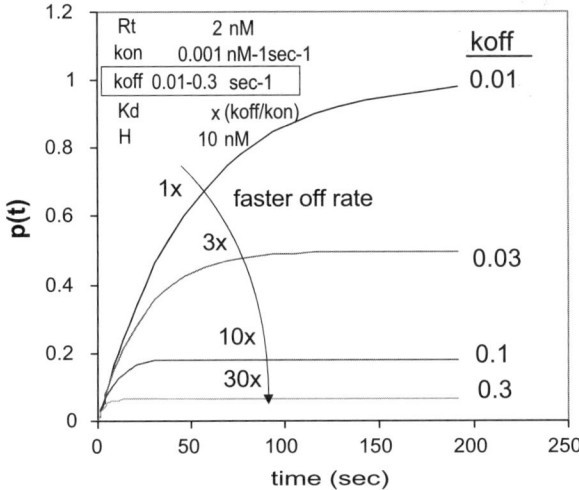
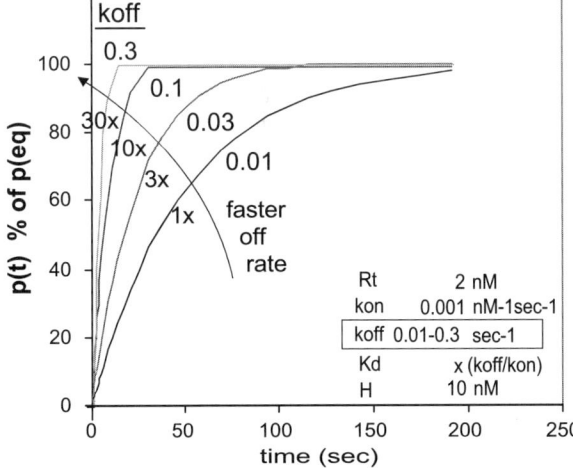

Figure 11 Approach to equilibrium of a reversible bimolecular reaction,

$$H + R \underset{koff}{\overset{kon}{\rightleftarrows}} HR$$

HR = 0 at t_o. These panels illustrate the effect of increasing the koff rate, equivalent to the effect of introducing a GTPase-activating RGS into a regulatory GTPase cycle.

active transducin will be formed, Mother Nature endowed the visual system with the highest known concentrations of receptor (rhodopsin; 40% of disk membrane protein) and regulated G protein (transducin; 10% of disc protein).

Other Forms of G-Protein Modulation: ASGs or Activators of G-Protein Signaling—the GoLoco domain

As described above, the free Gα-GTP and the Gβγ released upon G-protein activation, are both active regulators of effector systems. RGSs and effectors acting as GAPs shorten the lifetime of the active Gα with formation of Gα-GDP. Due to the high affinity of Gα-GDP for Gβγ, a GAP activity not only accelerates deactivation of Gα-regulated effectors, but also that of the Gβγ-regulated effectors from which Gβγ is sequestered by Gγ-GDP. AGSs, or activators of G-protein signaling, were uncovered in a search for molecules that potentiate the effect of α factor in baker's yeast [23]. α Factor acts by activating an αβγ heterotrimeric G protein and initiating a Gβγ-mediated cascade of reactions that leads to growth arrest and preparation for mating to the opposite mating type. One of the AGSs, AGS3, was found to potentiate Gβγ signaling by binding preferentially to Gα-GDP and to do so via a domain found in unrelated proteins and referred to as GoLoco. In mammalian systems, GoLoco domains act as they do in yeast. They bind preferentially to GDP-liganded Gα subunit of the Gi/Go type, thus prolonging Gβγ signaling. In addition to AGS3, proteins with a GoLoco domain include several RGSs and Rap-GAPII. This hints at a role that transcends its direct function of binding the GDP forms of Gi/Go and involves participation in integration of multicomponent signaling pathways (see reference [24] for recent review).

Even though they were identified in the same type of bioassay, AGS1, AGS2, and AGS3 differ in their mode of action. AGS1 (also Rasdex1) is a ras-related protein that appears to act as a GEF, whereas AGS2 interacts with Gβγ and AGS3 binds to α-GDP. Why a Gβγ-interacting protein (AGS2) would enhance the effectiveness of either a Gβγ or a Gα-GTP remains to be determined. Another Gβγ-interacting protein, phosducin, serves to attenuate the action of transducin in the retina. The existence of Gβγ- and Gα-interacting proteins (RGSs, GoLoco proteins, phosducin, and non-GoLoco AGSs) points to the fact that fine-tuning of the basic regulatory G-protein cycle is required for proper cell homeostasis. These fine-tuning mechanisms are therefore responsible for the ultimate ability of a cell to live a productive life that is in concert with the needs of the whole organism.

In Memoriam

This chapter is written 32 years after the first data on the GTP requirement in hormonal stimulation of liver adenylyl cyclase (then adenyl cyclase) were obtained in Martin Rodbell's laboratory where I was a postdoctoral fellow. It was then that we made the proposal that receptors may be acting on adenylyl cyclase through a signal transducer

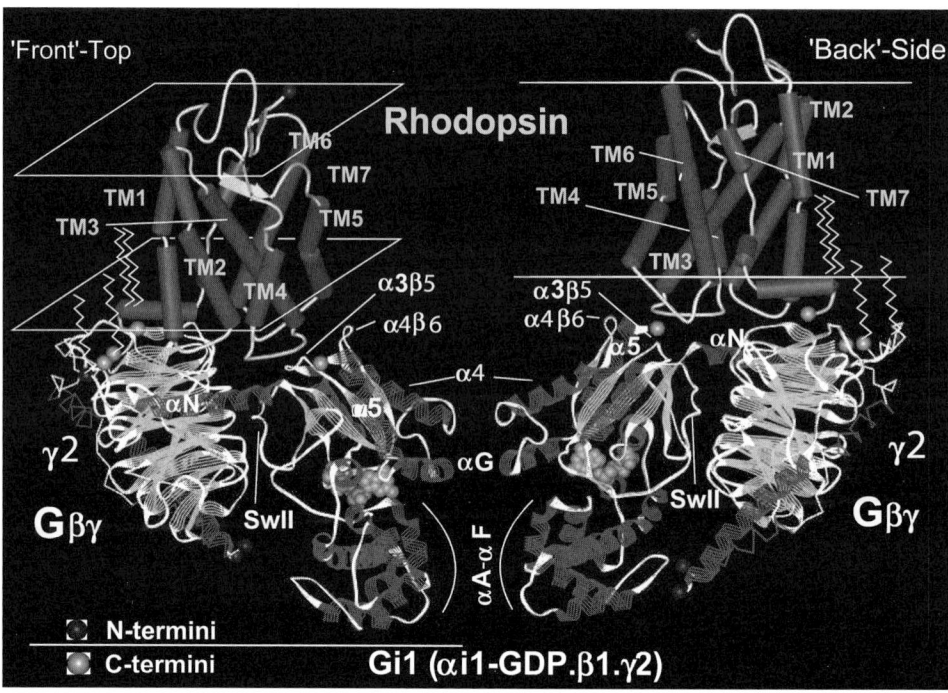

Figure 12 Model of orientation of a Gα subunit with respect to both the inner plane of the plasma membrane and a membrane-embedded GPCR. This is based on the crystal structures of bovine rhodopsin (PDB accession #1F88) and the GDP-occupied Gi1 trimer (PDB accession #1GG2). The molecules are shown in a near-docking situation with receptor in its "inactive" conformation and the G protein with GDP bound to it. Two views are shown and the main structural features are annotated. Vertical zigzag lines denote the α subunit N-terminal myristylation and/or palmitoylation and the γ subunit C-terminal polyisoprenylation. Positions of lipids are approximate because the modified N- and C-terminal amino acids were not resolved in the crystal structure of Gi1. Double zigzag approximately 13 amino acids after the TM7 of rhodopsin denotes double palmitoylation of the C terminus at that position. Single and double palmitoylation of GPCR C termini is a common but not universal feature of GPCRs.

driven by GTP—in some unknown way (Fig. 12). Two great thinking minds of that era are no longer with us. Martin Rodbell passed away in 1998 [25], as did Michael (Mickey) Schramm on June 8, 2002. Mickey visited often in the late 1960s, and fed us ideas that were incorporated into our thinking without us even realizing it. Seldom, if ever, have we properly given credit to Mickey's influence on our thinking. Better late than never, goes the saying, but I wish I would have done it earlier. While life is destined to end, our task as researchers dedicated to extracting nature's secrets never ends. Even though 32 years have passed, the mystery of signal transduction through G proteins has not been fully resolved, as the recent discoveries of RGS and AGS proteins demonstrates. While looking forward to new discoveries, we should also remember those that contributed in major ways to the way we think today. This chapter was written with the idea of introducing signaling through G proteins to the next generation of investigators. I hope not to have failed too badly.

References

1. Lodish, H., Berk, A., Zipursky, S. L., Matsudaira, P., Baltimore, D., and Darnell, J. (2000). *Molecular Cell Biology*, 14th Ed., pp 862–871, H. Freeman New York.
2. Rodbell, M., Birnbaumer, L., Pohl, S. L., and Krans, H. M. J. (1970). Properties of the adenyl cyclase systems in liver and adipose cells: The mode of action of hormones. *Acta Diabetolog. Latina* **7** (Suppl. 1), 9–57.
3. Kim, S.-H., Priveé, G. G., and Milburn, M. V. (1993). Conformational switch and structural basis for oncogenic mutations of *Ras* proteins, in Dickey, B. and Birnbaumer, L. Eds., GTPases in Biology" Handbook of Experimental Pharmacology Vol. 108/I, pp. 177–193, Springer Verlag, Heidelberg, Germany.
4. Dickey, B. and Birnbaumer, L., Eds. (1993). *GTPases in Biology, Handbook of Experimental Pharmacology*, Vol. 108/I. Springer-Verlag, Heidelberg, Germany.
5. Dickey, B. and Birnbaumer, L., Eds. (1993). *GTPases in Biology, Handbook of Experimental Pharmacology* Vol. 108/II. Springer-Verlag, Heidelberg, Germany.
6. Feig, L. A. (1993). Dominant inhibitory *Ras* mutants: Tools for elucidating *Ras* function, in Dickey, B. and L. Birnbaumer, Eds., *GTPases in Biology, Handbook of Experimental Pharmacology Vol. 108/I*, pp. 289–298. Springer-Verlag, Heidelberg, Germany.
7. Noel, J. P., Hamm, H. E., and Sigler, P. B.. (1993). The 2.2 Å crystal structure of transducin α-GTPγS. *Nature* **366**, 654–663.
8. Coleman, D. E., Berghuis, A. M., Lee, E., Lindner, M. E., Gilman, A. G., and Sprang, S. R. (1994). Structures of active conformations of $G_{i\alpha 1}$ and the mechanism of GTP hydrolysis. *Science* **265**, 1405–1412.
9. Higashijima, T., Ferguson, K. M., Sternweis, P. C., Smigel, M. D., and Gilman, A. G. (1987). Effects of Mg^{2+} and the βγ subunit complex on the interactions of guanine nucleotides with G proteins. *J. Biol. Chem.* **262**, 762–766.
10. Tesmer, J. J., Sunahara, R. K., Gilman, A. G., and Sprang, S. R. (1997). Crystal structure of the catalytic domains of adenylyl cyclase in a complex with Gsα. GTPγS. *Science* **278**, 1907–1916.

11. Lambright, D. G., Noel, J. P., Hamm, H. E., and Sigler, P. B. (1994). Structural determinants for the activation of the α subunit of a heterotrimeric G protein. *Nature* **239**, 621–628.
12. Wall, M. A., Coleman, D. E., Lee, E., Iniguez-Lluhi, J. A., Posner, B. A., Gilman, A. G., and Sprang, S. R.(1995). The structure of the G protein heterotrimer Giα1 β1 γ2. *Cell* **83**, 1047–1058.
13. Casey, P. J. (1995). Protein lipidation in cell signaling. *Science* **268**, 221–225.
14. Bourne, H. R. (1995). How receptors talk to trimeric G proteins. *Curr. Opin. Cell Biol.* **9**, 134–142.
15. Wess, J. (1998). Molecular basis of receptor/G protein coupling selectivity. *Pharmacol. Ther.* **80**, 231–264.
16. Grishina, G. and Berlot, C. H. (2000). A surface exposed region of $G_s\alpha$ in which substitutions decrease receptor mediated activation and increase receptor affinity. *Mol. Pharmacol.* **57**, 1081–1092.
17. Berlot, C. H. (2002). A highly effective dominant negative *ras* construct containing mutations that affect distinct functions inhibits multiple Gs-coupled receptor signalling pathways. *J. Biol. Chem.* **277**, 21080–21085.
18. Hamm, H. E. (2002). How activated receptors couple to G proteins. *Proc. Natl. Acad. Sci. USA* **98**, 4819–4821.
19. Rondard, P., Iiri, T., Srinivasan, S., Meng, E., Fujita, T., and Bourne, H. R. (2001). Mutant G protein α subunit activated by Gβγ: A model for receptor activation? *Proc. Natl. Acad. Sci. USA* **98**, 6150–6155.
20. Iyengar, R. and Birnbaumer, L. (1982). Hormone receptor modulates the regulatory component of adenylyl cyclases by reducing its requirement for Mg ion and enhancing its extent of activation by guanine nucleotides. *Proc. Natl. Acad. Sci. USA* **79**, 5179–5183.
21. DeVries, L., Zheng, B., Fischer, T., Elenko, E., and Farquhar, M. G. (2000). The regulator of G protein signaling family. *Annu. Rev. Pharmacol. Toxicol.* **40**, 235–271.
22. Chen, C. K., Burns, M. E., He, W., Wensel, T. G., Baylor, D. A., and Simon, M. I. (2000). Slowed recovery of rod photoresponse in mice lacking the GTPase accelerating protein RGS9–1. *Nature* **403**, 557–560.
23. Cismowski, M. J., Takesono, A., Bernard, M. L., Duzic, E., and Lanier, S. M. (2001). Receptor-independent activators of heterotrimeric G-proteins. *Life Sci.* **68**, 2301–2308.
24. Kimple, R. J., Willard, F. S., and Siderovski, D. R. (2002). The GoLoco motif: Heralding a new tango between G protein signaling and cell division. *Mol. Interventions* **2**, 88–100.
25. Birnbaumer, L. (1999). Martin Rodbell (1925–1998), *in memoriam. Science* **283**, 1656.
26. Wilkie, T. M., Gilbert, D. J., Olsen, A. S., Chen. X. N., Amatruda, T. T., Korenberg, J. R., Trask, B. J., de Jong, P., Reed, R. R., Simon, M. I., Jenkins, N. A., and Copeland, N. G. (1992). Chromosomal evolution of the G protein α subunit multigene family. *Nat. Genet.* **1**, 85–91.

CHAPTER 214

Genetic Analysis of Heterotrimeric G-Protein Function

Juergen A. Knoblich
Research Institute of Molecular Pathology (I.M.P.),
Vienna, Austria

Introduction

For a surprisingly long time, the yeast *Saccharomyces cerevisiae* has been the only organism in which signaling by heterotrimeric G proteins has been analyzed genetically. More recently, mutants affecting G-protein function have also been analyzed in *C. elegans* [1–3], *Dictyostelium* [4,5], and *Drosophila* [6] and have yielded some quite unexpected results. In this chapter, I want to first summarize the results obtained in yeast and then focus on more recent experiments that have revealed a function for heterotrimeric G-proteins in cell polarity in *Drosophila*. The beautiful experiments done in *C. elegans* and *Dictyostelium* are summarized in other chapters (see the contributions by Julie Ahringer and Peter Devreotes) and will not be reviewed here.

Signaling by Heterotrimeric G Proteins in Yeast

In response to the pheromone α-factor, yeast cells polarize their actin cytskeleton and extend a process called "shmoo." α-factor binds to a seven-transmembrane receptor in the plasma membrane and activates the only heterotrimeric G protein present in the yeast genome, Gpa1 [7]. G-protein activation leads to polarization of the actin cytoskeleton, arrest of the cell cycle, and a transcriptional response in the nucleus. In mutants affecting the β-subunit (Ste4) or the γ-subunit (Ste18) of the heterotrimeric G-protein complex, no response to α-factor can be detected, whereas in mutants affecting the α-subunit the signaling pathway is constitutively activated.

Thus, signal transduction occurs via the free Gβγ subunit, but the GTP-bound Gα subunit does not seem to play a role [7]. It is interesting that the defects observed in Gβ mutants can be rescued by a Gα-Gβ-fusion protein in which Gα and Gβγ are covalently linked but can still undergo conformational changes [8], indicating that the βγ-subunit does not have to physically separate from the α-subunit for signaling to occur.

The best studied signal transduction cascade that operates downstream of the heterotrimeric G protein involves the protein kinases Ste11, Ste7, and Fus3 (Fig. 1A). Upon G-protein activation, Ste11 phosphorylates Ste7, which in turn phosphorylates Fus3 and induces its translocation into the nucleus. Fus3 belongs to the MAP-kinase family of protein kinases and therefore this cascade of protein kinases is the prototype of a so-called MAP-kinase cascade. Activation of the MAP-kinase cascade occurs in two ways: First, Gβγ binds and activates the protein kinase Ste20 when released from the α-subunit and Ste20 initiates MAPK-signaling by phosphorylating its most upstream member, Ste11. Second, all kinases of the cascade bind to the scaffold protein Ste5. Ste5 can bind to free Gβγ and therefore G-protein activation recruits the whole kinase cascade to the site of receptor activation. Thus, G-protein signaling in yeast involves the relocalization of downstream signaling components to the site of G-protein activation by direct binding to free Gβγ.

Although the Map-kinase cascade transduces a signal into the nucleus, the most important downstream target for polarizing the actin cytoskeleton is the guanine exchange factor Cdc24 (Fig. 1A). Cdc24 binds free Gβγ through the

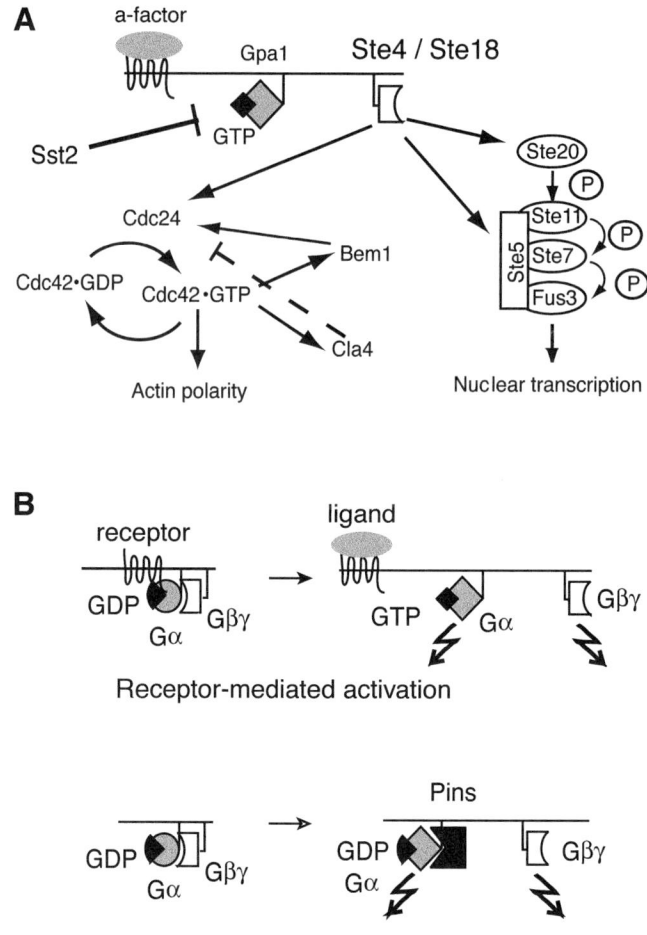

Figure 1 Outline of G-protein signaling in yeast (A) and Drosophila (B).

adaptor protein Far1 [9,10]. In the absence of α-factor, Far1 shuttles between the cytoplasm and the nucleus and sequesters Cdc24 in the nucleus [11,12]. Upon α-factor exposure, however, Far1 binds to free Gβγ and the trimeric complex is assembled, leading to an accumulation of Cdc24 at the site of receptor activation. Once it is localized, Cdc24 binds the adaptor protein Bem1 and this association stabilizes the cortical localization of the protein. Cdc24 is an exchange factor that catalyzes GDP/GTP exchange on the small G-protein Cdc42. Cdc42, in turn, is a well-characterized modulator of the actin cytoskeleton, and it is the local activation of Cdc42 by Cdc24 that is thought to polarize the actin cytoskeleton toward the site of a-factor exposure.

Two feedback loops have been shown to operate downstream of the G proteins. In its GTP bound form, Cdc42 can directly bind to Bem1 and recruit this adaptor protein to the cell cortex [13]. Bem1 in turn recruits Cdc24, which generates more Cdc42 and therefore amplifies the signal. Activated Cdc42 also recruits the protein kinase Cla4, which will phosphorylate Cdc24 at a site that is needed for it to bind to Bem1 [14]. This phosphorylation event will eventually terminate the signal and allow the signal transduction cascade to return to its inactive state. Thus, positive and negative feedback loops that operate downstream of the G-proteins both amplify and terminate the signal that is initially generated by receptor activation.

Regulation of signaling also occurs at the level of the G protein itself. The protein Sst2 is required for yeast cells to recover and reenter the cell cycle after exposure to α-factor. Sst2 directly binds to the G-protein α-subunit. It contains a so-called RGS (regulator of G-protein signaling) domain that enhances the rate of GTP hydrolysis and therefore increases the rate of G-protein inactivation. RGS proteins were also found in higher eukaryotes and represent an evolutionarily conserved protein family that acts as GAPs for heterotrimeric G-proteins [15].

Heterotrimeric G-Protein Function in *Drosophila*

A genetic analysis of heterotrimeric G-protein function in *Drosophila* has recently revealed another important protein family that regulates G-protein activity by direct interaction. *Drosophila* neural precursor cells called neuroblasts divide asymmetrically along the dorso-ventral axis. Before division, a protein called Inscuteable localizes to their apical cell cortex and establishes an axis of polarity that is used during mitosis for correct apical-basal spindle orientation and for asymmetric localization of protein determinants to the basal cell cortex [16]. Upon cytokinesis, these determinants are inherited by only one of the two daughter cells and ensure its correct cell fate. Inscuteable binds to a protein called Pins and recruits it to the apical cell cortex. Pins in turn recruits the heterotrimeric G protein Gαi through its C-terminal GoLoco domains, and both proteins are required for establishing an axis of polarity [17]. Thus, Inscuteable and Pins are adaptors that function by polarizing the distribution of heterotrimeric G-proteins.

Besides localizing the G proteins, Pins and its vertebrate homologs AGS-3 [18] and LGN [19] also seem to regulate their activity. Binding of Pins or just its GoLoco domains to Gαi induces the release of the βγ-subunit *in vitro* without the need for receptor activation [6]. βγ-release seems to occur in the GDP-bound form, since Pins does not induce nucleotide exchange and only binds the GDP-form of the G protein. In fact, GoLoco domains inhibit nucleotide exchange, suggesting they function as GDIs (GDP dissociation inhibitors) for heterotrimeric G-proteins [20]. The structural basis for these activities was recently revealed by X-ray crystallography [21]. GoLoco domains bind a region on Gα that overlaps the Gβγ-binding domain and induce a conformational change that precludes coincident Gβγ interaction. In addition, they interact with the α- and β-phosphates of Gα-bound GDP and inhibit its dissociation from the G protein. Thus, GoLoco domains represent a novel class of G-protein regulators that activate G proteins via a nonclassical mechanism that involves neither a seven transmembrane receptor nor GDP/GTP exchange on the α-subunit.

The function of heterotrimeric G-proteins in *Drosophila* has been addressed genetically. The *Drosophila* genome contains one of each of the major classes of Gα-subunits, Gαi, Gαo, and Gαs. Gαs mutants die for unknown reasons but do not show defects that can be associated with a lack of PKA activity or defects in any of the major developmental signaling pathways [22]. Mutants in Gαo have defects in heart development because epithelial polarity in cardiac precursors is not correctly established [23]. Mutants in Gαi have not been generated, but the function of this G protein was addressed by mutating the associated β-subunit [6]. Based on sequence similarity, Gβ13F is the common *Drosophila* homolog of vertebrate Gβ1-4. Gβ13F binds to Gαi, and in its absence Gαi can no longer be detected. Presumably, a homeostasis mechanism degrades Gαi in the absence of its associated β-subunit, ensures that cells contain equal amounts of Gα and Gβ, and prevents the accumulation of free Gαi in Gβ13F mutants. Besides the defects in asymmetric cell division mentioned above, these mutants have defects in gastrulation. This phenotype resembles the one described for *concertina*, the *Drosophila* homolog of Gα13. Whether Gα13 is also degraded in Gβ13F mutants cannot be addressed due to the lack of a good antibody, but this results suggests that both Gαi and Gα13 use Gβ13F for signal transduction in *Drosophila*.

Conclusions

Together with genetic results obtained in *C. elegans* and *Dictyostelium* (reviewed in the chapters by Devreotes and Ahringer in this volume), these experiments tell us that heterotrimeric G proteins—besides their well-studied function in hormone and neurotransmitter signal transduction—have an important role in cell polarity. To exert this role, G proteins may not always be activated by classical mechanisms involving seven-transmembrane receptors. Unconventional G-protein activators have been identified, and their further analysis may reveal more widespread functions that might change some of the classical textbook views on the role of this important protein family.

References

1. Jansen, G., Thijssen, K. L., Werner, P., van der Horst, M., Hazendonk, E., and Plasterk, R. H. (1999). *Nat. Genet.* **21**, 414–419.
2. Zwaal, R. R., Ahringer, J., van Luenen, H. G., Rushforth, A., Anderson, P., and Plasterk, R. H. (1996). *Cell* **86**, 619–629.
3. Gotta, M. and Ahringer, J. (2001). *Nat. Cell Biol.* **3**, 297–300.
4. Parent, C. A. and Devreotes, P. N. (1999). *Science* **284**, 765–770.
5. Janetopoulos, C., Jin, T., and Devreotes, P. (2001). *Science* **291**, 2408–2411.
6. Schaefer, M., Petronczki, M., Dorner, D., Forte, M., and Knoblich, J. A. (2001). *Cell* **107**, 183–194.
7. Dohlman, H. G. and Thorner, J. W. (2001). *Annu. Rev. Biochem.* **70**, 703–754.
8. Klein, S., Reuveni, H., and Levitzki, A. (2000). *Proc. Natl. Acad. Sci. USA* **97**, 3219–3223.
9. Butty, A. C., Pryciak, P. M., Huang, L. S., Herskowitz, I., and Peter, M. (1998). *Science* **282**, 1511–1516.
10. Nern, A. and Arkowitz, R. A. (1999). *J. Cell Biol.* **144**, 1187–1202.
11. Nern, A. and Arkowitz, R. A. (2000). *J. Cell Biol.* **148**, 1115–1122.
12. Shimada, Y., Gulli, M. P., and Peter, M. (2000). *Nat. Cell Biol.* **2**, 117–124.
13. Butty, A. C., Perrinjaquet, N., Petit, A., Jaquenoud, M., Segall, J. E., et al. (2002). *EMBO J.* **21**, 1565–1576.
14. Gulli, M. P., Jaquenoud, M., Shimada, Y., Niederhauser, G., Wiget, P., and Peter, M. (2000). *Mol. Cell* **6**, 1155–1167.
15. De Vries, L., Zheng, B., Fischer, T., Elenko, E., and Farquhar, M. G. (2000). *Annu. Rev. Pharmacol. Toxicol.* **40**, 235–271.
16. Knoblich, J. A. (2001). *Nature Rev. Mol. Cell Biol.* **2**, 11–20.
17. Knust, E. (2001). *Cell* **107**, 125–128.
18. Takesono, A., Cismowski, M. J., Ribas, C., Bernard, M., Chung, P. et al. (1999). *J. Biol. Chem.* **274**, 33202–33205.
19. Mochizuki, N., Cho, G., Wen, B., and Insel, P. A. (1996). *Gene* **181**, 39–43.
20. Natochin, M., Lester, B., Peterson, Y. K., Bernard, M. L., Lanier, S. M., and Artemyev, N. O. (2000). *J. Biol. Chem.* **275**, 40981–40985.
21. Kimple, R. J., Kimple, M. E., Betts, L., Sondek, J., and Siderovski, D. P. (2002). *Nature* **416**, 878–881.
22. Wolfgang, W. J., Hoskote, A., Roberts, I. J., Jackson, S., and Forte, M. (2001). *Genetics* **158**, 1189–1201.
23. Fremion, F., Astier, M., Zaffran, S., Guillen, A., Homburger, V., and Semeriva, M. (1999). *J. Cell Biol.* **145**, 1063–1076.

CHAPTER 215

Heterotrimeric G Protein Signaling at Atomic Resolution

David G. Lambright
Program in Molecular Medicine and Department of Biochemistry and Molecular Pharmacology, University of Massachusetts Medical School, Worcester, Massachusetts

Introduction

Heterotrimeric G proteins (subunits $G_{\alpha\beta\gamma}$) mediate a variety of physiological responses, including sensory perception, hormone action, polarization, chemotaxis, and growth control [1–3]. In the conventional paradigm for G-protein signaling, ligand-bound (or light-activated) heptahelical receptors catalyze release of GDP from the G_α subunit, resulting in a complex between the receptor and the nucleotide free $G_{\alpha\beta\gamma}$ heterotrimer. Association of GTP with G_α triggers release from the receptor and dissociation of G_αGTP and $G_{\beta\gamma}$. Depending on the particular signaling pathway, either G_αGTP or the released $G_{\beta\gamma}$ subunits interact with downstream effectors until the G_α subunit is deactivated by GTP hydrolysis. Although the G_α subunit possesses an intrinsic GTP hydrolytic activity, regulator of G protein signaling (RGS) domains, present in a variety of modular proteins, accelerate the rate of GTP hydrolysis [4].

Crystallographic studies of two different G-protein signaling pathways, involving the visual G protein transducin (G_t) as well as the hormone activated G proteins that stimulate (G_S) and inhibit (G_i) adenylyl cyclase (AC), reveal a highly conserved structural basis for heterotrimer assembly, activation by nucleotide exchange, and deactivation by GTP hydrolysis. The various structures explain a wealth of biochemical and cell biological data accumulated over the years and provide a springboard for mutational analyses aimed at dissecting structure-function relationships for the myriad diverse biological responses mediated by G proteins. The most salient observations are highlighted below. Interested readers are encouraged to consult the cited references for in-depth discussion of particular structures.

Architecture and Switching Mechanism of the G_α Subunits

As illustrated in Fig. 1A, the G_α subunits share a conserved architecture with a Ras-like domain, consisting of a core six stranded β sheet (β1–β6) surrounded by five helices (α1–α5), and a helical domain comprising a long helix (αA) enveloped on three side by five shorter helices (αB–αF) [5]. Compared with monomeric GTPases, the helical domain of the G_α subunits represents an insertion within the α1/β2 loop (known as the "effector binding loop" in Ras). Consequently, the Ras-like and helical domains are connected by two extended strands, one joining α1 of the Ras-like domain to αA of the helical domain and the other traversing from αF of the helical domain to β2 of the Ras-like domain. In all of the G_αGDP and G_αGTP structures determined to date, the αD/αE loop of the helical domain engages the Ras-like domain, thereby capping the nucleotide binding site [5–10]. In contrast, the ribose and phosphate moieties of the nucleotide are partially exposed in monomeric GTPases, particularly in the GDP-bound conformation in which the α1/β2 loop is highly flexible. Consistent with the structural observations,

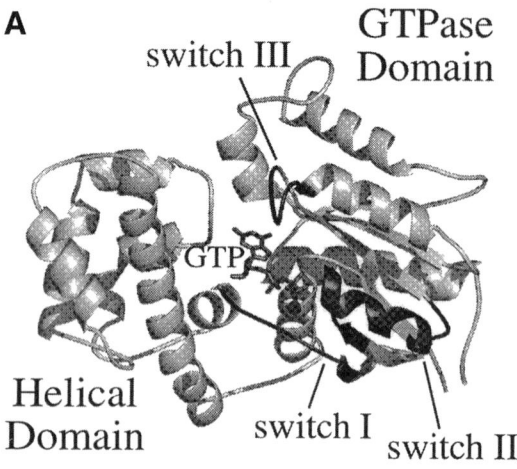

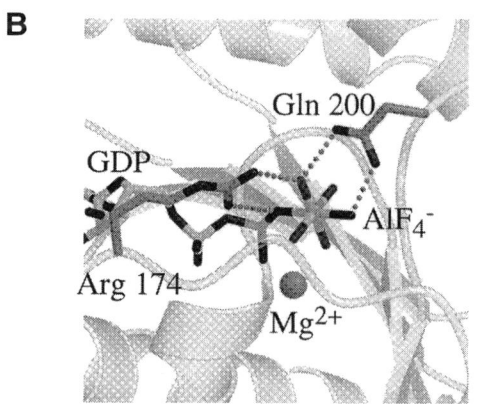

Figure 1 Active and transition states of a G_α subunit. (A) Structure of the active form of $G_{t\alpha}$ bound to GTPγS, a poorly hydrolyzeable GTP analog. The conformational switch regions, deduced by comparison with the inactive (GDP-bound) form, are highlighted in dark gray. The Mg^{2+} ion is represented as a dark gray sphere. (B) Structure of AlF_4^- activated $G_{t\alpha}$GDP.

the G_α subunits bind GDP with high affinity in the absence of Mg^{2+} whereas monomeric GTPases require Mg^{2+} for high-affinity binding, exhibiting rapid rates of nucleotide release in the presence of EDTA [11]. As an obvious consequence of the conserved domain architecture in the G_α subunits, intrinsic as well as receptor-catalyzed GDP release necessarily involves a transition to an "open" intermediate in which the helical domain has rotated away from the Ras-like domain.

Structural changes between the GDP-bound (inactive) and GTP-bound (active) conformations are localized to three nonconsecutive conformational switch regions corresponding to the first linker strand (switch I), a region extending from the C-terminus of β3 to the C-terminus of the α2/β4 loop (switch II), and the β4/α3 loop (switch III) [6,9]. In the GDP-bound structures, the switch regions are either disordered [9] or adopt a relaxed conformation stabilized by crystal contacts [6]. In the GTP-bound form, the γ phosphate is detected by direct hydrogen bonding interactions with the side chain hydroxyl of the invariant threonine residue in the RxxxT motif of switch I and the backbone NH group of the invariant glycine residue in the DxxGQ motif of switch II.

The conformation of switch III is indirectly coupled to the nucleotide state by a hydrogen-bonding interaction between the side chain carboxylate of an invariant glutamate residue in switch III and a main chain NH group in switch II. The active conformation is further stabilized through Mg^{2+} coordination by the side chain hydroxyl of the invariant threonine residue in switch I and through an extensive hydrophobic/ionic interface between the α2 helix of switch II and the α3 helix. Finally, the zippering of the β2 and β3 strands, resulting in two additional hydrogen bonds in the GTP-bound form, suggests cooperativity in the switching mechanism.

Insight into the GTP Hydrolytic Mechanism from an Unexpected Transition State Mimic

Aluminum fluoride (AlF_4^-) binds to G_αGDP near the binding site for the γ phosphate of GTP, inducing a conformational change that results in artificial activation of G_αGDP and dissociation from $G^{\beta\gamma}$ Crystal structures of the GDP·AlF_4^--bound forms of $G_{i\alpha}$ [7] and $G_{t\alpha}$ [8] revealed an unexpected finding. Although the conformation of all three switch regions closely resembles that of the GTP-bound form, AlF_4^- does not mimic the tetrahedral geometry of the γ phosphate but rather adopts an octahedral geometry with four fluoride ligands arranged in an equatorial plane and two axial ligands consisting of an oxygen from the β phosphate of GDP and a water molecule (Fig. 1B). The bound aluminum fluoride interacts with and orders the side chains of two critical residues previously implicated in GTP hydrolysis, namely the arginine of the switch I RxxxT motif and the glutamine of the switch II DxxGQ motif. These observations led to the hypothesis that AlF_4^- activates G_αGDP by approximating the expected stereochemistry of the pentavalent intermediate for GTP hydrolysis. This notion is strongly supported by the remarkable observation that AlF_4^- also binds to and stabilizes complexes of both Ras and Rho GTPases bound to their respective GAPs [12,13].

$G_{\beta\gamma}$ with and without G_α

Parallel crystallographic studies revealed the stunningly beautiful structure of the $G_{\beta\gamma}$ heterodimer alone [14] and in complex with G_αGDP [15,16]. The seven distinctive WD repeats of G_β fold into a seven bladed β propeller in which each blade consists of a four stranded antiparallel β sheet (Fig. 2A). It is interesting that the WD repeats do not coincide precisely with the individual blades of the propeller, but rather each repeat begins with the outer strand of one blade and extends through the inner three strands of the next blade. G_γ possesses an N-terminal helix, which forms a parallel coiled coil with the N-terminal helix of G_β, followed by an internal helix and a region of coil that extends across the bottom of the G_β propeller. Although G_γ has well-defined secondary structure, it is devoid of intramolecular interactions characteristic of tertiary structure and thus could not

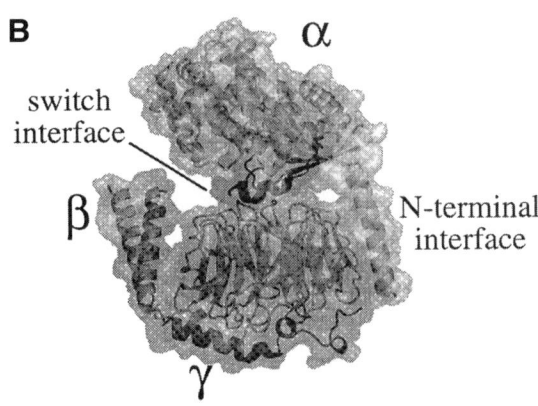

Figure 2 Structure of a heterotrimeric G protein. (A) Ribbon rendering of $G_{t\beta\gamma}$ with the γ subunit and one of the WD repeats in the β subunit shown in dark gray. (B) Heterotrimeric complex of a $G_{t\alpha}/G_{i\alpha}$ chimera and the unprenylated form of $G_{t\beta\gamma}$.

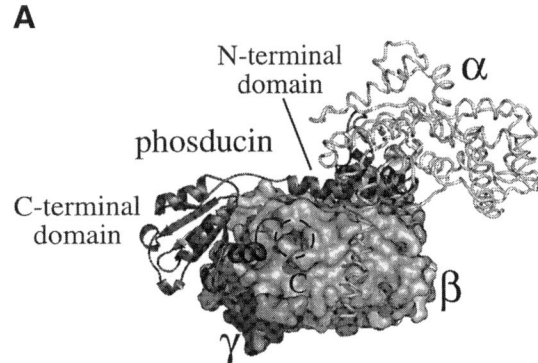

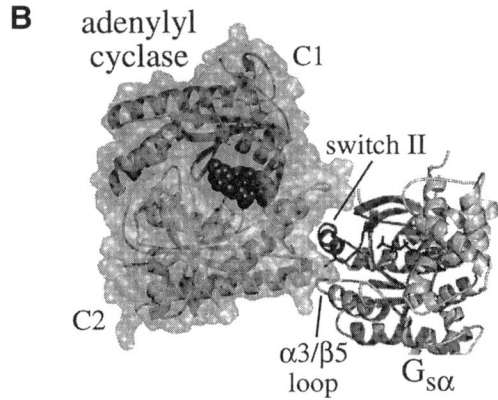

Figure 3 Regulatory and effector complexes with phosducin and adenylyl cyclase. (A) Structure of the unprenylated $G_{t\beta\gamma}$ (light surface) in complex with retinal phosducin (dark ribbon). For comparison, the $G_{t\alpha}/G_{i\alpha}$ chimera from Fig. 2B is overlaid as a light coil. A dashed circle denotes the location of the proposed binding cavity for the farnesyl moiety. (B) Structure of an active adenylyl cyclase C1-C2 heterodimer (ribbons and semitransparent surface) bound to forskolin (dark spheres) and $GS_\alpha GTP_\gamma S$ (ribbons). The active site is located to the left of the forskolin-binding site. Note that the interaction epitope of GS_α comprises the switch II region and the α3/β5 loop.

adopt a properly folded structure in the absence of G_β. Indeed, roughly half of the residues in G_γ are buried in an extensive hydrophobic interface with G_β, a finding that explains the unusually high stability of the $G_{\beta\gamma}$ heterodimer.

The interaction between $G_\alpha GDP$ and $G_{\beta\gamma}$ occurs at two distinct interfaces (Fig. 2B). The most extensive interface involves the switch I and II regions of G_α, which contact residues from the loops and turns at the top of the G_β propeller. The second interface forms between the N-terminal helix of G_α and the side of G_β. In the complex with $G_{\beta\gamma}$, the switch regions of G_α adopt a well-ordered conformation that is incompatible with the active conformation of $G_\alpha GTP$. In contrast, it appears that the interaction with the N-terminal helix of G_α would not be directly influenced by the state of the bound nucleotide, consistent with a residual low-affinity interaction between $G_\alpha GTP$ and the released $G_{\beta\gamma}$ subunits.

Phosducin and $G_{\beta\gamma}$

The first insight into how phosducin engages $G_{\beta\gamma}$ and promotes membrane dissociation came from the crystal structure of a phosducin complex with an unprenylated form of retinal $G_{\beta\gamma}$ [17]. Phosducin contains two domains, a small N-terminal helical domain composed primarily of hydrophilic residues and a C-terminal domain with a thioredoxin-like fold (Fig. 3A). The N-terminal domain interacts with the top of the G_β propeller, overlapping extensively with the epitope for interaction with the switch regions of G_α, whereas the thioredoxin domain contacts the side of G_β at a site distinct from the N-terminal epitope for G_α. These observations explain why the interaction of $G_{\beta\gamma}$ with phosducin and G_α is mutually exclusive. Furthermore, electrostatic calculations indicate that the presence of phosducin's thioredoxin-like domain introduces a substantial negative electrostatic potential near the prenylation site at the C-terminus of G_γ, thereby destabilizing the association with acidic membranes. Finally, the interaction with phosducin's N-terminal domain perturbs the conformation of three loops at the top of the G_β propeller. A subsequent structure of phosducin bound to farnesylated $G_{\beta\gamma}$ suggested that the conformational changes in G_β open a pocket of appropriate dimensions to accommodate the hydrophobic farnesyl group of G_γ [18].

$G_{S\alpha}$ and Adenylyl Cyclase

Adenylyl cyclase (AC) consists of two hexahelical transmembrane domains, each followed by a similar cytoplasmic domain referred to as C1 and C2. Expressed independently, the isolated C1 or C2 domains form soluble but inactive homodimers. When mixed, the C1 and C2 homodimers spontaneously equilibrate to form catalytically active heterodimers that retain the ability to be stimulated by $G_{S\alpha}$GTP and inhibited by $G_{i\alpha}$GTP [19]. The structure of a C1-C2 heterodimer in complex with the GTP-bound form $G_{S\alpha}$ and forskolin, a plant terpenoid that activates AC, provided the first glimpse of how a GTP-bound G_α subunit recognizes and activates a downstream effector [20]. The inactive C2 homodimer binds two molecules of forskolin at symmetrical sites located at the dimer interface [21]. In contrast, the pseudo-symmetrical C1-C2 heterodimer binds a single forskolin molecule at an analogous site (Fig. 3B). The ATP-binding site is located at a pseudo-symmetrical site analogous to the second forskolin site in the C2 homodimer. The switch II region and $\alpha 3/\beta 5$ loop of $G_{S\alpha}$ contact the C1-C2 domains at a location remote from the active site, thereby inducing a domain rotation that brings key catalytic and ATP-binding residues into register.

Filling in the GAP

RGS domains present in a variety of modular proteins accelerate GTP hydrolysis for $G\alpha$ subunits [4]. The underlying structural basis was established by the crystal structure of the helical RGS domain of RGS4 in complex with $G_{i\alpha}$GDP and AlF_4^- [22]. In contrast to GAPs (GTPase-activating proteins) for Ras and Rho GTPases, which supply an "arginine finger" analogous to the catalytic arginine in switch I of G_α, RGS proteins promote GTP hydrolysis by engaging the switch I and II regions so as to reorient the catalytic arginine and glutamine residues of $G_{i\alpha}$ to stabilize the pentavalent intermediate (Fig. 4A).

Visual Fidelity

Structures of the GTPγS-bound and AlF_4^- activated forms of $G_{t\alpha}$ in complex with RGS9 and/or the inhibitory subunit of the retinal phosphodiesterase (PDEγ) provided further insight into the cooperative mechanism of effector recognition and RGS stimulation of GTP hydrolysis in the visual system [23]. PDEγ forms a predominately hydrophobic interface with residues in the switch II/α3 cleft of $G_{t\alpha}$, consistent with mutational data. This interaction sequesters C-terminal residues of PDEγ implicated in PDEαβ inhibition. RGS9 engages the switch I and II regions of $G_{t\alpha}$ in a manner analogous to that observed for RGS4 and $G_{i\alpha}$. As shown in Fig. 4B, a small interface between PDEγ and a unique loop of RGS9, near the critical aspargine residue involved in positioning the catalytic glutamine of $G_{t\alpha}$,

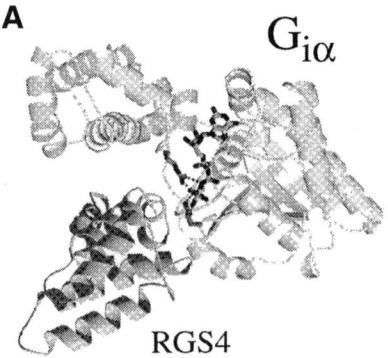

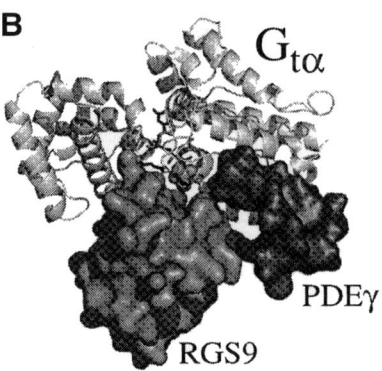

Figure 4 RGS and effector complexes with G_α subunits. (A) Structure of RGS4 (dark ribbon) bound to AlF_4^- activated $G_{i\alpha}$GDP (light ribbon). Also shown are the conserved arginine and glutatmine residues in the switch regions as well as GDP-AlF_4^-. (B) Structure of AlF_4^- activated $G_{t\alpha}$GDP (ribbons) bound to PDEγ (dark surface) and RGS9 (light surface). AlF_4^- is depicted as light spheres whereas GDP as well as the conserved arginine and threonine residues are shown as bonded cylinders.

couples the maximal GAP activity of RGS9 to the interaction of $G_{t\alpha}$ with PDEγ, thereby enhancing the fidelity of visual signal transduction.

What Structures May Follow

Clearly the most important unresolved structural question is how an activated receptor engages a heterotrimeric G protein so as to catalyze nucleotide exchange on the G_α subunit. The resolution of this question requires the crystal structure of a complex between a ligand-bound or light-activated receptor and the nucleotide free form of a G-protein heterotrimer. Only then can we claim to have glimpsed the conversion of extracellular signals into intracellular second messengers at atomic resolution.

References

1. Conklin, B. R. and Bourne, H. R. (1993). Structural elements of G_α subunits that interact with $G_{\beta\gamma}$ receptors, and effectors. *Cell* **73**, 631–641.
2. Neer, E. J. (1995). Heterotrimeric G proteins: organizers of transmembrane signals. *Cell* **80**, 249–257.

3. Hepler, J. R. and Gilman, A. G. (1992). G proteins. *Trends Biochem. Sci.* **17**, 383–387.
4. Ross, E. M. and Wilke, T. M. (2000). GTPase-activating proteins for heterotrimeric G proteins: regulators of G protein signaling (RGS) and RGS-like proteins. *Annu. Rev. Biochem.* **69**, 795–827.
5. Noel, J. P., Hamm, H. E., and Sigler, P. B. (1993). The 2.2 Å crystal structure of transducin-alpha complexed with GTPγS. *Nature* **366**, 654–663.
6. Lambright, D. G., Noel, J. P., Hamm, H. E., and Sigler, P. B. (1994). Structural determinants for activation of a G-protein α subunit. *Nature* **369**, 621–628.
7. Coleman, D. E., Berghuis, A. M., Lee, E., Linder, M. E., Gilman, A. G., and Sprang, S. R. (1994). Structures of active conformations of $G_{i\alpha 1}$ and the mechanism of GTP hydrolysis. *Science* **265**, 1405–1412.
8. Sondek, J., Lambright, D. G., Noel, J. P., Hamm, H. E., and Sigler, P. B. (1994). GTPase mechanism of G proteins from the 1.7 Å crystal structure of transducin α-GDP-AlF_4^-. *Nature* **372**, 276–279.
9. Mixon, M. B., Lee, E., Coleman, D. E., Berghuis, A. M., Gilman, A. G., and Sprang, S. R. (1995). Tertiary and quaternary structural changes in Gi alpha 1 induced by GTP hydrolysis. *Science* **270**, 954–960.
10. Sunahara, R. K., Tesmer, J. J., Gilman, A. G., and Sprang, S. R. (1997). Crystal structure of the adenylyl cyclase activator $G_{S\alpha}$. *Science* **278**, 1943–1947.
11. Sprang, S. R. (1997). G protein mechanisms: insights from structure analysis. *Annu. Rev. Biochem.* **66**, 639–78.
12. Scheffzek, K. *et al.* (1997). The Ras-RasGAP complex: structural basis for GTPase activation and its loss in oncogenic Ras mutants. *Science* **277**, 333–338.
13. Rittinger, K., Walker, P. A., Eccleston, J. F., Smerdon, S. J., and Gamblin, S. J. (1997). Structure at 1.65 Å of RhoA and its GTPase-activating protein in complex with a transition-state analogue. *Nature* **389**, 758–762.
14. Sondek, J., Bohm, A., Lambright, D. G., Hamm, H. E., and Sigler, P. B. (1996). Crystal structure of a G-protein βγ dimer at 2.1 Å resolution. *Nature* **379**, 369–374.
15. Wall, M. A., Coleman, D. E., Lee, E., Iniguez-Lluhi, J. A., Posner, B. A., Gilman, A. G., and Sprang, S. R. (1995). The structure of the G protein heterotrimer Giα1β1γ2. *Cell* **83**,1047–1058.
16. Lambright, D. G., Skiba, N., Hamm, H. E., and Sigler, P. B. (1996). The 2.0 Å structure of a heterotrimeric G-protein. *Nature* **379**, 311–316.
17. Gaudet, R., Bohm, A., and Sigler, P. B. (1996). Crystal structure at 2.4 Å resolution of the complex of transducin βγ and its regulator, phosducin. *Cell* **87**, 577–588.
18. Loew, A., Ho, Y. K., Blundell, T., and Bax, B. (1998). Phosducin induces a structural change in transducin beta gamma. *Structure* **6**, 1007–1019.
19. Tang, W. J. and Gilman, A. G. (1995). Construction of a soluble adenylyl cyclase activated by $G_{S\alpha}$ and forskolin. *Science* **268**, 1769–1772.
20. Tesmer, J. J., Sunahara, R. K., Gilman, A. G., and Sprang, S. R. (1997). Crystal structure of the catalytic domains of adenylyl cyclase in a complex with $G_{S\alpha}$·GTPγS. *Science* **278**, 1907–1916.
21. Zhang, G., Liu, Y., Ruoho, A. E., and Hurley, J. H. (1997). Structure of the adenylyl cyclase catalytic core. *Nature* **386**, 247–253.
22. Tesmer, J. J., Berman, D. M., Gilman, A. G., and Sprang, S. R. (1997). Structure of RGS4 bound to AlF4-activated Giα1: stabilization of the transition state for GTP hydrolysis. *Cell* **89**, 251–261.
23. Slep, K. C., Kercher, M. A., He, W., Cowan, C. W., Wensel, T. G., and Sigler, P. B. (2001). Structural determinants for regulation of phosphodiesterase by a G protein at 2.0 Å. *Nature* **409**,1071–1077.

In Vivo Functions of Heterotrimeric G Proteins

Stefan Offermanns
*Pharmakologisches Institut, Universität Heidelberg,
Heidelberg, Germany*

Introduction

The transmembrane signaling system, which uses heterotrimeric G proteins to couple heptahelical receptors to various effectors, operates in all cells of the mammalian organism and is involved in many physiological and pathological processes. The main properties of individual G proteins are determined by the identity of their α-subunits. To elucidate the role of G-protein-mediated signaling processes in the intact mammalian organism, almost all known genes encoding G-protein α-subunits have been inactivated by gene targeting in mice (Table I). So far, no mouse line was reported carrying targeted mutations of Gβ- or Gγ-genes. This short review summarizes the main phenotypical changes observed in mice lacking G protein α-subunits.

Development

Various Gα-deficient mouse models have pointed to the involvement of G-protein-mediated signaling pathways in certain developmental processes. For example, lack of $G\alpha_{13}$ results in embryonic lethality at about midgestation due to a defect in angiogenesis [1]. Mice deficient in both $G\alpha_q$ and $G\alpha_{11}$ suffer from a defect in heart development and die *in utero* (see below). In addition, signaling through G_q class members has also been implicated in the proliferation and/or migration of craniofacial neural crest cells [2]. The complete loss of $G\alpha_s$ in mice homozygous for an inactivating $G\alpha_s$ mutation leads to embryonic lethality before embryonic day 10 [3]. It is interesting that heterozygotes show varying phenotypes depending on the paternal origin of the intact allele; these are probably caused by genetic haploinsufficiency and/or tissue-specific imprinting of the maternal $G\alpha_s$ allele [4].

Central Nervous System

In the central nervous system (CNS), many mediators and neurotransmitters function through G-protein-coupled receptors to modulate neuronal activity or morphology. Neurotransmitters that induce an inhibitory modulation typically act on receptors that are coupled to members of the $G_{i/o}$ family, whereas G_q- and G_s-family members are primarily involved in excitatory responses.

The G-protein G_o is highly abundant in the mammalian nervous system and has been shown to mediate inhibition of neuronal (N-, P/Q-, R-type) voltage-dependent Ca^{2+} channels via its βγ-complex, thereby reducing the excitability of the cell. $G\alpha_o$-deficient mice suffer from tremors and have occasional seizures [5,6]; severely abnormal motor behavior can be observed in these animals (http://www.anes.ucla.edu/~lutzb/realmice.htm) as well. In addition, $G\alpha_o$-deficient mice appear to be hyperalgesic when tested in the hot plate assay [6]. The latter finding is consistent with the observation that opioid receptor-mediated inhibition of Ca^{2+} currents in dorsal root ganglia (DRG) from $G\alpha_o$-deficient animals was reduced by about 30 percent compared to those in wild type DRGs [6].

G_z, a member of the $G_{i/o}$-family of G proteins, shares with G_{i1}, G_{i2}, and G_{i3} the ability to inhibit adenylyl cyclases but has a rather limited pattern of expression, being found in brain, adrenal medulla, and platelets. $G\alpha_z$-deficient mice exhibit altered responses to a variety of psychoactive drugs. Cocaine-induced increases in locomotor activity were more pronounced, and short-term antinociceptive effects of

Table I Phenotypical Changes in Mice Lacking α-Subunits of Heterotrimeric G-Proteins

Family/Type		Gene	Expression	Effectors	Phenotype	Reference
$G\alpha_s$	$G\alpha_s{}^a$	Gnas	ubiquitous	AC (all types) ↑	embryonic lethal[d]	[3]
	$G\alpha_{olf}$	Gnal	olf. epithelium, brain	AC ↑	anosmia, hyperactivity	[9]
$G\alpha_{i/o}$	$G\alpha_{i1}$	Gnai1	widely distributed	AC ↓[e]	no obvious phenotype seen so far	[i]
	$G\alpha_{i2}$	Gnai2	ubiquitous	"	inflammatory bowel disease	[14]
	$G\alpha_{i3}$	Gnai3	widely distributed	"	no obvious phenotype seen so far	[i]
	$G\alpha_o{}^b$	Gnao	neuronal, neuroendocr.	Ca^{2+}-ch. ↓[f]	various CNS defects	[5,6,20]
	$G\alpha_z$	Gnaz	neuronal, platelets	AC ↓; ?	viable, increased bleeding time	[7,8]
	$G\alpha_{gust}$	Gnag	taste cells, brush cells	?	impaired bitter and sweet sensation	[21]
	$G\alpha_{t-r}$	Gnat1	retinal rods, taste cells	cGMP PDE ↑	mild retinal degeneration	[19]
	$G\alpha_{t-c}$	Gnat2	retinal cones	cGMP PDE ↑	no mouse mutant available	–
	$G\alpha_{i1} + G\alpha_{i3}$				no obvious phenotype seen so far	[i]
	$G\alpha_{i2} + G\alpha_{i3}$				lethal	[i]
$G\alpha_q$	$G\alpha_q$	Gnaq	Ubiquitous	PLC-β ↑[g]	ataxia, defective platelet activation	[12,23]
	$G\alpha_{11}$	Gna11	almost ubiquitous	"	no obvious phenotype seen so far	[2]
	$G\alpha_{14}$	Gna14	kidney, lung, spleen	"	no obvious phenotype seen so far	[j]
	$G\alpha_{15}{}^c$	Gna15	hematopoietic cells	"	no obvious phenotype seen so far	[26]
	$G\alpha_q + G\alpha_{11}$				myocardial hypoplasia (lethal e11)	[2]
					cardiomyocyte-restricted:	
					pressure overload induced hypertrophy ↓	[16]
	$G\alpha_q + G\alpha_{15}$				like $G\alpha_q$ (−/−)	[26]
$G\alpha_{12}$	$G\alpha_{12}$	Gna12	ubiquitous	?	no obvious phenotype seen so far	[k]
	$G\alpha_{13}$	Gna13	ubiquitous	?[h]	defective angiogenesis (lethal e9.5)	[1]
	$G\alpha_{12} + G\alpha_{13}$				embryonic lethal (e8.5)	[k]

[a] several splice variants
[b] 2 splice variants
[c] mouse form ($G\alpha_{16}$, human counterpart)
[d] parent of origin specific defects in heterozygotes
[e] adenylyl cyclase types I,V,VI
[f] N-,P/Q-type; effector is regulated through βγ-subunits
[g] β4;β3 ≥ β1 >> β2
[h] RhoGEF-proteins (p115RhoGEF)
[i] L. Birnbaumer, M. Jiang, G. Boulay, K. Spicher (personal communication);
[j] H. Jiang and M.I. Simon (personal communication)
[k] S. Müller, S.O., M.I. Simon (unpublished data). AC, adenylyl cyclase; Ca^{2+}-ch., Ca^{2+}-channel; cGMP PDE, cGMP-phosphodiesterase, PLC-β, β-isoforms of phospholipase C.

morphine were altered [7,8]. In addition, behavioral effects of catecholamine reuptake inhibitors were abolished in $G\alpha_z$-deficient mice [7], indicating that G_z is involved in signaling processes regulated by various neurotransmitters.

$G\alpha_{olf}$ is expressed in various regions of the CNS, including olfactory sensory neurons and basal ganglia. $G\alpha_{olf}$-deficient mice exhibit clear motoric abnormalities such as hypermotoric behavior [9]. Recent data indicate that G_{olf} is critically involved in dopamin(D_1)- and adenosine(A_{2A})-receptor-mediated effects in the striatum [10,11].

The two main members of the G_q family, G_q and G_{11}, are widely expressed in the central nervous system. Mice lacking $G\alpha_q$ develop an ataxia with clear signs of motor coordination deficits, and functional defects could be observed in the cerebellar cortex of $G\alpha_q$-deficient mice [12]. In addition, lack of $G\alpha_q$ resulted in defective cerebellar and hippocampal long-term depression (M. Kano et al., unpublished; [13]).

Immune System

Mice lacking $G\alpha_{i2}$ develop a lethal, diffuse inflammatory bowel disease that resembles in many aspects ulcerative colitis in humans [14]. In subsequent studies, dramatic changes in the phenotype and function of intestinal lymphocytes and epithelial cells have been described that are likely to be due to defective lymphocyte homing in enteric epithelia [15]. On a cellular level, G_{i2} may be involved in the regulation of T-cell function and trafficking. These processes are regulated through chemoattractant and chemokine receptors that show a predominant coupling to G_i-type G-proteins. In addition to the colitis, many $G\alpha_{i2}$-deficient mice develop colonic adenocarcinomas, which are probably secondary to colonic inflammation [14].

Heart

The $G\alpha_q/G\alpha_{11}$-mediated signaling pathway appears to play a pivotal role in the regulation of physiological myocardial growth during embryogenesis. This is demonstrated by the phenotype of $G\alpha_q/G\alpha_{11}$-double deficient mice that die at embryonic day 11 due to a severe thinning of the myocardial layer of the heart [2]. Adult cardiomyocytes are terminally differentiated post-mitotic cells that respond to stimulatory signals with cell growth rather than proliferation. Myocardial hypertrophy in the adult heart following mechanical stress depends on $G\alpha_q/G\alpha_{11}$-mediated signaling as demonstrated by the absence of a hypertrophic response in adult mice with cardiomyocyte-specific $G\alpha_q/G\alpha_{11}$ deficiency [16].

Inhibition of L-type Ca^{2+} channels in the heart through muscarinic M_2 receptors was found to be abrogated in hearts lacking $G\alpha_o$ as well as $G\alpha_{i2}$ [5,17]. This unexpected finding suggests that both G proteins may regulate this downstream signaling event in a complex fashion.

Sensory Systems

Odors, light, and many tastants act directly on G-protein-coupled receptors. The G protein G_{olf} is centrally involved in the transduction of odorant stimuli in olfactory cilia, and $G\alpha_{olf}$-deficient mice exhibit dramatically reduced electrophysiological responses to all odors tested [9]. Since nursing and mothering behavior in rodents is mediated a great deal by the olfactory system, most $G\alpha_{olf}$-deficient pups die a few days after birth due to insufficient feeding, and rare surviving mothers exhibit inadequate maternal behavior. In contrast to the olfactory epithelium, the vomeronasal organ, which detects pheromones, expresses receptors that are coupled to $G_{i/o}$. Absence of $G\alpha_o$ results in apoptotic death of receptor cells that usually express $G\alpha_o$ [18].

Rod-transducin (G_{t-r}) and cone-transducin (G_{t-c}) play well-established roles in the phototransduction cascade in the outer segments of retinal rods and cones, where they couple light receptors to cGMP-phosphodiesterase. In mice lacking $G\alpha_{t-r}$, the majority of retinal rods does not respond to light anymore, and these animals develop mild retinal degeneration with age [19]. The light response is transferred from the receptor cell to bipolar cells of the retina. In mice lacking $G\alpha_o$, modulation of ON bipolar cells in response to light is abrogated, indicating that G_o is critically involved in the tonic inhibition of these cells mediated by metabotropic glutamate (mGluR6) receptors [20].

Among the four taste qualities—sweet, bitter, sour, and salty—bitter and sweet tastes appear to signal through heterotrimeric G-proteins. Gustducin is a G protein mainly expressed in taste cells, and $G\alpha_{gust}$-deficient mice show impaired electrophysiological and behavioral responses to bitter and sweet agents [21]. The residual bitter and sweet taste responsiveness of $G\alpha_{gust}$-deficient mice could be further diminished by a dominant-negative mutant of gustducin-α, suggesting the involvement of other G proteins related to $G\alpha_{gust}$ [22].

Hemostasis

Hemostasis is a complex process involving platelet adhesion and aggregation as well as formation of fibrin through the coagulation cascade. Platelet activation results in a rapid shape-change reaction immediately followed by secretion of granule contents, as well as inside-out activation of the fibrinogen receptor, integrin $\alpha_{IIb}\beta_3$, leading to platelet aggregation. Most physiological platelet activators act through G-protein-coupled receptors, which in turn activate $G_{i2/3}$, G_q, G_{12}, and G_{13}. In platelets from $G\alpha_q$-deficient mice, the effect of various platelet stimuli on aggregation and degranulation was abrogated, demonstrating that $G\alpha_q$-mediated phospholipase C activation represents an essential event in platelet activation [23]. However, platelet shape change can still be induced in the absence of $G\alpha_q$, indicating that it is mediated by G proteins other than G_q, most likely G_{12}/G_{13} [24]. The defective activation of $G\alpha_q$-deficient platelets results in a primary hemostasis defect, and $G\alpha_q$ (−/−) mice are protected against platelet-dependent thromboembolism.

The role of G proteins of the $G_{i/o}$ family in platelet activation has recently been elucidated. Platelets contain at least three members of this class, G_{i2}, G_{i3}, and G_z. ADP, which is released from activated platelets and functions as a positive feedback mediator during platelet activation, induces platelet activation through the G_q-coupled $P2Y_1$ receptor as well as through the G_i-coupled $P2Y_{12}$ purinergic receptor. The general importance of the G_i-mediated pathway is indicated by the fact that responses to ADP but also to thrombin were markedly reduced in platelets lacking $G\alpha_{i2}$ [25]. In contrast to ADP or thrombin, epinephrine is not a full platelet activator *per se* in murine platelets. However, it is able to potentiate the effect of other platelet stimuli. In platelets from $G\alpha_z$-deficient mice, epinephrine's potentiating effects were clearly impaired, while the effects of other platelet activators appeared to be unaffected by the lack of $G\alpha_z$ [7]. Thus, members of the G-protein families G_q, G_{12}, and $G_{i/o}$ are involved in processes leading to platelet activation.

Conclusions

Mouse models lacking almost all known genes encoding G-protein α-subunits have been generated, and they provide a first insight into the biological roles of G-protein-mediated signaling pathways. To overcome embryonic lethality or complex phenotypes of some $G\alpha$ null mutations and to understand the degree of functional redundancy of closely related G proteins researchers have begun to cross individual mutants and to generate mouse lines that allow for the conditional inactivation of genes in a time- and tissue-specific manner. These approaches will soon provide more detailed views on the functions of G-protein-mediated signaling pathways in the developing and adult mammalian organism.

References

1. Offermanns, S., Mancino, V., Revel, J. P., and Simon, M. I. (1997). Vascular system defects and impaired cell chemokinesis as a result of $G\alpha_{13}$ deficiency. *Science* **275**, 533–536.
2. Offermanns, S., Zhao, L.-P., Gohla, A., Sarosi, I., Simon, M. I., and Wilkie, T. M. (1998). Embryonic cardiomyocyte hypoplasia and craniofacial defects in $G\alpha_q/G\alpha_{11}$ mutant mice. *EMBO J.* **17**, 4304–4312.
3. Yu, S., Yu, D., Lee, E., Eckhaus, M., Lee, R., Corria, Z., Accili, D., Westphal, H., and Weinstein, L. S. (1998). Variable and tissue-specific hormone resistance in heterotrimeric Gs protein alpha-subunit (Gsalpha) knockout mice is due to tissue-specific imprinting of the gsalpha gene. *Proc. Natl. Acad. Sci. USA* **95**, 8715–8720.
4. Weinstein, L. S. and Yu, S. (1999). The Role of genomic imprinting of Galpha in the pathogenesis of Albright hereditary osteodystrophy. *Trends Endocr. Sci.* **10**, 81–85.
5. Valenzuela, D., Han, X., Mende, U., Fankhauser, C., Mashimo, H., Huang, P., Pfeffer, J., Neer, E. J., and Fishman, M. C. (1997). G alpha(o) is necessary for muscarinic regulation of Ca^{2+} channels in mouse heart. *Proc. Natl. Acad. Sci. USA* **94**, 1727–1732.
6. Jiang, M., Gold, M. S., Boulay, G., Spicher, K., Peyton, M., Brabet, P., Srinivasan, Y., Rudolph, U., Ellison, G., and Birnbaumer, L. (1998). Multiple neurological abnormalities in mice deficient in the G protein Go. *Proc. Natl. Acad. Sci. USA* **95**, 3269–3274.
7. Yang, J., Wu, J., Kowalska, M. A., Dalvi, A., Prevost, N., O'Brien, P. J., Manning, D., Poncz, M., Lucki, I., Blendy, J. A., and Brass, L. F. (2000). Loss of signaling through the G protein, Gz, results in abnormal platelet activation and altered responses to psychoactive drugs. *Proc. Natl. Acad. Sci. USA* **97**, 9984–9989.
8. Hendry, I. A., Kelleher, K. L., Bartlett, S. E., Leck, K. J., Reynolds, A. J., Heydon, K., Mellick, A., Megirian, D., and Matthaei, K. I. (2000). Hypertolerance to morphine in G(z alpha)-deficient mice. *Brain Res.* **870**, 10–19.
9. Belluscio, L., Gold, G. H., Nemes, A., and Axel, R. (1998). Mice deficient in G(olf) are anosmic. *Neuron* **20**, 69–81.
10. Zhuang, X., Belluscio, L., and Hen, R. (2000). GOLFalpha mediates dopamine D1 receptor signaling. *J. Neurosci.* **20**, RC91.
11. Corvol, J. C., Studler, J. M., Schonn, J. S., Girault, J. A., and Hervé, D. (2001). Galpha(olf) is necessary for coupling D_1 and A_{2a} receptors to adenylyl cyclase in the striatum. *J. Neurochem.* **76**, 1585–1588.
12. Offermanns, S., Hashimoto, K., Watanabe, M., Sun, W., Kurihara, H., Thompson, R. F., Inoue, Y., Kano, M., and Simon, M. I. (1997). Impaired motor coordination and persistent multiple climbing fiber innervation of cerebellar Purkinje cells in mice lacking Galphaq. *Proc. Natl. Acad. Sci. USA* **94**, 14089–14094.
13. Kleppisch, T., Voigt, V., Allmann, R., and Offermanns, S. (2001). $G\alpha_q$-deficient mice lack metabotropic glutamate receptor-dependent long-term depression but show normal long-term potentiation in the hippocampal CA1 region. *J. Neurosci.* **21**, 4943–4948.
14. Rudolph, U., Finegold, M. J., Rich, S. S., Harriman, G. R., Srinivasan, Y., Brabet, P., Boulay, G., Bradley, A., and Birnbaumer, L. (1995). G_{i2} alpha protein deficiency: a model of inflammatory bowel disease. *Nat. Genet.* **10**, 143–150.
15. Hornquist, C. E., Lu, X., Rogers-Fani, P. M., Rudolph, U., Shappell, S., Birnbaumer, L., and Harriman, G. R. (1997). G(alpha)i2-deficient mice with colitis exhibit a local increase in memory CD^{4+} T cells and proinflammatory Th1-type cytokines. *J. Immunol.* **158**, 1068–1077.
16. Wettschureck, N., Rütten, H., Zywietz, A., Gehring, D., Wilkie, T., Chen, J., Chien, K. R., and Offermanns, S. (2001). Absence of pressure overload induced myocardial hypertrophy after conditional inactivation of Galphaq/Galpha11 in cardiomyocytes. *Nat. Med.* **7**, 1236–1240.
17. Chen, F., Spicher, K., Jiang, M., Birnbaumer, L., and Wetzel, G. T. (2001). Lack of muscarinic regulation of Ca(2+) channels in G(i2)alpha gene knockout mouse hearts. *Am. J. Physiol.* **280**, H1989–1995.
18. Tanaka, M., Treloar, H., Kalb, R. G., Greer, C. A., and Strittmatter, S. M. (1999). G(o) protein-dependent survival of primary accessory olfactory neurons. *Proc. Natl. Acad. Sci. USA* **96**, 14106–14111.
19. Calvert, P. D., Krasnoperova, N. V., Lyubarsky, A. L., Isayama, T., Nicolo, M., Kosaras, B., Wong, G., Gannon, K. S., Margolskee, R. F., Sidman, R. L., Pugh, E. N. Jr., Makino, C. L., and Lem, J. (2000). Phototransduction in transgenic mice after targeted deletion of the rod transducin alpha-subunit. *Proc. Natl. Acad. Sci. USA* **97**, 13913–13918.
20. Dhingra, A., Lyubarsky, A., Jiang, M., Pugh, E. N. Jr., Birnbaumer, L., Sterling, P., and Vardi, N. (2000). The light response of ON bipolar neurons requires G[alpha]o. *J. Neurosci.* **20**, 9053–9058.
21. Wong, G. T., Gannon, K. S., and Margolskee, R. F. (1996). Transduction of bitter and sweet taste by gustducin. *Nature* **381**, 796–800.
22. Ruiz-Avila, L., Wong, G. T., Damak, S., and Margolskee, R. F. (2001). Dominant loss of responsiveness to sweet and bitter compounds caused by a single mutation in alpha-gustducin. *Proc. Natl. Acad. Sci. USA* **98**, 8868–8873.
23. Offermanns, S., Toombs, C. F., Hu, Y. H., and Simon, M. I. (1997). Defective platelet activation in G alpha(q)-deficient mice. *Nature* **389**, 183–186.
24. Klages, B., Brandt, U., Simon, M. I., Schultz, G., and Offermanns, S. (1999). Activation of G12/G13 results in shape change and Rho/Rho-kinase-mediated myosin light chain phosphorylation in mouse platelets. *J. Cell Biol.* **144**, 745–754.
25. Jantzen, H. M., Milstone, D. S., Gousset, L., Conley, P. B., and Mortensen, R. M. (2001). Impaired activation of murine platelets lacking G alpha(i2). *J. Clin. Invest.* **108**, 477–483.
26. Davignon, I., Catalina, M. D., Smith, D., Montgomery, J., Swantek, J., Croy, J., Siegelman, M., and Wilkie, T. M. (2000). Normal hematopoiesis and inflammatory responses despite discrete signaling defects in Galpha15 knockout mice. *Mol. Cell. Biol.* **20**, 797–804.

Regulation of G Proteins by Covalent Modification

Jessica E. Smotrys and Maurine E. Linder
Department of Cell Biology and Physiology,
Washington University School of Medicine, St. Louis, Missouri

Introduction

G proteins are subject to a number of covalent modifications that affect their subcellular localization, protein-protein interactions, and regulation. The first modifications known were those catalyzed by bacterial toxins. The discovery of ADP-ribosylation of G-protein α subunits by cholera and pertussis toxins provided significant insights into G-protein function and is discussed in detail in another chapter in this volume (Chapter 100 by Di Girolamo and Corda). Regulation of G-protein activity by phosphorylation is also covered in a separate chapter (Chapter 99 by Luttrell and Luttrell). Other posttranslational modifications of G-protein subunits that have been reported include deamidation of $G_{o\alpha}$ [1,2] and ADP-ribosylation of G_β by a cellular ADP-ribosyltransferase [3]. Here we focus on the covalent modification of G proteins by lipids. All G_α subunits are fatty acylated by amide-linked myristate and/or thioester-linked palmitate. All G_γ subunits are modified at the C-terminus by a farnesyl (C15) or geranylgeranyl (C20) isoprenoid. Two themes have emerged concerning the role of lipid modifications in G-protein signaling. First, by conferring membrane affinity, they have a role in properly positioning G proteins at the plasma membrane, where they must reside in order to interact with receptors and effectors. Once the G protein is in place, lipids further affect G-protein signaling by modulating protein-protein interactions. This chapter will focus on the latter role, as subcellular localization is discussed elsewhere (Linder, Volume 1, Chapter 60).

Separating the roles of lipid modification in membrane affinity from protein affinity is challenging. *In vivo*, mutation of lipid modified residues often results in mislocalized G protein. Defects in signaling can be attributed to the mislocalization, making it difficult to assess additional roles of the lipid modification. Scientists have turned to *in vitro* methods to study effects of lipid modification on affinity between proteins. However, due to the hydrophobic nature of the proteins involved, detergents must often be present in the assays. Detergents at concentrations above the critical micelle concentration (CMC) form micelles. Lipid modifications mediate association with the micelles, concentrating proteins at the micelle surface. Thus, an apparent change in affinity could reflect the ability of lipid-modified proteins to cluster at the micelle surface. The use of soluble effector domains, lipidated peptides, and x-ray crystal structures have allowed more definitive analysis of lipid effects on protein interactions.

N-Terminal Acylation of G_α

As presented in Table I, fatty acids are found singly or in combination on G_α subunits. N-myristoylation is a cotranslational modification of G-protein subunits of the $G_{i\alpha}$ subfamily [4]. Myristic acid is added through an amide linkage to a glycine residue exposed after removal of the initiator methionine. The enzyme that catalyzes this modification, N-myristoyltransferase, is associated with ribosomes in mammalian cells [5].

All G-protein α subunits with the exception of $G_{t\alpha}$ and gustducin are substrates for palmitoylation [4]. Palmitate is linked through a labile thioester bond to one or more cysteine residues near the N-terminus of G_α. Palmitoylation of G-protein α subunits is believed to occur at the plasma membrane. In contrast to N-myristoylation, palmitoylation is a reversible and regulated modification. This topic is explored in a separate chapter (Chapter 106 by Wedegartner).

Acylation and Subunit Interactions

In the inactive state, G_α forms a high-affinity complex with $G_{\beta\gamma}$. The crystal structure of this complex reveals that the lipid-modified termini of G_α and G_γ are proximal to each other [6,7]. Although the crystallized proteins lacked lipid moieties, it is predicted that the lipids extend away from the protein complex and into nearby membrane. *In vitro*, myristoylated $G_{o\alpha}$ has higher affinity for $G_{\beta\gamma}$ than the nonmyristoylated form [8]. Studies with $G_{t\alpha}$ suggest that this apparent increase in affinity is due to interactions of the myristoylated subunit with detergent micelles, rather than direct interaction of the fatty acid with $G_{\beta\gamma}$ [9].

The N-terminal modifications found on $G_{s\alpha}$ are also important for subunit interactions. $G_{s\alpha}$ is palmitoylated on Cys3 and contains an additional unidentified hydrophobic moiety at the N-terminus [10]. $G_{s\alpha}$, which has both modifications (purified from liver or Sf9 insect cells), has higher affinity for $G_{\beta\gamma}$ than $G_{s\alpha}$, which lacks the modifications (Sf9 $G_{s\alpha}$ treated with palmitoyl esterase or purified from *E. coli*) [10,11]. The relative contributions of the modifications cannot be determined from these experiments. Detergent at concentrations above the CMC is present in the assays, and the effect of lipidation may be indirect.

Interaction of G_α with Effectors

Activated $G_{s\alpha}$ and $G_{i\alpha}$ work to stimulate and inhibit membrane-bound adenylyl cyclase (AC), respectively. The lipid modifications found on these subunits are important for this activity. The unidentified modification on $G_{s\alpha}$ supports high-affinity binding to AC, independent of the presence of palmitate [10,12]. It is not clear whether this modification directly binds AC or increases affinity by concentrating the G protein at membranes. It is clear, however, that the myristoyl group on $G_{i\alpha}$ directly affects affinity for AC independently of membrane localization [13]. Myristoylation was required for $G_{i\alpha}$ to interact with a soluble domain of AC in an assay free of micelles or membranes. It is unknown whether the myristoyl group has a binding site on AC or affects the conformation of $G_{i\alpha}$ to mediate its interaction with the effector [13]. In contrast to N-myristoylation, palmitoylation does not appear to have an important role in promoting G_α-effector interactions [10,14].

Palmitoylation and Signal Downregulation

Activation of receptors increases palmitate turnover on $G_{s\alpha}$ and $G_{i\alpha}$ [4]. This may serve to regulate the amount of G_α at the plasma membrane [15], although this has been controversial [16], and also to modify interactions with $G_{\beta\gamma}$ as mentioned previously in the section on acylation and subunit interactions. In addition, palmitoylation may be involved in signal down-regulation through effects on interactions with regulators of G-protein signaling (RGS), which stimulate the intrinsic GTPase activity of G_α subunits. Palmitoylated $G_{z\alpha}$ has reduced affinity for its RGS protein G_z GAP [17]. This same phenomenon was observed with other G_α-RGS pairs and is not due simply to the hydrophobicity of the palmitate moiety, since myristoylation of $G_{z\alpha}$ has the opposite effect and promotes interaction with G_z GAP. Thus, depalmitoylation after G-protein activation may be important for returning the G protein to the inactive state [17].

C-Terminal Modification of G_γ

All known G_γ subunits contain C-terminal CaaX motifs, where C is Cys, "a" is an aliphatic residue, and X is the C-terminal amino acid (Table I). These motifs direct prenylation of the cysteine residue, proteolysis of the three C-terminal residues (-aaX), and carboxylmethylation of the C-terminal prenylated cysteine [18]. G_γ is modified with farnesyl or geranylgeranyl depending on the identity of the C-terminal amino acid. Most G_γ subunits are geranylgeranylated; $G_{\gamma 1}$, $G_{\gamma c}$, and $G_{\gamma 11}$ are farnesylated. Prenylation occurs in the cytoplasm, and subsequent processing steps are likely to take place at the endoplasmic reticulum [19,20].

Prenylation and Subunit Interactions

Prenylation of G_γ is not required for the assembly of $G_{\beta\gamma}$ dimers [21]. However, like G_α fatty acylation, prenylation promotes interaction of $G_{\beta\gamma}$ with G_α [22]. The type of prenyl group attached affects the affinity of the interaction for many G_α subunits. Generally, G_γ modified with the more hydrophobic geranylgeranyl group displays higher apparent affinity for G_α. This is likely to be an indirect effect due to enhanced association with the membrane surface. For example, changing the prenyl group from farnesyl to geranylgeranyl on $G_{\gamma 1}$ increases affinity of $G_{\beta 1 \gamma 1}$ for both membranes and G_α [23].

Prenylation and Receptor Coupling

Prenylation is essential for receptor-G-protein coupling. Farnesylated peptides corresponding to the C-terminus of G_γ inhibit coupling to rhodopsin, suggesting that the prenyl group is directly involved in binding to receptor [24]. The type of prenyl moiety with which the peptides are modified influences the ability of the peptide to stabilize activated rhodopsin, a measure of affinity. Farnesylated peptides have greater affinity for rhodopsin than either geranylated (C10) or geranylgeranylated peptides [25]. Since the observed affinity does not correlate with hydrophobicity, this is suggestive of a specific binding site on rhodopsin that

Table I Lipid Modifications of Heterotrimeric G Proteins

Subunit	Lipid modifications	Modified sequence[a]
αi, αo, αz	N-Myr, S-palm	H$_2$N-M**G**C–
αt	N-Myr	H$_2$N-M**G**A–
αs	Unknown hydrophobic, S-Palm	H$_2$N-M**G**C–
αq	S-Palm	H$_2$N-MTLESIMA**CC**–
α12	S-Palm	H$_2$N-MSGVVRTLSR**C**–
α13	S-Palm	H$_2$N-MAD–^{14}**C**FPG**C**^{18}–
γ1, γc, γ11	Farnesyl	–Caa**S**-COOH[b]
γ2, γ3, γ4, γ5, γ7, γ8, γ10, γ12	Geranylgeranyl	–Caa**L**-COOH[b]

[a] modified residues are in boldface
[b] "a" is an aliphatic residue, additional processing includes proteolysis of the three C-terminal residues and carboxymethylation of the prenylated cysteine

preferentially recognizes farnesylated G$_\gamma$. However, in assays of rhodopsin-stimulated GTPγS binding, geranylgeranylated G$_{\beta\gamma}$ dimers are more effective than their farnesylated counterparts, arguing against a specific farnesyl binding site on rhodopsin [23,26]. How the prenyl moiety functions in receptor recognition requires further investigation.

Prenylation and Effector Interactions

Prenylation and primary sequence of G$_\gamma$ are both important determinants of G$_{\beta\gamma}$ effector activation. Similar to receptor coupling, the same G$_{\beta\gamma}$ complex is more effective in activating phospholipase C β (PLCβ) or AC when G$_\gamma$ is geranylgeranylated than when farnesylated [27]. In the case of PLCβ activation, there is evidence that the prenyl group is not simply promoting effector activation indirectly through membrane interactions, but is directly mediating protein-protein interactions. Peptides corresponding to the C-terminus of G$_{\gamma 2}$ inhibit activation of PLCβ by G$_{\beta\gamma}$ in a prenylation-dependent manner. A fluorescence-based binding assay demonstrated a direct interaction of the prenylated peptide with PLCβ$_2$ [28]. Precedence for prenyl-binding sites in proteins is provided by the structure of prenylated Cdc42, a Rho family GTPase, bound to Rho-GDI, a guanine nucleotide dissociation inhibitor, which regulates interaction of Rho family members with membranes. In this structure, the C-terminal geranylgeranyl group on Cdc42 inserts into a hydrophobic pocket formed by Rho-GDI [29].

A second mechanism for prenylation-dependent effector interactions is suggested by the structure of the G$_{\beta\gamma}$-phosducin complex. Phosducin binds tightly to free G$_{\beta\gamma}$, extracting it from membranes and preventing its reassociation with G$_\alpha$. Phosducin induces several local conformational changes in G$_\beta$ that are not seen in the structures of free G$_{\beta\gamma}$ or the heterotrimer, including the opening of a pocket between blades 6 and 7 [30,31]. Based on their structure of farnesylated G$_{\beta\gamma}$ with phosducin, Loew et al. proposed that the farnesyl group is sequestered in the crevice [30]. G$_{\beta\gamma}$ may undergo a similar conformational change when bound to effectors. Consistent with this model, mutations in G$_\beta$ that perturb the putative prenyl binding pocket exhibit reduced potency in effector activation assays [32].

Carboxymethylation of G$_\gamma$

Methylation of the C-terminal prenylated cysteine residue is of interest because it is the only step in the post-translational processing of G$_\gamma$ that is reversible and thus has the potential to be regulated. Methylation affects protein properties by neutralizing the negative charge on the C-terminal carboxylate ion. The effects of methylation on G$_{\beta\gamma}$ have only been studied for the farnesylated T$_{\beta\gamma}$ [33,34]. The most significant effect observed was that on effector interactions [34]. T$_{\beta\gamma}$ activation of phophoinositide-3-kinase and PLCβ, assayed in phospholipid/detergent micelles, was strongly dependent on methylation. Demethylated T$_{\beta\gamma}$ binds to phospholipid/cholate micelles, albeit less well than its methylated counterpart [9,34]. This suggests that the effect of methylation on effector interactions is not simply due to membrane affinity, but may involve direct protein contacts.

Conclusions

Lipid modifications found on G$_\alpha$ and G$_{\beta\gamma}$ modulate interactions with membranes and other proteins, playing an essential role in signal transduction. There is evidence for prenyl-binding sites on receptors and the effector PLCβ, and a direct role for myristate in binding of G$_{i\alpha}$ to AC. How these modifications mediate these protein interactions awaits additional structural studies with lipidated proteins. Novel roles for lipid modifications in protein interactions will surely be revealed in the future.

Note Added in Proof

The unknown hydrophobic modification on αs (Table I) has been identified as amide-linked palmitate at Gly2.

Kleuss, C. and Krause, E. (2003). Gαs is palmitoylated at the N-terminal glycine. *EMBO J.* **22**, 826–832.

References

1. McIntire, W., Schey, K., Knapp, D., and Hildebrandt, J. D. (1998). A major G protein αo isoform in bovine brain is deamidated at Asn346 and Asn347, residues involved in receptor coupling. *Biochemistry* **37**, 14651–14658.
2. Exner, T., Jensen, O. N., Mann, M., Kleuss, C., and Nurnberg, B. (1999). Posttranslational modification of Gαo1 generates Gαo3, an abundant G protein in brain. *Proc. Natl. Acad. Sci. USA* **96**, 1327–1332.
3. Lupi, R., Corda, D., and Di Girolamo, M. (2000). Endogenous ADP-ribosylation of the G protein β subunit prevents the inhibition of Type I adenylyl cyclase. *J. Biol. Chem.* **275**, 9418–9424.
4. Chen, C. and Manning, D. (2001). Regulation of G proteins by covalent modification. *Oncogene* **20**, 1643–1652.
5. Glover, C., Hartman, K., and Felsted, R. (1997). Human N-myristoyltransferase amino-terminal domain involved in targeting the enzyme to the ribosomal subcellular fraction. *J. Biol. Chem.* **272**, 28680–28689.
6. Lambright, D. G., Sondek, J., Bohm, A., Skiba, N. P., Hamm, H. E., and Sigler, P. B. (1996). The 2.0 A crystal structure of a heterotrimeric G protein. *Nature* **379**, 311–319.
7. Wall, M., Coleman, D., Lee, E., Iniguez-Lluhi, J., Posner, B., Gilman, A., and Sprang, S. (1995). The structure of the G protein heterotrimer Giα1β1γ2. *Cell* **83**, 1047–1058.
8. Linder, M. E., Pang, I. H., Duronio, R. J., Gordon, J. I., Sternweis, P. C., and Gilman, A. G. (1991). Lipid modifications of G protein subunits. Myristoylation of Gοα increases its affinity for βγ. *J. Biol. Chem.* **266**, 4654–4659.
9. Bigay, J., Faurobert, E., Franco, M., and Chabre, M. (1994). Roles of lipid modifications of transducin subunits in their GDP-dependent association and membrane binding. *Biochemistry* **33**, 14081–14090.
10. Kleuss, C. and Gilman, A. G. (1997). Gsα contains an unidentified covalent modification that increases its affinity for adenylyl cyclase. *Proc. Natl. Acad. Sci. USA* **94**, 6116–6120.
11. Iiri, T., Backlund, P. S. Jr., Jones, T. L., Wedegaertner, P. B., and Bourne, H. R. (1996). Reciprocal regulation of Gsα by palmitate and the βγ subunit. *Proc. Natl. Acad. Sci. USA* **93**, 14592–14597.
12. Graziano, M. P., Freissmuth, M., and Gilman, A. G. (1989). Expression of Gsα in *Escherichia coli*. Purification and properties of two forms of the protein. *J. Biol. Chem.* **264**, 409–418.
13. Dessauer, C. W., Tesmer, J. J., Sprang, S. R., and Gilman, A. G. (1998). Identification of a Giα binding site on type V adenylyl cyclase. *J. Biol. Chem.* **273**, 25831–25839.
14. Hepler, J. R., Biddlecome, G. H., Kleuss, C., Camp, L. A., Hofmann, S. L., Ross, E. M., and Gilman, A. G. (1996). Functional importance of the amino terminus of Gqα. *J. Biol. Chem.* **271**, 496–504.
15. Wedegaertner, P., Bourne, H., and von Zastrow, M. (1996). Activation-induced subcellular redistribution of Gsα. *Mol. Biol. Cell* **8**, 1225–1233.
16. Huang, C., Duncan, J. A., Gilman, A. G., and Mumby, S. M. (1999). Persistent membrane association of activated and depalmitoylated G protein α subunits. *Proc. Natl. Acad. Sci. USA* **96**, 412–417.
17. Tu, Y., Wang, J., and Ross, E. M. (1997). Inhibition of brain Gz GAP and other RGS proteins by palmitoylation of G protein α subunits. *Science* **278**, 1132–1135.
18. Zhang, F. L. and Casey, P. J. (1996). Protein prenylation: molecular mechanisms and functional consequences. *Annu. Rev. Biochem.* **65**, 241–269.
19. Schmidt, W. K., Tam, A., Fujimura-Kamada, K., and Michaelis, S. (1998). Endoplasmic reticulum membrane localization of Rce1p and Ste24p, yeast proteases involved in carboxyl-terminal CAAX protein processing and amino-terminal a-factor cleavage. *Proc. Natl. Acad. Sci. USA* **95**, 11175–11180.
20. Dai, Q., Choy, E., Chiu, V., Romano, J., Slivka, S. R., Steitz, S. A., Michaelis, S., and Philips, M. R. (1998). Mammalian prenylcysteine carboxyl methyltransferase is in the endoplasmic reticulum. *J. Biol. Chem.* **273**, 15030–15034.
21. Higgins, J. and Casey, P. (1994). In vitro processing of recombinant G protein γ subunits. *J. Biol. Chem.* **269**, 9067–9073.
22. Wedegaertner, P. B., Wilson, P. T., and Bourne, H. R. (1995). Lipid modifications of trimeric G proteins. *J. Biol. Chem.* **270**, 503–506.
23. Matsuda, T., Hashimoto, Y., Ueda, H., Asano, T., Matsuura, Y., Doi, T., Takao, T., Shimonishi, Y., and Fukada, Y. (1998). Specific isoprenyl group linked to transducin γ-subunit is a determinant of its unique signaling properties among G-proteins. *Biochemistry* **37**, 9843–9850.
24. Kisselev, O. G., Ermolaeva, M. V., and Gautam, N. (1994). A farnesylated domain in the G protein γ subunit is a specific determinant of receptor coupling. *J. Biol. Chem.* **269**, 21399–21402.
25. Kisselev, O., Ermolaeva, M., and Gautam, N. (1995). Efficient interaction with a receptor requires a specific type of prenyl group on the G protein γ subunit. *J. Biol. Chem.* **270**, 25356–25358.
26. Jian, X., Clark, W. A., Kowalak, J., Markey, S. P., Simonds, W. F., and Northup, J. K. (2001). Gβγ affinity for bovine rhodopsin is determined by the carboxyl-terminal sequences of the γ subunit. *J. Biol. Chem.* **276**, 48518–48525.
27. Myung, C. S., Yasuda, H., Liu, W. W., Harden, T. K., and Garrison, J. C. (1999). Role of isoprenoid lipids on the heterotrimeric G protein γ subunit in determining effector activation. *J. Biol. Chem.* **274**, 16595–16603.
28. Fogg, V. C., Azpiazu, I., Linder, M. E., Smrcka, A., Scarlata, S., and Gautam, N. (2001). Role of the γ subunit prenyl moiety in G protein βγ complex interaction with phospholipase Cβ. *J. Biol. Chem.* **276**, 41797–41802.
29. Hoffman, G. R., Nassar, N., and Cerione, R. A. (2000). Structure of the Rho family GTP-binding protein Cdc42 in complex with the multifunctional regulator RhoGDI. *Cell* **100**, 345–356.
30. Loew, A., Ho, Y. K., Blundell, T., and Bax, B. (1998). Phosducin induces a structural change in transducin βγ. *Structure* **6**, 1007–1019.
31. Gaudet, R., Bohm, A., and Sigler, P. B. (1996). Crystal structure at 2.4 A resolution of the complex of transducin βγ and its regulator, phosducin. *Cell* **87**, 577–588.
32. Myung, C.-S. and Garrison, J. (2000). Role of the C-terminal domains of the G protein β subunit in the activation of effectors. *Proc. Natl. Acad. Sci. USA* **97**, 9311–9316.
33. Fukada, Y., Matsuda, T., Kokame, K., Takao, T., Shimonishi, Y., Akino, T., and Yoshizawa, T. (1994). Effects of carboxyl methylation of photoreceptor G protein γ-subunit in visual transduction. *J. Biol. Chem.* **269**, 5163–5170.
34. Parish, C. A., Smrcka, A. V., and Rando, R. R. (1995). Functional significance of βγ-subunit carboxymethylation for the activation of phospholipase C and phosphoinositide 3-kinase. *Biochemistry* **34**, 7722–7727.

CHAPTER 218

G-Protein-Coupled Receptors, Cell Transformation, and Signal Fidelity

Hans Rosenfeldt, Maria Julia Marinissen, and J. Silvio Gutkind

Oral and Pharyngeal Cancer Branch,
National Institute of Dental and Craniofacial Research,
National Institutes of Health, Bethesda, Maryland

G protein-coupled receptors (GPCRs) represent the largest class of cell surface receptors and govern a multiplicity of intracellular signaling mechanisms, fulfilling a wide range of critical physiological and cell-specific actions. This vast signaling potential of GPCRs, however, leaves cell-specific transduction pathways vulnerable to profound alterations when G-protein coupled receptors are aberrantly expressed out of their cellular context or persistently activated by either mutatagenesis or excess availability of their ligands. This review covers some of the mechanisms that GPCRs modulate in order to stimulate normal and pathological cell proliferation, including the communication between heterotrimeric G proteins and guanine-nucleotide exchange factors for the Ras and Rho families of small GTPases, signaling cassettes regulating the activity of members of the MAP kinase superfamily, and more recent findings that identify a number of G-protein-independent effectors. We also explore the emerging role of scaffolding molecules in physically organizing the components of each transducing pathway downstream of GPCRs. These organizing molecules are likely to provide cell specificity to GPCR-driven signaling by preventing inappropriate cross-talk in physiological contexts, and may be one of the missing components that allow aberrantly expressed GPCRs or ectopic activation of their targets to cause cell transformation.

Introduction

The family of G-protein-coupled receptors (GPCRs) encompasses the largest group of transmembrane proteins implicated in signal transmission. Also known as heptahelical receptors, these molecules are sensitive to large variety of ligands, including peptide and nonpeptide neurotransmitters, hormones, growth factors, odorant molecules, and light. Such a wide spectrum of sensitivity is reflected in the number of genes that encode GPCRs in animal genomes, including *Drosophila* (1 percent of total genes), *Caenorhabditus elegans* (more than 5 percent of all genes), and even humans, where more than 2 percent of human genes are responsible for over 1,000 proteins with heptahelical structure [1,2]. Numerous GPCRs have been implicated in physiology and in the progression of hereditary diseases [3] and this link has transformed the investigation of new therapeutic drugs: GPCRs and the signaling pathways that they control have become a major focus among pharmaceutical firms [1,2,4].

Heptahelical receptors are called G-protein-coupled receptors because of their well-studied signaling relationship to heterotrimeric G proteins (α, β, and γ subunits). GPCRs become stabilized in an active conformation upon ligand-binding, allowing them to catalyze the exchange of GDP for GTP bound to the G-protein α subunit. Gα and Gβγ subunits subsequently activate effector proteins such as adenylyl and guanylyl cyclases, phosphodiesterases, phospholipase A_2 (PLA_2) and C (PLC), and phosphatidylinositol-3 kinases (PI3Ks). These downstream molecules modulate the synthesis of various second messengers such as cAMP, cGMP, diacylglycerol, IP_3, PIP_3, and arachidonic and phosphatidic acid and can also lead to increases in cytosolic $[Ca^{2+}]$ through the opening or closing of a variety of ion channels. Although much work has been focused on the signaling relationship between GPCRs and trimeric G proteins, recent findings suggest that heptahelical receptor activation can direct biochemical responses that are independent of heterotrimeric G proteins (reviewed in [5]).

In this review we will attempt to give a sense of the multiplicity of classical and novel signaling systems that transmit heptahelical receptor-induced changes in cell behavior. We emphasize that the complexity of known GPCR-initiated signals is likely not to exist in one single cell type; that the signaling pathways available to GPCRs in a particular tissue are much more limited. However, because of the vast signaling potential of GPCRs, these critical cell-specific transduction pathways are in danger of being overridden by disregulated mechanisms that are sensitive to heptahelical receptors. Thus, we will also focus on mechanisms that enforce GPCR signaling fidelity to physiological pathways as a barrier against GPCR-induced pathological outcomes such as cell transformation.

Heptahelical Receptors and Tumorigenesis

GPCRs can promote tumor formation in two different ways: through the stimulation of unregulated growth of cancer cells and by the recruitment of nascent blood vessels to tumor sites. Both tumor-promoting activities have received a great deal of attention since they suggest that drug-induced blockade of specific receptors might allow for the selective inhibition of cell growth or blood vessel formation in particular tumor types, and that such approaches would provide a mechanistic handle that could be exploited in the treatment of neoplastic diseases.

Growth factors such as thrombin, lysophosphatidic acid (LPA), bombesin, vasopressin, bradykinin, substance K, acetylcholine receptor agonists, angiotensin II, and many others induce cell division by binding to their cognate GPCRs in many different cell types (reviewed in [6–8]). The heptahelical protein encoded by the *mas* oncogene was the first clue of a connection between cancer and GPCRs. Unlike other oncogenes, the *mas* gene product does not contain any activating mutations when compared to other heptahelical receptors and requires ligand-binding for its transforming ability. This observation taken together with other findings showing that ectopic expression of serotonin 1C and muscarinic m1, m3, and m5 receptors transforms mouse cells in an agonist-dependent fashion [9,10] suggested that endogenous GPCRs can be tumorigenic in the presence of excess ligand and do not require to be mutated to be transforming.

The role of cellular GPCRs in tumorigenesis is an area of active research, and many heptahelical receptors, and their ligands, are coopted by tumors to support cell proliferation (Table I). For example, GPCR ligands such as bombesin, gastrin-releasing peptide (GRP), neuromedin B (NMB), bradykinin, cholecystokinin (CCK), galanin, neurotensin (NT), and vasopressin are secreted by small cell lung cancer cells (SCLC). These tumors also express the GPCRs sensitive to these agonists and thus use heptahelical receptors to stimulate their own proliferation in an autocrine or paracrine fashion (see [11] for an extensive review). A variety of neuropeptide receptors and their ligands play a role in the progression of colon adenomas and carcinomas, gastric hyperplasia and cancer, prostate cancer, and pancreatic hyperplasia and carcinoma (reviewed in [8,11]). Thus, ectopic expression of GPCRs allows tumor cells to override the heptahelical receptor-driven physiological pathways that are intrinsic to their original cell type and use the unchecked potential of GPCR-controlled signaling mechanisms to drive cell proliferation.

Ligand-dependence, however, is by no means an absolute requirement. Mutagenesis can cause GPCRs to become transforming even in an agonist-independent fashion, as is the case for α_{1b} adrenergic receptors [12]. Moreover, constitutively activating mutations do occur in nature. For example, 30 percent of hyperfunctioning human thyroid adenomas and a minority of differentiated thyroid carcinomas contained constitutively active TSH receptors [13], again linking GPCRs to human cancer. Hypermorphic mutations have also been detected in other GPCR such as LH receptors, which can cause hyperplastic growth of Leydig cells in a form of familial male precocious puberty [14], and Ca^{2+}-sensing G protein linked receptors, which can cause autosomal dominant hypercalcemia [15] and may be involved in certain cancers [16].

Genes that encode heptahelical receptors are also present in the genome of several DNA viruses, including the human cytomegalovirus (HCMV) [17], herpes virus saimiri (HVS) [18], and the Kaposi's sarcoma associated herpesvirus (KSHV) [19]. These receptors, such as the HCMV-encoded GPCRs, have a high degree of homology to chemokine receptors [20] and may help virally infected cells escape detection by the immune system. Moreover, viral GPCRs can also function in paracrine or autocrine fashion, encouraging inappropriate cell behaviors such as pathological cell proliferation. For example, the HVS-GPCR contributes to fatal lymphoproliferative diseases caused by HVS infection, such as leukemias and lymphomas, in several nonhuman primates [18].

Another way in which cellular and viral GPCRs can promote tumorgenesis is by promoting the development of blood vessels that support tumor growth. A variety of G-protein-coupled receptors, including those binding sphingosine-1 phosphate, LPA, PAF, thrombin, IL-8, GROα-γ, MCP-1, and SDF-1, have been implicated in tumor-induced angiogenesis and vasculogenesis [21]. The sphingosine-1-phosphate (S1P) receptor $S1P_1$/EDG-1 is a particularly interesting example

Table I G Proteins and G Protein-Coupled Receptors in Tumorigenesis

Activating mutations

a) G proteins

$G\alpha_s$	Thyroid toxic adenomas, thyroid carcinomas, growth hormone-secreting pituitary adenomas, McCune-Albright syndrome
$G\alpha_{i2}$	Ovarian sex cord tumors, adrenal cortical tumors

b) G protein-coupled receptors

TSH receptor	Thyroid adenoma, thyroid carcinoma
FSH receptor	Ovarian sex cord tumors, ovarian small cell carcinoma
LH receptor	Leydig cell hyperplasia, male precocious puberty
CCK-B receptor	Colorectal cancer
Ca^{2+}-sensing receptor	Autosomal-dominant hypocalcemia, neoplasms

Autocrine and paracrine activation

Neuromedin B receptor	Small cell lung carcinoma
Neurotensin receptor	Prostate cancer Small cell lung carcinoma
Gastrin receptor	Gastric cancer Small cell lung carcinoma
Cholecystokinin receptors	Pancreatic hyperplasia, pancreatic carcinoma, Gastrointestinal cancer Small cell lung carcinoma
Vasopressin receptors	Small cell lung carcinoma

Virally encoded G protein-coupled receptors

Kaposi's sarcoma associated herpesvirus (KSHV)	Kaposi's Sarcoma
Herpes virus saimiri (HVS)	Leukemias and lymphomas in non-human primates
Jaagsiekte sheep retrovirus (JSRV)	Ovine pulmonary carcinoma

because of its interactions with receptor tyrosine kinases [22]. This GPCR was originally cloned from endothelial cells and supports G_i-dependent cell migration and Rac activation in human embryonic kidney (HEK) cells [22,23] and mouse embryonic fibroblasts (MEF) [24]. It is interesting that MEF derived from EDG −/− animals not only exhibit deficits in S1P-directed Rac activation and cell migration but also in that elicited by other mitogens such as PDGF [25,26]. Previous data showing that the enzyme that makes S1P, sphingosine kinase, is stimulated by growth factors such as PDGF [27] led to a transactivation model in which S1P generated by PDGF receptor stimulation activated the EDG-1 receptor in a paracrine or autocrine way. Since cell migration is essential for blood vessel formation, this signaling relationship between the $S1P_1$/EDG-1 and the PDGF receptors might be a critical step in angiogenesis, including that promoted by tumors.

Viral GPCRs have also been implicated in pathological blood vessel formation. The KSHV-GPCR is a constitutively active G_q-coupled receptor and has been shown to be transforming when overexpressed in murine fibroblasts [19]. Recent work has shown that the KSHV-encoded GPCR can appropriate signaling pathways that are normally active in cell proliferation and use them to stimulate the inappropriate expression of VEGF [28,29] and promote cell survival [30], thus participating in the hyper-angiogenic response that characterizes Kaposi's sarcoma lesions.

G-Protein Signaling in Cancer

A minimum of ten of the seventeen $G\alpha$ subunits have been described to have transforming potential (reviewed in [31]), including members of the four trimeric G-protein classes: G_{12-13}, G_q, G_i, and G_s. In many cases, these proteins are similar to heptahelical receptors in that they stimulate carcinogenesis in their intact form when they are overexpressed outside their normal cellular context. However, mutations that inhibit the basal GTPase activity of two of these $G\alpha$ subunits, $G\alpha_s$ and $G\alpha_{i2}$, have been described in several tumors types.

Oncogenic mutations of $G\alpha_q$ family members have not been found in human cancers, and research carried out with laboratory-generated active forms of these proteins have yielded contradictory data. It seems that, depending on cell type, activated $G\alpha_q$ is transforming when expressed at low levels [32] but can lead to apoptosis when present at high levels [31]. It is interesting that highly transforming receptors, such as serotonin-1C, muscarinic m1 and α_1-adrenergic receptors, are coupled to G_q, and that the KSHV-GPCR, a constitutively active Gq-coupled receptor, has been implicated in Kaposi sarcoma progression [19], suggesting that parallel pathways emanating from Gq receptors may be necessary in addition to the activation of the $G\alpha_q$ subunit itself.

Examples of transforming G-proteins include the *gep*, *gip2*, and *gsp* oncogenes. The *gep* gene was simultaneously identified as an oncogenic sequence present in Ewing's sarcoma and as a transcript that induces strong transformation of NIH 3T3 cells [33,34]. *gep* turned out to be a wild-type $G\alpha_{12}$ subunit, belonging to the $G\alpha_{12/13}$ family. This result has been consistent with subsequent findings: increased expression of G_{12-13} subunits has been detected in many human cancers, but it is interesting that no mutations have been found. For example, breast, colon, and prostate adenocarcinoma-derived cell lines express elevated levels of wild-type $G\alpha_{12/13}$ (reviewed in [35]). The *gip2* oncogene is a constitutively active mutant of $G\alpha_{i2}$. This mutation has been found in human ovarian sex cord stromal tumors and adrenal cortical tumors [36], although how often $G\alpha_{i2}$ mutations occur in these cancer types remains a point of controversy. Fibroblast transformation resulting from transfection of *gip2* has been suggested to result from the derepression of the Ras-ERK1/2 pathway after cAMP/PKA inhibition [37]. However $G\alpha_{i2}$ can also stimulate the Ras-ERK1/2 cascade via Rap1 inhibition [38] or stimulation [39], depending on the cellular context, and by stimulating the release of βγ G-protein subunits [40]. The *gsp* oncogene codes for a GTPase-deficient mutant of $G\alpha_s$ and is found in thyroid toxic adenomas (30 percent), thyroid carcinomas (10 percent), growth hormone-secreting pituitary adenomas, and McCune-Albright syndrome. It is interesting that responses to *gsp* expression are cell-type specific. Increases in cAMP and activated PKA resulting from the presence of this oncogene can inhibit Raf1 and prevent transformation in some cells [41]. By contrast, the presence of *gsp* and the same downstream second messengers can inhibit cell growth in other cell types such as PC12 and thyroid cells [41–43].

gsp illustrates the importance of cellular context in regard to the transformation potential of GPCRs and G-proteins. Cell types that express the small GTPase Rap1 and the downstream kinase B-Raf, such as PC12 and thyroid cells, are predisposed to ERK activation in response to cAMP increases [42,43]. Thus the availability of the transforming ERK pathway to *gsp* depends on the organization of the signaling pathways present in a particular cell type. The cell type dependence of *gsp* transformation is only one specific case of a more general phenomenon. For example, the ERK pathway belongs to the MAP kinase group of related signaling cascades that are downstream of GPCRs, and many of these transduction pathways can be regulated as "signaling cassettes" to provide cell-type specific responses after heptahelical receptor stimulation.

A Matrix of MAPK Cassettes Links GPCRs to Biological Outcomes

Many intracellular cascades have been found to mediate GPCR cell growth effects. A major contributor to this function is a group of highly related proline-targeted serine-threonine kinases, generally known as MAP kinases (MAPKs). Although new members of this group have been recently discovered, we will focus on the three best-known classes of MAPKs: the extracellular signal-regulated kinases (ERKs), the c-Jun N-terminal kinases (JNK), and the p38 kinases. These molecules are phosphorylated by a family of proteins known as the MAP kinase kinases (MAPKKs), which are, in turn, themselves phosphorylated by the MAP kinase kinase kinase (MAPKKK) class of proteins. Under physiological conditions, there is great specificity among the MAP kinases in a particular cascade, with only a very limited number of proteins at each step. For example, in the ERK cascade the MAPKKK Raf-1 will not phosphorylate a MAPKK other than MEK1 or MEK2, two isotypes that perform the same function in that specific MAP Kinase pathway. Thus, MAP kinase molecules such as ERK are downstream of phosphorylation cascades, forming separate and parallel signaling cassettes.

G-protein coupled receptors can signal to a variety of MAPKKKs that are linked to MAP kinases, including those of the ERK, JNK, and p38 pathways. After activation, members of the MAPKs translocate to the nucleus, where they regulate the expression of genes that play a key role in physiological and pathological cell growth. These signaling molecules affect gene transcription by phosphorylating transcription factors that control the synthesis of these critical mRNAs [44]. Each MAP kinase signaling cassette has a different range of intracellular targets and is therefore able to induce different cellular responses such as cell proliferation, apoptosis, and migration. Much work has been done on the essential molecular events that GPCRs use to regulate the function of the MAPKs, and this research has provided insights into the underlying biochemistry that GPCRs use to govern a wide range of biological events.

ERK Cassette

Many GPCR agonists stimulate p42 and p44 MAPK (MAPK/ERK1/2) in multiple cell contexts (Fig. 1) (reviewed in [45]). Research focusing on the molecular mechanisms regulating this signaling cassette has revealed an assortment of cell-specific signaling cascades. These include the recognition of tyrosine kinases, PI-3 kinases, and PKC as possible downstream targets for GPCRs.

Tyrosine kinase inhibition reduces both the activation of ERK1/2 by GPCRs [46] and the rapid tyrosine phosphorylation of the adapter molecule Shc, a posttranslational modification that induces Shc-GRB2 complex formation [47]. These findings provided the first clue of a tyrosine kinase link between GPCRs and the Ras-ERK cassette. A subset of nonreceptor tyrosine kinases (NRTKs) and receptor tyrosine kinases (RTKs) have been suggested as mediators of this response. Src or Src-like kinases can phosphorylate Shc upon stimulation of β-adrenergic receptors, or βγ subunits [48]. NRTKs such as Csk, Lyn, Btk, Pyk2, and Fak have been implicated in signaling to mediate the ERK cassette by $G\alpha_i$ and $G\alpha_q$-coupled receptors in many cell contexts (reviewed in [49,50]). The variety of tyrosine kinases mediating the activation of the ERK cascade is a reflection of the multiplicity of cellular contexts in which this transduction

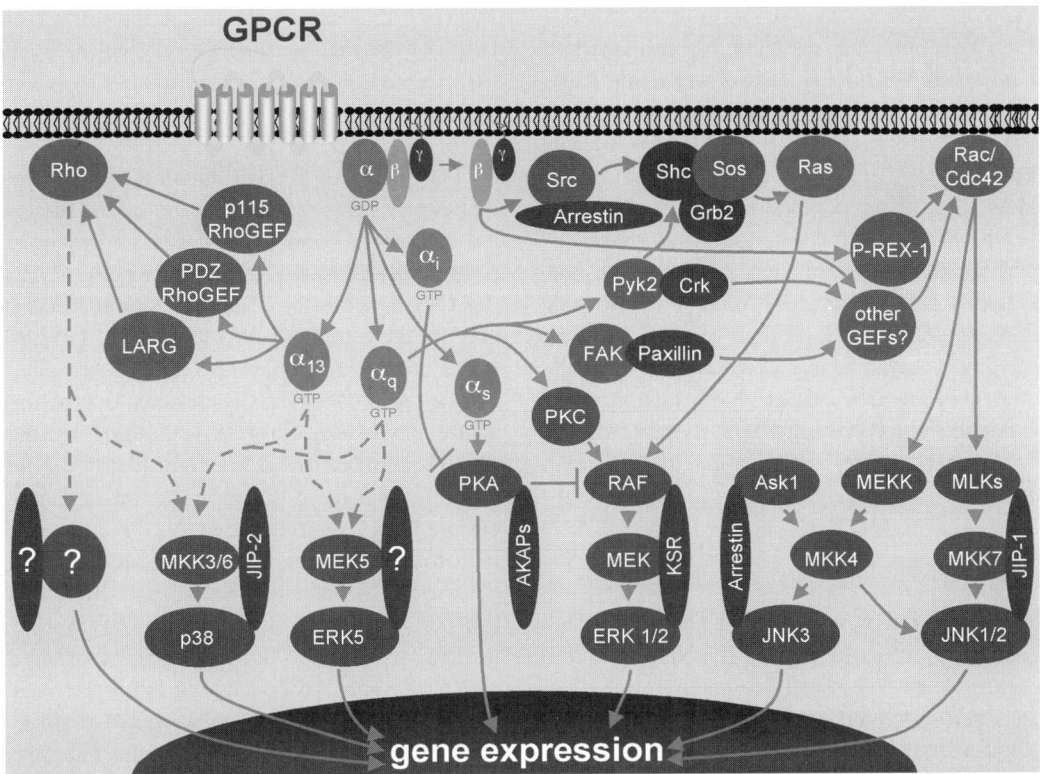

Figure 1 Signaling pathways connecting G-protein-coupled receptors to gene expression are embedded in scaffolding complexes. A multiplicity of signal transduction cascades can link G-protein-coupled receptor stimulation to nuclear events. Thus, aberrant expression of heptahelical receptors, or their cognate ligands, can lead to the activation of a large variety of biochemical routes that can promote neoplastic transformation. Under physiological conditions, scaffolding molecules ensure signal fidelity by physically linking multiple members of signaling cascades to proper subcellular locales and preventing inappropriate crosstalk between related pathways. Arrows represent positive stimulation; dashed lines, functional interactions where precise mechanisms are yet to be elucidated (see text for details).

pathway is involved and its modular nature. Thus, multiple tyrosine kinases in separate pathways may converge in the same signaling cascade to perform different functions.

The protein and lipid kinases of the PI3K family are also essential for GPCR signaling to ERKs. PI3Kγ activity is stimulated upon binding to Gβγ subunits and upon stimulation PI3Kγ activates by signaling to tyrosine kinases that phosphorylate Shc, leading to increases in ERK activity [51]. The PI3Kβ isoform can also mediate GPCR-directed signaling to the ERK cassette [52]. In this mechanism, it is possible that PI3K activates Rac and PAK in combination with Ras to stimulate Raf kinase activity [53]. The variety of pathways stimulated by PI3Kγ and β is an example of the complex mechanisms by which GPCRs can activate the ERK cascade. It is interesting that these pathways all converge on the ERK cassette, impinging on the MAPKKK Raf-1 and suggesting that although the GPCR-driven signals mediated by PI3 kinase and tyrosine kinases are complex, they are fundamentally ordered, as they all "plug-in" to a canonical mechanism leading to ERK activation.

$G\alpha_q$-coupled receptors can also use alternative mechanisms to activate the ERK pathway. These variations include signaling cascades that are mediated by protein kinase C, signals that are transmitted by Ras, and transduction pathways that require the function of both Ras and PKC. Which mechanism functions in a given case depends strongly on the cell type and the extracellular stimulus given. Examples of this specificity include the second messengers that are synthesized in response to G_q, such as diacylglycerol and elevated levels of intracellular Ca^{2+}, that can stimulate Ras through the guanine nucleotide exchange factors (GEFs) RasGRF and RasGRP (also called CalDGEF). These two GEFs are expressed only in certain tissues and stimulate only Ras and Ras-related GTPases (reviewed in [49]). By contrast, the mechanism by which PKC stimulates ERK activity is not yet completely elucidated, since direct phosphorylation of Raf does not seem to be sufficient to fully activate MEK and MAPK [54]. PKC may also modulate other molecules that regulate the interaction between Ras and Raf.

The consistent theme in GPCR signaling to ERK1/2 is that there is a complex matrix of transduction pathways that are ultimately integrated by a canonical set of three kinases (Raf, MEK, and ERK). This multiplicity of known signaling links that impinge on one cascade is unlikely to coexist in one single cell type but represents the summation of the different mechanisms by which GPCRs can connect to ERK in many cell types. In order to prevent pathological responses, it is critical for GPCRs to activate ERK in a controlled fashion, restricting its duration and intensity by cell-specific pathways that can ensure the fidelity of each signaling step. There is always potential, however, for GPCRs to inappropriately stimulate the ERK cassette via signal transduction pathways that are extraneous to cellular context. For example, overexpression of gastrin receptors in cancer cells leads to an activation of c-Src and, in turn, the ERK pathway. Thus, inappropriate c-Src activation may contribute to the transforming effects of gastrin receptors. When mechanisms of ERK activation are placed out of their normal cellular context, they are likely to be disregulated and, given the transforming potential of the ERK cascade, likely to contribute to tumor progression. Thus ensuring cell-specific signal transduction with respect to the ERK cassette is a critical aspect of proper GPCR physiology.

JNK Cassette

The detailed mechanisms by which GPCRs stimulate the MAP kinase cassette that terminates in the c-Jun NH2-terminal kinase (JNK) remain to be fully elucidated. This molecule, also known as stress-activated protein kinase (SAPK), has sequence similarity to ERK1/2, but is activated by GPCRs through distinct pathways. The most important difference in how these two MAP kinase cassettes are regulated is that whereas ERK1/2 stimulation often depends on Ras, the JNK cassette is downstream of the small G proteins Rac and Cdc42. Constitutively active mutants of Rac and Cdc42, for example, can stimulate JNK activity [55], and these two small G-proteins also mediate the activation of JNK by free G$\beta\gamma$ dimers and by Gα_{12}, Gα_{13}, Gq, and G$_i$ [49,50,56]. Little is known about how GPCRs activate JNK beyond these general constraints, but recent work greatly advanced the field by identifying the first GEF known to be responsive to G$\beta\gamma$ and PI-3-kinase, P-REX [57]. This GEF, however, was purified from neutrophils and it remains to be seen whether P-REX or similar GEFs stimulate Rac in response to G$\beta\gamma$ in other cell types.

Another area that remains unclear is how Gα_{12}/Gα_{13} stimulates Rac1/Cdc42 and the downstream JNK cassette. Current candidates that might mediate this effect include two Rac/Cdc42 GEFs, Tiam1 and Dbl, and two Ras GEFs, which may also catalyze Rac GTP exchange, Ras-GRF1 and Ras-GRF2. Further, the nonreceptor tyrosine kinases PYK2 and FAK that are stimulated by stress-fiber and focal complex formation can also stimulate the JNK cassette by interacting with the adaptor proteins Crk [58] or paxillin [59]. Crk and paxillin can, in turn, stimulate GEFs for Rac and Cdc42 (reviewed in [49]).

Still, many of the open questions concerning the activation of the JNK kinase pathway are in the end questions about how Rac and Cdc42 respond to GPCRs. The question of which Rac effectors stimulate the JNK pathway is also an important, outstanding issue. Here again, the presence of a variety of mechanisms impinging on the JNK cassette reflects the general theme of GPCR-signaling complexity and the exquisite cell specificity that is possible from such a wide range of alternative biochemical routes.

p38 Cassette

The p38 MAP kinases, like the JNK family, are stimulated by cellular stress and membrane-bound receptors [60]. There are presently four p38 MAPKs known: p38α(CSBP-1), p38β, p38γ (ERK6/SAPK3), and p38δ (SAPK4) [61].

Although the GPCRs and agonists that elicit increased activity from the p38 family of MAP kinases have been the focus of much investigation, there is no clear picture of the downstream mechanisms directly controlling p38 kinases. There have been some reports showing that Gα_q and $\beta\gamma$ dimers stimulate p38α [62] and that two NRTKs, BTK [63] and Src [64], are involved in this mechanism. Receptors that couple to Gα_q can also stimulate the p38α, p38γ, and p38δ isoforms [65]. Recent work indicates that Gα_q mediates p38 activation through the MKK3 and MKK6 MAPKKs [66]. Further, research using electrophysiological techniques suggest that p38 MAP kinases are downstream from G$_{13}$ [67], and it has been proposed that this heterotrimeric G protein can initiate the activity of the p38 pathway by stimulating Ask, a MAPKKK for this cascade [68]. It is expected that the use of novel techniques, such as RNA interference or the generation of knockout animals for molecules acting upstream of p38, will enable the molecular dissection of the mechanisms by which GPCRs and other cell surface receptors activate each member of the p38 family of MAPKs.

G-Protein-Independent Signaling

It has now become evident that the extent of heptahelical receptor signaling reaches beyond signal transduction pathways that are downstream of Gα and G$\beta\gamma$ subunits. Recent work suggests that GPCRs interact with a wide range of signaling molecules besides heterotrimeric G proteins. Molecules containing protein-protein interaction domains such as the PDZ, SH2, and SH3 motifs, as well as polyproline-containing regions, have been reported to directly interact with GPCRs (reviewed in [69]).

These specialized domains could serve GPCRs as cell-specific bypasses from trimeric G-proteins to activate intracellular signaling in some cellular contexts. Several signaling proteins containing these protein-protein interaction motifs have been shown to bind heptahelical receptors. For example, the PDZ domain interacts with proteins containing a C-terminal S/TxV(L/I) sequence common in GPCRs. SH2-containing molecules such as the adaptor Grb2 and the SH2-containing tyrosine phosphatase SHP (reviewed in [69]) have been reported to bind to the β_2-adrenergic and AT$_{1A}$ receptors, respectively. Other domains that have been reported to bind GPCRs are the polyproline-binding domains, such as SH3, WW, and EVH domains [5,70]. For example, metabotropic glutamate receptors (mGluRs) interact with a class of molecules that harbor Enabled/VASP homology (EVH)-like domains, such as Homer (1a-c, 2, and 3), which binds mGluRs through a C-terminal polyproline sequence (PPXXFP) (reviewed in [69]).

This expanded view of GPCR signaling, combined with emerging results showing that the *frizzled, smoothened,* and *Dictyostellium* cAMP receptors elicit biological responses that are independent of heterotrimeric G proteins, suggests a reevaluation of the "G protein-coupled receptor/heterotrimeric G protein associated effector" concept of heptahelical receptor function. Thus, some workers prefer terms other than "GPCR" in order to avoid using a designation that suggests a more limited range than the vast array of signaling cascades that these receptors actually control. Alternative terms include serpentine, seven-transmembrane, or heptahelical receptors [71]. Such a wide range of signaling possibilities requires strict organization among the signaling molecules that are downstream of GPCRs. One emerging example of how these downstream cascades can be physically organized intracellularly is a variety of scaffolding molecules that tether multiple components of signal transduction pathways in specific configurations.

GPCR Effectors Are Organized by Scaffolding Molecules

Heptahelical receptors stimulate physiological cell growth *via* a variety of tightly regulated signaling modules. In the past decade, scaffolding proteins have emerged as general regulatory mechanisms ensuring the fidelity of intracellular pathways. These proteins bind components of signaling pathways, physically organizing them to enable physiological responses. The prototypical signaling scaffold is the yeast protein Ste5p, which binds the components of the yeast MAP kinase cascade leading to the mating response after activation of the pheromone GPCR [72–75] and has been suggested to mantain the signaling fidelity of this cascade [76]. Ste5p is particularly important to the mating pathway of yeast because unlike multicellular organisms, MAP kinase cascades share most of their components such as the PAK-like kinase Ste20p and the MEKK-like protein Ste11p [77–80]. Like scaffolds in multicellular organisms, however, Ste5p is expressed in a specific cell type: the haploid yeast cell. Thus Ste5p allows a generalized MAP kinase module to elicit a specialized physiological response in a particular cellular context.

Several scaffolds binding a variety of signaling pathways have now been described in multicellular organisms. These include the various MAP kinase pathways organized by scaffolding proteins such as kinase suppressor of RAS (KSR) and c-Jun terminal kinase interacting protein family (JIPs), the cAMP dependent cascades that rely on the A-kinase anchoring proteins (AKAPs) for tethering, and the multiple cell signals downstream of proteins that bind molecules of the β-arrestin class. In many cases, overexpression of these scaffolding proteins will inhibit cell transformation, suggesting that these molecules play an important role in restricting the propagation of signaling events and preventing signal transduction cascades from causing pathological cell proliferation. Although there are very few published reports directly connecting G-protein-coupled receptors to these organizing molecules, we can expect that the relevance of scaffolding proteins to GPCR-induced transformation will become increasingly apparent in the foreseeable future because of their fundamental role in maintaining the integrity of biochemical routes connecting cell surface receptors to the nucleus.

KSR

KSR was first cloned in RAS suppressor screens by using *Caenorhabditis elegans* and *Drosophila* [81–83]. However, recent work has shown that KSR function is required for proper signaling between RAF-1 and ERK in both *Drosophila* [84,85] and *C. elegans* [86]. In these invertebrate systems, ablation of KSR by either RNAi [85] or mutation [86] prevented the efficient activation of MEK and ERK by constitutively active Ras. An intriguing finding is that overexpression of KSR has similar negative effects on preventing Ras-V12-induced tumorigenesis in *Drosophila* imaginal discs [84]. Roy *et al.* [85] have observed that wild-type KSR overexpression can only stimulate the ERK pathway if all other components binding KSR are concomitantly transfected; they suggest that the mechanism by which the overexpressed scaffolding protein prevents efficient signaling to ERK is through a stoichiometric excess of KSR isolating signaling components from each other.

The murine form of KSR was cloned concurrently with its invertebrate homologs [83,87]. Experiments in mammalian systems have paralleled results obtained in *Drosophila* and *C. elegans*. Transfection studies in NIH3T3 cells showed that KSR overexpression blocks RAS-induced transformation [88]. Moreover, KSR knockout mice are resistant to polyomavirus middle T-induced tumor formation [89], a finding that parallels results obtained with *Drosophila*. Mammalian KSR has been reported to bind βγ subunits, preventing ERK activation [90]. This result is particularly interesting because it is the first known link between the KSR scaffold and G-protein-coupled receptors. Further work remains to elucidate the physiological role of βγ subunit/KSR interactions and its effects on GPCR signaling to the ERK pathway.

JIP/IB Family of Scaffolding Molecules

The JIP/IB family of scaffolding proteins consists of three mammalian homologs JIP/IB-1, JIP/IB-1b, and JIP/IB-2 [91–93]. These molecules bind members of the mixed-lineage group of MAPKKK's (MLK) that have been reported to activate both JNK and p38 [94,95]. In addition, both IB1/JIP1 and IB2/JIP-2 have been found to bind the JNK MAPKK, MKK7 [93,96], and IB2/JIP-2 has also been reported to bind the p38 MAPKK MLK3 [97] and p38γ [98].

JIP-1 was originally described as an inhibitor of the JNK pathway that could block JNK-dependent cell growth and transformation [91]. Soon thereafter IB1, an alternatively transcribed isoform of JIP-1, was found in insulin-secreting cells [92]. Like KSR, the overexpression of JIP proteins inhibits the transmission of signals through the JNK pathway. However, the physiological function of this set of protein scaffolds seems to be to temper the activity of the JNK pathway, since cells derived from JIP-1 +/− heterozygous mice or cells depleted of endogenous JIP-1 with antisense RNA have augmented JNK activity [99–101]. These large increases lead to enhanced apoptosis in IL-1β/TNFα/IFN-treated pancreatic β cells. In fact, a mutation in the human *IB1/JIP-1* gene has recently been linked to adult onset diabetes [100]. An emerging property of these scaffolding proteins is their ability to bind certain GDP exchange factors that activate the small G-proteins Rho and Rac. The Rho exchange factor p190RhoGEF binds JIP-1 through its phosphotyrosine binding domain (PTB) in neuronal and PC12 cells [102]. Recent work describes the binding of IB2/JIP-2 to the RAC exchange factor TIAM and the RAC/RAS exchange factor RAS-GRF1, thereby physically linking known activators of RAC to components of the p38 pathway, which have previously been reported to be downstream of this small GTPase [97]. TIAM is phosphorylated in response to the G-protein-coupled receptor ligands LPA, endothelin-1, bombesin, and bradykinin [103], and this protein modification is required for the proper localization of TIAM after LPA stimulation [104]. Thus, the JIP scaffolding proteins might link G-protein-coupled receptors to the JNK and p38 pathways by physically interacting with the small GTPase exchange factors activated by heptahelical receptors.

AKAPs

cAMP signaling was the first second messenger described [105,106]. Production of this nucleotide is controlled by adenylyl cyclases, a class of enzymes that are mostly controlled by heptahelical receptors coupling to G proteins of the G_S class [107]. The best described target of cAMP is protein kinase A (PKA), a tetrameric enzyme that becomes activated upon binding to this nucleotide [108]. Activation of PKA does not occur randomly; instead a class of scaffolding proteins, the A kinase anchoring proteins (AKAPs), localizes PKA to specific sites of action [109]. These scaffolding proteins cannot be grouped together by homology, and are defined as AKAPs solely in terms of their capacity to bind PKA.

A common theme among AKAPs, besides their ability to bind PKA, is their multifunctional nature. A growing number of large, multivalent proteins that bind the PKA holoenzyme have been described. These proteins often serve as scaffolds for other signal transduction pathways or exhibit other functions in addition to binding PKA. For example, AKAPs now include proteins such as gravin, a protein that was first identified in patients with the autoimmune disease myasthemia gravis, and also binds protein kinase C (PKC) [110] and three orthologous members of the WASP family of proteins that interact with the ARP2/3 actin polymerization unit WAVE1-3 [111] and respond to RAC signals [112,113]. Moreover, recent evidence suggests that WAVE3 and gravin function as tumor suppressors [114–116], and it is probable that more of these PKA-binding proteins will be shown to function as tumor suppressors.

One reason AKAPs might serve as tumor suppressors is that the G_s/cAMP pathway often serves as an inhibitor of cell proliferation. For example, PKA phosphorylates RAF-1 at serines 43, 259, and 621 [117]. Phosphorylation at serine 259 is critical for PKA-dependent inhibition of RAF-1 kinase activity *in vitro* and blocks RAF-1-dependent *in vivo* activation of ERK in COS cells. In addition, mutation of serine 43 to alanine blocked the ability of RAF-1 to bind RAS in response to cAMP increases. Thus, proper organization of cAMP signaling to PKA in the correct subcellular compartment may be critical in preventing other mitogenic pathways such as the PKC or ERK cascades from becoming overactive and promoting inappropriate cell proliferation.

Arrestin

G-protein coupled receptors are phosphorylated by the G-protein-coupled receptor kinase (GRK) family of proteins after ligand binding [118]. The scaffolding protein arrestin subsequently interacts physically with heptahelical receptors. Although this molecule has been primarily implicated in targeting GPCRs for endocytosis, arrestin has been also shown to couple GPCRs to the activation of Src-like kinases. This process apparently involves the formation of large multiprotein complexes that can include components of the MAPK and JNK cassettes [119,120]. For example, an arrestin-tethered multimolecular complex has been described in the Rac/Cdc42 independent activation of the JNK cassette MAPKKK ASK1 by $G\alpha_{12}$ [68]. It is interesting that arrestin only binds the neural-specific JNK3 isoform, suggesting that this scaffold enforces cell-type specificity in the activation of the JNK pathway by heptahelical receptors.

Conclusion: GPCR Biology Requires Both Signal Integration and Separation

The essential molecular events that GPCRs use to govern such a wide range of biological events seem elusive because of their apparent complexity. Classical second messenger-generating systems are now understood to be only a subset of the mechanisms that GPCRs use in physiological and pathological contexts. At first glance, this "brave new world" of complexity looks like a highly interconnected meshwork wherein signals derived from a GPCR can travel to any of a wide variety of end-points (Fig. 1). From our current standpoint, we cannot predict with any confidence which signaling pathway, of the many possible routes, will be activated by a heptahelical receptor. Yet this situation is not the case in biology: usually GPCR stimulation in a given context will produce a repeatable result. Moreover, an emerging concept derived from recent studies with MAP kinases and other targets of GPCRs is that scaffold proteins are organizers and keepers of specificity. It may be as important to keep signals separate between closely related cascades, such as the MAP kinases, as it is to integrate them in a coordinated response.

Although physiological functions of GPCRs, including phenotypic differentiation and cell survival or death, most likely result from the integration of a complex network of signaling cassettes, it is probable that pathology induced by GPCRs, such as cancer or tissue hypertrophy, results from the breakdown of signal separation between targets downstream of these receptors. Recent advances in our understanding of GPCR-driven intracellular signaling networks and how they are organized by scaffolding proteins will provide a more global view of the general systems by which these receptors exert their numerous physiological roles and will elucidate their role in many pathological conditions. This new understanding may also point to novel approaches for pharmacological treatment of a variety of disease processes.

References

1. Flower, D. R. (1999). Modelling G-protein-coupled receptors for drug design. *Biochim. Biophys. Acta* **1422**(3), 207–234.
2. Attwood, T. K. and Findlay, J. B. (1994). Fingerprinting G-protein-coupled receptors. *Protein Eng.* **7**(2), 195–203.
3. Rohrer, D. K. and Kobilka, B. K. (1998). G protein-coupled receptors: functional and mechanistic insights through altered gene expression. *Physiol. Rev.* **78**(1), 35–52.
4. Schwarz, M. K. and Wells, T. N. (2002). New therapeutics that modulate chemokine networks. *Nat. Rev. Drug Discov.* **1**(5), 347–358.
5. Brzostowski, J. A. and Kimmel, A. R. (2001). Signaling at zero G: G-protein-independent functions for 7-TM receptors. *Trends Biochem. Sci.* **26**(5), 291–297.
6. Rozengurt, E. (1986). Early signals in the mitogenic response. *Science* **234**(4773), 161–166.
7. Young, D. et al. (1986). Isolation and characterization of a new cellular oncogene encoding a protein with multiple potential transmembrane domains. *Cell* **45**(5), 711–719.
8. Gutkind, J. S. (1998). Cell growth control by G protein-coupled receptors: from signal transduction to signal integration. *Oncogene* **17**(11 Reviews), 1331–1342.
9. Julius, D. et al. (1989). Ectopic expression of the serotonin 1c receptor and the triggering of malignant transformation. *Science* **244**(4908), 1057–1062.
10. Gutkind, J. S. et al. (1991). Muscarinic acetylcholine receptor subtypes as agonist-dependent oncogenes. *Proc. Natl. Acad. Sci. USA* **88**(11), 4703–4707.
11. Heasley, L. E. (2001). Autocrine and paracrine signaling through neuropeptide receptors in human cancer. *Oncogene* **20**(13), 1563–1569.
12. Allen, L. F. et al. (1991). G-protein-coupled receptor genes as protooncogenes: constitutively activating mutation of the alpha 1B-adrenergic receptor enhances mitogenesis and tumorigenicity. *Proc. Natl. Acad. Sci. USA* **88**(24), 11354–11358.
13. Parma, J. et al. (1993). Somatic mutations in the thyrotropin receptor gene cause hyperfunctioning thyroid adenomas [see comments]. *Nature* **365**(6447), 649–651.
14. Shenker, A. et al. (1993). A constitutively activating mutation of the luteinizing hormone receptor in familial male precocious puberty [see comments]. *Nature* **365**(6447), 652–654.
15. Spiegel, A. M. (1996). Mutations in G proteins and G protein-coupled receptors in endocrine disease. *J. Clin. Endocrinol. Metab.* **81**(7), 2434–2442.
16. Hoff, A. O. et al. (1999). Calcium-induced activation of a mutant G-protein-coupled receptor causes in vitro transformation of NIH/3T3 cells. *Neoplasia* **1**(6), 485–491.
17. Chee, M. S. et al. (1990). Human cytomegalovirus encodes three G protein-coupled receptor homologues. *Nature* **344**(6268), 774–777.
18. Nicholas, J., Cameron, K. R., and Honess, R. W. (1992). Herpesvirus saimiri encodes homologues of G protein-coupled receptors and cyclins. *Nature* **355**(6358), 362–365.
19. Arvanitakis, L. et al. (1997). Human herpesvirus KSHV encodes a constitutively active G-protein-coupled receptor linked to cell proliferation [see comments]. *Nature* **385**(6614), 347–450.
20. Ahuja, S. K. and Murphy, P. M. (1993). Molecular piracy of mammalian interleukin-8 receptor type B by herpesvirus saimiri. *J. Biol. Chem.* **268**(28), 20691–20694.
21. Moore, B. B. et al. (1998). CXC chemokine modulation of angiogenesis: the importance of balance between angiogenic and angiostatic members of the family. *J. Investig. Med.* **46**(4), 113–120.
22. Lee, M. J. et al. (1998). Sphingosine-1-phosphate as a ligand for the G protein-coupled receptor EDG-1. *Science* **279**(5356), 1552–1555.
23. Wang, F. et al. (1999). Sphingosine 1-phosphate stimulates cell migration through a G(i)-coupled cell surface receptor. Potential involvement in angiogenesis. *J. Biol. Chem.* **274**(50), 35343–35350.
24. Liu, Y. et al. (2000). Edg-1, the G-protein coupled receptor for sphingosine-1-phosphae, is essential for vascular maturation. *J. Clin. Invest.* **106**, 951–961.
25. Hobson, J. P. et al. (2001). Role of the sphingosine-1-phosphate receptor EDG-1 in PDGF-induced cell motility. *Science* **291**, 1800–1803.
26. Rosenfeldt, H. M. et al. (2001). EDG-1 links the PDGF receptor to Src and focal adhesion kinase activation leading to lamellipodia formation and cell migration. *FASEB J.* **15**(14), 2649–2659.
27. Olivera, A. and Spiegel, S. (1993). Sphingosine-1-phosphate as a second messenger in cell proliferation induced by PDGF and FCS mitogens. *Nature* **365**, 557–560.
28. Sodhi, A. et al. (2000). The Kaposi's sarcoma-associated herpes virus G protein-coupled receptor up-regulates vascular endothelial growth factor expression and secretion through mitogen-activated protein kinase and p38 pathways acting on hypoxia-inducible factor 1alpha. *Cancer Res.* **60**(17), 4873–4880.
29. Bais, C. et al. (1998). G-protein-coupled receptor of Kaposi's sarcoma-associated herpesvirus is a viral oncogene and angiogenesis activator [see comments] [published erratum appears in Nature 1998 Mar 12;392(6672):210]. *Nature* **391**(6662), 86–89.
30. Montaner, S. et al. (2001). The Kaposi's sarcoma-associated herpesvirus G protein-coupled receptor promotes endothelial cell survival through the activation of Akt/protein kinase B. *Cancer Res.* **61**(6), 2641–2648.
31. Dhanasekaran, N. et al. (1998). Regulation of cell proliferation by G proteins. *Oncogene* **17**(11 Reviews), 1383–1394.
32. Kalinec, G. et al. (1992). Mutated alpha subunit of the Gq protein induces malignant transformation in NIH 3T3 cells. *Mol. Cell Biol.* **12**(10), 4687–4693.
33. Xu, N. et al. (1993). A mutant alpha subunit of G12 potentiates the eicosanoid pathway and is highly oncogenic in NIH 3T3 cells. *Proc. Natl. Acad. Sci. USA* **90**(14), 6741–6745.
34. Chan, A. M. et al. (1993). Expression cDNA cloning of a transforming gene encoding the wild-type G alpha 12 gene product. *Mol. Cell Biol.* **13**(2), 762–768.
35. Gutkind, J. S., Coso, O. A., and Xu, N. (1998). In S. A. M., Ed., *G12 and G13 α Subunits of Heterotrimeric G Proteins: A novel Family of Oncogenes*, in *G Proteins, Receptors, and Disease*, pp. 101–117. Humana Press, Totowa, N.J.
36. Lyons, J. et al. (1998). Two G protein oncogenes in human endocrine tumors. *Science* **249**(4969), 655–659.
37. Miller, M. J. et al. (1998). Differential effects of protein kinase A on Ras effector pathways. *Mol. Cell Biol.* **18**(7), 3718–3726.

38. Mochizuki, N. et al. (1999). Activation of the ERK/MAPK pathway by an isoform of rap1GAP associated with G alpha(i) [see comments]. **400**(6747), 891–894.
39. Schmitt, J. M. and Stork, P. J. (2000). beta 2-adrenergic receptor activates extracellular signal-regulated kinases (ERKs) via the small G protein rap1 and the serine/threonine kinase B-Raf. *J. Biol. Chem.* **275**(33), 25342–25350.
40. Crespo, P. et al. (1994). Ras-dependent activation of MAP kinase pathway mediated by G-protein beta gamma subunits. *Nature* **369**(6479), 418–420.
41. Chen, J. and Iyengar, R. (1994). Suppression of Ras-induced transformation of NIH 3T3 cells by activated G alpha s. *Science* **263**(5151), 1278–1281.
42. Ehses, J. A. et al. (2002). Glucose-dependent insulinotropic polypeptide (GIP) activates the Raf-Mek 1/2-ERK 1/2 module via a cyclic AMP/PKA/Rap1-mediated pathway. *J. Biol. Chem.*, in press.
43. Erhardt, P. et al. (1995). Differential regulation of Raf-1 and B-Raf and Ras-dependent activation of mitogen-activated protein kinase by cyclic AMP in PC12 cells. *Mol. Cell Biol.* **15**(10), 5524–5530.
44. Davis, R. J. (1995). Transcriptional regulation by MAP kinases. *Mol. Reprod. Dev.* **42**(4), 459–467.
45. Gutkind, J. S. (1998). The pathways connecting G protein-coupled receptors to the nucleus through divergent mitogen-activated protein kinase cascades. *J. Biol. Chem.* **273**(4), 1839–1842.
46. Hordijk, P. L. et al. (1994). Protein tyrosine phosphorylation induced by lysophosphatidic acid in Rat-1 fibroblasts. Evidence that phosphorylation of map kinase is mediated by the Gi-p21ras pathway. *J. Biol. Chem.* **269**(1), 645–651.
47. van Biesen, T. et al. (1995). Receptor-tyrosine-kinase- and G beta gamma-mediated MAP kinase activation by a common signalling pathway [see comments]. *Nature* **376**(6543), 781–784.
48. Luttrell, L. M. et al. (1996). Role of c-Src tyrosine kinase in G protein-coupled receptor- and Gbetagamma subunit-mediated activation of mitogen-activated protein kinases. *J. Biol. Chem.* **271**(32), 19443–19450.
49. Gutkind, J. S. (2000). Regulation of Mitogen-Activated Protein Kinase signaling networks by G protein-coupled receptors. *Science's STKE* http://www.stke.org/cgi/content/full/OC_sigtrans;2000/40/re1.
50. Gudermann, T., Grosse, R., and Schultz, G. (2000). Contribution of receptor/G protein signaling to cell growth and transformation. *Naunyn Schmiedebergs Arch Pharmacol.* **361**(4), 345–362.
51. Lopez-Ilasaca, M. et al. (1997). Linkage of G protein-coupled receptors to the MAPK signaling pathway through PI 3-kinase gamma. *Science* **275**(5298), 394–397.
52. Murga, C., Fukuhara, S., and Gutkind, J. S. (2000). A novel role for phosphatidylinositol 3-kinase beta in signaling from G protein-coupled receptors to Akt. *J. Biol. Chem.* **275**(16), 12069–12073.
53. Sun, H. et al. (2000). Regulation of the protein kinase Raf-1 by oncogenic Ras through phosphatidylinositol 3-kinase, Cdc42/Rac and Pak. *Curr. Biol.* **10**(5), 281–284.
54. Macdonald, S. G. et al. (1993). Reconstitution of the Raf-1-MEK-ERK signal transduction pathway in vitro [published erratum appears in *Mol. Cell Biol.* (1994), **14**(3), 2223–2224]. *Mol. Cell Biol.* **13**(11), 6615–6620.
55. Coso, O. A. et al. (1995). The small GTP-binding proteins Rac1 and Cdc42 regulate the activity of the JNK/SAPK signaling pathway. *Cell* **81**(7), 1137–1146.
56. Yamauchi, J. et al. (2000). G(i)-dependent activation of c-Jun N-terminal kinase in human embryonal kidney 293 cells. *J. Biol. Chem.* **275**(11), 7633–7640.
57. Welch, H. C. et al. (2002). P-Rex1, a PtdIns(3,4,5)P3- and Gbetagamma-regulated guanine-nucleotide exchange factor for Rac. *Cell* **108**(6), 809–821.
58. Blaukat, A. et al. (1999). Adaptor proteins Grb2 and Crk couple Pyk2 with activation of specific mitogen-activated protein kinase cascades. *J. Biol. Chem.* **274**(21), 14893–148901.
59. Igishi, T. et al. (1999). Divergent signaling pathways link focal adhesion kinase to mitogen-activated protein kinase cascades. Evidence for a role of paxillin in c-jun nh(2)-terminal kinase activation [In Process Citation]. *J Biol Chem.* **274**(43), 30738–30746.
60. Davis, R. J. (2000). Signal transduction by the JNK group of MAP kinases. *Cell* **103**(2), 239–252.
61. Ono, K. and Han, J. (2000). The p38 signal transduction pathway: activation and function. *Cell Signal.* **12**(1), 1–13.
62. Yamauchi, J. et al. Activation of p38 mitogen-activated protein kinase by signaling through G protein-coupled receptors. Involvement of Gbetagamma and Galphaq/11 subunits. *J. Biol. Chem.* **272**(44), 27771–27777.
63. Bence, K. et al. (1997). Direct stimulation of Bruton's tyrosine kinase by G(q)-protein alpha-subunit. *Nature* **389**(6648), 296–299.
64. Nagao, M. et al. Involvement of protein kinase C and Src family tyrosine kinase in Galphaq/11-induced activation of c-Jun N-terminal kinase and p38 mitogen-activated protein kinase. *J. Biol. Chem.* **273**(36), 22892–22898.
65. Marinissen, M. J. et al. (1999). A network of mitogen-activated protein kinases links G protein-coupled receptors to the c-jun promoter: a role for c-Jun NH2-terminal kinase, p38s, and extracellular signal-regulated kinase 5. *Mol. Cell Biol.* **19**(6), 4289–4301.
66. Yamauchi, J. et al. (2001). Parallel regulation of mitogen-activated protein kinase kinase 3 (MKK3) and MKK6 in Gq-signaling cascade. *J. Biol. Chem.* **276**(26), 23362–23372.
67. Wilk-Blaszczak, M. A. et al. (1998). The mitogen-activated protein kinase p38-2 is necessary for the inhibition of N-type calcium current by bradykinin. *J. Neurosci.* **18**(1), 112–118.
68. Berestetskaya, Y. V. et al., Regulation of apoptosis by alpha-subunits of G12 and G13 proteins via apoptosis signal-regulating kinase-1. *J. Biol. Chem.* **273**(43), 27816–27823.
69. Bockaert, J. and Pin, J. P. (1999). Molecular tinkering of G protein-coupled receptors: an evolutionary success. *EMBO J.* **18**(7), 1723–1729.
70. Pawson, T. and Scott, J. D. (1997). Signaling through scaffold, anchoring, and adaptor proteins. *Science* **278**(5346), 2075–2080.
71. Hall, R. A., Premont, R. T., and Lefkowitz, R. J. Heptahelical receptor signaling: beyond the G protein paradigm. *J. Cell Biol.* **145**(5), 927–932.
72. Kranz, J. E., Satterberg, B., and Elion, E. A. (1994). The MAP kinase Fus3 associates with and phosphorylates the upstream signaling component Ste5. *Genes Dev.* **8**(3), 313–327.
73. Choi, K. Y. et al. (1994). Ste5 tethers multiple protein kinases in the MAP kinase cascade required for mating in *S. cerevisiae*. *Cell* **78**(3), 499–512.
74. Marcus, S. et al. (1994). Complexes between STE5 and components of the pheromone-responsive mitogen-activated protein kinase module. *Proc. Natl. Acad. Sci. USA* **91**(16), 7762–7766.
75. Printen, J. A. and Sprague, G. F. Jr. (1994). Protein-protein interactions in the yeast pheromone response pathway: Ste5p interacts with all members of the MAP kinase cascade. *Genetics* **138**(3), 609–619.
76. Elion, E. A. (2001). The Ste5p scaffold. *J. Cell Sci.* **114**(Pt 22), 3967–3978.
77. Gustin, M. C. et al. (1998). MAP kinase pathways in the yeast *Saccharomyces cerevisiae*. *Microbiol. Mol. Biol. Rev.* **62**(4), 1264–1300.
78. O'Rourke, S. M. and Herskowitz, I. (1998). The Hog1 MAPK prevents cross talk between the HOG and pheromone response MAPK pathways in *Saccharomyces cerevisiae*. *Genes Dev.* **12**(18), 2874–2886.
79. Liu, H., Styles, C. A., and Fink, G. R. (1993). Elements of the yeast pheromone response pathway required for filamentous growth of diploids. *Science* **262**(5140), 1741–1744.
80. Roberts, R. L. and Fink, G. R. (1994). Elements of a single MAP kinase cascade in *Saccharomyces cerevisiae* mediate two developmental programs in the same cell type: mating and invasive growth. *Genes Dev.* **8**(24), 2974–2985.
81. Kornfeld, K., Hom, D. B., and Horvitz, H. R. (1995). The ksr-1 gene encodes a novel protein kinase involved in Ras-mediated signaling in *C. elegans*. *Cell* **83**(6), 903–913.

82. Sundaram, M. and Han, M. (1995). The *C. elegans* ksr-1 gene encodes a novel Raf-related kinase involved in Ras-mediated signal transduction. *Cell* **83**(6), 889–901.
83. Therrien, M. *et al.* (1995). KSR, a novel protein kinase required for RAS signal transduction. *Cell* **83**(6), 879–888.
84. Karim, F. D. and Rubin, G. M. (1998). Ectopic expression of activated Ras1 induces hyperplastic growth and increased cell death in *Drosophila* imaginal tissues. *Development* **125**(1), 1–9.
85. Roy, F. *et al.* (2002). KSR is a scaffold required for activation of the ERK/MAPK module. *Genes Dev.* **16**(4), 427–438.
86. Ohmachi, M. *et al.* (2002). *C. elegans* ksr-1 and ksr-2 have both unique and redundant functions and are required for MPK-1 ERK phosphorylation. *Curr. Biol.* **12**(5), 427–433.
87. Nehls, M. *et al.* (1995). YAC/P1 contigs defining the location of 56 microsatellite markers and several genes across a 3.4-cM interval on mouse chromosome 11. *Mamm. Genome* **6**(5), 321–331.
88. Denouel-Galy, A. *et al.* (1998). Murine Ksr interacts with MEK and inhibits Ras-induced transformation. *Curr. Biol.* **8**(1), 46–55.
89. Nguyen, A. *et al.* (2002). Kinase suppressor of Ras (KSR) is a scaffold which facilitates mitogen-activated protein kinase activation in vivo. *Mol. Cell Biol.* **22**(9), 3035–3045.
90. Bell, B. *et al.* (1999). KSR-1 binds to G-protein betagamma subunits and inhibits beta gamma-induced mitogen-activated protein kinase activation. *J. Biol. Chem.* **274**(12), 7982–7986.
91. Dickens, M. *et al.* (1997). A cytoplasmic inhibitor of the JNK signal transduction pathway. *Science* **277**(5326), 693–696.
92. Bonny, C., Nicod, P., and Waeber, P. (1998). IB1, a JIP-1-related nuclear protein present in insulin-secreting cells. *J. Biol. Chem.* **273**(4), 1843–1846.
93. Yasuda, J. *et al.* (1999). The JIP group of mitogen-activated protein kinase scaffold proteins. *Mol. Cell Biol.* **19**(10), 7245–7254.
94. Rana, A. *et al.* (1996). The mixed lineage kinase SPRK phosphorylates and activates the stress-activated protein kinase activator, SEK-1. *J. Biol. Chem.* **271**(32), 19025–19028.
95. Tibbles, L. A. *et al.* (1996). MLK-3 activates the SAPK/JNK and p38/RK pathways via SEK1 and MKK3/6. *EMBO J.* **15**(24), 7026–7035.
96. Negri, S. *et al.* (2000). cDNA cloning and mapping of a novel islet-brain/JNK-interacting protein. *Genomics* **64**(3), 324–330.
97. Buchsbaum, R. J., Connolly, B. A., and Feig, L. A. (2002). Interaction of Rac exchange factors Tiam1 and Ras-GRF1 with a scaffold for the p38 mitogen-activated protein kinase cascade. *Mol. Cell Biol.* **22**(12), 4073–4085.
98. Schoorlemmer, J. and Goldfarb, M. (2001). Fibroblast growth factor homologous factors are intracellular signaling proteins. *Curr. Biol.* **11**(10), 793–797.
99. Bonny, C. *et al.* (2000). IB1 reduces cytokine-induced apoptosis of insulin-secreting cells. *J. Biol. Chem.* **275**(22), 16466–16472.
100. Waeber, G. *et al.* (2000). The gene MAPK8IP1, encoding islet-brain-1, is a candidate for type 2 diabetes. *Nat. Genet.* **24**(3), 291–295.
101. Tawadros, T. *et al.* (2002). The scaffold protein IB1/JIP-1 controls the activation of JNK in rat stressed urothelium. *J. Cell Sci.* **115**(Pt 2), 385–393.
102. Meyer, D., Liu, A., and Margolis, B. (1999). Interaction of c-Jun amino-terminal kinase interacting protein-1 with p190 rhoGEF and its localization in differentiated neurons. *J. Biol. Chem.* **274**(49), 35113–35118.
103. Fleming, I. N. *et al.* (1997). Lysophosphatidic acid induces threonine phosphorylation of Tiam1 in Swiss 3T3 fibroblasts via activation of protein kinase C. *J. Biol. Chem.* **272**(52), 33105–33110.
104. Buchanan, F. G. *et al.* (2000). Translocation of the Rac1 guanine nucleotide exchange factor Tiam1 induced by platelet-derived growth factor and lysophosphatidic acid. *J. Biol. Chem.* **275**(13), 9742–9748.
105. Sutherland, E. and Rall, T. (1957). The properties of an adenine ribonucleotide produced with cellular particles, ATP, Mg++, and epinephrine or glucagon. *J. Am. Chem. Soc.* **79**(13), 3608–3608.
106. Sutherland, E. W. (1972). Studies on the mechanism of hormone action. *Science* **177**(47), 401–408.
107. Simonds, W. F. (1999). G protein regulation of adenylate cyclase. *Trends Pharmacol. Sci.* **20**(2), 66–73.
108. Scott, J. D. (1991). Cyclic nucleotide-dependent protein kinases. *Pharmacol. Ther.* **50**(1), 123–145.
109. Colledge, M. and Scott, J. D. (1999). AKAPs: from structure to function. *Trends Cell Biol.* **9**(6), 216–221.
110. Gordon, T. *et al.* (1992). Molecular cloning and preliminary characterization of a novel cytoplasmic antigen recognized by myasthenia gravis sera. *J. Clin. Invest.* **90**(3), 992–999.
111. Westphal, R. S. *et al.* (2000). Scar/WAVE-1, a Wiskott-Aldrich syndrome protein, assembles an actin-associated multi-kinase scaffold. *EMBO J.* **19**(17), 4589–4600.
112. Mullins, R. D. and Machesky, L. M. (2000). Actin assembly mediated by Arp2/3 complex and WASP family proteins. *Methods Enzymol.* **325**, 214–237.
113. Diviani, D. and Scott, J. D. (2001). AKAP signaling complexes at the cytoskeleton. *J. Cell Sci.* **114**(Pt 8), 1431–1437.
114. Sossey-Alaoui, K. *et al.* (2002). WAVE3, an actin-polymerization gene, is truncated and inactivated as a result of a constitutional t(1;13)(q21;q12) chromosome translocation in a patient with ganglioneuroblastoma. *Oncogene* **21**(38), 5967–5974.
115. Gelman, I. H. (2002). The Role of SSeCKS/Gravin/AKAP12 scaffolding proteins in the spaciotemporal control of signaling pathways in oncogenesis and development. *Front. Biosci.* **7**, D1782–1797.
116. Wikman, H. *et al.* (2002). Identification of differentially expressed genes in pulmonary adenocarcinoma by using cDNA array. *Oncogene* **21**(37), 5804–5813.
117. Dhillon, A. S. *et al.* (2002). Cyclic AMP-dependent kinase regulates Raf-1 kinase mainly by phosphorylation of serine 259. *Mol. Cell Biol.* **22**(10), 3237–3246.
118. Ferguson, S. S. (2001). Evolving concepts in G protein-coupled receptor endocytosis: the role in receptor desensitization and signaling. *Pharmacol. Rev.* **53**(1), 1–24.
119. van Biesen, T. *et al.* (1996). Mitogenic signaling via G protein-coupled receptors. *Endocr. Rev.* **17**(6), 698–714.
120. McDonald, P. H. *et al.* (2000). Beta-arrestin 2: a receptor-regulated MAPK scaffold for the activation of JNK3. *Science* **290**(5496), 1574–1577.
121. Marinissen, M. J. and Gutkind, J. S. (2001). G-protein-coupled receptors and signaling networks: emerging paradigms. *Trends Pharmacol. Sci.* **22**(7), 368–376.

CHAPTER 219

Signaling through G_z

Jingwei Meng and Patrick J. Casey

*Department of Pharmacology and Cancer Biology,
Duke University Medical Center, Durham, North Carolina*

General Properties

Tissue Distribution

G_z is a member of the G_i subfamily that contains a 41 kDa α subunit $G\alpha_z$ that possesses several properties that set it apart from other $G\alpha_i$ proteins [1–4]. First, the tissue distribution of $G\alpha_z$ is quite restricted, being found primarily in brain, retina, adrenal medulla, and platelets; expression is virtually undetectable in other tissues [4,5]. In adult rat brain, $G\alpha_z$ is expressed mainly in the hippocampus and, at modest levels, in the cerebellum and neocortex. $G\alpha_z$ is expressed in large neurons, such as cholinergic interneurons, but not striatal neurons [6]. During development of the mouse peripheral nervous system, $G\alpha_z$ is expressed at high levels in the superior cervical ganglion, dorsal root ganglion, and trigeminal ganglion; however, expression appears to be downregulated when mice reach the age of three weeks, except in the superior cervical ganglion. In the mouse central nervous system, expression of $G\alpha_z$ peaks at around the third postnatal week in whole brain; however, in the cerebellum the peak is observed around birth [7]. These findings suggest a possible role for G_z signaling in development of the nervous system.

Biochemical Properties

One of the more intriguing properties of $G\alpha_z$ that was noted when the protein was first studied is that the intrinsic rate of GTP hydrolysis by $G\alpha_z$, 0.05 min^{-1}, is quite low compared with most other G-protein α subunits [3]. This relatively weak GTP hydrolysis activity may be a result of a Ser substitution of the second Gly in a conserved GAGES sequence among G-protein α subunits [8]. This slow rate of GTP hydrolysis by $G\alpha_z$ suggests that it may participate in longer-duration signaling events than other $G\alpha_i$ proteins.

In addition, RGS (regulator of G-protein signaling) or RGS-like molecules (discussed in the section on RGS proteins) may play crucial roles in regulating G_z signaling due to their ability to enhance the slow GTP hydrolysis rate. The rate of GDP dissociation from $G\alpha_z$ is extremely slow and almost completely suppressed at Mg^{2+} concentrations greater than 100 μM [3]. Arachidonate and other unsaturated fatty acids selectively inactivate $G\alpha_z$ *in vitro* via a mechanism involving binding of negatively charged acidic lipid micelles to the nucleotide-free form of $G\alpha_z$ [9]. However, the biological significance of this unique lipid effect remains to be established.

Covalent Modifications

Unlike other members of the $G\alpha_i$ subfamily, $G\alpha_z$ lacks a consensus Cys residue near the carboxyl-terminus that is the site of modification by PTX-catalyzed ADP-ribosylation [3,4,10]. This makes $G\alpha_z$ a candidate for PTX-insensitive signaling processes. Another distinct property of $G\alpha_z$ is that $G\alpha_z$ can be phosphorylated both *in vitro* and in cells. Treatment of platelets with phorbol 12-myristate 13-acetate, thrombin, or a thromboxane A_2 analogue (U46619) activates protein kinase C (PKC). PKC activation both *in vitro* and *in situ* via permeabilized and intact platelets promotes rapid and stoichiometric phosphorylation of $G\alpha_z$ but has no effect on other members of the G_i subfamily, including $G\alpha_{i1}$, $G\alpha_{i2}$, $G\alpha_{i3}$, and $G\alpha_o$, or members of G_s subfamily [11]. Site-directed mutagenesis of potential PKC consensus sites revealed that Ser^{16} and Ser^{27} of $G\alpha_z$ are responsible for nearly 80 percent of the total phosphorylation [12]. This PKC-mediated phosphorylation is effectively inhibited by the presence of βγ complex, and phosphorylated $G\alpha_z$ has markedly reduced its ability to interact with the βγ complex [13], suggesting that a βγ contact site is around the amino-terminus of $G\alpha_z$, near Ser^{27}. RGSZ1 (discussed in the section on RGS) is a G$_z$-selective RGS. Phosphorylation of Gα_z renders the α subunit much less susceptible to RGSZ1 action [14]. Taken together, PKC-mediated phosphorylation of Gα_z is thought to increase the strength and duration of G$_z$ signaling by both preventing RGSZ1-enhanced GTP hydrolysis and re-formation of an α-$\beta\gamma$ complex, although this hypothesis has not yet been directly tested.

Localization of Gα_z to the plasma membrane requires two lipid modifications: myristoylation Gly^2 during protein translation and palmitoylation on Cys^3. Prevention of myristoylation by substitution of Ala for Gly^2 decreases palmitoylation on Cys^3, but palmitoylation can be rescued by overexpression of $\beta\gamma$ complex, suggesting that membrane association triggers a palmitoylation event [15,16]. Substitution of Ala for Cys^3, which prevents palmitoylation, does not affect myristoylation during translation, suggesting that myristoylation, an irreversible modification, directly affects the cellular localization of Gα_z, while palmitoylation, a reversible modification, is more important for the interaction between Gα_z and its possible effectors and regulators [15,17].

Receptors That Couple to G$_z$

Many receptors that couple to G$_i$ proteins can also activate G$_z$ if the receptors are overexpressed in cells. In reconstituted lipid vesicles, nucleotide exchange of Gα_z is stimulated when m2-muscarinic receptors are activated [18]. In HEK293 cells, expression of either A$_1$-adenosine, α_2-adrenergic, or D2-dopamine receptors together with wild-type Gα_z confers PTX-insensitive inhibition of adenylyl cyclase in response to receptor activation [19,20]. Opioid receptors, including μ, δ, and κ opioid receptors, have also been reported to couple to several members of the G$_i$ subfamily, including Gα_z [21–23]. Clues to a receptor that preferentially couples to G$_z$ have recently come from the analysis of Gα_z knockout mice (discussed in the section on Gα_z knockout mice). Mice lacking Gα_z have decreased platelet aggregation and impaired inhibition of cAMP formation in response to epinephrine [24]. Since evidence indicates that the ability of epinephrine to promote platelet activation is mainly due to its ability to inhibit adenylyl cyclase via α_{2A}-adrenergic receptors [25], and the epinephrine response is lost even though the platelets still contain their normal complement of Gα_i proteins, it seems that α_{2A}-adrenergic receptors preferentially couple to G$_z$ in platelets. This preferential coupling was further confirmed in experiments with PC12 cells expressing wild-type Gα_z. When these cells are challenged with cAMP analogue or nerve growth factor (NGF), treatment with a specific agonist to α_{2A}-adrenergic receptors, UK14304, attenuates PC12 cell differentiation; such an effect is not observed in these cells in the absence of Gα_z expression [26]. This pathway is discussed further in the section on effectors of G$_z$ signaling.

Regulators of G$_z$ Signaling: RGS Proteins

RGS proteins act as negative regulators of G-protein signaling by binding to, and enhancing GTP hydrolysis of, G-protein α subunits. Several RGS proteins that have been identified (at least 20 are known in mammals) selectively act on members of the G$_i$ subfamily, including G$_z$. These include GAIP [27], RGS4 [28], and RGS10 [29]. In addition, an RGS protein identified through both biochemical and interaction-cloning approaches, termed RGSZ1, selectively acts on Gα_z but not on other members of G$_i$ subfamily [14,30]. It is not clear whether these specific RGS proteins function only as negative regulators of G-protein signaling, or whether they can also function as effectors of G proteins or as adaptors linking Gα_z to other signaling pathways. Hints at such a process have arisen from studies indicating that RGSZ1 binds to the stathmin family member SCG10 and inhibits its microtubule disassembly functions, suggesting that RGS proteins have a broader role in cellular signaling [31].

Effectors of G$_z$ Signaling

Although the cellular processes controlled by G$_z$ activation are not well understood, activated Gα_z does possess an ability to inhibit some subtypes of adenylyl cyclase, a property shared with other members of the G$_i$ subfamily [20,32]. Since expression levels of other members of the Gα_i subfamily are much higher than that of Gα_z, it is widely believed that the ability of Gα_z to inhibit adenylyl cyclase activity is largely masked and Gα_z has other unique, and presumably important, cellular functions. In this regard, stable expression of mutationally activated Gα_z can transform Swiss3T3 and NIH3T3 cells by stimulating mitogenic pathways. It is interesting that this stimulation is apparently unrelated to the ability of Gα_z to inhibit adenylyl cyclase [33].

Quite recently, potential clues to additional functions of Gα_z have emerged from yeast two-hybrid screens such as the one that resulted in the identification of RGSZ1 as a specific regulator of Gα_z. Two additional molecules were also identified in this screen that selectively interacted with mutationally activated Gα_z. These molecules were Rap1GAP, a GTPase-activating protein for the Ras-like monomeric G protein Rap1, and a transcriptional coactivator termed Eya2. There is emerging evidence that the interaction of Gα_z with each of these molecules has direct functional consequences. Eya2 interacts with transcription factors of Six family; this interaction results in a translocation of Eya2 from the cytosol to the nucleus. Expression of constitutively active Gα_z blocks this process by competing with Six4 in an activation-dependent manner [34].

The physical interaction between Gα_z and Rap1GAP blocks the ability of RGS proteins to stimulate GTP hydrolysis of the α subunit, and also attenuates the ability of activated Gα_z to inhibit adenylyl cyclase. In addition, co-precipitation assays revealed that Gα_z, Rap1GAP, and Rap1 could form a stable complex [35]. In cell-based studies,

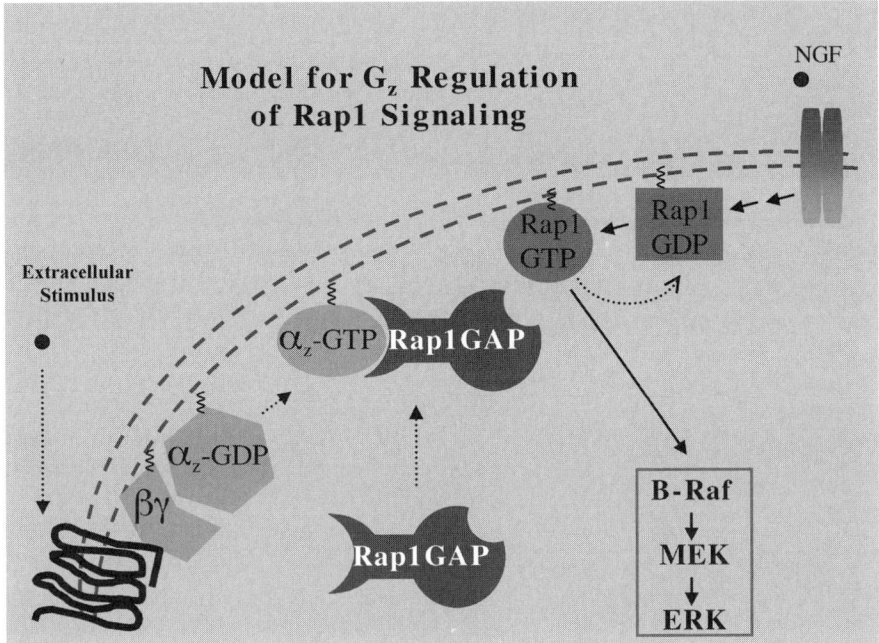

Figure 1 Model for Rap1GAP involvement in G_z signaling in neuronal cells. In the model shown, activation of Rap1 by elevated cAMP or NGF binding to its receptor leads to activation of a MAP kinase cascade that triggers cell differentiation. Concurrent activation of G_z by ligation of an appropriate G-protein-coupled receptor leads to recruitment of Rap1GAP to the membrane with subsequent downregulation of Rap1 signaling.

activated forms of $G\alpha_z$ were able to recruit Rap1GAP from a cytosolic location to the plasma membrane. Experiments in PC12 cells demonstrate that Rap1 activation, ERK phosphorylation, and cell differentiation induced by either cAMP analogue or NGF treatment were all blocked by either transfection of constitutively-activated $G\alpha_z$ or receptor-mediated G_z activation [26]. Based on these findings, a model has been proposed in which receptor-mediated activation of G_z results in recruitment of Rap1GAP to the plasma membrane, where it can effectively downregulate Rap1 signaling (Fig. 1).

$G\alpha_z$ Knockout Mice

$G\alpha_z$ knockout mice have been generated by two independent groups [24,36]. $G\alpha_z$ null mice are viable and no major phenotypes associated with loss of $G\alpha_z$ have yet been described. As already noted in the section on receptors that couple to G_z, however, one group has reported that platelet aggregation in response to challenge with epinephrine and collagen is impaired in the $G\alpha_z$ knockouts [24]. Furthermore, both groups have reported that $G\alpha_z$ knockout mice exhibit abnormal responses to certain psychoactive drugs, including a pronounced increase in locomotor activity in response to cocaine administration and a reduction in the analgesic effects of morphine.

One particularly intriguing phenotype of the $G\alpha_z$ null mice was a complete loss of the antidepressant effects of catecholamine reuptake inhibitors, suggesting that the signaling pathways invoked by antidepressant drugs of the norepinephrine reuptake inhibitor class are mediated primarily through G_z [24]. The physiological effects of this class of drugs are thought to be mediated at least in part through increased neurogenesis at the expense of cell differentiation [37,38]. The afore-mentioned findings that activation of G_z attenuates PC12 cell differentiation may provide a link between these observations. It is tempting to speculate that G_z signaling plays an important role in cell fate determination of neuronal cells. In this scenario, treatment with antidepressant drugs leads to increases in circulating norepinephrine that binds to G_z-coupled receptors, and subsequent activation of G_z leads to suppression of signaling pathways that promote cell differentiation and thereby allow cell proliferation associated with neurogenesis. In $G\alpha_z$ knockout mice, treatment with this class of antidepressant drugs cannot lead to activation of G_z and its downstream effectors, resulting in increased cell differentiation without further neurogenesis.

Summary

Although the precise role(s) of G_z in cellular signaling remain to be established, accumulating evidence points to the involvement of G_z in platelet function and, what is especially interesting, neurogenesis. It is intriguing to speculate that G_z may play a crucial role in determining cell fate during early brain development and later in neurogenesis in adulthood. Through subtle but extremely important actions,

G_z could participate in such critical brain functions as learning and memory [39,40].

References

1. Matsuoka, M., Itoh, H., Kozasa, T., and Kaziro, Y. (1988). Sequence analysis of cDNA and genomic DNA for a putative pertussis toxin-insensitive guanine nucleotide-binding regulatory protein α subunit. *Proc. Natl. Acad. Sci. USA* **85**, 5384–5388.
2. Fong, H. K., Yoshimoto, K. K., Eversole-Cire, P., and Simon, M. I. (1988). Identification of a GTP-binding protein α subunit that lacks an apparent ADP-ribosylation site for pertussis toxin. *Proc. Natl. Acad. Sci. USA* **85**, 3066–3070.
3. Casey, P. J., Fong, H. K., Simon, M. I., and Gilman, A. G. (1990). G_z, a guanine nucleotide-binding protein with unique biochemical properties. *J. Biol. Chem.* **265**, 2383–2390.
4. Gagnon, A. W., Manning, D. R., Catani, L., Gewirtz, A., Poncz, M., and Brass, L. F. (1991). Identification of G_z α as a pertussis toxin-insensitive G protein in human platelets and megakaryocytes. *Blood* **78**, 1247–1253.
5. Hinton, D. R., Blanks, J. C., Fong, H. K., Casey, P. J., Hildebrandt, E., and Simons, M. I. (1990). Novel localization of a G protein, G_z-α, in neurons of brain and retina. *J. Neurosci.* **10**, 2763–2770.
6. Friberg, I. K., Young, A. B., and Standaert, D. G. (1998). Differential localization of the mRNAs for the pertussis toxin insensitive G-protein α sub-units G_q, G_{11}, and G_z in the rat brain, and regulation of their expression after striatal deafferentation. *Brain Res. Mol. Brain Res.* **54**, 298–310.
7. Kelleher, K. L., Matthaei, K. I., Leck, K. J., and Hendry, I. A. (1998). Developmental expression of messenger RNA levels of the α subunit of the GTP-binding protein, G_z, in the mouse nervous system. *Brain Res. Dev. Brain Res.* **107**, 247–253.
8. Sprang, S. R. (1997). G protein mechanisms: insights from structural analysis. *Annu. Rev. Biochem.* **66**, 639–678.
9. Glick, J., Santoyo, G., and Casey, P. J. (1996). Arachidonate and related unsaturated fatty acids selectively inactivate the guanine nucleotide-binding regulatory protein, G_z. *J. Biol. Chem.* **271**, 2949–2954.
10. Fields, T. A. and Casey, P. J. (1997). Signalling functions and biochemical properties of pertussis toxin-resistant G-proteins. *Biochem. J.* **321**, 561–571.
11. Lounsbury, K. M., Casey, P. J., Brass, L. F., and Manning, D. R. (1991). Phosphorylation of G_z in human platelets. Selectivity and site of modification. *J. Biol. Chem.* **266**, 22051–22056.
12. Lounsbury, K. M., Schlegel, B., Poncz, M., Brass, L. F., and Manning, D. R. (1993). Analysis of G_z α by site-directed mutagenesis. Sites and specificity of protein kinase C-dependent phosphorylation. *J. Biol. Chem.* **268**, 3494–3498.
13. Fields, T. A. and Casey, P. J. (1995). Phosphorylation of G_z α by protein kinase C blocks interaction with the βγ complex. *J. Biol. Chem.* **270**, 23119–23125.
14. Glick, J. L., Meigs, T. E., Miron, A., and Casey, P. J. (1998). RGSZ1, a G_z-selective regulator of G protein signaling whose action is sensitive to the phosphorylation state of $G\alpha_z$. *J. Biol. Chem.* **273**, 26008–26013.
15. Wilson, P. T. and Bourne, H. R. (1995). Fatty acylation of α_z. Effects of palmitoylation and myristoylation on α_z signaling. *J. Biol. Chem.* **270**, 9667–9675.
16. Fishburn, C. S., Herzmark, P., Morales, J., and Bourne, H. R. (1999). Gβγ and palmitate target newly synthesized $G\alpha_z$ to the plasma membrane. *J. Biol. Chem.* **274**, 18793–18800.
17. Tu, Y., Wang, J., and Ross, E. M. (1997). Inhibition of brain G_z GAP and other RGS proteins by palmitoylation of G protein α subunits. *Science* **278**, 1132–1135.
18. Parker, E. M., Kameyama, K., Higashijima, T., and Ross, E. M. (1991). Reconstitutively active G protein-coupled receptors purified from baculovirus-infected insect cells. *J. Biol. Chem.* **266**, 519–527.
19. Tsu, R. C., Lai, H. W., Allen, R. A., and Wong, Y. H. (1995). Differential coupling of the formyl peptide receptor to adenylate cyclase and phospholipase C by the pertussis toxin-insensitive G_z protein. *Biochem. J.* **309**, 331–339.
20. Wong, Y. H., Conklin, B. R., and Bourne, H. R. (1992). G_z-mediated hormonal inhibition of cyclic AMP accumulation. *Science* **255**, 339–342.
21. Chan, J. S., Chiu, T. T., and Wong, Y. H. (1995). Activation of type II adenylyl cyclase by the cloned μ-opioid receptor: coupling to multiple G proteins. *J. Neurochem.* **65**, 2682–2689.
22. Lai, H. W., Minami, M., Satoh, M., and Wong, Y. H. (1995). G_z coupling to the rat κ-opioid receptor. *FEBS Lett.* **360**, 97–99.
23. Tsu, R. C., Chan, J. S., and Wong, Y. H. (1995). Regulation of multiple effectors by the cloned δ-opioid receptor: stimulation of phospholipase C and type II adenylyl cyclase. *J. Neurochem.* **64**, 2700–2707.
24. Yang, J., Wu, J., Kowalska, M. A., Dalvi, A., Prevost, N., O'Brien, P. J., Manning, D., Poncz, M., Lucki, I., Blendy, J. A., and Brass, L. F. (2000). Loss of signaling through the G protein, G_z, results in abnormal platelet activation and altered responses to psychoactive drugs. *Proc. Natl. Acad. Sci. USA* **97**, 9984–9989.
25. Hsu, C. Y., Knapp, D. R., and Halushka, P. V. (1979). The effects of α adrenergic agents on human platelet aggregation. *J. Pharmacol. Exp. Ther.* **208**, 366–370.
26. Meng, J. and Casey, P. J. (2002). Activation of G_z attenuates Rap1-mediated differentiation of PC12 cells. *J. Biol. Chem.* **277**, 43417–43424.
27. Woulfe, D. S. and Stadel, J. M. (1999). Structural basis for the selectivity of the RGS protein, GAIP, for $G\alpha_i$ family members. Identification of a single amino acid determinant for selective interaction of $G\alpha_i$ subunits with GAIP. *J. Biol. Chem.* **274**, 17718–17724.
28. Cavalli, A., Druey, K. M., and Milligan, G. (2000). The regulator of G protein signaling RGS4 selectively enhances α_{2A}-adreoreceptor stimulation of the GTPase activity of $G\alpha_{o1}$ and $G\alpha_{i2}$. *J. Biol. Chem.* **275**, 23693–23699.
29. Hunt, T. W., Fields, T. A., Casey, P. J., and Peralta, E. G. (1996). RGS10 is a selective activator of $G\alpha_i$ GTPase activity. *Nature* **383**, 175–177.
30. Wang, J., Tu, Y., Woodson, J., Song, X., and Ross, E. M. (1997). A GTPase-activating protein for the G protein $G\alpha_z$. Identification, purification, and mechanism of action. *J. Biol. Chem.* **272**, 5732–5740.
31. Nixon, A. B., Grenningloh, G., and Casey, P. J. (2002). The interaction of RGSZ1 with SCG10 attenuates the ability of SCG10 to promote microtubule disassembly. *J. Biol. Chem.* **277**, 18127–18133.
32. Taussig, R., Tang, W. J., Hepler, J. R., and Gilman, A. G. (1994). Distinct patterns of bidirectional regulation of mammalian adenylyl cyclases. *J. Biol. Chem.* **269**, 6093–6100.
33. Wong, Y. H., Chan, J. S., Yung, L. Y., and Bourne, H. R. (1995). Mutant α subunit of G_z transforms Swiss 3T3 cells. *Oncogene* **10**, 1927–1933.
34. Fan, X., Brass, L. F., Poncz, M., Spitz, F., Maire, P., and Manning, D. R. (2000). The α subunits of G_z and G_i interact with the eyes absent transcription cofactor Eya2, preventing its interaction with the six class of homeodomain-containing proteins. *J. Biol. Chem.* **275**, 32129–32134.
35. Meng, J., Glick, J. L., Polakis, P., and Casey, P. J. (1999). Functional interaction between $G\alpha_z$ and Rap1GAP suggests a novel form of cellular cross-talk. *J. Biol. Chem.* **274**, 36663–36669.
36. Hendry, I. A., Kelleher, K. L., Bartlett, S. E., Leck, K. J., Reynolds, A. J., Heydon, K., Mellick, A., Megirian, D., and Matthaei, K. I. (2000). Hypertolerance to morphine in $G\alpha_z$-deficient mice. *Brain Res.* **870**, 10–19.
37. Nibuya, M., Morinobu, S., and Duman, R. S. (1995). Regulation of BDNF and trkB mRNA in rat brain by chronic electroconvulsive seizure and antidepressant drug treatments. *J. Neurosci.* **15**, 7539–7547.
38. Malberg, J. E., Eisch, A. J., Nestler, E. J., and Duman, R. S. (2000). Chronic antidepressant treatment increases neurogenesis in adult rat hippocampus. *J. Neurosci.* **20**, 9104–9110.
39. Gould, E., Tanapat, P., Hastings, N. B., and Shors, T. J. (1999). Neurogenesis in adulthood: a possible role in learning. *Trends Cogn. Sci.* **3**, 186–192.
40. Cecchi, G. A., Petreanu, L. T., Alvarez-Buylla, A., and Magnasco, M. O. (2001). Unsupervised learning and adaptation in a model of adult neurogenesis. *J. Comput. Neurosci.* **11**, 175–182.

CHAPTER 220

Effectors of Gα$_o$ Signaling

Prahlad T. Ram, J. Dedrick Jordan, and Ravi Iyengar

Department of Pharmacology and Biological Chemistry,
Mount Sinai School of Medicine, New York, New York

Introduction

Heterotrimeric G proteins transduce signals through a wide variety of intracellular signaling pathways. Signaling pathways activated by G proteins vary depending on the identity of the specific alpha subunits. Direct effectors of the Gα$_o$ subunit are largely unknown, but recent work using yeast two-hybrid analysis [1] and screening a cDNA expression library [2] has yielded exciting new data and helped identify proteins that directly interact with Gα$_o$ and may serve as effectors. Some of the direct effectors identified by these methods include the GTPase activating protein for the small G protein Rap (Rap1-GAP), the GAP from the large G protein Gα$_z$ (G$_z$-GAP), the regulator of G-protein signaling 17 (RGS-17), and the G-protein regulated inducer of neurite outgrowth (GRIN). In addition to the direct effectors, some of the signaling pathways that are activated downstream from Gα$_o$ have also been identified. These include mitogen activated protein kinase 1,2 (MAPK 1,2) [3], the nonreceptor tyrosine kinase Src [4,5], signal transducer and activator of transcription 3 (Stat3) [5], and phospholipase C (PLC) [6]. In this chapter we will detail the direct effectors as well as the other downstream signaling pathways activated by Gα$_o$.

Heterotrimeric G proteins comprise α, β, and γ subunits. In the inactive state the Gα subunit is bound to GDP and the Gβγ dimer, and this trimeric complex is associated with the receptor. Upon activation there is an exchange of GDP for GTP on the Gα subunit and a disassociation from the βγ dimer. The GTP bound Gα is then able to bind and activate its effector molecules, as is the βγ dimmer in certain instances. The downstream signal specificity is dependent on the identity of the Gα subunit. There are currently four families of Gα subunits: Gα$_s$, which activates adenlyl cyclase (AC); Gα$_i$, which inhibits AC of which Gα$_o$ is a member; Gα$_{q/11}$, which activates phospholipase C; and finally Gα$_{12/13}$. Although the α subunits may activate some overlapping downstream signaling pathways, there are substantial differences between the individual members. Within the family of subunits, whereas the α subunits share common downstream targets they also associate with the common receptors that lead to their activation. A more detailed review of the richness of diversity among G-protein-coupled receptors, downstream targets, and physiological processes regulated by the various families of Gα subunits has recently been published [7].

Of all the members of the Gα subunits one of the more difficult to understand functionally has been Gα$_o$. Gα$_o$ was identified from bovine brain extract and is one of the most abundantly expressed proteins in the brain, making up almost 1 percent of the membrane proteins [8]. Gα$_o$ is present at high levels in the growth cones of neurons, and observations suggest a role for Gα$_o$ in regulation of neurite outgrowth [9,10]. Although Gα$_o$ is most abundantly expressed in the brain, other tissues have also been shown to express Gα$_o$, including the adenohypophysis, heart, kidney medulla, olfactory epithelium, and chromaffin cells [11].

Gα$_o$ is a member of the Gα$_i$ family of G proteins and has about an 80 percent homology to Gα$_{i1}$, but it does not appear to inhibit several forms adenylyl cyclase activity [12]. Early studies suggested a role for Gα$_o$ in coupling opiate and dopaminergic signals to the inhibition of voltage-sensitive calcium channels in neurons [13]. Gα$_o$ has also been to shown to couple the f-met-leu-phe (FMLP) receptor to PLC in HL-60 leukemia cells [14], the muscarinic receptor to PLC in *Xenopus* oocytes [15], and the α2 receptor to MAPK 1,2 in CHO cells [3]. Most Gα subunits contain an intrinsic GTPase domain that hydrolyses the GTP to GDP, thereby inactivating the GTP-bound Gα. Mutation of this intrinsic GTPase domain results in a constitutively active Gα subunit. The GTPase deficient Gα$_o$ subunit (Q-205-L)

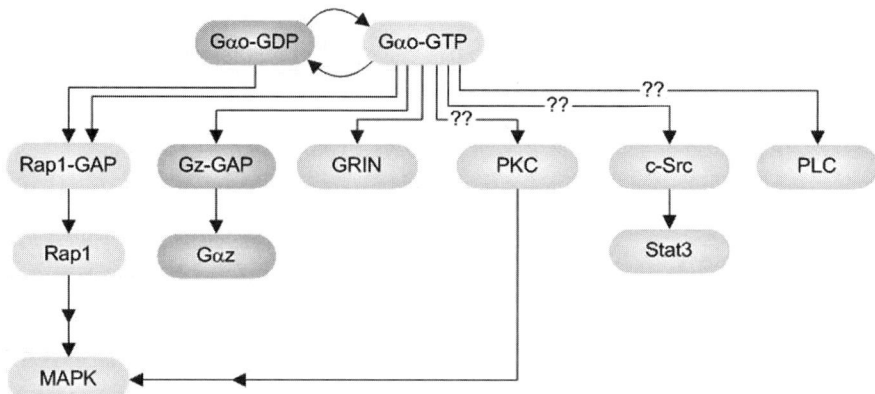

Figure 1 Effector pathways regulated by $G\alpha_o$. $G\alpha_o$ in the active GTP-bound or inactive GDP-bound state can regulate diverse signaling molecules. Not all pathways are likely to be present in all cell types. The regulation of multiple pathways allows for an increased complexity in the biological response to $G\alpha_o$ signaling.

has been used to identify physiological effects of $G\alpha_o$ (Q-L) expression in heterologous systems such as in *Xenopus* oocytes and mouse fibroblast cells [16,17]. This constitutively active mutant has also been a very useful tool to delineate and identify $G\alpha_o$-activated signaling pathways and putative direct effectors. The direct effectors and signaling pathways activated by $G\alpha_o$ are shown in Fig. 1.

By using yeast two hybrid we have identified proteins that can interact with $G\alpha_o$, and the concomitant use of either the inactive or the active $G\alpha_o$ subunit has allowed the identification of binding partners that can preferentially bind to either the inactive or the active $G\alpha_o$ subunit [1]. A chick dorsal root ganglia (DRG) neuron library was used for the two-hybrid analysis. The reason for using the chick DGR was that this is a model system in which the effects of activating $G\alpha_o$ have been well characterized [18]. The direct binding partners identified from the DRG neurons included Rap1GAP, Gz-GAP, RGS17, and GRIN3 [1]. One of the interesting aspects of this study was that Rap1-GAP binds the inactive or GDP-$G\alpha_o$ preferentially over the active or GTP-$G\alpha_o$. In addition, the interactions between $G\alpha_o$ and Rap1GAP lead to a change in Rap1 activity, whereby the binding of $G\alpha_o$ and Rap1GAP relieves the inhibitory effect of Rap1GAP on Rap1, leading to the activation of Rap1 [1]. The sequestration of a negative regulator to activate a third molecule was a novel finding in signaling via G-protein-coupled receptors.

Further studies in our lab have elucidated the mechanism for the negative regulation of Rap1GAP by $G\alpha_o$ [19]. The binding of $G\alpha_o$ to Rap1GAPII and RGS17 targets these proteins for degradation by the ubiquitin-dependent proteasome system. In rat hippocampal slices the induced degradation of Rap1GAPII could be stimulated by the addition of α_2 adrenergic agonist, which could be specifically blocked by inhibiting the proteasome. This novel mechanism for the regulation of a G-protein effector would allow for long-term modulation of signaling pathways. By selectively degrading negative regulators the intrinsic signaling properties would be altered, thus changing the required input for signal activation as well as signal duration.

A study to elucidate direct interactors for the $G\alpha_z$ subunit conducted by Kozasa and colleagues led to the identification of a protein that directly binds $G\alpha_o$ [2]. The approach taken in this study was to screen a mouse embryo expression library by using GTPγ-$G\alpha_o$. The novel $G\alpha_o$ interacting protein identified by this method was termed GRIN [2]. Expression of $G\alpha_o$ and GRIN in Neuro2A (N2A) mouse neuroblastoma cells leads to their differentiation marked by outgrowth of neurite-like fine processes. There are two isoforms of GRIN—GRIN1 and GRIN2—and both bind directly to $G\alpha_o$. One of the main differences between the two isoforms is their expression in different tissues: whereas GRIN1 is widely expressed in neuronal tissue with highest levels in the spinal cord, GRIN2 is only expressed in the cerebellum.

In addition to these direct interactors of $G\alpha_o$, several laboratories have identified signaling molecules and pathways that are activated by $G\alpha_o$. The signaling pathways that are activated by $G\alpha_o$ include c-Src, MAPK1,2, Stat3, and PLC. These pathways are also cell-type specific and are not activated in all cell types. Activation of MAPK 1,2 via ligand activation of $G\alpha_o$ occurs in CHO cells but not in COS-7 cells [3]. Further, this pathway is dependent on PKC activation, as depletion of PKC inhibited $G\alpha_o$-induced activation of MAPK 1,2. Experiments in chick DRG neurons show that the GABA-mediated inhibition of N-type calcium channels by $G\alpha_o$ occurs via activation of Src or a Src family kinase [4]. Src activation by $G\alpha_o$ has also been observed in NIH-3T3 mouse fibroblast cells [5]. Further characterization of the $G\alpha_o$ activation of Src showed that this leads to an increase in Stat3 tyrosine phosphorylation and activation [5].

Conclusions

It is still not clear how many signaling pathways are regulated by the $G\alpha_o$ subunit. However, what is increasingly clear is that in contrast to the $G\alpha_s$ and $G\alpha_q$ pathways, which are ubiquitous, the $G\alpha_o$ and $G\alpha_i$ pathways are likely to be cell-type dependent. Defining which pathways operate in which cell type is an immediate goal in many laboratories.

References

1. Jordan, J. D., Carey, K. D., Stork, P. J., and Iyengar, R. (1999). Modulation of rap activity by direct interaction of Galpha(o) with Rap1 GTPase-activating protein. *J. Biol. Chem.* **274**, 21507–21510.
2. Chen, L. T., Gilman, A. G., and Kozasa, T. A. (1999). A candidate target for G protein action in brain. *J. Biol. Chem.* **274**, 26931–26938.
3. van Biesen, T., Hawes, B. E., Raymond, J. R., Luttrell, L. M., Koch, W. J., and Lefkowitz, R. J. (1996). G(o)-protein alpha-subunits activate mitogen-activated protein kinase via a novel protein kinase C-dependent mechanism. *J. Biol. Chem.* **271**, 1266–1269.
4. Diverse-Pierluissi, M., Remmers, A. E., Neubig, R. R., and Dunlap, K. (1997). Novel form of crosstalk between G protein and tyrosine kinase pathways. *Proc. Natl. Acad. Sci. USA* **94**, 5417–5421.
5. Ram, P. T., Horvath, C. M., and Iyengar, R. (2000). Stat3-mediated transformation of NIH-3T3 cells by the constitutively active Q205L Galphao protein. *Science* **287**, 142–144.
6. Blitzer, R. D., Omri, G., De Vivo, M., Carty, D. J., Premont, R. T., Codina, J., Birnbaumer, L., Cotecchia, S., Caron, M. G., Lefkowitz, R. J. et al. (1993). Coupling of the expressed alpha 1B-adrenergic receptor to the phospholipase C pathway in *Xenopus* oocytes. The role of Go. *J. Biol. Chem.* **268**, 7532–7537.
7. Neves, S. R., Ram, P. T., and Iyengar, R. (2002). G protein pathways. *Science* **296**, 1636–1639.
8. Huff, R. M., Axton, J. M., and Neer, E. J. (1985). Physical and immunological characterization of a guanine nucleotide-binding protein purified from bovine cerebral cortex. *J. Biol. Chem.* **260**, 10864–10871.
9. Zubiaur, M. and Neer, E. J. (1993). Nerve growth factor changes G protein levels and localization in PC12 cells. *J. Neurosci. Res.* **35**, 207–217.
10. Strittmatter, S. M., Fishman, M. C., and Zhu, X. P. (1994). Activated mutants of the alpha subunit of G(o) promote an increased number of neurites per cell. *J. Neurosci* **14**, 2327–2338.
11. Bockaert, J., Brabet, P., Gabrion, J., Homburger, V., Rouot, B., and Toutant, M. (1990). Structural, immunobiological, and functional characterization of guanine nucleotide binding protein Go, in Iyengar, R. and Birnbaumer, L., Eds., *G Proteins*. Academic Press, San Deigo, CA.
12. Van Meurs, K. P., Angus, C. W., Lavu, S., Kung, H. F., Czarnecki, S. K., Moss, J., and Vaughan, M. (1987). Deduced amino acid sequence of bovine retinal Go alpha: similarities to other guanine nucleotide-binding proteins. *Proc. Natl. Acad. Sci. USA* **84**, 3107–3111.
13. Hescheler, J., Rosenthal, W., Trautwein, W., and Schultz, G. (1987). The GTP-binding protein, Go, regulates neuronal calcium channels. *Nature* **325**, 445–447.
14. Kikuchi, A., Kozawa, O., Kaibuchi, K., Katada, T., Ui, M., and Takai, Y. (1986). Direct evidence for involvement of a guanine nucleotide-binding protein in chemotactic peptide-stimulated formation of inositol bisphosphate and trisphosphate in differentiated human leukemic (HL-60) cells. Reconstitution with Gi or Go of the plasma membranes ADP-ribosylated by pertussis toxin. *J. Biol. Chem.* **261**, 11558–11562.
15. Moriarty, T. M., Padrell, E., Carty, D. J., Omri, G., Landau, E. M., and Iyengar, R. (1990). Go protein as signal transducer in the pertussis toxin-sensitive phosphatidylinositol pathway. *Nature* **343**, 79–82.
16. Kroll, S. D., Omri, G., Landau, E. M., and Iyengar, R. (1991). Activated alpha subunit of Go protein induces oocyte maturation. *Proc. Natl. Acad. Sci. USA* **88**, 5182–5186.
17. Kroll, S. D., Chen, J., De Vivo, M., Carty, D. J., Buku, A., Premont, R. T., and Iyengar, R. (1992). The Q205LGo-alpha subunit expressed in NIH-3T3 cells induces transformation. *J. Biol. Chem.* **267**, 23183–23188.
18. Diverse-Pierluissi, M., Goldsmith, P. K., and Dunlap, K. (1995). Transmitter-mediated inhibition of N-type calcium channels in sensory neurons involves multiple GTP-binding proteins and subunits. *Neuron* **14**, 191–200.
19. Jordan, D. J and Iyengar, R. Manuscript submitted.

CHAPTER 221

Phosphorylation of G Proteins

Louis M. Luttrell[1,2] and Deirdre K. Luttrell[3]

[1]The Geriatrics Research, Education and Clinical Center,
Durham Veterans Affairs Medical Center, Durham, North Carolina
[2]Department of Medicine, Duke University Medical Center,
Durham, North Carolina
[3]GlaxoSmithKline, Research Triangle Park, North Carolina

Introduction

In normal cells, heterotrimeric G-protein subunits are known to undergo several forms of co-translational or post-translational modification. Lipid modifications, including N-myristoylation and palmitoylation of Gα subunits and prenylation of Gγ subunits, are required for plasma membrane targeting and contribute to the interaction of Gα with Gβγ subunits, effectors, and regulators of G-protein signaling (RGS proteins) [1,2]. Gα subunits, including members of each of the four major families of Gα (Gαs, Gαi, Gαq, and Gα12), also undergo phosphorylation on serine or tyrosine residues. In addition, one Gγ subunit is serine phosphorylated. Here we review what is known about G-protein phosphoryation and its subsequent effects on G-protein function. (See Table I).

Serine Phosphorylation

Several heterotrimeric G-protein subunits are phosphorylated on serine residues *in vivo* and *in vitro*. The kinases responsible include various protein kinase C (PKC) isoforms, the p21-activated protein kinase (PAK), and cGMP-dependent protein kinase. In most cases, serine phosphoryation of Gα inhibits binding to Gβγ subunits and RGS proteins, which may lead either to prolonged Gα-subunit activation or to feedback inhibition of G-protein signaling.

Protein Kinase C

PKC-mediated phosphorylation of Gαi family (Gαz, Gαi, Gαt), Gαq family (Gα16), and Gα12 family (Gα12, Gα13) Gα subunits, along with one Gγ subunit (Gγ12), has been described. Of these, PKC phosphorylation of Gαz and Gα12 are the most thoroughly studied. In human platelets, exposure to phorbol ester, which directly activates classical PKC isoforms, or thrombin, which stimulates a Gq/11-coupled heptahelical receptor, results in rapid and nearly stoichiometric phosphorylation of endogenous Gαz [3,4]. Ser 27 has been identified as the major site of PKC phosphorylation both *in vivo* and *in vitro*, while Ser 16 serves as a secondary site [5,6]. The preferred substrate for PKC is the monomeric Gα subunit, since the addition of Gβγ *in vitro* strongly inhibits the phosphorylation [6–8]. The data are less clear as to whether the inactive GDP-bound or active GTP-bound form of Gαz is a preferred substrate. Since the Gα N-terminus contains determinants for the binding of the Gβγ subunit heterodimer, PKC phosphorylation of Gαz blocks Gβγ subunit binding [6–8]. Similarly, phosphorylation inhibits Gαz interaction with the RGS proteins RGSZ1, RET-RGS1, and GAIP [9,10]. The effect of PKC phosphorylation on Gαz function is unclear, however. Inhibition of Gβγ subunit and RGS protein binding might be expected to prolong Gαz activation *in vivo*, but phosphorylation has no effect on the ability of Gαz-GTPγS to inhibit adenylyl cyclase in a reconstituted system [8].

Conflicting data exist for the PKC-mediated phosphorylation of other Gαi family members. Some authors have reported PKC-mediated phosphorylation of Gαi or Gαt *in vitro* or phorbol ester-mediated phosphorylation *in vivo* [11–14]. Others, however, have failed to detect PKC-mediated phosphorylation of Gαi subunits [4,8]. In NG-108 cells, phorbol ester-induced phosphorylation of Gαi2 correlates with a loss of the inhibitory effect of Gαi on adenylyl cyclase activation [14], suggesting that Gαi phosphorylation

Table I Phosphorylation of Heterotrimeric G protein Subunits.

G protein subunit	Kinase	Phosphorylation site(s)	Functional effect	References
Serine phosphorylation				
Gαz	PKC	Ser27; Ser16	Inhibits Gβγ subunit and RGS protein binding	[4–10]
	PAK	Ser16	Inhibits Gβγ subunit and RGS protein binding	[6]
Gαi1, Gαi2, Gαi3	PKC	N.D.	Inhibits regulation of adenylyl cyclase	[11–14]
	CGMP-dependent Protein kinase	N.D.	Inhibits IGF-1 and IGF-2 stimulated Ca^{++} influx	[20]
Gα16	PKC	N.D.	Inhibits TRH-stimulated PI hydrolysis	[15]
Gα12, 13	PKC	N-terminal 50 amino acids	Inhibits Gβγ subunit binding	[8,16]
Gγ12	PKC	N-terminal SSK motif	Increases affinity for Gαo; Inhibits activation of type II adenyl cyclase; Increases binding to F-actin	[17–19]
Tyrosine Phosphorylation				
Gαs	c-Src	Tyr37; Tyr377	Increases GTP binding/hydrolysis; Enhances β2AR-stimulated cAMP production	[21–23]
	EGF receptor	N.D.	Enhances Gs activation	[29]
Gq/11	c-Src; c-Fyn; c-Yes; v-Src;	Tyr 356	Modulates R-G interaction; Enhances PI hydrolysis; Stimulates GLUT4 translocation	[24–26]
	Insulin Receptor	N.D.	Stimulates GLUT4 translocation	[28]
Gαi, Gαt	Insulin Receptor	N.D.	N.D.	[12,27]

might provide a mechanism for cross-talk between different G-protein pools.

When ectopically expressed in *Xenopus* oocytes, the Gαq family protein, Gα16, can be phosphorylated in response to phorbol ester treatment or stimulation of thyrotropin-releasing hormone (TRH) receptors [15]. Phosphorylation correlates with a loss of TRH responsiveness, suggesting Gα-subunit phosphorylation may represent part of a feedback inhibitory loop.

Gα12, like Gαz, undergoes PKC-mediated phosphorylation in platelets in response to phorbol ester or thrombin [8]. The site of phosphorylation has been mapped to the N-terminal 50 amino acids but has not been explicitly determined. As with Gαz, phosphorylation of Gα12 blocks Gβγ-subunit binding. Gα13 is also phosphorylated in platelets in response to phorbol ester [16]. However, in contrast to Gα12, attempts to phosphorylate Gα13 *in vitro* using purified PKCα, δ, ε, or ζ have been unsuccessful [8], suggesting that the effect of PKC on Gα13 may be indirect.

Gγ12 contains a Ser-Ser-Lys motif at its N-terminus that is not present in other Gγ subunits and that serves as a substrate for PKC *in vitro* [17,18]. Endogenous Gγ12 in Swiss 3T3 and aortic smooth muscle cells is phosphorylated in response to phorbol ester [17]. Phosphorylation of Gγ12 increases its affinity for Gαo, presumably allowing for more stable heterotrimer formation and allowing phosphorylated Gγ12 to enhance high-affinity agonist binding to A1 adenosine receptors [18]. Phosphorylation selectively affects the interaction of Gγ12 with effectors, impairing Gβγ subunit-dependent activation of type II adenylyl cyclase but having no effect on activation of phospholipase-Cβ. Phosphorylated Gγ12 also affects its interaction with F-actin, possibly allowing G-protein phosphorylation to regulate cell motility [19].

p21-Activated Protein Kinase

Gαz, but not Gαs, Gαi, Gαo, or Gαq, serves as a substrate for PAK1 *in vitro*, with a stoichiometry of phosphorylation of about 1:1 [6]. The preferred phosphorylation site is Ser 16, and the phosphorylation is independent of the activation state of Gαz. In transfected HEK-293 cells, Gαz is phosphorylated by coexpressed PAK1 to an extent similar to that observed with PKC. As with PKC phosphorylation of Ser27, PAK1 phosphorylation inhibits binding of Gβγ subunits and RGS proteins *in vitro*. The effect of PAK1-mediated phosphorylation on Gαz function *in vivo* remains to be determined but might be expected to resemble the effect of PKC phosphorylation.

cGMP-dependent Protein Kinase

In vitro, cGMP dependent protein kinase Iα phosphorylates the Gα subunits of heterotrimeric Gαi1, Gαi2, and Gαi3 to a stoichiometry of about 0.4:1 [20]. The site of phosphorylation is not known, but a potential cGMP kinase site (Arg-Lys-Asp-Thr-Lys) is present in the C-terminal effector region of the protein. In CHO cells, expression of cGMP kinase modestly increases phosphorylation of endogenous Gαi subunits and attenuates pertussis toxin-sensitive calcium influx following insulin-like growth factor stimulation, suggesting that phosphorylation of Gαi may interrupt the coupling of Gαi with effectors.

Tyrosine Phosphorylation

Tyrosine phosphorylation of Gα subunits may account for the ability of tyrosine kinases to enhance certain G-protein-mediated signals. Phosphorylation of Gαs, Gαi, and Gαq/11 by Src family nonreceptor tyrosine kinases, or in response to epidermal growth factor or insulin stimulation, has been reported. Specific phosphorylation sites have been determined only for Src-mediated phosphorylation of Gαs.

Src Family Tyrosine Kinases

Expression of avian c-Src in murine fibroblasts leads to a significant enhancement of adrenergic receptor-stimulated cAMP production, both in isolated plasma membranes and in intact cells [21]. *In vitro*, c-Src phosphorylates purified Gαs and Gαi subunits to a stoichiometry of 0.3–0.9:1 [22]. The inactive monomeric form of the Gα subunit appears to be the perferred substrate, since GTPγS and Gβγ subunits both inhibit phosphorylation. Phosphorylation of Gαs involves two tyrosine residues, Tyr37 and Tyr377 [23], and increases the rate of GTPγS binding and of β2 adrenergic receptor-stimulated GTP hydrolysis.

Similarly, v-*src* transformation of fibroblasts causes enhanced endothelin-1 receptor-stimulated phosphatidylinositol hydrolysis [24]. The effect correlates with tyrosine phosphorylation of Gαq/11 subunits. In a reconstituted system, Gαq/11 from v-*src* transformed cells exhibits increased AlF$_4^-$-stimulated phospholipase C activity. Although the phosphorylation *in vivo* was sensitive to the tyrosine kinase inhibitor herbimycin A, the authors were unable to demonstrate phosphorylation of Gαq/11 by v-Src *in vitro*. Tyrosine phosphorylation of Gαq/11 may also be mediated by the Src family kinase, c-Yes. In a recent study, endothelin 1 receptor-stimulated tyrosine phosphorylation of Gαq/11 in 3T3-L1 adipocytes was blocked by the Src inhibitor PP2 by microinjection of a dominant negative c-Src mutant and by microinjection of antibodies against c-Yes but not c-Src or c-Fyn [25]. Recruitment of a Src family kinase to the endothelin-1 receptor through the formation of β-arrestin1-Src kinase complexes was proposed as the mechanism underlying endothelin-stimulated Gαq/11 phosphorylation. The phosphorylation of Gαq/11 appears to play a role in the regulation of glucose transport by endothelin receptors, since inhibition of endothelin-stimulated Gαq/11 phosphorylation correlated with a marked decrease in of endothelin-stimulated translocation of the glucose transporter GLUT4.

Tyrosine phosphorylation of Gαq and Gα11 in response to M1 muscarinic receptor stimulation has also been reported [26]. In this case, the phosphorylation appears to involve Tyr 356, located in the C-terminus in a position analogous to Tyr 377 of Gαs. The responsible kinase has not been identified, but the effect can be mimicked by expression of c-Fyn, suggesting that it is mediated by a Src family kinase.

Insulin Receptor

Early work demonstrated that several Gα subunits, including Gαt, Gαi, and Gαo, can serve as substrates for the insulin receptor *in vitro* [12,27]. As is the case with Src kinases, the monomeric GDP-bound form of the Gα subunit is the preferred substrate.

Tyrosine phosphorylation of Gαq/11 subunits has been proposed to play a role in insulin-stimulated GLUT4 translocation [28]. Insulin stimulation is associated with a rapid and transient increase in tyrosine phosphorylation of Gαq/11 subunits in 3T3-L1 adiopcytes. In these cells, microinjection of RGS2 protein or antibodies against Gαq inhibited insulin-stimulated GLUT4 translocation, while expression of a consititutive!y active Gαq mutant stimulated GLUT4 translocation. Although the identity of the tyrosine kinase responsible for Gαq/11 phosphorylation was not determined in these studies, the data suggest that tyrosine phosphorylation of Gαq may play a role in regulating the translocation of glucose transporters to the plasma membrane.

Epidermal Growth Factor Receptor

Purified EGF receptors catalyze the tyrosine phosphorylation of Gαs subunits *in vitro*, with a stoichiometry of phosphorylation of about 2:1 [29]. In this case, the heterotrimeric form of the Gα subunit appears to be the preferred substrate. EGF receptor-phosphorylated Gαs exhibited increased GTPγS binding and GTPase activity and enhanced activation of adenylyl cyclase, suggesting that the phosphorylation leads to activation of Gαs.

Conclusions

There is convincing evidence that several heterotrimeric G-protein subunits are substrates for serine and/or tyrosine phosphorylation *in vitro* and in some cases, *in vivo*. The functional relevance of these phosphorylation events is less well established. Serine phosphorylation of Gα subunits by second messenger–dependent protein kinases is a potential mechanism of feedback regulation of G-protein-coupled receptor signaling. Tyrosine phosphorylation of Gα subunits may provide a mechanism for cross-talk between receptor and nonreceptor tyrosine kinases and G-protein-dependent signaling pathways.

References

1. Chen, C. A. and Manning, D. R. (2001). Regulation of G proteins by covalent modification. *Oncogene* **20**, 1643–1652.
2. Fu, H. W. and Casey, P. J. (1999). Enzymology and biology of CaaX protein prenylation. *Recent Prog. Hormone Res.* **54**, 315–342.
3. Carlson, K. E., Brass, L. F., and Manning, D. R. (1989). Thrombin and phorbol esters cause the selective phosphorylation of a G protein other than Gi in human platelets. *J. Biol. Chem.* **264**, 13298–13305.
4. Lounsbury, K. M., Casey, P. J., Brass, L. F., and Manning, D. R. (1991). Phosphorylation of Gz in human platelets. Selectivity and site of modification. *J. Biol. Chem.* **266**, 22051–22056.

5. Lounsbury, K. M., Schlegel, B., Poncz, M., Brass, L. F., and Manning, D. R. (1993). Analysis of Gzα by site-directed mutagenesis. Sites and specificity of protein kinase C-dependent phosphorylation. *J. Biol. Chem.* **268**, 3494–3498.
6. Wang, J., Frost, J. A., Cobb, M. H., and Ross, E. M. (1999). Reciprocal signaling between heterotrimeric G proteins and the p21-stimulated protein kinase. *J. Biol. Chem.* **274**, 31641–31647.
7. Fields, T. A. and Casey, P. J. (1995). Phosphorylation of Gzα by protein kinase C blocks interaction with the βγ complex. *J. Biol. Chem.* **270**, 23119–23125.
8. Kozasa, T. and Gilman, A. G. (1996). Protein kinase C phosphorylates G12α and inhibits its interaction with Gβγ. *J. Biol. Chem.* **271**, 12562–12567.
9. Glick, J. L., Meigs, T. E., Miron, A., and Casey, P. J. (1998). RGSZ1, a Gz-selective regulator of G protein signaling whose action is sensitive to the phosphorylation state of Gzα. *J. Biol. Chem.* **273**, 26008–26013.
10. Wang, J., Ducret, A., Tu, Y., Kozasa T., Aebersold, R., and Ross, E. M. (1998). RGSZ1, a Gz-selective RGS protein in brain. Structure, membrane association, regulation by Gαz phosphorylation, and relationship to a Gz GTPase-activating protein subfamily. *J. Biol. Chem.* **273**, 26014–26025.
11. Katada, T., Gilman, A. G., Watanabe, Y., Bauer, S., and Jakobs, K. H. (1985). Protein kinase C phosphorylates the inhibitory guanine-nucleotide-binding regulatory component and apparently suppresses its function in hormonal inhibition of adenylate cyclase. *Eur. J. Biochem.* **15**, 431–437.
12. Zick, Y., Sagi-Eisenberg, R., Pines, M., Gierschik, P., and Spiegel, A. M. (1986). Multisite phosphorylation of the α subunit of transducin by the insulin receptor kinase and protein kinase C. *Proc. Natl. Acad. Sci. USA* **83**, 9294–9297.
13. Bushfield, M., Murphy, G. J., Lavan, B. E., Parker, P. J., Hruby, V. J., Milligan, G., and Houslay, M. D. (1990). Hormonal regulation of Gi2 α-subunit phosphorylation in intact hepatocytes. *Biochem. J.* **268**, 449–457.
14. Strassheim, D. and Malbon, C. C. (1994). Phosphorylation of Giα2 attenuates inhibitory adenylyl cyclase in neuroblastoma/glioma hybrid (NG-108-15) cells. *J. Biol. Chem.* **269**, 14307–14313.
15. Aragay, A. M. and Quick, M. W. (1999). Functional regulation of Gα16 by protein kinase C. *J. Biol. Chem.* **274**, 4807–4815.
16. Offermanns, S., Hu, Y. H., and Simon, M. I. (1996). Gα12 and Gα13 are phosphorylated during platelet activation. *J. Biol. Chem.* **271**, 26044–26048.
17. Morishita, R., Nakayama, H., Isobe, T., Matsuda, T., Hashimoto, Y., Okano, T., Fukada, Y., Mizuno, K., Ohno, S., and Kozawa, O. (1995). Primary structure of a γ subunit of G protein, γ12, and its phosphorylation by protein kinase C. *J. Biol. Chem.* **270**, 29469–29475.
18. Yasuda, H., Lindorfer, M. A., Myung, C. S., and Garrison, J. C. (1998). Phosphorylation of the G protein γ12 subunit regulates effector specificity. *J. Biol. Chem.* **273**, 21958–21965.
19. Ueda, H., Yamauchi, J., Itoh, H., Morishita, R., Kaziro, Y., Kato, K., and Asano, T. (1999). Phosphorylation of F-actin-associating G protein γ12 subunit enhances fibroblast motility. *J. Biol. Chem.* **274**, 12124–12128.
20. Pfeifer, A., Nurnberg, B., Kamm, S., Uhde, M., Schultz, G., Ruth, P., and Hofmann, F. (1995). Cyclic GMP-dependent protein kinase blocks pertussis toxin-sensitive hormone receptor signaling pathways in Chinese hamster ovary cells. *J. Biol. Chem.* **270**, 9052–9059.
21. Bushman, W. A., Wilson, L. K., Luttrell, D. K., Moyers, J. S., and Parsons, S. J. (1990). Overexpression of c-src enhances β-adrenergic-induced cAMP accumulation. *Proc. Natl. Acad. Sci. USA* **87**, 7462–7466.
22. Hausdorff, W. P., Pitcher, J. A., Luttrell, D. K., Linder, M. E., Kurose, H., Parsons, S. J., Caron, M. G., and Lefkowitz, R. J. (1992). Tyrosine phosphorylation of G protein α subunits by pp60c-src. *Proc. Natl. Acad. Sci. USA* **89**, 5720–5724.
23. Moyers, J. S., Linder, M. E., Shannon, J. D., and Parsons, S. J. (1995). Identification of the in vitro phosphorylation sites on Gsα mediated by pp60c-src. *Biochem. J.* **305**, 411–417.
24. Liu, W. W., Mattingly, R. R., and Garrison, J. C. (1996). Transformation of Rat-1 fibroblasts with the v-src oncogene increases the tyrosine phosphorylation state and activity of the α subunit of Gq/G11. *Proc. Natl. Acad. Sci. U.S.A.* **93**, 8258–8263.
25. Imamura, T., Huang, J., Dalle, S., Ugi, S., Usui, I., Luttrell, L. M., Miller, W. E., Lefkowitz, R. J., and Olefsky, J. M. (2001). β-Arrestin-mediated recruitment of the Src family kinase Yes mediates endothelin-1-stimulated glucose transport. *J. Biol. Chem.* **276**, 43663–43667.
26. Umemori, H., Inoue, T., Kume, S., Sekiyama, N., Nagao, M., Itoh, H., Nakanishi, S., Mikoshiba, K., and Yamamoto, T. (1997). Activation of the G protein Gq/11 through tyrosine phosphorylation of the α subunit. *Science* **276**, 1878–1881.
27. Krupinski, J., Rajaram, R., Lakonishok, M., Benovic, J. L., and Cerione, R. A. (1988). Insulin-dependent phosphorylation of GTP-binding proteins in phospholipid vesicles. *J. Biol. Chem.* **263**, 12333–12341.
28. Imamura, T., Vollenweider, P., Egawa, K., Clodi, M., Ishibashi, K., Nakashima, N., Ugi S., Adams, J. W., Brown, J. H., and Olefsky, J. M. (1999). Gα-q/11 protein plays a key role in insulin-induced glucose transport in 3T3-L1 adipocytes. *Mol. Cell. Biol.* **19**, 6765–6774.
29. Poppleton, H., Sun, H., Fulgham, D., Bertics, P., and Patel, T. B. (1996). Activation of Gsα by the epidermal growth factor receptor involves phosphorylation. *J. Biol. Chem.* **271**, 6947–6951.

CHAPTER 222

Mono-ADP-Ribosylation of Heterotrimeric G Proteins

Maria Di Girolamo and Daniela Corda
Department of Cell Biology and Oncology,
Istituto di Ricerche Farmacologiche "Mario Negri," Consorzio Mario Negri Sud,
Santa Maria Imbaro, Chieti, Italy

Introduction

Heterotrimeric G proteins play a crucial role in determining the specificity of the cellular response to extracellular signals. Among the mechanisms controlling G-protein functions are a variety of covalent modifications, relevant in both normal and pathological conditions. The known post-translational modifications occurring on G protein α and βγ subunits include lipid modifications, which are required for targeting to the plasma membrane of α and βγ and for the interaction of α with either βγ or effectors or RGS proteins [1]; phosphorylation, which affects the interactions between α and βγ, and between G proteins with either receptors or effectors [2]; and finally mono-ADP-ribosylation, whose function has not yet been fully defined. In the following, we will focus on the enzymatic mono-ADP-ribosylation of the G protein subunits, catalyzed by both bacterial toxins and eukaryotic enzymes, and discuss the mechanisms and potential role of this reaction in mammals.

The Mono-ADP-Ribosylation Reaction

The mono-ADP-ribosylation reaction is catalyzed by mono-ADP-ribosyltransferases (EC 2.4.2.31) that transfer an ADP-ribose residue from βNAD$^+$ to a specific amino acid of the acceptor proteins, via N- or S-glycosidic linkages, with the release of nicotinamide (Fig. 1). This reaction is easily distinguished from that catalyzed by poly(ADP-ribose)polymerases (PARPs), which are nuclear proteins involved in DNA repair, cell differentiation, and apoptosis and are able to transfer multiple ADP-ribose residues, and even branched polymers of ADP-ribose linked by O-glycosidic linkage, onto target proteins [3]. The mono-ADP-ribosylation reaction is also different from the nonenzymatic binding of ADP-ribose to acceptor proteins that forms a thiazolidine, instead of thioglycoside (typical of the enzymatic reaction), bond [4]. ADP-ribose can be released from βNAD$^+$ by cellular NAD$^+$-glycohydrolases (NADases), of particular interest among which is CD38, an ectoenzyme that catalyzes the formation and hydrolysis of cyclic ADP-ribose, a potent intracellular Ca^{2+}-mobilizing agent [5]. It is interesting that CD38 also catalyzes both auto-ADP-ribosylation on cysteine and ADP-ribosylation of unidentified proteins on the cell surface [6], thus displaying both ADP-ribosyltransferase and NADase activities [7], as seen with other enzymes such as cholera toxin (CT) [8].

ADP-ribosyltransferase activities have been reported in different organisms, including viruses and prokaryotic and eukaryotic cells. The pionering studies by Rappuoli and colleagues [9] showed that ADP-ribosyltransferases fall into two groups: the first is characterized by a conserved histidine in the catalytic domain and includes diphtheria toxin, ART3, and PARPs; the second is characterized by a conserved arginine and includes pertussis toxin (PT), CT, heat-labile enterotoxins, and other ADP-ribosyltransferases.

The mono-ADP-ribosylation of proteins can be reverted by cellular ADP-ribosylhydrolases, which hydrolyze the protein-ADP-ribose linkage, thus releasing ADP-ribose [10,11]. So far three genes have been cloned from rat, mouse, and human, which encodes a soluble, intracellular protein that specifically hydrolyzes the ADP-ribose-arginine bond [12a].

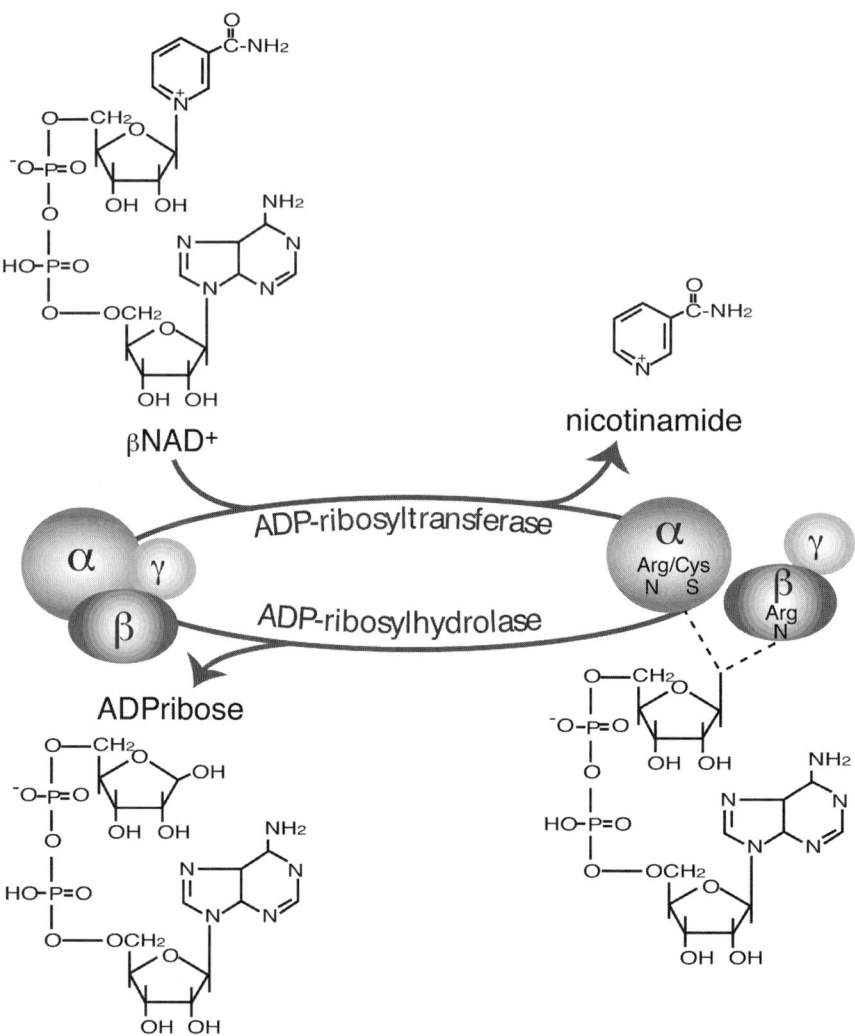

Figure 1 The mono-ADP-ribosylation cycle. The mono-ADP-ribosyltransferase transfers an ADP-ribose residue from βNAD⁺ to an arginine (αs and β) or to a cysteine (αi, αo, and αt) of the target G protein, via *N*- or *S*-glycosidic linkages, respectively, with the release of nicotinamide. Specific ADP-ribosylhydrolases hydrolyze the amino acid-ADP-ribose glycosidic linkage, thus regenerating free arginine or cysteine and releasing ADP-ribose.

The presence of ADP-ribosyltransferase and ADP-ribosylhydrolase activities in the cell is highly suggestive of a potential regulatory mechanism acting on the specific substrates of this reaction.

Bacterial Toxin-Induced ADP-Ribosylation

The mono-ADP-ribosylation of the G protein subunits was originally described as the mechanism of cell intoxication by bacteria, such as *Vibrio cholerae* (producing CT) and *Bordetella pertussis* (producing PT) [13,14]. These toxins, which specifically act on G proteins (see below), rapidly became important tools in the identification of these proteins and in investigations into their functional roles in cells (Table I). Other well-characterized bacterial ADP-ribosyltransferases are the diphtheria and clostridial toxins, which act by modifying crucial host cell proteins, such as the monomeric GTPases of the Rho family, monomeric actin, and elongation factor-2, permanently inactivating the cellular functions modulated by these proteins (reviewed by [15]).

The CT and heat-labile enterotoxins LT-1 and LT-2 from *Escherichia coli* are arginine-specific ADP-ribosyltransferases that are able to modify the α subunit of the stimulatory G protein (G_s) and to irreversibly inhibit its GTPase activity; this results in the activation of adenylyl cyclase and in the increase in intracellular cyclic AMP [13]. CT is an 84 kDa oligomeric protein that consists of the monomeric A subunit, a 29 kDa polypeptide that exhibits the ADP-ribosyltransferase activity, and the homopentamer B subunit, a complex of five 11 kDa polypeptides, which itself binds stoichiometrically with high affinity and specificity to five GM1 molecules on the plasma membrane [16]. To cause disease, both CT and LT coopt molecular machineries of the

Table I ADP-ribosylating/deribosylating Enzymes Acting on Heterotrimeric G Proteins

Enzyme/Source	Subunit structure/ Localization	Target/s	Effects	Ref.
Toxin ADPRT				
Cholera toxin	AB_5	G_s, G_t	Inhibition of G protein GTPase activity	13,18
Escherichia coli LT	AB_5	G_s, G_t	Inhibition of G protein GTPase activity	13
Pertussis toxin	AB_5	G_i, G_o, G_t	Uncoupling of G proteins from receptors	13,14
Cellular ADPRT				
Human erythrocytes	Membranes	G_i	Decrease in epinephrine-mediated AC inhibition	51,52
Human platelets	Membranes	G_i	Decrease in epinephrine-mediated AC inhibition	51,52
Rabbit ventricles	Membranes	G_s	Increase in AC activity	41
Human platelets	–	G_s	Increase in AC activity	42
Rat brain	Homogenate	G_s, G_o	–	43,44
Rabbit luteal	Membranes	G_s	Increase in AC activity	45
Chicken spleen	Membranes	G_s, actin	Increase in AC activity	46
NG108-15	Membranes	G_s	Increase in AC activity	47
CHO cells	Plasma Membranes	G_β	Inhibition of $\beta\gamma$-mediated function	55
Cellular Hydrolase				
Rat and human tissue	Cytosol	G_s	–	12
Human erythrocytes	Cytosol	G_i	–	11
CHO cells	Cytosol	G_β	–	55

This table lists the ADP-ribosylating activities (ADPRT) identified in different systems and related to G protein modification. The ADP-ribosylhydrolases (hydrolase) catalyzing the reverse reaction are also indicated. See the text for other enzymes for which the action on G proteins has not been proven directly, such as the mammalian ART family (including ART 1-5) and the mammalian Sirt family (including Sirtuin 1-7). AC, adenylyl cyclase; – not determined.

host cell; the toxin enters by endocytosis of the toxin-receptor complexes, follows retrograde transport to the Golgi cisternae or to the endoplasmic reticulum (ER), and finally translocates to its site of action on the inner surface of the plasma membrane [17]. The ADP-ribosyltransferase activity of the A subunit requires proteolysis for activation with the release of a smaller, carboxyl-terminus 7 kDa fragment (CTA2) and of a larger 22 kDa catalytically active fragment (CTA1) whose activity is promoted by a family of 20 kDa GTPases, the ADP-ribosylation factors (ARFs) [18].

PT, a cysteine-specific ADP-ribosyltransferase, ADP-ribosylates the α subunit of G_i, G_o, and G_t. Unlike CT, PT-catalyzed mono-ADP-ribosylation occurs when the α subunit is associated with βγ, thus generating a modified G protein unable to couple to activated receptors and to transduce signals [19]. PT is a 119 kDa toxin consisting of a 28 kDa monomeric A component, an S1 subunit expressing the ADP-ribosyltransferase activity, and a B component, a complex of five polypeptides (S2, S3, two S4s, S5) [20]. The function of the B component is to attach the native toxin to the cell-surface receptor, thus initiating toxin internalization by receptor-mediated endocytosis and its trafficking through early to late endosomes and the Golgi apparatus [21]. There is no evidence that the ADP-ribosyltransferase activity of the S1 subunit requires cofactors, such as the ARFs reported for CT.

Endogenous Mono-ADP-Ribosylation

Vertebrate ADP-ribosyltransferase activity was first detected in turkey erythrocytes [22,23], rat liver homogenates [24], and *Xenopus* tissues [25]. Specific enzymes have been cloned from different sources [26–31]. The family of mammalian ADP-ribosyltransferases includes five enzymes referred to as ART1 to 5 [32,33]. ART1, ART2, and most likely ART3 and ART4, are glycosylphosphatidylinositol (GPI)-anchored to the cell surface. ART5 possesses a hydrophobic N-terminal signal sequence and is likely to be a secretory protein but with an as yet unknown function [34]. ART1 and ART2 have roles in immune regulation: ART1 inhibits T-lymphocyte functions, such as their cytolytic activity and proliferation, by ADP-ribosylating arginines of cell surface molecules, such as the T-cell co-receptors [32,35]; ART2 is expressed on resting T cells and on natural killer cells, has not been found outside of the immune system, and exerts a regulatory role in rat models for autoimmune insulin-dependent diabetes mellitus, where its defective expression on T-cells has been associated with increased susceptibility to the disease [32,36]. ART3 is strongly expressed in human testis, and ART4 is preferentially expressed in human lymphatic tissue [31].

The catalytic domains of the mammalian ART enzymes are extracellular, and thus it is unlikely that members of this family are involved in mono-ADP-ribosylation of intracellular targets, like heterotrimeric G proteins. In principle, ARTs could be internalized from the cell surface and transported to the site of action, as with the bacterial toxins. However, the sorting mechanism reported for GPI-anchored proteins, which involves endocytic vesicles, would result in a lumenal localization of the catalytic domain, segregating it from cellular substrates such as the G proteins [37]. Thus, the possibility that a distinct family of intracellular enzymes that have ADP-ribosyltransferase activity should be considered. In support of this possibility, it has been suggested that the yeast silencing protein syr2 possesses ADP-ribosyltransferase activity [38]; moreover, the cloning and characterization of seven human proteins (sirtuins 1 to 7) that are able to metabolize NAD^+ and possess ADP-ribosyltransferase activity has recently been reported [39]. These sir2-like proteins do not share any obvious sequence homology with the well-characterized ART family, and represent good candidates for the intracellular ADP-ribosyltransferases.

Cellular ADP-ribosyltransferase activities able to catalyze the modification of G-protein arginine residues in a way similar to CT have been described in many cells and tissues (Table I) [40–49]. Although they present either the identification of the enzymatic activity or of the G-protein substrates, several of the initial reports do not fully demonstrate the mono-ADP-ribosylation reaction (no discrimination between the enzymatic and non-enzymatic ADP-ribose linkages, or use of α-NAD^+ rather than β-NAD^+, the preferred-substrate of ADP-ribosyltransferases, in the ADP-ribosylation assays); moreover, the identification of the G-protein substrates remains elusive, since these reports were only based either on the determination of the molecular mass or on the comigration with the bacterial toxin substrates by polyacrilamide gel electrophoresis [40–46].

A series of more recent reports has led to a better understanding of the endogenous mono-ADP-ribosylation machinery. It has been shown that in highly purified canine cardiac sarcolemma, isoproterenol produces the selective ADP-ribosylation of a single 45 kDa protein, identified as Gαs by immunoprecipitation. This effect was correlated with the ability of NAD^+ to increase cyclic AMP production [48]. Similarly, a 52 kDa protein, identified as an αs isoform, is ADP-ribosylated in smooth muscle cells from bovine coronary arteries [49]. In this case, the Gαs modification was related to the release of eicosanoids from the endothelium that stimulate the enzymatic reaction and thus activate K^+ channels [50].

Endogenous ADP-ribosylation of cysteine residues of G proteins that are comparable to the PT effect has been reported in human erythrocytes and platelets [51], where it attenuates the inhibition of adenylyl cyclase induced by epinephrine (Table I) [52]. Thus, the picture emerging from these observations is that the pathological modification of G proteins produced by bacterial toxins has an endogenous, possibly physiologically relevant, counterpart that could be looked upon as an additional mechanism of regulation of G-protein-mediated functions. In addition, there are indications in *in vitro* assays that not only the α but also the β subunit of G_t can be a substrate of an ADP-ribosyltransferase purified from the cytosol of turkey erythrocytes [53,54].

Recently, direct evidence of the functional, enzymatic modification of the G protein β subunit has been reported both in isolated plasma membranes and in intact cells [55]. The arginine-specific mono-ADP-ribosyltransferase that catalyzes this reaction is a plasma-membrane-associated, but not GPI-anchored, protein that acts intracellularly and specifically modifies residue 129 of the β subunit. It is interesting that the modified β subunit can be de-ADP-ribosylated by a cytosolic ADP-ribosylhydrolase, thus revealing a cellular ADP-ribosylation/de-ADP-ribosylation cycle that might parallel a functional activation-inactivation cycle of the $\beta\gamma$ dimer. It is worth mentioning that under resting conditions, the β subunit mono-ADP-ribosylation takes place also in intact cells, thus corroborating the potential physiological role of this reaction [55].

The modified β subunit loses its ability to modulate effector enzymes such as calmodulin-stimulated type 1 adenylyl cyclase and phospholipase C, indicating that the modification of arginine 129, a critical residue located in the β common-effector-binding surface [56], can indeed impair β subunit activities [55]. Mono-ADP-ribosylation can thus be considered to be a signal termination mechanism for $\beta\gamma$-mediated functions. In principle, the ADP-ribosylation of the β subunit could also affect the function of the α subunits, since it can lead to a sustained activation of α-subunit-dependent functions by sequestering the $\beta\gamma$ subunit from the signal cascade and preventing the reassociation of the heterotrimer.

Among the substrates of mono-ADP-ribosylating enzymes, there are other cellular proteins involved in signaling and cell organization, such as GRP78/BIP, p33, integrin α7, desmin, actin, and CtBP/BARS [57–64]. It is interesting, and in line with the data summarized above for the modification of the G-protein α and β subunits, that in all these cases ADP-ribosylation has been shown to cause protein inactivation.

Thus, although the regulation of mono-ADP-ribosylation in whole cells remains to be fully clarified, the information obtained to date indicates that this reaction has the characteristics of a fundamental regulatory process that is potentially relevant not only in G-protein-mediated signaling, but also in other cellular functions.

Acknowledgments

The authors would like to thank Dr. C. P. Berrie for editorial assistance and acknowledge the support of the Italian Association for Cancer Research (AIRC, Milano, Italy), Telethon (Italy) grant n. E.841, and the Italian National Research Council (CNR, Rome, Italy) Progetto Finalizzato "Biotecnologie" (ctr. n. 01.00027.PF49).

References

1. Chen, C. A. and Manning, D. R. (2001). Regulation of G proteins by covalent modification. *Oncogene* **20**, 1643–1652.
2. Morris, A. J. and Malbon, C. C. (1999). Physiological regulation of G protein-linked signaling. *Physiol. Rev.* **79**, 1373–1430.

3. Ueda, K. and Hayaishi, O. (1985). ADP-ribosylation. *Annu. Rev. Biochem.* **54**, 73–100.
4. McDonald, L. J., Wainschel, L. A., Oppenheimer, N. J., and Moss, J. (1992). Amino acid-specific ADP-ribosylation: structural characterization and chemical differentiation of ADP-ribose-cysteine adducts formed nonenzymatically and in a pertussis toxin-catalyzed reaction. *Biochemistry* **31**, 11881–11887.
5. Lee, H. C. (1994). Cyclic ADP-ribose: a calcium mobilizing metabolite of NAD^+. *Mol. Cell. Biochem.* **138**, 229–235.
6. Grimaldi, J. C., Balasubramanian, S., Kabra, N. H., Shanafelt, A., Bazan, J. F., Zurawski, G., and Howard, M. C. (1995). CD38-mediated ribosylation of proteins. *J. Immunol.* **155**, 811–817.
7. Berthelier, V., Tixier, J. M., Muller-Steffner, H., Schuber, F., and Deterre, P. (1998). Human CD38 is an authentic $NAD(P)^+$ glycohydrolase. *Biochem. J.* **330**, 1383–1390.
8. Moss, J., Stanley, S. J., and Lin, M. C. (1979). NAD glycohydrolase and ADP-ribosyltransferase activities are intrinsic to the A1 peptide of choleragen. *J. Biol. Chem.* **254**, 11993–11999.
9. Domenighini, M., Magagnoli, C., Pizza, M., and Rappuoli, R. (1994). Common features of the NAD-binding and catalytic site of ADP-ribosylating toxins. *Mol. Microbiol.* **14**, 41–50.
10. Moss, J., Tsai, S. C., Adamik, R., Chen, H. C., and Stanley, S. J. (1988). Purification and characterization of ADP-ribosylarginine hydrolase from turkey erythrocytes. *Biochemistry* **27**, 5819–5823.
11. Tanuma, S. and Endo, H. (1990). Identification in human erythrocytes of mono(ADP-ribosyl) protein hydrolase that cleaves a mono (ADP-ribosyl) Gi linkage. *FEBS Lett.* **261**, 381–384.
12. Takada, T., Iida, K., and Moss, J. (1993). Cloning and site-directed mutagenesis of human ADP-ribosylarginine hydrolase. *J. Biol. Chem.* **268**, 17837–17843.
12a. Glowacki, G., Braren, R., Firner, K., Nissen, M., Kuhl, M., Reche, P., Bazan, F., Cetkovic-Cvrlje, M., Leiter, E., Haag, F., and Koch-Nolte, F. (2002). The family of toxin related ecto-ADP-ribosyltransferases in humans and the mouse. *Protein Sci.* **11**, 1657–1670.
13. Moss, J. and Vaughan, M. (1988). ADP-ribosylation of guanyl nucleotide-binding regulatory proteins by bacterial toxins. *Adv. Enzymol. Relat. Areas Mol. Biol.* **61**, 303–379.
14. Katada, T. and Ui, M. (1982). Direct modification of the membrane adenylate cyclase system by islet-activating protein due to ADP-ribosylation of a membrane protein. *Proc. Natl. Acad. Sci. USA* **79**, 3129–3133.
15. Krueger, K. M. and Barbieri, J. T. (1995). The family of bacterial ADP-ribosylating exotoxins. *Clin. Microbiol. Rev.* **8**, 34–47.
16. Spangler, B. D. (1992). Structure and function of cholera toxin and the related *Escherichia coli* heat-labile enterotoxin. *Microbiol. Rev.* **56**, 622–647.
17. Lencer, W. I., Hirst, T. R., and Holmes, R. K. (1999). Membrane traffic and the cellular uptake of cholera toxin. *Biochim. Biophys. Acta* **1450**, 177–190.
18. Tsai, S. C., Noda, M., Adamik, R., Chang, P. P., Chen, H. C., Moss, J., and Vaughan, M. (1988). Stimulation of choleragen enzymatic activities by GTP and two soluble proteins purified from bovine brain. *J. Biol. Chem.* **263**, 1768–1772.
19. Hsia, J. A., Moss, J., Hewlett, E. L., and Vaughan, M. (1984). ADP-ribosylation of adenylate cyclase by pertussis toxin. Effects on inhibitory agonist binding. *J. Biol. Chem.* **259**, 1086–1090.
20. Tamura, M., Nogimori, K., Murai, S., Yajima, M., Ito, K., Katada, T., Ui, M., and Ishii, S. (1982). Subunit structure of islet-activating protein, pertussis toxin, in conformity with the A-B model. *Biochemistry* **21**, 5516–5522.
21. Xu, Y. and Barbieri, J. T. (1996). Pertussis toxin-catalyzed ADP-ribosylation of Gi-2 and Gi-3 in CHO cells is modulated by inhibitors of intracellular trafficking. *Infect. Immun.* **64**, 593–599.
22. Moss, J., Stanley, S. J., and Watkins, P. A. (1980). Isolation and properties of an NAD- and guanidine-dependent ADP-ribosyltransferase from turkey erythrocytes. *J. Biol. Chem.* **255**, 5838–5840.
23. West, R. E. Jr. and Moss, J. (1986). Amino acid specific ADP-ribosylation: specific NAD: arginine mono-ADP-ribosyltransferases associated with turkey erythrocyte nuclei and plasma membranes. *Biochemistry* **25**, 8057–8062.
24. Moss, J. and Stanley, S. J. (1981). Amino acid-specific ADP-ribosylation. Identification of an arginine-dependent ADP-ribosyltransferase in rat liver. *J. Biol. Chem.* **256**, 7830–7833.
25. Godeau, F. and Koide, S. S. (1983). *Xenopus* mono(adenosine diphosphate ribosyl) transferase: purification, assay, and properties. *Princess Takamatsu Symp.* **13**, 111–118.
26. Zolkiewska, A., Nightingale, M. S., and Moss, J. (1992). Molecular characterization of NAD:arginine ADP-ribosyltransferase from rabbit skeletal muscle. *Proc. Natl. Acad. Sci. USA* **89**, 11352–11356.
27. Davis, T. and Shall, S. (1995). Cloning of a chicken gene homologous to the rabbit mono-ADP- ribosyltransferase gene. *Biochem. Soc. Trans.* **23**, 207S.
28. Haag, F. A., Kuhlenbaumer, G., Koch-Nolte, F., Wingender, E., and Thiele, H. G. (1996). Structure of the gene encoding the rat T cell ecto-ADP-ribosyltransferase RT6. *J. Immunol.* **157**, 2022–2030.
29. Okazaki, I. J., Kim, H. J., and Moss, J. (1996). Cloning and characterization of a novel membrane-associated lymphocyte NAD:arginine ADP-ribosyltransferase. *J. Biol. Chem.* **271**, 22052–22057.
30. Shimoyama, M., Tsuchiya, M., Hara, N., Yamada, K., and Osago, H. (1997). Molecular cloning and characterization of arginine-specific ADP-ribosyltransferases from chicken bone marrow cells. *Adv. Exp. Med. Biol.* **419**, 137–144.
31. Koch-Nolte, F., Haag, F., Braren, R., Kuhl, M., Hoovers, J., Balasubramanian, S., Bazan, F., and Thiele, H. G. (1997). Two novel human members of an emerging mammalian gene family related to mono-ADP-ribosylating bacterial toxins. *Genomics* **39**, 370–376.
32. Haag, F. and Koch-Nolte, F. (1998). Endogenous relatives of ADP-ribosylating bacterial toxins in mice and men: potential regulators of immune cell function. *J. Biol. Regul. Homeost. Agents* **12**, 53–62.
33. Moss, J., Balducci, E., Cavanaugh, E., Kim, H. J., Konczalik, P., Lesma, E. A., Okazaki, I. J., Park, M., Shoemaker, M., Stevens, L. A., and Zolkiewska, A. (1999). Characterization of NAD:arginine ADP-ribosyltransferases. *Mol. Cell. Biochem.* **193**, 109–113.
34. Glowacki, G., Braren, R., Cetkovic-Cvrlje, M., Leiter, E. H., Haag, F., and Koch-Nolte, F. (2001). Structure, chromosomal localization, and expression of the gene for mouse ecto-mono(ADP-ribosyl)transferase ART5. *Gene* **275**, 267–277.
35. Wang, J., Nemoto, E., and Dennert, G. (1996). Regulation of CTL by ecto-nictinamide adenine dinucleotide (NAD) involves ADP-ribosylation of a p56lck-associated protein. *J. Immunol.* **156**, 2819–2827.
36. Whalen, B. J., Greiner, D. L., Mordes, J. P., and Rossini, A. A. (1994). Adoptive transfer of autoimmune diabetes mellitus to athymic rats: synergy of $CD4^+$ and $CD8^+$ T cells and prevention by $RT6^+$ T cells. *J. of Immunol.* **7**, 819–831.
37. Muniz, M. and Riezman, H. (2000). Intracellular transport of GPI-anchored proteins. *EMBO J.* **19**, 10–15.
38. Tanny, J. C., Dowd, G. J., Huang, J., Hilz, H., and Moazed, D. (1999). An enzymatic activity in the yeast Sir2 protein that is essential for gene silencing. *Cell* **99**, 735–745.
39. Frye, R. A. (2000). Phylogenetic classification of prokaryotic and eukaryotic Sir2-like proteins. *Biochem. Biophys. Res. Commun.* **273**, 793–798.
40. Reilly, T. M., Beckner, S., McHugh, E. M., and Blecher, M. (1981). Isoproterenol-induced ADP-ribosylation of a single plasma membrane protein of cultured differentiated RL-PR-C hepatocytes. *Biochem. Biophys. Res. Commun.* **98**, 1115–1120.
41. Feldman, A. M., Levine, M. A., Baughman, K. L., and Van Dop, C. (1987). NAD^+-mediated stimulation of adenylate cyclase in cardiac membranes. *Biochem. Biophys. Res. Commun.* **142**, 631–637.
42. Molina y Vedia, L., Nolan, R. D., and Lapetina, E. G. (1989). The effect of iloprost on the ADP-ribosylation of Gs alpha (the alpha-subunit of Gs). *Biochem. J.* **261**, 841–845.
43. Duman, R. S., Terwilliger, R. Z., and Nestler, E. J. (1991). Endogenous ADP-ribosylation in brain: initial characterization of substrate proteins. *J. Neurochem.* **57**, 2124–2132.

44. Matsuyama, S. and Tsuyama, S. (1991). Mono-ADP-ribosylation in brain: purification and characterization of ADP-ribosyltransferases affecting actin from rat brain. *J. Neurochem.* **57**, 1380–1387.
45. Abramowitz, J. and Jena, B. P. (1991). Evidence for a rabbit luteal ADP-ribosyltransferase activity which appears to be capable of activating adenylyl cyclase. *Int. J. Biochem.* **23**, 549–559.
46. Obara, S., Yamada, K., Yoshimura, Y., and Shimoyama, M. (1991). Evidence for the endogenous GTP-dependent ADP-ribosylation of the alpha-subunit of the stimulatory guanyl-nucleotide-binding protein concomitant with an increase in basal adenylyl cyclase activity in chicken spleen cell membrane. *Eur. J. Biochem.* **200**, 75–80.
47. Donnelly, L. E., Boyd, R. S., and MacDermot, J. (1992). Gs alpha is a substrate for mono(ADP-ribosyl)transferase of NG108-15 cells. ADP-ribosylation regulates Gs alpha activity and abundance. *Biochem. J.* **288**, 331–336.
48. Quist, E. E., Coyle, D. L., Vasan, R., Satumtira, N., Jacobson, E. L., and Jacobson, M. K. (1994). Modification of cardiac membrane adenylate cyclase activity and Gs alpha by NAD and endogenous ADP-ribosyltransferase. *J. Mol. Cell. Cardiol.* **26**, 251–260.
49. Li, P. L., Chen, C. L., Bortell, R., and Campbell, W. B. (1999). 11,12-Epoxyeicosatrienoic acid stimulates endogenous mono-ADP-ribosylation in bovine coronary arterial smooth muscle. *Circ. Res.* **85**, 349–356.
50. Li, P. L. and Campbell, W. B. (1997). Epoxyeicosatrienoic acids activate K^+ channels in coronary smooth muscle through a guanine nucleotide binding protein. *Circ. Res.* **80**, 877–884.
51. Tanuma, S., Kawashima, K., and Endo, H. (1988). Eukaryotic mono (ADP-ribosyl)transferase that ADP-ribosylates GTP-binding regulatory Gi protein. *J. Biol. Chem.* **263**, 5485–5489.
52. Tanuma, S. and Endo, H. (1989). Mono(ADP-ribosyl)ation of Gi by eukaryotic cysteine-specific mono(ADP- ribosyl) transferase attenuates inhibition of adenylate cyclase by epinephrine. *Biochim. Biophys. Acta* **1010**, 246–249.
53. Watkins, P. A., Kanaho, Y., and Moss, J. (1987). Inhibition of the GTPase activity of transducin by an NAD^+:arginine ADP-ribosyltransferase from turkey erythrocytes. *Biochem. J.* **248**, 749–754.
54. Ehret-Hilberer, S., Nullans, G., Aunis, D., and Virmaux, N. (1992). Mono ADP-ribosylation of transducin catalyzed by rod outer segment extract. *FEBS Lett.* **309**, 394–398.
55. Lupi, R., Corda, D., and Di Girolamo, M. (2000). Endogenous ADP-ribosylation of the G protein beta subunit prevents the inhibition of type 1 adenylyl cyclase. *J. Biol. Chem.* **275**, 9418–9424.
56. Chen, Y., Weng, G., Li, J., Harry, A., Pieroni, J., Dingus, J., Hildebrandt, J. D., Guarnieri, F., Weinstein, H., and Iyengar, R. (1997). A surface on the G protein β-subunit involved in interactions with adenylyl cyclases. *Proc. Natl. Acad. Sci. USA* **94**, 2711–2714.
57. Leno, G. H. and Ledford, B. E. (1989). ADP-ribosylation of the 78-kDa glucose-regulated protein during nutritional stress. *Eur. J. Biochem.* **186**, 205–211.
58. Mishima, K., Terashima, M., Obara, S., Yamada, K., Imai, K., and Shimoyama, M. (1991). Arginine-specific ADP-ribosyltransferase and its acceptor protein p33 in chicken polymorphonuclear cells: co-localization in the cell granules, partial characterization, and in situ mono(ADP-ribosyl)ation. *J. Biochem. (Tokyo)* **110**, 388–394.
59. Zolkiewska, A. and Moss, J. (1993). Integrin alpha 7 as substrate for a glycosylphosphatidylinositol-anchored ADP-ribosyltransferase on the surface of skeletal muscle cells. *J. Biol. Chem.* **268**, 25273–25276.
60. Huang, H. Y., Graves, D. J., Robson, R. M., and Huiatt, T. W. (1993). ADP-ribosylation of the intermediate filament protein desmin and inhibition of desmin assembly in vitro by muscle ADP-ribosyltransferase. *Biochem. Biophys. Res. Commun.* **197**, 570–577.
61. Fujita, H., Okamoto, H., and Tsuyama, S. (1995). ADP-ribosylation in adrenal glands: purification and characterization of mono-ADP-ribosyltransferases and ADP-ribosylhydrolase affecting cytoskeletal actin. *Int. J. Biochem. Cell Biol.* **27**, 1065–1078.
62. Di Girolamo, M., Silletta, M. G., De Matteis, M. A., Braca, A., Colanzi, A., Pawlak, D., Rasenick, M. M., Luini, A., and Corda, D. (1995). Evidence that the 50-kDa substrate of brefeldin A-dependent ADP-ribosylation binds GTP and is modulated by the G-protein beta gamma subunit complex. *Proc. Natl. Acad. Sci. USA* **92**, 7065–7069.
63. Weigert, R., Silletta, M. G., Spano, S., Turacchio, G., Cericola, C., Colanzi, A., Senatore, S., Mancini, R., Polishchuk, E. V., Salmona, M., Facchiano, F., Burger, K. N., Mironov, A., Luini, A., and Corda, D. (1999). CtBP/BARS induces fission of Golgi membranes by acylating lysophosphatidic acid. *Nature* **402**, 429–433.
64. Corda, D. and Di Girolamo, M. (2003). Functional aspects of the mono-ADP-ribosylation of cellular proteins. *EMBO J.* **22**, 1953–1958.

# CHAPTER 223

Using Receptor-G-Protein Chimeras to Screen for Drugs

Graeme Milligan, Richard J. Ward, Gui-Jie Feng,
Juan J. Carrillo, and Alison J. McLean

*Molecular Pharmacology Group, Division of Biochemistry and Molecular Biology,
University of Glasgow, Glasgow Scotland, United Kingdom*

Receptor-G-Protein Chimeras: An Introduction

Direct measures of the levels of expression of the α subunits of heterotrimeric G proteins indicate that they are generally in considerable excess over the levels of any particular G-protein-coupled receptor (GPCR) that might activate them. Despite this, an emerging strategy that has been used to address both basic questions on the details of GPCR-G protein interactions and to allow screens to be designed for agonist ligands at GPCRs has been to generate chimeric GPCR-G protein constructs that define a 1:1 stoichiometry of the two partner polypeptides. Such constructs provide a convenient means to assess the effects of mutations and polymorphisms in GPCRs on signal transduction effectiveness without altering the ratio of expression of GPCR to G protein. They have also allowed effective means of developing direct assays of GPCR-mediated guanine nucleotide exchange on G proteins of the G_s and G_q families that historically has been difficult to monitor. Such assays have allowed direct analysis of the extent of constitutive activity of different GPCRs and the detection of inverse agonists.

Construction of a GPCR-G protein chimera involves linking the 5′ end of a G protein α subunit cDNA to the 3′ end of a GPCR cDNA or DNA from which the stop codon has been eliminated. This results in the production of a single open reading frame containing the sequences of both partner proteins. Despite the artificial nature of such constructs all available data indicate that the expressed polypeptides are properly folded and functional. The α subunit in such chimeras is clearly able to interact with G protein β/γ complexes and the basic pharmacology of the GPCR is preserved. A number of recent reviews on the production and analysis of function of such chimeras are available [1–4] and should be consulted to expand on the information provided herein.

Defining the Signal

In the vast majority of cases analysis of GPCR-G protein chimeras takes place following expression in cell systems that also express endogenously the G protein of interest. A range of techniques has been developed to ensure that observed signals represent activation of the G protein within the chimera. For members of the G_i-family of pertussis toxin-sensitive G proteins, the cysteine residue that is the target for toxin catalyzed ADP-ribosylation is within a key receptor contact site but is not inherently required for function [5]. Amino acids with high hydrophobicity can substitute at least as effectively [5,6]. Thus, replacement of this cysteine in the GPCR-fused G protein ensures that the chimera is resistant to the actions of pertussis toxin. Prior treatment of cells with this toxin to modify the endogenously expressed G_i population thus prevents any possibility of agonist-induced signal deriving from interaction with these G proteins [7,8]. This idea has been adapted effectively to construct a ligand screen. The members of the edg family of GPCRs respond to lipid-derived products such as lysophosphatidic acid and sphingosine-1-phosphate [9]. However, at least eight GPCRs of this family are known and they display wide-ranging and complex patterns of expression. Virtually all cell lines respond to these ligands due to endogenously expressed edg family members, and such effects are generally transduced

Handbook of Cell Signaling, Volume 2 619 Copyright © 2003, Elsevier Science (USA).
All rights reserved.

via G_i family G proteins. A screening assay for ligands at the human edg2 receptor was thus established by expressing a GPCR-G protein chimera in which a pertussis toxin-resistant mutant of $G_{i1}\alpha$ was linked to this receptor [10]. Following pertussis toxin treatment, guanine nucleotide exchange assays provided a specific and direct assay for ligand activation of edg2 [10].

Because it has been well established that GPCRs within certain chimeras can also activate endogenously expressed G proteins [8], the functionality of GPCR-G protein chimeras has also been tested in cells that lack expression of certain G proteins. Elevation of intracellular $[Ca^{2+}]$ in cells derived from a mouse embryo in which the genes encoding both $G_q\alpha$ and $G_{11}\alpha$ were inactivated requires co-expression of both a GPCR and an appropriate G protein [11]. A chimeric α_{1b}-adrenoceptor-$G_{11}\alpha$ fusion protein is able to elevate $[Ca^{2+}]$ in response to agonist and thus must be functional [11]. This signal is in fact a monitor of release of β/γ complex from the chimeric protein and provides further confirmation that such GPCR-G protein chimeras function in the expected manner [11]. Introduction of mutations that block β/γ binding to the G protein α subunit or prevent α and β/γ dissociation also prevent agonist-induced elevation of $[Ca^{2+}]$ [12].

Guanine Nucleotide Exchange Assays

The initial steps in GPCR activation of a G protein are acceleration of GDP release from the G protein α subunit and its replacement with GTP. Deactivation requires hydrolysis of the terminal phosphate of the GTP. These processes provide conceptually easy assay endpoints. However, the low rates of basal guanine nucleotide exchange by G_s and G_q family G proteins has made this difficult to measure in membranes of mammalian cells because of the relatively high rates of exchange occurring on G_i family G proteins and on other non-signal transducing but GTP exchange-dependent polypeptides. Thus, in terms of assay screens for ligands, the signal-to-noise is generally too low to be acceptable. However, immunoprecipitation of such GPCR-G protein chimeras at the termination of a [^{35}S]GTPγS binding assay can produce data where signal-to-noise is between 20–40 to 1 [11–13]. This reflects that in the absence of agonist many GPCRs have a very low capacity to activate the G protein. Thus, when the chimeric protein is immunoprecipitated very little [^{35}S]GTPγS is associated with it. This is increased greatly by agonist occupation of the GPCR. Again, introduction of mutations that prevent guanine nucleotide exchange on the G protein can be used to confirm that the signal reflects activation of the G protein of the chimera [12,13]. Adaptation of this into a homogenous assay involving, for example, secondary antibody coated scintillation proximity assay beads offers an excellent opportunity to use this approach in ligand screening programs.

Although far less popular than [^{35}S]GTPγS binding assays, analysis of agonist-stimulated GTPase activity provides a highly attractive assay point for activation of GPCR-G protein chimeras. This is particularly so in two regards. First, the defined 1:1 stoichiometry of the two components means that combinations of analysis of the effects of agonist at Vmax and ligand binding studies means that direct measurement of GTP turnover and the catalytic efficiency of different ligands to activate the G protein can be obtained [7]. Second, the mechanisms of action of G protein interacting proteins can be deduced from their effects on enzyme kinetic parameters of GTP hydrolysis. These have been employed to examine the relative efficacy of different agonists and to demonstrate how regulators of G protein signaling (RGS) proteins function [14,15].

Constitutive Activity and Inverse Agonism

Many mutations have been described that introduce or enhance agonist-independent (also called constitutive) activity to GPCRs [16,17]. Such mutations may provide insights into the mechanisms of action of agonists and the conformational states of activated GPCRs. Such mutations have been introduced into GPCR-G protein chimeras. Using the [^{35}S]GTPγS binding assay in parallel with immunoprecipitation, direct measures of the extent of constitutive activity introduced by different mutations have been obtained for each of the α_{1b}-adrenoceptor when activating $G_{11}\alpha$ [13] and for the β_1-adrenoceptor to activate $G_s\alpha$ [18]. Equally, the very low levels of [^{35}S]GTPγS present in immunoprecipitates of the chimeras containing the wild-type forms of these GPCRs has allowed easy demonstration that a number of wild-type GPCRs display high levels of constitutive activity. For example when the melanocortin MC_4 receptor is fused to $G_s\alpha$ and a [^{35}S]GTPγS binding assay followed by immunoprecipitation, a large number of counts are present even in the absence of ligand. Addition of α MSH enhances this further but to a much smaller extent than produced by isoprenaline at an equivalent β_1-adrenoceptor-$G_s\alpha$ fusion. This is not inherently surprising because the agouti-related peptide acts as an endogenously expressed inverse agonist at the MC_4 receptor [19] and this ligand does indeed greatly reduce the levels of [^{35}S]GTPγS binding to the melanocortin MC_4 receptor-$G_s\alpha$ fusion protein. This approach is thus likely to be valuable in assessment of the extent of inherent constitutive activity in many GPCRs and could clearly be applied effectively to GPCRs encoded by viral genomes, at least some of which are thought to display high levels of ligand-independent activity [20]. It also offers excellent potential for development for screens for ligands that have inverse agonist activity. A limitation in the development of screens for ligands at orphan GPCRs has been to decide the G-protein coupling specificity of the orphan [21], but combinations of the use of so-called "universal" [22] and "chimeric" [23,24] G proteins may overcome this.

A second useful approach to improving the sensitivity of detection of ligands at GPCR-G protein chimeras takes advantage of constitutive activity but now monitors GTPase activity.

RGS proteins act as GTPase activating proteins for heterotrimeric G proteins [25]. Addition of a recombinant RGS to a GTPase assay in membranes expressing a chimera of the 5HT-1A receptor and $G_{o1}\alpha$ greatly enhances the responses to agonist ligands [26]. More usefully, however, the RGS protein also acts as a GAP for GTP loaded by the constitutive activity of the GPCR. As this increases the basal activity markedly, it greatly increases the dynamic range available for detection of decreased spontaneous loading of GTP produced by addition of an inverse agonist [26]. Adaptations of this whereby a GPCR that couples traditionally to $G_s\alpha$ is incorporated into a fusion protein containing a chimeric G protein that has the receptor recognition elements of $G_s\alpha$ and the guanine nucleotide exchange and RGS interaction sites of a $G_i\alpha$ [27] will allow this enhanced sensitivity to be used in screens for inverse agonists at $G_s\alpha$-coupled receptors.

Conclusions

Overall, GPCR-G protein chimeras display all the key properties of the two isolated polypeptides. However, due to the proximity produced by their fusion, they display enhanced effectiveness of interaction. Their defined stoichiometry also allows both detailed examination of basic research questions and the development of novel screens for GPCR ligands.

References

1. Seifert, R., Wenzel-Seifert, K., and Kobilka, B. K. (1999). GPCR-Gα fusion proteins: molecular analysis of receptor-G-protein coupling. *Trends Pharmacol. Sci.* **20**, 383–389.
2. Milligan, G. (2000). Insights into ligand pharmacology using receptor-G protein fusion proteins. *Trends Pharmacol. Sci.* **21**, 24–28.
3. Wurch, T. and Pauwels, P. J. (2001). Analytical pharmacology of G protein-coupled receptors by stoichiometric expression of the receptor and G(alpha) protein subunits. *J. Pharmacol. Toxicol. Methods* **45**, 3–16.
4. Milligan, G. (2002). Construction and analysis of function of G protein-coupled receptor-G protein fusion proteins. *Methods Enzymol.* **343**, 260–273.
5. Bahia, D. S., Wise, A., Fanelli, F., Lee, M., Rees, S., and Milligan, G. (1998). Hydrophobicity of residue^{351} of the G-protein $G_{i1}\alpha$ determines the extent of activation by the α2A-adrenoceptor. *Biochemistry* **37**, 11555–11562.
6. Waldhoer, M., Wise, A., Milligan, G., Freissmuth, M., and Nanoff, C. (1999). Kinetics of ternary complex formation with fusion proteins composed of the A1-adenosine receptor and G protein α-subunits. *J. Biol. Chem.* **274**, 30571–30579.
7. Wise, A., Carr, I. C., and Milligan, G. (1997). Measurement of agonist-induced guanine nucleotide turnover by the G protein $G_{i1}\alpha$ when constrained within an α_{2A}-adrenoceptor-$G_{i1}\alpha$ fusion protein. *Biochem. J.* **325**, 17–21.
8. Burt, A. R., Sautel, M., Wilson, M. A., Rees, S., Wise, A., and Milligan, G. (1998). Agonist-occupation of an α_{2A}-adrenoceptor-$G_{i1}\alpha$ fusion protein results in activation of both receptor-linked and endogenous G proteins: comparisons of their contributions to GTPase activity and signal transduction and analysis of receptor-G protein activation stoichiometry. *J. Biol. Chem.* **273**, 10367–10375.
9. Fukushima, N., Ishii, I., Contos, J. J. A., Weiner, J. A., and Chun, J. (2001). Lysophospholipid receptors. *Annu. Rev. Pharmacol. Toxicol.* **41**, 507–534.
10. McAllister, G., Stanton, J. A., Salim, K., Handford, E. J., and Beer, M. S. (2000). Edg2 receptor functionality: Giα1 coexpression and fusion protein studies. *Mol. Pharmacol.* **58**, 407–412.
11. Stevens, P. A., Pediani, J., Carrillo, J. J., and Milligan, G. (2001). Co-ordinated agonist-regulation of receptor and G protein palmitoylation and functional rescue of palmitoylation-deficient mutants of the G protein $G_{11}\alpha$ following fusion to the α_{1b}-adrenoceptor. Palmitoylation of $G_{11}\alpha$ is not required for interaction with β/γ complex. *J. Biol. Chem.* **276**, 35883–35890.
12. Liu, S., Carrillo, J. J., Pediani, J., and Milligan, G. (2002). Effective information transfer from the α_{1b}-adrenoceptor to $G_{11}\alpha$ requires both β/γ interactions and an aromatic group 4 amino acid from the C-terminus of the G protein. *J. Biol. Chem.* **277**, 25707–25714.
13. Carrillo, J. J., Stevens, P. A., and Milligan, G. (2002). Measurement of agonist-dependent and -independent signal initiation of α_{1b}-adrenoceptor mutants by direct analysis of guanine nucleotide exchange on the G protein Gα$_{11}$. *J. Pharm. Exper. Ther.* **302**, 1080–1088.
14. Cavalli, A., Druey, K. M., and Milligan, G. (2000). The regulator of G protein signaling RGS4 selectively enhances α_{2A}-adrenoreceptor stimulation of the GTPase activity of $G_{o1}\alpha$ and $G_{i2}\alpha$. *J. Biol. Chem.* **275**, 23693–23699.
15. Hoffmann, M., Ward, R. J., Cavalli, A., Carr, I. C., and Milligan, G. (2001). Differential capacities of the RGS1, RGS16 and RGS-GAIP regulators of G-protein signaling to enhance α_{2A}-adrenoreceptor agonist-stimulated GTPase activity of $G_{o1}\alpha$. *J. Neurochem.* **78**, 797–806.
16. Pauwels, P. J. and Wurch, T. (1998). Review; amino acids domains involved in constitutive activation of G protein coupled receptors. *Mol. Neurobiol.* **17**, 109–135.
17. Scheer, A. and Cotecchia, S. (1997). Constitutively active G protein-coupled receptors: potential mechanisms of receptor activation. *J. Recept. Signal Transduct. Res.* **17**, 57–73.
18. McLean, A. J., Zeng, F. Y., Behan, D., Chalmers, D., and Milligan, G. (2002). Generation and analysis of constitutively active and physically destabilized mutants of the human α_1-adrenoceptor. *Mol. Pharmacol.* **62**, 747–755.
19. Nijenhuis, W. A. J., Oosterom, J., and Adan, R. A. H. (2001). AgRP (83-132) acts as an inverse agonist on the human melanocortin-4 receptor. *Mol. Endocrinol.* **15**, 164–171.
20. Smit, M. J., Timmerman, H., Verzijl, D., and Leurs, R. (2000). Viral-encoded G-protein coupled receptors: new targets for drug research? *Pharm. Acta Helv.* **74**, 299–304.
21. Moller, S., Vilo, J., and Croning, M. D. (2001). Prediction of the coupling specificity of G protein coupled receptors to their G proteins. *Bioinformatics* **17** Suppl. 1, S174–181.
22. Kostenis, E. (2001). Is Galpha16 the optimal tool for fishing ligands of orphan G-protein-coupled receptors? *Trends Pharmacol. Sci.* **22**, 560–564.
23. Milligan, G. and Rees, S. (1999). Chimaeric G alpha proteins: their potential use in drug discovery. *Trends Pharmacol. Sci.* **20**, 118–124.
24. Cabrera-Vera, T. M., Thomas, T. O., Vanhauwe, J., Depree, K. M., Graber, S. G., and Hamm, H. E. (2002). Dissecting receptor-G protein specificity using G alpha chimeras. *Methods Enzymol.* **344**, 69–81.
25. Kozasa, T. (2001). Regulation of G protein-mediated signal transduction by RGS proteins. *Life Sci.* **68**, 2309–2317.
26. Welsby, P. J., Kellett, E., Wilkinson, G., and Milligan, G. (2002). Enhanced detection of receptor constitutive activity in the presence of regulators of G protein signalling: applications to the detection and analysis of inverse agonists and low efficacy partial agonists. *Mol. Pharmacol.* **61**, 1211–1221.
27. Feng, G. J., Cavalli, A., and Milligan, G. (2002). Engineering a V(2) vasopressin receptor agonist- and regulator of G-protein-signaling-sensitive G protein. *Anal. Biochem.* **300**, 212–220.

CHAPTER 224

Specificity of G Protein βγ Dimer Signaling

Janet D. Robishaw,[1] William F. Schwindinger,[1] and Carl A. Hansen[2]

[1]Weis Center for Research, Geisinger Clinic,
Danville, Pennsylvania and
[2]Department of Biological and Allied Health Sciences,
Bloomsburg University, Bloomsburg,
Pennsylvania

Introduction

The G protein βγ dimer interacts with multiple partners to perform numerous functions in the signal transduction process (for reviews see [1–4]). First, the βγ dimer interacts with the α subunit to form the G-protein heterotrimer. In this capacity, the βγ dimer functions to target the α subunit to the membrane and acts as a guanine nucleotide dissociation inhibitor for the α subunit. Second, the βγ dimer interacts with the receptor. In this capacity, the βγ dimer may be required for the receptor to activate the G protein; and the βγ dimer composition may contribute to recognition of the receptor. Third, the βγ dimer interacts with the effector, such as adenylyl cyclases (AC), phospholipases (PLC), ion channels, and kinases. In this capacity, the βγ dimer regulates the activity of the effector until re-association with the α subunit terminates the signal. Finally, the βγ dimer interacts with accessory proteins, including the cytoskeleton, phosducin, regulators of G protein signaling (RGS) proteins, and receptor kinases. In this capacity, the βγ dimer influences the magnitude and duration of the signal transduction process.

In the foregoing, the functions of "the βγ dimer" are discussed. In fact, there is the potential to generate a large number of different βγ dimers given the known existence of five β and twelve γ subtypes. However, their functional significance remains enigmatic. *In vitro* approaches have shown little specificity among the various βγ dimers, whereas *in vivo* approaches indicate greater specificity. Thus, a major question revolves around to what extent the G-protein βγ dimers are functionally interchangeable in the cell. The answer to this question has major implications for how these components are assembled into signaling pathways and how the fidelity of these signaling pathways are maintained. As discussed below, there is a growing recognition that the G-protein βγ dimers have specific functions *in vivo*.

Diversity of the β and γ Gene Families

The β-subunit family is highly conserved. At the genomic level, five genes encoding β subunits (GNB) have been identified [5]. Since their specific functions are not known, they have been named in order of cloning. The GNB1-GNB4 genes share a common intron-exon structure within the coding region, whereas the GNB5 gene does not [5]. The structure of the human GNB3 gene has been described in detail [6]. At the protein level, the β subunits display substantial sequence homology that is conserved across species.

Notably, the β_1 through β_4 subtypes share 78 to 88 percent amino acid identity, whereas the β_5 subtype shows only 51 to 53 percent amino acid identity with the other β subunits.

By contrast, the γ subunit family is more divergent. At the genomic level, 12 genes encoding γ subunits (GNG) have been identified [5,7]. Again, since their functions are not known, they have been named in order of cloning. As shown in Fig. 1, all GNG genes have two coding exons and share a common splice site in the coding region, with the exception of the GNG13 gene [8]. In addition, most GNG genes have one or more 5' noncoding exons, some of which are located more than 100 kb upstream of their coding regions.

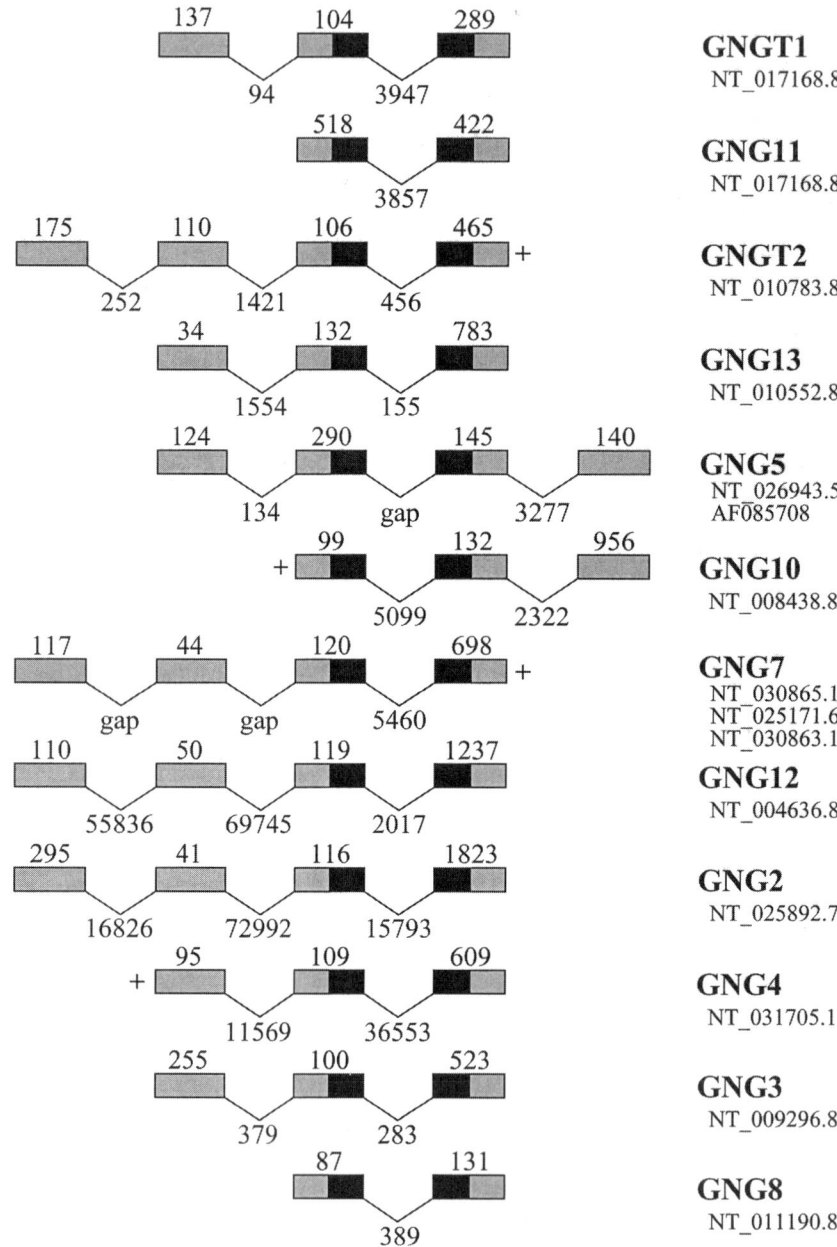

Figure 1 Structures of the genes encoding γ subunits, as predicted by comparing genomic sequences with mRNA and EST sequences via the data mining tools Evidence Viewer and Model Maker at the NCBI web site. Exons are diagrammed as shaded boxes, with size in bp indicated above each box. Dark shading indicates coding regions. Introns are diagrammed as V-shaped lines, with size in bp shown below each. The plus signs indicate the existence of ESTs with sequence beyond what is diagrammed, suggesting a possible additional exon or an extension of the indicated exon. At the right, the accession number for the working draft sequence containing each gene is shown below the gene name. Gaps were still present in the working draft sequences for the GNG5 gene and GNG7 genes. The coding portion of exon 2 of GNG5 was not identified in NT_026943.5 and was modeled based on AF085708. Two 5' noncoding exons of GNG7 were not identified in NT_030865.1, but were present in adjacent contigs.

Though the details remain to be worked out, these additional exons may allow for multiple promoters, coordinate regulation with other gene products, or alternate splicing of elements involved in regulation of translation. Finally, the GNG5 and GNG10 genes have one 3′ noncoding exon, which may be involved in mRNA stability or subcellular localization. This diversity in GNG gene structure suggests variability in gene expression that may underlie specificity in function. At the protein level, the γ subunits display a sequence diversity that is conserved across species, suggesting the individual subtypes have specific and distinct functions. Based on amino acid diversity, they can be arranged into five classes [2], with each group showing less than 50 percent homology to the other classes: (1) γ_7 and γ_{12} (76 percent identity); (2) γ_1, γ_{11}, and γ_{14} (62 to 73 percent identity); (3) γ_2, γ_3, γ_4, and γ_8 (56 to 75 percent identity); (4) γ_5 and γ_{10} (53 percent identity); and (5) γ_{13}. Analogous to the G-protein α subtypes, this pattern of structural diversity suggests that the γ subtypes may fall into functional classes that have yet to be characterized.

The genomic analysis can be extended to identify genes that are adjacent to GNG genes. Except for GNGT1 and GNG11, the genes encoding the γ subunits are widely dispersed throughout the genome [5,7]. The GNG2 gene is arranged head-to-head with an SH3 domain protein, NESH. The GNG3 gene is located head-to-head with the BSCL2 gene (murine Gng3-linked gene), which is responsible for Berardinelli-Seip congenital lipodystrophy [9]. Because of these arrangements, the promoter elements of these genes may overlap, suggesting the possibility of coordinate regulation. The bovine γ_5 subtype and chitobiase protein are alternatively spliced products of the same gene [10]; the human GNG5 and CTBS genes appear to be similarly linked.

Assembly of the βγ Dimer

Typically, the βγ subunits exist as a tightly associated complex that functions as a single unit. The only known exception is the β_5 subtype that interacts with certain RGS proteins [11]. Hence, an important step in deciphering which βγ subunit combinations can exist in intact cells is to understand the mechanics and determinants specifying their assembly. In this regard, there is some evidence suggesting that assembly of the βγ dimer occurs in a specific order: beginning with synthesis of β and γ in the cytosol, progressing to formation of the βγ dimer, to prenylation of the γ, to removal of the three terminal amino acids and carboxyl-methylation of the γ, and finally to translocation of the fully processed βγ dimer to the plasma membrane, perhaps in association with the α subunit [12,13]. Moreover, there is increasing evidence suggesting that spatial and structural constraints can contribute to selective assembly of specific βγ dimers.

Cellular Level

Just how β and γ subunits are brought together to form specific βγ dimers remains unresolved. At one level, the β and γ subtypes are expressed in cell-specific patterns [8,14–17], raising the possibility that each cell contains only a subset of all possible βγ dimers to participate in the signal transduction process at a given time. The mechanisms governing their expression have not been defined but may include differences in promoter function, mRNA stability, mRNA localization, translation efficiency, protein stability, or protein localization. Does this mean that βγ dimers assemble simply through random association in cells expressing multiple β and γ subtypes? *In vitro* studies indicate that structural constraints preclude a few βγ subunit combinations [18–20], most notably those involving β_2, β_3, and γ_1 subtypes. Moreover, *in vivo* studies suggest that spatial considerations contribute to the assembly of specific βγ subunit combinations [21–23]. For example, ribozyme-mediated suppression of the γ_7 subtype results in a coordinate reduction in the expression of the β_1 subtype but has no effect on the other three β subtypes expressed in human kidney cells [23]. Pulse-chase labeling studies show that there is a dramatic difference in the half-life of the β_1 monomer (20.8 min) compared to the $\beta_1\gamma_7$ dimer (14.2 hrs). Collectively, these results indicate that the β_1 protein is rapidly degraded when sufficient γ_7 protein is not available; and that other γ proteins are not able to randomly associate with the β_1 protein in these cells.

Subcellular Level

At another level, the γ subtypes exhibit distinct subcellular localizations [24,25], raising the possibility that they are involved in targeting the βγ dimers to specific locations within the cell membrane. For example, we have shown that the γ_7 subtype is present in the membrane overlying the actin fibers in the leading edge of the cell, the γ_5 subtype is highly enriched in the membrane overlying focal adhesions, and the γ_3 subtype is localized in a membrane site distinct from caveolae (J.D.R., unpublished; [24]). Other reports suggest that the γ subunit is not only associated with the membrane but also with intracellular components. For example, the γ_{12} subtype is present in a detergent-insoluble fraction in Swiss 3T3 cells in association with F-actin [25], and the βγ dimer associates with cytokeratin filaments in starfish oocytes [26]. These results suggest that differential subcellular localization of the γ subtypes may direct the specific assembly of βγ dimers and predicts that β subtypes may be found to share these patterns of expression.

The region(s) of the γ subunit responsible for their targeting includes the C-terminal tail and posttranslational modifications thereof. This raises the possibility that heterogeneity in their processing may account for their distinctive subcellular localizations. In this regard, the type of prenyl group may be important. The γ_1, γ_{11}, and γ_{14} subtypes are modified by the C15 farnesyl group, whereas the remaining subtypes are modified by the C20 geranylgeranyl group [17,19]. Though the prenyl group is important for targeting, the nature of this interaction is not clear. One possibility is that the prenyl group may interact as the result of a lipid-lipid interaction. In this case, the addition of the geranylgeranyl rather than the farnesyl group may confer additional hydrophobicity to the protein [17,27]. This is consistent with the finding that the geranylgeranyl moiety is sufficient to target the βγ dimer to the membrane, but farnesyl-dependent association of the βγ dimer with the membrane requires additional modifications, including carboxyl-methylation [27]. Another possibility is that the prenyl group may associate as the result of protein-protein interactions. In this case, the presence of the geranylgeranyl or farnesyl group may act as part of a recognition target for specific "docking proteins." This notion is consistent with recent results showing that the geranylgeranyl group of the γ subunit preferentially inserts into a hydrophobic binding pocket on PLC-β [28]. However, the type of prenyl group cannot explain the differences in subcellular localization among the γ_3, γ_5, and γ_7 subtypes noted above, since all three are modified with a geranylgeranyl group. This suggests that variable processing of the C-terminal tail and/or additional upstream regions may be important. In this regard, an alternatively processed form of the γ_5 subtype has been identified as the predominant species in brain [29]. Notably, this protein is prenylated but retains the three terminal amino acids. Whether this pattern of processing determines the unique localization of the γ_5 subtype in focal adhesions is the subject of ongoing studies. Also, there is some evidence that the N-terminal region of the γ protein could be important for subcellular localization based on its extensive sequence divergence [2] and the finding that phosphorylation within this region of the γ_{12} subtype alters the nature of its association with cytoskeleton [30]. Collectively, these results support a growing body of evidence that the G-protein-dependent signaling pathways may be segregated [31,32] and that this is a critical factor in maintaining the specificity of the βγ signal, as discussed below.

Specificity of G Protein βγ Dimer Signaling

Combinatorial association creates the potential to assemble 60 different βγ dimers that can be utilized to direct the fidelity of signaling. However, specificity has been difficult to show in transfected and reconstituted settings where there is substantial overlap in terms of their functions. This is most likely due to the absence of critical factors that provide specificity only in the context of the intact cell setting. For this reason, investigators have begun to use reverse genetic approaches to ascribe specific functions to the β and γ subtypes. Such strategies clearly support the notion that the composition of the βγ dimer, in particular the γ component, has important ramifications for the specificity of signaling that begins at the receptor level and continues to the effector level.

Structural Constraints on Specificity

Reconstitution studies of purified proteins show that the βγ dimer interacts directly with the receptor and certain types of the effector. To identify the contact sites, chimeric and mutagenesis approaches have been used to identify multiple sites on the βγ dimer: some regions appear to stabilize the interaction, other sites appear to modulate the interaction, and a few regions appear to specify the interaction. These regions will be discussed in the context of the three-dimensional structure of the βγ dimer [33,34]. In this structure, the β subunit is composed of two major domains—an N-terminal α helix and a seven-bladed β-propeller—whereas the γ subunit is composed of three major domains—an N-terminal α helix that forms a parallel coiled-coil structure with the corresponding region of the β subunit, a middle domain that forms extensive contacts along the bottom surface of the seven-bladed β-propeller, and a C-terminal domain, including the prenyl group.

βγ Dimer–Receptor

Recent evidence indicates that the βγ dimer directly interacts with the receptor [35], where it may actively participate in receptor-catalyzed nucleotide exchange of the α subunit [36]. Cross-linking studies confirm a receptor contact site within the C-terminal region of the β subunit [37], and reconstitution studies comparing different βγ subunit combinations reveal modest differences due to the β component [38–41]. Similar studies highlight even greater differences due to the γ component and point to the importance of the primary structure of the C-terminal domain and the type of prenyl group [38,41–45]. Collectively, these data indicate that the βγ-subunit composition contributes to the selectivity of receptor interaction. In addition, other results suggest that the βγ subunit composition may modulate the mechanism [46] and/or the duration [47] of receptor interaction. Nevertheless, the finding that the range of G protein βγ-receptor interactions *in vitro* is far greater than that observed *in vivo* suggests the presence of additional mechanisms for conferring specificity of signaling beyond direct receptor recognition of βγ subtypes.

βγ Dimer-Effector

Once released from the G-protein α subunit, the βγ dimer regulates a growing list of effectors. Multiple regions of the β subunit interact with the effector. One site lies within the N-terminal α-helix of the β subunit [48]. This region is exposed regardless of the activation state of the G-protein heterotrimer and hence may represent a region where the effector can remain bound to the βγ dimer even in the presence of the α subunit. Another region lies along the top surface of the β propeller of the β subunit (49). This region is exposed only upon activation of the G-protein heterotrimer and hence may be critical for modulation of the effector. Other regions include the sides of the β propeller, with each effector having its own characteristic set of contact points [49]. Despite the large number of sites on β, there is little evidence that any of these regions is involved in specifying the interaction with the effector consistent with the high homology of these domains among the β_1–β_4 subtypes. The apparent exception is the β_5 subtype [50,51].

Whereas the β subtypes are very similar, the γ subtypes are much more divergent, suggesting that functional specificity of different βγ subunit combinations is more likely due to the γ component. In this regard, comparison of βγ dimers differing only in γ reveal moderate differences in their abilities to activate effectors. The differences are due to both the primary structure of the N-terminal domain and the nature of the prenyl group on the C-terminal tail. It is remarkable that the relative contribution of these two regions appears to vary depending on the identity of the effector. For example, the amino acid sequence and/or charge of the N-terminal region of γ appears to be the major determinant for the AC activation. In this regard, the γ_1 and γ_{11} subtypes, which are negatively charged, are poor activators [17,52]. Likewise, the γ_{12} subtype, which is negatively charged as a result of phosphorylation, is a poor activator [53]. By contrast, the C-terminal domain and the type of prenyl group appear to be the more critical factors for activation of PLC-β [27,28,52]. (In one model, the prenyl group inserts into the β subunit, thereby producing a more active βγ dimer for regulation of PLC-β [54]. In the other model, the prenyl group inserts directly into PLC-β, thereby accounting for activation of PLC-β [28]. Supporting the latter model, fluorescence-based assays reveal that the prenylated γ peptide interacts directly with a site on PLC-β. Providing a possible basis for selectivity, this site prefers the geranylgeranylated to the farnesylated form of γ [28]. Finally, mutagenesis studies to alter the putative prenyl binding pocket of β [54] or the C-terminal tail of γ [28] reveal that all such mutants are less effective than wild type for activation of PLC-β. Thus, the position of the prenyl group on the βγ dimer has a profound impact on its ability to interact with PLC-β regardless of the model. However, there are still many unresolved questions. For instance, if the prenyl group of γ inserts into PLC-β, how does this fit with the crystal structure of phosducin-βγ dimer, which has the prenyl group of γ inserting into a hydrophobic pocket of the β subunit [55]? Also, how does this fit with the previous finding that the C-terminal tail and the prenyl group of γ bind directly to the receptor [45,46]?

Specific Functions of Individual β and γ Subtypes

The limited functional specificity that has been observed *in vitro* presents a conundrum. If all βγ dimers display substantial overlap in their interactions *in vitro*, how is specificity of the βγ signal maintained such that different classes of G-protein-coupled receptors have specific functions *in vivo*? For example, the β-adrenergic and muscarinic receptors activate G proteins to produce opposing effects on the rate and force of contraction of cardiac myocytes. This indicates that these two receptors cause release of βγ dimers that are not functionally interchangeable, and implies the existence of critical factors for specificity that can only be provided in the context of the cell. Increasingly, reverse genetic approaches provide the most direct and compelling evidence for specific functions of the β and γ subtypes. Analogous to the α subunit, loss of the β or γ subunit is likely to have major functional consequences by compromising the assembly of a specific G-protein αβγ trimer required for upstream interaction with the receptor and downstream regulation of effector(s).

RNA Suppression

Early studies relied on anti-sense oligonucleotides to suppress translation of the mRNAs encoding the β or γ subtypes. In rat pituitary cells, this approach was used to study regulation of voltage-dependent calcium channels following nuclear injection of anti-sense oligonucleotides against the various G-protein subunits. The intriguing results suggest a specific role for the $\beta_1\gamma_3$ dimer in the somatostatin receptor pathway, with a similarly selective requirement for the $\beta_3\gamma_4$ dimer in the muscarinic receptor pathway [56,57]. More recently, other studies in the same cells predict roles for the $\beta_2\gamma_2$ dimer in coupling the vasoactive intestinal peptide receptor to stimulation of AC, the $\beta_1\gamma_3$ dimer in linking the somatostatin receptor to inhibition of AC, and the $\beta_4\gamma_2$ dimer in coupling the thyroid-releasing hormone receptor to activation of PLC-β [58]. Finally, studies in mice suggest a role for the γ_2 subtype in antinociception by the opioid receptor [59]. Though intriguing, these studies do not show that such functional consequences are due to loss of the targeted protein(s). Compared to anti-sense RNA, ribozymes offer significant advantages by acting as site-specific nucleases [60]. In human kidney cells, we have used this approach for the first time to show a specific role for the $\beta_1\gamma_7$ dimer in coupling the β-adrenergic receptor to stimulation of AC [23,61]. Ongoing studies in mouse lung cells suggest a similarly selective requirement for γ_{11} subtype in control of cell growth. In both cases, these defects are linked to specific loss of the targeted protein(s).

Gene Targeting

More recently, we have begun analysis of mice or ES cells carrying targeted knockouts of the γ subunit genes to identify specific functions. Of the known γ subtypes, the γ_3 and γ_7 subtypes are closely related with respect to their primary structures and high expression levels in brain. Nevertheless, the γ_3 and γ_7 knockout mice exhibit two distinctive phenotypes, indicating they perform nonredundant roles in separate signaling pathways in the context of the whole animal. Collectively, the results of these reverse genetic approaches support the notion that receptors associate with particular combinations of G protein α and βγ subunits in the cell or whole animal setting; and that this provides a basis for the specificity of signaling through these pathways.

Conclusion

An intriguing, growing body of evidence suggests that polymorphisms in β_3 may be linked to disease [6]; and that sequestration of βγ may alter the course of diseases, including tumor invasiveness [62], vascular re-stenosis [63], and cardiac hypertrophy [64]. Thus, knowledge of which distinct subsets of βγ are involved in these processes will provide the rationale for more selective design of therapeutic strategies for these diseases.

Acknowledgment

This work was supported by National Institutes of Health Grant GM39864 awarded to J. D. R.

References

1. Clapham, D. and Neer, E. (1997). G protein βγ subunits. *Annu. Rev. Pharmacol. Toxicol.* **37**, 167–203.
2. Schwindinger, W. and Robishaw, J. (2002). Heterotrimeric G-protein βγ-dimers in growth and differentiation. *Oncogene* **20**, 1653–1660.
3. Azpiazu, I. and Gautam, N. (2002). Role of G protein βγ complex in receptor-G protein interactions. *Methods Enzymol.* **344**, 112–125.
4. Dell, E., Blackmer, T., Skiba, N., Daaka, Y., Shekter, L. et al. (2002). Defining G protein βγ specificity for effector recognition. *Methods Enzymol.* **344**, 421–434.
5. Hurowitz, E., Melnyk, J., Chen, Y., Kouros-Mehr, H., Simon, M., and Shizuya, H. (2000). Genomic characterization of the human heterotrimeric G protein α, β, γ subunit genes. *DNA Res.* **7**, 111–120.
6. Rosskopf, D., Busch, S., Manthey, I., and Siffert, W. (2000). G protein β_3 gene: structure, promoter, and additional polymorphims. *Hypertension* **36**, 33–41.
7. Downes, G. and Gautam, N. (1999). The G protein subunit gene families. *Genomics* **62**, 544–552.
8. Huang, L. Y., Shanker, J., Dubauskaite, J., Zheng, W., Yan. et al. (1999). G γ_{13} colocalizes with gustducin in taste receptor cells and mediates IP3 responses to bitter denatonium. *Nat. Neurosci.* **2**, 1055–1062.
9. Magre, J., Delepine, M., Khallouf, E., Gedde-Dahl, T. J., Van Malderger, L., et al. and BSCL Working Group (2002). Identification of the gene altered in Berardinelli-Seip congenital lipodystrophy on chromosome 11q13. *Nat. Genet.* **28**, 365–370.
10. Fisher, K. and Aronson, N. J. (1992). Characterization of the cDNA and genomic sequence of a G protein γ subunit (γ_5). *Mol. Cell Biol.* **12**, 1585–1591.
11. Siderovski, D., Snow, B., Chung, S., Brothers, G., Sondek, J., and Betts, L. (2002). Assays of complex formation between RGS protein G γ subunit-like domains and G β subunits. *Methods Enzymol.* **344**, 702–23.
12. Rehm, A. and Ploegh, H. (1997). Assembly and intracellular targeting of the βγ subunits of heterotrimeric G proteins. *Cell Biol.* **137**, 305–317.
13. Fu, H. and Casey, P. (1999). Enzymology and biology of CaaX protein prenylation. *Recent Prog. Horm. Res.* **54**, 315–342.
14. Peng, Y., Robishaw, J., Levine, M., and Yau, K. (1992). Retinal rods and cones have distinct G protein β and γ subunits. *Proc. Natl. Acad. Sci. USA* **89**, 10882–10886.
15. Betty, M., Harnish, S., Rhodes, K., and Cockett, M. (1998). Distribution of heterotrimeric G-protein β and γ subunit in the rat brain. *Neuroscience* **85**, 475–86.
16. Morishita, R., Shinohara, H., Ueda, H., Kato, K., and Asano, T. (1999). HIgh expression of the γ_5 isoform of G protein in neuroepithelial cells and its replacement of the γ_2 isoform during neuronal differentiation in the rat brain. *J. Neurochem.* **73**, 2369–2374.
17. Balcueva, E., Wang, Q., Hughes, H., Kunsch, C., Yu, Z., and Robishaw, J. (2000). Human G protein γ_{11} and γ_{14} subtypes define a new functional subclass. *Exp. Cell Res.* **257**, 310–319.
18. Mende, U., Schmidt, C., Yi, F., Spring, D., and Neer, E. (1995). The G protein γ subunit. Requirements for dimerization with β subunits. *J. Biol. Chem.* **270**, 15892–15898.
19. Ray, K., Kunsch, C., Bonner, L., and Robishaw, J. (1995). Isolation of cDNA clones encoding eight different human G protein γ subunits, including three novel forms designated the γ_4, γ_{10}, and γ_{11} subunits. *J. Biol. Chem.* **270**, 21765–21771.
20. Yan, K., Kalyanaraman, V., and Gautam, N. (1996). Differential ability to form the G protein βγ complex among members of the β and γ subunit families. *J. Biol. Chem.* **271**, 7141–7146.
21. Wilcox, M., Dingus, J., Balcueva, E., McIntire, W., Mehta, N., Schey, K., Robishaw, J., and Hildebrandt, J. (1995). Bovine brain Go isoforms have distinct γ subunit compositions. *J. Biol. Chem.* **270**, 4189–4192.
22. Asano, T., Morishita, R., Ueda, H., and Kato, K. (1999). Selective association of G protein β_4 with γ_5 and γ_{12} subunits in bovine tissues. *J. Biol. Chem.* **274**, 21425–21429.
23. Wang, Q., Mullah, B., and Robishaw, J. (1999). Ribozyme approach identifies a functional association between the G protein $\beta_1\gamma_7$ subunits in the β-adrenergic receptor signaling pathways. *J. Biol. Chem.* **274**, 17365–17371.
24. Hansen, C., Schroering, A., Carey, D., and Robishaw, J. (1994). Localization of a heterotrimeric G protein γ_5 subunit to focal adhesions and associated stress fibers. *J. Cell Biol.* **126**, 811–819.
25. Ueda, H., Saga, S., Shinohara, H., Morishita, R., Kato, K., and Asano, T. (1997). Association of the γ_{12} subunit of G proteins with actin filaments. *J. Cell Sci.* **110**, 1503–1511.
26. Chiba, K., Longo, F., Kontani, K., Katada, T., and Hoshi, M. (1995). A periodic network of G protein βγ subunits coexisting with cytokeratin filament in starfish oocytes. *Dev. Biol.* **169**, 415–20.
27. Matsuda, T., Hashimoto, Y., Ueda, H., Asano, T., Matsuura, Y. et al. (1998). Specific isoprenyl group linked to transducin γ-subunit is a determinant of its unique signaling properties among G-proteins. *Biochemistry* **37**, 9843–9850.
28. Akgoz, M., Azpiazu, I., Kalyanaraman, V., and Gautam, N. (2002). Role of the G protein γ subunit in βγ complex modulation of phospholipase C-β function. *J. Biol. Chem.* [epub ahead of print].
29. Cook, L., Schey, K., Wilcox, M., Dingus, J., and Hildebrandt, J. (1998). Heterogeneous processing of a G protein γ subunit at a site critical for protein and membrane interactions. *Biochemistry* **37**, 12280–12286.
30. Ueda, H., Yamauchi, J., Itoh, H., Morishita, R., Kaziro, Y., Kato, K., and Asano, T. (1999). Phosphorylation of F-actin-associating G protein γ_{12} subunit enhances fibroblast motility. *J. Biol. Chem.* **274**, 12124–12218.
31. Oh, P. and Schnitzer, J. (2001). Segregation of heterotrimeric G proteins in cell surface microdomains. G_q binds caveolin to concentrate in caveolae, whereas G_i and G_s target lipid rafts by default. *Mol. Biol. Cell* **12**, 685–698.
32. Davare, M., V., Avdonin, D., Hall, E., Peden, A., Burette, R. et al. (2001). A β_2 adrenergic receptor signaling complex assembled with the Ca^{2+} channel Cav1.2. *Science* **293**, 98–101.
33. Sondek, J., Bohm, A., Lambright, D., Hamm, H., and Sigler, P. (1996). Crystal structure of a G-protein βγ dimer at 2.1A resolution. *Nature* **379**, 369–374.
34. Wall, M., Coleman, D., Lee, E., Iniguez-Lluhi, J., Posner, B., Gilman, A., and Sprang, S. (1995). The structure of the G protein heterotrimer Gi $\alpha_1\beta_1\gamma_2$. *Cell* **83**, 1047–1058.
35. Phillips, W. and Cerione, R. (1992). Rhodopsin/transducin interactions. I. Characterization of the binding of the transducin-βγ subunit complex to rhodopsin using fluorescence spectroscopy. *J. Biol. Chem.* **267**, 17032–17039.
36. Rondard, P., Iiri, T., Srinivasan, S., Meng , E., Fujita, T., and Bourne, H. (2001). Mutant G protein α subunit activated by G βγ; a model for receptor activation?. *Proc. Natl. Acad. Sci. USA* **98**, 6150–6155.
37. Taylor, J., Jacob-Mosier, G., Lawton, R., VanDort, M., and Neubig, R. (1996). Receptor and membrane interaction sites on Gβ. A receptor-derived peptide binds to the carboxyl terminus. *J. Biol. Chem.* **271**, 3336–3339.
38. Richardson, M. and Robishaw, J. (1999). The α_{2A}-adrenergic receptor discriminates between Gi heterotrimers of different βγ subunit composition in Sf9 insect cell membranes. *J. Biol. Chem.* **274**, 13525–13533.
39. Hou, Y., Chang, V., Capper, A., Taussig, R., and Gautam, N. (2001). G protein β subunit types differentially interact with a muscarinic receptor but not adenylyl cyclase type II or phospholipase C-$\beta_{2/3}$. *J. Biol. Chem.* **276**, 19982–19988.
40. McIntire, W., MacCleery, G., and Garrison, J. (2001). The G protein β subunit is a determinant in the coupling of Gs to the β_1-adrenergic and A_{2a} adenosine receptors. *J. Biol. Chem.* **276**, 15801–15809.

41. Robillard, L., Ethier, N., Lachance, M., and Hebert, T. (2000). G βγ subunit combinations differentially modulate receptor and effector coupling in vivo. *Cell Signal* **12**, 673–682.
42. Hou, Y., Azpiazu, I., Smrcka, A., and Gautam, N. (2000). Selective role of G protein γ subunits is receptor interactions. *J. Biol. Chem.* **275**, 38961–38964.
43. Lim, W. K., Myung, C. S., Garrison, J. C., and Neubig, R. R. (2001). Receptor-G protein γ specificity: γ_{11} shows unique potency for A_1 adenosine and 5-HT_{1A} receptors. *Biochemistry* **40**, 10532–10541.
44. Yasuda, H., Lindorfer, M., Woodfork, K., Fletcher, J., and Garrison, J. (1996). Role of the prenyl group on the G protein γ subunit in coupling trimeric G proteins to A_1 adenosine receptors. *J. Biol. Chem.* **271**, 18588–18595.
45. Kisselev, O., Ermolaeva, M., and Gautam, N. (1995). Efficient interaction with a receptor requires a specific type of prenyl group on the G protein γ subunit. *J. Biol. Chem.* **270**, 25356–25358.
46. Azpiazu, I. and Gautam, N. (2001). G protein γ subunit interaction with a receptor regulates receptor-stimulated nucleotide exchange. *J. Biol. Chem.* **276**, 41742–41747.
47. Clark, W., Jian, X., Chen, L., and Northup, J. (2001). Independent and synergistic interaction of retinal G-protein subunits with bovine rhodopsin measured by surface plasmon resonance. *Biochem. J.* **358**, 389–397.
48. Yoshikawa, D., Bresciano, K., Hatwar, M., and Smrcka, A. (2001). Characterization of a phospholipase C β2-binding site near the amino-terminal coiled-coil of G protein βγ subunits. *J. Biol. Chem.* **276**, 11246–11251.
49. Ford, C. N., Skiba, H., Bae, Y., Daaka, E., Reuveny, L. *et al.* (1998). Molecular basis for interactions of G protein βγ subunits with effectors. *Science* **280**, 1271–1274.
50. Zhang, S., Coso, O., Lee, Gutkind, J., and Simonds, W. (1996). Selective activation of effector pathways by brain-specific G protein β_5. *J. Biol. Chem.* **271**, 33575–33579.
51. Lindorfer, M., Myung, C., Savino, Y., Yasuda, H., Khazan, R., and Garrison, J. (1998). Differential activity of the G protein $\beta_5\gamma_2$ subunit at receptors and effectors. *J. Biol. Chem.* **273**, 34429–34436.
52. Myung, C., Yasuda, H., Liu, W., Harden, T., and Garrison, J. (1999). Role of isoprenoid lipids on the heterotrimeric G protein γ subunit in determining effector activation. *J. Biol. Chem.* **274**, 16595–16603.
53. Yasuda, H., Lindorfer, M., Myung, C., and Garrison, J. (1998). Phosphorylation of the G protein γ_{12} subunit regulates effector specificity. *J. Biol. Chem.* **273**, 21958–21965.
54. Myung, C. and Garrison, J. (2000). Role of C-terminal domains of the G protein β subunit in the activation of effectors. *Proc. Natl. Acad. Sci. USA* **97**, 9311–9316.
55. Loew, A., Ho, Y., Blundell, T., and Bax, B. (1998). Phosducin induces a structural change in transducin βγ. *Structure* **6**, 1007–1019.
56. Kleuss, C., Scherubl, H., Hescheler, J., Schultz, G., and Wittig, B. (1992). Different beta-subunits determine G-protein interaction with transmembrane receptors. *Nature* **358**, 424–426.
57. Kleuss, C., Scherubl, H., Hescheler, J., Schultz, G., and Wittig, B. (1993). Selectivity in signal transduction determined by gamma subunits of heterotrimeric G proteins. *Science* **259**, 832–834.
58. Johansen, P., Lund, H., and Gordeladze, J. (2001). Specific combinations of G-protein subunits discriminate hormonal signalling in rat pituitary (GH_3) cells in culture. *Cell. Signal.* **13**, 251–256.
59. Hosohata, K., Logan, J., Varga, E., Burkey, T., Vanderah, T. *et al.* (2000). The role of the G protein γ_2 subunit in opioid antinociception in mice. *Eur. J Pharmacol.* **392**, R9–R11.
60. Robishaw, J., Wang, Q., and Schwindinger, W. F. (2002). Ribozyme-mediated suppression of G protein γ subunits. *Methods Enzymol.* **344**, 435–451.
61. Wang, Q., Mullah, B., Hansen, C., Asundi, J., and Robishaw, J. (1997). Ribozyme-mediated suppression of the G protein γ_7 subunit suggests a role in hormone regulation of adenylylcyclase activity. *J. Biol. Chem.* **272**, 26040–26048.
62. Faivre, S., Regnauld, K., Bruyneel, E., Nguyen, Q., Mareel, M., Emami, S., and Gespach, C. (2001). Suppression of cellular invasion by activated G-protein subunits $G\alpha_o$, $G\alpha_{i1}$, $G\alpha_{i2}$, and $G\alpha_{i3}$, and sequestration of Gβγ. *Mol. Pharmacol.* **60**, 363–372.
63. Iaccarino, G., Smithwick, L., Lefkowitz, R., and Koch, W. (1999). Targeting Gβγ signaling in arterial vascular smooth muscle proliferation: a novel strategy to limit restenosis. *Proc. Natl. Acad. Sci. USA* **96**, 3945–3950.
64. Naga Prasad, S., Esposito, G., Mao, L., Koch, W., and Rockman, H. (2000). Gβγ-dependent phosphoinositide 3-kinase activation in hearts with *in vivo* pressure overload hypertrophy. *J. Biol. Chem.* **275**, 4693–4698.

CHAPTER 225

The RGS Protein Superfamily

David P. Siderovski[1] and T. Kendall Harden
Department of Pharmacology, Lineberger Comprehensive Cancer Center, and
[1]*UNC Neuroscience Center, The University of North Carolina at Chapel Hill,*
Chapel Hill, North Carolina

Introduction

A large family of seven transmembrane–domain receptors for hormones, neurotransmitters, growth factors, chemoattractants, light, odorants, and other extracellular stimuli promote intracellular signaling responses by activation of heterotrimeric G proteins. Agonist-activated G protein-coupled receptors (GPCRs) selectively interact with one (or more) of approximately 20 different heterotrimeric G proteins, promoting exchange of GTP for bound GDP, dissociation of the G protein into GTP-bound Gα and free Gβγ subunits, and ensuing activation of downsteam effectors. Hydrolysis of GTP by GTPase activity intrinsic to each Gα subunit returns the heterotrimer to its ground state of Gα-GDP bound to Gβγ. Thus, signal amplitude of G protein-mediated responses is governed by the prevailing balance of receptor-promoted guanine nucleotide exchange and GTP hydrolysis.

Whereas activation of heterotrimeric G proteins by GPCRs has been understood in kinetic terms for almost two decades, understanding of the regulation of the deactivation process has accrued much more slowly. Purified G proteins hydrolyze GTP at a rate much lower than the very rapid deactivation of these proteins observed under physiological conditions. This discrepancy between the rates of GTP hydrolysis measured *in vitro* and *in vivo* presaged the existence of other classes of proteins that accelerate the deactivation of GTP-bound Gα-subunits. Observation of marked enhancement of the GTPase-activity of Gαq in the presence of its specific effector PLC-β1 [1] provided the first evidence for GTPase-activating proteins (GAPs) for heterotrimeric G proteins. More recently, a family of approximately 30 proteins was identified that contain a signature domain that selectively binds to GTP-bound Gα-subunits and stimulates their intrinsic GTPase activity up to several thousand-fold [2–8]. These "regulator of G protein signaling" (RGS) proteins promote the deactivation step in the heterotrimeric G protein regulatory cycle, catalyzing GTPase rates *in vitro* that are consistent with the rates of deactivation of G-protein signaling *in vivo*. Interaction of RGS proteins with GTP-bound Gα subunits confers to this large class of regulatory proteins the ability to modulate signaling response kinetics, amplitude, and specificity. The physiological functions of these proteins are currently under investigation and, in the case of many RGS proteins, include additional functionalities beyond their hallmark capacity to act as GAPs for Gα subunits.

The Signature RGS-Box as a Gα GAP

The defining feature of all RGS family proteins is an ~120 amino-acid region, the "RGS-box," that directly binds Gα-GTP subunits and markedly accelerates GTP hydrolysis. The RGS-box has a "modular" character like that of other recurrent protein functional modules such as PDZ (PSD-95/Dlg/ZO-1), PTB (phosphotyrosine-binding), and SH2 (Src homology-2) domains—that is, the minimal polypeptide sequence of the RGS-box is sufficient in isolation for full Gα-binding and GAP activity [9,10]. Soon after discovery of the RGS-box, Tesmer and colleagues [11] solved the first crystal structure of an RGS-box/Gα complex, namely RGS4 bound to the adenylyl cyclase-inhibitory Gα subunit Gαi1. The RGS-box of RGS4 folds as a compact bundle of nine-alpha helical segments [11]. Other structures of isolated RGS-boxes have since been solved by both NMR and X-ray diffraction methods [12–16], including a trimeric complex of RGS9 bound to both Gα-transducin (Gαt) and the gamma subunit of cGMP-phosphodiesterase [17]. Each of these

RGS-boxes presents the same global fold comprising predominantly alpha-helical secondary structure.

The atomic-resolution structures of RGS-boxes in complex with Gα partners [11,17] indicate that the RGS-box primarily interacts with the conformationally flexible "switch" regions in the GTP-binding domain of Gα subunits. RGS-box residues do not directly take part in the chemistry of the GTP hydrolytic cycle. Rather, GTP hydrolysis traverses through a bipyramidal transition state, and the crystal structure of the RGS4/Gαi1 complex revealed Gαi1 to exist in this transition state [11]. Gα-GDP subunits bound with the planar ion aluminum tetrafluoride (AlF_4^-) mimic the transition state of GTP hydrolysis, and therefore most RGS proteins bind most avidly to GDP/AlF_4^- complexed Gα subunits over the GTPγS- or GDP-bound states (for example Fig. 1A; [18]).

The multidomain structure of many RGS proteins confers a complexity to these proteins that makes it difficult to define unambiguously their biological activities *in situ* or *in vivo*. Nonetheless, their signature activity exists in a capacity to act as GAPs for heterotrimeric G proteins, and quantitation of the GTPase activities of Gα subunits is directly possible *in vitro* and is also feasible by indirect means *in vivo*.

Single-Turnover Assays of GAP Activity

The GAP activity of RGS proteins can be quantitated reliably in single-turnover assays with most purified Gα subunits and RGS proteins [19]. Such measurements require the binding of [$\gamma^{32}P$]GTP to Gα in sufficient stoichiometry and under conditions that minimize GTP hydrolysis. This is usually accomplished by incubation of purified Gα with [$\gamma^{32}P$]GTP in the absence of Mg^{2+}, followed by removal of free [$\gamma^{32}P$]GTP by gel filtration. Since GTP hydrolysis is Mg^{2+}-dependent, addition of Mg^{2+} allows quantitation of a single cycle of GTP hydrolysis in the basal state of the Gα subunit. Moreover, concomitant addition of an RGS protein with Mg^{2+} provides assessment of the extent of RGS protein-promoted increase in hydrolytic activity (for example, Fig. 1B). The monoexponential time courses for release of [$\gamma^{32}P$]Pi under these conditions allow calculation of the first-order rate constant for catalysis (k_{cat}) of the intrinisic GTPase activity and the rate constant (k_{gap}) for GAP-promoted hydrolysis. Bimolecular Michaelis-Menten enzyme kinetics are followed under assay conditions in which Gα-GTP concentration is in excess of RGS protein concentration.

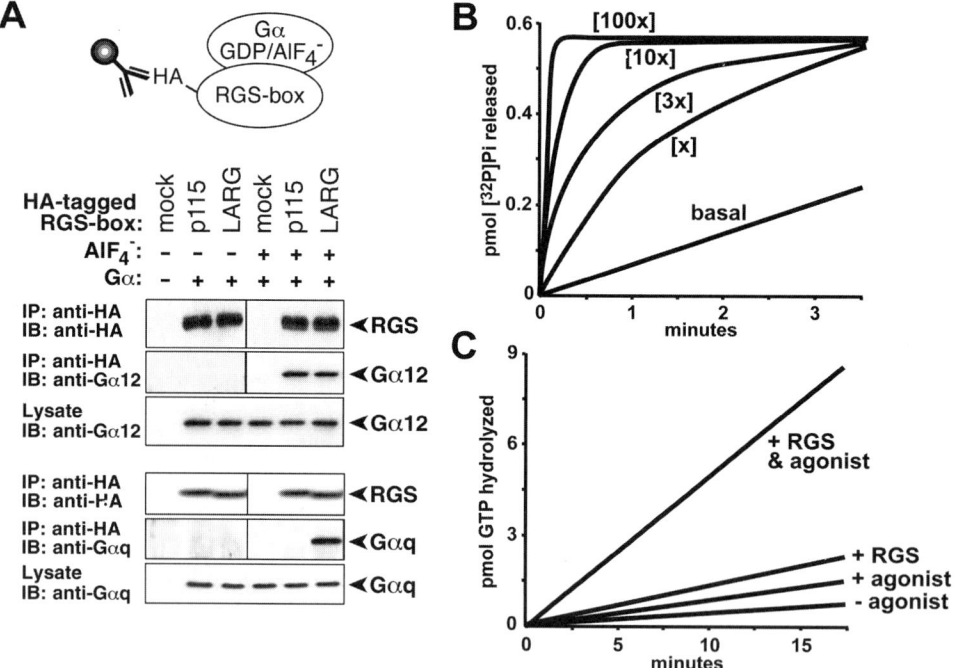

Figure 1 Representative functional data derived from assays of RGS-box activity. (A) Cellular co-transfection/co-immunoprecipitation assay to discern the Gα-binding specificity of hemagglutinin (HA)-epitope tagged RGS proteins is illustrated schematically above the immunoblot results. Binding of Gα subunits to the isolated RGS-boxes of p115-RhoGEF ("p115") or leukemia-associated RhoGEF ("LARG") is dependent on the addition to the cellular lysates of NaF and $AlCl_3$ (i.e., the components of aluminum tetrafluoride, AlF_4^-). IP, immunoprecipitation; IB, immunoblot. (B) Idealized results from a single-turnover assay measuring radioactive inorganic phosphate ([^{32}P]Pi) release (and, by inference, GTP hydrolysis) by [$\gamma^{32}P$]GTP-loaded Gα protein in the absence ("basal") or presence of increasing concentrations ([x] to [100x]) of RGS protein. (C) Idealized results from a steady-state GTP hydrolysis assay employing proteoliposomes reconstituted with purified GPCR, heterotrimeric G-protein Gα and Gβγ subunits, and incubated with [$\gamma^{32}P$]GTP in the absence or presence of agonist and/or RGS protein. As in part B, GTP hydrolysis activity is quantitated as the production of [^{32}P]Pi.

Such single-turnover GTPase assays are straightforward with Gα subunits that release GDP readily and therefore are readily labeled in solution by [γ^{32}P]GTP. Certain Gα subunits bind GDP tightly and are more difficult to label with GTP. This is particularly true with Gαq and Gα11. However, Ross and coworkers have experienced success in single-turnover assays with the Gαq mutant R183C [20], which exhibits reduced intrinsic GTPase activity and can thus be loaded (to relatively low stoichiometry) with [γ^{32}P]GTP by using prolonged incubations with [γ^{32}P]GTP.

A modification of the single-turnover assay takes advantage of the capacity of unlabeled Gα subunits to act as competitive inhibitors of the activity of RGS proteins against [γ^{32}P]GTP-bound Gα. That is, RGS protein-promoted increases in the rate of hydrolysis of [γ^{32}P]GTP-Gα are measured in the presence of AlF$_4^-$ and a given GDP-bound Gα-subunit (for example, [21]). Interaction of an RGS protein with the transition state Gα (that is, GDP and AlF$_4^-$ bound) will compete for the capacity of the RGS protein to act as a catalyst against the [γ^{32}P]GTP-bound Gα substrate. Thus, the selectivity of interaction of an RGS protein with a series of Gα subunits can be compared in measurements of their relative capacities to shift to the right the concentration effect curve for RGS protein-promoted release of [γ^{32}P]Pi.

Measurement of GTPase activity in single-turnover assays provides the most straightforward means of assessing RGS protein GAP activity. As discussed above, these assays are of limited use with Gα subunits that bind GDP tightly and are thus difficult to label with radioactive GTP. Since single-turnover GAP assays are carried out in solution, these measurements also ignore any role played by the lipid bilayer in concentrating an RGS protein in the milieu of its site of action. This is an increasingly important concern given the growing evidence that membrane-binding domains and important sites of posttranslational modification exist in many RGS proteins.

Steady-State Measurements of GAP Activity

GTPase activity of Gα subunits measured during multiple cycles of GTP hydrolysis depends on both GTP hydrolysis and guanine nucleotide exchange. Thus, steady-state GTPase activity can only be measured under assay conditions wherein GDP release and GTP binding are sufficiently rapid. In practice, it is the exchange rate that is usually rate-limiting, and increases in GTP catalysis by the presence of RGS proteins results in only small increases in measured GTPase activity (for example, Fig. 1C). Again, this also is particularly true for steady-state GTPase measurements with Gα subunits such as Gαq that bind GDP tightly in the basal state. In contrast, if guanine nucleotide exchange is promoted by GPCR activation, GTP hydrolysis now becomes rate-limiting and remarkably large increases in GTPase activity are observed due to the catalytic activity of RGS proteins (Fig. 1C). Indeed, up to 2000-fold stimulation of steady-state GTPase activity of Gαq has been observed in a system in which guanine nucleotide exchange was promoted by carbachol-mediated activation of the M1 muscarinic cholinergic receptor and GTP hydrolysis was accelerated by RGS4 [22].

Steady-state GTPase measurements have been effectively carried out with membrane preparations from various tissues [23]. Thus, activation of a native or heterologously expressed receptor with an appropriate agonist results in an increase in GTPase activity that may be enhanced by addition of a GAP for the involved G protein. Although such assays provide a readout for receptor activity, they are of limited value in assessing activities of RGS proteins per se. More elaborate assay systems have been developed that allow more quantitative analyses. For example, a useful test system is provided by heterologous expression of a GPCR in Sf9 insect cells (or a mammalian cell line) followed by membrane preparation and removal of endogenous G proteins by treatment with high concentrations of urea [24]. Purified G proteins then are reconstituted with the stripped membranes, and receptor-promoted G protein activation can be quantitated either by [^{35}S]GTPγS binding, which assesses guanine nucleotide exchange, or by steady-state GTPase assays. The activity of RGS proteins as GAPs for the reconstituted Gα subunit also can be assessed. Similarly, purified GPCRs and G proteins can be reconstituted in phospholipid vesicles so that steady-state GTPase activity can be measured as a function of receptor agonist and RGS protein concentration [22,25,26].

Steady-state GTPase activity of a Gα subunit under the concerted regulation by a GPCR and an RGS protein provides a complex signaling response that integrates many partial reactions. The work of Ross and coworkers in studying the GTPase activity of Gαq in proteolipomes reconstituted with the M1 muscarinic receptor, as a promoter of guanine nucleotide exchange, and PLC-β1, as a GAP for Gαq, provides an excellent example of the complexities that underlie such responses [27]. However, in its simplest application, such a system provides a reliable means to assess selectivity of RGS proteins (and other GAPs such as PLC-β isozymes) for Gα subunits. Thus, the relative activities of RGS proteins can be compared in a more physiologically relevant milieu of a lipid bilayer than is the case in single-turnover GAP assays in solution.

Heterologous overexpression of RGS proteins utilizing mammalian expression vectors results in attenuation of GPCR-promoted activation of signaling pathways. Although RGS protein-promoted inhibition of signaling through heterotrimeric G proteins is predictably a consequence of enhancing the GTPase activity of cellular Gα subunits, the physiological significance of such inhibition is often unclear. The Gα subunit selectivity of most RGS proteins *in vivo* is unknown and, under conditions of RGS protein overexpression, promiscuity of GAP activity against many, if not all, Gα subunits should be suspected. This may be a particular concern with overexpression of RGS protein constructs that include only the RGS-box. As discussed below, RGS proteins also exhibit many signaling modulatory activities, in addition to their Gα-GAP activity, that may influence their cellular actions. Moreover, certain RGS proteins directly inhibit heterotrimeric G-protein-mediated signal transduction by other means. For example, RGS2 acts both as a GAP for GTP-bound Gαq and as an "effector antagonist"—that is, by inhibiting GTP-Gαq promoted stimulation of PLC-β [28].

Gα GAP and Other Signaling Regulatory Activities of RGS Family Members

The function of RGS proteins as multifaceted signaling regulators, rather than simply as Gα GAPs, is reflected in the observed structures of these proteins—the RGS-box that embodies GAP activity is rarely found in the absence of other regulatory protein modules (Fig. 2). Even for those RGS proteins in subfamilies A (RZ) and B (R4) that lack defined modules beyond the RGS-box, recent work in the field has discerned roles for non-RGS-box components in their subcellular localization and/or receptor-selective GAP activity, as outlined below.

A- or RZ-Subfamily (RGS17, RGS19, RGS20)

The founding member of the A- or RZ-subfamily, GAIP (G-alpha interacting protein, now known as RGS19), was the first mammalian RGS protein identified via the yeast two-hybrid system, with Gαi3 serving as the "bait" [2]. Additional family members, RGSZ1 (RGS20) and RGSZ2 (RGS17), were also isolated in yeast two-hybrid screens with Gαz and Gαo baits, respectively [29,30]; in addition, RGSZ1 was independently identified by purification and protein sequencing of a bovine brain Gαz-specific GAP activity [31]. It remains unclear whether the RGS-boxes of these RGS proteins exhibit strict selectivity toward Gαz or are indeed active *in vivo* as GAPs for other Gα subunits. PKC-mediated phosphorylation of Gαz has been reported to block RGS-box interaction [29,31].

All three subfamily members share a cysteine-rich cluster or "cysteine string" that is N-terminal of the RGS-box (Fig. 2). This polypeptide is believed to be heavily palmitoylated, leading to tight membrane association [32]. Other RGS proteins outside the RZ-family, such as RGS4, RGS7, RGS10, and RGS16, are thought to be targetted to the membrane by palmitoylation of N-terminal cysteine residues, although they lack a formal cysteine string [33–35].

B- or R4-Subfamily (RGS1, RGS2, RGS3, RGS4, RGS5, RGS8, RGS13, RGS16, RGS18)

The largest group of RGS proteins, the B- or R4-subfamily, comprise some of the smallest proteins that possess

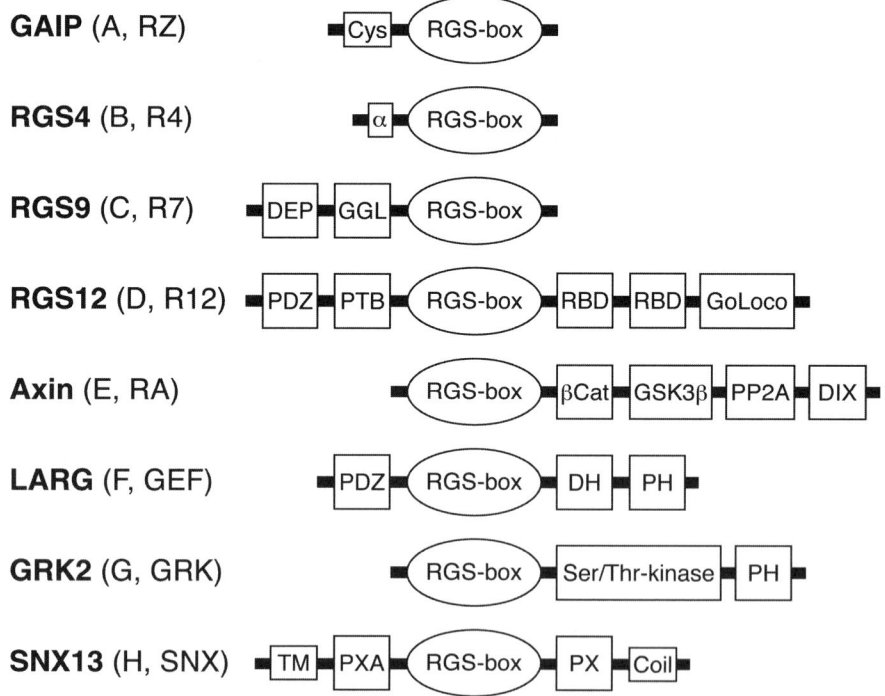

Figure 2 Protein domain architecture of representative members of the RGS protein subfamilies. Alternate nomenclatures for RGS protein subfamilies, as proposed independently by Zheng and Wilkie and coworkers [90,91], are denoted in brackets. Cys, cysteine-string motif; α, amphipathic alpha-helical region; DEP, dishevelled/Egl-10/pleckstrin domain; GGL, G-protein gamma subunit-like domain; PDZ, PSD-95/Dlg/ZO-1 domain; PTB, phosphotyrosine-binding domain; RBD, Ras-binding domain; GoLoco, Gαi/o-Loco interaction domain; βCat, β-catenin binding region; GSK3β, glycogen synthase kinase-3β binding region; PP2A, phosphatase PP2A binding region; DIX, domain of unknown function shared between Dishevelled and Axin proteins; DH, Dbl-homology domain; PH, pleckstrin-homology domain; Ser/Thr-kinase, protein kinase catalytic domain specific for serine and threonine residues; TM, putative transmembrane domain; PXA, "PX-associated" domain of unknown function; PX, Phox-homology domain; Coil, putative coiled-coil region.

RGS-boxes, with little in the way of non-RGS-box sequence.[1] Nevertheless, specific members of this subfamily exert receptor-selective inhibitory activity. Rat vascular smooth muscle cells, for example, express at least three R4-subfamily members capable of Gαq-directed GAP activity: RGS2, RGS3, and RGS5. However, ribozyme-mediated "knockdown" of the expression of each of these RGS proteins has a differential effect on cellular GPCR signaling coupled to Gαq activation. Neubig and colleagues [36] have observed that RGS3-ribozyme treatment selectively potentiates carbachol signaling via the M3 muscarinic receptor, whereas RGS5-ribozyme treatment only potentiates angiotensin II signaling via the AT1a receptor; RGS2 ribozyme treatment was without effect on either receptor signaling pathway.

Thus, determinants outside the RGS-box appear to play a role in targetting the Gα-GAP activity of these RGS proteins to particular receptors. In another example, RGS1, RGS4, and RGS16 exhibit large differences in their relative capacities to inhibit muscarinic-, bombesin-, or cholecystokinin-receptor-promoted Ca^{2+} signaling in permeabilized pancreatic acinar cells [37]. Since Ca^{2+} responses to all three GPCRs occur through Gq heterotrimers and all three RGS proteins exhibit similar activities *in vitro* as GAPs against Gαq, it was concluded that receptor-selective action of these RGS proteins contributes to their differential inhibitory activity *in vivo* [37]. Mutational analyses confirmed that selectivity did not involve the RGS-box per se, but rather, was contributed by sequences in a 33 amino-acid span of the RGS4 N-terminus [38]. This work has been extended by Wilkie, Muallem, and colleagues to propose a mechanism whereby RGS proteins play a central role in the Ca^{2+} oscillations that often accompany GPCR-promoted activation of phospholipase C [39]. Activation of Gq heterotrimers results in formation of GTP-Gαq, activation of phospholipase C-β, formation of $Ins(1,4,5)P_3$, and mobilization of intracellular Ca^{2+}. Gβγ also is released upon activation of Gq heterotrimers, which results in activation of phosphatidylinositol 3-kinase and formation of phosphatidylinositol-3,4,5-trisphosphate (PIP_3). PIP_3 binds to and inhibits the GAP activity of RGS4 [40]. Since Ca^{2+}/calmodulin reverses this PIP_3-mediated inhibition, Wilkie and coworkers have proposed that competition between PIP_3 and Ca^{2+}/calmodulin for inhibition and disinhibition of RGS4 may provide a feedback regulation that accounts for Ca^{2+} oscillations occurring in the constant presence of GPCR-activating agonists [39]. Kurachi and coworkers [41] recently proposed a similar role for PIP_3, Ca^{2+}/calmodulin, and RGS4 in the voltage-dependent regulation of the activation state of cardiac G_i heterotrimers, which in turn regulates potassium channel activity in rat heart.

---

[1]RGS3, while clearly a member of the R4-subfamily by virtue of its RGS-box sequence, is an outlier with respect to the overall length of its polypeptide sequence. RGS3 isoforms with considerable N-terminal extensions are known to exist; one extended RGS3 isoform contains an N-terminal PDZ domain that binds the ephrin-B cytoplasmic tail and mediates "reverse signaling" through these transmembrane-tethered ligands of Eph receptors (Schmucker, D., and Zipursky, S.L. (2001). Signaling downstream of Eph receptors and ephrin ligands. *Cell* **105**, 701–704.)

C- or R7-Subfamily (RGS6, RGS7, RGS9, RGS11)

The central structural features of the C- or R7-subfamily of RGS proteins are DEP and GGL domains N-terminal to the RGS-box (Fig. 2). The DEP domain [42] is a conserved ~80 amino-acid sequence found in *D*ishevelled (an intracellular component of Wnt/Frizzled signaling), *E*GL-10 (the first RGS protein discovered in *Caenorhabditis elegans* [3]), and *P*leckstrin (a major PKC substrate in platelets). DEP domains are also found in an N-terminal tandem repeat within Sst2, the archetypal RGS protein of the yeast *Saccharomyces cerevisiae* [43]. The role of the DEP domain within RGS proteins is assumed to be membrane localization [3,44], but the intracellular target(s) of the DEP domain remain undefined. A recent report by Dohlman and colleagues suggests that the N-terminal DEP domains of Sst2 modulate the yeast stress response pathway [45].

The discovery of the GGL or "G-protein gamma subunit-like" domain, present within all members of the R7-subfamily [21,46–48], has led to a radical departure from previous assumptions regarding the coupling of heptahelical receptors to heterotrimeric G-protein complexes [49]. In the standard model of heterotrimer assembly, conventional Gγ subunits exist as short, isoprenylated, alpha-helical polypeptides that form obligate heterodimers with the conventional Gβ subunits Gβ1 through Gβ4 [50]. In contrast, the GGL domains of R7-RGS proteins serve as obligate binding partners for the more distantly-related Gβ subunit, Gβ5 [21,46–48], thus supplanting the requirement for a conventional Gγ subunit [49]. This novel GGL/Gβ5 association implies a role for R7-RGS proteins in signaling functions akin to those well-characterized for conventional Gβγ subunits—namely, functional coupling of Gα subunits to GPCRs and modulation of downstream effectors [51]. With respect to the latter activity, screens of conventional Gβγ effects (for example, modulation of phospholipase C-β and adenylyl cyclase activity) using recombinant Gβ5/R7-RGS protein dimers have, to date, failed to identify any clear-cut effector activity by GGL/Gβ5 dimers [21,52]. However, a role for GGL/Gβ5 dimers in facilitating receptor/Gα coupling is suggested by recent findings that the Gβ5/RGS9 heterodimer can support agonist-stimulated guanine nucleotide exchange on Gαo by recombinant M2-muscarinic acetylcholine receptors (M2-mAChR) in proteoliposome reconstitution assays (Harden, T.K. *et al.*, unpublished observations). Such receptor-coupling activity could help to explain the accelerated, rather than attenuated, kinetics of GIRK channel activation by M2-mAChRs observed upon co-expression of Gβ5 with RGS7 or RGS9 in a *Xenopus* oocyte reconstitution system [53].

The best-characterized R7-RGS protein is RGS9-1, the retinal-specific isoform of RGS9 that represents the long-sought GTPase-activating protein for Gαt within the phototransduction cascade [54]. Inactivation of the mouse *Rgs9* gene leads to a greatly slowed inactivation of photon-induced signaling in both rod and cone photoreceptors [55,56]. RGS6, RGS7, and RGS11 all have been shown *in vitro* to accelerate GTP hydrolysis by Gαo subunits [21,52,57], but the physiological roles of these RGS/Gβ5 dimers remain to be elucidated.

D- or R12-Subfamily (RGS10, RGS12, RGS14)

All three of the D- or R12-subfamily RGS proteins, RGS10, RGS12, and RGS14, act as GAPs for Gαi-family Gα subunits *in vitro* [8, 58–60]. PKA-mediated phosphorylation of RGS10 at serine-168 attenuates its cellular function at the plasma membrane, not by reducing its intrinsic GAP activity, but by inducing its sequestration in the nucleus [61]. RGS10 belongs to the R12-subfamily given its RGS-box sequence similarity, but as a "small" RGS protein RGS10 is structurally similar to R4-subfamily members that lack discrete accessory modules.

In contrast, RGS12 and RGS14 share a multidomain organization C-terminal to the RGS-box that consists of tandem Ras-binding domains (RBDs) [62] and a GoLoco motif [63] (Fig. 2). Traver and colleagues [59] have identified the Ras-superfamily GTPases Rap1 and Rap2 as binding partners for the RBD region of RGS14; this interaction is dependent on the GTPases being in their GTP-bound or "activated" state, suggestive of a potential role for RGS14 as an effector for activated Rap proteins. The GoLoco motifs of RGS12 and RGS14 interact selectively with GDP-bound Gαi-family Gα subunits and prevent guanine nucleotide dissociation [60,64]. The X-ray crystal structure of the RGS14 GoLoco-motif region bound to Gαi1/GDP [65] indicates clear roles for both the Gα all-helical domain and polypeptide sequence C-terminal to the GoLoco motif in engendering binding specificity; guanine nucleotide dissociation inhibitor (GDI) activity is thought to result, in part, from direct contacts between a highly conserved arginine in the GoLoco motif and the phosphate groups of the bound GDP.

Unlike RGS14, RGS12 isoforms can also contain an N-terminal domain cassette consisting of a PDZ (PSD-95/Dlg/ZO-1) domain and a PTB (phosphotyrosine-binding) domain. The PDZ domain has a "group I" binding specificity (that is, C-terminal Ala/Ser-Thr-Xaa-Leu/Val motifs) and has been shown *in vitro* to interact with GPCR C-termini, including those of the interleukin-8 receptor CXCR2 [58] and the corticotrophin-releasing factor receptor CRF-R1 (D. P. Siderovski, *et al.*, unpublished observations). In chick dorsal root ganglia neurons, RGS12 controls desensitization of GABA$_B$-receptor coupling to calcium channel inhibition by virtue of its agonist-dependent recruitment to the phosphorylated N-type calcium channel via its PTB domain [66]. With the ability to bind phosphotyrosine-containing proteins, Gα subunits, and Ras superfamily GTPases, RGS12 appears to represent a signaling nexus that facilitates convergence and cross-regulation of receptor tyrosine-kinase, heterotrimeric G-protein, and Ras-superfamily GTPase signaling. Some as-yet undefined aspect of this cross-regulatory function must underlie the ability of RGS12 and RGS14 to inhibit G12/13-heterotrimer signaling; overexpression of either RGS12 or RGS14 is reported to block G12/13-dependent transcriptional readouts [67,68], yet neither RGS protein acts as a GAP for Gα12/13 subunits in solution-based assays [58,68].

E- or RA-Subfamily (Axin, Axil)

Axin ("*axis in*hibition") and Axil ("*axin-l*ike") proteins, negative regulators of Wnt signaling through Frizzled receptors, constitute the E- or RA-subfamily of RGS proteins. Axin and Axil are scaffold proteins that coordinately bind β-catenin, glycogen synthase kinase-3β (GSK3β), and APC (the *a*denomatous *p*olyposis *c*oli tumor-suppressor protein), facilitating the effects of GSK3β and APC on β-catenin destruction, which is antagonistic to Wnt-dependent transcriptional upregulation [69]. The RGS-boxes of Axin and Axil have, as yet, not been found to associate with Gα subunits; rather, the RGS-box represents the binding site for APC [70,71]. The crystal structure of Axin in complex with APC indicates that this interaction occurs on the opposing face of the RGS-box, leaving the Gα-interaction face unhindered [14]. Thus, the possibility exists that a Gα partner may yet be discovered for the RGS-boxes of Axin and Axil. Additional circumstantial evidence for the involvement of Gα subunits in the Wnt/Frizzled signaling pathway includes several reports suggesting that Frizzled receptors act as canonical GPCRs (for example, [72,73]) and the recent finding that activated Gα12 and Gα13 subunits can release β-catenin from cadherins [74].

F- or GEF-Subfamily (p115-RhoGEF, PDZ-RhoGEF, LARG)

The three members of the F- or GEF-subfamily (p115-RhoGEF/Lsc, PDZ-RhoGEF/GTRAP48, and leukemia-associated RhoGEF [LARG]) represent the clearest examples of RGS proteins as *positive* regulators of GPCR signaling—that is, as effectors coupling the activation of Gαq, Gα12, and Gα13 subunits to the activation of the small GTPase RhoA. All three RhoGEF proteins share a common structure of an N-terminal RGS-box and C-terminal Dbl-homology (DH) and pleckstrin-homology (PH) domains that collectively encode RhoA-specific guanine nucleotide exchange factor (GEF) activity [75]; the latter two proteins (PDZ-RhoGEF and LARG) also possess an N-terminal PDZ domain (Fig. 2). Kozasa and colleagues first demonstrated that the p115-RhoGEF RGS-box can accelerate GTP hydrolysis by Gα12 and Gα13 subunits [76]. Moreover, binding of activated Gα13, but not Gα12, to the RGS-box was shown to derepress the RhoGEF activity of the C-terminal DH/PH tandem [77]. LARG acts as a Gα-responsive RhoGEF not only for Gα12 and Gα13 subunits [78], but also for Gαq [18]. In addition, LARG associates with the C-terminal tail of the insulin-like growth factor (IGF)-1 receptor tyrosine kinase via its PDZ domain [79] and its RhoGEF activity can be modulated by tyrosine phosphorylation [80]. These recent findings suggest that LARG, and potentially other members of this subfamily, can participate in cross-talk between tyrosine kinase, GPCR, and monomeric GTPase signaling pathways.

G- or GRK-Subfamily (GRK1, GRK2, GRK3, GRK4, GRK5, GRK6, GRK7)

Among the original set of papers [2–5] heralding the discovery of the RGS superfamily, one report identified the conserved RGS-box sequence within a known family of GPCR signaling regulators, the G-protein-coupled receptor kinases or GRKs [4]. These serine/threonine kinases phosphorylate activated GPCRs, thereby allowing the binding of arrestin proteins, functional uncoupling from G proteins, and endocytosis of the phosphorylated receptor [81,82]. Three groups [83–85] have since shown that, at least for GRK2, the N-terminal RGS-box can act to inhibit Gαq signaling; this phosphorylation-independent inhibitory activity of GRK2 is thought not to result from RGS-mediated GAP activity (which is barely detectable *in vitro* [83]), but rather by sequestration of activated Gαq by the N-terminal RGS-box.

H- or SNX-Subfamily (SNX13, SNX14, SNX25)

The RGS-box sequence is also present in at least three of the sorting nexins or SNXs, a growing family of proteins involved in vesicular trafficking between cellular compartments [86]. Sorting nexins invariably contain Phox-homology (PX) domains that generally bind phosphatidylinositol-3-phosphate (PtdIns(3)P); the SNX-subfamily of RGS proteins also contain a poorly-defined PX-associated (PXA) domain N-terminal to the RGS-box and C-terminal to a putative transmembrane region (Fig. 2). The RGS-box of SNX13 was purported to have Gαs-directed GAP activity and inhibit Gαs-dependent signaling pathways upon overexpression [87], but the Gα specificities of the other two known subfamily members (SNX14 [88] and SNX25) have not been characterized to-date. SNX1, the founding member of the sorting nexins, is a binding partner of the epidermal growth factor receptor (EGF-R) and, upon overexpression, enhances EGF-R degradation [89]. Conversely, SNX13 overexpression delays EGF-R degradation [87]. A direct role for SNX13 or other SNX-subfamily RGS proteins in the trafficking of GPCRs and/or heterotrimeric G-protein subunits remains to be determined.

Acknowledgments

The authors thank M. Hains, R. Kimple, and F. Willard for critical appraisal of this manuscript. Work in the authors' laboratories is supported in part by P01 grant GM065533 from the National Institute of General Medical Sciences (NIH).

References

1. Berstein, G., Blank, J. L., Jhon, D.-Y., Exton, J. H., Rhee, S. G., and Ross, E. M. (1992). *Cell* **70**, 411–418.
2. De Vries, L., Mousli, M., Wurmser, A., and Farquhar, M. G. (1995). *Proc. Natl. Acad. Sci. USA* **92**, 11916–11920.
3. Koelle, M. R. and Horvitz, H. R. (1996). *Cell* **84**, 115–125.
4. Siderovski, D. P., Hessel, A., Chung, S., Mak, T. W., and Tyers, M. (1996). *Curr. Biol.* **6**, 211–212.
5. Druey, K. M., Blumer, K. J., Kang, V. H., and Kehrl, J. H. (1996). *Nature* **379**, 742–746.
6. Berman, D. M., Wilkie, T. M., and Gilman, A. G. (1996). *Cell* **86**, 445–452.
7. Watson, N., Linder, M. E., Druey, K. M., Kehrl, J. H., and Blumer, K. J. (1996). *Nature* **383**, 172–175.
8. Hunt, T. W., Fields, T. A., Casey, P. J., and Peralta, E. G. (1996). *Nature* **383**, 175–177.
9. Popov, S., Yu, K., Kozasa, T., and Wilkie, T. M. (1997). *Proc. Natl. Acad. Sci. USA* **94**, 7216–7220.
10. Faurobert, E. and Hurley, J. B. (1997). *Proc. Natl. Acad. Sci. USA* **94**, 2945–2950.
11. Tesmer, J. J., Berman, D. M., Gilman, A. G., and Sprang, S. R. (1997). *Cell* **89**, 251–261.
12. de Alba, E., De Vries, L., Farquhar, M. G., and Tjandra, N. (1999). *J. Mol. Biol.* **291**, 927–939.
13. Moy, F. J., Chanda, P. K., Cockett, M. I., Edris, W., Jones, P. G., Mason, K., Semus, S., and Powers, R. (2000). *Biochemistry* **39**, 7063–7073.
14. Spink, K. E., Polakis, P., and Weis, W. I. (2000). *EMBO J.* **19**, 2270–2279.
15. Longenecker, K. L., Lewis, M. E., Chikumi, H., Gutkind, J. S., and Derewenda, Z. S. (2001). *Structure (Cambridge)* **9**, 559–569.
16. Chen, Z., Wells, C. D., Sternweis, P. C., and Sprang, S. R. (2001). *Nat. Struct. Biol.* **8**, 805–809.
17. Slep, K. C., Kercher, M. A., He, W., Cowan, C. W., Wensel, T. G., and Sigler, P. B. (2001). *Nature* **409**, 1071–1077.
18. Booden, M. A., Siderovski, D. P., and Der, C. J. (2002). *Mol. Cell Biol.* **22**, 4053–4061.
19. Krumins, A. M. and Gilman, A. G. (2002). *Methods Enzymol.* **344**, 673–685.
20. Chidiac, P. and Ross, E. M. (1999). *J. Biol. Chem.* **274**, 19639–19643.
21. Snow, B. E., Krumins, A. M., Brothers, G. M., Lee, S.-F., Wall, M. A., Chung, S., Mangion, J., Arya, S., Gilman, A. G., and Siderovski, D. P. (1998). *Proc. Natl. Acad. Sci. USA* **95**, 13307–13312.
22. Mukhopadhyay, S. and Ross, E. M. (1999). *Proc. Natl. Acad. Sci. USA* **96**, 9539–9544.
23. Gierschik, P., Bouillon, T., and Jakobs, K. H. (1994). *Methods Enzymol.* **237**, 13–26.
24. McIntire, W. E., MacCleery, G., and Garrison, J. C. (2001). *J. Biol. Chem.* **276**, 15801–15809.
25. Ingi, T., Krumins, A. M., Chidiac, P., Brothers, G. M., Chung, S., Snow, B. E., Barnes, C. A., Lanahan, A. A., Siderovski, D. P., Ross, E. M., Gilman, A. G., and Worley, P. F. (1998). *J. Neurosci.* **18**, 7178–7188.
26. Posner, B. A., Mukhopadhyay, S., Tesmer, J. J., Gilman, A. G., and Ross, E. M. (1999). *Biochemistry* **38**, 7773–7779.
27. Biddlecome, G. H., Berstein, G., and Ross, E. M. (1996). *J. Biol. Chem.* **271**, 7999–8007.
28. Heximer, S. P., Watson, N., Linder, M. E., Blumer, K. J., and Hepler, J. R. (1997). *Proc. Natl. Acad. Sci. USA* **94**, 14389–14393.
29. Glick, J. L., Meigs, T. E., Miron, A., and Casey, P. J. (1998). *J. Biol. Chem.* **273**, 26008–26013.
30. Jordan, J. D., Carey, K. D., Stork, P. J., and Iyengar, R. (1999). *J. Biol. Chem.* **274**, 21507–21510.
31. Wang, J., Ducret, A., Tu, Y., Kozasa, T., Aebersold, R., and Ross, E. M. (1998). *J. Biol. Chem.* **273**, 26014–26025.
32. De Vries, L., Elenko, E., Hubler, L., Jones, T. L., and Farquhar, M. G. (1996). *Proc. Natl. Acad. Sci. USA* **93**, 15203–15208.
33. Tu, Y., Popov, S., Slaughter, C., and Ross, E. M. (1999). *J. Biol. Chem.* **274**, 38260–38267.
34. Druey, K. M., Ugur, O., Caron, J. M., Chen, C. K., Backlund, P. S., and Jones, T. L. (1999). *J. Biol. Chem.* **274**, 18836–18842.
35. Rose, J. J., Taylor, J. B., Shi, J., Cockett, M. I., Jones, P. G., and Hepler, J. R. (2000). *J. Neurochem.* **75**, 2103–2112.
36. Wang, Q., Liu, M., Mullah, B., Siderovski, D. P., and Neubig, R. R. (2002). *J. Biol. Chem.* **277**, 24949–24958.
37. Xu, X., Zeng, W., Popov, S., Berman, D. M., Davignon, I., Yu, K., Yowe, D., Offermanns, S., Muallem, S., and Wilkie, T. M. (1999). *J. Biol. Chem.* **274**, 3549–3556.

38. Zeng, W., Xu, X., Popov, S., Mukhopadhyay, S., Chidiac, P., Swistok, J., Danho, W., Yagaloff, K. A., Fisher, S. L., Ross, E. M., Muallem, S., and Wilkie, T. M. (1998). *J. Biol. Chem.* **273**, 34687–34691.
39. Luo, X., Popov, S., Bera, A. K., Wilkie, T. M., and Muallem, S. (2001). *Mol. Cell* **7**, 651–660.
40. Popov, S. G., Krishna, U. M., Falck, J. R., and Wilkie, T. M. (2000). *J. Biol. Chem.* **275**, 18962–18968.
41. Ishii, M., Inanobe, A., and Kurachi, Y. (2002). *Proc. Natl. Acad. Sci. USA* **99**, 4325–4330.
42. Ponting, C. and Bork, P. (1996). *Trends Biochem. Sci.* **21**, 245–246.
43. Dohlman, H. G., Song, J., Ma, D., Courchesne, W. E., and Thorner, J. (1996). *Mol. Cell Biol.* **16**, 5194–5209.
44. Hoffman, G. A., Garrison, T. R., and Dohlman, H. G. (2000). *J. Biol. Chem.* **275**, 37533–37541.
45. Burchett, S. A., Flanary, P., Aston, C., Jiang, L., Young, K. H., Uetz, P., Fields, S., and Dohlman, H. G. (2002). *J. Biol. Chem.* **8**, 8.
46. Snow, B. E., Betts, L., Mangion, J., Sondek, J., and Siderovski, D. P. (1999). *Proc. Natl. Acad. Sci. USA* **96**, 6489–6494.
47. Makino, E. R., Handy, J. W., Li, T., and Arshavsky, V. Y. (1999). *Proc. Natl. Acad. Sci. USA* **96**, 1947–1952.
48. Levay, K., Cabrera, J. L., Satpaev, D. K., and Slepak, V. Z. (1999). *Proc. Natl. Acad. Sci. USA* **96**, 2503–2507.
49. Sondek, J. and Siderovski, D. P. (2001). *Biochem. Pharmacol.* **61**, 1329–1337.
50. Sondek, J., Bohm, A., Lambright, D. G., Hamm, H. E., and Sigler, P. B. (1996). *Nature* **379**, 369–374.
51. Clapham, D. E. and Neer, E. J. (1997). *Annu. Rev. Pharmacol. Toxicol.* **37**, 167–203.
52. Posner, B. A., Gilman, A. G., and Harris, B. A. (1999). *J. Biol. Chem.* **274**, 31087–31093.
53. Kovoor, A., Chen, C. K., He, W., Wensel, T. G., Simon, M. I., and Lester, H. A. (2000). *J. Biol. Chem.* **275**, 3397–3402.
54. He, W., Cowan, C. W., and Wensel, T. G. (1998). *Neuron* **20**, 95–102.
55. Chen, C. K., Burns, M. E., He, W., Wensel, T. G., Baylor, D. A., and Simon, M. I. (2000). *Nature* **403**, 557–560.
56. Lyubarsky, A. L., Naarendorp, F., Zhang, X., Wensel, T., Simon, M. I., and Pugh, E. N. Jr. (2001). *Mol. Vis.* **7**, 71–78.
57. Lan, K. L., Zhong, H., Nanamori, M., and Neubig, R. R. (2000). *J. Biol. Chem.* **275**, 33497–33503.
58. Snow, B. E., Hall, R. A., Krumins, A. M., Brothers, G. M., Bouchard, D., Brothers, C. A., Chung, S., Mangion, J., Gilman, A. G., Lefkowitz, R. J., and Siderovski, D. P. (1998). *J. Biol. Chem.* **273**, 17749–17755.
59. Traver, S., Bidot, C., Spassky, N., Baltauss, T., De Tand, M. F., Thomas, J. L., Zalc, B., Janoueix-Lerosey, I., and Gunzburg, J. D. (2000). *Biochem. J.* **350**, 19–29.
60. Hollinger, S., Taylor, J. B., Goldman, E. H., and Hepler, J. R. (2001). *J. Neurochem.* **79**, 941–949.
61. Burgon, P. G., Lee, W. L., Nixon, A. B., Peralta, E. G., and Casey, P. J. (2001). *J. Biol. Chem.* **276**, 32828–32834.
62. Ponting, C. P. (1999). *J. Mol. Med.* **77**, 695–698.
63. Siderovski, D. P., Diversé-Pierluissi, M. A., and De Vries, L. (1999). *Trends Biochem. Sci.* **24**, 340–341.
64. Kimple, R. A., De Vries, L., Tronchere, H., Behe, C. I., Morris, R. A., Farquhar, M. G., and Siderovski, D. P. (2001). *J. Biol. Chem.* **276**, 29275–29281.
65. Kimple, R. J., Kimple, M. E., Betts, L., Sondek, J., and Siderovski, D. P. (2002). *Nature* **416**, 878–881.
66. Schiff, M. L., Siderovski, D. P., Jordan, J. D., Brothers, G., Snow, B., De Vries, L., Ortiz, D. F., and Diverse-Pierluissi, M. (2000). *Nature* **408**, 723–727.
67. Mao, J., Yuan, H., Xie, W., and Wu, D. (1998). *Proc. Natl. Acad. Sci. USA* **95**, 12973–12976.
68. Cho, H., Kozasa, T., Takekoshi, K., De Gunzburg, J., and Kehrl, J. H. (2000). *Mol. Pharmacol.* **58**, 569–576.
69. Barker, N., Morin, P. J., and Clevers, H. (2000). *Adv. Cancer Res.* **77**, 1–24.
70. Hart, M. J., de los Santos, R., Albert, I. N., Rubinfeld, B., and Polakis, P. (1998). *Curr. Biol.* **8**, 573–581.
71. Behrens, J., Jerchow, B.-A., Wuertele, M., Grimm, J., Asbrand, C., Wirtz, R., Kuehl, M., Wedlich, D., and Birchmeier, W. (1998). *Science* **280**, 596–599.
72. Liu, T., DeCostanzo, A. J., Liu, X., Wang, H., Hallagan, S., Moon, R. T., and Malbon, C. C. (2001). *Science* **292**, 1718–1722.
73. Sheldahl, L. C., Park, M., Malbon, C. C., and Moon, R. T. (1999). *Curr. Biol.* **9**, 695–698.
74. Meigs, T. E., Fields, T. A., McKee, D. D., and Casey, P. J. (2001). *Proc. Natl. Acad. Sci. USA* **98**, 519–524.
75. Snyder, J. T., Worthylake, D. K., Rossman, K. L., Betts, L., Pruitt, W. M., Siderovski, D. P., Der, C. J., and Sondek, J. (2002). *Nat. Struct. Biol.* **13**, 13.
76. Kozasa, T., Jiang, X., Hart, M. J., Sternweis, P. M., Singer, W. D., Gilman, A. G., Bollag, G., and Sternweis, P. C. (1998). *Science* **280**, 2109–2111.
77. Hart, M. J., Jiang, X., Kozasa, T., Roscoe, W., Singer, W. D., Gilman, A. G., Sternweis, P. C., and Bollag, G. (1998). *Science* **280**, 2112–2114.
78. Fukuhara, S., Chikumi, H., and Gutkind, J. S. (2000). *FEBS Lett.* **485**, 183–188.
79. Taya, S., Inagaki, N., Sengiku, H., Makino, H., Iwamatsu, A., Urakawa, I., Nagao, K., Kataoka, S., and Kaibuchi, K. (2001). *J. Cell Biol.* **155**, 809–820.
80. Chikumi, H., Fukuhara, S., and Gutkind, J. S. (2002). *J. Biol. Chem.* **277**, 12463–12473.
81. Krupnick, J. G. and Benovic, J. L. (1998). *Annu. Rev. Pharmacol. Toxicol.* **38**, 289–319.
82. Pitcher, J. A., Freedman, N. J., and Lefkowitz, R. J. (1998). *Annu. Rev. Biochem.* **67**, 653–692.
83. Carman, C. V., Parent, J. L., Day, P. W., Pronin, A. N., Sternweis, P. M., Wedegaertner, P. B., Gilman, A. G., Benovic, J. L., and Kozasa, T. (1999). *J. Biol. Chem.* **274**, 34483–34492.
84. Usui, H., Nishiyama, M., Moroi, K., Shibasaki, T., Zhou, J., Ishida, J., Fukamizu, A., Haga, T., Sekiya, S., and Kimura, S. (2000). *Int. J. Mol. Med.* **5**, 335–340.
85. Sallese, M., Mariggio, S., D'Urbano, E., Iacovelli, L., and De Blasi, A. (2000). *Mol. Pharmacol.* **57**, 826–831.
86. Teasdale, R. D., Loci, D., Houghton, F., Karlsson, L., and Gleeson, P. A. (2001). *Biochem. J.* **358**, 7–16.
87. Zheng, B., Ma, Y. C., Ostrom, R. S., Lavoie, C., Gill, G. N., Insel, P. A., Huang, X. Y., and Farquhar, M. G. (2001). *Science* **294**, 1939–1942.
88. Carroll, P., Renoncourt, Y., Gayet, O., De Bovis, B., and Alonso, S. (2001). *Dev. Dyn.* **221**, 431–442.
89. Kurten, R. C., Cadena, D. L., and Gill, G. N. (1996). *Science* **272**, 1008–1010.
90. Zheng, B., De Vries, L., and Gist Farquhar, M. (1999). *Trends Biochem. Sci.* **24**, 411–414.
91. Ross, E. M. and Wilkie, T. M. (2000). *Annu. Rev. Biochem.* **69**, 795–827.

CHAPTER 226

Mechanism of Gβγ Effector Interaction

Tohru Kozasa
Department of Pharmacology,
University of Illinois at Chicago, Chicago, Illinois

Introduction

The Gβγ subunit was initially recognized as an inactive subunit in G-protein-mediated signal transduction. Its main function was considered to form a high-affinity complex with GDP-bound Gα to generate an inactive heterotrimer. Only GTP-bound Gα subunits were thought to be responsible for the regulation of effectors. In 1987, Logothersis et al. presented the first clear evidence that βγ subunit free from Gα can activate K$^+$ channels in cardiac atrial cells [1]. Activated Gα subunit did not have an effect on K$^+$ channel activity. The following year, Whiteway et al. showed that βγ, but not Gα, is the responsible subunit to mediate mating-factor response in yeast [2]. Since then, various effectors or regulatory proteins have been identified as direct targets for the βγ subunit. It is now clear that receptor activation of G proteins generate two active signal mediators, GTP-bound Gα and free βγ subunit (reviewed in [3–5]).

Structure of βγ Subunits

cDNAs for five β subunits and 12 γ subunits have been isolated so far. Among five β subunits, β1, β2, β3, and β4 are highly homologous (about 80 percent identical to each other). β5 is distantly related (about 50 percent identical with other βs) and is expressed only in brain [6]. γ subunits are divergent in amino acid sequences and tissue distributions. The carboxyl terminus of the γ subunit contains a CAAX motif that directs prenylation of the molecule. Most of γ subunits are geranyl-geranylated except for γ1 and γ11, which are farnesylated. β1 to β4 subunits form a tightly complex with γs. They can only be separated under denaturing conditions. The interaction of β5 with γs is weak and sensitive to detergent condition [7].

The Gβ subunit is made up of two structurally distinct regions, an amino terminal α–helical segment, followed by seven WD-repeat sequences. The role of the WD repeat (about 40 amino acids in length) in cell signaling is not clearly understood, but many WD repeat-containing proteins participate in large macromolecular assemblies [8]. The crystal structures of βγ subunit, free or complexed with GDP-bound Gα, were solved recently [9–11]. The core WD repeat provides a rigid scaffold of a seven-bladed β-propeller structure. Each blade is made up of four anti-parallel β strands. The amino terminal region of γ forms an α-helical coiled-coil structure with the amino terminus of β. The rest of γ subunit extends along the wider bottom surface of the propeller through multiple interaction sites. Gα is mainly interacting with the opposite narrow surface of the propeller through its switch regions. The amino terminus of Gα is interacting with the outer strand of blade 2. It is important to note that the βγ subunit does not change its conformation when it dissociates from α subunit.

Effectors Interacting with βγ Subunits

GIRK Channel

G-protein-gated inwardly rectifying potassium channels (GIRKs) play an important role in regulating membrane excitability through G-protein-coupled receptors, especially in heart and brain [12, 13]. To date five mammalian cDNAs for GIRKs (GIRK1–5) have been isolated. GIRK functions as the heterotetramer, such as GIRK1 and 2 in heart and brain or GIRK1 and 4 in brain [12]. βγ subunits activate hetero- or homotetramer of GIRKs. It was shown that βγ subunits bind directly to the N- and C-terminal intracellular domains of GIRKs [14–16]. EC_{50} of β1γ2 for GIRK activation is about 30 nM [17,18]. βγ subunits with β1–β4 showed similar potency to activate GIRKs. It was shown recently that β5γ2 inhibited basal or agonist-induced GIRK1 and 4 activity by competing with β1γ2 [19]. The interaction of Giα with N-terminal region of GIRK was reported, but the regulatory function of this interaction is unknown [15].

PLCβ

Numerous G-protein-coupled receptors stimulate phosphoinositide phospholipaseCβ activity to induce intracellular Ca^{2+} release [20]. Both Gα subunits of Gq subfamily and βγ subunits can stimulate PLCβ activity in vitro and in vivo. Four isozymes of PLCβ are isolated, and all of them are activated by Gαq. The stimulation by βγ subunit was demonstrated for PLCβ1, β2, and β3 (with the potency of β3 > β2 > β1) but not for PLCβ4 [21–23]. Gqα and βγ directly bind to PLCβ with separate binding regions. In PLCβ2, the binding site for βγ has been mapped within the first half of the Y-domain of the catalytic region [24]. The stimulation of PLC activity by Gαq and βγ is additive for PLCβ3 but not for PLCβ2 [23,25]. The potency of βγ subunit to activate these PLCβs is much lower than Gαq. EC_{50} of βγ for PLCβ activation is 30–300 nM; in contrast, EC_{50} of Gαq is in the sub to low nanomolar range [23,25]. βγ-mediated PLCβ activation mainly contributes to pertussis-toxin sensitive PLC activation that is mediated through Gi/o. Recently, in vitro reconstitution experiments showed that β5γ2 could stimulate PLCβ1 and β2 but not PLCβ3 [26].

Adenylyl Cyclase

Isoforms of adenylyl cyclase (at least eight) are activated by Gαs and are differentially regulated by βγ subunits, Ca^{2+}, and phosphorylation [27]. AC-I is inhibited by βγ [28]. This inhibition by βγ is independent of Gαs stimulation. However, βγ activates AC-II and AC-IV synergistically with Gαs [28,29]. This activation of AC-II or AC-IV by βγ requires the activation by Gαs. Without Gαs stimulation, βγ has no effect on these adenylyl cyclases. Thus, in cells that are expressing AC-II or AC-IV, receptors that normally do not regulate cAMP production, such as receptors that couple to Gi or Gq, can further potentiate cAMP production. The site of βγ binding on adenylyl cyclase is not precisely known. For AC-I, the βγ binding region is separated from the Gαs binding region. The amino terminal part of the first large cytosolic domain of AC-I appears to include a βγ binding site [30]. For AC-II, a 14-amino acid peptide containing a QXXER motif derived from the second cytoplasmic domain blocks interaction with βγ [31]. It is interesting to note that the same peptide can block interaction of βγ with several other effectors, including PLCβ and GIRK channels. In reconstitution experiments, β5γ2 was markedly less effective than β1γ2 to activate AC-II (EC_{50} of ~700 nM versus 25 nM) [7].

N- and P/Q-type Calcium Channels

It has been shown that several neurotransmitters inhibit N-type and P/Q type calcium channels through pertussis-toxin sensitive G proteins. This effect is mediated by the direct binding of βγ subunit to calcium channels [32,33]. The βγ binding site was mapped at the intracellular loop and the C terminus of alpha1 subunit [34,35]. QXXER motif in loop I was shown to be important for the modulation by βγ subunit [35].

PI3K

Phosphoinositide 3-kinase (PI3K) catalyzes phosphorylation of phosphoinositide at the 3 position and plays a key role in a variety of cellular functions. It is stimulated by both G-protein-coupled receptors and receptor tyrosine kinases. PI3Kγ is specifically activated through the direct interaction with βγ subunit [36,37]. This activation involves the interaction of βγ subunit with both p110 catalytic subunit and p101 regulatory subunit (EC_{50} about 20 nM). The presence of p101 enhances the sensitivity of PI3Kγ to βγ subunit [38]. β5γ2 does not activate PI3Kγ but can bind to the molecule [26]. PI3Kβ is synergistically activated by βγ subunit and phosphotyrosine-containing peptide [39]. PI3K α and δ are activated by tyrosine kinase pathway but not by G-protein-mediated pathway.

MAP Kinase Pathway

It has been shown that the yeast homologue of βγ subunit (STE4 and STE18) is the active mediator for the pheromone-induced MAP kinase pathway to control cell cycle [2]. STE4 interacts with serine-threonine kinase STE20/PAK and scafolding protein STE5 [40,41]. These interactions are critical to mediate pheromone-mediated MAP kinase activation. G-protein-linked pathways can also initiate MAP kinase pathways in mammalian cells [42]. Some of these pathways are pertussis toxin-sensitive and some are pertussis toxin-insensitive. βγ subunit released from Gi or Go appears to initiate Ras-dependent MAP kinase activation in pertussis toxin-sensitive pathways. It was shown that βγ could activate several MAP kinase pathways, including the JNK/SAPK pathway and p38MAPK pathway [43,44]. Although the involvement of PI3K or nonreceptor tyrosine kinase has been suggested as a downstream component of βγ to MAPK activation, the precise biochemical mechanism to link βγ and MAP kinase activation has not been clearly understood. The complex of β5γ2 cannot activate either the MAPK or JNK pathway [45].

Other Effectors and Interacting Proteins

BTK FAMILY TYROSINE KINASES

Btk family tyrosine kinases are nonreceptor tyrosine kinases with a unique N-terminal extension that contain

PH and TH (Tec homology) domains. The regulation of these kinases in cell has not yet been well characterized. βγ subunit binds to the PH-TH domain and catalytic domains of Btk to stimulate the catalytic activity of the kinase [46]. Tsk, another member of this family, was also activated by co-transfection with βγ subunits [47].

PLCε

PLCε is a recently identified PI-PLC that is unique in that it contains a CDC25 domain and Ras-associating domains [48]. Involvement of Gα12 for the regulation of PLCε activity has been demonstrated. In addition, co-expression of β1γ2 but not β5γ2 with PLCε potently stimulated PLC activity [49]. Although an *in vitro* reconstitution experiment was unsuccessful, the βγ stimulatory effect on PLCε seems to be independent of Ras activation and PI3K activation, suggesting the direct stimulation by βγ.

GRK

βγ binds directly to the C-terminal region of the pleckstrin homology domain of β adrenergic receptor kinases (GRK2 and GRK3) and enhances their kinase activity [50,51]. It is not clearly understood whether βγ stimulates catalytic activity of GRKs or βγ facilitates membrane translocation of GRKs for activation. The C-terminal region of GRK2 can sequester βγ in cells without interacting with other α subunits. Thus, it is a useful tool to differentiate between βγ-mediated effects and Gα-mediated effects.

Phosducin

Phosducin is a phosphoprotein mainly expressed in retina or pineal gland. Phosducin regulates Gt-mediated phototransduction through high-affinity binding to βγ subunit and prevents the reassociation of βγ with Gαt [52]. Phosducin-like molecule has also been found in brain [53]. The crystal structure of phosducin and βγ complex revealed that phosducin interacts with the Gα binding surface of the βγ propeller and the outer surfaces of blades 1 and 7 [54]. The phosphorylation of phosducin inhibits its affinity to βγ. This phosphorylation site is not on the interacting surface on this structure, suggesting allosteric regulation of the interaction by phosphorylation.

In addition to these molecules, βγ subunits can also interact with other regulatory proteins such as Dbl, Raf-1 kinase, RasGEF, Rho family GTPases, or ARF [55–59]. The functional significance of these interactions is not clearly understood yet.

Specificity of the Interaction between βγ Subunit and Effectors

With multiple β and γ subunit cDNAs, various combinations of βγ subunit were tested to examine the specificity of their interaction with effectors. Most of them could form functional dimers. However, the degree of specificity in reconstitution assays was minimal in general. Most of βγ pairs showed almost the same potency and efficacy for multiple effectors. β1γ1, βγ for retinal transducin, showed lower potency for several effectors compared with other combination of βγs, while showing better interaction with rhodopsin and phosducin. It was also shown recently that β5, which is the most distantly related member of β subunits, differentially couples to several effectors; β5γ2 activates PLC β1 and β2 but not β3. β5γ2 does not activate GIRK, PI3Kγ, or MAP kinase pathways, although it can interact with GIRK or PI3Kγ [26,45]. In addition, several groups suggested that the physiological binding partner of β5 is not γ subunit but RGS proteins with a G-protein γ-subunit-like (GGL) domain (RGS6, 7, 9, and 11) [60,61]. Although the homologues of β5-RGS have a regulatory function in the egg-laying signaling pathway in *C. elegans,* the biochemical function of the β5-RGS complex in mammalian cells has not been understood yet [62].

βγ-effector interaction requires C-terminus prenylation of the γ subunit. Although lipid modification does not affect βγ subunit formation, it is essential for the interactions of βγs with effectors. It is likely that the difference of the lipid modification on γ1 (farnesyl) versus other γs (geranyl-geralyl) contributes to the difference of effector interaction mentioned above. Another common feature of βγ-effector interaction is that βγ has to dissociate from the heterotrimer in order to interact with effectors. This indicates that effectors share the top surface of βγ with the Gα subunit. This was confirmed by mutagenesis studies [63,64]. However, it also became evident that the different effectors overlap with the Gα surface in different ways.

The effector-interacting region of the β subunit is not restricted to the Gα interacting surface. The N-terminal α helical region of STE4 is involved in the interaction with STE20 to mediate phromone signaling [41]. It was also shown that the N-terminal 100 amino acid of β (including an α helical region and blade 1 and 2) interacts with GIRK1, AC-II, and PLCβ2 [65]. The importance of the C-terminal region of β for activation of PLC, AC-II, or MAP kinase pathways was also demonstrated [45,66,67]. The study with chimeras between β1 and β5 revealed that amino acids 52 to 143 of β1, which spans blade 2 and 3, is critical for the activation of GIRK [68]. The outer surface of the β-propeller also contributes to the effector interaction. The crystal structure revealed that the C-terminal part of phosducin interacts with the outer surface of blades 1 and 7 [54]. And a mutational study demonstrated that outer strands of blade 2 and 3 are differentially involved in GIRK or PLCβ2 activation [18,69].

It is now evident that βγ subunits of G protein play a variety of functions in cell signaling. In order to understand the physiological role of the βγ subunit, including subunit specificity, further investigation will be required using methods such as antisense RNA, RNAi, or gene knockout mice. Also, the determination of the crystal structure of the complex of βγ with different effectors will be critically important in understanding the effector activation mechanism by Gβγ subunits.

References

1. Logothersis, D. E., Kurachi, Y. et al. (1987). The βγ subunits of GTP-binding proteins activate the muscarinic K$^+$ channel in heart. *Nature* **325**, 321–326.
2. Whiteway, M., Hougan, L. et al. (1989). The STE4 and STE18 genes of yeast encode potential β and γ subunits of the mating factor receptor-coupled G protein. *Cell* **56**, 467–477.
3. Clapham, D. E. and Neer, E. J. (1997). G protein βγ sununits. *Annu. Rev. Pharmacol. Toxicol.* **37**, 167–203.
4. Hamm, H. E. (1998). The many faces of G protein signaling. *J. Biol. Chem.* **273**, 669–672.
5. Schwindinger, W. F., and Robishaw, J. D. (2001). Heterotrimeric G protein βγ-dimers in growth and differentiation. *Oncogene* **20**, 1653–1660.
6. Watson, J. A., Katz, A., and Simon, M. I. (1994). A fifth member of the mammalian G-protein β-subunit family. *J. Biol. Chem.* **269**, 22150–22156.
7. Lindorfer, M. A., Myung, C.-S. et al. (1998). Differential activity of the G protein β5γ2 subunit at receptors and effectors. *J. Biol. Chem.* **273**, 34429–34436.
8. Smith, T. F., Gaitatzes, C. et al. (1999). The WD repeat: a common architecture for diverse functions. *Trends Biochem. Sci.* **24**, 181–185.
9. Wall, M. A., Coleman, D. E. et al. (1995). The structure of the G protein heterotrimer Giα1β1γ2. *Cell* **83**, 1047–1058.
10. Lambright, D. G., Sondek, J. et al. (1996). The 2.0 A crystal structure of a heteritrimeric G protein. *Nature* **379**, 311–319.
11. Sondek, J., Bohm, A. et al. (1996). Crystal structure of a Gα protein βγ dimmer at 2.1 A resolution. *Nature* **379**, 369–374.
12. Sui, J.-L., Chan, K. W. et al. (1999). G protein-gated potassium channels. *Adv. Second Messenger Phosphoprotein Res.* **33**, 179–201.
13. Dascal, N. (2001). Ion-channel regulation by G proteins. *Trends Endocrinol. Metab.* **12**, 391–398.
14. Krapavinsky, G., Krapavinsky, L. et al. (1995). Gβγ binds directly to the G protein-gated K$^+$ channel, I$_{KACH}$. *J. Biol. Chem.* **270**, 29059–29062.
15. Huang, C. L., and Slesinger, P. A. (1995). Evidence that direct binding of Gβγ to the GIRK1, G protein-gated inwardly rectifying channel is important for channel activation. *Neuron* **15**, 1133–1143.
16. He, C. Yan, X. et al. (2002). Identification of critical residues controlling G protein-gated inwardly rectifying K$^+$ channel activity through interactions with the βγ subunits of G proteins. *J. Biol. Chem.* **277**, 6088–6096.
17. Wickmann, K. D., Iniguez-Lluhi, J. A. et al. (1994). Recombinant G protein βγ-subunit activate the muscarinic-gated atrial potassium channel. *Nature* **368**, 255–257.
18. Albsoul-Younes, A. M., Sternweis, P. M. et al. (2001). Interaction sites of the G protein β subunit with brain GIRK. *J. Biol. Chem.* **276**, 12712–12717.
19. Lei, Q., Jones, M. et al. (2000). Activation and inhibition of G protein-coupled inwardly rectifying potassium (Kir3) channels by G protein βγ subunits. *Proc. Natl. Acad. Sci. USA* **97**, 9771–9776.
20. Singer, W. D., Brown, A., and Sternweis, P. C. (1997). Regulation of eukaryotic phosphatidylinositol-specific phospholipase C and phospholipase D. *Annu. Rev. Biochem.* **66**, 475–509.
21. Park, D., Jhon, D.-Y. et al. (1993). Activation of phospholipase C isozymes by G protein βγ subunits. *J. Biol. Chem.* **268**, 4573–4576.
22. Jiang, H. P., Wu, D. Q. et al. (1994). Activation of phospholipaseC β4 by heterotrimeric GTP-binding proteins. *J. Biol. Chem.* **269**, 7593–7596.
23. Smrcka, A. V. and Sternweis, P. C. (1993). Regulation of purified subtypes of phosphatidylinositol-specific phospholipase Cβ by G protein α and βγ subunits. *J. Biol. Chem.* **268**, 9667–9674.
24. Kuang, Y., Wu, Y. et al. (1996). Identification of a phospholipase Cβ2 region that interacts with Gβγ. *Proc. Natl. Acad. Sci. USA* **93**, 2964–2968.
25. Hepler, J. R., Kozasa, T. et al. (1993). Purification from Sf9 cells and characterization of recombinant Gqα and G11α. *J. Biol. Chem.* **268** 14367–14375.
26. Maier, U., Babich, A. et al. (2000). Gβ5γ2 is a highly selective activator for phospholipid-dependent enzymes. *J. Biol. Chem.* **275**, 13746–13754.
27. Sunahara, R., Dessauer, C., and Gilman, A. G. (1996). Complexity and diversity of mammalian adenylyl cyclases. *Annu. Rev. Pharmacol. Toxicol.* **36**, 461–480.
28. Tang, W.-J. and Gilman, A. G. (1991). Type-specific regulation of adenylyl cyclase by G protein βγ subunits. *Science* **254**, 1500–1503.
29. Gao, B. N. and Gilman, A. G. (1991). Cloning and expression of a widely distributed (type IV) adenylyl cyclase. *Proc. Natl. Acad. Sci. USA* **88**, 10178–10182.
30. Chen, Y., Weng, G. et al. (1997). A surface on the G protein β subunit involved in interactions with adenylyl cyclases. *Proc. Natl. Acad. Sci. USA* **94**, 2711–2714.
31. Chen, J., DeVivo, M. et al. (1995). A region of adenylyl cyclase 2 critical for regulation by G protein βγ subunits. *Science* **268**, 1166–1169.
32. Ikeda, S. R. (1996). Voltage-dependent modulation of N-type calcium channels by G-protein βγ subunits. *Nature* **380**, 255–258.
33. Herlitze, S., Garcia, D. E. et al. (1996). Modulation of Ca^{2+} channels by G-protein βγ subunits. *Nature* **380**, 258–262.
34. Feng, Z.-P., Arnot, M. J. et al. (2001). Calcium channel β subunits differentially regulate the inhibition of N-type channels by individual Gβ isoforms. *J. Biol. Chem.* **276**, 45051–45058.
35. Herlitze, S., Hockerman, G. H. et al. (1997). Molecular determinants of inactivation and G protein modulation in the intracellular loop connecting domains I and II of the calcium channel α1A subunit. *Proc. Natl. Acad. Sci. USA* **94**, 1512–1516.
36. Stephens, L. R., Smrcka, A. et al. (1994). A novel phosphoinositide 3 kinase activity in myeloid-derived cells is activated by G protein βγ subunits. *Cell* **77**, 83–93.
37. Leopoldt, D., Hanck, T. et al. (1998). Gβγ stimulated phosphoinositide 3-kinase-γ by direct interaction with two domains of the catalytic p110 subunit. *J. Biol. Chem.* **273**, 7024–7029.
38. Krugmann, S., Hawkins, P. T. et al. (1999). Characterizing the interactions between the two subunits of the p101/p110γ phosphoinositide 3-kinase and their role in the activation of this enzyme by Gβγ subunits. *J. Biol. Chem.* **274**, 17152–17158.
39. Kurosu, H., Maehama, T. et al. (1997). Heterotrimeric phosphoinositide 3-kinase consisting of p85 and p110β is synergistically activated by the βγ subunits of G proteins and phosphotyrosyl peptide. *J. Biol. Chem.* **272**, 24252–24256.
40. Whiteway, M. S., Wu, C. et al. (1995). Association of the yeast pheromone response G protein βγ subunits with the MAP kinase scaffold Ste5p. *Science* **269**, 1572–1575.
41. Leeuw, T., Wu, C. et al. (1998). Interaction of a G-protein β-subunit with a conserved sequence in Ste20/PAK family protein kinases. *Nature* **391**, 191–194.
42. Gutkind, J. S. (1998). The pathways connecting G protein-coupled receptors to the nucleus through divergent mitogen-activated protein kinase cascade. *J. Biol. Chem.* **273**, 1839–1842.
43. Coso, O. A., Teramoto, H. et al. (1996). Signaling from G protein-coupled receptors to c-Jun kinase involves betagamma subunits of heterotrimeric G proteins acting on a Ras and Rac1-dependent pathway. *J. Biol. Chem.* **271**, 3963–3966.
44. Yamauchi, J., Kaziro, Y., and Itoh, H. (1997). C-terminal mutation of G protein β subunit affects differentially extracellular signal-regulated kinase and c-Jun N-terminal kinase pathways in human embryonal kidney 293 cells. *J. Biol. Chem.* **272**, 7602–7607.
45. Zhang, S., and Coso. O. A. (1996). Selective activation of effector pathways by brain-specific G protein β5. *J. Biol. Chem.* **271**, 33575–33579.
46. Lowry, W. E., and Huang, X.-Y. (2002). G protein βγ subunits act on the catalytic domain to stimulate Bruton's Agammaglobulinemia tyrosine kinase. *J. Biol. Chem.* **277**, 1488–1492.
47. Langhans-Rajasekaran, S. A. et al. (1995). Activation of Tsk and Btk tyrosine kinases by G protein βγ subunits. *Proc. Natl. Acad. Sci. USA* **92**, 8601–8605.
48. Lopez, I., Mak, E. C. et al. (2001). A novel bifunctional phospholipase C that is regulated by Gα12 and stimulates the Ras/MAP kinase pathway. *J. Biol. Chem.* **276**, 2758–2765.

49. Wing, M. R., Houston, D. *et al.* (2001). Activation of phospholipase C-ε by heterotrimeric G protein βγ-subunits. *J. Biol. Chem.* **276**, 48257–48261.
50. DebBurman, S. K., Ptasienski, J. *et al.* (1996). G protein-coupled receptor kinase GRK2 is a phospholipid-dependent enzyme that can be conditionally activated by G protein βγ subunits. *J. Biol. Chem.* **271**, 22552–22562.
51. Daaka, Y., Pitcher, J. A. *et al.* (1997). Receptor and Gβγ isoform-specific interactions with G protein-coupled receptor kinases. *Proc. Natl. Acad. Sci. USA* **94**, 2180–2185.
52. Yoshida, T., Willardson, B. M. *et al.* (1994). The phosphorylation state of phosducin determines its ability to block transducin subunit interactions and inhibit transducin binding to activated rhodopsin. *J. Biol. Chem.* **269**, 24050–24057.
53. Thilbault, C., Sganga, M. W., and Miles, M. F. (1997). Interaction of phosducin-like protein with G protein βγ subunits. *J. Biol. Chem.* **272**, 12253–12256.
54. Gaudet, R., Bohm, A., and Sigler, P. B. (1996). Crystal structure at 2.4 A resolution of the complex of transducin βγ and its regulator, phosducin. *Cell* **87**, 577–588.
55. Nishida, K., Kaziro, Y., and Satoh, T. (1999). Association of the proto-oncogene product Dbl with G protein βγ subunits. *FEBS Lett.* **459**, 186–190.
56. Slupsky, J. R., Quitterer, U. *et al.* (1999). Binding of Gβγ subunits to cRaf1 downregulated G protein coupled receptor signaling. *Curr. Biol.* **9**, 971–974.
57. Mattingly, R. R. and Macara, I. G. (1996). Phosphorylation-dependent activation of the Ras-GRF/CDC25Mm exchange factor by muscarinic receptors and G-protein βγ subunits. *Nature* **382**, 268–272.
58. Harhammer, R., Gohla, A., and Schultz, G. (1996). Interaction of G protein Gβγ dimmers with small GTP-binding proteins of the Rho family. *FEBS Lett.* **399**, 211–214.
59. Jaroma, C., Takizawa, P. A. *et al.* (1997). Regulation of Golgi structure through heterotrimeric G proteins. *Cell* **91**, 617–626.
60. Zhang, J. H. and Simonds, W. F. (2000). Copurification of brain G-protein β5 with RGS6 and RGS7. *J. Neurosci.* **20**, RC59.
61. Posner, B. A., Gilman, A. G., and Harris, B. A. (1999). Regulators of G protein signaling 6 and 7. Purification of complexes with Gβ5 and assessment of their effects on G protein-mediated signaling pathways. *J. Biol. Chem.* **274**, 31087–31093.
62. Hajdu-Cronin, Y. M. *et al.* (1999). Antagonism between Goα and Gqα in *C. elegans*: the RGS protein EAT-16 is necessary for Goα signaling and regulates Gqα activity. *Genes Devel.* **13**, 1780–1793.
63. Ford, C. E., Skiba, N. P. *et al.* (1998). Molecular basis for interactions of G protein βγ subunits with effectors. *Science* **280**, 1271–1274.
64. Li, Y., Sternweis, P. M. *et al.* (1998). Sites for Gα binding on the G protein β subunit overlap with sites for regulation of phospholipase Cβ and adenylyl cyclase. *J. Biol. Chem.* **273**, 16265–16272.
65. Yan, K., and Gautam, N. (1997). Structural determinants for interaction with three different effectors on the G protein β subunit. *J. Biol. Chem.* **272**, 2056–2059.
66. Myung, C.-S., and Garrison, J. C. (2000). Role of C-terminal domains of the G protein β subunit in the activation of effectors. *Proc. Natl. Acad. Sci. USA* **97**, 9311–9316.
67. Zhang, S., Coso, O. A. *et al.* (1996). A C-terminal mutant of the G protein β subunit deficient in the activation of phospholipase C-β. *J. Biol. Chem.* **271**, 20208–20212.
68. Mirshall, T., Robillard, L. *et al.* (2002). Gβ residues that do not interact with Gα underlie agonist-independent activity of K$^+$ channels. *J. Biol. Chem.* **277**, 7348–7355.
69. Panchenko, M. P., Saxena, K. *et al.* (1998). The sides of the G protein βγ subunit propeller structure contain regions important for phospholipase C β2 activation *J. Biol. Chem.* **273**, 28298–28304.

CHAPTER 227

βγ Signaling in Chemotaxis

Carol L. Manahan and Peter N. Devreotes
*Department of Cell Biology,
Johns Hopkins School of Medicine,
Baltimore, Maryland*

Introduction

Chemotaxis, the ability of a cell to sense and move up a chemical gradient, is important in various physiological processes and is involved in many disease states. Signal transduction during chemotaxis occurs through seven-transmembrane receptors that are coupled to heterotrimeric G proteins. The βγ subunits transmit the signal downstream, and although the direct effectors are unclear, increases in phosphoinositide 3,4,5-trisphosphate and phosphatidylinositide 3,4-bisphosphate occur at the leading edge of the migrating cell. The formation of these lipids results in the recruitment of a variety of signaling proteins. Recent work has implicated phosphatidylinositol 3-kinase and the lipid phosphatases, PTEN and SHIP in chemotaxis. This review discusses the role of G proteins in chemotaxis of *Dictyostelium discoideum* and neutrophils, with an emphasis on the possible role of proteins involved in lipid production and degradation.

Directed movement of cells toward soluble attractants (chemotaxis) is critical in immunity, wound healing, embryogenesis, and neuron guidance. Also, abnormalities in chemotaxis have been implicated in disease states such as tumor metastasis, atherosclerosis, arthritis, asthma, and multiple sclerosis [1]. The cell biology of chemotaxis has primarily been studied in two cell types, the social amoeba *Dictyostelium discoideum* and mammalian leukocytes. Due to the genetic manipulations available, *D. discoideum* is an excellent model system in which to study chemotactic signaling. Many observations originally made in these simple eukaryotic cells have subsequently held true in mammalian cells such as neutrophils.

Chemotaxis in amoebae and leukocytes is mediated through a heterotrimeric G protein that couples to a seven-transmembrane receptor on the plasma membrane. βγ dimers transduce the chemotactic signals downstream [2,3] to many known effectors ([reviewed in [4]). Recent work suggests that phosphatidylinositol-3-kinase (PI3K) is a critical effector involved in chemotaxis. βγ-signaling results in the accumulation of the lipids phosphoinositide 3, 4, 5-trisphosphate [PtdIns(3,4,5)P$_3$] and phosphatidylinositide 3,4-bisphosphate [PtdIns(3,4)P$_2$] at the leading edge of the cell (see below). Other downstream events include increases in actin polymerization, changes in cell shape, Ca^{2+} mobilization, myosin phosphorylation, and activation of adenylyl and guanylyl cyclases (Table I).

Evidence That G Proteins Are Involved in Chemotaxis

G-protein subunits were first implicated in chemotaxis by the generation of mutants in *D. discoideum*. Amoebae have at least 11 Gα subunits but only one β subunit and one γ subunit [5,6]. Mutation of the Gα2 or the Gα4 subunit [7–10] blocks chemotaxis to cAMP or folic acid, respectively. All chemotactic responses are defective in amoebae lacking the β subunit [11]. In mammalian systems, the scenario is more complicated, with many G protein subunits, receptors, and ligands involved (Table I). Early experiments showed that leukocyte chemotaxis was inhibited by pertussis toxin, indicating that many chemokine receptors couple to Gα subunits from the Gαi family. Studies using receptors that couple to Gαi, Gαq, or Gαs transfected into HEK293 cells or a

Table I Proteins Involved in Chemotaxis of Amoebae and Neutrophils

	D. discoideum	Neutrophils
Ligands	cAMP [28]	*Classes**
	folic acid [29]	chemokines [30]
	other nutrients	complement factors [31]
		N-formylated peptides [31]
		leukotrienes [32]
		lysophosphatidic acid [33]
		platelet-activating factor [34]
		prostaglandins [35]
Receptors (R)	cAR1 [36]	*Classes**
	cAR3 [37]	Chemokine receptors [39]
All are coupled to eterotrimeric G proteins	folic acid/pterin receptor [38]	Complement receptors
		N-formylated peptide R
		BLTR [40]
		Endothelial Differentiation Genes (Edg) [33]
		Platelet-Activating Factor R [40]
		Prostanoid R [35]
G proteins	Gα2, βγ (cAMP, [6,7,11])	Gαi family, βγ [2,3,12]
	Gα4, βγ (folic acid, [6,10,11])	
PI3Ks	PI3K1, PI3K2 [19,41,42]	PI3Kγ [21,22,23]
Signaling proteins found at leading edge	CRAC [14]	Akt/PKB [16]
	Akt/PKB [15]	Rac1 [43,44]
	PhdA [19]	PAK1 [45]
Signaling proteins on lateral edges or back of cell	Myosin II [46]	Myosin II [48]
	PAKa [47]	
	PTEN [25]	
Cell responses	↑ PtdIns(3,4)P_2 and PtdIns(3,4,5)P_3 [14,42]	↑ PtdIns(3,4)P_2 and PtdIns(3,4,5)P_3 [16]
	↑ actin polymerization	↑ actin polymerization
	↑ Ca^{++} [49]	↑ Ca^{++}
	Myosin phosphorylation [50,51]	Myosin phosphorylation
	↑ cAMP [52]	↑ inositol phosphates [54]
	↑ cGMP [53]	↑ cAMP [55]

*Due to space limitations the large number of ligands and receptors involved in neutrophil chemotaxis have been grouped into classes.

pre-B cell line have confirmed that Gαi is involved in chemotaxis [2,3]. In these same studies, chemotaxis was blocked when the βγ sequestering proteins (βARK or Gαt, transducin) were overexpressed, suggesting that the release of βγ is necessary for chemotaxis [2,3]. By expressing a chimeric Gα (which is activated by Gαi-coupled receptors but cannot inhibit adenylyl cyclase), Neptune et al. definitively showed that release of βγ subunits, not Gαi activation, was important for mediating the chemotactic signal downstream [12]. G-protein signaling, specifically the release of βγ, is critical for chemotaxis in amoebae and neutrophils.

Recently, it has been demonstrated that the G-protein heterotrimer disassociates upon chemoattractant stimulation in living cells. Using amoebae expressing Gα2 fused to cyan (CFP) and β fused to yellow (YFP) fluorescent proteins, Janetopoulous et al. observed fluorescence resonance energy transfer (FRET), which was lost when the cells were exposed to chemoattractant [13]. Upon removal of ligand, the heterotrimer quickly reassociated as indicated by the reappearance of the FRET signal. An important finding is that in the presence of continual stimulation, the FRET signal did not rebound, indicating that the heterotrimer remains disassociated in the presence of ligand and does not undergo desensitization [13]. Thus the key events that lead to desensitization are downstream or independent of G-protein activation.

PI3Ks—Role in Chemotaxis?

G-protein signaling activates the lipid kinase phosphatidylinositol-3-kinase (PI3K), which results in PtdIns(3,4,5)P_3

and PtdIns(3,4)P_2 being created at the leading edge. Studies on cytosolic regulator of adenylyl cyclase (CRAC), a pleckstrin-homology (PH) domain-containing protein in amoebae, provided the first clue to the involvement of phosphorylated lipids in the signaling pathway [14]. A class of PH-domains has been shown to associate with the lipids PtdIns(3,4)P_2 or PtdIns(3,4,5)P_3. CRAC (or just its PH domain) fused to green fluorescent protein (GFP) localized to the leading edge in chemotaxing amoebae [14], suggesting that these lipids are generated there. Subsequently, it has been shown that other PH-domain containing proteins, such as Akt/PKB, also localize to the leading edge in amoebae and neutrophils [15,16]. PI3K is undoubtedly involved in the formation of these lipids. However, studies using pharmacological inhibitors of PI3K, such as wortmannin and LY294002, have yielded conflicting results [17,18], with some studies showing effects on chemotaxis and others with no effect. *D. discoideum* have three genes encoding putative PI3Ks. Knockouts of these genes have been performed, and disruption of two of the PI3Ks result in impaired chemotaxis [19,20]. Disruption of all three PI3Ks in amoebae has not been completed. The conflicting genetic and pharmacological evidence indicates that the role of PI3K in chemotaxis is still unclear.

In mammalian systems, neutrophils from a PI3Kγ null mouse line have been evaluated. These neutrophils did not create detectable PtdIns(3,4,5)P_3 when stimulated with specific chemoattractants, suggesting that PI3Kγ is the sole PI3K that is coupled to chemoattractant receptors in these cells [21]. This phenotype could not be attributed to a global defect in G-protein signaling because other G-protein-mediated responses were functional. It is important that migration of the mutant neutrophils to chemokines *in vitro* (Boyden assay) and *in vivo* (infiltration assays) was reduced up to 50 percent [21–23], suggesting that PI3Kγ activity is required for proper chemotaxis.

Lipid Phosphatases, PTEN and SHIP

Since PI3K is involved in chemotaxis, one would predict that lipid phosphatases also play an important role. Previous work has suggested that the lipid phosphatase PTEN plays a negative role in chemotaxis. PTEN is a phosphatidylinositide 3-phosphatase. Fibroblasts from *pten*$^{-/-}$ mice exhibited faster motility and an increase in the number of pseudopods formed; however, directional sensing was not evaluated [24]. Recently, Iijima and Devreotes disrupted PTEN function in *D. discoideum* [25]. In the absence of chemoattractant, the mutant cells produced random pseudopods and some cells constitutively localized PH-GFP to the membrane [25]. This is consistent with PTEN playing a role in negatively regulating chemotaxis. However, the *pten*$^-$ cells do not undergo constitutive development, indicating that loss of PTEN is not sufficient to result in chemotaxis. In the presence of chemoattractant, *pten*$^-$ cells formed many pseudopods along the length of the cell, instead of only at the leading edge [25]. Therefore, the mutant cells took a slower, circuitous route up the gradient. Their hypothesis is that directional sensing requires a regulated balance between the activities of PI3K and PTEN [25].

Another lipid phosphatase, SHIP, which hydrolyzes the 5′-phosphate from PtdIns(3,4,5)P_3 and inositol 1,3,4,5-tetraphosphate seems to be involved in negatively regulating chemotactic signaling. Various hematopoietic cells from SHIP-deficient mice exhibited enhanced chemotaxis toward SDF-1 and B-lymphocyte chemoattractant (BLC) [26], but not to another chemokine, CKβ-ll. SHIP's role in chemotaxis may be dependent upon the cell type or chemokine involved. The enhanced chemotaxis of cells toward SDF-1 and BLC is consistent with the hypothesis that SHIP negatively regulates chemotaxis by hydrolyzing the PtdIns(3,4,5)P_3 created by PI3K. Consistent with this, decreasing the level of this lipid in B cells inhibited PKB/Akt activation [27]. Both lipid phosphatases PTEN and SHIP appear to be involved in chemotaxis signaling pathways.

Cells can sense and move up chemoattractant gradients that vary by less than 2 percent across their length. Understanding how cells sense and respond to these shallow gradients is important to understanding the basic mechanism of chemotaxis. Chemotaxis is mediated by G-protein βγ subunits and appears to be evolutionarily conserved from amoebae to mammalian neutrophils. Further study is necessary to determine the direct effector(s) of βγ and establish whether the key to directional sensing is in the regulation of the PtdIns levels in the migrating cell.

Acknowledgments

The authors would like to thank Chris Janetopoulous for critically reading the manuscript. We apologize to the authors whose work we could not cite due to space limitations. This work was funded by NRSA fellowship GM65644 to C.L.M. and NIH grants GM28007 and GM34922 to P.N.D.

References

1. Baggiolini, M. (2001). Chemokines in pathology and medicine. *J. Intern. Med.* **250**, 91–104.
2. Neptune, E. R. and Bourne, H. R. (1997). Receptors induce chemotaxis by releasing the βγ subunit of G_i, not by activating G_q or G_s. *Proc. Natl. Acad. Sci. USA* **94**, 14489–14494.
3. Arai, H., Tsou, C.-L., and Charo, I. F. (1997). Chemotaxis in a lymphocyte cell line transfected with C-C chemokine recptor 2B: Evidence that directed migration is mediated by βγ dimers released by activation of Gαi-coupled receptors. *Proc. Natl. Acad. Sci. USA* **94**, 14495–14499.
4. Rickert, P., Weiner, O. D., Wang, F., Bourne, H. R., and Servant, G. (2000). Leukocytes navigate by compass: roles of PI3Kγ and its lipid products. *T. Cell Biol.* **10**, 466–473.
5. van Es, S. and Devreotes, P. N. (1999). Molecular basis of localized responses during chemotaxis in amoebae and leukocytes. *Cell. Mol. Life Sci.* **55**, 1341–1351.
6. Zhang, N., Long, Y. and Devreotes, P. N. (2001). Gγ in *Dictyostelium*: Its role in localization of Gβγ to the membrane is required for chemotaxis in shallow gradients. *Mol. Biol. Cell* **12**, 3204–3213.
7. Kumagai, A., Pupillo, M., Gundersen, R., Miake-Lye, R., Devreotes, P. N., and Firtel, R. A. (1989). Regulation and function of G alpha protein subunits in *Dictyostelium*. *Cell* **57**, 265–275.
8. Pupillo, M., Kumagai, A., Pitt, G. S., Firtel, R. A., and Devreotes, P. N. (1989). Multiple α subunits of guanine nucleotide-binding proteins in *Dictyostelium*. *Proc. Natl. Acad. Sci. USA* **86**, 4892–4896.

9. Johnson, R. L., Gundersen, R., Lilly, P., Pitt, G. S., Pupillo, M., Sun, T. J., Vaghan, R. A., and Devretoes, P. N. (1989). G-protein-linked signal transduction systems control development in *Dictyostelium*. *Development* **107**, Suppl., 75–80.
10. Hadwiger, J. A., Lee, S., and Firtel, R. A. (1994). The G alpha subunit G alpha 4 couples to protein receptors and identifies a signaling pathway that is essential for multicellular development in *Dictyostelium*. *Proc. Natl. Acad. Sci. USA* **91**, 10566–10570.
11. Wu, L., Valkema, R., Van Haastert, P. J. M., and Devreotes, P. N. (1995). The G protein beta subunit is essential for multiple responses to chemoattractants in *Dictyostelium*. *J. Cell Biol.* **129**, 1667–1675.
12. Neptune, E. R., Iiri, T., and Bourne, H. R. (1999). $G\alpha_i$ is not required for chemotaxis mediated by G_i-coupled receptors. *J. Biol. Chem.* **274**, 2824–2828.
13. Janetopoulos, C., Jin, T., and Devreotes, P. (2001). Receptor-mediated activation of heterotrimeric G-proteins in living cells. *Science* **291**, 2408–2411.
14. Parent, C. A., Blacklock, B. J., Froehlich, W. M., Murphy, D. B., and Devreotes, P. N. (1998). G protein signaling events are activated at the leading edge of chemotactic cells. *Cell* **95**, 81–91.
15. Meili, R., Ellsworth, C., Lee, S., Reddy, T. B., Ma, H., and Firtel, R. A. (1999). Chemoattractant-mediated transient activation and membrane localization of Akt/PKB is required for efficient chemotaxis to cAMP in *Dictyostelium*. *EMBO J.* **18**, 2092–2105.
16. Servant, G., Weiner, O. D., Herzmark, P., Balla, T., Sedat, J. W., and Bourne, H. R. (2000). Polarization of chemoattractant receptor signaling during neutrophil chemotaxis. *Science* **287**, 1037–1040.
17. Wu, D., Huang, C. K., and Jiang, H. (2000). Roles of phospholipid signaling in chemoattractant-induced responses. *J. Cell Sci.* **113**, 2935–2940.
18. Wymann, M. P., Sozzani, S., Altruda, F., Mantovani, A., and Hirsch, E. (2000). Lipids on the move: phosphoinositide 3-kinases in leukocyte function. *Immunol. Today* **21**, 260–264.
19. Funamoto, S., Milan, K., Meili, R., and Firtel, R. A. (2001). Role of phosphatidylinositol 3' kinase and a downstream pleckstrin homology domain-containing protein in controlling chemotaxis in *Dictyostelium*. *J. Cell Biol.* **153**, 795–809.
20. Chung, C. Y., Funamoto, S., and Fitel, R. A. (2001). Signaling pathways controlling cell polarity and chemotaxis. *Trends Biochem. Sci.* **26**, 557–566.
21. Li, Z., Jiang, H., Xie, W., Zhang, Z., Smrcka, A. V., and Wu, D. (2000). Roles of PLC-β2 and -β3 and PI3Kγ in chemoattractant-mediated signal transduction. *Science* **287**, 1046–2049.
22. Hirsch, E., V. L., Katanaev, C., Garlanda, O., Azzolino, L., Pirola, L. *et al.* (2000). Central role for G protein-coupled phosphoinositide 3-kinase γ in inflammation. *Science* **287**, 1049–1053.
23. Sasaki, T., Irie-Sasaki, J., Jones, R. G., Oliveira-dos-Santos, A. J., Stanford, W. L. *et al.* (2000). Function of PI3Kγ in thymocyte development, T cell activation, and neutrophil migration. *Science* **287**, 1040–1046.
24. Liliental, J., Moon, S. Y., Lesche, R., Mamillapalli, R., Li, D. *et al.* (2000). Genetic deletion of the *Pten* tumor suppressor gene promotes cell motility by activation of Rac1 and Cdc42 GTPases. *Curr. Biol.* **10**, 401–404.
25. Iijima, M. and Devreotes, P. (2002). Tumor suppressor PTEN mediates sensing of chemoattractant gradients. *Cell*, in press.
26. Kim, C. H., Hangoc, G., Cooper, S., Helgason, C. D., Yew, S. *et al.* (1999). Altered responsiveness to chemokines due to targeted disruption of SHIP. *J. Clin. Invest.* **104**, 1751–1759.
27. Aman, M. J., Lamkin, T. D., Okada, H. Kurosaki, T., and Ravichandran, K. S. (1998). The inositol phosphatase SHIP inhibits Akt/PKB activation in B cells. *J. Biol. Chem.* **273**, 33922–33928.
28. Konijn, T. M., van de Meene, J. G., Chang, Y. Y., Barkley, D. S., and Bonner, J. T. (1969). Identification of adenosine-3', 5'-monophosphate as the bacterial attractant for myxamoebae of *Dictyostelium discoideum*. *J. Bacteriol.* **99**, 510–512.
29. Wurster, B. and Schubiger, K. (1977). Oscillations and cell development in *Dictyostelium discoideum* stimulated by folic acid pulses. *J. Cell Sci.* **27**, 105–114.
30. Fernandez, E. J. and Lolis, E. (2002). Structure, function, and inhibition of chemokines. *Annu. Rev. Pharmacol. Toxicol.* **42**, 469–499.
31. Yang, D., Chertov, O. and Oppenheim, J. J. (2001). The role of mammalian antimicrobial peptides and proteins in awakening of innate host defenses and adaptive immunity. *Cell Mol. Life Sci.* **58**, 978–989.
32. Leff, A. R. (2000). Role of leukotrienes in bronchial hyperresponsiveness and cellular responses in airways. *Thorax* **55**, Suppl. 32–37.
33. Goetzl, E. J. and An, S. (1998). Diversity of cellular receptors and functions for the lysophospholipid growth factors lysophosphatidic acid and sphingosine 1-phosphate. *FASEB J.* **12**, 1589–1598.
34. Snyder, F. (1990). Platelet-activating factor and related acetylated lipids as potent biologically active cellular mediators. *Am. J. Physiol.* **259**, C697–708.
35. Cirino, G. (1998). Multiple controls in inflammation. Extracellular and intracellular phospholipase A2, inducible and constitutive cyclooxygenase, and inducible nitric oxide synthase. *Biochem. Pharmacol.* **55**, 105–111.
36. Klein, P. S., Sun, T. J., Saxe, C. L., Kimmel, A. R., Johnson, R. L., and Devreotes, P. N. (1988). A chemoattractant receptor controls development in *Dictyostelium discoideum*. *Science* **241**, 1467–1472.
37. Johnson, R. L., Saxe, C. L., Gollop, R., Kimmel, A. R., and Devreotes, P. N. (1993). Identification and targeted gene disruption of cAR3, a cAMP receptor subtype expressed during multicellular stages of *Dictyostelium* development. *Genes Dev.* **7**, 273–282.
38. Van Dreil, R. (1981). Binding of the chemoattractant folic acid by *Dictyostelium discoideum* cells. *Eur. J. Biochem.* **115**, 391–395.
39. Horuk, R. (2001). Chemokine receptors. *Cytokine Growth Factor Rev.* **12**, 313–335.
40. Haribabu, B., Richardson, R. M., Verghese, M. W., Barr, A. J., Zhelev, D. V., and Snyderman, R. (2000). Function and regulation of chemoattractant receptors. *Immunol. Res.* **22**, 271–279.
41. Buczynski, G., Grove, B., Nomura, A., Kleve, M., Bush, J., Firtel, R. A., and Cardelli, J. (1997). Inactivation of two *Dictyostelium discoideum* genes, DdPIK1 and DdPIK2, encoding proteins related to mammalian phosphatidylinositide 3-kinases, results in defects in endocytosis, lysosome to postlysosome transport, and actin cytoskeleton organization. *J. Cell Biol.* **136**, 1271–1286.
42. Zhou, K., Pandol, S., Bokoch, G., and Traynor-Kaplan, A. E. (1998). Disruption of *Dictyostelium* PI3K genes reduces [32P] phosphatidylinositol 3,4 bisphosphate and [32P] phosphatidylinositol trisphosphate levels, alters F-actin distribution and impairs pinocytosis. *J. Cell Sci.* **111**, 283–294.
43. Kraynov, V. S., Chamberlain, C., Bokoch, G. M., Schwartz, M. A., Slabaugh, S., and Hahn, K. M. (2000). Localized Rac activation dynamics visualized in living cells. *Science* **290**, 333–337.
44. Cho, S. Y. and Klemke, R. L. (2002). Purification of pseudopodia from polarized cells reveals redistribution and activation of Rac through assembly of a CAS/Crk scaffold. *J. Cell Biol.* **156**, 725–736.
45. Dharmawardhane, S., Brownson, D., Lennartz, M., and Bokoch, G. (1999). Localization of p21-activated kinase 1 (PAK1) to pseudopodia, membrane ruffles, and phagocytic cups in activated human neutrophils. *J. Leukoc. Biol.* **66**, 521–527.
46. Clow, P. A. and McNally, J. G. (1999). *In vivo* observations of myosin II dynamics support a role in rear retraction. *Mol. Biol. Cell* **10**, 1309–1323.
47. Chung, C. Y., Poikyan, G., and Firtel, R. A. (2001). Control of cell polarity and chemotaxis by Akt/PKB and PI3 kinase through the regulation of PAKa. *Mol. Cell* **7**, 937–947.
48. del Pozo, M. A., Nieto, M., Serrador, J. M., Sancho, D., Vicente-Manzanares, M. *et al.* (1998). The two poles of the lymphocyte: specialized cell compartments for migration and recruitment. *Cell Adhes. Commun.* **6**, 125–133.
49. Milne, J. L. and Devreotes, P. N. (1993). The surface cyclic AMP receptors, cAR1, cAR2, and cAR3, promote Ca^{2+} influx in *Dictyostelium discoideum* by a G alpha 2-independent mechanism. *Mol. Biol. Cell* **4**, 283–292.
50. Malchow, D., Bohme, R., and Rahmsdorf, H. J. (1981). Regulation of phosphorylation of myosin heavy chain during the chemotactic response in *Dictyostelium* cells. *Eur. J. Biochem.* **117**, 213–218.

51. Zhang, H., Wessels, D., Fey, P., Daniels, K., Chisolm, R. L., and Soll, D. R. (2002). Phosphorylation of the myosin regulatory light chain plays a role in motility and polarity during *Dictyostelium* chemotaxis. *J. Cell Sci.* **115**, 1733–1747.
52. Shaffer, B. M. (1975). Secretion of cyclic AMP induced by cyclic AMP in the cellular slime mould Dictyostelium discoideum. *Nature* **255**, 765–775.
53. Wurster, B., Schubiger, K., Wick, U., and Gerisch, G. (1977). Cyclic GMP in *Dictyostelium discoideum*. Oscillations and pulses in response to folic acid and cyclic AMP signals. *FEBS Lett.* **76**, 141–144.
54. Snyderman, R. and Verghese, M. W. (1987). Leukocyte activation by chemoattractant receptors: roles of a guanine nucleotide regulatory protein and polyphosphoinositide metabolism. *Rev. Infect. Dis.* Suppl **5**, S562–569.
55. Verghese, M. W., Fox, K., McPhail, L. C., and Snyderman, R. (1985). Chemoattractant-elicited alterations of cAMP levels in human polymorphonuclear leukocytes require a Ca^{2+}-dependent mechanism which is independent of transmembrane activation of adenylate cyclase. *J. Biol. Chem.* **260**, 6769–6775.

CHAPTER 228

Reversible Palmitoylation in G-Protein Signaling

Philip Wedegaertner
Department of Microbiology and Immunology, Kimmel Cancer Institute,
Thomas Jefferson University, Philadelphia, Pennsylvania

Introduction

Numerous proteins involved in cellular signaling undergo reversible palmitoylation [1–3]. This review will focus on reversible palmitoylation of G-protein α subunits (Gα), briefly discuss some more general aspects of palmitoylation and depalmitoylation, and highlight recent evidence that many regulator of G protein signaling (RGS) proteins are palmitoylated.

Palmitoylation is a covalent lipid modification in which the saturated, 16-carbon fatty acid, palmitate, is linked through a thioester bond to a cysteine. Palmitoylation is often referred to as S-acylation or thioacylation to indicate that radiolabeled fatty acids of varying lengths can be incorporated into a protein. Very few studies have addressed the identity and stoichiometry of the endogenous fatty acid covalently attached to a particular cellular protein. In fact, it is mostly unknown if palmitoylated proteins exist as isoforms with different attached fatty acids, if specific cellular mechanisms exist to dictate a preference for palmitate, or if palmitate is primarily used merely because of its relatively high abundance in cells.

Much of the interest in palmitoylation stems from its dynamic nature and the resulting idea that this covalent modification is more than just a static membrane anchor. Numerous studies have demonstrated that palmitoylation is a reversible modification, i.e., attached palmitate often has a much shorter half-life than the protein it modifies. Such reversibility implies the potential for important regulatory roles for palmitoylation.

Sites of Palmitoylation in Gα and RGS Proteins

Besides the modified cysteine(s) itself, no primary amino acid consensus sequence has been identified for palmitoylation. The only consensus is that an additional hydrophobic modification or membrane-targeting motif is often a prerequisite for palmitoylation [1–4]. Table I lists identified sites of palmitoylation from representative Gα and RGS proteins. Based on lipid modification, Gα can be divided into two subfamilies (Table I)—ones that are palmitoylated only and ones that undergo both myristoylation and palmitoylation. Recently, it was demonstrated that the yeast *Saccharomyces cerevisiae* G-protein γ subunit is palmitoylated at a cysteine adjacent to the prenylated cysteine at the extreme C-terminus [5,6]. However, none of the 12 human Gγ subunits or the 2 Gγ of the model organism *Caenorhabditis elegans* contain a potential C-terminal palmitoylation site.

Palmitoylation has been identified in some members of the RGS family (Table I). Interestingly, palmitoylation of RGS proteins can occur at one or two cysteines within short N-terminal extensions [7,8], multiple cysteines in a cysteine string motif in the N-terminus [9], or at a cysteine within the RGS domain itself [10] (Table I). Additional palmitoylated proteins involved in G-protein signaling pathways include G-protein-coupled receptors (GPCR) [11], G-protein-coupled receptor kinases (GRK) [12], and small GTPases of the ras and rho superfamily [2,3].

Activation-Regulated Palmitoylation of Gα

If palmitoylation functions as a regulatory modification, then the expectation is that appropriate cellular stimuli will cause changes in the level of palmitoylation of a particular protein. Indeed, regulated changes in palmitoylation appear to be a general phenomenon for Gα [1]. Regulated palmitoylation was first demonstrated for α_s. Palmitate attached to

Table I Palmitoylation of Gα and RGS proteins[a]

Gα N-termini (myristoylated and palmitoylated)[b]

α_{i1}	M **G** *C* T L S A E D K A A V E R S K M I D
α_{o1}	M **G** *C* T L S A E E R A A L E R S K A I E
α_z	M **G** *C* R Q S S E E K E A A R R S R R I D
Gpa1	M **G** *C* T V S T Q T I G D E S D P F L Q N

Gα N-termini (non-myristoylated)[c]

α_s	M G *C* L G N S K T E D Q R N E E D A Q R
α_q	M T L E S I M A *C* *C* L S E E A K E A R R
α_{14}	M A G *C* *C* *C* L S A E E K E S Q R I S A E
α_{16}	M A R S L R W R *C* *C* P W *C* L T E D E K A
α_{12}	M S G V V R T L S R *C* L L P A E A G A R
α_{13}	M A D F L P S R S V L S V *C* F P G *C* V L

RGS N-termini[d]

RGS4	M *C* K G L A G L P A S *C* L R S A K D M K
RGS16	M *C* R T L A T F P N T *C* L E R A K E F K

RGS cysteine-string motif[e]

RGS-GAIP	35	S R N P *C* *C* L *C* W *C* *C* *C* *C* S *C* S W N Q E	54

RGS box palmitoylation[f]

RGS4	86	E E N I D F W I S *C* E E Y K K I K S P S	105
RGS10	57	E E N V L F W L A *C* E D F K K M Q D K T	76

[a]Palmitoylated (or potentially palmitoylated) cysteines are in italics.
[b]Myristoylated glycines at position 2 are in bold. Myristoylation of Gα has been well-reviewed [1–3].
[c]Potential sites of palmitoylation have not been addressed for α_{14} and α_{16}.
[d]Similar N-terminal sequence found in RGS5.
[e]Cysteine-string motifs are found in other members of the RZ sub-family [63].
[f]Cysteine present at similar position in RGS box of most RGS proteins.

α_s turns over much more rapidly after activation by the β-adrenergic receptor (β-AR) agonist isoproterenol, activation by cholera toxin, or activation by a constitutively activating mutation in α_s [13–15]. Time courses of palmitate incorporation and pulse-chase analyses are consistent with a model [4,16] in which activation leads to both more rapid depalmitoylation and more rapid subsequent repalmitoylation of α_s. For Gα other than α_s, a thorough recent study showed that agonist activation of stably transfected 5-HT$_{1A}$ receptors in CHO cells resulted in increased palmitate turnover on endogenous α_i [17]. In addition, stimulation of gonadotropin-releasing hormone (GnRH) receptors in pituitary cells caused increased palmitate incorporation into α_s, α_i, and α_q [18,19], stimulation of m1-muscarinic receptors in transfected cells increases the rate of palmitate incorporation into α_q (author's unpublished observations), stimulation of GPCRs in isolated membranes has been shown to increase palmitoylation of α_i, α_o, α_s, and α_q [18,20,21], and agonist-regulated palmitoylation has been observed for α_s and α_q even when they are directly fused to GPCRs [22,23]. A similar model of regulated palmitoylation/depalmitoylation in response to agonist activation has been well described for GPCRs [11,24,25]. Regulated changes in palmitoylation have not yet been described for RGS proteins.

Mechanisms of Reversible Palmitoylation

Regulated palmitoylation implies that changes in a protein's palmitoylation state are carried out by regulation of the palmitoylation and/or depalmitoylation machinery in the cell or by changes in the accessibility of the palmitoylated protein substrate to constitutive palmitoylating and depalmitoylating activities. Although intensely studied by many investigators, surprisingly little is understood regarding relevant enzymes and cellular pathways that regulate reversible palmitoylation.

Palmitoyl Transferases

A major unresolved question is whether palmitoylation occurs enzymatically or nonenzymatically. This controversial topic has been well discussed in recent reviews [2,3]. Compelling arguments for nonenzymatic palmitoylation are based on observations of transfer of palmitate to specific and appropriate cysteines in an *in vitro* reaction containing only palmitoyl CoA and a purified protein substrate, such as Gα [26]. On the other hand, others have argued that under physiological conditions, where most of a cell's palmitoyl CoA is bound to acyl-CoA binding proteins (ACBP), nonenzymatic palmitoylation would occur too slowly to be significant [27,28].

More importantly, the first convincing demonstrations of enzymatic palmitoylation have been described recently. Two different palmitoyl acyltransferases (PAT) were purified from yeast. One is composed of two proteins termed Erf2p and Erf4p and palmitoylates yeast Ras *in vitro* [29], while the other, Akr1p, was shown to palmitoylate the casein kinase Yck2p [30]. Both Erf2p and Akr1p contain a conserved Asp-His-His-Cys cysteine-rich domain (DHHC-CRD), and mutation of conserved residues in this domain abolishes PAT activity [29,30]. A large family of proteins exists that contains the DHHC-CRD domain, and clearly much research in the near future will be directed toward testing the possibility that the DHHC-CRD proteins are a family of PATs and identifying the relevant PAT(s) for heterotrimeric G-protein α subunits.

Palmitoyl Thioesterases

Recently, a candidate for a physiologically relevant palmitoyl thioesterase, termed an acyl-protein thioesterase (APT-I), was described [31]. Overexpression of APT in cultured cells increased the basal rate of depalmitoylation of co-expressed α_s. Further studies to substantiate a role for APT-I in depalmitoylation of signaling proteins are eagerly awaited.

Proteins and/or Pathways That Regulate Reversible Palmitoylation

G-protein βγ subunits appear to regulate both palmitoylation and depalmitoylation of Gα. βγ promotes palmitoylation of Gα *in vitro* [32] and allows palmitoylation of a non-myristoylated (G2A) mutant of α_i or α_z in transfected cells [33,34]. α_s or α_q containing mutations in N-terminal βγ interaction sites display greatly decreased palmitoylation [35,36], while a mutant α_o that has an increased affinity for βγ is palmitoylated to a higher level than α_o wild type [37]. One way in which $\beta\gamma$ can enhance palmitoylation of Gα is by promoting membrane targeting. Consistent with this, palmitoylation, and plasma membrane (PM) localization, of a $\beta\gamma$-binding-deficient α_q is recovered when it is engineered to undergo myristoylation [35], and $\beta\gamma$-binding mutations in α_o, a subunit that is normally myristoylated *and* palmitoylated (Table I), do not affect α_o palmitoylation [37]. Although myristoylation and binding to $\beta\gamma$ appear, in some cases, to function interchangeably as membrane-targeting signals for Gα, they may have additional, more specific roles in promoting palmitoylation [1,32,37].

$\beta\gamma$ can also inhibit depalmitoylation of α_s. A mutant α_s that binds tightly to $\beta\gamma$ is refractory to activation-induced rapid depalmitoylation in cultured cells [13,15], and purified $\beta\gamma$ inhibits depalmitoylation of α_s when assayed in cell extracts [15] or when using purified proteins [31,38]. These results suggest that activation-induced depalmitoylation of Gα is mediated, at least in part, by its dissociation from $\beta\gamma$.

Although studies of palmitate turnover on Gα and GPCRs suggest that palmitoylation and depalmitoylation are tightly coupled, a recent study utilizing a β-AR-α_s fusion protein was the first to separate these two activities [22]. The authors showed that palmitoylation occurred at appropriate sites in both the β-AR and α_s portions of the fusion protein, and isoproterenol induced rapid depalmitoylation, as seen with the separate proteins. However, isoproterenol-induced palmitoylation, likely re-palmitoylation, was blocked. This result suggests that, after rapid depalmitoylation, repalmitoylation requires dissociation of the G protein from the GPCR and/or later events that do not occur normally with this fusion protein, such as desensitization or internalization.

Tools for Studying Reversible Palmitoylation

As expected from our lack of knowledge of the proteins involved in palmitoylation and depalmitoylation, a dearth of tools exists for inhibiting these activities. Recently, the palmitate analog 2-bromopalmitate was used to inhibit palmitoylation of signaling proteins [39]. Another group synthesized analogs of cerulenin, a natural product inhibitor of fatty acid synthesis, and identified compounds that inhibited palmitoylation of H-ras without inhibiting a fatty acid synthase [40]. This interesting result suggests that not only might cerulenin analogs provide a powerful tool for studying palmitoylation of various proteins, but they might also provide a handle for identifying an elusive palmitoyl transferase [41].

Functions of Reversible Palmitoylation

Plasma Membrane Localization

An obvious role for palmitoylation is to tether a protein to cellular membranes, and thus regulation of this modification would allow changes in a protein's subcellular localization, either by dissociation off a cellular membrane or transfer to different membrane domains. Palmitoylation appears to function, in a poorly understood manner, as a specific membrane-targeting device that specifies localization to PM, and in some cases specialized PM microdomains [1–3]. This role of palmitoylation has been well documented and discussed in terms of a two-signal model for PM localization of signaling proteins [1–4, 42–44].

Although the consensus is that palmitoylation plays a critical role in PM localization of many proteins, including the α_i subfamily [4,34,45,46] of Gα (Table I), there is a surprising disagreement regarding a role for palmitoylation in the non-myristoylated Gα (Table I). Due to the lack of tools for perturbing palmitoylation in cells, virtually all localization studies have relied on preventing palmitoylation by mutating the relevant cysteines to serines or alanines. Subcellular fractionation and immunofluorescence localization of cells transiently or stably expressing nonpalmitoylated cysteine mutants of α_s, α_q, α_{11}, or α_{13} have demonstrated that such mutants are cytoplasmic and soluble [47–50], and, recently, these analyses were extended by the demonstration that a GFP-tagged, nonpalmitoylated mutant of α_q is cytoplasmic in living cells [51]. Moreover, overexpression of $\beta\gamma$ with palmitoylation-defective α_s or α_q cannot restore their PM localization [36], providing further support for the critical importance of palmitoylation. On the other hand, some researchers have observed that nonpalmitoylated mutants of α_s, α_q, or α_{12} remain in a particulate fraction after subcellular fractionation of transiently transfected cells [14,52–55]. However, it is important to note that in the latter cases [14,52–55] immunofluorescence localization of the mutants has not been reported, and thus it is unknown whether these mutants arrive correctly at the PM or are mistargeted to intracellular locations. Lastly, mutation of the N terminally palmitoylated cysteines in RGS4 and RGS16 (Table I) had little effect on their localization [7,8,56]. Possibly, additional membrane-targeting signals may be favored in different cells or experimental conditions.

Can Reversible Palmitoylation Regulate Changes in Subcellular Localization?

Although palmitoylation is reversible and plenty of evidence indicates that nonpalmitoylated mutants of various proteins are defective in PM localization and/or localization to PM microdomains, actually demonstrating a relationship between changes in a protein's palmitoylation and movement of that protein within a cell is a difficult problem. Nonetheless, α_s provides the best example of a correlation between reversible palmitoylation and reversible subcellular localization. As described above, activation of α_s induces rapid turnover of its attached palmitate. Similarly, activation of α_s by GPCRs, cholera toxin, or a constitutively activating mutation, can promote its redistribution from PM to cytoplasm [48,57–61]. Significantly, β-AR-induced redistribution of α_s appears to follow a similar time course as β-AR-induced depalmitoylation of α_s [15,48,57]. In addition, replacement of the N-terminal single site of palmitoylation of α_s with other membrane-targeting motifs results in mutant α_s subunits that are unable to translocate from PM to cytoplasm upon activation [62], consistent with depalmitoylation of a single palmitate playing a critical role in activation-induced subcellular redistribution of α_s. Others have speculated that regulated palmitoylation is more relevant for allowing reversible movement of Gα within PM microdomains [2,16,54]. Possibly, reversible palmitoylation could help to regulate the availability of G proteins at diverse subcellular locations.

Palmitoylation Affects RGS-Gα Interactions

One of the most interesting recent advances is evidence that palmitoylation of certain Gα influences their interaction with RGS proteins, and, vice versa, palmitoylation of certain RGS proteins affects their interactions with Gα. In vitro, palmitoylation of α_z or α_{i1} greatly inhibited their sensitivity to the GTP-hydrolysis-stimulating activity of several RGS proteins [63]. This result implies that reversible palmitoylation of Gα functions as a key switch to regulate the ability of RGS proteins to "turn-off" signaling.

Palmitoylation of RGS4 or RGS10 was shown to either inhibit or accelerate their ability to stimulate GTP hydrolysis of α_z or α_{i1} [10]. The positive or negative effect of palmitoylation differed depending upon the in vitro GAP assay used and whether palmitoylation occurred at N-terminal sites or cysteine sites in the conserved RGS box (Table I). Consistent with an important role for palmitoylation of RGS proteins, when nonpalmitoylated RGS16, containing N-terminal cysteine mutations (Table I), was expressed in cultured cells it failed to effectively inhibit signaling mediated by G_q or G_i [7]. It remains to be determined whether effects of palmitoylation on RGS-Gα functional interactions are mediated through direct binding or via binding to membranes, although for RGS4 in vitro lipid vesicle binding experiments are consistent with the proposal that palmitoylation, in addition to an N-terminal amphipathic helix, is important for membrane binding and properly orienting the RGS protein for optimal activity [64].

Conclusion

Palmitoylation plays an important role in membrane binding and regulating interactions of signaling proteins. Future challenges include defining cellular pathways and enzymes that regulate reversible palmitoylation, and understanding how changes in a protein's palmitoylation are translated into functional changes inside the cell.

References

1. Chen, C. A. and Manning, D. R. (2001). Regulation of G proteins by covalent modification. Oncogene 20, 1643–1652.
2. Dunphy, J. T. and Linder, M. E. (1998). Signalling functions of protein palmitoylation. Biochim. Biophys. Acta 1436, 245–261.
3. Resh, M. D. (1999). Fatty acylation of proteins: new insights into membrane targeting of myristoylated and palmitoylated proteins. Biochim. Biophys. Acta 1451, 1–16.
4. Wedegaertner, P. B. (1998). Lipid modifications and membrane targeting of Gα. Biol. Signal Recept. 7, 125–135.
5. Hirschman, J. E. and Jenness, D. D. (1999). Dual lipid modification of the yeast Gγ subunit Ste18p determines membrane localization of G$\beta\gamma$. Mol. Cell. Biol. 19, 7705–7711.
6. Manahan, C. L., Patnana, M., Blumer, K. J., and Linder, M. E. (2000). Dual lipid modification motifs in G(α) and G(γ) subunits are required for full activity of the pheromone response pathway in Saccharomyces cerevisiae. Mol. Biol. Cell 11, 957–968.
7. Druey, K. M., Ugur, O., Caron, J. M., Chen, C. K., Backlund, P. S., and Jones, T. L. (1999). Amino-terminal cysteine residues of RGS16 are required for palmitoylation and modulation of Gi- and Gq-mediated signaling. J. Biol. Chem. 274, 18836–18842.
8. Srinivasa, S. P., Bernstein, L. S., Blumer, K. J., and Linder, M. E. (1998). Plasma membrane localization is required for RGS4 function in Saccharomyces cerevisiae. Proc. Natl. Acad. Sci. USA 95, 5584–5589.
9. De Vries, L., Elenko, E., Hubler, L., Jones, T. L., and Farquhar, M. G. (1996). GAIP is membrane-anchored by palmitoylation and interacts with the activated (GTP-bound) form of Gα_i subunits. Proc. Natl. Acad. Sci. USA 93, 15203–15208.
10. Tu, Y., Popov, S., Slaughter, C., and Ross, E. M. (1999). Palmitoylation of a conserved cysteine in the regulator of G protein signaling (RGS) domain modulates the GTPase-activating activity of RGS4 and RGS10. J. Biol. Chem. 274, 38260–38267.
11. Morello, J. P. and Bouvier, M. (1996). Palmitoylation: a post-translational modification that regulates signalling from G-protein coupled receptors. Biochem. Cell Biol. 74, 449–457
12. Krupnick, J. G. and Benovic, J. L. (1998). The role of receptor kinases and arrestins in G protein-coupled receptor regulation. Annu. Rev. Pharmacol. 38, 289–319.
13. Degtyarev, M. Y., Spiegel, A. M., and Jones, T. L. (1993). Increased palmitoylation of the Gs protein alpha subunit after activation by the beta-adrenergic receptor or cholera toxin. J. Biol. Chem. 268, 23769–23772.
14. Mumby, S. M., Kleuss, C., and Gilman, A. G. (1994). Receptor regulation of G-protein palmitoylation. Proc. Natl. Acad. Sci. USA 91, 2800–2804.
15. Wedegaertner, P. B. and Bourne, H. R. (1994). Activation and depalmitoylation of Gsα. Cell 77, 1063–1070.
16. Mumby, S. M. (1997). Reversible palmitoylation of signaling proteins. Curr. Opin. Cell Biol. 9, 148–154.
17. Chen, C. A. and Manning, D. R. (2000). Regulation of Gα_i palmitoylation by activation of the 5-hydroxytryptamine-1A receptor. J. Biol. Chem. 275, 23516–23522.
18. Stanislaus, D., Janovick, J. A., Brothers, S., and Conn, P. M. (1997). Regulation of G(q/11)alpha by the gonadotropin-releasing hormone receptor. Mol. Endocrinol. 11, 738–746.
19. Stanislaus, D., Ponder, S., Ji, T. H., and Conn, P. M. (1998). Gonadotropin-releasing hormone receptor couples to multiple G proteins in rat gonadotrophs and in GGH3 cells: Evidence from palmitoylation and overexpression of G proteins. Biol. Reprod. 59, 579–586.
20. Gurdal, H., Seasholtz, T. M., Wang, H. Y., Brown, R. D., Johnson, M. D., and Friedman, E. (1997). Role of Gα_q or Gα_o proteins in alpha 1-adrenoceptor subtype-mediated responses in Fischer 344 rat aorta. Mol. Pharmacol. 52, 1064–1070.
21. Bhamre, S., Wang, H. Y., and Friedman, E. (1998). Serotonin-mediated palmitoylation and depalmitoylation of G alpha proteins in rat brain cortical membranes. J. Pharmacol. Exp. Ther. 286, 1482–1489.
22. Loisel, T. P., Ansanay, H., Adam, L., Marullo, S., Seifert, R., Lagace, M., and Bouvier, M. (1999). Activation of the beta(2)-adrenergic receptor-Gα(s) complex leads to rapid depalmitoylation and inhibition of repalmitoylation of both the receptor and Gα(s). J. Biol. Chem. 274, 31014–31019.
23. Stevens, P. A., Pediani, J., Carrillo, J. J., and Milligan, G. (2001). Coordinated agonist regulation of receptor and G protein palmitoylation and functional rescue of palmitoylation-deficient mutants of the

G protein G11alpha following fusion to the alpha1b-adrenoreceptor: palmitoylation of G11alpha is not required for interaction with beta*gamma complex. *J. Biol. Chem.* **276**, 35883–35890.
24. Mouillac, B., Caron, M., Bonin, H., Dennis, M., and Bouvier, M. (1992). Agonist-modulated palmitoylation of beta 2-adrenergic receptor in Sf9 cells. *J. Biol. Chem.* **267**, 21733–21737.
25. Loisel, T. P., Adam, L., Hebert, T. E., and Bouvier, M. (1996). Agonist stimulation increases the turnover rate of beta 2AR-bound palmitate and promotes receptor depalmitoylation. *Biochemistry* **35**, 15923–15932.
26. Duncan, J. A. and Gilman, A. G. (1996). Autoacylation of G protein alpha subunits. *J. Biol. Chem.* **271**, 23594–23600.
27. Leventis, R., Juel, G., Knudsen, J. K., and Silvius, J. R. (1997). Acyl-CoA binding proteins inhibit the nonenzymatic S-acylation of cysteinyl-containing peptide sequences by long-chain acyl-CoAs. *Biochemistry* **36**, 5546–5553.
28. Dunphy, J. T., Schroeder, H., Leventis, R., Greentree, W. K., Knudsen, J. K., Silvius, J. R., and Linder, M. E. (2000). Differential effects of acyl-CoA binding protein on enzymatic and non-enzymatic thioacylation of protein and peptide substrates. *Biochim. Biophys. Acta* **1485**, 185–198.
29. Lobo, S., Greentree, W. K., Linder, M. E., and Deschenes, R. J. (2002). Identification of a Ras palmitoyltransferase in *Saccharomyces cerevisiae*. *J. Biol. Chem.* **277**, 41268–41273.
30. Roth, A. F., Feng, Y., Chen, L., and Davis, N. G. (2002). The yeast DHHC cysteine-rich domain protein Akr1p is a palmitoyl transferase. *J. Cell Biol.* **159**, 23–28.
31. Duncan, J. A. and Gilman, A. G. (1998). A cytoplasmic acyl-protein thioesterase that removes palmitate from G protein alpha subunits and p21(RAS). *J. Biol. Chem.* **273**, 15830–15837.
32. Dunphy, J. T., Greentree, W. K., Manahan, C. L., and Linder, M. E. (1996). G-protein palmitoyltransferase activity is enriched in plasma membranes. *J. Biol. Chem.* **271**, 7154–7159.
33. Degtyarev, M. Y., Spiegel, A. M., and Jones, T. L. (1994). Palmitoylation of a G protein α_i subunit requires membrane localization not myristoylation. *J. Biol. Chem.* **269**, 30898–30903.
34. Morales, J., Fishburn, C. S., Wilson, P. T., and Bourne, H. R. (1998). Plasma membrane localization of $G\alpha_z$ requires two signals. *Mol. Biol. Cell* **9**, 1–14.
35. Evanko, D. S., Thiyagarajan, M. M., and Wedegaertner, P. B. (2000). Interaction with Gβγ is required for membrane targeting and palmitoylation of Gα(s) and Gα(q). *J. Biol. Chem.* **275**, 1327–1336.
36. Evanko, D. S., Thiyagarajan, M. M., Siderovski, D. P., and Wedegaertner, P. B. (2001). Gβγ isoforms selectively rescue plasma membrane localization and palmitoylation of mutant $G\alpha_s$ and $G\alpha_q$. *J. Biol. Chem.* **276**, 23945–23953.
37. Wang, Y., Windh, R. T., Chen, C. A., and Manning, D. R. (1999). N-Myristoylation and βγ play roles beyond anchorage in the palmitoylation of the G protein alpha(o) subunit. *J. Biol. Chem.* **274**, 37435–37442.
38. Iiri, T., Backlund, P. S., Jones, T. L., Wedegaertner, P. B., and Bourne, H. R. (1996). Reciprocal regulation of Gs α by palmitate and the β γ subunit. *Proc. Natl. Acad. Sci. USA* **93**, 14592–14597.
39. Webb, Y., Hermida-Matsumoto, L., and Resh, M. D. (2000). Inhibition of protein palmitoylation, raft localization, and T cell signaling by 2-bromopalmitate and polyunsaturated fatty acids. *J. Biol. Chem.* **275**, 261–270.
40. Lawrence, D. S., Zilfou, J. T., and Smith, C. D. (1999). Structure-activity studies of cerulenin analogues as protein palmitoylation inhibitors. *J. Med. Chem.* **42**, 4932–4941.
41. De Vos, M. L., Lawrence, D. S., and Smith, C. D. (2001). Cellular pharmacology of cerulenin analogs that inhibit protein palmitoylation. *Biochem. Pharmacol.* **62**, 985–995.
42. Shahinian, S. and Silvius, J. R. (1995). Doubly lipid-modified protein sequence motifs exhibit long-lived anchorage to lipid bilayer membranes. *Biochemistry* **34**, 3813–3822.
43. McLaughlin, S. and Aderem, A. (1995). The myristoyl-electrostatic switch: A modulator of reversible protein-membrane interactions. *Trends Biochem. Sci.* **20**, 272–280.
44. Cadwallader, K. A., Paterson, H., MacDonald, S. G., and Hancock, J. F. (1994). N-terminally myrisoylated ras proteins require palmitoylation or a polybasic domain for plasma membrane localization. *Mol. Cell. Biol.* **14**, 4722–4730.
45. Fishburn, C. S., Herzmark, P., Morales, J., and Bourne, H. R. (1999). Gβγ and palmitate target newly synthesized $G\alpha_z$ to the plasma membrane. *J. Biol. Chem.* **274**, 18793–18800.
46. Fishburn, C. S., Pollitt, S. K., and Bourne, H. R. (2000). Localization of a peripheral membrane protein: Gβγ targets Gα(z). *Proc. Natl. Acad. Sci. USA* **97**, 1085–1090.
47. Wedegaertner, P. B., Chu, D. H., Wilson, P. T., Levis, M. J., and Bourne, H. R. (1993). Palmitoylation is required for signaling functions and membrane attachment of Gq alpha and Gs alpha. *J. Biol. Chem.* **268**, 25001–25008.
48. Wedegaertner, P. B., Bourne, H. R., and von Zastrow, M. (1996). Activation-induced subcellular redistribution of $G_{s\alpha}$. *Mol. Biol. Cell* **7**, 1225–1233.
49. Wise, A., Parenti, M., and Milligan, G. (1997). Interaction of the G-protein G11alpha with receptors and phosphoinositidase C: The contribution of G-protein palmitoylation and membrane association. *FEBS Lett.* **407**, 257–260.
50. Bhattacharyya, R. and Wedegaertner, P. B. (2000). Gα 13 requires palmitoylation for plasma membrane localization, Rho-dependent signaling, and promotion of p115-RhoGEF membrane binding. *J. Biol. Chem.* **275**, 14992–14999.
51. Hughes, T. E., Zhang, H., Logothetis, D. E., and Berlot, C. H. (2001). Visualization of a functional Galpha q-green fluorescent protein fusion in living cells. Association with the plasma membrane is disrupted by mutational activation and by elimination of palmitoylation sites, but not by activation mediated by receptors or AlF4. *J. Biol. Chem.* **276**, 4227–4235.
52. Degtyarev, M. Y., Spiegel, A. M., and Jones, T. L. Z. (1993). The G protein α_s subunit incorporates [^{3}H]palmitic acid and mutation of cysteine-3 prevents this modification. *Biochemistry* **32**, 8057–8061.
53. Hepler, J. R., Biddlecome, G. H., Kleuss, C., Camp, L. A., Hofmann, S. L., Ross, E. M., and Gilman, A. G. (1996). Functional importance of the amino terminus of Gq α. *J. Biol. Chem.* **271**, 496–504.
54. Huang, C., Duncan, J. A., Gilman, A. G., and Mumby, S. M. (1999). Persistent membrane association of activated and depalmitoylated G protein α subunits. *Proc. Natl. Acad. Sci. USA* **96**, 412–417.
55. Jones, T. L. and Gutkind, J. S. (1998). Gα12 requires acylation for its transforming activity. *Biochemistry* **37**, 3196–3202.
56. Chen, C., Seow, K. T., Guo, K., Yaw, L. P., and Lin, S. C. (1999). The membrane association domain of RGS16 contains unique amphipathic features that are conserved in RGS4 and RGS5. *J. Biol. Chem.* **274**, 19799–19806.
57. Levis, M. J. and Bourne, H. R. (1992). Activation of the α subunit of G_s in intact cells alters its abundance, rate of degradation, and membrane avidity. *J. Cell Biol.* **119**, 1297–1307.
58. Hansen, S. H. and Casanova, J. E. (1994). Gsα stimulates transcytosis and apical secretion in MDCK cells through cAMP and protein kinase A. *J. Cell Biol.* **126**, 677–687.
59. Negishi, M., Hashimoto, H., and Ichikawa, A. (1992). Translocation of α subunits of stimulatory guanine nucleotide-binding proteins through stimulation of the prostacyclin receptor in mouse mastocytoma cells. *J. Biol. Chem.* **267**, 2367–2369.
60. Ransnäs, L. A., Svoboda, P., Jasper, J. R., and Insel, P. A. (1989). Stimulation of β-adrenergic receptors of S49 lymphoma cells redistributes the α subunit of the stimulatory G protein between cytosol and membranes. *Proc. Natl. Acad. Sci. USA* **86**, 7900–7903.
61. Yu, J.-Z. and Rasenick, M. M. (2002). Real-time visualization of a fluorescent Gαs: Dissociation of the activated G protein from plasma membrane. *Mol. Pharmacol.* **61**, 352–359.
62. Thiyagarajan, M. M., Bigras, E., Van Tol, H. H., Hebert, T. E., Evanko, D. S., and Wedegaertner, P. B. (2002). Activation-induced subcellular redistribution of Gα(s) is dependent upon its unique N-terminus. *Biochemistry* **41**, 9470–9484.

63. Tu, Y., Wang, J., and Ross, E. M. (1997). Inhibition of brain Gz GAP and other RGS proteins by palmitoylation of G protein α subunits. *Science* **278**, 1132–1135.
64. Tu, Y., Woodson, J., and Ross, E. M. (2001). Binding of regulator of G protein signaling (RGS) proteins to phospholipid bilayers. Contribution of location and/or orientation to GTPase-activating protein activity. *J. Biol. Chem.* **276**, 20160–20166.
65. Ross, E. M. and Wilkie, T. M. (2000). GTPase-activating proteins for heterotrimeric G proteins: regulators of G protein signaling (RGS) and RGS-like proteins. *Annu. Rev. Biochem.* **69**, 795–827.

CHAPTER 229

G Proteins Mediating Taste Transduction

Sami Damak[1] and Robert F. Margolskee[1,2]
[1]Department of Physiology and Biophysics,
[2]Howard Hughes Medical Institute,
The Mount Sinai School of Medicine,
New York, New York

Introduction

The sense of taste in humans is comprised of five primary qualities: sweet, bitter, sour, salty, and *umami* (the taste of glutamate) (reviewed in [1,2]). Taste sensation is initiated in primary taste sensory cells located in papillae at the surface of the tongue and in the soft palate. The papillae contain one to several taste buds, each comprising approximately 100 cells, including taste receptor cells (TRCs), precursor, and support cells [3]. TRCs are specialized epithelial cells that typically respond to one or more tastants by undergoing depolarization, elevating intracellular calcium, releasing a neurotransmitter, and activating the afferent neurons that they synapse upon.

Based on electrophysiological, genetic, and biochemical studies, several mechanisms have been identified for signal transduction by taste cells. For salty and sour compounds, the main signaling pathway is initiated by passive diffusion of ions through apical ion channels, amiloride-sensitive sodium channels, and acid-sensing channels [4–10]. For bitter, sweet, and *umami* compounds, the main signaling mechanisms depend on seven transmembrane-helix receptors, their coupled G proteins, and downstream effector enzymes (reviewed in [1,2]).

Various molecular techniques (e.g. molecular cloning, ribonuclease protection, RT-PCR, *in situ* hybridization, and immunohistochemistry) have shown that α-gustducin, $G\alpha_{i-2}$, $G\alpha_{i-3}$, $G\alpha_s$, $G\alpha_{14}$, $G\alpha_{15}$, α-transducin, $G\alpha q$, $G\beta_1$, $G\beta_3$, and $G\gamma_{13}$ are expressed in rodent TRCs [11–14].

α-Gustducin

α-Gustducin, a G-protein α subunit that shares 80 percent identity with rod α-transducin, is expressed in ~25 percent of TRCs [12] and has been implicated in responses of mice to bitter and sweet compounds [15]. Behavioral and nerve-recording studies showed that α-gustducin null mice have a marked reduction in their responses to the bitter compounds denatonium benzoate and quinine sulfate, and to the sweet compounds sucrose and SC45647, whereas their responses to salty and sour compounds were identical to those of their wild-type littermates [15].

Trypsin sensitivity and GTPγS binding assays showed that α-gustducin can be activated by native bovine taste receptors in the presence of a variety of bitter compounds [16]. Molecularly cloned bitter-responsive taste receptors (the T_{2r}/T_{rb} receptors) have been shown to couple to and activate heterotrimeric gustducin in preference to heterotrimers containing $G\alpha_s$, $G\alpha_i$, $G\alpha_o$, or $G\alpha_q$ [17]. Recombinant α-gustducin has been shown capable of activating retinal phosphodiesterase (PDE) [18]. Specific PDE subtypes (PDE1a) have been identified in bovine taste tissue that can be activated by rod α-transducin ([11]; Bakre *et al.* 2003 submitted) and by α-gustducin or a peptide derived from the "effector-interaction" region of α-gustducin ([19]; Lupi and R.F.M., unpublished). Rapid quench-flow experiments have shown that anti-α-gustducin antibodies block a decrease in cAMP and cGMP levels in taste tissues elicited by bitter compounds [20]. Based on the structural and biochemical similarity of α-gustducin and α-transducin, and the above observations, it appears that many bitter responses involve α-gustducin activation by heptahelical T2R/TRB taste receptors followed by α-gustducin activation of PDE1a in the taste cell.

α-Transducin

Rod α-transducin mRNA and protein has been shown to be expressed in rat taste tissue and TRCs [11,21]. The level of expression and the number of expressing cells were much lower than found with α-gustducin, suggesting a more limited role for α-transducin in taste signal transduction. The α subunit of cone transducin was also amplified by PCR from rat taste tissue RNA but was undetectable by ribonuclease protection, suggesting a very low level of expression and/or expression in only a small number of TRCs [11].

Rod α-transducin, like α-gustducin, can be activated *in vitro* by bitter-stimulated native bovine taste receptors [11,16]. A peptide that competitively inhibits activation of rod α-transducin by rhodopsin also inhibited activation of rod α-transducin by native taste receptors [11]. Taste PDE can be activated by aluminum fluoride-activated rod α-transducin or a peptide corresponding to the region of rod α-transducin that interacts with retinal PDE ([11]; Bakre *et al.* 2003 submitted). A small subset of frog taste cells respond to the sweeteners saccharin and NC-01 with the generation of an inward current [22]. Injection of a peptide derived from the effector-interaction region of rod α-transducin also induced this inward whole-cell current [22], suggesting that rod α-transducin, or the similar α-gustducin, might elicit this response *in vivo*.

To determine the role of rod α-transducin in taste signal transduction, behavioral tests with rod α-transducin single knockout mice and rod α-transducin/α-gustducin double knockout mice were carried out [23]. No differences were found in the responses to denatonium benzoate, quinine sulfate, sucrose, and SC45647 between rod α-transducin knockout and wild-type mice or between rod α-transducin/ α-gustducin double knockout and α-gustducin single knockout mice. However, the preference responses to monosodium glutamate (MSG) of rod α-transducin/α-gustducin double knockout mice were diminished compared to those of wild-type and α-gustducin single knockout mice [23]. At concentrations of MSG that are preferred strongly by wild-type mice, the double knockout mice were indifferent. These data show that rod α-transducin and α-gustducin are involved in *umami* signaling, but rod α-transducin, in contrast to α-gustducin, does not appear to contribute to sweet or bitter signaling.

Other G Protein α Subunits

The behavioral and gustatory nerve responses of α-gustducin knockout mice to bitter and sweet compounds were reduced, but not totally abolished, indicating that other G proteins and/or pathways are involved. Transgenic expression in the α-gustducin lineage of TRCs of a dominant-negative α-gustducin mutant that can bind to taste receptors and Gβγ subunits, but cannot be activated by receptors, further reduced the residual responses to sweet and bitter compounds of α-gustducin knockout mice [24]. This dominant-negative transgene also reduced responses to bitter and sweet compounds when introduced into α-gustducin-positive mice, demonstrating its effectiveness as a dominant-negative competitor [24]. $G\alpha_{i-2}$, $G\alpha_{i-3}$, $G\alpha_s$, $G\alpha_{14}$, and $G\alpha_{15}$ are possible candidates to mediate the residual responses in α-gustducin knockout mice; clearly these residual responses are not mediated by rod α-transducin based on results from the double knockouts. The precise role in taste transduction of G protein α subunits other than α-transducin and α-gustducin remains to be clarified.

βγ Subunits

A novel Gγ subunit, Gγ13, was found by screening cDNA libraries from individual α-gustducin-expressing versus nonexpressing TRCs [25]. Immunohistochemistry showed co-expression of Gγ13 and α-gustducin. Southern blot analysis of RT-PCR products from individual TRCs showed co-expression of α-gustducin and Gγ13 with Gβ3 (19/19 cells) and Gβ1 (15/19 cells) [25]. It was subsequently determined that α-gustducin is expressed in most, but not all, TRCs that express Gγ13, and that the Gγ13-expressing TRCs also express phospholipase Cβ2 and the type III inositol trisphosphate (IP_3) receptor [26–28]. The trypsin-sensitivity assay showed that Gγ13 can interact with α-gustducin and that α-gustducin/Gβ1/Gγ13 complexes can be activated by native taste receptors in the presence of the bitter compound denatonium benzoate [25]. Many bitter compounds elicit an increase in taste tissue levels of IP_3 and diacyl glycerol (DAG); this response was entirely blocked by antibodies directed against Gγ13, Gβ3 or PLCβ2 [14,25,29], implicating all three of these proteins in mediating this taste response. Control antibodies or antibodies directed against α-gustducin had no effect on IP_3 or DAG generation, while antibodies directed against Gβ1 had a slight effect. Gβ3Gγ13 appears to be the likely partner of α-gustducin and like α-gustducin is involved in TRC responses to bitter compounds.

G-Protein-Coupled Receptors

The T2r/TrB family of taste receptors contains approximately 25 members [30,31]. One member of this family, mT2r5, was shown *in vitro* to be activated by cycloheximide and to couple specifically to gustducin [17]; presumably the other T2r/TrB receptors are also bitter-responsive taste receptors. It was shown by *in situ* hybridization that multiple T2r/TrB receptors are co-expressed in a subset of the α-gustducin-expressing TRCs in mice and rats, indicating that most if not all T2r/TrB receptors are expressed in the same subset of TRCs and that they most likely couple to gustducin.

The T1rs constitute a three-member family of taste receptors expressed in TRCs [32–37]. T1r1 is predominantly expressed in TRCs of the fungiform and foliate papillae and palate, whereas T1r2 is expressed in the circumvallate and foliate papillae and palate, but not in the fungiform papillae [32]. T1r3 is expressed in both anterior and posterior papillae of the tongue as well as in the palate [36]. Calcium-imaging studies showed that HEK-293 cells expressing T1r3 and T1r2 responded to several sweeteners and D-amino-acids, whereas cells expressing T1r3 and T1r1 responded to L-amino-acids [36,38,39]. Transgenic expression of T1r3 from a highly sweet-preferring "taster" strain into the less sweet-preferring "nontaster" background converted the nontaster phenotype to that of the taster [36]. The G protein(s) that couple to these receptors are currently unknown, but probably include G_s for sugars based on biochemical and electrophysiological experiments, implicating adenylyl cyclase and cAMP in TRC responses to sugars.

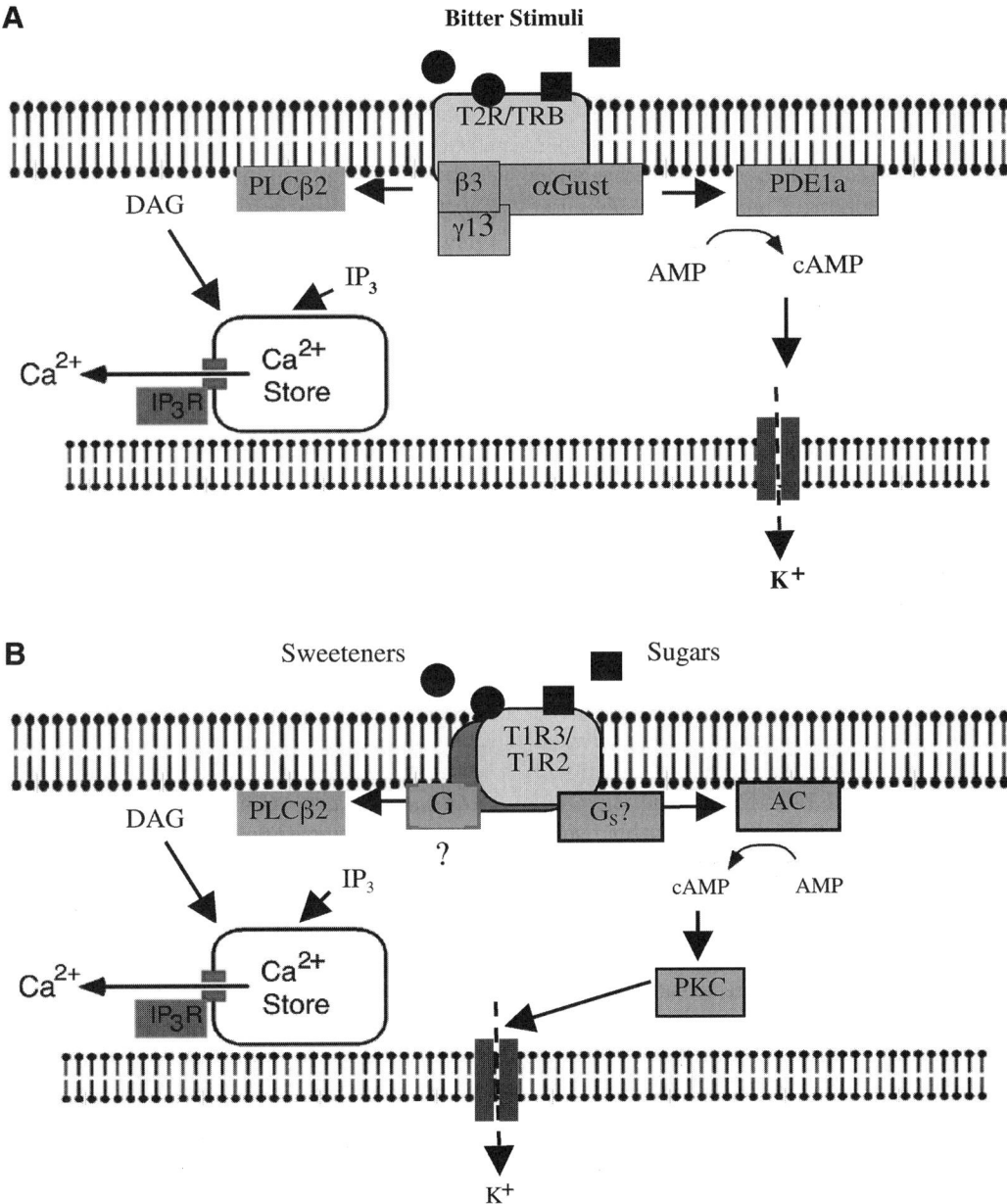

Figure 1 Schematic representation of the transduction pathways for bitter (A) and sweet (B) stimuli. One pathway, common to bitter compounds and artificial sweeteners, involves activation of PLCβ2, production of IP$_3$ and DAG, and release of Ca^{2+} from the ER. This pathway is activated by the β3γ13 subunits of gustducin for bitters and by unidentified G protein α and/or βγ subunits for sweeteners. The second main pathway for bitter compounds involves activation of PDE1a by the α subunit of gustducin and a drop in cAMP levels. For sugars the main pathway consists of activation of adenylyl cyclase (AC) by a G protein α-subunit, probably G_s, and a rise in cAMP.

Functional studies in heterologous cells, mRNA expression studies, and behavioral tests suggested that an mGluR4 variant, taste mGluR4, may function as an *umami* taste receptor [40,41]. It is currently not known whether taste mGluR4 couples to transducin and/or gustducin.

Second Messenger Pathways

The proposed main signaling pathways for bitter are

(1) T2r/TrB receptor → heterotrimeric gustducin → α-gustducin → activation of PDE1a → decrease in cAMP and cGMP → activation of a cyclic nucleotide suppressed cation channel → Ca^{2+} influx → neurotransmitter release;

(2) T2r/TrB receptor → heterotrimeric gustducin → βγ-gustducin (Gβ3/Gγ13) → PLC β2 → IP_3 + DAG → Ca^{2+} release from internal stores → neurotransmitter release (Figure 1a).

The two main pathways for sweetener signal transduction are believed to be

(1) T1r heterodimeric receptor → heterotrimeric G-protein (G_s?) → adenylyl cyclase → rise in [cAMP] → protein kinase phosphorlyation of K^+ channels → depolarization → Ca^{2+} influx → neurotransmitter release;

(2) T1r heterodimeric receptor → heterotrimeric G protein → PLC β2 → IP_3 + DAG → Ca^{2+} release from internal stores → neurotransmitter release. Experiments using rat TRCs showed that natural sweeteners use the cAMP/Ca^{2+} influx pathway whereas artificial sweeteners use the IP_3 + DAG/Ca^{2+} release pathway [42] (Figure 1b).

For *umami* signaling, biochemical experiments have implicated both a rise and a drop in [cAMP] and a rise in [IP_3] [43].

Conclusion

Heterotrimeric gustducin plays a central role in the signal transduction of bitter compounds by coupling to T2r/TrB receptors and activating PDE and PLC β2. It also plays a central role in the signal transduction of sweeteners, although its coupling to specific taste receptors has not been experimentally demonstrated. The transduction of *umami* signals appears to use both gustducin and transducin. The role of the other G proteins expressed in taste tissue is still unclear. The expression of these G proteins in a wide range of tissues limits the usefulness of the conventional mouse knockout as a tool for elucidating their role in taste transduction. Many answers will probably come from taste tissue specific knockouts using the Cre/LoxP system.

References

1. Kinnamon, S. C. and Margolskee, R. F. (1996). Mechanisms of taste transduction. *Curr. Opin. Neurobiol.* **6**, 506–513.

2. Gilbertson, T. A., Damak, S., and Margolskee, R. F. (2000). The molecular physiology of taste transduction. *Curr. Opin. Neurobiol.* **10**, 519–527.

3. Lindemann, B. (1996). Taste reception. *Physiol. Rev.* **76**, 718–766.

4. Avenet, P. and Lindemann, B. (1988). Amiloride-blockable sodium currents in isolated taste receptor cells. *J. Membr. Biol.* **105**, 245–255.

5. Avenet, P. and Lindemann, B. (1991). Noninvasive recording of receptor cell action potentials and sustained currents from single taste buds maintained in the tongue: the response to mucosal NaCl and amiloride. *J. Membr. Biol.* **124**, 33–41.

6. Heck, G. L., Mierson, S., and DeSimone, J. A. (1984). Salt taste transduction occurs through an amiloride-sensitive sodium transport pathway. *Science* **223**, 403–405.

7. Schiffman, S. S., Lockhead, E., and Maes, F. W. (1983). Amiloride reduces the taste intensity of Na^+ and Li^+ salts and sweeteners. *Proc. Natl. Acad. Sci. USA* **80**, 6136–6140.

8. Schiffman, S. S., Frey, A. E., Suggs, M. S., Cragoe, E. J. Jr., and Erickson, R. P. (1990). The effect of amiloride analogs on taste responses in gerbil. *Physiol. Behav.* **47**, 435–441.

9. Waldmann, R., Champigny, G., Bassilana, F., Heurteaux, C., and Lazdunski, M. (1997). A proton-gated cation channel involved in acid-sensing. *Nature* **386**, 173–177.

10. Liu, L. and Simon, S. A. (2001). Acidic stimuli activates two distinct pathways in taste receptor cells from rat fungiform papillae. *Brain. Res.* **923**, 58–70.

11. Ruiz-Avila, L., McLaughlin, S. K., Wildman, D., McKinnon, P. J., Robichon, A., Spickofsky, N., and Margolskee, R. F. (1995). Coupling of bitter receptor to phosphodiesterase through transducin in taste receptor cells. *Nature* **376**, 80–85.

12. McLaughlin, S. K., McKinnon, P. J., and Margolskee, R. F. (1992). Gustducin is a taste-cell-specific G protein closely related to the transducins. *Nature* **357**, 563–569.

13. Kusakabe, Y., Yamaguchi, E., Tanemura, K., Kameyama, K., Chiba, N., Arai, S., Emori, Y., and Abe, K. (1998). Identification of two alpha-subunit species of GTP-binding proteins, Galpha15 and Galphaq, expressed in rat taste buds. *Biochim. Biophys. Acta* **1403**, 265–272.

14. Rossler, P., Boekhoff, I., Tareilus, E., Beck, S., Breer, H., and Freitag, J. (2000). G protein betagamma complexes in circumvallate taste cells involved in bitter transduction. *Chem. Senses* **25**, 413–421.

15. Wong, G. T., Gannon, K. S., and Margolskee, R. F. (1996). Transduction of bitter and sweet taste by gustducin. *Nature* **381**, 796–800.

16. Ming, D., Ruiz-Avila, L., and Margolskee, R. F. (1998). Characterization and solubilization of bitter-responsive receptors that couple to gustducin. *Proc. Natl. Acad. Sci. USA* **95**, 8933–8938.

17. Chandrashekar, J., Mueller, K. L., Hoon, M. A., Adler, E., Feng, L., Guo, W., Zuker, C. S., and Ryba, N. J. (2000). T2Rs function as bitter taste receptors. *Cell* **100**, 703–711.

18. Hoon, M. A., Northup, J. K., Margolskee, R. F., and Ryba, N. J. (1995). Functional expression of the taste specific G-protein, alpha-gustducin. *Biochem. J.* **309**, 629–636.

19. Spickofsky, N., Robichon, A., Danho, W., Fry, D., Greeley, D., Graves, B., Madison, V., and Margolskee, R. F. (1994). Biochemical analysis of the transducin-phosphodiesterase interaction. *Nat. Struct. Biol.* **1**, 771–781.

20. Yan, W., Sunavala, G., Rosenzweig, S., Dasso, M., Brand, J. G., and Spielman, A. I. (2001). Bitter taste transduced by PLC-beta(2)-dependent rise in IP(3) and alpha-gustducin-dependent fall in cyclic nucleotides. *Am. J. Physiol. Cell. Physiol.* **280**, C742–C751.

21. Yang, H., Wanner, I. B., Roper, S. D., and Chaudhari, N. (1999). An optimized method for in situ hybridization with signal amplification that allows the detection of rare mRNAs. *J. Histochem. Cytochem.* **47**, 431–446.

22. Kolesnikov, S. S. and Margolskee, R. F. (1995). A cyclic-nucleotide-suppressible conductance activated by transducin in taste cells. *Nature* **376**, 85–88.

23. He, W., Margolskee, R. F., and Damak, S. (2002). "Signal transduction of umami taste by alpha-gustducin and alpha-transducin." XXIVth Am. Chem. Soc. meeting, Sarasota, Florida.

24. Ruiz-Avila, L., Wong, G. T., Damak, S., and Margolskee, R. F. (2001). Dominant loss of responsiveness to sweet and bitter compounds caused by a single mutation in alpha-gustducin. *Proc. Natl. Acad. Sci. USA* **98**, 8868–8873.
25. Huang, L., Shanker, Y. G., Dubauskaite, J., Zheng, J. Z., Yan, W., Rosenzweig, S., Spielman, A. I., Max, M., and Margolskee, R. F. (1999). Ggamma13 colocalizes with gustducin in taste receptor cells and mediates IP3 responses to bitter denatonium. *Nat. Neurosci.* **2**, 1055–1062.
26. Asano-Miyoshi, M., Abe, K., and Emori, Y. (2000). Co-expression of calcium signaling components in vertebrate taste bud cells. *Neurosci. Lett.* **283**, 61–64.
27. Kusakabe, Y., Yasuoka, A., Asano-Miyoshi, M., Iwabuchi, K., Matsumoto, I., Arai, S., Emori, Y., and Abe, K. (2000). Comprehensive study on G protein alpha-subunits in taste bud cells, with special reference to the occurrence of Galphai2 as a major Galpha species. *Chem. Senses* **25**, 525–531.
28. Clapp, T. R., Stone, L. M., Margolskee, R. F., and Kinnamon, S. C. (2001). Immunocytochemical evidence for co-expression of Type III IP3 receptor with signaling components of bitter taste transduction. *BMC Neurosci.* **2**, 6.
29. Rossler, P., Kroner, C., Freitag, J., Noe, J., and Breer, H. (1998). Identification of a phospholipase C beta subtype in rat taste cells. *Eur. J. Cell. Biol.* **77**, 253–261.
30. Adler, E., Hoon, M. A., Mueller, K. L., Chandrashekar, J., Ryba, N. J., and Zuker, C. S. (2000). A novel family of mammalian taste receptors. *Cell* **100**, 693–702.
31. Matsunami, H., Montmayeur, J. P., and Buck, L. B. (2000). A family of candidate taste receptors in human and mouse. *Nature* **404**, 601–604.
32. Hoon, M. A., Adler, E., Lindemeier, J., Battey, J. F., Ryba, N. J., and Zuker, C. S. (1999). Putative mammalian taste receptors: a class of taste-specific GPCRs with distinct topographic selectivity. *Cell* **96**, 541–551.
33. Kitagawa, M., Kusakabe, Y., Miura, H., Ninomiya, Y., and Hino, A. (2001). Molecular genetic identification of a candidate receptor gene for sweet taste. *Biochem. Biophys. Res. Commun.* **283**, 236–242.
34. Max, M., Shanker, Y. G., Huang, L., Rong, M., Liu, Z., Campagne, F., Weinstein, H., Damak, S., and Margolskee, R. F. (2001). Tas1r3, encoding a new candidate taste receptor, is allelic to the sweet responsiveness locus Sac. *Nat. Genet.* **28**, 58–63.
35. Montmayeur, J. P., Liberles, S. D., Matsunami, H., and Buck, L. B. (2001). A candidate taste receptor gene near a sweet taste locus. *Nat. Neurosci.* **4**, 492–498.
36. Nelson, G., Hoon, M. A., Chandrashekar, J., Zhang, Y., Ryba, N. J., and Zuker, C. S. (2001). Mammalian sweet taste receptors. *Cell* **106**, 381–390.
37. Sainz, E., Korley, J. N., Battey, J. F., and Sullivan, S. L. (2001). Identification of a novel member of the T1R family of putative taste receptors. *J. Neurochem.* **77**, 896–903.
38. Nelson, G., Chandrashekar, J., Hoon, M., Feng, L., Zhao, G., Ryba, N. J. P., and Zuker, C. S. (2002). An amino-acid taste receptor. *Nature* **416**, 199–202.
39. Li, X., Staszewski, L., Xu, H., Durick, K., Zoller, M., and Adler, E. (2002). Human receptors for sweet and umami taste. *Proc. Natl. Acad. Sci. USA* **26**, 4692–4696.
40. Chaudhari, N., Landin, A. M., and Roper, S. D. (2000). A metabotropic glutamate receptor variant functions as a taste receptor. *Nat. Neurosci.* **3**, 113–119.
41. Chaudhari, N., Yang, H., Lamp, C., Delay, E., Cartford, C., Than, T., and Roper, S. (1996). The taste of monosodium glutamate: membrane receptors in taste buds. *J. Neurosci.* **16**, 3817–3826.
42. Bernhardt, S. J., Naim, M., Zehavi, U., and Lindemann, B. (1996). Changes in IP3 and cytosolic Ca^{2+} in response to sugars and non-sugar sweeteners in transduction of sweet taste in the rat. *J. Physiol. (London)* **490**, 325–336.
43. Ninomiya, Y., Nakashima, K., Fukuda, A., Nishino, H., Sugimura, T., Hino, A., Danilova, V., and Hellekant, G. (2000). Responses to umami substances in taste bud cells innervated by the chorda tympani and glossopharyngeal nerves. *J. Nutr.* **130**, 950S–953S.

CHAPTER 230

Regulation of Synaptic Fusion by Heterotrimeric G Proteins

Simon Alford and Trillium Blackmer
Department of Biological Sciences, University of Illinois at Chicago, Chicago, Illinois

Introduction

Exocytotic release of a neurotransmitter is necessary for interneuronal communication. In neurons, exocytosis, or the fusion of a synaptic vesicle with a specialized area in the plasma membrane, is a tightly regulated process that must be activated with strict temporal control. Ca^{2+} entry to the terminal is the trigger for exocytosis and neurotransmitter release, but the release of neurotransmitter evoked by Ca^{2+} is modified by receptors located at the presynaptic terminal. These receptors include G-protein-coupled receptors (GPCRs).

The Vesicle Fusion Machinery

To understand how G proteins modulate the release of neurotransmitter we must understand some basic principles of exocytosis. The best-characterized proteins involved in exocytotic fusion are those that comprise the core complex, the formation of which is absolutely necessary for fusion to occur. The core complex, or SNARE (soluble NSF (N-ethylmaleimide-sensitive factor) attachment protein receptor), is a bundle of four α-helices, approximately 65 amino acids in length, which is thought to bridge the synaptic vesicle and plasma membranes [1]. These α-helices are donated by three different proteins; a family member from the syntaxin and SNAP-25 families, both located in the synaptic active zone, and a VAMP (vesicle associated membrane protein, also known as synaptobrevin) family member, located in the synaptic vesicular membrane. The syntaxin family consists of integral membrane proteins around 300 amino acids in length, which have been shown to bind to many regulatory proteins [2]. Of these regulatory proteins, synaptotagmin is the most likely candidate for the Ca^{2+} sensor in synchronous release of neurotransmitter [3,4] The core complex is sufficient to mediate fusion of lipid micelles *in vitro* [5], and fusion of the synaptic vesicle with the plasma membrane requires the interaction of syntaxin, SNAP-25, and VAMP.

Vesicles containing neurotransmitter must be able to fuse with the plasma membrane in microsecond time scales. Consequently, synaptic vesicles are located very near the point of fusion, the active zone, at the presynaptic terminal. This organization is termed *docking*. Vesicular recruitment and docking requires the actin assembly network and ATP [6]. Multistep fusion reactions are ruled out due to the speed of release. For this reason, it is believed that there is a pool of ready-to-fuse synaptic vesicles that have undergone a further maturation step referred to as *priming* [7]. Priming in large dense-core vesicle requires ATP, submicromolar Ca^{2+} concentrations, and alterations in membrane lipids by lipid transferases and kinases [8]. Furthermore, the possible role in priming of the N-ethylmaleimide-sensitive factor (NSF) in synaptic vesicles may indicate similar requirements [9]. Evoked synaptic fusion is thought to require high (hundreds of μM) local concentrations of Ca^{2+} [10] following action potential invasion of the nerve terminal, although more recent work suggests that low μM increases in Ca^{2+} concentrations (~ 10 μM) may activate fusion in some neurons [11].

G Protein-Coupled Receptor Mediated Modulation at the Presynaptic Terminal

Regulation of neurotransmitter release at the presynaptic terminal plays an important part in the plasticity of the nervous system [12]. Various neurotransmitters modulate release from presynaptic terminals, and many of these interactions involve the activation of a GPCR [13]. Modulation of exocytotic release by GPCRs is an important mechanism by which neurons are able to respond and adapt to changes in secretory requirements. Some GPCRs may couple to more than one G protein, while others show a great deal of specificity. Gβγ binding to Gα involves widespread contacts at two distinct interfaces. Following activation by a GPCR, the heterotrimeric G protein dissociates into an activated GαGTP subunit and a free Gβγ subunit [14]. Active Gα-GTP and free Gβγ may then activate many different signaling pathways [15].

Uncertainty over the mechanisms by which G proteins alter neurotransmitter release in part reflects the variety of G-protein effector targets and the difficulties in gaining experimental access to these small structures. Thus, most molecular studies of the detailed mechanisms come from either transfection of the relevant proteins into cultured cell lines and *Xenopus* oocytes or from electrophysiological measurements from neuronal cell bodies.

Possible Mechanisms of Presynaptic Inhibition by G Proteins

GPCRs that inhibit neurotransmitter release have perhaps been the mostly widely studied modulators of synaptic transmission. A consensus mechanism by which these transmitters may modulate synaptic transmitter release has been hypothesized to involve an alteration in action potential evoked Ca^{2+} entry to the presynaptic terminal.

1. This reduction in Ca^{2+} entry could occur by a direct action of G protein βγ subunit on the gating of voltage gated Ca^{2+} channels (VGCCs) [16,17]. If Gβγ inhibits VGCCs, less Ca^{2+} would enter the presynaptic terminal, and since neurotransmitter release is Ca^{2+} dependent, less neurotransmitter would be released. GPCR-mediated inhibition of release via a direct inhibition of VGCCs has been demonstrated at one presynaptic terminal [18] through multiple GPCRs [19,20]. Although, given the large number of neurotransmitters that inhibit release, many of which have very little effect on Ca^{2+} entry through VGCCs [20], it is unlikely that this is the only pathway involved in GPCR mediated inhibition at this terminal.

2. If G protein-coupled inwardly rectifying K^+ channels (GIRKs) were located at presynaptic terminals, activation by Gβγ could modulate action potential amplitudes, allowing fewer Ca^{2+} channels to open. Gβγ activates GIRKs in neuronal cell bodiestransfected cell lines, and *Xenopus* oocytes [21,22]. Although GIRKs have been histochemically localized to presynaptic terminals [23], physiological evidence for their presynaptic action is lacking. These channels are believed to be more important for postsynaptic modulation [24]. GPCR-mediated inhibition of voltage-gated K^+ channels at an autaptic presynaptic terminal has also been shown [25] to occur through activation of dopamine receptors, although it is not clear what G protein subunit, Gα or Gβγ, is responsible. At the reticulospinal-motoneuron synapse of the lamprey, both glutamate and 5-HT activate GPCRs, which modulate a K^+ current, although the channel subtype and G-protein subunit involved are unknown [26,27].

3. It has been suggested that G proteins may modulate voltage-gated Na^+ channels at the presynaptic terminal; however, no direct evidence has yet been presented [28]. Modulation of Na^+ channels could also indirectly affect the entry of Ca^{2+} into the presynaptic terminal.

These studies led to the idea that Gβγ-mediated inhibition of neurotransmission at the presynaptic terminal was through a direct or indirect effect on the amount of Ca^{2+} that enters the terminal during the action potential. However, there is growing evidence that Gβγ may also inhibit synaptic transmission by modulation distal to the point of Ca^{2+} entry. The ability of G proteins to inhibit neurotransmitter release by directly targeting the release apparatus was first demonstrated by Silinsky [29] in the neuromuscular junction. Spontaneous exocytotic events, where exocytosis occurs independently of Ca^{2+} entry, can be detected by recording miniature excitatory/inhibitory post-synaptic currents (mE/IPSCs). Measurements of mE/IPSCs allow the exocytotic modulatory processes that occur independently of Ca^{2+} entry to be isolated and studied, unlike evoked EPSCs. These mE/IPSCs have been shown to be regulated by many GPCRs [30].

In support of the hypothesis that G proteins inhibit vesicle fusion directly, G proteins can inhibit exocytosis after cell permeabilization, suggesting a role late in the exocytotic event [31]. In addition, exocytotic processes in pancreatic β cells, peritoneal mast cells, chromaffin cells, PC12 cells, and secretory granules are regulated independently of Ca^{2+} entry by G proteins [32]. Gβγ has been shown to interact directly with the fusion machinery in rat mast cells [33]. In the lamprey giant synapse 5-HT-mediated synaptic inhibition does not cause a reduction in Ca^{2+} entry to the synapse [27]. Furthermore, the actions of 5-HT at a GPCR are abolished by intracellular block of activated Gβγ [34]. A mechanism for a direct interaction between Gβγ and the core vesicle fusion machinery is suggested by the finding that Gβγ directly binds SNARE proteins syntaxin and SNAP-25 [34,35] as well as the cysteine string protein (CSP) [36].

Presynaptic Ca^{2+} Stores and Modulation of Neurotransmitter Release

Ca^{2+} release from internal stores located at the presynaptic terminal may lead to an enhancement of transmitter release. However, this Ca^{2+} may originate either from Ca^{2+} activated channels (that is, Ca^{2+}-induced Ca^{2+} release; CICRs) [37] or

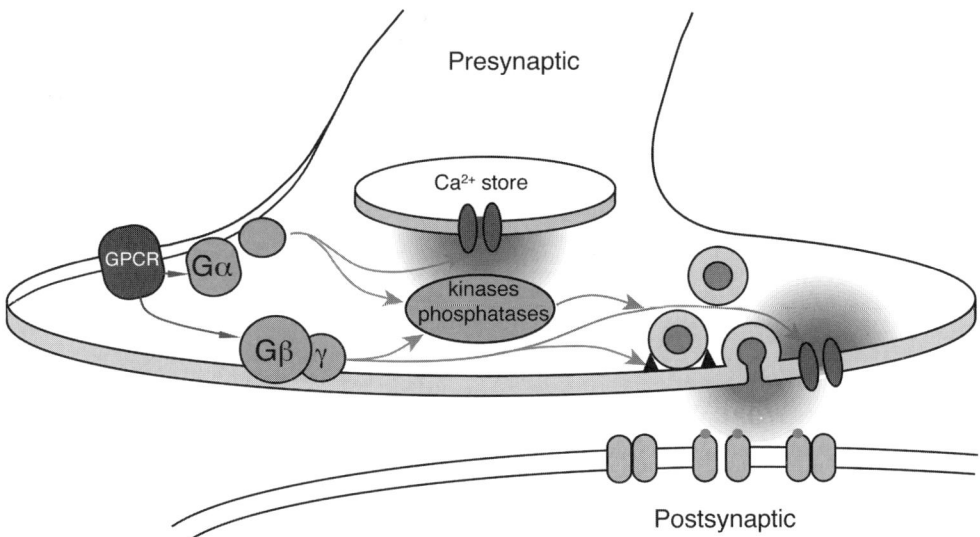

Figure 1 GPCRs at the presynaptic terminal (blue) may modulate transmission through either Gα or Gβγ. These pathways (green) may in turn alter the phosphorylation state of presynaptic proteins, cause the release of Ca^{2+} from presynaptic internal stores, or alter release through interactions with the release machinery or through actions at presynaptic ion channels.

following the activation of a presynaptic GPCR, leading to IP3 production and activation of presynaptic IP3 receptors [26,38]. It is also important to note that the activation of presynaptic receptors leading to IP3 production will also produce diacylglycerol in the nerve terminal, leading to the possible activation of PKC or direct effects on proteins associated with the release machinery, for example UNC13 [39].

G Proteins and Phosphorylation

G proteins may alter the efficacy of synaptic transmission either through phosphorylation or dephosphorylation of presynaptic components. Gβγ has been suggested to activate the calcineurin phosphatase pathway to inhibit release independent of Ca^{2+} entry [40] in neuroendocrine cells. In addition, the activation of both PKA and PKC have been implicated in the enhancement of synaptic transmission in the central nervous system. Indeed, tonic activation of PKA may be necessary for vesicle fusion to occur [41,42] and PKA phosphorylates CSP to alter its binding to syntaxin [43]. Metabotropic glutamate receptors in the mammalian CNS may activate either of these latter pathways [44,45].

Although relatively difficult to study, the presynaptic terminal may contain as rich an array of receptor-mediated mechanisms that modify information flow as has been identified at the postsynaptic side of the synapse. Figure 1 demonstrates some of the possible mechanisms by which presynaptic receptors may alter neurotransmitter release.

References

1. Sutton, R. B., Fasshauer, D., Jahn, R., and Brunger, A. T. (1998). Crystal structure of a SNARE complex involved in synaptic exocytosis at 2.4 A resolution. *Nature* **395**, 347–353.
2. Wu, M. N., Fergestad, T., Lloyd, T. E., He, Y., Broadie, K., and Bellen, H. J. (1999). Syntaxin 1A interacts with multiple exocytic proteins to regulate neurotransmitter release in vivo. *Neuron* **23**, 593–605.
3. Sugita, S., Han, W., Butz, S., Liu, X., Fernandez-Chacon, R., Lao, Y., and Sudhof, T. C. (2001). Synaptotagmin VII as a plasma membrane Ca(2+) sensor in exocytosis *Neuron* **30**, 459–473.
4. Davis, A. F., Bai, J., Fasshauer, D., Wolowick, M. J., Lewis, J. L., and Chapman, E. R. (1999). Kinetics of synaptotagmin responses to Ca^{2+} and assembly with the core SNARE complex onto membranes. *Neuron* **24**, 363–376.
5. Weber, T., Zemelman, B. V., McNew, J. A., Westermann, B., Gmachl, M., Parlati, F., Sollner, T. H., and Rothman, J. E. (1998). SNAREpins: minimal machinery for membrane fusion. *Cell* **92**, 759–772.
6. Iida, Y., Senda, T., Matsukawa, Y., Onoda, K., Miyazaki, J. I., Sakaguchi, H., Nimura, Y., Hidaka, H., and Niki, I. (1997). Myosin light-chain phosphorylation controls insulin secretion at a proximal step in the secretory cascade. *Am. J. Physiol.* **273**, E782–789.
7. Sudhof, T. C. (1995). The synaptic vesicle cycle: a cascade of protein-protein interactions. *Nature* **375**, 645–653.
8. Hay, J. C., Fisette, P. L., Jenkins, G. H., Fukami, K., Takenawa, T., Anderson, R. A., and Martin, T. F. (1995). ATP-dependent inositide phosphorylation required for Ca^{2+}-activated secretion. *Nature* **374**, 173–177.
9. Xu, J., Xu, Y., Ellis-Davies, G. C., Augustine, G. J., and Tse, F. W. (2002). Differential regulation of exocytosis by alpha- and beta-SNAPs. *J. Neurosci.* **22**, 53–61.
10. Augustine, G. J., Adler, E. M., and Charlton, M. P. (1991). The calcium signal for transmitter secretion from presynaptic nerve terminals. *Ann. NY Acad. Sci.* **635**, 365–381.
11. Bollmann, J. H., Sakmann, B., and Borst, J. G. (2000). Calcium sensitivity of glutamate release in a calyx-type terminal. *Science* **289**, 953–957.
12. Alford, S. and Grillner, S. (1991). The involvement of GABAB receptors and coupled G-proteins in spinal GABAergic presynaptic inhibition. *J. Neurosci.* **11**, 3718–3726.
13. Dutar, P. and Nicoll, R. A. (1988). A physiological role for GABAB receptors in the central nervous system *Nature* **332**, 156–158.
14. Stryer, L. and Bourne, H. R. (1986). G proteins: a family of signal transducers. *Annu. Rev. Cell Biol.* **2**, 391–419.
15. Hamm, H. E. (1998). The many faces of G protein signaling. *J. Biol. Chem.* **273**, 669–672.

16. Dolphin, A. C., Pearson, H. A., Menon-Johansson, A. S., Sweeney, M. I., Sutton, K., Huston, E., Cullen, G. P., and Scott, R. H. (1993). G protein modulation of voltage-dependent calcium channels and transmitter release. *Biochem. Soc. Trans.* **21**, 391–395.
17. Qin, N., Platano, D., Olcese, R., Stefani, E., and Birnbaumer, L. (1997). Direct interaction of Gβγ with a C-terminal gbetagamma-binding domain of the Ca^{2+} channel alpha1 subunit is responsible for channel inhibition by G protein-coupled receptors. *Proc. Natl. Acad. Sci. USA.* **94**, 8866–8871.
18. Takahashi, T., Forsythe, I. D., Tsujimoto, T., Barnes-Davies, M., and Onodera, K. (1996). Presynaptic calcium current modulation by a metabotropic glutamate receptor. *Science* **274**, 594–597.
19. Takahashi, T., Kajikawa, Y., and Tsujimoto, T. (1998). G-protein-coupled modulation of presynaptic calcium currents and transmitter release by a GABAB receptor. *J. Neurosci.* **18**, 3138–3146.
20. Mirotznik, R. R., Zheng, X., and Stanley, E. F. (2000). G-protein types involved in calcium channel inhibition at a presynaptic nerve terminal. *J. Neurosci.* **20**, 7614–7621.
21. Reuveny, E., Slesinger, P. A., Inglese, J., Morales, J. M., Iniguez-Lluhi, J. A., Lefkowitz, R. J., Bourne, H. R., Jan, Y. N., and Jan, L. Y. (1994). Activation of the cloned muscarinic potassium channel by G protein beta gamma subunits. *Nature* **370**, 143–146.
22. Huang, C. L., Slesinger, P. A., Casey, P. J., Jan, Y. N., and Jan, L. Y. (1995). Evidence that direct binding of G beta gamma to the GIRK1 G protein-gated inwardly rectifying K^+ channel is important for channel activation. *Neuron* **15**, 1133–1143.
23. Ponce, A., Bueno, E., Kentros, C., Vega-Saenz de Miera, E., Chow, A., Hillman, D., Chen, S., Zhu, L., Wu, M. B., Wu, X., Rudy, B., and Thornhill, W. B. (1996). G-protein-gated inward rectifier K^+ channel proteins (GIRK1) are present in the soma and dendrites as well as in nerve terminals of specific neurons in the brain. *J. Neurosci.* **16**, 1990–2001.
24. Dutar, P., Petrozzino, J. J., Vu, H. M., Schmidt, M. F., and Perkel, D. J. (2000). Slow synaptic inhibition mediated by metabotropic glutamate receptor activation of GIRK channels. *J. Neurophysiol.* **84**, 2284–2290.
25. Congar, P., Bergevin, A., and Trudeau, L. E. (2002). D2 receptors inhibit the secretory process downstream from calcium influx in dopaminergic neurons: implication of K^+ channels. *J. Neurophysiol.* **87**, 1046–1056.
26. Cochilla, A. J. and Alford, S. (1998). Metabotropic glutamate receptor-mediated control of neurotransmitter release. *Neuron* **20**, 1007–1016.
27. Takahashi, M., Freed, R., Blackmer, T., and Alford, S. (2001). Calcium influx-independent depression of transmitter release by 5-HT at lamprey spinal cord synapses. *J. Physiol. (London)* **532**, 323–336.
28. Ma, J. Y., Catterall, W. A., and Scheuer, T. (1997). Persistent sodium currents through brain sodium channels induced by G protein betagamma subunits. *Neuron* **19**, 443–452.
29. Silinsky, E. M. (1984). On the mechanism by which adenosine receptor activation inhibits the release of acetylcholine from motor nerve endings. *J. Physiol.* **346**, 243–256.
30. Scanziani, M., Gahwiler, B. H., and Thompson, S. M. (1995). Presynaptic inhibition of excitatory synaptic transmission by muscarinic and metabotropic glutamate receptor activation in the hippocampus: are Ca^{2+} channels involved? *Neuropharmacology* **34**, 1549–1557.
31. Luini, A. and De Matteis, M. A. (1990). Evidence that receptor-linked G protein inhibits exocytosis by a post-second-messenger mechanism in AtT-20 cells. *J. Neurochem.* **54**, 30–38.
32. Lang, J. (1999). Molecular mechanisms and regulation of insulin exocytosis as a paradigm of endocrine secretion [In Process Citation]. *Eur. J. Biochem.* **259**, 3–17.
33. Pinxteren, J. A., O'Sullivan, A. J., Tatham, P. E., and Gomperts, B. D. (1998). Regulation of exocytosis from rat peritoneal mast cells by G protein beta gamma-subunits. *EMBO J.* **17**, 6210–6218.
34. Blackmer, T., Larsen, E. C., Takahashi, M., Martin, T. F., Alford, S., and Hamm, H. E. (2001). G protein βγ subunit-mediated presynaptic inhibition: regulation of exocytotic fusion downstream of Ca^{2+} entry. *Science* **292**, 293–297.
35. Jarvis, S. E., Magga, J. M., Beedle, A. M., Braun, J. E., and Zamponi, G. W. (2000). G protein modulation of N-type calcium channels is facilitated by physical interactions between syntaxin 1A and Gbetagamma. *J. Biol. Chem.* **275**, 6388–6394.
36. Magga, J. M., Jarvis, S. E., Arnot, M. I., Zamponi, G. W., and Braun, J. E. (2000). Cysteine string protein regulates G protein modulation of N-type calcium channels. *Neuron* **28**, 195–204.
37. Peng, Y. (1996). Ryanodine-sensitive component of calcium transients evoked by nerve firing at presynaptic nerve terminals. *J. Neurosci.* **16**, 6703–6712.
38. Schwartz, N. E. and Alford, S. (2000). Physiological activation of presynaptic metabotropic glutamate receptors increases intracellular calcium and glutamate release. *J. Neurophysiol.* **84**, 415–427.
39. Nurrish, S., Segalat, L., and Kaplan, J. M. (1999). Serotonin inhibition of synaptic transmission: $G\alpha_o$ decreases the abundance of UNC-13 at release sites. *Neuron* **24**, 231–242.
40. Renstrom, E., Ding, W. G., Bokvist, K., and Rorsman, P. (1996). Neurotransmitter-induced inhibition of exocytosis in insulin-secreting beta cells by activation of calcineurin. *Neuron* **17**, 513–522.
41. Hilfiker, S., Czernik, A. J., Greengard, P., and Augustine, G. J. (2001). Tonically active protein kinase A regulates neurotransmitter release at the squid giant synapse. *J. Physiol.* **531**, 141–146.
42. Trudeau, L. E., Fang, Y., and Haydon, P. G. (1998). Modulation of an early step in the secretory machinery in hippocampal nerve terminals. *Proc. Natl. Acad. Sci. USA* **95**, 7163–7168.
43. Evans, G. J., Wilkinson, M. C., Graham, M. E., Turner, K. M., Chamberlain, L. H., Burgoyne, R. D., and Morgan, A. (2001). Phosphorylation of cysteine string protein by protein kinase A. Implications for the modulation of exocytosis. *J. Biol. Chem.* **276**, 47877–47885.
44. Kondo, S. and Marty, A. (1997). Protein kinase A-mediated enhancement of miniature IPSC frequency by noradrenaline in rat cerebellar stellate cells. *J. Physiol. (London)* **498**, 165–176.
45. Trudeau, L. E., Emery, D. G., and Haydon, P. G. (1996). Direct modulation of the secretory machinery underlies PKA-dependent synaptic facilitation in hippocampal neurons. *Neuron* **17**, 789–797.

CHAPTER 231

G Protein Regulation of Channels

Ofer Wiser and Lily Yeh Jan

*Howard Hughes Medical Institute,
Departments of Physiology and Biochemistry,
University of California San Francisco,
San Francisco, California*

Stimulation of G-protein coupled receptor (GPCR) facilitates GTP exchange for GDP to Gα, leading to dissociation of Gα^{GTP} from the Gβγ dimer and regulation of their effectors [1]. Ion channels as effectors allow neurotransmitters or hormones to elicit electric activity [2]. Measurements of ions that flow through the open channel pore by electrophysiological methods enable direct determination of the activity of a single channel protein at millisecond resolution *in vivo*. Different configurations of patch-clamp recording make it possible to control the composition of solutions on either side of the membrane.

There are numerous examples for modulation of channel activity by indirect means involving phosphorylation, second messengers (Ca^{2+}, cAMP), and regulators of G-protein activity such as RGS proteins [3]. In this chapter we will discuss modulation of channels due to their direct interaction with G proteins, and focus on the G-protein-gated inward rectifying potassium channels Kir3 (GIRK) [4], and the voltage-gated calcium channels [5,6]. This review is not comprehensive, as direct interaction of G proteins with other ion channels have been reported [7–10].

Interaction with K$^+$ Channels

The GIRK Channels

Opening of K channels allows outward flow of potassium ions down the concentration gradient, causing hyperpolarization. This way, the activation of GIRK channels by transmitter leads to calming of the heart rate and slowing of neuronal activity [4,11,12]. GIRK channels are tetramers of pore-lining subunits containing two transmembrane domains (TM1&2), and cytosolic N and C termini. The mammalian GIRK family has four members. They differ in the ability to form homotetrameric channels [13] and the traffic signals they possess [14]. Whereas midbrain dopamine neurons express GIRK2 channels, the brain and cortex contain primarily heterotetramers of GIRK1 and GIRK2 and the heart contains primarily GIRK1 and GIRK4 [15].

GIRK channels have low basal activity, and are activated by GPCR, due to their direct binding to the βγ subunit of the G protein [16,17]. Mutations in the transmembrane domain may rescue mutant GIRK channels constitutively active in the absence of Gβγ, indicating that channel gating triggered by Gβγ binding is likely to involve conformational changes within the membrane [18,19].

The Gβγ Interacting Domain of GIRK

Almost all Gβγ combinations can activate GIRK [17]. Two-hybrid analysis showed interaction between the Gβ1 and the N-terminal domain of GIRK1 [20], but only Gβγ dimers can activate GIRK channels [21]. A recent study compared Gβ1 with Gβ5, which does not activated GIRK, and identified Gβ1 residues (S67, S98, and T128) important for basal activity [22]. Another study showed the importance of residues close or within the Gα binding interface of Gβ indicating a possible Gα competition with GIRK [1].

Direct binding of Gβγ to both the N terminal (amino acid 34–86) and C terminal (amino acid 273–462) domains of GIRK1 shown by *in vitro* binding assay may be reduced by peptides from those GIRK1 domains, which also partially inhibit channel activation [23]. Peptides from GIRK4 C-terminal domain also affected Gβγ binding and channel activation [24]. A chimerical Kir2 channel gained sensitivity to Gβγ by introduction of the two GIRK1 termini [25]. Further experiments have localized Gβγ binding region in the

N- and C-terminal domains of GIRK1-4 [26]. Similar Gβγ activation has been shown for functional homotetramers of GIRK1 (F137S) and GIRK4 (S143T) [13]. Consistent with these findings, crucial GIRK1 (H57, L262) and GIRK4 (H64, L268) residues for Gβγ interaction have been identified at homologous sites [27].

Coupling of GIRK Activation to Specific Receptors

Remarkably, GIRK channels are activated specifically by Gi/o protein-coupled receptors *in vivo* and in mammalian HEK293 cells, even though GIRK channels are activated by Gβγ and in *Xenopus* oocytes, receptors coupled to different G proteins can all activate GIRK [28]. Chimeric Gα containing primarily Gαi except for the C-terminal residues allow Gs-coupled receptors to activate GIRK [29], indicating a pivotal role for Gα in G-protein/GIRK coupling. Conceivably, GIRK and Gi coupled receptors may be sequestered via scaffold proteins. If so, overexpression may saturate the scaffold protein that confers G-protein specificity *in vivo*. This may account for the observed GIRK activation by Gs coupled β1 adrenergic receptors [30]. It is also of interest to note that the receptor kinase GRK2 [31], which has Gβγ binding domains, has been implicated in mediating receptor specificity.

The kinetics of GIRK channel activation in mammalian cells *in vivo* are very fast (<1 s) as opposed to the much slower activation when receptors and GIRK channels are expressed in *Xenopus* oocytes. One possible explanation for this is that in the native cell GIRK channels and Gi/o-coupled receptors are colocalized [23,32]. Indeed, expression in *Xenopus* oocytes of fused muscarinic receptor m2R with the PTX resistant Gα$_Z$ resulted in acetylcholine (carbachol) stimulation of reduced currents with much faster activation kinetics [33]. The N-terminal domain of GIRK1 may also bind trimeric Gαβγ [23,34,35]. The physical association of Gα, though not needed for GIRK activation, may have a role in the coupling specificity to a certain GPCR, either by generating macro protein scaffold or by facilitating compartmentalization and hence rapid and specific channel activation by receptors. Interestingly, the N- and C-terminal domains of the GIRKs were found to interact with each other, including the N-terminal domain of GIRK1 to the C-terminal domain of GIRK4, and exhibited much enhanced binding of Gβγ [26]. Four Gβγ binding domains per GIRK tetramer have been implicated by cross-linking experiments [36], but any association between the channel and Gα remain to be examined.

The GIRK basal activity is due to free Gβγ without receptor stimulation. However, qualitative differences have emerged between the basal activity and the receptor-induced GIRK currents. Studies of chimeras between the GIRK4 (S143T, used for expression of functional homotetramer) and IRK1 channels have identified a single site, namely GIRK4 (L339E), that when mutated reduced binding to Gβγ and impaired agonist-induced activity [37]. What might be different between the basal activity and the receptor-induced activity? One can speculate that at resting, Gα is bound to the receptors and both GIRK domains are free to exhibit high affinity to the low level of cytosolic-free Gβγ [1,37]. Upon receptor activation, both Gα^{GTP} and Gβγ are separated from the receptors. Gα is then bound to the GIRK N terminal, reducing its affinity to Gβγ. Thus while the released Gβγ will bind GIRK and cause increase in GIRK currents above the basal activity, the reduction in GIRK affinity will result in a fast dissociation of Gβγ and hence fast reduction in channel activity (the apparent fast desensitization or deactivation of GIRK currents). In another scenario, the Gα^{GDP} alone or in a complex with βγ is bound to the N terminal and actively inhibiting the channel [35]. This inhibition will be relieved upon agonist application only if the receptor itself is in a complex with the same Gα^{GDP} and can facilitate the exchange to GTP. The mechanism of specificity of GIRK activation *in vivo* as well as the possible modulation of GIRK gating by Gα required further study.

Calcium Channel Interaction with G Proteins

High-voltage-activated (HVA) Ca^{2+} channels provide the coupling from the action potential to transmitter release in nerve terminals and endocrine cell, and of the excitation-contraction of muscle cells. The channel contains at least three subunits. The main subunit, α1 has four homologous repeats (I–IV) of six transmembrane (TM) domains (S1–S6). Each repeat contains the voltage-sensitive S4 domain, and the S5–S6 pore-forming domain. There are two auxiliary subunits with regulatory roles: α2/δ has extracellular domain and a single TM domain, and β is a cytosolic subunit [38]. There are three subfamilies of HVA Ca^{2+} channels named Ca_V1–3. Ca_V1.1–1.4 mediate the L-type currents; Ca_V2.1–2.3 mediate the P/Q-, N-, and R-type currents, respectively; Ca_V3.1–3.3 mediate the T-type currents. HVA Ca^{2+} channels are negatively regulated by G proteins in various neuronal preparations [2,39]. This response appears to be controlled by a membrane-delimited mechanism [40] via pertussis toxin (PTX)-sensitive G proteins [41].

G Protein Interacting Domains

Overexpression or injection of G-protein subunits in sympathetic neurons indicated that Gβγ mediates the Ca^{2+} channel inhibition [42,43]. Alanine mutations of Gβ1 residues 55 and 80, which reside at the N-terminal interface with Gα, had enhanced ability to inhibit current through Ca_V2.2, while mutations within the switch interface eliminated current inhibition [1], indicating a Gα role in regulating Gβγ interaction with Ca^{2+} channels.

The Gβγ Interacting Domain of HVA $Ca^{2\pm}$ Channels

Gi/Go coupled receptors selectively inhibited N-, P/Q- and R-type calcium currents [44], due to Gβγ interaction with the α1 subunit of Ca$_V$2.1–2.3, while the channel β subunit or strong depolarization (prepulse/voltage dependent facilitation), reduced this inhibition [45,46]. Multiple Gβγ interaction domains have been found, including the C-terminal region of CaV2 family [47–49], and the α1 cytoplasmic linker connecting S6 of the first repeat and S1 of the second repeat. Within this linker, Gβγ binding occurs both in the α1 interaction domain (AID), which also mediates the interaction with the channel β subunit, and in a second downstream sequence [49,50].

Modulation of Gβγ Inhibition

All channel β subunits antagonize Gβγ effect but not by direct displacement of Gβγ from the channel. The palmitoylation sites on subunit β2A are responsible for its unique modulation of the channel and the reduced competition with Gβγ [51]. Another putative modulator of Gβγ effect is the synaptic protein syntaxin 1A that may be involved in scaffolding Gβγ to the calcium channels [52]. RGS proteins interact with Gα and facilitate its GTPase activity. Such interaction may reduce calcium current inhibition by Gβγ [53,54].

Voltage-Independent G-protein-Mediated Inhibition of Calcium Channels

Functional studies in calyx-type nerve terminal [55] and in sympathetic neurons [56] have identified a rapid, PTX- and voltage-insensitive (VI) inhibition of calcium currents. The N terminus of Gα$_{q/11}$ is necessary for its binding to the C termini of Ca$_V$2.1–2.2 [49] and for mediating the VI inhibition [57]. Consistently, *in vitro* binding assay identified direct interaction of Gβγ with the N and C termini of Ca$_V$1.2, which mediates the L-type current. The mechanism of voltage-independent channel inhibition thus differs from that of voltage-dependent channel inhibition in several ways: instead of channel β subunit it involves Gα$_{q/11}$ binding to the same channel region as Gβγ, and it requires simultaneous interaction of calmodulin [58].

References

1. Ford, C. E., Skiba, N. P., Bae, H., Daaka, Y., Reuveny, E., Shekter, L. R., Rosal, R., Weng, G., Yang, C., Ivengar, R., Miller, R. J., Jan, L. Y., Lefkowitz, R. J., and Hamm, H. E. (1998). Molecular basis for interactions of G protein βγ subunits with effectors. *Science* **280**, 1271–1274.
2. Hille, B. (1994). Modulation of ion-channel function by G-protein-coupled receptors. *Trends Neurosci.* **17**, 531–536.
3. Ross E. M., and Wilkie, T. M. (2000). GTPase-activating proteins for heterotrimeric G proteins: Regulators of G protein signaling (RGS) and RGS-like proteins. *Annu. Rev. Biochem.* **69**, 795–827.
4. Dascal, N. (1997). Signaling via the G protein-activated K$^+$ channels. *Cell Signal* **9**, 551–573.
5. Catterall, W. A. (2000). Structure and regulation of voltage-gated Ca^{2+} channels. *Annu. Rev. Cell Dev. Biol.* **16**, 521–555.
6. Zamponi, G. W., and Snutch, T. P. (1998). Modulation of voltage-dependent calcium channels by G proteins. *Curr. Opin. Neurobiol.* **8**, 351–356.
7. Cohen, N. A., Sha, Q., Makhina, E. N., Lopatin, A. N., Linder, M. E., Snyder, S. H., and Nichols, C. G. (1996). Inhibition of an inward rectifier potassium channel (Kir2.3) by G-protein βγ subunits. *J. Biol. Chem.* **271**, 32301–32305.
8. Jing, J., Chikvashvili, D., Singer-Lahat, D., Thornhill, W. B., Reuveny, E., and Lotan, I. (1999). Fast inactivation of a brain K$^+$ channel composed of Kv1.1 and Kvbeta1.1 subunits modulated by G protein beta gamma subunits. *EMBO J.* **18**, 1245–1256.
9. Wada, Y., Yamashita, T., Imai, K., Miura, R., Takao, K., Nishi, M., Takeshima, H., Asano, T., Morishita, R., Nishizawa, K., Kokubun, S., and Nukada, T. (2000). A region of the sulfonylurea receptor critical for a modulation of ATP-sensitive K$^+$ channels by G-protein βγ-subunits. *EMBO J.* **19**, 4915–4925.
10. Reddy, M. M., Sun, D., and Quinton, P. M. (2001). Apical heterotrimeric G-proteins activate CFTR in the native sweat duct. *J. Membr. Biol.* **179**, 51–61.
11. Signorini, S., Liao, Y. J., Duncan, S. A., Jan, L. Y., and Stoffel, M. (1997). Normal cerebellar development but susceptibility to seizures in mice lacking G protein-coupled, inwardly rectifying K+ channel GIRK2. *Proc. Natl. Acad. Sci. USA* **94**, 923–927.
12. Kennedy, M. E., Nemec, J., Corey, S., Wickman, K., and Clapham, D. E. (1999). GIRK4 confers appropriate processing and cell surface localization to G-protein-gated potassium channels. *J. Biol. Chem.* **274**, 2571–2582.
13. Vivaudou, M., Chan, K. W., Sui, J., Jan, L. Y., Reuveny, E., and Logothetis, D. E. (1997). Probing the G-protein regulation of GIRK1 and GIRK4, the two subunits of the KACh channel, using functional homomeric mutants. *J. Biol. Chem.* **272**, 31553–31560.
14. Ma, D., Zerangue, N., Raab-Graham, K., Fried, S. R., Jan, Y. N., and Jan, L. Y. (2002). Diverse trafficking patterns due to multiple motifs in G protein-activated inwardly rectifying potassium channels from brain and heart. *Neuron* **33**, 715–729.
15. Krapivinsky, G., Gordon, E. A., Wickman, K., Velimirovic, B., Krapivin-sky, L., and Clapham, D. E. (1995). The G-protein-gated atrial K$^+$ channel IKACh is a heteromultimer of two inwardly rectifying K$^+$-channel proteins. *Nature* **374**, 135–141.
16. Reuveny, E., Slesinger, P. A., Inglese, J., Morales, J. M., Iniguez-Lluhi, J. A., Lefkowitz, R. J., Bourne, H. R., Jan, Y. N., and Jan, L. Y. (1994). Activation of the cloned muscarinic potassium channel by G protein beta gamma subunits. *Nature* **370**, 143–146.
17. Wickman, K. D., Iniguez-Lluhl, J. A., Davenport, P. A., Taussig, R., Krapivinsky, G. B., Linder, M. E., Gilman, A. G., and Clapham, D. E. (1994). Recombinant G-protein beta gamma-subunits activate the muscarinic-gated atrial potassium channel. *Nature* **368**, 255–257.
18. Yi, B. A., Lin, Y. F., Jan, Y. N., and Jan, L. Y. (2001). Yeast screen for constitutively active mutant G protein-activated potassium channels. *Neuron* **29**, 657–667.
19. Sadja, R, Smadja, K., Alagem, N., and Reuveny, E. (2001). Coupling Gβγ-dependent activation to channel opening via pore elements in inwardly rectifying potassium channels. *Neuron* **29**, 669–680.
20. Yan, K., and Gautam, N. (1997). Structural determinants for interaction with three different effectors on the G protein β subunit. *J. Biol. Chem.* **272**, 2056–2059.
21. Kawano, T., Chen, L., Watanabe, S. Y., Yamauchi, J., Kaziro, Y., Nakajima, Y., Nakajima, S., and Itoh, H. (1999). Importance of the G protein γ subunit in activating G protein-coupled inward rectifier K$^+$ channels. *FEBS Lett.* **463**, 355–359.
22. Mirshahi, T., Robillard, L., Zhang, H., Hebert, T. E., and Logothetis, D. E. (2002). Gβ residues that do not interact with Gα underlie agonist-independent activity of K$^+$ channels. *J. Biol. Chem.* **277**, 7348–7355.
23. Huang, C. L., Slesinger, P. A., Casey, P. J., Jan, Y. N., and Jan, L. Y. (1995). Evidence that direct binding of Gβγ to the GIRK1 G protein-gated inwardly rectifying K$^+$ channel is important for channel activation. *Neuron* **15**, 1133–1143.
24. Krapivinsky, G., Kennedy, M. E., Nemec, J., Medina, I., Krapivinsky, L., and Clapham, D. E. (1998). Gβγ Binding to GIRK4 subunit is critical for G protein-gated K$^+$ channel activation. *J. Biol. Chem.* **273**, 16946–16952.

25. Kunkel, M. T. and Peralta, E. G. (1995). Identification of domains conferring G protein regulation on inward rectifier potassium channels. *Cell* **83**, 443–449.
26. Huang, C. L., Jan, Y. N., and Jan, L. Y. (1997). Binding of the G protein βγ subunit to multiple regions of G protein-gated inward-rectifying K/channels. *FEBS Lett.* **405**, 291–298.
27. He, C., Yan, X., Zhang, H., Mirshahi, T., Jin, T., Huang, A., and Logothetis, D. E. (2001). Identification of critical residues controlling GIRK channel activity through interactions with the βγ subunits of G proteins. *J. Biol. Chem.* **277**, 6088–6096.
28. Leaney, J. L. and Tinker, A. (2000). The role of members of the pertussis toxin-sensitive family of G proteins in coupling receptors to the activation of the G protein-gated inwardly rectifying potassium channel. *Proc. Natl. Acad. Sci. USA* **97**, 5651–5656.
29. Leaney, J. L., Milligan, G., and Tinker, A. (2000). The G protein a subunit has a key role in determining the specificity of coupling to, but not the activation of, G protein-gated inwardly rectifying K^+ channels. *J. Biol. Chem.* **275**, 921–929.
30. Wellner-Kienitz, M. C., Bender, K., and Pott, L. (2001). Overexpression of β1 and β2 adrenergic receptors in rat atrial myocytes. Differential coupling to G protein-gated inward rectifier K^+ channels via Gs and Gi/o. *J. Biol. Chem.* **276**, 37347–37354.
31. Wellner-Kienitz, M. C., Bender, K., Brandts, B., Meyer, T., and Pott, L. (1999). Antisense oligonucleotides against receptor kinase GRK2 disrupt target selectivity of L-adrenergic receptors in atrial myocytes. *FEBS Lett.* **451**, 279–283.
32. Slesinger, P. A., Reuveny, E., Jan, Y. N., and Jan, L. Y. (1995). Identification of structural elements involved in G-protein gating of GIRK1 potassium channel. *Neuron* **15**, 1145–1156.
33. Vorobiov, D., Bera, A. K., Keren-Raifman, T., Barzilai, R., and Dascal, N. (2000). Coupling of the muscarinic m2 receptor to G protein-activated K^+ channels via Gα z and a receptor-Gαz fusion protein. Fusion between the receptor and Gαz eliminates catalytic (collision) coupling. *J. Biol. Chem.* **275**, 4166–4170.
34. Cohen, N. A., Sha, Q., Makhina, E, N., Lopatin, A. N., Linder, M. E., Snyder, S. H., and Nichols, C. G. (1996). Inhibition of an inward rectifier potassium channel (Kir2.3) by G-protein betagamma subunits. *J. Biol. Chem.* **271**, 32301–32305.
35. Peleg, S., Varon, D., Ivanina, T., Dessauer, C. W., and Dascal, N. (2002). Gαi controls the gating of the G protein-activated K^+ channel, GIRK. *Neuron.* **33**, 87–99.
36. Shawn Corey, S. and Clapham, D. E. (2001). The stoichiometry of Gβγ binding to G-protein-regulated inwardly rectifying K^+ channels (GIRKs). *J. Biol. Chem.* **276**, 11409–11413.
37. He, C., Zhang, H., Mirshahi, T., and Logothetis, D. E. (1999). Identification of a potassium channel site tha interacts with G protein βγ subunits to mediate agonist-induced signaling. *J. Biol. Chem.* **274**, 12517–12524.
38. Catterall, W. A. (1995). Structure and function of voltage-gated ion channels. *Annu. Rev. Biochem.* **65**, 493–531.
39. Hering, S., Berjukow, S., Sokolov, S., Marksteiner, R., Wei, R. G., Kraus, R., and Timin, E. N. (2000). Molecular determinants of inactivation in voltage-gated Ca^{2+} channels. *J. Physiol.* **528**, 237–249.
40. Forscher, P., Oxford, G. S., and Schulz, D. (1986). Noradrenaline modulates calcium channels in avian dorsal root ganglion cells through tight receptor-channel coupling. *J. Physiol.* **379**, 131–144.
41. Holz, G. G., 4th, Rane, S. G., and Dunlap, K. (1986). GTP-binding proteins mediate transmitter inhibition of voltage-dependent calcium channels. *Nature* **319**, 670–672.
42. Ikeda, S. R. (1996). Voltage-dependent modulation of N-type calcium channels by G-protein βγ subunits. *Nature* **380**, 255–258.
43. Herlitze, S., Garcia, D. E., Mackie, K., Hille, B., Scheuer, T., and Catterall, W. A. (1996). Modulation of Ca^{2+} channels by G-protein βγ subunits. *Nature* **380**, 258–262.
44. Jeong, S. W. and Wurster, R. D. (1997). Muscarinic receptor activation modulates Ca^{2+} channels in rat intracardiac neurons via a PTX- and voltage-sensitive pathway. *J. Neurophysiol.* **78**, 1476–1490.
45. Bourinet, E., Soong, T. W., Stea, A., and Snutch, T. P. (1996). Determinants of the G protein-dependent opioid modulation of neuronal calcium channels. *Proc. Natl. Acad. Sci. USA* **93**, 1486–1491.
46. Meza, U., and Adams, B. (1998). G-protein-dependent facilitation of neuronal α1A, α1B, and α1E Ca^{2+} channels. *J. Neurosci.* **18**, 5240–5252.
47. Zhang, J. F., Ellinor, P. T., Aldrich, R. W., and Tsien, R. W. (1996). Multiple structural elements in voltage-dependent Ca^{2+} channels support their inhibition by G proteins. *Neuron* **17**, 991–1003.
48. Qin, N., Platano, D., Olcese, R., Stefani, E., and Birnbaumer, L. (1997). Direct interaction of Gβγ with a C-terminal Gβγ-binding domain of the Ca^{2+} channel α1 subunit is responsible for channel inhibition by G protein-coupled receptors. *Proc. Natl. Acad. Sci. USA* **94**, 8866–8871.
49. Furukawa, T., Miura, R., Mori, Y., Strobeck, M., Suzuki, K., Ogihara, Y., Asano, T., Morishita, R., Hashii, M., Higashida, H., Yoshii, M., and Nukada, T. (1998). Differential Interactions of the C- terminus and the cytoplasmic I-II loop of neuronal Ca^{2+} channels with G-protein α and βγ subunits II. Evidence for direct binding. *J. Biol. Chem.* **273**, 17595–17603.
50. De Waard, M., Liu, H., Walker, D., Scott, V. E., Gurnett, C. A., and Campbell, K. P. (1997). Direct binding of G-protein βγ complex to voltage dependent calcium channels. *Nature* **385**, 446–450.
51. Cantí, C., Bogdanov, Y., and Dolphin, A. C. (2000). Interaction between G proteins and accessory β subunits in the regulation of α1B calcium channels in *Xenopus* oocytes. *J. Physiol.* **527**, 419–432.
52. Jarvis, S. E., Magga, J. M., Beedle, A. M., Braun, J. E. A., and Zamponi, G. W. (2000). G protein modulation of N-type calcium channels is facilitated by physical interactions between syntaxin 1A and Gβγ. *J. Biol. Chem.* **275**, 6388–6394.
53. Mark, M. D., Wittemann, S., and Herlitze, S. (2000). G protein modulation of recombinant P/Q-type calcium channels by regulators of G protein signalling proteins. *J. Physiol.* **528**, 65–77.
54. Zhou, J. Y., Siderovski, D. P., and Miller, R. J. (2000). Selective regulation of N-type Ca channels by different combinations of G-protein β/γ subunits and RGS proteins. *J. Neurosci.* **20**, 7143–7148.
55. Mirotznik, R. R., Zheng, X., and Stanley, E. F. (2000). G-protein types involved in calcium channel inhibition at a presynaptic nerve terminal. *J. Neurosci.* **20**, 7614–7621.
56. Kammermeier, P. J., Ruiz-Velasco, V., and Ikeda, S. R. (2000). A voltage-independent calcium current inhibitory pathway activated by muscarinic agonists in rat sympathetic neurons requires both Gα q/11 and βγ. *J. Neurosci.* **20**, 5623–5629.
57. Kinoshita, M., Nukada, T., Asano, T., Mori, Y., Akaike, A., Satoh, M., and Kaneko, S. (2001). Binding of Gαo N terminus is responsible for the voltage-resistant inhibition of α1A (P/Q-type, Cav2.1) Ca^{2+} channels. *J. Biol. Chem.* **276**, 28731–28738.
58. Ivanina, T., Blumenstein, Y., Shistik, E., Barzilai, R., and Dascal, N. (2000). Modulation of L-type Ca^{2+} channels by Gβγ and calmodulin via interactions with N and C termini of α1C. *J. Biol. Chem.* **275**, 39846–39854.

Ras and Cancer

Frank McCormick

*Cancer Research Institute,
University of California Comprehensive Cancer Center,
San Francisco, California*

Introduction: Ras Activation in Cancer

It has been 20 years since H-ras mutations were identified in DNA from the bladder cancer cell line T-24. Since this seminal observation, rates of mutation in H-ras, N-ras, and K-ras have been measured in most types of human cancers [1]. The clonal nature of these mutations in tumors strongly suggests a causal role, a suggestion that has been amply verified by mouse models of Ras-induced cancer. A striking result of this comprehensive survey is the considerable variation in frequency in Ras mutation between different types of cancer. In pancreatic carcinoma, K-ras is activated by point mutation in almost every case, whereas Ras mutations are hardly ever detected in mammary carcinomas, to cite two extreme examples. The biological or molecular basis of these observations is not yet understood. One interpretation is that alternative mechanisms of activating the Ras pathway (receptor amplification, activation of downstream pathways) also occur at varying frequencies. Another interpretation is that different types of cancer vary in their dependence on the Ras pathway. Another unresolved issue is the predominance of K-ras mutations over N-ras and H-ras: this may reflect different levels of expression of these genes in different tissues and different levels of dependence on each type. Mouse knockout experiments show that K-ras is essential [2], whereas N-ras and H-ras are not, consistent with K-ras being the most important form and therefore the most likely to be directly involved in carcinogenesis. However, other models must be considered: for example, each type of Ras may signal through a different set of downstream effectors, and K-ras happens to provide a repertoire of signals that is consistent with malignant progression. Although most evidence points toward shared effectors among all three types of Ras, evidence for discrimination among effectors also exists [3].

In addition to mutations in Ras genes, gains and losses of Ras genes have been reported in human tumors [4–9]. In mouse tumors double minute chromosomes encoding H-ras have been identified [10]. Also in mouse models, progress increase of copy numbers of H-ras mutants appears to drive malignant progression, along with selective loss of the wild-type allele [11].

Pathways Downstream of Ras

Figure 1 shows pathways regulated by Ras. In addition to the well-established pathways that Ras activates, the Raf-MAP kinase cascade and the PI 3′ kinase pathway, Ras activates RalGDS and possibly other effector pathways that are not well characterized [12]. Raf and PI 3′ kinase pathways act synergistically to mediate Ras transformation, suggesting that inhibitors of either pathway have profound effects on Ras transformation [13]. This is an important issue in the context of drug development based on Ras pathways.

The precise molecular basis of synergy between effector pathways is not fully understood. However, there are multiple elements of these pathways that intersect and could contribute to synergistic interaction. For example, the cyclin D1 gene is a transcriptional target of the Raf-MAP kinase pathway, and cyclin D1 protein is stabilized by the PI 3′ kinase pathway [14].

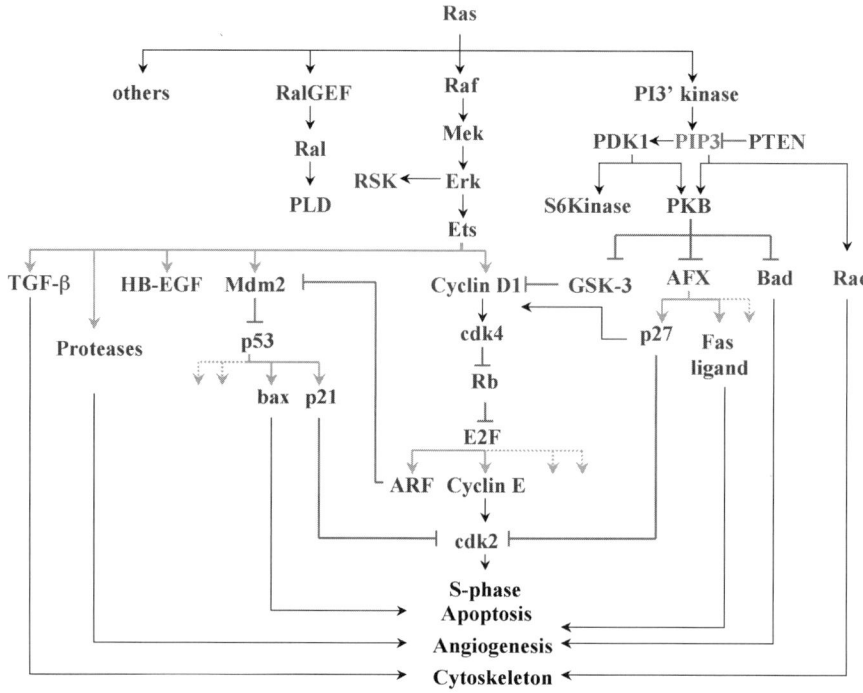

Figure 1

Until recently, there was little genetic evidence that the Raf-MAP kinase pathway is activated by mutation of gene copy number change in human cancer. However, the recent discovery that B-Raf is activated by mutation has changed this view dramatically. Two types of mutant have been described: one that renders B-ras independent of Ras and occurs in tumors in which Ras is wild type; another one that requires Ras interaction for full activity and occurs in tumors containing mutant Ras. It is conceivable that the high-throughput sequencing approach that identified these mutations may yet reveal other activating events in the Raf-MAP kinase pathway. Nonetheless, these new data imply that activation of the Raf effector pathway is the major selection for Ras mutation in these diseases.

In contrast, genetic changes activating the PI 3' kinase have been well documented and are considered of major importance in human cancer. Loss of PTEN is by far the most frequent event that activates this pathway, but increases in copy number of Akt/PKB have been documented and implicate this arm of the pathway in PTEN-deficient tumors. In endometrial and cutaneous melanoma cancers, loss of PTEN and Ras activation are mutually exclusive, suggesting that in these conditions, the major selection for Ras mutation is activation of the PI 3' kinase effector pathway [15,16].

Mouse Models of Cancer

Mouse models of cancer provide important clues relating to the role of Ras in cancer. Many have involved forced expression of mutant Ras proteins under tissue-specific promoters, revealing transforming power of the Ras oncogene in different physiological settings. An interesting aspect of these models is the sustained requirement for Ras expression even in advanced cancers: withdrawal of Ras expression causes complete regression of such tumors [17]. Recently a model has been developed in which mutant K-ras is activated sporadically: this appears to be an excellent model for sporadic human lung cancer [18].

Other informative rodent models have used mutagens to initiate cancers, followed by analysis of Ras activation and progression. The classic studies of Sukamar and Barbacid and coworkers proved conclusively that Ras mutation can be the initiating event in cancer and showed that mutations caused by an early chemical insult can persist in latent forms before progressing to cancer [19]. The skin cancer models of Balmain and coworkers have revealed a step-wise activation of H-ras during initiation and progression: mutant H-ras alleles created by exposure to carcinogen are selectively amplified in a step-wise manner as the tumors evolve. In parallel to increased ras activity, levels of cyclin D1 increase during progression. A role of cyclin D1 in Ras transformation in this model was confirmed by demonstration that tumors' progression is strongly retarded in mice lacking the cyclin D1 gene. Even more striking effects of cyclin D1 were demonstrated recently in a model of mammary carcinogenesis driven by Ras, erbB, wnt, or myc: the former two oncogenes were completely dependent on cyclin D1, whereas the latter were not [20]. This clear role of cyclin D1 points toward the importance of the Raf/MAP kinase effector pathway in Ras transformation, since this pathway activates transcription of cyclin D1 directly. However, a role of the PI kinase pathway cannot be ruled out, as this pathway stabilizes cyclin D1 through inhibition of GSK-3-mediated degradation.

Prospects for Cancer Therapy Based on Ras

Attempts to block Ras signaling in human cancers by inhibiting posttranslational farnesylation have been stalled by the fact that K-ras, the major form of ras involved in human cancer, can also be modified by geranylgeranylation. Thus allows continued K-ras activity in the presence of farnesyl transferase inhibitors. Such inhibitors may have clinical value through their action on other cellular targets, however [21]. More recent approaches to blocking Ras activity have targeted enzymes downstream of Ras. A Raf kinase inhibitor entered clinical trials recently [22], and a MEK inhibitor followed soon afterwards [23]. Attempts to block other enzymes downstream of Ras are also under way [24].

References

1. Bos, J. L. (1989). Ras oncogenes in human cancer: a review. *Cancer Res.* **49**, 4682–4689.
2. Johnson, L., Greenbaum, D., Cichowski, K., Mercer, K., Murphy, E., Schmitt, E., Bronson, R. T., Umanoff, H., Edelmann, W., Kucherlapati, R., and Jacks, T. (1997). K-ras is an essential gene in the mouse with partial functional overlap with N-ras. *Genes Dev.* **11**, 2468–2481.
3. Wolfman A. (2001). Ras isoform-specific signaling: location, location, location. *Sci. STKE* 2001, E2.
4. Kimura, E. and Armelin, H. A. (1988). Role of proto-oncogene c-Ki-ras amplification and overexpression in the malignancy of Y-1 adrenocortical tumor cells. *Brazil J. Med. Biol. Res.* **21**, 189–201.
5. Filmus, J., Trent, J. M., Pullano, R., and Buick, R. N. (1986). A cell line from a human ovarian carcinoma with amplification of the K-ras gene. *Cancer Res.* **46**, 5179–5182.
6. George, D. L., Scott, A. F., Trusko, S., Glick, B., Ford, E., and Dorney, D. J. (1985). Structure and expression of amplified cKi-ras gene sequences in Y1 mouse adrenal tumor cells. *EMBO J.* **4**, 1199–1203.
7. George, D. L., Scott, A. F., de Martinville, B., and Francke, U. (1984). Amplified DNA in Y1 mouse adrenal tumor cells: isolation of cDNAs complementary to an amplified c-Ki-ras gene and localization of homologous sequences to mouse chromosome 6. *Nucleic Acids Res.* **12**, 2731–2743.
8. Schwab, M., Alitalo, K., Klempnauer, K. H., Varmus, H. E., Bishop, J. M., Gilbert, F., Brodeur, G., Goldstein, M., and Trent, J. (1983). Amplified DNA with limited homology to myc cellular oncogene is shared by human neuroblastoma cell lines and a neuroblastoma tumour. *Nature* **305**, 245–248.
9. Schwab, M., Alitalo, K., Varmus, H. E., Bishop, J. M., and George, D. (1983). A cellular oncogene (c-Ki-ras) is amplified, overexpressed, and located within karyotypic abnormalities in mouse adrenocortical tumour cells. *Nature* **303**, 497–501.
10. Tanaka, K., Takechi, M., Nishimura, S., Oguma, N., and Kamada, N. (1993). Amplification of c-MYC oncogene and point mutation of N-RAS oncogene point mutation in acute myelocytic leukemias with double minute chromosomes. *Leukemia* **7**, 469–471.
11. Buchmann, A., Ruggeri, B., Klein-Szanto, A. J., and Balmain, A. (1991). Progression of squamous carcinoma cells to spindle carcinomas of mouse skin is associated with an imbalance of H-ras alleles on chromosome 7. *Cancer Res.* **51**, 4097–4101.
12. Campbell, S. L., Khosravi-Far, R., Rossman, K. L., Clark, G. J., and Der, C. J. (1998). Increasing complexity of Ras signaling. *Oncogene* **17**, 1395–1413.
13. Gille H. and Downward, J. (1999). Multiple ras effector pathways contribute to G(1) cell cycle progression. *J. Biol. Chem.* **274**, 22033–22040.
14. Diehl, J. A., Cheng, M., Roussel, M. F., and Sherr, C. J. (1998). Glycogen synthase kinase-3beta regulates cyclin D1 proteolysis and subcellular localization. *Genes Dev.* **12**, 3499–3511.
15. Ikeda, T., Yoshinaga, K., Suzuki, A., Sakurada, A., Ohmori, H., and Horii, A. (2000). Anticorresponding mutations of the KRAS and PTEN genes in human endometrial cancer. *Oncol. Rep* **7**, 567–570.
16. Tsao, H., Zhang, X., Fowlkes, K., and Haluska, F. G. (2000). Relative reciprocity of NRAS and PTEN/MMAC1 alterations in cutaneous melanoma cell lines. *Cancer Res.* **60**, 1800–1804.
17. Chin, L., Tam, A., Pomerantz, J., Wong, M., Holash, J., Bardeesy, N., Shen, Q., O'Hagan, R., Pantginis, J., Zhou, H., Horner, J. W. 2nd, Cordon-Cardo, C., Yancopoulos, G. D., and DePinho, R. A. (1999). Essential role for oncogenic Ras in tumour maintenance. *Nature* **400**, 468–472.
18. Johnson, L., Mercer, K., Greenbaum, D., Bronson, R. T., Crowley, D., Tuveson, D. A., and Jacks, T. (2001). Somatic activation of the K-ras oncogene causes early onset lung cancer in mice. *Nature* **410**, 1111–1116.
19. Sukumar, S., Notario, V., Martin-Zanca, D., and Barbacid, M. (1983). Induction of mammary carcinomas in rats by nitroso-methylurea involves malignant activation of H-ras-1 locus by single point mutations. *Nature* **306**, 658–661.
20. Yu, Q., Geng, Y., and Sicinski, P. (2001). Specific protection against breast cancers by cyclin D1 ablation. *Nature* **411**, 1017–1021.
21. Prendergast, G. C. and Rane, N. (2001). Farnesyltransferase inhibitors: mechanism and applications. *Expert Opin. Investig. Drugs* **10**, 2105–2116.
22. Lyons, J. F., Wilhelm, S., Hibner, B., and Bollag, G. (2001). Discovery of a novel Raf kinase inhibitor. *Endocr. Relat. Cancer* **8**, 219–225.
23. Sebolt-Leopold, J. S., Dudley, D. T., Herrera, R., Van Becelaere, K., Wiland, A., Gowan, R. C., Tecle, H., Barrett, S. D., Bridges, A., Przybranowski, S., Leopold, W. R., and Saltiel, A. R. (1999). Blockade of the MAP kinase pathway suppresses growth of colon tumors in vivo. *Nat. Med.* **5**, 810–816.
24. McCormick, F. (2000). Small-molecule inhibitors of cell signaling. *Curr. Opin. Biotechnol.* **11**, 593–597.

CHAPTER 233

The Influence of Cellular Location on Ras Function

Janice E. Buss[1], Michelle A. Booden[2], and John T. Stickney[3]

[1]Department of Biochemistry, Biophysics, and Molecular Biology, Iowa State University, Ames, Iowa
[2]Lineberger Comprehensive Cancer Center, University of North Carolina, Chapel Hill, North Carolina
[3]Department of Cell Biology, Neurobiology, and Anatomy, University of Cincinnati, Cincinnati, Ohio

Study of how cellular location influences the biological activity of signaling proteins is entering a second phase of important discoveries. Some of the very earliest studies on tyrosine kinases and Ras proteins recognized that these otherwise cytosolic proteins required membrane association in order to function. However, new work is providing a reminder that membrane structure is complicated and variable, and that the itineraries of Ras and other signaling proteins to and from membrane surfaces still contain mystery and controversy. The simple model of inactive cytosolic protein versus functional membrane-bound Ras has evolved into the greater challenge of mapping the topography of entire signaling pathways. As in the beginning, Ras proteins are teaching us many of these lessons.

Cytosolic Ras Is not Functional

The original work defining the membrane-binding domain of the vHRas protein discovered that if vHRas failed to achieve membrane binding it was completely and thoroughly transformation defective [1]. This crucial insight led to attempts to achieve similar potent interference with Ras function through pharmacological means, the results of which are described by A. D. Cox in another chapter of this volume. Further genetic and biochemical studies identified a farnesyl transferase (FTase) that attaches a farnesyl isoprenoid to the Ras C-terminus (see Fig. 1, Step 1; [2]). Two (and in most cases, three) additional modifications must also occur to convert Ras into a fully active protein at the plasma membrane [3]. However, the cellular site for these modifications came as a bit of a surprise. Both a unique farnesyl-directed protease and a farnesyl-cysteine methyl transferase were unexpectedly found to be integral proteins of the endoplasmic reticulum (ER) (Fig. 1, Steps 2 and 3;[4,5]). The speed and efficiency of these processing enzymes had kept the discrete visits of all four mammalian Ras proteins [6] to the endomembrane system from being noticed.

After Modifications by Endomembrane Enzymes, Ras Proteins Move Toward the Cell Surface

The KRas4B protein requires no further modifications, although it must contain a basic domain adjacent to its farnesylated cysteine to strengthen its eventual interaction with the plasma membrane [7]. The route through which KRas4B accesses the cell surface does not appear to be the traditional secretory pathway [8], but remains uncharacterized (Fig. 1, Step 5b). However, both HRas and NRas take similar journeys along the Golgi and onward to the cell surface, astride

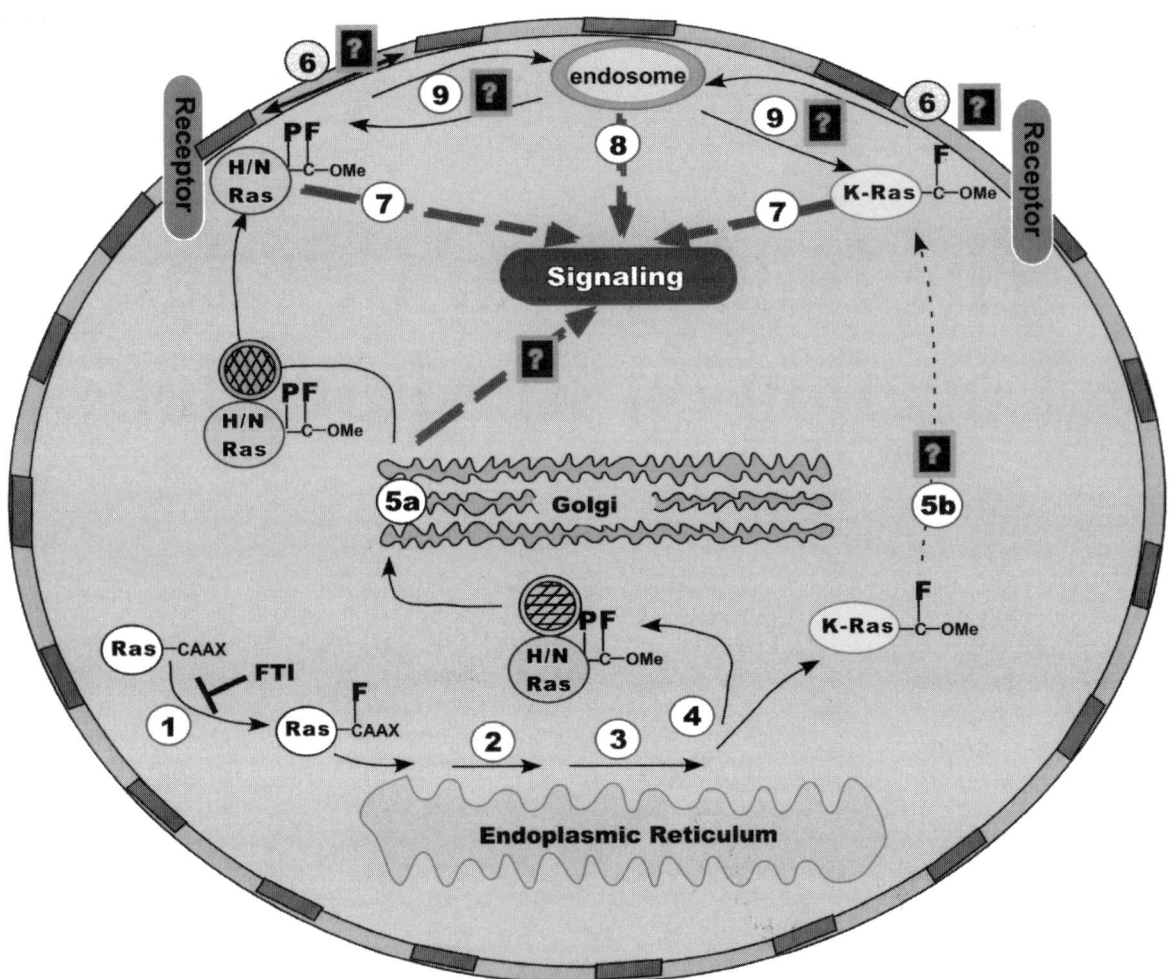

Figure 1 The life and times of Ras: a diagrammatic representation of Ras trafficking and localization. Newly synthesized Ras proteins are farnesylated in the cytoplasm (Step 1), then translocated to the cytoplasmic surface of the endoplasmic reticulum for further modification by a resident protease (Step 2) and carboxyl methyltransferase (Step 3). For H- and N-Ras, further modification occurs in the form of palmitoylation (Step 4). Processed Ras proteins then proceed to the plasma membrane and acquire functional activity. For H- and N-Ras, this occurs through vesicular transport through the endomembrane system (Step 5a). K-Ras4B is not palmitoylated and achieves plasma membrane localization via a separate, but undefined, mechanism (Step 5b). Once at the plasma membrane, H- and N-Ras (and perhaps K-Ras4B) partition into (and possibly out of) lipid rafts (shaded areas, Step 6). The primary events of Ras activation and signaling occur at the plasma membrane (Step 7). Endosomes (Step 8) may also be important sites for some Ras-mediated signaling events, but see text for questions on how Ras may be involved or recycled to the membrane (Step 9). Ras activity can be inhibited by preventing the first farnesylation step in this itinerary using farnesyl transferase inhibitors (FTI, Step 1). Steps where major unknowns remain in Ras trafficking and signaling are indicated with question marks.

intracellular vesicles [6]. Enroute both proteins claim a second modification—attachment of palmitoyl groups to C-terminal cysteines (Fig. 1, Step 4; [6,9]).

No enzyme for Ras palmitoylation has yet been identified. Two endomembrane proteins (ERF2 and ERF4/SHR5) that are required for palmitoylation of yeast Ras1 are known [10,11]. However, HRas palmitoylation also takes place, repeatedly, at the plasma membrane. A protein acyltransferase for G protein αsubunits has been partially purified from plasma membranes, but it acylates HRas poorly [12]. This leaves no current candidate for an enzyme with the crucial function of maintaining Ras plasma membrane attachment. Enzymes for Ras palmitate removal also await discovery. An acyl-protein thioesterase (APT1) that depalmitoylates G protein α subunits has been cloned [13] and can depalmitoylate HRas *in vitro*, but its activity toward HRas *in vivo* has not yet been confirmed.

Importantly, there are several studies showing that mutant HRas proteins that fail to be released from the ER or *cis*-Golgi [9,14] are almost completely transformation defective. A very new report suggests some intracellular HRas may be able to interact with Raf-1 even if endocytosis blocked [15]. Currently, there is almost nothing known about the fundamental processes or proteins that enable HRas and NRas to dock onto the surface of trafficking vesicles or move to the plasma membrane (Fig. 1, Step 5a). The first candidate is the PRA1 prenyl-dependent escort for Rab GTPases, which also interacts with Ras proteins in the Golgi [16]. Discovering these proteins will provide powerful new tools for controlling Ras activity.

Destination-Cell Surface: Ras Proteins Distribute Among Several Plasma Membrane Domains

A significant recent advance is the discovery that Ras proteins (and other signaling proteins as well) are not scattered evenly throughout the plasma membrane, but instead can reside in subdomains of the plasma membrane (Fig. 1, Step 6 [17]). These domains are popularly termed "rafts" and contain lipids that can coalesce into detergent-resistant, light density droplets within the more disordered phase of the general lipid bilayer [18,19]. Cells that express the protein caveolin have an additional raft subtype termed caveolae. It is notable that rafts are not a fixed entity. Microdomains can show dramatic variations in their lipid composition and size based on the lipids synthesized by a cell [20] and on rearrangements that occur during activation of the cell [21,22].

Because most immunofluorescence or gradient techniques utilize detergents or pH 11 carbonate buffers, it has proved difficult to study these dynamic, mobile microdomains without risking the release or intermixing of proteins that were separated in the native membrane. For proteins with a highly basic domain, such as KRas4B, pH 11 effectively neutralizes even the positive charge of lysines, and subsequent interactions may no longer reflect those initially present. For palmitoylated proteins, complete neutralization must occur rapidly to avoid hydrolysis of the delicate thioester-linked palmitates. The interaction of HRas with rafts appears to be tenuous, as the protein is easily dislodged by detergents [23,52], necessitating the use of nondetergent fractionation methods [24]. To complicate matters, although palmitates are readily accepted into the tightly packed core of a raft, farnesyl groups are poor inhabitants of raft-like domains [19,23,25]. HRas and NRas, which possess both palmitoyl and farnesyl modifications, therefore face a biophysical dilemma that is also a biological opportunity. The opposing effects of their lipids may poise these Ras proteins for easy transit into and out of raft domains. Such movement could then provide more than one location where regulatory interactions could occur. Given these caveats and speculations, optimism flavored with a good deal of skepticism should accompany current experimental conclusions.

Nevertheless, multiple techniques indicate that HRas proteins can reside graciously in raft domains [26,27]. Several reports indicate that some HRas is further localized within caveolae [28], where it may interact with caveolin-1 [29]. Whether this "some" represents a little or a lot of HRas is a matter of discussion. Discussions of how Ras proteins distribute among membrane domains are important, as HRas proteins with substitute membrane binding motifs can display aberrant signaling [30–32]. No consensus has yet been reached with the KRas4B protein. KRas4B has been reported to either avoid rafts [28,33], or to be concentrated in caveolin-rich (i.e., raft) domains [34]. Despite our poor understanding of the properties that control where or how tightly Ras proteins partition within the membrane, the underlying rationale that raft and nonraft membranes may each present unique environments for regulatory and signaling interactions is compelling.

Ras Proteins Finally Become Active at the Plasma Membrane

Normal Ras proteins gain functional competency at the plasma membrane. This is due largely to the requirement that in order for a cellular Ras protein to become GTPbound it must interact with a guanine nucleotide exchange factor (GEF) [35]. GEFs are themselves placed along the plasma membrane through interactions initiated by receptor-generated calcium signals, phosphatidyl inositols, and tyrosine phosphorylations. Thus coupling between Ras proteins and their GEFs may serve to (re)position *active* Ras at specific plasma membrane sites.

This possibility leads to two essential questions. Where are the active forms of each type of Ras located? Where do interactions of Ras-GTP with each of its effectors take place? A recent study suggests that HRas moves out of rafts when GTP-bound and that disrupting rafts diminishes the ability of HRas to activate Raf-1 [26,28]. In contrast, KRas4B is proposed to be in nonraft domains regardless of nucleotide state [28]. However, the question of whether Ras proteins are active when in raft domains (suggested by work in T and B cells and with EGF receptors in fibroblasts [21, 36–38]) or are inactive (especially if within caveolae) [27,39] remains an area of debate. One clear opportunity for discovery is to learn how GTP binding to HRas might be translated into lateral movement.

Although the mutations present in oncogenic Ras proteins cause constitutive elevations in GTP binding that occur prior to delivery to the cell surface, this does not seem to trigger premature signaling. In fact, in some situations, high levels of soluble oncogenic Ras proteins appear to dampen signaling [40]. An important unknown is whether membrane-bound oncogenic forms of Ras partition differently from the normal, cellular forms, or activate effectors in inappropriate locations.

Endocytosis—A New Stage for Ras Signaling

The plasma membrane has been considered the final destination for Ras, with the lipid-tethered protein undergoing repeated cycles of activation/inactivation and happily controlling signal passage through its territory. However, reports [41,42] that activation of the Raf-1/MEKK/ERK pathway may occur on early endosomes rather than the plasma membrane indicate that the tale of Ras location and signaling is not over [43,44]. During numerous signaling events, Raf-1 kinase is recruited to the plasma membrane where it interacts with GTP-bound Ras. However, this direct contact between Ras and Raf-1 does not seem to fully stimulate Raf-1 kinase activity [45]. Information from epidermal growth factor (EGF)- or insulin-treated cells indicates that within minutes of its membrane recruitment [24], Raf-1 departs the plasma membrane aboard early endosomes (Fig. 1, Step 7), and it is there that the further steps of MEK and ERK phosphorylation may take place [42]. This finding immediately raises a plethora of questions. Do the endosomes form from rafts, caveolae, or more traditional coated pits [46]? Does Ras-GTP actually leave the plasma membrane and recycle immediately or simply break off the relationship before Raf-1 departs and remain available for interaction with additional effectors? If Ras enters the endosome, how does it reestablish its association with the plasma membrane (Fig. 1, Step 8)? At what location does a Ras protein interact with a GAP protein and return to the GDP state? This latter point is important as this site will then define when and where Ras signaling terminates. In neurons a complete and functionally competent signaling pathway, from the ligand NGF to Ras to ERK kinases, has been detected on clathrin-coated vesicles moving from the axon back to the cell body [47].

Drugs that Affect Ras Membrane Binding

For more than a decade it has been known that preventing Ras membrane binding can inhibit its oncogenicity. This has spurred the development of potent inhibitors of the farnesyltransferase enzyme (Fig. 1, Step 1) as novel cancer chemotherapeutics [48]. FTase inhibitors are not without their difficulties, however. Multiple other farnesylated proteins (e.g., lamins of the nuclear envelope, retinal transducin γsubunits, Rho GTPases [49]) are also targets of these compounds and may cause impairments in other important pathways. A significant frustration is that both NRas and KRas4B can be modified by a distinct prenyl transferase, GGTase I, that is not susceptible to FTase inhibitors [50]. An alternative strategy of dislodging already prenylated Ras proteins from membranes and accelerating their degradation, currently being explored with the prototype compound farnesylthiosalicylic acid, also appears to diminish Ras signaling [51]. As these studies demonstrate, manipulating the location of Ras proteins is not only possible, it is a tactic that can control Ras biological activity. The recent discoveries of Ras intracellular trafficking, partitioning in or out of rafts, and the link to endocytosis identify several new sites where novel approaches for controlling Ras location and its function can now be envisioned.

References

1. Willumsen, B. M. et al. (1984). The p21ras C-terminus is required for transformation and membrane association. *Nature* **310**, 583–586.
2. Zhang, F. L. and Casey, P. J. (1996). Protein prenylation: Molecular mechanisms and functional consequences. *Annu. Rev. Biochem.* **65**, 241–269.
3. Gelb, M. H. et al. (1999). Protein prenylation: From discovery to prospects for cancer treatment. *Curr. Opin. Chem. Biol.* **2**, 40–48.
4. Otto, J. C. et al. (1999). Cloning and characterization of a mammalian prenyl protein specific protease. *J. Biol. Chem.* **274**, 8379–8382.
5. Dai, Q. et al. (1998). Mammalian prenylcysteine carboxyl methyltransferase is in the endoplasmic reticulum. *J. Biol. Chem.* **273**, 15030–15034.
6. Choy, E. et al. (1999). Endomembrane trafficking of Ras: The CaaX motif targets proteins to the ER and Golgi. *Cell* **98**, 69–80.
7. Hancock, J. F. et al. (1990). A polybasic domain or palmitoylation is required in addition to the CAAX motif to localize p21ras to the plasma membrane. *Cell* **63**, 133–139.
8. Apolloni, A. et al. (2000). H-ras but not K-ras traffics to the plasma membrane through the exocytic pathway. *Mol. Cell. Biol.* **20**, 2475–2487.
9. Willumsen, B. M. et al. (1996). Novel determinants of H-Ras plasma membrane localization and transformation. *Oncogene* **13**, 1901–1909.
10. Jung, V. et al. (1995). Mutations in the SHR5 gene of Saccharomyces cerevisiae suppress Ras function and block membrane attachment and palmitoylation of Ras proteins. *Mol. Cell. Biol.* **15**, 1333–1342.
11. Bartels, D. J. et al. (1999). Erf2, a novel gene product that affects the localization and palmitoylation of Ras2 in *Saccharomyces cerevisiae*. *Mol. Cell. Biol.* **19**, 6775.
12. Dunphy, J. T. et al. (1996). G-protein palmitoyltransferase activity is enriched in plasma membranes. *J. Biol. Chem.* **271**, 7154–7159.
13. Duncan, J. A. and Gilman, A. G. (1998). A cytoplasmic acyl-protein thioesterase that removes palmitate from G protein α subunits and p21ras. *J. Biol. Chem.* **273**, 15830–15837.
14. Hart, K. C. and Donoghue, D. J. (1997). Derivatives of activated H-Ras lacking C-terminal lipid modifications retain transforming ability if targeted to the correct subcellular location. *Oncogene* **14**, 945–953.
15. Chiu, V. K. et al. (2002). Ras signalling on the endoplasmic reticulum and the Golgi. *Nat. Cell Biol.* **4**, advance publication on-line.
16. Figueroa, C. et al. (2001). Prenylated Rab acceptor protein is a receptor for prenylated small GTPases. *J. Biol. Chem.* **276**, 28219–28225.
17. Simons, K. and Toomre, D. (2000). Lipid rafts and signal transduction. *Nat. Rev. Mol. Cell Biol.* **1**, 31–39.
18. Brown, D. A. and London, E. (1998). Functions of lipid rafts in biological membranes. *Annu. Rev. Cell Dev. Biol.* **14**, 111–136.
19. Wang, T.-Y. et al. (2001). Partitioning of lipidated peptide sequences into liquid-ordered lipid domains in model and biological membranes. *Biochemistry* **40**, 13031–13040.
20. Prinetti, A. et al. (2001). Changes in the lipid turnover, composition and organization of sphingolipid-enriched membrane domains in rat cerebellar granule cells developing in vitro. *J. Biol. Chem.* **276**, 21136–21145.
21. Pierce, S. K. (2002). Lipid rafts and B-cell activation. *Nat. Rev. Immunol.* **2**, 96–105.
22. Gomez-Mouton, C. et al. (2001). Segregation of leading-edge and uropod components into specific lipid rafts during T cell polarization. *Proc. Natl. Acad. Sci. USA* **98**, 9642–9647.
23. Melkonian, K. A. et al. (1999). Role of lipid modifications in targeting proteins to detergent-resistant membrane rafts: Many raft proteins are acylated, while few are prenylated. *J. Biol. Chem.* **274**, 3910–3917.
24. Mineo, C. et al. (1996). Localization of epidermal growth factor-stimulated Ras/Raf-1 interaction to caveolae membrane. *J. Biol. Chem.* **271**, 11930–11935.
25. Silvius, J. R. and l'Heureux, F. (1994). Fluorimetric evaluation of the affinities of isoprenylated peptides for lipid bilayers. *Biochemistry* **33**, 3014–3022.
26. Roy, S. et al. (1999). Dominant-negative caveolin inhibits H-Ras function by disrupting cholesterol-rich plasma membrane domains. *Nat. Cell Biol.* **1**, 97–105.
27. Galbiati, F. et al. (2001). Emerging themes in lipid rafts and caveolae. *Cell* **106**, 403–411.
28. Prior, I. A. et al. (2001). GTP-dependent segregation of H-ras from lipid rafts is required for biological activity. *Nat. Cell Biol.* **3**, 368–375.
29. Song, K. S. et al. (1996). Co-purification and direct interaction of Ras with caveolin, an integral membrane protein of caveolae microdomains. *J. Biol. Chem.* **271**, 9690–9697.
30. Booden, M. A. et al. (2000). Mutation of Ha-Ras C-terminus changes effector pathway utilization. *J. Biol. Chem.* **275**, 23559–23568.

31. Coats, S. G. et al. (1999). Transient palmitoylation supports H-Ras membrane binding but only partial biological activity. *Biochemistry* **38**, 12926–12934.
32. Buss, J. E. et al. (1989). Activation of the cellular proto-oncogene product p21ras by addition of a myristylation signal. *Science* **243**, 1600–1603.
33. Jaumot, M. et al. (2002). The linker domain of the Ha-Ras hypervariable region regulates interactions withexchange factors, Raf-1 and phosphoinositide 3-kinase. *J. Biol. Chem.* **277**, 272–278.
34. Kranenburg, O. et al. (2001). Regulating c-Ras function: cholesterol depletion affects caveolin association, GTP loading and signaling. *Curr. Biol.* **11**, 1880–1884.
35. Lowy, D. R. and Willumsen, B. M. (1993). Function and regulation of Ras. *Annu. Rev. Biochem.* **62**, 851–891.
36. Mineo, C. et al. (1999). Regulated migration of epidermal growth factor receptor from caveolae. *J. Biol. Chem.* **274**, 30363–30643.
37. Kabouridis, P. S. et al. (2000). Cholesterol depletion disrupts lipid rafts and modulates the activity of multiple signaling pathways in T lymphocytes. *Eur. J. Immunol.* **30**, 954–963.
38. Plyte, S. et al. (2000). Constitutive activation of the Ras/MAP kinase pathway and enhanced TCR signaling by targeting the Shc adaptor protein to membrane rafts. *Oncogene* **19**, 1529–1540.
39. Okamoto, T. et al. (1998). Caveolins, a family of scaffolding proteins for organizing "preassembled signalling complexes" at the plasma membrane. *J. Biol. Chem.* **273**, 5419–5422.
40. Fordalisi, J. J. et al. (2002). A distinct class of dominant negative Ras mutants: cytosolic GTP-bound Ras effector domain mutants that inhibit Ras signaling and transformation and enhance cell adhesion. *J. Biol. Chem.* **277**, 10813–10823.
41. Luttrell, L. M. et al. (2001). Activation and targeting of extracellular signal-regulated kinases by β-arrestin scaffolds. *Proc. Natl. Acad. Sci. USA* **98**, 2449–2454.
42. Rizzo, M. A. et al. (2001). Agonist-dependent traffic of raft-associated Ras and Raf-1 is required for activation of the MAPK cascade. *J. Biol. Chem.* **276**, 34928–34933.
43. DiFiore, P. P. and Gill, G. N. (1999). Endocytosis and mitogenic signaling. *Curr. Opin. Cell Biol.* **11**, 483–488.
44. DiFiore, P. P. and DeCamilli, P. (2001). Endocytosis and signaling: An inseparable partnership. *Cell* **106**, 1–4.
45. Morrison, D. K. and R. E. Cutler, J. (1997). The complexity of Raf-1 regulation. *Curr. Opin. Cell Biol.* **9**, 174–179.
46. Nichols, B. J. and Lippincott-Schwartz, J. (2001). Endocytosis without clathrin coats. *Trends Cell Biol.* **11**, 406–412.
47. Howe, C. L. et al. (2001.) NGF signaling from clathrin-coated vesicles: Evidence that signaling endosomes serve as a platform for the Ras-MAPK pathway. *Neuron* **32**, 801–814.
48. Cox, A. D. and Der, C. J. (1997). Farnesyltransferase inhibitors and cancer treatment: Targeting simply Ras? *BBA Rev. Cancer* **1333**, F51–F71.
49. Oliff, A. (1999). Farnesyltransferase inhibitors: Targeting the molecular basis of cancer. *Biochim. Biophys. Acta* **1423**, C19–C30.
50. Zhang, F. L. et al. (1997). Characterization of Ha-Ras, N-Ras, Ki-Ras4A, and Ki-Ras4B as in vitro substrates for farnesyl protein transferase and geranylgeranyl protein transferase type I. *J. Biol. Chem.* **272**, 10232–10239.
51. Haklai, R. et al. (1998). Dislodgement and accelerated degradation of ras. *Biochemistry* **37**, 1306–1314.
52. Baker, T. L., Zheng, H., Walker, J., Coloff, J. L., and Buss, J. E. (2003). Distinct rates of palmitate turnover on membrane-bound cellular and oncogenic H-Ras. *J. Biol. Chem.* **278**, 19242–19300.

CHAPTER 234

Role of R-Ras in Cell Growth

Gretchen A. Murphy, Adrienne D. Cox, and Channing J. Der

*Departments of Pharmacology and Radiation Oncology, and Lineberger Comprehensive Cancer Center,
University of North Carolina at Chapel Hill, Chapel Hill North Carolina*

Introduction

Ras proteins (H-Ras, K-Ras4A and 4B, and N-Ras) are regulators of signal transduction, mutated in 30% of human cancers, and targets for novel approaches for cancer treatment. Ras proteins are the founding members of a superfamily of small GTP binding and hydrolyzing proteins (GTPases). The small GTPases that share the greatest amino acid identity with Ras, such as R-Ras, Rap, and Ral, constitute members of the Ras family of proteins (Fig. 1). Within this family, the R-Ras subfamily proteins (R-Ras, TC21/R-Ras2, and M-Ras/R-Ras3) exhibit the strongest structural and biological similarities with Ras (Fig. 1). While studies in experimental model systems have shown that R-Ras proteins can promote oncogenic transformation, there is only limited evidence for aberrant R-Ras function in human cancers. Thus, are R-Ras proteins simply inferior versions of Ras proteins, or do they play distinct roles in normal cell physiology? In this review, we first summarize the general features of R-Ras proteins that are shared with Ras proteins, and then we highlight unique features of each R-Ras protein.

General Properties of R-Ras Proteins: Variations on Ras

Structure

R-Ras proteins are GTPases of approximately 200 amino acids that share significant primary, secondary, and tertiary structural characteristics with the 21-kDa Ras GTPases (Fig. 2). In addition, R-Ras proteins possess extended amino (10–26 residues) or carboxyl-terminal residues not present in Ras proteins that account for their larger size (approximately 25 kDa) and suggest functional differences with Ras proteins (Fig. 2). Therefore, the numbering of key R-Ras protein amino acid residues relative to Ras is by the addition of 26 (R-Ras), 11 (TC21), or 10 (M-Ras) to the Ras numbering system (Figs. 2 and 3).

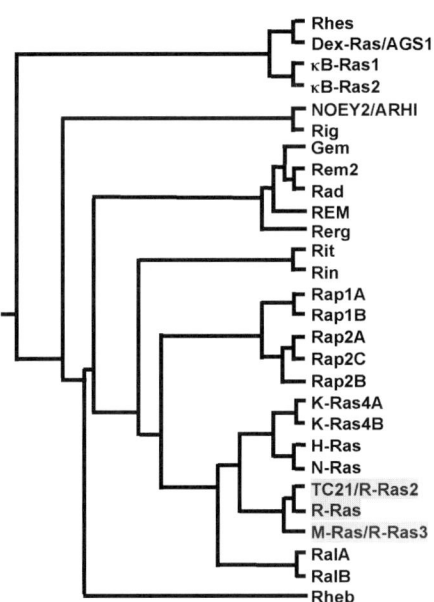

Figure 1 Ras family branch of the Ras superfamily. Small GTPases sharing strongest amino acid identity with Ras proteins are grouped together in the Ras branch. Among these, the three R-Ras proteins (highlighted) share the strongest overall amino acid identity with Ras. The dendrogram was generated by alignment of primary sequences of human proteins using ClustalW, a dynamic sequence alignment program. From the resulting multiple sequence alignment, a distance matrix was prepared and used to construct the dendrogram. Branch lengths are proportional to the estimated divergence along each branch.

```
· H-RAS   1   MT.......  .........  ........EY  KLVVVGAGGV  GKSALTIQLI
· R-RAS   1   MSSGAASGTG  RGRPRGGGPG  PGDPPPSETH  KLVVVGGGGV  GKSALTIQFI
· TC21    1   MAAA.....  .....GWRDG  SG....QEKY  RLVVVGGGGV  GKSALTIQFI
· M-RAS   1   MATSAVPSDN  L........  ........PTY KLVVVGDGGV  GKSALTIQFF

· H-RAS  25   QNHFV..DEY  DPTIEDSYRK  QVVID.GETC  LLDILDTAGQ  EEYSAMRDQY
· R-RAS  51   QSYFV..SDY  DPTIEDSYTK  ICSVD.GIPA  RLDILDTAGQ  EEFGAMREQY
· TC21   36   QSYFV..TDY  DPTIEDSYTK  QCVID.DRAA  RLDILDTAGQ  EEFGAMREQY
· M-RAS  35   QKIFV..PDY  DPTIEDSYLK  HTEID.NQWA  ILDVLDTAGQ  EEFSAMREQY

· H-RAS  72   MRTGEGFLCV  FAINNTKSFE  DIHQYREQIK  RVKDSDDVPM  VLVGNKCDLA
· R-RAS  98   MRAGHGFLLV  FAINDRQSFN  EVGKLFTQIL  RVKDRDDFPV  VLVGNKADLE
· TC21   83   MRTGEGFLLV  FSVTDRGSFE  EIYKFQRQIL  RVKDRDEFPM  ILIGNKADLD
· M-RAS  82   MRTGDGFLIV  YSVTDKASFE  HVDRFHQLIL  RVKDRESFPM  ILVANKVDLM

· H-RAS 122   A.RTVESRQA  QDLARSYGIP  YIETSAKT.R  QGVEDAFYTL  VREIRQHKLR
· R-RAS 148   SQRQVPRSEA  SAFGASHHVA  YFEASAK.LR  LNVDEAFEQL  VRAVRKYQEQ
· TC21  133   HQRQVTQEEG  QQLARQLKVT  YMEASAK.IR  MNVDQAFHEL  VRVIRKFQEQ
· M-RAS 132   HLRKITREQG  KEMATKHNIP  YIETSAKDPP  LNVDKAFHDL  VRVIRQQIPE

· H-RAS 170   KLNPP....  ...D....ESG  PGCMSCKCVL  S
· R-RAS 197   ELPPSP.PSA  PRK....KGG  .GCP...CVL  L
· TC21  182   ECPPSPEPTR  KEK....DKK  .GCH...CVI  F
· M-RAS 182   KSQKKKKKTK  WRG....DRA  TGTHKLQCVI  L
```

Figure 2 Amino acid sequence alignment of H-Ras and R-Ras proteins. R-Ras proteins are larger than Ras proteins due to additional amino (boxed) and carboxyl-terminal sequences. R-Ras proteins share complete sequence identity with the core Ras effector domain (H-Ras residues 32–40), but sequence differences in the extended effector domain (boxed; H-Ras residues 25–45) account for the different abilities of R-Ras proteins to interact with Ras effectors. Missense mutations at H-Ras residues 12, 13, and 61 (indicated by arrows) have been identified in human cancers and result in constitutively activated, transforming variants of Ras proteins. Mutation of the equivalent residues in R-Ras, TC21, or M-Ras results in the generation of constitutively activated mutants of the R-Ras proteins. The carboxyl-terminal CAAX motif (boxed) present in H-Ras and R-Ras proteins direct posttranslational modification by protein prenyltransferases. Whereas Ras proteins are modified by covalent addition of a C15 farnesyl isoprenoid by farnesyltransferase, R-Ras proteins are modified by covalent addition of a C20 geranylgeranyl isoprenoid by geranylgeranyltransferase I. This modification is followed by the proteolytic removal of the AAX residues and methylation of the now terminal isoprenylated cysteine residue. The CAAX-signaled modifications, together with specific sequences upstream of the CAAX motif, represent the two signals required to target Ras and R-Ras proteins to the plasma membrane. Like H-Ras, R-Ras and TC21 contain cysteine residues (bolded and underlined) that undergo posttranslational modification by the fatty acid palmitate which facilitates membrane association. Like K-Ras4B (not shown), M-Ras contains a lysine-rich sequence (bolded and underlined) in place of the palmitoylated cysteine that is important for membrane targeting.

Expression

The three *ras* genes are expressed ubiquitously, although at distinct tissue-specific levels. Similarly, the three R-*ras* genes are also expressed widely. R-*ras* transcripts showed a wide range of expression levels in a variety of cell types [1]. Human TC21 protein levels were found to be highest in adult human kidney, placenta, and ovary [2]. Interestingly, M-*ras* was isolated in two separate differential gene expression cloning strategies, reflecting its regulation at the level of gene expression [3,4]. M-*ras* gene expression is particularly high in brain and heart, with lower levels in skeletal muscle, ovary, and cells of hematopoietic origin [5–8].

Biochemistry

R-Ras proteins share 60–70% amino acid identity with each other and 50–60% amino acid identity to Ras, primarily in the consensus guanine nucleotide binding motifs and the switch I and II regions of Ras that alter conformation based on the bound guanine nucleotide (GTP or GDP), permitting function as GDP/GTP-regulated molecular switches (Fig. 2). These strong sequence similarities account for the fact that Ras and R-Ras proteins are regulated by an overlapping set of guanine nucleotide exchange factors (GEFs) and GTPase activating proteins (GAPs) (Table I). Like Ras, GTP-bound R-Ras is the active form, and mutations at positions analogous to those found in tumor-associated mutant Ras proteins (amino acids 12 and 61) also produce constitutively activated R-Ras GTPases (Figs. 2 and 3).

Like Ras, R-Ras proteins contain a carboxyl-terminal CAAX tetrapeptide motif that directs posttranslational modifications, including the covalent addition of an isoprenoid lipid, which promotes the association of R-Ras proteins with the inner face of the plasma membrane [9]. However, whereas Ras proteins are modified by a C15 farnesyl isoprenoid,

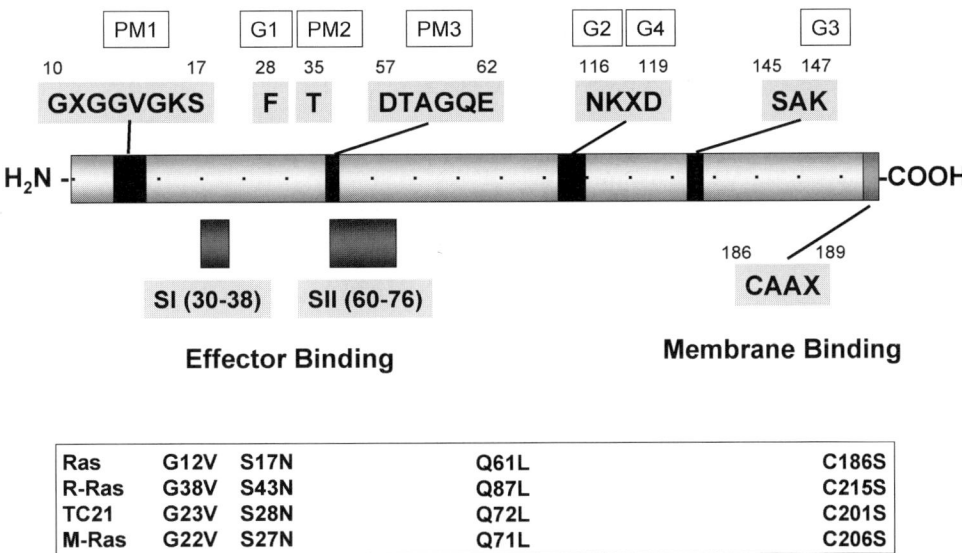

Figure 3 Functional amino acid sequence motifs of Ras and R-Ras proteins. R-Ras proteins possess a set of conserved sequence elements found in all small GTPases and are important for high-affinity binding of GDP and GTP guanine nucleotides, hydrolysis of GTP, or the conformational switch between the GDP- and GTP-bound states (numbers refer to positions in Ras proteins). These sequence elements are involved in binding to the phosphate and magnesium ions (PM) or to the guanine base (G) of the bound GDP or GTP nucleotide. By analogy to Ras proteins, the introduction of missense mutations in R-Ras proteins produces gain-of-function (e.g., Ras mutants G12V and Q61L), dominant-inhibitory (Ras mutant S17N), or loss-of-function cytosolic (Ras mutant C186S) mutants of R-Ras proteins that are useful reagents to study the signaling activities and biological functions of R-Ras proteins.

Table I Overlapping and Distinct Regulators and Effectors of R-Ras Proteins

GTPase	GAPs	GEFs[1]	Effectors[2]	Putative effectors[3]
Ras	P120 GAP, NF1-GAP, Gap1^m	Sos 1/2, RasGRF1/2, RasGRP1/2/3/4	Raf, PI3K, RalGEF, Rin1, PLCε	AF-6, Nore1, RASSF1, Rin1, PLCε
R-Ras	P120 GAP, NF1-GAP, Gap1^m, R-Ras GAP	RasGRF1, C3G, RasGRP1-3	PI3K	RalGEFs
TC21/R-Ras2	P120 GAP, NF1-GAP, Gap1^m, R-Ras GAP	Sos1, RasGRF1/2, RasGRP1-3, C3G	Raf*, PI3K, RalGEF*, PLCε	AF-6, Rin1
M-Ras/R-Ras3	P120 GAP, NF1-GAP, Gap1^m	Sos1, RasGRF1, RasGRP1/3	Raf, PI3K	RalGDS, RGL3, Nore1, AF-6, Rin1, RapGEFs

[1]RasGRP proteins are also called: Ca1DAG-GEFI (RasGRP2), Ca1DAG-GEFII (RasGRP), Ca1DAG-GEFIII (RasGRP3);
[2]Functional interactions
[3]Interactions determined, but functional activation not established
*conflicting reports

R-Ras proteins are modified by the more hydrophobic C20 geranylgeranyl isoprenoids [10]. Isoprenoid modification of Ras and R-Ras proteins is essential for biological function. Consequently, inhibitors (FTIs) of the enzyme farnesyltransferase which catalyzes the modification of Ras proteins, have been developed as inhibitors of oncogenic Ras function and are currently under clinical evaluation as anti-neoplastic drugs. Similarly, inhibitors (GGTIs) of the enzyme geranylgeranyltransferase I which catalyzes the modification of R-Ras proteins, have also been developed and have shown anti-tumor activity in preclinical studies [11].

Signal Transduction and Cell Biology

Like Ras, R-Ras proteins serve as relay switches that transmit signals initiated by diverse extracellular ligands to cytoplasmic signaling pathways. However, unlike Ras, the specific upstream signals which promote activation of endogenous R-Ras proteins are less well characterized, perhaps due in part to technical limitations and due in part to limited experimental analyses. Because Ras and R-Ras proteins share common GEFs, presumably many of the same signals that cause Ras activation also cause R-Ras protein family

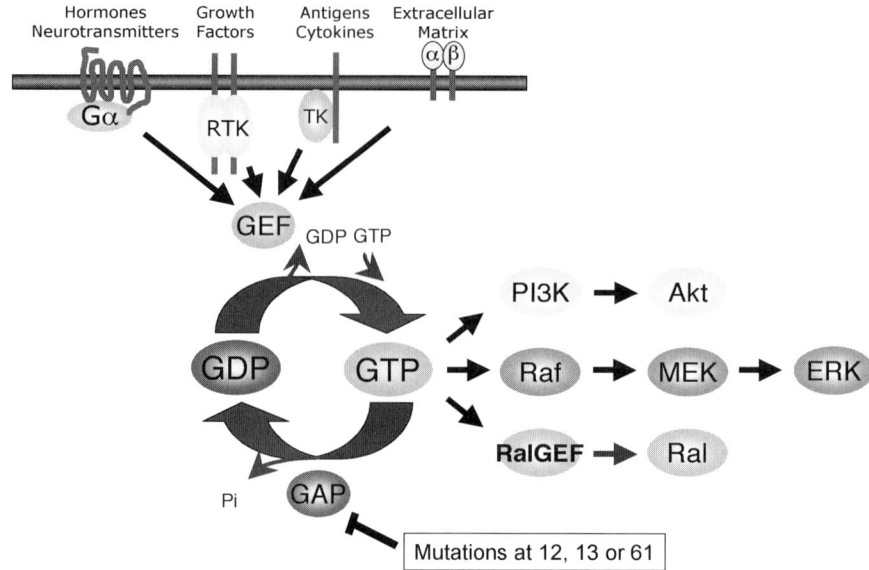

Figure 4 Ras family signaling pathways. Extracellular stimuli act on various cell surface receptors, including G-protein-coupled receptors, receptor tyrosine kinases (RTKs), tyrosine kinase (TK)-associated receptors, and integrins (α/β subunits). The activated receptors typically cause Ras and R-Ras GTP-binding via the activation of guanine nucleotide exchange factors (GEFs) specific for one or more Ras and/or R-Ras proteins. GTPase activating proteins (GAPs) stimulate the intrinsic GTPase activity of Ras and R-Ras proteins to return the proteins to the inactive, GDP-bound state. Mutations analogous to those at Ras residues 12, 13, or 61 render R-Ras proteins insensitive to GAP stimulation, and constitutively GTP-bound and active in the absence of extracellular stimuli. The GTP-bound protein preferentially binds to downstream effectors, such as phosphatidylinositol 3-kinases (PI3Ks), Raf serine/threonine kinases, or GEFs specific for Ral small GTPases. The PI3K lipid kinases phosphorylate phosphoinositides and promote formation of phosphatidylinositol 3,4,5-phosphate (PIP3), which in turn facilitates the activation of other cytoplasmic signaling proteins, including the serine/threonine kinase Akt/PKB. Raf phosphorylates the MEK1/2 dual specificity protein kinases, which in turn phosphorylate the ERK1/2 serine/threonine kinases.

activation (Fig. 4). Once activated, Ras protein signaling is mediated by interaction with a multitude of downstream effector targets (Table I). Among these, the Raf serine/threonine protein kinases, the phosphatidylinositol 3-kinase lipid kinases (PI3Ks), and the Ral small GTPase guanine nucleotide exchange factors (RalGEFs) have been established as key mediators of Ras-mediated transformation [12]. The contribution of each of these major Ras effectors to R-Ras family-mediated transformation is distinct, and will be discussed in the following sections. The core effector domain of Ras (amino acids 32–40) is critical for interactions with downstream signaling molecules, and R-Ras proteins share complete identity with this core effector domain. However, flanking residues within the extended effector domain (Ras amino acids 25–45) of R-Ras proteins differ more significantly, suggesting that R-Ras subfamily members may regulate a distinct set of effectors and cytoplasmic signaling pathways.

R-Ras subfamily GTPases are implicated in regulating biological functions both similar to and distinct from those controlled by Ras proteins. For example, like Ras, constitutively activated forms of all three R-Ras members have been shown to cause growth transformation of NIH 3T3 mouse fibroblasts. Similarly, both Ras and R-Ras family proteins regulate cell survival, actin cytoskeletal organization, and differentiation. However, R-Ras proteins also cause consequences opposing those of Ras, and the biological phenotypes of R-Ras function can be significantly different from those of TC21 and M-Ras. The following sections detail the distinct properties and roles of each R-Ras subfamily member in signal transduction and cell biology.

R-Ras

GDP/GTP Regulation

R-*ras* encodes a 218 amino acid protein that shares approximately 55% identity overall to Ras proteins (Fig. 2) [1]. While no GEFs have been shown to be specific for R-Ras, R-Ras GDP/GTP exchange is stimulated by GEFs for some Ras and/or Rap small GTPases (e.g., RasGRF, RasGRPs, C3G) (Table I) [13–19]. R-Ras interacts with some of the known Ras GAPs, and GTP hydrolysis is stimulated by p120GAP, GAP1[m] and R-Ras GAP (Table I) [20–22].

Signal Transduction

R-Ras fails to activate the Raf>MEK>ERK mitogen-activated protein kinase (MAPK) cascade, which is the key effector pathway required for Ras-mediated transformation of rodent fibroblasts (Table I). However, R-Ras strongly activates the PI3K/Akt signaling pathway, perhaps the second most critical effector pathway for Ras-mediated transformation and survival [23–25]. This effector pathway has also been shown to be required for R-Ras-mediated cell survival and transformation [24,26,27]. R-Ras apparently does not utilize the RalGEF pathway [26], which contributes significantly to Ras-mediated transformation of human cells [28]. Nevertheless, the GTP-bound form of each R-Ras family member can bind to isolated Ras binding domains of Raf and RalGDS, which enables the selective detection of activated R-Ras proteins [29].

Cell Biology

Although mutated R-*ras* genes have not been identified in human tumors, introduction of point mutations analogous to those that cause mutational activation of Ras (38V, 87L) stimulate R-Ras transforming potential. Unlike Ras, in which mutations at positions 12 and 61 are similar in their transforming potency, in members of the R-Ras family, activating mutations analogous to Ras codon 61 (e.g., 87L, 72L, or 71L) are significantly more potent than those analogous to Ras codon 12 [8,26,30,31]. Although constitutively activated mutants of R-Ras cause tumorigenic transformation of NIH3T3 mouse fibroblasts, the strong morphological transformation observed in Ras-transformed rodent fibroblasts is not observed with activated mutants of R-Ras [30,32]. Further, constitutively activated R-Ras fails to rescue the block in growth caused by expression of dominant negative Ras [33], suggesting that R-Ras induces transformation by regulation of signaling pathways distinct from those utilized by Ras. Hence, the normal biological role of R-Ras may lie in processes distinct from those that contribute to cell transformation.

R-Ras has been shown to control and promote integrin-mediated cellular adhesion [34,35]. For example, expression of dominant negative R-Ras (S43N) in adherent cells reduced cell spreading and expression of activated R-Ras in suspension cells promoted spreading and fibronectin assembly, suggesting that R-Ras is required for integrin-mediated cell adhesion [34]. R-Ras also promoted retinal neural outgrowth on laminin, a process dependent on integrin function [36]. R-Ras induced integrin-mediated migration and invasion of breast epithelial cells on collagen, by signaling to α2, but not α5 integrin receptors, suggesting that R-Ras induces selective activation of integrins [37]. R-Ras-mediated control of adhesion may be linked to both Src- and Raf-mediated pathways [38,39]. R-Ras has also been shown to control apoptosis [33], although the effect on the apoptotic response varies depending on cell type. For example, R-Ras promoted myeloid cell apoptosis in response to growth factor withdrawal, an effect which is abrogated by overexpression of Bcl-2 [40].

In contrast, a greater pool of evidence from several model cell systems suggests that R-Ras blocks the apoptotic response. For example, constitutively activated R-Ras induced protection from cell death following withdrawal of anchorage or growth factors in RIE-1 rat intestinal epithelial cells, C2C12 skeletal myoblasts, and BaF3 cells, an IL-3-dependent mouse pro-B-cell line [24,27,41].

In summary, R-Ras is both similar to and distinct from Ras. R-Ras utilizes only some of the same effectors and signaling pathways, such as PI3K/Akt, and regulates only some of the same cellular functions, such as cell proliferation and survival, similar to Ras. Conversely, R-Ras also functions in opposing cellular regulatory roles, such as in cell adhesion. Thus, R-Ras effector utilization is expected to be distinct from that of Ras.

TC21/R-Ras-2

GDP/GTP Regulation

Human TC21/R-Ras-2 is a 204 amino acid protein that shares approximately 70% amino acid identity to R-Ras and approximately 60% amino acid identity to H-Ras (Fig. 2) [32,42]. TC21 is regulated by a set of GEFs and GAPs similar to those that regulate R-Ras GDP/GTP cycling, but responds to more Ras regulators than does R-Ras (Table I) [17]. TC21 intrinsic GTP/GDP cycling activity is activated by SOS, RasGRF1/2, C3G, and RasGRP [2,17,43]. Hydrolysis of TC21-bound GTP is enhanced by known Ras GAPs (p120 GAP, Gap1^m, neurofibromin) (Table I) [2,17]. Hence, extracellular signals that regulate Ras GDP/GTP cycling are also likely to cause concurrent regulation of TC21 activity.

Signal Transduction

Whether Raf is an important mediator of TC21-induced transformation remains unclear (Table I). One study found that TC21 failed to interact with and activate Raf kinases *in vivo* [2,44]. Consistent with this failure to activate Raf is the observation that, unlike Ras, TC21, as well as R-Ras, failed to induce senescence in primary fibroblasts [45]. Ras-mediated induction of senescence has been attributed to activation of the Raf/ERK pathway. However, other studies determined that TC21 interacts with and activates Raf kinases, and activates the Raf/MEK/ERK signaling pathway in transformation [43,46]. The basis for these different observations is currently not known. The recent observation that association of 14-3-3 proteins with Raf-1 prevented R-Ras- or TC21-, but not Ras-induced, activation of Raf-1 suggests that the presence of other Raf-1 interacting proteins may account for these different observations [47].

Using various screening approaches for protein-protein interactions, TC21 was found to interact with RalGDS and RalGDS-like proteins (RGL, RGL2) [48,49]. However, whether TC21 activates RalGEFs is unclear (Table I).

One study showed that although TC21 interacted with RalGDS proteins, in contrast to Ras, TC21 did not promote activation of RalA [49]. Thus, similar to observations with R-Ras [50,51], TC21 can bind but not activate RalGDS, suggesting that RalGDS may not be a key effector for mediating TC21-induced growth transformation. In contrast, another study [52] determined that activated TC21 stimulated activation of Ral. These differences may be attributed to experimental, cell type, or cell strain type differences [53] or to the utilization of different expression vectors for ectopic expression of TC21 proteins [54].

Aside from the conflicting observations regarding whether Raf or RalGDS are involved in TC21-mediated transformation, there is a strong consensus that TC21 binds the p110 catalytic subunit of PI3K and stimulates PI3K and PI3K-dependent Akt activities (Table I) [49,52,55]. Inhibition of PI3K function with pharmacological inhibitors or genetic manipulation reversed TC21-mediated transformation of rodent fibroblasts [49,52,55]. However, activated PI3K alone does not cause transformation of fibroblasts [56], suggesting that TC21 must also utilize other effectors to mediate transformation. Like Ras, TC21 also activates phospholipase C epsilon (PLCε) (Table I) [49]. Further study is clearly necessary to determine whether PLCε-mediated second messenger activation is important for TC21-mediated transformation.

Cell Biology

Of the R-Ras subfamily members, TC21 regulates biological functions most similarly to Ras. Unlike other R-Ras subfamily members, activating mutations in *tc21* or overexpression of wild-type TC21 protein have been observed in several human tumors and tumor-derived cell lines [32,43,46,57–59]. Similar to activated forms of Ras, activated TC21 causes strong morphologic growth and neoplastic transformation in several cell types, including rodent fibroblasts and human breast epithelial cells [31,43,57]. In addition, unlike R-Ras, activated TC21 can overcome the growth inhibitory actions of dominant negative Ras, indicating that TC21 and Ras share some common functions important for normal cell proliferation [33]. However, in contrast to Ras, neither TC21 nor R-Ras caused premature senescence of primary rodent fibroblasts [45]. In addition to regulating cell proliferation and transformation, TC21 promotes cell survival of matrix-deprived intestinal epithelial cells and chemotherapeutic drug-treated NIH3T3 cells [41,55]. Like Ras, TC21 also promoted PC12 pheochromocytoma cell differentiation and growth cessation and blocked serum deprivation-induced differentiation of C2 myoblasts [44]. In addition, like R-Ras, activated TC21 has been shown to induce migration and invasion of breast epithelial cells [37].

In summary, of the three R-Ras subfamily proteins, TC21 is regulated and functions most similarly to Ras to control processes such as cell proliferation and transformation. While controversy exists over the mechanism by which TC21 causes transformation, it is clear that chronically activated versions are potent oncogenes, and contribute to various aspects of the transformation phenotype and to tumor formation in humans. Whether the three *ras* genes are functionally distinct or redundant remains an issue of considerable debate. Therefore, whether *tc21* is functionally distinct or redundant with *ras* genes is an issue that also remains to be elucidated.

M-Ras/R-Ras-3

GDP/GTP Regulation

Human M-Ras is a 209 amino acid protein that shares approximately 60% amino acid identity to R-Ras, TC21, and Ras. M-Ras GTP/GDP cycling is regulated more similarly to Ras than to R-Ras or TC21 (Table I). For example, while M-Ras is activated by RasGRF and RasGRP, unlike R-Ras and TC21, it is also stimulated by SOS [8,17]. In addition, like Ras but unlike TC21 and R-Ras, C3G failed to upregulate GTP-bound M-Ras [17]. Similarly, M-Ras GTPase activity is stimulated by the same set of GAPs, such as p120 Ras GAP that enhances GTP hydrolysis on Ras [8,17].

Signal Transduction

The effectors and pathways stimulated by M-Ras to cause transformation may involve both known Ras effectors and novel mechanisms, but are poorly characterized (Table I). Compared to Ras proteins, M-Ras interacts only weakly with Raf kinases [8], and activated M-Ras is only a weak stimulator of the Raf > MEK > ERK MAPK cascade [3,6,8,60]. However, inhibition of ERK activation by pharmacological inhibitors caused a reduction in M-Ras-mediated fibroblast transformation, and Raf-1 cooperated with M-Ras to induce transformation, suggesting that Raf is a key player in M-Ras-mediated transformation [8].

In addition to Raf, M-Ras associates, to varying extents, with other known Ras effectors, such as RalGEFs, PI3K, and Rin1 [8]. Like Ras and TC21, M-Ras was found to interact strongly with RGL3, a RalGDS-like protein [61]. In addition, M-Ras bound strongly to and activated PI3K in fibroblasts [60].

M-Ras-GTP also interacts with newly identified Rap GEFs, MR-GEF, and RA-GEF-2, which possess a Ras association (RA) domain [62,63]. Co-expression of constitutively activated M-Ras with RA-GEF-2 resulted in upregulation of GTP-bound, plasma membrane-bound Rap1 [63]. In contrast, activated M-Ras blocked MR-GEF-mediated GTP loading of Rap1A [62].

Cell Biology

Of the R-Ras subfamily proteins, the least is known regarding M-Ras biological function. Like R-Ras and TC21, M-Ras controls cell proliferation and survival and may also regulate actin cytoskeletal dynamics. M-Ras is only weakly transforming in NIH3T3 fibroblasts [6] and also induced transformation of melan-a immortalized mouse melanocytes [4].

Activated M-Ras also caused IL-3-independent proliferation of RX-6 mast cell/megakaryocyte or BaF3 mouse myeloid cells and blocked serum withdrawal induction of C2 myoblast differentiation [3,7,8]. Like Ras, R-Ras, and TC21, activated M-Ras also promoted cell survival in PC12 cells, by a mechanism that requires PI3K function [60]. It should be noted that these activities of M-Ras were revealed by overexpression of mutated M-Ras proteins, and hence, it is not clear whether they accurately reflect functions of endogenous M-Ras activated by extracellular stimuli.

Conclusions

In summary, the three R-Ras subfamily members, R-Ras, TC21/R-Ras2, and M-Ras/R-Ras3 mediate cell growth, division, differentiation, and death by utilizing both novel pathways and those regulated by Ras and other Ras-related proteins. Further, R-Ras subfamily members, especially R-Ras itself, clearly contribute to cellular processes such as integrin-mediated cell adhesion in a manner distinct from that of Ras. Deciphering the full contribution of R-Ras, TC21, and M-Ras to cellular growth control clearly awaits further study.

References

1. Lowe, D. G., Capon, D. J., Delwart, E., Sakaguchi, A. Y., Naylor, S. L., and Goeddel, D. V. (1987). Structure of the human and murine R-ras genes, novel genes closely related to ras proto-oncogenes. *Cell* **48**,137–146.
2. Graham, S. M., Vojtek, A. B., Huff, S. Y., Cox, A. D., Clark, G. J., Cooper, J. A., and Der, C. J. (1996). TC21 causes transformation by Raf-independent signaling pathways. *Mol. Cell. Biol.* **16**, 6132–6140.
3. Louahed, J., Grasso, L., De Smet, C., Van Roost, E., Wildmann, C., Nicolaides, N. C., Levitt, R. C., and Renauld, J. C. (1999). Interleukin-9-induced expression of M-Ras/R-Ras3 oncogene in T-helper clones. *Blood* **94**,1701–1710.
4. Wang, D., Yang, W., Du, J., Devalaraja, M. N., Liang, P., Matsumoto, K., Tsubakimoto, K., Endo, T., and Richmond, A. (2000). MGSA/GRO-mediated melanocyte transformation involves induction of Ras expression. *Oncogene* **19**, 4647–4659.
5. Matsumoto, K., Asano, T., and Endo, T. (1997). Novel small GTPase M-Ras participates in reorganization of actin cytoskeleton. *Oncogene* **15**, 2409–2417.
6. Kimmelman, A., Tolkacheva, T., Lorenzi, M. V., Osada, M., Chan, A. M. (1997). Identification and characterization of R-ras3: a novel member of the RAS gene family with a non-ubiquitous pattern of tissue distribution. *Oncogene* **15**, 2675–2685.
7. Ehrhardt, G. R., Leslie, K. B., Lee, F., Wieler, J. S., and Schrader, J. W. (1999). M-Ras, a widely expressed 29-kD homologue of p21 Ras: expression of a constitutively active mutant results in factor-independent growth of an interleukin-3-dependent cell line. *Blood* **94**, 2433–2444.
8. Quilliam, L. A., Castro, A. F., Rogers-Graham, K. S., Martin, C. B., Der, C. J., and Bi, C. (1999). M-Ras/R-Ras3, a transforming ras protein regulated by Sos1, GRF1, and p120 Ras GTPase-activating protein, interacts with the putative Ras effector AF6. *J. Biol. Chem.* **274**, 23850–23857.
9. Cox, A. D. and Der, C. J. (1997). Farnesyltransferase inhibitors and cancer treatment: targeting simply Ras? *Biochim. Biophys. Acta* **1333**, F51–F71.
10. Reuther, G. W. and Der, C. J. (2000). The Ras branch of small GTPases: Ras family members don't fall far from the tree. *Curr. Opin. Cell Biol.* **12**, 157–165.
11. Cox, A. D. and Der, C. J. (2002). Farnesyltransferase inhibitors: Promises and realities. *Curr. Opin. Pharmacol.* **2**, 388–393.
12. Shields, J. M., Pruitt, K., McFall, A., Shaub, A., and Der, C. J. (2000). Understanding Ras: "It ain't over 'til it's over.". *Trends Cell Biol.* **10**, 147–154.
13. Gotoh, T., Niino, Y., Tokuda, M., Hatase, O., Nakamura, S., Matsuda, M., and Hattori, S. (1997). Activation of R-Ras by Ras-guanine nucleotide-releasing factor. *J. Biol. Chem.* **272**, 18602–18607.
14. Gotoh, T., Tian, X., and Feig, L. A. (2001). Prenylation of target GTPases contributes to signaling specificity of Ras-guanine nucleotide exchange factors. *J. Biol. Chem.* **276**, 38029–38035.
15. Tian, X. and Feig, L. A. (2001). Basis for signaling specificity difference between Sos and Ras-GRF guanine nucleotide exchange factors. *J. Biol. Chem.* **276**, 47248–47256.
16. Kawasaki, H., Springett, G. M., Toki, S., Canales, J. J., Harlan, P., Blumenstiel, J. P., Chen, E. J., Bany, I. A., Mochizuki, N., Ashbacher, A., Matsuda, M., Housman, D. E., and Graybiel, A. M. (1998). A Rap guanine nucleotide exchange factor enriched highly in the basal ganglia. *Proc. Natl. Acad. Sci. USA* **95**, 13278–13283.
17. Ohba, Y., Mochizuki, N., Yamashita, S., Chan, A. M., Schrader, J. W., Hattori, S., Nagashima, K., and Matsuda, M. (2000). Regulatory proteins of R-Ras, TC21/R-Ras2, and M-Ras/R-Ras3. *J. Biol. Chem.* **275**, 20020–20026.
18. Gotoh, T., Cai, D., Tian, X., Feig, L. A., and Lerner, A. (2000). p130Cas regulates the activity of AND-34, a novel Ral, Rap1, and R-Ras guanine nucleotide exchange factor. *J. Biol. Chem.* **275**, 30118–30123.
19. Yamashita, S., Mochizuki, N., Ohba, Y., Tobiume, M., Okada, Y., Sawa, H., Nagashima, K., and Matsuda, M. (2000). CalDAG-GEFIII activation of Ras, R-ras, and Rap1. *J. Biol. Chem.* **275**, 25488–25493.
20. Rey, I., Taylor-Harris, P., van Erp, H., and Hall, A. (1994). R-ras interacts with rasGAP, neurofibromin and c-raf but does not regulate cell growth or differentiation. *Oncogene* **9**, 685–692.
21. Yamamoto, T., Matsui, T., Nakafuku, M., Iwamatsu, A., and Kaibuchi, K. (1995). A novel GTPase-activating protein for R-Ras. *J. Biol. Chem.* **270**, 30557–30561.
22. Li, S., Nakamura, S., and Hattori, S. (1997). Activation of R-Ras GTPase by GTPase-activating proteins for Ras, Gap1(m), and p120GAP. *J. Biol. Chem.* **272**, 19328–19332.
23. Marte, B. M., Rodriguez-Viciana, P., Wennstrom, S., Warne, P. H., and Downward, J. (1997). R-Ras can activate the phosphoinositide 3-kinase but not the MAP kinase arm of the Ras effector pathways. *Curr. Biol.* **7**, 63–70.
24. Suzuki, J., Kaziro, Y., and Koide, H. (1997). An activated mutant of R-Ras inhibits cell death caused by cytokine deprivation in BaF3 cells in the presence of IGF-I. *Oncogene* **15**, 1689–1697.
25. Suzuki, J., Kaziro, Y., and Koide, H. (1998). Synergistic action of R-Ras and IGF-1 on Bcl-xL expression and caspase-3 inhibition in BaF3 cells: R-Ras and IGF-1 control distinct anti-apoptotic kinase pathways. *FEBS Lett.* **437**, 112–116.
26. Osada, M., Tolkacheva, T., Li, W., Chan, T. O., Tsichlis, P. N., Saez, R., Kimmelman, A. C., and Chan, A. M. (1999). Differential roles of Akt, Rac, and Ral in R-Ras-mediated cellular transformation, adhesion, and survival. *Mol. Cell. Biol.* **19**, 6333–6344.
27. Suzuki, J., Kaziro, Y., and Koide, H. (2000). Positive regulation of skeletal myogenesis by R-Ras. *Oncogene* **19**, 1138–1146.
28. Hamad, N. M., Elconin, J., Karnoub, A. E., Bai, W., Rich, J. N., Der, C. J., Abraham, R. T., and Counter, C. M. (in press). The RalGEF pathway is critical for Ras-induced oncogenesis in human cells. *Genes Dev.*
29. Taylor, S. J., Resnick, R. J., and Shalloway, D. (2001). Nonradioactive determination of Ras-GTP levels using activated ras interaction assay. *Methods Enzymol* **333**, 333–342.
30. Cox, A. D., Brtva, T. R., Lowe, D. G., and Der, C. J. (1994). R-Ras induces malignant, but not morphologic, transformation of NIH3T3 cells. *Oncogene* **9**, 3281–3288.

31. Graham, S. M., Cox, A. D., Drivas, G., Rush, M. G., D'Eustachio, P., and Der, C. J. (1994). Aberrant function of the Ras-related protein TC21/R-Ras2 triggers malignant transformation. *Mol. Cell. Biol.* **14**, 4108–4115.
32. Chan, A. M., Miki, T., Meyers, K. A., and Aaronson, S. A. (1994). A human oncogene of the RAS superfamily unmasked by expression cDNA cloning. *Proc. Natl. Acad. Sci. USA* **91**, 7558–7562.
33. Huff, S. Y., Quilliam, L. A., Cox, A. D., and Der, C. J. (1997). R-Ras is regulated by activators and effectors distinct from those that control Ras function. *Oncogene* **14**, 133–143.
34. Zhang, Z., Vuori, K., Wang, H., Reed, J. C., and Ruoslahti, E. (1996). Integrin activation by R-ras. *Cell* **85**, 61–69.
35. Hughes, P. E., Renshaw, M. W., Pfaff, M., Forsyth, J., Keivens, V. M., Schwartz, M. A., and Ginsberg, M. H. (1997). Suppression of integrin activation: a novel function of a Ras/Raf-initiated MAP kinase pathway. *Cell* **88**, 521–530.
36. Ivins, J. K., Yurchenco, P. D., and Lander, A. D. (2000). Regulation of neurite outgrowth by integrin activation. *J. Neurosci.* **20**, 6551–6560.
37. Keely, P. J., Rusyn, E. V., Cox, A. D., and Parise, L. V. (1999). R-Ras signals through specific integrin alpha cytoplasmic domains to promote migration and invasion of breast epithelial cells. *J. Cell Biol.* **145**, 1077–1088.
38. Zou, J. X., Liu, Y., Pasquale, E. B., and Ruoslahti, E. (2002). Activated SRC oncogene phosphorylates R-ras and suppresses integrin activity. *J. Biol. Chem.* **277**, 1824–1827.
39. Sethi, T., Ginsberg, M. H., Downward, J., and Hughes, P. E. (1999). The small GTP-binding protein R-Ras can influence integrin activation by antagonizing a Ras/Raf-initiated integrin suppression pathway. *Mol. Biol. Cell* **10**, 1799–1809.
40. Wang, H. G., Millan, J. A., Cox, A. D., Der, C. J., Rapp, U. R., Beck, T., Zha, H., and Reed, J. C. (1995). R-Ras promotes apoptosis caused by growth factor deprivation via a Bcl-2 suppressible mechanism. *J. Cell Biol.* **129**, 1103–1114.
41. McFall, A., Ulku, A., Lambert, Q. T., Kusa, A., Rogers-Graham, K., and Der, C. J. (2001). Oncogenic Ras blocks anoikis by activation of a novel effector pathway independent of phosphatidylinositol 3-kinase. *Mol. Cell. Biol.* **21**, 5488–5499.
42. Drivas, G. T., Shih, A., Coutavas, E., Rush, M. G., and D'Eustachio, P. (1990). Characterization of four novel ras-like genes expressed in a human teratocarcinoma cell line. *Mol. Cell. Biol.* **10**, 1793–8.
43. Movilla, N., Crespo, P., and Bustelo, X. R. (1999). Signal transduction elements of TC21, an oncogenic member of the R-Ras subfamily of GTP-binding proteins. *Oncogene* **18**, 5860–5869.
44. Graham, S. M., Oldham, S. M., Martin, C. B., Drugan, J. K., Zohn, I. E, Campbell, S., and Der, C. J. (1999). TC21 and Ras share indistinguishable transforming and differentiating activities. *Oncogene* **18**, 2107–16.
45. Lin, A. W., Barradas, M., Stone, J. C., Van Aelst, L., Serrano, M., and Lowe, S. W. (1998). Premature senescence involving p53 and p16 is activated in response to constitutive MEK/MAPK mitogenic signaling. *Genes Dev.* **12**, 3008–3019.
46. Rosario, M., Paterson, H. F., and Marshall, C. J. (1999). Activation of the Raf/MAP kinase cascade by the Ras-related protein TC21 is required for the TC21-mediated transformation of NIH 3T3 cells. *EMBO J.* **18**, 1270–1279.
47. Light, Y., Paterson, H., and Marais, R. (2002). 14-3-3 antagonizes ras-mediated raf-1 recruitment to the plasma membrane to maintain signaling fidelity. *Mol. Cell. Biol.* **22**, 4984–4996.
48. Lopez-Barahona, M., Bustelo, X. R., and Barbacid, M. (1996). The TC21 oncoprotein interacts with the Ral guanosine nucleotide dissociation factor. *Oncogene* **12**, 463–470.
49. Murphy, G. A., Graham, S. M., Morita, S., Reks, S. E., Rogers-Graham, K., Vojtek, A., Kelley, G. G., and Der, C. J. (2002). Involvement of phosphatidylinositol 3-kinase, but not RalGDS, in TC21/R-Ras2-mediated transformation. *J. Biol. Chem.* **277**, 9966–9975.
50. Peterson, S. N., Trabalzini, L., Brtva, T. R., Fischer, T., Altschuler, D. L., Martelli, P., Lapetina, E. G., Der, C. J., and White, G. C. (1996). Identification of a novel RalGDS-related protein as a candidate effector for Ras and Rap1. *J. Biol. Chem.* **271**:29903–29908.
51. Spaargaren, M. and Bischoff, J. R. (1994). Identification of the guanine nucleotide dissociation stimulator for Ral as a putative effector molecule of R-ras, H-ras, K-ras, and Rap. *Proc. Natl. Acad. Sci. USA* **91**, 12609–12613.
52. Rosario, M., Paterson, H. F., and Marshall, C. J. (2001). Activation of the Ral and phosphatidylinositol 3′ kinase signaling pathways by the ras-related protein TC21. *Mol. Cell. Biol.* **21**, 3750–3762.
53. Khosravi-Far, R., White, M. A., Westwick, J. K., Solski, P. A., Chrzanowska-Wodnicka, M., Van Aelst, L., Wigler, M. H., and Der, C. J. (1996). Oncogenic Ras activation of Raf/mitogen-activated protein kinase-independent pathways is sufficient to cause tumorigenic transformation. *Mol. Cell. Biol.* **16**, 3923–3933.
54. Fiordalisi, J. J., Johnson, R. L., Ulku, A. S., Der, C. J., and Cox, A. D., (2001). Mammalian expression vectors for Ras family proteins: Generation and use of expression constructs to analyze Ras family function. *Methods Enzymol.* **332**, 3–36.
55. Rong, R., He, Q., Liu, Y., Sheikh, M. S., and Huang, Y. (2002). TC21 mediates transformation and cell survival via activation of phosphatidylinositol 3-kinase/Akt and NF-kappaB signaling pathway. *Oncogene* **21**, 1062–1070.
56. Rodriguez-Viciana, P., Warne, P. H., Khwaja, A., Marte, B. M., Pappin, D., Das, P., Waterfield, M. D., Ridley, A., and Downward, J. (1997). Role of phosphoinositide 3-OH kinase in cell transformation and control of the actin cytoskeleton by Ras. *Cell* **89**, 457–467.
57. Clark, G. J., Kinch, M. S., Gilmer, T. M., Burridge, K., and Der, C. J. (1996). Overexpression of the Ras-related TC21/R-Ras2 protein may contribute to the development of human breast cancers. *Oncogene* **12**, 169–176.
58. Huang, Y., Saez, R., Chao, L., Santos, E., Aaronson, S. A., and Chan, A. M. (1995). A novel insertional mutation in the TC21 gene activates its transforming activity in a human leiomyosarcoma cell line. *Oncogene* **11**, 1255–1260.
59. Barker, K. T. and Crompton, M. R. (1998). Ras-related TC21 is activated by mutation in a breast cancer cell line, but infrequently in breast carcinomas in vivo. *Br. J. Cancer* **78**, 296–300.
60. Kimmelman, A. C., Osada, M., and Chan, A. M. (2000). R-Ras3, a brain-specific Ras-related protein, activates Akt and promotes cell survival in PC12 cells. *Oncogene* **19**, 2014–2022.
61. Ehrhardt, G. R., Korherr, C., Wieler, J. S., Knaus, M., and Schrader, J. W. (2001). A novel potential effector of M-Ras and p21 Ras negatively regulates p21 Ras-mediated gene induction and cell growth. *Oncogene* **20**, 188–197.
62. Rebhun, J. F., Castro, A. F., and Quilliam, L. A. (2000). Identification of guanine nucleotide exchange factors for the Rap1 GTPase: regulation of MR-GEF by M-Ras-GTP interaction. *J. Biol. Chem.*
63. Gao, X., Satoh, T., Liao, Y., Song, C., Hu, C. D., Kariya, K. K., and Kataoka, T. (2001). Identification and characterization of RA-GEF-2, a Rap guanine nucleotide exchange factor that serves as a downstream target of M-Ras. *J. Biol. Chem.* **276**, 42219–42225.

Molecular and Structural Organization of Rab GTPase Trafficking Networks

Christelle Alory and William E. Balch

*Departments of Cell and Molecular Biology and
Institute for Childhood and Neglected Disease,
The Scripps Research Institute, La Jolla, California*

Introduction

Rab proteins are small GTPbinding proteins (20 to 25 kDa) that belong to the Ras superfamily. Rab proteins specifically direct membrane traffic in eukaryotic cells. Having been first discovered early in the 1980s in the budding yeast *Saccharomyces cerevisiae* and referred to as Ypt proteins, the name "Rab" comes from "the *Ra*s protein from *b*rain" as the first Rab (Rab3/Smg25) in mammalian cells was identified in bovine and rat brain [1]. The Rab family of GTPases is now the largest family of GTPbinding proteins in the cell [2]. They have been found in all eukaryotes studied and new members have been recently identified through the genome sequencing projects: *S. cerevisiae* contains 11 members, *Caenorhabditis elegans* has 29 members, *Drosophila melanogaster* has 26 members, and *Arabidopsis thaliana* has 57 members. *Homo sapiens* tops the list with 63 members. The fact that the number of Rab proteins is correlated with the complexity of the organism underlines their biological importance in membrane traffic. Rab proteins play a crucial role in regulating the intracellular protein and lipid trafficking between the different specialized organelles of a eukaryotic cell [3,4]. They are important in a myriad of vesicle-mediated transport pathways where they are found to be crucial in recruiting molecular motors, tethering factors, and SNAREs (SNAp REceptors) that guide, dock, and fuse the transport vesicle with target membrane compartments [5,6].

Rab Proteins are Recycling GTPases

Like all other GTPases, Rab proteins cycle between inactive GDP-bound and active GTP-bound forms. Because Rab

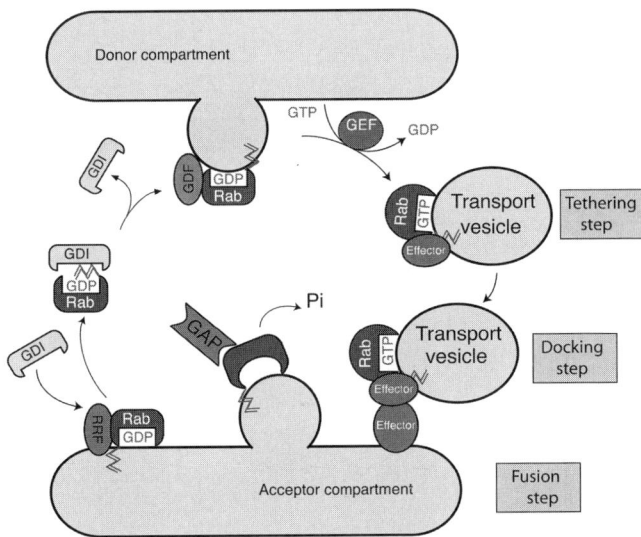

Figure 1 The general Rab GTPase cycle. Rab GTPases undergo a dynamic cycle of delivery and retrieval from the membrane. Delivery is initiated by the cytosolic form of Rab-GDP bound to GDI. Following recruitment, Rab is activated to the GTP-bound form by a Rab-specific GEF leading to recruitment of motors, tethers, and fusion factors. During or following fusion, Rab is converted back to the GDP-bound form and retrieved from the membrane by GDI through RRF.

proteins are prenylated and insoluble, they are maintained in the cytosol in their inactive GDP-bound state in a complex with a chaperone-like protein, the guanine-nucleotide dissociation inhibitor (GDI) (Fig. 1) [7,8]. During Rab recruitment to membranes, GDI is released and exchange of GDP for GTP is catalyzed by a guanine nucleotide exchange factor (GEF), activating the Rab protein. In the GTP-bound state, the Rab protein can recruit various effectors that direct the vesicle to the target membrane compartment and promote fusion [3–5,9]. Following hydrolysis by a GTPase-activating protein (GAP), the GDP-bound Rab protein is then extracted from the membrane and recycled back into the cytosol by GDI [7,8].

Rab Proteins: An Evolutionarily Conserved Family

Rab proteins define a diverse, but conserved functional family. They share ~20–95% overall amino acid identity within the mammalian family. Although more divergent between species, higher eukaryote homologs of several yeast proteins can complement function. For example, it has been shown that mammalian Rab1, involved in endoplasmic reticulum (ER) to Golgi transport, can replace the function of the yeast counterpart Ypt1p [10]. Sequence alignment of Rab proteins shows the presence of 8 conserved and 11 non-conserved regions (Fig. 2). Conserved domains include the universal 4 nucleotide binding regions (G1, G3, G4, G5) that are signature motifs found in nearly all Ras superfamily proteins, 3 phosphate/Mg^{2+} binding regions (PM1–PM3), and carboxyl-terminal cysteine motifs required for prenylation [11]. The variable regions are represented by the effector region G2 (also named switch I domain), the switch II domain, and the Rab family (RabF1–RabF5) and Rab subfamily (RabSF1–RabSF3) specific sequence motifs [2,12,13]. Based on the identification of these Rab-specific sequence motifs, Rab proteins have been clustered into 10 subfamilies (Rab1, 3–6, 8, 11, 22, 27, and 40). Based on the crystal structure of Rab proteins, Rab-specific motifs (RabF) are located in and around the switch domains and are thought to contribute to the binding of common regulatory molecules through such conserved regions. These have also been referred to as the complementarity determining regions (RabCDR) based on the structure of Rab3a bound to rabphilin3A (a Rab3A effector) [14]. Molecules that bind through RabF motifs may include GEFs, GAPs, GDIs, and Rab escort protein (REP) (also referred to as the choroideremia gene product; CHM [8]) involved in prenylation [8,15]. On the other hand, the RabSF regions, which show more variation, might allow specific binding of Rab proteins to a variety of divergent effector molecules involved in each of the different membrane trafficking pathways regulated by Rab [2,12].

Structural Organization of the Rab Proteins

Given their homology to other Ras superfamily GTPases, as early as the 1990s, the structural and biochemical information

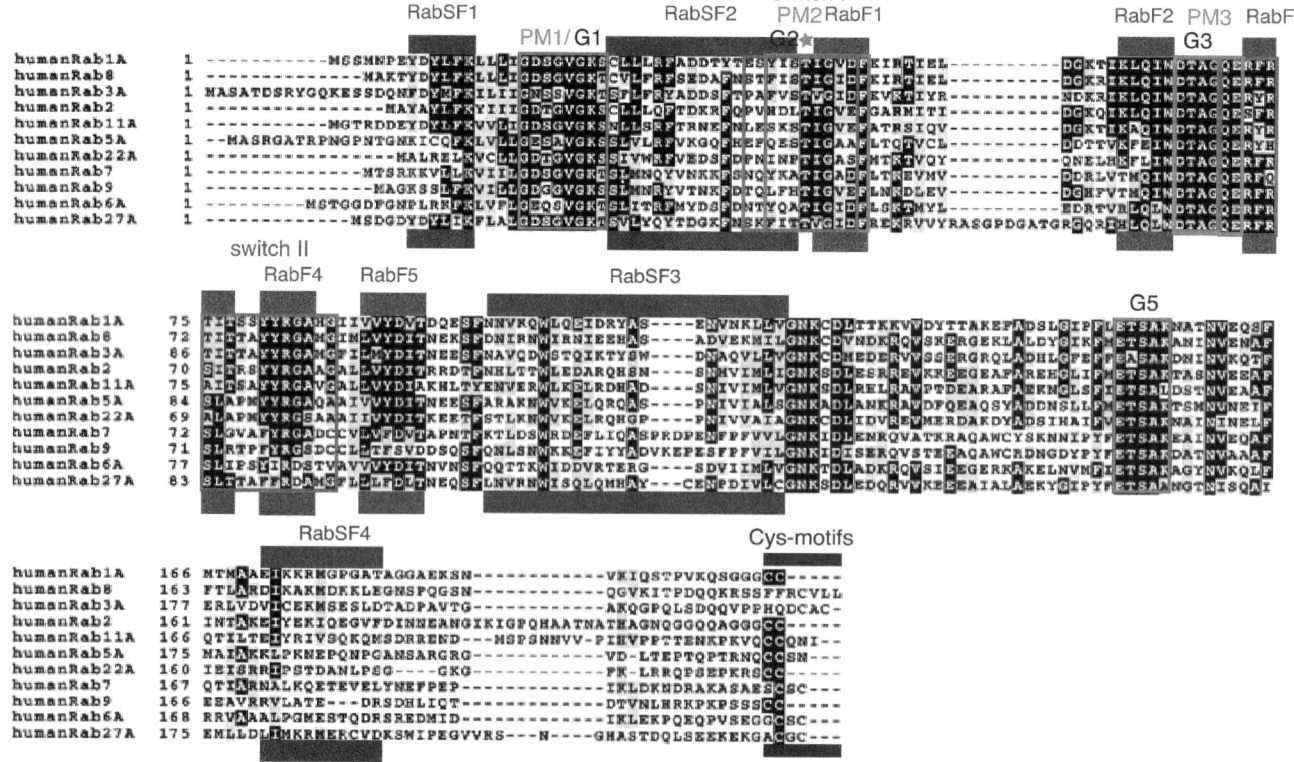

Figure 2 Rab-family and Rab subfamily domains direct Rab function. Illustrated are the location of the Rab family (RabF) and Rab subfamily (RabSF) domains in the context of nucleotide binding domains (G1–G5) and switch regions (switch I and II). These regions play an important role in defining the organization of the Rab gene family. (Adapted from Peirera-Leal and Seabra, *J. Mol. Biol.* 313, 889–901. With permission.)

available from the three-dimensional structure of the *ras* oncogene protein p21^{Ras} has been used to dissect Rab function [11]. This approach was confirmed by recent structures of Rab proteins. These now include Rab3A and Rab3A in complex with the effector Rabphilin-3A [13,14,16], Ypt51p (the Rab5 homolog) in its active GppNHp-bound conformation [17], and the yeast Sec4p (the Rab8 homolog) in both inactive GDP- and active GPPNHP-bound conformations [18]. Analysis of these structures reveals that the structure of Rab proteins is largely superimposable on the core structure *ras* oncogene despite its complete sequence divergence.

The structural core of the Rab proteins is comprised of six central β-strands in a relatively flat β-sheet that is sandwiched by three helices on either side, forming the guanine-nucleotide pocket. Comparison of the two Sec4 structures confirms that the protein undergoes a drastic conformational change in the regions called switch I and switch II which interact with effectors when Rab switches from the GDP- to the GTP-bound state [18]. Structurally, switch regions differ significantly from other Ras superfamily members. Consistent with this, mutagenesis studies have shown that the switch regions are crucial for the interaction of Rab proteins with regulatory proteins, such as GEF and GAP. Thus, Rab function recapitulates that classic *ras* oncogene GTPase paradigm.

Posttranslational Modification and Localization

To bind to membranes where they direct the assembly of protein complexes directing vesicle traffic, Rab GTPases need to be posttranslationally modified by the attachment of two 20-carbon geranylgeranyl moieties to cysteine residues located at their hypervariable carboxyl terminus [19,20]. Their consensus sequence for this posttranslational modification has been identified to be either -XXCC, -XCXC, or -CCXX where C is cysteine and X is any aliphatic amino acid [21]. The exceptions are Rab8 and Rab13, which contain only a single cysteine residue attached to a more traditional Ras-like -CAAX box. Deletion or mutation of the cysteine residues to serine eliminates the prenylation and inactivates all Rab species examined to date [22]. Prenylation requires the accessory component REP/CHM, and the enzyme complex geranylgeranyltransferase II (RabGGTase) composed of α and β subunits [23,24]. Unlike other members of the Ras superfamily that can be directly farnesylated by farnesyl transferase [19], newly synthesized Rab proteins are not recognized by the RabGGTase complex until they bind REP. Following prenylation, REP delivers the newly synthesized Rab proteins to the correct compartment through information found in their hypervariable 30–40 amino acid carboxyl terminus [25]. Indeed, each Rab protein has a very limited subcellular distribution diagnostic of its function [1,3,26]. For example, Rab1 is localized to the ER-Golgi interface in the exocytic pathway, Rab5 is found restricted to early endocytic compartments and clathrin-coated vesicles forming at the cell surface, Rab3A to synaptic vesicles, etc. The fact that 63 different Rab proteins have been identified to date suggests that there are at least this many different specialized trafficking pathways found in the differentiated cells of higher eukaryotes. As expected, many Rabs are associated with only a single tissue type, suggesting that the limited number of Rabs found in yeast (11 Rabs total) may define the rudimentary trafficking comprising the basic functions of the exocytic and endocytic pathways. In this regard, it is interesting to note that only 2 Rabs in yeast are essential proteins. Ypt1 is required for ER to Golgi transport, and Sec4 is required for Golgi to cell surface transport. We have proposed the "homing hypothesis" whereby each Rab dictates the formation of a effector complexes (the Rab interactome) that mediate a specific trafficking event in the cell, thereby ensuring high fidelity to membrane trafficking [6]. In addition, it was proposed that Rabs also control the mosaic organization of endocytic compartments by segregating cargo and transport factors into subdomains [27].

Effector Molecules: REP/CHM, GEF, Effectors (Motors/Tethers/Fusogens), GAP, and GDI

Rab proteins can interact with a diverse range of proteins [4,9]. GEFs are believed to play a critical role in the activation of Rab-GDP to Rab-GTP upon delivery to a specific membrane. Only a few exchange factors have been identified to date and they appear quite divergent indicating highly specialized function. However, it is likely that a conserved domain(s) promoting exchange will be a common feature of Rab GEFs. Once activated to the GTP-bound form and bound to membranes, Rab proteins recruit effector molecules [4,5,9]. While they are a highly heterogeneous group, they appear to fall into discrete categories in which they can be thought to function as (1) "tethers" that promote vesicle docking (such as p115, a Rab1 effector [28]; Rabphilin-3, RIM1 and 2, Rab3 effectors [29]; and EEA1, a Rab5 effector [5]), (2) SNAREs that direct vesicle fusion (SNAREs [30]), or molecular motors and their cognate receptor proteins that guide movement of membranes through the cell via the actin and microtubule cytoskeletons. A Rab6b effector is rabkinesin-6-involved microtubule binding during mitosis [31,32]. A Rab27a effector is the melanophilin/myosinVa complex involved in melanosome biogenesis and trafficking [33]. As many of the interactions between the Rab and various effector molecules could be indirect, this leads to the general principle that Rab GTPases serve as the focal point to regulate the assembly/disassembly of large macromolecular complexes to integrate the activity of these diverse components to move a vesicle from one cellular location to the next with speed and accuracy. While subfamily isoforms >80% homology (i.e., Rab3A–3D) may use evolutionarily related effectors; the high heterogeneity in effector function in more divergent Rabs emphasizes that they have become highly specialized in order to direct the specific function of a particular membrane transport step in the endocytic and exocytic pathways.

Following docking and fusion, conversion of Rab from the GTP- to the GDP-bound state is mediated by Rab-specific GAPs. Again, this appears to be a heterogeneous group,

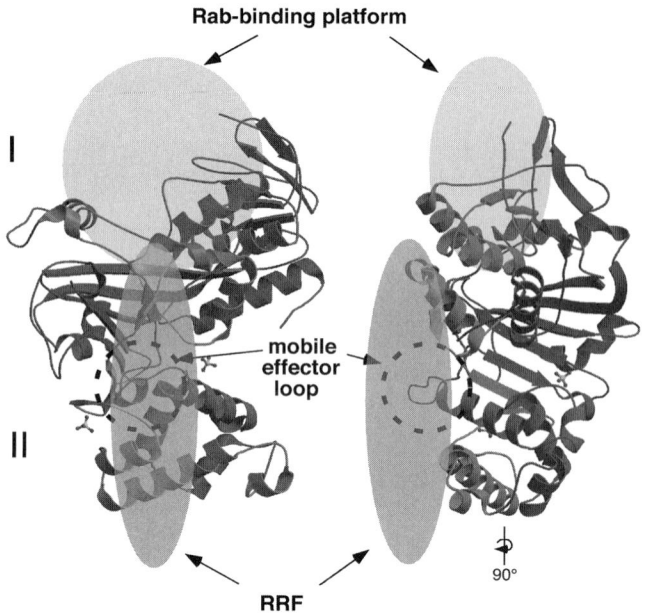

Figure 3 Structural organization of GDI, a Rab recycling factor. Shown is the two-domain organization of αGDI in which the upper domain (I) functions as a platform to bind Rab, whereas the lower domain (II) is involved in interaction with RRF, a membrane-associated recycling complex that assists GDI in removal of Rab from the membrane.

although only a few have been identified and their function in vesicle docking and fusion remains to be determined [9]. Following hydrolysis Rab-GDP must be recycled. This is the job of a common recycling protein, GDI, that interacts with all Rab GTPases studied to date [7,8]. The function of GDI has been extensively investigated using both structural and molecular approaches [34–38]. The structure of GDI shows that it is a two-domain protein—an upper domain I involved in Rab binding and a lower domain II involved in Rab extraction (Fig. 3). By binding of Rab through domain I in a reaction assisted by a membrane receptor (Rab recycling factor or RRF) which interacts with domain II [36], the prenyl groups of Rab are transferred from the lipid bilayer to the surface of GDI. Lipid binding triggers release of the Rab-GDI complex from the membrane. The cytosolic pool of the GDI-Rab complexes provides the principle source for re-recruitment of Rab to membrane. While the exact mechanism by which GDI releases Rab during delivery is unknown, it is believed to involve a guanine nucleotide dissociation factor (GDF) that displaces Rab from GDI in conjunction with interaction with the activity of Rab-specific GEFs found on the membrane [7,8].

Rab Dysfunction and Disease

Defects in intracellular trafficking can lead to a large variety of human diseases [39–41]. For example, mutation of the brain-specific guanine nucleotide dissociation inhibitor (αGDI), involved in Rab recycling, leads to X-linked mental retardation [42]. In contrast, mutation of the REP1/CHM involved in Rab prenylation leads to choroideremia, an X-linked recessive eye degeneration disease [43]. To date, only one Rab protein, Rab27, has been directly implicated in hereditary disease. Mutation of Rab27a results in Griscelli syndrome (GS) in humans and the *ashen* phenotype in mice [15]. Rab27a has been shown to play a critical role in the transport of secretory granules, such as melanosomes in melanocytes, lytic granules in cytotoxic T lymphocytes and platelet-dense granules [15]. Moreover, Rab27a might also be the trigger for the retinal degeneration in CHM, since it is present at high levels in the retinal pigment epithelium and choriocapillaris of the eye, and it is not prenylated in CHM patients where REP1/CHM is missing [44]. Because Rab proteins are the key regulators of all the membrane trafficking activity and dictate the function of highly specialized cellular trafficking pathways, we anticipate that a variety of other hereditary disorders are likely to be discovered that directly involve Rab dysfunction.

Perspective

Rab GTPases comprise a large gene family that dictates the organization of the exocytic and endocytic pathways of eukaryotic cells. Without Rabs, the subcellular organization typically found in eukaryotes would not exist. Thus, they provide a critical foundation for cell development and differentiation through their ability to organize divergent membrane trafficking complexes that dictate specificity to membrane interactions.

References

1. Nuoffer, C. and Balch, W. E. (1994). GTPases: Multifunctional molecular switches regulating vesicular traffic. *Annu. Rev. Biochem.* **63**, 949–90.
2. Peirera-Leal, J. B. and Seabra, M. (2001). Evolution of the Rab family of small GTP-binding proteins. *J. Mol. Biol.* **313**, 889–901.
3. Martinez, O. and Goud, B. (1998). Rab proteins. *Biochim. Biophys. Acta* **1404**, 101–12.
4. Pfeffer, S. R. (2001). Rab GTPases: specifying and deciphering organelle identity and function. *Trends Cell Biol.* **11**, 487–91.
5. Zerial, M. and McBride, H. (2001). Rab proteins as membrane organizers. *Mol. Cell. Biol.* **2**, 107–119.
6. Moyer, B. D., Allan, B. B., and Balch, W. E. (2001). Rab1 Interaction with a GM130 Effector Complex Regulates COPII Vesicle cis-Golgi Tethering. *Traffic* **2**, 268–76.
7. Wu, S. K., Zeng, K., Wilson, I. A., and Balch, W. E. (1996). Structural insights into the function of the Rab GDI superfamily. *Trends Biochem. Sci.* **21**, 472–6.
8. Alory, C. and Balch, W. E. (2001). Organization of the Rab-GDI/REP superfamily: functional basis for choroideremia disease. *Traffic* **2**, 532–543.
9. Segev, N. (2001). Ypt and Rab GTPases: insight into functions through novel interactions. *Curr. Opin. Cell Biol.* **13**, 500–11.
10. Haubruck, H., Prange, R., Vorgias, C., and Gallwitz, D. (1989). The ras-related mouse ypt1 protein can functionally replace the YPT1 gene product in yeast. *EMBO J.* **8**, 1427–1432.
11. Bourne, H. R., Sanders, D. A., and McCormick, F. (1991). The GTPase superfamily: conserved structure and molecular mechanism. *Nature* **349**, 117–127.

12. Pereira-Leal, J. B. and Seabra, M. C. (2000). The mammalian Rab family of small GTPases: definition of family and subfamily sequence motifs suggests a mechanism for functional specificity in the Ras superfamily. *J. Mol. Biol.* **301**, 1077–87.
13. Constantinescu, A. T., Rak, A., Alexandrov, K., Esters, H., Goody, R. S., and Scheidig, A. J. (2002). Rab-Subfamily-Specific Regions of Ypt7p Are Structurally Different from Other RabGTPases. *Structure (Camb.)* **10**, 569–79.
14. Ostermeier, C. and Brunger, A. T. (1999). Structural basis of Rab effector specificity: Crystal structure of the small G protein Rab3A complexed with the effector domain of rabphilin-3A. *Cell* **96**, 363–74.
15. Seabra, M. C., Mules, E. H., and Hume, A. N. (2002). Rab GTPases, intracellular traffic and disease. *Trends Mol. Med.* **8**, 23–30.
16. Dumas, J. J., Zhu, Z., Connolly, J. L., and Lambright, D. G. (1999). Structural basis of activation and GTP hydrolysis in Rab proteins. *Struct. Fold Des.* **7**, 413–23.
17. Esters, H., Alexandrov, K., Constantinescu, A. T., Goody, R. S., and Scheidig, A. J. (2000). High-resolution crystal structure of S. cerevisiae Ypt51(DeltaC15)- GppNHp, a small GTP-binding protein involved in regulation of endocytosis. *J. Mol. Biol.* **298**, 111–21.
18. Stroupe, C. and Brunger, A. T. (2000). Crystal structures of a Rab protein in its inactive and active conformations. *J. Mol. Biol.* **304**, 585–98.
19. Casey, P. J. and Seabra, M. C. (1996). Protein prenyltransferases. *J. Biol. Chem.* **271**, 5289–92.
20. Seabra, M. C. (1998). Membrane association and targeting of prenylated Ras-like GTPases. *Cell Signal.* **10**, 167–72.
21. Farnsworth, C. C., Seabra, M., Ericsson, L. H., Gelb, M. H., and Glomset, J. A. (1994). Rab geranylgeranyl transferase catalyzes the geranylgeranylation of adjacent proteins in the small GTPases Rab1A, Rab3A and Rab5A. *Proc. Natl. Acad. Sci. USA* **91**, 11963–11967.
22. Khosravi-Far, R., Lutz, R. J., Cox, A. D., Conroy, L., Bourne, J. R., Sinensky, M., Balch, W. E., Buss, J. E., and Der, C. J. (1991). Isoprenoid modification of rab proteins terminating in CC or CXC motifs. *Proc. Natl. Acad. Sci. USA* **88**, 6264–6268.
23. Seabra, M. C., Goldstein, J. L., Sudhof, T. C., and Brown, M. S. (1992). Rab geranylgeranyltransferase, a multisubunit enzyme that prenylates GTP-binding proteins terminating in cys-x-cys or cys-cys. *J. Biol. Chem.* **267**, 14497–14503.
24. Zhang, H., Seabra, M. C., and Deisenhofer, J. (2000). Crystal structure of Rab geranylgeranyltransferase at 2.0 A resolution. *Struct. Fold Des.* **8**, 241–51.
25. Chavrier, P., Gorvel, J. P., Stelzer, E., Simons, K., Gruenberg, J., and Zerial, M. (1991). Hypervariable C-terminal domain of rab proteins acts as a targeting signal. *Nature* **353**, 769–72.
26. Beraud-Dufour, S. and Balch, W. (2002). A journey through the exocytic pathway. *J. Cell Sci.* **115**, 1779–80.
27. Miaczynska, M. and Zerial, M. (2002). Mosaic organization of the endocytic pathway. *Exp. Cell Res.* **272**, 8–14.
28. Allan, B. B., Moyer, B. D., and Balch, W. E. (2000). Rab1 recruitment of p115 into a cis-SNARE complex: programming budding COPII vesicles for fusion [see comments]. *Science* **289**, 444–8.
29. Takai, Y., Sasaki, T., and Matozaki, T. (2001). Small GTP-binding proteins. *Physiol. Rev.* **81**, 153–208.
30. Chen, Y. A. and Scheller, R. H. (2001). SNARE-mediated membrane fusion. *Nat. Rev. Mol. Cell Biol.* **2**, 98–106.
31. Fontijn, R. D., Goud, B., Echard, A., Jollivet, F., van Marle, J., Pannekoek, H., and Horrevoets, A. J. (2001). The human kinesin-like protein RB6K is under tight cell cycle control and is essential for cytokinesis. *Mol. Cell. Biol.* **21**, 2944–55.
32. Echard, A., Jollivet, F., Martinez, O., Lacapere, J. J., Rousselet, A., Janoueix-Lerosey, I., and Goud, B. (1998). Interaction of a Golgi-associated kinesin-like protein with Rab6. *Science* **279**, 580–5.
33. Strom, M., Hume, A. N., Tarafder, A. K., Barkagianni, E., and Seabra, M. C. (2002). A family of Rab27-binding proteins: Melanophilin links Rab27a and myosin Va function in melanosome transport. *J. Biol. Chem.* **29**, 29.
34. Alory, C. and Balch, W. E. (2000). Molecular Basis for Rab Prenylation. *J. Cell Biol.* **150**, 89–103.
35. Luan, P., Balch, W. E., Emr, S. D., and Burd, C. G. (1999). Molecular dissection of guanine nucleotide dissociation inhibitor function in vivo. Rab-independent binding to membranes and role of Rab recycling factors. *J. Biol. Chem.* **274**, 14806–17.
36. Luan, P., Heine, A., Moyer, B. D., Greasely, S. E., Kuhn, P., Balch, W. E., and Wilson, I. A. (2000). A new functional domain of guanine nucleotide dissociation inhibitor (α-GDI) involved in Rab recycling. *Traffic* **1**, 270–281.
37. Schalk, I., Zeng, K., Wu, S.-K., Stura, E., Matteson, J., Huang, M., Tandon, A., Wilson, I., and Balch, W. E. (1996). Structure and mutational analysis of Rab GDP-dissociation inhibitor. *Nature* **381**, 42–48.
38. Wu, S. K., Luan, P., Matteson, J., Zeng, K., Nishimura, N., and Balch, W. E. (1998). Molecular role for the Rab binding platform of guanine nucleotide dissociation inhibitor in endoplasmic reticulum to Golgi transport. *J. Biol. Chem.* **273**, 26931–8.
39. Aridor, M. and Balch, W. E. (1999). Integration of endoplasmic reticulum signaling in health and disease. *Nat. Med.* **5**, 745–51.
40. Aridor, M. and Hannan, L. A. (2000). Traffic jam: a compendium of human diseases that affect intracellular transport processes. *Traffic* **1**, 836–51.
41. Olkkonen, V. M. and Ikonen, E. (2000). Genetic defects of intracellular-membrane transport. *N. Engl. J. Med.* **343**, 1095–104.
42. D'Adamo, P., Menegon, A., Lo Nigro, C., Grasso, M., Gulisano, M., Tamanini, F., Bienvenu, T., Gedeon, A. K., Oostra, B., Wu, S. K., Tandon, A., Valtorta, F., Balch, W. E., Chelly, J., and Toniolo, D. (1998). Mutations in GDI1 are responsible for X-linked non-specific mental retardation. *Nat. Genet.* **19**, 134–9.
43. Seabra, M. C., Brown, M. S., Slaughter, C. A., Sudhof, T. C., and Goldstein, J. L. (1992). Purification of component A of Rab geranylgeranyltransferase: Possible identity with the choroideremia gene product. *Cell* **70**, 1049–1057.
44. Seabra, M. C., Ho, Y. K., and Anant, J. S. (1995). Deficient geranylgeranylation of Ram/Rab27 in choroideremia. *J. Biol. Chem.* **270**, 24420–24427.

CHAPTER 236

Cellular Roles of the Ran GTPase

Jomon Joseph and Mary Dasso
*Laboratory of Gene Regulation, National Institute of Child Health and Human Development,
National Institutes of Health, Bethesda, Maryland*

Introduction

Ran is an abundant nuclear GTPase that has been extensively characterized for its role in nuclear transport. The Ran pathway is involved in spindle assembly and postmitotic nuclear assembly, independent of its role in nuclear transport. The distribution of Ran's nucleotide exchange factor (RCC1) and GTPase activating protein (RanGAP1) leads to the formation of gradients of Ran-GTP within the cell. These gradients provide spatial cues that direct Ran-dependent processes. These cues are monitored through a common set of Ran-GTP binding effector proteins that were originally characterized as nuclear transport receptors. The past year has seen a considerable deepening in our understanding of the molecular features of this pathway and of how gradients of Ran-GTP are built and utilized. In this chapter, we relate these new developments to Ran's role in cellular processes.

Introduction to the Ran Pathway

Ran's intrinsic rates of nucleotide hydrolysis and exchange are slow. These reactions are accelerated *in vivo* by RCC1, a guanine nucleotide exchange factor (GEF) [1], and by RanGAP, a GTPase activating protein [2]. A family of proteins that share homologous Ran-GTP-binding domains are also important *in vivo* regulators of Ran [3]. RanBP1 is the best-understood member of this family. RanBP1 does not have intrinsic GAP activity, but it increases the rate of RanGAP1-mediated nucleotide hydrolysis by roughly an order of magnitude *in vitro* [4]. Moreover, it is essential *in vivo* for RanGAP1-mediated hydrolysis of Ran-GTP bound to nuclear transport receptors [5].

There are four notable aspects of the Ran pathway that distinguish it from many other members of the Ras superfamily. First, Ran is not isoprenylated, and is thus freely soluble. Second, Ran is an extremely abundant protein (roughly 0.4% of total cellular protein [6]), as are many of its interacting partners. Third, there is an asymmetric distribution of Ran's regulators throughout interphase [7]. RCC1 is chromatin bound and hence nuclear, whereas RanGAP1 and RanBP1 are predominantly cytoplasmic. This causes a steep gradient of RanGTP across the nuclear membrane [8,9]. Fourth, as mentioned above, Ran-GTP binds to a family of proteins that have a common structural motif (RanBD) first found in RanBP1. RanBP2, another Ran binding protein with four RanBD domains, is localized to the cytoplasmic side of the nuclear pore complex (NPC) [10]. The role of RanBP2 in promoting nucleotide hydrolysis has not been clearly demonstrated. A third member of this family, RanBP3, acts as a cofactor for the formation of complexes destined for nuclear export [11,12].

Structural Analysis of Ran Pathway Components

Several points emerging from structural studies on Ran and its regulators are notable. First, the structure of Ran itself is dramatically regulated by nucleotide binding, particularly in the switch I region [13]. The orientation of the switch I region precludes association between Ran-GDP and transport receptors [14], providing a molecular rationale for the specificity for transport receptors in binding to Ran-GTP only. Second, the Ran pathway shows an interesting divergence from other Ras-like GTPases with respect to its mechanism of GAP-mediated hydrolysis [15]. RanGAP consists of 11 leucine rich repeats of 28 to 37 residues each, forming a crescent [16]. While it was widely assumed that GAP-mediated hydrolysis of all ras family GTPases would require an arginine residue provided by the GAP protein, this is not the case for Ran [15]. Structures of a Ran-RanBP1-RanGAP ternary complex in the ground-and in the transition-state show that RanGAP does not provide a catalytic arginine. Rather, Ran alone provides the machinery for GTP hydrolysis, with tyrosine 39 of Ran replacing the arginine provided by other GAPs.

Third, RanBP1 is essential *in vivo* because it dissociates Ran-GTP from transport receptors to permit access of RanGAP [5]. Transport receptor binding to Ran-GTP blocks RanGAP-mediated hydrolysis, because the interaction sites for RanGAP and receptors overlap [14,15]. However, binding of Ran to transport receptors does not inhibit association between Ran-GTP and RanBP1 family members because they bind on different surfaces of Ran [14]. RanBD domains make extensive contacts that have been described as a "molecular embrace" of Ran [14], causing a shift in the Ran's acidic C terminus, and promoting Ran's release from bound transport receptors. Although RanBP1 can activate GAP-mediated hydrolysis *in vitro* about an order of magnitude [4], RanBP1 does not directly contact RanGAP within the RanGAP-Ran-GTP-RanBP1 ternary complex [15]. It appears that RanBP1 exerts a positive influence on GAP-mediated hydrolysis by shifting the C terminus of Ran, thereby facilitating interactions between Ran and RanGAP.

Fourth, RCC1 is a seven-bladed propeller protein [17], similar in structure to the WD40 propeller structure of the GEF for the G-β heterotrimeric GTPase. RCC1's interaction interface with Ran resides in the loops between the propeller blades on one face of the protein [18]. RCC1 associates with chromatin through other proteins [19]. It has recently been demonstrated that the interaction of RCC1 with chromatin is mediated by histones [20]. RCC1 binds directly to mononucleosomes and to histones H2A and H2B [20], resulting in a modest stimulation of the catalytic activity of RCC1.

Ran's Role in Nuclear Transport

The requirement for Ran in nuclear transport has been extensively studied (Fig. 1) (reviewed in reference [7]). Ran-dependent nuclear transport is mediated by a family of transport receptors that are able to freely pass through the NPC in an energy-independent manner. Importin β was the first member of this family to be described, and all members of this family (both import and export receptors) bind to Ran-GTP. Import receptors bind to their cargo in the cytoplasm, where Ran-GTP is absent, and permit cargo translocation through the NPC. In the nucleus, RCC1-generated Ran-GTP binds to the import receptors and releases the cargo. Conversely, export receptors and their cargo associate within the nucleus as complexes containing Ran-GTP. After these complexes translocate to the cytoplasm, RanGAP1 induces GTP hydrolysis thereby causing the dissociation of the complexes. A small Ran-GDP-binding protein, NTF2, re-imports Ran-GDP back into the nucleus after each round of transport, where regeneration of Ran-GTP is achieved by RCC1.

Within this model, Ran-GTP concentration regulates receptor loading and unloading in a manner that is appropriate to the nucleus or cytosol. A number of interesting variations have been reported recently upon the way in which different receptors utilize the Ran-GTP gradient. First, while it is essential in this model that each transport receptor can only carry a particular cargo in one direction, there is no requirement that receptors cannot carry other cargos in the opposite direction. Indeed, bidirectional receptors have recently been reported in yeast [21] and mammals [22].

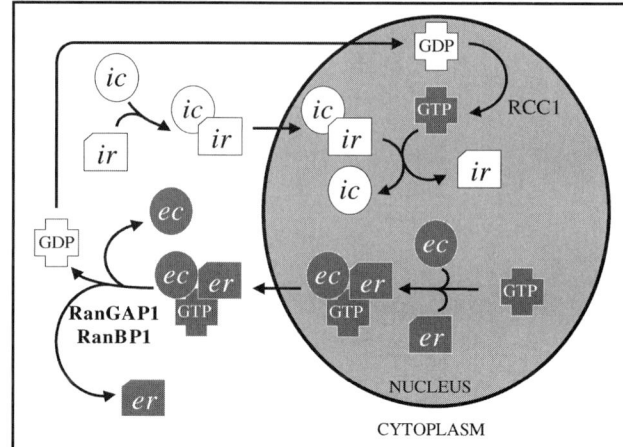

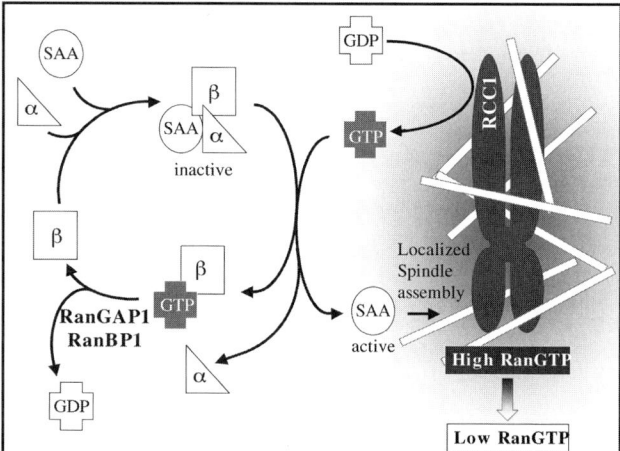

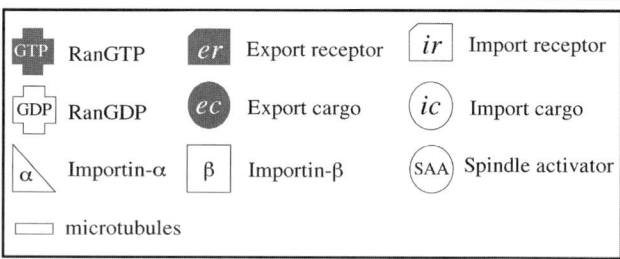

Figure 1 Ran regulates nuclear trafficking through the association and dissociation of transport complexes during interphase (upper panel). The compartmentalization of RanGAP1, RanBP1, and RCC1 cause an asymmetric distribution of Ran-GTP across the nuclear envelope. Import receptors and cargo form complexes in the cytosol and transit across the NPC. In the nucleus, Ran-GTP binds to the import receptors and causes cargo release. Import receptors return to the cytosol in association with Ran-GTP. RanGAP1, and RanBP1 hydrolyze Ran-GTP to Ran-GDP, promoting receptor recycling. Export receptors and Ran-GTP bind to export cargo in the nucleus. These complexes transit to the cytosol, where they dissociate after RanGAP1-mediated Ran-GTP hydrolysis. Elevated Ran-GTP levels near chromosomes promote mitotic spindle assembly (lower panel). Importin α and importin β inhibit spindle assembly activators (SAA) at low Ran-GTP concentrations in regions distant from chromosomes. Near chromosomes, Ran-GTP concentrations are higher. Ran-GTP binds to importin β, disrupting inhibitory complexes and allowing full SAA activity.

Second, many receptors interact with at least some of their cargo through adaptor proteins [7]. The best-characterized case of adaptor utilization is the import of substrates bearing classical nuclear localization signals (NLS) [7]. The Importin α adaptor protein binds to NLS sequences. Importin β associates with the importin α-NLS complex and promotes its transit through the NPC. Importin α is re-exported in association with Ran-GTP and CAS, an importin β-family export receptor. Two molecules of GTP are thus hydrolyzed during the import cycle of each NLS-bearing substrate, allowing substrates accumulation against a steeper nuclear-cytoplasmic concentration gradient through the expenditure of greater free energy.

Third, accessory proteins control the loading of cargo to receptors in many instances [7]. For example, the Crm1 export receptor binds RanBP3, a nuclear RanBD-containing protein [12,23]. RanBP3 does not directly bind to export substrates, but it increases the affinity of Crm1 for both Ran-GTP and export cargo. RanBP3 also associates with RCC1 in a manner that is stimulated by Ran, and activates RCC1s as Ran GEF [11]. Moreover, RanBP3 promotes the association of Crm1 and RCC1, perhaps acting as a scaffold to coordinate the loading of Ran-GTP onto Crm1. RanBP3 inhibits the association of unloaded Crm1 to the NPC in a manner that is relieved by Ran-GTP [12], suggesting that it permits Crm1 association to the NPC only after export complex assembly is complete.

Finally, import cargo unloading can be controlled by events in addition to Ran-GTP binding [24,25]. For example, DNA containing binding sites for the TBP transcription factor stimulates Ran-GTP-mediated dissociation of TBP from its import receptor (Kap114p) [24]. This stimulation suggests that TBP is released from Kap114p only when it finds its target DNA sequences within the nucleus. Import receptors may thus direct not only nuclear import but also intranuclear targeting or assembly of particular cargos into macromolecular complexes.

Ran's Function in Mitotic Progression

A number of observations have implicated Ran in regulating the onset of mitosis. When arrested in S phase and then shifted to the restrictive temperature, a mutant Hamster cell line with a temperature-sensitive allele of RCC1 (tsBN2 cells) progresses into mitosis prematurely, accompanied by nuclear envelope (NE) breakdown, precocious chromosome condensation, and activation of p34cdk1/cyclin B kinase [26]. These observations and others have shown that tsBN2 cells do not arrest appropriately in S phase in response to unreplicated DNA [27]. Moreover, mutant Ran proteins can block onset of mitosis in the presence or absence of nuclei in *Xenopus* cycling extracts, which would otherwise alternate between interphase and mitosis [28,29]. The molecular events whereby Ran regulates mitotic onset in metazoans remain unclear, and there is a lack of strong evidence from fission and budding yeast that Ran plays an analogous role for regulation of mitosis in either of those organisms [30].

Ran's Function in Spindle Assembly

Ran regulates spindle assembly in a manner that is independent of its nuclear transport function [31–35]. Spindle assembly is severely defective when Ran-GTP levels are lowered in *Xenopus* cytostatic factor (CSF) arrested egg extract, a mitotic system that is devoid of intact nuclei. Under these conditions, spindles are disorganized with low densities of microtubules (MTs) [33,35]. Conversely, increased levels of Ran-GTP in CSF extracts cause massive polymerization of MTs in a manner that does not require chromosomes or centrosomes [32,33,35]. It had been previously found that mitotic chromosomes can locally stabilize MTs, probably through the action of a diffusible MT-stabilizing factor produced by a chromatin-associated enzyme (reviewed in reference [36]). Since RCC1 binds to chromatin [37], it was natural to speculate from these results that Ran-GTP could play a role in the localized stabilization of mitotic MT by chromosomes. Consistent with this idea, it has recently been directly demonstrated through fluorescence resonance energy transfer (FRET) experiments that Ran-GTP concentrations are elevated in the vicinity of mitotic chromosomes [8].

The molecular mechanism by which Ran carries out its function in spindle assembly is closely related to its role in nuclear transport. Importin α/β bind and inhibit spindle assembly factors, such as the motor accessory proteins TPX2 and NuMA [38–40]. Ran-GTP near chromosomes destabilizes these inhibitory complexes, thereby allowing spindle assembly factors to function locally (Fig. 1). At a distance from chromosomes, Ran-GTP would presumably undergo nucleotide hydrolysis and inhibition would be restored. In reality, it is likely that the location of nucleotide hydrolysis may also be very important for spindle function, since vertebrate RanGAP1 is localized to the spindle in a highly regulated manner [41]. Another key aspect of this model is that importin α and β also promote the nuclear localization of spindle factors, ensuring that they are not inappropriately active on MT in interphase cytosol.

Since numerous other proteins involved in spindle assembly are nuclear during interphase, it is likely that additional targets are regulated in a similar manner. This notion is also supported by the fact that Ran regulates several different aspects of spindle assembly, such as the frequency of transition from shrinkage to growth of MTs (e.g., MT dynamics) [42,43], the capacity of centrosomes to nucleate MTs [42], and the behavior of other motor proteins or their accessory subunits [43,44]. Taken together, these data suggest that Ran may have acted in a multifaceted manner to coordinate assembly of spindles with respect to the mitotic chromosomes.

Ran's Role in Postmitotic Nuclear Assembly

Studies in *Xenopus* egg extracts have suggested that Ran is required for postmitotic NE assembly [45,46]. Nonhydrolyzable forms of Ran-GTP and loss of RCC1 activity or depletion of Ran all block early events of nuclear fusion during NE assembly in egg extracts [45]. Addition of Ran-GTP reverses this inhibition, leading to the conclusion that Ran has an important role in re-formation of NE at the end of mitosis. Both nucleotide exchange and hydrolysis on Ran are required for this function to be fulfilled [45]. Remarkably, GST-Ran bound to beads can assemble structures resembling NE in egg extracts [46]. These structures contain NPC and are capable of nuclear transport. RCC1-bound beads do not assemble NE in this assay. Zhang and Clarke [46,47] have argued that Ran actually precedes RCC1 in binding to chromatin and can act as an RCC1-independent signal for NE assembly.

In an exciting set of new reports [48], it has been shown that importin β is required for NE assembly induced by Ran on beads. Concentration of importin β on beads is sufficient to induce NE assembly in *Xenopus* egg extracts in a manner similar to Ran-coated beads, while beads bound with other transport receptors did not have this property. Moreover, the function of importin β in NE assembly is disrupted by a mutation that decreases the affinity of importin β for nucleoporins, but not by a mutation that disrupts the interactions of importin β with importin α. In contrast to the mechanisms of nuclear transport and spindle assembly, it thus appears that importin β functions in NE assembly by recruiting NPC components rather than by importin α-dependent interactions with cargo protein.

Conclusions

Ran regulates nuclear transport, cell cycle progression, spindle assembly, and postmitotic NE assembly, suggesting a wide role for Ran in coordinating events during the cell cycle. These facets of Ran are mechanistically linked in two ways. First, in all cases where we have a rudimentary molecular understanding of Ran function, it appears that Ran-GTP gradients provide spatial cues that serve to indicate the localization of the chromatin. This notion is particularly well supported for nuclear transport and spindle assembly. Second, gradients of Ran are interpreted through a common set of Ran-GTP-binding effectors, with the importin β protein playing a particularly prominent role.

References

1. Bischoff, F. R. and Ponstingl, H. (1991). *Nature* **354**(6348), 80–82.
2. Bischoff, F. R., Krebber, H., Kempf, T., Hermes, I., and Ponstingl, H. (1995). *Proc. Natl. Acad. Sci. USA* **92**(5), 1749–1753.
3. Beddow, A. L., Richards, S. A., Orem, N. R., and Macara, I. G. (1995). *Proc. Natl. Acad. Sci. USA* **92**(8), 3328–3332.
4. Bischoff, F. R., Krebber, H., Smirnova, E., Dong, W., and Ponstingl, H. (1995). *EMBO J.* **14**(4), 705–715.
5. Bischoff, F. R. and Gorlich, D. (1997). *FEBS Lett.* **419**(2–3), 249–254.
6. Bischoff, F. R. and Ponstingl, H. (1995). *Methods Enzymol.* **257**, 135–144.
7. Macara, I. G. (2001). *Microbiol. Mol. Biol. Rev.* **65**(4), 570–594, table of contents.
8. Kalab, P., Weis, K., and Heald, R. (2002). *Science* **295**(5564), 2452–2456.
9. Smith, A. E., Slepchenko, B. M., Schaff, J. C., Loew, L. M., and Macara, I. G. (2002). *Science* **295**(5554), 488–491.
10. Yokoyama, N., Hayashi, N., Seki, T., Pante, N., Ohba, T., Nishii, K., Kuma, K., Hayashida, T., Miyata, T., Aebi, U. *et al.* (1995). *Nature* **376**(6536), 184–188.
11. Nemergut, M. E., Lindsay, M. E., Brownawell, A. M., and Macara, I. G. (2002). *J. Biol. Chem.* **3**, 3.
12. Lindsay, M. E., Holaska, J. M., Welch, K., Paschal, B. M., and Macara, I. G. (2001). *J. Cell Biol.* **153**(7), 1391–1402.
13. Scheffzek, K., Klebe, C., Fritz-Wolf, K., Kabsch, W., and Wittinghofer, A. (1995). *Nature* **374**(6520), 378–381.
14. Vetter, I. R., Arndt, A., Kutay, U., Gorlich, D., and Wittinghofer, A. (1999). *Cell* **97**(5), 635–646.
15. Seewald, M. J., Korner, C., Wittinghofer, A., and Vetter, I. R. (2002). *Nature* **415**(6872), 662–666.
16. Hillig, R. C., Renault, L., Vetter, I. R., Drell, T. T., Wittinghofer, A., and Becker, J. (1999). *Mol. Cell* **3**(6), 781–791.
17. Renault, L., Nassar, N., Vetter, I., Becker, J., Klebe, C., Roth, M., and Wittinghofer, A. (1998). *Nature* **392**(6671), 97–101.
18. Azuma, Y., Renault, L., Garcia-Ranea, J. A., Valencia, A., Nishimoto, T., and Wittinghofer, A. (1999). *J. Mol. Biol.* **289**(4), 1119–1130.
19. Seino, H., Hisamoto, N., Uzawa, S., Sekiguchi, T., and Nishimoto, T. (1992). *J. Cell Sci.* **102**(Pt 3), 393–400.
20. Nemergut, M. E., Mizzen, C. A., Stukenberg, T., Allis, C. D., and Macara, I. G. (2001). *Science* **292**(5521), 1540–1543.
21. Yoshida, K. and Blobel, G. (2001). *J. Cell Biol.* **152**(4), 729–740.
22. Mingot, J. M., Kostka, S., Kraft, R., Hartmann, E., and Gorlich, D. (2001). *EMBO J.* **20**(14), 3685–3694.
23. Englmeier, L., Fornerod, M., Bischoff, F. R., Petosa, C., Mattaj, I. W., and Kutay, U. (2001). *EMBO Rep.* **2**(10), 926–932.
24. Pemberton, L. F., Rosenblum, J. S., and Blobel, G. (1999). *J. Cell Biol.* **145**(7), 1407–1417.
25. Senger, B., Simos, G., Bischoff, F. R., Podtelejnikov, A., Mann, M., and Hurt, E. (1998). *EMBO J.* **17**(8), 2196–2207.
26. Nishitani, H., Ohtsubo, M., Yamashita, K., Iida, H., Pines, J., Yasudo, H., Shibata, Y., Hunter, T., and Nishimoto, T. (1991). *EMBO J.* **10**(6), 1555–1564.
27. Nishijima, H., Seki, T., Nishitani, H., and Nishimoto, T. (2000). *Prog. Cell Cycle Res.* **4**, 145–156.
28. Kornbluth, S., Dasso, M., and Newport, J. (1994). *J. Cell Biol.* **125**(4), 705–719.
29. Clarke, P. R., Klebe, C., Wittinghofer, A., and Karsenti, E. (1995). *J. Cell Sci.* **108**, 1217–1225.
30. Sazer, S. and Dasso, M. (2000). *J. Cell Sci.* **113**, 1111–1118.
31. Zhang, C., Hughes, M., and Clarke, P. R. (1999). *J. Cell Sci.* **112**, 2453–61.
32. Wilde, A. and Zheng, Y. (1999). *Science* **284**(5418), 1359–1362.
33. Kalab, P., Pu, R. T., and Dasso, M. (1999). *Curr. Biol.* **9**(9), 481–484.
34. Carazo-Salas, R. E., Guarguaglini, G., Gruss, O. J., Segref, A., Karsenti, E., and Mattaj, I. W. (1999). *Nature* **400**(6740), 178–181.
35. Ohba, T., Nakamura, M., Nishitani, H., and Nishimoto, T. (1999). *Science* **284**(5418), 1356–1358.
36. Andersen, S. S. (1999). *Bioessays* **21**(1), 53–60.
37. Ohtsubo, M., Okazaki, H., and Nishimoto, T. (1989). *J. Cell Biol.* **109**, 1389–1397.
38. Wiese, C., Wilde, A., Moore, M. S., Adam, S. A., Merdes, A., and Zheng, Y. (2001). *Science* **291**(5504), 653–656.
39. Nachury, M. V., Maresca, T. J., Salmon, W. C., Waterman-Storer, C. M., Heald, R., and Weis, K. (2001). *Cell* **104**, 95–106.
40. Gruss, O. J., Carazo-Salas, R. E., Schatz, C. A., Guarguaglini, G., Kast, K., Wilm, M., Le Bot, N., Vernos, I., Karsenti, E., and Mattaj, I. W. (2001). *Cell* **104**, 83–93.
41. Joseph, J., Tan, S. H., Karpova, T. S., McNally, J. G., and Dasso, M. (2002). *J. Cell Biol.* **156**(4), 595–602.
42. Carazo-Salas, R. E., Gruss, O. J., Iain, W., Mattaj, I. W., and Karsenti, E. (2001). *Nat. Cell Biol.* **3**, 228–234.
43. Wilde, A., Lizarraga, S. B., Zhang, L., Wiese, C., Gliksman, N. R., Walczak, C. E., and Zheng, Y. (2001). *Nat. Cell Biol.* **3**(3), 221–227.

44. Ems-McClung, S. C., Zheng, Y., and Walczak, C. (2001). *Mol. Biol. Cell.* **12:S**(Suppl.), 181a.
45. Hetzer, M., Bilbao-Cortes, D., Walther, T. C., Gruss, O. J., and Mattaj, I. W. (2000). *Mol. Cell* **5**(6), 1013–1024.
46. Zhang, C. and Clarke, P. R. (2000). *Science* **288**(5470), 1429–1432.
47. Zhang, C. and Clarke, P. R. (2001). *Curr. Biol.* **11**(3), 208–212.
48. Zhang, C., Hutchins, J. R., Muhlhausser, P., Kutay, U., and Clarke, P. R. (2002). *Curr. Biol.* **12**(6), 498–502.

CHAPTER 237

Rho Proteins and Their Effects on the Actin Cytoskeleton

Anja Schmidt[1] and Alan Hall[2]
[1]MRC Laboratory for Molecular Cell Biology,
CRC Oncogene and Signal Transduction Group and
[2]Department of Biochemistry and Molecular Biology,
University College London, London, United Kingdom

Introduction

Rho GTPases form a discrete family within the Ras superfamily of small GTPases, which currently comprises 23 members, in mammalian cells [1]. They play an important role in controlling many cellular activities including transcription, cell cycle progression, membrane traffic, transformation, and apoptosis, but they are best known for regulating the organization of the actin cytoskeleton [2,3]. The actin cytoskeleton is a highly dynamic network of actin polymers that controls many aspects of cell behavior, including morphology, adhesion, migration, and phagocytosis. This chapter summarizes our current knowledge on the control of the actin cytoskeleton by Rho GTPases.

Effects of Rho GTPases on the Actin Cytoskeleton

The role of Rho, Rac, and Cdc42 in controlling actin cytoskeleton organization was first discovered in quiescent Swiss3T3 fibroblasts, where it was found that the activation of each GTPase induces distinct filamentous actin structures in response to extracellular stimuli (Fig. 1). Activation of Rho by lysophosphatidic acid (LPA), for example, causes the formation of stress fibers (contractile bundles of actin:myosin filaments traversing the cell and ending in focal adhesions) [4], activation of Rac by platelet-derived growth factor (PDGF) or insulin results in the formation of lamellipodia (membrane protrusions driven by actin polymerization that can detach and fall backward onto the cell body creating membrane ruffles) [5], and activation of Cdc42 by bradykinin leads to the formation of filopodia (thin finger-like protrusions consisting of actin filament bundles) [6,7]. In addition, crosstalk can occur between the three GTPases: Cdc42 can activate Rac and stimulate membrane ruffling, and Rac can activate Rho to form stress fibers [5–7].

The effects of Rho, Rac, and Cdc42 on the actin cytoskeleton are not, however, restricted to fibroblasts, and over the last decade Rho GTPases have been implicated in many actin-dependent events in a wide variety of different cell types. Often, Rho GTPases function in a coordinated manner to promote complex changes to the actin cytoskeleton, while in other cases Rho GTPases can have opposing effects with Rac and Cdc42 causing membrane extension and Rho causing membrane retraction.

Rho, Rac, and Cdc42 also control the assembly of structures that are intimately associated with the actin cytoskeleton like cell-matrix and cell-cell contacts. Rho, for example, is required for the assembly of classical focal adhesions, while Rac and Cdc42 induce so-called focal contacts found at the leading edge of migrating cells [4,7,8]. Rho, Rac, and Cdc42 have also been reported to participate in the formation of cadherin-dependent cell-cell adhesion complexes [9], as well as tight junction complexes in epithelial cells [10]. Whether these effects on surface adhesion complexes are direct or mediated through the actin cytoskeleton is not entirely clear.

Cell Migration

Rho GTPases generate the protrusive forces needed for forward movement of migrating cells. Rac, in particular, is essential for the protrusion of lamellipodia at the leading

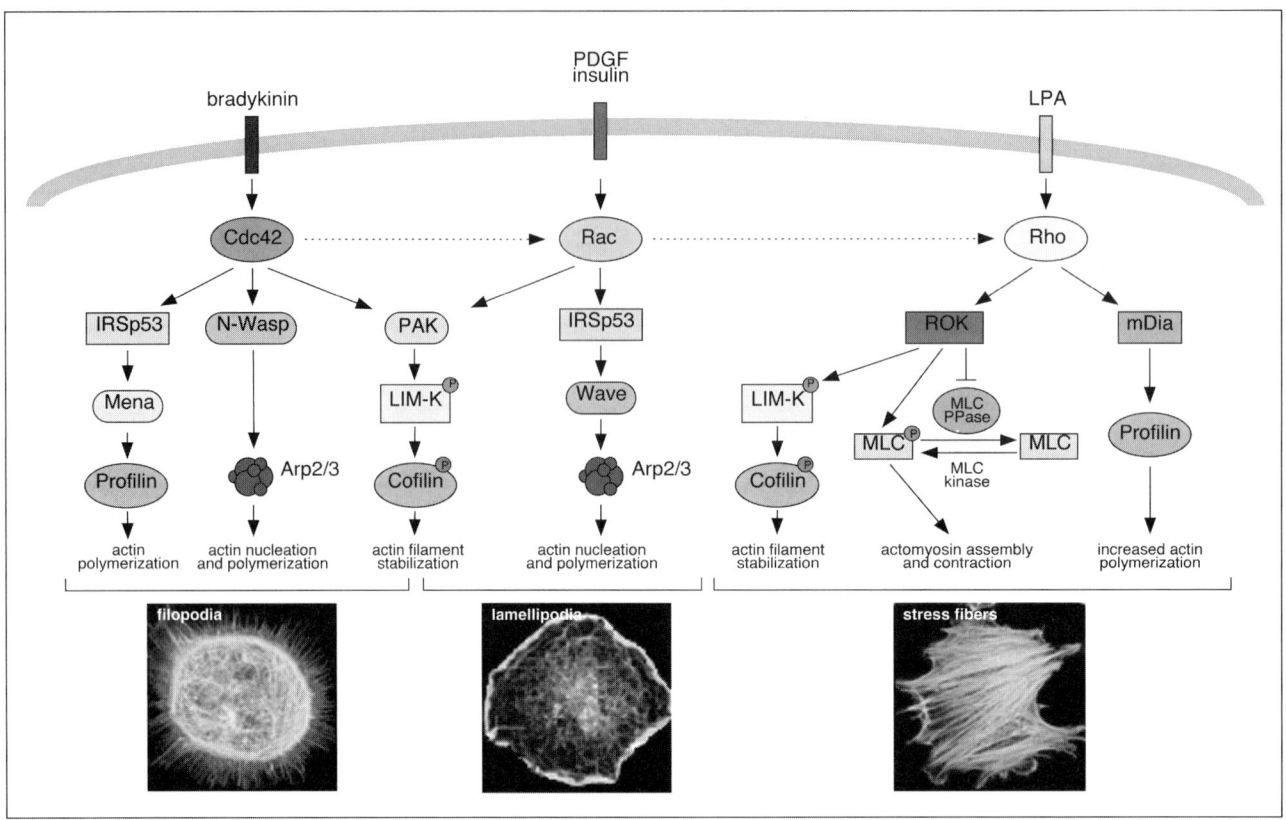

Figure 1 Schematic representation of Rho, Rac, and Cdc42 signaling pathways leading to organization of the actin cytoskeleton. Pictures of Swiss3T3 fibroblasts stained for filamentous actin structures are reprinted with permission from A. Hall (1998). *Science* **279**, 509–514, Copyright 1998 American Association for the Advancement of Science.

edge of a moving cell, while Cdc42-stimulated filopodia have been implicated in the sensing of chemotactic gradients and may be important for polarized movement [11,12]. By regulating focal adhesion sites, Rho, Rac, and Cdc42 can also control the transient loss and reactivation of substrate attachment required for traction during cell movement.

Neuronal Morphology and Guidance

Rho GTPases are important regulators of neurite outgrowth, guidance, and plasticity [13]. Rac and Cdc42 have been shown to stimulate neurite outgrowth and extension, dendritic branching and spine formation, and growth cone guidance toward attractive extracellular guidance cues. In contrast, Rho induces neurite retraction, prevents spine formation, and mediates growth cone collapse in response to repulsive cues.

Phagocytosis

Rho, Rac, and Cdc42 also drive cytoskeletal remodeling required for particle internalization during phagocytosis [14], and Rac and Cdc42 are required for uptake of fluids and particles during pinocytosis [15]. Interestingly, during bacterial invasion, pathogens have found ways to modulate the activity of Rho, Rac, and Cdc42 thereby eliciting changes in the actin cytoskeleton facilitating entry into the host cell [16].

Much less is known about the effects of other Rho GTPases on the actin cytoskeleton. The Cdc42-like protein TC10 induces filopodia [17], while two other Cdc42-like proteins, TCL and Chp, induce membrane ruffling [18,19]. Rnd1 and Rnd3 (RhoE), two apparently constitutively active Rho GTPases, cause disassembly of stress fibers and disruption of cortical actin filaments at least when overexpressed in cells [20]. RhoD stimulates the formation of actin-containing membrane protrusions and seems to antagonize Rho action by inducing stress fiber disassembly and blocking cytokinesis [21]. RhoG has been shown to induce Rac-dependent membrane ruffling and neurite outgrowth [22,23]. However, the physiological significance of these effects remains to be determined.

Signaling from Rho GTPases to the Actin Cytoskeleton

Signaling from Rho to the Actin Cytoskeleton

The pathways leading to Rho-induced actin:myosin filament assembly are relatively well understood and primarily involve two Rho-effectors, ROK and mDia (Fig. 1).

ROKα (Rho associated kinase, ROCKII) and its homolog ROKβ (ROCKI) are Ser/Thr kinases that bind activated Rho and are required for stress fiber formation and cytoskeletal contraction [24]. Constitutively active ROK stimulates the formation of stellate stress-fiber-like structures, whereas kinase-dead ROK or pharmacological inhibitors of ROK inhibit Rho-induced stress fibers. Several targets of ROK have been identified which are likely to be involved in stress fiber formation and contraction. ROK phosphorylates and thereby inhibits myosin-light chain (MLC) phosphatase and can also phosphorylate MLC directly. Together, this leads to an increase in phosphorylated MLC, activation of myosin motor function, and subsequent actin-myosin-based contraction. ROK can also phosphorylate and activate LIM kinase (LIM-K) which in turn phosphorylates and inhibits the actin-severing protein cofilin, leading to stabilization of actin-filaments.

The Rho effectors mDia1 and 2 are members of the formin-related protein family and contain three formin homology domains through which they interact with the actin monomer-binding protein profilin and stimulate actin polymerization [25]. Dominant-active mDia induces weak stress fiber formation and dominant-negative mDia inhibits Rho-induced stress fiber formation. Co-expression of active ROK and mDia induces stress fibers very reminiscent of those induced by activated Rho suggesting that the cooperative activity of these two target proteins is crucial for actin:myosin filament assembly [26].

Signaling Downstream of Cdc42 and Rac

Recently, considerable progress has been made in identifying downstream effectors of Cdc42 and Rac which may mediate filopodia and lamellipodia formation (Fig. 1). The Wiscott-Aldrich-syndrome protein (WASP) and its neuronal isoform N-Wasp have been implicated in filopodia formation downstream of Cdc42 [27]. Overexpression of N-WASP leads to the enhanced formation of filopodia, and Cdc42-induced filopodia can be blocked by anti-N-WASP antibodies or WASP-deletion mutants. Activated Cdc42 binds to WASP and synergizes with $PI4,5P_2$ to unfold the protein allowing it to interact with and activate the Arp2/3 complex. WASP also interacts with monomeric actin and profilin acting as a scaffold protein to bring together the machinery required for actin polymerization. The Arp2/3 complex is known to initiate actin nucleation and stimulate the formation of branched networks of actin as seen in lamellipodia [28]. However, it is not clear how Arp2/3 controls the formation and organization of the straight actin bundles seen in filopodia.

Another member of the WASP family, Wave/Scar, has been reported to mediate Rac-induced membrane ruffling by stimulating the Arp2/3 complex [27]. Rac activates Wave via the insulin receptor substrate p53 (IRSp53) which interacts both with GTP-bound Rac and Wave [29]. Interestingly, IRSp53 and its brain-enriched isoform IRSp58 also bind to activated Cdc42 and participate in filopodia formation and neurite outgrowth [30,31]. In this case, actin polymerization is mediated by binding of IRSp53 to Mena, a member of the Vasp/Ena family that interacts with profilin. Co-expression of Mena and IRSp53 strongly enhanced filopodia formation, though whether the interaction of Mena with profilin is needed in the formation of filopodia is unknown.

The PAK family of Ser/Thr kinases (PAK1, 2, and 3) also appears to promote actin assembly [32]. PAK binds to both Cdc42 and Rac and activated mutants of PAK1 have been reported to stimulate filopodia formation and membrane ruffling in fibroblasts, and induce neurite outgrowth in neuronal cells. Like ROK, PAK phosphorylates LIM-K leading to stabilization of actin filaments. PAK also phosphorylates and inactivates MLC kinase thereby reducing MLC phosphorylation and subsequent actomyosin assembly.

Conclusions

In the last decade it has emerged that Rho GTPases play a pivotal role in controlling actin filament assembly/disassembly during actin-based processes in all eukaryotic cells. A surprisingly large number of downstream effectors of Rho GTPases have been identified and some have been shown specifically to mediate signaling to the actin cytoskeleton. However, the detailed biochemical mechanisms by which this occurs are still unclear. The central role of Rho GTPases in regulating cell migration and bacterial invasion underscores their potential importance in the development of therapeutic agents to treat metastatic cancer or bacterial infections.

Acknowledgments

We thank John Connolly for comments on the manuscript, and the Cancer Research Campaign for their generous support.

References

1. Schultz, J., Milpetz, F., Bork, P., and Ponting, C. P. (1998). SMART, a simple modular architecture research tool: identification of signaling domains. *Proc. Natl. Acad. Sci. USA* **95**, 5857–5864.
2. Van Aelst, L. and D'Souza-Schorey, C. (1997). Rho GTPases and signaling networks. *Genes Dev.* **11**, 2295–2322.
3. Hall, A. (1998). Rho GTPases and the actin cytoskeleton. *Science* **279**, 509–514.
4. Ridley, A. J. and Hall, A. (1992). The small GTP-binding protein rho regulates the assembly of focal adhesions and actin stress fibers in response to growth factors. *Cell* **70**, 389–399.
5. Ridley, A. J., Paterson, H. F., Johnston, C. L., Diekmann, D., and Hall, A. (1992). The small GTP-binding protein rac regulates growth factor-induced membrane ruffling. *Cell* **70**, 401–410.
6. Kozma, R., Ahmed, S., Best, A., and Lim, L. (1995). The Ras-related protein Cdc42Hs and bradykinin promote formation of peripheral actin microspikes and filopodia in Swiss 3T3 fibroblasts. *Mol. Cell. Biol.* **15**, 1942–1952.
7. Nobes, C. D. and Hall, A. (1995). Rho, rac, and cdc42 GTPases regulate the assembly of multimolecular focal complexes associated with actin stress fibers, lamellipodia, and filopodia. *Cell* **81**, 53–62.
8. Hotchin, N. A. and Hall, A. (1995). The assembly of integrin adhesion complexes requires both extracellular matrix and intracellular rho/rac GTPases. *J. Cell Biol.* **131**, 1857–1865.

9. Braga, V. (2000). Epithelial cell shape: Cadherins and small GTPases. *Exp. Cell Res.* **261**, 83–90.
10. Nusrat, A., Giry, M., Turner, J. R., Colgan, S. P., Parkos, C. A., Carnes, D., Lemichez, E., Boquet, P., and Madara, J. L. (1995). Rho protein regulates tight junctions and perijunctional actin organization in polarized epithelia. *Proc. Natl. Acad. Sci. USA* **92**, 10629–10633.
11. Evers, E. E., Zondag, G. C., Malliri, A., Price, L. S., ten Klooster, J. P., van der Kammen, R. A., and Collard, J. G. (2000). Rho family proteins in cell adhesion and cell migration. *Eur. J. Cancer* **36**, 1269–1274.
12. Jones, G. E., Allen, W. E., and Ridley, A. J. (1998). The Rho GTPases in macrophage motility and chemotaxis. *Cell Adhes. Commun.* **6**, 237–245.
13. Luo, L. (2000). Rho GTPases in neuronal morphogenesis. *Nat. Rev. Neurosci.* **1**, 173–180.
14. Chimini, G. and Chavrier, P. (2000). Function of Rho family proteins in actin dynamics during phagocytosis and engulfment. *Nat. Cell Biol.* **2**, E191–E196.
15. Ridley, A. J. (2001). Rho proteins: linking signaling with membrane trafficking. *Traffic* **2**, 303–310.
16. Steele-Mortimer, O., Knodler, L. A., and Finlay, B. B. (2000). Poisons, ruffles and rockets: Bacterial pathogens and the host cell cytoskeleton. *Traffic* **1**, 107–118.
17. Murphy, G. A., Solski, P. A., Jillian, S. A., Perez de la Ossa, P., D'Eustachio, P., Der, C. J., and Rush, M. G. (1999). Cellular functions of TC10, a Rho family GTPase: Regulation of morphology, signal transduction and cell growth. *Oncogene* **18**, 3831–3845.
18. Vignal, E., De Toledo, M., Comunale, F., Ladopoulou, A., Gauthier-Rouviere, C., Blangy, A., and Fort, P. (2000). Characterization of TCL, a new GTPase of the rho family related to TC10 and Ccdc42. *J. Biol. Chem.* **275**, 36457–36464.
19. Aronheim, A., Broder, Y. C., Cohen, A., Fritsch, A., Belisle, B., and Abo, A. (1998). Chp, a homologue of the GTPase Cdc42Hs, activates the JNK pathway and is implicated in reorganizing the actin cytoskeleton. *Curr. Biol.* **8**, 1125–1128.
20. Chardin, P. (1999). Rnd proteins: a new family of Rho-related proteins that interfere with the assembly of filamentous actin structures and cell adhesion. *Prog. Mol. Subcell. Biol.* **22**, 39–50.
21. Tsubakimoto, K., Matsumoto, K., Abe, H., Ishii, J., Amano, M., Kaibuchi, K., and Endo, T. (1999). Small GTPase RhoD suppresses cell migration and cytokinesis. *Oncogene* **18**, 2431–2440.
22. Katoh, H., Yasui, H., Yamaguchi, Y., Aoki, J., Fujita, H., Mori, K., and Negishi, M. (2000). Small GTPase RhoG is a key regulator for neurite outgrowth in PC12 cells. *Mol. Cell. Biol.* **20**, 7378–7387.
23. Roux, P., Gauthier-Rouviere, C., Doucet-Brutin, S., and Fort, P. (1997). The small GTPases Cdc42Hs, Rac1 and RhoG delineate Raf-independent pathways that cooperate to transform NIH3T3 cells. *Curr. Biol.* **7**, 629–637.
24. Amano, M., Fukata, Y., and Kaibuchi, K. (2000). Regulation and functions of Rho-associated kinase. *Exp. Cell Res.* **261**, 44–51.
25. Wasserman, S. (1998). FH proteins as cytoskeletal organizers. *Trends Cell Biol.* **8**, 111–115.
26. Watanabe, N., Kato, T., Fujita, A., Ishizaki, T., and Narumiya, S. (1999). Cooperation between mDia1 and ROCK in Rho-induced actin reorganization. *Nat. Cell Biol.* **1**, 136–143.
27. Takenawa, T. and Miki, H. (2001). WASP and WAVE family proteins: Key molecules for rapid rearrangement of cortical actin filaments and cell movement. *J. Cell Sci.* **114**, 1801–1809.
28. May, R. C. (2001). The Arp2/3 complex: a central regulator of the actin cytoskeleton. *Cell. Mol. Life Sci.* **58**, 1607–1626.
29. Miki, H., Yamaguchi, H., Suetsugu, S., and Takenawa, T. (2000). IRSp53 is an essential intermediate between Rac and WAVE in the regulation of membrane ruffling. *Nature* **408**, 732–735.
30. Krugmann, S., Jordens, I., Gevaert, K., Driessens, M., Vandekerckhove, J., and Hall, A. (2001). Cdc42 induces filopodia by promoting the formation of an IRSp53:Mena complex. *Curr. Biol.* **11**, 1645–1655.
31. Govind, S., Kozma, R., Monfries, C., Lim, L., and Ahmed, S. (2001). Cdc42Hs facilitates cytoskeletal reorganization and neurite outgrowth by localizing the 58-kD insulin receptor substrate to filamentous actin. *J. Cell Biol.* **152**, 579–594.
32. Bagrodia, S. and Cerione, R. A. (1999). Pak to the future. *Trends Cell Biol.* **9**, 350–355.

CHAPTER 238

Regulation of the NADPH Oxidase by Rac GTPase

Becky A. Diebold and Gary M. Bokoch
Department of Immunology, The Scripps Research Institute, La Jolla, California

Phagocytic leukocytes play critical roles in the innate immune response to pathogens. An important component of this response is the ability of leukocytes to generate reactive oxygen species (ROS) via a membrane-associated NADPH oxidase [1,2]. This multicomponent enzyme utilizes electrons derived from intracellular NADPH to generate superoxide anion, which subsequently dismutates to H_2O_2 and other ROS that are used for host defense. The NADPH oxidase of phagocytic leukocytes was the first identified, and remains one of the best-characterized, Rho GTPase-regulated systems. Historically, the involvement of a GTP binding protein in the NADPH oxidase regulation was suspected when guanine nucleotide analogs such as GTPγS and Gpp(NH)p were found to enhance superoxide production in cell-free assays [3–5]. A requirement for a prenylated cytosolic component suggested the involvement of a small GTPase [6]. Subsequently, it was simultaneously shown that either Rac1 or Rac2 GTPase was required for oxidase activity in the cell-free system [7,8], with Rac2 being the predominantly active isoform in human neutrophils [9]. Several additional lines of evidence have established that Rac is an integral and required component of the NADPH oxidase. Rac antisense oligonucleotide inhibited superoxide production in Epstein-Barr virus (EBV)-transformed B lymphocytes [10]. In neutrophils of bcr-null transgenic mice, Rac activity was increased due to the loss of a GTPase-activating protein, Bcr, and superoxide production was concomitantly increased, suggesting that not only is Rac activation required, but that it may be rate-limiting for oxidase activity [11]. Finally, the generation of Rac2-null mice led to the demonstration that *rac2* −/− neutrophils had significantly reduced or absent superoxide production in response to various stimuli [12].

Components and Regulation of the NADPH Oxidase

The NADPH oxidase system of stimulated neutrophils catalyzes the one-electron reduction of oxygen to produce superoxide anion using NADPH as substrate (Fig. 1). In the unstimulated neutrophil, Rac exists as a complex with GDP dissociation inhibitor (GDI) in the cytosol [13]. When the phagocyte is activated through the action of soluble chemoattractants, chemokines, or phagocytic particles, Rac dissociates from GDI by an as yet undetermined mechanism. GDP is exchanged for GTP through the action of membrane-localized guanine nucleotide exchange factors (GEFs) [14] and Rac, now in its GTP-bound active form, becomes membrane associated [15]. Activation of the neutrophil also results in phosphorylation on multiple sites of cytosolic p47[phox], which exists in a complex with a third cytosolic component, p67[phox], followed by translocation of the p47[phox]/p67[phox] complex to the membrane [16,17]. Phosphorylation of p47[phox] is thought to lead to the disruption of an inhibitory, intramolecular interaction within p47[phox], allowing exposed SH3 domains in p47[phox] to interact with proline-rich regions of other NADPH oxidase components [reviewed in references 1, 2, and 18]. The translocation of the p47[phox]/p67[phox] complex occurs simultaneously and independently from the translocation of Rac GTPase [9,19]. p47[phox] is now known to be dispensable for NADPH oxidase activity under cell-free conditions, and appears to serve primarily as an adapter to facilitate membrane binding of p67[phox] [20]. The translocation of p47[phox], p67[phox], and Rac to the plasma membrane culminates in the formation of an active complex with integral flavocytochrome b_{558} (cyt *b*). Cyt *b* possesses an NADPH binding site, FAD, 2 hemes, and 2 subunits, gp91[phox] and p22[phox] [21]. The formation of this minimal

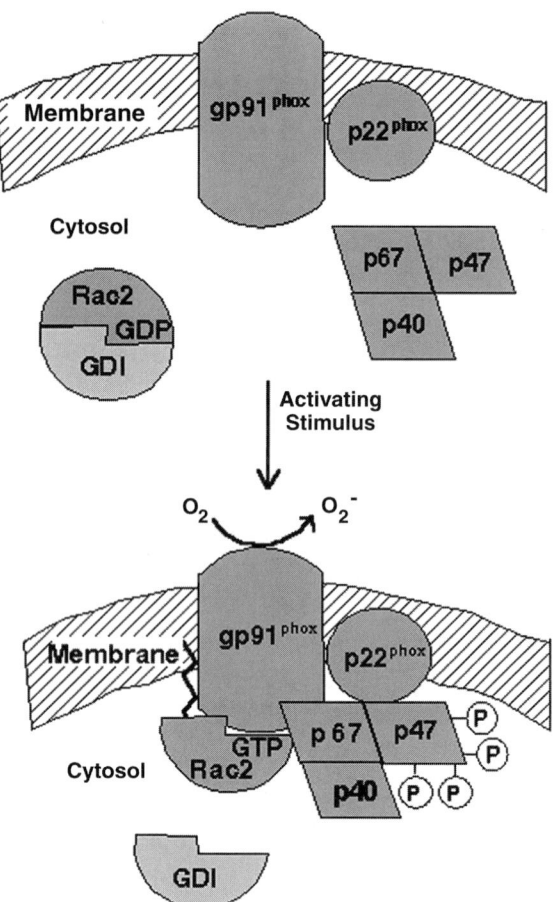

Figure 1 Formation of the NADPH oxidase complex. In the unstimulated neutrophil, cytosolic p47^{phox}, p67^{phox}, and p40^{phox} are believed to exist as a complex in the cytosol. The other cytosolic component, Rac1/2 GTPase in its GDP-bound form exists in a separate complex with Rho GDP dissociation inhibitor (RhoGDI). Upon activation of the neutrophil, Rac is released from GDI and nucleotide exchange occurs. Rac in its GTP-bound form translocates to the membrane. Also during activation, p47^{phox} is phosphorylated on multiple serines leading to the translocation of the p47/p67/(p40?) complex to the membrane; this event is separate from the translocation of Rac GTPase. At the membrane, the cytosolic components interact with cytochrome b_{558}, which is composed of two subunits, gp91^{phox} and p22^{phox}. gp91^{phox} contains the binding site for NADPH, FAD, and two heme groups. The interaction of the cytosolic components with cytochrome b_{558} allows electrons to flow from NADPH to FAD to the heme groups, and finally to the heme-bound oxygen to form superoxide anion.

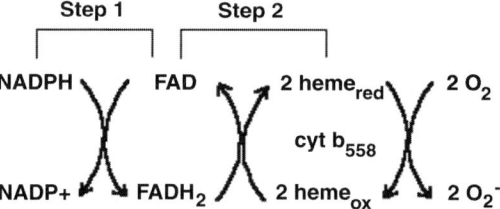

Figure 2 Electron transfer steps of the NADPH oxidase. Step 1 of electron transfer in the NADPH oxidase is defined as the pathway from the cytoplasmic electron donor NADPH to cytochrome b_{558}-bound FAD. Step 2 is defined as the pathway from FAD to the cytochrome b_{558}-associated heme groups.

complex allows electrons to flow from NADPH to FAD and from FAD to the heme of cyt b, and finally to the heme-bound oxygen whose reduction leads to the formation of superoxide anion. An additional cytosolic component, termed p40^{phox} [22], may play a role in regulating the response of the system to phosphatidylinositol-3-phosphate *in vivo* [23], but is not required for NADPH oxidase activity in the cell-free system.

The Role of Rac in NADPH Oxidase Regulation

Understanding the precise nature of the role of Rac in the oxidase regulation is facilitated by an examination of the specific domains of this GTPase. Most studies report that Rac is able to support oxidase activity only in its GTP-bound active form [24]. (There may be an exception to this—see reference [25]). This indicates that the switch I domain is important for Rac action on the oxidase (the switch II domain only undergoes minimal conformational changes upon GTP binding). Consistent with this, point mutations within the switch I or effector domain of Rac, including RacD38A, Y40K, A27K, and G30S, were unable to support oxidase activity [26–29]. The finding that Rac binds directly to p67^{phox} (but not p47^{phox}) via the switch I domain provided important insight into Rac function in the oxidase [26,30]. It was shown that the tetratricopeptide repeat (TPR) in the N terminus of p67^{phox} was the site of Rac binding [31], and this was confirmed upon determination of the crystal structure of the Rac-p67(TPR) complex [32]. The structure revealed specific stabilizing interactions between amino acids A27 and G30 of Rac and the TPR domain of p67^{phox}. At the same time, this structure revealed the availability of other Rac surface domains, particularly the insert domain present in members of the Rho GTPase subfamily, for possible protein interactions. Several prior investigations had suggested a requirement for the Rac insert domain in the activation of the NADPH oxidase. Peptide walking experiments indicated that blocking the insert domain of Rac abrogated NADPH oxidase activation [33]. Studies using insert domain deletion mutants of Rac have yielded conflicting results, however, concluding either that the insert domain was absolutely required [34,35] or unnecessary [36,37] for Rac oxidase activity.

The role of Rac in NADPH oxidase regulation had been generally supposed to be that of a docking protein, in which the prenylated C terminus of Rac bound to membrane phospholipids, while the switch I domain facilitated the binding of the p67^{phox}/p47^{phox} complex to cyt b. Our laboratory recently investigated the role of Rac in the electron transfer reactions of the NADPH oxidase using a purified cell-free system [34]. We determined that Rac2 was required for both electron transfer steps (from NADPH to FAD and from FAD to cyt b) (Fig. 2). In order to assess whether Rac2 had a functional role that was distinct from its interaction with p67^{phox}, we used Rac2 and p67 constructs that were mutated in domains involved in Rac/p67^{phox} interactions. We observed that the non-Rac-binding mutant, p67 Δ178-184, could still support electron transfer from NADPH to FAD (step 1), but not from FAD to cyt *b* (step 2). In a reciprocal experiment, we observed that a Rac2 mutant (Rac2 D38A) that is not able to bind p67phox supported step 1, but not step 2 activity. These data suggested that Rac2 and p67phox did not have to functionally interact for electrons to be transferred from NADPH to FAD, but that a Rac/p67phox interaction was required for completion of electron transfer from FADH$_2$ to cyt *b*. Furthermore, since Rac2 was operative in the absence of p47phox, this indicated that Rac2 must interact directly with cyt *b* to support the step 1 reaction. Based upon these observations, we proposed a two-step model for regulation of the NADPH oxidase by Rac2 GTPase (Fig. 3).

We hypothesized that the insert domain of Rac2 may play a role in binding to cyt *b* and regulation of electron transfer. Using an insert domain deletion mutant of Rac2 in our electron transfer assays, we found that the insert domain was critical for the step 1, and consequently, step 2 reactions. We furthermore demonstrated that the intensity of a fluorescent analog of mant-GppNHp bound to Rac2 increased in the presence of cyt *b*, indicating direct interaction between Rac2 and cyt *b*. In addition, using the insert domain deletion mutant of Rac2 in place of wild-type Rac2 eliminated this interaction, indicating that the insert domain of Rac2 is necessary for both the functional and physical interaction of Rac2 with cyt *b*. In support of the fluorescence data, we have recently demonstrated that glutathione-S-transferase (GST)-Rac2 can specifically bind cyt *b* purified from neutrophils in pulldown assays (Diebold and Bokoch, unpublished observations).

Current Models of Rac Function in NADPH Oxidase Regulation

Overall these data suggest a novel paradigm for NADPH oxidase regulation by Rac2 GTPase, showing that Rac2 can interact directly with cyt *b* and that Rac2 has a separate and required function in regulating catalytic activity of the NADPH oxidase (Fig. 4C). These two points differ from other models. The model of Lambeth *et al.* [2] (Fig. 4A) agrees with the first point that the insert domain of Rac may bind to cyt *b*. However, the only role of the insert domain of Rac according to this model is to facilitate binding of p67phox. p67phox, which has an activation domain and also binds cyt *b*, is proposed in the model of Lambeth *et al.* to be the only protein influencing the rate-limiting electron transfer step (step 1) of the NADPH oxidase (reviewed in [2]). Their model is based on the observation that a nonprenylated Rac1 mutant, which lacks the insert domain, decreased the affinity (EC$_{50}$) of Rac for the oxidase, but had no effect

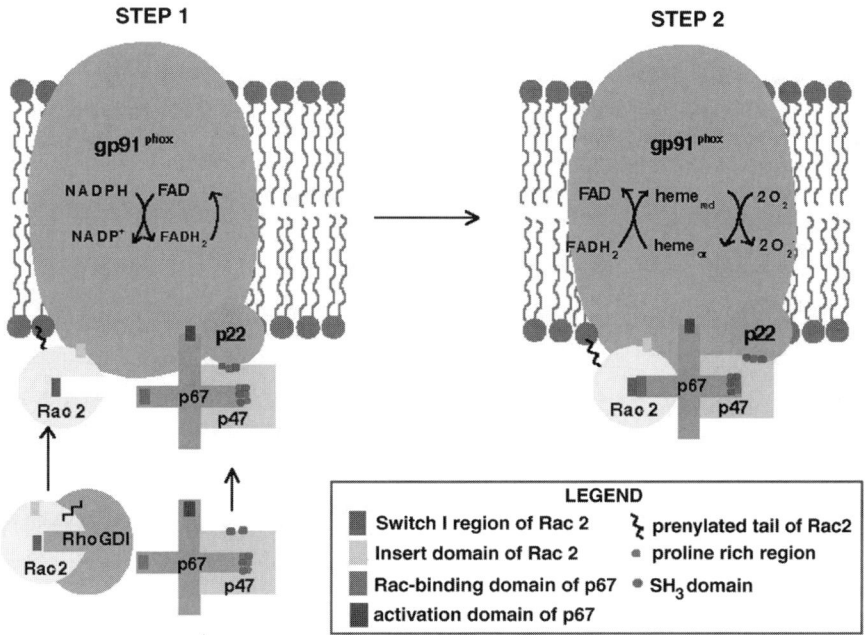

Figure 3 Two-step model for the regulation of the NADPH oxidase by Rac GTPase. Diebold and Bokoch [34] proposed a two-step model for the regulation of NADPH oxidase by Rac. In step 1, Rac translocates to the membrane and interacts with the phospholipid bilayer via its prenylated C terminus. In addition, Rac, via its insert domain, interacts with cytochrome b_{558} and contributes to the regulation of electron flow from NADPH to FAD without interacting with p67phox. p67phox is still required for electron flow to occur in step 1 and regulates electron flow via its activation domain. The interaction of the insert domain of Rac with cytochrome b_{558} may induce a conformational change in cytochrome b_{558} that improves the interaction of p67phox and cytochrome b_{558}. In step 2, the interaction between the switch I domain of Rac and the Rac-binding domain of p67phox is required for electrons to continue to flow from FAD to the heme groups of cytochrome b_{558}.

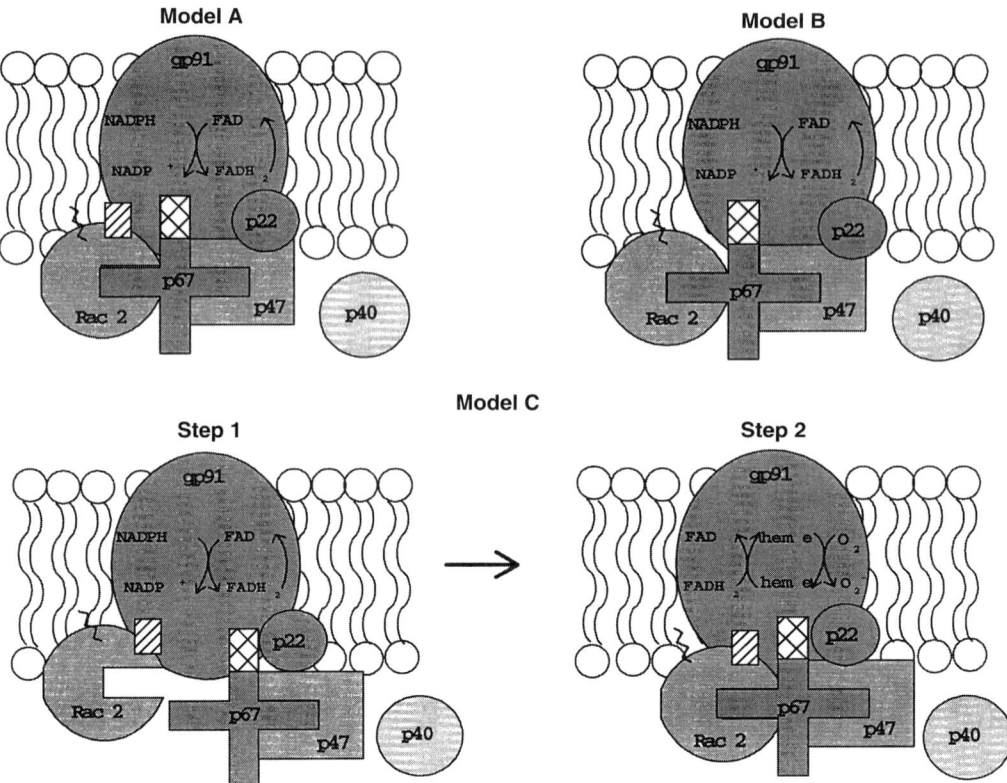

Figure 4 Comparison of proposed models of NADPH oxidase regulation by Rac. In each of the existing models, the switch I region of Rac (not indicated) interacts with p67phox, the prenylated tail of Rac (shown as a zigzag line) interacts with the membrane, and the activation domain of p67phox (a crosshatched section) interacts with cytochrome b_{558}. In the model of Diebold and Bokoch (C) [34] and that of Lambeth et al. (A) [2], the insert domain of Rac (hatched section) has direct protein interactions with cytochrome b_{558}. The Diebold and Bokoch model differs in that Rac contributes to the regulation of electron flow from NADPH to FAD, while Lambeth proposes that only the activation domain of p67phox regulates this electron transfer step. In the latter model, Rac and p47phox serve as adapters aiding in the interaction of p67phox with cytochrome b_{558}. In model B proposed by Pick et al [38], Rac interacts only with the phospholipids of the membrane via its prenylated C terminus and does not interact with cytochrome b_{558}. In this model, the insert domain is not involved in protein interactions or regulation of the NADPH oxidase. As in the model of Lambeth et al., p67phox is the only regulatory protein, while Rac and p47phox serve as adapters for p67phox.

on the maximal rate, V_{max}, of superoxide production [35]. Our laboratory, on the other hand, has observed a decrease in V_{max} when using a prenylated Rac2 version of this mutant [34].

The model of Pick et al. [38] (Fig. 4B) opposes the view that Rac interacts with cyt b. Instead, they propose that Rac interacts only with the phospholipids within the plasma membrane via its prenylated C terminus. This model is based upon observations that prenylated Rac can bind to phospholipid vesicles devoid of cyt b. (Phospholipids are used to re-lipidate cytochrome b_{558} purified from neutrophil membranes before it is used in cell-free assays). Like the previous model, this model portrays Rac and p47phox only as adapter molecules that aid p67phox in binding to cyt b. Interestingly, this group of investigators initially observed in a peptide walking study that peptides overlapping the insert domain of Rac inhibited superoxide production in the cell-free system [33]. Recently, however, this group used nonprenylated Rac1-p67phox chimeras in the cell-free system and showed that deletion of the insert domain of Rac1 did not affect the ability of this chimera to support superoxide production [37]. Consequently, based on these observation and observations by Lambeth's group that deletion of the insert domain does not affect V_{max} [35], the model of Pick et al. [38] does not support an interaction of the Rac insert domain with cyt b (Fig. 4B). At least part of the discrepancy between these models may be due to the use of prenylated Rac (Bokoch model) versus nonprenylated Rac (Lambeth and Pick models) in the cell-free system. The concentration of prenylated Rac required in the cell-free assay is at least 100-fold less than nonprenylated Rac. The use of such high concentrations of unprocessed GTPase may obscure relevant protein-protein interactions that normally occur at physiological concentration of reactants. Ultimately, a consensus model for the role of Rac in the oxidase will depend upon in vivo studies using neutrophils or cell lines bearing resemblance to neutrophils.

Rac GTPase—A More General Role in Regulating Oxidant-Based Signaling?

The finding that Rac can bind to cyt b and regulate electron transfer in the NADPH oxidase of neutrophils is relevant to other ROS-generating signaling pathways used by nonphagocytic cells. Recently, homologs of cyt b, called Nox, have been found in several tissues (reviewed in [39]). These new *NADPH ox*idases produce low levels of superoxide anion that appear to be used as signals for the control of cell growth and transformation. It has been known for some time that Ras and Rac contribute to the control of signaling pathways that are critical for mitogenesis and oncogenesis. In stimulated NIH3T3 cells, transient expression of a constitutively activated form of Ras leads to a significant increase in intracellular ROS, and expression of a dominant negative allele of Ras or Rac1 inhibited this rise in ROS production [40,41]. ROS production in NIH3T3 cells was suppressed by treatment with a diphenylene iodonium (DPI), a flavoprotein inhibitor of NADPH oxidase of phagocytes, indicating that a Nox protein may be involved [41]. Interestingly, it has been shown that the insert region of Rac1 was essential for mitogenesis and superoxide production in fibroblasts [42]. Rac1 also appears to be involved in the signaling pathway leading to reperfusion injury caused by ROS production during reoxygenation of vascular smooth muscle [43,44]. Recombinant adenoviral expression of a dominant negative Rac1 suppressed the reperfusion-induced injury in an *in vivo* model of mouse hepatic ischemia/reperfusion injury. This was also observed in mice deficient for the gp91^{phox} of phagocytic NADPH oxidase, suggesting that the Rac mutant inhibited ROS production by a Nox system rather than by one employing gp91^{phox} [44]. Thus, it appears that ROS production in nonphagocytic cells involves both a Nox protein and Rac GTPase. The possibility exists that Rac may directly regulate superoxide production by Nox proteins in nonphagocytic cells by a mechanism similar to its binding to cyt b and regulating oxidant production by phagocytes. Determining if Rac has a function in superoxide production by Nox may lead to a better understanding of redox signaling pathways in other cell types and potentially provide a means to therapeutically intervene in ROS-related pathological disease states.

References

1. Babior, B. M. (1999). NADPH oxidase: An update. *Blood* **93**, 1464–1476.
2. Lambeth, J. D. (2000). Regulation of the phagocyte respiratory burst oxidase by protein interactions. *Biochem. Mol. Biol.* **33**, 427–439.
3. Gabig, T. G., English, D., Akard, L. P., and Schell, M. J. (1987). Regulation of neutrophil NADPH oxidase activation in a cell-free system by guanine nucleotides and fluoride. Evidence for participation of a pertussis and cholera toxin-insensitive G protein. *J. Biol. Chem.* **262**, 1685–1690.
4. Doussiere, J., Pilloud, M. C., and Vignais, P. V. (1988). Activation of bovine neutrophil oxidase in a cell free system. GTP-dependent formation of a complex between a cytosolic factor and a membrane protein. *Biochem. Biophys. Res. Commun.* **152**, 993–1001.
5. Seifert, R., Rosenthal, W., and Schultz, G. (1986). Guanine nucleotides stimulate NADPH oxidase in membranes of human neutrophils. *FEBS Lett.* **205**, 161–165.
6. Bokoch, G. M. and Prossnitz, V. (1992). Isoprenoid metabolism is required for stimulation of the respiratory burst oxidase of HL-60 cells. *J. Clin. Invest.* **89**, 402–408.
7. Abo, A., Pick, E., Hall, A., Totty, N., Teahan, C. G., and Segal, A. W. (1991). Activation of the NADPH oxidase involves the small GTP-binding protein p21^{rac1}. *Nature* **353**, 668–670.
8. Knaus, R. G., Heyworth, P. G., Evans, T., Curnutte, J. T., and Bokoch, G. M. (1991). Regulation of phagocytic oxygen radical production by the GTP-binding protein Rac 2. *Science* **254**, 1512–1515.
9. Heyworth, P. G., Bohl, B. P., Bokoch, G. M., and Curnutte, J. T. (1994). Rac translocates independently of the neutrophil NADPH oxidase components p47^{phox} and p67^{phox}. *J. Biol. Chem.* **269**, 30749–30752.
10. Dorseuil, O., Vazquez, A., Lang, P., Bertoglio, A. J., Gacon, G., and Leca, G. (1992). Inhibition of superoxide in B lymphocytes by Rac antisense oligonucleotides. *J. Biol. Chem.* **267**, 20540–20542.
11. Voncken, J. W., van Schaik, H., Kaartinen, V., Deemer, K., Coates, T., Landing, B., Pattengale, P., Dorseuill, O., Bokoch, G. M., Groffen, J., and Heisterkamp, N. (1995). Increased neutrophil respiratory burst in Bcr null mutants. *Cell* **80**, 719–728.
12. Roberts, A. W., Kim, C., Zhen, L., Lowe, J. B., Kapur, R., Petryniak, B., Spaetti, A., Pollock, J. D., Borneo, J. B., Bradford, G. B., Atkinson, S. J., Dinauer, M. C., and Williams, D. A. (1999). Deficiency of the hematopoietic cell-specific Rho family GTPase Rac2 is characterized by abnormalities in neutrophil function and host defense. *Immunity* **10**, 183–196.
13. Chuang, T.-H., Bohl, G., and Bokoch, G. M. (1993). Biologically active lipids are regulators of Rac-GDI complexation. *J. Biol. Chem.* **268**, 26206–26211.
14. Bokoch, G. M., Bohl, B. P., and Chuang, T.-H. (1994). Guanine nucleotide exchange regulates membrane translocation of Rac/Rho GTP-binding proteins. *J. Biol. Chem.* **269**, 31674–31679.
15. Quinn, M. T., Evans, T., Loetterle, L. R., Jesaitis, A. J., and Bokoch, G. M. (1993). Translocation of Rac correlates with NADPH oxidase activation. *J. Biol. Chem.* **268**, 20983–20987.
16. Dusi, S., Della Bianca, V., Grzeskowiak, M., and Rossi, F. (1992). Relationship between phosphorylation and translocation to the plasma membrane of p47^{phox} and p67^{phox} and activation of the NADPH oxidase in normal and Ca^{2+}-depleted human neutrophils. *Biochem. J.* **290**, 173–178.
17. Rotrosen, D. and Leto, T. L. (1990). Phosphorylation of neutrophil 47-kDa cytosolic oxidase factor. Translocation to membrane is associated with distinct phosphorylation events. *J. Biol. Chem.* **265**, 19910–19915.
18. DeLeo, F. R. and Quinn, M. T. (1996). Assembly of the phagocyte NADPH oxidase: molecular interaction of oxidase proteins. *J. Leukocyte Biol.* **60**, 677–691.
19. Dorseuil, O., Quinn, M. T., and Bokoch, G. M. (1995). Dissociation of Rac translocation from p47^{phox}/p67^{phox} movements in human neutrophils by tyrosine kinase inhibitors. *J. Leukocyte Biol.* **58**, 108–113.
20. Freeman, J. L. and Lambeth, J. D. (1996). NADPH oxidase activity is independent of p47^{-phox} in vitro. *J. Biol. Chem.* **271**, 22578–22582.
21. Parkos, C. A., Allen, R. A., Cochrane, C. G., and Jesaitis, A. J. (1988). The quarternary structure of the plasma membrane b-type cytochrome of human granulocytes. *Biochim. Biophys. Acta* **932**, 71–83.
22. Wientjes, F. B., Panayotou, G., Reeves, E., and Segal. A. W. (1996). Interactions between cytosolic components of the NADPH oxidase: p40^{phox} interacts with both p67^{phox} and p47^{phox}. *Biochem. J.* **317**, 919–924.
23. Ellson, C. D., Gobert-Gosse, S., Anderson, K. E., Davidson, K., Erdjument-Bromage, H., Tempstt, P., Thuring, J. W., Cooper, M. A., Lim, Z. Y., Holmes, A. B., Piers, R. J., Gaffney, P. R. J., Coadwell, J., Chilvers, E. R., Hawkins, P. T., and Stephens, L. R. (2001). PtIns(3)P regulates the neutrophil oxidase complex by binding to the PX domain of p40^{phox}. *Nat. Cell Biol.* **3**, 670–683.

24. Bokoch, G. M. (1995). Regulation of the phagocyte respiratory burst by small GTP-binding proteins. *TCB* **5**, 109–113.
25. Di-Poi, N., Faure, J., Grizot, S., Molnar, G., Pick, E., and Dagher, M. C. (2001). Mechanism of NADPH oxidase activation by the Rac/Rho-GDI complex. *Biochemistry* **40**, 10014–10022.
26. Diekmann, D., Abo, A., Johnson, C., Segal, A., and Hall, A. (1994). Interaction of Rac with p67^{phox} and regulation of phagocytic NADPH oxidase activity. *Science* **265**, 531–533.
27. Freeman, J. L. R., Kreck, M. L., Uhlinger, D. J., and Lambeth, J. D. (1994). A Ras effector-homologue region on rac regulates protein associations in the neutrophil respiratory burst oxidase complex. *Biochemistry* **33**, 13431–13435.
28. Kwong, C. H., Adams, A. G., and Leto, T. L. (1995). Characterization of the effector-specifying domain of Rac involved in NADPH oxidase activation. *J. Biol. Chem.* **270**, 19868–19872.
29. Xu, S., Barry, D. C., Settleman, J., Schwartz, M. A., and Bokoch, G. M. (1994). Differing structural requirements for GTPase-activating protein responsiveness and NADPH oxidase activation by Rac. *J. Biol. Chem.* **269**, 23569–23574.
30. Nisimoto, Y., Freeman, J., Motalebi, S., Hirshberg, M., and Lambeth, J. D. (1997). Rac binding to p67^{phox}. *J. Biol. Chem.* **272**, 18834–18841.
31. Koga, H., Terasawa, H., Nunoi, H., Takeshige, K., Inagaki, F., and Sumimoto, H. (1999). Tetratricopeptide repeat (TPR) motifs of p67^{phox} participate in interaction with the small GTPase Rac and activation of the phagocyte NADPH oxidase. *J. Biol. Chem.* **274**, 25051–25060.
32. Lapouge, K., Smith, S. J. M., Walker, P. A., Gamblin, S. J., Serdon, S. J., and Rittinger, K. (2000). Structure of the TPR domain of p67^{phox} in complex with Rac-GTP. *Mol. Cell* **6**, 899–907.
33. Joseph, G. and Pick, E. (1995). "Peptide walking" is a novel method for mapping functional domains in proteins. *J. Biol. Chem.* **270**, 29079–29082.
34. Diebold, B. A. and Bokoch, G. M. (2001). Molecular basis for Rac2 regulation of the phagocyte NADPH oxidase. *Nat. Immun.* **2**, 211–215.
35. Freeman, J. L., Abo, A., and Lambeth, J. D. (1996). Rac "insert region" is a novel effector region that is implicated in the activation of NADPH oxidase, but not PAK65. *J. Biol. Chem.* **271**, 19794–19801.
36. Toporik, A., Gorzalczany, Hirshberg, Pick, E., and Lotan, O. (1998). Mutational analysis of novel effector domains in Rac1 involved in the activation of nicotinamide adenine dinucleotide phosphate (reduced) oxidase. *Biochemistry*, **37**, 7147–7156.
37. Alloul, N., Gorzalczany, Y., Itan, M., Sigal, N., and Pick, E. (2001). Activation of the superoxide-generating NADPH oxidase by chimeric proteins consisting of segments of the cytosolic component p67^{phox} and the small GTPase Rac1. *Biochemistry* **40**, 14557–14566.
38. Gorzalczany, Y., Sigal, N., Itan, M., Lotan, O., and Pick, E. (2000). Targeting of Rac1 to the phagocyte membrane is sufficient for the induction of NADPH oxidase assembly. *J. Biol. Chem.* **275**, 40073–40081.
39. Lambeth, J. D. (2002). Nox/Duox family of nicotinamide adenine dinucleotide (phosphate) oxidases. *Curr. Opin. Hematol.* **9**, 7–11.
40. Irani, K., Xia, Y., Zweier, J. L., Sollott, S. J., Der, C. J., Fearon, E. R., Sundaresan, M., Finkel, T., and Goldschmidt-Clermont, P. J. (1997). Mitogenic signaling mediated by oxidants in Ras-transformed fibroblasts. *Science*, **275**, 1649–1652.
41. Sundaresan, M., Yu, Z.-X., Ferrans, V. J., Sulciner, D. J., Gutkind, J. S., Irani, K., Goldschmidt-Clermont, P. J., and Finkel, T. (1996). Regulation of reactive-oxygen-species generation in fibroblasts by Rac1. *Biochem. J.* **318**, 379–382.
42. Joneson, T. and Bar-Sagi, D. (1998). A Rac1 effector site controlling mitogenesis through superoxide production, *J. Biol. Chem.* **273**, 17991–17994.
43. Kim, K.-S., Takeda, K., Sethl, R., Pracyk, J. B., Tanaka, K., Zhou, U. F., Yu, Z.-X., Ferrans, F. J., Bruder, J. T., Kovesdi, I., Irani, K., Goldschmidt-Clermont, P., and Finkel, T. (1998). Protection from reoxygenation injury by inhibition of rac1. *J. Clin. Invest.* **101**, 1821–1826.
44. Ozaki, M., Deshpande, S. S., Angkeow, P., Bellan, J., Lowenstein, C. J., Dinauer, M. C., Goldschmidt-Clermont, P. J., and Irani, K. (2000). Inhibition of the Rac1 GTPase protects against nonlethal ischemia/reperfusion-induced necrosis and apoptosis in vivo. *FASEB J.* **14**, 418–429.

The Role of Rac and Rho in Cell Cycle Progression

Laura J. Taylor and Dafna Bar-Sagi
*Department of Molecular Genetics and Microbiology,
State University of New York at Stony Brook, Stony Brook, New York*

Introduction

The ability of cells to progress through the cell cycle depends on the concerted action of mitogen- and anchorage-stimulated signal transduction pathways. The regulation of the cell cycle machinery is often disrupted in tumor cells, with the most common targets being proteins involved in G1 progression. Identifying the signaling pathways responsible for G1 progression is therefore important for increasing the understanding of control mechanisms that operate during normal cell proliferation and their subversion in tumor cells. One class of signaling proteins that has recently been recognized to play a significant role in G1 phase progression is the small GTP binding proteins of the Rho GTPase family. This chapter will focus on the role of two members of the Rho GTPase family, Rac and Rho, in G1 progression.

Regulation of G1 Progression

Cell cycle progression through G1 is a complex and tightly controlled process. It is regulated by stimulatory and inhibitory signals, both of which are targets of the Rho GTPases. Three activities are recognized to be important for progression through the G1 phase of the cell cycle: early-G1 transcriptional activation of immediate early genes, mid-G1 activation of cyclin D/cdk4/6, and late-G1 activation of cyclin E/cdk2 [1,2]. Mitogenic stimulation results in the induction of many immediate early genes containing the serum response element (SRE) in their promoter [1]. The SRE is activated by binding to a ternary complex containing the transcription factors, serum response factor (SRF) and ternary complex factor (TCF) [1]. Although activation of immediate early genes is necessary for early G1 progression, it is not sufficient for progression to S phase, and progression through later phases of G1 requires the activity of cyclin-dependent kinases (cdks).

Cdks are a group of serine/threonine kinases that are activated by binding to their respective cyclin partners and by phosphorylation [2]. Two main cdk activities play a role in G1, cyclin D/cdk4/6 functions in mid G1 and cyclin E/cdk2 functions in late G1. The major substrate of the G1 kinase complexes is the retinoblastoma protein (Rb). In its unphosphorylated form, Rb functions as an inhibitor of E2F, a transcription factor that controls the expression of genes required for G1 progression [3]. The inhibitory effect of Rb on E2F transcriptional activity is exerted by two mechanisms, one involving the direct binding to E2F and the other involving the recruitment of histone deacetylase (HDAC) [4]. Both inhibitory effects are antagonized by the coordinated and sequential phosphorylation of Rb by cyclin D/cdk4/6 and cyclin E/cdk2 which in turn allows the ordered expression of E2Fdependent genes [4,5]. Phosphorylation of Rb by cyclin D/cdk4/6 initially releases HDAC thereby alleviating transcriptional repression, and phosphorylation of Rb by cyclin E/cdk2 dissociates the Rb-E2F complex [6,7].

An important mechanism for regulation of cyclin/cdk activity involves inhibition by cyclin-dependent kinase inhibitors. The main inhibitors of cyclin D/cdk4/6 complexes are $p16^{ink4a}$ and $p21^{Cip1}$, whereas inhibition of cyclin E/cdk2 occurs by $p21^{Cip1}$ and $p27^{Kip1}$[2]. The levels of these inhibitors are regulated by multiple mechanisms. $p21^{Cip1}$ is regulated predominantly at the level of transcription [8] and mRNA stability [9]. $p27^{Kip1}$ can be regulated at multiple levels including transcriptional [10], translational [11,12], and posttranslational [13,14] mechanisms dependent on the cell type and the extracellular signal. However, the predominant mechanism by which p27^{Kip1} levels are controlled is through cyclin E/cdk2-dependent phosphorylation [15,16], which targets p27^{Kip1} for ubiquitination and proteolytic degradation [13].

The Function of Rac and Rho in Cell Cycle Progression and Transformation

The Rho family of GTPases functions as molecular switches by oscillating between an active GTP-bound form and inactive GDP-bound form. Activation of the Rho GTPases can be induced by soluble growth factor stimulation and cell adhesion to the extracellular matrix (ECM). Their biological effects are exerted through the activation of multiple effector pathways that control transcription, cytoskeleton organization, and changes in the redox state [17].

The importance of the Rho GTPases in cell cycle progression was initially illustrated through studies demonstrating their involvement in both growth-factor-induced proliferation and oncogenic transformation. Rac1 and RhoA are each required for transformation by Ras and co-expression of a constitutively active form of Raf, a Ras effector, with either constitutively active RhoA or Rac1 synergistically enhances focus-forming activity [18,19].

Rac and Rho have been shown to be necessary and sufficient for cell cycle progression. In Swiss 3T3 cells, introduction of dominant interfering forms of Rac1 and RhoA inhibits progression of growth-factor-induced cell cycle progression, while a dominant active form of each is sufficient to induce cell cycle progression [20]. However, the capacity of Rac1 and RhoA to promote cell cycle progression is cell-type specific. For example, in rat embryo fibroblasts,

G1 to S transition requires the synergistic activities of Rac and Raf [21]. Using partial loss of function mutants of Rac it has been shown that the contribution of Rac to cell cycle progression is dependent on two distinct effector functions, cytoskeleton rearrangements and superoxide production [21,22]. Rac-induced cytoskeleton rearrangements are mediated by the effector binding loop [21], a region spanning amino acids 26 to 40 that interacts with multiple downstream effector molecules. Rac-dependent superoxide generation is controlled by the insert region, a sequence of 11 amino acids common to all of the RhoGTPase family members, but not found in the Ras GTPase family members [23]. Significantly, superoxide generation has been shown to be essential for Ras-induced proliferation [24] indicating that Rac-mediated superoxide production might be functionally relevant to Ras-induced proliferation. It is noteworthy that the insert region of Rho also plays an important role with regard to cell cycle progression through the activation of the Rho effector, Rho kinase, which cooperates with activated Raf to promote transformation [25,26].

Cell Cycle Targets of Rac and Rho

Increasing evidence indicates that Rac and Rho influence cell cycle progression by targeting multiple regulatory steps throughout G1. A well-documented mechanism by which Rac and Rho affects early-G1 involves the activation of genes controlled by the SRE. Rac, through its effector PAK, promotes the phosphorylation of both Raf and MEK, two components of the signaling cascade leading to ERK activation [27,28]. Both PAK-mediated phosphorylation events act synergistically with the Ras pathway to promote full activation of ERK [27,29]. Subsequently, activated ERK phosphorylates and activates TCF thereby stimulating SRE-dependent transcription (Fig. 1). Rac activity has also been demonstrated

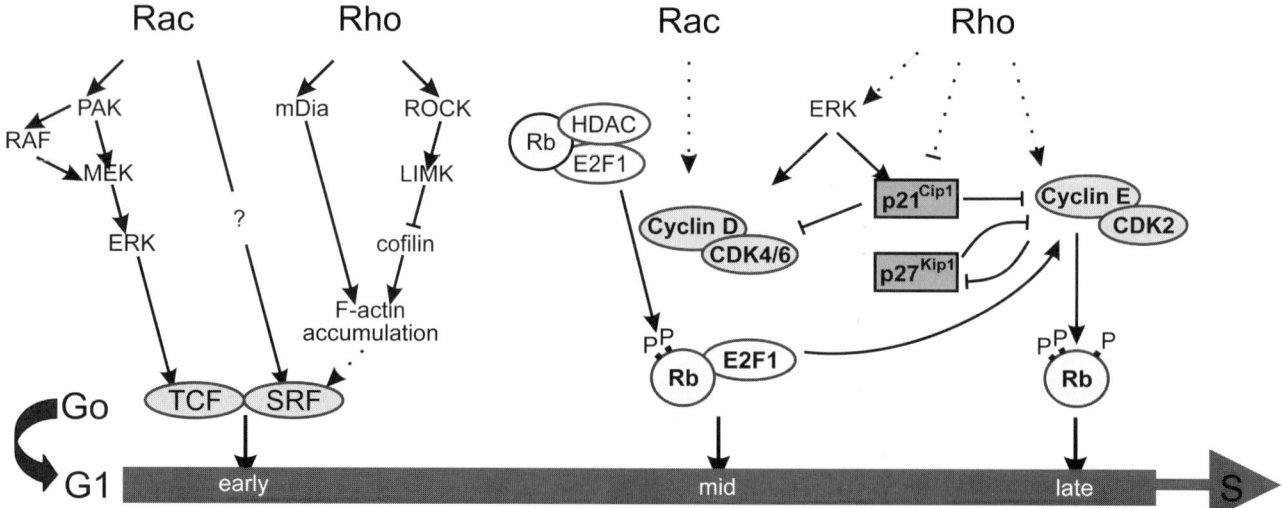

Figure 1 Integration of Rac and Rho signaling pathways with the cell cycle. Rac and Rho target multiple regulatory events during the G1 stage of the cell cycle. The contribution of Rac and Rho to TCF/SRF and cyclin/cdk complexes promotes G1-phase progression.

to potentiate SRF activity, but the signaling pathways regulating this response have not been identified [30,31].

RhoA activity is essential for mitogen-induced activation of SRF [30]. The ability of Rho to activate SRF is linked to its effect on actin cytoskeleton dynamics. This is indicated by studies showing that the Rho effectors LIMK and mDia potentiate SRF activity independently of extracellular signals [32–34]. Although the relative contribution of the Rho-dependent pathways to SRF activity seems to be cell-type dependent, both LIMK and mDia pathways contribute to F-actin accumulation, suggesting a causal role for F-actin levels in the activation of SRF (Fig. 1).

Progression through mid G1 of the cell cycle is dependent upon upregulation of cyclin D and formation of the cyclin D/cdk4/6 complex. Both Rac and Rho have been shown to contribute to the upregulation of cyclin D1 through ERK-dependent and -independent pathways (Fig. 1). Rac-mediated induction of cyclin D1 occurs in part through the Rac effector PAK [31], and is also dependent on NF-κB activation as evident from the findings that an intact NF-κB binding site in the cyclin D1 promoter is required for Rac-dependent cyclin D1 transcription [35].

Recent evidence by Welsh et al. demonstrates that Rho plays a central role in controlling adhesion- and mitogen-dependent cyclin D1 expression [36]. First, in early G1, Rho inhibits Rac-induced expression of cyclin D1 by antagonizing Rac through an unknown mechanism [36]. Second, in mid-G1 phase of the cell cycle, Rho promotes cyclin D1 induction by maintaining a sustained activation of ERK [36]. The mechanisms by which Rho might contribute to ERK activity are not well defined, but the Rho kinase pathway seems to be necessary for this effect [36]. Thus, Rho appears to have an important role in setting up the correct timing of cyclin D1 expression during cell cycle progression.

In addition to its role in the regulation of cyclin D1 expression, Rho regulates cyclin D/cdk4/6 activity by inhibiting the accumulation of the cdk inhibitor $p21^{Cip1}$ [37]. For example, in some cell types, high levels of Ras or Raf activities induce $p21^{Cip1}$ expression and cell cycle arrest [38–40], and this effect can be rescued by ectopic expression of activated Rho [37]. Furthermore, mouse embryo fibroblasts lacking $p21^{Cip1}$ do not require Rho for Ras-induced S-phase entry [37]. Rho can also be involved in the regulation of late-G1 progression by activating the cyclin E/cdk2 complex, which in turn promotes the degradation of the cdk inhibitor $p27^{Kip1}$ (Fig. 1) [41]. Together, the effects of Rho on the levels of cdk inhibitors are likely to contribute to the ability of cells to undergo G1 to S progression in response to proliferative signals.

Future Perspectives

Although, as outlined in this chapter, the involvement of Rac and Rho in regulating cell cycle progression is supported by many lines of evidence, the biochemical mechanisms that couple the signaling activities of these GTPases and the cell cycle machinery remain to be established. By virtue of their effects on the actin cytoskeleton, Rac and Rho play a key role in the regulation of cell shape changes that accompany adhesion and motility. It is well recognized that cell shape is an important determinant for the proliferative capacity of normal anchorage-dependent cells and loss of cell-shape-dependent growth control is a hallmark of oncogenically transformed cells. Thus, understanding the molecular basis for the involvement of Rac and Rho in cell cycle regulation should provide insights into the mechanisms by which alterations in cellular morphology can be sensed and converted to a growth response.

Acknowledgments

This work was supported by a grant from the National Institutes of Health (CA55360).

References

1. Treisman, R. (1990). The SRE: A growth factor responsive transcriptional regulator. *Semin. Cancer Biol.* **1**, 47–58.
2. Obaya, A. J. and Sedivy, J. M. (2002). Regulation of cyclin-Cdk activity in mammalian cells. *Cell. Mol. Life Sci.* **59**, 126–142.
3. Nevins, J. R. (2001). The Rb/E2F pathway and cancer. *Hum. Mol. Genet.* **10**, 699–703.
4. Harbour, J. W. and Dean, D. C. (2000). The Rb/E2F pathway: expanding roles and emerging paradigms. *Genes Dev.* **14**, 2393–2409.
5. Adams, P. D. (2001). Regulation of the retinoblastoma tumor suppressor protein by cyclin/cdks. *Biochim. Biophys. Acta* **1471**, M123–133.
6. Harbour, J. W., Luo, R. X., Dei Santi, A., Postigo, A. A., and Dean, D. C. (1999). Cdk phosphorylation triggers sequential intramolecular interactions that progressively block Rb functions as cells move through G1. *Cell* **98**, 859–869.
7. Zhang, H. S., Gavin, M., Dahiya, A., Postigo, A. A., Ma, D., Luo, R. X., Harbour, J. W., and Dean, D. C. (2000). Exit from G1 and S phase of the cell cycle is regulated by repressor complexes containing HDAC-Rb-hSWI/SNF and Rb-hSWI/SNF. *Cell* **101**, 79–89.
8. Gartel, A. L. and Tyner, A. L. (1999). Transcriptional regulation of the p21(WAF1/CIP1) gene. *Exp. Cell Res.* **246**, 280–289.
9. Macleod, K. F., Sherry, N., Hannon, G., Beach, D., Tokino, T., Kinzler, K., Vogelstein, B., and Jacks, T. (1995). p53-dependent and independent expression of p21 during cell growth, differentiation, and DNA damage. *Genes Dev.* **9**, 935–944.
10. Kolluri, S. K., Weiss, C., Koff, A., and Gottlicher, M. (1999). p27(Kip1) induction and inhibition of proliferation by the intracellular Ah receptor in developing thymus and hepatoma cells. *Genes Dev.* **13**, 1742–1753.
11. Hengst, L. and Reed, S. I. (1996). Translational control of p27Kip1 accumulation during the cell cycle. *Science* **271**, 1861–1864.
12. Millard, S. S., Yan, J. S., Nguyen, H., Pagano, M., Kiyokawa, H., and Koff, A. (1997). Enhanced ribosomal association of p27(Kip1) mRNA is a mechanism contributing to accumulation during growth arrest. *J. Biol. Chem.* **272**, 7093–7098.
13. Pagano, M., Tam, S. W., Theodoras, A. M., Beer-Romero, P., Del Sal, G., Chau, V., Yew, P. R., Draetta, G. F., and Rolfe, M. (1995). Role of the ubiquitin-proteasome pathway in regulating abundance of the cyclin-dependent kinase inhibitor p27. *Science* **269**, 682–685.
14. Nguyen, H., Gitig, D. M., and Koff, A. (1999). Cell-free degradation of p27(kip1), a G1 cyclin-dependent kinase inhibitor, is dependent on CDK2 activity and the proteasome. *Mol. Cell. Biol.* **19**, 1190–1201.
15. Sheaff, R. J., Groudine, M., Gordon, M., Roberts, J. M., and Clurman, B. E. (1997). Cyclin E-CDK2 is a regulator of p27Kip1. *Genes Dev.* **11**, 1464–1478.

16. Vlach, J., Hennecke, S., and Amati, B. (1997). Phosphorylation-dependent degradation of the cyclin-dependent kinase inhibitor p27. *EMBO J.* **16**, 5334–5344.
17. Van Aelst, L. and D'Souza-Schorey, C. (1997). Rho GTPases and signaling networks. *Genes Dev.* **11**, 2295–2322.
18. Qiu, R. G., Chen, J., Kirn, D., McCormick, F., and Symons, M. (1995). An essential role for Rac in Ras transformation. *Nature* **374**, 457–459.
19. Qiu, R. G., Chen, J., McCormick, F., and Symons, M. (1995). A role for Rho in Ras transformation. *Proc. Natl. Acad. Sci. USA* **92**, 11781–11785.
20. Olson, M. F., Ashworth, A., and Hall, A. (1995). An essential role for Rho, Rac, and Cdc42 GTPases in cell cycle progression through G1. *Science* **269**, 1270–1272.
21. Joneson, T., White, M. A., Wigler, M. H., and Bar-Sagi, D. (1996). Stimulation of membrane ruffling and MAP kinase activation by distinct effectors of RAS. *Science* **271**, 810–812.
22. Joneson, T. and Bar-Sagi, D. (1998). A Rac1 effector site controlling mitogenesis through superoxide production. *J. Biol. Chem.* **273**, 17991–17994.
23. Freeman, J. L., Abo, A., and Lambeth, J. D. (1996). Rac "insert region" is a novel effector region that is implicated in the activation of NADPH oxidase, but not PAK65. *J. Biol. Chem.* **271**, 19794–19801.
24. Irani, K., Xia, Y., Zweier, J. L., Sollott, S. J., Der, C. J., Fearon, E. R., Sundaresan, M., Finkel, T., and Goldschmidt-Clermont, P. J. (1997). Mitogenic signaling mediated by oxidants in Ras-transformed fibroblasts. *Science* **275**, 1649–1652.
25. Sahai, E., Ishizaki, T., Narumiya, S., and Treisman, R. (1999). Transformation mediated by RhoA requires activity of ROCK kinases. *Curr. Biol.* **9**, 136–145.
26. Zong, H., Kaibuchi, K., and Quilliam, L. A. (2001). The insert region of RhoA is essential for Rho kinase activation and cellular transformation. *Mol. Cell. Biol.* **21**, 5287–5298.
27. Frost, J. A., Steen, H., Shapiro, P., Lewis, T., Ahn, N., Shaw, P. E., and Cobb, M. H. (1997). Cross-cascade activation of ERKs and ternary complex factors by Rho family proteins. *EMBO J.* **16**, 6426–6438.
28. Sun, H., King, A. J., Diaz, H. B., and Marshall, M. S. (2000). Regulation of the protein kinase Raf-1 by oncogenic Ras through phosphatidylinositol 3-kinase, Cdc42/Rac and Pak. *Curr. Biol.* **10**, 281–284.
29. Chaudhary, A., King, W. G., Mattaliano, M. D., Frost, J. A., Diaz, B., Morrison, D. K., Cobb, M. H., Marshall, M. S., and Brugge, J. S. (2000). Phosphatidylinositol 3-kinase regulates Raf1 through Pak phosphorylation of serine 338. *Curr. Biol.* **10**, 551–554.
30. Hill, C. S., Wynne, J., and Treisman, R. (1995). The Rho family GTPases RhoA, Rac1, and CDC42Hs regulate transcriptional activation by SRF. *Cell* **81**, 1159–1170.
31. Westwick, J. K., Lambert, Q. T., Clark, G. J., Symons, M., Van Aelst, L., Pestell, R. G., and Der, C. J. (1997). Rac regulation of transformation, gene expression, and actin organization by multiple, PAK-independent pathways. *Mol. Cell. Biol.* **17**, 1324–1335.
32. Sotiropoulos, A., Gineitis, D., Copeland, J., and Treisman, R. (1999). Signal-regulated activation of serum response factor is mediated by changes in actin dynamics. *Cell* **98**, 159–169.
33. Tominaga, T., Sahai, E., Chardin, P., McCormick, F., Courtneidge, S. A., and Alberts, A. S. (2000). Diaphanous-related formins bridge Rho GTPase and Src tyrosine kinase signaling. *Mol. Cell* **5**, 13–25.
34. Geneste, O., Copeland, J. W., and Treisman, R. (2002). LIM kinase and Diaphanous cooperate to regulate serum response factor and actin dynamics. *J. Cell Biol.* **157**, 831–838.
35. Joyce, D., Bouzahzah, B., Fu, M., Albanese, C., D'Amico, M., Steer, J., Klein, J. U., Lee, R. J., Segall, J. E., Westwick, J. K., Der, C. J., and Pestell, R. G. (1999). Integration of Rac-dependent regulation of cyclin D1 transcription through a nuclear factor-kappaB-dependent pathway. *J. Biol. Chem.* **274**, 25245–25249.
36. Welsh, C. F., Roovers, K., Villanueva, J., Liu, Y., Schwartz, M. A., and Assoian, R. K. (2001). Timing of cyclin D1 expression within G1 phase is controlled by Rho. *Nat. Cell Biol.* **3**, 950–957.
37. Olson, M. F., Paterson, H. F., and Marshall, C. J. (1998). Signals from Ras and Rho GTPases interact to regulate expression of p21Waf1/Cip1. *Nature* **394**, 295–299.
38. Lloyd, A. C., Obermuller, F., Staddon, S., Barth, C. F., McMahon, M., and Land, H. (1997). Cooperating oncogenes converge to regulate cyclin/cdk complexes. *Genes Dev.* **11**, 663–677.
39. Pumiglia, K. M. and Decker, S. J. (1997). Cell cycle arrest mediated by the MEK/mitogen-activated protein kinase pathway. *Proc. Natl. Acad. Sci. USA* **94**, 448–452.
40. Sewing, A., Wiseman, B., Lloyd, A. C., and Land, H. (1997). High-intensity Raf signal causes cell cycle arrest mediated by p21Cip1. *Mol. Cell. Biol.* **17**, 5588–5597.
41. Hu, W., Bellone, C. J., and Baldassare, J. J. (1999). RhoA stimulates p27(Kip) degradation through its regulation of cyclin E/CDK2 activity. *J. Biol. Chem.* **274**, 3396–3401.

CHAPTER 240

Cdc42 and Its Cellular Functions

Wannian Yang and Richard A. Cerione
Department of Molecular Medicine, Cornell University, Ithaca, New York

Cdc42 is a member of the Rho subfamily of small GTP-binding (G) proteins. A variety of biochemical and cellular studies have shown that Cdc42 plays important roles in the regulation of cell growth, differentiation, programmed cell death, and in the establishment of cell polarity. The molecular mechanisms underlying these biological activities have emerged from recent breakthrough studies of the interactions of Cdc42 with its various downstream targets including Wiscott-Aldrich Syndrome Protein (WASP), γ-coatomer (γ–COP), and the p21-activated kinase (PAK), activated Cdc42-associated kinase (ACK) and partitioning-defective (PAR) proteins. Through these interactions, Cdc42 impacts two major cellular activities: F-actin polymerization and membrane vesicle trafficking. These findings now lead to the suggestion that Cdc42 serves as a convergence point for pathways that influence actin cytoskeletal architecture and intracellular trafficking.

Introduction

Cdc42 is a member of the Rho subfamily of Ras-related (small) GTP-binding proteins. The CDC42 gene was initially identified in *Saccharomyces cerevisiae* as being essential for polarized growth and assembly of the bud-site [1]. The human homolog, originally designated Cdc42Hs, was independently cloned following its identification as a potential participant in epidermal growth factor (EGF) receptor-coupled signaling [2,3]. Early microinjection studies in Swiss3T3 cells indicated that Cdc42 caused filopodium or microspike formation, thus linking it to actin cytoskeletal organization [4,5]. Now after a decade of studies, a variety of upstream signaling activators, which catalyze the guanine nucleotide exchange activity of Cdc42 (thus designated guanine nucleotide exchange factors or GEFs) have been identified including the prototype GEF, Dbl (diffuse B-cell lymphoma, as well as Cdc24, Fgd1 (facial genital dysplasia, intersectin-l/Ese1 (EH domain and SH3 domain regulator of endocytosis), hPEM (human homolog of ascidian protein posterior end mark-2), Brx (breast cancer nuclear hormone receptor auxiliary factor), and Clg (common-site lymphoma/leukemia GEF) [1, 6–8]. It is now felt that a variety of receptor-coupled signaling pathways feed into these GEFs, including those mediated by receptor tyrosine kinases, trimeric G-protein-coupled receptors, neurotrophic receptors, integrins, and cytokine receptors [12–18]. A number of targets or effectors suspected to function downstream of Cdc42 have also been discovered and characterized, including PAK, ACK, myotonic related Cdc42-associated kinase (MRCK), WASP, IQGAP, γ-COP, partitioning-defective protein 6 (PAR6) and binder of Rho GTPases (BORG) [1,9,10]. These discoveries seem to position Cdc42 in signaling pathways that regulate a vast array of cellular activities, ranging from cell growth and differentiation to apoptosis, as well as a number of fundamentally important molecular events including gene transcription, actin cytoskeletal organization, membrane vesicle endocytosis and trafficking, cell adhesion and migration, and RNA processing and transport [11]. In this review, we will focus on the molecular mechanisms underlying the ability of Cdc42 to mediate its cellular functions.

Biological Effects of Cdc42

Cell Growth Regulation

Cdc42 is essential for cell growth. Early genetic studies in *S. cerevisiae* showed that yeast cells containing function-defective mutants of Cdc42 were not viable [1], and more recently, it was shown that Cdc42-knockout mice were embryonic-lethal [19]. Overexpression of a GTPase-defective Cdc42 mutant, Cdc42(G12V), as well as a mutant capable of constitutive GTP-GDP exchange, Cdc42(F28L), induced the transformation of fibroblasts [20,21]. The dominant-negative mutant, Cdc42(T17N) blocked transformation by the oncogenic Ras protein, Ras(G12V), indicating that the activation of Cdc42 is somehow required for transforming signals of Ras. Mutations or truncations of various GEFs for Cdc42, including Dbl, Vav, Brx, and Clg, have been shown to give rise to transformation [8,22,23], further suggesting a connection between the activation of Cdc42 and cell growth regulation.

Differentiation and Development

Cdc42 has been shown to serve important functions in cellular differentiation and embryonic development, through roles in myogenesis, neurite outgrowth, monocyte differentiation, and embryogenesis [24–27]. For example, overexpression of the GTPase-defective Cdc42(G12V) mutant in L6 rat myoblasts blocked myotube formation [24]. Ectopic expression of Cdc42(G12V) in chick spinal cord neurons stimulated neurite outgrowth and enhanced the size of the growth cones and their number of filopodia [25]. In *Drosophila*, Cdc42 negatively regulates notch signaling during wing development [28].

Apoptosis

Cdc42 has been shown to regulate the apoptosis of neuronal cells and the survival of epithelial cells [29,30]. Mixed lineage kinase 3 (MLK3), a downstream effector of Cdc42, was shown to induce the apoptosis of neuronal cells through the activation of the c-Jun kinase (JNK [31]). However, it has also been suggested that Cdc42 provides a survival (antiapoptotic) signal through its activation of PAK [32], which in turn phosphorylates Bad [33]. Recent studies have shown that Cdc42 contains a caspase-substrate motif and is sensitive to caspases 3 and 7 [34]. Caspase-insensitive mutants of activated Cdc42 exhibit protective effects against Fas-induced apoptosis [34]. These data led to the suggestion that Cdc42 may somehow influence the timing of apoptotic signals, protecting against a full-scale apoptotic response, until a caspase-catalyzed degradation of Cdc42 occurs.

Cell Adhesion and Migration

Cell adhesion and migration require active cytoskeletal organization and rearrangements. Many lines of evidence indicate that Cdc42 participates in cell adhesion signaling pathways and promotes cell migration through its effects on actin cytoskeletal organization [35–37]. However, the mechanism of activation of Cdc42 by cell adhesion is still not clear.

Cell Polarity

Cell polarity encompasses asymmetrical properties of cells that are manifested during cell division, differentiation, morphogenesis, and embryogenesis. To generate or sustain cell polarity, many processes, including actin cytoskeletal organization and membrane vesicle trafficking, are required. Cdc42 has been shown to play a key role in cell polarity [1,38]. Two such polarity-dependent processes that are controlled by Cdc42 and its downstream effectors are bud-site assembly in *S. cerevisiae* and apical/basolateral protein transport in epithelial cells [1,39].

Molecular Mechanisms Underlying the Biological Activities of Cdc42

The recent identification and biochemical characterization of the downstream effectors of Cdc42 have provided important insights into the molecular mechanisms that underlie its many biological effects. Below, we describe the molecular mechanisms that are thought to be responsible for the effects of Cdc42 on actin cytoskeletal organization, membrane vesicle trafficking, and the establishment of cell polarity.

Cdc42/WASP/Arp2/3 Complexes in F-Actin Polymerization

Although a complete understanding of how Cdc42 influences actin cytoskeletal organization has not yet been achieved, it is becoming clear how Cdc42 stimulates F-actin polymerization. WASP is a specific downstream effector for Cdc42. The carboxyl terminus of WASP contains a VCA (verprolin-homology, cofilin-homology, and acidic) domain that serves as a binding site for the Arp2/3 complex to organize the actin nucleation core and initiate F-actin polymerization [40,41]. In its inactive conformation, the VCA domain is masked through an intramolecular interaction with its Cdc42/Rac-interactive binding (CRIB) domain. When activated Cdc42 binds to the CRIB domain on WASP, the VCA region is released and interacts with the Arp2/3 complex to form an actin nucleation core and initiate F-actin polymerization [41]. Cdc42-induced filopodia formation appears to be mediated through its interaction with WASP, as the overexpression or microinjection of an Arp2/3-binding defective mutant of WASP inhibits the ability of EGF or bradykinin to stimulate filopodia [41].

Cdc42/γ–COP Complexes in Cdc42-Induced Cellular Transformation and Intracellular Transport

Cdc42 is required for cell-cycle progression and DNA synthesis. It appears that the ability of Cdc42 to stimulate actin polymerization or produce actin filopodia is not required for its effects on cell growth [42]. A Cdc42 mutant, Cdc42 (Y40C), which excludes the binding of all CRIB-domain-containing effectors, only interfered with Cdc42-mediated cytoskeletal organization and JNK activation, but did not influence cell-cycle progression or DNA synthesis [42]. This implied that the effector responsible for mediating the effects of Cdc42 on cell-cycle progression is a non-CRIB-domain-containing protein. The γ subunit of the γ–COP, which lacks a CRIB domain, was found to be an essential effector for Cdc42-induced cellular transformation [10]. Because γ–COP is part of the COP1 complex, which is involved in vesicular trafficking, the interaction between Cdc42 and γ–COP suggests a direct link between Cdc42-mediated cell growth regulation and some type of a trafficking function.

Cdc42/WASP/Intersectin and Cdc42/ACK/Intersectin Complexes: Possible Roles in Receptor Endocytosis

Intersectin or Ese1 is a prototype for a family of endocytic proteins whose members contain two EH domains and five SH3 domains [43,44]. The long-form of Intersectin, designated Intersectin-l, contains a DH/PH domain and is a specific GEF for Cdc42 [6]. WASP interacts with the SH3 domains of Intesectin-l, activates its GEF activity, and influences receptor endocytosis [6,45]. It has been proposed that Cdc42/WASP/Intersectin complexes may play an important role in the regulation of endocytic vesicle transport by connecting endocytic vesicles to the actin cytoskeleton.

ACK is a nonreceptor tyrosine kinase that specifically interacts with activated forms of Cdc42 [46,47]. There are two isoforms of ACK that have very similar functional domains [46]. It has been shown that ACK contains a clathrin-binding motif and directly interacts with clathrin, an endocytic vesicle-coating protein [48,49]. Overexpression of ACK2 inhibits transferrin-receptor endocytosis via a competition with AP-2 for binding to clathrin [48]. Interestingly, both ACK2 and Intersectin-l are enriched in neuronal tissues and ACK2 interacts with the SH3 domains of Intersectin-l (Ese1) via its proline-rich domain 2 (PRD2) (Smith, W. *et al.*, unpublished data). This raises the interesting possibility that a ternary Cdc42/ACK/Intersectin-l may form in neuronal tissues, containing both a specific upstream activator and downstream target for Cdc42. Recently, it has been reported that ACK2 phosphorylates SH3PX1 (sorting nexin 9) in both *Drosophila* and mammalian cells [50,51] and facilitates the degradation of the EGF receptor [51], suggesting that the Cdc42/ACK/Intersectin-l complex may form a functional unit that directly influences receptor degradation.

Cdc42/PAR6/PKCζ Complexes in Cell Polarity

The PAR protein family has 6 members (PAR1–6) and were originally identified in *Caenorhabditis elegans* embryos as being essential for the establishment of polarity [52]. The PAR6 protein directly interacts with activated Cdc42, thus serving as a putative target/effector [53–55]. It has been shown that PAR6 binds to PAR3 through its PDZ domain [53] and that this heterodimer recruits both activated Cdc42 and PKCζ. The resultant tetrameric complex appears to function as a determinant for cell polarity as well as regulates tight junction structures in mammalian epithelial cells [54]. The Cdc42/PAR6 complex also stimulates PKCζ kinase activity and this activation has been reported to induce cellular transformation [55].

The Cdc42/PAR6/PKCζ complex apparently is not the only functional unit involved in Cdc42-mediated cell polarity. The Cdc42/γ–COP complex may also play an important role in cell polarity through the regulation of vesicular trafficking. Moreover, a ternary complex consisting of activated Cdc42, the putative Cdc42-effector IQGAP, and β–catenin has been reported to regulate cell-cell contact adhesion and cortical F-actin polymerization and to play an important role in embryogenesis and the establishment of epithelial cell polarity [56]. Recent studies have shown that Cdc42 regulates polarized exocytosis by directly interacting with Sec3p, a component of the exocyst complex [57].

Cdc42/PAK/Cool(Pix) Complexes in Cell Growth, Adhesion, and Arf6-Mediated Membrane Vesicle Trafficking

The PAK family has been studied extensively and its members have been reported to exhibit very broad effects on cell growth, differentiation, apoptosis, and actin cytoskeletal organization. However, the specific role of PAK in cells is not clear. It has been reported that PAK can mediate the Cdc42/Rac-stimulated activation of JNK, which leads to increased gene expression [58,59]. PAK activity has also been shown to be crucial for Ras-induced transformation, and PAK-catalyzed phosphorylation of Raf has been reported to be required for Raf activation [60]. PAK also phosphorylates the BAD protein to prevent cellular apoptosis [33]. There is a subfamily of Dbl-related proteins named Pix (PAK-interactive exchange factor) and Cool (cloned-out of library), whose members specifically interact with PAK [61,62]. In cells, Cdc42, PAK, and Cool(Pix), together with a Cool(Pix)-binding partner called Cat (Cool-associated tyrosine phosphosubstrate; also very similar if not identical to the G-protein-coupled receptor kinase interactor or GIT, and the Paxillin-kinase linker or PKL), form a stable complex [63]. The Cat protein has an Arf-GAP domain and specifically stimulates Arf6 GTPase activity, thus connecting Cdc42/Rac signaling to Arf6 signaling [64]. The Cat protein also directly interacts with Paxillin, an important component in focal adhesion complexes, thereby providing a possible link between Cdc42/Rac signaling and cell adhesion [65].

Conclusions

Cdc42 influences a broad range of biological and cellular activities that are mediated by multiple downstream effectors. How Cdc42 is able to discriminately interact with its many upstream activators, GEFs, and downstream targets to give rise to specific signals will be an important issue for further understanding how this GTP-binding protein mediates its multiple functions. One intriguing possibility is that specific signaling complexes are constructed that contain both a specific GEF and a specific target/effector for Cdc42.

Two such examples are the Cdc42/PAK/Cool(Pix) complex and the Cdc42/WASP/Intersectin-1 complex. At present, there appear to be two major cellular activities that are regulated by Cdc42: actin cytoskeletal organization and membrane vesicle trafficking. Whether these two functions are often intimately linked or more typically represent distinct activities under the control of Cdc42 remains to be established. There are some data implicating Cdc42 in the organization of F-actin cytoskeletal structures in vacuole membranes, as well as in T-cell receptor endocytic vesicles and in Golgi vesicles [66,67]. It is thus intriguing to hypothesize that Cdc42 organizes actin cytoskeleton-based vesicle "motors" that power and guide vesicle transport. Hopefully, in the not-too-distant future we will be able to determine the validity of this hypothesis.

References

1. Johnson, D. I. (1999). Cdc42: An essential Rho-type GTPase controlling eukaryotic cell polarity. *Microbiol. Mol. Biol. Rev.* **63**, 54–105.
2. Hart, M., Polakis, P., Evans, T., and Cerione, R. A. (1990). Identification and characterization of a low molecular weight GTP binding protein which is a phospho-substrate for the epidermal growth factor receptor/tyrosine kinase. *J. Biol. Chem.* **265**, 5990–6001.
3. Shinjo, K., Koland, J. G., Hart, M. J., Narasimhan, V., Johnson, D. I., Evans, T., and Cerione, R. A. (1990). Molecular cloning of the gene for the human placental GTP-binding protein Gp (G25K): Identification of this GTP-binding protein as the human homolog of the yeast cell-division-cycle protein CDC42. *Proc. Natl. Acad. Sci. USA* **87**, 9853–9857.
4. Nobes, C. D. and Hall, A. (1995). Rho, rac, and cdc42 GTPases regulate the assembly of multimolecular focal complexes associated with actin stress fibers, lamellipodia, and filopodia. *Cell* **81**, 53–62.
5. Nobes, C. D. and Hall, A. (1995). Rho, rac and cdc42 GTPases: regulators of actin structures, cell adhesion and motility. *Biochem. Soc. Trans.* **23**, 456–459.
6. Hussain, N. K., Jenna, S., Glogauer, M., Quinn, C. C., Wasiak, S., Guipponi, M., Antonarakis, S. E., Kay, B. K., Stossel, T. P., Lamarche-Vane, N., and McPherson, P. S. (2001). Endocytic protein intersectin-1 regulates actin assembly via Cdc42 and N-WASP. *Nat. Cell Biol.* **3**, 927–932.
7. Reid, T., Bathoorn, A., Ahmadian, M. R., and Collard, J. G. (1999). Identification and characterization of hPEM-2, a guanine nucleotide exchange factor specific for Cdc42. *J. Biol. Chem.* **274**, 33587–33593.
8. Himmel, K. L., Bi, F., Shen, H., Jenkins, N. A., Copeland, N. G., Zheng, Y., and Largaespada, D. A. (2002). Activation of Clg, a novel Dbl family guanine nucleotide exchange factor gene, by proviral insertion at Evi24, a common integration site in B cell and myeloid Leukemia. *J. Biol. Chem.* **277**, 13463–13472.
9. Joberty, G., Perlungher, R. R., and Macara, I. G. (1999). The Borgs, a new family of Cdc42 and TC10 GTPase-interacting proteins. *Mol. Cell. Biol.* **19**, 6585–6597.
10. Wu, W. J., Erickson, J. W., Lin, R., and Cerione, R. A. (2000). The gamma-subunit of the coatomer complex binds Cdc42 to mediate transformation. *Nature* **405**, 800–804.
11. Erickson, J. W. and Cerione, R. A. (2001). Multiple roles for Cdc42 in cell regulation. *Curr. Opin. Cell Biol.* **13**, 153–157.
12. Kozma, R., Ahmed, S., Best, A. and Lim, L. (1995). The Ras-related protein Cdc42Hs and bradykinin promote formation of peripheral actin microspikes and filopodia in Swiss 3T3 fibroblasts. *Mol. Cell. Biol.* **15**, 1942–1952.
13. Shekarabi, M. and Kennedy, T. E. (2002). The netrin-1 receptor DCC promotes filopodia formation and cell spreading by activating Cdc42 and Rac1. *Mol. Cell. Neurosci.* **19**, 1–17.
14. Wong, K., Ren, X. R., Huang, Y. Z., Xie, Y., Liu, G., Saito, H., Tang, H., Wen, L., Brady-Kalnay, S. M., Mei, L., Wu, J. Y., Xiong, W. C., and Rao, Y. (2001). Signal transduction in neuronal migration: roles of GTPase activating proteins and the small GTPase Cdc42 in the Slit-Robo pathway. *Cell* **107**, 209–221.
15. Driggers, P. H., Segars, J. H., and Rubino, D. M. (2001). The proto-oncoprotein Brx activates estrogen receptor beta by a p38 mitogen-activated protein kinase pathway. *J. Biol. Chem.* **276**, 46792–46797.
16. Ku, G. M., Yablonski, D., Manser, E., Lim, L., and Weiss, A. (2001). A PAK1-PIX-PKL complex is activated by the T-cell receptor independent of Nck, Slp-76 and LAT. *EMBO J.* **20**, 457–65.
17. Keely, P. J., Westwick, J. K., Whitehead, I. P., Der, C. J., and Parise, L. V. (1997). Cdc42 and Rac1 induce integrin-mediated cell motility and invasiveness through PI(3)K. *Nature* **390**, 632–636.
18. Puls, A., Eliopoulos, A. G., Nobes, C. D., Bridges, T., Young, L. S., and Hall, A. (1999). Activation of the small GTPase Cdc42 by the inflammatory cytokines TNF(alpha) and IL-1, and by the Epstein-Barr virus transforming protein LMP1. *J. Cell Sci.* **112**, 2983–2992.
19. Chen, F., Ma, L., Parrini, M. C., Mao, X., Lopez, M., Wu, C., Marks, P. W., Davidson, L., Kwiatkowski, D. J., Kirchhausen, T., Orkin, S. H., Rosen, F. S., Mayer, B. J., Kirschner, M. W., and Alt, F. W. (2000). Cdc42 is required for PIP(2)-induced actin polymerization and early development but not for cell viability. *Curr. Biol.* **10**, 758–765.
20. Qiu, R. G., Abo, A., McCormick, F., and Symons, M. (1997). Cdc42 regulates anchorage-independent growth and is necessary for Ras transformation. *Mol. Cell. Biol.* **17**, 3449–3458.
21. Lin, R., Bagrodia, S., Cerione, R., and Manor, D. (1997). A novel Cdc42Hs mutant induces cellular transformation. *Curr. Biol.* **7**, 794–797.
22. Driggers, P. H., Segars, J. H., and Rubino, D. M. (2001). The proto-oncoprotein Brx activates estrogen receptor beta by a p38 mitogen-activated protein kinase pathway. *J. Biol. Chem.* **276**, 46792–46797.
23. Zheng Y. (2001). Dbl family guanine nucleotide exchange factors. *Trends Biochem. Sci.* **26**, 724–732.
24. Meriane, M., Roux, P., Primig, M., Fort, P., and Gauthier-Rouviere, C. (2000). Critical activities of Rac1 and Cdc42Hs in skeletal myogenesis: Antagonistic effects of JNK and p38 pathways. *Mol. Biol. Cell* **11**, 2513–2528.
25. Brown, M. D., Cornejo, B. J., Kuhn, T. B., and Bamburg, J. R. (2000). Cdc42 stimulates neurite outgrowth and formation of growth cone filopodia and lamellipodia. *J. Neurobiol.* **43**, 352–364.
26. Aepfelbacher, M., Vauti, F., Weber, P. C., and Glomset, J. A. (1994). Spreading of differentiating human monocytes is associated with a major increase in membrane-bound CDC42. *Proc. Natl. Acad. Sci. USA* **91**, 4263–4267.
27. Crawford, J. M., Harden, N., Leung, T., Lim, L., and Kiehart, D. P. (1998). Cellularization in Drosophila melanogaster is disrupted by the inhibition of rho activity and the activation of Cdc42 function. *Dev. Biol.* **204**, 151–164.
28. Baron, M., O'Leary, V., Evans, D. A., Hicks, M., and Hudson, K. (2000). Multiple roles of the Dcdc42 GTPase during wing development in Drosophila melanogaster. *Mol. Gen. Genet.* **264**, 98–104.
29. Linseman, D. A., Laessig, T., Meintzer, M. K., McClure, M., Barth, H., Aktories, K., and Heidenreich, K. A. (2001). An essential role for Rac/Cdc42 GTPases in cerebellar granule neuron survival. *J. Biol. Chem.* **276**, 39123–39131.
30. Zugasti, O., Rul, W., Roux, P., Peyssonnaux, C., Eychene, A., Franke, T. F., Fort, P., and Hibner, U. (2001). Raf-MEK-Erk cascade in anoikis is controlled by Rac1 and Cdc42 via Akt. *Mol. Cell. Biol.* **21**, 6706–6717.
31. Xu, Z., Maroney, A. C., Dobrzanski, P., Kukekov, N. V., and Greene, L. A. (2001). The MLK family mediates c-Jun N-terminal kinase activation in neuronal apoptosis. *Mol. Cell. Biol.* **21**, 4713–4724.
32. Faure, S., Vigneron, S., Doree, M., and Morin, N. (1997). A member of the Ste20/PAK family of protein kinases is involved in both arrest of Xenopus oocytes at G2/prophase of the first meiotic cell cycle and in prevention of apoptosis. *EMBO J.* **16**, 5550–5561.

33. Schurmann, A., Mooney, A. F., Sanders, L. C., Sells, M. A., Wang, H. G., Reed, J. C., and Bokoch, G. M. (2000). p21-activated kinase 1 phosphorylates the death agonist bad and protects cells from apoptosis. *Mol. Cell. Biol.* **20**, 453–461.
34. Tu, S. and Cerione, R. A. (2001). Cdc42 is a substrate for caspases and influences Fas-induced apoptosis. *J. Biol. Chem.* **276**, 19656–19663.
35. Schmitz, A. A., Govek, E. E., Bottner, B., and Van Aelst, L. (2000). Rho GTPases: Signaling, migration, and invasion. *Exp Cell. Res.* **261**, 1–12.
36. Cox, E. A., Sastry, S. K., and Huttenlocher, A. (2001). Integrin-mediated adhesion regulates cell polarity and membrane protrusion through the Rho family of GTPases. *Mol. Biol. Cell* **12**, 265–277.
37. Weber, K. S., Klickstein, L. B., Weber, P. C., and Weber, C. (1998). Chemokine-induced monocyte transmigration requires cdc42-mediated cytoskeletal changes. *Eur. J. Immunol.* **28**, 2245–2251.
38. Chant, J. (1999). Cell polarity in yeast. *Annu. Rev. Cell Dev. Biol.* **15**, 365–391.
39. Cohen, D., Musch, A., and Rodriguez-Boulan, E. (2001). Selective control of basolateral membrane protein polarity by cdc42. *Traffic* **2**, 556–564.
40. Rohatgi, R., Ma, L., Miki, H., Lopez, M., Kirchhausen, T., Takenawa, T., and Kirschner, M. W. (1999). The interaction between N-WASP and the Arp2/3 complex links Cdc42-dependent signals to actin assembly. *Cell* **97**, 221–231.
41. Takenawa, T. and Miki, H. (2001). WASP and WAVE family proteins: key molecules for rapid rearrangement of cortical actin filaments and cell movement. *J. Cell Sci.* **114**, 1801–1809.
42. Lamarche, N., Tapon, N., Stowers, L., Burbelo, P. D., Aspenstrom, P., Bridges, T., Chant, J., and Hall, A. (1996). Rac and Cdc42 induce actin polymerization and G1 cell cycle progression independently of p65PAK and the JNK/SAPK MAP kinase cascade. *Cell* **87**, 519–529.
43. Hussain, N. K., Yamabhai, M., Ramjaun, A. R., Guy, A. M., Baranes, D., O'Bryan, J. P., Der, C. J., Kay, B. K., and McPherson, P. S. (1999). Splice variants of intersectin are components of the endocytic machinery in neurons and nonneuronal cells. *J. Biol. Chem.* **274**, 15671–15677.
44. Sengar, A. S., Wang, W., Bishay, J., Cohen, S., and Egan, S. E. (1999). The EH and SH3 domain Ese proteins regulate endocytosis by linking to dynamin and Eps15. *EMBO J.* **18**, 1159–1171.
45. McGavin, M. K., Badour, K., Hardy, L. A., Kubiseski, T. J., Zhang, J., and Siminovitch K. A. (2001). The intersectin 2 adaptor links Wiskott Aldrich Syndrome protein (WASp)-mediated actin polymerization to T cell antigen receptor endocytosis. *J. Exp. Med.* **194**, 1777–1787.
46. Yang, W. and Cerione, R. A. (1997). Cloning and characterization of a novel Cdc42-associated tyrosine kinase, ACK2, from bovine brain. *J. Biol. Chem.* **272**, 24819–24824.
47. Manser, E., Leung, T., Salihuddin, H., Tan, L., and Lim, L. (1993). A non-receptor tyrosine kinase that inhibits the GTPase activity of p21cdc42. *Nature* **363**, 364–367.
48. Yang, W., Lo, C. G., Despenza, T., and Cerione, R. A. (2001). ACK2 directly interacts with clathrin and inhibits AP-2 mediated receptor endocytosis. *J. Biol. Chem.* **276**, 17468–17473.
49. Teo, M., Tan, L., Lim, L., and Manser, E. (2001). The tyrosine kinase ACK1 associates with clathrin-coated vesicles through a binding motif shared by arrestin and other adaptors. *J. Biol. Chem.* **276**, 18392–18398.
50. Clemens, J. C., Worby, C. A., Simonson-Leff, N., Muda, M., Maehama, T., Hemmings, B. A., and Dixon, J. E. (2000). Use of double-stranded RNA interference in *Drosophila* cell lines to dissect signal transduction pathways. *Proc. Natl. Acad. Sci. USA* **97**, 6499–6503.
51. Lin, Q., Lo, C. G., Cerione, R. A., and Yang, W. (2002). The Cdc42-target ACK2 interacts with SH3PX1 (sorting nexin 9) to regulate EGF receptor degradation. *J. Biol. Chem.* **277**, 10134–10138.
52. Brazil, D. P. and Hemmings, B. A. (2000). Cell polarity: Scaffold proteins par excellence. *Curr. Biol.* **10**, R592–R594.
53. Joberty, G., Petersen, C., Gao, L., and Macara, I. G. (2000). The cell-polarity protein Par6 links Par3 and atypical protein kinase C to Cdc42. *Nat. Cell Biol.* **2**, 531–539.
54. Gao, L., Joberty, G., and Macara, I. G. (2002). Assembly of epithelial tight junctions is negatively regulated by Par6. *Curr. Biol.* **12**, 221–225.
55. Qiu, R. G., Abo, A., and Steven Martin, G. (2000). A human homolog of the C. elegans polarity determinant Par-6 links Rac and Cdc42 to PKCzeta signaling and cell transformation. *Curr. Biol.* **10**, 697–707.
56. Fukata, M. and Kaibuchi, K. (2001). Rho-family GTPases in cadherin-mediated cell-cell adhesion. *Nat. Rev. Mol. Cell Biol.* **2**, 887–897.
57. Zhang, X., Bi, E., Novick, P., Du, L., Kozminski, K. G., Lipschutz, J. H., and Guo, W. (2001). Cdc42 interacts with the exocyst and regulates polarized secretion. *J. Biol. Chem.* **276**, 46745–46750.
58. Bagrodia, S., Derijard, B., Davis, R. J., and Cerione, R. A. (1995). Cdc42 and PAK-mediated signaling leads to Jun kinase and p38 mitogen-activated protein kinase activation. *J. Biol. Chem.* **270**, 27995–27998.
59. Zhang, S., Han, J., Sells, M. A., Chernoff, J., Knaus, U. G., Ulevitch, R. J., and Bokoch, G. M. (1995). Rho family GTPases regulate p38 mitogen-activated protein kinase through the downstream mediator Pak1. *J. Biol. Chem.* **270**, 23934–23936.
60. King, A. J., Sun, H., Diaz, B., Barnard, D., Miao, W., Bagrodia, S., and Marshall, M. S. (1998). The protein kinase Pak3 positively regulates Raf-1 activity through phosphorylation of serine 338. *Nature* **396**, 180–183.
61. Manser, E., Loo, T. H., Koh, C. G., Zhao, Z. S., Chen, X. Q., Tan, L., Tan, I., Leung, T., and Lim, L. (1998). PAK kinases are directly coupled to the PIX family of nucleotide exchange factors. *Mol. Cell* **1**, 183–192.
62. Bagrodia, S., Taylor, S. J., Jordon, K. A., Van Aelst, L., and Cerione, R. A. (1998). A novel regulator of p21-activated kinases. *J. Biol. Chem.* **273**, 23633–23636.
63. Bagrodia, S., Bailey, D., Lenard, Z., Hart, M., Guan, J. L., Premont, R. T., Taylor, S. J., and Cerione, R. A. (1999). A tyrosine-phosphorylated protein that binds to an important regulatory region on the cool family of p21-activated kinase-binding proteins. *J. Biol. Chem.* **274**, 22393–22400.
64. Premont, R. T., Claing, A., Vitale, N., Freeman, J. L., Pitcher, J. A., Patton, W. A., Moss, J., Vaughan, M., and Lefkowitz, R. J. (1998). Beta2-Adrenergic receptor regulation by GIT1, a G protein-coupled receptor kinase-associated ADP ribosylation factor GTPase-activating protein. *Proc. Natl. Acad. Sci. USA* **95**, 14082–14087.
65. Turner, C. E., Brown, M. C., Perrotta, J. A., Riedy, M. C., Nikolopoulos, S. N., McDonald, A. R., Bagrodia, S., Thomas, S., and Leventhal, P. S. (1999). Paxillin LD4 motif binds PAK and PIX through a novel 95-kD ankyrin repeat, ARF-GAP protein: A role in cytoskeletal remodeling. *J. Cell Biol.* **145**, 851–863.
66. Muller, O., Johnson, D. I., and Mayer, A. (2001). Cdc42p functions at the docking stage of yeast vacuole membrane fusion. *EMBO J.* **20**, 5657–5665.
67. Fucini, R. V., Chen, J. L., Sharma, C., Kessels, M. M., and Stamnes, M. (2002). Golgi vesicle proteins are linked to the assembly of an actin complex defined by mAbp1. *Mol. Biol. Cell* **13**, 621–631.

CHAPTER 241

Tissue Transglutaminase: A Unique GTP-Binding/GTPase

Richard A. Cerione

*Department of Molecular Medicine and
Department of Chemistry and Chemical Biology,
Cornell University, Ithaca, New York*

Tissue transglutaminase (TGase) is capable of GTP-binding/GTPase activities like members of the large and small families of G proteins, as well as exhibiting an enzymatic activity that catalyzes the crosslinking of glutamine residues to primary amino groups or to the epsilon amino groups of protein lysine residues (transamidation). The binding of GTP to the TGase has a negative regulatory effect on transamidation, whereas Ca^{2+}, which is essential for transamidation activity, weakens the binding of guanine nucleotides. The recent determination of the three-dimensional structure for the guanosine diphosphate (GDP)-bound form of TGase sheds light on the molecular basis of the interplay between guanine binding and transamidation activity. Recent studies also point to an unexpected role for the GTP-binding activity, as well as transamidation activity, in the protection of cells against apoptosis.

Introduction

Tissue transglutaminase (TGase) is capable of multiple types of catalytic activity. Namely, it undergoes a GTP-binding/GTP hydrolytic cycle like members of the families of large heterotrimeric G proteins and small Ras-related G proteins [1–3], as well as it exhibits an ATP hydrolytic activity which appears to be noncompetitive with GTPase activity [4], and an acyl transferase activity [5,6] that results in the crosslinking of proteins (transamidation). Recent breakthroughs on the three-dimensional structure of TGase [7–10], and new insights into how TGase influences whether cells undergo differentiation versus apoptosis [11–13], provide interesting perspectives regarding the relationship between the GTP-binding/GTPase activities and the transamidation capability that resides on this single chain polypeptide. It is the relationship between these activities and how they might fit into new biological roles for the TGase that is the subject of this chapter.

General Overview

Transglutaminases are Ca^{2+}-dependent acyl transferases that catalyze the formation of an amide bond between the γ-carboxamide groups of peptide-bound glutamine residues (which serve as acceptor residues in the reaction) and the primary amino groups in various (donor) compounds, and in particular, the ε-amino group of lysine residues in proteins [5,6]. These enzymes contain an active site thiol group (6) within a Cys-His-Asp catalytic triad [7,10]. Six classes of transglutaminases have been identified and characterized in mammals. These are factor XIII-A and the keratinocyte (type I), tissue (type II or TGase), epidermal (type III), and prostate (type IV) transglutaminases, as well as TGX [6,14]. The erythrocyte membrane proteins Band 4.2 are also typically classified as a noncatalytic transglutaminase homologs [15]. It has been suggested that the different members of the transglutaminase family participate in diverse biological activities.

Tissue transglutaminase (TGase) is particularly interesting because of its ability to bind and hydrolyze GTP like traditional G proteins. There have been a number of suggestions that TGase functions as a signal transducer by mediating the stimulation of phospholipase C activity by α1-adrenergic receptors [16–21]. An important breakthrough came when it was appreciated that a GTP-binding protein, designated Gh, which was being studied for its ability to couple α adrenergic receptors to phosphoinositide lipid metabolism, was identical to TGase [18]. Other studies have shown that TGase binds most effectively to certain α1 adrenergic receptor subtypes (α1B and α1D but not α1A) [21,22], and that the target/effector for TGase is not phospholipase C-β, which is regulated by the heterotrimeric Gq protein [23], but rather phospholipase C-δ [16,24]. It remains to be established just how analogous a role TGase plays in α adrenergic-coupled phospholipase C activity compared to the well-characterized role of the Gq protein. The expectation is that some insight will come from comparisons of the structures of the GDP- and GTP-bound forms of TGase (see the following section), and whether there are specific conformationally sensitive regions on the TGase that are analogous to the switch I and II regions of large and small G proteins.

Recent Implications

Tissue transglutaminase appears to be ubiquitously distributed in mammalian tissues and exists in a variety of cellular locations. It has been generally found in cytosol, although TGase has also been purified from the nucleus [3], and appears in the plasma membrane upon stimulating the cells with certain factors (e.g., retinoic acid [25]). It has been suggested that TGase participates in a variety of biological activities including wound healing [26]), cell-matrix interactions [27], differentiation [25,28–30], neuronal degeneration [14,31], and programmed cell death [32–34]. In particular, possible roles for TGase in both the induction of and protection against apoptosis have received a good deal of attention over the past few years. A number of correlative observations have been made regarding TGase expression and transamidation activity and cellular apoptosis, among the most interesting being the TGase-catalyzed crosslinking of the retinoblastoma protein (Rb) [34]. Given the potential for transamidation to disrupt protein function, conditions that generate hyperactive forms of TGase could well have deleterious effects on cells and likely account for the tight regulatory control over TGase expression and activation that has been observed [25]. However, it is becoming increasingly attractive to consider that TGase is expressed and/or activated in response to factors that would normally cause some insult or stress to cells, and that even in some cases, TGase-catalyzed transamidation is directed toward protecting against these perturbations rather than exacerbating them (see the following sections). There are now reasons to believe that the GTP-binding activity and potential signaling function of the TGase, as well as its enzymatic transamidation activity, contribute to its protective (antiapoptotic) effects. This then makes it important to understand the molecular basis for the GTP-binding/GTP hydrolytic activity and the transamidation activity and their functional interplay.

TGase as a GTP-Binding/GTPase

Biochemical Characterizations

It has been realized for some time, starting with the initial characterizations of purified preparations of TGase, that there is a negative relationship between guanine-nucleotide-binding and transamidation activity [1–3]. In some studies, it has been shown that GTP is a much more potent inhibitor of transamidation activity, compared to GDP [1,2], thereby raising the provocative possibility of a direct coupling between the protein crosslinking activity of the TGase and its GTP-binding/GTPase cycle. However, in other studies, the differences in the abilities of GTP versus GDP to inhibit transamidation activity were minimal [4]. If true, this would then raise questions regarding the cellular conditions necessary for transamidation to occur, and in particular, whether those physiologically relevant transamidation substrates, upon binding to the TGase, weaken guanine nucleotide binding and stabilize a guanine-nucleotide-depleted state of the protein. Unlike the large and small G proteins, which show a high degree of preference for guanine nucleotides relative to adenine nucleotides, the TGase is able to bind and hydrolyze ATP. It is interesting that ATP binding and hydrolysis are not competitive with GTP binding and hydrolysis, suggesting that distinct binding sites exist on the TGase for these two nucleotides. At the present time, there is little known regarding the role of the ATP hydrolytic activity and further insight awaits the generation of TGase mutations that are highly specific for blocking ATP binding and hydrolysis without perturbing guanine nucleotide binding or GTP hydrolysis. There also have been questions regarding the affinity of guanine nucleotides for the TGase, as compared to guanine nucleotide binding to G proteins, as well as how rapidly GTP is hydrolyzed by the TGase. Some estimates have suggested that guanine nucleotides bind to the TGase with relatively weak affinity (dissociation constants in the micromolar range) and that GTP is hydrolyzed relatively slowly [4]. However, the turnover number for GTP hydrolysis catalyzed by the TGase purified from rabbit liver nuclei [3] was measured to be ~1 mol ^{32}Pi released per minute per mol protein, which is on the order of the intrinsic GTP hydrolytic activity for heterotrimeric G proteins. Moreover, the TGase has recently been crystallized with bound GDP (see the following sections), suggesting a high-affinity interaction similar to what has been demonstrated for members of both the large and small G-protein families.

Three-Dimensional Structure for the GDP-Bound TGase

The three-dimensional structure for the GDP-bound form of the human TGase has recently been determined to 2.8 Å resolution by x-ray crystallography [10]. The TGase organizes as three dimers within the unit cell. Each monomer contains a bound GDP molecule and is made up of four distinct domains, an amino terminal β-sandwich domain (Met1 to Phe139), a transamidation catalytic core (Ala147 to Asn460), and two carboxyl-terminal β-barrel domains (Gly472 to Tyr583 and Ile591 to Ala687, respectively). Cysteine 277 lies within the middle of a groove within the transamidation active site [thus the TGase(C277A) mutant is transamidation-defective] and is part of a catalytic triad, Cys277-His-335-Asp358. The transamidation active site and the general domain structure of TGase are similar to those reported for factor XIIIa and for TGase III [7,8].

The guanine-nucleotide-binding site lies in a cleft formed by the catalytic core domain and the first β-barrel domain [10]. Most of the residues contacting GDP come from the end of the first β-strand of the first β-barrel domain and the loop that connects it to the second β-strand. There are also two residues contributed by the catalytic core domain that are involved in binding the guanine ring moiety. In this regard, the overall architecture for the guanine-nucleotide-binding site on TGase differs significantly from the guanine-nucleotide-binding domains of the α subunits of heterotrimeric (large) G proteins or the Ras-related (small) G proteins. For both the large and small G proteins, the guanine-nucleotide-binding site consists of five helices surrounding a six-stranded β-sheet [35]. The α subunits of the large G proteins also contain a helical domain that is adjacent to (and in effect closes over) the guanine-nucleotide-binding site and enables the α subunits to bind GDP with high affinity. Ras and other small G proteins lack this helical domain and bind GDP with high affinity through the coordination of Mg^{2+}. Both large and small G proteins require Mg^{2+} for the GTP hydrolytic reaction and thus share conserved serine and threonine residues that bind to the β- and γ-phosphates of the guanine nucleotide. The TGase also requires Mg^{2+} for its GTP hydrolytic activity, however, the location of the Mg^{2+}-binding site is not obvious as the TGase lacks amino acids with either hydroxyl or carboxyl side chains in the immediate vicinity of the nucleotide phosphate groups. Mutation of serine 171 does inhibit the GTP-binding activity of the TGase, however, the position of this serine is closer to the guanine ring rather than to the nucleotide phosphate groups and thus would seem to be an unlikely participant in the GTP hydrolytic reaction.

The importance or location of the normally essential arginine (i.e., arginine finger) for the GTP hydrolytic reaction of the TGase is also not known at the present time. There are various positively charged groups that surround the phosphate moieties of the bound GDP on the TGase and could conceivably play a similar role as the essential arginine residues provided by the helical domains of the large G protein and by the GTPase-activating proteins (GAPs) of small G proteins. These include Arg 580 from the last β-strand of the first β-barrel domain of the TGase, which forms two ion-pairs with the α- and β-phosphate groups of the guanine nucleotide, and Arg 478 from the first β-strand, which is located close to the β-phosphate of GDP. Both Arg 478 and Arg 580 are conserved in all TGases (i.e., from different species) but are missing in other members of the transglutaminase family that are incapable of binding GTP. Because the β-phosphate is pointed toward Arg 478, the γ-phosphate of GTP would likely need to rotate to avoid clashing with the side chain of Arg 478 and its hydrogen-bonding partner Val 479. This would bring the γ-phosphate into the vicinity of the positively charged side chains of Lys 173 and Arg 476. Because the TGase lacks the conserved glutamine residue found in all large and small GTP-binding proteins, which functions to position water for nucleophilic attack during GTP hydrolysis, a plausible mechanism for the GTP hydrolytic activity of the TGase would be that a water molecule hydrogen bonded to either the side chain of Lys 173 or Arg 476 would serve as the attacking group. Consistent with this idea, mutation of Lys 173 yields a GTPase-impaired TGase molecule [36].

Another notable difference between the guanine-nucleotide-binding pockets of large and small G proteins and that of the TGase is the absence of the NKXD motif that is conserved in virtually every large and small G protein. The asparagine residue within this motif always forms a hydrogen bond with the N7 atom of the guanine moiety, and the aspartic acid residue hydrogen bonds with the N1 and N2 atoms. In the TGase, a main chain oxygen from Tyr 583 from the last β-strand of the first β-barrel domain forms hydrogen bonds with the N1 and N2 atoms of the guanine ring, and Ser 482 from the first β-strand forms an additional hydrogen bond with N2.

There is one interesting similarity regarding the guanine-nucleotide-binding site in the TGase, compared to that for Ras or other small G proteins like Cdc42. In the TGase, the guanine ring lies in a hydrophobic pocket that includes a phenylalanine residue (Phe 174), which appears to be positioned to stack with the guanine moiety. The α subunits of heterotrimeric G proteins lack such a phenylalanine residue, however, in Ras and other small GTP-binding proteins, a conserved phenylalanine approaches one side of the guanine ring at an ~90° angle and participates in π-π stacking interactions. This stabilization has been noteworthy, as mutation of the phenylalanine (Phe 28) to leucine in Ras and Cdc42 yields proteins that show an accelerated dissociation of tightly bound GDP and are capable of constitutive GTP-GDP exchange in cells, thereby giving rise to malignant transformation [37,38]. It will be interesting to see if the corresponding mutation in TGase yields a molecule capable of constitutive GDP-GTP exchange activity.

An examination of the TGase structure also provides some possible insight into the nature of the interplay between guanine nucleotide binding and transamidation activity. In the GDP-bound form of TGase, it appears that access to the enzymatic (transamidation) active site is restricted by a loop that connects the third and fourth β-strands, and a loop that connects the fifth and sixth β-strands of the first β-barrel domain. Tyrosine 516, a residue conserved in different members of the transglutaminase family, is located within the first loop and hydrogen bonds with the essential cysteine residue (Cys 277). It is expected that Tyr 516 needs to move to make the essential cysteine accessible to participate in catalysis, and it is easy to imagine how guanine nucleotide binding might stabilize the tyrosine residue and hinder the accessibility to the catalytic cysteine. The binding of GTP could further stabilize this catalytically compromised conformation thereby accounting for the enhanced inhibitory effects exhibited by GTP toward transamidation activity.

Based on comparisons of the structures of TGase, factor XIIIa, and type III transglutaminase [7–10], a major Ca^{2+}-binding site on the TGase is formed by the side chains of Asn 436, Asp 438, Glu 485, and Glu 490, as well as by the main chain oxygen from Ala 457. These are located near the end of the loop that connects the transamidation active site to the first β-barrel domain. In the TGase, Ile 416 and Ser 419 form a β-strand (anti-parallel) with Leu 577 and Glu 579 in a manner that maintains the first β-barrel domain and stabilizes the guanine-nucleotide-binding site [10]. It seems reasonable to expect that the binding of Ca^{2+} alters the position of Ile 416 and Ser 419 thus eliminating their stabilizing effects and weakening guanine nucleotide binding. Increasing Ca^{2+} would then potentially have a dual effect on transamidation activity by inducing a conformational change that is both essential for catalysis and weakens guanine nucleotide binding, which in turn reverses a negative regulatory effect normally imparted by bound guanine nucleotide.

New Links to Biological Function

It has been well documented that TGase activity is under the control of retinoic acid (RA) in a variety of cell types [11,25]. In some cases (e.g., the human leukemia cell line HL60), both the expression and activation of TGase are stimulated by RA treatment, whereas in HeLa cells, we have found that RA promotes the activation of TGase without having a significant effect on its expression. Because of its ability to trigger cell cycle arrest and cellular differentiation, RA has been examined as a possible therapeutic agent against human cancers. Whereas the natural retinoids have shown somewhat limited activity, the synthetic analog all-trans-N-(4-hydroxyphenyl) retinamide (HPR) has shown some promise in the treatment of breast and prostate cancers [39–041]. Interestingly, HPR has been consistently linked to apoptosis, while RA has been reported to give rise to cellular differentiation as well as apoptosis [42,43]. We have found that only HPR, and not RA, induces an apoptotic response in both NIH3T3 cells and HL60 cells. A major difference that has been detected in response to treatment with RA versus HPR is that RA stimulates the expression and/or activation of TGase, whereas HPR appears to inhibit its expression. It is interesting that pretreatment of cells with RA, prior to addition of HPR, induces TGase expression and activation and completely protects against HPR-induced apoptosis. The RA-mediated protection effect can be mimicked by the expression of wild-type TGase in NIH3T3 cells or by a TGase (C277S) mutant where the active site cysteine residue has been changed to serine. However, TGase mutants that are unable to bind GTP [e.g., TGase (S171E)] do not provide protection against HPR.

These results have led us to propose that the TGase may be functioning as a protection-factor that is activated by differentiation agents in order to ensure that cells remain viable and not susceptible to cell death programs during cellular differentiation. Apparently, the GTP-bound TGase can activate a signaling pathway that is essential for this protection effect. It will be interesting in the future to delineate this survival pathway and to identify the target that binds to GTP-bound TGase and mediates these effects. It should be noted that addition of the TGase competitive inhibitor, monodansylcadaverine (MDA), to RA-treated cells results in converting the RA response to one that mimics HPR treatment (i.e., apoptosis). This is interesting in light of our finding that mutation of the active site cysteine on the TGase did not abrogate its protection capability or give rise to apoptosis. Thus, the effects of MDA may be directed at altering the ability of TGase to bind GTP or preventing GTP-bound TGase from initiating signaling activities that can protect against apoptosis. However, recently, we have found that in Rb knock-out cells, which are sensitive to HPR, co-expression of wild-type TGase and Rb can overcome the apoptotic effects of HPR, whereas expression of TGase (C277A) together with Rb cannot [13]. Thus, there may be cell types where both the GTP-binding activity of the TGase and its transamidation activity are critical to trigger pathways that protect against programmed cell death.

Future Directions

A number of questions regarding the role of TGase, and in particular its GTP-binding/GTPase cycle, in various cellular functions await future studies. It will be important to know whether in fact the TGase undergoes significant conformational changes as an outcome of GDP-GTP exchange, analogous to large and small G proteins, and where those changes occur. How are such changes translated into the binding of target/effector proteins and what are the identities of these proteins? At the present time, phospholipase C appears to be a leading candidate, however, similar to the cases for other G proteins, there are likely to be additional cellular targets for the TGase. We also know very little regarding how the GDP-GTP exchange reaction for the TGase is regulated, and it will be important to determine whether guanine nucleotide exchange factors (GEFs) exist with functions analogous to the serpentine receptors (which stimulate the activation of heterotrimeric G proteins) and the GEFs for small G proteins. Likewise, nothing is known regarding the regulation of the GTP hydrolytic activity of the TGase. Are there specific GTPase-activating proteins? If so, how are they regulated, as GTP hydrolysis may be necessary to promote optimal transamidation activity? Can the TGase bind aluminum fluoride and will this induce a conformation that mimics the transition state for GTP hydrolysis similar to what has been observed in large and small G proteins?

Finally, what other binding partners exist for the TGase and how do these interactions give rise to survival or anti-apoptotic effects? We are at a much earlier stage in appreciating the functional importance of the GTP-binding/GTPase cycle of the TGase, compared to what we know about other more traditional G proteins, but there is every reason to believe that the TGase will prove to be an important, albeit somewhat unusual, type of G-protein switch.

References

1. Achyuthan, K. E. and Greenberg, C. S. (1987). Identification of a guanosinetriphosphate-binding site on guinea pig liver transglutaminase. Role of GTP and calcium ions in modulating activity. *J. Biol. Chem.* **262**, 1901–1906.
2. Takeuchi, Y., Birckbichler, P. J., Patterson Jr., M. K., and Lee, K. N. (1992). Putative nucleotide binding sites of guinea pig liver transglutaminase. *FEBS Lett.* **307**, 177–180.
3. Singh, U. S., Erickson, J. W., and Cerione, R. A. (1995). Identification and biochemical characterization of an 80 kDa GTP-binding/transglutaminase from rabbit liver nuclei. *Biochemistry* **34**, 15863–15871.
4. Lai, T. S., Slaughter, T. F., Peoples, K. A., Hettasch, J. M., and Greenberg, C. S. (1998). Regulation of human tissue transglutaminase function by magnesium-nucleotide complexes. Identification of distinct binding sites for Mg-GTP and Mg-ATP. *J. Biol. Chem.* **273**, 1776–1781.
5. Lorand, L. and Conrad, S. M. (1984). Transglutaminases. *Mol. Cell. Biochem.* **58**, 9–35.
6. Greenberg, C. S., Birckbichler, P. J., and Rice, R. H. (1991). Transglutaminases: Multifunctional cross-linking enzymes that stabilize tissues. *FASEB J.* **5**, 3071–3077.
7. Yee, V. C., Pedersen, L. C., Le Trong, L, Bishop, P. D., Stenkamp, R. E., and Teller, D. C. (1994). Three-dimensional structure of a transglutaminase: human blood coagulation factor XIII. *Proc. Natl. Acad. Sci. USA* **91**, 7296–7300.
8. Ahvazi, B., Kirn, H. C., Kee, S.-H., Nemes, Z., and Steinert, P. M. (2002). Three-dimensional structure of the human transglutaminase 3 enzyme: Binding of calcium ions changes structure for activation. *EMBO J.* **21**, 2055–2067.
9. Fox, B. A., Yee, V. C., Pedersen, L. C., Le Trong, L, Bishop, P. D., Stenkamp, R. E., and Teller, D. C. (1999). Identification of the calcium binding site and a novel ytterbium site in blood coagulation factor XIII by X-ray crystallography. *J. Biol. Chem.* **274**, 4917–4923.
10. Liu, S., Cerione, R. A., and Clardy, J. C. (2002). Structural basis for the guanine nucleotide-binding activity of tissue transglutaminase and its regulation of transamidation activity. *Proc. Nail. Acad. Sci.* **99**, 2743–2747.
11. Antonyak, M. A., Singh, U. S., Lee, D. A., Boehm, J. E., Combs, C, Zgola, M. M., Page, R. L., and Cerione, R. A. (2001). Effects of tissue transglutaminase on retinoic acid-induced cellular differentiation and protection against apoptosis. *J. Biol. Chem.* **276**, 33582–33587.
12. Antonyak, M. A., Boehm, J. E., and Cerione, R. A. (2002). Phosphoinositide-3 kinase activity is required for retinoic acid-induced expression and activation of the tissue transglutaminase. *J. Biol. Chem.* **277**, 14712–14716.
13. Boehm, J. E., Singh, U., Combs, C, Antonyak, M. A., and Cerione, R. A. (2002). Tissue Transglutaminase protects against apoptosis by modifying the tumor suppressor protein p110 Rb. *J. Biol. Chem.* **277**, 20127–20130.
14. Lesort, M., Chun, W., Johnson, G. V. W., and Ferrante, R. J. (1999). Tissue transglutaminase is increased in Huntington's disease brain. *J. Neurochem.* **73**, 2018–2027.
15. Korsgren, C. and Cohen, C. M. (1991). Organization of the gene for human erythrocyte membrane protein 4.2: structural similarities with the gene for the a subunit of factor XIII. *Proc. Natl. Acad. Sci. USA* **88**, 4840–4844.
16. Feng, J. F., Rhee, S. G., and Im, M. J. (1996). Evidence that phospholipase delta1 is the effector in the Gh (transglutaminase II)-mediated signaling. *J. Biol. Chem.* **271**, 16451–16454.
17. Nakaoka, H., Perez, D. M., Baek, K. J., Das, T., Husain, A., Misono, K., Im, M. J., and Graham, R. M. (1994). Gh: a GTP-binding protein with transglutaminase activity and receptor signaling function. *Science* **264**, 1593–1596.
18. Im, M. J. and Graham, R. M. (1990). A novel guanine nucleotide-binding protein coupled to alpha 1-adrenergic receptor. I. Identification by photolabeling or membrane and ternary complex preparation. *J. Biol. Chem.* **265**, 18944–18951.
19. Back, K. J., Das, T., Gray, C., Antar, S., Murugesan, G., and Im, M.-J. (1993). Evidence that the Gh protein is a signal mediator from alpha 1-adrenoceptor to a phospholipase C. I. Identification of alpha 1-adrenoceptor-coupled Gh family and purification of Gh7 from bovine heart. *J. Biol. Chem.* **268**, 27390–27397.
20. Hwang, K. C., Gray, C. D., Sivasubramanian, N., and Im, M. J. (1995). Interaction site of GTP binding Gh (transglutaminase II) with phospholipase C. *J. Biol. Chem.* **270**, 27058–27062.
21. Feng, J. F., Gray, C. D., and Im, M. J. (1999). α1B-adrenoceptor interacts with multiple sites of transglutaminase II: Characteristics of the interaction in binding and activation. *Biochemistry* **38**, 2224–2232.
22. Chen, S., Lin, F., Lismaa, S., Lee, K. N., Birckbichler, P. J., and Graham, R. M. (1996). Alpha 1-adrenergic receptor signaling via Gh is subtype specific and independent of its transglutaminase activity. *J. Biol. Chem.* **271**, 32385–32391.
23. Smrcka, A. V., Hepler, J. R., Brown, K. O., and Sternweis, P. C. (1991). Regulation of polyphosphoinositide-specific phospholipase C activity by purified Gq. *Science* **251**, 804–807.
24. Park, E. S., Won, J. H., Han, K. J., Suh, P. G., Ryu, S. H., Lee, H. S., Yun, H. Y., Kwon, N. S., and Back, K. J. (1998). Phospholipase C-delta1 and oxytocin receptor signalling: evidence of its role as an effector. *Biochem. J.* **331**, 283–289.
25. Singh, U. S. and, R. A. (1996). Biochemical effects of retinoic acid on GTP-binding protein/transglutaminases in HeLa Cells: Stimulation of GTP-binding and transglutaminase activity, membrane-association, and phosphatidylinositol lipid turnover. *J. Biol. Chem.* **271**, 27292–27298.
26. Upchurch, H. F., Conway, R, Patterson Jr., M. K., and Maxwell, M. D. (1991). Localization of cellular transglutaminase on the extracellular matrix after wounding: characteristics of the matrix bound enzyme. *J. Cell. Physiol.* **149**, 375–382.
27. Gentile, V., Thomazy, V., Piacentini, M., Fesus, L., and Davies, P. J. (1992). Expression of tissue transglutaminase in Balb-C 3T3 fibroblasts: effects on cellular morphology and adhesion *J. Cell Biol.* **119**, 463–474.
28. Birckbichler, P. J. and Patterson Jr., M. K. (1978). Cellular transglutaminase, growth, and transformation. *Ann. N. Y. Acad. Sci.* **312**, 354–365.
29. Byrd, J. C. and Lichti, U. (1987). Two types of transglutaminase in the PC12 pheochromocytoma cell line. Stimulation by sodium butyrate. *J. Biol. Chem.* **262**, 11699–11705.
30. Maccioni, R. B. and Seeds, N. W. (1986). Transglutaminase and neuronal differentiation. *Mol. Cell. Biochem.* **69**, 161–168.
31. Citron, B. A., Suo, Z., SantaCruz, K., Davies, P. J. A., Qin, R., and Festoff, B. W. (2002). Protein crosslinking, tissue transglutaminase, alternative splicing and neurodegeneration. *Neurochem. Int.* **40**, 69–78.
32. Piacentini, M., Fesus, L., Farrace, M. G., Ghibelli, L., Piredda, L., and Melino, G. (1991). The expression of tissue transglutaminase in two human cancer cell lines is related with programmed cell death (apoptosis). *Eur. J. Cell Biol.* **54**, 246–254.
33. Nemes, Z., Adany, R., Balazs, M., Boross, P., and Fesus, L. (1997). Identification of cytoplasmic actin as an abundant glutaminyl substrate for tissue transglutaminase in HL60 and U937 cells undergoing apoptosis. *J. Biol. Chem.* **272**, 20577–20583.
34. Oliverio, S. Amendola, A., DiSano, F., Farrace, M. G., Fesus, L., Nemes, Z., Piredda, L., Spinedi, A., and Piacentini, M. (1997). Tissue transglutaminase-dependent posttranslational modification of the retinoblastoma gene product in promonocytic cells undergoing apoptosis. *Mol. Cell. Biol.* **17**, 6040–6048.

35. Lambright, D. G., Noel, J. P., Hamm, H. E., and Sigler, P. B. (1994). Structural determinants for activation of the alpha-subunit of a heterotrirneric G protein. *Nature* **369**, 621–628.
36. Lismaa, S. E., Wu, M. J., Nanda, N., Church, W. B., and Graham, R. M. (2000). GTP binding and signaling by Gj/transglutaminase II involves distinct residues in a unique GTP-binding pocket. *J. Biol. Chem.* **275**, 18259–18265.
37. Reinstein, J., Shichting, I., French, M., Goody, R. S., and Wittinghofer, A. (1991). p21 with a phenylalanine 28 -» leucine mutation reacts normally with the GTPase activating protein GAP but nevertheless has transforming properties. *J. Biol. Chem.* **266**, 17700–17706.
38. Lin, R., Bagrodia, S., Cerione, R. A., and Manor, D. (1997). A novel Cdc42Hs mutant induces cellular transformation. *Curr. Biol.* **7**, 794–797.
39. Greenwald, P., Kramer, B., and Weed, D. (1993). Expanding horizons in breast and prostate cancer prevention and early detection. *J. Cancer Educ.* **8**, 91–107.
40. Costa, A., Formelli, F., Chiesa, F., Decensi, A., De Palo, G., and Veronesi, U. (1994). Prospects of chemoprevention of human cancers with the synthetic retinoid fenretinide. *Cancer Res.* **54**, 20328–20378.
41. Kienta, K. J., Esper, P. S., Zwas, F., Krzeminski, R., and Flaherty, L. E. (1997). Phase II chemoprevention trial of oral fenretinide in patients at risk for adenocarcinoma of the prostate. *Am. J. Clin. Oncol.* **20**, 36–39.
42. Kalemkerian, G. P., Slusher, R., Ramalingam, S., Gadgeel, S., and Mabry, M. (1995). Growth inhibition and induction of apoptosis by fenretinide in small-cell lung cancer cell lines. *J. Natl. Cancer Inst.* **87**, 1674–1680.
43. Zou, C. P., Kurie, J. M., Lotan, D., Zou C. C., Hong, W. K., and Lotan, R. (1998). Higher potency of N-(4-hydroxyphenyl)retinamide than all-trans-retinoic acid in induction of apoptosis in non-small cell lung cancer cell lines. *Clin. Cancer Res.* **4**, 1345–1355.

CHAPTER 242

The Role of ARF in Vesicular Membrane Traffic

Melissa M. McKay and Richard A. Kahn

*Department of Biochemistry,
Emory University School of Medicine,
Atlanta, Georgia*

The ARF Family of Regulatory GTPases

*ADP-r*ibosylation *f*actors (ARFs) are a family of highly conserved, ~20-kDa GTPases with a number of cellular activities; most notably the regulation of vesicular membrane traffic. ARFs are ubiquitous in eukaryotes with as many as 23 members of the ARF family expressed in mammalian cells. The ARF family includes both ARF and ARF-like (ARL) proteins, based upon the extent of sequence and functional identities; e.g., the 6 mammalian ARFs share > 60% identity to one another, while most ARLs share 40–60% identity to each other or to any ARF [1]. With this high level of structural conservation we expect that ARLs perform mechanistically and functionally homologous roles to those of ARFs, but these are much less well defined and will not be covered here. The ARF subfamily can be further divided into three classes. In mammals, class I includes ARF1–3, class II includes ARF4 and ARF5, and class III is ARF6 [2]. In contrast, *Caenorhabditis elegans* and *Drosophila melanogaster* each have single representatives of each ARF class. The results discussed in the following sections were obtained primarily from studies of ARF1. The extent to which ARF4 and ARF5 act as described below is unknown, while ARF6 clearly has some differences in location and presumed actions [3,4].

Like other regulatory GTPases, ARFs cycle between GDP- and GTP-bound conformations in a cycle whose rate is determined by binding to guanine nucleotide exchange factors (GEFs), GTPase-activating proteins (GAPs), and effectors. In addition to these functionally critical protein interactions, ARFs are unique among regulatory GTPases in that their affinity for biological membranes is affected by the nucleotide bound [5]. Specifically, ARFs are co-translationally modified by the addition of myristic acid to the N terminal glycine [6]. This hydrophobic N-terminal anchor functions coordinately with the amphipathic N terminus to orient the protein on the surface of the membrane when GTP is bound. ARFs are predominantly soluble when GDP is bound. Thus, activation of ARF (GTP binding) alters both its location and its affinity for effectors.

All of the currently identified actions of ARFs take place on membrane surfaces. These different functions can be divided into four different groups, (1) regulation of lipid metabolism, (2) regulation of vesicle transport, (3) cofactor for bacterial toxins [7,8], and (4) GTP-dependent binding to other miscellaneous proteins [9–11] with less well understood functions. Because of the likelihood of functional interplay, the rest of this review will focus only on the first two.

ARF as a Regulator of Membrane Traffic

The primary cellular roles for ARFs are the regulation of membrane traffic and lipid metabolism. Whether these represent one integrated or two distinct aspects of ARF biology is controversial. Most likely, ARF-dependent changes in lipid metabolism occur both during vesicle biogenesis and at other times. The observations that (1) activation of ARF is sensitive to its lipid environment, (2) specific lipids serve as cofactors in some ARF-dependent activities, (3) ARF can alter lipid metabolism, and (4) ARF is required for the biogenesis of a broad array of vesicles in cells (see Fig. 1),

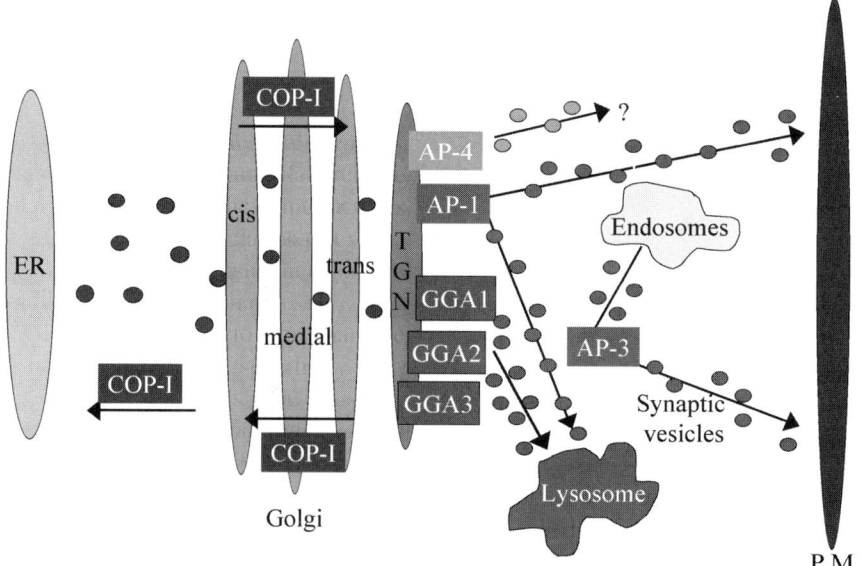

Figure 1 ARFs act to regulate the recruitment of coat proteins to a wide variety of intracellular membranes. The different compartments implicated as a source or destination of vesicles regulated by ARF proteins are shown. Coat/adaptor proteins or complexes are represented by shaded rectangles at the membranes at which they are thought to initiate vesicle budding. The direction and destination of the vesicles are indicated by arrows. Arrows emanating from AP-1 and GGA2 are intended only to indicate traffic from TGN to endosomes/lysosomes, and P.M. routing between different endosomal compartments is not shown. Note that the destination and possibly directionality of transport of each of the coated vesicles are still under investigation (particularly, e.g., AP-4). P.M. = Plasma Membrane.

implicate ARFs as both sensors and modulators of different lipids and membranes. The role of ARF as regulator of membrane traffic is very likely to be intimately linked to roles in lipid metabolism and signaling.

ARF is Activated on Membranes

The dependence on a hydrophobic environment for GTP binding was first described when ARF was purified and shown to bind guanine nucleotides [12]. The ability of a broad list of different lipids and detergents to satisfy this requirement reveals a nonspecific requirement for lipids in the activation process. This is likely an *in vitro* correlate of the GTP-stimulated translocation of ARF to membranes. ARFs are sufficiently hydrophobic to interact weakly with membranes in the GDP-bound state [13]. Once at the membrane they encounter ARF GEFs, which promote GTP binding and a more stable association with the membrane. In this way, specificity in membrane binding is likely tied to the localization of ARF GEFs, though we cannot exclude the lipid composition of the membrane from playing a role.

Specific Lipids are Cofactors for ARF-Stimulated Activities

The search for an ARF GEF led to the observation that acid phospholipids can stimulate the release of GDP from ARFs, and can stabilize the protein in the nucleotide-free state [14]. Phosphoinositides, specifically phosphatidylinositol 4,5-bisphosphate (PI(4,5)P_2), were the most active lipids in these studies. At about the same time, ARF was found to be a potent, direct, GTP-dependent activator of phospholipase D (PLD), which catalyzes the conversion of phosphatidylcholine to phosphatidic acid and choline [15,16]. Surprisingly, ARF-stimulated PLD activity was dependent on the addition of PI(4,5)P_2 [15]. Similarly, some ARF GAPs were found to be potently stimulated by PI(4,5)P_2 [17,18]. The product of PLD, phosphatidic acid, can also increase ARF GAP activity and alter affinity of ARF GAP for PI(4,5)P_2 [18]. Thus, there exist both nonspecific effects of lipids to stabilize GTP binding and specific effects of lipids to bind and modulate ARF, ARF effectors (PLD), or ARF GAPs. As seen in the following section, the interplay between ARF actions and PI(4,5)P_2 gets even more complicated.

ARF as Modifier of Lipid Metabolism

A role for ARF in lipid metabolism was first revealed by its role as co-activator of PLD [16,19]. This led to speculation that effects of ARF on vesicle traffic may result from changes in membrane lipid composition. This theory was supported by the observation that PLD was localized to Golgi membranes [20], a major site of ARF actions, and that a persistently active form of PLD can mimic many of the actions of activated ARF on coated vesicle formation *in vitro* [21]. More recently, ARF has been shown to recruit both PI 4-kinase and PI(4)P 5-kinase to membranes, [22] and is a direct activator of the latter [23]. The sequential actions of these two enzymes produce PI(4,5)P_2, the allosteric activator of PLD and some ARF GAPs and a molecule with widespread effects on cell signaling and morphology. Thus, ARF can increase PLD activity and PA levels both through direct activation of PLD and indirectly, through increased production of PI(4,5)P_2 via activation of PIP kinases. ARF activity can produce changes in local and perhaps cellular levels of phosphatidylcholine, phosphatidic acid, PI, PI(4)P, and PI(4,5)P_2.

Together these data reveal extensive interplay between ARF and phospholipids that may have profound effects on signal transduction, through the activation of specific enzyme effectors and generation of PI(4,5)P_2 and on changes in the lipid composition in microdomains that could impact the rate or ability to form transport vesicles.

The Role of ARF in Vesicle Biogenesis

Cell biological, genetic, and pharmacological data all pointed to intracellular membranes, and particularly the Golgi and *trans*-Golgi network (TGN), as important sites of ARF action. For example, electron microscopy and indirect immunofluorescence of cultured mammalian cells localized ARF to cis-Golgi [24] and TGN membranes [25]. Further, deletion of yeast *ARF1* resulted in defects in the processing of secreted proteins as they pass through the Golgi and genetic interactions with other Golgi proteins [24]. Later, brefeldin A was shown to progressively cause the release of proteins from the cytosolic surface of Golgi and TGN membranes [26], loss of Golgi integrity with fusion of Golgi elements with the endoplasmic reticulum (ER) [27,28], inhibition of protein secretion [29], and death of some cells [30]. All the cellular actions of brefeldin A result from inhibition of ARF GEFs [31], suggesting a role (though perhaps an indirect one) for ARFs in each.

A breakthrough occurred with the observation that the inhibition of intra-Golgi transport by GTPγS led to the accumulation of vesicles of uniform size and possessing an electron-dense coat. Purification of these vesicles allowed the identification of the coat proteins; consisting of the heptameric COP-I complex and ARF [32]. The brefeldin A sensitivity of both ARF and COP-I localization at Golgi membranes led to the proposal that binding to activated ARF is required for COP-I to bind Golgi membranes and promote the budding process [33]. Since that time the number of different coat proteins recruited by ARF has increased markedly, leading to the need for mechanistic details that can explain specificity in recruitment of cargo and coats into the nascent vesicle.

Thus far, seven coat complexes have been identified whose binding to ARF is GTP-dependent and to membranes is brefeldin A-sensitive. These include COP-I, the tetrameric adaptin complexes AP-1 [34], AP-3 [35], and AP-4 [36], and the three monomeric GGAs, GGA1-3 [37–39]. COP-I is required for retrograde transport from the Golgi to the ER and for transport between the Golgi stacks (see Fig. 1). The adaptin complexes are involved in vesicle traffic from the TGN (AP-1 and AP-4) and from endosomes and synaptic vesicles (AP-3). GGAs regulate vesicle traffic from the TGN to endosomes/lysosomes. Thus, ARF regulates the recruitment of coat complexes from a variety of membranes within a cell, including multiple coats at a single membrane, the TGN.

The initiation of vesicle budding requires a minimum of three components: ARF, the coat/adaptor[1] protein or complex, and a transmembrane anchor protein (see Fig. 2). For COP-I, these roles are filled by ARF, COP-I, and the p24 family of transmembrane proteins (Fig. 2A) [40]. The transmembrane proteins have cytosolic "tails" that carry sorting motifs, such as the acidic dileucine motif that bind GGAs (Fig. 2B) [41,42] or the tyrosine-based signals that bind adaptins (Fig. 2C) [43]. Once the initial three components are in place, the adaptor proteins recruit other components necessary for vesicle formation, transport, uncoating, and fusion. In only some cases does this include clathrin. It is not clear when ARF dissociates from the bud or vesicle but it is likely to be soon after initiation of the coating process. Thus, activation of ARF promotes vesicle budding through the recruitment of vesicle coat proteins which can act as scaffolding in the assembly of the components necessary for vesicular transport.

The correct matching of cargo with adaptor is required for appropriate targeting of the vesicle. An attractive model for how the three components act to "proofread" the coating process was provided by Goldberg [44] and was based on the observation that ARF GAP activity was increased by COP-I. When cargo with an incorrect sorting signal is present, GAP activity is stimulated by COP-I; leading to GTP hydrolysis on ARF, release of COP-I, and prevention of coat assembly and bud maturation. If the sorting motif on the cargo is appropriate to the adaptor, GTP hydrolysis is not stimulated, and vesicle coating proceeds. Though data supporting this model have been generated only with COP-I, we present a more general model for ARF-dependent vesicle biogenesis, showing the different sorting motifs and adaptors that result in the production of different coated vesicles (Fig. 2). Further experimentation is needed to support this three-component model for ARF action. An even more challenging goal is the integration of this protein-based model with the changes in membrane lipids that are known to occur in response to ARF activity. Such a holistic model will be required before detailed understanding of membrane traffic and signal transduction by ARFs is achieved.

---

[1]The terms coat protein and adaptor are used interchangeably here. The former name originates from the electron-dense coat first seen on the non-clathrin coated, COP-I vesicles. Such a coat can be seen in AP-1 bearing clathrin-coated vesicles but has not been described for other ARF coats. The term adaptor is more generic and intended to refer to the fact that whether monomeric or oligomeric, a key component in their biological functions is the ability to bind multiple proteins, including ARFs, transmembrane cargo, clathrin, and accessory proteins.

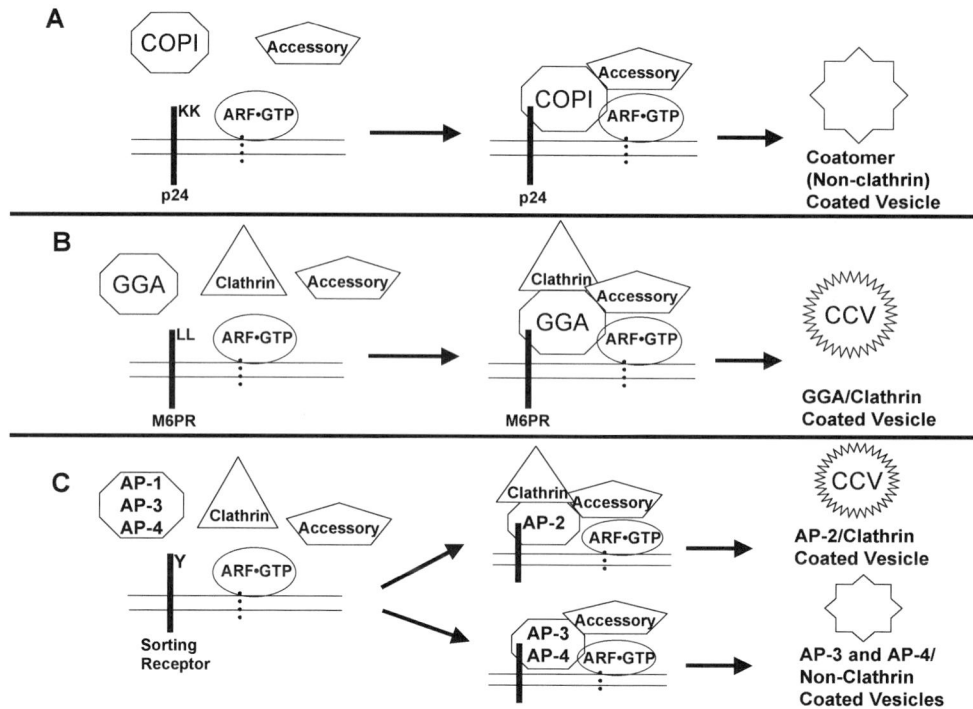

Figure 2 Three-component model for the initiation of ARF-mediated coated vesicle assembly. A minimum of three components are proposed as being critical to the assembly of coat proteins and initiation of vesicle budding; ARF-GTP, a coat/adaptor protein or complex, and a transmembrane anchor or sorting receptor. The latter carries sorting motifs which help match cargo and coat. The p24 family has a dilysine (KK) sorting motif, the mannose 6-phosphate receptors (M6PR) carry an acidic dileucine motif, and sorting receptors have a tyrosine-based signal. Once assembled the adaptor proteins are proposed to act as scaffolds to recruit clathrin (in some cases) and (poorly defined) accessory proteins. The minimal components required for coating of COP-I (panel A), GGA (panel B), and adaptin (panel C) vesicles are shown. Note that GGAs and AP-1 form clathrin-coated vesicles (CCV), while clathrin is not found on vesicles carrying COP-I, AP-3, or AP-4.

References

1. Boman, A. L. and Kahn, R. A. (1995). Arf proteins: the membrane traffic police? [see comments]. *Trends Biochem. Sci.* **20**, 147–150.
2. Lee, F. J., Stevens, L. A., Hall, L. M., Murtagh, J. J., Jr., Kao, Y. L., Moss, J., and Vaughan, M. (1994). Characterization of class II and class III ADP-ribosylation factor genes and proteins in Drosophila melanogaster. *J. Biol. Chem.* **269**, 21555–21560.
3. Turner, C. E. and Brown, M. C. (2001). Cell motility: ARNO and ARF6 at the cutting edge. *Curr. Biol.* **11**, R875–877.
4. Chavrier, P. and Goud, B. (1999). The role of ARF and Rab GTPases in membrane transport. *Curr. Opin. Cell Biol.* **11**, 466–475.
5. Regazzi, R., Ullrich, S., Kahn, R. A., and Wollheim, C. B. (1991). Redistribution of ADP-ribosylation factor during stimulation of permeabilized cells with GTP analogues. *Biochem. J.* **275**, 639–644.
6. Kahn, R. A., Goddard, C., and Newkirk, M. (1988). Chemical and immunological characterization of the 21-kDa ADP-ribosylation factor of adenylate cyclase. *J. Biol. Chem.* **263**, 8282–8287.
7. Vaughan, M. and Moss, J. (1997). Activation of toxin ADP-ribosyltransferases by the family of ADP-ribosylation factors. *Adv. Exp. Med. Biol.* **419**, 315–320.
8. Zhu, X. and Kahn, R. A. (2001). The Escherichia coli heat labile toxin binds to Golgi membranes and alters Golgi and cell morphologies using ADP-ribosylation factor-dependent processes. *J. Biol. Chem.* **276**, 25014–25021.
9. Kanoh, H., Williger, B. T., and Exton, J. H. (1997). Arfaptin 1, a putative cytosolic target protein of ADP-ribosylation factor, is recruited to Golgi membranes. *J. Biol. Chem.* **272**, 5421–5429.
10. Shin, O. H., Ross, A. H., Mihai, I., and Exton, J. H. (1999). Identification of arfophilin, a target protein for GTP-bound class II ADP-ribosylation factors. *J. Biol. Chem.* **274**, 36609–36615.
11. Boman, A., Kuai, J., Zhu, X., Chen, J., Kuriyama, R., and Kahn, R. (1999). Arf proteins bind to mitotic kinesin-like protein 1 (MKLP1) in a GTP-dependent fashion. *Cell Motil. Cytoskel.* **44**, 119–132.
12. Kahn, R. A. and Gilman, A. G. (1986). The protein cofactor necessary for ADP-ribosylation of Gs by cholera toxin is itself a GTP binding protein. *J. Biol. Chem.* **261**, 7906–7911.
13. Antonny, B., Beraud-Dufour, S., Chardin, P., and Chabre, M. (1997). N-terminal hydrophobic residues of the G-protein ADP-ribosylation factor-1 insert into membrane phospholipids upon GDP to GTP exchange. *Biochemistry* **36**, 4675–4684.
14. Terui, T., Kahn, R. A., and Randazzo, P. A. (1994). Effects of acid phospholipids on nucleotide exchange properties of ADP-ribosylation factor 1. Evidence for specific interaction with phosphatidylinositol 4,5-bisphosphate. *J. Biol. Chem.* **269**, 28130–28135.
15. Brown, H. A., Gutowski, S., Moomaw, C. R., Slaughter, C., and Sternweis, P. C. (1993). ADP-ribosylation factor, a small GTP-dependent regulatory protein, stimulates phospholipase D activity [see comments]. *Cell* **75**, 1137–1144.
16. Cockcroft, S., Thomas, G. M., Fensome, A., Geny, B., Cunningham, E., Gout, I., Hiles, I., Totty, N. F., Truong, O., and Hsuan, J. J. (1994). Phospholipase D: A downstream effector of ARF in granulocytes. *Science* **263**, 523–526.
17. Randazzo, P. A. (1997) Resolution of two ADP-ribosylation factor 1 GTPase-activating proteins from rat liver. *Biochem. J.* **324**, 413–419.
18. Randazzo, P. A. and Kahn, R. A. (1994). GTP hydrolysis by ADP-ribosylation factor is dependent on both an ADP- ribosylation factor

GTPase-activating protein and acid phospholipids [published erratum appears in *J. Biol. Chem.* 1994 Jun 10;**269**(23),16519]. *J. Biol. Chem.* **269**, 10758–10763.
19. Brown, H. A., Gutowski, S., Kahn, R. A., and Sternweis, P. C. (1995). Partial purification and characterization of Arf-sensitive phospholipase D from porcine brain. *J. Biol. Chem.* **270**, 14935–14943.
20. Ktistakis, N. T., Brown, H. A., Sternweis, P. C., and Roth, M. G. (1995). Phospholipase D is present on Golgi-enriched membranes and its activation by ADP ribosylation factor is sensitive to brefeldin A. *Proc. Natl. Acad. Sci. USA* **92**, 4952–4956.
21. Ktistakis, N. T., Brown, H. A., Waters, M. G., Sternweis, P. C., and Roth, M. G. (1996). Evidence that phospholipase D mediates ADP ribosylation factor-dependent formation of Golgi coated vesicles. *J. Cell Biol.* **134**, 295–306.
22. Godi, A., Pertile, P., Meyers, R., Marra, P., Di Tullio, G., Iurisci, C., Luini, A., Corda, D., and De Matteis, M. A. (1999). ARF mediates recruitment of PtdIns-4-OH kinase-beta and stimulates synthesis of PtdIns(4,5)P2 on the Golgi complex [see comments]. *Nat. Cell Biol.* **1**, 280–287.
23. Honda, A., Nogami, M., Yokozeki, T., Yamazaki, M., Nakamura, H., Watanabe, H., Kawamoto, K., Nakayama, K., Morris, A. J., Frohman, M. A., and Kanaho, Y. (1999). Phosphatidylinositol 4-phosphate 5-kinase alpha is a downstream effector of the small G protein ARF6 in membrane ruffle formation. *Cell* **99**, 521–532.
24. Stearns, T., Willingham, M. C., Botstein, D., and Kahn, R. A. (1990). ADP-ribosylation factor is functionally and physically associated with the Golgi complex. *Proc. Natl. Acad. Sci. USA* **87**, 1238–1242.
25. Traub, L. M., Ostrom, J. A., and Kornfeld, S. (1993). Biochemical dissection of AP-1 recruitment onto Golgi membranes. *J. Cell Biol.* **123**, 561–573.
26. Donaldson, J. G., Kahn, R. A., Lippincott-Schwartz, J., and Klausner, R. D. (1991). Binding of ARF and beta-COP to Golgi membranes: possible regulation by a trimeric G protein. *Science* **254**, 1197–1199.
27. Fujiwara, T., Oda, K., Yokota, S., Takatsuki, A., and Ikehara, Y. (1988). Brefeldin A causes disassembly of the Golgi complex and accumulation of secretory proteins in the endoplasmic reticulum. *J. Biol. Chem.* **263**, 18545–18552.
28. Lippincott-Schwartz, J., Yuan, L. C., Bonifacino, J. S., and Klausner, R. D. (1989). Rapid redistribution of Golgi proteins into the ER in cells treated with brefeldin A: Evidence for membrane cycling from Golgi to ER. *Cell* **56**, 801–813.
29. Misumi, Y., Miki, K., Takatsuki, A., Tamura, G., and Ikehara, Y. (1986). Novel blockade by brefeldin A of intracellular transport of secretory proteins in cultured rat hepatocytes. *J. Biol. Chem.* **261**, 11398–11403.
30. Ishii, S., Nagasawa, M., Kariya, Y., and Yamamoto, H. (1989). Selective cytotoxic activity of brefeldin A against human tumor cell lines. *J. Antibiot. (Tokyo)* **42**, 1877–1878.
31. Peyroche, A., Antonny, B., Robineau, S., Acker, J., Cherfils, J., and Jackson, C. L. (1999). Brefeldin A acts to stabilize an abortive ARF-GDP-Sec7 domain protein complex: involvement of specific residues of the Sec7 domain. *Mol. Cell* **3**, 275–285.
32. Serafini, T., Orci, L., Amherdt, M., Brunner, M., Kahn, R. A., and Rothman, J. E. (1991). ADP-ribosylation factor is a subunit of the coat of Golgi-derived COP-coated vesicles: a novel role for a GTP-binding protein. *Cell* **67**, 239–253.
33. Orci, L., Palmer, D. J., Amherdt, M., and Rothman, J. E. (1993). Coated vesicle assembly in the Golgi requires only coatomer and ARF proteins from the cytosol. *Nature* **364**, 732–734.
34. Stamnes, M. A. and Rothman, J. E. (1993). The binding of AP-1 clathrin adaptor particles to Golgi membranes requires ADP-ribosylation factor, a small GTP-binding protein. *Cell* **73**, 999–1005.
35. Faundez, V., Horng, J. T., and Kelly, R. B. (1998). A function for the AP3 coat complex in synaptic vesicle formation from endosomes. *Cell* **93**, 423–432.
36. Hirst, J., Bright, N. A., Rous, B., and Robinson, M. S. (1999). Characterization of a fourth adaptor-related protein complex. *Mol. Biol. Cell.* **10**, 2787–2802.
37. Boman, A. L., Zhang, C. J., Zhu, X., and Kahn, R. A. (2000). A family of ADP-ribosylation factor effectors that can alter membrane transport through the trans-Golgi. *Mol. Biol. Cell* **11**, 1241–1255.
38. Dell'Angelica, E. C., Puertollano, R., Mullins, C., Aguilar, R. C., Vargas, J. D., Hartnell, L. M., and Bonifacino, J. S. (2000). GGAs: a family of ADP ribosylation factor-binding proteins related to adaptors and associated with the Golgi complex. *J. Cell Biol.* **149**, 81–94.
39. Hirst, J., Lui, W. W., Bright, N. A., Totty, N., Seaman, M. N., and Robinson, M. S. (2000). A family of proteins with gamma-adaptin and VHS domains that facilitate trafficking between the trans-Golgi network and the vacuole/lysosome. *J. Cell Biol.* **149**, 67–80.
40. Stamnes, M. A., Craighead, M. W., Hoe, M. H., Lampen, N., Geromanos, S., Tempst, P., and Rothman, J. E. (1995). An integral membrane component of coatomer-coated transport vesicles defines a family of proteins involved in budding. *Proc. Natl. Acad. Sci. USA* **92**, 8011–8015.
41. Puertollano, R., Aguilar, R. C., Gorshkova, I., Crouch, R. J., and Bonifacino, J. S. (2001). Sorting of mannose 6-phosphate receptors mediated by the GGAs. *Science* **292**, 1712–1716.
42. Zhu, Y., Doray, B., Poussu, A., Lehto, V. P., and Kornfeld, S. (2001). Binding of GGA2 to the lysosomal enzyme sorting motif of the mannose 6-phosphate receptor. *Science* **292**, 1716–1718.
43. Ohno, H., Aguilar, R. C., Yeh, D., Taura, D., Saito, T., and Bonifacino, J. S. (1998). The medium subunits of adaptor complexes recognize distinct but overlapping sets of tyrosine-based sorting signals. *J. Biol. Chem.* **273**, 25915–25921.
44. Goldberg, J. (2000). Decoding of sorting signals by coatomer through a GTPase switch in the COPI coat complex. *Cell* **100**, 671–679.

CHAPTER 243

Yeast Small G Protein Function: Molecular Basis of Cell Polarity in Yeast

Hay-Oak Park[1] and Keith G. Kozminski[2]

[1]Department of Molecular Genetics,
The Ohio State University, Columbus, Ohio and
[2]Departments of Biology and Cell Biology, University of Virginia,
Charlottesville, Virginia

Introduction

Cell polarity is critical for the function of the many cell types involved in vectored processes such as nutrient transport, neuronal signaling, and motility. Development of a polarized cell, regardless of cell type, follows a common plan [1]. First, a spatial cue (landmark) marks the site of polarized growth. Signal transduction molecules then interpret the identity of the landmark and signal the establishment of polarity by an asymmetric organization of the cytoskeleton. Polarity is then reinforced with the secretory transport and deposition of molecules needed for growth at the site chosen for polarized growth.

The budding yeast *Saccharomyces cerevisiae* is an outstanding experimental organism with which to decipher the molecular mechanisms of polarity development by small G proteins (hereafter GTPases). GTPases make numerous molecular interactions with effectors and regulators and can be involved in multiple distinct signaling pathways in response to the internal and external cues. The facile experimental genetics of yeast affords the opportunity to study GTPases in individual signaling pathways under different physiological conditions [2]. Herein, the role of GTPases in linking spatial landmarks on the cell cortex to the reorganization of the cytoskeleton will be presented in the context of yeast budding.

The Rsr1/Bud1 Ras-Like GTPase Module Interprets Spatial Landmarks

Yeast cells undergo oriented cell division by selecting a specific site for polarized growth, the bud site, on their cell cortex. Haploid **a** and α cells bud in an axial pattern in which both mother and daughter cells select a bud site immediately adjacent to their previous division site. Diploid **a**/α cells bud in a bipolar pattern: mother cells select a bud site adjacent to their daughter or on the opposite end of the cell, whereas daughter cells always choose a bud site directed away from their mother [3–5] (Fig. 1). Thus, the choice of a bud site determines the axis for cell polarity and ultimately the cell division plane.

A GTPase module consisting of the Ras-like GTPase Rsr1/Bud1, its GDP-GTP exchange factor (GEF) Bud5, and its GTPase activating protein (GAP) Bud2, is essential for selecting the proper site for polarized growth in both haploid and diploid yeast cells [6–10]. Deletion of *RSR1/BUD1*, *BUD2*, or *BUD5* leads to random budding [6,7,10] as does the expression of $rsr1^{G12V}$ (predicted to be constitutively GTP-bound) or $rsr1^{K16N}$ (predicted to be constitutively GDP-bound or nucleotide-empty) [11], suggesting that Rsr1/Bud1 needs to be cycled between GTP- and GDP-bound states for proper bud-site selection. Rsr1/Bud1 localizes to the plasma membrane and then becomes

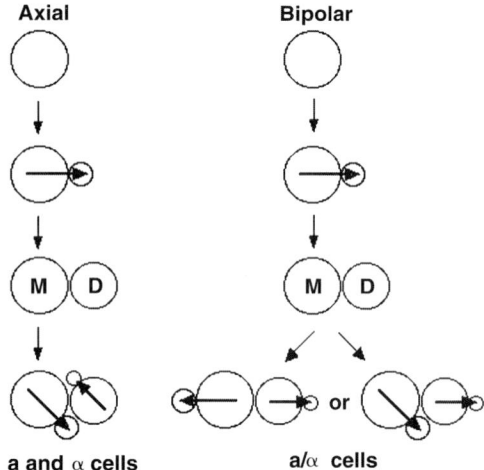

Figure 1 Wild-type patterns of bud site selection in yeast. In the axial pattern (exhibited by an **a** or α cell), the mother cell (M) buds immediately adjacent to its last daughter; the daughter cell (D) buds toward its mother. In the bipolar pattern (exhibited by an **a**/α cell), the mother cell can bud at or near either of its poles; the daughter cell buds away from its mother. The arrows within the cells indicate the axis of polarity. (Reprinted, with permission, from Chant, J. and Herskowitz, I. *Cell,* 65, 1203–1212, 1991. Copyright Cell Press).

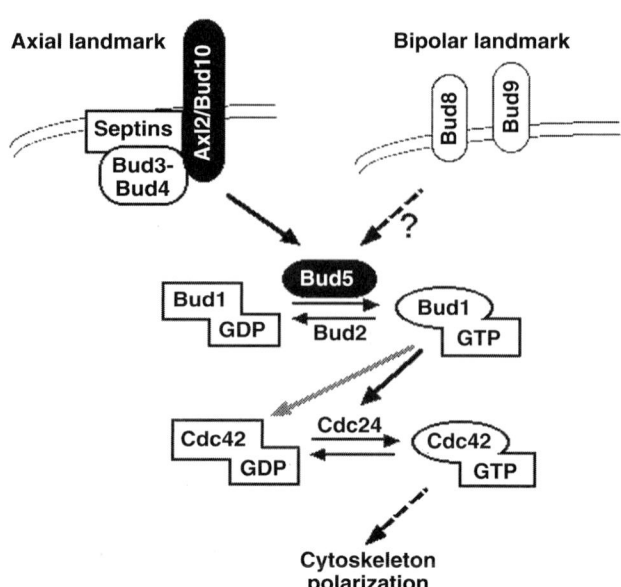

Figure 2 A model for spatial control of cell polarity during yeast budding. Bud5 is likely to be localized to the presumptive bud site in haploid **a** and α cells through the interaction with Axl2, a component of the axial landmark. Bud5 locally activates Bud1 to the GTP-bound state, which then associates with proteins necessary for bud-site assembly, such as Cdc24 and Cdc42. Bud5 may interact with Bud8 or Bud9, putative landmarks of diploid **a**/α cells, thus directing polarity establishment for bipolar budding pattern. (Adapted from Kang, P. J., Sanson, A., Lee, B., and Park, H.-O., *Science,* 292, 1376–1378, 2001. With permission.)

concentrated at the presumptive bud site in G1 of the cell cycle [11a]. Bud2 and Bud5 proteins are also found at the presumptive bud site in unbudded cells [12,13]. Although each protein of the Rsr1/Bud1 GTPase module can localize to the presumptive bud site independently of each other, localization of each protein to the proper bud site cannot be maintained in the absence of any other protein in the module. Thus the localized action of the Rsr1/Bud1 GTPase module is required for proper bud site selection [12,13].

Isolation of cell-type specific alleles of *BUD2* and *BUD5* suggests that each of the proteins encoded by these genes localizes to the presumptive bud site through interactions with cortical landmarks unique to haploid or diploid cells [13,14]. Recent studies on Bud5 have established that Bud5 interacts physically with a transmembrane protein, Axl2/Bud10 [15,16], a potential landmark for axial budding in haploid cells [13]. Bud5 mislocalizes in cells lacking Axl2/Bud10 [13,17]. Thus, the interpretation of a cortical landmark by a GTPase module occurs by the interaction of the landmark protein with the GEF of the module, which may result in the localized activation of module at the site of polarized growth. It is not known whether Bud5 interacts with potential bipolar landmarks such as Bud8 or Bud9 [14,18] to establish the bipolar budding pattern in diploid cells (Fig. 2).

The Rsr1/Bud1 module does not complete the link between cortical landmarks and polarity establishment; rather, it links the spatial signals from both axial and bipolar landmarks to the Cdc42 GTPase module (discussed in the following section), which is essential for the establishment of cell polarity.

The Cdc42p Rho-Like GTPase Module Regulates the Establishment of Cell Polarity

A GTPase module consisting of the Rho-like GTPase Cdc42, its GDP-GTP exchange factor (GEF) Cdc24, its GTPase-activating proteins (GAPs) Bem3, Rga1, and Rga2, and guanine nucleotide dissociation inhibitor (GDI) Rdi1, is essential for the establishment of cell polarity prior to budding (reviewed in [19]). Mutation of *CDC42* or *CDC24* prevents an asymmetric organization of the actin cytoskeleton at the bud site as well as bud formation [20,21]. Association of Cdc42 with the plasma membrane via posttranslational geranylgeranylation of its C terminus is essential for Cdc42 function [22]. Deletion of each of the GAP-encoding genes does not result in a *cdc42* phenotype, due perhaps to a redundancy of function among this class of proteins [23,24]. How the Cdc42 module effects the establishment of cell polarity at the site chosen for polarized growth is a question of expanding complexity because the proteins of the Cdc42 module make numerous physical and genetic interactions [19,25,26] that result not just in the polarization of the actin cytoskeleton (see the following section), but of other proteins involved in polarized growth as well (e.g., Sec3 of the exocyst complex) [27].

Coupling the Rsr1/Bud1 GTPase Module to the Cdc42 GTPase Module

A web of genetic interactions among the *BUD* genes and *CDC24* hinted at a linkage between the Rsr1/Bud1 and

Cdc42 GTPase modules (reviewed in [28]). Subsequent studies showed that Rsr1/Bud1 interacts with Cdc24 (as well as Cdc42 and the scaffold Bem1) in a guanine nucleotide-dependent manner [29,30] (Fig. 2). The scaffold protein Far1 sequesters Cdc24 in the nucleus [31–33]. In late G1, Cdc24 is exported into the cytoplasm upon cyclin activation of Cdc28 kinase [32,33]. In the absence of Rsr1/Bud1, Cdc24 localizes in a patch randomly located on the plasma membrane [11a], consistent with the proposed role of Rsr1/Bud1 GTPase module guiding bud-site assembly proteins to the proper bud site.

The Cdc42 GTPase Module Spatially Restricts Actin Assembly to the Site of Polarized Growth

Localized activation of Cdc42 at the bud site by Cdc24 leads to an asymmetric distribution of filamentous actin structures within the cell. In late G1, actin cables align along the cortex of the cell and cortical actin patches concentrate at the bud site (reviewed in [34]). Recent studies suggest that new Cdc42-dependent actin assembly and stabilization at the bud rather than the movement of existing actin structures to the bud causes the asymmetric distribution of these cytoskeletal structures within the cell [35–38].

A permeabilized cell assay provided the first evidence for Cdc42-dependent actin assembly at the site of polarized growth. Permeabilized cells with impaired Cdc42 function did not assemble exogenously added actin at the site of polarized growth, except with the prior addition of recombinant Cdc42 or an activated form of the Cdc42 effector Ste20, a PAK family kinase [39,40]. This latter result is consistent with the ability of GTP-bound Cdc42 to relieve PAK autoinhibition (reviewed in [41]), although a role for Ste20 in actin assembly remains undefined. Cla4, a PAK kinase that is partially redundant with Ste20, may, however, have a bona fide role in cortical actin assembly. Phosphorylation of class I myosins by Cla4 may activate the actin nucleating complex, Arp2/3, directly or indirectly through interaction with the Arp2/3 activator Las17/Bee1 [38]. To what extent this activity depends upon the kinase activity of Cla4 is unclear [42]. Interestingly, overexpression of *CLA4*, but not *STE20*, is lethal [2]. Phosphorylation of Cdc24 by Cla4 may release Cdc24 from a Bem1 scaffold that also contains Cdc42, ending polarized bud growth [43, 43a]. Thus, activation of Cla4 by Cdc42 may initiate actin assembly at the site of polarized growth and may result in the negative regulation of Cdc42 by acting on its GEF Cdc24.

Polarization of the actin cytoskeleton also depends upon two other Cdc42 effectors, the redundant formin family proteins Bni1 and Bnr1. Bni1 localization at the site of polarized growth depends upon Cdc42 activity and is enhanced by interactions with proteins of the polarisome and the Cdc42 effectors Gic1 and Gic2 [44–46]. In the absence of formin function, the actin cytoskeleton becomes depolarized. Upon recovery from formin inactivation, actin cables assemble from the site of polarized growth in an Arp2/3-independent manner [36,37]. Formins are also required for maintaining a polarized distribution of Las17 [38]. It is unclear whether recruitment of this activator of Arp2/3-dependent actin assembly depends upon Cdc42.

Furcation of Cdc42-dependent actin assembly pathways may allow for greater regulation of polarized actin assembly among growth conditions and cell types. For example, the Cdc42 effectors Gic1 and Gic2 may be more important in haploids and at higher temperatures in contrast to the putative Cdc42 effectors Msb3 and Msb4, which are suggested to be more important in diploids and at low temperatures [47].

Conclusion

Substantial progress has been made in deciphering the molecular basis of cell polarity in yeast. One important aspect, but still poorly understood, is the regulation of each component, in particular GEFs and GAPs, involved in polarity development. Some of the machinery is specific to yeast, but the general principles underlying cell polarity and the components in the signaling pathways appear highly conserved throughout evolution. Thus understanding the spatial and temporal control of the GTPases and their regulators in yeast is undoubtedly relevant to other eukaryotes. This article aimed to be a brief guide to the field. For in-depth discussion, we recommend recent reviews [25,34,48–51]. A guide to yeast genes can also be found in the *Saccharomyces* Genome Database (SGD) [52].

References

1. Drubin, D. G. and Nelson, W. J. (1996). Origins of cell polarity. *Cell* **84**, 335–344.
2. Kozminski, K. G., Chen, A. J., Rodal, A. A., and Drubin, D. G. (2000). Functions and functional domains of the GTPase Cdc42p. *Mol. Biol. Cell* **11**, 339–354.
3. Freifelder, D. (1960). Bud position in *Saccharomyces cerevisiae*. *J. Bacteriol.* **80**, 567–568.
4. Hicks, J. B., Strathern, J. N., and Herskowitz, I. (1977). Interconversion of yeast mating types. III. Action of the homothallism (*HO*) gene in cells homozygous for the mating type locus. *Genetics* **85**, 373–393.
5. Chant, J. and Pringle, J. R. (1995). Patterns of bud-site selection in the yeast *Saccharomyces cerevisiae*. *J. Cell Biol.* **129**, 751–765.
6. Bender, A. and Pringle, J. R. (1989). Multicopy suppression of the *cdc24* budding defect in yeast by *CDC42* and three newly identified genes including the *ras*-related gene *RSR1*. *Proc. Natl. Acad. Sci. USA* **86**, 9976–9980.
7. Chant, J. and Herskowitz, I. (1991). Genetic control of bud site selection in yeast by a set of gene products that constitute a morphogenetic pathway. *Cell* **65**, 1203–1212.
8. Chant, J., Corrado, K., Pringle, J. R., and Herskowitz, I. (1991). Yeast *BUD5*, encoding a putative GDP-GTP exchange factor, is necessary for bud site selection and interacts with bud formation gene *BEM1*. *Cell* **65**, 1213–1224.
9. Powers, S., Gonzales, E., Christensen, T., Cubert, J., and Broek, D. (1991). Functional cloning of *BUD5*, a *CDC25*-related gene from *S. cerevisiae* that can suppress a dominant-negative *RAS2* mutant. *Cell* **65**, 1225–1231.
10. Park, H.-O., Chant, J., and Herskowitz, I. (1993). *BUD2* encodes a GTPase-activating protein for Bud1/Rsr1 necessary for proper bud-site selection in yeast. *Nature* **365**, 269–274.

11. Ruggieri, R., Bender, A., Matsui, Y., Powers, S., Takai, Y., Pringle, J. R., and Matsumoto, K. (1992). *RSR1,* a ras-like gene homologous to Krev-1 (smg21A/rap1A): role in the development of cell polarity and interactions with the Ras pathway in *Saccharomyces cerevisiae. Mol. Cell Biol.* **12**, 758–766.
11a. Park, H.-O., Kang, P. J., and Rachfal, A. W. (2002). Localization of the Rsr1/Bud1 GTPase involved in selection of a proper growth site in yeast. *J. Biol. Chem.* **277**, 26721–26724.
12. Park, H.-O., Sanson, A., and Herskowitz, I. (1999). Localization of Bud2p, a GTPase-activating protein necessary for programming cell polarity in yeast, to the presumptive bud site. *Genes Dev.* **13**, 1912–1917.
13. Kang, P. J., Sanson, A., Lee, B., and Park, H.-O. (2001). A GDP/GTP exchange factor involved in linking a spatial landmark to cell polarity. *Science* **292**, 1376–1378.
14. Zahner, J., Harkins, H. I., and Pringle, J. R. (1996). Genetic analysis of the bipolar pattern of bud site selection in the yeast *Saccharomyces cerevisiae. Mol. Cell. Biol.* **16**, 1857–1870.
15. Roemer, T., Madden, K., Chang, J., and Snyder, M. (1996). Selection of axial growth sites in yeast requires Axl2p, a novel plasma membrane glycoprotein. *Genes Dev.* **10**, 777–793.
16. Halme, A., Michelitch, M., Mitchell, E. L., and Chant, J. (1996). Bud10p directs axial cell polarization in budding yeast and resembles a transmembrane receptor. *Curr. Biol.* **6**, 570–579.
17. Marston, A. L., Chen, T., Yang, M. C., Belhumeur, P., and Chant, J. (2001). A localized GTPase exchange factor, Bud5, determines the orientation of division axes in yeast. *Curr. Biol.* **11**, 803–807.
18. Harkins, H. A., Page, N., Schenkman, L. R., De Virgilio, C., Shaw, S., Bussey, H., and Pringle, J. R. (2001). Bud8p and Bud9p, proteins that may mark the sites for bipolar budding in yeast. *Mol. Biol. Cell* **12**, 2497–2518.
19. Johnson, D. I. (1999). Cdc42: An essential Rho-type GTPase controlling eukaryotic cell polarity. *Micro. Mol. Biol. Rev.* **63**, 54–105.
20. Sloat, B. F. and Pringle, J. R. (1978). A mutant of yeast defective in cellular morphogenesis. *Science* **200**, 1171–1173.
21. Adams, A. E., Johnson, D. I., Longnecker, R. M., Sloat, B. F., and Pringle, J. R. (1990). *CDC42* and *CDC43,* two additional genes involved in budding and the establishment of cell polarity in the yeast *Saccharomyces cerevisiae. J. Cell Biol.* **111**, 131–142.
22. Ziman, M., O'Brien, J. M., Ouelette, L. A., Church, W. R., and Johnson, D. I. (1991). Mutational analysis of *CDC42,* a *Saccharomyces cerevisiae* gene that encodes a putative GTP-binding protein involved in the control of cell polarity. *Mol. Cell. Biol.* **11**, 3537–3544.
23. Stevenson, B. J., Ferguson, B., De Virgilio, C., Bi, E., Pringle, J. R., Ammerer, G., and Sprague Jr., G. F. (1995). Mutation of *RGA1,* which encodes a putative GTPase-activating protein for the polarity-establishment protein Cdc42p, activates the pheromone-response pathway in the yeast *Saccharomyces cerevisiae. Genes Dev.* **9**, 2949–2963.
24. Zheng, Y., Cerione, R., and Bender, A. (1994). Control of the yeast bud-site assembly GTPase Cdc42: catalysis of guanine nucleotide exchange by Cdc24 and stimulation of GTPase activity by Bem3. *J. Biol. Chem.* **269**, 2369–2372.
25. Pruyne, D. and Bretscher, A. (2000). Polarization of cell growth in yeast. I. Establishment and maintenance of polarity states. *J. Cell Sci.* **113**, 365–375.
26. Drees, B. L. *et al.* (2001). A protein interaction map for cell polarity development. *J. Cell Biol.* **154**, 549–576.
27. Zhang, X., Bi, E., Novick, P., Du, L., Kozminski, K. G., Lipschutz, J. H., and Guo, W. (2001). Cdc42 interacts with the exocyst and regulates polarized secretion. *J. Biol. Chem.* **276**, 46745–46750.
28. Drubin, D. (1991). Development of cell polarity in budding yeast. *Cell* **65**, 1093–1096.
29. Zheng, Y., Bender, A., and Cerione, R. A. (1995). Interactions among proteins involved in bud-site selection and bud-site assembly in *Saccharomyces cerevisiae. J. Biol. Chem.* **270**, 626–630.
30. Park, H.-O., Bi, E., Pringle, J., and Herskowitz, I. (1997). Two active states of the Ras-related Bud1/Rsr1 protein bind to different effectors to determine yeast cell polarity. *Proc. Natl. Acad. Sci. USA* **94**, 4463–4468.
31. Toenjes, K. A., Sawyer, M. M., and Johnson, D. I. (1999). The guanine-nucleotide-exchange factor Cdc24p is targeted to the nucleus and polarized growth sites. *Curr. Biol.* **9**, 1183–1186.
32. Nern, A. and Arkowitz, R. A. (2000). Nucleocytoplasmic shuttling of the Cdc42p exchange factor Cdc24p. *J. Cell Biol.* **148**, 1115–1122.
33. Shimada, Y., Gulli, M.-P., and Peter, M. (2000). Nuclear sequestration of the exchange factor Cdc24p by Far1 regulates cell polarity during mating. *Nat. Cell Biol.* **2**, 117–124.
34. Pruyne, D. and Bretscher, A. (2000). Polarization of cell growth in yeast. II. The role of the cortical actin cytoskeleton. *J. Cell Sci.* **113**, 571–585.
35. Barral, Y., Mermall, V., Mooseker, M. S., and Snyder, M. (2000). Compartmentalization of the cell cortex by septins is required for maintenance of cell polarity in yeast. *Mol. Cell* **5**, 841–851.
36. Evangelista, M., Pruyne, D., Amberg, D. C., Boone, C., and Bretscher, A. (2002). Formins direct Arp2/3-independent actin filament assembly to polarize cell growth in yeast. *Nat. Cell Biol.* **4**, 32–41.
37. Sagot, I., Klee, S. K., and Pellman, D. (2002). Yeast formins regulate cell polarity by controlling the assembly of actin cables. *Nat. Cell Biol.* **4**, 42–50.
38. Lechler, T., Jonsdottir, G. A., Klee, S. K., Pellman, D., and Li, R. (2001). A two-tiered mechanism by which Cdc42 controls the localization and activation of an Arp2/3-activating motor complex in yeast. *J. Cell Biol.* **155**, 261–270.
39. Li, R., Zheng, Y., and Drubin, D. G. (1995). Regulation of cortical actin cytoskeleton assembly during polarized cell growth in budding yeast. *J. Cell Biol.* **128**, 599–615.
40. Eby, J. J., Holly, S. P., van Drogen, F., Grishin, A. V., Peter, M., Drubin, D. G., and Blumer, K. J. (1998). Actin cytoskeleton organization regulated by the PAK family of protein kinases. *Curr. Biol.* **8**, 967–970.
41. Bagrodia, S. and Cerione, R. A. (1999). Pak to the future. *Trends Cell Biol.* **9**, 350–355.
42. Weiss, E. L., Bishop, A. C., Shokat, K. M., and Drubin, D. G. (2000). Chemical genetic analysis of the budding-yeast p21-activated kinase Cla4p. *Nat. Cell Biol.* **2**, 677–685.
43. Gulli, M., Jaquenoud, M., Shimada, Y., Niederhauser, G., Wiget, P., and Peter, M. (2000). Phosphorylation of the Cdc42 exchange factor Cdc24 by the PAK-like kinase Cla4 may regulate polarized growth in yeast. *Mol. Cell* **6**, 1155–1167.
43a. Bose, I., Irazogui, J. E., Moskow, J. J., Bardes, E. S., Zyla, T. R., and Lew, D. J. (2001). Assembly of scaffold-mediated complexes containing Cdc42p, the exchange factor Cdc24p, and the effector Cla4p required for cell-cycle-regulated phosphorylation of Cdc24p. *J. Biol. Chem.* **276**, 7176–7186.
44. Sheu, Y. J., Santos, B., Fortin, N., Costigan, C., and Snyder, M. (1998). Spa2p interacts with cell polarity proteins and signaling components involved in yeast cell morphogenesis. *Mol. Cell. Biol.* **18**, 4053–4069.
45. Jaquenoud, M. and Peter, M. (2000). Gic2p may link activated Cdc42p to components involved in actin polarization, including Bni1p and Bud6p (Aip3p). *Mol. Cell. Biol.* **20**, 6244–6258.
46. Ozaki-Kuroda, K., Yamamoto, Y., Nohara, H., Kinoshita, M., Fujiwara, T., Irie, K., and Takai, Y. (2001). Dynamic localization and function of Bni1p at the sites of directed growth in *Saccharomyces cerevisiae. Mol. Cell. Biol.* **21**, 827–839.
47. Bi, E., Chiavetta, J. B., Chen, H., Chen, G. C., Chan, C. S., and Pringle, J. R. (2000). Identification of novel, evolutionarily conserved Cdc42p-interacting proteins and of redundant pathways linking Cdc24p and Cdc42p to actin polarization in yeast. *Mol. Biol. Cell* **11**, 773–793.
48. Chant, J. (1999). Cell polarity in yeast. *Annu. Rev. Cell Dev. Biol.* **15**, 365–391.
49. Gulli, M. P. and Peter, M. (2001). Temporal and spatial regulation of Rho-type guanine-nucleotide exchange factors: The yeast perspective. *Genes Dev.* **15**, 365–379.
50. Goode, B. L. and Rodal, A. A. (2001). Modular complexes that regulate actin assembly in budding yeast. *Curr. Opin. Microbiol.* **4**, 703–712.
51. Casamayor, A. and Snyder, M. (2002). Bud-site selection and cell polarity in budding yeast. *Curr. Opin. Microbiol.* **5**, 179–186.
52. http://www.yeastgenome.org/

Farnesyltransferase Inhibitors

James J. Fiordalisi and Adrienne D. Cox
*Departments of Radiation Oncology and
Pharmacology and the Lineberger Comprehensive Cancer Center,
University of North Carolina at Chapel Hill, Chapel Hill, North Carolina*

Introduction

Cell signaling requires proper localization of all the players involved. Many proteins are targeted to the appropriate subcellular location, and are directed to interact with their regulators and effectors, by lipid modification. Prenylation, the modification by a C15 farnesyl isoprenoid or a C20 geranylgeranyl isoprenoid, is required for the biological activity of small GTPases and other proteins containing specific C-terminal CAAX motifs. Inhibitors of farnesyltransferase (FTase), FTIs, have been designed for use as anti-Ras and anticancer drugs, but in fact they are selective for FTase, not for Ras. This distinction has important implications for their use as pharmacological tools to dissect signaling pathways. For example, FTIs can completely suppress the prenylation of H-Ras but not of the other Ras isoforms, so grouping all the Ras proteins together as FTI targets is an oversimplification that adversely affects the interpretation of results using FTIs to block Ras-mediated signaling events. Efforts are underway to identify the biological consequences of disrupting the farnesylation and geranylgeranylation of specific prenylated proteins, and it now seems clear that multiple proteins are involved in cellular responses to FTIs. For example, genetic evidence indicates that RhoB is required for the apoptotic response to FTIs, but not for the suppression of anchorage-independent growth. Understanding the contribution of specific targets to distinct consequences of FTI disruption will be important in identifying the roles of prenylated proteins in specific aspects of cell signaling.

Farnesylation and Protein Function

Like many other things in life, successful cell signaling requires accurate and specific localization of all the players.

For many proteins, the correct subcellular localization and biological activity requires the attachment of specific lipids that mediate both positioning within the appropriate cellular compartment and interaction with upstream regulators and downstream effectors. It has become apparent in the last decade or so that the modification of proteins by prenylation, that is, the attachment of isoprenoid lipids, plays a critical role in the signaling activity of G proteins both large and small. In particular, the finding that farnesylation, the attachment of a 15-carbon farnesyl isoprenoid group, to the small GTPase Ras is required for its biological activity, provoked massive efforts to develop inhibitors of (FTase), the enzyme responsible for this attachment. The resulting (FTI) compounds have helped to shed considerable light on the role of such modifications in cell signaling events, but have also introduced a few mysteries of their own.

Ras—the Prototype of Farnesylated Proteins

Ras is an important hub in the complex network of intracellular signaling pathways. Ras proteins receive signals from a wide range of cell surface receptors including receptor tyrosine kinases, G-protein-coupled receptors, and integrins [1]. In turn Ras transmits these signals through an ever-increasing list of effector proteins, the most well characterized of which are the serine/threonine kinase Raf, phosphatidylinositol-3-kinase (PI3-K) and Ral-guanine nucleotide dissociation stimulator (Ral-GDS). Ras signaling ultimately influences gene transcription to alter properties including cell survival, proliferation, differentiation, adhesion, motility, and morphology. Normally GDP-bound and inactive, Ras proteins are converted to the active, GTP-bound conformation by guanine nucleotide exchange factors (GEFs), and back again to the resting state by GTPase-activating proteins (GAPs) that stimulate the

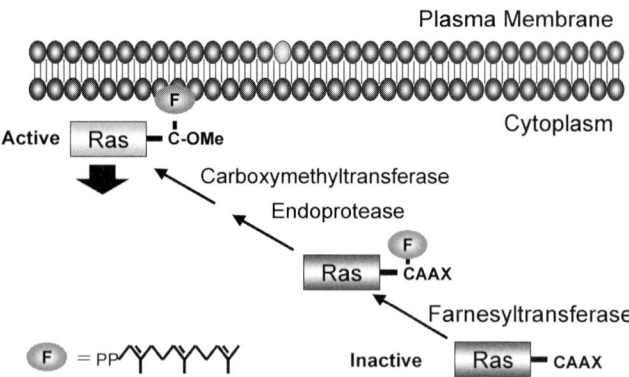

Figure 1 Posttranslational modifications are required for membrane localization and biological activity of prenylated proteins.

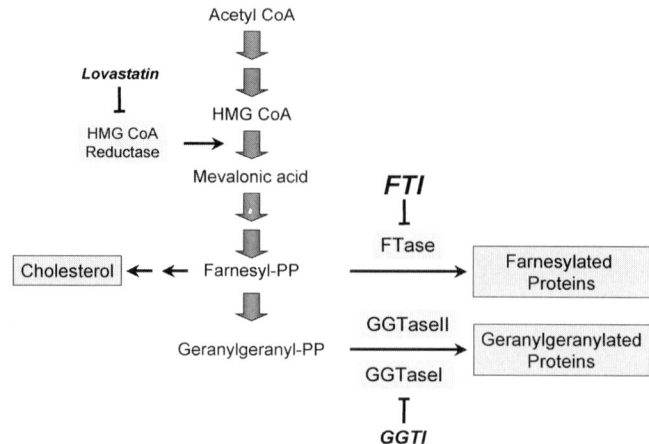

Figure 2 Farnesyl isoprenoids, obligate intermediates in cholesterol biosynthesis, are key elements of the mevalonate pathway.

intrinsic GTPase activity of Ras. When Ras is rendered constitutively GTP-bound and active by oncogenic mutations (typically at positions 12, 13, or 61), the resulting uncontrolled signaling leads to increased cell proliferation, survival in the absence of substratum, loss of contact inhibition, changes in adhesion, motility, and morphology, and the development of other aspects of the transformed phenotype. Alternative splicing results in the three *ras* genes encoding four Ras proteins, H-Ras, N-Ras, K-Ras4A, and K-Ras4B. Recent work suggests that these proteins, while highly similar, are not completely functionally equivalent. In keeping with this possibility, although activating Ras mutations in general are associated with approximately 30% of human cancers, the distribution of mutations of each Ras isoform differs among tumor types [2]. The high prevalence of mutated Ras in cancers has led to many attempts to develop compounds that block Ras signaling for cancer treatment.

Localization of Ras to the inner leaflet of the plasma membrane is an absolute requirement for both its normal and oncogenic functions, and is accomplished through a series of posttranslational modifications (Fig. 1) (reviewed in reference [3]). The first is the addition of a 15-carbon farnesyl isoprenoid to the cysteine of the C terminal four amino acid sequence called the "CAAX motif," where C = cysteine, A = aliphatic, and X = any amino acid. This prenylation step is then followed by proteolysis of the -AAX sequence, carboxymethylation of the resulting farnesylated C-terminal cysteine, and (in H-, N- and K-Ras4A) palmitoylation of either one or two cysteines upstream of the CAAX motif. The first and obligate step of all these posttranslational modifications is catalyzed by FTase. Mutants of Ras that lack a farnesylatable cysteine neither localize to membranes nor signal to downstream effectors. Thus, FTase has been an attractive target for anti-cancer drug development efforts directed toward inhibiting aspects of cellular transformation influenced by Ras. Farnesyl isoprenoids are obligate intermediates in the cholesterol biosynthetic pathway, but blocking FTase is far more selective than disrupting the entire pathway via statins, which are inhibitors of the mevalonic acid precursor (Fig.2). However, although FTIs were originally developed as anti-Ras drugs, the consequences of inhibiting FTase are more complicated than simply inhibiting the farnesylation of Ras. In this chapter we will describe the activities of FTIs, the current confusion surrounding the identity of the biologically relevant downstream targets of FTase, and the use of FTIs as pharmacological tools to disrupt cellular signaling events.

Identification and Development of FTIs

The first FTI was a simple synthetic tetrapeptide corresponding to the CAAX motif of K-Ras4B (CVIM), which was designed to act as a competitive inhibitor of full-length Ras for binding to FTase [4]. However, the instability and poor membrane permeability of simple peptides demanded that different chemical entities be developed. Thus, in rationally designed peptidomimetic FTIs, labile peptide bonds were replaced with more stable chemical moieties [5–7] to increase biological half-life, and modification of the negatively charged carboxyl terminus greatly improved cell permeability. Current FTIs are generally nonpeptidomimetics that have been identified by screening small molecule libraries generated by combinatorial chemistry [8]. Naturally occurring FTIs have also been discovered in screens of natural products from bacteria [9], fungi [10], and plants [11], although only the fungal FTI manumycin has been used extensively for research purposes. The mechanism and kinetics of Ras prenylation by the zinc-containing metalloenzyme FTase, and its inhibition by FTIs, have been studied extensively (reviewed in reference [12]). These studies have been facilitated by the determination of the crystal structure of FTase alone or in complex with Ras and/or farnesylpyrophosphate [13,14], and by the use of NMR to define the interactions between FTase and FTIs in solution. This work has further contributed to refining the structures of subsequent FTIs. Most FTIs developed to date have been competitive for the protein substrate of FTase, but prenyl-competitive [15,16] and dual-competitive [17] FTIs have also been developed successfully. Peptide-competitive FTIs with IC_{50}s in the micromolar to nanomolar range are commercially available for research purposes. Two of the best of these compounds for use in cell-based assays are FTI-277 [18] (Calbiochem, San Diego, CA) and L744,832 [19] (BioMol, Plymouth Meeting, PA). Other peptide-competitive FTIs with favorable pharmacokinetic and pharmacodynamic properties are currently in clinical trials. These include R115777 (Zarnestra, Johnson & Johnson), SCH66336 (Sarosar, Schering-Plough), BMS-214662 (Bristol Myers-Squibb), and CP-609754 (Pfizer). The status of clinical trials using FTIs is the subject of recent reviews [20,21].

FTI Activity in Cell Culture and Animal Models

Early FTIs showed great promise *in vitro* against purified FTase, with many FTIs inhibiting H-Ras farnesylation in the low nanomolar range [5,22], and demonstrating excellent specificity for FTase compared to the highly related enzyme geranylgeranyltransferase I (GGTase I) [23]. Further, FTIs inhibited the prenylation of H-Ras and other farnesylated proteins in a variety of cell culture systems, as demonstrated most often by gel shift analysis wherein the unprenylated protein migrates slower than the prenylated form. FTIs selectively inhibited transcriptional transactivation of reporters driven by Ras compared to Raf, and also caused reversion of many aspects of the transformed cell phenotype (reviewed in references [3,24,25]). FTIs inhibited the proliferation of H-Ras-transformed cell lines and human tumor-derived cell lines, reverted Ras-transformed morphology, re-established actin stress fibers lost upon Ras transformation, inhibited Ras-dependent anchorage-independent cell growth and migration, inhibited DNA synthesis, and affected the transcription of a variety of genes downstream of Ras-responsive promoters. Oddly, FTIs showed remarkably little toxicity toward normal cells, which is a highly desirable feature for potential drugs, but a surprise at the time given the inability of cells to cope with dominant negative Ras. Except for the relative lack of toxicity, these studies suggested that Ras was an important target of FTIs, and encouraged further studies in animal models.

In both mouse tumor xenograft models and transgenic mouse models, FTIs also produced impressive effects with little overt toxicity (reviewed in reference [26]). In nude mice, FTIs inhibited the growth of a wide variety of human tumor-derived cell lines and human tumor explants from the pancreas, lung, colon, blood, brain, prostate, and bladder. Inhibition of the growth of tumors derived from rodent epithelial and fibroblast cell lines transformed with specific oncogenes including Ras was also observed. Regression of tumors was rare in nude mouse models, consistent with a cytostatic mode of action. Transgenic mice told a different story. In transgenic animals harboring mutations in H-Ras [27], N-Ras [28], or K-Ras [29] giving rise to stochastic mammary, salivary, or lymphoid tumors, FTI treatment inhibited the growth of existing tumors and prevented the growth of new ones. But in H-Ras-driven transgenic tumors, rapid regression of existing tumors was also observed [27]. The basis for the different responses of xenografted and transgenic tumors to FTIs is still not understood.

Although FTIs were originally intended to be used against solid tumors harboring Ras mutations, there is increasing preclinical and clinical evidence that they are actually more effective against various forms of leukemia (reviewed in reference [30]), as well as gliomas and breast cancer (reviewed in reference [31]). These tumors are not typically associated with Ras mutations, but instead may have upregulated Ras signaling due to elevations in epidermal growth factor receptor (EGFR) family activity via overexpression or mutation [1]. Thus, inhibition of endogenous Ras function may play a role in this sensitivity. Whether such inhibition is direct or indirect is a matter of debate.

Studies in human tumor cell lines have also shown that FTIs are additive or synergistic in combination with a wide variety of other anti-cancer agents including [32], taxanes [33,34], aminobisphosphonates [35], cyclophosphamide [36], inhibitors of the MEK pathway [37], and ionizing radiation [38]. It is not clear why FTIs work particularly well with taxanes, but interactions with farnesylated microtubule binding proteins including CENP-E and K-Ras have been proposed. Clearly, many questions regarding the consequences of FTI inhibition still exist.

Alternative Prenylation in the Presence of FTIs

Early in the development of FTIs it was observed that, while prenylation of H-Ras was effectively blocked by FTIs, prenylation of N- and K-Ras was not [39,40]. Instead, when FTase is inhibited, N- and K-Ras were shown to be alternatively prenylated by GGTase I [41,42]. Sharing a common α subunit with FTase, GGTase I catalyzes the addition of a 20-carbon geranylgeranyl isoprenoid to proteins terminating in appropriate CAAX motifs. The canonical CAAX motif specifying geranylgeranylation terminates in leucine (L) [23]. However, in the absence of FTase activity, the C-terminal methionine (M) of N-Ras (CVVM) and K-Ras (CVIM) permits alternative prenylation while the serine (S) of H-Ras (CVLS) does not. Because of this, N- and K-Ras, but not H-Ras, were shown to bypass functional FTI inhibition. Unless farnesylated and geranylgeranylated N- and K-Ras proteins were functionally distinct, which does not appear to be the case, this end-run around FTI suggested that functional compensation by N- and K-Ras for FTI-inhibited H-Ras might explain the lack of toxic side effects of FTIs [40]. However, it also suggested that FTIs might not be effective against the majority of human tumors that harbor N- or K-Ras mutations, an assertion that is inconsistent with the observation that such tumors were sensitive to FTIs in mouse models. Alternative prenylation of N- and K-Ras, as well as of other farnesylated proteins with a C-terminal methionine, further complicates attempts to define clearly the appropriate downstream targets and mechanism of FTI action. Importantly, a critical role for non-Ras FTI targets is implied by these observations.

FTIs as Pharmacological Tools to Study Signaling and Biology

FTIs have been used to implicate Ras in numerous cellular processes including cell transformation, mitogenesis, neurite outgrowth, motility/migration, response to oxidative stress, transcriptional regulation, and protein secretion. Further, contributions of specific downstream Ras effectors, such as Raf/MEK/Erk, PI3-K/Akt, and upstream Ras activators such as receptor tyrosine kinases have also been implicated by using FTIs. However, the interpretation of such data can also be problematic. Despite their differing responses to FTI, H-, N-, and K-Ras proteins are often wrongly lumped together, and "Ras" is often wrongly presumed to be implicated in any process that is affected by FTI treatment. Thus, FTIs have been used as putative Ras inhibitors in ways analogous to inhibitors of other signaling intermediates such as the MEK inhibitors U0126 and PD98059, the PI3-K inhibitors LY294002 and wortmannin, and the p38 MAPK inhibitor SB203580. This is potentially problematic for two reasons. First, as was discussed in the above section, not all Ras proteins respond equally to FTIs (N- and K-Ras are resistant), and second, Ras proteins are neither the only farnesylated proteins affected by FTIs, nor necessarily the most sensitive. The mechanism of FTI action is still poorly understood because there are many farnesylated proteins other than Ras whose functions are affected by FTIs, making it difficult to conclude that a particular FTI-induced phenotype is Ras dependent. Other possible FTI targets will be discussed in the section about inhibition of signaling by FTIs.

In short, FTIs are literally FTase inhibitors, not Ras inhibitors. As will become apparent, only when the phenotype of interest is known to be H-Ras-dependent, is it really safe to use FTIs as Ras inhibitors. Instead, FTIs should be used as inhibitors of farnesylation, with the understanding that the processing, localization, and function of multiple farnesylated proteins are inhibited upon FTI treatment.

Targets of FTIs

During the early years of FTI research initial success was tempered by the increasing realization that the efficacy of FTIs was not simply the result of functional Ras inhibition, and that the original straightforward model of FTI-as-Ras-inhibitor was not the full story. FTI sensitivity does not correlate with the presence of activating Ras mutations in a series of human tumor cell lines [19], and FTIs inhibit N- and K-Ras-induced transformation, even though these isoforms are alternatively prenylated and therefore not rendered unprocessed and inactive by FTI. Together these observations support the notion that farnesylated proteins other than Ras proteins, whose prenylation and activity are also affected by FTIs, are also important targets of FTI action (reviewed in references [43] and [44]).

Our own search of the SwissProt database revealed approximately 250 human proteins containing a C terminus that resembles a CAAX motif. Some of these are shown in Table I. *In vitro* studies using synthetic or mutant CAAX sequences [4,45,46] demonstrated that many of these proteins are poor substrates for FTase and are not likely normally to be farnesylated. Of those that are potential FTase substrates, many, including numerous members of the Ras superfamily, are known to be farnesylated while others have not yet been evaluated. At present, several CAAX-containing, farnesylated human proteins are currently under investigation as mediators of FTI action.

The most intensively studied putative FTI target is RhoB [47], an immediate-early, inducible small GTPase that is highly related to RhoA and RhoC. Uniquely, RhoB exists in both a farnesylated (F) and a geranylgeranylated (GG) form [48]. The geranylgeranylated form of RhoB is growth inhibitory in both rodent fibroblasts and human carcinomas [49,50], and an increase in RhoB-GG is proposed to mediate the antineoplastic effects of FTIs [51]. It remains to be demonstrated whether a shift of RhoB-F to RhoB-GG or an induction of RhoB-GG correlates with FTI sensitivity in human tumor cells. Interestingly, RhoB may be upregulated as a consequence of FTI treatment, in keeping with its role as a stress detector. Genetic evidence from RhoB null MEFs shows that loss of RhoB impairs the apoptotic response to FTIs [51], suggesting that RhoB is required for apoptosis. By contrast, loss of RhoB does not impair the inhibition of anchorage-independent growth by FTIs [51], suggesting that RhoB is not the mediator of that response.

Other potentially critical FTI targets include the centromere binding protein CENP-E [52,53] which is involved in spindle formation and chromosome alignment during mitosis, the phosphatases PRL-1, -2, and -3, nuclear lamins, inositol 3,5-diphosphate phosphatases I and IV, kinases such as S/T kinase 11, and numerous small GTPases of the Ras (Rheb, Rap2) and Rho (RhoD, RhoE, Rho6, RhoN, and TC-10) families (reviewed in reference [43]). Proteins that have specialized expression patterns, such as the α and β subunits of skeletal muscle phosphorylase B kinase, may play important roles in specialized circumstances. CAAX-containing proteins whose prenylation status has not yet been explicitly determined, but whose CAAX sequences and homology to other farnesylated proteins imply that they too will be farnesylated, include the small GTPases RhoI, Ras-related protein 22 (RRP22), Rab28, and the putative tumor suppressor NOEY2/ARHI, as well as AGS1, a GEF for heterotrimeric G proteins. In addition, there are approximately 40 other human proteins unrelated to GTPases that contain CAAX motifs that could support farnesylation. Further complicating the issue is the observation that several potentially farnesylated proteins have a methionine-containing CAAX motif similar to N- and K-Ras, suggesting that they also might be resistant to FTI inhibition via alternative prenylation.

Table I CAAX-Containing Proteins as Potential FTI Targets

FTI targets	Accession no.	CAAX	Prenyl group
H-Ras	RASH_HUMAN	CVLS	F
hPRL-3	NP_116000	CCVM	F(=>GG?)
Lamin A/C	LAMA_HUMAN	CSIM	F(=>GG?)
Lamin B1	LAM1_HUMAN	CAIM	F(=>GG?)
Lamin B2	LAM2_HUMAN	CYVM	F(=>GG?)
CENP-E	CENE_HUMAN	CKTQ	F
hPRL-1 (PTPCAAX1)	NP_003454	CCIQ	F
hPRL-2 (PTPCAAX2/OV-1)	NP_003470	CCVQ	F
RhoB	RHOB_HUMAN	CKVL	F/GG
Known farnesylated			
DnaJ	DJA1_HUMAN	CQTS	F
gamma-Gt (Transducin)	GBG1_HUMAN	CVIS	F
gamma-T2	GBGU_HUMAN	CLIS	F
gamma11	GBGB_HUMAN	CVIS	F
G-protein-coupled recept.	GP41_HUMAN	CAES	F
InsP3 5-phosph.IV	AF187891	CSVS	F
Phosphorylase B kinase β	KPBB_HUMAN	CLIS	F
Rhodopsin kinase	RK_HUMAN	CLVS	F
K-Ras2A	RASK_HUMAN	CIIM	F=>GG
K-Ras2B	RASL_HUMAN	CVIM	F=>GG
N-Ras	RASN_HUMAN	CVVM	F=>GG
Peroxisomal protein (PxF)	PXF_HUMAN	CLIM	F(=>GG?)
Rheb	RHEB_HUMAN	CSVM	F(=>GG?)
Rho6(Rnd1)	RHO6_HUMAN	CSIM	F(=>GG?)
RhoE(Rho8/Rnd3)	RHOE_HUMAN	CTVM	F(=>GG?)
RhoN(Rho7/Rnd2)	RHON_HUMAN	CNLM	F(=>GG?)
RhoD	RHOD_HUMAN	CVVT	F
TC-10	NP_036381	CLIT	F
Hepatitis delta antigen	AANT_HDVAM	CRPQ	F
InsP3 5-phosph.I	I5P1_HUMAN	CVVQ	F
Phosphorylase B kinase α	KPB1_HUMAN	CAMQ	F
Rap2A	RAP2_HUMAN	CNIQ	F
Likely farnesylated			
AGS1	NP_057168	CVIS	F?
NOEY2/ARHI	U96750	CIIM	F(=>GG?)
RRP22	RR22_HUMAN	CSLM	F(=>GG?)
Rab28	RB28_HUMAN	CAVQ	F?
Serine/threonine kinase 11	ST11_HUMAN	CKQQ	F?

Also, the fact that RhoB (CKVL) is both farnesylated and geranylgeranylated, even though it ends with a leucine (L) that normally signals exclusively for geranylgeranylation, suggests that other proteins might be exceptions to the basic rules governing prenylation. Given the current confusion surrounding the identity of relevant FTI targets, it will be important to determine which of these proteins is farnesylated, affected by FTIs, and can account for the observed biological effects of FTIs. In any case, it is likely that FTIs will affect numerous farnesylated proteins to produce complex, cell-type-specific effects.

Inhibition of Signaling by FTIs

Why must farnesylated proteins be farnesylated? First, this lipid modification has been shown to be necessary for membrane localization. Both pharmacological inhibition with lovastatin and similar compounds and with FTIs prevent farnesylation and cause normally membrane-localized farnesylated proteins to remain cytosolic. Mutagenesis of the CAAX motif to a SAAX motif, which cannot be prenylated, also renders normally farnesylated proteins cytosolic. These results imply the existence of specific protein:lipid interaction sites at membranes, whether simply to increase hydrophobicity or for specific docking interactions. Second, farnesylation is important for protein:protein interactions. Farnesylated Ras proteins have long been known to be better substrates for GEFs such as SOS, to interact better with downstream effectors such as PI3-K p110δ and adenylyl cyclase [54] and to stimulate downstream signaling to kinases such as ERK2 as compared to their nonfarnesylated counterparts, even in cell-free systems [55]. The assembly of β/γ dimers of heterotrimeric G proteins depends upon appropriate prenylation of the γ subunit [56,57], as does cell-free activation of downstream effectors including PLC-β and adenylyl cyclase [57]. The enhancement of protein:protein interactions by lipid modification even in cell-free systems suggests that this is not simply a secondary consequence of necessary subcellular localization conferred by the isoprenoid.

If the critical consequences of FTI activity are due to cellular stress responses, then the most critical FTI targets may not necessarily even be farnesylated (viz. RhoB). But if the consequences of FTI activity are primarily due to blocking FTase-mediated farnesylation of target proteins, then the challenge is to determine the identities and relative affinity for FTase of all farnesylated proteins. FTIs can demonstrate the existence of farnesylated proteins in a particular pathway. Growth inhibition by FTIs is associated primarily with their induction of cell cycle arrest in either G0/G1 or G2/M, depending on p53 status. FTIs do not generally kill transformed cells except in conjunction with a second apoptotic signal such as serum starvation [58] or the absence of substratum [59], but exceptions have been observed. In some human tumor cell lines, sensitivity to FTI-induced apoptosis is dependent on the inhibition of the PI3-K/Akt/BAD pathway and a concomitant reduction in the level of phospho-Akt in those cells [60]. What farnesylated protein is critical to regulate this pathway is unknown at present, although it could be endogenous H-Ras; it has also been suggested that FTI does not inhibit Akt via blocking the processing and activity of a farnesylated protein, but rather by upregulating the expression of RhoB [61]. However, even the correlation between FTI-induced apoptosis and phospho-Akt suppression is not universal across all cell types, so additional mechanisms are likely. FTIs may also have an indirect effect on tumor growth by inhibiting angiogenesis [62,63], and this is known to be the consequence of a decrease in vascular endothelial growth factor (VEGF). However, it is unclear whether the farnesylated protein controlling VEGF that is inhibited by FTI to produce the anti-angiogenic effect is a Ras protein, which is known to regulate VEGF in a PI3-K-dependent manner, or is another protein altogether. Therefore, at this stage it is not possible to draw definitive causal connections between most of the observed cellular effects of FTIs and particular biochemical pathways that may mediate those effects. Making such connections is now a major goal for understanding what FTIs are doing in the context of both whole cells and animals.

Summary and Prospects

The development of FTIs during the past ten years constitutes a convoluted story. Originally designed to treat Ras-related cancers by inhibiting oncogenic Ras function, FTIs are now thought to be ineffective against N- and K-Ras due to alternative prenylation, and to exert their effects primarily through inhibition of the farnesylation of other substrate proteins or the induction of proteins such as RhoB. With so many potential FTI targets, each with a different cellular function and sensitivity to FTIs, the task of unraveling the mechanism of FTI action is complex. But even in the absence of an understanding of which specific substrates mediate which specific consequences, it is clear that FTIs do have utility as anti-cancer agents, and can be useful tools to demonstrate the contribution of farnesylated proteins to the control of a particular signaling pathway.

References

1. Shields, J. M. et al. (2000). Understanding Ras: 'it ain't over 'til it's over.' *Trends Cell Biol.* **10**(4), 147–154.
2. Bos, J. L. (1989). Ras oncogenes in human cancer: A review. *Cancer Res.* **49**(17), 4682–4689.
3. Cox, A. D. and Der, C. J. (1997). Farnesyltransferase inhibitors and cancer treatment: targeting simply Ras? *Biochim. Biophys. Acta* **1333**(1), F51–F71.
4. Reiss, Y. et al. (1990). Inhibition of purified p21ras farnesyl:protein transferase by Cys-AAX tetrapeptides. *Cell* **62**(1), 81–88.
5. James, G. L. et al. (1993). Benzodiazepine peptidomimetics: potent inhibitors of Ras farnesylation in animal cells. *Science* **260**(5116), 1937–1942.
6. Garcia, A. M. et al. (1993). Peptidomimetic inhibitors of Ras farnesylation and function in whole cells. *J. Biol. Chem.* **268**(25), 18415–18418.
7. Qian, Y. et al. (1994). Design and structural requirements of potent peptidomimetic inhibitors of p21ras farnesyltransferase. *J. Biol. Chem.* **269**(17), 12410–12413.
8. Wallace, A. et al. (1996). Selection of potent inhibitors of farnesyl-protein transferase from a synthetic tetrapeptide combinatorial library. *J. Biol. Chem.* **271**(49), 31306–31311.
9. Uchida, R. et al. (1996). Andrastin D, novel protein farnesyltransferase inhibitor produced by *Penicillium* sp. FO-3929. *J. Antibiot. (Tokyo)* **49**(12), 1278–1280.
10. Hara, M. et al. (1993). Identification of Ras farnesyltransferase inhibitors by microbial screening. *Proc. Natl. Acad. Sci. USA* **90**(6), 2281–2285.
11. Shaikenov, T. E. et al. (2001). Arglabin-DMA, a plant derived sesquiterpene, inhibits farnesyltransferase. *Oncol. Rep.* **8**(1), 173–179.
12. Fu, H. W. and P. J. Casey (1999). Enzymology and biology of CaaX protein prenylation. *Recent Prog. Horm. Res.* **54**, 315–342.
13. Strickland, C. L. et al. (1998). Crystal structure of farnesyl protein transferase complexed with a CaaX peptide and farnesyl diphosphate analogue. *Biochemistry* **37**(47), 16601–16611.
14. Long, S. B. et al. The crystal structure of human protein farnesyltransferase reveals the basis for inhibition by CaaX tetrapeptides and their mimetics. *Proc. Natl. Acad. Sci. USA* **98**(23), 12948–12953.

15. Aoyama, T. et al. (1998). A new class of highly potent farnesyl diphosphate-competitive inhibitors of farnesyltransferase. *J. Med. Chem.* **41**(2), 143–147.
16. Scholten, J. D. et al. (1997). Synergy between anions and farnesyldiphosphate competitive inhibitors of farnesyl:protein transferase. *J. Biol. Chem.* **272**(29), 18077–18081.
17. Manne, V. et al. (1995). Bisubstrate inhibitors of farnesyltransferase: A novel class of specific inhibitors of ras transformed cells. *Oncogene* **10**(9), 1763–1779.
18. Lerner, E. C. et al. (1995). Ras CAAX peptidomimetic FTI-277 selectively blocks oncogenic Ras signaling by inducing cytoplasmic accumulation of inactive Ras-Raf complexes. *J. Biol. Chem.* **270**(45), 26802–26806.
19. Sepp-Lorenzino, L. et al. (1995). A peptidomimetic inhibitor of farnesyl:protein transferase blocks the anchorage-dependent and -independent growth of human tumor cell lines. *Cancer Res.* **55**(22), 5302–5309.
20. Karp, J. E. et al. (2001). Current status of clinical trials of farnesyltransferase inhibitors. *Curr. Opin. Oncol.* **13**(6), 470–476.
21. Caponigro, F., Casale, M., Bryce, J. (2003). Farnesyl transferase inhibitors in clinical development. *Expert. Opin. Investig. Drugs* **12**(6), 943–954.
22. Vogt, A. et al. (1995). A non-peptide mimetic of Ras-CAAX: Selective inhibition of farnesyltransferase and Ras processing. *J. Biol. Chem.* **270**(2), 660–664.
23. Yokoyama, K., McGeady, P., and Gelb, M. H. (1995). Mammalian protein geranylgeranyltransferase-I: Substrate specificity, kinetic mechanism, metal requirements, and affinity labeling. *Biochemistry* **34**(4), 1344–1354.
24. Gibbs, J. B. and Oliff, A. (1997). The potential of farnesyltransferase inhibitors as cancer chemotherapeutics. *Annu. Rev. Pharmacol. Toxicol.* **37**, 143–166.
25. Gibbs, R. A., Zahn, T. J., and Sebolt-Leopold, J. S. (2001). Non-peptidic prenyltransferase inhibitors: diverse structural classes and surprising anti-cancer mechanisms. *Curr. Med. Chem.* **8**(12), 1437–1465.
26. Oliff, A. (1999). Farnesyltransferase inhibitors: Targeting the molecular basis of cancer. *Biochim. Biophys. Acta* **1423**(3), C19–C30.
27. Kohl, N. E. et al. (1995). Inhibition of farnesyltransferase induces regression of mammary and salivary carcinomas in ras transgenic mice. *Nat. Med.* **1**(8), 792–797.
28. Mangues, R. et al. (1998). Antitumor effect of a farnesyl protein transferase inhibitor in mammary and lymphoid tumors overexpressing N-ras in transgenic mice. *Cancer Res.* **58**(6), 1253–1259.
29. Omer, C. A. et al. (2000). Mouse mammary tumor virus-Ki-rasB transgenic mice develop mammary carcinomas that can be growth-inhibited by a farnesyl:protein transferase inhibitor. *Cancer Res.* **60**(10), 2680–2688.
30. Karp, J. E. (2001). Farnesyl protein transferase inhibitors as targeted therapies for hematologic malignancies. *Semin. Hematol.* **38**(3 Suppl. 7), 16–23.
31. Johnston, S. R. and Kelland, L. R. (2001). Farnesyl transferase inhibitors—A novel therapy for breast cancer. *Endocr. Relat. Cancer* **8**(3), 227–235.
32. Adjei, A. A. et al. (2000). A Phase I trial of the farnesyl transferase inhibitor SCH66336: Evidence for biological and clinical activity. *Cancer Res.* **60**(7), 1871–1877.
33. Moasser, M. M. et al. (1998). Farnesyl transferase inhibitors cause enhanced mitotic sensitivity to taxol and epothilones. *Proc. Natl. Acad. Sci. USA* **95**(4), 1369–1374.
34. Shi, B. et al. (2000). The farnesyl protein transferase inhibitor SCH66336 synergizes with taxanes *in vitro* and enhances their antitumor activity *in vivo*. *Cancer Chemother. Pharmacol.* **46**(5), 387–393.
35. Andela, V. B. et al. (2002). Synergism of aminobisphosphonates and farnesyl transferase inhibitors on tumor metastasis. *Clin. Orthop.* (397), 228–239.
36. Tahir, S. K. et al. (2000). Inhibition of farnesyltransferase with A-176120, a novel and potent farnesyl pyrophosphate analogue. *Eur. J. Cancer* **36**(9), 1161–1170.
37. Brassard, D. L. et al. (2002). Inhibitors of farnesyl protein transferase and MEK1,2 induce apoptosis in fibroblasts transformed with farnesylated but not geranylgeranylated H-Ras. *Exp. Cell Res.* **273**(2), 138–146.
38. Bernhard, E. J. et al. (1998). Inhibiting Ras prenylation increases the radiosensitivity of human tumor cell lines with activating mutations of ras oncogenes. *Cancer Res.* **58**(8), 1754–1761.
39. James, G. L., Goldstein, J. L., and Brown, M. S. (1995). Polylysine and CVIM sequences of K-RasB dictate specificity of prenylation and confer resistance to benzodiazepine peptidomimetic *in vitro*. *J. Biol. Chem.* **270**(11), 6221–6226.
40. James, G., Goldstein, J. L., and Brown, M. S. (1996). Resistance of K-RasBV12 proteins to farnesyltransferase inhibitors in Rat1 cells. *Proc. Natl. Acad. Sci. USA* **93**(9), 4454–4458.
41. Whyte, D. B. et al. (1997). K- and N-Ras are geranylgeranylated in cells treated with farnesyl protein transferase inhibitors. *J. Biol. Chem.* **272**(22), 14459–14464.
42. Rowell, C. A. et al. (1997). Direct demonstration of geranylgeranylation and farnesylation of Ki-Ras *in vivo*. *J. Biol. Chem.* **272**(22), 14093–14097.
43. Cox, A. D. and Der, C. J. (2000). Farnesyltransferase inhibitors: Anti-Ras or anti-cancer drugs?, in Gutkind, J. S., ed, *Signalling Networks and Cell Cycle Control: The Molecular Basis of Cancer and Other Diseases*, pp. 501–508. Humana Press: Totowa, NJ.
44. Tamanoi, F. et al. (2001). Farnesylated proteins and cell cycle progression. *J. Cell. Biochem.* **Suppl**(37), 64–70.
45. Moores, S. L. et al. (1991). Sequence dependence of protein isoprenylation. *J. Biol. Chem.* **266**(22), 14603–14610.
46. Kato, K. et al. (1992). Isoprenoid addition to Ras protein is the critical modification for its membrane association and transforming activity. *Proc. Natl. Acad. Sci. USA* **89**(14), 6403–6407.
47. Lebowitz, P. F. and Prendergast, C. C. (1998). Non-Ras targets of farnesyltransferase inhibitors: Focus on Rho. *Oncogene* **17**, 1439–1445.
48. Adamson, P. et al. (1992). Post-translational modifications of p21rho proteins. *J. Biol. Chem.* **267**(28), 20033–20038.
49. Chen, Z. et al. (2000). Both farnesylated and geranylgeranylated RhoB inhibit malignant transformation and suppress human tumor growth in nude mice. *J. Biol. Chem.* **275**(24), 17974–17978.
50. Du, W., Lebowitz, P. F., and Prendergast, G. C. (1999). Cell growth inhibition by farnesyltransferase inhibitors is mediated by gain of geranylgeranylated RhoB. *Mol. Cell. Biol.* **19**(3), 1831–1840.
51. Liu, A. et al. (2000). RhoB alteration is necessary for apoptotic and antineoplastic responses to farnesyltransferase inhibitors. *Mol. Cell. Biol.* **20**(16), 6105–6113.
52. Ashar, H. R. et al. (2000). Farnesyl transferase inhibitors block the farnesylation of CENP-E and CENP-F and alter the association of CENP-E with the microtubules. *J. Biol. Chem.* **275**(39), 30451–30457.
53. Crespo, N. C. et al. (2001). The farnesyltransferase inhibitor, FTI-2153, blocks bipolar spindle formation and chromosome alignment and causes prometaphase accumulation during mitosis of human lung cancer cells. *J. Biol. Chem.* **276**(19), 16161–16167.
54. Kuroda, Y., Suzuki, N., and Kataoka, T. (1993). The effect of post-translational modifications on the interaction of Ras2 with adenylyl cyclase. *Science* **259**(5095), 683–686.
55. Itoh, T. et al. (1993). The post-translational processing of ras p21 is critical for its stimulation of mitogen-activated protein kinase. *J. Biol. Chem.* **268**(5), 3025–3028.
56. Yasuda, H. et al. (1996). Role of the prenyl group on the G protein gamma subunit in coupling trimeric G proteins to A1 adenosine receptors. *J. Biol. Chem.* **271**(31), 18588–18595.
57. Myung, C. S. et al. (1999). Role of isoprenoid lipids on the heterotrimeric G protein gamma subunit in determining effector activation. *J. Biol. Chem.* **274**(23), 16595–16603.
58. Suzuki, N., Urano, J., and Tamanoi, F. (1998). Farnesyltransferase inhibitors induce cytochrome c release and caspase 3 activation preferentially in transformed cells. *Proc. Natl. Acad. Sci. USA* **95**(26), 15356–15361.

59. Lebowitz, P. F., Sakamuro, D., and Prendergast, G. C. (1997). Farnesyl transferase inhibitors induce apoptosis of Ras-transformed cells denied substratum attachment. *Cancer Res.* **57**(4), 708–713.
60. Jiang, K. *et al.* (2000). The phosphoinositide 3-OH kinase/AKT2 pathway as a critical target for farnesyltransferase inhibitor-induced apoptosis. *Mol. Cell. Biol.* **20**(1), 139–148.
61. Liu, A. and Prendergast, G. C. (2000). Geranylgeranylated RhoB is sufficient to mediate tissue-specific suppression of Akt kinase activity by farnesyltransferase inhibitors. *FEBS Lett.* **481**(3), 205–208.
62. Gu, W. Z. *et al.* (1999). Effect of novel CAAX peptidomimetic farnesyltransferase inhibitor on angiogenesis *in vitro* and *in vivo*. *Eur. J. Cancer* **35**(9), 1394–1401.
63. Kerbel, R. S. *et al.* (2000). "Accidental" anti-angiogenic drugs. Antioncogene directed signal transduction inhibitors and conventional chemotherapeutic agents as examples. *Eur. J. Cancer* **36**(10), 1248–1257.

CHAPTER 245

Structure of Rho Family Targets

Helen R. Mott and Darerca Owen
Department of Biochemistry, University of Cambridge, Cambridge, United Kingdom

The variety of cellular processes controlled by the Rho family proteins, of which Cdc42, Rac, and Rho are the most widely studied members, are mediated through a number of downstream effector proteins, many of which are still being characterized. The members of the Rho subfamily have a far greater number of effector proteins than other G-protein subfamilies. The effectors can be loosely divided into those that bind to Cdc42 and Rac and those that bind to Rho. The effectors are both functionally and structurally diverse, as is the nature of their interactions with the small G proteins.

Structurally, the Rho family proteins are distinguished from the Ras family by the presence of an extra pair of helices (the insert region) [1]. It has been reported that removal of this insert region in Rac expressed *in vivo* prevented both the formation of membrane ruffles [2] and transformation [3]. Thus it is expected that at least one target protein of the Rho family will interact with this region. The structures of six Rho family effectors and their complexes have been studied so far and will be described here.

CRIB Proteins

Many of the downstream effectors for the Rho family proteins Cdc42 and Rac contain a small (16 amino acid) consensus sequence known as the CRIB (Cdc42/Rac interactive binding), which is essential for mediating interactions with the G proteins [4]. In several studies it has been shown that additional residues, C terminal to the CRIB, are also necessary for tight binding, making the full G protein binding domain (GBD) 40–45 amino acids. All the CRIB proteins bind to Cdc42·GTP and some of them bind to Rac·GTP. Structural studies of the CRIB family proteins have addressed two fundamental questions: How do some CRIB proteins discriminate between the closely related Cdc42 and Rac, and how does binding a Rho family protein activate downstream events?

The solution structures of three different Cdc42/CRIB complexes have been solved: activated Cdc42 kinase (ACK), a tyrosine kinase, which has been implicated in integrin signaling and endocytosis [5]; Wiscott-Aldrich syndrome protein (WASP), which is thought to mediate interactions with the cytoskeleton [6], and p21 activated kinase (PAK), a serine/threonine kinase involved in JNK signaling and cytoskeletal rearrangements [7]. Comparison of these structures reveals interesting differences in the way that the effectors contact the G proteins (Fig. 1). In addition, since ACK and WASP are specific for Cdc42 while PAK binds to both Rac and Cdc42, the structures shed light on how the CRIB proteins may discriminate between two such similar molecules. In each structure, the CRIB consensus region binds in a similar manner, forming an intermolecular β-sheet with the β2-strand of Cdc42 and then interacting with switch I. This is reminiscent of the structure of Ras with its effectors Raf-1 [8] and Ral-GEF [9], both of which form a similar intermolecular β-sheet, although all the other features of these structures differ. The regions outside the CRIB consensus all interact with the same regions of Cdc42, helix α5, and switch II, although the details of the interactions are all different. ACK does not form any more secondary structure but wraps around the G protein, forming an irregular hairpin at the top of switch I. WASP and PAK both form a regular β-hairpin that interacts with switch I and switch II, followed by a short piece of α-helix which interacts with switch II. In WASP and PAK the relative orientation of this hairpin and α-helix are different, as is their orientation with respect to the switch II helix.

NMR studies on the free GBDs of these three proteins revealed no significant tertiary structure. In both PAK and WASP there was some evidence for the formation of the short section of α-helix that is seen in the complex with Cdc42 [7,10]. In ACK, where there is no secondary structure in the complex, none could be discerned in the free GBD [5].

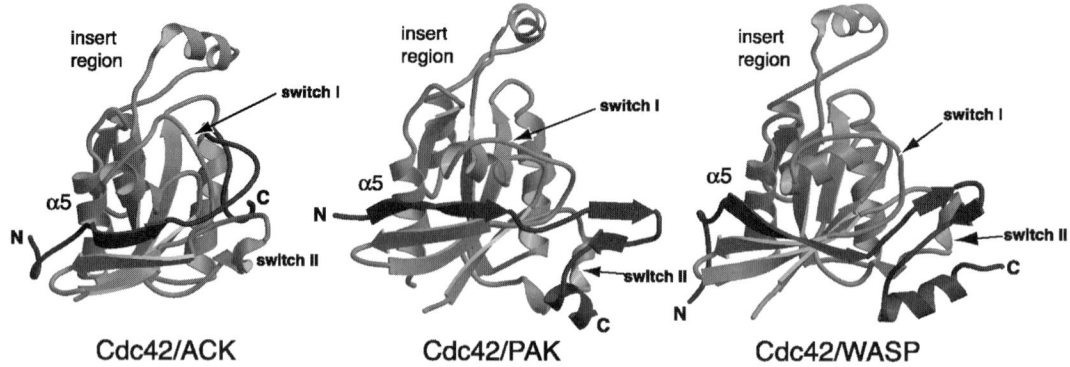

Figure 1 Comparison of CRIB/Cdc42 structures. In each case the G protein is shown in pale gray and the CRIB fragment in dark gray. The insert region that defines the Rho family is marked. The positions of the two switch regions that become fixed on effector binding are also shown, as is the position of the α5-helix, which interacts with the CRIB effector in all three cases and is the location of the mutation that specifically disrupts Cdc42/ACK and Cdc42/WASP binding.

Sequence alignments of Cdc42 and Rac reveal that switches I and II are almost completely conserved, the only difference being a single (conservative) substitution. It was therefore clear that the basis of the selectivity of CRIB effector proteins such as ACK and WASP for Cdc42 would lie outside these switches. Mutagenesis studies combined with measurements of K_d suggested that interactions with Leu-174 in helix α5 of Cdc42 contribute to binding of WASP and ACK to Cdc42 but do not contribute to PAK binding [11]. Position 174 is an Arg residue in Rac and thus it may represent one of the points of discrimination between G proteins. Analysis of single residue mutations may not lead us to a complete understanding of the discrimination between Cdc42 and Rac, because in all the Cdc42/effector complexes solved the buried surface area is large (2500–4000 Å^2). Also, since there is no Rac/CRIB fragment complex solved so far, the details of Rac/CRIB interactions are still unknown.

Insight into the activation mechanisms of the CRIB proteins came when the structures of both PAK and WASP in autoinhibited forms were solved [12,13]. WASP has no enzyme activity but it has a region at the C terminus that binds both the Arp2/3 complex and actin. It was found that a small region of the C terminus that is homologous to cofilin (CHR) interacts with residues within and C terminal to the GBD. The structure of a molecule comprising the CHR tethered by a flexible linker to the GBD extended at the C terminus by approximately 20 residues has been solved (Fig. 2). This structure showed that the β-hairpin and α-helix seen in the Cdc42 complex are still present but that they now interact with three extra α-helices C terminal to the GBD, mainly through hydrophobic contacts. These four helices form a hydrophobic surface, against which a helix from the CHR is packed. It is clear that in this form the protein cannot bind either to the Arp2/3 complex (via the CHR) or to Cdc42 (via the GBD). To bind Cdc42 the protein would have to undergo a conformational change that results in release of the CHR, allowing it to bind to other partners. The thermodynamic cost of this conformational change would be paid for by a lower binding affinity between the autoinhibited form of the protein and Cdc42 than between GBD fragments

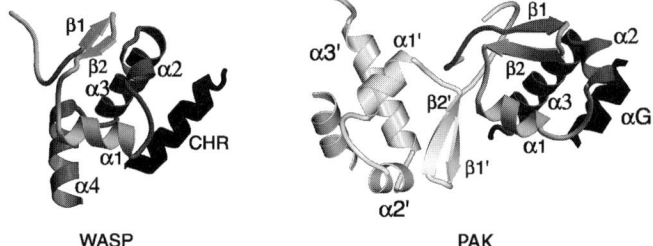

Figure 2 Structure of autoinhibited WASP and PAK fragments. The layers of the structures are shown in different shades. In both cases the β1/β2 and α1 elements are part of the GBD and are involved in the interaction with Cdc42; α2 and α3 are in the region immediately C terminal to the GBD that is involved in negative regulation of the C terminus of the molecule. In WASP, helix α4 follows α3; this helix has no counterpart in PAK. The CHR and αG are the cofilin homology region (WASP) and helix G from the kinase domain (PAK), respectively. The second PAK monomer is shown in pale gray and labeled β1′, α1′ etc.

and Cdc42 as is seen, with K_ds approximately 100-fold higher for the tethered construct [13].

It had been shown by yeast-two hybrid experiments with PAK that a region C terminal to the GBD, known as the kinase inhibitory (KI) domain could bind directly to its kinase domain, thus inhibiting its catalytic activity. An x-ray-derived structure has been solved of the GBD/KI bound to the kinase domain and reveals that the GBD/KI domain is strikingly similar to the equivalent region of WASP, although in PAK there are only three helices to WASP's four (Fig. 2). One of the helices from the kinase domain packs against the three helices of the KI domain in a manner closely resembling the CHR helix of WASP packing against its autoinhibitory domain. There is one striking difference in the PAK structure: it is a dimer that is held together by an interaction between the CRIB/KI regions. Given this structure it would seem to be impossible to form a Cdc42/PAK complex without breaking the dimer. This was shown in a later paper by the same authors [14]. However others have shown that PAK forms a stable dimer, even in the presence of Cdc42 [15].

The activation mechanism of ACK is not known at present but it is likely to involve a similar, intramolecular inhibition.

This is implied by the discovery that mutation of Leu-543 of ACK causes constitutive activation of the kinase [16]. This Leu interacts with switch II in the Cdc42/ACK complex [5] and mutation of the equivalent Leu residue in PAK (Leu-107) or WASP (Leu-270) also disrupts their autoinhibitory interactions [17,18]. The activation of ACK may, however, be more complicated since it appears that Cdc42 also disrupts an intramolecular interaction between the SH3 domain and a Pro-rich region [19]. An intramolecular SH3/Pro interaction is also thought to exist in another CRIB-containing protein, mixed lineage kinase 3 (MLK3) [20]. In both ACK and MLK3 the role of Cdc42 binding in activation is still unclear.

Non-CRIB Rac Effectors

p67^{phox}

p67^{phox} is one component of the multiprotein enzyme complex, NADPH oxidase. This complex, found in phagocytes, forms the principal defense mechanism against microbial infection in humans. Binding of Rac to p67^{phox} is a critical step in the activation of the latent NADPH oxidase complex. The Rac binding region of p67^{phox} had been localized to the N terminal 200 amino acids, which contains four copies of the tetratricopeptide repeat (TPR) motif. Structures of other TPR-containing proteins had shown that each TPR motif is composed of a pair of anti-parallel α-helices (A and B), and that these repeated units pack together to form an extended structure with an amphipathic groove on the A helix face of the domain (the TPR groove), which mediates interactions with other proteins [21,22]. The structure of Rac·GTP complexed with the N terminal 200 amino acids of p67^{phox} revealed both an effector GBD distinct from the CRIB family of effectors and also a different use of the TPR repeats as a binding motif [23]. The N terminus of p67^{phox} contains 9 α-helices: the first eight of these form four TPR motifs, while the ninth helix packs against the B helix of TPR4 (Fig. 3). The TPR groove is filled by a stretch of residues C terminal to the ninth helix, which binds in an extended conformation and thus it is not available for intermolecular interactions. Rather, contacts are made between Rac and one face of the TPR domain, which consist of a β–hairpin insertion between TPR3 and 4 and the loops connecting TPR1 with TRP2 and TPR2 with TPR3. Contacts on the G-protein side are also unusual and include residues from helix α1 and the following loop, residues from the N terminal end of switch I, and the loop between strand β5 and helix α5. The TPR domain does not contact all of switch I and no contacts are seen with switch II or the insert region. This is in contrast to the complexes between Cdc42 and the CRIB effectors, where extensive contacts are made with both switch I and II. The TPR/Rac complex also differs from the CRIB effector complexes in that no intramolecular β–sheet is formed. Still, 1170 Å^2 is buried in the TPR/Rac complex, less than that in the CRIB/Cdc42 complexes, possibly resulting in the lower affinity observed.

p67^{phox} binds specifically to Rac rather than Cdc42. Analysis of the residues involved in the interface showed that all were conserved between Rac and Cdc42, except Gly-30. Ala-27 and Gly-30 have been defined as critical residues for the specificity of the interaction between Rac and p67^{phox}: Mutation of these residues in Cdc42 to the corresponding residues in Rac results in a Cdc42 protein that binds p67^{phox} with a relatively high affinity [23]. Ala-27 does not directly contact the TPR domain but in Cdc42 this residue is a Lys, which would cause a steric clash, preventing binding.

Arfaptin

Arfaptin (or POR, partner of Rac) was identified independently as an effector for both the Rac and Arf small G proteins and, consequently, has been proposed to be a facilitator of crosstalk between signaling pathways. Arfaptin (residues 118–341) has been crystalized alone and in complex with both Rac·GDP and Rac·GMPPNP [24]. The Arfaptin domain consists of 3 α–helices (A–C) that form an anti-parallel α–helical bundle; 2 of these self-associate to give an elongated crescent-shaped dimer, with an overall length of 140Å, which binds to one Rac molecule. Rac sits on the concave surface of the Arfaptin crescent close to the dimer interface, and contacts are seen predominantly between switch I and II of Rac and monomer A of Arfaptin (Fig. 3). Switch I packs against helix αA while switch II interacts with helix αB. A single contact is seen to monomer B, at His 57 in helix αA′. This interaction is sufficient to preclude the binding of another Rac molecule to Arfaptin, thus accounting for the observed stoichiometry of 1 Rac:1 Arfaptin dimer. 1600Å^2 of solvent-accessible surface area is buried in the complex. Similar of Arfaptin affinities for both the GDP- and GTP-bound forms of Rac argues against its being a conventional effector for the G protein. In contrast, Arfaptin binds to both Arf1 and Arf6 in a GTP dependent manner. Examination of the Rac·GDP and Rac·GMPPNP molecules reveals crucial similarities between the two in complex with Arfaptin. Critically, Thr-35, which coordinates to the Mg^{2+} in G protein·GTP forms is instead in contact with Arfaptin, giving rise to a more GDP-like structure. It is predicted that a canonical Rac·GTP conformation could not be accommodated by Arfaptin [24]. Presumably, in the Arf/Arfaptin complex the G protein can take up its usual GTP-like conformation, thus allowing discrimination in that case. It is possible that Arfaptin's ability to bind both forms of Rac allows it to sequester Rac until Arf is activated, whereupon Arf displaces Rac freeing it to signal appropriately. In this model, the function of Arf is to allow coordinated activation of Rac and Arf.

Rho Effectors

Effectors for Rho include at least eleven proteins: DAG Kinase, PLD, PIP5-kinase, Kinectin, Rhotekin, Rhophilin, p140 Diaphanous, MBS, Citron, ROK and PRK. Within this

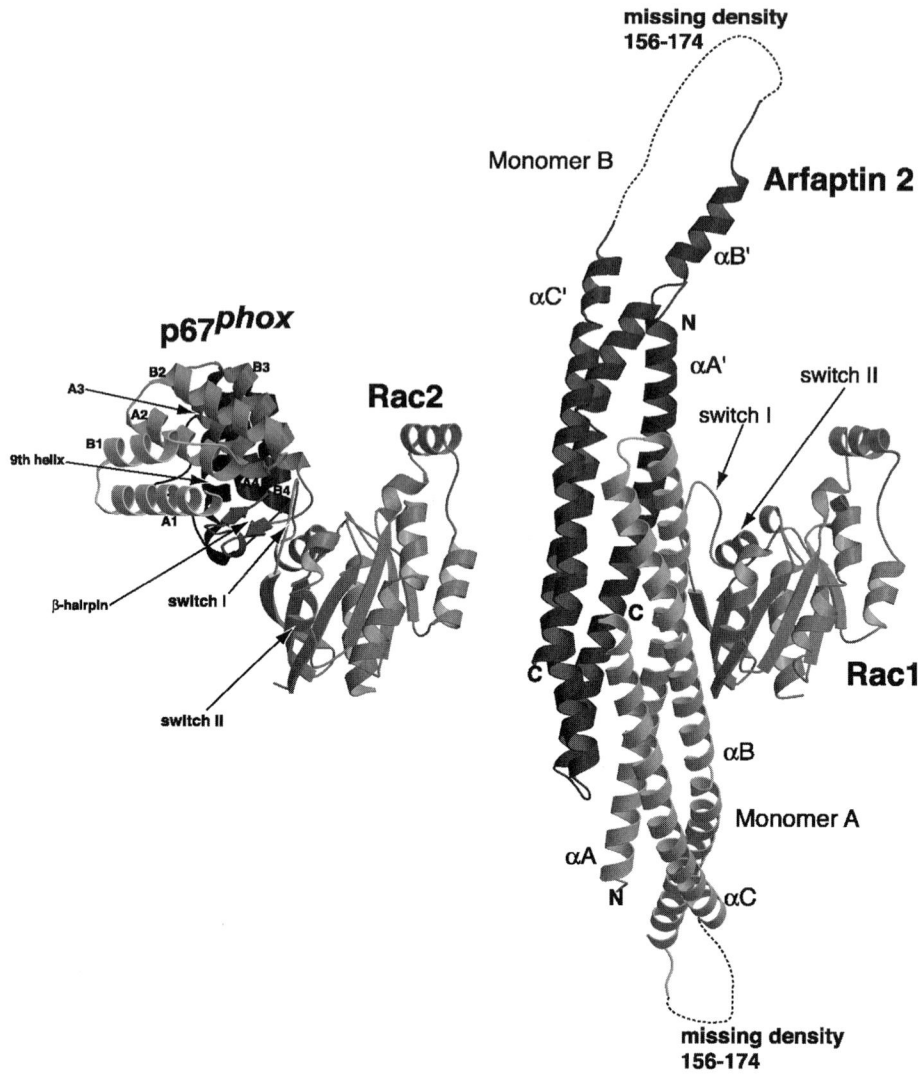

Figure 3 The structures of Rac2/p67^{phox} and Rac1/Arfaptin 2. The layers of the TPR repeats of the p67^{phox} become darker going from the N to the C terminus. The switch regions of Rac are shown in both structures. The two monomers in the Arfaptin 2 dimer are shown in different shades. In the Arfaptin structure the two regions where no electron density was observed are shown as dotted lines.

group of effector proteins there are at least two Rho binding motifs defined by sequence homology: REM proteins (or Class 1 Rho binding motif) include the PRKs, Rhophilin and Rhotekin, while RKH proteins (REM2 or Class 2 Rho binding motif) include the ROKs and Kinectin (reviewed in [25]). Structural information is limited at present to one of the REM proteins, PRK1, in complex with RhoA.

Protein Kinase C Related Kinases

PRK1 (PKN) and 2 are highly related serine/threonine kinases with a catalytic domain homologous to that of the protein kinase C family in their C termini and a unique regulatory domain in their N termini [26,27]. The N terminus of the PRK1 contains three HR1 repeats, one of which, HR1a, incorporates an inhibitory pseudosubstrate site [28]. Kinase activity is enhanced by binding of GTP-bound Rho or Rac [29–32].

The x-ray structure of RhoA in complex with the HR1a repeat of PRK1 describes the fold of the HR1a domain as an **a**nti-parallel **c**oiled-**c**oil (ACC) finger domain [33] (Fig. 4). The ACC finger domain is quite distinct from other G protein binding domains and known Rho family effectors. It does show limited similarity with the Rab binding domain of Rabphilin and the Rac/Arf effector Arfaptin, but the contacts that these effectors make with their G proteins are quite different [24,34]. The structure of the complex between HR1a and RhoA indicated two possible contacts sites on RhoA for HR1a. The major site, contact 1, has a buried surface area of 2080Å^2 and mainly involves hydrophilic interactions. This contact involves residues in the β2- and β3-strands of RhoA, the N-terminal part of helix α5 and residues at the ends of switch I. Contact 2, which involves more hydrophobic residues, buries a total surface area of 1640Å^2 and involves residues in switch I (Val38–Asn41),

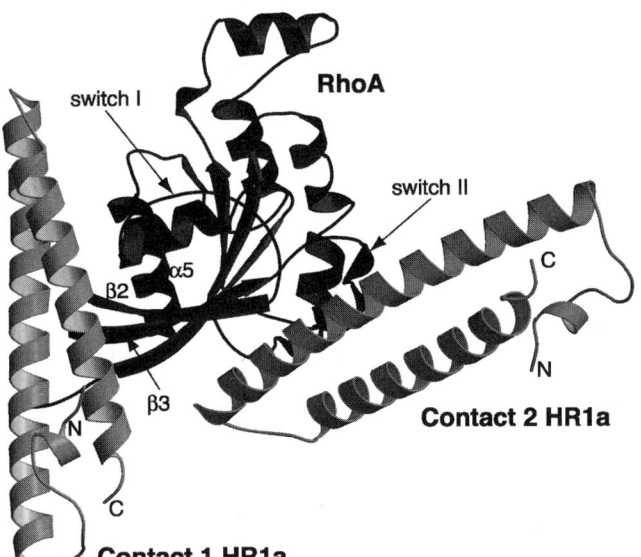

Figure 4 The structure of the RhoA/PRK1 HR1a complex. The two potential HR1a binding sites are shown as contact 1 and contact 2. Contact 1 was defined as the primary contact site on the basis of the buried surface area, although the HR1a in contact 2 makes more interactions with the switches.

strand β3, and switch II (Trp58 and Asp65–Asp76). The size of contact 2 means that it is unlikely to be an artifact of crystal packing. It is also noteworthy that it involves more residues in the switch regions of RhoA, which are likely to be involved in effector binding.

Concluding Remarks

We have attempted to summarize here the pertinent details of all the Rho family/effector structures determined. Several points emerge from this discussion. First, it is clear that the structural diversity in the Rho effectors is extensive. It is likely, for example, that the CRIB proteins, although they will all have some similarities in the way that they interact, will also differ in their details, particularly outside the short CRIB consensus sequence. The other three effector structures are also completely different both to each other and to the CRIBs, although it is likely that families will emerge whose members interact with the G proteins in a similar, but not identical manner. There are also, however, several other effector proteins with no sequence homology to Arfaptin, PRK, TPR domains, or CRIBs. It is likely that these proteins will adopt different structures and will interact with the Rho family proteins in novel ways.

The manner in which the effectors contact the Rho family G proteins is multifarious. They utilize β-strands (the CRIBs), β-hairpins (the CRIBs), α-helices (Arfaptin, PRK1, and the CRIBs), interhelical loops ($p67^{phox}$), and even a dimer interface (Arfaptin) to interact with the G protein. In some cases, such as the CRIB proteins, the effectors make an extensive set of contacts with the Rho family proteins, burying a large surface area (2500–4000 Å^2) while in others,

such as $p67^{phox}$, which binds with a lower affinity, the buried surface is only ~1200Å^2.

The region of the G proteins that interact with the downstream targets is also not conserved. In most cases, switch I is involved in the interactions, which is perhaps not a surprise given that effectors bind preferentially to the GTP-bound form of the G protein. It is usually the case that other regions of the G protein are also involved in binding to effectors; this may be in part to bring about specificity. The switches are relatively well conserved within the family while the diversity in the other regions is higher. In most of the structures discussed here switch II is also involved in the interaction; the exceptions being $p67^{phox}$ and PRK1, contact I. The other regions of the Rho family that are involved in effector binding are helix α1, the β2- and β3-strands, the β5-α5 loop, and the C-terminal helix, α5. In no case so far is the insert region involved in effector binding.

Structural information on the Rho family targets has moved at an exciting pace over the last three years. The next stage must be to determine how the binding of the Rho family protein then causes the downstream effects that are seen. This point has begun to be addressed with the elegant work on PAK and WASP activation. We can look forward to many more such breakthroughs in the future, where a combination of structural and mechanistic studies will help us to understand fully these complex systems.

References

1. Hirshberg, M., Stockley, R. W., Dodson, G., and Webb, M. R. (1997). The crystal structure of human rac1, a member of the rho-family complexed with a GTP analogue. *Nat. Struct. Biol.* **4**, 147–152.
2. Karnoub, A. E., Der, C. J., and Campbell, S. L. (2001). The insert region of Rac1 is essential for membrane ruffling but not cellular transformation. *Mol. Cell. Biol.* **21**, 2847–2857.
3. Joneson, T. and Bar-Sagi, D. (1998). A Rac1 effector site controlling mitogenesis through superoxide production. *J. Biol. Chem.* **273**, 17991–17994.
4. Burbelo, P. D., Drechsel, D., and Hall, A. (1995). A conserved binding motif defines numerous candidate target proteins for both Cdc42 and Rac Gtpases. *J. Biol. Chem.* **270**, 29071–29074.
5. Mott, H. R., Owen, D., Nietlispach, D., Lowe, P. N., Manser, E., Lim, L., and Laue, E. D. (1999). Structure of the small G protein Cdc42 bound to the GTPase-binding domain of ACK. *Nature* **399**, 384–388.
6. Abdul-Manan, N., Aghazadeh, B., Liu, G. A., Majumdar, A., Ouerfelli, O., Siminovitch, K. A., and Rosen, M. K. (1999). Structure of Cdc42 in complex with the GTPase-binding domain of the 'Wiskott-Aldrich syndrome' protein. *Nature* **399**, 379–383.
7. Morreale, A., Venkatesan, M., Mott, H. R., Owen, D., Nietlispach, D., Lowe, P. N., and Laue, E. D. (2000). Structure of Cdc42 bound to the GTPase binding domain of PAK. *Nat. Struct. Biol.* **7**, 384–388.
8. Nassar, M., Horn, G., Herrmann, C., Scherer, A., McCormick, F., and Wittinghofer, A. (1995). The 2.2-angstrom crystal-structure of the Ras-binding domain of the serine threonine kinase C-Raf1 in complex with Rap1a and a Gtp analog. *Nature* **375**, 554–560.
9. Vetter, I. R., Linnemann, T., Wohlgemuth, S., Geyer, M., Kalbitzer, H. R., Herrmann, C., and Wittinghofer, A. (1999). Structural and biochemical analysis of Ras-effector signaling via RalGDS. *FEBS Lett.* **451**, 175–180.
10. Rudolph, M. G., Bayer, P., Abo, A., Kuhlmann, J., Vetter, I. R., and Wittinghofer, A. (1998). The Cdc42/Rac interactive binding region motif of the Wiskott Aldrich syndrome protein (WASP) is necessary

but not sufficient for tight binding to Cdc42 and structure formation. *J. Biol. Chem.* **273**, 18067–18076.

11. Owen, D., Mott, H. R., Laue, E. D., and Lowe, P. N. (2000). Residues in Cdc42 that specify binding to individual CRIB effector proteins. *Biochemistry* **39**, 1243–1250.

12. Lei, M., Lu, W. G., Meng, W. Y., Parrini, M. C., Eck, M. J., Mayer, B. J., and Harrison, S. C. (2000). Structure of PAK1 in an autoinhibited conformation reveals a multistage activation switch. *Cell* **102**, 387–397.

13. Kim, A. S., Kakalis, L. T., Abdul-Manan, M., Liu, G. A., and Rosen, M. K. (2000). Autoinhibition and activation mechanisms of the Wiskott-Aldrich syndrome protein. *Nature* **404**, 151–158.

14. Parrini, M. C., Lei, M., Harrison, S. C., and Mayer, B. J. (2002). Pak1 kinase homodimers are autoinhibited in trans and dissociated upon activation by Cdc42 and Rac1. *Mol. Cell* **9**, 73–83.

15. Buchwald, G., Hostinova, E., Rudolph, M. G., Kraemer, A., Sickmann, A., Meyer, H. E., Scheffzek, K., and Wittinghofer, A. (2001). Conformational switch and role of phosphorylation in PAK activation. *Mol. Cell. Biol.* **21**, 5179–5189.

16. Kato, J., Kaziro, Y., and Satoh, T. (2000). Activation of the guanine nucleotide exchange factor Dbl following ACK1-dependent tyrosine phosphorylation. *Biochem. Biophys. Res. Commun.* **268**, 141–147.

17. Frost, J. A., Khokhlatcheva, A., Stippec, S., White, M. A., and Cobb, M. H. (1998). Differential effects of PAK1-activating mutations reveal activity-dependent and -independent effects on cytoskeletal regulation. *J. Biol. Chem.* **273**, 28191–28198

18. Devriendt, K., Kim, A. S., Mathijs, G., Frints, S. G. M., Schwartz, M., Van den Oord, J. J., Verhoef, G. E .G., Boogaerts, M. A., Fryns, J. P., You, D. Q., Rosen, M. K., and Vandenberghe, P. (2001). Constitutively activating mutation in WASP causes X-linked severe congenital neutropenia. *Nat. Genet.* **27**, 313–317.

19. Yang, W. N., Lin, Q., Guan, J. L., and Cerione, R. A. (1999). Activation of the Cdc42-associated tyrosine kinase-2 (ACK-2) by cell adhesion via integrin beta(1). *J. Biol. Chem.* **274**, 8524–8530.

20. Zhang, H. and Gallo, K. A. (2001). Autoinhibition of mixed lineage kinase 3 through its Src homology 3 domain. *J. Biol. Chem.* **276**, 45598–45603.

21. Das, A. K., Cohen, P. T. W., and Barford, D. (1998). The structure of the tetratricopeptide repeats of protein phosphatase 5: implications for TPR-mediated protein-protein interactions. *EMBO J.* **17**, 1192–1199.

22. Scheufler, C., Brinker, A., Bourenkov, G., Pegoraro, S., Moroder, L., Bartunik, H., Hartl, F. U., and Moarefi, I. (2000). Structure of TPR domain-peptide complexes: Critical elements in the assembly of the Hsp70-Hsp90 multichaperone machine. *Cell* **101**, 199–210.

23. Lapouge, K., Smith, S. J. M., Walker, P. A., Gamblin, S. J., Smerdon, S. J., and Rittinger, K. (2000). Structure of the TPR domain of p67(phox) in complex with Rac · GTP. *Mol. Cell* **6**, 899–907.

24. Tarricone, C., Xiao, B., Justin, N., Walker, P. A., Rittinger, K., Gamblin, S. J., and Smerdon, S. J. (2001). The structural basis of Arfaptin-mediated cross-talk between Rac and Arf signalling pathways. *Nature* **411**, 215–219.

25. Bishop, A. L. and Hall, A. (2000). Rho GTPases and their effector proteins. *Biochem. J.* **348**, 241–255.

26. Mukai, H. and Ono, Y. (1994). A novel protein-kinase with leucine zipper-like sequences—Its catalytic domain is highly homologous to that of protein-kinase-C. *Biochem. Biophys. Res. Commun.* **199**, 897–904.

27. Palmer, R. H., Ridden, J., and Parker, P. J. (1995). Cloning and expression patterns of 2 members of a novel protein-kinase-C-related kinase family. *Eur. J. Biochem.* **227**, 344–351.

28. Kitagawa, M., Shibata, H., Toshimori, M., Mukai, H., and Ono, Y. (1996). The role of the unique motifs in the amino-terminal region of PKN on its enzymatic activity. *Biochem. Biophys. Res. Commun.* **220**, 963–968.

29. Amano, M., Mukai, H., Ono, Y., Chihara, K., Matsui, T., Hamajima, Y., Okawa, K., Iwamatsu, A., and Kaibuchi, K. (1996). Identification of a putative target for Rho as the serine-threonine kinase protein kinase N. *Science* **271**, 648–650.

30. Lu, Y. and Settleman, J. (1999). The Drosophila Pkn protein kinase is a Rho Rac effector target required for dorsal closure during embryogenesis. *Genes Dev.* **13**, 1168–1180.

31. Vincent, S. and Settleman, J. (1997). The PRK2 kinase is a potential effector target of both Rho and Rac GTPases and regulates actin cytoskeletal organization. *Mol. Cell. Biol.* **17**, 2247–2256.

32. Watanabe, G., Saito, Y., Madaule, P., Ishizaki, T., Fujisawa, K., Morii, N., Mukai, H., Ono, Y., Kakizuka, A., and Narumiya, S. (1996). Protein kinase N (PKN) and PKN-related protein rhophilin as targets of small GTPase Rho. *Science* **271**, 645–648.

33. Maesaki, R., Ihara, K., Shimizu, T., Kuroda, S., Kaibuchi, K., and Hakoshima, T. (1999). The structural basis of Rho effector recognition revealed by the crystal structure of human RhoA complexed with the effector domain of PKN/PRK1. *Mol. Cell* **4**, 793–803.

34. Ostermeier, C. and Brunger, A. T. (1999). Structural basis of Rab effector specificity: Crystal structure of the small G protein Rab3A complexed with the effector domain of Rabphilin-3A. *Cell* **96**, 363–374.

CHAPTER 246

Structural Features of RhoGEFs

Jason T. Snyder, Kent L. Rossman, David K. Worthylake, and John Sondek

Department of Pharmacology, Department of Biochemistry and Biophysics, and Lineberger Comprehensive Cancer Center, The University of North Carolina at Chapel Hill, Chapel Hill, North Carolina

Introduction

Effective cellular signaling relies upon the tight control of the various proteins within a signal transduction pathway. For GTPases (guanosine triphosphatases), common mechanisms of regulating the activation and attenuation of these critical molecules have sustained throughout evolution. All GTPases cycle between two discrete states, an inactive guanosine diphosphate (GDP)-bound form, and an active guanosine triphosphate (GTP)-bound form. The removal of GDP nucleotide from an inactive GTPase allows subsequent loading of GTP, thereby triggering these "binary switches" to recognize downstream effectors. This critical process of GTPase activation is rigidly controlled by guanine nucleotide exchange factors (GEFs).

Dbl family proteins are the major recognized class of GEFs for the Rho family of small GTPases [1–3]. Rho GTPases have risen to prominence since a large body of work over the last ten years has implicated these ~25-kDa members of the Ras superfamily in controlling vital cellular functions, including organization of the actin cytoskeleton, progression through the cell cycle, and regulation of transcriptional activities [4–9]. Given that Rho GTPases manage various critical cellular processes, it is not surprising that these small GTPases, as well as their activators (RhoGEFs), promote oncogenesis when constitutively activated [10–16]. Membership within the Dbl family of RhoGEFs is solely dependent upon the possession of an ~300 amino acid segment containing a Dbl homology (DH) domain directly adjacent to a pleckstrin homology (PH) domain [17]. While PH domains exist in a multitude of signaling proteins, the DH domain is unique to these RhoGEFs, and accordingly constitutes the primary catalytic portion of a Dbl protein by supporting nucleotide exchange activity within a substrate Rho GTPase *in vitro* and *in vivo* [18,19].

Recent biophysical investigations into the function of Dbl-family proteins have revealed substantial insight into the means by which these RhoGEFs catalyze the removal of bound nucleotide from Rho proteins. Specifically, an understanding at atomic resolution of the roles of the conserved DH and PH domains found within all Dbl-related proteins is now available. Here we survey the structural features of RhoGEFs and highlight the key determinants responsible for dictating the activation of Rho GTPases.

Structural Accomplishments

The founding member of the Dbl family, Dbl, was first identified as a transforming factor from a human *d*iffuse *B*-cell *l*ymphoma [20,21]. Further analysis delimited the DH/PH domain segment as the minimal transforming portion of Dbl-related RhoGEFs [22]. While most Dbl proteins possess a plethora of protein modules, which most likely regulate GEF activity, including Src homology 2 (SH2) domains, Src homology 3 (SH3) domains, Ras-binding domains (RBD), and regulator of G-protein signaling (RGS) domains, the conserved tandem DH and PH domains have been clearly characterized as the region possessing biochemical guanine nucleotide exchange activity for Rho GTPase targets. This combined information has set the stage for an in-depth structural examination of the catalytic mechanism of Dbl proteins.

NMR structures of the isolated DH domains from the exchange factors Trio [23], βPIX [24], and Vav [25], as well

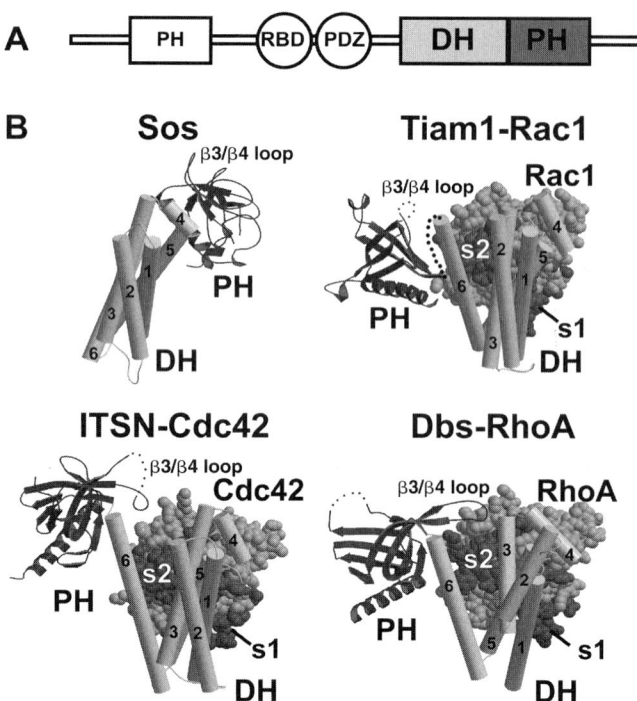

Figure 1 Structural comparison of the DH/PH portion of several Dbl exchange factors. (A) Domain architecture of Tiam1, a representative Dbl family protein, highlighting the conserved DH/PH domain cassette. Tiam1 also possesses an N-terminal PH domain, a Ras binding domain (RBD), and a PDZ domain. (B) The crystal structures of Sos DH/PH (PDB accession code #1DBH), Tiam1-Rac1 (PDB accession code #1FOE), ITSN-Cdc42 (PDB accession code #1KI1), and Dbs-RhoA (PDB accession code #1LB1) were aligned using the CR1 and CR3 regions of the DH domains. The identified helical segments of the DH domains (yellow) are depicted as cylinders, while the PH domains (blue) are presented as ribbon representations. Bound Rho GTPase substrates are portrayed as green CPK spheres, with the exception of the noted switch regions (s1, s2), highlighted in red. Black circles depict the linker region between the DH and PH domains of Tiam1, while blue circles illustrate PH domain loops that are disordered in the crystal structures. (A color representation of this figure is available on the CD version of the *Handbook of Cell Signaling*.)

as the crystal structure of the DH/PH portion of son of sevenless (Sos) [26], first presented the three-dimensional architecture of a DH domain (Fig. 1). In addition, significant insight has been gained by crystal structures of the DH/PH fragments of several RhoGEFs in complex with their cognate Rho GTPase substrates. Specifically, the initial structure of the DH/PH segment of Tiam1 bound to Rac1 (Tiam1-Rac1) [27], followed by subsequent similar structures, (Dbs-Cdc42, Dbs-RhoA, intersectin (ITSN)-Cdc42) [28,29] have explained several facets of the conserved mechanism of Dbl protein catalyzed nucleotide exchange.

DH Domain Features

DH domains form an elongated α-helical bundle composed of six major helical segments, with a unique fold unrelated to other nucleotide exchange factors (RCC1, Sec7 domain, EF-Ts, and the Cdc25 domain of Sos) for different G-protein families [30,31]. DH domains possess three evolutionarily *c*onserved *r*egions: CRs 1–3. Together CR1

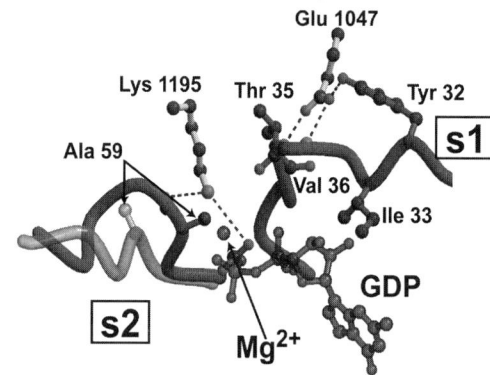

Figure 2 Mechanism of RhoGEF-induced release of GDP from a Rho GTPase. Close-up view of the switch regions of Rac1 (red) binding important Tiam1 residues (yellow) in the Tiam1-Rac1 complex. (Adapted from Worthylake, D. *et al.*, *Nature*, **408**, 682–688. With permission). GEF-GTPase interactions are depicted with dotted blue lines. GDP and magnesium ion (in semi-transparent blue) have been superimposed in their natural binding sites of a Rho GTPase based on the Cdc42-GDP structure (PDB accession code #1AN0). A portion of switch II is shown both before (pink), and after (red) engaging Tiam1, to highlight the dramatic rearrangement of Ala 59, which occludes magnesium binding, and therefore promotes removal of nucleotide. (A color representation of this figure is available on the CD version of the *Handbook of Cell Signaling*.)

(within helix 1), and CR3 (within helix 5), along with a small region of helix 6, create the primary binding surface for a substrate Rho protein (Fig. 1). In sharp contrast, CR2 (within helix 2), maps to the opposite face from the active site, and most likely functions to stabilize the helical bundle.

The Tiam1-Rac1 structure first revealed the critical DH domain positions that engage the nucleotide-binding "switch" regions of a Rho GTPase [27]. Specifically, well-conserved, solvent-exposed DH domain residues equivalent to Glu 1047 and Lys 1195 in Tiam1 make important contacts with Thr 35 and Ala 59 of Rac1, respectively (Fig. 2). Subsequent RhoGEF/Rho GTPase structures [28,29], and mutagenesis within the DH domains of Trio [23], and Dbl [32] support the fundamental importance of these positions in catalyzing the removal of bound GDP from a Rho GTPase. In all four RhoGEF/Rho GTPase structures [27–29], the DH domains bury significant solvent exposed surface area to stabilize the GTPase in a nucleotide-free transition between the inactive and active states.

DH-Associated PH Domains

Extensively characterized as lipid-binding membrane tethers, PH domains from many different proteins display a wide assortment of affinities and specificities for phosphoinositide components of the cell membrane [33–35]. The invariant organization of the DH and PH domains in RhoGEFs suggests a vital role of the PH domain in Dbl-catalyzed nucleotide exchange. Specifically, the PH domain has been implicated in anchoring a Dbl protein to the cellular membrane [36], allosterically regulating GEF activity via phosphoinositide interactions [37,38], and directly engaging the substrate GTPase [28].

Structurally, within Sos, the PH domain is connected to the DH domain by a flexible interdomain linker, while for Tiam1, Dbs and ITSN, the PH domain begins immediately after the terminal helix ($\alpha6$) of the DH domain. Overall, Dbl PH domains resemble the classical ~100 residue anti-parallel β-sandwich PH domain fold, composed of seven strands capped by a terminal α-helix. However, unlike PH domains from other proteins, DH-associated PH domains usually possess an additional short β-strand and a 3_{10} helix at the N terminus of the fold, that pack against the α6 helix of the DH domain using extensive hydrophobic interactions.

PH Domain Configurations

Comparison of the DH/PH structure of Sos with the Tiam1-Rac1, Dbs-Cdc42, Dbs-RhoA, and ITSN-Cdc42 structures provides significant insight into the dramatic conformational heterogeneity of the PH domain. Where the PH domain of Sos partially occludes the GTPase binding site of the DH domain, within the RhoGEF/Rho GTPase complexes, the PH domains are positioned to allow the Rho proteins access to the CR1/CR3 surface of the DH domain (Fig. 1). In addition, where the PH domains of Tiam1 and ITSN reside distant from Rac1 and Cdc42, respectively, the PH domain of Dbs is rotated about the α6 helix of the DH domain such that the β3/β4 loop of the PH domain now contacts the substrate GTPases, (Cdc42 and RhoA), with interactions shown to be functionally important for exchange activity [28]. Together, this structural data potentially reflect several of the configurations that DH and PH domains sample when engaging their target Rho GTPases. However, complicating this model is the fact that, *in vitro*, Dbs requires its PH domain for full catalytic activity when comparing DH to DH/PH elements [28], while the PH domain of ITSN is dispensable for robust exchange activity (our unpublished observations). Thus, it is quite possible that these structural snapshots depict two distinct classes of RhoGEFs, where Dbs and closely related family members can utilize their PH domains to assist in activating a Rho GTPase, while others cannot.

With regard to the functional role of the conserved PH domain, several reports document allosteric modulation of RhoGEF activity in response to specific phosphoinositides binding the PH domain [37–39]. In contrast, other studies find conflicting or no modulation of DH domain-catalyzed nucleotide exchange activity in response to phosphoinositides [40,41]. A large body of structural studies has revealed how PH domains specifically engage the head groups (inositol phosphates) of phosphoinositides, using a positively charged patch composed of variable length loops [33]. It is quite possible that at a cellular membrane, specific phosphoinositide-PH domain interactions could restrict the conformation of the PH domain leading to novel PH domain/GTPase, or PH domain/DH domain contacts that modulate exchange activity. However, studies involving prenylated Rho proteins, and using lipid bilayers will be necessary to dissect the complete functional roles of the DH-associated PH domain in activating Rho GTPases.

Mechanism of Nucleotide Exchange

An important feature revealed by the RhoGEF/Rho GTPase structures is the mechanism of release of bound GDP from the GTPase. As mentioned, all GTPases possess flexible switch regions that adopt distinct conformations depending upon the state of bound nucleotide [42]. Similar to other characterized GEFs [43], DH domains engage switches I and II of the Rho protein, and induce these regions to physically impede upon both magnesium and GDP binding sites. For example, significant rearrangement of switch II (residues 57–75) of Rac1 occurs in the Tiam1-Rac1 structure relative to the architecture of an inactive GTPase [27]. Importantly, Ala 59 of Rac1, stabilized by Lys 1195 of Tiam1 dramatically swings into the binding site of the Mg^{2+} ion normally associated with bound nucleotides (Fig. 2). The crystal structure of the Mg^{2+}-free form of RhoA-GDP documents the paramount consequence of the removal of magnesium in encouraging an open conformation of the switch regions of a Rho GTPase [44].

Additionally, other RhoGEF-induced rearrangements within the switch regions of a Rho GTPase help stabilize the nucleotide-free state of the protein. For instance, Glu 1047 in CR1 of Tiam1 binds several Rac1 residues of switch I (Tyr 32, Thr 35, Val 36) to effectively shift Ile 33 of Rac1 into the binding site of the bound GDP (Fig. 2). This reorientation of switch I destroys the positive interaction that Cys 18 of Rac1 normally makes with bound nucleotide in the Rho GTPase. Moreover, binding of the DH domain of Tiam1 is further secured by Tyr 64 of Rac1, which contacts a set of conserved Tiam1 residues [27]. In addition, Tiam1 binding prevents Phe 28 of Rac1 from making a positive contact with bound nucleotide. This observation is consistent with the finding that Rac1(F28L) exhibits a "fast-cycling" phenotype, showing an increased spontaneous nucleotide exchange rate, thereby mimicking a Dbl-induced Rho GTPase [16]. The combined network of DH domain-Rho GTPase interactions encourages release of nucleotide, and thus traps a Rho GTPase in a nucleotide-free state, where the large excess of GTP *in vivo* begets GTP loading and concomitant downstream effector activation. While the Tiam1-Rac1 structure, as well as the other RhoGEF/Rho GTPase complexes, serve as models to interpret the determinants responsible for nucleotide exchange in Rho proteins, a mechanistic study of the effect of these proposed critical residues is sorely lacking.

Molecular Recognition of Rho GTPase Substrates

The combined structural wealth of data has allowed a framework to interpret the determinants within RhoGEFs and Rho GTPases that dictate the proper pairing between these oncoproteins. To date, the Rho GTPase family possesses 21 members [45], while greater than 70 distinct mammalian Dbl proteins exist. Discrimination by RhoGEFs for the highly similar Rho GTPase substrates remains a paramount issue, given that individual Rho proteins stimulate diverse cellular events. For example, three well-studied Rho GTPases provoke dramatically different cytoskeletal morphologies, where Cdc42 stimulates filopodia production, Rac1 induces lamellipodia formation, and RhoA initiates actin-myosin filament assembly [5].

DH domains selectively recognize a nonconserved patch of positions between the switch regions of Rho GTPases, and utilize these residues for selecting the appropriate target. Structure-based mutagenesis of complementary residues within DH domains and Rho GTPases has demonstrated the basis for this superb specificity. For instance, mutation of Trp 56 in Rac1 to mimic the equivalent Cdc42 side chain (Phe 56) prohibits exchange activity catalyzed by the Rac1-specific GEF Tiam1, while simultaneously removing steric constraint to allow the Cdc42-specific GEF ITSN, to engage and exchange nucleotides within W56F Rac1 [46,47]. In addition, recent work using subtle point mutations within DH domains has remodeled the exchange profiles of ITSN and Dbs to alter their specificity for Rho GTPase substrates [29]. These studies have focused entirely upon Cdc42, Rac1, and RhoA, and must be extended to the entire Rho family to truly appreciate the basis for this selective molecular recognition.

External Regulation of the DH and PH Domains

While the tandem DH and PH domains effectively activate members of the Rho family, outside portions of Dbl proteins surely contribute to the regulation of nucleotide exchange activity. For example, several RhoGEFs (LARG, p115, and PDZ-RhoGEF) possess an N-terminal RGS domain, which binds and hydrolyzes GTP within Gα subunits. Accordingly, emerging studies suggest that modulation of RhoGEF activity is induced by Gα subunit binding to the RGS domain of a Dbl protein [48–50], however, further investigation must decipher this complex mode of regulation. In addition, the upregulation of ITSN-catalyzed exchange activity within Cdc42 by N-WASP binding to an SH3 domain of ITSN has been reported [51].

An exceptional example of external regulation of RhoGEF exchange activity was presented by Rosen and colleagues [25]. Using NMR spectroscopy, the authors determined the structure of the DH domain of Vav along with an N-terminal helical extension that folds back and occludes the conserved DH domain residue (equivalent to the previously mentioned Glu 1047 of Tiam1) at the GTPase binding surface. Interestingly, this amino terminal sequence adjacent to the DH domain contains an Src phosphorylation site (Tyr 174), and accordingly, phosphorylation of Tyr 174 disrupts the inhibitory contacts made on the catalytic surface of the DH domain. These structural data coincide well with previous findings reporting that phosphorylation stimulates the exchange activity of Vav homologs [52]. Thus, outside of the conserved DH/PH cassette, external autoregulation of exchange activity may prove a common theme in all Dbl proteins. Although, given the diverse protein modules present in these RhoGEFs, numerous means of regulating the capacity of the DH and PH domains are anticipated.

References

1. Cerione, R. A. and Zheng, Y. (1996). *Curr. Opin. Cell Biol.* **8**, 216–222.
2. Whitehead, I. P., Campbell, S., Rossman, K. L., and Der, C. J. (1997). *Biochim. Biophys. Acta* **1332**, F1–F23.
3. Zheng, Y. (2001). *Trends Biochem. Sci.* **26**, 724–732.
4. Hill, C. S., Wynne, J., and Treisman, R. (1995). *Cell* **81**, 1159–1170.
5. Symons, M. (1996). *Trends Biochem. Sci.* **21**, 178–181.
6. Van Aelst, L. and D'Souza-Schorey, C. (1997). *Genes Dev.* **11**, 2295–2322.
7. Ridley, A. J. (1997). *Biochem. Soc. Trans.* **25**, 1005–1010.
8. Mackay, D. J. and Hall, A. (1998). *J. Biol. Chem.* **273**, 20685–20688.
9. Hall, A. (1998). *Science* **279**, 509–514.
10. Khosravi-Far, R., Chrzanowska-Wodnicka, M., Solski, P. A., Eva, A., Burridge, K., and Der, C. J. (1994). *Mol. Cell. Biol.* **14**, 6848–6857.
11. Hart, M. J., Eva, A., Zangrilli, D., Aaronson, S. A., Evans, T., Cerione, R. A., and Zheng, Y. (1994). *J. Biol. Chem.* **269**, 62–65.
12. Michiels, F., Habets, G. G., Stam, J. C., van der Kammen, R. A., and Collard, J. G. (1995). *Nature* **375**, 338–340.
13. Olson, M. F. (1996). *J. Mol. Med.* **74**, 563–571.
14. Qiu, R. G., Abo, A., McCormick, F., and Symons, M. (1997). *Mol. Cell. Biol.* **17**, 3449–3458.
15. del Peso, L., Hernandez-Alcoceba, R., Embade, N., Carnero, A., Esteve, P., Paje, C., and Lacal, J. C. (1997). *Oncogene* **15**, 3047–3057.
16. Lin, R., Cerione, R. A., and Manor, D. (1999). *J. Biol. Chem.* **274**, 23633–23641.
17. Hoffman, G. R. and Cerione, R. A. (2002). *FEBS Lett.* **513**, 85–91.
18. Rossman, K. L. and Campbell, S. L. (2000). *Methods Enzymol.* **325**, 25–38.
19. Zheng, Y., Hart, M. J., and Cerione, R. A. (1995). *Methods Enzymol.* **256**, 77–84.
20. Eva, A., Vecchio, G., Rao, C. D., Tronick, S. R., and Aaronson, S. A. (1988). *Proc. Natl. Acad. Sci. USA* **85**, 2061–2065.
21. Eva, A. and Aaronson, S. A. (1985). *Nature* **316**, 273–275.
22. Ron, D., Zannini, M., Lewis, M., Wickner, R. B., Hunt, L. T., Graziani, G., Tronick, S. R., Aaronson, S. A., and Eva, A. (1991). *New Biol.* **3**, 372–379.
23. Liu, X., Wang, H., Eberstadt, M., Schnuchel, A., Olejniczak, E. T., Meadows, R. P., Schkeryantz, J. M., Janowick, D. A., Harlan, J. E., Harris, E. A., Staunton, D. E., and Fesik, S. W. (1998). *Cell* **95**, 269–277.
24. Aghazadeh, B., Zhu, K., Kubiseski, T. J., Liu, G. A., Pawson, T., Zheng, Y., and Rosen, M. K. (1998). *Nat. Struct. Biol.* **5**, 1098–1107.
25. Aghazadeh, B., Lowry, W. E., Huang, X. Y., and Rosen, M. K. (2000). *Cell* **102**, 625–633.
26. Soisson, S. M., Nimnual, A. S., Uy, M., Bar-Sagi, D., and Kuriyan, J. (1998). *Cell* **95**, 259–268.
27. Worthylake, D., Rossman, K., and Sondek, J. (2000). *Nature* **408**, 682–688.
28. Rossman, K. L., Worthylake, D. K., Snyder, J. T., Siderovski, D. P., Campbell, S. L., and Sondek, J. (2002). *EMBO J.* **21**, 1315–1326.
29. Snyder, J. T., Worthylake, D. K., Rossman, K. L., Betts, L., Pruitt, W. M., Siderovski, D. P., Der, C. J., and Sondek, J. (2002). *Nat. Struct. Biol.* **9**, 468–475.
30. Geyer, M. and Wittinghofer, A. (1997). *Curr. Opin. Struct. Biol.* **7**, 786–792.
31. Sprang, S. R. and Coleman, D. E. (1998). *Cell* **95**, 155–158.

32. Zhu, K., Debreceni, B., Li, R., and Zheng, Y. (2000). *J. Biol. Chem.* **275**, 25993–26001.
33. Lemmon, M. A. and Ferguson, K. M. (2000). *Biochem. J.* **350** Pt 1, 1–18.
34. Lemmon, M. A. and Ferguson, K. M. (1998). *Curr. Top. Microbiol. Immunol.* **228**, 39–74.
35. Kavran, J. M., Klein, D. E., Lee, A., Falasca, M., Isakoff, S. J., Skolnik, E. Y., and Lemmon, M. A. (1998). *J. Biol. Chem.* **273**, 30497–30508.
36. Olson, M. F., Sterpetti, P., Nagata, K., Toksoz, D., and Hall, A. (1997). *Oncogene* **15**, 2827–2831.
37. Russo, C., Gao, Y., Mancini, P., Vanni, C., Porotto, M., Falasca, M., Torrisi, M. R., Zheng, Y., and Eva, A. (2001). *J. Biol. Chem.* **276**, 19524–19531.
38. Han, J., Luby-Phelps, K., Das, B., Shu, X., Xia, Y., Mosteller, R. D., Krishna, U. M., Falck, J. R., White, M. A., and Broek, D. (1998). *Science* **279**, 558–560.
39. Crompton, A. M., Foley, L. H., Wood, A., Roscoe, W., Stokoe, D., McCormick, F., Symons, M., and Bollag, G. (2000). *J. Biol. Chem.* **275**, 25751–25759.
40. Snyder, J. T., Rossman, K. L., Baumeister, M. A., Pruitt, W. M., Siderovski, D. P., Der, C. J., Lemmon, M. A., and Sondek, J. (2001). *J. Biol. Chem.* **276**, 45868–45875.
41. Fleming, I. N., Gray, A., and Downes, C. P. (2000). *Biochem. J.* **351**, 173–182.
42. Sprang, S. R. (1997). *Annu. Rev. Biochem.* **66**, 639–678.
43. Cherfils, J. and Chardin, P. (1999). *TIBS* **24**, 306–311.
44. Shimizu, T., Ihara, K., Maesaki, R., Kuroda, S., Kaibuchi, K., and Hakoshima, T. (2000). *J. Biol. Chem.* **275**, 18311–18317.
45. Wherlock, M. and Mellor, H. (2002). *J. Cell Sci.* **115**, 239–240.
46. Gao, Y., Xing, J., Streuli, M., Leto, T. L., and Zheng, Y. (2001). *J. Biol. Chem.* **276**, 47530–47541.
47. Karnoub, A. E., Worthylake, D. K., Rossman, K. L., Pruitt, W. M., Campbell, S. L., Sondek, J., and Der, C. J. (2001). *Nat. Struct. Biol.* **8**, 1037–1041.
48. Wells, C. D., Liu, M. Y., Jackson, M., Gutowski, S., Sternweis, P. M., Rothstein, J. D., Kozasa, T., and Sternweis, P. C. (2002). *J. Biol. Chem.* **277**, 1174–1181.
49. Fukuhara, S., Chikumi, H., and Gutkind, J. S. (2000). *FEBS Lett.* **485**, 183–188.
50. Kozasa, T., Jiang, X., Hart, M. J., Sternweis, P. M., Singer, W. D., Gilman, A. G., Bollag, G., and Sternweis, P. C. (1998). *Science* **280**, 2109–2111.
51. Hussain, N. K., Jenna, S., Glogauer, M., Quinn, C. C., Wasiak, S., Guipponi, M., Antonarakis, S. E., Kay, B. K., Stossel, T. P., Lamarche-Vane, N., and McPherson, P. S. (2001). *Nat. Cell Biol.* **3**, 927–932.
52. Crespo, P., Schuebel, K. E., Ostrom, A. A., Gutkind, J. S., and Bustelo, X. R. (1997). *Nature* **385**, 169–172.

CHAPTER 247

Structural Considerations of Small GTP-Binding Proteins

Alfred Wittinghofer
*Max-Planck-Institut für molekulare Physiologie,
Dortmund, Germany*

Introduction

The basic switching apparatus of small GTP-binding proteins is the G domain. The G domain itself is a conserved structural unit with a canonical switch mechanism, with minor modifications in different members of the family. Their regulators and effectors show a large diversity in their structures and modes of interaction. Here we will try to define some underlying principles.

The G Domain Functional Unit

The basic structural unit of small GTP-binding proteins consists of about 160–170 residues. It is α,β fold typical for nucleotide binding proteins and is called the G domain (for earlier reviews and references, see references [1–3]). This G domain consists of a mixed six-stranded β-sheet and five helices on both sides. It contains four to five conserved sequence elements, which are lined up along the nucleotide binding site and mediate binding, GTP hydrolysis, and the switching apparatus (Fig. 1). The most important contributions to binding are first due to the interactions of the nucleotide base with the N/TKxD motif and an invariant Ala and second of the phosphates with the conserved P-loop GxxxxGKS/T motif [4]. Specificity is due to an Asp side chain forming a bifurcated H bond with the guanine ring, but also due to the Ala main chain interaction with the guanine oxygen, which for steric reasons does not allow replacement by the adenine amino group [5]. Other conserved sequence elements are a conserved Thr and a DxxG motif, both of which are involved in the conformational change.

Structures of several Ras-related proteins like Ras/Rap/Ral, Rac/Rho/Cdc42, Arf/Arl, Ran, and Rab have been solved. The easiest way to compare these structures is to consider the 166–171 residue long G domain of the Ras protein as the minimal switch unit, and to describe the others as variations of this canonical structure. In addition to the G domain, small GTP-binding proteins contain a C- and N-terminal extension of varying length and various insertions. Rho proteins Rho/Rac/Cdc42 contain a α-helical insertion of approximately 13 amino acids, Ran has an elongated C-terminal element very crucial for its function in nuclear transport, whereas Arf/Sar1 proteins contain a myristoylated N-terminal extension that is important for nucleotide-dependent insertion into the plasma membrane. The G domain with its conserved features is also found in multidomain GTP-binding proteins such as the heterotrimeric G proteins and the protein biosynthesis factors eIF5B, EF-Tu, and EF-G.

Whereas the small G proteins are structurally highly homologous, and the structural features of the switch mechanism are conserved, the regulators actually show a variety of shapes and the effectors have, in addition, a variety of function. Here we will concentrate on the major regulators GEF and GAP and the effectors.

The Conformational Switch

Requirements of the molecular switch can be defined from comparing GDP- and GTP-bound structures. Structural differences are mostly subtle and are primarily confined to two regions, which have first been observed in Ras and are called the switch regions [6]. These regions usually show an increased flexibility in X-ray structures and in magnetic

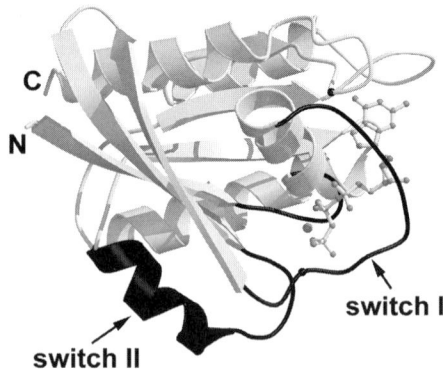

Figure 1 Structure of the G domain. Ribbon plot of the minimal G domain in gray, containing six β-strands and five α-helices, with the conserved sequence elements and the switch regions in black. GTP and Mg^{2+} ion are shown in ball and stick representation.

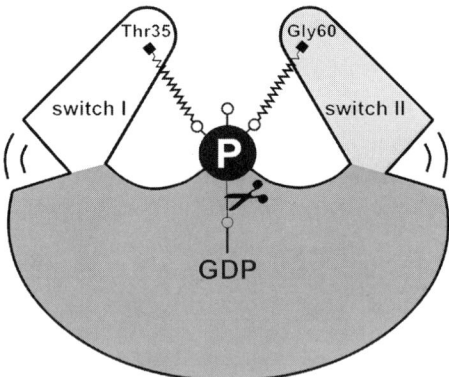

Figure 2 Schematic diagram of the universal switch mechanism where the switch I and II domains are bound to the γ-phosphate via the main chain NH groups of the invariant Thr and Gly residues, in what might be called a loaded spring mechanism. Release of the γ-phosphate after GTP hydrolysis allows the switch regions to relax into a different conformation.

resonance studies using NMR and electron paramagnetic resonance (EPR) [7,8]. Furthermore, whereas the GDP-bound proteins show a large variation of structural details, the GTP-bound forms of the G domain are remarkably similar [3]. Most important, the trigger for the conformational change is rather universal.

In the triphosphate form there are two hydrogen bonds from γ-phosphate oxygens to the main chain NH groups of the invariant Thr and Gly residues (Thr35/Gly60 in Ras) in switch I and II, respectively. The conformational change can best be described as a loaded-spring mechanism where release of the γ-phosphate after GTP hydrolysis allows the two switch regions to relax into the GDP-specific conformation (Fig. 2). The extent of the conformational change is different between the different proteins. In Ras the switch regions involve residues 32–38 for switch I and 59–57 for switch II.

The canonical switch mechanism is modified in many ways. Whereas Ras/Rap/Rho/Rac/Rab show minor changes involving only switch I and II, Ran experiences a large conformational change in switch I with an unfolding of an extra β-strand which in turn induces the long C-terminal extension, the so-called C-terminal switch, to drastically alter its location [9–11]. An even more dramatic change in switch I of Arf and Arl involves the change in register of two β-strands relative to the rest of the sheet and the detachment and subsequent membrane insertion of the N-terminal helix. This has been called the N-terminal switch [12,13].

Guanine Nucleotide Exchange Factors

Guanine nucleotide release from small GTP-binding proteins is slow and is stimulated by several orders of magnitude by guanine nucleotide exchange factors (GEFs). The mechanism of GEF action involves a series of fast kinetic steps, which lead from a binary protein-nucleotide complex via a trimeric complex to a binary nucleotide-free complex, which is stable in the absence of nucleotide. This series of reactions is reversed by rebinding of nucleotide, which is predominantly GTP due to its higher concentration in the cell. In principle, these reactions are fast and fully reversible such that GEF merely acts as a catalyst, which increases the rates at which equilibrium between the GDP- and GTP-bound form of the protein is reached.

Structures of GEFs are conserved within a given subfamily. GEFs for the Ras subfamily have a Cdc25 catalytic domain, whereas the GEFs for Rho-type proteins contain a DH (dibble-homology) domain (see chapter by John Sondek). Structures for the Ran-GEF RCC1, for the Ras-GEF Sos, for the Arf-GEFs Arno/Geα2p, the Rho-GEFs Sos, Trio, Pix, and Tiam and Dbs have been solved. To better understand the mechanism of nucleotide exchange structures of the nucleotide-free binary complexes Ras-Sos [14] Arf-Geα2 [12], of the Ran-RCC1 [15], Rac-Tiam [16], and Dbs-Cdc42 [17] complexes with a polyanion in the phosphate binding (P) loop have been solved. Although the details of the interactions are all different, arguing for a variety of kick-out mechanisms, the complexes have structural features in common suggesting mechanistic similarities. The GEFs do not act allosterically but directly interfere with nucleotide binding by inserting residues close to or into the P loop, the Mg^{2+} and/or guanine base binding area which create structural changes that are inhibitory for tight binding of nucleotide. This agrees with binding studies which show that the β-phosphate-P-loop interaction and the Mg^{2+} phosphate interaction are the most important elements for tight binding [5].

Effector B Via Switches and Others

Effectors for GTP-binding proteins are operationally defined as molecules interacting more tightly with the GTP- than with the GDP-bound form. This implies that effector binding involves the switch regions of G proteins, which is borne out by the structures. The Ras-binding domain (RBD) or Ras association domain (RA) has been identified in many different proteins [18], which may or may not be true effectors [19]. In any case many of the bona fide Ras effectors also bind to proteins of the Ras subfamily such as Rap, R-Ras, and TC21. The structures of several Ras effectors such as Raf, RalGDS, PI3kinase, and recently byr2 [20] either alone or in complex with Ras, Rap, and their mutants have shown that the RBD is a small, well-defined domain with a ubiquitin fold, which binds to Rap/Ras by forming a GTP-binding dependent interprotein β-sheet between both molecules [21]. In the case of the complex with the phosphatidylinositol-specific lipid kinase PI(3)Kγ, the contacts produce structural changes in PI3Kγ which are believed to allosterically influence the catalytic activity [22–24].

Binding of Rac/Cdc42 to its effectors, some of which contain a so-called CRIB (Cdc42/Rac interactive binding) region [25,26] induces major structural changes and has been shown to directly activate the protein kinases PAK (p21-activated kinase), activated Cdc42-associated kinase (ACK), and the scaffold protein WASP (Wiskott-Aldrich syndrome protein). Fragments containing the CRIB domain show no apparent three-dimensional structure, but are structured in complex with C-terminal fragments from the same protein in an autoinhibitory conformation [27,28] or with Cdc42 (see chapter by Mike Rosen) [29,30].

Other structures of effector complexes show a variety of interaction patterns (Fig. 3) the only common feature of which is to include the switch region(s) for the interaction, the exception being the complex between RhoA and an antiparallel coiled-coil fragment from the regulatory region of the protein kinase PKN, where it is not clear why the interaction should be GTP-specific [31]. The complex between Rab3A and its effector Rabphilin-3A involves two modes of binding [32], where one end of Rabphilin-3A contacts the switch region of Rab3A and thus senses its nucleotide status. A second site interacts with the complementarity-determining region (CDR), a site on the surface whose sequence varies among members of the Rab subfamily, and is thus responsible for specificity.

Small GTP-binding proteins such as Arf, Rab, or Ran are involved in a variety of transport processes. Ran·GTP binding to nuclear import factors-cargo complexes is necessary to release cargo on the nuclear side of the nuclear pore. Structures of the Ran complexes with importin-β and transportin show the switch I- and II-mediated interaction with the helical repeat motif of the factors [9,11]. Additional structural studies of a complex of importin with cargo [33] or with components from the nuclear pore [34] have suggested that the cargo-loaded transport factors can bind simultaneously to the nuclear pore but that interaction with Ran·GTP in the nucleus sterically interferes with binding of both, suggesting how Ran terminates the import reaction. The structure of the Arf-related protein Arl2 in complex with its effector PDEδ showed the latter to have a RhoGDI fold with an empty hydrophobic pocket [13], which together with biochemical studies suggested it to be involved in the transport of farnesylated proteins.

The only unifying principle that we find in the structures of effector complexes is that they involve the switch regions.

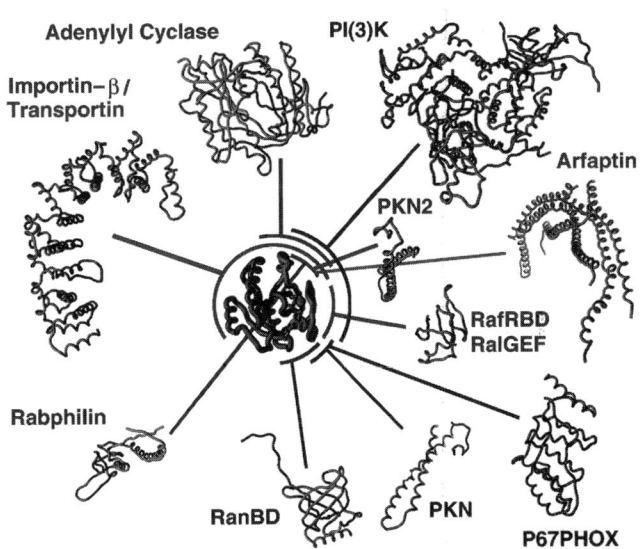

Figure 3 Interaction with effectors is mediated via different surface regions of GTP-binding proteins (see reference [3]). The structures were aligned on the G domain in the middle, and the effectors are removed from the interface in the direction indicated by the line. Involvement of the switch I and/or switch II region in the interface is the only common feature of these complexes. For PKN, the two variations found in the crystal structure are shown.

Other than that they show that the effectors have a variety of shapes of folds and interact in many different ways such that the different effectors cover almost the complete surface of the G domain (Fig. 3). We can distinguish between effectors, which contain a preformed binding domain and show no major structural change on binding, and those involving a large conformational change on binding to the GTP-binding protein. In the former case experimental evidence points toward (only) a recruiting function as the major signal transduction mechanism [24], whereas the other clearly involves (additional) allosteric regulation of the effector.

GAP Proteins and the GTPase Reaction

The GTPase reaction for most Ras-like proteins is slow and would not be suitable for most biological signal transduction processes where inactivation is complete within minutes after GTP loading. It is not surprising then that GTPase activating proteins (GAPs) have been discovered for all major forms of these proteins [35,36]. They increase GTP hydrolysis by several orders of magnitude. As with GEFs, the structures of GAPs for various (sub)family Ras proteins are different, although an evolutionary relationship between Ras- and RhoGAP has been suggested [37,38].

Biochemical experiments showed that the active site arginine of heterotrimeric Gα is supplied in "trans" by RasGAP [39]. Structures of Gα·GDP in the presence of aluminum fluoride had shown that the latter is in a planar conformation and thus seems to mimic the transferred phosphate of the reaction [40,41]. In those structures both the arginine and a conserved glutamine stabilize the conformation of the transition state mimic. The structures of RasGAP and RhoGAP in complex with their respective G proteins in the presence of AlF_x showed an intrinsic Gln and a so-called arginine finger supplied by GAP into the active site [42–44]. The structures also give an explanation for the inability of oncogenic mutants of Ras to hydrolyze GTP. The mechanism of inserting an arginine finger into the active site of Rho proteins is also used by some bacteria that insert these toxins into eukaryotic cells [45,46].

It has been discussed whether the mechanism of GAP-assisted GTP hydrolysis applies to all G-domain proteins [35]. This is clearly not the case. Rap proteins have a Thr, and protein synthesis factors have a histidine in place of the catalytic glutamine. For Arf, the presence or absence of an arginine of ArfGAP participating in catalysis is disputed [47,48]. Furthermore, in a recent structure of the trimeric complex between Ran, RanGAP, and a Ran-binding protein, no arginine was found to be participating in catalysis [49].

Conclusions

A large number of structural studies on small G proteins showed that there is a conserved module with a canonical structure and switch mechanism that can be considered as a tema con variazoni. The variations are derived from insertions into and additions to the canonical G domain and from a variety of regulators and effectors that are different for different types of G-binding proteins. The mechanisms by which GEFs and GAPs stimulate the otherwise slow intrinsic nucleotide dissociation and GTP hydrolysis have been worked out in some cases and suggest some underlying common principles in spite of the multitude of differences in detail. Interactions with effectors show a similar multitude of interactions. How these interactions generate the biological response of the particular system remains to be established in most cases with more structural studies on complexes of small GTP-binding proteins.

Acknowledgment

I thank Christina Kiel for Fig. 2 and Rita Schebaum for secretarial assistance.

References

1. Kjeldgaard, M., Nyborg, J., and Clark, B. F. (1996). The GTP binding motif: variations on a theme. *FASEB J.* **10**, 1347–1368.
2. Sprang, S. R. (1997). G protein mechanisms-insights from structural analysis. *Annu. Rev. Biochem.* **66**, 639–678.
3. Vetter, I. R. and Wittinghofer, A. (2001). Signal transduction—The guanine nucleotide-binding switch in three dimensions [Review]. *Science* **294**, 1299–1304.
4. Saraste, M., Sibbald, P. R., and Wittinghofer, A. (1990). The P-loop—A common motif in ATP- and GTP-binding proteins. *Trends Biochem. Sci.* **15**, 430–434.
5. Rensland, H., John, J., Linke, R., Simon, I., Schlichting, I., Wittinghofer, A., and Goody, R. S. (1995). Substrate and product structural requirements for binding of nucleotides to h-ras p21—the mechanism of discrimination between guanosine and adenosine nucleotides. *Biochemistry* **34**, 593–599.
6. Milburn, M. V., Tong, L., DeVos, A. M., Brünger, A., Yamaizumi, Z., Nishimura, S., and Kim, S.-H. (1990). Molecular switch for signal transduction: Structural differences between active and inactive forms of protooncogenic ras proteins. *Science* **247**, 939–945.
7. Farrar, C. T., Halkides, C. J., and Singel, D. J. (1997). The frozen solution structure of p21ras determined by Eseem spectroscopy reveals weak coordination of THR35 to the active site metal ion. *Structure* **5**, 1055–1066.
8. Morreale, A., Venkatesan, M., Mott, H. R., Owen, D., Nietlispach, D., Lowe, P. N., and Laue, E. D. (2000). Structure of Cdc42 bound to the GTPase binding domain of PAK. *Nat. Struct. Biol.* **7**, 384–388.
9. Vetter, I. R., Arndt, A., Kutay, U., Gorlich, D., and Wittinghofer, A. (1999). Structural view of the Ran-Importin beta interaction at 2.3 A resolution. *Cell* **97**, 635–646.
10. Vetter, I. R., Nowak, C., Nishimoto, T., Kuhlmann, J., and Wittinghofer, A. (1999). Structure of a Ran-binding domain complexed with Ran bound to a GTP analogue: implications for nuclear transport. *Nature* **398**, 39–46.
11. Chook, Y. M. and Blobel, G. (1999). Structure of the nuclear transport complex karyopherin-beta 2-Ran·GppNHp. *Nature* **399**, 230–237.
12. Goldberg, J. (1998). Structural basis for activation of arf gtpase—mechanisms of guanine nucleotide exchange and gtp-myristoyl switching. *Cell* **95**, 237–248.
13. Hanzal-Bayer, M., Renault, L., Roversi, P., Wittinghofer, A., and Hillig, R. C. (2002). The complex of Arl2-GTP and PDE delta: from structure to function. *EMBO J.* **21**, 2095–2106.
14. Boriack-Sjodin, P. A., Margarit, S. M., Barsagi, D., and Kuriyan, J. (1998). The structural basis of the activation of ras by sos. *Nature* **394**, 337–343.
15. Renault, L., Kuhlmann, J., Henkel, A., and Wittinghofer, A. (2001). Structural basis for guanine nucleotide exchange on Ran by the Regulator of Chromosome Condensation (RCC1). *Cell* **105**, 245–255.
16. Worthylake, D. K., Rossman, K. L., and Sondek, J. (2000). Crystal structure of Rac1 in complex with the guanine nucleotide exchange region of Tiam1. *Nature* **408**, 682–688.
17. Rossman, K. L., Worthylake, D. K., Snyder, J. T., Siderovski, D. P., Campbell, S. L., and Sondek, J. (2002). A crystallographic view of interactions between Dbs and Cdc42: PH domain-assisted guanine nucleotide exchange. *EMBO J.* **21**, 1315–1326.
18. Ponting, C. P. and Benjamin, D. R. (1996). A novel family of ras-binding domains. *Trends Biochem. Sci.* **21**, 422–425.
19. Kalhammer, G., Bahler, M., Schmitz, F., Jockel, J., and Block, C. (1997). Ras-binding domains—predicting function versus folding. *FEBS Lett.* **414**, 599–602.
20. Scheffzek, K., Grunewald, P., Wohlgemuth, S., Kabsch, W., Tu, H., Wigler, M., Wittinghofer, A., and Herrmann, C. (2001). The Ras-Byr2RBD complex: Structural basis for Ras effector recognition in yeast. *Structure* **9**, 1043–1050.
21. Nassar, N., Horn, G., Herrmann, C., Scherer, A., McCormick, F., and Wittinghofer, A. (1995). The 2.2 A crystal structure of the Ras-binding domain of the serine/threonine kinase c-Raf1 in complex with Rap1A and a GTP analogue. *Nature* **375**, 554–560.
22. Walker, E. H., Perisic, O., Ried, C., Stephens, L., and Williams, R. L. (1999). Structural insights into phosphoinositide 3-kinase catalysis and signalling. *Nature* **402**, 313–320.
23. Pacold, M. E., Suire, S., Perisic, O., Lara-Gonzalez, S., Davis, C. T., Walker, E. H., Hawkins, P. T., Stephens, L., Eccleston, J. F., and Williams, R. L. (2000). Crystal structure and functional analysis of Ras binding to its effector phosphoinositide 3-kinase gamma. *Cell* **103**, 931–943.
24. McCormick, F. and Wittinghofer, A. (1996). Interactions between Ras proteins and their effectors. *Curr. Opin. Biotechnol.* **7**, 449–456.
25. Burbelo, P. D., Drechsel, D., and Hall, A. (1995). A conserved binding motif defines numerous candidate target proteins for both cdc42 and rac gtpases. *J. Biol. Chem.* **270**, 29071–29074.
26. Symons, M., Derry, J. M. J., Karlak, B., Jiang, S., Lemahieu, V., McCormick, F., Francke, U., and Abo, A. (1996). Wiskott-Aldrich syndrome protein, a novel effector for the gtpase cdc42hs, is implicated in actin polymerization. *Cell* **84**, 723–734.

27. Kim, A. S., Kakalis, L. T., Abdul-Manan, M., Liu, G. A., and Rosen, M. K. (2000). Autoinhibition and activation mechanisms of the Wiskott-Aldrich syndrome protein." *Nature* **404**, 151–158.
28. Lei, M., Lu, W. G., Meng, W. Y., Parrini, M. C., Eck, M. J., Mayer, B. J., and Harrison, S. C. (2000). Structure of PAK1 in an autoinhibited conformation reveals a multistage activation switch. *Cell* **102**, 387–397.
29. Abdul-Manan, N., Aghazadeh, B., Liu, G. A., Majumdar, A., Ouerfelli, O., Siminovitch, K. A., and Rosen, M. K. (1999). Structure of Cdc42 in complex with the GTPase-binding domain of the "Wiskott-Aldrich syndrome" protein. *Nature* **399**, 379–383.
30. Mott, H. R., Owen, D., Nietlispach, D., Lowe, P. N., Manser, E., Lim, L., and Laue, E. D. (1999). Structure of the small G protein Cdc42 bound to the GTPase-binding domain of ACK. *Nature* **399**, 384–388.
31. Maesaki, R., Ihara, K., Shimizu, T., Kuroda, S., Kaibuchi, K., and Hakoshima, T. (1999). The structural basis of Rho effector recognition revealed by the crystal structure of human RhoA complexed with the effector domain of PKN/PRK1. *Mol. Cell* **4**, 793–803.
32. Ostermeier, M., Nixon, A. E., Shim, J. H., and Benkovic, S. J. (1999). Combinatorial protein engineering by incremental truncation. *Proc. Natl. Acad. Sci. USA* **96**, 3562–3567.
33. Cingolani, G., Petosa, C., Weis, K., and Muller, C. W. (1999). Structure of importin-beta bound to the IBB domain of importin-alpha. *Nature* **399**, 221–229.
34. Bayliss, R., Littlewood, T., and Stewart, M. (2000). Structural basis for the interaction between FxFG nucleoporin repeats and importin-beta in nuclear trafficking. *Cell* **102**, 99–108.
35. Scheffzek, K., Ahmadian, M. R., and Wittinghofer, A. (1998). GTPase-activating proteins: helping hands to complement an active site. *Trends Biochem. Sci.* **23**, 257–262.
36. Gamblin, S. J., and Smerdon, S. J. (1998). GTPase-activating proteins and their complexes. *Curr. Opin. Struct. Biol.* **8**, 195–201.
37. Bax, B. (1998). Domains of rasgap and rhogap are related. *Nature* **392**, 447–448.
38. Calmels, T. P. G., Callebaut, I., Leger, I., Durand, P., Bril, A., Mornon, J. P., and Souchet, M. (1998). Sequence and 3D structural relationships between mammalian ras- and rho-specific GTPase-activating proteins (GAPs)—the cradle fold. *FEBS Lett.* **426**, 205–211.
39. Mittal, R., Ahmadian, M. R., Goody, R. S., and Wittinghofer, A. (1996). Formation of a transition-state analog of the Ras GTPase reaction by Ras-GDP, tetrafluoroaluminate, and GTPase-activating proteins. *Science* **273**, 115–117.
40. Coleman, D. E., Berghuis, A. M., Lee, E., Linder, M. E., Gilman, A. G., and Sprang, S. R. (1994). Structures of active conformations of $G_i\alpha_1$ and the mechanism of GTP hydrolysis. *Science* **265**, 1405–1412.
41. Sondek, J., Lambright, D. G., Noel, J. P., Hamm, H. E., and Sigler, P. B. (1994). GTPase mechanism of G proteins from the 1.7-A crystal structure of transducin alpha-GDP-AIF-4. *Nature* **372**, 276–279.
42. Scheffzek, K., Ahmadian, M. R., Kabsch, W., Wiesmuller, L., Lautwein, A., Schmitz, F., and Wittinghofer, A. (1997). The Ras-RasGAP complex: structural basis for GTPase activation and its loss in oncogenic Ras mutants. *Science* **277**, 333–338.
43. Rittinger, K., Walker, P. A., Eccleston, J. F., Smerdon, S. J., and Gamblin, S. J. (1997). Structure at 1.65 Angstrom of RhoA and its GTPase-activating protein in complex with a transition-state analogue. *Nature* **389**, 758–762.
44. Nassar, N., Hoffman, G. R., Manor, D., Clardy, J. C., and Cerione, R. A. (1998). Structures of Cdc42 bound to the active and catalytically compromised forms of Cdc42GAP. *Nat. Struct. Biol.* **5**, 1047–1052.
45. Wurtele, M., Wolf, E., Pederson, K. J., Buchwald, G., Ahmadian, M. R., Barbieri, J. T., and Wittinghofer, A. (2001). How the Pseudomonas aeruginosa ExoS toxin downregulates Rac. *Nat. Struct. Biol.* **8**, 23–26.
46. Stebbins, C. E. and Galan, J. E. (2000). Modulation of host signaling by a bacterial mimic: structure of the Salmonella effector SptP bound to Rac1. *Mol. Cell* **6**, 1449–1460.
47. Goldberg, J. (1999). Structural and functional analysis of the ARF1-ARFGAP complex reveals a role for coatomer in GTP hydrolysis. *Cell* **96**, 893–902.
48. Mandiyan, V., Andreev, J., Schlessinger, J., and Hubbard, S. R. (1999). Crystal structure of the ARF-GAP domain and ankyrin repeats of PYK2-associated protein beta. *EMBO J.* **18**, 6890–6898.
49. Seewald, M. J., Korner, C., Wittinghofer, A., and Vetter, I. R. (2002). RanGAP mediates GTP hydrolysis without an arginine finger. *Nature* **415**, 662–666.

CHAPTER 248

Conventional and Unconventional Aspects of Dynamin GTPases

Sandra L. Schmid

*Department of Cell Biology,
The Scripps Research Institute, La Jolla, California*

Introduction

Dynamin is unique among GTPases for its large size (~96 kDa), its multidomain structure, its low affinity for GTP, its relatively high intrinsic rate of GTP hydrolysis and most strikingly, its ability to self-assemble into rings and helical stacks of rings and its very robust (≥ 100 min^{-1}) assembly-stimulated rates of GTP hydrolysis. These unique properties have led to the suggestion that dynamin functions, unlike any other GTPase superfamily member, as a mechanochemical enzyme that encircles the necks of deeply invaginated coated pits and generates the force necessary to drive scission. Recent structural data show that the GTPase domain of dynamin is, however, quite conventional. Moreover, dynamin overexpression affects many cellular processes, in addition to its effects on endocytosis, including actin dynamics, MAP kinase activation, transcriptional activation, and changes in cell morphology. Re-evaluation of the current data, from a new perspective, may suggest a broader role for the dynamin subfamily more akin to other regulatory GTPases.

Dynamin is one of the youngest members of the GTPase superfamily, having been discovered in the late 1980s. As a young upstart, it has proven to be unconventional and textbooks have rushed to include models for dynamin function as a fission GTPase—the first example of a GTPase functioning as a mechanochemical enzyme. However, more recent evidence suggests that, like signaling or regulatory GTPases, dynamin overexpression can effect many aspects of cell physiology, including actin dynamics, cell morphology, transcription, and even cell death. Dynamin's function in endocytosis has been extensively reviewed in references [1–3].

Here, with special emphasis on often overlooked aspects, I will review some of the features that set dynamin apart from its GTPase cousins, place it in the context of its siblings in the dynamin subfamily of GTPases, and describe dynamin effects that are not easily reconciled with its "textbook" function as a fission GTPase.

Common and Unique Features of Dynamin as a GTPase

Dynamin is a Multidomain Protein with a Conventional GTPase Domain

Atypically for GTPase family members, dynamin is large (~96 kDa) and has a modular domain structure (Fig. 1) consisting of an N-terminal GTPase domain, a middle domain

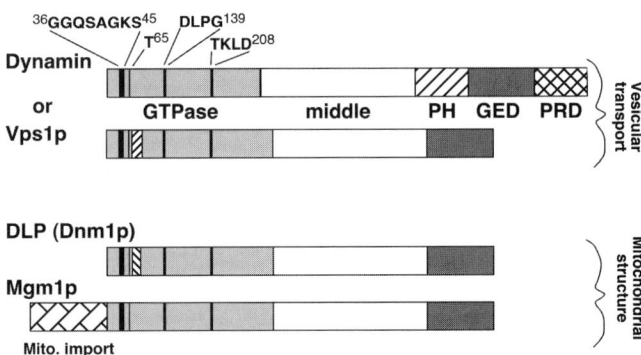

Figure 1 Domain structure of the only three evolutionarily conserved dynamin family members. The conserved elements of the GTPase binding site are indicated for dynamin-1.

of unknown function, a PH domain that binds PI4,5P$_2$, a GTPase effector domain (GED) that functions in dynamin self-assembly and assembly-stimulated GTPase activity, and a C-terminal proline and arginine-rich domain (PRD) that interacts with numerous SH3 domain-containing partners [4]. The recently solved three-dimensional structure of the 316 amino acid GTPase domain of *Dictyostelium* dynamin A [5], which is 60% identical to the GTPase domain of human dynamin-1, reveals it to have a typical, ras-like core GTPase domain fold (Fig. 2) with several distinguishing features. The first is an insertion, which is quite divergent among dynamin family members (see Fig. 1), that adds two additional strands to the core 6-stranded β-sheet and a variable length extended loop (Fig. 2). It is positioned immediately distal to T65 (in dynamin-1) which coordinates with γ-phosphate on bound GTP, at a region equivalent to switch I of other GTPases. Its position, its variability, and the high content of charged residues suggest that this loop could be involved in interactions with downstream partners, like other GTPases. The loop between β3 and α2 that constitutes the switch II region of the GTPase (element 3 of the GTP binding site), also undergoes GTP-dependent conformational changes and is considerably longer in dynamin than in ras. A third insert is located near the C terminus of the GTPase domain and constitutes two α-helices which extend out from and run perpendicular to the core β-sheet. The fourth distinguishing feature is that the N and C terminus of the GTPase domain come in close contact with each other through an additional N-terminal α-helix and an extended C-terminal α-helix. Interestingly, together these two terminal helices form a hydrophobic groove, which was occupied in the solved structure by hydrophobic residues from the N-terminal fusion partner used in the dynamin construct, the catalytic domain of *Dictyostelium* myosin II [5]. The author's speculation, that this groove might normally be a docking site for the GED domain of dynamin is consistent with results from the dynamin-related mx proteins which identified a hydrophobic residue within GED that is essential for self-assembly and assembly-stimulated GTPase activity [6], and results from two-hybrid analysis mapping the GED-GTPase interaction site in dynamin to the C-terminal regions of the GTPase domain [7]. Importantly, this model for GED-GTPase interaction is inconsistent with that based on the structure of GBP-1, another interferon-inducible GTPase [8]. However, comparison of the GTPase domain structures of GBP1 and dynamin A, together with their lack of sequence similarity would strongly

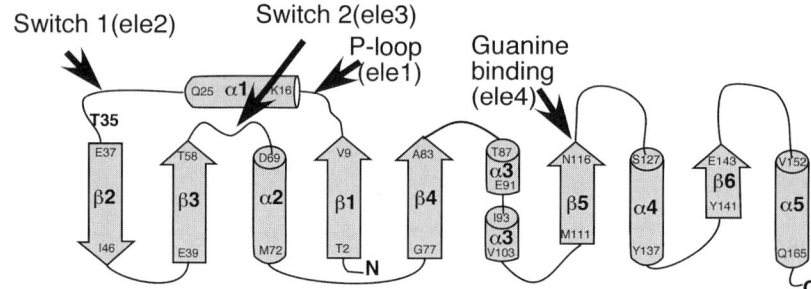

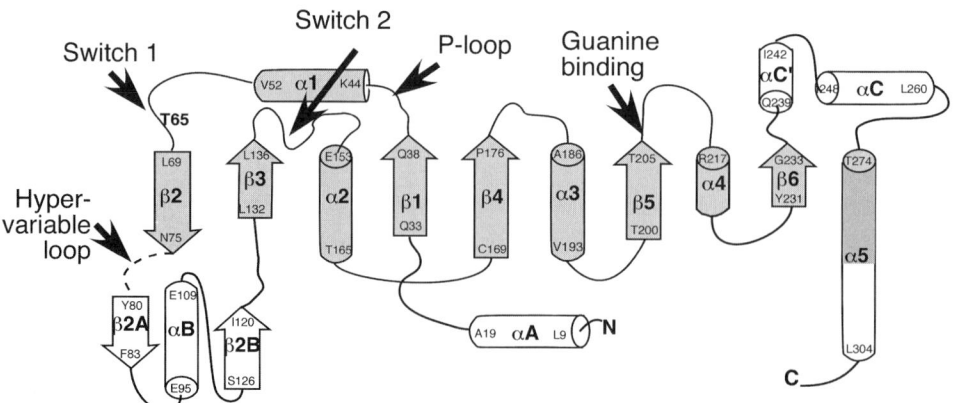

Figure 2 Topology of dynamin GTPase domain compared to Ras. The models for ras and dynamin are redrawn and modified from Wittinghofer, A., *GTPases*, Vol. 24, 244–310, Oxford University Press, New York, with permission and Niemann, H.H. *et al.*, *EMBO J.*, 20, 5813–5821, 2001, with permission, respectively. The location of the four GTPase binding site elements are indicated.

argue that they represent distinct GTPase subfamilies and, therefore, that the structure of distal domains of GBP-1 may not reflect those of dynamin-family members.

Dynamin's Unconventional GTPase Activity

Although dynamin's GTPase domain is structurally quite conventional, there are several important distinguishing features of its GTPase activity that set it apart from most GTPase family members. The first is dynamin's relatively low affinity for guanine nucleotides (2.5 µM for GTP and ~40-fold weaker for GDP) and the very rapid dissociation rates for both GTP ($2.1\ s^{-1}$) and for GDP ($93\ s^{-1}$) [9]. Thus, unlike most other GTPases, nucleotide exchange is not rate-limiting for dynamin's GTPase activity. The second is dynamin's relatively fast intrinsic rate of GTP hydrolysis. Reported measurements of this rate have varied considerably probably due to variable amounts of dynamin aggregation or self-assembly under different assay conditions. However, when assayed in high salt to inhibit self-assembly, more consistent rates (0.01–$0.03\ s^{-1}$) are reported [10,11]. This rapid intrinsic rate is comparable to that measured for trimeric G proteins; however, the steady-state GTPase activity of trimeric proteins is greatly limited by the very slow dissociation rates of GDP. Together, these unusual properties of dynamin's GTPase activity would suggest that *in vivo*, in the absence of as yet unidentified regulatory factors, dynamin would rapidly cycle between the GTP-bound and unoccupied state, occasionally hydrolyzing bound GTP.

Dynamin has an Intrinsic GAP Domain

As alluded to above, perhaps the most unusual property of dynamin is its ability to self-assemble into rings and helical stacks of rings. Depending on the template provided, self-assembly stimulates dynamin's GTPase activity >100-fold [9,12–14]. This assembly-stimulated GTPase activity is mediated by GED, which functions as an assembly-dependent GAP domain [15]. Thus, dynamin, unlike other GTPase family members, encodes its own GAP; however, like other GTPases, GED appears not to function intramolecularly, but is only activated by intermolecular interactions that occur between dynamin molecules upon self-assembly.

The mechanism of GED-stimulated GTPase activity is unknown. Mutagenesis studies had identified an arginine residue in GED (R725), which when mutated to alanine, appeared to specifically inhibit assembly-stimulated GTPase activity while not affecting dynamin's ability to self-assemble upon dilution into low-salt buffers or on a microtubule template [15]. These findings led to the suggestion that R725 might function as a catalytic residue akin to the Arg-finger in rasGAP. However, a more recent study has shown that dyn(R725A) is capable of near wild-type stimulated GTPase activity when assayed on lipid tubule assembly templates, and instead revealed an assembly defect in the R725 mutant [16]. Consistent with this result, we have found that dyn(R725K) exhibits wild-type self-assembly and GTPase activity (S. Sholly and S.L. Schmid., unpublished results), a finding inconsistent with a role for R725 in catalysis. Given that dynamin's intrinsic rate of GTP hydrolysis parallels that of trimeric G protein α subunits, it is possible that GED functions more in analogy to RGS-type GTP-activating proteins (GAPs) that stabilize and position catalytic residues already in place in the GTPase domain (reviewed in reference [17]).

Dynamin's Function in Endocytic Vesicle Formation

The rapid temperature-sensitive inhibition of endocytosis seen in *Drosophila* bearing mutations in *shibire*, the ortholog of dynamin, remains the strongest and most compelling evidence for dynamin's essential role in endocytosis [18]. In mammalian cells, endocytic clathrin-coated vesicle formation is inhibited by overexpression of dominant-negative mutants of dynamin. One striking example is the finding that HeLa cells overexpressing human dyn(G273D), corresponding to the shibire-ts1 allele, exhibit rapid and reversible temperature-sensitive inhibition of receptor-mediated endocytosis [19]. Several mutations have been made at conserved residues in the nucleotide-binding site within dynamin's GTPase domain (see Fig. 1), which by analogy to other GTPases are predicted to disrupt dynamin's ability to bind GTP. Overexpression of these "dominant-negative" mutants [e.g., dyn(K44A) or dyn(S45N)] potently inhibits receptor-mediated endocytosis and leads to the accumulation of invaginated clathrin-coated pits on the cell surface [16,20–22]. Other endocytic events, such as internalization of caveolae [23,24] and phagocytosis [25] are also inhibited, but the internalization of bulk fluid or membrane markers mediated by a constitutive, but poorly understood, clathrin-independent endocytic process is unaffected [20,26]. In stable HeLa cells induced to express dyn(K44A), the mutant also blocks the retrograde transport of the plant toxin ricin from late endosomes to the trans-Golgi network (TGN) [26]. Other membrane transport events, such as trafficking from the TGN to the cell surface, transport from the TGN to late endosomes or delivery of endocytic tracers to lysosomes are unaffected in these cells [20,27].

Two Models for Dynamin Function in Endocytic Vesicle Formation

While there exists considerable evidence establishing a critical role for dynamin in endocytic clathrin-coated vesicle formation, its exact function remains a matter of dispute [3]. One prevailing model (Fig. 3, model 1) suggests that after self-assembly around the necks of invaginating coated pits, dynamin functions as a mechanochemical enzyme to generate the force—either by constriction (model 1A [28]) or expansion (model 1B [14]) of the assembled ring driven by rapid GTP hydrolysis—that mediates membrane fission and vesicle detachment. This model, which is consistent with its unique enzymatic properties, places dynamin in a class of its

Dynamin as a mechanochemical GTPase

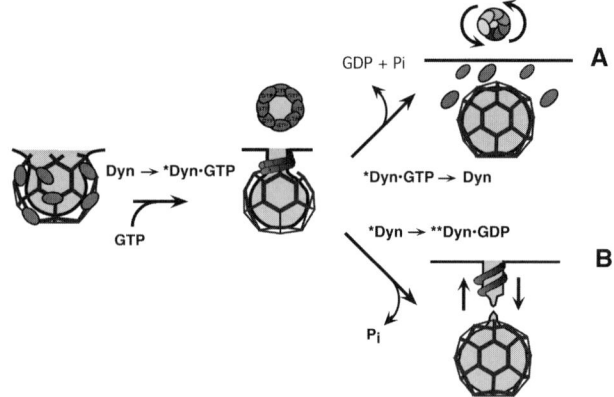

Dynamin as a regulatory GTPase

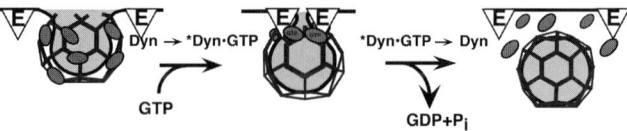

Figure 3 Two prevailing models for dynamin's function in endocytosis. Model 1 asserts that dynamin functions in a mechanochemical manner as either a "pinchase" or a "spring" to sever the necks of invaginated coated pits. Model 2 asserts that in its GTP-bound form dynamin recruits downstream effector molecules that mediated vesicle formation. In both models dynamin must bind and hydrolyze GTP (at least at its basal rate) and undergo GTP-dependent conformational changes (indicated by asterisks).

own as the first example of a mechanochemical GTPase. The second, more recently espoused model (Fig. 3, model 2), is more in keeping with dynamin's membership in the GTPase superfamily, and suggests that dynamin functions, in its GTP-bound form, to recruit downstream effectors, which themselves constitute the vesicle fission/detachment machinery [15]. In this model the collar functions as a geometric sensor and dynamin's GAP, when activated by self-assembly, functions by analogy with all other GTPase superfamily members, as a timer to turn off and recycle the activated GTPase. Naturally, these two models are not mutually exclusive and other models are possible that incorporate features of both extremes.

Testing Models for Dynamin Function *in vivo*: GTPase Domain Mutants

Good faith efforts have been made to test models for dynamin function *in vivo*, but as yet none of the experiments have unambiguously supported one model over the other. One approach is to ask whether GTP hydrolysis by dynamin is required for endocytosis. If dynamin functions like a regulatory GTPase one would predict that an "activating mutant" of dynamin locked in its GTP-bound form would support and perhaps stimulate endocytosis. In contrast, if dynamin functions as a mechanochemical enzyme, its high rates of GTP hydrolysis would be required for endocytosis. Based on analogy to other GTPases, Thr65 of dynamin, which is located in the switch I region of GTPases and is required to coordinate the Mg^{2+} ion essential for GTP hydrolysis, was mutated in an effort to generate such an activating mutant. Overexpression of dyn(T65A) inhibits receptor-mediated endocytosis, suggesting that GTP hydrolysis by dynamin is essential [16].

However, the results are not conclusive, as mutation of this conserved Thr in ras or trimeric G proteins reduces their affinity for GTP by >100-fold [17]. However, ras and Gα GTPases are still able to bind GTP with relatively high affinity and binding is not significantly impaired under *in vivo* conditions. By contrast, given dynamin's already low affinity for GTP, this mutation is more likely to perturb its ability to bind GTP *in vivo*. Moreover, the coordination of this conserved Thr to Mg^{2+} and the γ-phosphate in other GTPases is essential for the GTP-dependent conformational change in the effector loop necessary for interaction with downstream partners. Thus, it is unclear whether GTPases with this mutation are able to efficiently adopt the "active" conformation. Consistent with this interpretation, dyn(T65A) is unable to adopt the GTP-dependent conformation on lipid tubules [16].

Thus the issue of whether GTP hydrolysis by dynamin is essential for endocytosis remains unresolved. Ideally, one needs to mutate an essential catalytic residue, equivalent to Q61 in ras, that does not affect GTP binding or the GTP-dependent conformational change in order to the test whether GTP hydrolysis was required for dynamin function. Unfortunately, candidate residues based on other GTPases are not conserved in dynamin. However, the recently solved structures of the GTPase domain of dynamin A in the unoccupied and GDP-bound state [5] may suggest candidate residues for further mutagenesis.

Testing Models for Dynamin Function *in vivo*: GED Mutants

Given that dynamin encodes its own GAP, another approach to making activating mutants is to identify residues in GED that affect the assembly-dependent GAP activity. Two such mutations, dyn(K694A) and dyn(R725A), were shown to specifically perturb the assembly-stimulated GTPase activity of dynamin *in vitro* [15]. Importantly, these GED mutations do not alter the basal rate of GTP hydrolysis, nor do they effect GTP binding affinities. In contrast to the GTPase mutants of dynamin, overexpression of these GED mutants accelerated the rate-limiting step in clathrin-mediated endocytosis [15]. Dyn(K694A) and dyn(R725A) were overexpressed under the control of a tetracycline-regulatable promoter in stably-transfected HeLa cells and studied using stage-specific endocytosis assays that distinguish the formation of constricted coated pits, an intermediate in endocytosis, from subsequent membrane fission and coated vesicle release [29]. Dyn(K694A) overexpression increased the rate of formation of both constricted coated pits and coated vesicles. Consistent with these findings, expression in yeast of an analogous mutation (K705A) in Dnm1p, a dynamin family member, results in increased mitochondrial fission [30].

Given that the dyn(K694A) mutant was specifically defective in self-assembly, this result is difficult to reconcile with models for dynamin function as strictly a mechanochemical enzyme. Instead, these findings are most consistent with the interpretation that by impairing GAP activity and prolonging dynamin in its GTP-bound form, as yet unidentified, downstream effectors(s) are activated and in turn accelerate the rate of endocytosis. Importantly, dyn(K694A) was impaired but not incapable of self-assembly, and so these results do not rule out a requirement for dynamin self-assembly in vesicle formation. Moreover, results with dyn(R725A) were more difficult to interpret, because, unexpectedly, although overexpression of dyn(R72A) enhanced the rate of formation of constricted coated pits, it had no effect on or was slightly inhibitory for vesicle formation [29]. These results suggested that the dyn(R725A) mutant reduces the rate of membrane fission, while increasing the rate of constricted coated pit formation. Thus, dynamin assembly and assembly-stimulated GTPase activity may be required for the final stages in vesicle detachment. New mutants with more pronounced defects in assembly and assembly-stimulated GTPase activity will be needed to rigorously test models for dynamin function *in vivo*. Interestingly, mutations in mx protein that have abolished self-assembly and assembly-stimulated GTPase activity do not alter the protein's ability to confer viral resistance [6], suggesting that for this dynamin family member, self-assembly is not essential for function.

Dynamin's Siblings: The Dynamin Subfamily of GTPases

Although not the founding member, dynamin was born into a subfamily of GTPases that now bears its name. Members of the dynamin family of GTPases are functionally diverse, and not numerous. That is, unlike other components of the vesicle transport machinery (i.e., arfs, coats, SNAREs, rabs, etc.), there does not appear to be a dynamin analog functioning in every vesicle formation event within the cells. In the yeast, *Saccharomyces cerevisiae* there are only three dynamin family members. These include Vps1p, which is the only dynamin family member involved in vesicular transport in yeast and plays an as yet undefined role in protein trafficking between the *trans*-Golgi network (TGN) and the vacuole; Dnm1p, a cytosolic protein that binds to the outer mitochondrial membrane and plays a role in regulating the size and number of mitochondria; and, Mgm1p, an inner mitochondrial protein also involved in controlling mitochondrial division. These dynamin-related GTPases share high degrees of homology with dynamin's GTPase, middle and GED domains, but they lack the pleckstrin homology (PH) domain and the PRD of true dynamins (Fig. 1).

The *Drosophila* and *Caenorhabditis elegans* genomes also encode only three dynamin family members. These include Dnm1p and Mgm1p homologs, which appear to function like their *S. cerevisiae* counterparts in regulating mitochondrial membrane dynamics [31]. Surprisingly, there are no Vps1p homologs in either organism, instead they both encode a single gene for dynamin. Thus, Vps1p may be the yeast homolog of dynamin, given that both proteins function in vesicular trafficking, albeit at distinct sites. Vps1 mutations have no effect on endocytosis in yeast, whereas dynamin mutations in *C. elegans* and *Drosophila* specifically disrupt endocytosis, without affecting other trafficking pathways.

The issue is more complicated, however, in mammals, which in addition to a Dnm1p homolog (called DLP1 and also known as dymple, DRP1 or DVLP) and a Mgm1p homolog, both of which are associated with mitochondrial membrane dynamics, express three dynamin genes in a tissue-specific manner [31]. Dynamin-1 is exclusively expressed in neurons, dynamin-2 is ubiquitously expressed, and dynamin-3 is expressed primarily in testes, brain, and lung. There are also numerous splice variants of each gene product [2]. Gene duplication in mammals is not uncommon and the simplest interpretation is that the newly evolved isoforms perform specialized functions in specialized cells (e.g., rapid and regulated synaptic vesicle recycling at the synapse). In HeLa cells and MDCK cells, overexpression of dyn2(K44A) functions indistinguishably from dyn1(K44A) in membrane trafficking to specifically and potently inhibit receptor-mediated endocytosis without affecting other vesicular trafficking events [27]. There are conflicting data as to whether specific splice variants of dynamin-2 do [32] or do not [33] localize to the TGN and effect vesicle formation from this compartment. These discrepancies may reflect cell-type differences and/or they may reflect nonspecific effects or incomplete inhibition due to too much or too little overexpression. Knock-out experiments will be needed to unambiguously resolve this issue; however, given that a function for dynamin in the Golgi does not appear to be conserved in *C. elegans* or *Drosophila*, it seems unlikely that it is essential for vesicle formation at these sites.

Mammals also express a fourth, and original member of the dynamin subfamily, the α-interferon inducible mx proteins [34]. The mx proteins are localized in either the cytoplasm or the nucleus and confer cellular resistance to RNA viruses. Recent evidence suggests that one mx family member (mxA) does so by binding to and sequestering the viral nucleocapsid in large aggregates in the cytosol [35]. Thus, of the four dynamin family members in mammals, all of which self-assemble into rings and helical stacks of rings and all of which exhibit assembly-stimulated GTPase activity, only one is involved in vesicular trafficking. This latter point is important when considering dynamin's proposed function as a fission GTPase or vesicle "pinchase" that is integral to the fundamental mechanism of vesicle formation. One might expect such an elegant fission apparatus to be a ubiquitous component of the vesicle formation machinery

Dynamin as a Signaling Molecule

While the mechanisms remain to be established, a growing body of evidence suggests that dynamin (and dynamin

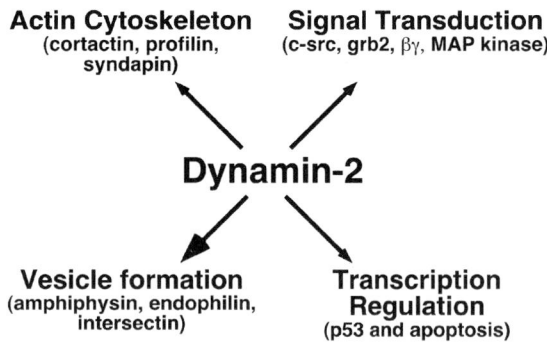

Figure 4 Dynamin as a signaling GTPase. There is now considerable evidence that dynamin can signal to multiple, functionally diverse downstream cellular processes. Some of these signaling pathways are specific for dynamin-2 versus dynamin-1.

family members) may instead, or in addition, function as signaling molecules to control multiple aspects of cell physiology (Fig. 4). For example, dynamin is known to interact with several actin regulatory proteins, including profilin, cortactin, syndapins, and Abp1 [36] and has recently been shown to be associated with sites of active actin assembly/disassembly, including membrane ruffles [37], podosomes [38], and actin comet tails [39,40]. Actin dynamics at these sites are disrupted or altered by overexpression of dyn2 (K44A) mutants suggesting a role for GTP binding and/or hydrolysis in regulating these events. Overexpression of dyn2(ΔPRD) in clone 9 cells causes dramatic changes in overall cell shape [37], again suggesting a more pleiotropic effect than one would expect from a defect in vesicle formation alone.

Dynamin can also effect other signaling pathways, directly or indirectly. Three studies have shown that dyn(K44A) overexpression perturbs G-protein-coupled receptor (GPCR) signaling to MAP kinase family members, independent of its effects on endocytosis [41–43]. Dyn(K44A) overexpression also inhibits high-affinity binding to epidermal growth factor (EGF) receptors [44] and causes reduced acidification of endosomes [45]. While the mechanisms underlying these effects are unknown, these observations are not easily reconciled with a model for dynamin as simply a vesicle pinchase.

Finally, low levels of overexpression of dynamin-2 induces p53-dependent apoptosis in several cell types including HeLa cells, rat-1 fibroblasts, and mouse embryo fibroblasts [46]. Several lines of evidence support the specificity of this effect: (1) <5-fold overexpression of dyn2 induces apoptosis, whereas >250-fold overexpression of the 80% identical dynamin-1 isoform has no cytotoxic effect; (2) dyn2-induced apoptosis requires GTP binding because dyn2(K44A), which inhibits endocytosis, does not induce apoptosis; (3) dyn2 at high levels of overexpression does not induce apoptosis in nondividing cells (e.g., contact inhibited MDCK cells or primary macrophages) or in p53-deficient mouse embryo fibroblasts. Interestingly, a recent report established that dominant-negative DRP1, a mitochondria-associated dynamin family member, can protect cells from staurosporine-induced apoptosis [47]. Together these data suggest that dynamin family members might play a more general role in monitoring membrane dynamics at critical cellular locations—the plasma membrane and the mitochondria—to control cellular homeostasis.

Conclusion and Perspectives

Consider the following re-ordering of the discoveries surrounding dynamin function: (1) dynamin-2 is cloned and found to be a GTPase superfamily member; (2) analysis of the complete genomes of *Drosophila* and *C. elegans* reveal that the only two other dynamin-related GTPases are mitochondrial—there is no "true" dynamin in yeast; (3) in mammals, a fourth dynamin related GTPase is identified as the α-interferon-inducible mx protein, which confers viral resistance; (4) overexpression of dyn2(wt) activates p53 and induces apoptosis, mutants unable to bind GTP do not; (5) overexpression of dyn2(K44A) inhibits MAP kinase activation and alters actin cytoskeleton dynamics; (6) all dynamin-related proteins, even those not involved in vesicular transport, are shown to self-assemble and to have assembly-stimulated GTPase activity; (7) mutants of mx incapable of either self-assembly or assembly-stimulated GTPase activity are fully functional in conferring viral resistance, although GTPase domain mutants are not; (8) dyn(K44A) mutants inhibit endocytosis and assembled dynamin accumulates in GTPγS-treated membranes. Given this set of facts, presented in this order, one might have reached a different conclusion regarding dynamin's cellular function, perhaps one in which dynamin functions more broadly and more in line with rho family regulatory GTPases. While images of "collared" pits are compelling, the function of dynamin GTPase's may turn out to be more than meets the eye.

References

1. Hinshaw, J. E. (2000). Dynamin and its role in membrane Fission. *Annu. Rev. Cell Dev. Biol.* **16**, 483–519.
2. McNiven, M. A., Cao, H., Pitts, K. R., and Yoon, Y. (2000). The dynamin family of mechanoenzymes: pinching in new places. *Trends Biochem. Sci.* **25**, 115–120.
3. Sever, S., Damke, H., and Schmid, S. L. (2000). Garrotes, springs, ratchets and whips: Putting dynamin models to the test. *Traffic* **1**, 385–392.
4. Schmid, S. L., McNiven, M. A., and De Camilli, P. (1998). Dynamin and its partners: A progress report. *Curr. Opin. Cell Biol.* **10**, 504–512.
5. Niemann, H. H., Knetsch, M. L., Scherer, A., Manstein, D. J., and Kull, F. J. (2001). Crystal structure of a dynamin GTPase domain in both nucleotide-free and GDP-bound forms. *EMBO J.* **20**, 5813–5821.
6. Janzen, C., Kochs, G., and Haller, O. (2000). A monomeric GTPase-negative MxA mutant with antiviral activity. *J. Virol.* **74**, 8202–8206.
7. Smirnova, E., Shurland, D. L., Newman-Smith, E. D., Pishvaee, B., and van der Bliek, A. M. (1999). A model for dynamin self-assembly based on binding between three different protein domains. *J. Biol. Chem.* **274**, 14942–14947.
8. Prakash, B., Praefcke, G. J., Renault, L., Wittinghofer, A., and Herrmann, C. (2000). Structure of human guanylate-binding protein 1 representing a unique class of GTP-binding proteins. *Nature* **403**, 567–571.
9. Binns, D. D., Barylko, B., Grichine, N., Adkinson, A. L., Helms, M. K., Jameson, D. M., Eccleston, J. F., and Albanesi, J. P. (1999). Correlation between self-association modes and GTPase activation of dynamin. *J. Protein Chem.* **18**, 277–290.

10. Binns, D. D., Helms, M. K., Barylko, B., Davis, C. T., Jameson, D. M., Albanesi, J. P., and Eccleston, J. F. (2000). The mechanism of GTP hydrolysis by dynamin II: a transient kinetic study. *Biochemistry* **39**, 7188–7196.
11. Warnock, D. E., Baba, T., and Schmid, S. L. (1997). Ubiquitously expressed dynamin-II has a higher intrinsic GTPase activity and a greater propensity for self-assembly than neuronal dynamin-I. *Mol. Biol. Cell* **8**, 2553–2562.
12. Tuma, P. L. and Collins, C. A. (1994). Activation of dynamin GTPase is a result of positive cooperativity. *J. Biol. Chem.* **269**, 30842–30847.
13. Warnock, D. E., Hinshaw, J. E., and Schmid, S. L. (1996). Dynamin self assembly stimulates its GTPase activity. *J. Biol. Chem.* **271**, 22310–22314.
14. Stowell, M. H. B., Marks, B., Wigge, P., and McMahon, H. T. (1999). Nucleotide-dependent conformational changes in dynamin: evidence for a mechanochemical molecular spring. *Nat. Cell Biol.* **1**, 27–32.
15. Sever, S., Muhlberg, A. B., and Schmid, S. L. (1999). Impairment of dynamin's GAP domain stimulates receptor-mediated endocytosis. *Nature* **398**, 481–486.
16. Marks, B., Stowell, M. H. B., Vallis, Y., Mills, I. G., Gibson, A., Hopkins, C. R., and McMahon, H. T. (2001). GTPase activity of dynamin and resulting conformation change are essential for endocytosis. *Nature* **410**, 231–235.
17. Wittinghofer, A. (2000). The functioning of molecular switches in three dimensions. in Hall, A., Eds., *GTPases Vol. 24*, pp. 244–310. Oxford University Press, New York.
18. Kosaka, T. and Ikeda, K. (1983). Reversible blockage of membrane retrieval and endocytosis in the garland cell of the temperature-sensitive mutant of *Drosophila melanogaster, shibire*[ts1]. *J. Cell Biol.* **97**, 499–507.
19. Damke, H., Baba, T., van der Bliek, A. M., and Schmid, S. L. (1995). Clathrin-independent pinocytosis is induced in cells overexpressing a temperature-sensitive mutant of dynamin. *J. Cell Biol.* **131**, 69–80.
20. Damke, H., Baba, T., Warnock, D. E., and Schmid, S. L. (1994). Induction of mutant dynamin specifically blocks endocytic coated vesicle formation. *J. Cell Biol.* **127**, 915–934.
21. Herskovits, J. S., Burgess, C. C., Obar, R. A., and Vallee, R. B. (1993). Effects of mutant rat dynamin on endocytosis. *J. Cell Biol.* **122**, 565–578.
22. Damke, H., Binns, D. D., Ueda, H., Schmid, S. L., and Baba, T. (2001). Dynamin GTPase domain mutants block endocytic vesicle formation at morphologically distinct stages. *Mol. Biol. Cell* **12**, 2578–2589.
23. Henley, J. R., Krueger, E. W., Oswald, B. J., and McNiven, M. A. (1998). Dynamin-mediated internalization of cavalla. *J. Cell Biol.* **141**, 85–99.
24. Oh, P., McIntosh, D. P., and Schnitzer, J. E. (1998). Dynamin at the neck of caveolae mediates their budding to form transport vesicles by GTP-driven fission from the plasma membrane of endothelium. *J. Cell Biol.* **141**, 101–114.
25. Gold, E. S., Underhill, D. M., Morrissette, N. S., Guo, J., McNiven, M. A., and Aderem, A. (1999). Dynamin 2 is required for phagocytosis in macrophages. *J. Exp. Med.* **190**, 1849–1856.
26. Llorente, A., Rapak, A., Schmid, S. L., van Deurs, B., and Sandvig, K. (1998). Expression of mutant dynamin inhibits toxicity and transport of endocytosed ricin to the Golgi apparatus. *J. Cell Biol.* **140**, 553–563.
27. Altschuler, Y., Barbas, S., Terlecky, L., Mostov, K., and Schmid, S. L. (1998). Common and distinct functions for dynamin-1 and dynamin-2 isoforms. *J. Cell Biol.* **143**, 1871–1881.
28. Hinshaw, J. E. and Schmid, S. L. (1995). Dynamin self assembles into rings suggesting a mechanism for coated vesicle budding. *Nature* **374**, 190–192.
29. Sever, S., Damke, H., and Schmid, S. L. (2000). Dynamin:GTP controls the formation of constricted coated pits, the rate limiting step in clathrin-mediated endocytosis. *J. Cell Biol.* **150**, 1137–1148.
30. Fukushima, N. H., Brisch, E., Keegan, B. R., Bleazard, W., and Shaw, J. M. (2001). The GTPase effector domain sequence of the Dnm1p GTPase regulates self-assembly and controls a rate-limiting step in mitochondrial fission. *Mol. Biol. Cell* **12**, 2756–2766.
31. van der Bliek, A. M. (1999). Functional diversity in the dynamin family. *Trends Cell Biol.* **9**, 96–102.
32. Cao, H., Thompson, H. M., Krueger, E. W., and McNiven, M. A. (2000). Disruption of Golgi structure and function in mammalian cells expressing a mutant dynamin. *J. Cell Sci.* **113**, 1993–2002.
33. Kasai, K., Shin, H. W., Shinotsuka, C., Murakami, K., and Nakayama, K. (1999). Dynamin II is involved in endocytosis but not in the formation of transport vesicles from the trans-Golgi network. *J. Biochem. (Tokyo).* **125**, 780–789.
34. Staeheli, P., Pitossi, F, and Pavlovic. J. (1993). Mx proteins: GTPases with antiviral activity. *Trends Cell Biol.* **3**, 268–272.
35. Kochs, G., Janzen, C., Hohenberg, H., and Haller, O. (2002). Antivirally active MxA protein sequesters La Crosse virus nucleocapsid protein into perinuclear complexes. *Proc. Natl. Acad. Sci. USA* **99**, 3153–3158.
36. Schafer, D. A. (2002). Coupling actin dynamics and membrane dynamics during endocytosis. *Curr. Opin. Cell Biol.* **14**, 76–81.
37. McNiven, M. A., Kim, L., Krueger, E. W., Orth, J. D., Cao, H., and Wong, T. W. (2000). Regulated interactions between dynamin and the actin-binding protein cortactin modulate cell shape. *J. Cell Biol.* **151**, 187–198.
38. Ochoa, G. C., Slepnev, V. I., Neff, L., Ringstad, N., Takei, K., Daniell, L., Kim, W., Cao, H., McNiven, M., Baron, R., and De Camilli, P. (2000). A functional link between dynamin and the actin cytoskeleton at podosomes. *J. Cell Biol.* **150**, 377–389.
39. Lee, E. and De Camilli, P. (2002). Dynamin at actin tails. *Proc. Natl. Acad. Sci. USA* **99**, 161–166.
40. Orth, J. D., Krueger, E. W., Cao, H., and McNiven, M. A. (2002). The large GTPase dynamin regulates actin comet formation and movement in living cells. *Proc. Natl. Acad. Sci. USA* **99**, 167–172.
41. Earnest, S., Khokhlatchev, A., Albanesi, J. P., and Barylko, B. (1996). Phosphorylation of dynamin by ERK2 inhibits the dynamin-microtubule interaction. *FEBS Lett.* **396**, 62–66.
42. Kranenburg, O., Verlaan, I., and Moolenaar, W. H. (1999). Dynamin is required for the activation of mitogen-activated protein (MAP) kinase by MAP kinase kinase. *J. Biol. Chem.* **274**, 35301–35304.
43. Hislop, J. N., Everest, H. M., Flynn, A., Harding, T., Uney, J. B., Troskie, B. E., Millar, R. P., and McArdle, C. A. (2001). Differential internalization of mammalian and non-mammalian gonadotropin-releasing hormone receptors. Uncoupling of dynamin-dependent internalization from mitogen-activated protein kinase signaling. *J. Biol. Chem.* **276**, 39685–39694.
44. Ringerike, T., Stang, E., Johannessen, L. E., Sandnes, D., Levy, F. O., and Madshus, I. H. (1998). High-affinity binding of epidermal growth factor (EGF) to EGF receptor is disrupted by overexpression of mutant dynamin (K44A). *J. Biol. Chem.* **273**, 16639–16642.
45. Huber, M., Brabec, M., Bayer, N., Blaas, D., and Fuchs, R. (2001). Elevated endosomal pH in HeLa cells overexpressing mutant dynamin can affect infection by pH-sensitive viruses. *Traffic* **2**, 727–736.
46. Fish, K. N., Schmid, S. L., and Damke, H. (2000). Evidence that dynamin-2 functions as a signal-transducing GTPase. *J. Cell Biol.* **150**, 145–154.
47. Frank, S., Gaume, B., Bergmann-Leitner, E. S., Leitner, W. W., Robert, E. G., Catez, F., Smith, C. L., and Youle, R. J. (2001). The role of dynamin-related protein 1, a mediator of mitochondrial fission, in apoptosis. *Dev. Cell* **1**, 515–525.

Mx Proteins: High Molecular Weight GTPases with Antiviral Activity

Georg Kochs, Othmar G. Engelhardt, and Otto Haller

*Abteilung Virologie,
Institut für Medizinische Mikrobiologie und Hygiene,
Universität Freiburg, Freiburg, Germany*

Mx proteins belong to a family of interferon-induced GTPases within the superfamily of high molecular weight GTPases. A unique property of some Mx GTPases is their antiviral activity against RNA viruses. Much progress has been made in better understanding the molecular mechanisms of Mx action over the years. Here we review recent insights into the workings of this fascinating class of large GTPases.

Antiviral Activity of Mx GTPases

Mx, for Myxovirus resistance, has first been described in mice, as an interferon-induced resistance phenomenon against influenza A virus (FLUAV) infection [1]. Subsequently, a variety of Mx proteins from different species were found to have antiviral activity, as shown by transfection experiments with Mx cDNA expression constructs [1,2]. Rodents express nuclear as well as cytoplasmic Mx proteins, whereas most other vertebrates express Mx proteins that accumulate exclusively in the cytoplasm. Interestingly, the antiviral specificity of rodent Mx proteins correlates with their subcellular localization. The nuclear forms (Mx1 in mouse and rat) confer resistance to orthomyxoviruses, FLUAV, and Thogoto virus (THOV), known to replicate in the cell nucleus. In contrast, cytoplasmic proteins (Mx2 in mouse and rat) inhibit multiplication of viruses that replicate in the cytoplasm, such as rhabdoviruses (e.g., vesicular stomatitis virus; VSV) and bunyaviruses (e.g., LaCrosse virus ;LACV) [1]. This suggests that rodents have evolved two distinct types of Mx GTPases to allow control of viruses with different intracellular replication sites.

Humans express a single antiviral Mx GTPase, called MxA. The cytoplasmic MxA has a rather wide antiviral spectrum against different classes of viruses, irrespective of their intracellular site of replication. MxA interferes with the multiplication of members of the bunyaviruses, orthomyxoviruses, paramyxoviruses, and rhabdoviruses [1]. Initially, Mx proteins were assumed to be active against negative-strand RNA viruses only, but recent studies showed inhibitory activity of MxA also against Semliki Forest virus and coxsackievirus B4, two plus-strand RNA viruses, as well as hepatitis B virus, a DNA virus with a genomic RNA intermediate [3–5]. The power of MxA as an intrinsic host defense mechanism is best illustrated in experiments with *MxA*-transgenic mice lacking interferon-α/β receptors. These animals survived an otherwise lethal infection with MxA-sensitive viruses, despite their interferon-nonresponsiveness [6]. Interestingly, Mx proteins without antiviral function also exist, such as human MxB or rat Mx3. It is presently not known whether these Mx GTPases are directed against a different set of microbial pathogens or serve other functions.

Mouse Mx1 Protein Blocks Primary Transcription of Orthomyxoviruses by Interfering with the Viral Polymerase Activity

The nuclear mouse Mx1 protein inhibits FLUAV and THOV multiplication at the level of primary transcription of the viral genome. It is noteworthy that primary transcription of these viruses occurs in the nucleus of infected cells and is catalyzed by the viral polymerase. The FLUAV polymerase consists of three subunits, PB1, PB2, and PA. Overexpression of the PB2 subunit abolished the antiviral effect of Mx1, suggesting that the PB2 subunit of FLUAV is an Mx1 target [7]. The subcellular localization of Mx GTPases is important for their antiviral activity, as demonstrated by artificial translocation of cytoplasmic Mx proteins into the nuclear compartment. When MxA or rat Mx2 was translocated into the nucleus with the help of a foreign nuclear translocation signal, both proteins gained an Mx1-like phenotype and blocked primary transcription of FLUAV and THOV [8,9].

Human MxA Interferes with Virus Multiplication by Missorting Viral Components

Studies in cell culture showed that cytoplasmic MxA inhibits primary transcription of VSV and measles virus, all of which are known to transcribe their genome in the cytoplasm [1]. Surprisingly, MxA also inhibited primary transcription of THOV which takes place in the nucleus [10]. In addition, MxA inhibits the multiplication of FLUAV, human parainfluenza virus type 3, and bunyaviruses, at later steps in the viral replication cycle [1]. In the case of THOV, we could demonstrate that MxA binds to the nucleocapsids and prevents their transport into the nucleus (Fig. 1A). As a consequence, THOV primary transcription is inhibited [10]. In the case of LACV, we could show that MxA binds to the viral nucleocapsid protein [11]. This interaction leads to the formation of MxA/nucleocapsid protein copolymers. A redistribution of the viral nucleocapsid protein takes place in MxA-expressing cells leading to the disappearance of nucleocapsid proteins from the putative viral replication sites near the Golgi apparatus and to accumulation in the perinuclear area (Fig. 1B). This indicates that the antiviral action of MxA against bunyaviruses is based on missorting of an essential virus component which thus becomes unavailable for the generation of new virus particles. In summary, our analysis of THOV and LACV, representing two viruses with different replication strategies, suggests that MxA acts by interfering with proper transport of critical viral components to their ultimate target compartments within the infected cell.

Mx Proteins Belong to the Superfamily of High Molecular Weight GTPases

Mx proteins are large GTPases that differ from small GTPases and heterotrimeric G proteins in their molecular weight of about 70,000 to 80,000, a relatively low affinity

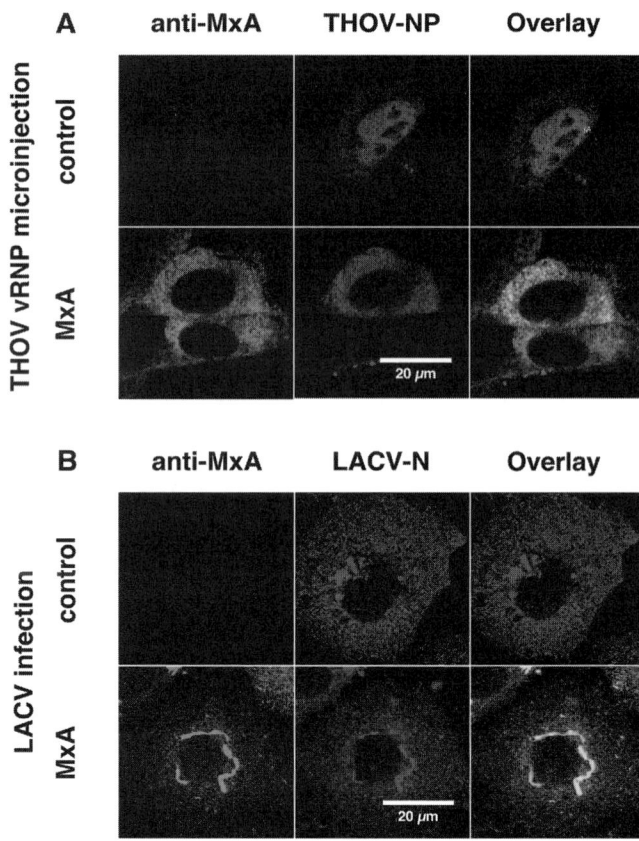

Figure 1 Inhibition of LACV and THOV multiplication by human MxA protein. (A) MxA-expressing or control cells were microinjected with purified viral nucleocapsids and incubated for 2 h. (B) Monolayers of the cells were infected with LACV for 12 h. Cells were fixed and analyzed for MxA and the viral nucleoproteins by double-immunofluorescence as described in references [10] and [11]. The right panels show the superimposition of the two images.

to GTP, and a high intrinsic rate of GTP hydrolysis [7]. Accordingly, Mx proteins belong to the superfamily of high molecular weight GTPases known to have mechanochemical properties. The prototype of this superfamily is the mammalian dynamin that is involved in endocytosis at the plasma membrane and other intracellular vesicle trafficking events (see previous chapter and reference[12]). Large GTPases share similarities in their overall domain structure. They contain a conserved tripartite GTP-binding motif within their N-terminal G domain and a less conserved C-terminal part that serves effector functions (Fig. 2A). GTP-binding and/or hydrolysis are necessary for large GTPases to display their force-generating functions [13]. Likewise, GTP binding and/or hydrolysis are required for the antiviral activity of Mx proteins [14].

The crystal structure of human guanylate-binding protein 1 (GBP-1) has been solved [15] and presently represents best the putative three-dimensional structure of high molecular weight GTPases. GBP-1 monomers consist of a globular G domain, a compact central core that is composed of two helix bundles and a long C-terminal helix, which interacts with the central core and the G domain. Figure 2B depicts a

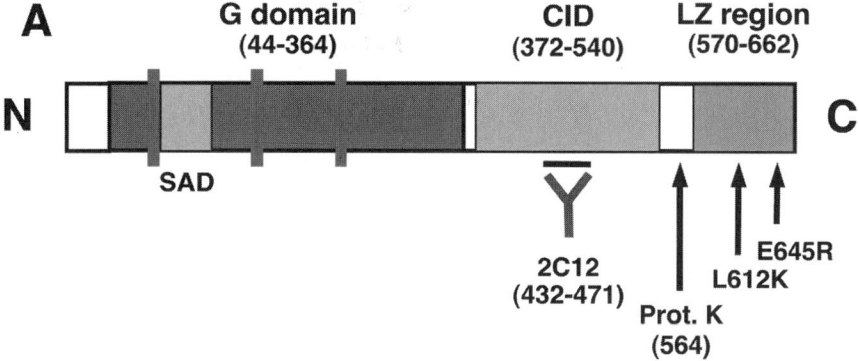

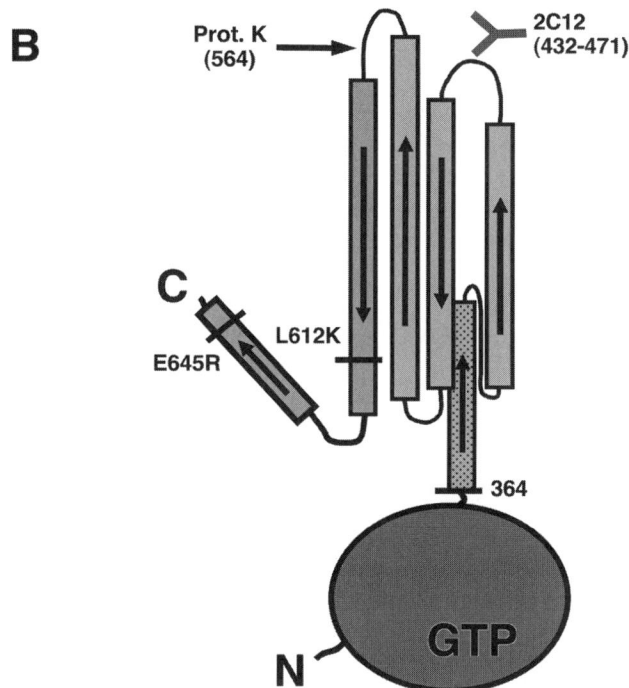

Figure 2 Functional domains of human MxA protein. (A) Schematic diagram of MxA domains and (B) a model of the putative structure according to the three-dimensional structure determined for GBP-1 [15]. The N-terminal globular G domain contains the tripartite GTP-binding element (red bars) and the "self-assembly domain," SAD (light blue box). The C-terminal effector domain contains the "central interactive domain." CID (yellow box), and the leucine zipper (LZ) region (green). The positions of the 2C12 antibody-binding site, the proteinase K cleavage site, and the critical amino acid substitutions L612K and E645R are indicated.

hypothetical model of the structure of human MxA, based on the structure of GBP-1. In this model, MxA consists of a large globular G domain and an extended helical C-terminal domain that controls self-assembly and association with other molecules, such as viral target structures or cellular binding partners.

Three different regions have been identified to be involved in molecular interactions of Mx proteins (Fig. 2A) [16]: (1) an N-terminal "self-assembly" sequence between GTP-binding elements 1 and 2 [17], (2) a "central interactive domain" (CID) [16], and (3) a leucine zipper motif (LZ) [18]. In our model, the central core containing the CID is composed of four helices which interact with the long C-terminal helix formed by the LZ region. This interaction was confirmed by biochemical studies and studies using cellular systems [19]. The CID and LZ region are connected by a putative hinge region that was identified by its accessibility to proteinase K cleavage [20]. The backfolding of the LZ region on CID results in increased GTPase activity, indicating that the C-terminal part acts as a "GTPase effector domain" [20]. Further biochemical studies with the C-terminal half indicated that the effector domain containing the CID and the LZ region is also crucial for binding to the viral target. Likewise, point mutations in the LZ region of MxA, like

MxA(E645R), and rat Mx2 identified the C-terminal effector domain as an important determinant for viral target recognition [8,11,21]. Other members of the superfamily of large GTPases have a comparable molecular structure which is also in agreement with recent reconstructions of electron microscopic pictures of human dynamin-1 [22] and dynamin A of *Dictyostelium discoideum* [23] (see also the previous chapter on dynamin).

What Is the Function of Mx Oligomerization?

Self-assembly of Mx proteins seems to be critical for GTPase activity, protein stability, and recognition of viral target structures. In cells, Mx proteins aggregate into punctate granula [24,25]. *In vitro*, purified Mx proteins form high molecular weight homo-oligomers and self-assemble into ring-like and helical structures [17,18,26]. Homo-oligomerization is essential for the biological function of other dynamin-like GTPases [13]. Dynamin, for example, self-assembles into ring-like and helical structures around membrane invaginations. Hydrolysis of guanosine triphosphate (GTP) to guanosine diphosphate (GDP) then induces conformational changes in these complexes, which generate the mechanochemical force necessary for membrane vesiculation.

A characteristic feature of Mx proteins and other large GTPases is their ability to hydrolyze GTP to GDP in the absence of additional factors like GTPase-activating proteins or nucleotide exchange factors [7]. Mutational analysis and proteolytic digestion of MxA showed that the association of the C-terminal LZ region with the CID is necessary for GTP binding and hydrolysis [20]. Accordingly, it has been demonstrated that oligomerization stimulates GTPase activity of MxA [27]. Homo-oligomerization most likely results from binding of the LZ region of one molecule to the CID of a second neighboring molecule [19]. Mutant MxA(L612K) carries an amino acid exchange from leucine to lysine at position 612 in the proximal part of the LZ region (Fig. 2A). MxA(L612K) has lost both its ability to self-assemble and its ability to hydrolyze GTP [28], indicating that oligomerization is the major regulator of GTPase activity.

Studies in cell culture showed that the monomeric MxA(L612K) is rapidly degraded [28]. In contrast, wild-type MxA is stable, with a half-life of over 24 h. Therefore, we propose that aggregation of Mx proteins prevents their rapid degradation and provides a storage form from which active molecules can be recruited for prolonged periods of time.

As already mentioned, biochemical studies demonstrated complex formation of MxA with the ribonucleoprotein complex of THOV and with the nucleocapsid protein of LACV [10,11]. This interaction only occurred when MxA was in the GTP-bound form, indicating that GTP-MxA is in a conformation able to recognize viral target structures. With LACV we demonstrated that binding of MxA to the viral nucleocapsid protein leads to the formation of protein complexes, most likely by copolymerization of MxA with the viral nucleocapsid protein [11]. We envisage that oligomerization of MxA occurs as a consequence of viral target recognition and is important in mediating the antiviral effect.

Cellular Interaction Partners of Mx GTPases

The antiviral activity of MxA against certain viruses was found to be cell-type specific. In the case of measles virus, MxA inhibited viral transcription in human glioblastoma cells, but not in a human monocytic cell line in which a later step, namely viral glycoprotein synthesis, was affected [29]. Also, the antiviral activity of MxA against Semliki Forest virus was detectable in human cells, but not in MxA-expressing mouse cells [3]. These findings suggested that unknown cellular factors are modulating the antiviral activity of MxA and influence its antiviral specificity for some viruses. Early work by Horisberger revealed an association of MxA with actin and tubulin [30], but the significance of this interaction remained unresolved. It is conceivable that the cytoskeleton plays a role in MxA-mediated relocalization of viral components as observed in infected cells [10,11]. Furthermore, nuclear factors may contribute to the antiviral activity of mouse Mx1 protein against influenza viruses [31]. Interestingly, rodent Mx1 proteins accumulate in distinct dots within the interchromatin nucleoplasm [24]. These dots are in close proximity to subnuclear structures, called promyelocytic leukemia protein nuclear bodies (PML NBs) (Fig. 3) [32,33]. Using a yeast-two-hybrid screen, we found that mouse Mx1 interacts with components of PML NBs, such as Sp100 and Daxx. Moreover, Mx1 interacted with components of the SUMO-1 protein modification system [33]. This is interesting because protein constituents of PML NBs are targets for SUMO-1 modification and SUMOylation of PML has been demonstrated to be crucial for PML NB assembly [34]. Again, the C-terminal effector domain of Mx seems to be responsible for at least some of these interactions (Trost, Kochs, Haller, unpublished results). PML NBs are involved in transcriptional regulation, cell cycle control,

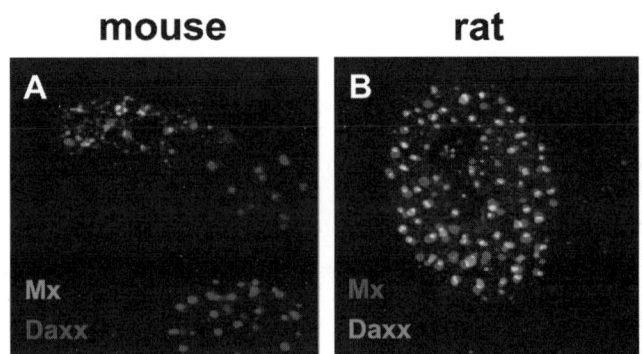

Figure 3 Mouse and rat Mx1 proteins form nuclear dots that are associated with PML NBs. Mouse embryo fibroblasts were transfected with cDNA expression constructs coding for mouse Mx1 (A) or rat Mx1 (B). The cells were fixed 24 h later and analyzed for the accumulation of Mx1 and endogenous Daxx by double-immunofluorescence. The pictures show the superimposition of the separately recorded images for Mx1 and Daxx.

apoptosis, and antiviral defense [34]. Whether these structures play a role for the antiviral activity of Mx1 is presently being investigated. It will be of interest to see whether the nuclear Mx1 dots are simply storage sites of Mx1, as discussed for MxA, or whether they represent, in fact, specialized nuclear compartments in which the inhibition of the viral polymerase takes place.

Acknowledgment

This work was supported by Grants HA 1582/3-1 and Ko 1579/1-4 of the Deutsche Forschungsgemeinschaft.

References

1. Haller, O., Frese, M., and Kochs, G. (1998). Mx proteins: mediators of innate resistance to RNA viruses. *Rev. Sci. Technol. OIE* **17**, 220–230.
2. Staeheli, P., Haller, O., Boll, W., Lindenmann, J., and Weissmann, C. (1986). Mx protein: Constitutive expression in 3T3 cells transformed with cloned Mx cDNA confers selective resistance to influenza virus. *Cell* **44**, 147–158.
3. Landis, H., Simon-Jödicke, A., Klöti, A., Di Paolo, C., Schnorr, J., Schneider-Schaulies, S., Hefti, H. P., and Pavlovic, J. (1998). Human MxA protein confers resistance to Semliki Forest virus and inhibits the amplification of a semliki forest virus-based replicon in the absence of viral structural proteins. *J. Virol.* **72**, 1516–1522.
4. Chieux, V., Chehadeh, W., Harvey, J., Haller, O., Wattre, P., and Hober, D. (2001). Inhibition of coxsackievirus B4 replication in stably transfected cells expressing human MxA protein. *Virology* **283**, 84–92.
5. Gordien, E., Rosmorduc, O., Peltekian, C., Garreau, F., Brechot, C., and Kremsdorf, D. (2001). Inhibition of Hepatitis B virus replication by the interferon-inducible MxA protein. *J. Virol.* **75**, 2684–2691.
6. Hefti, H. P., Frese, M., Landis, H., DiPaolo, C., Aguzzi, A., Haller, O., and Pavlovic, J. (1999). Human MxA protein protects mice lacking a functional alpha/beta interferon system against La Crosse virus and other lethal viral infections. *J. Virol.* **73**, 6984–6991.
7. Staeheli, P., Pitossi, F., and Pavlovic, J. (1993). Mx proteins: GTPases with antiviral activity. *Trends Cell Biol.* **3**, 268–272.
8. Zürcher, T., Pavlovic, J., and Staeheli, P. (1992). Mechanism of human MxA protein action: variants with changed antiviral properties. *EMBO J.* **11**, 1657–1661.
9. Johannes, L., Arnheiter, H., and Meier, E. (1993). Switch in antiviral specificity of a GTPase upon translocation from the cytoplasm to the nucleus. *J. Virol.* **67**, 1653–1657.
10. Kochs, G. and Haller, O. (1999). Interferon-induced human MxA GTPase blocks nuclear import of Thogoto virus nucleocapsids. *Proc. Natl. Acad. Sci. USA* **96**, 2082–2086.
11. Kochs, G., Janzen, C., Hohenberg, H., and Haller, O. (2002). Antivirally active MxA protein sequesters La Crosse virus nucleocapsid protein into perinuclear complexes. *Proc. Natl. Acad. Sci. USA*, in press.
12. Schmid, S. L., McNiven, M. A., and De Camilli, P. (1998). Dynamin and its partners: a progress report. *Curr. Opin. Cell Biol.* **10**, 504–512.
13. Sever, S., Damke, H., and Schmid, S. L. (2000). Garrotes, springs, ratchets, and whips: putting dynamin models to the test. *Traffic* **1**, 385–392.
14. Pitossi, F., Blank, A., Schröder, A., Schwarz, A., Hüssi, P., Schwemmle, M., Pavlovic, J., and Staeheli, P. (1993). A functional GTP binding motif is necessary for antiviral activity of Mx proteins. *J. Virol.* **67**, 6726–6732.
15. Prakash, B., Praefcke, G. J. K., Renault, L., Wittinghofer, A., and Herrmann, C. (2000). Structure of human guanylate-binding protein 1 representing a unique class of GTP-binding proteins. *Nature* **403**, 567–571.
16. Kochs, G., Trost, M., Janzen, C., and Haller, O. (1998). MxA GTPase: Oligomerization and GTP-dependent interaction with viral RNP target structures. *Methods: Compan. Methods Enzymol.* **15**, 255–263.
17. Nakayama, M., Yazaki, K., Kusano, A., Nagata, K., Hanai, N., and Ishihama, A. (1993). Structure of mouse Mx1 protein: molecular assembly and GTP-dependent conformational change. *J. Biol. Chem.* **268**, 15033–15038.
18. Melen, K., Ronni, T., Broni, B., Krug, R. M., Vonbonsdorff, C. H., and Julkunen, I. (1992). Interferon-induced Mx proteins form oligomers and contain a putative leucine zipper. *J. Biol. Chem.* **267**, 25898–25907.
19. Schumacher, B. and Staeheli, P. (1998). Domains mediating intramolecular folding and oligomerization of MxA GTPase. *J. Biol. Chem.* **273**, 28365–28370.
20. Schwemmle, M., Richter, M. F., Herrmann, C., Nassar, N., and Staeheli, P. (1995). Unexpected structural requirements for GTPase activity of the interferon-induced MxA protein. *J. Biol. Chem.* **270**, 13518–13523.
21. Johannes, L., Kambadur, R., Lee-Hellmich, H., Hodghinson, C. A., Arnheiter, H., and Meier, E. (1997). Antiviral determinants of rat Mx GTPases map to the carboxy-terminal half. *J. Virol.* **71**, 9792–9795.
22. Zhang, P. and Hinshaw, J. E. (2001). Three-dimensional reconstruction of dynamin in the constricted state. *Nat. Cell Biol.* **3**, 922–927.
23. Klockow, B., Tichelaar, W., Madden, D. R., Niemann, H. H., Akiba, T., Hirose, K., and Manstein, D. J. (2002). The dynamin A ring complex: Molecular organization and nucleotide-dependent conformational changes. *EMBO J.* **21**, 240–250.
24. Dreiding, P., Staeheli, P., and Haller, O. (1985). Interferon-induced protein Mx accumulates in nuclei of mouse cells expressing resistance to influenza viruses. *Virology* **140**, 192–196.
25. Staeheli, P. and Haller, O. (1985). Interferon-induced human protein with homology to protein Mx of influenza virus-resistant mice. *Mol. Cell. Biol.* **5**, 2150–2153.
26. Kochs, G., Haener, M., Aebi, U., and Haller, O. (2002). Self-Assembly of human MxA GTPase into highly-ordered dynamin-like oligomers. *J. Biol. Chem.*, in press.
27. Flohr, F., Schneider-Schaulies, S., Haller, O., and Kochs, G. (1999). The central interactive region of human MxA GTPase is involved in GTPase activation and interaction with viral target structures. *FEBS Lett.* **463**, 24–28.
28. Janzen, C., Kochs, G., and Haller, O. (2000). A monomeric GTPase-negative MxA mutant with antiviral activity. *J. Virol.* **74**, 8202–8206.
29. Schneider-Schaulies, S., Schneider-Schaulies, J., Schuster, A., Bayer, M., Pavlovic, J., and ter Meulen, V. (1994). Cell type-specific MxA mediated inhibition of measles virus transcription in human brain cells. *J. Virol.* **68**, 6910–6917.
30. Horisberger, M. A. (1992). Interferon-induced human protein MxA is a GTPase which binds transiently to cellular proteins. *J. Virol.* **66**, 4705–4709.
31. Zürcher, T., Pavlovic, J., and Staeheli, P. (1992). Nuclear localization of mouse Mx1 protein is necessary for inhibition of influenza virus. *J. Virol.* **66**, 5059–5066.
32. Chelbi-Alix, M. K., Pelicano, L., Quignon, F., Koken, M. H., Venturini, L., Stadler, M., Pavlovic, J., Degos, L., and de The, H. (1995). Induction of the PML protein by interferons in normal and APL cells. *Leukemia* **9**, 2027–2033.
33. Engelhardt, O. G., Ullrich, E., Kochs, G., and Haller, O. (2001). Interferon-induced antiviral Mx1 GTPase is associated with components of the SUMO-1 system and promyelocytic leukemia protein nuclear bodies. *Exp. Cell Res.* **271**, 286–295.
34. Salomoni, P. and Pandolfi, P. P. (2002). The role of PML in tumor suppression. *Cell* **108**, 165–170.

SECTION H

Developmental Signaling

Geraldine Weinmaster, Editor

Toll-Dorsal Signaling in Dorsal-Ventral Patterning and Innate Immunity

Ananya Bhattacharya and Ruth Steward
Waksman Institute, Department of Molecular Biology and Biochemistry, and Cancer Institute of New Jersey, Rutgers University, Piscataway, New Jersey

The Toll-Dorsal Pathway

The Toll-Dorsal pathway functions in *Drosophila* in the establishment of dorsal-ventral polarity in the early embryo. The same pathway also controls the innate humoral and cellular immune response in larvae and adults and, besides Dorsal, involves an additional NF-κB/Rel protein, Dif. The NF-κB/Rel pathway is conserved in flies and vertebrates. In mammals it functions in the immune and inflammatory responses, and it is critical for cell growth and survival. Moreover a large number of mammalian tumors are associated with misregulation of the NF-κB/Rel proteins. Dorsal, like all Rel proteins, is retained in an inactive state in the cytoplasm through direct interaction with a IκB protein, Cactus. The ventral signal, transmitted through the trans-membrane receptor Toll, destabilizes Cactus and controls the formation of a ventral-to-dorsal nuclear Dorsal gradient that, through the specific activation of zygotic genes, results in the formation of the dorsal-ventral axis.

Maturation of the Toll Ligand

Dorsal-ventral asymmetry is set-up in the egg chamber midway through oogenesis and is controlled mainly by EGF signaling [1]. This asymmetry is transmitted to the embryo through the interaction of two groups of genes. One group is expressed specifically on the ventral side of the follicle cells that surround the oocyte and secrete the egg-membranes. The other group is expressed in the oocyte and their products are secreted into the perivitelline space, between the egg or early embryo and the egg membranes. The interaction of the proteins encoded by the two groups of genes culminates in the maturation and cleavage of the ligand, Spatzle, which activates the Toll-Dorsal signaling pathway.

Spatzle is thought to diffuse and form a gradient in the perivitelline fluid. The extent of the Spätzle gradient is controlled by the upstream genes expressed in the ventral follicle cells and is refined by Spätzle autoregulation [2]. The Toll trans-membrane receptor is uniformly distributed in the plasma membrane of the early embryo and is asymmetrically activated by Spätzle.

Toll Signaling Establishes the Embryonic Dorsal Gradient

In early embryos Dorsal is uniformly retained in the cytoplasm bound to its inhibitor IκB protein, Cactus. The graded activation of the Toll receptor induces the nuclear translocation of the NF-κB-Rel homolog Dorsal in a gradient in the early embryo (Fig. 1A). The highest level of protein is present in the ventral nuclei and the concentration decreases laterally until on the dorsal side no Dorsal protein is present in the nuclei. The Dorsal gradient represents the first step in setting up the ultimate fate of each cell along the axis.

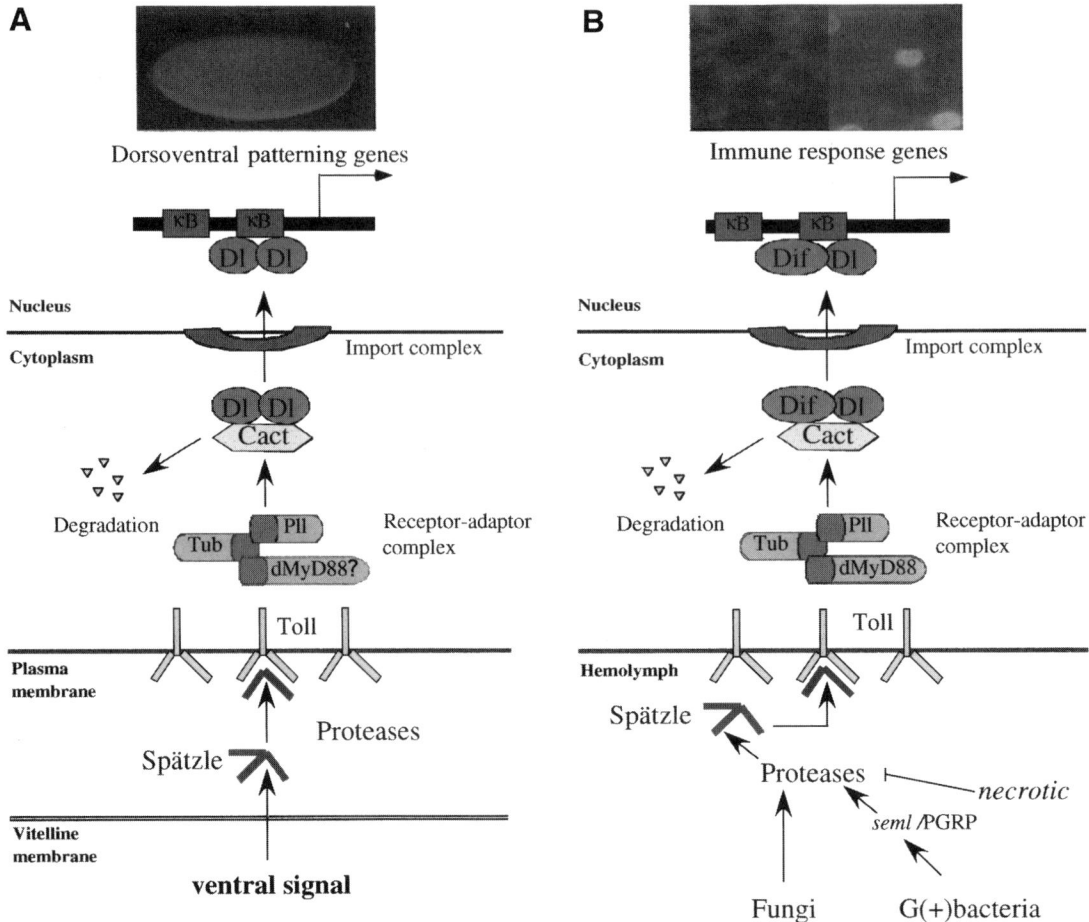

Figure 1 The Toll-Rel pathway in the early embryo and in immune response. (A) The top panel shows a blastoderm stage embryo, anterior to the left and dorsal side up, exhibiting a Dorsal-GFP gradient. Next, the steps in the Toll-Dorsal pathway occurring in the early embryo are depicted. Tub = Tube, Pll = Pelle, Cact = Cactus, Dl = Dorsal, κB = κB binding sites in DNA. (B) The top panel shows Dorsal-GFP in unchallenged (left) and bacterially challenged (right) larval fat body. Next, the steps in the Toll-Rel pathway occurring in the immune response are depicted. Dif = Dorsal-related immunity actor.

Cells with high levels of Dorsal protein will become mesoderm, while cells with no Dorsal protein will form the dorsal hypoderm, with ventral and lateral fates in-between.

The products of the *tube* and *pelle* genes relay the signal from Toll to the cytoplasmic Dorsal-Cactus complex. The Tube encodes a novel adaptor protein while Pelle functions as a serine-threonine kinase. Genetic and biochemical experiments suggest that the transmission of the Toll signal to the Dorsal-Cactus complex is mediated through the formation of a multimeric complex involving Tube, Pelle, Dorsal, Cactus, and possibly additional proteins [3]. Toll signaling results in the phosphorylation of Cactus and targets it for degradation. The same signal also controls Dorsal phosphorylation, essential for the high levels of nuclear protein and the establishment of the gradient [4]. The kinases involved in these phosphorylation events are not known.

In addition to the Toll-mediated degradation of Cactus, there is a signal-independent mechanism for degrading Cactus that also influences the Dorsal nuclear gradient. This degradation is probably controlled by an additional signaling pathway involving the maternally supplied TGFβ protein, Dpp [5]).

Dorsal Regulates the Function of Zygotic Genes

The Dorsal gradient subdivides the embryonic axis into distinct domains by setting the expression limits of key zygotic genes that are responsible for initiating the differentiation of various tissues. Dorsal interacts directly with κB binding sites and activates promoters of different genes. In the same cells it also functions as a repressor by recruiting co-repressors. The control of expression of genes in specific domains along the dorsal-ventral axis is controlled by the levels of Dorsal protein and the number and affinity of Dorsal binding sites in the target promoters [6].

The Intracellular Pathway Is Conserved in the *Drosophila* Immune Response

Fungal or gram-positive bacterial infections elicit a specific response in both adults and larvae that involves Spätzle, Toll, and the intracellular components of the pathway (Fig. 1B) [7].

Gram-positive and gram-negative bacterial infections are sensed by distinct pattern recognition proteins called peptidoglycan recognition proteins (PGRPs). PGRPs are highly conserved in organisms from insects to mammals. In the Toll-mediated immune response pathway Spätzle activation is under the control of a different set of genes than in the early embryo. The interaction between the microorganism and the PGRP is likely to occur in the circulating hemolymph and triggers a serine protease inhibitor (serpin)-controlled proteolytic cascade to activate Spätzle. Once activated, Toll interacts with death domain proteins: DmMyd88 and Tube, two adaptor proteins, and the kinase Pelle. This receptor adaptor complex relays the signal to the Cactus-Rel protein (Dif and/or Dorsal) complex. Dif is the predominant transactivator in the antifungal and anti-gram-positive bacterial infection. Dorsal acts redundantly with Dif in larvae but has no function in the adult immune response. Dif and Dorsal control the transcription of the antimicrobial peptide genes Drosomycin and Metchnikowin. In addition, a compound Rel protein, Relish, with an N-terminal Rel homology domain and a C-terminal IκB-like domain, is crucial for the response to gram-negative bacterial infection. While it is also inactive in the cytoplasm and translocated to the nucleus upon signaling, this signal is not transmitted through Toll.

Nuclear Import of Rel Proteins

Although the nuclear import of Rel proteins is a key step in the formation of the nuclear gradient and in the immune response, little is known about the factors involved. Proteins destined for nuclear import containing a nuclear localization sequence (NLS) are transported through nuclear pore complexes (NPCs) in a multistep process [8]. The NLS of Dorsal is essential and sufficient for nuclear import [9] and in conjunction with the signal-dependent phosphorylation of Dorsal it could provide a substrate for a specific import in α/β, which has not been identified thus far. Some of the nuclear pore proteins (Nups) forming the NPC have been shown to interact with importins. Mutant analysis has shown that both Nup88 or *mbo* and nuclear transport factor-2 (NTF-2) are involved in the import of the Rel proteins.

NTF-2 is an essential gene in yeast, *Caenorhabditis elegans*, and *Drosophila*. Dorsal, NTF-2, and Nup88 form a complex necessary for nuclear import [10,11]. The NLS in the NF-κB/Rel proteins and the proteins identified thus far as important for nuclear import in *Drosophila* are all highly conserved. It is therefore likely that the same mechanisms are functioning in vertebrates.

The NF-κB/Rel pathways are some of the best known signaling processes and play central roles in development, immune response, and cancer. Because of their high level of conservation from flies to humans, further study of the Toll-Dorsal (Rel) pathways in flies will continue to complement our understanding of events in vertebrates.

References

1. Nilson, L. A and Schupbach, T. (1999). EGF Receptor signaling in *Drosophila* oogenesis. *Curr. Top. Dev. Biol.* **44**, 203–243.
2. Morisato, D. (2001). Spatzle regulates the shape of the Dorsal gradient in the Drosophila embryo. *Development* **128**, 2309–2319.
3. Drier, E. A. and Steward, R. (1997). The Dorsoventral signal transduction pathway and the rel-like transcription factors in *Drosophila*. *Semin. Cancer Biol.* **8**, 83–92.
4. Drier, E. A., Huang, L. H., and Steward, R. (1999). Nuclear import of the *Drosophila* rel protein Dorsal is regulated by phosphorylation. *Genes Dev.* **13**, 556–568.
5. Araujo, H. and Bier, E. (2000). Sog and Dpp exert opposing maternal functions to modify toll signaling and pattern the dorsoventral axis of the *Drosophila* embryo. *Development* **127**, 3631–3644.
6. Rusch J. and Levine, M. (1996). Threshold responses to the Dorsal regulatory gradient and the subdivision of primary tissue territories in the *Drosophila* embryo. *Curr. Opin. Genet. Dev.* **6**, 416–423.
7. Hoffmann, J. A. and Reichhart, J.-M. (2002). *Drosophila* innate immunity: An evolutionary perspective. *Nat. Immun.* **3**(2), 121–126.
8. Jans, D. A., Xiao, C.-Y., and Lam, M. (2000). Nuclear targeting signal recognition: A key control point in nuclear transport? *BioEssays* **22**, 532–544.
9. Govind, S., Drier, E., Huang, L. H., and Steward, R. (1996). Regulated nuclear import of the *Drosophila* Rel protein Dorsal: Structure-function analysis. *Mol. Cell. Biol.* **16**, 1103–1114.
10. Bhattacharya, A. and Steward, R. (2002). The *Drosophila* homolog of NTF-2, the nuclear transport factor-2, is essential for immune response. *EMBO Rep.* **3**, 378–383.
11. Uv, A. E., Roth, P., Xylourgidis, N., Wickberg, A., Cantera, R., and Samakovlis, C. (2000). Members only encodes a *Drosophila* nucleoporin required for rel protein import and immune response activation. *Genes Dev.* **14**, 1945–1957.

CHAPTER 251

Developmental Signaling: JNK Pathway in *Drosophila* Morphogenesis

Beth E. Stronach[1] and Norbert Perrimon[2]

[1]*Department of Biological Sciences, University of Pittsburgh, Pittsburgh, Pennsylvania*
[2]*Department of Genetics, Harvard Medical School, Howard Hughes Medical Institute, Boston, Massachusetts*

Introduction

For a cell to be able to distinguish and respond appropriately to extracellular signals, intracellular information transfer must be precise and reliable yet versatile. One key mode of signal transduction is the use of sequential phosphorylation events mediated by modules of protein kinases. The mitogen-activated protein kinase (MAPK) cascade is a highly conserved linear module that threads information from the cell surface to the nucleus to modulate transcription factor activity and thus gene expression [1]. Several related but functionally distinct MAPK modules have evolved that transduce a variety of extracellular signals. Genetic and molecular studies in *Drosophila melanogaster* over the last several years have revealed that components of stress signaling MAPK pathways, namely the Jun NH2-terminal kinase (JNK) and p38 kinase signaling modules, are functionally conserved and participate in numerous processes during normal development [2].

Loss-of-function genetic studies using Drosophila have identified the JNK signal transduction cascade as one of the key modulators of morphogenetic events. The *Drosophila* model system is providing a number of new insights into the general organization of the JNK pathway, the identification of new components and their epistatic relationships, and the various biological roles of the signaling activity. In the context of normal development, JNK signaling has been linked to the movement and fusion of epidermal cell sheets during dorsal and thorax closure, and to the formation of epithelial tubes during oogenesis [3]. Besides the use of JNK signaling for developmental events, accumulating evidence suggests that activation of the JNK pathway can be induced by stress and injury, as it is in mammalian systems.

The Paradigm of JNK Signaling: Dorsal Closure

In a process that has been compared to mammalian wound healing, dorsal closure of the fly embryo occurs when hundreds of ectodermal cells undergo coordinated cell elongation and movement to close over the dorsal side of the embryo. [4]. The opening is covered initially by a squamous epithelial tissue called the amnioserosa. Coverage of the area occupied by the amnioserosa is achieved by contraction of amnioserosa cells concomitant with dramatic cell elongation within the ectoderm [5,6]. Upon completion of dorsal closure, the opposing ectodermal epithelia meet and fuse at the dorsal midline; this fusion serves to internalize the amnioserosa and to surround the embryo in a continuous protective epidermis [7].

JNK signaling is precisely regulated throughout dorsal closure. Prior to initiation of closure, JNK signaling is transiently active in a broad dorsal domain; this activity requires the product of the *u-shaped* (*ush*) gene, a protein related to mammalian FOG (friend of GATA), which appears to be permissive for JNK activation [8–10]. JNK signaling then becomes restricted to the ectoderm by the negative action of *hnt*, which downregulates JNK signaling specifically in the amnioserosa [8,9]. Thus, although the nature of the signal that activates the JNK pathway is still unknown, expression of transcriptional targets of the pathway reveal a spatial refinement of activation, by the start of dorsal closure, to the leading row of migrating cells at the epithelial front. There, JNK pathway activation mediates changes in gene expression, accumulation of a contractile band of actin and myosin, and movement of the epidermis toward the dorsal midline [5,11–13]. Loss of or failure to activate JNK signaling results in defective cell shape changes, failed closure, and lethality [14].

Figure 1 shows a current model of JNK signaling in dorsal closure. One of the molecules that is thought to be activated early in the JNK pathway is the small GTPase, Rac1 [15]. Genetic epistasis studies in the fly are consistent with the proposed biochemical mechanism: Rac1, together with a JNKKKK, *msn*, activates the JNKKK *slpr*, a mixed lineage kinase, through direct interaction with the Rac-binding CRIB domain [16–19]. MLK catalytic activity is stimulated, which by analogy, leads to phosphorylation of Hep (JNKK) and then Bsk (JNK) [19–22]. JNK substrates include, but are not limited to, dJun, a component of the jun/fos AP-1 transcription factor [23]. As a result of AP-1 activation, transcriptional targets of JNK signaling are expressed.

Among those targets are Dpp, a secreted morphogen related to the BMP/TGFβ family of signaling molecules, and Puc, a dual specificity phosphatase that provides crucial negative feedback regulation of JNK signaling via dephosphorylation of the JNK, Bsk [12,13,15,24,25]. The precise role of *dpp* expressed in leading cells is not well understood; however, Dpp signaling may control the speed and direction of leading edge cell motility through the GTPase, Cdc42 [3,26].

Given that loss-of-function mutations in any of the component kinases leads to a loss of target gene expression and a failure of dorsal closure, it has been suggested that JNK signaling regulates differentiation and function of the leading cells in which it is activated [9,27]. The special properties of these leading cells, where JNK signaling is highest, are essential for the cell biological mechanics that underlie the morphogenetic movements of dorsal closure. These properties include: (1) the assembly of a contractile actomyosin cable that provides tension for the epithelial front as it moves forward, (2) assembly of focal complexes presumed to regulate adhesion and possibly signal transduction, and (3) protrusion of actin-rich filopodia that may sense the correct segmental position of the approaching contralateral epithelium [5–8].

Until recently, it was not clear if the cytoskeletal changes observed in the leading ectoderm were controlled by JNK signaling directly or through a transcriptional mechanism.

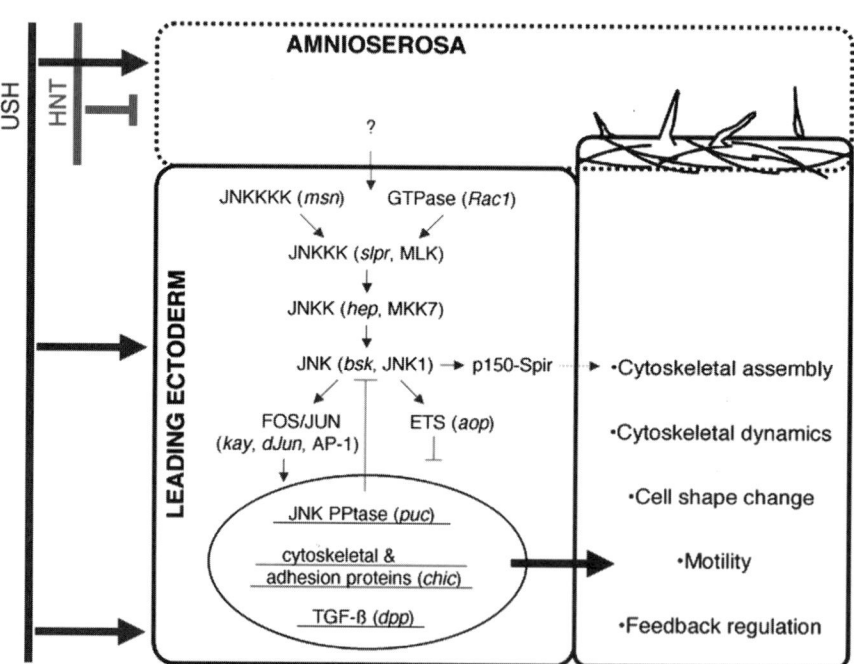

Figure 1 A model of JNK signaling during tissue morphogenesis in *Drosophila*. Prior to dorsal closure, JNK signaling is active in both the amnioserosa and the dorsal ectoderm, requiring the function of U-shaped (Ush), related to mammalian Friend of GATA (FOG) proteins. Hindsight (Hnt), a nuclear Zn-finger protein, participates in restricting JNK signaling to the leading cells of the ectoderm by repressing signaling specifically in the amnioserosa. In the leading ectoderm, JNK signaling is mediated by a kinase cascade, which results in activation of the FOS/JUN (AP-1) transcription factor and regulation of gene expression. The effectors of JNK signaling (both direct and via transcription) execute functions critical for tissue morphogenesis including cytoskeletal deployment and modulation, i.e., assembly of a contractile actomyosin network and extension of cell protrusions.

The expression of the Dpp morphogen in leading cells led to the speculation that Dpp signaling could mediate the majority of the downstream cellular events [13]. Transcript analysis by the serial analysis of gene expression (SAGE) method in embryos with either constitutive-active or blocked JNK signaling has provided compelling evidence that coordinated regulation of many genes encoding actin-binding proteins and cell adhesion molecules is essential for the execution of dorsal closure [28]. These data show that cytoskeletal reorganization driving morphogenetic movements can be mediated through JNK-dependent changes in gene expression. In addition though, biochemical methodology revealed a direct interaction between activated JNK and p150-Spir, a protein that contains an actin-binding motif with similarity to known cytoskeletal regulators of the WASP/WAVE family [29]. Thus, in some cases, the ability of JNK signaling to regulate cell shape change and motility may be direct, without a requirement for changes in transcription.

Thorax Closure

Adult tissues of the fly are generated from precursor cells (called imaginal cells or imaginal discs) that are set aside during embryogenesis and undergo proliferation and eventual large-scale morphogenetic movements during metamorphosis. Genetic analysis of semilethal *hep* (JNKK) and *kay* (dFos) zygotic mutant animals revealed an important role for JNK signaling in the spreading and fusion of sets of matched, laterally-positioned disks as they come together to meet dorsally [30,31]. In a scenario that parallels embryonic dorsal closure, Hep and dFos regulate Puc phosphatase expression in cells that adopt a position at the leading margin of the spreading discs. Ectopic puc expression, induced under conditions where imaginal cells are programmed to express an activated form of the Cdc42 GTPase, is also *hep*-dependent, suggesting that in this system Cdc42 may be the relevant upstream activator during disc morphogenesis [30]. In contrast to dorsal closure, *dpp* is not under transcriptional control by the JNK pathway, even though *dpp* and *puc* are co-expressed in a subset of disc cells, notably those at the leading margin [30,31].

Follicle Cell Morphogenesis

The somatic follicle cells that surround the germline consist of several distinct populations that, over the course of oogenesis, take on characteristic morphologies and migration patterns. The JNK signaling pathway has been implicated, through genetic mosaic analysis, in regulating a late-stage morphogenetic process whereby the respiratory appendages of the eggshell are built up from two populations of dorsal anterior follicle cells [32–34]. Dorsal appendage morphogenesis is impaired when mutant cell clones of *slpr*, *hep*, or *bsk* are induced prior to the convergence of the appendage primordia [34] (B. E. Stronach, unpublished). Short, thin appendages result from a failure of the clustered pre-appendage cells to fully elongate. Overexpression of the negative JNK regulator, Puc phosphatase, phenocopies the JNK loss-of-function effect on dorsal appendage outgrowth [34]. Together, these data suggest a model similar to that for the role of JNK signaling in dorsal closure; in both processes maintenance of the levels of JNK activity are critical to achieve and maintain proper epithelial cell shape change and migration [33]. Like thorax closure, but in contrast to dorsal closure, *dpp* is not under transcriptional control by Jun in follicle cells. During oogenesis Dpp does not appear to mediate the late morphogenetic events of dorsal appendage formation; in fact, Dpp acts before JNK signaling to regulate the A/P position of the appendage primordia [34].

A New Paradigm: Planar Cell Polarity

Epithelial planar polarity refers to the uniform orientation of cells or cell groups within the plane of an epithelium, observable in the adult fly as oriented bristles on the thorax, hairs on the wings, and facets of the compound eye [35]. Mutations in genes required for the development of planar polarity, such as the *frizzled* (*fz*) receptor, result in random orientation of bristles, hairs, or ommatidia [36]. Genetic mosaics and dominant interaction studies have uncovered the backbone of a signaling pathway required for establishment of planar polarity; this pathway shares several components with, yet is distinct from, the canonical Wg/Wnt pathway [37,38]. The emerging model places the JNK cascade downstream of Fz1 receptor activation. JNK activity is thought to be stimulated through a pathway that includes Disheveled (Dsh), a PDZ-containing adaptor molecule, Misshapen (Msn), related to the yeast Ste20 kinase, and the RhoA GTPase [3,35]. This linkage of Fz/Dsh signaling to JNK activation is also being investigated as a general mechanism underlying convergent extension in vertebrate embryos [39,40].

Cellular Stress Response and Wound Healing

To date, genetic analysis of the JNK pathway in *Drosophila* has revealed developmental requirements throughout the life cycle of the fly from embryogenesis to metamorphosis. Given the widespread use of the JNK signaling pathway in mammals in response to environmental and systemic stresses, such as an inflammatory response, it is important to ask if the *Drosophila* JNK pathway is similarly inducible by stress or injury. It has been suggested that the mechanics of dorsal closure movements resemble, to a large extent, the process of tissue repair during mammalian wound healing [30,41,42]. Recent investigations have probed the requirement for JNK pathway proteins during the response of flies to wounding or the response of fly cells to toxic stresses [43–45]. These studies reveal that *puc*, a transcriptional marker for activation of the JNK pathway, becomes upregulated at the site of tissue injury in the adult fly in a *Dfos*-dependent manner [44]. Moreover, the temporal pattern of *puc* expression is correlated with cell shape changes and movement of the wound margins to close the injury. In contrast to the noted requirement for Dpp in many of the processes that require JNK, adult tissue wounds do not appear to induce *dpp* marker expression, suggesting a Dpp-independent mechanism of healing. Whether one of the other TGFß-related ligands is upregulated upon injury remains an open question. The JNK-activating signal resulting from injury also remains elusive. Although the molecular identities of JNK-activating signals for numerous JNK-dependent processes are largely unknown, there must be a variety of developmental, inducible, and environmental signals, since markers of JNK signaling activity can be observed under various conditions, e.g., tissue closure, injury, infection, and chemical stress. Future work aims at identifying the relevant signals.

Perspectives

Most of what we have learned from studying JNK signaling in *Drosophila* comes from the dorsal closure paradigm. Further studies on the function of the pathway in other processes (polarity, stress signaling, immunity) are likely to reveal different routes of JNK activation and insights on signaling. Characterization of JNK signaling in various morphogenetic processes throughout development reveals several common themes. Cumulative evidence supports the notion that JNK signaling can modulate cytoskeletal organization and cell adhesion [46]. This would provide a basis to explain the observation that JNK signaling promotes the differentiation of margin cells to establish a boundary between columnar and squamous epithelia or to define the properties of a "free" motile edge [30]. Cooperation between the JNK pathway and the Dpp pathway is a predominant, but not obligate, theme in various morphogenetic events of *Drosophila* epithelia. The role of Dpp, whether or not its expression is transcriptionally coupled with JNK activity, seems to be involved in modulating the dynamic nature of cytoskeletal assemblages beneath the cell surface and in cell projections [46]. The uncoupling of JNK signaling and *dpp* expression in several instances, however, raises the interesting possibility that there may be tissue-specific regulators that act in combination with JNK-inducible factors to specify unique morphogenetic outcomes in a tissue- or context-dependent manner.

References

1. Chang, L. and Karin, M. (2001). Mammalian MAP kinase signalling cascades. *Nature* **410**, 37–40.
2. Stronach, B. E. and Perrimon, N. (1999). Stress signaling in Drosophila. *Oncogene* **18**, 6172–6182.
3. Noselli, S. and Agnes, F. (1999). Roles of the JNK signaling pathway in Drosophila morphogenesis. *Curr. Opin. Genet. Dev.* **9**, 466–472.
4. Jacinto, A. and Martin, P. (2001). Morphogenesis: Unravelling the cell biology of hole closure. *Curr. Biol.* **11**, R705–707.
5. Kiehart, D. P. et al. (2000). Multiple forces contribute to cell sheet morphogenesis for dorsal closure in Drosophila. *J. Cell Biol.* **149**, 471–490.
6. Young, P. E. et al. (1993). Morphogenesis in Drosophila requires nonmuscle myosin heavy chain function. *Genes Dev.* **7**, 29–41.
7. Jacinto, A. et al. (2000). Dynamic actin-based epithelial adhesion and cell matching during Drosophila dorsal closure. *Curr. Biol.* **10**, 1420–1426.
8. Reed, B. H., Wilk, R., and Lipshitz, H. D. (2001). Downregulation of Jun kinase signaling in the amnioserosa is essential for dorsal closure of the Drosophila embryo. *Curr. Biol.* **11**, 1098–1108.
9. Stronach, B. E. and Perrimon, N. (2001). Investigation of leading edge formation at the interface of amnioserosa and dorsal ectoderm in the Drosophila embryo. *Development* **128**, 2905–2913.
10. Cubadda, Y. et al. (1997). U-shaped encodes a zinc finger protein that regulates the proneural genes achaete and scute during the formation of bristles in Drosophila. *Genes Dev.* **11**, 3083–3095.
11. Glise, B., Bourbon, H., and Noselli, S. (1995). Hemipterous encodes a novel Drosophila MAP kinase kinase, required for epithelial cell sheet movement. *Cell* **83**, 451–461.
12. Hou, X. S., Goldstein, E. S., and Perrimon, N. (1997). Drosophila Jun relays the Jun amino-terminal kinase signal transduction pathway to the Decapentaplegic signal transduction pathway in regulating epithelial cell sheet movement. *Genes Dev.* **11**, 1728–1737.
13. Riesgo-Escovar, J. R. and Hafen, E. (1997). Drosophila Jun kinase regulates expression of decapentaplegic via the ETS-domain protein Aop and the AP-1 transcription factor DJun during dorsal closure. *Genes Dev.* **11**, 1717–1727.
14. Noselli, S. (1998). JNK signaling and morphogenesis in Drosophila. *Trends Genet.* **14**, 33–38.
15. Glise, B. and Noselli, S. (1997). Coupling of Jun amino-terminal kinase and Decapentaplegic signaling pathways in Drosophila morphogenesis. *Genes Dev.* **11**, 1738–1747.
16. Bock, B. C. et al. (2000). Cdc42-induced activation of the mixed-lineage kinase SPRK in vivo. Requirement of the Cdc42/Rac interactive binding motif and changes in phosphorylation. *J. Biol. Chem.* **275**, 14231–14241.
17. Burbelo, P. D., Drechsel, D., and Hall, A. (1995). A conserved binding motif defines numerous candidate target proteins for both Cdc42 and Rac GTPases. *J. Biol. Chem.* **270**, 29071–29074.
18. Teramoto, H. et al. (1996). Signaling from the small GTP-binding proteins Rac1 and Cdc42 to the c-Jun N-terminal kinase/stress-activated protein kinase pathway. A role for mixed lineage kinase 3/protein-tyrosine kinase 1, a novel member of the mixed lineage kinase family. *J. Biol. Chem.* **271**, 27225–27228.
19. Stronach, B. and Perrimon, N. (2002). Activation of the JNK pathway during dorsal closure in Drosophila requires the mixed lineage kinase, slipper. *Genes Dev.* **16**, 377–387.
20. Hirai, S. et al. (1997). MST/MLK2, a member of the mixed lineage kinase family, directly phosphorylates and activates SEK1, an activator of c-Jun N-terminal kinase/stress-activated protein kinase. *J. Biol. Chem.* **272**, 15167–15173.
21. Rana, A. et al. (1996). The mixed lineage kinase SPRK phosphorylates and activates the stress-activated protein kinase activator, SEK-1. *J. Biol. Chem.* **271**, 19025–19028.
22. Tibbles, L. A. et al. (1996). MLK-3 activates the SAPK/JNK and p38/RK pathways via SEK1 and MKK3/6. *EMBO J.* **15**, 7026–7035.
23. Kockel, L., Homsy, J. G., and Bohmann, D. (2001). Drosophila AP-1: lessons from an invertebrate. *Oncogene* **20**, 2347–2364.
24. Kockel, L. et al. (1997). Jun in Drosophila development: Redundant and nonredundant functions and regulation by two MAPK signal transduction pathways. *Genes Dev.* **11**, 1748–1758.
25. Martin-Blanco, E. et al. (1998). Puckered encodes a phosphatase that mediates a feedback loop regulating JNK activity during dorsal closure in Drosophila. *Genes Dev.* **12**, 557–570.
26. Ricos, M. G. et al. (1999). Dcdc42 acts in TGF-beta signaling during Drosophila morphogenesis: distinct roles for the Drac1/JNK and Dcdc42/TGF-beta cascades in cytoskeletal regulation. *J. Cell Sci.* **112**, 1225–1235.

27. Ring, J. M. and Martinez Arias, A. (1993). puckered, a gene involved in position-specific cell differentiation in the dorsal epidermis of the Drosophila larva. *Dev. Suppl.* 251–259.
28. Jasper, H. *et al.* (2001). The genomic response of the Drosophila embryo to JNK signaling. *Dev. Cell* **1**, 579–586.
29. Otto, I. M. *et al.* (2000). The p150-Spir protein provides a link between c-Jun N-terminal kinase function and actin reorganization. *Curr. Biol.* **10**, 345–348.
30. Agnes, F., Suzanne, M., and Noselli, S. (1999). The Drosophila JNK pathway controls the morphogenesis of imaginal discs during metamorphosis. *Development* **126**, 5453–5462.
31. Zeitlinger, J. and Bohmann, D. (1999). Thorax closure in Drosophila: involvement of Fos and the JNK pathway. *Development* **126**, 3947–3956.
32. Dequier, E. *et al.* (2001). Top-DER- and Dpp-dependent requirements for the Drosophila fos/kayak gene in follicular epithelium morphogenesis. *Mech. Dev.* **106**, 47–60.
33. Dobens, L. L. *et al.* (2001). Drosophila puckered regulates Fos/Jun levels during follicle cell morphogenesis. *Development* **128**, 1845–1856.
34. Suzanne, M., Perrimon, N., and Noselli, S. (2001). The Drosophila JNK pathway controls the morphogenesis of the egg dorsal appendages and micropyle. *Dev. Biol.* **237**, 282–294.
35. Mlodzik, M. (1999). Planar polarity in the Drosophila eye: a multifaceted view of signaling specificity and cross-talk. *EMBO J.* **18**, 6873–6879.
36. Adler, P. N. and Lee, H. (2001). Frizzled signaling and cell-cell interactions in planar polarity. *Curr. Opin. Cell Biol.* **13**, 635–640.
37. Axelrod, J. D. *et al.* (1998). Differential recruitment of Dishevelled provides signaling specificity in the planar cell polarity and Wingless signaling pathways. *Genes Dev.* **12**, 2610–2622.
38. Boutros, M. *et al.* (1998). Dishevelled activates JNK and discriminates between JNK pathways in planar polarity and wingless signaling. *Cell* **94**, 109–118.
39. McEwen, D. G. and Peifer, M. (2000). Wnt signaling: Moving in a new direction. *Curr. Biol.* **10**, R562–564.
40. Sokol, S. (2000). A role for Wnts in morphogenesis and tissue polarity. *Nat. Cell Biol.* **2**, E124–125.
41. Jacinto, A., Martinez-Arias, A., and Martin, P. (2001). Mechanisms of epithelial fusion and repair. *Nat. Cell Biol.* **3**, E117–123.
42. Kiehart, D. P. (1999). Wound healing: The power of the purse string. *Curr. Biol.* **9**, R602–605.
43. Botella, J. A. *et al.* (2001). The Drosophila cell shape regulator c-Jun N-terminal kinase also functions as a stress-activated protein kinase. *Insect Biochem. Mol. Biol.* **31**, 839–847.
44. Ramet, M. *et al.* (2002). JNK signaling pathway is required for efficient wound healing in Drosophila. *Dev. Biol.* **241**, 145–156.
45. Sluss, H. K. *et al.* (1996). A JNK signal transduction pathway that mediates morphogenesis and an immune response in Drosophila. *Genes Dev.* **10**, 2745–2758.
46. Martin-Blanco, E., Pastor-Pareja, J. C., and Garcia-Bellido, A. (2000). JNK and decapentaplegic signaling control adhesiveness and cytoskeleton dynamics during thorax closure in Drosophila. *Proc. Natl. Acad. Sci. USA* **97**, 7888–7893.

CHAPTER 252

Wnt Signaling in Development

Christian Wehrle, Heiko Lickert, and Rolf Kemler
Department of Molecular Embryology,
Max-Planck Institute of Immunobiology,
Freiburg, Germany

Introduction

Development is controlled by an interplay of many signaling molecules. Among these, Wnt proteins comprise a large family of secreted proteins with key roles in intercellular signaling throughout the animal kingdom. This chapter is dedicated to reviewing present knowledge about Wnt signaling in invertebrate and vertebrate development, beginning with a brief overview of the components of the Wnt signaling pathway. More details about this signaling cascade can be obtained from the Wnt homepage (www.stanford,edu/~rnusse/) and the references therein. Wnt genes were discovered from two different areas of research. In 1982, Nusse and Varmus [1] identified a proto-oncogene, called *Int-1*, by mouse mammary tumor virus integration into it in mammary tumors. The molecular cloning of the *Drosophila* segment polarity gene *wingless* (*wg*) revealed that *Int-1* and *wg* are orthologs [2], and both terms (W+nt) were combined in a common nomenclature of Wnt genes. By now, roughly 100 Wnt genes have been isolated in different species ranging from hydra to man. Information collected from genetic analysis in *Drosophila* and *Caenorhabditis elegans*, cell and molecular studies in mammalian cells, and functional overexpression experiments in *Xenopus*, as well as loss-of-function analysis in mice, has provided a detailed picture about the molecular components which transduce Wnt signals. In general, there is a high complexity of positive and negative regulation of Wnt signal transduction in each cellular compartment. At the cell membrane, binding of Wnts to their cognate receptors, Frizzled (Fz) proteins, is modulated by co-receptors or inhibitors. In the cytoplasm, Wnt signals can diverge into different molecular cascades, e.g., the best known canonical Wnt/β-catenin pathway, the planar polarity pathway, and the Wnt/Ca^{2+} pathway. The canonical Wnt/β-catenin signaling pathway provides by far the most comprehensive picture of Wnt signaling. Here the central component is β-catenin, which upon Wnt signaling enters the nucleus and functions as a transcriptional co-activator with members of the TCF/LEF-1 family of transcription factors. Wnt signaling results in the activation of specific target genes, which again involves an interplay of repressors (groucho, Sox) and activators (p300, TBP) and the coordinated action of other signaling pathways, e.g., TGFβ signaling (regulation of Twin in *Xenopus*), or retinoic acid receptor stimulation (regulation of cdx1 in mouse). The activation of Wnt/β-catenin target genes controls cell proliferation and cell fate decisions and regulates various morphogenetic events during the development of various organisms.

Wnt Signaling in Invertebrate Development

The *Drosophila wg* gene was discovered in the classical genetic screen for zygotic lethal mutations affecting the larval cuticle pattern [3]. Genome sequencing revealed a total of seven *wg* genes in the genome. The *wg* gene belongs to the class of segment polarity genes, which mediate intercellular interactions regulating parasegmental periodicity and cell fates in the larval cuticle [4]. Additionally, the adult appendages—legs and wings—are formed from sheets of epithelial cells in imaginal discs in which the dorsal-ventral (DV) and anterior-posterior (AP) axes are defined as the disc grows during larval development [5,6]. The AP and DV patterning of the imaginal discs is directed through organizing centers and morphogen gradients of the secreted signaling proteins Wingless, Hedgehog, and Decapentaplegic (reviewed in [7]).

Studies in *C. elegans* have shown that Wnt signaling acts very early in nematode development (at the four-cell stage) to induce endoderm formation. This is brought about by relieving transcriptional repression and partially by aligning the mitotic spindle in a responding cell along the AP body axis [reviewed in 8]. The outcome is the polarization of the bipotential precursor cell (EMS) producing one daughter that makes endoderm and one for mesoderm [9,10]. In the absence of the polarizing Wnt signal both daughters form mesoderm, a phenotype represented by seven different mutants, *mom-1-5* (for more mesoderm), *lit-1* (for loss of intestine), and *pop-1* (for posterior pharynx-defective). Cloning and sequencing of these genes revealed that they code for components of the Wnt signaling cascade [reviewed in 8]. There are 5 Wnt genes in *Caenorhabditis elegans*. In the invertebrate phylogenetic tree components of the Wnt pathway have been described in the diploblastic Cnidaria (e.g., Hydra) [11], and even in the slime mold *Dictyostelium* homologs to β-catenin (Aardvark) and GSK3β have been found (reviewed in [12]).

Wnt Signaling in Vertebrate Development

Our knowledge about the Wnt signaling cascade in vertebrate development began with pioneer work in *Xenopus laevis*. Injection of Wnt1 mRNA into ventral blastomeres of early *Xenopus* embryos led to duplication of the body axis, and this was subsequently used as a read-out to explore the details of this signaling cascade [13]. The siamese twin tadpoles obtained after ventral injection of mRNA for Wnt/β-catenin pathway members were strongly reminiscent of the experimental animals Hans Spemann and Hilde Mangold [14] had produced by transplanting the dorsal blastopore lip to the ventral side of a host newt embryo. It became apparent that the Spemann-Mangold organizer, which when transplanted had this remarkable ability to induce a secondary body axis, is induced by target genes of the Wnt/β-catenin signaling cascade (reviewed in [15]). This underlines the importance of the Wnt/β-catenin signaling cascade in the induction of the primary body axis. The endogenous Wnt molecule inducing the primary body axis is still not identified, but at least 16 Wnt genes have been described in *Xenopus*. Also in chicken and zebra fish the Wnt genes apparently have similar activities during early development, with 11 and 12 identified genes, respectively.

Conventional and conditional gene knockouts in the mouse have demonstrated the important function of the Wnt/β-catenin signaling cascade in higher vertebrate development in inducing the primary body axis and directing cell fate. Both Wnt3 and β-catenin have been functionally inactivated, leading to the arrest of the mutant embryos in gastrulation due to the inability to form a primitive streak and hence mesoderm and endoderm. Evidence that the Wnt/β-catenin signaling pathway is directly involved in Spemann-Mangold organizer formation in the mouse, here called the node, came from both gain- and loss-of-function experiments [16,17]. More recently, conditional gene inactivation of β-catenin supported the existence of a bipotential mesendodermal precursor cell in the mouse embryo similar to the EMS cell in *C. elegans* embryos [18]. In addition to these Wnt functions in early development, knock-out studies of different Wnt family members have revealed their involvement in organ formation, mid- and hindbrain patterning, and limb development [19–23]. Genome sequences from mouse and human revealed a total of 19 Wnt genes in each species.

Wnt/β-Catenin Target Genes

Based on the mechanisms that lead to gene activation, one can distinguish between direct and indirect target genes. Direct target genes are defined as genes with TCF/LEF-1 binding motifs in the gene regulatory regions, where the functional importance of these motifs in β-catenin-mediated gene activation has been demonstrated. The transcriptional activation of such direct target genes does not require *de novo* protein synthesis. In contrast, indirect target genes are not regulated by TCF/β-catenin complexes, and their activation depends on other factors and requires *de novo* protein synthesis.

Because the unphysiological activation of the Wnt/β-catenin pathway leads to tumor formation, much effort has been made to identify target genes of this pathway in tumorigenesis (reviewed in [24]). Expression profile analysis of cells with an activated Wnt/β-catenin pathway led to identifying direct target genes involved in cell cycle control, e.g., c-Myc [25], Cyclin D1 [26,27], transcription factors including Tcf-1 [28], c-jun and fra-1 [29], or cell surface components such as claudin-1 [30] and MMP-7 [31,32].

The first Wnt/β-catenin developmental target genes were identified via a candidate gene approach correlating gene expression temporo spatially with the known action of Wnt/β-catenin in morphogenetic processes. The *Xenopus* genes *siamois* and *Xnr3*, originally cloned in expression screens for dorsalizing genes, turned out to be direct target genes of the Wnt/β-catenin pathway [33–36]. Direct transcriptional regulation was also reported for *X. fibronectin* [37], *connexin43* [38], *twin* [39], and *engrailed-2* [40]. Developmental mutants proved very helpful in identifying direct Wnt/β-catenin target genes, e.g., *nacre* [41] and *bozozok* [42], in zebra fish or *Dpp* [43] and *Ubx* [44] in *Drosophila*. Although valuable, this candidate gene approach has obvious limitations, making a broader search for developmentally regulated target genes desirable.

In this respect the use of ES cells represents an interesting alternative [45]. It was shown that mouse ES cells express all the necessary components to respond to a Wnt signal. Hence, by seeding ES cells on a feeder layer of NIH3T3 fibroblasts expressing different Wnts, it was possible to screen for expression of Wnt/β-catenin target genes (Fig. 1). *T-brachyury* [45] and *Cdx1* [46] are examples of genes found to be directly regulated by the Wnt/β-catenin signaling

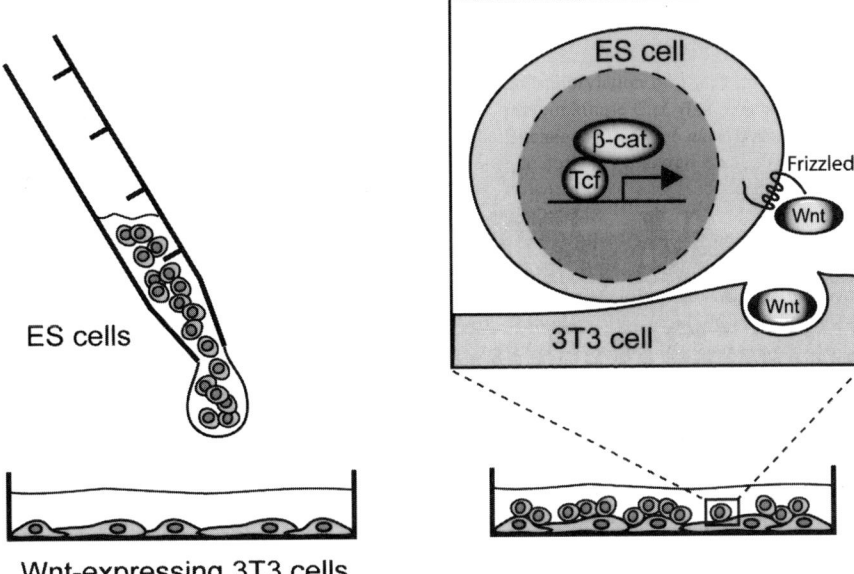

Figure 1 Scheme of the ES cell co-culture system suitable to search for Wnt/β-catenin target genes. (Adapted from Arnold, S. J. et al., *Mech. Dev.*, **91**, 249–258, (2000). With permission.) ES cells expressing a Wnt/β-catenin reporter gene (gfp driven by TCF-binding motifs) are seeded on 3T3 cells expressing various Wnts. ES cells receiving a Wnt signal as monitored by gfp expression can be separated and analyzed for Wnt/β-catenin target gene expression.

pathway, using this ES cell co-culture system. The T-box gene *T-brachyury* codes for a transcription factor involved in mesoderm formation [47–49]. Cdx1 is a homeodomain-containing transcription factor with important roles in anterior-posterior patterning [50] and likely in intestinal development [46,51]. The ES cell co-culture system in combination with microarray analysis should allow screening systematically for a large number of potential Wnt/β-catenin target gene candidates.

In conclusion, the Wnt/β-catenin signaling pathway is evolutionarily highly conserved on a biochemical level and regulates many morphogenetic processes during development. From the direct target genes found thus far, it is apparent that this pathway regulates gene expression of a large variety of cellular components, i.e., transcription factors, cytoplasmic and cell surface molecules, and growth factors.

Acknowledgments

We thank Rosemary Schneider for typing and Dr. Randy Cassada for critically reading the manuscript. C.W. is a Ph.D. student of the Faculty of Biology, University of Freiburg.

References

1. Nusse, R. and Varmus, H. E. (1982). Many tumors induced by the mouse mammary tumor virus contain a provirus integrated in the same region of the host genome. *Cell* **31**, 99–109.
2. Rijsewijk, F., Schuermann, M., Wagenaar, E., Parren, P., Weigel, D., and Nusse, R. (1987). The Drosophila homolog of the mouse mammary oncogene int-1 is identical to the segment polarity gene wingless. *Cell* **50**, 649–657.
3. Nüsslein-Volhard, C. and Wieschaus, E. (1980). Mutations affecting segment number and polarity in Drosophila. *Nature* **287**, 795–801.
4. Klingensmith, J. and Nusse, R. (1994). Signaling by wingless in Drosophila. *Dev. Biol.* **166**, 396–414.
5. Struhl, G. and Basler, K. (1993). Organizing activity of wingless protein in Drosophila. *Cell* **72**, 527–540.
6. Zecca, M., Basler, K., and Struhl, G. (1995). Sequential organizing activities of engrailed, hedgehog and decapentaplegic in the Drosophila wing. *Development* **121**, 2265–2278.
7. Strigini, M. and Cohen, S. M. (1999). Formation of morphogen gradients in the Drosophila wing. *Semin. Cell Dev. Biol.* **10**, 335–344.
8. Thorpe, C. J., Schlesinger, A., and Bowerman, B. (2000). Wnt signalling in *Caenorhabditis elegans*: Regulating repressors and polarizing the cytoskeleton. *Trends Cell Biol.* **10**, 7–10.
9. Goldstein, B. (1992). Induction of gut in *Caenorhabditis elegans* embryos. *Nature* **357**, 255–257.
10. Goldstein, B., Hird, S. N., and White, J. G. (1993). Cell polarity in early C. elegans development. *Dev.* (Suppl.), 279–287.
11. Hobmayer, B., Rentzsch, F., Kuhn, K., Happel, C. M., von Laue, C. C., Snyder, P., Rothbacher, U., and Holstein, T. W. (2000). WNT signalling molecules act in axis formation in the diploblastic metazoan Hydra. *Nature* **407**, 186–189.
12. Weeks, G. (2000). Signalling molecules involved in cellular differentiation during *Dictyostelium* morphogenesis. *Curr. Opin. Microbiol.* **3**, 625–630.
13. McMahon, A. P. and Moon, R. T. (1989). Ectopic expression of the proto-oncogene int-1 in *Xenopus* embryos leads to duplication of the embryonic axis. *Cell* **58**, 1075–1084.
14. Spemann, H. and Mangold, H. (1924). Über die Induktion von Embryonalanlagen durch Implantation artfremder Organisatoren. *Wilhelm Roux's Arch. Dev. Biol.* **100**, 599–638.
15. Moon, R. T., Brown, J. D., and Torres, M. (1997). WNTs modulate cell fate and behaviour during vertebrate development. *Trends Genet.* **13**, 157–162.
16. Popperl, H., Schmidt, C., Wilson, V., Hume, C. R., Dodd, J., Krumlauf, R., and Beddington, R. S. (1997). Misexpression of Cwnt8C

in the mouse induces an ectopic embryonic axis and causes a truncation of the anterior neuroectoderm. *Development* **124**, 2997–3005.

17. Zeng, L., Fagotto, F., Zhang, T., Hsu, W., Vasicek, T. J., Perry, W. L., III, Lee, J. J., Tilghman, S. M., Gumbiner, B. M., and Costantini, F. (1997). The mouse fused locus encodes axin, an inhibitor of the Wnt signaling pathway that regulates embryonic axis formation. *Cell* **90**, 181–192.

18. Lickert, H., Kutsch, S., Kanzler, B., Tamai, Y., Taketo, M. M., and Kemler, R. (2002). Formation of multiple hearts. *Dev. Cell* **3**, 171–181.

19. Stark, K., Vainio, S., Vassileva, G., and McMahon, A. P. (1994). Epithelial transformation of metanephric mesenchyme in the developing kidney regulated by Wnt-4. *Nature* **372**, 679–683.

20. McMahon, A. P. and Bradley, A. (1990). The Wnt-1 (int-1) proto-oncogene is required for development of a large region of the mouse brain. *Cell* **62**, 1073–1085.

21. Thomas, K. R. and Capecchi, M. R. (1990). Targeted disruption of the murine int-1 proto-oncogene resulting in severe abnormalities in midbrain and cerebellar development. *Nature* **346**, 847–850.

22. Brault, V., Moore, R., Kutsch, S., Ishibashi, M., Rowitch, D. H., McMahon, A. P., Sommer, L., Boussadia, O., and Kemler, R. (2001). Inactivation of the β-catenin gene by Wnt1-Cre-mediated deletion results in dramatic brain malformation and failure of craniofacial development. *Development* **128**, 1253–1264.

23. Parr, B. A. and McMahon, A. P. (1995). Dorsalizing signal Wnt-7a required for normal polarity of D-V and A-P axes of mouse limb. *Nature* **374**, 350–353.

24. Bienz, M. and Clevers, H. (2000). Linking colorectal cancer to Wnt signaling. *Cell* **103**, 311–320.

25. He, T. C., Sparks, A. B., Rago, C., Hermeking, H., Zawel, L., da Costa, L. T., Morin, P. J., Vogelstein, B., and Kinzler, K. W. (1998). Identification of c-MYC as a target of the APC pathway. *Science* **281**, 1509–1512.

26. Tetsu, O. and McCormick, F. (1999). β-catenin regulates expression of cyclin D1 in colon carcinoma cells. *Nature* **398**, 422–426.

27. Shtutman, M., Zhurinsky, J., Simcha, I., Albanese, C., D'Amico, M., Pestell, R., and Ben-Ze'ev, A. (1999). The cyclin D1 gene is a target of the β-catenin/LEF-1 pathway. *Proc. Natl. Acad. Sci. USA* **96**, 5522–5527.

28. Roose, J., Huls, G., van Beest, M., Moerer, P., van der Horn, K., Goldschmeding, R., Logtenberg, T., and Clevers, H. (1999). Synergy between tumor suppressor APC and the β-catenin-Tcf4 target Tcf1. *Science* **285**, 1923–1926.

29. Mann, B., Gelos, M., Siedow, A., Hanski, M. L., Gratchev, A., Ilyas, M., Bodmer, W. F., Moyer, M. P., Riecken, E. O., Buhr, H. J., and Hanski, C. (1999). Target genes of β-catenin-T cell-factor/lymphoid-enhancer-factor signaling in human colorectal carcinomas. *Proc. Natl. Acad. Sci. USA* **96**, 1603–1608.

30. Miwa, N., Furuse, M., Tsukita, S., Niikawa, N., Nakamura, Y., and Furukawa, Y. (2001). Involvement of claudin-1 in the beta-catenin/Tcf signaling pathway and its frequent upregulation in human colorectal cancers. *Oncol. Res.* **12**, 469–476.

31. Brabletz, T., Jung, A., Dag, S., Hlubek, F., and Kirchner, T. (1999). β-catenin regulates the expression of the matrix metalloproteinase-7 in human colorectal cancer. *Am. J. Pathol.* **155**, 1033–1038.

32. Crawford, H. C., Fingleton, B. M., Rudolph-Owen, L. A., Goss, K. J., Rubinfeld, B., Polakis, P., and Matrisian, L. M. (1999). The metalloproteinase matrilysin is a target of β-catenin transactivation in intestinal tumors. *Oncogene* **18**, 2883–2891.

33. Lemaire, P., Garrett, N., and Gurdon, J. B. (1995). Expression cloning of Siamois, a *Xenopus* homeobox gene expressed in dorsal-vegetal cells of blastulae and able to induce a complete secondary axis. *Cell* **81**, 85–94.

34. Smith, W. C., McKendry, R., Ribisi, S., Jr., and Harland, R. M. (1995). A nodal-related gene defines a physical and functional domain within the Spemann organizer. *Cell* **82**, 37–46.

35. Brannon, M., Gomperts, M., Sumoy, L., Moon, R. T., and Kimelman, D. (1997). A β-catenin/XTcf-3 complex binds to the siamois promoter to regulate dorsal axis specification in *Xenopus*. *Genes Dev.* **11**, 2359–2370.

36. McKendry, R., Hsu, S. C., Harland, R. M., and Grosschedl, R. (1997). LEF-1/TCF proteins mediate wnt-inducible transcription from the *Xenopus* nodal-related 3 promoter. *Dev. Biol.* **192**, 420–431.

37. Gradl, D., Kuhl, M., and Wedlich, D. (1999). The Wnt/Wg signal transducer β-catenin controls fibronectin expression. *Mol. Cell. Biol.* **19**, 5576–5587.

38. Van der Heyden, M. A., Rook, M. B., Hermans, M. M., Rijksen, G., Boonstra, J., Defize, L. H., and Destree, O. H. (1998). Identification of connexin43 as a functional target for Wnt signalling. *J. Cell Sci.* **111**, 1741–1749.

39. Laurent, M. N., Blitz, I. L., Hashimoto, C., Rothbacher, U., and Cho, K. W. (1997). The *Xenopus* homeobox gene twin mediates Wnt induction of goosecoid in establishment of Spemann's organizer. *Development* **124**, 4905–4916.

40. McGrew, L. L., Takemaru, K., Bates, R., and Moon, R. T. (1999). Direct regulation of the *Xenopus* engrailed-2 promoter by the Wnt signaling pathway, and a molecular screen for Wnt-responsive genes confirm a role for Wnt signaling during neural patterning in *Xenopus*. *Mech. Dev.* **87**, 21–32.

41. Dorsky, R. I., Raible, D. W., and Moon, R. T. (2000). Direct regulation of nacre, a zebrafish MITF homolog required for pigment cell formation, by the Wnt pathway. *Genes Dev.* **14**, 158–162.

42. Ryu, S. L., Fujii, R., Yamanaka, Y., Shimizu, T., Yabe, T., Hirata, T., Hibi, M., and Hirano, T. (2001). Regulation of dharma/bozozok by the Wnt pathway. *Dev. Biol.* **231**, 397–409.

43. Yang, X., van Beest, M., Clevers, H., Jones, T., Hursh, D. A., and Mortin, M. A. (2000). Decapentaplegic is a direct target of dTcf repression in the *Drosophila* visceral mesoderm. *Development* **127**, 3695–3702.

44. Riese, J., Yu, X., Munnerlyn, A., Eresh, S., Hsu, S. C., Grosschedl, R., and Bienz, M. (1997). LEF-1, a nuclear factor coordinating signaling inputs from wingless and decapentaplegic. *Cell* **88**, 777–787.

45. Arnold, S. J., Stappert, J., Bauer, A., Kispert, A., Herrmann, B. G., and Kemler, R. (2000). Brachyury is a target gene of the Wnt/β-catenin signaling pathway. *Mech. Dev.* **91**, 249–258.

46. Lickert, H., Domon, C., Huls, G., Wehrle, C., Duluc, I., Clevers, H., Meyer, B. I., Freund, J. N., and Kemler, R. (2000). Wnt/(beta)-catenin signaling regulates the expression of the homeobox gene Cdx1 in embryonic intestine. *Development* **127**, 3805–3813.

47. Wilkinson, D. G., Bhatt, S., and Herrmann, B. G. (1990). Expression pattern of the mouse T gene and its role in mesoderm formation. *Nature* **343**, 657–659.

48. Kispert, A., Koschorz, B., and Herrmann, B. G. (1995). The T protein encoded by Brachyury is a tissue-specific transcription factor. *EMBO J.* **14**, 4763–4772.

49. Yamaguchi, T. P., Takada, S., Yoshikawa, Y., Wu, N., and McMahon, A. P. (1999). T (Brachyury) is a direct target of Wnt3a during paraxial mesoderm specification. *Genes Dev.* **13**, 3185–3190.

50. Subramanian, V., Meyer, B. I., and Gruss, P. (1995). Disruption of the murine homeobox gene Cdx1 affects axial skeletal identities by altering the mesodermal expression domains of Hox genes. *Cell* **83**, 641–653.

51. Duprey, P., Chowdhury, K., Dressler, G. R., Balling, R., Simon, D., Guenet, J. L., and Gruss, P. (1988). A mouse gene homologous to the *Drosophila* gene caudal is expressed in epithelial cells from the embryonic intestine. *Genes Dev.* **2**, 1647–1654.

Hedgehog Signaling and Embryonic Development

Mark Merchant, Weilan Ye, and Frederic de Sauvage

*Department of Molecular Biology, Genentech, Inc.,
South San Francisco, California*

The Hedgehog Proteins: Generation and Distribution

The hedgehog mutation was originally identified in a *Drosophila* genetic screen as one of the segment-polarity genes important in fly development [1]. Cloning of the hedgehog gene (*hh*) revealed a secreted peptide whose expression is confined to Engrailed expressing cells and is dependent upon Wingless expression. It is now clear that Hh plays a vital role in the development of multiple organ systems in the fly and vertebrates [2]. In mammals there are three Hh proteins named Sonic Hh (Shh), Desert Hh (Dhh), and Indian Hh (Ihh). Dhh appears to be most closely related to *Drosophila* Hh, while Shh and Ihh are more closely related to one another. Production and diffusion of these factors in different tissues determines proper development of multiple organ systems.

Hh proteins are synthesized as precursor proteins of about 45 kDa. The C-terminal portion of the Hh precursor has autoproteolytic activity and cleaves Hh into a C-terminal peptide of about 25 kDa with no known function and an N-terminal fragment (Hh-N), which constitutes the biologically active portion of Hh [3]. During autoprocessing, a cholesterol moiety is coupled to the C-terminus of Hh-N, a form which is further denoted Hh-Np [4]. It is thought that the addition of cholesterol helps to retain Hh-Np to cell membranes, thus limiting the range of action of Hh activity. However, in mice engineered to express a form of Shh lacking cholesterol modification (N-Shh), short-range Hh signaling was maintained while long-range signaling was defective, resulting in loss of digits and proper patterning in the developing limb and suggesting differential requirements for cholesterol in Hh signaling [5]. Additional proteins involved in the secretion and diffusion of cholesterol-modified forms of Hh have been identified, such as Dispatched (Disp), which is required for release of Hh-Np from Hh-producing cells, and Tout velu (TTV), which is involved in the biosynthesis of a putative Hh-interacting proteoglycan [6–8].

Hh proteins are further modified by palmitoylation on a highly conserved N-terminal cysteine residue [9]. Mutation of the *sightless/skinny Hh* (*sig/ski*) gene, encoding a transmembrane (TM) acyl transferase, abrogates palmitoylation of Hh-N and results in an Hh-like phenotype, indicating that palmitoylation of Hh is required for some aspect of Hh function [10,11]. In some systems both modified and unmodified forms of Hh show equivalence, indicating that the importance of fatty acid modification may be context dependent [9,12]. Together these data indicate that the biological activity of the Hh proteins is finely tuned through posttranslational modification, affecting its activity and capacity to diffuse.

Transmitting the Hh Signal

Both *Drosophila* and mouse genetics indicate that the transmembrane protein, Smoothened (Smo), is required to transmit the Hh signal [13–15], and another TM protein, Patched (Ptc), negatively regulates Smo in the absence of Hh [16,17]. While it is widely accepted that Hh binds to Ptc [18,19], it is not clear how Hh binding results in downstream Smo activation and signal transduction.

In the conventional model, Ptc binds directly to Smo and represses its activity (Fig. 1A). Upon Hh binding the normal inhibition by Ptc is released and Smo initiates signaling (Fig. 1B). However, recent studies in *Drosophila* have

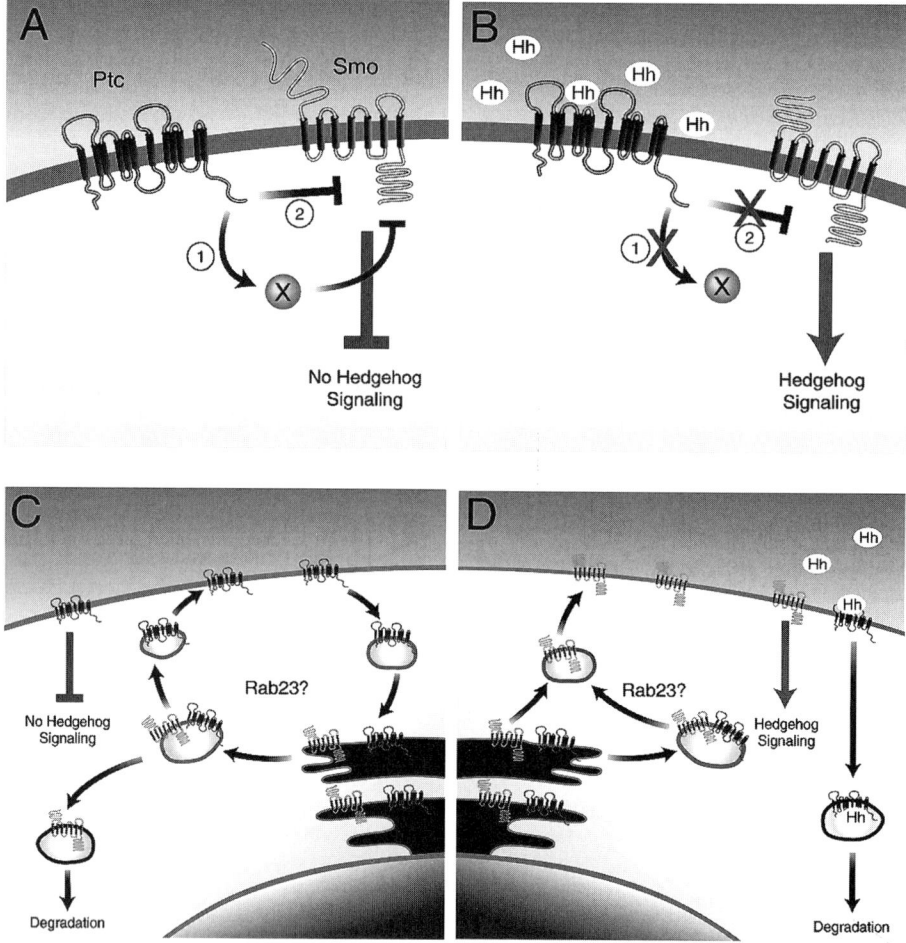

Figure 1 Two models for transmitting the Hh signal through Patched and Smoothened. (**A**) In the conventional model Patched (Ptc; red) inhibits Smoothened (Smo; green) function through (1) direct inhibition of Smo function or (2) an indirect effect, where Ptc acts upon an unknown factor (X) that subsequently inhibits Smo activity. (**B**) Upon Hedgehog (Hh; yellow) stimulation, the repressive function of Ptc is abrogated allowing for activation of Smo and transmission of the Hh signal to downstream targets. (**C**) In a more unconventional model, Ptc (red) acts to repress Smo vesicular trafficking. In the absence of Hh, Ptc and Smo are generated in the endoplasmic reticulum (ER) and are processed through the Golgi. Ptc traffics to and from the cell surface, while Smo may traffic to an unknown subcellular compartment, possibly resulting in Smo degradation, ultimately resulting in a lack of Hh signaling. Rab23 may act to coordinate some aspect of Ptc or Smo trafficking. (**D**) Upon Hh signaling, Ptc binds to Hh stimulation, the Ptc-Hh complex is internalized via endocytosis resulting in the degradation of Ptc. This allows for the transportation of Smo to the cell surface where it can mediate Hh signaling.

suggested that Ptc may not repress Smo activity through a direct interaction, but rather that Ptc inhibits Smo activity from a distance [20–24]. For example, regulation of Smo activity may occur through control of vesicular trafficking by Ptc. SCAP, a protein involved in cholesterol metabolism and trafficking, regulates the activity of SREBP by controlling its movement between the endoplasmic reticulum (ER) and Golgi via a mechanism involving use of a SSD in SCAP [25,26]. Ptc may regulate the activity of its cargo, Smo, in a similar fashion (Fig. 1C and D). Mutation within the Ptc SSD results in the abrogation of Ptc-mediated inhibition of Smo activity [27,28]. Further support for the regulation of Smo transport has come from the *open brain* (*opb*) mutation in the mouse, recently identified as the *Rab23* gene [29], a GTP-activating protein involved in vesiculotubular transport. A model involving the negative regulation of Smo movement to the cell surface by Rab23 has been hypothesized, however, it is also possible that Rab23 aids in the proper localization and function of Ptc.

The main target of Hh activity is a family of zinc finger transcription factors known as cubitus interruptus (Ci) in the fly and Gli in vertebrates. Gli activity is regulated at multiple levels, including nuclear export, proteosome-mediated degradation, and subcellular localization. In vertebrates there are three Ci orthologs; Gli1, Gli2, and Gli3, which may have retained aspects of Ci-155 and Ci-75 function (reviewed in references [2] and [30]). Knockout or transgenic mice of each Gli isoform have been generated and the

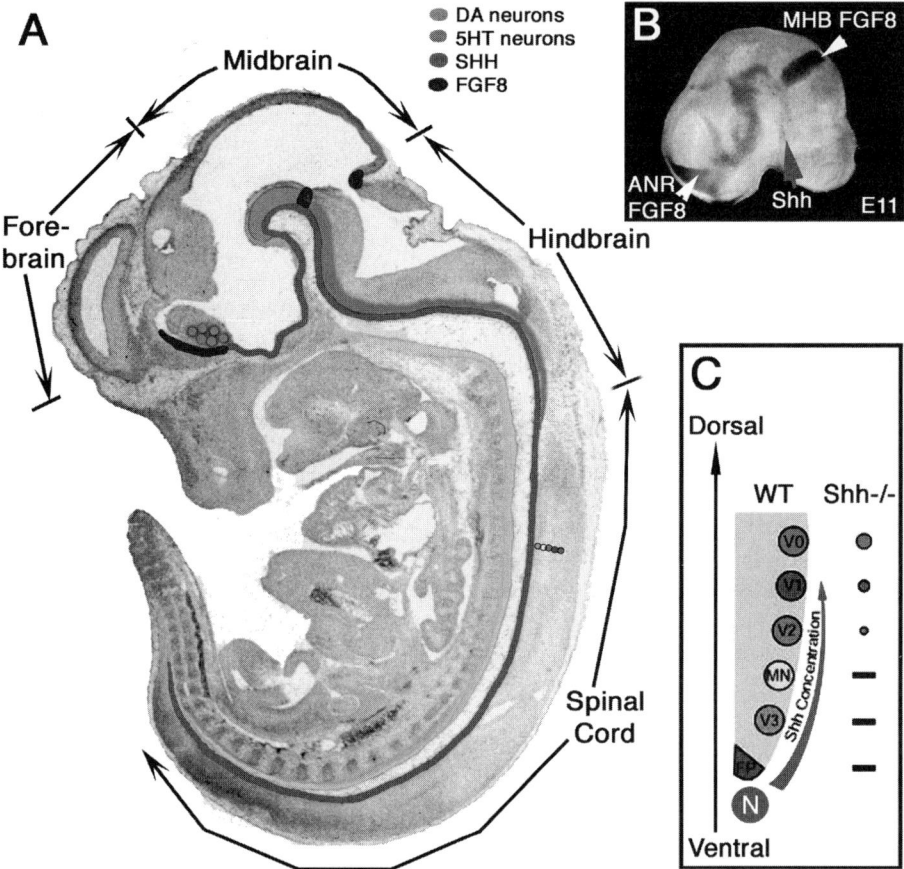

Figure 2 Shh in the developing nervous system and the cell types controlled by SHH. (**A**) Shh (red), FGF8 (dark blue), dopaminergic (DA) neurons (light blue), serotonergic (5HT) neurons (green), and six ventral spinal cord neurons (colored dots with black circles) are diagramed on a sagittal section of a 12-day-old mouse embryo. Shh in the notocord and floor plate is represented by a single line. (**B**) Whole mount *in situ* hybridization of Shh (red) and FGF8 (dark blue) in the brain of an 11-day-old embryo. Shh is expressed in the ventral midline (red arrow), FGF8 is expressed in two organizers—the mid-hindbrain junction (MHB) and the anterior neural ridge (ANR) (white arrow heads). (**C**) The position of six Shh-dependent cell types (color-filled circles) in the ventral spinal cord of wild-type (WT) and Shh mutant embryos (Shh$^{-/-}$), and their relationship to SHH concentration (red arrow). Different cell types are induced by different threshold concentrations of SHH. Higher concentration of SHH induces cells in progressively more ventral locations. FP = floor plate, N = notochord, MN = motor neuron, and V0, V1, V2, and V3 are four classes of interneurons.

phenotypes observed support the idea that Gli2 and Gli3 are critical for normal development, while Gli1 may be redundant for Gli2 and Gli3 function. The interplay between the activities of these isoforms increases the complexity of vertebrate systems dramatically making it difficult to conclusively delineate all of the functions for the different Gli proteins during normal development (reviewed in reference [2] and in Section III of this book).

Hh in Development and Disease

Studies of the normal functions for the Hhs in animal models have helped in our understanding of Hh-related diseases. Many studies have shown that Shh acts as a morphogen in the nervous system, where it is secreted from the notochord and later from the floor plate, patterning neurons along the dorsal-ventral axis of the neural tube in a dose-dependent manner [2] (Fig. 2). In the spinal cord and part of the hindbrain, a fine gradient of SHH with two- to three-fold incremental changes from the source (floor plate and notocord) delineates the ventral neural tube into six distinct domains along the dorsal-ventral axis. The expression of a set of homeodomain and bHLH transcription factors is tightly controlled by different SHH protein concentrations, thus generating six intricate combinatorial transcription factor codes in these domains [31,32]. These codes specify the identities of neural progenitors, which ultimately give rise to six cell types, including the floor plate (FP) cells, motor neurons (MN), and four classes of interneurons, V0, V1, V2, V3[33] (Fig. 2A and C). In the Shh$^{-/-}$ mouse embryos, FP cells, MNs, and V3 neurons are missing, and the numbers of V0, V1, and V2 neurons are greatly reduced [34,35] (Fig. 2C). In the brains of mouse embryos, SHH acts in concert with another organizer molecule FGF8 to create information grids, which serve as spatial cues for the specification of ventral cell types. Ventral neurons such as dopaminergic (DA) neurons and serotonergic (5HT) neurons are specified in locations where Shh and FGF8 intersect (Fig. 2A and B) [36,37]. Earlier patterning events dictate how neural stem cells respond to these organizer signals, so that DA neurons are induced by SHH and FGF8 in the fore- and midbrain, whereas 5HT neurons are induced by the same set of signals in the hindbrain, because the hindbrain progenitors have been prepatterned by another signal(s) from the posterior [36]. Shh controls cell fates not only by induction, but also by repression. It is found that SHH in the forebrain is required to repress Pax6 expression, thus resulting in the separation of a single eyefield into two retinal primordia [38,39]. This explains the cyclopic phenotype in the Shh-deficient embryos [34]. Late in development, Shh elicits other cellular responses than cell fate specification. For example, in the fetal cerebella, SHH induces proliferation of granule cells while inhibiting their differentiation [40–42].

Shh also acts to determine anterior-posterior (AP) patterning in the developing skeleton, limb bud, and gut tube [2,30]. Shh has recently been shown to act as an angiogenic factor leading to neovascularization and the proliferation of blood cells [43]. In some tissue types, such as the pancreas, Shh acts both as a positive and negative regulator, as its activity is needed for inhibition of pancreatic anlagen formation, but is also needed for specification of the pancreatic β cells [44,45]. Shh is also involved in the morphology of branching structures such as the lung [46]. The importance of Shh is highlighted by the phenotype of Shh knockout mice, which die at birth due to multiple defects, including cyclopia and holoprosencephaly (HPE), as well as other defects in limbs, brain, spinal cord, axial skeleton, and midline structures [34]. Overlapping roles of Shh and Ihh have been identified in heart development as well as specification of left/right (L/R) asymmetry, as observed in Shh/Ihh double knockout mice and Smo knockout mice [47]. The major impact of Ihh is in radial patterning of the gut as well as bone morphogenesis [48,49]. Loss of Ihh function results in a lack of chondrocyte proliferation and differentiation [49]. Dhh is predominantly involved in peripheral nerve sheath and germ cell development, particularly in the development of the male germline and maturation of the testes [50,51].

The patterning functions of Hh are also highlighted in humans where mutations have clearly been linked to developmental disorders, including spina bifida, neural tube defects, and skeletal deformations. For example, mutations in human Shh result in cyclopia and HPE [52]. Downstream of Hh, mutations within Gli3 are found in Grieg's cephalopolysyndactyly [53] and Pallister-Hall syndrome (PHS) [30,54], disorders associated with various abnormalities including polydactyly.

The first indication that the Hh pathway might be involved in tumor formation stemmed from the observation that human Ptc (PTCH) gene was mutated in individuals with Gorlin's syndrome (also known as basal-cell nevus syndrome; BCNS) [55], a familial inherited predisposition to the development of basal cell carcinomas (BCCs), medulloblastomas, and rhabdomyosarcomas [56,57]. Many sporadic BCCs have been associated with mutation within Ptc [58]. In all cases loss of Ptc function leads to constitutive Smo signaling and Hh pathway activation. In addition, point mutations in Smo, which result in constitutively active Ptc-insensitive forms of Smo, have been identified in sporadic BCC [59]. However, most of these tumors show upregulation of the Hh target genes, suggesting that other components of the Hh pathway may be mutated. One such example was identified in Smo where a point mutation, known as the Smo-M2 mutation (Smo-M2), results in a constitutively active form of Smo [59].

Interestingly, treatment of Hh-responsive cells with the teratogenic steroidal alkaloid cyclopamine blocks Hh signaling and Smo activation in cell lines and induces HPE in animal models [60,61]. Treatment of some glioma and medulloblastoma cell lines *in vitro* with cyclopamine has resulted in successful inhibition of cell growth, indicating that cyclopamine or other drugs that specifically block the Hh pathway may be effective in the treatment of Hh-associated tumors [62,63].

Acknowledgments

The authors would like to thank J. Peña and A. Bruce for their help in generating figures. The authors also apologize to the many authors whose work was not referenced in this chapter due to space constraints.

References

1. Nusslein-Volhard, C. and Wieschaus, E. (1980). Mutations affecting segment number and polarity in Drosophila. *Nature* **287**, 795–801.
2. Ingham, P. W. and McMahon, A. P. (2001). Hedgehog signaling in animal development: Paradigms and principles. *Genes Dev.* **15**, 3059–3087.
3. Lee, J. J., Ekker, S. C., von Kessler, D. P., Porter, J. A., Sun, B. I., and Beachy, P. A. (1994). Autoproteolysis in hedgehog protein biogenesis. *Science* **266**, 1528–37.
4. Porter, J. A., Young, K. E., and Beachy, P. A. (1996). Cholesterol modification of hedgehog signaling proteins in animal development. *Science* **274**, 255–9.
5. Lewis, P. M., Dunn, M. P., McMahon, J. A., Logan, M., Martin, J. F., St. Jacques, B., and McMahon, A. P. (2001). Cholesterol modification of sonic hedgehog is required for long-range signaling activity and effective modulation of signaling by Ptc1. *Cell* **105**, 599–612.
6. Bellaiche, Y., The, I., and Perrimon, N. (1998). Tout-velu is a *Drosophila* homologue of the putative tumour suppressor EXT-1 and is needed for Hh diffusion. *Nature* **394**, 85–8.
7. Thé, I., Bellaiche, Y., and Perrimon, N. (1999). Hedgehog movement is regulated through tout velu-dependent synthesis of a heparin sulfate proteoglycan. *Mol. Cell* **4**, 633–639.
8. Burke, R., Nellen, D., Bellotto, M., Hafen, E., Senti, K. A., Dickson, B. J., and Basler, K. (1999). Dispatched, a novel sterol-sensing domain protein dedicated to the release of cholesterol-modified hedgehog from signaling cells. *Cell* **99**, 803–815.
9. Pepinsky, R. B., Zeng, C., Wen, D., Rayhorn, P., Baker, D. P., Williams, K. P., Bixler, S. A., Ambrose, C. M., Garber, E. A., Miatkowski, K., Taylor, F. R., Wang, E. A., and Galdes, A. (1998). Identification of a palmitic acid-modified form of human Sonic hedgehog. *J. Biol. Chem.* **273**, 14037–14045.

10. Chamoun, Z., Mann, R. K., Nellen, D., von Kessler, D. P., Bellotto, M., Beachy, P. A., and Basler, K. (2001). Skinny hedgehog, an acyl transferase required for palmitoylation and activity of the hedgehog signal. *Science* **293**, 2080–4.
11. Amanai, K. and Jiang, J. (2001). Distinct roles of Central missing and Dispatched in sending the Hedgehog signal. *Dev.– Suppl.* **128**, 5119–5127.
12. Kohtz, J., Lee, H., Gaiano, N., Segal, J., Ng, E., Larson, T., Baker, D., Garber, E., Williams, K., and Fishell, G. (2001). N-terminal fatty-acylation of sonic hedgehog enhances the induction of rodent ventral forebrain neurons. *Development* **128**, 2351–2363.
13. Kalderon, D. (2000). Transducing the hedgehog signal (minireview). *Cell* **103**, 371–374.
14. Alcedo, J., Ayzenzon, M., Von Ohlen, T., Noll, M., and Hooper, J. E. (1996). The *Drosophila* smoothened gene encodes a seven-pass membrane protein, a putative receptor for the hedgehog signal. *Cell* **86**, 221–232.
15. van den Heuvel, M. and Ingham, P. W. (1996). Smoothened encodes a receptor-like serpentine protein required for hedgehog signalling. *Nature* **382**, 547–551.
16. Ingham, P. W., Taylor, A. M., and Nakano, Y. (1991). Role of the *Drosophila* patched gene in positional signalling. *Nature* **353**, 184–7.
17. Chen, Y. and Struhl, G. (1998). *In vivo* evidence that Patched and Smoothened constitute distinct binding and transducing components of a Hedgehog receptor complex. *Development* **125**, 4943–4948.
18. Marigo, V., Davey, R. A., Zuo, Y., Cunningham, J. M., and Tabin, C. J. (1996). Biochemical evidence that patched is the Hedgehog receptor. *Nature* **384**, 176–9.
19. Stone, D. M., Hynes, M., Armanini, M., Swanson, T. A., Gu, Q., Johnson, R. L., Scott, M. P., Pennica, D., Goddard, A., Phillips, H., Noll, M., Hooper, J. E., de Sauvage, F., and Rosenthal, A. (1996). The tumour-suppressor gene patched encodes a candidate receptor for Sonic hedgehog. *Nature* **384**, 129–34.
20. Chen, Y. and Struhl, G. (1996). Dual roles for patched in sequestering and transducing Hedgehog. *Cell* **87**, 553–63.
21. Johnson, R., Milenkovic, L., and Scott, M. P. (2000). *In vivo* functions of the Patched protein: Requirement of the C terminus for target gene inactivation but not Hedgehog sequestration. *Mol. Cell* **6**, 467–478.
22. Denef, N., Neubuser, D., Perez, L., and Cohen, S. M. (2000). Hedgehog induces opposite changes in turnover and subcellular localization of patched and smoothened. *Cell* **102**, 521–531.
23. Ingham, P. W., Nystedt, S., Nakano, Y., Brown, W., Stark, D., van den Heuvel, M., and Taylor, A. M. (2000). Patched represses the Hedgehog signalling pathway by promoting modification of the Smoothened protein. *Curr. Biol.* **10**, 1315–1318.
24. Alcedo, J., Zou, Y., and Noll, M. (2000). Posttranscriptional regulation of smoothened is part of a self-correcting mechanism in the Hedgehog signaling system. *Mol. Cell* **6**, 457–465.
25. DeBose-Boyd, R. A., Brown, M. S., Li, W. P., Nohturfft, A., Goldstein, J. L., and Espenshade, P. J. (1999). Transport-dependent proteolysis of SREBP: Relocation of site-1 protease from Golgi to ER obviates the need for SREBP transport to Golgi. *Cell* **99**, 703–712.
26. Nohturfft, A., DeBose-Boyd, R., Scheek, S., Goldstein, J., and Brown, M. (1999). Sterols regulate cycling of SREBP cleavage-activating protein (SCAP) between endoplasmic reticulum and Golgi. *Proc. Natl. Acad. Sci. USA* **96**, 11235–11240.
27. Martin, V., Carrillo, G., Torroja, C., and Guerrero, I. (2001). The sterol-sensing domain of Patched protein seems to control Smoothened activity through Patched vesicular trafficking. *Curr. Biol.* **11**, 601–607.
28. Strutt, H., Thomas, C., Nakano, Y., Stark, D., Neave, B., Taylor, A. M., and Ingham, P. W. (2001). Mutations in the sterol-sensing domain of Patched suggest a role for vesicular trafficking in Smoothened regulation. *Curr. Biol.* **11**, 608–613.
29. Eggenschwiler, J., Espinoza, E., and Anderson, K. (2001). Rab23 is an essential negative regulator of the mouse Sonic hedgehog signalling pathway. *Nature* **412**, 194–198.
30. Theil, T., Kaesler, S., Grotewold, L., Bose, J., and Ruther, U. (1999). Gli genes and limb development. *Cell Tissue Res.* **296**, 75–83.
31. Briscoe, J., Pierani, A., Jessell, T. M., and Ericson, J. (2000). A homeodomain protein code specifies progenitor cell identity and neuronal fate in the ventral neural tube. *Cell* **101**, 435–45.
32. Zhou, Q. and Anderson, D. J. (2002). The bHLH transcription factors OLIG2 and OLIG1 couple neuronal and glial subtype specification. *Cell* **109**, 61–73.
33. Stone, D. and Rosenthal, A. (2000). Achieving neuronal patterning by repression. *Nat. Neurosci.* **3**, 967–969.
34. Chiang, C., Litingtung, Y., Lee, E., Young, K. E., Corden, J. L., Westphal, H., and Beachy, P. A. (1996). Cyclopia and defective axial patterning in mice lacking Sonic hedgehog gene function. *Nature* **383**, 407–413.
35. Litingtung, Y. and Chiang, C. (2000). Specification of ventral neuron types is mediated by an antagonistic interaction between Shh and Gli3. *Nat. Neurosci.* **3**, 979–985.
36. Ye, W., Shimamura, K., Rubenstein, J., Hynes, M., and Rosenthal, A. (1998). FGF and Shh signals control dopaminergic and serotonergic cell fate in the anterior neural plate. *Cell* **93**, 755–766.
37. Matise, M. P., Epstein, D. J., Park, H. L., Platt, K. A., and Joyner, A. L. (1998). Gli2 is required for induction of floor plate and adjacent cells, but not most ventral neurons in the mouse central nervous system. *Development* **125**, 2759–2770.
38. Macdonald, R., Barth, K. A., Xu, Q., Holder, N., Mikkola, I., and Wilson, S. W. (1995). Midline signalling is required for Pax gene regulation and patterning of the eyes. *Development* **121**, 3267–3278.
39. Hallonet, M., Hollemann, T., Pieler, T., and Gruss, P. (1999). Vax1, a novel homeobox-containing gene, directs development of the basal forebrain and visual system. *Genes Dev.* **13**, 3106–3114.
40. Dahmane, N. and Ruiz y Altaba, A. (1999). Sonic hedgehog regulates the growth and patterning of the cerebellum. *Development* **126**, 3089–3100.
41. Wallace, V. (1999). Purkinje-cell-derived Sonic hedgehog regulates granule neuron precursor cell proliferation in the developing mouse cerebellum. *Curr. Biol.* **9**, 445–448.
42. Wechsler-Reya, R. J. and Scott, M. P. (1999). Control of neuronal precursor proliferation in the cerebellum by Sonic Hedgehog. *Neuron* **22**, 103–14.
43. Pola, R., Ling, L. E., Silver, M., Corbley, M. J., Kearney, M., Blake Pepinsky, R., Shapiro, R., Taylor, F. R., Baker, D. P., Asahara, T., and Isner, J. M. (2001). The morphogen Sonic hedgehog is an indirect angiogenic agent upregulating two families of angiogenic growth factors. *Nat. Med.* **7**, 706–711.
44. Hebrok, M., Kim, S., St. Jacques, B., McMahon, A. P., and Melton, D. (2000). Regulation of pancreas development by hedgehog signaling. *Development* **127**, 4905–4913.
45. diIorio, P. J., Moss, J. B., Sbrogna, J. L., Karlstrom, R. O., and Moss, L. G. (2002). Sonic hedgehog is required early in pancreatic islet development. *Dev. Biol.* **244**, 75–84.
46. Bellusci, S., Furuta, Y., Rush, M., Henderson, R., Winnier, G., and Hogan, B. (1997). Involvement of Sonic hedgehog (Shh) in mouse embryonic lung growth and morphogenesis. *Dev.– Suppl.* **124**, 53–63.
47. Zhang, X. M., Ramalho-Santos, M., and McMahon, A. P. (2001). Smoothened mutants reveal redundant roles for Shh and Ihh signaling including regulation of L/R asymmetry by the mouse node. *Cell* **105**, 781–792.
48. Lanske, B., Karaplis, A. C., Lee, K., Luz, A., Vortkamp, A., Pirro, A., Karperien, M., Defize, L. H. K., Ho, C., Mulligan, R. C., Abou-Samra, A. B., Juppner, H., Segre, G. V., and Kronenberg, H. M. (1996). PTH/PTHrP receptor in early development and Indian hedgehog-regulated bone growth. *Science* **273**, 663–666.
49. St. Jacques, B., Hammerschmidt, M., and McMahon, A. P. (1999). Indian hedgehog signaling regulates proliferation and differentiation of chondrocytes and is essential for bone formation. *Genes Dev.* **13**, 2072–2086.
50. Bitgood, M. J., Shen, L., and McMahon, A. P. (1996). Sertoli cell signaling by Desert hedgehog regulates the male germline. *Curr. Biol.* **6**, 298–304.

51. Yao, H. H., Whoriskey, W., and Capel, B. (2002). Desert Hedgehog/Patched 1 signaling specifies fetal Leydig cell fate in testis organogenesis. *Genes Dev.* **16**, 1433–1440.
52. Belloni, E., Muenke, M., Roessler, E., Traverso, G., Siegel-Bartelt, J., Frumkin, A., Mitchell, H. F., Donis-Keller, H., Helms, C., Hing, A. V., Heng, H. H., Koop, B., Martindale, D., Rommens, J. M., Tsui, L. C., and Scherer, S. W. (1996). Identification of Sonic hedgehog as a candidate gene responsible for holoprosencephaly. *Nat. Genet.* **14**, 353–356.
53. Hui, C. C. and Joyner, A. L. (1993). A mouse model of Greig cephalopolysyndactyly syndrome: The extra toes mutation contains an intragenic deletion of the Gli3 gene. *Nat. Genet.* **3**, 241–246.
54. Shin, S. H., Kogerman, P., Lindstrom, E., Toftgard, R., and Biesecker, L. G. (1999). GLI3 mutations in human disorders mimic *Drosophila* cubitus interruptus protein functions and localization. *Proc. Natl. Acad. Sci. USA* **96**, 2880–2884.
55. Gorlin, R. J. (1995). Nevoid basal cell carcinoma syndrome. *Dermatol. Clin.* **13**, 113–125.
56. Johnson, R. L., Rothman, A. L., Xie, J., Goodrich, L. V., Bare, J. W., Bonifas, J. M., Quinn, A. G., Myers, R. M., Cox, D. R., Epstein, E. H., Jr., and Scott, M. P. (1996). Human homolog of patched, a candidate gene for the basal cell nevus syndrome. *Science* **272**, 1668–71.
57. Hahn, H., Wicking, C., Zaphiropoulous, P. G., Gailani, M. R., Shanley, S., Chidambaram, A., Vorechovsky, I., Holmberg, E., Unden, A. B., Gillies, S., Negus, K., Smyth, I., Pressman, C., Leffell, D. J., Gerrard, B., Goldstein, A. M., Dean, M., Toftgard, R., Chenevix-Trench, G., Wainwright, B., and Bale, A. E. (1996). Mutations of the human homolog of *Drosophila* patched in the nevoid basal cell carcinoma syndrome. *Cell* **85**, 841–51.
58. Xie, J., Johnson, R. L., Zhang, X., Bare, J. W., Waldman, F. M., Cogen, P. H., Menon, A. G., Warren, R. S., Chen, L. C., Scott, M. P., and Epstein, E. H., Jr. (1997). Mutations of the PATCHED gene in several types of sporadic extracutaneous tumors. *Cancer Res.* **57**, 2369–2372.
59. Xie, J., Murone, M., Luoh, S. M., Ryan, A., Gu, Q., Zhang, C., Bonifas, J. M., Lam, C. W., Hynes, M., Goddard, A., Rosenthal, A., Epstein, E. H., Jr., and de Sauvage, F. J. (1998). Activating Smoothened mutations in sporadic basal-cell carcinoma. *Nature* **391**, 90–92.
60. Cooper, M. K., Porter, J. A., Young, K. E., and Beachy, P. A. (1998). Teratogen-mediated inhibition of target tissue response to Shh signaling. *Science* **280**, 1603–1607.
61. Incardona, J. P., Gaffield, W., Kapur, R. P., and Roelink, H. (1998). The teratogenic Veratrum alkaloid cyclopamine inhibits sonic hedgehog signal transduction. *Development* **125**, 3553–3562.
62. Taipale, J., Chen, J. K., Cooper, M. K., Wang, B., Mann, R. K., Milenkovic, L., Scott, M. P., and Beachy, P. A. (2000). Effects of oncogenic mutations in Smoothened and Patched can be reversed by cyclopamine. *Nature* **406**, 1005–1009.
63. Dahmane, N., Sánchez, P., Gitton, Y., Palma, V., Sun, T., Beyna, M., Weiner, H., and Ruiz i Altaba, A. (2001). The Sonic Hedgehog-Gli pathway regulates dorsal brain growth and tumorigenesis. *Development* **128**, 5201–5212.

CHAPTER 254

Control of Left-Right (L/R) Determination in Vertebrates by the Hedgehog Signaling Pathway

Javier Capdevila and Juan Carlos Izpisúa Belmonte

*The Salk Institute for Biological Studies,
Gene Expression Laboratory, La Jolla, California*

Introduction

Although the bodies of most vertebrate animals appear to be almost perfectly symmetrical, profound left-right (L/R) asymmetries exist that are clearly demonstrated by the specific disposition of internal organs (such as the heart to the left side of the chest cavity, or the liver to the right side of the abdomen, just to mention two examples). At another level, there is also asymmetry in the activity of the brain, with specific functions performed by each individual brain hemisphere, resulting in asymmetries in locomotor functions (among which being right- or left-handed is only one well-known manifestation). The normal disposition of organs is called *situs solitus*, and *situs inversus* refers to a complete or near-complete mirror-image reversal of organ disposition. Right or left isomerism usually refers to situations where individual organs are bilaterally symmetric.

Interestingly, this normal (that is, asymmetric) disposition of organs develops in embryos that are initially perfectly symmetrical, at least as far as we can tell from their morphological examination at very early stages of development. Therefore, the problem is to explain how asymmetric animals develop from (apparently) symmetric embryos, and to define the molecular mechanisms that control this process. That L/R development is under genetic control is clear from the existence (in humans as well as in other vertebrates) of congenital syndromes that result in a variety of laterality defects that can severely impair the function of the heart, lungs, liver, and other vital organs.

The problem of L/R specification has enticed biologists for many years, but until very recently a solid molecular entry point that could be used to successfully attack the problem had been missing. The situation changed in 1995, when the first molecular asymmetries were discovered in chick embryos (and later in other vertebrates). This discovery revealed that certain genes were expressed in asymmetric patterns in the developing embryo, and that these asymmetric patterns were somehow translated into asymmetric development of internal organs. Alterations in some of these genes have already been shown to be linked to some of the human syndromes that display laterality defects, further adding to our knowledge of the etiology of L/R defects in humans. Several comprehensive reviews published recently deal with multiple aspects of L/R development from a broader perspective [1–3].

One of the first genes shown to be involved in L/R determination in vertebrate embryos was *Sonic hedgehog* (*Shh*), a member of the *hedgehog* (*hh*) family of genes, which encode secreted factors that play key roles in multiple aspects of embryonic development in both vertebrate and invertebrate embryos. Since general aspects of HH signaling during embryonic development are reviewed by F. de Sauvage elsewhere in this volume, we will focus here only on the aspects of HH signaling relevant to the development of the embryonic L/R axis, especially in chick and mouse embryos. Further discussion on possible species-specific mechanisms can be found in the reviews mentioned above.

The Discovery of the First Molecular Asymmetries in Vertebrate Embryos and the Role of SHH

In 1995, a team of researchers led by Cliff Tabin and Claudio Stern published a seminal paper that brought into focus key aspects of the molecular basis of L/R determination [4]. The paper demonstrated that asymmetric gene expression did indeed exist in the early embryo, and that it played a key instructive role in the development of the L/R axis in vertebrates. Specifically, it was shown that the *Shh* gene was initially expressed in Hensen's node (the organizer of the chick embryo) without any apparent bias, but that shortly after, *Shh* expression became restricted to the left side of the node by an inferred Activin-like activity (most likely Activin ßB) coming from the right side of the node. Activins belong to the transforming growth factor ß (TGFß) super-family of secreted factors, another group of signaling factors that play fundamental roles in the development of vertebrate and invertebrate embryos and during tumorigenesis [5,6]. It was shown that application of purified Activin protein to the left side of the node resulted in repression of *Shh* expression, and that application of purified SHH protein (this time to the right side of the node) repressed transcription of the gene encoding the Activin receptor (*ActRIIa*), which depends on the Activin ßB ligand. Therefore, a mutually exclusive regulatory interaction between two secreted proteins with complementary patterns of expression resulted in the "specification" of a right side of the node (producing Activin ßB protein) and a left side (producing SHH protein; Fig. 1A).

Another gene encoding a TGFß factor, *Nodal*, was shown to be expressed on the left side of the chick node under the control of SHH. *Nodal* also has a second domain of expression in the left lateral plate mesoderm (LPM) of the embryo, including cells that eventually will give rise to the organ primordia (Fig. 1A). Importantly, left-specific expression of *Nodal* has been observed in all types of vertebrates examined thus far, correlating absolutely with the development of normal organ *situs*. For example, when *Nodal* is either completely absent or present on both sides, organ *situs* is randomized. When *Nodal* is present exclusively on the right side (which occurs naturally in the *inv* mutant mouse and in some human patients that suffer from specific ciliary dysfunctions), there is *situs inversus*. Therefore, left-specific expression of the SHH target *Nodal* appears to play an instructive role in determining organ *situs* and constitutes a key feature of L/R development that has been absolutely conserved during the evolution of vertebrates. It is important to point out that the correct (that is, left-specific) expression of *Nodal* is not required for the development of the organs, but only for the establishment of their correct L/R asymmetry. When *Nodal* expression is altered, organs still form, but with an altered L/R axis that can result in malformations incompatible with life.

Further work identified additional factors that interact with SHH during L/R determination (Fig. 1A). On the right side of the node, Activin ßB controls expression of another TGFß factor, bone morphogenetic protein-4 (BMP-4), which in turn induces fibroblast growth factor-4 and -8 (FGF-4, -8). Mutual antagonism between BMP-4 and SHH appears to be necessary and sufficient to maintain asymmetric *Shh* expression in the node [7]. Moreover, SHH was shown to induce *Nodal* expression in the left LPM through the control of Caronte (Car), a secreted factor that binds to BMPs and antagonizes their repressive effect on *Nodal* transcription in the left LPM. On the right side of the embryo, FGFs turn off expression of Car, and thus BMPs are free to signal, which keeps the *Nodal* gene repressed on the right [8–10]. *Nodal* expression on the left side of the embryo controls the expression of genes such as *Pitx2*, which direct asymmetric organogenesis by mechanisms that are still poorly understood (see the reviews mentioned before in references [1–3]).

The control of *Nodal* by the HH signal on the left side of the node is likely to involve the activity of the HH reception complex, composed of the transmembrane proteins Patched1 and Smoothened (see F. de Sauvage's review in Chapter 253). Recent results indicate that additional components of the HH signaling pathway also play key roles in L/R determination. For example, protein kinase A (PKA) had been previously shown to act as an antagonist of HH signaling in vertebrates, most likely by phosphorylating GLI proteins, which are key nuclear effectors of the pathway. However, PKA has been recently described to act as a positive regulator of the SHH target *Nodal* during L/R determination in the chick embryo [11]. Consistent with this, the endogenous PKA inhibitor, PKI is more strongly expressed on the right side of Hensen's node, where it inhibits PKA, thus restricting its *Nodal*-inducing effects to the left side of the node [11,12]. Interestingly, the induction of *Nodal* expression by PKA does not require *Car* activity, and thus PKA appears to control *Nodal* through a pathway which acts in parallel to that operated by SHH and Car. Other components of HH signaling pathways are also likely to operate during L/R determination, but their exact roles have not yet been determined.

Finally, it is important to point out that the exact nature of the event that breaks the initial symmetry of the chick embryo remains unknown, although an intriguing hypothesis has been proposed that stresses the role of gap junction communication in this process [13].

The Role of a Composite HH Signal during L/R Determination in the Mouse

In the context of L/R development in the chick embryo, the role of SHH could be defined as that of a "left determinant." That is, SHH becomes restricted to the left side of Hensen's node, it controls expression of *Nodal* in and around the left side of the node, and (through Car) it also controls the *Nodal/Pitx2* domains in the left LPM. Inhibition of SHH activity (achieved, for instance, by exposing the chick embryo to an anti-SHH blocking antibody), inhibits

CHAPTER 254 Control of Left-Right (L/R) Determination in Vertebrates by the Hedgehog Signaling Pathway

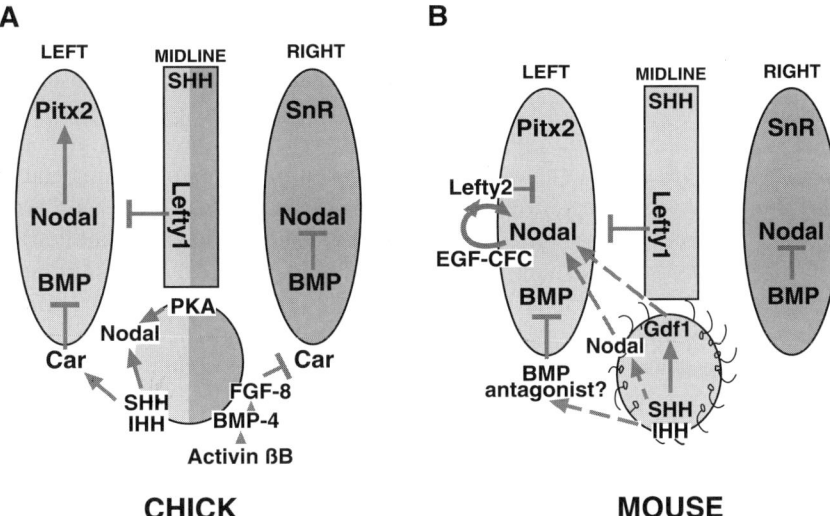

Figure 1 Comparison of the role of HH signaling during L/R determination in chick (A) and mouse embryos (B). At early stages of development of the chick embryo, breaking of the initial embryonic symmetry by mechanisms that are still unknown results in restriction of transcription of *Shh* (and most likely also of *Ihh*) to the left side of Hensen's node (the node is depicted as a circle, and the left side is shown in light gray). On the right side of the node (shown in darker gray), Activin ßB triggers a cascade that activates expression of the *Bmp-4* and *Fgf-8* genes (and also of *Fgf-4*). HH signals display an antagonistic regulatory interaction with both Activin ßB and BMP-4, which contributes to stabilize these side-specific patterns of transcription in the node. In and around the left side of the node, HH signaling mediated by the reception complex formed by the membrane proteins Patched1 (the receptor component) and Smoothened (the transducing component, not shown) results in activation of *Nodal* expression. PKA and other factors may act in parallel to ensure correct *Nodal* expression in the node area. HH signals also operate on the paraxial mesoderm cells adjacent to the node to maintain expression of the *Caronte* (*Car*) gene, which encodes an extracellular antagonist of BMP signaling. The presence of Car protein in the left side of the embryo antagonizes a repressive effect of the BMPs on transcription of the *Nodal* gene in the left LPM (the LPM is depicted as an ellipse), so that Nodal protein is produced and activates its target *Pitx2*. This gene encodes a bicoid-type homeoprotein that plays a key role in the transference of L/R positional information to the organ primordia. In contrast, on the right side of the embryo *Car* expression cannot be maintained, due to the absence of HH signaling and the presence of FGF-4 and/or -8, and thus BMP signals keep the *Nodal* gene repressed in the right LPM. This results in expression of *SnR*, a gene encoding a Snail-related zinc-finger transcription factor that also plays an important role in the determination of organ *situs*, in this case as a "right determinant" [19].

SHH (but not IHH) is also present in the midline (depicted as a rectangle), where it is required for the development of the floor plate, a structure in which the *Lefty1* gene is expressed. *Lefty1* encodes yet another TGFß member that has been proposed to function as a "midline barrier" to the diffusion of left-specific inducing signals from the left to the right side of the embryo (probably through direct interaction with Car and/or Nodal proteins). In cases where development of midline structures is compromised (such as in *Shh*-deficient embryos, for instance), bilateral expression of left-specific genes occurs, most likely due to secondary loss of *Lefty1* expression.

In the early mouse embryo, transcription of the *Shh*, *Ihh*, *Activin ßB*, and *Fgf-8* genes in the node is not restricted to the left side, and a leftward nodal cilia flow (which has been proposed to carry some still unidentified extracellular inducers) results in the transient asymmetrical expression of *Nodal* in the left side of the node. Despite the lack of left-specific restriction of HH signals, deficiency in both SHH and IHH proteins (or, alternatively, deficiency in Smo), completely abolishes *Nodal* and *Pitx2* expression in the left LPM, indicating that HH signaling also acts in the mouse as a left determinant. Interestingly, *Nodal* expression in the node is not abolished, but is nevertheless altered, suggesting that HH signaling does not control *Nodal* directly in the node, but is somehow involved in the specification of its normal domain. Both Gdf1 and Nodal itself play important roles downstream of HH signaling in controlling *Nodal* expression in the left LPM. Indeed, the Nodal protein, in the presence of extracellular cofactors of the EGF-CFC superfamily, has been shown to activate transcription from its own promoter. Exactly as in the chick embryo, Nodal expression in the left LPM activates *Pitx2* expression.

Thus far, no *Car* gene has been identified in vertebrates other than the chick, but it seems likely that a BMP antagonist (or an alternative mechanism of attenuation of BMP signaling) may also exist in the mouse embryo, as illustrated by the fact that mouse mutants deficient in BMP signaling display bilateral expression of *Nodal*. The nature of this putative BMP antagonist remains unknown, as is the chain of events that operate on the right side of the mouse embryo and its possible interactions with HH signaling in the node.

the expression of all these left-specific genes. Moreover, when SHH protein is applied ectopically to the right side of the embryo, it can ectopically trigger the whole program of "left expression."

In the mouse embryo, however, *Shh, Activin βB, ActRIIA*, and *Fgf-8* are all expressed in the node, but without any detectable asymmetry (Fig. 1B). *Nodal*, in contrast, displays a (transient) stronger expression on the left side of the node, besides its domain in the left LPM, further confirming it as a bona fide evolutionarily conserved left determinant. But does SHH act as a left determinant in the mouse embryo? Interestingly, recent studies in mouse embryos demonstrate that a second member of the HH family, named Indian Hedgehog (IHH), plays a role in L/R development that is partially redundant to that of SHH [14]. Indeed, the SHH and IHH proteins have been previously shown to display similar activities (and bind to the same receptor) in a variety of assays. Both SHH and IHH are expressed in the mouse node without apparent L/R bias, and *Shh*-null embryos still express IHH, which can perform part of the activities of SHH, thus obscuring the consequences of completely removing all HH signals from the node [15,16]. In fact, in *Shh/Ihh* double mutant embryos, *Pitx2* is completely absent from the LPM, which indicates that a compound HH signaling is absolutely required for the establishment of left-specific gene expression in the mouse embryo. Importantly, this phenotype is comparable to that of *Smo*-deficient embryos. The *Smoothened* (*Smo*) gene encodes the essential transducer of both SHH and IHH signals, and in the *Smo*-deficient embryos, *Nodal* (as well as other left-specific genes) is absent from the LPM. *Nodal* expression is, however, retained in the node, although with variable levels and patterns of expression, and the significance of this observation is still unclear.

These results demonstrate that HH signaling from the mouse node is absolutely required for the activation of left-specific genes in the LPM, exactly as in the chick embryo, even though expression is not restricted to the left side of the node. In contrast with the situation in the chick, however, ectopic HH signaling in the mouse fails to induce ectopic expression of *Nodal*, which suggests that the regulation of *Nodal* by HH is indirect. In this regard, the TGFβ family member Gdf1 and the EGF-CFC extracellular factor Cryptic (a cofactor for the Nodal protein) both appear to be targets of HH signaling, and may mediate the effects of HH proteins on the node and the LPM (Fig. 1B). In the mouse, there is also some evidence that the domain of *Nodal* expression in the node is required for expression of *Nodal* in the left LPM, where a positive autoregulatory loop is established. It is important to point out that the same situation of redundancy between SHH and IHH in the node is likely to occur in the chick (Fig. 1A), where the anti-SHH blocking antibody presumably blocks both SHH and IHH proteins.

Why isn't there left-specific expression of *Shh, Activin βB*, or *Fgfs* in the mouse node? It has been postulated that a leftward "nodal flow" driven by nodal cilia (possibly carrying some extracellular inducing factor), is the event that breaks the initial symmetry in the mouse embryo (and most likely also in other mammals), resulting in asymmetric expression of *Nodal* in and around the node [17]. The existence of this putative inducing factor carried by the nodal flow has not been demonstrated yet, but it is clear from the results indicated above that HH signaling in the node is required for the correct regionalization of *Nodal* expression in the mouse node. The reason why *Nodal* is transiently upregulated specifically on the left side of the mouse node is still unknown, and it is also important to point out that it is still unclear whether nodal cilia control the L/R axis in the chick embryo [18].

In conclusion, although the mechanisms that break the initial symmetry may be very different in chick and mouse embryos, the basic role of HH signaling as a left determinant appears to be conserved during evolution, as is the left-specific expression of the *Nodal* gene. Without a doubt, the characterization of additional components of HH signaling pathways and of their roles in the early vertebrate embryo will still provide very valuable information about the mechanisms that control L/R determination during early embryogenesis.

References

1. Hamada, H., Meno, C., Watanabe, D., and Saijoh, Y. (2002). Establishment of vertebrate left-right asymmetry. *Nat. Rev. Genet.* **3**, 103–113.
2. Mercola, M. and Levin, M. (2001). Left-right asymmetry determination in vertebrates. *Annu. Rev. Cell Dev. Biol.* **17**, 779–805.
3. Capdevila, J., Vogan, K. J., Tabin, C. J., and Izpisúa Belmonte, J. C. (2000). Mechanisms of left-right determination in vertebrates. *Cell* **101**, 9–21.
4. Levin, M., Johnson, R. L., Stern, C. D., Kuehn, M., and Tabin, C. (1995). A molecular pathway determining left-right asymmetry in chick embryogenesis. *Cell* **82**, 803–814.
5. Padgett, R. W. and Patterson, G. I. (2001). New developments for TGFβ. *Dev. Cell* **1**, 343–349.
6. Wakefield, L. M. and Roberts, A. B. (2002). TGF-ß signaling: positive and negative effects on tumorigenesis. *Curr. Opin. Genet. Dev.* **12**, 22–29.
7. Monsoro-Burq, A. and Le Douarin, N. M. (2001). BMP4 plays a key role in left-right patterning in chick embryos by maintaining Sonic Hedgehog asymmetry. *Mol. Cell* **7**, 789–799.
8. Rodríguez Esteban, C., Capdevila, J., Economides, A. N., Pascual, J., Ortiz, A., and Izpisúa Belmonte, J. C. (1999). The novel Cer-like protein Caronte mediates the establishment of embryonic left-right asymmetry. *Nature* **401**, 243–251.
9. Yokouchi, Y., Vogan, K. J., Pearse, R. V. II, and Tabin, C. J. (1999). Antagonistic signaling by Caronte, a novel Cerberus-related gene, establishes left-right asymmetric gene expression. *Cell* **98**, 573–583.
10. Zhu, L., Marvin, M. J., Gardiner, A., Lassar, A. B., Mercola, M., Stern, C. D., and Levin, M. (1999). Cerberus regulates left-right asymmetry of the embryonic head and heart. *Curr. Biol.* **9**, 931–938.
11. Rodríguez-Esteban, C., Capdevila, J., Kawakami, Y., and Izpisúa Belmonte, J. C. (2001). Wnt signaling and PKA control Nodal expression and left-right determination in the chick embryo. *Development* **128**, 3189–3195.
12. Kawakami, M. and Nakanishi, N. (2001). The role of an endogenous PKA inhibitor, PKIα, in organizing left-right axis formation. *Development* **128**, 2509–2515.
13. Levin, M. and Mercola, M. (1999). Gap junction-mediated transfer of left-right patterning signals in the early chick blastoderm is upstream of Shh asymmetry in the node. *Development* **126**, 4703–4714.
14. Zhang, X. M., Ramalho-Santos, M., and McMahon, A. P. (2001). Smoothened mutants reveal redundant roles for Shh and Ihh signaling

including regulation of L/R symmetry by the mouse node. *Cell* **105**, 781–792.

15. Meyers, E. N and Martin, G. R. (1999). Differences in left-right axis pathways in mouse and chick: functions of FGF8 and SHH. *Science* **285**, 403–406.

16. Tsukui, T., Capdevila, J., Tamura, K., Ruiz-Lozano, P., Rodríguez-Esteban, C., Yonei-Tamura, S., Magallón, J., Chandraratna, R. A., Chien, K., Blumberg, B., Evans, R. M., and Belmonte, J. C. (1999). Multiple left-right asymmetry defects in Shh(−/−) mutant mice unveil a convergence of the shh and retinoic acid pathways in the control of Lefty-1. *PNAS USA* **96**, 11376–11381.

17. Nonaka, S., Tanaka, Y., Okada, Y., Takeda, S., Harada, A., Kanai, Y., Kido, M., and Hirokawa, N. (1998). Randomization of left-right asymmetry due to loss of nodal cilia generating leftward flow of extraembryonic fluid in mice lacking KIF3B motor protein. *Cell* **95**, 829–837.

18. Essner, J. J., Vogan, K. J., Wagner, M. K., Tabin, C. J., Yost, H. J., and Brueckner, M. (2002). Conserved function for embryonic nodal cilia. *Nature* **418**, 37–38.

19. Isaac, A., Sargent, M. G., and Cooke, J. (1997). Control of vertebrate left-right asymmetry by a snail-related zinc finger gene. *Science* **275**, 1301–1304.

EGF-Receptor Signaling in *Caenorhabditis elegans* Vulval Development

Nadeem Moghal and Paul W. Sternberg

HHMI and Division of Biology, California Institute of Technology, Pasadena, California

A crucial aspect of *Caenorhabditis elegans* vulval development is induction of the vulval precursor cells by the anchor cell of the developing uterus. Precisely three of the six vulval precursor cells are induced by the anchor cell to generate vulval cells. As we shall discuss, this induction uses an epidermal growth factor (EGF)-receptor signaling pathway involving the *C. elegans* EGF-receptor homolog LET-23. There are a number of other cell-signaling events during vulval development including at least three uses of WNT [17,19,48,61], and the LIN-12 (Notch-type receptor) pathway. The LET-23 pathway is also used for a reciprocal induction by which a subset of the vulval cells signal back to the developing uterus [9]. Because vulval induction is relatively easy to observe, it can be used to study EGF-receptor signaling *in vivo*.

The Core LET-23 Signaling Pathway

Proteins necessary for LET-23 signaling during vulval induction were found by analyzing mutants with defective (vulvaless) or excessive (multivulva) vulval induction, or by suppressor mutations altering other pathway components. The order of action of these proteins was inferred from analysis of double mutant combinations, in the presence or absence of the inducing cell. The double mutant experiments relied on use of overexpressed EGF, activated EGF-receptor LET-23, or activated RAS. For example, a gain-of-function mutant version of LET-23 causes vulval development in the absence of the anchor cell. *lin-3* encodes multiple products sharing an EGF domain and a transmembrane domain, but differing in their amino termini and spacing of the EGF and transmembrane domain [27,40]. *let-23* encodes the *C. elegans* EGF receptor subfamily ortholog [1]. *sem-5* encodes a GRB2 ortholog with an SH2 domain and two SH3 domains [14]. *let-341* encodes an SOS ortholog, a Ras guanine nucleotide exchange factor [10,15]. *let-60* encodes RAS [23]. *lin-45* encodes RAF [24]. *mek-2* encodes MAP kinase kinase [35,36] and *mpk-1/sur-1* encodes MAP kinase [37,63]. The genetic data are consistent with a linear pathway from LET-23 to MAP kinase: LET-23 acts downstream or in parallel to LIN-3; SEM-5 acts downstream or in parallel to LET-23; LET-60 acts downstream or in parallel to SEM-5; LIN-45 acts downstream or in parallel to LET-60, etc. Downstream of MAP kinase the pathway is more complicated, with a number of positive- and negative-acting nuclear factors. The LIN-1 ETS-domain protein and the winged helix protein LIN-31 primarily act to inhibit vulval development, but direct targets are not yet known [2,46,58]. LIN-1 and LIN-31 can dimerize and affect each other's activity. Positive-acting transcription factors include LIN-25 and SUR-2 [54,60], which appear to act together. SUR-2 is homologous to a component of a human mediator complex [4], which in mammals and yeast links site-specific DNA binding proteins to RNA polymerase. The hox gene *lin-39* is a positive-acting nuclear factor, while the hox gene *mab-5* acts negatively on the two posterior vulval precursor cells [12,43].

Tissue Specificity

Analysis of the tyrosines in the carboxyl terminal, cytoplasmic domain of LET-23 revealed distinct positive functions in activation of the RAS pathway activity involved in cell fate specification and an inositol 1,4,5 trisphosphate (IP_3)

pathway involved in ovulation [6,13,39]. The three tyrosines sufficient for RAS pathway have the YXN consensus for GRB2 (SEM-5) binding, and a single tyrosine site can act via SEM-5 in the absence of the other two. By contrast, yet another tyrosine site stimulates the ovulation pathway not via RAS. The coupling of LET-23 to IP_3 is predicted to involve a phospholipase.

A number of cell fate choices during *C. elegans* development also depend on LET-23. For several of these choices, the outcome may depend on the particular hox gene the cells express. The vulval cells express *lin-39*, the male hook precursor cells express *mab-5*, and the P12 cell expresses *egl-5* [32,44,49].

Positive and Negative Regulators

Besides the core signaling components, additional positive and negative regulators of vulva development have been identified. Elimination of LIN-2, LIN-7, or LIN-10 results in a partial vulvaless phenotype identical to elimination of the carboxyl six amino acids of LET-23. The inferred signaling defect correlates with a shift in LET-23 localization from the basolateral to apical surface of vulval precursor cells, which is the side furthest away from the anchor cell [35,53]. Overexpression of LET-23 rescues the vulvaless phenotypes indicating that these genes facilitate signaling by concentrating LET-23 close to LIN-3 [53]. LIN-2, LIN-7, and LIN-10 all contain PDZ domains [29,53,62]. *In vitro* experiments indicate that the PDZ domain of LIN-7 binds the cytoplasmic tail of LET-23 [53], and that LIN-2 promotes ternary complex formation with LIN-7 and LIN-10 [34]. A truncated form of LIN-10 that lacks both PDZ domains has a relatively weak vulval phenotype indicating that at least for LET-23 signaling, LIN-10 PDZ domains are not crucial [62].

Genetic screens for mutations that suppress the multivulva phenotype conferred by a gain-of-function *let-60* allele *(n1046gf)* identified a number of modulators of vulval development. On their own, mutations in *ksr-1* [36,57], *sur-8/soc-2* [49,51], and *sur-6* [52] result in no obvious vulval phenotype. However, they cause *let-60(n1046gf)* animals to become wild-type, and produce highly penetrant vulvaless phenotypes in the presence of weak reduction of function mutations in core components such as *mpk-1* and *lin-45*. *ksr-1* encodes a predicted Ser/Thr kinase related to RAF, but lacks the CR1 and CR2 domains. *ksr-1* likely functions upstream of *lin-1*, and *in vitro* studies suggest that it may interact directly with MPK-1/SUR-1 through an FXFP site during vulval development [31]. Although some *ksr-1* alleles harbor missense mutations in the kinase domain, transgenes mutant in the ATP binding site or the catalytic aspartate have the same rescuing activity as wild-type DNA, indicating kinase activity is not crucial for its function [56]. *sur-8* encodes a protein consisting of 18 tandem Leu-rich repeats similar to yeast adenylate cyclases. *sur-8* mutations do not reduce the penetrance of the multivulva phenotype conferred by a gain-of-function LIN-45 RAF transgene, suggesting it works upstream of RAF activation. SUR-8 interacts with a different part of the LET-60 effector domain then LIN-45 RAF, and not all *sur-8* mutations affect this interaction [51]. *sur-6* encodes a regulatory β subunit of protein phosphatase 2A(PP2A). *sur-6* mutations do not suppress the multivulva phenotype conferred by a gain-of-function LIN-45 RAF transgene, suggesting it also works upstream of RAF activation. *sur-6* mutations produce a strong synthetic vulvaless phenotype with *sur-8*, but not with *ksr-1*, raising the possibility that SUR-6 function is intimately connected with KSR-1. dsRNA interference experiments against the A and C PP2A catalytic core components, also inhibit the multivulva phenotype of *let-60(n1046gf)* [51], suggesting that SUR-6 is a positive regulator of PP2A, and that the phosphatase activity of the complex promotes RAS-dependent vulva development.

Deletion of the SH2-domain containing protein tyrosine phosphatase, *ptp-2*, which is most closely related to SHP-2/corkscrew [21] does not perturb vulval development on its own, but reduces the multivulva phenotypes conferred by mutations in *lin-15* and activated alleles of *let-23* and *let-60*. In addition, a *ptp-2* mutation results in a synthetic vulvaless phenotype in the presence of a weak *sem-5* mutation.

Loci involved in negative regulation of vulva induction have been elucidated through three different approaches. Genetic screens for mutations, which suppress reduction-of-function mutations in the *let-23* pathway, led to the identification of *unc-101* [38], *sli-1* [33,65] *gap-1* [20], and *sur-5* [22]. UNC-101 is a functional homolog of the AP47 medium chain of the trans-Golgi clathrin-associated AP-1 complex. dsRNAi experiments indicate it works redundantly with a second AP47 homolog, *apm-1* [50]. An *sli-1* mutation suppresses a severe non-null allele of *let-23*, but not severe non-null alleles of *let-60*, suggesting it works upstream of RAS [66]. SLI-1 is related to the CBL family of adaptors/E3 ubiquitin ligases. Although both the PTB and RING finger domains are required for full inhibitory activity, SLI-1 displays some activity in the absence of the RING finger, indicating it may have some function independent of E3 ubiquitin ligase activity [66]. Recently, a tyrosine in the C-terminal tail of LET-23, in the sequence NSSRYKETP, was shown to be required for the genetic inhibition of *sli-1* on *sem-5*-dependent vulval induction [66], similar to a ZAP-70 binding site for the CBL PTB domain [42,45], and consistent with a model of direct binding of SLI-1 to LET-23. *gap-1* mutations suppress reduction-of-function mutations in *let-60*, but not *lin-45*, and its encoded protein is most similar to the Gap-1 and Gap-1m RAS GTPase activating proteins in *Drosophila* and vertebrates, respectively, strongly suggesting that it is a direct regulator of LET-60. *sur-5* was isolated as a suppressor of a dominant negative *let-60* allele. Unlike the previously described negative regulators, it has the unusual property of not being able to suppress any of the standard reduction-of-function mutations in the core components, including a weak allele of *let-23*. Moreover, it only suppresses a subset of dominant negative *let-60* alleles. SUR-5 has some identity to acetyl-CoA synthetases, but its mechanism of action is a mystery. The *cdf-1* gene also was identified as a suppressor of activated RAS [5]. Zinc ions inhibit RAS signaling in *C. elegans*, and the cation diffusion facilitator CDF-1, which lowers cytoplasmic zinc concentration, is thus a positive modulator of the pathway [5].

Genetic screens in sensitized, but phenotypically wild-type backgrounds have identified other negative regulators of vulva induction. Ferguson and Horvitz [18] first defined this approach by isolating mutations in two classes of loci, A and B, wherein any class A mutation caused a synthetic multivulva (synmuv) phenotype in combination with any class B mutation. Using class A or B sensitized backgrounds, a number of additional class A and B mutations have been isolated. *lin-15A* [16,30], *lin-15B* [16,30], and *lin-36* [59] encode novel proteins. LIN-35 is related to the retinoblastoma family of proteins [41], *lin-53* encodes the Rb-binding protein, RbAp48 [41], and DPL-1 is related to the DP family of transcription factors [8]. Reverse genetics indicate that *efl-1/2*, which are E2F-like genes, also have synmuv properties [8]. Genetic [8,30,41,59] and cell ablation [30,55] experiments indicate that the synmuv phenotype is strongly dependent on LET-23 and its downstream signaling components, but not LIN-3. Expression, transgenic, and mosaic analyses suggest that both the vulval precursor cells and the surrounding epidermal syncytium might be a source for an inhibitory signal by these genes [8,25,26,41,59].

Similar to the synmuv screens, *ark-1* was recovered as a mutation that caused a synthetic multivulva phenotype in the presence of a *sli-1* mutation [28]. An *ark-1* mutation also suppresses non-null mutations in the *let-23* pathway. *ark-1* encodes a putative tyrosine kinase most similar to Ack, and yeast two hybrid studies indicate that a C-terminal, proline-rich domain can bind to SEM-5, suggesting its function may be intimately connected to SEM-5. In addition, both SEM-5 and ARK-1 have an inhibitory effect on LET-23 signaling during ovulation, suggesting that SEM-5 might recruit ARK-1 to inhibit LET-23.

A loss-of-function mutation in *lip-1*, a gene predicted to encode a MAPK phosphatase, was generated based on its genomic sequence [3]. This mutation suppresses vulvaless defects conferred by loss-of-function mutations in the *let-23* pathway. LIP-1 transcription is induced by LIN-12 (Notch) signaling [3], and thus might be one way in which Notch signaling is antagonistic to LET-23 signaling [55].

Prospects

One key issue includes the identification of direct targets for the transcriptional response to the MAP kinase pathway. The *egl-17* gene, which encodes a fibroblast growth factor necessary for muscle precursor positioning, is induced by the LET-23 pathway [7], and is one candidate for an immediate target. The hox gene *lin-39* is upregulated by LET-23 signaling [43], and is another candidate. A long list of other candidates has been generated by DNA expression microarray experiments [47]. With the continued discovery of regulators and modulators of this core LET-23 pathway, studies in *C. elegans* will help place these many proteins into their physiological context, including other extracellular signals, tissue-specific effects, and environmental modulation.

References

1. Aroian, R. V., Koga, M., Mendel, J. E., Ohshima, Y., and Sternberg, P. W. (1990). The *let-23* gene necessary for *Caenorhabditis elegans* vulval induction encodes a tyrosine kinase of the EGF receptor subfamily. *Nature* **348**, 693–699.
2. Beitel, G. J., Tuck, S. P., Greenwald, I. S., and Horvitz, H. R. (1995). The *Caenorhabditis elegans* gene *lin-1* encodes an ETS-domain protein and defines a branch of the vulval induction pathway. *Genes Dev.* **9**, 3149–3162.
3. Berset, T., Hoier, E. F., Battu, G., Canevascini, S., and Hajnal, A. (2001). Notch Inhibition of RAS Signaling Through MAP Kinase Phosphatase LIP-1 During *C. elegans* Vulval Development. *Science* **291**, 1055–1058.
4. Boyer, T. G., Martin, M. E., Lees, E., Ricciardi, R. P., and Berk, A. J. (1999). Mammalian Srb/Mediator complex is targeted by adenovirus E1A protein. *Nature* **399**, 276–279.
5. Bruinsma, J. J., Jirakulaporn, T., Muslin, A. J., and Kornfeld, K. (2002). Zinc ions and cation diffusion facilitator proteins regulate Ras-mediated signaling. *Dev. Cell* **2**, 567–578.
6. Bui, Y. and Sternberg, P. W. (2002). *C. elegans* inositol 5-phosphatase homologue negatively regulates inositol 1,4,5-triphosphate signaling in ovulation. *Mol. Biol. Cell* **13**, 1641–1651.
7. Burdine, R. D., Branda, C. S., and Stern, M. J. (1998). EGL-17(FGF) expression coordinates the attraction of the migrating sex myoblasts with vulval induction in *C. elegans*. *Development* **125**, 1083–1093.
8. Ceol, C. J. and Horvitz, H. R. (2001). *dpl-1* DP and *efl-1* E2F act with *lin-35* Rb to antagonize Ras signaling in *C. elegans* vulval development. *Mol. Cell* **7**, 461–73.
9. Chang, C., Newman, A., and Sternberg, P. W. (1999). Reciprocal EGF signaling back to the uterus from the induced *C. elegans* vulva coordinates morphogenesis of epithelia. *Current Biol.* **9**, 237–246.
10. Chang, C., Hooper, N. A., and Sternberg, P. W. (2000). *C. elegans* SOS-1 is required for multiple RAS-dependent events and cooperates with a SOS-1-indipendent pathway during vulval induction. *EMBO J.* **19**, 3283–3293.
11. Church, D. L., Guan, K. L., and Lambie, E. J. (1995). Three genes of the MAP kinase cascade, *mek-2*, *mpk-1/sur-1* and *let-60* ras, are required for meiotic cell cycle progression in *Caenorhabditis elegans*. *Development* **121**, 2525–35.
12. Clandinin, T., Katz, W., and Sternberg, P. W. (1997). *Caenorhabditis elegans* HOM-C genes regulate the response of vulval precursor cells to inductive signal. *Dev. Biol.* **182**, 150–161.
13. Clandinin, T., DeModena, J., and Sternberg, P. W. (1998). Inositol triphosphate mediates a Ras-independent response to LET-23 receptor tyrosine kinase activation in *C. elegans*. *Cell* **92**, 523–533.
14. Clark, S. G., Stern, M. J., and Horvitz, H. R. (1992). *C. elegans* cell signaling gene *sem-5* encodes a protein with SH2 and SH3 domains. *Nature* **356**, 340–344.
15. Clark, S. G., Stern, M. J., and Horvitz, H. R. (1992). Genes involved in 2 *Caenorhabditis elegans* cell-signaling pathways. *Cold Spring Harbor Symp. Quant. Biol.* **57**, 363–373.
16. Clark, S., Lu, W., and Horvitz, H. (1994). The *Caenorhabditis elegans* locus *lin-15*, a negative regulator of a tyrosine kinase signaling pathway, encodes two different proteins. *Genetics* **137**, 987–997.
17. Eisenmann, D. M., Maloof, J. N., Simske, J. S., Kenyon, C. J., and Kim, S. K. (1998). The β-catenin homolog BAR-1 and LET-60 Ras coordinately regulate the Hox gene *lin-39* during *Caenorhabditis elegans* vulval development. *Development* **125**, 3667–3680.
18. Ferguson, E. and Horvitz, H. R. (1989). The multivulva phenotype of certain *Caenorhabditis elegans* mutants results from defects in two functionally redundant pathways. *Genetics* **123**, 109–121.
19. Gleason, J. E., Korswagen, H. C., and Eisenmann, D. M. (2002). Activation of Wnt signaling bypasses the requirement for RTK/Ras signaling during *C. elegans* vulval induction. *Genes Dev.* **16**, 1281–90.
20. Gu, T., Orita, S., and Han, M. (1998). *Caenorhabditis elegans* SUR-5, a novel but conserved protein, negatively regulates LET-60 ras activity during vulval induction. *Mol. Cell. Biol.* **18**, 4556–4564.
21. Gutch, M., Flint, A., Keller, J., Tonks, N., and Hengartner, M. (1998). The *Caenorhabditis elegans* SH2 domain-containing protein tyrosine

phosphatase PTP-2 participates in signal transduction during oogenesis and vulval development. *Genes Dev.* **12**, 571–585.

22. Hajnal, A., Whitfield, C., and Kim, S. (1997). Inhibition of *Caenorhabditis elegans* vulval induction by *gap-1* and by *let-23* receptor tyrosine kinase. *Genes Dev.* **11**, 2715–2728.

23. Han, M. and Sternberg, P. W. (1990). *let-60*, a gene that specifies cell fates during *C. elegans* vulval induction, encoedes a ras protein. *Cell* **63**, 921–931.

24. Han, M., Golden, A., Han, Y., and Sternberg, P. W. (1993). *C. elegans lin-45* raf gene participates in let-60 RAS-stimulated vulval differentiation. *Nature* **363**, 133–140.

25. Hedgecock, E. and Herman, R. (1995). The *ncl-1* gene and genetic mosaics of *Caenorhabditis elegans*. *Genetics* **141**, 989–1006.

26. Herman, R. and Hedgecock, E. (1990). Limitation of the size of the vulval primordium of *Caenorhabditis elegans* by *lin-15* expression in surrounding hypodermis. *Nature* **348**, 169–171.

27. Hill, R. and Sternberg, P. W. (1992). The gene *lin-3* encodes an inductive signal for vulval development in *C. elegans*. *Nature* **358**, 470–476.

28. Hopper, N. A., Lee, J., and Sternberg, P. W. (2000). ARK-1 inhibits EGFR signaling in *C. elegans*. *Mol. Cell* **6**, 65–75.

29. Hoskins, R., Hajnal, A., Harp, S., and Kim, S. (1996). The *C. elegans* vulval induction gene *lin-2* encodes a member of the MAGUK family of cell junction proteins. *Development* **122**, 97–111.

30. Huang, L., Tzou, P., and Sternberg, P. W. (1994). The *lin-15* locus encodes two negative regulators of Caenorhabditis elegans vulval development. *Mol. Biol. Cell* **5**, 395–411.

31. Jacobs, D., Glossip, D., Xing, H., Muslin, A. J., and Kornfeld, K. (1999). Multiple docking sites on substrate proteins form a modular system that mediates recognition by ERK MAP kinase. *Genes Dev.* **13**, 163–75.

32. Jiang, L. and Sternberg, P. W. (1998). Interactions of EGF, Wnt and Hom-C genes specify the P12 neuroectoblast fate in *C. elegans*. *Development* **125**, 2337–2347.

33. Jongeward, G., Clandinin, T., and Sternberg, P. W. (1995). *sli-1*, a negative regulator of *let-23*-mediated signaling in *C. elegans*. *Genetics* **139**, 1553–1566.

34. Kaech, S., Whitfield, C., and Kim, S. (1998). The LIN-2/LIN-7/LIN-10 complex mediates basolateral membrane localization of the *C. elegans* EGF receptor LET-23 in vulval epithelial cells. *Cell* **94**, 761–771.

35. Kornfeld, K., Guan, K. L., and Horvitz, H. R. (1995). The *Caenorhabditis elegans* gene *mek-2* is required for vulval induction and encodes a protein similar to the protein kinase MEK. *Genes Dev.* **9**, 756–68.

36. Kornfeld, K., Hom, D., and Horvitz, H. R. (1995). The *ksr-1* gene encodes a novel protein kinase involved in Ras-mediated signaling in *C. elegans*. *Cell* **83**, 903–913.

37. Lackner, M., Kornfeld, K., Miller, L., Horvitz, H., and Kim, S. (1994). A MAP kinase homolog, *mpk-1*, is involved in ras-mediated induction of vulval cell fates in *Caenorhabditis elegans*. *Genes Dev.* **8**, 160–173.

38. Lee, J., Jongeward, G., and Sternberg, P. W. (1994). *unc-101*, a gene required for many aspects of *Caenorhabditis elegans* development and behavior, encodes a clathrin associated protein. *Genes Dev.* **8**, 60–73.

39. Lesa, G. M. and Sternberg, P. W. (1997). Positive and negative tissue-specific signaling by a nematode EGF receptor. *Mol. Biol. Cell* **8**, 776–793.

40. Liu, J., Tzou, P., Hill, R., and Sternberg, P. W. (1999). Structural requirements for the tissue-specific and tissue-general functions of the *Caenorhabditis elegans* epidermal growth factor LIN-3. *Genetics* **153**, 1257–1269.

41. Lu, X. and Horvitz, H. R. (1998). *lin-35* and *lin-53*, two genes that antagonize a *C. elegans* Ras pathway, encode proteins similar to Rb and its binding protein RbAp48. *Cell* **95**, 981–991.

42. Lupher Jr., M. L., Songyang, Z., Shoelson, S. E., Cantley, L. C., and Band, H. (1997). The Cbl Phosphotyrosine-binding Domain Selects a D(N/D)XpY Motif and Binds to the Tyr292 Negative Regulatory Phosphorylation Site of ZAP-70. *J. Biol. Chem.* **272**, 33140–33144.

43. Maloof, J. and Kenyon, C. (1998). The Hox gene *lin-39* is required during *C. elegans* vulval induction to select the outcome of Ras signaling. *Development* **125**, 181–190.

44. Maloof, J., Whangbo, J., Harris, J., Jongeward, G., and Kenyon, C. (1999). A Wnt signaling pathway controls Hox gene expression and neuroblast migration in *C. elegans*. *Development* **126**, 37–49.

45. Meng, W., Sawasdikosol, S., Burakoff, S. J., and Eck, M. J. (1999). Structure of the amino-terminal domain of Cbl complexed to its binding site on ZAP-70 kinase. *Nature* **398**, 84–90.

46. Miller, L., Gallegos, M., Morisseau, B., and Kim, S. (1993). *lin-31*, a *Caenorhabditis elegans* HNF-3/fork head transcription factor homolog, specifies three alternative cell fates in vulval development. *Genes Dev.* **7**, 933–947.

47. Romagnolo, B., Jiang, M., Kiraly, M., Breton, C., Begley, R., Wang, J., Lund, J., and Kim, S. K. (2002). Downstream targets of *let-60* Ras in *Caenorhabditis elegans*. *Dev. Biol.* **247**, 127–136.

48. Sawa H., Lobel L., and Horvitz H. R. (1996). The *Caenorhabditis elegans* gene *lin-17*, which is required for certain asymmetric cell divisions, encodes a putative seven-transmembrane protein similar to the *Drosophila* frizzled protein. *Genes Dev.* **10**, 2189–2197.

49. Selfors, L., Schutzman, J., Borland, C., and Stern, M. J. (1998). *soc-2* encodes a leucine-rich repeat protein implicated in fibroblast growth factor receptor signaling. *Proc. Natl. Acad. Sci. (USA)* **95**, 6903–6908.

50. Shim, J., Sternberg, P. W., and Lee, J. (2000). Distinct and redundant functions of mu1 medium chains of the AP-1 clathrin-associated protein complex in the nematode *Caenorhabditis elegans*. *Mol. Biol. Cell* **11**, 2743–2756.

51. Sieburth, D., Sun, Q., and Han, M. (1998). SUR-8, a conserved Ras-binding protein with leucine-rich repeats, positively regulates Ras-mediated signaling in *C. elegans*. *Cell* **94**, 119–130.

52. Sieburth, D., Sundaram, M., Howard, R., and Han, M. (1999). A PP2A regulatory subunit positively regulates Ras-mediated signaling during *Caenorhabditis elegans* vulval induction. *Genes Dev.* **13**, 2562–2569.

53. Simske, J., Kaech, S., Harp, S., and Kim, S. (1996). LET-23 receptor localization by the cell junction protein LIN-7 during *C. elegans* vulval induction. *Cell* **85**, 195–204.

54. Singh, N. and Han, M. (1995). *sur-2*, a novel gene, functions late in the *let-60* ras-mediated signaling pathway during *Caenorhabditis elegans* vulval induction. *Genes Dev.* **9**, 2251–2265.

55. Sternberg, P. W. and Horvitz, H. R. (1989). The combined action of two intercellular signaling pathways specifies three cell fates during vulval induction in *C. elegans*. *Cell* **58**, 679–693.

56. Stewart, S., Sundaram, M. Zhang, Y., Lee, J., Han, M., and Guan, K. L. (1999). Kinase suppressor of Ras forms a multiprotein signaling complex and modulates MEK localization. *Mol. Cell. Biol.* **19**, 5523–5534.

57. Sundaram, M. and Han, M. (1995). The *C. elegans ksr-1* gene encodes a novel Raf-related kinase involved in Ras-mediated signal transduction. *Cell* **83**, 889–901.

58. Tan, P., Lackner, M. R., and Kim, S. K. (1998). MAP kinase signaling specificity mediated by the LIN-1 Ets/LIN-31 WH transcription factor complex during *C. elegans* vulval induction. *Cell* **93**, 569–580.

59. Thomas, J. and Horvitz, H. (1999). The *C. elegans* gene *lin-36* acts cell autonomously in the *lin-35* Rb pathway. *Development* **126**, 3449–3459.

60. Tuck, S. P. and Greenwald, I. S. (1995). *lin-25*, a gene required for vulval induction in *Caenorhabditis elegans*. *Genes Dev.* **9**, 341–357.

61. Wang, M. and Sternberg, P. W. (2000). Patterning of the *C. elegans* 1° lineage by RAS and Wnt pathways. *Development* **127**, 5047–5058.

62. Whitfield, C., Benard, C., Barnes, T., Hekimi, S., and Kim, S. (1999). Basolateral localization of the *Caenorhabditis elegans* epidermal growth factor receptor in epithelial cells by the PDZ protein LIN-10. *Mol. Biol. Cell* **10**, 2087–2100.

63. Wu, Y. and Han, M. (1994). Suppression of activated Let-60 Ras protein defines a role of *Caenorhabditis elegans* Sur-1 MAP kinase in vulval differentiation. *Genes Dev.* **8**, 147–159.

64. Wu, Y., Han, M., and Guan, K. L. (1995). MEK-2, a *Caenorhabditis elegans* MAP kinase kinase, functions in Ras-mediated vulval induction and other developmental events. *Genes Dev.* **9**, 742–755.

65. Yoon, C., Lee, J., Jongeward, G., and Sternberg, P. W. (1995). Similarity of *sli-1*, a regulator of vulval development in *C. elegans*, to the mammalian proto-oncogene c-cbl. *Science* **269**, 1102–1105.

66. Yoon, C. H., Chang, C., Hopper, N. A., Lesa, G. M., and Sternberg, P. W. (2000). Requirements of multiple domains of SLI-1, a *Caenorhabditis elegans* homologue of c-Cbl, and an inhibitory tyrosine in LET-23 in regulating vulval differentiation. *Mol. Biol. Cell* **11**, 4019–31.

Induction and Lateral Specification Mediated by LIN-12/Notch Proteins

Sophie Jarriault and Iva Greenwald
Howard Hughes Medical Institute, Department of Biochemistry and Molecular Biophysics, Columbia University, New York, New York

In this chapter, we describe general and conserved features of the LIN-12/Notch signal transduction pathway. We also describe features of two different binary cell fate decisions mediated by this pathway in *Caenorhabditis elegans*. Finally, we discuss whether the primary role of Notch is to mediate binary decisions between alternative cell fates or to block differentiation.

The LIN-12/Notch Pathway

Receptors of the LIN-12/Notch family are evolutionarily conserved single pass transmembrane proteins. A single *Notch* gene exists in *Drosophila*, whereas two, *lin-12 and glp-1*, have been described in *C. elegans* and four, *Notch-1* to *Notch-4*, in vertebrates (Table I). The ligands for LIN-12/Notch are single pass transmembrane proteins of the DSL family, after canonical proteins from *Drosophila* (Delta, Serrate) and *C. elegans* (Lag-2).

Genetic and biochemical experiments on invertebrates, mainly *C. elegans* and *Drosophila melanogaster*, and more recently on vertebrates, have defined members of the LIN-12/Notch signal transduction pathway (Table I), as well as an original model for signal transduction (Fig. 1). LIN-12/Notch proteins in vertebrates appear to be cleaved during their transit to the cell surface by the protease furin at site 1 (see Fig. 1). Upon activation of LIN-12/Notch by a DSL ligand, the receptor is cleaved both in its extracellular domain (site 2) and within its transmembrane domain (site 3), to produce an active intracellular fragment which translocates to the nucleus. The site 2 cleavage is likely to be mediated by a transmembrane metalloprotease of the ADAM family (see Table I).

The site 3 cleavage requires a complex that contains the membrane proteins Presenilin, APH-1, Nicastrin and PEN-2 (reviewed in references [1,2]).

In the nucleus, the intracellular part of LIN-12/Notch associates with a sequence-specific DNA binding protein (known as LAG-1 in *C. elegans*, Su(H) in *Drosophila*, and RBP-J or CBF1 in mammals [3]), with a glutamine-rich protein (SEL-8, for example, see Table I [4,5]), and with another protein called SKIP [6]. This complex, which is likely to include other proteins as well, binds to promoters of target genes, activating their transcription. In addition, the activation complex is believed to contribute to target gene activation by displacing a co-repressor complex previously associated with CBF1 [7].

A number of negative modulators of LIN-12/Notch signaling have been described, which influence the ability of a ligand to signal (Fringe, [8]), the ability of the receptor to transmit the signal (Numb, [9]), or the stability of the receptor (Sel-10, [10]). Crosstalk between the LIN-12/Notch signaling pathway and others, such as the frizzled pathway in *Drosophila* [11] or the Ras pathway in worms [12], has also been described.

Cell-Cell Interactions Mediated by the LIN-12/Notch Pathway

When equivalent cells interact with each other, that process has classically been termed "lateral inhibition," or more recently, "lateral specification" (see the following section). When the signal comes from a different cell type, that process

Table I Members of the LIN-12/Notch Pathway in *C. elegans*, Drosophila, and Mammals

Role	*C. elegans*	Drosophila	Mammals
Ligands (DSL family)	LAG-2, APX-1, others	Delta	Delta 1–4
		Serrate	Jagged 1–2
Receptors	LIN-12, GLP-1	Notch	Notch 1–4
Site 1 cleavage	(Furin)	(Furin)	Furin
Site 2 cleavage	SUP-17	Kuzbanian	TACE/ADAM-17
Site 3 cleavage	SEL-12, HOP-1	Presenilin	Presenilin
	APH-1	Aph-1	Aph-1
	APH-2	Nicastrin	Nicastrin
	PEN-2	PEN-2	PEN-2
Modulators	(NUMB)	Numb	(Numb)
	–	Fringe	Fringe
	SEL-10	(Sel-10)	Sel-10
Nuclear complex	LAG-1	Su(H)	RBP-J/CBF1
	SEL-8	Mastermind	Mastermind
	(SKIP)	(Skip)	SKIP
Selected targets	–	*E(spl)* complex	*HES-1,5*
	lip-1	–	–
	lin-12	(*Notch*)	*Notch-1*

Proteins drawn on the same line are equivalents in the three groups shown. Proteins shown in parentheses, e.g. (furin), exist in the organism, but have not been proven to function in that step. Fringe and genes of the E(spl) complex do not have clear orthologs in *C. elegans*; it is not known whether *lip-1* orthologs exist or are targets of the LIN-12/Notch pathway in other organisms. Note that SUP-17, Kuzbanian and TACE are metalloproteases belonging to the ADAM family.

has classically been termed "induction." LIN-12/Notch signaling has been shown to underlie both kinds of interactions in many different systems. The outcome is a binary decision, so that bipotential cells choose between one of two alternative fates depending on whether LIN-12/Notch activity is high or low. We will illustrate the basic principles here using two simple cell fate decisions from *C. elegans*.

Lateral Specification in the AC/VU Decision: A Decision between Two Equivalent Cells

During formation of the somatic gonad of the hermaphrodite, two cells, named Z1.ppp and Z4.aaa, have an equal probability to become either the anchor cell (AC) or a ventral uterine precursor cell (VU; Fig. 2A). In wild-type hermaphrodites, Z1.ppp and Z4.aaa interact with each other, so that only one of the two cells becomes the AC while the other becomes a VU [13,14]. Cell contact appears to be necessary for the interaction to occur correctly, because in mutants that disrupt contact between Z1.ppp and Z4.aaa, both cells become ACs [15].

Genetic studies of mutations in *lin-12* and other members of the LIN-12/Notch pathway have established that in Z1.ppp or Z4.aaa, a low level of *lin-12* activity results in the AC fate, whereas a high level results in the VU fate. Manipulation of the relative level of *lin-12* activity in genetic mosaics [14] and the detailed analysis of *lin-12* and its ligand *lag-2* transcriptional reporter gene expression during the course of the AC/VU decision [16] has revealed that *lin-12* and *lag-2* are initially expressed at comparable levels in Z1.ppp and Z4.aaa. A stochastic initial difference in *lin-12* or *lag-2* activity is further amplified by a feedback mechanism involving positive and negative autoregulatory loops. As a result, there is an increase of *lin-12* expression and a decrease in *lag-2* expression in one cell, and the reciprocal situation (decrease in *lin-12*, increase in *lag-2*) in the other. The cell in which *lin-12* activity is high becomes a VU and the other, in which *lin-12* activity is low, becomes an AC [14,16].

Lateral specification that includes transcription-based feedback loops also appears to operate in cell fate decisions in *Drosophila* and probably in vertebrates [17,18]. We prefer the term lateral specification to lateral inhibition to emphasize the bidirectional nature of the communication (see also reference [19]). Such simple circuits can also be modified by either intrinsic factors or external signaling events so that the element of randomness seen in the AC/VU decision is reduced or eliminated (see, for example, references [9] and [11]).

Inductive Signaling: Germline Proliferation and Meiotic Differentiation in *C. elegans*

The *C. elegans* gonad is organized as two symmetrical U-shaped arms extending distally from a common uterus (Fig. 2B). The distal-most germ cells divide mitotically, and

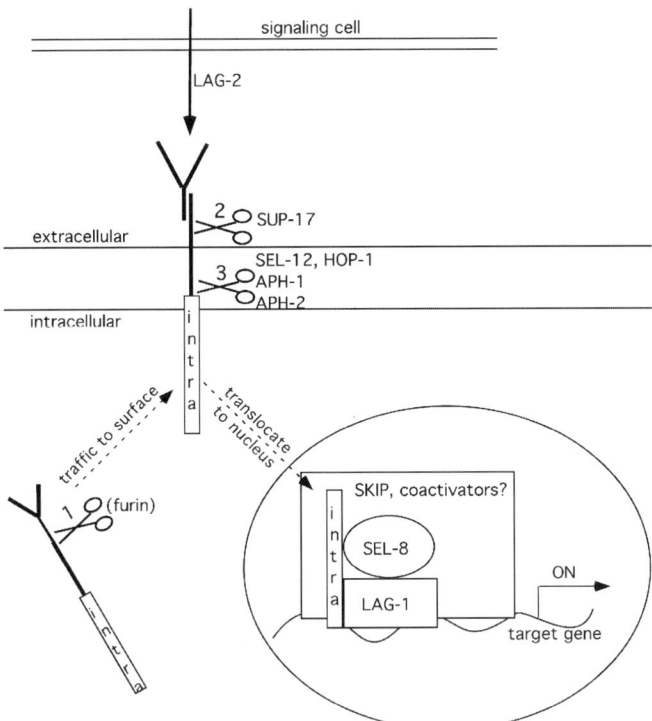

Figure 1 Simplified model of LIN-12/Notch signal transduction. The *C. elegans* names for components are used; see Table I for *Drosophila* and mammalian equivalents. A DSL ligand, such as LAG-2, on the surface of a neighboring cell activates LIN-12/Notch, triggering cleavage in the extracellular domain (site 2) and within its transmembrane domain (site 3). The released intracellular domain translocates to the nucleus, where in a complex with LAG-1, and probably SEL-8, SKIP, and other co-activators, it participates in the transcriptional activation of target genes. The protease furin has been shown to cleave LIN-12/Notch at site 1 during its transit to the surface in mammals [30]; however, this cleavage has not been addressed experimentally in *C. elegans* nor in *Drosophila*.

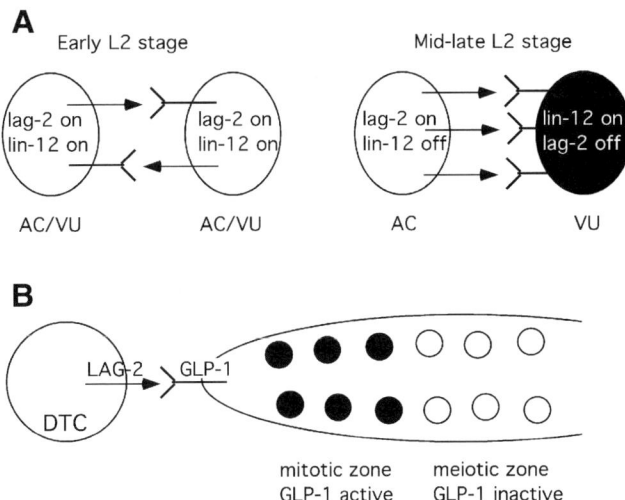

Figure 2 Lateral specification and inductive signaling events mediated by LIN-12/Notch. (A) Lateral specification, the AC/VU decision: Two equivalent cells each have the potential to adopt the AC or VU fates (AC/VU) in the early L2 stage. As the L2 stage progresses, a small variation in ligand or receptor activity is amplified by transcription-based feedback loops, so that activation of LIN-12 in one cell promotes *lin-12* expression and represses (or fails to maintain) *lag-2* expression. Failure to activate LIN-12 has the opposite effect. The presumptive AC is the cell that continues to express *lag-2*, and the presumptive VU is the cell that continues to express *lin-12*. (B) Induction, the mitosis/meiosis decision: The distal tip cell (DTC) of the somatic gonad expresses LAG-2, which activates GLP-1 in the distal portion of the germline syncytium. Distal germline nuclei therefore undergo mitosis, although as they move more proximally, they enter a zone in which GLP-1 is inactive and they begin to enter meiosis.

progress down the arm; at the bend, they enter a transition zone and proceed to meiosis. At the extremity of each arm, there is a somatic distal tip cell (DTC) that regulates the decision of the germ cells between mitosis and meiosis: If the DTC is ablated, the distal germ cell nuclei in that gonad arm enter meiosis precociously [20].

Genetic studies of *glp-1* have established that a low level of *glp-1* activity is required for germ cells to undergo meiosis, whereas high levels of *glp-1* activity promote mitosis [21–23]. The DTCs express the DSL gene *lag-2*, and the distal germ cells express the LIN-12/Notch protein GLP-1 [24,25]. Activation of GLP-1 in the germ cells that lie near the DTC causes them to remain proliferative.

The Role of LIN-12/Notch Proteins: Suppression of Differentiation versus Specification of Binary Cell Fate Decisions

Truncated forms of LIN-12/Notch proteins, such as the intracellular domain, mimic the natural cleavage product and behave like constitutively active receptors [26]. This approach has been important for studying the role and the molecular mechanism of the Notch pathway in vertebrates.

The idea that the role of Notch is to block differentiation was first proposed when activated Notch was thought to inhibit photoreceptor differentiation in *Drosophila* [27]. Furthermore, this view was reinforced by studies of proneural clusters in *Drosophila*, which showed that expression of the achaete-scute neural competence factors are lost in cells in which Notch has been activated [28]. The view that Notch activation inhibits differentiation in *Drosophila* strongly influenced the interpretation of phenotypes caused by expressing the Notch intracellular domain in vertebrates. However, recent work has established that Notch does not block differentiation of photoreceptors, and indeed, plays a direct role in photoreceptor differentiation (for example, see reference [29]). Thus, photoreceptor differentiation in the *Drosophila* eye, like the *C. elegans* examples illustrated above, suggest that activated Notch promotes a fate that is a normal outcome of a binary decision.

These considerations suggest that, while it may be that in some cases the effect of activated Notch is to "block differentiation" there may perhaps be an underlying binary cell fate decision; to differentiate or to remain a stem cell. Indeed, the continued mitotic proliferation of the germline when GLP-1 is constitutively activated in *C. elegans* may be considered a model for this kind of binary decision.

Acknowledgments

We are grateful to L. Johnston, C. Bais, X. Karp, and D. Shaye for critical reading of this manuscript.

References

1. Kopan, R. and Goate, A. (2002). Aph-2/Nicastrin: An essential component of gamma-secretase and regulator of Notch signaling and Presenilin localization. *Neuron* **33**, 321–324.
2. De Strooper, B. (2003). Aph-1, Pen-2 and Nicastrin with Presenilin generate an active γ-secretase complex. *Neuron* **38**, 9–12.
3. Jarriault, S. *et al.* (1995). Signalling downstream of activated mammalian Notch. *Nature* **377**, 355–358.
4. Doyle, T. G., Wen, C., and Greenwald, I. (2000). SEL-8, a nuclear protein required for LIN-12 and GLP-1 signaling in *Caenorhabditis elegans*. *Proc. Natl. Acad. Sci. USA* **97**, 7877–7881.
5. Petcherski, A. G. and Kimble, J. (2000). LAG-3 is a putative transcriptional activator in the *C. elegans* Notch pathway. *Nature* **405**, 364–368.
6. Zhou, S. *et al.* (2000). SKIP, a CBF1-associated protein, interacts with the ankyrin repeat domain of NotchIC To facilitate NotchIC function. *Mol. Cell. Biol.* **20**, 2400–2410.
7. Hsieh, J. J. *et al.* (1996). Truncated mammalian Notch1 activates CBF1/RBPJk-repressed genes by a mechanism resembling that of Epstein-Barr virus EBNA2. *Mol. Cell. Biol.* **16**, 952–959.
8. Irvine, K. D. (1999). Fringe, Notch, and making developmental boundaries. *Curr. Opin. Genet. Dev.* **9**, 434–441.
9. Guo, M., Jan, L. Y., and Jan, Y. N. (1996). Control of daughter cell fates during asymmetric division: interaction of Numb and Notch. *Neuron* **17**, 27–41.
10. Justice, N. J. and Jan, Y. N. (2002). Variations on the Notch pathway in neural development. *Curr. Opin. Neurobiol.* **12**, 64–70.
11. Blair, S. S. (1999). Eye development: Notch lends a handedness. *Curr. Biol.* **9**, R356–R360.
12. Wang, M. and Sternberg, P. W. (2001). Pattern formation during *C. elegans* vulval induction. *Curr. Top. Dev. Biol.* **51**, 189–220.
13. Kimble, J. (1981). Alterations in cell lineage following laser ablation of cells in the somatic gonad of *Caenorhabditis elegans*. *Dev. Biol.* **87**, 286–300.
14. Seydoux, G. and Greenwald, I. (1989). Cell autonomy of lin-12 function in a cell fate decision in *C. elegans*. *Cell* **57**, 1237–1245.
15. Hedgecock, E. M., Culotti, J. G., and Hall, D. H. (1990). The unc-5, unc-6, and unc-40 genes guide circumferential migrations of pioneer axons and mesodermal cells on the epidermis in *C. elegans*. *Neuron* **4**, 61–85.
16. Wilkinson, H. A., Fitzgerald, K., and Greenwald, I. (1994). Reciprocal changes in expression of the receptor lin-12 and its ligand lag-2 prior to commitment in a *C. elegans* cell fate decision. *Cell* **79**, 1187–1198.
17. Heitzler, P. *et al.* (1996). Genes of the Enhancer of split and achaete-scute complexes are required for a regulatory loop between Notch and Delta during lateral signalling in *Drosophila*. *Development* **122**, 161–171.
18. Robson MacDonald, H., Wilson, A., and Radtke, F. (2001). Notch1 and T-cell development: insights from conditional knockout mice. *Trends Immunol.* **22**, 155–160.
19. Greenwald, I. and Rubin, G. M. (1992). Making a difference: the role of cell-cell interactions in establishing separate identities for equivalent cells. *Cell* **68**, 271–281.
20. Kimble, J. E. and White, J. G. (1981). On the control of germ cell development in *Caenorhabditis elegans*. *Dev. Biol.* **81**, 208–219.
21. Austin, J. and Kimble, J. (1987). glp-1 is required in the germ line for regulation of the decision between mitosis and meiosis in *C. elegans*. *Cell* **51**, 589–599.
22. Berry, L. W., Westlund, B., and Schedl, T. (1997). Germ-line tumor formation caused by activation of glp-1, a Caenorhabditis elegans member of the Notch family of receptors. *Development* **124**, 925–936.
23. Priess, J. R., Schnabel, H., and Schnabel, R. (1987). The glp-1 locus and cellular interactions in early *C. elegans* embryos. *Cell* **51**, 601–611.
24. Henderson, S. T., Gao, D., Lambie, E. J., and Kimble, J. (1994). lag-2 may encode a signaling ligand for the GLP-1 and LIN-12 receptors of *C. elegans*. *Development* **120**, 2913–2924.
25. Crittenden, S. L., Troemel, E. R., Evans, T. C., and Kimble, J. (1994). GLP-1 is localized to the mitotic region of the C. elegans germ line. *Development* **120**, 2901–2911.
26. Struhl, G., Fitzgerald, K., and Greenwald, I. (1993). Intrinsic activity of the Lin-12 and Notch intracellular domains *in vivo*. *Cell* **74**, 331–345.
27. Artavanis-Tsakonas, S., Matsuno, K., and Fortini, M. E. (1995). Notch signaling. *Science* **268**, 225–232.
28. Culi, J. and Modolell, J. (1998). Proneural gene self-stimulation in neural precursors: an essential mechanism for sense organ development that is regulated by Notch signaling. *Genes Dev.* **12**, 2036–2047.
29. Tomlinson, A. and Struhl, G. (2001). Delta/Notch and Boss/Sevenless signals act combinatorially to specify the *Drosophila* R7 photoreceptor. *Mol. Cell* **7**, 487–495.
30. Logeat, F. *et al.* (1998). The Notch1 receptor is cleaved constitutively by a furin-like convertase. *Proc. Natl. Acad. Sci. USA* **95**, 8108–8112.

CHAPTER 257

Notch Signaling in Vertebrate Development

Chris Kintner

*The Salk Institute for Biological Studies,
La Jolla, California*

Introduction

The Notch signaling pathway was discovered in *Caenorhabditis elegans* and *Drosophila* by studying gene mutations that cause equivalent embryonic phenotypes, many of which entail alterations in cell fate [1]. Subsequently, homology cDNA cloning identified the vertebrate counterparts of these genes, paving the way for the discovery that Notch also plays a prominent role in the development of just about every organ system in the vertebrate embryo [2]. The first part of this chapter will describe the structural features of the receptors, ligands, and accessory proteins that mediate Notch signaling in vertebrates. The second part will describe how Notch signaling is used, in many cases in a similar manner, to regulate various aspects of cell fate determination and tissue patterning during vertebrate development.

Components Mediating Vertebrate Notch Signaling

Vertebrate Notch Receptors

The Notch receptors comprise a relatively small family of type I transmembrane proteins with just one member in *Drosophila*, two in *C. elegans*, and four in vertebrates [1]. As illustrated in Fig. 1A, the extracellular domains (ECN) of these receptors share a number of structural similarities, including multiple, tandem copies of a epidermal growth factor (EGF)-like motif that comprises most of the ectodomain as well as tandem copies of a second motif called the Notch/lin-12 repeat. The intracellular domain (NICD) contains a conserved juxtamembrane region called RAM23, six tandem ankyrin repeats (ANK), and a PEST (P) domain located at the carboxy terminus. The different receptors presumably arose during vertebrate evolution by several gene duplications starting from a receptor very similar to *Drosophila* Notch. Thus, mouse Notch1 and 2 contain the same number of EGF repeats as *Drosophila* (36) while Notch3 and Notch4 contain 34 and 29 repeats, respectively. The NICD of both Notch 3 and 4 is also shorter than that of Notch1 and 2. In addition, as expected for genes recently duplicated, the four Notch homologs show some degree of genetic redundancy based on the phenotypes produced by targeted mutations [3]. Nonetheless, the "knockout" phenotype of mouse Notch1 is more severe than that of the other Notch homologs, suggesting that the latter receptors have evolved to serve more specialized functions [4–6]. These specialized functions may have arisen by changes in their developmental expression pattern (for example, see [7,8]), in their preference for activation by the different ligands [9], or in the way that they signal. An example of the latter is Notch3, which may activate downstream targets differently than the other Notch receptors under at least some assay conditions [10].

The vertebrate Notch receptors were also identified as genes linked with several disease states. In certain forms of T-cell leukemia, a chromosomal translocation fuses the intracellular domain of the human Notch1 receptor to the extracellular domain of the β chain of the T-cell receptor [11]. Similarly, in MMTV-induced mammary tumors in the mouse, one of the viral integration sites (INT3) leads to expression of just the Notch4 intracellular domain [12]. Based on the known Notch signaling pathway (see the following section), these observations suggest that constitutive

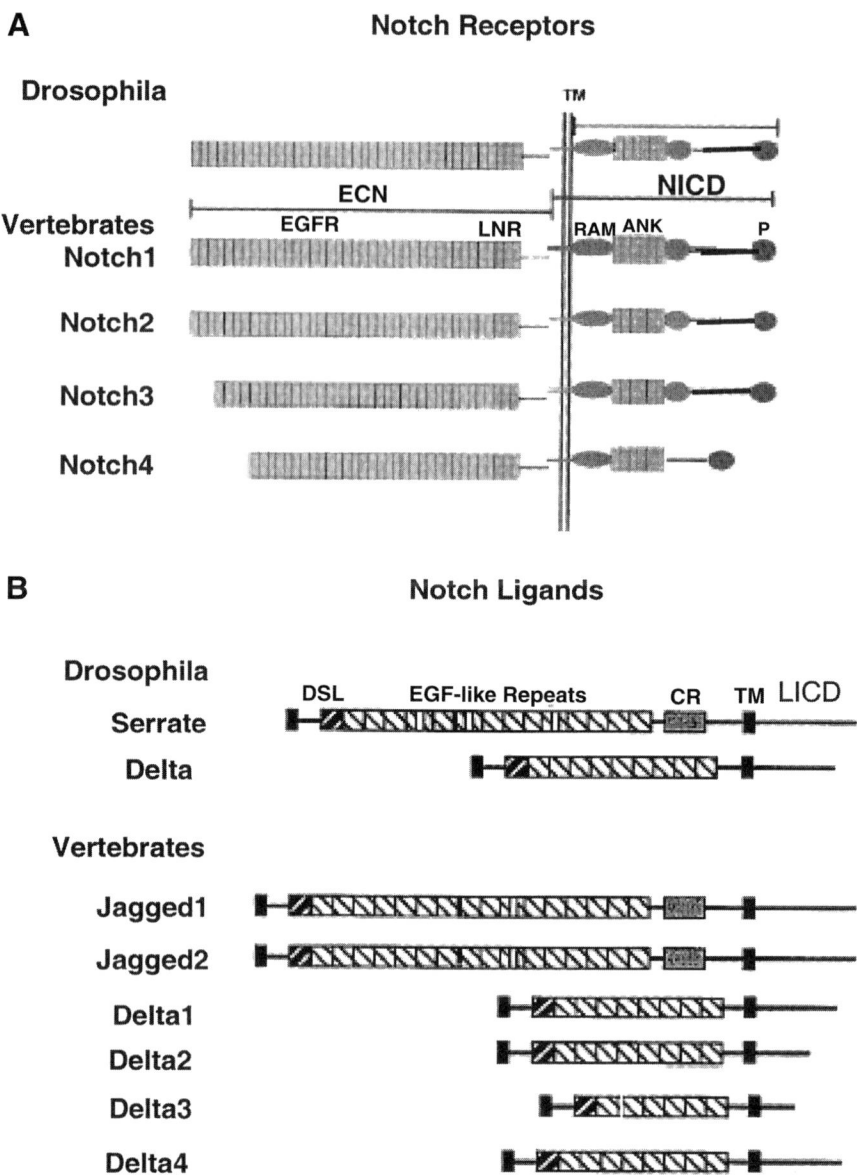

Figure 1 Diagram illustrating the structure of the Notch receptors (A) and ligands (B) in vertebrates relative to their *Drosophila* counterparts. See text for details.

Notch signaling contributes to tumor progression. Finally, an autosomal dominant disease resulting in stroke and dementia, called CADASIL, is associated with missense mutations in the human Notch3 receptor [13], although the consequences of these mutations on receptor activity are not fully known.

Vertebrate Notch Ligands

Ligands for the vertebrate Notch receptors are type I transmembrane proteins that are closely related in structure and sequence to the two ligands, Delta and Serrate, which bind and activate *Drosophila* Notch (Fig. 1B) [14]. The extracellular domains of these ligands are primarily made up of EGF-like repeats like the receptors but are distinguished by a signature divergent EGF-like repeat located at the amino terminus called the DSL domain (Delta, Serrate, and Lag-1). While the Delta-like ligands contain from 6–9 EGF-like repeats, the Serrate-like ligands (referred to mainly as the Jagged ligands) have 16 repeats, as well as a unique cysteine-rich (CR) domain located between the EGF repeats and the transmembrane domain. All vertebrate species have a Delta-like ligand, called Delta1, which seems to be the most closely related to *Drosophila* Delta in terms of structure and function. The other Delta-like ligands in vertebrates, such as Delta3 [15] and Delta4 [16] in the mouse or Delta2 in *Xenopus* [17], are either much more divergent in structure (Delta3), or more restricted in their expression pattern during development (Delta2 and 4). Vertebrates have at least two Serrate-like ligands (Jagged1 and Jagged2) [18,19] that contribute to the development of a number of tissues based on knockout phenotypes in the mouse [20,21]. Like the

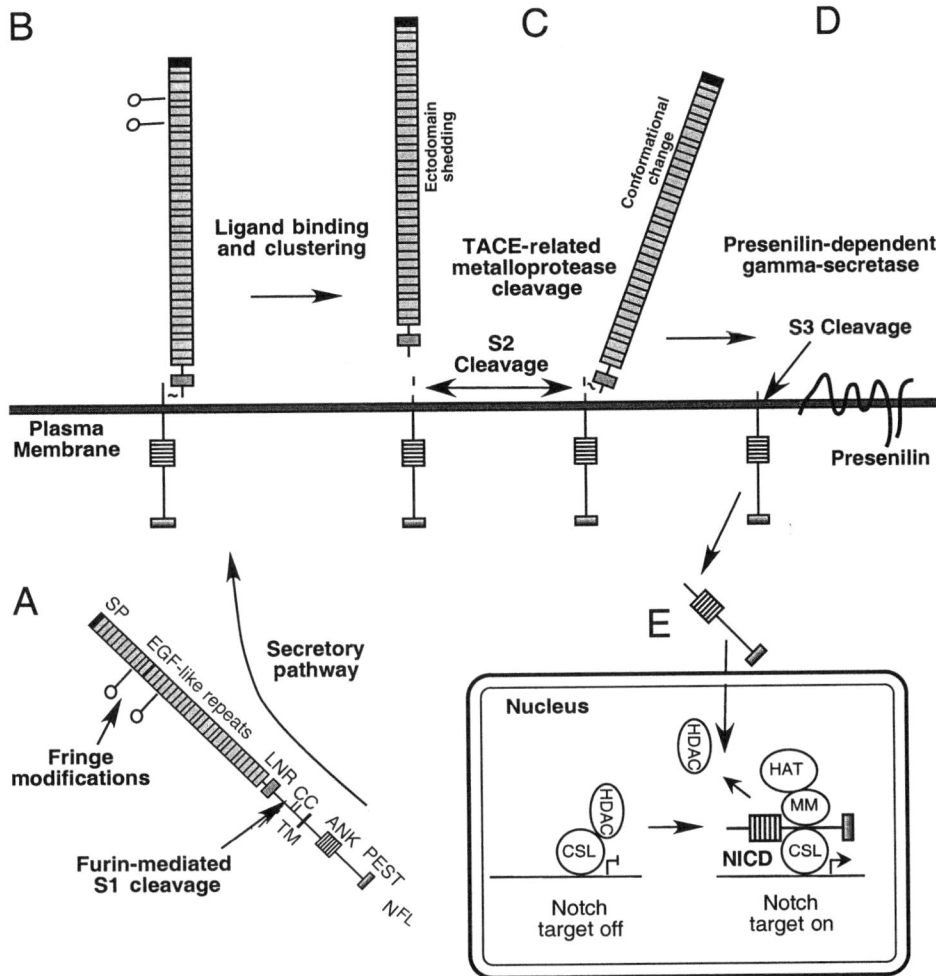

Figure 2 Signal transduction in the Notch pathway. (A) Receptors are modified during secretion by addition of sugar residues by the Fringe proteins as well as by a Furin cleavage that produces a heterodimeric receptor linked by metal bridge. (B) Ligand binding leads to receptor clustering which exposes an S2 cleavage site either by ectodomain shedding or a conformational change (C). Following S2 cleavage, the remaining receptor in the membrane is cleaved at S3 by a presenilin-dependent protease activity. (E) This cleavage releases the intracellular domain that then moves to the nucleus and associates with the CSL proteins to modulate transcription. In the absence of NICD, the CSL proteins repress transcription by associating with co-repressors with histone deacetylase (HDAC) activity. Binding of NICD to the CSL proteins displaces the co-repressor complexes and recruits co-activators including the mastermind (MM) proteins, which in turn recruit proteins such as p300 and CREB binding protein (CBP) with histone acetyltransferase (HAT) activity.

receptors, mutations in the ligands are associated with several disease states. Haploinsufficiency in the *Jagged1* gene is tightly associated with Alagille's syndrome; a human autosomal dominant, developmental disorder characterized by liver, heart, eye, skeletal, craniofacial, and kidney abnormalities [13]. A mutant *Delta3* gene is tightly linked with a skeletal abnormality spondylocostal dysplasia in humans, [22] and is responsible for the skeletal defects seen in the pudgy mouse [23].

Vertebrate Notch Signal Transduction

The mechanism used in vertebrates to transduce a Notch signal is largely the same as that used in *Drosophila* and *C. elegans*, and is one where the NICD is released from the plasma membrane, moves to the nucleus, and serves as a dedicated transcriptional co-activator for the CSL proteins (Fig. 2) [24–28]. CSL is the name given to essentially the same DNA binding protein found in different species (CBF1, vertebrates; SuH, *Drosophila*; Lag 2, *C elegans*), and this particular form of Notch signaling is termed the CSL-dependent pathway. Evidence for a CSL-independent form of Notch signaling exists in vertebrates [29], but the nature of this pathway remains poorly understood. In *Drosophila*, several examples of CSL-independent signaling by Notch have been described including one that is transduced by a cytoplasmic RING domain protein, called Deltex, [30,31], a second involving the abl tyrosine kinase during axon guidance [32], and a third involving the c-Jun N-terminal kinase (JNK) during dorsal closure [33].

Although the components involved in these pathways are found in vertebrates, including vertebrate homologs of Deltex [34–36], their contribution to Notch signaling during vertebrate development remains largely unexplored. These alternative pathways may gain further prominence as Notch signaling is studied in cases other than those involving cell fate choices, such as the reported affects of Notch signaling on the remodeling of neuronal dendrites [37,38].

The idea that NICD moves between the membrane and the nucleus was at first controversial [39] but became more generally accepted with the discovery of other dual-address transcription factors such as SREBP, which are activated by regulated intramembranous proteolysis (RIP) [40]. In the case of Notch, a key step in RIP (Fig. 2D) is carried out by a presenilin-dependent gamma-secretase activity that is also known for its role in processing the beta amyloid protein precursor (βAPP) into soluble peptides that potentially contribute to Alzheimer's disease [41,42]. Mutations in the presenilin homologs in *Drosophila* and *C elegans* disrupt Notch signaling, and subsequent biochemical experiments indicated that this activity cleaves the Notch receptor within the transmembrane domain at a site referred to as S3 [43]. By analogy with βAPP, the S3 cleavage of Notch is thought to occur constitutively upon the formation of a suitable substrate generated by other proteolytic processing events [44]. In the case of Notch, the substrate for S3 cleavage can be generated by cleaving the receptor by a TACE-related metalloprotease at a second site (S2), which lies extracellularly between the transmembrane domain and the conserved paired-cysteines (Fig. 2C) [45–47]. Thus the current model suggests that activation of the receptor induces changes that permit cleavage at S2, resulting in a substrate that is constitutively processed at S3, thereby releasing NICD from the membrane.

How ligand activation regulates cleavage of the receptor at S2 is not fully known, but one insight into this issue comes from the observation that forms of the receptor lacking an extracellular domain are constitutively processed at both S2 and S3 [45]. In the absence of ligand, the S2 cleavage site of the full-length receptor may be masked by interactions with another region in the extracellular domain, most likely the conserved Lin12/Notch repeats. The S2 cleavage site might be exposed during ligand binding through conformational changes that disrupt this inhibitory domain (Fig. 2B). Alternatively, for reasons discussed next, ligand binding might expose the S2 cleavage site by promoting the shedding of the receptor ectodomain (Fig. 2B). The Notch receptors could easily undergo ectodomain shedding because they reach the cell surface as heterodimers that are produced during secretion by cleavage with a Furin-like protease at a site called S1 (Fig. 2A) [48]. Disrupting the metal bridge that holds this heterodimer together exposes the S2 cleavage site [49], but whether or not this is ligand regulated is unknown. Regardless of whether the S2 cleavage is promoted by a conformational change or by ectodomain shedding, the proteolytic processing of the receptor is apparently inefficient, producing only small amounts of NICD that are difficult to detect biochemically, even though they are effective physiologically [50].

Upon release from the membrane, NICD associates with the CSL proteins, thereby converting these DNA binding proteins from transcriptional repressors into transcriptional activators (Fig. 2E) [27]. A prominent group of genes that are upregulated directly by Notch signaling encode basic helix-loop-helix (bHLH) transcriptional repressors [51], such as the HES and HRT genes in mouse [5,24,52], the HER/HRT genes in zebra fish [53,54], and the ESR genes in *Xenopus* [55]. Activation of these genes is dependent on tissue type and developmental context, indicating that Notch transcription in most cases acts in combination with additional transcriptional cofactors [56]. Additional complexity is likely to come from various proteins that are recruited by the Notch transcription complex that both positively and negatively regulate its activity. Further description of the mechanisms by which Notch regulates transcription of these genes via the CSL proteins are reviewed in more detail in Chapter 294.

Regulation of the Notch Ligands

One important feature of the Notch signaling pathway is that all of the known ligands presumably activate receptors on neighboring cells as membrane-bound proteins. Indeed, when secreted forms of the Notch ligands are engineered and expressed *in vivo*, they inhibit Notch signaling [57,58], suggesting that they bind but do not activate the Notch receptors. In cultured cell experiments, soluble ectodomains of the vertebrate ligands are much less effective at activating Notch signaling than membrane-bound ligands expressed on cells and used in co-cultivation assays [19]. Significantly, among the effective soluble Notch ligands are those engineered by fusing the ligand extracellular domain to the FC portion of antibodies, known as ligand-bodies. Ligand bodies activate Notch signaling but are the most effective when crosslinked extensively using FC antibodies or immobilized on plastic surfaces [59,60]. As a practical matter, these reagents have proved to be important tools for activating Notch signaling in many experimental systems [61]. Moreover their properties suggest that a simple monomeric interaction between the Notch ligands and receptors is not sufficient for receptor activation, but that clustering of the normally membrane-bound ligands may be required for signaling [62].

Even though the intracellular domains of the vertebrate Notch ligands (LICD) are not highly conserved in primary sequence, ligands lacking this domain are also potent dominant-negative inhibitors of Notch signaling [63,64]. Recent evidence suggests that one function of the LICD is to facilitate interactions between the ligands and the intracellular machinery that promote endocytosis into endosomes and perhaps other intracellular events that traffic ligands back to the cell surface or to destruction [65]. These trafficking events are regulated by ubiquitination and at least one E3 ubiquitin ligase, called neuralized, has been shown to promote the endocytosis of Notch ligands [66–68]. The function of ligand trafficking, and thus the LICD, in Notch signaling may be threefold. First, ligand endocytosis may be one means of promoting ligand clustering on the cell surface and thus receptor activation as described above [62]. Second, ligand trafficking could be used to produce asymmetric levels of ligand among populations of cells in cases where Notch signaling regulates cell fate by the process of lateral inhibition (see the following section). Finally, ligands lacking an intracellular domain may accumulate on the cell surface, resulting in a block of Notch signaling, presumably because ligand, at high levels, sequesters the Notch receptor into nonproductive cis-complexes [69]. Thus, ligand trafficking may be needed for the ligand to function in signal-emitting cells, as well as to keep ligand levels in a range that permits the Notch receptor to be activated in signal-receiving cells.

The ability of the ligands to bind and activate the Notch receptors is also regulated by a mechanism first found in *Drosophila* through the analysis of the *fringe* gene [70]. *Fringe* encodes a 1,3 *N*-acetylglucosamine transferase [71,72] that modifies an *O*-fucose linkage on a subset of the EGF-like repeats of Notch, thus enhancing activation of the receptor by Delta but inhibiting activation by Serrate [73]. The vertebrate genome contains three fringe homologs [74] that clearly interact at least in some cases with the Notch pathway [75–77]. Experiments in cultured cells have shown that mammalian lunatic fringe potentiates the ability of mammalian Notch1 to respond to signaling by Delta1 but inhibits the response to Jagged1 [9,78]. Significantly the change in ligand activation produced by lunatic fringe is not at the level of ligand binding, but rather at the level of receptor activation. One area where the fringes regulate Notch activity in vertebrates is during segmentation where they modulate signaling in a periodic fashion (see the following section).

Notch Signaling in Vertebrate Development

A large body of work has implicated the Notch pathway in the development of a variety of tissues in the vertebrate embryo. The ubiquitous contribution of Notch signaling to vertebrate development makes generalization difficult, particularly when, in many cases, the exact function of Notch signaling in a given developing tissue is still poorly understood. Nonetheless a few common themes are clear. First, the Notch pathway is commonly used to control the fate of progenitor cells, often through genetic interactions with the bHLH transcription factors. These interactions can take on many different forms, and examples are discussed in the following section where the dynamic output of Notch signaling is used to produce salt-and-pepper patterns of cell differentiation (neurogenesis) or sharp boundaries of differentiation between two groups of cells (segmentation). Second, during many developmental processes, rapid changes occur in the expression and/or activity of the Notch ligands or in the ability of cells to respond to Notch activation. Examples of this regulation are discussed in the context of lateral inhibition. Finally, Notch signaling is not likely to be limited to the regulation of cell fate, but may also have roles in cell proliferation, death, and morphology. A challenge for the future is to explore these other roles as the Notch pathway is examined in the context of developmental processes as hematopoiesis, vascularization, and limb development.

Notch Function during Neurogenesis

One function of Notch during vertebrate development is to regulate the activity of bHLH proteins that promote the differentiation of neural precursors into neurons. This regulation mirrors almost identically one that was first uncovered by genetic studies in *Drosophila*, where Notch signaling restricts the formation of neuroblasts or sensory organ precursors by regulating the activity of the proneural bHLH proteins [79]. In vertebrates, this interaction has been studied extensively in fish and frog embryos during primary neurogenesis where an early-born population of neurons is generated from neural progenitors located within discrete domains of the neural plate [63,80–82]. Within these domains, expression of bHLH proteins such as the neurogenins initially occurs in a relatively uniform pattern, thus establishing progenitor cells that can give rise to neurons [83,84]. While the neural bHLH proteins promote neuronal differentiation cell-autonomously, they inhibit neuronal differentiation non-cell-autonomously by promoting the expression of the Delta ligands [82,83], which activates Notch signaling in neighboring cells (Fig. 3A). Activation of Notch leads to the upregulation of genes encoding bHLH repressors, such as the ESR genes in *Xenopus* and the HER genes in zebra fish [53,55], which then act to antagonize the expression/activity of the neural bHLH proteins, both in terms of their ability to promote neuronal differentiation and to activate the expression of Notch ligands [51]. Blocking Notch signaling during primary neurogenesis results in a marked increase in the expression of the neural bHLH proteins within each proneural domain, with a corresponding increase in the number and density of neurons that subsequently differentiate. Conversely, increased Notch signaling in gain-of-function experiments represses the expression of the neural bHLH proteins and blocks neuronal differentiation.

Significantly, by restricting the number of progenitors that form neurons, Notch signaling keeps progenitor cells uncommitted, making them available for alternative neural fates, such as the neural crest [84]. Indeed, it is hard to argue that the raison d'être of Notch signaling is to regulate neuron number since neurons are usually overproduced in many regions of the developing nervous system during neurogenesis and then eliminated later on by cell death. It is also not likely that Notch exists to ensure that neurons differentiate in a precise spatial pattern since neurons undergo extensive migration to reach their final destination. Rather, Notch seems to be required to maintain a population of uncommitted precursors, which can contribute to later-born cell types that are generated over the protracted course of neural cell differentiation [85].

Studies of Notch function during secondary neurogenesis suggest that Notch signaling is used widely to maintain cells

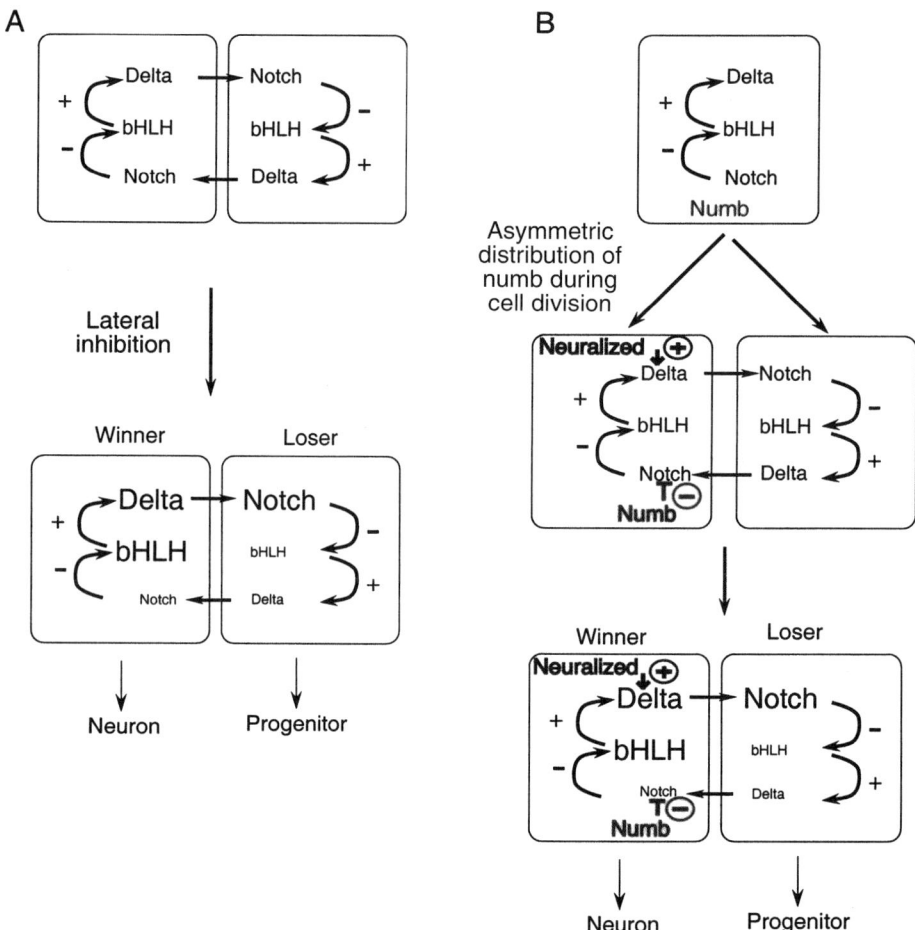

Figure 3 Lateral inhibition. (A) Genetic interactions between the Notch pathway genes and the differentiation bHLH proteins result in winners and losers during lateral inhibition. In this model, bHLH proteins promote neuronal differentiation cell autonomously but also inhibit neuronal differentiation in neighboring cells by activating the expression of ligand activating Notch, which inhibits the activity of the bHLH proteins. As a result, an unstable situation ensues in which some cells will tend to increase both bHLH activity and their ability to inhibit their neighbor (winner) while other cells will do the opposite (loser). (B) Example of mechanisms that preordain the outcome of lateral inhibition. In one case, the cytoplasmic determinate numb is asymmetrically localized to one daughter cell during division. Notch signaling is disabled in the cell that inherits numb, ensuring that this cell wins lateral inhibition while its sister loses. In another case, the protein neuralized is asymmetrically activated in one cell, and promotes Delta signaling by promoting its endocytosis in one cell, thus ensuring that it wins lateral inhibition.

in a progenitor state by preventing their differentiation en masse into neurons [86]. In the retina, loss of Notch function results in an excess production of early-born neuronal cell types at the expense of cell types that would be generated at later stages, while expression of activated forms of Notch activity in the retina represses neuronal differentiation [87–89]. Similarly, functional studies of Notch in neural precursors in the developing forebrain, the cerebellum, or in neural stem cells isolated from either the central or peripheral nervous system also point to a role for Notch in inhibiting neuronal differentiation [61,90–95]. Finally, Notch signaling negatively regulates the differentiation of hair cells in the inner ear as well as the differentiation of a mechanosensory neuron, the neuromast, in the developing lateral line of fish [20,96–98].

In cases where Notch inhibits neuronal differentiation, it seems to act by inducing the expression of the bHLH repressors that inhibit the activity/expression of the proneural bHLH proteins [5,99,100]. As a result, the role of Notch as a regulator of cell fate during neurogenesis has been viewed as permissive rather than instructive in nature. However, in some cases, expression of activated Notch in progenitor cells not only represses neuronal differentiation, but also promotes the differentiation of glial cells. For example, activated Notch promotes the differentiation of a later-born glial cell type, the Müller cell, in the retina [101], radial glial cells in the forebrain [90], myelin-producing glia in cultures of neural crest stem cells [61], or astrocytic glia in CNS neural stem cell cultures [95]. Nonetheless, it remains unclear whether Notch acts instructively in these cases since the targets of Notch required for glial differentiation are unknown [90].

Lateral Inhibition

During primary neurogenesis, the levels of RNA encoding the Notch ligand and to some extent the receptor change rapidly among progenitor cells. These changes have been attributed in part to the fact that the bHLH proteins not only promote neuronal differentiation but also promote the expression of the Delta-like ligands [82]. Activation of Notch therefore not only inhibits neuronal differentiation but also the ability of that cell to produce ligand and inhibit its neighbors (Fig. 3A). In principle, these interactions produce a dynamic negative feedback loop, called lateral inhibition, which amplifies differences in ligand expression between neighboring neural progenitor cells and generates winners that go on to form neurons (Fig. 3A)[102,103]. At one extreme, lateral inhibition has been proposed to act as a self-organizing selection process that uses the negative feedback loop to generate a salt-and-pepper pattern of ligand expression and differentiation from an initially equivalent population of neural progenitors (Fig. 3A). However, at the other extreme, lateral inhibition has been proposed to reinforce a bias that arose by another mechanism. An example of the latter has been described in *Drosophila*, whereby an asymmetrically localized, cytoplasmic determinant encoded by *Numb* has been shown to determine the fate of two daughter cells by regulating Notch signaling (Fig. 3B) [104,105]. In a similar manner, the vertebrate homologs of Numb, numb, and numb-like, may also determine cell fate by regulating Notch activity in the nervous system [106–109]. Another example is the protein neuralized which potentially promotes Delta activity, thus ensuring that certain cells consistently win lateral inhibition (Fig. 3B).

Notch and bHLH Factors

The role of Notch in regulating the differentiation of progenitor cells by targeting the activity of bHLH proteins is not restricted to the developing nervous system. Activated forms of Notch inhibit myogenesis in cultured myoblasts and this phenomenon has served as a model system for exploring potential inhibitory interactions between Notch and both myogenic bHLH proteins and associated transcription factors such as MEF2 [29,110–113]. Upregulation of MyoD and subsequent myogenic differentiation can also be blocked in chick embryos by Notch in gain-of-function experiments [114], but whether or not this regulation normally plays a physiological role *in vivo* is still unexplored. In the developing pancreas, there is strong genetic evidence showing that endodermal cells differentiate into an endocrine cell fate in response to a bHLH protein, neurogenin 3, and this differentiation step is inhibited by Notch signaling [115–117]. In a manner similar to that observed during neurogenesis, disabling Notch signaling, or the putative Notch target gene HES1, greatly increases the expression of neurogenin 3, resulting in a premature production of endocrine cells. Thus, the pancreas is another case where cell differentiation is promoted by a bHLH protein but inhibited by the Notch pathway [118].

Notch Function during Vertebrate Segmentation

During lateral inhibition, Notch signaling is used on a cell-by-cell basis to control cell fate. However, the study of Notch signaling in invertebrate embryos has also emphasized its role in patterning larger tissue domains such as when it is used to regulate the size of veins [119] or used to specify the margin in the developing *Drosophila* wing disc [120]. In an analogous manner, Notch signaling is also used in vertebrate embryos to pattern larger tissue domains, a particularly striking example of which occurs during segmentation of the mesoderm into somites.

Somites are the repeating, metameric building blocks of the vertebrate body plan that arise in early development from two bilateral sheets of cells, called the paraxial, or presomitic, mesoderm (PSM), which lie on both sides of the notochord [121]. Somites do not form en masse, but in an anterior to posterior progression, one unit arising at a regular interval every 30–90 minutes depending on the species. Because the process is progressive, a "differentiation wavefront" has been proposed that determines the point along the anterior-posterior (A/P) axis where paraxial cells mature and form a somite (Fig. 4A). However, to divide this differentiation process into a periodic pattern a segmental clock has been proposed which oscillates on and off during each segmental cycle, thus specifying the boundary between one somite and the next [121].

A role for the Notch pathway in somitogenesis was first evident from defects in somite morphology observed in mice with targeted mutations in the Notch1 gene or in the CSL protein, RBPjk [6,122]. Subsequently, the expression pattern of Notch pathway genes within the PSM revealed the formation of a segmental pattern of gene expression that prefigures the formation of somites. Indeed, based on this pattern, one could identify 1–3 segments, depending on the species, within the presomitic mesoderm (S0 to S-3 in Fig. 4A) [17]. Moreover, further analysis has produced a wealth of evidence indicating that Notch signaling is required for generating this segmental patterning, and that it contributes to this process somewhat as a component of the segmental clock [123]. First, the pioneering work of Pourquie and colleagues showed that genes encoding bHLH repressors oscillate in expression within the undifferentiated paraxial mesoderm with a periodicity corresponding to one segmental cycle (Fig. 4A) [124–127]. While it is not yet clear in all cases whether these genes are direct targets of Notch-mediated transcription, evidence from mouse and zebra fish indicate that their oscillatory expression is dependent on Notch signaling [127–129]. Secondly, Notch signaling seems to be activated synchronously within the paraxial cells during each segmental cycle. In fish and frog, the expression of RNA encoding the Notch ligand cycles off and on during

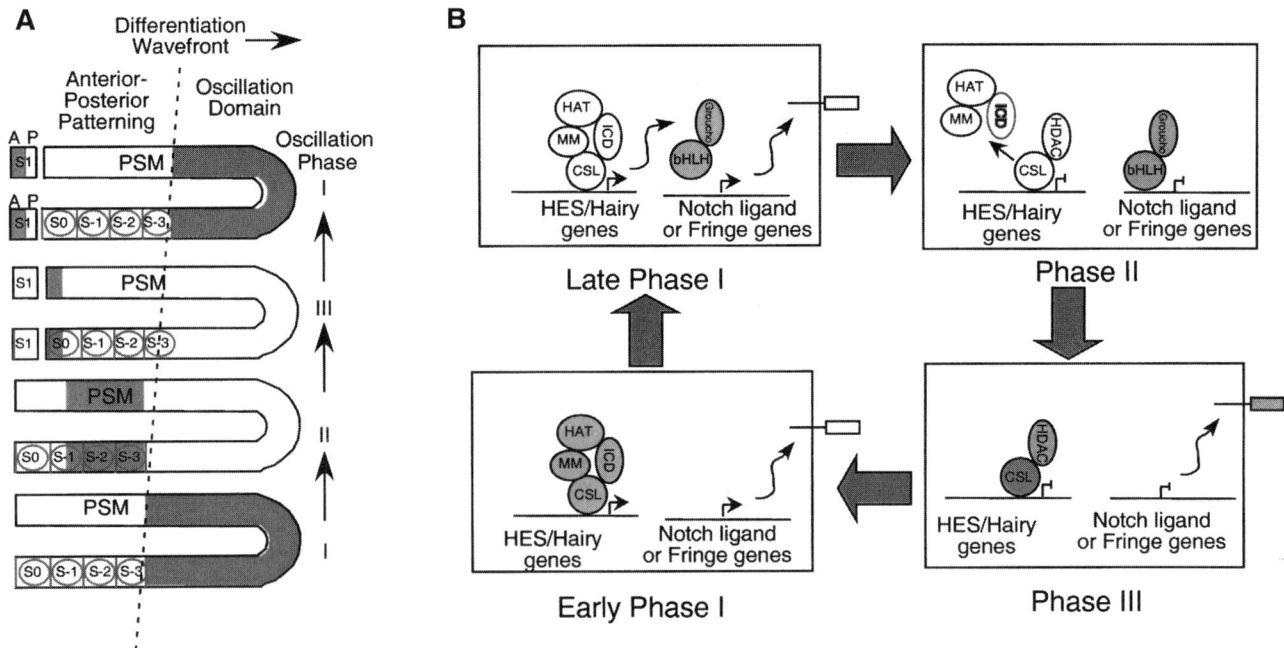

Figure 4 Model for Notch signaling during segmentation. (A) Somites form from the presomitic mesoderm (PSM) at regular intervals, with the most recently formed somite called S1. Each somite is divided into an anterior (A) and posterior (P) half (green shading). Notch signaling potentially plays two roles when segments (S0–S3) are generated within the PSM. In the more caudal region of the PSM (oscillatory domain), Notch signaling oscillates on and off during each segmental cycle (blue shading). This oscillation has been proposed to be part of the segmental clock that establishes segmental boundaries (purple lines). In the more rostral portion of the PSM (anterior-posterior patterning), Notch signaling may also act to subdivide segments into two halves by undergoing a dynamic refinement from a broad domain to a narrow half-segmental domain (red shading). (B) Model for the oscillation in Notch signaling that underlies the segmental clock. In this model, the bHLH repressors form a negative feedback loop that generates an oscillation much the same way the transcriptional repressors underlie oscillation in gene expression during circadian rhythms. A key element in the model shown is that the expression of a bHLH repressor (bHLH-groucho) is activated during each segmental cycle via Notch signaling, which then turns off the expression of genes required for Notch signaling, such as the Notch ligands or the fringe genes.

each segmental cycle within the undifferentiated paraxial mesoderm [130]. In contrast, ligand RNA expression is relatively uniform in the paraxial mesoderm of mouse and chick, but cycling is observed instead in the expression of RNA encoding the Notch modifier lunatic fringe [77,131]. Finally, mutations in these various components of the Notch pathway produce defects in somitic boundaries, indicating that the oscillating expression of these genes is critical for proper segmentation [76,77,126,129,132]. The exact role of Notch signaling in the segmental clock is still a matter of debate. One possibility is that the Notch pathway is the clock itself in which oscillations in gene expression are generated by Notch activation of the repressor bHLH proteins which then negatively feed back to shut down their own transcription by repressing the expression of the ligands or the fringe genes (Fig. 4B) [128]. Alternatively, Notch signaling may be downstream the target of a master clock, which drives the cyclic expression of components in the Notch pathway, perhaps as a means of synchronizing the phase of the clock among paraxial cells [130].

Establishment of boundaries during somitogenesis is tightly associated with the division of each somite into an anterior and posterior half, which is marked by the restricted expression of Notch ligands [121]. This A/P somite polarity has important functional consequences since it imposes segmental organization on the spinal cord and peripheral nervous system [133]. Loss of somite A/P polarity occurs in many cases where Notch signaling is disrupted during somitogenesis, indicating that Notch may also have a secondary role in polarizing somites along the A/P axis [121]. Significantly, the Notch pathway may contribute to A/P polarity by regulating the expression of the Mesp-like bHLH proteins [134,135]. Thus, the output of Notch signaling seems to specify domains of bHLH gene expression and thus patterns of differentiation during segmentation, in a manner that is similar to that seen during neurogenesis.

Notch Signaling in the Limb

The limb is another prominent example in vertebrate development where the contribution of Notch signaling to tissue patterning has been examined in some detail. In this case, Notch signaling contributes to the formation of a structure, called the apical epidermal ridge (AER), which forms at the boundary between the dorsal and ventral limb epidermis [120]. In chick limbs, Notch signaling may be modified during this process by a vertebrate homolog of *Drosophila* fringe, called *radical fringe*, which is expressed in the dorsal but not ventral limb epidermis [136–138]. When ectopically expressed ventrally using retrovirus vectors, radical fringe induces ectopic AER cells as does an activated form of Notch, while a dominant-negative form of the CSL proteins inhibits AER formation. It remains uncertain, however, whether these findings extend to the mouse where targeted mutations in *radical fringe* have no affect on limb development, although the situation is further complicated by potential genetic redundancy with the two other mammalian fringe genes, *lunatic* and *manic* [139]. Expression of the Notch ligand *Jagged2* is localized to the AER, and mice with targeted or naturally occurring mutations in this gene develop with syndactyly in both fore- and hindlimbs [140,141]. Additional work is needed to determine whether Notch signaling is a necessary factor in generating the AER, or whether it only has a secondary role in its maintenance and/or function.

Notch Signaling and Lymphoid Development

The hematopoietic system is an impressive example of cell diversification, and thus it is not surprising that the Notch pathway may have multiple roles in regulating cell fate as hematopoietic progenitors become progressively restricted in their developmental potential. The best evidence in this respect is for a role in Notch signaling in directing the differentiation of lymphoid-restricted progenitors into a T-cell versus a B-cell fate as these cells pass through the thymus and bone marrow, respectively. In mice where Notch1 is conditionally ablated in lymphocytic progenitors using the CRE-LOX system, T-cell development is inhibited [142] with a corresponding increase in intrathymic B cells [143]. Conversely, expressing activated Notch in bone marrow progenitors promotes extrathymic development of T cells and represses the development of early B cells [144]. A major unresolved issue is why Notch signaling is preferentially promoted in the thymus where T-cell development occurs [145] but repressed in the bone marrow during B-cell development. One likely target of Notch signaling within the T-cell lineage is the gene encoding one chain of the pre-TCR receptor whose expression then reinforces a T-cell fate [146], but the targets of Notch signaling that repress B-cell development are unknown.

Several lines of evidence indicate that Notch signaling influences the subsequent diversification within the T-cell lineage where progenitors are restricted further between a γδ or αβ fate or between the CD4 versus CD8 T-cell fate [147,148]. While most attention has been given to the role of Notch in the CD4/CD8 choice, how Notch affects this T-cell subdivision remains controversial [149–152]. The phenotypes reported for alterations in Notch signaling in this context may differ due to experimental differences in the forms of activated Notch used and how these forms were expressed in developing thymocytes. Nonetheless, one idea with far-reaching implications produced by this work is that Notch signaling may not be acting in the CD4/CD8 switch to influence cell fate per se, but rather to regulate such processes as lineage-specific cell death or differentiation. If this is indeed the case, then the targets of Notch signaling may be more extensive than previously appreciated, and inappropriate expression of such targets may potentially contribute to oncogenic transformation when the Notch pathway goes awry [153].

Notch Signaling during Vascular Development

Another venue where Notch signaling may control developmental processes other than simple cell fate choices is vascularization [154]. Indeed, one of the major reasons why mutations in the Notch pathway lead to embryonic lethality is defects in vascularization. These defects are likely to be due, in part, to a misspecification of angioblasts along an arterial or venous fate, thus producing profound defects in the how the vascular network is established. In zebra fish, loss of Notch signaling in fish embryos leads to a loss of artery-specific markers and a corresponding ectopic increase in cells that express venous markers [155]. Conversely, activated Notch signaling suppresses the expression of markers of venous differentiation. A potential direct target of Notch signaling during this process is a bHLH repressor encoded by the gene *gridlock*, which is also required genetically for establishing an arterial over a venous fate [54]. Further evidence for a role in vascular development comes from targeted mutations in various components of the Notch pathway in the mouse. Severe vascular defects associated with angiogenesis occur in mice that are double-mutant for both Notch1 and Notch4 [3]. Misexpressing the Notch4 intracellular domain specifically within the endothelial lineages produces a striking disorganization of the vascular network, indicating that Notch signaling acts in part by regulating some aspect of angioblast differentiation [156]. Both the Jagged1 and Delta1 knockout mice die at early embryonic stages with defects in vascular morphology [21,157] and a fourth ligand, Delta4, is expressed even earlier in the developing arterial system [3,16]. Thus, Notch signaling is likely to be required within the endothelial cells during the processes that elaborate the vascular network. Whether Notch functions simply as a regulator of cell differentiation or whether it has multiple roles during vascularization remains to be determined. Given that several Notch ligands and receptors are expressed with varying degrees of genetic overlap during vascular development, the role of Notch signaling during this process is likely to be quite complex.

Aspects of Notch Signaling in Other Organ Systems

The discussion thus far has focused mainly on the prominent examples of where Notch signaling has been shown to act during vertebrate development. However, this list is incomplete given the variety of tissue defects that have been described in various mouse mutants or combination of mutants in components in the Notch pathway. For example, mice with a hypomorphic mutation in Notch2 die from kidney defects that are most apparent within the developing glomeruli, and also show defects in various aspects of heart and eye development [158]. Various combinations of mutant forms of Notch2 and Jagged1 result in liver defects and other abnormalities that mimic those found in Alagille's syndrome [159]. In addition, the early lethality of the mouse mutants in many components of the Notch pathway has precluded the analysis of the pathway in tissues that develop at later stages, a problem that can be overcome by generating conditional knockouts. Based on both gene expression and functional studies, there is already a clear indication that Notch signaling has additional diverse roles in vertebrate development, including the patterning of the forebrain [160], the differentiation of oligodendrocytes into myelinating cells [161], the differentiation of epidermal stem cells into keratinocytes [162], and the formation of the lung [163].

Summary

Over the last ten years, the Notch pathway has emerged as one of the key signaling pathways operating during vertebrate development. In many cases, Notch signaling is used widely in vertebrate development to regulate cell fate often by activating the expression of bHLH repressors which in turn regulate the expression and/or activity of bHLH activators. These interactions can determine patterns of differentiation that range from salt-and-pepper to segmental. The role of Notch is not likely to be exclusive to cell fate decisions, but emerging evidence suggests that it might regulate a variety of cellular events ranging from differentiation, morphogenesis, and cell death to proliferation. A key goal of future studies is to identify the various gene targets of the CSL-dependent pathway that mediate these events, or to elucidate parallel pathways in which Notch signals by a CSL-independent mechanism if they exist. In addition, the diversity of Notch signaling is likely to depend in part on functional diversity of ligands and receptors, about which we still know very little. Finally, the Notch signaling pathway is likely to contribute to various human diseases, but more needs to be done to identify the nature of these lesions and how they alter cell physiology.

References

1. Greenwald, I. (1998). LIN-12/Notch signaling: Lessons from worms and flies. *Genes Dev.* **12**, 1751–1762.
2. Lewis, J. (1998). Notch signalling and the control of cell fate choices in vertebrates. *Semin. Cell. Dev. Biol.* **9**, 583–589.
3. Krebs, L. T., Xue, Y., Norton, C. R., Shutter, J. R., Maguire, M., Sundberg, J. P., Gallahan, D., Closson, V., Kitajewski, J., Callahan, R., Smith, G. H., Stark, K. L., and Gridley, T. (2000). Notch signaling is essential for vascular morphogenesis in mice. *Genes Dev.* **14**, 1343–1352.
4. Swiatek, P. J., Lindsell, C. E., del Amo, F. F., Weinmaster, G., and Gridley, T. (1994). Notch1 is essential for postimplantation development in mice. *Genes Dev.* **8**, 707–719.
5. de la Pompa, J. L., Wakeham, A., Correia, K. M., Samper, E., Brown, S., Aguilera, R. J., Nakano, T., Honjo, T., Mak, T. W., Rossant, J., and Conlon, R. A. (1997). Conservation of the Notch signalling pathway in mammalian neurogenesis. *Development* **124**, 1139–1148.
6. Conlon, R. A., Reaume, A. G., and Rossant, J. (1995). Notch1 is required for the coordinate segmentation of somites. *Development* **121**, 1533–1545.
7. Irvin, D. K., Zurcher, S. D., Nguyen, T., Weinmaster, G., and Kornblum, H. I. (2001). Expression patterns of Notch1, Notch2, and Notch3 suggest multiple functional roles for the Notch-DSL signaling system during brain development. *J. Comp. Neurol.* **436**, 167–181.
8. Lindsell, C. E., Boulter, J., diSibio, G., Gossler, A., and Weinmaster, G. (1996). Expression patterns of Jagged, Delta1, Notch1, Notch2, and Notch3 genes identify ligand-receptor pairs that may function in neural development. *Mol. Cell. Neurosci.* **8**, 14–27.
9. Hicks, C., Johnston, S. H., diSibio, G., Collazo, A., Vogt, T. F., and Weinmaster, G. (2000). Fringe differentially modulates Jagged1 and Delta1 signalling through Notch1 and Notch2. *Nat. Cell Biol.* **2**, 515–520.
10. Beatus, P., Lundkvist, J., Oberg, C., and Lendahl, U. (1993). The notch 3 intracellular domain represses notch 1-mediated activation through Hairy/Enhancer of split (HES) promoters. *Development* **126**, 3925–3935.
11. Aster, J., Pear, W., Hasserjian, R., Erba, H., Davi, F., Luo, B., Scott, M., Baltimore, D., and Sklar, J. (1994). Functional analysis of the TAN-1 gene, a human homolog of *Drosophila* notch. *Cold Spring Harbor Symp. Quant. Biol.* **59**, 125–136.
12. Jhappan, C., Gallahan, D., Stahle, C., Chu, E., Smith, G. H., Merlino, G., and Callahan, R. (1992). Expression of an activated Notch-related int-3 transgene interferes with cell differentiation and induces neoplastic transformation in mammary and salivary glands. *Genes Dev.* **6**, 345–355.
13. Joutel, A. and Tournier-Lasserve, E. (1998). Notch signalling pathway and human diseases. *Semin. Cell. Dev. Biol.* **9**, 619–625.
14. Weinmaster, G. (1997). The ins and outs of notch signaling. *Mol. Cell. Neurosci.* **9**, 91–102.
15. Dunwoodie, S. L., Henrique, D., Harrison, S. M., and Beddington, R. S. (1997). Mouse Dll3: a novel divergent Delta gene which may complement the function of other Delta homologues during early pattern formation in the mouse embryo. *Development* **124**, 3065–3076.
16. Shutter, J. R., Scully, S., Fan, W., Richards, W. G., Kitajewski, J., Deblandre, G. A., Kintner, C. R., and Stark, K. L. (2000). Dll4, a novel Notch ligand expressed in arterial endothelium. *Genes Dev.* **14**, 1313–1318.
17. Jen, W. C., Wettstein, D., Turner, D., Chitnis, A., and Kintner, C. (1997). The Notch ligand, X-Delta-2, mediates segmentation of the paraxial mesoderm in *Xenopus* embryos. *Development* **124**, 1169–1178.
18. Shawber, C., Boulter, J., Lindsell, C. E., and Weinmaster, G. (1996). Jagged2: A serrate-like gene expressed during rat embryogenesis. *Dev. Biol.* **180**, 370–376.
19. Lindsell, C. E., Shawber, C. J., Boulter, J., and Weinmaster, G. (1995). Jagged: A mammalian ligand that activates Notch1. *Cell* **80**, 909–917.
20. Lanford, P. J., Lan, Y., Jiang, R., Lindsell, C., Weinmaster, G., Gridley, T., and Kelley, M. W. (1999). Notch signalling pathway mediates hair cell development in mammalian cochlea. *Nat. Genet.* **21**, 289–292.
21. Xue, Y., Gao, X., Lindsell, C. E., Norton, C. R., Chang, B., Hicks, C., Gendron-Maguire, M. R., Weinmaster, G., and Gridley, T. (1999). Embryonic lethality and vascular defects in mice lacking the Notch ligand Jagged1. *Hum. Mol. Genet.* **8**, 723–730.
22. Pourquie, O. and Kusumi, K. (2001). When body segmentation goes wrong. *Clin. Genet.* **60**, 409–416.
23. Kusumi, K., Sun, E. S., Kerrebrock, A. W., Bronson, R. T., Chi, D. C., Bulotsky, M. S., Spencer, J. B., Birren, B. W., Frankel, W. N., and Lander, E. S. (1998). The mouse pudgy mutation disrupts Delta homologue Dll3 and initiation of early somite boundaries. *Nat. Genet.* **19**, 274–278.
24. Jarriault, S., Brou, C., Logeat, F., Schroeter, E. H., Kopan, R., and Israel, A. (1995). Signalling downstream of activated mammalian Notch. *Nature* **377**, 355–358.
25. Ebinu, J. O. and Yankner, B. A. (2002). A RIP tide in neuronal signal transduction. *Neuron* **34**, 499–502.
26. Weinmaster, G. (2000). Notch signal transduction: A real rip and more. *Curr. Opin. Genet. Dev.* **10**, 363–369.
27. Kadesch, T. (2000). Notch signaling: A dance of proteins changing partners. *Exp. Cell Res.* **260**, 1–8.

28. Mumm, J. S. and Kopan, R. (2000). Notch signaling: From the outside in. *Dev. Biol.*, in press.
29. Shawber, C., Nofziger, D., Hsieh, J. J., Lindsell, C., Bogler, O., Hayward, D., and Weinmaster, G. (1996). Notch signaling inhibits muscle cell differentiation through a CBF1-independent pathway. *Development* **122**, 3765–3773.
30. Matsuno, K., Diederich, R. J., Go, M. J., Blaumueller, C. M., and Artavanis-Tsakonas, S. (1995). Deltex acts as a positive regulator of Notch signaling through interactions with the Notch ankyrin repeats. *Development* **121**, 2633–2644.
31. Ramain, P., Khechumian, K., Seugnet, L., Arbogast, N., Ackermann, C., and Heitzler, P. (2001). Novel Notch alleles reveal a Deltex-dependent pathway repressing neural fate. *Curr. Biol.* **11**, 1729–1738.
32. Giniger, E. (1998). A role for Abl in Notch signaling. *Neuron* **20**, 667–681.
33. Zecchini, V., Brennan, K., and Martinez-Arias, A. (1999). An activity of Notch regulates JNK signalling and affects dorsal closure in Drosophila. *Curr. Biol.* **9**, 460–469.
34. Matsuno, K., Eastman, D., Mitsiades, T., Quinn, A. M., Carcanciu, M. L., Ordentlich, P., Kadesch, T., and Artavanis-Tsakonas, S. (1998). Human deltex is a conserved regulator of Notch signalling. *Nat. Genet.* **19**, 74–78.
35. Izon, D. J., Aster, J. C., He, Y., Weng, A., Karnell, F. G., Patriub, V., Xu, L., Bakkour, S., Rodriguez, C., Allman, D., and Pear, W. S. (2002). Deltex1 redirects lymphoid progenitors to the B cell lineage by antagonizing Notch1. *Immunity* **16**, 231–243.
36. Ordentlich, P., Lin, A., Shen, C. P., Blaumueller, C., Matsuno, K., Artavanis-Tsakonas, S., and Kadesch, T. (1998). Notch inhibition of E47 supports the existence of a novel signaling pathway. *Mol. Cell. Biol.* **18**, 2230–2239.
37. Redmond, L., Oh, S. R., Hicks, C., Weinmaster, G., and Ghosh, A. (2000). Nuclear Notch1 signaling and the regulation of dendritic development. *Nat. Neurosci.* **3**, 30–40.
38. Sestan, N., Artavanis-Tsakonas, S., and Rakic, P. (1999). Contact-dependent inhibition of cortical neurite growth mediated by notch signaling. *Science* **286**, 741–746.
39. Kopan, R., Schroeter, E. H., Weintraub, H., and Nye, J. S. (1996). Signal transduction by activated mNotch: importance of proteolytic processing and its regulation by the extracellular domain. *Proc. Natl. Acad. Sci. USA* **93**, 1683–1688.
40. Brown, M. S., Ye, J., Rawson, R. B., and Goldstein, J. L. (2000). Regulated intramembrane proteolysis: A control mechanism conserved from bacteria to humans. *Cell* **100**, 391–398.
41. Selkoe, D. J. (2001). Presenilin, Notch, and the genesis and treatment of Alzheimer's disease. *Proc. Natl. Acad. Sci. USA* **98**, 11039–11041.
42. Chan, Y. M. and Jan, Y. N. (1999). Presenilins, processing of beta-amyloid precursor protein, and notch signaling. *Neuron* **23**, 201–204.
43. De Strooper, B., Annaert, W., Cupers, P., Saftig, P., Craessaerts, K., Mumm, J. S., Schroeter, E. H., Schrijvers, V., Wolfe, M. S., Ray, W. J., Goate, A., and Kopan, R. (1999). A presenilin-1-dependent gamma-secretase-like protease mediates release of Notch intracellular domain. *Nature* **398**, 518–522.
44. Struhl, G. and Adachi, A. (2000). Requirements for presenilin-dependent cleavage of notch and other transmembrane proteins. *Mol. Cell* **6**, 625–636.
45. Mumm, J. S., Schroeter, E. H., Saxena, M. T., Griesemer, A., Tian, X., Pan, D. J., Ray, W. J., and Kopan, R. (2000). A ligand-induced extracellular cleavage regulates gamma-secretase-like proteolytic activation of Notch1. *Mol. Cell* **5**, 197–206.
46. Fortini, M. E. (2001). Notch and presenilin: A proteolytic mechanism emerges. *Curr. Opin. Cell Biol.* **13**, 627–634.
47. Brou, C., Logeat, F., Gupta, N., Bessia, C., LeBail, O., Doedens, J. R., Cumano, A., Roux, P., Black, R. A., and Israel, A. (2000). A novel proteolytic cleavage involved in Notch signaling: The role of the disintegrin-metalloprotease TACE. *Mol. Cell* **5**, 207–216.
48. Logeat, F., Bessia, C., Brou, C., LeBail, O., Jarriault, S., Seidah, N. G., and Israel, A. (1998). The Notch1 receptor is cleaved constitutively by a furin-like convertase. *Proc. Natl. Acad. Sci. USA* **95**, 8108–8112.
49. Rand, M. D., Grimm, L. M., Artavanis-Tsakonas, S., Patriub, V., Blacklow, S. C., Sklar, J., and Aster, J. C. (2000). Calcium depletion dissociates and activates heterodimeric notch receptors. *Mol. Cell. Biol.* **20**, 1825–1835.
50. Schroeter, E. H., Kisslinger, J. A., and Kopan, R. (1998). Notch-1 signalling requires ligand-induced proteolytic release of intracellular domain. *Nature* **393**, 382–386.
51. Davis, R. L. and Turner, D. L. (2001). Vertebrate hairy and Enhancer of split related proteins: transcriptional repressors regulating cellular differentiation and embryonic patterning. *Oncogene* **20**, 8342–8357.
52. Iso, T., Chung, G., Hamamori, Y., and Kedes, L. (2002). HERP1 is a cell type-specific primary target of Notch. *J. Biol. Chem.* **277**, 6598–6607.
53. Takke, C., Dornseifer, P., Weizsacker, E., and Campos-Ortega, J. A. (1999). her4, a zebrafish homologue of the Drosophila neurogenic gene E(spl), is a target of NOTCH signalling. *Development* **126**, 1811–1821.
54. Zhong, T. P., Childs, S., Leu, J. P., and Fishman, M. C. (2001). Gridlock signalling pathway fashions the first embryonic artery. *Nature* **414**, 216–220.
55. Wettstein, D. A., Turner, D. L., and Kintner, C. (1997). The Xenopus homolog of Drosophila Suppressor of Hairless mediates Notch signaling during primary neurogenesis. *Development* **124**, 693–702.
56. Barolo, S. and Posakony, J. W. (2002). Three habits of highly effective signaling pathways: principles of transcriptional control by developmental cell signaling. *Genes Dev.* **16**, 1167–1181.
57. Hukriede, N. A., Gu, Y., and Fleming, R. J. (1997). A dominant-negative form of Serrate acts as a general antagonist of Notch activation. *Development* **124**, 3427–3437.
58. Sun, X. and Artavanis-Tsakonas, S. (1997). Secreted forms of DELTA and SERRATE define antagonists of Notch signaling in Drosophila. *Development* **124**, 3439–3448.
59. Varnum-Finney, B., Wu, L., Yu, M., Brashem-Stein, C., Staats, S., Flowers, D., Griffin, J. D., and Bernstein, I. D. (2000). Immobilization of Notch ligand, Delta-1, is required for induction of notch signaling. *J. Cell Sci.* **113**, 4313–4318.
60. Shimizu, K., Chiba, S., Saito, T., Takahashi, T., Kumano, K., Hamada, Y., and Hirai, H. (2002). Integrity of intracellular domain of Notch ligand is indispensable for cleavage required for release of the Notch2 intracellular domain. *EMBO J.* **21**, 294–302.
61. Morrison, S. J., Perez, S. E., Qiao, Z., Verdi, J. M., Hicks, C., Weinmaster, G., and Anderson, D. J. (2000). Transient Notch activation initiates an irreversible switch from neurogenesis to gliogenesis by neural crest stem cells. *Cell* **101**, 499–510.
62. Parks, A. L., Klueg, K. M., Stout, J. R., and Muskavitch, M. A. (2000). Ligand endocytosis drives receptor dissociation and activation in the Notch pathway. *Development* **127**, 1373–1385.
63. Chitnis, A., Henrique, D., Lewis, J., Ish-Horowicz, D., and Kintner, C. (1995). Primary neurogenesis in Xenopus embryos regulated by a homologue of the Drosophila neurogenic gene Delta. *Nature* **375**, 761–766.
64. Sun, X. and Artavanis-Tsakonas, S. (1996). The intracellular deletions of Delta and Serrate define dominant negative forms of the Drosophila Notch ligands. *Development* **122**, 2465–2474.
65. Kramer, H. (2001). Neuralized: regulating notch by putting away delta. *Dev. Cell.* **1**, 725–726.
66. Deblandre, G. A., Lai, E. C., and Kintner, C. (2001). Xenopus neuralized is a ubiquitin ligase that interacts with XDelta1 and regulates Notch signaling. *Dev. Cell* **1**, 795–806.
67. Lai, E. C., Deblandre, G. A., Kintner, C., and Rubin, G. M. (2001). Drosophila neuralized is a ubiquitin ligase that promotes the internalization and degradation of delta. *Dev. Cell.* **1**, 783–794.
68. Pavlopoulos, E., Pitsouli, C., Klueg, K. M., Muskavitch, M. A., Moschonas, N. K., and Delidakis, C. (2001). Neuralized Encodes, a peripheral membrane protein involved in delta signaling and endocytosis. *Dev. Cell* **1**, 807–816.
69. Lai, E. C. (2002). Protein degradation: four E3s for the notch pathway. *Curr. Biol.* **12**, 74–R78.
70. Irvine, K. D. and Wieschaus, E. (1994). fringe, a Boundary-specific signaling molecule, mediates interactions between dorsal and ventral cells during Drosophila wing development. *Cell* **79**, 595–606.

71. Moloney, D. J., Panin, V. M., Johnston, S. H., Chen, J., Shao, L., Wilson, R., Wang, Y., Stanley, P., Irvine, K. D., Haltiwanger, R. S., and Vogt, T. F. (2000). Fringe is a glycosyltransferase that modifies Notch. *Nature* **406**, 369–375.

72. Bruckner, K., Perez, L., Clausen, H., and Cohen, S. (2000). Glycosyltransferase activity of Fringe modulates Notch-Delta interactions. *Nature* **406**, 411–415.

73. Fortini, M. E. (2000). Fringe benefits to carbohydrates. *Nature* **406**, 357–358.

74. Johnston, S. H., Rauskolb, C., Wilson, R., Prabhakaran, B., Irvine, K. D., and Vogt, T. F. (1997). A family of mammalian Fringe genes implicated in boundary determination and the Notch pathway. *Development* **124**, 2245–2254.

75. Zhang, N., Martin, G. V., Kelley, M. W., and Gridley, T. (2000). A mutation in the Lunatic fringe gene suppresses the effects of a Jagged2 mutation on inner hair cell development in the cochlea. *Curr. Biol.* **10**, 659–662.

76. Zhang, N. and Gridley, T. Defects in somite formation in lunatic fringe-deficient mice. *Nature* **394**, 374–377.

77. Evrard, Y. A., Lun, Y., Aulehla, A., Gan, L., and Johnson, R. L. (1998). The lunatic fringe is an essential mediator of somite segmentation and patterning. *Nature* **394**, 377–381.

78. Shimizu, K., Chiba, S., Saito, T., Kumano, K., Takahashi, T., and Hirai, H. (2001). Manic fringe and lunatic fringe modify different sites of the Notch2 extracellular region, resulting in different signaling modulation. *J. Biol. Chem.* **276**, 25753–25758.

79. Campos-Ortega, J. A. (1988). Cellular interactions during early neurogenesis of *Drosophila melanogaster*. *TINS* **11**.

80. Appel, B., Givan, L. A., and Eisen, J. S. (2001). Delta-Notch signaling and lateral inhibition in zebrafish spinal cord development. *BMC Dev. Biol.* **1**, 13.

81. Dornseifer, P., Takke, C., and Campos-Ortega, J. A. (1997). Overexpression of a zebrafish homologue of the *Drosophila* neurogenic gene Delta perturbs differentiation of primary neurons and somite development. *Mech. Dev.* **63**, 159–171.

82. Chitnis, A. and Kintner, C. (1996). Sensitivity of proneural genes to lateral inhibition affects the pattern of primary neurons in *Xenopus* embryos. *Development* **122**, 2295–2301.

83. Ma, Q., Kintner, C., and Anderson, D. J. (1996). Identification of neurogenin, a vertebrate neuronal determination gene. *Cell* **87**, 43–52.

84. Cornell, R. A. and Eisen, J. S. (2002). Delta/Notch signaling promotes formation of zebrafish neural crest by repressing Neurogenin 1 function. *Development* **129**, 2639–2648.

85. Lewis, J. (1996). Neurogenic genes and vertebrate neurogenesis. *Curr. Opin. Neurobiol.* **6**, 3–10.

86. Kintner, C. (2002). Neurogenesis in embryos and in adult neural stem cells. *J Neurosci.* **22**, 639–643.

87. Dorsky, R. I., Chang, W. S., Rapaport, D. H., and Harris, W. A. (1997). Regulation of neuronal diversity in the *Xenopus* retina by Delta signalling. *Nature* **385**, 67–70.

88. Henrique, D., Hirsinger, E., Adam, J., Le Roux, I., Pourquie, O., Ish-Horowicz, D., and Lewis, J. (1997). Maintenance of neuroepithelial progenitor cells by Delta-Notch signalling in the embryonic chick retina. *Curr. Biol.* **7**, 661–670.

89. Austin, C. P., Feldman, D. E., Ida, J. A., Jr., and Cepko, C. L. (1995). Vertebrate retinal ganglion cells are selected from competent progenitors by the action of Notch. *Development* **121**, 3637–3650.

90. Gaiano, N., Nye, J. S., and Fishell, G. (2000). Radial glial identity is promoted by Notch1 signaling in the murine forebrain. *Neuron* **26**, 395–404.

91. Lutolf, S., Radtke, F., Aguet, M., Suter, U., and Taylor, V. (2002). Notch1 is required for neuronal and glial differentiation in the cerebellum. *Development* **129**, 373–385.

92. Solecki, D. J., Liu, X. L., Tomoda, T., Fang, Y., and Hatten, M. E. (2001). Activated Notch2 signaling inhibits differentiation of cerebellar granule neuron precursors by maintaining proliferation. *Neuron* **31**, 557–568.

93. Yamamoto, S., Nagao, M., Sugimori, M., Kosako, H., Nakatomi, H., Yamamoto, N., Takebayashi, H., Nabeshima, Y., Kitamura, T., Weinmaster, G., Nakamura, K., and Nakafuku, M. (2001). Transcription factor expression and Notch-dependent regulation of neural progenitors in the adult rat spinal cord. *J. Neurosci.* **21**, 9814–9823.

94. Hitoshi, S., Alexson, T., Tropepe, V., Donoviel, D., Elia, A. J., Nye, J. S., Conlon, R. A., Mak, T. W., Bernstein, A., and van der Kooy, D. (2002). Notch pathway molecules are essential for the maintenance, but not the generation, of mammalian neural stem cells. *Genes Dev.* **16**, 846–858.

95. Tanigaki, K., Nogaki, F., Takahashi, J., Tashiro, K., Kurooka, H., and Honjo. T. (2001). Notch1 and Notch3 instructively restrict bFGF-responsive multipotent neural progenitor cells to an astroglial fate. *Neuron*. **29**, 45–55.

96. Haddon, C., Jiang, Y. J., Smithers, L., and Lewis, J. (1998). Delta-Notch signalling and the patterning of sensory cell differentiation in the zebrafish ear: Evidence from the mind bomb mutant. *Development* **125**, 4637–4644.

97. Riley, B. B., Chiang, M., Farmer, L., and Heck, R. (1999). The deltaA gene of zebrafish mediates lateral inhibition of hair cells in the inner ear and is regulated by pax2.1. *Development* **126**, 5669–5678.

98. Itoh, M. and Chitnis, A. B. (2001). Expression of proneural and neurogenic genes in the zebrafish lateral line primordium correlates with selection of hair cell fate in neuromasts. *Mech. Dev.* **102**, 263–266.

99. Kageyama, R. and Ohtsuka, T. (1999). The Notch-Hes pathway in mammalian neural development. *Cell Res.* **9**, 179–188.

100. Ohtsuka, T., Ishibashi, M., Gradwohl, G., Nakanishi, S., Guillemot, F., and Kageyama, R. (1999). Hes1 and Hes5 as notch effectors in mammalian neuronal differentiation. *EMBO J.* **18**, 2196–2207.

101. Vetter, M. L. and Moore, K. B. (2001). Becoming glial in the neural retina. *Dev. Dynam.* **221**, 146–153.

102. Wilkinson, H. A., Fitzgerald, K., and Greenwald, I. (1994). Reciprocal changes in expression of the receptor lin-12 and its ligand lag-2 prior to commitment in a *C. elegans* cell fate decision. *Cell* **79**, 1187–1198.

103. Heitzler, P. and Simpson, P. (1991). The choice of cell fate in the epidermis of *Drosophila*. *Cell* **64**, 1083–1092.

104. Spana, E. P. and Doe, C. Q. (1996). Numb antagonizes Notch signaling to specify sibling neuron cell fates. *Neuron* **17**, 21–26.

105. Guo, M., Jan, L. Y., and Jan, Y. N. (1996). Control of daughter cell fates during asymmetric division: interaction of Numb and Notch. *Neuron* **17**, 27–41.

106. Zhong, W., Feder, J. N., Jiang, M. M., Jan, L. Y., and Jan, Y. N. (1996). Asymmetric localization of a mammalian numb homolog during mouse cortical neurogenesis. *Neuron* **17**, 43–53.

107. Verdi, J. M., Bashirullah, A., Goldhawk, D. E., Kubu, C. J., Jamali, M., Meakin, S. O., and Lipshitz, H. D. (1999). Distinct human NUMB isoforms regulate differentiation vs. proliferation in the neuronal lineage. *Proc. Natl. Acad. Sci. USA* **96**, 10472–10476.

108. Wakamatsu, Y., Maynard, T. M., and Weston, J. A. (2000). Fate determination of neural crest cells by NOTCH-mediated lateral inhibition and asymmetrical cell division during gangliogenesis. *Development* **127**, 2811–2821.

109. Wakamatsu, Y., Maynard, T. M., Jones, S. U., and Weston, J. A. (1999). NUMB localizes in the basal cortex of mitotic avian neuroepithelial cells and modulates neuronal differentiation by binding to NOTCH-1. *Neuron* **23**, 71–81.

110. Nye, J. S., Kopan, R., and Axel, R. (1994). An activated Notch suppresses neurogenesis and myogenesis but not gliogenesis in mammalian cells. *Development* **120**, 2421–2430.

111. Kopan, R., Nye, J. S., and Weintraub, H. The intracellular domain of mouse Notch: a constitutively activated repressor of myogenesis directed at the basic helix-loop-helix region of MyoD. *Development* **120**, 2385–2396.

112. Wilson-Rawls, J., Molkentin, J. D., Black, B. L., and Olson, E. N. (1999). Activated notch inhibits myogenic activity of the MADS-Box transcription factor myocyte enhancer factor 2C. *Mol. Cell. Biol.* **19**, 2853–2862.

113. Kuroda, K., Tani, S., Tamura, K., Minoguchi, S., Kurooka, H., Honjo, T. (1999). Delta-induced Notch signaling mediated by RBP-J inhibits MyoD expression and myogenesis. *J. Biol. Chem.* **274**, 7238–7244.
114. Hirsinger, E., Malapert, P., Dubrulle, J., Delfini, M. C., Duprez, D., Henrique, D., Ish-Horowicz, D., and Pourquie, O. (2001). Notch signalling acts in postmitotic avian myogenic cells to control MyoD activation. *Development* **128**, 107–116.
115. Apelqvist, A., Li, H., Sommer, L., Beatus, P., Anderson, D. J., Honjo, T., Hrabe de Angelis, M., Lendahl, U., and Edlund, H. (1999). Notch signalling controls pancreatic cell differentiation. *Nature* **400**, 877–881.
116. Jensen, J., Pedersen, E. E., Galante, P., Hald, J., Heller, R. S., Ishibashi, M., Kageyama, R., Guillemot, F., Serup, P., and Madsen, O. D. (2000). Control of endodermal endocrine development by Hes-1. *Nat. Genet.* **24**, 36–44.
117. Jensen, J., Heller, R. S., Funder-Nielsen, T., Pedersen, E. E., Lindsell, C., Weinmaster, G., Madsen, O. D., and Serup, P. (2000). Independent development of pancreatic alpha- and beta-cells from neurogenin3-expressing precursors: a role for the notch pathway in repression of premature differentiation. *Diabetes* **49**, 163–176.
118. Skipper, M. and Lewis, J. (2000). Getting to the guts of enteroendocrine differentiation. *Nat. Genet.* **24**, 3–4.
119. Huppert, S. S., Jacobsen, T. L., and Muskavitch, M. A. (1997). Feedback regulation is central to Delta-Notch signalling required for *Drosophila* wing vein morphogenesis. *Development* **124**, 3283–3291.
120. Irvine, K. D. (1999). Fringe, Notch, and making developmental boundaries. *Curr. Opin. Genet. Dev.* **9**, 434–441.
121. Pourquie. O. (2001). Vertebrate somitogenesis. *Annu. Rev. Cell Dev. Biol.* **17**, 311–350.
122. Oka, C., Nakano, T., Wakeham, A., de la Pompa, J. L., Mori, C., Sakai, T., Okazaki, S., Kawaichi, M., Shiota, K., Mak, T. W., and Honjo, T. (1995). Disruption of the mouse RBP-J kappa gene results in early embryonic death. *Development* **121**, 3291–3301.
123. Pourquie, O. (1999). Notch around the clock. *Curr. Opin. Genet. Dev.* **9**, 559–565.
124. Palmeirim, I., Henrique, D., Ish-Horowicz, D., and Pourquie, O. (1997). Avian hairy gene expression identifies a molecular clock linked to vertebrate segmentation and somitogenesis. *Cell* **91**, 639–648.
125. Takke, C. and Campos-Ortega, J. A. (1999). her1, a zebrafish pair-rule like gene, acts downstream of notch signalling to control somite development. *Development* **126**, 3005–3014.
126. Bessho, Y., Sakata, R., Komatsu, S., Shiota, K., Yamada, S., and Kageyama, R. (2001). Dynamic expression and essential functions of Hes7 in somite segmentation. *Genes Dev.* **15**, 2642–2647.
127. Jouve, C., Palmeirim, I., Henrique, D., Beckers, J., Gossler, A., Ish-Horowicz, D., and Pourquie, O. (2000). Notch signalling is required for cyclic expression of the hairy-like gene HES1 in the presomitic mesoderm. *Development* **127**, 1421–1429.
128. Holley, S. A., Julich, D., Rauch, G. J., Geisler, R., and Nusslein-Volhard, C. (2002). her1 and the notch pathway function within the oscillator mechanism that regulates zebrafish somitogenesis. *Development* **129**, 1175–1183.
129. Oates, A. C. and Ho, R. K. (2002). Hairy/E(spl)-related (Her) genes are central components of the segmentation oscillator and display redundancy with the Delta/Notch signaling pathway in the formation of anterior segmental boundaries in the zebrafish. *Development* **129**, 2929–2946.
130. Jiang, Y. J., Aerne, B. L., Smithers, L., Haddon, C., Ish-Horowicz, D., and Lewis, J. (2000). Notch signalling and the synchronization of the somite segmentation clock. *Nature* **408**, 475–479.
131. McGrew, M. J., Dale, J. K., Fraboulet, S., and Pourquie, O. (1998). The lunatic fringe gene is a target of the molecular clock linked to somite segmentation in avian embryos. *Curr. Biol.* **8**, 979–982.
132. Holley, S. A., Geisler, R., and Nusslein-Volhard, C. (2000). Control of her1 expression during zebrafish somitogenesis by a delta-dependent oscillator and an independent wave-front activity. *Genes Dev.* **14**, 1678–1690.
133. Keynes, R. J. and Stern, C. D. (1984). Segmentation in the vertebrate nervous system. *Nature* **310**, 786–789.
134. Jen, W. C., Gawantka, V., Pollet, N., Niehrs, C., and Kintner, C. (1999). Periodic repression of Notch pathway genes governs the segmentation of *Xenopus* embryos. *Genes Dev.* **13**, 1486–1499.
135. Takahashi, Y., Koizumi, K., Takagi, A., Kitajima, S., Inoue, T., Koseki, H., and Saga, Y. (2000). Mesp2 initiates somite segmentation through the Notch signalling pathway. *Nat. Genet.* **25**, 390–396.
136. Irvine, K. D. and Vogt, T. F. (1997). Dorsal-ventral signaling in limb development. *Curr. Opin. Cell Biol.* **9**, 867–876.
137. Laufer, E., Dahn, R., Orozco, O. E., Yeo, C. Y., Pisenti, J., Henrique, D., Abbott, U. K., Fallon, J. F., and Tabin, C. (1997). Expression of Radical fringe in limb-bud ectoderm regulates apical ectodermal ridge formation. *Nature* **386**, 366–373.
138. Rodriguez-Esteban, C., Schwabe, J. W., De La Pena, J., Foys, B., Eshelman, B., and Belmonte, J. C. (1997). Radical fringe positions the apical ectodermal ridge at the dorsoventral boundary of the vertebrate limb. *Nature* **386**, 360–366.
139. Moran, J. L., Levorse, J. M., and Vogt, T. F. (1999). Limbs move beyond the radical fringe. *Nature* **399**, 742–743.
140. Sidow, A., Bulotsky, M. S., Kerrebrock, A. W., Bronson, R. T., Daly, M. J., Reeve, M. P., Hawkins, T. L., Birren, B. W., Jaenisch, R., and Lander, E. S. (1997). Serrate2 is disrupted in the mouse limb-development mutant syndactylism. *Nature* **389**, 722–725.
141. Jiang, R., Lan, Y., Chapman, H. D., Shawber, C., Norton, C. R., Serreze, D. V., Weinmaster, G., and Gridley, T. (1998). Defects in limb, craniofacial, and thymic development in Jagged2 mutant mice. *Genes Dev.* **12**, 1046–1057.
142. Radtke, F., Wilson, A., Stark, G., Bauer, M., van Meerwijk, J., MacDonald, H. R., and Aguet, M. (1999). Deficient T cell fate specification in mice with an induced inactivation of Notch1. *Immunity* **10**, 547–558.
143. Wilson, A., MacDonald, H. R., and Radtke, F. (2001). Notch 1-deficient common lymphoid precursors adopt a B cell fate in the thymus. *J. Exp. Med.* **194**, 1003–1012.
144. Pui, J. C., Allman, D., Xu, L., DeRocco, S., Karnell, F. G., Bakkour, S., Lee, J. Y., Kadesch, T., Hardy, R. R., Aster, J. C., and Pear, W. S. (1999). Notch1 expression in early lymphopoiesis influences B versus T lineage determination. *Immunity* **11**, 299–308.
145. Anderson, G., Pongracz, J., Parnell, S., and Jenkinson, E. J. (2001). Notch ligand-bearing thymic epithelial cells initiate and sustain Notch signaling in thymocytes independently of T cell receptor signaling. *Eur. J. Immunol.* **31**, 3349–3354.
146. Reizis, B. and Leder, P. (2002). Direct induction of T lymphocyte-specific gene expression by the mammalian Notch signaling pathway. *Genes Dev.* **16**, 295–300.
147. Robey, E., Chang, D., Itano, A., Cado, D., Alexander, H., Lans, D., Weinmaster, G., and Salmon, P. (1996). An activated form of Notch influences the choice between CD4 and CD8 T cell lineages. *Cell* **87**, 483–492.
148. Washburn, T., Schweighoffer, E., Gridley, T., Chang, D., Fowlkes, B. J., Cado, D., and Robey, E. (1997). Notch activity influences the alphabeta versus gammadelta T cell lineage decision. *Cell* **88**, 833–843.
149. Valdez, P. and Robey, E. (1999). Notch and the CD4 versus CD8 lineage decision. *Cold Spring Harbor Symp. Quant. Biol.* **64**, 27–31.
150. Deftos, M. L., Huang, E., Ojala, E. W., Forbush, K. A., and Bevan, M. J. (2000). Notch1 signaling promotes the maturation of CD4 and CD8 SP thymocytes. *Immunity* **13**, 73–84.
151. Izon, D. J., Punt, J. A., Xu, L., Karnell, F. G., Allman, D., Myung, P. S., Boerth, N. J., Pui, J. C., Koretzky, G. A., and Pear, W. S. (2001). Notch1 regulates maturation of CD4+ and CD8+ thymocytes by modulating TCR signal strength. *Immunity* **14**, 253–264.
152. Germain, R. N. (2002). T-cell development and the CD4-CD8 lineage decision. *Na. Rev. Immunol.* **2**, 309–322.
153. Aster, J. C. and Pear, W. S. (2001). Notch signaling in leukemia. *Curr. Opin. Hematol.* **8**, 237–244.
154. Gridley, T. (2001). Notch signaling during vascular development. *Proc. Natl. Acad. Sci. USA* **98**, 5377–5378.

155. Lawson, N. D., Scheer, N., Pham, V. N., Kim, C. H., Chitnis, A. B., Campos-Ortega, J. A., and Weinstein, B. M. (2001). Notch signaling is required for arterial-venous differentiation during embryonic vascular development. *Development* **128**, 3675–3683.
156. Uyttendaele, H., Ho, J., Rossant, J., and Kitajewski, J. (2001). Vascular patterning defects associated with expression of activated Notch4 in embryonic endothelium. *Proc. Natl. Acad. Sci. USA* **98**, 5643–5648.
157. Hrabe de Angelis, M., McIntyre, J., II, and Gossler, A. (1997). Maintenance of somite borders in mice requires the Delta homologue Dll1. *Nature* **386**, 717–721.
158. McCright, B., Gao, X., Shen, L., Lozier, J., Lan, Y., Maguire, M., Herzlinger, D., Weinmaster, G., Jiang, R., and Gridley, T. (2001). Defects in development of the kidney, heart and eye vasculature in mice homozygous for a hypomorphic Notch2 mutation. *Development* **128**, 491–502.
159. McCright, B., Lozier, J., and Gridley, T. (2002). A mouse model of Alagille syndrome: Notch2 as a genetic modifier of Jag1 haploinsufficiency. *Development* **129**, 1075–1082.
160. Zeltser, L. M., Larsen, C. W., and Lumsden, A. (2001). A new developmental compartment in the forebrain regulated by Lunatic fringe. *Nat. Neurosci.* **4**, 683–684.
161. Wang, S., Sdrulla, A. D., diSibio, G., Bush, G., Nofziger, D., Hicks, C., Weinmaster, G., and Barres, B. A. Notch receptor activation inhibits oligodendrocyte differentiation. *Neuron* **21**, 63–75.
162. Lowell, S., Jones, P., Le Roux, I., Dunne, J., and Watt, F. M. (2000). Stimulation of human epidermal differentiation by delta-notch signalling at the boundaries of stem-cell clusters. *Curr. Biol.* **10**, 491–500.
163. Post, L. C., Ternet, M., and Hogan, B. L. (2000). Notch/Delta expression in the developing mouse lung. *Mech. Dev.* **98**, 95–98.

CHAPTER 258

Reiterative and Concurrent Use of EGFR and Notch Signaling during *Drosophila* Eye Development

Raghavendra Nagaraj and Utpal Banerjee

*Departments of Molecular Cell and Developmental Biology,
Biological Chemistry, and Human Genetics,
Molecular Biology Institute, University of California at
Los Angeles, Los Angeles California*

Notch and epidermal growth factor receptor (EGFR) pathways are conserved throughout evolution and function together in many different developmental contexts. During the specification of the *Drosophila* eye, these two pathways either function in synergy or in opposition to achieve different outcomes. Similarly, in some instances, EGFR activation causes a serial activation of the Notch pathway through the control of the Notch ligand Delta, while in others, the two pathways act together on the same enhancer element in parallel to control target gene expression. We propose that the combinatorial effects of these two pathways have evolved as a network that generates a significant diversity in output with a small number of components used in a reiterative fashion.

Introduction

Intercellular signaling plays a critical role in the establishment of the body plan and cell fate during the development of multicellular organisms. The variation in the cell fates adopted is quite extensive; yet the number of conserved signaling pathways that mediate a majority of cell-cell interactions is extremely limited [1]. These few signaling pathways are used reiteratively and evoke specific responses depending upon the context in which they function. Understanding how this context is created within an equipotent group of cells is fundamental to our understanding of the role of cell-signaling in fate specification during development.

The *Drosophila* compound eye consists of a large number of ommatidia (facets), each containing a fixed number of cells: eight photoreceptor cells (R cells), four non-neuronal cone cells, three classes of pigment cells, and a bristle complex [2]. The fate of these cells is not derived by clonal mechanisms but through cell-cell communication [3]. The *Drosophila* eye develops from a sheet of epithelial tissue called the eye imaginal disc. In the third larval instar an indentation called the morphogenetic furrow (MF) initiates at the posterior tip and sweeps anteriorly across the disc. As cells emerge out of the furrow they attain the competence to respond to signaling pathways and initiate differentiation in a precise order (Fig. 1). The photoreceptors are the first cells to differentiate, followed by the cone cells and the pigment cells. This led to the hypothesis that unique signals from differentiated cells will sequentially induce the precursors of later developing cell types [4]. The molecular basis for such a combinatorial model is now becoming clear. The biggest surprise is that the signals involved are not very specific and that a small number of common signaling pathways and transcription factors can combine in different ways to generate a tremendous diversity of readouts [5–8]. The EGFR and Notch signaling pathways are prominently involved in this process. All components of these pathways, except the ligands, are ubiquitously expressed. The spatiotemporal

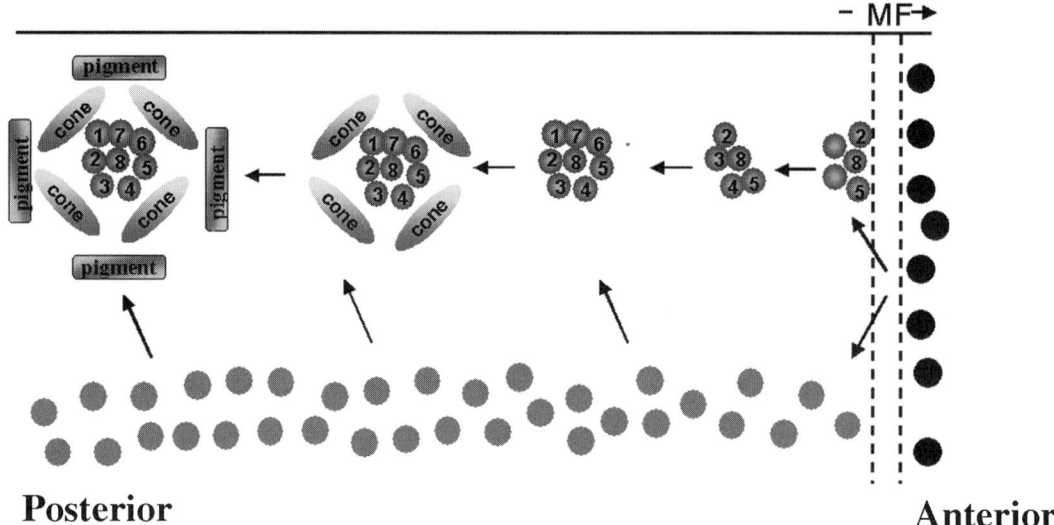

Figure 1 A schematic representation of cell fate specification during eye development. Cells ahead of (anterior to) the morphogenetic furrow (MF) are pluripotent and undergo extensive proliferation. The MF marks the onset of cell fate specification. The precluster cells (R2–R5, R8) differentiate first from cells ahead of the furrow. Cells that do not join the precluster undergo a single round of mitosis and become precursors to the rest of the cells in the ommatidial cluster (R1, R6, R7, Cone, and Pigment cells).

control of the ligand determines the cells in which a mosaic pattern of activated transcription factors is generated. In this chapter we briefly discuss steps in *Drosophila* eye development in which the Notch and EGFR pathways combine in different ways to generate unique outputs.

Establishment of the Eye Primordium

The *Drosophila* eye and antenna both arise from a single primordium at stages 14–15 of embryonic development [9], and only later the posterior region of this common primordium differentiates into the eye disc while the anterior region gives rise to the antenna. Expression of either a dominant-negative version of Notch or an activated version of EGFR in the common primordium converts the eye tissue into antenna [10]. This interplay of signals is important for the expression of Eyeless (DPax-6) in the aspect of the tissue that will give rise to the eye. Eyeless encodes a conserved transcription factor containing a Paired and a homeodomain and functions as a pivotal control point for a network of genes that establish eye fate [11]. Although the directness of the signaling control is not yet clear, it seems that Notch signaling is required for the expression of Eyeless, while EGFR antagonizes Notch-mediated Eyeless expression [10,12]. A balance between these two pathways is crucial for the specification of the eye primordium. It seems that such a balance is invoked repeatedly to create unique cell fates.

Proliferation and D/V Patterning

The eye disc primordium undergoes extensive proliferation during the later embryonic stages, as well as in the first and second larval instars. During late stages of the second larval instar, the eye disc primordium is patterned along the dorso-ventral (D/V) axis [13]. The segregation between the dorsal and ventral regions is not along traditional compartmental boundaries. Rather, the dorsal and ventral halves of the eye disc express distinct ligands for Notch. Cells in the ventral region express Serrate while those in the dorsal region express Delta (Dl) [14]. These ligands function differentially with the Fringe protein such that they activate Notch only at the boundary. This promotes the specification of a D/V organizer, which initiates cell proliferation and patterning along its edges [14]. Loss of EGFR at this stage will also cause a block in cell proliferation; however, its possible role in the specification of cell fate at this early stage remains to be established. In a comparable situation in the wing disc, EGFR functions with Notch during D/V boundary specification to control both proliferation and cell fate specification [15].

Morphogenetic Furrow and R8 Specification

Cells ahead of the morphogenetic furrow are undifferentiated and undergo several rounds of mitosis. These cells express all the genes that specify them as eye tissue and can adopt one of many possible differentiated fates upon the reception of an appropriate signal. Ectopic activation of EGFR and Notch in clones of cells ahead of the furrow causes them to differentiate as R cells, suggesting a readiness and capacity of these undifferentiated cells to respond to these signals [16,17].

The signaling pathways most closely linked to the initiation and progression of the morphogenetic furrow are Hh and Dpp (TGFβ) [18]. As the furrow progresses along the largely oval shaped disc it needs to be continuously reinitiated at the lateral edges so that it extends across the increasing width of the disc. This reinitiation requires Wingless (Wnt) signaling, and also the function of EGFR and the Notch pathways [19–21]. Loss of EGFR or Notch causes loss of furrow initiation at lateral edges and activation of EGFR and Notch at the lateral edges ahead of the furrow causes ectopic furrow initiation [21]. This is therefore an example of the EGFR and Notch pathways working cooperatively. In contrast, during the differentiation of the founder cell, R8, at the morphogenetic furrow, EGFR and Notch signals function in an antagonistic manner. EGFR has been proposed to be required for the establishment [22] and maintenance [23] of R8 cell fate, and these cells are lost in mutations that abolish the signal. In contrast, loss of Notch at the morphogenetic furrow causes overspecification of R8 cell types [24]. This is because Notch functions in a lateral inhibitory process at the furrow that helps restrict the number of R8 cells and generate regular spacing between the clusters. This lateral inhibition by Notch is mediated by the activation of repressor proteins of the E(spl) complex, which block the expression of the proneural gene *atonal* and the specification to an R8 cell type [25].

R-Cell Specification

Cells within the morphogenetic furrow are arrested in the G1 phase of the cell cycle. As they exit out of the morphogenetic furrow they follow one of two possible paths. A subset of these cells form "preclusters" consisting of the first five R cells (R2–R5, R8), and those that do not join the precluster undergo a terminal round of mitosis and continue to be pluripotent undifferentiated cells [26]. These cells express a distinct set of transcription factors (e.g., Lozenge) not found in the undifferentiated cells ahead of the furrow from which they are derived [27]. The remaining cell types within an ommatidium (R1, R6, R7, Cone, and Pigment cells) arise from the undifferentiated cells behind the morphogenetic furrow. This second round of recruitment completes the ommatidial cluster and is referred to as the second wave of morphogenesis in the eye [3]. The EGFR pathway is required for the G2/M transition during mitosis that gives rise to the precursors for the second wave [28]. Additionally, EGFR function is also required for maintaining viability and differentiation of all R cells. In contrast, during this later step of R-cell differentiation the function of Notch seems to be more specific. The Notch signal specifies the R4 cell and distinguishes it from R3, a process that is essential for the proper rotation of the ommatidial cluster with respect to the midline (equator) of the eye disc [29,30]. Furthermore, the Notch signal is required for the specification of the R7 cell fate as in its absence R7 cells are converted to an R1/6 type [31].

Sequential Linkage between Notch and EGFR Pathways

Although the Notch signal is required in only a subset of R cells for their proper specification, the Notch ligand Dl is expressed at high levels in all R cells [32]. The Dl protein functions in R cells to induce the later arising cone cells [8]. Interestingly, the expression of Dl in R cells is under the control of EGFR. A series of gain and loss-of-function studies indicate that loss of EGFR signaling causes a loss of Dl expression and ectopic activation of the EGFR pathway causes increase expression of Dl [33]. The EGFR-related activation of Dl requires the usual downstream components Ras, Raf, and MAPK but also two novel nuclear proteins Ebi and Strawberry notch (Sno) [33]. Ebi encodes an F-box/WD-40 repeat containing protein [34] and Sno codes for a novel nuclear protein [35]. The regulation of *Dl* in photoreceptor cells upon EGFR signaling requires the disruption of a Su(H)/SMRTER repression complex and involves the function of ubiquitin-mediated proteolysis and nuclear export of the co-repressor protein, SMRTER. This study highlights a mechanism whereby the EGFR and Notch pathways are linked in a serial manner. Activation of EGFR triggers a proteasome pathway leading to de-repression of *Dl*, which in turn can activate the Notch signal to be received by the neighboring cone cell precursors (Fig. 2).

Parallel Linkage between EGFR and Notch

The four non-neuronal cone cells are recruited to the ommatidial cluster after the R cells have joined. These cells express specific transcription factors such as D-Pax2, Prospero, and Cut, which are required for proper specification of their fate. The mechanisms that govern the specific expression of these factors have been uncovered in some detail. The expression of D-Pax2 and Prospero depend upon EGFR signaling, Notch signaling, and the transcription factor, Lozenge (Lz), expressed in the pool of undifferentiated cells behind the morphogenetic furrow [27]. Loss-of-function mutations in any one of these three components will cause a

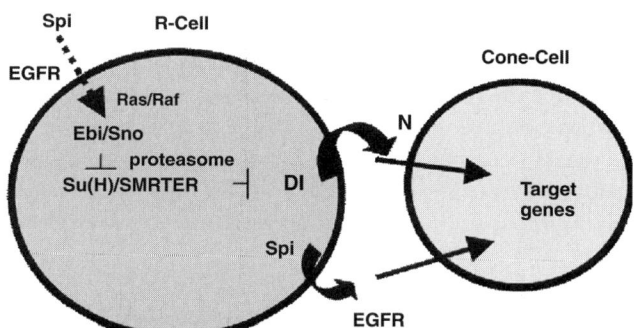

Figure 2 Sequential and parallel link between Notch and EGFR in cone cell specification. In R cells the EGFR and Notch pathways are linked sequentially. Activation of EGFR by its ligand Spi activates Ras/Raf/Sno/Ebi and leads to de-repression of the *Dl* locus by blocking its Su(H)/SMRTER-mediated repression. Dl protein expressed in R cells activates Notch signaling in cone cell precursors. These cells express the transcription factor, Lz, and also receive the EGFR signal from R cells. A combinatorial integration of the EGFR and Notch signals activates target genes required for cone cell specification.

complete loss of D-Pax2 and Prospero expression [8,36] (Nagaraj and Banerjee, unpublished). A 350 bp minimal enhancer element within the *D-Pax2* locus recapitulates its expression in the four cone cells in each cluster of the eye disc. Sequence analysis of this enhancer revealed functional binding sites for Lz and for Pointed and Su(H), the nuclear effectors of EGFR and Notch pathways, respectively. *In vivo* mutational analysis of these binding sites established that they are essential for the expression of D-Pax2 in the cone cells. Furthermore, ectopic activation of combinations of Notch, EGFR, and Lz using cell-specific drivers will cause D-Pax2 expression in different combinations of cell types in the eye disc that do not normally express this protein [8]. These results established that EGFR and Notch signals function combinatorially to specify cell fates among a group of pluripotent precursors.

Pigment Cell Differentiation and Apoptosis

The cooperativity and antagonism between EGFR and Notch is also evident in pigment cell differentiation. Cell ablation studies show that cone cells provide an inductive signal for the differentiation of the primary pigment cells [37]. In this function, Notch works combinatorially with EGFR as activation of EGFR is also needed for the specification of primary pigment cells. In contrast, the EGFR and the Notch pathways antagonize each other in promoting the survival of the secondary and tertiary pigment cells [37]. While EGFR promotes survival, Notch signaling in these cells has been proposed to promote cell death. The mechanisms by which these pathways function to regulate apoptosis remain to be established.

Conclusion

This review highlights how EGFR and Notch pathways function together in different ways within the same tissue and within small temporal windows to promote a diversity of cell fate specification events. These pathways can be linked to function in either a synergistic or an antagonistic manner; they can combine in series or in parallel to generate different outcomes. It is unlikely that these two pathways are chosen randomly to function together in so many different developmental decisions. Rather, this suggests the co-evolution of the two pathways to generate an efficient network involving possibilities for crosstalk. This intricate interplay between the EGFR and the Notch pathway can be seen from worms to man. During the induction of vulval fate in *C. elegans*, the anchor cell provides the EGF signal and induces the closest vulval precursors to differentiate into primary fate. This cell in turn uses the Notch signal to influence its neighbors to differentiate into secondary cells [38]. In mouse, differentiation of hematopoietic cells requires RTK and Notch signaling for proliferation and cell fate specification [39]. Finally, in tumor progression, Ras activation causes expression of Dl [40] similar to that observed in the developing *Drosophila* eye [33]. It is likely that the lessons learned from *Drosophila* will be useful in interpreting interactions between these pathways across species.

References

1. Hunter, T. (2000). Signaling—2000 and beyond. *Cell* **100**, 113–127.
2. Ready, D. F. (1989). A multifaceted approach to neural development. *Trends Neurosci.*, 102–110.
3. Ready, D. F., Hanson, T. E., and Benzer, S. (1976). Development of the *Drosophila* retina, a neurocrystalline lattice. *Dev. Biol.* **53**, 217–240.
4. Tomlinson, A. (1988). Cellular interactions in the developing *Drosophila* eye. *Development* **104**, 183–193.
5. Simon, M. A. (2000). Receptor tyrosine kinases: specific outcomes from general signals. *Cell* **103**, 13–15.
6. Ghazi, A. and VijayRaghavan, K. V. (2000). Developmental biology. Control by combinatorial codes. *Nature* **408**, 419–420.
7. Halfon, M. S., Carmena, A., Gisselbrecht, S., Sackerson, C. M., Jimenez, F., Baylies, M. K., and Michelson, A. M. (2000). Ras pathway specificity is determined by the integration of multiple signal-activated and tissue-restricted transcription factors. *Cell* **103**, 63–74.
8. Flores, G. V., Duan, H., Yan, H., Nagaraj, R., Fu, W., Zou, Y. M., Noll, M., and Banerjee, U. (2000). Combinatorial signaling in the specification of unique cell fates. *Cell* **103**, 75–85.
9. Younossi-Hartenstein, A., Nassif, C., Green, P., and Hartenstein, V. (1996). Early neurogenesis of the *Drosophila* brain. *J. Comp. Neurol.* **370**, 313–329.
10. Kumar, J. P. and Moses, K. (2001). EGF receptor and Notch signaling act upstream of Eyeless/Pax6 to control eye specification. *Cell* **104**, 687–697.
11. Callaerts, P., Halder, G., and Gehring, W. J. (1997). PAX-6 in development and evolution. *Annu. Rev. Neurosci.* **20**, 483–532.
12. Kurata, S., Go, M. J., Artavanis-Tsakonas, S., and Gehring, W. J. (2000). Notch signaling and the determination of appendage identity. *Proc. Natl. Acad. Sci. USA* **97**, 2117–2122.
13. Irvine, K. D. (1999). Fringe, Notch, and making developmental boundaries. *Curr. Opin. Genet. Dev.* **9**, 434–441.
14. Papayannopoulos, V., Tomlinson, A., Panin, V. M., Rauskolb, C., and Irvine, K. D. (1998). Dorsal-ventral signaling in the *Drosophila* eye. *Science* **281**, 2031–2034.
15. Nagaraj, R., Pickup, A. T., Howes, R., Moses, K., Freeman, M., and Banerjee, U. (1999). Role of the EGF receptor pathway in growth and patterning of the *Drosophila* wing through the regulation of vestigial. *Development* **126**, 975–985.
16. Dominguez, M., Wasserman, J. D., and Freeman, M. (1998). Multiple functions of the EGF receptor in Drosophila eye development. *Curr. Biol.* **8**, 1039–1048.
17. Baonza, A. and Freeman, M. (2001). Notch signalling and the initiation of neural development in the *Drosophila* eye. *Development* **128**, 3889–3898.
18. Heberlein, U. and Moses, K. (1995). Mechanisms of *Drosophila* retinal morphogenesis: The virtues of being progressive. *Cell* **81**, 987–990.
19. Treisman, J. E. and Rubin, G. M. (1995). Wingless inhibits morphogenetic furrow movement in the Drosophila eye disc. *Development* **121**, 3519–3527.
20. Ma, C. and Moses, K. (1995). Wingless and patched are negative regulators of the morphogenetic furrow and can affect tissue polarity in the developing *Drosophila* compound eye. *Development* **121**, 2279–2289.
21. Kumar, J. P. and Moses, K. (2001). The EGF receptor and notch signaling pathways control the initiation of the morphogenetic furrow during Drosophila eye development. *Development* **128**, 2689–2697.
22. Spencer, S. A., Powell, P. A., Miller, D. T., and Cagan, R. L. (1998). Regulation of EGF receptor signaling establishes pattern across the developing *Drosophila* retina. *Development* **125**, 4777–4790.

23. Kumar, J. P., Tio, M., Hsiung, F., Akopyan, S., Gabay, L., Seger, R., Shilo, B. Z., and Moses, K. (1998). Dissecting the roles of the *Drosophila* EGF receptor in eye development and MAP kinase activation. *Development* **125**, 3875–3885.
24. Cagan, R. L. and Ready, D. F. (1989). Notch is required for successive cell decisions in the developing *Drosophila* retina. *Genes Dev.* **3**, 1099–1112.
25. Artavanis-Tsakonas, S., Matsuno, K., and Fortini, M. E. (1995). Notch signaling. *Science* **268**, 225–232.
26. Banerjee, U. and Zipursky, S. L. (1990). The role of cell-cell interaction in the development of the *Drosophila* visual system. *Neuron* **4**, 177–187.
27. Flores, G. V., Daga, A., Kalhor, H. R., and Banerjee, U. (1998). Lozenge is expressed in pluripotent precursor cells and patterns multiple cell types in the *Drosophila* eye through the control of cell-specific transcription factors. *Development* **125**, 3681–3687.
28. Baker, N. E. and Yu, S. Y. (2001). The EGF receptor defines domains of cell cycle progression and survival to regulate cell number in the developing *Drosophila* eye. *Cell* **104**, 699–708.
29. Strutt, H. and Strutt, D. (1999). Polarity determination in the Drosophila eye. *Curr. Opin. Genet. Dev.* **9**, 442–446.
30. Fanto, M. and Mlodzik, M. (1999). Asymmetric Notch activation specifies photoreceptors R3 and R4 and planar polarity in the *Drosophila* eye. *Nature* **397**, 523–526.
31. Tomlinson, A. and Struhl, G. (2001). Delta/Notch and Boss/Sevenless signals act combinatorially to specify the *Drosophila* R7 photoreceptor. *Mol. Cell* **7**, 487–495.
32. Parks, A. L., Turner, F. R., and Muskavitch, M. A. (1995). Relationships between complex Delta expression and the specification of retinal cell fates during *Drosophila* eye development. *Mech. Dev.* **50**, 201–216.
33. Tsuda, L., Nagaraj, R., Zipursky, L., and Banerjee, U. (2002). An EGFR/Ebi/Sno pathway promotes Delta expression by inactivating Su(H)/SMRTER repression during inductive Notch signaling. *Cell* **110**, 625–637.
34. Dong, X., Tsuda, L., Zavitz, K. H., Lin, M., Li, S., Carthew, R. W., and Zipursky, S. L. (1999). ebi regulates epidermal growth factor receptor signaling pathways in *Drosophila*. *Genes Dev.* **13**, 954–965.
35. Majumdar, A., Nagaraj, R., and Banerjee, U. (1997). Strawberry notch encodes a conserved nuclear protein that functions downstream of Notch and regulates gene expression along the developing wing margin of *Drosophila*. *Genes Dev.* **11**, 1341–1353.
36. Xu, C., Kauffmann, R. C., Zhang, J., Kladny, S., and Carthew, R. W. (2000). Overlapping activators and repressors delimit transcriptional response to receptor tyrosine kinase signals in the *Drosophila* eye. *Cell* **103**, 87–97.
37. Miller, D. T. and Cagan, R. L. (1998). Local induction of patterning and programmed cell death in the developing *Drosophila* retina. *Development* **125**, 2327–2335.
38. Sternberg, P. W., Lesa, G., Lee, J., Katz, W. S., Yoon, C., Clandinin, T. R., Huang, L. S., Chamberlin, H. M., and Jongeward, G. (1995). LET-23-mediated signal transduction during *Caenorhabditis elegans* development. *Mol. Reprod. Dev.* **42**, 523–528.
39. Allman, D., Punt, J. A., Izon, D. J., Aster, J. C., and Pear, W. S. (2002). An invitation to T and more: notch signaling in lymphopoiesis. *Cell* **109**, Suppl., S1–11.
40. Weijzen, S., Rizzo, P., Braid, M., Vaishnav, R., Jonkheer, S., Zlobin, A., Osborne, B., Gottipati, A. S., Aster, J., Hahn, C. W., Rudolf, C. M., Siziopikou, K., Kast, M., and Miele, L. (2002). Activation of Notch-1 signaling maintains the neoplastic phenotype in human Ras-transformed cells. *Nat. Med.* **8**, 979–986.

BMPs in Development

Karen M. Lyons and Emmanuèle Délot*

Department of Molecular, Cell and Developmental Biology,
Department of Biological Chemistry, Department of Orthopaedic Surgery,
**Present Address, Department of Pediatrics*
University of California, Los Angeles, California

Introduction

Genes encoding bone morphogenetic proteins (BMPs) and their signaling components are found in many animal phyla, suggesting an ancient function as morphogens in metazoan development [1]. Where it has been possible to make comparisons (e.g., dorsal-ventral patterning and restriction of neuronal fate) there is a remarkable conservation of the roles of BMP signaling components in establishing and interpreting BMP gradients. A full discussion of the roles of BMPs in development is beyond our scope. Rather, we discuss recent concepts that have emerged from analysis of the BMP pathway in the fruit fly (*Drosophila*) embryo. We relate these concepts to the analysis of BMP function in vertebrates.

A fundamental task of the developing embryo is to inform individual cells of their relative positions through either direct cell-cell interactions, or soluble molecules that act as morphogens. To qualify as a morphogen, the factor must act directly on distant cells, generate a gradient of signaling activity, and elicit different patterns of gene expression at different concentrations [2,3]. Strong evidence for activity as a morphogen is available for only a few secreted proteins. BMP family members are among the best characterized examples [2–4].

Gradients of BMP Activity

BMPs satisfy the most critical requirement for morphogens: They act in a concentration-dependent manner to elicit differential gene expression [2,4]. For example, in *Drosophila*, neural cells form ventrally where BMP signaling is absent, ectoderm forms laterally where intermediate levels of BMP signaling take place, and peak levels induce the most dorsal tissue, the amnioserosa (Fig. 1). The most direct method to determine where BMP signaling pathways are active is the use of antibodies that specifically detect the active (phosphorylated) forms of Smads (pMAD and pSmad, in *Drosophila* and vertebrates, respectively). BMP activity gradients have been seen along the *Drosophila* and *Xenopus* dorso-ventral (D/V) axis, in the wing disc of the fly, and in the chick [5–9]. A common finding is that the shapes of the BMP ligand and activity gradients are considerably different. For example, along the *Drosophila* D/V axis, although pMAD levels form a gradient with highest levels dorsally, mRNAs for the ligands DPP and SCW are uniform dorsally (DPP) or are ubiquitously expressed (SCW) (Fig. 1) [8]. This finding underscores the substantial level of control over the shapes of BMP activity gradients that occurs extracellularly via ligand interactions with antagonists, and at the intracellular level by regulating BMP receptor and Smad degradation.

Establishing BMP Ligand Gradients

Many mechanisms are employed to establish BMP gradients, even within a single species. The propagation of endogenous BMP ligand gradients has not been studied directly in vertebrates, but in *Xenopus*, the TGFβ family member activin forms a gradient by diffusion [2]. On the other hand, BMP4 does not diffuse freely, perhaps due to its ability to bind proteoglycans [10], which are abundant in the extracellular matrix. In the *Drosophila* wing disc, the DPP gradient is controlled by receptor-mediated endocytosis [11]. DPP binds to its receptor (Thickveins, Tkv), and repeated cycles of endocytosis and re-secretion transport DPP from its source. DPP is also distributed extracellularly, and binding of DPP to Tkv limits its diffusion [6]. Little is

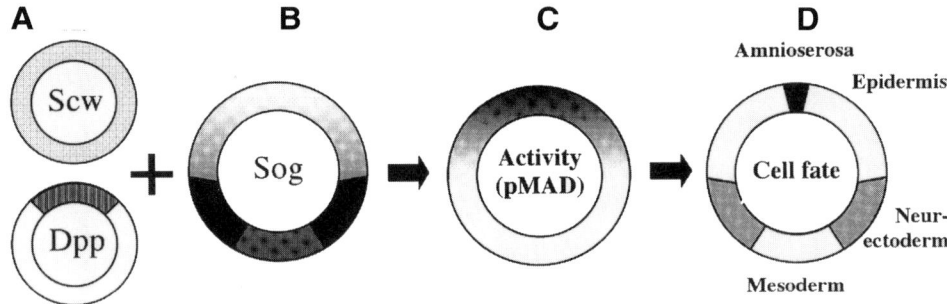

Figure 1 Formation of the BMP activity gradient along the *Drosophila* D/V axis. (A) mRNAs for the ligands are distributed uniformly (Scw) or broadly on the dorsal side of the embryo (Dpp). (B) Sog, the Scw/Dpp antagonist, is distributed in a ventral high-dorsal low gradient due to its expression in ventrolateral regions (black), and destruction in the dorsal region. (C) As a result of the Sog gradient, Dpp and Scw are antagonized ventrally, leading to a dorsal high-ventral low gradient of BMP signaling activity, as monitored by distribution of pMAD. (D) Different cell types are specified along the D/V axis. The amnioserosa is the most dorsal cell type.

known about how levels of BMP receptor expression are regulated, but BMPs can repress expression of their receptors [12], which limits the duration of responsiveness to BMP signals and prevents sequestration of BMPs close to the source of expression.

Extracellular Modifiers of BMP Activity

Antagonists that act by binding BMPs in the extracellular space play a pivotal role in shaping BMP gradients, and a large number of such antagonists have been identified [13]. The best characterized example, conserved through evolution, is the Short Gastrulation (Sog) protein, and its vertebrate homolog Chordin (Chd). They are the primary architects of the BMP activity gradients that specify the D/V axis in vertebrates and invertebrates (reviewed in references [13] and [14]).

Controlling the availability of these antagonists allows the embryo to create a BMP activity gradient despite uniform expression of the ligands. Sog/Chd is cleaved by the protease Tolloid (Xolloid in *Xenopus*). Along the *Drosophila* D/V axis, Sog is expressed in a ventral to dorsal gradient that opposes the gradient of DPP activity [9] (Fig. 1B). As a result of the Sog gradient, SCW is antagonized ventrally (where Sog levels are highest), but is active dorsally, where Sog is removed by endocytic degradation and cleavage by Tolloid [9]. Similar mechanisms are likely operational in vertebrates. For example, in zebrafish, the BMP2/4 homolog Swirl is not asymmetrically distributed, but it patterns the entire D/V axis through localized expression of zChd [15].

Twisted Gastrulation (Tsg), a protein that interacts with Sog/Chd and BMPs to affect the activity of Tolloid/Xolloid, brings an additional level of regulation. Tsg acts as an antagonist by forming a complex with Sog and SCW [16–19]. This promotes cleavage of Sog into a truncated form, Supersog, that can inhibit DPP in addition to SCW [16–19]. Thus, Tsg serves to steepen the Sog activity gradient, leading to more potent antagonism of BMP activity.

The consequences of this complex formation are not fully understood, and BMP binding proteins may not act solely as antagonists. For example, genetic studies indicate that Tsg and Sog have positive effects on BMP signaling, perhaps by mediating ligand diffusion (reviewed in references [9] and [14]). Moreover, in *Xenopus*, Tsg is multifunctional, acting as both a BMP agonist and antagonist depending on experimental conditions [18–20]. In addition, loss-of-function studies suggest that many antagonists have redundant functions that will be uncovered only through construction of double mutants. For example, although unrelated structurally, the antagonists Noggin [21] and Chd have a redundant function in induction of anterior neural tissues in mammals [22].

Interpreting the Gradient-Role of BMP Receptors

BMP signal transduction is initiated by formation of complexes of type I and type II receptors, leading to Smad phosphorylation. It is clear that different type I receptors activate common signaling pathways. How can such a simple system lead to differential gene expression? First, BMPs trigger different responses when different numbers of receptors are activated [2]. Second, different BMP receptors have different affinities for specific BMP ligands. Thus, two groups of cells might respond differently to a single BMP concentration because they express different type I BMP receptors.

A striking example of this latter mechanism is the synergy observed between some type I receptors. In *Drosophila*, Thickveins (Tkv) can act independently of the Saxophone (Sax) type I receptor, but signaling through Sax takes place only when Tkv is active. Moreover, Tkv and Sax strongly synergize when co-expressed, and this synergy is required for formation of the amnioserosa (reviewed in [4]). Evidence for a synergistic relationship has also been found in mammals [23]. BMPRIA and BMPRIB are structurally related to Tkv and function equivalently in that constitutively active (ca) forms induce chondrogenesis independently.

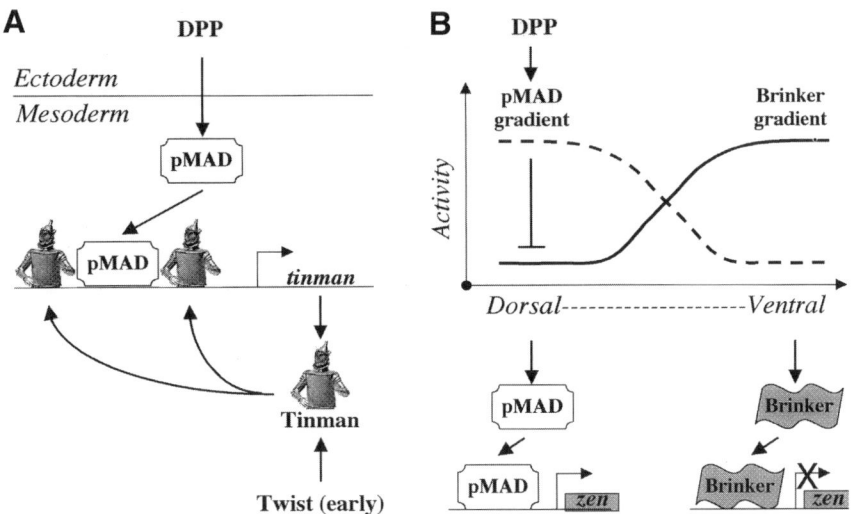

Figure 2 Regulation of target genes by BMP signaling. (A) The *Drosophila tinman* gene is expressed in dorsal mesoderm. At early stages, *tinman* is induced by Twist. The mesoderm subsequently migrates, and mesoderm located dorsally is exposed to high levels of DPP produced by the overlying ectoderm, resulting in high levels of pMAD activity. pMAD binds directly to the *tinman* promoter and cooperates with Tinman to maintain high levels of *tinman* expression. (B) Brinker is a transcriptional repressor of DPP target genes. *Brinker* expression is repressed by DPP. Hence, Brinker is present in areas of low pMAD activity. pMAD binds to the *zen* promoter to promote *zen* transcription. Brinker competes for binding and represses *zen* transcription. This mechanism would sharpen the boundary between *zen*-expressing and nonexpressing cells.

In contrast, ALK1 and ALK2 are structurally related to Sax, and ca forms of these receptors cannot induce chondrogenesis. However, ca BMPRIA or ca BMPRIB synergizes with ca ALK2 to promote chondrogenesis [23]. The mechanistic basis for the synergy is unknown, but may involve formation of a more stable receptor complex, differences in Smad activation, or effects on Smad cofactors.

Differential Gene Activity in Response to BMP Signal Transduction

The mechanisms by which changes in levels of activated Smads (pSmads) lead to changes in patterns of gene expression is an area of intensive investigation (reviewed in [24]). The ability of pSmads to interact with a myriad of transcriptional regulators is responsible for the dose-dependent effects of BMPs on gene expression [25]. The developmental history of the responding cell (i.e., its competence) determines which specific co-activators and co-repressors are present, thus defining the nature of the response. An example is the induction of *tinman* in dorsal mesoderm in *Drosophila* [26] (Fig. 2). The bHLH protein Twist activates *tinman* in the entire trunk mesoderm. During gastrulation, some of this mesoderm migrates to a position under the DPP-expressing dorsal ectoderm. The *tinman* promoter contains binding sites for Tinman and MAD, and in response to DPP, pMAD acts synergistically with Tinman to maintain *tinman* expression.

Different BMP target genes are activated by different levels of pSmads. One molecular explanation for this is that they have binding sites that recognize pSmad-containing complexes with different affinities. Another mechanism involves competitive binding of pSmads and repressors. For example, the *zen* promoter in *Drosophila* contains overlapping binding sites for pMAD and the repressor Brinker [27] (Fig. 2B). *zen* is active in the dorsal mesoderm, where the high levels of pMAD outcompete Brinker for binding to the *zen* promoter; DPP signaling also represses transcription of *Brinker*, leading to a sharp boundary of *zen* expression [28].

Intracellular Negative Regulation of BMP Signaling

BMP signaling pathways are subject to several levels of negative intracellular regulation that impact developmental outcomes. One of these involves inhibitory Smads (I-Smads) that stably associate with BMP receptors and/or other Smads to interfere with signaling. Expression of I-Smads is induced by BMP signaling, generating a feedback inhibition loop [25].

An additional level of modulation occurs at the level of Smad stability. Smurf1 and Smurf2 are Smad binding proteins that promote ubiquitin-mediated proteolysis of Smads and ligand-activated receptors (see references in [8]). The *Drosophila* cognate DSmurf is essential for establishing proper spatial control over DPP signaling, as *DSmurf* mutants exhibit expanded domains of DPP activity, leading to lethality [8]. In vertebrates, Smurfs may play an additional role; overexpression of Smurf1 not only decreases signaling through BMP pathways, but also enhances signaling through the TGFβ pathway [29]. This may occur through relief of competition for the common signal transduction component, Smad4 [30]. Competition between TGFβ and BMP pathways for limiting signal transduction components may be a widespread mechanism for modulating responsiveness to BMPs in vertebrates, as a number of examples of such competition have been described [31,32].

Lessons from Loss-of-Function Studies in Mammals

Over 15 different BMPs have been described in mammals, and expression of at least one is seen in almost every tissue type and stage of development, reflecting the universal importance of BMP pathways in vertebrates. Mutant strains have now been produced for most of the ligands, receptors, and Smads. Another useful approach has been to overexpress BMP antagonists or dominant-negative receptors in transgenic mice [33]. It has not been possible to uncover evidence that BMPs act as morphogens in mammals owing to the complexities of mammalian development and the lack of suitable markers. However, several important features have emerged from genetic analyses in mammals. First, they confirm widespread roles for BMP pathways in gastrulation, axis determination, and organogenesis. Second, the similar phenotypes of some mutant strains reveal the framework of the BMP pathways controlling specific developmental events. An example is the finding that mutations in the ligand nodal and the receptor ActRIIB are associated with defects in left/right asymmetry [34]. Another example is the nearly identical phenotype of mice lacking *Gdf5* or *BmprIB*, which suggests that GDF5 is a ligand for BMPRIB *in vivo* [35,36]. Third, mutant phenotypes are often more restricted than predicted from patterns of expression. One explanation that is consistent with analyses of double mutants is functional redundancy. For example, *Bmp7*–/– or *BmprIB*–/– mice have minor skeletal defects, but double mutants exhibit severe defects in long bones [35]. Hence, BMP7 and a different BMP (that acts through BMPRIB) have redundant functions in skeletal development. Surprisingly, ligands that act redundantly by genetic criteria are often not co-expressed in the affected tissues [37]. The molecular basis for the ability of ligands with nonoverlapping patterns of expression to compensate for each other genetically is not known, but may involve altered expression of other BMP ligands and/or signaling components, and reflects inductive interactions between adjacent tissues.

Conclusions

BMPs act as morphogens in both vertebrates and invertebrates. The distribution of BMP ligands often does not predict the shape of the BMP activity gradient. Clearly, ligand availability is highly regulated at the extracellular level by numerous BMP binding proteins. Many of these binding proteins can act both positively and negatively. How this occurs at a molecular level is not completely understood. Tissue-specific loss-of-function studies are needed to determine which BMP binding proteins act as positive and/or negative effectors of BMP signaling in particular developmental events.

Receptor synergy appears to play an important role in shaping BMP activity gradients in *Drosophila* and vertebrates. Additional studies will be required to delineate the mechanistic basis for this synergy, and to assess its role in specific developmental events in vertebrates. Our knowledge of how BMP signaling regulates the expression of specific target genes is increasing. It is clear from the large number of transcriptional regulators known to collaborate with Smads that multiple growth factor signaling pathways are integrated to control the nature of the response to the BMP signal. Understanding this integration on a molecular level will be challenging, but vital to our understanding of all developmental decisions in which BMPs participate.

References

1. Lelong, C., Mathieu, M., and Favrel, P. (2001). Identification of new bone morphogenetic protein-related members in invertebrates. *Biochemie*, 423–426.
2. Gurdon, J. and Bourillot, P.-Y. (2001). Morphogen gradient interpretation. *Nature* **413**, 797–803.
3. Teleman, A., Strigini, M., and Cohen, S. (2001). Shaping morphogen gradients. *Cell*, 559–562.
4. Podos, S. and Ferguson, E. (1999). Morphogen gradients: New insights from DPR. *Trends Genet.* **15**, 396–402.
5. Tanimoto, H., Itoh, S., ten Dijke, P., and Tabata, T. (2000). Hedgehog creates a gradient of DPP activity in *Drosophila* wing imaginal discs. *Mol. Cell* **5**, 59–71.
6. Teleman, A. and Cohen, S. (2000). Dpp gradient formation in the *Drosophila* wing imaginal disc. *Cell* **103**, 971–980.
7. Dorfman, R. and Shilo, B. (2001). Diphasic activation of the BMP pathway patterns the *Drosophila* embryonic dorsal region. *Development* **128**, 965–972.
8. Podos, S., Hanson, K., Wang, Y., and Ferguson, E. (2001). The DSmurf ubiquitin-protein ligase restricts BMP signaling spatially and temporally during *Drosophila* embryogenesis. *Dev. Cell* **1**, 567–578.
9. Srinivasan, S., Rashka, K., and Bier, E. (2002). Creation of a sog morphogen gradient in the *Drosophila* embryo. *Dev. Cell* **2**, 91–101.
10. Ohkawara, B., Lemura, S., ten Dijke, P., and Ueno, N. (2002). Action range of BMP is defined by its N terminal basic amino acid core. *Curr. Biol.* **12**, 205–209.
11. Entchev, E., Schwabedissen, A., and Gonzalez-Gaitan, M. (2000). Gradient formation of the TGF-β homolog Dpp. *Cell* **103**, 981–991.
12. Lecuit, T. and Cohen, S. (1998). Dpp receptor levels contribute to shaping the Dpp morphogen gradient in the *Drosophila* wing imaginal disc. *Development* **125**, 4901–4907.
13. De Robertis, E., Wessely, O., Oelgeschlager, M., Brizuela, B., Pera, E., Larrain, J., Abreu, J., and Bachiller, D. (2001). Molecular mechanisms of cell-cell signaling by the Spemann-Mangold organizer. *Int. J. Dev. Biol.* **45**, 189–197.
14. Ray, R. and Wharton, K. (2001). Twisted perspective: New insights into extracellular modulation of BMP signaling during development. *Cell* **104**, 801–804.
15. Hammerschmidt, M., Serbedzija, G., and McMahon, A. (1996). Genetic analysis of dorsoventral pattern formation in the zebrafish: Requirement of a BMP-like ventralizing activity and its dorsal repressor. *Genes Dev.* **10**, 2452–2461.
16. Yu, K., Srinivasan, S., Shimmi, O., Biehs, B., Rashka, K., Kimmelman, D., O'Connor, M., and Bier, E. (2000). Processing of the *Drosophila* Sog protein creates a novel BMP inhibitory activity. *Development* **127**, 2143–2154.

17. Ross, J., Shimmi, O., Vilmos, P., Petryk, A., Kim, H., Gaudenz, K., Hermanson, S., Ekker, S., O'Connor, M., and Marsh, J. (2001). Twisted gastrulation is a conserved extracellular BMP antagonist. *Nature* **410**, 479–483.
18. Scott, L,, Blitz, L,, Pappano, W., Maas, S., Cho, K., and Greenspan, D. (2001). Homologues of Twisted gastrulation are extracellular cofactors in antagonism of BMP signalling. *Nature* **410**, 475–478.
19. Chang, C, Holtzman, D., Chau, S., Chickering, T., Woolf, E., Holmgren, L., Bodorova, J., Gearing, D., Holmes, W., and Brivanlou, A. H. (2001). Twisted gastrulation can function as a BMP antagonist. *Nature* **410**, 483–487.
20. Larrain, J., Oelgeschlager, M., Ketpura, N., Reversade, B., Zakin, L., and De Robertis, E. (2001). Protcolytic cleavage of Chordin as a switch for the dual activities of Twisted gastrulation in BMP binding. *Development* **128**, 4439–4447.
21. Zimmerman, L. B., De Jesus-Escobar, and Harland, R. M. (1996). The Spemann organizer signal Noggin binds and inactivates bone morphogenetic protein 4. *Cell* **86**, 599–606.
22. Bachiller, D., Klingensmith, J., Kemp, I., Belo, J., Anderson, R., May, S., McMahon, J., McMahon, A., Harland, R., Rossant, J., and De Robertis, E. (2000). The organizer factors Chordin and Noggin are required for mouse forebrain development. *Nature* **403**, 658–661.
23. Aoki, H., Fujii, M., Imamura, T., Yagi, K., Takehara, K., Kato, M., and Miyazono, K. (2001). Synergistic effects of different bone morphogenetic protein type I receptors on alkaline phosphatase induction. *J. Cell Sci.* **114**, 1483–1489.
24. Massagué, J. and Wotton, D. (2000). Transcriptional control by the TGF-β/Smad signaling system. *EMBO J.* **19**, 1745–1754.
25. ten Dijke, P., Goumans, M.-J., Itoh, F., and Itoh, S. (2002). Regulation of cell proliferation by Smad proteins. *J. Cell. Physiol.* **191**, 1–16.
26. Xu, X., Yin, Z., Hudson, J., Ferguson, E., and Frasch, M. (1998). Smad proteins act in combination with synergistic and antagonistic regulators to target Dpp responses to the *Drosophila* mesoderm. *Genes Dev.* **12**, 2354–2370.
27. Affolter, M., Marty, T., Vigano, M., and Jazwinska, A. (2001). Nuclear interpretation of Dpp signaling in *Drosophila*. *EMBO J.* **20**, 3298–3305.
28. Rushlow, C., Colosimo, P., Lin, M., Xu, M., and Kirov, N. (2001). Transcriptional regulation of the *Drosophila* gene *zen* by competing Smad and Brinker inputs. *Genes Dev.* **15**, 340–351.
29. Zhu, H., Kavsak, P., Abdollah, S., Wrana, J., and Thomsen, G. (1999). A SMAD ubiquitin ligase targets the BMP pathway and affects embryonic pattern formation. *Nature* **400**, 687–693.
30. Candia, A. F., Watabe, T., Hawley, S. H. B., Onichtchouk, D., Zhang, Y., Derynck, R., Niehrs, C., and Cho, K. W. Y. (1997). Cellular interpretation of multiple TGF-β signals: intracellular antagonism between activin/BVgl and BMP-2/4 signaling mediated by Smads. *Development* **124**, 4467–4480.
31. Piek, E., Afrakhte, M., Sampath, K., van Zoelen, E. J., Heldin, C. H., and ten Dijke, P. (1999). Functional antagonism between activin and osteogenic protein-1 in human embryonal carcinoma cells. *J. Cell. Physiol.* **180**, 141–149.
32. Daluiski, A., Engstrand, T., Bahamonde, M. E., Gamer, L. A., Agius, E., Stevenson, S. L., Cox, K., Rosen, V., and Lyons, K. M. (2001). Bone morphogenetic protein 3 (BMP3) is a negative regulator of peak bone density. *Nat. Genet.* **27**, 84–88.
33. Weaver, M., Yingling, J. M., Dunn, N. R., Bellusci, S., and Hogan, B. L. M. (1999). Bmp signaling regulates proximal-distal differentiation of endoderm in mouse lung development. *Development* **126**, 4005–4015.
34. Whitman, M. and Mercola, M. (2001). TGF-β superfamily signaling and left-right asymmetry. *Science STKE* **64**, RE1.
35. Yi, S. E., Daluiski, A., Pederson, R., Rosen, V., and Lyons, K. M. (2000). The type I BMP receptor BMPRIB is required for chondrogenesis in the mouse limb. *Development* **127**, 621–630.
36. Baur, S. T., Mai, J. J., and Dymecki, S. M. (2000). Combinatorial signaling through BMP receptor IB and GDF5: Shaping of the distal mouse limb and the genetics of distal limb diversity. *Development* **127**, 605–619.
37. Storm, E. E. and Kingsley, D. M. (1996). Joint patterning defects caused by single and double mutations in members of the bone morphogenetic protein (BMP) family. *Development* **122**, 3969–3979.

Neurotrophin Signaling in Development

Albert H. Kim and Moses V. Chao

*Molecular Neurobiology Program, Skirball Institute of Biomolecular Medicine,
New York University School of Medicine,
New York, New York*

Introduction

The formation of the vertebrate nervous system is characterized by widespread programmed cell death, which determines cell number and appropriate target innervation during development. The neurotrophins, which include nerve growth factor (NGF), brain-derived growth factor (BDNF), NT-3, and NT-4, represent an important family of trophic factors that are essential for survival of selective populations of neurons during different developmental periods.

The neurotrophic hypothesis postulates that during nervous system development, neurons approaching the same final target vie for limited amounts of target-derived trophic factor [17]. In this way, the nervous system molds itself to maintain only the most competitive and appropriate connections. Competition among neurons for limiting amounts of neurotrophin molecules produced by target cells accounts for selective cell survival. Two predictions emanate from this hypothesis. First, the efficacy of neuronal survival will depend upon the amounts of trophic factors produced during development. Second, specific receptor expression in responsive cell populations will dictate neuronal responsiveness.

Neurotrophins exert their cellular effects through the actions of two different receptors, the tropomyosin-related kinase (Trk) receptor tyrosine kinase and the p75 neurotrophin receptor ($p75^{NTR}$), a member of the tumor necrosis factor (TNF) receptor superfamily. On one level, neurotrophins fit well with the neurotrophic hypothesis, as many peripheral neuronal subpopulations exhibit a predominant dependence on a specific neurotrophin during the period of naturally occurring cell death (Fig. 1). However, the biological reality appears much more complex. In the central nervous system, the overlapping expression of multiple neurotrophin receptors and their cognate ligands allows for the creation of diverse connectivity, which extends well into adulthood; even in the periphery complexities remain, such as the molecular mechanisms underlying the retrograde signal, a pathway that must efficiently transmit information over long distances, at times over a meter. The role of the neurotrophin system in development has been reviewed [12]. This chapter will focus upon new views concerning ligand-receptor interactions, signal transduction, and retrograde transport in the peripheral nervous system.

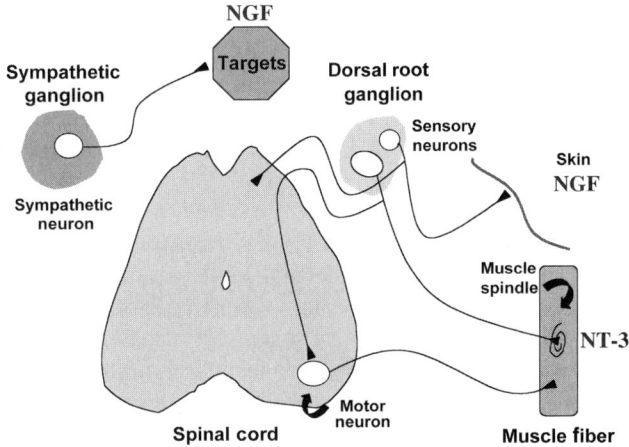

Figure 1 Neurotrophins serve as target-derived survival factors. NGF and NT-3 display specific survival and differentiative effects upon sympathetic and sensory neuron populations in the peripheral nervous system.

The Neurotrophin Ligands

The neurotrophins are initially synthesized as precursors or pro-neurotrophins that are cleaved to release the mature, active proteins. The mature proteins form stable, noncovalent dimers and are normally expressed at very low levels during development. Pro-neurotrophins are cleaved intracellularly by furin or pro-convertases utilizing a highly conserved dibasic amino acid cleavage site to release carboxy-terminal mature proteins of approximately 13 kDa. These extensively studied mature proteins mediate neurotrophin actions by selectively binding to members of the Trk family of receptor tyrosine kinases to regulate neuronal survival, differentiation, and synaptic plasticity. In addition, all mature neurotrophins interact with p75^{NTR}, which can modulate the affinity of Trk:neurotrophin associations.

Neurotrophins promote cell survival and differentiation during neural development. Paradoxically, they can also induce cell death. p75^{NTR} serves as a pro-apoptotic receptor during developmental cell death and after injury to the nervous system. Increases in p75^{NTR} expression are responsible for apoptosis in embryonic retina and sympathetic neurons during the period of naturally occurring neuronal death [3,7]. Whereas BDNF binding to p75^{NTR} in sympathetic neurons causes rapid cell death, NGF binding to the TrkA receptor on the same neurons provides a survival signal. In the context of neurotrophin processing, pro-neurotrophins are more effective than mature NGF in inducing p75^{NTR}-dependent apoptosis [16]. These results suggest that the biological action of the neurotrophins can be regulated by proteolytic cleavage, with pro-forms preferentially activating p75^{NTR} to mediate apoptosis and mature forms selectively activating Trk receptors to promote survival.

Neurotrophin Receptors

One way of generating more specificity during development is by imparting greater discrimination of ligands for the Trk receptors. NGF binds most specifically to TrkA; BDNF and NT-4 to TrkB; and NT-3 to TrkC receptors. The p75^{NTR} receptor can bind to each neurotrophin but has the additional capability of regulating a Trk's affinity for its cognate ligand. Trk and p75^{NTR} receptors have been referred to as high- and low-affinity receptors, respectively. However, this is not correct since TrkA and TrkB actually bind their ligands with an affinity of 10^{-9}–10^{-10} M, which is lower than the high-affinity site ($K_d = 10^{-11}$ M). Also, the precursor form of NGF displays high-affinity binding to p75^{NTR}. Trk-mediated responsiveness to low concentrations of NGF is dependent upon the relative levels of p75^{NTR} and TrkA receptors, and their combined ability to form high-affinity sites.

Although p75^{NTR} and Trk receptors do not appear to bind to each other directly, there is evidence that complexes form between the two receptors [5]. As a result of these interactions, increased ligand selectivity can be conferred onto Trks by p75^{NTR}. NGF and NT-3 both can bind to TrkA, but p75^{NTR} restricts signaling of TrkA to NGF and not to NT-3 [4]. Sympathetic neurons express TrkA and p75^{NTR} receptors and depend upon NGF for survival. In the absence of p75^{NTR}, NT-3 compensates for a lower level of NGF in vivo. Though NT-3 binds weakly to TrkA, the survival effects of NT-3 can be attributed to activation of TrkA receptors in the absence of p75^{NTR}. These observations support a role for p75^{NTR} in enhancing the specificity of Trk receptors for neurotrophins in vivo. Hence, p75^{NTR} and Trk receptors may interact to provide greater discrimination among different neurotrophins.

In addition to increasing ligand-receptor affinity and selectivity, p75^{NTR} contains a death domain sequence similar to the intracellular domains of the Fas and p55 TNF receptors [18]. Several examples of p75^{NTR}-mediated cell death now exist; these events appear to depend upon cell-specific interactions with adaptor proteins. It has been generally assumed that the mechanism of p75^{NTR}-mediated cell death is similar to apoptotic signaling by the TNF receptor and Fas. However, a number of experimental findings indicate that cell death by p75^{NTR} occurs quite differently from that of other TNF receptor family members [6,10]. First, regions of p75^{NTR} other than the death domain are responsible for apoptosis. Second, whereas recruitment of caspase-8 to the death domains of the p55 TNF and Fas receptors is critical for the initiation of apoptosis, other caspases are responsible for p75^{NTR}-mediated killing of oligodendrocytes.

What are the reasons for having a Trk receptor that mediates neuronal survival and a p75^{NTR} receptor that mediates apoptosis? One reason neurotrophins use a death receptor may be to prune neurons efficiently during periods of developmental cell death. In addition to competing for trophic support from the target, neurons must establish connections with the proper target. In the event of mistargeting, neurons may undergo apoptosis if the appropriate set of trophic factors is not encountered. In this case, a neurotrophin may not only fail to activate Trks but also bind to p75^{NTR} and eliminate cells by an active killing process [21]. For example, BDNF causes sympathetic neuronal death by binding to p75^{NTR} when TrkB is absent [3]. Likewise, NT-4 causes p75^{NTR}-mediated cell death in BDNF-dependent trigeminal neurons [1], due presumably to preferential p75^{NTR} rather than TrkB stimulation. Therefore, Trk and p75^{NTR} receptors can give opposite outcomes in the same cells. Cell death mediated by p75^{NTR} may be important for the refinement of correct target innervation during development.

Signaling Specificity during Development

Specific Trk receptor expression patterns determine the development of peripheral neuron populations. In the dorsal root ganglion, small-diameter unmyelinated neurons predominantly express TrkA whereas larger sized neurons express TrkC receptors. Many of the small diameter neurons are nociceptive and frequently terminate in the epidermis (Fig. 1). NGF is important for the development of these neurons during early postnatal periods. The large-diameter neurons are proprioceptive and are most responsive to NT-3. Consistent with the receptor expression, a lack of NGF leads to a lack of responsiveness to nociceptive stimuli, and a lack of NT-3 leads to a loss of muscle spindle afferents. Due to altered expression of specific Trk receptors, neurons change their dependency upon specific neurotrophins [27].

Trk receptors exhibit very high conservation in their intracellular domains, including the catalytic tyrosine kinase and the juxtamembrane NPXY motif that serves as the Shc binding site. However, several pronounced differences among the Trks exist. In a sympathetic neuronal background, TrkA relies predominantly upon phosphoinositide-3-kinase (PI3-K) activation for survival, whereas TrkB uses both PI3-K and ERK pathways [2]. Thus, each Trk receptor carries distinctive signaling properties. For example, TrkB may contain sequences that bind to factors that favor alternative pathways. Since there are now a number of different adaptor proteins and enzymatic functions associated with Trk receptors (Fig. 2), preferential interactions with these proteins must take place. Receptor utilization of substrates with differential association/dissociation kinetics, competition for binding among different substrates, or recruitment of unique target proteins, such as FRS-2, rAPs, and SH2-B for the Trk receptors, represent mechanisms by which each receptor may differentially utilize common substrates for signaling.

Alternatively, receptor processing or targeting into different membrane compartments may dictate function. A comparison of TrkA and TrkB receptors in neuronal cell lines has revealed a difference in turnover of each receptor. While NGF binding to TrkA does not lead to a significant downregulation of TrkA, BDNF binding to TrkB results in rapid turnover of TrkB receptors at the cell surface [28]. Additionally, the number of surface TrkB receptors is highly influenced by depolarization and levels of cAMP [22]. These observations hint at other receptor mechanisms that confer greater signaling specificities to the neurotrophins.

The Importance of Retrograde Transport

During development, neurotrophins are produced and released from the target tissues and become internalized into vesicles, which are then transported to the cell body. The biological effects of neurotrophins require that signals are conveyed over long distances from the nerve terminal to the cell body. Therefore, a central theme of the neurotrophic hypothesis is that neuronal survival and differentiation depend upon retrograde signaling of trophic factors produced at the target tissue.

Each neurotrophin binds to transmembrane receptors and undergoes internalization and transport from axon terminals to neuronal cell bodies [8,11,23]. Measurements of ^{125}I-NGF transport from distal axons to the cell body in compartment chambers indicate a rate from 3–10 mm/h. Both Trk and p75^{NTR} receptors undergo retrograde transport. The term "signaling endosome" has been coined to describe membrane vesicles that carry Trk, p75^{NTR}, and NGF [9].

A complex of NGF-TrkA has been found in clathrin-coated vesicles and endosomes, giving rise to the model that NGF and Trk are components of the retrograde signal. Several tyrosine phosphorylated proteins are associated with the TrkA receptor during transport, suggesting that signaling by neurotrophins persists following internalization of their receptors. Internalization of NGF from axon terminals is necessary for phosphorylation and activation of the

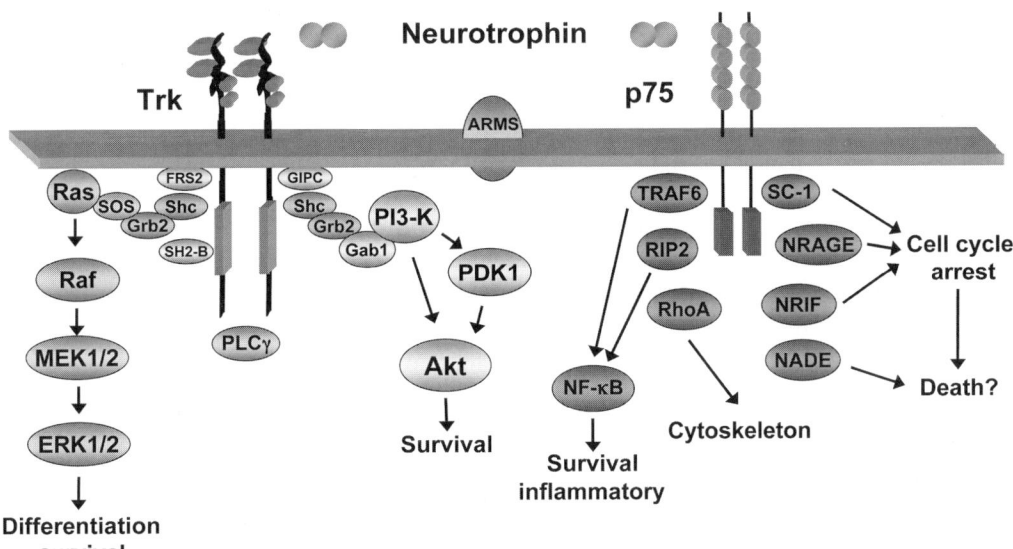

Figure 2 Neurotrophin receptors utilize multiple adaptor proteins. Each receptor undergoes ligand-induced dimerization that results in the recruitment of multiple cytoplasmic proteins, which serve to increase the activities of phospholipase Cγ, PI3-K, and MAP kinases. The p75 receptor is capable of initiating a cell death program in selected cells and signaling that leads to ceramide production and NF-κB or JNK activation.

CREB transcription factor, which leads to changes in gene expression and increased neuronal cell survival [25,30]. In addition, stimulation of Erk5 occurs in the cell body of DRG neurons after retrograde signaling by neurotrophins [29], as well as activation of PI3-K and Akt [14]. These events likely require internalization and transport of activated Trk receptors and result in a survival response. An alternative mechanism suggests that survival signals can be transmitted in the absence of NGF internalization and transport [20]. This explanation would account for the rapid appearance of phosphorylated TrkA in the cell body [26] and raises the possibility that a molecule other than endocytosed NGF may propagate a signal from distal axons to TrkA receptors at the cell body.

Interacting Proteins

The existence of neurotrophin-receptor complexes during their axonal transport suggests that other proteins may be associated with this complex. Neurotrophin receptors undergo ligand-induced dimerization that activates multiple signal transduction pathways. Neurotrophin binding to Trk family members produces biological responses through rapid increases in the phosphorylation of phospholipase C-γ and PI3-K. Increased *Ras* activity, a common signal from all tyrosine kinase receptors, results from the stimulation of guanine nucleotide exchange factors coupled to adaptor proteins which directly interact with Trk after ligand binding. These adaptor proteins include Shc, Grb2, SH2-B, and FRS-2 (Fig. 2).

A number of adaptor proteins also bind to p75^{NTR} (Fig. 2). Three different proteins, NRIF, NADE, and NRAGE contribute to apoptosis in immortalized cell lines or are correlated with neurotrophin-dependent cell death. Each protein binds to a separate sequence in the cytoplasmic domain of the p75^{NTR} [12]. Another protein that interacts with both p75^{NTR} and Trk receptors is ARMS, an ankyrin-rich transmembrane protein [13]. ARMS is rapidly tyrosine phosphorylated after binding of neurotrophins to Trk receptors. This protein may act as a scaffold to cluster proteins essential to neurotrophin signaling. Other proteins, including RhoA GTPase, SC-1, and NRAGE, exert nonapoptotic activities, such as neurite elongation and growth arrest. These proteins expand the functional scope of neurotrophins [15]. Still other proteins, such as cytoplasmic dynein and the PDZ-domain-containing GIPC protein may serve to target neurotrophin receptors intracellularly during important cellular processes such as internalization, retrograde transport, axonal and dendritic localization, and synapse formation.

Given the wide number of activities of neurotrophins and the small number of neurotrophins and neurotrophin receptor genes, it is likely other signaling systems are used. This includes ion channels such as TRP and glutamate receptors. How these signals are integrated to yield higher level neuronal functions, such as behavior, is unknown. But there are hints from knockout animals that lowering the levels of BDNF or its receptors TrkB and p75^{NTR} give rise to prominent aggressive behavior and abnormalities in eating and memory [19,24]. These findings bolster the notion that neurotrophins, in addition to their potent properties on the cellular level, possess abilities to influence cognitive functions.

References

1. Agerman, K., Baudet, C., Fundin, B., Willson, C., and Ernfors, P. (2000). Attenuation of a caspase-3 dependent cell death in NT-4 and p75-deficient embryonic sensory neurons. *Mol. Cell. Neurosci.* **16**, 258–268.
2. Atwal, J., Massie, B., Miller, F., and Kaplan, D. (2000). The TrkB-Shc site signals neuornal survival and local axon growth via MEK and PI3-kinase. *Neuron* **27**, 265–277.
3. Bamji, S., Majdan, M., Pozniak, C. D., Belliveau, D. J., Aloyz, R. J. K., Causing, C. G., and Miller, F. D. (1998). The p75 neurotrophin receptor mediates neuronal apoptosis and is essential for naturally occurring sympathetic neuron death. *J. Cell Biol.* **140**, 911–923.
4. Benedetti, M., Levi, A., and Chao, M. V. (1993). Differential expression of nerve growth factor receptors leads to altered binding affinity and neurotrophin responsiveness. *Proc. Natl. Acad. Sci. USA* **90**, 7859–7863.
5. Bibel, M., Hoppe, E., and Barde, Y. (1999). Biochemical and functional interactions between the neurotrophin receptors trk and p75NTR. *EMBO J.* **18**, 616–622.
6. Coulson, E. J., Reid, K., Baca, M., Shipham, K. A., Hulett, S. M., Kilpatrick, T. J., and Bartlett, P. F. (2000). Chopper, a new death domain of the p75 neurotrophin receptor that mediates rapid neuronal cell death. *J. Biol. Chem.* **275**, 30537–30545.
7. Frade, J. M., Rodriguez-Tebar, A., and Barde, Y.-A. (1996). Induction of cell death by endogenous nerve growth factor through its p75 receptor. *Nature* **383**, 166–168.
8. Ginty, D. and Segal, R. (2002). Retrograde neurotrophin signaling: Trk-ing along the axon, *Curr. Opin. Neurobiol.* **12**, 268–274.
9. Grimes, M., Beattie, E., and Mobley, W. (1997). A signaling organelle containing the nerve growth factor-activated receptor tyrosine kinase, TrkA. *Proc. Natl. Acad. Sci. USA* **94**, 9909–9914.
10. Gu, C., Casaccia-Bonnefil, P., Srinivaran, A., and Chao, M. (1999). Oligodendrocyte apoptosis mediated by caspase activation. *J. Neurosci.* **19**, 3043–3049.
11. Hendry, I., Stoeckel, K., Thoenen, H., and Iversen, L. (1974). The retrograde axonal transport of nerve growth factor. *Brain Res.* **68**, 103–121.
12. Huang, E. and Reichardt, L. (2001). Neurotrophins: Roles in neuronal development and function. *Annu. Rev. Neurosci.* **24**, 677–736.
13. Kong, H., Boulter, J., Weber, J., Lai, C., and Chao, M. (2001). An evolutionarily conserved transmembrane protein that is a novel downstream target of neurotrophin and ephrin receptors. *J. Neurosci.* **21**, 176–185.
14. Kuruvilla, R., Ye, H., and Ginty, D. (2000). Spatially and functionally distinct roles of PI3-K effector pathway during NGF signaling in sympathetic neurons. *Neuron* **27**, 499–512.
15. Lee, F., Kim, A., Khursigara, G., and Chao, M. (2001). The uniqueness of being a neurotrophin receptor. *Curr. Opin. Neurobiol.* **11**, 281–286.
16. Lee, R., Kermani, P., Teng, K., and Hempstead, B. (2001). Regulation of cell survival by secreted proneurotrophins. *Science* **294**, 1945–1948.
17. Levi-Montalcini, R. (1987). The nerve growth factor: Thirty-five years later. *Science* **237**, 1154–1164.
18. Liepinsh, E., Ilag, L. L., Otting, G., and Ibanez, C. F. (1997). NMR structure of the death domain of the p75 neurotrophin receptor. *EMBO J.* **16**, 4999–5005.
19. Lyons, W. E., Mamounas, L. A., Ricaurte, G. A., Coppola, V., Reid, S. W., Bora, S. H., Wihler, C., Koliatsos, V. E., and Tessarollo, L. (1999). Brain-derived neurotrophic factor-deficient mice develop aggressiveness and hyperphagia in conjunction with brain serotonergic abnormalities. *Proc. Natl. Acad. Sci. USA* **96**, 15239–44.

20. MacInnis, B. and Campenot, R. (2002). Retrograde support of neuronal survival without retrograde transport of nerve growth factor. *Science* **295**, 1536–1539.
21. Majdan, M. and Miller, F. (1999). Neuronal life and death decisions: Functional antagonism between the Trk and p75 neurotrophin receptors. *Int. J. Dev. Neurosci.* **17**, 153–161.
22. Meyer-Franke, A., Wilkinson, G., Kruttgen, A., Hu, M., Munro, E., Hanson, M., Reichardt, L., and Barres, B. (1998). Depolarization and cAMP rapidly recruit TrkB to the plasma membrane of CNS neurons. *Neuron* **21**, 681–693.
23. Neet, K. and Campenot, R. (2001). Receptor binding, internalization and retrograde transport of neurotrophic factors. *Cell. Mol. Life Sci.* **58**, 1021–1035.
24. Poo, M.-M. (2001). Neurotrophins as synaptic modulators. *Nat. Rev. Neurosci.* **2**, 24–31.
25. Riccio, A., Pierchala, B., Ciarallo, C., and Ginty, D. (1997). An NGF-TrkA-mediated retrograde signal to transcription factor CREB in sympathetic neurons. *Science* **277**, 1097–1100.
26. Senger, D. and Campenot, R. (1997). Rapid retrograde tyrosine phosphorylation of trkA and other proteins in rat sympathetic neurons in compartmented cultures. *J. Cell Biol.* **138**, 411–421.
27. Snider, W. D. (1994). Functions of the neurotrophins during nervous-system development—what the knockouts are teaching us. *Cell* **77**, 627–638.
28. Sommerfeld, M., Schweigreiter, R., Barde, Y., and Hoppe, E. (2000). Down-regulation of the neurotrophin TrkB following ligand binding. *J. Biol. Chem.* **275**, 8982–8990.
29. Watson, F., Heerssen, H., Bhattacharyya, A., Klesse, L., Lin, M., and Segal, R. (2001). Neurotrophins use the Erk5 pathway to mediate a retrograde survival response. *Nat. Neurosci.* **4**, 981–988.
30. Watson, F., Heerssen, H., Moheban, D., Lin, M., Sauvageot, C., Bhattacharyya, A., Pomeroy, S., and Segal, R. (1999). Rapid nuclear responses to target-derived neurotrophins require retrograde transport of ligand-receptor complex. *J. Neurosci.* **19**, 7889–7900.

CHAPTER 261

PDGF Receptor Signaling in Mouse Development

Richard A. Klinghoffer
CEPTYR, Inc., Bothell, Washington

Introduction

Highly defined characterization of the biochemical and cellular functions of platelet-derived growth factors (PDGFs) and their receptors (PDGFRs), set the stage to use this system as a genetic model to study receptor tyrosine kinase (RTK) signaling in mammalian development. Following binding to PDGF, PDGFRs autophosphorylate, enabling them to bind and then activate intracellular proteins that relay the receptor's signals throughout the cell. PDGF stimulates cellular functions including mitogenesis, survival, and chemotaxis (for a detailed review of PDGFR signal transduction see Volume 1, Chapter 70 and the references therein). Several features of the PDGFR system make it an appropriate model to investigate whether specific signals transmitted by RTKs translate into unique physiological functions *in vivo*. First, genetic analysis is facilitated as signaling is mediated by just two highly related yet distinct RTKs, the PDGFRs α and β (PDGFαR and PDGFβR) [1]. Second, the two PDGFRs bind a highly overlapping but nonidentical repertoire of signaling proteins [2]. Third, autophosphorylation site mutant PDGFRs, which lack the ability to activate specific intracellular signaling pathways, have been well characterized [3–5]. Finally, the phenotypes of the PDGFαR –/– mice and the PDGFβR –/– mice are easily distinguishable, demonstrating that each PDGFR mediates specific functions during embryonic development [6,7]. Since the phenotypes of PDGF and PDGFR-null mutants have recently been reviewed in detail [8], only the basics will be given here. This chapter focuses on what has been learned by analyzing mice bearing specific alterations in the intracellular signaling domains of the two PDGFRs.

PDGFβR Signaling *In Vivo*

PDGFβR signaling is required at late gestation (E16.5–E18.5), as deletion of the PDGFβR results in lost integrity of the developing microvasculature leading to capillary microaneurysm and perinatal lethality [7]. Abnormal capillary development is observed in multiple organs but is prominent at the kidney glomerulus, which completely lacks a well-formed capillary tuft. This is due to the absence of mesangial cells, matrix-secreting cells related to vascular smooth muscle cells (vSMCs), which coat and maintain the integrity of glomerular capillaries. The defects observed are strikingly similar to those displayed by mice harboring a null mutation of PDGFB [9]. This is not surprising, as dimeric PDGFBB is a predominant ligand for the PDGFβR [10]. Combined analysis of the two mutant lines revealed that PDGFB/PDGFβR signaling is required for proper communication between the two major vascular cell types: endothelial cells, which express PDGFB, and vSMCs/pericytes (vSMC/PC), which express the PDGFβR [11,12]. Consistent with identified roles of PDGFBB as a mitogen and chemoattractant for vSMCs, disruption of this signaling axis results in a failure of preexisting vSMC/PC to migrate to, and proliferate along angiogenic sprouts of the developing microvasculature [11,12].

PDGFαR Signaling *In Vivo*

While loss of PDGFβR signaling results in defects restricted primarily to vSMCs and related cells, loss of PDGFαR signaling affects multiple cell types. These include derivatives of sclerotome and non-neuronal neural crest cells [6]. Consequently, PDGFαR-null embryos exhibit a complex and variable phenotype. The hallmarks of this phenotype include a cleft face, defects of the axial and proximal appendicular skeleton, subepidermal blebbing, and hemorrhaging. Mice lacking the PDGFαR typically die at midgestation starting at E8.5 and very few embryos survive past E15.5. Consistent with a role for PDGFαR signaling in cell survival, the cranial defects appear to be due to increased apoptosis of migrating neural crest cells. The trunk defects, including skeletal malformations, appear to arise from a deficiency in myotome formation suggesting that PDGFαR signaling participates in somite patterning. In contrast to the high phenotypic similarity between PDGFB and PDGFβR-null mice, a substantial population of PDGFA-null mice (~20%) survives well beyond gestation, a stage never achieved by PDGFαR-null animals [13]. This is explained by the fact that the PDGFαR binds multiple PDGFs including homodimers of PDGFA, PDGFB, and PDGFC in addition to heterodimers of PDGFA and PDGFB [10,14]. Studies on postnatal PDGFA−/− mutants have revealed roles for PDGFαR signaling in development of oligodendrocytes, lung alveolar myofibroblasts, intestinal mesenchyme, and Leydig cells [8].

Specificity of PDGFR Signaling *In Vivo*

One important concern in signal transduction is whether individual RTKs transmit distinct biochemical signals that translate into the ability to direct unique functions *in vivo*. Different ligand-binding affinities, patterns of gene expression, and/or mechanisms of signal transduction could all account for the functional specificity displayed by the two PDGFRs *in vivo*. Although both PDGFRs bind to many of the same signaling proteins including PI3 kinase, PLCγ, SHP-2, and Src family kinases, each receptor also binds to some proteins in an exclusive fashion. Prominent examples include binding of RasGAP to the PDGFβR but not the PDGFαR, and binding of Crk family adaptor proteins to the PDGFαR but not the PDGFβR [15–17]. Whether such differences convey an intrinsic ability upon RTKs to transmit functionally distinct signals *in vivo* is a highly debated issue. This issue was addressed in part by analysis of two complementary lines of PDGFR knockin mice [18]. In each line, the intracellular signaling domains of one PDGFR were removed and replaced with those of the other PDGFR. Since the novel chimeric PDGFR retains the ligand binding capacity and the spatiotemporal expression of the replaced PDGFR, only differences in intracellular signaling should account for any abnormal phenotypes. Mice harboring PDGFβR intracellular domains in the place of the endogenous PDGFαR regions (called αβ) developed no overt defects as homozygotes (αβ/αβ) or hemizygotes (αβ/-). This suggests that signals transmitted by the two PDGFRs are interpreted in a highly redundant manner in cells normally expressing the PDGFαR. Furthermore, mice expressing the converse chimeric receptor at the PDGFβR locus (βα) are also viable and largely normal. However, suboptimal function of cells normally regulated by PDGFβR signaling in these animals is revealed by modest heart enlargement and an impaired mesangial cell response to induced glomerular injury. Furthermore, exacerbated systemic vascular defects are observed in βα/- hemizygotes. Consistent with loss of normal PDGFβR function, the defects are due to abnormal development of vSMC/PCs and subsequent failure to coat the microvasculature. The βα/- phenotype includes perinatal lethality (~40%), substantial heart enlargement, glomerulosclerosis, and retinopathy. The retinopathy, which results in retinal detachment and migration into the vitreous body, bears striking resemblance to retinopathy of diabetes. Diabetic retinopathy primarily affects the retinal capillaries and is characterized initially by loss of PCs. Indeed, marker analysis on βα/- animals clearly demonstrates that loss of retinal PCs is the underlying cause of the observed phenotype. Cellular analysis suggests that inability of PDGFαR-type signaling to sustain MAPK activation may contribute to the decreased function of βα-expressing vSMC/PCs. Taken together, these results indicate that while the functional specificity displayed by the two PDGFRs is largely due to differences in ligand affinities and/or patterns of gene expression, PDGFβR-specific signals are required for optimal function of vSMC/PCs. Further lines of knockin mice have since been generated in which the intracellular domains of the PDGFRs have been replaced with those of further divergent RTKs. Analysis of these lines clearly shows that increased divergence from PDGFR-type signaling reduces the ability to rescue the embryological functions of the PDGFRs [19].

Another major focus of signal transduction studies is to define the contributions of distinct intracellular signaling pathways to RTK function. Many RTK-associated signaling proteins contain phosphotyrosine-binding modules such as SH2 or PTB domains [20]. Mutation of receptor tyrosine residues can be used to uncouple RTK activation from activation of specific effector proteins. Knockin mice harboring such point mutations in the PDGFRs provided a means to test contributions of distinct signaling pathways to PDGFR function *in vivo*. Studies on cultured cells expressing PDGFβR mutants identified PI3 kinase and PLCγ as the major effectors of PDGFβR-triggered mitogenesis and chemotaxis [21,22]. These analyses also demonstrated that the functions promoted by these two effectors are highly redundant. Consistent with maintenance of this redundancy *in vivo*, homozygous mutant mice expressing a PDGFβR that cannot activate PI3K still develop normally [23]. More surprising is the finding that elimination of both PI3-K and PLCγ binding sites on the PDGFβR is still compatible with normal viability [24]. *In vivo* challenge assays and chimeric analysis does reveal that loss of these signals decreases vSMC/PC function. However, the relatively normal phenotype indicates that compensating mechanisms exist, which promote a threshold level of signaling permitting vascular integrity. Microarray experiments suggest that activation of other PDGFβR-associated signaling proteins compensate for loss of PI3-K and PLCγ signals [25]. However, PDGFβR signaling mutant mice have recently been generated that lack binding sites for SHP-2, RasGAP, Grb2, and Src family kinases, in addition to PI3-K and PLCγ, without loss of viability [26]. It is possible that a threshold of signaling required for viability is met by activating PDGFβR signaling pathways through a surrogate receptor. An obvious candidate is the PDGFαR, as heterodimerization between the two PDGFRs occurs in response to certain PDGF ligands. Thus far, PDGFαR signaling has not been shown to compensate for loss of PDGFβR signaling in cell types dependent upon PDGFβR function [24]. Perhaps interaction with less obvious cell surface receptors compensates for loss of direct PDGFβR binding to signaling proteins. Interactions between PDGFRs and integrins have been demonstrated and both activate many of the same intracellular signals [27–29]. Examination of mice harboring compound mutations in the PDGFβR and specific integrin subunits should shed light on this possibility.

In contrast to the apparent resilience of the PDGFβR to lost signals, mutations that uncouple the PDGFαR from distinct pathways cause developmental defects of varying severity. An allelic series of three PDGFαR signaling mutants has been analyzed [30]. Included in this series were mutants that prevented the PDGFαR from activating Src family kinases (α^{Src}), or PI3-K (α^{PI3-K}), or multiple pathways including the two previously mentioned in combination with PLCγ and SHP-2 (α^{F7}). α^{Src} animals are viable past birth but suffer from neurological-related defects including shaking, seizures, and decreased limb mobility. This is due to a specific defect in oligodendrocyte development and consequent hypomyelination of the central nervous system. Consistent with the interpretation that multiple PDGFαR signals are required for oligodendrocyte development, α^{PI3-K} mutants also displayed this phenotype. However, while the α^{Src} phenotype is restricted to oligodendrocytes, the α^{PI3-K} mutation affects multiple cell types. Most α^{PI3-K} mutants die perinatally and exhibit defects that are similar to, albeit less severe than, PDGFαR-null embryos. While α^{PI3-K} mutants do not exhibit an overt cleft face, they typically display a cleft palate. This suggests that PDGFαR signaling through PI3-K is required for optimal function of non-neuronal neural crest cells. Other skeletal abnormalities resulting from loss of PI3K signaling include spina bifida, misshapen cervical vertebrae, and malformations of the shoulder girdle. Furthermore, α^{PI3-K} mutants exhibit abnormal placental vascularization demonstrating a requirement for PDGFαR-initiated PI3-K signaling in extra-embryonic development [19]. While signaling through PI3-K is clearly essential for PDGFαR function *in vivo*, the receptor is still able to drive development to late embryogenesis. This is not likely due to compensatory signaling by other PDGFαR-associated proteins as the α^{F7} mutation essentially phenocopies α^{PI3-K}. Instead, it is likely that a low level of PI3-K signal, activated via heterodimer formation between mutant PDGFαRs and wild type PDGFβRs, is the compensating factor. This is based on the observation that double homozygous embryos harboring mutations in the PI3-K binding sites of both PDGFRs recapitulate the severity and the hallmark phenotypes of the PDGFαR-null mutants. These results indicate that PI3-K is the major effector of PDGFαR function *in vivo*, and that PDGFαR-initiated signals direct both specific and overlapping functions during mammalian development.

Building on work that established how PDGFRs transmit intracellular signals to carry out functions at the cellular level, genetic analysis in mice is beginning to demonstrate how these signals are utilized in the context of a living mammal. Studies are now beginning to reveal that most RTKs do not work alone, but instead participate as parts of a complex signaling matrix. Given that receptors such as integrins have been shown in cell culture systems to cooperate with PDGFRs for signal transmission, exploration of these interactions in genetic systems should be rewarding. Furthermore, while studies on mice have focused on signaling events that occur immediately following PDGFR activation, the significance of later events is largely unknown. To address this issue, gene trap screens in embryonic stem cells designed to find genes that exacerbate or ameliorate the PDGFR knockin mutant phenotypes are now underway. Such screens should provide a powerful means to identify novel players in PDGFR signal transduction *in vivo*.

References

1. Heldin, C.-H. and Westermark, B. (1999). Mechanism of action and *in vivo* role of platelet-derived growth factor. *Physiol. Rev.* **79**, 1283–1316.
2. Rosenkranz, S. and Kazlauskas, A. (1999). Evidence for distinct signaling properties and biological responses induced by the PDGF receptor alpha and beta subtypes. *Growth Factors* **16**, 201–216.
3. Kazlauskas, A. (1994). Receptor tyrosine kinases and their targets. *Curr. Opin. Genet. Dev.* **4**, 5–14.
4. Claesson-Welsh, L. (1994). Platelet-derived growth factor receptor signals. *J. Biol. Chem.* **269**, 32023–32026.
5. Rosenkranz, S., DeMali, K. A., Gelderloos, J. A., Bazenet, C., and Kazlauskas, A. (1999). Identification of the receptor-associated signaling enzymes that are required for platelet-derived growth factor-AA-dependent chemotaxis and DNA synthesis. *J. Biol. Chem.* **274**, 28335–28343.
6. Soriano, P. (1997). The PDGFα receptor is required for neural crest cell development and normal patterning of the somites. *Development* **124**, 2691–2700.
7. Soriano, P. (1994). Abnormal kidney development and hematological disorders in PDGF β-receptor mutant mice. *Genes Dev.* **8**, 1888–1896.
8. Betsholtz, C., Karlsson, L., and Lindahl, P. (2001). Developmental roles of platelet-derived growth factors. *BioEssays* **23**, 494–507.
9. Levéen, P., Pekny, M., Gebre-Medhin, S., Swolin, B., Larsson, E., and Betsholtz, C. (1994). Mice deficient for PDGF B show renal, cardiovascular, and hematological abnormalities. *Genes Dev.* **8**, 1875–1887.
10. Heldin, C.-H., Ostman, A., and Ronnstrand, L. (1998). Signal transduction via platelet-derived growth factor receptors. *Biochim. Biophys. Acta* **1378**, F79–F113.
11. Lindahl, P., Johansson, B. R., Leveen, P., and Betsholtz, C. (1997). Pericyte loss and microaneurysm formation in PDGF-B-deficient mice. *Science* **277**, 242–245.
12. Hellstrom, M., Kalen, M., Lindahl, P., Abramsson, A., and Betsholtz, C. (1999). Role of PDGF-B and PDGR-β in recruitment of vascular smooth muscle cells and pericytes during embryonic blood vessel formation in the mouse. *Development* **126**, 3047–3055.

13. Boström, H., Willetts, K., Pekny, M., Levéen, P., Lindahl, P., Hedstrand, H., Pekna, M., Hellström, M., Gebre-Medhin, S., Schalling, M. *et al.* (1996). PDGF-A signaling is a critical event in lung alveolar myofibroblast development and alveogenesis. *Cell* **85**, 863–873.
14. Li, X., Ponten, A., Aase, K., Karlsson, L., Abramsson, A., Uutela, M., Backstrom, G., Hellstrom, M., Bostrom, H., Li, H. *et al.* (2000). PDGF-C is a new protease-activated ligand for the PDGF α-receptor. *Nat. Cell Biol.* **2**, 302–309.
15. Heidaran, M. A., Beeler, J. F., Yu, J.-C., Ishibashi, T., LaRochelle, W. J., Pierce, J. H., and Aaronson, S. A. (1993). Differences in substrate specificities of α and β platelet-derived growth factor (PDGF) receptors. *J. Biol. Chem.* **268**, 9287–9295.
16. Bazenet, C. and Kazlauskas, A. (1994). The PDGF receptor alpha subunit activates p21ras and triggers DNA synthesis without interacting with rasGAP. *Oncogene* **9**, 517–525.
17. Yokote, K., Hellman, U., Ekman, S., Saito, Y., Ronnstrand, L., Saito, Y., Heldin, C.-H., and Mori, S. (1998). Identification of tyr-762 in the platelet-derived growth factor α-receptor as the binding site for crk proteins. *Oncogene* **16**, 1229–1239.
18. Klinghoffer, R. A., Mueting-Nelson, P. F., Faerman, A., Shani, M., and Soriano, P. (2001). The two PDGF receptors maintain conserved signaling *in vivo* despite divergent embryological functions. *Mol. Cell* **7**, 343–354.
19. Hamilton, T. G., Klinghoffer, R. A., Corrin, P. D., and Soriano, P. (2003). Evolutionary divergence of platelet-derived growth factor alpha receptor signaling mechanisms. *Mol. Cell. Biol.* **23**, 4013–4025.
20. Hunter, T. Signaling-2000 and beyond. (2000). *Cell* **100**, 113–127.
21. Valius, M. and Kazlauskas A. (1993). Phospholipase C-γ1 and phosphatidylinositol 3 kinase are the downstream mediators of the PDGF receptor's mitogenic signal. *Cell* **73**, 321–334.
22. Kundra, V., Escobedo, J. A., Kazlauskas, A., Kim, H. K., Rhee, S. G., Williams, L. T., and Zetter, B. R. (1994). Regulation of chemotaxis by the platelet-derived growth factor receptor-β. *Nature* **367**, 474–476.
23. Heuchel, R., Berg, A., Tallquist, M., Ahlen, K., Reed, R. K., Rubin, K., Claesson-Welsh, L., Heldin, C.-H., and Soriano, P. (1999). Platelet-derived growth factor β receptor regulates interstitial fluid homeostasis through phosphatidylinositol-3′ kinase signaling. *Proc. Natl. Acad. Sci. USA* **96**, 11410–11415.
24. Tallquist, M. D., Klinghoffer, R. A., Heuchel, R., Mueting-Nelson, P. F., Corrin, P. D., Heldin, C.-H., Johnson, R. J., and Soriano, P. (2000). Retention of PDGFR-β function in mice in the absence of phosphatidylinositol 3′-kinase and phospholipase Cγ signaling pathways. *Genes Dev.* **14**, 3179–3190.
25. Fambrough, D., McClure, K., Kazlauskas, A., and Lander, E. S. (1999). Diverse signaling pathways activated by growth factor receptors induce broadly overlapping, rather than independent, sets of genes. *Cell* **97**, 727–741.
26. Tallquist, M. and Soriano, P., personal communication.
27. Schneller, M., Vuori, K., and Ruoslahti, E. (1997). αvβ3 integrin associates with activated insulin and PDGFβ receptors and potentiates the biological activity of PDGF. *EMBO J.* **16**, 5600–5607.
28. Sundberg, C. and Rubin, K. (1996). Stimulation of β1 integrins on fibroblasts induces PDGF-independent tyrosine phosphorylation of PDGFβ-receptors. *J. Cell Biol.* **132**, 741–752.
29. DeMali, K. A., Balciunaite, E., and Kazlauskas, A. (1999). Integrins enhance platelet-derived growth factor (PDGF)-dependent responses by altering the signal relay enzymes that are recruited to the PDGFβ receptor. *J. Biol. Chem.* **274**, 19551–19558.
30. Klinghoffer, R. A., Hamilton, T. G., Hoch, R., and Soriano, P. (2002). An allelic series at the PDGFαR locus indicates unequal contributions of distinct signaling pathways during development. *Dev. Cell* **2**, 103–113.

VEGF and the Angiopoietins Activate Numerous Signaling Pathways that Govern Angiogenesis

Christopher Daly and Jocelyn Holash
Regeneron Pharmaceuticals, Inc., Tarrytown, New York

Angiogenesis is critical for both the growth and repair of tissues as well as for the advancement of many diseases. While there is great therapeutic potential in regulating angiogenesis, much is still to be learned about the processes that govern it. In order to design appropriate therapies to promote or inhibit blood vessel growth, we must identify factors that act selectively on the vasculature and define the mechanisms by which they act. Here we explore the signaling pathways regulated by members of two families of growth factors that act largely specifically on blood vessels: Vascular endothelial growth factor (VEGF) and the angiopoietins. These factors are involved in multiple steps of blood vessel growth where they play overlapping yet very distinct roles.

Introduction

While the great majority of endothelial cells (ECs) in the adult vasculature are quiescent, certain conditions can induce preexisting blood vessels to give rise to new vessels, a process termed angiogenesis. In some circumstances, such as wound healing, angiogenesis supports tissue repair, while in others, such as in tumors or diabetic retinopathy, angiogenesis is associated with the destruction of normal tissue. Although there are a number of processes that may contribute to angiogenesis, including vessel sprouting and branching, intussusceptive vessel growth and the recruitment of circulating endothelial cells into preexisting blood vessels, for this chapter we will focus on the expansion of the vascular tree that occurs as a consequence of the sprouting and branching of preexisting vessels.

Several events must occur in order for a vasculature to expand via sprouting angiogenesis. Endothelial cells must divide, migrate, and survive. Once a new vessel is formed, perivascular cells must be recruited to the vessel wall, and their interactions with ECs must be stabilized. Multiple factors regulate each of these events, and in some instances several factors appear to act redundantly; thus, there are numerous signaling events and interactive pathways that are associated with the growth of blood vessels. We have chosen to focus on the signaling pathways activated by VEGF and the Angiopoietins, Ang-1 and Ang-2, since these factors act far more specifically on the vasculature than other factors that regulate angiogenesis [1]. VEGF, for which both VEGFR-1 and VEGFR-2 act as receptors, appears to signal predominantly through VEGFR-2, and is required for both embryonic and adult angiogenesis [2]. Angiopoietins, which are ligands for the Tie 2 receptor, also appear to be involved in both embryonic and adult angiogenesis, although their functions are distinct from those of VEGF. Ang-1, in contrast to VEGF, is not required for the earliest stages of vascular formation in the embryo, but rather is necessary for the remodeling and stabilization of the primary vasculature [3–4]. Ang-2, which in some instances can antagonize the activation of Tie 2 by Ang-1[5], appears to play an important role in sprouting angiogenesis, possibly by destabilizing existing vessels [5–7]. Despite the cooperation between VEGF and the angiopoietins in vessel formation, these factors can also have opposing effects on vascular processes, for example, on permeability [8,9].

Endothelial Cell Proliferation

VEGF is highly expressed in angiogenic settings and drives EC proliferation [10]. Similar to other growth factors, induction of proliferation by VEGF requires activation of the ERK (MAP kinase) cascade. In cells that are normally adherent, such as endothelial cells, the ability of receptor tyrosine kinases (RTKs) to activate the ERK pathway and to induce proliferation is dependent on integrin-mediated adhesion to matrix [12,13]. Activation of ERK in ECs by VEGF and other mitogens is enabled by multiple integrins, suggesting a general requirement for cytoskeletal assembly [14,15]. Recent data suggest that colocalization of RTKs and integrins at sites of cell attachment results in assembly of signaling complexes that enhance the coupling of RTKs to downstream effectors [16]. The specific integrins with which VEGFR-2 cooperates may depend on the composition of the matrix in a particular vessel.

In the classical model of ERK activation by RTKs, the adaptor protein Grb2, in a complex with Sos (a Ras guanine nucleotide exchange factor or GEF), binds to an activated receptor [17]. Sos promotes GTP loading of Ras, which then triggers the Raf-MEK-ERK cascade. Activated ERK translocates to the nucleus where it induces transcription of immediate early genes, thereby promoting cell cycle progression [18]. Epidermal growth factor (EGF) and bFGF, which are potent mitogens for HUVECs, stimulate the Ras-ERK pathway in this fashion [19,20]. However, the Ras-ERK pathway can also be activated via protein kinase C (PKC), which can induce formation of a Ras/Raf complex through a mechanism that is distinct from the Grb2/Sos-dependent mechanism [21]. Strong evidence suggests that VEGF, in contrast to EGF and bFGF, activates Ras-ERK signaling primarily via the PLCγ-Ca^{2+}-PKC pathway [19,20,22,23], although this appears to be cell-type dependent [24,26]. VEGF signaling to ERK via the less conventional pathway ensures that the Ras-ERK and PLCγ-Ca^{2+} pathways are activated in concert (Fig. 1). Although the functional significance of linking Ca^{2+} mobilization and ERK activation in a single signaling cascade is unclear, since bFGF can stimulate EC proliferation *in vitro* without activating the PLCγ-Ca^{2+} pathway [20], it is possible that Ca^{2+} signaling helps to create an environment that is permissive for angiogenesis. *In vivo*, EC proliferation may require disruption of interendothelial junctions, a process which may be critically dependent on Ca^{2+} (see the following section).

Interestingly, Ang-1 does not induce EC proliferation *in vitro* [3]. The inability of Ang-1 to drive proliferation may reflect its relatively weak (as compared to VEGF) stimulation of the ERK pathway [27] or its failure to activate the PLCγ/Ca^{2+} pathway (C. Daly and J. Holash, unpublished data).

VEGF Promotes Vascular Permeability

VEGF rapidly and potently increases the permeability of vessels to plasma proteins (e.g., fibrinogen)[28]. Although it has been proposed that extravasated plasma proteins provide a scaffold for EC migration during angiogenesis [28], the precise role of VEGF-induced permeability is unclear. It is possible that the destabilizing effects of VEGF on interendothelial junctions, which lead to plasma protein leak, are also important in allowing ECs to adopt a proliferative and/or migratory phenotype.

Plasma protein extravasation in response to VEGF appears to occur via intercellular gaps [29], although a transcellular route dependent on vesicular transport has also been reported [30]. Studies of other permeability-inducing agents such as thrombin indicate that formation of actomyosin

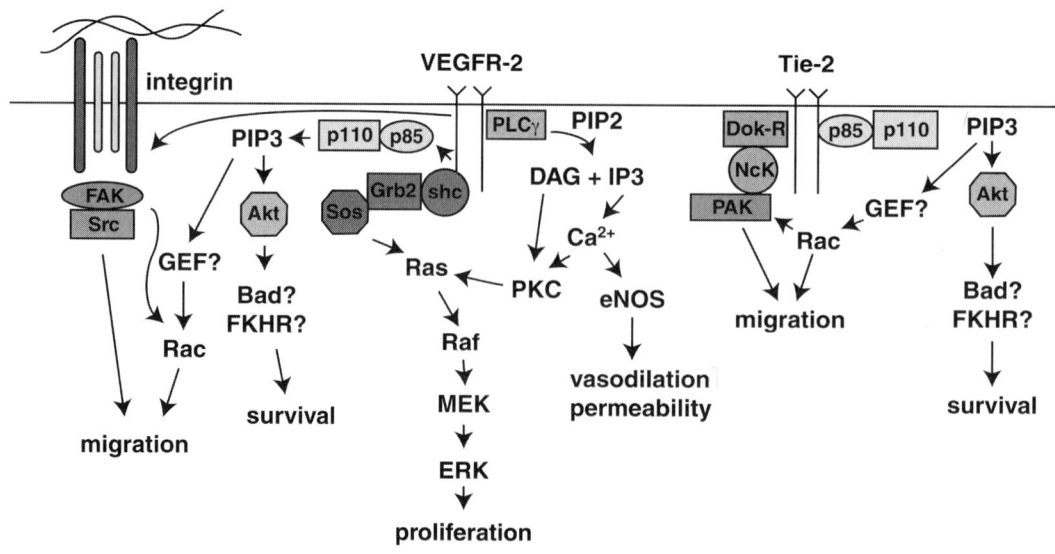

Figure 1 Major signaling pathways activated by VEGFR-2 and by Tie 2. The diagram is not meant to be comprehensive, but outlines some of the better characterized pathways employed by these two receptors. See the text for details.

stress fibers, which presumably create the tension required to form intercellular gaps, is important in controlling permeability[31]. Stress fiber formation is promoted by phosphorylation of myosin light chain, which can occur via Ca^{2+}-mediated activation of myosin light chain kinase or via activation of the small GTPase Rho [32]. Ca^{2+} mobilization is required for VEGF-induced permeability, consistent with a possible role for myosin light chain kinase [33]. In addition, VEGF appears to activate Rho and to promote stress fiber formation in HUVECs (J. Connolly and A. Hall, personal communication), although a requisite role for Rho in VEGF-induced leak remains to be demonstrated.

In addition to stress fiber formation, increases in permeability are likely to involve destabilization of the protein complexes at cell-cell junctions. VEGF has been reported to promote tyrosine phosphorylation of multiple components of EC adherens junctions, a modification which likely decreases junctional stability [34]. However, the precise mechanisms through which VEGF modulates cell-cell junctions remain to be elucidated.

eNOS, which is activated downstream of Ca^{2+} mobilization, is required for VEGF-induced permeability [35]. The generation of NO activates guanylate cyclase, resulting in cGMP production. cGMP appears to promote permeability, at least in part, by decreasing cAMP levels, although the mechanism(s) through which cAMP levels directly influence permeability are unknown [33].

Ang-1 Inhibits Vascular Permeability

Unlike VEGF, Ang-1 makes vessels resistant to leak. Transgenic overexpression of Ang-1 in the skin, or systemic administration of Ang-1, can block permeability induced by a number of inflammatory mediators including VEGF [8,9]. Intraocular injection of Ang-1 in diabetic mice reduces the edema that results from breakdown of the blood-retinal barrier, suggesting that it may be therapeutic in diabetic retinopathy [36]. The signaling mechanisms whereby Ang-1 blocks vascular permeability remain unknown, but might involve effects on the cytoskeleton and/or on cell-cell junctions.

Vessel Destabilization and EC Migration

Sprouting and migration of ECs requires the disruption of EC-perivascular cell interactions. Although signaling mechanisms controlling these interactions are not completely understood, Ang-1, as well as PDGFB [37] and TGFβ [38] appear to play important roles. The notion that Ang-1 contributes to vessel stabilization derives from the observation that Ang-1-null embryos exhibit disrupted endothelial-perivascular cell interactions [4]. Several studies have correlated Ang-2 expression in ECs with angiogenic sprouting [5,6,39], consistent with a role for Ang-2 (by antagonizing Ang-1 action [5]) in vessel destabilization [7]. However, the mechanisms through which Tie 2 signaling affects EC-perivascular cell interactions remain to be elucidated, and other factors are likely to contribute to vessel destabilization.

An essential aspect of cell migration is cytoskeletal rearrangement, a process in which Rho family GTPases play a critical role [40]. Rac, a member of the Rho family, is required for growth-factor-induced cell movement [40,41]. Through activation of a number of effector proteins, Rac promotes polymerization of actin in lamellipodia at the leading edge of motile cells [40]. VEGF has been shown to promote EC migration in a Rac-dependent fashion [42]. The pathway by which VEGF activates Rac is undefined, but may involve PI3-K-dependent stimulation of a Rac GEF [43]. Subsequent to Rac-induced lamellipodia formation, focal complexes containing clustered integrins are formed, presumably stabilizing the newly formed cell protrusion [44,46]. Focal complexes contain a variety of signaling molecules, including focal adhesion kinase (FAK), which is inducibly phosphorylated on tyrosine in response to integrin clustering and/or RTK activation [47]. VEGF has been shown to promote tyrosine phosphorylation of FAK [48], which in turn couples to multiple downstream signaling pathways, including Rac and the kinase Src [47]. Interestingly, Src appears to be required for VEGF-induced migration and angiogenesis [49,50].

Ang-1 has also been shown to induce EC migration *in vitro* [51,53], via a pathway involving Rac and the Rac effector PAK (Fig. 1)[54]. PAK, a kinase that inhibits actin depolymerization [55] and that is required for EC migration [56], is recruited to tyrosine-phosphorylated Tie 2 by the adaptors Dok-R and Nck [54,57], and subsequently becomes activated [54]. Understanding the significance of Ang-1-mediated activation of the PAK pathway will require further investigation, since the conditions under which Ang-1 is an important regulator of EC migration remain to be determined.

Regulation of EC Survival during Angiogenesis

ECs depend upon cell-cell and cell-matrix contacts for survival [58,59]. During angiogenesis, when matrix attachments and cell-cell contacts within stable vessels are disrupted in preparation for EC migration, the cells become dependent on VEGF for survival [60–63]. Once vessels mature and recruit pericytes, their dependence on VEGF is reduced [61,64], perhaps because pericytes elaborate additional survival factors or promote maturation of EC-matrix contacts.

Consistent with *in vivo* observations, VEGF promotes the survival of ECs *in vitro* under conditions of serum withdrawal [65,66]. The pro-survival effect of VEGF, like other growth factors, depends on activation of the PI3-K pathway [65,66]. Although VEGF induces tyrosine phosphorylation of the p85 subunit of PI3-K [67,68], it is unclear whether p85 interacts directly with VEGFR-2; it has been proposed that VEGF activates p85 indirectly, via a FAK dependent mechanism [69]. One critical component of the PI3-K pathway is AKT, a kinase which phosphorylates and inhibits several pro-apoptotic proteins, including Bad and the transcription factor FKHR [70]. Whether these AKT targets are relevant to VEGF signaling is unknown. In terms of pro-survival genes, VEGF induces the expression of bcl-2, survivin, Al, and xiap [71–74], by signaling pathways that remain to be characterized.

Ang-1 also promotes survival of ECs *in vitro* via the PI3-K/AKT pathway [75–77]. Interestingly, Ang-1 is a significantly more potent activator of the AKT pathway than is VEGF [78], suggesting that Ang-1 is likely to be an important regulator of EC survival *in vivo*. The AKT targets which are important in Ang-1 signaling remain to be identified. Like VEGF, Ang-1 increases survivin expression, although the underlying mechanism is unknown [77]. Recent data from partially-rescued Tie 2 knockout mice suggest that the Ang-1/Tie 2 signaling system is important for EC survival during vascular development [79]. Interestingly, activated Tie 2 can be detected in the quiescent vasculature of several adult tissues [80], suggesting that Ang-1 has a role in vascular maintenance, possibly including a pro-survival role.

Conclusion

The development of new blood vessels is an extraordinarily complex process. Although numerous growth factors have been evaluated for their angiogenic potential, very few factors that have specificity for the vasculature have been identified. It seems that most factors that are believed to have angiogenic activity also mediate numerous other processes, limiting the therapeutic potential of manipulating these systems. Thus, in this chapter we have focused on how VEGF and the angiopoietins, which are largely specific for the vasculature, may mediate a number of steps that are required for blood vessel formation.

Acknowledgments

The authors are grateful to Dr. George D. Yancopoulos for helpful insights and support and Vicki Lan for assistance with graphics.

References

1. Yancopoulos, G. D. *et al.* (2000). Vascular-specific growth factors and blood vessel formation. *Nature* **407**, 242–248.
2. Ferrara, N. and Davis-Smyth, T. (1997). The biology of vascular endothelial growth factor. *Endocr. Rev.* **18**, 4–25.
3. Davis, S. *et al.* (1996). Isolation of angiopoietin-1, a ligand for the TIE2 receptor, by secretion-trap expression cloning. *Cell* **87**, 1161–1169.
4. Suri, C. *et al.* (1996). Requisite role of angiopoietin-1, a ligand for the TIE2 receptor, during embryonic angiogenesis. *Cell* **87**, 1171–1180.
5. Maisonpierre, P. C. *et al.* (1997). Angiopoietin-2, a natural antagonist for Tie2 that disrupts *in vivo* angiogenesis. *Science* **277**, 55–60.
6. Holash, J. *et al.* (1999). Vessel cooption, regression, and growth in tumors mediated by angiopoietins and VEGF. *Science* **284**, 1994–1998.
7. Holash, J., Wiegand, S. J., and Yancopoulos, G. D. (1999). New model of tumorangiogenesis: Dynamic balance between vessel regression and growth mediated by angiopoietins and VEGF. *Oncogene* **18**, 5356–5362.
8. Thurston, G. *et al.* (1999). Leakage-resistant blood vessels in mice transgenically overexpressing angiopoietin-1. *Science* **286**, 2511–2514.
9. Thurston, G. *et al.* (2000). Angiopoietin-1 protects the adult vasculature against plasma leakage. *Nat. Med.* **6**, 460–463.
10. Leung, D. W., Cachianes, G., Kuang, W. J., Goeddel, D. V., and Ferrara, N. (1989). Vascular endothelial growth factor is a secreted angiogenic mitogen. *Science* **246**, 1306–1309.
11. Renshaw, M. W., Ren, X. D., and Schwartz, M. A. (1997). Growth factor activation of MAP kinase requires cell adhesion. *EMBO J.* **16**, 5592–5599.
12. Short, S. M., Talbott, G. A., and Juliano, R. L. (1998). Integrin-mediated signaling events in human endothelial cells. *Mol. Biol. Cell* **9**, 1969–1980.
13. Soldi, R. *et al.* Role of alphavbeta3 integrin in the activation of vascular endothelial growth factor receptor-2. *EMBO J.* **18**, 882–892.
14. Aplin, A. E., Short, S. M., and Juliano, R. L. (1999). Anchorage-dependent regulation of the mitogen-activated protein kinase cascade by growth factors is supported by a variety of integrin alpha chains. *J. Biol. Chem.* **274**, 31223–31228.
15. Senger, D. R. *et al.* The alpha(1)beta(1) and alpha(2)beta(1) integrins provide critical support for vascular endothelial growth factor signaling, endothelial cell migration, and tumor angiogenesis. *Am. J. Pathol.* **160**, 195–204.
16. Sieg, D. J. *et al.* (2000). FAK integrates growth-factor and integrin signals to promote cell migration. *Nat. Cell Biol.* **2**, 249–256.
17. Schlessinger, J. (2000). Cell signaling by receptor tyrosine kinases. *Cell* **103**, 211–225.
18. Hazzalin, C. A. and Mahadevan, L. C. (2002). MAPK-regulated transcription: A continuously variable gene switch? *Nat. Rev. Mol. Cell Biol.* **3**, 30–40.
19. Doanes, A. M. *et al.* (1999). VEGF stimulates MAPK through a pathway that is unique for receptor tyrosine kinases. *Biochem. Biophys. Res. Commun.* **255**, 545–548.
20. Wu, L. W. *et al.* (1998). Utilization of distinct signaling pathways by receptors for vascular endothelial cell growth factor and other mitogens in the induction of endothelial cell proliferation. *J. Biol. Chem.* **275**, 5096–5103.
21. Marais, R. *et al.* (1998). Requirement of Ras-GTP-Raf complexes for activation of Raf-1 by protein kinase C. *Science* **280**, 109–112.
22. Takahashi, T., Ueno, H., and Shibuya, M. (1999). VEGF activates protein kinase C-dependent, but Ras-independent Raf-MEK-MAP kinase pathway for DNA synthesis in primary endothelial cells. *Oncogene* **18**, 2221–2230.
23. Takahashi, T., Yamaguchi, S., Chida, K., and Shibuya, M. (2001). A single autophosphorylation site on KDR/Flk-1 is essential for VEGF-A-dependent activation of PLC-gamma and DNA synthesis in vascular endothelial cells. *EMBO J.* **720**, 2768–2778.
24. Kroll, J. and Waltenberger, J. (1997). The vascular endothelial growth factor receptor KDR activates multiple signal transduction pathways in porcine aortic endothelial cells. *J. Biol. Chem.* **272**, 32521–32527.
25. Yashima, R. *et al.* (2001). Heterogeneity of the signal transduction pathways for VEGF-induced MAPKs activation in human vascular endothelial cells. *J. Cell. Physiol.* **188**, 201–210.
26. Meadows, K. N., Bryant, P., and Pumiglia, K. (2001). Vascular endothelial growth factor induction of the angiogenic phenotype requires Ras activation. *J. Biol. Chem.* **276**, 49289–49298.
27. Kim, L. *et al.* (2002). EphB ligand, ephrinB2, suppresses the VEGF- and angiopoietin 1-induced Ras/mitogen-activated protein kinase pathway in venous endothelial cells. *FASEB J.* **16**, 1126–1128.
28. Dvorak, H. F. *et al.* (1995). Vascular permeability factor/vascular endothelial growth factor: an important mediator of angiogenesis in malignancy and inflammation. *Int. Arch. Allergy Immunol.* **107**, 233–235.
29. Roberts, W. G. and Palade, G. E. (1995). Increased microvascular permeability and endothelial fenestration induced by vascular endothelial growth factor. *J. Cell Sci.* **108** (Pt 6), 2369–79 (1995).
30. Dvorak, H. F. (2000). VPF/VEGF and the angiogenic response. *Semin. Perinatal* **24**, 75–78.
31. Stevens, T., Garcia, J. G., Shasby, D. M., Bhattacharya, J., and Malik, A. B. (2000). Mechanisms regulating endothelial cell barrier function. *Am. J. Physiol. Lung Cell Mol. Physiol.* **279**, L419–L422.

32. Ridley, A. J. (2001). Rho family proteins: coordinating cell responses. *Trends Cell Biol.* **11**, 471–477.
33. Michel, C. C. and Curry, F. E. (1999). Microvascular permeability. *Physiol. Rev.* **79**, 703–761.
34. Esser, S., Lampugnani, M. G., Corada, M, Dejana, E., and Risau, W. (1998). Vascular endothelial growth factor induces VE-cadherin tyrosine phosphorylation in endothelial cells. *J. Cell Sci.* **111** (Pt 13), 1853–1865.
35. Fukumura, D. *et al.* Predominant role of endothelial nitric oxide synthase in vascular endothelial growth factor-induced angiogenesis and vascular permeability. *Proc. Natl. Acad. Sci. USA* **98**, 2604–2609.
36. Joussen, A. M. *et al.* (2002). Suppression of diabetic retinopathy with angiopoietin-1. *Am. J. Pathol.* **160**, 1683–1693.
37. Hellstrom, M., Kal, N. M., Lindahl, P., Abramsson, A., and Betsholtz, C. (1999). Role of PDGF-B and PDGFR-beta in recruitment of vascular smooth muscle cells and pericytes during embryonic blood vessel formation in the mouse. *Development* **126**, 3047–3055.
38. Pepper, M. S. (1997). Transforming growth factor-beta: Vasculogenesis, angiogenesis, and vessel wall integrity. *Cytokine Growth Factor Rev.* **8**, 21–43.
39. Stratmann, A., Risau, W., and Plate, K. H. (1998). Cell type-specific expression of angiopoietin-1 and angiopoietin-2 suggests a role in glioblastoma angiogenesis. *Am. J. Pathol.* **153**, 1459–1466.
40. Hall, A. (1998). Rho GTPases and the actin cytoskeleton. *Science* **279**, 509–514.
41. Bar-Sagi, D. and Hall, A. (2000). Ras and Rho GTPases: a family reunion. *Cell* **103**, 227–238.
42. Soga, N. *et al.* (2001). Rho family GTPases regulate VEGF-stimulated endothelial cell motility. *Exp. Cell Res.* **269**, 73–87.
43. Scita, G. *et al.* Signaling from Ras to Rac and beyond: not just a matter of GEFs. *EMBO J.* **19**, 2393–2398.
44. Hotchin, N. A. and Hall, A. (1995). The assembly of integrin adhesion complexes requires both extracellular matrix and intracellular rho/rac GTPases. *J. Cell Biol.* **131**, 1857–1865.
45. Byzova, T. V. *et al.* (2000). A mechanism for modulation of cellular responses to VEGF: Activation of the integrins. *Mol. Cell* **6**, 851–860.
46. Kiosses, W. B., Shattil, S. J., Pampori, N., and Schwartz, M. A. (2001). Rac recruits high-affinity integrin alphavbeta5 to lamellipodia in endothelial cell migration. *Nat. Cell Biol.* **3**, 316–320.
47. Parsons, J. T., Martin, K. H., Slack, J. K., Taylor, J. M., and Weed, S. A. (2000). Focal adhesion kinase: A regulator of focal adhesion dynamics and cell movement. *Oncogene* **19**, 5606–5613.
48. Abedi, H. and Zachary, I. (1997). Vascular endothelial growth factor stimulates tyrosine phosphorylation and recruitment to new focal adhesions of focal adhesion kinase and paxillin in endothelial cells. *J. Biol. Chem.* **272**, 15442–15451.
49. Eliceiri, B. P. *et al.* (1999). Selective requirement for Src kinases during VEGF-induced angiogenesis and vascular permeability. *Mol. Cell* **4**, 915–924.
50. Abu-Ghazaleh, R., Kabir, J., Jia, H., Lobo, M., and Zachary, I. (2001). Src mediates stimulation by vascular endothelial growth factor of the phosphorylation of focal adhesion kinase at tyrosine 861, and migration and anti-apoptosis in endothelial cells. *Biochem. J.* **360**, 255–264.
51. Witzenbichler, B., Maisonpierre, P. C., Jones, P., Yancopoulos, G. D., and Isner, J. M. (1998). Chemotactic properties of angiopoietin-1 and -2, ligands for the endothelial-specific receptor tyrosine kinase Titil. *J. Biol. Chem.* **273**, 18514–18521.
52. Jones, N. *et al.* (1999). Identification of Tek/Tie2 binding partners. Binding to a multifunctional docking site mediates cell survival and migration. *J. Biol. Chem.* **274**, 30896–30905.
53. Kim, I. *et al.* (2000). Angiopoietin-1 induces endothelial cell sprouting through the activation of focal adhesion kinase and plasmin secretion. *Circ. Res.* **86**, 952–959.
54. Master, Z. *et al.* (2001). Dok-R plays a pivotal role in angiopoietin-1-dependent cell migration through recruitment and activation of Pak. *EMBO J.* 5919–5928.
55. Edwards, D. C., Sanders, L. C., Bokoch, G. M., and Gill, G. N. (1999). Activation of LIM-kinase by Pak1 couples Rac/Cdc42 GTPase signalling to actin cytoskeletal dynamics. *Nat. Cell Biol.* **1**, 253–259.
56. Kiosses, W. B., Daniels, R. H., Otey, C., Bokoch, G. M., and Schwartz, M. A. (1999). A role for p21-activated kinase in endothelial cell migration. *J. Cell Biol.* **147**, 831–844.
57. Jones, N. and Dumont, D. J. (1998). The Tek/Tie2 receptor signals through a novel Dok-related docking protein, Dok-R. *Oncogene* **17**, 1097–1108.
58. Meredith, J. E., Jr., Fazeli, B., and Schwartz, M. A. (1993). The extracellular matrix as a cell survival factor. *Mol. Biol. Cell.* **4**, 953–961.
59. Carmeliet, P. *et al.* (1999). Targeted deficiency or cytosolic truncation of the VE-cadherin gene in mice impairs VEGF-mediated endothelial survival and angiogenesis. *Cell* **98**, 147–157.
60. Benjamin, L. E. and Keshet, E. (1997). Conditional switching of vascular endothelial growth factor (VEGF) expression in tumors; induction of endothelial cell shedding and regression of hemangioblastoma-like vessels by VEGF withdrawal. *Proc. Natl. Acad. Sci. USA* **94**, 8761–8766.
61. Benjamin, L. E., Golijanin, D. M., Itin, A., Pode, D., and Keshet, E. (1999). Selective ablation of immature blood vessels in established human tumors follows vascular endothelial growth factor withdrawal. *J. Clin. Invest.* **103**, 159–165.
62. Alon, T. *et al.* (1995). Vascular endothelial growth factor acts as a survival factor for newly formed retinal vessels and has implications for retinopathy of prematurity. *Nat. Med.* **1**, 1024–1028.
63. Gerber, H. P. *et al.* (1999). VEGF is required for growth and survival in neonatal mice. *Development* **126**, 1149–1159.
64. Benjamin, L. E., Hemo, I., and Keshet, E. (1998). A plasticity window for blood vessel remodelling is defined by pericyte coverage of the preformed endothelial network and is regulated by PDGF-B and VEGF. *Development* **125**, 1591–1598.
65. Gerber, H. P. *et al.* (1998). Vascular endothelial growth factor regulates endothelial cell survival through the phosphatidylinositol 3'-kinase/Akt signal transduction pathway. Requirement for Flk-1/KDR activation. *J. Biol. Chem.* **273**, 30336–30343.
66. Fujio, Y. and Walsh, K. (1999). Akt mediates cytoprotection of endothelial cells by vascular endothelial growth factor in an anchorage-dependent manner. *J. Biol. Chem.* **274**, 16349–16354.
67. Guo, D., Jia, Q., Song, H. Y., Warren, R. S., and Donner, D. B. (1995). Vascular endothelial cell growth factor promotes tyrosine phosphorylation of mediators of signal transduction that contain SH2 domains. Association with endothelial cell proliferation. *J. Biol. Chem.* **270**, 6729–6733.
68. Thakker, G. D., Hajjar, D. P., Muller, W. A., and Rosengart, T. K. (1999). The role of phosphatidylinositol 3-kinase in vascular endothelial growth factor signaling. *J. Biol. Chem.* **274**, 10002–10007.
69. Qi, J. H. and Claesson-Welsh, L. (2001). VEGF-induced activation of phosphoinositide 3-kinase is dependent on focal adhesion kinase. *Exp. Cell Res.* **263**, 173–182.
70. Datta, S. R., Brunet, A., and Greenberg, M. E. (1999). Cellular survival: A play in three Akts. *Genes Dev.* **13**, 2905–2927.
71. Gerber, H. P., Dixit, V., and Ferrara, N. (1998). Vascular endothelial growth factor induces expression of the antiapoptotic proteins Bcl-2 and A1 in vascular endothelial cells. *J. Biol. Chem.* **273**, 13313–13316.
72. Nor, J. E., Christensen, J., Mooney, D. J., and Polverini, P. J. (1999). Vascular endothelial growth factor (VEGF)-mediated angiogenesis is associated with enhanced endothelial cell survival and induction of Bcl-2 expression. *Am. J. Pathol.* **154**, 375–384.
73. Tran, J. *et al.* (1999). Marked induction of the IAP family antiapoptotic proteins survivin and XIAP by VEGF in vascular endothelial cells. *Biochem. Biophys. Res. Commun.* **264**, 781–788.
74. O'Connor, D. S. *et al.* (2000). Control of apoptosis during angiogenesis by survivin expression in endothelial cells. *Am. J. Pathol.* **156**, 393–398.
75. Papapetropoulos, A. *et al.* (1999). Direct actions of angiopoietin-1 on human endothelium: evidence for network stabilization, cell survival, and interaction with other angiogenic growth factors. *Lab. Invest.* **79**, 213–223.
76. Kim, I. *et al.* (2000). Angiopoietin-1 regulates endothelial cell survival through the phosphatidylinositol 3'-Kinase/Akt signal transduction pathway. *Circ. Res.* **86**, 24–29.

77. Papapetropoulos, A. *et al.* (2000). Angiopoietin-1 inhibits endothelial cell apoptosis via the Akt/survivin pathway. *J. Biol. Chem.* **275**, 9102–9105.
78. Kim, L. *et al.* (2002). Angiopoietin-1 negatively regulates expression and activity of tissue factor in endothelial cells. *FASEB J.* **16**, 126–128.
79. Jones, N. *et al.* (2001). Rescue of the early vascular defects in Tek/Tie2 null mice reveals an essential survival function. *EMBO Rep.* **2**, 438–445.
80. Wong, A. L. *et al.* (1997). Tie2 expression and phosphorylation in angiogenic and quiescent adult tissues. *Circ. Res.* **81**, 567–574.

CHAPTER 263

Vascular Endothelial Growth Factors and their Receptors in Vasculogenesis, Angiogenesis, and Lymphangiogenesis

Marja K. Lohela and Kari Alitalo

Molecular/Cancer Biology Laboratory and Ludvig Institute for Cancer Research, Haartman Institute and Helsinki University Central Hospital, Biomedicum Helsinki, University of Helsinki, Helsinki, Finland

Vasculogenesis, Angiogenesis, and Lymphangiogenesis

During embryogenesis, blood and lymphatic vasculature develops through the regulated proliferation, migration, and differentiation of endothelial cells. Initially, a primary blood vascular plexus is formed in the process of vasculogenesis, where mesoderm-derived angioblasts differentiate into endothelial cells and assemble to form a network of uniformly sized primitive vessels [1]. The subsequent remodeling of the primary plexus to form a more complex treelike hierarchy of large and small vessels is called angiogenesis. Angiogenesis occurs by sprouting and branching of new vessels from the preexisting ones, followed by progressive pruning and remodeling of the vessels into a mature vascular system [1]. Periendothelial support cells, such as vascular pericytes and smooth muscle cells, are then recruited to the vessel wall and the extracellular matrix is reconstituted [2]. Vasculogenesis is probably restricted to early development, but physiological angiogenesis occurs in adult organisms during the female reproductive cycle and in wound healing. Pathological angiogenesis occurs in several diseases such as rheumatoid arthritis and diabetic retinopathy and during tumor growth [3]. The observation that solid tumors are dependent on neovascularization has brought endothelial cell signaling mechanisms and molecules into focus in the field of cancer research.

The lymphatic vessels arise from the embryonic veins through a process termed lymphangiogenesis. The lymph vessels become organized parallel to the blood vascular system and function to transport interstitial fluid, extravasated plasma proteins, and cells back into the blood circulation [4]. The lymphatic vessels also form a part of the immune system together with the lymphoid organs. Lymphatic capillaries consist of a thin, permeable layer of endothelial cells that are anchored to the surrounding connective tissue by elastic filaments called anchoring fibrils. The capillaries have a discontinuous basement membrane and only a few scattered support cells. The lymphatic capillaries transport the fluid into progressively larger collecting vessels consisting of endothelial, muscular, and adventitial layers, which ultimately drain into the venous circulation via the thoracic duct. Movement of lymph is brought about by the intrinsic contractility of the smooth muscle cells surrounding the larger vessels, and backflow is prevented by luminal valves [4]. Lymphedema, the accumulation of protein-rich fluid in interstitial tissues, occurs in humans either as hereditary disease or due to lymph vessel damage or removal. Because of the metastatic spread of tumor cells via the lymphatic system and tumor lymphangiogenesis, the lymphatic endothelial cell signaling is also of interest in cancer biology.

The Vascular Endothelial Growth Factors and their Receptors

Signaling by the vascular endothelial growth factor (VEGF) family members and their receptors (VEGFRs) is essential in vasculogenesis and angiogenesis, and has more recently been shown to be critical in lymphangiogenesis as well. The VEGF family consists of five members thus far, namely VEGF, VEGF-B, VEGF-C, VEGF-D, and placenta growth factor (PlGF). All family members are secreted glycoproteins that possess a VEGF homology domain belonging to the cysteine knot growth factor family, and they function as disulfide-linked anti-parallel dimers [5,6]. Out of the VEGF family, only VEGF itself is induced by hypoxia, via an intricate signal transduction mechanism involving prolyl hydroxylation and specific ubiquitin ligase complexes [7].

VEGFs transduce signals mainly via three known VEGFRs, VEGFR-1/Flt-1, VEGFR-2/Flk-1/KDR, and VEGFR-3/Flt-4. VEGFRs are high-affinity receptor tyrosine kinases, restricted largely to endothelial cells, and they are structurally and functionally related to the PlGF receptors [8].

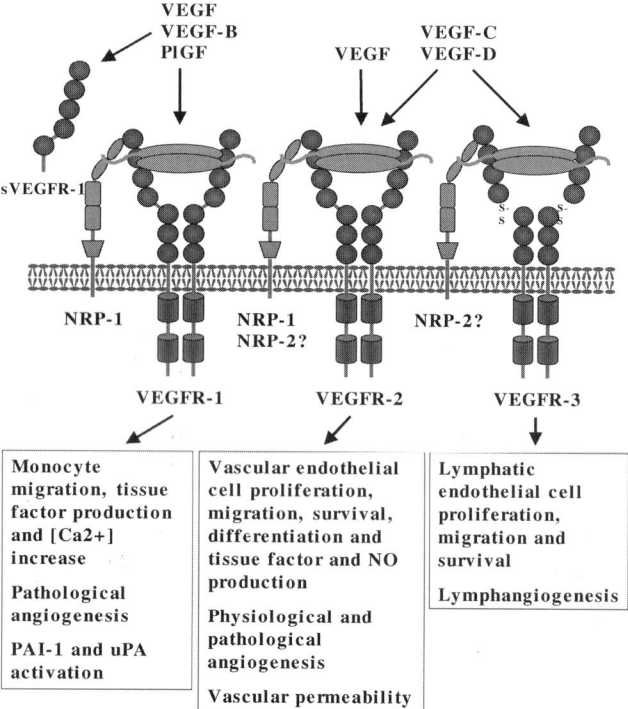

Figure 1 The VEGF receptors and their ligands. VEGFR-1 and VEGFR-2 have seven extracellular immunoglobulin homology domains (circles); in VEGFR-3 the fifth immunoglobulin domain is cleaved on receptor processing into two disulfide-linked subdomains. sVEGFR-1, soluble VEGFR-1; NRP, neuropilin; PAI-1, plasminogen activator inhibitor-1; uPA, urokinase-type plasminogen activator; NO, nitric oxide.

The VEGFRs have seven immunoglobulin homology domains in their extracellular part, and an intracellular tyrosine kinase domain split by a kinase insert sequence. VEGFR-1 exists also in a soluble form produced by alternative splicing. The VEGF family members have different binding specificities for the VEGFRs, schematically shown in Fig. 1. Neuropilins-1 and -2, which were originally characterized as receptors for semaphorins in the nervous system, also selectively bind members of the VEGF family in an isoform-specific manner and seem to act as co-receptors with VEGFRs, possibly acting by clustering of receptors (Fig. 1) [9,10]. Several excellent reviews have appeared that detail the structures of VEGFs and their receptors and their function in angiogenesis [5,8,11,12].

VEGF and VEGFR-1 and -2 are Essential for Vasculogenesis and Angiogenesis

Hemangioblasts, as yet ill-defined precursor cells for both the vascular endothelial and hematopoietic lineage [13,14], are thought to differentiate from pluripotent epiplastic precursor cells under the inductive influence of several factors, including the fibroblast growth factor [15]. VEGFR-2 is the earliest marker for hemangioblasts [16], and the importance of both VEGF and its receptors VEGFR-1 and VEGFR-2 in vasculogenesis and hematopoiesis is clear from the knockout mouse phenotypes. Inactivation of a single VEGF allele resulted in embryonic lethality between embryonic days 11 and 12 (E11–12), indicating that the effect of VEGF on vasculogenesis is dosage-dependent [17,18]. The VEGF$^{+/-}$ embryos appeared growth retarded and exhibited a number of developmental anomalies, including severe cardiovascular defects and reduced number of red blood cells. Mouse embryos lacking VEGFR-2 died at E8.5–9.5, and showed a total lack of vasculogenesis and hematopoiesis [19]. *In vitro* studies indicate that VEGFR-2 is not required for the formation of hemangioblasts, but is necessary for endothelial cell commitment [20]. Knockout studies have demonstrated the necessity of VEGFR-1 for vasculogenesis. VEGFR-1-null mice died at E8.5–9.5, and while their endothelial cells differentiated normally, they showed an increase in numbers and failed to organize into proper channels [21,22]. Disruption of VEGFR-3 also led to embryonic death due to defective remodeling of the vascular plexus and cardiovascular failure, but vasculogenesis appeared to occur normally [23].

VEGFR-2 appears to be the major receptor that conveys VEGF-induced signals in endothelial cells. In various endothelial cell types VEGF induces strong VEGFR-2 autophosphorylation, activation of the MAP kinase cascade, and the PI3-K-Akt pathway, cell survival, proliferation, and chemotaxis. The signal transduction via VEGFRs is summarized in Fig. 2; for reviews on VEGF signaling, see references [24] and [25]. Experiments with mice having the VEGFR-1 tyrosine kinase domain deleted have shown that the ligand-binding and transmembrane domains of the receptor are sufficient for normal vasculogenesis and angiogenesis [26].

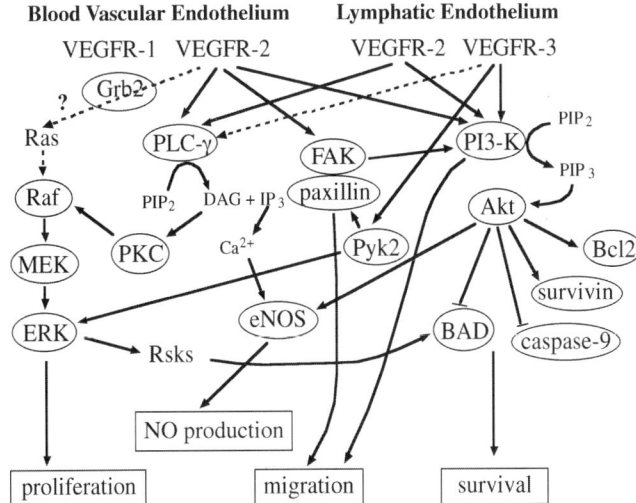

Figure 2 The major known VEGFR signaling pathways in blood, vascular, and lymphatic endothelium. PLC-γ, phospholipase C-γ; FAK, focal adhesion kinase; PI3-K, phosphatidylinositol 3′ kinase; PIP$_2$, phosphatidylinositol 4,5-bisphosphate; PIP$_3$, phosphatidylinositol 3,4,5-trisphosphate; PKC, protein kinase C; DAG, diacylglycerol; IP$_3$, inositol 1,4,5-trisphosphate; MEK, MAPK/ERK kinase; MAPK, mitogen-activated protein kinase; ERK, extracellular signal-regulated kinase; Rsks, ribosomal protein S6 kinases; eNOS, endothelial nitric oxide synthase; NO, nitric oxide. (Figure modified from Karkkainen, M. J. et al., Nat. Cell Biol., **4**, E2–E5. With permission. See also references [24] and [43]).

Thus it has been postulated that VEGFR-1 does not have a signaling role in these processes, but acts as a VEGF sink to negatively regulate VEGFR-2 signaling. Yet, PlGF, which only binds VEGFR-1 is essential for all forms of pathological angiogenesis in adult tissues [27].

Lymphangiogenesis is Regulated by VEGFR-3 and its Ligands VEGF-C and -D

VEGFR-3 is initially expressed in all embryonic endothelial cells, but becomes restricted to lymphatic vessels later in development [23,28]. VEGF-C and -D bind to and activate VEGFR-3 and in their fully mature, proteolytically processed forms, are also ligands for VEGFR-2 [29,30]. Both VEGF-C and -D have been shown to be lymphangiogenic. VEGF-C can induce lymphangiogenesis when administered as a recombinant protein onto the avian chorioallantoic membrane [31] or when overexpressed as a transgene in mouse skin [32]. A VEGFR-3-specific mutant form of VEGF-C also induced lymphangiogenesis in the skin [33], indicating that the lymphangiogenic effects of VEGF-C are transduced via VEGFR-3. Furthermore, a soluble form of VEGFR-3 expressed under the control of the same skin-specific promoter caused regression of developing lymphatic vessels by inducing endothelial cell apoptosis [34]. The dermal blood vessels were not affected, confirming the specificity of VEGFR-3 function in lymphatic endothelial cells in later development. In this model, lymphatic vessels were initially lost in several internal organs, but partially regenerated in adult mice [34].

Human early-onset primary lymphedema has been shown to be linked to mutations in the *VEGFR-3* gene that inactivate the tyrosine kinase [35,36]. Chy mice that have a similar mutation developed lymphedema of limbs due to lack of subcutaneous lymphatic vessels, but new lymphatics were induced to grow by VEGF-C therapy via viral gene delivery, suggesting that human lymphedema could also be treated with VEGF-C/D therapy [37]. Interestingly, VEGF-C and -D have also been shown to promote tumor lymphangiogenesis and lymphatic metastasis is mouse tumor models [38–42], and both soluble receptors and blocking antibodies could inhibit lymphatic metastasis [38,42].

Little is known about the signaling mechanisms involved in lymphangiogenesis. VEGFR-3 signaling has been investigated in isolated lymphatic endothelial cells utilizing the VEGFR-3-specific mutant form of VEGF-C. It was found that VEGFR-3 signaling can promote the growth, survival, and migration of lymphatic endothelial cells. VEGFR-3 phosphorylation leads to PI3-kinase-dependent Akt activation and protein kinase C-dependent activation of the p42/p44 MAPK (Fig. 2) [43].

Concluding Remarks

The complicated signaling networks regulating the growth and maintenance of blood– and lymphatic vasculature are only beginning to be elucidated. In addition to the VEGF/VEGFR system, two other receptor tyrosine kinase systems have been shown to be important in endothelial cells of vasculature (reviewed in [44]). Tie 1 and Tie 2/Tek receptors and the angiopoietins, which are Tie 2 ligands, seem to be especially important in regulating the stability of the vasculature. The Eph receptors and their Ephrin ligands, were originally discovered as regulators of the nervous system. There are interesting parallels between vascular and nervous development, and also the VEGF/VEGFR system may have a function in nervous development. The intricate interactions between the different signal transduction systems and the similarities and differences between the regulation of endothelial cell signaling in different parts of blood and lymphatic vasculature pose additional challenges in this field of research.

References

1. Risau, W. (1997). Mechanisms of ngiogenesis. *Nature* **386**, 671–674.
2. Folkman, J. and D'Amore, P. (1996). Blood vessel formation: What is its molecular basis? *Cell* **87**, 1153–1155.
3. Folkman, J. (1995). Angiogenesis in cancer, vascular, rheumatoid and other disease. *Nat. Med.* **1**, 27–31.
4. Witte, M. H., Way, D. L., Witte, C. L., and Bernas, M. (1997). In Regulation of Angiogenesis Rosen, I. D. G. A. E. M., Ed., pp. 65–112. Birkhauser Verlag, Basel.
5. Ferrara, N. (1999). Role of vascular endothelial growth factor in the regulation of angiogenesis. *Kidney Int.* **56**, 794–814.
6. Muller, Y. A., Li, B., Christinger, H. W., Wells, J. A., Cunningham, B. C., and de Vos, A. M. (1997). Vascular endothelial growth factor: Crystal structure and functional mapping of the kinase domain receptor binding site. *Proc. Natl. Acad. Sci. USA* **94**, 7192–7197.

7. Semenza, G. L. (2001). Hif-1, O_2, and the 3 Phds: How animal cells signal hypoxia to the nucleus. *Cell* **107**, 1–3.
8. Shibuya, M. (2001). Structure and function of Vegf/Vegf-receptor system involved in angiogenesis. *Cell Struct. Funct.* **26**, 25–35.
9. Gluzman-Poltorak, Z., Cohen, T., Herzog, Y., and Neufeld, G. (2000). Neuropilin-2 and europilin-1 are receptors for the 165-amino acid form of vascular endothelial growth factor (Vegf) and of placenta growth factor-2, but only neuropilin-2 functions as a receptor for the 145-amino acid form of Vegf. *J. Biol. Chem.* **275**, 18040–18045.
10. Soker, S., Takashima, S., Miao, H. Q., Neufeld, G., and Klagsbrun, M. (1998). Neuropilin-1 is expressed by endothelial and tumor cells as an isoform-specific receptor for vascular endothelial growth factor. *Cell* **92**, 735–745.
11. Veikkola, T., Karkkainen, M., Claesson-Welsh, L., and Alitalo, K. (2000). Regulation of angiogenesis via vascular endothelial growth factor receptors. *Cancer Res.* **60**, 203–212.
12. Carmeliet, P. and Collen, D. (1999). Role of vascular endothelial growth factor and vascular endothelial growth factor receptors in vascular development. *Curr. Top. Microbiol. Immunol.* **237**, 133–158.
13. Choi, K., Kennedy, M., Kazarov, A., Papadimitriou, J. C., and Keller, G. (1998). A common precursor for hematopoietic and endothelial cells. *Development* **125**, 725–732.
14. Eichmann, A., Corbel, C., Nataf, V., Vaigot, P., Breant, C., and Le Douarin, N. M. (1997). Ligand-dependent development of the endothelial and hemopoietic lineages from embryonic mesodermal cells expressing vascular endothelial growth factor receptor 2. *Proc. Natl. Acad. Sci. USA* **94**, 5141–5146.
15. Flamme, I., Frölich, T., and Risau, W. (1997). Molecular mechanisms of vasculogenesis and embryonic angiogenesis. *J. Cell. Physiol.* **173**, 206–210.
16. Yamaguchi, T., Dumont, D., Conion, R., Breitman, M., and Rossant, J. (1993). Flk-1, an Flt-related receptor tyrosine kinase is an early marker for endothelial cell precursors. *Development* **118**, 489–498.
17. Carmeliet, P., Ferreira, V., Breier, G., Pollefeyt, S., Kieckens, L., Gertsenstein, M., Fahrig, M., Vandenhoeck, A., Harpal, K., Ebenhardt, C., Declercq, C., Pawling, J., Moons, L., Collen, D., Risau, W., and Nagy, A. (1996). Abnormal blood vessel development and lethality in embryos lacking a single Vegf allele. *Nature* **380**, 435–439.
18. Ferrara, N., Carver-Moore, K., Chen, H., Dowd, M., Lu, L., O'Shea, K. S., Powell-Braxton, L., Hilan, K. J., and Moore, M. W. (1996). Heterozygous embryonic lethality induced by targeted inactivation of the Vegf gene. *Nature* **380**, 439–442.
19. Shalaby, F., Rossant, J., Yamaguchi, T. P., Gertsenstein, M., Wu, X. F., Breitman, M. L., and Schuh, A. C. (1995). Failure of blood island formation and vasculogenesis in Flk-1-deficient mice. *Nature* **376**, 62–66.
20. Schuh, A. C., Faloon, P., Hu, Q.-L., Bhimani, M., and Kyunghee, C. (1999). *In vitro* hematopoietic and endothelial potential of *Flk-1$^{-/-}$* embryonic stem cells and embryos. *Proc. Natl. Acad. Sci. USA*, **96**, 2159–2164.
21. Fong, G. H., Rossant, J., Gertsenstein, M., and Breitman, M. L. (1995). Role of the Flt-1 receptor tyrosine kinase in regulating the assembly of vascular endothelium. *Nature* **376**, 66–70.
22. Fong, G.-H., Zhang, L., Bryce, D.-M., and Peng, J. (1999). Increased hemangioblast commitment, not vascular disorganization, is the primary defect in *Flt-1* knock-out mice. *Development* **126**, 3015–3025.
23. Dumont, D., Jussila, L., Taipale, J., Mustonen, T., Pajusola, K., Breitman, M., and Alitalo, K. (1998). Cardiovascular failure in mouse embryos deficient in Vegf receptor-3. *Science* **282**, 946–949.
24. Matsumoto, T. and Claesson-Welsh, L. (2001). Vegf receptor signal transduction. *Sci. STKE* RE21.
25. Petrova, T. V., Mäkinen, T., and Alitalo, K. (1999). Signaling via vascular endothelial growth factor receptors. *Exp. Cell Res.* **253**, 117–130.
26. Hiratsuka, S., Minowa, O., Kuno, J., Noda, T., and Shibuya, M. (1998). Flt-1 lacking the tyrosine kinase domain is sufficient for normal development and angiogenesis in mice. *Proc. Natl. Acad. Sci. USA* **4**, 9349–9354.
27. Carmeliet, P., Moons, L., Luttun, A., Vincenti, V., Compernolle, V., De Mol, M., Wu, Y., Bono, F., Devy, L., Beck, H., Scholz, D., Acker, T., DiPalma, T., Dewerchin, M., Noel, A., Stalmans, I., Barra, A., Blacher, S., Vandendriessche, T., Ponten, A., Eriksson, U., Plate, K. H., Foidart, J.-M., Schaper, W., Charnock-Jones, D. S., Hicklin, D. J., Herbert, J.-M., Collen, D., and Persico, M. G. (2001). Synergism between vascular endothelial growth factor and placental growth factor contributes to angiogenesis and plasma extravasation in pathological conditions. *Nat. Med.* **7**, 575–583.
28. Kaipainen, A., Korhonen, J., Mustonen, T., van Hinsbergh, V. M., Fang, G.-H., Dumont, D., Breitman, M., and Alitalo, K. (1995). Expression of the Fms-like tyrosine kinase Flt4 gene becomes restricted to endothelium of lymphatic vessels during development. *Proc. Natl. Acad. Sci. USA* **92**, 3566–3570.
29. Joukov, V., Pajusola, K., Kaipainen, A., Chilov, D., Lahtinen, I., Kukk, E., Saksela, O., Kalkkinen, N., and Alitalo, K. (1996). A novel vascular endothelial growth factor, Vegf-C, is a ligand for the Flt4 (Vegfr-3) and Kdr (Vegfr-2) receptor tyrosine kinases. *EMBO J.*, **15**, 290–298.
30. Achen, M. G., Jeltsch, M., Kukk, E., Makinen, T., Vitali, A., Wilks, A. F., Alitalo, K., and Stacker, S. A. (1998). Vascular endothelial growth factor D (Vegf-D) is a ligand for the tyrosine kinases Vegf receptor 2 (Flk1) and Vegf receptor 3 (Flt4). *Proc. Natl. Acad. Sci. USA* **95**, 548–553.
31. Oh, S.-J., Jeltsch, M. M., Birkenhäger, R., McCarthy, J. E. G., Weich, H. A., Christ, B., Alitalo, K., and Wilting, J. (1998). Vegf and Vegf-C: Specific induction of angiogenesis and lymphangiogenesis in the differentiated avian chorioallantoic membrane. *Dev. Biol.* **188**, 96–109.
32. Jeltsch, M., Kaipainen, A., Joukov, V., Meng, X., Lakso, M., Rauvala, H., Swartz, M., Fukumura, D., Jain, R. K., and Alitalo, K. (1997). Hyperplasia of lymphatic vessels in Vegf-C transgenic mice. *Science* **276**, 1423–1425.
33. Veikkola, T., Jussila, L., Makinen, T., Karpanen, T., Jeltsch, M., Petrova, T. P., Kubo, H., Thurston, G., McDonald, D. M., Achen, M. G., Stacker, S. A., and Alitalo, K. (2001). Signalling via vascular endothelial growth factor receptor-3 is sufficient for lymphangiogenesis in transgenic mice. *EMBO J.* **20**, 1223–1231.
34. Mäkinen, T., Jussila, L., Veikkola, T., Karpanen, T., Kettunen, M. I., Pulkkanen, K. J., Kauppinen, R., Jackson, D. G., Thurston, G., McDonald, D., Kubo, H., Nishikawa, S.-I., Yla-Herttuala, S., and Alitalo, K. (2001). Inhibition of lymphangiogenesis with resulting lymphedema in transgenic mice expressing soluble Vegf receptor-3. *Nat. Med.* **7**, 199–205.
35. Kärkkäinen, M., Ferrell, R. E., Lawrence, E. C., Kimak, M. A., Levinson, K. L., McTigue, M. A., Alitalo, K., and Finegold, D. N. (2000). Missense mutations interfere with Vegfr-3 signaling in primary lymphedema. *Nat. Genet.* **25**, 153–159.
36. Irrthum, A., Kärkkäinen, M. J., Devrient, K., Alitalo, K., and Vikkula, M. (2000). Congenital hereditary lymphedema caused by a mutation that inactivates Vegfr-3 tyrosine kinase. *Am. J. Hum. Genet.* **67**, 295–301.
37. Kärkkäinen, M. J., Saaristo, A., Jussila, L., Karila, K. A., Lawrence, E. C., Pajusola, K., Bueler, H., Eichmann, A., Kauppinen, R., Kettunen, M. I., Ylä-Herttuala, S., Finegold, D. N., Ferrel, R. E., and Alitalo, K. (2001). A model for gene therapy of human hereditary lymphedema. *Proc. Natl. Acad. Sci. USA* **98**, 12677–12682.
38. Kärpänen, T., Egeblad, M., Kärkkäinen, M. J., Kubo, H., Jackson, D. G., S., Y.-H., Jäätteelä, M., and Alitalo, K. (2001). Vascular endothelial growth factor C promotes tumor lymphangiogenesis and intralymphatic tumor growth. *Cancer Res.* **61**, 1786–1790.
39. Mandriota, S. J., Jussila, L., Jeltsch, M., Compagni, A., Baetens, D., Prevo, R., Banerji, S., Huarte, J., Montesano, R., Jackson, D., Orci, L., Alitalo, K., Christofori, G., and Pepper, M. S. (2001). Vascular endothelial growth factor-C-mediated lymphangiogenesis promotes tumor metastasis. *EMBO J.* **20**, 672–682.
40. Skobe, M., Hamberg, L. M., Hawighorst, T., Schirner, M., Wolf, G. L., Alitalo, K., and Detmar, M. (2001). Concurrent induction of lymphangiogenesis, angiogenesis and macrophage recruitment by vascular endothelial growth factor-C in melanoma. *Am. J. Pathol.* **159**, 893–903.
41. Skobe, M., Hawighorst, T., Jackson, D. G., Prevo, R., Janes, L., Velasco, P., Riccardi, L., Alitalo, K., Claffey, K., and Detmar, M. (2001). Induction of tumor lymphangiogenesis by Vegf-C promotes breast cancer metastasis. *Nat. Med.* **7**, 192–198.

42. Stacker, S. A., Caesar, C., Baldwin, M. E., Thornton, G. E., Williams, R. A., Prevo, R., Jackson, D. G., Nishikawa, S., Kubo, H., and Achen, M. G. (2001). Vegf-D promotes the metastatic spread of tumor cells via the lymphatics. *Nat. Med.* **7**, 186–191.
43. Mäkinen, T., Veikkola, T., Mustjoki, S., Kärpänen, T., Catimel, B., Nice, E. C., Wise, L., Mercer, A., Kowalski, H., Kerjaschki, D., Stacker, S. A., Achen, M. G., and Alitalo, K. (2001). Isolated lymphatic endothelial cells transduce growth, survival and migratory signals via the Vegf-C/D receptor Vegfr-3. *EMBO J.* **20**, 4762–4773.
44. Gale, N. W. and Yancopoulos, G. D. (1999). Growth factors acting via endothelial cell-specific receptor tyrosine kinases: Vegfs, angiopoietins, and ephrins in vascular development. *Genes Dev.* **13**, 1055–1066.
45. Kärkkäinen, M. J., Mäkinen, T., and Alitalo, K. (2002). Lymphatic endothelium: A new frontier of metastasis research. *Nat. Cell Biol.* **4**, E2–E5.

Signaling from FGF Receptors in Development and Disease

Monica Kong-Beltran and Daniel J. Donoghue

*Department of Chemistry and Biochemistry, Center for Molecular Genetics,
University of California San Diego, La Jolla, California*

Introduction

Fibroblast growth factor receptors (FGFRs) are members of the receptor tyrosine kinase family which consists of four structurally related members: FGFR 1, 2, 3, 4, and their alternatively spliced variants. Each receptor contains two to three immunoglobulin (Ig)-like domains in the extracellular region, an acidic box between Ig-1 and Ig-2, a transmembrane domain, and a split intracellular tyrosine kinase domain (Fig. 1). Normal activation of FGFRs relies on interaction with their ligands: fibroblast growth factors (FGFs). At least 22 FGFs are known to date, and each FGFR member interacts with a specific subset of FGFs, as summarized in Fig. 1 (reviewed in [38] and reviewed in [11]). Heparan sulfate proteoglycans facilitate FGF binding with FGFR to induce FGFR dimerization and activation [44]. Crystal structure analyses have indicated that heparin binds to two FGFs such that the FGFs are not directly interacting with each other, and that each FGF is binding to an FGFR (reviewed in [19]). Another study has shown, however, that the heparin-FGF monomer alone can activate FGFR dimerization [39]. After ligand stimulation, FGFRs undergo dimerization and autotyrosine phosphorylation, resulting in kinase activation. The phosphorylated tyrosine residues may then serve as binding sites for secondary signaling molecules.

Expression of FGFR during Development

FGFRs are widely expressed during development and in adult tissues [18]. FGFR1 is expressed in the central nervous system and developing limbs during development. During early neurogenesis, FGFR1 expression is upregulated in the ventricular zone of the neural tube and is expressed in the mesenchyme of developing limbs [35,52]. At later stages of development, FGFR1 can be found in maturing neurons in the brain and in motor neurons in the spinal cord [17,35]. FGFR2 is also expressed in the brain, particularly in the developing mid- and hindbrain [35,47,52]. In addition, FGFR2 and FGFR3 expression occurs highly in the trophectoderm and to some degree in the mesenchyme [30,35,36]. Importantly, FGFR3 is expressed in the hypertrophic and proliferative zones of the bone growth plate, cochlea, brain, and spinal cord [6]. Expression levels of FGFR4 have not been examined as thoroughly as other FGFR family members, but FGFR4 has been shown to be expressed in definitive endoderm and skeletal muscle lineages, as well as the ventricular zone of developing spinal cord and dorsal root ganglia [27,34,45].

Role of FGFR in Development

Genetic alterations in mice reveal the importance of FGFR during development. In FGFR1 and FGFR2, null mutations result in embryonic lethality due to defects of mesodermal patterning or trophoblast cell proliferation [2,7,53,55]. FGFR1 chimeric mice further demonstrate that FGFR1 is necessary for neural tube and limb development [9], while FGFR2 chimeras indicate that FGFR2 is required for both limb outgrowth and branching morphogenesis of the lungs [2]. Additionally, FGFR2 has been shown to be important for keratinocyte differentiation [50]. FGFR3 null mice exhibit

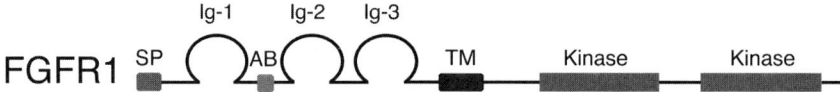

Member	Syndrome	FGF Interaction
FGFR1	Pfeiffer	FGF 1-6, 8, 10, 17
FGFR2	Apert Beare-Stevenson Crouzon Jackson-Weiss Pfeiffer	FGF 1-10, 17
FGFR3	Achondroplasia Hydpochondroplasia SADDAN TDI TDII CAN	FGF 1, 2, 4, 8, 9, 17
FGFR4	No known syndromes	FGF 1, 2, 4, 6, 8, 9, 17, 18, 19

Figure 1 Structure of FGFR1 is shown as representative of FGFR1, FGFR2, FGFR3, and FGFR4. SP, signal peptide; Ig, immunoglobulin-like domain; AB, acidic box; TM, transmembrane domain. Developmental syndromes associated with different FGFRs are summarized. SADDAN, severe achondroplasia with delayed development and acanthosis nigricans; TD, thanatophoric dysplasia; CAN, Crouzon syndrome with acanthosis nigricans.

bone defects and deafness, demonstrating that FGFR3 is important for normal skeletal and ear development [6,8]. The FGFR4 null mutant mice appear to be normal, but have depleted gallbladders and elevated liver bile acids [56].

Syndromes Associated with FGFRs

Activating mutations of FGFR1, FGFR2, and FGFR3 can lead to craniosynostosis syndromes of various degrees of severity as listed in Fig. 1. Craniosynostosis is characterized by the premature fusion of the cranial sutures between the developing flat bones of the skull, resulting in abnormal head shape, midface hypoplasia, and other skeletal defects. Related syndromes can be distinguished by examining the extremities [16,49,51]. For example, a mutation in FGFR1 which causes Pfeiffer syndrome, a classic form of craniosynostosis, is characterized by broad, sometimes medially deviated thumbs and big toes. Mutations in FGFR2 result in a broad group of craniosynostosis syndromes, including Pfeiffer, Apert, Beare-Stevenson, Crouzon, and Jackson-Weiss syndromes. Mutations in FGFR3 are generally responsible for skeletal dysplasia syndromes, including thanatophoric dysplasia (TD) types I and II, hypochondroplasia, achondroplasia, severe achondroplasia with developmental delay and acanthosis nigricans (SADDAN), and Crouzon syndrome with acanthosis nigricans (CAN). These FGFR3 syndromes range from a mild form of dwarfism (hypochondroplasia) to neonatal lethal dwarfisms (TD I and II). There are no known craniosynostoses or skeletal syndromes associated with mutations of FGFR4.

Signaling Pathways Mediated by FGFRs

FGFRs play key roles in cell proliferation, differentiation, migration, wound healing, survival, and angiogenesis. Ligand activation of FGFR results in kinase activity and autophosphorylation of tyrosine residues, thereby creating specific binding sites for downstream effector molecules that coordinate FGFR-mediated signaling. For example, phospholipase C-γ (PLC-γ) binds to activated FGFR1 and is phosphorylated by FGFR1, resulting in protein kinase C (PKC) activation (reviewed in [4]).

FGF receptor substrate 2 (FRS2), a membrane-associated docking protein, binds to the juxtamembrane of FGFR1 via its phosphotyrosine binding (PTB) domain [53]. The Grb2/Sos complex is recruited to the plasma membrane via FRS2 and activates Ras to trigger the mitogen-activated protein kinase (MAPK) pathway and expression of nuclear proteins including Fos and Myc [12,13,22,32]. In addition, activated FGFR1 results in FRS2 binding to Grb2 and leads to tyrosine

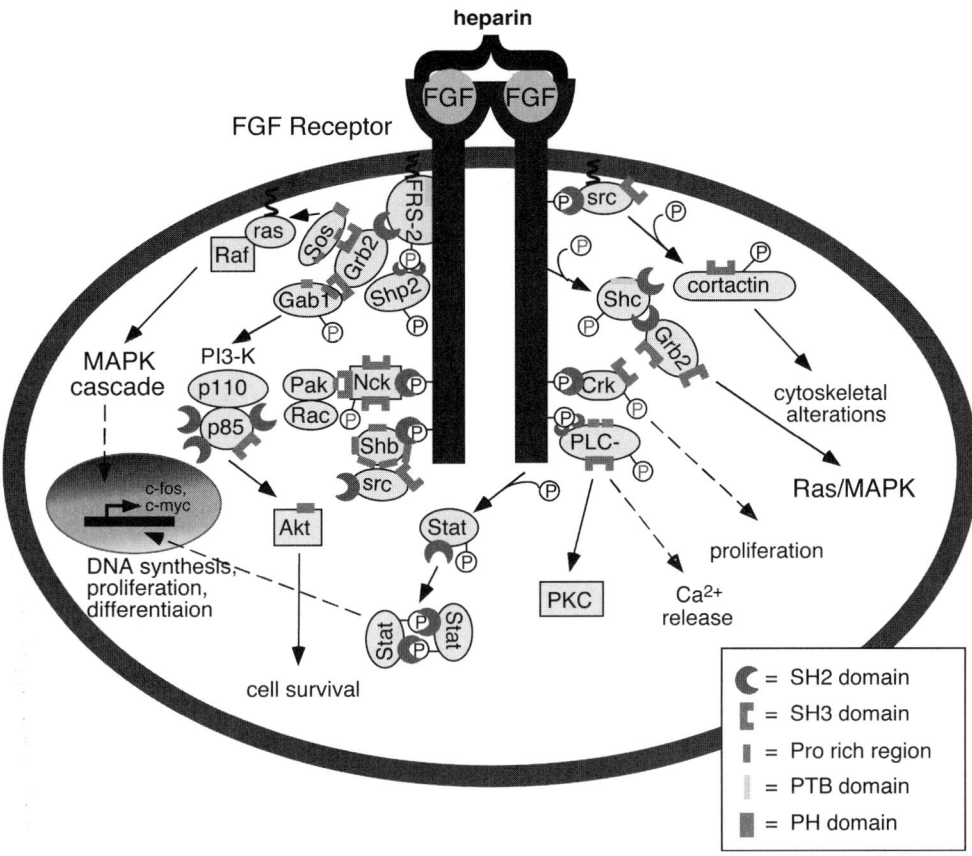

Figure 2 Intracellular signaling pathways activated by FGFRs are summarized. Tyrosine phosphorylation of receptor subunits facilitates recruitment and assembly of activated signaling complexes.

phosphorylation of a Grb2-associated docking protein named Gab1, which recruits and activates the phosphatidylinositol 3-kinase (PI3-kinase)/Akt pathway for cell survival [33]. Interestingly, Grb2 can interact with Sos and Gab1 simultaneously, allowing activation of both Ras/MAPK and PI3-kinase/Akt pathways [33]. Furthermore, the tyrosine phosphatase Shp2 becomes phosphorylated and interacts with the SH2 domain of Grb2 to form a complex with tyrosine phosphorylated FRS2, providing another link of activated FGFR to the Ras/MAPK signaling cascade [14,31,32,22]. Shp2 can also interact with Gab1 [31].

Adapter proteins play significant roles in FGFR-mediated signaling. Activated FGFR1 results in tyrosine phosphorylation of Shc. Shc binds to Grb2 and may link FGFR1, but not FGFR4, to the Ras signaling pathway [3,41,48]. The adapter protein Crk also interacts with phosphorylated FGFR1, becomes tyrosine phosphorylated, and results in cell proliferation [24]. In addition, Nck binds to activated FGFR1 and becomes tyrosine phosphorylated, potentially linking FGFR1 to the actin cytoskeleton since Nck facilitates the interaction between Pak and Rac [42], and reviewed in [26]. The adapter protein Shb also binds to FGFR1 and is tyrosine phosphorylated; Shb thus appears to be an adaptor protein linking SH3 domain proteins to tyrosine kinases or other tyrosine phosphorylated proteins [5,20]. Chimeric receptors composed of the extracellular domain of PDGFR-β and the transmembrane and intracellular domains of FGFR1, FGFR3, and FGFR4 were able to activate phospholipase Cγ, Shc, FRS2, and the mitogen-activated protein kinases, ERK1 and 2 [40]; in this study, however, these chimeric receptors did not appear to couple through Shc.

FGFRs are also implicated in other signaling pathways. Src is a non-receptor tyrosine kinase recruited by FGFRs and phosphorylates cortactin to affect cell migration; it remains unclear whether Src directly binds to FGFR1 [23,25,57]. Furthermore, activated FGFR 1, 3, and 4 promote Stat1 and Stat3 activation [15,46]. In the case of FGFR3, the adapter protein SH2-B has been found to activate Stat5 [21]. Most of these interactions and signaling pathways activated by FGFR mentioned above are summarized in Fig. 2, using FGFR1 as the model.

Summary

FGFRs play vital roles in growth, development, differentiation, and migration. Genetic analyses of FGFRs indicate their importance in organ and neuronal development. In addition, FGFs activate FGFRs to trigger multiple signaling pathways in the cell. Unfortunately, space has not permitted a discussion of the role of somatic FGFR mutations in human cancer. Many new pathways of FGFR signaling undoubtedly remain to be discovered that will be important in understanding mammalian development and also the control of abnormal cell proliferation in cancer.

References

1. Arman, E., Haffner-Krausz, R., Gorivodsky, M., and Lonai, P. (1999). Fgfr2 is required for limb outgrowth and lung-branching morphogenesis. *Proc. Natl. Acad. Sci. USA* **96**, 11895–11899.
2. Arman, E., Haffner-Krausz, R., Chen, Y., Heath, J. K., and Lonai, P. (1998). Targeted disruption of fibroblast growth factor (FGF) receptor 2 suggests a role for FGF signaling in pregastrulation mammalian development. *Proc. Natl. Acad. Sci. USA* **95**, 5082–5087.
3. Browaeys-Poly, E., Cailliau, K., and Vilain, J.-P. (2001). Transduction cascades initiated by fibroblast growth factor 1 on *Xenopus* oocytes expressing MDA-MB-231 mRNAs; role of Grb2, phosphatidylinositol 3-kinase, Src tyrosine kinase, and phospholipase Cγ. *Cell Signal* **13**, 363–368.
4. Carpenter, G. and Ji, Q. (1999). Phospholipase C-γ as a transducing element. *Exp. Cell Res.* **253**, 15–24.
5. Claesson-Welsh, L., Welsh, M., Ito, N., Anand-Apte, B., Soker, S., Zetter, B., O'Reilly, M., and Folkman, J. (1998). Angiostatin induces endothelial cell apoptosis and activation of focal adhesion kinase independently of the integrin-binding motif of RGD. *Proc. Natl. Acad. Sci. USA* **95**, 5579–5583.
6. Colvin, J. S., Bohne, B. A., Harding, G. W., McEwen, D. G., and Ornitz, D. M. (1996). Skeletal overgrowth and deafness in mice lacking fibroblast growth factor receptor 3. *Nat. Genet.* **12**, 390–397.
7. Deng, C., Wynshaw-Boris, A., Shen, M. M., Daugherty, C., Ornitz, D. M., and Leder, P. (1994). Murine FGFR-1 is required for early postimplantation growth and axial organization. *Genes Dev.* **8**, 3045–3057.
8. Deng, C., Wynshaw-Boris, A., Zhou, F., Kuo, A., and Leder, P. (1996). Fibroblast growth factor receptor 3 is a negative regulator of bone growth. *Cell* **84**, 911–921.
9. Deng, C., Bedford, M., Li, C., Xu, X., Yang, X., Dunmore, J., and Leder, P. (1997). Fibroblast growth factor receptor-1 (FGFR-1) is essential for normal neural tube and limb development. *Dev. Biol.* **185**, 42–54.
10. DiGabriele, A. D., Lax, I., Chen, D. I., Svahn, C. M., Jaye, M., Schlessinger, J., and Hendrickson, W. A. (1998). Structure of a heparin-linked biologically active dimer of fibroblast growth factor. *Nature* **398**, 812–817.
11. Ford-Perriss, M., Abud, H., and Murphy, M. (2001). Fibroblast growth factors in the developing central nerve system. *Clin. Exp. Pharmacol. Physiol.* **28**, 493–503.
12. Gillie, H., Sharrocks, A. D., and Shaw, P. E. (1992). Phosphorylation of transcription factor p26(TCF) by MAP kinase stimulates ternary complex formation at cfos promoter. *Nature* **358**, 414–417.
13. Gupta, S. and Davis, R. J. (1994). MAP kinase binds to the NH$_2$ terminal activation domain of cMyc. *FEBS Lett.* **353**, 281–285.
14. Hadari, Y. R., Kouhara, H., Lax, I., and Schlessinger, J. (1998). Binding of Shp2 tyrosine phosphatase to FRS2 is essential for fibroblast growth factor-induced PC12 cell differentiation. *Mol. Cell. Biol.* **18**, 3966–3973.
15. Hart, K. C., Robertson, S. C., Kanemitsu, M. Y., Meyer, A. N., Tynan, J. A., and Donoghue, D. J. (2000). Transformation and Stat activation by derivatives of FGFR1, FGFR2, and FGFR4. *Oncogene* **19**, 3309–3320.
16. Hehr, U. and Muenke, M. (1999). Craniosynostosis syndromes: from genes to premature fusion of skull bones. *Mol. Genet. Metab.* **68**, 139–151.
17. Heuer, J. G., von Bartheld, C. S., Kinoshita, Y., Evers, P. C., and Bothwell, M. (1990). Alternating phases of FGF receptor and NGF receptor expression in the developing chicken nervous system. *Neuron* **5**, 283–296.
18. Hughes, S. E. (1997). Differential expression of the fibroblast growth factor receptor (FGFR) multigene family in normal human adult tissues. *J. Histochem. Cytochem.* **45**, 1005–1019.
19. Kannan, K. and Givol, D. (2000). FGF receptor mutations: dimerization syndromes, cell growth suppression, and animal models. *IUBMB Life* **49**, 197–205.
20. Karlsson, T., Songyang, Z., Landgren, E., Lavergne, C., Di Fiore, P. P., Anafi, M., Pawson, T., Cantley, L. C., Claesson-Welsh, L., and Welshe, M. (1995). Molecular interactions of the Src homology 2 domain protein Shb with phosphotyrosine residues, tyrosine kinase receptors, and Src homology 3 domain proteins. *Oncogene* **10**, 1475–1483.
21. Kong, M., Wang, C. S., and Donoghue, D. J. (2002). Interaction of fibroblast growth factor receptor 3 and the adapter protein SH2-B: A role in Stat5 activation. *J. Biol. Chem.*, **277**, 15962–15970.
22. Kouhara, H., Hadari, Y. R., Spivak-Kroizman, T., Schilling, J., Bar-Sagi, D., Lax, I., and Schlessinger, J. (1997). A lipid-anchored Grb2-binding protein that links FGF-receptor activation to the Ras/MAPK signaling pathway. *Cell* **89**, 693–702.
23. Landgren, E., Blume-Jensen, P., Courtneidge, S. A., and Claesson-Welsh, L. (1995). Fibroblast growth factor receptor-1 regulation of Src family kinases. *Oncogene* **10**, 2027–2035.
24. Larsson, H., Klint, P., Landgren, E., and Claesson-Welsh, L. (1999). Fibroblast growth factor receptor-1-mediated endothelial cell proliferation is dependent on the Src Homology (SH) 2/SH3 domain-containing adaptor protein Crk. *J. Biol. Chem.* **274**, 25726–25734.
25. LaVallee, T. M., Prudovsky, I. A., McMahon, G. A., Hu, X., and Maciag, T. (1998). Activation of the MAP kinase pathway by FGF-1 correlates with cell proliferation induction while activation of the Src pathway correlates with migration. *J. Cell Biol.* **141**, 1647–1658.
26. Li, W., Fan, J., and Woodley, D. T. (2001). Nck/Dock: an adapter between cell surface receptors and the actin cytoskeleton. *Oncogene* **20**, 6403–6417.
27. Marcelle, C., Eichmann, A., Halevy, O., Breant, C., and Le Douarin, N. M. (1994). Distinct developmental expression of a new avian fibroblast growth factor receptor. *Development* **120**, 683–694.
28. Mohammadi, M., Honegger, A. M., Rotin, D., Fischer, R., Bellot, F., Li, W., Dionne, C. A., Jaye, M., Rubinstein, M., and Schlessinger, J. (1991). A tyrosine-phosphorylated carboxy-terminal peptide of the fibroblast growth factor receptor (Flg) is a binding site for the SH2 domain of phospholipase C-γ1. *Mol. Cell. Biol.* **11**, 5068–5078.
29. Mohammadi, M., Dionne, C. A., Li, W., Li, N., Spivak, T., Honegger, A. M., Jaye, M., and Schlessinger, J. (1992). Point mutation in FGF receptor eliminates phosphatidylinositol hydrolysis without affecting mitogenesis. *Nature* **358**, 681–684.
30. Noji, S., Koyama, E., Myokai, F., Nohno, T., Ohuchi, H., Nishikawa, K., and Taniguchi, S. (1993). Differential expression of three chick FGF receptor genes, FGFR1, FGFR2, and FGFR3, in limb and feather development. *Prog. Clin. Biol. Res.* **383**, 645–654.
31. Ong, S. H., Lim, Y. P., Low, B. C., and Guy, G. R. (1997). Shp2 associates directly with tyrosine phosphorylated p90 (SNT) protein in FGF-stimulated cells. *Biochem. Biophys. Res. Commun.* **238**, 261–266.
32. Ong, S. H., Guy, G. R., Hadari, Y. R., Laks, S., Gotoh, N., Schlessinger, J., and Lax, I. (2000). FRS2 proteins recruit intracellular signaling pathways by binding to diverse targets on fibroblast growth factor and nerve growth factor responses. *Mol. Cell. Biol.* **20**, 979–989.
33. Ong, S. H., Hadari, Y. R., Gotoh, N., Guy, G. R., Schlessinger, J., and Lax, I. (2001). Stimulation of phosphatidylinositol 3-kinase by fibroblast growth factor receptors is mediated by coordinated recruitment of multiple docking proteins. *Proc. Natl. Acad. Sci. USA* **98**, 6074–6079.
34. Ozawa, K., Uruno, T., Miyakawa, K., Seo, M., and Imamura, T. (1996). Expression of the fibroblast growth factor family and their receptor family genes during mouse brain development. *Brain Res. Mol. Brain Res.* **41**, 279–288.
35. Peters, K. G., Werner, S., Chen, G., and Williams, L. T. (1992). Two FGF receptor genes are differentially expressed in epithelial and mesenchymal tissues during limb formation and organogenesis in the mouse. *Development* **114**, 233–243.
36. Peters, K., Ornitz, D., Werner, S., and Williams, L. (1993). Unique expression pattern of the FGF receptor 3 gene during mouse organogenesis. *Dev. Biol.* **155**, 423–430.

37. Plotnikov, A. N., Schlessinger, J., Hubbard, S. R., and Mohammadi, M. (1999). Structural basis for FGF receptor dimerization and activation. *Cell* **98**, 641–650.
38. Powers, C. J. McLeskey, S. W., and Wellstein, A. (2000). Fibroblast growth factors, their receptors, and signaling. *Endo. Relat. Cancer* **7**, 165–197.
39. Pye, D. A. and Gallagher, J. T. (1999). Monomer complexes of basic fibroblast growth factor and heparan sulfate oligosaccharides are the minimal functional unit for cell activation. *J. Biol. Chem.* **274**, 13456–13461.
40. Raffioni, S., Thomas, D., Foehr, E. D., Thompson, L. M., and Bradshaw, R. A. (1999). Comparison of the intracellular signaling responses by three chimeric fibroblast growth factor receptors in PC12 cells. *Proc. Natl. Acad. Sci. USA* **96**, 7178–7183.
41. Rozakis-Adcock, M., McClade, J., Mbamalu, G., Pelicci, G., Daly, R. H. W., Batzer, A., Thomas, S., Brugge, J., Pelicci, P. G., Schlessinger, J., and Pawson, T. (1992). Association of the Shc and Grb2/Sem5 SH2-containing proteins is implicated in activation of the Ras pathway by tyrosine kinases. *Nature* **360**, 689–692.
42. Ryan, P. J. and Gillespie, L. L. (1994). Phosphorylation of phospholipase C gamma 1 and its association with the FGF receptor is developmentally regulated and occurs during mesoderm induction in *Xenopus laevis*. *Dev. Biol.* **166**, 101–111.
43. Ryan, P. J., Paterno, G. D., and Gillespie, L. L. (1998). Identification of phosphorylated proteins associated with the fibroblast growth factor receptor type I during early *Xenopus* development. *Biochem. Biophys. Res. Commun.* **244**, 763–767.
44. Spivak-Kroizman, T., Lemmon, M. A., Dikic, I., Ladbury, J. E., Pinchasi, D., Huang, J., Jaye, M., Crumley, G., Schlessinger, J., and Lax, I. (1994). Heparin-induced oligomerization of FGF molecules is responsible for FGF receptor dimerization, activation and cell proliferation. *Cell* **79**, 1015–1024.
45. Stark, K. L., McMahon, J. A., and McMahon, A. P. (1991). FGFR-4, a new member of the fibroblast growth factor receptor family, expressed in the definitive endoderm and skeletal muscle lineages of the mouse. *Development* **113**, 641–651.
46. Su, W. C., Kitagawa, M., Xue, N., Xie, B., Garofalo, S. Cho, J., Deng, C., Horton, W. A., and Fu, X. Y. (1997). Activation of Stat1 by mutant fibroblast growth-factor receptor in thanatophoric dysplasia type II dwarfism. *Nature* **386**, 288–292.
47. Walshe, J. and Mason, I. (2000). Expression of FGFR1, FGFR2, and FGFR3 during early neural development in the chick embryo. *Mech. Dev.* **90**, 103–110.
48. Wang, J. K., Gao, G., and Goldfarb, M. (1994). Fibroblast growth factor receptors have different signaling and mitogenic potentials. *Mol. Cell. Biol.* **14**, 181–188.
49. Webster, M. K. and Donoghue, D. J. (1997). FGFR activation in skeletal disorders: too much of a good thing. *Trends Genet.* **13**, 178–182.
50. Werner, S., Weinberg, W., Liao, X., Peters, K. G., Blessing, M., Yuspa, S. H., Weiner, R. L., and Williams, L. T. (1993). Targeted expression of a dominant-negative FGF receptor mutant in the epidermis of transgenic mice reveals a role of FGF in keratinocyte organization and differentiation. *EMBO J.* **12**, 2635–2643.
51. Wilke, A. (1996). Fibroblast growth factor receptor mutations and craniosynostosis: Three receptors, five syndromes. *Indian J. Pediatr.* **63**, 351–356.
52. Wilke, T. A., Gubbels, S., Schwartz, J., and Richman, J. M. (1997). Expression of fibroblast growth factor receptors (FGFR1, FGFR2, FGFR3) in the developing head and face. *Dev. Dynam.* **210**, 41–52.
53. Xu, H., Lee, K. W., and Goldfarb, M. (1998). Novel recognition motif on fibroblast growth factor receptor mediates direct association and activation of SNT adapter proteins. *J. Biol. Chem.* **273**, 17987–17990.
54. Xu, X., Weinstein, M., Li, C., Naski, M., Cohen, R. I., Ornitz, D. M., Leder, P., and Deng, C. (1998). Fibroblast growth factor receptor 2 (FGFR2)-mediated reciprocal regulation loop between FGF8 and FGF10 is essential for limb induction. *Development* **125**, 753–765.
55. Yamaguchi, T. P., Harpal, K., Henkemeyer, M., and Rossant, J. (1994). fgfr-1 is required for embryonic growth and mesodermal patterning during mouse gastrulation. *Genes Dev.* **8**, 3032–3044.
56. Yu, C., Wang, F., Kan, M., Jin, C., Jones, R. B., Weinstein, M., Deng, C.-X., and McKeehan, W. L. (2000). Elevated cholesterol metabolism and bile acid synthesis in mice lacking membrane tyrosine kinase receptor FGFR4. *J. Biol. Chem.* **275**, 15482–15489.
57. Zhan, X., Plourde, C., Hu, X., Friesel, R., and Maciag, T. (1994). Association of fibroblast growth factor receptor1 with cSrc correlates with association between cSrc and cortactin. *J. Biol. Chem.* **269**, 20221–20224.

CHAPTER 265

The Role of Receptor Protein Tyrosine Phosphatases in Axonal Pathfinding

Andrew W. Stoker

Neural Development Unit, Institute of Child Health,
University College London, London, United Kingdom

Introduction

One of the most impressive processes that occurs during development is the establishment of countless connections between neurons and their targets. Such precise connectivity requires long-distance growth and pathfinding by axons and short-range detection of target cells. This chapter reviews one family of molecules, the receptor-like protein tyrosine phosphatases (RPTPs), which direct axons in this astonishing feat. In the human genome there are around 22 RPTP genes, most of which have either orthologs or homologs in other species. Most RPTPs are strongly expressed, some times exclusively, in developing nervous systems, in particular within axons and their motile, pathfinding growth cones [1,2]. Evidence is reviewed here for RPTP roles in axon growth and guidance. Their potential signaling mechanisms are also briefly discussed. Due to space limitations, readers will, in most cases, be referred to two reviews and the references therein [1,2]. Figure 1 summarizes the axon growth and guidance events discussed in the following sections, and the RPTPs implicated in each. Other axonal receptor types are reviewed elsewhere in this volume.

RPTPs and the Visual System

Drosophila

The compound eye of the fly contains about 800 ommatidia, each with photoreceptor neurons R1 through R8. Axons of photoreceptors project to the optic lobe where they terminate either in the lamina (R1–R6) or in proximal layers (R8) or distal layers (R7) of the medulla. DPTP69D and DLAR control these axonal termination events [1,3,4]. If axons of R1–R6 are made DPTP69D-deficient, they will overshoot their target and terminate in the medulla. Loss of DPTP69D in R7 causes its axon to stop short in the R8 termination zone. DPTP69D thus appears to control the ability of growth cones to de-adhere (defasciculate) from R8 at correct navigational decision points. Interestingly, R7 axons that lack DLAR can reach and recognize medulla targets, but they later retract [3,4]. DLAR-deficient axons from R1–R6, however, terminate normally. DLAR may therefore be involved in specifically stabilising adhesion of R7 to its targets. DLAR mutants and cadherin mutants have similar phenotypes, suggesting that they may regulate similar adhesive signaling pathways [3]. The collective data also indicate that DPTP69D and DLAR function cell autonomously, although DLAR also shows evidence of nonautonomous function in R8. In contrast to their guidance roles, DRPTPs do not appear to be necessary for axon elongation in the visual system, unlike their vertebrate counterparts.

Vertebrate Retinotectal System

In vertebrate eyes, retinal ganglion cell (rgc) axons relay visual signals from the eye to the brain. Neighboring rgc axons establish precise topographic connectivity with neighboring neurons in the optic tectum. Several studies in cell culture demonstrate a role for RPTPs in rgc neurite growth.

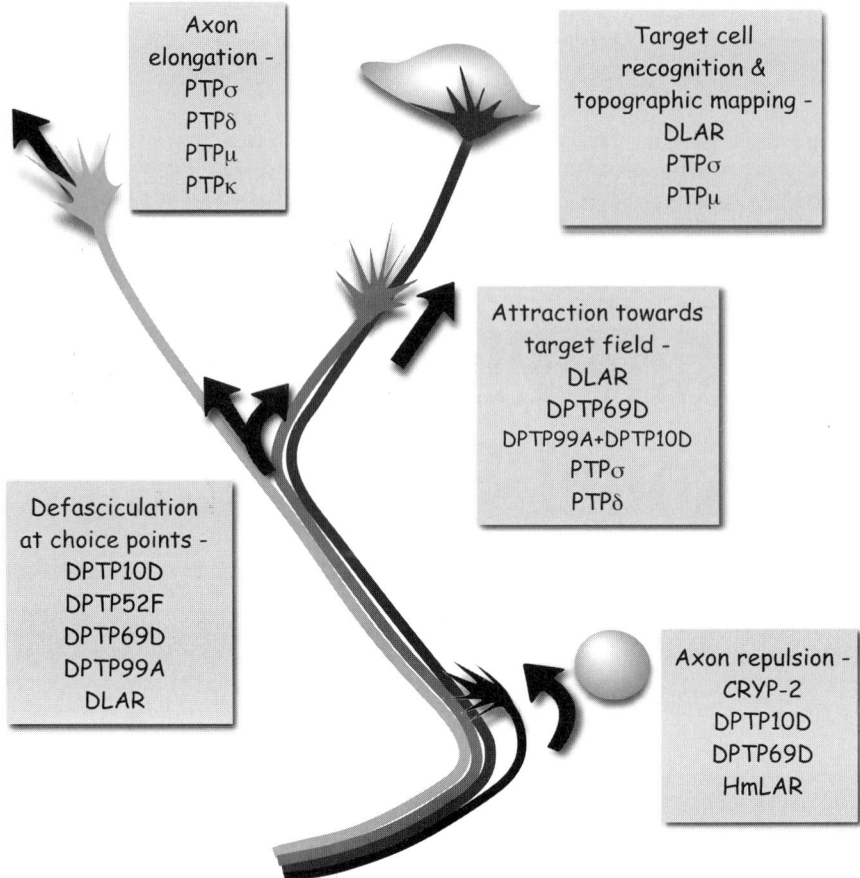

Figure 1 Schematic diagram showing the RPTPs implicated at different stages of axon growth and guidance. See text for details.

Signaling from chick PTPμ enhances cadherin-dependent retinal axon outgrowth [1]. Furthermore, PTPμ expression varies topographically across the retina and tectum and PTPμ has a selective, growth cone-collapsing function [5]. PTPμ may therefore differentially influence axon growth within the retinotectal projection. *Xenopus* PTPδ promotes both rgc axon growth along the optic tract *in vivo* and neurite growth on basement membranes in culture [6], while soluble ectodomains of human PTPδ can attract growth cones of cerebellar neurites [1]. Since both PTPδ and PTPμ bind homophilically, they may trigger signals directly between axons. Another retinal RPTP, CRYP-2 (encoded by avian *Ptpro*), is anti-adhesive in rgc cultures [7]. CRYP-2 may have an axon navigation role, since its ectodomain induces growth cone collapse and ectodomain gradients cause repulsive growth cone turning [7]. The interaction between chick PTPσ and a ligand(s) on basement membranes and glial endfeet maintains optimal retinal neurite outgrowth [1]. Perhaps counterintuitively, interference with intracellular signaling of *Xenopus* PTPσ causes faster neurite outgrowth in culture [6], suggesting a possible signaling model (see the section on ligands). The first evidence for RPTP function in axon targeting *in vivo* has come from chick PTPσ. Perturbation of the interactions between PTPσ and its ligands in the optic tectum causes retinal axon stalling and rostral mistargeting [8]. PTPσ may therefore function by maintaining retinal axon growth over the tectum and facilitating the recognition of correct target sites.

Neuromuscular System

Drosophila genetics has highlighted key RPTP functions during motor axon guidance. The segmental and intersegmental motor nerves ISN, ISNb, and SNa of the fly larva innervate body wall muscles in a highly stereotypical manner. Nerve defects arise after loss of function in *DLAR, DPTP69D, DPTP99A, DPTP10D,* and *DPTP52F* [2,9]. DPTP69D and DPTP99A are required for ISNb axons to defasciculate from the ISN at the correct choice point. Gene deficiency causes a "bypass" phenotype where axons fail to leave the ISN and thus travel past their targets. DLAR influences not only this defasciculation step, but also both the entry of axons into the muscle target field and synapse formation [10]. DPTP10D collaborates with other DRPTPs in guiding SNa, but antagonizes them during navigation of the ISN. Similarly, DLAR and DPTP99A antagonize each other within SNb axons. There is therefore a complex pattern of interaction between these RPTPs with "partial redundancy, competition, and collaboration" as described by Sun and co-workers [11].

Further Axon Growth and Guidance Roles

In the leech, LAR homolog HmLAR2 is expressed in growth cones of neurite-like processes of comb cells, where it controls the orderly outgrowth of these processes [1]. Evidence supports a homophilic interaction between HmLAR2 molecules, signaling a mutual repulsion between growth cones and neighboring processes.

In the ventral nerve cord of *Drosophila*, axon guidance across the midline is influenced by DLAR, DPTP99A, DPTP69D, and DPTP10D. The latter two in particular cooperate with Robo receptors to transduce repulsive signals from midline Slit protein [1]. How these receptors cooperate biochemically is unclear at present.

Several RPTP gene-deficiency models have been developed in mice. Loss of PTPσ function causes motor function deficits and hyposmia, as well as quite severe defects in sciatic nerve myelination and maturation [1]. Deficiency in PTPδ also causes milder motor defects as well as memory alterations, while loss of LAR causes a reduction in forebrain cholinergic neuron numbers and some mild defects in hippocampal innervation [1]. The developmental bases for all these neuronal and axonal defects have yet to be characterized.

Axonal Signaling by RPTPs

Instructive or Permissive?

Do axonal RPTPs send permissive or instructive signals during axon guidance? With DLAR, the fact that R7 growth cones reach targets, but then retract, supports an instructive role in securing adhesion to targets. DLAR may also control the instructive process of muscle cell recognition by motor axons [11]. For DPTP69D, the consensus is more in favor of a permissive role in controlling defasciculation rather than target recognition, although this remains under discussion [1]. In fact, the many complex interactions between *Drosophila* RPTPs may ultimately make simple instructive/permissive distinctions untenable. Vertebrate PTPδ and CRYP-2 may have instructive signaling roles given that they can force growth cone turning on otherwise permissive substrates [1,7]. In contrast, chick PTPσ may control a permissive event during axonal targeting in the tectum, given that its known ligands are uniformly distributed [8].

Ligands

PTPμ and PTPδ bind homophilically, with no evidence to date of heterophilic ligands. Ectodomains of these RPTPs can act as neurite growth-promoting substrates in culture, suggesting that the growth of fasciculated axons *in vivo* may be promoted by their homophilic action. Paradoxically, PTPμ ectodomains also have negative effects on growth cones [5], although we do not know yet if this signal is transduced by PTPμ receptors. The heparin-binding chemokine pleiotropin is a ligand for PTPζ, an RPTP expressed in both glia and some neurons. Pleiotropin can inhibit the PTPζ phosphatase and this leads to increased tyrosine phosphorylation of potential targets [12]. The heparan sulfate proteoglycans agrin and collagen XVIII are strong binding partners for chick PTPσ, although we await direct evidence for their roles in controlling PTPσ signaling in axons [13]. One of several models proposed also suggests that chick PTPσ ligands may inactivate the phosphatase, thereby facilitating neurite growth [6]. The effects of RPTP ligands are discussed further in part B of this Handbook. Although RPTP ectodomains have adhesive capacities, it is generally believed that their signaling roles require intact catalytic functions. For example, enzymatically active PTPμ is required for neurite outgrowth on cadherins [1]. Furthermore, genetic rescue studies with *Drosophila* RPTPs indicate that the rescuing genes must encode active phosphatases [1,3,4,11].

Downstream Signals

Figure 2 contains a summary of some of the known substrates and binding partners of neuronal RPTPs. Most of these impinge ultimately on the actin cytoskeleton, providing a logical handle on growth cone dynamics. DLAR interacts with several molecules including the tyrosine kinase Abl and its substrate Enabled (Ena, a VASP family member). Ena can be dephosphorylated by DLAR [1]. Dephosphorylation of Ena activates downstream signals that pass through profilin and on to actin. DLAR also interacts genetically with Trio, a large protein with two exchange factor domains for Rho family GTPases. *Drosophila* Trio also signals through the SH2–SH3 adaptor Dock and the p21-activated kinase Pak,

RPTPs	Interacting protein	Downstream effector	Refs.
LAR, PTPσ	> Trio >	Rac, Rho	1, 2
LAR, PTPδ, PTPσ	liprins		2, 10
DLAR	< Abl* <	ena*	1, 2
DLAR, DPTP69D	> ena* >	profilin	1
PTPα	> grb2 >		2
PTPα	> c-src* >		2
PTPμ	> RACK1 >	PKCδ	16
PTPμ, PTPκ, LAR family, PTPζ	< cadherins & catenins* <		15
DPTP10D, DPTP99A	gp150*		2

Figure 2 Table showing some of the proteins that interact with axonal RPTPs, and some of the predicted downstream effectors where known. Symbols > and < indicate either stimulation or repression of function by the RPTP, respectively. Asterisks indicate that the protein is a phosphatase substrate.

again impinging on the cytoskeleton. Mammalian Trio binds directly to LAR family members and signals through Rho family GTPases. Human Trio promotes neurite outgrowth in PC12 neurons, although it is not known yet if LAR RPTPs are involved in this event [14]. Catenins are a key target of several RPTPs, including PTPμ, PTPκ, PTPζ and LAR members [12,15]. These RPTPs may well antagonize cadherin/catenin-regulated cell adhesion by dephosphorylating β-catenin and p120(ctn), thereby directly influencing growth cone adhesion. PTPμ also binds to the adaptor protein RACK and in turn requires PKCδ to promote neurite outgrowth [16]. Other RPTP targets include the cell adhesion molecule-like gp150 and the tyrosine kinase c-src, but it is not clear yet if these are involved in RPTP signaling within axons. Finally, adaptor proteins of the liprin family bind to LAR family RPTPs and may be important for localizing these RPTPs in membranes and in forming complexes with roles in signaling, adhesion, and synapse function [2,10].

References

1. Stoker, A. W. (2001). Receptor tyrosine phosphatases in axon growth and guidance. *Curr. Opin. Neurobiol.* **11**, 95–102.
2. Van Vactor, D. (1998). Protein tyrosine phosphatases in the developing nervous system. *Curr. Opin. Cell Biol.* **10**, 174–81.
3. Clandinin, T. R., Lee, C. H., Herman, T., Lee, R. C., Yang, A. Y., Ovasapyan, S., and Zipursky, S. L. (2001). *Drosophila* LAR regulates R1-R6 and R7 target specificity in the visual system. *Neuron* **32**, 237–48.
4. Maurel-Zaffran, C., Suzuki, T., Gahmon, G., Treisman, J. E., and Dickson, B. J. (2001). Cell-autonomous and -nonautonomous functions of LAR in R7 photoreceptor axon targeting. *Neuron* **32**, 225–35.
5. Burden-Gulley, S. M., Ensslen, S. E., and Brady-Kalnay, S. M. (2002). Protein tyrosine phosphatase-mu differentially regulates neurite outgrowth of nasal and temporal neurons in the retina. *J. Neurosci.* **22**, 3615–27.
6. Johnson, K. G., McKinnell, I. W., Stoker, A. W., and Holt, C. E. (2001). Receptor protein tyrosine phosphatases regulate retinal ganglion cell axon outgrowth in the developing *Xenopus* visual system. *J. Neurobiol.* **49**, 99–117.
7. Stepanek, L., Sun, Q. L., Wang, J., Wang, C., and Bixby, J. L. (2001). CRYP-2/cPTPRO is a neurite inhibitory repulsive guidance cue for retinal neurons *in vitro*. *J. Cell Biol.* **154**, 867–78.
8. Rashid-Doubell, F., McKinnell, I., Aricescu, A. R., Sajnani, G., and Stoker, A. W. (2002). Chick PTPsigma regulates the targeting of retinal axons within the optic tectum. *J. Neurosci.*, **22**, 5024–5033.
9. Schindelholz, B., Knirr, M., Warrior, R., and Zinn, K. (2001). Regulation of CNS and motor axon guidance in Drosophila by the receptor tyrosine phosphatase DPTP52F. *Development* **128**, 4371–82.
10. Kaufmann, N., DeProto, J., Ranjan, R., Wan, H., and Van Vactor, D. (2002). *Drosophila* liprin-alpha and the receptor phosphatase Dlar control synapse morphogenesis. *Neuron* **34**, 27–38.
11. Sun, Q., Schindelholz, B., Knirr, M., Schmid, A., and Zinn, K. (2001). Complex genetic interactions among four receptor tyrosine phosphatases regulate axon guidance in *Drosophila*. *Mol. Cell. Neurosci.* **17**, 274–291.
12. Meng, K., Rodriguez-Pena, A., Dimitrov, T., Chen, W., Yamin, M., Noda, M., and Deuel, T. F. (2000). Pleiotrophin signals increased tyrosine phosphorylation of beta-catenin through inactivation of the intrinsic catalytic activity of the receptor-type protein tyrosine phosphatase beta/zeta. *Proc. Natl. Acad. Sci. USA* **97**, 2603–8.
13. Aricescu, A. R., McKinnell, I. W., Halfter, W., and Stoker, A. W. (2002). Heparan sulfate proteoglycans are ligands for receptor protein tyrosine phosphatase sigma. *Mol. Cell. Biol.* **22**, 1881–92.
14. Estrach, S., Schmidt, S., Diriong, S., Penna, A., Blangy, A., Fort, P., and Debant, A. (2002). The human rho-GEF Trio and its target GTPase rhoG are involved in the NGF pathway, leading to neurite outgrowth. *Curr. Biol.* **12**, 307–12.
15. Zondag, G. C., Reynolds, A. B., and Moolenaar, W. H. (2000). Receptor protein-tyrosine phosphatase RPTPmu binds to and dephosphorylates the catenin p120(ctn). *J. Biol. Chem.* **275**, 11264–9.
16. Rosdahl, J. A., Mourton, T. L., and Brady-Kalnay, S. M. (2002). Protein kinase C delta (PKCdelta) is required for protein tyrosine phosphatase mu (PTPmu)-dependent neurite outgrowth. *Mol. Cell. Neurosci.* **19**, 292–306.

CHAPTER 266

Attractive and Repulsive Signaling in Nerve Growth Cone Navigation

Guo-li Ming, and Mu-ming Poo
The Salk Institute, La Jolla, California
Division of Neurobiology, Department of Molecular and Cell Biology,
University of California, Berkeley, California

Introduction

The function of the nervous system depends on complex and precise connections between nerve cells [1]. The formation of specific connections during development often requires the growing axon to navigate over considerable distances to reach their final target cells. This long-range navigation is achieved by guidance factors within the developing tissue that regulate the motility or directionality of the growing tip of the axon, the growth cone [2]. During the last decade several families of guidance factors have been identified, including netrins, semaphorins, ephrins, and slits [3,4]. In addition, inhibitory factors associated with the myelin that exert repulsive actions on the navigation of regenerating axons have also been discovered. Different classes of membrane receptors for these factors have been identified and their intracellular signal transduction mechanisms are beginning to be elucidated.

How does a guidance factor affect the navigation of the growing axon? A general scheme of signal transduction cascades from the receptor activation to cytoskeletal rearrangements is shown in Fig. 1 [4]. It starts with the binding of the guidance factor with the receptor protein or protein complexes at the cell surface. Ligand-receptor binding in general stimulates the activities of the cytoplasmic domain of the receptor, which in turn interacts specifically with cytoplasmic adaptor proteins. These adaptors may then recruit or activate their downstream effectors to further mediate the guidance signal. The effectors (or mediators) can be enzymes or second messengers that activate or inhibit cytoskeleton-associated proteins, leading to polymerization or depolymerization of cytoskeletal structures and steering of the growth cone.

Two types of guidance signals may be distinguished: signals that convey a "stop or go" command regulating growth cone motility and signals that provide directional instructions to the growing axon, triggering turning responses of the growth cone. For nondirectional signals, mediators may simply alter the global cytoskeletal activity at the growth cone. For directional guidance signals, however, a gradient of cytoskeletal rearrangements must be created in order to induce directional motility. In the latter case, mediators must be activated or distributed in a gradient across the growth cone, and such a gradient may also need to be amplified in the cytoplasm in order to achieve a reliable directional response [4]. Although a number of cytoplasmic components have been implicated in such a scheme of signal transduction, definitive identification of signaling pathways are yet to be established for any one of the major families of guidance factors. This chapter summarizes some of the putative signaling pathways that have been shown to participate in growth cone navigation.

Netrin Signaling

Netrins are a family of secreted proteins and their receptors were identified to be DCC (deleted in colorectal cancer) and UNC-5 [5], two interacting transmembrane proteins that set the polarity of growth cone responses. Ectopic expression of UNC-5 in neurons converted netrin (UNC-6)-dependent

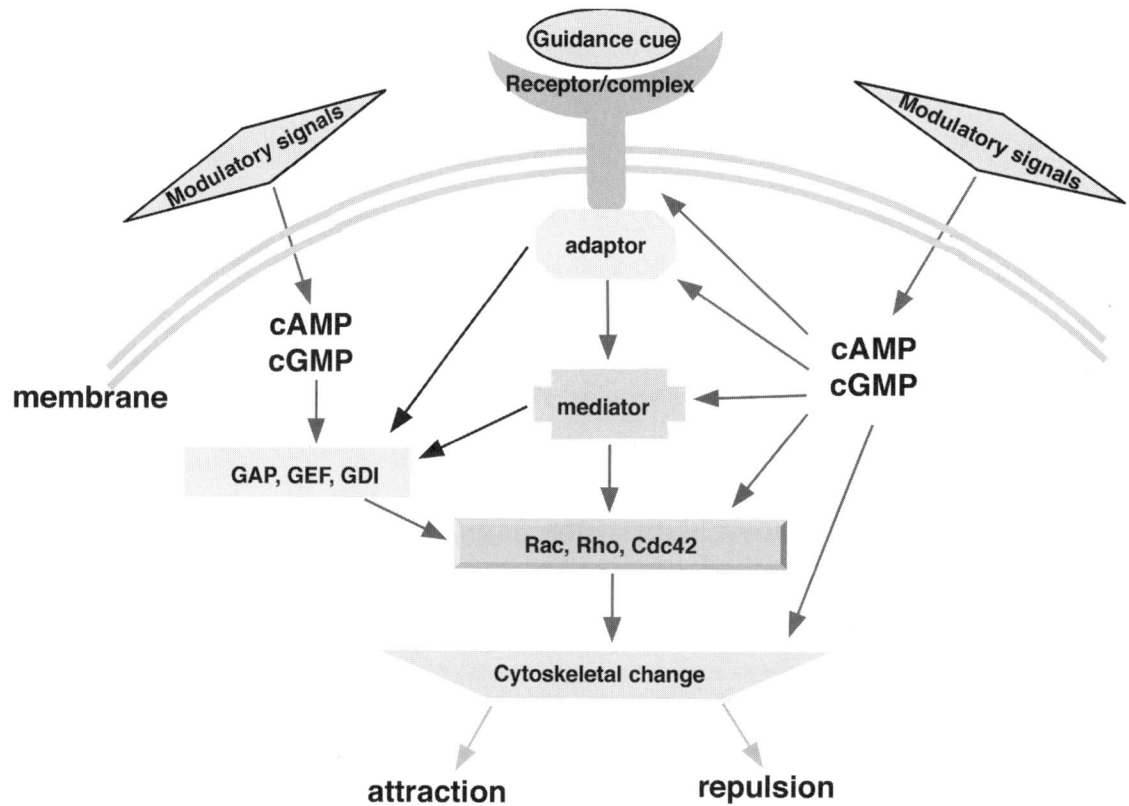

Figure 1 Signal transduction mechanism for nerve growth cone guidance.

chemoattraction to chemorepulsion in both *Caenorhabditis elegans* [6] and in dissociated *Xenopus* spinal neurons [7], a conversion that involves a netrin-dependent interaction between the cytoplasmic domain of DCC and UNC-5 [7]. The level of DCC can be actively regulated through degradation mediated by Sina/Siah protein [8] or through metalloprotease-mediated shedding [9], resulting in changes in the growth cone's sensitivity to netrin-1. Extracellular signal-regulated kinase1/2 (Erk 1/2) was found to be recruited to DCC receptor complex in rat commissural neurons [10] and activation of phosphoinositide 3-kinase (PI3-K) and phospholipase Cγ appear to mediate attractive turning of *Xenopus* spinal neurons induced by a gradient of netrin-1 [11]. More recently, it has been shown that activation of MAPK, local protein synthesis, and protein degradation are involved in the netrin-induced chemoattraction of *Xenopus* spinal and retinal neurons [13,14]. These downstream events were shown to be critical for adaptive changes of growth cone sensitivity to netrin-1 as the extracellular concentration of netrin-1 is increased [13]. How these cytoplasmic factors are linked to the cytoskeleton changes remains largely unclear. Two members of the Rho GTPase family, Rac1 and Cdc42, appear to be involved in netrin signaling [15,16], thus providing potential links to cytoskeletal regulation [17]. To serve for directional guidance signals, a mediator is not only required for the guidance responses, but must also be activated in a gradient across the growth cone. Furthermore, such a gradient should be sufficient to induce a turning response of the growth cone. None of the putative signaling components described above fulfill these criteria. Interestingly, the well-known second messenger Ca^{2+} appears to satisfy these criteria for netrin-1 signaling. Elevation of cytoplasmic Ca^{2+}, through both Ca^{2+} influx and release from internal stores, is required for netrin-1-induced turning responses and a netrin-1 gradient can trigger Ca^{2+} elevation and transient Ca^{2+} gradients across the growth cone [7,12]. Experimentally creating a gradient of Ca^{2+} across the growth cone in the absence of netrin-1 signals is sufficient to induce the turning of the growth cone [18]. However, it remains unclear how Ca^{2+} signals are linked to receptor activation upstream and cytoskeletal rearrangements downstream.

Semaphorin Signaling

The semaphorin family includes both membrane-bound and secreted molecules, thus it may work for both short- and long-range guidance [19,20]. Neuropilins were identified as semaphorin receptors and the plexin family of receptors was shown to be a co-receptor that transduces the signal. Several proteins have been shown to bind to neuropilins or plexins. These include a transmembrane protein OTK (off-track) [21], cytoplasmic protein NIP—a PSD-95/Dlg/ZO-1 domain—containing protein that may be involved in membrane trafficking [22], and MICAL, a flavoprotein oxidoreductase [23]. There is also evidence for the involvement of heterotrimeric G proteins [24]. The precise role of these receptor-interacting proteins and whether they mediate or modulate the signaling process remain to be determined. More is known about the downstream cascades that mediate cytoskeleton changes induced by semaphorins. Of particular interest is Rac1, a small Rho family GTPase. Introduction of dominant-negative Rac1 [25] or an inhibitory peptide for Rac1 or C3 transferase, a Rho GTPase inhibitor, blocks Sema3A-induced growth cone collapse in sensory neurons [26,27]. A major downstream mediator of Rac1 and Cdc42 is P21-associated kinase (PAK), and LIM-domain-containing kinase (LIM-kinase), a direct substrate of PAK, are necessary for Sema3A-induced growth cone collapse [28]. LIM-kinase is a serine-threonine kinase that inhibits cofilin's actin-severing function. Thus at least one of the mechanisms for semaphorin-mediated cytoskeletal changes is mediated through the activation of small GTPases and their targets PAK, which then regulates actin dynamics through a LIM-kinase- and cofilin-dependent pathway. Other small GTPases such as Rho and Rnd1 [29,30] and kinases such as GSK-3 (glycogen synthase kinase) [31] and Fes/Fps tyrosine kinase [32] have also been implicated in semaphorin signaling.

Slit Signaling

Slits are a family of secreted proteins that can exert short- and long-range guidance functions by activating their receptors, the roundabout (Robo) family of proteins [33]. Slits appear to act not only as directional guidance factors but also as stop signals through activation in a combinatorial manner of different Robo receptors expressed on the growth cone surface [34–36]. Both Abelson tyrosine kinase (Abl) and its substrate Enabled (Ena) can bind directly to the cytoplasmic domain of Robo and modulate its function [37]. Interfering with the binding between Ena and Robo partially impairs the Robo function, while a mutation in a conserved tyrosine residue that can be phosphorylated by Abl generates a hyperactive Robo. Small GTPases and their regulators also affect slit-Robo signaling. A slit-Robo-GTPase activating protein 1 (srGAP1) can bind to Robo and inactivate Cdc42, resulting in repulsion of growth cones [38], while GEF64C, a Dbl family guanine nucleotide exchange factor (GEF), can activate Rho and block Robo-induced repulsion [39]. Thus, Robo-mediated cytoskeleton changes also appear to be mediated by activation of GAP or GEF of small GTPases.

Ephrin Signaling

Ephrins and the Eph family of tyrosine kinase receptors are membrane-bound molecules that mediate short-range axon guidance via cell-cell contacts [40]. Ephrin-A ligands are attached to the plasma membrane via a glycophosphatidylinositol (GPI) linkage, whereas the ephrin-B ligand contains a transmembrane domain and a cytoplasmic tail [41]. Similar to slits, ephrins can function as either directional or nondirectional guidance factors. In addition, the signaling activated by ephrin-Eph binding is bidirectional [41], so that cytoplasmic activities are triggered in both interacting cells. A GEF, ephexin, binds to the kinase domain of EphA constitutively through its Dbl homology-pleckstrin homology (DH/PH) domain and activates both RhoA and Cdc42, thus regulating cytoskeletal structures [42]. In addition, focal adhesion kinase (FAK) and its downstream factor P130(cas) which are two proteins involved in actin reorganization, are also implicated in EphA-induced cytoskeletal changes [43].

Nogo and Myelin-Associated Glycoprotein Signaling

Two proteins associated with myelin, Nogo, and myelin-associated glycoprotein (MAG), have been identified as the major inhibitory factors that prevent axon regeneration after CNS injury [44]. Although the full length of these proteins are membrane-anchored, they can be released in a truncated form and function in repelling and inhibiting axon growth [45,46]. The receptors for Nogo (NogoR) [46] and for MAG (GD1a and GT1b) [47] have been identified. Interestingly, MAG also binds to NogoR [48]. Since NogoR is a GPI-anchored protein at the cell surface [46], and an as yet unidentified co-receptor(s) is required for transducing the cytoplasmic signal. The downstream signal cascade for NogoR signaling is largely unknown, although Ca^{2+} and PI3-K are required for MAG-induced repulsion of *Xenopus* spinal neurons [11,49]. Rho is activated by MAG through receptor GD1a and GT1b [47]. Inhibition of Rho activity can promote CNS axon regeneration, suggesting that Rho may also be involved in MAG-induced cytoskeletal rearrangement [50].

Critical Roles of Modulatory Signals

For a growth cone to make its navigational decisions, it must integrate information provided not only by the guidance factors, but also by other modulatory signals. *In vitro* studies have shown that cytoplasmic cyclic nucleotides play key roles in modulating signal transduction events triggered by most guidance factors identified thus far [4]. For example, the growth cone responses to netrin and MAG are modulated by a cAMP-dependent pathway, whereas Sema3A and slit signaling is modulated by a cGMP-dependent pathway. Elevating the cytoplasmic level of cyclic nucleotides favors attraction/growth, while lowering their level favors repulsion/collapse [4]. Many extracellular ligands, including neuromodulators, adhesion molecules, and extracellular matrix (ECM) components, may change the level of cyclic nucleotides within the cell, thus altering the growth cone behavior when they are present concurrently with the guidance signal [51]. For example, laminin, an abundant ECM protein, reduces the cAMP level in *Xenopus* retinal ganglion neurons and converts the growth cone response to a netrin-1 gradient from attraction to repulsion both *in vitro* and *in vivo* [52]. Conversely, the repulsive response of DRG and cortical axons induced by Sema3A can be converted to attraction by exposure to soluble L1-Fc chimeric molecules; activation of guanylate cyclase activity is required for the conversion [53]. The targets of PKA/PKG that are involved in regulating the polarity of growth cone turning responses remain to be identified.

Concluding Remarks

An emerging view of axon guidance factors is that they are multifunctional molecules capable of conferring attractive, repulsive, or stop signals. The precise behavior of a growth cone is determined by the nature of specific receptors and the status of cytoplasmic signal cascades, which are under the influence of a variety of extrinsic and intrinsic factors (Fig. 1). The combinatorial expression pattern of various receptors at the surface of a growth cone may trigger a differential downstream event [54]. In addition, the efficacy of receptor signaling across the plasma membrane can be modulated by various factors both extra- and intracellularly [4]. Recruitment of different adaptors and mediators in the cytoplasm can result in different growth cone behaviors. An area of interest for future studies is determining the mechanisms that control or modulate the recruitment and activation of these cytoplasmic factors and that amplify the signals conveyed by the receptors. Signal cascades triggered by all known guidance factors appear to eventually converge upon different members of the Rho family GTPases or their activators/inhibitors. It is of interest now to determine how spatiotemporal patterns of GTPase activation account for distinct navigational behaviors of the growth cone.

References

1. Kandel, E. R., Schwartz, J. H., and Jessell, T. M. (2000). *Principles of Neural Science*. McGraw-Hill, New York.
2. Tessier-Lavigne, M. and Goodman, C. S. (1996). The molecular biology of axon guidance. *Science* **274**, 1123–1133.
3. Mueller, B. K. (1999). Growth cone guidance: first steps towards a deeper understanding. *Annu. Rev. Neurosci.* **22**, 351–388.
4. Song, H. and Poo, M. (2001). The cell biology of neuronal navigation. *Nat. Cell Biol.* **3**, E81–E88.
5. Livesey, F. J. (1999). Netrins and netrin receptors. *Cell. Mol. Life Sci.* **56**, 62–68.
6. Hamelin, M., Zhou, Y., Su, M. W., Scott, I. M., and Culotti, J. G. (1993). Expression of the UNC-5 guidance receptor in the touch neurons of *C. elegans* steers their axons dorsally. *Nature* **364**, 327–330.
7. Hong, K, Nishiyama, M., Henley, J., Tessier-Lavigne, M., and Poo, M. (2000). Calcium signalling in the guidance of nerve growth by netrin-1. *Nature* **403**, 93–98.
8. Hu, G. and Fearon, E. R. (1999). Siah-1 N-terminal RING domain is required for proteolysis function, and C-terminal sequences regulate oligomerization and binding to target proteins. *Mol. Cell. Biol.* **19**, 724–732.
9. Galko, M. J. and Tessier-Lavigne, M. (2000). Function of an axonal chemoattractant modulated by metalloprotease activity. *Science* **289**, 1365–1367.
10. Forcet, C., Stein, E., Pays, L., Corset, V., Llambi, F., Tessier-Lavigne, M., and Mehlen, P. (2002). Netrin-1-mediated axon outgrowth requires deleted in colorectal cancer-dependent MAPK activation. *Nature* **417**, 443–447.
11. Ming, G., Song, H., Berninger, B., Inagaki, N., Tessier-Lavigne, M., and Poo, M. (1999). Phospholipase C-gamma and phosphoinositide 3-kinase mediate cytoplasmic signaling in nerve growth cone guidance. *Neuron* **23**, 139–148.
12. Ming, G.-L., Song, H.-J., Berninger, B., Holt, C. E., Tessier-Lavigne, M., and Poo, M. (1997). cAMP-dependent growth cone guidance by netrin-1. *Neuron* **19**, 1225–1235.
13. Ming, G. L., Wong, S. T., Henley, J., Yuan, X. B., Song, H. J., Spitzer, N. C., and Poo, M.-M. (2002). Adaptation in the chemotactic guidance of nerve growth cones. *Nature* **417**, 411–418.
14. Campbell, D. S. and Holt, C. E. (2001). Chemotropic responses of retinal growth cones mediated by rapid local protein synthesis and degradation. *Neuron* **32**, 1013–1026.
15. Li, X., Saint-Cyr-Proulx, E., Aktories, K., and Lamarche-Vane, N. (2002). Rac1 and Cdc42 but not RhoA or Rho kinase activities are required for neurite outgrowth induced by the Netrin-1 receptor DCC (deleted in colorectal cancer) in N1E-115 neuroblastoma cells. *J. Biol. Chem.* **277**, 15207–15214.
16. Shekarabi, M. and Kennedy, T. E. (2002). The netrin-1 receptor DCC promotes filopodia formation and cell spreading by activating Cdc42 and Rac1. *Mol. Cell. Neurosci.* **19**, 1–17.
17. Tapon, N. and Hall, A. (1997). Rho, Rac and Cdc42 GTPases regulate the organization of the actin cytoskeleton. *Curr. Opin. Cell. Biol.* **9**, 86–92.
18. Zheng, J. Q. (2000). Turning of nerve growth cones induced by localized increases in intracellular calcium ions. *Nature* **403**, 89–93.
19. He, Z., Wang, K. C., Koprivica, V., Ming, G., and Song, H.-J. (2002). Functions of semaphorins in the nervous system. *Science STKE* 119, RE1.
20. Nakamura, F., Kalb, R. G., and Strittmatter, S. M. (2000). Molecular basis of semaphorin-mediated axon guidance. *J. Neurobiol.* **44**, 219–229.
21. Winberg, M. L., Tamagnone, L., Bai, J., Comoglio, P. M., Montell, D., and Goodman, C. S. (2001). The transmembrane protein Off-track associates with Plexins and functions downstream of Semaphorin signaling during axon guidance. *Neuron* **32**, 53–62.
22. Cai, H. and Reed, R. R. (1999). Cloning and characterization of neuropilin-1-interacting protein: a PSD-95/Dlg/ZO-1 domain-containing protein that interacts with the cytoplasmic domain of neuropilin-1. *J. Neurosci.* **19**, 6519–6527.
23. Terman, J. R., Mao, T., Pasterkamp, R. J., Yu, H. H., and Kolodkin, A. L. (2002). MICALs, a amily of conserved flavoprotein oxidoreductases, function in plexin-mediated axonal repulsion. *Cell* **109**, 887–900.
24. Igarashi, M., Strittmatter, S. M., Vartanian, T., Fishman, M. C. (1993). Mediation by G proteins of signals that cause collapse of growth cones. *Science* **259**, 77–79.
25. Jin, Z. and Strittmatter, S. M. (1997). Rac1 mediates collapsin-1-induced growth cone collapse. *J. Neurosci.* **17**, 6256–6263.
26. Kuhn, T. B., Brown, M. D., Wilcox, C. L., Raper, J. A., and Bamburg, J. R. (1999). Myelin and collapsin-1 induce motor neuron growth cone collapse through different pathways: Inhibition of collapse by opposing mutants of rac1. *J. Neurosci.* **19**, 1965–1975.
27. Liu, B. P. and Strittmatter, S. M. (2001). Semaphorin-mediated axonal guidance via Rho-related G proteins. *Curr. Opin. Cell. Biol.* **13**, 619–626.
28. Aizawa, H., Wakatsuki, S., Ishii, A., Moriyama, K., Sasaki, Y., Ohashi, K., Sekine-Aizawa, Y., Sehara-Fujisawa, A., Mizuno, K., Goshima, Y., and Yahara, I. (2001). Phosphorylation of cofilin by LIM-kinase is necessary for semaphorin3A-induced growth cone collapse. *Nat. Neurosci.* **4**, 367–373.
29. Arimura, N. *et al.* (2000). Phosphorylation of collapsin response mediator protein-2 by Rho-kinase. Evidence for two separate signaling pathways for growth cone collapse. *J. Biol. Chem.* **275**, 23973–23980.
30. Zanata, S. M., Hovatta, I., Rohm, B., and Puschel, A. W. (2002). Antagonistic effects of Rnd1 and RhoD GTPases regulate receptor activity in Semaphorin3A-induced cytoskeletal collapse. *J. Neurosci.* **22**, 471–477.
31. Eickholt, B. J., Walsh, F. S., and Doherty, P. (2002). An inactive pool of GSK-3 at the leading edge of growth cones is implicated in Semaphorin3A signaling. *J. Cell Biol.* **157**, 211–217.

32. Mitsui, N., Inatome, R., Takahashi, S., Goshima, Y., Yamamura, H., and Yanagi, S. (2002). Involvement of Fes/Fps tyrosine kinase in semaphorin3A signaling. *EMBO J.* **21**, 3274–3285.
33. Guthrie, S. (2001). Axon guidance: Robos make the rules. *Curr. Biol.* **11**, R300–303.
34. Rajagopalan, S., Vivancos, V., Nicolas, E., and Dickson, B. J. (2000). Selecting a longitudinal pathway: Robo receptors specify the lateral position of axons in the *Drosophila* CNS. *Cell* **103**, 1033–1045.
35. Simpson, J. H., Bland, K. S., Fetter, R. D., and Goodman, C. S. (2000). Short-range and long-range guidance by Slit and its Robo receptors: A combinatorial code of Robo receptors controls lateral position. *Cell* **103**, 1019–1032.
36. Bagri, A., Marin, O., Plump, A. S., Mak, J., Pleasure, S. J., Rubenstein, J. L., and Tessier-Lavigne, M. (2002). Slit proteins prevent midline crossing and determine the dorsoventral position of major axonal pathways in the mammalian forebrain. *Neuron* **33**, 233–248.
37. Bashaw, G. J., Kidd, T., Murray, D., Pawson, T., and Goodman, C. S. (2000). Repulsive axon guidance: Abelson and Enabled play opposing roles downstream of the roundabout receptor. *Cell* **101**, 703–715.
38. Wong, K. *et al.* (2001). Signal transduction in neuronal migration: Roles of GTPase activating proteins and the small GTPase Cdc42 in the Slit-Robo pathway. *Cell* **107**, 209–221.
39. Bashaw, G. J., Hu, H., Nobes, C. D., and Goodman, C. S. (2001). A novel Dbl family RhoGEF promotes Rho-dependent axon attraction to the central nervous system midline in *Drosophila* and overcomes Robo repulsion. *J. Cell Biol.* **155**, 1117–1122.
40. Wilkinson, D. G. (2001). Multiple roles of EPH receptors and ephrins in neural development. *Nat. Rev. Neurosci.* **2**, 155–164.
41. Flanagan, J. G. and Vanderhaeghen, P. (1998). The ephrins and Eph receptors in neural development. *Annu. Rev. Neurosci.* **21**, 309–345.
42. Shamah, S. M., Lin, M. Z., Goldberg, J. L., Estrach, S., Sahin, M., Hu, L., Bazalakova, M., Neve, R. L., Corfas, G., Debant, A., and Greenberg, M. E. (2001). EphA receptors regulate growth cone dynamics through the novel guanine nucleotide exchange factor ephexin. *Cell* **105**, 233–244.
43. Carter, N., Nakamoto, T., Hirai, H., and Hunter, T. (2002). EphrinA1-induced cytoskeletal re-organization requires FAK and p130 (cas). *Nat. Cell Biol.* **22**, 565–573.
44. Fouad, K., Dietz, V., and Schwab, M. E. (2001). Improving axonal growth and functional recovery after experimental spinal cord injury by neutralizing myelin associated inhibitors. *Brain Res. Brain Res. Rev.* **36**, 204–212.
45. Tang, S., Qiu, J., Nikulina, E., and Filbin, M. T. (2001). Soluble myelin-associated glycoprotein released from damaged white matter inhibits axonal regeneration. *Mol. Cell. Neurosci.* **18**, 259–269.
46. Brittis, P. A. and Flanagan, J. G. (2001). Nogo domains and a Nogo receptor: Implications for axon regeneration. *Neuron* **30**, 11–14.
47. Vyas, A. A., Patel, H. V., Fromholt, S. E., Heffer-Lauc, M., Vyas, K. A., Dang, J., Schachner, M., and Schnaar, R. L. (2002). Gangliosides are functional nerve cell ligands for myelin-associated glycoprotein (MAG), an inhibitor of nerve regeneration. *Proc. Natl. Acad. Sci. USA* **99**, 8412–8417
48. Liu, B. P., Fournier, A., GrandPre, T., and Strittmatter, S. M. (2002). Myelin-associated glycoprotein as a functional ligand for the Nogo-66 receptor. *Science* **297**, 1190–1193.
49. Song, H., Ming, G., He, Z., Lehmann, M., McKerracher, L., Tessier-Lavigne, M., and Poo, M. (1998). Conversion of neuronal growth cone responses from repulsion to attraction by cyclic nucleotides. *Science* **281**, 1515–1518.
50. Lehmann, M., Fournier, A., Selles-Navarro, I., Dergham, P., Sebok, A., Leclerc, N., Tigyi, G., and McKerracher, L. (1999). Inactivation of Rho signaling pathway promotes CNS axon regeneration. *J. Neurosci.* **19**, 7537–7547.
51. Song, H. J. and Poo, M. M. (1999). Signal transduction underlying growth cone guidance by diffusible factors. *Curr. Opin. Neurobiol.* **9**, 355–363.
52. Hopker, V. H., Shewan, D., Tessier-Lavigne, M., Poo, M., and Holt, C. (1999). Growth-cone attraction to netrin-1 is converted to repulsion by laminin-1. *Nature* **401**, 69–73.
53. Castellani, V., Chedotal, A., Schachner, M., Faivre-Sarrailh, C., and Rougon, G. (2000). Analysis of the L1-deficient mouse phenotype reveals cross-talk between Sema3A and L1 signaling pathways in axonal guidance. *Neuron* **27**, 237–249.
54. Yu, T. W. and Bargmann, C. I. (2001). Dynamic regulation of axon guidance. *Nat. Neurosci.* **4** Suppl. 1169–1176.

CHAPTER 267

Semaphorins and their Receptors in Vertebrates and Invertebrates

Eric F. Schmidt, Hideaki Togashi, and Stephen M. Strittmatter
*Department of Neurology and Section of Neurobiology,
Yale University School of Medicine, New Haven, Connecticut*

The Semaphorin Family

Semaphorins are a large family of proteins originally identified as axon guidance factors of the developing nervous system. Over 30 family members are grouped into 8 classes based on structural and phylogenetic relationships (reviewed in [1]). Classes 1 and 2 are expressed in invertebrates, classes 3 through 7 are vertebrate semaphorins, and Class V is expressed in non-neurotropic DNA viruses. All semaphorins share a highly conserved 500 amino acid "Sema" domain at their amino terminus, but different classes possess divergent sequences in their carboxyl regions. Classes 1, 4, 5, and 6 are transmembrane proteins, class 7 has a GPI-anchor, and classes 2, 3, and V are secreted proteins. The presence of both membrane-bound and secreted semaphorins suggests that semaphorins act as both short- and long-range cues. In addition, the diversity of structural properties between the classes implies roles in a diversity of biological processes.

The best-documented function of semaphorins is their role in central nervous system (CNS) development. Semaphorins act as both repellents and attractants for growing axons. The first identified vertebrate semaphorin, Sema3A, causes retraction of axons and the collapse of the growth cone, a specialization at the tip of growing axons [2]. While repellents for certain neurons, Sema3C and 3F also serve as attractants for cortical and olfactory neurons, respectively. Within the large semaphorin family, certain semaphorins can exert antagonistic activity by competitively blocking the activity of other family members at certain receptors [3]. Semaphorins may guide growing dendrites as well as axons. Specifically, Sema3A attracts the apical dendrite of pyramidal neurons in the cerebral cortex toward the pial surface [4]. Many types of neurons are responsive to semaphorins, including dorsal root ganglion, sensory, motor, hippocampal, cortical, cerebellar, and olfactory. In addition to guiding axons and dendrites, semaphorins appear to play a role in fasciculation of nerve bundles, neuronal cell migration, axoplasmic transport, and apoptosis [5]. Like developing neurons, adult neurons of the regenerating CNS are responsive to semaphorins and semaphorin expression is upregulated after nerve injury [6].

Semaphorin signaling is not restricted to the CNS as evidenced by widespread expression throughout the embryo and adult tissue. Migrating non-neuronal cells are responsive to semaphorins, and cardiovascular abnormalities are observed when semaphorin signaling is disrupted [7]. In the immune system, expression of Sema4D (CD100) is regulated upon B- and T-lymphocyte activation and migrating monocytes are responsive to Sema3A and Sema4D [8]. Malignant lung cells show reduced levels and a cytoplasmic localization of semaphorins [9]. Taken together, it can be concluded that semaphorins act as guidance cues for many types of migrating cells in developmental, adult, and pathological tissue.

Receptors for Semaphorins

Neuropilins

Neuropilins are high-affinity transmembrane receptors for the secreted class 3 semaphorins in the CNS, but play no role in the activity of other semaphorins. A neuropilin family is composed of neuropilin-1 and several splice variants of neuropilin-2 [1]. Neuropilin-1 and -2 contain a number of conserved motifs on their extracellular domain including two CUB domains, FV/FVIII, and a MAM domain. The CUB domains are required for ligand binding and the MAM domain mediates neuropilin oligomerization [10]. Neuropilins have a short intracellular domain containing a PDZ binding motif that targets receptor localization to signaling components in the membrane of the cell [11]; however, the intracellular domain is not required to transduce the semaphorin signal [10]. The neuropilin isoforms bind differentially to various class 3 semaphorins [12], and the specificity of binding is determined by the CUB domains [10]. Although neuropilins are sufficient to bind class 3 semaphorins, the fact that the intracellular domain is not required for signaling suggests that a co-receptor transmits the semaphorin signal into the cell.

Outside of the CNS, neuropilins are found in the mesenchyme surrounding blood vessels and act as co-receptors for vascular endothelial growth factor (VEGF). Upon binding VEGF, neuropilins potentiate the kinase activity of the VEGF receptors flt-1 and KDR, resulting in endothelial cell migration [13]. There is some evidence to suggest that semaphorins and VEGF compete for neuropilin binding and a dysregulation of the competition may lead to pathological conditions [9].

Plexins

Plexins are the predominant receptors for membrane-bound, GPI-linked and viral semaphorins, and they bind to neuropilins to act as signaling co-receptors for the secreted class 3 semaphorins [14]. The initial discovery of Plexins as semaphorin receptors occurred with the identification of virus-encoded semaphorin protein receptor (VESPR; plexinC1) as a binding site for a class V semaphorin [15]. Currently, at least ten Plexins have been identified and are classified into four groups, Plexin A–D, which have different specificities for different semaphorins [1]. Plexins are distantly related to semaphorins since they possess the conserved Sema domain on their extracellular surface [16] and also share some sequence homology with the HGF receptor Met on their extracellular surface [17]. The intracellular domain of Plexin is highly conserved among family members, but is not significantly homologous to any known signaling motif. In their native state, Plexins are autoinhibited by their Sema domain and binding to semaphorin-neuropilin complexes or cleavage of the sema domain leads to activation of the protein and growth cone collapse in sensory neurons [18].

Intracellular Signaling Pathways

Actin Cytoskeleton and Monomeric GTPases

The actin cytoskeleton in growth cones undergoes dramatic rearrangement upon exposure to Sema3A. There is a relative decrease in F-actin within the lamellipodia [19] and actin colocalizes with neuropilin-1/PlexinA1 receptor complexes [20]. The actin reorganization is linked to increased endocytosis [20]. It was thought that semaphorins might regulate the actin cytoskeleton through monomeric G proteins due to the weak similarity of the conserved intracellular domain of Plexins to an R-Ras-GAP. However, no Plexin protein has been shown to possess GAP activity and semaphorin responses are not dependent on R-Ras. Instead, Rho family G proteins, namely Rac and Rho, seem to mediate the semaphorin response [21] (also reviewed in [22]). Active Rac binds directly to the intracellular domain of vertebrate and invertebrate PlexinB1, and this interaction is enhanced by the presence of ligand binding [23]. Activation of PlexinB1 appears to sequester active Rac from its endogenous substrate, p21-associated kinase, PAK [24–26]. RhoA is also activated as a result of PlexinB1, although it is not clear whether this is due to a direct or indirect action of PlexinB1 on RhoA [26], or is downstream of Rac-Plexin interactions [24]. Together, Rac sequestration and RhoA activation appear to mediate axon repulsion by PlexinB receptors.

Although the intracellular domain is highly conserved among all Plexin family members, it is not clear whether PlexinA functions in a similar fashion as PlexinB. PlexinA1 binds to both RhoD and Rnd1, and Rnd1 binding has been suggested to induce growth cone collapse [27,28], perhaps due to Rnd1-dependent inhibition of Rac [29]. Direct Rac-PlexinA interactions have not been demonstrated. It is possible that Plexins regulate monomeric GTPases indirectly by regulating Rho family GEFs and GAPs, factors that activate or inactivate monomeric GTPases, respectively.

A direct link between Sema3A and actin dynamics was recently demonstrated. Activated complexes of NP1 and PlexinA2 lead to the phosphorylation and deactivation of cofilin by LIM kinase [30]. Cofilin leads to F-actin turnover and plays a role in protrusion of lamellipodia and filopodia [31]. Further, LIM kinase is a substrate for both PAK and Rho kinase, which is consistent with the requirement for Rac and Rho, respectively, for Sema3A-induced collapse [31].

CRMP

Collapsin-response-mediator protein (CRMP) was identified in a *Xenopus* oocyte expression screen of mRNAs required for Sema3A responses [32]. The protein sequence of CRMP shares sequence homology with the *Caenorhabditis elegans* unc-33, a protein required for proper axonal pathfinding [33]. At least five isoforms of CRMP have been identified and they form heterotetramers *in vivo* [34]. Function blocking antibodies to CRMP block Sema3A-mediated growth cone collapse in chick DRG neurons [32], and CRMP is upregulated after axotomy of the sciatic [35] and olfactory [36] nerves. The mechanism of CRMP action is still unclear. Studies have shown that it is phosphorylated by Rho kinase, but this is not required for Sema3A-induced collapse [37]. The microtubule abnormalities of *unc-33* mutants and the observation that CRMP colocalizes with microtubules at certain stages of the cell cycle [38] suggest that Plexin/CRMP signaling may regulate microtubule dynamics. There is also evidence demonstrating CRMP binds to and inactivates phospholipase D2, an enzyme implicated in a variety of cell processes including actin dynamics, vesicle trafficking, and mitogenesis [39]. Finally, CRMP may act to mediate the cytoskeleton through Rho GTPases. Neuroblastoma cells overexpressing CRMP and constitutively active RhoA showed a Rac1-like morphology, whereas cells co-expressing CRMP and active Rac1 showed a RhoA-like morphology [40]. Thus, it seems likely that CRMP enhances the function of a Plexin-Rho family G-protein axis in axon repulsion.

Protein Phosphorylation

Receptors for several other axon guidance molecules, such as ephrins and neurotrophins, act via kinase cascades. Although semaphorin receptors themselves show no kinase activity, indirect evidence suggests that protein phosphorylation occurs and is required [41]. The involvement of PAK and LIM kinase downstream of Plexins is mentioned above. A recent study has shown that *Drosophila* PlexinA associates with the membrane-bound receptor tyrosine kinase-related protein Off-Track (Otk; [42]). In addition, two proteins with kinase activity have been co-purified with CRMP [38,43]. The serine/threonine kinase, glycogen synthase kinase (GSK)-3 is activated as a result of Sema3A in both neuronal cells and human breast cancer cells, and GSK-3 inhibitors prevent Sema3A-induced growth cone collapse [44].

Other Signaling Mechanisms

Another pathway that has been implicated in semaphorin signaling may utilize heterotrimeric G proteins. Much of the evidence for this has come from experiments with pertussis toxin (PTX), which blocks G-protein function [21,45]. Indeed, neuropilins were found to bind to a Gα-interacting protein (GIPC, SEMPCAP-1) that associates with a regulator of G-protein signaling (RGS) protein via its PDZ domains [11]. Interestingly, some transmembrane semaphorins interact with SEMCAP-1 as well, suggesting that semaphorins may act as receptors to transduce signals into the cell [46]. Semaphorin reverse signaling is further supported by the fact that Sema6B binds to Src both *in vitro* and *in vivo* [47], Sema4D interacts with a serine kinase [48], and Sema6A binds to the actin binding protein EVL [49].

Semaphorin-mediated signaling can be modulated by other pathways. Cyclic nucleotides can alter the response of growth cones to various signaling molecules [50,51]. Increasing levels of cGMP switches Sema3A responses from repulsion to attraction and decreasing cGMP potentiates the repulsive activity of Sema3A. Apical dendrites of cerebral cortical neurons are attracted to Sema3A while the axons of the same cells are repelled [4]. Remarkably, soluble guanylate cyclase (SGC) is asymmetrically localized to the dendrites of these cells, implicating an endogenous regulation of cGMP *in vivo*. The cell adhesion molecule, L1, is also able to modulate growth cone responses to Sema3A [52]. DRG neurons from L1-deficient mice show no response to Sema3A and soluble L1 protein switched repulsion to attraction. Finally, one or more of these pathways may impinge on protein synthesis and degradation within axons. Evidence indicates that local regulation of protein levels participates in multiple growth cone responses [53].

Semaphorin Signaling in the Immune System

Semaphorin signaling in activated lymphocytes does not rely on neuropilins and Plexins, but utilizes a different receptor called CD72 [54]. Under normal conditions, CD72 is phosphorylated on its intracellular domain by a Src tyrosine kinase. This phosphorylation leads to the recruitment of an SH2-domain-containing tyrosine phosphatase SHP-1, which then dephosphorylates and inactivates signaling proteins involved in lymphocyte activation [8]. Sema4D binding prevents the phosphorylation of CD72 therefore potentiating lymphocyte activation [54]. The migration of monocytes is inhibited by Sema4D and Sema3A; this effect is likely to be mediated via Plexins and neuropilins since monocytes do not express CD72 [55].

Summary and Future Directions

Many biological systems in the developing embryo and adult animal are dependent on semaphorin signaling. Although the importance of semaphorins in the developing CNS is well documented, their involvement in the immune response, the cardiovascular system, and in pathology is still being clarified. Most of the work to date has focused on the identification and classification of the various semaphorin families and their receptors, with less clarification of downstream signaling mechanisms. Regulation of the cytoskeleton is the most obvious effect of semaphorin signaling, and a number of studies have demonstrated a signaling connection of semaphorin receptors with actin filaments and microtubules. In particular, Rho family G proteins and CRMP appear to play major roles in this connection (see Fig. 1).

Acknowledgments

This work was supported by grants to Stephen M. Strittmatter from the NIH. Eric F. Schmidt is supported by an institutional NIH predoctoral training grant. Stephen M. Strittmatter is an Investigator of the Patrick and Catherine Weldon Donaghue Medical Research Foundation.

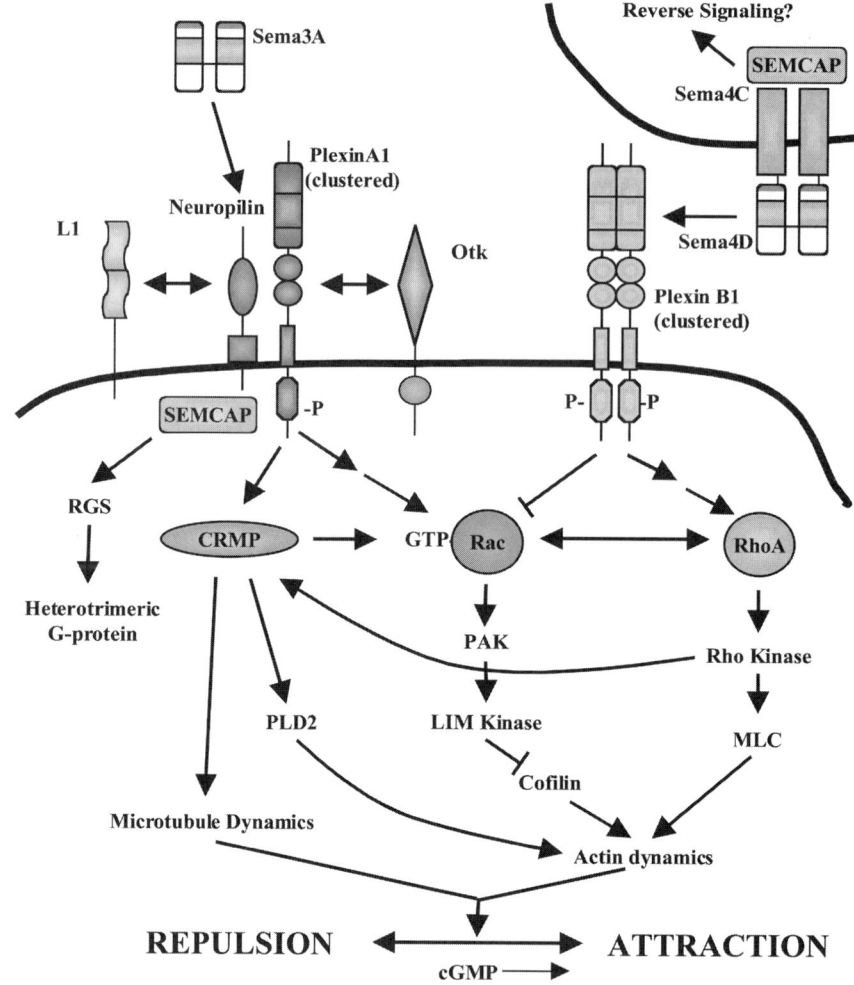

Figure 1 A schematic representation of semaphorin signaling in neuronal development. See text for details.

References

1. Nakamura, F., Kalb, R. G., and Strittmatter, S. M. (2000). Molecular basis of semaphorin-mediated axon guidance. *J. Neurobiol.* **44**, 219–229.
2. Luo, Y., Raible, D., and Raper, J. A. (1993). Collapsin: A protein in brain that induces the collapse and paralysis of neuronal growth cones. *Cell* **75**, 217–227.
3. Takahashi, T., Nakamura, F., Jin, Z., Kalb, R. G., and Strittmatter, S. M. (1998). Semaphorins A and E act as antagonists of neuropilin-1 and agonists of neuropilin-2 receptors. *Nat. Neurosci.* **1**, 487–493.
4. Polleux, F., Morrow, T., and Ghosh, A. (2000). Semaphorin 3A is a chemoattractant for cortical apical dendrites. *Nature* **404**, 567–573.
5. He, Z., Wang, K. C., Koprivica, V., Ming, G., and Song, H. J. (2002). Knowing how to navigate: mechanisms of semaphorin signaling in the nervous system. *Sci. STKE* **2002**, RE1.
6. Pasterkamp, R. J. and Verhaagen, J. (2001). Emerging roles for semaphorins in neural regeneration. *Brain Res. Brain Res. Rev.* **35**, 36–54.
7. Brown, C. B. *et al.* (2001). PlexinA2 and semaphorin signaling during cardiac neural crest development. *Development* **128**, 3071–3080.
8. Bismuth, G. and Boumsell, L. (2002). Controlling the immune system through semaphorins. *Sci. STKE* **2002**, RE4.
9. Brambilla, E., Constantin, B., Drabkin, H., and Roche, J. (2000). Semaphorin SEMA3F localization in malignant human lung and cell lines: A suggested role in cell adhesion and cell migration. *Am. J. Pathol.* **156**, 939–950.
10. Nakamura, F., Tanaka, M., Takahashi, T., Kalb, R. G., and Strittmatter, S. M. (1998). Neuropilin-1 extracellular domains mediate semaphorin D/III-induced growth cone collapse. *Neuron* **21**, 1093–1100.
11. Cai, H. and Reed, R. R. (1999). Cloning and characterization of neuropilin-1-interacting protein: A PSD-95/Dlg/ZO-1 domain-containing protein that interacts with the cytoplasmic domain of neuropilin-1. *J. Neurosci.* **19**, 6519–6527.
12. Feiner, L., Koppel, A. M., Kobayashi, H., and Raper, J. A. (1997). Secreted chick semaphorins bind recombinant neuropilin with similar affinities but bind different subsets of neurons in situ. *Neuron* **19**, 539–545.
13. Tamagnone, L. and Comoglio, P. M. (2000). Signalling by semaphorin receptors: cell guidance and beyond. *Trends Cell Biol.* **10**, 377–383.
14. Takahashi, T., Fournier, A., Nakamura, F., Wang, L. H., Murakami, Y., Kalb, R. G., Fujisawa, H., and Strittmatter, S. M. (1999). Plexin-neuropilin-1 complexes form functional semaphorin-3A receptors. *Cell* **99**, 59–69.
15. Comeau, M. R. *et al.* (1998). A poxvirus-encoded semaphorin induces cytokine production from monocytes and binds to a novel cellular semaphorin receptor, VESPR. *Immunity* **8**, 473–482.
16. Winberg, M. L., Noordermeer, J. N., Tamagnone, L., Comoglio, P. M., Spriggs, M. K., Tessier-Lavigne, M., and Goodman, C. S. (1998). Plexin A is a neuronal semaphorin receptor that controls axon guidance. *Cell* **95**, 903–916.
17. Bork, P., Doerks, T., Springer, T. A., and Snel, B. (1999). Domains in plexins: links to integrins and transcription factors. *Trends Biochem. Sci.* **24**, 261–3.

18. Takahashi, T. and Strittmatter, S. M. (2001). Plexina1 autoinhibition by the plexin sema domain. *Neuron* **29**, 429–439.
19. Fan, J., Mansfield, S. G., Redmond, T., Gordon-Weeks, P. R., and Raper, J. A. (1993). The organization of F-actin and microtubules in growth cones exposed to a brain-derived collapsing factor. *J. Cell Biol.* **121**, 867–878.
20. Fournier, A. E., Nakamura, F., Kawamoto, S., Goshima, Y., Kalb, R. G., and Strittmatter, S. M. (2000). Semaphorin3A enhances endocytosis at sites of receptor-F-actin colocalization during growth cone collapse. *J. Cell Biol.* **149**, 411–422.
21. Jin, Z. and Strittmatter, S. M. (1997). Rac1 mediates collapsin-1-induced growth cone collapse. *J. Neurosci.* **17**, 6256–6263.
22. Liu, B. P. and Strittmatter, S. M. (2001). Semaphorin-mediated axonal guidance via Rho-related G proteins. *Curr. Opin. Cell Biol.* **13**, 619–626.
23. Vikis, H. G., Li, W., He, Z., and Guan, K. L. (2000). The semaphorin receptor plexin-B1 specifically interacts with active Rac in a ligand-dependent manner. *Proc. Natl. Acad. Sci. USA* **97**, 12457–12462.
24. Driessens, M. H., Hu, H., Nobes, C. D., Self, A., Jordens, I., Goodman, C. S., and Hall, A. (2001). Plexin-B semaphorin receptors interact directly with active Rac and regulate the actin cytoskeleton by activating Rho. *Curr. Biol.* **11**, 339–344.
25. Vikis, H. G., Li, W., and Guan, K. L. (2002). The plexin-B1/Rac interaction inhibits PAK activation and enhances Sema4D ligand binding. *Genes Dev.* **16**, 836–845.
26. Hu, H., Marton, T. F., and Goodman, C. S. (2001). Plexin B mediates axon guidance in *Drosophila* by simultaneously inhibiting active Rac and enhancing RhoA signaling. *Neuron* **32**, 39–51.
27. Zanata, S. M., Hovatta, I., Rohm, B., and Puschel, A. W. (2002). Antagonistic effects of Rnd1 and RhoD GTPases regulate receptor activity in Semaphorin 3A-induced cytoskeletal collapse. *J. Neurosci.* **22**, 471–477.
28. Rohm, B., Rahim, B., Kleiber, B., Hovatta, I., and Puschel, A. W. (2000). The semaphorin 3A receptor may directly regulate the activity of small GTPases. *FEBS Lett.* **486**, 68–72.
29. Nobes, C. D., Lauritzen, I., Mattei, M. G., Paris, S., Hall, A., and Chardin, P. (1998). A new member of the Rho family, Rnd1, promotes disassembly of actin filament structures and loss of cell adhesion. *J. Cell Biol.* **141**, 187–197.
30. Aizawa, H. *et al.* (2001). Phosphorylation of cofilin by LIM-kinase is necessary for semaphorin 3A-induced growth cone collapse. *Nat. Neurosci.* **4**, 367–373.
31. Kuhn, T. B. *et al.* (2000). Regulating actin dynamics in neuronal growth cones by ADF/cofilin and rho family GTPases. *J. Neurobiol.* **44**, 126–144.
32. Goshima, Y., Nakamura, F., Strittmatter, P., and Strittmatter, S. M. (1995). Collapsin-induced growth cone collapse mediated by an intracellular protein related to UNC-33. *Nature* **376**, 509–514.
33. Li, W., Herman, R. K., and Shaw, J. E. (1992). Analysis of the *Caenorhabditis elegans* axonal guidance and outgrowth gene unc-33. *Genetics* **132**, 675–689.
34. Wang, L. H. and Strittmatter, S. M. (1997). Brain CRMP forms heterotetramers similar to liver dihydropyrimidinase. *J. Neurochem.* **69**, 2261–2269.
35. Minturn, J. E., Fryer, H. J., Geschwind, D. H., and Hockfield, S. (1995). TOAD-64, a gene expressed early in neuronal differentiation in the rat, is related to unc-33, a *C. elegans* gene involved in axon outgrowth. *J. Neurosci.* **15**, 6757–6766.
36. Pasterkamp, R. J., De Winter, F., Holtmaat, A. J., and Verhaagen, J. (1998). Evidence for a role of the chemorepellent semaphorin III and its receptor neuropilin-1 in the regeneration of primary olfactory axons. *J. Neurosci.* **18**, 9962–9976.
37. Arimura, N. *et al.* (2000). Phosphorylation of collapsin response mediator protein-2 by Rho-kinase. Evidence for two separate signaling pathways for growth cone collapse. *J. Biol. Chem.* **275**, 23973–23980.
38. Gu, Y. and Ihara, Y. (2000). Evidence that collapsin response mediator protein-2 is involved in the dynamics of microtubules. *J. Biol. Chem.* **275**, 17917–17920.
39. Lee, S. *et al.* (2002). Collapsin response mediator protein-2 inhibits neuronal phospholipase D(2) activity by direct interaction. *J. Biol. Chem.* **277**, 6542–6549.
40. Hall, C., Brown, M., Jacobs, T., Ferrari, G., Cann, N., Teo, M., Monfries, C., and Lim, L. (2001). Collapsin response mediator protein switches RhoA and Rac1 morphology in N1E-115 neuroblastoma cells and is regulated by Rho kinase. *J. Biol. Chem.* **276**, 43482–43486.
41. Tamagnone, L. *et al.* (1999). Plexins are a large family of receptors for transmembrane, secreted, and GPI-anchored semaphorins in vertebrates. *Cell* **99**, 71–80.
42. Winberg, M. L., Tamagnone, L., Bai, J., Comoglio, P. M., Montell, D., and Goodman, C. S. (2001). The transmembrane protein Off-track associates with Plexins and functions downstream of Semaphorin signaling during axon guidance. *Neuron* **32**, 53–62.
43. Inatome, R., Tsujimura, T., Hitomi, T., Mitsui, N., Hermann, P., Kuroda, S., Yamamura, H., and Yanagi, S. (2000). Identification of CRAM, a novel unc-33 gene family protein that associates with CRMP3 and protein-tyrosine kinase(s) in the developing rat brain. *J. Biol. Chem.* **275**, 27291–27302.
44. Eickholt, B. J., Walsh, F. S., and Doherty, P. (2002). An inactive pool of GSK-3 at the leading edge of growth cones is implicated in Semaphorin 3A signaling. *J. Cell Biol.* **157**, 211–217.
45. Igarashi, M., Strittmatter, S. M., Vartanian, T., and Fishman, M. C. (1993). Mediation by G proteins of signals that cause collapse of growth cones. *Science* **259**, 77–79.
46. Wang, L. H., Kalb, R. G., and Strittmatter, S. M. (1999). A PDZ protein regulates the distribution of the transmembrane semaphorin, M-SemF. *J. Biol. Chem.* **274**, 14137–14146.
47. Eckhardt, F., Behar, O., Calautti, E., Yonezawa, K., Nishimoto, I., and Fishman, M. C. (1997). A novel transmembrane semaphorin can bind c-src. *Mol. Cell. Neurosci.* **9**, 409–419.
48. Elhabazi, A., Lang, V., Herold, C., Freeman, G. J., Bensussan, A., Boumsell, L., and Bismuth, G. (1997). The human semaphorin-like leukocyte cell surface molecule CD100 associates with a serine kinase activity. *J. Biol. Chem.* **272**, 23515–23520.
49. Klostermann, A., Lutz, B., Gertler, F., and Behl, C. (2000). The orthologous human and murine semaphorin 6A-1 proteins (SEMA6A-1/Sema6A-1) bind to the enabled/vasodilator-stimulated phosphoprotein-like protein (EVL) via a novel carboxyl-terminal zyxin-like domain. *J. Biol. Chem.* **275**, 39647–39653.
50. Song, H., Ming, G., He, Z., Lehmann, M., McKerracher, L., Tessier-Lavigne, M., and Poo, M. (1998). Conversion of neuronal growth cone responses from repulsion to attraction by cyclic nucleotides. *Science* **281**, 1515–1518.
51. Ming, G., Henley, J., Tessier-Lavigne, M., Song, H., and Poo, M. (2001). Electrical activity modulates growth cone guidance by diffusible factors. *Neuron* **29**, 441–452.
52. Castellani, V., Chedotal, A., Schachner, M., Faivre-Sarrailh, C., and Rougon, G. (2000). Analysis of the L1-deficient mouse phenotype reveals cross-talk between Sema3A and L1 signaling pathways in axonal guidance. *Neuron* **27**, 237–249.
53. Campbell, D. S. and Holt, C. E. (2001). Chemotropic responses of retinal growth cones mediated by rapid local protein synthesis and degradation. *Neuron* **32**, 1013–1026.
54. Kumanogoh, A. *et al.* (2000). Identification of CD72 as a lymphocyte receptor for the class IV semaphorin CD100: a novel mechanism for regulating B cell signaling. *Immunity* **13**, 621–631.
55. Delaire, S., Billard, C., Tordjman, R., Chedotal, A., Elhabazi, A., Bensussan, A., and Boumsell, L. (2001). Biological activity of soluble CD100. II. Soluble CD100, similarly to H-SemaIII, inhibits immune cell migration. *J. Immunol.* **166**, 4348–4354.

CHAPTER 268

Signaling Pathways that Regulate Neuronal Specification in the Spinal Cord

Ann E. Leonard and Samuel L. Pfaff

Gene Expression Laboratory,
The Salk Institute for Biological Studies, La Jolla, California

A cascade of signaling events triggers the differentiation of specific neuronal and glial cell populations that comprise the central nervous system (CNS). Epidermal ectoderm deprived of bone morphogenic protein (BMP) signaling differentiates into "neural" ectoderm [1], the precursor of the CNS. These neural cells are multipotential and respond to signals in their environment in order to generate the appropriate types of neurons and glia at the correct positions. In this chapter we focus on the spinal cord, the most caudal region of the CNS, since it has served as a useful model in which to investigate signaling events that give rise to neuronal and glial populations within the developing neural tube.

Emerging from a combination of modern molecular studies and classical cellular studies, a central theme in spinal cord development is one in which inductive factors signal along the dorsoventral and rostrocaudal axes of the developing spinal cord to specify cell fate in a Cartesian-coordinate-like manner [2]. This signaling leads to the generation of dorsal spinal cord interneurons that process sensory information and relay it to the brain, while the ventral spinal cord forms interneurons and motor neurons involved in locomotor control (Fig. 1A). Along the rostrocaudal axis, discontinuous subclasses of motor neurons are generated in register with the peripheral targets that they innervate. In addition, numerous glial cell types are formed including the roof plate and floor plate, which act as organizing centers within the spinal cord, and astrocytes and oligodendrocytes, which support neuronal function and myelinate neurons, respectively.

Patterning along the Dorsoventral Axis

Two classes of factors play prominent roles in specifying distinct cell types along the dorsoventral axis of the spinal cord: members of the transforming growth factor β (TGFβ) superfamily acting dorsally, and Sonic hedgehog (Shh) ventrally (Fig. 1A) [3,4]. TGFβ signaling from the epidermal ectoderm flanking the dorsal neural tube leads to the differentiation of the roof plate [5], and Shh expression from the notochord below the neural tube triggers the formation of the floor plate [6]. These two glial structures in the spinal cord then express TGFβs dorsally and Shh ventrally. In this way, signals from the periphery are propagated into the spinal cord to control cell differentiation locally.

The dividing progenitor cells within the ventricular (medial) region of the spinal cord monitor the types and concentrations of TGFβs and Shh in order to determine their position, and consequently their fate, as they become postmitotic and migrate laterally into the mantle region (Fig. 1A). The signaling pathways triggered by these inductive factors lead to the activation of transcriptional networks that first define distinct domains along the dorsoventral axis of the ventricular zone, and ultimately lead to the expression of genes involved in controlling cell function (Table I) [7–10].

Dorsal Spinal Cord Development

In the embryonic dorsal spinal cord, four classes of interneurons (INs) termed D1–D4 arise in an orderly fashion

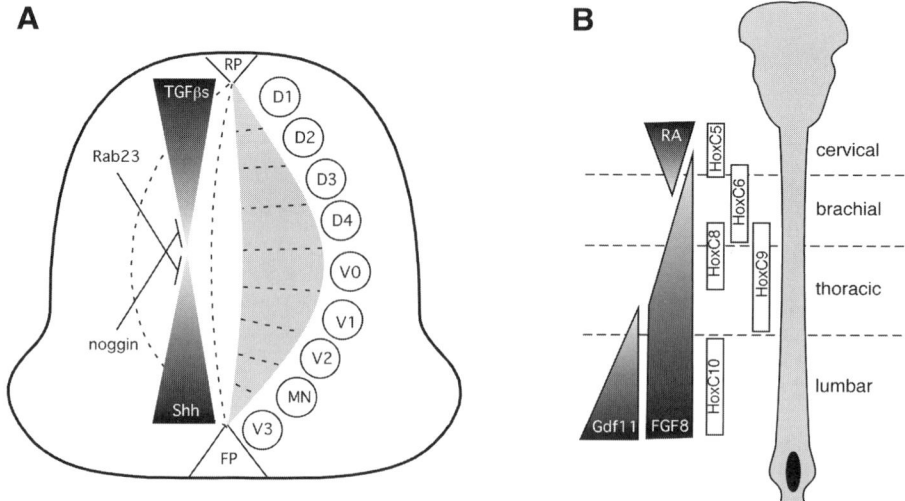

Figure 1 Patterning signals along the dorsoventral and rostrocaudal axes of the developing spinal cord. (A) Transverse view of the spinal cord in which the gray area represents the ventricular zone containing progenitor cells and the white area represents the mantle layer with mature cell types. FP, floor plate; RP, roof plate. (B) Rostrocaudal view of the spinal cord and the graded pattern of signals expressed by paraxial and axial mesoderm adjacent to the neural tube.

from specific regions of the progenitor zone (Fig. 1A). These INs, while not absolutely defined, consist of association and commissural cells that process and relay sensory information and depend upon the formation of the roof plate in order to develop [5,11]. The roof plate and adjacent neural epithelial cells express overlapping and nested combinations of several TGFβ members including BMP4/5/7, Gdf6/7, and Dsl1 [5,9]. How might the TGFβs produce different cell types in the spinal cord? Several strategies are likely to be involved including quantitative, qualitative, and timing differences in TGFβ activity. The expression pattern of the TGFβs suggests that a high-dorsal to low-ventral gradient of these proteins is present within the neural tube (Fig. 1A). Although more studies are needed to determine whether graded levels of TGFβs contribute to spinal cord patterning, *in vitro* experiments with neural explants have detected concentration-dependent activities for Activin A in the induction of D1 and D2 IN classes [5].

The clearest example of how TGFβs trigger the differentiation of specific IN types is based on the finding that individual members of this family have different qualitative activities [5]. The most convincing evidence for such a mechanism is found in the *Gdf7* mutant mice where a specific subpopulation of D1 INs fails to be generated [12]. These results suggest that the specificity of dorsal IN patterning is mediated, at least in part, by various TGFβ signaling molecules, some of which act directly to render class specificity (Table I). An additional mechanism for generating cellular diversity in the dorsal spinal cord involves a temporal switch in the way neuroepithelial cells respond to TGFβ signals. Early in development, progenitor cells produce neural crest cells when exposed to BMP4 or Activin A, but later, they give rise to dorsal INs in response to the same signals [5].

The basis for qualitative differences in TGFβ signaling and the mechanisms underlying the developmental switch in TGFβ responsiveness by neural epithelial cells remain important questions.

Shh signaling for ventral cell differentiation is attenuated by TGFβ signaling [13]. What limits the range of TGFβ activity to the appropriate areas of the developing spinal cord? Several TGFβ antagonists have been identified including noggin, chordin, and follistatin which bind directly to and sequester-specific TGFβs [14]. These antagonists are expressed by the somites and notochord near the ventral surface of the neural tube, and therefore are expected to limit the exposure of ventral cells to certain TGFβs. In *noggin* mutant mice, TGFβ signaling in the ventral neural tube is unmasked (Fig. 1A), which leads to a progressive loss of ventral cell differentiation [15].

The receptors of the TGFβs are serine/threonine kinases comprised of type I and type II dimers. These receptor complexes have not been well characterized in the spinal cord, but may select for different ligands and serve as the basis for the qualitative differences in cell differentiation induced by different TGFβ family members. The best known transducers of TGFβ signaling are the SMAD transcription factors [16], though the role of SMADs in spinal cord development also requires further characterization. Recently, a better understanding of the downstream targets of TFGβs has begun to emerge (Table I). For instance, it is now known that D1 INs, characterized postmitotically by the markers Lhx2/9, arise from progenitor cells that express the bHLH transcription factor mATH1 involved in establishing the fate of these cells [9,10]. Likewise, D3 IN progenitor cells marked by Lhx1/5 arise from progenitors that express the bHLH protein Ngn1. Thus, the identification of target genes activated by TGFβ

Table I Signaling Events in the Developing Spinal Cord

	Dorsal patterning Lateral ectoderm → Roof plate		Ventral patterning Notochord → Floor plate			Rostral/Caudal patterning Hensen's node, paraxial mesoderm			
Signal	TGFβ Superfamily BMP 4, 5, 6, 7 Gdf 6, 7 Activin B; Dsl1		Sonic Hedgehog			Retinoic Acid		FGF 8	Gdf 11
Inhibitors	Follistatin, Noggin, Chordin		Hedgehog-interacting Protein (Hip) Rab23 (vesicle transporter)			Cellular Retinoic Acid Binding Protein			Follistatin, Noggin, Chordin
Receptor(s)	Type I/II TGFβ receptors (serine/threonine receptor kinases)		Patched : ligand binding Smoothened : transducing			Nuclear Receptors RAR RXR		FGFR (receptor tyrosine kinase)	TGFβR (receptor serine/threonine kinase)
Signal transduction components	SMAD transcription factor family		Gli transcription factor family Gli 1, 2, 3			RAR RXR Cdx transcription factor family		Cdx transcription factor family	SMADs
Downstream transcription factors (ventricular zone)	*Progenitor Domain* pD1 pD3	*Transcription Factors* Math1 Ngn1	*Progenitor Domain* pV0 pV1 pV2 pMN pV3 pOlig	*Class I* Dbx1 Dbx2 Irx3 Pax6 – –	*Class II* ? Nkx6.2 Nkx6.1 Olig2 + MNR2 Nkx2.2 Nkx2.2 + Olig2	*Rostral/Caudal Level* hindbrain cervical brachial thoracic lumbar		*Transcription Factors* HoxB1/2/3/4 HoxC5 HoxC6/8 HoxC8/9 HoxC10	
Differentiated cell type (mantal zone)	*Cell Type* D1 IN D2 IN D3 IN D4 IN roof plate astrocytes	*Functional Markers* Lhx2/9 (Lh2A,B) Isl1 Lhx1/5 (Lim1, 2) Lmx1 Gdf7 GFAP	*Cell Type* V0 IN V1 IN V2 IN MN V3 IN floor plate oligodendrocytes astrocytes	*Functional Markers* Evx1/2 En1 Lhx3, Chox10 Isl1/2, HB9 Sim1 HNF3β PDGFRα GFAP		*MN Subtypes* columnar pool hindbrain visceral types		*Functional Markers* LIM transcription family ETS transcription family Phox transcription family	

signaling should help to work backward to characterize the signal transduction pathways.

Ventral Spinal Cord Development

Genetic studies as well as *in vitro* explant experiments have implicated Shh in the differentiation of ventral spinal cord cell types involved in locomotor control (V0–V3 INs and MNs), as well as oligodendrocytes (Fig. 1A) [3,8]. Unlike the nested combinations of TGFβ molecules in the dorsal spinal cord; however, only one hedgehog member appears to be involved in ventral spinal cord patterning in higher vertebrates. This raises the question of how different ventral cell types are induced by a single factor. Extensive studies with *in vitro* explants have shown that Shh concentration differences of approximately two- to threefold dramatically influence the types of cells that are triggered to differentiate. Decreasing concentrations of Shh progressively induce cell types found further from the ventral midline, recapitulating the normal organization of cells in the ventral spinal cord [17]. As with TGFβ signaling, there are also important temporal mechanisms that modify progenitor cell responses to Shh signaling during development. At early stages Shh acts on progenitor cells to trigger MN differentiation, but, later in development, oligodendrocytes are produced instead of MNs. The basis for this switch is not well understood but seems to involve the regulation of the transcription factor Nkx2.2 [18,19].

The active Shh signaling molecule is autoprocessed and cholesterol-modified, and binds to the patched/smoothened receptor complex [20]. In the absence of Shh, patched is thought to inhibit smoothened from signaling, and this inhibition is relieved when Shh binds patched. Many additional components of the Shh signaling pathway have been identified through genetic studies in *Drosophila*, including the downstream Gli family of zinc finger transcription factors [20]. Genetic studies of *Gli3* and *Gli3/Shh* compound mouse mutants indicate that this transcription factor is likely an intermediary in the Shh pathway, although it seems to function indirectly as a transcriptional repressor [21,22].

How might small gradations in the level of Shh signaling produce sharp progenitor cell domains that serve as the precursors for different ventral cell types? Studies of the factors regulated by Shh in the ventricular zone have uncovered a network of homeodomain proteins that mark distinct progenitor domains (Table I) [7]. The expression of these factors is controlled at two levels. First, Shh either represses (class I) or activates (class II) the expression of the homeodomain factors. If this were the only mechanism operating to control these factors, it might be expected that the interpretation of the fine Shh gradient would lead to imprecise boundaries of gene expression. However, the domains appear to be further refined by cross-repressive transcriptional interactions between factors from different domains. In this two-step manner, graded Shh leads to the activation of unique combinations of homeodomain transcription factors in precise progenitor cell domains [3,8]. The combinatorial activities of these homeodomain factors lead to the activation of downstream transcriptional regulators involved in cell specification and function (Table I) [23].

The opposing nature of the ventral Shh gradient meeting the dorsal TGFβ factors suggests that inhibitors of Shh activity might constrain its activity, much like the inhibitors of TGFβs. Hedgehog interacting protein (Hip) is a surface membrane protein that binds Shh and attenuates its activity [24]. In addition, characterization of the mouse *open brain* (*opb*) mutant, in which ventral cell types form inappropriately in the dorsal region of the spinal cord, has led to the identification of a member of the Rab family of vesicular transporters, Rab23, important in limiting the activity of Shh dorsally (Fig. 1A) [25]. Interestingly, mice deficient in both Shh and Rab23 regain many of the ventral cell types lost in Shh mutants. This, together with the observation that *Gli3/Shh* double mutants also regain many ventral cell types [22], suggests additional Shh-independent pathways might contribute to ventral spinal cord development. Studies to understand the basis for Shh-independent signaling have uncovered a parallel pathway involving retinoic acid (RA) expressed by paraxial mesoderm beside the neural tube [26]. It will be interesting to examine in more detail the interplay between Shh and RA signaling pathways in order to fully understand the molecular basis for neuronal specification along the dorsoventral axis of the neural tube.

Rostrocaudal Specification

The spinal cord can be subdivided into four broad, functional regions along the rostrocaudal axis: cervical, brachial, thoracic, and lumbosacral. The IN classes of the spinal cord extend continuously throughout these regions, while specific MN subclasses are found at each level [27]. Individual MN subclasses form discontinuous columns in register with their targets, such that MNs of the cervical region innervate axial muscles, brachial region MNs innervate the forelimb, MNs of the thoracic region innervate body wall muscle, and lumbar MNs innervate hindlimbs. Much like the initiation of dorsoventral patterning in the spinal cord, embryonic manipulations and *in vitro* explant studies suggest that members of several families of signaling molecules originating initially from sources outside the spinal cord contribute to the diversification process that leads to the generation of specific classes of MNs along the rostrocaudal axis [28–31].

Studies of the signals that control segmental identity along the rostrocaudal axis have used *Hox* gene expression patterns as downstream molecular correlates of the regional specification of cell identity (Table I). Furthermore, there is increasing functional data to suggest that *Hox* genes contribute to the proper development of MN subclasses [27,32,33]. As neuroepithelial cell identity is first established, it is thought to have a rostral identity which is then modified by "caudalizing" signals [4]. Hindbrain studies have found that increasing levels of RA activate more caudal-type *Hox* genes [34–36].

Likewise, the pattern of *Hox* gene expression in the cervical spinal cord is regulated by RA synthesized by the cervical paraxial mesoderm flanking the neural tube (Fig. 1B) [31].

However, at more caudal regions of the spinal cord, RA is insufficient to confer positional identity. A major source of additional regionalizing signals is detected in Hensen's node (HN), a precursor of the axial mesoderm that moves in a caudal direction below the nascent neural tube as development progresses. Interestingly, HN tissue taken from different stages (i.e., different rostrocaudal levels) is able to specify different regional values in neural explants [31]. Studies utilizing fibroblast growth factor (FGF) receptor antagonist SU5402 and expression of constitutively active FGF receptors are found to alter the Hox coding in neural cells, implicating FGF signaling as a mediator of HN activity. FGF8 is expressed by the HN and *in vitro* studies have found that this factor can act in a concentration-dependent manner to induce progressively more caudal positional values in neural explants. The Cdx family of transcription factors represents possible downstream mediators of both RA and FGF8 signaling in the regulation of Hox expression [36].

Taken together, these findings suggest that the signaling activity of FGF8 increases as the HN moves caudally (Fig. 1B) [31]. An additional mechanism that appears to contribute to the increased activity of FGF8 at more caudal positions is the involvement of accessory factors that enhance FGF signaling. One such example is the TGFβ superfamily member Gdf11. This factor is expressed in HN as it progresses through lumbosacral levels, where FGF signaling is expected to be highest (Fig. 1B). Unlike other TGFβ members, Gdf11 does not influence the dorsoventral pattern of the spinal cord, but rather acts to enhance FGF8 signaling activity. In this way, progressively more caudal regions of the spinal cord are defined by the composite functions of FGF8 and Gdf11 through the regulation of Hox codes involved in establishing regional levels of the spinal cord that will generate different MN subclasses (Table I).

References

1. Weinstein, D. C. and Hemmati-Brivanlou, A. (1999). Neural induction. *Annu. Rev. Cell. Dev. Biol.* **15**, 411–433.
2. Jessell, T. M. and Lumsden, A. (1997). Inductive signals and the assignment of cell fate in the spinal cord and hindbrain: An axial coordinate system for neural patterning, in Cowan, W. M., Jessell, T. M., and Zipursky, S. L., Eds., *Molecular and Cellular Approaches to Neural Development*, pp. 290–333. Oxford University Press, New York.
3. Jessell, T. M. (2000). Neuronal specification in the spinal cord: Inductive signals and transcriptional codes. *Nat. Rev. Genet.* **1**(1), 20–29.
4. Tanabe, Y. and Jessell, T. M. (1996). Diversity and pattern in the developing spinal cord. *Science* **274**(5290), 1115–1123.
5. Liem, K. F., Jr., Tremml, G., and Jessell, T. M. (1997). A role for the roof plate and its resident TGFbeta-related proteins in neuronal patterning in the dorsal spinal cord. *Cell* **91**(1), 127–138.
6. Placzek, M., Dodd, J., and Jessell, T. M. (2000). Discussion point. The case for floor plate induction by the notochord. *Curr. Opin. Neurobiol.* **10**(1), 15–22.
7. Briscoe, J. *et al.* (2000). A homeodomain protein code specifies progenitor cell identity and neuronal fate in the ventral neural tube. *Cell* **101**(4), 435–445.
8. Briscoe, J. and Ericson, J. (2001). Specification of neuronal fates in the ventral neural tube. *Curr. Opin. Neurobiol.* **11**(1), 43–49.
9. Lee, K. J. and Jessell, T. M. (1999). The specification of dorsal cell fates in the vertebrate central nervous system. *Annu. Rev. Neurosci.* **22**, 261–294.
10. Gowan, K. *et al.* (2001). Cross-inhibitory activities of Ngn1 and Math1 allow specification of distinct dorsal interneurons. *Neuron* **31**(2), 219–232.
11. Lee, K. J., Dietrich, P., and Jessell, T. M. (2000). Genetic ablation reveals that the roof plate is essential for dorsal interneuron specification. *Nature* **403**(6771), 734–740.
12. Lee, K. J., Mendelsohn, M., and Jessell, T. M. (1998). Neuronal patterning by BMPs: A requirement for GDF7 in the generation of a discrete class of commissural interneurons in the mouse spinal cord. *Genes Dev.* **12**(21), 3394–3407.
13. Liem, K. F., Jr., Jessell, T. M., and Briscoe, J., (2000). Regulation of the neural patterning activity of sonic hedgehog by secreted BMP inhibitors expressed by notochord and somites. *Development* **127**(22), 4855–4866.
14. Harland, R. (2000). Neural induction. *Curr. Opin. Genet. Dev.* **10**(4), 357–362.
15. McMahon, J. A. *et al.* (1998). Noggin-mediated antagonism of BMP signaling is required for growth and patterning of the neural tube and somite. *Genes Dev.* **12**(10), 1438–1452.
16. Massague, J. (2000). How cells read TGF-beta signals. *Nat. Rev. Mol. Cell Biol.* **1**(3), 169–178.
17. Ericson, J. *et al.* (1997). Pax6 controls progenitor cell identity and neuronal fate in response to graded Shh signaling. *Cell* **90**(1), 169–180.
18. Zhou, Q., Choi, G., and Anderson, D. J. (2001). The bHLH transcription factor Olig2 promotes oligodendrocyte differentiation in collaboration with Nkx2.2. *Neuron* **31**(5), 791–807.
19. Marquardt, T. and Pfaff, S. L. (2001). Cracking the transcriptional code for cell specification in the neural tube. *Cell* **106**(6), 651–654.
20. Ruiz, I. A. A., Palma, V., and Dahmane, N. (2002). Hedgehog-Gli signalling and the growth of the brain. *Nat. Rev. Neurosci.* **3**(1), 24–33.
21. Park, H. L. *et al.* (2000). Mouse Gli1 mutants are viable but have defects in SHH signaling in combination with a Gli2 mutation. *Development* **127**(8), 1593–1605.
22. Litingtung, Y. and Chiang, C. (2000). Specification of ventral neuron types is mediated by an antagonistic interaction between Shh and Gli3. *Nat. Neurosci.* **3**(10), 979–985.
23. Lee, S. K. and Pfaff, S. L. (2001). Transcriptional networks regulating neuronal identity in the developing spinal cord. *Nat. Neurosci.* **4** Suppl. 1.1183–1191.
24. Chuang, P. T. and McMahon, A. P. (1999). Vertebrate Hedgehog signalling modulated by induction of a Hedgehog-binding protein. *Nature* **397**(6720), 617–621.
25. Eggenschwiler, J. T., Espinoza, E., and Anderson, K. V. (2001). Rab23 is an essential negative regulator of the mouse Sonic hedgehog signalling pathway. *Nature* **412**(6843), 194–198.
26. Pierani, A. *et al.* (1999). A sonic hedgehog-independent, retinoid-activated pathway of neurogenesis in the ventral spinal cord. *Cell* **97**(7), 903–915.
27. Pfaff, S. and Kintner, C. (1998). Neuronal diversification: development of motor neuron subtypes. *Curr. Opin. Neurobiol.* **8**(1), 27–36.
28. Lance-Jones, C. *et al.* (2001). Hoxd10 induction and regionalization in the developing lumbosacral spinal cord. *Development* **128**(12), 2255–2268.
29. Matise, M. P. and Lance-Jones, C. (1996). A critical period for the specification of motor pools in the chick lumbosacral spinal cord. *Development* **122**(2), 659–669.
30. Ensini, M. *et al.* (1998). The control of rostrocaudal pattern in the developing spinal cord: specification of motor neuron subtype identity is initiated by signals from paraxial mesoderm. *Development* **125**(6), 969–982.
31. Liu, J. P., Laufer, E., and Jessell, T. M. (2001). Assigning the positional identity of spinal motor neurons. Rostrocaudal patterning of

Hox-c expression by FGFs, Gdf11, and retinoids. *Neuron* **32**(6), 997–1012.
32. Tiret, L. *et al.* (1998). Increased apoptosis of motoneurons and altered somatotropic maps in the brachial spinal cord of Hoxc-8-deficient mice. *Development* **125**(2), 279–291.
33. Studer, M. *et al.* (1996). Altered segmental identity and abnormal migration of motor neurons in mice lacking Hoxb-1. *Nature* **384**(6610), 630–634.
34. Itasaki, N. *et al.* (1996). Reprogramming Hox expression in the vertebrate hindbrain: Influence of paraxial mesoderm and rhombomere transposition. *Neuron* **16**(3), 487–500.
35. Marshall, H. *et al.* (1992). Retinoic acid alters hindbrain Hox code and induces transformation of rhombomeres 2/3 into a 4/5 identity. *Nature* **360**(6406), 737–741.
36. Gavalas, A. and Krumlauf, R. (2000). Retinoid signalling and hindbrain patterning. *Curr. Opin. Genet. Dev.* **10**(4), 380–386.

CHAPTER 269

Cadherins: Interactions and Regulation of Adhesivity

Barbara Ranscht
*The Burnham Institute, Neurobiology Program,
La Jolla, California*

Introduction

The ability of cells to distinguish between its neighbors and to recognize and associate with cells of specific subtypes is a prerequisite for the organization of the vertebrate and invertebrate body plan [1,2]. Cadherins are a class of transmembrane cell surface glycoproteins that regulate tissue morphogenesis and are necessary for maintaining the integrity of adult tissues [3–7]. The hallmark of cadherin molecules is their ability to confer calcium-dependent cell-to-cell adhesion, predominantly in a homophilic manner (e.g., molecules on one cell surface bind to molecules of the same molecule type on opposing cell surfaces [8]). This function plays profound roles in developing embryos as cells segregate from each other to form and distinguish specialized tissues from surrounding cells [9–12]. In the vertebrate nervous system, cadherins are implicated in the embryonic formation of specific brain subdivisions and the establishment of functional circuits [13]. Cadherins are clustered at synapses where they straddle the active zone [14,15] and regulate synaptic morphology and strength [16–18]. In epithelial tissue, cadherins are prominent components of adherens junctions [19,20]. In their junctional most adhesive configuration, they associate with the armadillo-repeat protein β-catenin (or plakoglobin) that connects the transmembrane cadherins via α-catenin to the actin-based cytoskeleton. Modulation of cadherin function is one of the central events that initiates cell spreading and migration during morphogenetic movements and cancer. Moreover, the transition from an epithelial to an invasive phenotype during growth and metastasis of malignant tumors is often associated with the loss of E-cadherin, and reintroduction of E-cadherin into metastatic cells can suppress the tumor phenotype [21,22]. Because of the recognized importance of cadherin cell adhesion molecules, major research efforts have concentrated on understanding the structural properties, dynamics, and signal transduction mechanisms by which cadherins regulate adhesive strength. Significant evidence has accumulated demonstrating that the fine-tuned balance of cellular signaling pathways that either enforce or reduce adhesion dynamically controls cadherin function. Moreover, it has become clear that cells regulate the degree of cadherin-mediated adhesion by diverse mechanisms. This chapter will summarize the mechanisms that contribute to the regulation of cadherin-mediated adhesivity.

The Members of the Family

Classical Cadherins

The vertebrate classical cadherins are transmembrane proteins composed of five tandemly arranged cadherin domains and a highly conserved cytoplasmic tail (Fig. 1). Cadherin domains (CD) are defined by a stretch of approximately 110 amino acids that contains characteristic motifs involved in calcium binding [3,23]. Classical cadherins or vertebrate type I cadherins, such as E-, P-, N-, and R-cadherin, display in the amino terminal CD1, a conserved HAV motif that has gained attention as a potential cell recognition sequence [24]. Although peptides containing this motif indeed inhibit cadherin-mediated functions [25,26], structural studies are inconsistent with this suggestion [27]. Atypical or vertebrate classical type II cadherins share the structural

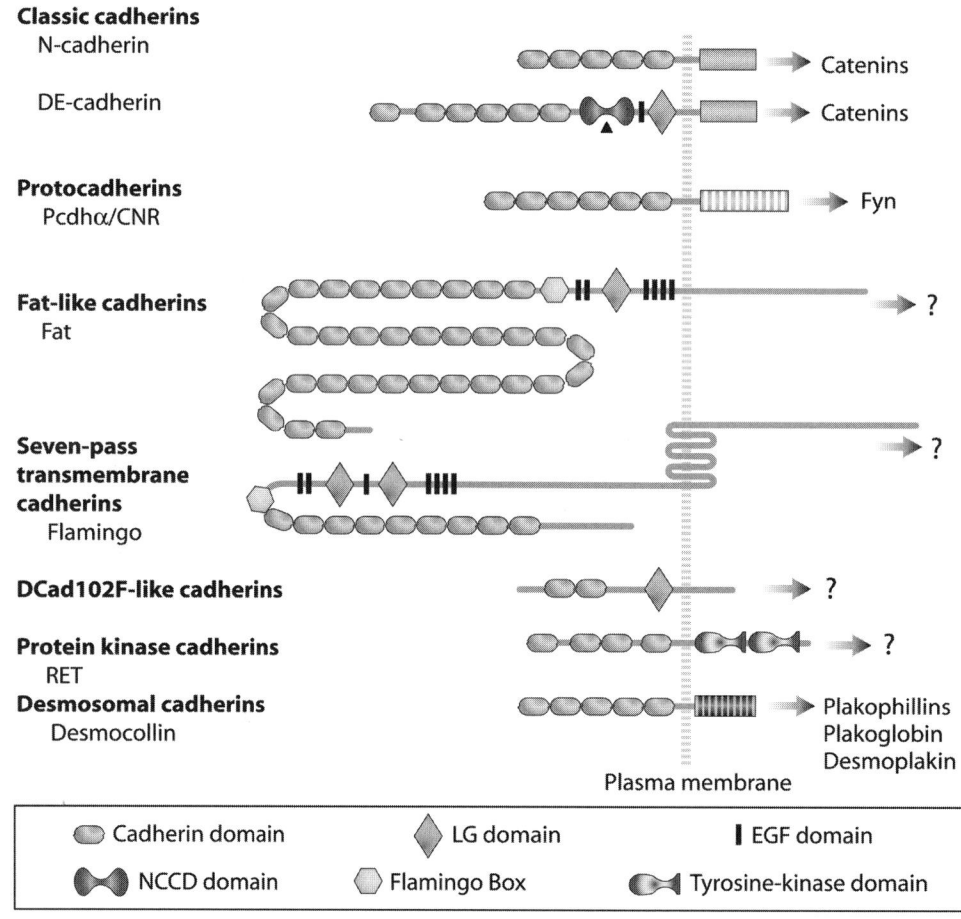

Figure 1 Molecular diversity of the cadherin family. The domain organization of representative members of each of the identified cadherin families is depicted. Arrows indicate the linkage with indicated intracellular molecules.

organization with the type I cadherins, but are less related based on sequence homology [28]. The intracellular domains of both groups are highly conserved. Transmembrane classical cadherins associate through interactions with the armadillo repeat family proteins β-catenin/plakoglobin with the actin cytoskeleton. This interaction is critical for dynamically regulating cadherin-mediated adhesivity. Classical cadherins exert profound functions in animal morphogenesis. Deletions of E- or N-cadherin gene expression in mice result in embryonic lethality respectively at the embryo preimplantation stage [29,30] and at embryonic day 10 due to defects in cardiac development [31].

Invertebrate classical cadherins including *Drosophila* DE- and DN-cadherin [32–34], *Caenorhabditis elegans* HMR-1 [35], and sea urchin LvG-cadherin [36] diverge in their extracellular domain organization from their vertebrate counterparts and contain additional CDs, cysteine-rich EGF repeats, laminin A G domains, and often a nonchordate classic cadherin domain (NCCD) [37] (Fig. 1). The cytoplasmic domain of invertebrate classical cadherins associates with *armadillo*, a molecule of the catenin family that can be linked to the actin cytoskeleton [5]. Mosaic analysis of mutations in the classical DE-cadherin-encoding *shotgun* (*shg*) gene provides evidence for homophilic binding activity [38]. Thus, despite structural divergence of their extracellular domains, the signaling mechanisms and principal functions of the classical cadherins are conserved between vertebrate and nonvertebrate species [39].

Cadherins with Divergent Structures

Bioinformatics has revealed CD-like structures in yeast and bacteria [40] suggesting that the CD represents an ancient structural motif. With the evolution of multicellular organisms, the diversity of proteins containing CDs has expanded enormously, and in humans more than 180 genes encoding proteins with cadherin domains have been reported (www.pfam.wustl.edu). Most, but not all of these genes are classified into distinct subgroups based on sequence homologies, domain organization, CD number, and genomic organization [5,28,39]. Representative members of the major subclasses are shown in Fig. 1.

PROTOCADHERINS

The largest subgroup of cadherin-related proteins is represented by the transmembrane protocadherins, which are composed of up to seven cadherin domains and contain divergent and distinct cytoplasmic regions [41,42]. Protocadherins are most abundant in the nervous system. Three gene clusters Pdchα, Pdchβ and Pdchγ located on human chromosome 5q31 are predicted to encode 52 cadherin-related molecules with a variable ectodomain and a characteristic cytoplasmic region [43]. Pdchα orthologs were independently identified as cadherin-related neuronal receptors (CNR) in mouse brain by their ability to interact with Fyn [44], an Src-related intracellular kinase that supports synaptic plasticity [45]. The observation that CNRs demarcate specific synapse populations [44] raises the possibility that these proteins regulate synaptic function or plasticity. Indeed, one of the synaptic protocadherins, Arcadlin (activity-regulated cadherin-like protein), was identified as a gene product induced by synaptic activity [46]. The functions of specific protocadherins during CNS development and in synaptic function are not yet known. Outside of the nervous system, the paraxial protocadherin (PAPC) was found to control cell movements during gastrulation [47] and establish segmental boundaries during somite formation [48]. These functions together with the demonstrated homophilic binding activity indicate that PAPC function resembles that of the classical cadherins [49,50], although its divergent cytoplasmic region indicates a different intracellular signaling mechanism.

FAT-LIKE CADHERINS

Drosophila Fat, the prototype of the fat-like cadherins, acts as a tumor suppressor as indicated by recessive lethal gene mutations that cause overgrowth of larval imaginal discs and other morphogenetic defects [51–54]. The hallmark of the fat-like transmembrane cadherins is their large extracellular domain that is composed of 17–34 tandemly arranged cadherin domains, EGF repeats, laminin A G domains, and, in some cases, a flamingo box [5,52]. *Drosophila fat* and the fat-like *dachsous* gene product are predominantly expressed in epithelial cells and contribute to planar polarity, the coordinated orientation of cells, in the eye and the wing [55,56]. Fat-like cadherins with functions in tissue morphogenesis are reported for the nematode *C. elegans* [57,58] and mammalian homologs have been identified [59,60], although their functions remain to be elucidated.

SEVEN-PASS TRANSMEMBRANE CADHERINS

This group of cadherins is anchored within the membrane by a seven-pass transmembrane domain similar to G-protein-coupled receptors. The extracellular domain of *Flamingo/starry night*, a *Drosophila* seven-pass transmembrane cadherin, consists of nine amino terminal cadherin domains, a flamingo box, EGF repeats and laminin A G domains [61]. *Flamingo/starry night* acts in concert with *fat*, *dachsous* and components of the *wnt/frizzled* pathway to establish planar polarity of hair cells in the *Drosophila* wing [53,56,61,62]. Seven-pass transmembrane cadherin receptors including mouse *Celsr1* [63], mouse *Fmi* [61], and rat *MEGF2* [64] have also been isolated from the vertebrate nervous system and remain to be assessed for their functions and molecular associations.

DCAD 102F-LIKE CADHERINS

Drosophila Dcad 102F and *C. elegans* CDH-11 represent a separate subclass of cadherin molecules with yet unknown functions. These transmembrane cadherins are composed of two cadherin domains, a laminin A G domain and a Glu-Ser-rich cytoplasmic domain region unrelated to other members of the cadherin family. They share significant homology with two proteins of unknown function from human brain (KIA0911 and KIAA0726) [65].

PROTEIN KINASE CADHERINS

The protein kinase cadherins are represented by Ret, which consists of four cadherin-like domains and an intracellular kinase region [66,67]. Ret is part of a tripartite receptor complex that is activated by interactions with neurotrophic factors of the glial cell-line-derived neurotrophic factor (GDNF) family. GDNF induces or stabilizes a complex between Ret and GPI-linked alpha receptors (GFR alpha 1–4) resulting in dimerization and activation of the Ret kinase [68]. Mutagenesis studies have shown that Ret, GFR alpha 1, and GDNF affect multiple developmental events including development of the enteric nervous system affected in Hirschsprung's disease [69–71].

DESMOSOMAL CADHERINS

The desmosomal cadherins comprise two separate subfamilies, the Desmocollins (DSC) and the Desmogleins (DSG) [72,73]. Each subfamily is represented by three members (DSC-1, -2, -3 and DSG-1, -2, -3) and each of these molecules displays a cell type- or differentiation-specific expression profile [74,75]. The extracellular domain of desmosomal cadherins is composed of five cadherin domains and confers homo- or heterophilic binding interactions with other members of the desmosomal cadherin family. DSCs and DSGs contain characteristic intracellular domains that diverge from those of the classical cadherins and interact with either of the armadillo family proteins plakoglobin and plakophilin. The latter provide a link (via desmoplakin) to intermediate filaments. The complex constituted by desmosomal cadherins and associated intracellular proteins is essential for the assembly of desmosomal plaques [76–79]. Several autoimmune skin blistering diseases (such as pemphigus vulgaris or foliaceus) are caused by desmoglein autoantibodies that act on disrupting desmosomes within epidermal layers [80].

Cadherin Structure-Function Relationships

Cadherin activity is typically measured by the ability of transfected cells to confer calcium-dependent cell-to-cell adhesion and is well established for the classical cadherins. Accordingly, analyses probing cadherin structure and function have focused on classical cadherin subtypes. Earlier work has established that the active binding configuration of these cadherins is only accomplished in the presence of calcium [3]. Mutation of one of the calcium-binding sites renders E-cadherin inactive in conferring homophilic binding [81].

Calcium binds to conserved motifs (DXD, DRE, and DXNDNXPXF) at the interface between cadherin domains and provides for the structural rigidity and dimerization of the extracellular cadherin domain. In the absence of calcium, E-cadherin polypeptides appear collapsed and disorganized by rotary shadowing and electron microscopy, while the addition of calcium at low concentration (50 μM) enforces the formation of rigid rod-like structures [82,83]. An increase in the calcium concentration (500 μM) supports *cis* association of adjacent cadherin extracellular domains, and concentrations above >1mM drive *trans* interactions between opposing *cis* strand dimers [83–85]. The strongest adhesive force between cadherin molecules is exerted in the *trans* dimer configuration. These and cell biological studies have corrected the previous concept that the classical cadherins are inactive without their linkage to the actin cytoskeleton [86–90]. It is now accepted that the cadherin extracellular domain alone suffices to induce cell-to-cell adhesion and the interaction of the full-length molecules with the cytoskeleton is required for compaction. Thus, dimerization is a critical parameter in regulating the strength of cadherin-mediated adhesivity.

Initial structural studies of the specificity-determining cadherin extracellular domain CD1 [91] revealed a β-strand organization similar to the immunoglobulin fold [92,93]. The crystal structures suggest that calcium plays an important role in cadherin dimerization [92,94–96]. Dimerization can occur between identical cadherin molecules (homodimerization) and heterodimers [97]. One of the suggested major features for cadherin dimerization is the association of Trp-2 of mature cadherin polypeptides with Ala-80, which is presented within a hydrophobic acceptor pocket formed around the HAV sequence [27,94,96–98].

Several experimental approaches provide evidence for multiple binding sites along the cadherin extracellular domain. First, biophysical studies demonstrate that the distance between opposing cadherin-covered lipid bilayers (250 Å) corresponds to the length of the cadherin extracellular domain [99,100]. Force application results in the stepwise increase of the intermolecular distance between cadherin molecules and suggests additional binding sites along the molecule [99,100]. Second, analyses of the binding properties of C-cadherin domain constructs show that the highest homophilic binding activity is conferred by the entire extracellular domain, while CD1 polypeptides exhibit only low binding activity [101]. Third, the crystal structure of the full C-cadherin ectodomain displays *cis*- and *trans* associations of the CD1 domain with multiple sites along the cadherin extracellular domain [27].

These cumulative data on cadherin structure-function relationships are consistent with a model in which adhesion occurs in multiple steps (Fig. 2). First, in the presence of calcium, cadherins adopt a rigid structure that is able to undergo *cis* dimerization. Lateral cadherin clustering into strand dimers enables subsequent engagement with *cis* dimers from opposing cell surfaces. One may speculate that the initial interactions occur between opposing CD1 domains.

Upon favorable conditions, the adhesive bonds tighten and the cadherin extracellular domains fully interdigitate to bring adjacent cell surfaces into tight contact. Despite these seminal insights into the structure and interactions of the cadherin extracellular domains in these models, interactions *in vivo* may be more complex as cadherins were recently shown to also communicate at adherens junctions with another adhesion system, the actin-linked nectin-afadin-ponsin complex [102,103]. Studies of the molecular and structural interactions between these combinatorially expressed protein complexes may reveal new insights into the adherens junction assembly.

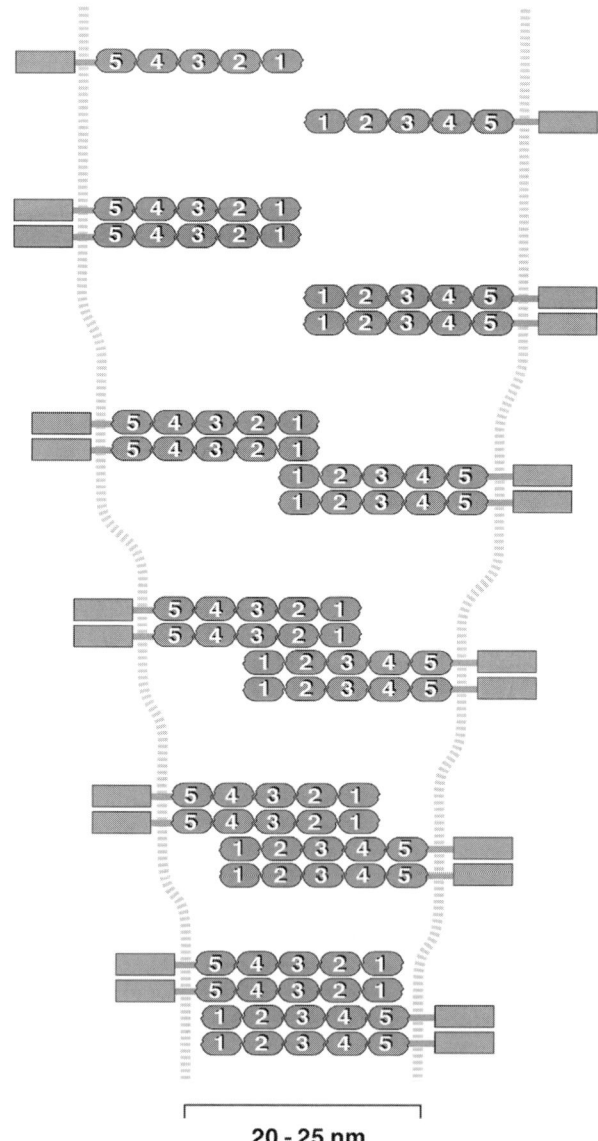

Figure 2 Stages of cadherin-mediated adhesive interactions. Cadherin monomers associate into strand dimers that subsequently engage with *cis* dimers from opposing cell surfaces. Initial interactions may occur between opposing CD1 domains. Upon favorable conditions, the adhesive bonds tighten and the cadherin extracellular domains fully interdigitate to bring adjacent cell surfaces into close contact.

Multiple Modes for Regulating Cadherin Adhesive Activity

During tissue morphogenesis, cells undergo dynamic rearrangements that require the spatial and temporal regulation of adhesive strength. The dynamic expression patterns of cadherins in developing embryos, and the functional correlation of cadherin-mediated adhesion with tissue reorganization during development suggest that regulation of cadherin function is critical for morphogenetic processes. In their most activate state, cadherins contribute to the formation and maintenance of tight adhesive contacts between adjacent cells. In this configuration, cadherin clusters at the contact site link adjacent cell surfaces to the filamentous actin cytoskeleton [104,105]. Cell migration requires the loosening of strong adhesive bonds and the downregulation of cadherin function. For example, neural crest cells modulate N-cadherin-mediated adhesion prior to migrating and ectopic cadherin expression in crest cells prevents migration [106]. The demonstrated role of N-cadherin in neurite extension *in vitro* [107,108] and axon guidance *in vivo* [109,110] requires low stringency adhesive interactions to allow cell motility. A myriad of studies have established that cadherin function is regulated at multiple levels, including gene transcription, proteolysis, endocytosis, and association with intracellular proteins. The highly conserved cadherin cytoplasmic domain provides selective target sites for regulating the cadherin-actin association that in turn influences the adhesive cell properties (Fig. 3).

Association with Intracellular Proteins

β-CATENIN/PLAKOGLOBIN

The central player in regulating the function of the classical cadherins is the armadillo repeat protein β-catenin (or the closely related plakoglobin) that binds to a conserved sequence within the cadherin carboxy-terminal cytoplasmic domain [3,81]. Deletion of the catenin-binding domain renders the mutant cadherin polypeptides inactive in promoting cell-to-cell aggregation [86,87]. The central status of β-catenin is derived from its ability to associate in a mutually exclusive manner with different cellular signaling pathways that regulate the formation of adherens junctions, gene transcription, and protein degradation [111–113]. In association with the cadherin cytoplasmic tail, β-catenin complexes with the actin-binding protein α-catenin to regulate cadherin association with actin filaments [86,114]. The small GTPases Rac, Rho, and Cdc42 are well known for regulating and specifying membrane interactions with cortical actin filaments [115], and thus it is not surprising that cadherin function is modified by such proteins. IQGAP1, an effector of the Rho family GTPases Rac and Cdc42, can modify the association of the cadherin cytoplasmic region with actin filaments [116,117]. At cell-to-cell contact sites, IQGAP1 is associated with active GTP-bound Rac and Cdc42 which prohibits binding to β-catenin. In their inactive GDP-bound form, Rac and Cdc42 do not associate with IQGAP1, which can then bind to β-catenin and dissociate the cadherin-β-catenin complex from α-catenin and the actin cytoskeleton thereby downregulating adhesivity [118]. The association of β-catenin with the cadherin cytoplasmic domain is also regulated by the Gα subunit of heterotrimeric G proteins. Binding of activated GTP-bound Gα12/13 to the cadherin cytoplasmic region dissociates β-catenin from the complex and negates adhesive interactions [119,120]. These and many other studies provide evidence that cadherin-mediated adhesive interactions are controlled to a large extent by the available pool of β-catenin in binding configuration to interact with cadherin cytoplasmic tails.

The pool of intracellular β-catenin that is not bound to cadherins can associate with a large protein complex that contains the adenomatous polyposis coli (APC) tumor suppressor gene product, axin, conductin, and the glycogen synthase kinase-3 (APC complex). In association with this complex, β-catenin becomes rapidly phosphorylated and is subsequently targeted for ubiquitination and degradation in the proteosome [121]. An effective way to counteract β-catenin phosphorylation and degradation is by activation of the Wnt signaling pathway. Wnt plays an important role in cell fate determination during embryonic development and in cancers [121,122]. Wnt-binding to a receptor of the Frizzled family antagonizes β-catenin phosphorylation and degradation. Free unphosphorylated β-catenin can then bind to unoccupied cadherin cytoplasmic domains and enforce adhesive interactions through the cytoskeletal link. When the free β-catenin pool exceeds the number of available cadherin binding sites, it accumulates in the nucleus where it can bind to DNA binding proteins of the T-cell specific factor (TCF)/lymphoid enhancer binding factor (LEF) family. β-catenin binding to TCF/LEF transcription factors enhances expression of new genes that contribute to regulating the cell cycle [121]. Thus, β-catenin serves as a molecular switch for regulating cell behavior. The pool of β-catenin, its posttranslational modifications, and its association with the cytoskeleton are central to the regulation of cadherin-mediated adhesive activity.

p120ctn

An earlier study [9] attributed functional importance to the cadherin membrane proximal cytoplasmic domain, and it has now become clear that this site is a target for a complex regulation of cadherin-mediated adhesivity. The membrane proximal cadherin cytoplasmic region contains the binding site for the armadillo repeat protein, p120ctn. Similar to β-catenin, it is dynamically regulated by phosphorylation [90,123,124]. Phosphorylated p120ctn does not interact with the binding site on cadherins, and other proteins to regulate cadherin-mediated adhesivity can occupy the binding region (see Fig. 3). P120ctn function is also affected by Rho family GTPases which act on the actin cytoskeleton to increase or decrease cell motility [118,125,126]. As p120ctn per se seems dispensable for adhesive interactions, it may

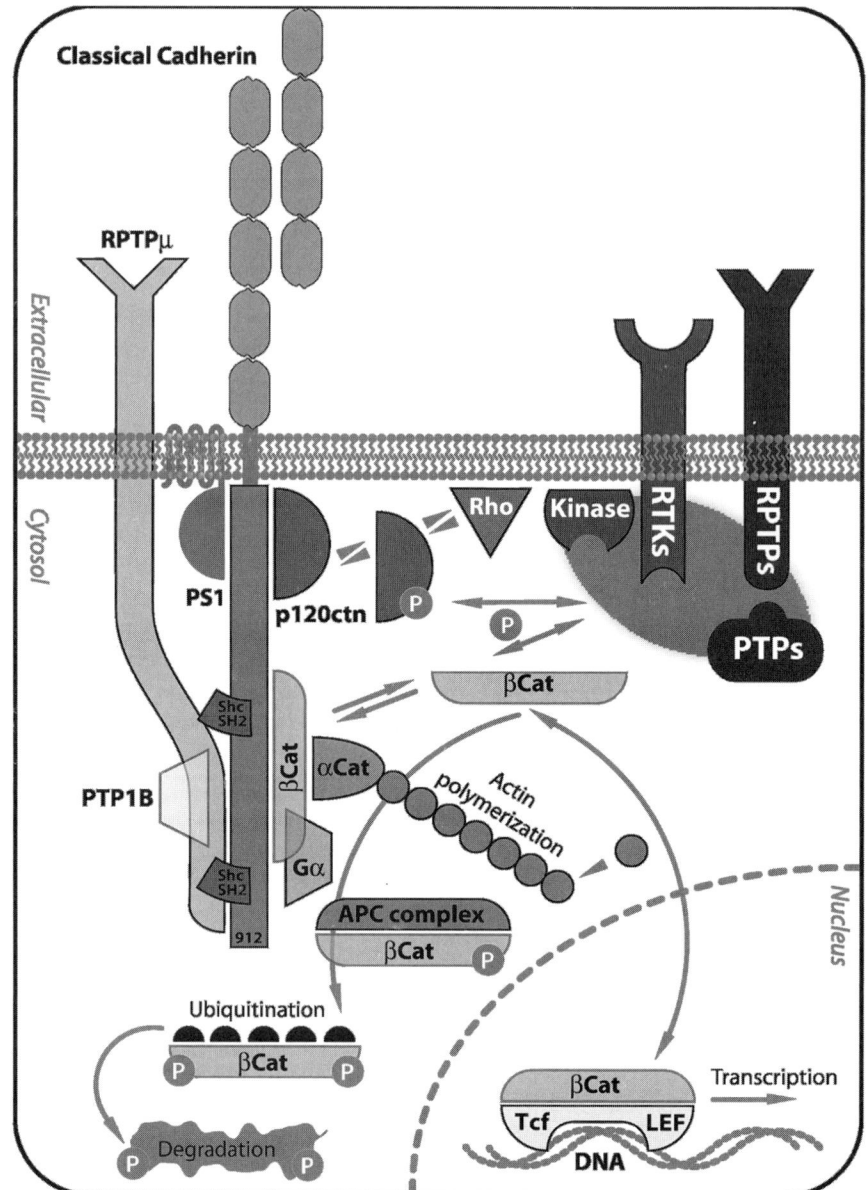

Figure 3 Regulation of cadherin function by associated intracellular proteins. The conserved cytoplasmic region of the classical cadherins interacts with multiple proteins that regulate adhesivity. Cadherin association with β-catenin and linkage of the complex to actin filaments (via α-catenin) plat a central role in this regulation. Receptor protein kinases (RTKs), receptor tyrosine phosphatases (RPTPs), ad intracellular phosphatases (PTPs) balance cadherin, β-catenin, and p120ctn phosphorylation. PS1 = Presenilin-1, βcat = β-catenin, αcat = α-catenin, and RPTPμ = receptor tyrosine phosphatase α.

serve as a molecular switch that can promote or prevent adhesion depending on its state of phosphorylation and the cellular context [90,124,127].

Protein Phosphorylation and Dephosphorylation

Structural studies of the interacting cadherin-β-catenin domains show that the cadherin domain is structured only when it is phosphorylated on serine residues [128]. The cadherin cytoplasmic region displays serine phosphorylation consensus sites for casein kinase II and glycogen synthase. Serine phosphorylation seems to strengthen the association between the cadherin cytoplasmic region and β-catenin and strengthen adhesion [129]. In contrast, tyrosine phosphorylation of the cadherin cytoplasmic tail and β-catenin dramatically decreases the interaction between these proteins and weakens adhesive bonds [130–132]. Both nonreceptor- and receptor-type kinases regulate phosphorylation of the cadherin cytoplasmic tail and β-catenin. Overexpression of the nonreceptor tyrosine kinases Src or Fer in cultured cells induces phosphorylation of tyrosine residues on cadherins and β-catenin [130,133,134] and decreases adhesive interactions in favor of a motile phenotype. Interestingly, in cancer cells, the lipid phosphatase activity of PTEN can counteract Src-induced cell scattering and invasiveness and stabilize the E-cadherin junctional complex through an as yet unknown mechanism [135]. Similarly, E-cadherin and β-catenin tyrosine phosphorylation by receptor kinase type growth factor receptors results in cell scattering [136,137]. Tyrosine-phosphorylated cadherin can bind the adaptor protein Shc, which participates in stimulating mitogenic signaling pathways by growth factor activation of Ras [138]. An alternate pathway for tyrosine-phosphorylated E-cadherin is ubiquitination and rapid endocytosis. Recent work has identified Hakai, an E3 ubiquitin-ligase that can interact with E-cadherin in a tyrosine-phosphorylation-dependent manner and stimulate endocytosis [139]. Cell surface removal of E-cadherin through this pathway decreases adhesion and enhances cell scattering. Thus, cadherin function is controlled by multiple cell signaling pathways, which regulate the availability of cadherins on the cell surface and balance the cadherin association with β-catenin and hence with the cytoskeleton.

Formation of the cadherin-catenin complex is also fine-tuned by the balance between protein kinase and phosphatase activity. Several phosphatases associate with and stabilize the cadherin-β-catenin complex, presumably by preventing phosphorylation. The nonreceptor-type phosphatase PTP1B is targeted to the cadherin complex where it interacts with sequences partially overlapping with the binding site for β-catenin [140]. The receptor tyrosine phosphatase μ (PTPμ) can interact with a largely overlapping site [141], indicating that the contributions of these proteins are mutually exclusive. Downregulation of either PTP1B or PTPμ suppresses N-cadherin-mediated neurite extension [142,143] by an as yet unresolved mechanism. Also, the nature of the interactions between phosphatases and β-catenin during cell adhesion and the process of neurite growth has not been determined. Other phosphatases including LAR, receptor tyrosine phosphatase β/ζ, and the Meprin/A5/Mu domain receptors κ and γ do not associate directly with cadherins, but may regulate cadherin function through β-catenin modifications.

Regulation by Proteolytic Cleavage

An effective strategy for regulating cadherin activity is cleavage by extra- and intracellular proteases. Mature classical cadherins are derived from precursor proteins that are cleaved at the RKQR sequence in transit to or at the cell surface [144]. Mutation of the cleavage site renders the mutant E-cadherin protein inactive in conferring cellular aggregation [144]. The proprotein convertase furin can exert proprotein cleavage of E-cadherin *in vitro* [145]; however, the expression and activity of furin *in vivo* have not been established.

The cadherin extracellular domain is subject to cleavage by metalloproteases. A 90-kDa soluble extracellular fragment of N-cadherin is generated during retinal development and may partake in modulating retinal axon guidance [146,147]. Numerous studies have implicated the loss of E-cadherin with tumor cell growth and metastasis [21].

E-cadherin is proteolytically cleaved in noncancerous mammary epithelial cells by ectopically expressed metalloprotease stromelysin-1 [148]. E-cadherin cleavage triggers the progressive conversion of the epithelial into a mesenchymal-invasive phenotype characterized by the disappearance of E-cadherin and β-catenin from cell-cell contacts and induction and activation of growth factors and endogenous metalloproteases [148]. Stromolysin, however, could not be detected in cancer cells or in the embryo, and the tissue-endogenous metalloproteases cleaving E-cadherin [149] remain to be defined.

Cellular responses to apoptotic signals are characterized by the disruption of cell-to-cell and cell-to-extracellular matrix contacts and cytoskeletal reorganization. During programmed cell death, adherens junctions disintegrate due to the actions of both metalloproteases and caspases on cadherin and β-catenin/plakoglobin molecules [150,151]. A metalloprotease activity releases most of the E-cadherin extracellular domain, while caspase-3 cleaves at an intracellular membrane proximal site [150]. These data enforce the suggestion that the structural integrity of cadherins, their assembly within adherens junctions, and their linkage to the actin filament network are critical for cell survival [152].

Lastly, in response to apoptotic stimuli, the γ-secretase activity of presenilin-1, a protein associated with Alzheimer's disease, can cleave E-cadherin at the membrane-cytoplasm interface [153]. The cleavage releases the cadherin intracellular domain, increases the intracellular pool of β-catenin, and facilitates the disassembly of adherens junctions by disconnecting cadherins from the cytoskeleton. However, under conditions that favor cell-to-cell adhesion, presenilin-1 binding stabilizes the junctional complex [154]. Such dual functions have also been reported for p120ctn which competes for the presenilin-1 binding site on E-cadherin in a mutually exclusive manner [154]. Presenilin-1 has also been demonstrated to bind to and regulate β-catenin function and trafficking thereby providing an additional mechanism for regulating adhesivity [155–157]. As presenilin-1 can be recruited to sites of synaptic contact [158] and synaptic morphology and function are regulated by the cadherin-catenin system [16–18], the cadherin-catenin-presenilin-1 interaction may favor the loss of synaptic structures at an early stage of Alzheimer's disease and increase vulnerability to neuronal apoptosis.

Conclusions and Perspectives

Over the years, it has become clear that the adhesive function of the classical cadherins is dynamically regulated. While beginning to grasp some of the principal mechanisms of this regulation, we are faced with new challenges. First, a large number of new cadherin-like molecules with cytoplasmic sequences different from those of the classical cadherins have been revealed. Little is known about the distribution, function, and modes of signaling of these molecules. Second, recent work suggests that cadherins may be far more promiscuous in their binding specificities than previously assumed [159]. N-cadherin-deficient mutant mice die of defects in heart development [31]. This phenotype can be rescued by the cardiac-specific ectopic expression of E-cadherin [160]. suggesting that cadherin-mediated adhesivity but not adhesive specificity is required. Moreover, recent work has demonstrated that cadherin function is required for the sorting of motor neurons into specific pools [161]. While cadherin-mediated sorting has been attributed to the binding specificity of one cadherin type, motor neuron pools are defined by the combinatorial expression of multiple cadherins. Overexpression of one of these cadherins, MN cadherin, disrupts pool sorting. Although the mechanism for MN-cadherin function remains to be determined, there is good evidence to suggest that the cadherin homophilic binding activity is not required as ectopic expression of E-cadherin in motor neurons has no effect on pool sorting [161]. The myriad of data that speak to the regulation of cadherin function have made it clear that cadherin-mediated adhesivity is a multifaceted issue. The diversity and magnitude of the cadherin family adds a new level of complexity to understanding the cellular interactions conferred by the combinatorial cadherin expression during development and in adult organisms.

Acknowledgments

I thank Dr. Chris Kintner for critical reading of this manuscript and Kosi Gramatikoff for preparing the illustrations. Cadherin research in our laboratory is supported by NIH Grant HD 25938.

References

1. Townes, P. L. a. Holtfretter, J. (1955). Directed movements and selective adhesion of embryonic amphibian cells. *J. Exp. Zool.* **128**, 53–120.
2. Steinberg, M. S. (1963). Reconstruction of tissue by dissociated cells. *Science* **141**, 401–408.
3. Takeichi, M. (1990). Cadherins: A molecular family important in selective cell-cell adhesion. *Annu. Rev. Biochem.* **59**, 237–252.
4. Ranscht, B. (1994). Cadherins and catenins: Interactions and functions in embryonic development. *Curr. Opin. Cell Biol.* **6**, 740–746.
5. Tepass, U., Truong, K., Godt, D., Ikura, M., and Peifer, M. (2000). Cadherins in embryonic and neural morphogenesis. *Nat. Rev. Mol. Cell Biol.* **1**, 91–100.
6. Yagi, T. and Takeichi, M. (2000). Cadherin superfamily genes: Functions, genomic organization, and neurologic diversity. *Genes Dev.* **14**, 1169–1180.
7. Redies, C. (2000). Cadherins in the central nervous system. *Prog. Neurobiol.* **61**, 611–648.
8. Takeichi, M., Nakagawa, S., Aono, S., Usui, T., and Uemura, T. (2000). Patterning of cell assemblies regulated by adhesion receptors of the cadherin superfamily. *Philos. Trans. R. Soc. Lond. B Biol. Sci.* **355**, 885–890.
9. Kintner, C. (1992). Regulation of embryonic cell adhesion by the cadherin cytoplasmic domain. *Cell* **69**, 225–236.
10. Inoue, T. *et al.* (2001). Role of cadherins in maintaining the compartment boundary between the cortex and striatum during development. *Development* **128**, 561–569.
11. Detrick, R. J., Dickey, D., and Kintner, C. R. (1990). The effects of N-cadherin misexpression on morphogenesis in *Xenopus* embryos. *Neuron* **4**, 493–506.
12. Fujimori, T., Miyatani, S., and Takeichi, M. (1990). Ectopic expression of N-cadherin perturbs histogenesis in *Xenopus* embryos. *Development* **110**, 97–104.
13. Redies, C. and Puelles, L. (2001). Modularity in vertebrate brain development and evolution. *Bioessays* **23**, 1100–1111.
14. Uchida, N., Honjo, Y., Johnson, K. R., Wheelock, M. J., and Takeichi, M. (1996). The catenin/cadherin adhesion system is localized in synaptic junctions bordering transmitter release zones. *J. Cell Biol.* **135**, 767–779.
15. Fannon, A. M. and Colman, D. R. (1996). A model for central synaptic junctional complex formation based on the differential adhesive specificities of the cadherins. *Neuron* **17**, 423–434.
16. Togashi, H. *et al.* (2002). Cadherin regulates dendritic spine morphogenesis. *Neuron* **35**, 77–89.
17. Murase, S., Mosser, E., and Schuman, E. M. (2002). Depolarization drives beta-Catenin into neuronal spines promoting changes in synaptic structure and function. *Neuron* **35**, 91–105.
18. Bruses, J. L. (2000). Cadherin-mediated adhesion at the interneuronal synapse. *Curr. Opin. Cell Biol.* **12**, 593–597.
19. Yap, A. S., Brieher, W. M., and Gumbiner, B. M. (1997). Molecular and functional analysis of cadherin-based adherens junctions. *Ann. Rev. Cell Dev. Biol.* **13**, 119–146.
20. Adams, C. L. and Nelson, W. J. (1998). Cytomechanics of cadherin-mediated cell-cell adhesion. *Curr. Opin. Cell Biol.* **10**, 572–577.
21. Birchmeier, C., Birchmeier, W., and Brand-Saberi, B. (1996). Epithelial-mesenchymal transitions in cancer progression. *Acta Anat.* **156**, 217–226.
22. Christofori, G. and Semb, H. (1999). The role of the cell-adhesion molecule E-cadherin as a tumour-suppressor gene. *Trends Biochem. Sci.* **24**, 73–76.
23. Ozawa, M., Engel, J., and Kemler, R. (1990). Single amino acid substitutions in one Ca^{2+} binding site of uvomorulin abolish the adhesive function. *Cell* **63**, 1033–1038.
24. Blaschuk, O. W., Sullivan, R., David, S., and Pouliot, Y. (1990). Identification of a cadherin cell adhesion recognition sequence. *Dev. Biol.* **139**, 227–229.
25. Doherty, P., Rowett, L. H., Moore, S. E., Mann, D. A., and Walsh, F. S. (1991). Neurite outgrowth in response to transfected N-CAM and N-cadherin reveals fundamental differences in neuronal responsiveness to CAMs. *Neuron* **6**, 247–258.
26. Tang, L., Hung, C. P., and Schuman, E. M. (1998). A role for the cadherin family of cell adhesion molecules in hippocampal long-term potentiation. *Neuron* **20**, 1165–1175.
27. Boggon, T. J. *et al.* (2002). C-cadherin ectodomain structure and implications for cell adhesion mechanisms. *Science* **296**, 1308–1313.
28. Nollet, F., Kools, P., and van Roy, F. (2000). Phylogenetic analysis of the cadherin superfamily allows identification of six major subfamilies besides several solitary members. *J. Mol. Biol.* **299**, 551–572.
29. Larue, L., Ohsugi, M., Hirchenhain, J., and Kemler, R. (1994). E-cadherin null mutant embryos fail to form a trophectoderm epithelium. *Proc. Natl. Acad. Sci. USA* **91**, 8263–8267.
30. Riethmacher, D., Brinkmann, V., and Birchmeier, C. (1995). A targeted mutation in the mouse E-cadherin gene results in defective preimplantation development. *Proc. Natl. Acad. Sci. USA* **92**, 855–859.
31. Radice, G. L. *et al.* (1997). Developmental defects in mouse embryos lacking N-cadherin. *Dev. Biol.* **181**, 64–78.
32. Tepass, U. *et al.* (1996). Shotgun encodes Drosophila E-cadherin and is preferentially required during cell rearrangement in the neuroectoderm and other morphogenetically active epithelia. *Genes Dev.* **10**, 672–685.
33. Uemura, T. *et al.* (1996). Zygotic *Drosophila* E-cadherin expression is required for processes of dynamic epithelial cell rearrangement in the *Drosophila* embryo. *Genes Dev.* **10**, 659–671.
34. Iwai, Y. *et al.* (1997). Axon patterning requires DN-cadherin, a novel neuronal adhesion receptor, in the *Drosophila* embryonic CNS. *Neuron* **19**, 77–89.
35. Costa, M. *et al.* (1998). A putative catenin-cadherin system mediates morphogenesis of the *Caenorhabditis elegans* embryo. *J. Cell Biol.* **141**, 297–308.
36. Miller, J. R. and McClay, D. R. (1997). Characterization of the role of cadherin in regulating cell adhesion during sea urchin development. *Dev. Biol.* **192**, 323–339.

37. Oda, H. and Tsukita, S. (1999). Nonchordate classic cadherins have a structurally and functionally unique domain that is absent from chordate classic cadherins. *Dev. Biol.* **216**, 406–422.
38. Niewiadomska, P., Godt, D., and Tepass, U. (1999). DE-Cadherin is required for intercellular motility during *Drosophila* oogenesis. *J. Cell Biol.* **144**, 533–547.
39. Tepass, U. (1999). Genetic analysis of cadherin function in animal morphogenesis. *Curr. Opin. Cell Biol.* **11**, 540–548.
40. Dickens, N. J., Beatson, S., and Ponting, C. P. (2002). Cadherin-like domains in alpha-dystroglycan, alpha/varepsilon-sarcoglycan and yeast and bacterial proteins. *Curr. Biol.* **12**, R197–R199.
41. Sano, K. *et al.* (1993). Protocadherins: A large family of cadherin-related molecules in central nervous system. *EMBO J.* **12**, 2249–2256.
42. Hamada, S. and Yagi, T. (2001). The cadherin-related neuronal receptor family: A novel diversified cadherin family at the synapse. *Neurosci. Res.* **41**, 207–215.
43. Wu, Q. and Maniatis, T. (1999). A striking organization of a large family of human neural cadherin-like cell adhesion genes. *Cell* **97**, 779–790.
44. Kohmura, N. *et al.* (1998). Diversity revealed by a novel family of cadherins expressed in neurons at a synaptic complex. *Neuron* **20**, 1137–1151.
45. Grant, S. G. *et al.* (1992). Impaired long-term potentiation, spatial learning, and hippocampal development in fyn mutant mice [see comments]. *Science* **258**, 1903–1910.
46. Yamagata, K. *et al.* (1999). Arcadlin is a neural activity-regulated cadherin involved in long term potentiation. *J. Biol. Chem.* **274**, 19473–19479.
47. Kim, S. H., Yamamoto, A., Bouwmeester, T., Agius, E., and Robertis, E. M. (1998). The role of paraxial protocadherin in selective adhesion and cell movements of the mesoderm during *Xenopus* gastrulation. *Development* **125**, 4681–4690.
48. Kim, S. H., Jen, W. C., De Robertis, E. M., and Kintner, C. (2000). The protocadherin PAPC establishes segmental boundaries during somitogenesis in *Xenopus* embryos. *Curr. Biol.* **10**, 821–830.
49. Lee, C. H. and Gumbiner, B. M. (1995). Disruption of gastrulation movements in *Xenopus* by a dominant-negative mutant for C-cadherin. *Dev. Biol.* **171**, 363–373.
50. Fagotto, F. and Gumbiner, B. M. (1994). Beta-catenin localization during *Xenopus* embryogenesis: Accumulation at tissue and somite boundaries. *Development* **120**, 3667–3679.
51. Bryant, P. J., Huettner, B., Held, L. I., Jr., Ryerse, J., and Szidonya, J. (1988). Mutations at the fat locus interfere with cell proliferation control and epithelial morphogenesis in *Drosophila*. *Dev. Biol.* **129**, 541–554.
52. Mahoney, P. A. *et al.* (1991). The fat tumor suppressor gene in *Drosophila* encodes a novel member of the cadherin gene superfamily. *Cell* **67**, 853–868.
53. Clark, H. F. *et al.* (1995). Dachsous encodes a member of the cadherin superfamily that controls imaginal disc morphogenesis in *Drosophila*. *Genes Dev.* **9**, 1530–1542.
54. Buratovich, M. A. and Bryant, P. J. (1997). Enhancement of overgrowth by gene interactions in lethal(2)giant discs imaginal discs from *Drosophila melanogaster*. *Genetics* **147**, 657–670.
55. Yang, C. H., Axelrod, J. D., and Simon, M. A. (2002). Regulation of Frizzled by fat-like cadherins during planar polarity signaling in the *Drosophila* compound eye. *Cell* **108**, 675–688.
56. Adler, P. N., Charlton, J., and Liu, J. (1998). Mutations in the cadherin superfamily member gene dachsous cause a tissue polarity phenotype by altering frizzled signaling. *Development* **125**, 959–968.
57. Pettitt, J., Wood, W. B., and Plasterk, R. H. (1996). cdh-3, a gene encoding a member of the cadherin superfamily, functions in epithelial cell morphogenesis in *Caenorhabditis elegans*. *Development* **122**, 4149–4157.
58. Hill, E., Broadbent, I. D., Chothia, C., and Pettitt, J. (2001). Cadherin superfamily proteins in *Caenorhabditis elegans* and *Drosophila melanogaster*. *J. Mol. Biol.* **305**, 1011–1024.
59. Dunne, J. *et al.* (1995). Molecular cloning and tissue expression of FAT, the human homologue of the *Drosophila* fat gene that is located on chromosome 4q34-q35 and encodes a putative adhesion molecule. *Genomics* **30**, 207–223.
60. Ponassi, M., Jacques, T. S., Ciani, L., and ffrench Constant, C. (1999). Expression of the rat homologue of the *Drosophila* fat tumour suppressor gene. *Mech. Dev.* **80**, 207–212.
61. Usui, T. *et al.* (1999). Flamingo, a seven-pass transmembrane cadherin, regulates planar cell polarity under the control of Frizzled. *Cell* **98**, 585–595.
62. Chae, J. *et al.* (1999). The *Drosophila* tissue polarity gene starry night encodes a member of the protocadherin family. *Development* **126**, 5421–5429.
63. Hadjantonakis, A. K., Formstone, C. J., and Little, P. F. (1998). mCelsr1 is an evolutionarily conserved seven-pass transmembrane receptor and is expressed during mouse embryonic development. *Mech. Dev.* **78**, 91–95.
64. Nakayama, M. *et al.* (1998). Identification of high-molecular-weight proteins with multiple EGF-like motifs by motif-trap screening. *Genomics* **51**, 27–34.
65. Nagase, T. *et al.* (1998). Prediction of the coding sequences of unidentified human genes. XI. The complete sequences of 100 new cDNA clones from brain which code for large proteins *in vitro*. *DNA Res.* **5**, 277–286.
66. Takahashi, M. and Cooper, G. M. (1987). ret transforming gene encodes a fusion protein homologous to tyrosine kinases. *Mol. Cell. Biol.* **7**, 1378–1385.
67. Anders, J., Kjar, S., and Ibanez, C. F. (2001). Molecular modeling of the extracellular domain of the RET receptor tyrosine kinase reveals multiple cadherin-like domains and a calcium-binding site. *J. Biol. Chem.* **276**, 35808–35817.
68. Airaksinen, M. S., Titievsky, A., and Saarma, M. (1999). GDNF family neurotrophic factor signaling: four masters, one servant? *Mol. Cell. Neurosci.* **13**, 313–325.
69. Schuchardt, A., D'Agati, V., Larsson-Blomberg, L., Costantini, F., and Pachnis, V. (1994). Defects in the kidney and enteric nervous system of mice lacking the tyrosine kinase receptor Ret. *Nature* **367**, 380–383.
70. Romeo, G. *et al.* (1994). Point mutations affecting the tyrosine kinase domain of the RET proto-oncogene in Hirschsprung's disease. *Nature* **367**, 377–378.
71. Edery, P. *et al.* (1994). Mutations of the RET proto-oncogene in Hirschsprung's disease. *Nature* **367**, 378–380.
72. Koch, P. J. *et al.* (1990). Identification of desmoglein, a constitutive desmosomal glycoprotein, as a member of the cadherin family of cell adhesion molecules. *Eur. J. Cell Biol.* **53**, 1–12.
73. Parker, A. E. *et al.* (1991). Desmosomal glycoproteins II and III. Cadherin-like junctional molecules generated by alternative splicing. *J. Biol. Chem.* **266**, 10438–10445.
74. Green, K. J. and Gaudry, C. A. (2000). Are desmosomes more than tethers for intermediate filaments? *Nat. Rev. Mol. Cell Biol.* **1**, 208–216.
75. Ishii, K. and Green, K. J. (2001). Cadherin function: Breaking the barrier. *Curr. Biol.* **11**, R569–R572.
76. Allen, E., Yu, Q. C., and Fuchs, E. (1996). Mice expressing a mutant desmosomal cadherin exhibit abnormalities in desmosomes, proliferation, and epidermal differentiation. *J. Cell Biol.* **133**, 1367–1382.
77. Koch, P. J. *et al.* (1997). Targeted disruption of the pemphigus vulgaris antigen (desmoglein 3) gene in mice causes loss of keratinocyte cell adhesion with a phenotype similar to pemphigus vulgaris. *J. Cell Biol.* **137**, 1091–1102.
78. Roberts, G. A. *et al.* (1998). Antisense expression of a desmocollin gene in MDCK cells alters desmosome plaque assembly but does not affect desmoglein expression. *Eur. J. Cell Biol.* **76**, 192–203.
79. Serpente, N. *et al.* (2000). Extracellularly truncated desmoglein 1 compromises desmosomes in MDCK cells. *Mol. Membr. Biol.* **17**, 175–183.
80. Stanley, J. R. (1995). Autoantibodies against adhesion molecules and structures in blistering skin diseases. *J. Exp. Med.* **181**, 1–4.
81. Ozawa, M., Ringwald, M., and Kemler, R. (1990). Uvomorulin-catenin complex formation is regulated by a specific domain in the cytoplasmic region of the cell adhesion molecule. *Proc. Natl. Acad. Sci. USA* **87**, 4246–4250.

82. Pokutta, S., Herrenknecht, K., Kemler, R., and Engel, J. (1994). Conformational changes of the recombinant extracellular domain of E-cadherin upon calcium binding. *Eur. J. Biochem.* **223**, 1019–1026.
83. Pertz, O. *et al.* (1999). A new crystal structure, Ca^{2+} dependence and mutational analysis reveal molecular details of E-cadherin homoassociation. *EMBO J.* **18**, 1738–1747.
84. Tomschy, A., Fauser, C., Landwehr, R., and Engel, J. (1996). Homophilic adhesion of E-cadherin occurs by a co-operative two-step interaction of N-terminal domains. *EMBO J.* **15**, 3507–3514.
85. Koch, A. W., Pokutta, S., Lustig, A., and Engel, J. (1997). Calcium binding and homoassociation of E-cadherin domains. *Biochemistry* **36**, 7697–7705.
86. Nagafuchi, A. and Takeichi, M. (1988). Cell binding function of E-cadherin is regulated by the cytoplasmic domain. *EMBO J.* **7**, 3679–3684.
87. Ozawa, M., Baribault, H., and Kemler, R. (1989). The cytoplasmic domain of the cell adhesion molecule uvomorulin associates with three independent proteins structurally related in different species. *EMBO J.* **8**, 1711–1717.
88. Vestal, D. J. and Ranscht, B. (1992). Glycosyl phosphatidylinositol-anchored T-cadherin mediates calcium-dependent, homophilic cell adhesion. *J. Cell Biol.* **119**, 451–461.
89. Brieher, W. M., Yap, A. S., and Gumbiner, B. M. (1996). Lateral dimerization is required for the homophilic binding activity of C-cadherin. *J. Cell Biol.* **135**, 487–496.
90. Ozawa, M. and Kemler, R. (1998). The membrane-proximal region of the E-cadherin cytoplasmic domain prevents dimerization and negatively regulates adhesion activity. *J. Cell Biol.* **142**, 1605–1613.
91. Nose, A., Tsuji, K., and Takeichi, M. (1990). Localization of specificity determining sites in cadherin cell adhesion molecules. *Cell* **61**, 147–155.
92. Overduin, M. *et al.* (1995). Solution structure of the epithelial cadherin domain responsible for selective cell adhesion. *Science* **267**, 386–389.
93. Shapiro, L., Kwong, P. D., Fannon, A. M., Colman, D. R., and Hendrickson, W. A. (1995). Considerations on the folding topology and evolutionary origin of cadherin domains. *Proc. Natl. Acad. Sci. USA* **92**, 6793–6797.
94. Shapiro, L. *et al.* (1995). Structural basis of cell-cell adhesion by cadherins. *Nature* **374**, 327–337.
95. Nagar, B., Overduin, M., Ikura, M., and Rini, J. M. (1996). Structural basis of calcium-induced E-cadherin rigidification and dimerization. *Nature* **380**, 360–364.
96. Tamura, K., Shan, W. S., Hendrickson, W. A., Colman, D. R., and Shapiro, L. (1998). Structure-function analysis of cell adhesion by neural (N-) cadherin. *Neuron* **20**, 1153–1163.
97. Shan, W. S. *et al.* (2000). Functional cis-heterodimers of N- and R-cadherins. *J. Cell Biol.* **148**, 579–590.
98. Kitagawa, M. *et al.* (2000). Mutation analysis of cadherin-4 reveals amino acid residues of EC1 important for the structure and function. *Biochem. Biophys. Res. Commun.* **271**, 358–363.
99. Sivasankar, S., Brieher, W., Lavrik, N., Gumbiner, B., and Leckband, D. (1999). Direct molecular force measurements of multiple adhesive interactions between cadherin ectodomains. *Proc. Natl. Acad. Sci. USA* **96**, 11820–11824.
100. Sivasankar, S., Gumbiner, B., and Leckband, D. (2001). Direct measurements of multiple adhesive alignments and unbinding trajectories between cadherin extracellular domains. *Biophys. J.* **80**, 1758–1768.
101. Chappuis-Flament, S., Wong, E., Hicks, L. D., Kay, C. M., and Gumbiner, B. M. (2001). Multiple cadherin extracellular repeats mediate homophilic binding and adhesion. *J. Cell Biol.* **154**, 231–243.
102. Tachibana, K. *et al.* (2000). Two cell adhesion molecules, nectin and cadherin, interact through their cytoplasmic domain-associated proteins. *J. Cell Biol.* **150**, 1161–1176.
103. Pokutta, S. and Weis, W. I. (2002). The cytoplasmic face of cell contact sites. *Curr. Opin. Struct. Biol.* **12**, 255–262.
104. Adams, C. L., Nelson, W. J., and Smith, S. J. (1996). Quantitative analysis of cadherin-catenin-actin reorganization during development of cell-cell adhesion. *J. Cell Biol.* **135**, 1899–1911.
105. Gumbiner, B., Stevenson, B., and Grimaldi, A. (1988). The role of the cell adhesion molecule uvomorulin in the formation and maintenance of the epithelial junctional complex. *J. Cell Biol.* **107**, 1575–1587.
106. Nakagawa, S. and Takeichi, M. (1998). Neural crest emigration from the neural tube depends on regulated cadherin expression. *Development* **125**, 2963–2971.
107. Bixby, J. L. and Zhang, R. (1990). Purified N-cadherin is a potent substrate for the rapid induction of neurite outgrowth. *J. Cell Biol.* **110**, 1253–1260.
108. Matsunaga, M., Hatta, K., Nagafuchi, A., and Takeichi, M. (1988). Guidance of optic nerve fibres by N-cadherin adhesion molecules. *Nature* **334**, 62–64.
109. Riehl, R. *et al.* (1996). Cadherin function is required for axon outgrowth in retinal ganglion cells *in vivo*. *Neuron* **17**, 837–848.
110. Lee, C. H., Herman, T., Clandinin, T. R., Lee, R., and Zipursky, S. L. (2001). N-cadherin regulates target specificity in the *Drosophila* visual system. *Neuron* **30**, 437–450.
111. Hulsken, J., Birchmeier, W., and Behrens, J. (1994). E-cadherin and APC compete for the interaction with beta-catenin and the cytoskeleton. *J. Cell Biol.* **127**, 2061–2069.
112. von Kries, J. P. *et al.* (2000). Hot spots in beta-catenin for interactions with LEF-1, conductin and APC. *Nat. Struct. Biol.* **7**, 800–807.
113. Gottardi, C. J. and Gumbiner, B. M. (2001). Adhesion signaling: How beta-catenin interacts with its partners. *Curr. Biol.* **11**, R792–R794.
114. Itoh, M., Morita, K., and Tsukita, S. (1999). Characterization of ZO-2 as a MAGUK family member associated with tight as well as adherens junctions with a binding affinity to occludin and alpha catenin. *J. Biol. Chem.* **274**, 5981–5986.
115. Hall, A. (1998). Rho GTPases and the actin cytoskeleton. *Science* **279**, 509–514.
116. Kuroda, S. *et al.* (1998). Role of IQGAP1, a target of the small GTPases Cdc42 and Rac1, in regulation of E-cadherin-mediated cell-cell adhesion. *Science* **281**, 832–835.
117. Fukata, M. *et al.* (1999). Cdc42 and Rac1 regulate the interaction of IQGAP1 with beta-catenin. *J. Biol. Chem.* **274**, 26044–26050.
118. Fukata, M. and Kaibuchi, K. (2001). Rho-family GTPases in cadherin-mediated cell-cell adhesion. *Nat. Rev. Mol. Cell Biol.* **2**, 887–897.
119. Meigs, T. E., Fedor-Chaiken, M., Kaplan, D. D., Brackenbury, R., and Casey, P. J. (2002). Gα 12 and gα 13 negatively regulate the adhesive functions of cadherin. *J. Biol. Chem.* **277**, 24594–24600.
120. Kaplan, D. D., Meigs, T. E., and Casey, P. J. (2001). Distinct regions of the cadherin cytoplasmic domain are essential for functional interaction with Galpha 12 and beta-catenin. *J. Biol. Chem.* **276**, 44037–44043.
121. Polakis, P. (2000). Wnt signaling and cancer. *Genes Dev.* **14**, 1837–1851.
122. Wodarz, A. and Nusse, R. (1998). Mechanisms of Wnt signaling in development. *Annu. Rev. Cell Dev. Biol.* **14**, 59–88.
123. Anastasiadis, P. Z. and Reynolds, A. B. (2000). The p120 catenin family: complex roles in adhesion, signaling and cancer. *J. Cell Sci.* **113**, 1319–1334.
124. Yap, A. S., Niessen, C. M., and Gumbiner, B. M. (1998). The juxtamembrane region of the cadherin cytoplasmic tail supports lateral clustering, adhesive strengthening, and interaction with p120ctn. *J. Cell Biol.* **141**, 779–789.
125. Grosheva, I., Shtutman, M., Elbaum, M., and Bershadsky, A. D. (2001). p120 catenin affects cell motility via modulation of activity of Rho-family GTPases: A link between cell-cell contact formation and regulation of cell locomotion. *J. Cell Sci.* **114**, 695–707.
126. Anastasiadis, P. Z. *et al.* (2000). Inhibition of RhoA by p120 catenin. *Nat. Cell Biol.* **2**, 637–644.
127. Aono, S., Nakagawa, S., Reynolds, A. B., and Takeichi, M. (1999). p120(ctn) acts as an inhibitory regulator of cadherin function in colon carcinoma cells. *J. Cell Biol.* **145**, 551–562.
128. Huber, A. H. and Weis, W. I. (2001). The structure of the beta-catenin/E-cadherin complex and the molecular basis of diverse ligand recognition by beta-catenin. *Cell* **105**, 391–402.

129. Lickert, H., Bauer, A., Kemler, R., and Stappert, J. (2000). Casein kinase II phosphorylation of E-cadherin increases E-cadherin/beta-catenin interaction and strengthens cell-cell adhesion. *J. Biol. Chem.* **275**, 5090–5095.
130. Hamaguchi, M. *et al.* (1993). p60v-src causes tyrosine phosphorylation and inactivation of the N-cadherin-catenin cell adhesion system. *EMBO J.* **12**, 307–314.
131. Roura, S., Miravet, S., Piedra, J., Garcia de Herreros, A., and Dunach, M. (1999). Regulation of E-cadherin/Catenin association by tyrosine phosphorylation. *J. Biol. Chem.* **274**, 36734–36740.
132. Piedra, J. *et al.* (2001). Regulation of beta-catenin structure and activity by tyrosine phosphorylation. *J. Biol. Chem.* **276**, 20436–20443.
133. Behrens, J. *et al.* (1993). Loss of epithelial differentiation and gain of invasiveness correlates with tyrosine phosphorylation of the E-cadherin/beta-catenin complex in cells transformed with a temperature-sensitive v-SRC gene. *J. Cell Biol.* **120**, 757–766.
134. Rosato, R., Veltmaat, J. M., Groffen, J., and Heisterkamp, N. (1998). Involvement of the tyrosine kinase fer in cell adhesion. *Mol. Cell. Biol.* **18**, 5762–5770.
135. Kotelevets, L. *et al.* (2001). The lipid phosphatase activity of PTEN is critical for stabilizing intercellular junctions and reverting invasiveness. *J. Cell Biol.* **155**, 1129–1135.
136. Hoschuetzky, H., Aberle, H., and Kemler, R. (1994). Beta-catenin mediates the interaction of the cadherin-catenin complex with epidermal growth factor receptor. *J. Cell Biol.* **127**, 1375–1380.
137. Hazan, R. B. and Norton, L. (1998). The epidermal growth factor receptor modulates the interaction of E-cadherin with the actin cytoskeleton. *J. Biol. Chem.* **273**, 9078–9084.
138. Xu, Y., Guo, D. F., Davidson, M., Inagami, T., and Carpenter, G. (1997). Interaction of the adaptor protein Shc and the adhesion molecule cadherin. *J. Biol. Chem.* **272**, 13463–13466.
139. Fujita, Y. *et al.* (2002). Hakai, a c-Cbl-like protein, ubiquitinates and induces endocytosis of the E-cadherin complex. *Nat. Cell Biol.* **4**, 222–231.
140. Lilien, J., Balsamo, J., Arregui, C., and Xu, G. (2002). Turn-off, drop-out: Functional state switching of cadherins. *Dev. Dynam.* **224**, 18–29.
141. Brady-Kalnay, S. M. *et al.* (1998). Dynamic interaction of PTPmu with multiple cadherins *in vivo*. *J. Cell Biol.* **141**, 287–296.
142. Pathre, P. *et al.* (2001). PTP1B regulates neurite extension mediated by cell-cell and cell-matrix adhesion molecules. *J. Neurosci. Res.* **63**, 143–150.
143. Burden-Gulley, S. M. and Brady-Kalnay, S. M. (1999). PTPmu regulates N-cadherin-dependent neurite outgrowth. *J. Cell Biol.* **144**, 1323–1336.
144. Ozawa, M. and Kemler, R. (1990). Correct proteolytic cleavage is required for the cell adhesive function of uvomorulin. *J. Cell Biol.* **111**, 1645–1650.
145. Posthaus, H. *et al.* (1998). Proprotein cleavage of E-cadherin by furin in baculovirus over-expression system: potential role of other convertases in mammalian cells. *FEBS Lett.* **438**, 306–310.
146. Paradies, N. E. and Grunwald, G. B. (1993). Purification and characterization of NCAD90, a soluble endogenous form of N-cadherin, which is generated by proteolysis during retinal development and retains adhesive and neurite-promoting function. *J. Neurosci. Res.* **36**, 33–45.
147. Roark, E. F., Paradies, N. E., Lagunowich, L. A., and Grunwald, G. B. (1992). Evidence for endogenous proteases, mRNA level and insulin as multiple mechanisms of N-cadherin down-regulation during retinal development. *Development* **114**, 973–984.
148. Lochter, A. *et al.* (1997). Matrix metalloproteinase stromelysin-1 triggers a cascade of molecular alterations that leads to stable epithelial-to-mesenchymal conversion and a premalignant phenotype in mammary epithelial cells. *J. Cell Biol.* **139**, 1861–1872.
149. Ito, K. *et al.* (1999). Calcium influx triggers the sequential proteolysis of extracellular and cytoplasmic domains of E-cadherin, leading to loss of beta-catenin from cell-cell contacts. *Oncogene* **18**, 7080–7090.
150. Steinhusen, U. *et al.* (2001). Cleavage and shedding of E-cadherin after induction of apoptosis. *J. Biol. Chem.* **276**, 4972–4980.
151. Herren, B., Levkau, B., Raines, E. W., and Ross, R. (1998). Cleavage of beta-catenin and plakoglobin and shedding of VE-cadherin during endothelial apoptosis: evidence for a role for caspases and metalloproteinases. *Mol. Biol. Cell* **9**, 1589–1601.
152. Carmeliet, P. *et al.* (1999). Targeted deficiency or cytosolic truncation of the VE-cadherin gene in mice impairs VEGF-mediated endothelial survival and angiogenesis. *Cell* **98**, 147–157.
153. Marambaud, P. *et al.* (2002). A presenilin-1/gamma-secretase cleavage releases the E-cadherin intracellular domain and regulates disassembly of adherens junctions. *EMBO J.* **21**, 1948–1956.
154. Baki, L. *et al.* (2001). Presenilin-1 binds cytoplasmic epithelial cadherin, inhibits cadherin/p120 association, and regulates stability and function of the cadherin/catenin adhesion complex. *Proc. Natl. Acad. Sci. USA* **98**, 2381–2386.
155. Kang, D. E. *et al.* (1999). Presenilin 1 facilitates the constitutive turnover of beta-catenin: differential activity of Alzheimer's disease-linked PS1 mutants in the beta-catenin-signaling pathway. *J. Neurosci.* **19**, 4229–4237.
156. Nishimura, M. *et al.* (1999). Presenilin mutations associated with Alzheimer disease cause defective intracellular trafficking of beta-catenin, a component of the presenilin protein complex. *Nat. Med.* **5**, 164–169.
157. Soriano, S. *et al.* (2001). Presenilin 1 negatively regulates beta-catenin/T cell factor/lymphoid enhancer factor-1 signaling independently of beta-amyloid precursor protein and notch processing. *J. Cell Biol.* **152**, 785–794.
158. Georgakopoulos, A. *et al.* (1999). Presenilin-1 forms complexes with the cadherin/catenin cell-cell adhesion system and is recruited to intercellular and synaptic contacts. *Mol. Cell* **4**, 893–902.
159. Niessen, C. M. and Gumbiner, B. M. (2002). Cadherin-mediated cell sorting not determined by binding or adhesion specificity. *J. Cell Biol.* **156**, 389–399.
160. Luo, Y. *et al.* (2001). Rescuing the N-cadherin knockout by cardiac-specific expression of N- or E-cadherin. *Development* **128**, 459–469.
161. Price, S. R., De Marco Garcia, N. V., Ranscht, B., and Jessell, T. M. (2002). Regulation of motor neuron pool sorting by differential expression of type II cadherins. *Cell* **109**, 205–216.

ISBN 0-12-124548-9